KF   Bierlein, Lawrence W.
3945
.B5   Red Book on Trans-
1988   portation of Hazard-
          ous Materials

DATE DUE			
SEP. 15.1994			

LEARNING RESOURCES CENTER
NH VOC-TECH COLLEGE
1066 FRONT STREET
MANCHESTER, NH   03102

RED BOOK on
Transportation of Hazardous Materials

Red Book on Transportation of Hazardous Materials

SECOND EDITION

Lawrence W. Bierlein

 VAN NOSTRAND REINHOLD COMPANY

New York

Copyright © 1988 by Van Nostrand Reinhold Company Inc.

Library of Congress Catalog Card Number: 86-32558
ISBN 0-442-21044-2

Printed in the United States of America

Published by Van Nostrand Reinhold Company Inc.
115 Fifth Avenue
New York, New York 10003

Van Nostrand Reinhold Company Limited
Molly Millars Lane
Wokingham, Berkshire RG11 2PY, England

Van Nostrand Reinhold
480 Latrobe Street
Melbourne, Victoria 3000, Australia

Macmillan of Canada
Division of Canada Publishing Corporation
164 Commander Boulevard
Agincourt, Ontario MIS 3C7, Canada

15 14 13 12 11 10 9 8 7 6 5 4 3 2 1

Library of Congress Cataloging-in-Publication Data

Bierlein, Lawrence W., 1942–
    Red book on transportation of hazardous materials.

    Includes index.
    1. Hazardous substances—Transportation—Law and
legislation—United States. I. Title.
KF3945.B5 1987   347.73'093   86-32558
ISBN 0-442-21044-2   347.30393

To Katherine, Kirsten and Anna, for the inspiration of a new generation

INTRODUCTION

The first edition of the *Red Book on Transportation of Hazardous Materials* was written in 1976. At that time, DOT had just completed a major rule-making action in Docket Nos. HM-103 and HM-112, pulling the water and air regulations into the same CFR volume with the rail and highway rules, and making many more substantive changes. The ORM classifications were created, and shipping documents and other hazard communications were streamlined. The hazardous materials program has had a long history, and the first edition of the *Red Book* collected many years of interpretations of the regulations, along with an explanation of the rules which then had just been adopted.

A great deal has happened in the past decade to warrant a second edition of the *Red Book*. As you will note, there are major new elements of the program in the form of controls on the transportation of environmental hazards, and there has been a dramatically increased influence on domestic commerce from the international regulations. In the intervening years, DOT also has made changes to the details covered in the first edition. The second edition, therefore, is a complete rewrite. Much of the material in it is new.

DOT is on the verge of yet another massive change in the regulations, through rule-making Docket No. HM-181. The effect of this action will be to give even greater emphasis to the international regulatory codes developed by the UN and related organizations. As of the printing of the second edition of the *Red Book*, DOT is just embarking on the public notice phase of this rule making. Certain aspects of it are bound to be controversial. Others will generate a great deal of comment because unanswered issues are raised for that kind of discussion. In short, while we anticipate additional changes in the DOT regulatory program, it is unlikely to be completed much before 1990. Even when it is completed, the existing system will continue to remain operative through grandfather provisions of the new regulations.

A third edition of the *Red Book* will be forthcoming when DOT completes its rule making in Docket No. HM-181, in approximately three years. The second edition of the *Red Book*, however, has been written with these changes in mind, based upon the existing provisions of international codes, so the reader should find the second edition a helpful text for many years to come.

LAWRENCE W. BIERLEIN
Washington, DC

ACKNOWLEDGMENTS

As noted on several occasions in this edition of the *Red Book,* the regulatory and liability issues associated with the transportation of hazardous materials have become so complex that they are beyond the knowledge of any single person. Gordon Rousseau and I have worked closely together for more years than I like to count, dealing on a daily basis with the various regulations of domestic and international agencies involved in hazardous materials transportation.

I said in the first edition that Gordon's input was invaluable, and it is even more true in this edition. Working effectively with regulations goes far beyond simply finding them and reading them. You have to have a sense of them, a feel that goes beyond easy description. Gordon has this patient sensitivity, and much of that quality of his is incorporated into the *Red Book.* I am deeply grateful for his support in this endeavor, as in all the rest of our professional ventures.

CONTENTS

CHAPTER 1

Using the Red Book

The *Red Book on Transportation of Hazardous Materials* is written as a guide for a wide range of people concerned with the transportation of hazardous materials, including hazardous substances and hazardous wastes designated by the Environmental Protection Agency (EPA). It is designed to enable people to better understand the nature and scope of the federal transportation regulations issued and enforced by the Department of Transportation (DOT), EPA, and related State and international agencies.

Federal safety regulations and, indeed, all regulations, are written with an unfortunate common flaw. The regulations are not written with the user or the subject matter in mind but, rather, with the legal enforcement of those regulations as the dominant factor affecting their format and phrasing. Although federal officials have long sought a complete and comprehensible series of hazardous materials regulations and although some improvement definitely can be achieved, those regulations will never reach the simple level of expression used in conversational English. The regulations must be legally complete, and they must apply to everyone involved with transportation, anticipating every foreseeable and unforeseeable circumstance, and this dooms any effort to clarify and simplify the rules. For every simple statement there is an exception. Listing those exceptions and plugging loopholes is done by the regulations, and simple communication is the victim.

The needs of safety and a concern for liability demand some mechanism to simplify the process of compliance with these regulations, and the *Red Book* has been prepared to meet that need. The *Red Book* is not a substitute for the regulations, however; use of this book calls for and depends upon frequent reference back and forth between the *Red Book* and the DOT and EPA regulations.

The government regulations, explained at length in the following text, are available from the U.S. Superintendent of Documents. Chapter 3, explaining "The What, Who, Where, When, and Why of Hazardous Materials Regulation," tells you which documents to get. Prices will vary from volume to volume and year to year. All requests should be addressed to:

Superintendent of Documents
U.S. Government Printing Office
Washington, D.C. 20402

1

Each request should identify the title and parts of the regulations being ordered, as these are explained in later chapters.

The federal hazardous materials regulations are also available from private sources. Two tariffs in the form of republications of title 49 of the Code of Federal Regulations are supplied on a subscription basis. They are available from:

Association of American Railroads
50 F Street, N.W.
Washington, D.C. 20001

American Trucking Associations, Inc.
2200 Mill Road
Alexandria, VA 22314

In addition, several publishers market looseleaf editions of the regulations, and provide revision sheets to keep them current.

The *Red Book* is user-oriented for the full range of people dealing with hazardous materials regulation, including shippers, carriers, container manufacturers, freight forwarders, lawyers, educators, legislators, and regulators. Since the basic responsibilities under the regulations are assigned to the shipper, however, the shipper will derive the greatest benefit from the *Red Book*. Under the law it is the responsibility of the shipper to identify, classify, name, pack, mark, label, and placard his shipment properly. He must tender it to the carrier with proper documentation and the shipping papers must certify compliance with the regulations. Accordingly, the following chapters are placed in the order of responsibilities listed above.

A persistent problem in achieving compliance and in understanding the regulations, and a problem as well in the writing of the *Red Book*, is the fact that the body of regulations is always in flux. At this period of regulatory history the flux is even greater, and the extent of the changes being proposed and adopted is enormous. Insofar as possible, the *Red Book* has been written to pull all requirements together and to express them in a way that will be least affected by the process of change. Obviously, this is not entirely possible and, with the implementation and phase-in of new requirements, reading the *Red Book* by itself could be misleading. Again, this book must be read in conjunction with a copy of the most current government regulations.

Any text addressing such a complex subject necessarily makes heavy reading. It is recommended that the *Red Book* be read and reread in short units. Effort has been made to provide a helpful discussion in a concise and compact style. Every statement in the *Red Book* has been written and rewritten, often with hours spent on a single page. It is anticipated, therefore, that periodically rereading part of the *Red Book* will be helpful and probably essential to a full understanding of the subject matter.

The lengthy and tedious treatment of hazardous materials in the regulations, and somewhat in the *Red Book*, is the natural result of having to discuss all materials shipped by all persons, in all modes of transportation, and in all quantities throughout the world. Few companies have needs as broad as those addressed in the regulations. For any single company it is worthwhile to extract the regulations applicable to that company's limited number of products, in the quantities shipped, and to the destinations and via the modes of transportation customarily used. This distillation of the requirements will make a much more handy document which, when coupled with the general explanations found in the *Red Book*, will facilitate compliance with the law.

CHAPTER 2

Increasing Complexity in the Transportation of Hazardous Materials, Wastes, and Substances

In recent memory the regulation of hazardous materials transportation was a relatively simple matter, generally involving a single definition of materials, administered by a single agency under a single statute.

In 1980, this field began a dramatic expansion, and consequent complication that has been maintained until today and promises to continue. For the hazardous materials shipper, carrier, and container manufacturer, and the lawyer or practitioner advising them, the task of knowing what to do and the penalties for failure to do it has become far more difficult than it had been.

The reasons for this are several, but generally it is because the purposes of regulation have become different and more complex, the population of regulated materials and companies has been enlarged, and the force of involved government officials has been expanding at the same time as the number of corporate and legal professionals familiar with transportation laws and regulations has been shrinking.

In analyzing this growth of regulatory complexity and associated confusion among the regulated industries, it will be helpful to review the pertinent statutory additions and other sources of influence bringing it about.

STATUTORY BASES FOR TRANSPORTATION REGULATION

(1) HAZARDOUS MATERIALS TRANSPORTATION ACT (HMTA), 49 U.S. CODE 1801, *ET SEQ*

From 1977 to 1980, generally the only law in this field with which one had to be familiar was the HMTA. If the problem involved the transport of radioactive materials, then one also needed some familiarity with the Atomic Energy Act of 1954, 42 U.S. Code 2011.

As administered by the U.S. Department of Transportation (DOT), the HMTA built upon a base of regulation that had been developed over the preceding half century under the since-repealed Explosives & Other Dangerous Articles Act, 18

3

U.S. Code 831-835. The key purpose of this early law, and the regulations issued under it first by the Interstate Commerce Commission and thereafter DOT, was the protection of transportation employees and the public from acute personal injuries that might be suffered as a result of the mishandling of certain chemical products in transportation. The types or "classes" of hazard addressed in this law were explosion, flammability, toxicity, corrosion, human infectious disease, and radioactivity. In other words, there was no concern for potential damage to the environment and, except for some forms of radioactivity, little or no concern for most long-term health hazards.

Although DOT can regulate any traffic that "affects" interstate or foreign commerce, the agency has chosen to limit the scope of its coverage to shipments via carriers engaged in interstate commerce. While in the rail, air, and water modes this is not very significant, in the highway mode of transportation a tremendous amount of intrastate commerce is not covered by the traditional DOT regulations. The agency did say it would reserve the authority to expand coverage to intrastate commerce on a case by case basis and, as discussed below, that has happened with regard to environmentally designated materials.

Related to the HMTA are specific modal provisions in the Federal Aviation Act (49 U.S. Code 1301, 1472(h)), the Motor Carrier Acts of 1980 and 1984 (P.L. 96–296 and P.L. 98–554), the Federal Railroad Safety Act (45 U.S. Code 421), and certain laws governing water transportation and administered by the U.S. Coast Guard (46 U.S. Code 170, 391a).

(2) CLEAN WATER ACT

A major addition to the DOT regulations were materials identified as "hazardous substances" by the Environmental Protection Agency (EPA) under the Clean Water Act, 33 U.S. Code 1251, 1321. This statute imposes a reporting burden on any facility that might be involved in the spill of a "reportable quantity" of a material that might be harmful to aquatic life, including spills from vessels or vehicles operated by any type of carrier in any mode of transportation. The task undertaken by DOT, at the urging of EPA and the transportation community, was development of a regulatory mechanism by which the shipper of a reportable quantity (RQ) of an EPA-designated hazardous substance could advise the carrier of the presence of the material, thus enabling the carrier to meet its criminally-enforced reporting obligations under the Clean Water Act. Unless the carrier, especially a common carrier, were so advised, EPA reasoned that no enforcement action could be brought for a "knowing" violation of the spill reporting requirements.

This original effort involved approximately 300 specifically named materials, some of which were already listed by DOT, others of which were regulated but only in generic (n.o.s.) categories, and still others which had not been DOT-regulated before because they lacked any acute human injury potential. In 1980, these specific materials were listed by name in the DOT regulations with a new "RQ" (reportable quantity) marking and shipping paper requirement. For materials which had not been formerly regulated by DOT, a new classification of Other Regulated Materials Group E, or "ORM-E," was created. (See DOT Docket No. Hm-145B, 45 Fed. Reg. 34560, May 22, 1980.)

The new DOT regulations on these environmental hazards were extended to intrastate as well as interstate commerce, thereby bringing into the DOT system a great number of people who may not otherwise have been familiar with it.

(3) RESOURCE CONSERVATION & RECOVERY ACT

In 1976, Congress enacted the basic hazardous waste law entitled the Resource Conservation & Recovery Act (RCRA), 42 U.S. Code 6901, *et seq.* As implemented by EPA through regulations finalized in 1980, this law is intended to establish cradle-to-grave identification and management of hazardous wastes, including waste materials in transportation by any mode.

Consistent with the conclusion reached with regard to environmental hazards under the Clean Water Act, DOT and EPA determined that they would add the new hazardous waste transportation controls on top of the existing hazardous materials transportation regulatory system. (See DOT Docket No. HM-145A, 45 Fed. Reg. 34560, May 22, 1980.)

This involved a variety of changes to the DOT regulations, as well as the adoption of corresponding EPA transportation rules. In terms of identification of materials regulated, the existing DOT regulations were used, including the new ORM-E classification for materials that had not been regulated by DOT previously. This effort was complicated substantially by EPA's use of definitional terminology that was similar to DOT's, but definitely not the same.

The primary differences between EPA, RCRA, and DOT coverage of materials are:

(a) EPA's "ignitable" characteristic is much broader than DOT's "flammable." (This has been proposed for change by DOT.)

(b) EPA does not regulate nonflammable gases, but DOT does.

(c) EPA's "corrosive" characteristic uses pH as a measure, whereas DOT's does not.

(d) EPA regulates corrosive liquids only, whereas DOT regulates both liquids and solids.

(e) EPA's "reactive" characteristic includes a number of DOT explosives, dangerous-when-wet materials, forbidden materials, and flammable solids.

(f) EPA's "extraction procedure (EP) toxicity" does not correlate with DOT's poison definitions.

(g) EPA's approach of identifying regulated wastes by listing their processes or industry sources differs greatly from DOT's approach of using simpler, more generic terms (e.g., EPA's "F026 Wastes (except wastewater and spent carbon from hydrogen chloride purification) from the production of materials on equipment previously used for the manufacturing use (as a reactant, chemical intermediate, or component in a formulating process) of tetra-, penta-, or hexachlorobenzene under alkaline conditions" and, "K071 Brine purification muds from the mercury cell process in chlorine production, where separately prepurified brine is not used"), as opposed to simply naming specific chemicals or hazard characteristics.

These definitional differences have stimulated confusion primarily among shippers of wastes (called "generators" by EPA), who must rely on technical data to determine proper packaging, marking, labeling, etc.

The most visible addition to the DOT regulations was a requirement to use the Uniform Hazardous Waste Manifest developed by the two agencies as a shipping document for hazardous wastes. Whereas DOT has always had shipping paper requirements, those papers never required the name and address of the shipper or consignee, or signatures by the carrier or consignee. In addition, although prohibiting certain stylistic deviations such as the use of coded symbols, DOT does not otherwise prescribe the format of a hazardous materials shipping paper. The signature blocks on a waste manifest for the hazardous waste carrier (called a "transporter" by EPA), and for the recipient EPA-permitted hazardous waste treatment, storage, or disposal facility, are unique and are part of the RCRA waste tracking system.

Generally speaking, EPA tells a waste generator to follow the DOT regulations for packaging of wastes, and to meet the related DOT package marking and labeling requirements in 49 CFR Part 172. In addition, however, a separate hazardous waste marking is prescribed by EPA in 40 CFR 262.32.

Several facets of the hazardous waste controls were not seen in the hazardous materials transportation regulations before 1980. The primary one is the tracking and sequential signature responsibility for wastes from point of generation to point of treatment or disposal, including intervening transportation. The second is the concern for environmental damage that might result from spillage of the wastes in transit. One might say that, whether a material is identified as a waste or not, the environmental impact of a spill is the same and so the regulatory controls should be the same. This type of logic is being pursued in current rule making at both agencies.

Compounding confusion caused by the new regulations imposed by DOT on hazardous waste is the fact that most companies do not use the same employees to prepare wastes for transport and products for shipment. Common experience is that compliance with the old DOT regulations fell in the traffic department's realm, while hazardous waste shipments are part of plant operations or environmental control. Thus, even though the company may have been DOT-regulated before, new people are having to cope with the amended DOT regulations.

Transporters of hazardous wastes are addressed specifically by 40 CFR Part 263 of the EPA regulations, with the key feature being the obligation to apply for and receive an identification number from EPA as a waste transporter. This is a relatively passive system at the federal level, but there is a strong thrust in the regulations to shift the daily responsibility for hazardous waste regulations to the States. While most States have followed the federal scheme in terms of basic waste identification and generator requirements, there is wide variation with regard to transporter controls. Generally speaking, the States have been more formalistic in this area than federal regulators, often requiring State certification of the carrier in proceedings similar to applications for operating authority having to show convenience and necessity for the proposed service. State certification often may also involve detailed identification of the principals involved in the company, the rates charged, and the carrier's financial responsibility or insurance available to government and third party claimants in the event of an incident.

Like the rules adopted by DOT with regard to hazardous substances under the Clean Water Act, the DOT hazardous waste rules apply in intrastate as well as interstate commerce. As noted earlier, especially in the highway mode, there is a considerable amount of intrastate commerce that has not been exposed to the DOT hazardous materials regulations, i.e., there is a broader population of people regulated whose familiarity with the DOT rules is limited at best.

In rule making implemented in 1985 and 1986, EPA dramatically increased the population of substantively regulated waste generators by lowering its coverage from those who generate 1000 kilos per month or more, to those who produce only 100 kilos per month. Several hundred thousand so-called "small quantity generators" were brought within the total hazardous waste regulatory system, including the DOT hazardous materials regulations, and the likelihood is high that these generators have little or no existing familiarity with either the EPA or DOT systems.

(4) TOXIC SUBSTANCES CONTROL ACT (TSCA), 15 U.S. CODE 2601, *ET SEQ*

TSCA is another very broad statute, giving EPA wide latitude to regulate materials including their handling in transportation. As an example, materials specifically regulated under this statute include polychlorinated byphenyls (PCBs). PCBs were not historically regulated by DOT, nor are they considered hazardous wastes under RCRA even when discarded (although regulation under RCRA has been discussed and may be implemented). PCBs came into the DOT regulations because they were listed under the Clean Water Act, discussed above, but the most detailed federal regulations are imposed by EPA under TSCA in 40 CFR Part 761. These regulations address the storage and disposal of PCBs, and specifically prescribe use of DOT specification packaging for these purposes, although there is no clear mention of transportation. DOT imposes some requirements when there is a reportable quantity (RQ) of PCBs in a package, but not otherwise.

The net effect of TSCA on transportation to date has been to add a further element of confusion. The breadth of the statute, however, and the fact that it would enable EPA to extend regulatory controls beyond wastes and into the production and pre-production phases of chemical manufacture, leads me to conclude that it is a likely statutory vehicle for future transportation regulations.

(5) COMPREHENSIVE ENVIRONMENTAL RESPONSE, COMPENSATION AND LIABILITY ACT (CERCLA, OR SUPERFUND), 42 U.S. CODE 9601, *ET SEQ*

In 1980, Congress adopted Superfund. As the title states, this law is comprehensive, and addresses the release or threatened release of "hazardous substances" from any vessel or facility, including transportation equipment.

The term "hazardous substance" used in Superfund is far broader than the same term used in the Clean Water Act. At 42 U.S. Code 9601(14), the term is defined to include those same water pollutants, plus other water pollutants listed under 33 U.S. Code 1317(a), all hazardous wastes under RCRA (42 U.S. Code 3001), hazardous air pollutants listed under section 112 of the Clean Air Act (42 U.S. Code 7412), imminently hazardous chemical substances or mixtures subject to action under section 7 of TSCA (15 U.S. Code 2606), as well as "any element, compound, mixture, solution, or substance" designated by EPA under section 102 of Superfund (42 U.S. Code 9602).

Like the Clean Water Act, Superfund calls for spill reporting when a "reportable quantity" (RQ) of material is involved. Congress assigned an RQ of 1 pound to all newly regulated materials but, frankly, no one paid much attention to that stringent level. EPA has been engaged in a lengthy and ongoing effort to assign more realistic RQs to Superfund materials, and has published several final rules and notices of proposed rule making on this subject.

As amended by the Superfund Amendments and Reauthorization Act of 1986 (SARA), P.L. 99–499, Section 306(a) of Superfund declares that a hazardous substance listed or designated under this law "shall, within ninety days ... be listed and regulated as a hazardous material under the Hazardous Materials Transportation Act." When Superfund was first passed in 1980, DOT had to list the materials but not regulate them. With the SARA amendments, DOT was compelled to list and regulate them. DOT has done this in rule making establishing package marking and shipping document requirements. Most of these requirements had an effective date of July 1, 1987.

SARA also included several major new requirements pertaining to emergency planning and community right-to-know. These have expanded industry and government record-keeping efforts, and have stimulated additional confusion because of definitional differences. Section 302 of SARA compelled EPA to publish a "list of extremely hazardous substances." This list, published on November 17, 1986, included 402 materials plus "threshold planning quantities" as well as reportable quantities for each. This list does not coincide with any other regulated list of materials.

Section 311 of SARA added yet more complexity by requiring the submission of material safety data sheets (MSDSs) or a list of materials as regulated by the Occupational Safety & Health Administration (OSHA) under their hazard communication requirements. The OSHA program is discussed more fully below. These materials are more inclusive than chemicals on other lists, but only affect people who are subject to the OSHA rules. At this time, those are chemical manufacturers, importers and distributors offering chemicals to businesses in the manufacturing SIC Codes.

Under Section 312 of SARA there is a discussion of hazardous chemical inventory forms. The chemicals affected are those subject to the OSHA standard, but certain information can be requested by local governments about other materials.

The confusion resulting from these added layers of information and regulatory responsibility is more likely to frustrate effective emergency response planning than facilitate it.

An extremely important element of the EPA effort, which will be reflected in the DOT regulations, is their conclusion that if a material would be subject to hazardous waste regulations under RCRA when it is discarded, then that material falls within the Superfund "hazardous substance" definition *before* it is discarded. In other words, the two agencies have discussed imposing substantive transportation requirements on nonwastes, in anticipation of those materials becoming wastes at some time through a spill or otherwise. The logical extension of this rationale will result in the broader waste regulations developed by EPA superseding the traditional hazardous materials regulations administered by DOT.

(6) OCCUPATIONAL SAFETY & HEALTH ACT, 29 U.S. CODE 651, *ET SEQ*

The Occupational Safety & Health Administration (OSHA) of the U.S. Department of Labor, operating under this statute, implemented far-reaching regulations on the communication of hazards to employees. These rules are found in 29 CFR 1910.1200 and, at least initially, were directed to manufacturing employers in the Standard Industrial Classifications (SIC) 20–39.

Through this standard, the manufacturers or importers of chemicals are obligated to develop a Material Safety Data Sheet (MSDS) for each product, and to pass that document to distributors and customers in the manufacturing SIC classifications. In addition, each chemical container must be labeled to advise employees of the hazards of the contents.

The OSHA field manual for inspectors specifically declares that compliance with the DOT system of marking, labeling and placarding for hazardous materials does not conclusively determine whether the hazard is adequately communicated to employees. In addition, it is clear that a DOT shipping paper would not satisfy the OSHA MSDS requirements.

One of the results of the OSHA requirements, therefore, is that the shipper of a chemical is likely to augment his package labeling, and may also tender freight with an MSDS (although there is no requirement that the MSDS move with the cargo in transportation). DOT has a regulation saying that no label or warning may be used on a package that might diminish or confuse the DOT hazard warning system. Satisfaction of the requirements of both agencies is a delicate and confusing task, at best.

GROWING INFLUENCE OF INTERNATIONAL TRANSPORTATION REGULATIONS

(1) THE UN "ORANGE BOOK"

The United States participates as a member of the United Nations Committee of Experts on the Transport of Dangerous Goods. The result of the deliberations of this committee, and a parallel committee on the transportation of explosives, is the UN "Orange Book" of recommendations on the transport of dangerous goods. These recommendations are a compendium of detail on the classification, identification, packaging, marking, labeling, documentation, placarding and handling of hazardous materials in transportation. Although the U.S. has been a participant in this process, the meetings generally are heavily dominated by European countries, and the system that has been developed differs from that currently followed in domestic commerce in the U.S.

One of the most important results of these deliberations has been the development of performance standards for packaging, found in Chapter 9 of the Orange Book. Completed a few years ago, this effort is leading to changes in packaging in nearly all international and national hazardous materials regulatory codes. See Chapter 25 on the international requirements.

These packaging recommendations are the culmination of years of effort by the UN and, although they are expected to be controversial and, in many instances are alleged to be inadequate, the latitude of any one country to deviate from them or to change them will be extremely limited.

(2) THE INTERNATIONAL MARITIME ORGANIZATION (IMO)

The UN recommendations are just that—recommendations that lack legal force and effect until promulgated by some other regulatory body. The broadest mechanism through which the UN recommendations are implemented is the International Maritime Dangerous Goods (IMDG) Code, published by the International Maritime Organization in London (formerly called IMCO). IMO's rules still are just recommendations, but often are adopted as the law of the land and, as a practical matter, are required commercially by most international water carriers.

As the name advises, the IMO IMDG Code is designed for the marine mode, and generally is followed or required by most international water carriers. It is adopted as law in many countries, and is a viable option for compliance in the water mode in the U.S. (with some exceptions). See 49 CFR 171.12.

Because the IMDG Code is amended periodically to implement the UN recommendations, a significant lag time appears between UN action on an item and practical implementation of that action in daily commerce. While this provides some benefit to companies who follow the UN, in terms of offering advance warning of requirements, it also can work to the detriment of anyone seeking a change, because changes occur very slowly.

The IMO's implementation of new packaging requirements based upon Chapter 9 of the UN Orange Book appears in Annex I of the IMDG Code, with a mandatory effective date of January 1, 1990. Grandfather clauses generally permit the continued use of existing packaging until then.

Another significant new element of the IMDG Code will be the identification and regulation of marine pollutants under the International Convention for the Prevention of Pollution from Ships (MARPOL 73/78). Just as EPA found the existing DOT regulations to be a convenient vehicle for the implementation of new environmental requirements, the IMO has found the IMDG Code to be an available mechanism for implementation of environmental regulations on marine pollutants. These will be phased in over the next few years, but will involve changes in documentation, packaging, package markings, and other handling controls for both bulk and packaged volumes of these goods. Mixtures and solutions containing listed ingredients also will be affected seriously.

(3) THE INTERNATIONAL CIVIL AVIATION ORGANIZATION (ICAO)

Under various international treaties involving air transportation, ICAO has adopted an annex and technical instructions on the preparation of dangerous goods for air transport. This code, too, is based upon the UN structure, and it has been in place as an option in the United States since 1983. See 49 CFR 171.11. The mandatory date to follow these requirements in air transport coincides with the maritime effective date for UN—packaging January 1, 1990.

With substantial deregulation of the air mode, air carriers have exercised the flexibility to adopt their own regulations, incorporating and adding to one or more government regulatory codes. International air carriers, through the International Air Transport Association (IATA) have done this, and the IATA dangerous goods (also often called "restricted articles") regulations are in common use in both domestic and international commerce. The rules are based upon ICAO, but generally are more restrictive in certain technical respects. Although in the U.S. an air carrier has the option to follow either the DOT regulations in 49 CFR or ICAO, most air carriers have chosen to use only ICAO, as implemented by IATA.

(4) DOT DOCKET NO. HM-181

This docket, opened by DOT in 1982 as an advance notice to discuss the advisability of adopting the UN recommendations as law in the United States, is still open while the *Red Book* goes to publication. The docket proposes implementation in domestic commerce of the classification, identification, and packaging schemes developed under the UN Hazardous materials labeling and placarding in the U.S. already are based upon the UN system.

The conclusion of this rule making action seems foregone, because of the general trend in other countries, including Canada, to shift to this system. It has been argued that the UN system, particularly for packaging, offers less public safety than the requirements currently enforced by DOT. Only time will tell whether the UN system is adopted, or is adopted with unique U.S. variations, and whether the predicted negative impact on public safety will occur.

STATE AND LOCAL REGULATION OF HAZARDOUS MATERIALS TRANSPORTATION

Section 112 of the HMTA, 49 U.S. Code 1811, generally preempts any State or local regulation of hazardous materials transportation that is "inconsistent" with the purposes of Congress in enacting the HMTA or the regulations promulgated thereunder.

Litigation and administrative discussions to date have focused on the meaning of the term "inconsistent," with the result that technical matters such as classification, identification, packaging, marking, labeling, documentation and placarding have been found to be federally exclusive areas. Operational matters such as the routing of vehicles have been held to be within the nonfederal province, so long as the nonfederal decision is technically well-founded and does not have the effect of simply shifting the traffic from one jurisdiction to another. Less clear subjects such as permits, fees, registration, curfews, and the like have been judged upon the unique fact circumstances involved.

There is a procedure in 49 CFR Part 107 for obtaining a DOT ruling on the inconsistency of a particular nonfederal enactment, but it is a slow process. Approximately 20 uncodified "inconsistency rulings" have been issued by DOT's Research & Special Programs Administration. DOT has chosen to limit its consideration of the specific local regulation relative to the DOT regulations, and explicitly ignores questions of constitutionality, such as the burden the local rule may impose on the flow of interstate commerce. Consequently, to date, even when there has been an administrative proceeding, there often has been a parallel judicial proceeding as well.

If an enactment is adjudged or admitted to be inconsistent, DOT has a procedure whereby preemption may be waived upon a specific showing by the affected State government. See 49 U.S. Code 1811 (b) and 49 CFR Part 107 on "nonpreemption determinations." This is a little-used procedure with no helpful precedent.

Under the Motor Carrier Act of 1980, *supra*, and the Surface Transportation Assistance Act of 1982 (P.L. 97–424), a grant program was created by which State and local governments may receive federal funding to administer federal truck safety and hazardous materials requirements.

The net result is that more local officials are getting involved in the application, interpretation and enforcement of the federal hazardous materials regulations. As mentioned earlier, this is occurring just as the federal program is rapidly expanding to cover new materials and companies. The combination of increased complexity and numbers of people for whom the regulatory system is a novelty, does not bode well for any shipper or carrier of regulated materials.

Another area of State and local activity that is having an impact on transportation are so-called "right-to-know" laws. These vary greatly and, in part, may be preempted either under the OSHA standard in 29 CFR 1910.1200, or the HMTA. Nonetheless, an impact is being felt, particularly in the carrier industry which may operate terminals which handle a variety of materials governed by nonfederal right-to-know laws.

CONCLUSION

What historically was a technical but relatively straightforward area of regulation is rapidly becoming so complex that no one understands it all. The federal programs involving the designation of regulated materials have enlarged, and the reasons for designation have become more varied. This is happening not only at the federal level in the U.S., but in international contexts where U.S. companies will be affected.

Despite increasing complexity, compliance by hazardous materials shippers, carriers, and container manufacturers remains enforceable by severe civil and criminal penalties. The force of inspection and enforcement personnel at federal, State and local levels is increasing dramatically.

CHAPTER 3

The What, Who, Where, When, and Why of Hazardous Materials Transportation Regulations

WHAT ARE HAZARDOUS MATERIALS REGULATIONS?

Several federal administrative agencies have the authority to issue regulations regarding questions of safety of hazardous materials. In the *Red Book*, however, the focus is on transportation, and the primary agency authorized to regulate the transportation of hazardous materials is the U.S. Department of Transportation (DOT), in Washington, D.C. See Chapter 29 on DOT organization and identification of the offices directly involved with the issuance and enforcement of these regulations.

DOT's regulations are issued under authority given to the agency by Congress in a statute signed by the President. The regulations issued by the agency under that statute are part of the law. They contain affirmative requirements as well as prohibitions affecting all aspects of the transportation (and related storage) of hazardous materials and people involved in that transportation. These aspects include the identification, classification, description, packing, marking, labeling, placarding, and shipping documentation for the materials, as well as the handling of the materials in transit by carriers. Carrier requirements touch upon placarding of motor vehicles, rail cars, and containerization units and tanks, as well as mandatory guidelines on documentation, loading, and storage of materials in the possession of the carrier. Vehicle safety standards are also keyed to the presence of the hazardous materials, as are the qualifications of drivers of motor vehicles carrying such materials.

Specific rules tell carriers what to do in case of an accident, spillage, or damage to a hazardous materials container, or if there is unusual delay in transit or an inability to effectuate delivery of the material. Included as well are accident and incident reporting requirements for all carriers in the event of an unintentional spillage or release of a hazardous material from its package. Motor carriers of hazardous materials also must meet specific financial responsibility requirements.

The hazardous materials regulations have evolved over more than a century, beginning with a law to control the transportation of explosives by rail after the Civil War. Generally speaking, the authority to issue these safety regulations was an ancil-

13

lary power given to agencies that had economic regulation of the carrier industry as a primary mission. Thus, the regulations have an inherent orientation toward particular modes of transportation that has lasted under DOT. This agency, born from the aggregation of other existing agencies and laws, maintained the regulation of hazardous materials along exclusively modal lines until 1975. This regulatory authority was vested in the Federal Aviation Administration for shipments by aircraft, the Federal Highway Administration for shipments by motor carrier, the Federal Railroad Administration for shipments by rail carrier, and the United States Coast Guard for all shipments by water.

With the enactment of the Hazardous Materials Transportation Act (49 U.S.C. 1801, *et seq.*) in early 1975, DOT was called upon to consolidate the multi-modal regulatory power into unified offices, not oriented toward any particular mode of transportation. Although the regulations pertaining exclusively to a single mode of transportation are still under some degree of control by the appropriate modal DOT operating administration, the bulk of the authority for issuing the hazardous materials regulations is now the province of the DOT Research and Special Programs Administration (RSPA). Within RSPA are two offices, one pertaining to pipeline safety and another devoted specifically to hazardous materials, called the Office of Hazardous Materials Transportation (OHMT).

The regulations naturally evolved along the lines of the administrative groups that were handling them. Thus, until a massive regulatory consolidation accomplished in DOT's 1976 rule making Docket No. HM-112 (see Appendix C), the regulations for air shipment and water shipment appeared in different places than those for rail and highway shipments. The regulations of the federal government generally are found in a series of volumes called the Code of Federal Regulations (CFRs). The CFRs are divided into titles, or subject groupings, with a total of fifty titles in all.

Title 49 of these titles addresses "Transportation." This title is divided into subtitles and chapters, each of which contain a designated series of "Parts," as follows:

Subtitle A Office of the Secretary of Transportation (Parts 0–99).

Subtitle B Other Regulations Relating to Transportation:

Chap. I Research & Special Programs Administration (Parts 100–199).

Chap. II Federal Railroad Administration, DOT (Parts 200–299).

Chap. III Federal Highway Administration, DOT (Parts 300–399).

Chap. IV Coast Guard, DOT (Parts 400–499).

Chap. V National Highway Traffic Safety Administration (Parts 500–599).

Chap. VI Urban Mass Transportation Administration (Parts 600–699).

Chap. VII National Railroad Passenger Corporation (Amtrak) (Parts 700–799).

Chap. VIII National Transportation Safety Board (Parts 800–899).

Chap. IX United States Railway Association (Parts 900–999).

Chap. X Interstate Commerce Commission (Parts 1000–1399).

The DOT regulations of particular concern in the *Red Book* are found in Chapter I of the DOT regulations in title 49 CFR Parts 100–199. Throughout the *Red Book*, one must realize that there is no consideration given to federal requirements governing the bulk transportation of hazardous materials by water, i.e., transportation of materials loaded in a vessel's integral tanks without benefit of any portable container or packaging.

Before outlining the format of the title 49 CFR regulations in detail one should note several other titles of the CFRs pertinent to hazardous materials transportation:

Title 10 Energy.
  Chap. I Nuclear Regulatory Commission (Parts 0–199).
Title 14 Aeronautics & Space.
  Chap. I Federal Aviation Administration, DOT (Parts 0–199).
Title 16 Commercial Practices.
  Chap. II Consumer Product Safety Commission (Parts 1000–1799).
Title 27 Alcohol, Tobacco Products and Firearms.
  Chap. I Bureau of Alcohol, Tobacco, and Firearms, Dept. of the Treasury (Parts 0–99).
Title 29 Labor; Subtitle B Regulations Relating to Labor.
  Chap. XVII Occupational Safety & Health Administration, DOT (Parts 1900–1999).
  Chap. XX Occupational Safety & Health Review Commission (Parts 2200–2299).
Title 33 Navigation & Navigable Waters.

  Chap. I Coast Guard, DOT (Parts 1–199, specifically Parts 124 & 126, pertaining to Control Over Movement of Vessels, and Handling Explosives or Other Dangerous Cargoes Within or Contiguous to Waterfront Facilities, respectively).
Title 39 Postal Service.
  Chap. I United States Postal Service (Parts 0–999, specifically Part 124, Nonmailable Matter—Articles & Substances; Special Mailing Rules).
Title 40 Protection of Environment.
  Chap. I Environmental Protection Agency (Parts 1–799).
Title 42 Public Health.
  Chap. I Public Health Service, Dept. of Health and Human Services (Parts 1–199, specifically regarding etiologic agents).
Title 46 Shipping.
  Chap. I Coast Guard, DOT (Parts 1–199).

Within 49 CFR one finds the DOT hazardous materials regulations that have been applicable to rail and highway transportation for years. In rule making Docket No. HM-112 (see Appendix C), DOT consolidated into 49 CFR those requirements that formerly had appeared in 14 CFR Part 103, pertaining to air shipments, and 46 CFR Part 146, pertaining to water shipments.

The hazardous materials regulations in title 49 CFR are organized as follows:

Title 49 Transportation.
Subtitle B Other Regulations Relating to Transportation.
Chapter I Research & Special Programs Administration.
Subchapter B Hazardous Materials Transportation and Pipeline Safety.

Part 106 Rulemaking procedures.
Part 107 Hazardous materials program procedures (non–preemption determinations, exemptions, and approvals, and enforcement sanctions).
Subchapter C Hazardous Materials Regulations.

Part 171 General information, regulations, and definitions.
Part 172 Hazardous materials tables and hazardous materials communications regulations.

    Subpart A General.
    Subpart B Tables of hazardous materials, their description, proper shipping names, class, label, packaging, and other requirements.
    Subpart C Shipping papers.
    Subpart D Marking (of packages).
    Subpart E Labeling (of packages).
    Subpart F Placarding (of containers and vehicles).

Part 173 Shippers.

    Subpart A General.
    Subpart B Preparation of hazardous materials for transportation.
    Subpart C Explosives and blasting agents; definitions and preparation.
    Subpart D Flammable, combustible, and pyrophoric liquids; definitions and preparation.
    Subpart E Flammable solids, oxidizers, and organic peroxides; definitions and preparation.

    Subpart F Corrosive materials; definition and preparation.
    Subpart G Gases; definition and preparation.
    Subpart H Poisonous materials, etiologic agents, and radioactive materials; definition and preparation.
    Subpart I Radioactive materials.
    Subpart J Other regulated material; definition and preparation.
    Subpart K Other regulated material; ORM-A.
    Subpart L Other regulated material; ORM-B.
    Subpart M Other regulated material; ORM-C.
    Subpart N Other regulated material; ORM-D.
    Subpart O Other regulated material; ORM-E.

Part 174 Carriage by rail.
Part 175 Carriage by aircraft.
Part 176 Carriage by vessel.
Part 177 Carriage by public highway.
Part 178 Shipping container specifications.
Part 179 Specifications for tank cars.

Other titles of the Code of Federal Regulations also directly regulate hazardous materials transportation, including:

Title 14 Aeronautics & Space.
Chapter I Federal Aviation Administration.
Subchapter F Air Traffic & General Operating Rules.
    Part 103 (49 CFR Part 175, then Part 135) Transportation of dangerous articles and magnetized materials by aircraft.
Title 40 Protection of the Environment.

Chapter I Environmental Protection Agency.

Subchapter I Solid Wastes.
    Part 260 Hazardous waste management system (definitions).
    Part 261 Identification and listing of hazardous wastes.
    Part 262 Standards applicable to generators of hazardous waste.
    Part 263 Standards applicable to transporters of hazardous waste.

Title 46 Shipping.

Chapter I Coast Guard, DOT.

Subchapter N Dangerous Cargoes.

Part 146 Transportation or storage of explosives or other dangerous articles or substances, and combustible liquids aboard vessels.

WHO IS REGULATED BY DOT UNDER THE HAZARDOUS MATERIALS REGULATIONS?

There is no simple answer to this question and any answer that looks simple should be regarded with suspicion. It is quick and easy to say the regulations apply to shippers, carriers, and packaging manufacturers, but the exceptions and additions to this simple phrase are critical to an understanding of federal regulatory obligations.

1. SHIPPERS

The Hazardous Materials Transportation Act (HMTA), 49 U.S.C. 1801 (see Chapter 30 and Appendix D), does not define "shipper" as a term. This lack of a definition was a conscious effort to implement what might be called a functional definition of this person, that is, defining the shipper by his actions rather than by any name tags.

Use of the "shipper" name tag led to great confusion and ambiguity in the application of these regulations for years. For a time many thought that if one person was a shipper, then all others performing some aspect of the shipper's role were not shippers and, since most of them were not operating a vehicle, they were not considered carriers either. So, it seemed, a broad range of intermediate persons such as freight forwarders, contract packagers, and the like were falling through the regulatory cracks. Many people handling a large volume of hazardous materials believed they had no obligations under the DOT regulations.

To disabuse them of this idea, the general counsel of DOT prepared an interpretation of the word "shipper," which was proposed but not adopted as a definition in rule making Docket No. HM-112 (see Appendix C):

> Shipper means any person who performs any function assigned by the regulations ... to a shipper. Performance of any function by one individual as a shipper does not exclude another person from also being considered a shipper. For example, a warehouseman who presents hazardous materials to a carrier may be subject to the regulations as a shipper or as the agent of a shipper, and the person who packed, marked, classified, and labeled the shipment initially may also be considered a shipper. [39 F.R. 3037, January 24, 1974.]

This functional approach was picked up in the subsequent legislation, where Section 105 of the Hazardous Materials Transportation Act (HMTA) declares that the DOT regulations "shall be applicable to any person who ... causes to be transported or shipped, a hazardous material." Causing the transportation of a hazardous material includes more than arranging for the transportation and covers all aspects of the regulatory preparation of the material for transport. These steps encompass identification of the material as hazardous, classification of its hazard, determination of the proper DOT shipping description, selection of packaging and proper packing of the material, application of required markings and labels to that package, preparation of shipping documents and signing the shipper's certificate of compliance on those papers, and the actual tender of the package and paperwork to a carrier.

In many instances more than one person performs each of these steps, and often more than one company is involved. DOT's consideration of any one of these as a shipper does not exclude consideration of the others as shippers as well. In an enforcement action the question will narrow to what the regulation specifically required and who should have met that requirement. Often, in ambiguous circumstances, it is the signer of the shipper's certificate who is held responsible for overall compliance, especially when it cannot be determined who else should have performed the step in question. In an enforcement action, the extraction of a penalty from one person or company does not preclude extraction of a penalty from any one else, if that other person or company was involved in the particular violation.

Accordingly, any persons performing any step in the chain of causing a hazardous material to be transported will be subject to the regulations and penalties for violation of those regulations. So, the contract packer, the freight forwarder, and the pier broker will each bear some responsibility and should be aware of the scope of their potential liability. Of course, those who act on behalf of a company as its agent or under a power of attorney are the responsibility of that company, and their acts may be attributed to that company.

The law and the regulations use the term "person" to describe the applicability of the regulations. Person includes individuals, partnerships, corporations, and other business entities. The applicability of the regulations to other government agencies is less clear but is generally acknowledged if the agency is acting in a quasi-commercial capacity. Individuals are only rarely subjected to DOT enforcement action, but it is clear that such an action is authorized by the statute and criminal penalties have been imposed, especially upon drivers of motor vehicles.

In the hazardous waste regulations issued by the Environmental Protection Agency (EPA) under the Resource Conservation & Recovery Act, 42 U.S. Code 6901, there is frequent use of the term "generator." Although the terms "generator" and "shipper" are used interchangeably in transportation, the generator is the party who is responsible for a waste material coming into existence. If the material is not shipped off site, that party remains a generator but never becomes a shipper. The EPA regulations do not use the word shipper, but they do impose specific requirements upon generators who ship hazardous wastes off the site of generation. See 40 CFR Part 262.

EPA adopts the DOT regulatory system by reference in 40 CFR Part 262, and the DOT regulations have been revised to reflect hazardous waste rules and the requirement to use the Uniform Hazardous Waste Manifest. See Chapter 14 on hazardous wastes and substances.

The focus of DOT regulatory responsibility is on that person or group of persons who initially identify, classify, and describe the material to be transported under the regulations, and thereafter who select the packaging and prepare the package for eventual offer to the operator of a vehicle, aircraft, or vessel. The bulk of the initial discussion in the *Red Book* pertains to the shipper, as a reflection of DOT's focus on the shipper under the regulations. Historically, despite this direction of regulations toward the shipper, enforcement actions have tended to be directed more toward the carrier industry. It is unfair to say that if the shipper performs his functions properly, safety is assured, but it is certainly likely that if the shipper performs his functions improperly, safety is endangered.

The steps to follow in shipping any hazardous material may be summarized as follows:

1. Identify the chemistry and properties of the material.
2. Determine whether the material as shipped contains a reportable quantity (RQ) of any chemicals listed in 49 CPR 172.101, Appendix
3. Classify that material according to the DOT regulations.
4. After classification, select the proper DOT shipping name and identification number.
5. Select a packaging authorized for that shipping name.
6. Mark and label that package for shipment.
7. Prepare shipping documents to accompany the package.
8. Certify the propriety of the shipment and tender the package and certified paperwork to the carrier.
9. Determine proper placarding and supply placards to the motor carrier, or apply them to rail cars or freight containers loaded by the shipper.

The initial inquiry as to the potential applicability of federal regulations to the transportation of a material is predicated upon the great likelihood that there will be regulation at some level of every material that is dangerous. Proper classification is something that generally requires a technical background and long experience and, accordingly, is usually not one for the traffic department. The United States Supreme Court said in a hazardous materials transportation case, " ... (W)here ... dangerous or deleterious devices or products or obnoxious waste materials are involved, the probability of regulation is so great that anyone who is aware that he is in possession of them or dealing with them must be presumed to be aware of the regulation." *United States v. International Minerals & Chemical Corp.*, 402 U.S. 558 (1970).

Clearly, the manufacturer of the material is the most qualified to know and identify the dangerous properties of that material, if any, and it is the manufacturer of the product on whom the government reposes the most reliance in making these determinations. If the shipper is not the maker of the material, then it is wise for him to contact the maker to ascertain the true identity of the product and its potential hazards.

Although DOT looks to the maker of the material as the most knowledgeable in the identification of its characteristics, penalties also may apply to those who simply put the material into transportation, so it behooves the shipper who is not the maker of the commodity to obtain the maker's assurance of its properties.

2. CARRIERS

All types of carriers are within the scope of the DOT hazardous materials regulations, including private, contract, and common carriers. Included as well are private motor vehicles used for commercial purposes and every civil aircraft is subject to the regulations. The original statutory base for years of hazardous materials regulation offered a definition of carrier, but the more recent Hazardous Materials Transportation Act declines to define the term. As in the case of shippers, the

HMTA refers instead to the functions of a carrier under the regulations. Thus, the HMTA uses the phrase "person who transports" to encompass all types of carriers.

Docket No. HM-112 (see Appendix C) offered the following new definition:

> Carrier means any person who performs any function assigned by the regulations . . . to a carrier, and includes the owner, charterer, agent, master, pilot, driver, or any person in charge of a transport vehicle or vessel and other carrier employees. Consideration of one individual as a carrier does not exclude another person from also being considered a carrier for an assigned function unless the regulations specifically provide that one party is to be responsible. [39 F.R. 3036, January 24, 1974.]

The carrier definition actually adopted was much more brief:

> Carrier means a person engaged in the transportation of passengers and property by—(1) land or water, as a common, contract, or private carrier, or (2) civil aircraft. [41 F.R. 15994, April 15, 1976.]

A contract packer who operates his own vehicle in a pickup and delivery service is a carrier for DOT purposes when he is operating that motor vehicle. An air carrier who also operates motor vehicles in pickup and delivery service is a motor carrier while operating that vehicle. The fact that the air carrier is certificated as an air carrier does not remove its motor carrier liability when a truck is being operated. So, too, when an air carrier hauls its own supplies from one station to another, if those supplies are hazardous materials, the operator of the aircraft is both the shipper and the carrier, much the same as when the private carrier by highway operates a motor vehicle to haul his own property. Regulation by one agency as an economic entity does not affect regulation of the same company by DOT as a different type of entity for safety purposes.

The Federal Aviation Regulations pertaining to hazardous materials are binding upon all operators of civil (non-public) aircraft in the United States, and upon all civil aircraft of U.S. registry anywhere in the world, except aircraft of U.S. registry under lease to and operated solely by foreign nationals outside the United States.

The water carrier regulations in 46 CFR Part 176 and 49 CFR Part 176 apply to all domestic and foreign vessels when in the navigable waters of the United States, regardless of character, tonnage, size, service, whether self-propelled or not, whether arriving or departing, underway, moored, anchored, aground, or in dry-dock. With the exception of vessels carrying high explosives, the following vessels are exempt from the hazardous materials regulations:

1. Any public vessel not engaged in commercial service;
2. Any vessel constructed or converted for the principal purpose of carrying flammable or combustible liquid cargo in bulk in its own tanks (regulated under 46 U.S.C. 391a);
3. Any vessel specifically exempted from these regulations by statute, including:
   a. Vessels not exceeding fifteen gross tons when not engaged in carrying passengers for hire;

    b. Vessels used exclusively for pleasure;

    c. Vessels not exceeding five hundred gross tons while engaged in the fisheries;

    d. Tugs or towing vessels, except as to fire protection and extinguishing requirements;

    e. Cable vessels, dredges, elevator vessels, fireboats, icebreakers, pile drivers, pilot boats, welding vessels, salvage and wrecking vessels.

The water regulations apply to all shippers making shipments of hazardous materials via any vessels within the regulations.

The HMTA significantly broadened the scope of coverage of the DOT regulations through a redefinition of "commerce." The regulations apply to all shippers and carriers of hazardous materials in commerce, which is defined as follows:

> Commerce means trade, traffic, commerce, or transportation, within the jurisdiction of the United States, (A) between a place in a State and any place outside of such State, or (B) which affects trade, traffic, commerce, or transportation described in clause (A). [49 U.S. Code 1802(1).]

Translated into different terms, this means that only the hazardous materials transportation that has no effect on interstate or foreign trade, traffic, commerce, or transportation falls outside the realm of this definition. As one can imagine, not much falls outside DOT's available jurisdiction. No longer would the carrier have to be engaged in interstate or foreign commerce or have cargo on board that is destined to travel in interstate or foreign commerce. Essentially all transportation is within DOT's reach at all times, and the exceptions are almost too narrow and infrequent to matter. DOT has voluntarily limited the scope of its own activity, however, so the scope of the regulation is no broader than it was before the HMTA, i.e., covering carriers engaged in interstate or foreign commerce and shippers using the services of such carriers, except for (1) shipments of EPA-designated hazardous wastes, (2) hazardous substances, and (3) flammable cryogenic liquids in bulk.

EPA's hazardous waste regulations apply to carriers of such materials, but for some reason use the word "transporter." A hazardous waste transporter must be specifically identified by EPA or one of the approved State environmental agencies and, in addition, many States also impose more stringent certification and licensing provisions upon waste transporters.

The EPA regulations most directly applicable to carriers are in 40 CFR Part 263. The DOT regulations are adopted by reference by EPA. In addition, the carrier has specific obligations with regard to signing the Uniform Hazardous Waste Manifest and taking waste materials to the permitted hazardous waste treatment, storage or disposal facility designated by the shipper/generator. See Chapter 14 for more detail on hazardous waste regulations.

3. PACKAGING MANUFACTURERS

This term also inadequately describes the full scope of DOT's jurisdiction over the packaging industry. At the outset it will be helpful to note that in DOT parlance, "packaging" means the assembly of the container and any other outer components necessary to assure compliance with prescribed packaging requirements, while "package" means the packaging plus its hazardous materials contents as offered for transportation.

Before passage of the HMTA, the packaging industry fell outside DOT's jurisdiction. The regulations were generally binding upon shippers and carriers, but a packaging manufacturer's failure to properly build a container to DOT specifications was not an act subject to DOT penalties or enforcement action. The HMTA, therefore, as the first statement of this power, defines the applicability of DOT regulations to this industry. Section 105 of the HMTA, 49 U.S.C. 1804, states:

> The Secretary [of DOT] may issue . . . regulations for the safe transportation in commerce of hazardous materials. Such regulations shall be applicable to any person . . . who manufactures, fabricates, marks, maintains, reconditions, repairs, or tests a container which is represented, marked, certified, or sold by such person for use in the transportation in commerce of certain hazardous materials.

See Chapter 30 for a discussion of the HMTA and its detailed individual provisions.

DOT for many years prescribed construction and quality control requirements for the making of DOT specification containers. Although DOT's request to Congress for authority to regulate the packaging industry evidently had only DOT specification containers in mind, it is clear that Congress expanded this coverage to include all packaging used for hazardous materials or that is sold or otherwise represented as being proper for use in shipping hazardous materials. It is essential to note that the regulations are applicable and enforceable against the packaging industry without the government having to show that the packaging actually was in transit containing hazardous materials.

Not only is the maker of the original container covered, but the repairer, reconditioner, or person who tests a container, such as a compressed gas cylinder on its periodic retest, is covered as well. One who maintains a packaging, such as the person who maintains cargo tanks and the related valving and relief devices for such tanks, also is covered by this regulatory power. This is a broad statement of applicability, drafted by Congress with great breadth in mind.

EPA prescribes the use of DOT packaging for hazardous wastes in 40 CFR 262.30, but that agency does not exercise any jurisdiction over companies because they manufacture such packaging.

WHERE DO THE REGULATIONS APPLY?

The HMTA definition of "commerce" says that all commerce in or out of a State is covered, whether to or from another State or a foreign nation. All transportation affecting interstate commerce is included in DOT coverage if that transportation involves hazardous wastes, hazardous substances, or cryogenic flammable liquids in bulk.

In a substantive broadening of regulatory applicability, "State" was redefined in the HMTA to mean any State of the United States, the District of Columbia, the Commonwealth of Puerto Rico, the Virgin Islands, American Samoa, and Guam.

In addition, as noted above in the discussion of carriers under the regulations, all civil aircraft in the United States, including all the States as defined above, and all civil aircraft of U.S. registry wherever they may be in the world, are covered.

The only exception is for aircraft of U.S. registry that are under lease to and are operated solely by foreign nationals outside the United States.

The water regulations apply to all vessels in the navigable waters of the United States including its territories and possessions, with the sole exception of the Panama Canal Zone.

WHEN DO THE REGULATIONS BEGIN TO APPLY?

The question often is asked when transportation begins or ends as far as the reach of DOT is concerned. DOT has long defined transportation as including more than the actual movement of the cargo and even more than the time the cargo is aboard the transport vehicle or vessel. This interpretation is codified in Section 103 of the HMTA, which declares that "transportation" means any movement of property by any mode of transportation and any loading, unloading, or storage of that property incidental to its transportation.

The regulations of the Occupational Safety & Health Administration of the Department of Labor (OSHA) apply where another federal agency has not exercised its authority to prescribe requirements for the safety and health of employees in that workplace. Since the perimeter of OSHA authority is defined by DOT's exercised authority, the two agencies spent a great deal of time trying to draw the line between them. It is unquestionable that the breadth of the DOT definition of transportation begins to overrun OSHA's activity relating to in-plant safety, when it comes to examining the storage of hazardous materials incident to transportation. There is no clear answer, and one can only address the issue on a case-by-case basis.

It should be noted that the authority over the packaging industry is not predicated upon transportation either beginning or ending, but is keyed to acts that may be described as representing, marking, certifying, or selling the packaging for the purpose of shipping regulated hazardous materials in transportation. Here, therefore, the "when" question can only be answered by looking to the acts of the person to determine when such representations, marks, or certifications took place.

The "when" question also arises in considering import shipments from nations with different or nonexistent standards. Although U.S. jurisdiction over the shipment itself may not begin until the shipment is in U.S. waters, or on an aircraft of U.S. registry, regulations apply to importers of those materials before that moment arrives. Specifically, with the exception of unique provisions applicable to Canadian shipments, the importer must furnish with the order to the foreign shipper, and also to the forwarding agent at the port of entry, full and complete information as to the packing, marking, labeling, and other DOT requirements, since these requirements apply to all hazardous materials transported within the United States outside the port area.

WHY DOT ISSUES HAZARDOUS MATERIALS REGULATIONS?

Under law there must be a specifically defined purpose for any regulatory requirements. In the case of DOT's hazardous materials responsibility, that purpose is described in Section 105 of the HMTA, 49 U.S.C. 1804, which authorizes the issuance of regulations in the first place:

> The Secretary may issue . . . regulations for *the safe transportation in commerce of hazardous materials* . . . .

Such regulations may govern any *safety* aspect of the transportation of hazardous materials which the Secretary deems necessary or appropriate, including, but not limited to, the packing, repacking, handling, labeling, marking, placarding, and routing (other than with respect to pipelines) of hazardous materials, and the manufacture, fabrication, marking, maintenance, reconditioning, repairing, or testing of a package or container which is represented, marked, certified, or sold by such person for use in the transportation of certain hazardous materials. [Emphasis added.]

DOT is a safety agency and, unless its regulations can be supported on the basis of their relation to safety in transportation, those regulations fall outside DOT's lawful scope of regulatory authority. See Chapter 26, "The Administrative Procedure Act," and the provisions for judicial review of rule making actions.

Strict environmental considerations may not truly be called safety problems. So, too, economic issues such as the balance of international trade cannot serve as the basis for DOT regulation, such as those that could serve as non-tariff barriers to imported hazardous materials or packaging. The fact that a product may be hazardous in use does not mean that DOT may regulate it in transportation, if the product is in a quantity and form that does not "pose an unreasonable risk to health and safety or property when transported in commerce." (See 49 U.S. Code 1802(2).)

Development of property-protective regulations appears to be authorized to a certain extent but, in the context of the HMTA, it seems that the property to be so protected should be preserved from the type of violent destruction that potentially could be a threat to persons as well. An example might be the regulation of aluminum-corrosive materials aboard aircraft, where the destruction of aircraft structural members is not itself violent, but the resulting failure of those structural members is. In qualifying the definition of hazardous material as that which poses an unreasonable risk to property, it would appear logical to assume that mere damage to property to the extent that is adequately compensated by the loss and damage claims structure of the relationship between shippers and carriers is not the type of property damage that would justify HMTA controls.

It is important to recognize, however, that under Section 306 of Superfund, 42 U.S. Code 9656, DOT is obligated to "list and regulate" as hazardous materials in transportation any substances designated under Superfund or by EPA under its Superfund authority. Many of these materials bear an environmental risk that is not combined with a risk to human health and safety. See 49 CFR 172.101, Appendix.

CHAPTER 4

DOT-Regulated Hazardous Materials

The federal Hazardous Materials Regulations are published by DOT in 40 CFR Parts 100–189. The regulations are issued under the authority of the Hazardous Materials Transportation Act (HMTA), 49 U.S. Code 1801, 1804. They may be applied to all carriers and shippers of hazardous materials, and all makers of packaging sold for hazardous materials, moving in commerce "affecting" interstate or foreign commerce.

Section 1803 of the HMTA, describing DOT's authority to designate hazardous materials, authorizes controls over any materials including "but not limited to" explosives, radioactive materials, etiologic agents, flammable liquids or solids, combustible liquids or solids, poisons, oxidizing or corrosive materials, and compressed gases. The regulations to date have only been expanded significantly beyond those enumerated hazards to accommodate EPA-designated hazardous substances and hazardous wastes, described more fully below and in Chapter 14.

Under 49 U.S. Code 1803, in order to designate a material as hazardous, the Secretary of Transportation must find it is in a quantity and form that may pose an unreasonable risk to health and safety or property in transportation. There is no separate procedural mechanism for making findings to carry out the designation authority, i.e., to identify new materials to be brought under transportation regulatory controls. To date, to the extent designation has occurred under the HMTA, it has taken place in the informal and loosely structured dialogue of agency rulemaking. The regulations currently appearing in 49 CFR were reissued as HMTA regulations with an effective date of January 3, 1977, but the rules did not come into being at that time. They have been in place for many decades, administered firstby the Interstate Commerce Commission and then DOT under authority cntained in the Explosives and Other Dangerous Articles Act, 18 U.S. Code 831–835 (since repealed).

Neither the facts necessary for designation of new hazardous materials under the HMTA nor the quantitative criteria upon which a designation decision would be based are published, and perhaps they are not even known. DOT has developed no systematic method for risk assessment or evaluation. There is no consistent consideration of the quantity or form of materials in making a risk assessment. There is no known "level of safety" in the regulations. There is no assurance of consistency in controls, from commodity to commodity, hazard to hazard, mode to mode, quantity to quantity, or day to day.

The system, such as it is, has evolved piecemeal over the past three-quarters of a century. It is a development in which all judgment has been subjective and all requirements negotiated. To fail to recognize this is to seek more substance and rationality in the DOT regulations than exists. Today there are many inconsistencies and illogical elements to the DOT hazardous materials regulations. This is particularly true when reviewing parallel regulations implemented and administered by the Coast Guard in 46 CFR.

A greater degree of rationality has been proposed in Docket No. AM-18, May 5, 1987.

The system of regulation administered by DOT requires the *classification* of a material to provide a "handle" for the regulatory mechanism. In other words, to be subject to substantive requirements under the DOT regulations, a material must first be categorized into one of the existing class definitions of hazardous materials. DOT defines the classes and it is the shipper's burden and opportunity to describe his materials according to the most appropriate definition.

The issue of proper classification of materials having multiple hazards is addressed in 49 CFR 173.2, where a "pecking order" of classification is offered. For certain materials such as explosives and blasting agents, however, the classification function is considered so important that it is not relegated to the shipper alone, but is reserved to specified agencies or laboratories. For other materials listed by name in 49 CFR 172.101 with a "+" mark, the classification is mandated by the table.   For the most part, however, it is up to the shipper to identify his materials, to determine the correct DOT classification for them, to select the proper shipping name from the list in 49 CFR 172.101, and then to comply with all requirements applicable to that description. While DOT may request data on how a given shipper reached his conclusions, there is no obligation to file this information with any agency as a matter of routine, nor is there an obligation to have such data on file. The classification decision, which bears so heavily on the requirements related to a given material, is one of the most complex a shipper makes and is the one on which he is given little guidance by the DOT regulations.

DOT has the authority to expand and revise the classifications and has done so in recent years with the creation of the Blasting agent and Other Regulated Materials (ORM-A, B, C, D, and E) classes. While many ORM-A, B, C, and D materials had been subject to some controls in transportation, the rule changes creating the ORM classes made applicability of the federal regulations more clear.

The reader should recognize that the United Nations also has developed a similar system of hazardous materials classification, in the UN Recommendations on the Transport of Dangerous Goods. The UN system of classification has been implemented in Europe, in international air and water transportation, and in Canada. In Docket No. HM-181, DOT has proposed that the UN system be implemented, at least as an option, in domestic transportation in the United States.

Currently regulated DOT classifications and their 49 CFR definition sections are as follows:

> Explosives, Class A, B and C (173.50, 173.53, 173.88, and 173.100)
> Blasting agents (173.114a)
> Flammable and combustible liquids (173.115)
> Flammable solids (173.150)
> Oxidizers (173.151)
> Organic peroxides (173.151a)
> Corrosive materials (liquid and solid) (173.240)

Compressed gases (flammable and non-flammable) (173.300)
Poisons, Class A and B (173.326 and 173.343)
Irritating materials (173.381)
Etiologic agents (173.386)
Radioactive materials (173.389)
ORM-A, B, C, D and E (173.500)

Although not a classification provision *per se*, there is a new definition of materials toxic by inhalation that must be labeled and otherwise identified as poisons, regardless of their primary classification. See 49 CFR 173.3a.

In virtually every classification, there is a mechanism whereby hazardous materials that do not "fit" the class may be relegated to that class nonetheless. In addition, for some cases there is a method to remove the material from the class despite the fact that it does fit technically. The legality of this awkward process has never been challenged. As the various classifications are described below, those paragraphs by which DOT adapts the fit of the classification will be highlighted. In general these are little used mechanisms with almost no procedural precedent, and they are unlikely to be used extensively in the future. The above classifications will each be described in turn.

EXPLOSIVES AND BLASTING AGENTS

An *explosive* (whether Class A, B, or C) is defined as "any chemical compound, mixture, or device, the primary or common purpose of which is to function by explosion, i.e., with substantially instantaneous release of gas and heat, unless such compound, mixture, or device is otherwise specifically classified" elsewhere in the DOT regulations. The explosives classification is an oddity. It is the one DOT classification (along with the related Blasting agent class) *not* left to the discretion of the shipper of the product. The shipper of an article that may be an explosive first must submit that material to the Bureau of Explosives, a technical arm of the Association of American Railroads. This is a private, non-governmental laboratory that for over 75 years has performed this function. After testing, the Bureau of Explosives will classify the material and, if it is assigned to the explosives class, will prescribe the A, B or C designation. Under 49 CFR 173.86, the classification of commercial explosives is reserved to the Bureau of Explosives or the U.S. Bureau of Mines of the Department of the Interior. Explosives developed for the Energy Research and Development Administration or the military may be classified and approved by the Department of Energy or the Department of Defense. While there is no shipper discretion to classify explosives, the shipper does have the obligation to know that there are materials described in 49 CFR 173.51 that are *forbidden* in transportation. These materials are so unstable that they may not be transported by any mode at any time.

Class A explosives are defined by eight types and by such generic descriptions as "ammunition for cannon," "explosive projectiles," "grenades," and the like. The type definitions of Class A explosives relate to the sensitivity of the material stated in terms of its sensitivity to testing with varying size blasting caps, or to dropping or exposure to heat.

Class B explosives are those that "deflagrate," or function by rapid combustion rather than by detonation. These include explosive devices such as special fireworks, flash powders, some pyrotechnic signal devices, and liquid or solid propellant explosives. Here again, materials in the Class B explosives classification are assigned to that class by the Bureau of Explosives, the Bureau of Mines, or the Departments of Energy or Defense. They fit within specific descriptions given in 49 CFR 173.88, which purports to define this class. This in fact is not a definition section but, rather, an assignment of shipping names to articles categorized as Class B explosives.

Class C explosives are the least hazardous materials in the explosives classification. These are defined in 49 CFR 173.100 as manufactured articles that contain Class A or Class B explosives or both as components, but in restricted quantities. Class C explosives include fireworks and small arms ammunition. Section 173.100 of 49 CFR, while nominally a definition section, like the sections defining Classes A and B is made up of descriptions of materials assigned to these classes by the Bureau of Explosives rather than definitions, *per se.*

Blasting agents have been those designated by DOT as a new classification. The DOT blasting agent rule making gave two reasons for the new class. The creation of a new class contributes to increased safety in transportation because "some materials now shipped as nitro carbo nitrates (oxidizing materials) present a potential explosive hazard. Establishing a blasting agent class will bring the DOT regulations into closer conformity with Mining Enforcement and Safety Administration (MESA) and Bureau of Alcohol, Tobacco and Firearms (BATF) regulations which now incorporate definitions of blasting agents." (43 Fed. Reg. 57898, Dec. 11, 1978.) With creation of the new class there were simultaneous creations of new shipping names, labels and placarding requirements for blasting agents.

A blasting agent is defined in 49 CFR 173.114a as, "a material designed for blasting which has been tested in accordance with (prescribed tests in this section) and found to be so insensitive that there is very little probability of accidental initiation to explosion or of transition from deflagration to detonation." The definition includes tests that must be passed, including a blasting cap sensitivity test, a differential thermal analysis test, a thermal stability test, an electrostatic sensitivity test, an impact sensitivity test, and a fire test. After performance of the tests the manufacturer in addition must receive DOT approval for assignment of the blasting agent classification, unless specified laboratories or agencies have written that there are no significant differences in hazard characteristics relative to a blasting agent already approved by DOT. The examination and tentative classification of new blasting agents must be done by the Bureau of Explosives, the Bureau of Mines, or the Departments of Energy or Defense.

While this classification is tied to the explosives group as opposed to the oxidizer group, it is not an explosive A, B or C, but is an independent classification. The label on the blasting agents looks similar in color to an explosives label, but does not carry the "exploding bomb" UN symbol. DOT said in its final amendment that no symbol is displayed "due to the fact that blasting agents present a much lower level of hazard in terms of their detonation potential than Class A and B explosives for which an exploding bomb is displayed on the label."

As noted above, for many years the Bureau of Explosives had the primary responsibility for assigning articles to the explosives class. The work of the Bureau of Explosives has evolved since the 19th Century and has been criticized for a lack of procedural consistency or quantification. Much of the classification decision is based upon the experience of the person running the test and his subjective reaction to the performance of materials in the test, relative to the performance of other materials in other tests. The Department of Transportation has long had a low priority project to revise the explosives classification to more quantitatively and consistently define materials in that class, but the conclusion of this rulemaking may be years away. It also is unclear whether the Bureau of Explosives will continue in this role due to high liability insurance considerations.

Another factor that must be noted about this class is that the classification is inextricably linked to the "primary or common purpose" of the material to function by explosion, as stated in the definition. Literally, this means that a material which may explode, but does not have the *purpose* of exploding, does not fit the DOT definition. While this factor has been raised in criticism of the definition, it has never been judicially resolved.

It should be noted that the "reactive" category in the EPA regulations under Section 3001 of the Resource Conservation and Recovery Act will include many explosives that are discarded, but that category does not overlap the DOT explosives classification exactly.

FLAMMABLE AND COMBUSTIBLE LIQUIDS

Section 173.115 of 49 CFR declares that a DOT-regulated *flammable liquid* is one having a closed-cup flash point below 37.8°C (100°F), with certain specific exceptions. The exceptions basically are those which fall within the flammable liquid class by the closed-cup flash point tester, but by logic do not fit it. These include gases and materials having minor amounts of certain ingredients that give extremely low or high flash point readings in the closed tester, without regard to their true fire hazard.

A *combustible liquid* is one posing less of a flammability hazard, having a closed-cup flash point at or above 37.8°C (100°F) and below 93.3°C (200°F). The combustible liquid class also has certain exceptions based upon illogical results of test methods. For example, an aqueous solution containing 24% or less alcohol by volume is "considered to have" a flash point no less than 37.8°C (100°F), i.e., is a combustible liquid, if the remainder of the solution does not meet the definition of another hazardous material. Such liquids are assigned to the combustible liquid class despite the fact they otherwise meet the *flammable* liquid definition, because of misleading test results caused by the closed-cup test method when applied to alcohol/water mixtures.

"*Pyroforic*" liquids also are regulated by DOT and are described in 49 CFR 173.115, the class definition section, although they do not form a separate classification. For DOT purposes a pyroforic liquid is any that ignites spontaneously in dry or moist air at or below 54.5°C (130°F).

The flammable and combustible liquid classes were revised by DOT over a period of several years in the 1970's. The primary purposes of the redefinition of the flammable and combustible liquid classes were: (1) to raise the definition limit for flammable liquids from 80°F to 100°F, and (2) to change the basic methodology from the open to the closed-cup tester. The first change was *not* justified on the basis of transportation safety but, rather, on administrative consistency with similar definitions under federal Occupational Safety and Health (OSHA) Regulations (29 CFR 1910.106). A similar purpose was cited above, for creation of the new "Blasting agent" class. The primary rationale for changing to the closed-cup tester was a parallel drawn between a closed transportation vehicle and the closed-cup tester, in the trapping of fumes from a liquid that might not give a reading of flammability in an open tester. The only other major federal definition of flammability at that time was administered by the Food and Drug Administration but was prescribed by statute at 80°F.

The 100–200°F combustible liquid flash point definition was taken from OSHA regulations as they reflected National Fire Protection Association (NFPA) standards. This is a classification of materials generally not subjected to substantive DOT regulations. *No* regulation applies to unit quantities smaller than 110 gallons (unless these materials are wastes or hazardous substances), and even in bulk land transportation the requirements are only of a communications nature—shipping papers and vehicle placards. The primary concern with creation of the combustible liquid class was proper notification to fire personnel when a bulk unit of such materials was involved in an accident. Too many very burnable materials had been moving in tank trucks bearing a "Nonflammable" placard.

A suggestion in the flammable liquids rulemaking, that the class be defined by an upper flash point limit of 140°F like the U.N. definition, was not adopted because there was no contention that liquids in this flash point range pose a safety hazard in transportation. As noted earlier, the shift from 80° to 100°F also was unsupported on safety grounds, but was adopted because of the consistency thereby derived with the Occupational Safety and Health Regulations. EPA's adoption of an upper limit of 140°F for its ignitable waste definition is not based upon transportation concerns, but upon disposal site considerations. Thus, an EPA ignitable waste with a flash point of 120°F will not be regulated as a flammable liquid under DOT, but will be consigned to the combustible liquids class.

Section 173.115(g) of the DOT regulations is one of the unique provisions mentioned earlier, providing DOT with a designation flexibility. It reads: "If experience or other data indicate that the hazard of a material is greater or less than indicated by the criteria specified (in the flammable liquid definition), the Department may revise its classification or make the material subject to the requirements" of the hazardous materials regulations. A specific product of concern to DOT at the time of the flammable liquids rulemaking was nitromethane, which does not meet the flash point definition of flammable liquid and yet had been involved in three major transportation explosions. When the flammable liquid class was redefined, this paragraph was inserted to provide a published mechanism by which a material not otherwise fitting the class could be forced into that class based upon experience or other data. Presumably a material could also be removed from the class, but to date this has not been done via this paragraph.

FLAMMABLE SOLIDS

The flammable solid class is described only generally in the DOT regulations. The term includes, "any solid material, other than one classed as an explosive, which, under conditions normally incident to transportation is liable to cause fires through friction, retained heat from manufacturing or processing, or which can be ignited readily and when ignited burns so vigorously and persistently as to create a serious transportation hazard." This class includes spontaneously combustible solids and water-reactive materials.

As can be seen from this definition, there are several subjective and amorphous qualifications relevant to placement of a material in this class. Unlike the explosives classification, the assignment of a material to this class is entirely at the discretion of the *shipper* of the material. Of course, however, if the material is classed by the

Bureau of Explosives or the Bureau of Mines as an explosive, it definitely cannot also be in the flammable solid class. Under DOT interpretation, "conditions normally incident to transportation" do not include accident circumstances, but do include the vibration and abuse a material will receive in usual cargo transportation. In air transport these conditions also include variations in pressure, temperature, and humidity due to rapid changes in altitude. In land transportation they include all environmental conditions likely to be encountered, including temperatures up to 130°F.

There are three alternative elements to the definition. The material under normal conditions must pose the danger of *causing* fires through (1) friction, (2) retained heat, or (3) ready ignition accompanied by vigorous and persistent burning. *None* of these elements of the definition has been quantified.

The specific inclusion of water-reactive materials in the flammable solid class is caused by the DOT system of requiring a material to fall within an existing classification before it may be regulated. Water-reactive materials are defined in 49 CFR 171.8 as those which through interaction with water are likely to become spontaneously flammable or to give off flammable or *toxic* gases in dangerous quantities. Although a solid material that gives off a toxic gas when wet does not pose a flammability hazard, for administrative convenience it has been assigned to the flammable solid classification. An additional rationale for this assignment is that the most likely source of significant quantities of water in transportation is a fire hose, and warning fire personnel that the material becomes more hazardous when wet is the primary purpose of the regulation of water reactive materials. Through DOT's four-digit "UN/NA" hazard information system, emergency response personnel are advised of the specific hazards of the material (flammability or toxicity) in an Emergency Response Guide, and in the interim they will not have sprayed it with water.

The flammable solid classification, because of its subjective nature and its nonspecific, unquantified terms, has been one of the most difficult to administer for DOT and shippers. It is presumed generally that a shipper in doubt will classify his material as a flammable solid if for no other reason than to avoid the civil liability that might attach in the event of an accident injuring third parties.

Many people have sought to quantify this definition for their own purposes by having testing done by various independent laboratories. For several years, DOT has had a low priority project underway to quantify this definition, and this class is the subject of continuing discussions in the UN, but implementation of changes is years away.

OXIDIZERS

DOT defines an oxidizer as a chlorate, permanganate, inorganic peroxide or nitrate that yields oxygen readily to stimulate the combustion of organic matter. Here again, while the definition is somewhat more specific than the flammable solid definition, the terms still are subject to varying interpretation. For example, it is unclear what level the oxygen yield must be to stimulate the combustion of organic matter to pose a danger in transportation.

With both the flammable solid and oxidizer classes, many materials recognized by DOT as being within these classifications are listed as such by name in 49 CFR 172.101. When a material is not so listed, the shipper must determine on his own data and at his peril whether the definition and classification fit his material.

ORGANIC PEROXIDES

DOT defines organic peroxides in 49 CFR 173.151(a). The definition used to be linked directly with that of flammable solids and oxidizers but later was broken out as a specific classification to recognize the unique properties of these materials. An organic peroxide is, "An organic compound containing the bivalent –O–O– structure which may be considered a derivative of hydrogen peroxide where one or more of the hydrogen atoms have been replaced by organic radicals." Exceptions to this definition are materials that have been classed as explosives, materials forbidden in transportation because of their instability, and materials meeting the definition of flammable liquids not specifically assigned to the organic peroxide class in 49 CFR 172.101.

In this classification, there is another provision giving DOT flexibility in their designation function. This one is in 49 CFR 173.151(a)(4), saying that any material meeting the definition of an organic peroxide must be so classed unless "according to data on file it has been determined that the material does not present a hazard in transportation." This paragraph, unlike the one discussed above with regard to flammable liquids, does not give the agency authority to put materials *into* the class that do not otherwise fit. It only provides a mechanism to take materials *out* of the class. The phrasing of this is also different from many of the other paragraphs on the same subject, since it only requires data to be on file at DOT and does not appear to require an agency response or finding to remove the material from the classification.

CORROSIVE MATERIALS

Corrosive liquids and solids are defined for DOT purposes in 49 CFR 173.240. A corrosive material is a liquid or solid that "causes a visible destruction or irreversible alterations in human skin tissue at the site of contact, or in the case of leakage from its packaging a liquid that has a severe corrosion rate on steel." These general descriptions are quantified by a specific 4-hour rabbit skin test and a corrosion rate test on steel giving results in inches per year (0.250 IPY).

This classification has undergone substantive revision. This docket, concluded in the early 1970s, examined the quantification of the definition in light of specific tests to be run. The proceeding took the form of advanced notices of proposed rulemaking and several notices followed by an amendment and several corrective amendments, each making significant adjustments in the definition.

The rabbit skin test is described in Appendix A to Part 173 of 49 CFR. It is similar to, but not the same as, a Food and Drug Administration (now Consumer Product Safety Commission) test methodology for skin irritants. The prominent factor in the DOT test is that the skin need not be abraded, and consequently many materials that might be found regulated under the other tests are not regulated by DOT. Tissue at the site of contact must be destroyed or changed irreversibly, and this has been clarified by DOT to mean actual necrosis of the tissue. Reddening or soreness or irritation that will heal does not fall within this definition.

Under the rabbit skin test an exposure period of four hours or less is prescribed. Longer periods were proposed in the early rulemaking, but it was determined by the agency that the likelihood of a transportation worker keeping spilled material on his hands without washing for a period in excess of four hours was so small as to not warrant regulation. The four-hour period is keyed to the typical transportation worker's day, where it is likely the worker will wash midway through the day prior to eating.

In the process of quantification of the corrosive classification, DOT considered and rejected a pH test as being a criterion not sufficiently broad to cover all materials that have a destructive effect on tissue. Notably, pH criteria exclude purely organic chemicals (such as amines), and also do not relate well on the basic (as opposed to acidic) side of the scale to materials corrosion. The pH characteristic has been adopted by EPA, however, in defining corrosive hazardous waste liquids in 40 CFR Part 261.

For both DOT and EPA purposes, the corrosion rate on steel is determined by a recognized test of the National Association of Corrosion Engineers (NACE). DOT also proposed that materials corrosive to aluminum be regulated in land transportation but, for a variety of reasons including the particular concern with air transportation, the aluminum corrosive qualification was not adopted for all modes.

The corrosive materials classification, as many of those already discussed, includes a flexibility paragraph reading: "If human experience or other data indicate that the hazard of a material is greater or less than indicated by the results of the tests specified in [49 CFR 173.240(a)], the Department may revise its classification or make the material subject to the requirements" of the regulations. This paragraph, similar to that found under the flammable liquids classification, allows DOT on the basis of data other than prescribed test results to put a material in or take it out of the classification. Unlike the comparable organic peroxide paragraph, the statement clearly requires some action on the part of DOT in response to the submission of information on human experience, or other data.

Regulation of corrosives, particularly those that are not damaging to human tissue, is one of the few areas where DOT clearly controls materials because of a threat to *property*. Historically it is understood that the concern with property in transportation focuses on damage to transportation equipment and on other packaged freight moving with the regulated commodity. There is no historical record that the transportation regulatory concern with corrosive materials has ever extended to damage to the environment, i.e., damage to natural resources that could not be considered private property.

Even in its regulation of property-corrosive materials, DOT's primary concern has been the impact on *transportation safety* that might be caused through destruction of adjacent packaging containing other hazardous materials, or the destruction of a structural element of a transport vehicle leading to a transportation safety hazard. Mere damage to adjacent *unregulated* articles is a matter more appropriately left to commercial loss and damage claims procedures and has not received a great deal of consideration in the evolution of DOT's corrosive classification. Without question, development of the corrosive materials classification occurred without any consideration of impact on the environment caused by the spill of any materials.

COMPRESSED GASES (FLAMMABLE AND NONFLAMMABLE)

DOT defines *compressed gas* in 49 CFR 173.300. The term includes "any material or mixture having in the container an absolute pressure exceeding 40 p.s.i. at 70°F, or regardless of the pressure at 70°F, having an absolute pressure exceeding 104 p.s.i. at 130°F; or any liquid flammable material having a vapor pressure exceeding 40 p.s.i. absolute at 100°F as determined by ASTM Test T-323."

The concern addressed historically by the DOT regulations is safety in transportation. As a result, the gas definition is stated in terms of pressure that might be encountered under conditions normally incident to transportation. As noted earlier, this includes temperatures up to 130°F. Within this definition are materials otherwise not considered gases, but flammable liquids having a high vapor pressure when exposed to temperatures likely to be encountered in transportation. Of course, any flammable fluid not having a vapor pressure sufficiently high to meet the gas definition is regulated under the flammable liquids class.

Despite the fact that the section defines "compressed gas," this term is not a classification. A gas is classed either as a "Flammable gas" or "Nonflammable gas." Flammable gases are defined. All materials meeting the definition of a gas and that are not flammable are considered Nonflammable gases (with the exception of some toxic gases in the Class A Poison classification). See important changes to the gas classifications in Docket HM-181, May 5, 1987.

The *flammable gas* determination relates to the potential for forming a flammable mixture in air. For solutions of materials and gas, such as in aerosols, any one of three mandatory tests (described in 49 CFR 173.300(b)) might lead to a flammable gas classification: (1) having the material project a flame more than 18 inches beyond an ignition source with the container valve open fully, or with a flame flashing back and burning at the valve with any degree of valve opening; (2) having any significant propogation of flame away from the ignition source in the Bureau of Explosives open drum apparatus; or (3) having any explosion of the vapor air mixture using the Bureau of Explosives closed-drum apparatus. These tests and apparatus are described in documents available from the Bureau of Explosives.

The only significant difficulty in implementation of this definition relates to a lack of data involving vapor pressure curves of flammable liquids (see Appendix A) and other data on the properties of gases at various temperatures. While the classification is within the discretion of the shipper, the distinction between flammable and nonflammable gases makes heavy use of Bureau of Explosives testing methods.

POISONS

Materials regulated by DOT as poisons are those posing a very significant threat to human health when transported. Materials that might be regulated by other agencies because of toxic properties in the home or in the workplace are not necessarily regulated by DOT, since the acute hazard to the industrial transportation worker may not be present. Within the poison definition there are three classes: Poison A, Poison B, and Irritating materials. Under a non-classification definition, DOT also recognizes the inhalation toxicity of materials, in 49 CFR 173.3a.

Poison A is defined in 49 CFR 173.326 and, again, the definition is not really a definition at all. The section declares that Class A poisons are "Poisonous gases or liquids of such nature that a very small amount of the gas, or vapor of the liquid, mixed with air is dangerous to life." Like many of the DOT definition sections, the true definition is given by example rather than by quantitative criteria. In 49 CFR 173.326 there are eleven specifically named materials that are designated poisons, Class A.

Poisons that pose a lesser hazard are in the *Poison B* group, defined in 49 CFR 173.343 as, "Those substances, liquid or solid (including pastes and semi-solids), other than Class A poisons or irritating materials, which are known to be so toxic to man as to afford a hazard to health during transportation; or which, in the absence of adequate data on human toxicity, are presumed to be toxic to man because they fall within any one of (three prescribed) categories when tested on laboratory animals." The DOT definition thereafter gives the following quantitative test methodologies for determining oral, inhalation, and skin absorption toxicity:

Oral Toxicity. Those which produce death within 48 hours in half or more than half of a group of 10 or more white laboratory rats weighing 200 to 300 grams at a single dose of 50 milligrams or less per kilogram of body weight, when administered orally.

Toxicity on inhalation. Those which produce death within 48 hours in half or more than half of a group of 10 or more white laboratory rats weighing 200 to 300 grams, when inhaled continuously for a period of one hour or less at a concentration of 2 milligrams or less per liter of vapor, mist, or dust, provided such concentration is likely to be encountered by man when the chemical product is used in any reasonable foreseeable manner.

Toxicity by skin absorption. Those which produce death within 48 hours in half or more than half of a group of 10 or more rabbits tested at a dosage of 200 milligrams or less per kilogram body weight, when administered by continuous contact with the bare skin for 24 hours or less.

Once again, a paragraph is inserted in this definition to give regulatory flexibility to the Class B poison classification. This one reads: "The foregoing categories (of Class B poison descriptions) shall not apply if the physical characteristics or the probable hazards to humans as shown by experience indicate a substance will not cause serious illness or death. Neither the display of danger or warning labels pertaining to use nor the toxicity tests set forth above shall prejudice or inhibit the exemption of any substances from the provisions" of the DOT regulations. An example of material that may be relieved from regulation despite its technical inclusion within the Class B poison classification is one shipped in units so small that a hazardous dose for the most foreseeable human exposure never is present in a single shipment. In this instance, compliance of the full scope of regulations has been considered unwarranted, and regulation is not applied.

In contrast to similar paragraphs having to do with regulatory flexibility, this paragraph does not appear to require the filing of any material with DOT, nor any form of DOT response. This paragraph gives discretion to the shipper of the material to determine whether his product is regulated or not, regardless of test results. Due to potential for civil liability, however, it is not believed that this mechanism has been used frequently by shippers of toxic materials without filing supportive documents with the agency.

An *irritating material* is defined in 49 CFR 173.381 as "A liquid or solid substance which upon contact with fire or when exposed to air gives off dangerous or intensely irritating fumes," such as certain specifically named materials but not including any Class A poisons. This is a very limited class dealing primarily with tear gas, chemical ammunition, and other materials infrequently encountered in transportation.

After the tragic accident in Bhopal, India, in which the release of methyl isocyanate (classed by DOT as a flammable liquid) killed and injured thousands of people, DOT adopted new requirements on the *inhalation toxicity* of materials. The emphasis was on those materials that have high vapor pressures and that are likely to have a rapid and wide impact in the event of a spill. The new rules were placed in 49 CFR 173.3a, and define materials "having a saturated vapor concentration at 20°C (68°F) equal to or greater than ten times its LC_{50} (vapor) value if the LC_{50} value is 1000 parts per million (ppm) or less."

Inhalation toxicity can be determined by animal testing, existing animal data, or in some circumstances by calculation. For purposes of determining inhalation toxicity, 49 CFR 173.3a(c) says:

(1) LC_{50} means the concentration of vapor that, when administered by continuous inhalation of both male and female young albino rats for one hour, is most likely to cause death within 14 days to one half of the animals tested. The result is expressed in milliliters per cubic meter of air (ppm). Wherever practicable, the test should be conducted in accordance with the procedure described in the Organization for Economic Cooperation and Development (OECD) for Acute Inhalation Toxicity except that the periods of exposure shall be one hour instead of four hours.

(2) Saturated vapor concentration (SVC) means the concentration of vapor at equilibrium with the liquid phase at 20°C (68°F) and standard atmospheric pressure expressed in milliliters per cubic meter (expressed in ppm). This concentration may be calculated from the vapor pressure (VP) of the liquid at 20°C (68°F). The general formula is the vapor pressure divided by the standard atmospheric pressure and multiplied by a million. If the vapor pressure is expressed in millimeters (mm) of mercury the calculation would be

$$\frac{VP \text{ (in mm Hg)}}{760} \times 10^6 = SVC \text{ (in ppm)}$$

(3) If LC_{50} data are available based on other than a one hour exposure, a factor may be used to determine an acceptable one hour value for the purposes of this section. If the only value available is for a 4 hour exposure, that value is multiplied by 2. This method of estimating a LC_{50} value may not be used when a material causes death by direct pulmonary effect, i.e., by destruction of lung tissue as opposed to systemic poisoning. For these corrosive poisons, the exposure period must be one hour.

(4) LC_{50} data published in scientific and technical handbooks, journals and texts may be used in place of new tests using animals to determine

compliance with this section. Where different values for the LC_{50} of a material are found, the most credible value must be used. The Registry of Toxic Effects of Chemical Substances (RTECS) published by NIOSH is a recommended source of these data.

(5) Limit test. As an alternative to determine a LC_{50} value, the following procedure may be used to determine whether a material is subject to this section: The saturated vapor concentration at 20°C (68°F) is determined as in subparagraph (2), above. This then is divided by 10 and the resulting concentration used to test 10 animals in accordance with the OECD procedure noted in subparagraph (1) of this section, with a one hour exposure period. If 5 or more animals die during the 14 day observation period, the material is subject to this section. For example: If a liquid has a saturated vapor concentration of 500 ppm at 20°C, the concentration used in the test outlined in this paragraph would be 50 ppm.

The 14-day observation period following 1 hour exposure was not put into effect with the rest of the new rules on inhalation toxicity, but was postponed until January 1, 1988. Until then data based upon only a 2-day observation period may be used. It is likely that two things will result near the January 1, 1988 date—most people will not have developed the data, and those who have will find a substantially wider population of materials affected by this rule. In other words, continuing controversy is expected to surround this subject.

A key factor to note is that the test animals need not die of systemic poisoning, but can die of any cause during the observation period. Thus, many corrosive products which cause tissue damage may be impacted.

Although this definition was drawn from the international one adopted by the International Civil Aviation Organization (ICAO), and later by the International Maritime Organization (IMO), the full implementation of it is most likely to occur first in the United States because the U.S. rule applies to all modes of transportation.

ETIOLOGIC AGENTS

An etiologic or disease-producing agent is defined by DOT as "A viable microorganism, or its toxin, which causes or may cause human disease." This definition is limited to specifically listed agents identified by the Department of Health and Human Services in 42 CFR 72.25(c). Etiologic agents were not regulated in transportation prior to 1960, when amendments to the Explosives & Other Dangerous Articles Act (18 U.S. Code 831-835) specifically authorized direct regulation of such materials. DOT regulation of etiologic agents to date has not been substantive, but health hazards generally are a likely area for regulatory expansion in the future, either by DOT directly or by their reference to some other regulatory agency's controls.

While the etiologic agent definition includes many articles, the exceptions to substantive regulation tend to limit the scope of applicability. Diagnostic specimens, biological products, and cultures of etiologic agents of 50 ml or less in a single outside package are fully relieved from DOT regulation, but this 50 ml provision does not exist in air transportation under ICAO/IATA.

RADIOACTIVE MATERIALS

In 49 CFR 173.401 of the DOT regulations, "radioactive material" is defined as "any material having a specific activity greater than 0.002 microcuries per gram of material (μCi/g)." Within this classification "fissile radioactive materials" are defined to include plutonium-238, plutonium-239, plutonium-241, uranium-233, and uranium-235, and any material containing any of these elements. Fissile radioactive materials are grouped into Classes I, II, and III, with Class III posing the greatest hazard in transportation.

"Low specific activity material" is defined as meaning any of a specified list of types of radioactive materials. "Normal form radioactive material" means radioactive materials that are not "special form." Special form radioactive materials are those which, are either a single solid piece or contained in a sealed capsule that can be opened only by destroying the capsule, and the piece or capsule has at least one dimension of 5 mm (0.197 inch) or more, and it satisfies specific impact, percussion, bending, and heat tests prescribed in 49 CFR 173.469.

The DOT rules on radioactive materials are essentially the same as those adopted by the International Atomic Energy Agency (IAEA). The definitions, other than the bare definition of radioactive materials quoted above, take the form of specifically named materials or activity levels.

Prior to amendment of the Explosives and Other Dangerous Articles Act in 1960, the Interstate Commerce Commission (now DOT) regulated radioactive materials as Class D poisons. The poison classification did not smoothly fit the nature of the hazard and authority was granted by Congress to clearly and directly regulate radioactive materials in transit.

The authority of the Department of Transportation significantly overlaps that of the Nuclear Regulatory Commission (NRC). This overlap is addressed in a memorandum of understanding between the two agencies. In general, while direct regulatory authority exists within DOT to govern these materials in transit, the function of designating these materials is performed by nuclear experts in NRC and in international regulatory bodies.

The regulation of radioactive materials is one of the few examples of DOT concern extending beyond the threat of immediate or acute damage from materials in transit. While long-term health effects are not discussed in the DOT regulations, clearly these effects have been considered by the technical personnel in NRC and IAEA who have defined many of the terms and developed the regulatory concepts.

OTHER REGULATED MATERIAL (ORM)

An Other Regulated Material is one that is deemed to pose some hazard in transportation, but that does not meet the definition of any other classification of hazardous material, and is specified in 49 CFR 172.101 as an ORM material or that possesses one or more of the characteristics described in the ORM-A, B, C, D, or E classifications.

Most of the ORM classifications were created in 1976. The classifications were to accommodate the historical nature of the DOT system, which requires a class before materials may be regulated. Because all other classes are devoted to specific *general* hazards, there was a need for miscellaneous hazard classes. For the most part, these hazards are specific to given modes of transportation. In other words, a material that is harmless in one mode but dangerous in another, such as an aluminum-corrosive liquid, is regulated in an ORM class so the regulation can be focused on the mode in question. Being a miscellaneous hazard aggregation, the ORM classifications are readily susceptible to expansion, as evidenced by creation of the ORM-E class.

The suffix letters of the ORM classes do not convey any meaning, but are merely in alphabetical sequence. In other words, the ORM-E class having to do with environmental hazards has no correlation with the fact that "environment" begins with the letter "e". It was merely the group next in line after ORM-D.

ORM-A is a material which has "an anesthetic, irritating, noxious, toxic, or other similar property which can cause extreme annoyance or discomfort to passengers and crew in the event of leakage during transportation." These materials are particularly hazardous in a cargo transportation unit that also encloses people, passengers or crew, such as a vessel or an aircraft. ORM-A materials, therefore, are almost never subject to regulation in truck or rail transport, unless they are being regulated because of their *environmental* impact as hazardous substances, or as hazardous wastes defined under Section 3001 of RCRA. See Chapter 14 on hazardous substances and hazardous wastes.

The majority of materials in the ORM-A classification would impair breathing or vision and are taken from a list of materials regulated for years in international air transportation by IATA (International Air Transport Association). Under the IATA regulations, these materials were called ORA (Other Regulated Articles). Some materials formerly listed by the Coast Guard as "hazardous articles" also have come into the ORM-A class. DOT examined all such IATA and Coast Guard regulated materials for their hazard potential in air or water transportation. Not all listings were adopted by DOT.

Materials included in the ORM-A classification are specifically listed by name in 49 CFR 173.605 through 173.655, or fall within an "ORM-A, n.o.s." shipping description. "ORM-A, n.o.s.," items are not specifically listed but pose the type of hazard described in the basic ORM-A definition. The ORM-A definition is highly subjective and depends heavily on the judgment and caution of the shipper rather than on quantitative criteria.

ORM-B is defined in 49 CFR 173.500 as "a material (including a solid, when wet with water) capable of causing significant damage to a transport vehicle from leakage during transportation." Materials in the ORM-B class either are specifically designated by chemical name in 49 CFR 172.101, or fall within the class based upon the results of a given NACE aluminum-corrosion test. If a *liquid* has a corrosion rate in excess of 0.250 inches per year on aluminum under this test, and exhibits no other regulated hazards, it falls within the ORM-B classification. This classification is limited almost exclusively to substantive restrictions in air transportation, and is keyed directly to the potential to damage to the aircraft if the material were to spill and attack aluminum or steel structural members. Calcium oxide (quicklime) is the only ORM-B regulated in water transportation. While the concern in this designation is protection of *property*, it is a property-protective concern linked directly to transportation safety, since destruction of any structural member of an aircraft has an obvious safety impact.

An *ORM-C* is "a material which has other inherent characteristics not described as an ORM-A or ORM-B, but which make it unsuitable for shipment, unless properly identified and prepared for transportation." Each ORM-C material is specifically named in 49 CFR 172.101. This classification does not have an n.o.s. shipping name. To a large extent, these specifically named materials had been regulated under Coast Guard authority under the former term "hazardous articles." Understandable confusion between the terms "hazardous materials" and "hazardous articles" was one of the reasons for changing the name of the latter.

One of the purposes of DOT's major rule making completed in 1976 (Docket No. HM-112) was to bring air, water, and land transportation regulations into a single volume of the Code of Federal Regulations. The air rules had appeared in Title 14, the water rules in Title 46, and the rail and highway rules in Title 49 CFR. Upon consolidation of all modes into Title 49, with certain exceptions air-regulated materials became ORM-A, and water-regulated materials became ORM-B and ORM-C.

The hazard common to most ORM-C entries is spontaneous combustion. In other words, these are solid materials such as vegetable fibers, tankage, metal borings, and oiled materials that have a strong potential for spontaneous heating and ignition, especially when enclosed in a vessel in bulk in the presence of humidity. In other modes of transportation, unit quantities are smaller and the environment is different and does not warrant the regulation of ORM-C materials.

Generally speaking these materials are regulated only in the water mode. Exceptions are asbestos, calcium cyanamide, life rafts, lithium batteries, and motor vehicles.

Another exception to these modal limitations occurs if an ORM-C material *also* is a hazardous substance or hazardous waste under RCRA definitions. In that case, it is regulated in all modes of transportation because of its environmental hazard, but for consistency it remains regulated under the ORM-C classification.

ORM-D is a novel entry in the DOT regulations. This classification did not, in fact, have a predecessor class but was a new grouping of materials created in the Docket HM-112 rulemaking. An ORM-D is defined in 49 CFR 173.500 as, "A material such as a consumer commodity which, though otherwise subject to the regulations of this subchapter, presents a limited hazard during transportation due to its form, quantity, and packaging." In other words, products in the ORM-D class are those which fall within definitions of *other* general DOT classifications, but do not warrant substantive regulation. Many people contend that these materials do not warrant any regulation in transportation.

The only articles currently in the new ORM-D classification are consumer commodities and small arms ammunition. "Consumer commodity" is defined in 49 CFR 171.8 as, "A material that is packaged and distributed in a form intended or suitable for sale through retail sales agencies or instrumentalities for consumption by individuals for purposes of personal care or household use. This term also includes drugs and medicines."

Not every consumer product falls within the ORM-D classification. The consumer product must be one for which general exceptions are permitted as listed in 49 CFR 172.101. It also must be within packaging limitations for the consumer market as well as specific quantity limits and packaging types prescribed in 49 CFR 173.1200. See 49 CFR 173.1201 for limitations on the kinds of ammunition that may be reclassed as ORM-D.

In determining whether an article may be shipped as a "consumer commodity" in the ORM-D class, one must first look to its hazard, for example, flammability. If the article is (1) a flammable liquid, (2) a consumer commodity (as defined in 49

CFR 171.8), (3) is granted exceptions due to its small size in various "limited quantity" paragraphs of the regulations (e.g., 173.118(d)), and (4) is in specifically described packages less than 65 lbs. gross weight, it may be *re*named "Consumer commodity" and may be *re*classified from the Flammable liquid class into the ORM-D class.

For small arms ammunition, the article must first fit that shipping description and be classed as a Class C explosive. If it meets the further descriptive and packaging limitations of 49 CFR 173.1201, it may be *re*classed as ORM-D, but it retains the original shipping name "Small arms ammunition."

ORM-D in effect is a secondary classification for materials that primarily meet basic definitions but, due to small size packaging and regulation imposed by other agencies (such as the Consumer Product Safety Commission), need not be subjected to substantive regulation DOT in transportation. Most general DOT regulations do *not* apply to ORM-D materials and, to the extent there is regulation, it is distinctly related to air transportation, not other modes.

HAZARDOUS WASTES, HAZARDOUS SUBSTANCES, AND ORM-E

Hazardous wastes are defined by EPA under Section 3001 of the Resource and Conservation Recovery Act (RCRA), 42 U.S. Code 6901, *et seq.* EPA definitions implementing Section 3001 appear in 40 CFR Part 261. It is clear from RCRA that control of hazardous wastes in transportation was envisioned, and that the EPA requirements on such transportation should be consistent with those adopted by the Department of Transportation. Implementing this intent, DOT in conjunction with EPA proposed regulation of hazardous wastes in transportation. After a significant comment period, new rules on hazardous wastes and hazardous substances were adopted by the Department of Transportation on May 22, 1980 (45 Fed. Reg. 34560).

With the determination that DOT would administer the transportation regulations for hazardous wastes defined by EPA, it was necessary to create a mechanism for this regulation. To simply establish a new classification would have been unduly confusing, since many EPA-designated wastes were covered already by other classifications under the DOT regulations. Basically, the approach followed by DOT was to overlay the existing DOT classification definitions with the EPA definitions. As a result, for example, any flammable liquid that was already regulated by DOT continued to be regulated in the Flammable liquid class, and its properties as a waste are highlighted by container markings and by the Uniform Hazardous Waste Manifest.

Since definitions applied by both agencies are not identical, there are places where the EPA regulations apply and no DOT rules were in place. In overlaying the DOT classes with the EPA waste definitions, there was extension beyond the purview of the existing DOT rules at certain points. For materials in this extended area that were unregulated prior to 1980, DOT created a *new* classification—ORM-E. This classification is discussed more fully below but, in short, is made up of articles that are environmentally threatening, but that do not pose an acute safety threat in transportation under traditional DOT definitions.

Hazardous substances are also regulated by DOT. Amendments adopting hazardous waste and hazardous substance rules were developed concurrently and consequently are somewhat coordinated. Like hazardous wastes, the original list of approximately 300 hazardous substances developed by EPA under Section 311(b) of the Clean Water Act encompassed many items already regulated by DOT for their safety hazards. These included materials in all of the traditional classifications as well as many in the ORM-A, B, C and D classes.

For practicality, DOT adapted EPA's definition of hazardous substance to fit transportation principles. By statute, the term does not include petroleum products that are lubricants or fuels. Under 49 CFR 171.8, "hazardous substance" for DOT purposes means a quantity of material offered for transportation in one package, or transport vehicle when the material is not packaged, that—(a) equals or exceeds the reportable quantity (RQ) specified for the material in EPA regulations at 40 CFR Parts 116 and 117, and (b) when in a mixture or solution in any package offered for transportation is in a concentration which equals or exceeds the following (based on the RQ weight quantities specified in 49 CFR 172.101):

RQ, pounds	RQ, kilograms	Concentration by Weight	
		percent	PPM
5000	2270	10	100,000
1000	454	2	20,000
100	45.4	0.2	2,000
10	4.54	0.02	200
1	0.45	0.002	20

The key element of this definition is that, at a minimum, one does not have a hazardous substance until the reportable quantity for that substance is contained *within a single package*. By this mechanism, DOT avoids the applicability of unnecessary regulation to hazardous substances that may be present but not in significant amounts. Whereas it is conceivable that trace amounts of hazardous substances in a full transportation unit, dumped in its entirety in a navigable waterway, might in the aggregate reach a reportable quantity threshold for spill reporting, DOT found that practical experience in transportation accidents made this circumstance unlikely. It is not impossible, but the extreme burden upon carriers in particular, who would have to arithematically aggregate all trace quantities on one vehicle before moving, was not felt worthwhile in light of the goals to be achieved. In a subsequent Federal Register publication, EPA concurred in this interpretation for transportation purposes (45 Fed. Reg. 61617, Sept. 17, 1980).

As with the hazardous waste definition, the hazardous substance definition developed by DOT was laid over the hazard class definitions traditionally contained in the hazardous materials regulations. Requirements for materials already regulated by DOT *by name* were adjusted to include a reference to the reportable quantity (RQ), to appear on shipping documents and package markings. Requirements for materials regulated by DOT but not by their chemical names were modified so documents and package markings would include the EPA-listed name. Rules applicable to materials in the ORM-A, B, C, and D classifications were modified so that a material that is a hazardous substance or a hazardous waste is regulated in *all* modes of transportation, regardless of the fact that in its non-hazardous substance or non-waste form it is only regulated in air or water commerce. An ORM-A, for example, that is covered by the hazardous substance definition now is regulated as an ORM-A in all modes of transportation and must be marked and documented as an ORM-A hazardous substance.

As with the hazardous waste definition, the hazardous substance definition did not fit neatly over the DOT regulated materials. There were several materials that did not fit any historic DOT hazard definitions. For these DOT assigned the ORM-E classification. This is the same class applied to hazardous wastes that do not meet another DOT definition. ORM-E is discussed more fully below.

ORM-E is a recent addition to the DOT list of classifications. It was adopted on May 22, 1980 (45 Fed. Reg. 34560). An ORM-E is a material subject to DOT regulation that is not included in any other DOT hazard class. Materials in the ORM-E class now include hazardous wastes (as defined by EPA under Section 3001 of RCRA) and hazardous substances (as designated by EPA but defined by DOT in 49 CFR 171.8 of the DOT Regulations). The class is structured to accommodate more materials in the future, e.g., "hazardous substances" regulated under Superfund, 42 U.S. Code 9601, P.L. 96–510.

Creation of the ORM-E class and other hazardous waste and substance regulations were the first clear entry by DOT into regulation of materials that are *environmentally* hazardous, but that may not pose direct threats to people or transportation property. A question of the agency's authority under the HMTA to issue such regulations was raised in comments. DOT's answer, published on May 22, read: "The pertinent language in Section 311 of the Clean Water Act (33 U.S. Code 1321), EPA's authority for designating hazardous substances, is contained in paragraph (b)(4), which provides for determination of 'those quantities of . . . any hazardous substances the discharge of which may be harmful to the public health or welfare . . . '" Clearly, many of the risks involved in the transportation of hazardous materials relate to the possibility of unintentional release, and such releases may involve discharges into the navigable waters of the United States. To the extent that EPA has designated certain substances in specific quantities as potentially harmful, it is appropriate for (DOT) to designate those quantities of those materials as hazardous materials under the HMTA. Moreover, should (DOT) not take this action, it would be left to EPA to fill the void covering the transportation of those hazardous substances not reached by DOT regulations. Such a split in regulatory coverage would be inefficient and a hinderance to all concerned." (45 Fed. Reg. 34569)

DOT's response to a question of their authority does little to dispel the contention that there was little basis in the HMTA, at the time these rules were adopted, for DOT to regulate environmentally hazardous substances. The designation concerns under the Clean Water Act and the HMTA are totally different, and the Constitutional bases of the two statutes are different. The claim of authority based upon administrative efficiency carries no legal weight. A hazardous material in the DOT statute is defined as one posing an unreasonable risk to health or safety and property, and that definition has not been changed by Congress.

Section 306 of the Comprehensive Environmental Response, Compensation, & Liability Act of 1980 (Superfund), 42 U.S. Code 9601, Public Law 96–510, Dec. 11, 1980, compels the Secretary of Transportation to list and regulate as a hazardous material under the HMTA, any material designated as a hazardous substance in the new statute. The term hazardous substance in Superfund, of course, includes more than merely Section 311(b) Clean Water Act materials customarily described by that term. The term now includes water pollutants, air pollutants, materials receiving attention under TSCA, hazardous wastes, and other materials that might later be designated as hazardous substances by the Administrator of EPA. Listing a hazardous substance in the DOT regulations for purposes of documentation and notice, however, may not be the same as an alteration of the HMTA statutory definition of hazardous material. In other words, the order in Superfund for DOT to list EPA-designated hazardous substances does not necessarily give *DOT* the direct authority to *initiate* the designation of materials that endanger the environment, but pose no threat to human health or safety or commercial property in transportation, as a hazardous material.

CONCLUSION

Under the HMTA the Department of Transportation has clear authority to designate materials as hazardous when transported in commerce. That authority has not been used, except in rare instances, to designate new materials. Rather, DOT used its authority to take materials designated under a predecessor statute in a blanket adoption of those designations and requirements for those materials as "new law" regulations. (41 Fed. Reg. 38175, Sept. 9, 1976.) There is no established mechanism or quantitative basis upon which DOT acts in designation of materials. The materials must be classed before being regulated, and many of the classes are ambiguous, subjective, inconsistent, and not susceptible to ready definition or quantification. There is no basic mission statement or policy position guiding the agency or the public on DOT actions under the HMTA. The process is subjective and informal and must be viewed with a certain degree of skepticism by those who would prefer a quantitative, consistent approach to controls.

The primary value of the DOT system is that it is there, and it is familiar to people who are regulated by it. Without question, the early thrust of regulation of materials in transit dealt with the acute hazard potential of the materials. Now that DOT has undertaken the designation (or at least the listing) of materials that pose environmental hazards or long term health hazards, it is unclear what limit there is to the exercise of HMTA authority.

U.S. COAST GUARD HAZARDOUS MATERIALS REGULATIONS

The Dangerous Cargo Act, 46 U.S. Code 170, was developed to augment the statutory authority in the Explosives and Other Dangerous Articles Act, 18 U.S. Code 831–835. The latter statute applied to shippers by all modes including water, but needed to be statutorily extended to water carriers. The authority under the Dangerous Cargo Act is exercised by the United States Coast Guard, a part of the Department of Transportation.

The Dangerous Cargo Act required the Coast Guard, in exercising this authority, to accept and adopt the "definitions, descriptions, descriptive names and classifications" adopted under the Explosives and Other Dangerous Articles Act. Designation of hazardous materials now occurs under the successor statute, the Hazardous Materials Transportation Act (HMTA), implemented by the Department of Transportation with input from the Coast Guard.

In DOT rulemaking completed in 1976, regulations applicable to water transportation of most packaged hazardous materials were relocated from 46 CFR into 49 CFR Part 176. Many of the provisions unique to water transportation were removed to that Part and also are reflected in the creation of the ORM-A, ORM-B and ORM-C classifications discussed above. Not all Coast Guard regulations of materials, however, are addressed in 49 CFR. The Dangerous Cargo Act still supports regulations appearing elsewhere in the Code of Federal Regulations.

I. MILITARY EXPLOSIVES

One of these still active areas appears in 46 CFR Part 146 entitled, "Transportation or storage of military explosives on board vessels." In Subpart 146.20, the Coast Guard sets out detailed regulations governing explosives. Explosives here are defined as they are in 49 CFR.

This coincidence of classification definition, however, is somewhat misleading in the regulations on handling military explosives. It was found that the commercial regulations for ocean shipment of explosives and other hazardous materials were too restrictive for vessels that were devoted to transportation of ammunition, such as those operated for the military. The 49 CFR regulations simply do not provide enough alternatives for the loading and stowage of explosives to satisfy the needs of a vessel for military explosives.

Consequently, in 46 CFR 146.29–100, there is an entirely different system of "classification" of explosives. These classifications are for the purpose of segregating materials to assure compatibility and to designate stowage locations aboard ammunition-carrying vessels. Thus, for example, one finds Class I made up of small arms ammunition (without explosive bullets), detonating fuses, DOT Class C mechanical time fuses, and like items. Class II-A is made up of bulk propellants such as ballistite, cordite, propellant charges, and "made-up" bag charges in outside shipping containers. Class IV includes fixed and semi-fixed ammunition with explosive loading projectiles. Class VIII includes various mine fuses, blasting caps and detonators. Class X-C includes guided missiles and solid propellant motors packed with or without warheads. Class XI-B includes non-lethal chemical ammunition. Class XI-D covers oxidizers in containers for guided missiles and rockets.

These various reclassifications, done for the sake of specialized military transportation, may or may not be consistent with the original Class A, B and C classifications developed in 49 CFR and referenced in 46 CFR. One finds A, B and C explosives all included in several of the military classifications. In Class XI-D mentioned above, the "oxidizers" are classed under 49 CFR as corrosive liquids, nonflammable gases and oxidizing materials, without *any* of them being classed as 49 CFR explosives for commercial purposes.

This is an esoteric area of classification developed for the specialized use of military shipments of ammunition. It is oriented toward the loading and stowage of materials to minimize incidents in transportation where the vessel is filled with a variety of explosives. Because of its limited applicability, it is not discussed in greater depth here.

II. SHIPS' STORES AND SUPPLIES

Section 170 of 46 U.S. Code authorizes the Coast Guard to develop regulations to control the *use* of hazardous materials aboard vessels. These regulations are published in 46 CFR Part 147 and are entitled, "Regulations governing use of dangerous articles as ships' stores and supplies on board vessels."

Ships' stores and supplies are defined in 46 CFR 147.02–1 as "any article or substance which is used on board a vessel subject to the regulations of (the Coast Guard) for the upkeep and maintenance of the vessel; or for the safety or comfort of the vessel, its passengers or crew; or for the operation or navigation of the vessel (except fuel for its own machinery); except shipboard fumigation" regulated elsewhere. Section 147.02–2 of 46 CFR defines ships' stores "of a dangerous nature." For the purpose of these regulations, materials with "such characteristic properties as will cause the substance to properly classify as either an explosive, inflammable liquid, inflammable solid, oxidizing material, corrosive liquid, compressed gas, poisonous article, hazardous article or combustible liquid in accordance with the definitions for such substance as contained in the regulations in this subchapter are defined as ships' stores and supplies of a dangerous nature."

Because of a series of rulemaking actions elsewhere over the past few years, this definition is incomprehensible. For example, at the latter portion of the quote where it speaks of definitions of substances as contained "in this subchapter," one finds there are *no* such definitions in the subchapter containing these regulations today. This is a vestigial reference to a series of definitions moved to 49 CFR in 1976. The words *in*flammable liquid and solid are now outdated. In 49 CFR, oxidizing materials are called oxidizers. "Hazardous articles" are those that became ORM-A, ORM-B or ORM-C in 49 CFR. Combustible liquids at the time this regulation was developed meant liquids with a flash point of 80°F to 150°F, whereas today in 49 CFR that definition begins at 100°F and goes up to 200°F. There is no reference whatsoever to the new ORM classifications. In short, the regulations on ships' stores and supplies are fatally defective and give little illustration of what is required.

In practice, it is understood that these rules apply to all articles classed as hazardous materials in 49 CFR. This is not written anywhere. It is merely known in the regulated community and by the Coast Guard.

These regulations are more akin to occupational safety and health rules than to hazardous materials transportation rules. The reference to the DOT regulations, therefore, perhaps indicates a deficiency in these rules. There are many hazards encountered in the use of a product that are not present in transportation and, therefore, are not addressed in the DOT rules in 49 CFR.

The handling of various regulated materials that do not need to be specially "certificated" or approved by the Coast Guard is addressed in 46 CFR 147.05–100. The materials are described and by their various properties are limited in use. These properties may or may not coincide with elements of the 49 CFR classifications. For example, the definition of combustible liquid in 49 CFR extends from 100° to 200°F. End use regulations in 46 CFR, however, often refer to goods with flash points of 150°F. Hazardous materials not covered by this table are intended to be identified as in 49 CFR 172.101 and require *special* Coast Guard certification before they may be used on board vessels. Some materials are not permitted for use, but this is not written in the regulations.

III. BULK SOLID HAZARDOUS MATERIALS BY VESSEL

The Dangerous Cargo Act includes authority for the Coast Guard to regulate shipments of bulk solid hazardous materials by vessel. Section 170(6)(a) of 46 U.S. Code prohibits transportation of bulk solid hazardous materials on board any vessel unless authorized by Coast Guard regulation or other special authorization. The term "hazardous materials" for these purposes is defined in 46 CFR 148.01–1 to mean hazardous materials as defined in 49 CFR Parts 170–189. Materials listed in 46 CFR 148.01–7 by the Coast Guard as authorized for carriage include a variety of flammable solids, oxidizing materials and ORM materials, as well as low specific activity radioactive materials. In addition to the specific listed articles permitted by regulation, there is a procedure for seeking special permission to carry bulk solids. This exemption procedure is outlined in 46 CFR 148.01–9 and 148.01–11.

IV. BULK LIQUID DANGEROUS CARGOES BY VESSEL

Coast Guard regulation of hazardous materials in bulk tankers and barges occurs under rules commonly referenced as "Subchapter D" and "Subchapter O" regulations. Subchapter D, made up of Parts 30–40 of 46 CFR, addresses liquids and liquefied gases bearing the hazards of flammability or combustibility, but without additional hazards. Generally these are hydrocarbons. Subchapter O, made up of Parts 151–154 of 46 CFR, addresses chemicals bearing other hazards that may include flammability and combustibility as well. Each subchapter prescribes requirements, but there is heavy cross-referencing between subchapters for definitions and other requirements.

Within Subchapter D, Part 30 contains general requirements; Part 31, general inspection and certification rules for tankers; Part 32, special equipment, machinery, and hull requirements; Parts 33, lifesaving equipment; Part 34, firefighting equipment; Part 35, tanker operations; Part 36, elevated temperature cargoes; Part 37, special construction, arrangement and other provisions for nuclear vessels carrying flammable or combustible cargo; Part 38, liquefied and compressed flammable gases; and, Part 40, special construction arrangement, and other provisions for carrying vinyl chloride in bulk.

Within Subchapter O, Part 151 addresses unmanned barges carrying certain bulk dangerous cargoes; Part 153, safety rules for self-propelled vessels carrying hazardous liquids; and Part 154, safety standards for self-propelled vessels carrying bulk liquefied gases.

Regulations on unmanned barges are issued by the Coast Guard under the authority of the Dangerous Cargo Act, 46 U.S. Code 170, as well as 46 U.S. Code 391a. The unmanned barge regulations are set forth in 46 CFR Part 151 and contain uniform minimum requirements for transportation of liquids or liquefied gases in bulk (other than flammables and combustibles regulated under Subchapter D).

As with the regulations on ships' stores and supplies, the regulations on unmanned barges also suffer from inconsistencies and inaccuracies related to rulemaking changes elsewhere. For example, in 46 CFR 151.01–1 there are lists of cargoes affected by these regulations and the list includes hazardous articles now properly classified as ORM-A, B and C materials. There are many sectional cross-references to regulations that no longer exist—predecessors to hazard definitions now appearing in 49 CFR. The requirements on unmanned barges do not involve the designation of hazards, but cross-reference designations made elsewhere in 46 CFR and in 49 CFR.

Coast Guard regulations in 46 CFR Part 153 on bulk chemicals aboard self-propelled vessels, issued jointly under 46 U.S. Code 170 and 391a, also adopt the hazardous materials classification definitions published in 49 CFR. Exceptions to this are definitions of flammable and combustible liquids that appear in 46 CFR 153.2. There, a flammable liquid is defined as it is in Subchapter D, in 46 CFR 30.10-22, and a combustible liquid as it is defined in 30.10-15.

The referenced definition sections are published as part of the Subchapter D. The definitions in Subchapter D predate changes that took place in the early 1970s and continue to draw distinctions and limitations from that earlier period. These are the definitions applicable to bulk tankers (including manned and unmanned barges). Section 30.10–22 of 46 CFR defines a flammable liquid as, "any liquid which gives off flammable vapors (as determined by flash point from an *open* cup tester as used for test of burning oils) at or below a temperature of 80°F." Flammable liquids are referred to by grades as follows: (a) Grade A, any flammable liquid having a Reid vapor pressure of 14 lbs or more; (b) Grade B, any flammable liquid having a Reid vapor pressure under 14 lbs and over 8½ lbs; and (c) Grade C, any flammable liquid having a Reid vapor pressure of 8½ lbs or less and a flash point of 80°F or below.

Section 30.10–15 of 46 CFR defines combustible liquid to mean, "any liquid having a flash point above 80°F (as determined from an open cup tester, as used by test of burning oils)." In Subchapter D, combustible liquids also are referred to by grades, as follows: Grade D, any combustible liquid having a flash point below 150°F and above 80°F; and Grade E, any combustible liquid having a flash point of 150°F or above.

The flammable liquid definition here stops at 80°F, as it used to for land regulations before rule making in the early 1970s. The combustible liquid definition is interesting because of the classification of combustibles below 150°F, rather than 200°F as in 49 CFR. Even more importantly, regulation as a combustible liquid occurs with any material with a flash point from 150°F to *infinity*. This means, for example, that molasses in a bulk tanker is shipped as a combustible liquid.

Materials listed in 46 CFR Parts 30 and 153 may be carried in bulk (when the vessel is certificated for such carriage by the Coast Guard) as long as the published rules are met. Materials not listed by name in either part must be made the subject of a *special* application for a Coast Guard certificate, in order to be carried in bulk aboard any self-propelled vessel or tank ship. There are no published criteria upon which such a certificate is granted or denied.

Coast Guard rules on self-propelled vessels carrying bulk liquefied gases are issued under the authority of 46 U.S. Code 391a, and are published in 46 CFR Part 154. In 46 CFR 154.3, a liquefied gas is defined as a cargo having a vapor pressure of 25 psia or more at 37.8°C (100°F). The regulations provide for carriage of specific listed gases. This list was developed over time from known gases for which a great need was felt or gases that were specifically requested by individual applicants to be moved in bulk vessel operations. The list is in Table 4 of Part 154.

The term "flammable cargoes" is defined in 46 CFR 154.3 as including the following gases from Table 4:

Acetaldehyde	Ethyl chloride	Methyl chloride
Butadiene	Ethylene	Propane
Butane	Ethylene oxide	Propylene
Butylene	Methane (LNG)	Vinyl chloride
Dimethylamine	Methyl acetylene–	
Ethane	propadiene mixture	
Ethylamine	Methyl bromide	

"Toxic cargoes" are defined to mean the following liquefied gases from Table 4.

Acetaldehyde	Ethylene oxide
Anhydrous ammonia	Methyl bromide
Dimethylamine	Methyl chloride
Ethylamine	Sulfur dioxide
Ethyl chloride	Vinyl chloride

Materials regulated in bulk include *all* those falling within the definition of a hazardous material. For the shipment of a hazardous material or a mixture of hazardous materials that is not authorized specifically by name in these various parts of 46 CFR, it is necessary to go through a *special* approval process before the vessel may be certificated by the Coast Guard to carry the material. The process may or may not result in the issuance of a certificate. Generally speaking the applicant for certification is subjected to a variety of requirements that are keyed to the given material. It has occurred, however, that certification has been denied for materials too dangerous to be carried in bulk. There is no assurance that the conditions attached to issuance of a certificate would be within the realm of economic feasibility. It does not appear that in making these determinations, the Coast Guard uses any published criteria, although related standards under IMO are used for reference. Since the Coast Guard participates heavily in the development of these standards, most requirements are becoming completely consistent. Each decision is a case unto itself, and decisions are made on an application-by-application basis.

For ships constructed before July 1, 1986, the IMO Code for the Construction and Equipment of Ships Carrying Dangerous Chemicals in Bulk is followed. This source is often cited as the "BCH Code." For ships constructed after that date, a different IMO International Code for the Construction and Equipment of Ships Carrying Dangerous Chemicals in Bulk is used. This latter code is often called the International Bulk Chemicals Code, or the "IBC" Code.

V. DANGEROUS CARGOES ON WATER FRONT FACILITIES

Under the Ports and Waterways Safety Act, 33 U.S. Code 1225, the Coast Guard has issued regulations published in 33 CFR Part 126 pertaining to the handling of explosives or other dangerous cargoes within or contiguous to water front facilities. Section 126.07 of 33 CFR defines dangerous cargo to mean all explosives and other hazardous materials or cargo covered by regulations of the Coast Guard in 46 CFR Parts 33-38, 146, and 148. In addition, the definition includes hazardous materials except those preceded by an "A" in Title 49 CFR Parts 170-179. Those preceded by an "A" in the hazardous materials table, of course, are only regulated by air (unless they are hazardous wastes or hazardous substances).

Section 126.10 of 33 CFR lists "cargo of particular hazard," noting those of high hazard such as explosives and others that pose a significant danger in bulk quantities. Under these regulations, the primary concern is safety of the water front facility. Judging by the materials regulated under 33 CFR Part 126, it is clear that the acute injury potential caused by fire and explosion are the primary concerns in establishment of these regulations. The mechanism of control is the issuance of permits to people engaged in activities involving hazardous materials coming to or being held at water front facilities.

VI. CONTROL OF POLLUTION BY OIL AND HAZARDOUS SUBSTANCES, DISCHARGE REMOVAL

These Coast Guard regulations are published in 33 CFR Part 153 and are issued under the authority of 33 U.S. Code Section 1321(a)(1) and (b)(2).

Section 153.103 of 33 CFR defines oil to mean oil "of any kind or in any form, including but not limited to petroleum, fuel oil, sludge, oil refuse and oil mixed with wastes other than dredged spoil." The term hazardous substance is defined as any substance designated by EPA pursuant to Section 311(b) of the Clean Water

Act. There is no independent Coast Guard authority to designate materials for regulation under these portions. Both oil and hazardous substances are defined in the statute and discretion to designate hazardous substances lies in the hands of EPA.

NUCLEAR REGULATORY COMMISSION REGULATIONS

The regulations of the Nuclear Regulatory Commission are issued under the authority of the Atomic Energy Act of 1954, 42 U.S. Code 2011. The regulations are published in 10 CFR Parts 0–170.

Part 20 of 10 CFR prescribes standards for protection against radiation. Section 20.3 within this Part offers several definitions pertinent to the designation of radioactive materials. Radioactive material is defined as, "any such material whether or not subject to licensing control by the (Nuclear Regulatory) Commission." Radiation is defined in this section as "any or all of the following: alpha rays, beta rays, gamma rays, X-rays, neutrons, high-speed electrons, high-speed protons, and other atomic particles; but not sound or radio waves, or visible, infrared, or ultraviolet light."

Part 20 of 10 CFR prescribes standards in terms of "permissible dose levels and concentrations" of radiation exposure for workers and the public. Various precautionary procedures are prescribed. The basic controls of the Nuclear Regulatory Commission are carried out through various licensing programs for source material, special nuclear material, and by-product material. These are regulated in Parts 40, 70 and 30, respectively of 10 CFR. Part 40 of 10 CFR defines source material as, "(1) uranium or thorium, or any combination thereof, in any physical or chemical form or (2) ores which contain by weight one-twentieth of one percent (0.05%) or more of: (i) uranium, (ii) thorium or (iii) any combination thereof. Source material does not include special nuclear material."

Special nuclear material is defined in 10 CFR 70.4(m) as "(1) plutonium, uranium 233, uranium enriched in the isotope 233 or in the isotope 235, and any other material which the Commission, pursuant to the provisions of Section 51 of the Act, determines to be special nuclear material, but does not include source material; or (2) any material artificially enriched by any of the foregoing but does not include source material."

By-product material is defined in 10 CFR 30.4(d) as, "any radioactive material (except special nuclear material) yielded in or made radioactive by exposure to the radiation incident to the process of producing or utilizing special nuclear material." The same term is given a somewhat different definition in 10 CFR 40.4(a–1). There, by-product material is defined as, "the tailings or wastes produced by the extraction or concentration of uranium or thorium from any ore processed, primarily for its source material content, including discrete surface wastes resulting from uranium solution extraction processes. Underground ore bodies depleted by such solution extraction operations do not constitute 'by-product' material within this definition." It should be noted that the definition of by-product material in 10 CFR 40.4 is limited to the use of that term in Part 40, whereas the definition in 10 CFR 30.4 is for use in Parts 30–35.

Whereas the definitions of basically regulated materials are virtually identical to definitions for these terms in the Atomic Energy Act, there is an exercise of discretion in application of regulatory requirements to these materials. In addition, there appears to be significant use of exemption authority to relieve certain defined materials from either certain regulations or all regulations.

Throughout the operative parts of the NRC regulations in 10 CFR, there are frequent references to exemptions and "specific exemptions." The exemptions include exemptions by rule, which generally relieve low-hazard materials or people governed under other regulations, or people who have only a peripheral contractual involvement in the handling of radioactive materials. Section 30.14 of 10 CFR, for example, is entitled "Unimportant Quantities of Source Material" and describes low-risk materials and exempts people handling those materials from various requirements, generally including exemption from licensing requirements. Throughout the regulations one also finds reference to specific exemptions. Section 30.11 of 10 CFR says that "The Commission may, upon application of any interested person or upon its own initiative, grant such exemptions from the requirements of the regulation in this part as it determines or authorized by law and will not endanger life or property or the common defense and security and are otherwise in the public interest." In its deliberations upon an application for exemption, the NRC must consider and "balance" the following factors: "(1) whether conduct of the proposed activities will give rise to a significant adverse impact on the environment and the nature and extent of such impact, if any; (2) whether redress of any adverse environmental impact or conduct of the proposed activities can reasonably be effected should such redress be necessary; (3) whether conduct of the proposed activities would foreclose subsequent adoption of alternatives; and (4) the effect of delay in conducting such activities in the public interrest. During the period of any exemption granted pursuant to this paragraph, any activities conducted shall be carried out in such a manner as will minimize or reduce their environmental impact."

Exemption alternatives such as that just quoted appear in the following sections: 10 CFR 20.501, 21.7, 30.11, 34.51, 40.14, 50.12, 51.4, 55.7, 70.14, 71.6, 110.10(a), and 140.8.

As can be seen from the proliferation of exemption sections, in addition to exemptions by rule, the NRC's designation authority takes the form of *non*-designation of hazardous materials. In other words, the Atomic Energy Act of 1954 defines materials to be regulated, and this statutory rigidity is ameliorated by the exemption process through which specific people, materials, and functions are relieved from the full impact of regulatory requirements, primarily licensing. Although the regulations do discuss the generic categories of concern, such as the concern for the public interest and the environment, these matters are not in any form quantified nor is the applicant for an exemption advised by the regulations of the details that will prevail in an exemption application.

CHAPTER 5

The Process
of Determining
Classification

See Chapter 4 for detailed definitions of the classes of materials regulated by DOT.

Experience indicates that most people shipping hazardous materials deal with flammables, corrosives, poisons, gases, and some of the other broad classes, while a distinct minority are involved in the transportation of high explosives, radioactive materials, etiologic agents, and organic peroxides. The *Red Book*, therefore, will put the accent on the more commonly encountered transportation by focusing attention on the former classifications rather than on the latter.

For shippers, classification of the material to be shipped is the single most important task under the regulations, for on the basis of this one decision all the regulatory requirements and safeguards will be determined. From this task will flow all decisions regarding packaging, marking, labeling, documentation, and carrier requirements and limitations. The responsive action taken in the event of an incident involving the material also will key to the classification decision. This chapter, therefore, will point out major elements that need to be considered in making this decision.

The process is made more difficult because the decision is not a pure scientific conclusion. It is not based solely on precisely developed technical test data, or the physical and chemical properties of the material. The classification of a product must include an understanding of the framework of the DOT regulations and their somewhat systematic approach to the achievement of safety.

The classification decision must combine science, regulations, experience, common sense, and good judgment in order to be a correct decision. For these reasons, classification is actually more of an art than a science, and it is entirely possible for two knowledgeable people to arrive at different conclusions. As long as there is proper rationale for each of them, it is possible that neither would be wrong or in violation of the regulations.

Given the nature of classification, it is advisable to involve more than one person in the process, drawing upon talent within the company that is familiar with the scientific, regulatory, safety, packaging, and transportation areas. Although the final decision should probably be vested in one person, the cross-pollination of ideas that such a group offers will improve the quality of that decision. Since this is such a critical decision, affecting the costs of transportation and the liability of the company, it should be made at a responsible level within the company, and involvement of counsel to the company is often advisable as well.

This chapter of the *Red Book* will not make anyone expert on classification. It is important that any shipper tendering a hazardous material to any carrier recognize his own limitations in the area of classification. If there is a need for knowledge about the product that is not satisfied from within the company, the answers should be sought from the original producers of the material, or from outside testing facilities. Perhaps other shippers of the same material can share in their knowledge of the proper safety precautions to take in classifying the product, and trade associations may have helpful information on the subject, too.

In addition to these outside sources, there are a number of texts that describe the properties of materials and the hazards associated with their manufacture and handling. No text purports to be complete in its treatment of materials for purposes of DOT classification and, accordingly, none should be relied upon as the sole source of classification information, but technical treatises and texts are an invaluable part of the classification process.

THE DOT COMMODITY LIST

The alphabetical table of regulated materials in 49 CFR 172.101 is the key to compliance. Without a thorough familiarity with the structure of this columnar table and the fund of information available in it, no one can hope to develop any facility in reading the regulations or can ever be assured of being in compliance.

THE STEPS OF CLASSIFICATION

The best way to illustrate the steps of classification and to provide a working familiarity with the commodity list is to take out your copy of the DOT regulations and, by example, learn.

FIRST EXAMPLE

The task is to determine the classification of a material known to you as "phosphorus tribromide."

Step 1. Look in the commodity list at the column of alphabetically arranged article names. There is a specific entry in the list for phosphorus tribromide. Do not be misled by entries of a similar nature or spelling. The entry in the list for phosphorus trichloride or phosphorus trisulfide, for example, is for a different material that has different properties and is subject to very different requirements on packaging and other significant details. Chemical names often are spelled like one another, or the name you are looking for may be incorporated in the longer name of another material. Use great care in reading the list, and do not assume that a similar appearance of the words indicates any similarity in properties or regulatory requirements.

Step 2. Now that the proper alphabetical listing for phosphorous tribromide has been found, look to the column to the right of this name, which is headed "Hazard class." One sees that this commodity is classed as a Corrosive material. This is the proper classification of this material for DOT regulatory purposes. The terminology is very precise, and, therefore, when preparing shipping papers for this material, it is essential to note the proper classification, among other things, without abbreviation.

SECOND EXAMPLE

Determine the proper DOT hazard classification for a material called diethylaminoethyl chloride, which is not found very frequently in commerce.

Step 1. Knowing the chemical name of the material, first examine the alphabetical commodity list in the regulations. The material does not appear in the list by name.

Step 2. With no listing of the material by name, the determination of classification calls for a technical evaluation at this point. This is not a job for the typical traffic manager or packaging engineer, unless he also happens to have a solid background in chemistry.

In the technical evaluation the first things to look for are whether the material is highly reactive (unstable) or significantly sensitive to shock or heat, or whether it is radioactive. Generally speaking, the kind of instability DOT means is relatively easily detectable through a review of technical literature. Under 49 CFR 173.21 one may not ship any material that under conditions normally incident to transportation may polymerize or decompose so as to cause a dangerous evolution of heat or gas. This means that under ordinary transportation conditions—taking into account the shocks and vibrations, temperature, and confinement in closed areas under any atmospheric condition—no reaction should be triggered. For known materials such reactivity is readily detectable and a chemist will usually be aware of these hazards in handling the material. Radioactivity also is easily detected through the literature or through use of detection devices.

The next level of reactivity is more difficult to detect, viz., significant sensitivity to shock or high heat. Experienced chemical know-how is a must to make this determination. If there is any suspicion that the material may be reactive in this fashion, the regulations require an assessment of the hazards of the material by the Bureau of Explosives or the Bureau of Mines of the Department of the Interior. No one may ship such a product until a sample of it has been specifically evaluated by one of these to determine if it is an explosive. See 49 CFR 173.51 and 173.86.

With the material in this example, diethylaminoethyl chloride, there is no hazard of reactivity or radioactivity.

The next technical evaluation is to determine whether it is a disease-bearing material, called an etiologic agent. This determination is relatively simple, and in this case the material is not etiologic.

Thus, in this step, three potential DOT classifications have been eliminated: explosives, radioactive materials, and etiologic agents.

Step 3. Ascertain the physical form of the material to determine whether it is a solid, liquid, or gas. This will have been developed in the preceding technical review. The remaining potential DOT classifications are: flammable or combustible liquid, flammable solid, oxidizing material (liquid or solid), corrosive material (liquid or solid), gas (flammable and nonflammable), and poison (liquid, solid, or gaseous).

Diethylaminoethyl chloride is a liquid. This determination eliminates all the potential solid and gaseous classifications from consideration, leaving four: flammable or combustible liquid, liquid oxidizing material, liquid corrosive material, and liquid poison.

In 49 CFR 173.2, DOT published its ranking of classes in the order of their classification priority in transportation (other than explosives that must be classified by the Bureau of Explosives, not the shipper, and other than specifically identified etiologic agents and oxidizers). This ranking, in declining order, is as follows:

1. Radioactive material (except limited quantities).
2. Poison A.
3. Flammable gas.
4. Nonflammable gas.
5. Flammable liquid.
6. Oxidizer.
7. Flammable solid.
8. Corrosive material (liquid).
9. Poison B.
10. Corrosive material (solid).
11. Irritating material.
12. Combustible liquid in packaging over 110 gallons capacity.
13. ORM-B.
14. ORM-A.
15. Combustible liquid (in containers of capacities of 110 gallons or less).
16. ORM-E.

This indication of DOT's ranking of classes offers guidance to the shipper endeavoring to classify a material that may have multiple hazard characteristics.

Proceeding from the top of this rank, having eliminated the radioactivity or gaseous nature of this material, one looks to its toxicity. Examination of the toxic hazards of diethylaminoethyl chloride by testing in accordance with procedures specified by DOT or by examination of the literature reveals that it is quite toxic, but not nearly as toxic as the Poison A materials listed in the regulation devoted to great toxicity, 49 CFR 173.326(a). One therefore proceeds down the hazard ranking list.

Testing or review of the literature would show that this commodity does not meet the flammable liquid or oxidizer definitions. It is not a solid, so the next point of examination is its corrosive property.

Very little can be found in the literature regarding any material's corrosive properties as measured by the tests prescribed by DOT. In most instances, therefore, actual testing will have to be done on the product and, in this case, testing would show that diethylaminoethyl chloride meets the corrosive materials definition. This is not the end of the road, however, for certain requirements apply to multihazard products and, therefore, each definition must be checked.

Step 4. Having noted in the examination of the toxicity of this material that it is quite toxic, one may see that it meets the criteria to classify it as a Class B poison. In this example the material is both a corrosive material and a Poison B, and the shipper is required to class it as a corrosive under the sequence in 49 CFR 173.2.

CHAPTER 6

Determination of Proper Dot Shipping Name

The term "proper shipping name" in the context of the *Red Book* and the DOT regulations is the specific name given in 49 CFR 172.101 (or in some international commerce, 49 CFR 172.102) for that material. Note that in Docket No. Hm-181, May 5, 1987, DOT has proposed to consolidate these two tables. The shipping name is not necessarily the brand name of the product has as a brand name, or the name for freight classification (rating) purposes, or even the chemical name of the ingredients of the product. In fact, the DOT shipping name often will be different from all of these. The proper shipping name must be determined, however, for compliance with packaging, marking, documentation, incident reporting, and other regulatory considerations in the shipment of hazardous materials, hazardous wastes, and hazardous substances.

The proper shipping name is the most specific entry in Roman type appearing in the DOT commodity list (49 CFR 172.101) that most accurately describes the material. Look at the alphabetical list of hazardous materials published in the front of the DOT regulations in 49 CFR 172.101. This should not be confused with 49 CFR 172.102, which offers international shipping names that may be used under some circumstances covered below.

The names appearing in the column farthest to the left of the page are in alphabetical order. Those in Roman (non-italic) type are authorized shipping names. The words in italics are not proper shipping names but are merely finding aids for the reader. Remember, no entry in italics is a shipping name that satisfies the DOT regulations.

A plus (+) mark placed before the shipping name for a commodity indicates that this particular material is regulated by DOT under the shipping name and classification shown. The shipper has no option but to use this description.

Materials listed without a plus (+) mark may not have the properties that would fit the definition of the DOT classification and should not be given that classification or shipping name. Look, for example, at the shipping name "Medicines, n.o.s. . . . Flammable liquid." The lack of a plus mark tells the reader that the flash point of some medicines may be sufficiently low to meet the definition of a flammable liquid, but that other medicines have flash points which fall above the limits of that definition. The shipper should match his product with the definition of the classification shown to determine whether it is regulated and should be given this shipping name.

It is important to note at this time whether the material is preceded in the commodity list in 49 CFR 172.101 by an A or W. The A signifies that the material is regulated only in air transportation, while the W signifies applicability of regulations only when that material is shipped by water. Of course, the presence of both letters means it is regulated in both modes of transportation, but not in land transportation.

The regulations compel selection of the most specific and accurate name for the product. There are roughly five levels of specificity, given here in declining order:

1. Specific listing by chemical name.
2. Less specific listing by n.o.s. chemical family.
3. Generic listing by end-use description.
4. Generic listing by n.o.s. end-use description.
5. Generic listing by n.o.s. classification description.

SPECIFIC LISTING BY CHEMICAL NAME

An example of a hazardous material that appears in the commodity list by its chemical name is sulfuryl fluoride, which is classed as a nonflammable gas. This type of entry means that the article being shipped is sulfuryl fluoride only. Any mixture of this chemical with another chemical or chemicals would make this shipping name inaccurate and improper for DOT purposes.

The safety interest is in the material as it is shipped and, although a review of the regulations to examine the nature of each of the ingredients in a mixture may be informative, the chemical name of each of those ingredients is not the proper shipping name for that mixture, nor are the classification, packaging, marking, labeling, and handling requirements for that mixture necessarily the same as they would be for each of the components of that mixture individually. The nature of the components is a clue to the nature of the total mixture, but the individual names should not be used for any other purposes with the exception for certain marking requirements imposed on n.o.s. commodities shipped by water, or hazardous substances. These are discussed in more detail in Chapter 9 on marking, Chapter 24 on hazardous substances, and Chapter 18 on unique requirements applicable to shipments by water.

If the commodity to be shipped appears in Roman type by its chemical name, the rest of the job is greatly simplified. This simple situation arises all too infrequently, however.

LESS SPECIFIC LISTING BY N.O.S. CHEMICAL FAMILY

If the article to be shipped is not specifically listed by name, it may be listed according to the family of chemicals to which it belongs. Examples of this n.o.s. listing are "Alcohol, n.o.s.," and "Perchlorate, n.o.s."

GENERIC LISTING BY END-USE DESCRIPTION

A generic name by definition is less specific than the chemical name or chemical family of the material, much as the term "fireman" is less specific than giving the name of an individual fireman or his family's surname. There are many generic shipping names in the DOT commodity list that describe materials by their functional purpose or end use. The reader will see such examples in the commodity list as "Insecticide, liquid," which describes the purpose to which this product will be put rather than its individual chemical ingredients. Other examples include "Compound, cleaning, liquid," "Igniter, rocket motor," "Motor fuel antiknock compound," and "Refrigerating machine."

If the article to be shipped is not a straight chemical listed by name or by the name of its chemical family, then the next level to examine is this level of generic descriptions based upon common parlance or the shipper's intended end-use of the product. The proper tariff description for economic rating purposes may provide a clue to the applicable generic description of the material. If no such description can be found, then the review of the commodity list should continue, proceeding to the next level of generality.

GENERIC LISTING BY N.O.S. END-USE DESCRIPTION

By its very nature, a generic n.o.s. description is less specific than a generic description without the n.o.s. qualifier. Examples of these generic n.o.s. names are "Cosmetics, n.o.s." and "Medicines, n.o.s."

GENERIC LISTING BY N.O.S. CLASSIFICATION DESCRIPTION

This is the broadest type of shipping name, to be used only when none of the above types applies to the material to be shipped. Although this type of shipping name reveals little about the specific characteristics of the material or its common function, it does reveal the major hazard posed by that material in transportation. Examples of shipping names of this type are "Flammable liquid, n.o.s.," "Corrosive liquid, n.o.s.," and "Poisonous liquid, n.o.s."

FIRST EXAMPLE

The article to be shipped is a liquid cleaning agent composed of phosphoric acid, acetic acid, and water. Testing shows it to be a corrosive material, and that is its classification for DOT purposes. The task is to determine the proper shipping name for this material.

Phosphoric acid and acetic acid appear individually in the commodity list. A solution of phosphoric acid is also listed by name. A mixture of phosphoric acid and other chemicals, however, cannot be properly described by just calling it phosphoric acid or by calling it a phosphoric acid solution. The specific chemical name is not available as a proper shipping name, and checking the next level shows that no chemical family n.o.s. entry fits either.

Having no success at these two levels of specificity, it is necessary to go to the next level to check for a generic name. Because the material is intended to serve as a cleaning agent, a possible listing would be "cleaning agent," but no such entry appears in the commodity list.

Further scanning of the commodity list for likely descriptions reveals a number of entries under "Compounds, cleaning." (Note: use of the term "compound" in the commodity list does not necessarily mean what that term means in chemistry.) They are as follows:

Compound, cleaning, liquid	**Corrosive**
Compound, cleaning, liquid	**Flammable liquid**
Compound, cleaning, liquid	**Combustible liquid**
Compound, cleaning, liquid (*containing hydrochloric (muriatic) acid*)	**Corrosive**
Compound, cleaning, liquid (*containing hydrofluoric acid*)	**Corrosive**
Compound, cleaning, liquid (*containing phosphoric acid, acetic acid, sodium or potassium hydroxide*).	**Corrosive**

The second and third entries can be eliminated as possible shipping names immediately because it is known that the mixture as shipped is not a flammable or combustible liquid.

Keeping in mind the requirement that the most specific and accurate name is the one to use, look at the other names listed. "Compound, cleaning, liquid" would be a possibility, since this product is a liquid cleaning compound and the classification is proper. Looking further down the list, however, one finds a more specific designation, actually indicating a cleaning compound that would have phosphoric acid and acetic acid in it. This is the proper shipping name for this product. Since italicized information is not part of the name but is just a guide for readers, the shipping name to be noted in markings and documentation is "Compound, cleaning, liquid."

What, one might ask, would result from using this entry rather than an even less specific entry with the same words "Compound, cleaning, liquid"? The answer is found in looking across the page to the various entries for packaging references and quantity limitations in certain transportation. Thus, for example, the maximum amount allowed per package for cargo aircraft under the more specific listing for mixtures containing phosphoric and acetic acid is only one quart, whereas the broader listing for other cleaning compounds allows up to ten gallons per package in cargo aircraft transportation.

SECOND EXAMPLE

The product to be shipped is cyclopentene, a product intended by the shipper for use in organic synthesis, which has a closed-cup flash point below 0° F. The task is to determine the proper shipping name for this material.

Step 1. From a reading of the chapter on classification and a familiarity with the basic definitions of the various classifications, it is known that a material with a flash point near 0° F definitely qualifies as a flammable liquid. Moderate research into the other properties of the material reveals no higher classification priorities as DOT ranks them in 49 CFR 173.2 (see the preceding chapter on classification). In other words, it is not explosive, radioactive, extremely toxic, or gaseous. Accordingly, the classification for cyclopentene is "Flammable liquid."

Step 2. Having looked at the DOT commodity list, one will note that cyclopentene is not listed specifically by name. There are listings for cyclopent*a*ne with an "a," but not cyclopent*e*ne with an "e." Do not assume chemicals with similar spelling have similar properties. Use of the different spelling would not be a proper shipping name for this material.

The next level of exploration through the commodity list is for a chemical family or a generic category or, failing this, a generic n.o.s. listing for a chemical used in organic synthesis. The reader will find that, although most materials will have a purpose that is described by one of the generic end-use descriptions, the one in this example does not.

With no listing specifically by name, and no generic end-use descriptions, the last possibility is a generic n.o.s. description by classification. Looking again at the commodity list for the alphabetical reference to the known classification, one finds "Flammable liquid, n.o.s." as a shipping name in Roman type. This is the proper shipping name for cyclopentene.

In Chapter 11 on shipping papers, note that when the n.o.s. shipping name is the same as the description of its classification, there is no need to repeat the two, so that shipping papers for this commodity can read: "Flammable liquid, n.o.s.," rather than "Flammable liquid, n.o.s., Flammable liquid". In order to familiarize yourself with the selection of the proper shipping name of materials, it is recommended that you develop a few examples and determine their proper DOT shipping name.

Having covered the basic rationale for shipping names, the following narrative in the form of a flow chart should help you select shipping names in the future.

GUIDE TO DOMESTIC PROPER SHIPPING PAPER DESCRIPTIONS FOR PACKAGED HAZARDOUS MATERIALS

In a series of changes to its Hazardous Materials Regulations published in the Federal Register during 1980 and 1981, the Department of Transportation (DOT) altered virtually all shipping paper descriptions for hazardous materials. In continuing rule changes affecting shipping names and classifications, DOT incorporated new hazardous substance shipping name regulations, the latest publication being the February 17, 1987 correction in Docket No. HM-145F. This guide has been prepared to provide assistance in making necessary changes in existing company shipping descriptions.

The DOT rules for "building" a proper shipping description have become very complex because of seemingly endless special circumstances and varieties of shipping name qualifiers. The assignment of shipping descriptions needs to be carefully managed to avoid oversights. To assist you, this guide presents a checklist. It is an attempt to address each description element in a logical seqence, to build a complete description. The guide also can be used after the fact to identify omissions in descriptions already in use.

There is considerable flexibility in the choice of acceptable DOT shipping paper descriptions and shipping name package markings. Frequently, there may be four or five different, all legally correct, ways to describe a product. It is very important to realize that if one arrives at a different description than one determined by using the methodology in this guide, it should not be assumed that there is an error.

Both may be correct. The guide sets forth an order of priority between five generic shipping name categories that are referenced ambiguously in the DOT rules (Section 172.101(c)(13)(ii)). The DOT rules require use of the most specific description without clearly setting forth priorities beyond technical nomenclature and non-chemical terms. On the basis of knowledge, experience and history, the authors of the guide have set forth a series of descriptive categories in descending order of priority.

In using this guide, you should note the following points regarding the DOT hazardous materials table:

- Shipping names may be singular or plural.
- Italics in the DOT list are not a required part of the shipping name.
- N.O.I., N.O.I.B.N., and N.O.S. are interchangeable.
- Word sequences within shipping names are optional.
- Concentration of chemicals may be expressed as the range in the table or as an exact concentration.
- Use of the prefix "mono" is optional in chemical shipping names.

The guide should *not* be used for the following:

- Consumer Commodity, ORM-D.
- Empty containers.
- Bulk shipments via tank car or cargo tank.
- Organic peroxides.
- Explosive material or blasting agents.
- Radioactive material.
- Forbidden material.
- Etiologic agents.
- Export shipments.
- HM-196 materials

To use this guide, you *must* know:

- The DOT classification definitions.
- EPA 40 CFR Part 261 as it defines or lists hazardous wastes.
- EPA and DOT regulations as they cover hazardous substances.
- Whether the product is one chemical, or a solution or mixture.
- Whether the ingredients of a solution or mixture meet any DOT hazard class.
- Whether the solution or mixture contains any hazardous substance.

This guide requires you first to catalog your material in one of seven categories which are found below in Items I through VII, respectively. But first, it requires you to ascertain if your material is an EPA hazardous substance. For ease of reference, they may be summarized as follows:

I. Material is a straight chemical, specifically named in Section 172.101. See page 63.

II. Material is a mixture or solution containing only one hazardous material specifically named in Section 172.101. See page 65.

III. Material is a mixture or solution containing more than one hazardous material specifically named as a mixture or solution in Section 172.101. See page 67.

IV. Material is not specifically named in Section 172.101, nor is it a named mixture or solution, but is identified in the table by a chemical family or grouping, n.o.s. See page 70.

V. Material does not fit any of the above categories, but is covered in Section 172.101 by a generic description which is not an n.o.s. See page 72.

VI. Material is none of the above but is covered by an end-use n.o.s. description. See page 74.

VII. Material is none of the above and is covered by a DOT classification n.o.s. description. See page 76.

For your convenience, reproduced at page 79 is 49 CFR 173.2, the section of the DOT regulations that prescribes the priority to be followed in classifying a material that meets more than one DOT classification definition.

INTRODUCTORY DETERMINATIONS

1. Is the material named specifically, or is it a listed or unlisted hazardous waste, in the Appendix to 49 CFR 172.101?

   No—Go to Item I, page 63.
   Yes—Go to next question.

2. Does the package contain a reportable quantity of the material described in the Appendix?

   No—Go to Item I, page 63.
   Yes—Go to next question.

3. If there is a reportable quantity in the package, does the concentration in the dilution chart in the definition of a hazardous substance in 49 CFR 171.8 exceed the number related to the material's RQ?

   No—Go to Item I, page 63.
   Yes—Go to to next question.

4. Is the name of the material in the Appendix followed by an asterisk?

   No   Give it the proper shipping name "Hazardous Substance, liquid (*or solid*), n.o.s." followed by the chemical name in parentheses, the classification "ORM-E" and the ID number "NA9188." The letters "RQ" must precede or follow this description. If it is a hazardous waste, immediately precede the proper shipping name with the word "Waste." In addition, following the proper shipping name or the complete description, include in parentheses the EPA Waste number or the name of the material.

Yes   Go to the DOT Table in 172.101 and using the proper shipping name given in the Hazardous Materials Table, follow the appropriate Item I through VII in this guide that corresponds to the make-up of the product named in Section 172.101.

ITEM I: SPECIFICALLY NAMED MATERIAL

1. Is the material *specifically* named in 49 CFR 172.101? (i.e., actual chemical name, not a mixture or solution, except that in the case of corrosive liquids water is not considered an added ingredient.)

    No—Go to Item II, page 65.
    Yes—Go to next question.

2. Does name in Section 172.101 have + mark?

    No—Go to next question.
    Yes—Go to question 4.

3. Does the classification given in Section 172.101 for the named material agree with its DOT hazard property? (Note that if the specifically named material has more than one DOT hazard, the classification assigned must apply. If, however, it has only one DOT hazard and that hazard is different from the class shown, and the entry is not preceded by a + mark, the shipper *must* reclassify according to actual hazard.)

    No   Assign proper classification and give it proper shipping name listed in Section 172.101, according to the following sequence:
        a. Chemical family or chemical grouping n.o.s. Go to Item IV, page 70.
        b. Generic description which is not class or n.o.s. Go to Item V, page 72.
        c. Generic description which is an n.o.s. Go to Item VI, page 74.
        d. DOT class n.o.s. description. Go to Item VII, page 76.
    Yes—Go to next question.

4. Does name have a UN or NA number?

    No   #1 Use name and classification shown. Go to next question.
    Yes   #2 Use name, classification and number shown (Section 172.202(a)). *Example*, "Ethyl alcohol, Flammable liquid, UN 1170." Go to next question.

5. Is it a hazardous waste?

> No—Go to next question.
>
> Yes   Use #1 or #2 from Question 4, preceded by word "Waste" (Section 172.101(c)(10)). *Example* (per #2, above), "Waste ethyl alcohol, Flammable liquid, UN 1170." Go to next question.

6. Is the material a hazardous substance, or a listed hazardous waste in a reportable quantity, or an unlisted hazardous waste (ICR) in a quantity of 100 pounds or more in one package? (It is a hazardous substance if you answered "Yes" to questions 1 through 3 of the "Introductory Determinations.")

> No—Go to next question.
>
> Yes   Use the above applicable input and precede or follow the complete description with the letters "RQ." If it is a hazardous waste, add the EPA Waste number, or the material's name in parentheses following the proper shipping name. For EPA ICR wastes, you may add the letters "EPA" followed by the word "ignitability," "corrosivity" or "reactivity," as appropriate, instead of the EPA waste ID number. Go to next question.

7. Is the material preceded by an "A" and/or a "W" symbol in column 1 of Section 172.101?

> No—Go to next question.
>
> Yes   If the material is not a hazardous waste and does not contain any hazardous substance, and shipment is not by air or water, this product is not DOT regulated and the DOT shipping description rules do not apply. For shipment by air (A) or water (W), if letter before name corresponds to mode being used, go to the next question.

8. Is it DOT poisonous (systemic)?

> No—Go to next question.
>
> Yes   Use above applicable input followed by the word "Poison" if the product is not classed in Section 172.101 as a poison A or B (Section 172.203(k)(2)). Go to next question.

9. Is it Dangerous When Wet?

> No—Go to next question.
>
> Yes   Use above applicable input followed by

the words "Dangerous When Wet"(Section 172.203(j)). Go to next question.

10. Is it a limited quantity?

    No—Go to next question.

    Yes   Use above applicable input followed by phrase "Ltd. Qty." or "Limited Quantity" (Section 172.203(b)). *Example*, "Ethyl alcohol, Flammable liquid, UN 1170, Ltd. Qty." Go to next question.

11. Is it covered by an exemption?

    No—Go to next question.

    Yes   Use above applicable input followed by "DOT-E"and the four-digit number assigned by DOT (Section 172.203(a)). *Example*, "Ethyl alcohol, Flammable liquid, UN 1170, Ltd. Qty, DOT E1234." Go to next question.

12. Is it being shipped by air?

    No   Use description as concluded from above input.

    Yes—Go to next question.

13. Is it prohibited on passenger aircraft?

    No   Use description as concluded from above input.

    Yes   Add phrase: "Cargo aircraft only" to description from above.

ITEM II: SPECIFICALLY NAMED MATERIAL IN MIXTURE, SINGLE HAZARDOUS INGREDIENT

1. Is the material a mixture or solution containing one hazardous material named as a straight chemical in Section 172.101 and other ingredients that are non-hazardous by DOT standards for the modes of transportation involved?

    No—Go to Item III, page 67.
    Yes—Go to next question.

2. Is the classification of the solution or mixture the same as the classification of the named ingredient?

    No—Go to Item IV, page 70.
    Yes—Go to next question.

3. Does the name have a UN or NA number?

No   #1 Use name shown with the additional word "mixture" or "solution" as appropriate followed by classification (Section 172.101(c) (11)). Go to next question.

Yes   #2 Use name shown with the additional word "mixture" or "solution" as appropriate followed by classification and number (Section 172.101(c)(11)). *Example*, "Ethyl enediamine solution, Corrosive material, UN 1604." Go to next question.

4. Is it a hazardous waste?

No—Go to next question.

Yes   Use #1 or #2 above preceded by word "Waste" (Section 172.101(c)(10)). *Example*, "Waste ethylenediamine solution, Corrosive material, UN 1604." Go to next question.

5. Does the material contain a hazardous substance, or is it a listed hazardous waste in a reportable quantity, or an unlisted hazardous waste (ICR) in a quantity of 100 pounds or more in one package? (It is a hazardous substance if you answered "Yes" to questions 1 through 3 of the "Introductory Determinations.")

No—Go to next question

Yes   Use the above applicable input and precede or follow the complete description with the letters "RQ." If it is a hazardous waste, add the EPA Waste number, or the material's name in parentheses following the proper shipping name. For EPA ICR wastes, you may add the letters "EPA" followed by the word "ignitability," "corrosivity" or "reactivity," as appropriate, instead of the EPA waste ID number. Go to next question.

6. Is the material preceded by an "A" and/or a "W" symbol in column 1 of Section 172.101?

No—Go to next question.

Yes   If the material is not a hazardous waste or does not contain any hazardous substance and shipment is not by air or water, this product is not DOT regulated and the DOT shipping description rules do not apply. For shipment by air (A) or water (W), if letter before name corresponds to mode being used, go to the next question.

7. Is it DOT poisonous (systemic)?

  No—Go to next question.

  Yes Use above applicable input followed by the word "Poison" if the product is not classed in Section 172.101 as a poison A or B (Section 172.203(k)(2)). Go to next question.

8. Is it Dangerous When Wet?

  No—Go to next question.

  Yes Use above applicable input followed by the words "Dangerous When Wet" (Section 172.203(j)). Go to next question.

9. Is it a limited quantity?

  No—Go to next question.

  Yes Use above applicable input followed by phrase "Ltd. Qty." or "Limited Quantity" (Section 172.203(b)). Go to next question.

10. Is it covered by an exemption?

  No—Go to next question.

  Yes Use above applicable input followed by "DOT-E" and the four-digit number assigned by DOT (Section 172.203(a)). Go to next question.

11. Is it being shipped by air?

  No Use description as concluded from above input.

  Yes—Go to next question.

12. Is it prohibited on passenger aircraft?

  No Use description as concluded from above input.

  Yes Add phrase: "Cargo aircraft only" to description concluded from above.

ITEM III: SPECIFICALLY NAMED MIXTURE OR SOLUTION

1. Is the material a mixture or solution containing more than one hazardous material and is it specifically named as a mixture or solution in Section 172.101?

  No—Go to Item IV, page 70.

  Yes

2. Does the name listed in Section 172.101 have a + mark?

> No—Go to next question.
> Yes—Go to question 4.

3. Does the material have only one DOT hazard, and the hazard is different from the classification shown in Section 172.101? (Note that if the specifically named material has more than one DOT hazard, the classification assigned must apply. If, however, it has only one DOT hazard, and that hazard is different from the class shown and the entry is not preceded by a + mark, the shipper *must* reclassify according to the actual hazard. See 49 CFR 173.2(b)(1) and 172.101(c)(13).)

> No—Go to next question.
>
> Yes   Assign proper classification and select the proper shipping name listed in Section 172.101, according to the following sequence:
>
> > a. Chemical family or chemical grouping, n.o.s. **Go to Item IV,** page 12.
> >
> > b. Generic description which is not the class or an n.o.s. **Go to Item V,** page 15.
> >
> > c. Generic description which is an n.o.s. **Go to Item VI,** page 18.
> >
> > d. DOT classification, n.o.s., description. **Go to Item VII,** page 22.

4. Does the name have a UN or NA number?

> No   #1 Use the name and classification shown. Go to next question.
>
> Yes   #2 Use the name, classification and number (Section 172.202(a)). *Example,* "Sodium methylate, alcohol mixture, Flammable liquid, NA 1289." Go to next question.

5. Is it a hazardous waste?

> No—Go to next question.
>
> Yes   Use #1 or #2 above preceded by word "Waste" (Section 172.101(c)(10).

6. Does the material contain a hazardous substance, or is it a listed hazardous waste in a reportable quantity, or an unlisted hazardous waste (ICR) in a quantity or 100 pounds or more in one package? (Is it a hazardous substance if you answered "Yes" to questions 1 through 3 of the "Introductory Determinations.")

> No—Go to next question.
>
> Yes   Use the above applicable input and precede or follow the complete description with the letters "RQ." If it is a hazardous waste, add the EPA Waste number, or the

material's name in parentheses following the proper shipping name. For EPA ICR wastes, you may add the letters "EPA" followed by the word "ignitability," "corrosivity" or "reactivity," as appropriate, instead of the EPA waste ID number. Go to next question.

7. Is the material preceded by an "A" and/or a "W" symbol in column 1 of Section 172.101?

No—Go to next question.

Yes   If the material is not a hazardous waste or does not contain any hazardous substance and shipment is not by air or water, this product is not DOT regulated and the DOT shipping description rules do not apply. For shipment by air (A) or water (W), if letter before name corresponds to mode being used, go to the next question.

8. Is it DOT poisonous (systemic)?

No—Go to next question.

Yes   Use above applicable input followed by the word "Poison" if the product is not classed in Section 172.101 as a poison A or B (Section 172.203(k)(2)). Also add the technical name of the principal constituent that causes the material to meet the definition of a DOT poison, if this name is not already included in the description. Go to next question.

9. Is it Dangerous When Wet?

No—Go to next question.

Yes   Use above applicable input followed by the words "Dangerous When Wet" (Section 172.203(j)). Go to next question.

10. Is it a limited quantity?

No—Go to next question.

Yes   Use above applicable input followed by phrase "Ltd. Qty." or "Limited Quantity" (Section 172.203(b)). Go to next question.

11. Is it covered by an exemption?

No—Go to next question.

Yes   Use above applicable input followed by "DOT-E" and the four-digit number assigned by DOT (Section 172.203(a)). Go to next question.

12. Is it being shipped by air?

>No Use description as concluded from above input.
>
>Yes—Go to next question.

13. Is it prohibited on passenger aircraft?

>No Use description as concluded from above input.
>
>Yes Add phrase: "Cargo aircraft only" to description concluded from above.

ITEM IV: CHEMICAL FAMILY, N.O.S., DESCRIPTION

1. If the material is not specifically named in Section 172.101, and it is not a mixture or solution specifically identified in Section 172.101, is it noted in Section 172.101 as a chemical family or grouping, n.o.s., together with the correct classification?

>No—Go to Item V, page 72.
>
>Yes

2. Does the name have a UN or NA number?

>No #1 Use the name and classification shown. Go to next question.
>
>Yes #2 Use the name, classification and number shown (Section 172.202(a)). *Example*, "Chlorate, n.o.s., Oxidizer, UN 1461." Go to next question.

3. Is it a hazardous waste?

>No—Go to next question.
>
>Yes Use #1 or #2 above preceded by word "Waste" (Section 172.101(c)(10)). Go to next question.

4. Does the material contain a hazardous substance, or is it a listed hazardous waste in a reportable quantity, or an unlisted hazardous waste (ICR) in a quantity or 100 pounds or more in one package? (It is a hazardous substance if you answered "Yes" to questions 1 through 3 of the "Introductory Determinations.")

>No—Go to next question.
>
>Yes Use the above applicable input and precede or follow the complete description with the letters "RQ." If it is a hazardous waste, add the EPA Waste number, or the material's name in parentheses following

the proper shipping name. For EPA ICR wastes, you may add the letters "EPA" followed by the word "ignitability," "corrosivity" or "reactivity," as appropriate, instead of the EPA waste ID number. Go to next question.

5. Is the material preceded by an "A" and/or a "W" symbol in column 1 of Section 172.101?

    No—Go to next question.

    Yes   If the material is not a hazardous waste or does not contain any hazardous substance and shipment is not by air or water, this product is not DOT regulated and the DOT shipping description rules do not apply. For shipment by air (A) or water (W), if letter before name corresponds to mode being used, go to the next question.

6. Is it DOT poisonous (systemic)?

    No—Go to next question.

    Yes   Use above applicable input followed by the word "Poison" if the product is not classed in Section 172.101 as a poison A or B (Section 172.203(k)(2)). Also add the technical name of the principal constituent that causes the material to meet the definition of a DOT poison if this name is not included in the description. Go to next question.

7. Is it Dangerous When Wet?

    No—Go to next question.

    Yes   Use above applicable input followed by the words "Dangerous When Wet" (Section 172.203(j)). Go to next question.

8. Is it a limited quantity?

    No—Go to next question.

    Yes   Use above applicable input followed by phrase "Ltd. Qty." or "Limited Quantity" (Section 172.203(b)). Go to next question.

9. Is it covered by an exemption?

    No—Go to next question.

    Yes   Use above applicable input followed by "DOT-E" and the four-digit number assigned by DOT (Section 172.203(a)). Go to next question.

10. Is it being shipped by air?

> No　　Use description as concluded from above input.
>
> Yes—Go to next question.

11. Is it prohibited on passenger aircraft?

> No　　Use description as concluded from above input.
>
> Yes　Add phrase: "Cargo aircraft only" to description concluded from above.

ITEM V: NON-N.O.S. GENERIC DESCRIPTIONS

1. If (a) the material is not specifically named in Section 172.101, (b) it is not a mixture or solution specifically identified in Section 172.101, and (c) it is not covered by a chemical family or chemical grouping n.o.s., is it described in Section 172.101 by a generic description which is not a hazard class or n.o.s., and which is paired with the proper classification?

> No—Go to Item VI, page 74.
>
> Yes

2. Does the name have a UN or NA number?

> No　　#1 Use name and classification shown. Go to next question.
>
> Yes　#2 Use name, classification and number shown (Section 172.202(a)). *Example*, "Coating solution, Flammable liquid, UN 1139." Go to next question.

3. Is it a hazardous waste?

> No—Go to next question.
>
> Yes　Use #1 or #2 above preceded by word "Waste" (Section 172.101(c)(10). Go to next question.

4. Does the material contain a hazardous substance, or is it a listed hazardous waste in a reportable quantity, or an unlisted hazardous waste (ICR) in a quantity or 100 pounds or more in one package? (It is a hazardous substance if you answered "Yes" to questions 1 through 3 of the "Introductory" Determinations.")

> No—Go to next question.
>
> Yes　Use the above applicable input and precede or follow the complete description with the letters "RQ." If it is a hazardous waste, add the EPA Waste number, or the

material's name in parentheses following the proper shipping name. For EPA ICR wastes, you may add the letters "EPA" followed by the word "ignitability," "corrosivity" or "reactivity," as appropriate, instead of the EPA waste ID number. Go to next question.

5. Is the material preceded by an "A" and/or a "W" symbol in column 1 of Section 172.101?

No—Go to next question.

Yes   If the material is not a hazardous waste or does not contain any hazardous substance and shipment is not by air or water, this product is not DOT regulated and the DOT shipping description rules do not apply. For shipment by air (A) or water (W), if letter before name corresponds to mode being used, go to the next question.

6. Is it DOT poisonous (systemic)?

No—Go to next question.

Yes   Use above applicable input followed by the word "Poison" if the product is not classed in Section 172.101 as a poison A or B (Section 172.203(k)(2)). Also add the technical name of the principal constituent that causes the material to meet the definition of a DOT poison if this name is not included in the description. Go to next question.

7. Is it Dangerous When Wet?

No—Go to next question.

Yes   Use above applicable input followed by the words "Dangerous When Wet" (Section 172.203(j)). Go to next question.

8. Is it a limited quantity?

No—Go to next question.

Yes   Use above applicable input followed by phrase "Ltd. Qty." or "Limited Quantity" (Section 172.203(b)). Go to next question.

9. Is it covered by an exemption?

No—Go to next question.

Yes   Use above applicable input followed by "E-" and the four-digit number assigned by DOT (Section 172.203(a)). Go to next question.

10. Is it being shipped by air?

    No   Use description as concluded from above
            input.

    Yes—Go to next question.

11. Is it prohibited on passenger aircraft?

    No   Use description as concluded from above
            input.

    Yes   Add phrase: "Cargo aircraft only" to de-
            scription concluded from above.

ITEM VI: N.O.S. GENERIC DESCRIPTIONS

1. If (a) a material is not specifically named in Section 172.101, (b) it is not a
mixture or solution specifically identified in Section 172.101, (c) it is not described in Section 172.101 by a chemical family or chemical grouping, n.o.s.,
and (d) it is not covered by a generic non-n.o.s. description, is it described in
Section 172.101 by an end use n.o.s. description with the correct classification?*

    No—Go to Item VII, page 76.

    Yes

2. Does this material have a UN or NA number?

    No   #1 Use name and classification shown.
            Go to next question.

    Yes   #2 Use name, classification and number
            shown (Section 172.202(a)). Go to next
            question.

3. Is it a hazardous waste?

    No—Go to next question.

    Yes   Use #1 or #2 above preceded by word
            "Waste" (Section 172.101(c)(10)).

4. Does the material contain a hazardous substance, or is it a listed hazardous
waste in a reportable quantity, or an unlisted hazardous waste (ICR) in a quantity or 100 pounds or more in one package? (It is a hazardous substance if you
answered "Yes" to questions 1 through 3 of the "Introductory Determinations.")

    No—Go to next question.

    Yes   Use the above applicable input and pre-
            cede or follow the complete description

---

*For "Hazardous waste, liquid or solid, n.o.s."   n.o.s.," "Corrosive liquid n.o.s." and "Corrosive
and "Hazardous substance, liquid or solid,   solid n.o.s" see Item VII page 76.

with the letters "RQ." If it is a hazardous waste, add the EPA Waste number, or the material's name in parentheses following the proper shipping name. For EPA ICR wastes, you may add the letters "EPA" followed by the word "ignitability," "coorosivity" or "reactivity," as appropriate instead of the EPA waste ID number. Go to the next question.

5. Is the material preceded by an "A" and/or a "W" symbol in column 1 of Section 172.101?

No—Go to next question.

Yes   If the material is not a hazardous waste or does not contain any hazardous substance and shipment is not by air or water, this product is not DOT regulated and the DOT shipping description rules do not apply. For shipment by air (A) or water (W), if letter before name corresponds to mode being used, go to the next question.

6. Is it DOT poisonous (systemic)?

No—Go to next question.

Yes   Use above applicable input followed by the word "Poison" if the product is not classed in Section 172.101 as a poison A or B (Section 172.203(k)(2)). Also add the technical name of the principal constituent that causes the material to meet the definition of a DOT poison if this name is not included in the description. Go to next question.

7. Is it Dangerous When Wet?

No—Go to next question.

Yes   Use above applicable input followed by the words "Dangerous When Wet" (Section 172.203(j)). Go to next question.

8. Is it a limited quantity?

No—Go to next question.

Yes   Use the above applicable input followed by phrase "Ltd. Qty." or "Limited Quantity" (Section 172.203(b)). Go to next question.

9. Is it covered by an exemption?

No—Go to next question.

Yes   Use above applicable input followed by

"DOT-E" and the four-digit number assigned by DOT (Section 172.203(a)). Go to next question.

10. Is it being shipped by air?

No   Use description as concluded from above input.

Yes—Go to next question.

11. Is it prohibited on passenger aircraft?

No   Use description as concluded from above input.

Yes   Add phrase: "Cargo aircraft only" to description concluded from above.

ITEM VII: Hazard Class N.O.S. Descriptions

1. If (a) the material is not specifically named in Section 172.101, (b) it is not a mixture or solution specifically identified in Section 172.101, (c) it is not covered by a chemical family or chemical grouping, n.o.s., (d) it is not covered by a generic non-n.o.s. description, and (e) it is not covered by an end use n.o.s. description, is it covered by a DOT hazard classification n.o.s. description?*

No   Seek specialized advice on how material should be handled.

Yes—Go to next question.

2. Does name in Section 172.101 have a + mark?

No   If it has more than one DOT hazard, is not an etiologic agent, and is not covered by one of the n.o.s. descriptions in Section 172.101, assign it the classification per Section 173.2. Otherwise go to the next question.

Yes   Use the name shown and go to the next question.

3. Does this material have a UN or NA number?**

No   Use name shown. Go to next question.

Yes   Use name and number shown (Section 172.202(a)). If the material is Hazardous

---

*For the purposes of this guide, "Hazardous substance, liquid or solid, n.o.s." is covered in the "Introductory Determinations." "Hazardous waste, liquid or solid, n.o.s." is considered a DOT classification n.o.s. description, although in fact this materials are classified as "ORM-E."

**If a material is properly described as a "Corrosive liquid, n.o.s." or "Corrosive solid, n.o.s." and the material is corrosive only to skin, the phrase "skin corrosive only" may be added to the basic description. This authorizes stowage below deck on vessels.

Substance n.o.s. or Hazardous Waste
n.o.s., or the name contains two DOT
class names, then also show the classification from the column 3 of Section
172.101, before the UN or NA number.
Go to the next question.

4. Is it a hazardous waste?

No—Go to next question.

Yes   Use material from questions above as applicable, preceded by the word "Waste"
(Section 172.101(c)(10)) unless the description is "Hazardous waste, liquid or
solid, n.o.s.". Go to next question.

5. Does the material contain a hazardous substance, or is it a listed hazardous
waste in a reportable quantity, or an unlisted hazardous waste (ICR) in a quantity or 100 pounds or more in one package? (It is a hazardous substance if you
answered "Yes" to questions 1 through 3 of the "Introductory Determinations.")

No—Go to next question.

Yes   Use the above applicable input and precede or follow the complete description
with the letters "RQ." If it is a hazardous
waste, add the EPA Waste number, or the
material's name in parentheses following
the proper shipping name. For EPA ICR
wastes, you may add the letters "EPA" followed by the word "ignitability," "coorosivity" or "reactivity," as appropriate instead of the EPA waste ID number. Go to
the next question.

6. Is the material preceded by an "A" and/or a "W" symbol in column 1 of Section
172.101?

No—Go to next question.

Yes   If the material is not a hazardous waste or
does not contain any hazardous substance
and shipment is not by air or water, this
product is not DOT regulated and the
DOT shipping description rules do not
apply. For shipment by air (A) or water
(W), if letter before name corresponds to
mode being used, go to next question.

7. Is it DOT poisonous (systemic)?

No—Go to next question.

Yes   Use above applicable input followed by
the word "Poison" if the product is not
classed in Section 172.101 as a poison A

or B (Section 172.203(k)(2)). Also add the technical name of the principal constituent that causes the material to meet the definition of a DOT poison if this name is not included in the description. Go to next question.

8. Is it Dangerous When Wet?

No—Go to next question.

Yes   Use above applicable input followed by the words "Dangerous When Wet" (Section 172.203(j)). Go to next question.

9. Is it a limited quantity?

No—Go to next question.

Yes   Use above applicable input followed by phrase "Ltd. Qty." or "Limited Quantity" (Section 172.203(b)). Go to next question.

10. Is it covered by an exemption?

No—Go to next question.

Yes   Use above applicable input followed by "DOT-E" and the four-digit number assigned by DOT (Section 172.203(a)). Go to next question.

11. Is it being shipped by air?

No   Use description as concluded from above input.

Yes—Go to next question.

12. Is it prohibited on passenger aircraft?

No   Use description as concluded from above input.

Yes   Add phrase: "Cargo aircraft only" to description concluded from above.

APPENDIX. 49 CFR SECTION 173.2, PRIORITY OF CLASSIFICATION

173.2 CLASSIFICATION OF MATERIAL

(a) Classification of material having more than one hazard as defined in this part. Except as provided in paragraph (b) of this section, a hazardous material, having more than one hazard as defined in (Part 173), must be classed according to the following order of hazards:

(1) Radioactive material (except a limited quantity).
(2) Poison A.
(3) Flammable gas.
(4) Non-flammable gas.
(5) Flammable liquid.
(6) Oxidizer.
(7) Flammable solid.
(8) Corrosive material (liquid).
(9) Poison B.
(10) Corrosive material (solid).
(11) Irritating materials.
(12) Combustible liquid (in containers having capacities exceeding 110 gallons).
(13) ORM-B.
(14) ORM-A.
(15) Combustible liquid (in containers having capacities of 110 gallons or less.)
(16) ORM-E.

(b) Exceptions. Paragraph (a) of this section does not apply to—(1) a material specifically identified in Section 172.101 of this subchapter;

(2) An explosive required to be classed and approved under Section 173.86, or a blasting agent required to be classed and approved under Section 173.114a.

(3) An etiologic agent identified in Section 173.386 as those materials listed in 42 CFR 72.3; or

(4) An organic peroxide. See Sections 172.101 and 173.151a of this subchapter.

(5) A limited quantity radioactive material that also meets the definition of another hazard class (see Section 173.421–2).

CHAPTER 7

Hazardous Materials Packaging in General

SUMMARY

Packaging for hazardous materials in transportation is one of the major elements of the Department of Transportation's regulatory system. This is not a new system, but one that has evolved over decades.

In general the packaging used for hazardous materials transportation is effective. There are major changes underway that could affect this system. Most countries in the world are in the process of authorizing hazardous materials packaging (other than bulk, or packaging for explosives or gases) constructed in accordance with United Nations (UN) performance standards. The U.S. also is engaged in this transition. It has been contended rather persuasively that these standards may in fact result in or stimulate a reduction in the quality of packaging that has been required by DOT for many years.

INTRODUCTION

This chapter pertains to packaging authorized by DOT for use in the transportation of hazardous materials in the United States. See Chapter 8 on how to select proper packaging for your hazardous material.

Hazardous materials have been regulated in U.S. transportation for more than 70 years, initially under the authority of the Explosives & Other Dangerous Articles Act, 18 U.S. Code 831–835 (since repealed). To a great extent, early regulations codified the responsibilities of shippers and rail common carriers as they had evolved under the common law. These obligations generally were for the shipper to know the hazards of the product, to package it safely, and to advise the carrier of those hazards. The common carrier's principal obligation was to provide service upon reasonable request, and to assure the product was delivered without damage to the consignee. The common carrier functioned basically as an insurer of the property he carried.

The original focus was on explosives, and then on what were called "other dangerous articles." These articles included flammable, corrosive, and poisonous materials, as well as compressed gases. The emphasis throughout the first fifty years of this regulatory program has been on these hazard classes, i.e., on materials that pose the significant potential to cause substantial injury if they are released from their packaging while in transportation.

Although the regulations became more complex through the years, these basic elements remain. The highway, water and air modes of transportation were added to the regulatory scope, and materials bearing other hazards (such as radiation or potential for environmental damage) were encompassed. Special requirements have been implemented for hazardous wastes. See Chapter 14.

The core of the DOT regulations, however, remains effective containment of materials that are likely to cause immediate injury in the event of a leak in transit, and the communication of the presence and nature of the hazard.

SIGNIFICANT TERMS AND CONCEPTS
Certain basic concepts in the hazardous materials packaging field are helpful to understand before beginning a discussion of the general topic.

A. SOLIDS, LIQUIDS, AND GASES
The goal of the packaging regulations is effective containment of the materials under conditions normally incident to transportation. Generally speaking, the steps that must be taken to contain solids are simpler than those for liquids, which are simpler than those for gases. Packaging required for corrosive solids, for example, may be stated only as "a metal drum"; corrosive liquids, however, would have to be packaged in a DOT *specification* drum built to detailed requirements; corrosive gases would have to be in DOT cylinders constructed in rigid compliance with detailed engineering specifications, with that construction observed by independent inspectors who prepare and maintain significant records on the steel, test results, etc.

B. SPECIFICATION AND NONSPECIFICATION PACKAGING
For higher hazards, or more difficult containment, DOT prescribes details of construction in the form of "specification" packagings in 49 CFR Parts 178 and 179. For materials of lower hazard, or that are easier to contain, the DOT regulations may authorize use of general types of "nonspecification" packaging without prescribing how that packaging must be constructed in any particular detail. It is understood that the marketplace (including requirements imposed by insurance carriers and transportation carriers, plus the shipper's and the consignee's desire to deliver product without loss or damage) will result in use of adequate nonspecification packaging. Generally speaking, specification packaging is significantly more expensive than nonspecification packaging.

C. BULK AND NONBULK PACKAGING
DOT's regulations apply to any size hazardous materials container. Some regulatory concepts apply equally to all packaging (such as the shipper's obligation to assure the compatibility of the hazardous material with the container), but most of the requirements are distinctly different depending on whether the material is shipped in bulk or not in bulk. As a general matter, the dividing line between the two terms is 110 gallons or 1,000 pounds water capacity. Bulk packagings include tank cars, tank trucks (called cargo tanks in the regulations), and portable tanks. Nonbulk packagings include drums, cylinders, boxes, cans, and bags.

D. REUSE OF PACKAGING
The more sturdy (and costly) a container, the more likely that it will be used more than once. Thus, virtually all of the bulk containers are intended to be

used over a period of years. Some smaller, heavy containers such as cylinders may be in use for many decades. Packaging that only is intended to be used once is marked "STC" (single-trip container) or "NRC" (nonreusable container). There are provisions for the reuse of some single-trip containers after proper reconditioning. Any container in use must be equivalent to a new container, so that periodic maintenance and repair of reusable containers is an important part of compliance.

E. REGULATIONS, EXEMPTIONS, AND APPROVALS

Regulations are published in 49 CFR. They have general applicability to everyone. Some general regulations require individual government approval. For example, a specification tank may be generally authorized, but anyone wanting to use that tank for hydrogen peroxide must have safety relief devices for it individually approved by DOT. Exemptions on the other hand are specific authorizations of alternative practices or packages, and these are issued in a public process to individual applicants. There often may be little difference between approvals and exemptions, other than the public nature of the procedures for the latter.

F. HAZARDOUS MATERIALS, HAZARDOUS WASTES, AND HAZARDOUS SUBSTANCES

See Chapters 3, 4, and 14 for more on what DOT regulates generally, and on hazardous wastes and substances specifically. The broad term "hazardous materials" encompasses wastes and hazardous substances. These latter concepts have evolved in environmental legislation (the Resource Conservation & Recovery Act, the Clean Water Act, and the Comprehensive Environmental Response, Compensation, & Liability Act). Generally, if a material poses the threat of acute injury to people in transportation, then it must be in DOT specification packaging. If the hazard is limited to the environment, or would require longer-term exposure than is commonly encountered in transportation, nonspecification packaging may be authorized and DOT requirements are limited to general packaging rules, hazard communications, and spill reporting.

G. RADIOACTIVITY AND OTHER HAZARDOUS MATERIALS

Rules for radioactive materials are of more recent origin than those for most other hazardous materials. Because of the unique nature of the public perception of radioactivity, and the specialized nature of the nuclear industry, packaging for radioactive materials is based on different concepts. The most fundamental difference is that for other hazards, packaging is designed for normal conditions of transportation with some exposure to abuse, while most radioactive materials packaging must be designed to survive transportation accidents. The findings and conclusions in this chapter should not be read as pertaining to radioactive materials.

HISTORICAL ORIGINS OF HAZARDOUS MATERIALS PACKAGING

As indicated above, early regulations generally codified the legal roles of shippers and rail carriers. Rules for other modes came later. A specialized element in the Association of American Railroads (AAR), called the Bureau of Explosives, managed the development of regulatory requirements before government regulation began. This effort was a relatively informal process, reaching accommodation between the interests of the shipper in moving his product cost-effectively, the interests of the railroads in minimizing damage to their equipment, other freight (for which they are an insurer), and their employees, and everyone's desire to avoid injury to the public.

Early packaging decisions were only as broad as the shipper's interest at the time. If the shipper were interested in using only a particular tank car, the decision-making process generally would not extend to other cars, trucks, or types of packaging. This pattern has continued, so that today's packaging authorizations have a spotty, somewhat erratic character to them. The fact that a particular type of packaging is not authorized for a material probably only reflects the fact that no shipper was interested in using it, not a conscious decision of regulators to limit the scope of available containers. DOT is planning to reduce this somewhat erratic characteristic in rule making in Docket No. HM-181, discussed more fully below.

In the Explosives & Other Dangerous Articles Act of 1908, 18 U.S. Code 831–835, the regulatory responsibility was entrusted to the Interstate Commerce Commission, but a specific provision in the law authorized continued reliance upon the Bureau of Explosives. The process generally involved a shipper or several similarly situated shippers, meeting with the Bureau of Explosives and perhaps other affected parties to determine how the interests of all parties could be satisfied. The conclusions of this effort were introduced as a proposal before the Interstate Commerce Commission and, barring unforeseen objections, the proposal became a legally binding regulation. Because most of the detailed aspects of these deliberations occurred before presentation to the ICC, and were not on the formal record, there is little evidence available upon which to determine today why a particular packaging may have been authorized several decades ago. This is not to say, however, that the shipper and carrier did not evaluate the package adequately—only that the record does not exist.

When DOT was created in 1966, the safety functions of the ICC and several other transportation agencies were transferred to DOT, including those pertaining to hazardous materials transportation in all modes. Since that time, the role of the Bureau of Explosives has diminished substantially, but the basic regulatory framework and the levels of safety established in packaging for many materials were already established.

GOVERNMENTAL PACKAGING STANDARDS

In 1966, DOT received the hazardous materials regulatory authority that had been exercised by the Interstate Commerce Commission, the Federal Aviation Agency, and the United States Coast Guard. Virtually all packaging decisions had been made by the ICC (using the work of the Bureau of Explosives), for all modes.

General packaging issues are now managed by DOT's Research & Special Programs Administration. Packaging issues that are limited to a single mode, such as tank truck, tank car, or tank vessel specifications, are delegated to the Federal Highway Administration, the Federal Railroad Administration, and the Coast Guard, respectively. There still is a modal dependence upon the technical expertise of the Research & Special Programs Administration, however. Despite the delegation, particularly in matters involving cargo tanks, this is a shared function.

NONGOVERNMENTAL STANDARDS

A variety of private groups promulgate standards or specifications for packaging in addition to the Department of Transportation. The most common of these groups are rail or highway carriers. By tariff, rail (Uniform Freight Classification (UFC)) and truck (National Motor Freight Classification (NMFC)) common carriers establish packaging standards which are applicable particularly to nonhazardous materials. Other carriers, notably United Parcel Service, have established packaging standards in conjunction with the National Safe Transit Association (NSTA). The UFC, NMFC, and NSTA specifications are related specifically to the capability of shippers to successfully file claims for loss or damage to their freight. The carrier's liability for claims generally is contingent upon the shipper having used carrier-recommended packaging. These carrier standards form a baseline of quality in packaging for the transportation of all materials, including hazardous materials.

Other nongovernmental bodies have packaging standards, or standards relating to packaging, including the Association of American Railroads (AAR), the American National Standards Institute (ANSI), the American Society for Testing Materials (ASTM), and the American Society of Mechanical Engineers (ASME). Their standards may be incorporated by reference in other requirements, contracts, and governmental codes.

HAZARDOUS MATERIALS TRANSPORTATION ACT

This enabling legislation, (49 U.S. Code 1801, *et seq.*), referred to as the HMTA, supports the DOT hazardous materials regulatory program. See Chapter 30. It was enacted in 1975 and has superseded the Explosives & Other Dangerous Articles Act.

Among other things, the HMTA expanded DOT's scope of regulatory authority to include those who manufacture, mark, or sell packaging represented to be proper for the transportation of hazardous materials (49 U.S. Code 1804). Included within this population of regulated industry are those who repair, recondition, test, retest, or maintain packaging. Prior to the adoption of regulations in 1977 implementing the HMTA, container makers were not regulated directly by DOT. The only governmental controls were indirect, through regulatory obligations on shippers to use packaging that had been properly fabricated.

The container specifications published in Part 178 originally were developed through the Bureau of Explosives processes. These included input from the container makers, shippers, and carriers. The Bureau of Explosives had several field inspectors who frequently made compliance inspections of container makers' facilities. For many years, at least in the packaging industry, the Bureau of Explosives was synonymous with the government.

Vestiges of their role are seen in the DOT approvals program. In many cases, while basic design features of a container would be published, certain aspects were left to individual approvals by the Bureau of Explosives inspectors. When counsel for the ICC and then DOT advised that these approvals were unlawful delegations of authority to a nongovernmental agency, the process of bringing the approvals back into the government was begun. Many of these involve DOT review of Bureau of Explosives decisions (see 49 CFR 173.21(d) and 173.86), while others vest all approval authority within DOT. Except for certain functions related to tank car and vessel construction, delegations of authority from DOT to nongovernmental bodies have been withdrawn.

DIRECT REGULATION OF CONTAINER MAKERS

Under 49 CFR 173.22, a shipper may take at face value the specification markings on packaging. Thus, although the shipper is responsible for using the proper packaging, that responsibility does not extend beyond assuring that the packaging is represented by its maker as being proper.

The inability of DOT to directly regulate package makers, who might represent that containers are proper for hazardous materials that are not in fact proper, led DOT to request the authority under the HMTA to regulate container makers directly.

The existing container specifications were shifted from industry guidelines that were not binding on the container makers to law that could be enforced criminally, without any significant adjustments in the language. There were those who expressed concern at the time that there were ambiguities and uncertainties in the regulations that became critical in light of criminal sanctions. DOT replied:

> Container manufacturers' own past practices under the "shipper" regulations with regard to manufacture and testing provide guidance as to the purpose and meaning of the regulations governing the design and construction of containers;

> Container manufacturers' long exposure to the past practices of the DOT with regard to interpretations and enforcement of manufacture and testing specifications with respect to shippers also provide guidance as to future DOT practices with regard to manufacturers;

> As in the case of any Federal compliance or enforcement action, any container manufacturer who may be accused of non-conforming to a regulation which he considers vague or ambiguous is entitled to assert that as a defense; and

> Container manufacturers have the same rights and opportunities to request interpretations of and changes to the regulations as are currently exercised with great frequency by the shipper and carriers who have long been directly subject to the Federal hazardous materials transportation regulations.

Docket No. HM-138, 41 Fed. Reg. 38175, Sept. 9, 1976.

"REPRESENTATION" UNDER THE HMTA

Section 1804(c) of the HMTA says that, "No person shall, by marking or otherwise, represent that a container or package for the transportation of hazardous materials is safe, certified, or in compliance with the requirements of this Act, unless it meets the requirements of all applicable regulations issued under this Act." This is the only statutory duty placed upon any regulated party under the HMTA.

The marking of a container as being proper for use constitutes a certification of compliance, under 49 CFR 178.0–2(b) and 178.0–3. Although the rule says the marking is the responsibility of the container maker, by interpretation DOT has indicated that anyone can apply the markings but, by so doing, takes on the regulatory liability of a container maker.

DOT is given the authority to "register" any regulated party under 49 U.S. Code 1805(b). The legislative history of the HMTA clearly confirms that this is not a licensing authority. Registration has been used infrequently by DOT, and usually with regard to the makers of packaging. See 49 CFR 178.651–4(d), for example.

BASIC PACKAGING CRITERIA

All packaging for hazardous materials, whether specification or nonspecification, must meet the general requirements of 49 CFR 173.24. A packaging must be "so designed and constructed, and its contents so limited, that under conditions normally incident to transportation

> (1) There will be no significant release of the hazardous materials to the environment;
> (2) The effectiveness of the packaging will not be substantially reduced, and
> (3) There will be no mixture of gases or vapors in the package which could, through any credible spontaneous increase of heat or pressure, or through an explosion, significantly reduce the effectiveness of the packaging.

Packaging materials and contents shall be such that there will be no significant chemical or galvanic reaction among any of the materials in the package. Closures must prevent leakage and use gaskets that will not be deteriorated by the contents. There are specific requirements for compatibility and permeability of polyethylene packaging.

DOT did issue several contracts a few years ago to study the transportation environment, portions of which have been incorporated in part into the regulations.

CURRENT NONBULK PACKAGING

All types of packaging are used in transportation. These include combination packaging, such as bottles in fiberboard boxes, composite packaging such as a plastic liner in a steel drum, as well as free standing single units such as steel drums and cylinders for gases. Materials of construction include fiberboard, plastic, wood, glass (fiberglass), paper, metal and combinations of these.

Most containers can be used for a multitude of products, although certain types of packaging are designed for a particular commodity. In addition to compatibility with the contents, factors which heavily influence container design relate to the transportation system. For example, under federal statutes, maximum vehicle sizes on the federal system have been established. Packaging to go on these trucks should be designed to facilitate loading, unloading, and an efficient use of vehicle space.

The basic size and configuration of rail cars as well is determined by the width of track, and size of tunnels, bridges, and the like through which the car must pass. Given a uniform car configuration, the packaging configurations that will work well in that space become more limited.

Other factors are involved in packaging design, such as the type of handling equipment that will be used. The steel drum, for example, has been called "man-sized" packaging, because it is about the largest unit that can be handled by a single person and will fit through a regular doorway. Products that are only used in small unit quantities generally are only transported in those quantities.

CURRENT BULK PACKAGING

Bulk packaging includes tank cars, cargo tanks (tank trucks), and portable tanks. These packagings differ significantly from one another, as well as from nonbulk packagings.

A. TANK CARS

DOT prescribes tank car specifications in 49 CFR Part 179. After construction, the qualification, maintenance, and use of tank cars is governed by 49 CFR 173.31, as well as individual commodity sections of the regulations.

Generally speaking, the tank car specifications have been developed from industry standards adopted by the Tank Car Committee of the Association of American Railroads. This committee prepares specifications for nonhazardous materials cars, and is heavily involved in the hazardous materials car specifications as well. In addition, the Mechanical Division of the Association of American Railroads is involved in the approval of tank car designs, materials and construction, as well as the conversion or alteration of tank cars. See 49 CFR 179.3.

Tank cars are not owned by railroads but, rather, are commonly owned by shippers or lessors who lease them on a long term basis to shippers.

Several years ago, a series of derailments involved the puncture of flammable gas tank cars by the couplers of adjoining cars. The ignited material venting from the punctured car impinged upon other flammable gas cars in the derailment, simultaneously heating and expanding their contents beyond the capacity of safety relief devices, and weakening the tank shell at that location.

DOT rule making led to the installation of "head shields" on many of these cars, thereby lessening the potential for coupler impact with the tank. The agency also mandated installation of couplers recommended by industry that would be less likely to disengage, particularly in a vertical direction that would propel the coupler into the next car. This rule making also called for the installation of insulation on these flammable gas cars.

Since the implementation of these remedial measures and others developed by the AAR, the Tank Car Committee, the Federal Railroad Administration, and the Research & Special Programs Administration, with the assistance of the Railway Progress Institute, the accident picture for hazardous materials in the rail mode has improved dramatically.

B. TANK CAR TANKS

Certain multimodal pressure vessels (Specifications 106 and 110) are meant to be lifted on and off vehicles that commonly are called ton tanks. Specifications for these are published as part of the tank car rules in 49 CFR Part 179. These are some of the most effective containers in commerce.

C. CARGO TANKS

In the highway mode, there is no corresponding group to that in rail, to deal with issues involving tank truck shipments of hazardous or other materials, although the Truck Trailer Manufacturers Assocation has a committee dealing with tank trucks. Within the railroad industry, there are a relatively unified and limited number of carriers, who by interchange agreements have established effective national standards of acceptability for rail cars. No parallel organization effectively represents the multitude of people operating trucks, or even those operating only tank trucks. The National Tank Truck Carriers, Inc., a conference of the American Trucking Associations does a great deal of work in this area, but primarily with larger interstate motor common carriers. Deregulation of many aspects of the trucking industry has made the formation of a cohesive group even more unlikely.

The AAR-related efforts govern each shipment of a tank car. Each accepting carrier is obligated to perform inspection functions upon receipt of the car, and undertakes certain liabilities as a result of that inspection.

There is no corresponding inspection force for the tank truck industry. Many cargo tanks are owned by the carrier, unlike tank cars which are seldom if ever owned by the rail carrier. Although there is a self-inspection requirement in the Motor Carrier Safety Regulations, 49 CFR Part 396, it does not have the same legal import as the rail carriers' inspection upon the receipt of a tank car from a shipper.

Periodic inspections and tests are necessary for cylinders and for bulk equipment, because those containers generally are reused without reconditioning. A tank car or truck may be in use for many years. Maintenance is required for cargo tanks in accordance with 49 CFR 173.33 of the regulations. There is no motor carrier industry interchange agreement, however, to police compliance with these requirements, and so enforcement is a government rather than a commercial matter.

In addition to the large number of diverse owners of cargo tanks, a significant complicating factor is the applicability of the federal truck regulations almost exclusively to carriers engaged in interstate commerce. This was the scope of regulatory coverage under the predecessor statute to the Hazardous Materials Transportation Act, and it was maintained as a voluntary limitation on the agency's regulations in DOT rule making.

As a result, a significant volume of commerce in tank trucks occurs outside the exercised jurisdiction of the Department of Transportation. Although several States have adopted the DOT regulations, and under federal grant programs more States are being encouraged to do so, there are locations where there are no packaging regulations for the bulk delivery of hazardous materials by highway.

This has many impacts, one of which is that it appears that intrastate carriers have become the market for used equipment that no longer meets federal standards. There also are new tanks built for the intrastate trade that do not meet DOT specifications, because those specifications are inapplicable. Such tanks also do not need to meet the periodic retest and maintenance requirements prescribed in the federal rules. The noncompliance of these tanks with federal standards has caused severe administrative problems for States implementing the federal rules. DOT has addressed this problem to a limited extent in rule making (see 49 CFR 173.315(m)), and through exemptions.

Cargo tank specifications generally have been developed in some isolation from the rest of the motor vehicle unit. In other words, while studies have shown that much of a tank truck's stability and resistance to rollover is dependent upon factors such as the suspension, fifth wheel, and tires of the vehicle, these factors are not taken into account in the DOT tank specifications.

A major DOT rule making on cargo tanks is in process, and this is expected to address some but not all of the difficulties that have been and are being encountered.

D. PORTABLE TANKS

A portable tank is a bulk unit (other than a ton tank described above) that is meant to be lifted off transportation equipment, unlike tank cars and cargo tanks that are permanently attached to or because of their size are loaded and unloaded without being removed from the transportation vehicle.

Portable tanks may be designed for use in a single mode of transportation, but are mostly intermodal. Only one series of tank specifications is designated Inter-Modal (IM), but most portable tanks are actually intermodal containers by usage if not by DOT title. In recent years this confusion in use of terms has caused serious difficulties in understanding the DOT regulations and complying with them.

Because of significant labor and material cost differentials, most intermodal portable tanks in use today are manufactured outside the United States. They may be constructed to U.S. specifications, however, and the builders may have to meet ASME, AAR, and other nongovernmental construction and inspection codes. There are a limited number of builders of portable tanks in the U.S.

DOT-designated IM tanks are used in multimodal service, that is, they can be loaded aboard ship and, upon arrival in the port area, can be transferred to rail cars or truck chassis. The IM tank specifications have been influenced substantially by international concerns expressed through the recommendations of the International Maritime Organization (IMO). The IMO tank designs have been adapted in the DOT specifications IM 101 and IM 102, for liquids and solids. IMO has also developed a specification for a portable tank for gases.

Like tank cars, these portable tanks are owned or leased by shippers. For rail transportation, the AAR has developed a portable tank standard (AAR 600) to set certain requirements above DOT IM tank minimum requirements.

The qualification, maintenance and use of portable tanks other than IM tanks is prescribed in 49 CFR 173.32. Under 49 CFR 173.32a, IM tanks must also be "approved" by third-party agencies designated pursuant to 49 CFR Part 107, Subpart E. Their periodic testing, inspection and use are governed by 49 CFR 173.32b and 173.32c.

EXEMPTIONS

As noted earlier, hazardous materials regulation is an old field. It has grown piecemeal, and much of the structure and original packaging specifications date back several decades. These generally are considered design specifications, prescribing in detail the features the packaging should include, in contrast to performance standards discussed more extensively below.

Because of the preciseness of the specifications, there is no substantial latitude for variation or innovation. While this has brought a measure of beneficial uniformity and familiarity to hazardous materials packaging, it also has been criticized as an unnecessary impediment to new product development.

For many years, new packaging ideas have been developed by container makers and shippers, and then they have been discussed with the regulatory agencies. If the new packaging appeared to be well designed, and to have been successfully tested, the agencies would approve its use, but seldom would immediately authorize general use of that packaging by everyone. There are many reasons for this, but the primary one is that no amount of testing or calculation can anticipate the variations that will be encountered in actual experience. Thus, the people involved in the development of the container would obtain an authorization to use that packaging. This was not so much a matter of experimentation, as of verification of conclusions already justified by tests and calculations. In effect, this was a type of registration program, identifying the people involved and having periodic contact with them, particularly in the event of adverse experience.

These individual authorizations used to be called special permits. In the HMTA, their issuance came to be more substantially controlled and limited, and they then came to be called exemptions. Each is issued for a specific period of time, up to two years, and upon a showing of successful experience, an exemption can be renewed. The exemption holder must report any adverse experience under the exemption as soon as practicable. This reporting is in addition to any other incident reporting requirements. There is an on-going rule making (Docket No. HM-139) in which successful exemptions are incorporated into the rules of general applicability.

There is an opportunity for exemptions to become administrative problems for both industry and the government, and that occurs when rule making is too slow to keep up with the demand for this type of innovation. Under 49 CFR Part 107, the exemption is issued to the original applicant. Other persons can become "parties to" that exemption. In addition, the exemption might be issued to a container maker who is authorized to "manufacture, mark, and sell" a container, and the users of the container need only have a copy of the exemption at each shipping location. See 49 CFR 173.22a.

PACKAGING FOR HAZARDOUS WASTE

Hazardous wastes are designated by the Environmental Protection Agency (EPA), which says in 40 CFR 262.30 that waste generators should employ packaging authorized by DOT in 49 CFR Part 173. See Chapter 14 on hazardous waste.

Many, but by no means all, hazardous wastes were regulated by DOT under the original hazardous materials transportation regulations. The definitions of hazard used by EPA are broader in some respects, particularly with regard to flammability and corrosivity. EPA's concerns, of course, extend beyond transportation and include waste generation and disposal, whereas DOT's interest is limited to transportation.

Wastes that meet DOT's hazard classifications, as they existed before creation of EPA's program in 1980, must be packaged the same as if they were products having commercial value. A more limited number of wastes that fall outside the DOT hazard definitions due to the broader EPA definitions are generally classed as "Other

Regulated Materials, Group E," or simply ORM-E. One must first assess the properties of a waste, and then compare them with the DOT hazard classification definitions, to determine whether one of those classes applies. If none does, then the proper class will be ORM-E. Nonspecification packaging is authorized for ORM-E materials.

DOT has two particular packaging rules for waste shipments that do not apply to commercial products. The first is 49 CFR 171.3(e), which authorizes an open head steel drum even though a closed-head would be prescribed for that hazard, if "the hazardous waste contains solids or semisolids that would make its placement in a closed head drum impracticable."

The second is 49 CFR 173.28(p), which authorizes the shipment of wastes in used packaging that has not been reconditioned or retested, under certain qualifications including a limitation to transportation by highway. This packaging must be held for 24 hours after loading, but it need not be leak tested as other reconditioned packaging must be.

The general packaging rules apply to all hazardous wastes, so that the shipper is responsible to assure the compatibility of the waste with the packaging used, including cargo tanks that may be supplied by the carrier. There is a presumption in the DOT regulations that the shipper of a product is the one most familiar with its properties. This presumption may be valid for the manufacturer of the product, but it is less so for the user of it, who must put the used and perhaps contaminated product back into transportation as a waste.

There have been many recorded instances, particularly with corrosive materials, where the packaging has been detrimentally affected by the waste because of corrosion, galvanic reaction, or the generation of heat or gas. This is an area that warrants greater examination by DOT and EPA, and a reevaluation of the presumption of the shipper having sufficient knowledge to select packaging for hazardous waste.

PERFORMANCE STANDARDS

The concept of performance standards has been advocated by most people ever since there has been regulation of packaging. Performance standards say what a packaging must do, not how it must be constructed.

The difficulty with performance standards is that either no one has documented the transportation environment sufficiently, to know what a packaging must do, or at least that agreement cannot be reached on what the tests representing that environment must be.

The Airlie House recommendations of 1969 pushed the concept of performance standards for packaging. The incentives for such standards are (1) elimination of much of the government program granting exemptions, approvals, or specific rule making petitions on packaging, (2) elimination of much of the existing volume of complex regulations, and (3) opening the door to new packaging product development.

Contracts were let by DOT in an unsuccessful or at least incomplete attempt to define the transportation environment. Other nongovernmental groups, such as United Parcel Service (and the National Safe Transit Association) have defined their own transportation environment, but their experience may not be readily transferrable to other modes.

While everyone recognizes the benefit of performance standards, they have not been well developed to date. As discussed below, in the portion of this chapter on international influences, many factors other than actual transportation experience enter into the development of performance standards.

Performance standards remain theoretically viable, but will probably be eclipsed by the adoption of international standards.

INTERNATIONAL INFLUENCES

The United States has participated in the development of international hazardous materials transportation regulations for more than 20 years. The U.S. has chaired the UN Committee of Experts on the Transportation of Dangerous Goods for many of those years. The working groups reporting to the Committee are the Group of Rapporteurs and the Committee of Experts on Explosives, on which the U.S. has played a continuing role.

The Committees and the Group are made up of the representatives of several nations, including the United States. The U.S. has one vote. The vast majority of the representatives are members of the European Common Market. Obviously, their votes significantly influence these decisions. Within Europe, it has been contended that the transportation system is substantially better in many respects than it is in the U.S., and the distances travelled often are shorter than in the U.S. or in much overseas transportation. Most distribution of European products can occur by rail, highway, or inland waterway.

The original goal of the Committee of Experts was to develop pure performance standards. Again, this project was undertaken without much information on the nature of the transportation environment. These standards had to be usable in nonindustrialized countries, and so the investment in technical equipment had to be minimized.

As each portion of the proposed standard was suggested, it appears that those portions that involved controversy or complexity were postponed or eliminated. Substantive revisions to the UN packaging rules were accelerated in the 1980s for a variety of reasons, including the creation of a new aviation code under the International Civil Aviation Organization (ICAO), and planned revisions to packaging rules within Europe. The streamlined performance standards as we now know them were completed for nonbulk and non-gas packaging with the UN adoption of performance-oriented revisions to Chapter 9 of their "Orange Book" of recommendations on the international transportation of dangerous goods, in December 1982. (See Appendix B.)

Chapter 9 describes generic types of packaging (e.g., 1A1 closed-head steel drums, 4G fibre drums), and subjects those types to drop testing, hydrostatic and "leakproofness" testing (for liquids), and stacking tests. There is no vibration test, and there is no test of puncture resistance. There is no firm basis upon which to correlate these performance tests with the transportation environment in any country.

ICAO was the first to implement revised Chapter 9, through their Technical Instructions on the transport of dangerous goods authorized December 31, 1983, and mandatory by 1990. The European road and rail regulatory conventions (RID/ADR) have implemented the system for flammable liquids, corrosives, and Class 6.1 poisons, but by grandfather provisions allow other packaging until 1990. The International Maritime Organization (IMO) will mandate its amendments in response to revised Chapter 9 by 1990.

There is significant pressure, therefore, particularly on and through multinational shippers, to conform the U.S. system to that being implemented elsewhere in the world. On behalf of the U.S. government, DOT issued an advance notice of proposed rule making on this subject in Docket No. HM-181, in 1982. A notice was published in 1987. The notice proposes at least reciprocity with international standards, if not wholesale adoption of them and abandonment of the historic DOT system.

Docket No. HM-181 includes the UN system of packaging, as well as UN materials classification, labeling, and shipping descriptions. There may be some U.S. exceptions to this because of inadequacies in the international standard that were highlighted in comments on the advanced notice. For example, DOT has proposed a vibration test capability in a further effort to mirror the as-yet-unquantified transportation environment in the performance standards.

Commenters on the advance notice generally noted a favorable reaction to the concept of performance standards, and the removal of unnecessary impediments to the flow of international commerce. Everyone also favors anything that can make the DOT exemption process either unnecessary, or at least less cumbersome.

Commenters who appeared to be skeptical of the proposal raised the question of whether the Chapter 9 standards alone will result in packaging that is adequate to withstand the rigors of international commerce. It has been contended that it is possible to make packaging that will withstand the UN performance tests but that will lose its contents quickly after being put into actual transportation. The DOT proposal addresses these concerns, but there will be pressure to avoid additional testing requirements for packaging used in the U.S. because they could become barriers to the free flow of commerce which was one of the basic reasons for the UN standards in the first place.

Another question raised in the discussion of UN standards is the fate of current U.S. container specifications. While the performance standards tend to encourage innovation, it is recognized that innovation will be a costly process, and that at least initially only the larger companies are likely to take advantage of the process. Innovation not only means substantial research, design and testing, but it also entails a greater measure of liability for failures. Thus, it is anticipated that smaller container manufacturers and shippers may need to limit themselves to proven packaging described in current 49 CFR.

One of the rationales upon which Docket No. HM-181 was founded was the proposed elimination of many pages of the Code of Federal Regulations, since the UN standards are shorter than the current specifications. To retain and publish both sets of rules would defeat this purpose.

It has been persuasively suggested that the existing standards be "frozen" as of a particular date, and that they then be authorized for continued use in that state for a decade or more. Anyone wanting to employ a variation on these old specifications would have to follow the new UN performance-oriented procedures. This authorization for continued use of the design specifications would not have to entail the concomitant obligation on the part of the government to print the specifications. There are several examples of past standards that continue to be authorized in a single sentence of federal regulation, but that are printed by and at the expense of trade associations and private publishers. This approach would preserve the existing specifications for smaller businesses and others who choose to use packaging of a known quality, and yet would permit the elimination of many CFR pages and much of the exemptions program.

INNOVATION IN PACKAGING

As indicated throughout this chapter, today's hazardous materials packaging was developed by container makers, shippers, and carriers as a cost-effective means of distributing hazardous materials without major risk to employees or the public. Included in the design influences were the anticipated rigors of transportation, and the size and configuration of transport equipment. The final product of these combined factors was presented to the regulatory agencies and, if the expected level of safety was provided, the packaging was authorized.

There is little indication that removal of regulatory constraints would bring about any significant change in packaging. The factors which led to the design were not regulatory, and remain relatively fixed. Developing a system of performance standards may smooth the authorization process, which has become cumbersome, but it alone is unlikely to stimulate greater innovation. It only has the potential of getting innovations implemented sooner than they are today.

It also is unlikely that totally new technologies (for example, packaging for cryogenic materials adopted into rules in Docket No. HM-115) would be accommodated in performance standards in any case. These technologies have encountered considerable delays in receiving general federal authorization. For example, completion of Docket No. HM-115 took seventeen years after the initial petition for rule making was filed. It is unquestionable that this delay inhibited innovation in that field. Adoption of performance standards alone will not improve this process. Thus, in addition to developing performance standards, DOT should give more expeditious consideration to major innovations and new technologies not addressed in those performance standards.

THIRD-PARTY CERTIFICATION

Under the UN performance standards, the design of a container must be tested, and successful passage of the tests noted by a certification. For the most part, this testing and certification within Europe has been done by government owned or otherwise designated testing laboratories. It is unclear how much the actions of these testing and certification facilities result in quality containers beyond simple passage of the few tests prescribed, but it is believed they have a positive influence on quality.

Under the U.S. packaging rules in 49 CFR, the marking of the specification (e.g., DOT 17E) constitutes a certification of compliance with that specification, including any prescribed tests. In this country, however, those tests and that marking and certification are usually done by the maker of the packaging.

With the implementation of UN Chapter 9 standards in Europe, through the European road and rail regulatory conventions (RID and ADR), it will be necessary for U.S. packaging shipped to Europe to comply with those standards. This includes UN marking and certification. What is unclear is the extent to which European nations will accept a manufacturers self-certification of compliance, when manufacturers within their own countries must get independent, third-party certification of their products from a government approved laboratory.

If a U.S. container maker wanted to get third-party certification from a government approved laboratory, the only way to do that was to ship the empty packaging to one of the laboratories in Europe. Several container makers did this, and found the process time-consuming and expensive.

DOT established a mechanism by which to grant government approval to third-party testing facilities. Use of these facilities is not required by the U.S. government. If a container maker or shipper wants third-party certification by a government recognized facility, however, it now is possible to obtain it within the United States.

Third-party inspection and testing is required for some U.S. containers, notably high pressure compressed gas cylinders. It also is required for intermodal portable tank construction. The AAR Mechanical Division is involved in tank car construction. For pressure vessels in all modes except rail, construction usually has to be done in accordance with rigid procedures of the American Society of Mechanical Engineers (ASME). There is no third-party role in the construction of nonpressure tank trucks.

Although self-certification is broadly advocated for smaller packaging, no one has suggested that third-party involvement be eliminated for larger units.

CHAPTER 8

Selecting Hazardous Materials Packaging

This chapter is devoted to how a shipper selects an appropriate packaging for regulated hazardous materials. See Chapter 7 for a general discussion of packaging under the DOT regulations.

This chapter addresses the use of DOT specification and nonspecification packaging under 49 CFR as it is in effect at the end of 1987. The reader should be aware of a general trend throughout the world to shift to UN performance oriented packaging standards. These are published in Chapter 9 of the UN Recommendations on the Transport of Dangerous Goods (Orange Book), Appendix B.

The UN recommendations in Chapter 9 of the Orange Book are also addressed in Chapter 25 of the *Red Book*. As noted there, the UN recommendations are implemented in other codes that may apply to a given shipment. Examples would be Annex I of the International Maritime Organization (IMO) International Maritime Dangerous Goods (IMDG) Code for international water shipments, or the Technical Instructions issued by the International Civil Aviation Organization (ICAO). These latter two codes have mandatory effective dates for UN-type packaging of January 1, 1990, and so one should expect to see UN-marked containers more frequently as that date approaches.

DOT has proposed the authorization of UN packaging in domestic as well as international commerce in the United States, in rule making Docket No. HM-181. The same has been done in Canada. Because of grandfather provisions, however, it is likely that the U.S. system of specification packaging described in 49 CFR in 1987 will remain in effect for many more years to come.

Without any doubt a fundamental key to safe transportation of hazardous materials is the selection of the right packaging for those materials. (As an aside, please note that in regulatory parlance, "package" is defined as the packaging material plus its content of hazardous materials as presented for transportation, while "packaging" refers to the assembly of the container and any other components of that packaging but does not include the intended contents.) The selection of a packaging for hazardous materials to transport those materials safely, legally, and successfully to the customer must represent the conclusion of a long and careful thought and experience process.

That package, bearing the name of the shipper or his logo, can create almost as many good or bad impressions of that company as the company sales force or its advertising. Experience shows that the public, customers, emergency personnel, carrier personnel, government inspectors and others involved in any hazardous materials accident often remember the name of the shipper far longer and more clearly than the chemical name of the product.

This situation, coupled with public safety considerations and the potential liabilities involved with non-compliance under federal safety regulations on packaging, should provide ample motivation to place the responsibility for packaging selection high on the corporate ladder. For years it appears that many firms have lived dangerously, delegating responsibility for packaging as a collateral function to an employee often greatly burdened with other responsibilities to which he necessarily may give his primary allegiance. The Hazardous Materials Transportation Act (see Chapter 30) and escalating liability have made such a practice an example of corporate irresponsibility.

The potential criminal and civil liability of corporate officers, and the enormous civil liabilities experienced by companies involved in personal injury lawsuits, compel recognition of the importance of the packaging field and demand the assignment of strong, knowledgeable, and experienced people to the packaging function in the company. This field is much too complex for a company that ships a variety of products to rely entirely on one person to perform the entire job, and often a team approach is warranted. In a large firm this team may be drawn almost entirely from within the company, while small firms may have to give their packaging staff the leeway to call upon the available services of packaging suppliers, trade associations, consultants, etc.

The individual responsible for packaging and compliance with federal safety regulations on packaging, including OSHA and the Consumer Product Safety Commission as well as DOT, faces a dilemma in his job. That dilemma is often encountered in this field, and it is the conflict between short-term economic interests and achievement of ideal safety in package shipment. The principles underlying acceptability or unacceptability of a given loss are very dissimilar in the areas of safety and economics. Economic returns may be measured against what may be termed "acceptable losses." In other words, there is a point at which it stops being economically feasible to expend additional energy and money in further reducing losses. The economic breakpoint, however, does not necessarily coincide with the government's viewpoint on environmental damage and personal injuries or the potential for such damages.

This inherent dissimilarity is the source of the communications problem so frequently encountered by corporate packaging employees in talking with government people in a federal safety context. The corporate employee bears a responsibility to shave costs, while the government would like him to use a packaging that may cost far more than the product it contains. The packaging engineer may be familiar with the materials of package construction and the physical and chemical properties of the materials to be shipped in that package, but he may not be thinking in terms of the hazard to be contained by that packaging and the parties at risk in the transportation of the material.

This brief introduction is for the purpose of advising that frustration in communications between government and industry packaging people is common and is not a sign of abnormality in the relationship, or that the government is out to get that particular company. It is essential to the resolution of this communications problem that the company view hazardous materials from the government's perspective as well as its own. The responsibility for packaging must not only lead to the shipment of the company's product to the customer in an undamaged and attractive condition, but that shipment must also occur with a minimum of risk to the safety of people and property exposed to that package in transportation, including company personnel, employees of the carrier, those of the consignee, and the public at large. What may seem to be uneconomic in the short run may be a cheaper way to proceed given the long-term impact on the company and on persons exposed to that package if there is an unintentional release of a hazardous material into the environment.

Often the responsibility for safety in packaging will run afoul of the economic considerations of the production and marketing groups, and for this reason such a responsibility should be borne by a strong office that is capable of opposing those groups if necessary.

HOW TO SELECT PACKAGING

Various factors go into the process of selecting proper packaging for a shipment of hazardous materials. Several of these are addressed in 49 CFR 173.24, having to do with basic packaging integrity and compatibility of that packaging with the lading that will be put into it. Another factor is the overall performance of the container—even if it is authorized in the regulations, if it arrives at your customer's location in damaged condition, or if it is the subject of numerous claims, then common sense dictates selecting an improved alternative packaging.

Yet another factor that has become influential with the growth of liability and environmental laws is consideration of the final disposal of the packaging after it has been used. Containers that may only be used once become a liability to the customer who then must get rid of them. This is a serious difficulty, since few disposal facilities or scrap dealers will accept containers that have not first been cleaned. Often they want them crushed and shredded as well. Under EPA solid waste regulations, an empty container cannot be placed in a landfill without first being crushed, because the agency anticipates the filled area will subside when containers eventually corrode and collapse, leading to cracking of impermeable caps on the landfill.

Thus, in making the original selection of a packaging, the shipper should consider the eventual disposal of the emptied unit, planning either to take it back from the customer, or to use a sound container that can be reconditioned effectively and therefore has a resale or salvage value.

Before consideration of the range of these factors, however, it is of primary importance to determine the limited number of alternative packagings offered under the DOT regulations. The fact that a customer asks for a product in a certain packaging, or that marketing feels it would be more attractive in a certain packaging, or that production feels it would be easier to fill certain kinds of packaging, or that the comptroller feels certain packaging would be cheaper, cannot be the exclusively determining factors. Unless the packaging is specifically authorized in the regulations, it must not be used. (If there is a strong desire to use a different packaging, one may seek an exemption from the DOT regulations. See Chapter 28 on exemptions.)

The determination of what packaging is authorized goes back to the discussion in earlier chapters on classification and determination of the proper shipping name for the material. Until the material is accurately classified and named in DOT terms, it is impossible to select packaging with any assurance of compliance.

In the front of the DOT regulations is an alphabetical commodity list, 49 CFR

172.101. After determination of the proper classification and shipping name of the article (see Chapters 5 and 6), look again at the list for that article. In the columns entitled "Exceptions" and "Packaging requirements," one will see the reference to section numbers offering exceptions to the regulations and others providing the particular range of packaging authorized for that commodity. The matter of exceptions is treated separately in Chapter 12 on limited quantities and will not be given greater coverage here. The following discussion presumes that the article is of such a nature that exceptions are not offered or that the package quantity restrictions to qualify for those exceptions are not suitable for the shipments in question.

A few examples may be helpful:

Shipping Name	Exception	Packaging
Ethyl benzene	173.118	173.119
Ethyl methyl ether	None	173.119
Flammable liquid, n.o.s.	173.118	173.119
Nicotine sulfate, liquid	173.345	173.346
High explosive, liquid	None	173.62
Corrosive liquid, n.o.s.	173.244	173.245, 173.245a
Potassium hydroxide, liquid	173.244	173.249

The first example above is an article properly described by the shipping name "ethyl benzene." In the commodity list it is classed as a flammable liquid, and 49 CFR 173.118 prescribes limited quantity exceptions to specification packaging requirements. Assuming the quantity the shipper desires to package is greater than that for which a partial exception is granted, it is necessary to turn to 49 CFR 173.119 to determine the only authorized packaging. That section is quite lengthy and groups flammable liquids above and below a flash point line drawn at 20°F with vapor pressures not over 16 p.s.i.a. at 100°F. (*Note:* A handy source for essential DOT oriented vapor pressure curve data on listed flammable liquids is found in Appendix A to the *Red Book*.)

Section 173.119 offers a variety of sizes of specification packaging, defined as that DOT-prescribed packaging meeting the detailed construction specifications in Parts 178 and 179 of the regulations. Each authorization of a DOT specification in 49 CFR 173.119 is combined with a sectional reference to the detailed package construction requirements in Part 178 of the regulations. Read the desired paragraph of 49 CFR 173.119 carefully for any limitations placed on that packaging that may be in addition to those appearing in a description of the specification itself. For example, 49 CFR 173.119(a)(3) authorizes DOT Specification 17E drums, but further prescribes that if those drums are between five and thirty gallons rated capacity, they must be constructed of 19-gauge body and head sheets. It is the shipper's responsibility to order the proper specification from the packaging supplier and in this case just asking for a Specification 17E drum may well lead to a violation of the regulations, for which the shipper will be liable because the body and head sheets of that general specification may be constructed of 22-gauge steel unless the shipper orders the required 19-gauge steel.

The next example is ethyl methyl ether which, because of its secondary hazards, is denied any exceptions. Authorized DOT specification packaging is set forth in 49 CFR 173.119, just as it was for ethyl benzene. The shipper would determine the packaging size he wanted, selecting from among those specifications authorized for that material given its flash point and vapor pressure.

"Flammable liquid, n.o.s." is an entry keying to the same regulatory sections as those referenced for ethyl benzene. One should read them the same way and the packaging should be selected in the same manner.

"Nicotine sulfate, liquid" is regulated as a poison, and partial packaging exceptions are described in 49 CFR 173.345. Specification packaging, if the excepted sizes are unsuitable to the shipper, is authorized in 49 CFR 173.346.

"High explosive, liquid" is an article obviously regulated as an explosive, for which no exceptions are offered. The only authorized packaging for this article is set forth in 49 CFR 173.62. Any shipper of such materials is cautioned to review the general regulations on explosives set out in the introductory sections on explosives in 49 CFR Part 173, including the description of a forbidden explosive that may not be transported under any circumstances (49 CFR 173.21) and requirements on the classification of new explosives (49 CFR 173.86).

A material properly described as "corrosive liquid, n.o.s." may qualify for limited quantity exceptions set forth in 49 CFR 173.244. If it does not, the DOT specification packaging listed in 49 CFR 173.245 and 173.245a is the only packaging authorized. "Potassium hydroxide, liquid" also is regulated as a corrosive material and the shipper is guided to 49 CFR 173.244 for possible limited quantity exceptions. If the desired packaging is not qualified for an exception, however, the shipper is referred to 49 CFR 173.249 to find the authorized specification packaging for this article. As with many specific listings authorizing certain packaging, 49 CFR 173.249 authorizes those packagings that would be authorized for a corrosive liquid, n.o.s., but gives several additional alternatives as well. Each entry should be read carefully.

The DOT regulations on the packaging of hazardous materials are very specific and do not provide a great deal of discretion to the maker of that packaging or the shipper in varying these specifics. The regulations, as we have seen, key the packaging to the proper shipping name of the article and its hazard classification. The entries in the alphabetical commodity list lead to the specific sections referencing authorized packaging. For a hazardous material, if a particular packaging is not specifically authorized in this manner, it must not be used without an exemption (Chapter 28).

DOT specification packaging, i.e., that packaging to which DOT specification numbers are assigned and for which there are exceedingly detailed construction and quality control requirements, are set forth in 49 CFR Part 178, and Part 179 for tank cars.

The specifications detail what is minimally (not nominally) required in the construction and arrangement of given packaging. For some products, particularly in smaller quantities, no detailed specifications are prescribed in the regulations. Note, however, that all packaging for hazardous materials must satisfy the general packaging requirements contained in 49 CFR 173.24, entitled "Standard Requirements for All Packages." There have been many enforcement cases brought under this section, which pertains to such things as the compatibility of the product with the packaging materials, legibility of markings, general integrity of the packaging, common marking requirements, etc. No shipper should overlook this section.

Part 64 of 46 CFR contains pertinent requirements on the packaging of combustible liquids shipped by sea in units of a capacity greater than 110 gallons each. Part 64 requirements are not applicable to combustible liquids shipped by rail, highway, or air. There are no requirements on packaging for combustible liquids shipped in quantities of 110 gallons or less.

For the sake of easy reference and to provide a summary of all DOT specification packaging, the following table is provided to give the DOT specification number, the type of container covered by that specification, and a reference to the current section of the DOT regulations in title 49 CFR where that specification may be found.

Spec. No.	Type Section	Description
CARBOY		
1A	carboy, boxed (178.1)	Glass boxed carboy, 5 to 13 gal. rated capacity.
1D	carboy, boxed (178.4)	Glass boxed carboy, up to 6.5 gal. nominal capacity.
1X	carboy, boxed (178.5)	Glass boxed carboy 5 to 6.5 gal. rated capacity, authorized only for export; single-trip.
1EX	carboy, plywood boxed (178.6)	Glass boxed carboy, 5 to 6.5 gal. rated capacity; single-trip.
1H	carboy, metal-enclosed (178.13)	Polyethylene carboy, within a metal crate, 5 to 13 gal. rated capacity.
1K	carboy, boxed (178.14)	Glass carboy, within expandable polystyrene in a wirebound wooden box, 5 to 13 gal. rated capacity.
1M	carboy, expanded polystyrene (178.17)	NRC glass carboy in NRC expandable polystyrene packaging.
INSIDE CONTAINERS		
2C	carton, fibreboard (178.22)	Corrugated fibreboard box.
2D	bag, paper (178.23)	Duplex paper bag not over 25 pounds capacity.
2E	bottle (178.24a)	Polyethylene container, up to 5 quart rated capacity.
INSIDE CONTAINERS		
2G	can (178.26)	Fibre (may have metal top and bottom), up to 6 pounds capacity.
2N	can (178.32)	Metal can not over 14 pounds water capacity.
2P	can (178.33)	Non-refillable aerosol metal can, not over 27.7 fl. oz. capacity.
2Q	can (178.33a)	Heavy duty, non-refillable aerosol metal can, not over 27.7 fl. oz. capacity.
2R	vessel, containment (178.34)	Metal or equivalent tube, maximum 12 inches diameter by 72 inches length.

(cont.)

Spec. No.	Type Section	Description
INSIDE LINERS		
2T	liner (178.21)	Polyethylene, heavy duty liner, 5 to 13 gal. rated capacity.
2U	liner (178.24)	Polyethylene, light duty liner, not over 55 gal. rated capacity.
2F	liner (178.25)	Tin plate sealed liner.
2TL	liner (178.27)	Polyethylene, light duty liner, up to 14 gal. marked capacity.
2J	bag liner (178.28)	Fitted waterproof creped paper liner, in a bag configuration.
2K	bag liner (178.29)	Fitted paper liner in a bag configuration.
2L	liner or bag (178.30)	Waterproof paper or other material lining or bag for box.
2M	liner (178.31)	Waterproof paper lining for box.
2S	liner (178.35)	Polyethylene, heavy duty liner, up to 55 gal. marked capacity.
2SL	liner (178.35a)	Polyethylene, medium weight liner, up to 55 gal. marked capacity.
CYLINDERS		
3A, 3AX	cylinder (178.36)	Steel, seamless, high pressure, not over 1,000 pounds water capacity.
3AA, 3AAX	cylinder (178.37)	Steel, seamless, high pressure, over 1,000 pounds water capacity.
3B	cylinder (178.38)	Steel, seamless, low pressure, not over 1,000 pounds water capacity.
3BN	cylinder (178.39)	Nickel, seamless, low pressure, not over 125 pounds water capacity.
3E	cylinder (178.42)	Steel, seamless, 1800 p.s.i. service pressure with maximum diameter of 2 inches and length of 24 inches.
3HT	cylinder (178.44)	Steel, seamless, high pressure, not over 150 pounds water capacity; inside packaging—overpack required.
3T	truck trailer tube (178.45)	Steel, seamless, high pressure, 1,000 pounds or over, water capacity.
3AL	cylinder (178.46)	Alumninum, seamless, high pressure, not over 1,000 pounds water capacity.
4DS	sphere (178.47)	Steel, welded, low pressure, not over 100 pounds water capacity; inside packaging—overpack required.

Spec No.	Type Section	Description
4B	cylinder (178.50)	Steel, welded or brazed, low pressure, not over 1,000 pounds water capacity.
4BA	cylinder (178.51)	Steel, welded or brazed, low pressure, not over 1,000 pounds water capacity.
4D	sphere (178.53)	Steel, welded, low pressure, not over 40 pounds water capacity; inside packaging—overpack required.
4B240-FLW	cylinder (178.54)	Steel, welded or brazed, 240 p.s.i. service pressure, not over 240 pounds water capacity.
4B240-ET	cylinder (178.55)	Steel, welded or brazed, 240 p.s.i. service pressure, not over 12 pounds water capacity.
4AA-480	cylinder (178.56)	Steel, welded, 480 p.s.i. service pressure, not over 1,000 pounds water capacity.
4L	cylinder, insulated (178.57)	Steel, welded, low pressure, not over 1,000 pounds water capacity; cryogenic container.
4DA	sphere (178.58)	Steel, welded, low pressure, not over 100 pounds water capacity; inside packaging—overpack required.
8	cylinder (178.59)	Steel, seamless or welded, 250 p.s.i. service pressure, porous filling (acetylene).
8AL	cylinder (178.60)	Steel, seamless or welded, 250 p.s.i. service pressure, porous filling (acetylene).
4BW	cylinder (178.61)	Steel, welded, low pressure, not over 1,000 pounds water capacity.
39	cylinder or sphere (178.65)	Steel or aluminum, not over 55 pounds water capacity; non-refillable; inside packaging—overpack required.
4E	cylinder (178.68)	Aluminum, welded, low pressure, not over 1,000 pounds water capacity.

DRUMS

Spec No.	Type Section	Description
5	drum, open or tighthead (178.80)	Steel, heavy duty, reusable container, up to 110 gal. rated capacity.
5A	drum, tighthead (178.81)	Steel, heavy duty, reusable container, up to 110 gal. rated capacity.
5B	drum, open or tighthead (178.82)	Steel, heavy duty, reusable container up to 110 gal. rated capacity.

(cont.)

Spec. No.	Type Section	Description
DRUMS (*cont'd.*)		
5C	drum, tighthead (178.83)	Stainless steel, heavy duty, reusable container, up to 110 gal. rated capacity.
5K	drum, tighthead (178.88)	Nickel, heavy duty, reusable container, not over 110 gal. rated capacity.
5L	drum, tighthead (178.89)	Steel, rectangular-shaped reusable drum, not over 5 gal. rated capacity.
5M	drum, tighthead (178.90)	Monel, heavy duty, reusable container, not over 55 gal. rated capacity.
6B	drum, open or tighthead (178.98)	Steel, medium duty, reusable container, not over 110 gal. rated capacity.
6C	drum, open or tighthead (178.99)	Steel, medium duty, reusable container, not over 110 gal. rated capacity.
6D	overpack, for inside plastic container (178.102)	Steel, reusable container with plastic inside container, not over 55 gal. rated capacity.
6J	drum, open or tighthead (178.100)	Steel, medium duty, reusable container, not over 110 gal. rated capacity.
6L	drum, open-head (178.103)	Steel, reusable container, 55 to 110 gal. rated capacity; used for radioactive materials only.
6M	drum, open-head (178.104)	Steel, reusable container, 10 to 110 gal. rated capacity; used for radioactive materials only.
42B	drum, tighthead (178.107)	Aluminum, medium duty, reusable container, 5 to 110 gal. rated capacity.
42D	drum, tighthead (178.109)	Aluminum, light duty, reusable container, not over 110 gal. rated capacity.
17C	drum, open or tighthead (178.115)	Steel, heavy duty, single-trip container not over 55 gal. rated capacity.
17E	drum, tighthead (178.116)	Steel, light duty, single-trip container, not over 55 gal. rated capacity.
17F	drum, tighthead (178.117)	Steel, heavy duty, single-trip container, not over 55 gal. rated capacity.
37K	drum, open or tighthead (178.130)	Steel, light duty, single-trip container, not over 55 gal. rated capacity, and 275 pounds gross weight.
37A	drum, open-head (178.131)	Steel, light duty, single-trip container, not over 55 gal. rated capacity and 490 pounds gross weight.

Spec No.	Type Section	Description
37B	drum, tighthead (178.132)	Steel, light duty, single-trip container, not over 55 gal. rated capacity and 650 pounds gross weight.
37P	drum, open-head (178.133)	Steel, light weight, nonreusable container, to be used only with 10-mil polyethylene liner not over 15 gal. rated capacity.
17H	drum, open-head (178.118)	Steel, light duty, single-trip container, not over 55 gal. rated capacity.
37M	overpack, for inside plastic container (178.134)	Steel, light weight, nonreusable container, not over 55 gal. rated capacity.
37C	drum, open-head (178.135)	Steel, light duty, nonreusable container, not over 5 gal. rated capacity and 80 pounds gross weight.
37D	drum, tighthead (178.137)	Steel, light weight, nonreusable container, ribbed or beaded construction, 55 gal. rated capacity.
13	keg, tighthead (178.140)	Steel, reusable container, not over 150 pounds gross weight.
13A	drum, tighthead (178.141)	Steel, light weight, top seamed on head after filling.
34	drum, tighthead (178.19)	Polyethylene, reusable container, not over 30 gal. rated capacity.
21C	drum, open-head (178.224)	Fibre, not over 75 gal. and 400 pounds gross weight.
21P	drum, open-head (178.225)	Fibre, with inside plastic container, not over 55 gal. rated capacity and 600 pounds gross weight.

CASES

32A	case, film type, hinged (178.146)	Sheet metal, riveted or lockseamed.
32B	case. film type, hinged (178.147)	Sheet metal, riveted or welded.
32C	trunk, film type (178.148)	Metal.
32D	box, film (178.149)	Steel, riveted or welded.
33A	case (178.150)	Expandable polystyrene, not over 60 pounds gross weight, nonreusable container.

WOODEN DRUMS, BARRELS, AND KEGS

10B	barrel or keg, tighthead (178.156)	Wood, not over 50 gal. rated capacity.

(cont.)

Spec. No.	Type Section	Description
WOODEN DRUMS, BARRELS, AND KEGS (*cont'd*)		
22A	drum, tighthead (178.196)	Plywood, authorized net weight not over 200 pounds.
22B	drum, tighthead (178.197)	Plywood, authorized net weight not over 200 pounds.
22C	drum, open-head (178.198)	Plywood, with inside plastic container, not over 14 gal. marked capacity.
WOODEN BOXES		
14	box, nailed (178.165)	Wood, not over 140 pounds gross weight.
15A	box, nailed (178.168)	Wood, not over 500 pounds gross weight.
15B	box, nailed (178.169)	Wood, not over 500 pounds gross weight.
15C	box, nailed (178.170)	Wood, not over 500 pounds gross weight.
15D	box, nailed (178.171)	Wood, not over 400 pounds gross weight.
15E	box, nailed, screwed, or stapled (178.172)	Wood, fibreboard lined, not over 550 pounds gross weight.
15L	box, brass screws (178.176)	Wood, inside metal and rubber container, not over 10 quarts.
15M	box, brass screws (178,177)	Wood, inside metal and rubber containers, with no more than 6 inside metal containers of a nominal capacity of 10 quarts each.
15X	box, nailed (178.181)	Wood, for two, inside 5 gal. cans.
15P	box, with inside container (178.182)	Wood or plywood.
16A	box, wirebound (178.185)	Wood or plywood, not over 400 pounds gross weight.
16B	box, wirebound (178.186)	Wood, not over 500 pounds gross weight.
16D	box, wirebound, overwrap (178.187)	Wood, with inside liquid container, not over 55 gal. marked capacity.
19A	box, nailed (178.190)	Wood or plywood, not over 400 pounds gross weight.
19B	box, nailed (178.191)	Wood or plywood, not over 150 pounds gross weight.
18B	kit, hooped (178.193)	Wood, not over 40 pounds gross weight.

Spec No.	Type Section	Description
FIBREBOARD BOXES		
12B	box (178.205)	Solid or corrugated board, not over 65 pounds gross weight.
12C	box (178.206)	Solid or corrugated board, not over 65 pounds gross weight.
12D	box (178.207)	Corrugated board, double-wall not over 75 pounds gross weight.
12E	box (178.208)	Corrugated board, for up to 2 inside metal containers not over 5 gal. each; not over 110 pounds gross weight.
12H	box (178.209)	Corrugated board, not over 65 pounds gross weight, except one type authorized to 103 pounds.
12A	box (178.210)	Corrugated board, not over 80 pounds gross weight.
12P	box, for inside plastic container(s) (178.211)	Solid or corrugated, nonreusable container, not over 80 pounds gross weight; plastic inside container(s) over 1 gal. capacity.
12R	box (178.212)	Paper-faced, expanded polystyrene board, not over 75 pounds gross weight.
23F	box (178.214)	Solid or corrugated board, not over 65 pounds gross weight.
23G	box, cylindrical (178.218)	Solid board, tubular, not over 12 inches diameter and 65 pounds gross weight.
23H	box (178.219)	Solid board, telescoping type, not over 65 pounds gross weight.
29	box, mailing tube (178.226)	Solid board with metal caps or ends.
BAGS		
36A	bag, cloth & paper (178.230)	Paper cloth-lined or cloth paper-lined, not over 100 pounds net weight.
36B	bag, burlap (178.233)	Burlap paper-lined, not over 100 pounds net weight.
36C	bag, burlap (178.234)	Burlap paper-lined, not over 100 pounds net weight.
44B	bag, paper (178.236)	Minimum double-wall plain or extensible plain shipping sack kraft paper, or asphalt laminated or polyethyene coated.
44C	bag, paper (178.237)	Minimum four-wall plain or extensible plain shipping sack kraft paper, or asphalt laminated or polyethylene coated.

(cont.)

Spec. No.	Type Section	Description
BAGS (*cont'd*)		
44D	bag, paper (178.238)	Minimum five-wall plain or extensible plain shipping sack kraft paper, or asphalt laminated or polyethylene coated.
44E	bag, paper (178.239)	Minimum triple-wall plain or extensible plain shipping sack kraft paper, or asphalt laminated or polyethylene coated; not over 100 pounds net weight.
45B	bag, paper (178.240)	Paper cloth-lined bag, not over 100 pounds net weight.
44P	bag, plastic (178.241)	Low-density polyethylene, not over 81 pounds net weight.
OVERPACKS (FOR RADIOACTIVE MATERIALS)		
20PF	overpack, metal (178.120)	Phenolic foam insulated with inner wood bracing supports for inside steel container, not over 1000 pounds gross weight.
21PF	overpack, metal (178.121)	Phenolic foam insulated with inner wood bracing supports for inside steel container, not over 8200 pounds gross weight.
20WC	overpack, wood (178.194)	Stacked plywood or hardwood disks assembled with steel rods, with a cavity for a steel container, not over 6000 pounds gross weight.
21WC	overpack, wood (178.195)	Nested wood boxes, configuration with outer steel strapping framework, not over 3000 pounds gross weight.
7A	Type A, general package (178.350)	Must meet requirements of Sections 173.403 and 173.465.
PORTABLE TANKS		
51	tank, steel (178.245)	Minimum 100 psig, maximum 500 psig design ASME steel tank, over 1000 pounds water capacity.
56	tank, metal (178.251, 178.252)	Steel, aluminum, or magnesium alloy tank (for solids), not over 7000 pounds gross weight.
57	tank, metal (178.251, 178.253)	Steel, aluminum, or magnesium alloy tank (for liquids), 110 or more gal., but not more than 660 gal.
60	tank, steel (178.255)	Steel, ASME Code vessel, minimum 1/4 inch shell thickness.
IM 101	tank, steel (178.270, 178.271)	Steel tank (for liquids), intermodal, 25.4 p.s.i.g. to less than 100 p.s.i.g.
IM 102	tank, steel (178.270, 178.272)	Steel tank (for liquids), intermodal, 14.5 p.s.i.g. to less than 25.4 p.s.i.g.

Spec No.	Type Section	Description
TANK TRUCKS		
MC 331	tank steel (178.337)	Truck tank, ASME Code vessel, minimum 100 psig/maximum 500 psig design.
MC 306	tank, metal (178.340, 178.341)	Steel or aluminum truck tank, non-pressure type.
MC 307	tank, metal (178.340, 178.342	Steel or aluminum truck tank, low pressure type.
MC 312	tank, metal (178.340, 178.343)	Steel or aluminum truck tank, non-pressure or pressure type with bulkhead, baffles, or ring stiffeners required.
MC 338	tank, metal (178.338)	Insulated steel or aluminum truck tank, pressure type 25.3 p.s.i.g. to not over 500 p.s.i.g.

NOTE: For tank truck specifications MC 300, MC 302, MC 303, MC 304, MC 305, MC 310, MC 311, and MC 330, see Appendix B.

Spec No.	Type Section	Description
SPECIALIZED MOTOR VEHICLES FOR EXPLOSIVES		
MC 200	178.315	Container for liquid nitroglycerin or diethylene glycol dinitrate.
MC 201	178.318	Container for blasting caps and percussion caps.
TANK CARS		
103W	non-pressure (179.201)	Insulated or uninsulated tank car with an expansion dome.
104W	non-pressure (179.201)	Insulated steel tank car with an expansion dome.
105A	pressure (179.101)	Insulated steel tank car with top loading facilities only, and no expansion dome.
105J	pressure (179.106)	Insulated steel tank car equipped with coupler restraint, head puncture resistance and thermal protection systems.
105S	pressure (179.106)	Insulated steel tank car equipped with coupler restraint and head puncture resistance system.
106A	pressure (179.301)	Uninsulated small steel lateral tanks designed to be removed from a multi-unit car; commonly called "ton tanks."
107A	pressure (179.500)	Uninsulated high pressure seamless tube-type tanks permanently mounted longitudinally in a cluster on a rail car.
109A	pressure (179.101)	Insulated or uninsulated steel tank car with top loading only and no expansion dome.

(cont.)

Spec. No.	Type Section	Description
TANK CARS (*cont'd*)		
110A	pressure (179.301)	Uninsulated small steel lateral tanks designed to be removed from a multi-unit car; commonly called "ton tanks."
111A	non-pressure (179.201)	Insulated or uninsulated tank car without an expansion dome.
111J	non-pressure (179.203)	Steel or aluminum having coupler restraint, thermal protection and head puncture resistance systems.
112A	pressure (179.101)	Uninsulated steel tank car with top loading facilities only and no expansion dome.
112J	pressure (179.105)	Uninsulated steel tank car with thermal protection system in metal jacket.
112S	pressure (179.105)	Uninsulated steel tank car with tank puncture resistance system.
112T	pressure (179.105)	Steel tank car with non-jacketed thermal protection system.
113W	pressure, low (179.400)	Insulated steel tank car having 30-day design holding time; designed for specific cold temperature loading.
114A	pressure (179.103)	Uninsulated steel tank car with top loading facilities only, no expansion dome, and optional non-circular cross section.
114J	pressure (179.105)	Uninsulated steel tank car with thermal protection system in metal jacket.
114S	pressure (179.105)	Uninsulated steel tank car with tank puncture resistance system.
114T	pressure (179.105)	Steel tank car with non-jacketed thermal protection system.
115A	non-pressure (179.220)	Insulated tank car with inner container suspended in foamed-in-place polyurethane.
ARA-III	non-pressure	Insulated or uninsulated riveted tank car with expansion dome, superseded by DOT-103W.
ARA-IV	non-pressure	Insulated riveted tank car with an expansion dome.
ARA-IV-A	non-pressure	Insulated riveted tank car with an expansion dome, without bottom outlet; superseded by DOT-104W.
ARA-V	pressure	Insulated steel tank car, forge welded, with top loading facilities only, superseded by DOT-105A.

CARBOYS

With regard to hazardous materials, the Federal Standard, "Glossary of Packaging Terms," defines a carboy as:

> A bottle or similar container made of glass, earthenware, clay, stoneware, plastics or metal, having a capacity of 3 to 13 gallons. Where the carboy is used for shipping purposes, principally for carrying corrosive liquids, chemicals, distilled spirits, and the like, it is usually designed to be encased in a rigid protective outer container for shipment, often with the use of cushioning materials prescribed by D.O.T. specifications, particularly when the carboy contains dangerous liquids.

DOT has viewed carboys for several years with some suspicion. In fact, a few years ago the Coast Guard actually proposed to ban carboys from transportation aboard vessels, but strong comments in opposition led to a withdrawal of this proposal.

Another problem for carboys relates to UN Dangerous Goods recommendations and philosophies of performance-oriented standards that have been favored by DOT, since carboys are relatively fragile in transportation. All of these factors should be considered in making plans for packaging of hazardous materials in carboys. It is clear, however, that there are some materials that can only be packaged effectively in carboys.

INSIDE CONTAINERS

Although 49 CFR Part 178 lists nine inside containers other than liners, the frequency with which these specifications are mentioned in the body of the regulatory text is low.

Reference to Specifications 2D and 2G, other than in a few regulations on poisons, is found only in one other section. Specification 2N is familiar mostly to shippers of hydrocyanic acid, while only the aerosol and LP-gas industries are generally aware of Specifications 2P and 2Q. DOT Specification 2R is a container used exclusively by the nuclear industry. Specification 2C is only mentioned in a few sections. The 2E Specification is the most frequently referenced of the inside container specifications, but it too is only prescribed a few times compared to the number of non-specification polyethylene bottle references.

Terms used in describing the DOT inside containers often are used as well for describing non-DOT packagings. As a consequence, one should know what is intended by the use of the terms in the DOT regulations. If the term is not defined in 49 CFR 171.8, a good source for this information is the Federal Standard, "Glossary of Packaging Terms," which described these terms as follows:

> **bag** A performed container made of flexible material, generally enclosed on all sides except one which forms an opening that may or may not be sealed after filling. May be made of any flexible material, or multiple plies of flexible materials, or a combination of two or more materials, such as paper, metal foil, cellulose and plastic films, tex-

tiles, etc., any of which may be coated, laminated or treated in other ways to provide the property required for the packaging, storing and distribution of a product. Although often used as a synonym for bag, the term "sack" generally refers to the heavier duty or shipping bags.

The four basic standard *styles* of bags are: (1) automatic self-opening bag, (2) satchel-bottom bag, (3) flat bag, (4) square bag.

The five basic standard *types* of bags are: (1) grocery bag, (2) merchandise paper, (3) industrial, (4) textile, (5) paper shipping sack.

bottle (1) As used in packaging, a container having a round neck of relatively smaller diameter than the body and an opening capable of holding a closure for retention of the contents. The cross section of the bottle may be round, oval, square, oblong, or a combination of these. Generally made of glass, but also of polyethylene or other plastics, earthenware, metal, etc. (2) Generally, any glass container capable of holding a closure (not including ampules which are sealable by fusion of the opening, or shell vials). Includes a large variety of glass containers of various sizes, shapes and finishes, such as: jar, demijohn, carboy, lask, flagon, magnum, etc. (3) Specifically, a narrow-neck container as compared with a jar, or widemouth container. [Note: This definition is similar to, but more detailed than, DOT's definition in Section 171.8.]

can, composite (1) a rigid container with the body made of fibreboard and one or both ends of metal, plastic or other material. (2) A rigid container constructed with the body of fibreboard or fibreboard in combination with other materials such as metal foils or plastics. One or both ends may be made of metal, plastic or other meterials.

can, fibre or fibreboard A rigid container constructed almost completely of lightweight fibre stock, which may be lined, treated, or coated to achieve desired chemical and physical characteristics; with ends of paperboard, or of metal (composite can), or of other material. The body may be spirally wound, convolutely wound, or it may be laminated or lap-seamed from an appropriate number of plies built up to the desired strength characteristics, and may be round, oval or rectangular. The end may incorporate pouring spouts or other fittings. A fibre can is an interior packaging container as compared with a fibre drum that is designed to be shipped without further packaging. Does not refer to paper milk container.

carton Folding boxes generally made from boxboard, for merchandising consumer quantities of products (e.g., shelf packages or prime packages). In domestic commerce, the item *carton* is generally recognized as the acceptable designation for folding paperboard boxes—never for a shipping container—although in maritime and export usage the term *carton* refers to a corrugated or solid fibre shipping container. [The term *carton* conveys several meanings to different people, particularly when incorrectly used to designate a shipping container. Unless it is properly qualified, the term *carton*, standing alone, can lead to misunderstandings.]

These definitions are useful as a starting point to determine the meaning of a DOT regulation. The complete answer, however, may not be apparent without referring to the historical evolution of the regulation, the regulator's intent, or to persons fluent in DOT jargon. Whenever in doubt, one should check to determine the proper meaning of the term. Unfortunately, DOT itself does not publish a glossary of its terms. It also may be helpful to check the *Glossary of Packagings* in Part II of Annex I of the IMO IMDG Code.

For international shipments, it is helpful to consult the *Illustrations of Selected Packagings* in Part II of Annex I of the IMO IMDG Code.

INSIDE LINERS

Although the DOT regulations make little reference to specification inside containers, inside specification liners are often prescribed.

In some instances, DOT uses the term containers to describe these units, but upon examination of the type of use for which these units are authorized, it is clear that they are more akin to a lining or bladder. In fact, even when DOT prescribes test requirements, such tests are always to be conducted with the liner inside its overpack, which is also a DOT specification container. Specifications 2F, 2K, 2L, and 2M liners are usually in a box configuration, while specifications 2T, 2TL, 2U, 2S, and 2SL are in a drum arrangement. One notable exception to this is the 12P fibreboard box, which holds a cubical Specification 2U "liner-container."

Many of these units are used in arrangements that are not declared to be either non-reusable or single-trip packagings. Because of this silence, the question of reuse is a recurring one. If the DOT regulations do not limit a packaging by declaring it non-reusable or single-trip, as a matter of law reuse may be authorized subject to compliance with 49 CFR 173.28. What may be legal, however, is not guaranteed to be safe without some further evaluation. This general responsibility is placed squarely upon the shipper by 49 CFR 173.22, 173.24, and 173.28.

In a combination or composite packaging unit, such as represented by these liners, the condition and suitability of the package is ascertainable only with difficulty without removal of the inner unit and full knowledge of all previous contents that have been contained by that liner. Honest compliance with 49 CFR 173.24 and 173.28 cannot be achieved without a thorough inspection program, usually requiring dismantling of the packages. Complete retest is normally out of the question because of the destructive nature of some of the testing, but serious thought should be given to use of any non-destructive testing prescribed for the original new unit as a means of assuring compliance with 49 CFR 173.28. The DOT regulations do not require the container to be good as new, because aging and use always bring some wear. The regulations do require, however, that the package "comply in all respects with the prescribed (specification) requirements" for the original container. It seems, therefore, that something more than a visual examination of the package is necessary.

This requirement also indicates that if the original new container was built to "bare bones" minimum requirements, reuse of it must be verified much more certainly than for a unit originally built with a more substantial margin for compliance.

Accordingly, in order to reuse packaging with confidence, one should know his packaging supplier and how his manufacturing is geared to compliance with DOT regulations in terms of his awareness of those regulations, his recognition of the serious responsibilities assumed by a shipper of hazardous materials, and his willingness to open his compliance program to customer scrutiny.

Most DOT specification containers are required to be marked with a specific DOT identification code. In some cases, all or part of this marking is required to be repeated on any overpack. This is particularly true of so-called liners, such as those addressed here. The shipper should always check Part 178 of the regulations for any specification marking requirements for his shipment to verify that the packaging he is using bears the correct and complete marking.

Many specifications require marking the identity of the packaging manufacturer on the package. This can be done in several ways. The permissible means of marking are spelled out in each specification. Not all specifications have consistent marking requirements, so each must be examined to determine the proper marking to be applied for that specification.

The three most common methods of prescribed container manufacturer identification are:

1. Marking the full name and address of the manufacturer on the package;
2. Marking a symbol of that manufacturer on the package that has been registered by DOT's Research & Special Progams Administration, Washington, DC 20590; or
3. Marking a registration number on the package that has been obtained from DOT's Research & Special Programs Administration, Washington, DC 20590.

Whichever marking scheme is used must be specifically authorized under the regulations for the specification in question.

CYLINDERS

Included in this category is a varied assortment of compressed gas pressure vessels made of steel or aluminum. These containers may be seamless, brazed or welded, cylindrical or spherical, refillable or nonrefillable, according to the particular specification involved. They further may be described as high or low pressure, the latter being those vessels having service pressure of less than 900 pounds per square inch. Specifications 3AX, 3AAX, and 3T are large cylinders made for mounting on a truck trailer chassis, commonly called tube trailers.

It should be noted that cylinders are authorized as packaging for many more classes of hazardous materials than gases. These include flammable liquids, corrosive liquids, radioactive materials, and poisons, as indicated by the specific regulatory sections for those materials. Gases authorized may be liquefied or non-lique-fied. The regulations are very specific on how much gas may be filled into a cylinder, (see 49 CFR 173.301 through 173.316). Most cylinders are subject to very specific retesting requirements. Cylinders once used in corrosive liquid service are subject to strict retesting requirements before being authorized for reuse in gas service. When compared to other DOT containers, reuse of cylinders is subject to much more specific regulation than other packaging.

Valve protection on cylinders is an important requirement under the DOT reg-ulations. Even though such protection generally is required only on cylinders filled with flammable, corrosive, or toxic gases, it is a good practice to use valve protec-tion at all times if possible.

Safety relief devices, generally required on cylinders containing gases, must be evaluated in relation to each gas with which they are used. In some cases, safety relief devices actually are prohibited, particularly in the case of extremely hazard-ous materials or materials that pose an unusual containment problem due to their corrosive or toxic properties.

Testing apparatus used for testing cylinders must be specifically approved both as to type and operation by DOT. Inspectors used by manufacturers of high-pres-sure cylinders may not be employees of that manufacturer but must be third-party, disinterested inspectors approved by DOT.

One unusual requirement in the manufacture of cylinders is that unless other-wise specifically approved by DOT, the chemical analyses of the steel and any tests prescribed must be performed within the limits of the United States. This require-ment can have a serious impact on any person considering fabrication of cylinders outside the United States. Note that fabrication outside the United States is not prohibited, just made more difficult because of the requirement of specific analysis and testing being performed in the United States or otherwise approved by DOT.

DRUMS

The largest group of DOT specification containers are drums. Drums may be made out of steel, fibre, aluminum, nickel, monel, plastic, or rub-ber. Some are reusable, others are non-reusable, while others are designated as singletrip. Most are designed to contain liquids or solids directly, but some are for use as an overpack for an inner container or liner such as DOT Specifications 2U and 2SL polyethylene units.

The most frequently discussed ambiguity in DOT terminology regarding drums is that agency's distinction between a non-reusable drum and a single-trip drum. The two terms sound synonymous, but DOT draws a fine distinction.

A non-reusable container (marked NRC) is a packaging that generally may not be used a second time for shipment of hazardous materials. (The second use is not considered to be merely the second time the packaging is put into transportation but, rather, when it is put into transportation after having been emptied and re-filled or having its contents significantly altered following the initial shipment.) Examples of nonreusable drums are DOT Specifications 37C, 37M, and 37P.

A single-trip container (marked STC) may be used again for subsequent ship-ments of flammable liquids, flammable solids, oxidizing materials, radioactive ma-terials, and certain corrosive liquids covered by 49 CFR 173.249 and 173.249a. Reuse, however, is predicated upon very specific regulations governing recondi-tioning and retesting of those containers before subsequent hazardous materials use (except for hazardous wastes, see Chapter 14). Note that not all hazardous materials may be shipped in a reconditioned single-trip drum, and that reuse requires careful reconditioning of the drum before each reuse. See 49 CFR 173.28. Of course, subsequent reuse of either type of packaging for non-hazardous materials outside DOT's jurisdiction is not prohibited by DOT's regulations, but particular care should be taken with the residual effects of the original contents in mind. A single-trip packaging also may be reused for any corrosive solid, ORM-A, B, or C, or any material not required to be shipped in a DOT specification container.

There are particular provisions on the reuse of packaging for the shipment of hazardous wastes. See 49 CFR 173.28(p) and Chapter 14 on hazardous wastes.

Reconditioners of drums for hazardous materials must be registered with DOT's Office of Hazardous Materials Transportation and must mark the identification number issued to them on each drum they recondition. Each plant for each reconditioner must be separately registered, to facilitate follow-up if the container fails in transportation. Any person, including a shipper, may be a reconditioner, but registration is mandatory for all.

Some drums are really cylindrical overpacks for other packaging, since DOT regulations do not authorize their use without an inside plastic liner. These are DOT Specifications 6D, 37M, and 21P. These containers may be used with a variety of DOT polyethylene liner-containers.

In limiting shippers to use of tight-head drums for certain hazardous materials, two regulatory mechanisms are used:

1. By listing specifications under which tight-head containers are the only type authorized, or
2. By using the phrase "with openings not over 2.3 inches in diameter" to indicate the largest permissible opening in a drum, thereby precluding use of any open-head drum that may be included within the same specification.

It is important to note this dual practice of prescribing a tight-head drum. It means that although a regulation such as 49 CFR 173.119(a)(2) authorizes a Specification 5B drum, which by specification may be either open or tight-head, the qualifier of "openings not over 2.3 inches in diameter" limits the authorization to a tight-head Specification 5B. Use of an open-head Specification 5B drum for a flammable liquid with a flash point below 20°F would constitute a violation of the regulations for which the shipper would be liable. It is essential, therefore, to note both the shipper regulations under the commodity authorization and the packaging specifications in ordering packaging from a supplier. In the example, simply ordering a Specification 5B without also noting the fact that it must be a tight-head container could have led to an expensive error.

A frequent question pertains to the use of safety relief devices on drums of flammable liquids. DOT neither prohibits nor requires such devices. If a safety relief device is used, however, the drum on which it is installed must be capable of passing all tests (except hydrostatic) prescribed for that drum specification. Typically these would be the four-foot drop test and air leakage tests. In addition, any general packaging requirements, e.g., compatibility of the container with its contents, will apply to the relief device as well as the rest of the drum. DOT does point out that the device must not leak under any normal conditions of transportation. To DOT, this means no leakage of liquid or vapor in an ambient temperature up to 130°F.

There are only two DOT specification plastic drums as of this writing. One is the

DOT Specification 34 drum, a tight-head polyethylene unit that is authorized in sizes up to fifty-five gallons rated capacity and is permitted for many hazardous materials. These drums are considered reusable by DOT. The general packaging regulations require compatibility of the hazardous material with the polyethylene and the integrity of the closure. In addition, these plastic containers as well as any other bottles, liners, etc., must meet the more specific compatibility requirements of 49 CFR 173.24(d) and Appendix B to 49 CFR Part 173.

The only other polyethylene plastic drum authorized is DOT Specification 35, a non-reuseable, open-head plastic pail of not over 7-gallons capacity. The top head may be made of polyethylene, 26-gauge steel, or any combination thereof. This drum is only authorized for solids, although several exemptions have been issued authorizing a very similar drum for certain liquids. When containing liquids under these exemptions, the pail must pass a 5 p.s.i.g. hydrostatic pressure test, which is notably lower than what is required for DOT liquid containers other than fibre.

The issues involving plastic drums are illustrative of DOT's general philosophy of giving a new industry wide latitude initially, depending upon self-policing and industry standards to achieve safety. Upon a showing of any inability to maintain safe controls, however, DOT initiates federal controls that invariably are more stringent than the industry standards. Only after the fact have several industries realized that a little greater care and self-control initially may have prevented such adverse federal intervention.

DOT provides two basic types of fibre drums, one for solids (Specification 21C) and another for liquids (Specification 21P). Rules covering these drums are very similar to those applicable to the steel overpacks that have a plastic liner insert such as DOT-2S, 2SL, and 2U. The fibre units for solids often do not have separate linings, although they may have an integral coating or lining, if some concern exists. The liquid units typically incorporate the same liners described in the discussion of steel composite units above (DOT-6D and 37M).

Testing for these packagings also is similar to that for composite steel overpacks, in that no leakage or hydrostatic test must be performed. Similar to the steel units, fibre drums have to meet specified drop test requirements but DOT-21C tests include a special "beam" test, consisting of a 4-foot side drop onto a 2″ x 6″ timber with the 6″ side vertical. Fibre drums must meet certain atmospheric (temperature and humidity) conditioning criteria and must be capable of sustaining given compression levels prescribed according to size. No vibration testing is prescribed as it is for steel composites. The drop tests for DOT-21C (solids) must be conducted in various orientations, no drum being required to pass more than one test although the manufacturer is not precluded from using a single drum for multiple tests.

Fibre drums offer an interesting feature in today's world of concern with protection of the environment. This container is being used now to package goods for incineration. Being combustible, the drums become part of the fuel used to destroy the waste materials. Fibre drums to date have not been authorized for wide usage for liquids under the DOT regulations, although the opportunity to use them in conjunction with waste incineration may lead to the development and greater recognition of their liquid containment capability.

FIBREBOARD BOXES

Fibreboard boxes play an enormous role in the transportation of hazardous materials. The problem of widespread regulatory unfamiliarity is coupled with the fact that the fibreboard box regulations are antiquated in terms of the technological advances in packaging. Many plants no longer even have the equipment to build or close fibreboard boxes in the manner prescribed by these regulations. Examples of obsolete requirements may be seen in the specification for the DOT Specification 12B box, which requires types of tape no longer in common use, or authorization for glued closures on slotted fibreboard boxes only if the adhesive covers the entire contact surface of the closing flaps. Although out-of-date, these requirements are the ones on the books, and, unless industry effort to update the requirements is undertaken or the box specifications are rendered moot by rule making in Docket No. Hm-181, a multitude of operations will remain in technical noncompliance and subject to substantial civil penalties.

CONCLUSION

Until passage of the Hazardous Materials Transportation Act in 1975 (see Chapter 30), manufacturers and reconditioners of packaging for hazardous materials fell outside the direct regulatory jurisdiction of the Department of Transportation. Those persons are now as subject to the regulations as shippers and carriers are.

This long-term freedom from direct regulation, coupled with a provision in the regulations releasing the shipper from responsibility if the package bore the right specification markings, led to the evolution of a universe of compliance problems that is still being faced in the packaging field. For the maker or reconditioner of the packaging, as well as the shipper using that packaging, the future holds a great deal in store in terms of regulatory revisions and enforcement of federal requirements.

As noted in the opening of this chapter, the responsibilty for selecting and maintaining a shipper's packaging is a critical link in a company's efforts to minimize federal liability. With the shift to performance-oriented standards, the shipper's liability will increase dramatically, because there will be more latitude for choice and correspondingly more liability for errors in making those choices. As such packaging should be a responsibility delegated to a level in the company that is attuned to the dimension of the problems in terms of staffing and authority.

CHAPTER 9

Required Markings on Hazardous Materials Packages

Marking, as it is used in the *Red Book* and in the DOT regulations, means those letters, numbers, words, and symbols that DOT requires to be marked on hazardous materials packages, portable tanks, cargo tanks, and tank cars. There are no DOT hazardous materials marking requirements for aircraft, although there is some written signing aboard vessels.

The term does not include DOT labeling, which is discussed in the following chapter. These two concepts must be kept separate and distinct in reading the DOT regulations. Placarding, the application of large warning signs to motor vehicles and other bulk containers, occasionally has been called marking, but in this chapter the term does not include placarding. See Chapter 12 on placarding.

Marking as used in this chapter also does not refer to those marks on containers that are applied by a shipper or a carrier for his own purposes, or other government regulations, as opposed to markings made in response to regulatory requirements imposed by DOT. The shipper's logo or other advertising is not part of the marking discussed in the regulations (except insofar as it may obscure or confuse DOT-required marks).

Several types of markings must appear on most hazardous materials containers, including (1) markings to be applied by a packaging manufacturer or reconditioner, (2) the shipper's marking of the name of contents (proper shipping name), (3) an indication that a "reportable quantity" or "RQ" of a hazardous substance is contained in the package, (4) special markings on portable tanks, (5) cargo tanks, (6) tank cars, and (7) certain miscellaneous series of advisory messages, e.g., "This Side Up," and other information such as the name and address of the consignee under certain circumstances. All marking must be in English, must be legible, must be placed on a contrasting background, and may not be near any other marking that could reduce its effectiveness substantially. Each type of marking will be discussed in turn.

1. MARKINGS RELATING TO SPECIFICATION PACKAGING

Specification packaging means packaging designed and constructed to the detailed specifications published in the DOT regulations in 49 CFR Parts 178 and 179. Packaging specifications are also prescribed by the United

Nations (UN) and other regulatory bodies, and these usually include a specification packaging marking that may appear alone or with the DOT specification designation.

Almost all, but not every, DOT specification packaging must bear the appropriate letters and numerals identifying the particular specification, e.g., DOT-6L, DOT MC 306, or DOT-105A200-F. The particular letters and numerals are set forth in the applicable specification, which prescribes the format and location for that marking.

If the regulations require the use of a specification packaging and there is a requirement to mark that packaging with the specification numerals and letters, the use of a packaging without the prescribed specification numerals and letters marked on it may be a violation of the regulations by the packaging manufacturer (for having represented that this was a specification package without applying the necessary markings), by the shipper (for using an improperly marked container), and perhaps by the carrier (for accepting a hazardous material for transportation that required specification packaging but that did not bear the marks showing that it was in fact a specification package). Thus, although these markings must be applied by the maker or reconditioner of the container, it is of regulatory concern to all people in the transportation system that those markings be applied and applied correctly.

Specification packaging usually must also show the name and address, registered symbol or registration number of the manufacturer or reconditioner of that packaging. Symbols must be registered with the Department of Transportation and cannot duplicate a symbol already being used by another manufacturer. Some specifications require the container manufacturer to obtain a registration number from DOT. This number is used in place of that company's name and address, and it must appear on the package as described in the particular specification. Look at the sections of the regulations pertaining to the specification to determine the nature and location of the markings of the maker's identity.

Specification packaging markings must be stamped, embossed, burned, printed, or otherwise durably marked on the packaging to provide adequate accessibility and permanence, and they must contrast so as to be readily understood and apparent. The specification frequently identifies how and where the marking must be applied.

Putting specification markings or a DOT registration number on a packaging that does not meet the specifications violates both the statute and the regulations. See 49 U. S. Code 1804(c), 49 CFR 173.24(c)(1)(v), and 178.0-2(b).

DOT specification containers that are authorized for reuse under the regulations for the shipment of hazardous materials must have the markings for that specification maintained in a legible condition. If, because of painting or any other reason, the markings prescribed for any container cannot be kept plain and legible, the regulations permit fastening a metal plate to the container with a reproduction of the prescribed markings plainly stamped on it.

Single-trip Specifications 17C, 17E, and 17H steel drums, if reconditioned in accordance with the regulations applicable to reuse of packaging, must have all previous test markings, commodity identification markings, and labels removed. The outside of each drum properly qualified for reuse must be marked with an indication of the test or inspection date, as well as the DOT registration number issued to the reconditioner of that drum. If the specification is altered by the reconditioner, the regulations require specific markings showing that fact. See 49 CFR

173.28. This section does permit certain containers to be reused for the shipment of hazardous wastes without reconditioning. See Chapter 14 on hazardous waste for more detail.

The marking of DOT specification numerals and letters or registration number on a packaging is a representation and certification that such packaging complies with all specification requirements. The maker of the packaging, and any person other than the maker who applies the DOT markings to packaging, is subject to the same civil and criminal penalty provisions of the law as any shipper or carrier might be. Accordingly, the act of placing DOT specification markings on any container or packaging is not an act to be taken lightly, and any person applying such markings should be fully aware of the substantial liability involved.

No simple summary of specification package marking is complete, and the reader is urged to examine the details for each specification in question to determine the correct markings and their size and location on the packaging or container.

2. MARKING THE NAME OF CONTENTS ON THE PACKAGE

When DOT uses the phrase "name of contents" in the regulations, it means the proper DOT shipping name as it appears in Roman type (not italics) in the alphabetical commodity list at the front of the regulations. See 49 CFR 172.101 and Chapter 6 on determining the proper shipping name.

Shipping names may be used in the singular or plural. When qualifying words are used as part of the shipping name, the sequence of words as they appear on the package is optional, e.g., "lithium amide, powdered" may be printed "powdered lithium amide." The italicized word "or" between two terms or names in the commodity list indicates that any one term in the sequence may be used to complete the shipping name. It is not required that each term in the sequence be used, e.g., "Caustic sode, flake" may be used in place of "Caustic soda, dry, solid, flake, bead, *or* granular," as it appears in the commodity list.

Common marking requirements (but not all marking requirements) are set out in 49 CFR 172.300-330. Each package of hazardous materials, unless otherwise provided for the specific quantity or material in question, must be marked with the name of contents. Except for use of a "W" to mean "with" and "W/O" to mean "without" on ammunition packages, "ORM" for Other Regulated Materials, "n.o.s." or its synonyms as part of the proper shipping name, "RQ" for "reportable quantity," or as otherwise specifically authorized by DOT, no abbreviation may be used in marking the name of contents on hazardous materials packages.

These markings must satisfy the same general requirements applicable to all marking, including legibility, durability, and contrast with the background for that marking.

Each package containing a hazardous material offered for export by water and described by an n.o.s. entry in the alphabetical commodity list must have the technical name of the material added in parentheses immediately following the shipping name. See 49 CFR 172.302.

Commodities in the "Other Regulated Materials" or "ORM" classes, in addition to the shipping name, must be marked with the ORM-A, B, C, D or E designation in a rectangle that is approximately ¼ inch (6.3 mm) larger on each side than the reporting requirements.

3. REPORTABLE QUANTITIES OF HAZARDOUS SUBSTANCES

Each hazardous substance present in a reportable quantity in one package (smaller than 110 gallons) must be marked on the package if the shipping name does not already identify the hazardous substance(s). See Chapter 14 on hazardous substances, the Appendix to 49 CFR 172.101, and 49 CFR 172.324.

In addition, these packages must also be marked near the shipping name with the initials "RQ" for "reportable quantity." The purpose of this marking is to alert carrier personnel and others to the environmental threat posed by leaks from any of these packages, and to facilitate their compliance with hazardous substance spill reporting requirements.

4. MARKINGS ON PORTABLE TANKS

Portable tanks must be marked on opposing sides in letters at least 2 inches (50.8 mm) high with the name of contents. The tank must also bear the name of the owner or lessee, if appropriate. As with smaller containers, shipments of n.o.s. materials for export by water must include the technical (chemical) name of the material following the DOT shipping name.

Of course, once the name of a hazardous material is marked on a portable tank, no other material may be shipped in that tank unless the markings are removed or appropriately changed to reflect the new contents. The tank must continue to bear these markings, however, even when it is emptied and only contains residue of the former contents. After it is filled with other material, or cleaned of its residues and purged of their vapors, the markings must be removed or covered.

See the particular requirements for the material being shipped for additional marking requirements (if any), and also look at the specification for the tank for the DOT specification numerals and letters. Portable tanks under exemptions must bear the "DOT E" marking of the exemption number as required in the exemption and Appendix B of Subpart B of 49 CFR Part 107.

5. MARKING ON CARGO TANKS

Cargo tanks are tanks permanently mounted on motor vehicles, or any bulk liquid or gas packaging not permanently attached which, by reason of its size, construction, or attachment to a motor vehicle, is loaded or unloaded without first being removed from the motor vehicle. This does not include packaging fabricated to gas cylinder specifications. Each cargo tank used for the shipment of gases (including cryogenic liquids) must be marked with the shipping name of the contents, although it is permitted to mark such tanks with an appropriate common name for the material such as "Refrigerant Gas."

Letters must be at least 2 inches (50.8 mm) high with a one-fourth inch (6.3 mm) stroke in the prescribed colors. This marking is in addition to the placarding required for cargo tanks transporting gases. As with portable tanks, marking the name of one material on the cargo tank precludes its use with another material unless the first marking is removed or appropriately changed to reflect the current contents.

Each DOT specification MC 330 and MC 331 cargo tank must be appropriately marked "QT" or "NQT," to indicate whether it is constructed of quenched and tempered steel or of some other material. Cargo tanks, of course, must bear the appropriate specification markings, described in detail in the specifications for each tank. The "QT" or "NQT" must be marked near this specification marking.

Cargo tanks authorized under DOT exemptions must bear the "DOT E" marking of the exemption number as required in the exemption and Appendix B of Subpart B of 49 CFR Part 107.

6. MARKING ON TANK CARS

Tank cars containing certain materials must be marked with the shipping name or a common name of the material authorized in the regulations, in 4-inch (101.6 mm) high letters. See Section 172.330. In addition, throughout the shipper regulations and in 49 CFR Part 179 there are specific stenciling and marking requirements for certain commodities shipped in tank cars. Tank cars authorized under exemptions must bear the "DOT E" marking of the exemption number as required in the exemption and Appendix B of Subpart B of 49 CFR Part 107.

7. MISCELLANEOUS ADDITIONAL MARKINGS

a. "This Side Up" or "This End Up" must be marked on the outside of packages containing liquid hazardous materials, with certain exceptions provided for flammable liquids in inside packages of one quart or less. See 49 CFR 172.312. When packages so marked are placed inside another overpack, that overpack must then be marked with the same wording, indicating the proper positioning of those packages. See 49 CFR 173.25(a)(3).

b. Whenever outside specification packaging is overpacked in another container that would preclude seeing the proper specification markings on that packaging, the overpack must be marked with the prescribed shipping name that appears on the overpacked packages, as well as with the message that "Inside Packages Comply With Prescribed Specifications." A similar message must appear on shipping cartons of aerosol cans other than those classed as ORM-D. If the specification packaging that is overpacked is not outside packaging but is designated in the regulations as specification inside packaging, this requirement does not apply (see 49 CFR 173.306(a)(3)(vi)). The overpack must also show the labels that have been obscured. These requirements are applicable to crates and similar overpacks but not to containerizing units of 640 cubic feet or more. See 49 CFR 173.25.

c. To qualify for certain less restrictive regulations, outside packaging for flammable liquids with closed-cup flash points between 73°F and 100°F must be marked with an indication of the flash point, which may be a specific number or may say that the flash point is in the range from 73° to 100°F, or may merely indicate that the flash point is over 73°F. See 49 CFR 173.118. The following guide may prove helpful in applying this unique requirement. The shipper does *not* need to mark the flash point or an indication of the flash point on the outside package when:

1. The material has a flash point of less than 73°F.
2. The material has a flash point of 100°F or more.
3. The material has a flash point of 73°–100°F and is inside metal containers which each have a capacity of 1 quart or less, or inside nonmetal containers which each have a capacity of 1 pint or 16 ounces net weight, or less.
4. The product is an alcoholic beverage (wine and distilled spirits as defined in the regulations of the Treasury Department in 27 CFR 4.10

and 5.11), and is inside containers with a rated capacity of 1 gallon or less.

5. The material is an aqueous solution containing 24 percent or less alcohol by volume, if the remainder of the solution does not meet the definition of a hazardous material in the DOT regulations, e.g., the remainder is not poisonous, corrosive, etc.

6. The material has a flash point of 73°–100°F, and is in non-metal inside containers exceeding 1 pint or 16 ounces net weight, or is in metal containers exceeding 1 quart capacity, and the shipper prefers to ship the product in DOT specification packaging with "Flammable" red labels applied and the shipping name of the material marked on the shipping case. In other words, if the shipper chooses not to take advantage of the larger packaging offered in the flammable liquid exception for materials in the 73°–100°F range, and he wants to comply with the full scope of DOT regulation, then the flash point marking is not necessary.

7. The material has a flash point of 73°–100°F and is in individual packages exceeding 110 gallons capacity.

8. The material is properly reclassified and marked as ORM-D. See Chapter 13.

In other words, under DOT regulations the shipper need only mark the flash point of the material or an indication of the flash point range when: the material has a flash point of 73°–100°F, does not fall within the descriptions set out as items 4, 5, or 8, above, and is shipped in:

- Inside metal containers with a capacity over 1 quart but not over 1 gallon, overpacked in strong outside non-specification packaging;
- Inside non-metal containers with a capacity between 1 pint or 16 ounces net weight and 1 gallon, overpacked in strong outside non-specification packaging; or,
- Outside non-specification packaging with a capacity between 1 gallon and 110 gallons.

d. If the material happens to fit within exceptions drawn particularly for materials with that shipping name, the applicability of the general exception section may vary. Check the particular provisions involved. Also see Chapter 13 on limited quantity exceptions to DOT requirements.

Some specific regulations on certain materials include additional marking requirements for the outside packaging. See, for example, 49 CFR 173.266 for marking "Keep This End Up" or "Keep Plug Up To Prevent Spillage," on certain drums containing hydrogen peroxide.

e. Each package of hazardous materials must be marked with the name and address of the consignee or consignor, except when shipped in carloads and truckloads or in less-than-truckloads when handled by a motor vehicle not requiring transfer from one motor carrier to another.

f. Certain materials require the application of a rectangular advisory message that is not properly called a DOT hazard label but in common parlance is more like a label than anything else. See Chapter 10 on labeling for a discussion of the "Bung" label, "Etiologic Agent" label, and others.

CHAPTER 10

Hazardous Materials Package Labeling

Hazard labeling is a common governmental approach to warning workers and the public about that hazard, and a multitude of federal agencies employ one form of labeling or another. Among these agencies are the Food and Drug Administration, the Consumer Product Safety Commission, the Center for Disease Control of HHS, the Environmental Protection Agency (EPA), the Department of Agriculture, and the Occupational Safety and Health Administration of the Department of Labor (OSHA). The fact that one agency prescribes labeling for a material does not preclude the imposition of additional labeling requirements by other agencies if that material also falls within their jurisdiction.

The only agency purposely established with the view of avoiding overlap with the safety and health authority of other federal agencies is OSHA, but that goal has been frustrated by expansion of OSHA into other areas, particularly into "hazard communication" to employees under 29 CFR 1910.1200. This incursion into other regulated areas includes transportation, and so in transportation one often finds material safety data sheets (MSDSs) and package labeling that supplements the DOT hazard labeling.

In complying with labeling requirements from other agencies, or those developed privately, one should keep in mind the DOT rule in 49 CFR 172.401, which says that no one may offer or transport a package "bearing any marking or label which by its color, design, or shape could be confused with or conflict with" a DOT label.

With so many agencies dealing with labeling, the term "label" has varied meanings depending on its context. The philosophy behind imposition of labeling requirements is equally varied. For the purposes of the *Red Book* and DOT, the term "label" only means the 4-inch-by-4-inch colored diamond with a pictorial symbol and word describing the hazard, which is applied according to DOT requirements to the outside shipping container for hazardous materials. (There is provision made in 49 CFR 172.400 for smaller labels for compressed gas cylinders under certain circumstances.)

The term label does not include words of warning or other advisory material that must appear on the outside of the container, which are covered in Chapter 9 on marking. It also does not include the warning words applied to inside containers under the requirements of other agencies and local fire departments.

Shippers must furnish and attach hazardous materials labels, as prescribed by DOT, for the packages being shipped. The label should be conspicuous, and it must be placed on the package near the DOT-required marking of the name of contents of that package. The label definitely should not be applied to the bottom of the package where it will probably go unseen.

The specifications for the appearance of the labels are very particular and include the size, shape, hazard symbology, color, and written legend. Unless there is a clear instruction to the contrary, each label must be at least four inches square and should be oriented on the package with the point up, giving the square a diamond configuration. Each label must have a solid line border at least 3½ inches long on each side. Except when having an outer border consisting of a dotted line, each label on a package must be on a background of contrasting color.

Unless labeling exceptions are provided for the specific material or the quantity being shipped, the DOT label is required for all shipments within the United States. It is compatible with the UN labeling system and therefore may be used for export shipments by air and water as well, as long as the single digit UN hazard number appears in the lower point of the label and the UN class of the material is the same as its U.S. classification.

The basic elements of the label are: the color, which is carefully prescribed in the regulations and is also part of the international labeling system; the symbol, which pictorially displays the primary hazard of the product; and the verbal message or legend, which gives the name of the primary hazard for which the material is classified. In other nations, the language portion of the label will be different and may be absent, but the dimensions, color, and symbol will be the same in all countries. For import shipments, but only import shipments, the verbal hazard description that meets the requirements of the country of origin will be accepted in the United States, assuming the dimension, symbol, and color are the same as the UN requirements.

The UN hazard classification numeral, mentioned above, may appear on the label and should be at least ¼-inch high, located in the bottom point of the label.

The terms "UN class" and "UN division" are not synonymous, but in literature describing UN terminology for shipping papers, labels, and package markings, it often is difficult to grasp the distinction made between the terms since it is seldom explained. The following table should be helpful in understanding the UN class and division relationship to DOT hazard classifications. Although the relationship is not perfectly equivalent, as the table perhaps would indicate, understanding the relationship in this general manner can serve as a guide to the reader.

At one time the hazardous materials labels were referred to by their colors. Thus,

DOT Classification	U.N. Class	U.N. Division
Explosive, Class A	1	1.1 or 1.2
Explosive, Class B	1	1.3
Explosive, Class C	1	1.4
Flammable compressed gas	2	2.0 or just 2
Nonflammable compressed gas	2	2.0 or just 2
Flammable liquid (except pyrophoric)	3	3.0 or just 3
Combustible liquid	3	3.0 or just 3 (UN only covers up to 141°F. cc)

DOT Classification	U.N. Class	U.N. Division
Flammable solid	4	4.1
Flammable liquid (pyrophoric)	3	4.2 (UN—substance liable to spontaneous combustion)
Flammable solid (only certain ones)	4	4.3 (UN—substance which emits flammable gas on contact with water)
Oxidizer (except organic peroxides	5	5.1
Oxidizing material (organic peroxides)	5	5.2
Oxidizing material (non-explosive blasting agents)	1	1.5
Poison, Class A	6	6.1
Poison, Class B	6	6.1
Etiologic agent	6	6.2
Irritating material	6	6.1
Radioactive material	7	7
Corrosive material	8	8.0 or just 8
___ ___ ___	9 (Miscellaneous dangerous substances)	-9.0 or just 9

a red label product was a flammable, a white label product a corrosive, etc. In fact, in some circles, the term "red label" is used to denote *any* hazardous material, regardless of class. DOT has abandoned this manner of reference to the labels by color and the proper name for the labels is now the full name of the particular classification involved, e.g., "Corrosive label." One often will find current reference in non-DOT sources to labels by their color, however, and it is worthwhile to be familiar with this traditional practice.

The proper names and corresponding colors of today's labels are as follows:

Label Name	Colors
Explosive A, B, or C	Orange, with bomb-burst symbol, inscription, and border in black.
Nonflammable Gas	Green, with cylinder symbol, inscription, and border in black.
Flammable Gas	Red, with flame symbol, inscription, and border in black.
Flammable Liquid	Red, with flame symbol, inscription, and border in black.
Flammable Solid	White with vertical "candy stripes" in red, and with flame symbol, inscription, and border in black. The words "Flammable Solid" must not touch any red stripe.
Oxidizer	Yellow, with "flaming doughnut" symbol, inscription, and border in black.
Organic Peroxide	Yellow, with "flaming doughnut" symbol, inscription, and border in black.

(cont.)

Label Name	Colors
Poison Gas	White, with skull and crossbones symbol, inscription, and border in black.
Poison	White, with skull and crossbones symbol, inscription, and borders in black.
Irritant	White, with no symbol, but an incription in red and border in black; for import and export, the "Irritant" label may be white, with a skull and crossbones symbol and border in black.
Radioactive— White I	White, with radioactivity symbol, inscription, and border in black, and with a single overprinted vertical bar in red.
Radioactive— Yellow II	Upper half yellow and lower half white, with radioactivity symbol, inscription, and border in black, and with overprinted double vertical bars in red.
Radioactive— Yellow III	Upper half yellow and lower half white, with radioactivity symbol, inscription, and border in black, and with overprinted triple vertical bars in red.
Corrosive	Upper half white, with "eaten hand" and metal bar symbols in black; lower 3-½-inch dimension in black, with inscription in white and outer ¼-inch in white; dotted-line border in black.
Spontaneously Combustible	Upper half white and lower half red, with flame symbol, inscription, and border in black.
Dangerous When Wet	Blue, with flame symbol, inscription, and border in black.

Label colors must follow the recommended specifications by their Munsell notations, given in the appendix to the labeling regulations, 49 CFR Part 172, Appendix A.

Combustible liquids are regulated by DOT only in bulk unit quantities of 110 gallons or more, unless they are also hazardous wastes or hazardous substances. There is no DOT combustible liquid label, even if they do meet these environmental descriptions.

At least one label must be affixed to each package (except radioactive materials, which must bear two labels on opposite sides of the package), based upon the classification of the material being shipped. If the package contains different individual articles that are of varying DOT classifications, the outside of the package must bear a label for each hazard present in that package. In other words, if a shipper has a flammable liquid and a flammable compressed gas in two separate inside containers that are placed together in one outside container, that outside container must have two labels, Flammable Liquid and Flammable Gas. (As an aside, shippers putting more than one type of hazardous material in the same outside container should pay careful heed to regulations on compatibility of materials in such packaging.)

Two labels are required to be displayed on each end or each side of any package over 64 cubic feet in volume, or any freight container of up to 640 cubic feet. On a freight container, one of the labels must be near the closure. If a freight container of less than 640 cubic feet is shipped by air, one placard may be used instead of the two labels.

If the same material meets more than one definition of a regulated classification, one of which is a Class A explosive, Class A poison, or radioactive material, then the outside package for that single material must be labeled to reflect each hazard of that material. For example, a radioactive material that is also a flammable liquid will have three labels on the package, two Radioactive Material labels and one Flammable Liquid label. See Chapter 4 for a discussion of the classification of products of multiple hazard.

Each package containing a material classed as a flammable solid, flammable liquid, corrosive, or oxidizer, which also meets the definition of a Poison B, must be labeled as required for the class of material and as a Poison B. Flammable liquids and solids, and oxidizers that are also corrosive must be labeled for each hazard.

If a material is classed as one hazard but also exhibits the characteristics of inhalation toxicity defined in 49 CFR 173.3a, then it must bear the poison label as well as the classification label. In addition, specific entries in 49 CFR 172.101 prescribe multiple labels.

Materials, once classified, are labeled according to their classification and, except as noted above, only one label is affixed to each outside container. For export, however, a shipper may apply additional hazard labels of the same class or different classes if so required by the destination country. It is interesting to note here for export purposes that, according to regulations for rail and road in Europe, many liquid hazardous materials, if packed in inner fragile packaging (they say glass, porcelain, stoneware, or similar) of over 5 litres capacity (1.3 gallon), must bear two labels on the package.

Over-labeling is discouraged by the DOT regulations. For example, unless regulated by the foreign destination country, no label should be applied to a product that does not have the hazard depicted by that label, although it need not actually meet the definition of that DOT hazard class. See 49 CFR 172.401. Also, if a liquid meets the classification for two hazards, only one classification will apply, in accordance with Section 173.2. As noted above, in some combinations DOT will prescribe two different labels for that package.

DOT labels can be paste-on, self-adhesive, preprinted, or incorporated into the shipper's identifying labeling on the package. As noted earlier, shippers are prohibited from using any other label that might by its design, shape, or color be confused with the DOT warning label. The label may contain form identification information, including the name of its maker, if that information is printed outside of the solid line border in no larger than 10-point type.

The label should be applied so that it looks like a diamond, not laying down on its side to look like a square, although this is not a requirement. If the package is too small to accommodate the 4″ × 4″ label, the label can be attached to a tag on the package, or may be shown "by other suitable means."

No person may offer or accept any package for transportation that bears a DOT label that has been obscured, altered, or obliterated by additional marks or writing or that is affixed to a package containing material that is not regulated by DOT. A label advising of a hazard that is not present dilutes the effect of required labels and may stimulate emergency response actions that are contrary to everyone's best interest in the event of an incident or accident involving the package.

As noted above, throughout the discussion of labeling the term "label" as used in the *Red Book* and in the DOT regulations means the diamond-shaped colored label. To confuse the issue somewhat, it is important to note that there are other visual signals required to be applied to certain products that in common parlance would also be called labels. Except for the etiologic agent label, which in the UN is diamond-shaped, these labels are rectangular and are more in the nature of advisory labels than hazard alert labels: Bung, Empty, Magnetized Material, Etiologic Agent, and Danger. Their description is as follows:

Label Name	Description and Use
Bung label	White, 5 inch by 3 inch rectangle, offering caution advisory language, to be applied to metal barrels and drums containing flammable liquids with vapor pressures exceeding 16 p.s.i.a.
Empty Label	White, 6 inches square, with "EMPTY" in black letters at least 1 inch high. To be used on certain empty radioactive materials packages or may be used to obliterate hazard labels on emptied hazardous materials packages.
Magnetized Materials	Blue and white rectangle measuring 3 9/16 inches (90 mm.) high and 4 5/16 inches (110 mm.) wide. For use on air shipments.
Etiologic Agent	White background with predominantly red printing on rectangle measuring 2 inches high and 4 inches long; prescribed by HEW regulations in 42 CFR 72.25(c)(4) for each package of etiologic agents.
Danger	Also called the Cargo Aircraft Only label; orange background with black symbol and inscription, on a rectangle measuring 4 11/32 inches (110 mm.) by 4 3/4 inches (120 mm.). To be used on all packages authorized for shipment by cargo aircraft, but which are prohibited aboard passenger-carrying aircraft.

CHAPTER 11

Shipping Paper Requirements

Shipping papers, the documentation required by DOT for hazardous materials, have been one of the most controversial and topical aspects of those regulations. Although each shipping paper in itself may not pose sufficient difficulty to generate such controversy, the large volume shipped by many companies, the detail of the requirements, the relatively low responsibility level of the persons filling out the papers, and the advent of computerization of shipping practices have served to magnify the difficulties and to compound the costs inherent in compliance.

Virtually all shipments of regulated hazardous materials, even those which otherwise are granted limited quantity exceptions from specification packaging, marking, and labeling requirements, must be accompanied by shipping papers that are given to the carrier. ORM-A, B, and C materials only necessitate shipping papers when shipped by air or water, as noted in Section 172.101. ORM-D materials only need shipping papers if destined for shipment by air. Even the ORM materials, however, need shipping papers in every mode if those materials meet the definition of a hazardous waste or hazardous substance (see Chapters 13 and 14).

The carrier, in turn, is obligated to have certain information at hand regarding the nature of the hazardous materials aboard the transport vehicle, whether that vehicle is an aircraft, a vessel, or a surface mode of transportation.

DOT went through special rule making to make the shipping paper a well-defined, exactly formulated document. DOT requires that it be a paper (any piece of paper will do, unless the shipment is a hazardous waste; see Chapter 14 on waste manifests) that states the following information, as a minimum, with the description in this exact sequence:

1. The proper shipping name of each different hazardous material in the shipment, as prescribed in 49 CFR 172.101 or 172.102. (If a concentration or percentage of ingredient range is given in Roman type in the shipping name, the actual percentage may be shown on shipping papers, e.g., "Hypochlorite solution, 5% available chlorine" instead of the Section 172.101 entry "Hypochlorite solution containing not more than 7% available chlorine.")

2. The classification prescribed for each different hazardous material, as it is paired with the proper shipping name in the applicable column of the alphabetical commodity list.

3. The United Nations (UN) or North American (NA) identification number for this shipping name and class, as given in the commodity list.

4. The total quantity of each different hazardous material in the shipment by weight, volume, or some other appropriate means. (The quantity may precede the shipping name.)

5. The initials "RQ" either preceding or following the shipping description, if the shipment contains a reportable quantity of a hazardous substance. See Chapter 14 for specific requirements on hazardous substances.

6. It must also include a signed *certificate* worded as follows: "This is to certify that the above-named (or 'herein-named') materials are properly classified, described, packaged, marked, and labeled, and are in proper condition for transportation, according to the applicable regulations of the Department of Transportation."

For shipments by air, the certificate must add the following sentence to the basic certificate: "This shipment is within the limitations prescribed for passenger aircraft/cargo-only aircraft (delete nonapplicable)."

Also, one may conclude the certificate by saying that the shipment is prepared "for transport by (mode) according to applicable international and national governmental regulations," in lieu of the closing phrase of the basic certificate that mentions DOT. These variations facilitate shipments made under the provisions of the International Civil Aviation Organization (ICAO) or the International Maritime Organization (IMO), or Transport Canada.

The words "each different hazardous material," as they are used above, refer to materials having different shipping names in the alphabetical commodity list of materials in the front of the volume of regulations. For example, if one were shipping 3500 pounds of acetone and 2000 pounds of hexane, those articles would be described on the shipping papers as follows:

Acetone, Flammable liquid, UN1990, 3500 pounds
Hexane, Flammable liquid, UN1208, 2000 pounds

If, on the other hand, one were shipping 3500 pounds of ethyl butylamine, a flammable liquid, and 2,000 pounds of propyl benzene, another flammable liquid, the description could be as follows because neither article is listed by its specific chemical name:

Flammable liquids, n.o.s., Flammable liquid,
UN1993, 5500 pounds

Since both of these articles use the shipping name "Flammable liquids, n.o.s.," the single entry suffices. One could, of course, list them separately if that method was a personal preference, but it is not required.

This shipping paper description may, in some instances, be contracted further. The regulations do not compel use of repetitive or duplicative language where the shipping name is identical to the classification. Thus, in the last example the shipping paper description for that shipment could have been written as follows:

Flammable liquids, n.o.s., UN1993, 5500 pounds

One may use this particular means of contraction only with the following shipping names:

Flammable liquid, n.o.s.

Flammable solid, n.o.s.

Combustible liquid, n.o.s.

Poison B liquid, n.o.s.

Poison B solid, n.o.s.

Corrosive liquid, n.o.s.

Corrosive solid, n.o.s.

Oxidizer, n.o.s.

Flammable gas, n.o.s.

Nonflammable gas, n.o.s.

Organic peroxide, liquid *or* solution, n.o.s.

Organic peroxide, solid, n.o.s.

Etiologic agent, n.o.s.

Radioactive material, n.o.s.

ORM-A, n.o.s.

ORM-B, n.o.s.

In all other cases, both the shipping name and the classification of the material must appear on the shipping paper, because the shipping names and classifications are not repetitive in this way.

In rail operations when the shipping paper is a switching ticket prepared by the shipper, the document must also bear the appropriate placard endorsement as it appears below.

Also in rail operations, an intermediate shipper or carrier who tenders any trailer, semi-trailer, or freight container to a rail carrier must show a description of the vehicle or freight container and the kind of rail placards that are affixed.

Hazardous material or class	Placard endorsement	Placard notation*
Explosives, class A	Explosives	EXPLOSIVES A
Explosive chemical ammunition containing class A poison gas	Explosives and Poison Gas	EXPLOSIVES A and POISON GAS
Explosives, class B	Dangerous	EXPLOSIVES B
Explosives, class C	Dangerous	FLAMMABLE
Blasting agents	Dangerous	BLASTING AGENTS
Flammable liquids	Dangerous	FLAMMABLE
Flammable solids	Dangerous	FLAMMABLE SOLID
Oxidizers	Dangerous	OXIDIZER
Corrosive materials	Dangerous	CORROSIVE
Nonflammable gases	Dangerous	NON-FLAMMABLE GAS
Flammable gases	Dangerous	FLAMMABLE GAS
Poisonous gases or liquids, class A	Poison Gas	POISON GAS

(cont.)

Hazardous material or class	Placard endorsement	Placard notation*
Poisons, class B	Dangerous	POISON
Irritating materials	Dangerous	DANGEROUS
Organic peroxides	Dangerous	ORGANIC PEROXIDE
Radioactive materials with radioactive yellow-III label	Radoactive Material	RADIOACTIVE
Combustible liquids	None	COMBUSTIBLE
Chlorine	Dangerous	CHLORINE
Fluorine	Dangerous	POISON
Oxygen, cryogenic liquid	Dangerous	OXYGEN
Empty tank cars last containing hazardous material other than a combustible liquid	Dangerous	See 49 CFR 174.25(c)

*The word "Placarded" must precede the notation on the switching tickets except when using the last notation ("EMPTY").

The certificate on the shipping paper must be signed by the shipper. The signature does not need to be placed in immediate proximity to the certificate. A signature in any location on the document, which generally applies to all of the information given on that document, is sufficient to comply with the DOT requirements.

The exact nature of the signature has been a frequent source of questions and controversy. DOT has issued an interpretation to the effect that DOT will accept any endorsement that is recognized as a binding signature in commercial law. As a result, the signature may be manual, typewritten, computer-printed, facsimile stamped, or applied by other mechanical means. (In air transportation under IATA, however, a typewritten signature is not acceptable.) No signature flexibility is available when completing the Uniform Hazardous Waste Manifest, which must actually be signed by the generator, the waste transporter, and the receiving treatment, storage or disposal facility. See Chapter 14 on hazardous waste transportation.

The signature must refer to a person, however, as an identifiable individual within the company. The company name alone is insufficient. So too, mere initials of an individual are not acceptable.

The person whose name appears as the signature must be fundamentally knowledgeable of the hazardous materials regulations, to the end that his certificate has some real meaning. Of course, that person must depend upon others in the chain of classification, packaging, and preparation of the articles for shipment. The validity and purpose of the certificate and specifically the signature often have been questioned, but the requirement remains in the regulations.

It is common practice for compliance inspectors to seek out the person whose signature appears on the certificate to inquire as to details of the shipment, its packaging, labeling, and description. Many shippers have received punitive citations for not having instructed this person on the regulations. As in any investigation, if this initial contact reveals an unfamiliarity with the regulatory requirements, it serves as a stimulus to pursue the investigation in more thorough detail.

Remember that errors, omissions, or other flaws in documentation historically have been the item most responsible for attracting compliance inspectors to a shipper's facility, for the documentation with a shipment is not only the key to the nature of that shipment but also can reveal a shipper's unfamiliarity with the details of safety regulations that could prove to be the source of a serious accident in transportation.

DOT specifically prohibits use of alphanumeric coding on shipping papers to convey required information describing the hazards of the shipment. When a hazardous material and nonhazardous material are described on the same shipping paper, the hazardous material must be entered first, or must be entered in a contrasting color, or must be identified by an "X" placed before the proper shipping name in a column captioned "HM." Entry of hazard information on the back of the shipping paper is also forbidden. The primary purpose of the DOT shipping paper requirement is to readily communicate the specific nature of the hazardous materials to firemen, police, and other emergency response crews. Legibility and clarity in that communication are essential.

See Chapter 13 for special shipping paper notations for limited quantity shipments that are granted certain exceptions and must be identified on shipping papers by the words "Limited quantities" or "Ltd. Qty." immediately following the proper shipping name and material classification.

See Chapter 14 for identification of hazardous substances that are shipped in a reportable quantity, or "RQ," and special instructions on completion of the Uniform Hazardous Waste Manifest.

CHAPTER 12

Placarding of Vehicles, Rail Cars, Tanks and Containers

Placards are the large, diamond-shaped, color-coded signs used on the outside of motor vehicles, rail cars, portable and intermodal tanks, and freight containers to convey the hazard of their contents. Placards look like enlarged labels, but they are not exactly the same as labels. The appearance of placards, and rigid details as to color and other tolerances, are found in 49 CFR 172.519–172.558, and in Appendices A and B to 49 CFR Part 172.

In adopting the current DOT placarding system, the agency noted that the prior "communications requirements of the regulations (1) generally are not addressed to more than one hazard; (2) do not in all instances require disclosure of the presence of hazardous materials in transport vehicles; (3) are not addressed to the different hazard characteristics of a mixed load of hazardous materials; (4) do not provide sufficient information whereby fire fighting and other emergency response personnel can acquire adequate immediate information to handle emergency situations; and (5) are inconsistent in their application to the different modes of transport." (Docket No. HM-103, 37 Fed. Reg. 12660, June 27, 1972.) A substantial effort was made to address these problems through the adoption of multiple placarding rules, coupled with the display of the UN/NA identification number of the material on the outside of bulk tanks and transport vehicles.

In transportation by highway, it is the shipper's obligation to offer correct placards to the motor carrier for his particular shipment, even though the carrier may already have other hazardous materials on board. If the placards should be changed because of materials already on the vehicle, or additional materials the carrier may pick up later, this change is the responsibility of the motor carrier. The shipper also need not provide placards to a motor carrier whose vehicle is already properly placarded for his shipment. The motor carrier who consolidates freight at his own terminal remains responsible for providing and affixing proper placards. No DOT placarding is prescribed for the exterior of vessels or aircraft.

There is a basic requirement to placard each side and each end of each motor vehicle, rail car, larger portable tank, or freight container holding a hazardous material in any quantity *unless* the regulations provide an exception to the contrary. Much of the discussion of placarding focuses on the exceptions to placarding and when they apply.

No placarding is required for etiologic agents, for materials classed as ORM-A, B, C, D, or E, or for limited quantities (49 CFR 172.500).

NONBULK PACKAGING

Placarding is not required for vehicles carrying packaging containing residues of hazardous materials listed in 49 CFR 172.504 Table 2, and having a capacity of 110 gallons or less (49 CFR 172.504(d) and 173.29(a)(3)(i)), nor is placarding required for vehicles transporting combustible liquids in unit quantities of 110 gallons or less.

Two tables in 49 CFR 172.504 distinguish between materials that require placarding in any quantity, and those which are relieved from placarding in highway or rail transportation unless carried in packages having a total gross weight of 1000 pounds or more.

Table 1 materials are Class A and B explosives, Class A poisons, flammable solids required to bear a "Dangerous-When-Wet" label, radioactive materials required to bear a Yellow III label, and certain low specific activity radioactive materials shipped under 49 CFR 173.425(b). A material that is toxic by inhalation, as that concept is defined in 49 CFR 173.3a, must be "Poison" placarded in any quantity whether listed in Table 2 or not.

Table 2 materials include virtually all hazardous materials except those in Table 1 or that are excluded from placarding altogether. The 1000-pound threshold below which placarding is not required generally means 1000 pounds gross weight of hazardous materials in unit sizes that do not qualify for the "limited quantity" exceptions provided for each class. For example, if there were a shipment of 200 pounds of labeled corrosive liquids, 700 pounds of labeled flammable liquids, and 300 pounds of flammable liquids in limited quantities (unlabeled), then no placarding would be required.

Mixed loads of hazardous materials may be placarded for each of the classes, or may be placarded "Dangerous." The exception to this is that when 5000 pounds of a single class of material is loaded at one location, the placard for that class must be shown along with any others. For example, a load of 300 pounds of labeled corrosives and 800 pounds of labeled flammable liquids could be placarded "Corrosive" and "Flammable," or just "Dangerous." A load of 800 pounds of corrosive liquids and 6000 pounds of flammable liquids loaded at one time would be placarded "Flammable" and "Corrosive," or "Flammable" and "Dangerous." A load of 6000 pounds of corrosive liquids and 7000 pounds of flammable liquids loaded at one location *must* be placarded both "Corrosive" and "Flammable."

This threshold level was developed after careful consideration by DOT:

> While disclosure of the presence of hazardous materials in any quantity is considered desirable, it appears that to require disclosure in any quantity in the highway area is totally impractical. For example, while not pointed out specifically in the comments received, it must be recognized that the "Dangerous" placard would have been required on thousands of vehicles operated by utility companies, construction companies, and others who carry small quantities of flammable liquids and

compressed gases to perform their functions, particularly in the private carriage area. Also, considerable difficulty would be involved in placarding the thousands of vehicles used to transport small parcel shipments. [DOT] believes the benefit of such a requirement would be outweighed by the diminishing effect it would have on other placards that would convey information on the potential hazards of materials in significant quantities. Therefore, the 1000 pound rule presently used in the placarding regulations for highway vehicles is being continued, except for Class A and B explosives, highly or extremely toxic gases, thermally unstable or self reactive materials, water reactive materials, certain organic peroxides, extremely toxic poisons, and certain radioactive materials. [Docket No. HM-103, 39 Fed. Reg. 3164, Jan. 24, 1974.]

BULK PACKAGING

Cargo tanks, portable tanks, and tank cars must be placarded for their contents unless they have been cleaned of residues and purged of hazardous vapors, or refilled with an unregulated product. In other words, a tank that is emptied but continues to contain residues of its former hazardous materials contents must continue to bear the appropriate placard for that residue, unlike smaller packaging.

The person who offers a rail car, cargo tank or portable tank for transportation is responsible for affixing the proper placards to that unit. A portable tank having a rated capacity of less than 1,000 gallons that may be labeled under 49 CFR 172.406(e)(4) may also be placarded instead. If it is placarded, the placards need only appear on two opposite sides.

There is an obligation to identify the UN/NA number of hazardous materials shipped in bulk. See the special provision on these requirements below.

Proper placards must be kept on loaded cars and vehicles, and should remain visible and not obscured by appurtenances, other markings, or dirt.

SPECIAL PLACARDING RULES

There is an assortment of special rules, which may be summarized as follows:

1. Poison-inhalation hazards described in 49 CFR 173.3a must be placarded with "Poison" placards in addition to any other placards that might be on the vehicle, car or tank.
2. Highway route controlled quantities of radioactive materials need to have their "Radioactive" placard on a square white background with a black border.
3. A square background also is required for each "Explosive A," "Poison Gas" and "Poison Gas-Empty" placard in rail.
4. Every domed tank car containing a flammable liquid having a vapor pressure exceeding 16 p.s.i. at 100°F (37.8°C) must have a "Dome" placard.

5. Each tank car containing a hazardous material residue must bear special RESIDUE placards corresponding to the placards that were required for the material when the car was loaded.

6. Each container or car in the rail mode that has been fumigated or treated with poisonous liquid, solid or gas must be placarded "Fumigation" on or near each door.

UN/NA IDENTIFICATION NUMBER MARKINGS

Portable tanks, cargo tanks, and tank cars must have the UN or NA (North American) identification number for the contents displayed on the unit. There are two ways to do this: either by use of a separate orange panel, or by incorporating the numbers into the placards used.

The orange panels must be 16 cm (6.25 in) by 40 cm (15.75 in), with a 15 mm (9/16 in) black border. The numbers must be displayed in black digits 4 inches high. Other details are prescribed in 49 CFR 172.332.

Numbers incorporated into the placards must appear in the center area of the placard in 3-inch digits on a white background which is 4½ inches high.

No identification number may be displayed on a "Poison Gas," "Radioactive," or "Explosives" placard. No identification number may be placed on, or remain on, a freight container or transport vehicle that does not contain that material. No identification numbers are required on the ends of multi-compartment tanks if different materials are in the tanks; they should appear on the sides of the tanks, however, in the same sequence as the compartments in which those materials are carried. If a gasoline or fuel oil truck is marked "Gasoline" or "Fuel Oil" on each side and rear, then no numbers are required. For distillate fuels in compartmented tanks, the number may be the proper one for the compartment containing the lowest flash point material.

If one chooses to use the orange panel on a vehicle that also is placarded, the two must be displayed in close proximity to one another. If no placards are required, but one also does not want to use the orange panel, a plain white diamond placard configuration may be used, showing the number as if it were on a placard.

EMERGENCY RESPONSE GUIDEBOOK

DOT has published a guidebook for hazardous materials incidents, identified as DOT P 5800.3. Enough of these books were printed to have one in each emergency response vehicle in the country that might respond to an incident.

The book describes the placarding and, in particular, the UN/NA identification numbering system. By having the shipping name of the material, one is referred to the UN/NA number; or having only the UN/NA number one is referred to the proper shipping name. Either method of looking up a material guides the reader to one of a limited number of response guide pages.

These few pages describe the fire, explosion and health hazards of the material, and give brief guidance on what emergency actions to take, including advice on what to do with fires, spills, and injuries.

Obviously, if the wrong numbers are displayed, or no numbers are displayed when they are required, then the actions of emergency response personnel are impaired—perhaps dangerously. Thus, the proper use of placards, and of identification numbers, is critical for transportation safety.

CHAPTER 13

Limited Quantities and ORM-D Shipments

Before getting into the subject of this chapter, it is essential to distinguish between exemptions discussed in Chapter 28, and those instances of partial relief from regulation discussed here. Exemptions in Chapter 28 are true exemptions or waivers, issued to a specific applicant for the exemption and relieving him from a specific regulatory requirement, as described procedurally in 49 CFR Part 107. The exemptions or, more properly, "exceptions" covered by this chapter are distinctly different and are simply opportunities for lesser regulation offered to all shippers, based on the classification of the materials and the quantity shipped in each inside container. These exceptions, which relieve all shippers of such articles from some of the more onerous requirements, are set forth in detail in the regulations of title 49 CFR. It is not necessary for the shipper to file an application or have any contact with DOT to take advantage of the partial relief they offer.

It must be stressed that the relief is partial, since certain requirements are maintained and enforced. Qualifying for an an exception does not remove the material or the shipper from all responsibility, or put him outside the reach of DOT jurisdiction. Exceptions are instances of partial relief, not total relief, from regulatory requirements.

The various special exceptions, also called limited quantity exceptions, vary depending on DOT hazard classification of the material and the mode of transportation by which it will be shipped. Accordingly, the discussion in this chapter is subdivided along these lines. Initially, however, a general summary will be helpful.

Larger quantities of materials bearing a relatively higher hazard are subject to the full spectrum of DOT regulatory requirements. Generally speaking, these include identification and classification of the material, selection and use of the proper DOT shipping name, use of DOT specification packaging, marking the package with the DOT shipping name and other information, application of a DOT warning label to the package, preparation of shipping documents, certification of compliance on those documents, placarding of freight containers, railroad cars, and motor vehicles, and distinct limitations on stowage, storage, and handling of the materials. A partial or limited quantity exception relieves shippers and carriers from some of these basic requirements, but not from all of them.

Other than ORM-D discussed further below, most hazardous materials, regardless of quantity or mode of transportation, must be accompanied by proper DOT shipping documents. With limited exceptions, every rail car carrying any hazardous material, regardless of the quantity, must be placarded. Other than ORM, most hazardous materials shipped by air must be labeled. Regardless of quantity or classification, most packages of hazardous materials regulated under title 49 CFR, including ORM, must be marked with the name of the contents, i.e., the DOT shipping name of the material. Exceptions to these general statements are found in 49 CFR 173.4 (very small quantities), 173.5 and 173.315(m) (certain agricultural operations), and 173.421 (limited quantities of radioactive materials).

Because of the partial nature of the limited quantity exceptions, the regulatory paragraph describing each exception must be read with great care. Do not read relief into that paragraph that is not specifically provided there.

Relief from specification packaging means that the shipper need not incur the expense of using the packaging described in detail in 49 CFR Part 178. Specification packaging is described at greater length in Chapter 8. Relief from specification packaging does not mean that the shipper may use just any type of packaging for his product, no matter how leaky or flimsy, since the shipper still is covered by the general packaging responsibility paragraphs of 49 CFR 173.24. For air shipments, despite relief from specification packaging, there are general, additional requirements set out in new Section 173.6.

Marking, as that term is used in regulations, refers to all those words and symbols generally described in 49 CFR Part 172. Those sections do not include all marking requirements, however, so look at Chapter 9 on package marking to determine the full requirements. In any case, each exception section should be read carefully to establish just how broad the marking relief is, if any, since there are some unique requirements maintained for certain materials. Note that broad relief from marking the shipping name on small packages transported by land and air was eliminated in 1976.

Labeling, as mentioned in the limited quantity exceptions, is labeling which otherwise would be required for the particular material, as described in 49 CFR 172.400-450. A label, further discussed in Chapter 10 on labeling, is a diamond-shaped color-coded label, at least four inches square, which depicts symbolically and in words what the primary hazard of the material is in transportation.

The location of the limited quantity exceptions is determined by using the alphabetical commodity list in 49 CFR 172.101. Once the shipper has determined the proper shipping name for the material he has classified, the columnar tables in the commodity list will lead him to other requirements. The column headed "Packaging Exceptions" is the one of particular interest here. In a few rare instances only a single regulatory reference entry will provide both an exception and the specification packaging requirements for that material. If no general limited quantity exception is given in the regulations, the first entry will read "None". Even when such an indicator appears in the commodity list, however, it is possible that the particular packaging section for that material in 49 CFR Part 173 will offer some exceptions to certain packaging, labeling, or other requirements.

SHIPMENTS BY RAIL AND HIGHWAY

EXPLOSIVES

The commodity list indicates that no exceptions are offered for shipments of explosives. The shipper, however, should review the specific packaging requirements for his commodity, since there are certain less onerous variations on packaging and labeling offered for certain explosive commodities in certain quantities.

FLAMMABLE AND COMBUSTIBLE LIQUIDS

In view of the frequency with which shippers will need to refer to the limited quantity exception section for this classification, and the fact that it can serve as an illustrative model for limited quantity exceptions for other classes of hazard, the flammable liquids exception section is reproduced here in its entirety, and with annotations explaining each paragraph.

SECTION 173.118 Limited quantities of flammable liquids.

(a) Limited quantities of flammable liquids that do not meet the definition of another hazard class in [these regulations] and for which exceptions are permitted as noted by reference to this section in Section 172.101 ... , are excepted from labelling [except when offered for transportation by air] and specification packaging requirements ... when packed according to the following paragraphs. In addition, shipments are not subject to placarding requirements of Part 172 ... , to Part 174 ... except Section 174.24 and to Part 177 ... except Section 177.817.

(1) In metal containers not over 1 quart capacity each, packed in strong outside containers,

(2) In containers having a capacity not over 1 pint or 16 ounces by weight each, packed in strong outside containers, or

(3) In inside containers having a rated capacity of one gallon or less when packed in strong outside containers. The provisions of this partial exemption apply only if the flash point of the material is 73F° or higher and the flash point, or an indication that the flash point is 73F° or higher is marked on the outside package.

(b) A flammable liquid that does not meet the definition of another hazard class and has a flash point of 73F° or higher is not subject to the specification packaging requirements of [these regulations] when in packagings of 110 gallons or less. The provisions of this paragraph apply only if the flash point, or an indication that its flash point is 73F° or higher, is marked on the outside package. Notwithstanding Section 172.101 of [these regulations], the net quantity limitation for flammable liquids meeting the conditions of this paragraph is one gallon per package for carriage aboard passenger-carrying aircraft or railcar and 55 gallons per package for carriage aboard cargo aircraft only.

(c) Alcoholic beverages [wine and distilled spirits are defined in 27 CFR 4.10 and 5.11] in containers having a rated capacity of one gallon or less are

not subject to the requirements of [these regula-
tions].

(d) Special exceptions for shipment of certain
flammable liquids in the ORM-D class are provided in Subpart N of [these regulations].

This flammable liquids exception section evolved over many years, with bits and
pieces added or deleted like patchwork. In part, that accounts for the general lack
of readability of the section. In order to comprehend the full meaning of the relief
granted or not granted in this section, and as a guide to proper understanding of
exceptions sections for other classes of hazardous material, a detailed review of
each paragraph will be helpful.

(a) Limited quantities of flammable liquids that do
not meet the definition of another hazard class in
(these regulations) and for which exceptions are
permitted as noted by reference to this section
172.101 ... , are excepted from labeling [except
when offered for transportation by air and specification packaging requirements of [the regulations]
when packed according to the following paragraphs. In addition, shipments are not subject to
(placarding requirements) of Part 172, to Part
174 ... except Section 174.24 and to Part 177 ...
except Section (tab)177.817.

To determine whether a shipment qualifies for relief under this paragraph, one
must read it in conjunction with the three subparagraphs that follow it. For the
moment, however, it is important to know what that relief may or may not be, and
the qualifying subparagraphs will be covered in detail later.

See 49 CFR 173.115 and Chapter 5 on identification and classification of hazardous materials if you are unclear on what constitutes a flammable or combustible
liquid. The opening paragraph of this limited quantity exception confirms that
relief is not available if the material meets more than one DOT definition of hazardous class or, if the commodity list in 49 CFR 172.101 shows "None" for that material.
The more dangerous of the regulated materials fall within this non-
excepted category. Among the flammable liquids for which no exceptions are provided, for example, are butyl mercaptan, ethyl chloride, ethylene oxide, and ethyl
ether.

Materials qualifying for the limited quantity exception set forth in paragraph
(a) of Section 173.118 are given relief from specification packaging and labelling
requirements. Remember that the full breadth of this exception is only available
when shipping by rail, water, or highway. If air transportation will be used the
package must be labeled. This will be discussed more below.

Paragraph (a) contains the following sentence: "In addition, shipments are not
subject to [placarding requirements] of Part 172, to Part 174 ... except Section
174.24 and to Part 177 [of the regulations] except Section 177.817." This sentence
says a lot. Parts 174 and 177 of the hazardous materials regulations pertain to requirements for rail and motor carriers. Those parts, therefore, include requirements on receiving, loading, handling, storing, and unloading hazardous materials,
as well as certain other operational requirements. When a rail or motor carrier
(including a private carrier) is hauling materials that qualify for the limited quan-

tity exception, he is relieved from compliance with Parts 174 and 177 for that shipment and thus need not concern himself with loading compatibility and other operational requirements. Placarding requirements also do not apply.

For all modes of transportation the regulations specifically require shipping papers for all materials except ORMs-A, B, C and D (unless they are hazardous wastes or hazardous substances). Shipping papers are required for ORM-A, B, and C only when they are offered or intended for transportation by the air or water modes in which they are regulated, as indicated by the "A" or "W" preceding the shipping name of the material in Section 172.101. An extremely limited number of these ORMs are not preceded by an "A" or a "W" and, therefore, are regulated in all modes, for example, asbestos (ORM-C). Shipping papers are required for ORM-D only if material is offered or intended for transportation by aircraft, or if the material is a hazardous waste or hazardous substance. The shipper requirements to prepare shipping papers are matched by parallel sections applicable to carriers such as 49 CFR 174.24, 175.30(a)(2), 176.24, and 177.817.

Shipping papers are required even when a material qualifies for the limited quantity exception. The only distinction for limited quantity shipments on the shipping papers is the requirement that those papers bear the notation "Limited Quantities" or "Ltd. Qty." immediately following the shipping name and classification description of the material. But for this special limited quantity notation, the shipping paper is identical to that which would be prepared for the same material in a large quantity. It must show the shipping name of the material, its UN/NA number, its hazard classification, and the total quantity of material of that description in the shipment. Below are some examples of shipping paper descriptions for flammable liquids in limited quantities:

> Paint, Flammable liquid, UN1263, Limited Quantity 750 pounds
>
> Methyl acetate, Flammable liquid, UN1231, Ltd.Qty., 10 gallons
>
> Flammable liquid, n.o.s., UN1993, Limited Quantity, 20 gallons

See Chapter 11 for specific requirements on shipping papers. Except in private carriage and certain tank car and tank truck shipments, every shipping paper, including those for limited quantities, must bear the shipper's certificate of compliance.

Historically, the limited quantity exceptions also used to provide relief from package markings. This was eliminated in DOT rule making, so that the full marking requirements now apply even to smaller units of regulated materials, including ORMs. See Chapter 9 on marking.

To summarize, qualifying for the limited quantity exception granted for flammable liquids relieves the shipper of specification packaging and labeling (other than air) requirements. Shipping papers, however, are mandatory.

The criteria for the limited quantity exceptions for flammable liquids and other classifications are very specific, relating generally to the size or weight of the inner units being shipped. The variety of packaging dicussed in the flammable liquids exception simply reflects DOT's belief that a metal container is stronger than plastic or glass, and that materials with high flash points are less hazardous than those with lower flash points.

Before certain DOT rule making amendments several years ago, the upper limit on DOT's definition of "flammable liquid" was 80°F, as determined by the open-cup tester. DOT theoretically determined that the change in test method from open to closed-cup would ordinarily result in about a seven degree shift in the flash point of a given material. In other words, a material that had a flash point of 57°F by the open-cup tester would register a flash point of about 50°F by the closed-cup method. Pursuing this theory to its logical end, a material at the former upper definitional limit of 80°F by the open-cup tester would have a flash point of 73°F by the closed tester. Thus, the flash point range of materials that had not been regulated as flammables before but were considered regulated flammable liquids after that rule making, runs from 73° to 100°F. With this history in mind, perhaps some of the otherwise inexplicable number breakpoints will be more understandable.

A flammable liquid having no secondary hazard, that is not listed with a "None" entry in the exception column of 49 CFR 172.101 qualifies for the limited quantity exception from specification packaging and labeling (unless shipped by air) when packed:

1. In metal inside containers that each have a rated capacity of 1 quart or less, with all of the inside containers overpacked in a strong outside container. Note that the quantity limitation is on the size of the inside container, not the case as packed for shipment. In limited quantity exception sections for certain other classes and materials there may also be a limit on the size or weight of the filled outer shipping case. Though each inside container for flammable liquids may not exceed a quart, there is no limit on the number of metal containers in a single overpack. The overpack may be fibreboard, wood, or other material that makes up a "strong outside container," although this phrase may need some interpretation in unusual packaging configurations. For example, DOT generally has stated that shrink film wrapping does not constitute a strong outside container for the purposes of these regulations. A reasonable guide is to determine what packaging is acceptable according to published freight tariff requirements such as those published by the National Classification Committee or the Uniform Freight Classification Committee, although this, like every other general statement in this regulatory area, has its exceptions.

2. In inside non-metal containers each having a rated capacity of 1 pint or 16 ounces weight or less, overpacked in a strong outside container. The capacity may be either 16 fluid ounces (1 pint) or 16 net weight ounces, but it does not have to be both, so that a pint of material that weighs 20 ounces still qualifies for the exception.

3. In inside containers of any construction, with a rated capacity of one gallon or less, when the material as shipped has a flash point of 73°F or higher, and the containers are overpacked in a strong outside container. In order to take advantage of this larger packaging, the shipper must mark the flash point of the material on the outside shipping case, or at least an indication that the flash point is above 73°F.

These are the three basic categories into which a flammable liquid must fit to qualify for the broad limited quantity exception to the specification packaging and labeling requirements. The regulations offer yet another type of exception, but this exception only gives relief from specification packaging requirements (not labeling). The shipping units must be marked and labeled as if the full scope of regulations applied and, of course, shipping papers must be prepared. Flammable liquids without a secondary hazard and having a flash point in the range of 73°–100°F may be packaged in a non-specification outside container if that packaging has a rated capacity of 110 gallons or less. In other words, a nonspecification 55-gallon drum may be used. Once again, since this exception is limited to materials in a specific flash point range, the flash point, or an indication that it is above 73°F, must be marked on the outside shipping unit. There are no specific requirements on size or location of this marking, but it should be very clear and near the other required markings and label.

Before leaving the classification of flammable and combustible liquids, there are a few items of particular note with regard to limited quantity exceptions:

1. Combustible liquids are those with closed-cup flash points at or above 100°F and up to 200°F. There are no requirements imposed on such liquids in unit quantities of less than 110 gallons (unless they are hazardous wastes or reportable quantities of hazardous substances), and thus no need for limited quantity exceptions. Even for hazardous wastes and substances, the requirements would be for markings and shipping papers as a combustible liquid, i.e., there is no combustible liquid label and nonspecification packaging could be used.

2. For the purposes of the DOT regulations, a distilled spirit of 140 proof or less is considered to have a flash point no lower than 73°F, thus qualifying for the larger capacity containers provided under the exception for materials having flash points of 73°–100°F.

3. Alcoholic beverages (wine and distilled spirits as defined in the Treasury Department regulations in 27 CFR 4.10 and 5.11) are not subject to any DOT regulations when shipped in containers having a rated capacity of 1 gallon or less.

4. An aqueous solution containing 24 percent or less alcohol by volume is considered to have a flash point no less than 100°F if the remainder of the solution does not meet the definition of a hazardous material, e.g., if the rest of the solution is not poisonous, corrosive, etc. Under this provision, such aqueous solutions are declared to be above 100°F and, if their flash point is below 200°F, they will be properly classed as "combustible liquids." As noted above, no regulations apply to packages of combustible liquids of 110 gallons or less (unless they are hazardous wastes or substances).

Paragraph (d) of the flammable liquids limited quantity exception section advises that special exceptions for shipment of certain flammable liquids are provided in Subpart N of Part 173 if those liquids may be shipped under the classification of ORM-D.

Subpart N of Part 173 opens with 49 CFR 173.1200, entitled "Consumer Commodity." Consumer commodities and certain small arms ammunition are the only materials in the ORM-D classification, but the structure of the regulations has been arranged to facilitate the entry of more items into this special exception in later rule making. According to the definitions in 49 CFR 171.8:

> "Consumer commodity" means a material that is
> packaged and distributed in a form intended or
> suitable for sale through retail sales agencies or in-
> strumentalities for consumption by individuals for
> purposes of personal care or household use. This
> term also includes drugs and medicines.

Section 173.1200 declares that in order for a material to be transported under the shipping name "consumer commodity," it must in fact meet this quoted defini- tion, being both a material that would be available by law to consumers and in consumer-type packaging.

A consumer commodity that otherwise fits the description of a flammable liquid may be reclassed and offered for shipment as ORM-D material provided that a special exception like the one in 49 CFR 173.118(d) is authorized in the specific limited quantity regulations applicable to that material, and further provided that it is packaged as prescribed in 49 CFR 173.1200 based upon its initial classification.

For example, a household cleaning agent that falls within the definition of a flammable liquid in 49 CFR 173.115 would initially be classed as a flammable liq- uid, and a review of the commodity list in 49 CFR 172.101 might show the most accurate shipping name for the material to be "Compound, cleaning, liquid." Look- ing at that entry in the commodity list, the reader scans across the table to the "exception" column (5)(a), which refers him to 49 CFR 173.118.

Once at 49 CFR 173.118, one is led to paragraph (d) and then to Subpart N and the ORM-D regulations. Section 173.1200 begins Subpart N and says that if a mate- rial is a consumer commodity, and if 49 CFR 172.101 originally sent the reader to an exception section that contained a special reference to ORM-D, and if it is within the general packaging limits of 49 CFR 173.1200, it may be renamed "Con- sumer commodity" instead of "Compound, cleaning, liquid" and it may be re- classed as ORM-D instead of flammable liquid.

Section 173.1200 includes a parenthetical reference to 49 CFR 173.500 where the broad definition of ORM-D is found in paragraph (a)(4):

> An ORM-D material is a material such as a con-
> sumer commodity which, though otherwise subject
> to regulations . . . ,presents a limited hazard during
> transportation due to its form, quantity and pack-
> aging. They must be materials for which excep-
> tions are provided in 49 CFR 172.101. . . . A ship-
> ping description applicable to each ORM-D
> material or category of ORM-D materials is found
> in 49 CFR 172.101. . . .

The packaging criteria for ORM-D are found in 49 CFR 173.510 and in 49 CFR 173.1200. In the example of the household cleaning compound that was initally classed as a flammable liquid, the pertinent paragraphs under 49 CFR 173.1200 are the following ones:

> (1) Flammable Liquids must be:
>
> (i) In inside metal containers, each having a
> rated capacity of 1 quart or less, packed in strong
> outside packagings;

(ii) In inside containers, each having a rated capacity of 1 pint or less, packed in strong outside
packagings;

(iii) In inside containers, each having a rated capacity of one gallon or less, packed in strong outside packagings. The provisions of this exception
apply only if the flash point of the material is 73°F
or higher.

These three criteria obviously track the smaller packagings of the basic limited
quantity exception for flammable liquids in 49 CFR 173.118, but with the one difference that use of the larger packaging in subparagraph (iii) does *not* entail marking
the flash point or an indication of the flash point range on the outer container as
one would have to do under the limited quantity provisions of 49 CFR 173.118.

Qualifying for reclassification as ORM-D represents a lot more regulatory relief
than just this flash point marking requirement, however. As noted throughout the
remainder of the *Red Book*, many requirements that apply to fully regulated materials and to materials under the limited quantity exceptions do not apply to
ORM-D shipments. For example, there are no specification packaging requirements, no labeling requirements even in air transportation [49 CFR 172.400(b)(8)],
no shipping paper requirements except when offered or intended for air transportation [49 CFR 172.200(b)(3)] or when the material is a hazardous waste or hazardous substance, no placarding requirements [49 CFR 172.500(b)], no individual package inspection requirements when containerized and shipped by air by one
consignor [49 CFR 175.30(c)], no overall weight limitations aboard aircraft [49 CFR
175.75(b)], and no carrier loading or stowage requirements for shipments by water
[49 CFR 176.11(e)]. ORM-D is not relieved from marking requirements, however,
and each outside package must be marked with the shipping name and the classification of "ORM-D" in a rectangle. If the package meets the additional criteria of
49 CFR 173.6 it may be shipped by air, but the marking has to be adjusted to show
the class in the rectangle as "ORM-D-AIR." See the general marking requirements
beginning at 49 CFR 172.300, paying particular heed to 49 CFR 172.316 on marking
of ORM units.

FLAMMABLE SOLIDS, OXIDIZERS, AND ORGANIC PEROXIDES

See 49 CFR 173.150, 173.151, and 173.151a, and Chapter 4 for the specific
definitions of each of these classifications. The limited quantity exception section
for these classifications is similiar in structure to other such exception sections and
is found in 49 CFR 173.153. Except for flammable solids and oxidizers for which
"None" is noted in the exception column of 49 CFR 172.101 list of shipping names,
these materials are given relief from specification packaging and labeling (unless
shipped by air) when shipped in inside containers weighing no more than 1 pound
net weight each, overpacked in an outside container that does not exceed 25
pounds net weight. These shipments also are excepted from the rail and highway
requirements in 49 CFR Parts 174, 177 and 397, except for the shipping paper
requirements in 49 CFR 174.24 and 177.817. As with all exceptions, the shipper
should look not only at the exception section to which he is referred in the commodity list but also to the specific regulatory section applicable to that commodity.
In these classifications, for example, there are specific provisions relieving safety
matches and certain motion picture film from compliance with any of the regulations.

A similar limited quantity exception is provided for organic peroxides, when packed:

1. In strong outside containers having no more than one pint or one pound net weight of organic peroxide in any one outside package, with inside containers securely packed and cushioned with noncombustible cushioning material. Such cushioning material is not necessary when liquid organic peroxides are contained in strong, securely closed plastic containers that each do not exceed one fluid ounce capacity, if those containers are properly packed to prevent breakage or leakage.

2. Strong outside containers having no more than twenty-four inside fibreboard containers, which each in turn contain no more than seventy chemically resistant closed plastic tubes no larger than one-sixth fluid ounce capacity each, all securely packed in incombustible cushioning material.

These limited quantity exception sections also refer to special exceptions for flammable solids, oxidizers and organic peroxides that are reclassed as ORM-D, described in Subpart N beginning with 49 CFR 173.1200. This reference is used just as it was in the flammable liquid section, but 49 CFR 173.1200 provides different basic quantity limits for ORM-D packages that were initially classed as flammable solids, oxidizers, and organic peroxides than it does for flammable liquids. A special exception appears in 49 CFR 173.1200 authorizing ORM-D classification for charcoal briquets in packages not exceeding 65 pounds each.

A shipper of nitrates should specifically examine the regulatory sections applicable to those materials to determine the exception that may be offered.

CORROSIVE MATERIALS

The definition of this class of materials is found in 49 CFR 173.240. Limited quantity exceptions for corrosive materials are presented in 49 CFR 173.244.

Corrosive liquids, except those for which there is a "None" notation in the exception column of 49 CFR 172.101, are offered relief from specification packaging and labeling requirements (unless shipped by air), and are relieved of all Parts 174 and 177 except for the sections on shipping papers, when packed:

1. In inside bottles with a rated capacity no greater than 16 fluid ounces, each enclosed in a metal can, overpacked in an outside packaging. Several of these cans may be packed together.

2. In inside metal or plastic containers with a rated capacity no greater than 16 fluid ounces, overpacked in a strong outside packaging.

3. In glass containers having a rated capacity of not over 8 fluid ounces in strong outside packaging, and cushioned with sufficient absorbent material to completely absorb the liquid contents in the event of breakage, and which cannot react chemically with the corrosive liquid.

Other exceptions for corrosive liquids will be found in the specific regulatory sections of 49 CFR Part 173 devoted to those particular liquids.

Corrosive solids are offered a limited quantity exception to specification packaging and labeling requirements (unless shipped by air), when packed:

1. In inside earthenware, glass, plastic, or paper receptacles having a capacity of no more than 5 pounds each, overpacked in metal, wooden, or fibreboard outside containers, and not exceeding 25 pounds net weight.

2. In inside metal, rigid fibre, or composition cans or cartons, or in rigid plastic receptacles, having a shipping capacity of no more than 10 pounds each, overpacked in metal, wooden, or fibreboard outside containers, and not exceeding 25 pounds net weight.

Section 173.244 of 49 CFR includes a reference to the special exceptions for corrosive materials that qualify for reclassification as ORM-D. The concept of ORM-D works the same for these materials as it does for the example of a flammable liquid discussed above. Special provisions are made in 49 CFR 173.1200 for corrosive liquids and solids that are consumer commodities. Somewhat larger packaging may be used under the ORM-D concept for corrosive liquids, providing the mixture contains 15 percent or less corrosive material and the remainder of the mixture does not in itself meet the definition of a hazardous material, and for corrosive solids provided the solid mixture contains 10 percent or less corrosive material and the remainder of the mixture does not meet the definition of another DOT classification.

COMPRESSED GASES

The DOT regulations provide limited quantity exceptions for compressed gases that are not Class A poisons and are not accompanied by a "None" entry in the exception column of 49 CFR 172.101. The definition of the class of compressed gas, and the detailed distinction between flammable and nonflammable gases, are found in 49 CFR 173.300.

The exception sections for qualified compressed gases are particularly convoluted, having evolved over a considerable period of time with each paragraph being styled to fit a specific need at the moment. As with the other limited quantity exceptions, qualified gases (being neither poisonous or otherwise ineligible) are given relief from specification packaging and labeling requirements (unless shipped by air), and also are not subject to the rail and motor carrier requirements of Parts 174 and 177 except for shipping papers, when packaged as follows:

1. In containers of any material of construction that have a water (overflow) capacity not exceeding 4 fluid ounces (7.22 in.^{3}) except cigarette lighters.

2. In metal containers filled with nondangerous material to no more than 90 percent capacity at 70°F, then charged with a nonflammable, nonliquefied gas. Each such container must be tested to three times the gas pressure at 70°F and, when the container is to be refilled it must be retested to three times the gas pressure at 70°F, provided:
   a. Containers not over 1 quart capacity are not charged to more than 170 p.s.i.g. at 70°F.
   b. Containers not over 30 gallons capacity are not charged to more than 75 p.s.i.g. at 70°F.

3. In inside nonrefillable metal containers charged with a nonpoisonous solution of materials and compressed gas, provided each of the following conditions is met:

a. The overflow water capacity of the container is no more than 50 in.³ (27.7 fl.oz.).

b. Pressure in the container does not exceed 140 p.s.i.g. at 130°F.* In any case, the metal container must be capable of withstanding, without bursting, a pressure of one and one-half times the equilibrium pressure of the contents of 130°F.

c. Liquid content of the material and gas must not completely fill the container at 130°F.

d. Each and every container, as completed and filled for shipment, must be heated until the pressure in the container is equivalent to the equilibrium pressure of the contents at 130°F without evidence of any leakage, container distortion, or other defect. Note that this is a 100 percent test requirement, not a periodic sampling. Compliance with this step is often accomplished through use of a hot water bath in the aerosol industry.

e. Each outside packaging must be marked "INSIDE CONTAINERS COMPLY WITH PRESCRIBED REGULATIONS," unless a DOT 2P or 2Q specification container is required as explained in the note, in which case the marking indicating specification packaging is to be used on the outside case.

Section 173.306 of 49 CFR also provides a limited quantity exception for foodstuffs, soap, beverages, biologicals, electronic tubes, and audible fire alarm systems. Relief from specification packaging and labeling requirements (unless shipped by air) is offered for:

1. Foodstuffs or soaps in nonrefillable metal containers not exceeding 50 in.³ (27.7 fl.oz.) overflow capacity, with soluble or emulsified compressed gas, provided the pressure at 130°F does not exceed 140 p.s.i.g. The metal container must be capable of withstanding, without bursting, an internal pressure of one and one-half times the equilibrium pressure of the contents at 130°F. No testing is required. The outside container must be marked "INSIDE CONTAINERS COMPLY WITH PRESCRIBED REGULATIONS."

2. Cream may be packaged in nonrefillable metal containers with soluble or emulsified compressed gas. Containers must have a design capability to hold pressure up to 375 p.s.i.g. without deformation, and must be equipped with a safety relief device. This particular exemption is only authorized for shipment of these containers by refrigerated motor vehicles. No testing is required.

---

*If the container has an internal pressure at 130°F that exceeds 140 p.s.i.g. but does not exceed 160 p.s.i.g., then the full breadth of the exception is not available and the shipper must use a DOT specification 2P container as the inside container. If the internal pressure at 130°F exceeds 160 p.s.i.g., the shipper must use a DOT specification 2Q container as the inside container. Under no circumstances may the internal pressure at 130°F exceed 180 p.s.i.g., unless the container is not over 4 fl.oz. overflow capacity. In all cases, the metal container must be capable of withstanding, without bursting, a pressure of one and one-half times the equilibrium pressure of the contents at 130°F. Although at these higher pressures one must use a DOT specification can, the outer shipping case does not have to be a specification container itself, but it must be marked "INSIDE CONTAINERS COMPLY WITH PRESCRIBED SPECIFICATIONS," as well as with the proper shipping name of the material. It need not be labeled unless shipped by air.

3. Biological products or medical preparations that will deteriorate with heat, in solution with compressed gas, may be packed in inside nonrefillable metal containers, but the capacity may be no more than 35 in.[3] (19.3 fl.oz.). Pressure in the container must not exceed 140 p.s.i.g. at 130°F, and the container must not go liquid full at 130°F. One container out of each lot of 500 such containers or less, completed as for shipment, must be heated until the pressure in the container is equivalent to the equilibrium pressure of the contents at 130°F without any evidence of leakage, distortion, or other defects. Note here, where the contents are heat-sensitive, 100 percent heat testing is not required as it is for many other aerosol containers.

Shippers of electronic tubes, audible fire alarm systems powered by compressed gas, fire extinguishers, refrigerating units, auto airbag systems, cigarette lighters, and hydraulic accumulators should check the regulations for specific criteria applicable to those articles to qualify for limited quantity exceptions.

Several paragraphs of the limited quantity exception for compressed gases make reference to certain special additional exceptions granted for certain compressed gases transported under ORM-D classification. Each paragraph must be read carefully, however, since this reference does not appear uniformly throughout 49 CFR 173.306. The "beer keg" containers described in paragraph (a)(2) of the regulation, for example, do not benefit from a reference to ORM-D provisions.

As noted earlier ORM-D is limited exclusively to consumer commodities and small arms ammunition. Nonconsumer materials under the compressed gas limited quantity exception, such as fire extinguishers, may not qualify for ORM-D due to the narrower packaging authorized in 49 CFR 173.1200 for compressed gases and due to the non-consumer nature of many fire extinguishers. Although the description of packaging in 49 CFR 173.1200 under compressed gases looks a lot like 49 CFR 173.306, there are very significant differences and the reader is cautioned to examine 49 CFR 173.1200 very carefully.

POISONS

Extremely dangerous poisonous gases and liquids, classed as Class A poisons in 49 CFR 173.326, are not authorized for shipment under any exception provisions. Less toxic materials, as Class B poisons, do qualify for limited quantity exceptions that include relief from specification packaging, placarding, and 49 CFR Parts 174 and 177 except for shipping papers and prohibitions against co-loading with foodstuffs.

In 49 CFR 173.345, Class B poisonous liquids, except those assigned a "None" entry in the exception column of 49 CFR 172.101 or as other specifically provided under sections applicable to the particular material, are relieved from specification packaging requirements (not from labeling) when packed as follows:

1. In glass containers having a rated capacity not exceeding one quart each, or in metal containers or polyethylene bottles having a rated capacity not exceeding one gallon each, overpacked in strong, outside steel or wooden boxes, barrels, or drums.
2. In glass containers having a rated capacity not exceeding 1 pint each, or in metal or polyethylene containers having a rated capacity not exceeding 1 quart each, overpacked in a strong, outside fibreboard box or molded expanded polystyrene case.

In 49 CFR 173.364, Class B poisonous solids, except for a series of specifically named materials for which minimal individual exceptions are offered, are given relief from specification packaging requirements (not labeling) when packed in tightly closed inside containers, securely cushioned when necessary to prevent breakage:

1. In inside glass, earthenware, or composition bottles or jars, or metal containers, or lock corner sliding lid wooden boxes, having a capacity not exceeding 5 pounds, or in inside chipboard, pasteboard, or fibre cartons, cans, boxes, or tightly closed strong plastic bags or bottles compatible with the contents, each having a capacity not exceeding 1 pound, overpacked in outside wooden or fibreboard box, or wooden barrel or keg, or molded expanded polystyrene case. The net weight of the outside package may not exceed one hundred pounds.
2. In inside plastic bottles or jars, chipboard, pasteboard, or fibre cartons, cans or boxes, each having a capacity not exceeding 5 pounds, overpacked in outside fibreboard or wooden boxes. No more than six of these cartons shall be packed in any one outside container.

The reader should check here, as with other classifications, for specific regulatory sections applicable to the material being shipped, since many of these specifc sections contain partial exceptions for the materials they cover.

The limited quantity exceptions for poison B materials do offer a narrow special exception for shipment as ORM-D for drugs and medicines that qualify for that treatment under 49 CFR 173.1200. Since ORM-D reclassification relieves these drugs and medicines from shipping paper requirements (unless intended for air shipment) and labeling, ORM-D drugs and medicines are not subject to the prohibition against co-loading poisons and foodstuffs. After reclassification, these drugs are not classed as poisons, but as ORM-D.

IRRITATING MATERIALS
These hazardous materials, in a classification defined in 49 CFR 173.381, are not eligible for any exceptions.

ETIOLOGIC AGENTS
This classification is defined in 49 CFR 173.386. Because of the unique nature of the DOT requirements, which supplement the packaging, marking, and labeling regulations of the Department of Health & Human Services, the reader is urged to examine those sections appplicable to them to determine the extent of the DOT requirements. Because almost all air carriers use the ICAO/IATA Dangerous Goods Code, it should be noted that the important 50 ml exception in 49 CFR 173.386(d) cannot be used except with specialized air carriers who accept shipments under 49 CFR.

RADIOACTIVE MATERIALS
DOT defines "limited quantity of radioactive material" to mean a quantity of a solid, liquid, or gas that does not exceed the activity limits shown in Table 7 in 49 CFR 173.423, and that is packed to meet 49 CFR 173.421, 173.421-1 or 173.421-2.

When the activity of the total package does not exceed the Table 7 limits, these materials are excepted from specification packaging, shipping paper and certification requirements, and labeling. In fact, excepted radioactive materials prepared in accordance with the requirements below are not otherwise subject to the DOT regulations except for incident reporting. See Chapter 22 on incident reporting.

The packaging requirements may be summarized as follows:

1. Packaging must be such that it will not leak material under normal conditions of transportation.
2. The radiation level at any external point on the package must not exceed 0.5 millirem/hour.
3. Any removable external radioactive contamination must not exceed the limits in 49 CFR 173.443(a).
4. The outside of the inner package (or the outside package if there is no inner package) must be marked "Radioactive."
5. The package does not contain more than 15 grams of uranium-235 (unless it is natural or depleted uranium as a manufactured article enclosed in an inactive metal or otherwise durable sheath packaged in accordance with 49 CFR 173.421-1). Note that there are separate limited quantity exceptions for manufactured articles in which the sold content is natural or depleted uranium or natural thorium.

If the radioactive material also has other hazards regulated by DOT, for example, flammable or corrosive liquid, then the additional provisions of 49 CFR 173.421-2 apply.

In addition, DOT provides similar exceptions for instruments and articles which do not exceed the activity limits of Table 7 in 49 CFR 173.423 and which are prepared for shipment in accordance with 49 CFR 173.421-1, if:

1. All the conditions given above are met.
2. The radiation level at 10 cm (4 inches) from any external surface point of the unpackaged device does not exceed 10 millirem/hour.
3. The radiation level for any exclusive use shipment does not exceed 2 millirem/hour.

There also are special transportation conditions that apply to the shipment of low specific activity (LSA) radioactive materials (49 CFR 173.425) and empty radioactive materials packagings (49 CFR 173.427).

Materials prepared for shipment under 49 CFR 173.421, .422, .425, or .427 must be certified as being acceptable for transportation by enclosing a notice in, or attaching it to the package as follows:

(According to the contents)

This package conforms to the conditions and limitations specified in 49 CFR 173.421 for excepted radioactive material, limited quantity, n.o.s., UN2910;

This package conforms to the conditions and limitations specified in 49 CFR 173.422 for excepted radioactive material, instruments and articles, UN2911;

This package conforms to the conditions and limitations specified in 49 CFR 173.424 for excepted radioactive material, articles manufactured from natural *or* depleted uranium *or* natural thorium, UN2909;

This package conforms to the conditions and limitations specified in 49 CFR 173.427 for excepted radioactive material, empty packages, UN2908.

SHIPMENTS BY WATER

Because of the international nature of most shipping by water, most shipments must be prepared in accordance with the International Maritime Organization (IMO) International Maritime Dangerous Goods (IMDG) Code. See Chapters 18, 19 and 25 for specific information about IMO and the IMDG Code. The Code has been amended recently to provide for the shipment of consumer commodities (ORM-D) and certain other hazardous materials in limited quantities.

Requirements found in the current edition of the IMO IMDG Code in Section 18 include a limited quantity system that is so alien to the U.S. regulations that it has been completely unusable for U.S. shippers. It has been used to a very limited extent by shippers in other countries. Most people, including most regulators and the IMO Secretariat, readily admitted an inability to understand the provisions. The 37th Session of the IMO Sub-committee on the Carriage of Dangerous Goods adopted a completely revised limited quantity section which was approved by the Maritime Safety Committee. Because of delays in administrative processing, this important change will not appear in the IMO IMDG Code before July 1988. The revised text of Section 18 appears at the end of this chapter.

In the meantime, shippers and carriers by water should be aware that DOT has been issuing a Competent Authority Certificate for many years that is almost identical to these new IMO provisions. Shippers wanting to make shipments under this waiver should apply to DOT for authorization. The certificate is commonly referred to as the ORM-D Competent Authority Certificate, and does not cover other limited quantities.

The DOT regulations governing shipment of limited quantity hazardous materials by water require marking the shipping name of a hazardous material on the outside shipping case, just as it must be marked for other modes.

Keeping account of the Optional Table in 49 CFR 172.102, limited quantity shipments may be made by water described and marked as for highway shipment, except for n.o.s. items. For water shipments other than domestic shipments, where the proper shipping name of the commodity is an n.o.s. entry in the commodity list, this marking on the outside container shall be qualified by the technical name of the commodity in parenthesis, e.g., "Corrosive liquid, n.o.s. (caprylyl chloride)." This requirement applies to n.o.s. materials in the new limited quantity exceptions as well as to fully regulated articles, but it does not apply to ORM-D because the only shipping names in that class are not n.o.s. items.

As noted throughout this chapter on limited quantity exceptions and ORM-D, a shipping paper is required in every mode of transportation for every hazardous material other than an ORM. ORM-A, B, or C require shipping papers by air and water if regulated in those modes. (A few ORM-Cs are regulated in all modes, such as asbestos.) ORM-D only requires a shipping paper by air. The shipping paper for water transportation, with specific notations as to the shipping name, classification, and quantity of material in the shipment, followed by the words "Limited Quantities" or "Ltd. Qty." for limited quantity shipments, is the same as it would be for rail or highway transportation. The shipping name of an n.o.s. item must be followed by the technical name of the material, however, just like the marking. Of course, the shipping paper must bear the signed shipper's certification of compliance. See Chapter 11 on shipping papers generally.

Limited quantity shipments (other than Class B poisons) and all ORM materials do not require placarding. Any quantity of the remainder of the range of hazardous materials has to be placarded when shipped by water. Placarding is required for limited quantity shipments made under the IMO IMDG Code until revised Section 18, described above, is published.

SHIPMENTS BY AIR

As you have noticed, hazardous materials are given unique treatment when transported by air, and the treatment of limited quantity articles is also unique in this mode of transportation.

The domestic aviation regulations formerly were found in 14 CFR Part 103, but were consolidated into title 49 CFR as Part 175. Historically these modal regulations were the last to evolve and, consequently, there used to be a great deal of cross referencing in Part 103 to the regulations in title 49 CFR. Now the commodity list in 49 CFR 172.101 includes a column devoted specifically to air transportation.

The U.S. regulations applicable to air commerce specifically declare that every hazardous material must bear the appropriate label for its classification, despite the fact that regulations for other modes of transportation may relieve that same package from labeling. Thus, when reading title 49 CFR, the aspect of the exception sections that offers relief from labeling is only available for land and water transportation. It is not available for air transportation of these articles. There are no labeling requirements for ORMs in any mode, however.

The proper shipping paper entry to accompany a limited quantity shipment by air is "Limited Quantities" or "Ltd. Qty." immediately following the description of the material. The shipper's certification of compliance is also different for air transportation. See Chapter 11 on shipping papers and Chapter 20 on shipments by air generally.

To this time, in international air transportation there are no limited quantity exceptions provided by ICAO or IATA. Thus far, ICAO has included a provision for "transitional packaging" which accommodates materials which do not require specification packaging under the DOT regulations because of limited quantity relief. All other requirements, however, apply to these shipments, including shipping papers, marking, and labeling. For more information on transitional packaging, see Chapter 20.

Transitional packaging is scheduled to be phased out on January 1, 1990. On that date both IMO and ICAO are scheduled to require across-the-board use of UN performance standard packaging discussed elsewhere. IMO has acted to relieve some of the undue burden by adopting the limited quantity/consumer commodity system described above. ICAO has been much more restrained and has provided exceptions for only very limited shipments of certain materials in extremely small quantities. These provisions are based on the small quantity exceptions of 49 CFR

173.4, the so-called "30 ml rule." Although perhaps ICAO will alter this situation before 1990, there is no assurance that this will occur.

REVISED IMO SECTION 18

THE CARRIAGE OF DANGEROUS GOODS IN LIMITED QUANTITIES

18.1. The requirements of this section concern the transport of dangerous goods of certain classes in limited quantitites. The quantity limitations are specified in 18.3, but are subject to the exceptions listed in 18.2. The full requirements of this Code apply equally to limited quantities except as provided elsewhere in this section.

18.2. The requirements contained in this section do not apply to:

.1 Explosives of class 1;

.2 Gases of class 2 which have a subsidiary risk such as flammable, corrosive, oxidizing or toxic;

.3 Self-reactive substances of class 4.1;

.4 Substances which are liable to spontaneous combustion of class 4.2;

.5 Organic peroxides of class 5.2, with the exception of test kits, repair kits or similar mixed packages that may contain small quantities of these substances;

.6 Infectious substances of class 6.2;

.7 Radioactive materials of class 7;

.8 Aerosols included in class 9; and

.9 Dangerous goods to which packaging group I has been assigned.

18.3 QUANTITY LIMITATIONS

Class	Packaging Group	State	Maximum Quantity per Inner Packaging
2	—	gas	120 ml
3	II	liquid	1 l(metal) 500 ml (glass or plastics)
3	III	liquid	5 l
4.1	II	solid	500 g
4.1	III	solid	3 kg
4.3	II	liquid solid	25 ml 100 g
4.3	III	liquid or solid	1 kg
5.1	II	liquid or solid	500 g
5.1	III	liquid or solid	1 kg
5.2[a]	II	solid	100 g
5.2[a]	II	liquid	25 ml

(cont.)

Class	Packaging Group	State	Maximum Quantity per Inner Packaging
6.1	II	solid	500 g
6.1	II	liquid	100 ml
6.1	III	solid	3 kg
6.1	III	liquid	1 l
8	II	solid	1 kg
8	II	liquid	500 ml[b]
8	III	solid	2 kg
8	III	liquid	1 l

[a]See 18.2.5.

[b]Glass, porcelain or stoneware inner packagings should be enclosed in a compatible and rigid intermediate packaging.

18.4. Dangerous goods transported according to these special requirements should be packaged only in inner packagings placed in suitable outer packagings that would be capable of meeting the requirements for packaging group III. The total gross weight of a package should not exceed 30 kg and should in no case exceed that permitted in the individual schedules for the substances concerned.

18.5. Different dangerous goods in limited quantities may be packaged in the same outer packaging, provided the segregation requirements of the individual schedules are taken into account and the goods will not interact dangerously in the event of leakage.

18.6. The segregation requirements of section 15 are not applicable for packagings containing dangerous goods in limited quantities.

CHAPTER 14

Transportation of Hazardous Wastes and Hazardous Substances

Hazardous wastes and hazardous substances are groups of materials identified by the U.S. Environmental Protection Agency (EPA). Wastes are designated by EPA under the Resource Conservation & Recovery Act (RCRA), 42 U.S. Code 6901. Hazardous substances are designated under the Clean Water Act, 33 U.S. Code 1251, and the Comprehensive Environmental Restoration, Compensation & Liability Act (CERCLA, or Superfund), 42 U.S. Code 9601. All of these statutes encompass all facilities that handle such materials, including fixed facilities, rolling stock, vessels, vehicles, and other transportation facilities.

DOT's regulation of hazardous materials in transportation predated these environmental programs by many years. For the sake of administrative and industry convenience, DOT and EPA agreed that most (but not all) of the transportation restrictions on these environmentally designated materials would be published by DOT in 49 CFR as part of the hazardous materials regulations. The DOT rules, in turn, are referenced in the EPA regulations.

At first this meant only a modest addition to the DOT regulations, but the continuing expansion of the environmental programs led to a larger and larger portion of the DOT regulations being devoted to these materials.

The basic thrust of the traditional DOT regulations was to prescribe effective packaging and hazard communication (shipping papers, markings, labels and placards) to warn people of the presence of materials that could cause acute personal injury if they were released unintentionally during transportation. A large proportion of materials specifically identified by EPA as hazardous wastes and hazardous substances also fall within the traditional DOT hazard classifications, and for these materials there are simply supplemental requirements because of the EPA designation. For other materials, however, which were designated by EPA because of a perceived hazard not recognized historically by DOT, a new classification of "Other Regulated Material" (ORM) was created—"ORM-E."

It is extremely important to recognize that the full burden of DOT packaging and other regulations applies to hazardous wastes and substances that fit the traditional DOT hazard classes. The few remaining materials assigned to the ORM-E class generally are far less hazardous in transportation and, therefore, are subject to substantially less control under the regulations. It would be a serious and perhaps deadly mistake to assume that every hazardous waste or hazardous substance is in the ORM-E class—in fact, the greater likelihood is that it meets one of the more closely regulated DOT classifications of explosives, blasting agents, flammable liquids, combustible liquids, flammable solids, organic peroxides, oxidizers, corrosive materials, flammable gases, nonflammable gases, Poison A's, Poison B's, irritating materials, etiologic agents, radioactive materials, or ORMs-A, B, C or D. Each of these classes should be examined and determined *not* to apply before assigning an EPA-designated material to the ORM-E class.

HAZARDOUS WASTES

See Chapter 4 on the materials regulated by DOT for a more specific discussion of the technical aspects of waste definitions. The summary here will be more conceptual than detailed.

Basic but critical information is that EPA and its various statutes use the term "solid waste" to describe regulated materials, and a subgroup of solid wastes are hazardous wastes. Do not be misled by the term "solid" into thinking that liquid wastes are not covered. "Solid waste" incorporates a number of wastes "including solid, liquid, semisolid, or contained gaseous material" (42 U.S. Code 6903(27)).

At this point, at the risk of totally confusing the reader, it is important to mention that every hazardous waste also is a hazardous substance under Superfund. Hazardous substances have specific DOT shipping paper description and marking requirements described more fully below, but such additional DOT requirements only come into play when the amounts in one package exceed what EPA has established as the "reportable quantity" or "RQ" for the particular material. In 49 CFR 172.101, Appendix, DOT publishes the RQs established by EPA.

The only way to understand all of the DOT requirements applicable to hazardous wastes is to read *both* the hazardous waste and hazardous substance portions of this chapter, preferably more than once.

WHEN IS A MATERIAL A WASTE?

The critical initial determination in considering whether a substance is a *hazardous* waste is whether it is a waste at all. By its very nature, a waste usually is a material that originally was not a waste but became one only when someone decided that it should be discarded or recycled. In other words, the regulatory existence of many wastes begins somewhere on the continuum of that material's physical existence. For the sake of EPA and DOT compliance, it is important to determine where that point is. In a corporate context, this often is when the material is first consigned in writing for discard or disposal. This decision may be phrased in different ways on company paperwork and, although the decision on what to do with the material may have been made in someone's mind at an earlier time, in a practical sense it is most important to find when company paperwork first writes the material off as scrap or waste or assigns it for transfer to a recycling operation.

Determining this moment is vital to knowing when the material became a regulated waste. For one thing, under EPA's 40 CFR 262.34, this start date must be marked on the container of waste to record the beginning of the 90-day grace period that is allowed for accumulation of waste without an EPA permit. Because the EPA regulations commence their applicability at this time, it may be possible to defer the decision to discard or recycle a material to a more convenient time or location—that is, to delay making the decision that a material is a waste until that decision is unavoidable.

See 40 CFR Parts 260, 261 and 266 to determine when a material is a hazardous waste, either because it is being destroyed or discarded, or because it is being re-cycled or reclaimed.

IDENTIFICATION AND LISTING OF WASTES

As noted in greater detail in Chapter 4, EPA characterizes and identifies wastes in a manner similar to but different from DOT. The shipper of a waste must know both systems in order to be in compliance.

The primary resource for waste identification and characterization is 40 CFR Part 261. Some wastes are identified by their industry or process sources such as "Quenching waste water treatment sludges from metal heat treating operations where cyanides are used in the process" (40 CFR 261.31), or "Bottom sediment sludge from the treatment of wastewaters from wood preserving processes that use creosote and/or pentachlorophenol" (40 CFR 261.32). These source listings also in-clude waste streams made up of more than 10% of certain solvents.

In addition to generic waste streams, EPA lists certain discarded commercial chemical products, off-specification products, manufacturing chemical intermedi-ates, container residues, and spill residues in 40 CFR 261.33. Some of these are routine hazardous wastes (40 CFR 261.33(f), called the "U" list), and some are iden-tified as "acute hazardous wastes" (40 CFR 261.33(e), called the "P" list) which are subject to much more stringent EPA and DOT regulation. At the time of publica-tion of this edition of the *Red Book*, these lists are limited to the straight materials named in them. They do not include products which may incorporate these mate-rials as ingredients. Rule making is underway at EPA exploring ways to cover mix-tures and solutions of these materials. Whenever using the *Red Book*, be sure to use it in conjunction with the latest version of the DOT and EPA regulations and, especially on this question, review the regulations to determine the current EPA posture on mixtures and solutions of materials listed in 40 CFR 261.33.

Beyond wastes from various sources, and those identified by name as "listed" hazardous wastes, EPA includes many wastes on the basis of their generic hazard characteristics. The four EPA characteristics are "ignitability" (40 CFR 261.21), "cor-rosivity" (40 CFR 261.22), "reactivity" (40 CFR 261.23), and "EP toxicity" (40 CFR 261.24). These characteristic wastes are not identified by name in the EPA regula-tions and often are referenced as "ICR" or "ICRE" wastes. Generally speaking, al-though the DOT and EPA definitions are similar-sounding, there are critical differ-ences. See Chapter 4 on materials regulated by DOT for a much more detailed discussion of each of these characteristics, particularly as they relate to the hazard class definitions applied by DOT.

DESCRIPTION OF HAZARDOUS WASTES IN TRANSPORTATION

Nothing is simple, and that is true in discussing this topic as well. The EPA regulations list some materials and regulate others by generic characteristics. Each of these entries is combined with an EPA "Hazardous Waste Number"—a letter followed by three digits. For example, a waste exhibiting the characteristic of EPA ignitability is assigned the EPA Hazardous Waste Number of "D001," and a listed chemical example would be formaldehyde with the EPA Hazardous Waste Number "U122." See 40 CFR Part 261.

This information is essential to the hazardous waste generator for purposes of completing the annual report due every two years. It also will be of interest to the receiving facility for the shipment of hazardous waste, for their recordkeeping. In transportation, however, this waste number is part of the DOT shipping description of some, but not all wastes. In addition, on the Uniform Hazardous Waste Manifest certain information that may be demanded by States, and this includes the EPA waste number.

In transportation, it is essential that the generator develop the proper DOT shipping description for the material, including the shipping name, the appropriate UN/NA identification number (which is not related to and should not be confused with the EPA Hazardous Waste Number), and DOT hazard classification. If this material would have been regulated by DOT before it became a waste, then the shipping name is that original material shipping name, preceded by the word "waste." If the material is now a mixture of a variety of materials, the DOT classification should be determined as described in Chapter 5, and its shipping name and UN/NA number should be determined as shown in Chapter 6. The DOT description may also include a reference to the waste's reportable quantity, if such a quantity occurs in a single package. See the description discussion for hazardous substances, below.

As noted both at the introduction to this chapter and in the discussion below on hazardous substances, the terms tend to overlap, but not perfectly so. A hazardous waste is also a hazardous substance under Superfund. Under the DOT regulations, however, the unique hazardous substance marking and shipping paper entries do not become applicable until there is a "reportable quantity" or "RQ" of the material in a single package. It is possible, especially with high RQ levels such as 5000 pounds, to have a hazardous waste that is not regulated by DOT as a hazardous substance. For the sake of the examples below, the first description for the example assumes that an RQ is present.

If the material does not fit any DOT hazard class definitions other than ORM-E, then its shipping name is "Hazardous waste, liquid *or* solid, n.o.s., NA9189," and its classification is "ORM-E." The description of materials that fit no other DOT classification, and which have a shipping name that does not reveal the name of the regulated constituent, also must include in parentheses the EPA name of the material (found in DOT's 49 CFR 172.101, Appendix) or, if there is no chemical name for the material in the Appendix, by the EPA waste stream number or in the case of ICRE materials by the parenthetical "(EPA ignitability)" or corrosivity, or reactivity or EP-toxicity, whichever is the case. Characteristic wastes also can be identified parenthetically by number, such as "D001" for ignitability, "D002" for corrosivity, or "D003" for reactivity. The EPA number for EP-toxic materials varies with the constituent that gives it the toxicity. See 40 CFR 261.24(b).

For example:

1. The description for discarded or recycled acetone would be "RQ, Waste acetone, Flammable liquid, UN1090." If less than the reportable quantity for acetone of 5000 pounds (2270 kilos) were in one package or tank, the description would not include the initials "RQ".

2. The description for discarded or recycled paint having a flash point under 100°F would be "RQ, Waste paint, Flammable liquid, UN1263 (EPA ignitability) *or* (D001)." If less than 100 pounds of this material were in one package, the material would be a hazardous waste but

not a hazardous substance and the description would be "Waste paint, Flammable liquid, UN1263."

3. The description of discarded or recycled paint having a flash point over 100°F but under 140°F would be "RQ, Waste paint, Combustible liquid, UN1263 (EPA ignitability) *or* (D001)." As with the other paint example, if less than 100 pounds were in one package, then the description would be "Waste paint, Combustible liquid, UN1263." Likewise, if the discarded or recycled paint had a flash point above 140°F, but below 200°F, it would only be regulated by DOT in unit quantities of 110 gallons or more, and would be described as "Waste paint, Combustible liquid, UN1263." Materials at this higher flash point level are above the EPA definitions, so no EPA regulations or description would apply.

4. If the discarded or recycled material met the description of EPA's listed waste bottom sediment sludge from the treatment of wastewaters from wood preserving processes that use creosote and/or pentachlorophenol, and this sludge did not meet any other DOT hazard class definition, it would be described on the Uniform Hazardous Waste Manifest as "RQ, Hazardous waste, n.o.s., ORM-E, NA9189 (K001)." At the time of publication of the *Red Book*, the reportable quantity for this waste stream is only 1 pound in each package.

5. If the discarded or recycled material were in a package containing a reportable quantity of the hazardous substance furan, its DOT description would be "RQ, Waste hazardous substance, liquid, n.o.s., ORM-E, NA 9188 (Furan)." If less than 5000 pounds of furan were in each container, the description would be "Hazardous waste, n.o.s., ORM-E, NA 9189."

6. If the material was being discarded or recycled, and it met no DOT hazard classification definitions other than ORM-E but it did meet one of the EPA characteristics, for example a non-DOT corrosive sludge with a pH of 2 in a reportable quantity of over 100 pounds, then its description would be "RQ, Hazardous waste, liquid, n.o.s., ORM-E, NA 9189, (EPA corrosivity) *or* (D002)."

UNIFORM HAZARDOUS WASTE MANIFEST

Virtually every DOT-regulated material must be accompanied by a shipping document. For hazardous wastes designated by EPA, this requirement is satisfied by the Uniform Hazardous Waste Manifest developed by the two agencies. An example of the manifest appears at the end of this chapter.

Every generator should obtain an identification number from EPA or the approved State environmental agency. This number and the name of the generator appear on the manifest, along with the name and identification numbers of the transporter and the receiving treatment or disposal facility selected by the generator.

Every hazardous waste must be examined in light of the DOT hazard classifications and shipping names, and a proper description developed as described above. This description must be entered on the manifest, along with a description of the number and type of containers and the quantity of material in each description. Shaded areas of the manifest identify additional items that may be required by the States, but are not federally mandated. Check your State rules and those of the receiving State.

Although most DOT shipping papers must include a shipper's certificate of compliance, it need not be signed manually. A Uniform Hazardous Waste Manifest, however, must be signed by hand by the generator. In addition, each transporter as well as the receiving facility must sign. The idea is to have a paper chain of custody and responsibility.

PACKAGING OF HAZARDOUS WASTES

If the material otherwise meets one of the DOT hazard classes (other than ORM-E), see the requirements for packaging that class and shipping description. Also see Chapter 8 on selecting a proper packaging.

A hazardous waste in the ORM-E class is subject only to general, non-DOT-specification packaging requirements. See the general provisions in 49 CFR 173.24 as well as those in 49 CFR 173.510.

Three unique DOT packaging requirements for hazardous wastes are not available for other materials. The first is in 49 CFR 171.3(e) where it says that, even though a material otherwise may be required to be shipped in a closed-head DOT specification drum (e.g., low flash point flammable liquids), an "equivalent specification" open head drum may be used if the material is a waste containing solids or semisolids that cannot pass through the small bung opening on a closed-head drum. How one determines equivalency is unclear.

The second provision has to do with reuse of packaging, in 49 CFR 173.28(p). A container that otherwise would not be reusable, or that would have to be reconditioned before reuse, may be reused without reconditioning for the shipment of hazardous wastes, provided the following conditions are met:

1. The container otherwise has to be the proper one for shipment of a material of this description;
2. Transportation can be by highway only;
3. It must sit for at least 24 hours after filling and before transportation, and must be inspected for leaks just before it is shipped;
4. The motor carrier may not load or unload the container (unless the motor carrier is a private or contract carrier); and,
5. This may only be done once with a packaging, after which it must be discarded or reconditioned before use again in shipping any hazardous materials, including hazardous wastes.

The leeway this provision offers may be more than a prudent shipper would want to take. Leaks from hazardous waste packaging pose a much more expensive liability threat than the modest cost of using a new or reconditioned packaging. In addition to leaks, people using this provision should be extremely careful to assure that the waste being put into the container is compatible with any residues of former contents remaining in that container. Of course, all prior markings and labels on used packaging must be proper for the new waste contents, or must be obliterated or removed.

The third unique regulation pertains to so-called "lab packs," consolidations of a variety of smaller containers of hazardous wastes that otherwise could not be shipped under a single shipping description or in the same package. See 49 CFR 173.12 for the detailed limitations on lab packs.

LABELING HAZARDOUS WASTES

If the waste otherwise meets one or more of the DOT hazard classes requiring labeling, then those labels must be used. See Chapter 10 on labeling. In addition, note the EPA "label" described below under "marking."

MARKING HAZARDOUS WASTE PACKAGES

All aspects of the DOT marking regulations for other hazardous materials apply as well to hazardous wastes. See Chapter 9 on package markings.

In addition, there are two special EPA markings for hazardous wastes. The first is prescribed in 40 CFR 262.32, and reads as follows:

> HAZARDOUS WASTE—Federal Law Prohibits Improper Disposal. If found, contact the nearest police or public safety authority or the U.S. Environmental Protection Agency.
>
> Generator's Name and Address _____.
>
> Manifest Document Number _____.

Under 40 CFR 262.34, a generator of hazardous waste may accumulate that waste on site for a 90-day period without having to obtain a treatment, storage or disposal facility permit. This section is contingent, among other things, upon the following marking requirement: "The date upon which each period of accumulation begins is clearly marked and visible for inspection on each container."

PLACARDING HAZARDOUS WASTES

If the waste meets any of the DOT hazard class definitions that require placarding under 49 CFR 172.504, then the generator must offer the transporter the proper DOT placards for that shipment. See Chapter 12 for more on placarding generally. There are no unique federal rules for marking hazardous waste transport vehicles, although you might check your State hazardous waste transporter rules for any nonfederal provisions.

EXPORT OF HAZARDOUS WASTES

Most hazardous wastes meet the definition of "dangerous goods" regulated by the U.S. DOT and the International Maritime Organization (IMO). These address the details of identifying, packaging, and documenting chemical shipments.

The export of any material involves certain federal permits and licenses but, in addition, hazardous waste exports are governed by a 1984 amendment to RCRA (42 U.S. Code 6938). This law is implemented by regulations in 40 CFR Part 262, Subpart E.

Under these regulations, no export of an EPA-regulated hazardous waste may occur without EPA's acknowledgment of the consent of the destination country to receive the waste. Here is how this is supposed to work.

Someone, usually the generator of the waste although someone else could do it for the generator, advises EPA of the intention to export a hazardous waste, giving the details of who, what, where, how and when. This is supposed to occur at least 60 days before the intended date of export, but that time period is unrealistically short. The exporter should notify EPA as soon as the details are known.

EPA, through the State Department, then transmits the exporter's notification to the receiving country and all other countries through which the material would move, and asks the destination country for their consent or objection. The U.S. embassy in the receiving country actually communicates this information to the appropriate environmental agencies of that country. The destination country as well as all countries to be crossed are advised how the particular waste stream is managed under the EPA regulations in the U.S., along with a notice that the shipment cannot leave the U.S. unless the receiving country consents.

Not only may the shipment not leave the U.S., but EPA's hazardous waste transporter rules in 40 CFR Part 263 prohibit any transporter from accepting a shipment with a foreign destination unless the manifest is accompanied by the acknowledgment of consent. Thus, the shipment with a foreign destination may not leave the generator's site even to get to the port area in anticipation of export.

The response from the receiving country is passed back to the State Department through the U.S. embassy, and in turn to EPA. Assuming consent, presumably with some conditions, that consent cable is acknowledged by EPA, and EPA's acknowledgment of the receiving country's consent is given to the exporter, to be attached to the manifest for each shipment. State Department personnel provide necessary translation services.

The notification and consent process can cover a number of different shipments over a year, but has to be repeated at least annually or whenever there is any significant change.

As a practical matter, this system cannot work at all unless the exporter has made his own thorough preparations overseas in advance, so that when these complicated communications are received from U.S., someone at the foreign destination knows how to react to them.

Through the UN and affiliated organizations, international recommendations on the movement of hazardous wastes are being developed. These may or may not have the force of law in the United States, depending upon how it is done.

Despite the delay and paperwork involved, there are good reasons why the export of hazardous waste will continue to have market appeal. First, because of high permit and insurance costs for domestic disposal facilities, getting rid of waste in the U.S. always will be more expensive than sending it elsewhere. Second, Superfund's perpetual generator liability for the clean-up of disposal sites does not apply outside the U.S., and this long-term liability far exceeds the immediate costs of disposal.

HAZARDOUS SUBSTANCES

CLEAN WATER ACT

DOT initially ventured into the regulation of hazardous substances by adding approximately 300 specifically named materials under the Clean Water Act to the table of hazardous materials in 49 CFR 172.101.

Because any facility having a spill of these materials in a reportable quantity had to report that spill, and because most carriers would have no way of knowing whether their cargo required such a report, DOT established two communications requirements for hazardous substances. These are a notation on shipping documents of the presence of a reportable quantity of material in one package, along with an obligation to mark packaging. If the shipping name of the material did not include the name of the hazardous substance, such as an n.o.s. name, then that specific name had to appear in parentheses on shipping documents and package markings. No marking requirements were imposed for bulk equipment.

SUPERFUND

The Comprehensive Environmental Response, Compensation & Liability Act (CERCLA or Superfund), 42 U.S.C. 9601, also uses the term "hazardous substance." See Chapter 4 for the details of all definitions of materials regulated under DOT.

The Superfund hazardous substance definition includes not only the original Clean Water Act materials, but certain others designated under other provisions of the Clean Water Act, the Clean Air Act, the Toxic Substances Control Act, and RCRA. EPA also may add to these lists.

Superfund, as amended in 1986, required DOT to "list and regulate" all hazardous substances as hazardous materials within 30 days of their designation by EPA (42 U.S.C. 9656).

DOT has done this. Certain hazardous substances, but not all, that already met the traditional hazard class definitions, were added by name to the commodity list in 49 CFR 172.101. All of these, plus all others under Superfund, are listed in an Appendix to 49 CFR 172.101, along with their specific "reportable quantity" or RQ in pounds and kilograms. In addition, the shipping name "Hazardous substance, liquid *or* solid, n.o.s." appears in the DOT list of proper shipping names.

One of the terribly confusing things about these rules is that the term "hazardous substance" under Superfund includes hazardous wastes. It is easy to think the two are synonymous. Some hazardous substances, however, are never hazardous wastes, such as PCBs regulated under the Toxic Substances Control Act but not under RCRA.

Hazardous substances are regulated by DOT even when they are not wastes, but only when a reportable quantity is in each package. While Superfund liability for the release of hazardous substances extends to spills in any quantity, under the DOT regulations no additional regulation of hazardous substances applies until the reportable quantity threshold is exceeded in each package. The term "package" includes bulk equipment.

Some people have criticized the DOT definition, only imposing more stringent requirements when a reportable quantity is in each package. One criticism is that a motor carrier having a spill of material in less than a reportable quantity may be liable for damage to natural resources caused by that spill, despite the fact that he had no way of knowing that the material was a hazardous substance. Another criticism speculates that a carrier could spill material from two or more unidentified packages which, in the aggregate, constitute a reportable quantity of a hazardous substance without the carrier knowing that a spill report should be made. EPA and DOT have addressed this issue, saying that prosecution for failures to report a spill will be predicated upon the carrier's knowledge of the existence of a reportable quantity of the material. Thus, in ignorance there is freedom from prosecution, or at least freedom from conviction. While this protection may exist for common carriers, it is not available for private carriers who legally are held to have the same knowledge as their owner, the shipper. The liability of contract carriers is a factual matter in each case, but the assumption would be that the contract carrier has the same knowledge as a private carrier.

Even when there is a reportable quantity in each package, DOT regulations may not apply if the material is very dilute. A dilution table in 49 CFR 171.8 says that if the concentration of the hazardous substance is less than that shown below, then the DOT regulations will not apply despite the fact that there is enough of the substance in the package to otherwise make up a reportable quantity:

RQ pounds	RQ kilograms	Concentration by weight	
		%	kilos
5000	2270.0	10.0	100000
1000	454.0	2.0	20000
100	45.4	0.2	2000
10	4.54	0.02	200
1	0.45	0.002	20

Hazardous wastes, on the other hand, are regulated by DOT and EPA in any quantity, although because they are part of the hazardous substance definition they also have reportable quantities. The RQ here is pertinent for two reasons—it necessitates additional markings and shipping paper notations, and it requires immediate incident reporting in case of a spill.

As a result of the system that has been established, you could have a waste hazardous substance that is not a hazardous waste (e.g., discarded PCBs), and hazardous wastes that are not hazardous substances (e.g., wastes below their RQ in one package). One would hope that this needless confusion would be straightened out, but clarification is not coming soon.

SHIPPING DESCRIPTIONS FOR HAZARDOUS SUBSTANCES

The first thing to do is to determine if the package of material being shipped contains a hazardous substance listed by name in the Appendix to 49 CFR 172.101, in a quantity equal to or exceeding the reportable quantity given in the Appendix. (Even if the quantity exceeds this level, in particularly dilute concentrations the material may still fail to be regulated as a hazardous substance, based upon the dilution table discussed above and in the definition of a hazardous substance in 49 CFR 171.8).

If a hazardous substance is present in a regulated reportable quantity per package, and the material being shipped meets any DOT hazard class definition other than ORM-E, then it must be classed according to that definition, and a shipping name that is paired with that class must be used. The shipping name that most accurately describes the material as shipped is the one that must be selected.

For example, barium cyanide is listed in 49 CFR 172.101 as a Poison B. It also is in the Appendix as a hazardous substance with an RQ of 10 pounds (4.54 kilos). The proper description for packages containing more than 10 pounds of this material would be "RQ, Barium cyanide, solid, Poison B, UN1565."

If the proper shipping name does not include the hazardous substance name as it appears in the Appendix then that specific name must be added to the description, in parentheses. For example, if there were drums containing cleaning materials including ammonia (which is in the Appendix as a hazardous substance with an RQ of 100 pounds (45.4 kilos)), and the mixture met the definition of DOT's combustible liquid class, and there were more than 100 pounds of ammonia in each drum, the proper description would be "RQ, Compound, cleaning, liquid, Combustible liquid, NA 1993, (Ammonia)."

Some waste streams regulated by EPA under RCRA have very long names. Others are not designated by EPA by name, but by the EPA characteristics of ignitability, corrosivity, reactivity, and extraction-procedure (EP) toxicity. When the DOT shipping name does not reveal the reason such a material may be regulated as a hazardous substance, one has to put something in parentheses.

Rather than use the cumbersome descriptions of waste streams listed under 40 CFR 261.31 and 261.32, DOT allows the parenthetical reference to be simply the EPA waste stream number. For example, one waste stream is "Wastes (except wastewater and spent carbon from hydrogen chloride purification) from the production or manufacturing use (as a reactant, chemical intermediate, or component in a formulating process) of pentachlorophenol, or of intermediates used to produce its derivatives." This waste stream has the EPA Hazardous Waste number "F021." Under the DOT regulations, a shipment of this material that did not meet any DOT classification except ORM-E, and that was in package quantities of 100 pounds or more, would be described on the Uniform Hazardous Waste Manifest as: "RQ, Hazardous waste, liquid, n.o.s., ORM-E, NA 9189 (F021)." If there were less than 100 pounds in the package, it still would be a hazardous waste but would not be a hazardous substance for DOT purposes because it was below the RQ level for this shipping name. In this latter case, the manifest description would be "Hazardous waste, liquid, n.o.s., ORM-E, NA 9189."

If the waste were not listed by EPA, either by its source or by a specific chemical name, but fell within EPA's rules because it meets one of the four EPA characteristics, then that characteristic or the EPA number for that characteristic must appear in the parentheses. For example, if a sludge were not listed in 40 CFR Part 261, but it exhibited the characteristic of corrosivity and more than 100 pounds were in each package, then a proper DOT description would be "RQ, Hazardous waste, solid, n.o.s., ORM-E, NA 9189, (EPA corrosivity) *or* (D002)." If less than 100 pounds were in each package, the DOT description for this material would be "Hazardous waste, solid, n.o.s., ORM-E, NA 9189."

PACKAGING HAZARDOUS SUBSTANCES

No special DOT packaging rules apply to hazardous substances. If the material otherwise meets a DOT hazard class definition, then it should be packaged as prescribed for that class and shipping name. If it is in the ORM-E class, then the only packaging rules are general provisions in 49 CFR 173.24 and 173.510. Also, shipments of any regulated material by air must meet 49 CFR 173.6.

MARKING OF PACKAGES OF HAZARDOUS SUBSTANCES

Every package of 110 gallons capacity or less that contains a reportable quantity of a hazardous substance must be marked with the DOT shipping name, UN/NA number, a parenthetical indication of what the substance is if that information is not already part of the shipping name, and the initials "RQ." No hazardous substance marking requirements apply to portable tanks, cargo tanks, or tank cars.

LABELING HAZARDOUS SUBSTANCE PACKAGES

There are no special labels for hazardous substances. DOT labels for other classifications would be required if the material met one of those classes and was in a quantity requiring labeling.

PLACARDING HAZARDOUS SUBSTANCES

There are no special placarding rules for hazardous substances.

INCIDENT REPORTING FOR HAZARDOUS WASTE AND HAZARDOUS SUBSTANCE RELEASES AND SPILLS

Hazardous wastes and substances are a subgroup of hazardous materials generally, and the general incident reporting requirements prescribed by DOT in 49 CFR 171.15 (immediate notice for certain serious accidents) and 171.16 (follow-up written notification within 15 days) are applicable. See Chapter 22 for specific information on these requirements.

UNIFORM HAZARDOUS WASTE MANIFEST	1. Generator's US EPA ID No.	Manifest Document No.	2. Page 1 of	Information in the shaded areas is not required by Federal law.

3. Generator's Name and Mailing Address

A. State Manifest Document Number

B. State Generator's ID

4. Generator's Phone (   )

5. Transporter 1 Company Name	6. US EPA ID Number	C. State Transporter's ID
		D. Transporter's Phone

7. Transporter 2 Company Name	8. US EPA ID Number	E. State Transporter's ID
		F. Transporter's Phone

9. Designated Facility Name and Site Address	10. US EPA ID Number	G. State Facility's ID
		H. Facility's Phone

11. US DOT Description *(Including Proper Shipping Name, Hazard Class, and ID Number)*	12. Containers No.	Type	13 Total Quantity	14 Unit Wt/Vol	I. Waste No.
a.					
b.					
c.					
d.					

J. Additional Descriptions for Materials Listed Above

K. Handling Codes for Wastes Listed Above

15. Special Handling Instructions and Additional Information

16. GENERATOR'S CERTIFICATION: I hereby declare that the contents of this consignment are fully and accurately described above by proper shipping name and are classified, packed, marked, and labeled, and are in all respects in proper condition for transport by highway according to applicable international and national government regulations.

Unless I am a small quantity generator who has been exempted by statute or regulation from the duty to make a waste minimization certification under Section 3002(b) of RCRA, I also certify that I have a program in place to reduce the volume and toxicity of waste generated to the degree I have determined to be economically practicable and I have selected the method of treatment, storage, or disposal currently available to me which minimizes the present and future threat to human health and the environment.

Printed/Typed Name	Signature	Month Day Year

17. Transporter 1 Acknowledgement of Receipt of Materials

Printed/Typed Name	Signature	Month Day Year

18. Transporter 2 Acknowledgement of Receipt of Materials

Printed/Typed Name	Signature	Month Day Year

19. Discrepancy Indication Space

20. Facility Owner or Operator: Certification of receipt of hazardous materials covered by this manifest except as noted in Item 19.

Printed/Typed Name	Signature	Month Day Year

EPA Form 8700-22 (Rev. 4-85) Previous edition is obsolete.

CHAPTER 15

Motor Carrier Hazardous Materials Requirements

The U.S. trucking industry is regulated in a multitude of ways by various components of the Department of Transportation. All safety aspects of the vehicle, its driver, and the roadway the vehicle runs on are affected by DOT regulations. So too, with regard to hazardous materials cargo moving by highway, DOT regulations apply. There are two sets of pertinent regulations: the Motor Carrier Safety Regulations, particularly 49 CFR Parts 387 and 397, and the hazardous materials regulations, particularly 49 CFR Parts 172 and 177. These will be discussed in turn.

APPLICABILITY

The rules cover contract and private carriers as well as common carriers.

A shipper using his own vehicles or leased vehicles in a private carrier operation must be aware of the applicability of the Motor Carrier Safety Regulations to his operations.

The hazardous materials aspect of motor carrier regulation, before enactment of the Hazardous Materials Transportation Act (HMTA), 49 U.S. Code 1801, was supported by the Explosives & Other Dangerous Articles Act, 18 U.S. Code 831–835, since repealed. (See Appendix D for these laws.) With passage of the HMTA, the obvious legislative desire was to have all future hazardous materials regulatory activities carried out under that law.

Under 18 U.S. Code 831–835, first the Interstate Commerce Commission and then DOT were authorized to issue hazardous materials regulations "binding upon all carriers engaged in interstate or foreign commerce which transport explosives or other dangerous articles by land" and upon all shippers using the services of such regulated carriers.

"Carrier" was defined as any person engaged in the transportation of passengers or property, by land, as a common, contract, or private carrier, or freight forwarder as those terms are used in the Interstate Commerce Act, and officers, agents, and employees of such carriers. Since the general applicability of this law depended upon the nature of the carrier being used, these words received a great deal of attention and were subject to frequent interpretation over several decades.

The consistent interpretation is that a motor carrier is engaged in interstate or foreign commerce if any shipment aboard the particular vehicle originates from outside the State or is consigned to a destination outside the State. The motor carrier is considered to be engaged in interstate and foreign commerce at all times if his vehicles cross a State line with any frequency. DOT has approached unclear cases with a case-by-case determination, analysing the degree of common management and direction between the carrier in question and known interstate or foreign operations, the extent of involvement of vehicles and drivers in both operations, and similar factors. If the carrier is found to be engaged in interstate or foreign commerce, then all hazardous materials aboard his vehicles at all times are considered regulated by DOT, even though some articles may be moving only in intrastate operations or that particular segment of transportation is only within a single State. The legislative history of the law confirms that DOT's jurisdiction under the Explosives & Other Dangerous Articles Act could encompass intrastate shipments.

Under the HMTA, discussed at length in Chapter 30, DOT's regulatory jurisdiction was extended far beyond that described in 18 U.S. Code 831–835. In fact, carriers are not a defined entity under the HMTA, which simply makes the DOT regulations binding upon any person who transports hazardous materials in commerce. "Transports" is defined as any movement of property by any mode, and any loading, unloading, or storage incident thereto. "Commerce" is broadly defined to mean any trade, traffic, commerce, or transportation in the United States that crosses a State line or that affects trade, traffic, commerce, or transportation crossing State lines. Thus, although 18 U.S. Code 831–835 occasionally covered intrastate shipments, under the HMTA it is rare that such coverage would not exist. In other words, intrastate and local cartage could be covered by any regulations issued under the HMTA, but DOT has voluntarily limited the scope of its regulations to shippers and motor carriers formerly covered under 18 U.S. Code 831–835. There are a limited number of exceptions to this general statement, including motor carriers of (1) hazardous wastes, (2) hazardous substances, and (3) flammable cryogenic gases in bulk.

By refraining from a specific definition of "carrier," the HMTA broadened the regulatory coverage to any person who transports whether he regularly deals in the business of carriage for hire or not. An air carrier operating trucks in hazardous materials pickup operations has to comply with the motor carrier aspects of DOT regulations for those operations. The businessman making deliveries in his own vehicles also is covered.

The Motor Carrier Safety Regulations are published in 49 CFR Parts 350–397, addressed to the following subjects:

Part 350	Commercial Motor Carrier Safety Assistance Program
Part 385	Safety Ratings
Part 387	Minimum Levels of Financial Responsibility
Part 389	Rulemaking Procedures
Part 390	General Provisions
Part 391	Qualifications of Drivers
Part 392	Driving of Motor Vehicles
Part 393	Parts and Accessories
Part 394	Reporting of Accidents
Part 395	Hours of Service, Drivers
Part 396	Inspection and Maintenance
Part 397	Explosives and Dangerous Articles (defined as those hazardous materials regulated in title 49 CFR)

49 CFR PART 385

Motor carrier safety ratings are carried out under this part of the regulations. The Federal Highway Administration, through its field offices, headquarters, and information exchange with State officials, determines the carrier's safety rating.

A carrier's rating is either Satisfactory, Unsatisfactory, Conditional, or there is insufficient information to give a rating. A significant effort is underway to give every carrier a rating.

The carrier's rating is communicated to the Interstate Commerce Commission when they need it. It also is available to anyone else who asks by writing to the Office of Motor Carrier Field Operations, HFO-1, Federal Highway Administration, Washington, DC 20590. The motor carrier is not advised of its rating unless it files a request in this manner.

An Unsatisfactory rating is a very serious matter and can result in the carrier being compelled to cease all operations. There are sketchy review procedures if a carrier believes its rating is inaccurate.

49 CFR PART 387

This part of the regulations should be examined carefully. It prescribes minimum levels of financial responsibility for most motor carriers, including most carriers of hazardous materials. Hazardous materials carriers not covered are those operating vehicles with a gross vehicle weight rating of less than 10,000 pounds, unless that carrier is hauling Class A or B explosives, poison gas, or highway route controlled quantities of radioactive materials.

The rules also are inapplicable to the intrastate carriage of oil, hazardous substances, hazardous wastes, or other hazardous materials (except highway route controlled quantities of radioactive materials) unless those materials are carried in bulk.

The form of this financial responsibility is prescribed in great detail in Part 387, as is the verification of its existence. For hazardous materials carriers, the amounts of coverage are $5,000,000 for bulk (3500+ gallons) hazardous substances (Chapter 14), liquefied petroleum gas or compressed gas, or any quantity of explosives A and B, Poison A, or highway route controlled quantities of radioactive materials.

A lower level of $1,000,000 is set for DOT-regulated oil, hazardous waste, hazardous substances, and hazardous materials not subject to the $5,000,000 requirements.

49 CFR PART 397

The provisions of Part 397 are issued jointly under the authority in the HMTA and the Interstate Commerce Act. Part 397 does not apply to drivers or vehicles wholly engaged in exempt intracity operations. This term means a vehicle or driver used wholly within a municipality or the commercial zone thereof, as defined by the Interstate Commerce Commission on October 1, 1975, transporting:

1. Passengers or property, or both, for which no placarding is required, or
2. Property consisting of hazardous materials of a type and quantity that requires placarding but that weigh less than 2,500 pounds in the case

of one hazard class or less than 5,000 pounds in the case of more than one class.

The Motor Carrier Safety Regulations refer to the definition of a commercial zone as it appeared in the ICC regulations in 49 CFR Part 1048, as that term was defined in 1975. The reference is so dated because after that date, the ICC began to expand that definition and DOT, not wishing to expand the relief that is provided from safety regulations, froze the definition as it existed then. Note, however, that at the time of its freezing of the old definition, DOT declared that it would issue a notice of proposed rule making "in the very near future to establish exempt intracity areas as they relate to the safe operation of vehicles in interstate commerce within those areas. There is considerable concern that many commercial zone areas, as they exist today have been expanded beyond the limits originally intended when the commercial zone exemption was established." (41 F.R. 8175, February 25, 1976.) That change has not been forthcoming, but still is expected. If you need to know the extent of the 1975 definition, contact the Federal Highway Administration or a law library that would maintain back issues of the Code of Federal Regulations.

Part 397 applies to motor carriers engaged in the transportation of hazardous materials by a motor vehicle that must be placarded and to each officer or employee of the carrier who performs supervisory duties related to the transportation of hazardous materials, as well as to each person who operates or who is in charge of a motor vehicle containing hazardous materials. Each of these people must know and obey the rules in Part 397.

If the operation is an exempt intracity operation, as defined above, it is not covered by Part 397. So too, if no placards need be applied to the motor vehicle, then part 397 does not apply. Every motor carrier or other person to whom Part 397 does apply must meet all of the rules set forth in Parts 390 through 397, inclusive.

Part 397 prescribes vehicle attendance and surveillance requirements for carriers hauling Class A or B explosives as well as detailed regulations on the parking of such vehicles. Unless there is no practicable alternative, a motor vehicle that contains hazardous materials must be operated over routes that do not go through or near heavily populated areas, places where crowds are assembled, tunnels, narrow streets, or alleys. Operating convenience is not a basis for determining whether it is practicable to operate a vehicle in accordance with this requirement. There are more specific requirements on the route to be followed by haulers of Class A and B explosives.

This provision on avoiding populated areas has been the subject of considerable discussion in the context of local routing requirements for hazardous materials vehicles. These requirements are examined in light of Section 112 of the HMTA, 49 U.S. Code 1811, which preempts nonfederal restrictions that are deemed "inconsistent" with the DOT hazardous materials program. In short, DOT and the courts have found that this provision of the Motor Carrier Safety Regulations does permit local routing restrictions, but those restrictions must (1) result in an improvement in public safety for all citizens, not just those in the enacting jurisdiction, and (2) must be coordinated with other communities that would be affected by a change in truck routes.

A motor vehicle subject to Part 397 containing any hazardous material must not be operated near can open fire unless the driver has first taken precautions to ascertain that the vehicle can pass that fire safely without stopping. No such motor vehicle may be parked within three hundred feet of an open fire.

No person may smoke or carry a lighted smoking article within twenty-five feet of any motor vehicle containing explosives, oxidizers, or flammable materials, or an empty cargo tank which has been used to transport flammable liquids or gases and when so used had to be placarded. A motor vehicle subject to Part 397 containing hazardous materials must not have its engine in operation during fueling, and someone must be in control of the fueling process where the fuel tank is filled.

If a placarded motor vehicle is equipped with dual tires on any axle, the driver must stop the vehicle in a safe location at least once during each two hours or one hundred miles of travel, whichever is less, and he must examine the tires. He also must examine the tires at the beginning of the trip and each time he parks the vehicle. Of course, if he finds problems with the tires, appropriate responsive action should be taken, as prescribed in Part 397.

In addition to the shipping paper requirements for all hazardous materials described below, the driver of a motor vehicle carrying Class A or B explosives must have certain other instructions and documents noted in Part 397.

Any private carrier operating a motor vehicle under its own power, which must be placarded for the hazardous materials it is carrying, must also mark the following information on that vehicle:

1. The name or trade name of the private carrier operating the vehicle.
2. The city or community in which the private carrier maintains its principal office or in which the vehicle is customarily based.

These markings must be on both sides of the truck, readily legible from fifty feet while the truck is stationary.

Under the Motor Carrier Safety Act of 1984, Congress declared that DOT "shall not eliminate or modify any existing motor carrier safety rule pertaining exclusively to the maintenance, equipment, loading or operation (including routing requirements)" of hazardous materials trucks "unless and until an equivalent or more stringent regulation has been promulgated under the Hazardous Materials Transportation Act." See Chapter 30 on this statute. As of the printing of the second edition of the *Red Book*, no revisions to Part 397 have been proposed under any statute, but such a project is expected and the reader is encouraged to verify the current provisions of this Part.

49 CFR PARTS 172 AND 177

DOT publishes general motor carrier requirements related to hazardous materials in title 49 CFR Part 177 with the rest of the hazardous materials regulations. Under changes adopted in rule making Docket Nos. HM-103 and HM-112 several years ago, many communications requirements pertaining to shipping papers, labels, and placards were put with such requirements for other modes of transportation in a consolidated 49 CFR Part 172.

Part 177 is intended to complement the requirements applicable to shippers of hazardous materials. For example, while the shipper requirements prescribe the details for shipping papers and compel the shipper to prepare papers and tender them to the carrier, the carrier sections also prescribe certain details of shipping papers and include a requirement that the carrier receive those papers with the hazardous materials. The carrier parts are not all repetition, however, so the private carrier should be cognizant of both sides of his responsibility as a shipper/carrier.

ACCEPTANCE OF HAZARDOUS MATERIALS

The motor carrier is not only obligated to assure that the materials received from the shipper comply with regulatory requirements, but he must assure that those shipments received from or transferred to connecting carriers are in compliance as well. So too, any hazardous material or supplies belonging to the carrier are just as regulated as materials belonging to the shipper.

Most hazard classifications defined by DOT include a limited quantity exception. An example would be that for flammable liquids in 49 CFR 173.118, when packed in limited size, inside containers. Such exceptions and the ORM-D class are discussed at considerable length in Chapter 13. A common element in most of these limited quantity exceptions is specific relief from the requirements of Part 177 except Section 177.817 dealing with shipping papers.

A shipment under a limited quantity exception relieves the motor carrier from compliance and concern with the entirety of Part 177 except for the paperwork. Placarding requirements apply to all hazardous materials except ORMs and limited quantities (other than Class B poisons). There are no highway shipping paper requirements for ORM-D. Without placards, there is no applicability of Part 397 either. This should be kept in mind while reading the remainder of this chapter, devoted to the specific requirements of Part 177.

Part 177 requires that the diamond-shaped, color-coded DOT label appear on packages provided by the shipper, unless those packages are excepted from labeling and the material is an ORM or the labeling exception is appropriately noted on the shipping papers by the words "Limited Quantities" or "Ltd. Qty." See Chapter 13 for a detailed discussion of hazard labels and how they are to be applied. Motor carriers must keep a supply of labels on hand and must replace lost or detached labels from the information given on the waybill, manifest, or other shipping paper. In obtaining these labels, the carrier may have his name and stationery form-number printed on the labels, if it is in small type in the border of the label.

Each package of hazardous materials also must be marked with the name and address of the consignee or consignor, unless the transportation is to be exclusively by motor vehicle between points during which the packages will not be transferred from one motor carrier to another, in either truckload or LTL lots, or unless it is part of a truckload lot or freight container load and the entire contents of the truck or freight container are tendered from one consignor to one consignee.

SHIPPING PAPERS

Motor carriers must not accept hazardous materials for transportation or transport them without a description of the materials on the shipping papers. This description must be by DOT shipping name (Chapter 6) and classification (Chapter 5). Further description not inconsistent with these names is allowed. Abbreviations of the DOT shipping name or classification (other than ORM) must not be used on shipping papers, i.e., calling sulfuric acid "sul. acid" is a violation, as is writing of "f.l." for flammable liquid or "cor." for a corrosive material. Abbreviations may be used to describe the type of packaging and weight or volume. The shipping paper must also note the total quantity of each shipping name and class by weight, volume, or as otherwise appropriate.

Limited quantity packages of articles excepted from labeling by the shipper sections of the regulations should not be labeled, and the exception must be shown on shipping papers by the words "Limited Quantities" or "Ltd. Qty." following the shipping description of the material.

The regulations go into considerable detail requiring that the driver of the vehicle have a copy of the shipping papers in his possession in the cab of the truck showing the information regarding hazardous materials that must appear on such papers. This, among other things, is to facilitate communication of the hazard to emergency personnel who may need to know the nature of the cargo in a hurry. The paperwork need not be the original certified paper received from the shipper.

Upon offering or delivering any loaded motor vehicle, trailer, freight container, or semitrailer containing hazardous materials to a rail carrier for further transportation, the motor carrier must show the material by shipping name, classification, and quantity, must describe the vehicle or container, and must show the kind of placards affixed to that vehicle or container.

Before leaving the subject of shipping papers, it should be emphasized that every shipping paper must bear a shipper's certificate of compliance (except for shipments by private carrier that are not to be reshipped or transferred from one carrier to another, or for bulk shipments in cargo tanks supplied by the carrier). The shipper's certificate should read as follows:

> This is to certify that the above-named materials are properly classified, described, packaged, marked, and labeled, and are in proper condition for transportation, according to the applicable regulations of the Department of Transportation.

This certification must be signed, but an actual handwritten signature is not required. It must be the name of a person, however, as an employee or agent of the shipper company. The signature may be typewritten or otherwise machine-printed. Any signature that is binding for commercial purposes will satisfy the DOT regulations. Initials, however, are insufficient.

PLACARDING. Placarding is the required notation of certain hazard information on the outside of vehicles. This subject is addressed separately in Chapter 12.

LOADING AND UNLOADING. Several general requirements concern the loading and unloading of hazardous materials from motor vehicles:

1. Any tank, barrel, drum, or cylinder not designed to be permanently attached to the motor vehicle that contains any flammable liquid, compressed gas, corrosive material, or poison must be reasonably secured from movement within the vehicle during transportation.
2. No hazardous material may be put on or transported via any pole trailer.
3. No person may smoke near any vehicle while loading or unloading any explosive, flammable liquid, flammable solid, oxidizing material, or flammable compressed gas. Extreme care should be taken during the loading or unloading of these materials and all fire and flame should be kept away.
4. The handbrake must be set when loading or unloading any hazardous material and all other reasonable precautions should be taken to prevent movement of the vehicle during the process.

5. The carrier must not use any tools likely to damage the closure or otherwise adversely affect the integrity of any package.

6. Packages containing explosives, flammable liquids, flammable solids, oxidizing materials, corrosive materials, compressed gases, or poisonous liquids or gases must be braced to prevent motion relative to the vehicle during transportation. Containers having valves or other fittings must be loaded to minimize the likelihood of damage to them in transit.

7. Reasonable care should be taken to avoid any dangerous rise in the temperature of the lading in transit. No one should tamper with any hazardous materials container or its contents between the points of origin and destination, nor may there be any discharge of the contents of any container other than a cargo tank prior to the removal of that container from the trailer.

8. It is absolutely prohibited to leave a cargo tank unattended during loading or unloading of a hazardous material. A person attending the loading or unloading must be qualified, must be awake, must maintain an unobstructed view of the tank, and remain within 25 feet (7.62 meters) of the cargo tank. A person is considered qualified if he has been made aware of the nature of the hazardous material he is handling, he has been instructed in emergency procedures, and he has both the authority and the means to move the tank if he has to do so.

9. Use of cargo heaters with hazardous materials is carefully regulated: (1) When transporting explosives any cargo heater must be rendered inoperative by draining or removing the heater fuel tank or disconnecting the power source for the heater, and (b) when transporting labeled flammable liquids or flammable compressed gas a combustion cargo heater may only be used if it is a flameless, catalytic heater, which is not ignited in the vehicle and which has a surface temperature that cannot exceed 130°F (54°C) either by thermostatic controls or other controls when the outside or ambient temperature is 60°F (15.6°C) or less. The manufacturer of such a catalytic heater must certify and mark the heater indicating that it "Meets DOT requirements for catalytic heaters used with flammable liquid and gas." Such a heater also must be marked "Do not load into or use in cargo compartments containing flammable liquid or gas if flame is visible on catalyst or in heater."

10. An automatic, cargo-space-heating, temperature control device may be used when transporting labeled flammable liquid or gas only if (a) electrical apparatus in the cargo compartment is nonsparking or explosion proof, (b) there is no combustion apparatus in the cargo compartment, (c) there is no connection or return of air from the cargo compartment to the combustion apparatus, and (d) the heating system will not elevate the temperature of any part of the lading above 130° F. (54° C). If these restrictions cannot be met, the heating device must be rendered inoperative through emptying or removing the fuel tank. (Each LPG fuel tank for automatic temperature control equipment must have its discharge valve closed and fuel feed line disconnected in lieu of actual emptying or removal of the tank.)

11. So-called ton tanks meeting DOT specifications 106A or 110A, au-

thorized to carry certain hazardous materials by highway, must be securely chocked or clamped to prevent any shifting. Equipment suitable for handling such tanks must be provided at any point of loading or unloading.

12. Specification 56 or 57 portable tanks may not be stacked on top of one another, nor may other freight be stacked on them during transportation.

13. No motor carrier may load any article bearing a poison label into any vehicle with material marked as, or known to be, foodstuff, feed, or any material intended for oral consumption by humans or animals.

In addition to these general regulatory requirements on loading and unloading of hazardous materials, Part 177 contains specific requirements regarding the loading and unloading of articles under headings for each hazard classification. The reader is urged to examine these in detail for the class of materials to be carried.

There also are particular sections on the highway routing of certain radioactive materials and the registration of carriers of bulk flammable cryogenic materials. See Part 177, including the Appendix to that part, for details.

Section 177.848 of 49 CFR is a tabular chart of the loading and storage requirements, pairing each classification with potential loads of other classes. Motor carriers handling explosives should examine that chart in detail to assure safe and compatible loading and storage.

The loading and storage chart for materials other than explosives may be summarized as follows:

1. Labeled flammable liquids and flammable compressed gases must not be loaded or stored with Class A poisons.

2. Labeled flammable solids, oxidizers, or organic peroxides must not be loaded or stored with labeled corrosive liquids or Class A poisons.

3. Labeled corrosive liquids must not be loaded or stored with labeled flammable solids, oxidizers, organic peroxides, or Class A poisons, cyanides or cyanide mixtures.

4. Nitric acid, when loaded in the same motor vehicle with other acids or other corrosive liquids in carboys must be separated from the other carboys as prescribed in the notes to the loading and storage chart.

NOTE: Shippers loading truckload shipments of corrosive liquids and flammable solids or oxidizing materials packages and who have obtained prior approval from DOT may load such materials together when it is known that the mixture of contents would not cause a dangerous evolution of heat.

TRANSPORTATION AND DELIVERY

All shipments of hazardous materials must be transported without unnecessary delay, including delay in loading and unloading. The key word here is "unnecessary." DOT and the courts have noted with frequency that certain nonfederal restrictions that delay the movement of hazardous materials vehicles constitute an unnecessary delay and, therefore, are inconsistent with this regulation. As such, those State and local restrictions have been preempted under the HMTA, 49 U.S.C. 1811(a).

Shipments of hazardous materials that are refused by the consignee or that cannot be delivered within 48 hours after arrival at destination promptly must be (1) returned to the shipper if in proper shipping condition, (2) stored, assuming a suitable storage place for such articles is available, (3) sold, or (4) destroyed when necessary to safety. (Charged electric storage batteries may be held for 30 days after arrival, pending delivery.) This rule, however, must be read in the context of the legal relationship between a shipper and carrier with regard to this property. Selling the cargo, for example, although consistent with the hazardous materials regulations, may constitute an unlawful conversion of someone else's property. Accordingly, this rule should be read and applied with care.

Motor carriers may only deliver hazardous materials to persons authorized to receive those shipments, but explosive shipments may be delivered to a magazine that is locked immediately thereafter.

DISABLED VEHICLES AND BROKEN OR LEAKING PACKAGES

Of course, whenever a motor vehicle carrying hazardous materials breaks down on the highway, the driver should take necessary precautions to minimize the risk of damage to the vehicle or its lading. Special effort must be made to move the vehicle to a place where the hazards of the materials will be minimized.

When a leaking package is discovered in transit, the carrier should dispose of that package by the safest practical means among the following:

1. Packages may be repaired, in accordance with the best and safest practice known and available.
2. A repaired package may only be transported further to the nearest place for safe disposal, and that package must be safe for transportation, must not lead to contamination or mixture of contents with other lading, and must be marked with the name and address of the consignee.
3. If the package is not safe to transport, it should be stored pending proper disposal.

One also should note a special provision for using overpacks (so-called salvage drums) to handle damaged or leaking packaging, found in 49 CFR 173.3(c).

A motor vehicle other than a cargo tank used to carry flammable liquids or flammable compressed gases or other vehicle carrying Class A or B explosives must be marked with flares (pot torches), fusees, red electric lanterns, red emergency reflectors and red flags in accordance with 49 CFR Part 392 when it is stopped on the traveled portion of any highway or shoulder for reasons other than traffic conditions. Cargo tanks used to carry flammable liquids or flammable compressed gases or other motor vehicles carrying Class A or B explosives in these circumstances should not be marked with flares (pot torches) or fusees. No repair of any vehicle carrying hazardous materials should be attempted unless it is safe to do so, and no such vehicle should be repaired in a closed garage.

No one may make any repair of any cargo tank or compartment thereof used to carry any flammable or poisonous liquid, nor may anyone make a repair of any container for any fuel of any nature on such a vehicle by any method employing a flame, arc, or other means of welding unless the tank or compartment is first made gas-free.

Part 177 also contains detailed requirements on actions to take in the event of an accident involving any motor vehicle containing hazardous materials. Persons responsible for the operation of motor vehicles should refer directly to those sections of 49 CFR Part 177, segregated by class of hazardous materials, and should be familiar with the details of those requirements. These measures are too detailed and too important to paraphrase in the *Red Book*.

PASSENGER-CARRYING FOR-HIRE MOTOR VEHICLES

Where other practicable means of transportation is available, the DOT regulations do not permit the carriage of labeled hazardous materials by for-hire passenger-carrying motor vehicles except for small-arms ammunition, emergency shipments of drugs, chemicals and hospital supplies, and munitions of war accompanying the armed forces. Note that this restriction applies only to labeled articles and that materials excepted from labeling and 49 CFR 177 (see Chapter 13) are not limited by this restriction.

When hazardous materials are carried, no explosives other than small-arms ammunition may be carried in the passenger space of the vehicle. No more than one hundred pounds aggregate gross weight of explosives permitted to be transported by rail express may be carried in any passenger vehicle. Certain more flexible provisions are made for samples of explosives being shipped for laboratory examination.

The gross weight of any given class of hazardous materials other than explosives, when allowed on passenger vehicles at all, must not exceed one hundred pounds, and the aggregate gross weight of all such materials must not exceed five hundred pounds.

No motor carrier operating any bus transporting passengers may carry any Class A poison, any liquid Class B poison, any irritating material, or any paranitraniline. Nor may he carry any non-liquid Class B poison above an aggregate gross weight of one hundred pounds.

Packages of labeled radioactive materials that may be carried may only be stored in the trunk or baggage compartment of the vehicle, not in the passenger space.

CHAPTER 16

Motor Carrier Safety Act of 1984

In October 1984, Congress passed a significant new law entitled the Motor Carrier Act of 1984.

There is a justified concern with the impact this law will have on future DOT operations, particularly those of the Federal Highway Administration. It is cumbersome, heavily procedural, and tends to enhance the likelihood of litigation. It is a significant congressional statement endorsing local initiatives in transportation safety regulation. Of great concern is the potential that procedures described in this law will be mirrored in future Hazardous Materials Transportation Act (HMTA) amendments.

The law addresses commercial truck safety. Affected are trucks (1) having a gross vehicle weight rating over 10,000 pounds, (2) designed to transport more than 15 passengers, or (3) carrying hazardous materials requiring DOT placarding.

Section 203 is the only part of the legislation that appears to put a value on uniformity, setting forth the finding that "improved, more uniform commercial motor vehicle safety measures and strengthened enforcement would reduce the number of fatalities and injuries and the level of property damage related to commercial motor vehicle operations."

The value this finding places on uniformity, however, is more than countered by the phrase in Section 206(c)(2) that DOT, in issuing motor carrier regulations, should consider "State laws and regulations pertaining to commercial motor vehicle safety in order to minimize unnecessary preemption of such State laws and regulations under this Act."

In addition, with regard to the approach toward uniformity reflected in this law, Congressman Snyder of Kentucky made the following remarks on the floor of the House of Representatives:

> We start with the premise that, to a degree, greater uniformity in safety requirements can lead to less uncertainty and confusion and to improved enforcement and better performance on the part of the carriers. We also recognize that differing safety considerations among the various States can at times justify requirements in addition to, or more stringent than, a minimum level of Federal requirements. Hence, our approach consists of a limited Federal preemption of State laws and regulations governing interstate truck and bus

operations, based on extensive analysis of such requirements and comparison of them with Federal
regulations. We build in the presumption that the
State laws and regulations shall not be preempted
unless the Secretary so determines. [Congressional
Record—House, Oct. 11, 1984, H 12228.]

Section 206 calls for federal regulations. In issuing any new truck and bus safety regulations, DOT must consider "costs and benefits" and, as noted above, must avoid "unnecessary preemption."

Section 206(e) declares that if DOT fails to issue new truck safety regulations, the existing motor carrier safety regulations shall be deemed "new law" regulations. This will be the case because DOT was given insufficient time to get out truly new regulations.

Section 206(b) declares that the Secretary "shall not eliminate or modify any existing motor carrier safety rule pertaining *exclusively* to the maintenance, equipment, loading or operation (including routing requirements)" of vehicles carrying hazardous materials, "unless and until an equivalent or more stringent regulation has been promulgated under the Hazardous Materials Transportation Act."

The only motor carrier safety regulations pertaining exclusively to hazardous materials now appear in 49 CFR Part 397. This includes Section 397.9, which orders carriers to avoid populated areas to the extent practicable, without consideration of operating convenience. The Second Circuit Court of Appeals held that Section 397.9 supported New York City's ban on through shipments of liquefied petroleum gas, in litigation brought against that ban. It is also the section cited by Boston, Cincinnati, Columbus and other cities enacting local restrictions.

For reasons that remain obscure, several years ago, in adopting existing motor carrier safety regulations under the authority of the Hazardous Materials Transportation Act, DOT chose *not* to include Section 397.9. With passage of this new law, no one in DOT may alter or even clarify the section unless they do so through HMTA rule making adopting an equivalent or more stringent provision.

Section 209 creates a Commercial Motor Vehicle Safety Regulatory Review Panel, generally called the Safety Panel. The actions of this group will be nonbinding, but are expected to be very persuasive. DOT, in fact, must give "great weight" to their decisions in deciding preemption questions.

The panel is imbalanced. There are fifteen members, including the Secretary. Seven represent State and local government. The other seven "represent the interests of business, consumer, labor, and safety groups." Thus, business is likely to have a maximum of four representatives, with a strong likelihood that the opinions of the other ten panel members frequently will coincide on preemption issues.

Under Section 208, the Safety Panel must review all State and local laws and regulations to identify those that "pertain" to commercial motor vehicle safety. The effect of each of these then must be analyzed, categorizing that effect as (1) the same as the federal rules, (2) less stringent than the federal rules, or (3) in addition to or more stringent than the federal rules. As a practical matter, the panel will be identifying individual provisions of these laws, since few laws could be fairly described in their entirety by a single characterization.

The panel must decide whether nonfederal rules found to be in addition to or more stringent than the federal rules—(a) have *no* safety benefit, (b) are "incompatible" with the federal motor carrier safety regulations, or (c) impose an undue burden on interstate commerce. There are obvious difficulties here. First, one must assume that virtually all local rules originally enacted with a safety intent have some safety benefit. The fact that the benefit may be minor or may be outweighed by other factors does not seem relevant. Second, legislative invention of the undefined term "incompatible" is an automatic stimulus for litigation. We may have begun to know what "inconsistent" meant under the Hazardous Materials Transportation Act (see Chapter 30), but it will take the courts to define "incompatible."

Although Section 207 implies that States have the burden of coming forward with their laws and their own analyses, paragraph (e) of that section undermines the obligation by saying, "If any State fails to submit any State law or regulation pertaining to commercial motor vehicle safety in accordance with this section, the Safety Panel shall analyze the laws and regulations of such State and determine which of such State's laws and regulations pertain to commercial motor vehicle safety." In other words, if a State does not undertake the obligation, the panel must do it for them. Other than speeding the process, which is *not* in the State's interest because it may hasten preemption, there is no incentive to relieve the panel of the job.

The panel is made up of unpaid appointees. The real work has fallen to DOT, therefore, since Section 209(e) requires the Secretary to provide DOT offices and personnel "to assist the Safety Panel in carrying out its duties."

The Safety Panel does analytical work and then makes its findings available to the Secretary.

Under Section 208(c), DOT must repeat the process. Independent of the Safety Panel, and despite the likelihood that DOT people actually did the panel's work, the Secretary "shall review each State law and regulation pertaining to motor vehicle safety." This must be done in the context of a rule making. The same determinations must be made, as to whether the individual provisions of nonfederal law have the same effect as, are less stringent than, or are in addition to or more stringent than the federal rules.

In the final rule issued in this rule making, DOT must find the following types of nonfederal laws to be not in effect or capable of being enforced, i.e., preempted—(a) those that are less stringent, and (b) those that are in addition to or more stringent than the federal rules that also have no safety benefit, are incompatible with the federal rules, or unduly burden interstate commerce.

No such preemption will occur sooner than five years from enactment, and DOT may (and is likely to) extend this period by another year.

Once the final rule is issued announcing the Secretary's determination of preemption, "any person" may seek a waiver of that determination. The waiver must be granted if granting it would not be "contrary to the public interest and is consistent with the safe operation of commercial motor vehicles." It is difficult to see how the Secretary can take all pertinent factors into account in the initial rule making action, and yet grant a waiver in the public interest.

The waiver process must include an opportunity for the petitioner to be heard on the record, i.e., in a hearing conducted in quasi-judicial manner, probably by an administrative law judge.

Even after the administrative rule making and consideration of any waivers, the process continues through the availability of judicial review in the courts of appeals, at the request of "any person (including any State) adversely affected" by the decision.

DOT's action in support of the panel, and in conducting an independent review of all provisions of all State and local laws, will consume an enormous percentage of their regulatory resources. While some personnel may be shifted from the enforcement mode into rule making, it still seems unlikely that the task could be handled efficiently.

Section 210 establishes a new federal truck inspection procedure, and calls for each affected truck to pass the inspection at least annually. Recognition is given to inspection programs implemented by Commercial Vehicle Safety Alliance (CVSA) member States. Section 211 authorizes DOT to delegate enforcement authority to nonfederal jurisdictions receiving grants under Section 402 of the Surface Transportation Assistance Act.

Federal enforcement is stimulated by a new provision in Section 212, calling for investigation of complaints, particularly employee complaints. Except under narrow and controlled circumstances, DOT may not disclose the identity of the complainant.

Civil and criminal penalty procedures and dollar figures were adopted in Section 213. These closely parallel those in the HMTA, but the dollar amounts are substantially lower. Experience under the HMTA shows that civil penalties are employed heavily by the agency, almost to the exclusion of criminal sanctions. The civil penalty provisions in the new law also anticipate compromise of the penalty originally assessed, and compel DOT to take into account "the nature, circumstances, extent, and gravity of the violation committed and, with respect to the violator, the degree of culpability, history of prior offenses, ability to pay, effect on ability to continue to do business, and such other matters as justice and public safety may require. In each case, the assessment shall be calculated to induce further compliance." All fines and penalties are to be deposited in the Highway Trust Fund.

Section 215 formalizes what was an informal process at DOT and the ICC to make findings of safety fitness of owners and operators, "including persons seeking new or additional operating authority." An application for new authority must be denied if the applicant is found unfit under standards DOT is instructed to adopt.

Studies are mandated on heavy truck operating characteristics, crash protection for truck occupants, commercial truck safety performance, emergency warning devices, and health hazards to carrier employees. This last study could involve carrier employees hauling hazardous materials or wastes.

Mexican and Canadian carriers became obligated to assure compliance with DOT financial responsibility requirements, and there also were new provisions for registering foreign motor carriers.

CONCLUSION

This law is having a significant impact upon Federal Highway Administration. Required analyses of State and local laws, if done properly, will be a nearly overwhelming task.

Although there is a call for new truck safety regulations, it is uncertain that all of the new rules will be issued in the first five years.

The activities of the Safety Panel are likely to run counter to the interests of business with considerable frequency. It remains to be seen whether the recommendations of this panel and the actions of the Secretary diverge significantly.

Undoubtedly more latitude will be granted to local truck safety initiatives, and enforcement of requirements should increase somewhat. Vehicle inspection procedures are expected to have a significant impact in any States where such inspections are not already mandated.

CHAPTER 17

Rail Carrier Hazardous Materials Requirements

The hazardous materials regulations had their origin before the turn of the century in self-imposed rules developed by the railroad industry; it is not surprising, therefore, that the railroad regulations contain some of the oldest provisions in 49 CFR.

The hazardous materials regulations applicable to motor carriers in 49 CFR Part 177 were drawn to a large extent from these rail carrier provisions in the 1930s. As a consequence, many sections of the rail regulations and the highway regulations used to be identical, and others follow a very similar structure. The reader will find a great deal of similarity between these carrier parts.

Although DOT made extensive changes in the format of the rail regulations the same degree of change has not yet been implemented for Part 177. Given the substantial differences in these modes of transportation, it is expected that more careful distinction between the two will generally improve the quality of the regulations.

Part 174 of 49 CFR contains the standard requirements applicable to all modes of transportation, i.e., acceptable articles, forbidden explosives, carriers' materials and supplies, Canadian shipments, hazardous materials incident reporting, shipping papers, loading and unloading, stowage compatibility, and the handling of cars. Those provisions of particular interest that have been emphasized in DOT inspections are highlighted in this chapter.

Section 174.45 requires the carrier's preparation of a standard hazardous materials incident report in the event of an unintentional release or spillage of any hazardous material from its package. See Chapter 22 on the details of preparing and filing DOT form F.5800.1, the hazardous materials incident report.

Section 174.24 requires a member of the crew of each train transporting hazardous materials in any amount to have with him the required DOT shipping paper information but not necessarily the shipper certified copy. Shipping papers for ORMs are not required unless these are hazardous substances or hazardous wastes. In addition to the standard information that would be required for highway transportation, discussed in detail in Chapter 11 on shipping papers, 49 CFR 174.25 requires waybills, switching orders, or other billing to show specific placard endorsements and placard notations according to the hazard classification of the materials on the train. This requirement applies to shippers' documentation when the initial movement is a switching operation. In this case, the switching order, switching receipt, or switching ticket (and any copies) must bear the shipper's certificate of compliance described in 49 CFR 172.204.

A large number of special loading and unloading requirements are provided in 49 CFR 174.55 through 174.81. These are binding upon any shipper loading a rail car but, because of their location in the carrier portion of the regulations, they may be overlooked by such shippers. Truck bodies on flat cars are covered by 49 CFR 174.61. Note that cargo tanks (tank truck trailers) may not be transported in TOFC service except under specific conditions approved by the Federal Railroad Administrator within DOT. Section 174.63 prescribes special requirements for the transportation of dry freight containers and portable tanks. Section 174.81 sets forth a compatibility table between hazardous materials very similar to that found in 49 CFR 177.848 of the motor carrier regulations.

Sections 172.500–558 contain the placarding regulations for rail cars as well as for other transport vehicles and freight containers and portable tanks. For specific information on placarding, see Chapter 12.

The rail placards must be attached to both sides and both ends of a car. The regulations prescribe how the placards should be mounted but not the specific location on the sides or ends. Placards on container cars must be similarly applied, or placards may be attached to each end of the car and to the side of any container on the car.

The central regulatory idea to maintain in placarding a trailer on flat car (TOFC) or a container on flat car (COFC) is that one is placarding the car, not the trailers or containers on it. In accordance with general rail car placarding rules, it is required that placards appear at each end and both sides of a TOFC or COFC car. For example, a two-trailer flat car, in which the forward trailer contains hazardous materials and the rear does not, could have the forward trailer placarded on its nose and the rear trailer placarded on both sides and its rear. Obviously, it is important to remove all rail placards from trailers or containers when they are off-loaded to prevent confusion in highway transportation caused by a situation like the one in this example.

Sections 174.55 through 174.81 prescribe a number of requirements regarding handling hazardous materials, unloading cars, documentation, inspection, leaks, fires, cleaning, etc. Unfortunately several of the requirements in this area are old and obsolete. For example, the tank car unloading provisions do not take much of today's car equipment into account.

One particular subject that pertains to a great extent to shippers but is submerged in the carrier requirements is the one dealing with track arrangement of a tank car for unloading. Any consignor handling tank cars must be familiar with the following rules in 49 CFR 174.67:

1. Be sure that plant personnel unloading the car have specific instructions, know what they are handling (DOT hazard), and are generally aware of the DOT hazardous materials regulations, especially 49 CFR Part 174.

2. Car brakes must be set and wheels blocked on a car before any hose is connected to it.

3. Before a discharge hose is connected, signs must be placed on the track or car so persons approaching the car from open ends or ends of the siding can see them. These signs are not to be removed until the discharge connection is disconnected. Signs must be made of metal or other durable material. Words that must appear are:

STOP
Tank Car Connected

or

STOP
Men at Work

The words on these signs must be white on a blue background. The sign itself must be 12 inches by 15 inches overall size, the "STOP" must be at least 4 inches high, and the other lettering must be at least 2 inches high.

A key group of rail regulations from the perspective of rail carriers is set forth in Subpart D of Part 174, entitled "Handling of Placarded Cars." In these sections are the details of placement of cars, handling of cars during switching operations, certain notifications to train crews, and the location of cars containing personnel.

DOT published specific provisions for cargo-only aircraft in rule making that did away with the rail express regulations in Part 175. Some attention was given in consideration of personal baggage and small package shipments via Amtrak. These are quantity limits only and are the same quantity limits as those applicable to passenger aircraft.

One additional area of hazardous materials regulation in the railroad industry is the broad authority granted to the Federal Railroad Administrator within DOT under Section 203 of the Federal Railroad Safety Act, 49 U.S.C. 432. This statute authorizes the administrator to issue emergency orders compelling railroads to take certain actions or to stop the performance of certain acts.

CHAPTER 18

Shipment of Hazardous Materials by Water

While the regulations found in 49 CFR Part 176 are oriented strongly to water carriers, interspersed among those carrier requirements are requirements binding upon shippers. This chapter will deal with the unique aspects of those "hidden" shipper obligations. It will also address other parts of 49 CFR that are unique to water transportation.

All hazardous materials must be prepared for shipment according to the specific classification, description, marking, labeling, and packaging requirements of 49 CFR. It is important to note certain exceptions to these requirements provided for shipments by water:

1. A material being imported or exported, or passing through the United States (neither originating from nor being destined for delivery into the U.S.), may be shipped and transported according to the classification and labeling requirements of the International Maritime Organization (IMO) International Maritime Dangerous Goods Code (IMO IMDG Code) if it otherwise meets the DOT regulations on shipping name, packaging, and other shipper and carrier requirements. This provision applies to the segment of the movement within the U.S. that takes place by land. Even the packaging may be relieved from compliance with U.S. requirements if it meets the IMO IMDG Code when the shipment arrives at a U.S. seaport and is not destined for further transportation outside the port area. Note, however, that this packaging exception specifically does not apply to Class A and B explosives or to radioactive materials.

2. DOT specification package markings need not be applied or affixed in accordance with 49 CFR Part 178 to import or export shipments or those just passing through the U.S., but those markings may be displayed instead on decals, labels, or tags securely attached to the package. This provision is not available for gas cylinders or packagings of a capacity greater than 110 gallons.

3. IMO regulations may be used for import or export shipments or those just passing through the U.S. when transported by motor vehicle or vessel within a single port area, provided the shipping paper format and placarding requirements of 49 CFR Part 172 are met.

4. When any import or export shipment will not be transported by rail,

highway, or air, a shipper may certify that the hazardous material is properly classed, described, marked, packaged, and labeled according to the IMO IMDG Code in lieu of the required U.S. certification.

The carrier requirements applicable to shipment of hazardous materials by water found in 49 CFR 176 used to be published in 46 CFR Part 146. This explains some of the vestiges of ambiguous or erroneous references that continue to be found in Part 176 and 46 CFR Part 146 to this day. The shipper requirements formerly found in 46 CFR Part 146 have been spread throughout 49 CFR Parts 172 and 173. Combustible liquids that used to be listed in the Coast Guard's Dangerous Cargo Regulations in 46 CFR have been added to 49 CFR 172.101, and "hazardous articles" as the Coast Guard once termed them have been absorbed into 49 CFR as ORM-A, B, or C materials. These are indicated in 49 CFR 172.101 by the "W" which precedes them in the first column of that table.

Combustible liquids are not regulated aboard any vessels when in packagings of less than 110 gallons capacity each. The regulations generally specify where one must stow tanks of combustible liquids in excess of 110 gallons that are acceptable for transportation. Special carrier requirements for these tanks also are indicated. One should refer directly to the regulations in 49 CFR to determine applicable requirements for shipment of these combustible liquids by water.

Combustible liquids in tank vehicles or portable tanks on board vessels are subject to 49 CFR 173.118a. Shipping paper and placarding requirements for these shipments appear in 49 CFR Part 172, and special carrier requirements are given in 49 CFR 176.74 and 176.76. Additional carrier restrictions are found in 49 CFR 176.305–325. There are *no* specification packaging requirements applicable to these tanks under 49 CFR Part 100–199, but there are certain construction and shipper requirements (packagings authorized) in 46 CFR. A reference to these is found in 46 CFR 176.340.

A unique group of regulated materials are those classed as ORM-A, B, and C. These replace the former Coast Guard hazardous articles class and also include certain materials posing peculiar threats to transportation aboard aircraft. Most ORM-A, B, and C materials are not considered a significant threat to safety in other modes of transportation and accordingly are not regulated by DOT in highway and rail transportation unless they are hazardous wastes or hazardous substances, or are offered or intended for transportation by air or water, in which case shipping papers must be prepared. Those articles in these classes subject to regulation only when shipped by water (some also by air) are listed below for the convenience of the reader.

*Ammonium hydroxide (less than 12% ammonia): ORM-A

Ammonium sulfate nitrate: ORM-C

Asphalt shipped at or above its flash point: ORM-C

Battery parts (plates, grids, etc., unwashed, exhausted): ORM-C

Bleaching powder (not over 39% available chlorine): ORM-C

Calcium cyanamide, not hydrated (0.1% or more calcium carbide): ORM-C

---

*Above specified quantities, these materials are also hazardous substances, hence, classed as shown by all modes of transportation if that quantity or more is in one package. See 49 CFR 172.101, Appendix.

Calcium oxide: ORM-B

Camphene: ORM-A

Carbon dioxide, solid (dry ice): ORM-A

*Carbon tetrachloride: ORM-A

Castor beans: ORM-C

Castor pomace: ORM-C

Chlorinated lime: ORM-C

*Chloroform: ORM-A

Coconut meal pellets (6–13% mixture, not over 10% fat): ORM-C

Copra: ORM-C

Cotton: ORM-C

1,2-Dibromethane: ORM-A

*Ethylene dibromide: ORM-A

Exothermic ferrochrome, ferromanganese, or silicon-chrome: ORM-C

Ferrophosphorus: ORM-A

Ferrosilicon: ORM-A

Fibers, jute, hemp, flax, sisal, coir, kapok, etc.: ORM-C

Fish scrap or meal: ORM-C

Formaldehyde: ORM-A (see 172.101 entry)

Hexachloroethane: ORM-A

Lead dross or scrap: ORM-C

*Maleic anhydride: ORM-A

Mercaptan mixture, aliphatic (see 172.101 entry)

Metal borings or shavings: ORM-C

*Naphthalene: ORM-A

*Paraformaldehyde: ORM-A

Pesticides, water reactive: ORM-C

Petroleum coke: ORM-C

Sulfur: ORM-C

Tetrachloroethane: ORM-A

For detail on specific requirements applicable to empty packagings, see 49 CFR 172.101 and 173.29.

For export water shipments, when a material is properly described by an n.o.s. entry in the commodity list, e.g., "Flammable liquids, n.o.s.," this shipping name on shipping papers and package markings must include as well by the technical name of the commodity in parentheses. The DOT regulations give the following example: "Corrosive liquid, n.o.s. (caprylyl chloride)."

When loading freight containers for eventual shipment by water, one must keep in mind that the Coast Guard has special compatibility requirements for each class of regulated materials, as well as specific product entries affecting compatibility and stowage. These requirements are phrased as carrier requirements, but when a shipper loads an ocean container, the carrier is not permitted to accept that container unless the shipper has met all of the compatibility requirements.

To determine the compatibility rules, one must check the particular entries in 49 CFR 172.101 column (7)(c), 176.83, and Subparts G through O of Part 176 according to the class of the material. There are certain special compatibility requirements by individual product, as well as by classification.

One area of common confusion among shippers is an international maritime code called the IMO International Maritime Dangerous Goods Code, or IMDG Code. IMO stands for International Maritime Organization. Many countries have adopted the IMO recommendations as their own law or are using it as the basis for their own regulations. See Chapter 25 for a list of countries that have adopted the IMO IMDG Code.

The IMO IMDG Code is comprised of five volumes, covering shipper, packaging manufacturer, and carrier requirements. It is amended approximately every 18 months. See Chapter 25 for a further discussion on the origin and purposes of IMO and other international regulatory bodies.

A shipper preparing packages for export by sea should be familiar with this international code. By shipping hazardous materials in compliance with it, as well as with the U.S. requirements, the shipper assures a smooth transition to the overseas destination. Since most U.S. shipments are made via foreign flag vessels, compliance with IMO will minimize the frustration of shipments at the dockside before loading.

Some useful key items with respect to the international aspects of shipping by sea are the following:

1. Even though DOT (Coast Guard) regulations may exempt a particular shipment from labeling, the shipper may apply hazardous materials warning labels if such labels are required by the IMDG Code, or the foreign country of origin or destination.
2. For import shipments, labels may contain inscriptions required by the country of origin.
3. The UN hazard class or division number may be overprinted on DOT labels for both import and export shipments.
4. IMO compatibility code numbers may be displayed on the labels.

There is one additional international requirement to which the attention of U.S. shippers should be drawn. The UN has broken poisons down into three groups. The so-called "Group III" toxic materials do not fall within the DOT definition for Class B poisons. Under some countries' requirements these materials require a special label of the same dimensions and orientation as that prescribed for the DOT hazardous materials labels. The label is white with black inscriptions and symbol, portrayed as follows:

The criteria for the three routes of exposure for a so-called UN Group III toxic material are as follows:

Oral toxicity		
LD_{50}	Solids	50–200
(mg/kg)	Liquids	50–500
Dermal toxicity		
LD_{50}	200–1,000	
(mg/kg)		
Inhalation toxicity	Vapors	
LC_{50}		[1]$V - 1/5\ LC_{50}$ and
		$LC_{50} - 5.00\ ml/m^3$.[2]

[1]V = saturated vapor concentration in ml/m3 of air at 20°C and standard atmospheric pressure.
[2]ml/m^3 is equivalent to ppm (parts per million).

One final area should be given as a precautionary note. Most, but not all the packagings listed in the DOT surface regulations in 49 CFR are acceptable for ocean transport. Accordingly, when a shipper selects a DOT specification packaging for export use, he must closely check the commodity section in 49 CFR Part 173 for the article listed in the tables to verify that the packaging is indeed authorized for water transportation. See 49 CFR 173.204, for example, to illustrate such a water restriction. Also see 49 CFR 176.76(c)(5) on the prohibition against use of aluminum and magnesium tanks.

Classification, labeling, and shipping papers for water shipments where the IMO IMDG Code applies can be very complex because of differences in classification between shipments which are transported over land in conjunction with export or import by water requires review of the Optional Table, 49 CFR 172.102. The introduction to the table explains how it is to be used.

A major difficulty is that the Optional Table, originally designed to reflect the IMO IMDG recommendations, has not been kept up to date by DOT. The result is that the Optional Table is different from both the DOT rules and the current IMDG Code. Rule making in Docket No. HM-181 may reduce this problem.

Generally speaking, therefore, it is best to use the Optional Table to see if there are any DOT restrictions against using a description or class being considered and, finding none, to consult 49 CFR 172.101 together with the entry in the IMDG Code to discover any variations. DOT indicates unacceptable IMO descriptions and classifications by using the letter "N" in the first column next to the entry in the 49 CFR 172.102 Optional Table. The Optional Table also does not contain information consistent with the hazardous waste and hazardous substance rules. See Chapter 14 on these materials.

As indicated earlier in connection with import and export shipments transported outside the port area, except for Class A and B explosives and radioactive materials, and provided DOT shipping paper descriptions, placarding, and packaging are used overland, the shipper is free to classify, mark, and label his goods as provided under IMO. Physically changing the markings and labels on each package is not required provided the accompanying domestic shipping papers meet 49 CFR domestic requirements for the product. This solution, however, requires intimate and current knowledge of both systems, a difficult feat for anyone.

The introduction to the Optional Table in 49 CFR 172.102 contains helpful information giving the counterpart classification to the corresponding DOT classification which permits correct description and placarding for the overland segment.

CHAPTER 19

Water Carrier Hazardous Materials Requirements

The hazardous materials regulations applicable to water carriers are found in 49 CFR Parts 172 and 176, and Part 171 provides general requirements applicable to all persons under the DOT regulations. The water carrier requirements used to be found in 46 CFR Part 146, but most of that portion of the regulations was abolished after the substantive requirements were shifted into 49 CFR Part 176 in rule making several years ago. The alphabetical commodity list in 49 CFR 172.101 gives the proper DOT shipping name for materials, their classification, labeling, packaging, stowage on board ship and certain compatibility restrictions. Part 172 of 49 CFR also contains the requirements on shipping papers, labeling, and placarding.

The shipper requirements are found in the shipper sections of 49 CFR, discussed in Chapter 18. This chapter will not reiterate those shipper regulations but will concentrate on requirements binding upon water carriers transporting packages of hazardous materials. Title 49 CFR does not cover water carrier requirements for bulk shipments in a vessel's integral tanks and this chapter will not address those requirements either. Specialized vessels such as tank ships, oceanographic vessels, oil rigs, etc., are covered by special sections of 46 CFR and are administered directly by the Coast Guard. All of these 46 CFR rules reference 49 CFR when addressing small packages and portable tanks, and generally prescribe more strict regulations for handling, stowage and sometimes even design. This is because the Coast Guard believes that the long voyages and isolation from shore for personnel on these carriers dictates stricter standards.

In addition to the regulations found in 49 CFR Parts 100–199, it is necessary to mention the strong and long-standing tendency for local Coast Guard port officials (usually called Captain of the Port) to establish local practices that have the effect of regulation. These local regulations are not processed under the Administrative Procedure Act (Chapter 26) or other DOT rule making procedures (Chapter 27). Local Coast Guard authorities are precluded from routinely modifying, adjusting, or otherwise changing any of the requirements in 49 CFR. Any variation on interpretations of the regulations in this volume are inappropriate and local authorities should be so advised. The only official sources of interpretation of any of the regulations applicable to water are Coast Guard Headquarters and the Research and Special Programs Administration (Office of Hazardous Materials Transportation), both in Washington, D.C. The limitation on the authority of local officers pertains only to 49 CFR, however, and does not extend to port safety regulations in 33 CFR. These latter rules tend to be more operational in nature and do not affect such matters as classification, labeling, packaging, but dwell on the physical handling and storing of hazardous materials in the port area.

It is necessary to highlight this situation because experience shows that many ports have different interpretations that are forced upon shippers and water carriers, often resulting in unfortunate and time-consuming incidents.

In the enforcement area, shippers and water carriers should both be aware that the Coast Guard has developed a computer data bank, collecting data on alleged shipper and carrier violations, written up in Coast Guard citations. Those citations may not have led to actual prosecution, however. This data bank gives the Coast Guard a compliance profile of each business entity under its jurisdiction.

Given this record-keeping effort, as well as substantial liability for violation of the hazardous materials regulations, there is little room for ambiguous rulings or subjective interpretations by individual Captains of the Port. Shippers and water carriers should know the full extent of their obligations in order to conduct efficient operations with a minimum of unnecessary delay.

The penalty for violation of these regulations may be either civil or criminal. For a discussion of the significance of the distinction between these penalties, see that portion of Chapter 23 devoted to the penalty provisions of the Hazardous Materials Transportation Act, 49 U.S. Code 1801 et seq. The civil penalty administered by the Coast Guard may be as high as $10,000 for each day of each violation of the regulations by a shipper or a water carrier. The criminal penalty, involving the Department of Justice and the court system, can be as high as $25,000 fine or five years in jail, or both. This area of regulation definitely has teeth in it, and should be accorded serious consideration.

IMPORT AND EXPORT SHIPMENTS

A material being imported or exported, or passing through the United States (neither originating from or being delivered into the U.S., may be transported according to the classification and labeling requirements of International Maritime Organization (IMO) International Dangerous Goods (IMDG) Code if it otherwise meets the DOT regulations, i.e., it is covered by documentation meeting 49 CFR domestic requirements, packed in 49 CFR Parts 173 and 178 packaging, and all surface carrier handling, loading, and placarding requirements are met. See Chapter 18 for more on import/export shipments by water. In addition, other than Class A and B explosives or radioactive materials, any material arriving at a domestic port not destined for further transportation outside the port area may be *packaged* according to the IMO IMDG Code, in lieu of compliance with U.S. packaging requirements.

The DOT specification identification package markings need not be applied as prescribed in Part 178 but may be displayed on decals, labels, or tags secured to the package. This provision, however, does not apply to gas cylinders or packagings with a capacity greater than 110 gallons.

When an import or export shipment will be transported by highway, the water carrier may accept the shipper's certification to the effect that the hazardous materials is properly classed, described, marked, packaged, and labeled according to the IMO IMDG Code, in place of the certification of compliance with DOT regulations.

DAMAGED PACKAGING

Section 176.50 of 49 CFR prohibits the acceptance of certain damaged packaging aboard a vessel, and 49 CFR 176.48 advises the master of the vessel what to do with packages that are damaged while on board or jettisoned or lost from the vessel. In addition to the acts the master must take, and the reports he must prepare, there are special hazardous materials incident reports to be filed as described in 49 CFR 171.15 and 171.16 and discussed at length in Chapter 22 of the *Red Book*.

VESSEL REPAIRS

Any repairs involving welding or burning are strictly controlled. While in port, Coast Guard authorities control these activities by a permit system administered through the local Coast Guard Captain of the Port, under the authority of 33 CFR 126.

EXEMPTIONS

Any waiver, deviation, or alternative practice under the DOT regulations is governed by the procedural requirements for DOT hazardous materials regulations generally, published in 49 CFR Part 107. See Chapter 28 for more on those exemption procedures. There are situations, however, that are more interpretive in nature than a true waiver or deviation from general requirements. An example of such an instance would be the applicability of these regulations to novel vessels, which did not exist at the time these regulations were written. A certain flexibility in the application of the rules is necessary to accommodate novel circumstances, and this occasionally may be achieved by the Coast Guard under certain other maritime regulations, outside the scope of the general DOT exemption procedures.

DEFINITIONS

There are two basic locations in the regulations that should be checked in determining the meaning of any term in those regulations. Section 171.8 of 49 CFR contains general definitions while 49 CFR Part 176 describes certain terms and abbreviations used in water transportation of hazardous materials. For example, see 49 CFR 176.63, defining "on deck" and "under deck."

GENERAL SHIPPER REQUIREMENTS

These are set out in Parts 171, 172, and 173 of 49 CFR. For details of special water requirements on shipper preparation of shipping papers and package marking, see Chapters 11 and 9, respectively.

Section 171.12 establishes the responsibility of an importer to furnish the foreign shipper and the forwarding agent at the port of entry with detailed information from U.S. regulations regarding the material he is shipping to the United States. This requirement is intended to assure that hazardous materials arriving in the United States already meet domestic requirements, thereby facilitating subsequent transportation within the United States. If such a shipment does not meet U.S. requirements, DOT will hold the importer responsible for a portion of this liability, unless he can show he discharged his obligations under 49 CFR 171.12. It is insufficient, however, for the importer to just tell the foreign shipper that the "shipment must comply with 49 CFR Parts 100–199." The information he passes to the foreign shipper must be in detail on the specific requirements applicable to the commodity being imported.

DANGEROUS CARGO MANIFEST

By far some of the most important sections of 49 CFR Part 176 are those relating to the Dangerous Cargo Manifest. This is the document to which the Coast Guard enforcement officer keys his inspection. In view of today's enforcement practices and the dimension of the penalties involved, it cannot be sufficiently emphasized that the manifest is the single most important step in any water carrier's compliance program. A carrier must be thoroughly familiar with 49 CFR 176.30 and 176.36 on this subject, for more violations have been recorded on these sections of the Dangerous Cargo Regulations than all others in 49 CFR Part 176 and its precursor 46 CFR 146, combined. This does not necessarily mean that in terms of safety the paperwork is most important or that more violations actually existed here, but it does illustrate that this is a central requirement that is easy to check and that the Coast Guard accentuates its importance.

The information required to appear on the manifest is the information actually furnished to the vessel by the shipper of the hazardous materials on his bill of lading or other shipping paper. The person under whose supervision the actual preparation of the manifest is made shall cause the information required on it to be correctly transcribed. The person preparing or supervising the preparation of the manifest shall certify to the truth and accuracy of the information on it to the best of his knowledge and belief by his signature and a notation of the date the manifest is prepared. The master, or a licensed deck officer designated by the master and attached to the vessel, or the person in charge of a barge, shall by his signature acknowledge the correctness of the dangerous cargo manifest, although this signature requirement does not apply to unmanned barges.

The dangerous cargo manifest, list, or stowage plan generally shall show:

1. The name of the vessel and official number (or international radio call sign if there is no official number).
2. The nationality of the vessel.
3. The true shipping name and identification number of all hazardous materials as it appears in the alphabetical commodity lists in the U.S. regulations or the IMO IMDG Code. For export shipments, when the shipping name of a commodity is an n.o.s. entry in the particular table, this entry must be qualified by the technical name of the commodity in parentheses, e.g., "Corrosive liquid, n.o.s. (caprylyl chloride)."
4. The number and description of packages (such as barrels, drums, cylinders, boxes, etc.) and the gross weight for each type of packaging.
5. The hazard classification of the materials under the DOT regulations (such as class A explosive, flammable liquid, nonflammable compressed gas), or in accordance with the IMO IMDG Code. There are specific instructions on how the classification information is to be shown.
6. Additional descriptive information as required by 49 CFR 172.203.

A requirement has been added to specify that the Dangerous Cargo Manifest must be kept in a *designated* holder on or near the vessel's bridge. If a material is being transported by vessel under the authority of an exemption and the exemption requires that a copy be on board the vessel, this copy of the exemption must be kept with the Dangerous Cargo Manifest.

TRANSPORTING FREIGHT CONTAINERS

Section 176.83 of 49 CFR, the stowage compatibility table, has great importance because it prescribes requirements on the transportation of freight containers and LASH barges by cargo vessel. This section also covers water transportation of highway and railroad vehicles aboard non-ferry cargo vessels. Since railroad vehicle carriage is such a specialized operation, it will not be given further coverage here. Section 176.83 of 49 CFR supposes a degree of adaptation to apply these rules effective to RO-RO operations. Such vessels, however, are outside the scope of the *Red Book*, and no further coverage will be given to that type of operation here. The discussion regarding this section is limited to containers, containerships, and compatibility.

While this section relates generally to vessel operations, there are several paragraphs that govern the acts of persons loading or stuffing freight containers. Unfortunately, many shippers engaged in loading containers are unaware that these requirements apply to them. As a rule of thumb, one should apply a given regulation to a shipper, carrier, or third party on the basis of who is performing the acts covered by that regulation.

Familiarity with DOT highway carrier regulations is not enough information for people who fill containers for water transportation. The highway compatibility rules are given in chart form in 49 CFR 177.848, but the Coast Guard rules on compatibility are very different and must be met by the person who fills the container. The general compatibility regulations are arranged according to hazard classification, in 49 CFR 176.83. Any shipper offering containers or vehicles for cargo vessel carriage, as opposed to ferry vessel transportation, must be fully acquainted with this section of 49 CFR Part 176. The water carrier should be alert to potential ignorance of this rule on the shipper's part and the unintended noncompliance that may result.

Once the general compatibility is established, individual items must be looked up in the alphabetical commodity list in 49 CFR 172.101. What may be permissible according to the general rules may be prohibited by a specific rule for that commodity. For example, there is no prohibition against loading sodium cyanide solution with other corrosive liquids under the general rules, but in 49 CFR 172.101 it is stated that these solutions must be stowed away from acids. This means that this material may not be in the same freight container with acids. (See 49 CFR 176.83(d)(1).)

For containerized shipments, a material required to be stowed "away from" another may not be stowed in the same container. They may be stowed in adjacent closed containers, however. The term "adjacent" as used in this context means that a side or end surface directly faces another side or end of another container. They cannot be in the same vertical line unless separated by a deck resistant to fire and liquid. When a *solid* substance must be stored "away from," it may be stowed above the other provided at least 8 feet intervene. There are different requirements for "open" freight containers.

Different requirements apply to railroad and highway vehicles transported on ferry vessels. These are found in 49 CFR 176.11(d) and Subpart E of Part 176. In 49 CFR 172.101, column 7(b) on passenger vessels applies to passenger ferry vessels and indicates the materials that may be transported by passenger ferry vessel.

This subpart is an unfortunate and classic example of unusable federal regulations. Highway vehicles are loaded by shippers or motor carriers for the most part. Neither group is known to be familiar, in detail, with the Coast Guard ferry vessel regulations. In addition, a shipper loading a vehicle for pickup by a carrier may not even be aware of that carrier's use of ferry vessels. Interline shipments present the same problem. The ferry vessel operator may handle dozens of trucks per hour through his ticket booth. It is unlikely that he can check each hazardous material in the DOT regulations for compliance with the ferry vessel regulations. In addition, much of the material permitted on passenger-carrying ferry vessels is permitted on deck stowage only, a form of stowage not available on passenger ferries. There is no way for the ferry vessel operator to verify compliance with the regulations, short of carefully reviewing each shipping paper. This is an unreasonable burden on a passenger ferry ticket booth operator in a busy metropolitan area.

Section 176.89 of 49 CFR sets forth special requirements on the driver in his operation of a vehicle containing hazardous materials. Many ferry vessel operators display large signs paraphrasing this regulation and apply it to all vehicles.

Dry freight barges are the subject of Subpart F of 49 CFR Part 176. The transportation of packaged hazardous materials on these barges is such a small part of commerce relative to overall tons shipped that the requirements will not be covered in detail here.

Section 176.78 of 49 CFR is a very important section for carriers, since it contains all the basic requirements for use of power-operated industrial trucks (e.g., forklifts) on board vessels, including particular requirements applicable for each hazard classification.

Key general sections a water carrier should be sure to master include:

172.101	Explanation of hazardous materials list.
	Hazardous materials list.
172.102	Explanation of optional hazardous materials list.
	Optional hazardous materials list.
172.504	Placarding of containers, portable tanks and vehicles.
176.30	Dangerous Cargo Manifest.
176.63	Stowage locations.
176.74	On deck stowage.
176.76	Stowage of vehicles, freight containers, portable tanks.
176.83	Compatibility and segregation requirements.

Notwithstanding all the requirements in the other portions of 49 CFR, each class of hazardous material has certain additional, special requirements when carried by water:

176.100–177	Explosives.
176.200–230	Compressed gases.
176.305–340	Flammable and combustible liquids.
176.400–419	Flammable solids, oxidizers, organic peroxides, and blasting agents.
176.600–605	Poisons and irritating materials.
176.700–715	Radioactive materials.

176.800–805   Corrosive materials.
176.900–906   ORM materials.

One particularly significant change brought about by the amendments adopted in DOT rule making a few years ago was to convert what had been known as the Coast Guard "hazardous articles" class into ORM-A, B, or C. Generally speaking the same articles are regulated in somewhat the same fashion as they had been, but they have been subdivided into these new classifications.

Many of the regulations in 49 CFR Part 176 do not apply to ORM-A, B, or C, but one must examine the particular rule to determine this inapplicability with any assurance. It also is necessary to determine whether a material otherwise granted one of these excepted classifications is a hazardous substance or hazardous waste (Chapter 24), because these environmentally hazardous materials are regulated at all times in all modes of transportation. Representative sections that exclude ORM from substantive requirements are 49 CFR 176.11(e), 176.69(a) and 176.83(b). ORM-A, B, and C materials regulated in water and not in land transportation are listed in 49 CFR 172.101 and are preceded by a "W" in column (1) of that section.

A significant class of hazardous materials is called ORM-D. This class is discussed at length in Chapter 13 on limited quantities and ORM-D. It currently *only* applies to consumer commodities in certain small size consumer packaging described in 49 CFR 173.1200 and small arms ammunition. ORM-D is completely unregulated in water transportation (assuming it is not a hazardous substance or hazardous waste) except for the outside packaging marking "Consumer Commodity, ORM-D" and other markings such as "This Side Up." Section 172.200 of 49 CFR declares that there are no shipping papers for ORM-D transported by water, hence there is no entry required for ORM-D on the vessel's Dangerous Cargo Manifest. Stowage and compatibility requirements are also inapplicable to ORM-D. (See 49 CFR 176.11(e).) Note that under the IMO IMDG Code, these materials are not so exempted. See Chapter 13 regarding revised IMO limited quantity provisions and competent authority certificates.

"Limited quantities" is a term in 49 CFR that is pertinent to a discussion of 49 CFR Part 176. Special stowage provisions used to apply to limited quantity shipments in the form of certain relief from on deck stowage limitations. This stowage provision no longer appears in the regulations, however, so that no matter what the quantity of the hazardous material (other than ORM-D) the stowage location is prescribed in column (7) of the table in 49 CFR 172.101.

As before, however, limited quantity shipments are not subject to the compatibility requirements of 49 CFR Part 176, but they are subject to the compatibility limitations of 49 CFR 172.101.

The Coast Guard regulations are based historically and in part today on a different statute than the other DOT hazardous materials regulations. See Appendix D for all statutes. The applicable statute in this case is called the Dangerous Cargo Act and is cited as 46 U.S. Code 170, as amended. The major amendments in the past have been in 1940 and again in 1975, with the passage of the Hazardous Materials Transportation Act.

PORT SAFETY REGULATIONS

For many years, pursuant to President Truman's Executive Order (E.O.) 10173, October 20, 1950, the Coast Guard has exercised port security/safety jurisdiction in port areas. Since the Port and Waterways Safety Act (PWSA) of 1972, 33 U.S. Code 1221 et seq., the Coast Guard administratively has been shifting the emphasis of these regulations from port security, which is oriented toward military defense, to port safety generally. This shift in emphasis actually began some time after the Korean conflict and consequently is well established as a part of the Coast Guard law enforcement effort. Until the PWSA in 1972, however, the Coast Guard's entire authority rested on President Truman's Executive Order, which was very much oriented toward defense security in ports.

The regulations promulgated under the Executive Order are found for the most part in 33 CFR Parts 121 through 126. Readers of the *Red Book* will be particularly concerned with 33 CFR Parts 126 and 161.

Subpart P of 33 CFR is entitled "Ports and Waterways Safety" and specifies certain notification requirements for most vessels to follow before they arrive at U.S. ports. It consists of Parts 160 through 165. In Part 161, 33 CFR 161.11 (Notice of arrival; vessels carrying certain dangerous cargo) and 33 CFR 161.13 (Notice of departure; vessels carrying certain dangerous cargo) are of particular interest to the vessel operator who is carrying hazardous materials. Such a vessel, whether foreign or domestic, is generally required to give at least 24-hours advance notice of arrival and departure to the local Coast Guard Captain of the Port at each port or place in the United States where the vessel will call, giving the amount and stowage location of specified hazardous materials.

Packaged hazardous materials affected by this notification requirement are:

1. Class A explosives.
2. Oxidizers and blasting agents that require a Coast Guard permit for loading or discharging according to 49 CFR 176.415.
3. Large quantity radioactive materials or Fissile Class III shipments.
4. Each cargo under Table I of 46 CFR Part 153 when carried in bulk.
5. Certain other hazardous materials in bulk. For flammable and combustible liquids the Coast Guard interprets any liquid container of over 110 gallons capacity as bulk. For other classifications, bulk would be materials delivered without benefit of packaging, and handled without mark or count. These materials are:

Acetaldehyde	Ethylene	Methyl chloride
Ammonia, anhydrous	Ethylene oxide	Phosphorous,
Butadiene	Methane	elemental
Butane	Methyl acetylene,	Propane
Butene	propadiene mixture,	Propylene
Butylene oxide	stabilized	Sulfur dioxide
Chlorine	Methyl bromide	Vinyl chloride
Ethane		

The other major regulations on port safety are published in 33 CFR Part 126, and are entitled "Handling of Explosives or Other Dangerous Cargoes Within or Contiguous to Waterfront Facilities." Dangerous cargoes are quite strictly controlled on the waterfront. The Coast Guard requires that hazardous materials be handled only on designated waterfront facilities. For these facilities it prescribes conditions for security guards, smoking, welding or hot work, vehicles, pier automotive equipment, cleanliness, stores and supplies, electric wiring, heating, fire extinguishing equipment, first aid appliance, lighting, arrangement of cargo, and handling of cargo. Noncompliance with any of these regulations subjects the pier operator to revocation of a general permit issued to him under the regulations for the operation of a pier handling hazardous materials.

The pier operator must notify the Captain of the Port if he handles at one time:

- Class B explosives, in excess of one net ton.
- Class C explosives, in excess of 10 net tons.
- Flammable liquids, in excess of 10 net tons.
- Flammable solids, in excess of 100 net tons.
- Oxidizers, in excess of 100 net tons.
- Flammable gases, in excess of 10 net tons.
- Class A poisons.
- Bulk shipment of any cargo in the chemicals listing above.
- Bulk liquids as named in 46 CFR Part 153.

In addition, the pier operator must abide by certain compatibility regulations on the pier:

1. Maximum separation is called for between flammable gases and liquids on the one hand, and corrosive liquids and combustible materials on the other.
2. Flammable solids and oxidizers must be protected from being wetted.
3. Corrosives must be handled to avoid contact with organic cargo.
4. Poisons must be stored so as to prevent contact with corrosive or flammable liquids, and flammable solids.

Waterfront facilities obviously are under a multitude of regulations when storing hazardous materials. This leads to special storage sites away from the facility proper and often causes refusal to accept hazardous materials except on sailing days. This, in turn, creates logistic problems for the shipper. As DOT and EPA pursue a course of increasing the population of regulated articles, this confusing and disruptive situation should be receiving more attention.

Finally, under 33 CFR Part 126 there are two sections that give the local Coast Guard Captain of the Port broad authority to control the handling of hazardous materials in port areas under his jurisdiction. These sections have no counterpart in any other DOT regulations and are worth quoting in part:

> 33 CFR 126.29 Supervision and control of dangerous cargo.
>
> (a) Authority. The Captain of the Port is authorized to require that any transaction of handling, storing, stowing, loading, discharging, or transporting the dangerous cargo covered by (these regulations) shall be undertaken and continued only under the immediate supervision and control of the Captain of the Port or his duly authorized representative. In case the Captain of the Port exercises such authority, all direction, instructions, and orders of the Captain of the port or his representative, not inconsistent with [these regulations], with respect to such handling, storing, loading, discharging, and transporting; with respect to the operation of the waterfront facility; with respect to

vessels handling, stowing, loading, or discharging of dangerous cargo at anchorages when the operations are under the immediate control and supervision of the Captain of the Port or his duly authorized representative, with respect to the ingress and egress of persons, articles, and things and to their presence on the waterfront facility or vessel; and with respect to vessels approaching, moored at, and departing from the waterfront facility, shall be promptly obeyed.

33 CFR 126.31 Termination or suspension of general permit.

The Captain of the Port is hereby authorized to terminate or to suspend the general permit granted by Section 125.27 in respect to any particular designated waterfront facility whenever he deems that the security or safety of the port or vessels or waterfront facilities therein so requires. Confirmation of such termination or suspension shall be given to the permittee in writing. After such termination, the general permit may be revived by the District Commander with respect to such particular waterfront facility upon a finding by him that the cause of termination no longer exists and is unlikely to recur. After such suspension, the general permit shall be revived by the Captain of the Port with respect to such particular waterfront facility when the cause of suspension no longer exists, and he shall so advise the permittee in writing.

In the first section quoted, note that as long as the Captain of the Port's requirements are not inconsistent with the regulations, he has authority to take complete charge of all aspects of a cargo operation. The rules are not conditional, i.e., related to a safety cause/effect situation for protection of the public. When this section is used, however, Captains of the Port always give public safety as their reason for acting.

Note that the exercise of this authority does not require compliance with the Administrative Procedure Act, and Coast Guard officers using this authority have seldom given any opportunity for public participation and comment before promulgating a rule. They simply have published the rule as a military commander would publish orders to his base personnel.

The second quoted section relates to the general permit granted to pier operators to handle dangerous cargo on their facilities if they comply with certain regulations generally described above. Again, for cause the Coast Guard is empowered to order a facility to cease handling hazardous materials immediately. This termination is not necessarily required to be related to noncompliance with existing regulations. Rather, it is based on a general Coast Guard determination that the security or safety of the port requires the action.

The termination of a permit is not an action frequently taken by Captains of the Port. Under 33 CFR 126.29, however, myriads of local Coast Guard restrictions have been enacted that vary greatly from port to port. These regulations cause a great deal of frustration and delay in handling shipments when a shipper or carrier is unaware of them. Anyone who has not handled or shipped a hazardous material through a particular port should find out from the Coast Guard what the local requirements may be. With more frequent shipments through a port, one adjusts to working with these unique regulations.

Not only the Captain of the Port, but the fire departments and port authorities in the area also may be involved in promulgating regulations on hazardous materials in the port. The local regulatory picture, therefore, is not static.

There is a serious need for DOT to take more of these inefficiencies and local variables into account in administering a system of national commerce. For the moment, however, it is each man for himself in determining the nature of and complying with local port regulations.

CHAPTER 20

Shipment of Hazardous Materials by Air

The federal regulation of hazardous materials, as noted on many occasions in this book, was not conceived as a unified total package of governmental controls but evolved in bits and pieces through sporadic revisions to various transportation statutes. This evolution was unmistakably linked to the mode of transportation covered by the particular statute being amended, rather than to the nature of the products, the hazards, or the people involved in the shipment of these materials. The regulatory authority was designed initially to cope with problems confronting commerce and the public at the time, and thus we find that the earliest group of regulations were those having their origin in railroad carriage of explosives. As each mode of carriage became more efficient and the range and volume of hazardous materials shipments became greater, laws affecting rail, highway, water, and air transportation were enacted to reflect this development.

The history of air regulation evolved somewhat differently and it is important to the understanding of how two sets of standards currently co-exist governing air transportation of hazardous materials. Thus a bit of history is in order.

Safety in air transportation, the most recent of the four modes to be developed, was not federally regulated until the Civil Aeronautics Act of 1938. This act had a specific section on explosives and other dangerous articles, which was the forerunner of the section of the Federal Aviation Act of 1958 that underlies the air regulations on hazardous materials today found in 49 CFR.

The air regulations on hazardous materials nominally appeared in Part 103 of title 14 CFR for many years. Actually Part 103 was a selective cross-reference to the more conservative provisions in 49 CFR—those applicable to partially exempt small quantity packages and those governing shipments by rail express. In those days, the air regulations in 14 CFR had to be read in conjunction with 49 CFR. In 1976, in rulemaking Docket HM-112, DOT shifted the air requirements from 14 CFR to Part 175 of 49 CFR.

While a large portion of the air regulations found in 49 CFR Part 175 are requirements binding upon the air carrier, certain shipper requirements are limited to air transportation, and these are covered in this chapter. See Chapter 21 for requirements binding on air carriers.

Another aspect of hazardous materials shipments by air needs to be addressed here. In other modes of transportation, regulations that the government developed became the norm for operation. In air transportation a particularly active group of carriers dissatisfied with what they perceived to be the safety criteria established by the government, undertook to prepare their own rules by tariff. This dissatisfaction existed not only in this country. In fact, at that time, few air carrier requirements existed outside the United States. This international group of air carriers became the International Air Transport Association (IATA) and developed, among other things, international rules for air transportation of hazardous materials, based on the existing U.S. air carriers' tariff system.

The first government rules were based on regulations relating to rail express and limited quantity shipments. This indirect approach necessitated referring back and forth between different volumes of the Code of Federal Regulations. This did not encourage use of the regulations. The American Air Transport Association (ATA), under the auspices of Tariff 6-D, paraphrased the federal regulations in an entirely different format using a commodity-by-commodity approach, specifically describing the types of packages and the limits on quantities in each package permitted aboard aircraft. Aircraft were separated into cargo-only aircraft and passenger aircraft.

While developing these standards, airlines began to change the governmental system and to impose special requirements and restrictions not found in the federal rules. No one protested the tariff filings, and the air carriers' restrictions took on the form of law because the tariff was filed with the Civil Aeronautics Board, a government agency.

Two bodies of "commercial regulations," the IATA Restricted Articles Regulations and the U.S. Tariff 6-D became the embodiment of air shipping practices and the de facto rules that were followed. Both publications were relatively well-coordinated, but neither were the federally enforced regulations.

This lead to a degree of confusion since, from a practical standpoint, shippers and carriers found the tariff format easier to use than government regulations, and yet they were not the same.

In the mid-1970s, the DOT rules were consolidated into a single volume of the Code of Federal Regulations. At the same time, many of the carrier-developed concepts were incorporated into the federal rules. About the same time, hazardous materials regulations in the form of tariffs began to be protested by shipper groups, and the Civil Aeronautics Board agreed that the proper place for restrictions was in the DOT regulations, not the tariffs. Before any conclusive action could be taken, however, commercial air transportation was deregulated, tariffs were eliminated, and the role of the Civil Aeronautics Board diminished.

In the late 1970s, simultaneously with the deregulation of the air industry in this country, two developments occurred. These were based on the correct perception of the airlines that some other mechanism had to be found to express their view of safety in the transportation of hazardous materials. In the U.S., with the demise of the Civil Aeronautics Board and tariffs, "Circular 6-D" was developed, almost completely copying the federal regulations. The airlines continued, however, to insist on a few particular rules of their own.

It is important to note here that the U.S. airlines had always exercised a particularly pervasive influence even in international transport because of their unique hazardous materials air Tariff 6-D. The IATA regulations were patterned after this tariff and the IATA system became widely used throughout the world.

A special organization under the auspices of the United Nations called the Committee of Experts on the Transport of Dangerous Goods had been developing an international standard which was slowly being adopted into the rail, highway and water modes of transportation. Elsewhere in this book we cover the international recommendations for transportation of hazardous materials by water, called the International Maritime Organization (IMO) IMDG Code. Governments began planning a similar international code for air transportation and this resulted in a corresponding set of international standards based not on the U.S. DOT rules, nor Tariff 6-D, nor on IATA, but on the UN Recommendations.

The International Maritime Organization (IMO) had a counterpart for air transportation known as the International Civil Aviation Organization (ICAO). Both organizations are specialized agencies of the United Nations. See Chapter 25 on international regulations.

A Dangerous Goods Panel was formed under ICAO for the purpose of developing an international dangerous goods code. As explained in the chapter on international organizations, the basic principles governing the international transport of dangerous goods in air were already found in Annex 18 to the Convention on International Civil Aviation—"The Safe Transport of Dangerous Goods by Air." It remained for the dangerous goods panel to develop a detailed code based on the UN standards.

It had been expected that the work of the UN Committee of Experts on the Transport of Dangerous Goods would serve as the core for a unified set of international standards to govern the transportation of hazardous materials by all modes of transportation throughout the world.

ICAO already had a system of detailed standards in other areas called "Technical Instructions." These were used to elaborate on more basic provisions of various annexes of the Chicago Convention governing air travel. It was natural that the detailed requirements for dangerous goods should take this form, in implementing Annex 18.

During the early days of the project to convert the UN recommendations into an aviation code, IATA made a strong bid to become the coordinator, if not promulgator, of standards for the international air transportation of hazardous materials. Throughout the world, many countries had adopted the IATA Restricted Articles Regulations as national regulations. The air carriers' earlier work in developing a comprehensive, detailed set of recommendations had made it easy for many governments to fill safety regulatory voids by simply adopting air carrier standards. The U.S. was one of the governments that had resisted this approach and had maintained independence in administering its safety program.

During the early meetings of the ICAO Dangerous Goods Panel, IATA fought to retain the status quo by assuring the governments that they could do the job. The pressure was great on these governments since many of them had relied for so many years on the air carriers to set their standards. Only a handful of governments had safety expertise on their staffs to permit a reasonable assessment of what was needed in regulating the air transportation of hazardous materials.

ICAO itself, however, is an international organization of governments and is a strong link connecting governments throughout the world administering a complex air traffic system. These governments were accustomed to dealing with each other in setting international standards for the safe passage of aircraft, and many thought that they could do as well in setting standards for the transportation of dangerous goods. The panel felt that the role of government was to make the basic safety decisions and not turn it over to a non-governmental enterprise.

The discussions went so far as to have the International Air Transport Association consider withdrawal from the activity and then offer to become the editor and publisher of the ICAO's standards to the exclusion of an independent publication by ICAO. ICAO resisted this offer and decided that they would go it alone. Recognizing the potential for significant input from the airlines, IATA was accorded a voting position on the Dangerous Goods Panel, a decision that some still question considering the governmental nature of ICAO. IATA continues to sit on the panel exercising a potent influence on the final decisions of the group, although affected shippers of dangerous goods do not have an equivalent opportunity to participate.

This history leads to the fact that now there are two major international publications governing the air transportation of dangerous goods. IATA having lost the battle to retain control over international air standards undertook to set the format for the publication of the international governmental standards. ICAO did not want to present the appearance of an IATA controlled body. IATA, on the other hand, disagreed with what they viewed to be a bureaucratic appearance of the standards being developed and often mentioned during the development phases that the ICAO publication would not be practical to use.

It was no surprise, therefore, that upon ICAO's completion of the standards IATA announced it would continue to publish its own regulations in a format more suitable for the using public. Interestingly enough, IATA also announced that although their version would be essentially identical in substance to the ICAO requirements, variations would occur where air carriers felt that the governments had not accommodated the carriers' concerns. IATA announced that while the government would have its exceptions for countries, IATA also would have exceptions for air operators.

As a result, today there is the official governmental publication produced by ICAO called "The Technical Instructions for the Safe Transport of Dangerous Goods by Air." This publication was produced annually from 1980 to 1986, but now will be produced only biennially. The ICAO publication contains so-called State (national) exceptions, and even prints carrier exceptions in an Attachment. State exceptions are where a nation, perhaps because of internal compatibility problems with other modes of transportation or because of dissatisfaction with the handling of a particular subject under ICAO, files its own requirements for the handling of specified hazardous materials. These State exceptions are relatively few in number and generally do not present any significant threat to international coordination.

The second publication is the IATA Dangerous Goods Regulations, published annually. Both publications are revised as of January 1, annually or biennially, as indicated above.

The world of commerce is a practical place. The necessity to deliver goods using a cooperating carrier is necessary. Air carriers presented a unified front in maintaining independence in assessing safety in the transportation of hazardous materials. They promote use of the IATA Dangerous Goods Regulations. It is clearly a practical matter that communications taking place between shipper and carrier refer to the contents of the IATA regulations rather than the ICAO technical instructions. It is a fact, however, that airlines and shippers are held accountable by governments for compliance with the ICAO Technical Instructions, not the IATA regulations.

In some ways, the IATA regulations do not appear so different from the ICAO standards but upon closer review it is clear that much material has been edited, reorganized, and restructured in the IATA format. Many people have great difficulty working with the two volumes and relating requirements from one to the other. From the commercial aspect, despite not favoring a unilaterally prepared nongovernmental document, people generally refer only to the IATA regulations.

IATA has taken steps to assure that its regulations are at least compatible with the ICAO Technical Instructions. It is intended that IATA's rules will meet the minimum of the ICAO Technical Instructions. If the IATA Dangerous Goods Regulations differ intentionally, they differ only in being more conservative by not authorizing what is authorized under the ICAO Technical Instructions.

It is an enormous undertaking that both organizations have been involved in and both publications are relatively inflexible. Neither has provisions for interim changes or amendments. Under ICAO one can only change regulations every other year. Under IATA it can be done annually. Neither system has an efficient method whereby an exception can be granted from the rule because of significant technological developments or emergency needs. ICAO purports to have an emergency system allowing a shipper or carrier to have problems addressed. In most instances, however, the shipper or carrier is expected to get the agreement of the nation of origin of the shipment, the nation of destination and in most instances the countries of overflight. It is not a practical mechanism for dealing with emergencies or other situations that may warrant a change in the rules.

Under IATA the air carriers do not provide for emergency transportation. A provision under IATA mentions the validity of exceptions issued by governments but IATA compromises the operation by saying that the acceptance of dangerous goods offered under State exemptions or approvals is at the discretion of the air operator. In other words, even though the shipper has obtained all the approvals, in last analysis the air carrier may refuse the shipment.

In the United States certain air carriers continued to provide transportation of hazardous materials according to the DOT regulations found in 49 CFR. A limited number of other carriers provide service according to the U.S. regulations but with a few exceptions. These airlines publish their exceptions and restrictions in Circular 6-D.

In DOT rule making Docket No. HM-181, DOT has proposed substitution of the international standards for transportation of hazardous materials by rail and highway in the U.S. For shipments involving air transportation this has already happened because land movement of a shipment that will have an aviation segment may follow the international rules. See 49 CFR 171.11.

Before going into a detailed discussion of the ICAO Technical Instructions/IATA Dangerous Goods Regulations, a few general comments are in order.

SHIPPING PAPERS

Shipping papers of the type described in 49 CFR Part 172 and Chapter 11 of the *Red Book* are required for virtually all shipments of hazardous materials by air. Shipping papers for air must be prepared in duplicate. One copy travels with the shipment and the other stays with the originating air carrier and must be retained for at least 90 days. For shipments on passenger-carrying aircraft the shipper must add the following words to his certification of the propriety of the shipment: "This shipment is within the limitations prescribed for passenger aircraft/cargo aircraft only (delete nonapplicable)." One also may note that the shipment is compatible with certain international air requirements by adjusting the domestic certificate to close with the statement that the shipment is "in proper condition for carriage by air according to applicable national governmental regulations," in lieu of the domestic certificate's reference to DOT.

The use of shipping papers by air has become institutionalized. A specific carrier-prescribed form must be used, but no international governmental standard or any domestic governmental regulation requires use of a special form. This requirement has emanated from the airlines and continues to be only an airline requirement.

PASSENGER-CARRYING AIRCRAFT DISTINGUISHED FROM CARGO AIRCRAFT

It is essential that the shipper and air carrier distinguish between passenger-carrying and cargo-only aircraft in the transportation of hazardous materials.

For DOT purposes, a passenger-carrying aircraft is one that carries any person other than a crewmember, company employee, an authorized representative of the United States (FAA), or a person accompanying the shipment. Even if the only non-crewmember on the flight is the spouse of one of the crew, the quantity and other limitations applicable to hazardous materials aboard passenger aircraft come into effect.

Materials authorized aboard passenger-carrying aircraft are more conservative generally than those allowed on cargo aircraft, while restrictions as to cargo aircraft are more conservative than others applicable to rail, highway or water transportation.

The regulations in 49 CFR specially relieve a number of products in small quantity packages classed as Other Regulated Materials (ORM) from DOT specification packaging and labeling when those materials are shipped by water, rail, and highway. See Chapter 13 for a complete discussion of these exceptions. ORM packages and DOT limited quantity shipments make up the major segment of the materials permitted aboard passenger-carrying aircraft. In the alphabetical commodity list in 49 CFR 172.101, in the column headed "exceptions," a number of materials show the phrase "None." Most of these materials are prohibited aboard passenger aircraft.

All materials authorized aboard passenger-carrying aircraft may be carried on cargo aircraft. Additional materials that may be carried aboard cargo planes are found by referring to 49 CFR 172.101. In this reference, attention should be directed to that column pertaining to cargo-only aircraft. A hazardous material, in addition to those materials authorized aboard passenger-carrying aircraft, is authorized for carriage aboard cargo aircraft provided:

1. It is permitted in transportation by cargo aircraft according to the cargo aircraft column in the alphabetical commodity list in 49 CFR; and
2. It is packed, marked, and labeled, including quantity limitations, as required by 49 CFR Parts 172 and 173 for shipment on cargo aircraft.

Quantity limitations with regard to inaccessible cargo compartments hold true for cargo aircraft as well as passenger aircraft. Under the regulations, accessible means that a crew member can see, handle, and when the size and weight permit, can separate the package from other cargo during flight. On a cargo aircraft, accessible cargo locations are limited in quantity only by the per package limitations applicable to cargo aircraft. The number of packages is not limited in these locations, provided accessibility is maintained. On cargo aircraft, one may place up to 50 pounds of material that is authorized on passenger aircraft in each freight container in an accessible cargo compartment even though the freight containers may not themselves be accessible to the crew. Materials authorized only for cargo flights must be accessible to the crew at all times and, therefore, must be in an accessible cargo compartment and may not be containerized.

The need to provide accessibility has come under broad criticism from governmental as well as nongovernmental sources, because it presumes the articles must be accessible for some purpose that may no longer be valid considering today's commercial aircraft. DOT has revised the restriction, noting that the following hazardous materials on cargo aircraft may be carried in a location which is not accessible to the crew member, and they are not subject to the weight limitations in 49 CFR 175.75(a)(2):

1. Radioactive materials, ICAO Class 7.
2. Poison B liquids and solids, except those labeled flammable, ICAO 6.1.
3. Irritating materials, ICAO Class 6.1.
4. Etiologic Agents, ICAO Class 6.2.
5. Flammable liquids with a flashpoint above 73°F that do not meet the definition of any other hazard class.
6. Materials in ICAO Class 9, or ORM-A, C, D or E materials.

A different set of hazardous materials regulations apply in the case of shipments via small, single-pilot aircraft where other means of transportation are not available or practicable. These regulations, designed with remote areas such as Alaska in mind, have limited applicability.

The regulations applicable to hazardous materials in air commerce apply to everyone, including the passengers aboard such aircraft. Personal baggage, whether checked or carry-on, is governed by the same restrictions as other cargo. In recognition of the fact that many personal care products fall within the regulations, DOT provided a total regulatory relief for "medicinal and toilet articles" in the baggage of passengers and crew. This exception is limited, however, to a total capacity of 75 ounces for all containers of such articles, and no non-aerosol container may exceed 16 fluid ounces or 1 pound of material.

PACKAGING FOR AIR SHIPMENT

Regulations generally require that liquid hazardous materials be packed in containers strong enough to prevent leakage or distortion of the containers from temperature or pressure changes during transportation on the aircraft.

Containers usually are filled at or near sea level so that the air space in the container is likely to be near 14.7 psi. Aboard an aircraft the altitude and pressure vary considerably from sea level. Modern aircraft cargo compartments generally are pressurized to the equivalent of about 8,000 feet above sea level, irrespective of the actual flight altitude, which often may exceed 40,000 feet.

This means there may be up to a 3.8 psi differential between the atmospheric conditions at the time of filling of the container and the cargo hold pressures encountered. In extreme circumstances and in many older aircraft, this pressure differential may be as high as 8 psi. A container closure that is adequate at sea level could be deficient if the pressure outside the closure were reduced substantially, resulting in leakage of a hazardous liquid. This problem must be avoided through development of satisfactory inspection and test programs by an air shipper to assure that packaging can withstand the environmental conditions found in air shipment without leakage.

Temperatures in the hold of an aircraft tend to drop significantly and may go below 0°F. On the ground that same aircraft hold may have temperatures as elevated as 130°F. The air shipper must foresee such ambient temperatures and must guard against their effects, particularly with regard to expansion of liquids. Sufficient ullage (outage) must be provided to accommodate potential expansion of the liquid through elevated temperature or freezing at reduced temperature.

Some glass, metal, or plastic containers which are perfectly satisfactory for surface transportation will not withstand the rigors of the aircraft environment. Even if the containers appear to withstand these problems initially, testing procedures should anticipate successive pressure and temperature changes that may affect closures, gaskets, container seams, soldered joints, etc.

Low temperatures also can affect materials such as plastics, resulting in contraction of necks or shoulders of bottles and jars. Contents of containers have disappeared without a trace during transportation as a result of low temperature contraction, pressure differentials, and high volatility. Failure of friction lids on metal containers of paints and other products has been reported as well. Ballooning of plastic bags because of entrapped air is another problem that must be foreseen and guarded against.

Federal safety regulations on packaging, particularly in 49 CFR 173.6, address some of these problems, and those regulations place the burden for safe transportation on the shipper of the regulated materials. Air transportation is a specialized mode of commerce, and packaging for air shipments should reflect the unique aspects of this mode.

LABELING FOR AIR SHIPMENT

Air transportation of hazardous materials is unique in that almost all hazardous materials except ORM must bear a DOT warning label when transported by air. Even though a limited quantity material may be specifically relieved from labeling as a limited quantity when shipped by rail, highway, or water, it must be labeled when shipped by air.

CIRCULAR NO. 6-D

Commercial air transportation, like other modes of transportation, at one time was defined by tariffs prepared by carriers and required by law to be kept on file with the government and to be made available to the public. The federal regulatory body entrusted with the responsibility for all economic matters affecting air carriage, including rates, routes, ownership, mergers, etc., used to be the Civil Aeronautics Board in Washington, D.C. The CAB's economic functions were eliminated to a great extent with deregulation of air transportation, and those remaining functions were transferred to DOT.

One of these tariffs was Restricted Articles Tariff No. 6-D, C.A.B. No. 82, which reiterated the DOT regulations on hazardous materials in air transportation along with carrier restrictions.

The regulatory portion of the Restricted Articles Tariff has become Circular 6-D, an airline directive to customers. It is published by the Airline Tariff Publishers, Inc., located at Dulles Airport, as agent on behalf of participating carriers. It is available in a subscription form.

IATA DANGEROUS GOODS REGULATIONS

IATA is the International Air Transport Association, a nongovernmental body of participating international air carriers. The IATA Dangerous Goods Regulations are edited and reissued, generally on an annual basis, by the IATA Restricted Articles Board pursuant to authority of the IATA Cargo Traffic Conferences. This Board is made up of members and technical advisors representing the wider range of international air carriers participating in the tariff.

IATA's publication, like domestic Circular No. 6-D, sets forth the various requirements and limitations on the carriage of hazardous materials by air. IATA includes restrictions that may be encountered throughout the world, however, not just those in the United States. The IATA regulations do not of themselves have the force of law. Those legally binding regulations are found in the ICAO Technical Instructions for the Safe Transport of Dangerous Goods by Air.

ICAO TECHNICAL INSTRUCTIONS

The ICAO Technical Instructions for the Safe Transport of Dangerous Goods by Air are the applicable international and often are the domestic rules as well for the shipment of hazardous materials via aircraft. Traffic departments and air operators do not generally reference these ICAO standards but use the IATA Dangerous Goods Regulations instead. There is a need when ascertaining compliance to reference the ICAO standards rather than the IATA regulations. Many people have great difficulty doing this because although both publications contain the same essential information, it is organized in a very different manner.

At the end of this chapter is a side-by-side comparison chart indicating where in the ICAO Technical Instructions one can find the corresponding IATA requirement. For example, one may want to know an individual country's governmental restrictions on the shipment of a particular item, or at least to check if one exists. These government variations are in Section 1.8 of the IATA regulations. The same information is found at the end of the ICAO Technical Instructions in Chapter 1 of Attachment 3. Similarly, IATA sets forth information on transitional packing instructions in Section 5.10, but in ICAO one must look to Part 3, Chapter 1 for this information. Variations in the structure of the two publications, therefore, are of concern.

The IATA regulations are organized by sections. Section 1 is information regarding general applicability of the regulations. Section 2 is general information concerning definitions, abbreviations and units of measurement. Section 3 is the classification of dangerous goods giving all the definitions for the various classes and special rules, for example, for classification of materials with more than one hazard. Section 4 is the dangerous goods list. Each section is divided into subsections identified as follows:

Section 1, General Applicability
Section 1.2, Application
Section 1.3, General Shipper's Responsibilities
Section 1.4, Air Operator's Responsibilities, etc.

ICAO, on the other hand, is organized by Parts and there are seven such parts. In addition, there are four Attachments. All of these are bound together in one book. Each part is divided into chapters. For example, one looking for general information on the scope and applicability of ICAO would look to Part 1, Chapter 1. One seeking general shipper's information would turn to Part 4, Chapter 1. To find the dangerous goods list turn to Part 2, Chapter 11. Thumbing through the ICAO volume, note the page numbering system at the outside top of each page. It is helpful in identifying where one is within the structure.

If you have the ICAO book, turn to page 2-7-9. Two (2) indicates the Part, 7 indicates the Chapter, and 9 is the page within that chapter. It is relatively easy, therefore, to find the Part since there are only 7 of them. Locating the Chapter is relatively simple too, once you are in the correct Part, but until this sytem of numbering is obvious, it is confusing to try to locate regulations in the ICAO volume. With the information above, you can turn to the comparative table at the end of this chapter and understand how to locate an item in either publication.

Both publications use the United Nation's Dangerous Goods List which is amended regularly by the United Nation's Committee of Experts on the Transport of Dangerous Goods. Sometimes it is necessary to locate a material by its UN number. Again, both publications have numerical cross-references permitting you to search for a commodity in this manner. Using the comparison chart at the end of the chapter, look to Section 4, subsection 4.3 in the IATA column and see how easy it is to locate the same numerical cross-reference list in the ICAO volume. Imagine the difficulty when comparing the location of these two items of information without such a comparative chart if you are unfamiliar with one of the publications.

The basic information in both volumes is derived from the United Nations, but the very extensive packing notes referenced in the Dangerous Goods List are ICAO's own. One of the elements missing in the UN Transport of Dangerous Goods Code, except for certain classes such as explosives and organic peroxides, is the assignment of specific packaging to given materials. The assignment of an authorized packaging to a chemical is a function of the ICAO Dangerous Goods Panel and not the United Nations. On the other hand, assignment of a chemical to a Hazard Class and its labeling are done by the United Nations and, thereafter, they are adopted by the ICAO Panel. For example, when looking at the Dangerous Goods Listing in the IATA regulations, columns A, B, C, D, E and F are derived directly from the United Nations; columns G, H, I, J and K are ICAO's as modified by the airlines.

Another thing to note are the cryptic alphanumeric notations throughout the ICAO publication. Under the heading for Section 5.1, Packing Instructions, Class 1, Explosive, one finds the following listing in the January 1, 1987 ICAO regulations. "State and operator variations (see Subsection 1.8) AN-03, BEG-02, FM-01, GBG-01, HKG-03, IC-01," etc. These are the State (country) and operator variations which you can locate in either publication by scanning the comparative table at the end of this chapter. These entries begin with either two letters or three letters. Three-letter entries are governmental variations and would be found in that section covering the government's exceptions. Two-letter entries are air operator exceptions and would be found in the listing of airlines showing restrictions they apply to specific hazardous materials.

The discussion of international transportation of hazardous materials and its packaging would not be complete without a summary of what is commonly referred to as transitional packing.

When the ICAO Technical Instructions were adopted and the use of UN packaging prescribed, a period had to be afforded the industry to permit an adjustment in packaging and marking designations. The approach that was followed by ICAO was to adopt United Nations packaging but to allow continued use of current packaging until a specific termination date. For a United States company these are packages that were authorized for use by DOT as of January 1, 1983. These packages continue to be authorized for use until December 31, 1989, but there are five conditions:

1. The specific material must be permitted in air transport by ICAO now, and by the U.S. rules that existed before January 1, 1983.
2. Packages authorized for air transport of the specific item must bear any specification packaging identification that was required by DOT.
3. The material must be packaged in accordance with the requirements of those earlier U.S. rules.
4. The packages must meet all DOT specification requirements that they reference by their markings.
5. The maximum net quantity in a specification packaging may not exceed the quantity specified *today* in ICAO. The maximum net quantity in a non-DOT specification packaging must be the quantity that was permitted by the U.S. rule before January 1, 1983, or the quantity specified in the current ICAO rules, whichever is *less*.

When such packagings are being used in place of UN specification packaging, the shipping paper must include the word "transitional" or "transitional packaging." It is not acceptable simply to precede the packing instruction number by the letter "T" to identify this fact. This latter point has caused confusion.

In the IATA regulations a whole series of packagings are covered by so-called "T" notes. These T notes are found only in the IATA regulations. They appear first in the Dangerous Goods List as one of the authorized packagings shown for many items and, second, they are described at the end of Section 5 in a series of T-packing notes. International rules require that packing notes be identified on documentation. Thus, the use of T-packing instruction numbers, for example, "T960," has led to the confusion noted above.

Although on documentation the packing note preceded by the letter T might seem to indicate that transitional packaging is being used, because the T-note system is not part of the ICAO standards it is not recognized by the U.S. Department of Transportation. DOT requires the use of indicative wording *such as* "transitional," or "transitional packaging." This is an example of a governmental variation. In this case, under government variations in ICAO see the note "USG-13." "USG" indicate U.S. government variations. As of January 1, 1987, there were 33 such variations, by far the largest number filed by any country under ICAO.

By way of comparison only one other country has near this number of notes and that is Japan with 22. Otherwise, most countries if they have filed at all (and most have not), have filed a half dozen or fewer variations. Likewise, when reviewing the air operator variations, note that the majority of exceptions have been filed by U.S. airlines.

The DOT regulations (49 CFR 171.11) say that any hazardous material transported by aircraft under ICAO may be transported before or afterward by motor vehicle in accordance with the ICAO technical instructions, provided, however, certain conditions are met. It is important to note, in contrast with requirements for shipments by sea in 49 CFR 171.12, that the required use of an aircraft to qualify for land use of the international rules is not limited to import or export traffic. In other words, the international standards for the transportation of hazardous materials by air and related highway transportation may be followed within the United States for purely domestic shipments. Several basic requirements must be met by the shipper who elects to follow the ICAO technical instructions rather than 49 CFR:

1. The material must be packaged, marked, labeled, classified, described and certified on a shipping paper as required by the ICAO Technical Instructions.
2. The material must be within quantity limits prescribed for transportation aboard aircraft by ICAO.
3. The material may not be forbidden or the package itself forbidden, according to 49 CFR 173.21 or Column 3 of 49 CFR 172.101.

In addition, a series of special requirements will apply depending upon whether the material has peculiarities of regulation here in the United States:

- If the material is a hazardous substance or waste, the U.S. regulations applying to these categories of goods must be completely followed.
- If a material is not subject to the requirements of ICAO and yet under the U.S. rules is a hazardous material, then the U.S. rules must be followed.

A series of requirements apply to the transportation of the ICAO shipment by highway:

1. Under ICAO hazard classes are referred to by number. DOT has determined that in the U.S. an English name description should be used. Therefore, with the exception of hazardous material in ICAO Class 6.1 packing group III, and materials in ICAO Class 9, the name of the DOT hazard class that most closely corresponds to the ICAO Class must be used unless the shipping name contains the key word or words of the hazard class.
2. The designation "ORM-E" must be used in association with the basic description for a material in ICAO 6.1 packing group III, or ICAO Class 9, that is also a hazardous substance.
3. The words "Dangerous When Wet" must be used in association with the basic description for Class 4, Division 4.3, labeled items.
4. An indication should appear on shipping papers that the shipment is being made under ICAO. Simply adding the letters "ICAO" at the end of the description is a good way of accomplishing this.
5. If the material being shipped meets the definition of a poison accord-

ing to the DOT rules, and it is not clear from the shipping name or the class entry that the material is a poison, then the word "Poison" must be entered on the shipping paper in association with the description.

6. Several special requirements pertaining to the movement of radioactive materials relate specifically to highway route controlled quantities, competent authority certification for type B packaging.

Following is a comparative IATA/ICAO table showing the relative location of all the rules:

IATA	ICAO
Section 1 **APPLICABILITY**	**Part 1, Chapter 1** **SCOPE AND APPLICABILTY**
1.2 Dangerous Goods/Air Mail	1.4 Dangerous Goods/Air Mail
1.3 Shipper's responsibilities	Part 4, Chapter 1 Shipper's general responsibilities
1.4 Operator's responsibilities	Part 5, Chapter 1 General procedures
1.5 Training requirements	Part 6, Chapter 1 Establishing training programs
Training curricula	Part 6, Chapter 2 Training curricula
1.6 Limitations on dangerous goods on aircraft	Part 1, Chapter 2 Limitations on dangerous goods on aircraft
Table 1.6.A List of dangerous goods forbidden	No counterpart
1.7 Exceptions for operator's dangerous goods	Part 1, Chapter 2 Exceptions for the operator's dangerous goods
Passenger or crew's dangerous goods exceptions	Passenger or crew's dangerous goods exceptions
1.8 Table 1.8.A Government variations	Attachment 3 (End of Vol.) Chapter 1 Government variations
Table 1.8.B Air operator variations	Attachment 3 (End of Vol.) Chapter 2 Air operator variations
Section 2 **GENERAL INFORMATION**	**Part 1, Chapter 3** **GENERAL INFORMATION**
2.1 Definitions	3.1 Definitions
2.2 Abbreviations and reference marks	Foreword, Abbreviations and symbols

(cont.)

IATA	ICAO
Section 3 **CLASSIFICATION OF** **DANGEROUS GOODS**	**Part 2** **CLASSIFICATION AND LIST OF** **DANGEROUS GOODS**
3.1–3.9 Explosives through Class 9, Miscellaneous Dangerous Goods	Part 2, Chapters 1–9 Explosives through Class 9, Miscellaneous Dangerous Goods
3.10 Classification of materials with multiple hazards	Part 2, Chapter 10 Classification of materials with multiple hazards
Section 4 **DANGEROUS GOODS LIST**	**Part 2, Chapter 11** **DANGEROUS GOODS LIST**
4.3 UN numerical cross-reference of listed materials	Attachment 1, (End of Vol.) List of UN numbers with associated ship- ping names
4.4 Explanation of terms	Attachment 2, (End of Vol.) Explanation of terms in Dangerous Goods List
4.5 Special provisions (notes)	Part 2, Chapter 12 Special provisions (notes)
Section 5 **PACKING INSTRUCTION**	**Part 3** **PACKING INSTRUCTIONS**
5.0 General packing requirements and transitional packaging	Part 3, Chapter 1 General Packing Instructions and transitional packaging
5.1–5.9 Explosives through Class 9, Miscellaneous Dangerous Goods packing notes	Part 3, Chapter, 3–11 Explosives through Class 9, Miscellaneous Dangerous Goods packing notes
5.10 Transitional packing notes (T-Notes)	No counterpart
Section 6 **PACKAGING SPECIFICATION** **AND PERFORMANCE TESTS**	**Part 7** **PACKAGING, NOMENCLATURE,** **MARKING AND TEST** **REQUREMENTS**
6.0 General provisions	Part 7, Chapter 1 General rules
Definitions and Codes	Part 7, Chapter 1 Packaging codes
Specification markings	Part 7, Chapter 2 Specification markings
6.1 Specifications—Inner specifications shown first, followed by outer specifications	Part 7, Chapter 3, Specifications—Outer specifications shown first, followed by in- ner specifications—
6.2 Performance tests	Part 7, Chapter 4 Performance tests
6.3 Radioactive materials packaging	Part 7, Chapter 7 Radioactive materials packaging

IATA	ICAO
6.4 Radioactive materials packaging performance tests	Part 7, Chapter 7 Radioactive materials packaging performance tests
6.5 Packaging for cold gases	Part 7, Chapter 5 Packaging for cold gases
6.6 Infectious substances packaging performance tests	Part 7, Chapter 6 Testing of infectious substances packaging
6.7 Transitional specifications and performance tests	No counterpart
6.8 Transitional specification package marking	No counterpart
Section 7 MARKING AND LABELING	**Part 4, Chapters 2–3**
7.1 Marking	Part 4, Chapter 2 Package markings
7.2 Labeling	Part 4, Chapter 3 Labeling
7.3 Label specifications	Part 4, Chapter 3 3.4 Label specifications
Section 8 DOCUMENTATION	**Part 4, Chapter 4 DOCUMENTATION**
Section 9 AIR OPERATOR HANDLING	**Part 5, Chapters 1–4 AIR OPERATOR HANDLING**
PACKAGE PERFORMANCE TESTING FACILITIES	No counterpart

CHAPTER 21

Air Carrier Requirements

DOT's regulations applicable to air carriers in their transportation of hazardous materials correspond to the requirements on shippers who offer such materials for transportation by air. The shipper requirements are discussed at greater length in Chapter 20, and this chapter will address the regulations from the perspective of the air carrier operating under U.S. DOT rules (not ICAO).

Part 175 of 49 CFR prescribes regulations for loading and carrying what used to be called "dangerous articles and magnetized materials" in any civil aircraft in the United States, and in civil aircraft of U.S. registry anywhere in the world. For most purposes, "dangerous articles" can be read as being synonymous with "hazardous materials" as that term is used in the rest of the DOT regulations and the *Red Book*. Magnetized materials, which can pose a threat to delicate navigational instruments, are forbidden under 49 CFR 173.21 unless their magnetic field is less than 0.00525 gauss when measured at 15 feet from any package or article (if not packaged) surface.

Every civil (nongovernmental) aircraft in the U.S. is covered by the DOT regulations *in* the United States. This includes all commercial airliners as well as corporate and private aircraft in general aviation operations. This is unique, for although the land regulations would pertain to a private automobile if it were engaged in a commercial trip on a for hire basis, the land regulations are not interpreted generally as applying to personal transportation. A personal aircraft, on the other hand, is subject to the aviation regulations on hazardous materials.

Note that every civil aircraft in the "United States" is covered, including aircraft of foreign airlines. The United States are defined in the statute to encompass the States, the District of Columbia, Puerto Rico and the U.S. possessions, including the territorial waters, and the airspace of those areas. Operators of aircraft should also note that civil aircraft of U.S. registry are covered by 49 CFR Part 175 wherever they may be flying in the world. This does not include aircraft of U.S. registry under lease to and operated solely by foreign nationals outside the United States.

Part 175 contains rules on air transportation of hazardous materials but sets forth certain items that are specifically not regulated:

1. Aviation fuel and oil in tanks that comply with FAA installation requirements in 14 CFR Chapter 1.
2. Small arms ammunition for personal use carried by a crew member or passenger in his baggage (excluding carry-on baggage) if securely packed in fibre, wood, or metal boxes.

3. Hazardous materials required aboard an aircraft in accordance with FAA airworthiness requirements and operating regulations. Unless otherwise approved by DOT, items of replacement must be transported in accordance with 49 CFR, except:
   a. In place of required packagings, packagings deemed equivalent that are specially designed for transport of air craft, spares and supplies may be used;
   b. Aircraft batteries are not subject to the 50 pound quantity limitation or to those specified in the table in 49 CFR 172.101;
   c. Tire assemblies with servicable tires are not subject to the regulations provided the tire is not inflated beyond the maximum rated pressure for the tire.
   d. Aerosol dispensers;
   e. Distilled spirits;
   f. Hydraulic accumulators;
   g. Non-spillable batteries;
   h. First-aid kits;
   i. Signaling devices;
   j. Tires; and,
   k. Items of replacement for the foregoing items, except that batteries, aerosol dispensers, and signaling devices must be packed in strong outside containers, and tires must be deflated to a pressure no greater than 100 p.s.i.g.
4. Hazardous materials loaded and carried in hoppers or tanks of aircraft certified for use in aerial seeding, dusting, spraying, fertilizing, crop improvement, or pest control, to be dispensed during such an operation.
5. Medicinal and toilet articles carried by a crewmember or passenger in his baggage (including carry-on baggage) when:
   a. The total quantity of all the containers used by a crewmember or passenger does not exceed seventy-five ounces (net weight and fluid ounces); and
   b. The capacity of each container other than an aerosol container (which is regulated insofar as capacity is concerned under other provisions of the hazardous materials regulations) does not exceed sixteen fluid ounces or one pound of material.
6. Oxygen, or any dangerous article used for the generation of oxygen, carried for medical use by a passenger in accordance with 14 CFR 121.574 or 135.114.
7. Human beings and animals with an implanted medical device, such as a heart pacemaker, that contains radioactive material or with radio pharmaceuticals that have been injected or ingested.
8. Smoke grenades, flares, or similar devices carried only for use during a sport parachute jumping activity.
9. Personal smoking materials intended for use by any individual when carried on his person except cigarette lighters with flammable liquid reservoirs and containers of lighter fluid for use in refilling lighters.
10. Smoke grenades, flares and other pyrotechnic devices that are affixed to an aircraft that is carrying no person other than a required flight member during exercises conducted at or as part of a scheduled air

show or exhibition of air skills. However, these fixed installations must be approved by the FAA.

12. Hazardous materials which are loaded and carried on a cargo aircraft and which are to be dispensed during flight for weather control, forest preservation and protection, flood control or avalanche control purposes, when the following requirements are met:
    a. These operations cannot be conducted over densely populated areas in a congested airway or near any airport where air carrier passenger operations take place.
    b. Currently maintained manuals containing operational guidelines and handling procedures must be prepared and kept available for the use and guidance of flight maintenance and ground personnel.
    c. No one other than a required flight crew member, an FAA inspector or person necessary for handling the hazardous material can be on the aircraft.
    d. The operator of the aircraft must have advanced permission from the owner of the airport that will be used in the operation.
    e. When explosives are carried for avalanche control flights, these explosives must be under the control at all times of a blaster who is licensed under a state or local authority that has been identified in writing to the FAA office that is responsible for that operators overall aviation security program.

13. Carbon dioxide solid (dry ice) in quantities not exceeding 5 pounds per package, packed as prescribed by 49 CFR 173.615(a) and used as a refrigerant for the contents of the package. This package must be marked with the name of the contents being cooled, the net weight of the dry ice or an indication that the net weight is 5 pounds or less and also marked "Carbon Dioxide, Solid," or "Dry Ice."

14. A transport incubator unit necessary to protect life or an organ preservation unit necessary to protect human organs on the condition that:
    a. The compressed gas used to operate the unit is marked, labeled, filled, maintained in an authorized DOT specification container;
    b. Any battery used in the operation of the unit is of the non-spillable type;
    c. The unit is constructed so that valves, fittings, and gauges are protected from damage;
    d. The pilot has been advised of the unit being aboard and when it is intended for use;
    e. The unit is accompanied by a qualified operator;
    f. The unit is secured in the aircraft so as not to restrict access to any emergency equipment or regular exit or aisle in the passenger compartment; and,
    g. There will be no smoking within 10 feet of the unit.

15. Alcoholic beverages, perfumes and colognes carried aboard a passenger aircraft for use or sale on the aircraft.

16. Alcoholic beverages, perfumes and colognes purchased through duty-free sales carried by passengers or crew as carry-on baggage.

17. Carbon dioxide, solid (dry ice) intended for use in food and beverage

services aboard the aircraft. And dry ice in quantities not exceeding 4 pounds per passenger when used to pack perishables in carry-on baggage provided the package is so designed as to permit release of carbon dioxide gas.

18. Carbon dioxide cylinders worn by passengers for the operation of mechanical lens and spare cylinders of a similar size for this only purpose in sufficient quanitites to assure adequate supply for the duration of the journey.

19. Wheelchairs with non-spillable batteries as checked baggage provided that the battery is disconnected, the battery terminals are insulated and the battery is securely attached to the wheelchair.

20. Wheelchairs with spillable batteries as checked baggage, provided that the wheelchair can be loaded, stowed, secured and unloaded always in an upright position. In this case, the battery must be disconnected, the terminals insulated and the battery securely attached to the wheelchair. The pilot must be advised of the location of the wheelchair aboard the aircraft prior to departure. If the wheelchair cannot be loaded stowed, secured, loaded and unloaded in an upright position, the battery must be removed, and then the wheelchair can be carried as checked baggage without restriction. Special conditions apply to the packaging of this removed battery. These are covered in 49 CFR 175.10(a)(20).

21. Catalytic hair curlers containing hydrocarbon gas as checked baggage provided not more than 1 per passenger or crew member and provided that the safety cover is securely fitted over the heat element. Gas refills for such curlers are not permitted in checked or carry on baggage.

22. Mercury barometers carried as carry-on baggage only by a representative of a government weather bureau or similar agency provided that the individual advises the operator of the presence of the barometer in its package. Special packaging requirements apply to this barometer.

For the purposes of applying the regulations in 49 CFR Part 175, "radioactive materials" means any materials or combination of materials which spontaneously emit ionizing radiation. The term does not include materials in which the estimated specific activity is no greater than 0.002 microcuries per gram of material, and the radiation is distributed in an essentially uniform manner. (This definition is drawn from Section 108 of the Hazardous Materials Transportation Act, 49 U.S. Code 1807.) "Research" means investigation or experimentation aimed at the discovery of new theories or laws and the discovery and interpretation of facts or revision of accepted theories or laws in the light of new facts; it is not limited to medical research.

CERTIFICATION AND SHIPPING PAPERS

No one may offer any hazardous material for air shipment unless that shipment is accompanied by a clear and visible statement that it complies with the content, quantity, packaging, marking, and labeling requirements of 49 CFR Part 175. The shipper or his authorized agent shall sign the statement or it may be mechanically signed with his signature. Note that under the IATA interpretation of the international rules, air carriers will not accept a mechanically signed document.

Those sections of 49 CFR pertaining to shipping papers and the shipper's certificate of compliance govern these documents as well. See Chapter 11 on the preparation of shipping papers by all modes, including air. The wording of the shipper's certificate for air transportation is as follows:

> This is to certify that the above-named materials are properly classified, described, packaged, marked, and labeled, and are in proper condition for transportation, according to the applicable regulations of the Department of Transportation. This shipment is within the limitations prescribed for passenger aircraft/cargo-only aircraft (delete nonapplicable).

In order to meet the certification requirements of ICAO/IATA, the shipper also may add the following instead of that portion of his certification indicating compliance with the DOT regulations: ". . . and in proper condition for carriage by air according to applicable national governmental regulations."

In addition to these requirements, no shipper may offer and no operator of an aircraft may knowingly accept any radioactive material subject to 49 CFR Part 175 for shipment on a passenger aircraft unless the shipper's certificate also declares that the shipment contains radioactive materials intended for use in, or incident to research, medical diagnosis, or treatment, and meets the regulatory requirements of 49 CFR Part 175 for shipment on passenger-carrying aircraft.

PASSENGER-CARRYING AND CARGO AIRCRAFT DISTINGUISHED

The hazardous materials regulations make a great distinction between passenger-carrying and cargo-only aircraft. Both the shipper and the operator of the aircraft must be very sensitive to this distinction, because many substantive safety requirements are keyed to it. Section 171.8 of 49 CFR declares that a passenger-carrying aircraft is an aircraft that carries any person other than a crewmember, company employee, an authorized representative of the United States (such as an FAA inspector), or a person accompanying the shipment. A person accompanying the shipment need not be there to guard or handle the material in order to qualify for the last designation. Thus, in aerial pipeline inspection operations, for example, where small aircraft are used and carry some hazardous materials related to the pipeline inspection function as well as the inspector of the pipeline, the inspector has been interpreted to be accompanying the shipment of hazardous materials.

A "cargo-only aircraft" is any aircraft that is not a passenger-carrying aircraft and is used to transport cargo.

LIMITATIONS ON PASSENGER-CARRYING AIRCRAFT

The materials authorized for carriage aboard a passenger aircraft are very limited. As a general category they can best be described as hazardous materials which are specified in 49 CFR Part 173 as excepted from the specification packaging and labeling requirements of 49 CFR, when those materials are prepared as required for exception and as shown in 49 CFR 172.101. See Chapter 13 on limited quantity exceptions to regulations under 49 CFR. It should be noted that although an exception to labeling requirements is given for land and water transportation of these materials, when shipped by air they must be labeled.

LIMITATIONS ON CARGO AIRCRAFT

One may carry any article on a cargo aircraft that would be permitted on a passenger-carrying aircraft, as well as any article specified in 49 CFR 172.101 as acceptable for carriage aboard cargo-only aircraft.

PACKAGING AND MARKING REQUIREMENTS

For a discussion of packaging, especially packaging selected for air transportation, see Chapter 8 on hazardous materials packaging, and 49 CFR 173.6. For marking, see Chapter 9.

LABELING

A label as it is addressed here is the diamond-shaped and color-coded hazard label discussed at length in Chapter 10. For the purposes of this discussion, it also includes rectangular etiologic agent and the cargo aircraft only label (danger label).

Almost every hazardous materials package subject to the air regulations other than ORM must be labeled. This includes limited quantities otherwise excepted from labeling for surface transportation. Although this discrepancy between the modes of transportation has been the topic of frequent criticism and although some change was adopted in this area in rule making Docket Nos. HM-103 and HM-112 (see Appendix C), the requirement still may stimulate some confusion in air transportation.

QUANTITY LIMITATIONS

Quantity limits are expressed in these regulations in two ways: by package and by aircraft compartment. Except under specific regulations governing circumstances in which small, single pilot, cargo-only aircraft are used when other means of transportation are not available or are impracticable, no carrier may put more than 50 pounds net weight of hazardous materials authorized on passenger aircraft plus 150 pounds net weight of any authorized nonflammable gas in an inaccessible cargo compartment on any aircraft.

All materials that are only acceptable on cargo aircraft, not passenger aircraft, must be accessible to the crew, and, therefore, must be in an accessible cargo compartment and may not be containerized. One may place no more than 50 pounds hazardous materials authorized for passenger aircraft (plus 150 pounds on nonflammable gas) in each freight container as long as the cargo compartment is accessible.

No aircraft operator may carry more packages of radioactive materials covered by 49 CFR Part 175 on passenger aircraft than will exceed a combined transport index number of fifty. On cargo aircraft this limit is increased to 200. There are special requirements for fissile radioactive materials covered in Section 175.702. On cargo aircraft, however, no single package may exceed an index of 10. One additional concept uses the "total transport index" which is covered below. The total index number is determined by adding the transport indices shown on the labels of each package.

Except as noted above, no carrier may put more than 50 pounds net weight of any hazardous materials authorized on passenger aircraft (other than nonflammable gases or radioactive materials) in any inaccessible cargo compartments in any aircraft. This 50-pound limit means the combined weight of all regulated materials, not the weight of each article, or even of each class of articles. One may carry 150 pounds of nonflammable gas in addition to the 50 pounds of other material. No limitation applies, however, to the number of ORM packages aboard an aircraft. No ORM-D package may exceed 65 pounds gross weight according to 49 CFR 173.1200.

INSPECTION REQUIREMENTS

Every aircraft operator, before placing any hazardous material on his aircraft, must inspect the package or the outside container prepared in accordance with 49 CFR 173.25 in which that article is packed, and must determine:

1. That the container has no dents, holes, leakage or other indication that the product-holding effectiveness of the container has been impaired and, for radioactive materials, that the package seal has not been broken.
2. That the labeling and marking of the container complies with the regulations. Most packages regulated under 49 CFR Part 175 must be labeled and marked with name of contents, except ORM that are not subject to labeling at any time.
3. For radioactive materials to be carried aboard a passenger aircraft, accompanying the shipment is a clear and visible statement, signed or stamped by the shipper or his agent, that the shipment contains radioactive materials intended for use in, or incident to, research or medical diagnosis or treatment and meets the requirements of the regulations for carriage aboard passenger-carrying aircraft.
4. The requirements of this inspection regulation do not apply to ORM-D materials packed in a freight container and offered for transportation by one consignor, or to Dry Ice.

TRAINING REQUIREMENTS

Aviation is unique, in that the Federal Aviation Regulations prescribe training manual requirements for air carriers. Section 121.135 of title 14 CFR describes the contents of required personnel training manuals. It declares that each manual must include procedures and instructions relating to the handling of dangerous articles if these materials are to be carried, stored, or handled, including:

1. Procedures for determining the proper shipper certification required under 49 CFR 172, as well as the proper packaging, marking, labeling, shipping documents, compatibility of articles, and instructions on the loading, storage, and handling of those articles.

2. Notification procedures for reporting incidents of the spillage or other unintentional release of a hazardous material from its packaging. See Chapter 22 on the details of these requirements.

3. Instructions and procedures for the notification of the pilot in command when there are hazardous materials aboard his aircraft.

A similar manual requirement appears in 14 CFR Part 135 of the Federal Aviation Regulations.

NOTIFICATION OF THE PILOT IN COMMAND

Whenever a hazardous material is carried in an aircraft, the pilot in command must be given a copy of the shipping papers (or the information from the shipping papers) for each hazardous material on board, and must be told in writing the location where the hazardous material is stowed on the aircraft, and the results of the package inspection.

DAMAGE TO HAZARDOUS MATERIALS PACKAGES

Any hazardous materials package that appears to be damaged or leaking must be removed from the aircraft, and that package may not be carried in any aircraft until it is determined that it meets the requirements of 49 CFR Part 175 of the regulations.

REPORTING HAZARDOUS MATERIALS INCIDENTS

Every carrier, by every mode of transportation, must file a report with DOT whenever that carrier is aware of any incident in which a hazardous material has been released or spilled unintentionally from its package. See Chapter 22 on the detailed requirements, copies of the form, instructions on completing the form, and the location of the offices to which it should be sent.

CARGO LOCATION

Hazardous materials generally may not be carried in the passenger space of any aircraft.

A special table has been devised by the DOT to govern the stowage compatibility of cargo aboard an aircraft. There are specific compatibility requirements relating to the stowage of explosives, blasting agents, compressed gases, flammable liquids, flammable solids, oxidizers, organic peroxides, and corrosive materials.

A hazardous materials package marked "This Side Up" or "This End Up," or marked with arrows indicating the proper orientation of the package, must be loaded aboard an aircraft in accordance with those markings, and must be secured in a manner that will prevent any movement that would change the proper orientation of that package. (Note that under international requirements a very specific arrow label is prescribed that must be used to comply with those requirements.)

No operator of an aircraft may carry hazardous materials without guarding against the hazards of shifting cargo. For packages bearing Radioactive Yellow-II or Radioactive Yellow-III labels, this safeguarding action must prevent movement that would permit the package of radioactive materials to be closer to a space that is occupied by a person or an animal than is permitted by 49 CFR 175.701.

Except in the case of a small, single pilot aircraft being used where other means of transportation are not available or are impracticable, addressed in detail in Section 175.320, each operator of an aircraft carrying materials acceptable only on all-cargo aircraft must carry those hazardous materials in a location accessible to a crewmember during flight. When materials acceptable for cargo-only aircraft are carried on a small single pilot cargo-only aircraft being used where other means of transportation are not available or are impracticable, the materials may be carried in a location that is not accessible to the pilot, subject to certain specific conditions:

1. No one may be carried on that aircraft other than the pilot, an FAA inspector, the shipper or consignee or their representative (so designated in writing), or a person necessary for handling the hazardous materials.

2. The pilot of this aircraft must be provided with written instructions on the characteristics of the material and the proper means of handling it. Any change in the pilot requires that the new pilot be briefed under a hand-to-hand signature service provided by the operator of the aircraft.

TRANSPORTATION OF FLAMMABLE LIQUID FUELS IN SMALL, PASSENGER-CARRYING AIRCRAFT

A small aircraft or helicopter operated entirely within Alaska or in a remote area elsewhere in the United States may carry, in other than scheduled passenger operations, up to twenty gallons of flammable liquid fuels under carefully prescribed conditions set forth in 49 CFR 175.310. Persons to whom such a section would apply are encouraged to review the specific requirements in the regulations.

SPECIAL REQUIREMENTS FOR POISONS AND ETIOLOGIC AGENTS

No operator of an aircraft may carry material marked as or known to be Class A or B poison, or etiologic agent in the same cargo compartment of an aircraft with material that is marked as or known to be foodstuffs, feeds, or any other edible material intended for consumption by humans or animals. (Note that except for Gas Identification Sets, Class A poisons are never authorized for carriage aboard any aircraft, whether passenger or cargo.) When either the poisons and etiologic agents or the foodstuffs, feed or other edible materials are loaded in unit load devices which when stowed on the aircraft are not adjacent to each other, this compartment restriction does not apply. Alternatively, if the poisons or etiologic agents are loaded in one closed unit load device and the foodstuffs, feed or other edible materials are in another, the separate compartment rule does not apply.

The operator of an aircraft used to transport material marked as or known to be Class A or B poison must, upon removal of the poison, inspect the compartments in which it was carried for leakage, spillage, or other contamination. All contamination that may be discovered must be isolated or removed from the aircraft. Until such contamination is cleaned up, no foodstuffs, feeds, or any other edible material intended for human or animal consumption may be put in the compartment.

SPECIAL REQUIREMENTS FOR ASBESTOS

A general regulation requires that asbestos is to be loaded, handled, and unloaded and any asbestos contamination removed in a manner that will minimize occupational exposure to airborn asbestos particles that might be released incident to its transportation. Although not stated here, this obviously relates only to those types of asbestos that are regulated under the DOT hazardous materials regulations (see 49 CFR 173.1090).

SPECIAL REQUIREMENTS ON CARRIAGE
OF RADIOACTIVE MATERIALS BY AIR

In recent years there has been an intensity of concern with the transportation of radioactive materials by all modes, and it has manifested itself in aviation with a series of statutory restrictions and regulatory outgrowths of those statutes. The reader affected by this area of regulation is encouraged to check on current requirements at the time of shipment. See Section 108 of the Hazardous Materials Transportation Act, 49 U.S.C. 1807, in Appendix D and also discussed in Chapter 30.

No radioactive Yellow-II or Yellow-III labeled material may be carried on board a passenger aircraft unless (1) the Yellow-II labeled package has a transport index not exceeding 1.0; (2) the Yellow-III labeled package has a transport index not exceeding 3.0; (3) the package is carried on the floor of the cargo compartment or freight container; and, (4) the package conforms to specific requirements described in 49 CFR 175.701 (separation distances) and 175.703(c)(2).

There is a special provision in 49 CFR 175.702 relating to the carriage of packages containing radioactive materials when these are carried in cargo aircraft only. A concept in this section refers to a "Group of Packages" as a mechanism for controlling stowage. The term means packages that are separated from each other in an aircraft by a distance of 20 feet or less.

Yellow-II or Yellow-III labeled packages may not be carried on cargo aircraft unless the total transport index for all the packages does not exceed 50 and the prescribed separation distances aboard passenger carrying aircraft are met. As an alternative, this rule also permits use of the group of packages concept wherein the transport index for any group does not exceed 50 and the separation distance between the surfaces of the radioactive materials packages and the surfaces bounding the space occupied by persons or animals is at least 30 feet. (Note that in this case each group of packages is required to be separated from every other group in the aircraft by not less than 20 feet.)

In this last case the total transport index for all packages containing fissile radioactive materials cannot exceed 50. There are other special requirements for the acceptance and carriage of radioactive materials, including stowage in relation to undeveloped film and radioactive materials which are not Yellow-II or Yellow-III labeled. In addition, there are restrictions against the air transport of Type B packaging with an accessible surface temperature in excess of 120°F, continuously vented Type B packages which require external cooling by an ancillary cooling system, or any packages subject to operational controls during transport.

Radioactive materials which contain or are a liquid pyroforic are prohibited. Finally, packages with radiation levels at the package surface or that have a transport index in excess of limits specified in 49 CFR 173.441(a) cannot be transported by aircraft except under special arrangements approved by DOT.

INSPECTION OF AIRCRAFT FOR CONTAMINATION BY RADIOACTIVE MATERIALS

Aircraft used routinely for the carriage of radioactive materials must be periodically checked for contamination. There is no prescribed frequency for these checks except that it must relate to the likelihood of contamination and the extent to which radioactive materials are carried.

Part 175 of 49 CFR contains a detailed table of distances that must divide packages of radioactive materials bearing Radioactive Yellow-II or Yellow-III labels from any space that may be continuously occupied by people, animals, or marked undeveloped film. Any operator of an aircraft carrying these radioactive materials should be familiar with that table, which is based upon the transport index of the packages.

Chapter 22 addresses carrier reporting requirements for hazardous materials incidents. In addition to these requirements, the carrier must also notify any radioactive materials shipper following any incident in which there has been breakage, spillage, or suspected contamination of his shipment. Aircraft in which radioactive materials have spilled may not be placed in service or routinely occupied again until the radiation dose rate at any accessible surface is less than 0.5 millirem per hour, and there is no significant removable radioactive surface contamination as that term is defined in 49 CFR 173.443. In these instances, the package should be segregated as far as practicable from personnel. If radiological advice is needed, the Department of Energy should be contacted, as described in detail in Chapter 22. See that chapter for regions and telephone numbers. In case of obvious leakage or if it appears likely that the inside container may be damaged, care should be taken to avoid inhalation, ingestion, or contact with the radioactive materials. Any loose radioactive materials should be left in a segregated area pending proper disposal.

Fissile Class III radioactive materials may be carried only aboard certain all-cargo aircraft on a dedicated use basis. See the specific regulations in 49 CFR Part 175 if such materials are to be carried.

CHAPTER 22

Hazardous Materials Incident Reporting

The DOT regulations require all carriers, including private and contract carriers, to file reports of any accident or incident involving the spillage or unintentional release of any hazardous material from its packaging. Although the requirements are binding upon carriers, any person may submit such a report or supplement or rebut a report filed by any other person.

The incident reporting system, as it is commonly called, was developed in response to a recommendation made by the National Transportation Safety Board (NTSB; see Chapter 32) to the Secretary of Transportation in 1969. Criticizing the lack of a system at that time, the NTSB said:

> A unified data system, based on uniform definitions of terms, utilizing a common reporting form to be submitted by carriers, with a flow of reports and supplemental information designed to be channeled to a common data center, and with the processed data (and results of special studies) being made available to all Administrations, would be a logical and necessary prerequisite toward solving many of the problems now confronting all Administrations in the transport of hazardous materials. The increase in traffic, the increase in demand for materials classified as hazardous, and the increasing need for intermodal coordination make this essential not only as an economic necessity, but for the safety of all concerned.

DOT established an incident reporting system in 1971. The system covers two primary areas. The first is a requirement that all carriers make immediate reports to DOT by telephone when incidents of a specified severity occur. (The word "incident" is used in the regulations to cover all reportable occurrences that involve hazardous materials.) The second part of the system is a routine reporting requirement governing the submission of written reports in a uniform format in those instances when an immediate report is required and also in any case where there has been an unintentional release of hazardous materials from a package. The detailed written report must be filed within fifteen days of the date of discovery. If the information at that time is incomplete, the carrier should file within fifteen days and complete it later.

The reports on file with DOT are public information, as are the accumulated totals and other compilations prepared by the DOT staff from these reports. Any person may obtain a copy of any report, and no information is considered confidential. The tide of ambulance chasers that many persons feared would take unfair advantage of the information in DOT's files has never appeared.

A detailed written incident report is required for every unintentional spillage or release of a product. There is no quantity threshold that would preclude reporting small amounts of unintentional spills, although no report is needed for normal occurrences such as spills from hoses during normal disconnections. As DOT said in adopting the system:

> The (Hazardous Materials Regulations) Board does not feel it is in a position at this time to determine whether there are insignificant unintentional releases of hazardous materials that do not warrant the filing of a written report. While it may be true that under the amendment the Board will receive reports of unintentional releases of hazardous materials that may prove to be insignificant, the Board does not have any criteria at this time on which it could draw a line betweeen those releases that should be reported and those that should not. As experience is gained under this incident report program, the program will be subject to continuing review. If it is found that the present criteria is [sic] putting an undue burden on carriers and that the Board is receiving unusable or irrelevant incident reports, the Board will not hesitate to review the reporting requirements and to take future rulemaking action.

The matter of reporting every incident, no matter how small, was the subject of frequent critical comment and internal frustration and, in a subsequent rule making, DOT eliminated the obligation to report minor incidents involving consumer commodities, wet storage batteries, and paint and related materials. These exclusions do not apply if there is a major incident requiring an immediate report, or if the incident involves hazardous waste, or if it occurs aboard an aircraft.

The specific requirements are as follows:

IMMEDIATE TELEPHONIC HAZARDOUS MATERIALS INCIDENT REPORTS

> (a) At the earliest practicable moment, each carrier who transports hazardous materials shall give notice in accordance with paragraph (b), below, after each incident that occurs during the course of transportation (including loading, unloading, and temporary storage) in which as a direct result of hazardous materials:
> (1) A person is killed;
> (2) A person receives injuries requiring his admission to a hospital;
> (3) Estimated carrier or other property damage exceeds $50,000;
> (4) Fire, breakage, spillage, or suspected radioactive contamination occurs involving a

shipment of radioactive material or etiologic agents.

[In the case of radioactive materials, in addition to the immediate and detailed written reporting requirements, the carrier must also notify the shipper at the earliest practicable moment following any incident in which there has been breakage, spillage, or suspected radioactive contamination involving that radioactive materials shipment. Vehicles, building, areas, or equipment in which radioactive materials have been spilled may not be again placed in service or routinely occupied until the radiation dose rate at any accessible surface is less than 0.5 millirem per hour and there is no significant removable radioactive surface contamination, as that term is defined in the shipper regulations on radioactive materials contamination control. In these instances, the radioactive materials package or materials should be segregated as far as practicable from personnel contact. If radiological advice or assistance is needed, the U.S. Department of Energy should also be notified, at the offices listed below. In case of obvious leakage, or if it appears likely that the inside container may have been damaged, care should be taken to avoid inhalation, ingestion, or contact with the radioactive material. Any loose radioactive materials should be left in a segregated area and held pending disposal instructions from qualified persons. Details involving the handling of radioactive materials in the event of an accident can be found in Bureau of Explosives Pamphlets 1 and 2.]

(5) Fire, breakage, spillage, or suspected contamination occurs involving shipment of etiologic agents; or

(6) A situation exists of such a nature that, in the judgment of the carrier, it should be reported by Immediate Telephonic Report, even though it does not meet the criteria listed in this paragraph, for example, a continuing danger to life exists at the scene of the incident.

(b) Each notice required by paragraph (a), above, shall be given to DOT by *telephone* at (800) 424-8802. Notice involving etiologic agents may be given to the Director, Center for Disease Control, U.S. Public Health Service, Atlanta, Georgia, at area code (404) 633-5313, instead of giving notice to DOT. Each immediate telephone notification must include the following information:

(1) The name of the person calling.

(2) The name and address of the carrier on whose behalf the call is being made.

(3) A telephone number where the caller can be reached.

(4) he date, time, and location of the incident.

(5) The extent of injuries, if there are any.

(6) The classification, name, and quantity of hazardous materials involved, if this information is available at the time of the call.

(7) The type of incident and the nature of the involvement of hazardous materials in that incident, and whether a continuing danger to life exists at the scene of the incident.

(c) Each carrier making an immediate telephone report of an (tab)incident must also make a detailed written report as described below.

DETAILED HAZARDOUS MATERIALS INCIDENT REPORT

(a) Each carrier who transports hazardous materials shall report in writing, in duplicate, on DOT Form F 5800.1, illustrated in Figure 22.1 below, to DOT *within 15 days* of the date of discovery, each incident that occurs during the course of transportation (including loading, unloading, or temporary storage) in which, as a direct result of the hazardous materials, any of the circumstances warranting an immediate telephone report have occurred or there has been any unintentional release of hazardous materials from a package (including a tank).

(b) Each carrier making such a detailed written report shall send it to the Information Systems Manager, Research & Special Programs Administration, Department of Transportation, Washington, D.C. 20590.

AIR CARRIERS

With regard to an air carrier's immediate telephone report of a hazardous materials incident, required under the criteria discussed for such reports above, there are different persons specified to whom that air carrier's call should be made. Each operator of an aircraft that transports hazardous materials must report to the nearest ACDO, FSDO, GADO, or other FAA facility, except that a certificate holder under 14 CFR Parts 121, 127, or 135 may report instead to the FAA District Office holding that carrier's operating certificate and charged with overall inspection of its operations, by telephone at the earliest practicable moment after an incident as described above for immediate telephone notices. The caller should be prepared to give the information described above in paragraph (b) describing these immediate telephone reporting requirements.

RADIOLOGICAL ASSISTANCE

The map in Figure 22.1 illustrates the location, address, and telephone numbers for the Regional Coordinating Offices of the Department of Energy in the event the carrier needs radiological assistance or advice.

DEPARTMENT OF ENERGY
REGIONAL COORDINATING OFFICES
for
RADIOLOGICAL ASSISTANCE
AND
GEOGRAPHICAL AREAS
OF RESPONSIBILITY

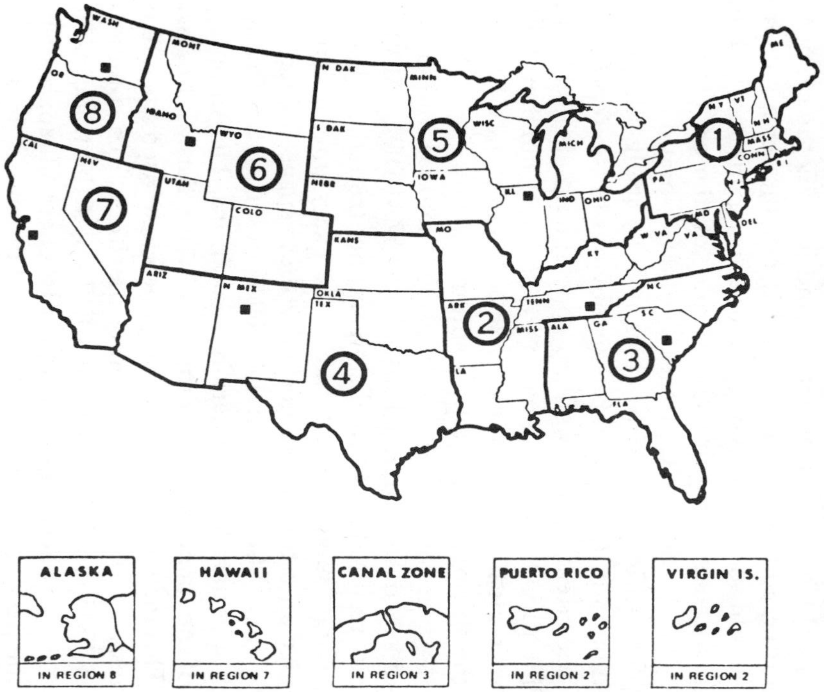

REGIONAL COORDINATING OFFICE	POST OFFICE ADDRESS	TELEPHONE for ASSISTANCE
① BROOKHAVEN AREA OFFICE	UPTON, L. I. NEW YORK 11973	(516) 282-2200
② OAK RIDGE OPERATIONS OFFICE	P.O. BOX E, OAK RIDGE, TENNESSEE 37830	(615) 576-1005 OR (615) 525-7885
③ SAVANNAH RIVER OPERATIONS OFFICE	P.O. BOX A, AIKEN, S.C. 29601	(803) 725-3333
④ ALBUQUERQUE OPERATIONS OFFICE	P.O. BOX 5400, ALBUQUERQUE, NEW MEXICO 87115	(505) 844-4667
⑤ CHICAGO OPERATIONS OFFICE	9800 S. CASS AVE. ARGONNE, ILLINOIS 60439	DUTY HRS. (312) 972-4800 OFF HRS. (312) 972-5731
⑥ IDAHO OPERATIONS OFFICE	550 SECOND ST. IDAHO FALLS, IDAHO 83401	(208) 526-1515
⑦ SAN FRANCISCO OPERATIONS OFFICE	1333 BROADWAY OAKLAND, CALIFORNIA 94612	(415) 273-4237
⑧ RICHLAND OPERATIONS OFFICE	P.O. BOX 550 RICHLAND, WASHINGTON 99362	(509) 373-3800

VIRGIN IS. IN REGION 2
PUERTO RICO IN REGION 2
CANAL ZONE IN REGION 3
HAWAII IN REGION 7
ALASKA IN REGION 8

Figure 22.1

DOT FORM 5800.1: WHAT IT LOOKS LIKE
AND HOW TO FILL IT OUT

Figure 22.2 illustrates what the form looks like for filing Detailed Hazardous Materials Incident Reports. Before going on, review the form and the types of information requested from the carrier. You will note that the form is divided into several sections, each given an alphabetical designation. Within each section are items describing the details for that section. The following instructions may prove helpful in illustrating how one should complete this form.

INSTRUCTIONS

Fill in all blanks. Use N/A when not applicable. If there are none, state "No marking on container," "No label applied," "No symbols," "No serial numbers," etc. as the case may be.

Section A: If items 1.1 through 1.5 do not apply, insert at 6 your operational area: Manufacturer, Warehouse, etc. For items A2 and A3, if the actual date and location are not known, give the date and location of discovery. Do not include terms such as "on trailer 376" or "between New York and Philadelphia."

Section B: Item B4 should indicate the complete company name. Do not use abbreviations. If the report is submitted by someone other than the carrier involved in the incident, please indicate your connection with the incident such as "J & J Chemicals-Consignee" and identify the carrier. Item B5 should be the main office address of the company, not the terminal preparing the report. Item B6 should specify the type of vehicle or facility in which the unintentional release took place: tank car, van trailer, trailer on flat car (TOFC), storage warehouse, etc.

Section C: Items C7 and C8 should include the complete company name. "Scientific Div.—AHS" does not, by itself, identify the shipper or consignee although it may be completely obvious to the reporter as "American Hotel Supply." The street address and zip code should also be included. Item C9 should clearly identify the shipping papers. A series of numbers without any identification is not very meaningful. An example of "Other" in C10 would be the broker or agent of the shipper on an import shipment.

Section D: For items D11 and D12 enter the number of persons injured or killed AS A RESULT OF THE HAZARDOUS MATERIALS INVOLVED. If a casualty resulted from a collision and not from the release of a hazardous material, then "none" should be entered. If the exact amounts for D13 and D14 are not known, give an estimate. Do not leave these spaces blank.

Section E: In item E15 enter the classification of the commodity as shown in the hazardous materials regulations. The shipping name in E16 MUST be one of the names shown in the commodity lists of the hazardous materials regulations. This may or may not be the same name used for rate or billing purposes. Nevertheless, the regulations are quite specific as to a commodity's proper hazardous material shipping name. In item E17 enter the trade name if any.

Section F: In item F18 check all spaces which may have contributed to the package failure. An "External Puncture" may have been caused by "Other Conditions" such as a traffic collision. Do not make any mark in item 19.

Section G: Columns 1, 2, and 3 may be used to convey a variety of information. You may report details of three different types of containers from which hazardous materials escaped, or three containers of the same type but of different capacities, or three containers of the same type and size but made by three different container manufacturers. In the example below, Columns 1 and 2 have been used

DEPARTMENT OF TRANSPORTATION

Form Approved OMB No. 04-5613

HAZARDOUS MATERIALS INCIDENT REPORT

NSTRUCTIONS: Submit this report in duplicate to the Secretary, Hazardous Materials Regulations Board, Department of Transportation, Washington, D.C. 20590, (ATTN: Op. Div.). If space provided for any item is inadequate, complete that item under Section H, ''Remarks'', keying to the entry number being completed. Copies of this form, in limited quantities, may be obtained from the Secretary, Hazardous Materials Regulations Board. Additional copies in this prescribed format may be reproduced and used, if on the same size and kind of paper.

A | INCIDENT

1. TYPE OF OPERATION

1 ☐ AIR 2 ☒ HIGHWAY 3 ☐ RAIL 4 ☐ WATER 5 ☐ FREIGHT FORWARDER 6 ☐ OTHER (Identify) ____

2. DATE AND TIME OF INCIDENT (Month - Day - Year)	3. LOCATION OF INCIDENT
March 7, 1972 11:30 a.m. / p.m.	Exit 3 on I-495 near Alexandria, Va.

B | REPORTING CARRIER, COMPANY OR INDIVIDUAL

4. FULL NAME	5. ADDRESS (Number, Street, City, State and Zip Code)
ABC Trucking Company, Inc.	204 Post Avenue Fayetteville, North Carolina 28301

6. TYPE OF VEHICLE OR FACILITY

Tractor - Van Trailer

C | SHIPMENT INFORMATION

7. NAME AND ADDRESS OF SHIPPER (Origin address)	8. NAME AND ADDRESS OF CONSIGNEE (Destination address)
XYZ Chemical Company 1101 South Peachtree Street Atlanta, Ga. 30303	J & J Chemicals 1506 Wayne Street Alexandria, Va. 22301
9. SHIPPING PAPER IDENTIFICATION NO. Shipper's B/L: FNC 12345 Carrier's Pro: 98765	10. SHIPPING PAPERS ISSUED BY ☒ CARRIER ☐ SHIPPER ☐ OTHER (Identify)

D | DEATHS, INJURIES, LOSS AND DAMAGE

DUE TO HAZARDOUS MATERIALS INVOLVED

11. NUMBER PERSONS INJURED	12. NUMBER PERSONS KILLED	13. ESTIMATED AMOUNT OF LOSS AND/OR PROPERTY DAMAGE INCLUDING COST OF DECONTAMINATION (Round off in dollars)
-1-	-0-	
14. ESTIMATED TOTAL QUANTITY OF HAZARDOUS MATERIALS RELEASED 45 gals.		$ 1,000.00

E | HAZARDOUS MATERIALS INVOLVED

15. CLASSIFICATION (Sec. 172.4)	16. SHIPPING NAME (Sec. 172.5)	17. TRADE NAME
Corrosive Liquid	Formic Acid	None

F | NATURE OF PACKAGING FAILURE

18. (Check all applicable boxes)

(1) DROPPED IN HANDLING	x	(2) EXTERNAL PUNCTURE	x	(3) DAMAGE BY OTHER FREIGHT
(4) WATER DAMAGE		(5) DAMAGE FROM OTHER LIQUID		(6) FREEZING
(7) EXTERNAL HEAT		(8) INTERNAL PRESSURE		(9) CORROSION OR RUST
(10) DEFECTIVE FITTINGS, VALVES, OR CLOSURES		(11) LOOSE FITTINGS, VALVES OR CLOSURES		(12) FAILURE OF INNER RECEPTACLES
(13) BOTTOM FAILURE		(14) BODY OR SIDE FAILURE		(15) WELD FAILURE
(16) CHIME FAILURE	x	(17) OTHER CONDITIONS (Identify) Traffic Collision	19. SPACE FOR DOT USE ONLY	

Form DOT F 5800.1 (10-70)

Figure 22.2

G PACKAGING INFORMATION - *If more than one size or type packaging is involved in loss of material show packaging information separately for each. If more space is needed, use Section H "Remarks" below keying to the item number.*				
ITEM		**#1**	**#2**	**#3**
20	TYPE OF PACKAGING INCLUDING INNER RECEPTACLES (*Steel drums, wooden box, cylinder, etc.*)	(Inner) Plastic Liner	(Outer) Steel Drum	
21	CAPACITY OR WEIGHT PER UNIT (*55 gallons, 65 lbs., etc.*)	55 gals.	55 gals.	
22	NUMBER OF PACKAGES FROM WHICH MATERIAL ESCAPED	1	1	
23	NUMBER OF PACKAGES OF SAME TYPE IN SHIPMENT	72	72	
24	DOT SPECIFICATION NUMBER(S) ON PACKAGES (*21P, 17E, 3AA, etc., or none*)	DOT 2SL	DOT 17H	
25	SHOW ALL OTHER DOT PACKAGING MARKINGS (*Part 178*)	55-12-71	STC 18/16-55-70	
26	NAME, SYMBOL, OR REGISTRATION NUMBER OF PACKAGING MANUFACTURER	AAA	FUBAR	
27	SHOW SERIAL NUMBER OF CYLINDERS, CARGO TANKS, TANK CARS, PORTABLE TANKS	N/A	N/A	
28	TYPE DOT LABEL(S) APPLIED	N/A	Corrosive Liquid	
29	IF RECONDITIONED OR REQUALIFIED, SHOW — A REGISTRATION NO. OR SYMBOL	N/A	DOT R1000	
	B DATE OF LAST TEST OF INSPECTION	N/A	Tested 2/72	
30	IF SHIPMENT IS UNDER DOT OR USCG SPECIAL PERMIT, ENTER PERMIT NO.	None	None	

H   REMARKS - Describe essential facts of incident including but not limited to defects, damage, probable cause, stowage, action taken at the time discovered, and action taken to prevent future incidents. Include any recommendations to improve packaging, handling, or transportation of hazardous materials. Photographs and diagrams should be submitted when necessary for clarification.

Our vehicle was involved in a minor traffic accident which caused the load to shift and puncture one of the drums. The leaking drum was removed by the consignee to their disposal area and buried. The vehicle was taken to our Alexandria terminal and cleaned (washed down and steamed). A Highway Patrolman on the scene had some of the spilled liquid splash on his hand. He was taken to a local hospital where he was treated and released.

31. NAME OF PERSON PREPARING REPORT (*Type or print*) Ira Jeopard	32. SIGNATURE
33. TELEPHONE NO. (*Include Area Code*) (202) 143-0510	34. DATE REPORT PREPARED March 15, 1972

Reverse of Form DOT F 5800.1 (10-70)

to separate the details of inner and outer containers. If columns 1, 2, and 3 are not adequate, a separate sheet may be attached to the report, or you may utilize the space in the "Remarks."

Additional examples for G20 are "Carboys" and "Fibreboard Box," and for G21, the capacity of a tank trailer or tank car. G22 and G23— In the example, the report clearly indicates that hazardous materials escaped from 1 drum and 1 liner out of 72 lined drums in the shipment. When the inner and outer containers are of a different capacity or nomenclature, the report should clearly state this fact. For example: 2 glass bottles out of 4 glass bottles in a carton were broken. If there were 10 such cartons in the shipment, then the report should state that hazardous materials escaped from 2 bottles out of 40 bottles in the shipment and from 1 carton out of 10 cartons. There should be no doubt that the 40 bottles were the inner containers of 10 outer containers in one shipment.

In G24 show all of the markings related to the container. "12B" is not the complete marking for a fibreboard box. It should be "12B40," or "12B60," etc. If the container bears no DOT specification marking, enter "NONE" in the space. DO NOT leave G24 blank.

G25—The hazardous materials regulations also require additional markings in some cases, such as: "HIGH EXPLOSIVES—DANGEROUS" or "HANDLE CAREFULLY."

In G26 enter the name of the container manufacturer. Keep in mind that some manufacturers use initials, abbreviations, symbols and combinations of letters and symbols.

For G27 the serial number of a cylinder should appear just below the cylinder neck ring. A tank car serial number might be similar to "GUTX98765."

Enter in G28 "Flammable Liquid," "Compressed Gas," etc. If no label appears on the package, state "NONE."

G29A—Include symbols and registration numbers, e.g., R 100, M 100, etc.

G29B—Show periodic test dates for containers which require same (e.g., cylinders, tank vehicles, reconditioned drums).

G30—Include DOT Exemption numbers (e.g., DOT SP 9999, E 2414, SP99–72).

Section H: In addition to the information requested following "remarks" on the form, this section should be used to include any information which the reporter feels is pertinent. For instance, if there was a spill of a flammable liquid and the driver was burned, and you did not indicate "fire" in F17 (Other Conditions), then Section H should clearly explain that there was a fire involving the flammable cargo, the origin of the fire, etc. In instances of contamination of a vehicle or freight, the method of decontamination and disposition of the contaminated freight should be explained.

COMPILED DATA

Although little has been done with the body of data collected by DOT thus far, the original NTSB recommendations called for a flow of reports "through a central 'clearinghouse' where such data would be collected and evaluated to determine whether greater emphasis should be directed to shipper and carrier compliance with existing requirements, or to the need for change in containers, in hazardous classifications, or in handling requirements." The Secretary of DOT responded to this recommendation with agreement "as to the importance of accident and incident data in evaluating the effectiveness of existing regulations and in developing new hazardous materials regulations."

At this time, the detailed information for the reports is lodged in computer banks, and it is possible to retrieve compiled information according to each of the major subject categories on form DOT F 5800.1. In other words, incident data can be retrieved from the full system by carrier name, shipper name, container specification, incident location, material classification, product shipping name, etc. This information is available to the public in accordance with 49 CFR Part 7 of the DOT regulations that prescribes the manner in which to request it and the costs of providing it. Since the reporting carrier is not obligated to give notice to a shipper when he files a report on the unintentional release of that shipper's product, it is possible that several reports indicating defects in packaging, etc., may be on file in that shipper's name without his awareness. The existence of such materials has been mentioned in connection with compliance inspections of shippers' facilities and, therefore, it behooves the prudent shipper to comb these files regularly for data relating to his products.

CHAPTER 23

DOT Compliance Inspections and Enforcement

RECOMMENDED IMMEDIATE ACTIONS

A procedure should be established for dealing with visits to a company site by any regulatory inspector. At a minimum, when an inspector arrives do the following immediately:

1. Identify the inspector. Ask to see credentials, and write down the relevant information including the inspector's name, agency affiliation, address, telephone number, and statutory authority under which the inspection is being conducted.

2. Designate the appropriate company official to interact with the inspector.

3. Determine the scope of the inspection. Ask the inspector what areas of company activity are of interest and what has triggered this inspection.

4. Advise counsel of the presence of the inspector.

5. Take notes on what is seen, what is said, who is spoken to, and any samples or copies taken.

6. When in doubt on any question, do not answer. Ask the inspector to put it in writing, addressed to counsel for the company.

7. Prepare a memorandum of the visit as soon as the inspector leaves.

See below for more detail on these basic steps, some background information on the agency, the scope of their authority, what to expect, and what to do about it.

INTRODUCTION

This chapter describes the process of hazardous materials transportation regulatory enforcement from the perspective of the industry subject to that enforcement. It describes what is being enforced, who is authorized to do it, how the process works, and what the company should do to protect its rights throughout the process.

In addition to the federal government, most States and many city governments also have adopted all or part of the DOT hazardous materials regulations. While the requirements may be virtually the same, usually the enforcement process is not. In most States and cities, this type of regulatory enforcement is adapted to fit a pre-existing enforcement scheme, often one comparable to traffic offenses. The

243

description of the federal procedure in this chapter will give some guidance on dealing with State and local enforcement of the same regulations, but if the prosecution is being conducted by nonfederal authorities, it is essential that the company also obtain a copy of the State or local laws to determine their rights during the process.

WHO IS SUBJECT TO ENFORCEMENT?

The DOT regulations apply to hazardous materials shippers, carriers, and container makers. The term "shipper" incorporates anyone who does anything to prepare a hazardous material for transportation (including classification, description, packaging, marking, labeling, and documentation), or who causes it to be transported by tendering it to a carrier or otherwise getting the material into transportation.

The term "carrier" means anyone who moves the material, by any mode of transportation, whether as a private, contract, or common carrier. Although the regulations generally only apply to carriers engaged in interstate or foreign commerce (and their shipper customers), this distinction means little to carriers or shippers of any substantial size.

The term "container maker" includes anyone engaged in the fabrication of any hazardous materials container or its subparts, or in its reconditioning, testing, retesting, or marking, when that container is sold or represented as being appropriate for hazardous materials shipment.

Many companies will fall within the DOT hazardous materials regulations because they ship or carry hazardous wastes or hazardous substances, as those terms are defined by the U.S. Environmental Protection Agency (EPA). See Chapter 14. EPA generally has agreed with DOT, through a memorandum of understanding, that DOT will be the primary agency involved in the enforcement of rules applicable to the transportation of these materials.

Generally speaking, the regulations are divided into parts which are assigned to particular types of operations—for example, virtually all requirements in 49 CFR Part 173 are shipper obligations, Part 174 rail carriers, and Part 178 packaging manufacturers. It is important to be sure that the respondent in the particular enforcement action is the one who is obligated to satisfy the rule the agency said was violated.

WHO ENFORCES THE RULES?

The hazardous materials transportation regulations may be enforced by DOT personnel, or by State or local government authorities. DOT is made up of five relevant modal subgroups, each of which is vested with some regulatory enforcement authority:

- The Federal Railroad Administration deals with rail carriers, tank car manufacturers, and shippers primarily of tank car quantities of hazardous material.
- The United States Coast Guard deals with water carriers and shippers of hazardous materials by water, both in bulk and in packaged form.
- The Federal Aviation Administration deals with operators of aircraft, including corporate aircraft, and with anyone who has anything to do

with offering a hazardous material for air transport or carrying such material onto an airplane, including passengers doing so in their baggage.

- The Federal Highway Administration enforces the regulations applicable to motor carriers of all types and sizes, the makers of cargo tanks, and companies shipping hazardous materials by highway. The Federal Highway Administration also adopts and enforces the Federal Motor Carrier Safety Regulations, 49 CFR Parts 387–397.
- The Research & Special Programs Administration enforces regulations applicable to the makers, reconditioners, and retesters of packaging. They also may have some residual authority to enforce rules against shippers who have used multiple modes of transportation.

Each of these enforcement bodies, although implementing the same words of the same statute, does so in somewhat different ways and follows different procedures, and so it is essential to discuss them separately.

FEDERAL RAILROAD ADMINISTRATION

This group within DOT, also known as FRA, uses a field staff to enforce a wide range of railroad safety requirements adopted over many years under a variety of statutes, including the Railroad Safety Act of 1970. Hazardous materials enforcement is only one part of their role. Their enforcement actions against rail carriers, therefore, may have multiple statutory bases. A typical hazardous materials infraction by a rail carrier may relate to improper car placement in a train, such as when hazardous materials cars are placed too close to the engine, caboose, or certain other hazardous materials cars.

Historically the FRA has focused its shipper enforcement of hazardous materials requirements upon shippers of tank cars. Typical violations relate to using cars that are out of retest date, or that have improper or inadequately secured closures.

Inspectors are based regionally, and spend their inspection time visiting railyards and shipper facilities where tank cars may be awaiting filling or filled cars may be in transit or awaiting rail carrier pick-up.

The field inspector who finds violations often will take the first step of advising the staff person of the company in writing of his findings. This advice is, in essence, a warning letter. It means at least two things—first, the warning is going into the company's file for potential enforcement action later and, second, it means the inspector will return to determine whether corrections have been made. If they have not, enforcement action usually is commenced. Cases are generally well prepared by FRA inspectors, and will include copies of earlier correspondence, photographs, and other documentation of the violations.

All FRA hazardous materials enforcement actions are processed through the FRA Chief Counsel's office in Washington, DC.

UNITED STATES COAST GUARD

The Coast Guard's hazardous materials enforcement responsibilities are exercised in conjunction with other laws and regulations governing safety at sea. Inspectors are assigned to the Captain of the Port in the area in question, and they focus on the cargo and on the water carrier's handling of that cargo.

Because the Coast Guard is a military organization, and cycles its personnel on a three-year basis, the violations noted tend to be those that are relatively obvious.

The inexperience of personnel may lead to citations for shipping practices that are not in fact violations.

It is unlikely that Coast Guard personnel will visit a shipper's facility if it is outside the port area. Shipper cases usually involve correspondence following the Coast Guard's examination of the shipper's freight in the port area or aboard a vessel.

Enforcement is conducted through the offices of the Captain of the Port, and Washington headquarters is seldom involved.

FEDERAL AVIATION ADMINISTRATION

FAA inspectors are part of the agency's cargo security force. They are located at principal airports throughout the country, and conduct their examination of air carriers' operations and shippers' cargo that may be present. They also will visit air freight forwarder facilities on the airport grounds. FAA visits to shipper facilities off the airport property are extremely rare.

Hazardous materials cases are developed in the regions, but are prosecuted through the FAA chief counsel's office in Washington.

FEDERAL HIGHWAY ADMINISTRATION

FHWA operates regionally, in enforcing hazardous materials regulations and motor carrier safety regulations. As with other modes, inspectors have broad carrier safety responsibilities as well as hazardous materials compliance. There generally is one hazardous materials specialist for each region, but each region covers multiple states. Hazardous materials specialists spend a great deal of time in an educational effort, assisting shippers and carriers on how to achieve compliance, and are less frequently involved in direct enforcement.

FHWA inspectors are concerned primarily with truck safety, so they will spend the most time at common carrier terminals, and the facilities of larger private carriers. Violations observed involving shipper's freight may be tracked back to the shipper, but usually not if the shipper is domiciled outside the inspector's region.

FHWA inspectors have no authority to stop a moving vehicle, so they often conduct inspections at stopping points such as weigh stations. Very frequently their inspections are performed in conjunction with State law enforcement personnel, such as the highway patrol.

With the growing number of State and local personnel engaged in motor carrier safety and hazardous materials regulatory enforcement, in the highway mode the inspector often will not be a federal employee. The inspector may be acting under the provisions of a cooperative agreement with FHWA, or under the authority of specific State or local law that adopts the DOT rules.

The FHWA enforcement of hazardous materials may be handled through the regional counsels' offices, or the FHWA chief counsel's office in Washington.

RESEARCH & SPECIAL PROGRAMS ADMINISTRATION

The RSPA's enforcement staff is devoted almost exclusively to packaging makers, reconditioners, and retest facilities. These include companies dealing with fibreboard boxes, drums, cylinders, and portable tanks.

RSPA inspectors are based in Washington, D.C., and generally will conduct periodic campaigns involving a particular type of packaging. In their travels in the country they often attempt to visit every container maker of a particular type in the locale being visited.

All enforcement actions by the RSPA are handled through the RSPA in Washington, and if contested will involve the chief counsel of the RSPA.

CIVIL VERSUS CRIMINAL PROSECUTION

Section 110 of the HMTA sets forth the enforcement mechanism for hazardous materials violations. The term "civil" is used in this context to distinguish this type of enforcement from the criminal provisions discussed below.

DOT's civil penalty or civil forfeiture program is administered entirely within the agency, without the involvement of the court system or the Justice Department unless it is necessary for the government to initiate a collection action. The penalties are applicable to those persons who "have knowingly committed an act which is a violation of a provision of this title or of a regulation issued under this title. . . ." These words have been carefully drafted to avoid saying that one must knowingly violate the law. Note that the adverb "knowingly" modifies the commission of an act, not the matter of violation. Thus, if a person is in possession of his senses, it is likely that he is aware of his direct acts. If those acts happen to violate the law, liability ensues. Arguments over knowledge of the law generally are fruitless. If, however, the person did not know they were dealing with a hazardous material (such as a carrier who has not been so advised by a shipper), this lack of knowledge may be a valid defense.

The second and third sentences in Section 110(a) look the same at first, but actually have a significant difference. The second sentence states that shippers and carriers are subject to civil penalty liability of up to $10,000 for each violation, "and if any such violation is a continuing one, each day of violation constitutes a separate offense." The third sentence is applicable only to the packaging industry, but does *not* contain a continuing violation provision. It would appear that Congress did not intend the potential economic destruction of a packaging manufacturer because of the repetitive nature of his business and production. Each day and each packaging is like the last, and Congress apparently felt that a continuing violation would be too onerous for such a repetitive operation.

Under a civil penalty procedure, the agency determines the amount of the penalty that *could* be assessed (up to $10,000 per violation), and in compromise proceedings with the company involved, effort is made to agree on the figure that *will* be assessed. In determining the amount of the penalty the agency must take into account the nature, circumstances, extent, and gravity of the violations committed and, with respect to the person found to have committed such violation, the degree of culpability, any history of prior offenses, ability to pay, effect on ability to continue to do business, and such other matters as justice may require.

It is unusual that the amount assessed originally by the agency would be the final amount paid—usually a compromise figure is reached that is accepted with some reluctance by both the government and the alleged violator of the regulations. How much of a reduction can be achieved varies with the modal administration involved and the particular facts of the case. It is extremely unlikely, however, that the agency will drop the penalty to a negligible amount once the proceeding has gotten underway, since to dramatically reduce the penalty would reflect poorly on the quality of the original preparation of the case. Frequently one can get specific counts dropped upon a factual showing. Even if a hearing is requested, a responding party can ask for an opportunity to discuss settlement. Upon completion of the respondent's recitation of everything he believes is in his favor, it often speeds the process if a realistic compromise penalty is suggested first by the respondent.

Before getting to this stage, it is essential to consider at the outset whether one should request a hearing or forever lose that right. (See Hearings, below.) Whether handling the matter informally, or engaging in settlement or compromise discussions with agency personnel, several subject areas should be covered:

1. Any factual discrepancies. Is the allegation accurate? If not, provide proof of the inaccuracy.

2. Any interpretive discrepancies. Does the inspector read the requirement in a manner different than the company? If so, the company should take the steps to verify the soundness of their position. This may involve discussions with trade associations and other companies in the business. Expert advice may be necessary. The company's efforts also may involve discussions with the drafters of the regulation within the agency (who are separate and distinct from those who enforce it). Legislative interpretation is very much a historical function, and a legal one, and a review of the originating documents for the rule in question can be very revealing.

3. Any mitigating factors. This can include actions by third parties, unauthorized actions by employees or agents, weather, or other unusual circumstances unlikely to be repeated.

4. Past record. If this is the first alleged offense by the company, this factor should be brought out. A good safety record and program should be stressed as well.

5. Financial hardship. The law compels DOT to consider the economic impact of the penalty, but as a factual matter it is difficult for DOT to assess such matters independently. It is in the interest of any alleged violator to bring financial hardship to the attention of the enforcing agency as early as possible, in an effort to reduce or eliminate the penalty.

6. Corrective action. Here is where great emphasis should be placed. Regardless of mitigating circumstances and the like, the most effective means to achieve a reasonable compromise is to assure the agency that the infraction will not be repeated. This involves a safety program which provides a level of confidence of adequate quality controls to preclude mistakes from getting out the door. The better this aspect of the presentation, and the more forcefully it is presented by high officials in the company authorized and committed to making it effective, the more successful it will be.

If, after consideration of any mitigating circumstances and the negotiations with the alleged violator, no mutually acceptable penalty figure is determined and the company refuses to pay the amount demanded by the agency, the agency may ask the Justice Department, through the U.S. attorneys, to bring a collection action in federal district court. Traditionally, the U.S. attorney also is capable of compromising the penalty, but the agency discourages any great flexibility at this stage since it tends to make alleged violators less inclined to settle at the DOT level, in anticipation of getting a better deal from the Justice Department.

The final determination of a penalty, either through litigation or compromise, may be deducted from sums owed by the United States to the person charged, such as income tax refunds.

HEARINGS

Under the civil penalty system, the alleged violator is given the opportunity for a hearing. Despite the common language of the statute, the various DOT enforcement agencies take different approaches in their procedures and this is particularly evident with regard to hearing rights. The Federal Railroad Administration, the Federal Highway Administration, the Federal Aviation Administration, and the Research and Special Programs Administration all will schedule a hearing before an administrative law judge hired from a federal pool of such judges. The Coast Guard, on the other hand, conducts a more informal hearing before a hearing officer who is a member of the Coast Guard.

In the FRA and FAA, one can ask for a hearing at virtually any stage of discussions with the agency. In FHWA, one is encouraged to ask for a hearing at the outset, but it is possible to undertake settlement discussions first and, if those do not resolve the matter, still pursue a hearing. In the Coast Guard, a hearing must be requested at the outset. Only in RSPA is it explicitly stated that starting informal settlement discussions with the RSPA *waives* any right to a hearing. Before the RSPA, therefore, it is essential to request a hearing whenever there is any doubt about the agency's or the company's perception or interpretation of the facts or the regulation.

A hearing is a right of the alleged violator. Once requested, that request may be withdrawn by the requesting party, without penalty.

The various agencies differ in level of formality in their hearings. Generally speaking, it is up the the administrative law judge who is hired for the proceeding to set forth the procedures that will be followed. It is essential to review the rules of procedure carefully, at the outset of the process and frequently thereafter, to avoid losing any rights that are granted in those procedures. Note particularly any requirements that claims must be stated in the pleadings, that documents must be served in any particular manner, that discovery may be conducted or not, and how motions are handled.

Generally hearings take place at the most convenient location for the parties, usually in the area where the alleged violator resides.

The decision of an administrative law judge usually is appealable within the agency. If there is a possibility of conflict of interest, disinterested agency counsel and officials may be requested to hear the appeal.

CRIMINAL PENALTIES

The HMTA also authorizes criminal sanctions of up to $25,000 or five years, or both. The criminal penalties are carefully circumscribed, however, by making them applicable only to violations *willfully* committed. This phrasing, indicating that there must be a willful violation as opposed to a knowing act that happens to be a violation, involves a more substantial burden on the part of the prosecution in showing intent to violate the law.

Criminal penalties, since they involve use of grand juries, the federal criminal courts, and the Justice Department, are more time consuming and difficult for the government to prosecute than civil penalties, which are imposed by DOT itself. In addition, the accused in a criminal procedure benefits from a presumption of innocence. The scales are tipped in favor of the accused under the Federal Rules of Criminal Procedure, which compel proof of guilt beyond a reasonable doubt. In one of the few criminal prosecutions sought by DOT, for the forgery of a DOT exemption paper, the U.S. attorney chose to pursue conviction for mail fraud for having mailed the forged document, rather than attempt a trial on the DOT regulations.

Under the civil forfeiture system, in contrast, the scales are unquestionably tipped against the interest of the alleged violator. Accordingly, the agency most often will choose to enforce regulations through the civil penalty procedure, although the agency has the discretion to impose both types of penalties on the same person for the same offense.

COMPLIANCE ORDERS, SUBPOENAS, ETC.

Under 49 U.S. Code 1808, the agencies within DOT are authorized to perform many administrative and quasi-judicial functions that relate to enforcement, including issuing subpoenas, conducting inspections, holding hearings, taking depositions, and requiring production of documents and property.

The agency also can issue "compliance orders," after notice and an opportunity for hearing. These orders are enforceable through the district courts upon request by the attorney general to whom DOT would refer the matter for enforcement. To date, this is a little used power under the HMTA, but it is available if circumstances would so warrant.

Particularly in cases involving the Federal Highway Administration, civil penalty cases may be resolved with the payment of a penalty combined with a "consent order" imposing requirements on the party that may be in addition to the regulations. For example, a carrier might be obliged to establish a specific safety program, with periodic progress reports to be filed with the agency.

SPECIFIC RELIEF

Section 111 of the HMTA, 49 U.S. Code 1810, authorizes the Justice Department, at the request of DOT, to bring suit in an appropriate federal district court for equitable relief to redress a violation by any person of a provision of the HMTA or an order or regulation issued under the HMTA. The district courts are authorized to grant such relief as is necessary or appropriate, including mandatory or prohibitive injunctive relief, interim equitable relief, and punitive damages.

It is unclear when such equitable relief would be appropriate, or the manner in which one might provoke DOT to initiate such a request. Although this equitable authority complements the assortment of powers given to DOT under this statute, this power will not be exercised with any frequency.

If DOT has reason to believe that an imminent hazard exists, the agency may go directly to the federal district courts, or may ask the Justice Department to do so, to petition for an order suspending or restricting the transportation of the hazardous materials responsible for that imminent hazard. The HMTA defines "imminent hazard" as a substantial likelihood that serious harm will occur prior to the completion of an administrative hearing or other formal proceeding initiated to abate the risk of that harm. "Serious harm" is defined in Section 103 of the HMTA, 49 U.S. Code 1802, as meaning "death, serious illness, or severe personal injury." It clearly does not include property damage or other economic injury.

ENFORCEMENT PROCEDURES

FRA's enforcement procedures are published in 49 CFR Part 209.

The Coast Guard's procedures are in 33 CFR Part 1.

The FAA's enforcement procedures are in 14 CFR Part 13.

The FHWA's procedures are in 49 CFR Part 386.

The RSPA's procedures are in 49 CFR Part 107.

Each of the administrations has its own enforcement procedures, and these should be examined in detail, for there are substantial differences between them.

WHAT TO DO ABOUT HAZARDOUS MATERIALS ENFORCEMENT

The best thing to do about any enforcement of any regulations is to expect it to happen sometime, and to prepare for it. The best preparation is to establish a written procedure that is made part of the operational manual for the site and is incorporated in the company's training program for all new employees.

The likelihood of a hazardous materials inspection is small. With the growing number of State and local officials getting involved with enforcement, however, that likelihood is growing. In addition, many federal enforcement programs are similar, and effective preparation for one often will be helpful for another.

There should be a procedural manual to guide employees on what to do. The company's procedural manual should be consistent, for all sites and all inspectors. A firm company-wide decision should be made, and followed, on at least the following points:

1. Who will see the inspector? This should be a supervisory level employee. If none is present, the inspector should be asked to return at another time, or to await the arrival of the supervisor. As long as this request is valid and is made politely, it usually will be accommodated. Many companies also have their supervisor contact legal counsel for the company before the inspection begins, or at least while the inspector is there, although the lawyer need not be present.

2. Will photographs be allowed? Especially if there are trade secrets that may be revealed through photographs, this policy should be considered very carefully.

3. Will the inspector be asked to get a warrant before being admitted? Under most regulatory enforcement programs, inspectors may be forced to obtain a search warrant before entering the premises. They will not always be able to do so, and in any case getting the warrant will consume time. Careful thought should be given to the balance between deterring the inspection, however, and inciting the inspector to be particularly thorough upon his return. In any case, the policy should be consistent company-wide, so the individual asking the inspector to get the warrant is not perceived as having personally obstructed the inspection.

4. Will inspectors be permitted to speak with nonsupervisory employees? This generally should be discouraged.

5. Will tape recordings of conversations with company personnel be permitted? This also should be discouraged.

Every inspector should be asked to show proper credentials, and the information from these credentials should be written down. It should include the inspector's name, agency, and office within that agency. One also should request the telephone number and address of the inspector, and a reference to the regulatory or statutory authority by which the inspection is being conducted. An inspector who refuses to show credentials should be refused admittance.

At the outset, the company supervisor should determine the scope and purpose of the inspector's visit, and what has caused it (e.g., employee complaint, customer complaint, incident report). An inspector wishing to see one aspect of the facility should be shown no more than that. Broad guided tours should not be volunteered.

Similarly, any direct questions should be answered, politely, but no additional information should be volunteered.

At no time should any employee sign any document or statement, although signing to acknowledge receipt of a copy of the inspector's report is not itself incriminating.

If the inspector asks any questions or asks for copies of any information about which the supervisor is uncertain, the supervisor should express this uncertainty, and should politely ask the inspector to make the request to counsel for the company, in writing. The inspector should be given counsel's name and address. The supervisor later should advise counsel of this action, so that counsel may anticipate the request.

The supervisor should take notes during the inspection, in order to prepare a subsequent memo about the visit. These notes and the follow-up memo should describe all parts of the facility seen by the inspector, all people who spoke to the inspector and a general description of the topic of conversation, all areas that seemed of particular interest to the inspector, any photographs taken, any samples taken, any documents copied, and any reference by the inspector to potential violations during the inspection or in a closing conference. The followup memo also should attach duplicates of all such photos, samples, or documents. If the inspector leaves any paperwork, this too should be attached.

The memo should be kept by the supervisor, and in the files at the facility. Copies should be sent immediately to other affected supervisory personnel and to legal counsel for the company.

WHAT TO DO AFTER THE INSPECTION

The initial inspection should stimulate several actions within the company. To do nothing is a serious mistake.

1. Confer with legal counsel to determine what the next step may be on the agency's part, and the timing of that step. Did the inspector say he was returning? When? Enforcement actions usually take months to commence, but a return visit could occur at any moment.

2. Did the inspector reveal his discovery of any violations? These should be verified and corrected immediately, and appropriate memos to the file made to document the correction. It also is helpful to write the inspector advising him of the correction.

3. Are there broader implications? Is it possible this infraction is present in other company facilities or operations? If so, broader corrective action should be undertaken. If the problem seems to be industry-wide, perhaps a trade association should be brought into the picture.

4. Counsel should rough out the defense to an enforcement action, even if one will not commence for many months. Preparation when the events are fresh is invaluable. This outline also will reveal areas where more information may be necessary. If it would be helpful to get any regulatory requirements interpreted, realize that this is a time-consuming process and counsel should act on it immediately. Interpretations often are essential if the inspector is nonfederal, and is interpreting the federal regulations. Any interpretation from any government agency should be sought through counsel, who should take pains to

determine the answer he is likely to get to the question before asking his question in writing.

5. If the company believes it was incorrectly advised of a potential violation, it is worthwhile to bring that matter to the attention of the inspector and/or his supervisor. It is always more difficult to stop an enforcement action once it gets underway, than to head it off before it gets going.

A compliance inspection involving any government agency should serve as a learning experience for the company. Counsel and management should critique the written company procedures, to determine whether they need revision, or whether additional training of personnel may be appropriate. No one likes surprises during an inspection, and all efforts should be made following an inspection that involves surprises to revise procedures to minimize that potential in the future.

CONCLUSION

The best preparation for any inspection is a good compliance program.

This truly is where the effort should begin. A company that is in compliance generally need not be concerned with preparing a defense for an enforcement action.

CHAPTER 24

Liability Associated with the Shipment and Transportation of Hazardous Materials

INTRODUCTION

This chapter describes various liabilities associated with the shipment and carriage of hazardous materials in domestic transportation in the U.S.

The term liability is meant to describe a judicially enforceable obligation to pay money to someone for injuries they may have suffered. Although the report discusses various statutory and regulatory obligations as those impact the duties and standards of care to which parties may be held in determining their liability for damages, it does not discuss the civil or criminal enforcement mechanisms which are part of virtually every statute or regulation. These can be determined by reference to the particular statutes cited in this report. See Chapter 23 on compliance inspections and enforcement.

The negligence theory of tort liability requires an injured plaintiff to prove fault on the part of the defendant. Absolute or strict liability, on the other hand, is imposed without the plaintiff having to prove fault or negligence and, therefore, is a legal theory much more favorable to plaintiffs and unfavorable to defendants.

Within the concept of strict liability are two pertinent theories, discussed further below. The first is liability for injuries resulting from having engaged in "ultrahazardous" or "abnormally dangerous" activity. The second is "enterprise liability" as a means of risk distribution, imposed in some States as a matter of policy to shift the loss of unavoidable accidents to the broadest range of enterprises that, in turn, can accommodate that liability in the cost of their goods or services.

HAZARDOUS MATERIALS

Hazardous materials described in the *Red Book* are those regulated by the U.S. Department of Transportation (DOT). A court, however, is not limited to existing statutory or regulatory definitions in assigning liability.

For this chapter, a hazardous material includes not only those products that can cause immediate personal injury, but also hazardous wastes and pollutants that are damaging to the environment but not necessarily to people. Materials designated by EPA to date include few long-term health hazards. A major expansion of the population of regulated materials is expected to be caused by heightened public and congressional perceptions of toxicity, as well as the chronic health effects of exposure to materials.

Although there is some question whether the number of chemicals being transported is increasing significantly, there is no doubt that the number of them deemed "hazardous materials" is increasing because of expanding regulatory definitions. The increased population of regulated materials means an increased population of regulated transportation equipment and carriers, and an increased likelihood of the handling of that equipment being considered an "ultrahazardous activity," as that term is explained more fully below.

COMMON CARRIAGE

The federal government's original hazardous materials transportation program was substantially the same as that established by and for the railroad industry. The purpose of the rules was to compel shippers to use sufficient packaging and to communicate hazards to the carriers in the form of markings, signs (placards), labels, and documents. Packaging requirements imposed on shippers include proper periodic inspection, maintenance, and retest of containers such as tank cars.

The shift of this program from the private sector to government had at least two general effects favored by the railroad carriers—(1) it made violation of the regulations a criminal offense and, (2) it provided a measure of national uniformity beyond the scope and influence of the Association of American Railroads. As other modes of transportation developed, the regulatory agencies created to govern those modes dealt with hazardous materials transportation by simple cross reference to the original railroad regulations then administered by the ICC. The impact of the railroad industry on all hazardous materials regulation, as expressed through the AAR's Bureau of Explosives and the AAR Tank Car Committee, is felt to this day.

The railroads who developed these regulations were common carriers. Common carriers are characterized by an obligation to accept, transport, and deliver cargo to its destination upon reasonable request for that service, and without unjust discrimination. There is a public interest in transporting many difficult materials, that may hold little appeal for a common carrier. These include hazardous materials.

The concept that evolved under English common law was that the common carrier received a public trust, and was protected from destructive competition, but in turn had the duty to provide service to all who asked, including hazardous materials shippers. The carrier became the insurer of the goods it carried, because of the perceived public benefit in protecting the shipper in the event the carrier lost, damaged, or stole the freight.

HAZARDOUS MATERIALS TRANSPORTATION ACT (HMTA)

For greater detail on the HMTA, see Chapter 30. The HMTA superseded the Explosives and Other Dangerous Articles Act and was implemented by Department of Transportation (DOT) regulations in 1977. Under the HMTA, DOT issues regulations that are binding upon "any person who transports, or causes to be transported or shipped, a hazardous material, or who manufactures, fabricates, marks, maintains, reconditions, repairs, or tests a package or container which is represented, marked, certified, or sold by such person for use in the transportation in commerce of certain hazardous materials." (49 U.S. Code 1804(a).)

The statute does not use the terms shipper, carrier, or container manufacturer. Statutory and regulatory obligations, rather, are functionally assigned to anyone who might be considered to have transported or caused the transportation of a hazardous material, or who performed any task with regard to packaging and who represented that packaging to be appropriate for use in transporting hazardous materials.

As noted above, DOT regulations refrain from defining regulated parties. See 49 CFR 171.2. Generally speaking, however, the DOT regulations are allocated into shipper, type of carrier, and type of packaging manufacturer "parts." Thus, as a general proposition Part 173 is for shippers, Part 174 for rail carriers, Part 175 for air carriers, Part 176 for water carriers, Part 177 for highway carriers, Part 178 for various packaging makers, and Part 179 for tank car makers.

The DOT rules, however, rarely assign responsibility to any particular person. Section 173.1(c), for example, says that "when a person other than the person preparing a hazardous material for shipment performed a function required by this part, that person shall perform the function in accordance with this part." Part 173 sets forth various shipper requirements, including obligations to maintain and retest cylinders, portable tanks, cargo tanks and tank cars.

Parts 178 and 179 of 49 CFR, setting forth DOT packaging and tank car specifications, is preceded by a similar section which says, "Any person who performs a function prescribed in this part, shall perform that function in accordance with this part."

It is assumed by the regulations that responsibility for individual shipper and equipment or packaging construction tasks will be assigned to specific parties through contractual or other commercial understandings. If the contractual documents between parties are silent with regard to performance of a specific functions, for example, tank car maintenance, DOT appears to have the leeway in an enforcement action to decide whether the car owner or the car user should have performed it. Enforcement liability can be shared, so that a DOT penalty levied against one party does not preclude a similar sanction being levied against another for the same violation.

Making a container or building bulk transportation equipment and then providing it to the shipper are two functions that can be divided, and again DOT appears to have the option of going against the container maker or equipment builder, or the container or equipment owner, or both, if that container or equipment does not in fact meet the specification it is represented to meet. (See 49 U.S. Code 1804(c).) The owner's responsibilities and thus liabilities will be larger if the owner participates in determining which container or transportation equipment will be provided for specific hazardous materials.

Part 174 responsibilities, including car inspection, are more clearly assigned to rail carriers. Section 174.7 says that, unless otherwise indicated in the regulations, "each carrier, including a connecting carrier, shall perform the duties specified and comply with each applicable requirement of this part, and shall instruct its employees in relation thereto."

Further on the issue of training and instruction, Section 173.1(b) declares: "It is the duty of each person who offers hazardous materials for transportation to instruct each of its officers, agents, and employees having any responsibility for preparing hazardous materials for shipment as to applicable regulations in this subchapter."

Insofar as any shipper, carrier or container maker performs any function regulated under the HMTA, employees should have been trained in the proper conduct of that work, and there should be records of that training.

NEGLIGENCE

By statute in most jurisdictions, as well as by common law, anyone who breaches a duty to act toward another person in a prudent manner is liable for the foreseeable results of that breach. This is true of motor carriers, railroads, shippers, container makers and anyone else. The emphasis in negligence litigation, therefore, is on whether the defendant was at fault in causing the injury.

In many jurisdictions, a defendant's violation of an applicable safety statute or regulation constitutes negligence *per se*, if the plaintiff is within the class of people intended to be protected by that statute or rule. In other words, proving the violation in those jurisdictions satisfies the requirement of proving negligence. The corollary, that compliance with the DOT hazardous materials regulations proves a lack of negligence, is not equally true. *Howell* v. *Lehigh Valley R. Co.*, 109 A. 309 (N.J. 1920); *Lehigh Valley R. Co.* v. *Allied Machinery Co. of America*, 271 F. 900 (2d Cir. 1921).

If a derailment and release of hazardous material from a cargo tank or tank car occurs because the highway or rail carrier was negligent in maintaining its equipment, track or roadbed, or in the actions of its employees, then the carrier will be held liable for resulting damages. The right of a railroad to run its engines and trains over its tracks is coupled with the duty so to operate them as not to injure the property of others near the track.

If a shipper or provider of equipment, in providing equipment or more often in managing that equipment, is *negligent* in the performance of its duties, that company will be held liable for the damages that result. The more actions anyone takes with the shipment or the equipment, clearly the more opportunities there are for negligence in the performance of those actions.

RAIL CARRIER INSPECTION OF CARS

A question of allocating liability frequently arises when a defective car is tendered to a railroad, and the railroad accepts it for service on its line. When a railroad company hauls a car over its lines, it thereby, in theory, adopts it as part of its own equipment, irrespective of ownership or the source from which it may have been received. (*Dominices* v. *Monongahela Connecting R. Co.*, 195 A. 747, 749 (Pa. 1938).)

Carrier inspection of cars, and thereby the accrual of liability for failure to inspect or to detect defects during that inspection, have been the subject of much litigation.

DOT assigns the inspection task to the railroad for hazardous materials tank cars in 49 CFR 174.9. Each loaded car must be inspected by the carrier before acceptance at the originating point, and when received in interchange, to see that it is not leaking and that the air and hand brakes, journal boxes, and trucks are in proper condition for service. The regulation goes on to note that manhole covers, and outlet valves, caps and plugs must be secured before transportation of empty cars containing residues of hazardous materials.

The Association of American Railroads (AAR) also indicates that the carriers will perform inspections of cars upon receipt and at interchange, to assure that the cars are in good operating condition.

The detail of this inspection is not prescribed by DOT, nor have the courts generally imposed detailed inspection obligations on rail carriers.

> In Pennsylvania, as in other jurisdictions, it is settled that a railroad company, before hauling freight cars over its lines, must subject them to an inspection sufficiently thorough to ascertain whether there is any fairly obvious defect in their construction or state of repair which constitutes a likely source of danger. [*Dominices, supra,* at 748.]

This case involved an acid tank car with a defective valve cover, and the rail carrier was held not liable for damages despite its failure to spot the defect during inspection:

> The inspection which the [railroad] company is required to make of a foreign car tendered to it by another company for transportation over its lines is not merely a formal one, but it should be made with reasonable diligence, so that its employees will not be exposed to perils, which reasonable care would have guarded against. The company receiving a foreign car can be held responsible by an employee who sustains an injury from its defects, only for failure to furnish a competent inspector, or for failure of the inspector to exercise due care in making the inspection. It is not, however, to be held responsible for hidden defects, which could not be discovered by such an inspection as the exigencies of traffic will permit. [*Railway Co.* v. *Fry,* 131 Ind. 319, 28 N.E. 989.]

Colorado & S. Ry. Co. v. *Rowe,* 238 S.W. 908, 912 (Tex. 1922); *Ambrose* v. *Western Maryland Ry. Co.,* 81 A.2d 895 (Pa. 1951); *Erie R. Co.* v. *Murphy,* 108 F.2d 817 (6th Cir.).

In *Vondergoltz* v. *Oil & Chemical Products,* 280 S.W.2d 774 (Tex. 1955), the court reached a similar conclusion, finding that injury caused by a corroded hand rail on an acid car "could not have been discovered from the ground." The court refrained from describing the inspection in detail, but said:

> The type of examination or inspection used by the Railroad was this: Some eleven hundred to fourteen hundred cars a day pass through the Railroad's yards. There are some sixteen inspectors to perform the inspection of them. There is an inspector on each side of the track when the cars are drawn by. These inspectors have about one minute to inspect each car for defects. Such inspection is made from the ground. If some defect is discovered which requires it, the inspector will get on the car for a more minute inspection.

A requirement for more effective inspection was implied, however, in *Rylander* v. *Chicago Short Line Railway Co.*, 153 N.E.2d 225 (Ill. 1958), where the railroad failed to detect a creosote-covered inoperative lug bolt on a tank car dome cover, despite the fact that it probably could not have been inspected from the ground.

It is unlikely that all defects in cars would be found by carrier inspectors. This would raise the question of whether the car owner or shipper was negligent in providing a defective car, or whether the defect was such that strict liability would be imposed upon owner or the car builder as the provider of the car, under product liability legal theories. Strict liability also could be imposed under the theories discussed below.

ULTRAHAZARDOUS (OR ABNORMALLY DANGEROUS) ACTIVITY AND STRICT LIABILITY

"One who carries on an abnormally dangerous activity is subject to liability for harm to the person, land, or chattels of another resulting from the activity, although he has exercised the utmost care to prevent the harm." (Restatement of Torts, Second, Section 519.)

This concept was codified by Prosser as follows:

> This ... policy frequently has found expression where the defendant's activity is unusual and abnormal in the community, and the danger which it threatens to others is unduly great—and particularly where the danger will be great even though the enterprise is conducted with every possible precaution. The basis of the liability is the defendant's intentional behavior in exposing those in his vicinity to such a risk. The conduct which is dealt with here occupies something of a middle ground. It is conduct which does not so far depart from social standards as to fall within the traditional boundaries of negligence—usually because the advantages which it offers to the defendant and to the community outweigh even the abnormal risk; but which is still so far socially unreasonable that the defendant is not allowed to carry it on without making good any actual harm which it does to his neighbors.

This doctrine had its origin in *Rylands* v. *Fletcher*, L.R.1 Ex. 265 (1866), affirmed, House of Lords, 3 H.L. 330 (1868), where a landowner who had impounded water in a reservoir was held liable for the damage caused to neighbors when it escaped.

Although the *Rylands* doctrine has not been applied to all materials, it certainly has been applied to many *hazardous* materials. Whether the handling of a given hazardous material constitutes ultrahazardous or abnormally dangerous activity is a question for the court. Decisions have found the handling of explosives, propane, and flammable liquids in populated areas to be ultrahazardous activities. The cases are mixed on whether it applies to gasoline in bulk.

It should be assumed for purposes of this report that the transportation of bulk quantities of flammable gases and poisons would be considered engaging in an ultrahazardous activity. These materials would include chlorine and anhydrous ammonia. Whether the concept would apply to transportation of other materials, including those that have been regulated only because of their environmental impact, is more difficult to determine. As noted below in the discussion on strict liability under Superfund, such a determination is largely moot.

ULTRAHAZARDOUS ACTIVITY AND COMMON CARRIAGE

Section 521 of the Restatement says: "The rules as to strict liability for abnormally dangerous activities do not apply if the activity is carried on in pursuance of a public duty imposed on the actor as a public officer or employee *or as a common carrier.*"

This statement is based upon several cases, one of the earliest of which was *Acties-selskabet Ingrid* v. *Central R. Co. of New Jersey*, 216 F. 72, 78 (2d Cir. 1914):

> At the time of the explosion the dynamite was in the course of transportation. A common carrier must transport freight of this character over its line, and the doctrine of *Rylands* v. *Fletcher*, if applicable at all, cannot be applied to cases of this nature. We think there can be no doubt, so far as a common carrier is concerned, that such danger as necessarily results to others from the performance of its duty, without negligence, must be borne by them as an unavoidable incident of the lawful performance of legitimate business. . . .
>
> It certainly would be an extraordinary doctrine for courts of justice to promulgate to say that a common carrier is under legal obligation to transport dynamite and is an insurer against any damage which may result in the course of transportation, even though it has been guilty of no negligence which occasioned the explosion which caused the injury.

This is an old concept, but it still is viable in most jurisdictions in the country. In a recent suit by a town against both the railroad and the shipper of carbolic acid that was spilled in a derailment and polluted the town's water supply, it was held that in Wisconsin a "common carrier is not subject to strict liability for the transportation of goods which it is required by law to undertake." *Town of East Troy* v. *Soo Line Railroad Co.*, 409 F. Supp. 326 (E.D.Wis. 1976); see also, *Pecan Shoppe, etc.* v. *Tri-State Motor Transit Co.*, 573 S.W.2d 431, 434 (Mo. 1978).

"A common carrier, in the transportation of high explosives by any proper type of conveyance in the absence of conduct which constitutes a nuisance, is not an insurer against or absolutely liable for damages caused by an explosion of such substances, but is liable only for such damages as are caused by its negligence." *Pope* v. *Edward M. Rude Carrier Corp.*, 75 S.E.2d 584, 597 (W.Va. 1953); *accord, Christ Church Parish* v. *Cadet Chemical Corp.*, 199 A.2d 707, 708 (Conn. 1964).

Although the Restatement position represents the majority view, there have been a limited number of significant departures from it. One of the most notable was *Siegler* v. *Kuhlman*, 502 P.2d 1181 (Wash. 1972), cert. denied 411 U.S. 783 (1972). In this case, gasoline leaked from a tank truck involved in a single vehicle accident. A passing motorist drove through the pool of gasoline, which caught fire and fatally injured the motorist. The Washington State Supreme Court held:

> In many respects, hauling gasoline as freight is no more unusual but more dangerous, than collecting water. When gasoline is carried as cargo—as distinguished from fuel for the carrier vehicle—it takes on uniquely hazardous characteristics as does water impounded in large quantities. Dangerous in itself, gasoline develops an even greater potential for harm when carried as freight—extraordinary dangers deriving from sheer quantity, bulk and weight, which enormously multiply its hazardous properties and the very hazards inhering from the size of the load, its bulk or quantity and its movement along the highways present another reason for application of the *Fletcher* v. *Rylands* rule not present in impounding large quantities of water—the likely destruction of cogent evidence from which negligence or the want of it may be proved or disproved. It is quite probable that the most important ingredients of proof will be lost in a gasoline explosion and fire.

ENTERPRISE LIABILITY

The majority in the *Siegler* case rested on the principle of *Rylands*, which broadly imposed liability for the harm resulting from the escape of any material which had been accumulated. The concurring opinion by Judge Rosellini, however, has had a greater impact. He based his finding of carrier liability, and rejection of the Restatement position, on the following view: "In my opinion, a good reason to apply these principles, which is not mentioned in the majority opinion, is that the commercial transporter can spread the loss among his customers—who benefit from this ultrahazardous use of the highways." *Siegler, supra*, at 1187.

One of the leading cases on the risk distribution doctrine, favorably citing the views of Judge Rosellini, is *Chavez* v. *Southern Pacific Transp. Co.*, 413 F. Supp. 1203, 1208 (E.D.Cal. 1976). The district court, determining what California law would be on the subject, concluded that there should be strict liability for ultrahazardous activity, but *without* an exception for common carriers. The case involved the explosion of an ammunition train in Roseville, California, and the carrier moved to dismiss the case based upon the exception from strict liability for common carriers set forth in the Restatement:

> Notwithstanding [the defendant's] protestations to the contrary, one public policy now recognized in California as justifying the imposition of strict liability for the miscarriage of an ultrahazardous activity is the social and economic desirability of distributing the losses, resulting from such activity, among the general public. . . .
>
> There is no logical reason for creating a "public duty" exception when the rationale for subjecting

the carrier to absolute liability is the carrier's ability to distribute the loss to the public. Whether the carrier is free to reject or bound to take the explosive cargo, the plaintiffs are equally defenseless. Bound or not, Southern Pacific is in a position to pass along the loss to the public. Bound or not, the social and economic benefits which are ordinarily derived from imposing strict liability are achieved. Those which benefit from the dangerous activity bear the inherent costs. The harsh impact of inevitable disasters is softened by spreading the cost among the greater population and over a larger time period. A more efficient allocation of resources results.

A good law review article on this topic is "The Enterprise Liability Theory of Torts," by Howard Klemme, 47 Colo. Law Review 153 (Winter '76). He offers the following statement of purpose for the shift in some jurisdictions toward enterprise liability:

> [The cost accounting and cost distribution criteria of the enterprise liability theory] are expected to achieve four interrelated purposes: (1) preventing as many tort-like losses as economically feasible; (2) distributing as fairly as possible among various segments of the consuming public the cost of such prevention or alternatively, the cost of insuring against the tort-like losses which will nonetheless occur; (3) encouraging individual members of the community to make decisions about the use of their personally available resources as rationally as possible; and (4) avoiding the creation of distortions in the use of the marketplace as a tool for otherwise "best" allocating the community's total limited resources. . . . [Using a] criterion of normal expectations. . . , the ultimate burden of an otherwise compensable tort loss is assignable to that particular enterprise which, among all the infinite number of "but for [this, the loss would not have occurred]" enterprises which may have caused the loss, failed to function as most people would normally have expected and thereby disturbed the normally-to-be-expected status quo. Because of the basic expectation that those carrying on an enterprise or activity should and will seek to avoid causing tort-like losses to themselves and others, the enterprise which is identified as the one which failed to function as would normally be expected will usually, if not always, be the one in which the most effective preventive action could probably have been taken to avoid the loss, if any such action could have been taken at all. For the enterprise to which the ultimate burden of a tort loss is thus assigned under this criterion, the imperative of the enterprise liability theory is: take appropriate preventive action to avoid similar losses in the future and in addition, or in the alternative, insure against those losses which nonetheless will occur.

Klemme's analysis is of some assistance in answering the inevitable question of which of several potential defendants should bear the ultimate cost. Among those whose actions may not have met "normal expectations" may be several business enterprises, each capable of spreading the cost to a larger number of people through the price asked for their goods or services.

> The application of this criterion of normal expectations also determines which group in society ought most fairly, ultimately, to bear tort-like losses. Ideally, that group will be the segment of the consuming public which purchases the goods or services being provided by the enterprise. Should market conditions be such that these losses as costs cannot be passed on entirely in the marketplace to purchasers, the economic burden is likely to fall or be shared with employees or investors in the enterprise. In any event, the economic burden is placed on those who derive or are seeking to derive some fairly direct economic gain or benefit from carrying on the enterprise which proved to be disruptive of the normally-to-be-expected status quo.
>
> The superior risk bearer in any case will be that person who is a member of that group of participants which is probably in the most effective position to bring about two results within the enterprise: (1) cause more preventive action to be taken to avoid similar losses in the future, if possible, or to seek to discover more effective means of avoiding such losses, and (2) cause the alternative cost of prevention or insurance to be passed on most efficiently, economically, to the purchasing consumers of the enterprise.
>
> The theory of enterprise liability simply acknowledges that rules of tort liability and nonliability do function in today's society as cost accounting, cost distribution rules. It is unrealistic any longer to hold the simplistic view that how the tort law resolves a dispute between the immediate parties will have no significant impact on other people. Thus, while the enterprise liability theory, like its predecessors, is concerned about the community's limited resources and seeks to conserve them, it recognizes that people are human, "accidents will happen," losses will occur, and when they do occur, someone will have to pay for them. The "most effective cost distributor" to "the economic beneficiaries of that enterprise which failed to meet normal expectations" are the criteria by which the law, through the enterprise liability theory, seeks to distribute among society's various members, in a fairly acceptable, rational way, those never-to-be-completely-avoided losses.

The trend seen in *Siegler, Chavez,* and the Klemme article was recently felt in an Iowa case involving damages to private property from propane tank cars involved in a derailment. The court in *National Steel Service Center, Inc.* v. *Gibbons* (trustee in bankruptcy for the Rock Island railroad), 319 N.W.2d 269, 272 (Iowa 1982), rejected the carrier's claim of the common carrier exception to strict liability for ultrahazardous activity. It pointed out that the risk of loss could be included in the carrier's tariff rates, citing *Akron, Canton & Youngstown Railroad Company* v. *Interstate Commerce Commission*, 611 F.2d 1162, 1170 (6th Cir. 1979), cert. denied 449 U.S. 830 (1980). The court endorsed enterprise liability as a means of risk distribution:

> Here we have two parties without fault. One of them, the carrier, engaged in an abnormally dangerous activity under compulsion of public duty. The other, who was injured, was wholly innocent. The carrier was part of the dangerous enterprise, and the victim was not. The carrier was in a better position to investigate and identify the cause of the accident. When an accident destroys the evidence of causation, it is fairer for the carrier to bear the cost of that fortuity. Apart from the risk distribution concept, the carrier is also in a better position than the ordinary victim to evaluate and guard against the risk financially.
>
> Furthermore, the carrier is in a superior position to develop safety technology to prevent such accidents, and assessment of safety costs is one means of inducing such developments. . . .

Also see Posner's *Economic Analysis of Law*, Section 6.11, 2nd Ed., at 141.

The court in the *Akron* case went on to say that its views on strict liability for all enterprises, not just carriers, went beyond the Restatement, and paralleled the holding in *Rylands* v. *Fletcher* in applying liability for activity in addition to that which is ultrahazardous.

Although courts have addressed the issue of common carrier's liabilities in most of these cases, that is primarily because the carriers were the ones sued. The courts' discussion of a carrier's ability to distribute losses to a greater population of beneficiaries over a longer time may be applied equally well, however, to other parties to the overall transportation enterprise, including hazardous materials equipment owners and operators.

An extremely pertinent case is one brought by an injured railroad employee against a railroad, an acrylonitrile shipper, and the lessor of a tank car from which acrylonitrile spilled. The complaint, among other things, alleges liability based upon the conduct of ultrahazardous activity.

The shipper moved to dismiss the strict liability count for failure to state a cause of action under Illinois law. That motion was denied in *Indiana Harbor Belt R. Co.* v. *American Cyanamid Co.*, 517 F. Supp. 314 (N.D. Ill. 1981). The court engaged in a discussion of strict liability for engaging in ultrahazardous activity, citing both *Siegler* and *Chavez, supra,* as "two opinions which this court believes provide persuasive rationales for finding liability in analogous circumstances."

Quoting *Chavez*, at 1209, the Illinois court said,

> The risk distribution justification for imposing strict liability is well suited to claims arising out of the conduct of ultrahazardous activity. The victims

of such activity are defenseless. Due to the very nature of the activity, the losses suffered as a result
of such activity are likely to be substantial—an
"overwhelming misfortune to the person injured".
. . . By indirectly imposing liability on those that
benefit from the dangerous activity, risk distribution benefits the social-economic body in two ways:
(1) the adverse impact of any particular misfortune
is lessened by spreading its cost over a longer time
period, and (2) social and economic resources can
be more efficiently allocated when the actual costs
of goods and services (including the losses they entail) are reflected in their price to the consumer.

Generally speaking, under the Restatement type of rationale, a defendant must
have engaged in an ultrahazardous "activity," which was the proximate cause of
injury through some foreseeable mechanism. Thus, under such a rationale, it would
appear to be difficult for the strict liability to extend to a passive owner of transportation equipment, since the activity of investing and owning is not itself ultrahazardous.

The more active the equipment owner's role, however, the more a plaintiff might
successfully characterize that activity as ultrahazardous. For example, in short-term
management of the equipment, including responsibility for periodic maintenance,
the company may allocate certain specification equipment to shippers of certain
hazardous materials. Especially if an incident were alleged to have involved a problem with the equipment, there would seem to be a basis for finding strict liability
because (1) the equipment owner may have been in the best position to have taken
preventive measures to preclude such incidents, and (2) through its lease the equipment owner may have been in the best position to allocate the anticipated costs of
that liability to the greatest number of enterprises benefitting from it.

The opportunity for imposition for liability, under negligence or strict liability
theories, grows substantially with an increase in control over the equipment. Where
the equipment owner not only manages the equipment, but actually performs repairs and maintenance functions in the owner's shops, there is the greatest exposure.

Despite the judicial and academic analyses of rationales that have been followed
in imposing strict liability, it appears that what may determine the "best" party to
bear the ultimate loss in some cases is the financial capability of such a party to do
so. This is most likely in jurisdictions intent upon distribution of the risk to enterprises which profit from the activity.

It should be anticipated that in a major incident, particularly if the common
carrier is found exempt from strict liability, and no other parties to the transaction
have sufficient assets to cover the loss, the equipment owners and/or operators may
be held liable. The greater the managerial control exercised in the allocation of
equipment, as well as its maintenance and repair, the more likely this liability will
attach.

STRICT LIABILITY UNDER SUPERFUND
Superfund, also known as the Comprehensive Environmental Response,
Compensation, and Liability Act (CERCLA), 42 U.S. Code 9601, 9607,
created a federal cause of action for recovery of response costs, and for damage to

natural resources. The new liability is strict, without fault having to be shown, and it is joint and several. It also operates retroactively. In other words, of all the potential defendants, only one needs to be sued for the entire cost of clean-up of a hazardous substance release.

Many products regulated as hazardous materials under the HMTA are hazardous substances under Superfund. In addition, after spillage of lading from transportation equipment, once those in charge of the clean-up decide to discard the spilled material, it becomes a hazardous waste and, in turn, a Superfund hazardous substance. A hazardous waste under RCRA is a hazardous substance under Superfund.

If residual lading is removed from transportation equipment in owner-operated repair shops, EPA deems the shop to be the "generator" of a hazardous waste under the Resource Conservation & Recovery Act (RCRA). See 40 CFR 261.4(c).

Superfund liability applies to the owner or operator of a facility from which there is a release, or a threatened release of a hazardous substance which causes the incurrence of response costs. Liability includes costs of removal and remedial action incurred by the government, necessary response costs incurred by any other person consistent with the National Contingency Plan, and damages for injury to natural resources. It is liability that cannot be transferred to any other person, although arrangements for indemnification can be made.

It is difficult to detach oneself from this liability once it attaches. Selling the property alone is insufficient. In addition, in some jurisdictions (notably New Jersey), no property that might be contaminated may be sold before the seller completes clean-up of the site. "Owner or operator" is defined in 42 U.S. Code 9601(20)(A) to *exclude* "a person who, without participating in the management of a vessel or facility, holds indicia of ownership primarily to protect his security interest in the vessel or facility." Thus, it is unlikely that the equipment owner's passive investor role by itself would result in Superfund liability. Where a more active managerial role is taken and in assigning hazardous materials equipment to particular shippers and commodities, this statutory exclusion would seem to be unavailable.

As mentioned above, Superfund liability is joint and several. In any incident involving the release of a hazardous substance which would include a large percentage of the ladings carried in bulk transportation equipment, Superfund liability will attach. The entire liability for the incident *could* be assigned to the equipment owner or operator, with the company left to seek contribution from other "owners and operators" involved in the incident.

The availability of insurance to cover this liability is very limited, and is growing more so.

Under Section 9607(c) there is a limitation on Superfund liability of $50,000,000, but this limitation would be removed if the spill were (a) the result of the defendant's willful misconduct or willful negligence, (b) the primary cause was the violation of an applicable safety, construction, or operating standard or regulation, or (c) if the defendant were deemed to have been insufficiently cooperative in the clean-up activity.

Defenses are limited to an act of God, act of war, or act of a third party other than an agent, employee, or someone under contract to the defendant "except where the sole contractual arrangement arises from a published tariff and acceptance for carriage by a common carrier by rail." This third-party defense is predicated upon the defendant's having exercised "due care" taking into account the hazards of the substance, and having taken "precautions against foreseeable acts or omissions of any such third party" and the consequences of those acts or omissions.

As noted earlier, Superfund liability is limited to response costs, and damages to natural resources. Several elected representatives have indicated they would legislate a private cause of action for personal injury or property damage. If Congress or a State legislature passes such "toxic tort" legislation, it may be expected to increase every hazardous materials shipper and carrier's potential liability substantially, by offering private plaintiffs the opportunity to sue the owner or operator of a transportation equipment under theories of strict, as well as joint and several liability.

CONCLUSIONS

A carrier who is negligent in causing injuries in the release of materials from transportation equipment will be liable for those injuries. A shipper, carrier, or packaging maker will be liable for its own negligent acts that result in injury.

Even without negligence, if the transportation equipment is defective, the one who provides that equipment might be held liable under strict product liability theories. Imposition of this liability will be more likely if the owner has performed maintenance and repair on the equipment.

The handling of bulk quantities of many hazardous materials would be held to be an ultrahazardous activity. Under the Restatement position, those who engage in an ultrahazardous activity may be held strictly liable. There is a widely recognized exception to this rule for common carriers.

Under risk distribution or enterprise liability theories which are gaining support in several jurisdictions, those who are in the best position (1) to avoid similar incidents, and (2) to distribute the costs widely among those who benefit from the activity, may be held liable. Especially if other parties to the transportation have insufficient assets, a solvent shipper or equipment owner could be exposed to liability under these theories. Here, too, the greater the management role with regard to the equipment, the greater the liability potential.

Superfund imposes liability upon owners and operators of facilities from which hazardous substances are released. This term would encompass many products carried in transportation equipment. If the ownership is coupled with management of the equipment, Superfund's strict as well as joint and several liability would apply. This liability also is retroactive, encompassing sites at which spills may have occurred prior to the passage of Superfund. While currently limited to $50 million of clean-up costs and damage to natural resources, legislation could expand this liability to encompass personal injuries and damage to private property.

CHAPTER 25

International Regulatory Bodies

Apart from the international organizations within the United Nations, two main bodies are concerned directly with the transportation of hazardous materials abroad, and they are highlighted here merely to assist the reader who may encounter them. The most commonly known of the two to the Western shipper is the Central Office for International Railway Transport (OCTI), established on October 14, 1890 by the first International Convention Concerning the Carriage of Goods by Rail (CIM). Members of OCTI are: in Europe—Austria, Belgium, Bulgaria, Czechoslovakia, Denmark, Finland, France, German Democratic Republic, Federal Republic of Germany, Ireland, Italy, Liechtenstein, Luxembourg, Netherlands, Norway, Poland, Portugal, Romania, Spain, Sweden, Switzerland, Turkey, the United Kingdom, and Yugoslavia; in Africa—Algeria, Morocco, and Tunisia; and in Asia—Iran, Iraq, and Syria. OCTI provides an international forum for rail transportation requirements between member countries.

OCTI's counterpart in Eastern Europe is the Organization for the Collaboration of Railways (OSJD). Its operations began much later, in September 1957. Members of OSJD are: in the Western Hemisphere—Cuba; in Europe—Albania, Bulgaria, Czechoslovakia, German Democratic Republic, Hungary, Poland, Romania, and the Union of Soviet Socialist Republics; and in Asia—People's Republic of China, Democratic People's Republic of Korea, Mongolia, and the Democratic Republic of Vietnam. The aims of OSJD are to develop international traffic and technical operations in the rail area. Since 1959, highway considerations have also come under the aegis of OSJD.

Any person involved directly in the regulations of these bodies, or any of the UN bodies discussed below, is directed to those specific regulations for proper guidance on compliance. This chapter will be too abbreviated to serve as a substitute for the regulations themselves. In addition, such organizations as the ADN (inland waterways generally) and the ADNR (Rhine River transport) are not discussed here despite their links with the United Nations, because they are considered too narrow and peripheral to the interests of the typical *Red Book* user. The principal coverage of this chapter is on the organizations and their interactions, as indicated by the following accumulation of acronyms: ECOSOC, ECE, the Committee of Experts, IMO (formerly IMCO), IAEA, RID, ADR, ICAO, and IATA. This grouping is not to imply that these organizations are of equal stature, but merely that they are tied together via UN structure.

ECOSOC

The main policy body in the United Nations concerned with hazardous
materials transportation is the Economic and Social Council (ECOSOC),
which reports to the UN General Assembly in New York. The General Assembly is
the main body of the United Nations and includes representatives of all the member nations. The General Assembly, for the sake of practical functioning, is broken
into several main subgroups of which ECOSOC is one. ECOSOC has over fifty members.

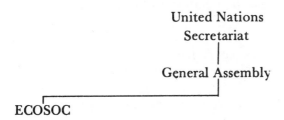

ECOSOC works through commissions, committees, and various other subsidiary
bodies. In hazardous materials the groups of primary concern are the Economic
Commission for Europe (ECE) and two committees called the Committee of Experts
on the Transport of Dangerous Goods and the Committee of Experts on Explosives.
These committees are made up of representatives of twelve nations: Canada, Federal Republic of Germany, France, Holland, Italy, Japan, Norway, United Kingdom,
United States, Poland, Sweden, the Union of Soviet Socialist Republics, the Netherlands and Sweden. Belgium attends regularly but only as an observer. Australia,
Thailand, Iran and Iraq have been invited to join but have made no commitment.
Chile and China both attended individual sessions to evaluate belonging and, to
date, have declined to participate. Finland, Switzerland and Spain also have been
discussed as potential participants but no action has been taken on this.

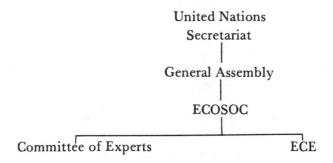

The Group of Rapporteurs on the Transport of Dangerous Goods is a working
group under the Committee of Experts on the Transport of Dangerous Goods for
the purpose of working out the details of positions to be formally considered and
adopted or rejected by the committee. For explosives, a separate body called the
Group of Rapporteurs on Explosives likewise operates under the same Committee.
Membership on the Groups of Rapporteurs is open to any members on the parent
Committee. Except for Poland, the other countries regularly are involved. Aside
from these committees, commission, and groups, as direct descendant bodies deriv-

ing their existence from ECOSOC, there are other so-called consultative groups, which may be either governmental or nongovernmental in their makeup.

There are a number of intergovernmental consultative agencies that are separate, autonomous organizations related to the United Nations through special agreements. They work with the United Nations through ECOSOC. Well-known acronyms of some of these are ILO, UNESCO, WHO, World Bank, FAO, ICAO, and IMO. These groups are known as specialized agencies, and they report annually to ECOSOC. Another well-known body of this type is the International Atomic Energy Agency (IAEA), which reports annually directly to the UN General Assembly. In the hazardous materials field, one should be interested particularly in ILO, IAEA, WHO, and IMO.

The International Labor Organization, (ILO) was established on April 11, 1919 and became the United Nations' first specialized agency in 1946. It frequently becomes involved in the work of the Committee of Experts, when pursuing its concerns with hazards in the workplace.

The World Health Organization (WHO), established in 1948, has participated actively in the past few years in the work of the Committee of Experts, specifically on matters relating to pesticides and other toxic materials.

The International Atomic Energy Agency (IAEA) was established in 1957. In 1959, ECOSOC asked the UN Secretary General to inform IAEA of ECOSOC's desire to have that agency draft recommendations for the transportation of radioactive materials that would be consistent with the framework and principles of the Committee of Experts' recommendations (Transport of Dangerous Goods—1970). As a result of this request, the IAEA Regulations for the Safe Transport of Radioactive Materials-1985 Edition, were drafted and were adopted by IAEA. Those regulations as they pertain to shipping names, in turn have become part of the UN recommendations. ICAO (air) and IMO (water) have begun the process of revising their codes as well to incorporate the revised standard, but this work is not expected to be completed until 1989 or 1990.

In the maritime area, general guidance has long existed regarding the ocean carriage of hazardous materials. As another step in the development of international standards for transportation by sea, the 1960 International Conference on Safety of Life at Sea (SOLAS) asked the Inter-governmental Maritime Consultative Organization (now the International Maritime Organization or IMO) to undertake a study designed to establish a unified code of recommendations for sea transport of hazardous materials in packages including portable tanks and various vehicles (ferry vessels). The IMO Dangerous Goods Code was the result of this study, conducted by IMO's Sub-Committee on the Transport of Dangerous Goods, which is a subsidiary of IMO's main technical body, the Maritime Safety Committee.

The agencies discussed above are all governmental in nature. As mentioned initially, however, certain nongovernmental agencies have a role in the United Nations by special agreement. One such body with a distinct impact on the hazardous materials field is the International Air Transport Association (IATA). IATA is made up of air carriers who develop their Restricted Articles Regulations through a subgroup called the Restricted Articles Board. The IATA regulations are not true regulations that may be enforced against shippers but are more closely aligned with carrier-developed tariffs on the subject. Under the UN rules of operation, IATA may be consulted by ECOSOC on matters of hazardous materials transportation by air.

ICAO, the International Civil Aviation Organization, is a body of governmental representatives that has taken over much of the international air regulation exercised in part by IATA. This is a broad overview of the ICAO structure. It is not a complete recitation of the organization but, rather, identifies the bodies the shipper and carrier of hazardous materials is likely to encounter in publications and actions of this group.

There are 156 countries that are members of ICAO. One of its chief activities is standardization by the establishment of international standards which govern many aspects of modern aviation and ground support for air transportation.

The constitution of ICAO is the Convention of International Civil Aviation that was developed by a conference in Chicago in 1944. The organization is made up of an Assembly, a Council of limited membership with various subordinate bodies, and a Secretariat. The Assembly is composed of all member states and is the sovereign body which sets policy. It meets every three years. The Council is the governing body which is selected by the Assembly, consisting of 33 members selected for a term of three years.

Standards and recommended practices are adopted in the Council and are incorporated as Annexes to the Convention on International Civil Aviation. The Council is assisted by the Air Navigation Commission and two committees.

Basic principles governing the transportation of hazardous materials by air are found in Annex 18 to the Chicago Convention. The specific and detailed standards by which shippers and carriers are governed in daily operations are published in the ICAO "Technical Instructions for the Safe Transport of Dangerous Goods by Air." The first edition appeared in 1982 and, until 1986, it was published annually. In 1987, on a trial basis, a two-year interim publication period was adopted.

The detailed standards are developed by a Dangerous Goods Panel, assisted by its subsidiary body, the Hazardous Materials Working Group, which keeps the standards up to date and revised according to industrial developments taking place. The Panel recommends changes to the Air Navigation Commission which in turn presents them to the Council for regular approval and publication authorization. See Chapter 20 for additional information on these standards.

ECOSOC generally recognizes that these organizations should have the opportunity to express their views, and that they often possess particular expertise that is of great value in ECOSOC's work. Organizations that have been given consultative status may send observers to public meetings of ECOSOC and its subsidiary bodies, and they may submit written statements relelvant to the work of ECOSOC.

The above organizations are the broad groups with a worldwide perspective and participation. It is helpful to consider at this point those smaller European intergovernmental organizations, examining how they interact and where they fit in the overall UN structure.

In 1954, when ECOSOC's Committee of Experts met for the first time, there were three basic codes the group considered as relevant starting points. These were:

1. The International Regulations concerning the carriage of dangerous goods by rail (RID, related to OCTI).
2. The U.S. Interstate Commerce Commission's Explosives and Other Dangerous Articles Regulations, now administered by DOT and largely used at the time by IATA.
3. The British "Blue Book," which was quite widely recognized in maritime circles.

COUNTRY	Date of last reply	Adoption being considered	Partially adopted	Fully adopted	Effective from
ALGERIA	15.7.66	Yes	—	—	—
ARGENTINA	17.2.86	—	—	Yes	—
AUSTRALIA	6.8.85	—	—	Yes[1]	20.3.68
BAHAMAS	15.3.78	—	—	Yes	1976
BELGIUM	27.2.81	—	—	Yes	6.5.66
BRAZIL	24.5.83	—	—	Yes	4.10.72
BULGARIA	29.12.84	—	—	Yes	1.1.83
CANADA	6.12.84	—	—	Yes[2]	24.11.84
CHILE	20.11.78	—	—	Yes	24.10.78
CHINA	17.12.84	—	—	Yes	1.10.82
DENMARK	18.1.82	—	—	Yes	1.2.72
EGYPT	24.7.75	Yes	—	—	—
FINLAND	10.8.81	—	—	Yes	1.10.80
FRANCE	10.9.85	—	—	Yes	8.8.68
GERMAN DEMOCRATIC REPUBLIC	8.12.81	—	—	Yes	1.12.81[3]
GERMANY, FEDERAL REPUBLIC OF	5.5.72	—	—	Yes	7.4.72
GREECE	2.9.85	—	—	Yes	4.2.74
ICELAND	6.12.84	—	—	Yes	20.9.82
INDIA	12.6.80	—	—	Yes[4]	4.11.78
IRELAND	19.9.85	—	—	Yes	1968
ISRAEL	24.7.85	—	—	Yes	11.11.72
ITALY	27.2.81	—	—	Yes[5]	1.8.69
JAPAN	29.6.84	—	Yes[6]	—	1.9.84
LIBERIA	27.2.81	—	—	Yes	1.3.79
MEXICO	4.1.83	Yes	—	—	—
MOROCCO	15.2.85	Yes[17]	Yes	—	—
NETHERLANDS	7.2.86	—	—	Yes[7]	1.1.74
NEW ZEALAND	18.2.81	—	—	Yes[8]	9.2.79
NORWAY	27.2.81	—	—	Yes	1971
PAKISTAN	9.8.84	—	—	Yes[16]	1.9.73
PANAMA	1.4.86	—	—	Yes[18]	1.3.86
PAPUA NEW GUINEA	28.3.85	Yes	—	—	—
PERU	19.3.85	—	—	Yes	1.1.70
PHILIPPINES	9.7.81	Yes[9]	—	Yes[9]	5.1.81[9]
POLAND	11.21.81	—	—	Yes	9.1.74
PORTUGAL	14.4.86	—	—	Yes[18]	—
REP. OF KOREA	28.5.81	—	—	Yes	1.7.79
SAUDI ARABIA	22.11.76	—	—	Yes	9.11.75
SINGAPORE	6.1.83	—	—	Yes	16.6.81
SOUTH AFRICA	1.7.74	Yes	—	—	—
SPAIN	19.4.83	—	—	Yes[10]	25.5.80

COUNTRY	Date of last reply	Adoption being considered	Partially adopted	Fully adopted	Effective from
SWEDEN	9.10.85	—	—	Yes	1.8.81[11]
SWITZERLAND	11.12.81	Yes	—	Yes	1.2.73
USSR	10.9.85	—	—	Yes[12]	1.1.69
UNITED KINGDOM	17.9.85	—	—	Yes[13]	29.12.78
UNITED STATES	13.6.83	—	—	Yes[14]	10.11.80
Associate Member					
HONG KONG	6.12.85	—	—	Yes[15]	29.12.78

1. As of 1 January 1980, the national legislation for the sea transport of dangerous goods incorporates the 1977 edition of the IMDG Code, including all amendments in force as approved by the Maritime Safety Committee. The Code is amplified by Part II of the 1979 edition of the Australian Supplement to the IMDG Code, as amended from time to time, which is published by the Australian Government Publishing Service.

2. Canadian Dangerous Shipping Regulations including Amendments Nos. 19, 20 and 21 to the IMDG Code (SOR 1981–951, SOR 1982–1053, SOR 1983–482 and SOR 1984–936). Some modifications have been made with respect to domestic trade.

3. "Ordnung vom 17.8.1981 Uber den Seetransport und Hafenumschlag gefahrlicher Guter—OSHG, Anlage 1." The national edition of the IMDG Code corresponds to the 1977 IMO edition of the IMDG Code with the exception of section 13 on portable tanks and road tank vehicles, and Annex I on packaging recommendations.

    The application of the IMDG Code is obligatory on all ships flying the flag of the German Democratic Republic (GDR) under supervision of the Board of Navigation and Maritime Affairs of the GDR and carrying packaged dangerous goods as well as for ships flying foreign flags operating in the waterways of the GDR and handling dangerous goods in GDR ports.

4. Merchant Shipping (Carriage of Dangerous Goods) Rules 1978.

5. Regolamento per l'imbarco, il Transporto per Mare, lo Sbarco e il Trasbordo delle Merci Pericolose in Colli.

6. The 1982 Edition of the IMDG Code (incorporating all amendments up to and including No. 20–82) with the exception of Class 1 and Annex I to the IMDG Code. Shipments of dangerous goods may be permitted in the case of an international voyage, provided that packing/packaging and labeling are in accordance with the provisions of Amendment No. 20–82 from 1 September 1984.

7. The 1977 Edition of the IMDG Code incorporating all amendments up to and including No. 22–84 (Amendment No. 22–84 effective from 1 July 1986 with the exception of Section 22 of the Introduction). Revised Annex I (Packing Recommendations) to the IMDG Code has been adopted with the proviso that existing packagings should comply with the recommendations therein from 1 January 1990.

8. Shipping (Dangerous Goods) Rules 1980.

9. Fully adopted for the Port of Manila, effective 5.1.81. In some other Government ports the adoption of the IMDG Code is still being considered.

10. Although there is at present no legal instrument covering this matter, the IMDG Code is being applied in all Spanish ports and to all Spanish ships for the carriage of dangerous goods by sea.

11. The 1977 Edition of the IMDG Code and Amendments up to and including No. 21–84 effective from 1.2.1985 applicable to Swedish ships and foreign ships in Swedish territorial waters (Ordinages SJOFS 1984:24).

12. Except for Government Explosives. Including all amendments and supplements in force as approved by the Maritime Safety Committee.

13. Full adoption—United Kingdom 1978 "Blue Book", revised (Fourth) Edition 1984, with Amendment No. 1 (1985).

14. United States Regulations allow the use of the IMDG Code for hazardous materials being imported or exported from the United States by vessel, with the exception of Classes A and B Explosives and Radioactive Substances (see US Title 49, Code of Federal Regulations Parts 100–177).

15. Merchant Shipping (Safety) Ordinance and the Merchant Shipping (Dangerous Goods) Regulations.

16. The IMDG Code and all amendments to it as approved by the Maritime Safety Committee and recommended for implementation.

17. The Moroccan maritime authorities use the provisions of the IMDG Code as a reference in order to authorize or to prohibit the carriage of dangerous goods by sea. The Ministry of Fishing and Merchant Navy is preparing a Maritime code which will incorporate provisions of the IMDG Code.

Each of these codes was based on a different approach to materials classification and labeling. Within Europe, efforts were being made for standardization. In the European Agreement Concerning the International Carriage of Dangerous Goods by Road (ADR), concluded in 1957, the provisions of RID established for rail were generally followed. Thus, within Europe, a compatible RID/ADR system came into being through the efforts of a Working Party designated "WP 15." At the start of its work, therefore, the Committee of Experts was faced with established RID/ADR, U.S./IATA, and United Kingdom maritime systems.

The result of the UN work is the UN Recommendations for the Transport of Dangerous Goods (also called the "Orange Book") which in effect is a fourth system that is different from each of the three systems from which the work began. Added to this and supporting it is the IMO Internatonal Maritime Dangerous Goods (IMDG) Code, based entirely on the UN recommendations.

Numerous countries have adopted the IMO's IMDG Code, and the current list as of the time of publication of the *Red Book* appears on pp. 272–273.

As a result of the disparate regulations and recommendations that are extant:

1. A shipper by sea must follow UN recommendations under IMO, which are different from requirements on either side of the Atlantic, where the United States is under the DOT regulations and the European continent follows RID/ADR.
2. A shipper by air usually will be required to abide by ICAO/IATA, but may also meet U.S. requirements while the shipment is in the United States or aboard any U.S.-registered air carrier; in Europe the shipment will have to satisfy RID/ADR requirements for surface movement.

There are provisions under RID/ADR recognizing that if a shipment arrives in Europe under IMO or ICAO, it may move to the delivery destination under the international standard being followed without application of RID/ADR. Further distribution, if necessary, is subject to RID/ADR.

In addition to these shipper headaches, one must consider the applicability of local port restrictions, which exist to some degree in every port. The shipper to the Far East and Africa must also acquaint himself with the requirements of those destinations, which are not insubstantial.

The picture is not clear but the outlook is encouraging in the sense of growing international regulatory harmony. Often, however, it seems that this harmony is being achieved by reducing existing standards to the lowest common denominator, and concern has been expressed that the uniform standards, particularly in packaging, may be inadequate.

IATA

The "IATA Restricted Articles Regulations" are published annually in the Fall and are available in a single volume from:

Traffic Services
International Air Transport Association
2000 Peel Street
Montreal, Quebec
Canada H3A 2R4

One should consult the bibliography in the back of the IATA regulations for other publications that may be helpful, such as a slide program and training materials on the regulations. As a matter of agreement with other air carriers, any U.S. carrier who participates in IATA has agreed to demand preparation of all hazardous materials tendered to that carrier, whether for international shipment or not, in accordance with the IATA regulations. The IATA requirements, however, are based upon the ICAO Technical Instructions and are not the same as U.S. air regulations in 49 CFR. The most expeditious way to cope with the differences is to satisfy both sets. Compliance with ICAO is authorized in U.S. air transportation, including domestic flights, as well as the highway movement of materials that may move by air. It is important for the shipper to determine whether to use the system in 49 CFR or ICAO, and then to stick with the system that has been selected. One may not use portions of each.

IMO

The IMO "International Maritime Dangerous Goods Code" is published with looseleaf supplements about every 18 months. The code consists of five volumes. These are available from:

International Maritime Organization
4 Albert Embankment
London, SE1 7SR
United Kingdom

New York Nautical Instrument & Service Corporation
140 West Broadway
New York, NY 10013

or

American Labelmark Co., Inc.
5724 N. Pulaski Rd.
Chicago, Ill. 60646

The reader is offered the names and addresses of participants in IMO, and it is recommended that these be the first points of inquiry for questions to local port requirements and other restrictions on export of hazardous materials by water to their respective countries.

**List of Contact Names and Addresses of the Offices
of Designated National Competent Authorities***

ARGENTINA	Prefectura Naval Argentina (Argentine Coast Guard)
	Direccion de Policia de Segiridad de la Navegacion
	Departmento de Contaminacion y Mercancias Peligrasas
	Avda. Eduardo Madero 235, 4° piso, Oficina 4.15
	Buenos Aires (1106)
	Republica Argentina
	Tel. No. 34–1633
	Telex: 18581 PREFEC AR
AUSTRALIA	First Assistant Secretary
	Maritime Safety Division
	Department of Transport
	G.P.O. Box 594
	Canberra ACT 2601
	Australia

(cont.)

**List of Contact Names and Addresses of the Offices
of Designated National Competent Authorities*** (*continued*)

Tel. No. 61–62–687799
Telex: AA61680

BELGIUM

Head Office:
Administration de la Marine et de la Navigation Interieure
104 Rue d'Arlon
Bruxelles
Belgique
Tel. No. 02.233.12.11

Antwerp Office:
Inspection Maritime
Quai Tavernier, 37
Anvers
Belgique
Tel. No. 031.33.12.75
Telex: 35028B

BRAZIL

Directoria de Portos e Costas
Departmento do Material da Marinha Mercante (DPC-2)
Rua 1° de marco, 118, 16° andar
200100 Rio de Janeiro RJ
Brazil
Tel. No. (21) 253–7386
Telex: 021–215–3210
        021–215–2178

BULGARIA

Main Office:
State Shipping Inspectorate
Ministry of Transport
Levski Str. 9/11
1000 SOFIA
Tel. No. 88–55–29

Sections:
1. State Shipping Inspectorate
   Chervenoarmejski Blvd. 1
   VARNA
   Tel. No. 2–54–09
2. State Shipping Inspectorate
   Burgas—port
   BURGAS
   Tel. No. 4–31–40

CANADA

The Chairman
Board of Steamship Inspection
Canadian Coast Guard
Tower "A"
Ottawa KIA 0N7
Canada
Tel. No. (613) 992–0242
Telex: 0533128

CHILE

Direccion General del Territorio Maritimo y de Marina Mer-
cante
Errazuriz 537
Correo Naval
Valparaiso
Chile
Tel. No. 58091–6

Telex: DIRECTEMAR
034 30443 CTCV CL

CHINA
The Bureau of Harbour Superintendency of the People's Republic of China
10 Fu Xing Road
Yan Fang Dien
Beijing
China
Tel. No. 366184
Telex: 22462 COMCT CN

DENMARK
Government Ships Inspection Service
Snorresgade 19
DK-2300 Copenhagen S
Denmark
Tel. No. (01) 547131
Telex: 31141 stsk dk

FINLAND
Board of Navigation
P.O. Box 158
SF-00141 Helsinki 14
Finland
Tel. No. (90)–18081
Telex: 12–1471

FRANCE
Secretariat d'Etat charge de la Mer
Bureau du Controle des Navires
3 Place de Fontenoy
F 75700 Paris
France
Tel. No. 1–273.55.05
Telex: 250823 MIMER F

GERMANY, FEDERAL REPUBLIC
Ministry of Transport
Postfach 200100
D-5300 Bonn 2
Federal Republic of Germany
Tel. No. (0228) 3001/300 2492/300 2495
Telex: 885700 BMV

GREECE
Ministry of Mercantile Marine
Safety of Navigation Division
Gr. Lambraki Avenue
185 18 Piraeus
Greece
Telex: 021–2022, 2273 YEN GR

ICELAND
Directorate of Shipping
Hringbraut 121
P.O. Box 484
Reykjavik
Tel. No. (1)–25844
Telex: 2307 ISINFO

INDIA
The Directorate General of Shipping
Jahz Bhawan
Walchand Hirachand Marg
Bombay—400 001
India
Tel. No. 263651

(cont.)

**List of Contact Names and Addresses of the Offices
of Designated National Competent Authorities* (*continued*)**

	Telex: DEGESHIP 2813—BOMBAY
IRELAND	The Chief Surveyor Marine Survey Office 27 Eden Quay Dublin 2 Eire Tel. Nos. 744900, 722045, 743021 Telex: 33358 MSO EI
ISRAEL	Technical Services Department Shipping and Ports Administration 102 Haatzmaut Road Haifa Tel. No. 972–4–526210 Telex: 46632
ITALY	Ministero Della Marina Mercantile Viale Asia-eur 00144 Roma Italy Tel.No. 5908 Telex: 62153
JAMAICA	Harbour Master Harbour Master's Department P.O. Box 116 Myers Wharf Kingston 15 Jamaica Tel. No. 923–9774 922–22222
JAPAN	Inspection and Measurement Division Marine Technology and Safety Bureau Ministry of Transport 2–1–3 Kasumigaseki, Chiyoda-ku Tokyo Japan Tel. No. (03) 580–3111
LIBERIA	National Port Authority Monrovia Liberia Tel. No. 221 306 Telex: 4275
NETHERLANDS	Directorate-General Shipping and Maritime Affairs *Mailing address:* P.O. Box 5817 NL-2280 HV Rijswijk Netherlands *Office:* Bordewijkstraat 4 NL-2288 EB Rijswijk Netherlands Tel. No. (070) 949420 Telefax: (070) 996274 Telex: 31040 DGSM NL

NEW ZEALAND	The Ministry of Transport Marine Division Private Bag Wellington 1 New Zealand Tel. No. 721–253 Telex: NZ 31524 Telegram: DIRMARINE
NORWAY	Maritime Directorate P.O. Box 8123 Dept. N-Oslo 1 Norway Tel. No. (02) 350250 Telex: 76997 SDIR N
PAKISTAN	Mercantile Marine Department 70/4, Timber Hard N.M. Reclamation Keamari, Post Box No. 4534 Karachi Pakistan Tel. Nos. 270117, 270118, 270119, 270289 Telex: 2733 NSC KAR, 2833, 2683, 2765
PANAMA	Direccion General Consular y Naves Apartado Postal 5245 Panama 5 Rep. de Panama
PAPUA NEW GUINEA	First Assistant Secretary Department of Transport Division of Marine P.O. Box 457 Konedobu Papua New Guinea (PNG) Tel. No. 211866 Telex: 22203
PERU	1. Direccion General de Capitanias y Guardacostas Plaza Grau s/n Callao 1 Peru Tel. Nos. 29–0693, 24–4692, 29–7278 Telex: 26071 2. Empresa Nacional de Puertos Terminal Maritimos del Callao s/n Callao 1 Peru Tel. Nos. 29–9210, 29–0355 Telex: 26010
PHILIPPINES	Philippine Ports Authority Port of Manila Safety Staff P.A. 193, Port Area Manila, 2803 Philippines Tel. Nos. 47–34–41 to 49
POLAND	Office of Maritime Economy ul. Hoza 20

(cont.)

**List of Contact Names and Addresses of the Offices
of Designated National Competent Authorities* (*continued*)**

	00–521 Warszawa Poland Tel. No. 284071, 284081 Telex: 812681 GOMO PL, 813407, 817421
REPUBLIC OF KOREA	Inspection and Measurement Division Seafarers and Ship Bureau Korea Maritime and Port Administration 263, Yeungi-dong, Jongro-Ku Seoul Korea Tel. No. 763–8972 Telex: KPA 26528
SINGAPORE	Director of Maritime Marine Department 1 Maritime Square No. 09–66 Singapore 0409 Republic of Singapore Tel. No. 2785611 Telex: RS 50287 MARDEP
SPAIN	Direccion General de la Marina Mercante Ministerio de Transportes, Turismo y Communicaciones Ruiz de Alarcon No. 1 Madrid 14 Spain Tel. No. 232–84–20 or 232–85–20 Telex: 27298 MAMER EO, 43579 MAMER EO
SWEDEN	The National Swedish Administration of Shipping and Navigation Division of Dangerous Goods and Marine Environment S–601 78 Norrkoping Sweden Tel. No. 011–191000 Telex: 64380 SHIPADMS Telefax: 011–101949
SWITZERLAND	Office suisse de la navigation maritime Elisabethenstrasse 31 4002 Bale Switzerland Tel. No. (061) 23.53.33 Telex: 62073 SSA
USSR	Main Department for Shipping and Port Operations (GLAVFLOT) Ministry of Merchant Marine Zhdanov Street, 1/4 Moscow USSR Tel. No. 228–38–82 Telex: 411197 MORFLOT
UNITED KINGDOM	Department of Transport Marine Directorate Sunley House 90/93 High Holborn

London WCIV 6LP
United Kingdom
Tel. No. 01-405-6911
Telefax: 01-831-2508
Telex: 264084 MARBOT G

UNITED STATES

Office of Hazardous Materials Transportation
U.S. Department of Transportation
Attn: Exemption and Approvals Division (DMT/23)
Washington, D.C. 20590
USA
Tel. No. (202) 366-4516
Telex: 892427 DOT WASH D.C.

Associate Member
HONG KONG

The Director of Marine
Marine Department
Harbour Building
38, Pier Road
Hong Kong
Tel. No. 5-8523085
Telex: 64553 MARHQ/HX

---

*For National Competent Authorities responsible for approval and authorizations in respect of the transport of Radioactive Material see also the International Atomic Energy Agency's (IAEA) National Competent Authorities List No. 16 of October 1984.

RID

The RID regulations are not structured in the same way as U.S. regulations and at first may appear to be confusing. This confusion is amplified in that RID is in a state of transition between the former RID regulations and the new UN standards. Studying the table of contents and then examining one of the classes of regulated articles, however, will show that these requirements are in quite a usable format. The contents of each section, one to a class, vary according to whether it has been converted to the UN style. Classes 3 (flammable liquids), 6.1 (poisons), and 8 (corrosive materials) have been so converted. The other classes remain in the old RID format and content of the original regulations. Each class of hazardous materials is broken down in its chapter as follows:

I. List of substances.
II. Provisions (conditions of carriage).
    A. Packages.
        1. General conditions.
        2. Packing of individual substances.
        3. Mixed packing.
        4. Marking and danger labels.
    B. Particulars in the consignment note (shipping papers).
    C. Empty packagings.
    D. Transport equipment.
        1. Conditions for wagons and their loading.
        2. Markings and danger labels on wagons and small containers.
    E. Prohibitions on mixed loading.
    F. Empty packages.
    G. Other regulations.

This structure is very similar for each class of materials and greatly simplifies use of the book once one is familiar with it. In the index at the back of the book is a complete listing of the hazardous materials similar to the DOT commodity list found in 49 CFR 172.101.

In the European system each class of hazardous materials is now identified by using the basic UN system. Awareness of these designations will further simplify the task of understanding these regulations, since Europeans regularly use these code designations instead of the common name for the class of materials.

While the packaging for classes other than 3, 6.1, and 8 continue to refer to the old RID packaging standards, and the descriptions are found generally within the class in the text there, the UN classes are referenced for packaging to an Appendix A.V.

This appendix is now found in an Annex 1, which is similar to UN Chapter 9. These rules differ from the UN recommendations in two major respects. First, they provide for some light gauge metal packaging for liquids commonly used in the paint industry. Second, they provide a detailed evaluation system for determining the compatibility of the lading with plastic packaging.

It is beyond the scope of this book to go into detail on this special evaluation system. It is sufficient here to note that it exists in RID and as well in ADR, and shippers may be called upon to describe the acceptability of packaging used for various chemicals in these "Appendix A.V." terms. Application of these standards is very uneven outside continental Europe, and even within Europe for incoming packaging, so that it only becomes a serious matter if a U.S. shipper has an extensive distribution system in Europe that is separate from the direct imports.

ADR

The structure of the ADR regulations for highway transport within Europe is essentially the same for those of RID, pertaining to rail transportation. The class designations are identical and refer to the UN system. Once one becomes familiar with either the ADR or the RID, the other becomes very simple to use. Both publications are available in English from:

Bernan Associates
10033 Martin Luther King Highway
Lanham, MD 20706

The cost of ADR or RID is about $20.00 each.

The same countries are parties to both the RID and the ADR. They are as follows:

Austria	Netherlands
Belgium	Norway
Denmark	Poland
Finland	Portugal
France	Spain
Fed. Republic of Germany	Sweden
German Dem. Republic	Switzerland
Hungary	United Kingdom
Italy	Yugoslavia
Luxembourg	

CHAPTER 26

The Administrative Procedure Act

GENERAL

A federal administrative agency, although under the direct authority of the White House insofar as management is concerned, is a creation of Congress. Congress dervies its power to rule from the United States Constitution. In today's society, however, problems on a daily basis are so complex and require such expertise to effectively cope with them, that Congress delegates its regulatory power to administrative agencies such as the Department of Transportation and the Environmental Protection Agency.

As a creation of Congress, these agencies have no greater authority than Congress is able to give, and only have that authority specifically granted to them in the form of statutes. The typical regulatory agency today is created by a statute that gives the agency the power to regulate in a specific area. This authority, under early principles of constitutional law, had to be carefully defined with legislative guidance from the Congress, on the theory that the agency was really just an agent of the Congress in carrying out the congressional will to regulate. Especially in early court cases, the functions of the agency had to be truly ministerial, with a minimum of discretion lodged in the agency to make law.

The concept of detailed and specific statutory guidelines defining and limiting the power of administrative agencies has waned, however, and since 1970 statutes have reposed enormous discretion in administrative agencies. The statutory bases for the Occupational Safety & Health Administration, the Consumer Product Safety Commission, and the Environmental Protection Agency are all very broad and conceptual. In an endeavor to remove the opportunity for loopholes in the law, Congress has only vaguely defined the problems confronting the country and has granted authority to the agencies to more specifically define the problems, and to make policy decisions in selecting a solution to problems the agencies define.

ADMINISTRATIVE PROCEDURE ACT

One of the consistent standards by which all regulatory agencies are guided in their actions is the Administrative Procedure Act, 5 U.S. Code 500, *et seq*. Not only must the actions of a regulatory agency be within the specific statutory grant of power that created the agency, but the Administrative Procedure Act (APA) must be satisfied as well. The policing body to assure satisfaction of APA requirements by agencies is the federal court system.

283

It is worthwhile knowing the boundaries to a regulatory agency's sphere of action, and for this reason this chapter on the APA is included in the *Red Book*. Examination of a portion of the table of contents of the APA is helpful:

Section 551. Definitions. This section contains helpful definitions of such terms as "rule" (to be read synonymously with regulation), "order," "license," "adjudication," and "sanction."

Section 552. Public Information. This section, commonly referred to individually as the Freedom of Information Act (FOIA), deals with the limited ability of an agency to withhold information and documents from the general public. It touches upon trade secrets, proprietary information, and confidentiality of information submitted to an agency.

Section 553. Rule Making. This section will be discussed in greater detail below, because it is the statutory guide for agencies engaged in informal rule making, such as DOT and EPA.

Section 554. Adjudications. These adjudications, or administrative trials, are the formal processes followed whenever a regulating body is engaged in the issuance of licenses, certificates of public convenience and necessity, etc. It prescribes formal proceedings, on the record, such as one might find in a rate hearing at the ICC.

Judicial review of administrative rule making is governed by Chapter 7 of the APA, discussed below.

RULE MAKING

For the benefit of users of the *Red Book*, those portions of the APA informal rule making section that are most pertinent to DOT regulatory procedures are reproduced here:

Section 553. Rule Making.

(b) General notice of proposed rule making shall be published in the *Federal Register*, unless persons subject thereto are named and either personally served or otherwise have actual notice thereof in accordance with law.

The notice shall include—
  (1) a statement of the time, place, and nature of public rule making proceedings;
  (2) reference to the legal authority under which the rule is proposed, and
  (3) either the terms of the substance of the proposed rule or a description of the subject and issues involved.

Except when notice or hearing is required by statute this subsection does not apply—

(A) to interpretative rules, general statements of policy, or rules of agency organization, procedure, or practice; or

(B) when the agency for good cause finds (and incorporates the finding and a brief statement of reasons therefore in the rules issued) that notice and public procedure thereon are impracticable, unnecessary, or contrary to the public interest.

(c) After notice required by this section, the agency shall give interested persons an opportunity to participate in the rule making through submission of written data, views, or arguments with or without opportunity for oral presentation. After consideration of the relevant matter presented, the agency shall incorporate in the rules adopted a concise general statement of their basis and purpose ...

(d) The required publication or service of a substantive rule shall be made not less than 30 days before its effective date, except—

(1) a substantive role which grants or recognizes an exemption or relieves a restriction;

(2) interpretative rules and statements of policy; or

(3) as otherwise provided by the agency for good cause found and published with the rule.

(e) Each agency shall give an interested person the right to petition for the issuance, amendment, or repeal of a rule.

The important aspects of the preceding statutory quote are those that compel the publication of a notice of proposed rule making describing the proposed rule so people might offer intelligent comment, and publication of a final rule after consideration of the relevant matter presented by commenters. That final publication must be accompanied by a concise general statement of the basis and purpose of the rule change. No substantive change may be given an effective date of less than thirty days from the time of publication unless it meets one of the specified exceptions to that requirement. Any interested person must have the opportunity to ask for new rules, changes in old rules, or the abolition of old rules.

JUDICIAL REVIEW OF RULE MAKING

The courts are the forum for final review of administrative action. They are not the forum of first resort, however, for most regulatory procedures of most agencies provide some opportunity to petition for reconsideration of the final rule published by the agency. DOT has such provisions as well. Generally speaking, courts will not regard an agency decision as a final order unless the complaining party has exhausted his administrative remedies by petitioning the agency for reconsideration.

Judicial review of rule making is addressed in Chapter 7 of the APA. For the benefit of readers of the *Red Book*, the pertinent portions of these statutory sections are reproduced here:

Section 701. Application; definitions.

(a) This chapter applies, according to the provisions thereof, except to that extent that—

    (1) statutes preclude judicial review; or
    (2) agency action is committed to agency discretion by law.

Section 702. Right of review.

A person suffering a legal wrong because of agency action, or adversely affected or aggrieved by agency action within the meaning of a relevant statute, is entitled to judicial review thereof.

Section 705. Relief pending review.

When an agency finds that justice so requires, it may postpone the effective date of action taken by it, pending judicial review. On such conditions as may be required and to the extent necessary to prevent irreparable injury, the reviewing court, including the court to which a case may be taken on appeal from or on application for certiorari or other writ to a reviewing court, may issue all necessary and appropriate process to postpone the effective date of an agency action or to preserve status or rights pending conclusion of the review proceedings.

Section 706. Scope of review.

To the extent necessary to decision and when presented, the reviewing court shall decide all relevant questions of law, interpret constitutional and statutory provisions, and determine the meaning or applicability of the terms of an agency action. The reviewing court shall—

    (1) compel agency action unlawfully withheld or unreasonably delayed; and
    (2) hold unlawful and set aside agency action, findings, and conclusions found to be—
        (A) arbitrary, capricious, an abuse of discretion, or otherwise not in accordance with law;
        (B) contrary to constitutional right, power, privilege, or immunity;
        (C) in excess of statutory jurisdiction, authority, or limitations, or short of statutory right;
        (D) without observance of procedure required by law;
        (E) unwarranted by the facts to the extent that the facts are subject to trial de novo by the reviewing court.

In making the foregoing determinations, the court
shall review the whole record or those parts of it
cited by a party, and due account shall be taken of
the rule of prejudicial error.

These quoted words describing the scope of judicial review of informal rule making, such as that conducted by DOT and EPA, should be the measure of the quality
and propriety of DOT or EPA action. Although it is clear that courts will defer to
the expertise of technical agencies, especially in the safety field, it is equally clear
that the agency has an obligation to establish a rational basis for rules that are
adopted.

If a rule change is unreasonably delayed, or if the action taken by the agency is
arbitrary, capricious, an abuse of the discretion given the agency by Congress, or
otherwise is not in accordance with the law, it will be held unlawful by the courts.
If the procedural requirements of Section 553, quoted in part above, are not fully
satisfied, then the court under (D), above, must hold that agency action to be unlawful and must set it aside.

Occasionally these boundaries to administrative action are overlooked by officials in charge of regulatory programs. It does not hurt to remind an agency of its
obligations under the APA, nor is court action unwise if the agency action is in
violation of the provisions of the APA.

CHAPTER 27

Rule Making
on Hazardous Materials

In order to comprehend the various steps DOT follows in developing regulations, it is helpful to know the guidelines set for them by the Congress under the Administrative Procedure Act (APA). The reader who is unfamiliar with the particular provisions of the APA is urged to review Chapter 26 before reading this chapter.

Like every other function of DOT as an administrative agency, there is a statutory provision underlying DOT's power to issue regulations. Throughout the years that authority has appeared in various statutes in the United States Code, but the critical provision is now found in title 49 U.S. Code 1804, enacted originally as Section 105 of the Hazardous Materials Transportation Act (HMTA) on January 3, 1975. (49 U.S.C. 1801 et seq., P.L. 93–633, 88 Stat. 2156; see Chapter 30 for detailed information on the HMTA.)

Section 105 of the HMTA generally declares that DOT may issue, in accordance with the informal rule making sections of the APA, regulations for the safe transportation of hazardous materials in commerce. This provision also confirms that rule making on hazardous materials shall include "an opportunity for informal oral presentation." Any person may initiate changes in DOT regulations. It is therefore essential for every person subject to the regulations to know how to go about achieving the changes believed to be necessary.

WHERE TO FIND RULE MAKING PROCEDURES
The rule making procedures of DOT's Research & Special Programs Administration (RSPA), including the Office of Hazardous Materials Transportation (OHMT), are in 49 CFR Part 106.

WHAT IS RULE MAKING?
Under the APA, rule making means the federal agency process for formulating, amending, or repealing a rule. A rule, which can be read as synonymous with regulation for DOT purposes, means the whole or a part of an agency statement of general applicability to the public and of future effect in controlling coming events, designed to implement, interpret, or prescribe law or policy. It also may be the type of statement that describes the organization, procedure, or practice requirements of the agency. These rules, or regulations, are issued by the agency

under the authority of a statute written by Congress. Statutes and regulations are both considered law, so merely referring to the law does not distinguish between regulations and statutes. For the most part, a statute is broad and general, giving an agency the power to issue regulations that are narrow and specific, carrying out the overall purposes of the statute.

So, in the case of DOT the statute is the HMTA, and the regulations are found in various titles (or subject groupings) of the Code of Federal Regulations (CFRs). Title 49 CFR pertains to transportation generally and contains many of the regulations of the various operating administrations within DOT. The bulk, although not all, of the regulations pertaining to transportation of hazardous materials are found in the volumes of title 49 CFR containing Parts 100–199.

PETITIONS TO INSTITUTE RULE MAKING

As indicated above, any person may ask DOT to issue, change, or repeal a regulation. That person does not have to be subject to the regulations in order to ask for rule making but can be any member of the public. To refer to this broadest classification of people and companies, DOT uses the phrase "any interested person."

Petition is just a legalistic word for anything written that seeks some action from DOT. There is no prescribed format, paper size, length, or limitation on who may prepare a petition. The only requirements are those dictated by common sense, that the petition say clearly and concisely, with a minimum of argument and a maximum of fact, what the petition seeks to have done. Petitions for rule making should be addressed to:

Administrator, Research & Special Programs Administration
U.S. Department of Transportation
400 Seventh Street, S.W.
Washington, D.C. 20590

Each petition to establish, amend, or repeal a regulation must:

1. Set forth the text or substance of the regulation or amendment proposed, or specify the regulation that the petitioner seeks to have repealed, as the case may be.
2. Explain the petitioner's interest in the action requested; and,
3. Contain any information and arguments available to the petitioner to support the action he seeks.

No specific number of copies are required, but it is always good to provide at least two copies of any correspondence with DOT.

PROCESSING OF PETITIONS FOR RULE MAKING

It is unlikely that there will be any public hearing, argument, or other proceeding on any petition for rule making before it stimulates some action of a positive nature or is denied. Because of perennial backlogs of pending petitions, it is unlikely the petitioner will see anything more than a post card receipt at the time the petition is sent to DOT.

If DOT determines, eventually, that the petition contains adequate justification, rule making action will be initiated in accordance with the procedures in 49 CFR Part 106. If, however, it is determined that the petition does not justify rule making, the petition is denied. Whenever DOT determines that a petition should be granted or denied, the Office of the Chief Counsel of RSPA prepares a notice of that grant or denial for issuance to the petitioner by the RSPA.

RULE MAKING IN THE ABSENCE OF A PETITION

Rule making may be initiated by DOT without having an outside stimulus in the form of a petition. In addition, other documents may serve as a basis for initiating rule making, such as a request for an exemption that more appropriately should be treated as a general matter than as an exemption limited to one or a few exemption holders. In fact, the bulk of rule making in the past few years has been stimulated by internal forces seeking change, or has been based upon successful experience with novel concepts under exemptions.

THE NOTICE OF PROPOSED RULE MAKING

Under the APA, with a few exceptions discussed further below, in making or changing a rule DOT must publish a notice of proposed rule making (NPRM) in the *Federal Register*. This notice advises the public that a new rule or revision of a rule is under consideration and affirmatively solicits the public's response to that idea. No notice is required for interpretive rules, general statements of policy, or rules of agency organization, procedure, or practice, or when the agency for good cause finds and states that notice and public procedure are impracticable, unnecessary, or contrary to the public interest. Each notice of proposed rule making is published in the *Federal Register*, although this publication may be omitted if all persons subject to the proposed rule are personally served with a copy of the proposal. Whether published or personally served, each NPRM includes:

1. A statement of the time, place, and nature of the proposed proceeding;
2. A reference to the statutory authority under which it is issued;
3. A description of the subjects and issues involved, or the terms of the proposed regulation;
4. A statement of the time within which written comments on the proposal must be submitted; and
5. A statement of how and to what extent persons may participate in the proceeding. In response to every notice of proposed rule making, as a minimum interested persons may participate by submitting written comments containing information, views, or arguments.

WRITTEN COMMENTS: HOW AND WHERE TO FILE

The APA is unique in world law, giving the regulated public in the U.S. an opportunity to express opinions and views on proposed requirements before those requirements are actually imposed. Any affected person is urged to participate in the process, for every comment is read and any one could change the course of the proposal.

All written comments must be in English. It is requested by DOT, but not required, that five copies be submitted. A commenter making a statement of fact should submit all the material the commenter considers relevant to back up that fact. Incorporation by reference, i.e., referring to material in another source by just mentioning that other source without repeating the material, is discouraged. If DOT does not have access to that other source of information, the comment is virtually worthless. If incorporation by reference is unavoidable, the proper citation and title to the source should be given, as well as the page at which the referenced material appears.

The NPRM will give the name and address of the official to whom any comments on that NPRM should be sent. Every rule making action is given a docket number to identify it. As indicated in the NPRM, the proper reference to the docket should appear on the comments to assure filing in the proper public file.

Experience shows that comments on rule making are usually too wordy, consuming many pages to say what could be said in a few sentences. Commenters are often compelled, however, to include lengthy references and descriptions of the company or group that they represent, and often the comments are lengthy in order to adequately explain the point to persons who otherwise may be unfamiliar with the rule making or regulations in question. It is more effective to provide an executive summary to precede the body of the comment. Although every comment is considered, ones that are more easily, quickly, and clearly understood will be more persuasive than wordy, rambling, or vague comments. Persuasion, after all, is the purpose of the comment, so keep it short and to the point.

TIME TO COMMENT

The time given to comment, called the comment period, varies with the size and complexity of the proposal. The NPRM will give the date on which the comment period will close. There is little benefit derived from waiting until the final day to submit the comments, and since other members of the public review the files before submitting their own comments, early filing may help to convince other commenters to support a position taken in those first comments. Seldom is any comment period shorter than thirty days, and sixty to ninety days is more common.

If anyone desires more time in which to prepare their comments, provision is made for petitions for an extension of the comment period. Any petition for an extension of time to comment must be received no later than ten days before the close of the comment period sought to be extended. Asking for more time does not guarantee that more will be given. In fact, a petition for more time is granted only if the petitoner shows good cause for the extension and if the extension is consistent with the public interest. Any extension that is granted will be available to all commenters, who will be so notified by a publication in the *Federal Register*.

PUBLIC DOCKET

Rule making is a public process, and all information and data DOT considers relevant to the rule making, including the NPRM, comments received in response to the notice, petitions for rule making, petitions for reconsideration, denials of such petitions, records of meetings or additional rule making proceedings, and final regulations adopted are maintained in the public files or dockets. A new docket is opened for each rule making and is assigned a number.

When the DOT was formed in 1967, rule making dockets were assigned sequential numbers preceded by the letters "HM," for hazardous materials. Thus, the first rule making was denominated Docket No. HM-1. By the end of 1986, DOT was well beyond Docket No. HM-197. The reader will note that once a docket number is assigned it is maintained and, although new notices or subsequent amendments will be given numbers distinguishing them from one another, within the same docket they remain always under the same umbrella number.

The public docket files are kept at the Materials Transportation Bureau, 400 7th Street, S.W., Washington, DC 20590. Any person may examine any docket material at these offices during regular business hours and may obtain a copy upon payment of the customary fee.

The docket not only contains active rule making projects but those which have been completed as well. It does not contain internal staff memoranda or internal reactions to the comments. It includes all public comments, however, and since no person commenting is obliged to provide copies of those comments to anyone else, the docket in Washington is the only source of all the public comment that goes into any rule making decision. Especially in a controversial action, interested commenters owe it to themselves to periodically review the public file to find out what others are saying. In a controversial rule making, this review should take place every few weeks.

Not only are comments placed in the file, but if any official in the decision-making chain meets with any nongovernment person and discusses an open rule making, that official is obliged to provide a memorandum to that docket file summarizing what was said and what commitments, if any, were made. Again, review of the docket itself in Washington is the only way to determine that such memoranda exist and to identify them sufficiently well to obtain a copy.

ADDITIONAL RULE MAKING PROCEEDINGS

A mechanism used occasionally by DOT to raise a subject for public consideration, short of actually proposing specific action, is the "advance notice of proposed rule making." This is not a notice that can be followed in turn by an amendment but is merely an agency announcement of a need for additional information in an area where an NPRM eventually may be issued. The concepts and ideas set forth in an advance notice often change dramatically before they are published as an NPRM and there is no assurance the idea will even get that far. Public participation at this time is particularly valuable, since the concepts are at their most malleable stage and no element of the staff will worry about saving face if the ideas are dropped entirely. This kind of resistance to change has been observed on occasion in many agencies once ideas have gone so far as to have appeared in the *Federal Register* as an NPRM.

There have been other hybrid publications that defy description other than to say they are requests for information or public participation. Typically these matters relate to the germ of an idea within DOT that is coupled with the recognition that there is no information within the agency with which to assess that idea; so help is sought on the outside via the *Federal Register*. These publications do not necessarily lead to anything, nor do they fall within any of the aspects of the APA. They are simply communications between the agency and the public, with the *Federal Register* being the medium of communication.

A standard rule making involves an NPRM with a preamble explaining the purpose of the proposal, followed by a written comment period and eventual publication of a final amendment, this time with a preamble responding to the written comments and explaining DOT's position on those comments. DOT may initiate any further rule making proceedings, however, if it is found by DOT to be necessary or desirable. For example, interested persons could be invited to make oral arguments, to participate in conferences with the DOT staff in which minutes are kept for filing with the docket, to appear at informal hearings at which a transcript is kept and later placed in the docket, or to participate in any other proceeding to assure informed administrative action and to protect the public interest.

HEARINGS

The HMTA provides an opportunity to the public to make an informal oral presentation during DOT rule making. On occasion the NPRM will indicate that a hearing will be held and will provide details as to time and place. If the NPRM does not provide for a hearing, any interested person may petition the RSPA for an informal hearing. The petition must be received by the director no later than twenty days before the close of the comment period stated in the NPRM.

A petition requesting an informal hearing (which is the only kind DOT holds) does not automatically result in the scheduling of a hearing, despite what the statute apparently compels. Under the procedural regulations a petition seeking a public hearing is granted only if the petitioner shows good cause for a hearing. If the petition is granted, DOT will publish a notice in the *Federal Register* giving the subject matter, as well as the time and place of the hearing.

By "informal" is meant that the hearing is non-adversary in nature, and is more like a legislative fact-finding procedure at which there are no formal pleadings or adverse parties, and no opportunity for cross-examination of witnesses. Testimony is given by persons desiring to give it and, although a transcript is kept, the testimony is not given under oath and there are no rules of evidence prescribing what may or may not be said. Furthermore, the decision ultimately issued in the rule making need not be based exclusively on the hearing record. Usually the hearing is held in an auditorium and is conducted by a DOT representative with counsel. Members of the public do most of the talking, not DOT officials, and so, while there is some cross-pollination of ideas among the public, there is no true opportunity to plumb the DOT staff ideas behind the proposed rule.

ADOPTION OF FINAL RULES AND AMENDMENTS

Upon consideration of all the comments filed in response to the NPRM, the office dealing with the specific docket in question prepares a draft final rule. This is reviewed by the RSPA Chief Counsel's office and passed on to the Director of the Office of Hazardous Materials Transportation. Note that the person authorized to sign the final decision is the Administrator of the RSPA, not the director of OHMT who issues the NPRM. If the RSPA adopts the regulation, it is published in the *Federal Register* unless all persons subject to it are named and served with a copy, which is an unlikely event.

Each final rule or amendment of an existing regulation will be published with a preamble explaining the comments received and the basis for the DOT decision on the issues raised by those comments. It is DOT policy to respond to every major comment or group of similar comments. Although the rationale may be vague or questionable on occasion, the preamble to the change or new rule provides an invaluable insight into the application of that regulation in the future and the interpretation the agency will give it in specific future situations. The new or revised regulation will appear in subsequent compilations of the regulations in the CFR's, but the preamble is not republished.

Interpretation of regulations necessarily is a historical function, since the intent at the time of adoption of the regulation governs its meaning thereafter. As of this writing, for approximately twenty years DOT has been publishing preambles to notices explaining the original intent of a proposal and preambles to amendments explaining and responding to comments, thereby elaborating on this intent. Outside of DOT, however, it is unlikely that copies of the significant dockets exist. The regulatory preambles of what are believed to be the key rule making actions appear in Appendix C. Appendix C is accompanied by an introduction explaining the various features of the notices and amendments, as well as a list of contents of the subject matter in those proceedings. In addition, each docket is preceded by a brief summary of the significance of that rule making. This compilation can provide a wealth of information and explanatory background material, and the reader is urged to use it and become familiar with its contents.

Under the APA, no rule can be made effective in less than thirty days from publication except (1) a substantive rule that grants or recognizes an exemption or relieves a restriction, (2) interpretive rules and statements of policy, or (3) when found by the agency for good cause that is published with the rule. The effective date is keyed to publication not signing.

PETITIONS FOR RECONSIDERATION: STAY OF EFFECTIVE DATE

Any person may petition the RSPA for reconsideration of any amendment or final rule. It is requested, though not required, that the petition be submitted in triplicate. A petition for reconsideration must be received no later than thirty days after publication of the amendment or final rule in the *Federal Register*. Few companies can obtain a copy of the amendment or final rule, assess its impact on their operations, develp a position on that impact, and communicate that position in time to be received by DOT within thirty days of the initial publication. There is no room for delay.

One way to cope with this is to remember that no final rule or amendment may be substantively broader than the NPRM. No rule may be imposed under the APA unless adequate notice has been given and an opportunity provided for public participation. If a final rule goes beyond the scope of the NPRM, that extra material has been adopted without notice and the agency has violated the APA. Careful examination of the NPRM, therefore, will tell the reader far in advance just how far the agency might go in the final rule. Affected companies could decide, at that time, what their position would be if DOT were to follow through with the proposal. Then, if that happens, there need not be such a scramble to determine the impact on the company and to develop a position in the form of a petition for reconsideration. It demands participation at each stage of the process, however, for a company to keep attuned to potential impacts.

Any petition for reconsideration filed later than thirty days after publication of the final rule or amendment in the *Federal Register* is not treated as a petition for reconsideration at all, but is relegated to the level of a petition for rule making to start the process anew under 49 CFR 106.31. In this circumstance, no additional revision could take place without issuance of a new notice of proposed rule making.

The petition for reconsideration must contain a brief statement of who the petitioner is and the nature of the complaint. The petition further must include an explanation as to why compliance with the new rule is not practicable, is unreasonable, or is not in the public interest. If the petitioner feels the rule was adopted in violation of a law, such as the APA, this would fall within the sphere of items not in the public interest. If the petitioner requests consideration of additional facts, he must state why those facts were not presented in a more timely fashion during the comment period on the NPRM. DOT will not consider repetitious petitions.

Unless DOT decides differently at the time, the filing of a petition for reconsideration does not in itself act to stay the effective date of the new rule. In any case, if the petitioner believes a stay is in order, that belief should be clearly expressed in the petition and should be well supported.

RSPA may grant or deny, in whole or in part, any petition for reconsideration. In the event they determine to reconsider any regulation, RSPA may issue a final decision on reconsideration without further proceedings, providing that final decision would have been within the scope of the original NPRM if it had been the decision reached the first time. RSPA also may reopen the matter, providing such opportunity to submit additional comments, information, and data deemed appropriate.

As with petitions for rule making, whenever it is decided that a petition for reconsideration should be granted or denied, the RSPA's Chief Counsel prepares a notice to that effect for issuance to the petitioner over the signature of the Administrator of the RSPA.

It is DOT policy to issue this notive of the action taken on a petition for reconsideration within ninety days after the date of *Federal Register* publication of the regulation in question, unless it is found impracticable to take action within that time. If there will be delay beyond this ninety day period and that delay is expected to be substantial, notice to that effect and a date by which action will be taken will be issued to the petitioner and will be published in the *Federal Register* as well.

CHAPTER 28

Exemptions
From Regulation

As used in this chapter the term exemption means a waiver of a requirement in the form of a DOT authorization of an alternative packaging or
handling practice. It does not include exceptions to the full impact of the
regulations accorded to limited quantity and low hazard materials like ORMs.
Those situations, in which less than the full weight of regulation is levied on the
transportation of certain materials under certain circumstances, are not true
exemptions, but merely examples of lesser regulation in recognition of unique and
lesser hazards, and are called "exceptions."

The Random House Dictionary of the English Language, Unabridged Edition, offers as
a primary definition of the verb "exempt" the following guidance: "1. to free from
an obligation or liability to which others are subject; release." This is the sense of
the word as used in this chapter, for that is the sense of the process described in
the Hazardous Materials Transportation Act (HMTA), 49 U.S.C. 1801, 1806. Thus,
when referring here to an exemption, what is meant is a procedure by which a
person subject to a DOT regulation seeks to be relieved from the obligation to
comply with that regulation, while other people similarly situated must continue to
comply. This may be done by waiving the regulation entirely (which is rare) or by
application of alternative restrictions.

No discussion of how to get an exemption from the DOT regulations would be
complete without setting out the full text of the exemption section of the HMTA,
for many of the DOT procedures do not make sense unless it is realized that the
HMTA compels DOT to take certain steps.

Sec. 107. (a) General. The Secretary [of DOT], in
accordance with procedures prescribed by regulation, is authorized to issue or renew, to any person
subject to the requirements of this title, an exemption from the provisions of this title, and from regulations issued under Section 105 of this title, if
such person transports or causes to be transported
or shipped hazardous materials in a manner so as
to achieve a level of safety (1) which is equal to or
exceeds that level of safety which would be required in the absence of such exemption, or (2)
which would be consistent with the public interest
and the policy of this title in the event there is no
existing level of safety established. The maximum

296

period of an exemption issued or renewed under this section shall not exceed 2 years, but any such exemption may be renewed upon application to the Secretary. Each person applying for such an exemption or renewal shall, upon application, provide a safety analysis as prescribed by the Secretary to justify the grant of such exemption. A notice of an application for issuance or renewal of such exemption shall be published in the *Federal Register*. The Secretary shall afford access to any such safety analysis and an opportunity for public comment on any such application, except that nothing in this sentence shall be deemed to require the release of any information described by subsection (b) of section 552 of title 5, United States Code, or which is otherwise protected by law from disclosure to the public.

* * *

(d) Limitation on Authority. Except when the Secretary determines that an emergency exists, exemptions or renewals granted pursuant to this section shall be the only means by which a person subject to the requirements of this title may be exempted from or relieved of the obligation to meet any requirements imposed under this title.

For a full discussion on the language and background of this and other specific sections of the HMTA, see Chapter 30. For the HMTA itself, see Appendix D.

WHERE TO FIND EXEMPTION PROCEDURES
Detailed procedures describing how one applies for an exemption and all related matters are found in 49 CFR Part 107.

HOW TO APPLY
Any person subject to a DOT hazardous materials regulation may apply for relief from that regulation. As DOT itself said in the preamble to the rule making adopting exemption procedures, "A person's right to petition an agency for relief from a regulation of that agency which directly affects that person is so well established as to be beyond question."

There is no special format for an exemption application and no special legalistic phrases that must be included. No lawyer is necessary to prepare an application or present it to DOT, though consultation with counsel is always advisable before you have any contact with any federal agency on any subject.

The application must be submitted in triplicate but does not have to be a sworn statement or be signed in the presence of a notary. It should be mailed or otherwise sent to:

Director, Office of Hazardous Materials Transportation
Research & Special Programs Administration
U.S. Department of Transportation
Washington, D.C. 20590
Attn: Exemptions Branch

WHAT TO INCLUDE

Each application for an exemption from the DOT hazardous materials regulations must include the following information. Failure to include all of this information or a reason why it is not included will only result in a return of the application for incompleteness, thereby adding more delay to a process which is already time-consuming.

1. Set forth the text or substance of the regulation from which the exemption is sought. This should refer not only to the substantive requirement but should include the proper citation to the specific regulation in 49 CFR. This portion of the application need not be lengthy, since the question is not what you want relief from but why and on what basis.

2. State the name, address, and telephone number of the applicant. If the application seeks an exemption for use by several divisions within a company doing business in their own name, be sure to include any such names along with the appropriate addresses.

3. Provide a detailed description of the proposal, including when appropriate any drawings, plans, calculations, procedures, test results, previous exemptions on the subject, a list of DOT specification packaging to be used, if any, a list of modified DOT specification packaging to be used, if any, and a description of the modifications. Include as much supporting data for the exemption as is available. Save the argument for later in the application. For confidentiality of data, see below.

4. State the chemical name, DOT shipping name, common name, DOT hazard classification, form, quantity, properties, and characteristics of the material covered by the application, including the composition and percentage (specify if by volume or weight) of each chemical, if a solution or mixture. Giving the proper DOT shipping name (see Chapter 6) and classification (see Chapter 5) is essential. Determination of the proper shipping name and classification is a shipper's function, and asking DOT to name and classify the material for you usually is a futile gesture. For confidential treatment of any of this information, see below.

5. Describe all relevant shipping and accident experience. If this is a totally new product, obviously there will be no shipping or accident experience and, if that is the case, be sure to say so. If there is relevant shipping and accident experience with the same packaging (perhaps used in shipping an unregulated product) or a product with similar characteristics shipped in the same manner as proposed, it is helpful to mention it. DOT compiles data on accident experience that is filed in computer banks and that may be retrieved. See Chapter 21 on hazardous materials incident reports. If you have no direct shipping and accident experience, perhaps the DOT database can supply some information that will be helpful.

6. Specify the proposed mode of transportation, identify any increased risks that are likely to result if the exemption is granted, and specify the safety control measures that you consider necessary or appropriate to compensate for those increased risks. If you only seek to ship

or carry a material by one mode of transportation, such as by motor carrier, it simplifies the process to say so. The more modes of transportation involved, the more coordination and review that will be done before issuance; so do not apply whimsically for authority that will not be used. The delay alone should discourage shotgun requests unless all modes are necessary to distribution of the product. If cargo or passenger aircraft authority is requested, the safety review is more intense, and you should only ask for air transportation authority if you genuinely need it. Requesting authorization to ship by water also adds significantly to DOT processing time. Do not ignore the requirement of identifying any increased risks under the proposal, and be specific both in identifying such risks and the measures to counter them.

7. Specify the proposed duration or describe the proposed schedule of events for which the exemption is sought. If the exemption is only to make one shipment or is only necessary pending some coming event, say so.   Remember from the statute that 2 years is the longest period that can be granted, although this can be renewed.

8. State why you believe the proposal, including any safety control measures you specify, will achieve a level of safety that
   a. Is at least equal to that specified in the regulation from which the exemption is sought, or
   b. Will be consistent with the public interest and will adequately protect against the risks of life and property that are inherent in the transportation of hazardous materials in commerce if the regulations do not contain a specified level of safety, and they generally do not.

   This is a difficult area of the application but a critical one, in which you pull together all the detail requested in the first part of the application with your reasons for wanting the exemption, and all the bases for your belief that safety will not be impaired by a grant of the exemption. If an affirmative requirement is imposed by a regulation and it has an obvious safety purpose, it is unlikely the exemption will be granted unless you can positively show that your proposal achieves the same purpose. For example, if a regulation requires a threaded closure on a container and you want to use a friction closure but cannot show through tests that it is of equivalent integrity, and you are not proposing additional measures to counter the removal of the threads, you probably will not get the exemption.

9. If you want to have the application processed on a priority basis, set forth the supporting facts and reasons. By priority the procedural regulations merely mean expeditious handling ahead of other applications that ordinarily are handled in the order of receipt. This almost never happens and, therefore, is usually a waste of time to request. Priority handling does not skip any of the public procedures or *Federal Register* publication, but only refers to going to the head of the line of applications that go through the full processing cycle. For situations in which the public procedures may be omitted, see emergencies, below. A priority situation is not subject to exact definition but would include those instances in which there are compelling rea-

sons for expeditious handling, but not so compelling as to justify declaring an emergency. Getting or maintaining a competitive advantage would not justify priority handling, but if an operation were left in economic limbo awaiting a shipment that could only come by exemption, perhaps a case could be made that priority treatment was warranted. To reiterate, this does not omit any element of the processes; it merely puts the applicant to the head of the line if DOT is so inclined, rather than making him wait his turn, but you will not know how DOT actually is handling your application.

10. If the applicant is not a resident of the United States, the application should include a designation of a permanent resident of the United States as his agent for service of process. In other words, if the exemption holder gets sued, someone has to be in this country to receive notice of that suit. Whether that person also can be held liable is a question that has never been tested.

TIMING

By looking at the detailed steps through which DOT must go in processing exemptions, it is clear that it will be a slow process. Section 107.103(c) states that, "to permit timely consideration, an application *should* be submitted at least 120 days before the requested effective date." DOT has stated it may not consume the entire 120-day period, and that original applications will be processed as expeditiously as possible. This could take less than 120 days, but applicants should be aware that it also may take more than 120 days. File the application as soon as it is prepared, as far in advance of the desired effective date as possible.

Once received, an original application is subject to administrative review to determine its completeness and conformity with the procedural requirements of 49 CFR Part 107. This review, according to 49 CFR 107.107, will take place within 30 days of receipt. If the application is not returned for further work in about 30 days, the applicant may assume it has begun the actual processing steps on the way to grant or denial. History has shown that the bulk of time wasted in the processing of any request for rule making or exemption has been the lack of completeness of the application or the inadequacy of the grounds offered in support of the proposal. Extra effort is warranted at the beginning of the process to prepare the most effective application, because it will bring ample returns in smoothing the processing.

DOT holds the exemption applications received in a given month until the end of that month, and then publishes a list of applications received, in the Federal Register in mid-month. Thus, if you have an opportunity to file on the 27th of the month, or on the first of next month, taking the latter date usually will cost at least an additional 30 days in processing. Also, if you are close to the end of the month, have the application hand-delivered, return-receipt-requested, and do not entrust your application to the mail room.

RENEWALS OF EXEMPTIONS: HOW TO APPLY

Quick reference to the statutory section quoted at the beginning of this chapter will show that Congress spoke of renewals of exemptions in the same terms as original exemptions. Further, that reference will show that no exemption may be granted for a period longer than two years, at which time the exemption will expire by its terms unless the holder of the exemption properly applies for renewal.

As with original applications, the papers must be filed in triplicate to the same address given above. The application for renewal should include the following:

1. The identity of the exemption, by exemption number as well as by its principle effect, e.g., Exemption No. _____ authorizing the shipment of commodity XYZ in non-specification steel drums.
2. State the name, address, and telephone number of the applicant.
3. Include:
   a. A certification that the descriptions, technical information, and safety assessment submitted in the original application, or as may have been updated by any subsequent correspondence or application for renewal, remain accurate and correct, or
   b. Such amendments to the previously submitted descriptions, technical information and safety assessment as may be necessary to update them and assure their accuracy and correctness.

   Experience shows that the people in the company who were involved in the original application are seldom the same as those who deal with the renewal. The DOT files on the exemption should be reviewed for current accuracy of the information in them. Do not rely on memory or company files alone.
4. Provide a statement describing all relevant shipping and accident experience that has occurred in connection with the exemption since its issuance or most recent renewal or, if no accidents have been experienced, a certification to that effect. This statement must include the approximate number of shipments made or packages shipped, and the number of shipments or packages involved in any loss of contents, including loss by venting when transporting a compressed or cold temperature gas. The applicant for renewal should not take this portion of the application lightly, for renewal is predicated upon successful experience under the original exemption. The fact that the matter covered by the exemption may have been involved in an accident is not necessarily detrimental, if it can be shown that under accident conditions there was no release of hazardous materials or personal injury attributable to those materials. In offering a certificate swearing that no accidents have occurred, the wise applicant will check the DOT files for any incidents involving the exemption that may have been reported to DOT by others. See Chapter 21 on the incident reporting system. As for the giving of the number of shipments, reasonably accurate approximations are sufficient. If this commercial data is considered confidential, see the discussion of confidentiality, below.

TIMING OF RENEWAL APPLICATIONS
DOT advises that applications for renewal, which will involve somewhat less evaluation of the supporting data than for the initial exemption, should be filed at least sixty days before the date of expiration appearing on the exemption. The prudent applicant will file within sixty days of that expiration,

because if such a timely application is made and DOT does not complete its processing before the expiration date, the exemption will not lapse but will remain viable until processing is complete. This provision for preservation of exemptions on renewal is totally dependent upon filing at least sixty days before the expiration date. The applicant who waits beyond this filing period does so as his peril, since there is no guarantee of processing within the time left to DOT, and no shipments may take place under an exemption that has been allowed to expire. Shipments under a dead exemption could result in greater liability, in fact, than unlawful shipments not involving any exemption.

DOT provides a 15-day comment period in the Federal Register for on renewal applications. Before this, DOT reviews the renewal application to determine its completeness and its conformity with the procedural requirements. If the applicant does not get his application back for further work prior to publication, it is reasonable to assume that the application has entered the actual processing machinery.

FEDERAL REGISTER PUBLICATION

As required by Section 107 of the HMTA, quoted in the beginning of this chapter, notice regarding all applications for original exemptions and renewals must be published in the *Federal Register*. The application itself is not published but merely a brief indication of the identity of the applicant and the nature of the request. Any interested member of the public may go to DOT's offices to see or may write for copies of any information relevant to the application, including the application and supporting data, memoranda of any informal meetings of DOT staff with the applicant, and the DOT grant or denial of the application if that stage has been reached.

An unspecified period, averaging thirty days for new exemptions and fifteen days for renewals, will be allowed after publication of the notice in the *Federal Register* to permit interested members of the public to come to DOT to review the file and to submit comments on the applicant's proposal. Such commenters are not obligated to provide copies of their comments to the applicant, nor will DOT on its own motion advise an applicant that comments have been received. Any such comments, as well as DOT memoranda of any meetings relating to the application, will be in the public files for that application in Washington. The applicant and any other person may review those files and obtain copies of any documents in the files under the procedures outlined in 49 CFR Part 7, which are DOT's regulations implementing the Freedom of Information Act. Any applicant would be well advised to review his own file on the chance that someone, including public interest groups and competitors, may have opposed issuance of the requested exemption. The applicant may submit additional information to counter any such opposition, and this too will go into the public file.

After grant or denial of an application for exemption, DOT again will publish a notice in the *Federal Register*, although this publication is not required by statute. In other words, any application accepted for processing will be highlighted twice in the *Federal Register*, once upon receipt and again upon its grant or denial.

REQUESTS FOR AUTHORITY TO USE EXISTING EXEMPTIONS

Any exemption will be issued to the original applicant only. If any other person wishes to take advantage of the provisions of that exemption or to join in the initial application, that person may apply to become a "party to" an exemption, at the same address noted above. This process is covered by 49 CFR 107.111. Such an application must include the following information:

1. The specific identity of the current exemption, or exemption application if the exemption is not yet issued, to which the applicant wishes to become a party. This should give the name of the current exemption holder or applicant for exemption, a brief description of the nature of the exemption issued or sought, and any numerical reference to the exemption if it has already been issued.
2. The name, address, and telephone number of the applicant to be a party to an exemption. If the applicant is not a resident of the United States, the application should include a designation of a permanent resident of the United States as his agent for service of process. If authority is sought to use the exemption in the name of several divisions or aspects of the applicant company, each of these should be named and an address given for each. The applicant should also give a clear showing that he is subject to the regulations and is eligible to seek an exemption from the requirement in question. An example of ineligibility would be a request from a motor carrier to be exempted from certain packaging requirements, since those packaging requirements are regulations binding upon the shipper of the materials, not the motor carrier of them. Another example would be a motor carrier seeking relief from a regulation binding only upon air carriers.

If the applicant is eligible, and there is need for the exemption because of the continuing nature of the matter covered by the exemption, and if the current exemption or application for exemption is not dependent upon confidential information to which the applicant to become a party is not privy, the application to become a party is granted.

The *Federal Register* will contain a notice of the fact that someone has applied to be a party to an exemption as well as a notice of the grant or denial of the "party to" application. Public comment on this application is invited.

By filing an application to become a party to an exemption the applicant constructively adopts as his own the technical and safety information submitted by the original applicant and, if he is granted the status of a party to the exemption, he is bound by the same limitations and conditions that apply to the initial holder of the exemption, and he will be identified separately as a holder in his own name on exemption documents issued to him. If an exemption authorizes use of a packaging for the shipment or transportation of a hazardous material by any person or class of persons other than or in addition to the holder of the exemption, that person or a member of that class of persons may use the packaging under that exemption, but must maintain a copy of the exemption at each facility where the packaging is being used in connection with the shipment or transportation of the hazardous materials concerned. See 49 CFR 173.22a. Copies of exemptions may be obtained from the Office of Hazardous Materials Transportation, U.S. Department of Transportation, Washington, D.C. 20590, Attention: Docket Section.

In subsequent renewals, a party to an exemption must go through the same renewal application process described above, and the *Federal Register* will contain the necessary notices of application and grant or denial of the renewal application. If a party to an exemption fails to apply for renewal, and the original exemption holder has that exemption renewed, the former party to that exemption cannot continue to ship under that exemption after its original expiration date.

TIMING OF APPLICATION TO BECOME A PARTY

No specific time limits are set, but the application should be filed at least sixty days before the desired inclusion as a party. See above on timing of renewal applications.

VOLUNTARY WITHDRAWAL OF AN APPLICATION

Any applicant may withdraw his application from consideration any time before a final decision on that application has been issued. Withdrawal of the application, however, does not remove the paperwork from the public files, except in the case of confidential information discussed below. Any applicant asking for withdrawal after the *Federal Register* has contained a notice of the application should expect a corrective notice in the *Federal Register* indicating that withdrawal has taken place.

REQUESTS FOR EMERGENCY EXEMPTIONS

Referring back once again to the section of the statute quoted in the beginning of this chapter, one sees that the full public process, including *Federal Register* publication and an opportunity for public comment, must be followed in every case of application for issuance or renewal of an exemption "except when the Secretary determines that an emergency exists."

Congress did not define emergency, preferring to leave the widest latitude to the Secretary of DOT to determine in his discretion when the public processes should be sacrificed to higher needs. In the preamble to the exemption procedure amendments, DOT kindly quoted the author of the *Red Book* in justifying its decision to avoid a clear definition of emergency in the regulations. That quoted comment read:

> The regulations implementing the exemption power should . . . refrain from specific definition of an "emergency", for the very nature of emergencies is their unforeseen timing and character. The need for expedited treatment as an emergency matter is best left to the judgment and discretion of the Materials Transportation Bureau and its staff, to determine on a case-by-case basis as each situation arises. Attempts at definition of an indefinable concept will only serve to frustrate the equitable exercise of this power, by boxing it into criteria that fail to accommodate every situation that will be encountered.

In an emergency, unless you have handled an application like this before, the first step is to contact DOT by telephone, because the written application to be submitted must include "such supporting information with respect to each of the topics [found in a routine application] as the receiving Department of Transportation official considers necessary for processing the [emergency] application." (49

CFR 107.113(b).) The Washington offices of the RSPA are not open on a twenty-four hour basis, but the modal offices should be contacted by telephone at the earliest opportunity. The initial contact should be to the representatives of the modal administration for the mode of transportation that will be involved, as follows:

1. *Transportation by certificate holding aircraft operators.* Telephone the Federal Aviation Administration Civil Aviation Security Office that serves the place where the flight(s) concerned will originate or that is responsible for overall inspection of the air carrier's operations. Those who are unaware of the location and number for these offices should get this information from the air carrier or aircraft operator that will perform the transportation involved.

2. *Transportation by noncertificate holding aircraft operators* (operating under Part 91 of the Federal Aviation Regulations). Telephone the Federal Aviation Administration Civil Aviation Security Office that serves the place where the flight(s) will originate. Those who are unaware of the location and number for these offices should get this information from the operator of the aircraft who will perform the transportation involved or from the FAA Duty Officer, day or night, at (202) 426-3333.

3. *Transportation by motor carrier.* Telephone the office of the Chief of the Office of Motor Carrier Standards, Federal Highway Administration, Department of Transportation, Washington, D.C. 20590. Day (202) 366-2989 and Night (202) 267-2100.

4. *Transportation by rail carrier.* Telephone the office of the Associate Administrator for Safety, Federal Railroad Administration, Department of Transportation, Washington, DC 20590. Day (202) 366-9178 and night (202) 267-2100.

5. *Transportation by water carrier.* Telephone the office of the Chief, Hazardous Materials Branch, Marine Technical & Hazardous Materials Division, United States Coast Guard, Washington, DC 20593. Day (202) 267-1577 or night (202) 267-2100.

The modal offices will contact the RSPA, but a telephone call to the Exemptions Branch of RSPA's Office of Hazardous Materials Transportation from the applicant, indicating the nature of the emergency and alerting the office to expect to hear from the modal administration, will serve to smooth the process: (202) 366-4513.

The telephonic inquiries should identify the caller, the nature of the problem, and should inquire as to the specific information DOT will need to process the request for emergency exemption. The initial evaluation will be made by the modal administration with final action being taken by RSPA. The determination of an existing emergency shall be made if, on the basis of the information provided, RSPA finds that:

1. Existing conditions require the hazardous material concerned to be transported in commerce for the protection of life or property (other than the hazardous materials itself); and

2. The protection of life or property to be provided by the hazardous material would not be possible if the application is processed on a routine (more time-consuming) basis.

Not all emergency situations involve the protection of life or property. The regulations give RSPA the discretion to deal with these economic crises. The rules state that RSPA may determine that an emergency exists if:

1. Existing conditions require the hazardous material concerned to be transported in commerce to prevent or minimize serious economic loss; and
2. The prevention and minimizing of serious economic loss to be provided by the hazardous material would not be possible if the application is processed on a routine basis.

In determining what constitutes serious economic loss in this discretionary area, DOT will consider the nature and extent of the expected loss. Experience has shown that if the economic injury is to the applicant alone, this will not justify emergency processing; the applicant must show significant economic injury to other parties such as the customer who is waiting for the shipment.

For economic pressures that call for speedy action but do not rise to the level of an emergency, see the discussion of priority handling above, but realize that priority handling is rarely given. Either a problem qualifies as an emergency, or it gets handled routinely.

The emergency provisions do not give DOT the leeway to allow any transportation that would be unsafe or that would otherwise endanger the public. These provisions only pertain to the time of processing and the advisibility of delaying the responsive exemption pending *Federal Register* publication of a notice of application, allowing a period for public comment on the proposed exemption and consideration of those comments. In other words, the emergency provisions of the HMTA and the regulations short-cut processing time, not safety. DOT will publish a notice in the *Federal Register* advising the public of the grant or denial of any exemption, including an emergency application.

TERMINATION OF EXEMPTIONS

Every exemption has a stated expiration date not more than two years from date of issuance. In addition, RSPA may suspend any exemption, thereby temporarily prohibiting shipments under that exemption, if it is determined that:

1. Any activity under the exemption is not being performed in accordance with the terms of the exemption. Exemptions are granted based upon satisfaction of certain conditions set out in the exemption itself and in the appendices to 49 CFR Part 107, and failure to satisfy those conditions could result in a lifting of the exemption.
2. Or, on the basis of information not available at the time the exemption was granted, an amendment to the terms of that exemption is necessary to adequately protect against risks to life and property. In other words, additional requirements can be imposed if experience or new information shows them to be necessary.

DOT may terminate, or cancel, any exemption if it is determined that:

1. The exemption is no longer consistent with the public interest. This ground for termination is incapable of explicit definition but presum-

ably, since DOT is a safety agency, the public interest must be synony-mous with safety in transportation. In other words, if the practice authorized under the exemption is found to be unsafe, that clearly would make any continuation of that exemption inconsistent with the public interest.

2. The exemption is no longer necessary because of an amendment to the regulations. Obviously, if an exemption is needed to relieve a person from a specific regulation, and that regulation is later deleted or amended in a rule making action, the exemption becomes superfluous and is terminated.

3. The exemption was granted on the basis of false, fraudulent, or misleading representations or information.

Unless DOT believes that immediate suspension or termination is necessary to abate the risk of an imminent hazard, the holder of that exemption will be given notice in writing of the reasons for the proposed suspension or termination and will be given an opportunity to show cause why that action should not be taken before such action is taken. Under the HMTA an imminent hazard exists if there is substantial likelihood that serious harm will occur prior to the completion of an administrative hearing or other proceeding initiated to abate the risk of such harm. Serious harm, in turn, is defined as meaning death, serious illness, or severe personal injury.

APPEALS FROM DECISIONS AFFECTING EXEMPTIONS

Any applicant for an original exemption or the renewal of an existing exemption who believes himself aggrieved by a DOT decision or any person whose exemption is suspended or terminated by the RSPA may appeal that decision to the Administrator of the RSPA, who sits as the supervisor of the director of OHMT. Any such appeal must be filed within thirty days of the decision or action. The administrative action with regard to this matter is not to be considered final until a decision on the appeal has been issued in the form of an order from the RSPA Administrator. In other words, no court will consider hearing a case regarding this matter until the RSPA rules on the appeal, and any petition to a court for review will be returned unanswered due to the petitioner's failure to exhaust his administrative remedies. An appeal could cover the denial of an application, the partial grant of an application coupled with a partial denial, etc.

CONFIDENTIALITY OF APPLICATION INFORMATION

The Freedom of Information Act, as it is popularly called, is published in 5 U.S. Code 552. It gives broad guidelines on the types of information that a federal agency may withhold from public disclosure. The act itself does not compel the withholding of any information but, rather, compels federal agencies to disclose information upon request unless that information falls within certain specified exceptions to the general rule. Like other statutes, it is implemented within each federal agency through regulations issued by that agency.

The DOT regulations governing the availability of information in all the offices and administrations of DOT are found in 49 CFR Part 7. As a general rule, the types of information that may be withheld are limited to narrowly interpreted exceptions to the general law that compels disclosure. Even if an exception to the rule of disclosure is available, DOT as a matter of longstanding policy will disclose the information unless the facts of the particular situation warrant nondisclosure.

Section 107.5 of the RSPA's rules is addressed directly to the matter of confidentiality of information submitted to DOT with an application for exemption or any other matter. Note that this section is of general applicability, not just applicability to exemption procedures.

If any person filing any document with RSPA claims that some or all of the information in that document is exempt from the mandatory public disclosure requirements of the law, and if that person asks RSPA not to disclose the information to the public, that person must file a second document with the first from which the confidential information has been deleted. He must indicate in the full original document that it is confidential or contains confidential information and he may file a statement specifying the justification for which confidential treatment is claimed. If the grounds for the claim are that the information is a trade secret or is commercial or confidential information protected under the Freedom of Information Act, some statement must be included in the document indicating how the exceptions to that act apply and why the information is privileged or confidential. If the person filing a document does not submit a second copy of the document with the confidential information deleted, RSPA will assume there is no objection to public disclosure of the document in its entirety.

Even if a second document is provided, and specific reasons are given in justification of why the material should not be disclosed, RSPA still retains the right to make its own determination with regard to any claim of confidentiality. Notice of an RSPA decision to deny the claim of confidentiality, in whole or in part, and an opportunity to respond shall be given to the person making the claim no less than five days before RSPA will disclose the information. If the decision on the overall application has not yet been made, the applicant may withdraw any portion of it for which confidentiality was claimed but denied. If, however, the documents have served as the basis for completed official action, they may not be removed from the public file even though confidentiality was claimed and denied. Other aspects of the documents may be voluntarily withdrawn from official consideration, but those papers will remain in the public file unless they were the subject of a specific claim of confidentiality that was denied before official action was taken on the application.

The Freedom of Information Act has been the subject of voluminous litigation, with each decision adding to the vast body of interpretive case law under the statute. The law itself is written vaguely and the regulations are not much better. It is rare in dealing with the hazardous materials offices of DOT that legal advice is essential, but in the case of a claim of confidentiality the advice of counsel is a must.

SERVICE OF PROCESS ON NONRESIDENTS OF THE UNITED STATES

When an applicant for an exemption or to become party to an exemption is not a resident of the United States, he is required by the regulations to designate a permanent resident of the United States to serve as his agent upon whom service of process may be made for him and on his behalf. The agent may be an individual, a firm, or a domestic corporation. Any number of principals may designate the same person as agent. A designation is binding on the nonresident principal even if it is technically not in compliance with the detailed designation procedures, unless and until that designation is rejected by RSPA. A designated agent may not assign performance of his functions under the designation to another person.

The designation must be in writing and must be dated. It has to be in the legal form necessary to make it valid and binding on the principal under the laws, corporate by-laws, or other requirements governing the making of the designation by the principal at the place and time where it is made, and the person or persons signing the designation must certify that it is so made. The designation must include the principal's full legal name, name of business, and mailing address. It must also provide that it shall remain in effect until withdrawn or replaced by the principal. Of course, the designation must include the legal name and mailing address of the agent, and must bear a declaration of acceptance of the responsibility signed by that agent.

Service of any process, notice, order, decision, or requirement of RSPA may be made by registered or certified mail addressed to the agent with return receipt requested, or in any other manner authorized by law. If service cannot be effected because the agent has died (or, if a firm or corporation, has ceased to exist), or moved, or otherwise does not receive correctly addressed mail, service may be made by publication in the *Federal Register*.

It has never been resolved whether the responsibility taken on by an agent for service of process might extend to additional liability for injuries or costs involved in a legal action brought by DOT or any other party.

STANDARD CONDITIONS APPLICABLE TO ALL EXEMPTIONS

Exemptions from the regulations governing packages, containers, and the preparation and offering of hazardous materials for shipment are subject to these standard conditions:

1. The outside of each package must be plainly and durably marked "DOT-E" followed by the exemption number assigned. On portable tanks, cargo tanks and tank cars, the marking must be in letters at least two inches high on a contrasting background.
2. Each shipping paper issued in connection with any shipment made under an exemption must bear the notation "DOT-E" followed by the exemption number assigned in addition to the various shipping paper entries required by 49 CFR Part 172.
3. When an exemption issued to a shipper includes special carrier requirements, the shipper must furnish a copy of the exemption to the carrier before or at the time of tender of each shipment under the exemption.

FLIGHTS OF CIVIL AIRCRAFT

Exemptions from the regulations governing the transportation of hazardous materials on civil aircraft are subject to standard limitations set forth in detail in Appendix B to the Part 107 procedural requirements. See those specific requirements for shipments under exemptions applicable to these aircraft.

CONCLUSION

The preceding pages describing the extensive regulatory procedures governing application for and issuance of exemptions probably have left the reader wondering whether it is worth it or whether the extra effort to satisfy the general regulations might not be less costly and time-consuming in the long run. The legislative history of the HMTA reveals that bewilderment and frustration were intended objectives. The drafters of Section 107 intended to discourage the exemption process, to urge the applicants and the government to avoid exemptions, and to move instead more quickly into rule making of general applicability.

Unfortunately, however, there are many instances in which an exemption is the only practical way to resolve a transportation problem. In these instances, the applicant will be frustrated needlessly by the tremendous amounts of time consumed in the process, while the staff implementing the program will face even greater frustration in touching all the bases set out by the Congress for each exemption and renewal.

If an application for an exemption concerns a matter of such general applicability and future effect as to warrant being made the subject of rule making, RSPA may initiate rule making under 49 CFR Part 106 in addition to or in lieu of granting or denying the application. In other words, DOT may accept an application for exemption and treat it as a petition for rule making, perhaps issuing the exemption in the interim pending completion of the rule making process. This is the eventual goal of all exemptions, to be accepted into the ranks of general regulations in 49 CFR, and it may be that the exemption process will become so cumbersome as to make rule making a viable alternative to the exemption. As of the writing of the second edition of the *Red Book*, through Docket Nos. HM-139 the process of conversion of exemptions to general rules has accelerated. Still, rule making often is a laboriously slow process consuming years of effort. Both exemptions and rule making are time-consuming, and good company planning should incorporate delays into the schedule.

For more on hazardous materials rule making processes, see Chapter 27.

APPROVALS

Throughout the DOT regulations one occasionally finds a reference to an approval that must be obtained from the agency. Although similar in many respects to exemptions, approvals are processed differently. The most notable difference is the lack of a public procedure, including any Federal Register notice or opportunity to comment. In fact, there are no procedural rules governing approval applications, their grant or denial, appeals or anything else. Perhaps DOT will adopt procedural regulations for approvals in the near future, since the lack of procedures confuses the regulated public and opens the agency to criticism for being arbitrary.

If you need to obtain an approval, you might consider using the same approach in an application as it outlined for exemptions. Before doing anything, however, contact the Exemptions & Approvals Division, Office of Hazardous Materials Transportation, Research & Special Programs Administration, Department of Transportation, Washington, DC 20590, and determine what in particular need be included in your application.

CHAPTER 29

DOT Organization
and Hazardous Materials

The Department of Transportation (DOT) is a cabinet-level, executive department reporting directly to the White House. It was created by the Department of Transportation Act of 1967, 49 U.S. Code 1651, *et seq.*, largely through the accumulation of transportation functions previously performed by other agencies of the federal government under other provisions of law. To a great extent, the people performing these functions at that time also were transferred into DOT, including those who had been directly responsible for the regulation of hazardous materials in transportation.

DOT generally consists of a varied series of operating administrations under a superimposed layer of policy and administrative offices. It may be fairly said that most line functions are within the operating administrations, while policy decisions involving national and intermodal transportation issues are the responsibility of the overall layer, called the Office of the Secretary, or OST. Operating administrations pertaining to a single mode of transportation (for instance, the Federal Aviation Administration) have become commonly referred to in DOT as the "modes." Thus, although DOT includes such agencies as the Saint Lawrence Seaway Development Corporation and the National Highway Traffic Safety Administration, the term "modes" has come to mean only four of the operating administrations: the Federal Aviation Administration (FAA), Federal Highway Administration (FHWA), Federal Railroad Administration (FRA), and the Coast Guard (USCG).

As indicated above, DOT's original development was less a process of legislative creation than legislative accumulation. The real novelty was having all of these functions under a single umbrella, but most of the modes had been in operation for years in one form or another. As far as hazardous materials regulations are concerned, the regulations now found in 49 CFR as they applied to shippers and carriers by land and shippers by water had been developed and enforced by the Interstate Commerce Commission. The Coast Guard, as part of the Department of Treasury, had developed and enforced regulations applicable to carriers by water, and were statutorily required to draw heavily upon the work of the ICC. The Federal Aviation Agency, as an independent agency, had regulated shippers and carries by air, again with heavy reliance on the regulations developed by the ICC.

Although these functions were all brought within DOT, by statute the hazardous materials regulatory authority was split among the four modes, with the Secretary of Transportation and the people in OST having no substantive authority. The Hazardous Materials Regulations Board (HMRB, since abolished) was created as a mechanism to minimize inconsistencies and conflicts in the regulations between the modes of transportation. The HMRB consisted of four representatives, one from each of the modes, who exercised the decision-making and signature powers for the creation and amendment of regulations. A representative of the Secretary of Transportation served as non-voting chairman of the HMRB, while a representative of OST's general counsel sat as legal advisor.

The Hazardous Materials Transportation Act (HMTA), 49 U.S. Code 1801, *et seq.*, brought a reorganization of these functions. (See Chapter 30 for more detail on the HMTA.) By delegating regulatory power directly to the Secretary of Transportation for the Secretary to redelegate elsewhere or not, the HMTA removed the mandatory delegations of authority to the modes. The most immediate and visible impact of this was to put the authority to sign notices of proposed rule making, amendments, and exemptions into one hand rather than four.

The Secretary of DOT chose to implement this new power by creating a new operating arm with a rank equivalent to the modes but not allied with any specific mode of transportation itself. This body now is called the Research & Special Programs Administration (RSPA) and the Administrator of the RSPA reports to the Secretary of Transportation and Deputy Secretary, not to any of the modal administrations.

The RSPA in its hazardous materials role is by nature a line organization, appropriately placed outside the policy, evaluation, and related staff functions that are the primary responsibilities of the Office of the Secretary.

RESEARCH & SPECIAL PROGRAMS ADMINISTRATION

RSPA includes within it the DOT responsibilities and duties relating to gas and liquid pipeline safety as well as those relating to hazardous materials transportation via land, water, and air carriers. Hazardous materials statutory responsibility exclusive to a single mode of transportation is still the responsibility of the appropriate operating administration. All activity with respect to intermodal, operational hazardous materials functions, however, is the responsibility of the RSPA. The discussion of the RSPA in the *Red Book* will only be directed to the hazardous materials functions, not those pertaining to pipeline safety or the Office of Pipeline Safety Operations.

According to the internal notice which created the organization, its hazardous materials responsibilities include:

1. Developing, approving, issuing, and enforcing compliance with intermodal hazardous materials regulations, including hazardous materials container manufacturers and shippers, except for those activities exclusive to a single mode. Functioning as publication agent for hazardous materials regulations developed by modal administrations.
2. Collecting, compiling, and analyzing data.
3. Performing intermodal training and education.
4. Conducting research and development except with respect to bulk vehicles peculiar to a single mode.
5. Approving and issuing exemptions and approvals after coordination with modal administrations, as necessary.
6. Managing the selective registration of persons who transport or cause

to be transported or shipped in commerce hazardous materials or who are engaged in packaging or container activities related thereto.

Responsibility is assigned to the respective modal administrations—the Coast Guard, FAA, FHWA, and FRA—for all hazardous materials functions pertaining exclusively to their individual modes. Operating administrations developing hazardous materials regulations that pertain exclusively to their mode of transport coordinate such regulations with other administrations and the RSPA as required prior to final approval by the administration and publication by the RSPA. For actual delegations of authority from the Secretary to each of these groups, see 49 CFR Part 1.

OFFICE OF HAZARDOUS MATERIALS TRANSPORTATION

OHMT is an operating group made up of an office director and several specialized divisions relating to such functions as technical materials evaluation, packaging evaluation, training, regulations development and revision, exemptions, approvals, enforcement and compliance, etc. The director of OHMT is authorized by specific delegation to conduct all rule making proceedings, except the issuance of final rules and the grant or denial of petitions for reconsideration. Most applications for exemptions and approvals, and grants or denials of exemptions of any nature, are handled through OHMT with coordination with modal representatives as may be necessary to the processing of that exemption.

MODAL ADMINISTRATIONS

The Federal Highway Administration, the Federal Railroad Administration, and the U.S. Coast Guard retain direct regulatory authority with regard to cargo tanks, tank cars, and tank vessels, respectively. In addition, these modal administrations as well as the Federal Aviation Administration are involved in compliance inspections and enforcement of the hazardous materials regulations against shippers and carriers in those modes. See Chapter 23 on inspections and enforcement. In addition to Washington headquarters' staff involved with hazardous materials, each of the modal groups has an extensive field force in regional offices throughout the country.

U.S. DEPARTMENT OF TRANSPORTATION

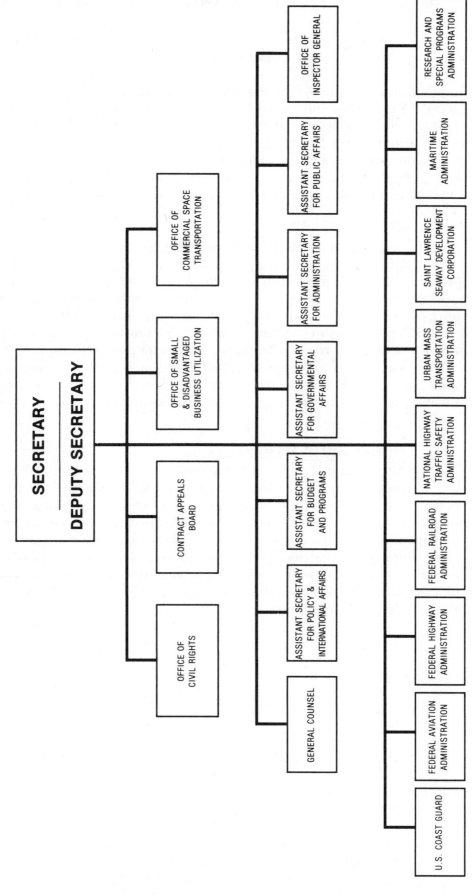

CHAPTER 30

Hazardous Materials Transportation Act

In 1975, President Ford signed the Transportation Safety Act of 1974 into law, title I of which individually was named the Hazardous Materials Transportation Act (HMTA). Much of this new law was designed to cure longstanding problems in the administration of the former statutes on this subject, but it also was an expression of congressional discontent with the weakness of those earlier laws and the lack of governmental enforcement of them. It was a substantial change in the course of regulatory events in the hazardous materials field and, therefore, this chapter is devoted to a section-by-section analysis of the more important provisions of the HMTA. Those provisions are repeated in the text of the *Red Book* for the sake of easy reference, and the entire statute appears in Appendix D.

The proper legal citation for the HMTA is 49 U.S.C. 1801 *et seq.*, 88 Stat. 2156, P. L. 93–633.

> **Sec. 102.** It is declared to be the policy of Congress in this title to improve the regulatory and enforcement authority of the Secretary of Transportation to protect the Nation adequately against the risks to life and property which are inherent in the transportation of hazardous materials in commerce.

This opening Declaration of Policy should be read as DOT reads it: that Congress was dissatisfied with administration of this function as it had been accomplished before 1975 and that substantial changes in the regulatory practices had to be made. This congressional declaration meant there will be more rule making and more enforcement of existing regulations than had been historically true in this area.

DEFINITIONS

> **Sec. 103.** As used in this title, the term—"commerce" means trade, traffic, commerce, or transportation, within the jurisdiction of the United States, (A) between a place in a State and any place outside of such State, or (B) which affects trade, traffic, commerce, or transportation described in clause (A).

This definition of commerce, although difficult to read, could have brought about one of the major changes in this regulatory field. Under earlier hazardous materials law, regulations were made binding upon carriers engaged in interstate or foreign commerce and upon the shippers using the services of such carriers. Although this occasionally covered intrastate shipments aboard a carrier otherwise engaged in interstate commerce, generally speaking the federal regulations were limited to interstate or foreign commerce. Under the HMTA definition of commerce, however, any trade, commerce, or transportation that affects interstate or foreign trade, traffic, commerce, or transportation could be covered. Laws drafted to cover interstate activities or those which affect interstate or foreign commerce, like this one and like the Occupational Safety and Health Act of 1970, are as broad as Congress can make them based upon their authority under the Commerce Clause of the U.S. Constitution. Court decisions interpreting this use of the word "affect" commerce have held that virtually all trade, traffic, commerce, or transportation has some effect on interstate or foreign commerce.

The net result is that almost all intrastate commerce under this law came within the regulatory sphere of DOT for the first time. Although essentially all commerce is within DOT's reach, in a rule making to implement the HMTA, DOT voluntarily limited the scope of its regulatory control over shippers and carriers to those who were covered under the jurisdiction of the old law. The only exceptions to this are the transportation of hazardous substances and hazardous wastes, and flammable cryogenic gases in bulk.

> "Hazardous material" means a substance or material in a quantity and form which may pose an unreasonable risk to health and safety or property when transported in commerce.

"Hazardous material" was not a defined term before the HMTA was enacted. The law providing the basic foundation for the body of regulations developed to that date described the materials to be regulated by example, not by any common characteristics or generic descriptions. "Quantity and form" are words of limitation that recognize that the smaller sizes and dilute forms of some materials that are otherwise regulated by name do not pose the degree of hazard that would warrant federal regulation.

"Unreasonable risk" is an unwieldy concept that DOT never has defined. There have been many contracts and studies on risk analysis and risk assessment, but all have foundered on the lack of actual data on the transportation environment.

"Health and safety" is a combination of terms that would appear to overlap some of the jurisdiction of the Occupational Safety and Health Administration, but here it is limited to those risks to health that occur when that material is transported in commerce. Nothing under OSHA applies to the working conditions of employees with respect to which agencies such as DOT "exercise statutory authority to prescribe or enforce standards or regulations affecting occupational safety or health." Thus, DOT's regulations could presumably preempt those of OSHA in the transportation workplace and, in some cases, this has happened. In hazard communication regulations under OSHA, however, the package labeling and material safety data sheet requirements adopted under 29 CFR 1910.1200 have extended into transportation and product distribution.

"When" is a time word that must be read in conjunction with "health and safety."

The legislative intent was to cover those hazards to health and safety that are present to an unreasonable degree during transportation. The fact that intense exposure to a material over a period of several weeks might make someone sick does not of itself bring that material within the definition of a hazardous material under HMTA. The danger or threat during transportation is too minimal to be considered unreasonable. A more difficult case is a material that, upon repeated short exposure, may lead to illness. Perhaps DOT or OSHA both would regulate such a material due to the repeated exposure of transport workers to the hazard.

> "Serious harm" means death, serious illness, or severe personal injury.

This term must be read in conjunction with the provisions on imminent hazard, discussed below.

DESIGNATION OF HAZARDOUS MATERIALS

> **Sec. 104.** Upon a finding by the Secretary, in his discretion, that the transportation of a particular quantity and form of material in commerce may pose an unreasonable risk to health and safety or property, he shall designate such quantity and form of material or group or class of such materials as a hazardous material. The materials so designated may include, but are not limited to, explosives, radioactive materials, etiologic agents, flammable liquids or solids, combustible liquids or solids, poisons, oxidizing or corrosive materials, and compressed gases.

This is the prime section of the HMTA determining what materials are covered by DOT regulations. It, combined with the hazardous material definition discussed above, gives DOT broad authority to regulate many classes of articles, including but not limited to those classifications that they regulated historically. Earlier controversies as to whether DOT could extend beyond the enumerated examples into such areas as combustibles were laid to rest by this amendment to the law.

REGULATIONS GOVERNING TRANSPORTATION OF HAZARDOUS MATERIALS

> **SEC. 105.** (a) General—The Secretary may issue, in accordance with the provisions of section 553 of title 5, United States Code, including an opportunity for informal oral presentation, regulations for the safe transportation in commerce of hazardous materials.

The first sentence forms the core of this statute, as it impacts the regulated industries, since it gives the Secretary of Transportation the power to issue regulations. This entire section is important and will be dissected sentence-by-sentence for the benefit of the reader.

"In accordance with the provisions of section 553 of title 5, United States Code" refers to the informal rule making provisions of the Administrative Procedure Act. See Chapter 26. The reference to the "opportunity for an informal oral presentation" is consistent with that section of the APA. By informal is meant that statements are not under oath, there is no opportunity for cross-examination, there are no rules of evidence, and generally there are no findings, conclusions, or rights of appeal from findings and conclusions. See Chapter 27 on rule making procedures at DOT. Note that an opportunity for hearing is not interpreted by most courts or by DOT as meaning an automatic hearing or even a guarantee of a hearing if requested. DOT requires the requesting party to show good cause for holding a hearing.

"Safe transportation in commerce" is a phrase of limitation, to the end that DOT may not directly regulate economic or other non-safety conditions of transportation. A similar provision in earlier statutes formed the basis of DOT's refusal to support non-tariff barriers to imports in the form of discriminatory, hazardous materials regulations.

> Such regulations shall be applicable to any person who transports, or causes to be transported or shipped, a hazardous material, or who manufactures, fabricates, marks, maintains, reconditions, repairs, or tests a package or container which is represented, marked, certified, or sold by such person for use in the transportation in commerce of certain hazardous materials.

"Any person who transports" is a wordy phrase meaning carrier. The statute specifically and intentionally declines to define either a carrier or a shipper, primarily due to difficulties encountered under other definitions when it came to categorizing various intermediate parties such as freight forwarders, pier brokers, warehousemen, contract packagers, etc. Thus, "any person who transports," without specifically saying so, covers all private, common, and contract carriers by any mode of transportation, including the pickup and delivery services of contract packagers.

"Causes to be transported or shipped" is another long phrase deliberately chosen to avoid use of the term "shipper." Again, the difficulty inherent in categorizing intermediate persons who are not agents of the primary shipper is the basis for selecting this functional description of the persons to whom regulations are applicable. It is worthwhile to note, in this regard, the definition of a shipper proposed by DOT in rule making:

> "Shipper" means any person who performs any function assigned by the regulations in this subchapter to a shipper. Performance of any function by one individual as a shipper does not exclude another person from also being considered a shipper. For example, a warehouseman who presents hazardous materials to a carrier may be subject to the regulations as a shipper or as the agent of a shipper, and the person who packed, marked, classified, and labeled the shipment initially may also be considered a shipper. [Docket No. HM-112, 39 F.R. 3037, January 24, 1974.]

The phrase "who manufactures, fabricates, marks, maintains, reconditions, repairs, or tests a package or container" applied the DOT regulatory power for the first time directly to the maker of packaging for hazardous materials. Until this provision was created, the DOT regulations were only binding upon carriers and shippers. Occasionally problems related to the manufacture, reconditioning, and repair of packaging for hazardous materials had gone unresolved because of the lack of direct federal regulatory jurisdiction to apply or enforce the hazardous materials regulations with regard to the persons involved. Again, the terms define the applicability not by tagging a label on the people involved but by describing the functions they perform. As with shippers discussed above, more than one person could be subject to DOT jurisdiction for any single container. So, for example, the maker of component parts such as gaskets, the assembler of those parts, and the independent lab that determines that the completed packaging meets the DOT specifications, all could be liable for DOT penalties if there were a failure of the container.

"A package or container which is represented, marked, certified, or sold by such person for use in the transportation in commerce of certain hazardous materials" is limiting language that defines the packages and containers of concern. This applicability is unique in its focus on the marketing department of a container manufacturer through the use of the words "represented" and "sold." If the manufacturer does not intend his product to be used for hazardous materials but his salesman tells customers that the product is good for that use, the manufacturer is drawn within the scope of applicability of the regulations and may be penalized for any violation of the regulations. This applicability extends not only to DOT specification containers, but to all containers that are represented, marked, certified, or sold for the purpose of use in hazardous materials transportation.

HANDLING OF HAZARDOUS MATERIALS

Sec. 106. (a) Criteria—The Secretary is authorized to establish criteria for handling hazardous materials. Such criteria may include, but need not be limited to, a minimum number of personnel; a minimum level of training and qualification for such personnel; type and frequency of inspection; equipment to be used for detection, warning, and control of risks posed by such materials; specifications regarding the use of equipment and facilities used in the handling and transportation of such materials; and a system of monitoring safety assurance procedures for the transportation of such materials. The Secretary may revise such criteria as required.

This laundry list of criteria is self-explanatory, but it should be read as an indicator of the focus of congressional attention on quality control and training in the total range of activities leading to the transportation of hazardous materials.

(b) Registration.—Each person who transports or causes to be transported or shipped in commerce hazardous materials or who manufactures, fabricates, marks, maintains, reconditions, repairs, or

tests packages or containers which are represented, marked, certified, or sold by such person
for use in the transportation in commerce of certain hazardous materials (designated by the Secretary) may be required by the Secretary to prepare
and submit to the Secretary a registration statement not more often than once every 2 years.

This provision of Section 106 of the HMTA stirred a great deal of concern among the legal staffs of regulated industries, but it must be stressed that as used in this section, "registration" does not mean licensing. Early drafts of this section as it was considered in Congress clearly show that, although licensing was seriously consid-ered at one time and a licensing provision did pass the Senate, the concept was abandoned in conference as a time-consuming and expensive paperwork burden on regulated industry and government alike, and that the other punitive powers given to the Secretary made a license revocation power unnecessary. The registra-tion described in this section is for the sake of identification, not for the sake of licensing. Thus, it declares that the registration shall include, but need not be lim-ited to, the registrant's name, principal place of business, the location of each activ-ity handling such hazardous materials, a complete list of all such hazardous mate-rials handled, and an averment that he is in compliance with all the applicable criteria prescribed by DOT.

This is not a mandatory provision compelling DOT to register all persons in this field, but a discretionary provision permitting DOT to register the persons dealing with those particular hazardous materials for which this registration or identifica-tion is considered by DOT to be necessary. Obviously, one would imagine that the materials selected will be those in the higher range of hazard, including all the shippers, carriers, and makers of packaging for the transportation of those mate-rials. Historically, DOT has required registration of certain container manufac-turers and reconditioners, such as drum reconditioners and makers of Specification 39 nonreusable gas cylinders.

Periodically various factions have encouraged DOT to engage in licensing of shippers, carriers or others. This necessarily would require a legislative change, since the registration authority itself does not permit DOT to assess the capabilities of an applicant, or to withdraw the registration. To date, none of those who have suggested a legislative change to give DOT licensing authority have been able to justify the major administrative burden this task would put on the agency.

The averment of compliance may cause some legal staffs to pause, as will the following statement in Section 106:

(c) Requirement—No person required to file a registration statement under subsection (b) of this
section may transport or cause to be transported
or shipped hazardous materials, or manufacture,
fabricate, mark, maintain, recondition, repair, or
test packages or containers for use in the transportation of extremely hazardous materials, unless he
has on file a registration statement.

Although this registration is not a true licensing scheme, it is also not a pure registration system either, since one must have a registration statement on file be-fore being able to lawfully conduct the business described by that statement.

EXEMPTIONS

Sec. 107. (a) General—The Secretary, in accordance with procedures prescribed by regulation, is authorized to issue or renew, to any person subject to the requirements of this title, an exemption from the provisions of this title, and from regulations issued under section 105 of this title, if such person transports or causes to be transported or shipped hazardous materials in a manner so as to achieve a level of safety (1) which is equal to or exceeds that level of safety which would be required in the absence of such exemption, or (2) which would be consistent with the public interest and the policy of this title in the event there is no existing level of safety established.

This sentence of Section 107 of the HMTA stimulated the first litigation under the new law. The law was signed on January 3, 1975 and as a legal document came into effect. As a matter of practicality, however, DOT required a reasonable period of time to phase in the new procedures and requirements. For other requirements, this was clearly foreseen in Section 114 of the HMTA, discussed below. For the exemption section, DOT also delayed the phase-in of procedures, since the first sentence directly states that exemptions must be issued "in accordance with procedures prescribed by regulation." DOT interpreted this to mean that new regulations had to be adopted under Section 105 of HMTA and that the business of issuing waivers, exemptions, and special permits would continue as usual until such regulations were developed. In the interim, however, a consumer group took the issue to court, where the judge disagreed with DOT and declared all exemptions and special permits issued from January 3, 1974, until the phase-in of Section 107 procedures to be null and void. This matter is now moot, since the adoption of final procedural regulations implementing this section effective October 16, 1975.

A flaw in the exemption section is its failure to make any explicit provision to issue exemptions to packaging manufacturers and reconditioners. The regulatory power in Section 105 clearly encompasses regulations binding upon such persons, but the exemption section does not mention them. In the preamble to the exemption procedures rule making, DOT addressed this matter declaring: "A person's right to petition an agency for relief from a regulation of that agency which directly affects that person is so well established as to be beyond question."

Each person applying for such an exemption or renewal shall, upon application, provide a safety analysis as prescribed by the Secretary to justify the grant of such an exemption.

Although this sentence calls for a safety analysis, the initial regulations to implement this requirement do not use that term. (See Chapter 28 for more information on exemptions.)

A notice of an application for issuance or renewal of such exemption shall be published in the Federal Register. The Secretary shall afford access to any such safety analysis and an opportunity for

> public comment on any such application, except that nothing in this sentence shall be deemed to require the release of any information described by subsection (b) of section 552 of title 5, United States Code, or which is otherwise protected by law from disclosure to the public.

The provision calling for *Federal Register* notification to the public on the application for an exemption was a novelty for DOT but has been in the environmental, occupational safety and health, and other federal regulatory systems. This requirement does not compel the publication of the application for exemption itself, but merely the fact that such an application has been filed and by whom. The public—including public interest groups, labor groups, and competitors—has an opportunity to review the application and safety analysis, within certain bounds. Section 552 of title 5 U.S.C. is commonly called the Freedom of Information Act. Any claim of confidentiality, trade secrets, or proprietary data that demands nondisclosure should be prepared by someone familiar with the Freedom of Information Act, as amended, and the regulations of DOT implementing that Act, found in 49 CFR Part 7 and Section 107.5. In any event, no request for nondisclosure of information guarantees that DOT will indeed withhold that information from the public. Upon request, however, DOT will render a decision and, if it is adverse, will allow the applicant to withdraw the material if no decision based on that material has yet been made.

> (d) Except when the Secretary determines that an emergency exists, exemptions or renewals granted pursuant to this section shall be the only means by which a person subject to the requirements of this title may be exempted from or relieved of the obligation to meet any requirements imposed under this title.

The first part of this paragraph from Section 107 opens the door for DOT to skip the time-consuming processes of an ordinary exemption when the Secretary determines that an emergency exists. An emergency is not defined in the statute, but under the procedural regulations in 49 CFR Part 107 it definitely includes situations other than straight life-and-death emergencies. The regulations implementing this provision compel a finding of emergency in a life-and-death circumstance but also reserve the power to the Secretary to make an emergency exemption to avoid or minimize substantial economic harm.

All exemptions in the hazardous materials field must be processed through this section, including situations formerly covered by special permits, Coast Guard portable tank special permits, and Federal Aviation Administration waivers and deviations.

TRANSPORTATION OF RADIOACTIVE MATERIALS ON PASSENGER-CARRYING AIRCRAFT

This section called for enactment of new regulations with respect to the transportation of radioactive materials on any passenger-carrying aircraft. These regulations prohibit the transportation of radioactive materials on any such aircraft unless the radioactive materials are intended for use in, or incident to, research, or medical diagnosis or treatment, so long as such materials, as prepared for and during transportation, do not pose an unreasonable hazard to health and safety. The research mentioned in this section need not be medical research.

POWERS AND DUTIES OF THE SECRETARY

Section 109 of the HMTA provides a laundry list of powers given the Secretary to carry out responsibilities under the HMTA. One interesting aspect of the general powers clause is the following sentence:

> The Secretary is further authorized after notice and an opportunity for a hearing, to issue orders directing compliance with this title or regulations issued under this title; the district courts of the United States shall have jurisdiction, upon petition by the Attorney General, to enforce such orders by appropriate means.

It is unclear to what extent this power will or could be exercised by the Secretary of Transportation, especially in light of those powers generally described in Section 111 of the HMTA, 49 U.S.C. 1810, ambiguously entitled "Specific Relief." The essential elements of this power are the indirect manner of enforcement through the courts via the Attorney General, and the requirement that any Secretarial orders directing compliance be issued "after notice and an opportunity for a hearing." These aspects of the regulatory authority tend to deter action by the agency and it is more likely, therefore, that the general provisions of the agency-administered enforcement program will continue to be the active statutory sections.

DOT has always maintained certain record-keeping requirements, and paragraph (b) of Section 109 codifies the power to continue to impose such requirements.

Under Section 109(c), DOT's officers, employees, or agents may enter upon, inspect, and examine, at reasonable times and in a reasonable manner, the records and properties of persons subject to the HMTA, including shippers, carriers, and packaging manufacturers. This inspection power is specifically limited, however, to the extent to which such records and properties relate to the hazardous materials function of that person. Any such inspector shall, upon request, display proper credentials and the person receiving such an inspector would be prudent to make such a request.

Although the section authorizes the taking of depositions and the issuance of subpoenas, it does not call for the taking of signed statements from regulated persons or their employees during the course of an investigation, and regulated persons should be wary of providing such signatures or in allowing their employees to sign statements.

Paragraph (d) of Section 109 calls for specific risk evaluation capability within the federal government, which should be seen as a more flexible requirement than the maintenance of such a capability within DOT itself. The central reporting system called for in this provision is maintained and carriers of hazardous materials are required to report any unintentional release or spill of hazardous materials. See Chapter 21 as to reportable quantities and proper forms for preparing hazardous materials incident reports.

The Annual Report to the Congress mentioned in Section 109(e) was originally established by the Hazardous Materials Control Act of 1970, 49 U.S.C. 1761–1762, which was repealed by HMTA. These reports serve as a compilation of DOT activities and may provide interesting statistical background.

PENALTIES
Also see Chapter 23 on enforcement of the hazardous materials regulations.

CIVIL PENALTIES
The term "civil" is used to distinguish this enforcement program from the criminal provisions discussed below. The legislative history reveals that this broad enforcement power, sought by DOT and eagerly given by the Congress, is one of the critical sections of the HMTA. Since vigorous enforcement was contemplated, it is worthwhile to dissect this power carefully.

"Any person (except an employee who acts without knowledge). . . . " This opening would appear to relieve an untrained employee from personal liability for the civil penalties imposed by DOT. In theory, this situation should not be encountered, because all the applicable regulations are preceded by a requirement that the regulated company acquaint its employees with the regulations, and those employees are supposed to become familiar with them. In the real world, high turnover rates at the lower employment levels of a company as well as the complexity of the training function may bring about the frequent instance of the employee who acts without knowledge. Providing a training course for such employees and keeping accurate records of those who attend such courses would seem to refute any claim of lack of knowledge on the part of any person attending those courses.

" . . . [Who] is determined by the Secretary, after notice and an opportunity for a hearing. . . . " This indicates that the Secretary makes the initial determination of the existence of a violation and the persons responsible for that violation of the statute or regulations.

The "notice and opportunity for a hearing" are not very formal requirements, and in other contexts they have been interpreted by DOT to mean that the alleged violator receives a letter advising him of the Secretary's initial conclusions of liability and offering an opportunity to meet with someone within DOT to talk about it. Although explanations of how the alleged event took place are of some interest, the true focus of attention by the agency will be upon the steps that have been taken subsequent to the incident to preclude recurrence. Internal disciplinary action, development of new procedures and manuals, implementation of corrective policies, and the like, are of critical importance in the Secretary's determination of his willingness to compromise on the severity of the penalty to be assessed for that violation.

" . . . [To] have knowingly committed an act which is a violation of a provision of this title or of a regulation issued under this title. . . . " These words have been carefully drafted to avoid saying that one must knowingly violate the law. Note that the adverb "knowingly" modifies the commission of an act, not the matter of violation. Thus, if a person is in possession of his senses, it is likely that he is aware of his direct acts. If those acts happen to violate the law, liability ensues. As the United States Supreme Court held in *United States v. International Minerals & Chemical Corp.*, 402 U.S. 558 (1971), ignorance of the law is no excuse. Grammatically, however, this requirement of knowing violations would appear to be contradictory to the opening pardon given to those employees who act without knowledge. Even the president of a corporation is an employee of that corporation.

The second and third sentences in Section 110(a) look the same at first, but actually have significant differences. The first sentence states that shippers and carriers are subject to civil penalty liability up to $10,000 for each violation, "and if any such violation is a continuing one, each day of violation constitutes a separate offense." The latter sentence is applicable to the packaging industry, but does not contain a continuing violation provision. Congress did not intend the potential economic destruction of a packaging manufacturer because of the repetitive nature of his business and production. Each day is like the last, and it appears Congress felt that a continuing violation would be too onerous for such a repetitive operation. As a matter of reality, there may not be that much difference in shipper, carrier, and packaging manufacturer's operations to justify this difference in the statute, but it was a distinction drawn by Congress and it remains to the benefit of the packaging industry.

Under a civil penalty procedure, the agency determines the amount of the penalty that could be assessed, and in proceedings and discussions with the company involved, an effort is made to mutually agree on a compromise figure that will be assessed. In determining the amount of the penalty that could be assessed, the Secretary shall take into account the nature, circumstances, extent, and gravity of the violations committed and, with respect to the person found to have committed such violation, the degree of culpability, any history of prior offenses, ability to pay, effect on ability to continue to do business, and such other matters as justice may require. These mitigating factors must be considered at the outset in determining the maximum penalty liability.

If, after consideration of the mitigating circumstances and the negotiations with the alleged violator, no mutually acceptable penalty figure is determined, the Secretary may ask the Attorney General to collect the penalty through an action in the district courts. Traditionally, the Attorney General is also capable of compromising the penalty, but the agency discourages any great flexiblity at this stage, because it tends to make alleged violators less inclined to settle at the DOT level in anticipation of getting a better deal from the Attorney General.

The final determination of a penalty, either through litigation or compromise, may be deducted from sums owed by the United States to the person charged, such as tax refunds. All moneys collected are paid into the U.S. Treasury as miscellaneous receipts, to discourage the development of a self-funding enforcement program that would expand agency coffers through unduly hard positions or refusals to compromise penalties.

CRIMINAL PENALTIES

The HMTA did not eliminate criminal penalties and in fact raised the previous dollar figures and prison terms to a maximum of $25,000 and five years, or both. The criminal penalties are carefully circumscribed, however, by making them applicable only to violations that are willfully committed. This phrasing, indicating that there must be a willful violation as opposed to a knowing act that happens to be a violation, involves a more substantial burden of proof on the part of the prosecution in showing knowledge of the law and intent to violate that law despite this knowledge.

Criminal penalties, since they involve use of grand juries, the federal courts, and representatives of the Justice Department, are more time consuming and difficult to prosecute than civil penalties, which usually are compromised by DOT itself. In addition, the accused in a criminal procedure benefits from a presumption of innocence with the scales tipped in his favor under the Federal Rules of Criminal Procedure, which compel proof of guilt beyond a reasonable doubt. The scales are unquestionably tipped against the interest of the person charged in a civil penalty proceeding, however. Accordingly, the administrative agency most often will choose to enforce regulations through the civil penalty procedure, although it is possible to impose both types of penalties on the same person for the same offense.

SPECIFIC RELIEF

Section 111 of the HMTA, 49 U.S.C. 1810, authorizes the Attorney General of the United States, at the request of the Secretary of Transportation, to bring a legal action in an appropriate federal district court for equitable relief to redress a violation by any person of a provision of the HMTA or an order or regulation issued under the HMTA. The district courts are authorized to grant such relief as is necessary or appropriate, including mandatory or prohibitive injunctive relief, interim equitable relief, and punitive damages.

It is unclear when such equitable relief would be appropriate, or the manner in which one might stimulate DOT to initiate such a request. Although this rounds out the collection of powers given to the Secretary of DOT under this statute, it is not anticipated that this power will be exercised with any frequency, if at all.

If the Secretary of DOT has reason to believe that an imminent hazard exists, he may go directly to the federal district courts, or may ask the Attorney General to do so, to petition for an order suspending or restricting the transportation of the hazardous material responsible for that imminent hazard. The HMTA defines "imminent hazard" as existing if there is a substantial likelihood that serious harm will occur prior to the completion of an administrative hearing or other formal proceeding initiated to abate the risk of that harm. "Serious harm" is defined in Section 103 of the HMTA, 49 U.S.C. 1802, as meaning "death, serious illness, or severe personal injury." It clearly does not include property damage or other economic injury.

ENFORCEMENT PROCEDURES

Subpart D of 49 CFR Part 107 describes the investigative procedures and enforcement authorities of DOT's Research & Special Programs Administration (RSPA) with respect to its assigned area of responsibility for enforcement of the hazardous materials regulations. It also sets forth the RSPA's procedures governing the imposition of sanctions with respect to violations of those regulations.

Section 107.301 of 49 CFR describes the division of enforcement responsibilities among the operating or modal elements of the Department of Transportation, including RSPA. The RSPA exercises its hazardous materials enforcement responsibility through the Office of Hazardous Materials Transportation (OHMT). Section 107.305 describes the investigative procedures followed by OHMTO inspectors.

Under separate headings the remainder of Subpart D describes the various enforcement sanctions available and, with respect to those which are primarily administrative to nature (i.e., compliance orders and civil penalties), sets forth procedures governing individual actions. The choice of which enforcement action or combination of actions to be initiated in a given case is, of course, a matter of administrative discretion with the OHMT acting under the policy direction of the RSPA. The four types of enforcement sanctions available are compliance orders, injunctive actions, civil penalties, and criminal penalties.

Sections 107.307 through 107.331 set forth the procedures applicable to the issuance of compliance orders and civil penalties under Sections 109(a) and 110(a) of the HMTA. In many respects of the compliance order and civil penalty procedures are the same. Both are initiated by the issuance of a notice of probable violation with the respondent having 30 days in which to reply. Both provide the respondent with an opportunity to present a rebuttal to the allegations, in writing or orally, or a combination of both. They also provide means for settlement through informal compromise. In addition, both provide the respondent with options as to the degree of formality with which the matter is to be processed.

In a situation where an ongoing or impending violation poses a risk requiring corrective steps be taken for the protection of public health and safety without delay, an order directing immediate compliance can be issued under Section 107.339. Although prior notice and opportunity for a hearing procedure do not apply in such a situation, the order is subject to administrative appeal.

Section 107.337 reflects the authority contained in HMTA Section 111 (a) and (b). These provisions provide for injunctive relief and punitive damages for violations of the hazardous materials regulations and for the enforcement of compliance orders. When time permits the Department of Justice will bring an action on behalf of the RSPA in the appropriate U.S. District Court. However, when the situation involves an imminent hazard RSPA may bring the action on its own motion in the appropriate U.S. District Court.

When a respondent receives a notice of probable violation indicating that the OHMT is considering assessing a civil penalty, the respondent may, within 30 days of service, (1) pay the amount of the preliminary assessment, (2) make an informal response denying the allegations in whole or in part and offering explanatory information, or (3) request a hearing. The filing of an informal response provides the opportunity for informal conference and possible compromise of the case. Taking an informal route means that an opportunity for a formal hearing is waived—not a step any company should take lightly. If a hearing is requested it will be conducted before an administrative law judge who may dismiss the case or issue an order assessing a civil penalty. His order may be appealed within 20 days after service. In any case in which a civil penalty is assessed, the factors listed in Section 107.359 are considered.

At any time after the issuance of a notice of probable violation in a civil penalty case and before it is referred to the Department of Justice for collection, the civil penalty can be compromised and settled by payment of the amount agreed upon in compromise.

RELATIONSHIP TO OTHER LAWS—PREEMPTION
Section 112 of the HMTA, 49 U.S.C. 1811, addresses the overlap between federal, state and local requirements by noting that any state or local governmental requirement that is "inconsistent" with any DOT requirement under this statute is preempted. The section goes on, however, to declare that even inconsistent state and local regulations may have validity if, upon the application of an appropriate state agency, the Secretary of DOT determines that the inconsistent requirement affords an equal or greater level of protection to the public and that it does not unreasonably burden commerce. To the extent that this inconsistent

requirement is administered and enforced, DOT has the discretion to determine that it is not preempted. Preemption procedures are published in 49 CFR 107.201–107.225

A substantial amount of litigation has occurred under this section, and one should check to see the latest details of any decisions. Look not only at court decisions, but at "inconsistency rulings" issued by the RSPA. The latter decisions are difficult to find because they are not codified, but contact with the RSPA Chief Counsel's office should provide some guidance on what rulings may be pertinent to a given situation.

Most proceedings to date have involved State or local limitations in the highway mode of transportation, but the same issues can arise in rail and the other modes. Generally speaking, DOT and the courts have held that technical aspects of hazardous materials regulatory compliance, including all packaging, classification, definitions, shipping names, and the like, are an exclusively federal province and that State or local initiatives that deviate from the federal rules are inconsistent and preempted. The same is true for hazard communications, including shipping papers, labels, markings and placards. Anything that unnecessarily delays the transportation of a hazardous material also is inconsistent, with the arguments on this point revolving around what is necessary.

In less technical, more operational areas, such as vehicle routing and time-of-day restrictions (which are considered a form of routing), several local enactments have been upheld. Those that have been allowed have include two major elements: (1) an objective assessment of the impact of the local enactment, having considered such things as the quality of the road, traffic conditions, proximity to population centers, accident history, and selection of the safest alternative for all citizens, and (2) effective political coordination with all other affected jurisdictions, primarily those which would bear the brunt of the traffic shift. It seems clear that unilateral action by a single community, that is designed primarily to shift traffic to its neighbors, is inconsistent with the purposes of Congress in enacting the HMTA and is preempted.

More difficult questions arise when hazardous materials restrictions are tied to nonsafety issues such as burdens on commerce, permits and fees, or to other regulations such as the Federal Highway Administration's Motor Carrier Safety Regulations, 49 CFR 387–397.

CHAPTER 31

Chemtrec: an Emergency Response System

**CALL CHEMTREC—(800) 424–9300
FOR CHEMICAL EMERGENCIES IN TRANSPORTATION**

CHEMTREC, the Chemical Transportation Emergency Center, is a public service of the Chemical Manufacturers Association (CMA), Washington, D.C. CHEMTREC provides immediate information and assistance to those at the scene of emergencies involving chemicals, and promptly contacts the shipper of the chemicals involved for more detailed assistance and appropriate follow-up.

CHEMTREC operates around the clock—twenty-four hours a day, seven days a week—to receive direct-dial toll-free calls through a wide area telephone service number: (800) 424–9300 (or 483–7616 for calls originating within the District of Columbia; (202) 483–7616 for calls from outside the continental U.S.).

The Chemtrec number is circulated widely in professional literature and to emergency response personnel, carriers, the chemical industry, government agencies, associations, and others who may have need to use it. If there are questions about chemicals of a non-emergency nature, one should telephone CMA's Chemical Referral Center (CRC) at 800–CMA–8200 (887–1315 in the District of Columbia or call collect to (202) 887–1315 from outside the continental U.S.).

CHEMTREC's primary mission is to help in transportation incidents, but it also provides support in chemical and hazardous materials emergencies in nontransportation situations.

Shippers participating in the system, principally CMA member companies, are notified through pre-established telephone contacts, providing twenty-four hours' accessibility. These chemical shippers also are urged to include the following directive on their shipping papers: "For Chemical Emergency—Spill, Leak, Fire, Exposure or Accident, Call CHEMTREC 800–424–9300 day or night."

An emergency reported to CHEMTREC is received by the Communicator on duty. Recording detail in writing, or on a video screen as well as by tape recorder, the Communicator will question the caller to determine as much essential information as possible. This information enables CHEMTREC to provide the best initial information on the chemical(s) reported to be involved, giving the caller specific indication of hazards, what to do, or what not to do in case of spills, fire or exposure.

Having advised the caller, CHEMTREC's Communicator proceeds immediately to notify the shipper of the material involved in the incident. The known particulars of the emergency are relayed, and responsibility for further guidance, including dispatching personnel to the scene or other measures, passes to the shipper.

Shipper notification is accomplished either by data transmission by computer link or by voice communication. Companies receiving data transmission and voice communication have at their disposal the CHEMTREC teleconference bridge to bring on line any needed additional chemical experts or personnel at the scene for expert advice. This teleconference system also is available to medical personnel treating patients exposed to chemicals. Consultation is provided by chemical company medical staff when the manufacturer is known.

The second stage of assistance to the scene is more difficult when the shipper of the material is unknown to CHEMTREC. The Communicator can turn to other resources, such as the U.S. Department of Energy and State Radiological Emergency response plans if radioactive materials were involved, as an example.

CHEMTREC's Communicators are not scientists. They are chosen for their ability to remain calm under emergency stresses. To preclude unfounded personal speculation regarding a reported emergency, they are under instructions to abide strictly by the information prepared by technical experts for their use.

Because chemicals find so many uses and have such a wide range of characteristics, there is much need for information about them: composition and purity, physical and chemical properties, effects on people and the environment, sources of supply, etc. It is important for all to understand that CHEMTREC is not intended and is not equipped to function as a general information source but by design is confined to dealing with chemical transportation emergencies. Do not call the emergency number if you do not have an emergency involving chemicals. As noted earlier, for this purpose call CMA's Chemical Referral Center.

Mutual aid programs exist for many products, whereby one producer will service field emergencies involving another producer's product. Initial referral may be in accord with the applicable mutual aid plan rather than direct to the shipper. Arrangements of this sort are established for emergencies involving chlorine, pesticides, vinyl chloride, hydrogen cyanide, hydrogen fluoride and liquefied petroleum gas. CHEMTREC serves as a communications link for these programs. In addition, a number of individual companies have information and service networks for their own products.

CHEMTREC is not a replacement or substitute for these individual programs but collaborates with them and endeavors to enhance their effectiveness through use of CHEMTREC's single telephone number.

CHEMTREC is a private sector program, but its capabilities have been recognized for many years by the government, and under the Hazardous Materials Transportation Act a close working relationship is maintained between the U.S. Department of Transportation and CHEMTREC. See 49 U.S. Code 1808. Through a formal relationship established by the two groups, the U.S. Coast Guard's National Response Center receives notification of all significant incidents.

Inquiries for further particulars about CHEMTREC are welcomed by CMA. Contact CHEMTREC, c/o CMA, 2501 M Street, N.W., Washington, D.C. 20037. The telephone number for such informational contacts is (202) 887-1255. Do not use the CHEMTREC emergency number for such calls.

For incidents involving the release or leakage of any hazardous material or hazardous substance from its packaging, also see the applicable reporting requirements in 49 CFR 171.15–171.17, discussed in Chapter 22.

CHAPTER 32

National Transportation Safety Board

The National Transportation Safety Board (NTSB) should not be confused with the former Civil Aeronautics Board or the former Hazardous Materials Regulations Board. The NTSB was created by the Department of Transportation Act of 1966, which simultaneously established the Department of Transportation. The NTSB is not a part of DOT, however.

Title III of the Transportation Safety Act of 1974, 49 U.S. Code 1901, was named the Independent Safety Board Act of 1974. This statute had three titles, the first of which is the Hazardous Materials Transportation Act, addressed specifically in Chapter 30. The second title pertained to railroad safety, while the third was entitled the "Independent Safety Board Act of 1974" and attempted to divorce the NTSB from any political influence or operational regulatory body.

The Congress found and declared:

> Proper conduct of the responsibilities assigned to this Board requires rigorous investigation of accidents involving transportation modes regulated by other agencies of Government; demands continual review, appraisal, and assessment of the operating practices and regulations of such agencies; and calls for the making of conclusions and recommendations that may be critical of or adverse to any such agency or its officials. No Federal agency can properly perform such functions unless it is totally separate and independent from any other department, bureau, commission, or agency of the United States.

The NTSB is relevant to the hazardous materials field not because of any impact from direct NTSB regulation, but because persuasive recommendations the Board provides to other government agencies that stimulate direct regulation by those agencies. The Safety Board is charged with:

1. Conducting a continuing across-the-board review of safety in all modes of transportation;
2. Determining cause or probable cause of transportation accidents and reporting the facts, conditions, and circumstances relating to such accidents;

3. Making recommendations to the Secretary of the Department of Transportation or Administrators of the operating administrations of the Department of Transportation to promote safety;

4. Conducting special studies and investigations on matters pertaining to safety in transportation and the prevention of accidents;

5. Assessing techniques and methods of accident investigation and publishing recommended procedures;

6. Evaluating the adequacy of safeguards and procedures concerning the transportation of hazardous materials and the performance of government agencies charged with regulating this transportation; and,

7. Making public its findings, reports, and recommendations.

The NTSB is made up of five members, appointed by the President, by and with the advice and consent of the Senate. No more than three members of the Board shall be of the same political party. At any given time, no less than three members of the Board must be individuals who have been appointed upon the basis of technical qualification, professional standing and demonstrated knowledge in the fields of accident reconstruction, safety engineering, human factors, transportation safety, or transportation regulation. Besides offices that specifically examine air, highway, railroad and pipeline transportation, the NTSB inquires into accidents in marine transportation and is obligated to establish and maintain an office to investigate and report on the safe transportation of hazardous materials.

The NTSB is equipped with the full powers of an investigative body, including the power to administer oaths, issue subpoenas and compel testimony. Any employee of the Board, upon presenting appropriate credentials and a written notice of inspection authority, may enter any property where a transportation accident has occurred or wreckage is located. In addition, the Board's investigators may inspect, at reasonable times, records, files, papers, processes, controls, and facilities relevant to the investigation of a transportation accident. The Board may also order the performance of an autopsy.

Any investigation of an accident conducted by the Safety Board has priority over all other investigations of such accidents conducted by other federal agencies, although the Board will provide for participation by those agencies.

The NTSB, in making its findings and recommendations public, does so in the form of reports that may be purchased from the National Technical Information Service, Springfield, Virginia 22151. These reports are the distillation of the evidence gathered, often through a public inquest, plus the conclusions as to probable cause of the incident and the recommendations to be drawn therefrom. These inquests involve the accumulation of facts and testimony presented under oath in response to questions from the Board and its staff as well as from representatives of all directly interested parties to the incident. Unfortunately, the forum often provides an opportunity for parties to indulge in fishing expeditions for evidence to be used in later civil lawsuits among themselves.

It should be noted, however, that no part of any report of the NTSB, relating to any accident or the investigation thereof, may be admitted as evidence or used in any suit or action for damages growing out of any matter mentioned in such report or reports. See 49 U.S. Code 1903(c).

Although the recommendations of the Safety Board are not mandatory directions, the Secretary of Transportation is obligated to respond to each recommendation formally and in writing, not later than ninety days after receipt of the recommendation. The response must indicate an intention:

1. To initiate and conduct procedures for adopting the recommendation in full, pursuant to a proposed timetable;
2. To initiate and conduct procedures for adopting the recommendation in part, pursuant to a proposed timetable, indicating in detail the reasons for not adopting the remainder of the recommendation; or
3. To refuse to initiate or conduct procedures for adopting such recommendation, indicating in detail the reasons for the refusal.

Notice of the Safety Board's recommendations and the DOT's responses appear periodically in the Federal Register and, in addition, are in DOT's annual report to Congress.

The National Transportation Safety Board is answerable only to the Congress, and it has been used as an investigative body by Congress to delve into specific issues on occasion. One instance of this, for example, is the "Special Study of the Carriage of Radioactive Materials By Air," noted above. Through such reports and testimony offered before congressional committees as well as its report to Congress the NTSB stimulates responsive regulations and legislation. Although the Board is criticized occasionally for its findings, and although some of this criticism is justified, the fact remains that the NTSB has stimulated legislatures and agencies to take action that otherwise might not have been taken.

One who wants to maintain a familiarity with the hazardous materials field, especially to speculate on future regulatory restrictions, should follow the actions of the NTSB through its investigations and reports. One should contact the National Transportation Safety Board, Publications Branch, Washington, D.C. 20591, asking for a list of publications and for copies of NTSB reports when they are highlighted in the Federal Register. Future planning of any activity involving hazardous materials should take the DOT regulations into account, and no assessment of the future impact of those regulations would be complete without including the NTSB and its recommendations.

In the hazardous materials field, several critical accident investigations and special studies have been issued, although not all of the recommendations in these reports have been followed. Some of the more important NTSB hazardous materials reports include:

- "Study of Uniform Reporting System for All Modes of Transportation in Reporting Incidents and Accidents Involving the Shipment of Hazardous Materials," adopted 3/21/69.
- "Marine Board of Investigation, M/V *Thorstream* Explosion and Fire with Loss of Life, Buffalo, New York, June 2, 1967," adopted 5/17/68.
- "Pennsylvania Railroad Train PR 11A, Extra 2210 West and Train SW-6, Extra 2217 East, Derailment and Collision, Dunreith, Indiana, Jan. 1, 1968," NTSB-RAR-68-3.
- "Southern Railway Company Train 154 Derailment With Fire and Explosion, Laurel, Mississippi, Jan. 25, 1969," NTSB-RAR-69-1.
- "Chicago, Burlington, and Quincy Railroad Company Train 64 and

Train 824 Derailment and Collision, With Tank Car Explosion, Crete, Nebraska, Feb. 18, 1969," NTSB-RAR-71-2.

- "Fire and Explosions on Tank Barge MOS 106 at La Grange, Missouri on Mississippi River, May 12, 1969," adopted Jan. 7, 1971.
- "Illinois Central Railroad Company Train Second 76 Derailment at Glendora, Mississippi, Sept. 11, 1969," NTSB-RAR-70-2.
- "SS Badger State Explosion Aboard and Eventual Sinking in the North Pacific Ocean, Dec. 26, 1969," adopted Oct. 21, 1971.
- "Illinois Central Railroad Company Train No. 1 Collision With Gasoline Tank Truck at South Second Street Grade Crossing, Loda, Illinois, Jan. 24, 1970," NTSB-RHR-71-1.
- "Liquefied Oxygen Tank Explosion Followed by Fires in Brooklyn, New York, May 30, 1970," NSTB-HAR-71-6.
- "Derailment of Toledo, Peoria & Western Railroad Company's Train No. 20, With Resultant Fire and Tank Ruptures, Crescent City, Illinois, June 21, 1970," NSTB-RAR-72-2.
- "Penn Central Transportation Company Freight Train Derailment/ Passenger Train Collision with Hazardous Material Car, Sound View, Connecticut, Oct. 8, 1970," NTSB-RAR-72-1.
- "Special Study, Risk Concepts in Dangerous Goods Transportation Regulations," Jan. 27, 1971, NTSB-STS-71-1.
- "Accidental Mixing of Incompatible Chemicals, Followed by Multiple Fatalities, During a Bulk Delivery, Berwick, Maine, April 2, 1971," NTSB-HAR-71-7.
- "Automobile-Truck Collision Followed by Fire and Explosion of Dynamite Cargo on U.S. Highway 78, near Waco, Georgia, June 4, 1971," NTSB-HAR-72-5.
- "Truck-Automobile Collision Involving Spilled Methyl Bromide on U.S. 90 near Gretna, Florida, Aug. 8, 1971," NTSB-HAR-72-3.
- "Tank-Truck Combination Overturn Onto Volkswagon Microbus Followed by Fire: U.S. Route 611, Moscow, Pennsylvania, Sept. 5, 1971," NTSB-HAR-72-6.
- "Derailment of Missouri Pacific Railroad Company's Train 94 at Houston, Texas, Oct. 19, 1971," NTSB-RAR-72-6.
- "SS V.A. Fogg, Sinking in the Gulf of Mexico on 1 February 1972 with Loss of Life," NSTB-MAR-74-8.
- "Analysis of the Safety of Transportation of Hazardous Materials on the Navigable Waters of the United States," March 1972, NTSB-MSS-72-2.
- "Special Study of the Carriage of Radioactive Materials by Air," April 1972, NTSB-AAS-72-4.
- "Hazardous Materials Railroad Accident in the Alton & Southern Gateway Yard in East St. Louis, Illinois, Jan. 22, 1972," NTSB-RAR-73-1.
- "Propane Tractor-Semitrailer Overturn and Fire, U.S. Route 501, Lynchburg, Virginia, March 9, 1972," NTSB-HAR-73-3.
- "Multiple-Vehicle Collision Followed by Propylene Cargo-Tank Explosion, New Jersey Turnpike, Exit 8, Sept. 21, 1972," NTSB-HAR-73-4.
- "Southern Pacific Transportation Co. Freight Train 2nd BSM 22

Munitions Explosion, Benson, Arizona, May 24, 1973," NTSB-RAR-75-2.
- "SS C.V. Sea Witch—SS Esso Brussels (Belgium); Collision and Fire in New York Harbor, June 2, 1973, with Loss of Life," NSTB-MAR-75-6.
- "Hoppy's Oil Service, Inc., Truck Overturn and Fire, State Route 128, Braintree, Massachusetts, Oct. 18, 1973," NTSB-HAR-74-4.
- "Pan American World Airways, Inc. Boeing 707–321C, N458PA, Boston, Massachusetts, Nov. 3, 1973," NTSB-AAR-74-16.
- "Derailment and Subsequent Burning of Delaware & Hudson Railway Freight Train at Oneonta, New York, Feb. 12, 1974," NTSB-RAR-74-4.
- "Surtigas, S.A., Tank-Semitrailer Overturn, Explosion and Fire Near Eagle Pass, Texas, April 29, 1975," NSTB-HAR-76-4.
- "Hazardous Materials Accident in the Railroad Yard of the Norfolk & Western Railway at Decatur, Illinois, July 19, 1974," NTSB-RAR-75-4.
- "Burlington Northern, Inc., Monomethylamine Nitrate Explosion at Wenatchee, Washington, Aug. 6, 1974," NTSB-RAR-76-1.
- "Hazardous Materials Accident at the Southern Pacific Transportation Company's Englewood Yard, Houston, Texas, Sept. 21, 1974," NTSB-RAR-75-7.
- "Derailment of Tank Cars with Subsequent Fire and Explosion on Chicago, Rock Island and Pacific Railroad Company near Des Moines, Iowa, Sept. 1, 1975," NSTB-RAR-76-8.
- "Union Oil Company of California, Tank Truck and Full Trailer Overturn and Fire, Seattle, Washington, Dec. 4, 1975," NTSB-HAR-76-7.
- "Transport Company of Texas, Tractor-Semitrailer (Tank) Collision with Bridge Column and Sudden Dispersal of Anhydous Ammonia Cargo, I-610 at Southwest Freeway, Houston, Texas, May 11, 1976," NSTB-HAR-77-1.
- "Chicago and North Western Transportation Company, Freight Train Derailments and Collision, Glen Ellyn, Illinois, May 16, 1976," NSTB-RAR-77-2.
- "Derailment of a Burlington Northern Freight Train at Belt, Montana, Nov. 26, 1976," NSTB-RAR-77-7.
- "Usher Transport, Inc., Tractor-Cargo Tank-Semitrailer Overturn and Fire, State Route 11, Beattyville, Kentucky, Sept. 24, 1977," NTSB-HAR-78-4.
- "Louisville and Nashville Railroad Company Freight Train Derailment and Puncture of Anhydrous Ammonia Tank Cars at Pensacola, Florida, Nov. 9, 1977," NSTB-RAR-78-4.
- "Collision of a L.A. & R.K. Railway Freight Train with a Rhymes Tractor-Semitrailer, Goldonna, Louisiana, Dec. 28, 1977," NSTB-RHR-78-1.
- "Derailment of Louisville & Nashville Railroad Company's Train No. 584 and Subsequent Rupture of Tank Car Containing Liquefied Petroleum Gas, Waverly, Tennessee, Feb. 22, 1978," NSTB-RAR-79-1.
- "Overview of Bulk Gasoline Delivery Fire and Explosions," Feb. 24, 1978, NTSB-HZM-78-1.
- "Derailment of Atlanta and Saint Andrews Bay Railway Company Freight Train, Youngstown, Florida, Feb. 26, 1978," NTSB-RAR-78-7.
- "St. Louis Southwestern Railroad Company Freight Train Derailment

and Rupture of Vinyl Chloride Tank Car, Lewisville, Arkansas, March 29, 1978," NSTB-RAR-78-8.

- "Analysis of Proceedings of the National Transportation Safety Board into Derailments and Hazardous Materials, April 4–6, 1978," NSTB-SEE-78-2.
- "Head-end Collision of Louisville and Nashville Railroad Local Freight Train and Yard Train at Florence, Alabama, Sept. 18, 1978," NTSB-RAR-79-2.
- "Safety Effectiveness Evaluation of the Federal Railroad Administration's Hazardous Materials and Track Safety Programs," March 8, 1979, NSTB-SEE-79-2.
- "Louisville and Nashville Railroad Company Freight Train Derailment and Puncture of Hazardous Materials Tank Cars, Crestview, Florida, April 8, 1979," NSTB-RAR-79-11.
- "Standardized Hazardous Materials Accident Maps," May 3, 1979, NTSB-HSM-79-1. See, also, NSTB-HZM-79-2, 79-3, 80-1, 80-2, 80-3, 80-4, 80-7, 80-8, 81-1, 81-2, 81-3, 81-4, 81-5, 81-6, and 83-2.
- "Noncompliance with Hazardous Materials Safety Regulations, Special Study," Aug. 3, 1979, NTSB-HZM-79-2.
- "Collision of Peruvian Freighter M/V Inca Tupac Yupanqui and U.S. Butane Barge Panama City, Good Hope, Louisiana, Aug. 30, 1979," NSTB-MAR-80-7.
- "Onscene Coordination Among Agencies at Hazardous Materials Accidents," Sept. 13, 1979, NSTB-HZM-79-3.
- "Survival in Hazardous Materials Transportation Accidents," Dec. 6, 1979, NSTB-HZM-79-4.
- "Multiple Vehicle Collision and Fire, U.S. Route 101, Los Angeles, California, March 3, 1980," NSTB-HAR-80-5.
- "The Accident Performance of Tank Car Safeguards," March 8, 1980, NSTB-HZM-80-1.
- "Phosphorus Trichloride Release in Boston & Maine Yard During Switching Operations, Somerville, MA, April 3, 1980," NSTB-HZM-81-1.
- "Illinois Central Gulf Railroad Company Freight Train Derailment, Hazardous Material Release and Evacuation, Muldraugh, Kentucky, July 26, 1980," NTSB-RAR-81-1.
- "Tank Car Structural Integrity After Derailment," Oct. 16, 1980, NTSB-SIR-80-1.
- "Illinois Central Gulf Railroad Freight Train and Mobil Oil Company Tractor-Cargo Tank Semitrailer Collision and Fire, Kenner, Louisiana, Nov. 25, 1980," NSTB-RHR-81-1.
- "Direct Transit Lines, Inc., Tractor-Semitrailer Multiple Vehicle Collision and Fire, U.S. Route 40, Frostburg, Maryland, Feb. 18, 1981," NSTB-HAR-81-3.
- "Safety Effectiveness Evaluation—Federal and State Enforcement Efforts in Hazardous Materials Transportation by Truck," Feb. 19, 1981, NTSB-SEE-81-2.
- "Derailment of Southern Pacific Transportation Company Freight Train Extra 9164 West at Surf, California, May 22, 1981," NSTB-RAR-81-8.

- "Miller Transporters, Inc., Tractor Cargo Tank-Semitrailer/Southern Railway System Freight Train Collision and Fire, Huntsville, Alabama, Sept. 15, 1981," NSTB-TSR-RHR-82-1.
- "Railroad/Highway Grade Crossing Accidents Involving Trucks Transporting Hazardous Materials," Sept. 24, 1981, NSTB-HZM-81-2.
- "Status of the Department of Transportation's Hazardous Materials Regulatory Program," Sept. 29, 1981, NTSB-SR-81-2.
- "Pacific Intermountain Express Tractor Cargo Tank Semitrailer Eagle, F.B. Truck Lines, Inc., Tractor Lowboy Semitrailer Collision and Fire, U.S. Route 50, Near Canon City, Colorado, Nov. 14, 1981," NSTB-HAR-82-3.
- "Collision of a Southeastern Pennsylvania Transportation Authority Commuter Train No. 114 with a Gasoline Truck, Southampton, Pennsylvania, Jan. 2, 1982," NTSB-TSR-RHR-82-3.
- "Derailment of Southern Pacific Transportation Company 01-BSMFF-05 Carrying Radioactive Material, Thermal, California, Jan. 7, 1982," NSTB-RAR-83-1.
- "Multiple Vehicle Collisions and Fire, Caldecott Tunnel Near Oakland, California, April 7, 1982," NSTB-HAR-83-1.
- "Derailment of Seaboard Coast Line Railroad Train No. 120 at Colonial Heights, Virginia, May 31, 1982," NTSB-RAR-83-4.
- "Derailment of Illinois Central Gulf Railroad Freight Train Extra 9629 East (GS-2-28) and Release of Hazardous Materials at Livingston, Louisiana, Sept. 28, 1982," NTSB-RAR-83-5.
- "United Air Lines Flight 2885, N8053U, McDonnell Douglas DC-8-54F (radioactive materials), Detroit, Michigan, Jan. 11, 1983," NSTB-AAR-83-7.
- "Illinois Central Gulf Railroad Company Freight Train Derailment, Fort Knox, Kentucky, March 22, 1983," NTSB-RAR-83-7.
- "Ramming of the Poplar Street Bridge by the Towboat M/V City of Greenville and Its Four-Barge Tow, St. Louis, Missouri, April 2, 1983," NTSB-MAR-83-10.
- "Denver & Rio Grande Western Railroad Company Train Yard Accident Involving Punctured Tank Car, Nitric Acid and Vapor Cloud, and Evacuation, Denver, Colorado, April 3, 1983," NTSB-RAR-85-10.
- "Derailment of Burlington Northern Railroad Company Freight Train No. MTC-0718 near Crystal City, Missouri, July 18, 1983," NTSB-RAR-84-1.
- "Vinyl Chloride Monomer Release from a Railroad Tank Car and Fire, Formosa Plastics Corporation Plant, Baton Rouge, Louisiana, July 30, 1983," NSTB-RAR-85-8.
- "Samuel Coraluzzo Co., Inc., Tractor Cargo Tank-Semitrailer, Mechanical Failure, Overturn and Fire, I-76 (Schuylkill Expressway), Philadelphia, Pennsylvania, Oct. 7, 1983," NSTB-HAR-84-2.
- "Release of Waste Acid from Cargo Tank Truck, Orange County, Florida, March 6, 1984," NSTB-HZM-85-1.
- "Seaboard System Railroad Freight Train FERHL Derailment and Fire, Marshville, North Carolina, April 10, 1984," NSTB-RAR-85-5.
- "Overturn of a Tractor-Semitrailer Transporting Torpedoes, Denver, Colorado, Aug. 1, 1984," NSTB-HZM-85-2.

- "Hazardous Materials Release, Missouri Pacific Railroad Corporation's North Little Rock, Arkansas, Railroad Yard, Dec. 31, 1984," NSTB-SIR-85-3.
- "Safety Effectiveness Evaluation of the Federal Railroad Administration's Hazardous Materials and Track Safety Programs."
- "Anhydrous Hydrogen Fluoride Release from NATX 9408, Train No. BNEL3Y at Contrail's Receiving Yard, Elkhart, Indiana, Feb. 4, 1985," NSTB-HZM-85-3.
- "Hazardous Materials Special Investigation Report, Release of Oleum during Wreckage Clearing Following Derailment of Seaboard System Railroad Train Extra 8294, North Clay, Kentucky, Feb. 5, 1984," NTSB-SIR-85-1.
- "Derailment of Seaboard System Railroad Train No. F-690 with Hazardous Material Release, Jackson, South Carolina, Feb. 23, 1985," and "Collision of Train No. F-481 with Standing Cars, Robbins, South Carolina, Feb. 25, 1985," NSTB-RAR-85-12.
- "Railroad Yard Safety: Hazardous Materials and Emergency Preparedness," April 30, 1985, NTSB-SIR-85-2.
- "Derailment of St. Louis Southwestern Railway Company (Cottonbelt) Freight Train Extra 4835 North, and Release of Hazardous Materials near Pine Bluff, Arkansas, June 9, 1985," NTSB-RAR-86-4.
- "Failure of Cargo Tank Transporting Hazardous Waste on the Washington, D.C., Beltway, I-95, Fairfax County, Virginia, Aug. 12, 1985," NTSB-SIR-86-2.
- "Collision Between a Tractor-Semitrailer Transporting Bombs and an Automobile, Resulting in Fire and Explosions, Checotah, Oklahoma, Sept. 4, 1985," NSTB-SIR-87.
- "Explosions and Fire on U.S. Chemical Tank Ship Puerto Rican in the Pacific Ocean Near St. Francis County, California, Oct. 31, 1986," NSTB-MAR-86-5.

APPENDICES

APPENDIX A

Vapor Pressure Curves for Selected Flammable Liquids

prepared by Carl A. Bierlein

Note: All flash points shown are by closed cup tester unless specifically marked open cup (oc).

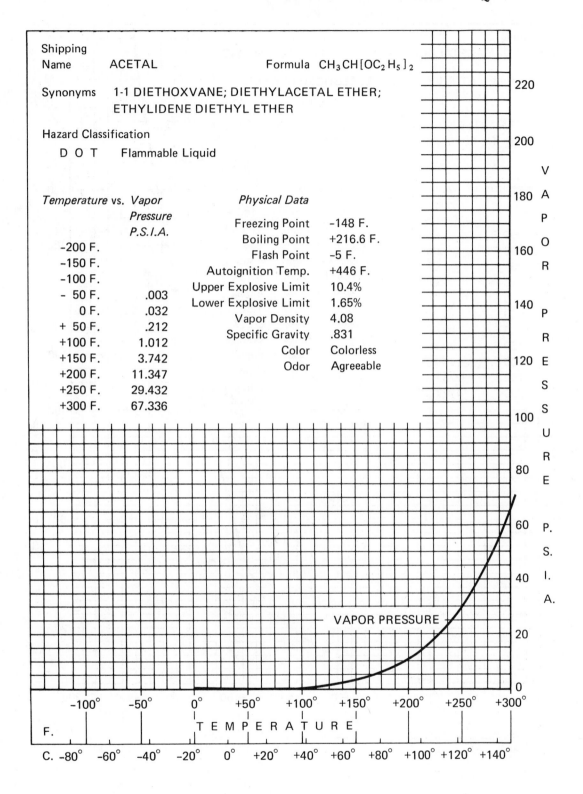

Shipping
Name   ACETAL   Formula $CH_3CH[OC_2H_5]_2$

Synonyms   1-1 DIETHOXVANE; DIETHYLACETAL ETHER;
           ETHYLIDENE DIETHYL ETHER

Hazard Classification
    D O T   Flammable Liquid

Temperature vs. *Vapor*
                Pressure
                P.S.I.A.

Temperature	Vapor Pressure P.S.I.A.
-200 F.	
-150 F.	
-100 F.	
- 50 F.	.003
0 F.	.032
+ 50 F.	.212
+100 F.	1.012
+150 F.	3.742
+200 F.	11.347
+250 F.	29.432
+300 F.	67.336

Physical Data

Freezing Point	-148 F.
Boiling Point	+216.6 F.
Flash Point	-5 F.
Autoignition Temp.	+446 F.
Upper Explosive Limit	10.4%
Lower Explosive Limit	1.65%
Vapor Density	4.08
Specific Gravity	.831
Color	Colorless
Odor	Agreeable

VAPOR PRESSURE

VAPOR PRESSURE P.S.I.A.

TEMPERATURE

F.
C.

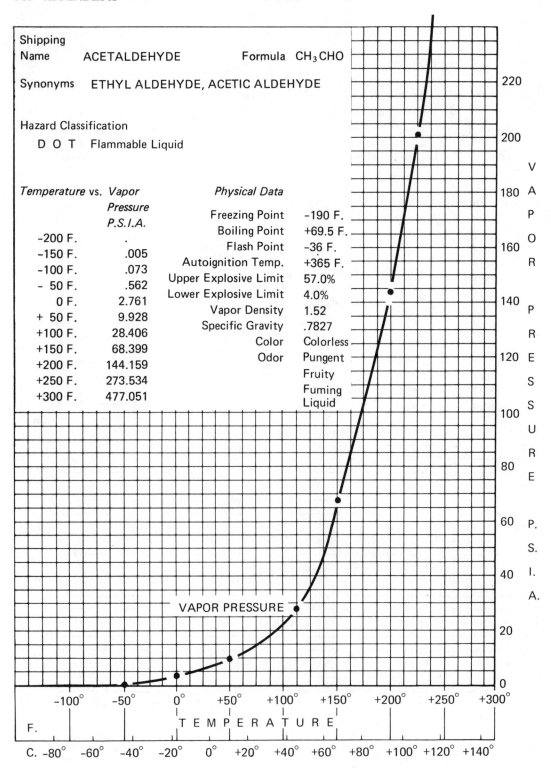

Shipping
Name   ACETALDEHYDE   Formula   CH₃CHO

Synonyms   ETHYL ALDEHYDE, ACETIC ALDEHYDE

Hazard Classification
   D O T   Flammable Liquid

Temperature vs. Vapor
                 Pressure
                 P.S.I.A.

Temperature	P.S.I.A.
-200 F.	.
-150 F.	.005
-100 F.	.073
- 50 F.	.562
0 F.	2.761
+ 50 F.	9.928
+100 F.	28.406
+150 F.	68.399
+200 F.	144.159
+250 F.	273.534
+300 F.	477.051

Physical Data

Freezing Point	-190 F.
Boiling Point	+69.5 F.
Flash Point	-36 F.
Autoignition Temp.	+365 F.
Upper Explosive Limit	57.0%
Lower Explosive Limit	4.0%
Vapor Density	1.52
Specific Gravity	.7827
Color	Colorless
Odor	Pungent
	Fruity
	Fuming
	Liquid

VAPOR PRESSURE

VAPOR PRESSURE P.S.I.A.

TEMPERATURE

F.   -100°   -50°   0°   +50°   +100°   +150°   +200°   +250°   +300°

C. -80° -60° -40° -20° 0° +20° +40° +60° +80° +100° +120° +140°

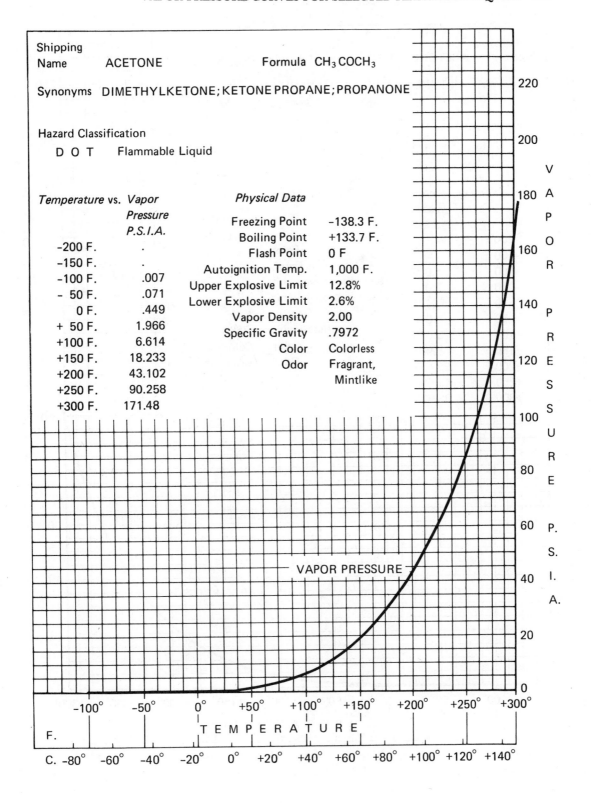

Shipping
Name   ACETONE   Formula CH_3COCH_3

Synonyms   DIMETHYLKETONE; KETONE PROPANE; PROPANONE

Hazard Classification
    D O T   Flammable Liquid

Temperature vs. Vapor
                  Pressure
                  P.S.I.A.

Temperature	P.S.I.A.
-200 F.	.
-150 F.	.
-100 F.	.007
- 50 F.	.071
0 F.	.449
+ 50 F.	1.966
+100 F.	6.614
+150 F.	18.233
+200 F.	43.102
+250 F.	90.258
+300 F.	171.48

Physical Data

Freezing Point	-138.3 F.
Boiling Point	+133.7 F.
Flash Point	0 F
Autoignition Temp.	1,000 F.
Upper Explosive Limit	12.8%
Lower Explosive Limit	2.6%
Vapor Density	2.00
Specific Gravity	.7972
Color	Colorless
Odor	Fragrant, Mintlike

VAPOR PRESSURE

VAPOR PRESSURE P.S.I.A.

TEMPERATURE

F.
C.

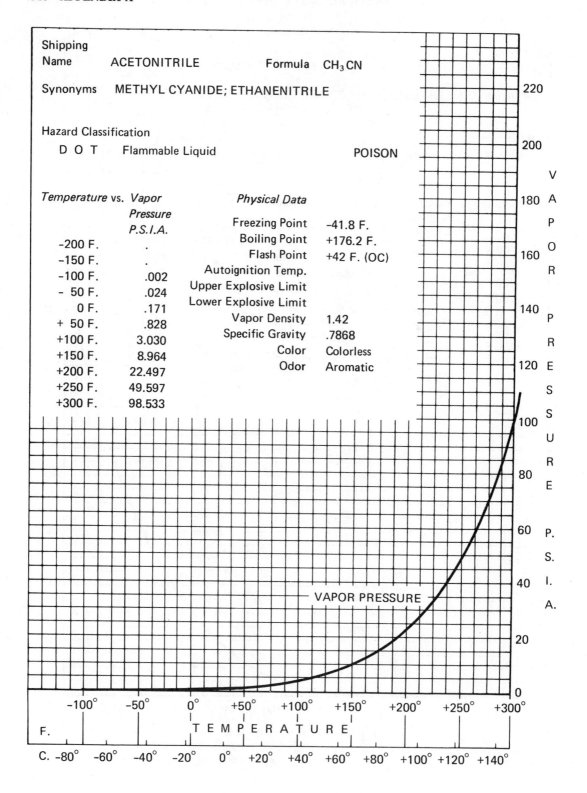

Shipping
Name   ACETONITRILE   Formula CH_3CN

Synonyms   METHYL CYANIDE; ETHANENITRILE

Hazard Classification
   D O T   Flammable Liquid   POISON

Temperature vs. Vapor Pressure P.S.I.A.

Temperature	P.S.I.A.
-200 F.	.
-150 F.	.
-100 F.	.002
- 50 F.	.024
0 F.	.171
+ 50 F.	.828
+100 F.	3.030
+150 F.	8.964
+200 F.	22.497
+250 F.	49.597
+300 F.	98.533

Physical Data

Freezing Point	-41.8 F.
Boiling Point	+176.2 F.
Flash Point	+42 F. (OC)
Autoignition Temp.	
Upper Explosive Limit	
Lower Explosive Limit	
Vapor Density	1.42
Specific Gravity	.7868
Color	Colorless
Odor	Aromatic

VAPOR PRESSURE

V A P O R   P R E S S U R E   P. S. I. A.

TEMPERATURE

F.   -100°   -50°   0°   +50°   +100°   +150°   +200°   +250°   +300°

C. -80° -60° -40° -20° 0° +20° +40° +60° +80° +100° +120° +140°

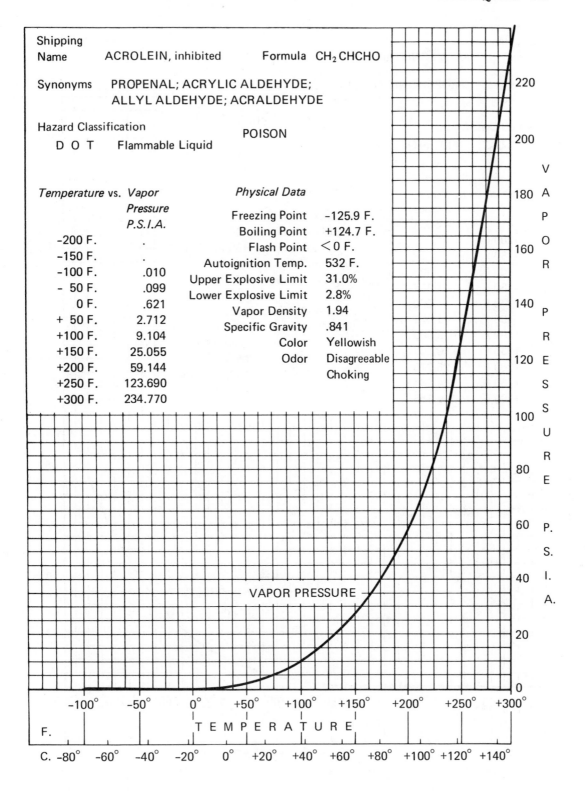

Shipping
Name   ACROLEIN, inhibited   Formula CH₂CHCHO

Synonyms   PROPENAL; ACRYLIC ALDEHYDE;
          ALLYL ALDEHYDE; ACRALDEHYDE

Hazard Classification
          D O T   Flammable Liquid   POISON

Temperature vs. Vapor
                  Pressure
                  P.S.I.A.

Temperature	Pressure P.S.I.A.
-200 F.	.
-150 F.	.
-100 F.	.010
- 50 F.	.099
0 F.	.621
+ 50 F.	2.712
+100 F.	9.104
+150 F.	25.055
+200 F.	59.144
+250 F.	123.690
+300 F.	234.770

Physical Data

Freezing Point	-125.9 F.
Boiling Point	+124.7 F.
Flash Point	< 0 F.
Autoignition Temp.	532 F.
Upper Explosive Limit	31.0%
Lower Explosive Limit	2.8%
Vapor Density	1.94
Specific Gravity	.841
Color	Yellowish
Odor	Disagreeable Choking

VAPOR PRESSURE

VAPOR PRESSURE P.S.I.A.

TEMPERATURE
F.
C.

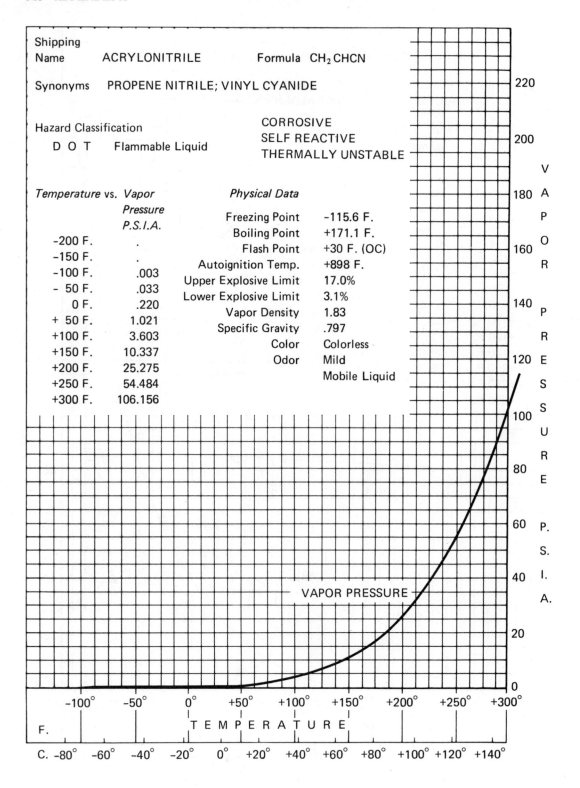

Shipping
Name   ACRYLONITRILE   Formula CH_2CHCN

Synonyms   PROPENE NITRILE; VINYL CYANIDE

Hazard Classification   CORROSIVE
   D O T   Flammable Liquid   SELF REACTIVE
                                 THERMALLY UNSTABLE

Temperature vs. *Vapor Pressure P.S.I.A.*		*Physical Data*	
-200 F.	.	Freezing Point	-115.6 F.
-150 F.	.	Boiling Point	+171.1 F.
-100 F.	.003	Flash Point	+30 F. (OC)
- 50 F.	.033	Autoignition Temp.	+898 F.
0 F.	.220	Upper Explosive Limit	17.0%
+ 50 F.	1.021	Lower Explosive Limit	3.1%
+100 F.	3.603	Vapor Density	1.83
+150 F.	10.337	Specific Gravity	.797
+200 F.	25.275	Color	Colorless
+250 F.	54.484	Odor	Mild
+300 F.	106.156		Mobile Liquid

VAPOR PRESSURE

TEMPERATURE

F.   -100°   -50°   0°   +50°   +100°   +150°   +200°   +250°   +300°

C. -80° -60° -40° -20° 0° +20° +40° +60° +80° +100° +120° +140°

V A P O R   P R E S S U R E   P. S. I. A.

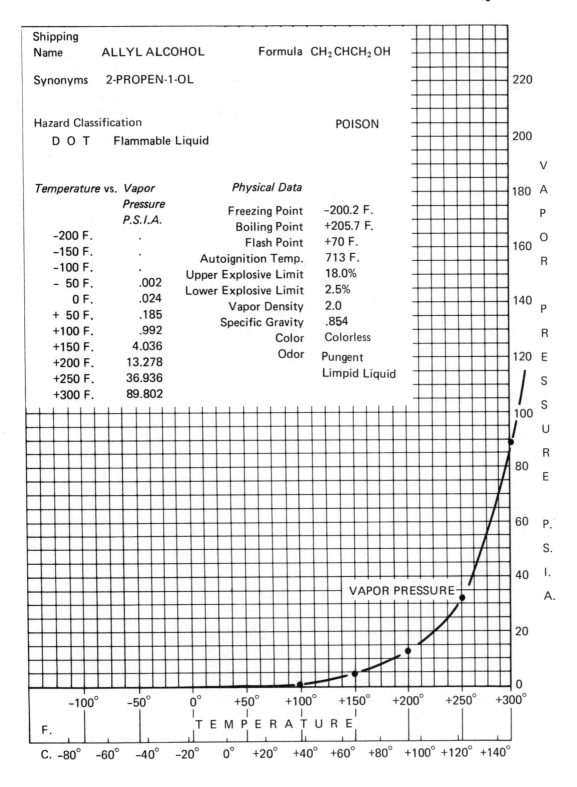

Shipping
Name   ALLYL ALCOHOL   Formula CH_2CHCH_2OH

Synonyms   2-PROPEN-1-OL

Hazard Classification   POISON
    D O T   Flammable Liquid

Temperature vs. Vapor Pressure P.S.I.A.

Temperature	Vapor Pressure P.S.I.A.
–200 F.	.
–150 F.	.
–100 F.	.
– 50 F.	.002
0 F.	.024
+ 50 F.	.185
+100 F.	.992
+150 F.	4.036
+200 F.	13.278
+250 F.	36.936
+300 F.	89.802

Physical Data

Freezing Point	–200.2 F.
Boiling Point	+205.7 F.
Flash Point	+70 F.
Autoignition Temp.	713 F.
Upper Explosive Limit	18.0%
Lower Explosive Limit	2.5%
Vapor Density	2.0
Specific Gravity	.854
Color	Colorless
Odor	Pungent Limpid Liquid

VAPOR PRESSURE

V A P O R   P R E S S U R E   P. S. I. A.

TEMPERATURE

F.   –100° –50° 0° +50° +100° +150° +200° +250° +300°

C. –80° –60° –40° –20° 0° +20° +40° +60° +80° +100° +120° +140°

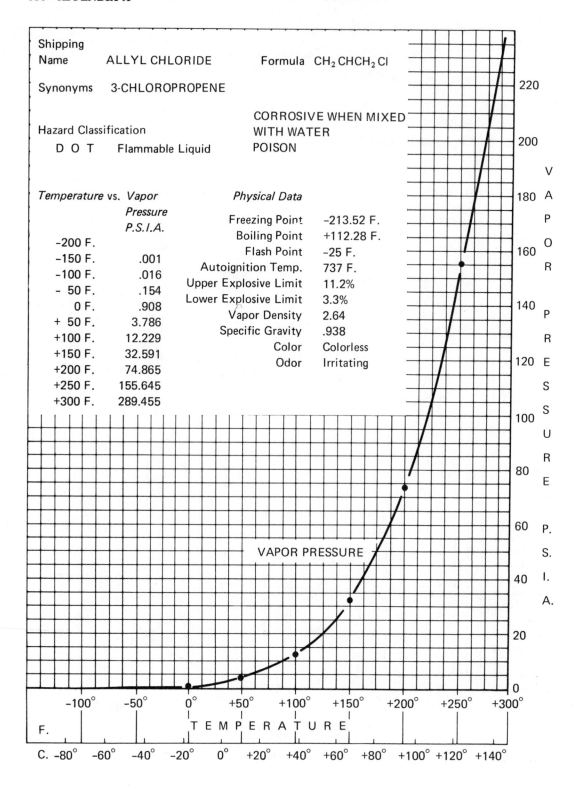

Shipping
Name   ALLYL CHLORIDE   Formula $CH_2 CHCH_2 Cl$

Synonyms   3-CHLOROPROPENE

Hazard Classification

    D O T   Flammable Liquid

CORROSIVE WHEN MIXED
WITH WATER
POISON

Temperature vs. *Vapor Pressure P.S.I.A.*

Temperature	P.S.I.A.
−200 F.	
−150 F.	.001
−100 F.	.016
− 50 F.	.154
0 F.	.908
+ 50 F.	3.786
+100 F.	12.229
+150 F.	32.591
+200 F.	74.865
+250 F.	155.645
+300 F.	289.455

Physical Data

Freezing Point	−213.52 F.
Boiling Point	+112.28 F.
Flash Point	−25 F.
Autoignition Temp.	737 F.
Upper Explosive Limit	11.2%
Lower Explosive Limit	3.3%
Vapor Density	2.64
Specific Gravity	.938
Color	Colorless
Odor	Irritating

VAPOR PRESSURE

VAPOR PRESSURE P.S.I.A.

TEMPERATURE

F.   −100°   −50°   0°   +50°   +100°   +150°   +200°   +250°   +300°

C. −80° −60° −40° −20°   0°   +20° +40° +60° +80° +100° +120° +140°

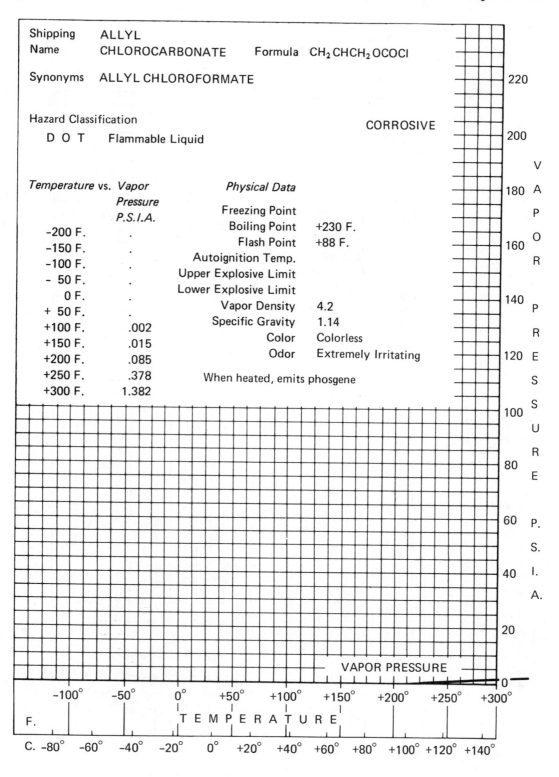

Shipping Name   ALLYL CHLOROCARBONATE   Formula CH_2CHCH_2OCOCl

Synonyms   ALLYL CHLOROFORMATE

Hazard Classification   CORROSIVE
   D O T   Flammable Liquid

Temperature vs. Vapor Pressure P.S.I.A.

Temperature	Vapor Pressure P.S.I.A.
−200 F.	.
−150 F.	.
−100 F.	.
− 50 F.	.
0 F.	.
+ 50 F.	.
+100 F.	.002
+150 F.	.015
+200 F.	.085
+250 F.	.378
+300 F.	1.382

Physical Data

Freezing Point	
Boiling Point	+230 F.
Flash Point	+88 F.
Autoignition Temp.	
Upper Explosive Limit	
Lower Explosive Limit	
Vapor Density	4.2
Specific Gravity	1.14
Color	Colorless
Odor	Extremely Irritating

When heated, emits phosgene

VAPOR PRESSURE

VAPOR PRESSURE P.S.I.A.

TEMPERATURE

F.   −100°   −50°   0°   +50°   +100°   +150°   +200°   +250°   +300°

C. −80°   −60°   −40°   −20°   0°   +20°   +40°   +60°   +80°   +100°   +120°   +140°

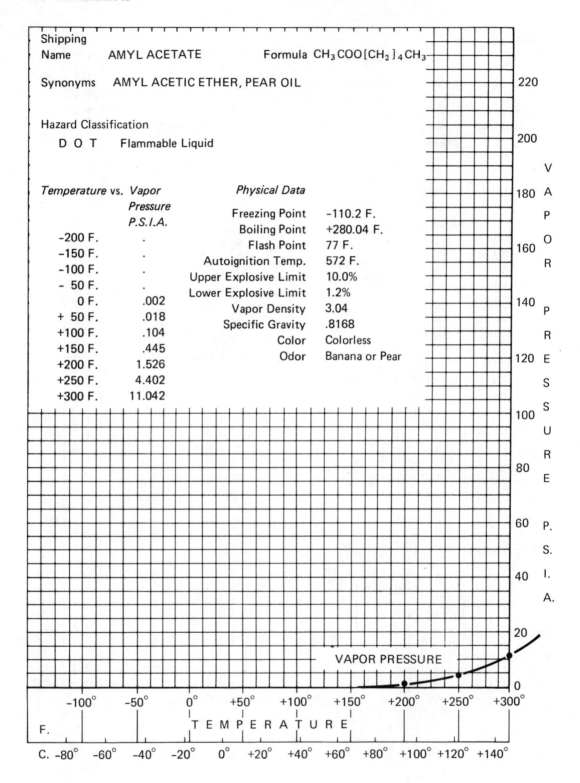

Shipping
Name   AMYL ACETATE   Formula $CH_3COO[CH_2]_4CH_3$

Synonyms   AMYL ACETIC ETHER, PEAR OIL

Hazard Classification
    D O T   Flammable Liquid

Temperature vs. _Vapor_   _Physical Data_
                Pressure
                _P.S.I.A._   Freezing Point   -110.2 F.
                                   Boiling Point   +280.04 F.
    -200 F.   .   Flash Point   77 F.
    -150 F.   .   Autoignition Temp. 572 F.
    -100 F.   .   Upper Explosive Limit   10.0%
    - 50 F.   .   Lower Explosive Limit   1.2%
       0 F.   .002   Vapor Density   3.04
    + 50 F.   .018   Specific Gravity   .8168
    +100 F.   .104   Color   Colorless
    +150 F.   .445   Odor   Banana or Pear
    +200 F.   1.526
    +250 F.   4.402
    +300 F.   11.042

V A P O R   P R E S S U R E   P. S. I. A.

VAPOR PRESSURE

220
200
180
160
140
120
100
80
60
40
20
0

-100°   -50°   0°   +50°   +100°   +150°   +200°   +250°   +300°

F.

T E M P E R A T U R E

C. -80° -60° -40° -20° 0° +20° +40° +60° +80° +100° +120° +140°

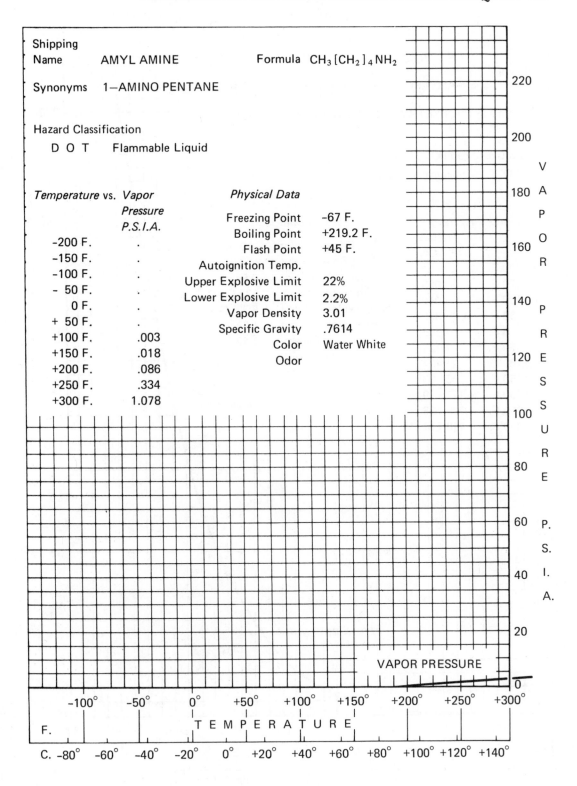

Shipping
Name   AMYL AMINE   Formula $CH_3[CH_2]_4NH_2$

Synonyms   1—AMINO PENTANE

Hazard Classification
   D O T   Flammable Liquid

Temperature vs. *Vapor Pressure P.S.I.A.*

Temperature	Vapor Pressure P.S.I.A.
-200 F.	.
-150 F.	.
-100 F.	.
- 50 F.	.
0 F.	.
+ 50 F.	.
+100 F.	.003
+150 F.	.018
+200 F.	.086
+250 F.	.334
+300 F.	1.078

Physical Data

Freezing Point	-67 F.
Boiling Point	+219.2 F.
Flash Point	+45 F.
Autoignition Temp.	
Upper Explosive Limit	22%
Lower Explosive Limit	2.2%
Vapor Density	3.01
Specific Gravity	.7614
Color	Water White
Odor	

VAPOR PRESSURE

VAPOR PRESSURE P.S.I.A.

TEMPERATURE

F. -100° -50° 0° +50° +100° +150° +200° +250° +300°

C. -80° -60° -40° -20° 0° +20° +40° +60° +80° +100° +120° +140°

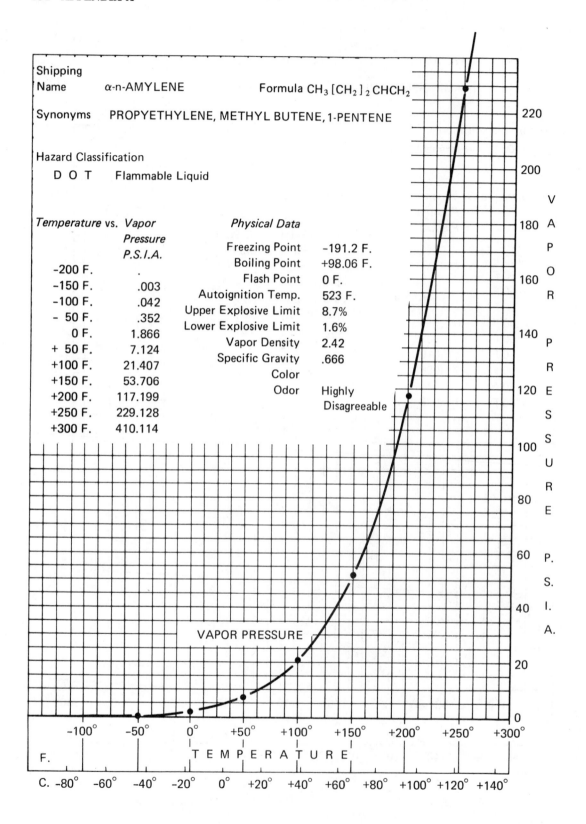

Shipping
Name   α-n-AMYLENE   Formula CH₃[CH₂]₂CHCH₂

Synonyms   PROPYETHYLENE, METHYL BUTENE, 1-PENTENE

Hazard Classification
   D O T   Flammable Liquid

Temperature vs. Vapor   Physical Data
               Pressure
               P.S.I.A.   Freezing Point   −191.2 F.
                               Boiling Point   +98.06 F.
   −200 F.   .   Flash Point   0 F.
   −150 F.   .003   Autoignition Temp. 523 F.
   −100 F.   .042   Upper Explosive Limit 8.7%
   − 50 F.   .352   Lower Explosive Limit 1.6%
     0 F.   1.866   Vapor Density   2.42
   + 50 F.   7.124   Specific Gravity   .666
   +100 F.   21.407   Color
   +150 F.   53.706   Odor   Highly
   +200 F.   117.199   Disagreeable
   +250 F.   229.128
   +300 F.   410.114

VAPOR PRESSURE

VAPOR PRESSURE P.S.I.A.

TEMPERATURE

F.   −100°   −50°   0°   +50°   +100°   +150°   +200°   +250°   +300°

C. −80° −60° −40° −20°   0°   +20°   +40°   +60°   +80° +100° +120° +140°

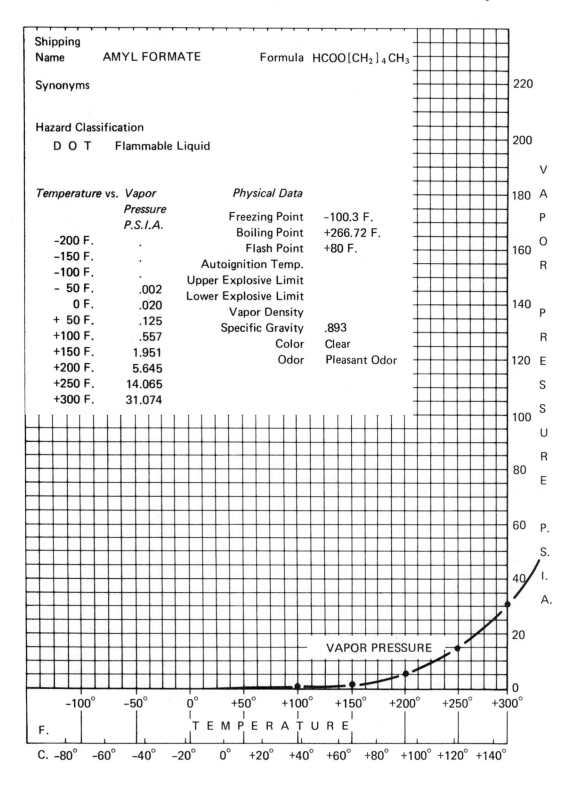

Shipping
Name   AMYL FORMATE   Formula   HCOO[CH₂]₄CH₃

Synonyms

Hazard Classification
   D O T   Flammable Liquid

Temperature vs. Vapor
                    Pressure
                    P.S.I.A.

Temperature	P.S.I.A.
-200 F.	.
-150 F.	.
-100 F.	.
- 50 F.	.002
0 F.	.020
+ 50 F.	.125
+100 F.	.557
+150 F.	1.951
+200 F.	5.645
+250 F.	14.065
+300 F.	31.074

Physical Data

Freezing Point	-100.3 F.
Boiling Point	+266.72 F.
Flash Point	+80 F.
Autoignition Temp.	
Upper Explosive Limit	
Lower Explosive Limit	
Vapor Density	
Specific Gravity	.893
Color	Clear
Odor	Pleasant Odor

VAPOR PRESSURE

VAPOR PRESSURE P.S.I.A.

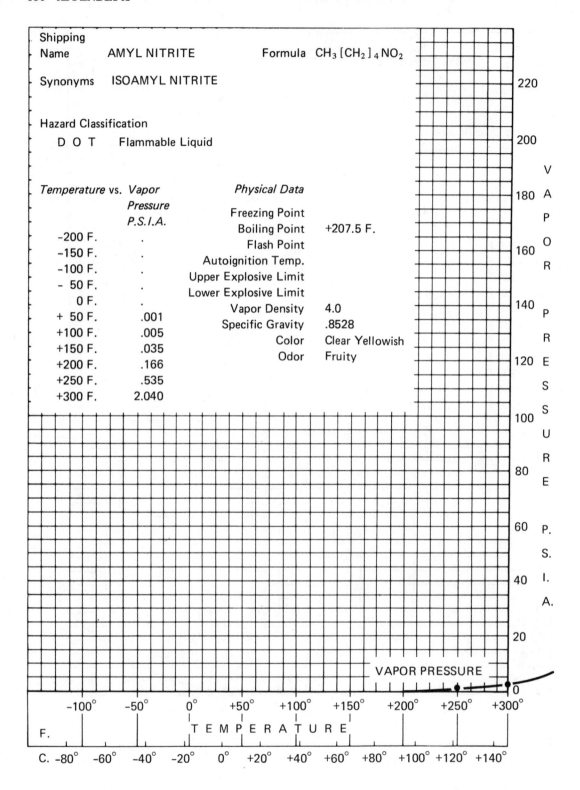

Shipping
Name   AMYL NITRITE   Formula   CH$_3$[CH$_2$]$_4$NO$_2$

Synonyms   ISOAMYL NITRITE

Hazard Classification
    D O T   Flammable Liquid

Temperature vs. Vapor Pressure P.S.I.A. | Physical Data
| |
-200 F.   .   Freezing Point
-150 F.   .   Boiling Point   +207.5 F.
-100 F.   .   Flash Point
- 50 F.   .   Autoignition Temp.
   0 F.   .   Upper Explosive Limit
+ 50 F.   .001   Lower Explosive Limit
+100 F.   .005   Vapor Density   4.0
+150 F.   .035   Specific Gravity   .8528
+200 F.   .166   Color   Clear Yellowish
+250 F.   .535   Odor   Fruity
+300 F.   2.040

VAPOR PRESSURE

V A P O R   P R E S S U R E   P. S. I. A.

220
200
180
160
140
120
100
80
60
40
20
0

T E M P E R A T U R E

-100°   -50°   0°   +50°   +100°   +150°   +200°   +250°   +300°

F.

C. -80° -60° -40° -20° 0° +20° +40° +60° +80° +100° +120° +140°

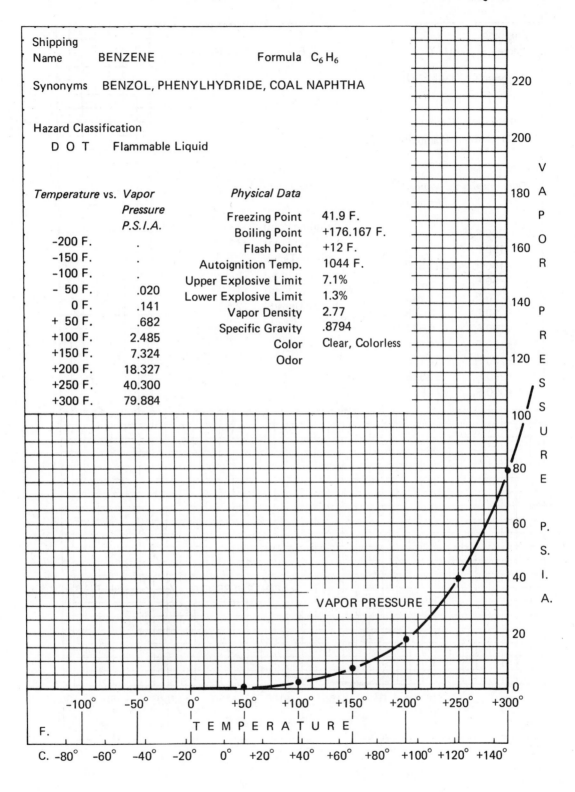

Shipping
Name   BENZENE   Formula C_6H_6

Synonyms   BENZOL, PHENYLHYDRIDE, COAL NAPHTHA

Hazard Classification
   D O T   Flammable Liquid

Temperature vs. *Vapor Pressure P.S.I.A.*

Temperature	Vapor Pressure P.S.I.A.
-200 F.	.
-150 F.	.
-100 F.	.
- 50 F.	.020
0 F.	.141
+ 50 F.	.682
+100 F.	2.485
+150 F.	7.324
+200 F.	18.327
+250 F.	40.300
+300 F.	79.884

Physical Data

Freezing Point	41.9 F.
Boiling Point	+176.167 F.
Flash Point	+12 F.
Autoignition Temp.	1044 F.
Upper Explosive Limit	7.1%
Lower Explosive Limit	1.3%
Vapor Density	2.77
Specific Gravity	.8794
Color	Clear, Colorless
Odor	

VAPOR PRESSURE

V A P O R   P R E S S U R E   P. S. I. A.

T E M P E R A T U R E

F.

C. -80° -60° -40° -20° 0° +20° +40° +60° +80° +100° +120° +140°

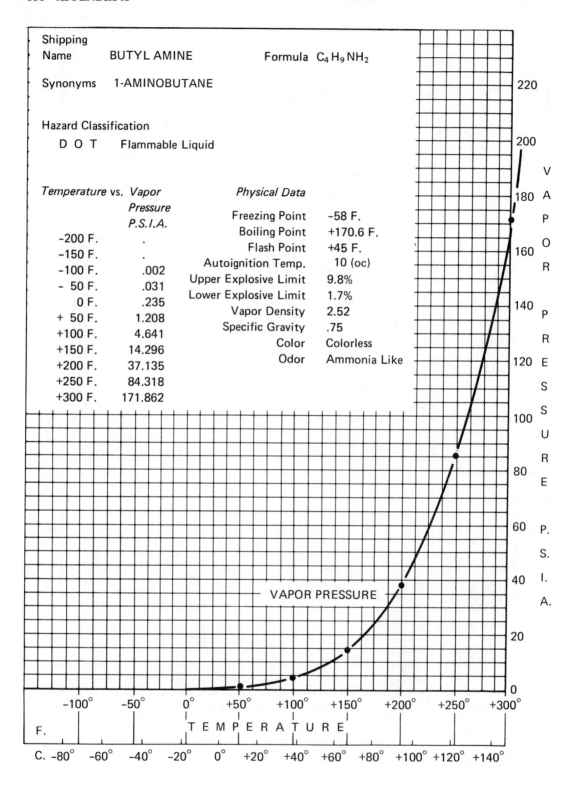

Shipping
Name   BUTYL AMINE   Formula $C_4H_9NH_2$

Synonyms   1-AMINOBUTANE

Hazard Classification
   D O T   Flammable Liquid

Temperature vs. Vapor
                 Pressure
                 P.S.I.A.

Temperature	P.S.I.A.
-200 F.	.
-150 F.	.
-100 F.	.002
- 50 F.	.031
0 F.	.235
+ 50 F.	1.208
+100 F.	4.641
+150 F.	14.296
+200 F.	37.135
+250 F.	84.318
+300 F.	171.862

Physical Data

Freezing Point	-58 F.
Boiling Point	+170.6 F.
Flash Point	+45 F.
Autoignition Temp.	10 (oc)
Upper Explosive Limit	9.8%
Lower Explosive Limit	1.7%
Vapor Density	2.52
Specific Gravity	.75
Color	Colorless
Odor	Ammonia Like

VAPOR PRESSURE

VAPOR PRESSURE P.S.I.A.

TEMPERATURE

F.   -100°   -50°   0°   +50°   +100°   +150°   +200°   +250°   +300°

C. -80° -60° -40° -20°   0°   +20° +40° +60° +80° +100° +120° +140°

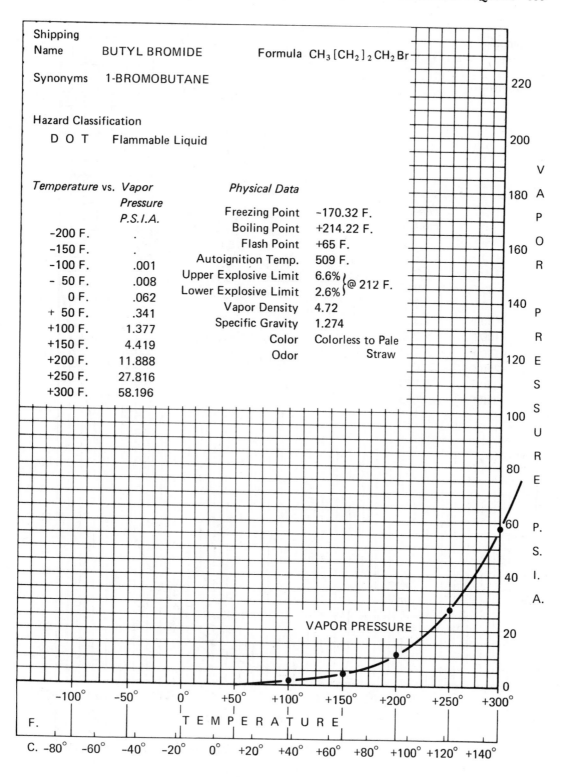

Shipping
Name   BUTYL BROMIDE   Formula $CH_3[CH_2]_2CH_2Br$

Synonyms   1-BROMOBUTANE

Hazard Classification
   D O T   Flammable Liquid

Temperature vs. *Vapor*
             Pressure
             P.S.I.A.

		Physical Data
-200 F.	.	Freezing Point -170.32 F.
-150 F.	.	Boiling Point +214.22 F.
-100 F.	.001	Flash Point +65 F.
- 50 F.	.008	Autoignition Temp. 509 F.
0 F.	.062	Upper Explosive Limit 6.6%
+ 50 F.	.341	Lower Explosive Limit 2.6% } @ 212 F.
+100 F.	1.377	Vapor Density 4.72
+150 F.	4.419	Specific Gravity 1.274
+200 F.	11.888	Color Colorless to Pale
+250 F.	27.816	Odor Straw
+300 F.	58.196	

VAPOR PRESSURE

V A P O R   P R E S S U R E   P. S. I. A.

T E M P E R A T U R E

F.

C. -80° -60° -40° -20° 0° +20° +40° +60° +80° +100° +120° +140°

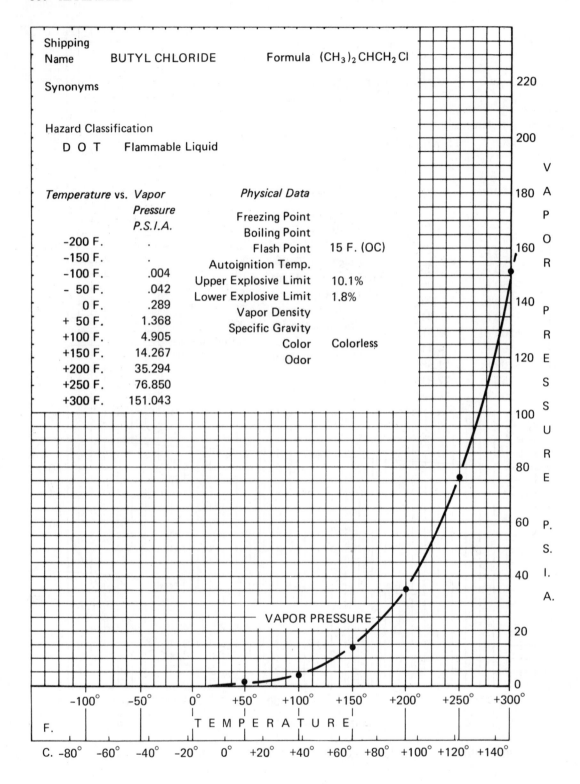

Shipping
Name   BUTYL CHLORIDE   Formula $(CH_3)_2 CHCH_2 Cl$

Synonyms

Hazard Classification
    D O T   Flammable Liquid

Temperature vs. *Vapor Pressure P.S.I.A.*

Temperature	Pressure P.S.I.A.
–200 F.	.
–150 F.	.
–100 F.	.004
– 50 F.	.042
0 F.	.289
+ 50 F.	1.368
+100 F.	4.905
+150 F.	14.267
+200 F.	35.294
+250 F.	76.850
+300 F.	151.043

Physical Data

Freezing Point	
Boiling Point	
Flash Point	15 F. (OC)
Autoignition Temp.	
Upper Explosive Limit	10.1%
Lower Explosive Limit	1.8%
Vapor Density	
Specific Gravity	
Color	Colorless
Odor	

VAPOR PRESSURE

VAPOR PRESSURE P.S.I.A.

TEMPERATURE

F.

C.

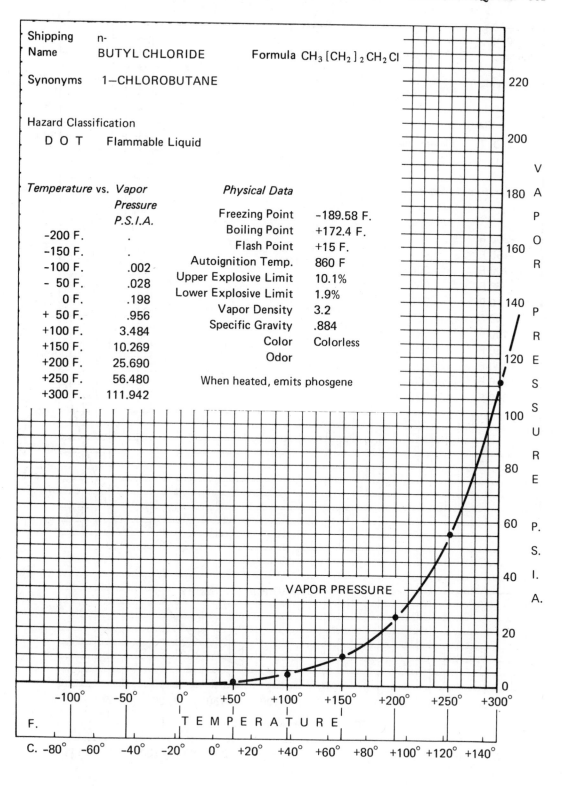

Shipping n-
Name BUTYL CHLORIDE   Formula $CH_3[CH_2]_2CH_2Cl$

Synonyms   1—CHLOROBUTANE

Hazard Classification
    D O T   Flammable Liquid

Temperature vs. Vapor
                     Pressure
                     P.S.I.A.

-200 F.	.
-150 F.	.
-100 F.	.002
- 50 F.	.028
0 F.	.198
+ 50 F.	.956
+100 F.	3.484
+150 F.	10.269
+200 F.	25.690
+250 F.	56.480
+300 F.	111.942

Physical Data

Freezing Point	-189.58 F.
Boiling Point	+172.4 F.
Flash Point	+15 F.
Autoignition Temp.	860 F
Upper Explosive Limit	10.1%
Lower Explosive Limit	1.9%
Vapor Density	3.2
Specific Gravity	.884
Color	Colorless
Odor	

When heated, emits phosgene

VAPOR PRESSURE

VAPOR PRESSURE P.S.I.A.

TEMPERATURE

F.

C. -80° -60° -40° -20° 0° +20° +40° +60° +80° +100° +120° +140°

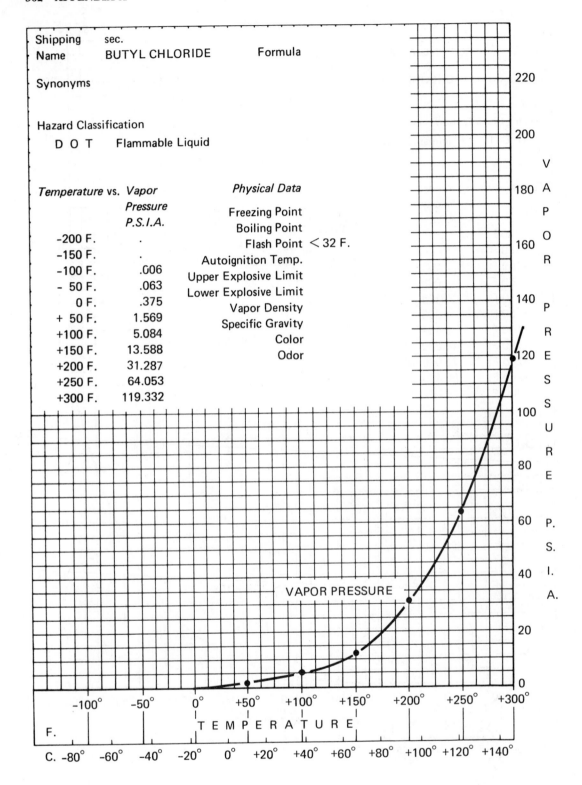

Shipping sec.
Name   BUTYL CHLORIDE   Formula

Synonyms

Hazard Classification
  D O T   Flammable Liquid

Temperature vs.	Vapor Pressure P.S.I.A.
-200 F.	.
-150 F.	.
-100 F.	.006
- 50 F.	.063
0 F.	.375
+ 50 F.	1.569
+100 F.	5.084
+150 F.	13.588
+200 F.	31.287
+250 F.	64.053
+300 F.	119.332

Physical Data

Freezing Point
Boiling Point
Flash Point < 32 F.
Autoignition Temp.
Upper Explosive Limit
Lower Explosive Limit
Vapor Density
Specific Gravity
Color
Odor

VAPOR PRESSURE

VAPOR PRESSURE P.S.I.A.

TEMPERATURE

F.

C. -80° -60° -40° -20° 0° +20° +40° +60° +80° +100° +120° +140°

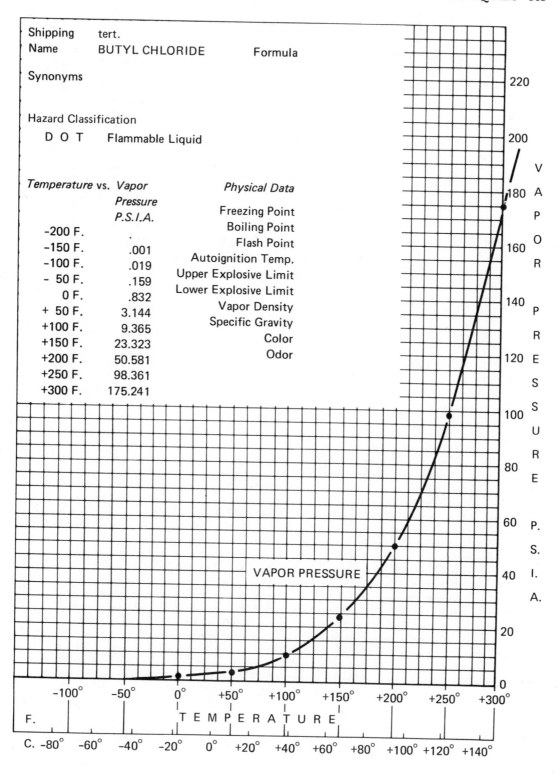

Shipping Name: tert. BUTYL CHLORIDE

Formula

Synonyms

Hazard Classification

D O T   Flammable Liquid

Temperature vs. Vapor Pressure P.S.I.A.

Temperature	Vapor Pressure P.S.I.A.
-200 F.	.
-150 F.	.001
-100 F.	.019
- 50 F.	.159
0 F.	.832
+ 50 F.	3.144
+100 F.	9.365
+150 F.	23.323
+200 F.	50.581
+250 F.	98.361
+300 F.	175.241

Physical Data

Freezing Point
Boiling Point
Flash Point
Autoignition Temp.
Upper Explosive Limit
Lower Explosive Limit
Vapor Density
Specific Gravity
Color
Odor

VAPOR PRESSURE

VAPOR PRESSURE P.S.I.A.

F.   T E M P E R A T U R E

-100° -50° 0° +50° +100° +150° +200° +250° +300°

C. -80° -60° -40° -20° 0° +20° +40° +60° +80° +100° +120° +140°

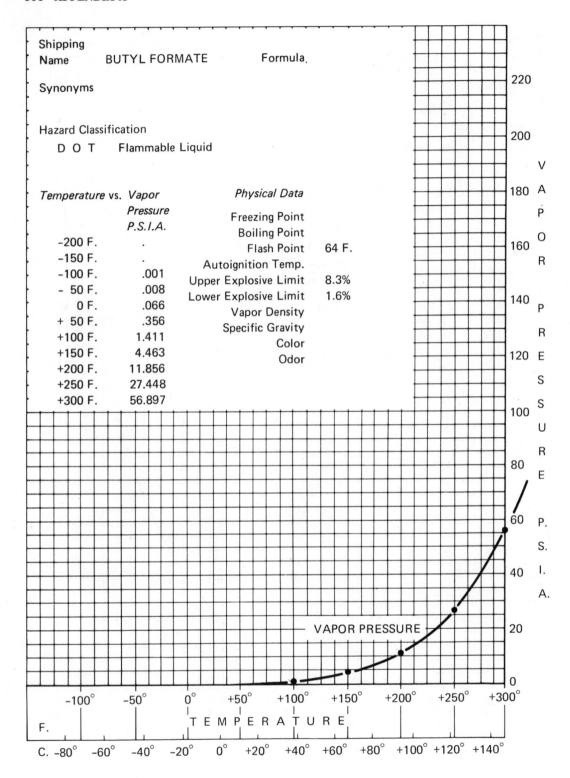

Shipping
Name   BUTYL FORMATE   Formula.

Synonyms

Hazard Classification
   D O T   Flammable Liquid

Temperature vs. Vapor
              Pressure
              P.S.I.A.

-200 F.	.
-150 F.	.
-100 F.	.001
- 50 F.	.008
0 F.	.066
+ 50 F.	.356
+100 F.	1.411
+150 F.	4.463
+200 F.	11.856
+250 F.	27.448
+300 F.	56.897

Physical Data

Freezing Point	
Boiling Point	
Flash Point	64 F.
Autoignition Temp.	
Upper Explosive Limit	8.3%
Lower Explosive Limit	1.6%
Vapor Density	
Specific Gravity	
Color	
Odor	

VAPOR PRESSURE

VAPOR PRESSURE P.S.I.A.

220
200
180
160
140
120
100
80
60
40
20
0

TEMPERATURE

F.   -100°   -50°   0°   +50°   +100°   +150°   +200°   +250°   +300°

C. -80° -60° -40° -20°   0°   +20° +40° +60° +80° +100° +120° +140°

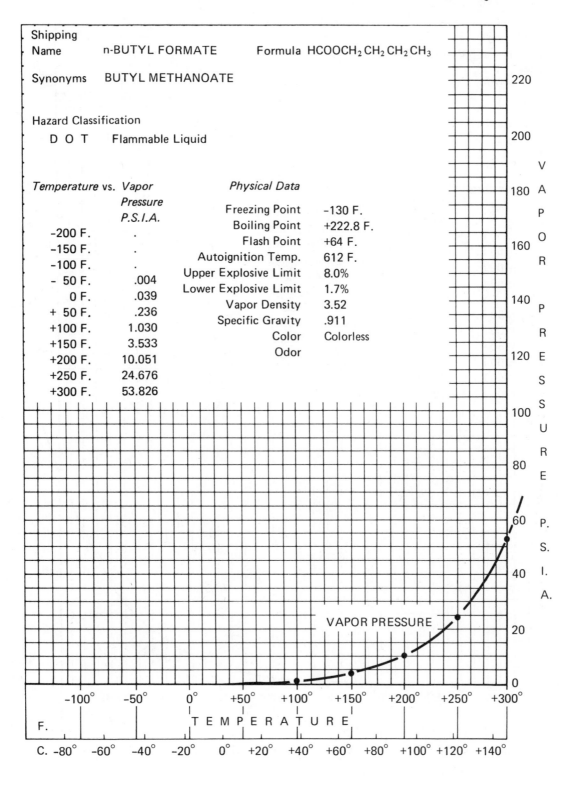

Shipping
Name   n-BUTYL FORMATE   Formula $HCOOCH_2 CH_2 CH_2 CH_3$

Synonyms   BUTYL METHANOATE

Hazard Classification

   D O T   Flammable Liquid

Temperature vs. *Vapor Pressure P.S.I.A.*

Temperature	Vapor Pressure P.S.I.A.
−200 F.	.
−150 F.	.
−100 F.	.
− 50 F.	.004
0 F.	.039
+ 50 F.	.236
+100 F.	1.030
+150 F.	3.533
+200 F.	10.051
+250 F.	24.676
+300 F.	53.826

Physical Data

Freezing Point	−130 F.
Boiling Point	+222.8 F.
Flash Point	+64 F.
Autoignition Temp.	612 F.
Upper Explosive Limit	8.0%
Lower Explosive Limit	1.7%
Vapor Density	3.52
Specific Gravity	.911
Color	Colorless
Odor	

VAPOR PRESSURE

VAPOR PRESSURE P.S.I.A.

TEMPERATURE

F.   −100°   −50°   0°   +50°   +100°   +150°   +200°   +250°   +300°

C. −80° −60° −40° −20° 0° +20° +40° +60° +80° +100° +120° +140°

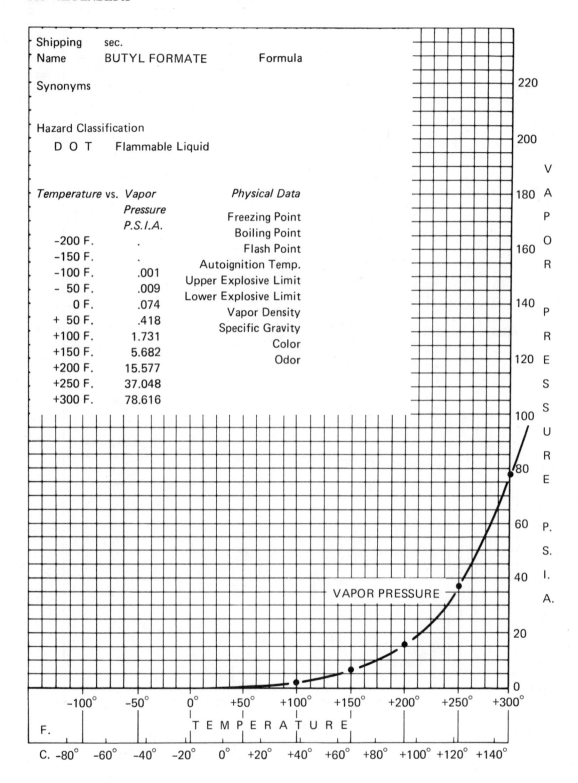

Shipping   sec.
Name   BUTYL FORMATE   Formula

Synonyms

Hazard Classification
   D O T   Flammable Liquid

Temperature vs. *Vapor Pressure P.S.I.A.*

–200 F.	.
–150 F.	.
–100 F.	.001
– 50 F.	.009
0 F.	.074
+ 50 F.	.418
+100 F.	1.731
+150 F.	5.682
+200 F.	15.577
+250 F.	37.048
+300 F.	78.616

Physical Data

Freezing Point
Boiling Point
Flash Point
Autoignition Temp.
Upper Explosive Limit
Lower Explosive Limit
Vapor Density
Specific Gravity
Color
Odor

VAPOR PRESSURE

V A P O R   P R E S S U R E   P.S.I.A.

220
200
180
160
140
120
100
80
60
40
20
0

TEMPERATURE

F.   –100°   –50°   0°   +50°   +100°   +150°   +200°   +250°   +300°

C. –80°   –60°   –40°   –20°   0°   +20°   +40°   +60°   +80°   +100°   +120°   +140°

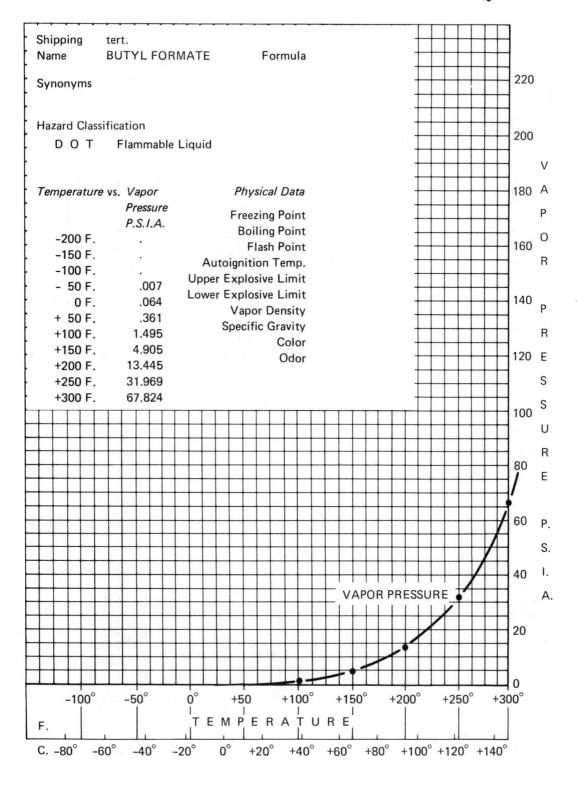

Shipping   tert.
Name   BUTYL FORMATE   Formula

Synonyms

Hazard Classification
    D O T   Flammable Liquid

Temperature vs. *Vapor*   *Physical Data*
                  Pressure
                  P.S.I.A.   Freezing Point
    -200 F.   ·   Boiling Point
    -150 F.   ·   Flash Point
    -100 F.   ·   Autoignition Temp.
    - 50 F.   .007   Upper Explosive Limit
      0 F.   .064   Lower Explosive Limit
    + 50 F.   .361   Vapor Density
    +100 F.   1.495   Specific Gravity
    +150 F.   4.905   Color
    +200 F.   13.445   Odor
    +250 F.   31.969
    +300 F.   67.824

VAPOR PRESSURE

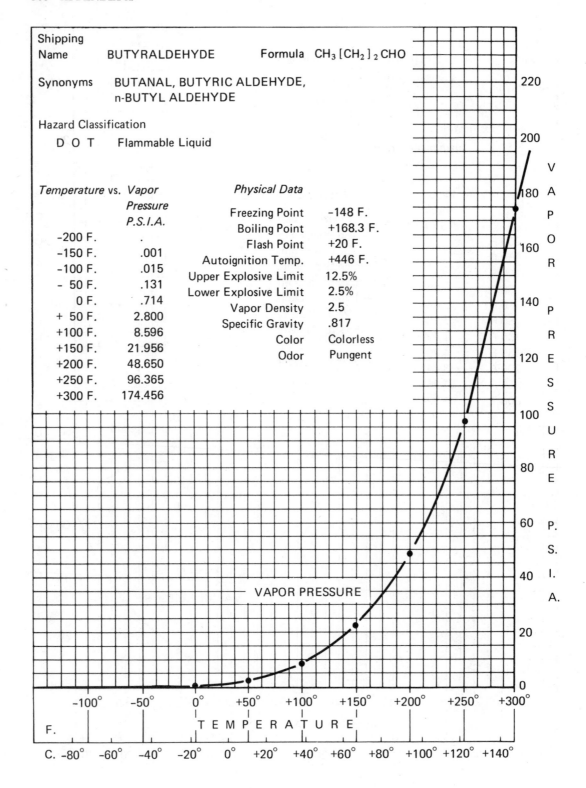

Shipping
Name   BUTYRALDEHYDE   Formula $CH_3[CH_2]_2CHO$

Synonyms   BUTANAL, BUTYRIC ALDEHYDE,
           n-BUTYL ALDEHYDE

Hazard Classification
    D O T   Flammable Liquid

Temperature vs. Vapor Pressure P.S.I.A.

Temperature	Vapor Pressure P.S.I.A.
-200 F.	.
-150 F.	.001
-100 F.	.015
- 50 F.	.131
0 F.	.714
+ 50 F.	2.800
+100 F.	8.596
+150 F.	21.956
+200 F.	48.650
+250 F.	96.365
+300 F.	174.456

Physical Data

Freezing Point	-148 F.
Boiling Point	+168.3 F.
Flash Point	+20 F.
Autoignition Temp.	+446 F.
Upper Explosive Limit	12.5%
Lower Explosive Limit	2.5%
Vapor Density	2.5
Specific Gravity	.817
Color	Colorless
Odor	Pungent

VAPOR PRESSURE

VAPOR PRESSURE P.S.I.A.

TEMPERATURE

F. -100° -50° 0° +50° +100° +150° +200° +250° +300°

C. -80° -60° -40° -20° 0° +20° +40° +60° +80° +100° +120° +140°

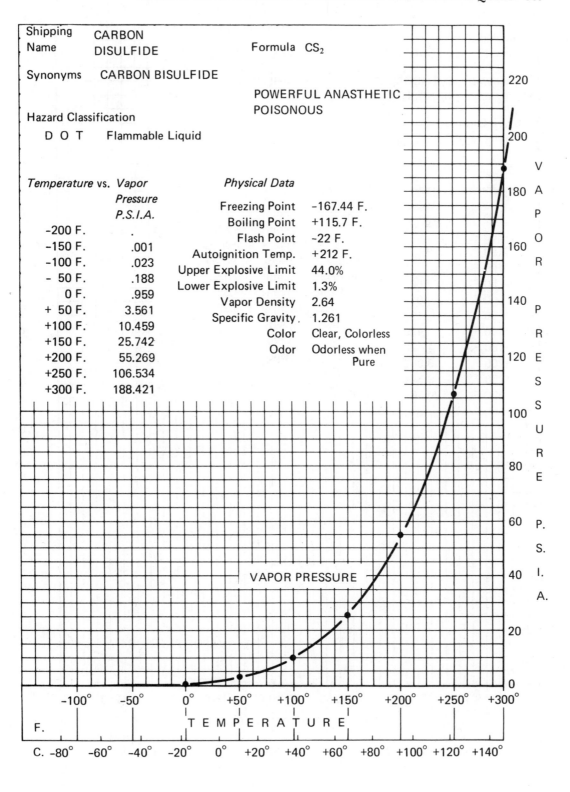

Shipping Name: CARBON DISULFIDE

Formula CS$_2$

Synonyms: CARBON BISULFIDE

POWERFUL ANASTHETIC
POISONOUS

Hazard Classification

DOT Flammable Liquid

Temperature vs. Vapor Pressure P.S.I.A.

Temperature	Vapor Pressure P.S.I.A.
-200 F.	.
-150 F.	.001
-100 F.	.023
- 50 F.	.188
0 F.	.959
+ 50 F.	3.561
+100 F.	10.459
+150 F.	25.742
+200 F.	55.269
+250 F.	106.534
+300 F.	188.421

Physical Data

Freezing Point	-167.44 F.
Boiling Point	+115.7 F.
Flash Point	-22 F.
Autoignition Temp.	+212 F.
Upper Explosive Limit	44.0%
Lower Explosive Limit	1.3%
Vapor Density	2.64
Specific Gravity	1.261
Color	Clear, Colorless
Odor	Odorless when Pure

VAPOR PRESSURE

VAPOR PRESSURE P.S.I.A.

TEMPERATURE

F.

C. -80° -60° -40° -20° 0° +20° +40° +60° +80° +100° +120° +140°

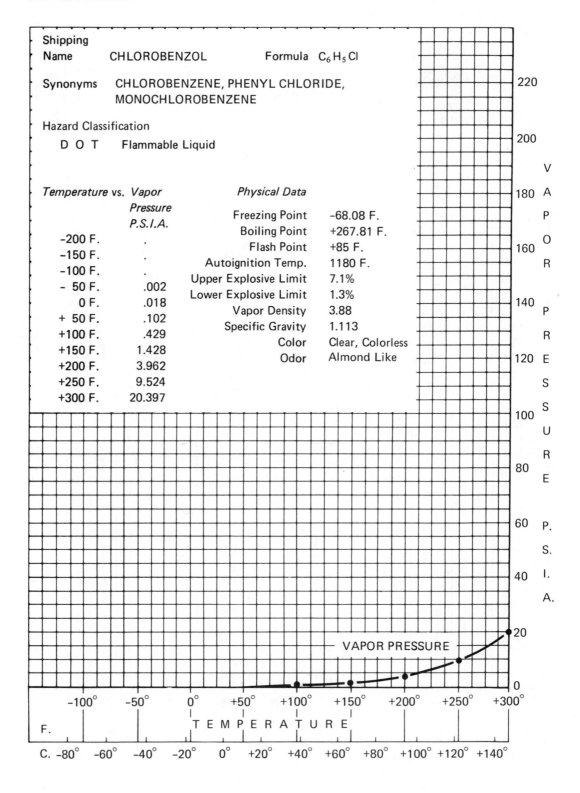

Shipping
Name   CHLOROBENZOL   Formula C_6H_5Cl

Synonyms   CHLOROBENZENE, PHENYL CHLORIDE,
            MONOCHLOROBENZENE

Hazard Classification
     D O T   Flammable Liquid

Temperature vs. Vapor
            Pressure
            P.S.I.A.

Temperature	Vapor Pressure P.S.I.A.
-200 F.	.
-150 F.	.
-100 F.	.
- 50 F.	.002
0 F.	.018
+ 50 F.	.102
+100 F.	.429
+150 F.	1.428
+200 F.	3.962
+250 F.	9.524
+300 F.	20.397

Physical Data

Freezing Point	-68.08 F.
Boiling Point	+267.81 F.
Flash Point	+85 F.
Autoignition Temp.	1180 F.
Upper Explosive Limit	7.1%
Lower Explosive Limit	1.3%
Vapor Density	3.88
Specific Gravity	1.113
Color	Clear, Colorless
Odor	Almond Like

VAPOR PRESSURE P.S.I.A.

VAPOR PRESSURE

TEMPERATURE

F.   -100°   -50°   0°   +50°   +100°   +150°   +200°   +250°   +300°

C. -80° -60° -40° -20° 0° +20° +40° +60° +80° +100° +120° +140°

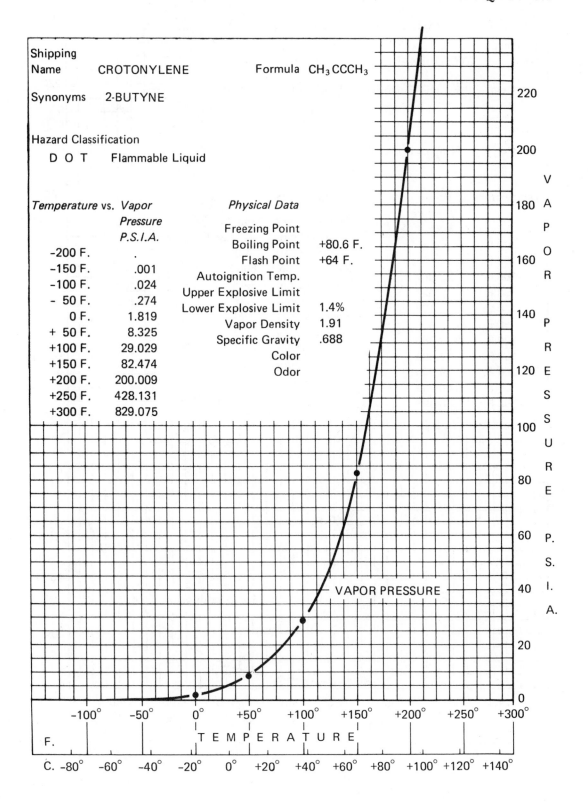

Shipping
Name   CROTONYLENE   Formula CH_3CCCH_3

Synonyms   2-BUTYNE

Hazard Classification
    D O T   Flammable Liquid

Temperature vs. Vapor
                    Pressure
                    P.S.I.A.
    -200 F.   .
    -150 F.   .001
    -100 F.   .024
    - 50 F.   .274
       0 F.   1.819
    + 50 F.   8.325
    +100 F.   29.029
    +150 F.   82.474
    +200 F.   200.009
    +250 F.   428.131
    +300 F.   829.075

Physical Data
    Freezing Point
    Boiling Point   +80.6 F.
    Flash Point   +64 F.
    Autoignition Temp.
    Upper Explosive Limit
    Lower Explosive Limit   1.4%
    Vapor Density   1.91
    Specific Gravity   .688
    Color
    Odor

VAPOR PRESSURE

VAPOR PRESSURE P.S.I.A.

TEMPERATURE

F.
C.

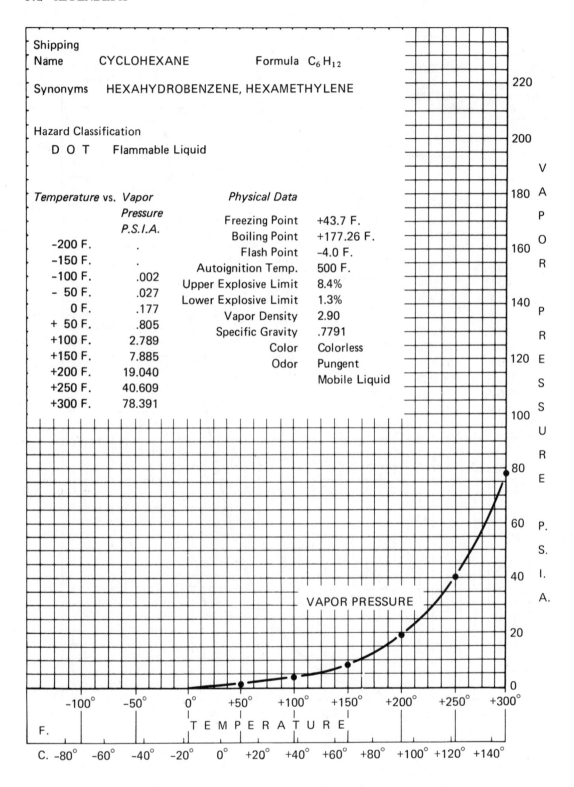

Shipping
Name   CYCLOHEXANE   Formula C_6H_{12}

Synonyms   HEXAHYDROBENZENE, HEXAMETHYLENE

Hazard Classification
    D O T   Flammable Liquid

Temperature vs. Vapor
                   Pressure
                   P.S.I.A.

Temperature	Pressure P.S.I.A.
-200 F.	.
-150 F.	.
-100 F.	.002
- 50 F.	.027
0 F.	.177
+ 50 F.	.805
+100 F.	2.789
+150 F.	7.885
+200 F.	19.040
+250 F.	40.609
+300 F.	78.391

Physical Data

Freezing Point	+43.7 F.
Boiling Point	+177.26 F.
Flash Point	-4.0 F.
Autoignition Temp.	500 F.
Upper Explosive Limit	8.4%
Lower Explosive Limit	1.3%
Vapor Density	2.90
Specific Gravity	.7791
Color	Colorless
Odor	Pungent
	Mobile Liquid

VAPOR PRESSURE

VAPOR PRESSURE P.S.I.A.

TEMPERATURE

F.   -100°   -50°   0°   +50°   +100°   +150°   +200°   +250°   +300°

C. -80° -60° -40° -20° 0° +20° +40° +60° +80° +100° +120° +140°

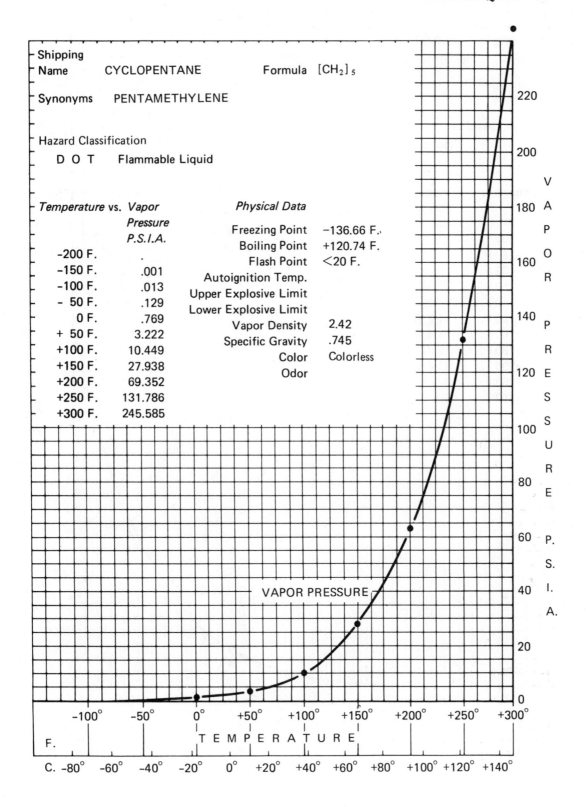

Shipping
Name   CYCLOPENTANE   Formula $[CH_2]_5$

Synonyms   PENTAMETHYLENE

Hazard Classification
    D O T   Flammable Liquid

Temperature vs. Vapor
                Pressure
                P.S.I.A.
    -200 F.   .
    -150 F.   .001
    -100 F.   .013
    - 50 F.   .129
      0 F.   .769
    + 50 F.   3.222
    +100 F.   10.449
    +150 F.   27.938
    +200 F.   69.352
    +250 F.   131.786
    +300 F.   245.585

Physical Data
    Freezing Point   −136.66 F.
    Boiling Point   +120.74 F.
    Flash Point   <20 F.
    Autoignition Temp.
    Upper Explosive Limit
    Lower Explosive Limit
    Vapor Density   2.42
    Specific Gravity   .745
    Color   Colorless
    Odor

VAPOR PRESSURE

VAPOR PRESSURE P.S.I.A.

TEMPERATURE

F.   −100°   −50°   0°   +50°   +100°   +150°   +200°   +250°   +300°

C. −80°   −60°   −40°   −20°   0°   +20°   +40°   +60°   +80°   +100°   +120°   +140°

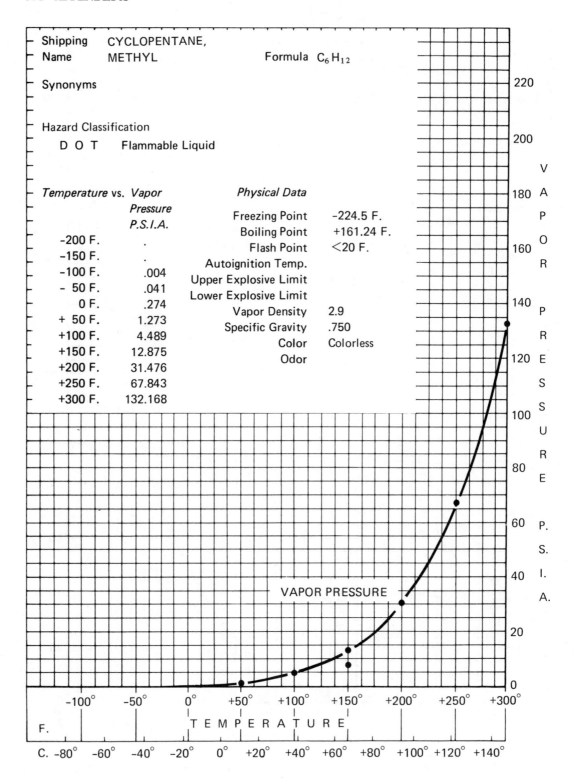

Shipping Name   CYCLOPENTANE, METHYL   Formula C_6H_{12}

Synonyms

Hazard Classification
     D O T   Flammable Liquid

Temperature vs. Vapor Pressure P.S.I.A.

Temperature	P.S.I.A.
-200 F.	.
-150 F.	.
-100 F.	.004
- 50 F.	.041
0 F.	.274
+ 50 F.	1.273
+100 F.	4.489
+150 F.	12.875
+200 F.	31.476
+250 F.	67.843
+300 F.	132.168

Physical Data

Freezing Point	-224.5 F.
Boiling Point	+161.24 F.
Flash Point	<20 F.
Autoignition Temp.	
Upper Explosive Limit	
Lower Explosive Limit	
Vapor Density	2.9
Specific Gravity	.750
Color	Colorless
Odor	

VAPOR PRESSURE

V A P O R   P R E S S U R E   P. S. I. A.

TEMPERATURE
F.   -100°   -50°   0°   +50°   +100°   +150°   +200°   +250°   +300°
C. -80° -60° -40° -20° 0° +20° +40° +60° +80° +100° +120° +140°

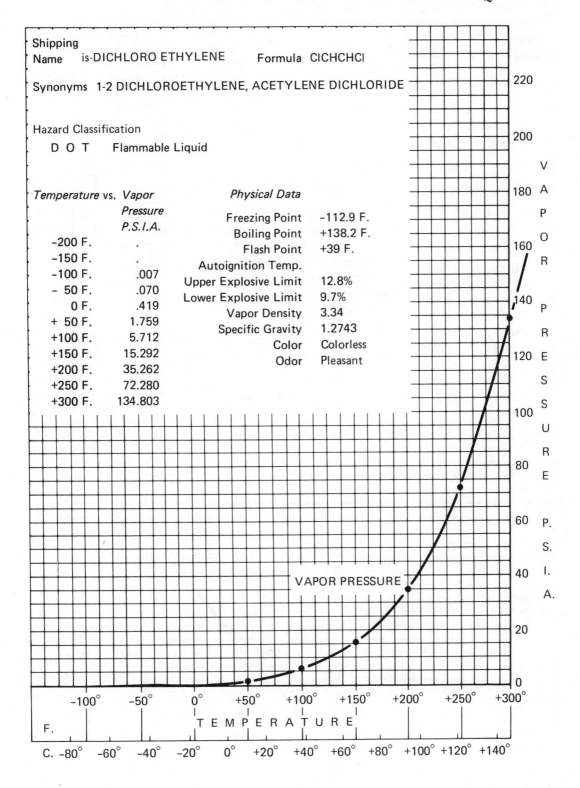

Shipping
Name   is-DICHLORO ETHYLENE   Formula CICHCHCI

Synonyms 1-2 DICHLOROETHYLENE, ACETYLENE DICHLORIDE

Hazard Classification
   D O T   Flammable Liquid

Temperature vs. Vapor Pressure P.S.I.A.

Temperature	P.S.I.A.
-200 F.	.
-150 F.	.
-100 F.	.007
- 50 F.	.070
0 F.	.419
+ 50 F.	1.759
+100 F.	5.712
+150 F.	15.292
+200 F.	35.262
+250 F.	72.280
+300 F.	134.803

Physical Data

Freezing Point	-112.9 F.
Boiling Point	+138.2 F.
Flash Point	+39 F.
Autoignition Temp.	
Upper Explosive Limit	12.8%
Lower Explosive Limit	9.7%
Vapor Density	3.34
Specific Gravity	1.2743
Color	Colorless
Odor	Pleasant

VAPOR PRESSURE

VAPOR PRESSURE P.S.I.A.

TEMPERATURE

F.

C.

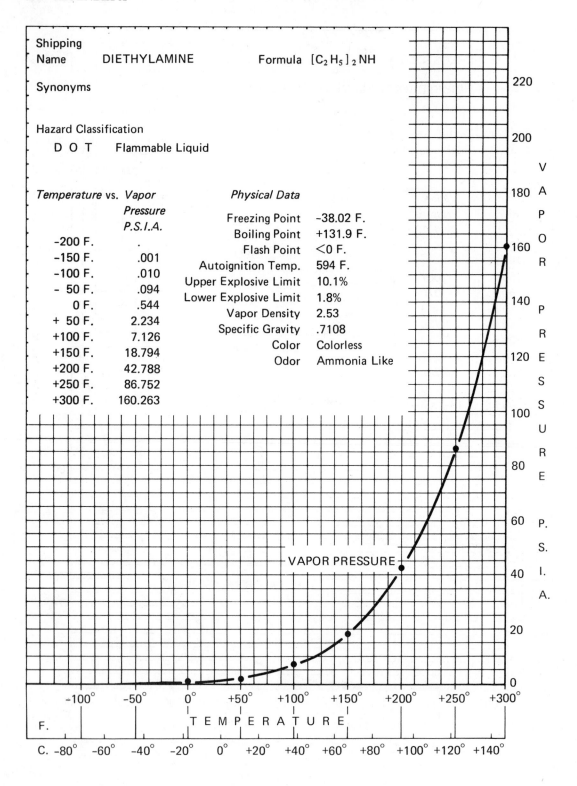

Shipping
Name   DIETHYLAMINE   Formula   [C₂H₅]₂NH

Synonyms

Hazard Classification
   D O T   Flammable Liquid

Temperature vs. Vapor
Pressure
P.S.I.A.

Temperature	P.S.I.A.
-200 F.	.
-150 F.	.001
-100 F.	.010
- 50 F.	.094
0 F.	.544
+ 50 F.	2.234
+100 F.	7.126
+150 F.	18.794
+200 F.	42.788
+250 F.	86.752
+300 F.	160.263

Physical Data

Freezing Point	-38.02 F.
Boiling Point	+131.9 F.
Flash Point	<0 F.
Autoignition Temp.	594 F.
Upper Explosive Limit	10.1%
Lower Explosive Limit	1.8%
Vapor Density	2.53
Specific Gravity	.7108
Color	Colorless
Odor	Ammonia Like

VAPOR PRESSURE

VAPOR PRESSURE P.S.I.A.

TEMPERATURE

F.

C.

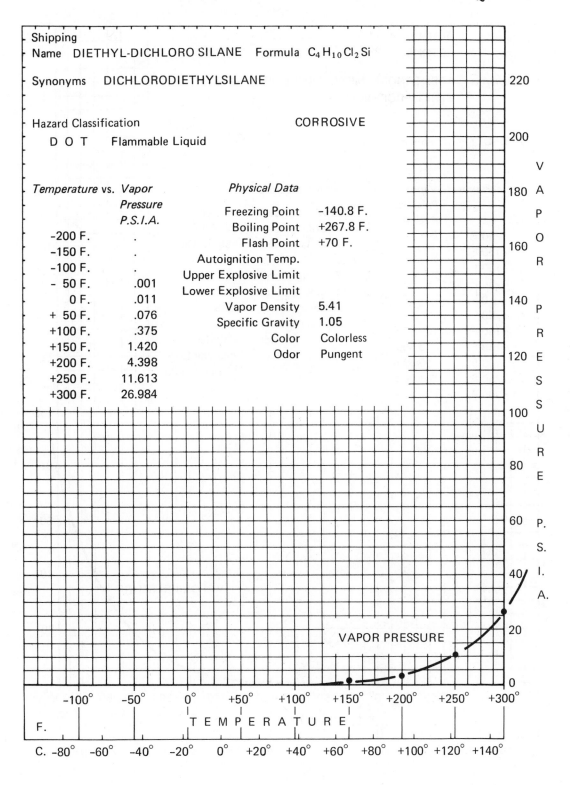

Shipping
Name   DIETHYL-DICHLORO SILANE   Formula $C_4H_{10}Cl_2Si$

Synonyms   DICHLORODIETHYLSILANE

Hazard Classification   CORROSIVE
   D O T   Flammable Liquid

Temperature vs. Vapor
                    Pressure
                    P.S.I.A.

Temperature	Pressure P.S.I.A.
-200 F.	.
-150 F.	.
-100 F.	.
- 50 F.	.001
0 F.	.011
+ 50 F.	.076
+100 F.	.375
+150 F.	1.420
+200 F.	4.398
+250 F.	11.613
+300 F.	26.984

Physical Data

Freezing Point	-140.8 F.
Boiling Point	+267.8 F.
Flash Point	+70 F.
Autoignition Temp.	
Upper Explosive Limit	
Lower Explosive Limit	
Vapor Density	5.41
Specific Gravity	1.05
Color	Colorless
Odor	Pungent

VAPOR PRESSURE

VAPOR PRESSURE P.S.I.A.

TEMPERATURE

F.

C.

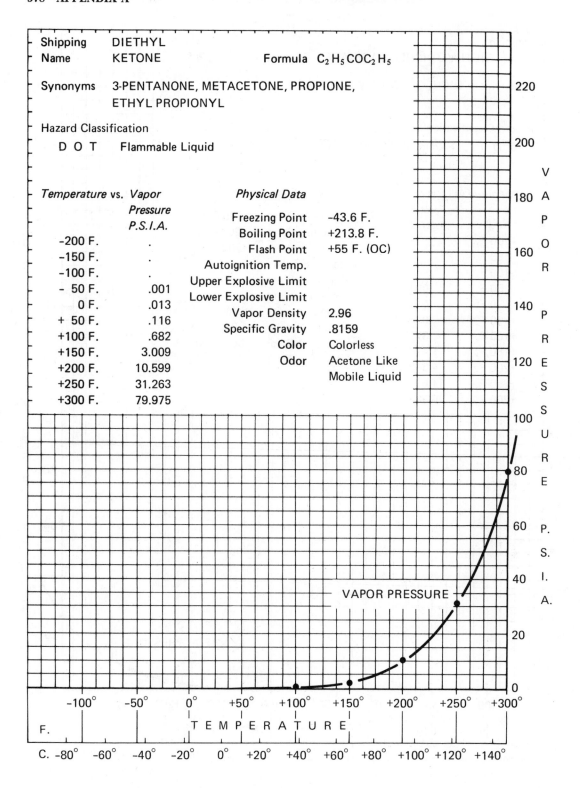

Shipping Name DIETHYL KETONE Formula $C_2H_5COC_2H_5$

Synonyms 3-PENTANONE, METACETONE, PROPIONE, ETHYL PROPIONYL

Hazard Classification

D O T Flammable Liquid

Temperature vs. *Vapor Pressure P.S.I.A.*

Temperature	P.S.I.A.
-200 F.	.
-150 F.	.
-100 F.	.
- 50 F.	.001
0 F.	.013
+ 50 F.	.116
+100 F.	.682
+150 F.	3.009
+200 F.	10.599
+250 F.	31.263
+300 F.	79.975

Physical Data

Freezing Point	-43.6 F.
Boiling Point	+213.8 F.
Flash Point	+55 F. (OC)
Autoignition Temp.	
Upper Explosive Limit	
Lower Explosive Limit	
Vapor Density	2.96
Specific Gravity	.8159
Color	Colorless
Odor	Acetone Like
	Mobile Liquid

VAPOR PRESSURE

VAPOR PRESSURE P. S. I. A.

TEMPERATURE

F. -100° -50° 0° +50° +100° +150° +200° +250° +300°

C. -80° -60° -40° -20° 0° +20° +40° +60° +80° +100° +120° +140°

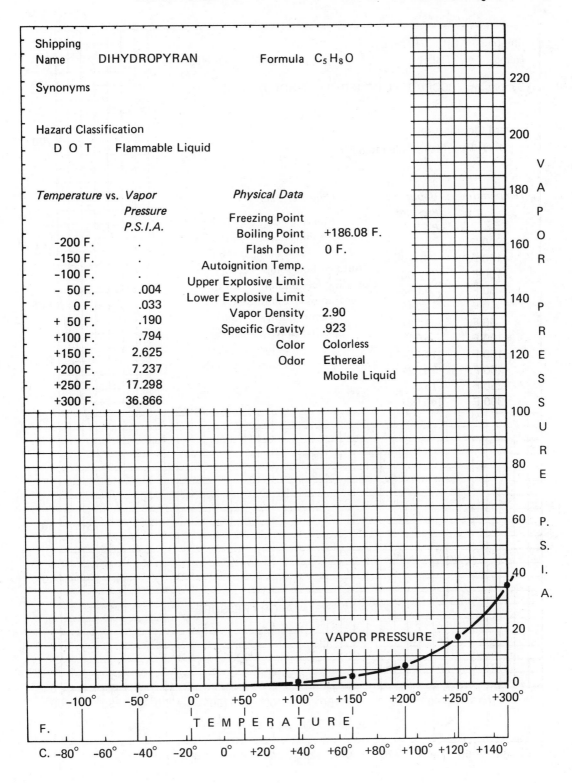

Shipping
Name   DIHYDROPYRAN   Formula C_5H_8O

Synonyms

Hazard Classification
   D O T   Flammable Liquid

Temperature vs. Vapor
                  Pressure
                  P.S.I.A.

Temperature	P.S.I.A.
-200 F.	.
-150 F.	.
-100 F.	.
- 50 F.	.004
0 F.	.033
+ 50 F.	.190
+100 F.	.794
+150 F.	2.625
+200 F.	7.237
+250 F.	17.298
+300 F.	36.866

Physical Data

Freezing Point	
Boiling Point	+186.08 F.
Flash Point	0 F.
Autoignition Temp.	
Upper Explosive Limit	
Lower Explosive Limit	
Vapor Density	2.90
Specific Gravity	.923
Color	Colorless
Odor	Ethereal
	Mobile Liquid

VAPOR PRESSURE

VAPOR PRESSURE P.S.I.A.

TEMPERATURE

F.

-100° -50° 0° +50° +100° +150° +200° +250° +300°

C. -80° -60° -40° -20° 0° +20° +40° +60° +80° +100° +120° +140°

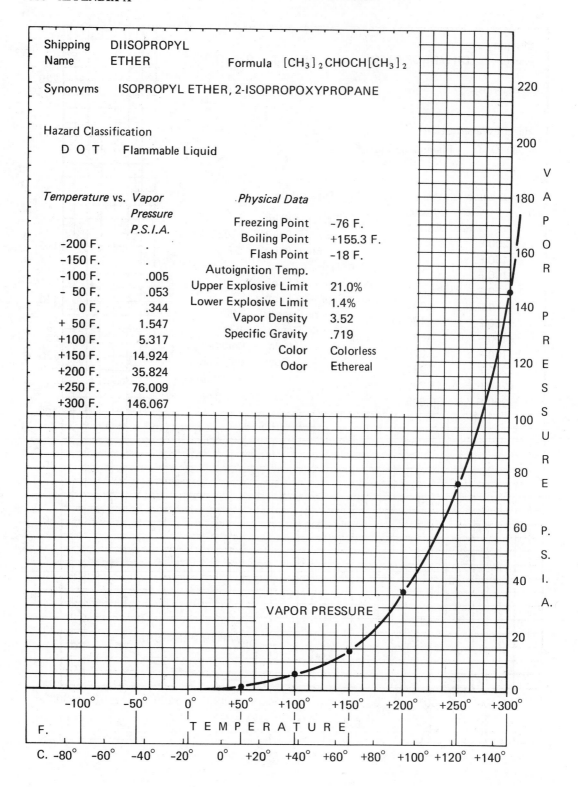

Shipping Name   DIISOPROPYL ETHER

Formula $[CH_3]_2CHOCH[CH_3]_2$

Synonyms   ISOPROPYL ETHER, 2-ISOPROPOXYPROPANE

Hazard Classification

  D O T   Flammable Liquid

Temperature vs. *Vapor Pressure P.S.I.A.*

-200 F.	.
-150 F.	.
-100 F.	.005
- 50 F.	.053
0 F.	.344
+ 50 F.	1.547
+100 F.	5.317
+150 F.	14.924
+200 F.	35.824
+250 F.	76.009
+300 F.	146.067

Physical Data

Freezing Point	-76 F.
Boiling Point	+155.3 F.
Flash Point	-18 F.
Autoignition Temp.	
Upper Explosive Limit	21.0%
Lower Explosive Limit	1.4%
Vapor Density	3.52
Specific Gravity	.719
Color	Colorless
Odor	Ethereal

VAPOR PRESSURE

VAPOR PRESSURE P.S.I.A.

TEMPERATURE

F.   -100°   -50°   0°   +50°   +100°   +150°   +200°   +250°   +300°

C. -80° -60° -40° -20° 0° +20° +40° +60° +80° +100° +120° +140°

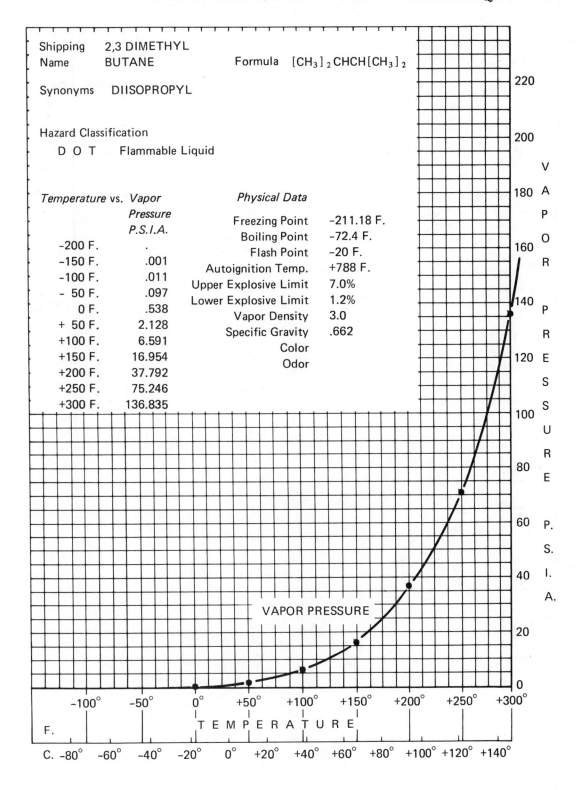

Shipping Name   2,3 DIMETHYL BUTANE

Formula $[CH_3]_2CHCH[CH_3]_2$

Synonyms   DIISOPROPYL

Hazard Classification
  D O T   Flammable Liquid

Temperature vs. Vapor Pressure P.S.I.A.

Temperature	Vapor Pressure P.S.I.A.
-200 F.	.
-150 F.	.001
-100 F.	.011
- 50 F.	.097
0 F.	.538
+ 50 F.	2.128
+100 F.	6.591
+150 F.	16.954
+200 F.	37.792
+250 F.	75.246
+300 F.	136.835

Physical Data

Freezing Point	-211.18 F.
Boiling Point	-72.4 F.
Flash Point	-20 F.
Autoignition Temp.	+788 F.
Upper Explosive Limit	7.0%
Lower Explosive Limit	1.2%
Vapor Density	3.0
Specific Gravity	.662
Color	
Odor	

VAPOR PRESSURE

VAPOR PRESSURE P.S.I.A.

TEMPERATURE

F.
C.

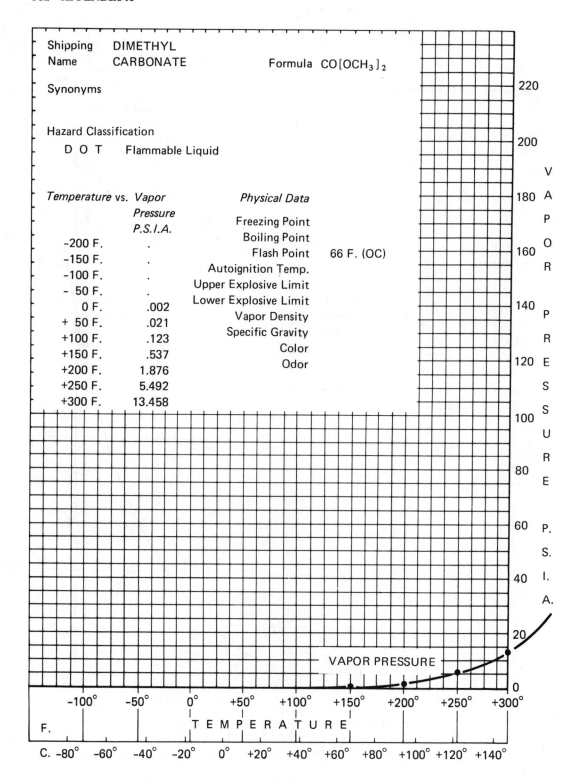

Shipping Name   DIMETHYL CARBONATE   Formula $CO[OCH_3]_2$

Synonyms

Hazard Classification
   D O T   Flammable Liquid

Temperature vs. Vapor Pressure P.S.I.A.		Physical Data	
-200 F.	.	Freezing Point	
-150 F.	.	Boiling Point	
-100 F.	.	Flash Point	66 F. (OC)
- 50 F.	.	Autoignition Temp.	
0 F.	.002	Upper Explosive Limit	
+ 50 F.	.021	Lower Explosive Limit	
+100 F.	.123	Vapor Density	
+150 F.	.537	Specific Gravity	
+200 F.	1.876	Color	
+250 F.	5.492	Odor	
+300 F.	13.458		

VAPOR PRESSURE

VAPOR PRESSURE P.S.I.A.

220
200
180
160
140
120
100
80
60
40
20
0

TEMPERATURE

F. -100° -50° 0° +50° +100° +150° +200° +250° +300°

C. -80° -60° -40° -20° 0° +20° +40° +60° +80° +100° +120° +140°

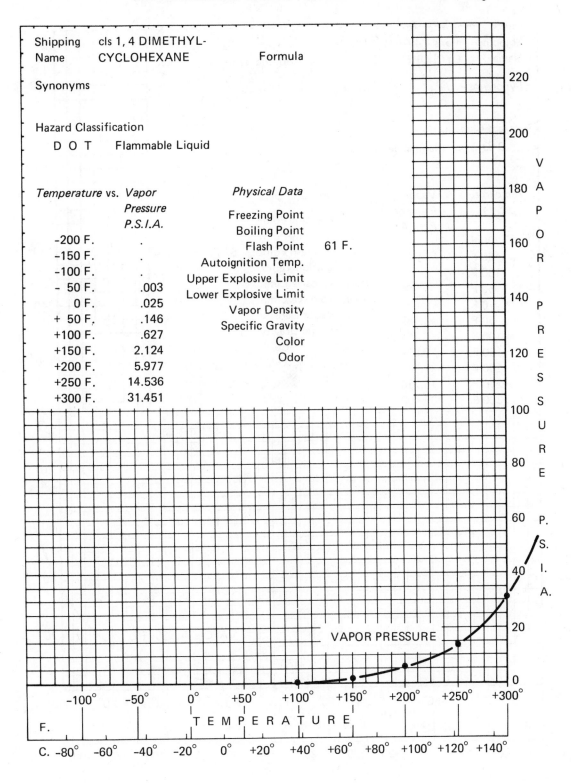

Shipping Name   cls 1, 4 DIMETHYL-CYCLOHEXANE

Formula

Synonyms

Hazard Classification
   D O T   Flammable Liquid

Temperature vs. Vapor Pressure P.S.I.A.

Temperature	Vapor Pressure P.S.I.A.
-200 F.	.
-150 F.	.
-100 F.	.
- 50 F.	.003
0 F.	.025
+ 50 F.	.146
+100 F.	.627
+150 F.	2.124
+200 F.	5.977
+250 F.	14.536
+300 F.	31.451

Physical Data

Freezing Point
Boiling Point
Flash Point   61 F.
Autoignition Temp.
Upper Explosive Limit
Lower Explosive Limit
Vapor Density
Specific Gravity
Color
Odor

VAPOR PRESSURE

VAPOR PRESSURE P.S.I.A.

TEMPERATURE

F.   -100°   -50°   0°   +50°   +100°   +150°   +200°   +250°   +300°

C. -80°   -60°   -40°   -20°   0°   +20°   +40°   +60°   +80°   +100°   +120°   +140°

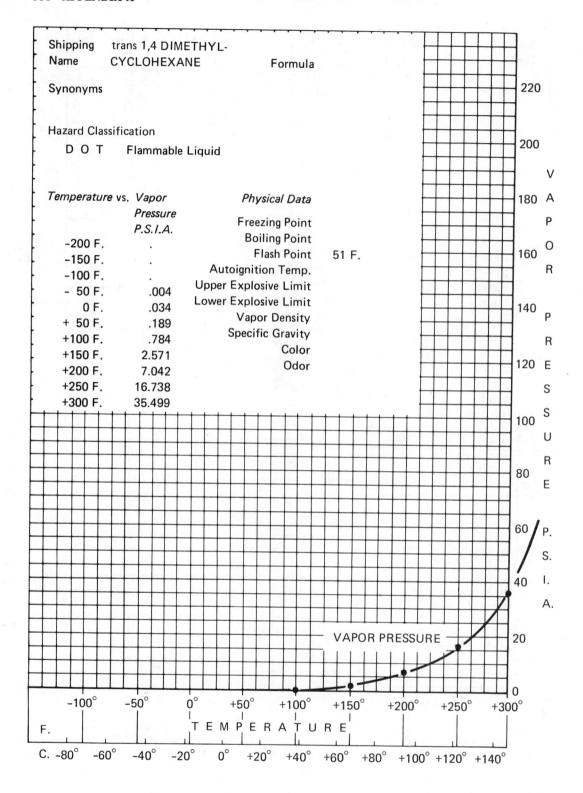

Shipping Name: trans 1,4 DIMETHYL-CYCLOHEXANE

Formula

Synonyms

Hazard Classification

D O T   Flammable Liquid

Temperature vs. Vapor Pressure P.S.I.A.

Temperature	Vapor Pressure P.S.I.A.
-200 F.	.
-150 F.	.
-100 F.	.
- 50 F.	.004
0 F.	.034
+ 50 F.	.189
+100 F.	.784
+150 F.	2.571
+200 F.	7.042
+250 F.	16.738
+300 F.	35.499

Physical Data

Freezing Point
Boiling Point
Flash Point   51 F.
Autoignition Temp.
Upper Explosive Limit
Lower Explosive Limit
Vapor Density
Specific Gravity
Color
Odor

VAPOR PRESSURE P.S.I.A.

VAPOR PRESSURE

TEMPERATURE

F. -100° -50° 0° +50° +100° +150° +200° +250° +300°

C. -80° -60° -40° -20° 0° +20° +40° +60° +80° +100° +120° +140°

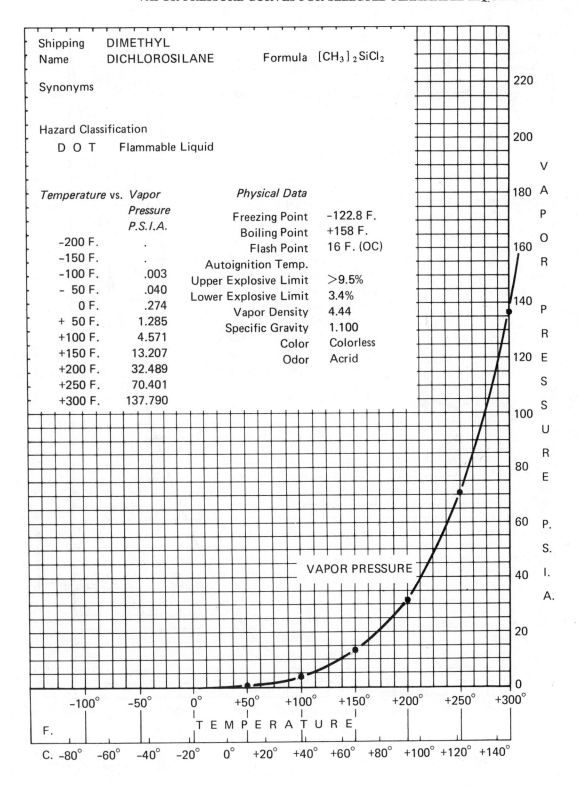

Shipping Name   DIMETHYL DICHLOROSILANE

Formula $[CH_3]_2 SiCl_2$

Synonyms

Hazard Classification
    D O T   Flammable Liquid

Temperature vs. *Vapor Pressure P.S.I.A.*

Temperature	Vapor Pressure P.S.I.A.
-200 F.	.
-150 F.	.
-100 F.	.003
- 50 F.	.040
0 F.	.274
+ 50 F.	1.285
+100 F.	4.571
+150 F.	13.207
+200 F.	32.489
+250 F.	70.401
+300 F.	137.790

Physical Data

Freezing Point	-122.8 F.
Boiling Point	+158 F.
Flash Point	16 F. (OC)
Autoignition Temp.	
Upper Explosive Limit	>9.5%
Lower Explosive Limit	3.4%
Vapor Density	4.44
Specific Gravity	1.100
Color	Colorless
Odor	Acrid

VAPOR PRESSURE

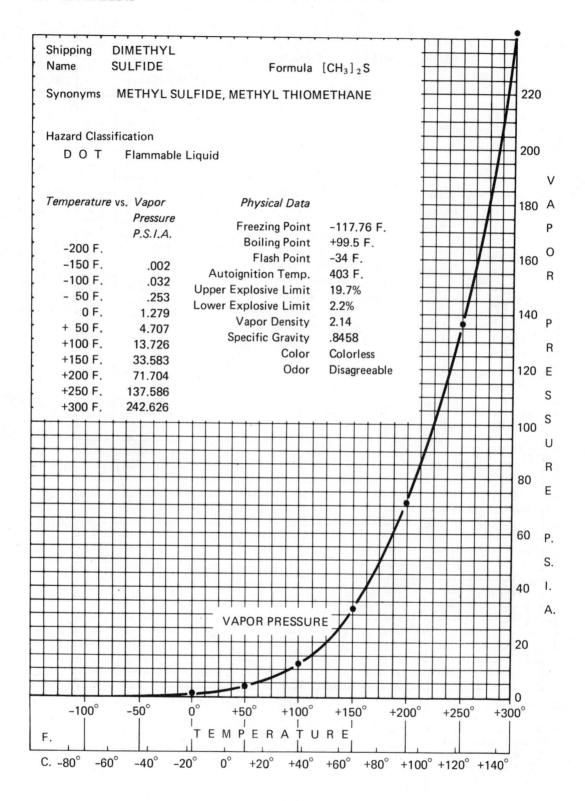

Shipping Name   DIMETHYL SULFIDE   Formula   [CH₃]₂S

Synonyms   METHYL SULFIDE, METHYL THIOMETHANE

Hazard Classification
  D O T   Flammable Liquid

Temperature vs. Vapor Pressure P.S.I.A.

Temperature	Vapor Pressure P.S.I.A.
-200 F.	
-150 F.	.002
-100 F.	.032
- 50 F.	.253
0 F.	1.279
+ 50 F.	4.707
+100 F.	13.726
+150 F.	33.583
+200 F.	71.704
+250 F.	137.586
+300 F.	242.626

Physical Data

Freezing Point	-117.76 F.
Boiling Point	+99.5 F.
Flash Point	-34 F.
Autoignition Temp.	403 F.
Upper Explosive Limit	19.7%
Lower Explosive Limit	2.2%
Vapor Density	2.14
Specific Gravity	.8458
Color	Colorless
Odor	Disagreeable

VAPOR PRESSURE

VAPOR PRESSURE P.S.I.A.

TEMPERATURE

F. -100° -50° 0° +50° +100° +150° +200° +250° +300°

C. -80° -60° -40° -20° 0° +20° +40° +60° +80° +100° +120° +140°

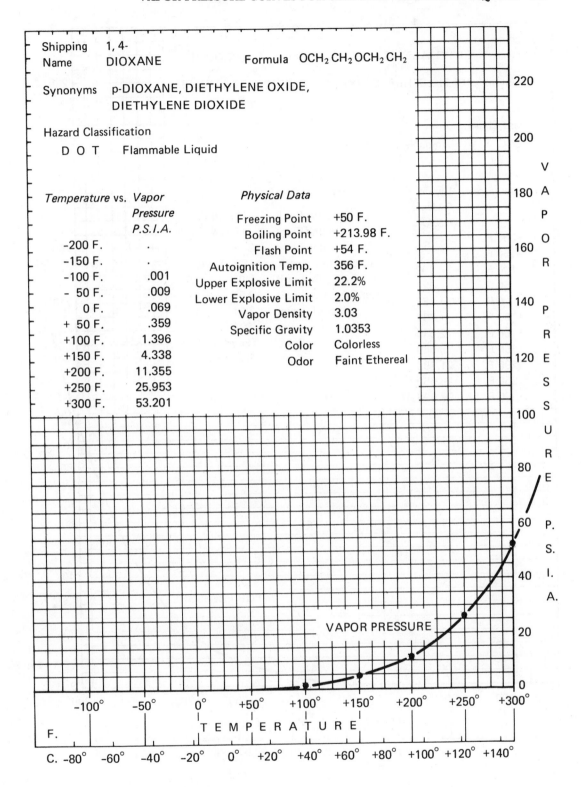

Shipping Name   1, 4- DIOXANE

Formula $OCH_2CH_2OCH_2CH_2$

Synonyms   p-DIOXANE, DIETHYLENE OXIDE, DIETHYLENE DIOXIDE

Hazard Classification
D O T   Flammable Liquid

Temperature vs. Vapor Pressure P.S.I.A.

Temperature	Vapor Pressure P.S.I.A.
-200 F.	.
-150 F.	.
-100 F.	.001
- 50 F.	.009
0 F.	.069
+ 50 F.	.359
+100 F.	1.396
+150 F.	4.338
+200 F.	11.355
+250 F.	25.953
+300 F.	53.201

Physical Data

Freezing Point	+50 F.
Boiling Point	+213.98 F.
Flash Point	+54 F.
Autoignition Temp.	356 F.
Upper Explosive Limit	22.2%
Lower Explosive Limit	2.0%
Vapor Density	3.03
Specific Gravity	1.0353
Color	Colorless
Odor	Faint Ethereal

VAPOR PRESSURE

VAPOR PRESSURE P.S.I.A.

TEMPERATURE
F.
C.

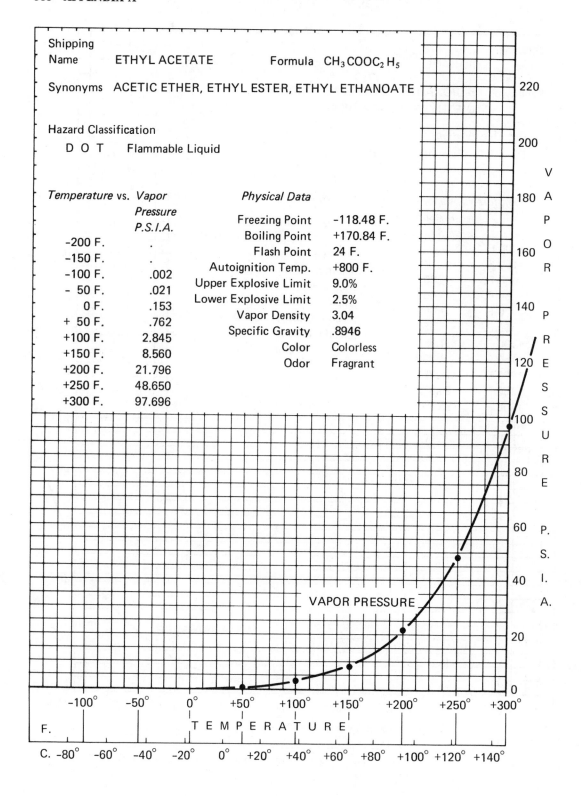

Shipping
Name   ETHYL ACETATE   Formula $CH_3COOC_2H_5$

Synonyms   ACETIC ETHER, ETHYL ESTER, ETHYL ETHANOATE

Hazard Classification
    D O T   Flammable Liquid

Temperature vs. *Vapor Pressure P.S.I.A.*

Temperature	Vapor Pressure P.S.I.A.
−200 F.	.
−150 F.	.
−100 F.	.002
− 50 F.	.021
0 F.	.153
+ 50 F.	.762
+100 F.	2.845
+150 F.	8.560
+200 F.	21.796
+250 F.	48.650
+300 F.	97.696

Physical Data

Freezing Point	−118.48 F.
Boiling Point	+170.84 F.
Flash Point	24 F.
Autoignition Temp.	+800 F.
Upper Explosive Limit	9.0%
Lower Explosive Limit	2.5%
Vapor Density	3.04
Specific Gravity	.8946
Color	Colorless
Odor	Fragrant

VAPOR PRESSURE

VAPOR PRESSURE P.S.I.A.

TEMPERATURE

F.   −100°   −50°   0°   +50°   +100°   +150°   +200°   +250°   +300°

C. −80° −60° −40° −20° 0° +20° +40° +60° +80° +100° +120° +140°

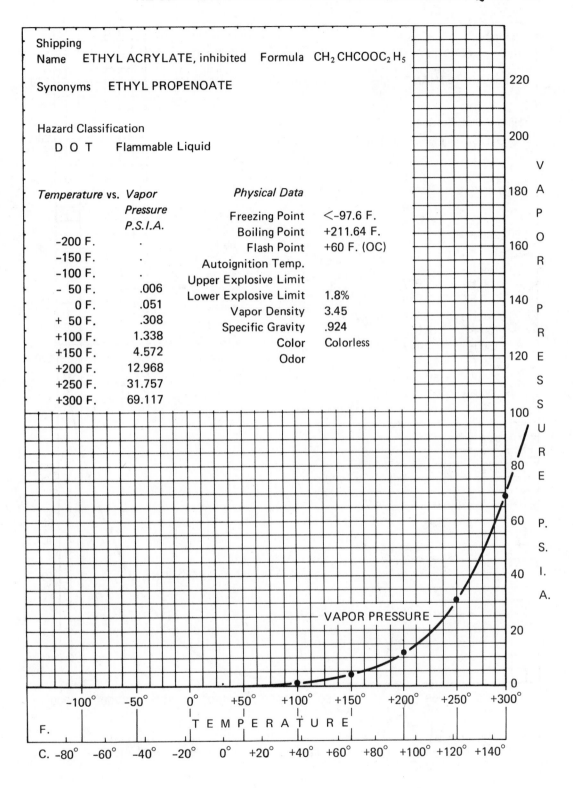

Shipping
Name   ETHYL ACRYLATE, inhibited   Formula $CH_2 CHCOOC_2 H_5$

Synonyms   ETHYL PROPENOATE

Hazard Classification

  D O T   Flammable Liquid

Temperature vs.	Vapor Pressure P.S.I.A.
-200 F.	.
-150 F.	.
-100 F.	.
- 50 F.	.006
0 F.	.051
+ 50 F.	.308
+100 F.	1.338
+150 F.	4.572
+200 F.	12.968
+250 F.	31.757
+300 F.	69.117

Physical Data

Freezing Point	<-97.6 F.
Boiling Point	+211.64 F.
Flash Point	+60 F. (OC)
Autoignition Temp.	
Upper Explosive Limit	
Lower Explosive Limit	1.8%
Vapor Density	3.45
Specific Gravity	.924
Color	Colorless
Odor	

VAPOR PRESSURE P.S.I.A.

VAPOR PRESSURE

TEMPERATURE

F.   -100°   -50°   0°   +50°   +100°   +150°   +200°   +250°   +300°

C. -80°   -60°   -40°   -20°   0°   +20°   +40°   +60°   +80°   +100°   +120°   +140°

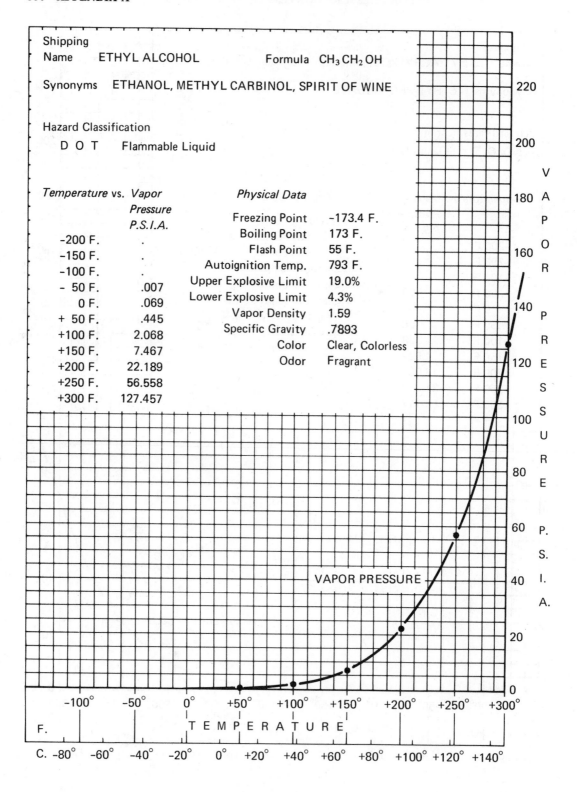

Shipping
Name   ETHYL ALCOHOL   Formula CH_3CH_2OH

Synonyms   ETHANOL, METHYL CARBINOL, SPIRIT OF WINE

Hazard Classification
    D O T   Flammable Liquid

Temperature vs. *Vapor Pressure P.S.I.A.*

-200 F.	.
-150 F.	.
-100 F.	.
- 50 F.	.007
0 F.	.069
+ 50 F.	.445
+100 F.	2.068
+150 F.	7.467
+200 F.	22.189
+250 F.	56.558
+300 F.	127.457

Physical Data

Freezing Point	-173.4 F.
Boiling Point	173 F.
Flash Point	55 F.
Autoignition Temp.	793 F.
Upper Explosive Limit	19.0%
Lower Explosive Limit	4.3%
Vapor Density	1.59
Specific Gravity	.7893
Color	Clear, Colorless
Odor	Fragrant

VAPOR PRESSURE

VAPOR PRESSURE P.S.I.A.

TEMPERATURE

F.   -100°   -50°   0°   +50°   +100°   +150°   +200°   +250°   +300°

C. -80°   -60°   -40°   -20°   0°   +20°   +40°   +60°   +80°   +100°   +120°   +140°

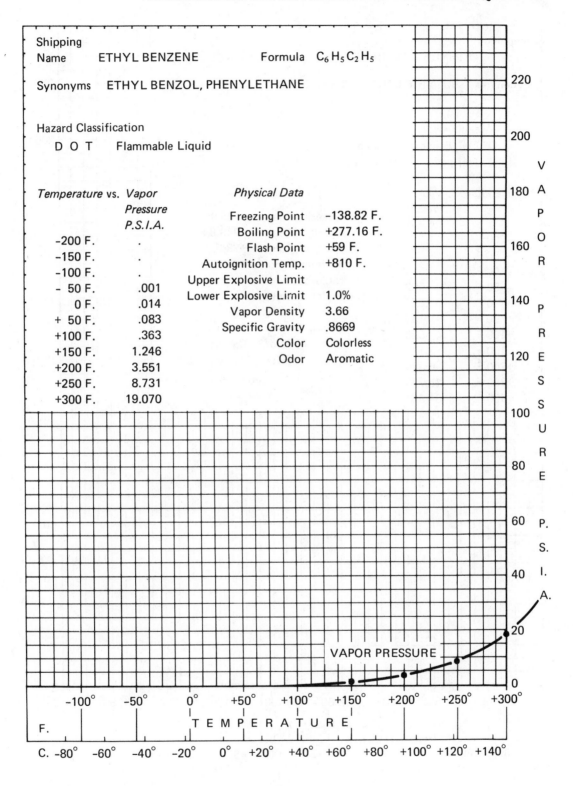

Shipping
Name   ETHYL BENZENE   Formula $C_6H_5C_2H_5$

Synonyms   ETHYL BENZOL, PHENYLETHANE

Hazard Classification
    D O T   Flammable Liquid

Temperature vs. Vapor Pressure P.S.I.A.

Temperature	Vapor Pressure P.S.I.A.
-200 F.	.
-150 F.	.
-100 F.	.
- 50 F.	.001
0 F.	.014
+ 50 F.	.083
+100 F.	.363
+150 F.	1.246
+200 F.	3.551
+250 F.	8.731
+300 F.	19.070

Physical Data

Freezing Point	-138.82 F.
Boiling Point	+277.16 F.
Flash Point	+59 F.
Autoignition Temp.	+810 F.
Upper Explosive Limit	
Lower Explosive Limit	1.0%
Vapor Density	3.66
Specific Gravity	.8669
Color	Colorless
Odor	Aromatic

VAPOR PRESSURE

VAPOR PRESSURE P.S.I.A.

TEMPERATURE
F.
-100° -50° 0° +50° +100° +150° +200° +250° +300°
C. -80° -60° -40° -20° 0° +20° +40° +60° +80° +100° +120° +140°

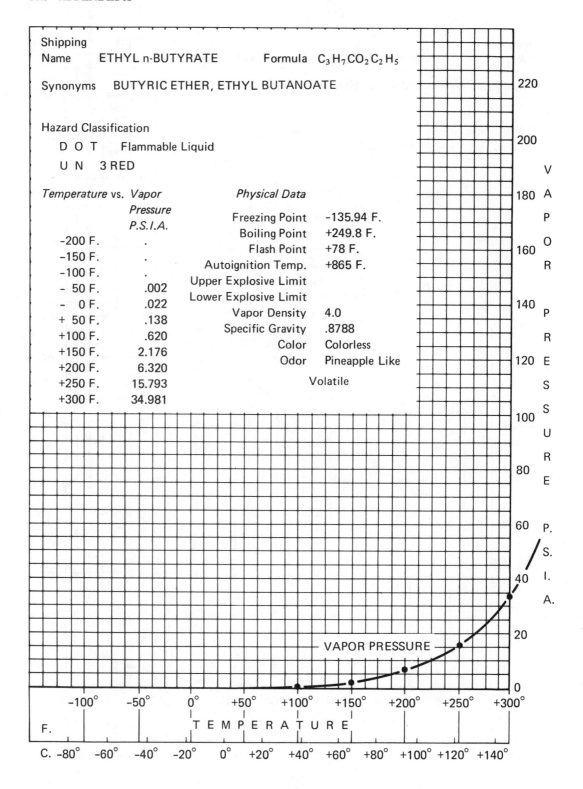

Shipping
Name ETHYL n-BUTYRATE Formula $C_3H_7CO_2C_2H_5$

Synonyms BUTYRIC ETHER, ETHYL BUTANOATE

Hazard Classification
  D O T Flammable Liquid
  U N 3 RED

Temperature vs. *Vapor Pressure P.S.I.A.*

Temperature	Vapor Pressure P.S.I.A.
-200 F.	.
-150 F.	.
-100 F.	.
- 50 F.	.002
- 0 F.	.022
+ 50 F.	.138
+100 F.	.620
+150 F.	2.176
+200 F.	6.320
+250 F.	15.793
+300 F.	34.981

Physical Data

Freezing Point	-135.94 F.
Boiling Point	+249.8 F.
Flash Point	+78 F.
Autoignition Temp.	+865 F.
Upper Explosive Limit	
Lower Explosive Limit	
Vapor Density	4.0
Specific Gravity	.8788
Color	Colorless
Odor	Pineapple Like

Volatile

VAPOR PRESSURE P.S.I.A.

VAPOR PRESSURE

TEMPERATURE

F. -100° -50° 0° +50° +100° +150° +200° +250° +300°

C. -80° -60° -40° -20° 0° +20° +40° +60° +80° +100° +120° +140°

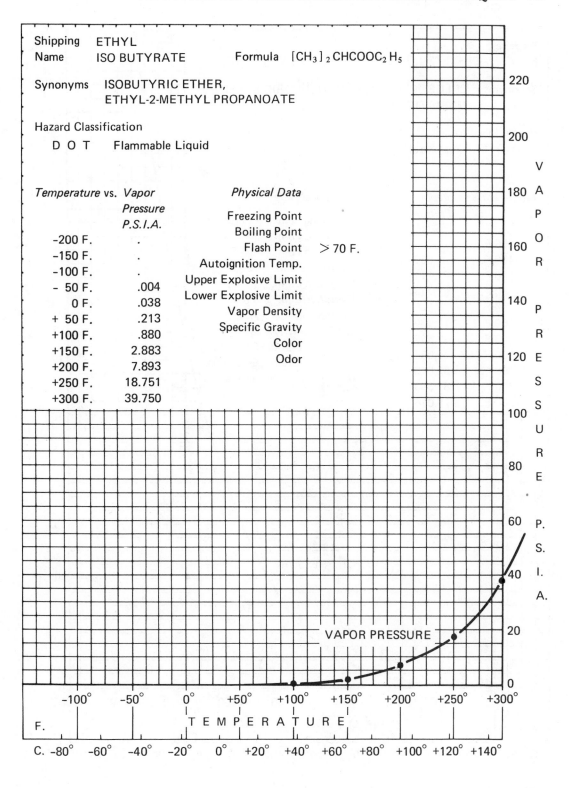

Shipping Name   ETHYL ISO BUTYRATE

Formula $[CH_3]_2 CHCOOC_2 H_5$

Synonyms   ISOBUTYRIC ETHER,
ETHYL-2-METHYL PROPANOATE

Hazard Classification
D O T   Flammable Liquid

Temperature vs. Vapor Pressure P.S.I.A.

Temperature	Vapor Pressure P.S.I.A.
-200 F.	.
-150 F.	.
-100 F.	.
- 50 F.	.004
0 F.	.038
+ 50 F.	.213
+100 F.	.880
+150 F.	2.883
+200 F.	7.893
+250 F.	18.751
+300 F.	39.750

Physical Data

Freezing Point
Boiling Point
Flash Point   > 70 F.
Autoignition Temp.
Upper Explosive Limit
Lower Explosive Limit
Vapor Density
Specific Gravity
Color
Odor

VAPOR PRESSURE

VAPOR PRESSURE P.S.I.A.

TEMPERATURE

F.
C.

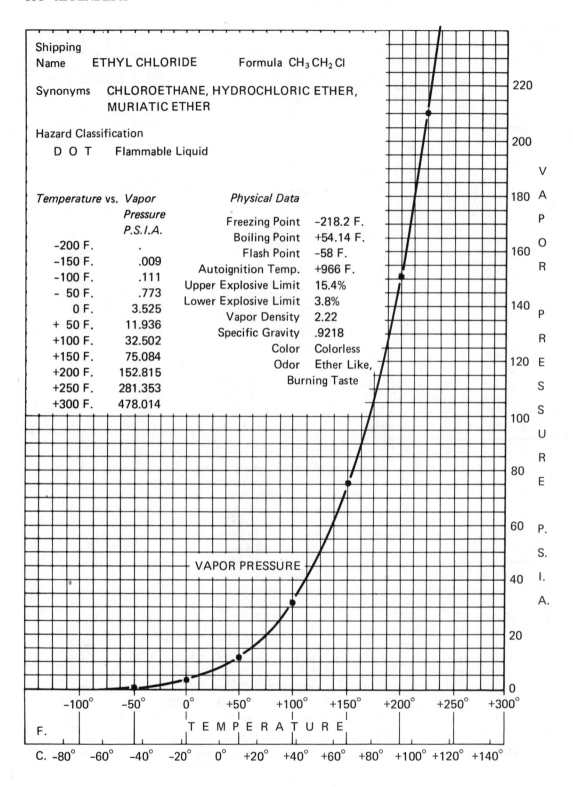

Shipping
Name   ETHYL CHLORIDE   Formula CH_3CH_2Cl

Synonyms   CHLOROETHANE, HYDROCHLORIC ETHER,
MURIATIC ETHER

Hazard Classification
D O T   Flammable Liquid

Temperature vs. *Vapor Pressure P.S.I.A.*

Temperature	Vapor Pressure P.S.I.A.
-200 F.	.
-150 F.	.009
-100 F.	.111
- 50 F.	.773
0 F.	3.525
+ 50 F.	11.936
+100 F.	32.502
+150 F.	75.084
+200 F.	152.815
+250 F.	281.353
+300 F.	478.014

Physical Data

Freezing Point	-218.2 F.
Boiling Point	+54.14 F.
Flash Point	-58 F.
Autoignition Temp.	+966 F.
Upper Explosive Limit	15.4%
Lower Explosive Limit	3.8%
Vapor Density	2.22
Specific Gravity	.9218
Color	Colorless
Odor	Ether Like, Burning Taste

VAPOR PRESSURE

VAPOR PRESSURE P.S.I.A.

TEMPERATURE

F.   -100°   -50°   0°   +50°   +100°   +150°   +200°   +250°   +300°

C. -80° -60° -40° -20° 0° +20° +40° +60° +80° +100° +120° +140°

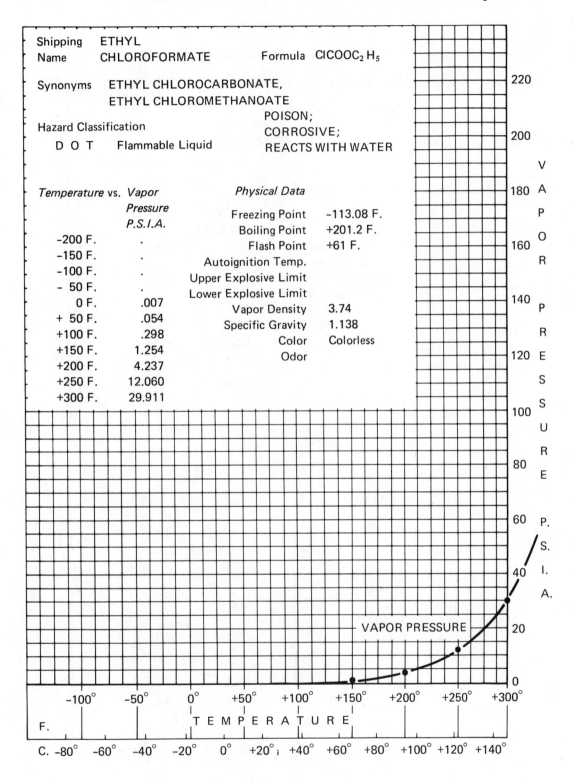

Shipping Name: **ETHYL CHLOROFORMATE**

Formula $ClCOOC_2H_5$

Synonyms   ETHYL CHLOROCARBONATE, ETHYL CHLOROMETHANOATE

Hazard Classification
DOT   Flammable Liquid

POISON;
CORROSIVE;
REACTS WITH WATER

Temperature vs. Vapor Pressure P.S.I.A.

Temperature	P.S.I.A.
-200 F.	.
-150 F.	.
-100 F.	.
- 50 F.	.
0 F.	.007
+ 50 F.	.054
+100 F.	.298
+150 F.	1.254
+200 F.	4.237
+250 F.	12.060
+300 F.	29.911

Physical Data

Freezing Point	-113.08 F.
Boiling Point	+201.2 F.
Flash Point	+61 F.
Autoignition Temp.	
Upper Explosive Limit	
Lower Explosive Limit	
Vapor Density	3.74
Specific Gravity	1.138
Color	Colorless
Odor	

VAPOR PRESSURE

VAPOR PRESSURE P.S.I.A.

TEMPERATURE
F.
C.

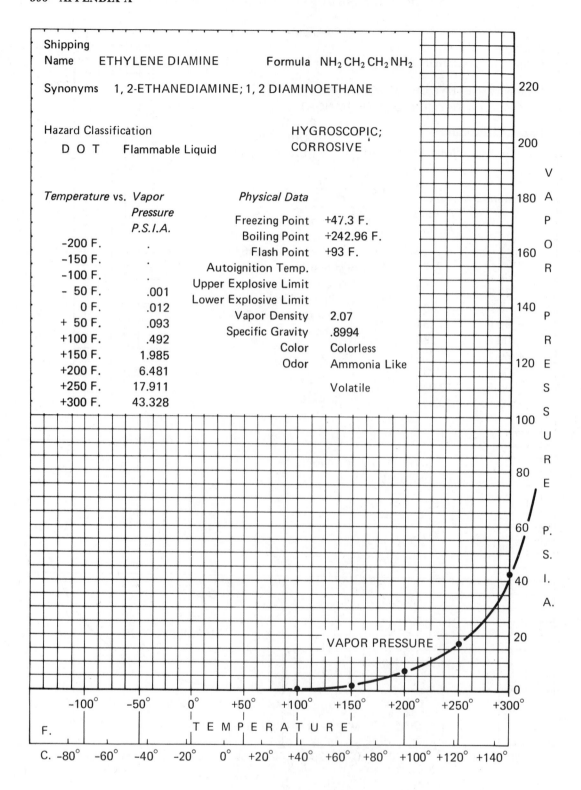

Shipping Name ETHYLENE DIAMINE   Formula $NH_2 CH_2 CH_2 NH_2$

Synonyms 1, 2-ETHANEDIAMINE; 1, 2 DIAMINOETHANE

Hazard Classification   HYGROSCOPIC;
   D O T   Flammable Liquid   CORROSIVE

Temperature vs. *Vapor Pressure P.S.I.A.*

Temperature	Vapor Pressure P.S.I.A.
-200 F.	.
-150 F.	.
-100 F.	.
- 50 F.	.001
0 F.	.012
+ 50 F.	.093
+100 F.	.492
+150 F.	1.985
+200 F.	6.481
+250 F.	17.911
+300 F.	43.328

Physical Data

Freezing Point	+47.3 F.
Boiling Point	+242.96 F.
Flash Point	+93 F.
Autoignition Temp.	
Upper Explosive Limit	
Lower Explosive Limit	
Vapor Density	2.07
Specific Gravity	.8994
Color	Colorless
Odor	Ammonia Like
	Volatile

VAPOR PRESSURE

VAPOR PRESSURE P.S.I.A.

TEMPERATURE

F. -100° -50° 0° +50° +100° +150° +200° +250° +300°

C. -80° -60° -40° -20° 0° +20° +40° +60° +80° +100° +120° +140°

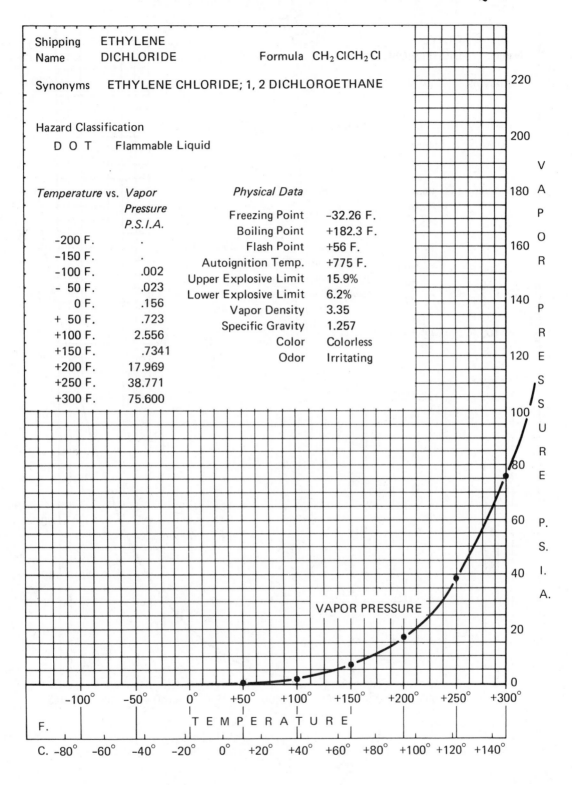

Shipping Name: ETHYLENE DICHLORIDE

Formula CH_2ClCH_2Cl

Synonyms: ETHYLENE CHLORIDE; 1, 2 DICHLOROETHANE

Hazard Classification

D O T   Flammable Liquid

Temperature vs. Vapor Pressure P.S.I.A.

Temperature	P.S.I.A.
-200 F.	.
-150 F.	.
-100 F.	.002
- 50 F.	.023
0 F.	.156
+ 50 F.	.723
+100 F.	2.556
+150 F.	.7341
+200 F.	17.969
+250 F.	38.771
+300 F.	75.600

Physical Data

Freezing Point	-32.26 F.
Boiling Point	+182.3 F.
Flash Point	+56 F.
Autoignition Temp.	+775 F.
Upper Explosive Limit	15.9%
Lower Explosive Limit	6.2%
Vapor Density	3.35
Specific Gravity	1.257
Color	Colorless
Odor	Irritating

VAPOR PRESSURE

VAPOR PRESSURE P.S.I.A.

TEMPERATURE

F. -100° -50° 0° +50° +100° +150° +200° +250° +300°

C. -80° -60° -40° -20° 0° +20° +40° +60° +80° +100° +120° +140°

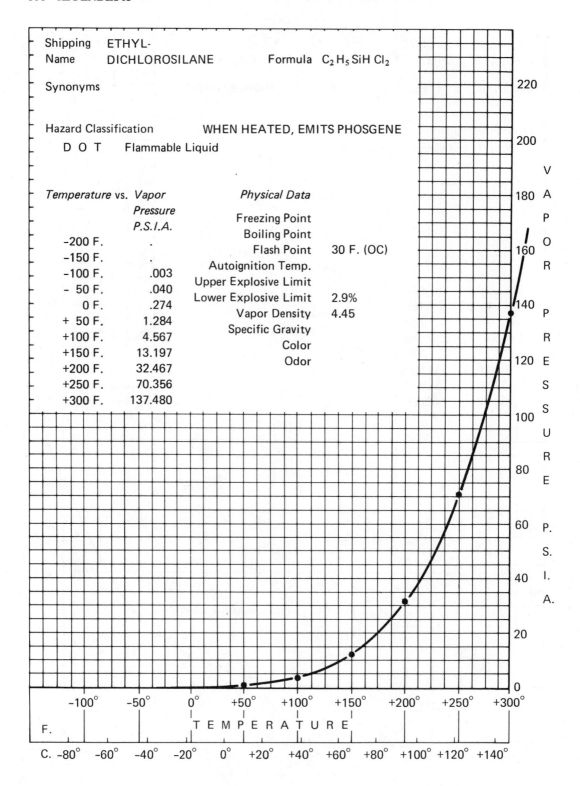

Shipping Name ETHYL-DICHLOROSILANE   Formula $C_2H_5SiHCl_2$

Synonyms

Hazard Classification   WHEN HEATED, EMITS PHOSGENE
    D O T   Flammable Liquid

Temperature vs. Vapor Pressure P.S.I.A.

Temperature	Vapor Pressure P.S.I.A.
-200 F.	.
-150 F.	.
-100 F.	.003
- 50 F.	.040
0 F.	.274
+ 50 F.	1.284
+100 F.	4.567
+150 F.	13.197
+200 F.	32.467
+250 F.	70.356
+300 F.	137.480

Physical Data

Freezing Point	
Boiling Point	
Flash Point	30 F. (OC)
Autoignition Temp.	
Upper Explosive Limit	
Lower Explosive Limit	2.9%
Vapor Density	4.45
Specific Gravity	
Color	
Odor	

VAPOR PRESSURE P.S.I.A.

F.
-100°   -50°   0°   +50°   +100°   +150°   +200°   +250°   +300°
TEMPERATURE

C. -80° -60° -40° -20° 0° +20° +40° +60° +80° +100° +120° +140°

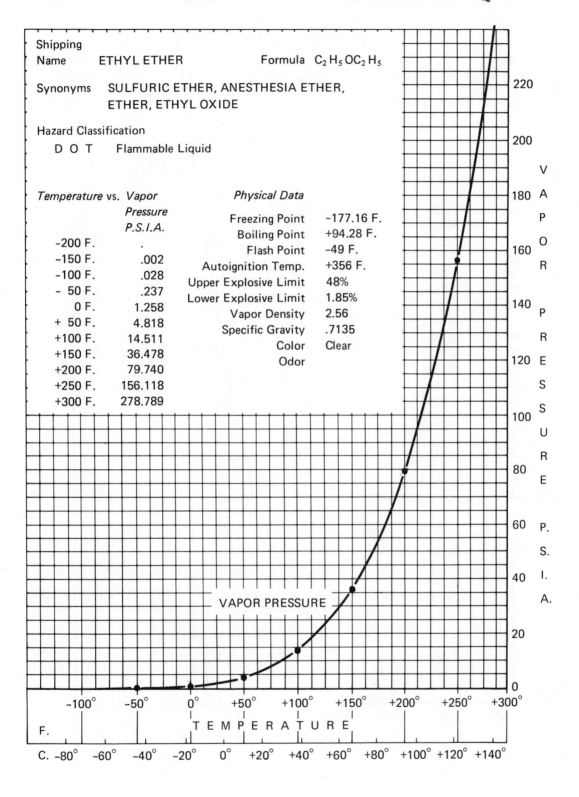

Shipping
Name   ETHYL ETHER   Formula $C_2H_5OC_2H_5$

Synonyms   SULFURIC ETHER, ANESTHESIA ETHER,
           ETHER, ETHYL OXIDE

Hazard Classification
    D O T   Flammable Liquid

Temperature vs. *Vapor*
                Pressure
                P.S.I.A.

Temperature	Pressure P.S.I.A.
−200 F.	.
−150 F.	.002
−100 F.	.028
− 50 F.	.237
0 F.	1.258
+ 50 F.	4.818
+100 F.	14.511
+150 F.	36.478
+200 F.	79.740
+250 F.	156.118
+300 F.	278.789

Physical Data

Freezing Point	−177.16 F.
Boiling Point	+94.28 F.
Flash Point	−49 F.
Autoignition Temp.	+356 F.
Upper Explosive Limit	48%
Lower Explosive Limit	1.85%
Vapor Density	2.56
Specific Gravity	.7135
Color	Clear
Odor	

VAPOR PRESSURE

VAPOR PRESSURE P.S.I.A.

TEMPERATURE

F.

C.

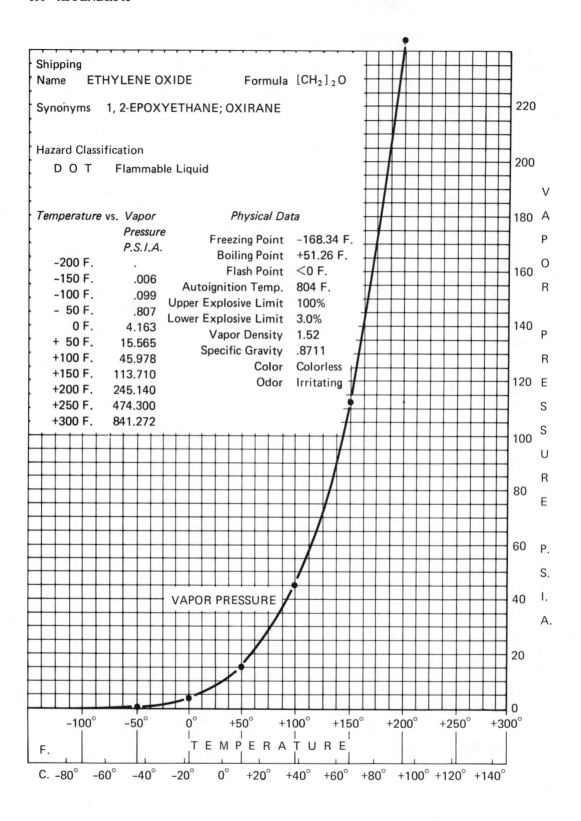

Shipping
Name   ETHYLENE OXIDE   Formula $[CH_2]_2O$

Synonyms   1, 2-EPOXYETHANE; OXIRANE

Hazard Classification
   D O T   Flammable Liquid

Temperature vs. Vapor Pressure P.S.I.A.

Temperature	Pressure P.S.I.A.
-200 F.	.
-150 F.	.006
-100 F.	.099
- 50 F.	.807
0 F.	4.163
+ 50 F.	15.565
+100 F.	45.978
+150 F.	113.710
+200 F.	245.140
+250 F.	474.300
+300 F.	841.272

Physical Data

Freezing Point	-168.34 F.
Boiling Point	+51.26 F.
Flash Point	<0 F.
Autoignition Temp.	804 F.
Upper Explosive Limit	100%
Lower Explosive Limit	3.0%
Vapor Density	1.52
Specific Gravity	.8711
Color	Colorless
Odor	Irritating

VAPOR PRESSURE

VAPOR PRESSURE P.S.I.A.

TEMPERATURE

F.

C.

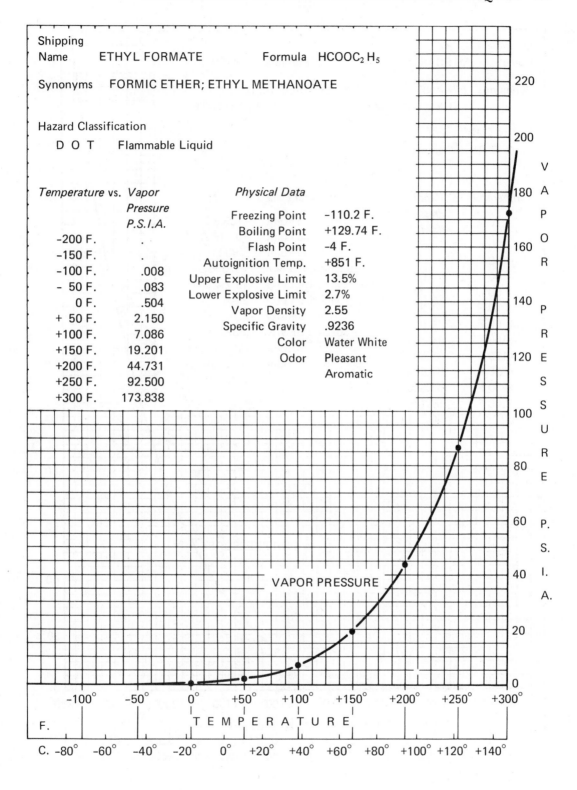

Shipping
Name   ETHYL FORMATE   Formula $HCOOC_2H_5$

Synonyms   FORMIC ETHER; ETHYL METHANOATE

Hazard Classification
    D O T   Flammable Liquid

Temperature vs. Vapor Pressure P.S.I.A.

Temperature	Vapor Pressure P.S.I.A.
−200 F.	.
−150 F.	.
−100 F.	.008
− 50 F.	.083
0 F.	.504
+ 50 F.	2.150
+100 F.	7.086
+150 F.	19.201
+200 F.	44.731
+250 F.	92.500
+300 F.	173.838

Physical Data

Freezing Point	−110.2 F.
Boiling Point	+129.74 F.
Flash Point	−4 F.
Autoignition Temp.	+851 F.
Upper Explosive Limit	13.5%
Lower Explosive Limit	2.7%
Vapor Density	2.55
Specific Gravity	.9236
Color	Water White
Odor	Pleasant Aromatic

VAPOR PRESSURE

VAPOR PRESSURE P.S.I.A.

TEMPERATURE

F.
C.

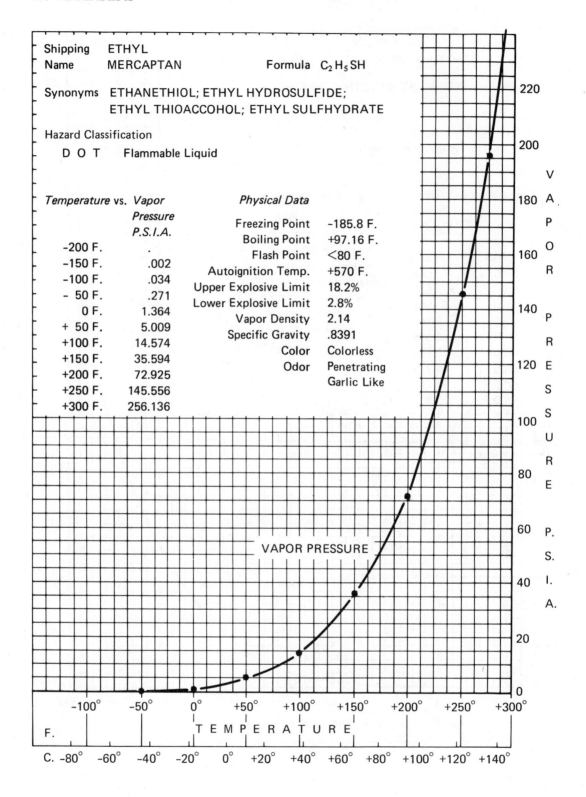

Shipping Name   ETHYL MERCAPTAN

Formula C_2H_5SH

Synonyms   ETHANETHIOL; ETHYL HYDROSULFIDE; ETHYL THIOACCOHOL; ETHYL SULFHYDRATE

Hazard Classification

D O T   Flammable Liquid

Temperature vs. Vapor Pressure P.S.I.A.

Temperature	Vapor Pressure P.S.I.A.
-200 F.	.
-150 F.	.002
-100 F.	.034
- 50 F.	.271
0 F.	1.364
+ 50 F.	5.009
+100 F.	14.574
+150 F.	35.594
+200 F.	72.925
+250 F.	145.556
+300 F.	256.136

Physical Data

Freezing Point	-185.8 F.
Boiling Point	+97.16 F.
Flash Point	<80 F.
Autoignition Temp.	+570 F.
Upper Explosive Limit	18.2%
Lower Explosive Limit	2.8%
Vapor Density	2.14
Specific Gravity	.8391
Color	Colorless
Odor	Penetrating Garlic Like

VAPOR PRESSURE

VAPOR PRESSURE P.S.I.A.

TEMPERATURE

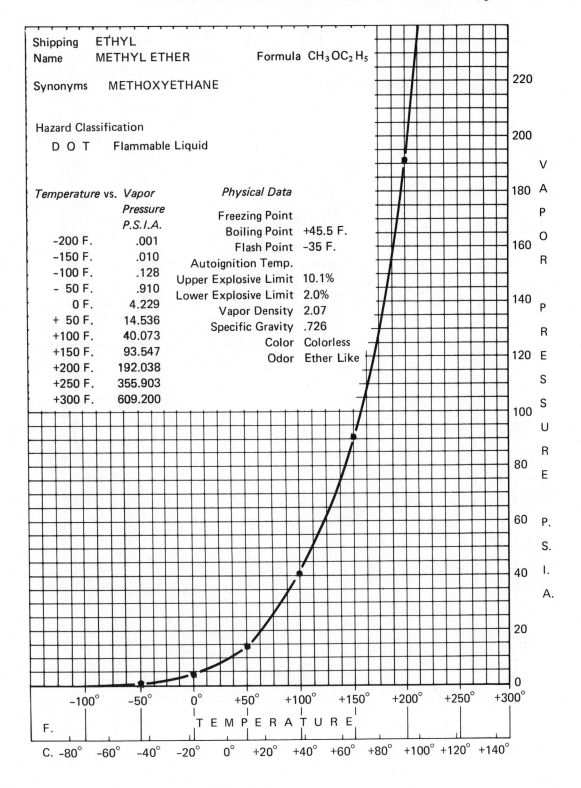

Shipping Name   ETHYL METHYL ETHER

Formula $CH_3OC_2H_5$

Synonyms   METHOXYETHANE

Hazard Classification
   D O T   Flammable Liquid

Temperature vs. Vapor Pressure P.S.I.A.

Temperature	Vapor Pressure P.S.I.A.
-200 F.	.001
-150 F.	.010
-100 F.	.128
- 50 F.	.910
0 F.	4.229
+ 50 F.	14.536
+100 F.	40.073
+150 F.	93.547
+200 F.	192.038
+250 F.	355.903
+300 F.	609.200

Physical Data

Freezing Point	
Boiling Point	+45.5 F.
Flash Point	-35 F.
Autoignition Temp.	
Upper Explosive Limit	10.1%
Lower Explosive Limit	2.0%
Vapor Density	2.07
Specific Gravity	.726
Color	Colorless
Odor	Ether Like

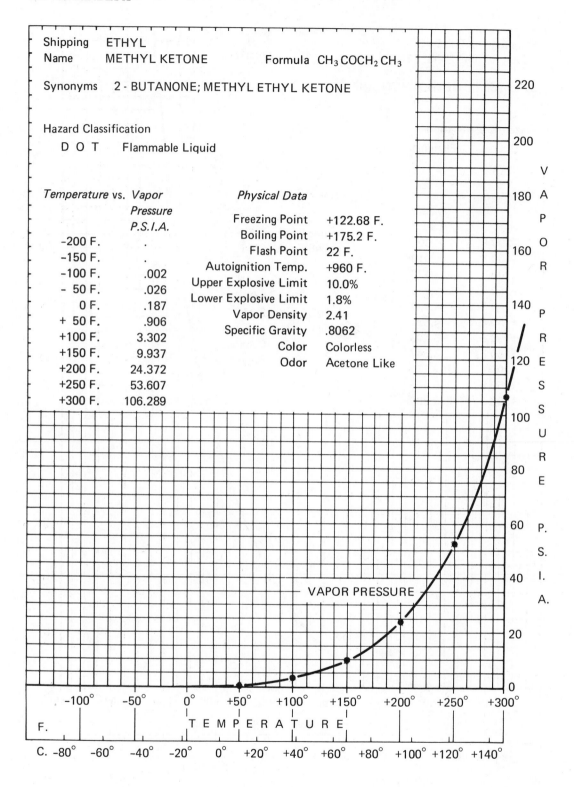

Shipping Name ETHYL METHYL KETONE

Formula $CH_3COCH_2CH_3$

Synonyms 2 - BUTANONE; METHYL ETHYL KETONE

Hazard Classification
   D O T   Flammable Liquid

Temperature vs. *Vapor Pressure P.S.I.A.*

Temperature	Vapor Pressure P.S.I.A.
-200 F.	.
-150 F.	.
-100 F.	.002
- 50 F.	.026
0 F.	.187
+ 50 F.	.906
+100 F.	3.302
+150 F.	9.937
+200 F.	24.372
+250 F.	53.607
+300 F.	106.289

Physical Data

Freezing Point	+122.68 F.
Boiling Point	+175.2 F.
Flash Point	22 F.
Autoignition Temp.	+960 F.
Upper Explosive Limit	10.0%
Lower Explosive Limit	1.8%
Vapor Density	2.41
Specific Gravity	.8062
Color	Colorless
Odor	Acetone Like

VAPOR PRESSURE

V A P O R   P R E S S U R E   P. S. I. A.

TEMPERATURE

F. -100° -50° 0° +50° +100° +150° +200° +250° +300°

C. -80° -60° -40° -20° 0° +20° +40° +60° +80° +100° +120° +140°

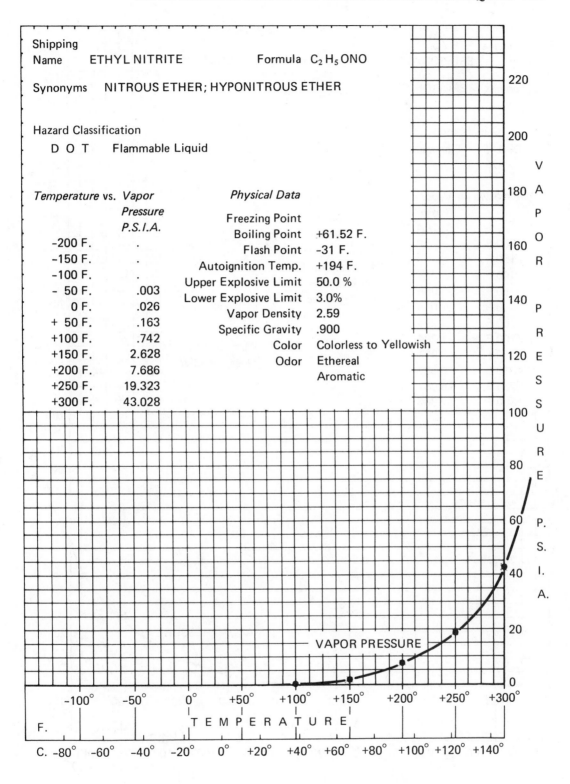

Shipping
Name   ETHYL NITRITE   Formula C_2H_5ONO

Synonyms   NITROUS ETHER; HYPONITROUS ETHER

Hazard Classification
    D O T   Flammable Liquid

Temperature vs. *Vapor Pressure P.S.I.A.*

Temperature	Vapor Pressure P.S.I.A.
-200 F.	.
-150 F.	.
-100 F.	.
- 50 F.	.003
0 F.	.026
+ 50 F.	.163
+100 F.	.742
+150 F.	2.628
+200 F.	7.686
+250 F.	19.323
+300 F.	43.028

Physical Data

Freezing Point	
Boiling Point	+61.52 F.
Flash Point	-31 F.
Autoignition Temp.	+194 F.
Upper Explosive Limit	50.0 %
Lower Explosive Limit	3.0%
Vapor Density	2.59
Specific Gravity	.900
Color	Colorless to Yellowish
Odor	Ethereal Aromatic

VAPOR PRESSURE

VAPOR PRESSURE P.S.I.A.

TEMPERATURE

F.

C.

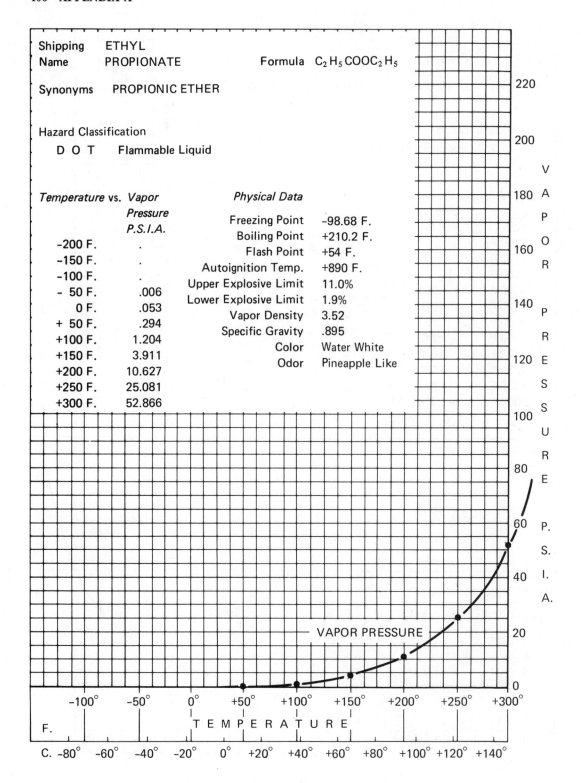

Shipping Name   ETHYL PROPIONATE

Formula $C_2H_5COOC_2H_5$

Synonyms   PROPIONIC ETHER

Hazard Classification
    D O T   Flammable Liquid

Temperature vs. Vapor Pressure P.S.I.A.

Temperature	Vapor Pressure P.S.I.A.
-200 F.	.
-150 F.	.
-100 F.	.
- 50 F.	.006
0 F.	.053
+ 50 F.	.294
+100 F.	1.204
+150 F.	3.911
+200 F.	10.627
+250 F.	25.081
+300 F.	52.866

Physical Data

Freezing Point	-98.68 F.
Boiling Point	+210.2 F.
Flash Point	+54 F.
Autoignition Temp.	+890 F.
Upper Explosive Limit	11.0%
Lower Explosive Limit	1.9%
Vapor Density	3.52
Specific Gravity	.895
Color	Water White
Odor	Pineapple Like

VAPOR PRESSURE P.S.I.A.

VAPOR PRESSURE

TEMPERATURE

F.   -100°   -50°   0°   +50°   +100°   +150°   +200°   +250°   +300°

C. -80° -60° -40° -20° 0° +20° +40° +60° +80° +100° +120° +140°

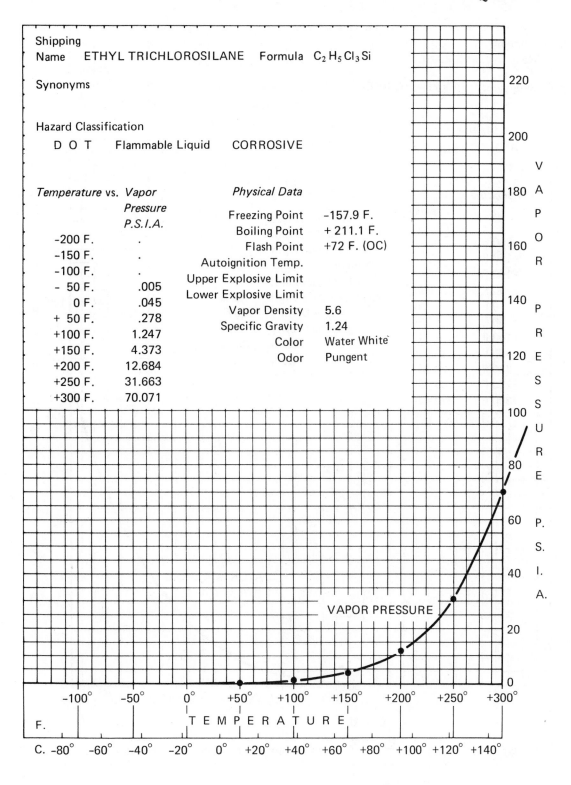

Shipping
Name   ETHYL TRICHLOROSILANE   Formula $C_2H_5Cl_3Si$

Synonyms

Hazard Classification
    D O T   Flammable Liquid   CORROSIVE

Temperature vs. Vapor Pressure P.S.I.A.

Temperature	Vapor Pressure P.S.I.A.
-200 F.	.
-150 F.	.
-100 F.	.
- 50 F.	.005
0 F.	.045
+ 50 F.	.278
+100 F.	1.247
+150 F.	4.373
+200 F.	12.684
+250 F.	31.663
+300 F.	70.071

Physical Data

Freezing Point	-157.9 F.
Boiling Point	+ 211.1 F.
Flash Point	+72 F. (OC)
Autoignition Temp.	
Upper Explosive Limit	
Lower Explosive Limit	
Vapor Density	5.6
Specific Gravity	1.24
Color	Water White
Odor	Pungent

VAPOR PRESSURE

VAPOR PRESSURE P.S.I.A.

TEMPERATURE

F.

C.

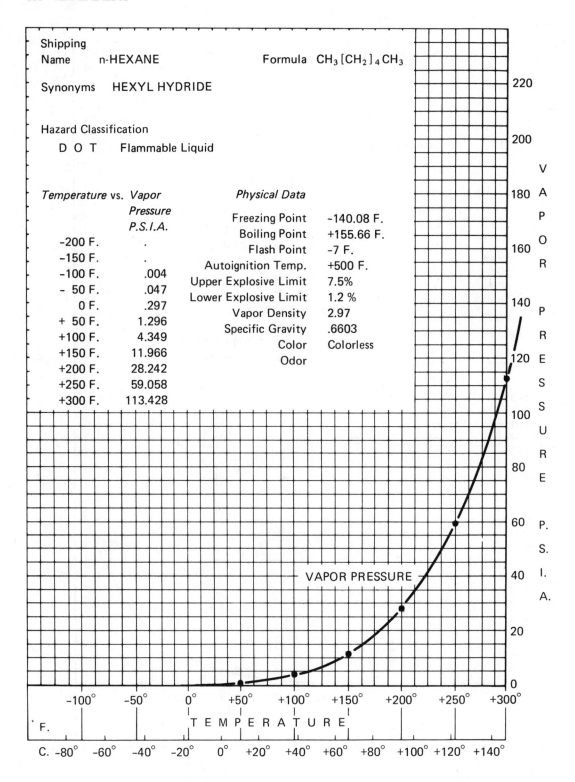

Shipping
Name   n-HEXANE   Formula $CH_3[CH_2]_4CH_3$

Synonyms   HEXYL HYDRIDE

Hazard Classification
    D O T   Flammable Liquid

Temperature vs. Vapor   Physical Data
                    Pressure
                    P.S.I.A.   Freezing Point   -140.08 F.
    -200 F.   .   Boiling Point   +155.66 F.
    -150 F.   .   Flash Point   -7 F.
    -100 F.   .004   Autoignition Temp.   +500 F.
    - 50 F.   .047   Upper Explosive Limit 7.5%
      0 F.   .297   Lower Explosive Limit 1.2 %
    + 50 F.   1.296   Vapor Density   2.97
    +100 F.   4.349   Specific Gravity   .6603
    +150 F.   11.966   Color   Colorless
    +200 F.   28.242   Odor
    +250 F.   59.058
    +300 F.   113.428

VAPOR PRESSURE

VAPOR PRESSURE P.S.I.A.

TEMPERATURE
F.
-100° -50° 0° +50° +100° +150° +200° +250° +300°
C. -80° -60° -40° -20° 0° +20° +40° +60° +80° +100° +120° +140°

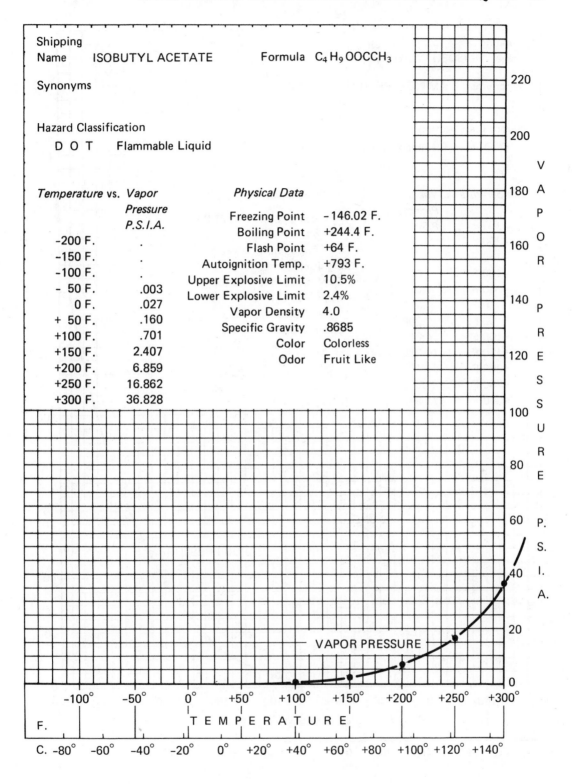

Shipping
Name   ISOBUTYL ACETATE   Formula $C_4H_9OOCCH_3$

Synonyms

Hazard Classification
    D O T   Flammable Liquid

Temperature vs. Vapor Pressure P.S.I.A.	
-200 F.	.
-150 F.	.
-100 F.	.
- 50 F.	.003
0 F.	.027
+ 50 F.	.160
+100 F.	.701
+150 F.	2.407
+200 F.	6.859
+250 F.	16.862
+300 F.	36.828

Physical Data

Freezing Point	- 146.02 F.
Boiling Point	+244.4 F.
Flash Point	+64 F.
Autoignition Temp.	+793 F.
Upper Explosive Limit	10.5%
Lower Explosive Limit	2.4%
Vapor Density	4.0
Specific Gravity	.8685
Color	Colorless
Odor	Fruit Like

VAPOR PRESSURE

VAPOR PRESSURE P.S.I.A.

TEMPERATURE

F.

C. -80° -60° -40° -20° 0° +20° +40° +60° +80° +100° +120° +140°

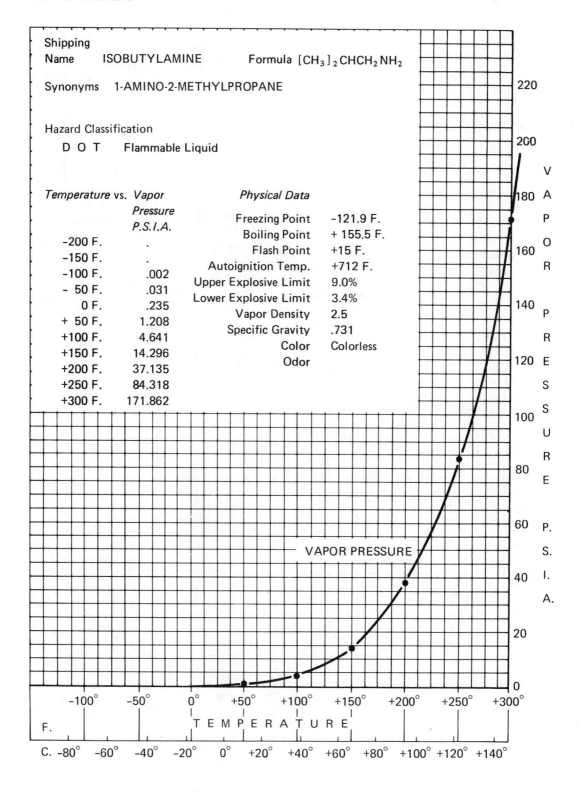

Shipping
Name   ISOBUTYLAMINE   Formula $[CH_3]_2 CHCH_2 NH_2$

Synonyms   1-AMINO-2-METHYLPROPANE

Hazard Classification
    D O T   Flammable Liquid

Temperature vs. *Vapor*   *Physical Data*
                  Pressure
                  P.S.I.A.   Freezing Point   -121.9 F.
    -200 F.   .   Boiling Point   + 155.5 F.
    -150 F.   .   Flash Point   +15 F.
    -100 F.   .002   Autoignition Temp. +712 F.
    - 50 F.   .031   Upper Explosive Limit 9.0%
      0 F.   .235   Lower Explosive Limit 3.4%
    + 50 F.   1.208   Vapor Density   2.5
    +100 F.   4.641   Specific Gravity   .731
    +150 F.   14.296   Color   Colorless
    +200 F.   37.135   Odor
    +250 F.   84.318
    +300 F.   171.862

VAPOR PRESSURE

VAPOR PRESSURE P.S.I.A.

TEMPERATURE

F.   -100°   -50°   0°   +50° +100° +150° +200° +250° +300°

C. -80° -60° -40° -20°   0°   +20° +40° +60° +80° +100° +120° +140°

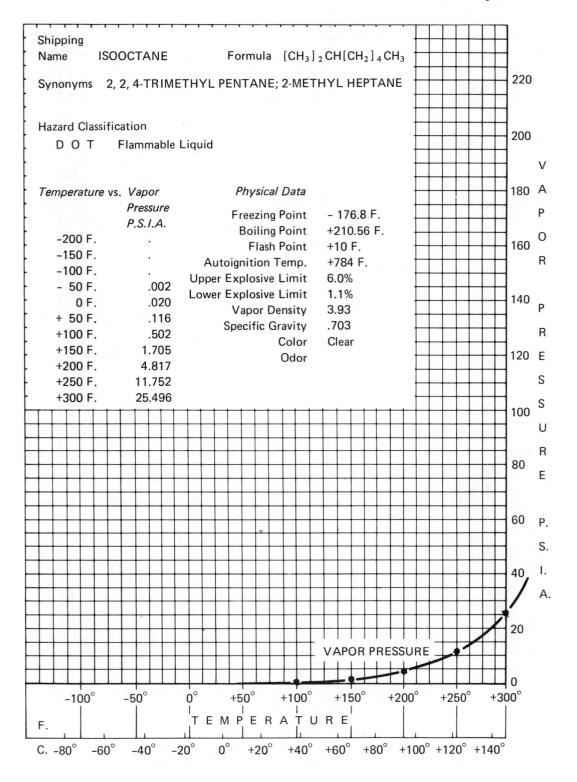

Shipping
Name   ISOOCTANE   Formula $[CH_3]_2 CH[CH_2]_4 CH_3$

Synonyms   2, 2, 4-TRIMETHYL PENTANE; 2-METHYL HEPTANE

Hazard Classification
   D O T   Flammable Liquid

Temperature vs. *Vapor*
                    Pressure
                    P.S.I.A.

Temperature	Vapor Pressure P.S.I.A.
-200 F.	.
-150 F.	.
-100 F.	.
- 50 F.	.002
0 F.	.020
+ 50 F.	.116
+100 F.	.502
+150 F.	1.705
+200 F.	4.817
+250 F.	11.752
+300 F.	25.496

Physical Data

Freezing Point	- 176.8 F.
Boiling Point	+210.56 F.
Flash Point	+10 F.
Autoignition Temp.	+784 F.
Upper Explosive Limit	6.0%
Lower Explosive Limit	1.1%
Vapor Density	3.93
Specific Gravity	.703
Color	Clear
Odor	

VAPOR PRESSURE

VAPOR PRESSURE P.S.I.A.

TEMPERATURE

F.   -100°   -50°   0°   +50°   +100°   +150°   +200°   +250°   +300°

C. -80° -60° -40° -20°   0°   +20° +40° +60° +80° +100° +120° +140°

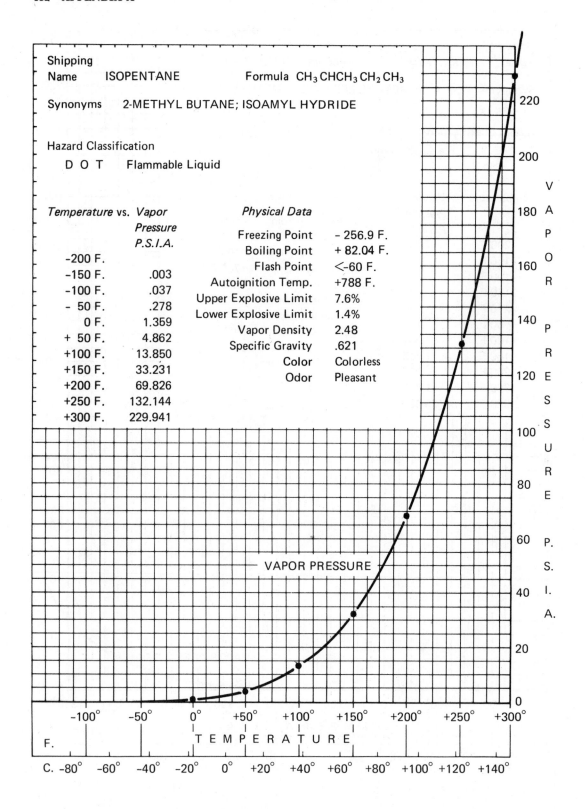

Shipping
Name   ISOPENTANE   Formula $CH_3 CHCH_3 CH_2 CH_3$

Synonyms   2-METHYL BUTANE; ISOAMYL HYDRIDE

Hazard Classification
    D O T   Flammable Liquid

Temperature vs. Vapor Pressure P.S.I.A.

Temperature	Vapor Pressure P.S.I.A.
−200 F.	
−150 F.	.003
−100 F.	.037
− 50 F.	.278
0 F.	1.359
+ 50 F.	4.862
+100 F.	13.850
+150 F.	33.231
+200 F.	69.826
+250 F.	132.144
+300 F.	229.941

Physical Data

Freezing Point	− 256.9 F.
Boiling Point	+ 82.04 F.
Flash Point	<−60 F.
Autoignition Temp.	+788 F.
Upper Explosive Limit	7.6%
Lower Explosive Limit	1.4%
Vapor Density	2.48
Specific Gravity	.621
Color	Colorless
Odor	Pleasant

VAPOR PRESSURE

VAPOR PRESSURE P.S.I.A.

TEMPERATURE

F.   −100°   −50°   0°   +50°   +100°   +150°   +200°   +250°   +300°

C. −80°   −60°   −40°   −20°   0°   +20°   +40°   +60°   +80°   +100°   +120°   +140°

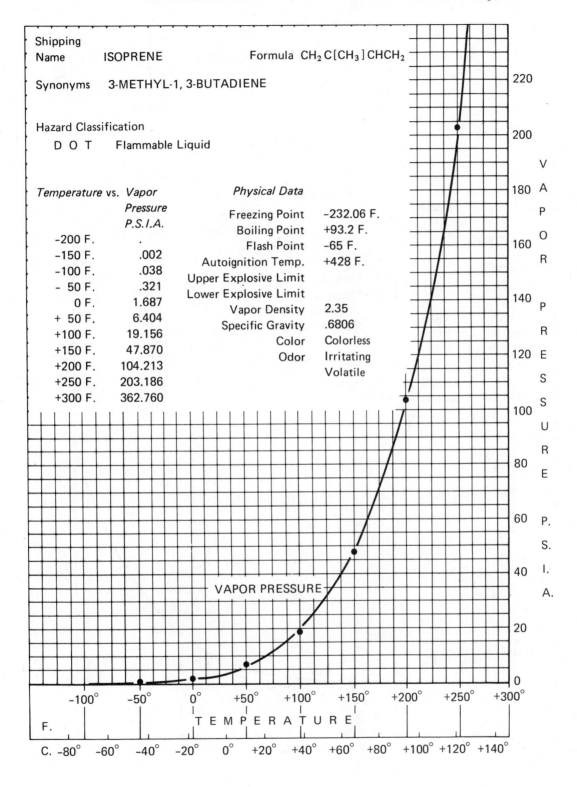

Shipping
Name   ISOPRENE   Formula $CH_2 C[CH_3] CHCH_2$

Synonyms   3-METHYL-1, 3-BUTADIENE

Hazard Classification
   D O T   Flammable Liquid

Temperature vs. Vapor
                  Pressure
                  P.S.I.A.

-200 F.	.
-150 F.	.002
-100 F.	.038
- 50 F.	.321
0 F.	1.687
+ 50 F.	6.404
+100 F.	19.156
+150 F.	47.870
+200 F.	104.213
+250 F.	203.186
+300 F.	362.760

Physical Data

Freezing Point	-232.06 F.
Boiling Point	+93.2 F.
Flash Point	-65 F.
Autoignition Temp.	+428 F.
Upper Explosive Limit	
Lower Explosive Limit	
Vapor Density	2.35
Specific Gravity	.6806
Color	Colorless
Odor	Irritating
	Volatile

VAPOR PRESSURE

VAPOR PRESSURE P.S.I.A.

TEMPERATURE

F.

C. -80° -60° -40° -20° 0° +20° +40° +60° +80° +100° +120° +140°

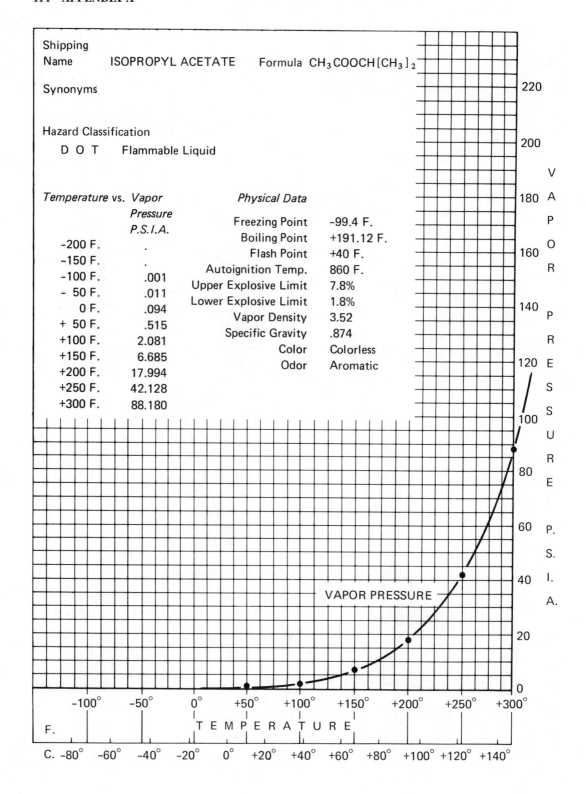

Shipping
Name   ISOPROPYL ACETATE   Formula $CH_3COOCH[CH_3]_2$

Synonyms

Hazard Classification
    D O T   Flammable Liquid

Temperature vs. Vapor Pressure P.S.I.A.

Temperature	P.S.I.A.
-200 F.	.
-150 F.	.
-100 F.	.001
- 50 F.	.011
0 F.	.094
+ 50 F.	.515
+100 F.	2.081
+150 F.	6.685
+200 F.	17.994
+250 F.	42.128
+300 F.	88.180

Physical Data

Freezing Point	-99.4 F.
Boiling Point	+191.12 F.
Flash Point	+40 F.
Autoignition Temp.	860 F.
Upper Explosive Limit	7.8%
Lower Explosive Limit	1.8%
Vapor Density	3.52
Specific Gravity	.874
Color	Colorless
Odor	Aromatic

VAPOR PRESSURE

VAPOR PRESSURE P.S.I.A.

TEMPERATURE

F.   -100°   -50°   0°   +50°   +100°   +150°   +200°   +250°   +300°

C. -80° -60° -40° -20° 0° +20° +40° +60° +80° +100° +120° +140°

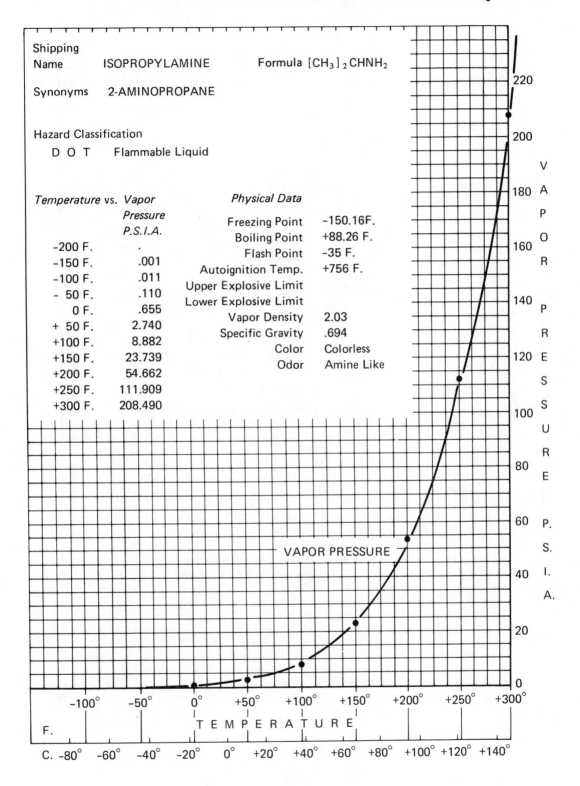

Shipping
Name   ISOPROPYLAMINE   Formula $[CH_3]_2 CHNH_2$

Synonyms   2-AMINOPROPANE

Hazard Classification
    D O T   Flammable Liquid

Temperature vs. *Vapor Pressure P.S.I.A.*

Temperature	Pressure P.S.I.A.
-200 F.	.
-150 F.	.001
-100 F.	.011
- 50 F.	.110
0 F.	.655
+ 50 F.	2.740
+100 F.	8.882
+150 F.	23.739
+200 F.	54.662
+250 F.	111.909
+300 F.	208.490

Physical Data

Freezing Point	-150.16F.
Boiling Point	+88.26 F.
Flash Point	-35 F.
Autoignition Temp.	+756 F.
Upper Explosive Limit	
Lower Explosive Limit	
Vapor Density	2.03
Specific Gravity	.694
Color	Colorless
Odor	Amine Like

VAPOR PRESSURE

VAPOR PRESSURE P.S.I.A.

TEMPERATURE

F.   -100° -50° 0° +50° +100° +150° +200° +250° +300°

C. -80° -60° -40° -20° 0° +20° +40° +60° +80° +100° +120° +140°

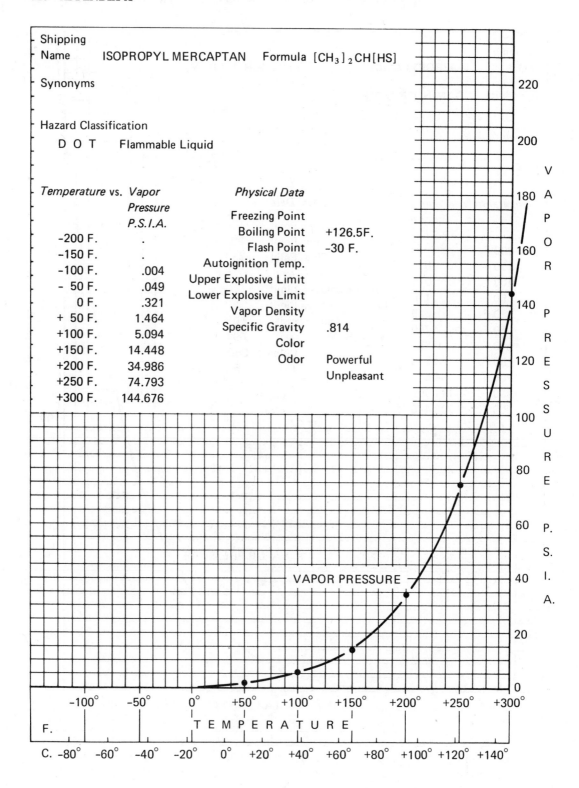

Shipping
Name   ISOPROPYL MERCAPTAN   Formula [CH₃]₂CH[HS]

Synonyms

Hazard Classification
    D O T   Flammable Liquid

Temperature vs.	Vapor Pressure P.S.I.A.	Physical Data	
-200 F.	.	Freezing Point	
-150 F.	.	Boiling Point	+126.5F.
-100 F.	.004	Flash Point	-30 F.
- 50 F.	.049	Autoignition Temp.	
0 F.	.321	Upper Explosive Limit	
+ 50 F.	1.464	Lower Explosive Limit	
+100 F.	5.094	Vapor Density	
+150 F.	14.448	Specific Gravity	.814
+200 F.	34.986	Color	
+250 F.	74.793	Odor	Powerful
+300 F.	144.676		Unpleasant

VAPOR PRESSURE

V A P O R   P R E S S U R E   P. S. I. A.

TEMPERATURE

F.

C.

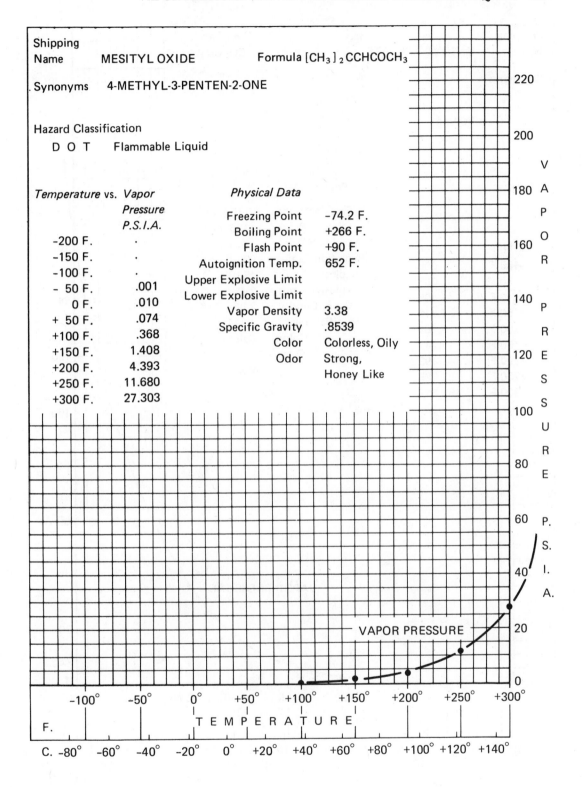

Shipping
Name   MESITYL OXIDE   Formula $[CH_3]_2CCHCOCH_3$

Synonyms   4-METHYL-3-PENTEN-2-ONE

Hazard Classification
    D O T   Flammable Liquid

Temperature vs. Vapor
                    Pressure
                    P.S.I.A.
    -200 F.   .
    -150 F.   .
    -100 F.   .
    - 50 F.   .001
      0 F.   .010
    + 50 F.   .074
    +100 F.   .368
    +150 F.   1.408
    +200 F.   4.393
    +250 F.   11.680
    +300 F.   27.303

Physical Data
    Freezing Point   -74.2 F.
    Boiling Point   +266 F.
    Flash Point   +90 F.
    Autoignition Temp.   652 F.
    Upper Explosive Limit
    Lower Explosive Limit
    Vapor Density   3.38
    Specific Gravity   .8539
    Color   Colorless, Oily
    Odor   Strong,
                          Honey Like

VAPOR PRESSURE P.S.I.A.

VAPOR PRESSURE

TEMPERATURE

F. -100° -50° 0° +50° +100° +150° +200° +250° +300°

C. -80° -60° -40° -20° 0° +20° +40° +60° +80° +100° +120° +140°

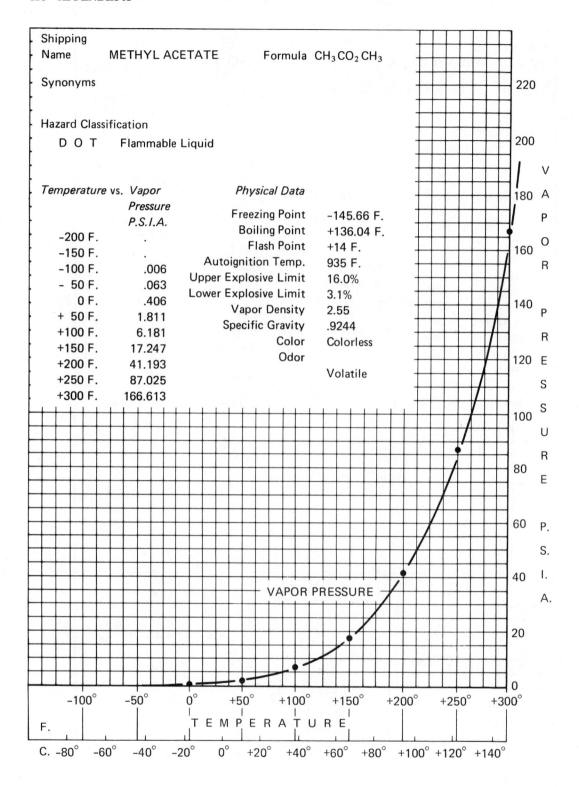

Shipping
Name   METHYL ACETATE   Formula $CH_3CO_2CH_3$

Synonyms

Hazard Classification
   D O T   Flammable Liquid

Temperature vs. *Vapor*
                  Pressure
                  P.S.I.A.

Temperature	Vapor Pressure P.S.I.A.
-200 F.	.
-150 F.	.
-100 F.	.006
- 50 F.	.063
0 F.	.406
+ 50 F.	1.811
+100 F.	6.181
+150 F.	17.247
+200 F.	41.193
+250 F.	87.025
+300 F.	166.613

Physical Data

Freezing Point	-145.66 F.
Boiling Point	+136.04 F.
Flash Point	+14 F.
Autoignition Temp.	935 F.
Upper Explosive Limit	16.0%
Lower Explosive Limit	3.1%
Vapor Density	2.55
Specific Gravity	.9244
Color	Colorless
Odor	Volatile

VAPOR PRESSURE

VAPOR PRESSURE P.S.I.A.

TEMPERATURE

F.   -100°   -50°   0°   +50°   +100°   +150°   +200°   +250°   +300°

C. -80° -60° -40° -20° 0° +20° +40° +60° +80° +100° +120° +140°

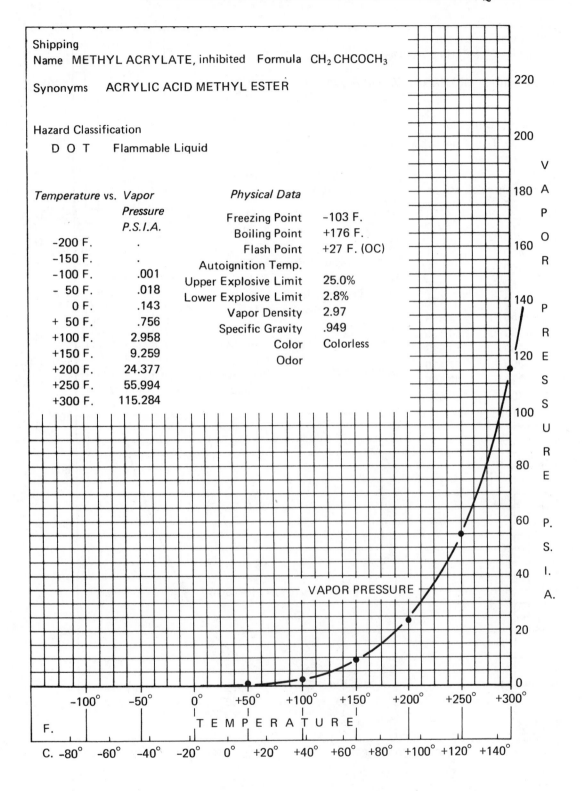

Shipping
Name METHYL ACRYLATE, inhibited   Formula $CH_2CHCOCH_3$

Synonyms   ACRYLIC ACID METHYL ESTER

Hazard Classification
   D O T   Flammable Liquid

Temperature vs. *Vapor Pressure P.S.I.A.*

Temperature	P.S.I.A.
−200 F.	.
−150 F.	.
−100 F.	.001
− 50 F.	.018
0 F.	.143
+ 50 F.	.756
+100 F.	2.958
+150 F.	9.259
+200 F.	24.377
+250 F.	55.994
+300 F.	115.284

Physical Data

Freezing Point	−103 F.
Boiling Point	+176 F.
Flash Point	+27 F. (OC)
Autoignition Temp.	
Upper Explosive Limit	25.0%
Lower Explosive Limit	2.8%
Vapor Density	2.97
Specific Gravity	.949
Color	Colorless
Odor	

VAPOR PRESSURE

VAPOR PRESSURE P.S.I.A.

TEMPERATURE

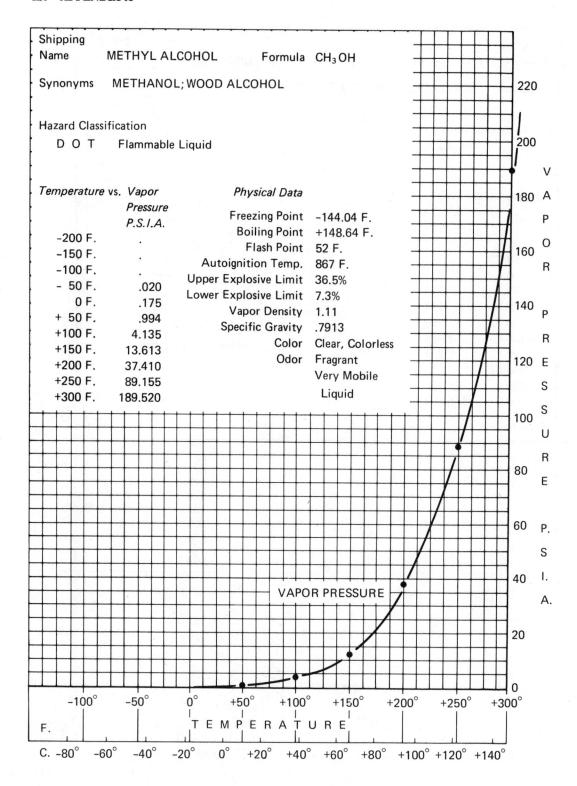

Shipping
Name   METHYL ALCOHOL   Formula   CH₃OH

Synonyms   METHANOL; WOOD ALCOHOL

Hazard Classification
   D O T   Flammable Liquid

Temperature vs. Vapor Pressure P.S.I.A.

Temperature	Vapor Pressure P.S.I.A.
-200 F.	.
-150 F.	.
-100 F.	.
- 50 F.	.020
0 F.	.175
+ 50 F.	.994
+100 F.	4.135
+150 F.	13.613
+200 F.	37.410
+250 F.	89.155
+300 F.	189.520

Physical Data

Freezing Point	-144.04 F.
Boiling Point	+148.64 F.
Flash Point	52 F.
Autoignition Temp.	867 F.
Upper Explosive Limit	36.5%
Lower Explosive Limit	7.3%
Vapor Density	1.11
Specific Gravity	.7913
Color	Clear, Colorless
Odor	Fragrant
	Very Mobile
	Liquid

VAPOR PRESSURE

VAPOR PRESSURE P.S.I.A.

TEMPERATURE

F.   -100°   -50°   0°   +50°   +100°   +150°   +200°   +250°   +300°

C. -80° -60° -40° -20° 0° +20° +40° +60° +80° +100° +120° +140°

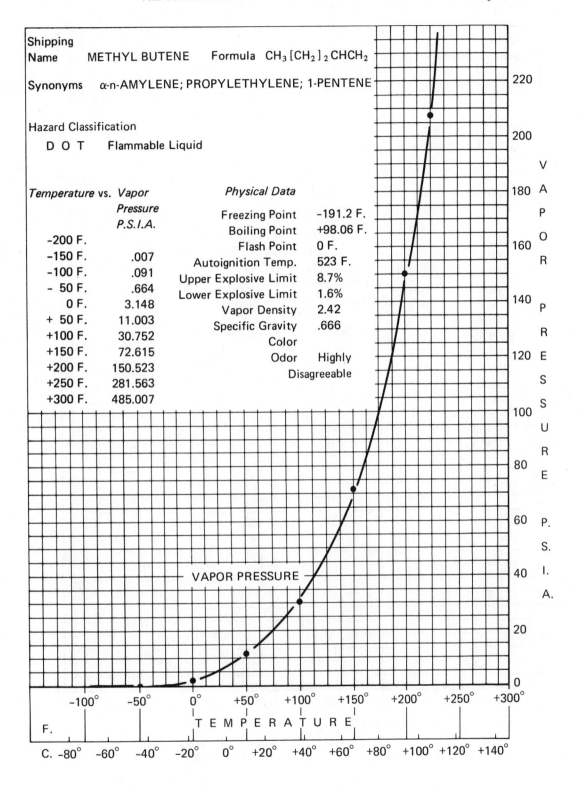

Shipping
Name   METHYL BUTENE   Formula $CH_3[CH_2]_2CHCH_2$

Synonyms   α-n-AMYLENE; PROPYLETHYLENE; 1-PENTENE

Hazard Classification
   D O T   Flammable Liquid

Temperature vs. *Vapor Pressure P.S.I.A.*

Temperature	Vapor Pressure P.S.I.A.
-200 F.	
-150 F.	.007
-100 F.	.091
- 50 F.	.664
0 F.	3.148
+ 50 F.	11.003
+100 F.	30.752
+150 F.	72.615
+200 F.	150.523
+250 F.	281.563
+300 F.	485.007

Physical Data

Freezing Point	-191.2 F.
Boiling Point	+98.06 F.
Flash Point	0 F.
Autoignition Temp.	523 F.
Upper Explosive Limit	8.7%
Lower Explosive Limit	1.6%
Vapor Density	2.42
Specific Gravity	.666
Color	
Odor	Highly Disagreeable

VAPOR PRESSURE

VAPOR PRESSURE P.S.I.A.

TEMPERATURE

F.

C.

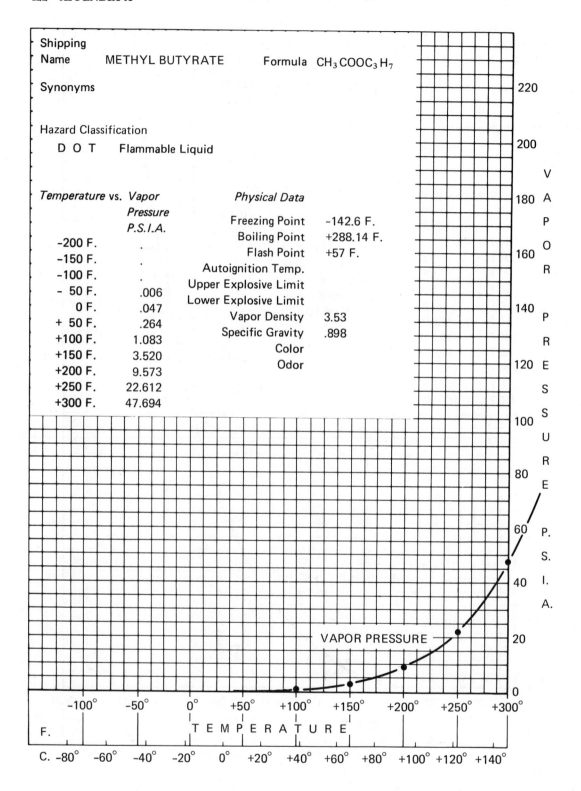

Shipping
Name   METHYL BUTYRATE   Formula $CH_3COOC_3H_7$

Synonyms

Hazard Classification

   D O T   Flammable Liquid

Temperature vs. *Vapor*   *Physical Data*
                Pressure
                P.S.I.A.   Freezing Point   -142.6 F.
   -200 F.   .   Boiling Point   +288.14 F.
   -150 F.   .   Flash Point   +57 F.
   -100 F.   Autoignition Temp.
   - 50 F.   .006   Upper Explosive Limit
     0 F.   .047   Lower Explosive Limit
   + 50 F.   .264   Vapor Density   3.53
   +100 F.   1.083   Specific Gravity   .898
   +150 F.   3.520   Color
   +200 F.   9.573   Odor
   +250 F.   22.612
   +300 F.   47.694

VAPOR PRESSURE P.S.I.A.

VAPOR PRESSURE

TEMPERATURE

F.   -100°   -50°   0°   +50° +100° +150° +200° +250° +300°

C. -80° -60° -40° -20°   0°   +20° +40° +60° +80° +100° +120° +140°

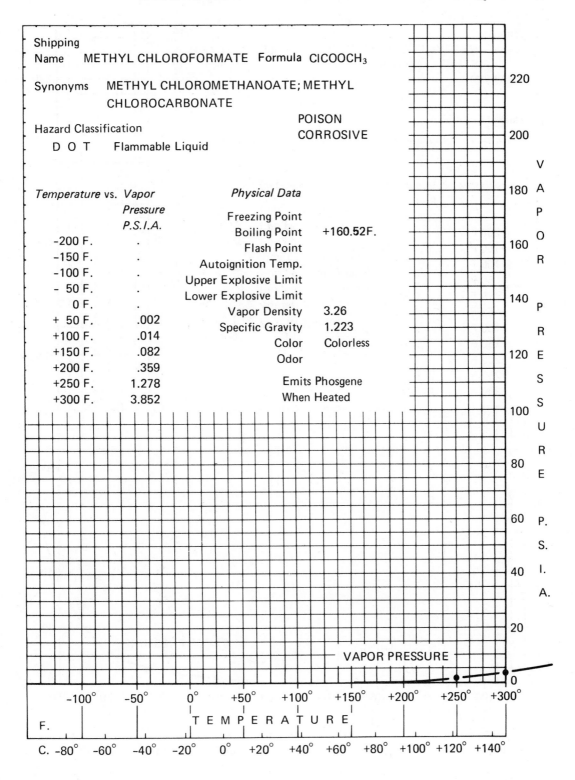

Shipping
Name   METHYL CHLOROFORMATE Formula ClCOOCH$_3$

Synonyms   METHYL CHLOROMETHANOATE; METHYL
            CHLOROCARBONATE

Hazard Classification   POISON
    D O T   Flammable Liquid   CORROSIVE

Temperature vs. Vapor   Physical Data
                   Pressure
                   P.S.I.A.   Freezing Point
    -200 F.   .   Boiling Point   +160.52F.
    -150 F.   .   Flash Point
    -100 F.   .   Autoignition Temp.
    - 50 F.   .   Upper Explosive Limit
      0 F.   .   Lower Explosive Limit
    + 50 F.   .002   Vapor Density   3.26
    +100 F.   .014   Specific Gravity   1.223
    +150 F.   .082   Color   Colorless
    +200 F.   .359   Odor
    +250 F.   1.278   Emits Phosgene
    +300 F.   3.852   When Heated

VAPOR PRESSURE

VAPOR PRESSURE P.S.I.A.

TEMPERATURE

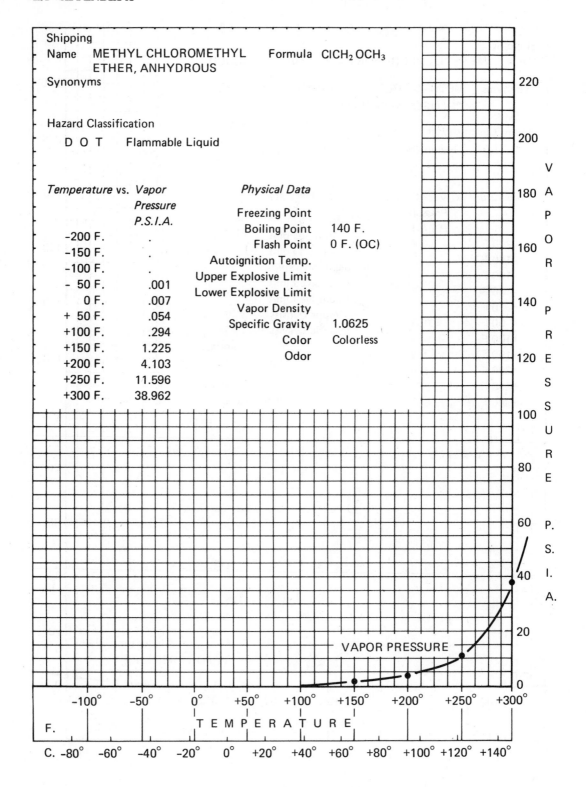

Shipping

Name   METHYL CHLOROMETHYL   Formula   CICH$_2$OCH$_3$
         ETHER, ANHYDROUS

Synonyms

Hazard Classification

   D O T   Flammable Liquid

Temperature vs. *Vapor Pressure P.S.I.A.*

Temperature	Vapor Pressure P.S.I.A.
-200 F.	.
-150 F.	.
-100 F.	.
- 50 F.	.001
0 F.	.007
+ 50 F.	.054
+100 F.	.294
+150 F.	1.225
+200 F.	4.103
+250 F.	11.596
+300 F.	38.962

Physical Data

Freezing Point	
Boiling Point	140 F.
Flash Point	0 F. (OC)
Autoignition Temp.	
Upper Explosive Limit	
Lower Explosive Limit	
Vapor Density	
Specific Gravity	1.0625
Color	Colorless
Odor	

VAPOR PRESSURE P.S.I.A.

VAPOR PRESSURE

TEMPERATURE

F.   -100°   -50°   0°   +50°   +100°   +150°   +200°   +250°   +300°

C. -80° -60° -40° -20° 0° +20° +40° +60° +80° +100° +120° +140°

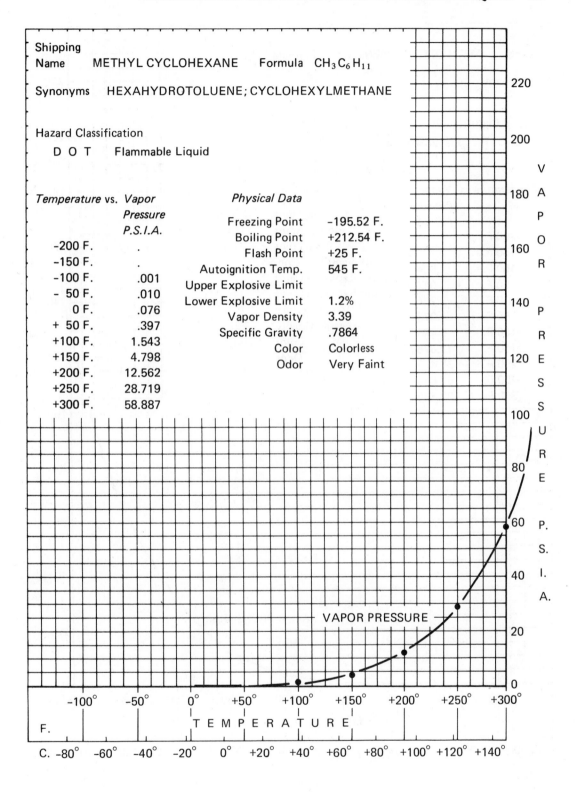

Shipping
Name   METHYL CYCLOHEXANE   Formula $CH_3C_6H_{11}$

Synonyms   HEXAHYDROTOLUENE; CYCLOHEXYLMETHANE

Hazard Classification
    D O T   Flammable Liquid

Temperature vs. Vapor
                    Pressure
                    P.S.I.A.
    -200 F.   .
    -150 F.   .
    -100 F.   .001
    - 50 F.   .010
       0 F.   .076
    + 50 F.   .397
    +100 F.   1.543
    +150 F.   4.798
    +200 F.   12.562
    +250 F.   28.719
    +300 F.   58.887

Physical Data

Freezing Point	-195.52 F.
Boiling Point	+212.54 F.
Flash Point	+25 F.
Autoignition Temp.	545 F.
Upper Explosive Limit	
Lower Explosive Limit	1.2%
Vapor Density	3.39
Specific Gravity	.7864
Color	Colorless
Odor	Very Faint

VAPOR PRESSURE

VAPOR PRESSURE P.S.I.A.

TEMPERATURE

F.

C. -80° -60° -40° -20° 0° +20° +40° +60° +80° +100° +120° +140°

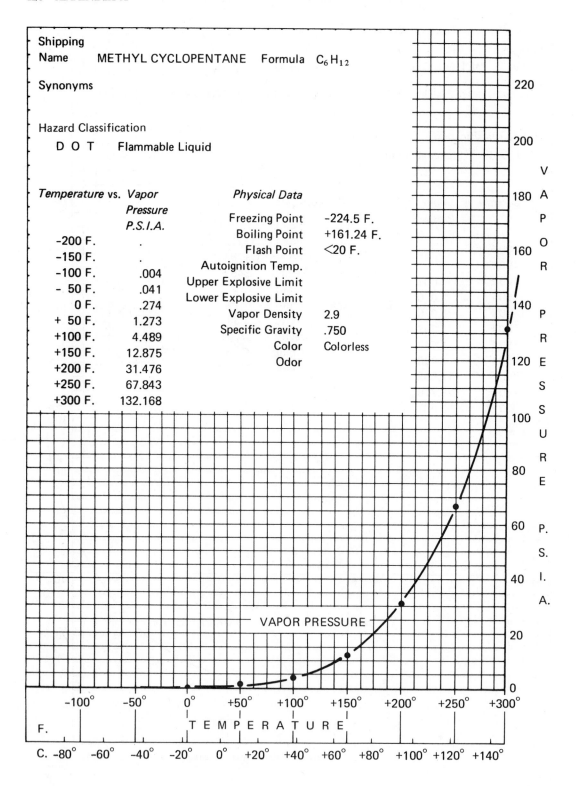

Shipping
Name   METHYL CYCLOPENTANE   Formula C_6H_{12}

Synonyms

Hazard Classification
    D O T   Flammable Liquid

Temperature vs. Vapor Pressure P.S.I.A.

Temperature	Vapor Pressure P.S.I.A.
-200 F.	.
-150 F.	.
-100 F.	.004
- 50 F.	.041
0 F.	.274
+ 50 F.	1.273
+100 F.	4.489
+150 F.	12.875
+200 F.	31.476
+250 F.	67.843
+300 F.	132.168

Physical Data

Freezing Point	-224.5 F.
Boiling Point	+161.24 F.
Flash Point	<20 F.
Autoignition Temp.	
Upper Explosive Limit	
Lower Explosive Limit	
Vapor Density	2.9
Specific Gravity	.750
Color	Colorless
Odor	

VAPOR PRESSURE

VAPOR PRESSURE P.S.I.A.

TEMPERATURE

F.   -100°   -50°   0°   +50°   +100°   +150°   +200°   +250°   +300°

C. -80° -60° -40° -20° 0° +20° +40° +60° +80° +100° +120° +140°

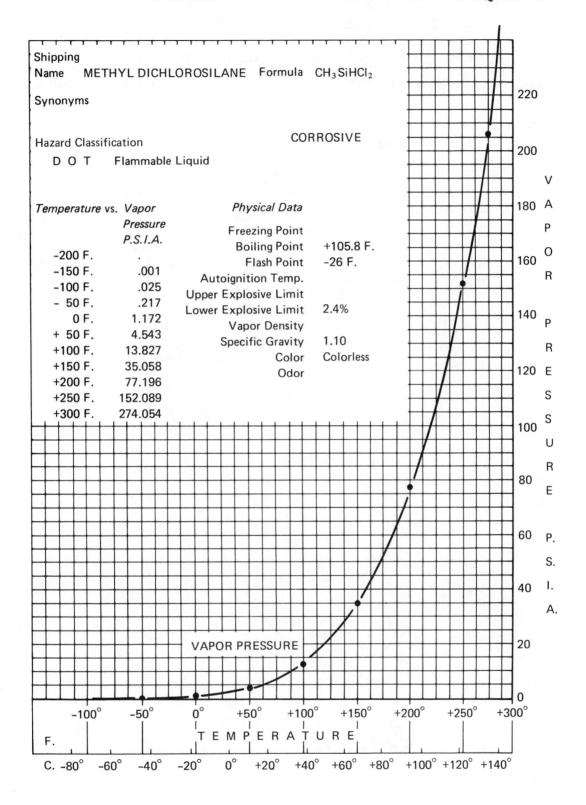

Shipping
Name   METHYL DICHLOROSILANE   Formula CH_3SiHCl_2

Synonyms

Hazard Classification   CORROSIVE

   D O T   Flammable Liquid

Temperature vs. Vapor   Physical Data
                Pressure
                P.S.I.A.   Freezing Point
   -200 F.   .   Boiling Point   +105.8 F.
   -150 F.   .001   Flash Point   -26 F.
   -100 F.   .025   Autoignition Temp.
   - 50 F.   .217   Upper Explosive Limit
      0 F.   1.172   Lower Explosive Limit   2.4%
   + 50 F.   4.543   Vapor Density
   +100 F.   13.827   Specific Gravity   1.10
   +150 F.   35.058   Color   Colorless
   +200 F.   77.196   Odor
   +250 F.   152.089
   +300 F.   274.054

VAPOR PRESSURE

V A P O R   P R E S S U R E   P. S. I. A.

220
200
180
160
140
120
100
80
60
40
20
0

-100°   -50°   0°   +50°   +100°   +150°   +200°   +250°   +300°

F.   T E M P E R A T U R E

C. -80° -60° -40° -20° 0° +20° +40° +60° +80° +100° +120° +140°

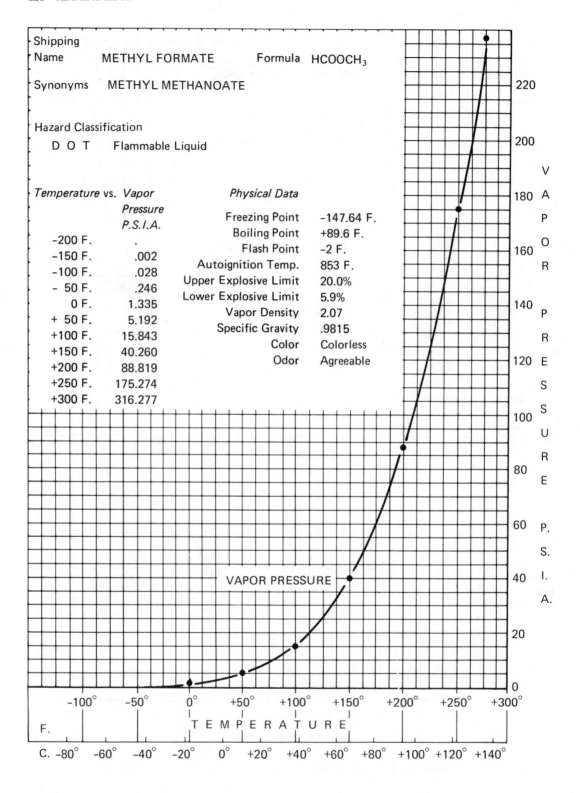

Shipping
Name   METHYL FORMATE   Formula   HCOOCH₃

Synonyms   METHYL METHANOATE

Hazard Classification
   D O T   Flammable Liquid

Temperature vs. *Vapor*
                Pressure
                P.S.I.A.

-200 F.	.
-150 F.	.002
-100 F.	.028
- 50 F.	.246
0 F.	1.335
+ 50 F.	5.192
+100 F.	15.843
+150 F.	40.260
+200 F.	88.819
+250 F.	175.274
+300 F.	316.277

Physical Data

Freezing Point	-147.64 F.
Boiling Point	+89.6 F.
Flash Point	-2 F.
Autoignition Temp.	853 F.
Upper Explosive Limit	20.0%
Lower Explosive Limit	5.9%
Vapor Density	2.07
Specific Gravity	.9815
Color	Colorless
Odor	Agreeable

VAPOR PRESSURE

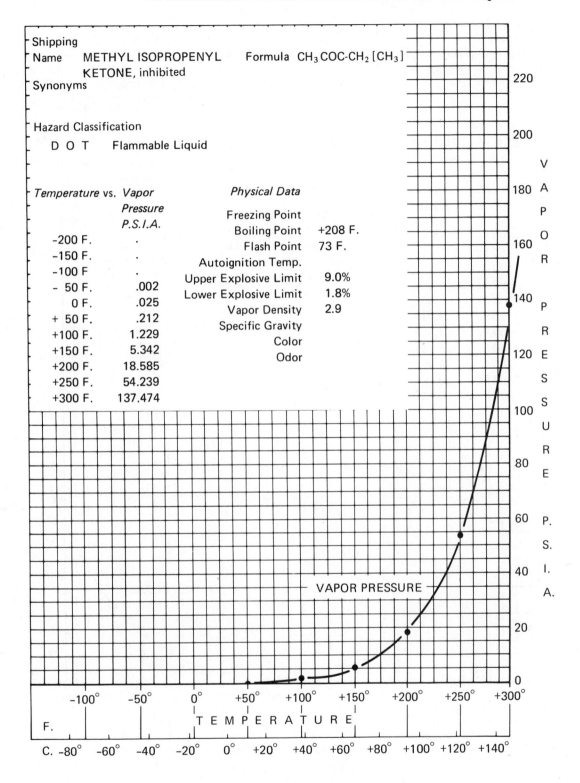

Shipping

Name   METHYL ISOPROPENYL   Formula $CH_3 COC\text{-}CH_2 [CH_3]$
        KETONE, inhibited

Synonyms

Hazard Classification

  D O T   Flammable Liquid

Temperature vs. *Vapor Pressure P.S.I.A.*

Temperature	Vapor Pressure P.S.I.A.
-200 F.	.
-150 F.	.
-100 F	.
- 50 F.	.002
0 F.	.025
+ 50 F.	.212
+100 F.	1.229
+150 F.	5.342
+200 F.	18.585
+250 F.	54.239
+300 F.	137.474

Physical Data

Freezing Point	
Boiling Point	+208 F.
Flash Point	73 F.
Autoignition Temp.	
Upper Explosive Limit	9.0%
Lower Explosive Limit	1.8%
Vapor Density	2.9
Specific Gravity	
Color	
Odor	

VAPOR PRESSURE

VAPOR PRESSURE P.S.I.A.

TEMPERATURE

F. -100° -50° 0° +50° +100° +150° +200° +250° +300°

C. -80° -60° -40° -20° 0° +20° +40° +60° +80° +100° +120° +140°

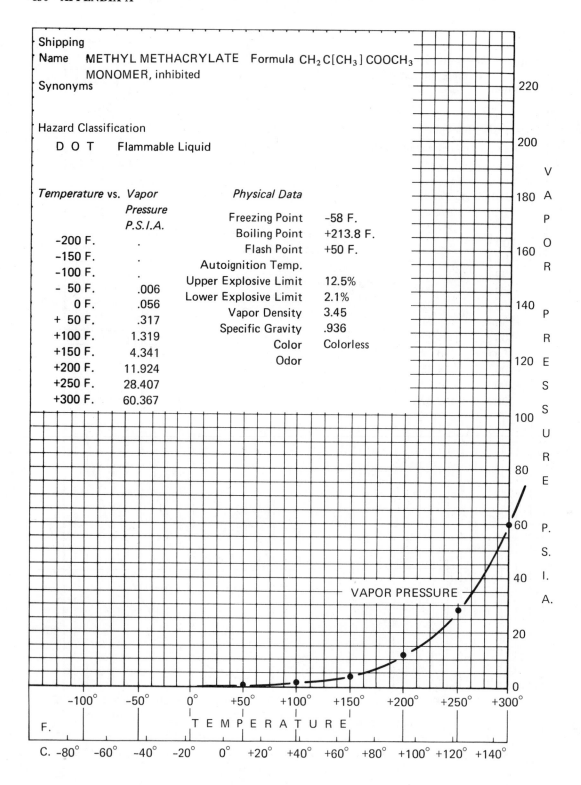

Shipping
Name   METHYL METHACRYLATE   Formula $CH_2 C[CH_3] COOCH_3$
       MONOMER, inhibited
Synonyms

Hazard Classification
     D O T   Flammable Liquid

Temperature vs. Vapor
                 Pressure
                 P.S.I.A.

Temperature	P.S.I.A.
-200 F.	.
-150 F.	.
-100 F.	.
- 50 F.	.006
0 F.	.056
+ 50 F.	.317
+100 F.	1.319
+150 F.	4.341
+200 F.	11.924
+250 F.	28.407
+300 F.	60.367

Physical Data

Freezing Point	-58 F.
Boiling Point	+213.8 F.
Flash Point	+50 F.
Autoignition Temp.	
Upper Explosive Limit	12.5%
Lower Explosive Limit	2.1%
Vapor Density	3.45
Specific Gravity	.936
Color	Colorless
Odor	

VAPOR PRESSURE

VAPOR PRESSURE P.S.I.A.

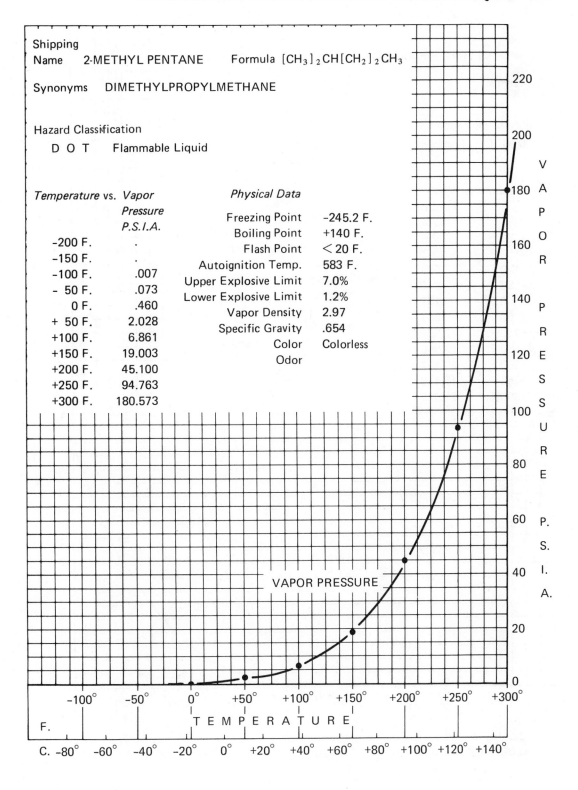

Shipping
Name   2-METHYL PENTANE   Formula $[CH_3]_2CH[CH_2]_2CH_3$

Synonyms   DIMETHYLPROPYLMETHANE

Hazard Classification
    D O T   Flammable Liquid

Temperature vs. *Vapor*
                    Pressure
                    P.S.I.A.

Temperature	Pressure P.S.I.A.
-200 F.	.
-150 F.	.
-100 F.	.007
- 50 F.	.073
0 F.	.460
+ 50 F.	2.028
+100 F.	6.861
+150 F.	19.003
+200 F.	45.100
+250 F.	94.763
+300 F.	180.573

Physical Data

Freezing Point	-245.2 F.
Boiling Point	+140 F.
Flash Point	< 20 F.
Autoignition Temp.	583 F.
Upper Explosive Limit	7.0%
Lower Explosive Limit	1.2%
Vapor Density	2.97
Specific Gravity	.654
Color	Colorless
Odor	

VAPOR PRESSURE

VAPOR PRESSURE P.S.I.A.

TEMPERATURE

F.   -100°   -50°   0°   +50°   +100°   +150°   +200°   +250°   +300°

C. -80° -60° -40° -20° 0° +20° +40° +60° +80° +100° +120° +140°

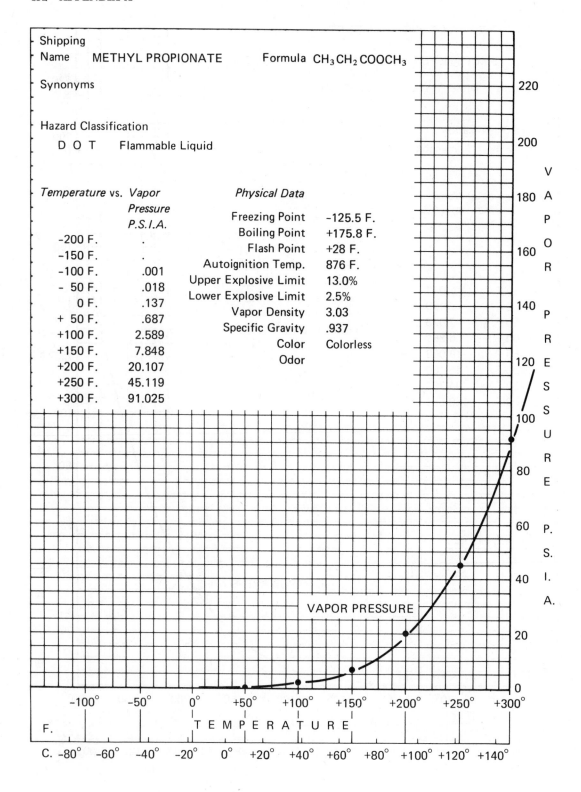

Shipping
Name   METHYL PROPIONATE   Formula $CH_3CH_2COOCH_3$

Synonyms

Hazard Classification

   D O T   Flammable Liquid

Temperature vs. *Vapor*
                Pressure
                P.S.I.A.

Temperature	P.S.I.A.
-200 F.	.
-150 F.	.
-100 F.	.001
- 50 F.	.018
0 F.	.137
+ 50 F.	.687
+100 F.	2.589
+150 F.	7.848
+200 F.	20.107
+250 F.	45.119
+300 F.	91.025

Physical Data

Freezing Point	-125.5 F.
Boiling Point	+175.8 F.
Flash Point	+28 F.
Autoignition Temp.	876 F.
Upper Explosive Limit	13.0%
Lower Explosive Limit	2.5%
Vapor Density	3.03
Specific Gravity	.937
Color	Colorless
Odor	

VAPOR PRESSURE

V A P O R   P R E S S U R E   P. S. I. A.

TEMPERATURE

F. -100° -50° 0° +50° +100° +150° +200° +250° +300°

C. -80° -60° -40° -20° 0° +20° +40° +60° +80° +100° +120° +140°

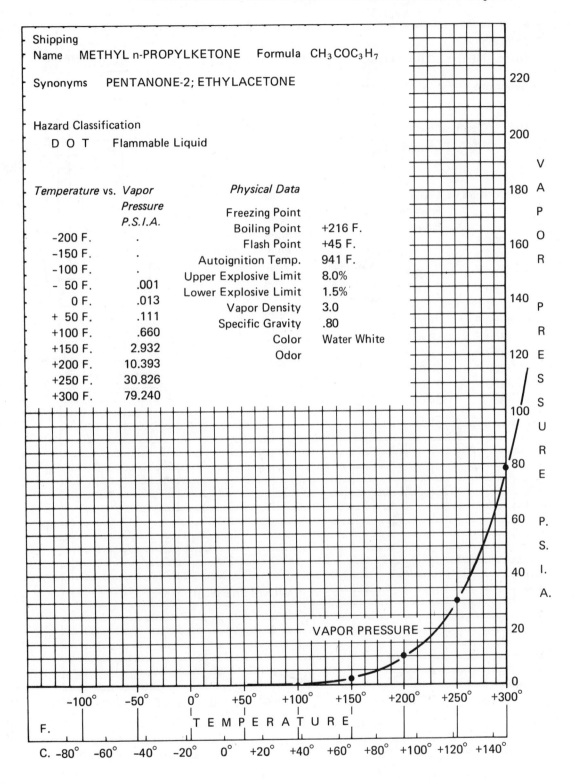

Shipping
Name   METHYL n-PROPYLKETONE   Formula $CH_3COC_3H_7$

Synonyms   PENTANONE-2; ETHYLACETONE

Hazard Classification
   D O T   Flammable Liquid

Temperature vs. Vapor
                  Pressure
                  P.S.I.A.

Temperature	Pressure P.S.I.A.
-200 F.	.
-150 F.	.
-100 F.	.
- 50 F.	.001
0 F.	.013
+ 50 F.	.111
+100 F.	.660
+150 F.	2.932
+200 F.	10.393
+250 F.	30.826
+300 F.	79.240

Physical Data

Physical Data	
Freezing Point	
Boiling Point	+216 F.
Flash Point	+45 F.
Autoignition Temp.	941 F.
Upper Explosive Limit	8.0%
Lower Explosive Limit	1.5%
Vapor Density	3.0
Specific Gravity	.80
Color	Water White
Odor	

VAPOR PRESSURE

VAPOR PRESSURE P.S.I.A.

TEMPERATURE

F.   -100° -50° 0° +50° +100° +150° +200° +250° +300°

C. -80° -60° -40° -20° 0° +20° +40° +60° +80° +100° +120° +140°

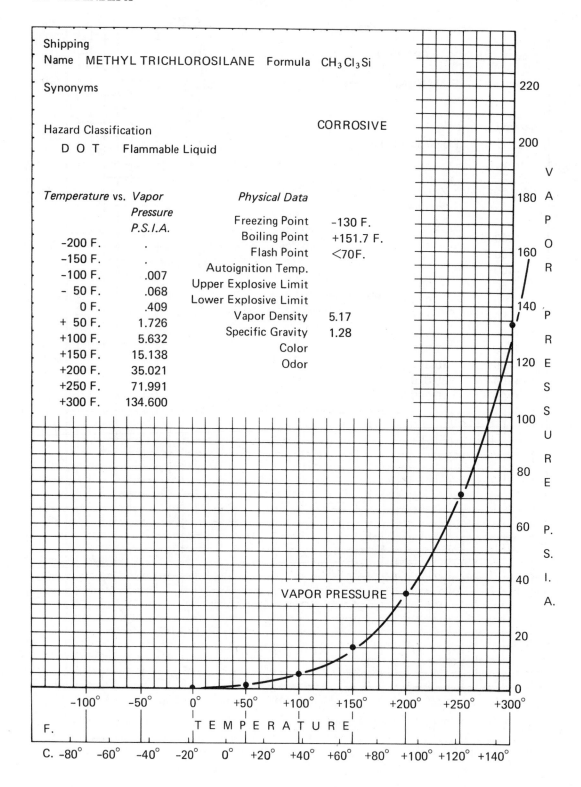

Shipping
Name   METHYL TRICHLOROSILANE   Formula CH_3Cl_3Si

Synonyms

Hazard Classification   CORROSIVE
       D O T   Flammable Liquid

Temperature vs. Vapor   Physical Data
                  Pressure
                  P.S.I.A.   Freezing Point   –130 F.
    –200 F.   .   Boiling Point   +151.7 F.
    –150 F.   .   Flash Point   <70F.
    –100 F.   .007   Autoignition Temp.
    – 50 F.   .068   Upper Explosive Limit
       0 F.   .409   Lower Explosive Limit
    + 50 F.   1.726   Vapor Density   5.17
    +100 F.   5.632   Specific Gravity   1.28
    +150 F.   15.138   Color
    +200 F.   35.021   Odor
    +250 F.   71.991
    +300 F.   134.600

VAPOR PRESSURE

V A P O R   P R E S S U R E   P. S. I. A.

TEMPERATURE

F.   –100°   –50°   0°   +50°   +100°   +150°   +200°   +250°   +300°

C. –80° –60° –40° –20°   0°   +20° +40° +60° +80° +100° +120° +140°

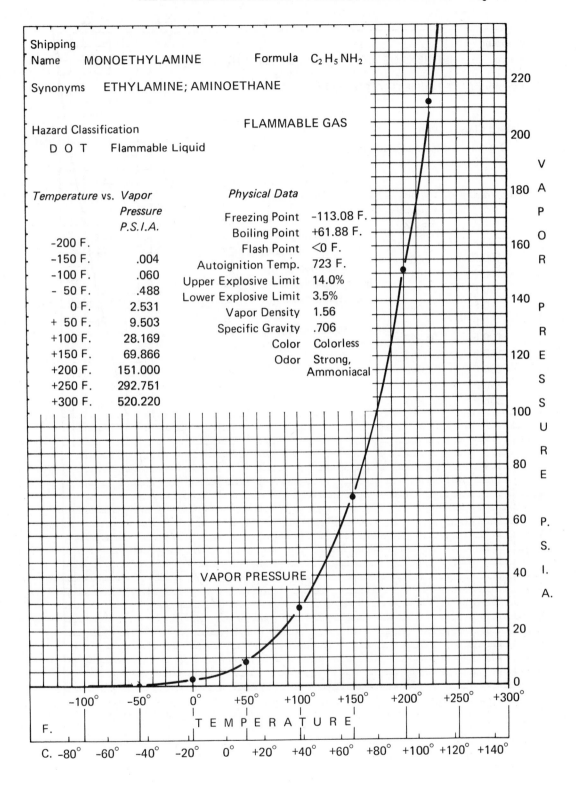

Shipping
Name   MONOETHYLAMINE   Formula $C_2H_5NH_2$

Synonyms   ETHYLAMINE; AMINOETHANE

Hazard Classification   FLAMMABLE GAS
     D O T   Flammable Liquid

Temperature vs. *Vapor*
                Pressure   *Physical Data*
                P.S.I.A.   Freezing Point   -113.08 F.
 -200 F.   Boiling Point   +61.88 F.
 -150 F.   .004   Flash Point   <0 F.
 -100 F.   .060   Autoignition Temp.   723 F.
 - 50 F.   .488   Upper Explosive Limit   14.0%
   0 F.   2.531   Lower Explosive Limit   3.5%
 + 50 F.   9.503   Vapor Density   1.56
 +100 F.   28.169   Specific Gravity .706
 +150 F.   69.866   Color   Colorless
 +200 F.   151.000   Odor   Strong,
 +250 F.   292.751   Ammoniacal
 +300 F.   520.220

VAPOR PRESSURE

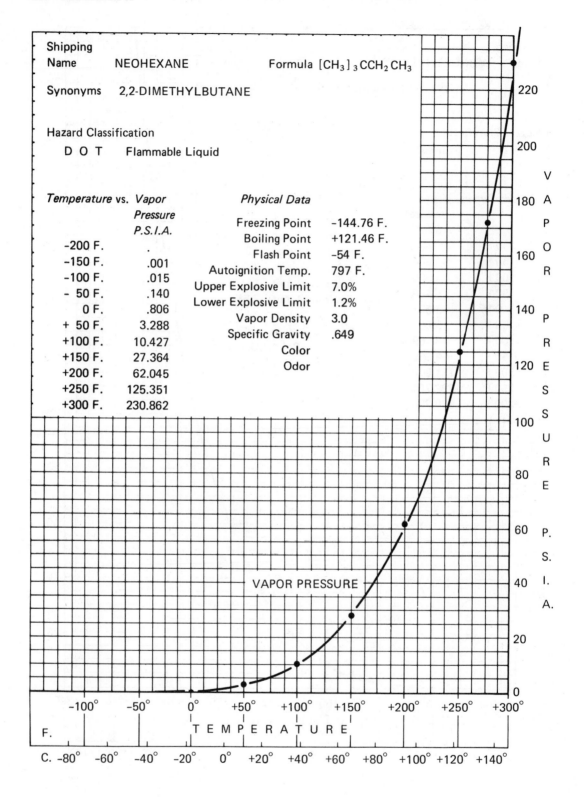

Shipping
Name   NEOHEXANE   Formula [CH₃]₃CCH₂CH₃

Synonyms   2,2-DIMETHYLBUTANE

Hazard Classification
    D O T   Flammable Liquid

Temperature vs. Vapor   Physical Data
                   Pressure
                   P.S.I.A.   Freezing Point   −144.76 F.
    −200 F.   .   Boiling Point   +121.46 F.
    −150 F.   .001   Flash Point   −54 F.
    −100 F.   .015   Autoignition Temp. 797 F.
    − 50 F.   .140   Upper Explosive Limit   7.0%
      0 F.   .806   Lower Explosive Limit   1.2%
    + 50 F.   3.288   Vapor Density   3.0
    +100 F.   10.427   Specific Gravity   .649
    +150 F.   27.364   Color
    +200 F.   62.045   Odor
    +250 F.   125.351
    +300 F.   230.862

VAPOR PRESSURE

V A P O R   P R E S S U R E   P. S. I. A.

−100° −50° 0° +50° +100° +150° +200° +250° +300°
F.   T E M P E R A T U R E
C. −80° −60° −40° −20° 0° +20° +40° +60° +80° +100° +120° +140°

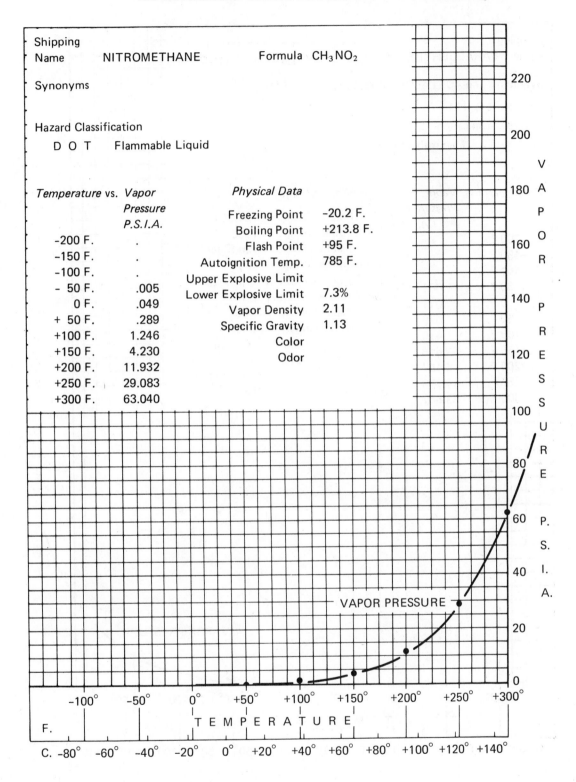

Shipping
Name   NITROMETHANE   Formula CH_3NO_2

Synonyms

Hazard Classification
    D O T   Flammable Liquid

Temperature vs. Vapor Pressure P.S.I.A.	
-200 F.	.
-150 F.	.
-100 F.	.
- 50 F.	.005
0 F.	.049
+ 50 F.	.289
+100 F.	1.246
+150 F.	4.230
+200 F.	11.932
+250 F.	29.083
+300 F.	63.040

Physical Data

Freezing Point	-20.2 F.
Boiling Point	+213.8 F.
Flash Point	+95 F.
Autoignition Temp.	785 F.
Upper Explosive Limit	
Lower Explosive Limit	7.3%
Vapor Density	2.11
Specific Gravity	1.13
Color	
Odor	

VAPOR PRESSURE

V A P O R   P R E S S U R E   P.S.I.A.

TEMPERATURE

F.

C. -80° -60° -40° -20° 0° +20° +40° +60° +80° +100° +120° +140°

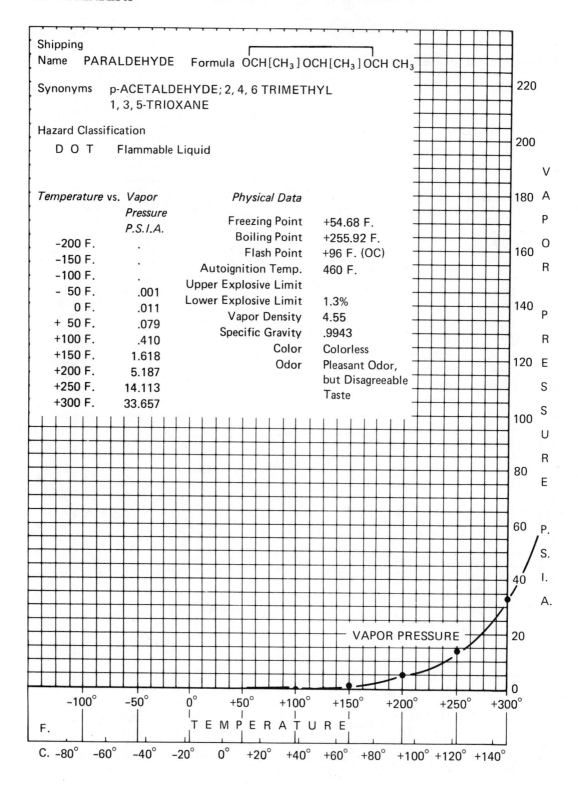

Shipping
Name   PARALDEHYDE   Formula $OCH[CH_3]OCH[CH_3]OCH CH_3$

Synonyms   p-ACETALDEHYDE; 2, 4, 6 TRIMETHYL
1, 3, 5-TRIOXANE

Hazard Classification
    D O T   Flammable Liquid

Temperature vs. Vapor Pressure P.S.I.A.

Temperature	P.S.I.A.
-200 F.	.
-150 F.	.
-100 F.	.
- 50 F.	.001
0 F.	.011
+ 50 F.	.079
+100 F.	.410
+150 F.	1.618
+200 F.	5.187
+250 F.	14.113
+300 F.	33.657

Physical Data

Freezing Point	+54.68 F.
Boiling Point	+255.92 F.
Flash Point	+96 F. (OC)
Autoignition Temp.	460 F.
Upper Explosive Limit	
Lower Explosive Limit	1.3%
Vapor Density	4.55
Specific Gravity	.9943
Color	Colorless
Odor	Pleasant Odor, but Disagreeable Taste

VAPOR PRESSURE P.S.I.A.

VAPOR PRESSURE

TEMPERATURE
F.
C.

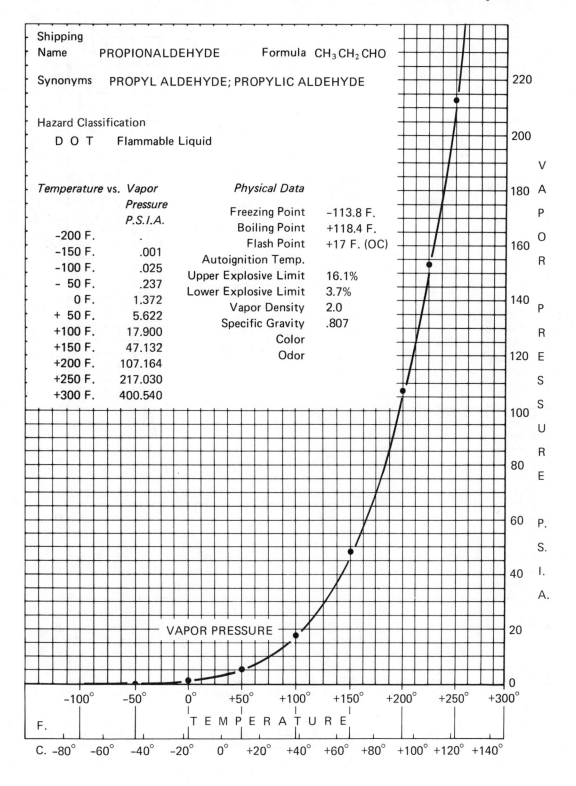

Shipping
Name   PROPIONALDEHYDE   Formula CH_3CH_2CHO

Synonyms   PROPYL ALDEHYDE; PROPYLIC ALDEHYDE

Hazard Classification

   D O T   Flammable Liquid

Temperature vs. Vapor   Physical Data
                  Pressure
                  P.S.I.A.   Freezing Point   -113.8 F.
   -200 F.   .   Boiling Point   +118.4 F.
   -150 F.   .001   Flash Point   +17 F. (OC)
   -100 F.   .025   Autoignition Temp.
   - 50 F.   .237   Upper Explosive Limit   16.1%
     0 F.   1.372   Lower Explosive Limit   3.7%
   + 50 F.   5.622   Vapor Density   2.0
   +100 F.   17.900   Specific Gravity   .807
   +150 F.   47.132   Color
   +200 F.   107.164   Odor
   +250 F.   217.030
   +300 F.   400.540

VAPOR PRESSURE

VAPOR PRESSURE P.S.I.A.

TEMPERATURE

F.

C.

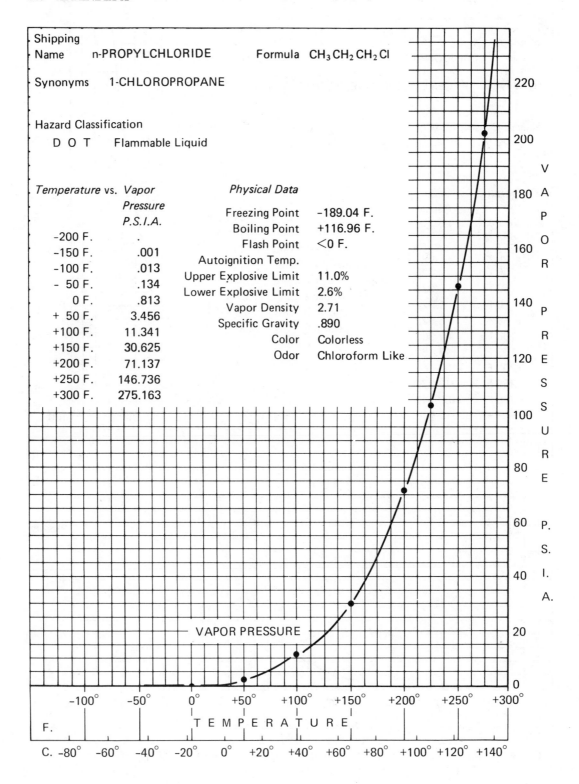

Shipping Name   n-PROPYLCHLORIDE   Formula $CH_3CH_2CH_2Cl$

Synonyms   1-CHLOROPROPANE

Hazard Classification
    D O T   Flammable Liquid

Temperature vs. *Vapor Pressure P.S.I.A.*

Temperature	P.S.I.A.
-200 F.	.
-150 F.	.001
-100 F.	.013
- 50 F.	.134
0 F.	.813
+ 50 F.	3.456
+100 F.	11.341
+150 F.	30.625
+200 F.	71.137
+250 F.	146.736
+300 F.	275.163

Physical Data

Freezing Point	-189.04 F.
Boiling Point	+116.96 F.
Flash Point	$<$0 F.
Autoignition Temp.	
Upper Explosive Limit	11.0%
Lower Explosive Limit	2.6%
Vapor Density	2.71
Specific Gravity	.890
Color	Colorless
Odor	Chloroform Like

VAPOR PRESSURE

VAPOR PRESSURE P.S.I.A.

TEMPERATURE

F.

C.

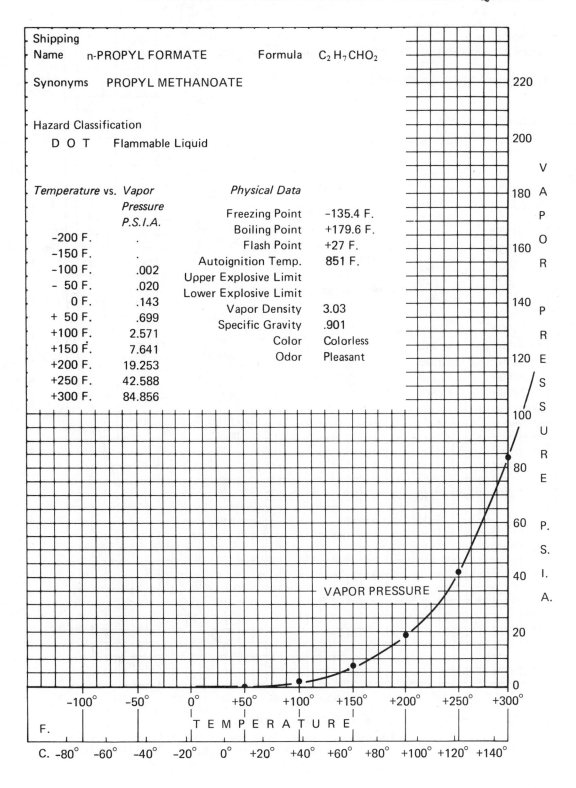

Shipping
Name   n-PROPYL FORMATE   Formula $C_2H_7CHO_2$

Synonyms   PROPYL METHANOATE

Hazard Classification
    D O T   Flammable Liquid

Temperature vs. Vapor
                Pressure
                P.S.I.A.
    -200 F.   .
    -150 F.   .
    -100 F.   .002
    - 50 F.   .020
       0 F.   .143
    + 50 F.   .699
    +100 F.   2.571
    +150 F.   7.641
    +200 F.   19.253
    +250 F.   42.588
    +300 F.   84.856

Physical Data
    Freezing Point   -135.4 F.
    Boiling Point   +179.6 F.
    Flash Point   +27 F.
    Autoignition Temp.   851 F.
    Upper Explosive Limit
    Lower Explosive Limit
    Vapor Density   3.03
    Specific Gravity   .901
    Color   Colorless
    Odor   Pleasant

VAPOR PRESSURE

VAPOR PRESSURE P.S.I.A.

TEMPERATURE

F.
-100°   -50°   0°   +50°   +100°   +150°   +200°   +250°   +300°

C. -80° -60° -40° -20° 0° +20° +40° +60° +80° +100° +120° +140°

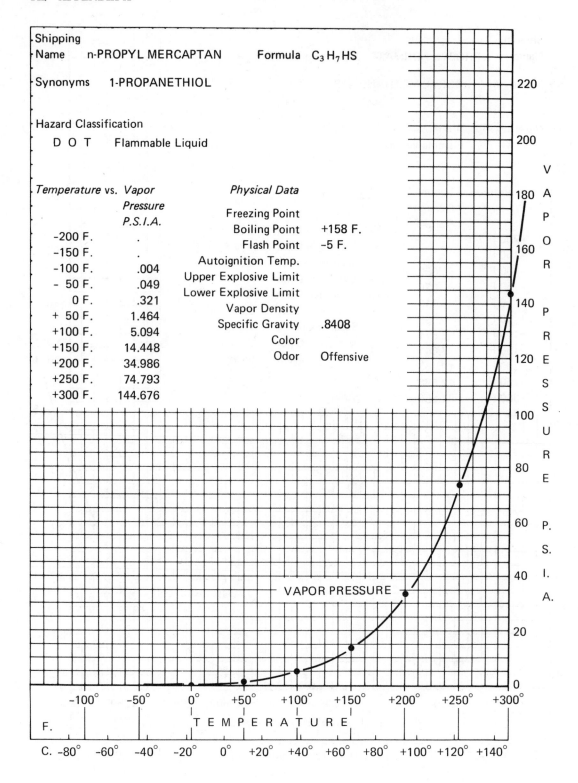

Shipping
Name   n-PROPYL MERCAPTAN   Formula C_3H_7HS

Synonyms   1-PROPANETHIOL

Hazard Classification
   D O T   Flammable Liquid

Temperature vs. *Vapor Pressure P.S.I.A.*

Temperature	P.S.I.A.
-200 F.	.
-150 F.	.
-100 F.	.004
- 50 F.	.049
0 F.	.321
+ 50 F.	1.464
+100 F.	5.094
+150 F.	14.448
+200 F.	34.986
+250 F.	74.793
+300 F.	144.676

Physical Data

Freezing Point	
Boiling Point	+158 F.
Flash Point	-5 F.
Autoignition Temp.	
Upper Explosive Limit	
Lower Explosive Limit	
Vapor Density	
Specific Gravity	.8408
Color	
Odor	Offensive

VAPOR PRESSURE

VAPOR PRESSURE P.S.I.A.

TEMPERATURE

F.   -100°   -50°   0°   +50°   +100°   +150°   +200°   +250°   +300°

C.   -80°   -60°   -40°   -20°   0°   +20°   +40°   +60°   +80°   +100°   +120°   +140°

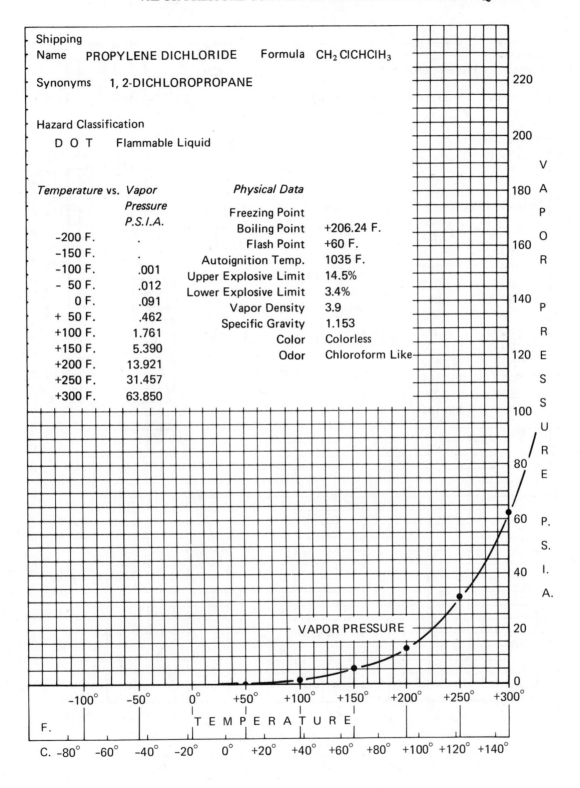

Shipping
Name   PROPYLENE DICHLORIDE   Formula $CH_2CICHCIH_3$

Synonyms   1, 2-DICHLOROPROPANE

Hazard Classification

   D O T   Flammable Liquid

Temperature vs. *Vapor*
                    Pressure
                    P.S.I.A.

Temperature	Vapor Pressure P.S.I.A.
−200 F.	.
−150 F.	.
−100 F.	.001
− 50 F.	.012
0 F.	.091
+ 50 F.	.462
+100 F.	1.761
+150 F.	5.390
+200 F.	13.921
+250 F.	31.457
+300 F.	63.850

Physical Data

Freezing Point	
Boiling Point	+206.24 F.
Flash Point	+60 F.
Autoignition Temp.	1035 F.
Upper Explosive Limit	14.5%
Lower Explosive Limit	3.4%
Vapor Density	3.9
Specific Gravity	1.153
Color	Colorless
Odor	Chloroform Like

VAPOR PRESSURE P.S.I.A.

VAPOR PRESSURE

TEMPERATURE

F.

C.

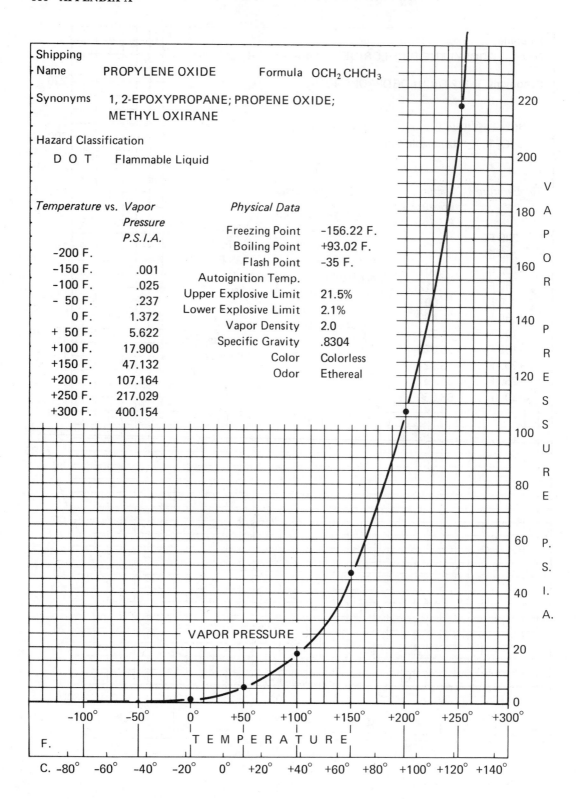

Shipping
Name   PROPYLENE OXIDE   Formula OCH_2CHCH_3

Synonyms   1, 2-EPOXYPROPANE; PROPENE OXIDE;
            METHYL OXIRANE

Hazard Classification

    D O T   Flammable Liquid

Temperature vs. *Vapor*
                Pressure
                P.S.I.A.

    -200 F.
    -150 F.   .001
    -100 F.   .025
    - 50 F.   .237
       0 F.   1.372
    + 50 F.   5.622
    +100 F.   17.900
    +150 F.   47.132
    +200 F.   107.164
    +250 F.   217.029
    +300 F.   400.154

Physical Data

Freezing Point	-156.22 F.
Boiling Point	+93.02 F.
Flash Point	-35 F.
Autoignition Temp.	
Upper Explosive Limit	21.5%
Lower Explosive Limit	2.1%
Vapor Density	2.0
Specific Gravity	.8304
Color	Colorless
Odor	Ethereal

VAPOR PRESSURE

VAPOR PRESSURE P.S.I.A.

TEMPERATURE

F. -100° -50° 0° +50° +100° +150° +200° +250° +300°

C. -80° -60° -40° -20° 0° +20° +40° +60° +80° +100° +120° +140°

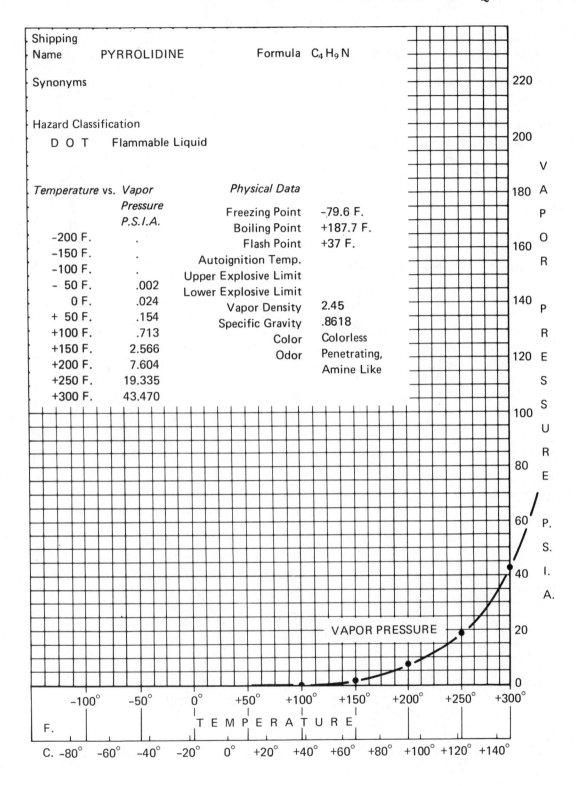

Shipping
Name   PYRROLIDINE   Formula C_4H_9N

Synonyms

Hazard Classification
   D O T   Flammable Liquid

Temperature vs. *Vapor Pressure P.S.I.A.*

Temperature	P.S.I.A.
-200 F.	.
-150 F.	.
-100 F.	.
- 50 F.	.002
0 F.	.024
+ 50 F.	.154
+100 F.	.713
+150 F.	2.566
+200 F.	7.604
+250 F.	19.335
+300 F.	43.470

Physical Data

Freezing Point	-79.6 F.
Boiling Point	+187.7 F.
Flash Point	+37 F.
Autoignition Temp.	
Upper Explosive Limit	
Lower Explosive Limit	
Vapor Density	2.45
Specific Gravity	.8618
Color	Colorless
Odor	Penetrating, Amine Like

VAPOR PRESSURE

V A P O R   P R E S S U R E

P. S. I. A.

TEMPERATURE

F.

C. -80° -60° -40° -20° 0° +20° +40° +60° +80° +100° +120° +140°

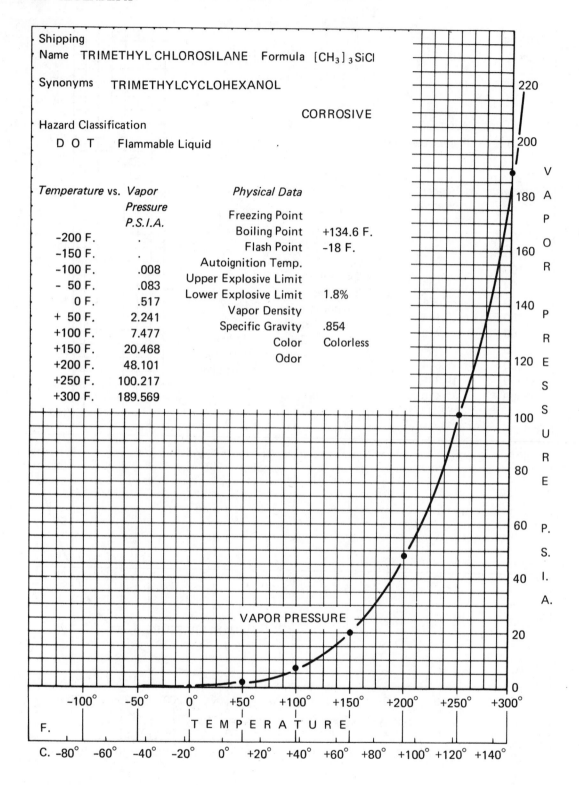

Shipping
Name   TRIMETHYL CHLOROSILANE   Formula $[CH_3]_3SiCl$

Synonyms   TRIMETHYLCYCLOHEXANOL

CORROSIVE

Hazard Classification
   D O T   Flammable Liquid

Temperature vs.	Vapor Pressure P.S.I.A.
-200 F.	.
-150 F.	.
-100 F.	.008
- 50 F.	.083
0 F.	.517
+ 50 F.	2.241
+100 F.	7.477
+150 F.	20.468
+200 F.	48.101
+250 F.	100.217
+300 F.	189.569

Physical Data

Freezing Point	
Boiling Point	+134.6 F.
Flash Point	-18 F.
Autoignition Temp.	
Upper Explosive Limit	
Lower Explosive Limit	1.8%
Vapor Density	
Specific Gravity	.854
Color	Colorless
Odor	

VAPOR PRESSURE

VAPOR PRESSURE P.S.I.A.

T E M P E R A T U R E

F.
C.

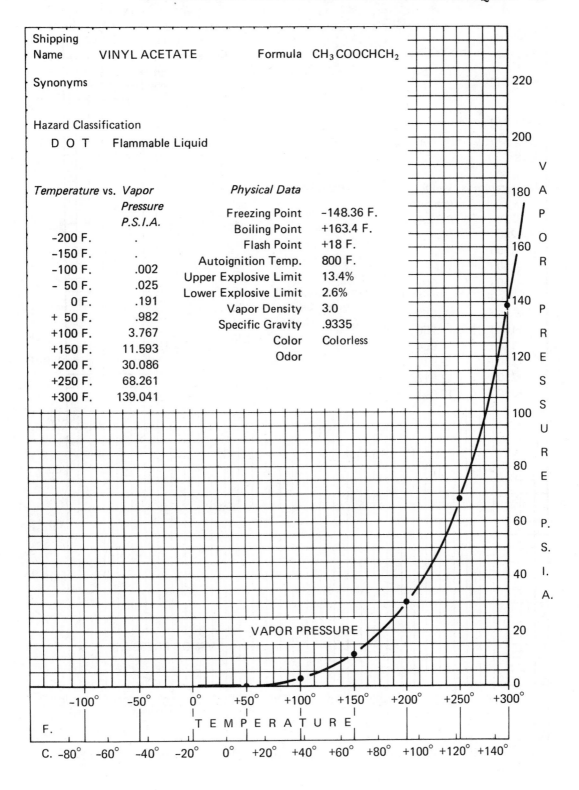

Shipping
Name   VINYL ACETATE   Formula $CH_3COOCHCH_2$

Synonyms

Hazard Classification
   D O T   Flammable Liquid

Temperature vs. *Vapor Pressure P.S.I.A.*

Temperature	Vapor Pressure P.S.I.A.
-200 F.	.
-150 F.	.
-100 F.	.002
- 50 F.	.025
0 F.	.191
+ 50 F.	.982
+100 F.	3.767
+150 F.	11.593
+200 F.	30.086
+250 F.	68.261
+300 F.	139.041

Physical Data

Freezing Point	-148.36 F.
Boiling Point	+163.4 F.
Flash Point	+18 F.
Autoignition Temp.	800 F.
Upper Explosive Limit	13.4%
Lower Explosive Limit	2.6%
Vapor Density	3.0
Specific Gravity	.9335
Color	Colorless
Odor	

VAPOR PRESSURE

TEMPERATURE

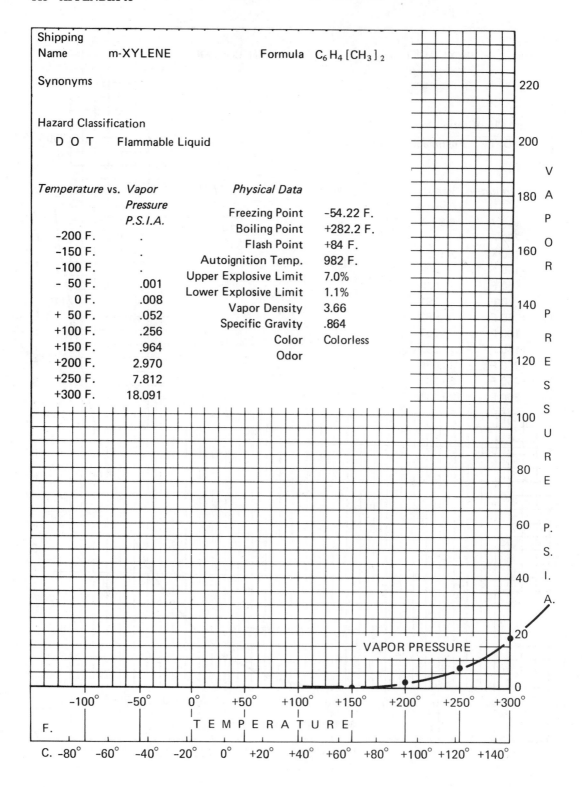

Shipping
Name   m-XYLENE   Formula $C_6H_4[CH_3]_2$

Synonyms

Hazard Classification
     D O T   Flammable Liquid

Temperature vs. Vapor
                   Pressure
                   P.S.I.A.

-200 F.	.
-150 F.	.
-100 F.	.
- 50 F.	.001
0 F.	.008
+ 50 F.	.052
+100 F.	.256
+150 F.	.964
+200 F.	2.970
+250 F.	7.812
+300 F.	18.091

Physical Data

Freezing Point	-54.22 F.
Boiling Point	+282.2 F.
Flash Point	+84 F.
Autoignition Temp.	982 F.
Upper Explosive Limit	7.0%
Lower Explosive Limit	1.1%
Vapor Density	3.66
Specific Gravity	.864
Color	Colorless
Odor	

VAPOR PRESSURE

V A P O R   P R E S S U R E   P.S.I.A.

TEMPERATURE

F.
C.

APPENDIX B

Chapter 9 of the United Nations "Orange Book"— Recommendations on the Transport of Dangerous Goods

Chapter 9 of the UN recommendations contains performance-oriented packaging standards being adopted by various national and international bodies, including the RID/ADR in Europe, the International Maritime Organization, the International Civil Aviation Organization, and Transport Canada. It is proposed for adoption as a replacement for 49 CFR Part 178 in DOT's rule making Docket No. HM-181.

General Recommendations on Packing

9.1 GENERAL NOTES

9.1.1 The recommendations on the packing of dangerous goods are based, in the main, on existing international and national regulations. Account is also taken of a prevailing trend to replace the detailed specifications of packagings, which may vary considerably from one country to another, by tests designed to ensure that packages containing dangerous goods can withstand normal conditions of transport and to ensure the desirable level of safety. When drafting the recommendations sight was not lost of improvements and changes that may occur as a result of progress in science and technology. So provisions are made for the use of packagings which, while not complying exactly with the specifications set out in the recommendations, would be nevertheless as satisfactory in every respect as those that do, and would successfully pass the recommended tests when prepared for transport.

9.1.2 The Recommendations Do Not Cover:

(a) packages containing radioactive material, which should comply with the Regulations of the International Atomic Energy Agency; except that radioactive materials possessing other dangerous properties (subsidiary risks) should also comply with special provision 172;

(b) cylinders and other receptacles for gas;

(c) packages whose net mass exceeds 400 kg;

(d) packagings with a capacity exceeding 450 litres.

9.1.3 Dangerous goods of all classes other than Classes 1, 2, 6.2 and 7 have for packing purposes been divided among three groups according to the degree of danger they present, i.e. great danger—Packing Group I; medium danger—Packing Group II; and minor danger—Packing Group III. The Packing Group to which a substance is assigned is listed in chapter 2.

9.1.4 Because of the special nature of explosives, the varying degree of hazard they present according to the manner in which they are packed, and the desirability of improving the uniformity of their packing, detailed recommendations are included on the way in which individual explosive substances and articles, or groups thereof, should be packed (see UN chapter 10). Unless specific provision to the contrary is made in the individual recommendations, the packagings used for goods of Class I should comply with the requirements for the medium danger (Packing Group II) category mentioned in 9.1.3.

9.1.5 For similar reasons, recommendations are included on the way in which organic peroxides and some self-reactive substances should be packed, the maximum quantities, the indication of the subsidiary risk of explosion and, for those which should be carried at a control temperature, that temperature (see UN chapters 11 and 14).

9.1.6 The only provisions of this chapter which apply to packagings for infectious substances are those in 9.2 and 9.3 (except in 9.3.3 and 9.3.9 to 9.3.13). Provisions for packing and test procedures for packagings for infectious substances are given in chapter 6 of these recommendations.

9.1.7 When incorporating these recommendations in international and national regulations, consideration should be given to providing a transitional period of a few years for packagings not conforming to the provisions of this chapter but which have been found safe for use or are acceptable according to existing regulations.

9.1.8 It is recognized that certain substances, for example flammable liquids with a high viscosity, are transported in packagings which do not meet all the performance tests in this chapter. It may be appropriate for provisions to be made for such substances and/or packagings in relation to individual modes of transport on a regional basis.

9.2 TERMS AND DEFINITIONS

9.2.1 For the Purpose of the Recommendations in this Chapter:

- *Bags* are flexible packagings made of paper, plastics film, textiles, woven material or other suitable materials;
- *Boxes* are packagings with complete rectangular or polygonal faces, made of metal, wood, plywood, reconstituted wood, fibreboard, plastics or other suitable material;
- *Closures* are devices which close an opening in a receptacle;
- *Combination packagings* are a combination of packagings for transport purposes, consisting of one or more inner packagings secured in an outer packaging in accordance with 9.3.5;
- *Composite packagings* are packagings consisting of an outer packaging and an inner receptacle so constructed that the inner receptacle and the outer packaging form an integral packaging. Once assembled it remains thereafter an integrated single unit; it is filled, stored, transported and emptied as such;
- *Crates* are outer packagings with incomplete surfaces;
- *Drums* are flat-ended or convex-ended cylindrical packagings made of metal, fibreboard, plastics, plywood or other suitable materials. This definition also includes packagings of other shapes made of metal or plastics, e.g. round taper-necked packagings, or pail-shaped packagings. Wooden barrels or jerricans are not covered by this definition;
- *Inner packagings* are packagings for which an outer packaging is required for transport;
- *Inner receptacles* are receptacles which require an outer packaging in order to perform their containment function;
- *Jerricans* are metal or plastics packagings of rectangular or polygonal cross-section;
- *Maximum capacity* as used in 9.6 is the maximum inner volume of receptacles or packagings expressed in litres;
- *Maximum net mass* is the maximum net mass of contents in a single packaging or maximum combined mass of inner packagings and the contents thereof and is expressed in kg;
- *Outer packaging* is the outer protection of a composite or combination packaging together with any absorbent materials, cushioning and any other components necessary to contain and protect inner receptacles or inner packagings;
- *Packages* are the complete product of the packing operation, consisting of the packaging and its contents prepared for transport;
- *Packagings* are receptacles and any other components or materials necessary for the receptacle to perform its containment function;
- *Receptacles* are containment vessels for receiving and holding substances or articles, including any means of closing;
- *Wooden barrels* are packagings made of natural wood, of round cross-section, having convex walls, consisting of staves and heads and fitted with hoops.

9.2.2 The following explanations and examples are meant to assist in clarifying the use of the terms defined in 9.2.1.

9.2.2.1 The definitions in 9.2.1 are consistent with the use of the defined terms throughout the recommendations. However, many national and international regulations, based on the recommendations, currently use some of the defined terms in other ways. This is particularly evident in respect of the term "inner receptacle" which has often been used to describe the "inners" of a combination packaging.

9.2.2.2 The "inners" of "combination packagings" are always termed "inner packagings" not "inner receptacles". A glass bottle is an example of such an "inner packaging".

9.2.2.3 The "inners" of "composite packagings" are normally termed "inner receptacles". For example, the "inner" of a 6HA1 composite packaging (plastics material) is such an "inner receptacle" since it is normally not designed to perform a containment function without its "outer packaging" and is not therefore an "inner packaging".

9.3 GENERAL PACKING REQUIREMENTS APPLICABLE TO THE PACKING OF DANGEROUS GOODS OF ALL CLASSES OTHER THAN CLASS 2 AND CLASS 7

9.3.1 Dangerous goods should be packed in packagings of good quality which should be constructed and closed so as to prevent the package as prepared for shipment from any leakage which might be caused under normal conditions of transport by changes in temperature, humidity or pressure (resulting from altitude, for example). No harmful quantity of a dangerous substance should adhere to the outside of packages. These provisions apply both to new and to reused packagings.

9.3.2 Parts of packagings which are in direct contact with dangerous substances should not be affected by chemical or other action of those substances. Where necessary, they should be provided with a suitable inner coating or treatment. Such parts of packagings should not incorporate constitutents liable to react dangerously with the contents so as to form hazardous products, or to weaken them significantly.

9.3.3 Each packaging, except inner packagings of combination packagings, should conform to a design type successfully tested in accordance with the requirements laid down in 9.7.

9.3.4 When filling packagings with liquids sufficient ullage (outage) should be left to ensure that neither leakage nor permanent distortion of the packaging occurs as a result of an expansion of the liquid caused by temperatures likely to occur during transport. Unless specific requirements are prescribed in national or international rules, agreements or recommendations, liquids should not completely fill a packaging at a temperature of 55°C.

9.3.4.1 Packagings for air transport should be able to withstand, without leakage, an internal pressure test designated by the competent authority.

9.3.5 Inner packagings should be packed in an outer packaging in such a way that, under normal conditions of transport, they cannot break, be punctured or leak their contents into the outer packaging. Inner packagings that are liable to break or be punctured easily, such as those made of glass, porcelain or stoneware or of certain plastics materials, etc., should be secured in outer packagings with suitable cushioning material. Any leakage of the contents should not substantially impair the protective properties of the cushioning material or of the outer packaging.

9.3.6 Inner packagings containing different substances which may react dangerously with one another should not be placed in the same outer packaging.

9.3.7 The closures of packagings containing wetted or diluted substances should be such that the percentage of liquid (water, solvent or phlegmatizer) does not fall below the prescribed limits during transport.

9.3.8 Where pressure may develop in a package by the emission of gas from the contents (as a result of temperature increase or other cause), the packaging may be fitted with a vent provided that the gas emitted will not cause danger on account of its toxicity, its inflammability, the quantity released, etc. The vent should be so designed that, when the packaging is in the attitude in which it is intended to be transported, leakages of liquid and the penetration of foreign substances are prevented under normal conditions of transport. Venting of the package should not be permitted for air transport.

9.3.9 New, reused or reconditioned packagings should be capable of passing the tests prescribed in 9.7. Before being filled and handed over for transport, every packaging should be inspected to ensure that it is free from corrosion, contamination or other damage. Any packaging which shows signs of reduced strength as compared with the approved design type should no longer be used or should be so reconditioned that it is able to withstand the design-type tests.

9.3.10 Liquids should be filled only into packagings which have an appropriate resistance to the internal pressure that may be developed under normal conditions of transport. Packagings marked with the hydraulic test pressure prescribed in 9.5.1 (d) should be filled only with a liquid having a vapour pressure:

(a) such that the total gauge pressure in the packaging (i.e. the vapour pressure of the filling substance plus the partial pressure of air of other inert gases, less 100 kPa) at 55°C, determined on the basis of a maximum degree of filling in accordance with 9.3.4 and a filling temperature of 15°C, will not exceed two-thirds of the marked test pressure; or

(b) at 50°C less than four-sevenths of the sum of the marked test pressure plus 100 kPa; or

(c) at 55°C less than two-thirds of the sum of the marked test pressure plus 100 kPa.

9.3.11 An empty packaging that has contained a dangerous substance should be treated in the same manner as is required by these recommendations for a filled packaging until it has been purged of the residue of that dangerous substance.

9.3.12 Every packaging intended to contain liquids should undergo the leakproofness test prescribed in 9.7.4.3 to 9.7.4.5:

(a) before it is first used for transport;

(b) after reconditioning, before it is reused for transport.

This test is not necessary for inner packagings of combination packagings.

9.3.13 The requirements for packagings in 9.6 are based on packagings currently used. In order to take into account progress in science and technology, there is no objection to the use of packagings having specifications different from those in 9.6, provided that they are equally effective, acceptable to the competent authority and able successfully to withstand the tests described in 9.3.12 and 9.7. Methods of testing other than those described in the recommendations are acceptable, provided they are equivalent.

9.4 CODE FOR DESIGNATING TYPES OF PACKAGINGS

9.4.1 The code should consist of:

- an Arabic numeral indicating the type of packaging, e.g. drum, jerrican, etc., followed by
- a capital letter(s) in Latin characters indicating the nature of the material, e.g., steel, wood, etc., followed where necessary by
- an Arabic numeral indicating the category of packaging within the type to which the packaging belongs.

9.4.2 In the case of composite packagings, two capital letters in Latin characters should be used in sequence in the second position of the code. The first should indicate the material of the inner receptacle and the second that of the outer packaging.

9.4.3 In the case of combination packagings, only the code number for the outer packaging should be used.

9.4.4 The letter "W" may follow the packaging code signifying that the packaging, although of the same type indicated by the code, is manufactured to a specification different to that in 9.6 and is considered equivalent under the provisions of 9.3.13.

9.4.5 The following numerals should be used for the types of packaging:

1. Drum
2. Wooden barrel
3. Jerrican
4. Box
5. Bag
6. Composite packaging
7. Pressure receptacle

9.4.6 The following capital letters should be used for the types of material:

A. Steel (all types and surface treatments)
B. Aluminium
C. Natural wood
D. Plywood
F. Reconstituted wood
G. Fibrebroad
H. Plastics material
L. Textile
M. Paper, multiwall
N. Metal (other than steel or aluminium)
P. Glass, porcelain or stoneware

9.4.7 The following types and codes of packaging are assigned:

TYPE	MATERIAL	CATEGORY	CODE	PARAGRAPH
1. Drums	A. Steel	non-removable head	1A1	9.6.1
		removable head	1A2	
	B. Aluminium	non-removable head	1B1	9.6.2
		removable head	1B2	
	D. Plywood		1D	9.6.4
	G. Fibre		1G	9.6.6
	H. Plastics	non-removable head	1H1	9.6.7
		removable head	1H2	
2. Barrels	C. Wooden	bung type	2C1	9.6.5
		removable head	2C2	
3. Jerricans	A. Steel	non-removable head	3A1	9.6.3
		removable head	3A2	

TYPE	MATERIAL	CATEGORY	CODE	PARAGRAPH
	H. Plastics	non-removable head	3H1	9.6.7
		removable head	3H2	
4. Boxes	A. Steel	—	4A1	9.6.13
		with inner liner or coating	4A2	
	B. Aluminium	—	4B1	9.6.13
		with inner liner or coating	4B2	
	C. Natural wood	ordinary	4C1	9.6.8
		with sift-proof walls	4C2	
	D. Plywood	—	4D	9.6.9
	F. Reconstituted wood	—	4F	9.6.10
	G. Fibreboard	—	4G	9.6.11
	H. Plastics	expanded	4H1	9.6.12
		solid	4H2	
5. Bags	H. Woven plastics	without inner lining or coating	5H1	9.6.15
		sift-proof	5H2	
		water resistant	5H3	
	H. Plastics film	—	5H4	9.6.16
	L. Textile	without inner lining or coating	5L1	9.6.14
		sift-proof	5L2	
		water resistant	5L3	9.6.14
	M. Paper	multiwall	5M1	9.6.17
		multiwall, water resistant	5M2	
6. Composite packagings	H. Plastice receptacle	in steel drum	6HA1	9.6.18
		in steel crate or box	6HA2	9.6.18
		in aluminium drum	6HB1	9.6.18
		in aluminium crate or box	6HB2	9.6.18
		in wooden box	6HC	9.6.18
		in plywood drum	6HD1	9.6.18
		in plywood box	6HD2	9.6.18
		in fibre drum	6HG1	9.6.18
		in fibreboard box	6HG2	9.6.18
		in plastics drum	6HH	9.6.18

(cont.)

TYPE	MATERIAL	CATEGORY	CODE	PARAGRAPH
	P. Glass, porcelain or stoneware receptacle	in steel drum	6PA1	9.6.19
		in steel crate or box	6PA2	9.6.19
		in aluminium drum	6PB1	9.6.19
		in aluminium crate or box	6PB2	9.6.19
		in wooden box	6PC	9.6.19
		in plywood drum	6PD1	9.6.19
		in wickerwork hamper	6PD2	9.6.19
		in fibre drum	6PG1	9.6.19
		in fibreboard box	6PG2	9.6.19
		in expanded plastics packaging	6PH1	9.6.19
		in solid plastics packaging	6PH2	9.6.19

9.5 MARKING

Introductory Notes

The marking indicates that the packaging which bears it corresponds to a successfully tested design type and that it complies with the provisions of this chapter which are related to the manufacture, but not to the use, of the packaging. In itself, therefore, the mark does not necessarily confirm that the packaging may be used for any substance: generally the type of packaging (e.g. steel drum), its maximum capacity and/or mass, and any special requirements are specified for each substance in the regulations for each mode of transport.

The marking is intended to be of assistance to packaging manufacturers, reconditioners, packaging users, carriers and regulatory authorities. In relation to the use of a new packaging, the original marking is a means for its manufacturer(s) to identify the type and to indicate those performance test recommendations that have been met.

The marking does not always provide full details of the test levels, etc. and these may need to be taken further into account, e.g. by reference to a test certificate, test reports or register of successfully tested packagings. For example, a packaging having an X or Y marking may be used for substances to which a packing group having a lesser degree of danger has been assigned with the relevant maximum permissible value of the relative density */ determined by taking into account the factor 1.5 or 2.25 indicated in the packaging test requirements in 9.7 as appropriate, i.e. Group I packaging tested for products of relative density 1.2 could be used as a Group II packaging for products of relative density 1.8 or a Group III packaging of relative density 2.7, provided of course that all the performance criteria can still be met with the higher relative density product.

9.5.1 Each packaging intended for use according to these recommendations should bear durable and legible markings showing:

(a) The United Nations packaging symbol

This must not be used for any purpose other than certifying that a packaging complies with the relevant recommendations in this chapter. For embossed metal packagings the capital letters "UN" may be applied as the symbol;

(b) The code number designating the type of packaging according to 9.4.
(c) A code in two parts:
    (i) a latter designating the packing group(s) for which the design type has been successfully tested:

        X for packing groups I, II and III
        Y for packing groups II and III
        Z for packing group III only

*/Relative density (d) is considered to be synonymous with Specific Gravity (SG) and is used throughout this text.

(ii) the relative density, rounded off to the first decimal, for which the design type has been tested for packagings without inner packaging: intended to contain liquids; this may be omitted when the relative density does not exceed 1.2. For packagings intended to contain solids or inner packagings, the maximum gross mass in kilograms;

(d) Either a letter "S" denoting that the packaging is intended for the transport of solids or inner packagings, or, where a hydraulic pressure test has been successfully passed, the test pressure in kPa rounded off to the nearest 10 kPa;

(e) The last two digits of the year during which the packaging was manufactured. Packagings of types 1H and 3H should also be appropriately marked with the month of manufacture; this may be marked on the packaging in a different place from the remainder of the marking. An appropriate method is:

(f) The State authorizing the allocation of the mark, indicated by the distinguishing sign for motor vehicles in international traffic;

(g) The name of the manufacturer or other identification of the packaging specified by the competent authority.

9.5.2 Every reusable packaging liable to undergo a reconditioning process which might obliterate the packaging markings should bear the marks indicated in 9.5.1 (a) to (e), in a permanent form (e.g. embossed) able to withstand the reconditioning process.

9.5.3 Marking should be applied in the sequence of the subparagraphs in 9.5.1; for examples, see 9.5.6. Any additional markings authorized by a competent authority must still enable the parts of the mark to be correctly identified with reference to 9.5.1.

9.5.4 After reconditioning a packaging, the reconditioner should apply to it, in sequence, a durable marking showing:

(h) The State in which the reconditioning was carried out, indicated by the distinguishing sign for motor vehicles in international traffic;

(i) The name or authorized symbol of the reconditioner;

(j) The year of reconditioning; the letter "R"; and, for every packaging successfully passing the leak-proofness test in 9.3.12, the additional letter "L."

9.5.5. The marking referred to in 9.5.4 should be applied near the marking referred to in 9.5.1, and may replace that of 9.5.1 (f) and (g) or be in addition to that marking.

9.5.6 Examples of markings for NEW packagings:

⊕	4G/Y145/S/83 NL/VL823	as in 9.5.1 (a), (b), (c), (d) and (e) as in 9.5.1 (f) and (g)	For a new fibreboard box
⊕	1A1/Y1.4/150/83 NL/VL824	as in 9.5.1 (a), (b), (c), (d) and (e) as in 9.5.1 (f) and (g)	For a new steel drum to contain liquids
⊕	1A2/Y150/S/83 NL/VL825	as in 9.5.1 (a), (b), (c), (d) and (e) as in 9.5.1 (f) and (g)	For a new steel drum to contain solids, or inner packagings
⊕	4HW/Y136/S/83 NL/VL826	as in 9.5.1 (a), (b), (c), (d) and (e) as in 9.5.1 (f) and (g)	For a new plastics box of equivalent specification.

9.5.7 Examples of Markings for RECONDITIONED Packagings

(symbol) 1A1/Y1.4/150/83   as in 9.5.1 (a), (b), (c), (d) and (e)
NL/RB/85 RL   as in 9.5.4 (h), (i) and (j)

(symbol) 1A1/Y1.4/150/83   as in 9.5.1 (a), (b), (c), (d) and (e)
NL/VL824   as in 9.5.1 (f) and (g)
NL/RB/85 RL   as in 9.5.4 (h), (i) and (j)

(symbol) 1A2/Y150/S/83   as in 9.5.1 (a), (b), (c), (d) and (e)
USA/RB/85 R   as in 9.5.4 (h), (i) and (j).

9.6 REQUIREMENTS FOR PACKAGINGS

9.6.1 *Steel drums:* 1A1 non-removable head; 1A2 removable head

9.6.1.1 Body and heads should be constructed of steel sheet of suitable type and adequate thickness in relation to the capacity of the drum and the intended use.

9.6.1.2 Body seams should be welded on drums intended to contain more than 4 litres of liquid. Body seams should be mechanically seamed or welded on drums intended to contain solids or 40 litres or less of liquids.

9.6.1.3 Chimes should be mechanically seamed or welded. Separate reinforcing rings may be applied.

9.6.1.4 The body of a drum of a capacity greater than 60 litres should, in general, have at least two expanded rolling hoops or, alternatively, at least two separate rolling hoops. If there are separate rolling hoops they should be fitted tightly on the body and so secured that they cannot shift. Rolling hoops should not be spot welded.

9.6.1.5 Openings for filling, emptying and venting in the bodies or heads of non-removable head (1A1) drums should not exceed 7 cm in diameter. Drums with larger openings are considered to be of the removable head type (1A2). Closures for openings in the bodies and heads of drums should be so designed and applied that they will remain secure and leakproof under normal conditions of transport. Closure flanges may be mechanically seamed or welded in place. Gaskets or other sealing elements should be used with closures, unless the closure is inherently leakproof.

9.6.1.6 Closure devices for removable head drums should be so designed and applied that they will remain secure and drums will remain leakproof under normal conditions of transport. Gaskets or other sealing elements should be used with all removable heads.

9.6.1.7 If materials used for body, heads, closures and fittings are not in themselves compatible with the contents to be transported, suitable internal protective coatings or treatments should be applied. These coatings or treatments should retain their protective properties under normal conditions of transport.

9.6.1.8 Maximum capacity of drum: 450 litres

9.6.1.9 Maximum net mass: 400 kg.

9.6.2 *Aluminium drums:* 1B1 non-removable head; 1B2 removable head

9.6.2.1 Body and heads should be constructed of aluminium at least 99% pure or of an aluminium base alloy. Material should be of a suitable type and of adequate thickness in relation to the capacity of the drum and the intended use.

9.6.2.2 All seams should be welded. Chime seams, if any, should be reinforced by the application of separate reinforcing rings.

9.6.2.3 The body of a drum of a capacity greater than 60 litres should, in general, have at least two expanded rolling hoops or, alternatively, at least two separate rolling hoops. If there are separate rolling hoops they should be fitted tightly on the body and so secured that they cannot shift. Rolling hoops should not be spot welded.

9.6.2.4 Openings for filling, emptying and venting in the bodies or heads of non-removable head (1B1) drums should not exceed 7 cm in diameter. Drums with larger openings are considered to be of the removable head type (1B2). Closures for openings in the bodies and heads of drums should be so designed and applied that they will remain secure and leakproof under normal conditions of transport. Closure flanges should be welded in place so that the weld provides a leakproof seam. Gaskets or other sealing elements should be used with closures, unless the closure is inherently leakproof.

9.6.2.5 Closure devices for removable head drums should be so designed and applied that they will remain secure and drums will remain leakproof under normal conditions of transport. Gaskets or other sealing elements should be used with all removable heads.

9.6.2.6 Maximum capacity of drum: 450 litres

9.6.2.7 Maximum net mass: 400 kg.

9.6.3 *Steel jerricans:* 3A1 non-removable head; 3A2 removable head

9.6.3.1 Body and heads should be constructed of steel sheet of a suitable type and of adequate thickness in relation to the capacity of the jerrican and intended use.

9.6.3.2 Chimes of all jerricans should be mechanically seamed or welded. Body seams of jerricans intended to contain more than 40 litres of liquid should be welded. Body seams of jerricans intended to carry 40 litres or less should be mechanically seamed or welded.

9.6.3.3 Openings in jerricans (3A1) should not exceed 7 cm in diameter. Jerricans with larger openings are considered to be of the removable head type (3A2). Closures should be so designed that they will remain secure and leakproof under normal conditions of transport. Gaskets or other sealing elements should be used with closures, unless the closure is inherently leakproof.

9.6.3.4 If materials used for body, heads, closures and fittings are not in themselves compatible with the contents to be transported, suitable internal protective coatings or treatments should be applied. These coatings or treatments should retain their protective properties under normal conditions of transport.

9.6.3.5 Maximum capacity of jerrican: 60 litres

9.6.3.6 Maximum net mass: 120 kg.

9.6.4 *Plywood drums:* 1D

9.6.4.1 The wood used should be well seasoned, commercially dry and free from any defect likely to lessen the effectiveness of the drum for the purpose intended. If a material other than plywood is used for the manufacture of the heads, it should be of a quality equivalent to the plywood.

9.6.4.2 At least two-ply plywood should be used for the body and at least three-ply plywood for the heads; the plies should be firmly glued together by a water resistant adhesive with their grain crosswise.

9.6.4.3 The body and heads of the drum and their joins should be of a design appropriate to the capacity of the drum and intended use.

9.6.4.4 In order to prevent sifting of the contents, lids should be lined with kraft paper or some other equivalent material which should be securely fastened to the lid and extend to the outside along its full circumference.

9.6.4.5 Maximum capacity of drum: 250 litres

9.6.4.6 Maximum net mass: 400 kg.

9.6.5 *Wooden barrels:* 2C1 bung type; 2C2 removable head

9.6.5.1 The wood used should be of good quality, straight grained, well seasoned and free from knots, bark, rotten wood, sapwood or other defects likely to lessen the effectiveness of the barrel for the purpose intended.

9.6.5.2 The body and heads should be of a design appropriate to the capacity of the barrel and intended use.

9.6.5.3 Staves and heads should be sawn or cleft with the grain so that no annual ring extends over more than half the thickness of a stave or head.

9.6.5.4 Barrel hoops should be of steel or iron of good quality. The hoops of 2C2 barrels may be of a suitable hardwood.

9.6.5.5 Wooden barrels 2C1: the diameter of the bunghole should not exceed half the width of the stave in which it is placed.

9.6.5.6 Wooden barrels 2C2: heads should fit tightly into the crozes.

9.6.5.7 Maximum capacity of barrel: 250 litres

9.6.5.8 Maximum net mass: 400 kg.

9.6.6 *Fibre drums:* 1G

9.6.6.1 The body of the drum should consist of multiple plies of heavy paper or fibreboard (without corrugations) firmly glued or laminated together and may include one or more protective layers of bitumen, waxed kraft paper, metal foil, plastics material, etc.

9.6.6.2 Heads should be of natural wood, fibreboard, metal, plywood or plastics material and may include one or more protective layers of bitumen, waxed kraft paper, metal foil, plastics material, etc.

9.6.6.3 The body and heads of the drum and their joins should be of a design appropriate to the capacity of the drum and intended use.

9.6.6.4 The assembled packaging should be sufficiently water resistant so as not to delaminate under normal conditions of transport.

9.6.6.5 Maximum capacity of drum: 450 litres

9.6.6.6 Maximum net mass: 400 kg.

9.6.7 *Plastics drums and jerricans*

1H1	drums, non-removable head	3H1	jerricans, non-removable head
1H2	drums, removable head	3H2	jerricans, removable head

9.6.7.1 The packaging should be manufactured from suitable plastics material and be of adequate strength in relation to its capacity and intended use. No used material other than production residues or regrind from the same manufacturing process may be used. The packaging should be adequately resistant to ageing and to degradation caused either by the substance contained or by ultra-violet radiation. Any permeation of the substance contained should not constitute a danger under normal conditions of transport.

9.6.7.2 Unless otherwise approved by the competent authority, the period of use permitted for the transport of dangerous substances should be five years from the date of manufacture of the packaging except where a shorter period of use is prescribed because of the nature of the substance to be transported.

9.6.7.3 If protection against ultra-violet radiation is required, it should be provided by the addition of carbon black or other suitable pigments or inhibitors. These additives should be compatible with the contents and remain effective throughout the life of the packaging. Where use is made of carbon black, pigments or inhibitors other than those used in the manufacture of the tested design type, retesting may be waived if the carbon black content does not exceed 2% by mass or if the pigment content does not exceed 3% by mass; the content of inhibitors of ultra-violet radiation is not limited.

9.6.7.4 Additives serving purposes other than protection against ultra-violet radiation may be included in the composition of the plastics material provided that they do not adversely affect the chemical and physical properties of the material of the packaging. In such circumstances, retesting may be waived.

9.6.7.5 The wall thickness at every point of the packaging should be appropriate to its capacity and intended use, taking into account the stresses to which each point is liable to be exposed.

9.6.7.6 Openings for filling, emptying and venting in the bodies or heads of non-removable head drums (1H1) and jerricans (3H1) should not exceed 7 cm in diameter. Drums and jerricans with larger openings are considered to be of the removable head type (1H2 and 3H2). Closures for openings in the bodies or heads of drums and jerricans should be so designed and applied that they will remain secure and leakproof under normal conditions of transport. Gaskets or other sealing elements should be used with closures unless the closure is inherently leakproof.

9.6.7.7 Closure devices for removable head drums and jerricans should be so designed and applied that they will remain secure and leakproof under normal conditions of transport. Gaskets should be used with all removable heads unless the drum or jerrican design is such that, where the removable head is properly secured, the drum or jerrican is inherently leakproof.

9.6.7.8 Maximum capacity of drums and jerricans: 1H1, 1H2: 450 litres
3H1, 3H2: 60 litres.

9.6.7.9 Maximum net mass: 1H1, 1H2: 400 kg
3H1, 3H2: 120 kg.

9.6.8 *Boxes of natural wood:* 4C1 ordinary; 4C2 with sift-proof walls

9.6.8.1 The wood used should be well seasoned, commercially dry and free from defects that would materially lessen the strength of any part of the box. The strength of the material used and the method of construction should be appropriate to the capacity and intended use of the box. The tops and bottoms may be made of water resistant reconstituted wood such as hardboard, particle board or other suitable type.

9.6.8.2 Box 4C2: each part should consist of one piece or be equivalent thereto. Parts are considered equivalent to one piece when one of the following methods of glued assembly is used: Lindermann joint, tongue and groove joint, ship lap or rabbet joint or butt joint with at least two corrugated metal fasteners at each joint.

9.6.8.3 Maximum net mass: 400 kg.

9.6.9 *Plywood boxes:* 4D

9.6.9.1 Plywood used should be at least 3-ply. It should be made from well seasoned rotary cut, sliced or sawn veneer, commercially dry and free from defects that would materially lessen the strength of the box. The strength of the material used and the method of construction should be appropriate to the capacity and intended use of the box. All adjacent plies should be glued with water resistant adhesive. Other suitable materials may be used together with plywood in the construction of boxes. Boxes should be firmly nailed or secured to corner posts or ends or be assembled by equally suitable devices.

9.6.9.2 Maximum net mass: 400 kg.

9.6.10 *Reconstituted wood boxes:* 4F

9.6.10.1 The walls of boxes should be made of water resistant reconstituted wood such as hardboard, particle board or other suitable type. The strength of the material used and the method of construction should be appropriate to the capacity of the boxes and their intended use.

9.6.10.2 Other parts of the boxes may be made of other suitable material.

9.6.10.3 Boxes should be securely assembled by means of suitable devices.

9.6.10.4 Maximum net mass: 400 kg.

9.6.11 *Fibreboard boxes:* 4G

9.6.11.1 Strong and good quality solid or double-faced corrugated fibreboard (single or multiwall) should be used, appropriate to the capacity of the box and to its intended use. The water resistance of the outer surface shall be such that the increase in mass, as determined in a test carried out over a priod of 30 minutes by the Cobb method of determining water absorption, is not greater than 155 g/m^2—see ISO International Standard 535—1976 (E). It should have proper bending qualities. Fibreboard should be cut, creased without scoring, and slotted so as to permit assembly without cracking, surface breaks or undue bending. The fluting of corrugated fibreboard should be firmly glued to the facings.

9.6.11.2 The ends of boxes may have a wooden frame or be entirely of wood. Reinforcements of wooden battens may be used.

9.6.11.3 Manufacturing joins in the body of boxes should be taped, lapped and glued or lapped and stitched with metal staples. Lapped joins should have an appropriate overlap. Where closing is effected by gluing or taping a water resistant adhesive should be used.

9.6.11.4 Boxes should be designed so as to provide a good fit to the contents.

9.6.11.5 Maximum net mass: 400 kg.

9.6.12 *Plastics boxes:*

4H1 expanded plastics boxes
4H2 solid plastics boxes

9.6.12.1 The box should be manufactured from suitable plastics material and be of adequate strength in relation to its capacity and intended use. The box should be adequately resistant to ageing and to degradation caused either by the substance contained or by ultra-violet radiation.

9.6.12.2 An expanded plastics box should comprise two parts made of a moulded expanded plastics material, a bottom section containing cavities for the inner packagings and a top section covering and interlocking with the bottom section. The top and bottom sections should be designed so that the inner packagings fit snugly. The closure cap for any inner packaging should not be in contact with the inside of the top section of this box.

9.6.12.3 For dispatch, an expanded box should be closed with a self-adhesive tape having sufficient tensile strength to prevent the box from opening. The adhesive tape should be weather resistant and its adhesive compatible with the expanded plastics material of the box. Other closing devices at least equally effective may be used.

9.6.12.4 For solid plastics boxes, protection against ultra-violet radiation, if required, should be provided by the addition of carbon black or other suitable pigments or inhibitors. These additives should be compatible with the contents and remain effective throughout the life of the box. Where use is made of carbon black, pigments or inhibitors other than those used in the manufacture of the tested design type, retesting may be waived if the carbon black content does not exceed 2% by mass or if the pigment content does not exceed 3% by mass; the content of inhibitors of ultra-violet radiation is not limited.

9.6.12.5 Additives serving purposes other than protection against ultra-violet radiation may be included in the composition of the plastics material provided that they do not adversely affect the chemical or physical properties of the material of the box. In such circumstances, retesting may be waived.

9.6.12.6 Solid plastics boxes should have closure devices made of a suitable material of adequate strength and so designed as to prevent the box from unintentional opening.

9.6.12.7 Maximum net mass 4H1: 60 kg; 4H2: 400 kg

9.6.13 *Steel or aluminium boxes*

4A1 steel	4B1 aluminium
4A2 steel, with inner liner or coating	4B2 aluminium, with inner liner or coating

9.6.13.1 The strength of the metal and the construction of the box should be appropriate to the capacity of the box and to its intended use.

9.6.13.2 Boxes 4A2 and 4B2: should be lined with fibreboard or felt packing pieces as required or should have an inner liner or coating of suitable material. If a double seamed metal liner is used, steps should be taken to prevent the ingress of substances, particularly explosives, into the recesses of the seams.

9.6.13.3 Closures may be of any suitable type, they should remain secured under normal conditions of transport.

9.6.13.4 Maximum net mass: 400 kg.

9.6.14 *Textile bags:*

5L1 without inner liner or coating
5L2 sift-proof
5L3 water resistant

9.6.14.1 The textiles used should be of good quality. The strength of the fabric and the construction of the bag should be appropriate to the capacity of the bag and intended use.

9.6.14.2 Bags, sift-proof, 5L2: the bag should be made sift-proof, for example by the use of:

- paper bonded to the inner surface of the bag by a water resistant adhesive such as bitumen; or
- plastics film bonded to the inner surface of the bag; or
- one or more inner liners made of paper or plastics material.

9.6.14.3 Bags, water resistant, 5L3: to prevent the entry of moisture the bag should be made waterproof, for example by the use of:

- separate inner liners of water resistant paper (e.g. waxed kraft paper, tarred paper or plastics-coated kraft paper); or
- plastics film bonded to the inner surface of the bag; or
- one or more inner liners made of plastics material.

9.6.14.4 Maximum net mass: 50 kg

9.6.15 *Woven plastics bags:*
5H1 without inner liner or coating
5H2 sift-proof
5H3 water resistant

9.6.15.1 Bags should be made from stretched tapes or monofilaments of a suitable plastics material. The strength of the material used and the construction of the bag should be appropriate to the capacity of the bag and intended use.

9.6.15.2 If the fabric is woven flat, the bags should be made by sewing or some other method ensuring closure of the bottom and one side. If the fabric is tubular, the bag should be closed by sewing, weaving or some other equally strong method of closure.

9.6.15.3 Bags, sift-proof, 5H2: the bag should be made sift-proof, for example by means of: paper or a plastics film bonded to the inner surface of the bag; or one or more separate inner liners made of paper or plastics material.

9.6.15.4 Bags, water resistant, 5H3: to prevent the entry of moisture, the bag should be made waterproof, for example by means of:

- separate inner liners of water resistant paper (e.g. waxed kraft paper, double-tarred kraft paper or plastics-coated kraft paper); or
- plastics film bonded to the inner or outer surface of the bag; or
- one or more inner plastics liners.

9.6.15.5 Maximum net mass: 50 kg.

9.6.16 *Plastics film bags:* 5H4

9.6.16.1 Bags should be made of a suitable plastics material. The strength of the material used and the construction of the bag should be appropriate to the capacity of the bag and the intended use. Joins and closures should withstand pressures and impacts liable to occur under normal conditions of transport.

9.6.16.2 Maximum net mass: 50 kg.

9.6.17 *Paper bags:* 5M1 multiwall; 5M2 multiwall, water resistant

9.6.17.1 Bags should be made of a suitable kraft paper or of an equivalent paper with at least three plies. The strength of the paper and the construction of the bags should be appropriate to the capacity of the bag and intended use. Joins and closures should be sift-proof.

9.6.17.2 Bags 5M2: to prevent the entry of moisture, a bag of four plies or more should be made waterproof by the use of either a water resistant ply as one of the two outermost plies or a water resistant barrier made of a suitable protective material between the two outermost plies; a bag of three plies should be made waterproof by the use of a water resistant ply as the outermost ply. Where there is a danger of the contained substance reacting with moisture or where it is packed damp, a water resistant ply or barrier should also be placed next to the substance. Joins and closures should be waterproof.

9.6.17.3 Maximum net mass: 50 kg.

9.6.18 *Composite packagings* (plastic material):

6HA1 plastics receptacle with outer steel drum

6HA2 plastics receptacle with outer steel crate or box

6HB1 plastics receptacle with outer aluminium drum

6HB2 plastics receptacle with outer aluminium crate or box

6HC plastics receptacle with outer wooden box

6HD1 plastics receptacle with outer plywood drum

6HD2 plastics receptacle with outer plywood box

6HG1 plastics receptacle with outer fibre drum

6HG2 plastics receptacle with outer fibreboard box

6HH plastics receptacle with outer plastics drum

9.6.18.1 Inner receptacle

9.6.18.1.1 The provisions of 9.6.7.1 and 9.6.7.4 to 9.6.7.7 should apply to inner plastics receptacles.

9.6.18.1.2 The inner plastics receptacle should fit snugly inside the outer packaging, which should be free of any projection that might abrade the plastics material.

9.6.18.1.3 Maximum capacity of inner receptacle:

6HA1, 6HB1, 6HD1, 6HG1, 6HH: 250 litres

6HA2, 6HB2, 6HC, 6HD2, 6HG2: 60 litres.

9.6.18.1.4 Maximum net mass:

6HA1, 6HB1, 6HD1, 6HG1, 6HH: 400 kg

6HA2, 6HB2, 6HC, 6HD2, 6HG2: 75 kg.

9.6.18.2 Outer packaging

9.6.18.2.1 Plastics receptacle with outer steel or aluminium drum 6HA1 or 6HB1; the relevant provisions of 9.6.1 or 9.6.2, as appropriate, should apply to the construction of the outer packaging.

9.6.18.2.2 Plastics receptacle with outer steel or aluminium, crate or box 6HA2 or 6HB2; the relevant provisions of 9.6.13 should apply to the construction of the outer packaging.

9.6.18.2.3 Plastics receptacle with outer wooden box 6HC; the relevant provisions of 9.6.8 should apply to the construction of the outer packaging.

9.6.18.2.4 Plastics receptacle with outer plywood drum 6HD1; the relevant provisions of 9.6.4 should apply to the construction of the outer packaging.

9.6.18.2.5 Plastics receptacle with outer plywood box 6HD2; the relevant provisions of 9.6.9 should apply to the construction of the outer packaging.

9.6.18.2.6 Plastics receptacle with outer fibre drum 6HG1; the provisions of 9.6.6.1 to 9.6.6.4 should apply to the construction of the outer packaging.

9.6.18.2.7 Plastics receptacle with outer fibreboard box 6HG2; the relevant provisions of 9.6.11 should apply to the construction of the outer packaging.

9.6.18.2.8 Plastics receptacle with outer plastics drum 6HH; the provisions of 9.6.7.1 and 9.6.7.3 to 9.6.7.7 should apply to the construction of the outer packaging.

9.6.19 *Composite packagings* (glass, procelain or stoneware):

6PA1 receptacle with outer steel drum

6PA2 receptacle with outer steel crate or box

6PB1 receptacle with outer aluminium drum

6PB2 receptacle with outer aluminium crate or box

6PC receptacle with outer wooden box

6PD1 receptacle with outer plywood drum

6PD2 receptacle with outer wickerwork hamper

6PG1 receptacle with outer fibre drum

6PG2 receptacle with outer fibreboard box

6PH1 receptacle with outer expanded plastics packaging

6PH2 receptacle with outer solid plastics packaging

9.6.19.1 Inner receptacle

9.6.19.1.1 Receptacles should be of a suitable form (cylindrical or pear-shaped) and be made of good quality material free from any defect that could impair their strength. The walls should be sufficiently thick at every point.

9.6.19.1.2 Screw-threaded plastics closures, ground glass stoppers or closures at least equally effective should be used as closures for receptacles. Any part of the closure likely to come into contact with the contents of the receptacle should be resistant to those contents. Care should be taken to ensure that the closures are so fitted as to be leakproof and are suitably secured to prevent any loosening during transport. If vented closures are necessary, they should comply with 9.3.8.

9.6.19.1.3 The receptacle should be firmly secured in the outer packaging by means of cushioning and/or absorbent materials.

9.6.19.1.4 Maximum capacity of receptacle: 60 litres

9.6.19.1.5 Maximum net mass: 75 kg.

9.6.19.2 Outer packaging

9.6.19.2.1 Receptacle with outer steel drum 6PA1; the relevant provisions of 9.6.1 should apply to the construction of the outer packaging. The removable lid required for this type of packaging may nevertheless be in the form of a cap.

9.6.19.2.2 Receptacle with outer steel crate or box 6PA2; the relevant provisions of 9.6.13 should apply to the construction of the outer packaging. For cylindrical receptacles the outer packaging should, when upright, rise above the receptacle and its closure. If the crate surrounds a pearshaped receptacle and is of matching shape, the outer packaging should be fitted with a protective cover (cap).

9.6.19.2.3 Receptacle with outer aluminium drum 6PB1; the relevant provisions of 9.6.2 should apply to the construction of the outer packaging.

9.6.19.2.4 Receptacle with outer aluminium crate or box 6PB2; the relevant provisions of 9.6.13 should apply to the construction of the outer packaging.

9.6.19.2.5 Receptacle with outer wooden box 6PC; the relevant provisions of 9.6.8 should apply to the construction of the outer packaging.

9.6.19.2.6 Receptacle with outer plywood drum 6PD1; the relevant provisions of 9.6.4 should apply to the construction of the outer packaging.

9.6.19.2.7 Receptacle with outer wickerwork hamper 6PD2. The wickerwork hamper should be properly made with material of good quality. It should be fitted with a protective cover (cap) so as to prevent damage to the receptacle.

9.6.19.2.8 Receptacle with outer fibre drum 6PG1; the relevant provisions of 9.6.6.1 to 9.6.6.4 should apply to the body of the outer packaging.

9.6.19.2.9 Receptacle with outer fibreboard box 6PG2; the relevant provisions of 9.6.11 should apply to the construction of the outer packaging.

9.6.19.2.10 Receptacle with outer expanded plastics or solid plastics packaging (6PH1 or 6PH2); the materials of both outer packagings should meet the relevant provisions of 9.6.12. Solid plastics packaging should be manufactured from high density polyethylene or some other comparable plastics material. The removable lid for this type for packaging may nevertheless be in the form of a cap.

9.7 TEST REQUIREMENTS FOR PACKAGINGS

9.7.1 *Performance and frequency of tests*

9.7.1.1 The design type of each packaging should be tested as provided in 9.7 in accordance with procedures established by the competent authority.

9.7.1.2 Tests should be successfully performed on each packaging design type before such packaging is used. A packaging design type is defined by the design, size, material and thickness, manner of construction and packing, but may include various surface treatments. It also includes packagings which differ from the design type only in their lesser design height.

9.7.1.3 Tests should be repeated on production samples at intervals established by the competent authority. For such tests on paper or fibreboard packagings, preparation at ambient conditions is considered equivalent to the provisions of 9.7.2.3.

9.7.1.4 Tests should also be repeated after each modification which alters the design, material or manner of construction of a packaging.

9.7.1.5 The competent authority may permit the selective testing of packagings that differ only in minor respects from a tested type, e.g. smaller sizes of inner packagings or inner packagings of lower net mass; and packagings such as drums, bags and boxes which are produced with small reductions in external dimension(s).

9.7.1.6 Where an outer packaging or a combination packaging has been successfully tested with different types of inner packagings, a variety of such different inner packagings may also be assembled in this outer packaging.

9.7.1.7 The competent authority may at any time require proof, by tests in accordance with this section, that serially-produced packagings meet the requirements of the design type tests.

9.7.1.8. If an inner treatment or coating is required for safety reasons, it should retain its protective properties even after the tests.

9.7.2 *Preparation of packagings for testing*

9.7.2.1 Tests should be carried out on packagings prepared as for transport including inner packagings of combination packagings. Inner or single receptacles or packagings should be filled to not less than 95% of their capacity for solids or 98% for liquids. The substances to be transported in the packaging may be replaced by other substances except where this would invalidate the results of the tests. For solids, when another substance is used it should have the same physical characteristics (mass, grain size, etc.) as the substance to be carried. It is permissible to use additives, such as bags of lead shot, to achieve the requisite total package mass, so long as they are placed so that the test results are not affected.

9.7.2.2 In the drop tests for liquids, when another substance is used, it should be of similar relative density and viscosity to those of the substance being transported. Water may also be used for the liquid drop test under the conditions in 9.7.3.4.

9.7.2.3 Paper or fibreboard packagings should be conditioned for at least 24 hours in an atmosphere having a controlled temperature and relative humidity (r.h.). There are three options, one of which should be chosen. The preferred atmosphere is 23° ± 2°C and 50% ± 2% r.h. The two other options are 20° ± 2°C and 65% ± 2% r.h. or 27° ± 2°C and 65% ± 2% r.h.

9.7.2.4 Bung-type barrels made of natural wood should be left filled with water for at least 24 hours before the tests.

9.7.2.5 Steps should be taken to ascertain that the plastics material used in the manufacture of plastics drums, plastics jerricans and composite packagings (plastics material) complies with the provisions in 9.3.2, 9.6.7.1 and 9.6.7.4. This may be done, for example, by submitting sample receptacles or packagings to a preliminary test extending over a long period, for example six months, during which the samples would remain filled with the substances they are intended to contain, and after which the samples should be submitted to the applicable tests listed in 9.7.3, 9.7.4, 9.7.5 and 9.7.6. For substances which may cause stress-cracking or weakening in plastics drums or jerricans, the sample, filled with the substance or another substance that is known to have at least as severe a stress-cracking influence on the plastic materials in question, should be subjected to a superimposed load equivalent to the total mass of identical packages which might be stacked on it during transport. The minimum height of the stack, including the test sample which should be considered, is 3 metres.

PACKAGING	NO. OF TEST SAMPLES	DROP ORIENTATION
Steel drums Aluminium drums Steel jerricans Plywood drums Wooden barrels Fibre drums Plastics drums and jerricans Composite packagings which are in the shape of a drum	Six (three for each drop)	First drop (using three samples): the packaging should strike the target diagonally on the chime or, if the packaging has no chime, on a circumferential seam or an edge. Second drop (using the other three samples): the packaging should strike the target on the weakest part not tested by the first drop, for example a closure or, for some cylindrical drums, the welded longitudinal seam of the drum body
Boxes of natural wood Plywoood boxes Reconstituted wood boxes Fibreboard boxes Plastics boxes	Five (one for each drop)	First drop: flat on the bottom Second drop: flat on the top Third drop: flat on the long side Fourth drop: flat on the short side
Steel or aluminium boxes Composite packagings which are in the shape of a box		Fifth drop: on a corner
Bags—single-ply with a side seam	Three (three drops per bag)	First drop: flat on a wide face Second drop: flat on a narrow face Third drop: on an end of the bag
Bags—single-ply without a side seam, or multi-ply	Three (two drops per bag)	First drop: flat on a wide face Second drop: on an end of the bag

9.7.3 *Drop test*

9.7.3.1 Number of test samples (per design type and manufacturer) and drop orientation. For other than flat drops the centre of gravity should be vertically over the point of impact.

9.7.3.2 Special preparation of test samples for the drop test:

Testing of plastics drums, jerricans and boxes (see 9.6.7 and 9.6.12), of composite packagings (plastics material) (see 9.6.18) and of combination packagings with inner plastics packagings—with the exception of bags and expanded polystyrene boxes—should be carried out when the temperature of the test sample and its contents has been reduced to −18°C or lower; where test samples are prepared in this way the conditioning specified in 9.7.2.3 may be waived. Test liquids should be kept in the liquid state, if necessary by the addition of anti-freeze.

9.7.3.3 Target. The target should be a rigid, non-resilient, flat and horizontal surface.

9.7.3.4 Drop height. For solids and liquids, if the test is performed with the solid or liquid to be carried or with another substance having essentially the same physical characteristics:

Packing Group 1	Packing Group II	Packing Group III
1.8 m	1.2 m	0.8 m

For liquids if the test is performed with water:

(a) where the substances to be transported have a relative density not exceeding 1.2

Packing Group 1	Packing Group II	Packing Group III
1.8 m	1.2 m	0.8 m

(b) where the substances to be transported have a relative density exceeding 1.2, the drop height should be calculated on the basis of the relative density (d) of the substance to be carried, rounded up to the first decimal, as follows:

Packing Group 1	Packing Group II	Packing Group III
d × 1.5 (m)	d × 1.0 (m)	d × 0.67 (m)

9.7.3.5 Criteria for passing the test:

9.7.3.5.1 Each packaging containing liquid should be leakproof when equilibrium has been reached between the internal and external pressures, except for inner packagings of combination packagings when it is not necessary that the pressures be equalized.

9.7.3.5.2 Where a packaging for solids undergoes a drop test and its upper face strikes the target, the test sample passes the test if the entire contents are retained by an inner packaging or inner receptacle (e.g. a plastics bag), even if the closure is no longer sift-proof.

9.7.3.5.3 The packaging or outer packaging of a composite or combination packaging should not exhibit any damage liable to affect safety during transport. There should be no leakage of the filling substance from the inner receptacle or inner packaging(s).

9.7.3.5.4 Neither the outermost ply of a bag nor an outer packaging should exhibit any damage liable to affect safety during transport.

9.7.3.5.5 A slight discharge from the closure(s) upon impact should not be considered to be a failure of the packaging provided that no further leakage occurs.

9.7.3.5.6 No rupture is permitted in packagings for explosives.

9.7.4 *Leakproofness test*

The leakproofness test should be performed on all design types of packagings intended to contain liquids; however, this test is not required for the inner packagings of combination packagings.

9.7.4.1 Number of test samples: three test samples per design type and manufacturer.

9.7.4.2 Special preparation of test samples for the test: either vented closures should be replaced by similar non-vented closures or the vent should be sealed.

9.7.4.3 Test method and pressure to be applied: for design type tests the packagings including their closures should be restrained under water while an internal air pressure is applied; the method of restraint should not affect the results of the test. Other methods at least equally effective may be used. The air pressure (gauge) to be applied should be:

Packing Group 1	Packing Group II	Packing Group III
Not less than 30 kPa (0.3 bar)	Not less than 20 kPa (0.2 bar)	Not less than 20 kPa (0.2 bar)

9.7.4.4 For the leakproofness test specified in 9.3.12 the packagings need not have their own closures fitted. Each packaging should be tested as specified in 9.7.4.3.

9.7.4.5 Criterion for passing the test: there should be no leakage.

9.7.5 *Internal pressure (hydraulic) test*

9.7.5.1 Packaging to be tested: the internal pressure (hydraulic) test should be carried out on all metal, plastics and composite packagings intended to contain liquids. Except for air transport, this test is not required for inner packagings of combination packagings.

9.7.5.2 Number of test samples: three test samples per design type and manufacturer.

9.7.5.3 Special preparation of packagings for testing: either vented closures should be replaced by similar non-vented closures or the vent should be sealed.

9.7.5.4 Test method and pressure to be applied: metal packagings and composite packagings (glass, porcelain or stoneware) including their closures should be subjected to the test pressure for 5 minutes. Plastics packagings and composite packagings (plastics material) including their closures should be subjected to the test pressure for 30 minutes. This pressure is the one to be included in the marking required by 9.5.1 (d). The manner in which the packagings are supported should not invalidate the test. The test pressure should be applied continuously and evenly; it should be kept constant throughout the test period. The hydraulic pressure (gauge) applied, as determined by any one of the following methods, should be:

(a) not less than the total gauge pressure measured in the packaging (i.e. the vapour pressure of the filling substance and the partial pressure of the air or other inert gases, minus 100 kPa) at 55°C, multiplied by a safety factor of 1.5; this total gauge pressure should be determined on the basis of a maximum degree of filling in accordance with 9.3.4 and a filling temperature of 15°C;

(b) not less than 1.75 times the vapour pressure at 50°C of the substance to be transported, minus 100 kPa but with a minimum test pressure of 100 kPa;

(c) not less than 1.5 times the vapour pressure at 55°C of the substance to be transported, minus 100 kPa but with a minimum test pressure of 100 kPa.

9.7.5.5 In addition, packagings intended to contain substances of Packing Group I should be tested to a minimum test pressure of 250 kPa (gauge) for a test period of 5 or 30 minutes depending upon the material of construction of the packaging.

9.7.5.6 The special requirements for air transport, including minimum test pressures, may not be covered in 9.7.5.4.

9.7.5.7 Criterion for passing the test: no packaging should leak.

9.7.6 *Stacking test:*

All packagings other than bags should be subjected to a stacking test.

9.7.6.1 Number of test samples: three test samples per design type and manufacturer.

9.7.6.2 Test method: the test sample should be subjected to a force applied to the top surface of the test sample equivalent to the total weight of identical packages which might be stacked on it during transport; where the contents of the test sample are non-dangerous liquids with relative density different from that of the liquid to be transported, the force should be calculated in relation to the latter. The minimum height of the stack including the test sample should be 3 metres. The duration of the test should be 24 hours except that plastics drums, jerricans, and composite packagings 6HH intended for liquids should be subjected to the stacking test for a period of 28 days at a temperature of not less than 40°C.

9.7.6.3 Criterion for passing the test: no test sample should leak. In composite packagings or combination packagings, there should be no leakage of the filling substance from the inner receptacle or inner packaging. No test sample should show any deterioration which could adversely affect transport safety or any distortion liable to reduce its strength or cause instability in stacks of packages. In instances (such as guided load tests of drums and jerricans) where stacking stability is assessed after completion of the test, this may be considered sufficient when two filled packagings of the same type placed on each test sample maintain their position for one hour. Plastics packagings should be cooled to ambient temperature before the assessment.

9.7.7 *Cooperage test for bung type wooden barrels*

9.7.7.1 Number of samples: one barrel.

9.7.7.2 Method of testing: remove all hoops above the bilge of an empty barrel at least two days old.

9.7.7.3 Criterion for passing the test: the diameter of the cross-section of the upper part of the barrel should not increase by more than 10%.

APPENDIX C

Selected DOT Rule-Making Notices and Amendments Docket Nos. HM-1 to HM-198

LIST OF DOCKETS

471

INTRODUCTION

Under the Administrative Procedure Act and the Hazardous Materials Transportation Act, both discussed at length earlier in the Red Book, an administrative agency such as the Department of Transportation (DOT) issues regulations describing in detail what regulated people must or must not do. This issuance is broadly described as rule making and, although purists may find some differences between rules and regulations, for our purposes the terms are synonymous.

Appendix C contains the preambles to all of the rule making issuances by DOT's Hazardous Materials Regulations Board and its successor, the Materials Transportation Bureau, since the creation of DOT in 1967, from Docket No. HM-1 through Docket No. HM-198.

An "advance notice of proposed rule making" is a mechanism through which DOT requests the filing of information or other comments on an embryonic concept that has not taken sufficient form yet to constitute a notice of proposed rule making.

A "notice of proposed rule making" is the publication in which DOT advises the public that a specific change is being considered for adoption, requesting public comments on the advisability of that change.

An "amendment" is the publication of the change that was first announced in the notice of proposed rule making. It is issued after consideration of the comments received in response to the notice.

A "spot amendment" is one in which an immediate change is adopted without a notice of proposed rule making, under circumstances where the need for change is so important that matters cannot wait while a notice is published and comments are received, or is so unimportant that public comment would not be worthwhile, such as with a change to an internal DOT procedure.

The "effective date" of the amendment is that date after which the change comes into effect as a matter of law. The "mandatory effective date" is a date that may follow the ordinary effective date, thereby providing an interim period within which regulated people may phase-in newly-adopted changes. Under this concept, people may begin to comply after the effective date but must be in compliance by the mandatory effective date.

DOT rule making is "informal" in nature and, despite the holding of some hearings, in the past has been an exchange of ideas and views in writing. DOT's proposals and responses are published in the daily Federal Register, which is the newspaper of the government, and the public's comments are submitted in writing to DOT.

With few exceptions, each advance notice or notice of proposed rule making is preceded by a "preamble," outlining the basis for and intent of the proposed changes. Each amendment is preceded by a preamble that should respond to the comments that have been received, and that should identify any change in the agency's thinking or the final rule that has taken place since the publication of the notice, either because new information has become available or in response to the comments that have been submitted by the public.

Although the final rule is eventually published in the bound volumes of the Code of Federal Regulations (CFRs), the CFRs do not contain any advance notice or notice of proposed rule making, or the preambles to such notices, nor do the CFRs contain the explanatory preamble to the final rule that was adopted in the amendment. In other words, the material that fleshes out the skeleton of the rule by giving the background and purpose of that rule as well as alternative solutions that may have been rejected, is lost.

The author of the Red Book, engaged in a daily legal practice devoted exclusively to interpretation and application of hazardous materials regulations, has found a complete set of all preambles to be an invaluable tool. Accordingly, preambles to significant notices and amendments have been reproduced in Appendix C, in a larger typeface than that encountered in the Federal Register, reading across the page rather than in Federal Register columns.

An explanation of the detailed elements of these publications will be helpful:

"Docket" is a term applied to the particular file opened by DOT each time a specific subject will be considered in the rule-making processes. The publications are kept in a public docket room at DOT, which also will include copies of the public comments that are filed in response to the publications.

Each docket is assigned its own number and, to indicate that the subject is hazardous materials, the docket number is preceded by the letters "HM," e.g., "HM-102." This example merely means that a docket has been opened for a new subject matter and that it is the 102nd docket opened since the inception of DOT in 1967. Dockets are numbered sequentially, so that the first docket was HM-1 and the last in Appendix C is Docket No. HM-198.

It is likely that there will be several individual publications within the same docket, or subject file. In most instances, for example, there will be at least two publications, a notice and an amendment. In the more controversial dockets, there have been a multitude of notices and amendments. For notices, each is identified by a reference to the year of its publication, and the sequence in which it has been published among all the notices for that year. For example, "Notice 74–10" means that document was the tenth notice published in 1974. For amendments, individual identification is provided by the CFR Part that is affected, and the sequence of changes that have taken place in that Part since DOT began. Thus, "Amendment No. 173–91" signifies that the amendment is to Part 173 and that this amendment is the ninety-first time DOT has amended Part 173.

Generally speaking, each publication will be introduced by a reference to earlier related publications. These are not only identified by a broad descriptive term, but also will be identified by the date, volume, and page number of the Federal Register in which the earlier publication appeared. The reference to the Federal Register, or citation, will appear with the annual volume number of that year's Federal Register and the page number for that year's volume. Since Federal Register pagination begins with "1" at the beginning of the year and is cumulative throughout the year, by the end of the year, the page numbers are quite high. The most proper form of reference is to give not only the volume and page number, but also the date of the particular issue in which the document appeared, e.g., "40 F.R. 24902, June 11, 1975" means that the particular material was published in the Federal Register for that date, that it is in volume 40 of the Federal Register, and that the document begins on page 24902, although it may continue for several pages thereafter.

The proper citation for each publication in Appendix C appears in parentheses following the publication.

In summary, the regulatory people at DOT have been more or less consistent in their approach to problems and their regulatory resolution of those problems. The preambles to notices and amendments provide an invaluable insight into past approaches and considerations, and this resource should be drawn upon in predicting the interpretation DOT will make of its own rules or in assessing the likelihood of the agency taking certain directions in the future.

Appendix C does not constitute light reading and is not intended to. It does provide a rich vein of information, however, that may be mined whenever difficulties arise in the interpretation of a particular rule or in the contemplation of asking DOT to take action in a given area in the future. It is a research tool that should serve the reader well.

Office of the Secretary
HAZARDOUS MATERIALS REGULATIONS BOARD
Notice of Establishment

The Department of Transportation Act (80 Stat. 931) transferred to the Secretary of Transportation the regulatory authority in various sections of law relating to the shipment and transportation of hazardous materials by all civil modes of transportation. In accordance with the Act and pursuant to delegation by the Secretary, the functions of the Secretary relating to the regulation of the shipment and transportation of hazardous materials by air, water, highway, and railroad and pipeline are now exercised by the Federal Aviation Administrator, the Commandant, U.S. Coast Guard, the Federal Highway Administrator, and the Federal Railroad Administrator, respectively. (See sections 6(c) (1) and 9(e) of the Department of Transportation Act and Part 1 of the Regulations of the Secretary (49 CFR Part 1).) Shipments of hazardous materials may move through all or several of the modes of transportation over which the heads of the operating administrations of the Department of Transportation have cognizance. The Department considers it to be essential that shippers and carriers be able to refer to a cohesive set of authoritative regulations upon which they may rely in preparing, shipping, and transporting materials regardless of the mode of transportation.

By Departmental Order 1100.11, dated July 27, 1967, there was established a Hazardous Materials Regulations Board. The Board is composed of the Assistant Secretary for Research and Technology as Chairman; and the Commandant, U.S. Coast Guard, Federal Aviation Administrator, Federal Highway Administrator, and Federal Railroad Administrator, or their designees, as members. The General Counsel of the Department is the legal advisor to the Board and the Director of the Office of Hazardous Materials is the secretary to the Board.

The function of the Board is to handle all matters relating to regulations (including special permits for waiver or exemption) issued under Title 18 U.S.C. 831–835 and Title VI and section 902(h) of the Federal Aviation Act of 1958 (49 U.S.C. 1421–1430, and 1472(h)), for the shipment and transportation of hazardous materials.

Regulations, other than special permits, developed by the Board for the shipment and transportation of hazardous materials, will be published in the FEDERAL REGISTER and in the Code of Federal Regulations. To the extent practicable these regulations will be the same for all modes of transportation, will be adopted under the same procedures, and will be published in the same document or series of documents. Regulations now contained in Parts

171–190 of Title 49 CFR and in Part 103 of Title 14 CFR will henceforth be designated as the "Hazardous Materials Regulations of the Department of Transportation".

(32 F.R. 145659, October 19, 1967)

SUMMARY • **Docket No. HM-1** • *Establishment of the Hazardous Materials Regulations Board and the Board's Rules of Procedure for handling rule changes and issuance of special permits.*

[49 CFR Part 170]
[Docket No. HM-1; Notice 67–1]
RULES OF PROCEDURE
Notice of Proposed Rule Making

On October 19, 1967 (32 F.R. 14569), the Department published a notice of the creation of the Hazardous Materials Regulations Board, and stated its composition and jurisdiction.

The function of the Board is to handle all matters relating to regulations (including special permits for waiver or exemption, issued under Title 18, U.S.C. 831–835 and title IV and section 902(h) of the Federal Aviation Act of 1958 (49 U.S.C. 1421–1430, and 1472(h)), for the shipment and transportation of hazardous materials.

Regulations of the Board, other than special permits, for the shipment and transportation of hazardous materials will be published in the FEDERAL REGISTER and in the Code of Federal Regulations. Regulations now contained in Parts 171–190 of Title 49 CFR and in Part 103 of the Federal Aviation Regulations (14 CFR Part 103) are now designated as the "Hazardous Materials Regulations of the Department of Transportation".

At the present time, regulations under Title 18, U.S.C. 831–835 are issued under procedures established by the Interstate Commerce Commission which was vested with this regulatory function before the effective date of the Department of Transportation Act. Similarly, regulations in Part 103 of the Federal Aviation Regulations are currently issued under Part 11 thereof (14 CFR Part 11). Section 12 of the Department of Transportation Act provides that regulations (including procedural regulations) issued in the exercise of powers, functions, and duties transferred by the Act shall continue in effect until modified, terminated, or superseded in the exercise of authority under that Act.

The purpose of this notice is to request public comment on procedures proposed for use in prescribing all hazardous materials regulations under the cited authority. The proposed rule would supersede the applicable procedural rules of the Interstate Commerce Commission and, so far as Part 11 of the Federal Aviation Regulations relates to Hazardous Materials Regulations, would supersede that part. The proposal would establish one procedure for the issuance of all of those regulations.

Sections 556 and 557 of title 5, United States Code (formerly sections 7 and 8 of the Administrative Procedure Act) relating to the conduct of hearings on the record, would not apply to rule making under the proposed part. Therefore, hearings are not a required part of the rule-making procedure. However, hearings may be held, in the discretion of the Board, as a supplementary factfinding procedure. Any hearing held would be nonadversary, with no formal pleadings, and any resultant rule would not necessarily be based exclusively on the record of the hearing.

(32 F.R. 16437, November 30, 1967)

[Docket No. HM-1]
PART 170—RULE-MAKING PROCEDURES OF THE HAZARDOUS MATERIALS REGULATIONS BOARD

The purpose of this amendment is to add a new Part 170 "Rule-making Procedures of the Hazardous Materials Regulations Board" to Title 49 of the Code of Federal Regulations.

A notice of proposed rule making (Notice 67–1) regarding this action was published in the FEDERAL REGISTER on November 30, 1967 (32 F.R. 16437). A supplemental notice (Notice 68–3) was published on February 28, 1968 (33 F.R. 3439), and a public hearing on the entire proposal was held on March 19, 1968.

Interested persons were then afforded an opportunity to participate in the rule making through the submission of oral and written comments. Due consideration has been given to all relevant matter presented.

Historical background. The shipment and carriage of hazardous materials by carriers subject to the Interstate Commerce Act (rail, motor carrier, pipeline, inland waterway) is governed by sections 831–835 of Title 18, United States Code. Before April 1, 1967, this authority was exercised by the Interstate Commerce Commission. Since that date, the authority has been exercised by the Department of Transportation, pursuant to the Department of Transportation Act. The shipment and carriage of hazardous materials by air has since 1958 been governed by section 902(h) of the Federal Aviation Act of 1958, which authorizes the Federal Aviation Administrator to issue his own regulations or to adopt, in whole or in part, the regulations issued under Title 18 by U.S.C. by the Interstate Commerce Commission (ICC). The FAA has, in the past, for the most part adopted the regulations issued by the ICC. The shipment and carriage of hazardous materials by vessel is governed by section 4472 of the Revised Statutes (46 U.S.C. 170) which authorizes the Commandant of the Coast Guard to define, describe, name, and classify hazardous materials and establish regulations for their carriage. It also requires him to "accept and adopt" for these purposes, the ICC regulations so far as they apply to shippers by common carriers by water.

Thus, the vast majority of the regulations applicable to hazardous materials were those adopted by the ICC in Title 49 of the CFR's.

(1) In adopting those regulations the ICC followed the normal requirements of the Administrative Procedure Act for notice to the public of proposed rules, time for public comment, evaluation of comments and issuance of final rules. These procedures were not, however, completely set forth in the published procedural rules.

(2) While public hearings were specifically authorized in § 171.7 of Title 49, in the Administrative Procedure Act, and in the basic statutory authority, such hearings were not mandatory and, in fact, were rarely held.

(3) Within the ICC the rule-making authority with respect to hazardous materials was delegated to an employee board of the Commission called the Explosives and Other Dangerous Articles Board. This Board issued the notices, evaluated the comments and adopted final amendments for the Commission. Actions of this Board could be appealed to Division 3 of the Commission which consisted of three commissioners who reviewed the appeals. A second level of appeal could, in the discretion of the Commission, be made to the entire Commission after which the only recourse was to the courts.

(4) In the course of the rule-making procedure the ICC's section of Explosives and Other Dangerous Articles held frequent informal meetings and discussions with representatives of affected industry groups.

(5) Section 17(8) of the Interstate Commerce Act provided that the effective date of any amendment would, in the case of an employee Board regulation or Division regulation, be suspended upon the request by an interested person for reconsideration until further action by the Commission. This section of the Interstate Commerce Act was not, however, made applicable to the Department of Transportation by the Department of Transportation Act. Only those rules and regulation of the Interstate Commerce Commission that were effective before the Department's effective date were continued in effect. Thus, the procedural rules issued by the ICC but not the statutory rules were continued in effect by section 12 of the Department of Transportation Act, until changed and superseded by the rules adopted by the Department.

Upon establishment of the Department, effective April 1, 1967, the above described powers and duties of the ICC, FAA, and Coast Guard with respect to the transportation of hazardous materials were transferred to the Department of Transportation. As was stated in the original notice announcing the establishment of the Hazardous Materials Regulations Board (32 F.R. 14569), the Department considers it to be essential that shippers and carriers be able to refer to a cohesive set of authoritative regulations upon which they may rely in preparing, shipping, and transporting materials regardless of the mode of transportation. To this end the Secretary, by Departmental Order 1100.11 dated July 27, 1967, established a Hazardous Materials Regulations Board. One of the first acts of this Board was to issue Notice 67–1, proposed rules of procedure, on November 30, 1967 (32 F.R. 16437), for public comment.

Analysis of comments. In general, the bulk of comments received on the notices and at the hearing dealt with four major areas—(1) Special permit requirements; (2) the treatment of confidential and proprietary information received by the Board; (3) procedures for administrative appeal and automatic stay of rules; and (4) mandatory hearings.

(1) *Special permits.* Many of the comments received on notice 67–1 were, so far as they related to special permits, reflected in the restatements of proposed §§ 170.13 and 170.15 contained in supplemental Notice 68–3 and were fully discussed in that notice. Other comments on this subject related to the providing of notice to the public, the Bureau of Explosives, and certain carriers of the granting of special permits. The Board has determined, as an administrative matter, that it will periodically publish, in the FEDERAL REGISTER, a listing of permits issued. The Board will also continue its policy of close cooperation with the Bureau of Explosives in this regard. The Board has determined, however, that it cannot accept a requirement that it furnish copies of special permits to carriers. In many cases special permits are issued without knowledge of the specific carrier or carriers to be used. Therefore, even if it adopted such a role, the Board could not comply with it. It is the duty of the holder of the permit to notify the carriers concerned of the provisions of the special permit, and no carrier is authorized or required to operate in ignorance of the terms of a special permit. As a result, under the prevailing situation the Board has elected to continue the practice that has been followed in the past in this regard, both by it and the ICC.

Several comments suggested deletion of the phrase "why the public interest would be served by the proposal" in § 170.13(b) (2). It should be made clear that the proposed language does not state that the granting of a permit depends upon a showing of a public interest. It merely requires that the petitioner submit information on this point. If there is no information that can be submitted, the petitioner may so state. It is the Board's position that the required information, or the absence of it, is pertinent to its considerations and will be of use to it in making its determinations, particularly in cases where there is some question as to the other justifications for the permit.

Finally, several comments stated that § 170.13(b) (4) should be narrowed to delete the requirement that the "composition and percentage (specified by volume or weight) of each chemical, if a solution or mixture" be submitted with each applicable petition for a permit. The Board considers this information to be relevant to its consideration in the applicable cases. So far as this requirement relates to confidential information, the Board will accord it the desired protection (see discussion below).

(2) Confidential and proprietary information. In the written comments and at the public hearing, several persons questioned the protection to be afforded confidential or proprietary information submitted with petitions for rule making or special permits. Under the "Freedom of Information Act" (section 552 of Title 5, United States Code), all records of each Government agency are open to public inspection except certain records that the statute specifically authorizes to be exempted. The regulations of the Department of Transportation relating to the public availability of information are set forth in Part 7 of Title 49 CFR. The general policy of the Department as stated in §7.51 is that it will release a record that is otherwise authorized to be withheld under the statutory exemption "unless it determines that the release of that record would be inconsistent with a purpose of the section concerned."

In Notice 68–3 the restatement of §170.15 stated that "The treatment of confidential or proprietary material furnished by any petitioner is governed by §7.59 of this title". Section 7.59(a) states that the types of information that are within the statutory exemption include "Information furnished by any person, to the extent that the person furnishing the information would not customarily release it to the public" and "Information furnished and accepted in confidence". Section 7.59(b) states "The purpose of this section is to authorize the protection of records that are customarily privileged or are appropriately given to the Department in confidence * * *. In any case in which the Department has obligated itself not to disclose information it receives, this section authorizes the Department to honor that obligation".

Section 7.59 therefore provides persons submitting confidential or proprietary information to the Hazardous Materials Regulations Board the same protection that is provided to all persons submitting these types of information to the Department. This protection is considered adequate and where applicable the affected materials will not be placed in the public docket.

(3) Procedures for administrative appeal and automatic stay of rules. As indicated previously, under the ICC, actions of the employee board that issued the notices and adopted the final amendments could automatically be appealed to a designated division of the Commission and the Commission could in its discretion grant a second level of appeal. Under section 17(8) of the Interstate Commerce Act, whenever, under certain circumstances, a final rule was appealed before its effective date, that effective date was stayed until final disposition of the appeal. Several comments indicated that the Board's procedural rules should provide both an administrative appeal procedure within the Department of Transportation and an automatic stay provision.

As the Hazardous Materials Regulations Board is presently constituted within the Department of Transportation an administrative appeal procedure is impractical and would appear to be unnecessary. It should be kept in mind that since the Board is made up for the most part of the Administrators of the various operating administrations within the Department, the initial issuance of regulations will be from a higher source, and after more consideration, than was the case of employee board issued regulations of the ICC. To provide for a formal appeal procedure to the same Board (Administrators) that issued the rule in question would not appear to satisfy the desires of those persons who requested a "second chance" within the Department. In this connection, § 170.35 as proposed, provided for petitions for rehearing or reconsideration of rules. In cases where facts are submitted before the effective date of a rule that warrants postponement of the effective date, the Board will not hesitate to provide for a postponement. Furthermore, even where there is no postponement, if after evaluation of the comments, the Board determines that a revision of the rule is warranted, it will not hesitate to make or propose changes to the adopted rule.

With respect to "staying" the effect of adopted rules, proposed §170.35 specifically provided that "Unless the Board otherwise provides, the filing of a petition under this section does not stay the effectiveness of a rule". In the absence of a procedure providing an appeal to a separate body within the Department, automatic stay would appear to serve no practical purpose except delay. Nor is the Board convinced that an automatic stay is an appropriate procedure for safety regulations even where a formal appeal is provided.

(4) Hearings. The proposed rules of procedure did not provide in every case for a public hearing. Proposed §170.31 did provide that, whenever the Board should elect to hold a hearing, the hearing would be a nonadversary, factfinding type of hearing. Several of the comments were addressed to the lack of provision for a mandatory hearing, upon request, and to the lack of provision for a formal evidentiary type of hearing. As is stated in §170.31, sections 556 and 557 of Title 5, United States Code (relating to the conduct of hearings required to be on the record) have no application to hearings held for the purposes of proposed rule-making action under the pertinent statutes. The purpose of any hearing held by the Board would be to provide the Board with information, views, and arguments relative to proposed rule-making action. The Board is not and should not be required to act solely on the narrowly developed information that would be produced at a formal hearing. It is only in the area of economic rule making such as the ICC, CAB, FTC, and FCC that evidentiary hearings are generally used. Seldom, if ever, are

evidentiary hearings held in safety matters. The FAA, for example, and the Coast Guard have traditionally used the legislative type of hearing in safety rule making. Therefore, it would be inappropriate and unduly burdensome for the Board to attempt to provide formal evidentiary hearings.

Nor does the Board consider it necessary or appropriate that a hearing be required in every case where there is a request for one. In appropriate cases the Board will not hesitate to hold an informal public hearing of the type prescribed in these rules. However, to require the Board to hold such a hearing whenever one interested party requests, would be to unnecessarily burden the Board and tie its hands administratively. In every proposed rule-making action of a substantive nature adequate time for public comment will be provided and the Board will consider all comments received. It is not necessary or appropriate in every case that interested parties be permitted to state their comments orally. However, if in any case, it would be helpful, the Board will not hesitate to call a hearing on any matter before it.

Miscellaneous comments. In addition to the above major areas of comment, comments were received on the following matters.

Several comments objected to the statement in proposed §170.21(b) that "Unless the Board determines that notice and public rule-making proceedings are desirable, interpretive rules, general statements of policy, and rules relating to organizations, procedure, or practice are prescribed as final without notice or other public rule-making procedures". It must be noted that this authority is strictly limited to non-substantive rules, and not those affecting the rights, privileges, and duties of the regulated public and is therefore completely consistent with specific provisions of the Administrative Procedure Act (see 5 U.S.C. 553). Furthermore, the Board intends, wherever time permits, to provide for public participation in the adoption of rules, even these nonsubstantive rules, regardless of the fact that it is not required to do so.

Several comments suggested the addition of language to provide that the effective date for all regulations be at least 30 days after issue. In view of the fact that section 835(d) of Title 18, United States Code, specifies that all regulations take effect 90 days after issue "unless a shorter time is specified" and in view of section 553(d) of Title 5 United States Code which requires (except in certain enumerated cases) at least 30 days between issue and effective date, the Board does not consider it necessary to add this provision.

It was suggested that the rules provide for specific service of notices of proposed rule making on all parties. This comment must be rejected on practical grounds as it would be impossible for the Board to ascertain all interested parties in each case. However, the Board will furnish copies of notices to any person who requests to have his name placed on a mailing list for this purpose. Interested persons are invited to take advantage of this opportunity.

Notice 68–3 stated that the FAA was considering inclusion in Part 170 of the authority to grant deviations now in §103.5 of the Federal Aviation Regulations. On the basis of comments received, the FAA has decided not to make the transfer at this time.

Miscellaneous changes based on comments. It was suggested that §170.11(b) (4) be amended to delete the requirement that persons petitioning the Board for the issue, amendment, or repeal of a rule be required to submit "arguments" to support the petition. The Board accepts this comment on the basis that "information" to support the request will suffice for the purpose.

Several comments suggested that §170.21(a) be amended to provide, under normal circumstances, a minimum of 30 days for public comment on notices of proposed rule making issued by the Board. It has been the Board's policy, in all except emergency situations, to allow more than 30 days for comment on each notice. Therefore, the Board has accepted this comment and §170.21(a) is amended accordingly.

Several comments requested that §170.21(a) be amended to require the Board to incorporate a brief statement of findings as to why notice and public procedure are dispensed with in any case involving a substantive rule. Since this practice would be followed in any event, the Board has no objection to its incorporation in the regulation. The section has therefore been amended accordingly.

Several comments indicated a misunderstanding that §170.29, which relates to additional rule-making procedures, would be used to initiate rule making without notice or other public procedure. Since §170.21 provides for notice in all appropriate cases, this was not the intended meaning. To obviate any question in this regard, however, §170.29 is amended by inserting the words "After issuing a notice of proposed rule making in any particular case," at the beginning of the section.

Several comments called attention to the fact that proposed §170.35 would have required all petitions for reconsideration to be filed more than 10 days before the effective date of the rule. This would effectively prohibit the filing of such a petition in the case of any rule issued within 10 or less days before the effective date. In order to allow petitions for reconsideration of these rules of an emergency nature, a sentence has been added to allow petitions to be filed at any time before the effective date, if the effective date is less than 15 days after the date of issuance.

A comment requested information as to the meaning of proposed §170.35(c) which stated that the Board will not consider repetitious petitions. It was the Board's intention that after it had denied a petition it would not consider repetitious petitions on the same subject for the same purpose. However, on reconsideration this would not appear to be a problem of great magnitude and the language could possibly be misconstrued to prevent multiple filings by different persons or groups. For these reasons the provision has been deleted.

All other comments suggesting changes, additions, or deletions were carefully considered. In the opinion of the Board, all comments that merited acceptance have been accepted.

This part applies to all rule-making activities of the Board in existence on or after its effective date.

(33 F.R. 8277, June 4, 1968)

SUMMARY • Docket No. HM-2 • *Complete revision of the radioactive materials regulations to harmonize the domestic U.S. requirements with those of the International Atomic Energy Agency (IAEA).*

[49 CFR Parts 170–190; 14 CFR Part 103]
[Docket No. HM-2; Notice No. 68–1]
RADIOACTIVE MATERIALS
Notice of Proposed Rule Making

On April 1, 1963, the Interstate Commerce Commission (ICC) published its Notice No. 58 in Docket No. 3666. The notice proposed to modify the ICC Regulations for transporting radioactive materials to bring them into accord with the recommended regulations of the International Atomic Energy Agency (IAEA). Based upon the comments received pursuant to that notice of proposed rule making and after discussion with representatives of the U.S. Atomic Energy Commission (USAEC), it became apparent that it would not be in the public interest to adopt those amendments at that time. This area of regulation was transferred to the Department of Transportation by the Department of Transportation Act (80 Stat. 931).

Since that time this Department, the ICC, and the Atomic Energy Commission have worked toward the preparation of a revision to the radioactive materials regulations. Many meetings have been held between industry and Government representatives. Several significant "enabling" regulatory amendments have been adopted which now make it practical to propose a revised major revision of these regulations. In 1966, the USAEC published its packaging standards in Part 71 of Title 10, CFR. At the same time, the ICC published Order No. 70 relating to transportation of fissile radioactive materials. Early in 1967, the ICC also published Order No. 74 which made further modifications regarding radioactive materials.

During the past 18 months a task force comprised of representatives of the USAEC and its contractors prepared a series of draft regulatory changes designed to incorporate the principles of the recommended regulations of the IAEA into the regulations as amended by Orders 70 and 74. These drafts were further modified as a result of participation by representatives of the ICC, Federal Aviation Administration, U.S. Coast Guard, and various atomic energy and transportation industry personnel. The results of all of these reviews and discussions are reflected in this notice of proposed rule making.

This notice includes proposed amendments to the Hazardous Materials Regulations of the Department of Transportation (49 CFR Parts 171–178) (formerly a part of the ICC Regulations) and Part 103 of the Federal Aviation Regulations (14 CFR Part 103). The purpose of this notice is to request public comment on procedures proposed for the transportation of radioactive materials. Interested persons are invited to participate in the making of proposed rules by submitting such written data, views, or arguments as they may desire. Communications should identify the regulatory docket and notice number and be submitted in duplicate to the Secretary, Hazardous Materials Regulations Board, Department of Transportation, 400 Sixth Street SW., Washington, D.C. 20570.

Communications received before April 15, 1968, would be considered by the Board before taking final action on the notice. All comments will be available for examination by interested persons at the Office of the Secretary of the Board, both before and after the closing date for comments. The proposals contained in this notice may be changed in light of comments received.

Several references are made in the proposed regulatory amendments to authorizations issued under Part 170 of Title 49, CFR. Part 170 has been reserved for the Rules of Procedure for the Hazardous Materials Regulations Board. Part 170 has been published as a notice of proposed rule making but has not yet been adopted. It is expected that Part 170 will be in effect before the expiration of the comment period for this notice of proposed rule making. Part 170 will include the procedures for general rule making as well as those for handling applications for special permits.

The basic consideration in the transportation of radioactive materials is that they may present radiation and contamination hazards to transportation workers, passengers, and the general public. In addition, radiation exposure may damage other materials in transport, such as undeveloped photographic film. The proposed regulatory amendments will provide for the control of these potential hazards by considering the three basic factors of (1) relative hazard potential (2) packaging performance and (3) the transportation environment. The existing regulations place the primary emphasis on the packaging requirements for the normal conditions of transportation. The proposed revisions will provide a system of allowing sufficient emphasis to be placed not only on the normal conditions of transportation, but also on the environmental conditions which a package of radioactive materials might encounter in an accident.

This notice of proposed rule making establishes a separate hazard classification category for radioactive materials, apart from the poisonous category. Radioactive materials would be classified as radioactive materials and not as Class D poisons as they currently are. Appropriate changes are being proposed to the commodity list in Part 172.

Several provisions which are presently contained in the regulations of the U.S. Atomic Energy Commission, title 10 CFR Part 71, have been incorporated into this proposed revision. Examples are the definitions of "special form," "normal form," and "large quantities" of radioactive materials.

A major change is proposed in the method of hazard identification of radioactive materials. Assignments of hazard categories which are based solely upon the type of radiation emanating from the package is not truly representative of the transportation hazards to be considered. The proposed system is based instead upon the radiotoxicity of the isotope concerned. The hazard potential of radioactive materials is defined by consideration of radiotoxicity and physical form, and by assigning each radionuclide to an appropriate "transport group." In addition, some special classes of materials are considered, such as very large or very small quantities, low specific activity materials, and fissile materials. This system is presently prescribed in regulations of the Atomic Energy Commission, 10 CFR Part 71.

Another major area of change is in package identification. A new labeling system is proposed to conform to the recommendations of the United Nations and the IAEA. The labels will also be used to determine the need for placarding of vehicles. A later regulatory proposal will incorporate the remainder of the U.N. labels for other hazardous materials.

The proposed regulatory change will provide more types of specification packaging, increased flexibility for the shipper in terms of new package development, and a clearer definition of the criteria which the Department will be using to evaluate the adequacy of various packaging methods.

A further change would allow an increase in the amount of radioactive materials that may be carried aboard a vehicle from 40 units to 50 units. This revision would also change the name for the term "radiation unit" to "transport index."

Proposed new § 173.393 contains a number of general packaging requirements, many of which are in the existing regulations. Sections 173.394 and 173.395 contain the particular packaging requirements for special form and normal form radioactive materials. These two sections could be combined into a single section but there have been indications from industry sources of the desirability of separation. Section 173.396 proposes specific packaging requirements for fissile material. This section is essentially unchanged from the present regulations except for some additional flexibility in the packaging of small amounts of fissile materials. Sections 173.396a and 173.397 incorporate the provisions included in the existing § 173.392 for "exempt quantities," and also make an additional provision for the transportation of contaminated items and bulk low specific activity materials. Section 173.398 prescribes the special test conditions for special form material and for the hypothetical accident conditions of transportation. These provisions are presently contained in Part 71 of the USAEC regulations. Section 173.399 prescribes new labeling requirements. Section 173.399a consolidates and updates the general contamination control requirements.

Appropriate changes are proposed for Parts 174, 175, and 177 to incorporate the new placarding requirements, to increase the transport index from 40 to 50, to delete certain consignee requirements that are not within the jurisdiction of these regulations, and to provide for more comprehensive distance—time handling provisions.

In Part 178 revisions are made to specifications 6L and 12B, and two new specifications are being proposed. Specification 6L is being modified to provide a wider flexibility in drum size and centering mechanisms. Tests have shown the inadequacy of the present closure requirements and the specification is being modified to require higher strength locking rings. A new specification 6M metal package is being proposed for both fissile and non-fissile radioactive materials. The special specification 12B fiberboard box for radioactive materials would be deleted since the requirements contained therein would now be included in § 173.393. A new specification 7A general package is being proposed for radioactive materials. Specification 7A provides for performance criteria rather than detailed engineering design requirements. The shipper would be given a great deal of flexibility in the exact design of his specification 7A package.

A number of editorial changes are being proposed in this Notice which do not directly bear on substantive requirements for the transportation of radioactive materials, but are being made in related provisions as a part of the general updating of the regulations. Examples are in the changes being proposed for §§ 173.22, 173.23, 173.24, and 173.28.

In Part 103 of 14 CFR appropriate amendments are being proposed to incorporate the provisions of the general revision into the Hazardous Materials Regulations applicable to aviation. At the same time, § 103.3 is being amended to reflect Amendment No. 75 regarding shipping paper requirements. Several other minor changes are being proposed to provide consistency between Parts 174–177 and Part 103.

Since the Federal Aviation Administration does not exercise jurisdiction over the handling and storage of hazardous materials in air freight terminals or other storage locations outside of aircraft, the provisions for handling, storage, and accidents are limited to aircraft only. However, the Department is considering the need for providing similar safeguards in connection with the storage and handling of radioactive materials at all times once they have entered into the realm of air transportation.

Paragraph (d) of § 103.23 would be deleted from Part 103 under this proposed amendment. This provision makes the shipper and the carrier jointly responsible for providing personnel monitoring devices. There are no similar requirements for rail, highway, and water, and the experience of the transportation industry has been that none are required in these regulations. The Atomic Energy Commission, the Department of Labor, and the Department of Health, Education, and Welfare already have established standards for exposure control of people. Removal of the requirement does not, of course, preclude the carrier or the shipper from fulfilling his responsibilities in this area.

This amendment is proposed under the authority of Title 18, United States Code, sections 831–835, section 9 of the Department of Transportation Ace (49 U.S.C. 1657), and Title VI and section 902(h) of the Federal Aviation Act of 1958 (49 U.S.C. 1421–1430 and 1472(h)).

(33 F.R. 750, January 20, 1968)

Title 49—Chapter I
[Docket No. HM-2; Amdts. 171–1, 172–1, 173–3, 174–1, 175–1, 177–3, 178–1

RADIOACTIVE MATERIALS AND OTHER MISCELLANEOUS AMENDMENTS

On January 20, 1968, the Hazardous Materials Regulations Board published Docket No. HM-2; Notice No. 68–1 (33 F.R. 750), which proposed amendments to the Department's Hazardous Materials Regulations (49 CFR Parts 170–190 and 14 CFR Part 103). These proposals dealt with a major revision to the regulations for the transportation of radioactive materials, along with a number of other general packaging modifications. The public was given 90 days for comment. Numerous comments were filed and have been studied by the Department staff. Several meetings and discussions were held with staff personnel of the U.S. Atomic Energy Commission (AEC), as required by 18 U.S.C. 834(b), and the amendments reflect the results of those discussions. All other comments suggesting changes, additions, or deletions were carefully considered.

One of the most controversial items in the notice of proposed rule making involved the proposed changes in the regulations for the transportation of radioactive materials by air and bus. Restrictions on shipments of radioactive liquids and shipments of packages having significant external radiation levels had been proposed. After consideration of the comments received, and after evaluation of the impact of the proposal on the atomic energy industry, particularly with regard to the use of radiopharmaceuticals, those proposed restrictions have been deleted, and the present provisions for such shipments have been retained. No regulatory restrictions on shipment of large quantities of radioactive materials are considered necessary since each such shipment is covered by a Department special permit. Each situation can then be analyzed on its own merits, and appropriate restrictions can be imposed in the language of the permit.

Numerous comments were received regarding suggested changes to the Department's proposed labeling system for radioactive materials packages. The proposed system was in harmony with the regulations of the International Atomic Energy Agency (IAEA), and the proposed regulations of the United Nations. Certain parties in the United States felt that those international standards are not appropriate in all respects for U.S. usage, and asked that the use of the labels be modified accordingly. However, the Department believes that the interests of international harmony in this area are overriding, and has retained the IAEA-type labels and labeling criteria. The Department will pursue the item further with the IAEA to determine if changes could be made in the international standards which would reflect the total United States interests.

Many of the modifications in these amendments will require parallel changes in the AEC regulations (10 CFR Part 71) to assure harmony between the two complementary sets of regulations. The AEC has indicated that it expects to be able to publish the necessary amendments to its Part 71 prior to the effective date of these amendments.

Many of the new procedures prescribed in these amendments have been previously authorized by Departmental special permits. Special Permit No. 5000 authorized the use of a drum-type birdcage now listed as the Specification 6M package. Special Permit No. 5300 authorized the use of a type of packaging now listed as the Specification 7A package. Special Permit No. 5400 provided for the shipment of enriched uranium under the terms of § 71.6 of the AEC regulations, and the terms of that permit are now included in § 173.396 of these regulations. Special Permit No. 5417 provided for the transportation of radioactively contaminated items, and the terms of that permit are not included in the low specific activity provisions of § 173.392. Accordingly, those special permits are no longer appropriate, and are hereby terminated. Several of the carriers objected to increasing the transport index from the present limit of 40 to a new limit of 50. Although this increase means that more radioactive materials could be carried aboard a vehicle, it does not present a significant increase in hazard. The extra packages may only be carried under additional transport controls for segregation of packages from passengers, transportation workers, and film. The increase is not mandatory, but only allows more packages to be carried. Each carrier is still free to load his vehicles as he sees fit within the overall regulatory limitations. Studies have shown that the previous limit of 40 was greatly overconservative, and that the new limit of 50 would still provide adequate safety in transportation. The new limit will also provide a higher degree of consistency with the international regulations which already provide for a transport index of 50.

A large number of special permits have been issued for the transportation of fissile radioactive materials under Fissile Class II conditions. Because these amendments reflect the international standard of a transport index limit of 50, rather than the 40 radiation unit maximum presently prescribed in the U.S. regulations, a modification of these permits must be made in order that the number of Fissile Class II packages per vehicle remains the same. Therefore, for all special permits issued prior to September 26, 1968, the allowable transport index listed for Fissile Class II packages is increased by a factor of 1.25; i.e., an increase of 25 percent over the present assignments. All holders of such permits will receive individual notification of this change. Future special permits and revisions to existing permits will reflect the new criteria in making transport index assignments.

The notice of proposed rule making did not utilize the Type A-Type B quantity provisions of the IAEA regulations, but instead referred only to specified quantities of radioactive materials for the various categories of packaging. This was done at the request of a number of interested parties in the atomic energy field. These parties felt that there was a certain stigma attached to these terms as a result of previous unsuccessful rule making efforts by the Interstate Commerce Commission. However, the comments received indicated that the use of those IAEA terms would be not only acceptable but would clarify and simplify the packaging provisions. Accordingly, those terms are defined and used in these amendments.

On February 28, 1969, all existing Bureau of Explosives (Association of American Railroads) permits for radioactive materials packages will expire. Many comments indicated that the regulations were not sufficiently clear as to whether those previously authorized containers could ever be used again. The acceptability of these containers after February 28, 1969, will be a function of their ability to meet the prescribed structural integrity, shielding, and thermal resistance criteria. In each case, the shipper should examine the design and construction details of his container and compare them to the new regulations. If the container does not fit within one of the prescribed categories or usages, he may not use the container after that date without first having secured a Department special permit. The Department's safety evaluation of each of those containers will be based upon the criteria in these amendments. The detailed procedures for petitioning the Department for a special permit are prescribed in Part 170 of these regulations. Part 170 was published in the FEDERAL REGISTER on June 3, 1968 (33 F.R. 68–6562). Copies of Part 170 may be obtained by writing the Secretary, Hazardous Materials Regulations Board, 400 Sixth Street SW., Washington, D.C. 20590; there is no charge.

The present regulations in § 173.393, mention that containers authorized by the Interstate Commerce Commission (now the Department) under special permit may be used for the transportation of radioactive materials. In the light of the recent publication of Part 170, those statements are extraneous, and have been deleted. This does not mean to imply, however, that special permits are no longer available. Any person may petition the Department to use a container which is not prescribed in the regulations, whether for radioactive materials or any of the other hazardous materials covered in the regulations.

A number of comments were received, primarily from carrier interests, objecting to the shifting of responsibility for vehicle monitoring from the consignee to the carrier. They stated that they had neither the trained personnel nor the equipment to perform such services. The Transportation of Explosives Act (18 U.S.C. 831–835), which gives the Department the responsibility for developing and administering regulations for the safe transportation of hazardous materials, limits the Department's jurisdiction to shippers and carriers. The Department cannot impose requirements on consignees since it has no jurisdiction over them. Carriers have historically been responsible for cleaning up spills of other hazardous materials in their vehicles or on their property. Their responsibility with respect to radioactive materials is no different. The amended regulations prescribe performance standards for monitoring and cleanup of spills. The carrier may utilize the services of any qualified person, including the consignee, in performing the required functions. The present regulations often refer to actions to be carried out by the shipper or his authorized agent. Since 18 U.S.C. 831 includes a shipper's authorized agent in the definition of a shipper, the use of both terms in the same regulatory provision is redundant. Where the term shipper appears in the regulations, it is implied that the term includes his authorized agent. Accordingly, several of the sections have been modified to delete the reference to the authorized agent.

The Department acts as the U.S. competent authority as that term is used in the IAEA regulations. In issuing special permits for radioactive materials packages, the Department is often asked to provide the certificate required of competent authorities in the IAEA regulations. The details of these certificates are outlined in Marginal C-6 of those regulations. In order to provide this information, it will be necessary for the petitioner for the special permit to certify in his petition that his packaging, and the contents (particularly with respect to the special form criteria), meet all of the standards prescribed in the IAEA regulations. Although these amendments will bring the U.S. regulations more in harmony with the international standards, there are still some significant differences that will be dealt with in future rule making actions. It is the shipper's responsibility, as prescribed in § 173.393, to make the determination that his package meets all of the requirements of the foreign countries as well as the United States, and the shipper must certify to the Department that he has made that determination. He must present to the Department the basis of his evaluation that those standards have been met. The Department will review the petitioner's data and, if it is satisfied that the petitioner has in fact made a proper determination, it will issue the necessary IAEA certificate as a part of the special permit.

Several comments indicated that there will be difficulty in complying with the placarding requirements unless there was some indication on the shipping papers as to the type of label required for the packages being shipped. The Board agrees that the shipping paper should contain adequate information from which the placarding require ment can be determined and has therefore amended § 173.427 to require that the shipping paper description include the type of label required.

A number of additional editorial changes have been made throughout the regulations to correct such items as references to radioactive materials as Class D poisons, correction of paragraph references, and incorrect format.

At the request of a number of interested parties, the order of presentation of the radioactive materials packaging criteria in Part 173 has been modified to clarify the applicability of certain requirements, and to simplify the use of the regulations. This modified order of presentation is also more in harmony with the regulations for transportation of other hazardous materials.

In addition to the general changes discussed above, a number of specific changes to the notice of proposed rule making are worthy of highlighting.

Proposed § 173.22 has been modified to separate the subject of shipper's responsibility from the types of packages authorized under "grandfather clauses." The latter have been included in a new § 173.23. In § 173.23, two additional months have been provided for continued use of packages operating under permits from the Bureau of Explosives. The expiration date of the B of E permits is now February 28, 1969.

A table of steel thicknesses has been added to the general construction standards in § 173.24. The general prohibi tion against vented packages has been deleted.

In § 173.29, the "Empty" label is now required to be affixed to empty radioactive materials packagings.

In § 173.389, the definition of "fissile materials" has been clarified so that it agrees with the current definition in 10 CFR Part 71 of the USAEC Regulations. The use of the transport index numbers has also been clarified. New definitions for "large quantity" radioactive materials, "Type A" and "Type B" quantities, and "Type A" and "Type B" packaging have been included to obviate the need for repetitive definitions throughout the packaging regulations.

In § 173.390, an additional transport group, Group VII, has been added to conform with the IAEA regulations, and to obviate the need for descriptive limits throughout the packaging regulations. The provisions for determining the transport group of unknown mixtures have been expanded to conform with the IAEA definition.

In § 173.391, a total package limit has been placed on the amount of tritium which may be shipped under the exemption. The permissible contamination limits for the exempt packages has been changed from "detectable" to "significant removable." The requirement for the marking "Radioactive" on exempt devices has been deleted, and the maximum radioactivity content of each such device has been modified to conform with the IAEA regulations. An exemption has been added to provide for packaging in which natural or depleted uranium (such as shipping casks) is incorporated into the packaging.

Proposed § 173.393 has been modified to provide for a security seal, similar to the present special permit require ments, and in accord with the IAEA provisions. Section 173.393(d) has been clarified with regard to the requirements for internal bracing. Section 173.393 now includes restrictions on the surface temperatures in order to prevent injury to employees and to reduce the fire hazard to other cargo. The temperature restrictions are those commonly provided in special permits. Special permits are required for all shipments involving high internal decay heat, so this addition represents no change from present practice. Pyrophoric liquids are not authorized for air transporta tion under either the IATA or the IAEA regulations, and that restriction has been noted in § 173.393(f). Section 173.393(g) has been modified to remove the requirement that the inner container be made of metal. Section 173.393(j) has been reworded for clarification as to its applicability. The radiation level restrictions for occupied positions in private vehicles have been removed, since radiation exposures to personnel operating or riding in those vehicles are adequately controlled by existing regulations of the AEC and the Department of Labor.

Proposed §§ 173.394 and 173.395 have been modified to provide for delivery of IAEA Type A packages to their destination in the United States without need for special permit. Type B packages, other than Spec. 55 or 6M, will require Departmental approval in every case at the present time due to a lack of specification containers for Type B quantities.

Proposed § 173.396 has been modified to provide for package limits for the Spec. 6L and 6M packages. The limits are presently specified in Part 178.

Proposed § 173.398 has been modified to prescribe the criteria for Type A packages (normal conditions incident to transportation) as well as the previously prescribed criteria for Type B packages (hypothetical accident condi tions). The allowable release of radioactive material from packages under the Type B tests, and the test conditions themselves, have been clarified to conform with the present requirements of 10 CFR Part 71 of the AEC or the IAEA regulations.

In proposed § 173.399, the reference to a zero transport index for the white label has been deleted. An additional example of dual labeling requirements is shown for radioactive materials containing nitric acid. Provisions have been included in § 173.402 to require two radioactive materials labels on opposite sides of each package, and to allow the use of foreign labels which conform to the IAEA regulations. Labels for other hazardous materials which are required for air transportation are authorized for surface transportation as well.

The proposed change in the package marking requirements for full-load shipments of all hazardous materials has been retained. These requirements have been in effect for all shipments by water and for Department of Defense shipments for many years.

The provisions of paragraph (b) have been modified to reflect the shipping paper requirements of § 173.427, which itself has been changed to include informational material required on the shipping papers for radioactive materials shipments. These informational modifications conform to the IAEA regulations. Section 173.430 has been modified to allow for the use of an optional reference to IATA regulations for air shipments.

Proposed § 177.870(g) has been modified to allow transportation of radioactive materials on buses under essentially the same conditions as presently provided for. Storage and loading restrictions have been prescribed in place of the proposed prohibitions for Category II and III packages.

Specification 2R, in § 178.34, has been modified to provide for reduced size of the letters of identification.

A number of cylinder specifications have been corrected to reflect the proper cross-references to Part 173.

Specification 6L, in § 178.103, has been modified to provide for additional types of spacers ("spiders"). The total quantity of required vermiculite has been deleted as extraneous because the required density provides automatically for the total weight control. Marking requirements have been modified to conform with other steel drum requirements. Closure requirements have been modified to require a specified metal thickness and locking ring attachment. Recent accident tests demonstrated the inadequacy of the more common lightweight locking rings. Loading capacity limitations have been relocated to § 173.396.

Section 103.31 of Title 14 has been modified to clarify the identification of certain labels used on mixed cargoes.

Because of the complex nature of these amendments, and the impact that they will have on the transportation of radioactive materials, and to allow a reasonable time for compliance with the changes made herein, the effective date of the amendments is December 31, 1968. However, compliance with these amendments is authorized on and after the date of publication in the FEDERAL REGISTER.

(33 F.R. 14918, October 4, 1968)

[49 CFR Part 173]
[Docket No. HM-2, Notice 69–14]

<h1 style="text-align:center">SHIPPERS</h1>

Packaging of Certain Radioactive Materials

The Hazardous Materials Regulations Board is considering amending §§ 173.394 and 173.395 of the Hazardous Materials Regulations to simplify the procedures for obtaining approval of packaging for Type B quantities and large quantities of both special form and normal form radioactive materials.

Interested persons are invited to participate in the making of these proposed rules by submitting written data, views, or arguments as they may desire. Communications should identify regulatory docket and notice number and be submitted in duplicate to the Secretary, Hazardous Materials Regulations Board, Department of Transportation, 400 Sixth Street SW., Washington, D.C. 20590. Communications received before July 22, 1969, will be considered by the Board before taking final action on the notice. All comments will be available for examination by interested persons at the Office of the Secretary of the Board both before and after the closing date for comments. The proposals contained in this notice may be changed in light of comments received.

Amendment 173–3 (Docket HM-2) contained a major revision of the requirements for packaging of radioactive materials. One part of that revision was the adoption of what are in effect performance type packaging requirements for certain types of radioactive materials. Section 173.398(c) contains the "Type B" packaging requirements. However, notwithstanding the adoption of these performance type requirements, the Board required that before Type B quantities or large quantities of special form or normal form radioactive materials could be shipped, the packaging must be approved by the Department. In the case of large quantities the Type B packaging must also meet either the U.S. Atomic Energy Commission standards or the 1967 regulations of the International Atomic Energy Agency. The regulations further specify that the Departmental approval of large quantity packaging must be issued under Part 170 of the Department's regulations (i.e., a special permit). As a matter of practice all approvals, whether for Type B quantities or for large quantities, have been treated as requests for special permits.

Upon a review of the procedure for handling the above described approvals, the Board has concluded that it is inappropriate to continue treating requests of this type as requests for special permits since in fact no exception from the Hazardous Materials Regulations is involved. What is involved is a review by the Department's Office of Hazardous Materials (in cooperation with the U.S. Atomic Energy Commission) of whether in fact the proposed packaging complies with the performance specification requirements of § 178.398. The document issued by the Department is in effect a statement to the effect that the Department agrees that a certain described packaging is for certain stated contents consistent with the Department's regulations. In view of the recommended International Atomic Energy Agency requirement that for international shipments the "national competent authority" certify (with an identifying symbol) as to the packaging adequacy and that the package be so marked, the Board believes that the required approval should take the form of a "Certificate of Approval" with an identifying number. Thus persons who obtain such approval will be assured that they can comply with international requirements in those countries that have adopted the above described recommended practice.

In addition, certain requirements now routinely included in special permits would be made a part of the regulations, such as prior notifications by the shipper and special unloading instructions.

(34 F.R. 8244, May 28, 1969)

SUMMARY • **Docket No. HM-4** • *Development of the prohibition against the loading and transportation of Class A or B poisons with foodstuffs, feed, or other material intended for human or animal consumption. See Docket No. HM-112.*

[Docket No. HM-4; Amdt. No. 67–1]

PARTS 171–190—HAZARDOUS MATERIALS REGULATIONS OF THE DEPARTMENT OF TRANSPORTATION
Miscellaneous Restrictions Against Loading and Transporting Class B Poisonous Liquids or Solids With Foodstuffs

There have been several recent instances of food poisoning attributed to the consumption of food which had become contaminated by a poisonous insecticide or pesticide during the course of transportation. The poisons involved were liquids or solids, of the types defined and described as class B in section 173.343 of the Hazardous Materials Regulations of this Department. These incidents have caused the death of several persons. While none of the incidents occurred in the United States, there has been, within the past year, a number of container leakages, adverse handling experiences, and motor vehicle accidents involving shipments of poisonous liquids or solids, class B. Therefore, it is possible for the conditions which caused the deaths in other countries to arise in this country.

Investigations of the leakages and other accidents in the United States have not yet developed all of the information which would indicate conclusively the need for a change in the specification packaging requirements for poisons. However, a review of all of the incidents concerned clearly shows that there is good cause to restrict mixed shipments of poisonous liquids or solids, class B, and foodstuffs, feeds, and other materials intended for consumption by humans or animals, which are not packaged in air tight nonpermeable containers to minimize the possibility of food poisoning that could be caused by inadvertent contamination during transportation. Also, because of the multiple uses of transportation equipment, it is considered necessary to place a restriction on the reuse of transportation equipment which has been contaminated by the leakage of poisonous liquids or solids, class B, until the contamination has been removed, to preclude injury to transportation personnel and contamination of subsequent shipments.

As a situation exists which demands immediate adoption of this regulation in the interests of public safety, it is found that notice and public procedure hereon are impractical and good cause exists for making this amendment effective in less than 30 days.

(32 F.R. 20982, December 29, 1967)

[49 CFR Parts 170–189]
[Docket No. HM-4]

TRANSPORTATION OF PESTICIDES
Advance Notice of Proposed Rule Making; Request for Public Advice

The use of liquid class B poisons (particularly pesticides) is increasing rapidly throughout the United States. From reported incidents, we believe the leakage of liquid pesticides during transportation is also increasing rapidly. Accordingly, it is important that the present regulations be reviewed to determine whether they provide an adequate level of safety for the transportation of these poisons.

This advance notice of proposed rule making invites the public to advise us on the reasons for the leakage, the resulting safety hazard, and appropriate regulatory action. We invite advice only on liquid poisons in this proceeding.

Recent regulatory action. On December 29, 1967, the Department published Amendment No. 67–1 (32 F.R. 20982) in Docket No. HM-4, Miscellaneous Restrictions Against Loading and Transporting Class B Poisonous Liquids or Solids with Foodstuffs. This regulation restricted transportation of any Class B poison in the same vehicle with foodstuffs, feeds, or any other material intended for consumption by humans or animals. The amendment also provided for inspection and decontamination of vehicles.

This regulation sought to minimize harm resulting from leaks, but it did nothing to prevent leaks. Further, it applies only to foodstuffs and feeds, not to clothing, cosmetics, and other consumer items capable of transmitting the poisons.

Numerous petitions and complaints were filed with the Board since issuance of Amendment 67–1. The Board has issued a notice of proposed rule making (Notice No. 69–12, Docket No. HM-4, 34 F.R. 7456) to resolve some of these problems by clarifying the language, adopting additional restrictions on the commingling of poisons and foodstuffs and making the rule also applicable to Class A poisons.

Facts. Our knowledge of the number of leaks and the quantities of poisons which escaped is limited to major spills. We do not have a system for collecting information on all accidents and incidents. We are developing such a system, but it is of no help in this instance.

From the limited information we have, we know that hundreds of containers of class B poisons leaked last year. This leads us to believe that thousands of leaks actually occurred during transportation. A substantial number of the leaks were of the more toxic class B poisons.

We do not know of any deaths in the United States resulting from these leaks, but there have been hundreds of deaths abroad from similar leaks.

The following examples of recent poison leaks were selected to show that leaks occur (i) in more than one container on some shipments, (ii) in drums of different sizes, built to different specifications, (iii) in seams, chimes, heads, and closures, and (iv) in both truck and train transport. The containers were made by different manufacturers, filled with different poisons by different shippers, and shipped via different carriers.

Safety problems. During the past year we have worked with shippers, carriers, container manufacturers, and Federal and State government officials, seeking the precise causes of the leaks. The number of leaks indicate a need for regulation, but we need more precise information to determine what regulation is needed. The first step is to define the safety problems: the causes of the leaks.

The principal safety problems appear to be inadequacy of containers and carelessness of shippers. Theoretically, the authorized containers are adequate, if the manufacturers, shippers, and carriers carefully follow all regulatory requirements. Actually, many of the containers leak during transit. It follows that our safety standards are not high enough; they are not people-proof; they do not provide a margin for predictable error.

More particularly, these are areas of inquiry to define the safety problems:

1. Whether the authorized containers, such as Specification 17E and possibly others of the Specification 17 series, are adequate for the transportation of the more toxic materials. This inquiry covers everything which contributes to container integrity, such as gauge and quality of the steel, quality and resilience of lining material, the manufacturing process, inspection and quality control, and testing of finished containers.

2. Whether the leaks result from improper filling and closing of containers. If so, is it because the regulations are inadequate or because the regulations are not followed? If they are not followed, is it because of practical or other problems of complying with the regulations?

3. Whether the leaks result from damage in transit. If so, is it because the regulations are inadequate or because the regulations are not followed? If they are not followed, is it because of practical or other problems of complying with the regulations?

4. Whether regulatory standards should be higher for the more toxic class B poisons.

Possible solutions. As we have been defining the safety problems, we have been considering possible solutions. We have received specific recommendations from the California State Health Department, the National Agricultural Chemicals Association, and the Steel Shipping Container Institute. These are some of the regulatory solutions which we are considering:

1. Require shippers (where appropriate, this term includes the person who fills the container) to use containers produced by manufacturers approved by the Hazardous Materials Regulations Board. Provide for the Board to withdraw approval from manufacturers who do not meet regulatory standards.

2. Prohibit use of Specification 17E and possibly others of the Specification 17 series.

3. Improve integrity of presently authorized containers (for example, by raising the specification standards for gauge of steel or quality of steel, or both) with particular attention to drum heads.

4. Require manufacturing procedures which will not unduly stress the steel.

5. Require comprehensive nondestructive testing of each container and complete destructive testing of frequent random samples, relating test procedures to the actual use for which the container is built.

6. Establish quality standards for lining material, including sufficient resilience to withstand transportation shocks without cracking.

7. Require quality control procedures which will ensure that the manufacturer meets regulatory standards.

8. Require shippers to inspect each container before filling, to ensure that it has not been damaged in transit to him; prohibit use of damaged containers.

9. Require shippers to leave enough outage after filling so that container can be closed without overflow.

10. Require shippers to use fail-safe closure devices and attachment procedures.

11. Require shippers to inspect and clean each container after filling.

12. Require shippers to observe containers in both the upright and inverted positions long enough to detect leaks.

13. Require shippers to palletize or crate (bottom, side, and top protection) all shipments of small containers.

14. Require shippers to inspect each container after storage and before shipment.

15. Require shipper to furnish, and carrier to have, precise chemical name and emergency instructions with each shipment.

16. Prescribe stowage rules, including vertical bulkheads between poisons and other freight, horizontal partitions between layers of containers, and stack height limitations.

17. Prohibit trailer-on-flat-car carriage.

18. Require "poison" label on each package, even in truckload or carload lots, and placard on each truck, even when the amount of poison is less than 1,000 pounds.

19. Impose routing and stop-over restrictions, to limit extent of public exposure.

20. Require shipment in fully enclosed vehicles, to lessen chance of loss of containers.

21. Prohibit shipment on vehicles which have wooden floors, because of difficulty of decontamination after a leak.

22. Prescribe rules for handling contaminated freight and decontamination vehicles.

Scope of notice. This is not a proposal to change the regulations. It is an effort to obtain public participation early in the rule-making process. It is an effort to develop facts upon which to base rational rule making. We invite the general public to advise us on all aspects of this subject.

We invite interested persons to give us their views by July 21, 1969. Advice (identifying the docket number) should be submitted in duplicate to the Secretary, Hazardous Materials Regulations Board, Department of Transportation, 400 Sixth Street SW., Washington, D.C. 20590.

(34 F.R. 7545, May 9, 1969)

[49 CFR Parts 174, 175, 177]
[Docket No. HM-4; Notice No. 69-12]
LOADING AND TRANSPORTING POISONS, CLASS A OR B WITH FOODSTUFFS
Proposed Restrictions

The Hazardous Materials Regulations Board is considering amending Parts 174, 175, and 177 of Title 49 and Part 103 of Title 14 of the Department's Hazardous Materials Regulations (1) to clarify a previous amendment (Amendment 67-1, Docket HM-4) concerned with the carriage of poisons and foodstuffs and (2) to propose further amendments to the current loading and carriage requirements. An advance notice of proposed rule making will soon be published which will request public advice on the reasons for leakage of packages, the resulting safety hazards, and appropriate regulatory action.

Interested persons are invited to participate in the making of the proposed rule by submitting such written data, views, or arguments as they may desire. Communications should identify the docket and notice number and be submitted in duplicate to the Secretary, Hazardous Materials Regulations Board, 400 Sixth Street SW., Washington, D.C. 20590. Communications received on or before June 10, 1969, will be considered before final action is taken on the proposal. All comments received will be available for examination by interested persons at the Office of the Secretary, Hazardous Materials Regulations Board, both before and after the closing date for comments.

Amendment 67-1 was issued on December 21, 1967 (32 F.R. 20982, 12-29-67) after a number of deaths occurred outside the United States due to eating food that was contaminated by an insecticide or pesticide. Many of the contamination incidents resulted from shipments originating in the United States. A number of injuries occurred in the United States from similar contamination incidents. Numerous petitions and complaints have been filed with the Board since issuance of Amendment 67-1 pointing out certain problems caused by the rule. In this notice the Board proposes to resolve some of these problems by making certain clarifying changes in language. In addition, the Board proposes to adopt additional restrictions on the commingling of shipments of poisons and foodstuffs.

One major difficulty in attempting to comply with the amendment was a lack of certainty as to how far a carrier was required to go in identifying foodstuffs, feeds, or any other material intended for consumption by humans or animals under the provisions of the amendment. The Board recognizes the difficulties inherent in attempting to segregate packages of foodstuffs under normal cargo handling procedures. It is proposed to clarify and relax this requirement. Only those foodstuffs and feed, etc., which are clearly marked as or are known to be such need be considered in applying these regulations.

Carriers have had difficulty in understanding the meaning of the terms "airtight and nonpermeable." Although the normal dictionary meaning of these terms was intended, the amount of confusion that has ensued warrants reconsideration of this requirement. In many cases, the mechanics of making on-the-spot field determinations of whether a particular packaging for foodstuffs was airtight or nonpermeable were impracticable. The practical result often was an operational restriction by carriers against loading any poisons with any foodstuffs, regardless of packaging. Several carrier representatives have suggested that the regulations be amended to that effect to reflect the practicalities of transportation.

Although the regulations in question at present deal only with poison, class B, several class A poisons are shipped as liquids. The Board therefore proposes to include all class A poisons in the restriction against loading with foodstuffs, feeds, or any other material intended for consumption by humans or animals. Accordingly, the Board proposes to place a total restriction on the transportation of all packages of poisons, class A or B, with packages of foodstuffs, feed, or any other material intended for consumption by humans or animals as discussed above.

Another clarifying change that is proposed would limit post-transportation inspection on aircraft to the compartment in which the poisonous material was carried. Aircraft have a number of isolated cargo compartments with no reasonable way for materials to leak from one compartment to another. Therefore, it is unreasonable to require inspection of an entire aircraft in every case.

(34 F.R. 7456, May 8, 1969)

[Docket No. HM-4, Amdt. 174-5, 175-3, 177-9]

PART 174—CARRIERS BY RAIL FREIGHT
PART 175—CARRIERS BY RAIL EXPRESS
PART 177—SHIPMENTS MADE BY WAY OF COMMON, CONTRACT, OR PRIVATE CARRIERS BY PUBLIC HIGHWAY

Miscellaneous Restrictions Against Loading and Transporting Poisons (Class A or B) With Foodstuffs

The purpose of these amendments to the Hazardous Materials Regulations is to modify certain restrictions on the loading and transporting of poisons (class A or B) with foodstuffs, feed, or other material intended for consumption by humans or animals.

On May 8, 1969, the Hazardous Materials Regulations Board published a notice of proposed rule making, Docket No. HM-4; Notice No. 69-12 (34 F.R. 7456) proposing to modify the existing restrictions against commingling poisons and foodstuffs during shipment. The comments in response to the notice for the most part supported the proposed changes. Several comments raised questions that indicated that the intent of both the present and proposed requirements could be further clarified. One commenter raised numerous objections to both the present and proposed restrictions, most of which had been previously submitted to, and considered by, the Board. The most significant comments and changes to the regulations are as follows:

One commenter indicated that there was still some confusion as to the intent of the words "foodstuffs, feeds, or any other material intended for consumption by humans or animals". This commenter questioned whether these words could be interpreted to cover any materials that normally might come in contact with the human body or did they exclude " . . . clothing, cosmetics, and other consumer items capable of transmitting poisons," as indicated in an advance notice of proposed rule making published by the Director of the Office of Hazardous Materials on May 9, 1969 (34 F.R. 7545). The intent of these words is to cover edibles and the language of the regulation has been clarified in this regard. The Board recognizes that there are many items such as clothing that could become a hazard to human life if contaminated by certain poisons. However, these items were not included in the original amendment adopted in December of 1967 or in the proposal upon which this amendment is based. The need for further rule making in this regard is still being considered based on the response to the aforementioned advance notice of proposed rule making.

Two commenters suggested that the Board require foodstuffs to be marked as such. The Board recognizes that the marking of foodstuffs would greatly enhance both the problems of compliance and the safety benefits resulting from the subject regulation. However, the Hazardous Materials Regulations Board has no jurisdiction directly over the transportation of foodstuffs. Similarly, the Board does not have the authority, as suggested by one commenter, to prescribe requirements for the safe disposal of the poisonous products resulting from any decontamination.

One commenter suggested that with respect to the car cleaning requirements it would be helpful if acceptable contamination levels could be prescribed. The Department of Health, Education, and Welfare (Public Health Service and Food and Drug Administration) is presently working actively in this field and the Board intends to make use in future regulatory changes of any significant information developed from these studies.

Two commenters suggested that foodstuffs and poisons could be shipped in the same vehicle provided they are separated by airtight and nonpermeable partitions. Such a provision would appear to have all of the inherent problems and confusion that arose from the use of the terms "airtight" and "nonpermeable" in the original amendment. For rail cars and highway vehicles, the Board believes that it is not too much of a burden, considering the potential dangers, to make the prohibition against commingling apply to each car.

One commenter suggested that the inspection of aircraft cargo compartments should only be carried out if a package has been found to be leaking or damaged. The Board has been concerned, however, with the number of instances of contamination of other goods by poisons when it was not immediately known that a package had leaked. By the time the package leakage was noticed, the other freight had been transshipped in many different directions. Therefore, the Board does not consider it appropriate to limit the inspection requirement to cases of known leakage.

One commenter protested the application of the prohibition against mingling to all classes A and B poisons. This commenter indicated that the Board should single out and limit the restriction only to the most dangerous items, such as parathion and other organic phosphates. The Board does not agree. While there is necessarily a difference in the degree of hazard among classes A and B poisons, the Board believes that the leakage of any class A or B poison on edible foodstuffs is so potentially hazardous that no effort should be made at this time to determine "safe" class A or B poisons insofar as shipments with foodstuffs are concerned. This commenter also suggested that the restriction should be imposed only on liquid poisons and not on solid poisons. The Board believes that the restriction should be total regardless of state of the poisonous material. Many food products are shipped in packagings that could be penetrated by dry materials, or which could retain deposits of dry material in or near discharge openings.

One commenter pointed out an inconsistency existing in the present regulations that results from the present exemptions for small quantities of class B poisons when transported by highway. The Board recognizes that there is no justification for permitting a small quantity of a class B poison in a motor vehicle carrying foodstuffs while the same quantity of class B poison would be barred from a railroad car carrying foodstuffs. The Board is presently reconsidering in toto the authorized small quantity exemptions and intends to include this term for consideration in that study.

One commenter suggested that the rule could be averted by shippers who fail to mark the edible nature of the contents on the package and by originating carriers who fail to carry forward the "poison" notation on the interchange forms. The Board recognizes that these restrictions are not foolproof and that the effectiveness of its regulation is directly related to the ease of identifying foodstuffs. However, as indicated above, solution of this problem is to some extent outside the scope of the Board's authority. This rule, as is any rule, is effective only if complied with. That persons may render regulations ineffective by ignoring them is not valid reason against regulating.

One commenter stated that both the present rule and the proposed changes would disrupt the marketing and distribution of class B poisons by grocery warehouses that reship along with edibles. This commenter made this same point earlier with respect to the Board's first action in this regard in December 1967. The present regulation has been in effect now for over 18 months and the Board has received no evidence that commerce of any kind has been adversely affected thereby to any significant degree. Nor has the Board received any specific evidence that this amendment will have such an effect. Therefore, the Board must conclude that neither the present rule nor the changes adopted herein will adversely affect the grocery industry in the United States.

(34 F.R. 18553, November 21, 1969)

SUMMARY • Docket No. HM-5 • *Requirement for special test procedures and future control methods to prevent stress corrosion cracking in DOT MC 330 and 331 pressure tank trucks used for transporting anhydrous ammonia or liquified petroleum gas.*

[Docket No. HM-5; Amdts. 173–1, 177–1]
PART 173—SHIPPERS
PART 177—SHIPMENTS MADE BY WAY OF COMMON, CONTRACT, OR PRIVATE CARRIERS BY PUBLIC HIGHWAY
Stress Corrosion in MC 330 and MC 331 Cargo Tanks

Transporters of anhydrous ammonia in MC 330 and MC 331 cargo tanks (constructed of quenched and tempered steels) have recently experienced numerous occurrences of a phenomenon known as stress corrosion cracking. Stress corrosion cracking is the spontaneous failure of a metal resulting from the combined effects of corrosion and stress.

Fortunately, the stress corrosion cracking experienced in MC 330 and MC 331 cargo tanks to date has been such that only minor leaks have occurred with sufficient warning being provided to the operator to facilitate corrective action. However, stress corrosion cracking in the pressure vessels concerned could result in a much more drastic failure wherein an entire section of the vessel could fail and thereby permit a catastrophic escape of the vessel's contents.

Under § 173.315 of the hazardous materials regulations the MC 330 and MC 331 quenched and tempered steel cargo tanks concerned are authorized for the transportation of 23 different kinds of gas. All of these materials are transported under high pressure. Many of these commodities are toxic or flammable.

The purpose of this amendment is to require the immediate inspection of MC 330 and MC 331 quenched and tempered cargo tanks to determine the need for repair, and to insure the product retention integrity of the vessels involved. The priority of inspection is based on the commodities transported in the past and the existing or potential use of the vessel. Tests already conducted have resulted in the withdrawal from service of a large number of defective tanks.

Recognizing the potential hazard of anhydrous ammonia so far as it contributes to the stress corrosion cracking of quenched and tempered steel MC 330 and MC 331 cargo tanks this amendment simultaneously requires inhibiting and preventive measures to reduce the stress corrosion cracking in future operations. The introduction of a minimum of 0.2 percent water by weight to the anhydrous ammonia shipped after January 31, 1968, is being required to inhibit the corrosive action of anhydrous ammonia. Purging of cargo tanks before loading with anhydrous ammonia is being required to remove the air in the tank, since this procedure has also been proved to be an effective deterrent against stress corrosion cracking. (The presence of carbon dioxide and oxygen in the tanks under pressure have been shown to be contributory to the corrosion cracking.)

Research by the affected industries has also indicated that metallurgical grade anhydrous ammonia (i.e., at least 99.995 percent pure) will not cause stress corrosion cracking if the vessel has been properly purged of air. This amendment makes appropriate provisions for this special use.

This amendment is addressed to known stress corrosion conditions. There are still some important operating conditions involving potential stress corrosion cracking which may be uncovered during the required inspection. There is reason to suspect that the sulfides, which may be found in "sour" liquified petroleum gas, are potential contributors to stress corrosion cracking.

Reference is made in the amendments to Part 177 to the "ASME Code, 1965 Edition." Copies of this code are available from the American Society of Mechanical Engineers, 345 East 47th Street, New York, N.Y. 10017, for a nominal charge.

As a situation exists which demands immediate adoption of this regulation in the interests of a public safety, it is found that notice and public procedure hereon are impractical and good cause exists for making this amendment effective without notice and in less than 30 days.

In accordance with the Federal Reports Act of 1942, the reporting requirements contained in Amendment 177-1 have been approved by the Bureau of the Budget under Docket No. 4S68001.

(33 F.R. 2389, January 31, 1968)

[Docket No. HM-5; Amdts. 173–2, 177–2]
PART 173—SHIPPERS
PART 177—SHIPMENTS MADE BY WAY OF COMMON, CONTRACT, OR PRIVATE CARRIERS BY PUBLIC HIGHWAY
Hazardous Materials Regulations Board; Stress Corrosion in MC 330 and MC 331 Cargo Tanks

On January 31, 1968, the Federal Highway Administrator published Docket HM-5; Amendments 173-1, 177-1 (33 F.R. 2389), containing amendments to the Hazardous Materials Regulations to require the prompt inspection of MC 330 and MC 331 cargo tanks made of quenched and tempered steels to determine the need for repair and to insure the product retention integrity of vessels involved. This action was predicated on the relatively recent occurrences of stress corrosion cracking being experienced and reported to the Department by transporters of anhydrous ammonia. The amendment was addressed to known stress corrosion conditions and contained requirements for inspections and repairs based in the main on the recommendations of affected shippers and tank motor carriers. The amendments were also made applicable to shipments of liquified petroleum gas because there was good reason to suspect that the sulfides which may be found in "sour" liquefied petroleum gas are potential contributors to stress corrosion cracking.

After extensive consultations with persons knowledgeable in the manufacture of quenched and tempered steels and in the design and fabrication of cargo tanks, it was determined that stress corrosion cracking could be precluded by purging air from cargo tanks before loading with anhydrous ammonia and by requiring anhydrous ammonia to be inhibited with 0.2 percent water by weight or be 99.995 percent pure. Requirements for these aspects were included in the initial order.

The Compressed Gas Association (CGA) has now requested the FHWA to give further consideration to those aspects of Docket HM-5 with respect to: (1) The validity of postweld heat treatment of welded repair areas, and (2) internal areas to be inspected.

CGA states that the reason for requiring welded repair areas to be postweld heat treated was that the steel companies recommended that quenched and tempered steel cargo tanks be postweld heat treated as a precaution against stress corrosion cracking resulting from transportation of anhydrous ammonia or other contaminated lading.

CGA now states that postweld heat treatment is not necessary since the lading conditions that contribute to stress corrosion cracking are being controlled. Also, they state that certain producers of quenched and tempered steels indicate that pressure vessels fabricated of these steels are better in the as-welded condition than if they are postweld heat treated. Two major producers of the steel in question have confirmed this. Also, these producers have advised that while postweld heat treatment may in some cases reduce the likelihood of stress corrosion cracking, postweld heat treatment will not guarantee that stress corrosion cracking will not occur.

On the basis of this new information and other supporting evidence recently submitted pertaining to the effect of postweld heat treatment on quenched and tempered steels of the type and thicknesses used to construct MC 330 and MC 331 cargo tanks, § 177.824(f) (5) is amended to eliminate the requirement for postweld heat treatment after welded repairs are made.

CGA has requested modification of the requirement for internal inspection opposite external welds so that inspection of areas opposite nonloadbearing supports such as lighting brackets, ladders, etc., is not required. This request is based on the premise that stress corrosion cracks opposite these small external welds have been found only in cases where cracks have also been found in the more critical external weld areas of the tank. Although the Administrator does not have any information to indicate that this premise is incorrect, the Administrator does not believe that this information alone justifies modification of the internal inspection requirement to the extent requested.

However, the Administrator has determined that inspection of internal areas opposite exterior welds that are visibly discernible on the interior of the tank will provide the level of testing desired and necessary. Accordingly, an appropriate modification of § 177.824(f) (2) is provided.

In § 173.315(a) (1) table, Note 14, the words "for metallurgical grade" have been deleted to avoid any possible confusion that could result from use of the concerned grade of ammonia for other than metallurgical purposes.

After completion of the test required therein, the concluding clause of § 177.824(f) (3) (ii) prohibited the use of quenched and tempered steel MC 330 and MC 331 cargo tanks in the carriage of liquified petroleum gases except those that meet the requirements of National Gas Processors Association specification 2140 (1962 edition). As drafted, this limitation did not apply to quenched and tempered steel tanks that were subject to § 177.824(f) (3) (i), although the safety justification for this requirement applies equally to those tanks previously used for both anhydrous ammonia and liquified petroleum gases. To remove this inconsistency, and also to place this continuing prohibition in a more appropriate portion of the regulations a new Note 15 is being added to the table in paragraph (a) (1) of § 173.315. This note makes it clear that after December 1, 1968, or after completion of the test required by § 177.824(f) (3) (i) and (ii), whichever occurs first, quenched and tempered steel MC 330 and MC 331 cargo tanks may only carry liquefied petroleum gas if it meets NGPA 2140 (1962 edition).

In addition, consistent with Note 15 of the table in § 173.315(a) (1), §§ 173.427, and 177.817 are amended by requiring shipping papers to show the notation "NGPA 2140" indicating suitability of the liquefied petroleum gas to be transported in MC 330 or MC 331 cargo tanks constructed of quenched and tempered steels. In § 177.824, paragraph (h) has been amended to require cargo tanks tested in conformance with the requirements of § 177.824 (f) to be marked with the letters "WF" to indicate the wet fluorescent magnetic particle tests have been completed. The period between publication of this amendment and the effective date is considered to be sufficiently long to allow for the marking of cargo tanks that have already completed this test.

To the extent that these amendments are other than clarifying they are for the most part relaxatory in nature and consistent with the changes requested by, and discussed with, representatives of the affected interested groups. Therefore, in order to best serve the purposes set forth above, I find that notice and public procedure is impractical and unnecessary.

To allow a reasonable time for compliance with the changes made herein, these amendments are not being made effective upon issuance. However, compliance with these amendments is authorized on and after the date of publication in the FEDERAL REGISTER.

(33 F.R. 7493, May 21, 1968)

SUMMARY • Docket No. HM-7 • *This was a broad conceptual docket covering the Board's long-range planning on revision of the regulations and setting forth a portion of the Board's regulatory philosophy.*

[Docket No. HM-7; Notice No. 68–5]

TRANSPORTATION OF HAZARDOUS MATERIALS
Notice of Plan to Revise Regulations

The Hazardous Materials Regulations Board plans to revise the regulations governing the transportation of hazardous materials, casting the regulations in general terms and eliminating much of the detail. This notice states the reasons for the revision and requests public comment on the general principles which the Board proposes to follow.

Background. The regulations reflect a commodity-by-commodity and package-by-package approach. As a result, the regulations focus on commodities instead of hazards. For example, (i) the present classification of hazards does not identify spontaneous combustion and (ii) excepting class A poisons and radioactive materials, only one classification (the greatest hazard) may be identified, even though the material may present more than one serious hazard. As another reason, the packaging regulations are repetitious how-to-do-it instructions, rather than general performance standards. For example, Part 178—Shipping Container Specifications consists almost entirely of specifications and tests for individual packages already developed; it does not set standards for the development of new packages. Recent regulatory actions have sought to synthesize the specifics, but the bulk of the regulations still deal in detail.

Different authorities developed the safety regulations for the air, land, and water modes of transportation. As a result, the hazardous materials regulations differ in many particulars between the modes; some differences are of form, others of substance. The differences impose burdens on shippers and carriers in intermodal shipments. One of the functions of the Board is to make the regulations uniform, to the extent that uniformity is consistent with the differences inherent in the modes.

Scope. The Board plans to issue notices of proposed rule making in at least these areas: Classification and labels. Handling and Stowing. Placards and Emergency Procedures. Packages.

The Board's initial emphasis will be on intermodal shipments of packaged materials, but the classification, placards, and emergency procedures notices will also cover shipments by portable tank, tank car, and tank truck. The Board does not have jurisdiction over bulk shipments of hazardous materials by water under Title 46—Shipping.

Uniformity. The regulations should be uniform for all modes of transportation, differing only where the inherent characteristics of an individual mode require a difference. The regulations should be consistent with international standards, differing only where our national needs require a difference.

Simplicity. The regulations should serve a practical safety purpose. They should be written so that those who handle the materials can understand and apply them. Throughout the regulations there will be a compromise between the complexity required to cover all contingencies and the simplicity required by human limitations.

Classifications and labels. Classifications should be based on the hazards involved in transporting the materials. Each classification should group together materials which require similar packaging and handling during transportation, or handling after an accident. Materials which pose a similar threat to safety should be classed together, without regard to historic classifications.

Labels should reflect the classifications. Labels should (i) give notice of the hazard potential of the material in the package and (ii) call attention to the need for special handling and stowing.

Handling and stowing. Instructions for in-transit handling should be on the label and should be written so that all persons involved can understand and apply them. One way of doing this might be to have a simple color-coded scheme for separating incompatible materials. Since many thousands of people, of varying levels of competence and training, are involved in in-transit handling, there will be some difficulty in finding the proper balance between flexibility (which increases complexity) and simplicity.

Placards and emergency procedures. Placards required to be posted on portable tanks, tank cars, tank motor vehicles, transport vehicles, and containers should parallel the label requirement. Placards should (i) give notice of the hazard potential of the material being transported, (ii) call attention to the need for special treatment by the carrier, and (iii) give notice of the need for special care after an accident. Since many thousands of people (cargo handlers, longshoremen, policemen, firemen, ambulance attendants) are concerned with handling emergencies involving hazardous materials, some means should be devised for giving these persons instructions on handling these emergencies. This could be done by putting code numbers on the placards and distributing booklets with emergency instructions keyed to the placard code numbers.

Packaging. Packaging requirements should relate to the classifications, the quantity of material involved, and the transport environment. Packaging requirements should be stated in terms of performance standards, rather than manufacturing specifications. The regulations should prescribe tests to determine whether the packages meet the requirements.

Cost/benefit. Each safety requirement should be subject to a cost-benefit analysis in which (i) cost is the direct cost of buying and maintaining special equipment and the indirect cost of special handling and stowing and (ii) benefit is the decrease in the hazard to the public. The public is endangered by the transportation of many materials, such as fertilizers and pesticides for agriculture, chemicals for industry, and explosives for the military. Despite the danger, the national interest requires the transportation of these materials. The regulations should minimize the hazard to the public, within the limits of economic feasibility.

Interested persons are invited to participate in developing these basic regulatory principles to guide the Board in revising the regulations. Comments (identifying the docket or notice number) should be submitted in duplicate to the Secretary, Hazardous Materials Regulations Board, Department of Transportation, 400 Sixth Street SW., Washington, D.C. 20590.

Comments received before October 15, 1968, will be considered by the Board as it proceeds with the phased implementation of the proposed revision. The principles proposed in this notice may be changed in the light of comments received. All comments will be available for examination by interested persons at the Office of the Secretary, Hazardous Materials Regulations Board, both before and after the closing date for comments.

(33 F.R. 11862, August 21, 1968)

SUMMARY • **Docket No. HM-8** • *Conversion of the U.S. hazardous materials labeling system to the United Nations recommended graphic, or symbol method of communicating hazard. See also Docket No. HM-103.*

[Docket No. HM-8]

TRANSPORTATION OF HAZARDOUS MATERIALS
Request for Public Advice on Labels and Classification

On August 21, 1968 (33 F.R. 11862), the Hazardous Materials Regulations Board announced a plan to revise the regulations governing the transportation of hazardous materials. That document announced the intention to issue notices of proposed rule making in at least four areas, including "classification and labels", and invited public help in developing the basic regulatory principles to guide the Board in revising the regulations.

The Board is planning to consider, in the near future, a proposal for classification and labels. To assist the Board in that consideration, the public is invited to express its views on the labeling system recommended by the United Nations Committee of Experts on the Transport of Dangerous Goods, and on the correlative classifications.

This document is not a proposal to change the regulations. It is an effort to get public participation early in the rule-making process.

The United Nations labels are color coded and identify each kind of hazard with a pictorial symbol and classification number.

Each label represents a class of hazard, so the first thing to consider is classification. To limit the number of classes, each class must be a compromise between (i) the detail which would be necessary to cover the infinite number of materials which may be transported, and (ii) the simplicity which is necessary for the regulations to have any practical value. The changes in classification proposed below are such a compromise.

The following table compares the U.N. classes and the proposed classes. Although they are not shown, most of the U.N. classes have subdivisions which are similar to the subdivisions of the proposed classes:

U.N. classes	*Proposed classes (classification)*
Class 1—Explosives	Class 1—Explosives. 　Explosives A. 　Explosives B. 　Explosives C.
Class 2—Gases: Compressed, liquified, or dissolved under pressure.	Class 2—Compressed gases (including liquified gases): 　Nonflammable. 　Flammable.
Class 3—Inflammable liquids. Liquids which give off an inflammable vapor at or below 150° F. open test. (The word "Inflammable" in the U.N. system has the same meaning as "Flammable.")	Class 3—Flammable liquids; combustible liquids: 　Flammable liquids—flash point at or below 80° F. (open-cup) or 73° F. (closed-cup). 　Combustible liquids—flash point at 81° to 150° F. (open-cup) or 74° F to 141° F. (closed-cup). 　Note: This is an alternative to the proposal in Docket No. HM-3; Notice No. 68-2, to raise the definition of flammable liquids to 110° F. (open-cup) or 100° F. (closed-cup).
Class 4—Inflammable solids; substances liable to spontaneous combustion; substances which, on contact with water, emit inflammable gases.	Class 4—Flammable solids; spontaneously combustible materials; water reactive materials.
Class 5—Oxidizing substances; organic peroxides.	Class 5—Oxidizing materials; organic peroxides.

Class 6—Poisonous (toxic) and infectious substances.

Class 6—Poisonous (toxic) and Etiologic agents:
Poison gas.
Poison A (solid or liquid).
Poison B (solid or liquid).
Tear Gas.
Etiologic agents.

Class 7—Radioactive substances
Class 8—Corrosives

Class 7—Radioactive materials.
Class 8—Corrosives:
Corrosive liquids.
Corrosive solids.

Class 9—Miscellaneous dangerous substances. (This is an undefined catchall.)

Definitions of some of the proposed classifications are not contained in the Hazardous Materials Regulations and some of the classifications in the regulations may be inadequately defined. We ask for discussion of the adequacy of the proposed classifications using a common sense definition of the terms and assuming that they can be defined satisfactorily at a later date. For example, commenters should assume that poison A represents a greater hazard than poison B and that the line of demarcation will be drawn in a later rulemaking procedure.

We are considering new labels, shown in the appendix, to represent the classes of hazards discussed above. The proposed labels are consistent with the present recommended U.N. labeling system. This will result in one labeling system serving both domestic and international shipments. The illustrations have been somewhat simplified to indicate the general approach under consideration, and the recommendations relate essentially to danger labels. This system permits the addition of handling instructions and other safety information, such as "Keep Away From Fire and Heat" and "Do Not Drop".

The proposed colors are as indicated in the appendix. For compressed gases which also have flammable or poisonous characteristics, a second pictorial symbol—a gas cylinder—has been added to the bottom half of the flammable or poisonous label to denote additional hazard. This additional symbol complements the use of the U.N. class numeral and provides a better identification of the compressed gas.

Before the U.N. labeling system could be adopted, numerous changes would be required in the Commodity List in § 172.5 of the regulations. For example, flammable solids now require a "yellow" label, as do all oxidizing materials. Under the proposed system, each flammable solid and oxidizing material must be evaluated to determine which of the flammable or oxidizing materials labels would be required, and an appropriate change must be made in the Commodity List entry for that item.

When one label does not represent all of the serious hazards presented by a material, the U.N. system provides for the additional hazards to be represented by additional labels without the class number. For example, chlorine trifluoride would have a combination of labels to show that it is a corrosive (class 8), poisonous (class 6), compressed gas (class 2), and also an oxidizing material (class 5). We are not now considering the adoption of this part of the U.N. system. The regulations now provide that materials which present more than one hazard must be classified and labeled according to the greatest hazard, except that those materials which are also class A poisons or radioactive materials must be classified and labeled to show both hazards. The matter of dual (or maybe multiple) labels may be considered at a later date.

Public views are specifically requested on these basic questions: (1) Do the proposed classes identify the hazard potential related to transportation? (2) Do the proposed classes identify the hazards which require special packaging, special handling during transportation, or special handling after an accident? (3) Would each label give notice of the hazard potential of the material in the package to which it is attached? (4) Would each label call attention to the need for special handling and stowing of the container?

Interested persons are invited to give us their views, before December 1, 1968, on whether the U.N. labeling system would be appropriate for our regulations. Correspondence (identifying the docket number) should be submitted in duplicate to the Secretary, Hazardous Materials Regulations Board, Department of Transportation, 400 Sixth Street SW., Washington, D.C. 20590. All responses to this request will be available for examination by interested persons at the Office of the Secretary, Hazardous Materials Regulations Board.

(33 F.R. 15347, October 16, 1968)

[49 CFR Parts 172, 173, 174, 175, 176, 177]
[Docket No. HM-8; Notice No. 70-13]
CLASSIFICATION AND LABELING OF HAZARDOUS MATERIALS
Transportation of Hazardous Materials

On October 16, 1968 (33 F.R. 15347), the Department of Transportation sought public advice and comment on adoption of the United Nations' hazardous materials classification and labeling system. Comments generally favored adopting the U.N. system. This notice proposes to (1) convert the existing labeling requirements to the U.N. system (with minor exceptions), (2) to completely revise the commodity list, and (3) to revise some of the hazard classification categories.

ANALYSIS OF COMMENTS

Most of the comments indicated that the question of redefinition of flammable and combustible liquids remains highly controversial. Most of the controversy stemmed from a concern that new packaging requirements would be imposed for combustible liquid shipments, and from the continuing argument as to whether the closed cup or the open cup method should be prescribed. The open cup method is being retained for the present, keeping the limit at 80° F. for flammable liquids. Most nonshipper comments on combustible liquids indicated a need for identification of vehicles carrying combustible liquids. The question of combustible liquids is being covered in a separate rule-making action. Accordingly, this notice will be generally limited to those changes necessary to effect a meaningful conversion from the present labels to the U.N. labels.

Concern was expressed by numerous commenters that previously established packaging procedures would be changed by the changes in classification. To prevent this, the same packaging references are used for those reclassified items in almost all cases. If rearrangement is necessary at a later time, it will be the subject of a separate rule-making action.

A number of commenters suggested that the labels could be blown up to a 10″ × 10″ size for use as placards on rail cars and on container vans for water shipment. This notice deals primarily with labels for packages. Rail and highway placarding generally will be the subject of separate rule-making action.

The only new hazard classification items covered in this notice are etiologic agents, and a separation of the present flammable liquids and solids categories into flammable liquids, flammable solids, water-reactive materials, and spontaneously combustible materials. The Board recognizes the continuing need for benchmark tests for these materials, and is undertaking to develop those benchmarks.

The present term "tear gas, poison class C" would be replaced by the term "irritant". This is to bring the Department's terminology more in line with the terminology used by the Departments of Defense, Agriculture, and Health, Education, and Welfare. At present, users of these materials refer to the materials by one name during use or storage, and a different name for transportation.

About half of the comments pertained to the provisions for etiologic materials. Many commenters objected to including etiologic agents as "poisonous materials". Others suggested that the transportation of these materials are already adequately regulated by other government agencies. The Department has coordinated this proposal with the Departments of Defense, Agriculture, and Health, Education, and Welfare. The Transportation of Explosives Act (18 U.S.C. 831-835) clearly assigns this jurisdiction to the Department of Transportation. The inclusion of etiologic agents as poisonous materials is in conformance with the U.N. system. Therefore, the proposed classification of etiologic agents is retained in this notice.

There was almost unanimous disagreement with the use of the skull-and-crossbones symbol for etiologic agents. Since the U.N. system does not prescribe a label for etiologic agents, the Board is proposing use of the symbol developed by the National Cancer Institute of the Public Health Service.

Several commenters objected to the use of the U.N. numbering system on the labels. Although a need for the use of these numbers in domestic shipments has not yet been clearly defined, the numbers are important in international shipments. This notice proposes as an alternate method, to leave the numbers off the labels for routine domestic use, but permit them to be overstamped on the labels for all export shipments. Import shipments may have the numbered labels, of course. Further comments would be useful on this matter.

The Board has decided to postpone the inclusion of corrosive solids until a later rule-making action.

There was adverse comment to the requirement that classes A, B., and C explosives all carry the same type of label. While there is merit in the objection, we are keeping the labels consistent with the U.N. system, but are providing exemptions to labeling for certain class C explosives of minimal risk.

Several commenters suggested that explosive (class C) and organic peroxides should really be classified as flammable solids and explosives, respectively. While those comments have some merit, their adoption would not be appropriate in this notice. This may be considered later when setting up benchmarks and tests to be used in classification of materials.

Many commenters expressed concern about the U.N. system of multiple labels. This notice retains the present single hazard labeling concept, with the existing exceptions. If multiple labeling is necessary for safety, it will be proposed in a separate rule-making action.

OTHER PROPOSED CHANGES

It appears that several basic types of information are needed by those who may be exposed to hazardous materials between the producer and the consumer. These include:

1. General warning. This should indicate the presence and nature of the hazard. This warning would be provided

by the color, hazard symbol, and hazard name, as described in the request for public advice, and retained in this proposal.

2. Special handling precautions. These should consist of "alert" words or short phrases. Long, "small print" statements are likely to go unread. The Board proposes to include the "DANGER/WARNING/CAUTION" system used by the Departments of Agriculture, and Health, Education, and Welfare in order to provide great uniformity with other governmental requirements. Specific comments are requested on the usefulness of this system.

3. Emergency procedures. Information is needed, but we doubt whether this detailed information should be included on the primary warning label.

With regard to the special handling precautions, a number of suggestions were made to the Board as to the exact words to be used. These suggestions included the following listed approximately in order of priority:

Explosive A:
 *Keep away from heat, sparks, and open flame
 Handle carefully
 Do not drop
Explosive B:
 *Keep away from heat, sparks, and open flame
 Handle carefully
Explosive C:
 *Keep away from heat, sparks, and open flame
Nonflammable Compressed Gas:
 *Keep away from heat and fire
 Do not drop
 Contents under pressure
Flammable Compressed Gas:
 *Keep away from heat, sparks, and open flame
 *Handle carefully
 Do not drop
 *Ventilated storage only
 Contents under pressure
 *Do not apply water to contents
 *Keep dry
Oxidizing Material:
 *Do not store near combustible materials
Organic Peroxide:
 *Keep away from fire and open flame
 *Do not store near combustible materials
 *Handle carefully
 Ventilated storage only
 Remove damaged packages to safe place
Poison A:
 *Keep away from food products
 *Wash thoroughly after handling
 *Harmful or fatal if taken into body
 If leaking: Do not touch contents or breath fumes
 Handle carefully

Remove damaged packages to safe place
Poison Gas:
 *Keep away from heat and fire
 *Handle carefully
 *Do not drop
 Avoid breathing gas
 Beware of fumes—avoid inhalation
 Ventilated storage only
 Contents under pressure
Flammable Liquid:
 *Keep away from heat, sparks, and open flame
 *Ventilated storage only
 Remove leaking packages to safe place
Flammable Solid:
 *Keep away from heat and open flame
Spontaneously Combustible Material:
 *Keep away from heat and open flame
 *Ventilated storage only
Water-Reactive Material:
Poison B:
 *Keep away from food products
 *Wash thoroughly after handling
 *Harmful or fatal if taken into body
 If leaking: Do not touch contents
 Handle carefully
Irritant:
 *Wash thoroughly after handling
 *Avoid irritating fumes
Etiologic:
 *If leaking or broken: Evacuate area, do not touch
 *Handle carefully
 *Keep away from food products
 Wash thoroughly after handling
Corrosive Liquid:
 *Wash thoroughly after handling
 *Flush spillage with water
 Remove leaking packages to safe place

*The Board proposes to require that the phrases marked with an asterisk (*) appear on each label.

Some of this suggested label information is duplicative, and in some cases it is probably not possible to include all of the informational items. Commenters who believe that a different priority would be more appropriate should indicate their preference.

Although the list of commodities (§ 172.5) would be completely revised to reflect the new labels, no previously unregulated specific commodities have been added to the list.

Additional columns would be added to the commodity list for cargo aircraft and passenger aircraft limitations. The quantity limits specified for air are taken from the Air Transport Association Tariff 6-D, supplemented as appropriate by data from the International Air Transport Association regulations.

The terms "mark", "label", and "placard" have often been misused and misunderstood. Definitions of those terms have proposed to help in the proper application of these terms. The terms "not accepted" and "forbidden" would be clarified to reflect current usage and needs.

In Subpart D, new definitions are proposed for flammable solids, spontaneously combustible materials, and water-reactive materials. For the most part, the redefinitions are only a rearrangement of the terms of the present definitions for flammable solids.

The Board hopes to consolidate all of the definitions, which are presently scattered throughout the regulations into Part 171. Although this notice proposes to place the new and revised definitions in the format of the present regulations, the Board intends in the final rule to make that consolidation.

The general marking and labeling requirements in §§173.402 through 173.414 would be revised to provide for the new labeling usage. Numerous editorial changes would be made in the final rule to incorporate the proper names and section references for the new labels. Since these changes are not substantive in nature, they are not listed specifically herein. Examples are §§173.25, 173.29, and 173.134.

If the proposed labeling system is adopted, numerous other changes relating to shipments of hazardous materials may also be desirable, and would be proposed at a later date.

[Docket No. HM-8; Amdts. 172–19, 173–70, 174–17, 175–10, 176–5, 177–24]

LABELING OF HAZARDOUS MATERIALS

The purpose of these amendments to the hazardous materials regulations of the Department of Transportation is to adopt new labels and specify new regulations pertaining to their use on packages containing hazardous materials.

On July 22, 1970, the Hazardous Materials Regulations Board published a notice of proposed rule making, Docket No. HM-8; Notice 70–13 (35 FR 11742), which proposed to change its labeling requirements to be consistent with those contained in the United Nations recommendations. Interested persons were afforded an opportunity to participate in this rule making. Prior to Notice No. 70–13, the Board published a request for public advice on labels and classification in the FEDERAL REGISTER on October 16, 1968 (33 FR 15347), setting out the basis of the United Nations Labeling System. Subsequent to Notice No. 70–13, the Board published Notice No. 71–13 in the FEDERAL REGISTER on May 25, 1971 (36 FR 9449), setting forth additional proposals pertaining to the implementation of the new labeling system. The amendments published herein, however, are derived primarily from the proposals made under Notice No. 70–13 to the extent necessary to adopt the new labels. The other matters proposed under Docket No. HM-8 will be handled in future rule making actions.

The reason for the Board's delay in acting on this docket is the consideration it has been giving to the implementation of a hazard information system, such as was proposed in Docket No. HM-103 published in the FEDERAL REGISTER on June 27, 1972 (37 FR 12660, advanced notice of proposed rule making). Since it was proposed to modify the labels that were proposed in Docket No. HM-8 by the addition of "hazard information blocks" in the lower portion of most of the labels, the Board did not want to impose one change immediately upon another change in an area that would involve considerable effort and expense on the part of regulated industries. However, recent events have convinced the Board that it should no longer delay the implementation of the labeling requirements previously proposed.

The principal reason for immediate adoption of the new labeling requirements now is the fact that a number of foreign governments have adopted the United Nations Labeling System and have made the use of United Nations' labels mandatory. The Board was recently advised by one large shipper that it was having difficulty making shipments into the Federal Republic of Germany since that country has adopted the new labels. The situation has been temporarily alleviated by the granting of an extension to March 31, 1973. Other countries have advised that the use of the new labels will be made mandatory by January 1, 1974. It has been reported that the United States exports chemicals worth in excess of $4 billion. This does not include explosives, petroleum products, and consumer materials. A substantial portion of the chemicals exported are classed as hazardous materials. With this in mind, and since the Board has decided to adopt the new labeling requirements, the Board believes it should no longer delay action in this area because of the proposals it has made in Docket No. HM-103. This does not mean the Board is abandoning the proposals it made in that docket; however, it will have to consider appropriate timing for implementation of changes if it decides to adopt the proposals made in Docket No. HM-103.

A number of commenters responding to Notice No. 70–13 indicated their support of the basic purpose of the proposal, i.e., to adopt the United Nations' labels. Many comments were made pertaining to matters outside the scope of the proposal and others to particular matters concerning the implementation of the proposal.

Concerning the content of labels, several commenters requested that no language pertaining to special handling precautions be included on labels, such as the words "Danger," "Warning," or "Caution." One commenter pointed out that the proposed signal words would be inconsistent with the requirements of the Food and Drug Administration in some cases. Concerning the Board's statement that it doubts whether this detailed information should be included on a primary warning label, no commenters presented sufficient justification to include this type of information on the labels. Therefore, the Board has not adopted any of the suggested language that was set forth in the notice. The Board recognizes the limitations on placing detailed handling and emergency procedure information on labels and has decided that the information should not be included on the labels due to space limitations precluding the use of such information in a beneficial manner.

Several commenters raised objections to the use of a skull and crossbones symbol in the irritant label pointing out that, historically, the use of the skull and crossbones has been limited to those materials which are highly or extremely toxic. The Board agrees that it is not appropriate to require the use of this symbol on materials that are intended for use on people and animals. However, this symbol is required on the United Nations' label. Therefore, the Board is adopting two labels for irritants—one containing the skull and crossbones which will be authorized for import and export traffic, and the other for domestic shipments. The Board intends to propose that some other symbol, such as a tearing eye, be adopted in the United Nations recommendations to correct this matter in the future.

A comment was made concerning the proposed change to § 173.86 that would permit shippers to make tentative classifications of explosives shipped for laboratory examination. The Board agrees that this would permit the exercise of judgment by persons who may not be properly qualified to make explosives classifications. Therefore, the Board has decided that the determination of the classification and appropriate labels to be applied to packages containing these samples will be made by specified elements of the Department of Defense for military explosives, and the Bureau of Explosives for commercial explosives.

Several commenters objected to the requirements for the labeling of Class C explosives. The Board has decided to adopt the exemption for small arms ammunition as proposed in the notice. As to the comments raised concerning primers and percussion caps, no proposal was made to change § 173.107. Paragraph (h) of that section does not require the labeling of these materials. Concerning the Board's proposal that other Class C explosives be labeled, no convincing argument was presented to alter the Board's position that most Class C explosives be labeled to identify their hazards. Future exemptions from labeling in this area will be handled on a case-by-case basis in accordance with the rule making procedures (see § 170.11).

Several commenters questioned the entry "No exemption" in the commodity list, § 172.5, following the entries for certain Class C explosives. The heading of the column is "Exemptions and packing." When there are no small quantity partial exemptions provided for an individual class, or when the nature of the material is such that no small quantity exemption is warranted, the entry "No exemption," followed by section reference means the shipper must look to the referenced section for the requirements that pertain to the material. For example, § 173.101 specifies the packaging and marking requirements for small arms ammunition. There is no small quantity exemption for these materials. Therefore, a "No exemption" entry is entered in the list of hazardous materials. This does not mean that the individual section does not contain an exemption for labeling or certain other requirements.

The notice contained proposals to adopt two new classes—"Water reactive materials" and "Spontaneously combustible materials." The Board has decided to withdraw these proposals for further study. It is not necessary they be adopted to implement the new labeling requirements. Use of the "Dangerous When Wet" and "Spontaneously Combustible" labels in conjunction with the specified labels is permitted by this amendment.

A commenter recommended that the Board adopt specifications for the colors on labels. Such a proposal was made by the Board in Docket No. HM-103. The Board agrees that the different colors should be specified due to the wide variations observed in the colors of the labels presently in use. However, since specific color requirements were not proposed in the notice, color specifications are only being included in this amendment by the addition of an appendix to Part 173 with a statement in § 173.404(d) to indicate that the colors specified in the appendix "should" be used. Inclusion of the word "should" indicates that the specifications in the appendix are not mandatory.

The list of hazardous materials set forth in § 172.5 is not being changed other than to list "Irritant" in place of "Poison, Class C," to list names of the new labels in place of the labels previously specified, to add new label entries for explosives, and to make minor editorial changes. The Board has decided that it is no longer appropriate to describe labels by their colors and that the required inscription is the best method to specify the labels to be used. Also, there are many paragraphs of regulations set forth in these amendments which only change references to labels or change references from Poison C or Tear Gas to Irritant or irritating material.

The Board believes the labels it is adopting in the amendment are a significant improvement in the Department's hazardous materials communications requirements. With their adoption, there will be a single, intermodal, labeling system harmonious with the regulations of other countries. It should be noted that new paragraph (h) in § 173.404 authorizes continued use of the labels presently required (except for explosives) until January 1, 1975. However, exporters are advised that their shipments may not be accepted for transportation by water or air to foreign destinations if they bear the labels required before this amendment since many countries are implementing labeling requirements conforming to the United Nations Recommendations.

(38 F.R. 5292, February 27, 1973)

SUMMARY · Docket No. HM-10 · *A proposal to completely up-date the DOT tank car specifications in order to incorporate earlier established Association of American Railroads standards. This proposal was not adopted as an amendment. A number of the proposals were covered in a later docket. See Docket No. HM-90.*

[49 CFR Parts 173, 179]
[Docket No. HM-10; Notice No. 68–8]
TRANSPORTATION OF HAZARDOUS MATERIALS
Tank Car Specifications

The Hazardous Materials Regulations Board is considering amending Parts 173 and 179 of the Hazardous Materials Regulations to authorize the use of additional tank cars and to include tank car specifications for these tank cars. These proposed amendments are almost entirely based on requests for the adoption of additional tank car specifications submitted by the Association of American Railroads (AAR). The tank car proposals submitted by the Committee of the AAR were based largely on experience gained under a number of outstanding special permits.

Interested persons are invited to participate in the making of these proposed rules by submitting written data, views, or arguments as they may desire. Communications should identify the regulatory docket and notice number and be submitted in duplicate to the Secretary, Hazardous Materials Regulations Board, Department of Transportation, 400 Sixth Street SW., Washington, D.C. 20590. Communications received before February 18, 1969, will be considered by the Board before taking final action on the notice. All comments will be available for examination by interested persons at the Office of the Secretary of the Board both before and after the closing date for comments. The proposals contained in this notice may be changed in light of comments received.

These proposals are set forth in the type of detailed specifications that have been used for many years in this field. As has been announced previously (see for example the proposed oil pipeline regulation (33 F.R. 10213) and the notice published in 33 F.R. 11862), it is the Board's intention in the future to prescribe minimum performance type requirements rather than detailed specification type requirements. However, in view of the substantial amount of time and effort of both the industry and of the Federal Railroad Administration that can be saved through the elimination of many special permits, it is desirable to issue this proposal in the traditional specification form rather than to let the existing situation stand while performance specifications are developed.

The special commodity requirements presently shown in §§ 179.102 and 179.202 relating to pressure tank car tanks and non-pressure tank car tanks, respectively, have for the most part been relocated in appropriate sections of Part 173 applicable to the commodity. This is for consistency of format since the subject material is primarily applicable to shipper requisites rather than to specifications for the construction of tank car tanks. Those requirements presently in §§ 179.102 and 179.202 pertinent to the construction of the tank have not been altered. This relocation affects §§ 173.124, 173.264, 173.268, and 173.314 in Part 173.

Section 179.2(a) (4) identifies "DOT" and "Department" as the Department of Transportation. Similar editorial adjustments are reflected in §§ 179.3, 179.4, and 179.5. In § 179.6, reference is made to Appendix R of the AAR Specifications for Tank Cars. Section 179.12–2(b) would permit a 20 percent reduction in carbon steel pipe, if welded, and would reflect provisions that have been incorporated in other tank car standards for many years.

Section 179.100, in addition to updating section cross-references and incorporating references to various appendices of the AAR Specifications for Tank Cars, requires, at subsection 4, that the jacket covering insulation be of a thickness not less than eleven gauge. At subsection 6, more stringent requirements are added to assure that tank plates are not reduced during forming below specification requirements. Also the welded joint efficiency factor change to 1.0 recognizes improvements in welding and weld inspection over the years in addition to the fact that all such tank welds are 100 percent radiographed. At subsection 7 the tables are deleted and in place thereof reference is made to Appendix M of the AAR Specifications for Tank Cars which includes all approved materials including specifications on high tensile strength steel, and on cladding. Subsection 14 would define more clearly the limitation on extreme projection of bottom outlets on cars. Paragraph (a) (5) would require a screw plug on closure which has been standard approved practice, but never specified heretofore. Subsection 15(a) is to be consistent with § 179.100–1 which includes valve flow rating pressures for safety-relief valves. Subsection 15(c) would require the safety-relief valve on specification 105A500W tanks to be set for a start-to-discharge pressure of 360 p.s.i. in keeping with current practice.

Section 179.101 provides for the construction of tank cars to new specifications 114A400W, 117A340W, 120A300W, and 112A400F. At the same time, provisions are made for specifying valve flow rating pressure, and for bottom outlets on certain cars.

Section 179.102 has been altered as previously mentioned. In addition, provisions for liquefied carbon dioxide and chlorine tanks to accommodate the use of new steels are made. Provisions for the alternate setting of safety-relief valves on butadiene and vinyl chloride tanks are made. Additionally, provisions have been made for hydrogen sulfide tanks built to specification 105A600W.

Section 179.103 adds special requirements for class 117AW uninsulated tank cars used in the dual service of compressed gases and flammable liquids; and 120AW insulated tank cars used in the dual service of compressed gases and flammable liquids. Subsection 5 provides specific requirements for approved bottom outlet valves.

Section 179.200, general specifications applicable to nonpressure tank cars tanks, makes substantive changes in subsections 3, 4, 6, 7, 15, 16, 17, 22, and 24. Subsection 4 requires insulation on tanks to be covered with a metal jacket not less than 11 gauge. Subsection 6 is clarified to assure that thickness of tank plate is not reduced during forming below specification requirements. Also the weld joint efficiency factor is to be 1.0 for seamless heads of all tanks. Subsection 7 omits reference to ASTM specifications in preference to adoption of Appendix M of the AAR Specifications for Tank Cars which appendix has been updated in keeping with current industry practices. Subsection 15 has been amended to delete reference to AAR specification M-402, Grade 35018, malleable iron castings. Subsection 16 requires the application of shutoff valves at specific locations on the tank when top loading and discharge devices are installed. Subsection 17 would define more clearly the limitation on extreme projection of bottom outlets on cars with truck centers less than or greater than 60 feet 6 inches. Subsection 22 is addressed to all lined tanks instead of rubber-lined tanks only. Subsection 24 adds paragraph (b) to provide an abbreviated marking for class 111A tank car tanks by omitting the suffix numeral.

Section 179.201-1 provides for the construction of tank cars to new specifications 111A100ALW, 111A60W2, 111A60W5, 117A340W, 120A300W, 111A60W1, 111A60ALW2, 111A100ALW1, 111A100ALW2, and 111A60W7. Subsection 3 makes distinct the requirements applying to rubber-lined tanks and tanks lined with material other than rubber. Subsections 4, 5, and 6 adopt by reference the requirements specified in Appendix M of AAR Specifications for Tank Cars. Subsection 7 requires safety relief devices to comply with § 179.200-18.

Section 179.202 has been altered as previously mentioned. In addition, reference to hydrofluoric acid, and nitric acid has been included.

Sections 179.220 and 179.221 have been added to provide general specifications applicable to class 115AW tank cars consisting of an inner container supported within an outer shell.

In §§ 179.300-6 and 179.300-8 more stringent requirements are added to assure that tank plates are not reduced during forming below specification requirements. In subsection 7 reference to ASTM specifications has been deleted in preference to adopting certain material specifications by reference to Appendix M of the AAR Specifications for Tank Cars. In subsection 17 the tests for frangible discs of safety vents are required to comply with Appendix A of the AAR Specifications for Tank Cars.

In § 179.301, a correction is made to the values prescribed for the safety relief devices applicable to 110A1000W tanks. The change to § 179.302 rearranges the commodities in alphabetical order and consolidates the family of aluminum alkyls (pyroforic materials) under the generic description, "pyroforic liquids, n.o.s."

Sections 179.400 and 179.401 have been expanded to provide for the construction of class 113AW tank cars for liquefied ethylene service, and to provide for advancements in engineering and design of tank cars for the transportation of liquefied hydrogen.

In § 179.500, editorial changes have been made to subsections 3, 6, 8, 10, 12, and 17. Subsection 4 has been modified to assure that the cylinder shell is not reduced during forming below specification requirements. Subsection 7 has been retitled and provisions inserted to rely upon the requirements for test specimens prepared in accordance with Appendix W of the AAR Specifications for Tank Cars. In subsection 17, the size of letters has been reduced to 1 ½ inches to be uniform with other marking requirements. Throughout Part 179, editorial changes have been made to the marking requirements for tank cars to substitute the letters "DOT" for "ICC" and "Department" for "Commission."

This amendment is proposed under the authority of sections 831–835, title 18, United States Code, section 9 of the Department of Transportation Act (49 U.S.C. 1657), and § 1.4(d)(6) of the regulations of the Office of the Secretary of Transportation.

(33 F.R. 17246, November 21, 1968)

SUMMARY • Docket No. HM-12 • *Authorization of the transportation of anhydrous hydrazine in spec DOT-42B, 42C, and 42D aluminum drums.*

[49 CFR Part 178]
[Docket No. HM-12; Notice No. 68–9]

TRANSPORTATION OF HAZARDOUS MATERIALS
First Public Notice on a Petition for Special Permit

The Department receives scores of requests for special permits each month. Most of the requested permits fall into three categories:

(1) Permits for one-of-a-kind, emergency, or military shipments. (2) Experimental or developmental permits, which

develop information for future regulatory action. (3) General interest permits, which are based on existing knowledge.

A special permit is a special regulation, a waiver or exemption from some provision of the general regulations. A petition for a special permit is usually evaluated on the basis of information submitted with the petition (49 CFR 170.3) without the benefit of public comment. The Department issues a special permit when it appears that the waiver or exemption will be in the public interest and will result in an appropriate level of safety.

Special permits can be issued more quickly than the regulations can be amended under normal procedures. As a result, applicants have come to petition for special permits, rather than for amendments to the regulations. Further, industry has come to expect the Department to give priority treatment of petitions for special permits, at the expense of the general regulatory program.

A special permit is usually issued to a single company, giving that company the right to do something which the regulations prevent other companies from doing. If the permit is of general interest, it may give the holder a competitive advantage over similarly situated companies. As competing companies find out about the special permit, they individually petition for the same waiver or exemption. Once the first petition has been evaluated and the permit issued, the Department routinely issues general interest permits to similarly situated companies.

Regulation by special permit gives the first petitioner quicker service than he could get through a change in the regulations. But competing companies do not fare as well. By the time they find out about the special permit and get special permits of their own, usually more time has passed than would have been required to amend the regulation in the first place.

A general interest permit, when issued to all similarly situated companies, is really a disguised amendment to the regulations. This method of regulation has these disadvantages:

1. Safety standards are changed without an opportunity for public comment on the change.

2. Changes in safety standards, issued to individual companies, are not codified and published as a part of the regulations.

3. The procedure wastes industry and government time and manpower.

The Department believes that the disadvantages of regulation by special permit outweigh the advantages. Accordingly, the Department proposes to treat as requests for rule making those petitions for special permits which are clearly within the general interest class. If a petition is without merit, the Department will deny it. If a petition appears to have merit, the Department will issue a notice of proposed rule making, usually with a 30-day comment period, and then, after evaluating the comments, either amend the regulations or deny the petition.

Special permits for experimental, developmental, one-of-a-kind, emergency, and military shipments, would continue to be issued under present procedures. Special permits for radioactive materials and for cryogenic compressed gases would also be handled under the present procedures for the time being.

This is the first such notice under this procedure. Commentors should address themselves to the procedure itself as well as to the merits of this individual proposal.

(34 F.R. 1175, January 13, 1969)

SUMMARY • Docket No. HM-15 • *Removal of the 20 degree Fahrenheit minimum flash point restriction applicable to the transportation of aerosols.*

[49 CFR Part 173]
[Docket No. HM-15; Notice No. 69–3]
TRANSPORTATION OF HAZARDOUS MATERIALS
Notice of Proposed Rule Making

The purpose of this notice is to request public comment on a proposed amendment to § 173.306 of the Hazardous Materials Regulations (49 CFR 170–189) to eliminate § 173.306(a) (3) (iv). This subparagraph now provides that the flammable contents of inside nonrefillable metal cans charged with solutions of compressed gas or gases must not have a flash point of less than 20° F. in order to be eligible for shipment under the exemption provisions of this section. A petitioner, Chemical Specialties Manufacturing Association, has indicated that the deletion of the subparagraph would have the effect of allowing the "exempt shipment" of aerosol products which are flammable without regard to flash point, but still under all of the other restrictions, such as use of metal cans only, 50-cubic-inch size limit, pressure limit of contents in relation to can strength, adequate head space, and a test to 130° F. of each complete can filled for shipment. In support of its petition, the Association states that . . . "a test of flash point is not a proper test to be applied to a compressed gas including aerosol products, which are formulated products under pressure, usually with a liquefied gas in solution with the other ingredients."

Petitioner further points out that over a period of time numerous special permits have been issued by the Department authorizing the shipment of large quantities of aerosol products having flash points lower than 20° F. and that these shipments have been entirely successful. In requesting a change in the regulations, petitioner stated that in view of the remaining requirements in the regulations as listed above, it saw no need for incorporating either specification container requirements or weight limitations such as were included in the experimental special permits.

The Board believes that there is merit in the petitioner's proposal and that adoption of this change would not adversely affect safety.

Interested persons are invited to give their views on this proposal. Communications should identify the docket number and be submitted in duplicate to the Secretary, Hazardous Materials Regulations Board, Department of Transportation, 400 Sixth Street SW., Washington, D.C. 20590. Communications received on or before April 15, 1969, will be considered before final action is taken on the proposal. All comments received will be available for examination by interested persons at the Office of the Secretary, Hazardous Materials Regulations Board, both before and after the closing date for comments.

(34 F.R. 5112, March 12, 1969)

[Docket No. HM-15; Amdt. 173–7]
PART 173—SHIPPERS
Aerosol Flash Point Restriction

The purpose of this amendment to the Hazardous Materials Regulations (49 CFR 170–189) is to delete § 173.306(a) (3) (iv) and thereby remove the flash point restriction applicable to aerosol products packaged in compliance with the "exempt shipment" provisions otherwise specified.

On March 12, 1969, the Hazardous Materials Regulations Board published a notice of proposed rule making, Docket No. HM-15; Notice No. 69–3 (34 F.R. 5112) to eliminate § 173.306(a) (3) (iv). The deletion of the subparagraph would have the effect of allowing the "exempt shipment" of aerosol products which are flammable by § 173.300(b) criteria without regard to flash point. However, all the other restrictions specified such as use of metal containers only, 50 cubic inches capacity limitation, pressure limitation of contents in relation to container strength, adequate head space, and a heat test of 130° F. of each complete container filled for shipment, would be retained. Interested persons were afforded an opportunity to participate in this rule making.

A number of comments were received in response to this notice. The large majority of comments concurred with the proposal. One commenter objected to removal of the 20° F. limitation for aerosol products shipped by air or water in the belief that, should leakage occur in a confined cargo space, a flash fire could result from a spark or other source of heat. Another commenter expressed the opinion that the flashpoint limitation should be retained, asserting that materials used to pressurize these containers could explode and splash burning materials over a wide range, thus propagating fire over a wider range. It is true that there is some possibility that these incidents could occur but the same is true for materials having flashpoints higher than 20° F. when they are expelled under pressure.

No adverse experience has been reported to the Department concerning the transportation of several million cans shipped under special permits issued by the Department wherein the flashpoint restriction had been removed. Further, the integrity of each can will be adequately preserved by compliance with the requirements retained in § 173.306(a) (3). In view of the previous experience under special permits and of the requirements retained, and considering the conditions normally incident to transportation, deletion of § 173.306(a) (3) (iv) is considered reasonable and justified.

In consideration of the foregoing, 49 CFR Part 173 is amended, effective September 3, 1969, by canceling § 173.306(a) (3) (iv). However, compliance with the regulations as amended herein is authorized immediately.

(34 FR. 9869, June 26, 1969)

SUMMARY • Docket No. HM-27 • *Established requirements for the reconditioning of 17C, 17E, and 17H specification drums and provided authorization for their use to transport certain materials.*

[49 CFR Part 173]
[Docket No. HM-27; Notice 69–19]
TRANSPORTATION OF HAZARDOUS MATERIALS
Reuse of Spec. 17 Series Steel Drums

The Hazardous Materials Regulations Board is considering amending § 173.28 of the Department's Hazardous Materials Regulations to specify the standards which must be met in order for shippers to reuse certain DOT specification 17-series drums for the shipment of specified classes of hazardous materials.

Interested persons are invited to give their views on this proposal. Communications should identify the docket number and be submitted in duplicate to the Secretary, Hazardous Materials Regulations Board, Department of Transportation, 400 Sixth Street SW., Washington, D.C. 20590. Communications received on or before September

23, 1969, will be considered before final action is taken on the proposal. All comments received will be available for examination by interested persons at the Office of the Secretary, Hazardous Materials Regulations Board, both before and after the closing date for comments.

A review of reports of incidents involving leaking steel drums has revealed that many of the "leakers" reported upon were "single-trip" specification 17-series drums that had been reconditioned or repaired and reused. Follow-up investigations on several incidents and inspections of drum reconditioning facilities have revealed that such operations do not always produce reconditioned drums which would be of such a quality as to lend themselves to compatibility with established minimum safety standards for new drums. Yet, the reconditioned drums are used under essentially the same transportation conditions as new drums. Deficiencies noted included attempts to repair badly damaged drums, removal of parent metal of a drum during reconditioning with resultant unacceptable reduction in wall thickness, and inadequate inspection and testing of the reconditioned drums prior to reuse for the shipment of hazardous materials.

The Department's regulations now provide that single-trip drums may be reconditioned and reused only under conditions approved by the Bureau of Explosives of the Association of American Railroads. The regulations do not prescribe standards against which the reconditioning operations might be evaluated. The Board believes that it is the responsibility of Government to clearly set forth safety standards which it expects industry to meet.

This proposal would amend § 173.28(h) and add paragraphs (m) and (n) to prescribe the conditions under which single-trip drums may be reused for shipment of certain hazardous materials. General requirements for cleaning, reconditioning, inspection, and testing of drums are proposed. A drum marking system is proposed which would include identification of the drum reconditioner through a DOT code number. Procedures would also be prescribed for drum alteration in a new § 173.28(n).

(34 F.R. 12187, July 23, 1969)

[Docket No. HM-27; Amdt. 173–31]

PART 173—SHIPPERS
Reuse of Specification 17 Series Steel Drums

The purpose of this amendment to the Hazardous Materials Regulations of the Department of Transportation is to prescribe standards for the use of reconditioned and converted steel drums for the shipment of hazardous materials.

On July 23, 1969, the Hazardous Materials Regulations Board issued a notice of proposed rule making, Docket No. HM-27; Notice No. 69-19 (34 F.R. 12187), requesting public comment on a proposal to amend the Hazardous Materials Regulations to prescribe drum reconditioning standards. The proposed standards were based largely on those that had been used by the Bureau of Explosives of the Association of American Railroads for many years.

The general nature of the comments was in support of the intent of the proposal to prescribe the standards in the regulations rather than relying upon individual case by case subjective evaluations. Most of the comments received dealt with only a very few basic points of the proposal. Most of these comments have been incorporated in one form or another in this amendment. Changes have been made in several sections of the proposed rule as a result of comments received:

(1) The requirement for inspection and replacement of closure devices (including gaskets) has been modified to simplify the language and to state the requirements in more general terms. This revised wording reflects the propriety of methods and techniques presently in use for this purpose.

(2) The procedure for conducting the internal air pressure test of each drum has been modified to allow other testing methods which are at least equivalent to the proposed tests.

(3) The prohibition against repairs of drums has been deleted. The Board believes that the performance standards themselves provide adequate control over any repairs to be made. The limitations in § 173.28(a) and (m) (1), as proposed, already preclude any major repairs. The Board agrees that minor repairs should be allowed as long as the drum is still capable of meeting the prescribed standards.

(4) The marking requirements have been simplified. Based on the comments received the Board agrees that the listing of the test pressure is unnecessary. In incorporating a registration number system for drum reconditioners, the Board believes that the registration number itself will provide adequate identification of the locale of the drum reconditioner's plant, so the specific marking of the location is unnecessary.

(5) One commenter indicated that the "type test" referred to in proposed § 173.28(n) (1) required destructive testing of each drum. This was not intended. Since the requirement that the converted drum meet the new specifications provides adequate assurance that the converted drum would be capable of meeting the type test, the specific provision is unnecessary.

(6) A provision has been made in § 173.28(n) (2) to require that the means of attachment of the metal plate be of such a nature as to not adversely affect the integrity of the drum. Methods of attachment such as welding, epoxy

bonding, or brazing would be allowed under this provision, so long as they do not adversely affect the integrity of the drum.

Several commenters requested that the reconditioned specification 17 series drums be authorized for shipments of extremely flammable liquids (flash point below 20° F.) and poisons. However, the Board believes that the degree of potential hazard of these materials is sufficiently great that the authorization of used drums would not be in the public interest. On May 9, 1969, in Docket HM-4, the Board pointed out a number of problems involved in the shipping of poisons in light weight steel drums. That matter has still not been resolved, and the Board considers it inappropriate to authorize these high hazard materials in second-hand light weight drums.

Several commenters objected to the requirement that cleaning processes not remove parent metal from the drums. The Board believes that for the specification 17 series drums, there is insufficient allowance for significant reduction of parent metal thickness without a resultant unacceptable loss of integrity of the drum. The provision has therefore been retained.

One commenter objected to application of a DOT registration number other than the reconditioner's trade symbol. Nothing in this amendment would preclude a reconditioner from marking the drums with his own symbol in addition to the required markings. Furthermore, this amendment does not establish a licensing or certification scheme for directly controlling the industry. The registration is a merely ministerial function to facilitate identification of drum reconditioners. While the Board is considering licensing in this area, and others, such substantive regulation will not take place without separate rule making action. That same commenter recommended that manufacturers be required to emboss the gauge thickness on each removable head rather than requiring this to be done by the reconditioner. The Board agrees that this might be a desirable procedure but believes it would not be appropriate to include in this amendment. The Board will consider that specific recommendation in a future rule making action involving steel drum specifications.

The question of applicability of these standards to all steel drums (e.g., Specifications 5 and 6 series) arose during the comment period. The Board recognizes that these same reconditioning and testing standards would be appropriate for those drums as well. Another question arose regarding the need for a temperature limitation on the burning process used to remove residue from the drums. The Board considers both of these points to be beyond the scope of this rule-making action. They will be handled later as appropriate.

One commenter pointed out that the process of converting a closed head drum to an open head drum is subject to differences in quality control due to inherent design differences in different drums. The Board is investigating this situation.

Interested persons were afforded an opportunity to participate in this rule-making action and due consideration has been given to all relevant matter presented.

(35 F.R. 12275, July 31, 1970)

[Docket No. HM-27; Amdt. No. 173–31A]

PART 173—SHIPPERS
Reuse of Specification 17 Series Steel Drums

On July 31, 1970, an amendment to the Hazardous Materials Regulations of the Department of Transportation was published in the FEDERAL REGISTER (Docket No. HM-27); Amendment 173-31 (35 F.R. 12275) concerning the reuse of specification 17 series steel drums. Since the publication date, a large number of petitions have been received by the Hazardous Materials Regulations Board requesting reconsideration of certain of the requirements contained in that amendment, and requesting an extension of its December 31, 1970, effective date.

Upon consideration of the comments and new information provided by the petitioners, the Board believes modification of certain of the requirements of the amendment is warranted. In addition, the Board is making several clarifying changes. The changes to § 173.28 as set forth in Amendment 173–31, and the reasons therefore, are as follows:

1. The words "as prescribed in this part," are being added in paragraph (m) to make it clear that this section does not authorize the use of specification 17 drums beyond those authorizations contained in the commodities sections of Part 173.

2. A number of petitioners requested the Board reconsider the restriction against flammable liquids having flash points of 20° F. or lower. One petitioner referred to 29 years of experience in using reconditioned drums and stated he is presently shipping more than 100,000 drums annually, filled with below 20° F. flash point flammable liquids. He stated he has had excellent experience in the use of reconditioned drums over the 29-year period. Other petitioners stated that the Bureau of Explosives has been authorizing such usage under its present delegation of authority in § 173.28(h).

The Board has decided to remove the flash point restriction based on the good experience reported by the petitioners and lack of evidence to the contrary. The Board further believes experience should be gained under the new standards promulgated by this amendment before requiring a costly adjustment in industry practices. Paragraph

(m) of § 173.28 is changed to authorize flammable liquids without a flash point restriction. Shippers are reminded of the steel gage limitations specified in § 173.119(a) (3) for the filling of drums with flammable liquids having flash points of 20° F. or lower.

3. One petitioner objected to the language in paragraph (m) (1), which states that any drum which shows evidence of significant reduction in parent metal thickness due to cleaning processes does not qualify for reuse. The intent of the requirement was to preclude the use of any cleaning process that would cause a deteriorative effect to the integrity of a drum. This could include the use of concentrated acidic solutions, certain abrasives, or even a hammer and chisel. It does not include use of those methods that do not cause significant reduction in parent metal thickness. Upon full consideration of the petition for modification of paragraph (m) (1) the Board has decided it should be and is hereby denied.

4. Several petitioners raised strong objection to the pressure test requirements for open-head drums in paragraph (m) (2), pointing out the limited availability of equipment to perform such tests. A representative of the drum reconditioning industry indicated that the necessary equipment is being purchased, but that it would not be available to all reconditioners by December 31, 1970. He requested an extension of the compliance date for testing open-head drums to October 1971. The Board agrees that such an extension is warranted, and in anticipation of further delays in obtaining equipment believes an extension of 1 year should be provided. Further, in considering the practicality of the testing requirement for open-head drums, the Board believes the pressure test requirement should apply to side walls, bottom chimes, and bottom heads only and is therefore excluding the removable head and adjacent chime area from the pressure testing requirements. Otherwise the Board believes the testing requirement has merit and petitions for its cancellation are hereby denied.

5. A number of petitions have been received concerning the marking requirements. There was objection to (1) all of the marking requirements; (2) marking the word "Tested"; (3) marking the month of test; (4) marking a DOT registration number; (5) marking the thickness of removable heads; (6) marking on a metal plate information concerning conversion from a closed-head drum to an open-head drum specification; and (7) marking the new rated capacity when capacity is reduced by more than 2 percent.

The Board has carefully considered all the petitions concerning the marking requirements and has decided that certain changes are warranted. The changes provided are: (1) Addition of alternative means for making the month and year of test (or inspection); (2) deletion of the requirement for the marking of gage on removable heads (such a requirement may be considered as part of a separate rule-making action dealing with specifications for drums); (3) deletion of the requirement for attachment of metal plates to converted drums, in order to permit marking by stencil, decal, or other means prior to each reuse; (4) revision of the marking requirements for converted drums so that only the conversion marking (e.g., 17E/17H) need to be added to those markings required by paragraph (m); and (5) deletion of the reduced capacity marking requirement, thus relying on requirements such as those specified in § 173.116 and paragraph (a) of § 173.24. The revised requirements make it clear that the original specification markings must be retained on a converted drum. They do not have to be made part of the markings on the body, as could have been construed from the language in Amendment 173–31.

Except for the changes made herein, petitions for modification or deletion of the marking requirements are hereby denied.

6. Several petitioners pointed out serious difficulty with meeting the effective date of the amendment due, in part, to the labor difficulties in the automobile industry. Many drums, refilled with paint for new automobiles, could not be shipped and will not be shipped prior to the effective date of this amendment. In each instance, a petitioner stated that such drums are authorized for reuse by the Bureau of Explosives under the provisions of § 173.28(h) in effect at the present time. The Board agrees that 6 months should be provided to allow shipment of drums reconditioned in accordance with the regulations in effect prior to the effective date of these new regulations. A provision is being added to the amendment accordingly. Except as provided by the revised regulations, petitions for extension of the effective date are hereby denied.

(35 F.R. 19021, December 16, 1970)

SUMMARY • Docket No. HM-28 • *Removed the labeling exemption for certain truckload and carload shipments.*

[49 CFR Part 173]
[Docket No. HM-28; Notice 69–20]
TRANSPORTATION OF HAZARDOUS MATERIALS
Removal of Label Exemption

The Hazardous Materials Regulations Board is considering amending paragraph (c) and canceling paragraph (e) of § 173.402 of the Department's Hazardous Materials Regulations to remove certain exemptions from requirements for the labeling of packages containing specified classes of hazardous materials when they are transported in carload or truckload lots. The Board is also planning to cancel § 173.404(h) since the provision therein is no longer necessary.

Interested persons are invited to give their views on this proposal. Communications should identify the docket number and be submitted in duplicate to the Secretary, Hazardous Materials Regulations Board, Department of Transportation, 400 Sixth Street SW., Washington, D.C. 20590. Communications received on or before September 23, 1969, will be considered before final action is taken on the proposal. All comments received will be available for examination by interested persons at the office of the Secretary, Hazardous Materials Regulations Board, both before and after the closing date for comments.

Carload and truckload shipments of hazardous materials, except classes A or C poisons, etiologic agents, and radioactive materials, are presently exempt from labeling requirements when such shipments are loaded by the shipper and are unloaded by the consignee from the transport vehicle in which originally loaded. In addition, carload and truckload shipments of classes A or C poisons, etiologic agents, and radioactive materials made by, for, or to the Department of Defense are presently exempt from the labeling requirements if loaded by the shipper and unloaded by the consignee from the transport vehicle in which originally loaded when accompanied by qualified personnel who are supplied with equipment to repair leaks or other container failures which will permit escape of contents.

These labeling exemptions were provided over 30 years ago for rail shipments. The exemptions were later extended to truckload shipments when transported by highway. In either case a car or motor vehicle containing carload or truckload shipments is required to be placarded or marked as prescribed for the hazardous materials contained therein. The placard (or marking) has about the same relationship to the rail car or motor vehicle as the label has to the package. Basically, the label provides precautionary information to the handler of the package and governs the loading or storage of the package while in the custody of the carrier. The placard (or marking) governs the placement of the rail car in a train, is a warning to train crews and operating personnel, and provides precautionary information to persons responding to the scene of an accident. Essentially the same type of safeguards apply to a placarded motor vehicle.

Packages of hazardous materials often are not confined within transport vehicles as a result of collisions, derailments, and overturns. These packages may or may not be intact. Persons engaged in firefighting cleanup operations, enforcement, and the general public should be afforded sufficient warning of the potential hazards of the materials in packages. Prescribed labels on packages are a means of informing persons of the hazards involved.

There are occurrences when handling personnel, other than those employed by consignees, would come into contact with these hazardous materials even though such events are not contemplated at the time of shipment. Such occurrences are mechanical failure of transport equipment, shipments reconsigned to more than one destination, and the placement of shipments temporarily in storage are not uncommon.

The Board believes that the absence of labels from certain packages of hazardous materials even when carried in carload or truckload shipments is no longer justified except for shipments of the Department of Defense which are loaded and unloaded under its supervision and which are escorted by its personnel.

(34 F.R. 12188, July 23, 1969)

[Docket No. HM-28; Amdts. 173-41, 177-15]

PART 173—SHIPPERS
PART 177—SHIPMENTS MADE BY WAY OF COMMON, CONTRACT, OR PRIVATE CARRIERS BY PUBLIC HIGHWAY
Removal of Label Exemption

The purpose of this amendment to the Hazardous Materials Regulations of the Department of Transportation is to remove certain exemptions from the labeling requirements in § 173.402 and to make corresponding changes in § 177.815.

On July 23, 1969, the Hazardous Materials Regulations Board published a notice of proposed rule making, Docket No. HM-28; Notice 69-20 (34 F.R. 12188), which proposed to remove certain exemptions from the requirements for labeling of packages containing specified classes of hazardous materials. The Board also proposed to cancel § 173.404(h) since the provision therein is no longer necessary.

As a basis for removing the exemptions the Board said:

Carload and truckload shipments of hazardous materials, except classes A or C poisons, etiologic agents, and radioactive materials, are presently exempt from labeling requirements when such shipments are loaded by the shipper and are unloaded by the consignee from the transport vehicle in which originally loaded. In addition, carload and truckload shipments of classes A or C poisons, etiologic agents, and radioactive materials made by, for, or to the Department of Defense are presently exempt from the labeling requirements if loaded by the shipper and unloaded by the consignee from the transport vehicle in which originally loaded when accompanied by qualified personnel who are supplied with equipment to repair leaks or other container failures which will permit escape of contents.

These labeling exemptions were provided over 30 years ago for rail shipments. The exemptions were later extended to truckload shipments when transported by highway. In either case a car or motor vehicle containing carload or truckload shipments is required to be placarded or marked as prescribed for the hazardous materials contained therein. The placard (or marking) has about the same relationship to the rail car or motor vehicle as the label has to the package. Basically, the label provides precautionary information to the handler of the package and governs the loading or storage of the package while in the custody of the carrier. The placard (or marking) governs the placement of the rail car in a train, is a warning to train crews and operating personnel, and provides precautionary information to persons responding to the scene of an accident. Essentially the same type of safeguards apply to a placarded motor vehicle.

Packages of hazardous materials often are not confined within transport vehicles as a result of collisions, derailments, and overturns. These package may or may not be intact. Persons engaged in firefighting, cleanup operations, enforcement, and the general public should be afforded sufficient warning of the potential hazards of the materials in packages. Prescribed labels on packages are a means of informing persons of the hazards involved.

There are occurrences when handling personnel, other than those employed by consignees, would come into contact with these hazardous materials even though such events are not contemplated at the time of shipment. Such occurrences as mechanical failure of transport equipment, shipments reconsigned to more than one destination and the placement of shipments temporarily in storage are not uncommon.

The Board believes that the absence of labels from certain packages of hazardous materials even when carried in carload or truckload shipments is no longer justified except for shipments of the Department of Defense which are loaded and unloaded under its supervision and which are escorted by its personnel.

The Board received comments from persons involved in one or more of five areas of interest: (1) The transportation industry; (2) manufacturers of chemicals and explosives; (3) manufacturers and distributors of compressed gases; (4) fire prevention organizations; and (5) government agencies.

Respondents from the transportation industry, government agencies (with one exception) and fire prevention agencies fully supported the proposal. The General Services Administration said, in its supporting statement, "By requiring the individual shipping containers to be labeled with the required hazardous label for the commodity at the initial point of origin or manufacture, inadvertent omission of required label(s) in further shipment of the commodity in less than car or truckload quantities would be eliminated to a substantial degree, further contributing to the safe handling of hazardous commodities in transportation."

The Military Traffic Management and Terminal Service (MTMTS) of the Department of Defense (DOD) objected to the proposal pointing out that many of its shipments are not normally escorted. MTMTS requested that DOD be exempt from labeling requirements for carload or truckload lots either with or without escorts. If not exempted, a considerable number of military items would require labels. The Board was aware that large numbers of packages not presently subject to the labeling requirements would have to be labeled in proposing to remove most of the exemptions. This included shipments by DOD since its shipments are subject to the same types of exposure during transportation. The proposal to maintain the label exemption for shipments being escorted by DOD personnel in separate vehicles is adopted. Otherwise, DOD shipments must be labeled in the same manner as those of commercial shippers, and for the same reasons.

A number of municipal fire departments and fire prevention organizations submitted comments indicating their strong support for the proposal. One fire department recommended adoption of labels as specified in the National Fire Protection Associations's Pamphlet No. 704-M. The content of labels is being handled in another rule making action—Docket No. HM-8. The manufacturers of chemicals (other than compressed gases) and explosives, with one exception, indicated their support for the proposal. One manufacturer, commenting in opposition to the proposal, said "Although we agree with the reasoning given that packages are not necessarily confined within a transport vehicle as a result of a collision, derailment, or overturn, it is highly questionable in our mind that such labeling is going to provide any additional help over and above that provided in placarding to persons engaged in firefighting, cleanup operations, enforcement, and the general public. The cost to many people to apply labels, where they are now exempt from the labeling requirements, such as those shipping cylinders and drummed products, would be extensive without sufficient benefit to the general public." The Board believes that removal of the exemption will be of benefit to the public for the reasons stated in the notice. For shipments by highway, there are occasions when trucks carry certain classes of hazardous materials in quantities of less than 1,000 pounds gross weight. In such cases, exterior markings or placarding is not required on motor vehicles. The label on a package, in many instances, will be the only communication of the potential hazard of the material it contains. The marking of the name of contents on the outside of a package, in compliance with § 173.401, will not always provide immediate communication as to the type of hazard involved. Also, as was stated in the notice, events occur that are not contemplated at the time of shipment such as reconsignment, and transfers due to mechanical failure. The Board is concerned not only with events following collisions, derailments, and overturns, but with such matters as material compatibility. For example, packages of acids bearing white labels must not be loaded near packages of oxidizing materials bearing yellow labels (see §§ 174.538 and 177.848). Labels affixed to packages are one means of effecting compliance with certain of the compatibility requirements.

In general, manufacturers, distributors, and associations of the compressed gas industry indicated strong objection to the proposal. Their principal comments were: (1) The requirement would be an unnecessary duplication of decals presently employed by the industry; (2) it will place an undue burden on shippers; (3) there will be considerable additional expense incurred in shipping millions of cylinders annually; (4) no purpose will be accomplished and no benefit will be provided to the public.

Concerning the first comment, the Board does not agree that the label requirement will be an unnecessary duplication of industry decals. From a review of cylinder shipments, the Board found that cylinders usually did not bear any decal. Further, several of the decals that were observed were not legible. The use of such decals, often containing proprietary legends, are not acceptable as a substitute for the universally accepted diamond-shaped warning label of specified size and color that can be recognized from a considerable distance.

Concerning the second comment, the Board is aware that safety requirements are often a burden. However, the Board believes that it is not an undue burden to require that packages containing hazardous materials be labeled to communicate the nature of the hazard, unless the hazard is minimized by some other means.

Concerning the third comment, the Board knows that the new safety requirements will cost money in material and labor. Of the large number of comments received, two commenters responded in detail as to cost. One commenter indicated the material and labor cost per cylinder would be approximately 2.4 cents. He indicated that the new requirement would cost his company in excess of $200,000 per year. The other commenter estimated a cost of 2 cents per label, and an added cost to the liquefied petroleum gas industry of at least 2.8 million dollars annually. The Board has considered the cost of labeling each package of hazardous materials and believes the benefit to the public exceeds the cost.

Concerning the fourth comment, the Board has stated previously what it believes to be the purpose and benefit of the rule change. One commenter provided considerable justification with his proposal that cylinders transported by contract and private carriers be exempt from the labeling requirements. He also stated, "There is probably no container that is more suggestive of its contents or more familiar to the general public than the compressed gas cylinder. This is probably the result of ubiquitous presence of such cylinders at hospitals, building and construction sites, welding shops, auto and body shops, auto repair stations, suburban heating systems, and other locations. Indeed, so well known is the bottle configuration of the compressed gas cylinder that a picture of such a cylinder, without more, is used to represent compressed gas in the United Nations label recommendations and in Federal Aviation Administration regulations for air transportation. Moreover, the Board presently has under consideration, in Docket No. HM-8, a proposal to adopt the same method for representing compressed gas. In view of the foregoing, we are entirely confident that the absence of labels on compressed gas cylinders, when moved in private and contract carriage, would not in any manner diminish or impair the safety of such transportation or increase the hazard to emergency crews and the general public." The Board agrees with the commenter that a compressed gas cylinder itself signifies to some degree the presence of a hazard and acknowledges that the silhouette of a cylinder is being proposed for inclusion on the new label for compressed gases. The Board does not agree that by granting an exemption for all cylinders, as proposed by the commenter, the degree of the hazard will be indicated; therefore, the amendment will provide an exemption only for compressed gases classed as nonflammable when carried by private or contract motor carriers. As a further condition of the exemption, it will apply only to those cylinders that are not overpacked so that the cylinder configuration will be readily visible in place of the communication of hazard by a warning label.

In reviewing the comments made in response to the ;notice, the Board believes it necessary to remind all shippers that the requirements of § 173.401(a) apply to all shipments of hazardous materials unless they are specifically exempt from the requirement. Some of the sample decals that were submitted with comments do not conform to the requirements of this section in that the commodity name, as listed in § 172.5, was not included.

(36 F.R. 1473, January 30, 1971)

[Docket No. HM-28; Amdts. 173.41A, 177–15A]

PART 173—SHIPPERS
PART 177—SHIPMENTS MADE BY WAY OF COMMON, CONTRACT, OR PRIVATE CARRIERS BY PUBLIC HIGHWAY
Removal of Label Exemption

On January 30, 1971, amendments to the Hazardous Materials Regulations of the Department of Transportation were published in the FEDERAL REGISTER (Docket No. HM-28; Amendments Nos. 173–41, 177–15, 36 F.R. 1473), concerning the removal of certain exemptions from the requirements for labeling of packages containing specified classes of hazardous materials. The preamble to these amendments explained the Board's reasons for removing the exemptions.

In accordance with 49 CFR 170.35, the Compressed Gas Association and the National LP-Gas Association, representing more than 2,000 member companies, have petitioned the Board for reconsideration of the amendments as they pertain to shipments of compressed gases by motor vehicle. In their petitions, the Associations recommended that the Board adopt an alternate proposal for shipments of compressed gases by highway. The proposals appear to have considerable merit as they would pertain to contract and private motor carriage.

Essentially, the Associations proposed a standardized marking system that would consist of a red or green diamond centered within a marking area having a contrasting white background. The marking area to the left would be designated for marking the name of contents as listed in § 172.5. The marking area to the right, designated and outlined by a dotted line would be reserved for proprietary and precautionary information as desired by the shipper and would not be a mandatory part of the standard.

Included in the diamond would be the word "flammable" or "nonflammable", as appropriate. The most practical manner to apply the marking would be the use of decals of durable quality. The Compressed Gas Association has identified its standard as "CGA Pamphlet C-7, Appendix A," dated May 15, 1971.

The Board member representing the Federal Highway Administration believes that provision for this alternative system is warranted in the private and contract motor carriage sector, where personnel are more experienced in handling the materials they carry than are the handling personnel and drivers of common carriers. In the common carrier sector, even when truckload lots are involved, it is essential that a uniform and consistent labeling system be maintained in order to assure its maximum effectiveness. Also, there are occurrences which necessitate the "breaking up" of truckload lots even though not intended at the time of shipment. The petitioners' request that truckload quantities by common carriers be included in this amendment are hereby denied.

The alternative marking system is responsive to the statement made in the earlier amendment concerning the degree of hazard of the gas contained in the cylinder. The Board stated that the cylinder itself signifies to some degree the presence of a hazard. This system distinguishes between flammable and non flammable compressed gases. Another convincing contention is the likelihood that the alternative markings will be maintained on cylinders more than labels following delivery, and will thus improve safety in many sectors perhaps not subject to the Department's regulations.

Additional time will be needed to apply the alternative markings to millions of cylinders. Therefore, the effective date of these amendments as they apply to the transportation of compressed gases classes as "flammable" or "nonflammable" by contract and private motor carriage, is extended to October 1, 1971. Otherwise, the amendments are effective on June 10, 1971, as specified in the earlier amendment. Only those portions of amendments 173–41 and 177–15 pertaining to transportation of compressed gases by contract and private motor carrier are being amended. However, to avoid confusion, all amendments made under Docket No. HM-28 are set forth in this document.

(36 F.R. 9068, May 19, 1971)

[Docket No. HM-28; Amdts. 173–41A, 177–15A]

PART 173–SHIPPERS
PART 177—SHIPMENTS MADE BY WAY OF COMMON, CONTRACT, OR PRIVATE CARRIER BY PUBLIC HIGHWAY
Removal of Label Exemption

On May 19, 1971, the Hazardous Materials Regulations Board published Docket No. HM-28, Amendments Nos. 173–41A and 177.15A (36 F.R. 9068), concerning the removal of certain exemptions from labeling requirements for packages containing specified classes of hazardous materials. The effective date for alternate use of marking for compressed gas cylinders was stated as October 1, 1971.

On the basis of a petition filed by the Compressed Gas Association, the Board understands that makers of the decals have not been able to deliver sufficient numbers of decals to shippers to effect compliance with the marking requirements.

In consideration of this matter, the effective date for these amendments as they apply to the marking of cylinders containing compressed gases classed as "flammable" or "nonflammable," when transported by contract and private motor carriers, is extended to December 31, 1971.

(36 F.R. 19370, October 5, 1971)

SUMMARY • Docket No. HM-35 • *Required emergency discharge controls for motor vehicle tank trucks transporting flammable liquids; authorized DOT-MC 330 and 331 tank trucks for a number of flammable liquids.*

[49 CFR Part 173]
[Docket No. HM-35; Notice NO. 69–28]
TRANSPORTATION OF HAZARDOUS MATERIALS
Flammable Liquids in MC 330 and MC 331 Cargo Tanks; Emergency
Discharge Controls for Cargo Tanks

The Hazardous Materials Regulations Board is considering amending the Department's Hazardous Materials Regulations to (1) authorize specification MC 330 and MC 331 cargo tanks for transporting certain flammable liquids for which those tanks are not presently authorized, (2) require the bottom outlets on MC 330 and MC 331 cargo tanks currently prescribed in §§ 173.123, 173.134, 173.136, and 173.141 to be equipped with valves conforming with § 178.337–11(c), and (3) require bottom outlets on MC 304 cargo tanks in flammable liquid service to be equipped with valves conforming with § 178.342–5(a).

For several years MC 330 cargo tanks, used primarily in compressed gas service, have been authorized for the transportation of ethyl chloride, pyroforic liquids, methyl dichlorosilane, trichlorosilane, and various mercaptans. On the basis of the satisfactory experience in the service of these flammable liquids and on the basis of a special permit for the transportation of flammable liquids, n.o.s., in MC 330 cargo tanks, it appears that MC 330 and MC 331 cargo tanks are suitable for the general category of flammable liquids not otherwise specified.

Specification MC 330 and MC 331 cargo tanks currently authorized in §§ 173.123, 173.134, 173.136, and 173.141 are required to be equipped with suitable automatic excess flow valves or automatic quick closing internal valve but are not required to have emergency discharge controls for operation in the event of fire or other accident. To provide the level of safety generally provided for cargo tanks that are presently used primarily in flammable liquid service the board believes that bottom outlets on MC 330 and MC 331 cargo tanks should be equipped with valves employing a secondary closing means remote from tank filling or discharge openings.

In addition, the Board believes that MC 304 cargo tanks currently in flammable liquid service should employ emergency discharge controls similar to those used by other types of specification cargo tanks used primarily in the service of flammable liquids. Accordingly, it is proposed to require the bottom outlets on these cargo tanks to be equipped with valves conforming with the emergency discharge controls specified for MC 307 cargo tanks.

(34 F.R. 15660, October 9, 1969)

SUMMARY • Docket No. HM-36 • *Established the DOT hazardous materials incident reporting system and report form (Form DOT F 5800.1).*

[49 CFR Parts 171, 173, 174, 175, 176,177, 180]
[Docket No. HM-36; Notice No. 69–29]
TRANSPORTATION OF HAZARDOUS MATERIALS
Reports of Hazardous Materials Incidents

The Hazardous Materials Regulations Board is considering amending the Department's Hazardous Materials Regulations to include (1) a requirement for the immediate reporting of serious incidents involving hazardous materials and (2) a requirement for the reporting of certain information concerning all hazardous materials incidents whether or not an immediate notification is required.

On April 1, 1969, the National Transportation Safety Board submitted to the Secretary of Transportation a study titled, "Uniform Reporting System for All Modes of Transportation in Reporting Incidents and Accidents Involving the Shipment of Hazardous Materials" (copies may be obtained from the National Transportation Safety Board, 1626 K Street NW., Washington, D.C. 20591). In setting forth the background for its study the NTSB stated:

In the process of reviewing a number of hazardous materials accidents in the past year, the National Transportation Safety Board has become aware of the need for a centralized and coordinated system within the Department of Transportation to collect, process, and disseminate information among the modes pertaining to improving the safety of hazardous materials transport in all its phases in all modes. In its background summary, the Board further stated:

The effective use of data is of prime importance in developing hazardous materials regulations—data from both accidents and incidents (which require definition)—but under existing reporting requirements and procedures it is virtually impossible to consolidate and compare analytically such information on a cross-modal basis. Each mode is essentially concerned only with itself, and hazardous materials accident or incident, reports, where submitted, do not contain information appropriate in character, depth, or detail to have much value in preventing hazardous materials accidents in other modes. There is no satisfactory system now for collecting and analyzing such data so that all modes may to a maximum degree benefit from the experience of each.

After reviewing the present reporting requirements (or lack thereof) in each of this Department's operating administrations, the NTSB concluded:

A unified data system, based on uniform definitions of terms, utilizing a common reporting form to be submitted by carriers, with a flow of reports and supplemental information designed to be channeled to a common data center, and with the processed data (and results of special studies) being made available to all Administrations, would be a logical and necessary prerequisite toward solving many of the problems now confronting all Administrations in the transport of hazardous materials. The increase in traffic, the increase in demand for materials classified as hazardous, and the increasing need for intermodal coordination make this essential not only as an economic necessity, but for the safety of all concerned.

The National Transportation Safety Board then made the following recommendations to the Secretary of Transportation:

1. That the term "hazardous materials incident" be defined in regulations governing transport of such materials in all transportation modes, and that the definitions of hazardous materials accidents, now established independently for each mode, be revised for greater standardization across all transportation modes.

2. That a uniform, cross-modal reporting form be developed, appropriate for automatic data processing purposes, for hazardous materials incidents and accidents.

3. That a centralized reporting system be established within the Department of Transportation, coordinating the handling of reports of all hazardous materials incidents and accidents by carriers to the Administrations and the Coast Guard (as applicable), to operate through a central "clearinghouse" where such data would be collected and evaluated to determine whether greater emphasis should be directed to shipper and carrier compliance with existing requirements, or to the need for change in containers, in hazardous classifications, or in handling requirements.

4. That the Department's Hazardous Materials Regulations Board expedite its action to amend or to revise existing Federal Regulations. It should develop uniform regulations for all modes of transport relating to the shipment and carriage of hazardous materials, as may be necessary to assure substantial uniformity among all modes as to reporting requirements, and processing of incident and accident reports involving hazardous materials, so that a centralized and unified hazardous materials reporting system and clearinghouse might function effectively. If this cannot be done within the existing statutory framework, consideration should then be given to seeking legislation which would authorize the issuance of one regulation applicable to all modes by the Secretary, following appropriate consultation with the Administration and the Coast Guard.

The Hazardous Materials Regulations Board has for sometime been working on a centralized and unified system for collecting information about accidents and incidents involving hazardous materials. The Board agrees with the NTSB's statements as to the importance of accident and incident data in evaluating the effectiveness of existing regulations and in developing new hazardous materials regulations. At present, the little information which is being received is often inaccurate or incomplete; therefore, it is difficult for the Department to properly respond to such occurrences.

This proposal covers two primary areas. The first is a requirement that carriers (including private carriers) make immediate reports to the Department by telephone when incidents of a specified severity occur (the word "incident" is used in the proposed regulations to cover all reportable occurrences that involve hazardous materials). The single telephone number to be provided in the regulations will be for a telephone that is attended by personnel of the Department on a 24-hour basis. While notification by telephone is specified in this notice, comments are requested on the feasibility of using other means of communication to accomplish the required notification with the immediacy that the Board considers necessary. The immediate report would cover the essential items of information necessary for the operating administrations of the Department and the National Transportation Safety Board to determine what immediate action should be taken by them, if any. The immediate notification requirement proposed would also apply to the transportation of liquids by pipeline now covered by Part 180 of this chapter.

The second part of the proposal is a routine reporting requirement that would require the submission of reports in a prescribed format to the Office of Hazardous Materials in those instances where an immediate report is required and also in any case where there has been an unintentional release of hazardous materials from a package. The proposed report (copies of which may be obtained from the Secretary, Hazardous Materials Regulations Board at the address set forth below), which is to be submitted within 15 days of occurrence of discovery of an incident, would provide information and data such as: Hazardous materials involved, consequences, packaging information, probable cause of packaging failure, shipper and consignee identification, and a narrative account explaining the incident. The information derived from these reports will be used by the Department: (1) As an aid in evaluating the effectiveness of the existing regulations; (2) to assist in determining the need for regulatory changes to cover changing transportation safety problems; and (3) to determine the major problem areas so that the attention of the Department may be more suitably directed to those areas.

This centralized reporting system would amend or replace the existing reporting requirements presently provided for in §§ 173.11, 174.506, 174.508, 174.565, 174.588, 175.660, 176.707, 177.807, 177.814, 177.861, and add an immediate reporting requirement to Part 180.

One effect of this proposal would be to eliminate the requirement that is now contained in several of these sections that the Bureau of Explosives of the Association of American Railroads must be notified in certain circumstances. Elimination of this requirement would not prevent carriers from voluntarily notifying the bureau of Explosives but this notification would no longer be required by Federal regulation.

Interested persons are invited to give their views on the proposal discussed herein. Communications should identify the docket number and be submitted in duplicate to the Secretary, Hazardous Materials Regulations Board, Department of Transportation, 400 Sixth Street SW., Washington, D.C. 20590. Communications received on or before January 12, 1970, will be considered before final action is taken on the proposal. All comments received will be available for examination by interested persons at the Office of the Secretary, Hazardous Materials Regulations Board, both before and after the closing date for comments.

(34 F.R. 17450, October 29, 1969)

[Docket No. HM-36; Amdts. Nos. 171–7, 173–39, 174–7, 175–5, 176–3, 177–14]
REPORTS OF HAZARDOUS MATERIALS INCIDENTS

The purpose of this amendment to the Department's Hazardous Materials Regulations is to establish uniform requirements for (1) immediate telephonic reports of serious incidents involving hazardous materials; and (2) written reports containing detailed information for certain hazardous materials incidents.

This amendment is based on a notice of proposed rule making published in the FEDERAL REGISTER October 29, 1969, Docket No. HM-36; Notice No. 69-29 (34 F.R. 17450). (Separate notices of proposed rule making were issued by the U.S. Coast Guard and the Federal Aviation Administration and were published in the same issue of the FEDERAL REGISTER. Final action on these notices is announced in documents published at pages 16829 and 16832 of this issue.) A number of comments were received in response to that notice and all of the comments were carefully considered. The most significant comments and the changes that were made in this amendment as a result of the comments are discussed below.

A number of comments recommended that the reports to be submitted to the Department on hazardous materials incidents should be classified as confidential and should not be made available to the general public. The commenters suggested a number of reasons for the requested confidentiality. For example, several indicated that the reporters would be much more candid if the comments were not to be available to the general public.

Most of the arguments for classifying hazardous materials incident reports as confidential are necessarily speculative. After considering and analyzing all the comments, the Board concludes that they do not contain any argument substantial enough to require that the reports be kept confidential.

It is the policy of the Department of Transportation to make information available to the public to the greatest extent possible in keeping with the spirit of the Freedom of Information Act (5 U.S.C. 552). In the light of that statute, a refusal to permit the public access to accident reports would be contrary to sound policy. The public is better served by not keeping such reports confidential.

The only statutory exceptions to the basic requirement of disclosure are set out in section 552(b). None of these exceptions provides confidentiality for the reports under consideration here. Section 552(b) (4) excepts "trade secrets and commercial or financial information obtained from a person and privileged or confidential". However, the legislative history indicates that this exception refers to instances where privileged information (not required by law, and that would not customarily be released to the public) is voluntarily furnished and received in confidence. Examples are commercial or financial information submitted with loan applications, or information voluntarily given to the Government in confidence for the purpose of compiling statistics which are then published in the aggregate.

Moreover, in promulgating the regulations by which the Department implemented the Freedom of Information Act (49 CFR Part 7), the Secretary announced that "the policy of the Department will be to make all information available to the public except that which must not be disclosed in the national interest, to protect the right of an individual to personal privacy, or to insure the effective conduct of public business. To this end, the (regulation) provides that information will be made available to the public even if it falls within one of the exemptions set forth in section 552(b), unless the release of that information would be inconsistent with the purpose of the exemption" (32 F.R. 9287 (1967)).

The exemption of documents from mandatory public disclosure merely authorizes the Secretary to withhold them, it does not compel him to do so.

Section 7.51 of the Department's regulations provides that, even though a record is exempt from public inspection, nevertheless the Department will release it, "unless it determines that the release of that record would be inconsistent with a purpose of" the particular exemption.

A number of commenters suggested that the Department should require carriers to furnish the concerned shipper with a copy of each hazardous materials incident report. The basis for this suggestion was that the shipper should have an opportunity to review the carriers description of the incident so that the shipper could file a supplementary report if he felt that the carriers report did not state the facts fairly. While the Department does not agree that a carrier should be required to file a copy of each incident report with the concerned shipper, this does not mean that the Department is not interested in obtaining any supplemental information that a shipper may wish to provide concerning a hazardous materials incident. Since the incident reports will be available to the general public and since it is likely that shippers will be appraised of hazardous materials incidents of interest to them, any shipper is free to review a carrier's report relating to a specific incident and to file supplemental information on that incident with the Department. After the incident reporting system has been in effect for a period of time, the Department will review its effectiveness and, if it is found necessary, additional rule making action could be taken to formalize shipper input on hazardous materials incidents.

The most significant comments made by a large number of commenters with respect to the immediate telephonic reporting requirement was that, under proposed criteria, the Department would be receiving telephonic notification in many instances where it was not clear that the incident would be of such significance to require any immediate action by anyone within the Department of Transportation. Upon reviewing the criteria proposed for telephonic notification and the experience of the Office of Pipeline Safety under its immediate report requirement (which has been in effect for approximately 8 months), it has been concluded that there is justification for further limiting the criteria for immediate notification. This has been done by (1) increasing the property damage from $5,000 to $50,000; (2) eliminating the requirement for a telephonic notification where it is estimated that the resumption of normal transportation facilities involved would be prevented for 2 hours or more; and (3) by establishing an overall judgmental requirement that carriers should notify the Department when they believe that the incident is of such significance as to warrant a telephonic notification even though it does not involve a fatality, serious injury, or property damage in excess of $50,000. Incidents involving a significant continuing danger to life would be one type that would fall into this last category.

A number of commenters indicated that the 15-day reporting requirement for the written report might in many cases be difficult to comply with. It is recognized that in some cases a carrier may find it difficult to furnish all of the information required in the incident reporting form within 15 days of the incident. However, the Board believes that in the vast majority of cases this information should be available within a few days of the incident and that the clerical work involved in completing the form should not delay the submission for longer than 15 days. In the event the carrier is not able to obtain all the necessary information within the 15-day period, it may submit the report and file a supplementary report when the additional information becomes available.

A number of commenters objected to the requirement that a detailed written report must be filed in every case where there "has been an unintentional release of hazardous materials from a package." Many commenters felt that the Department would as a result of this requirement be flooded with numerous incident reports relating to the release of insignificant amounts of hazardous materials. These commenters pointed out that this would place a substantial and apparently unnecessary paper work burden on both the carriers and the Department. The Board does not feel that it is in a position at this time to determine whether there are insignificant unintentional releases of hazardous materials that do not warrant the filing of a written report. While it may be true that under the amendment the Board will receive reports of unintentional releases of hazardous materials that may prove to be insignificant, the Board does not have any criteria at this time on which it could draw a line between those releases that should be reported and those that should not. As experience is gained under this incident report program, the program will be subject to continuing review. If it is found that the present criteria is putting an undue burden on carriers and that the Board is receiving unusable or irrelevant incident reports, the Board will not hesitate to review the reporting requirements and to take future rule-making action.

A number of commenters made specific suggestions as to detailed requirements of the incident report. Many of these comments were considered warranted and a number of changes have been made in the report form. For example, many commenters pointed out that Item C of the proposed report was entitled probable causes while many of the items listed thereunder were not in actuality "probable causes". The most significant overall comment by a number of commenters on the report form was that it was too detailed and that it would result in the Board receiving much more information than is necessary for the purposes for which the report is required. Each of the items in the report form has been carefully reviewed in the light of this criticism. The Board does not agree that the report form is unduly detailed nor does the Board believe the completion of the form will place any undue burden on carriers. Nevertheless, as indicated above, this is the first comprehensive hazardous materials incident reporting system for all modes of transportation and the Board intends to continually review the requirements adopted in this amendment in the light of the information received under its requirements. As both the carriers and the operating administration of the Department of Transportation acquire experience under the new incident reporting system, it may well be that the Board will wish to review some of the requirements in the form presently adopted. The Board will be interested in hearing of actual experience from carriers in completing the forms and their recommendations for further changes in the forms, whether deletions therefrom or additions thereto.

Due to the time required to prepare, print, and distribute adequate supplies for use, the printed forms may not be available at the time this regulations becomes effective. In that event, a small supply of temporary forms will be distributed for use until receipt of permanent forms. These temporary forms may be obtained in limited quantities from the Office of Hazardous Materials, Department of Transportation, Washington, D.C. 20590. They may be reproduced by any company if additional copies are needed in the interim period.

(35 F.R. 16836, October 31, 1970)

[49 CFR Parts 171, 174, 175, 177]
TRANSPORTATION OF HAZARDOUS MATERIALS
[Docket No. HM-36; Notice No. 71–22]
Reports on Incidents Involving Radioactive Materials

On October 31, 1970, the Hazardous Materials Regulations Board published new hazardous materials incident reporting requirements under Docket No. HM-36 (35 F.R. 16836 and 16837). The purpose of the amendments to the Department's Hazardous Materials Regulations was to establish uniform reporting requirements for incidents occurring as a direct result of hazardous materials in transportation.

Regulation for the reporting of incidents involving shipments of radioactive materials have been in effect for more than 2 years. requiring immediate reports by the carrier to the shipper and to the Department. The purpose of these proposed amendments to §§ 171.15, 174.588, 175.655, and 177.861 is to make the reporting requirements for incidents involving radioactive materials consistent with the more recent general reporting requirements established.

The Board is also of the opinion that the requirement for reporting based on "unusual delay" involving radioactive material shipments should be deleted from §§ 174.588(c) (1), 175.655(j) (3), and 177.861(a). Experience and comments from carriers have indicated that the term lacks precision and is subject to serious variance in interpretation. The Board believes that the requirements in § 171.15(a) (4) which provide for reporting on a judgment basis, are sufficient to include situations involving unusual delay.

A new reporting criteria is proposed to be added to §§ 171.15(a), 174.588(c) (1), 175.655(j) (3), and 177.861(a) relating to incidents involving "suspected radioactive contamination". When compared to other hazardous material hazards, radiation hazards present a special problem in detection. Radiation cannot be detected by the senses, but must be observed by the means of special measuring or detection instruments. For this reason, the Board believes that immediate notification should be made whenever a carrier suspects radioactive contamination.

(36 F.R. 12913, July 9, 1971)

[Docket No. HM-36; Amdt. Nos. 171–13, 174–11, 175–6, 177–18]
RADIOACTIVE MATERIALS
Reporting Requirements

The purpose of this amendment to the Hazardous Materials Regulations of the Department of Transportation is to make reporting requirements for incidents involving radioactive materials consistent with reporting requirements applicable to other hazardous materials.

On July 9, 1971, the Hazardous Materials Regulations Board published a notice of proposed rule making, Docket No. HM-36; Notice No. 71–22 (36 F.R. 12913), which proposed this amendment. Interested persons were invited to give their views and several comments were received by the Board. Most comments were in support of the proposal.

Several commenters correctly noted that reference to § 173.399 in §§ 174.588, 175.655, and 177.861 should be to § 173.397. These changes have been made.

One commenter was concerned that difficulties might arise with a requirement to report when "suspected" radioactive contamination is involved. This commenter stated that the requirement for reporting should be conditioned upon observed leakage only and cited as a reason the lack of availability of radiation detection equipment to many carriers. The Board is particularly concerned with failures to report because of this lack of detection devices. It recognizes that when radioactive contamination occurs, detailed investigation, extensive tracing, and many contacts with potentially exposed persons may become necessary. The time, effort, and cost incurred in such cases dictate that incidents be reported as quickly as possible to minimize the extent and effect of any possible radioactive contamination. Therefore, suspect situations warrant a carrier ascertaining the facts and reporting them to the Department. He may not have the means to do this without enlisting aid from knowledgeable and properly equipped organizations. Past history has shown that many carriers are aware of the potential seriousness of contamination to persons, equipment, or facilities, and have requested radiological assistance when in doubt about the integrity of a package. Since this reaction is desired, the Board has concluded that the conditions suggested by the commenter are not warranted.

One commenter recommended that notification be required within 48 hours following a radioactive materials incident, instead of at the "earliest practicable moment." The "earliest practicable moment" policy is consistent with

§ 171–15(a). By this amendment, the Board is establishing the radioactive material incident described in §§ § 174.588, 175.655, and 177.861 as immediately reportable, rather than only reportable within some longer time frame. The nature of a radioactive contamination problem indicates to the Board that immediate reporting to the Department and the shipper involved is particularly essential and entirely appropriate with respect to the public's interest.

It could be emphasized that the reporting requirements for radioactive materials transportation incidents as established by the Board in this amendment are in no way intended to replace or impair the existing national system for obtaining emergency radiological assistance in the event of incidents involving radioactive materials. That system, which is intended to provide response capabilities for radiological monitoring in emergencies, is administered by the U.S. Atomic Energy Commission and supported by the capabilities of other Federal and State organizations. Further information on that system may be obtained from the U.S. Atomic Energy Commission, Division of Operational Safety, Washington, D.C. 20545.

(36 F.R. 21200, November 4, 1971)

49 CFR part 171
[Docket No. HM-36A; Notice No. 80–5]
Elimination of Certain Reporting Requirements

This notice proposes to eliminate the detailed hazardous materials incident reports required by § 171.16 of the Hazardous Materials Regulations for materials being transported under the following proper shipping names, "Consumer commodity"; "Battery, electric storage, wet"; or "Paint, enamel, lacquer, stain, shellac or varnish; aluminum, bronze, gold, wood filler liquid or lacquer base, liquid." With respect to paint and related materials, this exception would apply only when shipped in quantities of five gallons or less.

The MTB has analyzed the hazardous materials incident data base and believes that continued reporting of incidents involving these materials would be of minimal value when weighed against the burden placed upon the carriers who are required to prepare and submit incident reports.

In 1979, approximately 18,000 incident report were filed with the MTB involving all hazardous materials. Of the incident reports submitted between January 1971 and October 1979 (excluding 1977), paint and related materials represented 21.4 percent of the incidents reported and batteries represented 8.1 percent of the incidents reported. Of the paint reports, including related materials, 86 percent involved packagings of five gallons or less. The major cause of the incidents were reported as resulting from packages "dropped in handling" and "damage by other freight." If this proposal is adopted, it is expected that there would be a 30 percent reduction in the number of hazardous materials incident reports submitted during the first complete reporting year. This would result in substantial savings for the carriers and MTB believes that information which is already contained in its data base is sufficient for its purposes.

The severity level of incidents involving paint and related materials, when shipped in quantities of five gallons or less, shows that there have been no deaths; a few minor injuries and property damage averaging $95 per report. The severity level of incidents involving batteries (electric storage, wet) shows no deaths, a few injuries, mostly minor in nature, and property damage averaging $157 per report. Reports pertaining to consumer commodities, ORM-D, represent less than one percent of the incident reports received by the MTB. For incidents involving these materials, no deaths or injuries were reported and property damage per incident averaged less than $11.

MTB estimates that a combined total of 9,000 man hours of effort will be eliminated by this change in the reporting requirements if it is adopted. This includes the time taken to obtain the information necessary to complete incident reports, the time consumed preparing incident reports, and the review and processing time consumed by MTB and its contractors in compiling the data base after receipt of the incident reports.

This notice does not propose to change the requirement in § 171.16 which calls for a detailed written report for serious incidents as provided in § 171.15 or to change any of the reporting requirements relative to transportation by air. The MTB believes that continued reports of these materials being transported by air would be of value for further analysis because of the potential for serious consequences from incidents aboard aircraft which would otherwise be minor when they occur elsewhere.

(45 F.R. 40628, June 16, 1980)

49 CFR Part 171
[Docket No. HM-36A; Amdt. No. 171–56]
Elimination of Certain Reporting Requirements

On June 16, 1980, the MTB published a notice of proposed rulemaking, Docket HM-36A; Notice No. 80–5 (45 FR 40628) which proposed the elimination of certain requirements for the submission of hazardous materials incident reports presently required by § 171.16 of the Hazardous Materials Regulations. Subject to this action are materials being transported under the following proper shipping names "Consumer Commodity," "Battery, electric storage, wet", or "Paint, Enamel Lacquer, Stain, Shellac or Varnish; Aluminum, Bronze, Gold, Wood filler liquid or Lacquer base, liquid." With respect to paint and related materials, this exception applies only when shipped in packagings of five gallons or less. The reasons for this action were stated in Notice 80–5.

The MTB received twenty-four comments in response to the proposed regulation and twenty-three of them, including one from the National Transportation Safety Board, were substantive in nature. All comments were in favor of the proposal.

Two commenters requested that all paperwork requirements be eliminated relative to the materials covered by this action. These comments are clearly outside the scope of this action. One commenter suggested that all reporting requirements for hazardous materials incidents other than those required under § 171.15(a) be eliminated. The MTB believes that written incident reports are valuable in determining problem areas which should be addressed by rulemaking. However, as stated in the notice of proposed rulemaking, the MTB believes that sufficient data has been obtained on the specified hazardous materials to justify eliminating the reporting requirements for these materials. The preamble to Docket No. HM-36; Amendments Nos. 171–7, 173–39, 174–7, 175–5, 176–3, 177–14, (35 FR 16836), published October 31, 1970, addressed the question of whether the incident reports would be of value. The preamble stated that as experience was gained under the incident report system, it would be easier to analyze which reports were of significant value and which were not. The MTB believes this rulemaking takes a step in determining those reports of significant value.

Several commenters suggested that the MTB consider eliminating reporting requirements for additional commodities. Further analysis will determine whether additional exceptions to the reporting requirements would be warranted and if so, additional notices of proposed rulemaking will be published.

Two commenters requested that wording of the proposed amendment in § 171.16(c)(3) be changed from "... when shipped in quantities of five gallons or less", to include multiple packagings of five gallon or less. The MTB agrees with this comment since the analysis upon which the notice was based involved incident reports of package failures, most of which were multiple packagings. This comment has been incorporated in § 171.16(c)(3).

Although not mentioned in the notice, § 171.16(b) is being revised to reflect a change in operational procedures in processing incident reports. Because this action does not impose any burden on the general public, public proceedings are unnecessary.

(45 F.R. 73682, November 6, 1980)

49 CFR Part 171
[Docket No. HM-36B; Notice 84–1]
Detailed Hazardous Materials Incident Reports

Background. The Materials Transportation Bureau (MTB) is reviewing the requirements of § 171.16 that each carrier who transports hazardous materials submit to the Department a hazardous materials (hazmat) incident report (DOT Form F 5800.1) for each incident that occurs during the course of transportation (including loading/unloading, or temporary storage). The review was conducted in accordance with Executive Order 12291 as a part of MTB's program to evaluate existing regulations for clarity and to revoke or revise those that are not achieving their intended purpose; or can achieve their intended purpose in a more effective and efficient manner. To accomplish this purpose, a review team consisting of the personnel of several offices of MTB was established.

The review is also consistent with the final rule under Docket HM-36 (35 FR 16836, October 31, 1970) which established the current reporting requirements for hazardous materials incidents. In that docket it was noted that, after a period of time, the Department would evaluate the effectiveness of the incident reporting system and, as appropriate, take further rulemaking action to incorporate additional input on the reporting of hazardous materials incidents.

One of the major objections raised in Docket HM-36 referred to the requirement that a detailed, written report be filed in every case where there "has been an unintentional release of hazardous materials from a package." Many commenters believed that the Department would be flooded with numerous incident reports relating to the release of insignificant amounts of hazardous materials. In response to these comments, the Hazardous Materials Regulations Board (the predecessor of the present Materials Transportation Bureau) stated that it was not in a position to determine whether there are insignificant unintentional releases of hazardous materials that do not warrant the filing of a written report and that it lacked criteria to establish a line between those releases that should and those that should not be reported.

This Advanced Notice of Proposed Rulemaking is intended to assist MTB in determining what these criteria should be, in light of the more than 130,000 hazmat incident reports submitted to MTB over the past 12 years.

The Present Reporting Requirements. The present reporting requirements of § 171.16 are triggered by the following criteria (Reporting criteria (A) and (B) below also require telephonic reports, as required by § 171.15.):

(A) All releases of a hazardous material, which as a direct result of the hazardous material, result in:

A fatality;

An injury requiring hospitalization;

Estimated carrier or other property damage exceeding $50,000.

(B) All incidents, whether or not there is an actual release of a hazardous material, in which:

A fire, breakage, spillage, or suspected contamination occurs involving shipment of radioactive materials;

A fire, breakage, spillage, or suspected contamination occurs involving shipment of etiologic agents;

A situation exists of such a nature that in the judgment of the carrier, it should be reported, e.g., a continuing danger to life exists at the scene of the incident.

(C) All unintentional releases of hazardous materials from a package (including a tank) or any quantity of a hazardous waste during transportation, except for the following hazardous materials (except aboard aircraft):

Consumer commodity; Battery, electric storage, wet, filled with acid or alkali; Paint and paint related materials when shipped in packagings of five gallons or less.

Under these criteria, an average of 7,900 incidents per year have been reported to MTB over the last two years. The vast majority of these reports pertain to criterion (C)—that is, they do not involve a death, an injury, damage exceeding $50,000, etc., and are primarily associated with incidents involving small packages, such as drums, bottles, cans, boxes, bags, etc. Approximately 79 percent of all incident reports involve small packages.

Nature and Extent of the Existing Hazmat Incident Reporting Data Base. At the beginning of 1983, there were approximately 130,000 hazmat incident reports (DOT Form F 5800.1) in the hazmat incident computerized data base. These reports span the 12-year period 1971–1982. During the two-year period 1981–1982, the data base increased by an average of 7,900 reports per year. Each report (see DOT Form F 5800.1 at the end of this document) contains approximately 30 primary data fields (e.g., date of incident, mode, name of carrier, name of shipper, commodity released, etc.). The data fields are further broken down by various codes including the following:

11,700	Companies (name, duns number, address, etc.)
1,400	Specific hazardous materials (e.g., gasoline)
328	Package types and specifications
27	Failure codes (e.g., dropped in handling)—of which only 15 actually appear on the report (the other 12 codes being inferred and assigned by MTB personnel)
25	Violation codes (e.g., driver not in attendance)
25	Significance codes (e.g., incidents involving 1-10 injuries)
35	Placard codes (e.g., empty)
21	Miscellaneous codes (e.g., vandalism suspected)
12	Restriction codes (e.g., removable head not authorized)
8	Type of Record codes (e.g., generic container type)

The 30 primary data fields on the incident report, plus the detailed and extensive data codes that have been applied to the reports, lead to an extremely vast and varied data base (e.g., the 30 primary data fields alone can be combined in 2.6×10^{32} or 260 million, trillion trillion ways). Even if a minute fraction of such combinations were analytically useful or meaningful, any attempt to analyze all of them would be very difficult, probably impossible, and in any case, enormously costly.

Several salient aspects of the existing hazmat data base are the following:

There were a total of 282 fatalities and 7,150 minor to severe injuries associated with the approximately 130,000 hazmat incident reports in the data base as of the beginning of 1983. Over the last three years, an annual average of 7,154 incidents, 8 fatalities and 172 injuries have been reported to the MTB.

Twenty-one percent of the 130,000 incident reports pertain to bulk packagings (e.g., cargo tanks, rail tank cars). Over the 12-year period, 1971–1982, hazmat incidents involving these containers resulted in 270 fatalities (96 percent of the total of all hazmat fatalities) and 4,305 injuries (60 percent of the total of all hazmat injuries).

Seventy-nine percent, or 102,700 of the 130,000 incident reports in the data base, pertain to small packages, such as bags, boxes, and drums. Of these 102,700 incidents, 84 percent are accounted for by only five DOT drum specifications, and seven generic or general purpose packages (e.g., cans, jugs, and bottles) which can be used to transport hazardous materials not requiring a DOT specification package. Over the 1971–1982 period, reported incidents involving these small packages resulted in 12 deaths and 2,845 injuries.

Seventy-six percent of all fatalities and 50 percent of all injuries have involved the following 12 selected hazardous materials.

Hazardous material	Percent total fatalities	Percent total injuries
Gasoline	40.2	4.9
LP-Gas	18.6	8.8
Anhydrous Am (NH$_3$)	5.7	6.5
Corrosive liquid NOS	2.7	2.9
Chlorine	2.7	5.5
Explosive, A	2.7	0.5
Flammable liquid, NOS	1.7	3.7
Sulfuric acid	0.7	8.5
Sodium hydroxide	0.7	3.0
Hydrochloric acid	0.3	2.1
Compound cleaning liquid		1.7
Poison liquid, NOS		1.5

It is MTB's belief that the continued augmentation of the existing data base under current requirements for incident reporting will not significantly increase an understanding of the causes, the nature, and the consequences associated with hazmat incidents. These incidents primarily pertain to incidents involving small packages.

This belief is based on, (1) the vast amount of data on small packages/containers already in the 12-year data base, (2) the diminishing marginal utility associated with the continued growth in the data base, rather than selective and judicious increases in the data base, in terms of the 30 primary data fields contained in the current incident report form, and (3) given the underlying millions of shipments, vehicle transit miles, and the varied nation-wide transportation environment, the fact that incidents involving small package/container of hazardous materials have been largely low consequences events.

Development of New Reporting Criteria. MTB has sought to develop alternatives to the current reporting criteria in terms of the following set of factors.

(A) Characterization of Hazmat Accident/Incident Event

Type of Event (e.g., in-transit, loading/unloading)

Type of Package (e.g., bulk/non-bulk)

Type of Hazmat (e.g., flammable liquid, explosives, etc.)

Mode (e.g., rail, highway, air, etc.)

Severity of Event

Frequency of Event

(B) Definition of Users

DOT/MTB

Other Federal Agencies

State and Local Governments

Public Interest Groups Industry

(C) Objectives of Users

Public Safety

Product/Container Performance

Research and Development

Determination of Liability

(D) User Data Requirements

Analytic Purposes (e.g., human factor analysis, cause-consequence analysis, fault-tree analysis, procedures analysis, cost/benefit/risk analysis)

Programmatic and Policy Analysis (e.g., enforcement and compliance, regulatory development, package performance)

(E) Nature of Data Requirements To Meet Purpose (e.g., essential/non-essential, level of detail, usefulness, i.e., multiple/single purpose applications, utilization, i.e., actual/potential, non-duplicative)

(F) Methods of Data Collection (e.g., routine reporting, special studies/surveys, other data sources)

(G) Costs Incurred in Data Collection (e.g., industry, government)

The above factors are all interrelated and entail a large number of considerations. The following summarizes the review team's major findings concerning them.

In terms of the characterization of a hazmat accident/incident event; clearly, an event involving a hazmat accident/incident—e.g., a cargo tank spill during loading/unloading operations—can be described in an extremely large number of ways, and can serve to generate an enormous array of data such as time of day, weather conditions, age of driver, type of truck, type of valve, manufacturer of valve, age of valve, design characteristics of valve, location of incident, type of hazmat released, amount released, etc.

Further distinctions characterizing a hazmat accident/incident event are also possible and useful. One can distinguish between events in which a hazardous material is actually spilled and events in which a hazardous material package is involved, but no spillage occurs. The current reporting requirements of 171.16, for the most part, pertain to events involving the actual spillage of a hazardous material. An event of this kind is termed an "incident." An event involving a hazmat package (e.g., a gasoline cargo tank overturning) but not involving a spillage of a hazardous material is not required to be reported to MTB. It should be noted, however, that this does not necessarily mean that such an event is not reported to the Department since, in the case of a cargo tank, it may be reported to the Bureau of Motor Carrier Safety as a motor vehicle accident.

Two further and related distinctions concern the "severity" and "frequency" of hazmat accident/incident events. These distinctions lead to the four-fold typology of:

High consequence—high frequency events.

High consequence—low frequency events.

Low consequence—high frequency events.

Low consequence—low frequency events.

Of the four types of events, the first two are considered of greater inherent concern, even though the remaining two types cannot be completely ignored, because certain types of low consequence events may have the potential for producing very high consequences under certain circumstances.

Just what type or arrays of data are to be generated is a function of the objectives of the users of the data, their data requirements, and a host of other considerations, including the cost associated with collecting, storing, and analyzing the data. Cost is a particularly important consideration, since many people find that certain data are "essential" for their purposes, only so long as they do not bear the cost of obtaining and maintaining the data.

With respect to the users of hazmat data and their objectives, the review team found, perhaps not surprisingly, that MTB is now and will continue to be the "primary" user of such data; that its program data requirements have priority over other user requirements; and that, although other user requirements should be accommodated to the extent possible, the hazmat data base cannot be all things to all possible users.

With respect to the primary objectives to be served from the vast array of data that can be generated by a hazmat accident/incident event, the review team found that these data should (1) serve as an aid in evaluating the effectiveness of the existing regulations, (2) assist in determining the need for regulatory changes to cover transportation safety problems, and (3) determine the major problem areas in hazmat transportation so that the attention of the Department may be more suitably directed to those areas.

To accomplish these objectives, the general nature of the data to be reported to MTB should have the following characteristics:

(A) The data should be essential, not merely "desirable." Indeed, the essential nature of the data is implied by the term "requirement."

(B) Because the data are essential, they must to be collected on a routine basis, rather than on a one-time, or intermittent, basis.

(C) Because it is a routine procedure, it is not practical that each separate report be of an extremely detailed nature. Extremely detailed data should be obtained through special studies or surveys (follow ups). The data should, therefore, be general purpose data, which maximizes their usefulness and actual (as against potential) utilization.

(D) The data should be nonduplicative with respect to the existence of other data of the same or similar nature and with respect to the volume of data (e.g., 100 incident reports may provide as much information as 1,000 reports) if each report contains essentially the same data.

Change Under Consideration. On the basis of the foregoing discussion, MTB is considering changing § 171.16 with respect to criteria for reporting incidents and the content and format of the report form (DOT Form F 5800.1). Under this change, carriers would be required to submit detailed written reports for incidents having the following characteristics:

(A) All incidents involving telephonic notifications required under § 171.15.

(B) All incidents involving bulk packagings.

(C) All incidents involving transportation aboard aircraft.

(D) All incidents involving property damage from the incident, including cleanup and decontamination, resulting in costs equal to or in excess of $1,000, incurred or anticipated to be incurred within 15 days of the incident.

(E) All incidents involving the evacuation of people.

(F) All incidents involving materials or packages shipped under MTB's exemption program.

(G) All incidents involving the release of hazardous waste.

Under this approach, § 171.16(a) would read as follows:

(a) Each carrier who transports hazardous materials shall report in writing, in duplicate, on DOT Form F 5800.1 to the Department within 15 days of the discovery, each incident that occurs during the course of transportation (including loading, unloading, or temporary storage) in which, as a direct result of the hazardous materials, any of the circumstances set forth in § 171.15(a) occurs; and all unintentional releases of hazardous materials involving:

(1) Bulk packagings;

(2) Shipments aboard aircraft or in air terminals;

(3) Property damage, including cleanup and decontamination, resulting in costs equal to or in excess of $1,000 incurred or reasonably anticipated to be incurred within 15 days of the incident;

(4) The evacuation of people;

(5) Packages or hazardous materials shipped under an exemption; and

(6) Any quantity of hazardous waste that has been discharged during transportation.

The current § 171.16(a)(1) and 171.16(a)(2) requirements pertaining to hazardous waste would be retained and redesignated as §§ 171.16(a)(6)(i), and 171.16(a)(6)(ii).

To assist in the selection of appropriate criteria for the submission of detailed, written reports on hazmat incidents, MTB invites interested persons to participate in this rulemaking. In particular, MTB requests comments addressed to the following questions and submission of any substantiating information:

1. In terms of the foregoing discussion and proposed reporting criteria identified herein, are there other criteria that should be considered for purposes of submitting detailed written reports on accidents or incidents involving hazardous materials? If so, what are they?

2. Does the current DOT incident report form (DOT Form F 5800.1) provide an adequate basis for:

a. Identifying major safety performance trends in the transportation of hazardous materials?

b. Providing a source of data for small packages and bulk packages safety design information and optimization in the transportation environment?

3. Should a separate incident report form be developed to focus exclusively on small package failure mechanisms in the transportation environment (in contrast to the present report form, DOT Form F 5800.1, which is used to describe hazmat incident data involving both packages, e.g., cargo tanks and small package incidents)? What data fields or failure mechanisms might such a report form include?

4. Is a $1,000 damage figure an adequate criterion for determining a threshold for reporting hazmat incidents that are otherwise without consequence? What is an appropriate property damage reporting criterion? Should an environmental damage criterion be included?

5. If no other formal proposal is made to the present incident reporting system, what changes do you recommend to the format and content of the present incident report form (DOT Form F 5800.1)?

6. Does your organization report, or is your organization now required to report hazardous material or hazardous waste, or hazardous substance accidents/incidents to another organization (e.g., insurance company, state or local government, other federal agency)? What are the criteria for reporting such accidents/incidents? Is there a standard form to be filled out? (Please attach a copy of such form, if appropriate.)

7. To what extent does your organization utilize hazmat incident data? Does your organization collect hazmat incident data? If so, what is the source and nature of these data? How often are such data collected (routinely, special surveys, etc.)? If any standardized forms are utilized in the collection of such data, we would appreciate receiving a copy of them.

(53 F.R. 10042, March 16, 1984)

SUMMARY • Docket No. HM-38 • *Established a tank car capacity and gross weight limitation; required certain couplers for tank cars, and their approval by the Federal Railroad Administrator.*

[49 CFR Part 179]
[Docket No. HM-38; Notice 69–31]
TRANSPORTATION OF HAZARDOUS MATERIALS
Interlocking Couplers and Restriction of Capacity of Tank Cars

The Hazardous Materials Regulations Board is considering amending Part 179 of the Hazardous Materials Regulations (1) to require interlocking couplers on all new and rebuilt tank cars transporting hazardous materials, and (2) to restrict the capacity of new and rebuilt tank cars used to transport hazardous materials.

The Board has become concerned with the increasing number of railroad accidents involving tank cars transporting hazardous materials in which the tank released its contents, either because of a puncture or because of a rupture, causing a dangerous condition. In many instances this release of hazardous materials resulted in personal injury and substantial property damage; and in two instances it resulted in multiple fatalities. According to the records compiled by the Federal Railroad Administration, since January 1, 1968, there have been 43 instances in which tank

HAZARDOUS MATERIALS INCIDENT REPORT

INSTRUCTIONS: Submit this report in duplicate to the Director, Office of Hazardous Materials Operations, Materials Transportation Bureau, Department of Transportation, Washington, D.C. 20590, (ATTN: Op. Div.). If space provided for any item is inadequate, complete that item under Section H, "Remarks", keying to the entry number being completed. Copies of this form, in limited quantities, may be obtained from the Director, Office of Hazardous Materials Operations. Additional copies in this prescribed format may be reproduced and used, if on the same size and kind of paper.

A | INCIDENT

1. TYPE OF OPERATION

1 ☐ AIR   2 ☐ HIGHWAY   3 ☐ RAIL   4 ☐ WATER   5 ☐ FREIGHT FORWARDER   6 ☐ OTHER (Identify) _____

2. DATE AND TIME OF INCIDENT *(Month - Day - Year)*

_____ a.m.
_____ p.m.

3. LOCATION OF INCIDENT

B | REPORTING CARRIER, COMPANY OR INDIVIDUAL

4. FULL NAME

5. ADDRESS *(Number, Street, City, State and Zip Code)*

6. TYPE OF VEHICLE OR FACILITY

C | SHIPMENT INFORMATION

7. NAME AND ADDRESS OF SHIPPER *(Origin address)*

8. NAME AND ADDRESS OF CONSIGNEE *(Destination address)*

9. SHIPPING PAPER IDENTIFICATION NO.

10. SHIPPING PAPERS ISSUED BY

☐ CARRIER   ☐ SHIPPER

☐ OTHER
(Identify) _____

D | DEATHS, INJURIES, LOSS AND DAMAGE

DUE TO HAZARDOUS MATERIALS INVOLVED

11. NUMBER PERSONS INJURED

12. NUMBER PERSONS KILLED

13. ESTIMATED AMOUNT OF LOSS AND OR PROPERTY DAMAGE INCLUDING COST OF DECONTAMINATION *(Round off in dollars)*

14. ESTIMATED TOTAL QUANTITY OF HAZARDOUS MATERIALS RELEASED

$

E | HAZARDOUS MATERIALS INVOLVED

15. HAZARD CLASS (*Sec. 172.101, Col. 3)	16. SHIPPING NAME (*Sec. 172.101, Col. 2)	17. TRADE NAME

F | NATURE OF PACKAGING FAILURE

18. *(Check all applicable boxes)*

(1) DROPPED IN HANDLING	(2) EXTERNAL PUNCTURE	(3) DAMAGE BY OTHER FREIGHT
(4) WATER DAMAGE	(5) DAMAGE FROM OTHER LIQUID	(6) FREEZING
(7) EXTERNAL HEAT	(8) INTERNAL PRESSURE	(9) CORROSION OR RUST
(10) DEFECTIVE FITTINGS, VALVES, OR CLOSURES	(11) LOOSE FITTINGS, VALVES OR CLOSURES	(12) FAILURE OF INNER RECEPTACLES
(13) BOTTOM FAILURE	(14) BODY OR SIDE FAILURE	(15) WELD FAILURE
(16) CHIME FAILURE	(17) OTHER CONDITIONS *(Identify)*	19. *SPACE FOR DOT USE ONLY*

Form DOT F 5800.1 (10-70) (9/1/76)
*-Editorial change to incorporate redesignation per HM-112.

PACKAGING INFORMATION - *If more than one size or type packaging is involved in loss of material show packaging information. separately for each. If more space is needed, use Section H "Remarks" below keying to the item numbers.*

	ITEM		#1	#2	#3
20	TYPE OF PACKAGING INCLUDING INNER RECEPTACLES (*Steel drums, wooden box, cylinder, etc.*):				
21	CAPACITY OR WEIGHT PER UNIT (*55 gallons, 65 lbs., etc.*)				
22	NUMBER OF PACKAGES FROM WHICH MATERIAL ESCAPED				
23	NUMBER OF PACKAGES OF SAME TYPE IN SHIPMENT				
24	DOT SPECIFICATION NUMBER(S) ON PACKAGES (*21P, 17E, 3AA, etc., or none*)				
25	SHOW ALL OTHER DOT PACKAGING MARKINGS (*Part 178*)				
26	NAME, SYMBOL, OR REGISTRATION NUMBER OF PACKAGING MANUFACTURER				
27	SHOW SERIAL NUMBER OF CYLINDERS, CARGO TANKS, TANK CARS, PORTABLE TANKS				
28	TYPE DOT LABEL(S) APPLIED				
29	IF RECONDITIONED OR REQUALIFIED, SHOW	A REGISTRATION NO. OR SYMBOL			
		B DATE OF LAST TEST OF INSPECTION			
30	IF SHIPMENT IS UNDER DOT OR USCG SPECIAL PERMIT OR EXEMPTION, ENTER PERMIT OR EXEMPTION NO.				

H REMARKS - Describe essential facts of incident including but not limited to defects, damage, probable cause, stowage, action taken at the time discovered, and action taken to prevent future incidents. Include any recommendations to improve packaging, handling, or transportation of hazardous materials. Photographs and diagrams should be submitted when necessary for clarification.

31. NAME OF PERSON PREPARING REPORT (*Type or print*)	32. SIGNATURE
33. TELEPHONE NO. (*Include Area Code*)	34. DATE REPORT PREPARED

cars released their contents as the result of a rail accident. In 22 instances, evacuation of the surrounding population was undertaken as a safety measure. Although the records do not indicate the amount of property damage which resulted from these occurrences, conservative estimates indicate that the loss exceeds $15 million.

The Department is presently reviewing the design of these tank cars in an effort to determine the relationship between the capacity of tank cars and the likelihood of the release of contents in an accident. However, pending the completion of these studies, the Board believes that some action should be taken to reduce the likelihood that new or rebuilt tank cars will release their hazardous lading when involved in a rail accident. Therefore, the Board proposes to require that all tank cars built or rebuilt after June 30, 1970, that are used to transport hazardous materials, must be equipped with interlocking couplers that have been approved by the Federal Railroad Administration. These couplers should reduce the incidence of couplers disengaging in an accident and puncturing the head of a tank car. Since puncture of the tank car tank head is the primary cause of release of product, this requirement would reduce the number of head punctures in tank heads by a considerable amount. In addition, the Board proposes to limit the capacity of tank cars used to transport hazardous materials that are built after June 30, 1970, to those not over 34,500 gallons capacity or not over 263,000 pounds gross weight on rail.

If the above-referenced studies warrant, the Board will in future rule making actions apply these requirements and limitations and any others found to be necessary to existing tank cars used to transport hazardous materials. These and other actions will be discussed further in an advance notice of proposed rulemaking to be issued in the near future.

Interested persons are invited to give their views on this proposal. Communications should identify the docket number and be submitted in duplicate to the Secretary, Hazardous Materials Regulations Board, Department of Transportation, 400 Sixth Street SW., Washington, D.C. 20590. Communications received on or before February 10, 1970, will be considered before final action is taken on the proposal. All comments received will be available for examination by interested persons at the Office of the Secretary, Hazardous Materials Regulations Board, both before and after the closing date for comments.

(34 F.R. 19553, December 11, 1969)

[Docket No. HM-38; Amendment No. 179-4]
PART 179—SPECIFICATIONS FOR TANK CARS
Restriction of Capacity of Tank Cars and Interlocking Couplers

The purpose of this amendment to the Hazardous Materials Regulations of the Department of Transportation is to restrict the gross weight and volume capacity of, and require interlocking couplers on all new tank cars used to transport hazardous materials.

On December 11, 1969, the Hazardous Materials Regulations Board published Docket No. HM-38; Notice No. 69-31 (34 F.R. 19553) proposing to amend Part 179 of the Hazardous Materials Regulations as indicated above. In that notice, the Board stated its concern with the increasing number of railroad accidents involving tank cars transporting hazardous materials in which the tank released its contents, through either puncture or rupture. Reference was made to the mounting death and personal injury rate resulting from these accidents, as well as the property loss. Interested persons were afforded an opportunity to participate in this rule making.

Regarding the imposition of a capacity limitation of 34,500 gallons, many respondents noted that large capacity tank cars tended to reduce the hazard to the public by reducing the number of cars required for a given volume movement. No consideration was expressed for the fact that increased capacity will result in a greater hazard in the event that the tank car is punctured or ruptured in a derailment. Large capacity tank cars also increase the hazard of soil, water and air pollution.

Many responses were addressed to the question of limiting the total gross weight on rail to 263,000 pounds. Some of the data discussed the validity of a weight limitation as a control measure to improve railroad safety, focusing primarily on weight-related causative accident factors and the effects on kinetic energy of the tank car.

Causative accident factors show that stress failures in track and car parts account for approximately 50 percent of all rail accidents. The Board believes that the relationship between such stress failures and car weight is direct.

In every example offered citing rail loads in excess of the proposed limit, particular mention was made of the special routing clearances and controls exercised over the movement of these cars. Such special measures are not present in normal tank car movement, which is the situation to which the Board must address itself. Only one response offered design data which showed that due consideration had been given to overbuilding a tank and running gear to obtain the margin of safety which is required by good engineering practice.

Weight related stress failures are known to have occurred in existing "100 ton" capacity, 263,000 pounds gross weight tank cars which have been in service for a period of years. "Fix" programs to correct buckling and fatigue cracking at both ends of stub sills on underframeless cars have been underway for several years. It is necessary to have an upgrading of the present tank car fleet in order to withstand the rigors of the normal railroad environment over the expected life of the tank cars. This upgrading must be accomplished before considering allowing increase of the stress loads on equipment and the rail plant caused by heavier cars.

One respondent addressed himself to the influence of weight on kinetic energy of the tank car and mentioned the ability of a larger mass to absorb a larger amount of kinetic energy. Increasing the weight of the tank car produces a linear increase in its kinetic energy at equal velocity. This increased kinetic energy increases the likelihood that the tank will be punctured or will rupture in an accident. Therefore, the Board believes that limiting the maximum weight of a tank car will reduce incidents of puncture and rupture.

Inadequate consideration has been given in current design practice to the selection of material thicknesses to compensate for greater kinetic energy levels encountered as tank car weight increases. As train operating speeds increase, this kinetic energy increases exponentially.

Sill design has been held nearly constant despite change in tank car weight and capacity, and shell thickness has varied only as a function of the tensile strength of materials and tank diameter. It is apparent that the weight (stress) related elements have not been strengthened as a direct function of capacity. The Board believes that this, in effect, results in a lower factor of safety in larger capacity tank cars as related to smaller capacity cars.

Virtually all respondents mentioned the economic impact of the proposed weight-capacity limitations. It must be recognized that the cost of accidents is also a part of the national distribution costs and is reflected in freight rates.

In order to accurately determine the economic effect of this rule making, the Board retained an independent expert to analyze the overall costs of "large capacity" tank cars as related to "smaller capacity" tank cars. The following table summarizes his findings:

	Dollars per ton	Cents per gallon
500-mile movement:		
70-ton capacity	8.58	2.0151
100-ton capacity	7.22	1.6957
125-ton capacity	6.71	1.5757
140-ton capacity	7.65	1.7983
1,000-mile movement:		
70-ton capacity	13.52	3.1777
100-ton capacity	11.54	2.7111
125-ton capacity	10.84	2.5465
140-ton capacity	11.98	2.8165
1,500-mile movement:		
70-ton capacity	18.47	4.3403
100-ton capacity	15.86	3.7265
125-ton capacity	14.97	3.5173
140-ton capacity	16.32	3.8347

The table indicates that costs involved in utilizing the "100 ton" capacity tank car differ little from those costs involved in utilizing the "125 ton" capacity tank car. The "100 ton" capacity tank car actually offers some cost savings over the "140 ton" capacity tank car. The Board believes public safety warrants the slight reduction in economic efficiency which results from utilizing "100 ton" capacity tank cars in place of "125 ton" capacity tank cars.

For the above reasons, the Board concludes that the proposed restrictions on tank car weight-capacity are in the public interest. Until the present problems involved in using the "100 ton" capacity tank cars are resolved and until evidence is presented to show that increased stress levels associated with higher unit loadings on the rail plant and tank car equipment at prevailing speeds have been adequately compensated for, this will remain the Board's conclusion.

The Board further believes that the application of interlocking automatic couplers on all new tank cars will materially improve safety by reducing the incidence of tank head puncture and tank car pileup.

Since the date of Notice No. 69–31, there have been 19 accidents involving tank cars transporting hazardous materials in which the contents have been released causing severe hazard. One such accident occurred at Crescent City, Ill., on June 21, 1970. The continuing occurrence of accidents of this nature makes evident the need for action. The Hazardous Materials Regulations Board is aware that research efforts are being made by the affected industries, and that the Federal Railroad Administration has entered into contracts to study certain aspects of tank car design and accident behavior. It is hoped that these efforts will develop improved tank designs and methods of construction, including specialized hardware, which will enable all newly built tank cars to be able to safely transport hazardous materials. Until the results of these research activities are known, the Board believes that the proposed steps must be taken to prevent proliferation of the problems resulting from the continued construction of large capacity tank cars exceeding 34,500 gallons. While the Board recognizes that the Crescent City accident involved tank cars having capacities in the 30,000-gallon range, it believes that larger capacity cars would have released much greater quantities of hazardous materials, with consequently increased fire hazard and property damage. In addition, the added weight on rail would have increased the impact forces in the derailment and might well have resulted in additional punctures, fires, and violent ruptures.

Several responses noted the lack of a readily acceptable definition of the term "rebuilt tank car." This term has been deleted from the amendment pending the Board's further review.

The Board believes that by requiring installation of interlocking couplers that will resist car telescoping and jackknifing in derailments and emergency stops, the incidence of tank head and side puncture will be markedly reduced. At Crescent City, a tank head puncture caused the eventual conflagration and violent ruptures.

(35 F.R. 14216, September 9, 1970)

[Docket No. HM-38; Amdt. No. 179–4]
CERTAIN RESTRICTIONS ON NEW TANK CARS USED TO TRANSPORT HAZARDOUS MATERIALS
Petitions for Reconsideration

On September 9, 1970, the Hazardous Materials Regulations Board of the Department of Transportation published Docket No. HM-38; Amendment No. 179–4 (35 F.R. 14216) restricting the gross weight and volume capacity of, and requiring interlocking couplers on, all new tank cars used to transport hazardous materials.

The Board subsequently has received several timely petitions for reconsideration pursuant to the provisions of §170.35 of the Hazardous Materials Regulations. The points raised by petitioners and the Board's responses follow.

One petition noted that the preamble to the amendment for the first time made certain information public in justification of the rule. Petitioner contended that the public did not have an opportunity to comment upon this information and data. Information appearing in the preamble that had not appeared in the notice (34 F.R. 19553) was offered in response to specific comments made on that notice. It did not form the primary basis of the Board's decision in HM-38. The Board is of the opinion that to offer an opportunity for public comment on all information offered in response to specific comments on the notice of proposed rule making could give rise to an impractical and unnecessary dialogue.

It was also pointed out that the Federal Railroad Safety Act of 1970, 84 Stat. 971, has been enacted subsequent to the publication of the notice, and that this Act grants more comprehensive regulatory authority to the Department of Transportation in the field of railroad safety. While this new legislation may enable the Department to cope more effectively with the breadth of problems involved with today's railroads, the enactment in itself does not lead the Board to alter the conclusions expressed in HM-38.

A petitioner cited a recently completed study made for the U.S. Coast Guard by the National Academy of Sciences, relating to the factors involved in cargo size limitations. The Board is aware of the study, but as it was directed to bulk shipments by water and involved dimensions and transportation factors not encountered in tank car service, it was considered to be inapplicable to the subject matter of HM-38.

It is contended that Docket No. HM-60, Request for Public Advice on Speed Restriction on Tank Cars (35 F.R. 16180), is a proceeding interconnected with the issues involved in HM-38, and that therefore the public ought to have the opportunity to comment on the integrated package of regulations. Docket No. HM-60 is limited in its applicability to DOT Specifications 112A and 114A tank cars transporting liquefied flammable gases and, in the Board's opinion, is not so related to the matters involved in HM-38 as to require delay of the effective date of the amendment.

Apparent confusion was noted regarding the term "built", as it appears in the amendment. The Board believes this term to be one of common usage in the tank car construction industry, and that it is reflected in the "built" date presently stenciled on all tank cars. For the sake of clarity, the Board may initiate rule making to provide a comprehensive definition of the term "built", but the Board is not of the opinion that sufficient confusion exists at the present time to warrant extension of the effective date of HM-38.

Docket No. HM-38 requires installation of "approved" interlocking automatic couplers, but petitioners noted that as yet, no couplers had received approval from the Federal Railroad Administrator. On November 13, 1970, the Board published Docket No. HM-38; Amendment No. 179–5 (35 F.R. 17418), amending new §179.14 to list those interlocking couplers approved as of that date. In order to provide adequate time to assure compliance with the new section, the amendment also extended the date for required installation of approved couplers to January 1, 1971.

Certain research is being conducted to further analyze difficulties encountered in tank car operations, but the Board is of the opinion that it is not in the public interest to defer the effective date of HM-38 to await receipt of tangible results from those studies. If such research reveals evidence in addition or contrary to the present conclusions of the Board, appropriate rule-making proceedings may be initiated at that time.

The Board concludes that, except for the above-noted amendment to §179.14, the petitions for reconsideration of HM-38 should be and are hereby denied.

(35 F.R. 18009, November 24, 1970)

SUMMARY • **Docket No. HM-42** • *Proposed the development of a new materials classification called "Combustible liquids". See Docket Nos. HM-67, HM-102, and HM-112.*

[49 CFR Parts 172, 173, 174, 177]
[Docket No. HM-42; Notice No. 70–3]
TRANSPORTATION OF HAZARDOUS MATERIALS
Combustible Liquids

The Department's Hazardous Materials Requirements Regulations presently define a flammable liquid as any liquid having a flash point of 80° F. or lower. Liquids having flash points higher than 80° F. are not now within the scope of the Department's Hazardous Materials Regulations. Liquids in this higher flash point range include kerosene, fuel oil, turpentine, and certain alcohols, all of which present fire hazards during transportation. These liquids are often referred to by the generic name, combustible liquids, which normally refers to liquids having flash points between 80° F. and 200° F. These materials are routinely transported in tank cars, tank trucks, and portable tanks with no requirement that these tanks be identified during transportation as containing a material having a fire hazard.

Fire, police, and rescue personnel are generally trained to deal with fuel oil and kerosene accidents in the same manner as they deal with gasoline accidents. In order to be able to do their job, they must have immediate information regarding the contents of these tanks. Without this information, the emergency personnel might well be misled into believing that the tanks contained some innocuous commodity such as milk or molasses. Their attention might, therefore, be misdirected away from this significant potential hazard. The Board believes that it will be in the public interest to require that tanks containing combustible liquids be marked or placarded to properly reflect the hazard present and is proposing in this notice to adopt an identification system.

Compounding the problem of lack of information as to hazards is the fact that many tank truck operators are transporting combustible liquids in tanks which bear the placard "Non-Flammable". This is apparently done in order to be able to permanently mark the word "Flammable" on tanks which are used interchangeably in shipping flammable or combustible liquids. In that way, the carrier need only to add a small tag or plate with the word "Non" on it rather than having to constantly remove and replace a larger placard having the word "Flammable". Placarding of this type is a gross misrepresentation of the actual hazard that would be present should such vehicles be involved in accidents, parked or stopped near fires, or otherwise placed in jeopardy.

A second related problem involves the transportation of liquids which have flash points above 200°F. but which are transported at temperatures at or exceeding their flash points. If tanks containing these hot liquids fail during transportation resulting in rupture or leakage, the overall public hazard may be just as great as flammable or combustible liquids with lower flash points.

The Board believes that these two closely related problem areas can best be resolved by including in the regulations a new hazard classification for combustible liquids (flash points between 80° F. and 200° F.) and other liquids (flash points above 200° F.) which are transported at a temperature higher than their flash points.

The 200° F. upper limit is one commonly used by industry and Government. The National Fire Protection Association, in their "Fire Protection Guide on Hazardous Materials," second edition, 1967, uses a 200° F. break-point in flash point. The Federal Highway Administration does the same in its "Motor Carrier Safety Regulations" (49 CFR 392). The 80°–200° F. range will include almost all of the commonly transported combustible fuel oils which comprise the major portion of bulk shipments. The Board recognizes that for certain other purposes a cutoff of 150° F. has been used. The Board therefore, while proposing a cutoff at 200° F., requests specific comment on whether a 150° F. cutoff would be warranted. Reasons for recommending either cutoff point or for any cutoff within this range should be given.

The Board is proposing to require that shipments of combustible liquids in tanks be clearly identified by the same type of vehicle placards used for identifying other rail and highway shipments. A new "Combustible" placard is proposed for highway use as an alternative to the "Flammable" placard, and the existing "Dangerous" placard is proposed for shipments by rail.

Because of the significantly lower degree of potential hazard involved in shipments of combustible liquids in smaller containers, the Board is proposing to apply these new rules only to shipments in containers having a volume greater than 110 gallons.

(35 F.R. 3298, February 21, 1970)

SUMMARY • **Docket No. HM-45** • *Prohibited the transportation of cargo tanks by rail in trailer-on-flatcar (TOFC) service, except under conditions approved by the Federal Railroad Administrator.*

[49 CFR Part 174]
[Docket No. HM-45; Notice No. 70–6]
TRANSPORTATION OF HAZARDOUS MATERIALS
Cargo Tanks in Trailer-on-Flatcar Service

The Hazardous Materials Regulations Board is considering a proposal to amend §174.533 of the Department's hazardous materials regulations to prohibit the transportation of cargo tanks containing hazardous materials by rail in trailer-on-flatcar service except under conditions approved by the Federal Railroad Administrator.

Interested persons are invited to give their views on the proposal discussed herein. Communications should identify the docket number and be submitted in duplicate to the Secretary, Hazardous Materials Regulations Board, Department of Transportation, 400 Sixth Street SW., Washington, D.C. 20590. Communications received on or before June 16, 1970 will be considered before final action is taken on the proposal. All comments received will be available for examination by interested persons at the Office of the Secretary, Hazardous Materials Regulations Board, both before and after the closing date for comments.

The Board believes that the transportation of hazardous materials in cargo tanks of certain designs and construction in trailer-on-flatcar service can present an unacceptable hazard and may not be in the public interest. The Board considered the following factors in reaching this conclusion:

(1) Because of their design and construction, cargo tanks are capable of withstanding the dynamic loadings experienced during the normal course of highway transportation. The design criteria for specification cargo tanks for highway use require a resistance to forces of 2G. Railroad tankcars are designed to withstand forces of 7G longitudinally and 3G vertically and transversely. Since tankcars are designed to maintain their integrity in impacts and derailments with a certain degree of safety, it follows that cargo tanks would be much more susceptible to damage or failure in the same transportation environments.

(2) When a loaded cargo tank vehicle is placed on a flatcar, the center of gravity of the vehicle is approximately 120 inches or more above the top of the rail. The Association of American Railroads has specified acceptable safe limits of 98 inches for loaded railcars. This, combined with lateral instability resulting from the flexibility in motor vehicle suspensions and tires, creates a hazardous situation when not compensated for by additional design and structural requirements for cargo tanks.

(3) In rail accidents, particularly derailments (5,487 in 1968), there is little probability that any cargo tank, unless of extraordinary design and construction, could sustain the rail accident environment without failure and resulting discharge of its contents.

Section 174.533 (c) presently authorizes cargo tanks that are mounted on truck bodies or trailer chassis and that contain hazardous materials to be transported by rail only under conditions approved by the Bureau of Explosives. The Board proposes to withdraw this delegation of authority and to prohibit such transportation except under conditions approved by the Federal Railroad Administrator.

(35 F.R. 6151, April 15, 1970)

[Docket No. HM-45; Amdt. No. 174–8]
PART 174—CARRIERS BY RAIL FREIGHT
Cargo Tanks in Trailer-on-Flat-Car Service

The purpose of this amendment to the Hazardous Materials Regulations of the Department of Transportation is to make transportation of cargo tanks containing hazardous materials in trailer-on-flat-car service (TOFC) subject to conditions approved by the Federal Railroad Administrator. Such approval authority is presently exercised by the Bureau of Explosives (AAR).

On April 15, 1970, the Hazardous Materials Regulations Board published a notice of proposed rule making, Docket No. HM-45; Notice No. 70–6 (35 F.R. 6151), proposing the amendment described above. In addition, the notice expressed certain of the Board's views regarding the hazards involved in transportation of cargo tanks in trailer-on-flat-car service.

Interested persons were invited to give their views on the proposal and on the opinions expressed by the Board. Comments generally favored the proposed vesting of approval authority in the Federal Railroad Administrator. Many helpful comments and data were received relating to the hazards involved in TOFC service. These will be given full consideration by the Federal Railroad Administrator in the development of criteria for conditions under which TOFC service for transportation of hazardous materials may be approved.

It should be emphasized that the proposed amendment does no more at this time than change the approving authority from that of a non-governmental agency to a designated official of the cognizant Federal agency. In evaluating conditions submitted for approval, the Federal Railroad Administrator will avail himself of the information and advice of the Bureau of Explosives (AAR), and will take into consideration the Bureau of Explosives' criteria in granting approvals heretofore.

The Board concludes that it is in the public interest to adopt the amendment as proposed. In order to facilitate compliance with the amendment, it is suggested that during the 90-day period preceding the effective date, shippers and railroad carriers presently conducting this type of operation under existing Bureau of Explosives' approval apply to the Federal Railroad Administrator for his approval of their continued operations.

(35 F.R. 17950, November 21, 1970)

SUMMARY • **Docket No. HM-51** • *Several advance notices were issued in this docket requesting public comment on a new DOT system for classifying poisonous materials. See also Docket No. HM-112.*

[49 CFR Parts 170–189]
[Docket No. HM-51]
CLASSIFICATION OF CERTAIN HAZARDOUS MATERIALS ON BASIS OF THEIR HEALTH HAZARDS
Advance Notice of Proposed Rule Making

On August 21, 1968 (33 F.R. 11862), the Hazardous Materials Regulations Board announced a plan to revise the regulations governing the transportation of hazardous materials. That document announced the intention to issue notices of proposed rule making in at least four areas, including, "classification and labels", and invited public help in developing the basic regulatory principles to guide the Board in revising the regulations.

The Board is planning to consider, in the near future, a proposal for classification tests for poisonous materials. To assist the Board in that consideration, the public is invited to express its views on the health hazard classification tests proposed herein. This document is not a proposal to change the regulations. It is an effort to get public participation early in the rule-making process.

The present definitions of poisonous materials contain specific testing criteria only in the case of class B poisons. There are no criteria now provided for class A poisons or irritating materials (including tear gases). As a result, the public cannot practically rely upon those definitions to determine when the Federal regulations apply. In order to correct that situation, the Department proposes to adopt testing criteria for those latter two categories.

The National Research Council-National Academy of Sciences assisted the Department in developing these test criteria. In addition, the testing procedures and benchmarks used by the Departments of Agriculture and Health, Education, and Welfare have also been considered to ensure harmony between the regulatory standards of the several Federal agencies having jurisdiction in this area (see, for example, §191.1 of the regulations of the Department of Health, Education, and Welfare, 21 CFR Part 191, and §362.116 of the regulations of the Department of Agriculture, 7 CFR Part 362).

Types of health hazards. The health hazards of materials being transported are characterized by their acute effects on human health. Hazards to be considered are: Systemic hazards. Contact hazards. Irritant hazards.

Systemic hazards exist when materials are capable of causing harmful effects through inhalation, ingestion, or absorption through the skin. Contact hazards exist when materials are capable of causing destruction of living tissue or tissue reaction by thermal or chemical action at the site of contact. Irritant hazards exist when the materials are capable of causing local irritating effects on eyes, nose, or throat, or are capable of causing severe lachrimation.

Hazard degrees. Degrees of hazard are ranked according to the potential severity of the hazard to people. The establishment of hazard degrees is necessary in order to establish packaging criteria reflecting the potential severity of the damage if a product should escape from its packaging during transportation. This potential must be taken into account in the design and integrity of packaging used in the shipment of the toxic products. The major categories and criteria are as follows:

Extremely dangerous poisons. Materials would be classified as extremely dangerous poisons if, on short exposure, they could cause deaths or major residual injury to humans. In the absence of adequate data on human toxicity, a material would be presumed to be extremely poisonous to humans if it falls within any one of the following categories when tested on laboratory animals:

(1) Ingestion (oral). Any material that has a single dose LD_{50}[1] of 5 milligrams or less per kilogram of body weight when administered orally to both male and female rats, each weighing between 200 and 300 grams, and which have been fasted for a period of 24 hours.

(2) Inhalation. Any material that has an LC_{50}[1] of 75 parts per million by volume or less or 0.75 milligrams per liter by volume or less of vapor, mist or dust when administered by continuous inhalation for 1 hour or less to both male and female rats, each weighing between 200 and 300 grams. If the material is administered to the animals as a dust or mist, more than 90 percent of the particles available for inhalation in the test must have a diameter of 10 microns or less.

(3) Skin contact. Any material that has an LC_{50} of 100 milligrams or lower per kilogram of body weight when administered by continuous contact for 1 hour with the bare skin of rabbits, each weighing between 2.3 and 3.0 kilograms, according to test procedures described in §191.10 of the regulations of the Department of Health, Educa-

tion, and Welfare (21 CFR Part 191). (This test procedure is also listed in NAS-NRC Publication 1138, "Principles and Procedures for Establishing the Toxicity of Household Substances", available from the Printing and Publishing Office, National Academy of Sciences, Washington, D.C. 20408 at $1.50.)

Toxic materials. Materials would be classified as toxic if on short exposure they could cause serious temporary or residual injury to humans. In the absence of adequate data on human toxicity, a material would be presumed to be toxic to humans if it falls within any one of the following categories when tested on laboratory animals:

(1) Ingestion (oral). Any material that has a single dose LD_{50} of more than 5 milligrams but not more than 50 milligrams per kilogram of body weight when orally administered to both male and female rats, each weighing between 200 and 300 grams, and which have been fasted for a period of 24 hours.

(2) Inhalation. Any material that has an LC_{50} of more than 75 parts per million by volume but not more than 200 parts per million or more than 0.75 milligram but not more than 2 milligrams per liter of vapor, mist, or dust when administered by continuous inhalation for 1 hour or less to both male and female rats, each weighing between 200 and 300 grams. If the product is administered to the animals as a dust or mist, more than 90 percent of the particles available for inhalation in the test must have a diameter of 10 microns or less.

(3) Skin contact. Any material that has an LD_{50} of greater than 100 milligrams but not more than 200 milligrams per kilogram of body weight when administered by continuous contact for 1 hour with the bare skin of rabbits, each weighing between 2.3 and 3.0 kilograms, according to the test procedure described in §191.10 of 21 CFR Part 191.

Irritating materials. Materials would be classified as irritants if they would cause reversible local irritant effects on eyes, nose, or throat, or cause slight irritation to the skin of humans. In the absence of adequate data on human reaction, they would be presumed to be irritating if they fall within either of the following categories:

(1) Skin irritation. Any material with an average irritation score of 4 or more, but less than 6, according to the test procedures described in §191.11 of 21 CFR Part 191. (If the score is 6 or more, the material would be classified as corrosive.)

(2) Eye irritation. Any material which exhibits a reaction with an average score of 2 or more according to the test for eye irritants described in § 191.12 of 21 CFR Part 191, and the "Illustrated Guide for Grading Eye Irritation by Hazardous Substances" prepared by the U.S. Department of Health, Education, and Welfare, and available from the U.S. Government Printing Office, Washington, D.C.

If human experience or other data indicate that the hazard of a given material that may be encountered during an accidental exposure in transportation is greater or lesser than indicated by the data from the specified animal tests, the classification for that specific material would be revised upward or downward.

If these classifications are adopted, appropriate changes will be required in the labels required to be applied to packages.

Interested persons are invited to give their views as to whether this approach is a reasonable and practical one. Alternative approaches, along with supporting data, will be welcome. Comments (identifying the docket number) should be submitted prior to September 4, 1970, in duplicate to the Secretary, Hazardous Materials Regulations Board, Department of Transportation, 400 Sixth Street SW., Washington, D.C. 20590. All comments received will be available for examination by interested persons at the Office of the Secretary, Hazardous Materials Regulations Board, both before and after the closing date for comments.

[1] LD_{50}, LC_{50}: That dose (LD) or concentration (LC) which will cause death within 14 days to at least one-half of the test animals.

(35 F.R. 8831, June 6, 1970)

[49 CFR Parts 170–189]
[Docket No. HM-51]

CLASSIFICATION OF CERTAIN HAZARDOUS MATERIALS ON BASIS OF THEIR HEALTH HAZARDS
Second Advanced Notice of Proposed Rule Making

On June 6, 1970, the Hazardous Materials Regulations Board published an Advance Notice of Proposed Rule Making Docket No. HM-51 (35 F.R. 8831), inviting public assistance in developing regulatory principles for the classification of certain hazardous materials on the basis of their health hazards.

The comments received generally related to toxicity test procedures, classification, and degrees of toxicity.

Toxicity test procedures. Most commenters agreed that toxicity test procedures should be uniform among regulatory agencies, noting even minor variations by DOT could be confusing. Apprehension was displayed concerning the use of tests and other criteria which were not developed specifically for the transportation environment.

Classification. Many commenters contended that materials having only minor temporary and reversible irritating effects should not be regulated. The potential safety hazard of these materials during transportation, they claimed, is not considered so severe as to justify regulation. Others suggested that materials causing irreversible destruction of living tissue should be covered under a corrosive classification, such as discussed in Docket No. HM-57. Some commenters suggested that the Poison C category be retained for lachrymatory and respiratory irritation.

Degrees of toxicity. There was no common opinion expressed in this area. One group of commenters suggested retaining only one toxic category as Poison B, leaving the Poison A category for gases only, and possibly placing some quantitative benchmarks on this category. Others agreed in principle with the designation of various degrees but suggested modifications.

A second advance notice is being used to give the public ample opportunity to contribute to the development of the matter before the Board, prior to publication of a notice of proposed rule making. On the basis of comments received on this advance notice, the Board is of the opinion it will be able to prepare a more complete and meaningful notice.

The present definitions of poisonous materials only contain specific testing criteria or guidelines for Class B poisons. There are no criteria or sufficiently descriptive guidelines for Class A or Class C poisons. Consequently, the public may encounter difficulty in relying solely on those definitions to determine the applicability of the regulations. In order to improve this situation, the Board proposes to adopt testing criteria wherever possible and better descriptive guidelines for all toxic materials covered by the Department's regulations.

The National Research Council-National Academy of Sciences assisted the Department in developing these test criteria. In addition, the testing procedures and hazard degrees used by the Departments of Agriculture and Health, Education, and Welfare were considered to insure harmony among the regulatory standards of Federal agencies having jurisdiction with respect to health hazards of chemicals.

The health hazards of materials being transported are proposed to be characterized by their acute effects on human health. The hazards considered are systemic hazards and irritant hazards. Systemic or internal hazards exist when materials, if inhaled, ingested, or absorbed through the skin can have harmful effects on organs and tissues other than at the site of contact. Irritant hazards exist when materials such as gases, vapors, or mist can have local irritating effects on eyes, nose, or throat temporarily impairing the person's ability to function to the degree that he cannot take necessary emergency action.

Materials which otherwise produce reversible injury to the tissues of the skin are not proposed to be regulated. Materials which cause destruction of these tissues by chemical action would be considered under the "Corrosive" classification discussed in Docket No. HM-57, Classification of Corrosive Hazards; advance notice of proposed rule making, published September 4, 1970 (35 F.R. 14090).

Degrees of hazard would be ranked according to the potential severity of the hazard to people. The establishment of hazard degrees is necessary in order to establish packaging criteria reflecting the potential severity of the damage if a product should escape from its packaging during transportation. The major categories and criteria which would be proposed are as follows:

Extremely toxic substances. Materials would be classified as extremely toxic substances if, on short exposure, they could cause death or major residual injury to humans. In the absence of adequate data on human toxicity, a material would be presumed to be extremely toxic to humans if it fell within any one of the following categories when tested on laboratory animals, according to the U.S. Department of Agriculture test procedures described under Title 7, Chapter 3, §362.8 of the Federal Regulations.

(1) Ingestion (oral): Any material that has a single dose LD_{50}[1] of 5 milligrams or less per kilogram of body weight when administered orally to both male and female rats (young adults).

(2) Inhalation: Any material that has an LC_{50}[1] of 50 parts per million or less by volume of a gas or vapor, or 0.50 milligrams or less of mist or dust per liter of air when administered by continuous inhalation for 1 hour to both male and female white rats (young adults). If the material is administered to the animals as a dust or mist, more than 90 percent of the particles available for inhalation in the test must have a diameter of 10 microns or less, provided the Department finds it reasonably foreseeable that such concentrations could be encountered by man.

(3) Skin absorption. Any material that has an LD_{50} of 20 milligrams or less per kilogram of body weight when administered by continuous contact for 24 hours with the bare skin of rabbits, according to test procedures described in Title 21, §191.10 of the Code of Federal Regulations.

Highly toxic materials. Materials would be classified as highly toxic if, on short exposure, they could cause serious temporary or residual injury to humans. In the absence of adequate data on human toxicity, a material would be presumed to be highly toxic to humans if it fell within any one of the following categories when tested on laboratory animals, according to the U.S. Department of Agriculture test procedures described under Title 7, Chapter 3, §362.8 of the Code of Federal Regulations.

(1) Ingestion (oral): Any material that has a single dose LD_{50} of more than 5 milligrams but not more than 50 milligrams per kilogram of body weight when orally administered to both male and female white rats (young adults).

(2) Inhalation: Any material that has an LC_{50} of more than 50 parts per million by volume of gas or vapor but not more than 200 parts per million or more than 0.50 milligram, but not more than 2 milligrams of mist or dust per liter of air when administered by continuous inhalation for 1 hour or less to both male and female white rats (young adults). If the product is administered to the animals as a dust or mist, more than 90 percent of the particles available for inhalation in the test must have a diameter of 10 microns or less provided the Department finds that it is reasonably foreseeable that such concentrations could be encountered by man.

(3) Skin absorption: Any material that has an LD_{50} of greater than 20 milligrams but not more than 200 milligrams per kilogram of body weight when administered by continuous contact for 24 hours with the bare skin of rabbits, according to the test procedures described in Title 21, §191.10 of the Code of Federal Regulations.

Tear gas or irritating substances. Materials would be classified as tear gas or irritating substances if they cause reversible local irritant effects on eyes, nose, or throat temporarily impairing a person's ability to function to the degree that he cannot take necessary emergency action. Military and police tear gases and riot control agents would fall under this category. It is planned to include a list of materials falling under this category in the notice of proposed rule making.

If human experience or other data indicate that the hazard of a given material encountered during an accidental exposure in transportation is greater or less than indicated by the data from the specified animal tests, the classification for the specific material could be revised to reflect this data.

If these classifications are adopted, appropriate changes will be required in the labels required to be applied to packages.

Interested persons are invited to give their views on this notice. Communications should identify the docket number and be submitted in duplicate to the Secretary, Hazardous Materials Regulations Board, Department of Transportation, 400 Sixth Street SW., Washington, DC 20590. Communications received on or before April 27, 1971, will be considered before final action is taken on this subject. All comments received will be available for examination by interested persons at the Office of the Secretary, Hazardous Materials Regulations Board, both before and after the closing date for comments.

[1]LD$_{50}$, LC$_{50}$: that dose (LD) or concentration (LC) which will cause death within 14 days to one half of the test animals.

(36 F.R. 2934, February 12, 1971)

SUMMARY • **Docket No. HM-53** • *Amended Part 177 of the Regulations to clarify its applicability to private motor carriers. See also Docket No. HM-122.*

[Docket No. HM-53; Amendment No. 177–12]
PART 177—SHIPMENTS MADE BY WAY OF COMMON, CONTRACT, OR PRIVATE CARRIERS BY PUBLIC HIGHWAY
Application to Private Carriers by Motor Vehicle

Effective September 3, 1969, the Hazardous Materials Regulations of the Department of Transportation were amended to make it clear that private carriers of hazardous materials by motor vehicle are subject to those regulations (34 F.R. 7162, May 1, 1969). Although §§177.800 and 177.801 were amended to add private carriers to the carriers within the purpose and scope of the regulations, the note at the beginning of the part (pertaining to the applicability of the part to private carriers) which would no longer be needed because of the specific amendments to the part, was inadvertently left in the part, and the words "private carrier" were not added to §177.802(a). In addition, a comma after the word "private" in §§177.800 and 177.801 was omitted in the printing of Title 49 of the 1970 edition of the Code of Federal Regulations, although it did appear in the FEDERAL REGISTER printing of the amendment.

(35 F.R. 9284, June 13, 1970)

SUMMARY • **Docket No. HM-57** • *Controversial redefinition of classification of corrosive materials. See Docket No. HM-112 with regard to aluminum corrosives.*

[49 CFR Parts 170–189]
[Docket No. HM-57]
TRANSPORTATION OF HAZARDOUS MATERIALS
Classification of Corrosive Hazards; Advance Notice of Proposed Rule Making

On August 21, 1968 (33 F.R. 11862), the Hazardous Materials Regulations Board announced a plan to revise the regulations governing the transportation of hazardous materials. That document announced the intention to issue notices of proposed rule making in at least four areas, including "classification and labels," and invited public help in developing the basic regulatory principles to guide the Board in revising the regulations.

The Board is planning to consider, in the near future, a proposal for classification tests for corrosive materials. The present definition of corrosive materials does not make any distinction between damage to living tissue and damage to materials. That definition also contains no specific testing criteria. As a result, the public cannot practically rely upon the definition to determine when the Federal regulations apply. In order to correct this situation, the Department is developing benchmarks and testing procedures to be used in determining exactly when a material is to be considered sufficiently corrosive as to present a significant transportation hazard. That material would then be subject to the Department's regulations.

To assist the Board in considering a definitive classification scheme for corrosive materials, the public is invited to express its views on the corrosive hazard classification system outlined herein. This document is an effort to get public participation early in the rule making process. In developing test criteria, the Department's Office of Hazardous Materials has had preliminary discussions and communications with a number of knowledgeable persons, including the National Academy of Sciences (NAS), National Association of Corrosion Engineers (NACE), American

Institute of Chemical Engineers (AICHE), the Manufacturing Chemists Association (MCA), the Bureau of Explosives of the Association of American Railroads (B. of E.), and other industry representatives. In order to insure harmony between the regulatory standards of the several Federal agencies having jurisdiction in this area, the Department is considering the testing procedures used by the Departments of Agriculture, and Health, Education, and Welfare (see, for example, §191.1 of the regulations of the Department of Health, Education, and Welfare, 21 CFR Part 191, and §362.116 of the regulations of the Department of Agriculture, 7 CFR Part 362).

At the present time, §173.240 of the Department's regulations defines corrosives as follows:

Corrosive liquids . . . are those acids, alkaline caustic liquids and other corrosive liquids which, when in contact with living tissue, will cause severe damage of such tissue, by chemical action; or in case of leakage, will materially damage or destroy other freight by chemical action; or are liable to cause fire when in contact with organic matter or with certain chemicals.

The Board is considering revising the definition as follows:

Corrosive materials are those substances which, by chemical action, can (1) cause severe damage when in contact with living tissue, or (2) materially damage or destroy property.

The Board believes the reference to causing fires when in contact with organic matter more properly belongs under the classification of oxidizing materials.

This document is not concerned with materials which are mildly irritating, but only with those that cause significant, irreversible damage. Materials which are only irritating would be covered under the proposed new scheme for classifying health hazards (Docket No. HM-51, 35 F.R. 8831, June 6, 1970).

Degrees of hazard would be ranked as follows:

Class A: Potential damage to living tissue.

Class B: Potential damage to property, without damage to living tissue.

Establishment of hazard degree is required in order to specify packaging criteria reflecting the severity of potential damage. The degree of severity must be taken into account through the design and integrity of the packaging used for shipment of corrosive materials.

In order to classify products according to their degree of corrosiveness on living tissue, and on other materials, specific standard corrosion criteria must be established. A product would be considered corrosive to the living tissue if, when tested on the intact skin of the albino rabbit, the structure of the tissue at the site of contact is destroyed or changed irreversibly in 24 hours or less or, the primary test score is 6 or higher according to the test procedure described in §191.11 of the regulations of the Department of Health, Education, and Welfare (10 CFR Part 191). This test procedure is also listed in NAS-NRC Publication 1138, "Principle and Procedures for Establishing the Toxicity of Household Substances" (available from the Printing and Publishing Office, National Academy of Sciences, Washington, D.C. 20418 at $1.50 a copy).

The toxicity of these materials as determined by ingestion, inhalation, and skin penetration (absorption) is a systemic problem, rather than one of local damage, and is covered under the health hazards discussed in Docket No. HM-51.

In addition to the corrosiveness on living tissue, the corrosive effect on commonly used materials, e.g., steel, stainless steel, and aluminum should be considered. Also, plastic, wood, paper, and other nonmetallic materials are often used in packaging. The deterioration of these nonmetallics perhaps cannot be expressed in the same manner as the corrosion of metals. However, practical experience and a review of the literature indicate that, with few minor exceptions, corrosive materials which severely attack nonmetallics would also be corrosive to one or more of the metals mentioned above.

In establishing standard criteria we must realize that the concentration of a liquid may have a significant effect on its corrosive properties. It is necessary to consider the possible aqueous dilution of the liquid in the course of transportation, as well as its concentration at the time of shipment. The concentration having the highest corrosion rate would always serve as the basis for determination of the proper classification.

Another important factor is temperature. A maximum temperature of 130° F. is assumed as the basis for corrosion rate determination, and tests should be conducted at that temperature. This temperature has often been used to represent the upper limit encountered under normal transport conditions.

In addition to corrosive liquids, we must also consider any solid product such as sodium hydroxide pellets, chromic fluoride, aluminum chloride, etc. When in contact with living tissue or property, in the presence of moisture, these products may act as a concentrated solution of the product. The solid would be classified according to the more severe effect of either the concentrated or diluted solution.

In order to obtain exact corrosion values on presently regulated products, considerable experimental work would be required. Few data are available in the literature or from the manufacturers of the products covering the desired dilution and temperature range as well as corrosion attack rates. The general data presented in "Corrosion Data Survey," 1967 edition, published by the National Association of Corrosion Engineers, has been used as a basis for suggested corrosion criteria. This survey considers corrosion rates within the following ranges:

Less than 0.002 inches per year (ipy)

Less than 0.020 (ipy)

0.020 to 0.050 (ipy)
Greater than 0.050 (ipy)

Many factors besides concentration and temperature influence corrosion rates. For example, velocity, aeration, heat flux, changes in the metal due to welding, and the presence of oxidizing agents and other chemical contaminants can either increase or decrease the corrosion rate. The general information presented in "Corrosion Data Survey" appears to cover these points satisfactorily, but public comment is specifically invited here as to the need to consider these points in more detail.

The class B hazard degree would include materials which in concentrated or diluted form may severely damage or destroy property on contact. Damage to property is to be considered severe if the materials' corrosion rate on steel, stainless steel, or aluminum exceeds 0.050 inches per year. An acceptable corrosion test is described in NACE Standard TM-01-69, "Laboratory Corrosion Testing of Metals for the Process Industries", March 1969, available from the National Association of Corrosion Engineers, 2400 West Loop South, Houston, Tex. 77027.

The Board requests advice as to whether the test criteria should be based on a corrosion rate of 0.020 ipy. It appears that materials having a corrosion rate of less than 0.020 ipy are not sufficiently corrosive to be regulated as class B corrosives.

The Board is also in need of advice on corrosion test methods and benchmarks for certain organic materials which are corrosive or destructive to plastics, paper, etc., but not to metals. Also, stress corrosion is not presently covered in the regulations, and public comment is invited as to whether and how it should be.

As an example of the practical application of the proposed system, listed in the appendix hereto are representative chemicals now classified as corrosive materials under DOT or U.N. regulations, coupled with the proposed hazard degree class.

The Board may change the assigned hazard degree if human experience or other data indicate that the hazard that may be encountered during an accidental exposure is greater or less than that indicated by the specified tests.

If these classifications are adopted, appropriate changes will be required in the labels proposed in Notice No. 70-13, Docket No. HM-8 (35 F.R. 11742).

Interested persons are invited to give their views prior to December 2, 1970, as to whether this approach is a reasonable and practical one.

Comments (identifying the docket number) should be submitted in duplicate to the Secretary, Hazardous Materials Regulations Board, Department of Transportation, 400 Sixth Street SW., Washington, D.C. 20590. All comments received will be available for examination by interested persons at the Office of the Secretary, Hazardous Materials Regulations Board, both before and after the closing date for comments.

(35 F.R. 14090, August 31, 1970)

[49 CFR Parts 171, 172, 173, 174, 175, 177]
[Docket No. HM-57; Notice No. 71-17]

TRANSPORTATION OF HAZARDOUS MATERIALS
Proposed Classification of Corrosive Hazards

On August 31, 1970, the Director, Office of Hazardous Materials, published Docket No. HM-57, an advance notice of proposed rule making on classification of corrosive hazards (35 F.R. 14090). That document noted that the present definition for corrosives does not prescribe any testing criteria for damage to living tissue or to materials, and that present regulations may be subject to varied interpretation.

The Director suggested establishment of two corrosive classifications, Class A and Class B, with definite benchmarks for each classification. The notice suggested that the Class A designation cover potential damage to living tissue, and that the Class B designation cover potential damage to other material.

The Hazardous Materials Regulations Board has thoroughly reviewed all of the comments received and concludes that the following changes in the classification system presented in the advance notice have merit.

Hazard degrees. The Board recognizes that some materials are considerably more corrosive than others, e.g., hydrofluoric acid as compared to formic acid. A differentiation in hazard degree classification would appear justified. However, the Board agrees with many of the commenters that a dual classification system for corrosives would include only a relatively small number of compounds in the suggested Class B category, and would appear to complicate the regulations unnecessarily without a significant contribution to the improvement of safety. The Board also believes that some hazardous materials now classed as corrosive liquids should be otherwise classified. They will be the subject of future regulatory proposals.

Test methods—1. Damage to living tissue. The Board concludes that the proposed 24-hour duration rabbit test, specified by §191.11 of the HEW Regulations, may be too stringent. However, exposure times of from 10 minutes to 4 hours as suggested by some commenters would not provide a sufficient safety factor. The corrosive effect on the skin may not be apparent immediately, and an exposed individual may delay washing off the corrosive material. Based upon these considerations, the Board is proposing to reduce the specified HEW test duration from 24 to 8 hours. The Board also concludes that the suggested scoring number criteria are not sufficiently reliable and should be eliminated as an evaluation of skin destruction. The Board proposes, therefore, to adopt the definition of tissue destruction given in 21 CFR 191.1(h), but with the 24-hour exposure prescribed in that section reduced to 8 hours.

2. Damage to materials. The Board does not agree with some of the comments that claimed the suggested corrosion rate of 0.05 inch per year (IPY) is too stringent and went so far as to suggest a minimum regulated rate of 0.05 inch per week. The Board considers a minimum corrosion rate of 0.05 IPY necessary in order to provide sufficient flexibility to account for the many and varied factors encountered under transportation conditions. For example, one of the lighter steel drums used as a shipping container has a wall thickness of 0.024 inch. Small steel pails can be as thin as 0.015 inch. A corrosion rate of 0.050 IPY would result in a 0.008 inch loss during a 60-day transportation period. This is a severe decrease in wall thickness of such containers. Allowing a corrosion rate greater than 0.050 IPY would be unduly hazardous for materials shipped in these thin-walled containers. Therefore, 0.050 IPY is being proposed as the corrosion rate benchmark.

In view of these considerations, the Board is proposing new definition and test criteria for the corrosive effects of materials on human tissue and on two basic packaging materials—steel and aluminum.

The adoption of these criteria and standard test methods will enable a shipper to determine the proper classification for his product much more readily than is now possible. This new definition and the subsequently outlined criteria are not intended in themselves to alter the present packaging requirements as described in the Department's Hazardous Materials Regulations. The Board does recognize, however, that some materials formerly not considered subject to the regulations may come under these packaging requirements; therefore, it might be necessary for different packagings to be introduced by a shipper for certain products. Substantive changes to packaging requirements, where appropriate, may be the subject of future rule making. The only exception is packaging for solid materials for which there are no present requirements, and this is being covered in this notice.

The Board believes the effect of corrosion on commonly used packaging materials, steel and aluminum, also must be considered. The Board recognizes that plastics, wood, and paper composites are also used in the packaging of hazardous materials. The deterioration of these nonmetallics cannot be expressed in the same manner as the corrosion of metals. However, practical experience indicates that with the exception of organic solvents, products which severely attack the nonmetallics generally would be corrosive to aluminum or steel. The Board concludes that present regulatory test criteria relating to the definition should be limited to the corrosion data determined for these two commonly used metals. Primary concern involves the severe effect of a product leaking from its shipping container onto other containers which are not resistant to its corrosiveness, thus causing damage within the limited period of transportation.

A maximum temperature of 130° F. was selected as the basis for corrosion rate determination. This temperature is often used to represent the nominal upper limit of the normal transport environment.

In addition to corrosive liquids, the Board also considered the potential corrosiveness of solid products such as sodium hydroxide pellets, chromic fluoride, and aluminum chloride. When spilled from a container on human tissue or other packaging or lading, and in the presence of moisture, solid corrosive products may have an effect similar to that of a concentrated solution of the product. For this reason, the Board believes that corrosive solids should be classified according to the most severe effect of a concentrated solution of the commodity.

In order for the Department to obtain exact corrosion values of regulated commodities on steel and aluminum, considerable experimental work would be required. Few data are available in the literature or from the manufacturers of the products, especially data covering the desired concentration and temperature range as well as severe corrosion attack rates. However, the data presented in "Corrosion Data Survey," 1967 edition, published by the National Association of Corrosion Engineers, provides a useful basis for suggested corrosion criteria. The information presented in "Corrosion Data Survey" appears to cover the general conditions of a relatively severe attack during the period of transportation.

In Docket No. HM-8; Notice No. 70–13 (35 F.R. 11742), on page 11764, a new label was proposed for corrosive liquids. If the regulations proposed herein are adopted, the Board would delete the word "Liquid" on the label.

(36 F.R. 11304, June 11, 1971)

[49 CFR Part 173]
[Docket No. HM-57; Notice No. 71–17]
TRANSPORTATION OF HAZARDOUS MATERIALS
Proposed Classification of Corrosive Liquids

On June 11, 1971, the Hazardous Materials Regulations Board published Docket No. HM-57; Notice No. 71–17 (36 F.R. 11304), relating to the classification of corrosive hazards. The purpose of this supplemental notice is to clarify §173.245b, as that section was proposed in Notice No. 71–17.

Notice 71–17 did not specifically prescribe in §173.245b that fiber drums would be an authorized packaging for corrosive solids. A comment recently filed with the Board contends that this omission may be interpreted as an indication of intention not to allow this packaging for corrosive solids. This was not the Board's intention.

The packaging proposed in Notice 71–17 was intended to reflect packaging known by the Board to be currently in use for corrosive solids. On the basis of comments to be received on Notice 71–17, the Board may add other packaging demonstrated to be in use for these materials. Of course, the Board expects comments of this nature to state that the packaging has proven to be satisfactory and to provide adequate support for that statement. This intent may not have been stated clearly in Notice 71–17 and has evidently caused difficulties for some container manufacturers including suppliers of fiber drums. The Board does not propose at this time to make substantive changes to packaging currently in use.

(36 F.R. 15762, August 18, 1971)

[Docket No. HM-57, Amdts. 171–14, 172–14, 173–14, 173–61, 174–14, 175–7, 177–21]

CLASSIFICATION OF CORROSIVE HAZARDS

The purpose of this amendment to the Hazardous Materials Regulations is to amend §173.240 to provide a quantitative definition for corrosive materials and to change several sections of the regulations now applicable only to acids and other corrosive liquids to make them applicable to shipments of corrosive solids.

On June 11, 1971, the Hazardous Materials Regulations Board published a notice of proposed rule making, Docket No. HM-57; Notice No. 71–17 (36 F.R. 11304), which proposed this amendment. This notice was amended by a supplemental notice in the same docket published on August 18, 1971 (36 F.R. 15762). Interested persons were invited to give their views and many comments were received by the Board.

The very large majority of the comments were directed to the criteria in the definition for corrosive materials and to the proposal to include regulatory controls on shipments of corrosive solids.

Several persons objected to the proposed 8-hour duration period for the skin contact test. After reviewing the various arguments presented, mostly for a shorter duration period, the Board agrees that 8 hours duration is too lengthy an exposure to be reasonably related to transportation conditions. On the other hand, the Board believes that a 1-hour test as proposed by some commenters, although supportable in some aspects of actual expected exposure, would not provide a sufficient safety factor to account for variations in test conditions or the transportation environment. A few commenters claimed that the proposed test procedure would distort the results by the requirement for an impervious occlusion of the test material on the skin of a rabbit, thereby preventing evaporation of any volatile solvents. Upon review of the procedure, however, the Board finds that this claim is not correct. Impervious occlusion of the test material is not required. The test patches are held in position in such a manner that evaporation of the solvent may be retarded, but it is not prevented. Therefore, the Board concludes that a 4-hour exposure time, using the rabbit test as described in 21 CFR 191.11, is necessary.

The other area significantly commented upon was the proposed corrosion rate of 0.050 inch per year (IPY) on steel and aluminum. Commenters demonstrated that this portion of the proposed definition would result in the regulation of many commodities that are not regulated at present which do not present hazards that require regulations in the transportation environment. After careful consideration, the Board concludes that the corrosion rate proposed was too stringent. Therefore, the Board has decided to specify a corrosion rate criterion in excess of 0.250 inch per year (IPY) on a specific steel and aluminum. Also, under the same evaluation parameters, the Board concluded that dry materials could be adequately regulated by referring only to the contact hazard of the solid on skin. Using a solution for testing and the resultant data for classifying solids would result in the regulation of materials that do not have a significant hazard in transportation. Therefore, the Board has prescribed testing only for the material in the "as shipped" condition. Also, solids are not required to be evaluated in solution for skin contact, and are not required to be evaluated for compatibility with materials of construction for other than the packaging material as now prescribed under section 173.24, or under more specific sections of the regulations.

The Board received comments requesting that a qualifying term be added to the basic definition to cover those circumstances where animal testing does not significantly represent the results of human exposure. The Board agrees with this observation and has added a qualifying statement.

General observations and editorial suggestions were made on the applicability of §§173.245b, 173.286 (b) and (c), and 174.597(a). The Board has incorporated many of the suggested changes as improvements in language or safety and within its intent in this docket.

Although the Board proposed multi-wall paper bags that would have to be tested by six drops, the Board agrees with several comments received that the current standard four-drop test should continue to be used for basic guidance in bag evaluation. The Board realizes that on the basis of studies now being conducted, this reference point may undergo change. But in line with its basic approach not to radically change packaging standards for corrosive solids, and to maintain a satisfactory packaging level, the Board has specified four 4-foot drops for bagged corrosive materials.

Objections were raised regarding compatibility requirements for corrosive solids with other hazardous materials as covered by §§174.538 and 177.848. The Board agrees that standards of compatibility for solids should be more closely evaluated for possible differences from corrosive liquid loading and storage standards. Applying the same standards could very likely result in overregulation. Therefore, the Board is not making any change with respect to solids in these two sections of the regulations at this time.

A comment was made as to the correct classification for solid uranium hexafluoride. Future rule making will consider the problem of dual classifications. Meanwhile, present requirements for the packaging of this commodity are prescribed under the radioactive materials classification. However, as specified in §§173.2 and 173.402(a) (15) herein, both classifications and their appropriate label requirements apply.

(37 F.R. 5946, March 23, 1972)

[Docket No. HM-57; Amdts. 171-14, 172-14, 173-61, 174-14, 175-7, 177-21]
Classification of Corrosive Hazards; Notice of Board Action to Authorize Immediate Compliance

On March 23, 1972, the Hazardous Materials Regulations Board published Docket No. HM-57; Amendments Nos. 171-14, 172-14, 173-61, 174-14, 175-7, 177-21 (37 F.R. 5946) prescribing new regulations for the classification, packaging, marking, labeling, and transportation of corrosive materials. It was stated that these amendments became effective December 31, 1972, however, compliance was authorized as of October 1, 1972.

The Board has been advised that the 3-month period authorized effectively limits the conversion period in a manner that precludes efficient changeover, particularly where leadtimes in packaging orders and disposal of existing stock are highly variable. It has been suggested that compliance be authorized immediately to permit the needed latitude. The Board Members for the Federal Aviation Administration, the Federal Highway Administration, and the Federal Railroad Administration have agreed to expand the implementation period and to advise the public by notice thereof in the FEDERAL REGISTER effective with the publication of this notice.

Accordingly, all amendments in Docket No. HM-57 continue to be effective December 31, 1972; however, compliance with the regulations as amended in this docket is authorized immediately.

(37 F.R. 8383, April 26, 1972)

[Docket No. HM-57; Amdts. 171-14, 172-14, 173-61, 174-14, 175-7, 177-21]
Classification of Corrosive Hazards; Postponement of Mandatory Effective Date

On March 23, 1972, and on April 26, 1972, the Hazardous Materials Regulations Board published Docket No. HM-57; Amendments Nos. 171-14, 172-14, 173-61, 174-14, 175-7, 177-21 (37 F.R. 5946 and 8383) prescribing new regulations for the classification, packaging, marking, labeling, and transportation of corrosive materials. Compliance with these amendments has been authorized as of April 21, 1972. However, the mandatory effective date was specified as December 31, 1972.

The Hazardous Materials Regulations Board has received several petitions concerning difficulties with completing testing of materials now being shipped, in time to meet the mandatory effective date. Also several petitioners have requested the Board to consider authorizing additional packaging, particularly for materials shipped in bulk quantities. The Board considers that the petitions have merit. They contain justification to postpone the effective date in this docket and to further study the required packaging. In view of these petitions, the Board hereby gives notice that the mandatory effective date for Amendments Nos. 171-14, 172-14, 173-61, 174-14, 175-7, 177-21 is changed from December 31, 1972, to June 30, 1973. However, voluntary compliance continues to be authorized.

In addition, the Board intends to publish at a later date additional regulations pertaining to authorized packaging of corrosive materials, particularly liquids.

Several petitions for reconsideration were received listing many chemicals heretofore not covered by the Hazardous Materials Regulations. However, the petitioners failed to indicate by which test each material met the corrosive materials definition. The Board is requesting that supplemental information be furnished by the petitioners to permit the Board to evaluate separately those corrosive substances that are corrosive only by the steel or aluminum test criteria, or both. The Board requests that this information be submitted before November 30, 1972.

(37 F.R. 18918, September 16, 1972)

[Docket No. HM-57; Notice No. 73-1]
CLASSIFICATION AND PACKAGING OF CORROSIVE MATERIALS
Notice of Proposed Rule Making

On March 23, 1972, April 26, 1972, and September 16, 1972, the Hazardous Materials Regulations Board published Docket No. HM-57; Amendment Nos. 171-14, 172-14, 173-61, 174-14, 175-7, 177-21 (37 FR 5946, 8383, and 18918) prescribing new regulations for the classification, packaging, marking, labeling, and transportation of corrosive materials. Compliance with these amendments has been authorized as of April 21, 1972. The mandatory effective date is specified as June 30, 1973.

The Hazardous Materials Regulations Board has received numerous petitions to provide additional packagings for shipment of corrosive materials. In accordance with its statement of intent, the Board has reviewed these petitions and this notice contains the Board's proposal for the packagings and transportation of such materials.

In addition, the Board is proposing to authorize bottom unloading of cargo tanks containing corrosive liquids. This is based on (1) satisfactory experience reported by a number of petitioners transporting corrosives not previously considered to be covered by the corrosive liquid definition; (2) several special permits that authorize bottom discharge outlets for corrosive liquids; and (3) the absence of any unsatisfactory experience reported to the Board through the Department's incident reporting system, as applicable to the use of bottom outlets on cargo tanks transporting corrosive liquids. Also, the Board is proposing that §178.343-5 be revised to incorporate protective safeguards in the design and construction of these outlets. This is consistent with the regulations applicable to other hazardous materials when such outlets are authorized.

The Board has not yet finalized its rule making on this same subject with regard to tank cars. The subject of bottom outlets on tank cars was introduced in Docket No. HM-90 (36 FR 16680, 18873, and 21343) and as stated in that docket on November 6, 1971 (36 FR 21343), it remains an issue for future rule making action. The Board intends, shortly, to present its proposal to the public in this matter. Meanwhile, the Board is proposing to provide temporary authorization to ship certain corrosive liquids in tank cars with bottom outlets to provide limited relief because of the numbers of these materials presently being so shipped.

As requested, the Board is proposing to list solid caustic soda and phosphoric acid by name in the regulations. In addition, it proposes to add a new packaging reference to crotonaldehyde, diethylamine, ethyl dichloride, and propylene oxide to adequately cover these materials under paragraph 173.119(m).

Disagreement in the industry with respect to classification has arisen regarding the following chemicals: maleic anhydride, phthalic anhydride, alphapicoline, gamma-picoline, ethyl morpholine, nonyl phenol, phenol sulphonic acid, and sodium bisulfite. The Department is studying these products and based on the results of its findings will publish a notice later if it concludes that these materials meet the criteria of §173.240.

The Board is proposing to revise §173.240 to clarify the intent of the definition and scope. Also, it has included proposed changes to incorporate the proposed revision to 21 CFR 191.11 which was published by the Commissioner of Food and Drugs, Food and Drug Administration, in the FEDERAL REGISTER of December 19, 1972 (37 FR 27635). Based on the statement that he (the Commissioner) "does not expect the new test procedure, if adopted, to materially alter the classification of more severe dermal irritants and corrosives to the skin" and that rather "[i]t will aid in classifying corrosives by clarifying certain troublesome language, and will resolve situations where borderline irritation is an issue," the Board does not consider it necessary for transportation purposes to retest any materials previously tested according to the definition in the present amendment in Docket HM-57. The results obtained under the existing procedure may be considered acceptable by shippers. Final action on these definition changes would be taken on the basis of the final rule published by the Food and Drug Administration.

Some petitioners requested that the regulations be amended to authorize the use of Specification MC 300, MC 301, MC 302, MC 303, MC 305, and MC 306 cargo tanks for transporting corrosive liquids. The Board is concerned about the adequacy of these tanks because some of them may be very old (MC 301 cargo tanks have not been authorized for construction since June 11, 1961) and because they have not been designed in accordance with standard requirements for corrosive liquids. The Board finds that the specification requirements for MC 304 and MC 307 call for a higher design integrity. The Board does not find comparable factors concerning the cargo tanks listed above. The use of tanks as described above, except MC 304 and MC 307, has therefore not been proposed in this notice.

(38 F.R. 4270, February 12, 1973)

[Docket No. HM-57; Amdt. Nos. 171–14, 172–14, 173–61,
174–14, 175–7, 177–21]
**Classification of Corrosive Hazards;
Postponement of Mandatory Effective Date**

On March 23, 1972, April 26, 1972, and on September 16, 1972, the Hazardous Materials Regulations Board published docket No. HM-57; amendment Nos. 171–14, 172–14, 173–61, 174–14, 175–7, 177–21 (37 FR 5946, 8383, 18918) prescribing new regulations for the classification, packaging, marking, labeling, and transportation of corrosive materials. Compliance with the amendments has been authorized as of April 21, 1972. However, the mandatory effective date is specified as June 30, 1973. Also, on February 12, 1973, the Board published Notice 73–1 (38 FR 4270) under this docket. The comment period closed on April 10, 1973. However, the Board has not completed its study of the comments and therefore has yet to publish an amendment relative to this notice.

The Hazardous Materials Regulations Board has been developing what it considers final resolutions in several areas relating to such matters as open-head fiber drums, materials corrosive only to metals, viscosity of certain corrosive liquids, bottom unloading of cargo tanks, and handling of corrosives in the distribution systems that are not packed, marked, and identified as would be required by HM-57.

The Board is of the opinion that to make HM-57 effective June 30, 1973, with most of these matters unresolved would cause much confusion and administrative difficulties and an undue burden on carriers and shippers. The matters mentioned above, and others should be fully resolved within the next 90 days. The Board, therefore, considers it reasonable and necessary to delay the mandatory effective date. Accordingly, the Board hereby gives notice that the mandatory effective date for amendment Nos. 171–14, 172–14, 173–61, 174–14, 175–7, 177–21 is changed from June 30, 1973, to December 31, 1973. Voluntary compliance continues to be authorized. Note that because of the wording in docket HM-8; amendment No. 173–70 (38 FR 5292) paragraph 173.404(h), the previously authorized corrosive labels will thereby be authorized until January 1, 1975.

The Board also requests additional information as follows from those persons concerned about the omission of a provision for open-head fiber drums in corrosive liquid service.

1. Specific identification of these liquids by maximum and minimum percentages of corrosive ingredient in their formulations;

2. Identification of what criterion indicates these materials to be corrosive in the percentages identified, that is, skin corrosive, or metal corrosive only, and to what metal.

3. Identification of the minimum and maximum viscosity, in centipoises measured at 20° C., for the formulations proposed to be shipped in these drums.

The Board cannot finalize its action in this area without this information. The Board requests that this information be submitted before June 30, 1973, to the Secretary, Hazardous Materials Regulations Board, Department of Transportation, Washington, D.C. 20590.

(38 F.R. 12807, May 16, 1973)

[Docket No. HM-57; Amdts. 172–20, 173–74, 178–26]
CLASSIFICATION AND PACKAGING OF CORROSIVE MATERIALS

The purpose of these amendments to the Hazardous Materials Regulations is to amend §§172.5, 173.119, 173.245, 173.245b, and 178.343–5 to specify certain shipping names and to add authorized packagings for the shipment of corrosive materials.

On February 12, 1973, the Hazardous Materials Regulations Board published a notice of proposed rulemaking, Docket No. HM-57; Notice No. 73-1 (38 FR 4270), which proposed these amendments. Interested persons were invited to give their views and many comments were received by the Board.

1. Shipping descriptions. Several commenters requested the inclusion of additional shipping names and packagings which were not the subject of Notice No. 73–1. Since these were not proposed to be added to the regulations, it would be more appropriate for the Board to handle these requests in a separate notice of proposed rule making to allow for adequate comment. The Board intends to publish such a notice.

2. Drum size limitation for corrosive solids. One commenter noted, in a petition for reconsideration in Docket HM-57, Amendment No. 173–61, that the 55-gallon capacity limitation for corrosive solids in §173.245(a) (4) was too restrictive in view of their experience in shipping drums of larger size. In view of the Board's previously stated position in packaging requirements for solids, (See 36 FR 11304 and 15762, 37 FR 5946) and on the basis of the reported satisfactory experience, the Board agrees with the commenter and has removed the size restriction in the subject paragraph.

3. Inside PE container for corrosive liquids. One commenter stated that § 173.245(a) (16), as proposed, should be changed to provide for a certain high density, non-DOT specification, 55-gallon capacity polyethylene inner container for shipping corrosive liquids, n.o.s. The commenter cites satisfactory experience with the use of such an inner container for materials previously not considered regulated. Such a change was not proposed in the notice. The Board, however, is considering a proposal to amend the inner container specifications. Meanwhile, it believes such packaging should be approved only under the terms of a special permit so that it can study the experience factor more closely. Since similar permits now exist, this commenter should petition the Board for a special permit so that the Board may add his experience to that being obtained under other permits.

4. Corrosive material definition. Several comments were made on the Board's proposal to adjust the corrosive material definition (§173.240) to coincide with certain proposals by the Commissioner of Food and Drugs (FDA) (37 FR 27635). The Board's proposals were mainly related to handling of test animals and detail on preparation of the test material. Before acting on this proposal, the Board is awaiting the outcome of the Food and Drug Administration's rule making. However, as stated in the preamble of Notice No. 73–1, the Board continues to believe that any changes it proposed will not materially alter the classification of materials severely corrosive (that cause destruction or irreversible alteration) to human skin. When the FDA rule making action is completed, the Board will then consider all the comments received on the definition in §173.240 and the FDA decision.

5. Test animals. One commenter pointed out the difficulties in relating results of tests on rabbit skin to the results on human skin contact, particularly with respect to materials proven not corrosive to human skin although giving positive results under rabbit skin testing. This commenter went on to state that such a substance need not be labeled under the provisions of the Federal Hazardous Substances Act. The Board points out that this was one of the primary reasons for the adoption of §173.240(b). This paragraph specifically alludes to human experience. There is a mecha-

nism, established by the Board, whereby materials may be "declassified" by its action. It should be noted, however, that the Board, under the same provision, may "classify" a material because of human experience without being strictly limited to the quantitative rabbit skin test results. In conclusion on this matter, the Board chose rabbit skin testing for its criteria because it wanted to use recognized methods and not establish completely new criteria. To have established new criteria would have posed a different series of difficulties for manufacturers.

6. Ethylene dichloride. Based on a comment the Board has removed ethylene dichloride from the list (§172.5) of flammable liquids which are also corrosive. The Board will study this matter in more depth.

7. Cargo tank density limitation. Several comments were made objecting to the proposed lading density limitation of 10 pounds per gallon when transporting a corrosive material in a DOT specification MC 304 or MC 307 cargo tank. The regulation has been amended to remove the restriction for MC 307 cargo tanks since the weight of lading is adequately controlled by §178.340-10(b) (1) wherein such a cargo tank has a maximum product load limitation established by regulation. This maximum product load must be marked in pounds on the metal certification plate attached to each MC 307 cargo tank. Also, the proposed regulation has been amended to remove the reference to a lading weight restriction for MC 304 based on the comments and reports of operation and experience received indicating that such a restriction is not justified.

8. Tank car bottom outlets. Objections were made to the "interim" authorization for bottom outlets on tank cars. The Board recognizes that this situation creates some serious difficulties with respect to program planning, cost allocations, leasing, long-term operations, forecasts, etc. It does, however, have comments on file questioning the safety of permitting bottom outlets on tank cars containing certain corrosive materials. While the Board has not completed its evaluation of the reported problems, it nevertheless believes that it would be misleading to provide for bottom outlets in §173.245 without an appropriate warning that such authorization is provided temporarily pending a final decision. On the other hand, the Board does not consider it has progressed far enough in its evaluation to justify any curtailment of such cars in the services where they are used. Therefore, the Board is using an "interim" solution which it considers to be appropriate under these circumstances.

9. Emergency discharge controls. Several comments were received objecting to a requirement for heat actuated and remote control mechanisms in bottom outlet valves as proposed for MC 304, MC 310, MC 311, and MC 312 cargo tanks used in the transportation of corrosive liquids. The Board agrees that the requirement for a heat-actuated mechanism, in view of the nature of the lading, is not justifiable under conditions of transportation as known today. The Board, however, believes that the remote control mechanism is necessary to provide adequate safety in handling corrosive materials. The comments that objected to the need for the remote control shut-off argued that such a shut-off was unnecessary. But duplication with respect to control is known to exist in several plants handling these materials. The concern for an adequately safe operation in unloading the tank truck should be of the same order of importance as the plant handling of the material. It is of interest to note that the National Transportation Safety Board (NTSB) in its report NTSB-HAR-72-3, a truck-automobile collision involving spilled methyl bromide, observed that "substantial differences exist between methyl bromide hazard control measures exercised in manufacturing plants and those found in transportation in this accident." The NTSB concluded: "Inplant hazard control procedures should be considered by the regulators in developing future changes in transportation regulations." The Board (HMRB) believes such an analogy exists here. The requirement for a second station in the event of incident involving the discharge point station is consistent with safety engineering practices in use and is not overburdening when considering the plant safety procedures in use for handling corrosive materials. Therefore, the Board is requiring use of a remote station except when liquids contain solids in suspension in sufficient quantity that settling may form a layer of solid material that may interfere with sealing of the valve seat.

10. "Grandfather" clause. Several commenters pointed out the difficulty in handling materials which were in the distribution system before the effective date of these amendments. The Board considers these comments valid and has provided a limited "grandfather" clause. See Note 1 to §173.240.

11. Inside plastic receptacles for corrosive solids. In Notice No. 71-17, the Board proposed to authorize plastic receptacles as inside packagings for shipment of corrosive solids covered by §173.244(c). This type of packaging was inadvertently omitted from Amendment No. 173-61. This oversight has been corrected by amending §173.244(c).

(38 F.R. 20837, August 3, 1973)

[49 CFR Part 172-173]
[Docket No. HM-57; Notice No. 73-6]
CLASSIFICATION AND PACKAGING
OF CORROSIVE MATERIALS
Notice of Proposed Rule Making

On March 23, 1972, April 26, 1972, September 16, 1972, May 16, 1973, and August 3, 1973, the Hazardous Materials Regulations Board (the Board) published Docket No. HM-57, Amendment Nos. 171-14, 172-14, 172-20, 173-61, 173-74, 174-14, 175-7, 177-21, 178-26 (37 FR 5946, 8383, and 18918, 38 FR 12807 and 20837) prescribing new regulations for the classification, packaging, marking, labeling, and transportation of corrosive materials. Compliance with these amendments has been authorized as of April 21, 1972. The mandatory effective date is December 31, 1973.

When the Board extended the mandatory effective date to December 31, 1973 (38 FR 12807), it stated that several problems remained to be resolved. On August 3, 1973, the Board published Amendment Nos. 172–20, 173–74, and 178–26 in this docket which resolved several issues. This notice proposes changes to resolve the remaining difficulties except for the one involving materials corrosive only to metals. The Board expects to propose a final resolution in this matter in the near future. In this document, the Hazardous Materials Regulations Board is considering amendment of Parts 172 and 173, to add a new §173.249a, and to amend §§172.5, 173.28, 173.119, 173.244, and 173.245.

Tariff 6D.—The proposed changes to §172.5 are based on several different sources of information. The primary source is the Air Transport Restricted Articles Tariff No. 6D which is the tariff used by many airlines as their public disclosure on the materials authorized aboard aircraft and the conditions under which they will be transported. The materials listed in Tariff 6D are identified in §172.5 of this document by abbreviations following the name of the material, i.e., "(Cor)" for corrosive liquid, "(B)" for ORA B. "ORA B" means "Other Restricted Article, Group B" and is defined in Tariff 6D as "a solid material which, when wet, becomes strongly corrosive so as to be capable of causing damage to aircraft structure." These abbreviations represent the classification given the subject materials in Tariff 6D before July 28, 1973. In a recent revision to Tariff 6D, the classification of a number of materials was changed from ORA B to "corrosive material." These changes have highlighted the existence of discrepancies between the DOT Hazardous Materials Regulations and Tariff 6D regarding corrosive materials. The Board has information regarding certain materials that indicates they may not meet the corrosive material definition. Both publications use the same definition.

Based on the evaluation of the information available to the Board, it is of the opinion that many of the corrosive materials listed in the tariff are properly classed. In an effort to ascertain that the materials listed in §172.5 as proposed herein meet the corrosive material definition, the Board requests that additional data be made available to it.

The Board is primarily interested in reviewing data representing results of the tests prescribed in §173.240. It also requests that persons submitting data only on metals testing indicate that the material is not skin corrosive according to the test in §173.240(a) if this is the case. They should also indicate if the material is corrosive to steel or aluminum, or both. According to the items that are finally added to §172.5, the Board will make corresponding changes in Part 173 in the final amendment.

As an ancillary action, after the data has been reviewed and classification determinations have been completed through the rule making process, the Department will undertake to notify the Civil Aeronautics Board of all discrepancies. In this manner, it is hoped that Tariff 6D and the DOT Hazardous Materials Regulations will be made more uniform thereby facilitating compliance with all regulations. The Board intends to follow this procedure for all classes of hazardous materials as it proceeds with the updating of its own regulations.

Additional shipping names. The Board has information on approximately 40 additional materials which could be corrosive, and are not named in this notice. However, on the basis of the information submitted, the Board believes that some of the materials possibly should be classed as flammable liquids or Class B poisons, and is reviewing this matter in more depth. If a material requested by a petitioner to be added to §172.5 has a Tagliabue Open-Cup (T.O.C.) flash point slightly above 100° F. or between 80° F. and 100° F., the Board did not include the material in the list because of other pending rule making in HM-102. Any person observing that the Board has not provided for a material in this list as he requested, may request further consideration by providing data indicating that the material is not a Class B poison or a flammable liquid (open and closed cup flash point). There is no need to provide such data if these persons anticipate shipping these materials as Class B poisons, n.o.s. or corrosive liquids, n.o.s., as they deem appropriate. If the T.O.C. flash point is 80° F. or below and the material meets the definition in §173.343, the material would be classed as a flammable liquid or a Class B poison, not a corrosive material.

Additional packagings.—Some persons petitioned the Board to add certain packagings in §§173.119(m) and 173.245. On the basis of these petitions and the satisfactory experience gained with the use of these packagings to transport materials similar to those now covered by the regulations, §§173.119(m) and 173.245 are proposed to be amended.

Reconditioned drums.—Several petitions were received to amend §173.28(h) to provide for the use of series 17 reconditioned drums. This proposal includes a provision for limited use of these drums in a service where they have been successfully used under special permit or for materials not considered corrosive prior to revision of §173.240.

Small quantities and exemptions.—The Council for Safe Transport of Hazardous Articles (COSTHA) has petitioned the Board for reconsideration of Amendment 173–61 in Docket HM-57 to amend the small quantities exemption provision of §173.244 to relieve shippers from what it considers an unwarranted burden of regulation when packaging and shipping small quantities of corrosive liquids. The Board believes that the portion of COSTHA's petition relating to exemption from specification packaging requirements when corrosive liquids are packed in inside metal or plastic packagings each not over 32 ounces by volume or weight has merit. It proposes to exempt corrosive liquids when so packaged from specification packaging, by amending §173.244.

"Low hazard" corrosive liquids.—Several petitions were received requesting that the use of certain non-DOT specification packagings be authorized for corrosive liquids not considered corrosive by shippers prior to revision of §173.240. These materials have been identified to the Board as presenting a lesser degree of hazard. In this proposal, the Board has agreed with several of the petitioners. However, it has modified the description of the material as presented by petitioners, in an effort to assure that corrosive liquids having a higher degree of hazard would not be shipped under these provisions. A new §173.249a is proposed to prescribe packaging for these materials. Meanwhile, the Board has been inquiring into the strength of some of the packaging referenced to determine if upgrading of the packaging is warranted and feasible. Depending on the outcome of the Board's review and the records of experience that it gathers regarding the shipment of materials in these packagings, the Board may propose further rule making in this area.

Previously submitted data.—Some data has been provided to the Board indicating that certain materials proposed to be classified as corrosive materials do not meet §173.240. For example, some aluminum chloride solutions apparently do not meet §173.240. While the Board does not dispute this data and this data has been made part of the docket, it desires to obtain the advice of persons who possess information either supporting or opposing the proposed classification of the materials as corrosive. However, persons who have submitted any data on materials listed in this notice should reaffirm their conclusions by simply referring to the data previously submitted. It is not necessary to resubmit the data. This procedure will permit the Board to have the benefit of the latest data available and it will avoid any oversights caused by the extremely large volume of data that has been developed in this docket.

(38 F.R. 24915, September 11, 1973)

[Docket No. HM-57; Amdts. 172–22, 173–77, 178–30]
Classification and Packaging of Corrosive Materials

The purpose of these amendments to the Hazardous Materials Regulations is to amend §§172.5, 173.28, 173.119, 173.244, and 173.245 and to add §173.249a to identify specifically a number of corrosive materials that would have been shipped as "corrosive liquids, n.o.s." or "corrosive solids, n.o.s." pursuant to the amendment dated March 23, 1972 (37 FR 5946), and to authorize the use of certain packagings for which the Hazardous Materials Regulations Board ("the Board") has received numerous letters citing satisfactory experience.

On September 11, 1973, the Board published a notice of proposed rule making, Docket No. HM-57, Notice 73–6 (38 FR 24915), which proposed these amendments. Interested persons were invited to give their views and many comments were received by the Board.

1. Editorial. Several commenters suggested modifications to the list of hazardous materials (§172.5), including provisions for use of alternate designations for certain products, all oriented to increasing the utility of the list and improving its quality. The Board appreciates the show of concern and has incorporated most of the recommendations.

2. List of hazardous materials. Several commenters presented data indicating that certain materials should not be entered in the list as corrosive materials. On the basis of this data, the following entries have been deleted: Aluminum chloride anhydrous, solid; Aluminum chloride solution; Aminoethylethanolamine; Ammonium hydrogen fluoride, solid; Dichloropropene-dichloropropane mixture; Ferric chloride, solid; Ferric chloride solution; Methyl chloroform; Nonyl phenol; Phenolsulfonic acid, liquid; Sodium hydrogen sulfate, solid; Sodium hydrogen sulfite, solid; Tetra-ethylenepentamine; and 1,1,2-trichloro-ethane. The Board has concluded that the data it has on these materials is either positively determinative that the materials are not corrosive or that, in some cases, the available data was not sufficiently conclusive to include the materials in the list.

3. Reconditioned drums. A commenter objected to the proposed restriction relating to re-use of containers in §173.28 to corrosive materials containing caustic compounds because he believed that such solutions of over 15 percent concentration would not be permitted to be shipped in reconditioned drums. This commenter pointed out that a special permit (No. 6683) had been issued to authorize the shipment of caustic soda, liquid, alkaline caustic liquids, n.o.s., and alkaline corrosive liquids without regard to the 15 percent caustic limitation in §173.249a. The Board has reviewed the permit and finds that the amendment to §173.28, as proposed, encompasses all materials covered by Special Permit No. 6683 and that this permit did not go beyond the scope of §§173.249 and 173.249a. It should be noted that caustic and alkaline materials subject to the packaging in §173.249 are not subject to the 15 percent limitation of §173.249a. The basis for the restriction in §173.249a is due to the nature of the packaging authorized by that section. Because of the provisions of §173.28, reconditioned drums are authorized for any strength of liquid caustic soda, alkaline caustic liquids, n.o.s., and alkaline corrosive liquids, n.o.s., that are subject to the packaging provisions of §173.249.

4. Flammable liquids which are also corrosive. In Notice No. 73–6, the Board proposed to authorize DOT specification 6D or 37M cylindrical steel overpacks with an inside DOT specification 2S or 2SL polyethylene container. One commenter requested that the DOT specification 2U inside polyethylene container be authorized as well. The commenter argued that existing regulations authorize the use of inside polyethylene containers that are less substantial than the DOT 2U in shipping containers less substantial than the DOT 6D or 37M, thus supporting his case for authorizing the DOT 2U in a DOT 6D or 37M overpack. A review of these regulations indicates that the commenter's statement is correct provided the comparison is limited to not more than 5-gallon capacity containers. Therefore, the Board agrees with the commenter's position if the specification DOT 2U packaging is limited to a maximum

capacity of 5 gallons. Consequently it is providing for this shipping container by adding a new paragraph §173.119 (m) (17).

5. Exemptions. Several comments were received generally supporting the expansion of the existing exemption from specification packaging up to metal and plastic packagings of not over 32-ounce capacity, proposed in Notice No. 73–6 in §173.244. However, the Bureau of Explosives objected to any expansion in the size of packaging for exemptions and stated that any "increase in the small package exemption . . . should only be made after convincing proof on an individual commodity basis." The Bureau further stated that "some transportation experience should be gained through special permits." Although the Board disagrees with the Bureau's reasoning as being an impractical approach in all cases, it does believe that the matter of increasing exempt quantities for corrosives should be studied further in the light of the materials that are presently permitted to be exempt from specification packaging. Therefore, to assure an adequate level of safety in the shipment of corrosive materials, the Board is withdrawing the proposed change in §173.244 as regards the 32-ounce exemption provision for further study.

However, exemptions from specification packaging are provided for 16-ounce glass, metal, and plastic packaging. The special requirement for overpacking of a glass bottle in a metal can remains unchanged.

6. "Low hazard" corrosive liquids. Several comments were received supporting the provision to allow low hazard corrosives to be shipped in non-specification packaging. However, one commenter, the Bureau of Explosives, objected strongly to the proposal stating: "If the commodities proposed for the new §173.249a are of such a slight hazard that they can be shipped in non-specification packaging they should not be regulated at all." The Board finds that the present regulations contain many requirements for materials described according to their end use. Almost always these materials are permitted in packaging less restrictive (including non-specification packaging) than permitted for other materials not identified by end use. Also, the Hazardous Materials Regulations require these materials to be identified on shipping papers and, often, by labeling and marking. Therefore, the contention that when materials "can be shipped in non-specification packaging then they should not be regulated at all" is not reflected by the existing regulations. In several instances, the regulations provide for some form of identification without applying the full scope of all the regulations. That is what the Board is doing in this instance. Further, the Board does not deny that there can be excesses practiced under such a system and has proposals under development that will attempt to control some shipping practices which appear to be designed to thwart the intent of particular safety requirements. Nevertheless, the record has shown that, to be reasonable, and not unduly burdensome, end use descriptions can be used in the interests of safety. The Board agrees that it would prefer a better system and is attempting to reach such a goal. Meanwhile, the available record does not support that §173.249a would result in unsafe practices. It is important to note that the descriptions set forth in §173.249a will permit the Board to monitor the adequacy of the packagings authorized in this section through its hazardous materials incident reporting system.

The Board agrees with several commenters that the description "liquid acid chloride compound" is too vague as proposed and should further be defined if it is to be included in the regulations. Therefore, it has deleted this entry from §§172.5 and 173.249a. Further, it agrees with a commenter that the nature of the products covered by §173.249a could nevertheless pose more serious problems aboard aircraft and has added a restriction against use of non-specification packaging for transportation of these materials by aircraft.

7. Withdrawal of HM-57. The Bureau of Explosives requested that the Board withdraw HM-57. They stated that this action was justified by the problems discussed in items 5 and 6 above and for the following reason:

The rail transportation industry is concerned about the manner in which this docket is being handled. Not only has HMRB found it necessary to publish six different effective dates, but parts of the original proposed rules have been placed into effect in a piecemeal, optionally observable fashion with the result that, for instance, a car of phosphoric acid may or may not be subject to placarding, train placement and switching requirements depending on the election of the shipper to follow or not to follow the newly promulgated regulations. The confusion and potential for transportation errors and delays which result from such inconsistencies are so easily imaginable that the point need not be further discussed.

The Board rejects this request for withdrawal because it does not consider the matters discussed in items 5 and 6 stemming from a Notice of Proposed Rule Making to justify such a radical action. Further with respect to the final point, i.e., delays of effective date and optional compliance with the regulations, the Board again believes there is no justification for withdrawal. The rule was published effective March 23, 1972. Approximately 20 months have been provided for compliance. The options open to the Board following publication of any rule are: (1) To adjust the regulations by further rule making, (2) to withdraw the rule or portions thereof, (3) to delay the effective date of the rule or (4) to maintain the effective date of the rule and entertain petitions for special permits during necessary periods of adjustment.

With regard to the first option, the Board has made changes since March 26, 1972 (the original publication date of the amendments in this docket) which were primarily based on justifications submitted by the petitioners. This amendment is the latest in a series of such adjustments.

Concerning the second option, the Board does not agree that the "optional compliance period" is a justifiable reason for withdrawal of the Amendment made under Docket HM-57. The Board knows of no means it could use to avoid the optional period. It is a practical impossibility for instance, that as of a given moment, all the labels, markings, placards, or whatever be changed. In any complex rule making action, there is necessarily a long period in which adjustments have to be made and optional compliance must be permitted.

In respect to the third option, the Board has not been presented with any reasons nor is it aware of any circumstances that would justifiably support further delay of the effective date of the amendments published under this docket, except those matters pertaining to materials corrosive only to aluminum.

Concerning the fourth option, past history, covering many years, indicates the Bureau formerly used the special permit approach in petitioning the Interstate Commerce Commission for developing further proposals for a rule change. The Board does not agree that this is a preferred approach.

Accordingly, the petition of the Bureau of Explosives seeking withdrawal of the amendments adopted under this docket is hereby denied.

It was also brought to the Board's attention that it had inadvertently omitted the term "companion flange" in §178.343-5(b) (2) (i). This oversight has been rectified.

(38 F.R. 35467, December 28, 1973)

[Docket No. HM-57; Amdts. Nos. 171–14, 172–14, 172–20, 173–61, 173–74, 174–14, 175–7, 177–21, 178–26]
CLASSIFICATION OF CORROSIVE HAZARDS
Postponement of Mandatory Effective Date

On October 12, 1973 (38 FR 28292), the Hazardous Materials Regulations Board ("the Board") published a revised text to the next to the last paragraph of FR Doc. 73–1594 (39 FR 20837) which read as follows:

Amendments Nos. 171–14, 172–14, 172–20, 173–61, 173–74, 174–14, 175–7, 177–21, 178–26, except as they pertain to materials corrosive only to aluminum, are effective December 31, 1973, and are effective in their entirety on September 30, 1974. However, compliance with the regulations, as amended therein, is authorized immediately.

The Council for Safe Transportation of Hazardous Articles (COSTHA) has petitioned for a further extension of the effective date on materials corrosive only to aluminum pending the resolution of matters in Docket HM-112. In that Docket (39 FR 3022), the Board proposed reclassification of these materials as Other Regulated Materials, Group B (ORM-B), which would have certain marking requirements but would otherwise not be substantively regulated unless shipped by aircraft.

COSTHA points out that:

The original date for close of the comment period of Docket No. HM-112 has been extended from May 28, 1974 to August 31, 1974. It is unlikely that there could be any administrative determination on the status of regulation of aluminum corrosives before September 30, 1974. Furthermore, necessary packaging changes and alteration in shipping practices require a lead time of at least 6 months for many shippers.

The Board agrees with this evaluation of the circumstances and hereby changes the mandatory effective date of Amendments Nos. 171–14, 172–14, 172–20, 173–61, 173–74, 174–14, 175–7, 177–21, 178–26 as they pertain to materials corrosive only to aluminum from September 30, 1974 to September 30, 1975. However, voluntary compliance with these regulations as they pertain to aluminum is authorized.

(39 F.R. 32615, September 10, 1974)

[Docket No. HM-57; Amdts. Nos. 171–14, 172–14, 172–20, 173–61, 173–74, 174–14, 175–7, 177–21, 178–26]
PART 171—GENERAL INFORMATION AND REGULATIONS
PART 172—COMMODITY LIST
PART 173—SHIPPERS
PART 174—CARRIERS BY RAIL FREIGHT
PART 175—CARRIERS BY RAIL EXPRESS
PART 177—SHIPMENTS MADE BY WAY OF COMMON, CONTRACT, OR PRIVATE CARRIERS BY PUBLIC HIGHWAY
PART 178—SHIPPING CONTAINER SPECIFICATIONS
Classification of Corrosive Hazards;
Postponement of Mandatory Effective Date; Correction

On March 23, 1972; April 26, 1972; September 16, 1972; May 16, 1973; and August 3, 1973, the Hazardous Materials Regulations Board ("the Board") published Amendment Nos. 171–14, 172–14, 172–20, 173–61, 173–74, 174–14, 175–7, 177–21, 178–26 (37 FR 5946, 8383, and 18918; 38 FR 12807 and 20837) under Docket No. HM-57 prescribing new regulations for the classification, packaging, marking, labeling, and transportation of corrosive materials. Compliance with these amendments has been authorized as of April 21, 1972. The mandatory effective date was specified as December 31, 1973.

Included in the new criteria for defining corrosive material is a corrosion rate on aluminum. The Board has received a number of comments relative to this specified test criteria and the imposition of new regulations applicable to more than transportation by aircraft. In its comments on the amendments, the Counsel for Safe Transport of Hazardous Materials (COSTHA) pointed out that "It is COSTHA's view that the aluminum testing criterion was determined essential to the definition because of the aluminum shell and structural members of today's aircraft. This safety concern is realistic and not questioned by COSTHA. What is questioned, however, is the need to apply the full scheme of regulation to these products when shipped by rail or highway." Based on these comments and others similar in nature, the Board is giving further consideration to the regulation of materials corrosive only to aluminum, therefore it has extended the effective date of the amendments under Docket No. HM-57 as they pertain to materials that are corrosive only to aluminum.

(38 F.R. 28292, October 12, 1973)

SUMMARY • Docket No. HM-59 • *Authorized the shipment of Class A poisons in DOT-3A, 3AA, and 3E1800 cylinders; established new requirements covering the filling and performance adequacy of any cylinder used in this service.*

[49 CFR Part 173]
[Docket No. HM-59; Notice 70–18]
CLASS A POISONS IN CYLINDERS
Notice of Proposed Rule Making

The Hazardous Materials Regulations Board is considering amending the Department's Hazardous Materials Regulations to provide for the use of specification DOT 3A, 3AA, and 3E1800 cylinders for the transportation of certain class A poisonous liquids or gases.

This proposal is based upon a petition for rule making and the satisfactory experience gained under the terms of several special permits in existence for several years. These cylinders would be prescribed for class A poisons not otherwise specifically provided for (§173.328). At the same time they would automatically be authorized for certain specifically named poisonous materials covered in the sections that adopt the packaging requirements of §173.328 by reference to that section. Simultaneously, criteria for adequate valve protection would be more clearly defined in §173.327.

The packaging and handling requirements for class A poisons, in general, as well as the need for specifically naming in §172.5 certain class A poisonous materials currently described under the generic term "Poisonous liquid or gas, n.o.s." is under review by the Board. However, the Board believes that the changes proposed in this notice are significant enough to warrant separate rule-making action in the interim.

(35 F.R. 16005, October 10, 1970)

[Docket No. HM-59; Amdt. 173–42]
PART 173—SHIPPERS
Class A Poisons in Cylinders

The purpose of this amendment is to provide for the use of specification DOT-3A, 3AA, and 3E1800 cylinders for the transportation of certain class A poisonous liquids or gases.

On October 10, 1970, the Hazardous Materials Regulations Board published a notice of proposed rule making, Docket No. HM-59; Notice No. 70–18 (35 F.R. 16005), proposing to amend the regulations as stated above. No objections were received to the basic proposal, but two commenters objected to certain elements. Several others suggested some editorial changes, of which most were adopted.

One commenter objected to the specificity of the requirement for taper-threaded connections of valves to cylinders, noting the Board's announced intention of striving for performance standards. Another objected to the limitation for the gas pressure at 130°F. to not exceed the service pressure of the cylinder, and to the 6-foot boxed drop test. Both commenters referred to good experience in the shipment of gases in general, under conditions other than were proposed to be required for class A poison liquids and gases. The Board acknowledges that the proposals it made impose more specific and limiting conditions for class A poison shipments. It did so on the basis that the hazard level of a material should be a primary factor in specifying packaging. Consistent with this philosophy, its proposal intended to provide a better package than currently required for nonpoisonous gases or liquids that are otherwise regulated. Also, the proposal generally reflected packaging standards that have been in use for several years under the terms of many special permits. The permit experience has been completely satisfactory and, according to the comments received by one holder of such a permit and an association representing the majority of shippers of class A poisons who hold such permits, implementation of the proposal was recommended.

Editorial changes are being made to insure completeness and consistency between §§173.34, 173.327, and 173.328. For example, the sentence "Safety relief devices are forbidden" is being deleted in §173.327(a), since this is already covered by §173.34(d) (3); reference to §173.301(g) is being deleted, as §173.327 is complete in itself.

(36 F.R. 1472, January 30, 1971)

SUMMARY • Docket No. HM-60 • *Requested public advice on speed restrictions for trains hauling liquefied flammable gases in DOT-112A or 114A tank cars. No further action has ever been taken in this docket.*

[49 CFR Parts 173, 174]
[Docket No. HM-60]
TRANSPORTATION OF HAZARDOUS MATERIALS
Request for Public Advice on Speed Restriction on Tank Cars

The Hazardous Materials Regulations Board is considering amending Parts 173 and 174 of the Department's Hazardous Materials Regulations to cope with the increasing number of railroad accidents involving DOT Specifications 112A and 114A tank cars transporting liquefied flammable gas. Although Docket No. HM-38; Amendment No. 179-4 limited the capacity of tank cars built after November 30, 1970, to 34,500 gallons, existing cars carrying liquefied flammable gas with a capacity of over 25,000 gallons appear to have caused a major portion of the more serious difficulties encountered to date.

The sequence of events in accidents such as occurred at Laurel, Miss., and Crescent City, Ill., may be briefly summarized. Following derailment of the train, one or more tank cars became punctured, releasing their contents which, in most cases, then burst into flame. The volume in these cars is such that a fire of considerable magnitude develops which heats and burns other cars in the immediate area. Often liquefied flammable gas moves in groups of more than five tank cars coupled together. When punctured and ignited, the fire impinges on the uninsulated shells of adjacent tank cars, causing them to heat, lose tensile strength and explode. It is usually a very short time between derailment and explosion of these adjacent tank cars. Such explosions occur despite the safety relief devices with which these cars are equipped.

The Board believes that placing certain restrictions on speed of trains in which liquefied flammable gas tank cars are included will significantly reduce the likelihood of derailment and potential impact on the cars. The Board is therefore considering proposing, among other things, that the speed of trains carrying liquefied flammable gas in DOT Specifications 112A and 114A tank cars having a capacity of 25,000 gallons or more, be limited to a maximum of 25 miles per hour when moving through incorporated communities, unless one of the following provides qualified mechanical or visual surveillance of the train within 35 miles of the limits of such communities:

  a. Hotbox detector;
  b. Initial terminal interchange or 500-mile inspection point;
  c. Roll-by inspection at crew change point; or
  d. Qualified railroad personnel adjacent to the track.

A positive indication of satisfactory surveillance or inspection of the train would have to be received by the train crew in the caboose, and that information would then have to be transmitted to the engineman.

The Board is also considering proposing that shippers, when offering carriers loaded tank cars of 25,000 gallons or greater capacity be required to designate on shipping orders for inclusion in way-bills that such tank cars are loaded with liquefied flammable gas and that the carrier should institute the special handling provisions which would be necessary in moving those cars.

Interested persons are invited to give their views on the speed restriction the Board is considering proposing as part of the solution to a problem it considers too complex to be solved by a single change to the regulations. Therefore, public advice is also specifically requested as to additional means for coping with the increasing number of railroad accidents involving these tank cars, including lower filling densities, addition of insulation, etc., to be considered as part of an overall system aimed at significantly reducing the loss of hazardous materials in such accidents. Comments, identifying the docket number, should be submitted in duplicate to the Secretary, Hazardous Materials Regulations Board, Department of Transportation, 400 Sixth Street SW., Washington, D.C. 20590, prior to December 16, 1970. All comments received will be available for examination by interested persons at the Office of the Secretary, Hazardous Materials Regulations Board, both before and after the closing date for comments.

(35 F.R. 16180, October 15, 1970)

SUMMARY • Docket No. HM-63 • *Proposed removal of authorizations for higher safety relief device discharge settings on DOT-112A and 114A tank cars transporting liquefied petroleum gas; proposed cancellation of special permits authorizing use of DOT-112A and 114A tank cars constructed with a weld joint efficiency of E=1.0. This controversial proposal was withdrawn.*

[49 CFR Part 179]
[Docket No. HM-63; Notice No. 70-20]
TRANSPORTATION OF HAZARDOUS MATERIALS
Tank Car Specifications

The Hazardous Materials Regulations Board is considering amending Part 179 of the Department's Hazardous Materials Regulations to remove the authorization for higher discharge safety valve settings on DOT Specifications 112A and 114A tank cars transporting liquefied petroleum gas and anhydrous ammonia. The Board also proposes to cease issuance and renewal of special permits authorizing these higher discharge safety valve settings on Specifications 112A and 114A tank cars, and special permits authorizing use of specifications 112A and 114A tank cars constructed with a welded joint efficiency of E=1.0. Upon adoption of this proposal, the Board would withdraw existing special permits of these two types.

As of this publication date, all outstanding special permits authorizing a discharge safety valve setting of 280.5 p.s.i.g. on specification 112A340W and 114A340W, or 330 p.s.i.g. on specification 112A400W tank cars, are hereby revised to preclude addition of new tank cars to such service. No further requests to record tank cars under special permits authorizing these higher safety valve settings will be accepted. Requests to record tank cars constructed with a weld joint efficiency of E = 1.0, and built after November 25, 1970, will not be accepted.

A continuing series of major accidents resulting in deaths, personal injuries, and massive property losses have involved tank cars of the above descriptions.

Section 179.102-11 presently allows discharge safety relief valve settings to prevent buildup of pressure in excess of 90 percent of tank test pressure. The Board is of the opinion that this percentage does not afford an adequate margin of safety in specifications 112A and 114A tank cars, and the Board is therefore proposing to remove that authorization for those cars. The highest allowable setting would be that specified in section 179.101-1. Consistent with this proposal, the Board would cease to issue or renew special permits authorizing such higher safety valve settings, and the Board is further proposing withdrawal of those special permits outstanding. The effective date for this withdrawal would be June 1, 1971.

Section 179.100-6 requires that the welded joint efficiency of pressure tank cars be E=0.9. This figure is inserted in the formula for determination of tank car plate thickness. All other factors remaining constant, an increase in the welded joint efficiency figure leads to a decrease in the required tank plate thickness. The Board has concluded that this reduction in plate thickness on specifications 112A and 114A pressure tanks is not in the interest of safety, and therefore its authorization by special permit should be discontinued. The Board is accordingly proposing to cease issuance and renewal of special permits granting such authorization, and upon adoption of this proposal would withdraw outstanding permits for E=1.0 on specifications 112A and 114A tank cars, as of June 1, 1971.

Return to service of cars affected by the above permit withdrawals may be considered by the Board upon addition of adequate modifications or operational controls. Comments suggesting appropriate modifications of cars or operational controls will be appreciated.

(35 F.R. 16741, October 29, 1970)

[49 CFR Part 179]
[Docket No. HM-63; Notice 70-20]
TRANSPORTATION OF HAZARDOUS MATERIALS
Tank Car Specifications

On October 29, 1970, the Hazardous Materials Regulations Board of the Department of Transportation published Docket No. HM-63; Notice No. 70-20 (35 F.R. 16741), entitled Tank Car Specifications.

The second paragraph of the preamble to that document precluded the addition of new tank cars to service under existing special permits which authorize a discharge safety relief valve setting 280.5 p.s.i.g. on specifications 112A340W and 114A340W tank cars, and 330 p.s.i.g. on specification 112A400W tank cars, and which authorize a welded joint efficiency of E=1.0 on either specification 112A or 114A tank cars constructed after November 25, 1970.

Based upon consideration of petitions received, the Board concludes that implementation of this paragraph will work an undue hardship upon both tank car builders and users, and the public. Therefore, Docket No. HM-63; Notice No. 70-20 is hereby revised to delete the second paragraph of the preamble in its entirety.

(35 F.R. 17790, November 19, 1970)

[49 CFR Part 179]
[Docket No. HM-63; Notice 70–20]

TRANSPORTATION OF HAZARDOUS MATERIALS
Tank Car Specifications; Withdrawal of Notice of Proposed Rule Making

On October 29, 1970, the Hazardous Materials Regulations Board published Docket No. HM-63; Notice No. 70-20 (35 F.R. 16741), and modified that notice on November 19, 1970 (35 F.R. 17790). The Board proposed to amend §179.102–11 of the Department's Hazardous Materials Regulations to remove the authorization for higher discharge safety relief valve settings on DOT Specifications 112A and 114A tank cars transporting liquefied petroleum gas and anhydrous ammonia. The Board also proposed to cease issuance and renewal of special permits authorizing these higher discharge safety valve settings on Specifications 112A and 114A tank cars and special permits authorizing use of Specifications 112A and 114A tank cars constructed with a welded joint efficiency of E = 1.0, as that figure is used in the regulations to determine tank plate thickness. The proposed date for special permit withdrawals was set at June 1, 1971.

Interested persons were given an opportunity to comment on the Board's proposal. On the basis of information and data received in response to the proposal, and available for examination at the Office of the Secretary of the Hazardous Materials Regulations Board, the Board concludes that it would be inadvisable to carry through with the proposal at this time. Should research by the Government or by industry indicate a need for further action on this subject, the matter may be opened for re-examination in its entirety.

In consideration of the foregoing, Docket No. HM-63; Notice No. 70–20 is withdrawn.

(36 F.R. 3376, February 23, 1971)

SUMMARY • Docket No. HM-67 • *Proposed a new flash point definition (closed-cup tester, 100 degrees F.) for flammable liquid class. See Docket Nos. HM-42, HM-102, and HM-112.*

[49 CFR Part 173]
[Docket No. HM-67; Notice 70–23]

TRANSPORTATION OF HAZARDOUS MATERIALS
Flash Points of Flammable Liquids

The Hazardous Materials Regulations Board is considering amending §§173.115 and 173.119 of the Department's Hazardous Materials Regulations to specify use of the Tagliabue (Tag) closed-cup tester (ASTM D 56–70) to determine flash points of flammable liquids, instead of the Tagliabue (Tag) open-cup tester (ASTM D 1310–67), presently specified.

The flash point is generally accepted as a useful means to determine the flammability of flammable liquids, and therefore their potential fire hazard during transportation. The Tagliabue open-cup testing method, which has been in use with only minor modification for many years, lacks the precision, reliability, and reproducibility necessary to properly estimate the flammability hazard that may be encountered during transportation.

This notice is not intended to change the present established classification ranges or packaging of flammable liquids. Its purpose is to propose adoption of a more accurate method for determining flash points than the Tag open cup presently affords.

As part of the Department's overall review of the Hazardous Materials Regulations, the Board and the staff of the Office of Hazardous Materials (OHM) have been evaluating methods used for classification of materials according to the hazard presented during transportation. OHM contracted with the Safety Research Center, U.S. Bureau of Mines, to examine the limitations of the available flash point testers and to recommend the best method for adoption by DOT.

In reaching their conclusions, the Bureau of Mines measured the present state of the art against the following criteria:

1. Repeatability (data obtained by the same analyst in several determinations, using the same equipment and the same sample).

2. Reproducibility (data obtained by several analysts, each using a different piece of equipment of the same type, and using the same sample).

3. Reliability in assessing the fire or explosion hazard.

In addition, the Bureau of Mines considered and evaluated all comments received in response to that part of a prior notice of proposed rule making (NPRM)[1] dealing with definitions of flammable liquid, flashpoint, open-cup tester, and closed-cup tester. The results and recommendations of the Bureau's study have been reported.[2]

The Bureau's report recommends that the Tag closed-cup method be used to determine flashpoints of flammable liquids for purposes of the DOT Hazardous Materials Regulations. The conclusions, proposing adoption of the closed-cup method, may be summarized as follows:

1. The closed-cup method is more precise and reliable than the open-cup method, gives more reproducible data, and provides a more conservative estimate of the hazard presented by the formation of flammable vapor-air mixtures under either confined or unconfined conditions.

2. "It is often proposed that an open-cup more nearly approximates the geometry of a spill situation than does a closed-cup. In our judgment, this is a trivial consideration in choosing among the variations of existing apparatus. The actual likelihood of ignition of a spill depends heavily upon factors which are beyond the scale of laboratory apparatus, such as the cooling of the liquid surface by evaporation or the gustiness of the atmosphere."[3]

"The greatest explosion hazard results from leakage or spillage into surroundings that provide some confinement, such as a railroad box car, a van-type truck, or the hold of a ship. In this situation, convection currents aid the formation of homogeneous vapor-air mixtures and the magnitude of overpressures in confined combustion is usually greatest with homogeneous mixtures. Here again, the closed-cup gives the best definition of a hazard."[4] Experience shows that spills and leaks in confinement are common accident situations and must be considered in the development of safety criteria.

3. Due to its greater reliability, the closed-cup method has been accepted by the National Fire Protection Association, the National Academy of Sciences, the United Nations Intergovernmental Maritime Consultative Organization (IMCO), and many western European industrial countries, including Great Britain, France, West Germany, Sweden, and the Netherlands.

Additional reasons supporting the closed-cup method may be found in a review of various technical publications and comments received on a prior notice of rule making.[5] The following is quoted from the International Chamber of Shipping's statement which was attached to the IMCO October 15, 1969, communication to the sixth session of the Committee of Experts on the Transport of Dangerous Goods:

The closed-cup method of testing should be used rather than the open-cup method in view of the former's much better precision.[6]

Proponents of the open-cup method point out that improvement in technique in recent years has resulted in increased precision and reproducibility of data. It is agreed that refinement of test methods has brought some improvement. However, in spite of this improvement, the Board believes that the open cup is still not equal to the closed-cup method for overall transportation safety purposes. For example, the report of Technical Subcommittee No. II of the Chicago Society for Paint Technology[7] summarizes the testing done during 1968 with six different types of flashpoint testers and 27 solvents having flash points ranging from 20° F. to 190° F. The report concluded that, "All closed-cups were considerably more reliable and easier to work with than the other cups. . . . "

Some comments received on Docket HM-3; Notice No. 68–2 stated that a closed-cup is not responsive to mixtures that contain low-volatility nonflammable components; it is, on the other hand, far too stringent for mixtures containing very small (less than 0.2 percent) amounts of highly volatile flammable compounds. During the test of a mixture, the closed-cup can concentrate nonflammable vapors as readily as flammable vapors. These nonflammable vapors can have a suppressant effect upon the flammability of the sample, thereby raising the flash point beyond the limit prescribed in the regulations for flammable liquids. In an open-cup, part or all of the vapors can escape, thus reducing this suppressant effect. On the other hand, comments noted that a non-flammable anti-knock compound containing less than 0.2 percent of dissolved hydrocarbon, because of trapping of the hydrocarbon traces in the vapor space of the apparatus, had a closed-cup flash point of 58°–73° F., compared to an open-cup flash point of 180°–245° F.

The Board realizes that none of the presently available test methods accurately applies to all mixtures. To cover the unusual behavior of certain mixtures, the Board can issue the necessary rulings. For example, the Board could classify such mixtures according to the flash point of their major component. There may be alternative means to cover certain mixtures which do not lend themselves to the proposed testing procedure, and the Board welcomes any suggestions in this regard. The decision as to proper classification of exceptions could be based upon other data or experience showing that the liquid is more or less hazardous than the flash point data indicate. The exceptions should not govern the general rule, however, and the Board is concerned with covering the great majority of substances by a single test method.

In defining flammable liquid,[8] the United Nations Organization recognizes both the open- and closed-cup methods. It is the Board's understanding that the U.N. included the open-cup method principally to accommodate the United States' regulations.

The United Nations Committee of Experts on the Transport of Dangerous Goods, in arriving at the value of 73° F. for the closed-cup as being equivalent to 80° F. in the open-cup test, considered all available information on the subject. The Paint Technology Report shows an average difference of 7° F. between the Tag open- and closed-cup methods.[9] A review of the pertinent literature confirms this relationship. Therefore, the Board intends to substitute 73° F. for 80° F. in the Hazardous Materials Regulations as the upper limit for flash points of regulated flammable liquids in implementing the change from the open-cup to the closed-cup method. The Board does realize that for a few materials the difference between methods may be much more. It is important to emphasize that this change is in no way an attempt to change the classification of the existing flammable liquids. It is recognized, however, that there may be some isolated cases where the classification would change based upon closed-cup test results. The Board would appreciate receiving advice on how to deal with such situations so as to minimize the hardship on industry.

Upon adoption of this proposal, all references in the Department's Hazardous Materials Regulations will be changed from open-cup to closed-cup.

The Board intends to retain the lower flash point limit of 20° F., as prescribed in §173.119, open-cup for the closed-cup method. The corresponding flash point difference between the open- and closed-cup methods at this temperature range generally would be very slight, and therefore a change in the lower limit is considered unnecessary.

In the event that the new classification, "Combustible liquids," is established pursuant to proposed rule making,[10] the test method proposed herein, conducted at an appropriately reduced heating rate, would be prescribed in place of the open-cup test. The equivalent closed-cup temperatures would be substituted for the adopted "Combustible liquid" open-cup temperatures.

Interested persons are invited to give their views on the amendment proposed herein. Communications should identify the docket number and be submitted in duplicate to the Secretary, Hazardous Materials Regulations Board, Department of Transportation, 400 Sixth Street SW., Washington, DC 20590. Communications received on or before March 2, 1971, will be considered before final action is taken on the proposal. All comments received will be available for examination by interested persons at the Office of the Secretary, Hazardous Materials Regulations Board, both before and after the closing date for comments.

[1]Docket No. HM-3; Notice No. 68–2 (33 F.R. 3382, Feb. 27, 1968).

[2]Kuchta, Joseph M. and Burgess, David, Report No. S 4131, Apr. 29, 1970, Safety Research Center, U.S. Bureau of Mines. This document is available from the Clearing House for Federal Scientific and Technical Information. National Bureau of Standards, U.S. Department of Commerce, Springfield, Va. 22151 at a cost of $3 per copy, or Microfiche copy at 65 cents.

[3]Kuchta, Joseph M. and Burgess, David, Report No. S 4131, Apr. 29, 1970, Safety Research Center, U.S. Bureau of Mines, p. 5.

[4]Ibid., p. 6.

[5]Docket No. HM-3; Notice No. 68–2 (33 F.R. 3382, Feb. 27, 1968).

[6]United Nations Economic and Social Council, E/CN.2/CONF.5/R.198.

[7]Probst, K.G., Correlation of Apparatus for Measuring Flash Point of Solvents, J. of Paint Technology, Vol. 40, No. 527, pp. 576–81 (December 1968).

[8]United Nations, Transport of Dangerous Goods (1966), Vol I, p. 5, ST/ECA/81/Rev.1, E/CN 2/Conf. 5/10/Rev. 1.

[9]Probst, K.G., Correlation of Apparatus for Measuring Flash Point of Solvents, J. of Paint Technology, Vol. 40, No. 527, pp. 576–81 (December 1968).

[10]Docket No. HM-42; Notice No. 70–3 (35 F.R. 3298, Feb. 21, 1970).

(35 F.R. 18534, December 5, 1970)

SUMMARY • Docket No. HM-68 • *Established new specification DOT-56 and 57 metal portable tanks; deleted authorization for construction of specification DOT-52 and 53 portable tanks.*

[49 CFR Parts 173, 178]

[Docket No. HM-68; Notice 70–24]

TRANSPORTATION OF HAZARDOUS MATERIALS
Portable Tank Specification

The purpose of this notice of proposed rule making is to request public comment on a proposed amendment to Parts 173 and 178 of the Hazardous Materials Regulations (49 CFR Parts 170–189) to prescribe a new specification for portable tanks (bins) and the uses of that new tank. The existing specifications 52 and 53 portable tanks (§§178.246, 178.247) no longer meet the needs of industry, and many new portable tanks are being shipped under the terms of DOT Special Permits. The specification 56 tanks proposed in this notice would replace the specifications 52 and 53 tanks, although continued use of the latter tanks would be allowed, and would eliminate the need for future special permits of this type.

The proposed specification is based primarily on performance standards, rather than detailed engineering design requirements. The Board announced in the FEDERAL REGISTER on August 21, 1968 (33 F.R. 11862), its intention to convert the regulations from design specifications to performance standards. Public comment on that announcement was for the most part favorable, and several performance-type standards have already been issued by the Board.

The existing specifications 52 and 53 for portable tanks are deficient in that they do not provide for the use of steel, and they do not allow the sizes of tanks now being shipped. The proposed specification provides for the new aluminum alloys, for combinations of metals, and for metal quality testing. New standards are also being proposed for fitting protection, venting capacity, and stacking, mounting, and tie-down provisions, similar to those provided for in the general cargo tank specifications for MC 306, MC 307, and MC 312 (§178.340). No limitation is proposed in the specification for the size of openings, although the Board, when authorizing use of specifications 56A and 56B in Part 173, may include opening limitations according to the particular product being authorized. As written, the proposal does provide for hopper-type and side-opening doors. Specific comment is requested on this aspect of the proposal. A series of performance tests on the completed tanks would take the place of a specific metal wall thickness requirement.

The proposed specification provides for two types of tank specifications—56A and 56B. Specification 56A would be authorized in Part 173 for appropriate dry flowable or solid materials, but would not be authorized for liquids. Specification 56B would be authorized for liquids having in the container an absolute vapor pressure not exceeding 16 p.s.i. at 100° F. The specifications are written in such a way that they could be expanded to cover higher vapor pressure materials—up to perhaps 25 p.s.i.a. Public comment is specifically invited as to how this might be done in a later rule-making action.

The primary benefit to shippers and manufacturers from this type of specification is that it provides for maximum design and construction latitude. The choice of materials, appurtenances, and design configuration is placed in the hands of shippers and the tank designers. The minimum performance criteria must be met by specific proof testing procedures or analyses.

During the development of this specification, the question arose regarding minimum wall thickness. The Board prefers to specify performance rather than design, but recognizes that meaningful puncture tests are not yet developed. Since a minimum wall thickness itself implies a certain puncture resistance, it may be necessary to prescribe a minimum. For example, the present specification 53 (§178.247) prescribes 0.25-inch thick aluminum for the bottom of the tank and 0.09-inch thick aluminum for the sides and top. The Board believes that the design and testing requirements of this proposed specification will automatically provide sufficient wall thickness to include puncture resistance, but invites specific public comment on this point.

Consistent with the proposed new tank specification, the appropriate paragraphs of §173.32 relating to retest of portable tanks would be revised to consolidate the retest requirements and to provide for the new tanks.

Appropriate changes would also be made (although not specifically listed herein) to Subparts C, D, E, and G of Part 173 to authorize the appropriate type of specification 56 tank for hazardous materials now authorized in specification 52 or 53 tanks. For example, §173.128(a) (3) authorizes the specification 52 tank for paints; the specification 56B tank would also be authorized; the same would be done for cements in §173.132(a) (2).

(35 F.R. 18919, December 12, 1970)

[Docket No. HM-68; Amdts. 173–60, etc.]

PORTABLE TANK SPECIFICATIONS

The purpose of these amendments to the Department's Hazardous Materials Regulations is to provide two new specifications for portable tanks and to prescribe uses for these tanks. Existing authorization for fabrication of DOT specifications 52 and 53 portable tanks will terminate on May 31, 1972.

On December 12, 1970, the Hazardous Materials Regulations Board published a notice of proposed rule making, Docket No. HM-68; Notice No. 70–24 (35 F.R. 18919) which proposed these amendments. The Board received several comments regarding these changes.

Several comments were addressed to the material specifications for aluminum and the difficulties of expressing permitted aluminum alloys in a format that would not be unnecessarily restrictive and that would not require frequent updating to be kept current with advancing technology. Also, the Board was requested to treat aluminum in a similar manner as steel. After reviewing all the comments on this matter, the Board concluded that a table similar in principle and format to that for steel would be the most effective solution. This table is included in the amendment.

Many comments were made regarding the criteria for venting of liquids and the inadequacy of the proposal as it was written. The Board agrees with these comments and has included provisions for the use of frangible discs, adjustment of the total emergency venting capacity table, a requirement for total containment under conditions of transportation which include temperatures as high as 130° F., allowance for minimum venting capacity calculated from a tank pressure of 3 p.s.i.g. to 5 p.s.i.g., and clarification that spring-loaded devices are not mandatory.

In response to several comments regarding materials of construction, the Board has clearly noted that pressure relief devices are not required to be made of metal.

The Board received many comments on the written format of the specifications and has adopted the format for which almost all commenters expressed a preference; viz., a general requirement section for all portable tanks followed by separate specification requirements for tanks for dry and liquid products. Under this format the Board further recognized a difference between requirements for containers for liquid and containers for solids. This was in consideration of the overall difference in hazard between the liquid and solid state of a material in the event of a leak.

Several comments were made concerning the adequacy, desirability, and effect of the proposed tests and how they relate to the transportation environment. Although the testing criteria now set forth in the Hazardous Materials Regulations may not be fully responsive, in consideration of results these criteria have produced with respect to the adequacy of present hazardous materials packaging, the Board prefers to retain the conventional drop, vibration, and pressure testing now found in the regulations. The Department has a study in progress regarding the relation of testing to the environment and will await the conclusions of this study before embarking on any extensive changes in testing philosophy. A particular concern in testing was the requirement for vibration testing, especially as it involves solids. Here also, the Board has chosen to retain existing test criteria. With respect to the tanks to be used for solids, it is the opinion of the Board that use of a fine, dry, powdered filling material in the vibration test will be helpful. It should also permit evaluation of closures located low in the tank if the dry powdered material is of sufficiently fine mesh size. However, the Board has chosen not to prescribe the actual mesh size, but will rely on the clear intent of the rule describing testing and the fact that design and testing information must be available to the shipper to assure compliance with §173.24 of the Hazardous Materials Regulations.

The retest periods remain as proposed because of the exposure these portable tanks receive. The Board believes that a longer retest period would not be responsive to the type of service these smaller portable tanks will see.

Some objection against the tie-down requirements was expressed. Most of these comments did not contain explanatory or supporting data. Consequently, the Board has made adjustments only as it considered were necessary but otherwise maintained the requirement.

Several commenters objected to the joint efficiency requirements as expressed and referred the Board to the current requirements for cargo tanks as a better form for the rule. The Board agrees with these suggestions and has added requirements similar to §178.340-5 of the Hazardous Materials Regulations.

On the basis of several comments and re-evaluation, the Board has determined that existing requirements and tests for fittings, closures, guards, and supports assure the adequacy of the specifications without the need for specific mounting pad requirements. This proposed section, therefore, is not included in this amendment.

Several comments related to the difficulty in expressing tank failure. The Board originally proposed terms such as "no stress in excess of yield strength" and other similar expressions. The difficulty in describing inadequate strength was demonstrated by several commenters. The Board has changed the rule to refer to the term "significant deformation" or similar expressions, being of the opinion that the term "significant" relating to the projected adequacy of a container to hold product is capable of being determined.

On the basis of some of the comments received, it became obvious to the Board that the purpose of the term "Maximum gross weight" in §178.251-7(a) was not fully understood. In performance oriented specifications such as these, testing must of necessity be closely allied to the packaging in the "as used" condition in transportation. The maximum gross weight limitation can refer only to a rated weight based on design and testing. The Board noted, however, that although its performance oriented philosophy was expressed in notice No. 70-24, some of the rules applying its philosophy could be clearer. Accordingly, the term "Maximum gross weight" in §178.251-7(a) has been changed to "Rated gross weight" and §§178.251-1(b), 178.252-3(a), and 178.253-5(a) incorporate language relating testing more specifically to the gross weight marking on the certification plate.

Currently, there are at least 25 outstanding permits authorizing use of modified specifications 52 or 53 portable tanks. These are commonly known in the trade as "Tote Tanks", "Tote Bins", "Liqua-Bins", etc. The Board is hereby authorizing their continued usage until April 30, 1974, under special permit, each permit subject to cancellation for cause. This authorization is only in effect for the commodities and modes of transportation specified in each permit. Each permit holder must examine the tanks in use to determine compliance with specifications 56 or 57 and if possible modify, re-rate, and re-mark the tanks according to the applicable specification. If the tank covered by a special permit cannot be brought within specification, the permit holder must advise the Board before July 1, 1973, and explain why the tank can not be brought within specification. The Board will then make individual determinations regarding additional rule making or continuance of the permit.

Those special permit holders having specification 52, 53, 56 or 57 portable tanks or similar modified tanks in services not covered by this amendment must also petition the Board for additional rule making or continuance of the permit under the same conditions.

Permits outstanding for portable tanks related to MC specifications or specifications 51 and 60 are not affected by this amendment.

(37 F.R. 2885, February 9, 1972)

SUMMARY • Docket No. HM-69 • *Established new DOT-39 non-reusable compressed gas cylinder specification and provided for its use.*

[49 CFR Parts 171, 173, 178]
[Docket No. HM-69; Notice 70–25]
TRANSPORTATION OF HAZARDOUS MATERIALS
Cylinder Specifications

The Hazardous Materials Regulations Board is considering amending the Department's Hazardous Materials Regulations to provide a specification for a nonreusable (non-refillable) cylinder for certain compressed gases and to eliminate existing specifications 9, 40 and 41. Also, a new paragraph (k) is proposed for addition to §173.301 to specify that certain cylinders, including the new specification proposed herein, must be shipped in outside packagings with their valves protected.

Interested persons are invited to give their views on this proposal. Communications should identify the docket number and be submitted in duplicate to the Secretary, Hazardous Materials Regulations Board, Department of Transportation, 400 Sixth Street SW., Washington, DC 20590. Communications received on or before March 9, 1971, will be considered before final action is taken on the proposal. All comments received will be available for examination by interested persons at the Office of the Secretary, Hazardous Materials Regulations Board, both before and after the closing date for comments.

The basis for this proposal is a petition of the Compressed Gas Association, Inc., and more than 100 special permits issued during the last 12 years. In its petition, the Compressed Gas Association submitted a draft of a proposed specification designated by it as DOT-42. Since the number 42 is designated for aluminum drums, the specification designation has been changed to DOT-39.

In reviewing the proposal, the Board has concluded that adoption of this specification will eliminate any need for specifications 9, 40, and 41. Comments are invited as to continuing need for manufacture of these specifications. Existing cylinders would be authorized until there would no longer be a need for their authorization.

The proposed new specification is primarily performance oriented. There are proposed limitations as to the type of metal to be used. Design for the cylinder would be based upon the pressure of the intended contents at 130° F., described as the "test pressure". The performance pressure (burst) test, would be a function of the test pressure. The marked service pressure would be limited to a maximum of 80 percent of the test pressure. A cylinder would be marked to indicate both service and test pressures. Rather than designating these pressures according to the different types of gases authorized, the regulations in paragraphs (e) and (f) of §173.301 would apply.

Consistent with the terms of most of the outstanding special permits, the specification would be limited to 55 pounds water capacity for service pressures of 500 p.s.i. or less, and 10 pounds water capacity for service pressures in excess of 500 p.s.i. For flammable gases, cylinder size would be limited to 75 cubic inches. The CGA petition proposed certain exemptions for safety relief devices. The Board is not proposing any exemption for relief devices for cylinders made under this specification. Further, the Board believes safety valves should be required for cylinders intended for liquefied flammable gases to reduce the hazard of total release of contents should a cylinder accidentally be overfilled or exposed to high temperatures of short duration.

There are two major performance tests proposed; a "burst" test and a flattening (crush) test. The emphasis is placed on the "burst" test since the Board believes it the more significant and reliable of the two. A minimum burst criteria is specified plus restrictions on the allowable manner of failure. The manner of failure at burst is a basic safety consideration. In developing the testing requirements, it became evident that the imposition of test requirements on a production line from the beginning-of-forming operations could cause manufacturers some difficulty. Such a limitation is not being proposed at this time for inclusion in the definition of a lot. However, if it is determined that the proposed definition of a lot is inadequate, an appropriate change will be made to insure that tests are truly representative of the production of all cylinders.

A manufacturer's registration system is proposed as part of this Notice. At this time, it is not proposed to establish a licensing or certification scheme. As stated in §173.24, the specification identification marking of a packaging is a certification that it complies with all specification requirements. The proposed issuance of registration numbers to manufacturers would be a ministerial function to facilitate the identification of manufacturers. Although the Board is considering licensing in this area, and others, such substantive regulations will not take place without separate rule-making action.

The Board is placing emphasis on the nonreusable limitation of this specification by proposing a penalty statement as part of the specification marking requirement. The regulations concerned with the marking "NRC" is found presently in paragraph (i) of §173.28.

The Board agrees with the CGA proposal, as modified herein, since the proposed cylinder will be: (1) Overpacked at all times during transportation, (2) limited to noncorrosive gases, (3) nonreusable and not subject to cyclic stresses resulting from refilling, (4) equipped with safety devices, (5) made of ductile materials, and (6) the subject of continuous performance tests.

(35 F.R. 18879, December 11, 1970)

[Docket No. HM-69; Amdt. Nos. 171–11, 173–53, 178–20]
CYLINDER SPECIFICATIONS

The purpose of these amendments to the Department's Hazardous Materials Regulations is to provide a new specification for a nonrefillable cylinder to be used for certain compressed gases, to remove existing authorization for fabrication of Department of Transportation specifications 9, 40 and 41 cylinders, and to require that certain cylinders be shipped in outside protective packagings.

On December 11, 1970, the Hazardous Materials Regulations Board published a notice of proposed rule making, Docket No. HM-69; Notice No. 70–25 (35 F.R. 18879) which proposed these changes.

Several commenters objected to Note 1 of §173.34(d) (1). Some objections were addressed to the requirement for use of a safety relief valve on specifications 39 cylinders containing liquefied flammable gases, thereby prohibiting the use of other type safety relief devices. It was stated that CGA pamphlet S-1.1 is a standard accepted and used by the industry and that such a standard should be referenced. This standard allows other safety relief devices to be used. The Board notes that although a standard exists, this does not mean its provisions are followed by industry in all respects. It cannot be concluded that this standard is the decisive factor in the safe history of transportation of gases. Although the industry claims a safe history, the actual practices followed are the standards to consider, rather than a permissive standard which may not be fully utilized. Safety relief valves are the devices routinely used by industry on cylinders containing flammable liquefied gases. Other objections related to specifying metal as the construction material for the valve. Since the Board has determined that a valve is desirable in place of a fusible device for liquefied flammable gases, then the valve should be capable of functioning as a valve, i.e., opening and closing. Use of a valve made of materials that would lose their shape, or act as a fusible device after short exposure to elevated temperatures would defeat the purpose of the original requirement for a valve. Objections also were made to the prohibition against use of fusible safety relief devices. The Board has re-examined its position in this matter, including consideration of experience under special permits, and has concluded that fusible safety relief devices are not desirable for use with liquefied compressed gases. However, the Board believes that fusible safety relief devices are adequate for use with non-liquefied gases and the rule has been amended accordingly.

Comments made on §173.301(k) indicated a definite need for clarification of the objective for the outside packaging. The basic design for specification 39 was predicated on this cylinder's transportation in an outside container. Section 173.301(k) (1) as it was proposed, did not clearly express this fact and therefore has been modified. Specification 3E is included in this paragraph because of its specified size limitation.

Objection was raised concerning the 75-cubic-inch container limitation for flammable gases. This limitation is based on extensive experience under special permits and the consideration that, in transportation, nonreusable cylinders of larger sizes would be used in place of higher-integrity reusable cylinders now used in flammable gas service. The Board believes that this usage would degrade the level of transportation safety now established for the shipment of flammable gases.

Requests were made to add products to the table in §173.304(a) (2), but since these items would be new commodities to be covered by name in the Hazardous Materials Regulations, a separate rule-making action would be required.

Several of the comments concerned the specification itself. One commenter suggested an upper design pressure limit of 3,000 p.s.i., stating that the Board should not exceed that limit unless a need and justification were demonstrated. Need and justification for pressures higher than 3,000 p.s.i. have been demonstrated and several permits are now in effect for pressures above 3,000 p.s.i. The Board considers the design parameters of the specification to be adequate and finds no justification for setting a specified upper design pressure limit.

Some commenters observed that, as written, §178.65-2 (c) and (d) would require remarking of cylinders for different service and test pressures. However, §173.34(a) (2) provides for using higher service pressure marked units than are required by the regulations and applies in this circumstance. To avoid ambiguity because of the requirements of §178.65-14 relating to markings, §178.65-2(d) has been changed by the addition of the word "maximum".

Comments on §178.65-3 concerned foreign chemical analyses and test provisions and the requirements for disinterested inspectors for cylinders having a marked service pressure over 900 p.s.i.g. The question of the need for chemical analyses and testing to be performed in the United States is the subject of Docket No. HM-74 (36 F.R. 11224). The manner in which the question is resolved in that docket will have a bearing on the final wording of this section. Meanwhile, the rule is being published consistent with existing requirements. Based on visits to plant sites by Department personnel, the Board believes there is need for disinterested inspectors. There appears to be a higher degree of compliance with Department of Transportation specification requirements where disinterested inspection is required or used even though not required. Therefore, on the basis of available information, the Board is not authorizing the use of interested inspectors for cylinders having a marked service pressure over 900 p.s.i.g.

Two commenters objected to §178.654(c) (1), stating that complete internal and external inspection, particularly if by a disinterested inspector, would present extreme difficulty if high capacity production lines were used. The Board agrees with this observation, and for this inspection only, is authorizing use of an interested inspector. This permits use of qualified production personnel under the overall supervision of assigned inspectors.

Some commenters noted that the steel material specifications would be improved in the interest of safety by a modification of §178.65-5(a) to guard against age hardening and loss of ductility in a completed cylinder. The Board acknowledges the correctness of this observation and has amended the rule accordingly.

Objection was raised to the pressure limitation proposed to be placed on cylinders fabricated from aluminum. The Department's knowledge of experience with high pressure aluminum cylinders is very limited. At the present time these cylinders have to be considered experimental. The Board does not consider it appropriate to authorize aluminum cylinders of higher service pressure.

It was pointed out by several commenters that the carbon content of steel should be increased to provide for the use of seamless steel tubing. The Board agrees with this comment and has so provided.

One commenter objected to §178.65-6(b) (7), which states that welded joints must have a strength equal to or greater than the minimum strength of the shell material in the completed cylinder. This requirement is consistent with §178.65-11(b) (2), which requires that the entire lot must be rejected if a failure during testing initiates in a weld or its heat affected zone. The objective of the rule is to assure that the weld area is not the weakest point of a completed cylinder. The Board concludes that the rule, as it was proposed, is valid but has modified it to clearly express this objective.

Two commenters objected to the requirement that openings be permitted in heads only. The Board is of the opinion that side openings create greater stress raisers than end openings. The stresses present in a side would be higher than in an ellipsoidal, spherical or flat head. Another commenter objected to the limitation that the diameter of a head opening not exceed 80 percent of the outside diameter of the cylinder. Requests were also made to authorize welded steel tubing in cylinder fabrication and to authorize additional aluminum alloys. The Board has not had the opportunity to fully evaluate these suggested designs and materials and does not have sufficient knowledge of experience in their use. Therefore, it considers these designs and fabrication to be experimental, and will deal with them on an individual rule-making basis. However, in this amendment, the Board is authorizing use of longitudinal or helical welded cylinders up to a 500 p.s.i.g. service pressure on the basis of existing cylinder specifications.

An objection was received on the lot size to be used for testing, suggesting a larger lot size. In a performance oriented specification adequate test requirements are paramount. The Board is concerned that testing be meaningful. To positively relate various tests to the end product, tests must be conducted at a frequency to assure that the unit tested is representative. No data have been presented to the Board to support the contention that a larger lot size could be used with assurance that there would be adequate monitoring of production for this type of packaging.

Objection was raised regarding the severity of the flattening test when compared to stationary vessel standards. The Board recognizes that the flattening test may be more restrictive, but these vessels will be exposed under pressure to a varied transportation environment and greater abuse. The test proposed is similar to testing requirements for other Department of Transportation cylinder specifications.

A commenter objected to the proposed wording of §178.65-12(a) (2) regarding the inclusion of the weld in the crush test. He observed that the metal adjacent to the weld could be annealed by the welding process and thus would not be representative of the base metal. Since the intent of the crush test is to assure that the base metal remains ductile in the finished condition, the Board agrees with this observation and the rule has been modified accordingly.

A question was raised concerning the need for marking both service and test pressure on the cylinder. Under the concept of this specification, the test pressure is specifically related to the maximum pressure of the contents at 130° F., and the filler needs this information. Therefore, the marking requirement is being retained as proposed.

One commenter considered the statement required by §178.65-14(b) (8) to be too lengthy. The Board agrees that the statement is longer than any similar statement previously required, but believes that by the emphasis placed on the penalties prescribed by law, the statement will serve as a better deterrent to refilling. The proposed statement is a prime requisite to the safe applicability of this specification in transportation and therefore is being retained.

The Board received several comments on various aspects of the use and disposition of this specification cylinder, and while many of these appeared to have merit, they did not sufficiently relate to transportation safety to be an appropriate matter for the Board's consideration.

(36 F.R. 16579, August 24, 1971)

SUMMARY · **Docket No. HM-70** · *Established conditions for special permits which had been authorizing transportation of hydrogen sulfide in tank cars and tank trucks.*

[49 CFR Parts 170–189]
[Docket No. HM-70]
TRANSPORTATION OF HAZARDOUS MATERIALS
Hydrogen Sulfide Gas in Cargo Tank Trucks and Tank Cars

The Hazardous Materials Regulations Board of the Department of Transportation is proposing to take action with regard to the transport of hydrogen sulfide gas (H_2S) in cargo tank trucks and tank cars.

Hydrogen sulfide is presently classified under §172.5 of the Hazardous Materials Regulations as a flammable gas, and is authorized for transport in certain compressed gas cylinders and DOT spec. 106A800X tanks. Transport of hydrogen sulfide in spec. MC 330 and MC 331 tank trucks and DOT spec. 105A600W tank cars is also authorized under certain special permits. Additional petitions for special permit authorization for such transport of large quantity shipments of hydrogen sulfide have been received by the Board.

Hydrogen sulfide, though presently classified as a flammable gas, is also a severe irritant and is highly toxic. This gas has the further insidious property of causing olfactory fatigue, and dangerous concentrations cannot be smelled after short exposure. The gas is heavier than air and thus tends to "pool" rather than to dissipate into the atmosphere to harmless concentrations. Obviously, the greater the quantity of this dangerous gas present, the greater will be the transportation hazard inherent in its movement, and the greater the number of injuries or fatalities suffered in case of accidental release of the gas.

It is the conclusion of the Board that the public interest precludes amendment of the regulations to provide a general authorization for transport of hydrogen sulfide in cargo tank trucks or tank cars. The Board further concludes that this limitation on authorization should not be weakened by issuance of additional special permits for such transport. The Board does not consider it appropriate to deny new petitions for cargo tank truck and tank car shipments of hydrogen sulfide while continuing the existence of those special permits already issued. For this reason, the Board is hereby proposing to rescind such existing permits.

Interested persons are invited to comment on the Board's proposed action in this rescission. The Board is also requesting advice regarding alternative large quantity packaging which would provide the public with a margin of safety equivalent to that offered by the DOT spec. 106A800X tank. Comments, identifying the docket number, should be submitted in duplicate to the Secretary, Hazardous Materials Regulations Board, Department of Transportation, 400 Sixth Street SW., Washington, D.C. 20590. Communications received on or before February 10, 1971 will be considered before final action is taken on this notice. All comments received will be available for examination by interested persons at the Office of the Secretary, Hazardous Materials Regulations Board, both before and after the closing date for comments.

(35 F.R. 18919, December 12, 1970)

[49 CFR Parts 170–189]
[Docket No. HM-70]
TRANSPORTATION OF HAZARDOUS MATERIALS
Hydrogen Sulfide Gas in Cargo Tank Trucks and Tank Cars;
Notice of Board Action

On December 12, 1970, the Hazardous Materials Regulations Board published a notice of proposed action, Docket No. HM-70 (35 F.R. 18919), with respect to the transportation of hydrogen sulfide gas in cargo tank trucks and tank cars.

In that notice, the Board expressed its view that the tank trucks and tank cars authorized by DOT special permits for bulk shipment of hydrogen sulfide did not provide a margin of transportation safety equivalent to the container which is prescribed by the regulations for that commodity, viz, the DOT Specification 106A800X tank. Based upon this view, the Board proposed rescission of those special permits and requested public comment on the proposal, including a specific request for alternative methods by which to ship bulk quantities of hydrogen sulfide.

Many comments were received on the proposal and two meetings were held with the permit holders and other interested persons. All comments and a summary of the meetings are available for inspection and copying in the public files of the Secretary of the Hazardous Materials Regulations Board.

The Board has concluded, on the basis of the comments received and technical reevaluation of the containers in question, that the permits for shipment of hydrogen sulfide in the DOT Specification 105A600W tank car should be continued as originally written, and that the permits for shipment of the gas in DOT Specification MC 331 tank trucks may be continued with significant revision of the provisions of those permits. The authorization for use of MC 330 tank trucks will be terminated October 1, 1971. The bases for the Board's conclusions follow:

The 105A600W tank car actually in use under permit, while basically conforming to the DOT specifications for that car, exceeds the requirements of that specification in several respects. It is constructed to an Association of American Railroads' standard especially designed for hydrogen sulfide. In addition, the car actually in use has a minimum tank shell thickness of 1-1/8 inches, which is thicker than the AAR specification. Other factors required by the AAR hydrogen sulfide car specification and considered by the Board to be particularly significant are the insulation of the car, the controlled hardness of the steel used in the tank, the two semiannual and then annual visual inspections of the inside of the tank, the good low-temperature impact properties of the tank material, and the quality control exercised during construction of the tank.

To elaborate, the Board believes that insulation of the tank shell provides a safeguard against heat transfer to the tank in an accident situation. This aids in keeping the commodity at a temperature low enough to preclude or significantly delay undesired relief valve venting or possible tank rupture due to increased vapor pressure, as could occur in an unprotected, uninsulated tank exposed to intense heat. The hardness of the steel of the shell is carefully controlled in construction of the tank in order to provide ductility and to avoid the phenomenon of hydrogen embrittlement, more frequently encountered in harder steels. All tank steel is normalized, and the tank is subjected to post-weld heat treating to assure evenness of the stresses present in the steel. More frequent inspections of the interior of the completed tank are required in order to detect any impairment of the integrity of the tank. Quality control during construction of the tank requires that all specimens tested meet the standard of the specification and not just the average of those tested, as is usually the case. The puncture resistance of the tank is enhanced by the insulation, the increased tank shell thickness, and the excellent impact-resistance properties of the tank material. The Department is unaware of any instance of puncture of the DOT Specification 105A600W tank car, and no failure of these specially built cars has occurred, despite exposure to an accident environment.

The DOT Specification MC 331 tank trucks presently used in hydrogen sulfide service also have a greater tank thickness than required by the specification and consequently have a higher design pressure, 500 p.s.i. versus 460 p.s.i. required by the specification. This gives a test pressure comparable to that of the DOT Specification 106A800X tank. Like the tank car, only normalized steel is used in the cargo tank and the tank is subjected to postweld heat treatment. A hydrogen probe is installed to detect conditions that would lead to hydrogen embrittlement of the tank steel.

The MC 331 cargo tanks already authorized are being continued in use, but must be retrofitted with insulation comparable to the tank car by January 1, 1973. In addition, the permits have been revised to include the following new immediate requirements:

1. Insulation on all new tanks.
2. Two drivers on each trip.
3. Vehicles must be equipped with a self-contained breathing apparatus for each driver.
4. Compliance with revised Part 397 of the Motor Carrier Safety Regulations (36 F.R. 4874 and 9780), as that part pertains to the transportation of Class A explosives. This includes requirements relating, among other things, to attendance and surveillance of motor vehicles, parking locations, selection of routes, periodic tire inspection, and the documents and instructions to be given to the drivers.
5. A speed limit of 55 m.p.h. or that posted, whichever is lower.

The Board is of the opinion that these additional requirements will serve to provide the public with a margin of safety comparable to that of the tank car and the Specification 106A800X tank.

The tank wall thickness of the two Specification MC 330 cargo tanks authorized under special permits is the minimum prescribed by that tank specification for the pressure involved. In addition, new construction of cargo tanks under this specification is no longer authorized. It is the Board's position that these tanks would present an unacceptable hazard potential over a long period of time. Therefore, these MC 330 cargo tanks will be taken out of hydrogen sulfide service after October 1, 1971.

One commenter supported the proposed rescission of the special permits, noting as did the Board that hydrogen sulfide is a commodity of extreme hazard, and that great danger would be involved in the unintentional release of a bulk quantity of the gas. After weighing this aspect of the proposal with particular care, giving full consideration to the actual hazard potential, the Board has concluded, in this instance, that the greater number of smaller containers that would be necessary to move the same volume of hydrogen sulfide would provide a greater hazard to the public than a lesser number of carefully controlled bulk shipments in these specially designed cargo tank trucks and tank cars. With bulk shipments, the number of containers exposed to the environment and to the public is less, the number of loading-unloading operations which are so frequently dangerous is less, the number of valves which could leak is less, and the likelihood of accident of abuse leading to a failure of the container is less because there are fewer individual containers involved. Thus, given the need to move hydrogen sulfide in interstate commerce, the Board is continuing authorization to ship it under the controlled conditions of special permits.

This action is taken under the authority of 18 U.S.C. 831-835 and section 9 of the Department of Transportation Act (49 U.S.C. 1657).

(36 F.R. 13792, July 24, 1971)

SUMMARY • **Docket No. HM-73** • *Provided for a more efficient approval system between DOT and AEC for radioactive materials packages.*

[49 CFR Part 173]
[Docket No. HM-73; Notice 71–1]
TRANSPORTATION OF HAZARDOUS MATERIALS
Design Approvals for Radioactive Materials Packages

The Hazardous Materials Regulations Board is considering amending the Department's Hazardous Materials Regulations to transfer the administrative requirements for approvals of radioactive materials packages from the Department to the U.S. Atomic Energy Commission (USAEC).

At the present time, the Department issues special permits for radioactive materials packages which meet the Department's performance standards but which are not among the limited number of specification packagings prescribed for radioactive materials in the regulations. In most instances, the USAEC performs detailed safety evaluations to determine whether a petitioner's package design does in fact meet the Department's performance standards. As a result, the Department's special permits are often only duplicative paperwork incorporating the USAEC approval by reference. The Board has concluded that it would be in the public interest to eliminate this duplicative ministerial requirement. This would reduce the administrative burden on both the nuclear industry and the Department, without adversely affecting the safe transport of radioactive materials. The USAEC has agreed to this procedure.

Under this change in procedure, special permits would only be issued by the Department in those cases where it is necessary and appropriate to provide an exemption or waiver of the regulations. Petitioners for routine package approvals would apply directly to the USAEC for package review, evaluation, and approval. The procedures for obtaining USAEC approval are contained in the regulations of the USAEC (10 CFR Part 71) for USAEC licensees, and in USAEC Manual Chapter 0529 for USAEC prime operating contractors. Specific procedures for USAEC approval of Type B packages are being developed at this time. Provisions would be made for continuation of the existing special permits until their date of expiration as well as for establishment of a package numbering identification system. The Office of Hazardous Materials of the Department would continue to issue certificates of competent authority under IAEA regulations.

The Board believes that this proposal would effectively reduce delays in granting package approval without adversely affecting safety in transportation.

(36 F.R. 292, January 8, 1971)

[49 CFR Parts 173, 174, 175, 177]
[Docket No. HM-73; Notice No. 71–30]
TRANSPORTATION OF HAZARDOUS MATERIALS
Design Approvals for Radioactive Materials Packages

On January 8, 1971, the Hazardous Materials Regulations Board published Docket HM-73; Notice 71–1 (36 F.R. 292), proposing to transfer the administrative requirements for the approvals of certain radioactive materials packages from the Department of Transportation to the U.S. Atomic Energy Commission (USAEC). That notice further stated that it would be in the public interest and would not adversely affect safety in transportation to eliminate the present duplicative ministerial procedure of the issuance of special permits for packages which have been reviewed and approved by the U.S. Atomic Energy Commission.

Several comments were received by the Board, including those of the USAEC. The majority of the comments expressed support for the proposal. Several commenters qualified that support, requesting that additional information regarding concurrent changes be made to the USAEC regulations in conjunction with the DOT proposal.

One commenter indicated an objection to the proposal and questioned the advisability and need for changing the present division of responsibility between the USAEC and DOT. The commenter further stated that the proposal appears to be inconsistent with the intent of the Congress in amending the Transportation of Explosives Act of 1960. The amendment placed responsibility for regulating the transport of radioactive materials with the ICC. (That responsibility has since been transferred to the Department.) The Board has concluded that the proposal is not inconsistent with that amendment. Only the ministerial system for issuance of certain package design approvals is involved in the proposal. The Department would continue to be responsible for its regulatory role as directed by the Transportation of Explosives Act.

After further discussion with the USAEC, it has become apparent that several additional sections of the Hazardous Materials Regulations will be needed to fully implement a transition from the present system of "special permit" issuances to the "AEC approvals." The related requirements of the International Atomic Energy Agency's (IAEA) regulations concerning competent authority certifications of package design also need clarification.

Under the regulations proposed in this second notice, petitioners for TYPE B, Fissile, and large quantity packages would apply directly to the USAEC for package review, evaluation, and approval. The standards and requirements which must be met to obtain AEC approval are published in the USAEC Regulations (10 CFR Part 71). In a separate document on page 22184 of this issue of the FEDERAL REGISTER, the AEC is publishing a notice of proposed rule making to add procedures for review and approval of Type B packages to 10 CFR Part 71.

This proposal would require non-AEC licensees, including licensees of agreement states to which the AEC has transferred certain regulatory authority over radioactive material by formal agreement, and radium shippers in nonagreement states, to apply to the AEC for approval of Type B, fissile material, and large quantity packages. Non-AEC licensees, other than license-exempt contractors of the AEC, would apply to the AEC in the same manner as AEC licensees. These persons would be required to submit an application containing the information set forth in 10 CFR 71.21, 71.22, 71.23, and 71.24, to the Director, Division of Materials Licensing, U.S. Atomic Energy Commission, Washington, D.C. 20545. License-exempt AEC contractors would apply to the appropriate AEC field office.

The DOT requirements for the AEC approval proposed in this notice would, therefore, be satisfied by:

1. A lincense or linse amendment issued by the AEC under 10 CFR Part 71; or
2. an administrative approval issued by the AEC's Division of Materials Licensing; or
3. an administrative approval issued to AEC license-exempt contractors by AEC Field Offices under AEC Manual Chapter 0529.

With the adoption of these proposed amendments, a fissile class III shipment would no longer require a special permit by regulation. However, the Board is concerned that the absence of special permits for fissile class III shipments may contribute or create an inability to identify these shipments by carrier personnel who must handle and stow them. A normal means of control by carriers of radioactive materials packages is through the transport index entry on the package label. Since fissile class III packages require special stowage controls based on the number of packages to be transported together, and since the package label does not convey this information, the Board believes that the additional warning statement proposed in §173.427(a) must be added to the shipping papers for fissile class III shipments, to warn the carrier that he is to exercise special stowage controls on that shipment. This statement would therefore effectively replace the language formerly conveyed by certain provisions of the special permit.

This notice also contains a proposal to change the section of the regulations applicable to transport of fissile class III radioactive materials by aircraft. The proposed changes to §173.396(g) are in response to a recent petition by the Air Transport Association of America. The petition stated that the present wording of §173.396(g) (2) is impracticable for transportation of fissile class III shipments by air, and suggested that a new §173.396(g) be added which would be applicable to fissile class III shipments by air. The Board agrees with the Association's petition and accordingly has included proposed changes in the appropriate sections of the shipper and air carrier sections for fissile class III shipments.

In 14 CFR Part 103, the Board is also proposing to limit the transport of radioactive materials packages on passenger-carrying aircraft to those packages containing less than a large quantity of radioactive material (§173.389(b)), except as specifically approved on an individual shipment basis. This proposal is consistent with the present limitation appearing in the International Atomic Energy Agency's Regulations, the IATA Restricted Articles Regulations, and in the Air Tariff No. 6D.

Notice 71-1 is superseded by this second notice of proposed rule making.

(36 F.R. 22181, November 20, 1971)

[14 CFR Part 103]
[Docket No. 11558; Notice No. 71-39]
TRANSPORTATION OF HAZARDOUS MATERIALS
Design Approvals for Radioactive Materials Packages

On January 8, 1971, the Hazardous Materials Regulations Board published Docket HM-73; Notice 71-1 (36 F.R. 292), proposing to transfer the administrative requirements for the approvals of certain radioactive materials packages from the Department of Transportation to the U.S. Atomic Energy Commission (USAEC). That notice further stated that it would be in the public interest and would not adversely affect safety in transportation to eliminate the present duplicative ministerial procedure of the issuance of special permits for packages which have been reviewed and approved by the U.S. Atomic Energy Commission.

As a result of the comments received and the discussion on page 22181 of this issue of the FEDERAL REGISTER, the Hazardous Materials Regulations Board is publishing a second notice of proposed rule making in Docket No.

HM-73 incorporating several additional proposed changes needed to fully implement a transition from the present system of "special permit" issuances to the "AEC approvals".

(36 F.R. 22181, November 20, 1971)

[Docket No. HM-73; Amdts. 171–17, 173–69, 174–16, 175–9, 177–23]
DESIGN APPROVALS FOR
RADIOACTIVE MATERIALS PACKAGES

The purpose of these amendments to the Hazardous Materials Regulations is to eliminate the present duplicative procedure for the issuance of special permits by DOT for type B, fissile, and large quantity radioactive materials packages which have been reviewed and approved by the U.S. Atomic Energy Commission (USAEC). The present regulations for these materials requires that the packages be reviewed and approved by the USAEC, prior to the issuance of a special permit by the DOT. This amendment will eliminate this second step. These amendments also establish and clarify the Department's procedures for issuance of competent authority certificates with respect to the regulations for the safe transport of radioactive materials, as established by the International Atomic Energy Agency (IAEA), hereinafter referred to as the IAEA Regulations. Essentially the IAEA Regulations provide for certain administrative approvals of radioactive materials package designs by the "competent national authority" of countries when the export-import of such packages is involved. These approvals are known as competent authority certificates. The IAEA regulations also prescribe the information required to be submitted to the competent authority by petitioners, as well as the content of the certificates, and the requirements for exchange of such certificates between the countries involved.

On January 8, 1971, the Hazardous Materials Regulations Board published a notice of proposed rule making, Docket No. HM-73; notice No. 71-1 (36 FR 292), which initially proposed these amendments. After further consideration, it became apparent that several additional sections of the regulations needed clarification in order to fully implement a transition from the "Special Permit" issuances for radioactive materials to the "AEC approvals," as well as to describe the related requirements pursuant to IAEA competent authority certificate issuances. Accordingly, notice No. 71-1 was superseded by a second notice of proposed rule making in the same docket, HM-73 published in the FEDERAL REGISTER on November 20, 1971, as notice No. 71-30 (36 FR 22181). Concurrent proposed revisions to Title 10 CFR Part 71 were also published by the U.S. Atomic Energy Commission on the same date. Interested persons were invited to give their views and several comments were received by the Board. These comments were generally in support of the proposed amendments as were those on the original notice.

One commenter stated that the proposed package approval procedures would lead to a duplication of effort and diverse interpretation of requirements by establishing various review groups within the USAEC. In fact, more than one package review and approval group already exists within the USAEC and these amendments recognize the approvals of any one of those USAEC groups. These amendments, however, reduce an existing duplication of effort by eliminating the requirements for issuance of a special permit by the Department. Amendments to the USAEC regulations (10 CFR Part 71) to be published soon in the FEDERAL REGISTER as well as changes to be made in that Commission's own operating rules (USAEC manual), will describe the implementation by the USAEC of the package approvals by the appropriate USAEC review group.

In these amendments, several minor changes have been made regarding the revalidation and furnishing of the IAEA competent authority certificates and the shipping paper information for export shipments of radioactive materials packages. As proposed in §§173.393a(a) (4) and 173.393(a) (4), each shipper of a package described by an IAEA certificate would have been required to furnish a copy of the certificate to the competent authority of each country prior to the first export shipment. After reconsideration, the Board has modified this requirement to the effect that the shipper simply must insure that the proper certificate has been furnished, i.e., if the applicable certificate has already been supplied by another shipper, it will not be necessary for an additional copy to be furnished. Also, a requirement for the competent authority certificate to accompany each shipment of fissile class III (§173.393b(a) (5)) is redundant and inconsistent especially in view of the requirement in §173.427(a) (5) (v) for certain specific shipping paper information on each fissile class III shipment. Accordingly, §173.393(a) (5) has been modified to require that shippers shall furnish competent authority certificates to carriers upon request. Concurrent with this, the shipping paper requirement in §173.427(a) (5) (vi) is deleted.

A further change has been made with respect to the requirement for revalidation of foreign made and certified packages. As proposed in each earlier notice, the revalidation of foreign-made packages which had been certified by a foreign competent authority, and which contained less than a large quantity of nonfissile radioactive material, would not have been required by the U.S. competent authority. The principal requirement would have been for the furnishing of the applicable foreign certificate to the Department and a registration by each shipper using such certificate. Upon further reconsideration, the Board feels that it would be desirable to maintain the degree of administrative control over foreign made and certified packages at the present level. Accordingly, a modification has been made to §173.393b(a) to require that the certificate for all foreign-made packages be revalidated by the U.S. competent authority.

Several comments were received with respect to the proposed changes to 14 CFR Part 103. One commenter recommended that mention of required "Special arrangements" between shipper and carrier as made in proposed §103.24(a) (1) should also be made in §103.24(a) (2). The Board agrees and has amended §103.24(a) (2), accordingly. Another commenter objected to the proposed prohibition of large quantities (also referred to as large radioactive sources) for carriage on passenger-carrying aircraft. The commenter noted that protection to the public is afforded by package integrity, and that a properly packaged large quantity shipment should present no more hazard than a properly packaged type A or B package. Although this comment has some merit in principle, the Board is of the opinion that the large radioactive source threshold is a practical and logical maximum delineating point for establishment of a quantity limit on passenger-carrying aircraft. It should be also noted in support of this opinion, that in the regulations of both IAEA and the International Air Transport Association (IATA), as well as in the domestic air tariff No. 6D provisions, the routine carriage of large sources of radioactivity on passenger-carrying aircraft is not permitted. Therefore, 14 CFR 103.19(d) has been added to impose this same requirement in the Federal Aviation Regulations.

Upon the effective date of these amendments, the Department will cease the issuance of new or amended special permits for those packages which have become subject to the USAEC approval requirements of §173.393a. However, all existing DOT Special Permits will continue in effect until their stated expiration date. Prior to the expiration of any permit containing an IAEA certification clause, this clause will have to be replaced by a separate IAEA certification to be issued by the Department pursuant to §173.393b. All persons concerned should submit petitions for such replacement certificates at least 45 days prior to the expiration date of the special permit.

(38 F.R. 4396, February 14, 1973)

SUMMARY • **Docket No. HM-74** • *Proposed a system to provide for the authorization to use cylinders for which testing and chemical analysis have been conducted outside the United States.*

[Docket No. HM-74]
NOTICE OF PUBLIC HEARING
DOT Specifications 3A, 3AA, and 39 Cylinders Manufactured
Outside United States

Section 173.301(i) of the Department's Hazardous Materials Regulations prohibits the transportation of charged compressed gas cylinders unless they have been made in accordance with applicable DOT specifications and unless the tests required by the specifications under which they are manufactured have been conducted within the United States. Sections 178.36 and 178.37 of Title 49, CFR, prescribe DOT specifications 3A and 3AA, respectively, and require chemical analyses and tests as specified in those sections to be made within the limits of the United States. A similar requirement appears in the proposed new specification 39 high-pressure disposable cylinder provisions, published December 11, 1970 (35 F.R. 18879).

These cylinders are used to transport various compressed gases, such as carbon dioxide, oxygen, helium, argon, and nitrogen for industrial and medical purposes. In addition, new National Highway Safety Bureau motor vehicle Standard No. 208 (35 F.R. 16927) specifies occupant crash protection requirements for certain motor vehicles manufactured on or after July 1, 1973, including those of foreign manufacture sold in the United States. One major type of passive restraint system contemplated employs a high pressure gas cylinder, which would also be subject to the requirement that analyses and tests be performed within the United States. In view of the desire to import foreign-made cylinders for industrial and medical gas service and the future difficulties which will evolve from passive restraint systems being incorporated into foreign manufactured automobiles, the Board is considering the necessity for continuing to require the prescribed chemical analyses and tests to be performed within the United States. The Board believes that public participation and comments would be particularly helpful and therefore it will conduct a public hearing on this matter at 10 a.m. on February 23, 1971, in Room 10430 Nassif Building, 400 Seventh Street SW., Washington, DC.

The Board will appreciate comments addressed to the following principle area of consideration: Are the regulatory controls imposed on manufacturers of specifications 3A, 3AA, and other cylinders made in the United States suitable for controlling manufacturers outside the United States? For example, if the Bureau of Explosives is capable of qualifying competent and disinterested inspectors for cylinder manufacturing operations in the United States, would it also have the capability to approve, in a satisfactory manner, competent and disinterested inspectors performing their duties outside the United States? Also, in what manner is safety enhanced by the requirement that tests and analyses be performed in the United States? Could the same inspector witness tests performed outside the United States?

The hearing will be an informal one conducted by the Board. It will not be a judicial or evidentiary type hearing. There will be no cross-examination of persons presenting statements. A representative of the Board will make an opening statement outlining the scope of the hearing. Statements should focus on the issue raised by this notice. After all initial statements have been completed, those persons who wish to make rebuttal statements will be given the opportunity to do so in the same order in which they made their initial statements. Additional procedures for the conduct of the hearing will be announced at the hearing.

Interested persons are invited to attend the hearing and present oral or written statements on the matter set for hearing. These statements will be a matter of public record. Any person who wishes to make an oral statement at the hearing should notify the Secretary of the Hazardous Materials Regulations Board by February 17, 1971, stating the approximate amount of time required for his initial statement. The Board will also receive written comments until March 9, 1971.

All communications concerning the hearing should be addressed to the Secretary, Hazardous Materials Regulations Board, Department of Transportation, 400 Sixth Street SW., Washington, DC 20590.

This notice is issued under the authority of sections 831–835 of title 18, United States Code, section 9 of the Department of Transportation Act (49 U.S.C. 1657).

(36 F.R. 838, January 19, 1971)

[49 CFR Parts 173, 178]
[Docket No. HM-74; Notice No. 71–16]
TRANSPORTATION OF HAZARDOUS MATERIALS
Cylinders Manufactured Outside United States

The Hazardous Materials Regulations Board is considering amendment of Parts 173 and 178 of the Department's Hazardous Materials Regulations to authorize the performance, outside the limits of the United States, of chemical analyses and tests prescribed for DOT specification compressed gas cylinders, under conditions approved by the Department. In addition, the Board is proposing to require Departmental approval of disinterested inspectors and inspection procedures prescribed for all DOT specification cylinders, whether they are made inside or outside the United States.

This proposal is based on petitions from foreign compressed gas cylinder manufacturers, received over a period of several years, requesting relief from the provisions of the regulations requiring specified chemical analyses and tests to be performed within the United States (see, for example, 49 CFR 178.36-3). In response to these petitions and in an endeavor to gather information on the necessity for continuing to require the prescribed analyses and tests to be performed within the United States, the Board sought public participation in its publication of a notice of public hearing (Docket No. HM-74, 36 F.R. 838, 35 F.R. 3836), which was held on February 23 and March 16, 1971. An additional item prompting resolution of this question was the publication by the National Highway Safety Bureau (now the National Highway Traffic Safety Administration) of motor vehicle Standard No. 208 (35 F.R. 16927), specifying occupant crash protection requirements for certain motor vehicles manufactured after July 1, 1973, including those of foreign manufacture sold in the United States. One major type of passive restraint system contemplated employs a high pressure gas cylinder, which would be subject to the requirement that chemical analyses and tests be performed within the United States.

The record of the hearing, available for inspection in the public file of the Secretary of the Board, confirms the need for greater flexibility in the regulations for those foreign manufacturers who can assure the Department of their competence and ability to produce compressed gas cylinders meeting U.S. safety standards. Questions raised in the hearing regarding the need for more effective approval and inspection procedures for domestic production of cylinders have been carefully noted, and will be treated in later rule making action. The Board is proposing, however, to withdraw the authority presently vested in the Bureau of Explosives to approve inspectors in the United States and would place within the Department the authority for approval of both domestic and foreign inspectors.

It is the Board's conclusion, on the basis of its investigations and the public record, that approval to perform specified chemical analyses and tests outside the United States may be granted to foreign manufacturers upon favorable consideration of several matters, including the acceptance of quality of production materials, manufacturing procedures, testing methods, inspection methods, and the inspectors. In addition, each foreign manufacturer requesting approval would be required to specify an agent, domiciled within the United States, upon whom service of process effectively could be made.

(36 F.R. 11224, June 10, 1971)

[49 CFR Parts 173, 178]
[Docket No. HM-74; Notice No. 71-16]
TRANSPORTATION OF HAZARDOUS MATERIALS
Cylinders Manufactured Outside United States;
Reopening for Additional Comments

On June 10, 1971, the Hazardous Materials Regulations Board published Docket No. HM-74; Notice No. 71-16 (36 F.R. 11224), Cylinders Manufactured Outside the United States. In response to a petition, the Board extended the period for comments on this notice to November 1, 1971 (36 F.R. 13793).

Upon review of the comments filed, the Board is of the opinion that additional information would be beneficial in assisting it to arrive at a proper determination in this matter. Accordingly, the comment period is reopened for the purpose of receiving comments on the following questions:

1. Are changes to the specifications for cylinders in Part 178 of the Department's Hazardous Materials Regulations required to insure the safe quality of cylinders transported in the United States regardless of their origin? If so, what specific changes are required?

2. In the June 10, 1971 notice, it was proposed that the Board withdraw the authority presently vested in the Bureau of Explosives to approve inspectors in the United States and place within the Department the sole authority to approve both domestic and foreign inspectors. In light of that proposal, what specific qualifications and requirements should cylinder inspectors be required to meet in order to obtain and maintain Departmental approval?

To contribute to the information that will be available to the Board regarding these questions, representatives of the Office of Hazardous Materials will continue their study and examination of the specifications and inspection requirements. A full report of the examinations and the observations of Office of Hazardous Materials representatives will be placed in the docket.

Interested persons are invited to give their views on the questions posed in this notice. Communications should identify the docket number and be submitted in duplicate to the Secretary, Hazardous Materials Regulations Board, Department of Transportation, 400 Sixth Street SW., Washington DC 20590. Communications received on or before June 1, 1972, will be considered before final action is taken on the proposal. All comments received will be available for examination by interested persons at the Office of the Secretary, Hazardous Materials Regulations Board, both before and after the closing date for comments.

This proposal is made under the authority of sections 831-835 of title 18, United States Code, section 9, Department of Transportation Act (49 U.S.C. 1657), and title VI and section 902(h) of the Federal Aviation Act of 1958 (49 U.S.C. 1421-1430 and 1472(h)).

(37 F.R. 2588, February 3, 1972)

[49 CFR Parts 173, 178]
[Docket No. HM-74; Notice 76-1]
CYLINDERS MANUFACTURED OUTSIDE UNITED STATES
Revised Notice of Proposed Rulemaking

This docket was established in January 1971, when the Hazardous Materials Regulations Board published a notice of public hearing (36 FR 838) on the subject of DOT Specifications 3A, 3AA, and 39 cylinders manufactured outside the United States. The hearing convened on February 23, 1971, and because of the amount of public interest shown, a second session was scheduled and the public comment period was extended (36 FR 3836). On June 10, 1971, the Board proposed a series of amendments to authorize chemical analyses and tests for DOT specification compressed gas cylinders to be performed outside the United States under conditions to be approved on a case-by-case basis by the Department of Transportation, and to require Departmental approval of all disinterested inspectors of DOT specification cylinders, whether manufactured within or outside the United States (36 FR 112). The close of the public comment period on that proposal, which was first set for September 14, 1971, was extended to November 1, 1971 (36 FR 13793). The Board later reopened the Docket for the purpose of receiving additional comments on possible changes to the existing cylinder specifications and recommendation as to the qualification standards for cylinder inspectors (37 FR 2588). In connection with seeking this additional information, the Board announced that it would continue its study and examination of the specifications and inspection requirements. The Docket subsequently received comments in response to that request, as well as supplemental staff materials bearing on the subject. At the time the Hazardous Materials Regulations Board was abolished on July 7, 1975, and authority to issue regulations transferred to the newly created Materials Transportation Bureau (40 FR 30821), the Board had not taken any further action on the proposal.

The Bureau believes to be sound the premises underlying the Board's proposal of June 10, 1971 (36 FR 3836), to make all "disinterested" (i.e., independent) inspectors of cylinders directly subject to a DOT-administered approval system and to then recognize their performance of cylinder inspections and verifications as required by DOT regulations without regard to geographic limitations. However, as indicated by its February 1972, decision to continue its study of the matter, the Board was of the view that the specifics of its proposal could afford a degree of refinement. It was also conscious of questions raised during the course of the proceeding as to the quality of the existing domestic system of inspections, particularly with regard to so-called low pressure cylinders as carried out through "interested" inspectors.

As the Board's successor in this effort, the Bureau is now prepared to receive public comments on a modified version of a series of amendments which would implement the basic changes earlier proposed by the Board and provide for the independent inspection of low pressure cylinders. Specifically, the modified procedures provide for the following:

1. The term "independent inspection agency" is used to describe what is now covered by the term "competent and disinterested inspector." Except for persons currently holding Bureau of Explosives' approval as "competent and disinterested inspectors", applicants (foreign or domestic) would be evaluated and, if qualified, approved by DOT. Provision is made for converting Bureau of Explosives approved holders. (Proposed § 173.300a.)

2. Foreign cylinder manufacturers and U.S. manufacturers who may manufacture cylinders outside the United States could apply to DOT for authorization to have their cylinders inspected and tests verified outside the United States by an approved independent inspection agency. (Proposed § 173.300b.)

3. To ensure continuing DOT jurisdiction over approvals issued to non-U.S. domiciled inspection agencies and manufacturers, it would be necessary for them to establish a U.S. domiciled agent. (Proposed § 173.300c.)

4. Commencing on the effective date of the proposed amendments, cylinders manufactured, inspected and tested outside the United States in accordance with all the appropriate requirements of the cylinder regulations (49 CFR, Part 178) would be acceptable for use in transportation within the United States. (Proposed § 173.301(i)). Foreign manufactured cylinders which do not meet all of the requirements would only be acceptable for export. (Proposed amendment to § 173.300(j)).

5. The inspection provisions of each DOT cylinder specification would be revised to reflect the change from "disinterested inspector" to "independent inspection agency" and to recognize the method whereby approval could be obtained for inspections to be performed and chemical analyses to be verified outside the United States by approved independent inspection agencies. In addition, those specifications which now authorize inspection by "interested" inspectors would be changed to require all inspection and test verrifications (both within and outside the United States) to be performed by approved independent inspection agencies. (Proposed amendments to §§ 178.36–3(a) through 178.68–3(a).)

6. In the case of those cylinders domestically manufactured in accordance with DOT specifications in which authorization for interested inspectors is to be dropped, the manufacturers could continue to use interested inspectors until January 1, 1978.

7. The Bureau's out-of-pocket costs for any inspections or tests it must perform in approving an inspection or manufacturing facility located outside the United States would be borne by the applicant.

The proposed effective date is May 1, 1976.

Interested persons are invited to submit views and comments on this revised proposal. Comments should refer to the docket number and be submitted to: Docket Section, Materials Transportation Bureau, U.S. Department of Transportation, Trans Point Building, Washington, D.C. 20590. All comments received before the close of business on March 15, 1976, will be considered, and will be available in the docket for examination both before and after the closing date. Comments received after the closing date and too late for consideration will be treated as suggestions for future rulemaking.

(41 F.R. 1919, January 3, 1976)

[Docket No. HM-74; Amdt. Nos. 173–97; 178–39]
PART 173—SHIPPERS
PART 178—SHIPPING CONTAINER SPECIFICATIONS
Inspection and Testing Requirements for Cylinders Manufactured Outside
the United States

This docket was opened on January 19, 1971, when the Hazardous Materials Regulations Board announced that it was considering the necessity for continuing the domestic analysis and test rule (36 FR 838) and that a public hearing had been scheduled for that purpose. The domestic analysis and test rule, found in the compressed gas cylinder specifications of Part 178, requires that an analysis of metal to be used in making a cylinder, as well as tests on the finished or partially finished product, be conducted within the United States, regardless of where that cylinder is manufactured. On June 10, 1971, following a two-day public hearing, the Board announced, based on information then available, that it had concluded that analyses and tests could be performed outside the United States under appropriately controlled manufacturing procedures. The Board at the same time also proposed amendments it believed would establish that control (36 FR 11224).

The 1971 proposals would have—

1. Required all disinterested inspectors to be approved by DOT rather than by the Bureau of Explosives of the American Association of Railroads, as is the current practice;

2. Required disinterested inspection of all foreign-made cylinders, while continuing to allow interested inspection of domestic-made low pressure cylinders (inspection by an employee of the cylinder manufacturer); and

3. Allowed, for the first time, analyses and tests to be made outside the United States, but only upon DOT manufacturing approval, and only in conjunction with DOT-approved disinterested inspection.

The docket remained open for public comment until November 1971. It was reopened February 3, 1972, to consider what additional changes to the cylinder specifications of Part 178, if any, might be necessary to the transportation safety of compressed gas cylinders. The February 1972 notice also sought comment on what specific qualifications and requirements cylinder inspectors should be required to meet before being approved by DOT. The docket remained open for comment until October 3, 1972.

On January 13, 1976 (41 FR 1919), after thorough consideration of the contents of the docket, a revised notice of proposed rulemaking was published, which essentially repeated the 1971 proposals. In addition, the revised notice also proposed—

1. A substitution of the term "independent inspection agency" for "disinterested inspector";

2. A specific process by which a person could apply for approval as an independent inspection agency, a similar process by which a manufacturer could apply for approval to conduct analyses and tests outside the United States, and the information necessary to support such applications (including designation of an agent for service of process for nonresident applicants);

3. The discontinuance of authority for domestic manufacturers of low pressure cylinders to use interested inspectors in favor of independent inspectors (a proposal which has since been severed from this docket and is presently being considered under Docket HM-74A, 41 FR 11179, March 17, 1976).

Well over 300 comments have been received on this rulemaking since it was first opened, about 30 of which have been received since publication of the revised notice early this year. Interest has been expressed by domestic cylinder users, domestic steel suppliers, and both foreign and domestic cylinder manufacturers, trade associations and inspection agencies. References herein are to comments received on the January 1976 revised notice. Those comments, however, are generally representative of comments on earlier docket publications.

THE DOMESTIC ANALYSIS AND TEST RULE

The domestic analysis and test rule dates to 1922 and was originally intended to protect American citizens against gas cylinders of uncertain pedigree. At a time preceding rapid transoceanic travel and communication, the necessity for the rule was clear.

The nature of that necessity has gradually altered. A substantial exchange of complex industrial and scientific information now occurs among Europe, the United States and elsewhere, and it is presently possible for the Department to perform an inspection at a foreign location almost as quickly as at a domestic location. The MTB believes it is practical to establish a properly supervised alternate method involving analysis and testing outside the United States, by which a foreign cylinder manufacturer can comply with the Department's gas cylinder regulations.

It was apparent early in this docket that some domestic users of compressed gas cylinders, as well as some foreign manufacturers, consider themselves unnecessarily burdened by the domestic analysis and test rule. To enter the American cylinder market, a foreign manufacturer must not only adjust his usual testing and manufacturing cycle to meet DOT requirements, he must also face additional costs and manufacturing delays resulting from the domestic analysis and test rule.

Some domestic cylinder users believe that price and supply in the domestic cylinder market reflect a lack of competition and attribute that condition to the rule, perceiving in it a non-tariff trade barrier that effectively prevents the entry of quality foreign-made cylinders. A representative of the Department of Justice Antitrust Division, in the March 16, 1971 public hearing which is part of this docket, observed similaritites between trade restraints intended to be remedied by an antitrust suit filed against the American Society of Mechanical Engineers and the claimed trade barrier effects of the domestic analysis and test rule. In a separate action as late as last year, the Justice Department obtained a consent decree effectively reversing the acquisition of Pressed Steel Tank Company by Norris Industries, the second and fourth largest producers of high pressure cylinders in the United States.

In light of a docket which extends back to 1971, the MTB has concluded that domestic analysis and testing are not any more conducive to safety than properly supervised analysis and testing occurring elsewhere. Moreover, the MTB recognizes the obvious difficulties that the domestic analysis and test rule imposes on foreign cylinder manufacturers and the possibility that those difficulties may be reflected in the domestic cylinder market. Continuance of the Department's reliance on the domestic analysis and test rule as the exclusive means by which foreign-made cylinders can be manufactured in compliance with safety regulations may be tantamount to regulating transportation safety by effectively prohibiting importation of most foreign-made cylinders without regard to quality.

The domestic analysis and test rule was never intended to prohibit the importation of foreign-made compressed gas cylinders but to insure that those imported are safe. The amendments are intended to provide a more reliable and economically less burdensome means of distinguishing between good and bad cylinders.

In defense of retaining the domestic analysis and test rule, the American Cylinder Manufacturers Committee (ACMC), commenting on other materials found in the docket, states that—

> [t]estimony ... which seeks to establish that the current safety regulations are a non-tariff trade barrier or provide the domestic cylinder manufacturers with a monopoly in the domestic cylinder market or limit the supply of cylinders available in this country is irrelevant to this proceeding and invalid.... [T]he only information which OHMO may consider in its evaluation of the issues raised by HM-74 is information relevant to the safety of compressed gas cylinders introduced into interstate commerce.

The ACMC is generally correct. The statutory responsibility of the Department is transportation safety. On that basis, the new amendments are an improvement over the existing regulations. The amendments are expected to increase the control and supervision exercised by DOT over foreign manufacturers, as well as over many domestic manufacturers. The amendments accomplish this by requiring all independent cylinder inspectors to be approved by DOT, by requiring that all foreign-made cylinders and domestic-made high pressure cylinders be subjected to independent inspection, and by requiring DOT manufacturing approval in any case where analyses and tests are to be performed outside the United States.

An additional consideration is the fact that retention of the domestic analysis and test rule, absent some justification in transportation safety, wrongly places the Department in the position of preemptively regulating an aspect of national economic policy and foreign trade which is properly addressed by Congress and other Federal agencies. In short, although the new amendments promise greater transportation safety, even if they did not, there would still remain a legitimate question of whether the existing regulations achieve safety in an efficient manner.

CONDITIONS OF FOREIGN CYLINDER MANUFACTURE

Many of the comments addressed to foreign manufacturers as a group, asserting that foreign manufacturers have in the past fallen short of meeting DOT specifications, do not now manufacture to DOT specifications, lack adequate testing and inspection procedures and have poor quality control. The conclusion apparently urged is that until all identifiable foreign manufacturers have been evaluated as part of that group, there is not any single manufacturer who can be said to be competent to manufacture gas cyclinders to DOT specifications.

An attempt to exhaustively evaluate all foreign manufacturers before approving any one of them would be wasteful and would produce results of questionable value. Comments from both foreign and domestic interests recognize that foreign cylinder manufacturers constitute a diverse group which unquestionably includes a great many concerns that will never seek entry into the U.S. cylinder market, as well as concerns that will not or cannot comply with DOT regulatory standards. The amendments are therefore structured to provide an individual evaluation of each foreign inspection agency and foreign manufacturer who seeks DOT approval.

Several other comments expressed the view that foreign cylinder manufacturers will have an unfair price advantage because of the availability of cheap labor, or because ineffective regulatory supervision will allow production of defective and thus less expensive cylinders than the quality product of a domestic manufacturer. Cheap labor, to the extent it does exist in countries sufficiently advanced technologically to manufacture cylinders, may indeed result in low manufacturing costs. Foreign producers, however, are also subject to a 5% or 7½% tariff, additional transportation costs, and DOT inspection costs that are not faced by their domestic counterparts. There exist outside the DOT appropriate means of dealing with unfair import competition.

With regard to the possibility of lax regulatory enforcement, it is the intent of the Department that regulatory compliance by foreign manufacturers will be as complete as compliance by domestic manufacturers.

REGULATION OF FOREIGN MANUFACTURERS AND INSPECTORS

A number of comments expressed the view that regulating foreign cylinder manufacturers and inspectors is difficult, expensive and beyond the capacity of DOT. One comment suggested that unannounced inspection of foreign manufacturers would be "impractical, if not impossible". DOT inspection of foreign facilities may in some cases be more difficult than inspection of domestic facilities, but it is practical and will be used in essentially the same fashion as it is used domestically. The amendments require the cost of foreign inspection by the Office of Hazardous Materials Operations to be borne by the manufacturer or inspection agency seeking DOT approval as a condition of that approval. The intention is to recover "out-of-pocket" costs to the United States Government for foreign inspections considered necessary to evaluate an approval application, or necessary to monitor an approval holder, but not to recover salary of OHMO personnel.

Another series of comments suggested that the regulations governing cylinder manufacture are so vague that only the domestic industry, with its record of safety, common regulatory experience and common language can be relied upon for comprehension and compliance. It is clear that some foreign manufacturers are capable of making cylinders to DOT specifications and that the regulatory provisions governing cylinder manufacture are capable of communication outside the United States. Differences between domestic and foreign manufacturers can be evaluated in the course of considering approval applications and monitoring approval holders.

Finally, a number of commenters addressed problems foreseen in making civil or criminal penalties effective against a foreign cylinder manufacturer or inspection agency, or collecting from him a tort judgment. A nonresident manufacturer who chooses to conduct analyses and tests outside the United States, or a nonresident inspection agency, must designate a domestic agent for service of process before DOT approval will be granted. Service on that agent will be sufficient for purposes of civil or criminal action under the Hazardous Materials Transportation Act of 1974 (Pub. L. 93–633, 49 U.S.C. 1801 et seq.) when the necessary implementation of the Act's relevant provisions is completed (see Docket HM-134, 41 FR 9188, March 3, 1976). Actual enforcement of any such action is in any event backed by withdrawal of Departmental approvals. In the case of a civil suit, the MTB recognizes that reaching assets located outside the United States may be more difficult than reaching domestic assets. The concern of the MTB in this matter is that some products liability exposure exist to provide additional motivation for a cylinder producer to avoid manufacturing errors. Distinctions between national jurisdictions as to proof of liability or manner of recovery are marginal to this concern.

THE APPROVAL PROCESS

A criticism made by several commenters dealt with what is perceived as a lack of specificity in the criteria to be used in determining whether to grant approval to a foreign manufacturer or inspector. One commenter addressing the approval process in particularly useful detail was Union Carbide Corporation. Certain of the Union Carbide comments regarding clarity of the proposed rules have been incorporated into the final rules, and others are addressed here.

The term "person" used in the amendments is defined at 49 CFR 171.8 (41 FR 15995, April 15, 1976) as an individual, firm, co-partnership, corporation, company, association, joint stock association, or trustee, receiver, assignee or personal representative of the foregoing.

Among the items of information necessary to support an inspection agency application, new § 173.300a(b) (6) requires identification and qualifications of those inspectors responsible for certifying inspection and test results (certifying inspectors). Certifying inspectors are responsible for the proper performance of inspection duties. Certifying inspectors may witness or perform tests themselves, or supervise others in such activity. In the latter case, new section 173.300a(b) (7) requires a method by which such supervised inspectors may be individually identified. Supervised inspectors may not certify inspection or test results. They are answerable as part of the independent inspection agency, cannot be an employee of the cylinder manufacturer, and cannot delegate their functions. The certifying inspector cannot delegate his certification functions. Actual organizational arrangements must be specified in the application and must meet the circumstances of manufacture.

From applicant inspection agencies, the amendments also require identification and description of testing facilities, a description of the agency's ability to perform duties imposed by Part 178, a description of ownership interests in the agency, and for nonresident agencies, a designation of agent for service of process.

From applicant manufacturers, the amendments require identification and description of each facility at which cylinders are to be manufactured or where analyses and tests will occur. Complete details on each specification cylinder for which manufacturing approval is sought must be provided, and the independent inspection agency to be used must be identified. Nonresident manufacturers must designate an agent for service of process.

The MTB believes that the level of specificity in the new amendments is sufficient to give notice as to how the approval process is expected to operate. A great number of factors, such as experience, credentials, training, available equipment and other resources, as well as (for inspection agencies) independence, are involved in each approval decision. To attempt to enumerate each factor and identify a constant relationship it may bear to any final approval action would suggest absolutes that do not exist and might tend to rule out concerns that may prove to be important. It is the intent of the amendments that the Director retain substantial discretion in approval decisions. Additional information may be sought for any approval application or in the course of monitoring an approval holders activities.

The effect of an approval issued to either an independent inspection agency or a foreign manufacturer is limited by the operation of any terms or conditions considered necessary by the Director, OHMO, and specified therein.

An approval issued either a manufacturer or an inspection agency may be terminated for fraud, noncompliance with Subchapter C, nonsatisfaction of Federal civil or criminal enforcement action, or if continuation of the approval is not consistent with the requirements of transportation safety. The latter category could encompass nonsatisfaction of a final judgment involving a tort claim related to cylinder manufacturing or inspection deficiencies; other circumstances indicating the practical nonexistence of an approval holders' exposure to product safety tort liability; or, a loss of independence by an approved inspection agency.

Prior to approval termination, the approval holder will be notified of the basis for that action and given an opportunity to show why the approval should not be terminated.

Provision has been made for any domestic inspection agency, which the Bureau of Explosives has designated as a competent and disinterested inspector prior to May 1, 1976, upon timely application and presentation of credentials, to be approved as a domestic independent inspection agency. Such agencies will be limited by the terms of such an approval to activities within the United States, for which reason they may choose to submit a full application for DOT approval subsequent to or instead of presentation of Bureau of Explosives credentials. Submission of Bureau of Explosives credentials must be made by July 15, 1976. Until August 15, 1976, Bureau of Explosives designation is acceptable as DOT approval. Following that date, such designation will not be recognized for any purpose.

(41 F.R. 18412, May 4, 1976)

SUMMARY • Docket No. HM-80 • *Classified phosphorus pentasulfide as a flammable solid and prescribed specific packaging.*

[49 CFR Parts 172, 173]
[Docket No. HM-80; Notice 71–7]
TRANSPORTATION OF HAZARDOUS MATERIALS
Phosphorus Pentasulfide

The Hazardous Materials Regulations Board is considering amending § § 172.5 and 173.225 to identify phosphorus pentasulfide by name as a flammable solid and to prescribe general packaging conditions for its transportation.

The question of regulating this commodity has been raised several times and confusion may exist regarding its classification, i.e., whether or not the product is subject to the Department's Hazardous Materials Regulations.

In view of the properties of this material and its accident history, the Board believes that the product should be identified and classed as a flammable solid and hazard warnings should be provided during transportation. The Board also considers the additional placarding described in § 177.823(c) appropriate. On the basis of the information the Board has available, it appears that any tight, sift-proof packaging meeting the requirements of § 173.24 would be adequate.

(36 F.R. 4626, March 10, 1971)

[49 CFR Parts 172, 173]
[Docket No. HM-80; Notice No. 72–8]
PHOSPHORUS PENTASULFIDE
Proposed Packaging Specifications

On March 10, 1971, the Hazardous Materials Regulations Board of the Department of Transportation published Notice No. 71–7 Phosphorus Pentasulfide in Docket No. HM-80 (36 F.R. 4626). Several comments were received.

Some commenters objected to the classification of the material but in view of accidents with the product that have continued to occur during the rule making proceeding, and the basic properties of the material necessitating certain responses to an incident, the Board continues to believe that the material should be classed as a flammable solid. In addition, on the basis of its properties and the experience developed from these incidents, it further believes that specification packaging should be specified. In the notice of March 10, 1971, DOT specification packaging was not a subject of the rule making action. Therefore, the Board is republishing the notice to propose that specification packaging be required except when the material will be fused into one solid mass for transportation.

Some commenters stated that the same requirements applicable to phosphorus sesquisulfide should be applicable to phosphorous pentasulfide. Upon examination of the literature describing the physical and chemical properties of the two compounds, the Board considers phosphorus sesquisulfide to be of a higher level of hazard, and believes that less stringent packaging limitations for phosphorus pentasulfide are warranted.

This notice supersedes Notice No. 71–7 in Docket HM-80.

(37 F.R. 14239, July 18, 1972)

[Docket No. HM-80; Amdt. 172–18, 173–68]
PART 172—COMMODITY LIST
PART 173—SHIPPERS
Phosphorus Pentasulfide

The purpose of this amendment to the Department's Hazardous Materials Regulations is to identify phosphorus pentasulfide by name as a flammable solid and to prescribe packaging requirements for its transportation.

On March 10, 1971, the Hazardous Materials Regulations Board published a notice of proposed rule making, Docket No. HM-80; Notice 71-7 (36 FR 4626) which proposed to classify phosphorus pentasulfide as a flammable solid. On July 18, 1972, the Board published another notice of proposed rule making in Docket No. HM-80, superseding Notice 71-7 (Notice 72-8 (37 FR 14239)). This later notice repeated the proposal to classify the material as a flammable solid and, in addition, proposed specification packaging. Several comments were received on this later notice.

One commenter objected to classification of the subject material as a flammable solid. In his comments, he referred to Docket No. HM-103, an advance notice of proposed rule making on a DOT Hazardous Information System (37 FR 12660), and cited that document as support for classification as a "Dangerous When Wet" material. The advance proposal, as set forth in Docket No. HM-103, does not establish "Dangerous When Wet" as a separate hazard classification, but rather establishes that the material has an additional hazard that is proposed to be specifically identified. It is contemplated that such materials will be classed and labeled as flammable solids. Therefore, since the Board does not intend to propose this additional hazard as a classification, the Board has classed this material as a flammable solid.

Some commenters pointed out that costly modification of certain bin-type packagings would be necessary if the Board pursued its proposed action. Six incidents involving phosphorus pentasulfide have been reported, three of which involved fire. The Board has considered the cost/benefit aspects of the rule and, in view of the incidents that have occurred and the experiences noted in the development of DOT specification 56 and the experience with present packaging, it has concluded that such modifications are warranted. Strengthening of the side-opening door closure in these types of packaging has been a concern of the Board for some time. This conclusion is consistent with its decision in the amendments to Docket HM-68 (37 FR 2885 and 3524) which were based on observed deficiencies in the closure which resulted in sifting of certain materials. The Board acknowledges that these modifications will require a certain period of time in order that the flow of traffic of this material will not be disrupted and is of the opinion that an 8-month modification period is reasonable if no further incidents occur. In this respect, the Board advises shippers to exercise special precautions in preparing these packagings for shipment during the interim period. Some commenters requested that the regulations provide for nonspecification (DOT) packaging. However, other commenters objected to the use of any packaging other than DOT specification packaging. After considering the incidents involving this material, and considering the nature of the product, the Board does not find justification for authorizing other than DOT specification packaging except for material that is fused solid.

One commenter suggested that the packaging for phosphorus pentasulfide be limited to the packaging authorized for phosphorus sesquisulfide. Adoption of such packaging criteria would severely restrict available packaging options for shipment and would in fact entail a substantial deviation from present practices for packaging this chemical. After reviewing the properties of the material and transportation experience with packaging now in use, the Board is of the opinion that DOT specification packagings should be prescribed, but that the restriction to use only packagings now listed in § 173.225 is not warranted. However, the Board does agree that the packaging used must be water-tight, i.e., a packaging design to prohibit the ingress of water, and has provided such a requirement in this amendment.

The note for specification 53 portable tanks indicating that new construction is not authorized has not been included in § 173.225(b) (2) because this specification is no longer shown in Part 178 and continued use is controlled by the requirements of § 173.32(d). The note would be redundant.

(38 F.R. 1507, January 15, 1973)

[Docket No. HM-80; Amdt. No. 173–68]

PART 173—SHIPPERS
Phosphorus Pentasulfide

On January 15, 1973, the Hazardous Materials Regulations Board published Docket No. HM-80; Amendment Nos. 172–18 and 173–68 (38 FR 1507) identifying phosphorus pentasulfide by name as a flammable solid and prescribing packaging requirements for its transportation.

The Board has received petitions for reconsideration of Note 1 following § 173.225(b) (2) as published in the amendment. This note authorizes non-DOT specification metal portable tanks similar to specification 53 or 56 to be used until August 30, 1973. After this date, DOT specificaiton tanks would be required. The petitioners stated that due to the following conditions, more time was needed to obtain, or convert tanks to DOT specification 56 portable tanks:

1. Larger than anticipated numbers of tanks are being removed from service since they are being considered as unacceptable by the petitioners themselves for conversion to DOT-56.

2. The necessity to replace these unacceptable tanks with new ones has required new capital appropriations with attendant delays.

3. Modification of large numbers of tanks has placed an overwhelming demand on the supplier's conversion hardware and replacement parts so that delivery dates have had to be extended resulting in the extension of work schedules.

The Board indivdiually contacted the petitioners in writing and asked, as a condition of extension of the August 30, 1973 date, a written commitment that every reasonable effort would be made to have tanks modified as early as practicable under the circumstances. The Board had received such commitments from the petitioners and believes an eight-month extension is reasonable and justified.

One petitioner requested "grandfathering" of the tanks not meeting the DOT-56 specification until all bins are modified. The petitioner did not establish that an unlimited extension of the date in Note 1 to § 173.225(b)(2) would be justified; therefore that portion of his petition is hereby denied.

(38 F.R. 16875, June 27, 1973)

SUMMARY • Docket No. HM-88 • *Established June 30, 1974, as the expiration date for all special permits authorizing the transportation of propylene having a pressure over 255 psi at 110 degrees F., in tank cars not otherwise authorized by regulation for this commodity pressure.*

[Docket No. HM-88]
TRANSPORTATION OF HAZARDOUS MATERIALS
Notice of Proposed Board Action

The Hazardous Materials Regulations Board is proposing taking action with respect to several special permits which authorize the bulk rail shipment of propylene in DOT Specification 112A340W and 114A340W tank cars.

As a part of the course of its usual regulatory activity, the Board evaluates special permits which have been outstanding for a period of several years, with the view to incorporating successful provisions into the Hazardous Materials Regulations by amendment. The Board has evaluated the permits noted above, authorizing use of the Specifications 112A340W and 114A340W tank cars for shipment of propylene. Existing regulatory provisions for propylene only permit use of a tank having a test pressure equal to or greater than that of the Specification 112A400W car.

Although there has been no accident history accompanying the permits in question, the Board believes that the standard design relationship between a pressure car's test pressure and its authorized operating pressure should be preserved as a principle of transportation safety. Section 173.314(d) of the Hazardous Materials Regulations prescribes that, "The gas pressure . . . at 115° F. in any uninsulated tank car of the DOT-112AW and 114AW class . . . must not exceed three-fourths times the prescribed retest pressure of the tank. . . ." The vapor pressure of propylene at 115° F is at least 259 p.s.i.g., and exceeds the upper limit of 255 p.s.i.g. prescribed for the Specifications 112A340W and 114A340W tank cars. The Board is of the opinion that this exceeds the intended design safety factor of the Specifications 112A340W and 114A340W tank cars, and should not be authorized in the regulations by amendment or continued under special permits.

The Board is proposing, therefore, that all permits authorizing the use of the Specification 112A340W and 114A340W tank cars for propylene be terminated within a reasonable period of time. After that termination date, proposed to be June 30, 1972, bulk rail shipment of propylene would only be authorized in tank cars having at least the test pressure of the DOT Specification 112A400W car. New permits would be issued until that date and existing permits would be renewed until that date, but all permits would be terminated simultaneously on June 30, 1972.

Information and advice are specifically requested regarding the Board's position that the 75 percent relationship between tank test pressure and operating pressure should be carefully preserved. In addition, data relating to the direct impact of this proposal on the permit holders and the public, including cost analyses and statistics on the availability of Specification 112A400W cars, would be very helpful to the Board in reaching its final decision in this matter.

Interested persons are invited to give their views on this proposal. Communications should identify the docket number and be submitted in duplicate to the Secretary, Hazardous Materials Regulations Board, Department of Transportation, 400 Sixth Street SW., Washington, DC 20590. Communications received on or before October 12, 1971, will be considered before final action is taken on the proposal. All comments received will be available for examination by interested persons at the Office of the Secretary, Hazardous Materials Regulations Board, both before and after the closing date for comments.

This proposal is made under the authority of sections 831–835 of title 18, United States Code, and section 9 of the Department of Transportation Act (49 U.S.C. 1657).

(36 F.R. 13173, July 15, 1971)

[Docket No. HM-88]
TRANSPORTATION OF HAZARDOUS MATERIALS
Notice of Board Action

On July 15, 1971, the Hazardous Materials Regulations Board published a notice of proposed action, Docket No. HM-88 (36 F.R. 13173) with respect to bulk rail shipment of propylene in DOT Specification 112A340W and 114A340W tank cars.

In that notice, the Board expressed its opinion that the standard design relationship betwen a tank car's test pressure and its maximum authorized operating pressure should be preserved as a principle of transportation safety. It observed that there were several special permits outstanding for propylene shipment in cars not in accord with that regulatory principle. The vapor pressure of propylene at 115° F. is at least 259 p.s.i.g. This exceeds the upper limit of 255 p.s.i.g. prescribed for Specification 112A340W and 114A340W tank cars. Based on these considerations, the Board proposed rescission of these special permits.

Persons objecting to the Board's proposed rescission of the permits did not present engineering or design data that would support continuance of the present departures from the current regulatory principles. The Board further notes that DOT Specification 114A400W was authorized for use in propylene service by a recent amendment to the Hazardous Materials Regulations thereby providing another specification car for use by shippers.

In consideration of the foregoing, the Hazardous Materials Regulations Board hereby establishes June 30, 1972, as the expiration date for all special permits issued to authorize the transportation of propylene having a pressure in the tank car over 255 p.s.i.g but not over 250 p.s.i.g. at 115° F. New permits may be issued until that date, and existing permits will be renewed if necessary, but all permits will terminate simultaneously on June 30, 1972.

(37 F.R. 149, January 6, 1972)

SUMMARY · Docket No. HM-89 · *Established a new DOT specification 115 insulated tank car and provided for its use to transport flammable liquids, n.o.s.*

[49 CFR Parts 173, 179]
[Docket No. HM-89; Notice No. 71–23]
TRANSPORTATION OF HAZARDOUS MATERIALS
Specifications for Tank Cars

The Hazardous Materials Regulations Board of the Department of Transportation is considering amending the Hazardous Materials Regulations to authorize a new DOT Specification 115 tank car for shipment of flammable liquids not otherwise specified (n.o.s.).

This proposal is based on more than 5 years of satisfactory experience under special permit authorizations for this type of car in flammable liquid service. The proposed car is basically an inner container supported within an outer tank shell. The construction involves a different design concept from the ordinary tank car, in that the inner container is suspended within the outer tank shell solely by an insulation material such as polyurethane, which is foamed in place, without any metal supports. Without metal supports for the inner tank, heat transfer from an outside source would be reduced significantly, and thus the proposed specification provides a highly efficient thermal insulation of the inner container and the commodity being shipped. The insulating suspension system is designed to withstand prescribed acceleration loads and has been proved under actual tests. The outer tank shell serves as the structural member and transfers the load forces to the conventional tank car body structure. The specification provides for conventional loading and unloading fittings, and appropriate safety relief devices.

(36 F.R. 13405, July 21, 1971)

[Docket No. HM-89; Amdts. Nos. 173–58, 179–9]
PART 173—SHIPPERS PART 179—SPECIFICATIONS
FOR TANK CAR
Provisions for Shipment of Flammable Liquids

The purpose of this amendment to the Department's Hazardous Materials Regulations is to amend § 173.119(a)(12) and to add new § § 179.220 and 179.221 to provide for the use of a new DOT specification 115 tank cars for shipment of flammable liquids not otherwise specified (n.o.s.).

On July 21, 1971, the Hazardous Materials Regulations Board published Docket No. HM-89; Notice No. 71–23 (36 F.R. 13405) proposing these changes. Several comments were received.

One commenter requested inclusion of additional requirements to provide for the use of specification 115 cars in low temperature service (service below −20° F.). Since this was not included as subject matter in the present rulemaking action and it would introduce substantial changes, the Board is of the opinion that these comments should be subject to a separate rule-making action.

The commenter also noted that the Board apparently was proposing to reject tanks having metal support members extending from inner tank to outer shell. The Board did not intend to restrict the design in this manner. Section 179.220–15 does not contain this limitation. The regulation that "the inner container must be thermally isolated from the outer shell to the best practical extent" is a rule referring to a support system of "approved design." There is no limitation on such a system provided it complies with the specification requirements. The phrase "without metal supports" found in the preamble of Notice 71–23 was meant to be descriptive of the manner in which these specification cars are now being built, and was not intended to be a restriction in future design.

Several comments were received suggesting clarification of various texts and most of these suggestions were adopted. One change relating to § 179.220–18(a) (3), location of bottom unloading valves, was set aside as a matter for separate rule-making. Another suggestion related to the requirement for a locking arrangement on valve-operating mechanisms. This requirement does not mean that exterior actuating mechanisms must be physically connected by means of linkages to the actual seating mechanism of the valve. On valves that might otherwise be open during transportation, some securing device or arrangement must be supplied to insure that positive closure is maintained during transportation and that the valve cannot accidently be opened.

The Bureau of Explosives, Association of American Railroads, objected to the addition of DOT specification 115A type cars to § 173.119(a) (12) on the following bases:

(1) Specification 115A cars were designed primarily for transportation of "time temperature" products as defined in DOT Special Permits for dichlorobutene, methyl methacrylate monomer. For this service there are restrictions imposed on both the cars and the products which are not contained in the proposal.

(2) Under the special permit system the special restrictions mentioned above can be and are imposed; but the blanket authorization proposed in the instant notice would eliminate all of them, even those demonstrably necessary for safety.

(3) New-time-temperature-sensitive products may be developed that require other and different restrictions for safe transportation. These restrictions likewise could not be imposed under the proposal.

The Board does not believe the 115A car should be singled out for restriction. Under the present regulations, transportation of dichlorobutene and methyl methacrylate is authorized in many specification cars that are not required to be insulated. However, this transportation is always subject to § 173.21, which states in part that,

"The offering for transportation of any package or container of any liquid, solid, or gaseous material which under conditions incident to transportation may polymerize (combine or react with itself) or decompose so as to cause dangerous evolution of heat or gas is prohibited. Such materials may be offered for transportation when properly stabilized or inhibited. Refrigeration may be used as a means of stabilization only when approved by the Bureau of Explosives."

The effect of this rule would not be changed, i.e., when a time-temperature relationship is involved in safely transporting a product, this means of stabilization may be used only when approved by the Bureau of Explosives. Therefore, if dichlorobutene not otherwise safely inhibited is proposed to be shipped safely under temperature control, such as by the use of a type 115A tank car or any other insulated car now authorized by § 173.119(a) (12), this method of stabilization of inhibition continues to be subject to the Bureau's approval and any conditions it may impose. This includes use of the car in this service. This section of the regulations presently controls nearly all shipments requiring time-temperature controls due to instability. Any permits now outstanding that involve § 173.119(c) (12) and that include time-temperature relationship involving safety would not be obviated by this amendment. Rather, control would revert back to the Bureau by virtue of § 173.21(b). Therefore, the Board is of the opinion that the amendment as proposed should be sustained.

(36 F.R. 21339, November 6, 1971)

SUMMARY • **Docket No. HM-90** • *This is a continuing or open docket which has provided for updating many of the requirements proposed in Docket No. HM-10; See Docket No.*

[Docket No. HM-90; Notice No. 71–24]
TRANSPORTATION OF HAZARDOUS MATERIALS
Specifications for Tank Cars

The Hazardous Materials Regulations Board is considering amendment of Part 179 of the Hazardous Materials Regulations to update tank car specification requirements and to add new tank car specifications. Suitable corresponding changes would be made to Part 173 for commodities when the amendment is published. These proposed amendments are almost entirely based on requests submitted by the Association of American Railroads (AAR). These requests are based largely on experience gained under a very significant number of outstanding special permits.

These proposals are set forth in the type of detailed specifications that have been used for many years in this field. As has been mentioned frequently, it is the Board's intention in the future to prescribe minimum performance type requirements rather than detailed specification requirements. However, in view of the substantial amount of time and effort of both the industry and the Federal Railroad Administration that can be saved through the elimination of many special permits, it is desirable to issue this proposal in the traditional specification form rather than to let the existing situation stand while performance specifications are developed.

The major changes proposed are discussed in this preamble. Numerous editorial changes relating to clarification of text and updating of references are not specifically highlighted.

Section 179.2(a) (4) would be amended to indicate that "DOT" identifies the Department of Transportation. Section 179.5 would be revised to assure completeness of Bureau of Explosives and AAR Mechanical Division files. The Department does not maintain such records.

In § 179.6 reference would be made to Appendix R of the AAR Specifications for Tank Cars. Section 179.12–2(b) would permit a 20 percent reduction in carbon steel pipe, if welded, and would reflect provisions that have been incorporated in other tank car standards for many years.

Section 179.100, Pressure Tank Car Tanks, in addition to updating section cross-references and incorporating references to various appendices of the AAR Specifications for Tank Cars, would require, at subsection 4, that the jacket covering insulation be of a thickness not less than 11 gage. At subsection 6, more stringent requirements would be added to assure that tank plates are not reduced during forming below specification requirements. Also the welded joint efficiency factor change to 1.0 would recognize improvements in welding and weld inspection over the years in addition to the fact that all such tank welds are 100 percent radiographed. In subsection 7, the tables would be revised to update the requirements by deleting obsolete material specifications and by adding new specifications currently in use under special permits. In subsection 10, the phrase "postweld heat treatment" would be substituted for the phrase "stress relieving" to agree with terminology used in other codes such as the ASME Code. This change is also repeated later in the text. In subsection 12, minimum requirements would be added for the size and number of studs and the minimum thickness of material for dome housings. Subsection 13 would be amended by providing for the addition of a sump or siphon to the bottom of a tank. Subsection 14 would define more clearly the limitation on extreme projection of bottom outlets on cars. Paragraph (a) (5) would require a screw plug on closure which has been standard approved practice, but never specified heretofore. Subsection 15(a) would be changed to be consistent with § 179.100–1 which includes valve flow rating pressures for safety-relief valves. Subsection 15(b) would require the safety-relief valve on specification 105A500W tanks to be set for a start-to-discharge pressure of 360 p.s.i. in keeping with current practice. Subsection 16 would specify use of reinforcing pads to distribute stresses and to prevent punctures and tearing of a tank by attachments. Subsection 21 would require a water capacity stencil on a pressure tank to provide additional infomration to loading personnel to prevent overfilling.

Section 179.101 would provide for the construction of tank cars to new specification 114A400W. At the same time, provisions would be made for specifying valve flow rating pressure, for use of white paint in place of "reflecting paint", and for bottom outlets on certain cars.

Section 179.102 would be amended to include new requirements for cars in service transporting liquefied carbon dioxide, liquefied flammable gases, vinyl chloride, inhibited vinyl methyl ether, ethylene oxide, anhydrous hydrofluoric acid, inhibited acrolein, metallic sodium, stabilized sulfur trioxide, flammable liquids, n.o.s., and unsymmetrical dimethyl hydrazine. Provisions for the alternate setting of safety-relief valves on butadiene and vinyl chloride tank cars, for flow rating data, and for gasketing are included.

Section 179.103 is proposed to be amended to specify minimum requirements for a protective housing with cover, to eliminate the requirement for excess flow valves when the lading is nonflammable, and to broaden the application of § 179.103–4 to safety-relief devices and pressure regulators. Subsection 5 would be added to provide specific requirements for approved bottom outlet valves.

Section 179.200, general specifications applicable to nonpressure tank car tanks, would be amended to make substantive changes in subsections 3, 4, 6, 7, 8, 10, 15, 16, 17, 19, and 24. Subsection 4 would require insulation on tanks to be covered with a metal jacket not less than 11 gage. Subsection 6 would be clarified to assure that thickness of tank plate is not reduced during forming below specification requirements. Also the weld joint efficiency factor would be 1.0 for seamless heads of all tanks. In subsection 7, the tables would be revised to update the requirements by deleting obsolete material specifications and by adding new specifications currently in use under special permits. Subsection 8 would be amended to require 2:1 ellipsoidal heads for class 111A tank car tanks, except for internal compartment heads. Subsection 10 would specify that welding to certain fittings is not authorized. Subsection 15 would be amended to delete reference to AAR specification M-402, Grade 35018, malleable iron castings. Subsection 16 would require the application of shutoff valves at specific locations on the tank when top loading and discharge devices are installed, would set design parameters for sumps and siphon bowls, would provide protection for certain fittings, and would specify minimum requirements for protective housings. Subsection 17 would define more clearly the design criteria and a limitation on extreme projection of bottom outlets on cars with truck centers less than or greater than 60 feet 6 inches. Subsection 19 would specify use of reinforcing pads to distribute stresses and to prevent punctures and tearing of a tank by attachments. Subsection 24 would add a paragraph (b) to provide an abbreviated marking for class 111A tank car tanks by omitting the suffix numeral.

Section 179.201-1 would be amended to provide for the construction of tank cars to new specifications 111A100-ALW, 11A60W2, 111A60W5, 111A60W1, 111A60ALW2, 111A100ALW1, 111A100ALW2, and 111A60W7. Subsection 3 would be changed to make distinct the requirements applying to rubber-lined tanks and tanks lined with material other than rubber. Subsections 4, 5, and 6 would be changed to adopt by reference the requirements specified in Appendix M of AAR Specifications for Tank Cars. Subsection 7 would require safety relief devices to comply with § 179.200-18. Subsection 9 would be amended to indicate that an excess flow valve is not needed if the gaging device does not allow passage of the lading when the device is in operation.

Section 179.202 is also proposed to be changed. Most revisions would be editorial in nature. Subsection 9 would be amended to add requirements for sodium chlorite. Subsections 12 would provide for a frangible disc to be made from other materials than lead. In subsection 14, authorization for use of Type 347 stainless steel would be deleted and addition of special safety valve requirements would be made. Subsection 18 would be amended to prohibit the use of copper or copper bearing alloys. Subsections 20, 21, and 22 would be added to provide special commodity requirements for hydrofluoric acid, nitric acid, and mixed acids.

In § § 179.300-6 and 179.300-8 more stringent requirements would be added to assure that tank plates are not reduced during forming below specification requirements. In subsection 7 the table would be revised to update the requirements by deleting obsolete material specifications and by adding new specifications currently in use under special permits. In subsection 17 the tests for frangible discs of safety vents would be required to comply with Appendix A of the AAR Specifications for Tank Cars.

The change to § 179.302 would rearrange the commodities in alphabetical order and would consolidate the family of aluminum alkyls (pyroforic materials) under the generic description "pyroforic liquid, n.o.s."

(36 F.R. 16680, August 25, 1971)

[Docket No. HM-90; Amdts. Nos. 173-59, 179-10]
PART 173—SHIPPERS
PART 179—SPECIFICATIONS FOR TANK CARS
Miscellaneous Amendments

The purpose of this amendment to the Department's Hazardous Materials Regulations is to update tank car specification requirements and to add new tank car specifications.

On August 25, 1971, the Hazardous Materials Regulations Board published Docket No. HM-90; Notice No. 71-24 (36 F.R. 16680) proposing these changes. A supplemental notice in this docket was published on September 23, 1971 (36 F.R. 18873), concerning bottom outlets on class 114A tank cars. Several comments were received.

Many of the commenters suggested editorial changes to improve on the language or the format proposed by the Board. These comments were very helpful in clarifying the intent of the regulations. The Board adopted a large portion of them.

Several comments appeared to be presented as editorial in nature but actually were not because the changes suggested would have resulted in substantial revision of material or construction requirements. Changes suggested but not adopted were in § § 179.100-7(a), 197.200-7(b), 179.200-7(f), and 179.300-7(a) concerning elongation criteria for ASTM materials, § 179.102-1 concerning examination of certain fillet welds, § 179.102-4 concerning vinyl fluoride, § 179.102-18 concerning hydrogen chloride, and § 179.200-22(c) concerning testing on other than rubber-lined tanks.

Other commenters suggested changes that were not only substantial but actually introduced new factors for evaluation on which the public should be afforded the opportunity to comment. Such changes concerned § § 179.101-1(a) and 179.201-1(a) regarding steel and aluminum equivalencies, § 179.202-15 regarding certain tank cars in formic acid service, and § 179.300-7(a) regarding the addition of new materials. If after review of these amendments commenters are of the opinion that further consideration should be given to their comments for inclusion in future rule making, they should advise the Board.

Several comments were received regarding the Board's proposal to remove bottom outlets on class 114A tank cars. These outlets are now authorized in the regulations on DOT-114A340W tank cars. The Board now considers this a matter for separate rule making and intends to reopen the issue in a future rule making action. This will enable the Board to clearly present its reasons and conclusions to the public. Therefore, since the Board wants to be certain that proper consideration is given this subject, it is not changing the present construction requirements for the class 114A tank car.

One commenter specifically proposed that the markings "Ethylene Oxide and Propylene Oxide Only" be authorized as well as the individual marking "Ethylene Oxide Only." The Board has not amended the rule as requested but will give consideration to such a petition upon the submission of detailed data on the safe interchangeability of these products including information on cleaning between changes, if such is the practice.

Another commenter requested to know if cold ammonia would continue to be authorized in class 119A (interim proposed designation for a DOT tank car specification) tank cars under special permit since specification 119A was not covered by this docket. This class car was not a subject of this rule making action. By these amendments the Board does not intend to preclude consideration of class 119A tank cars in ammonia service.

(36 F.R. 21343, November 6, 1971)

[49 CFR Parts 173, 179]
[Docket No. HM-90; Notice No. 74-2]
BOTTOM OUTLETS ON FLAMMABLE COMPRESSED GAS TANK CARS
Notice of Proposed Rule Making

The Hazardous Materials Regulations Board is considering amendments to §§ 173.314, 179.102-3, and 179.103-5 of the Department's Hazardous Materials Regulations pertaining to prohibiting bottom outlet devices on DOT Class 114A tank car tanks used for the transportation of liquefied flammable compressed gases.

The Board on August 25, 1971, in Docket Number HM-90, Notice 71-24 (36 FR 16680) proposed amendments to § 179.103-5 dealing with bottom outlet devices on DOT class 114A tank cars. By a supplemental notice published on September 23, 1971 (36 FR 18873), this proposed amendment was deleted and withdrawn from the docket after the Board determined from comments received that the amendment proposed should be the subject of a separate rule-making proceeding.

There are approximately 175,000 tank cars approved for the rail transportation of hazardous materials and other non-regulated commodities. Of this total there are approximately 650 specification 114A-W tank cars, equipped with bottom fittings used for the transport of compressed gases.

The Railway Progress Institute/Association of American Railroads "Tank Car Safety Research and Test Project" study of railroad accidents during the period of 1965-1972 indicates that there were 2,624 reported accidents involving some 4,385 tank cars. The lading loss (monetary) from tank cars attributable to accidents during 1965-1970 amounted to more than $300,000; not including resultant property and other damages, etc.

The Board, in the interim, has become increasingly aware of the poor safety performance record of bottom fittings currently authorized on DOT Class 114A tank cars. Incident reports received by the Board indicate that there has been an increase in the number of liquefied flammable compressed gas leaks occurring through tank car bottom outlet devices even when the tank car has not been involved in any accident.

There have been eight reported leakage incidents of flammable compressed gas from specification 114A-W type tank cars, six of which occurred during 1972. Two leakage incidents were attributable to derailments. There is currently an industry study of the feasibility, as well as the design features, for bottom outlet devices and fittings for tank cars used to transport hazardous materials.

The Board's review of the leakage incident reports indicates that in large part, the failure of these devices is due to their inadequate maintenance or a deficiency in their design. As a result of these findings and in view of the serious threat to the public safety, the Board proposes to prohibit bottom outlets on DOT Class 114A tank cars. The proposed changes would not become effective (mandatory) until 6 months after issuance of the final rules.

Pursuant to the provisions of Section 102(2) (c) of the National Environmental Policy Act (42 U.S.C. 4321 et seq.), the Board has considered the requirements of the Act concerning Environmental Impact Statements and has determined that the amendments proposed in this notice would not have a significant impact upon the environment. Accordingly, an Environmental Impact Statement is not necessary and will not be issued with respect to the proposed amendments.

(39 F.R. 7432, February 26, 1974)

SUMMARY · Docket No. HM-91 · *This proposal to establish new tank car specification for the transportation of cold (cryogenic) compressed gases was withdrawn in Docket No. HM-115.*

[49 CFR Parts 172, 173, 179]
[Docket No. HM-91; Notice No. 71-25]
TRANSPORTATION OF HAZARDOUS MATERIALS
Cold Compressed Gases in Tank Cars

The Hazardous Materials Regulations Board is considering amendment of the Department's Hazardous Materials Regulations to provide for the shipment of ethylene, hydrogen, methane, natural gas, and inhibited vinyl fluoride in a cold liquefied gas state in certain tank cars.

The movement of several liquefied compressed gases at low temperatures in insulated tank cars equipped with safety relief valves to permit venting, if the commodity remains in the tank for an extended period of time, has proved to be a safe method of transportation. Under the provisions of § 173.314, inhibited vinyl fluoride is authorized to be so shipped in certain tank cars with start-of-shipment commodity temperature of not over 0°F. Each tank is required to be insulated to insure that at least 30 days will pass before the pressure of the contents reaches the level of the safety relief valve setting.

For several years, these commodities have been transported under similar conditions, authorized by special permits. The experience reported by the special permit holders relating to the changes proposed has been satisfactory.

This proposal is based on a petition to incorporate the terms of these permits into the regulations. The proposal also would amend the regulations relating to conditions for transportation of inhibited vinyl fluoride, and those relating to present specification requirements for certain insulated tank cars. In addition, new tank car specifications 113C120W and 113D120W would be added to the regulations.

The petition for rule change also included a request to amend the regulations to allow the shipment of cold anhydrous ammonia in a proposed specifiation 119A60W tank car, and to provide proposed new specifications 113B60W, 113C60W, and 113D60W. The small amount of experience obtained under special permits thus far for these items is considered too limited to serve as a basis for rule change. In addition, actual operating difficulties encountered with the proposed 113D60W specification tank car have raised serious doubts regarding use of the "60-pound car". The Board, therefore, is not proposing these requested changes at this time.

(36 F.R. 20166, October 16, 1971)

SUMMARY • Docket No. HM-93 • *Prohibited the shipment of Class B propellant explosives in DOT-21C drums via rail transportation except when transported in trailer-on-flat car service.*

[49 CFR Part 173]
[Docket No. HM-93; Notice No. 71–28]
TRANSPORTATION OF HAZARDOUS MATERIALS
Class B Propellant Explosives in Fiber Drums

The Hazardous Materials Regulations Board is considering amendment of § 173.93 of the Hazardous Materials Regulations to prohibit the shipment of Class B propellant explosives in specification 21C fiber drums by rail freight transportation.

Several instances of serious accidents and gross container failures involving Class B propellant explosives packed in DOT-21C fiber drums have been brought to the Board's attention. These instances appear to be directly related to transportation handling practices. Although these problems are reported to have existed in rail and highway transportation, the Board has no evidence that shippers by highway transportation are now experiencing difficulties. Rail carrier representatives together with involved shippers and Federal officials have been unable to resolve the difficulties associated with the rail transportation of this commodity in the packaging. Significantly, on June 8, 1971, the Department of Defense, a major shipper of this Class B propellant explosives, issued an order to all U.S. Army Ammunition Depots to discontinue such shipments by rail box car. Therefore, in view of the continuing seriousness of the situation, the Board is proposing to prohibit the shipment of Class B propellant explosives in DOT-21C fiber drums in rail transportation.

(36 F.R. 21360, November 6, 1971)

[Docket No. HM-93, Amdt. 173–63]
PART 173—SHIPPERS
Class B Propellant Explosives in Fiber Drums

The purpose of this amendment to the Department's Hazardous Materials Regulations is to prohibit the shipment of Class B propellant explosives in specification 21C fiber drums in rail boxcars and in container-on-flat-car service.

On November 6, 1971, the Hazardous Materials Regulations Board published Docket No. HM-93; Notice No. 71–28 (36 F.R. 21360) proposing the change described above. On December 16, 1971 (36 F.R. 23921), the comment period in this docket was extended from January 4 to February 22, 1972 to permit test results to be presented to the Board by the petitioner. Again, on February 10, 1972, the same petitioner requested further extension of the comment period for 60 days, for the same reason. This later petition for extension was denied since no additional information was submitted to the Board showing good cause and public interest. To date, the results of the tests referred to above have not been submitted to the Board.

Several comments were received. The objectors took issue with the Board's proposed solution to the problem. Commenters did not question the existence of the problem cited, namely, serious failures in rail transportation of DOT-21C fiber drums filled with Class B propellant explosives. The Board admits that if conditions in rail transportation were changed, the package could be satisfactory in this service. But, given the existing problem, which the parties involved have been unable to solve, the Board must take this action to prevent undue public exposure to hazards. No evidence has been presented that would indicate that the Board could satisfactorily resolve this issue which shippers and carriers have also been unable to solve.

Several commenters correctly noted that the basis for the Board's decision did not include unsatisfactory experience reports in container-on-flat-car or trailer-on-flat-car service, but by its proposed rule this transportation by rail was precluded. The Board agrees that the incidents reported do not directly support excluding such transportation. However, as far as is known to the Board, no shipments have been made by COFC and in view of the problems in boxcar shipments the Board has no information on which to base a determination that COFC shipments could safely be made. Safe shipments are being made in TOFC service. Therefore the rule has been changed to authorize TOFC service.

(37 F.R. 12494, June 24, 1972)

SUMMARY • **Docket No. HM-95** • *Prohibited the transportation of blasting caps in the same motor vehicle with other explosives (other than Class C explosives), unless approved by DOT.*

[49 CFR Part 177]
[Docket No. HM-95; Notice 71–31]

TRANSPORTATION OF BLASTING CAPS WITH OTHER EXPLOSIVES BY MOTOR VEHICLE
Notice of Proposed Rule Making

The Hazardous Materials Regulations Board of the Department of Transportation is considering amending § 177.835 of the Hazardous Materials Regulations to remove authorization for the transportation of blasting caps in the same motor vehicle with other explosives, except under certain stowage conditions. The Board proposes to continue the authorization for use of specification MC 201 containers for all types of blasting caps. It further proposes to provide for use of another type of container for electric blasting caps.

The major concern with the transportation of blasting caps in the same vehicle with other explosives is the detonation of the other explosives should the blasting caps be subjected to heat, heavy shock, or other potential initiation sources.

The Board has two basic choices to consider in addressing itself to this matter. It could prohibit the transportation of blasting caps in the same vehicle with other explosives, or it could recognize containment of the blasting caps in a manner so that they would not initiate other explosives under evaluated credible circumstances. A total prohibition does not appear justifiable if a reasonable and safe alternative is available. Under a prohibition, a separate vehicle would be required for a small quantity of blasting caps on a move to a blast site possibly several hundred miles distant. The costs of blasting operations, a vital function of the construction industry, would be greatly increased.

In considering a possible alternative to prohibition, explosives experts were contacted to obtain information and suggestions on methods that would provide for increased safety in the transportation of blasting caps with explosives.

Two major problems presented for solution were heat transfer and blast penetration. Blasting caps initiate at relatively low temperatures and their blast effects cause penetration of most ordinary packaging materials. Various persons were asked to examine different methods whereby a container could be constructed to protect packages containing blasting caps from high rates of heat input in a fashion that would preclude penetration by a cap or caps should they be initiated.

A container has been developed and tested. It is constructed of ½-inch plywood, ½-inch gypsum wallboard, ⅛-inch low carbon steel and ¼-inch plywood put together in sandwich fashion which hereafter is described as a barrier lamination. All sides, the bottom, and the lid are constructed of this material. On October 8, 1971, a fire test on the container was witnessed by representatives of the Department. A test container was loaded with electric blasting caps and placed in a wood fire for 62 minutes before the first blasting cap initiated. During the next 20 minutes there were many detonations but the structure of the steel shell of the container was not affected. There was no indication of penetration of the steel shell whatsoever. Later the same day, a second test was performed on another type container presently in use. The results were nearly identical—66 minutes before first detonation and no penetration. Complete reports of the tests, including photographs and 8-mm. movies, are available for examination in the Board's public docket room.

The Institute of Makers of Explosives (IME) has prepared a document entitled "IME Standard for the Transportation of Electric Blasting Caps in the Same Vehicle With Other Explosives," for distribution by the IME Safety Library. It contains specific requirements for construction of containers and compartments with barrier laminate material, and diagrams to illustrate vehicle configuration. The Board is proposing to adopt the IME Standard by reference.

Copies are available upon request from the Secretary, Hazardous Materials Regulations Board, or from the Institute of Makers of Explosives, 420 Lexington Avenue, New York, NY 10017.

The Board believes that, by adoption of this proposed amendment, the safe transportation of blasting caps in the same vehicle with other explosives would be better assured if the proposed method of containment is used and that adoption of a complete prohibition would not be necessary.

(36 F.R. 24125, December 21, 1971)

[Docket No. HM-95; Amdt. 177-22]

PART 177—SHIPMENTS MADE BY WAY OF COMMON, CONTRACT, OR PRIVATE CARRIERS BY PUBLIC HIGHWAY
Transportation of Blasting Caps With Other Explosives

The purpose of this amendment to the Hazardous Materials Regulations of the Department of Transportation is to prohibit the transportation of blasting caps in the same motor vehicle with other explosives, except Class C explosives, unless approved by the Department.

On December 21, 1971, the Hazardous Materials Regulations Board published Docket No. HM-95; Notice No. 71-31 (36 F.R. 24126) which proposed regulations consistent with this amendment. Interested persons were invited to give their view and several comments were received by the Board.

Several commenters objected to the proposal indicating many years of successful experience transporting blasting caps in accordance with the present regulations. The Board wishes to reiterate its concern as expressed in the preamble of the notice:

The major concern with the transportation of blasting caps in the same vehicle with other explosives is the detonation of the other explosives should the blasting caps be subjected to heat, heavy shock, or other potential initiation sources.

At the present time, § 177.835(g) authorizes the transportation of blasting caps in fiberboard boxes in the same vehicle with high explosives. In 1964, there was a serious explosion following a fire in a semitrailer. The vehicle contained blasting caps packed in fiberboard boxes placed immediately adjacent to dynamite packed in fiberboard boxes. No one is in a position to say that the blasting caps initiated the dynamite due to their penetrating effect or that the dynamite was initiated due to elevated temperature. The fact remains that blasting caps, when placed in certain orientations to high explosives, could cause the initiation of high explosives even though they are properly packaged in accordance with Part 173 of the Hazardous Materials Regulations.

Other commenters made recommendations that the Board impose additional requirements most of which are outside the scope of this rule making action. Others suggested requirements that are already in existence in 49 CFR Part 397.

Several commenters objected to the regulation being expanded to apply when blasting caps are transported with class C explosives, such as oil well jet perforating guns. The Board agrees with the commenters that the amendment should be restricted to classes A and B explosives. The suggested change has been incorporated in this amendment.

In the notice, the Board proposed to adopt by reference a standard prepared by the Institute of Makers of Explosives (IME) which contained specific requirements for the construction of containers and compartments using a barrier laminated material which was tested by IME and witnessed by representatives of the Department. The Board now believes that adoption of the IME standard would be too restrictive and would prevent the use of other barrier material that may provide equal penetration and heat transfer resistance. However, the Board believes that the IME standard contains sufficient and adequate criteria to serve as a model for approval of other containers and blasting caps on a comparative basis. In using the standard, the Board believes that containers and compartments constructed of other materials, that will perform in fire tests as well as the containers covered by the Standard, can be used to transport electric blasting caps or other types of blasting caps in the same vehicle with Class A and B explosives.

Several commenters were concerned that there might be a conflict between this amendment and the Occupational Safety and Health Administration (OSHA) standards of the Department of Labor which totally prohibit the transportation of blasting caps in the same vehicle with explosives. There is no conflict because the OSHA and the Department of Transportation regulations relate to different areas. Section 4(b) (2) of the William-Steiger Occupational Safety and Health Act of 1970 provides that the Act does not apply to working conditions with respect to which other Federal agencies exercise their statuatory authority to prescribe regulations affecting occupational safety. The Department of Transportation has such statutory authority under Title 18 U.S.C. 831-835 to prescribe regulations for shippers and carriers involved in interstate commerce. Therefore, with respect to the statuatory authority of the Department of Transportation relative to interstate commerce, the Department of Transportation regulations would apply and not the OSHA standard.

One commenter, representing the for-hire motor carrier industry in the transportation of munitions and explosives, objected to the proposed amendment being placed in Part 177 of Title 49, thus making the regulation applicable to motor carriers instead of shippers. This amendment relates to the loading as well as transportation of blasting caps in the same transport vehicle with other explosives. Section 173.30 specifies that any person who loads shipments of hazardous materials into transport vehicles shall comply with the applicable loading provisions, in this case with §§ 177.834 through 177.848. If a shipper loads a vehicle subject to the rules in this amendment, he must comply with these requirements.

(37 F.R. 21531, October 12, 1972)

SUMMARY • Docket No. HM-96 • *Established the classification of etiologic agents and the requirements for the transportation of these materials.*

[49 CFR Parts 172, 173]
[Docket No. HM-96; Notice No. 71–32]
TRANSPORTATION OF HAZARDOUS MATERIALS
Proposed Requirements for Shipment of Etiologic Agents

The Hazardous Materials Regulations Board is considering amendment of Parts 172 and 173 to add new §§ 173.386, 173.387, and 173.416 and to amend § 172.5, to provide for the shipment of etiologic agents.

Under the provisions of Public Law 86–710, the Interstate Commerce Commission was directed to formulate regulations for the safe transportation of etiologic agents. Upon the establishment of the Department of Transportation in 1967, the safety functions of the ICC were transferred to the Department.

This notice adopts by reference existing Department of Health, Education, and Welfare standards for classification and its current general packaging requirements of etiologic agents. In the FEDERAL REGISTER of May 13, 1971 (36 F.R. 8815), DHEW has proposed certain changes to these regulations. This Department would accept by reference any of the applicable changes made by those proposals. This notice also proposes application of performance test criteria for packagings for etiologic agents. In addition, a new hazard warning label has been proposed for packages of etiologic agents. This label would incorporate the biological hazards warning symbol proposed by the National Institute of Health's National Cancer Institute.

Docket No. HM-8; Notice No. 71–13 (36 F.R. 9449) and Notice No. 70–13 (35 F.R. 11742), insofar as they apply to etiologic agents, are superseded by this notice of proposed rule making. The comments made on those notices were considered in preparing this notice. There was some objection to the symbol on the label, but it is being retained because it is a symbol in use. The Board proposes to rely on education of the public, as was originally done with the radioactive materials "trefoil".

36 F.R. 25163, December 29, 1971)

[49 CFR Part 171]
[Docket No. HM-96; Notice No. 72–9]
ETIOLOGIC AGENTS
Proposed Requirements for Shipment

On December 29, 1971, the Hazardous Materials Regulations Board published Docket No. HM-96; Notice No. 71–32 (36 F.R. 25163), proposed requirements for shipment of etiologic agents.

Since the proposal was published, the Department has been discussing the coordination of these proposed regulations with the Center for Disease Control, U.S. Public Health Service, Atlanta, Ga. The Center commented on the Board's proposal.

It is the opinion of the Board that, by utilizing the expertise of the Center for Disease Control, safer transportation and a higher level of public protection will result if there is more direct contact with the Center by a carrier in the event of an incident. The Center has an organization, manned on a 24-hour basis, capable of providing response information and guidance in the event of leakage in transportation of an etiologic material package.

The newly adopted Department of Health, Education, and Welfare label (37 F.R. 12915) would provide on-the-scene information regarding a telephone number to call for emergency information or assistance. Section 171.15, as proposed herein, would also contain this information. Immediate reporting of an incident involving etiologic agents would be made, not to the Department of Transportation, but to the Center for Disease Control, of the Department of Health, Education, and Welfare. Section 171.16 would be unchanged thereby requiring that the subsequent written report be made to the Hazardous Materials Regulations Board, Department of Transportation.

(37 F.R. 14728, July 22, 1972)

[Docket No. HM-96, Amdts. 172-17 and 173-67]

PART 172—COMMODITY LIST OF HAZARDOUS MATERIALS CONTAINING THE SHIPPING NAME OR DESCRIPTION OF ALL ARTICLES SUBJECT TO PARTS 170–189 OF THIS CHAPTER
PART 173—SHIPPERS
Etiologic Agents

The purpose of this amendment to the Hazardous Materials Regulations of the Department of Transportation is to provide for the shipment of etiologic agents.

On December 29, 1971, the Hazardous Materials Regulations Board published a notice of proposed rule making, Docket No. HM-96; No. 71-32 (36 F.R. 25163), which proposed these amendments. Interested persons were invited to give their views.

The International Air Transport Association (IATA) and the Air Transport Association (ATA), representatives for many aircraft operators, submitted several similar comments generally objecting to the use of the proposed label, to some of the requirements for better packaging than is now being used by the industry (there are no requirements in the present DOT regulations), and to the prohibition against the shipment of other than exempt etiologic agents via passenger-carrying aircraft.

They objected to the use of the symbol on the label. The proposed symbol was adopted and published as a national standard (USAS Z35.1-1968) by those persons most familiar with these materials and most interested in their safe handling. These comments were filed with the Department of Health, Education, and Welfare (HEW), and the Hazardous Materials Regulations Board, and considered before the publication of the new HEW label in 37 F.R. 12915. The Board cannot justify a requirement for two different labels on a shipping container for etiologic agents. The Department of Health, Education, and Welfare also advised the Board that the skull and crossbones symbol proposed by these commenters was unacceptable. The Board has determined that since a nationally acceptable symbol already exists, the transportation interests should undertake to educate themselves on its meaning and use rather than to develop another symbol.

The associations voiced serious reservations to some of the requirements for better packagings on the basis of past experience and the fact that the packaging would ... "be difficult and expensive to manufacture and demonstrate" ... for compliance with the regulations. The comment was not supported by data. They stated that this was primarily a matter for shippers to resolve. The Air Line Pilots Association (ALPA) stated that the ... "testing procedures proposed are good, but they are the least that should be required for this type of commodity." Inasmuch as no person objected to the packaging requirements, and since the regulations are acceptable to an association which represents persons who must work in an environment involving these packages, the Board believes that the regulations are necessary.

Both IATA and ATA suggested that the proposed reduced pressure requirements did not provide an adequate safety margin under severe depressurization conditions. The requirements of 0.5 atmospheric (absolute) are based on safety margin under severe depressurization conditions. The requirements of 0.5 atmospheric (absolute) are based on existing performance criteria in the Hazardous Materials Regulations. This value is also consistent with present IAEA and IATA regulations. Before making any change the Board wants to evaluate the performance criteria as they presently exist for other materials and consider the need for change wherever the criteria are used. Also, additional consideration is being given by the Board to general pressure requirements for packaging. Regulations for solid carbon dioxide (dry ice) that will apply to air transportation will be covered in a future rule making notice.

The air carrier associations (IATA and ATA) and the representative for the pilots (ALPA) expressed opposing opinions regarding the presence of these materials on passenger aircraft. The carrier associations objected that ... "very considerable hardship and transport difficulties" ... would arise ... "for research and medical facilities" ... since there is much air freight traffic with these materials and ... "many cities are not served by all-cargo aircraft." No data was furnished to support this position. The pilots' association (ALPA), in the opposite view, declared that ... "we are reviewing the possibility of the association adopting a policy in strong opposition to the carriage of etiologic agents by public air transport." The Board received no other indication that its restriction against carriage on passenger-carrying aircraft would cause any hardship. In its deliberations, the Board's prime motivation must be the public interest. After weighing the curtailment of certain shipments in passenger-carrying aircraft against the lack of any demonstrated public interest need for such transportation, the Board concluded that the published regulations are in the best interest of the public.

Several other comments were received regarding definitions and scope of the proposed etiologic agent category. Since these comments related to technical matters with which the Department of Health, Education, and Welfare is more familiar, and since the Board is relying on that Department for adequate identification of what constitutes an etiologic agent and the degree of control necessary for each material, the Board has adopted the decisions of that Department as published in its regulations on June 30, 1972, in the FEDERAL REGISTER (37 F.R. 12915).

(37 F.R. 20554, September 30, 1972)

[Docket No. HM-96; Notice 72-13]
ETIOLOGIC AGENTS
Shipment on Passenger-Carrying Aircraft

On September 30, 1972, the Hazardous Materials Regulations Board published Amendments 172-17 and 173-67 in Docket No. HM-96 establishing requirements for the shipment of etiologic agents.

The regulation permits the shipment of diagnostic specimens and biological products on passenger-carrying aircraft. However, it does not permit the shipment of cultures of etiologic agents on such aircraft. These cultures are presently being transported on passenger-carrying aircraft and prior to the above amendment they were not prohibited.

The Board has received many petitions for reconsideration indicating that the proposed change in transportation conditions would seriously and detrimentally affect the timely response and diagnostic capability of many laboratories involved in the protection of the public health. This position was expressed to the Board by numerous State health agencies, by the American Society of Clinical Pathologists, the American Type Culture Collection, the College of American Pathologists, the Association of State and Territorial Laboratory Directors, The Mycological Society of America, the Institute for Medical Research, The American Association of Immunologists, the American Society for Medical Technology, the American Society for Microbiology, the Infectious Diseases Society of America, the Association of Schools of Public Health, Inc., the American Association of Bioanalysts, a large number of hospitals, clinics, Federal health agencies, and several individual members of the medical profession.

One commenter, the Center for Disease Control, Health Services and Mental Health Administration, U.S. Department of Health, Education, and Welfare, summarized part of the problem by stating that " . . . [a]s an example, the physicians who live in areas not served by cargo-only carriers will be forced to rely on surface transportation to carry cultures to laboratories for determination of antibiotic resistance of cultured bacterial isolates—knowledge which is essential for correct treatment. Loss of time due to slower surface transportation delays the treatment of the patient. In addition, some agents are so sensitive that they may perish if their arrival is delayed in any way. Other problems such as changes in the required degree of acidity, etc., which already cause difficulties in the shipment of microbiologic cultures, will be increased as time between shipment and receipt is lengthened. . . . " This type of concern and statements that the level of protection for public health would be seriously affected permeated the dozens of comments received by the Board.

The Center for Disease Control petitioned that the regulations be amended to permit cultures of etiologic agents in volumes of less than 50 milliliters (1.666 fluid ounces) to be transported on passenger-carrying aircraft. They stated that " . . . [b]ased on our past experience with over 100,000 shipments of etiologic agents annually, our scientific knowledge of these agents, and the public health need for their rapid movement, you are assured that undelayed shipments of cultures of etiologic agents in quantities less than 50 ml. are in the interest of public health, and that the hazard to passengers or crews of aircraft is infinitesimal. . . . "

Throughout this proceeding, the Board has relied on information supplied by the Center for Disease Control because of its expertise in the knowledge and handling of these agents. Based on the information it now has, the Board proposes to modify its regulations and to adopt the proposal of the U.S. Health Services and Mental Health Department. This action would have no effect on the present Department of Health, Education, and Welfare regulations on etiologic agents which continue to apply to the packaging of these substances.

(37 F.R. 25243, November 29, 1972)

[Docket No. HM-96; Amdt. Nos. 171-18, 173-72]
PART 171—GENERAL INFORMATION AND REGULATIONS
PART 173—SHIPPERS
Etiologic Agents

On July 22, 1972, and November 29, 1972, the Hazardous Materials Regulations Board published two notices, (1) 72-9 and (2) 72-13, respectively, in Docket No. HM-96 (37 FR 14728 and 25243). Interested persons were invited to comment on the proposals they contained.

1. Notice 72-9 proposed to require direct reporting to the Center for Disease Control (CDC) of the Department of Health, Education, and Welfare in the case of fire, breakage, spillage, or suspected contamination involving etiologic agents. This report was proposed to replace the immediate report to this Department required by § 171.15 for certain hazardous materials incidents.

Comments were about equally divided in their position for or against the proposal. However, those commenters voicing their opinion against the proposal based their objections on the difficulties for carriers to maintian separate emergency telephone numbers for incidents involving different hazardous materials. Because of a higher probability of confusion, safety in transportation of hazardous materials could suffer. In addition, the public should not be required to make two phone calls, reporting the same matter.

The Board finds that the objections are valid and that, as much as possible, it should not establish rules that would cause proliferation of telephone numbers. Therefore, the amendment provides that reports must be made either to the Department of Transportation or to the CDC. The Board still recognizes the importance of quickly informing the CDC should a report be made to DOT and not CDC. Consequently, the Board has made arrangements to assure that any immediate reports it receives on etiologic agents will be promptly relayed to the Center for Disease Control. Accordingly, the rule has been changed to required reporting to DOT as specified for other hazardous materials. However, any immediate report made directly to CDC will constitute compliance with the regulations without the requirement for an additional call to the Department.

2. Notice 72–13 proposed to authorize that quantities of etiologic agents of less than 50 ml in one outside packaging be exempt from the Hazardous Materials Regulations. Several comments were received on the proposal and each commenter agreed except one, the Atomic Energy Commission. That Commission stated in part: "[h]owever, to exempt them from the regulations in toto by listing them in § 173.386(d) seems to be contrary to the interest of public safety. They should be controlled and regulated, not exempted. They should be allowed on passenger-carrying aircraft by specific provisions for their safe transportation, not by deleting all requirements for their safe packaging and labeling."

The Board has carefully considered the comments from the AEC, which objected in part to the proposed amendment, and all the other comments from other authorities supporting the amendment, and has determined in the interest of public safety, to promulgate the amendment with the exemption. It is to be noted, however, that exempt quantities of etiologic agents will still be regulated by other agencies. The Board specifically stated in the preamble to Notice 72–13 that "[t]his action would have no effect on the present Department of Health, Education, and Welfare regulations on etiologic agents which continue to apply to the packaging of these substances." The petition of CDC and the Board's proposal were based on the fact that the packaging and labeling requirements of 42 CFR 72.25 will continue to apply. No comments or objections were received regarding these packaging requirements.

The Food and Drug Administration of the Department of Health, Eduation, and Welfare noted that it is of critical importance to the public health that it be permitted to ship samples for analysis, including suspect food products and quality assurance samples, on passenger-carrying aircraft. It requested that § 173.386(d) be amended to list "samples for analysis" as an additional exemption. This request is outside the scope of the present rulemaking and will be covered in a separate notice of proposed rulemaking.

(38 F.R. 8161, March 29, 1973)

SUMMARY • **Docket No. HM-98** • *Increased the requirements for preshipment preparation of radioactive materials packages to include certain examinations and tests.*

[49 CFR Part 173]
[Docket No. HM-98; Notice 72–2]
RADIOACTIVE MATERIALS
Preparation of Packages for Shipment

The Hazardous Materials Regulations Board is considering amending §§ 173.389 and 173.393 of the Department's Hazardous Materials Regulations to specify requirements to be fulfilled before each shipment of radioactive materials packages.

As a result of past experience arising from the very few incidents involving leakage from radioactive materials packages in transportation, it has become increasingly evident that more attention and emphasis should be placed by the shippers on the proper preparation of their packages for shipment. Although the regulatory standards are considered quite adequate in terms of the containment, radiation, and contamination limitations, the Board believes the regulatory language in the general packaging requirements, which effectively relates to quality control, is not sufficient. The reasons behind improper quality control or failure to properly prepare packages for shipment usually involve a breakdown in administrative controls, rather than a faulty design or inadequate packaging standards, per se.

The U.S. Atomic Energy Commission has recommended to the Board that certain wording be added to the general packaging requirements for radioactive materials. The Board agrees that such wording would be helpful, not only in clarifying the quality control and package preparation requirements, but also would improve the enforceability of the regulations which may be necessitated due to faulty package preparation. Further, such regulatory language is generally consistent with the present standards, as well as those proposed by the International Atomic Energy Agency (IAEA).

(35 F.R. 5641, March 17, 1972)

[Docket No. HM-98; Amt. No. 173–66]
PART 173—SHIPPERS
Radioactive Materials; Preparation of
Packages for Shipment

The purpose of this amendment to the Department's Hazardous Materials Regulations is to improve the requirements for preshipment preparation of radioactive materials packages by prescribing certain examinations and test procedures.

On March 17, 1972, the Hazardous Materials Regulations Board published Docket HM-98; Notice 72–2 (37 F.R. 5641) proposing certain additions to §§ 173.389 and 173.393 of the Hazardous Materials Regulations. Interested persons were given an opportunity to comment on the proposed changes.

Numerous comments were received from persons representing shippers, carriers, labor unions, medical isotope users, and military. The great majority of comments were in favor of the amendment, although several changes were recommended. On the basis of these suggestions, the Board has made several changes in this amendment.

The introductory language of § 173.393(m) has been clarified to more appropriately reflect and emphasize the quality control and administrative procedural requirements which follow.

A significant number of commenters, including the Atomic Industrial Forum and the Atomic Energy Commission, recommended that for air shipment § 173.393(n) require a special preshipment leakage test on each package containing liquid radioactive material exceeding a Type A quantity. Such a test provides confirmation of the existing performance requirement in § 173.398(b) (2) (iii). In view of these comments and the circumstances peculiar to air transportation, particularly those involving a package containing a greater quantity of radioactive material, the Board has changed § 173.393(n) to require this preshipment leak test. The Board has provided that the test may be conducted either on the entire containment system as a unit, or on any receptacle or vessel within the containment system.

Several comments were received from manufacturers of irradiated fuel casks regarding an inconsistency between the definition of "maximum normal operating pressure" (MNOP) in § 173.389(n) and its use in § 173.393(n) (8). The inconsistency involved the 1-year period specified for pressure buildup in the MNOP definition when compared to the proposed requirement that the internal pressure of the containment system must not exceed the MNOP during the "anticipated period of transport." Commenters stated that § 173.389(n) does not recognize the possiblity of the MNOP being below atmospheric. Also, some commenters considered the word "normal" in the MNOP definition inappropriate since it ignores the provisions which might be made for venting or for ancillary cooling systems. The Board acknowledges these difficulties presented by the proposal.

The MNOP definition was taken from the draft of proposed changes to the International Atomic Energy Agency Regulations (IAEA) which are now expected to be published late in 1972. The IAEA definition of MNOP sets forth the maximum pressure that could develop under credible conditions of transportation. It assumes such things as malfunction of a pressure relief device, misplacement of a package for as long as 1 year, and lack of provision for external mechanical cooling or administrative shipment controls. It establishes an idealized design benchmark. However, within the context of the IAEA Regulations, it will be qualified by several technical requirements not presently in the Hazardous Materials Regulations. These will relate to the criteria for international competent authority certifications of package design in contrast to the "unilateral" or "multilateral" approval concepts.

The Board acknowledges that in the development of § 173.393(n) (8), it did not consider the fact that in many package designs, the MNOP would never be reached because of pressure relief devices, external cooling systems, etc. Further, for domestic transport purposes, the period of transport of large irradiated fuel packages is not likely to exceed 2 months. For these reasons, it is neither necessary nor appropriate to require that the containment system be designed to meet the MNOP definition.

Therefore, § 173.393(n) (8) is modified to require that for any package likely to develop a significant internal pressure, such pressure may not exceed the "design pressure" at any time during transportation.

One commenter suggested modifying the requirement in § 173.393(n) (2) that packagings be in "unimpaired physical condition." He stated that this rule could be interpreted in an unnecessarily restrictive manner and preclude repeated shipments of containers which have only superficial damage such as scratches or surface oxidation. The Board does not agree with this comment. However, it recognizes that a certain amount of judgment is involved to determine if a package is impaired to safety in transportation.

The source of the values in § 173.393(a) (7) concerning pressure differential and any future contemplated changes are of particular interest to some persons according to comments the Board has received in this docket and other dockets relating to other classes of hazardous materials. The value of 0.5 atmosphere (absolute) (7.3 p.s.i. or 0.5

kg/cm²) in § 173.393(a) (7) is based on the existing performance criteria in § 173.398(b) (2) (iii). That value is also consistent with existing IAEA and IATA Regulations. However, it is foreseen that, with the expected revisions to the IAEA Regulations later in 1972 and subsequent revisions to the IATA Regulations in 1973, this value may be changed to 0.25 atmospheric absolute (3.56 p.s.i. or 0.25 kg/cm²). Therefore, the Board advises that it may propose such a change at some future date, together with other changes, to make the Hazardous Materials Regulations as compatible as possible with international standards.

(37 F.R. 17969, September 2, 1972)

SUMMARY • Docket No. HM-99 • *Established a new DOT specification 3T cylinder (tube trailers), and provided for its use; made miscellaneous changes regarding the use of DOT 3AX and 3AAX cylinders.*

[49 CFR Parts 171, 173, 178]

[Docket No. HM-99; Notice 72–3]

TRANSPORTATION OF HAZARDOUS MATERIALS
Specifications 3AX, 3AAX, and 3T Cylinders

The Hazardous Materials Regulations Board is considering amendments to §§ 171.7, 173.34, 173.301, 173.302, and 173.304, and addition of a new § 178.45, to provide for the shipment of certain gases in large cylinders or tubes mounted on a motor vehicle.

These proposed changes are based on several outstanding permits and two petitions for rule change from the Compressed Gas Association, Inc. Nineteen shippers, including most major gas suppliers, are involved in the special permits which have been outstanding for up to 4 years. Consequently, the Board has evaluated considerable experience data. It has found that, without exception, all reports have cited satisfactory experience, without loss of any product.

Although not specifically covered by petitions, the Board specified several special conditions of transportation in the permits and is proposing them in this rule making action. These special conditions are covered essentially by the proposed new paragraph (1) to § 173.301.

Also covered by this notice are matters relating to manufacturer's registration and method of maintaining cylinder reports. However, the purposes of the Board regarding these new items have been the subject of similar rule making actions in Dockets HM-27 and HM-69, and, therefore, are not discussed specifically in this docket.

(37 F.R. 6747, April 4, 1972)

[Docket No. HM-99; Amdts. Nos. 171–21, 173–75, 178–27]

SPECIFICATIONS 3AX, 3AAX, and 3T CYLINDERS

The purpose of this amendment to the Hazardous Materials Regulations is to amend §§ 171.7, 173.34, 173.301, 173.302 and 173.304, and to add a new section 178.45, DOT specification 3T, to provide for the shipment of certain gases in large cylinders or tubes mounted on a motor vehicle.

On April 4, 1972, the Hazardous Materials Regulations Board published a notice of proposed rulemaking. Docket No. HM-99; Notice No. 72–3 (37 FR 6747), which proposed this amendment. Interested persons were invited to give their views and several comments were received by the Board.

1. Hydrogen chloride. A commenter stated that the proposed regulations omitted the authorization for hydrogen chloride to be shipped in DOT-3T cylinders. This authorization was purposely omitted because the Board has not completed its evaluation of DOT-3T cylinders in hydrogen chloride service.

2. Nonliquefied natural gas. Another commenter requested that the regulations be amended to permit nonliquefied natural gas to be shipped in DOT-3AX, 3AAX, and 3T cylindrs. The Board has not authorized nonliquefied natural gas to be shipped in these cylinders because it has not been demonstrated that impurities which may be present in the gas would not affect the structure of the particular steel used in the manufacture of these cylinders.

3. Design criteria for cylinders. The proposed regulations provide design criteria for cylinders of 1000 pounds or more water capacity. A commenter has stated that this same design criteria should be applicable for cylinders of less than 1000 pounds water capacity and has requested that, wherever appropriate, the proposed regulations be amended to include the design criteria for the smaller water capacity cylinders. As this request is not within the scope of the present rule making, the Board did not address itself to the comment in this amendment.

4. Cylinder attachment to vehicle. Another commenter objected to the proposal which would require the rear end of the cylinders to be affixed to the vehicle. This commenter stated that adequate expansion provisions can be incorporated in the cylinders when they are affixed to the vehicle at the front end of the cylinders. The Board agrees that it need not specify rear end rigid attachment and has authorized the cylinders to be affixed at either end with thermal expansion provisions required at the opposite end.

5. Editorial. Section 178.45–11(d) is reworded to more clearly state the intent of the regulation. As previously worded, the words "the test may be repeated" did not clearly indicate the other option of reheat treatment and subsequent compliance with all the prescribed tests. Consequently, the objective of the use of the word "may" could be misunderstood.

(38 F.R. 21989, August 15, 1973)

SUMMARY • **Docket No. HM-100** • *Improved the requirements for the shipment of ethylene oxide in tank cars and deleted authorizations for ARA-IV-A, ARA-IV, DOT-104, and 104W tank cars in this service; provided for use of DOT-51 portable tanks.*

[49 CFR Parts 173, 179]
[Docket No. HM-100; Notice 72–4]
TRANSPORTATION OF HAZARDOUS MATERIALS
Ethylene Oxide; Opening in Tank Car Heads

The Hazardous Materials Regulations Board is considering amendment of §§ 173.124, 179.102, 179.201, and 179.202 of the Department's Hazardous Materials Regulations to authorize the shipment of ethylene oxide in insulated portable tanks and to upgrade the specifications of tank cars authorized for ethylene oxide. In addition, the Board proposes to remove authorization for the use of certain other tank cars in this service.

The Board has received petitions to make these changes to the regulations. Support for the petition to permit ethylene oxide to be transported in specially modified Specification 51 portable tanks is based on favorable experience data reported to the Board on shipments moving since 1964 under special permit.

The proposed changes for tank cars are based on recommendations by the Manufacturing Chemists Association, Inc. Its petition indicates that the present ethylene oxide tank car spacifications warrant revisions for improved safety performance, that some currently authorized tank cars for this service are obsolete, and that others are inadequate for safe rail transportation.

The Board believes that the adoption of this proposal would provide greater safety in the transportation of ethylene oxide. Also, the Board requests advice on the need for continuing the authorization for "Openings in tank heads to facilitate application of lining" which is found in § 173.124(a) (5) and numerous other sections such as §§ 173.119(a) (12), (e) (2), and (f) (3), 173.314(c) Note 16, 173.354(a) (4), 179.102–12, 179.102–17, 179.102–20, 179.102–6(a) (3), 179.202–1, and 179.202–18. The Board believes that this is an obsolete requirement and is no longer needed.

The Board is developing improved identification requirements for tank cars containing certain hazardous materials such as ethylene oxide. As part of this development, the Board will propose changes to the present marking requirements for tank cars in a separate notice. Any changes resulting from that notice of proposed rule making would be reflected in those sections dealing with marking of ethylene oxide tank cars.

(37 F.R. 6871, April 5, 1972)

[Docket No. HM-100; Amdt. Nos. 173–82, 179–14]
PART 173—SHIPPPERS
PART 179—SPECIFICATIONS FOR TANK CARS
Ethylene Oxide; Opening in Tank Car Heads

On April 5, 1972, the Hazardous Materials Regulations Board ("the Board") published a notice of proposed rulemaking, Docket No. HM-100; Notice No. 72–4 (37 FR 6871), which proposed these amendments. The reasons for all these amendments were discussed in that notice of proposed rulemaking. Interested persons were invited to give their views and several comments were received by the Board.

The purpose of these amendments is to authorize the shipment of ethylene oxide in insulated portable tanks and to upgrade the requirements of filling and the specifications of tank cars authorized for ethylene oxide. In addition, Association of American Railroads specification ARA-IV-A and ARA-IV, and DOT specification 104 and 104W, tank cars are no longer authorized to be used for the transportation of ethylene oxide.

Gas padding. Several comments were received regarding the adequacy of the requirements proposed by the Board in Notice No. 72–4 covering gas padding. Several commenters pointed out that there are a number of gases that are suitable for padding. However, because of differences in solubility of these gases in liquid ethylene oxide, differences in loading temperatures and other factors, different amounts of padding gas (and hence different charging pressures) are required to provide the same level of safety for these different loading conditions. The Board agrees that the proposed pressure range of 35 to 60 p.s.i.g. at 70° F. does not completely reference the range of padding pressures producing safe loading and transportation conditions. The Board also agrees with one commenter's suggestion that it is more appropriate to state the padding gas requirements in the form of performance criteria. As a consequence, §§ 173.124(a) (5) and (a) (6) have been changed from what was proposed in accordance with these recommendations.

Fusible plugs on portable tanks. In Notice No. 72–4, the Board proposed requirements for portable tanks to be used in the transportation of ethylene oxide. The proposal was based on extensive experience gained by one shipper under special permit. Essentially, the proposal was based on the conditions existing in the special permit. One commenter, the special permit holder, suggested increasing the proposed thickness of the tank shell material since permit experience was with heavier gauge tanks. The Board agrees with this suggestion and has so provided in § 173.124(a) (6). With respect to these portable tanks, the Board proposed to require that the tanks be equipped with safety relief valves or frangible discs. Under the special permit, these tanks have been equipped with fusible plugs, and not safety relief valves or frangible discs. The same commenter requested that fusible plugs, in addition to safety relief valves or frangible discs, be permitted for such small tanks (maximum 300 gallons) designed according to the same basic criteria as are DOT-5P insulated drums which are used for liquefied petroleum gases. The commenter pointed out that the DOT-5P drum specification was established based on extensive fire testing to determine the proper sizing of the fuse plugs to prevent rupture of the container under fire exposure and that the surface to volume ratio of his portable tanks is the same as that of a DOT-5P drum. The Board agrees that, with more specific requirements applicable to the fusible plug and to performance criteria for the insulation, fusible plugs may be permitted in this size packaging. Section 173.124(a) (6) specifies the requirements accordingly.

Gaskets. The Manufacturing Chemists Association, Inc., the submitter of the petition upon which Docket No. HM-100 was based, pointed out that it did not recommend Teflon, or interwoven stainless steel and Teflon, as the only acceptable resilient materials for gaskets in ethylene oxide service. It stated that it wished to merely identify these materials as being among those suitable for ethylene oxide transportaton service. It noted that to list all of the suitable materials in the regulations would not be practicable nor desirable since to do so could place unnecessary restraints on the development of technology in this application. It stated that it believed that it would be more appropriate to specify the materials that are not suitable (neoprene, natural rubber, and asbestos) and to state the requirements for gaskets on a performance criterion basis. In a related observation, the Board notes that the Coast Guard has successfully applied this philosophy to gasketing materials for use on ethylene oxide barges. (See 46 CFR 40.05–40, valves, fittings, and accessories.) In consideration of these factors, the requirements concerning gaskets have been restated in performance terms. Certain undesirable materials, including asbestos, have been prohibited. One commenter objected to the absence of approval for compressed asbestos gaskets stating that such gaskets have been in use and he has found no evidence to support their prohibition. The Board has been advised that asbestos is not suitable because it significantly lowers the autoignition temperature of ethylene oxide. In addition, neoprene and natural rubber are considered unsuitable for use because of their undesirable reaction to ethylene oxide. Another commenter stated that the prohibition of certain materials and the establishment of performance criteria for gaskets were fully justifiable but concluded that the effective date of these requirements should be delayed for 18 months after the effective date of the amendments in this docket. The Board does not agree with such a delay. Since the hazards of neoprene, natural rubber, and asbestos with ethylene oxide have been recognized, there is no justification for delaying their replacement for what would be over two years.

(39 F.R. 24909, July 8, 1974)

SUMMARY • **Docket No. HM-101** • *Proposed special marking requirements on tank cars transporting certain hazardous materials. Action remanded to Docket Nos. HM-103 and HM-112.*

[49 CFR Parts 173, 179]
[Docket No. HM-101; Notice 72–5]
TRANSPORTATION OF HAZARDOUS MATERIALS
Markings on Tank Cars

The Hazardous Materials Regulations Board is considering amendment of several sections in Parts 173 and 179 of the Department's Hazardous Materials Regulations. These amendments would require the name of certain hazardous materials being transported to be marked on the sides of tank cars in compliance with specific detailed marking requirements.

Based on the Department's observations during accident situations, and on recommendations from fire, safety and police personnel, the Board believes that existing regulations covering special commodity markings on tank cars are not adequate. Identification by name of certain materials in tank cars by improved markings designed for greater visibility is an area of the regulations which needs upgrading. Improvements in marking would contribute to overall safety in transportation of these hazardous materials by tank car.

Several flammable liquefied compressed gases have been frequently involved in accidents. The present regulations do not require the names of these materials to be identified on tank cars. This proposal is to require the name of any flammable liquefied compressed gas being transported to appear on the tank car. In addition, the proposal is to change the marking requirements for other hazardous materials that are now required to be identified by marking on a tank car. These same changes would be made to cars covered under Docket No. HM-91; Notice No. 71-25 (36 F.R. 20166).

The proposal states that the commodity marking on a tank car must be in compliance with certain detailed requirements. These requirements are found in proposed § 173.31(a) (6) and specify the height of the letters, the size of the marking stroke, the spacing of the letters, and the use of contrasting colors.

Editorial changes are proposed in § 173.31 (a) (4) for clarification purposes.

(37 F.R. 7104, April 8, 1972)

SUMMARY • **Docket No. HM-102** • *Established expanded definitions for flammable, combustible and pyroforic liquids, as determined by use of the closed-cup tester. See also Docket Nos. HM-42, HM-67, HM-112, and HM-133.*

[49 CFR Part 173]
[Docket No. HM-102; Notice 72–7]
TRANSPORTATION OF HAZARDOUS MATERIALS
Flammable, Combustible, and Pyroforic Liquids; Definitions

The Hazardous Materials Regulations Board in considering an amendment to § 173.115 of the Department's Hazardous Materials Regulations to specify a new definition for the class of materials identified as "Flammable Liquid" and to create and define a new class of materials identified as "Combustible Liquid." Also, it is proposing to modify the definition for pyroforic liquids within the "Flammable Liquid" class.

On February 27, 1968, the Board published a notice of proposed rule making, Docket No. HM-3 (33 F.R. 3382) proposing a new definition for "Flammable Liquid." On February 21, 1970, the Board published a notice of proposed rule making, Docket No. HM-42; Notice No. 70-3 (33 F.R. 3298) proposing to create and define a new class of materials identified as "Combustible Liquid." On December 5, 1970, the Board published a notice of proposed rule making, Docket No. HM-67; Notice No. 70-23 (35 F.R. 18534) proposing to change the method for determining the flashpoint of materials from the Tagliabue open-cup test method to the Tagliabue closed-cup test Materials Regulations. The matters proposed in those dockets, hereafter referred to as 3, 42, and 67, are hereby consolidated within this docket and the reasons and justifications, except as modified herein, given in their preambles are made a part of this rule-making proposal.

The proposals made in 3, 42, and 67 raised considerable controversy. Comments were addressed to the need for change, the degree of change, the method specified for testing, and the lack of uniformity in defining flammable and combustible materials.

Combustible liquids. Much interest was expressed in the proposal to regulate "combustible liquids." The Board notes that while virtually all commenters acknowledged the problem the rule making was designed to solve, there was considerable divergence of views on the proposed solution. In 42, the Board described the problem as follows:

Combustible liquids are routinely transported in tank cars, tank trucks, and portable tanks with no requirement that these tanks be identified during transportation as containing a material having a fire hazard.

Fire, police, and rescue personnel are generally trained to deal with fuel oil and kerosene accidents in the same manner as they deal with gasoline accidents. In order to be able to do their job, they must have immediate information regarding the contents of these tanks. Without this information, the emergency personnel might well be misled into believing that the tanks contained some innocuous commodity such as milk or molasses. Their attention might, therefore, be misdirected away from this significant potential hazard. . . .

Compounding the problem of lack of informatoin as to hazards is the fact that many tank truck operators are transporting combustible liquids in tanks which bear the placard "Non-Flammable." This is apparently done in order to be able to permanently mark the word "Flammable" on tanks which are used interchangeably in shipping flammable or combustible liquids. In that way, the carrier need only to add a small tag or plate with the word "Non" on it rather than having to constantly remove and replace a larger placard having the word "Flammable." Placarding of this type is a gross misrepresentation of the actual hazard that would be present should such vehicles be involved in accidents, parked or stopped near fires, or otherwise placed in jeopardy.

No one questioned the basis for the Board's concern. In fact, several commenters, including State governments, agreed that a problem existed that required solution for the public's protection. Rather than question the need for the new classification, most commenters addressed themselves to the details of scope and implementation.

One commenter noted that 18 U.S.C. 834 directs the Department "to formulate regulations for the safe transportation within the United States of explosives and other dangerous articles, including radioactive materials, etiologic agents, flammable liquids, flammable solids, oxidizing materials, corrosive liquids, compressed gases, and poisonous substances." He contended that the word "including" tended to limit the Department's jurisdiction to regulation of the listed items, thereby excluding "combustibles." As a common practice in legal drafting, utilized throughout the

United States Code, the term "including" serves to introduce examples of a broad class of items in order to provide a partial definition of that class. The Board believes this to have been the intent of Congress in enacting the Explosives and Combustibles Act of 1908, and is of the opinion that the contention of lack of jurisdiction is without merit.

A very large percentage of the commenters on 42 addressed themselves to what temperature level, 150° F. or 200° F., is the more justified upper limit. The same sources were cited in certain instances to support either the 150° or the 200° break point. This depended on the approach they considered in citing the reference.

No convincing argument was presented to support the 150° F. cutoff. The Board believes that it must not ignore the significant number of materials having flashpoints between 150° and 200° F. being transported. To do so would not accomplish the stated objective of its proposal. The Board is aware of the fact that these materials have flashpoints higher than credible ambient temperatures, and that they are less likely to ignite than the lower flashpoint materials. Their vapors, however, can ignite when exposed to elevated temperatures caused by other than normal ambient conditions. Several commenters suggested the Board had no adequate accident data in the area of higher flashpoint materials. It is true that such data is limited due to the fact that these materials have never been covered by a hazardous materials incident reporting procedure. There are, however, accident reports on file with the Bureau of Motor Carrier Safety, Federal Highway Administration, that relate the facts of accidents involving fires fueled by "combustible liquid" cargoes. For those who question the potential of these materials to cause or contribute to harm, the Board urges reading of the National Transportation Safety Board report, dated March 7, 1968, on the railroad-highway grade crossing accident in Everett, Mass., on December 29, 1966. The tank motor vehicle involved in that particular accident was transporting fuel oil. Thirteen people were killed " . . . due to thermal burns and smoke inhalation . . . "

One concern expressed by several commenters was the need for establishing a new classification. This is necessary because of the structure of the Hazardous Materials Regulations. Before a material is regulated as a hazardous material, it must be classed as a hazardous material. Although the Board has several rule making actions and studies in progress concerning test and definition criteria for the classification of materials, it does not contemplate any change from a classificaton type of system. It is necessary, therefore, to establish a "Combustible Liquid" classification.

Method of test for flashpoint. In its proposal to convert from the open-cup to the closed-cup test method in 67, the Board said:

The flash point is generally accepted as a useful means to determine the flammability of flammable liquids, and therefore their potential fire hazard during transportation. The Tagliabue open-cup testing method, which has been in use with only minor modification for many years, lacks the precision, reliability, and reproducibility necessary to properly estimate the flammability hazard that may be encountered during transportation . . . .

As part of the Department's overall review of the Hazardous Materials Regulations, the Board and the staff of the Office of Hazardous Materials (OHM) have been evaluating methods used for classification of materials according to the hazard presented during transportation. OHM contracted with the Safety Research Center, U.S. Bureau of Mines, to examine the limitations of the available flash point testers and to recommend the best method for adoption by DOT.

In reaching their conclusions, the Bureau of Mines measured the present state of the art against the following criteria:

1. Repeatibility (data obtained by the same analyst in several determinations, using the same equipment and the same sample).

2. Reproducibility (data obtained by several analysts, each using a different piece of equipment of the same type, and using the same sample).

3. Reliability in assessing the fire or explosion hazard.

In addition, the Bureau of Mines considered and evaluated all comments received in response to that part of a prior notice of proposed rulemaking (NPRM)[1] dealing with definitions of flammable liquid, flashpoint, open-cup tester, and closed-cup tester. The results and recommendations of the Bureau's study have been reported.[2]

The Bureau's report recommends that the Tag closed-cup method be used to determine flashpoints of flammable liquids for purposes of the DOT Hazardous Materials Regulations. The conclusions, proposing adoption of the closed-cup method, may be summarized as follows:

1. The closed-cup method is more precise and reliable than the open-cup method, gives more reproducible data, and provides a more conservative estimate of the hazard presented by the formation of flammable vapor-air mixtures under either confined or unconfined conditions.

---

[1]Docket No. HM-3; Notice No. 68–2 (33 F.R. 3882, Feb. 27, 1968).

[2]Kuchta, Joseph M. and Burgess, David, Report No. S. 4131, Apr. 29, 1970, Safety Research Center, U.S. Bureau of Mines. This document is available from the Clearing House for Federal Scientific and Technical Information, National Bureau of Standards, U.S. Department of Commerce, Springfield, Va. 22151, at a cost of $3 per copy, or Microfiche copy at 65 cents.

2. It is often proposed that an open-cup more nearly approximates the geometry of a spill situation than does a closed-cup. In our judgment, this is a trivial consideration in choosing among the variations of existing apparatus. The actual likelihood of ignition of a spill depends heavily upon factors which are beyond the scale of laboratory apparatus, such as the cooling of the liquid surface by evaporation or the gustiness of the atmosphere.

The greatest explosion hazard results from leakage or spillage into surroundings that provide some confinement, such as a railroad boxcar, a van-type truck, or the hold of a ship. In this situation, convection currents aid the formation of homogeneous vapor-air mixtures and the magnitude of overpressures in confined combustion is usually greatest with homogeneous mixtures. Here again, the closed-cup gives the best definition of hazard. Experience shows that spills and leaks in confinement are common accident situations and must be considered in the development of safety criteria.

3. Due to its greater reliability, the closed-cup method has been accepted by the National Fire Protection Association, the National Academy of Sciences, the United Nations Intergovernmental Maritime Consultative Organization (IMCO), and many western European industrial countries, including Great Britain, France, West Germany, Sweden, and the Netherlands.

Additional reasons supporting the closed-cup method may be found in a review of various technical publications and comments received on a prior notice of rule making. The following is quoted from the International Chamber of Shipping's statement which was attached to the IMCO October 15, 1969, communication to the sixth session of the Committee of Experts on the Transport of Dangerous Goods:

The closed-cup method of testing should be used rather than the open-cup method in view of the former's much better precision.[3]

Proponents of the open-cup method point out that improvement in technique in recent years has resulted in increased precision and reproducibility of data. It is agreed that refinement of test methods has brought some improvement. However, in spite of this improvement, the Board believes that the open-cup is still not equal to the closed-cup method for overall transportation safety purposes. For example, the report of Technical Subcommittee No. II of the Chicago Society for Paint Technology[4] summarizes the testing done during 1968 with six different types of flashpoint testers and 27 solvents having flash points ranging from 20° F. to 190° F. The report concluded that, "All closed-cups were considerably more reliable and easier to work with than the other cups. . . ."

Some comments received on Docket HM-3; Notice No. 68-2 stated that a closed-cup is not responsive to mixtures that contain low-volatility nonflammable components; it is, on the other and, far too stringent for mixtures containing very small (less than 0.2 percent) amounts of highly volatile flammable compounds. During the test of a mixture, the closed-cup can concentrate nonflammable vapors as readily as flammable vapors. These nonflammable vapors can have a suppressant effect upon the flammability of the sample, thereby raising the flashpoint beyond the limit prescribed in the regulations for flammable liquids. In an open-cup, part or all of the vapors can escape, thus reducing this suppressant effect. On the other hand, comments noted that a nonflammable anti-knock compound containing less than 0.2 percent of dissolved hydrocarbon, because of trapping of the hydrocarbon traces in the vapor space of the apparatus, had a closed-cup flashpoint of 58°–73° F., compared to an open-cup flashpoint of 180°–245° F.

The Board realizes that none of the presently available test methods accurately applies to all mixtures. To cover the unusual behavior of certain mixtures, the Board can issue the necessary rulings. For example, the Board could classify such mixtures according to the flash point of their major component. There may be alternative means to cover certain mixtures which do not lend themselves to the proposed testing procedure, and the Board welcomes any suggestions in this regard. The decision as to proper classification of exceptions could be based upon other data or experience showing that the liquid is more or less hazardous than the flashpoint data indicate. The exceptions should not govern the general rule, however, and the Board is concerned with covering the great majority of substances by a single test method. . . .

The other principal matter in the preamble of 67 dealt with the Board's intent to not change "the present established classification ranges or packaging of flammable liquids," a position that has been modified and which will be discussed later in this preamble.

The comments made in response to 67 were rather diverse, ranging from full support of hte proposal to being totally against it in all respects. Since that proposal is being modified by this notice, no attempt will be made to respond to all of the arguments presented, only to those that relate to changes made in response to comments.

Several commenters again pointed out that small quantities of volatile nonflammable materials in mixtures could mask the danger of "flammable" materials. The Board agrees and is proposing that tests be conducted on partially evaporated samples of mixtures. Conversely, another commenter pointed out that very small amounts of dissolved

[3]United Nations Economic and Social Council, E/CN.2/CONF.5/R.198.

[4]Probst, K. G., Correlation of Apparatus for Measuring Flash Point of Solvents, J. of Paint Technology, Vol. 40, No. 527, pp. 576–81 (Dec. 1968).

hydrocarbons (in his case less than 0.2 percent) in a mixture could cause an anomalously low closed-cup flash point. The Board agrees that very small quantities of materials, meeting a proposed definition, should not have the effect of making 99 percent or more of a mixture subject to the requirements pertaining to that definition. Therefore, it is proposing exceptions to the two definitions. If tests on a material prove positive, the shipper will be afforded the opportunity of analyzing his material to determine if 99 percent or more of its components, when tested, do not meet either or both of the proposed definitions. Several commenters pointed out that the Tag closed-cup method is not appropriate for viscous materials and liquids which tend to form a surface film under test conditions, such as most paint products. The Board agrees and its proposing use of the Pensky-Martens Closed Tester (ASTM D93–71) for these materials as well as liquids that contain suspended solids.

The Board is proposing a modification of the proposal it made in 67 by raising the flashpoint for "flammable liquids" to (but not including) 100° F. closed cup. Also, it is proposing to change the upper limit for "combustible liquids" from 200° F. open cup to 200° F. closed cup with the same test criteria applicable to both definitions. The two principal reasons for these proposed modifications are: (1) To more properly reflect credible ambient temperatures in defining "flammable liquids," and (2) uniformity.

Ambient temperatures. A report entitled "A Survey of Environmental Conditions Incident to the Transportation of Materials"[5] was recently prepared for the Department. In the "Summary of Conclusions" portion of the report, the following statement pertaining to temperature is presented:

4.7 Temperature. From the results of storage temperatures reported in the western desert, northern cold regions (Maine, Alaska, Washington), various other storage areas in the continental United States, Puerto Rico, and Hawaii, it is seen that temperature criteria for military equipment are too severe. A more accurate, but still conservative criterion is to apply extremes of local air temperatures. While this appears to neglect the results of solar thermal radiation, which for desert areas in summer is great, the data indicate that the thermal inertia and insulation of storage structures is sufficiently great such that attenuation of the swings in air temperature inside storage chambers results. The recorded extremes in air temperature in storage areas over the entire range of localities and structure types is −9° F. to 119° F., a much narrower range than the −65° F. to 160° F. expected values stated in MIL-STD-210A.

A limited amount of data for truck and rail transport also indicate that the cargo material undergoes swings in temperature which are greatly diminished from that of the forcing functions, the outdoor air temperature and solar thermal radiation.

The referenced 119° F. was arrived at from a report on the occurrence of higher temperatures in standing boxcars in which the highest measured temperature was 119° F. Similarly, in another study made under extreme temperature conditions in Death Valley, an overall maximum skin temperature and temperature within the cargo under test in a truck was 116° F. in response to a 130° F. maximum outside temperature on the day of the test. The Board concludes that it can reasonably assume that the temperature of cargo in transport vehicles can and often will reach or exceed 100° F. under conditions normally incident to transportation. This view is further supported by dry-bulb air temperatures for a 10-year period for 91 stations operated by the U.S. Weather Bureau. Temperature maximums for 10 representative locations were as follows:

Weather Bureau station	Period of record	Dry-Bulb Temperature Maximum
Chicago, Ill	January 1949–December 1958.	104
El Paso, Tex	January 1950–December 1959.	106
Los Angeles, Calif	January 1949–December 1958.	107
Miami, Fla	January 1948–December 1958.	98
Montgomery, Ala	January 1949–December 1958.	105
New Orleans, La	do	100
Phoenix, Ariz	do	117
San Antonio, Tex	do	
Seattle, Wash	do	97
Washington, D.C.	do	102

[5]"A Survey of Environmental Conditions Incident to the Transportation of Materials, October 1971, PB-204-442" prepared by General American Research Division of GATX. This document is available from the Clearing House for Federal Scientific and Technical Information, National Bureau of Standards, U.S. Department of Commerce, Springfield, Va. 22151 at a cost of $3 per copy or Microfiche copy at 95 cents.

The above data do not reflect the effects of radiation on transport vehicles and storage facilities used during the course of transportation.

The Board believes the regulations that apply to flammable liquids as they are defined at present should be made applicable to materials meeting its proposed new definition. However, the Board will consider providing additional packagings for these materials newly covered by the regulations if it adopts this proposal as an amendment.

Uniformity. One type of comment repeated often in 3, 42, and 67 was a need for uniformity among the different regulatory agencies and other organizations having an effect on the manner in which shippers and carriers ship, store, and handle flammable and combustible liquids. Following publication of 67, this situation was further compounded by publication of new regulations by the Occupational Safety and Health Administration, U.S. Department of Labor, on May 29, 1971 (36 F.R. 10529) defining a flammable liquid as any liquid having a flashpoint below 140° F. (closed cup) and a combustible liquid as any liquid having a flashpoint at or above 140° F. and below 200° F. The Board agrees with the commenters who voiced their concern over the lack of uniformity and believes the area of greatest concern is the interface between transportation and nontransportation activities under the jurisdiction of the Department of Transportation and the Department of Labor, respectively. Another agency, the Food and Drug Administration, Department of Heath, Education, and Welfare, has definitions for these materials defined by statute that are not consistent with the proposals herein. However, since the regulations of FDA are addressed to consumer-type packages that are primarily inside packages during transportation, the Board believes its most immediate concern should be the development of regulations compatible with those of the Department of Labor. The Assistant Secretary of Labor for Occupational Safety and Health agrees that there is a need for uniformity and his proposal for the modification of definitions set forth in 29 CFR 1910.106(a) are published at page 11901 of this issue of the FEDERAL REGISTER.

The Board will continue to seek adoption of the definitions proposed herein by all agencies in the United States, both State and Federal, and will also seek their adoption internationally.

Implementation. Some commenters requested that sufficient time be provided for re-evaluation of materials under the test method that was proposed in 67. The Board believes that approximately 1 year should be provided to permit testing and other necessary adjustments to accomplish compliance with the regulations under the new definitions. However, compliance should be authorized at an early date to permit adherence to the regulations of the Occupational Safety and Health Administration.

There are no proposals in this docket pertaining to placarding or marking of vehicles and portable tanks as proposed in 42. The Board will be making proposals in this area in the near future in a separate notice. Also, that portion of 42 pertaining to materials transported at temperatures higher than their flash points is not proposed in this docket as a mandatory requirement but in advisory language pertaining to materials that have flash points of 200° F. or higher.

(37 F.R. 11898, June 15, 1972)

[Docket No. HM-102; Amdt. Nos. 172–73, 173–78, 174–19, 177–29]

FLAMMABLE, COMBUSTIBLE, AND PYROPHORIC LIQUIDS, DEFINITIONS

The purpose of this amendment to the Department's Hazardous Materials Regulations is to:
1. Specify a new definition for the class of materials identified as "Flammable liquid";
2. Create and define a new class of materials identified as "Combustible liquid";
3. Modify the definition for Pyrophoric liquids within the flammable liquid class; and
4. Set for the requirements for the materials that are covered by the new definitions.

On June 15, 1972, the Board published a notice of proposed rule making, Docket No. HM-102 (37 FR 11898), proposing the new definitions. Consolidated within the notice were matters previously covered by Docket Nos. HM-3 (33 FR 3382), HM-42 (35 FR 3298), and HM-67 (35 FR 18534). Interested persons were invited to participate in this rule making proceeding and all comments received have been given full consideration by the Board before it decided on the amendments made herein.

A large number of comments were received by the Board. Many commenters, principally from fire service organizations, requested that the Board take no action that would have an adverse effect on the standards set forth in National Fire Protection Association (NFPA) Standard No. 30. The NFPA's definition for flammable liquids included those liquids having flash points below 140° F. (closed cup) for several years. The Board assumes that the commenters were not aware that, to the present time, the Department's regulations in 49 CFR 170–189 do not cover liquids having flash points higher than 80° F. (open cup), and that the proposal in Notice 72–7 was to raise the definition from 80° F. (open cup) to 100° F. (closed cup). Also the Board has been advised that the NFPA has modified its definition of flammable liquids to be consistent with the proposals made by the Department of Transportation and the Department of Labor.

Other commenters presented diverse viewpoints ranging from full support of the proposal to total objection. The Board agrees that several of the objections made are valid and adjustments have been made accordingly. Particular areas which the Board believes should be discussed in detail are covered in the following paragraphs:

1. A commenter recommended that the new definition for flammable liquid specify that materials defined as compressed gases are not included. This has been done, and an appropriate change has been made to recognize the division between the definition in § 173.115 and the definition specified in § 173.300.

2. Several commenters expressed their views that the Board should give special consideration to materials completely miscible with water. Particular concern was addressed to the effect the new definition would have on materials such as distilled spirits and a number of household products. The Board believes it is making proper provision of these materials within the partial exemption provisions, including marking, labeling, and specification packaging, when their flash points are 73° F. (closed cup) or higher.

3. When the conversion to 73° F. closed cup was selected, the Board believed that there were very few materials having open-cup flash points higher than 80° F. that have closed-cup flash points lower than 73° F. The closed cup flash point of 140 proof spirits, which involves several millions of gallons stored and transported in barrels, is close to 73° F.; therefore, the Board believes it should specify the flash point demarcation for these materials due to the confusion that could exist because of very slight deviations in test results from laboratory to laboratory.

4. A number of commenters expressed concern over the testing requirements specified for materials having a viscosity of 45 S.U.S. or more at 100° F. They pointed out that, in some instances, use of the Pensky-Martens Closed Tester would not be appropriate. The Board agrees that there are some materials for which the Pensky-Martens Tester would be inappropriate and believes that some flexibility should be provided in this area to allow alternate test methods. Therefore, the requirement has been modified to permit use of other test methods as authorized in Note 1 of ASTM Standard D93–71. Further, the Board has been advised that a number of testing procedures are presently under review by experts involved in ASTM standardization procedures. The Board will give full consideration to any petition for a change in its reference to ASTM D93 if it is modified in the future.

5. Several commenters questioned the double testing of mixtures having different volatility and flash points. It was pointed out that a second test should be unnecessary when the first test discloses a material has a flash point in the lowest regulated range. Since the lowest regulated range is applied to materials having flash points of 20° F. or lower, the requirement has been changed to make it clear that once a material has been tested and found to have a flash point of 20° F. or lower a second test is not required. It was also pointed out that there is no need to specify the temperature at which a material is evaporated and all that is necessary is to specify that a certain portion of a material shall be evaporated. The Board agrees and a change has been made accordingly.

As stated earlier in this preamble, this rule making proceeding incorporates matters covered by three other rule making proceedings. With the proposals made in Docket Nos. HM-67 and HM-102 pertaining to modifications of § 173.119(b) and (L) and the proposals made in Notice 70–3 (HM-42), the application of the new requirements for flammable and combustible liquids have been proposed and are being adopted by the Board in this amendment with certain exceptions. The application of these regulations based on the size of packaging and flash point is as follows:

1. A provision is added to § 173.118 to provide partial exemption for materials having flash points between 73° and 100° F. when in packagings having a rated capacity of 1 gallon or less.

2. Specification packaging is not being required for a flammable liquid having a flash point of 73° F. or higher closed cup when in a packaging having a capacity of 110 gallons or less.

3. For transportation by highway and rail, a combustible liquid will not be subject to the regulations when in a packaging having a rated capacity of 110 gallons or less.

Interested persons should note that there is another rulemaking action, Docket No. HM-112, in progress which relates to this amendment. (See page 3022 of this issue of the FEDERAL REGISTER.) That notice refers to the application of the regulations pertaining to flammable and combustible liquids aboard aircraft and vessels. These regulations will not apply to water transportation because the Coast Guard has determined that its statutory authority does not allow this change in flash point. The Coast Guard is considering an amendment to 46 U.S.C. 170 to remove the restrictions which prevent the Coast Guard from adjusting the flash point.

On April 24, 1973, the Board published a notice entitled "Request for Information for Environmental Impact Statement" (38 FR 10118) in response to a request from the Association of American Railroads (AAR) that the Board issue an environmental impact statement. Besides comments from the AAR, four other comments were received, one supporting the comments of the AAR and three presenting opposing views. After carefully considering the issue and after reviewing the comments received, the Board believes that no statement is necessary because the adoption of this amendment will not have a significant impact upon the environment. The AAR argued (1) that the adoption of this amendment would "drastically" increase the number of commodities subject to regulation; (2) that the amendment would expose these newly regulated commodities to "more restrictive railroad operating rules . . . ," (3) that restrictive railroad operating procedures would cause a shift of the usual means of transportation of these newly regulated materials from rail to highway; and (4) that this shift would cause an increase in pollution and fuel use.

The AAR also argued that this amendment would cause an increase in the number of required rail cars, increased fuel consumption and potential for accidents, and longer trains.

The Board believes the AAR's comments were addressed primarily to the proposed combustible liquids class. The Board shares the AAR's concern for the environment and appreciates the extensive efforts that went into the AAR presentation; however, the Board does not agree that regulations on combustible liquids it is adopting for rail freight will have a significant impact on the environment. The requirements proposed for rail and highway were essentially the same, i.e., marking, placarding, shipping papers, and accident reporting, and it is difficult to see how such requirements would cause a shift of traffic from rail to highway thereby causing an increase in pollution.

It is also difficult to see how the amendment as it was proposed would bring the United States to the "brink . . . of bankruptcy" with respect to its international trade as the second commenter favoring the AAR position claimed. As to the regulations being adopted in this rule making, the Board simply wishes to state that regulation of combustible liquids and certain additional flammable liquids will primarily involve requirements for the identification of the potential hazards of flammable liquids and combustible liquids. Also, flammable liquids with flash points of 73° F. or higher in packages of less than 110 gallons and combustible liquids are exempt from specification packaging.

As stated in the preamble of the Notice of Proposed Rule Making for this Docket, the Board is providing approximately one year to accomplish compliance with the requirements of this amendment with a provision that immediate compliance is authorized except as they pertain to placarding of tank cars.

(39 F.R. 2768, January 24, 1974)

[Docket No. HM-102; Amdt. No. 173-78A]

PART 173—SHIPPERS
Flammable, Combustible, and Pyrophoric Liquids; Definitions

The purpose of these revisions to the amendment made under Docket HM-102; Amdt. 173-78 is to:

(1) Permit the use of an additional test method for the determination of flash points,

(2) Modify the evaporation requirements for mixtures,

(3) Permit the marking of the flash point range of materials as an alternative to the marking of a specific flash point on the outside package.

(4) Specify the minimum flash point of certain aqueous solutions containing alcohol,

(5) Grant a special exemption for alcoholic beverages, and

(6) Extend the mandatory compliance date of amendments made under Docket HM-102.

On January 24, 1974, the Board published amendments 172-23, 173-78, 174-19, and 177-29 (39 FR 2768) to: (1) specify a new definition for the class of materials identified as "Flammable liquids", (2) create and define a new class of materials identified as "Combustible liquid"; (3) modify the definition for pyrophoric liquids within the flammable liquid class; and (4) set forth the requirements for the materials that are covered by the new definitions. Since publication of the amendments, a large number of petitions have been received by the Board pertaining to the amendments and the Board has decided that many of the petitions have merit.

Setaflash tester. Several petitions have been received for the addition of the Setaflash closed tester for flash point testing. This apparatus has been subjected to extensive testing within the United States and Europe. Standard test procedures for its use have been issued by the American Society for Testing and Materials (ASTM), The International Standards Organization (ISO), and the British Standards Institution (BSI). The Setaflash tester has been presented as suitable for highly viscous materials such as pastes. Special advantages of the Setaflash tester include: (1) increased safety due to small sample size required (2 milliliters vs. 50 to 70 milliliters for Tag or Pensky-Martens); (2) a decrease in the time necessary to run a test due to the small sample size, with a corresponding reduction in the heating time at different temperatures; and (3) data comparing the Setaflash tester with other flash point testers indicates that the Setaflash test gives better reproducibility and repeatability because it produces liquid-vapor equilibrium test results. Based on these considerations, the Board has decided to specify two standard test methods, using the Setaflash tester, as alternate test methods for the determination of flash points.

Evaporation procedures. A number of questions concerning the validity of the evaporation requirement and the double testing of mixtures having components of widely differing volatilities and flash points have been raised. The Board believes safety considerations require that provisions be made for the most hazardous situation that could reasonably be expected to occur and, since flash point is the sole criterion specified for the classification of flammable and combustible liquids, it is necessary to classify a material according to the lowest flash point that could reasonably be expected to occur. The requirement for the evaporation of 90% of the original volume before retesting is not feasible for mixtures containing 10% or less volatiles. Therefore, the Board has changed the requirements to specify an evaporation time as an alternative to the 90% evaporation requirement. Also, no evaporation temperature was specified in the amendment; therefore, in many cases it would have been possible to selectively remove any component desired. By specifying an evaporation temperature in conjunction with an evaporation time or volume decrease, the Board believes these concerns will be alleviated.

Specification of flash point vs. flash point range. In order to use the partial exemptions specified in § 173.118(a) (3) and (b), for flammable liquids with a flash point between 73° F. and 100° F., amendment 173–78 required that the flash point must be marked on the outside of the package. It is not necessary to show the exact flash point of a flammable liquid as long as marking indicates that the flash point is 73° F. or higher. Therefore, changes are being made to permit an alternative marking.

One petitioner questioned the need for marking flash points on containers in order to qualify for these partial exemptions since no similar requirement exists for such a marking on smaller quantity packagings. The reason for the marking is to provide a visual affirmation that the material in the package has a flash point high enough to qualify for partial exemptions when in quantities of more than one quart. The small quantity partial exemption presently in effect differentiates between metal containers (up to one quart) and other kinds of containers (up to one pint). The new partial exemption (up to and including one gallon) applies to flammable liquids having flash points from 73° F. to 100° F. and does not specify the material of construction for their packaging. Therefore, the marking will be an indication that the flammable liquid (73–100° F.) is authorized under the partial exemption where the absence of such a marking will be an indication that specification packaging, marking, and labeling is required. The same applies to packaging having capacities of more than 1 gallon which would not be subject to the specification packaging requirements if they are marked to indicate that the flash point is 73° F. or higher. Therefore, the petition for removal of the flash point marking requirement is hereby denied.

Alcohol water solutions. The Board has decided to make a declaration in the regulations that the flash point of dilute alcohol water solutions, containing no more than 24 percent alcohol by volume, are considered to have flash points of 100° F. or higher because the Board believes the combustible liquid classification is more appropriate for such materials (if they have a flash point below 200° F.). The declaration is applicable only when the remainder of the solution does not meet any definition of a hazardous material as defined in Title 49 CFR including those of a flammable or a combustible liquid. This decision is based in part on petitions for reconsideration received relative to wine and certain consumer commodities. The maximum alcohol content specified is based on the maximum alcohol content for wine as defined in 27 CFR 4.10 which is 24% by volume alcohol. Rather than restrict this revision to the wine industry, it has been expanded to include other alcohol water solutions.

Exemption for alcoholic beverages. A special provision for alcoholic beverages is included in this amendment to provide complete exemption from the Department's Hazardous Materials Regulations when they are shipped in containers having a rated capacity of one gallon or less. These beverages are restricted to those defined in 27 CFR 4.10 and 5.11 as wine and distilled spirits. Due to the controls exercised by the Department of the Treasury pertaining to these products, the Board believes that it is unnecessary, from a transportation safety standpoint, to make them subject to the Department's regulations.

Environmental impact. Pursuant to the provisions of section 102(2) (c) of the National Environmental Policy (42 U.S.C. 4321 et seq.), the Board has considered the requirements of that Act concerning Environmental Impact Statements and has determined that this amendment would not have a significant impact upon the environment. Accordingly, an Envrionmental Impact Statement is not necessary and will not be issued with respect to this amendment.

For purposes of clarity, several editorial changes have been made to §§ 173.115 and 173.118, and the two sections are set forth in this amendment in their entirety.

(40 F.R. 22263, May 22, 1975)

[Docket No. HM-102; Amdt. No. 174–19]
PART 174—CARRIERS BY RAIL FREIGHT
Placarding of Tank Cars Containing Combustible Liquids

On January 24, 1974, the Hazardous Materials Regulations Board (the Board) published Amendment 174–19 (39 FR 2768) setting forth in Part 174 of the Department's Hazardous Materials Regulations certain new requirements pertaining to the placarding of tank cars and portable tanks (when transported by rail freight) containing combustible liquids. In response to petitions for reconsideration, Amendment No. 173–78A was published (40 FR 22263) on May 22, 1975, containing revisions to regulations published on January 24, 1974. In addition, this latter amendment and a subsequent publication on June 12, 1975 (40 FR 25024), extended the effective date of the new regulations, including those pertaining to the placarding of tank cars and portable tanks containing combustible liquids, to January 1, 1976. In each case, although immediate voluntary compliance was authorized with respect to other matters covered by the new regulations it was not authorized relative to the placarding of tank cars containing combustible liquids.

One portion of these new regulations which has generated concern is the requirement to placard tank cars containing combustible liquids with "Dangerous" placards. There are a number of operating requirements in Part 174 pertaining to tank cars placarded "Dangerous" that are not intended to apply to tank cars containing combustible liquids. Until a distinctive "Dangerous" placard for combustible liquids is prescribed, implementation of this requirement could cause confusion. In consideration of the fact that an unnecessary burden could be imposed if the placarding requirements pertaining to combustible liquids transported by railroad become effective on January 1, 1976, and that action to establish distinctive "Dangerous" placards for tank cars containing combustible liquids will be taken by the Bureau in amendments to be published under Dockets HM-103 and HM-112, the Bureau believes that the compliance date of Amendment No. 174–19 should be deferred.

In consideration of the foregoing, the effective date for compliance with §§ 174.541 and 174.584 under Amendment No. 174–19 as they pertain to the documentation and placarding of cars containing combustible liquids is changed from January 1, 1976, to January 1, 1977. Use of "Dangerous" placards on cars containing materials classed combustible is not authorized until January 1, 1977, or such earlier date as the Bureau may designate in connection with regulations to be issued under Dockets HM-103 and HM-112.

(40 F.R. 57433, December 10, 1975)

SUMMARY • **Docket No. HM-103** • *Proposed a new verbal, pictorial, and numerical Hazard Information (HI) system for communicating the primary, secondary, and tertiary hazards of materials through labels, placards, and shipping papers. Digital system subsequently withdrawn but amendments issued pertaining to labels, placards, and shipping papers in conjunction with Docket No. HM-112. Also see Docket No. HM-126.*

[49 CFR Part 172]
[Docket No. HM-103]
HAZARD INFORMATION SYSTEM
Advance Notice of Proposed Rule Making

The Hazardous Materials Regulations Board is considering the adoption of regulations that would provide for more complete identification of the hazards of materials in transportation. Included in this consideration is some modification of the present hazard communications requirements and their relocation in one part of the regulations.

The need for improved hazard communications has been the subject of considerable controversy and debate during recent years. It has been pointed out that the communications requirements of the regulations (1) generally are not addressed to more than one hazard; (2) do not in all instances require disclosure of the presence of hazardous materials in transport vehicles; (3) are not addressed to the different hazard characteristics of a mixed load of hazardous materials; (4) do not provide sufficient information whereby fire fighting and other emergency response personnel can acquire adequate immediate information to handle emergency situations; and (5) are inconsistent in their application to the different modes of transport. The Board believes there are deficiencies in this area and that certain changes are necessary to provide for the adequate communication of hazards for materials in transportation. However, the Board also believes that it must consider the complexity of any regulations it adopts in this area and what is to be imposed on the personnel who will be required to follow them.

Considerable time has been spent on the development of this proposed system thereby causing a delay of final rules resulting from the proposals in Docket No. HM-8. The Board believes that it is time to place this proposal before the public for consideration and constructive comment in advance of a complete notice of proposed rule making. The complete notice would encompass all of the proposed changes and modifications necessary to implement this system which the Board has identified as the Department of Transportation Hazard Information System.

The Hazard Information System encompasses all hazardous materials regulated by the Department of Transportation except those specifically exempted, such as small quantities of certain classes of materials. Each shipping paper, label, and placard would, in addition to the classification identification bear a hazard information number consisting of two digits. The first digit represents a base hazard and, with few exceptions, would be the same as the United National class number. The second digit will convey (1) there are no additional significant hazards, or (2) that a material has one or more additional significant hazards. Associated with each of the 59 hazard information numbers designated so far is a hazard information action "card." Each card contains information on potential hazards of a material and, in the event of an incident, recommended action for the handling of the material. With certain exceptions, the cards have been developed on a generic basis so that different materials having similar characteristics would be covered by one set of instructions. The system is designed so that, in most instances, shippers may determine the hazard information numbers of their materials without "preclearance" by the Department. It is estimated that the exceptions to this procedure will involve only a small percentage of the different materials offered for transportation.

The Board is proposing to consolidate all of the hazard communications regulations into Part 172 except those requirements peculiar to a mode of transportation. Significant changes being proposed are as follows:

1. Assignment of hazard information number, where possible, to each hazardous material listed in a new § 172.101 (not contained herein);

2. Designation of each label by class wording, rather than color, in the list of hazardous materials and elsewhere in the regulations;

3. A requirement that the description of a hazardous material be readily identifiable on shipping papers;

4. A requirement that hazardous materials be listed first when other materials not subject to the regulations are described on the same shipping paper;

5. A requirement that the hazard information number for a material be placed immediately after the classification identification on a shipping paper;

6. A proviso that the class name need not be reentered when it is the same as the shipping name, except for the entry "n.o.s.";

7. A clarification of the requirements relating to the signing of certificates;

8. A requirement that a hazard information number be placed on labels in the hazard information block specified;

9. A table specifying the placarding requirements for highway and rail transport vehicles, based on hazard information numbers;

10. A provision for "Dangerous" placards in place of specific identification placards for not more than 1,000 pounds of certain materials;

11. Requirements pertaining to the giving and affixing of placards with alternatives provided to allow flexibility in the arrangements made between shippers and carriers;

12. A distinction between large packages and small packages to provide a break point where labeling stops and placarding begins;

13. A clarification of the placarding requirements for "Empty" cargo tank motor vehicles and a special proviso for gasoline and fuel oil;

14. Detailed requirements and provisions for the attaching of placards to transport vehicles;

15. Specifications for new placards;

16. Specifications for new labels;

17. The method for derivation of a hazard information number when it is not assigned in the list of hazardous materials; and

18. Proposed definitions for hazard characteristics of materials to be used for derivation of hazard information numbers.

Aside from the basic concept of the use of a two-digit number to convey information on one or more hazards of a material, the Board believes the most significant concern of persons who would be affected by these proposed requirements is placarding. Except for those materials which would be exempt from these requirements, this proposal is designed to provide for the disclosure of the presence of a hazardous material in a transport vehicle regardless of quantity. Specific placards identifying certain types of hazardous materials would be required regardless of quantity. For other materials, specific identification of hazards would not be required until a single consignor offers for transportation (or transports as a private carrier) more than 1,000 pounds of a material bearing the same hazard information number in one vehicle.

Subpart F of Part 172 contains the proposed revisions pertaining to placarding. A placarding table is specified in § 172.502. Column A of the table specifies the list of hazard information numbers, column B, specifies the placard for each hazard information number, and column C contains the exceptions to specific placarding when another required placard would bear a number that would provide the same hazard information for that material as for another material having similar but also additional and more significant characteristics. Paragraph (b) of that section provides for an optional "Dangerous" placard for certain materials bearing the same hazard information number received from one consignor and not exceeding 1,000 pounds. There is no requirement that a driver must add up the amount of material bearing the same hazard information number when received from different consignors in LTL motor carrier operations. This will substantially reduce the burden on pickup and delivery drivers. Subparagraph (2) of that paragraph lists hazard information numbers that could require specific placards regardless of the amount of material offered for transportation unless excepted by other placards in accordance with column C of the placarding table. This would include certain highly or extremely toxic compressed gases, water reactive materials, organic perioxides of high sensitivity or requiring refrigeration, and certain radioactive materials, extremely toxic liquids or solids and such materials that are highly toxic by skin absorption. The Board believes that the hazards of these materials are of such significance as to require specific identification regardless of the quantity contained in a transport vehicle.

Section 172.503 contains proposed requirements applicable to the giving and affixing of placards. This section is designed to set forth the prime responsibility for the giving of placards while providing for flexibility in the arrangements made between shippers and carriers. No distinction is made between rail and highway vehicles in these proposed requirements. The Board believes that the only practicable method to accomplish the specific placarding in this proposal is for the shipper to supply the four placards pertaining to his material. The shipper either knows the characteristics of the material he offers for transportation or has access to the person who manufactured it to obtain information concerning its characteristics. It does not appear appropriate to require rail and highway carriers to carry 59 different sets of placards in anticipation of receiving a hazardous materials shipment. However, in the case of dedicated equipment used repeatedly for the same service, such as a cargo tank used to transport gasoline, it would not be necessary for the shipper to provide placards for each trip. The proposal also recognizes the optional "Dangerous" placard which could be provided as a permanent type of device by motor carriers and which could be used for many LTL shipments of less than 1,000 pounds both prior to and following terminal transfer operations.

A new series of placards is being proposed which will be the same in their application for shipments by rail and highway. They also would be affixed to portable tanks, large packages, and reusable transport containers. Each placard would communicate through its color, shape, symbol, keyword, and hazard information number. While the letters on the placards are smaller than those presently prescribed for motor vehicles, the placards will be visible from a greater distance due to their form and presentation and could eventually become a "trade-mark" identifying vehicles containing hazardous materials much the same as the diamond shaped label is for smaller packages today.

A number of different methods for attachment of placards to vehicles are set forth in this proposal. These methods were drawn mainly from the existing regulations for the attachment of placards to rail cars. The Board is particularly interested in comments addresssed to these various methods and in any alternative suggestions. The placard holder specified in Appendix A to Subpart F appears to be a practical means for affixing placards to vehicles, particularly vehicles frequently used to transport different materials. The proposal requires placards to be affixed to each end and side of a rail or highway transport vehicle as is required at present. A semitrailer by definition is a transport vehicle and no provision has been made in this proposal to permit the attachment of placards to truck tractors. Since it is proposed to require that placards be maintained on semitrailers at all times they contain hazardous materials subject to placarding, the Board believes the four placards should be displayed on them to preclude violation of the regulations when they are dropped at pickup or delivery points. Also, it is contemplated that a placarded semitrailer will be considered sufficient for the placarding of a railcar when trailer-on-flatcar operations are involved. The Board is aware that the effectiveness of a placard could be somewhat reduced when it is affixed to the front end of a semitrailer and partially obscured by the towing vehicle. Specific comments concerning the merits of one method over the other are requested.

The labels proposed in Subpart E are essentially those proposed in Docket No. HM-8 with the exception that a hazard information block, which would contain the hazard information number pertaining to the material in the package, has been added at the bottom of the label. Also, it is proposed to identify all labels by key words rather than colors to prevent confusion.

The Board is proposing to specify the colors to be used on labels and placards. Color specifications are proposed in Appendix A to Part 172 of this proposal. Each color and tolerance is identified in the Munsell Notation System. This will permit a label and placard manufacturer to use any suitable color base to make the colors for the labels and placards providing the resulting colors match the specified standard or are within the tolerances specified for each standard. Color standards have been identified in Munsell notations for several years in such standards as:

1. USAS Z53.1, "USA Standard Safety Color Code for Marking Physical Hazards,"
2. USAS Z35.1, "USA Standard Specifications for Accident Prevention Signs," and
3. USAS Z35.2, "USA Standard Specifications for Accident Prevention Tags."

In addition to the Munsell notations, chips for each color standard and allowable tolerances will be made available to facilitate production of labels and placards with specified colors.

Subpart G to Part 172 of this proposal covers the procedure for derivation of hazard information numbers. A step-by-step approach is set forth in §172.602 which will provide for derivation of hazard information numbers for most materials shipped. The exceptions to the derivation procedure involve those materials that the Board believes should be reviewed by the Department prior to designation of their hazard information numbers. As mentioned earlier, the hazard information numbers will be designated in the list of hazardous materials for those materials listed by name. Many mixtures and "n.o.s." materials will not have hazard information numbers assigned them in the list.

Concerning the proposed definitions in Appendices A and B to Subpart G, the Board was faced with setting forth either specific definition criteria for the characteristics of materials, knowing that some of the proposed definitions could be controversial, or setting frth criteria lacking in sufficiently specific definition that could lead to inconsistent derivation of hazard information numbers. Since classification projects are in progress concerning these definitions and will be handled by separate rule making, the Board believes the better course is to propose, so far as practicable, tentative definitions for derivation purposes during the interim. Also, consideration is being given to a requirement (not proposed herein) that shippers notify the Department of their assignment of a hazard information number to a material on an after-the-fact basis in order to maintain an overview of consistency in the derivation of hazard information numbers. It is not contemplated that such a requirement would be on a preshipment basis.

The 59 hazard information cards presented at the end of this publication were developed with the assistance of knowledgeable persons both within and outside the Department. The Board wishes to express its appreciation to those persons who expended time and effort in this endeavor. The cards are not presented in their final form, but are presented to indicate the style of their presentation and the kind of information that would be provided in relation to the hazard information numbers assigned different materials. It is planned to combine the cards into a booklet or manual that would be distributed through the Government Printing Office and possibly by any interested organization. It has been decided that the cards presented with this publication would not be suitable for use in the marine environment and that a separate publication would have to be prepared for use aboard vessels. The Board is interested in receiving comments concerning the content and format of the cards that will lead to their improvement. One factor that must be recognized is that a card must contain information that is applicable to all materials assigned the same number as that of the card. This is the principal difficulty with this sytem or any system that is generic in nature and addressed to thousands of different materials.

The information and instructions given on the hazard information cards should be considered advisory in nature. The intent of the cards is to give fire-fighters and other emergency personnel sufficient information to enable them to make informed judgments in the handling of emergency situations during their initial phases. The cards are not intended to perform the same service that could be provided by a centralized information agency having ready access to large amounts of data. The Board wishes to emphasize that, by implementation of this system, it does not intend to usurp the prerogatives of any jurisdiction in its handling of emergency situations.

This proposal does not include Subparts A, B, or D to Part 172. They will be included in a future notice of proposed rule making under this docket. Subpart A will contain general regulations pertaining to the application of Part 172; Subpart B, the list of hazardous materials; and Subpart D, the pakage marking requirements. To permit a review of contemplated assignments of hazard information numbers to materials, a sample list illustrating the use of the definitions in Appendices A and B to Subpart G of Part 172 in assigning numbers is provided as follows:

30—Acetone.
67—Acetone cyanohydrin.
36—Acrolein, inhibited.
60—Aldrin mixtures, liquid, with more than 60 . percent aldrin.
32—Allyl alcohol.
21—Ammonia, anhydrous.
54—Ammonium perchlorate.
35—Aniline oil, liquid.
20—Argon.
62—Arsenic chloride (arsenous), liquid.
60—Arsenic pentoxide, solid.
30—Benzene (benzol).
59—Benzoyl peroxide.
60—Calcium arsenate, solid.
26—Chlorine.
53—Chlorine trifluoride.
05—Chloroacetophenone, gas, liquid or solid.
82—Chlorosulfonic acid.
20—Dichlorodifluoromethane.
57—Dicumyl peroxide, solid.
83—Diethyl dichlorosilane.
67—Dimethyl sulfate.
23—Ethylene.
35—Ethylene oxide.
19—Explosive mine.
15—Explosive power device, class B.

11—Explosive rivets.
19—Explosive torpedo.
27—Fluorine.
80—Formic acid.
30—Gasoline (including casing head and natural).
64—Hexaethyl tetraphosphate, liquid.
32—Hydrazine, anhydrous.
29—Hydrocyanic acid, liquefied.
29—Hydrogen sulfide.
23—Liquefied petroleum gas.
53—Nitrogen tetroxide, liquid.
22—Oxygen, liquefied.
64—Parathion, liquid.
57—Peracetic acid.
54—Potassium perchlorate.
50—Potassium permanganate.
46—Sodium aluminum hydride.
60—Sodium cyanide, solid.
46—Sodium, metallic.
26—Sulfur dioxide.
84—Sulfuric acid (oil of vitriol).
64—Tetraethyl dithio pyrophosphate, liquid.
64—Tetraethyl lead, liquid.
81—Thionyl chloride.
23—Trifluorochloroethylene.
23—Vinyl chloride.

The list above also will permit association of the proposed hazard information cards with different hazardous materials, and serves to point out one other advantage of the system. Note that Acetone has the hazard information number 30, and Acetone cyanohydrin, 67. Here we have two materials that would require different kinds of response if spilled in an accident. It is not difficult to visualize that the word cyanohydrin could be lost in communications; however, incomplete communications of this type could be discovered if hazard information numbers are communicated with the names of materials. When verbal communications via radio, telephone, or other means are necessary to obtain further assistance beyond that provided by a hazard information card, the number will serve as an authenticator of the word being transmitted.

(37 F.R. 12660, June 27, 1972)

[49 CFR Part 172]
[Docket No. HM-103; Notice No. 73–10]
TRANSPORTATION OF HAZARDOUS MATERIALS
Hazard Information System and Miscellaneous Proposals; Notice of Proposed
Rule Making

The Hazardous Materials Regulations Board is considering the adoption of amendments to the Department's Hazardous Materials Regulations (1) to adopt a hazard information system; (2) to consolidate the Hazardous Materials communications regulations; and (3) to make miscellaneous changes to the documentation, marking, labeling, and placarding requirements.

On June 27, 1972, the Board published an advance notice of proposed rule making under Docket No. HM-103 (37 FR 12660) proposing a hazard information system and miscellaneous changes to the communications regulations. Interested persons were invited to comment on the proposals made by submission of written comments. A large number of comments were received by the Board for its review and consideration.

Many of the comments contained constructive suggestions and criticism while others contained only expressions of support or non-support for the advance proposals. All comments received have been reviewed and considered. No attempt is being made in this preamble to address all of the comments and suggestions submitted; therefore, commenters should study the proposals made in this Notice as compared to the proposals made in the Advance Notice and resubmit their comments in the manner specified at the end of this Notice if they believe their earlier comments were possibly ignored, misinterpreted or overlooked.

The Board is aware of the significance and far reaching ramifications of the proposals made in this Notice. In the preamble of the Advance Notice, the Board stated the following:

"The need for improved hazard communications has been the subject of considerable controversy and debate during recent years. It has been pointed out that the communications requirements of the regulations (1) generally are not addressed to more than one hazard; (2) do not in all instances require disclosure of the presence of hazardous materials in transport vehicles; (3) are not addressed to the different hazard characteristics of a mixed load of hazardous materials; (4) do not provide sufficient information whereby fire fighting and other emergency response personnel can acquire adequate immediate information to handle emergency situations; and (5) are inconsistent in their application to the different modes of transport. The Board believes there are deficiencies in this area and that certain changes are necessary to provide for the adequate communication of hazards for materials in transportation. However, the Board also believes that it must consider the complexity of any regulations it adopts in this area and what is to be imposed on the personnel who will be required to follow them."

In accordance with the last sentence above, the Board has made adjustments in this proposal, based on its own investigation and the comments received, to simplify the requirements pertaining to the placarding of transport vehicles while not defeating its original purpose to any significant degree.

A number of commenters stated that the Hazard Information System is too complex. However, they did not address themselves to the basic derivation procedure for hazard information numbers as being an area of difficulty. The complexity of the system, as pointed out by some commenters, would be in the area of its application during transportation operations. These views were expressed primarily by persons who transport hazardous materials by motor vehicle and were pointed primarily to the proposed placarding requirements which will be discussed later in this preamble.

Several commenters took exception to the use of numerical identifiers (hazard information numbers) as a mechanism for conveying the hazards and multiple hazards of different materials. Two commenters suggested use of words and symbols as the primary means of conveying information. The Board does not agree with this approach and believes, contrary to the views of these commenters, that such a method would be more confusing and, possibly, misapplied. There would be no continuity in the system if such an approach were followed. There would be words on shipping papers (presumably for each kind of hazard) and symbols and words on placards and labels. It was not shown how this proposed approach would provide communication of hazards such as thermal instability, toxicity by skin absorption, and others—factors the Board believes should be considered. On the other hand, with great simplicity the use of the hazard information numbers would serve to tie together and authenticate communications of the hazards of materials on shipping papers, labels, placards, the hazard information cards, and verbal communications to assisting personnel who may or may not go to the site of an incident.

Many comments were received from individuals and organizations recommending that the Board adopt regulations that would incorporate "A Recommended System for the Identification of the Fire Hazard of Materials" commonly referred to as NFPA 704M. The Board wishes to point out that it respects the efforts and objectives of the many individuals who participated in the development of the 704M System. Their efforts are to be commended even though the 704M System is not applicable for transportation purposes as has been clearly expressed in the following statement which appears in the foreword to NFPA No. 704M.

"As originally conceived, the purpose of the guide is to safeguard the lives of those individuals who may be concerned with fires occurring in an industrial plant or storage location where the fire hazards of materials may not be readily apparent. It does not envision possible application to other situations, such as chemical laboratories, rail or truck transportation, lumber and coal storage yards and tobacco warehouses."

The Board agrees with the above statement and wishes to emphasize that it is not denying the possible value of the 704M System for purposes other than transportation.

In its comments, the International Association of Fire Chiefs (IAFC) " . . . strongly advocates that any new system of hazardous materials information must positively identify the NATURE of the hazard and the potential degree of severity." They further state that they must be given adequate instant information to allow emergency personnel " . . . to undertake actions designed to protect victims and bystanders from possible injury or death. Denying the emergency forces total instant information may be regarded as little short of criminal . . . "

In its comments, the IAFC indicated that the 704M System satisfies these requirements. The Board does not agree. On the other hand the Board is of the opinion that the proposed Hazard Information System meets the essential requirements pointed out by the IAFC.

Upon examining the 704M hazard number assignments pertaining to health, the Board found that many dissimilar materials were assigned the same numerical degree of severity. For example, chlorine, oxygen (liquid), sodium hydroxide (lye), hydrogen sulfide, and tetraethyl-lead carry the health hazard classification of 3. The required response of an accidental spill for any of these materials is quite different. The Hazard Information System recognizes these differences by assignment of the appropriate numbers to these compounds. The hazard information cards will give specific, simple, and concise information to emergency personnel. The 704M health hazard number 3 does not convey any response information aside from indicating that these materials are extremely hazardous to health. Several commenters and the IAFC were critical of the Board's intent to place reliance on a "secondary source" of information to assist emergency response personnel. After careful consideration of this aspect, the Board is convinced that no simplistic symbol, placarding or number system like 704M can possibly convey the necessary response information for the different hazard characteristics of the thousands of commodities shipped without the addition of at least a minimum of short, concise, descriptive information. The reading of this information, as presented on a hazard information card, can be readily accomplished in less than one minute.

In its comments, the IAFC also stated "The fire department and rescue squads of America, which we represent, have an inherent responsibility to function wisely and judiciously in the public welfare in emergency conditions." The fundamental purpose of the Board's proposal herein is to provide them and other emergency personnel information whereby they can make better informed judgments on how to handle incidents involving hazardous materials during the course of transportation. It is not intended to supplant the exercise of good judgment nor to provide information that could be misconstrued or misleading. Considerable effort has been expended to make the proposed placards recognizable and distinguished from a distance by their format and color. For example, the difference between the "Explosives" placard and the "Radioactive" placard can be determined at a considerable distance from a vehicle to quickly inform persons who need this information of the kind of potential hazard they are approaching. However, the instructions contained in the proposed manual are intended to assist in handling a situation, not merely to inform that a situation exists.

Finally, the IAFC states that the descriptive legend on the hazard information card may not adequately cover the hazard of a specific material. It is recognized that in some instances more definite information might be desirable for a specific product. However, this shortcoming lies in the nature of any generic approach. The only alternative would be individual cards for each product which, obviously, would be impractical and certainly would not meet the requirements postulated by IAFC. The Board believes that the generic hazard information cards will be adequate for the vast majority of the hazardous materials shipped, and will provide the required instant information. However, it also plans to distribute spill manuals for selected materials which will complement the Hazard Information System proposed for implementation through the Hazardous Materials Regulations.

The Board has studied the users need for the hazard information manual and believes that it should aim for the vehicles used by emergency response personnel in making an initial distribution. It has been estimated that there are 392,000 police and fire vehicles used in the United States. Most responses to hazardous materials emergencies would involve the use of these types of vehicles. Also, ambulance operators, police and fire dispatchers, and others involved in emergency response activities should receive copies. Therefore, the Board plans to have 500,000 copies available for distribution, if it decides to adopt the proposals it is making in this Notice.

PROPOSED NEW PART 172

Subpart A contains a section pertaining to the applicability of the regulations in Part 172 and a section containing proposed definitions including one for a transport container.

SUBPART B—LIST OF HAZARDOUS MATERIALS

Subpart B will contain the list of hazardous materials and appears in this issue of the FEDERAL REGISTER under Docket No. HM-112.

SUBPART C—SHIPPING PAPERS

Subpart C will contain virtually all of the Department's regulations pertaining to the preparation of shipping papers for hazardous materials. Note that it does not apply to an "Other Regulated Material" (ORM) unless it is intended for transportation by air or water and then only when the Hazardous Materials Regulations apply to the material in one or both of these modes. Section 172.201 contains the requirements pertaining to the placement of the required description on a shipping paper. A number of commenters expressed concern over the Board's proposal to require hazardous materials to be described first when other materials are described on the same shipping paper. The principal concern expressed was the proposal's effect on stocks of computerized shipping papers. No explanation was provided by these commenters as to the technical feasibility of adjusting computerized programs to comply with the proposed requirements. The Board believes that sufficient and adequate information, for purposes of safety, must be conveyed on shipping papers in a uniform and readable manner. Therefore, the Board is again proposing the requirement that hazardous materials be described first for two principal reasons; (1) that the information pertaining to hazardous materials be available quickly when needed in an emergency, and (2) that the information on hazardous materials be stated first in order that carriers may more readily comply with the placarding requirements and compatibility restrictions.

In consideration of the fact that the Board is proposing to place greater emphasis on the value of shipping documents by expanding use of the DANGEROUS 01 placard, the Board believes that the hazardous material description should be shown before other descriptive information for a material on a shipping paper. This position has been strengthened by examination of some of the documents that were submitted by commenters to illustrate their difficulties if the proposal in the Advance Notice were adopted. Again, it is the Board's position that the description of hazardous materials on shipping papers be presented in a manner that will quickly give a potential user the information he must know. Therefore, the Board is proposing to require that the hazardous material description be placed before other information that is not required by the Department's safety regulations.

Recognizing that it will be necessary to allow sufficient time for shippers to adjust their documentation procedures, the Board is proposing to allow sufficient time for procedures to be adjusted.

The Board agrees with the commenter who suggested that placing the hazard information number after the classification would cause confusion with other numerical information even though paragraph (d) of § 172.201 in the Advance Notice stated that the number must be clearly separated from any other numbers. Accordingly, the Board has revised the proposal to require that the hazard information number be printed between the shipping name and classification prescribed for a material.

Several changes are proposed pertaining to the shipping paper certification requirements. The Board is proposing to authorize use of the certificates that has been used for a number of years by those persons subject to the IATA (International Air Transport Association) "Restricted Articles Regulations." Use of this certification would be limited to those materials offered or intended for transportation by air and, in such instances, could serve in place of the certificate required for surface movements. The Board understands the desirability of having a single form of certification that will be considered acceptable for all modes of transportation, both domestically and internationally. However, it does not contemplate resolution of the differences within the time frame of this rulemaking action.

A number of commenters pointed out that they have procedures whereby a computer prepares and executes an entire shipping document including the certification. The Board is proposing to permit certificates to be signed by mechanical means.

SUBPART D—MARKING

Proposed subpart D, pertaining to marking, was not proposed in the Advance Notice of Proposed Rule Making. It contains the basic marking requirements for packages, portable tanks, cargo tanks and tank cars. However, it does not contain all the marking requirements for packages and it will still be necessary for shippers to determine particular marking requirements that are required by Part 173.

It is proposed in § 172.305 to require all portable tanks to be marked in 2 inch letters on two sides with the shipping name of the material it contains. While marking on two sides is proposed, it should be noted that the size of lettering has been standardized and would no longer be relative to the tank diameter as is the present requirement in § 173.401(a). The requirements for marking of cargo tanks are essentially the same as the requirements contained at present in § 177.823. The main difference is the proposal to require that each side and the rear of a cargo tank be so marked. Also included in the proposed new subpart is the proposal made in Docket HM-101 (37 FR 7104) to add a new § 173.31(a) (6) pertaining to the marking of tank cars.

SUBPART E—LABELING

Proposed subpart E would contain practically all the Department's regulations pertaining to the labeling of hazardous materials.

Section 172.400 contains the labeling exemptions while section 172.401 contains the labeling prohibitions. To clarify the intent of the labeling prohibition in present § 173.404(b), the Board is proposing a provision to authorize the labeling of a package containing a material (except explosives) that has been tentatively classified pending the outcome of the tests or the development of data to permit appropriate classification.

No date is provide in § 172.400(a) (2) (iii) to indicate the Board's further adoption of CGA Pamphlet C-7, Appendix A. This regulation authorizes use of special markings on compressed gas cylinders in place of labels for highway transportation only. The Board will await the proposals it anticipates will be submited by the CGA, or others, before it decides on whether to continue recognition of the standard.

Section 172.402 sets forth the proposed regulations for labeling. They are essentially the same as those proposed in the Advance Notice of Proposed Rule Making with the exception of a requirement for multiple labeling of poisons. Several commenters suggested that a poison label be required for materials that are highly toxic when they are in another classification. The Board agrees and has proposed such a requirement in § 172.402.

Section 172.403 contains detailed requirements for labeling radioactive materials. In order to resolve the conflict created by the existence of a transport index block on the lower half of RADIOACTIVE-YELLOW labels with the proposed hazard information number block, the Board is proposing herein to modify the labels so as to require the application of the transport index on a "third line" as is presently required on the label of the regulations of the International Atomic Energy Agency. This allows for the use of the block for the application of the hazard information number. Another noteworthy change to the label format involves the removal of all of the text adjacent to the word "RADIOACTIVE" and the slight shifting of the vertical red bars so as to eliminate interference with the legibility of the label entries. None of these changes are expected to create any significant incompatibility of the United States label format with that of the International Atomic Energy Agency.

Section 172.405 clarifies the requirements for one or more labels on a package depending on its size.

Section 172.406 contains detailed requirements pertaining to the hazard information number on labels. In response to suggestions that the proposed requirements for a material requiring two labels on a package be clarified, the Board is proposing that the hazard information number for such a material be entered on each label.

§ 172.407 contains the detailed specifications for labels. These labels are substantially the same as those adopted under Docket No. HM-8 (38 FR 5292) with the addition of the hazard information number block. It is proposed that some of the names on the labels would be printed in white to improve visibility. Increasing the visibility of the hazardous materials labels is considered by the Board to be a significant improvement in hazard communication; however, in order to maintain a harmonious international labeling system, the Board will withdraw this portion of the proposal before final rule making if the United Nations Recommendations are not changed to accommodate the visibility improvements. Similar consideration is being given to the color of symbols. The Board recognizes that, if it adopts the label modifications it is proposing, sufficient time should be allowed to permit conversion.

It has been suggested that specific labels be provided for certain materials such as Oxygen and Chlorine. While the Board had been attempting to limit the number of different labels it would require, further comments concerning the need for additional labels are solicited.

SUBPART F—PLACARDING

This proposed subpart will contain the Department's placarding requirements pertaining to transportation of hazardous materials by water, rail, highway, and air.

Section 172.502 contains specific requirements for placarding highway and rail transport vehicles and, to simplify the procedure for determining the appropriate placard, the requirements are presented in three tables.

While disclosure of the presence of hazardous materials in any quantity is considered desirable, it appears that to require disclosure in any quantity in the highway area is totally impractical. For example, while not pointed out specifically in the comments received, it must be recognized that the DANGEROUS placard would have been required on thousands of vehicles operated by utility companies, construction companies, and others who carry small quantities of flammable liquids and compressed gases to perform their functions, particularly in the private carriage area. Also, considerable difficulty would be involved in placarding the thousands of vehicles used to transport small parcel shipments. The Board believes the benefit of such a requirement would be outweighed by the diminishing effect it would have on other placards that would convey information on the potential hazards of materials in significant quantities. Therefore, the 1,000 pound rule presently used in the placarding regulations for highway vehicles is being continued, except for Class A and B explosives, highly or extremely toxic gases, thermally unstable or self reactive materials, water reactive materials, certain organic peroxides, extremely toxic poisons, and certain radioactive materials.

Another part of the complexity problem relates to the specific identity of materials. It was proposed in the Advance Notice to identify the specific hazards of certain materials when 1,000 pounds or more were loaded at any one loading point. Again, this would cause difficulty in LTL operations due to the nature of placarding requirements proposed. Recognizing the kinds of materials that have been included in Table I of Subpart F, and the information content on hazard information card 01 which would be conveyed through the DANGEROUS placard, the specific placarding criteria has been raised to 5,000 pounds for the materials covered by Table I.

At the end of Table (3) in § 172.502, there is a footnote authorizing use of a flammable placard in place of combustible placard on tank motor vehicles. A similar provision was adopted in Docket HM-102. Further comments on the merits of continuing this provision are solicited for the Board's consideration.

A provision has been added to § 172.502(b) to authorize use of Oxygen 22 placards on vehicles transporting liquefied oxygen that, as carried, does not meet the definition of § 173.300.

Section 172.503, pertaining to the giving and affixing of placards, has been modified to specify more clearly who is required to perform each function. As a matter of practicality, the Board visualizes that motor carriers who handle LTL shipments will find it necessary to have DANGEROUS 01 placards mounted on a permanent basis for required display when they load different shipments of hazardous materials covered by Table (1) totalling more than 1000 pounds gross weight.

Section 172.508 establishes requirements for placement of placards on highway and rail transport vehicles. Not included are the requirements previously proposed pertaining to the number of tacks and staples which must be used to attach placards. The Board is proposing in Paragraph (c) that the front placard must be attached to the front of a truck tractor in certain cases.

Section 172.511 contains proposed minimum strength and durability requirements for placards. Nothing in these proposed requirements prohibits the use of more durable placards or permanent placards made from metal, plastic, etc.

In Section 172.512 it is proposed that, if more than one placard is required for the same material, such as for hazard information numbers 26, 27, 28, and 29, the hazard information number for the material will be required on both placards.

Several comments were received on the merits of having all the placards of a standard shape and size. One commenter submitted a redesign of the placards for uniformity and increased visibility which the Board has made a part of this proposal. Section 172.513 contains the proposed EXPLOSIVES placards. One is of a diamond shape so that all highway transport vehicle placards will be of the same size and shape to permit uniformity in placard holders.

Section 172.520 and Appendices B and C to Subpart F contain the specifications for the diamond shaped placards. Additional specifications are provided for each placard when it differs from the general specifications.

SUBPART G—HAZARD INFORMATION NUMBERS

This proposed subpart contains detailed procedures for determining the source of the hazard information number for a material before it is offered for transportation. In most cases, the derivation procedure is dependent upon quantitative definition criteria. Only partial definitions for flammable solids, oxidizing materials, and self reactive or thermally unstable materials are presently available. Quantitative criteria for these hazards are presently under development and will be proposed in future rule making actions.

HAZARD INFORMATION NUMBERS FOR RADIOACTIVE MATERIALS

In the derivation of hazard information numbers for radioactive materials, as proposed herein, a number of different conceptual approaches were identified and are considered relevant to any approach. For example:

(1) Radioactive materials labels are now unique in that a transport index "block" is already present on the lower half of two of the three labels, obviously creating a conflict with the use of such block for a hazard information number.

(2) Placarding of transport vehicles carrying radioactive packages is presently based on the presence of only the RADIOACTIVE-YELLOW III category label, i.e., vehicles carrying RADIOACTIVE-WHITE I or YELLOW II labels are not required to be placarded; and

(3) In the proposed Hazard Information System, the purpose and intent of the vehicle placard takes on greatly expanded importance with respect to providing meaningful information in the event of a transport incident.

After careful consideration, the Board has concluded that placarding of a transport vehicle carrying all categories of labeled radioactive packages is not justified on a safety basis. Therefore, in the derivation of hazard information number assignments in § 172.101 for the "generic" proper shipping name entries of radioactive materials, a distinction between the hazard information numbers 70 and 71 is being proposed relative to placarding. No placard will be required for a transport vehicle carrying packages bearing hazard information number 70, and placards will be required if any quantity of packages are present bearing a hazard information number of 71. In determining which hazard information number to assign to a package, basically, the shipper would make this distinction on the basis of the quantity of radioactivity contained in the package, i.e., if a Type A quantity is present, hazard information number 70 will be assigned, and if more than a Type A quantity is present, hazard information number 71 will be assigned. One important exception to this rule-of-thumb would prevail, however, and a package bearing a RADIOACTIVE-YELLOW III label will require a hazard information number 71 regardless of the quantity of radioactivity thereby continuing the placarding requirements in the same manner as the present regulations for these packages.

Therefore, the content of the hazard information instructions have been developed on the basis that hazard information number 70 (no placards required) represents a hazard of relatively lower order than hazard information number 71.

It is pointed out that in deriving the hazard information numbers, the numbers 70 and 71 represent only those materials in which radioactivity is the only hazard, whereas, numbers 72, 73, 74, 78 and 79 are for those specific "non-n.o.s." materials which are listed by name in § 172.101 which possess other hazards such as corrosivity, pyrophoricity, flammability, etc., in addition to radioactivity.

It should be noted that those certain small quantities of radioactive materials and small radioactive devices which are exempt from specification packaging, marking, and labeling would not be required to have hazard information numbers assigned.

THE HAZARD INFORMATION CARDS

The proposed hazard information cards or pages, which will constitute the major portion of the Hazard Information Manual presently under development, are included with this Notice for review and comments. It should be noted that the text of each card has been shortened as was recommended by several commenters and several experts contacted by the Board's staff. Any person who desires to participate in final preparation of the entire manual should notify the Secretary, Hazardous Materials Regulations Board.

ENVIRONMENTAL CONSIDERATIONS

In response to comments from the Association of American Railroads (AAR) requesting the Board to issue an environmental impact statement, the Board issued a Request for Information on April 24, 1973 (38 FR 10117), asking the public to comment on the need for such a statement. A similar request was made for Docket HM-102 and a request for Information was also issued. The Board determined that it would not issue an environmental impact statement for HM-102 because there would not be a significant impact on the environment, and the Board has determined that it will not issue an environmental impact statement with respect to this Notice because there will not be a significant impact upon the environment.

The AAR submitted comments to the two Requests for Information in a single document, and the Board has already discussed some of the points raised in its adoption of regulations under Docket HM-102. With respect to HM-103, the AAR said the greatest single impact upon the environment will be the "blossoming of placards" and that such an increase in the number of placards will cause a "dilution" in the effect of the hazardous materials warning system. The AAR argues that studies have shown that the "titillation factor" for each of a series of pornographic photographs decreases in direct proportion to the number of such photos shown, and, by analogy, that too many hazardous materials placards will mean less safety observance and not more.

The Board believes that the AAR has not shown that the adoption of the hazard information system will cause any significant impact on the environment. The argument that the adoption of the hazard information system will dilute the effect of the existing system is without basis and evidence. In addition, the Board is well aware, as must be the AAR, of the inadequate warning system that presently exists. The public docket for this amendment is replete with comments about the need to implement a new hazard warning system, and the inadequacy of the present system. To talk of "diluting" it appears premature at best.

In order to assist interested persons in their efforts to understand fully the proposals made in this Notice, the Board has scheduled a meeting which will be open to the public. The meeting will be for the purpose of answering questions and providing explanations of the proposals contained in this Notice. Statements relative to the merits of the proposals in this Notice may not be made at the meeting, but must be submitted in writing.

The meeting will be held on February 14, 1974, in the Departmental Auditorium located on Constitution Avenue between 12th and 14th Streets, NW, Washington, D.C. beginning at 10 a.m.

Interested persons are invited to give their views on these proposals. Communications should identify the docket number and be submitted in duplicate to the Secretary, Hazardous Materials Regulations Board, Department of Transportation, Washington, D.C. 20590. Communications received on or before May 28, 1974, will be considered before final action is taken on these proposals. All comments received will be available for examination by interested persons at the Office of the Secretary, Hazardous Materials Regulations Board, room 6215 Buzzards Point Building, Second and V Streets, SW, Washington, D.C., both before and after the closing date for comments.

Commenters are requested to make their comments in a manner that will clearly identify the particular matters on which they are commenting. Unless comments are general in nature pertaining to the entire Notice, it is requested that each paragraph of comments be identified in the following manner:

"Subpart C—Shipping Papers—We think . . . " or

"Section 172.503—We believe . . . "

Also, those commenters submitting more than two pages of comments are requested to submit six copies in order to facilitate their handling by the Board.

(39 F.R. 3164, January 24, 1974)

[Docket No. HM-103; Notice No. 73–10A]
TRANSPORTATON OF HAZARDOUS MATERIALS
Notice of Public Hearing Hazard Information System
and Miscellaneous Proposals

On January 24, 1974, the Hazardous Materials Regulations Board ("the Board") issued a notice of proposed rule making in the FEDERAL REGISTER (39 FR 3164; Notice No. 73–10) under Docket HM-103 entitled "Hazard Information System and Miscellaneous Proposals." The final date for filing written comments was specified as October 3, 1974, following publication of two notices granting extensions of time to prepare and submit original or additional comments.

The comments received by the Board on its proposals, and placed in the Public Docket, represent many diverse points of view indicating support, partial support, or in some cases essentially total nonsupport for the proposals contained in the Notice. Two major areas are the subject of the majority of the comments—the proposed placarding requirements and the proposed shipping paper requirements.

Placarding. Many commenters object to the placarding requirements proposed to implement the Hazard Information System. In general, they claim the proposal (1) is too complex; (2) will increase the likelihood of the commission of errors by persons required to carry out the proposed requirements; (3) is unjustified based on experience; (4) does not take into account the cost of implementation; (5) does not fulfill the needs of the fire service as well as the system identified as NFPA 704M would; and (6) is insufficient because reliance is placed on a secondary source for complete information on the hazards presented.

Shipping Papers. Many commenters also object to the Board's proposals pertaining to the preparation of shipping papers. The majority of the comments fall into four basic categories: (1) interference with the use of such documents for economic purposes such as the entry of a freight classification description; (2) conflict with automated billing and document processing system; (3) the proposed requirements are confusing and unnecessary; and (4) the requirements pertaining to the listing of hazardous materials first should be set forth in the regulations as a "recommended practice" and not as a mandatory requirement.

While the Board has not considered in full detail all of the voluminous comments submitted, and has not made any determination as to its acceptance or nonacceptance of the points raised in opposition to its proposals, several suggestions have raised a sufficient concern at this stage of the rulemaking process to warrant further input by interested persons on alternate proposals.

Following his statement of support for the comments submitted by the American Trucking Associations, Inc., the safety director of a motor carrier stated the following:

As an individual, I would like to comment further on Docket HM-103 part 172.200 (shipping papers). I strongly believe that the simplest and most errorless system for dealing with hazardous materials would be to require a separate and individual bill of lading for hazardous material shipments. The bill of lading should be titled "Hazardous Materials Bill of Lading", be distinct in color, and be used solely for hazardous materials shipments. I am aware that these could present problems in other areas of the industry regarding rates and tariffs, etc., however, from a safety point of view, I strongly believe it would be the best method.

This comment and other comments, such as those submitted by the Association of American Railroads pertaining to the information obtainable from shipping papers, has led the Board to consider the following which will be considered a part of this rulemaking proceeding under Docket HM-103.

The Board proposes publication of a standardized form which would be entitled "Hazardous Materials Manifest" (HMM) or some similar title. Under consideration will be one HMM for hazardous materials generally, and one for radioactive materials due to the specialized information required. Required entries on the form would be:

(1) The name and address of the shipper.
(2) Shipping point.
(3) Destination.
(4) Consignee.

(5) Quantity of material(s) (as appropriate).
(6) Shipping name of material(s) as specified in §172.101.
(7) Classification(s).

(8) Hazard information reference number.

(9) Other information required by the regulations.

(10) Shipper's certificate.

(11) Signature of shipper or his representative (agent).

Other entries to be considered although possibly not as mandatory entries, except in certain cases are:

(1) A cross reference to other shipping documents (if any).

(2) Vehicle identification number (possibly for bulk shipments only).

(3) Name of carrier(s).

(4) Signature of recipient.

(5) Date and time of delivery.

On the reverse of the HMM would be a listing of the hazard information reference numbers in sequence and an indication of their meaning. Also, the phone number of Chemtrec would be conspiciously displayed to facilitate the obtaining of further assistance when needed. Other information or instructional material could be provided by shippers by their utilization of certain numbers (such as 90–99) on an optional basis or as required by another agency: i.e., "Cancer-Suspect Agent." The border of the HMM would be highlighted by diagonal red stripes to emphasize its presence. It is proposed that the HMM would be prepared in triplicate, a copy to be retained by the shipper, a copy to be retained by the originating carrier, and a copy to accompany the shipment as proposed in Notice 73–10.

A number of factors will be considered anew or considered further by the Board including (1) the advantages to be gained or lost in placing greater emphasis on documents moving with hazardous materials with no requirement that hazard information numbers be displayed on labels and placards, except possibly for bulk shipments (more than 110 gallons per container); (2) the potential for elimination of confusion and conflicts at the shipper-carrier interface; and (3) the potential for improving the communication of different hazards to both transportation workers and emergency personnel by implementation of a standardized documentation system intended for safety purposes only.

The Board believes there is merit in placing greater emphasis on a document having a standardized format since (1) there would be no confusion or conflict with entries on documents which are used for commercial purposes and to comply with the economic requirements of other agencies such as the Interstate Commerce Commission; (2) its purpose would be for compliance with safety regulations and the communication of safety information (3) it would be prepared by the shipper—the person who must know the characteristics of the materials he is shipping; (4) potential errors would be minimized since little or no information would have to be transferred to other documents; (5) hazard information reference information would be preprinted on the reverse of the HMM being always present on the document when needed. The Board is aware that this proposal involves increased cost to shippers since they would be required to purchase and execute forms that do not exist at present. This factor will be fully considered by the Board before the conclusion of the rulemaking action under this Docket.

To develop a full and complete record concerning the proposals made in Notice 73–10 and above, the Board will hold an informal public hearing beginning at 9:30 a.m. on February 11, 1975 in Washington, D.C. Any person desiring to make an oral presentation at the hearing is requested to notify the Board's Secretary on or before January 8, 1975 stating the approximate time he will need to make his presentation. Additional information about the conduct of the hearing and its location will be announced in the FEDERAL REGISTER during the last week of January 27, 1975.

The Board will also accept written or oral comments from those persons who are the proponents of other methods pertaining to the communication of hazards. If the proponents intend to make a presentation at the hearing, it is requested that they illustrate how they would have their proposals integrated into the regulatory structure of the Department's hazardous materials regulations.

The Board proposes to adopt all or part of the draft revision of Appendix A to Compressed Gas Association, Inc. (CGA) Pamphlet C–7 dated July 1974. The draft revision, entitled "CGA Marking System for Compressed Gas Cylinders," was submitted to the Board for its consideration on August 26, 1974 in response to the statement in the preamble of Notice 73–10 as to why no date was provided in the incorporation reference in proposed § 172.400(a) (2) (iii). The Board believes the draft revision submitted by CGA warrants adoption, except for those adjustments necessary for consistency with other requirements adopted by the Board under this rulemaking proceeding. A copy of the draft revision is available for examination at the address specified below and at the Offices of CGA located at 500 Fifth Avenue, New York City, New York.

Interested persons not desiring to make oral presentations are invited to give their views in writing on the original proposals as amended by this Notice. Communications should identify the docket number and be submitted in duplicate to the Secretary, Hazardous Materials Regulations Board, Department of Transportation, Washington, D.C. 20590. Communications received on or before March 7, 1975 will be considered before final action is taken on these proposals. All comments received will be available for examination by interested persons at the Office of the Secretary, Hazardous Materials Regulations Board, room 6215 Trans Point Building, Second and V Streets SW., Washington, D.C., both before and after the closing date for comments.

(39 F.R. 43091, December 10, 1974)

Amendments in Docket No. HM-103 were published on April 14, 1976 in conjunction with and as part of amendments in Docket No. HM-112. Please refer to the preamble for that latter docket for discussion of amendments in Docket No. HM-103.

SUMMARY • Docket No. HM-109 • *Require installation of tank car head sheilds on DOT-112A and 114A cars used for the transportation of compressed gases; shields are required on all new DOT-112A and 114A tank cars. Partially withdrawn. See Docket Nos. HM.*

[49 CFR Part 179]
[Docket No. HM-109; Notice No. 73–4]
TANK CAR TANK HEAD SHEILDS
Notice of Proposed Rulemaking

The Hazardous Materials Regulations Board (HMRB), is considering an amendment to § 179.100–8, of the hazardous materials regulations to require a protective shield for uninsulated pressure tank car tank heads.

As a result of the growing concern with tank car accidents involving uninsulated pressure tank cars, the Federal Railroad Administration commissioned the railroad tank car safety research and test project (a cooperative program of the Railway Progress Institute and the Association of American Railroads), to study the design of a railroad tank car head protective device, which would reduce the frequency of head puncture in accidents. This study was undertaken under contract No. DOT FRA 00035 and the final report, entitled "Hazardous Materials Tank Cars—Tank Head Protective Shield or Bumper Design," was completed in August 1971. The study showed that for uninsulated pressure tank cars conforming to DOT specifications 112A and 114A, most punctures occur on the lower portion of the tank head. In addition, the study indicated that there was merit in terms of cost/benefit in applying head protection to the lower portion of specifications 112A and 114A tank car tank heads.

Subsequent to the issuance of the report on protective head shields, an accident occurred in the East St. Louis railroad yard of the Alton and Southern Railway. As a result of an impact, the lower portion of the head of a specification 112A tank car was punctured, releasing a vapor cloud of liquefied petroleum gas which exploded. More than 230 persons were injured and property damage was estimated at $7½ million. A complete analysis of this accident was published by the National Transportation Safety Board in Report No. NTSB-RAR-73-1, adopted January 31, 1973. The HMRB believes that rulemaking action is necesary to prevent repetition of such an accident.

Based on the Federal Railroad Administration's studies and analysis of the various accidents, including the one in East St. Louis, the HMRB is proposing to amend § 179.100–8 to include a requirement for a protective shield for each tank head on the specifications 112A and 114A uninsulated tank car tanks. Protective shields would be installed on all newly constructed tank cars of these specifications effective January 1, 1974, although existing tank cars would not have to be so equipped until January 1, 1978. The Board considers this proposal to be more practical and effective than that made in the HM-60 advance notice of proposed rulemaking (35 FR 16180) which proposed speed restrictions for certain hazardous materials trains or wayside inspections and checking of the trains by hotbox detectors or dragging equipment detectors.

The HMRB also believes that even though interlocking couplers are now required on all new tank cars, their presence does not obviate the need for protective end shielding. Shielding is necesary due to the possibility of tank cars being coupled to cars having couplers which are not of the interlocking type, or whose shanks may break, allowing puncture of tank heads in derailments.

(38 F.R. 14112, May 29, 1973)

[Docket No. HM-109; Amdt. Nos. 173–83, 179–15]
PART 173—SHIPPERS
PART 179—SPECIFICATIONS FOR TANK CARS
Tank Car Tank Head Shields

This amendment establishes a requirement for a protective shield for certain uninsulated tank car heads. The amendment was proposed on May 29, 1973, in Docket No. HM-109, Notice No. 73-4 (38 FR 14112). In that notice the Board stated that it believed this requirement would materially reduce the number of head punctures on tank cars carrying liquefied flammable compressed gases and thereby increase safety to the public and railroad employees.

Interested pesons were invited to participate in this rulemaking proceeding and all comments received have been given full consideration by the Board. There were nineteen commenters on the Notice including representatives of the railroad industry and shippers. The interest shown and the comments expressed are appreciated by the Board.

All of the respondents were of the opinion that a regulation calling for head shields is premature and that a modified coupler design with a more positive means of preventing vertical displacement of freight cars during impact would be preferable. The Board does not agree with this position for the following reasons:

1. Statistical evidence already exists through testing that a head shield would be both effective in reducing tank head punctures and would also be cost beneficial. There have been three studies on tank car head shields. Results of these studies are as follows:

(a) The first study, Railroad Tank Car Safety Research and Test Project, was conducted by the Railway Progress Institute (RPI) and the Association of American Railroads (AAR) under an FRA contract. This report was submitted in August 1971. Damage data in the report were based on tank head punctures for the period 1965–1970. Benefits were based on the head shield being 77 percent efficient. The cost of application used in the report was developed by the tank car manufacturers. The average costs of application used in this report were $280 for a new car and $335 for an existing car. The present value benefit of the head shield was computed in this study as the resultant of investing the annual per car damage savings for a thirty year period at an interest rate of 10 percent. The report stated that the net economic value of the head shield was $105 on new cars and $50 on existing cars.

(b) The Association of American Railroads submitted a report in November 1972 on tank car head shields. The same data base and statistical approach as used in the RPI/AAR report was employed. The AAR assumed that the head shield would be only 50 percent effective and estimated the cost to be $272 for new cars and $474 for existing cars. On this basis, the net economic value was negative. On new cars the economic loss was stated as $8 and on existing cars the economic loss was $210.

(c) Examination of the two reports by the FRA and the Calspan Corporation revealed that the separation of tank car head punctures from other tank shell intrusions accompanying or resulting from a head puncture may have caused bias in the data base discussed in (a) and (b) above. FRA totaled all shell puncture damage and assigned the portion to head punctures based on the percentage of incidents originating from a head puncture. Application costs were based on the highest estimates from both head shield reports. On this basis, the net economic value of the head shield is $395 on new cars and $201 on existing cars. The following table shows a comparison of the three reports.

2. The modified coupler design which consists of a standard coupler with top and bottom shelf has had little testing and there is no basis for assuming that it is superior to the head shield as a puncture preventative. In the event that the modified coupler design also proves cost beneficial, the head shield can serve as back up system and increase the total effectiveness of both. Some commenters were concerned about the 500,000 pounds dynamic force strength requirement. The Board concurs with their recommendation that the shield be designed to pass the normal impact test required for all tank cars. The regulation has been revised to reflect this change. For the purposes of clarity a new paragraph Head Shields (179.100–23) has been introduced rather than amend the paragraph captioned Tank Heads (179.100–8).

In developing the final rule in this proceeding, the Board seriously considered reducing by one or two years the proposed period for retrofitting the more than 18,000 existing DOT specification 112A and 114A tank cars with head shields. However, upon further consideration, it was determined that this task is of such a magnitude that it cannot be completed before December 31, 1977. Reducing the retrofit period by one or two years would only result in removal of many of these cars from service thereby futher intensifying the energy crisis and severely restricting the rail movement of fuels, fertilizers, chemicals and liquefied compressed gases vital to the nation's economy. The Board believes that prompt action must be taken by tank car owners to ensure that all existing 112A and 114A tank cars are equipped with head shields by the end of 1977. Accordingly, the Board requests that each owner of these tank cars file with the Federal Railroad Administrator, Washington, D.C. 20590, by September 1, 1974, its head shield retrofit program or schedule, followed by annual progress reports to be filed by September 1 each year and a final report when the program is completed. The Board expects each owner to retrofit all of its tank cars with head shields as soon as possible and will not be receptive to petitions to extend the retrofit program completion date.

(39 F.R. 27572, July 30, 1974)

[Docket No. HM-109; Amdt. Nos. 173–83, 179–15]
PART 173—SHIPPERS
PART 179—SPECIFICATIONS FOR TANK CARS
Tank Car Tank Head Shields; Denial of Petitions fo Reconsideration

On July 23, 1974 (39 FR 27572, July 30, 1974), the Hazardous Materials Regulations Board issued Amendments Numbered 173-83 and 179-15 under Docket Number HM-109. These amendments require a tank head protection device (head shield) to be affixed to each end of all specification DOT-112A and 114A tank cars built after August 30, 1974, used for the transportation of compressed gases be equipped with protective head shields by January 1, 1978. In developing these amendments, the Board analyzed economic, research and accident data and concluded that the head shield specified in these amendments was cost beneficial and would be effective in reducing tank head punctures.

Subsequent to the issuance of thse amendments, the following persons submitted Petitions for Reconsideration under provision of Title 49 of the Code Federal Regulations, §170.35:

The Compressed Gas Association.

The Railway Progress Institute on behalf of ACF
   Industries, Inc., General American
   Transportation Corporation, North American
   Car Corporation, and Trans Union
Corporation.

The Association of American Railroads.

Phillips Petroleum Company.

Cities Service Oil Company.

Ethyl Corporation.

Pennwalt Corporation.

Cities Service Pipe Line Company.

Amoco Oil Company.

The Manufacturing Chemists Association.

Additionally, the following petitions were filed subsequent to ten days prior to the effective date of the rule, but prior to the effective date of the rule:

American Petroleum Institute.

FMC Corporation.

Continental Oil Company.

Republic Car Line Inc.

Although these four petitions were filed late, the Board has decided to consider them.

Most of the petitions endorsed the petition submitted by the Railway Progress Institute. In that petition, the Institute stated three allegations which are discussed separately.

I. The Board has misconceived and misconstrued the material persented to it by respondents; reconsideration is urgently required to permit appropriate analysis of this critically relevant data.

To the contrary, the Board stated in the preamble to these amendments that all respondents believed that imposition of a requirement for head shields was premature; however, the Board noted that:

Statistical evidence already exists through testing that a head shield would be both effective in reducing tank head punctures and would also be cost beneficial.

Results of several studies were summarized in that preamble.

The petitioners state that the effectiveness of the prescribed head shield is based essentially upon laboratory tests and assumptions based on analysis of prior accident information. The Board has relied upon such data and information in developing this regulation and it believes that due to the potential tragic consequences of head punctures in liquefied compressed gas laden specification DOT-112A and 114A tank cars, prompt action to apply this data and information so as to upgrade the safety of this equipment is essential.

The petitioners state that the Board failed to consider the relative merits of a standard coupler with top and bottom shelves vis-a-vis the specified head shield. The Board considered the information submitted both prior to and subsequent to the closing date for the filing of comments to the Notice of Proposed Rule Making. In addition, the Board analyzed the results of studies, tests and accident investigatory reports in developing these amendments. In its expert opinion, the Board does not believe that there is likelihood that the top and bottom shelved coupler will prevent head punctures as effectively as the specified head shield.

II. The material which has become available since the terminal date for filing statements in Docket No. HM-109 (September 4, 1973), clearly establishes and confirms the superirity of the E coupler with top and bottom shelves as a head puncture preventative; reconsideration of the issued rule should be granted and a substitute rule involving the shelf-type coupler should be proposed for adoption.

In claiming that new information is now available to the Board which clearly establishes and confirms the superiority of the E coupler with top and bottom shelves as a head puncture preventative, Railroad Tank Car Safety Research and Test Project Reports RA-10-3-25 concerning a July 1, 1973, railroad accident and RA-10-4-28 concerning a February 9, 1974, railroad accident are submitted as "new material." The Board had received copies of both reports and included that information in its analysis of the effectiveness of a shelf coupler prior to issuing these amendments. Also, the Board had the benefit of a field investigatory report on the accident that occurred near Romney, Kentucky, which is referred to by the petitioners as RA-10-3-25. After its review of those reports, the Board determined that the information contained provided no reason to conclude that such couplers have the effectiveness that the specified head shields have in preventing tank head punctures.

Likewise, the report by Siniat, Helliesan & Eichner, Inc., dated June 12, 1974, had been analyzed by the Board prior to issuance of these amendments and has again been analyzed prior to the issuance of this denial. The Board does not concur with their conclusion:

"We further conclude that the evidence bearing on the E type coupler modified with top and bottom shelves, compared with either the RPI/AAR-designator or the DOT-design head shield indicates the preferability of the coupler solution under all circumstances."

As a result of analysis of head punctures in these cars from January 1970 through June 1974, the Board has determined that of the 29 cars which received head punctures 18 cars (or 62 percent) were punctured by couplers, while 11 cars (or 38 percent) were punctured by rail or other objects. The shelved E coupler would not have protected the tank head from punctures caused by "rail or other objects" while the head shield would have provided protection against such punctures.

III. Rescission of the issued rule and promulgation of a proposed rule involving the shelf-type coupler would be in the public interest.

The petitioners state that the application of head shields is complicated, costly and time-consuming whereas the installation of the shelf coupler could be accomplished in a relatively short period of time. The Board was and is aware of the fact that installation of protective head shields will cause tank cars to be out-of-service. However, the Board has weighed public safety against inconvenience and believes that safety demands the installation of shields in order to prevent tank head punctures.

The petitioners indicate that the application of head shields will increse the light weight of each tank car by approximately 1,400 pounds. However, the Association of American Railroads under contract DOT-FR-00035 stated: "increase (in) the light weight of the cars in some cases will and in other cases will not, have an effect on the commodity carrying capacity of the car." The Board was and is aware that the majority of the tank cars affected by these amendments are designed to transport both liquefied petroleum gas and anhydrous ammonia. Since liquefied petroleum gas (LPG) has a lower density than anhydrous ammonia, the LPG capacity of such dual service designed tank cars would normally not be reduced due to the application of these head shields. In those cases where the product capacity of these tank cars will be reduced by the added weight of the shield, the Board had concluded that public safety considerations override this loss of lading capacity.

One petitioner expressed concern that the head shield might not remain attached to the car under actual operating conditions. The Board knows that competent engineering and use of good car construction practices can assure that the shields will remain affixed to the car during transportation. Inspection of the head shield can be performed at each originating and interchange point (as required by 49 CFR 173.596) to assure that it is properly secured to the car.

Another petitioner stated:

"The smaller diameters of tanks of cars of less than 1,500 gallons capacity and the greater curvature of tank heads may well result in such cars being far less susceptible to punctures of the type which the head shield is intended to prevent."

If the petitioner has technical data and test results to support this position, it may submit a petition for special permit under the provisions of 49 CR 170.13.

Petitioners have requested the Board to consider the use of a newly developed E coupler with a top and bottom shelf on tank cars. The Board has awarded contract numbered DOT-OS-40106 to Washington University, St. Louis, Missouri to "perform a study of criteria and technology for the design of shelf couplers." Upon completion of this contract and publication of the report, the Board will give further consideration to this coupler arrangement.

Not included in the "Economic Evaluation of Tank Car Shield" cited in the amendments was accident damage occurring after 1970. Between January 1, 1971, and May 31, 1974, there were 17 incidents involving 19 head punctures to these tank cars reported to the Board. Estimated damage caused by these punctures exceeded $15,000,000. On June 19, 1974, at the Norfolk and Western Railway Yard in Decatur, Illinois, a specification DOT-114A tank car tank head was punctured by a coupler on a box car. The escaping liquefied petroleum gas ignited and the resulting explosion and fires killed seven railroad employees, injured eight others and caused property damage in excess of $20,000,000. It is the opinion of the Board that had the car been equipped with protective head shields, as specified in these amendments, the tank head would not have been punctured and the Decatur catastrophe would not have occurred. When these losses are considered, the economic "cost/benefit" of applying head shields in accordance with these amendments becomes more advantageous.

The Hazardous Materials Regulations Board has evaluated the Petitions for Reconsideration and it has determined that the information submitted does not justify reconsideration; therefore, these petitions are denied.

(39 F.R. 32912, September 12, 1974)

SUMMARY • **Docket No. HM-110** • *This is a miscellaneous docket combining several unrelated changes to requirements imposed on motor carriers engaged in transporting hazardous materials. See individual headings for subject matter.*

[49 CFR Parts 173, 177]
[Docket No. HM-110; Notice No. 73-5]
HANDLING OF HAZARDOUS MATERIALS ON MOTOR VEHICLES
Miscellaneous Amendments

The Hazardous Materials Regulations Board is considering amendment of several sections of the Department's Hazardous Materials regulations. Commenters need only identify the particular proposal on which they wish to comment when responding. The proposals covered in this document are:

A. Emergency discharge controls on MC 330 cargo tanks.

B. Cargo tank certificate retention.

C. Hydrostatic and pneumatic testing of cargo tanks.

D. Cargo heaters with explosives and flammable commodities.

E. Attendance of tank vehicles during loading and unloading.

F. Openings on cargo tank to be closed during transportation.

G. Repairs and maintenance to vehicles in closed garages.

H. Waning devices on vehicles containing hazardous materials.

PROPOSAL A
EMERGENCY DISCHARGE CONTROLS ON MC 330 CARGO TANKS

The Hazardous Materials Regulations Board is considering amendment of § 173.33 of the Department's Hazardous Materials regulations to require that all Specification MC 330 cargo tanks used for the transportation of flammable compressed gases and anhydrous ammonia be equipped with emergency discharge controls as is now required on Specification MC 331 cargo tanks. At the time Specification MC 331 cargo tank specification was adopted, the Interstate Commerce Commission decided not to require an updating of the standards for Specification MC 330 cargo tanks.

The Board believes that it is now necessary to require that these tanks conform to the same emergency discharge control standards as are required for MC 331 cargo tanks to assure the same degree of safety. A recent accident involving an MC 330 cargo tank has demonstrated the need to require that these tanks be retrofitted with remote controlled internal shutoff valves. In this accident, the propane from an MC 330 cargo tank provided fuel to a fire which resulted from the accidental rupture of a manifolded storage tank intake line into which the cargo tank was unloading. The escaping propane from the cargo tank was not discharging at a rate high enough to activate the excess flow valve. The fire was directed to another cargo tank that eventually exploded. As a result of the fire and explosion, one person was killed and over $200,000 in property damage occurred. There is little doubt that if an internal valve, as specified in section 178.337-11(c) had been installed on the cargo tank, the flow of propane could have been shutoff by manual means, if not automatically by the melting of the fusible element.

In order not to impose an undue burden on tank owners, the Board is proposing that the emergency discharge controls may be installed when the cargo tanks are scheduled for the 5-year retest required in § 173.33.

PROPOSAL B
CARGO TANK CERTIFICATE RETENTION

The Hazardous Materials Regulations Board is considering an amendment to clarify the requirement for retention of the manufacturer's certificate for specification cargo tanks.

The Board has found that many motor carriers are not aware of a requirement that the manufacturer's certificate for a specification cargo tank must be retained by the motor carrier for as long as the tank is in service and for 1 year thereafter. The confusion may be caused by the fact that the certificate-retention requirements are presently contained in the cargo tank specifications in Part 178, and many of the specifications no longer appear in the published codification of the regulations, although the tanks may be continued in use. The Board believes that a general retention provision is needed in Part 177 to resolve this problem. In addition, the Board is proposing to require carriers to retain all retest and inspection reports in the same file with the manufacturer's certificate.

PROPOSAL C
HYDROSTATIC AND PNEUMATIC TESTING OF CARGO TANKS

The Hazardous Materials Regulations Board is considering amendment of § 177.824 of the Department's Hazardous Materials Regulations to clarify the hydrostatic and pneumatic testing procedures for cargo tanks.

Present requirements specify hydrostatic or pneumatic testing of cargo tanks under certain conditions. But the procedures for pneumatic testing are not contained in the regulations. Therefore, the Board is proposing that these procedures be incorporated into the regulations.

PROPOSAL D
CARGO HEATERS WITH EXPLOSIVES AND FLAMMABLE COMMODITIES

The Hazardous Materials Regulations Board is considering amendment of § 177.834 of the Department's Hazardous Materials regulations, to clarify the prohibition against the use of certain heaters in a transport vehicle which is loaded with explosives or flammable commodities.

There has been much confusion concerning the use of catalytic cargo heaters in vehicles transporting flammable liquids and flammable gases. The Federal Highway Administration has taken the position that, for the purposes of these regulations, a catalytic heater is a combustion heater. This proposal would specifically state that catalytic heaters are considered as such, and not permitted in vehicles transporting flammable materials. However, commenters are invited to submit test data and other evidence in support of the use of catalytic heaters as a safe means of heating the cargo spaces of motor vehicles.

The Board also feels that the precautions against the loading of explosives into transport vehicles containing a heater should be the same as the pertaining to flammables, and that the specific provision should be included under general requirements.

PROPOSAL E
ATTENDANCE OF TANK MOTOR VEHICLES DURING LOADING AND UNLOADING OPERATIONS

The Hazardous Materials Regulations Board is considering amendment of § 177.834 to clarify the meaning of "attendance" as it pertains to a tank motor vehicle being loaded or unloaded.

The Board has found that several dangerous incidents have occurred during the loading or unloading of tank motor vehicles which could have been avoided, if there had been someone near the cargo tank to take corrective or precautionary action. The Board feels that there may be some confusion as to the intent of the term "attendance" as it is used in § 177.834(i).

PROPOSAL F
OPENINGS ON CARGO TANKS TO BE CLOSED DURING TRANSPORTATION

The Hazardous Materials Regulations Board is considering amendment of § 177.839 by adding a requirement that internal valves and manholes be in a closed and secured position during transportation. A similar provision was added during recent rule making pertaining to the transportation of compressed gases, and the Board now proposes to add similar requirements for the transportation of flammable liquids, poisons and corrosive materials in cargo tanks.

PROPOSAL G
REPAIRS AND MAINTENANCE TO MOTOR VEHICLES CONTAINING HAZARDOUS MATERIALS

The Hazardous Materials Regulations Board is considering amendment of § 177.854 of the Department's Hazardous Materials Regulations to authorize repairs to a motor vehicle containing hazardous materials in a closed garage.

This proposal is based, in part, on a petition from Consolidated Freightways Corp. of Delaware. Petitioner states that " . . . minor repairs as adjustment of brakes, changing of tires, replacing burned out lamps, etc., would necessarily have to be made in the open or under a shed-type building . . . this rule causes undue hardship on our, or any motor carrier's operation, especially during the hours of darkness and also inclement weather.

The Board has concluded that the petitioner's request may have merit. A carrier faced with the prohibition in § 177.854(g) might well choose to delay making necessary, though perhaps minor, repairs until after the vehicle has reached its destination and has been emptied. By doing so, a greater hazard may exist than the potential hazard posed by the making of repairs in a garage, especially since many repairs do not involve any particular risk of explosion or fire.

However, the Board believes that the prohibition should be removed only if adequate safeguards against explosion are maintained. The safeguards propsed include protecting the vehicle from open flames or welding devices in use, and requiring every vehicle to have a means of motive power while it is in the garage.

PROPOSAL H
WARNING DEVICES FOR STOPPED VEHICLES

The Hazardous Materials Regulations Board is considering editorial changes to §§ 177.854, 177.856, and 177.859 to reflect recent changes to the Motor Carrier Safety regulations (49 CFR Parts 390–397) pertaining to warning devices for stopped vehicles.

(38 F.R. 22901, August 27, 1973)

[Docket No. HM-110; Amdts. Nos. 173–87, 177–31]
PART 173—SHIPPERS
PART 177—SHIPMENTS MADE BY WAY
OF COMMON, CONTRACT, OR PRIVATE CARRIERS
BY PUBLIC HIGHWAY
Handling of Hazardous Materials on Motor Vehicles;
Miscellaneous Amendments

The purpose of these amendments to the hazardous materials regulations of the Department of Transportation is to (a) require the installation of liquid discharge controls on certain MC 330 cargo tanks; (b) specify location requirements for retention of manufacturer's certificates for specification cargo tanks; (c) clarify the hydrostatic and pneumatic testing procedures for cargo tanks; (d) provide for the use of certain heaters in vehicles loaded with flammable commodities; (e) define "attendance" as it pertains to a tank motor vehicle being loaded or unloaded; (f) require that internal valves and manholes on cargo tanks be in a closed and secured position during transportation; (g) authorize certain repairs to a motor vehicle containing hazardous materials in a closed garage; and, (h) make several editorial changes to reflect recent amendments to the Federal motor carrier safety regulations (49 CFR 390–397) pertaining to warning devices for stopped vehicles.

On August 27, 1973, the Hazardous Materials Regulations Board published a notice of proposed rule making, Docket No. HM-110; Notice No. 73–5 (38 FR 22901), which proposed these amendments. Interested persons were invited to give their views.

Several commenters pointed out that because many MC 330 cargo tanks are not equipped with manholes, especially those of less than 3500-gallon water capacity, it would be difficult to retrofit these tanks with internal emergency valves at vapor discharge openings. The Board recognizes this problem and has decided to make this rule making applicable only to liquid discharge openings. The Board will consider further the problems raised concerning retrofitting vapor discharge openings in future rule making. Also, this decision is based on the fact that the Board inadvertently failed to include MC 331 cargo tanks constructed before March 31, 1974, in order to be consistent with amendments made in HM-86 (38 FR 27595).

A number of commenters objected to the cargo tank certificate retention proposal. One commenter stated "[it] is an unrealistic and unnecessary requirement," and "could significantly increase paperwork for many carriers. . . ." The Board reiterates that this proposal is merely intended to clarify existing record keeping requirements. Presently, each cargo tank specification requires that the carrier procure and retain a manufacturer's certificate. Sections 173.33 and 177.824 require the retention of periodic retest and inspection records. Part 396 of the Federal motor carrier safety regulations (49 CFR 390–397) requires the retention of inspection and maintenance records for all vehicles. The only "new" aspect of the proposal is that the records be kept together in the same file to form a composite history of the cargo tank. The Board recognizes that some record keeping duplication may result, in that part of the inspection and maintenance records required by Part 396 may be duplicated by this amendment depending upon each carrier's record keeping system. However, only those records pertaining to inspection and maintenance of the tank vessel need be retained with the manufacturer's certificate. Recognizing the burden this would place on single trip leased cargo tanks, the Board has not made the requirement applicable to cargo tanks leased for less than 30 days when the required records are retained by the lessor.

In its proposal to clarify the procedures for pneumatic testing in § 177.824(d) (2), the Board said the test pressure must be increased by a pressure equivalent to the static head in the tank when fully loaded. A commenter pointed out that in some cases, the total test pressure may exceed the maximum design pressure of the cargo tank. The Board agrees and has dropped this provision. Another commenter suggested that the provisions for hydrostatic and pneumatic testing be listed separately. The Board agrees and had made editorial changes accordingly.

In its proposal to prohibit the use of catalytic heaters in vehicles transporting flammable liquids and flammable gases, the Board stated the rule making action was merely clarifying the position of the Federal Highway Administration that catalytic heaters are combustion heaters. The Board Member for Federal Highway Administration has not changed that position. However, based on experience data submitted in response to the notice, the Board Member for Federal Highway Administration believes catalytic heaters may be a safe means to heat the cargo space of motor vehicles.

The original position of the Board Member for the Federal Highway Administration was based on fires reported to the Bureau of Motor Carrier Safety under the Federal Motor Carrier Safety Regulations' accident reporting system. Recent reports submitted to the Bureau have indicated that the majority of such fires were caused by cargo being too close to the heaters. Based on this data, the Board believes catalytic heaters can safely be used with flammable liquids and flammable gases provided adequate guards are installed to prevent lading from being too close to the heating elements. Therefore, § 177.834 has been changed to require guards to keep lading at least 12 inches from any catalytic heater.

That part of the proposal pertaining to the use of catalytic heaters in vehicles containing explosives is being retained as proposed.

Section 177.834(i) has been modified to clearly distinguish who is responsible for ensuring the attandance of cargo tanks during loading and unloading. For example, the carrier's responsibility will depend upon whether its contractual obligation to transport the hazardous material includes loading or unloading.

Many commenters pointed out that shipper or consignee personnel often perform the actual loading or unloading of the cargo tank even though the carrier's driver is present. The proposed definition of "qualified representative" would not allow attendance by such persons. The section has been changed to permit attendance by a "qualified person" without the requirement that he be a representative of a motor carrier.

Editorial changes have been made to the proposal that all openings on a tank vehicle be closed to clarify application of the rule.

The Board proposed to change § 177.845(g) to allow repairs to vehicles containing hazardous materials in closed garages with certain restrictions. Several commenters objected to the restriction pertaining to presence of welding or flame producing devices in the same enclosed area and suggested a distance factor instead of the blanket prohibition. While the prohibition may cause some inconveniences, accident experience has shown that distance is not always an adequate safeguard, especially considering the nature of some gases and flammable vapors. The Board does not believe the restrictions are overbearing when considering the alternative of the complete prohibition against any repairs in a closed garage. Editorial changes have been made in § 177.854(g) (1) to include a cross-reference to paragraph (h) and to limit application to those materials having a definite fire or explosion hazard.

The Board is cancelling § 177.815(g) which was inadvertently overlooked during rule making in Docket No. HM-8 (38 FR 5292).

(39 F.R. 41741, December 2, 1974)

[49 CFR Part 177]
[Docket No. HM-110; Notice No. 75–5]
USE OF CATALYTIC HEATERS IN CERTAIN MOTOR VEHICLES, AND REPAIRS TO VEHICLES IN CLOSED BUILDINGS
Proposed Rule Making

On December 2, 1974, the Hazardous Materials Regulations Board (The Board) published Amendments Nos. 173–87 and 177–31 under Docket HM-110 (39 FR 41741). One portion of the amendment pertained to the use of catalytic heaters in the cargo compartment of a motor vehicle transporting flammable liquids or flammble compressed gases. The revision authorizes use of catalytic heaters in these motor vehicles if guards are installed to keep the cargo at least one foot away from the heater. The amendment was to become effective April 1, 1975, but those sections covered by this notice were postponed until October 1, 1975 (40 FR 12269).

After the revision was issued, a petition for reconsideration was received from Cargo Safe, Inc., a manufacturer of catalytic heaters, containing the following statement:

"We believe the wording as it currently exists does not deal adequately with the temperature problem and can allow for dangerous catalytic heaters to be produced even though they conform to the present requirements."

Specifically, the petitioner seeks the addition of a limitation upon the temperature that may be reached on the outside surface of a catalytic heater used in the cargo compartment of a vehicle transporting flammable liquids or flammable compressed gases. Petitioner cited tests of a prototype heater, with guards 12 inches away from the heater, which allowed the outside skin temperature of the guard to reach 284° F.

After reviewing the data submitted in support of the petition, the Board has concluded that the petitioner's contention has merit. Accordingly, the Board proposes to add to paragraph (1) of § 177.834 a limit to the maximum temperature permitted on the outside surface of a catalytic heater which is used in the cargo compartment of motor vehicles transporting flammable liquids or flammable gases. However, the Board has not removed the requirement for the guard on the heater as the petitioner suggested because the guard affords protection against damage to packages coming in contact with the heater. The Board further believes that a catalytic heater needs to be marked so carriers will know if the heater complies with these requirements. Therefore, the Board is proposing that "Meets DOT

Requirements" be marked on the heaters as a manufacturer's certification that the heater complies with the require-ments of § 177.834 (1).

Upon further consideration, the Board believes that catalytic heaters should not be lighted or used in the cargo compartment of a motor vehicle containing flammable liquids or flammable gases if any flame is present on the catalyst or visible anywhere in the heater. Such a restriction is proposed to be added to § 177.834 (1) along with a requirement that catalytic heater manufacturers place a sign on each heater warning of this danger.

The Board also received a letter from Phillips Petroleum Company concerning the § 177.854(g) amendment being different from what was proposed in the notice. After further review, it appears that the preamble did not clearly explain the rationale behind the final amendment, and that § 177.854(g) as amended does not clearly state the Board's decision and a proposed editorial revision of the section.

Prior to this rule making, § 177.854(g) contained a blanket prohibition against a vehicle containing any hazardous material being in a closed garage for repairs. In response to a petition, the Board proposed that this prohibition be lifted provided certain conditions existed in the closed garage. The proposed restriction was without regard to any particular class of hazardous material. Following a review of comments submitted, the Board decided that vehicles containing material which posed an inherent fire or explosion danger, i.e., explosives, flammable liquids or gases, should not be in a closed garage for repairs or maintenance regardless of the added conditions. Therefore, the amendment was written to allow vehicles containing hazardous materials, with certain exceptions to be in a closed garage for repairs. This decision was a compromise between the original blanket prohibition and the proposed relaxation without regard to type of material on the vehicle. The Board is also proposing a definition for a closed building for purposes of § 177.854(g).

(40 F.R. 21485, May 16, 1975)

[49 CFR Part 177]
[Docket No. HM-110; Reference Notice 75-5]
REPAIRS TO VEHICLES IN CLOSED BUILDINGS
Withdrawal of Notice of Proposed Rulemaking

This notice withdraws that part of the notice of proposed rulemaking, 75-5 (40 FR 21485), which proposed a revision to the regulation pertaining to repair and maintenance, in closed buildings, of vehicles containing flamma-ble liquid or gas.

Of the comments received in response to this proposal, most objected for several reasons. Upon review of the proposal, in the light of these objections, the Materials Transportation Bureau found a need for additional research and consideration.

Withdrawal of this proposal also shows the need to delay the effective date of a recent amendment, which ad-dresses the same matter. (Amendment No. 177-31 (39 FR 41741).) It would otherwise become effective on October 1, 1975. A notice to postpone the effective date appears on page 44821.

The Bureau expects to issue another notice of proposed rulemaking after further consideration. The views of interested parties, which are currently on file, will be reconsidered in the development of the new proposal.

(40 F.R. 44842, September 30, 1975)

[Docket No. HM-110; Amdt. 177-34]
PART 177—SHIPMENTS MADE BY WAY
OF COMMON, CONTRACT, OR PRIVATE CARRIERS,
BY PUBLIC HIGHWAY
Use of Catalytic Heaters

The purpose of this amendment to Part 177 of the Hazardous Materials Regulations is to permit the use of catalytic heaters in motor vehicle cargo compartments during the transportation of flammable liquid or flammable gas.

On May 16, 1975, the Hazardous Materials Regulations Board published a notice of proposed rulemaking, Docket No. HM-110; Notice No. 75-5 (40 FR 21485), which proposed this amendment. It proposed:

(1) To establish a maximum allowable temperature of 130° F. (54° C.) for all surfaces of the heater which might contact the cargo of flammable liquid or gas;

(2) To require installation of a guard on a heater to prevent the hazardous material cargo from being closer than 12 inches (30.05 cm) to the heater, as a means of preventing exposure of the cargo to more than 130° F. (54° C.);

(3) To prohibit heater ignition before the vehicle is loaded;

(4) To prohibit flame, either in the catalyst or in any part of the heater, and to require that notice to this effect be marked on the heater;

(5) To require the use of a heater that has been certified by the manufacturer that the heater meets all the Department's requirements by properly marking the heater with the certification; and

(6) To clarify the restrictions on the use of automatic cargo-space-heating temperature control equipment with explosives or flammable liquid and gas.

Interested persons were invited to give their views on these proposals. Of the comments received, no objection was made to the proposals as outlined in items (3), (4), and (6) above. Rewording these proposals in this amendment clarifies, rather than changes, substance.

There was one comment, a significant one, regarding the manufacturer's certification-marking requirement. As proposed, the wording was so similar to marking requirements under the Federal Motor Carrier Safety Regulations that confusion, threatening a compromise of safety, was fostered. Consequently, the certification-marking language for catalytic heaters is changed.

All other comments addressed the proposed maximum allowable temperature and guard requirements.

Several commenters objected that the requirement to install a heater guard was unnecessary in view of the proposed maximum temperature standard and the high cost resulting from installation, retrofitting, and loss of revenue producing cargo space.

There was no significant objection to approaching the problem of assuring safety in the use of catalytic heaters by establishing a maximum temperature for heater surfaces. Nor was there any significant objection to establishing 130° F. (54° C.) as the standard. One objection, based upon the premise that a higher maximum was proper, was withdrawn after close examination.

The Materials Transportation Bureau, which is now vested with the authority to issue hazardous materials regulations, concluded that assuring safety in the use of catalytic cargo heaters for transporting flammable liquid or gas is properly approached by establishing a maximum allowable temperature for heater surfaces of 130° F. (54° C.). It was further concluded that the use of a 12 inch heater guard, as the means of keeping heater surface temperatures within the temperature standard, is not required.

In support of these conclusions, the evidence indicated the following:

(1) 130° F. (54° C.) is the maximum temperature at which human skin may be subjected for short duration without causing non-reversable tissue damage. See, for example, the Calspan Corporation's study for the Consumer Products Safety Commission, entitled, "Investigation of Safety Standards for Flame-fired Furnaces, Hot-Water Heaters, Clothes Dryers and Ranges."

(2) 130° F. (54° C.) permits a wide and comfortable safety factor before auto-ignition occurs with properly packaged flammable liquid and gas that is listed under 49 CFR 172.5 as hazardous material.

(3) Two of the major manufacturers of commercial catalytic heaters came forth to assure the Bureau that heaters may be readily designed to operate within the 130° F. (54° C.) maximum, in an outside or ambient temperature range beyond which there is no reasonable expectation to find a heater in operation.

(4) It is reasonable to expect that a heater would not be operated when the outside or ambient temperature is above 60° F. (15.6° C.).

(5) Heaters with 12 inch-guards, if the guards were installed on some of the heaters currently available, could reach a surface temperature as high and as dangerous as 284° F. (140° C.).

(40 F.R. 46106, October 6, 1975)

[Docket No. HM-110; Amdt. No. 177–42]
PART 177—CARRIAGE BY PUBLIC HIGHWAY
Repair and Maintenance of Vehicles

I. PREVIOUS NOTICES AND AMENDMENTS IN THIS DOCKET

On August 27, 1973 (38 FR 22901), the Department of Transportation (DOT) published a notice that, among other things, proposed to modify an existing requirement (49 CFR 177.854(g)) which prohibited repair and maintenance work on vehicles containing hazardous materials inside a building, or whenever that work could not be performed without hazard. The notice was prompted by a rulemaking petition from a motor carrier who observed that the existing requirement prohibited minor but important repairs, such as brake adjustments and tire and lamp replacement, from being performed indoors on vehicles carrying hazardous materials. The carrier asserted that such minor repair work could be performed safely indoors. The notice resulted in an amendment (39 FR 41741, Dec. 2, 1974) with a 6-month delay in effective date. The effective date was extended another 6 months (40 FR 12269, Mar. 18, 1975) during which time the original proposal was modified (40 FR 21485, May 16, 1975) and then the modification withdrawn (40 FR 44842, Sept. 30, 1975). The effective date was again extended until July 1, 1976 (40 FR 44821, Sept. 30, 1975). Finally, § 177.854(g) as it originally read was confirmed until further amendment (41 FR 27968, July 8, 1976). On October 12, 1976 (41 FR 44712), a new notice was published upon which this amendment is based.

II. PURPOSE OF THIS AMENDMENT

This amendment is intended to reduce the possibility of an accident involving significant quantities of hazardous materials as a result of maintenance or repair work on the motor vehicle in which the materials are contained. Commenters have suggested that a complete examination of the work area might be necessary to control possible sources of accidents. The MTB agrees and has prescribed only general conditions in this amendment to deal with hazards arising from the materials a cargo vehicle may carry, and with conditions that threaten the integrity of the system on the vehicle that contains the materials. This amendment does not exhaustively address conditions in the work area. Because of the variety of possible circumstances involved, reliance must be placed on existing local, State, and Federal laws concerning motor vehicle repair work, and on the sound judgment and experience of those persons performing the work.

Paragraph (g)(1) of this amendment prohibits hot work on the cargo or fuel containment system of a vehicle carrying a sufficient quantity of a hazardous material to necessitate placarding. Vehicles carrying combustible liquids are excluded from this prohibition.

Under paragraph (g)(2) of the amendment, repair or maintenance inside a building on any placarded vehicle, except one placarded COMBUSTIBLE, is prohibited unless the cargo and fuel containment systems are free from leaks, the vehicle can be quickly removed from the enclosed area if necessary in an emergency, and the vehicle is removed from the enclosed area upon completion of repair or maintenance work. Both a person capable of operating the vehicle, and a means of moving the vehicle (in the event the vehicle cannot or should not be moved under its own power) must be available if removing the vehicle becomes necessary in an emergency. Conditions under which the vehicle should be removed, and the location to which it should be removed, are not prescribed in the amendment. Those conditions, which may vary, are left to be determined by the carrier or repair facility. In some circumstances it may be advisable not to attempt to remove the vehicle until further precautions are taken. To preclude the potential dangers that may develop, the vehicle must be removed from the enclosed area upon completion of repair or maintenance work. This condition was proposed in the notice of proposed rulemaking and generated very few comments because from a practical safety standpoint it appears to be followed by motor carrier shops and repair facilities.

In addition to these three restrictions on inside repair to a placarded vehicle, two more conditions apply if the vehicle contains explosives A or B (any quantity of which necessitates placarding) or contains a flammable liquid or flammable gas (note that cargo tanks and portable tanks containing even residues of these materials necessitate placarding, unless cleaned and purged). The MTB recognizes that hazardous materials other than those prescribed in paragraph (g)(2)(iv) may pose fire hazards.

Such hazards are commensurate with hazards posed by the presence of gasoline as motor fuel in vehicles subject to customary repair practices. However, although it may be common repair practice to allow some ignition sources near a vehicle being repaired in an enclosed area, the additional and potentially much more extensive hazards, and accident consequences, involving explosives or significant quantities of flammable liquids or flammable gases warrant special precautions. Consequently, a placarded vehicle containing those materials may be repaired inside a building only if ignition sources are eliminated. The method used to eliminate ignition sources is not prescribed in the amendment but is left to the sound judgment of the repair facility, subject to applicable local, State, and Federal laws concerning those facilities.

Some of the differences between this amendment and the notice proposed on October 16, 1976, should be reiterated.

1. The proposed prohibition against use of heat, flame or spark outdoors within 100 feet of a vehicle containing certain hazardous materials has been dropped in favor of a general prohibition on the use of heat, flame, or spark on the cargo or fuel containment system of a placarded vehicle. Judgment on the safe use of welding equipment, outside of a building, on other components of a placarded vehicle is left to the repair facility.

2. Except for a vehicle carrying combustible liquids, any vehicle required to be placarded is subject to this amendment. The previous proposal would have applied to any vehicle containing any quantity of a flammable liquid, combustible liquid, flammable gas, poisonous liquid, oxidizer, or explosive. Excluded from the amendment's coverage is any vehicle not required to be placarded (however, note that materials, especially explosives A or B, listed in table 1 at § 172.504, as well as cargo tanks and portable tanks that have not been cleaned and purged, necessitate placarding regardless of quantities).

3. Repair work performed indoors on placarded vehicles containing cargoes of explosives A or B, flammable liquids, or flammable gases is permitted, provided exposed sources of ignition are eliminated. Compliance with National Fire Protection Association Pamphlet No. 70, "National Electrical Code," although proposed in the last notice for electrical equipment, is not required by the amendment. In addition to that publication, there exist various requirements imposed by local, State, and Federal agencies pertaining to sources of ignition. Rather than requiring compliance with a particular method of protecting exposed ignition sources, the amendment allows use of any reliable method that would prevent the ignition of an explosive fuel-air mixture.

4. Provided that the requirements of paragraph (g)(2) of the amendment are met, cold repairs to closed fuel

and cargo containment systems may be performed indoors. The vehicle's motor and electrical system may remain operational to the extent necessary to accomplish maintenance or repair.

5. The previous proposal to delete paragraph (h) (concerning hot repair to cargo tanks and fuel containers) has been dropped. Paragraph (h) will be retained in this amendment since repair of a cargo tank or compartment thereof, formerly containing certain hazardous materials, can be safely performed by using flame, arc, or other means of welding if the tank or compartment has been made gas free prior to the repair. In response to the notice, several comments indicated that paragraph (h) should be retained since it reflects long standing safety practices set forth by safety-oriented organizations, such as NFPA, and which should be adhered to by all motor carriers and repair shops that engage in such work.

III. IMPACT

This amendment may still necessitate the performance of some kinds of repair work outdoors, such as hot work on vehicles carrying flammable or explosive cargoes. The MTB has attempted to balance the hazards of working in an enclosed area on such vehicles against the disadvantages and possible hazards of requiring the work to be performed outside. A concern raised in comment, with which MTB agrees, is that a rule that is too stringent may result in deferral of repair work, failure to correct vehicle deficiencies, and the continued operation of vehicles in an unsafe condition. On the other hand, hot work in an enclosed area on vehicles carrying explosives or large volumes of liquids, gases or vapors which are flammable presents substantial, inherent risks. Reliance on the repair worker's sense of smell to detect cargo or fuel leakage is not a sufficient safeguard against the possibility of fire or explosion.

This amendment is a relaxation of existing requirements found in § 177.854(g). Moreover, the restrictions imposed by the amendment are similar to restrictions resulting from existing fire codes and insurance practices. Comments indicate that cleaning and purging a cargo tank can cost $100 or more and that many repair facilities lack tankage to receive hazardous materials that this amendment may require to be removed from a vehicle on which work is to be performed. The incremental safety thereby achieved is worth the cost of cleaning and purging a cargo tank. For facilities that lack receiving tanks, the carrier seeking the repairs will have to arrange to remove the vehicle's contents before hot repair may be undertaken.

(43 F.R. 41401, September 18, 1978)

SUMMARY • Docket No. HM-111 • *Amended regulations to improve on radioactive materials requirements, based upon the first three years experience in using the International Atomic Energy Agency's recommendations which had been adopted in Docket No. HM-2.*

[49 CFR Parts 171, 173, 174, 175, 177, 178]
[Docket No. HM-111; Notice No. 73-7]
TRANSPORTATION OF HAZARDOUS MATERIALS
Miscellaneous Proposals Relating to Radioactive Materials

On October 4, 1968, the Hazardous Materials Regulations Board (the Board) published extensive amendments to the Department's Hazardous Materials Regulations relating to radioactive materials (33 FR 14918). Those amendments were instrumental in bringing the transportation regulations of this country into substantial conformity with the international standards for the transportation of radioactive materials. The internationl standards were promulgated by the International Atomic Energy Agency (IAEA) and published in its "Regulations for the Safe Transportation of Radioactive Materials," 1967 edition.

With over three years of experience in the application of those amendments by shippers, carriers, and the Department, the Board now believes there is need for a number of changes, many of which would be minor or editorial in nature. Many of these proposed changes have been suggested by various persons or have been developed by the Board in response to formal and informal comments it has received. Several of the proposed changes have also been recommended by the U.S. Atomic Energy Commission. A number of these proposed changes would incorporate into the regulations various provisions which are now authorized under general type special permits held by many shippers, including several new specification packagings, which have been in use under special permits for several years with satisfactory experience.

The Board also wishes to note that at a future date, it will consider other, more substantive changes. Although the regulations relating to radioactive materials as amended October 4, 1968, conformed substantially at that time to the IAEA standards, the IAEA has been developing comprehensive revisions to its standards. In 1970 and 1971, regulatory review panels of the IAEA met. The membership on these panels comprised all major IAEA member states and international bodies, including the United States. The revisions to IAEA Safety Series No. 6 resulting from those panels have recently been published. The Board emphasizes that the substance of the changes proposed in this notice are not based on the most recent changes to the IAEA standards. However, to maintain United States' standards in as close conformity to international standards as possible, the Board intends to propose future changes as necessary; these proposed changes will be the subject of separate rule making.

Several additions to § 171.7 relating to matter incorporated by reference are proposed. These additions generally relate to certain sections of the new specification packagings proposed to be added in Part 178.

In § 173.23, the reference to radioactive materials packagings previously authorized by the Bureau of Explosives would be deleted since such packagings have not been authorized for shipment since February 28, 1969.

Two of the proposed changes herein (§ § 173.202 and 173.206) would permit the transportation of small quantities of certain metals or alloys (also classed as flammable solids) when such materials are a component part of fissile or large quantity radioactive materials. The current packaging requirements for radioactive materials are substantially superior to those presently required in the regulations for these metals or alloys and the former are the requirements that are proposed. Also, many years of successful experience with shipments authorized on a special permit basis have been acquired for the proposed packaging.

The revision of Note 1 in § 173.69(a) and the addition of a footnote in § 173.226 would clarify that Class A detonating fuzes with radioactive components, and thorium metal also are subject to the provisions for radioactive materials.

In § 173.389, two additional definitions are proposed, i.e., "full load" and "closed transport vehicle." Based on its experience, the Board believes that these definitions are needed to clarify the administrative requirements for control of certain types of radioactive materials shipments.

The proposal in § 173.391(c) would change the requirement from no "detectable" radioactive surface contamination on certain packages of exempt manufactured articles containing natural uranium or depleted uranium to no "significant" contamination as provided for in § 173.397. Another proposed change to § 173.391(c) would add natural thorium and its alloys to the classes of material which could qualify as a certain type of exempt manufactured radioactive material.

Several changes to the provisions of § 173.392 are proposed to clarify the requirements for advance special arrangements and written instructions between the shipper and carrier for the transportation of "full loads" of low specific activity radioactive materials.

In § 173.393, the section heading would be amended to include general shipment as well as general packaging requirements. Section 173.393(j) (3) would also be amended to clarify the method of measuring the radiation level at six feet from a vehicle containing a "full load" of radioactive materials; § 173.393(L) would be amended for clarification and to make reference to the general requirements of packages destined for export; § 173.393(o) would be added to require that packages not be offered for transport until the temperature of the packaging system has reached equilibrium; § 173.393(k) would be added to require that the inner containment vessel which comprises a separate unit of any portion of a packaging must be securely closed by a positive fastening device which is independent of any other part of the packaging. (This provision is intended to make it clear that a piece of masking tape is not sufficient to hold a lead plug in place.) Also in § 173.393(d), a significant change is proposed in conjunction with the authorized Type A packaging changes which are being proposed in § § 173.394 and 173.395, i.e., the statement that the presently listed specification Type A packages " . . . may be assumed to meet those standards. . ." would be deleted.

A further change is proposed to require that packages containing liquid radioactive materials meet the conditions of both § 173.393(g) (1) and (2). Such a change has been recommended by the National Transportation Safety Board in its "Special Study of the Transportation of Radioactive Materials by Air." It is felt that the double precaution provided by this approach will significantly improve the safety of such packages.

In § § 173.394 and 173.395, the most significant proposed change relates to the change to the "performance criteria" concept for Type A packages. The present listing of the various authorized DOT specifications would be deleted. Instead, complete reliance would be placed (except for the DOT Spec. 55) on the use of the DOT-7A, Type A, general packaging specification. Also, one very significant requirement would be added, i.e., that each user of a Spec. 7A package would be required to document and maintain on file for one year after the latest shipment a written record of his determination of compliance with the DOT Spec. 7A performance requirement for the specific package design. The Board's experience indicates that the present method of listing DOT detailed design specifications is somewhat misleading when the general packaging requirements for radioactive materials are considered, i.e., liquid packaging requirements, shielding requirements, inner vessel closure, etc. The Board believes that the present method has resulted in a system which is misleading since it specifies only the "outer" packaging in most cases. It is possible that in some cases a shipper might mistakenly consider only the outer packaging requirements without properly taking into account the additional general requirement such as illustrated above. Concurrently with these changes, DOT Specification 55 is proposed to be deleted from Part 178. However, DOT Spec. 55 packaging would continue to be authorized for use in Part 173 but would be limited to packagings constructed prior to the effective date of the amendments in this docket. The Board intends to phase out this specification as a "limited Type B" packaging (up to 300 curies of special form material) at some later date.

Further additions to §§173.394 and 173.395 would prescribe the quantities of radioactive materials authorized in the new specification Type B DOT Spec. 20WC and 21WC packagings (DOT Special Permit Nos. 5684, 5800, 6008, and 5725) and would clarify that any approved Type B packaging may be used for a shipment of a Type A quantity of radioactive material.

Numerous changes also are proposed to the packaging provisions for fissile radioactive materials in §173.396 and to the design requirements for the two specification packagings for fissile materials, i.e., the DOT Spec. 6L (§178.103 and Spec. 6M (§178.104). These proposed changes deal with modifications relating to the permitted radioactive material content, quantity, and physical details of construction of each packaging. Based on a recommendation received from the USAEC, DOT Spec. 2R (§178.34) is proposed to be extensively revised to provide more definitive requirements for flanged closures. These changes would also affect the design requirements for DOT Spec. 6L and Spec. 6M packagings, each of which utilizes a Spec. 2R inner containment vessel. Another change (§173.396 (f) would provide for the shipment of Fissile Class II packages under Fissile Class III type controls, thus allowing commingling of Fissile Class II and III packages by a specific consignor. This provision is presently authorized under a special permit (SP 5908) issued to many shippers. Another provision, §173.396(b) (7) would add a useful general package loading authorization for small amounts of fissile radioactive material as limited quantities of Uranium-235, in standard DOT specification steel drums (DOT SP 5021).

Also §173.396(b) (6) and (c) (5) would prescribe the quantities of fissile uranium hexafluoride (UF_6) that could be transported in the new DOT specification packagings 20PF and 21PF (new §§178.120 and 178.121). These are inner metal cylinders within certain types of phenolic-foam insulated steel protective overpacks (DOT SP 4909). Shipments of UF_6 have been performed routinely and successfully under this special permit for many years. In §173.396(b) (8), the Board proposes to provide for the shipment of a limited quantity of uranium hexafluoride as a residual "heel" in a cylinder. These shipments, as Fissile Class I, would be permitted in a bare cylinder without overpack.

In §173.397(a), changes are proposed to clarify the determination of the allowable amount of removable (non-fixed) radioactive surface contamination, in terms of quantified "significant removable contamination." The present provisions in §173.397 are a modified version of the 1967 IAEA Standards in Marginal C-3.3 and Table IV, Annex I. The Hazardous Materials Regulations specify a method for determination of external removable surface contamination based on the activity on the "wipe" sample. That limit is set at 10 percent of the IAEA values on the surface itself.

Many questions have arisen with respect to the "averaging" of multiple wipe samples. The IAEA standards clearly allow for "averaging" of contamination over any area of 300 cm^2 of any part of the surface. The proposed revision of §173.397(a) would provide that "averaging" is only allowable over any one 300 cm^2 area of any part of the surface and it is not allowed to average wipe samples from several 300 cm^2 areas. However, an exception is made for somewhat higher levels of contamination on packages consigned for "full load" shipments in §173.397(b). In §173.397(c), another change is proposed which would require each vehicle to be monitored after having been used for any "full load" shipment of radioactive material, and not only for a bulk shipment of low specific activity material, as is required by the present provisions of §173.397(b).

Section 173.398(a) is proposed to be changed to require that each shipper of special form radioactive material document and maintain on file for one year after the latest shipment a certification and supporting safety analysis demonstrating the method of determination that the special form test requirements were met. Also, requirements are proposed which outline the information to be submitted to the Department in petitions for certifications of special form designs when foreign shipments of these materials are intended. These requirements are based on provisions of the IAEA regulations. In this regard it should be noted that in some countries, competent authority certification is required domestically for all special form materials.

In §173.399, the reference to the labeling requirement for packages previously approved by the Bureau of Explosives would be deleted since it is no longer appropriate. In §173.401, a requirement to mark the gross weight on packages exceeding 110 pounds would be imposed to achieve consistency with an equivalent international requirement. A further clarifying provision would also be added to require the external marking of any Type A or Type B package, as appropriate, including the letters "USA", if foreign shipments are involved. Section 173.404(a) would be amended to make it clear that the blank spaces on the package labels must be filled in as appropriate. Further, a new §173.416(d) would be added to provide more precise guidance in completing the label entries; there has been some degree of confusion on this matter during the past few years.

Two significant changes are being proposed to the rail and motor vehicle carrier requirements in Parts 174, 175, and 177. One change would provide for the controlled spacing of groups of packages at 20-foot distances, when more than one group containing 50 transport indices or less is present in any single storage area. This provision is presently contained in the IAEA regulations as well as in the U.S. Coast Guard regulations in 46 CFR Parts 146–149. The other change which would apply to carriers would clearly specify that in preparing their manifests, waybills, etc., for radioactive materials shipments, the carrier must transpose all of the applicable shipping paper information as it has been supplied by the shipper pursuant to § 173.427(a) (5). Under the present carrier regulations in Parts 174, 175, and 177, the carrier is only required to include on his shipping papers the proper shipping name and the classification of the material, with the result that the information being supplied by the shipper in many cases does not accompany the shipment during transportation.

In Part 178, major revisions to existing requirements for DOT Specification 2R, 6L, and 6M are proposed. New specifications DOT 20PF, 21PF, 20WC, and 21WC are also proposed.

(38 F.R. 29483, October 25, 1973)

[Docket No. HM-111; Amdt. Nos. 171–28, 173–90, 174–25, 175–12, 177–32, 178–35]

RADIOACTIVE MATERIALS
Miscellaneous Amendments

On October 25, 1973, the Hazardous Materials Regulations Board ("the Board") published extensive proposals to amend the Department's Hazardous Materials regulations relating to radioactive materials (38 FR 29483). The reasons for the various miscellaneous proposals were explained in detail in the preamble to that notice. Interested persons were invited to participate in this rule-making proceeding and all comments received have been given full consideration by the Board before it decided on the amendments made herein.

As was pointed out in the notice, the board wishes to reiterate that the substantial matters covered by these amendments are not based on the 1973 revisions by the International Atomic Energy Agency (IAEA) to its Safety Series No. 6, "Regulations for the Safe Transport of Radioactive Materials." The Board intends to propose such changes in the very near future as the subject of a separate rule making action.

Comments were received by the Board on this notice from many organizations including government agencies, members of the Nuclear Industry, and the Air Transport Association of America. The U.S. Atomic Energy Commission submitted several comments. On the basis of the information submitted by these commenters, a number of modifications to the regulations proposed in the notice were made which are reflected in these amendments. These changes are explained as follows:

1. Section 173.206(a) (10) and (11). A commenter suggested that the 25-pound material limit that was proposed to be specified in § 173.206(a) (11) was not necessary in view of the technical evaluation which is performed on each individual Type B package. He further questioned the reason for the limited applicability of § 173.206(a) (10) as proposed to combinations of such alkali materials with radioactive materials, in view of the authority included in § 173.206(a) (11) for such materials. The Board agrees with these comments and has modified these two paragraphs accordingly.

2. Section 173.389(o). One commenter stated that the proposed "full load" definition was contrary to the present meaning of that term as it is used in transportation practice in the United States. It was suggested that the term "exclusive use" be used for the definition, in place of "full load," since "exclusive use" of a vehicle as it is presently defined in several carrier freight tariffs means in part "A service offered to shippers who require segregation of their freight from the freight of other shippers for protection against scrutiny, pilferage, or any other reason." The Board agrees with this comment and has adopted the term "exclusive use." However, a parenthetical reference has been added to the definition to clarify that "exclusive use" is also referred to as "sole use" and may be identified with the term "full load" as it is used in the IAEA regulations.

3. Section 173.389(p). One commenter suggested that a definition of the term "radioactive device" be included in the regulations, since it is one of the proper shipping names listed in § 172.5. The Board agrees with this suggestion and has therefore included a new definition as an editorial addition based on the existing requirements of § 173.391(b).

4. Section 173.392. Several comments were received on this paragraph covering low specific activity material. An editorial change has been made to clarify the relationship of the various changes which were proposed in this paragraph, as well as to correct an inadvertency in the notice which would have permitted low specific activity materials in less than exclusive use shipments to be in nonspecification containers.

5. Section 173.393(g). A number of comments were received on the proposal to strengthen the requirements for packaging of radioactive liquids. Several of the commenters suggested that the specific requirements not be applicable to Type B and fissile material packages, in view of the fact that such packages, including their containment systems, are subject to individual governmental review and approval. The Board agrees with this comment and has modified the regulation accordingly, since the original intent of the proposal was that it be applicable only to Type A packages and not to Type B and fissile packages. Several commenters stated that, although they supported the basic proposal to strengthen the liquid packaging requirements, the proposal should be modified to provide another option to the absorbent material provision. They requested that use of a secondary containment vessel be permitted,

enclosing the primary inner liquid-containing vessel so as to provide overall containment of the liquid, assuming the failure of the innermost vessel. The Board also agrees with these suggestions and has added § 173.393(g) (3) permitting this additional option.

6. Section 173.393(k). Several commenters suggested certain changes in the wording of the proposed requirement for closure of certain inner containment systems by positive fastening devices independent of the other packaging. In view of the fact that this proposed regulation is contained, with other related provisions, in the 1973 Revised IAEA regulations, the proposal to add this requirement is deleted from this docket. It will be included in another notice of proposed rule making to be published in the near future that will propose revisions specifically based on the 1973 Revised IAEA regulations.

7. Section 173.393(o). One commenter suggested that each shipper, prior to the first use of a package, be required to subject the package to an appropriate thermal test to demonstrate the thermal performance of the package under normal conditions of transport with its design decay heat load. Although this comment has merit, it is only one of several acceptable methods of confirming compliance with the performance criteria specified in the amendment. For this reason the Board believes that the regulations should not specify one method to be used.

8. Sections 173.394 and 173.395. The proposals to eliminate the various "hardware oriented" DOT specifications for Type A packages received several comments. Several of the commenters questioned whether a need for such a proposal is justified on the basis of adverse shipping experience. The Board continues to believe that the proposal is justified, particularly on the basis of observations of packaging compliance in field surveillance activities, and the reasons stated in the preamble to the notice. The Board recognizes that the elimination of listed packaging specifications will require that some existing packages be modified and that some be tested against the performance standards. In view of the manpower and effort that may be involved and in order to not cause any unreasonable hardship, a transition period of one year has been provided before compliance with this requirement becomes mandatory. Further, the U.S. Atomic Energy Commission who supported the proposal, has informed this Department that it intends to provide a consolidated testing program in support of AEC contractor operations to develop and certify packaging designs against the DOT 7A Specification. This program is expected to yield data and results which, when appropriate, will be useful to others in establishing their supporting package safety evaluation and certification. It is expected to reduce the duplicative effort that might otherwise be required by persons in the nuclear industry who use the same or similar packaging. Further questions on this program should be directed to the USAEC, Division of Waste Management and Transportation, Washington, D.C. 20545.

The Board emphasizes that this amendment is not intended to preclude the use of certain existing DOT specification packagings as a component of the Specifiation 7A package.

In response to several comments, the introductory headings of § § 173.394(a) and 173.395(a) have been editorially revised to call the attention of the shipper to the need for proper consideration of the other applicable general packaging requirements.

9. Section 137.396. In response to technical recommendations from the USAEC, a number of substantive revisions have been made to this section, as follows:

a. Section 173.396(b) (6). A note has been added to the table specifying that the maximum H/U ratio of 0.088 applies only to 30-inch cylinders and not to the other sizes of cylinders.

b. Section 173.396(b) (7). This paragraph has been changed to specify clearly that the inner package must meet the Type A, Spec. 7A package requirements, including the liquid packaging provisions.

c. Section 173.396(c) (1). Requirements have been added to specify a 5-watt limit on decay energy of contents and also to specify that large quantity radioactive material in normal form in the DOT-6L must be packaged in one or more sealed and leak tight cans or polyethylene bottles within the Spec. 2R containment vessel, a requirement which is consistent with a similar requirement of the DOT-6M (§ 178.104). In addition, the proposal to add fissile Class I authorized loadings in the Spec. 6L has been deleted because the USAEC has not completed its detailed nuclear safety analysis of the proposed loadings. An appropriate proposal will be the subeject of future rule making. In the table of authorized contents, a footnote has been added specifying that plutonium solutions are not authorized in the Spec. 6L.

d. Section 173.396(c) (2). The footnote limiting the maximum U-235 enrichment for contents in the Spec. 6M has been retained. The available data indicates that removal of the 93 weight percent limitation would increase the reactivity by about two percent. Persons shipping enriched uranium exceeding 93 weight percent will be required to petition the USAEC for specific approvals of such shipments with lowered material quantities.

e. Section 173.396(c) (2) (iii). An inadvertency has been corrected to specify that each Fissile Class III rather than Class II shipment is subject to § 173.396(g). Also in the table of authorized contents the column headings "H/X equals 3" have been changed to read "HX ≥ 3".

f. Section 173.396(f) (1) and (2). The requirements in these two paragraphs have been modified to reflect more appropriately the nuclear safety philosophy and criteria used in limiting Fissile Class II and III shipments. Upon the effective date of these amendments, DOT Special Permit 5908 presently authorizing shipments under similar provisions will be canceled with individual notification of the cancellation to be sent to each permit registrant.

10. Section 173.397. Several editorial changes have been made to clarify the requirements of this paragraph. One commenter noted that the allowable contamination levels of § 173.397(a) are significantly higher than those which would be recommended in a forthcoming ANSI standard which is being prepared for contamination on equipment and facilities to be released for uncontrolled use. The Board emphasizes that the allowable contamination levels in § 173.397(a) are not being changed in this rule making, only those applicable to "exclusive use" shipments pursuant to § 173.397(a), which have been raised by a factor of 10. In § 173.397(a) (1), the discussion on methodology for assessment of removable surface contamination has been modified to incorporate one commenter's suggestion that other measurement methods of equal or greater efficiency than the cited "wipe test" method may be utilized.

11. Section 173.398(a) (4). A statement delaying the effective date of this paragraph for one year from the publication of these amendments has been added to Note 1 in the paragraph. This is necessary to permit a reasonable transition period for compliance with the new requirement for the certification and supporting safety analysis to be maintained on file by shippers of special form radioactive material.

12. Section 173.416. In § 173.416(d) (1), a change has been made to clarify that when symbols are used on label entries, such symbols must conform to established radiation protection terminology which utilizes a superscript designating the atomic mass on the left side of the chemical symbol for the radionuclide.

In § 173.416(d) (2), a clarification has been added to provide that for fissile radioactive materials, the insertion of the weight in grams or kilograms of the fissile radioisotope in the "number of curies" entry on the label is optional.

13. Sections 174.586(h) (2), 175.655(j) (2), and 177.842(b). In each of these paragraphs a clarification has been added to the effect that when groups of packages are stored in a single location, the required separation of 20 feet between adjacent groups is measured from edge to edge between the groups. One commenter noted that the provisions being added to these paragraphs would have the effect of allowing an increase in radiation dose rate to transport workers, and that such an increase is not justified under the recent concepts of limiting radiation to "as low as practicable" levels. The Board wishes to emphasize that these amendments are not intended to increase radiation levels to transport workers. The required segregation distances of packages from areas occupied by persons are not being changed. The provision for situations where there are more than one group of packages with 50 transport indexes is intended solely to cover an inadvertent gap in the surface regulations (Highway and Rail) for carriers, which does not exist in either vessel (46 CFR Part 146) regulations or the IAEA standards. The Board also wishes to point out that the broad area of radiation exposure to transport workers as a result of handling radioactive packages is currently under study by this Department, in cooperation with the USAEC and several States. It is possible that changes to the carrier requirements for handling and stowing radioactive packages may be proposed later as a result of the findings in these studies.

14. Section 178.34. A change in the required temperature rating of the luting compounds from 250° F. to 300° F. has been made to make this requirement compatible with the limitation on decomposition characteristics of the authorized contents of the Spec. DOT 6L (§ 173.396(c) (1)), as well as to achieve compatibility with the operational requirements of the DOT-6L (§ 178.103) and DOT-6M (§ 178.104) specifications.

15. Section 178.103. In § 178.103-4, a clause has been added to specify that the requirement for increased fire resistance of welded joints applies only to the added spacer rods as prescribed for compliance with § 178.103-3(c) (1). Further, a statement has been added to § 178.103-3(c) (1) providing that compliance with the new requirement for four additional welded spacer rods is not mandatory for existing packagings until one year after the date of publication of these amendments.

16. Section 178.104. Section 178.104-3(a) has been changed to provide for the welding together of different capacity drums. Also, in § 178.104-3(a) (2), provision has been made for the optional utilization of a layer of porous refractory fiber beneath the pressure-relief vent holes. Several commenters suggested that such an option be added to the specification. The intended purpose of the refractory layer is to preclude smoldering of the insulation media after exposure to the accidental fire test condition. Since the presence of the refractory fiber layer has not been demonstrated to be necessary for the package to meet the accident damage test sequence, its utilization has been made optional. In § 178.104-3(b), limitations on the material of construction of the Spec. 2R have been added. An editorial reorganization of § 178.104-3(c) has also been made.

17. Section 178.195. An illustrative sketch showing typical assembly detail for this specification has been added, as well as several editorial clarifications.

18. Metric/English Units. Throughout these amendments, units have been stated in metric units with equivalent English units in parentheses.

In accordance with section 102 of the National Environmental Policy Act (Pub. L. 91–90, (42 U.S.C. 4231 et seq.)) the Board has considered the environmental impact of these amendments. It has determined that the changes made in these amendments would not have a significant impact on the environment. Accordingly, it considers that an Environmental Impact Statement is not necessary and has not issued such a statement with respect to these amendments.

(39 F.R. 45238, December 31, 1974)

Note: The following pages, a blend of amendments in Docket Nos. HM-103 and 112, have been shifted from their position after HM-103 to this more appropritae location in Docket No. HM-112.

SUMMARY • Docket No. HM-112 • *A major rule making effort to consolidate the hazardous materials regulations for rail and highway (49 CFR 170–189) with such regulations for air (14 CFR 103) and water (46 CFR 146); includes several other miscellaneous changes, some of which were related to Docket Nos. HM-4, HM-8, HM-51, HM-57, HM-101, HM-102, and HM-103.*

[14 CFR Part 103]
[46 CFR Part 146]
[49 CFR Parts 170, 171, 172, 173, 174, 175, 176, 177]
[Docket No. HM-112; Notice No. 73–9]
HAZARDOUS MATERIALS REGULATIONS
Proposed Consolidation

Since the inception of the Department of Transportation, it has been the goal of the Hazardous Materials Regulations Board to make the Department's Hazardous Materials Regulations as uniform and clearly stated as possible. One major difficulty has stood in the way of progress towards this goal: The regulations are found in three different Codes of Federal Regulations according to the mode of transportation involved and much referencing between them exists. Also, the manner of form and presentation of their contents vary and could be improved. In this notice, as a first step, the Board is proposing to consolidate the air, water, and surface transportation Hazardous Materials Regulations of the Department into one volume of the Code of Federal Regulations, 49 CFR Parts 170 to 189. These regulations would not include the bulk regulations of the U.S. Coast Guard for transportation by water. The Board is of the opinion that after this step is taken, the second step of restating the regulations would become more manageable.

This document necessarily encompasses a wide breadth of subject matter, some of which is substantive and much of which is merely editorial. To single out the substantive changes proposed as completely as possible, the Board is going to great lengths in presenting a detailed preamble to facilitate review by the public. Matters of particular interest in the consolidation of the Hazardous Materials Regulations are summarized as follows:

1. The Board is proposing to create four new classifications of hazardous materials to handle the problem of materials which pose a significant hazard when transported by air or water, but do not present such a hazard by surface transportation. These are identified as "Other Regulated Materials".

2. The Board is proposing a new regulatory mechanism to regulate small quantities of hazardous materials considered to pose a significant hazard only when transported in aircraft. These are identified as "Other Regulated Materials, Group D," abbreviated "ORM-D".

3. The Board is proposing to resolve certain issues that remain outstanding from other dockets because these issues relate to matters involved in this proposed consolidation of regulations or to a companion docket published elsewhere in this issue of the FEDERAL REGISTER, Docket HM-103. Some of the Dockets affected by this Notice are: HM-4, 8, 51, 57, 101, 102, and 103.

4. The Board is proposing a new format for the list of hazardous materials presently found in 49 CFR 172.5 which will recognize the recommendations of such organizations as the United Nations Committee of Experts on the Transport of Dangerous Goods and the Dangerous Goods Code of the Intergovernmental Maritime Consultative Organization (IMCO).

5. The Board is proposing to adopt a very large portion of the recommendations of the International Air Transport Association Restricted Articles Regulations as also published in C.A.B. No. 82, Official Air Transport Restricted Articles Tariff No. 6-D.

6. The Board is proposing to eliminate exemptions from specification packaging for certain hazardous materials it believes should not be so exempted.

7. The Board is proposing to prohibit certain hazardous materials from being transported on passenger-carrying aircraft which may now be so transported.

8. The Board is proposing to clearly regulate hazardous materials being transported in a passenger rail car in baggage. This is an interim rule making proposal while more complete regulations are under development. The present regulations apply only to rail carriers in baggage service.

9. The Board is proposing to change stowage requirements aboard vessels to align them completely with the International Dangerous Goods Code of IMCO.

10. The Board is proposing to change the present poison classifications from Class A & B poisons to Extremely Toxic and Highly Toxic classifications. This has been the subject of two Advance Notices of Proposed Rulemaking under Docket HM-51.

11. The Board is proposing to change the classification of a limited number of hazardous materials to more clearly reflect the potential hazard they pose in transportation.

12. The Board is proposing to consolidate present Parts 174, 175 and 176 governing carriers by rail into a new Part 174 and update some of the requirements therein.

A specific discussion in more detail follows:

I. GENERAL

The Board recognizes that this document could require an inordinate amount of time to review. Inordinate, in the sense that because of the volume, commenters could conceivably request delay after delay in asking for more time to study the docket. Also, the Board realizes that during the period that this document is under review, it would be confusing to publish too many additional proposed changes. Therefore, the Board has taken these factors into consideration and to assist the public as much as possible in reviewing this document in a timely and efficient manner, it has developed a very detailed preamble. This preamble is designed to set forth in such detail the changes the Board proposes that review by the commenters should be greatly assisted without lengthy and time consuming studies aimed at trying to "discover" what is being changed.

To again facilitate review, this preamble has been divided into various sections representing the major proposed changes and the different parts in the Code of Federal Regulations. In the conclusion is the presentation of a plan on how the Board proposes to change the present arrangement of Parts 170 to 189 in the final amendment to permit better organization of the Hazardous Materials Regulations.

In accordance with section 102 of the National Environmental Policy Act (Pub. L. 91–190, 42 U.S.C. 4321 et seq.) the Board has considered the environmental impact of its proposed consolidation of the Department's Hazardous Materials Regulations. Although there are included in the proposed consolidation, some changes in or additions to the substantive requirements of certain existing regulations, the Board is of the opinion that the proposed consolidation, even when viewed cumulatively is not a Federal action that has a significant impact on the environment. The primary purpose of the consolidation is to achieve editorial clarity and uniformity in existing regulations. The substantive changes are basically related to matters of classification, form, and format, and will not have any adverse impact upon the environment. The Board also notes that a separate assessment of the environmental impact of the substantive proposals contained in Dockets HM-102 and 103 is set forth in the preamble to the notice of proposed rulemaking for those proposals, published elsewhere in this edition of the FEDERAL REGISTER.

II. NEW CLASSIFICATION—OTHER REGULATED MATERIALS (ORM)

The Board's effort to consolidate virtually all of the Department's Hazardous Materials Regulations into one volume has made it necessary to develop a regulatory mechanism whereby the application of the regulations may be clearly established for each of the modes of transportation. This poses no difficulty concerning those materials already included in existing classifications; however, for certain materials, it will be necessary to make the regulations applicable only to transportation by air or water or both, due to the kinds of potential hazards presented when transported in those transportation environments. These materials are discussed in § 170.10 and defined in proposed § 173.500. These materials are now described as "hazardous articles" when transported by water (46 CFR 146.27), "other restricted articles" when transported by air (Tariff 6-D, C.A.B. No. 82 and the IATA Restricted Articles Regulations), or partially "exempt" materials when transported by highway or rail (49 CFR Part 173). The Board believes the best way to handle all these materials is to create an "Other Regulated Material" (ORM) category which would be divided into four classifications. By the addition of these new classifications, and including the letters A, B, C, or D to identify them, a method would be provided to identify the four types of materials covered and the application of various regulations to them when they are transported by air, highway, rail, or water. A general description of the four classes of ORM materials are as follows:

An ORM-A matterial is a substance which has an anesthetic, irritating, noxious, toxic, or other similar property and which can cause extreme annoyance or discomfort to passengers and driver or crew in the event of leakage during transportation.

An ORM-B material is a substance capable of causing significant damage to a transport vehicle or vessel from leakage during transportation.

An ORM-C material is a substance which has other inherent characteristics not described as an ORM-A or ORM-B but which make it unsuitable for shipment, unless properly identified and prepared for transportation. Each ORM-C material is specifically named in § 172.101.

An ORM-D material is a material that is classed as a flammable liquid, corrosive material, flammable compressed gas, nonflammable compressed gas, flammable solid, oxidizing material, or organic peroxide, that, due to its limited quantity in a package, may be described and shipped as an ORM material. Shippers of ORM-D materials would be required to comply with the regulations pertaining to such materials regardless of the transportation mode utilized.

However, no carrier operating regulations would apply, such as the carrying of shipping papers unless the ORM-D material is to be transported by aircraft.

On January 6, 1972 the Director of the Office of Hazardous Materials published a request for public participation entitled "Exemptions" in the FEDERAL REGISTER (37 FR 149). He explained the various aspects of the existing requirements pertaining to partial exemptions. Interested persons were invited to participate in the project by submission of supported information that would lead to the development of an appropriate rulemaking proposal. Many letters were received indicating interest in the project. However, very few submissions contained any form of supported information.

One organization, the Council for Safe Transportation of Hazardous Articles, hereafter referred to as COSTHA, submitted a petition for rulemaking in response to the project. COSTHA represents more than 1,000 companies engaged in manufacturing of end-use products packaged primarily for consumption in the home. The petition urges the Board to initiate rulemaking to delete the shipping paper requirements for small-quantity shipments of partially exempt hazardous materials which meet the existing quantity limitations, when overpacked in strong outside containers during transportation. As an alternative, COSTHA requested exemption for materials identified as consumer commodities and provided supporting data in the form of references to statutes pertaining to consumer protection. The Board does not agree with this latter approach since many of the other materials subject to its regulations are virtually identical in composition to materials that could be characterized as consumer commodities. However, they are not classed as such since they are intended for purposes other than consumption by consumers. For example, a container of a liquid that is packaged and marked "Nail Polish Remover" is intended for consumer use. Another package containing essentially the same material, acetone, is intended for some industrial purpose. If the Board were to accept the Council's proposed definition, it would be involved in discrimination against certain kinds of materials that present the same limited potential hazard as those identified as consumer commodities. Further, it would base certain hazardous materials regulations on vague and open-ended terminology.

Two other comments providing substantial comment in response to the request were received. The President of the Uniformed Fire Fighters Association of Greater New York expressed his view that shipping papers for materials covered by the partial exemptions in the regulations serve little or no benefit to fire fighters. The General Manager International Association of Fire Chiefs stated in part as follows:

"Specifically our comments are directed to that requirement, as referred to in your release, that each driver of a motor vehicle have in his possession copies of the shipping papers describing the articles transported by their technically prescribed shipping nomenclature and classification. Regardless of quantity involved, we refer now, not to articles that by their nature are so hazardous as to warrant application of your total safety requirements, but only to those classified as hazardous but designated partially exempt by reason of their relatively low-risk potential. In this partially exempt category, there are found ordinary every day household products such as insecticides, room fresheners, a multitude of toilet articles and medicines.

Under existing regulations, the driver is expected to carry the shipping documents for each of the many shipments in his vehicle either on his person or in the cab of the truck. The purpose—and we consider it laudable—is to transmit to firemen, in an emergency, knowledge of the contents of the vehicle so that they may deal effectively with the situation without undue danger to their persons. However, in a fire situation where the materials mentioned above are in limited quantities and are mixed in transit with ordinary combustible materials, it is unlikely that a fire officer would attempt to examine shipping papers prior to extinguishing the fire. The hazard does not warrant the delay involved in gaining control of the fire.

It is our opinion, therefore, that when all these limitations are considered, the shipping papers notations serve no practical purpose. They could be eliminated for all those commodities which, because of their relatively low hazard, have been exempted from virtually all other requirements of the regulations."

The Board agrees with the above statement and believes that shipper compliance with conditions specified for ORM-D materials will preclude the necessity for adherence to any carrier regulations or shipping paper requirements except for transportation by aircraft.

II. OTHER DOCKETS

Matters discussed in other dockets that were left outstanding namely HM-4, HM-8, HM-51, HM-57, HM-101, HM-102, and HM-103, are considered proposed for resolution in the combined HM-103 and this docket, e.g., foodstuffs with poisons, pyrophoric solids, water reactive materials, toxicity definitions, materials which are corrosive only to aluminum, special marking of tank cars, packaging for flammable and combustible liquids. Any person who is of the opinion that these combined dockets do not satisfactorily resolve the pending issues in these other dockets should comment on the issues and indicate to the Board what other equivalently safe alternatives are available.

IV. LOADING POISONS WITH FOODSTUFFS

The Board is proposing revisions to §§ 174.632(m) (See § 174.680 in this Notice) and 177.841 to specify that packages bearing poison labels may not be loaded in the same transport vehicle with foodstuffs, feeds, etc. The present restrictions refer to poisons Class A or B which will no longer exist as classifications if the proposals contained in this Notice pertaining to toxic materials are adopted. Consistent with those proposals, and in recognition of the need for ready identification of materials subject to the loading restriction, the Board believes the best approach is to relate the restriction to labels on packages which would be easily recognized at the time of loading. The Board is aware that additional materials will be made subject to the loading restriction if the proposals made herein are adopted.

The matter of commingling of poisons and foodstuffs was the subject of Docket HM-4. The resulting action by the Board was to totally preclude loading within the same rail or highway transport vehicle. The Council for the Safe Transportation of Hazardous Articles (COSTHA), in its comments on HM-103 has requested that the Board consider segregation requirements in place of total restrictions. The basis for the request was recognition of the proposed hazard information requirements for additional highly toxic materials and the fact that the Committee of Experts on the Transport of Dangerous Goods (United Nations) has recognized a segregation requirement for Toxic Materials Group II which corresponds to the Highly Toxic Classification proposed herein. The recommendations adopted by the UN Group were based on an approval procedure adopted by a competent authority which for the United States would be the Department's Hazardous Materials Regulations Board. The Board is not prepared at this time to relax the restriction until workable and easily understood segregation requirements are proposed for consideration. The only apparent solutions at this time do not appear practical e.g., separate compartments and specialized containers for containment of packages. However, the Board invites comments on whether or not there should be some relaxation in this area and, if so, how it should be accomplished.

V. PART 170—GENERAL INFORMATION, REGULATIONS, AND DEFINITIONS

In essence, this part is proposed to be comprised of the sections formerly found in Part 171. Changes are briefly explained as follows:

(a) Section 170.1 is a new section citing the sources for the Hazardous Materials Regulations and was formerly § 171.2. This section is expanded now that the Hazardous Materials Regulations will include regulations promulgated not only under the authority of (18 U.S.C. 831–835), but regulations promulgated under similar authorities for the Federal Aviation Administrator and the Commandant, U.S. Coast Guard.

(b) Section 170.2 is the former § 171.1, expanded in an effort to make the orgnization of the regulations clearer. It includes subject matter formerly covered in §§ 171.9 and 171.10.

(c) Section 170.6 is basically self-explanatory within its own text, as are §§ 170.7, 170.12, 170.14, 170.15 and 170.16.

(d) Section 170.8 is a new section proposing to clarify the applicability of the Hazardous Materials Regulations to combustible liquids.

(e) Section 170.10 is a new section proposing to explain the applicability of the regulations to ORM—A, B, C and D materials.

(f) Section 170.30 is a new section proposing to encompass all of the definitions formerly found in § 171.8 in alphabetical order in a "dictionary" format so that the definitions can be more easily located and so that new definitions can be easily added. Definitions for the following words are proposed which were not found heretofore in § 171.8: Air commerce, Approved, c.c., Cargo-only aircraft, cargo tank (new definition) Carrier, Civil aircraft, Express rail car, Flash point, Flight crewmember, Gross weight, Hazardous materials (new definition), Hermetically sealed, Includes, May, Net weight, No person may, Passenger-carrying aircraft, Passenger vessel, Person who offers for transportation, Public aircraft, Shipper, Solid, S.U.S., U.F.C., Viscous. It should be noted that a number of these definitions are now found in 14 or 46 CFR.

VI. PART 171—RULEMAKING PROCEDURES OF THE HAZARDOUS MATERIALS REGULATIONS BOARD.

Essentially, this part is proposed to be comprised of the sections formerly located in Part 170.

(a) Section 171.1 Editorial.

(b) Section 171.3 Updates former § 170.1. Editorially provides for practices that are presently being followed by the Board.

(c) Sections 171.5, 171.7, 171.9, editorial 171.11. Editorial. Proposes to require that petitions for rulemaking be submitted in English.

(d) Section 171.13 Editorial. Also, proposes to limit petitions to persons over whom the Board has jurisdiction, i.e., shippers and carriers. Proposes to require that petitions for waivers or exemptions be submitted in English.

(e) Section 171.15 Editorial. Also sets forth manner of processing of petitions by the Board, or any individual member of the Board.

(f) Section 171.23 Proposes to make reference to hearings to be held by the Commandant, U.S. Coast Guard.

(g) Section 171.31 Proposes to refer to hearings to be held by the Commandant, U.S. Coast Guard to the extent under (46 U.S.C. 170(a)), as amended.

(h) Section 171.35 Proposes to require that petition for rehearing or reconsideration of a rule be submitted at least 20 days before its effective date. The specified time is not 10 days. Also, proposes to provide that in any case in which a rule becomes effective in less than 30 days after issuance, such a petition may be filed at any time before the effective date. The specified time is now 15 days.

VII. PART 172—LIST OF HAZARDOUS MATERIALS AND HAZARDOUS MATERIALS COMMUNICATIONS REGULATIONS.

This part, except for Subpart B which is comprised of §§ 172.100 and 172.101, is covered in Docket HM-103, under which a notice is published in this issue of the FEDERAL REGISTER. For any explanations regarding this Part, see that document. However, Subpart B is covered in this Docket and the following comments are pertinent.

1. Format. The list of hazardous materials is presented in a new format. Basically, § 172.100 in the text gives a complete explanation of the new table.

2. New entries. To the present list, 341 new entries are proposed to be added. To clearly single out these new entries, a complete listing of them is given below together with the source of the entry. "6D" means the entry is found in C.A.B. No. 82, Air Transport Restricted Articles Tariff and the International Air Transport Association Restricted Articles Regulations. "USCG" means the item is found in the Dangerous Cargo Regulations of the U.S. Coast Guard, 46 CFR 146.04–5. "HM-57" means the item is proposed to be added as a result of the rule making completed in that docket. "Petition" means that the Board has received a request to list the subject material classed as shown. In the list below, some of the entries are followed by the words "air", "water", or "air and water" in parentheses. This is to indicate that it is proposed to regulate these materials only by these modes of transportation. The list below does not include new organic peroxide entries that are given in § 172.101. These are being shown for information only and to make § 172.101 as complete as possible. Some organic peroxide entries in the present § 172.5 list have been deleted. All these peroxide entries will be handled by separate rule making to be introduced in a notice before action is completed in this docket. Also, based on petitions received and the data submitted, lead arsenate, sodium cacodylate, and cacodylic acid are proposed to be removed from the listing. "Chemicals, n.o.s." is proposed to be deleted as an authorized description. This description has caused much confusion in the past. Based on another petition, the listings covering motion picture film are proposed to be greatly simplified. The list of new entries is as follows:

Acetal: 6D

Acetaldehyde ammonia: 6D

Acetic acid (aqueous solution): 6D, HM-57

Acetic acid, glacial: 6D, HM-57

Acetyl bromide: 6D, HM-57

Acetylene tetrabromide: 6D (air)

Acetyl iodide: 6D, HM-57

Acid butyl phosphate: 6D, HM-57

Acrylic acid: HM-57

Ammonium hydrogen sulfate: 6D (air)

Ammonium hydrosulfide solution: 6D (air)

Ammonium polysulfide solution: 6D (air)

Ammonium sulfate nitrate: USCG (water)

Ammonium sulfide solution: 6D (air)

Amyl acid phosphate: HM-57

Amylamine: 6D

Amylene: 6D

Amyl formate: 6D

Antimony lactate: 6D (air)

Aldrin, cast solid: 6D (air)

Aldrin mixture, dry, with 65% or less Aldrin: 6D (air)

Alkanesulfonic acid: 6D, HM-57

Aluminum bromide, anhydrous: 6D, HM-57

Aluminum chloride, anhydrous: 6D (air)

Aluminum chloride solution (air)

Aluminum hydride: 6D

Aluminum metallic, powder: 6D, USCG (air and water)

Aluminum phosphate: HM-57

Aluminum phosphide: 6D

Aminoethyl ethanolamine: HM-57

2-(2-aminoethoxy) ethanol: HM-57

n-aminoethyl piperazine: HM-57

Aminopropyl diethanolamine: HM-57

n-aminopropyl piperazine: HM-57

bis-aminopropyl piperazine: HM-57

Ammonium chlorate: 6D

Ammonium fluoride: 6D

Ammonium hydrogen fluoride: 6D, HM-57 (air)

Ammonium hydrogen fluoride solution: 6D, HM-57

Antimony potassium tartrate: 6D (air)

Antimony sulfide: 6D (air)

Antimony trichloride solution: 6D HM-57

Argon, cryogenic liquid: Notice on cryogenic liquids soon to be issued by the Board

Antimony trichloride, solid: 6D, HM-57

Antimony trichloride solution: 6D, HM-57

Argon, cryogenic liquid: Notice on cryogenic liquids soon to be issued by the Board

Arsine: 6D, petition

Asphalt, at or above its flash point: USCG (water)

Barium oxide: 6D (air)

Battery parts: USCG (water)

Benzaldehyde: USCG

Benzine: USCG

Bleaching powder: 6D, USCG (water)

Bone oil: 6D (air)

Boron trifluoride-acetic acid complex: 6D, HM-57

Box toe board: USCG (water)

Box toe gum: USCG

Bromoacetic acid, solid: 6D, HM-57

Bromoacetic acid solution: 6D, HM-57

Bromobenzene: USCG

Bromochloromethane: 6D (air)

Burlap bags, new: USCG (water)

Burlap bags, used, not cleaned: 6D, USCG (air and water)

Burlap bags, cleaned: USCG (water)

Burlap cloth: USCG (water)

Butylamine: 6D

Butyl bromide: 6D

Butyl ether: 6D

Butyl chloride: 6D

Butyl formate: 6D

Butyric acid: HM-57

Calcium carbide: 6D, USCG (air and water)

Calcium cyanamide: 6D, USCG (air and water)

Calcium hydrogen sulfite solution: 6D, HM-57

Calcium oxide: 6D, USCG (air and water)

Camphene: 6D, USCG (air and water)

Camphor oil: USCG

Carbaryl: 6D

Carbon dioxide, solid: 6D, USCG (air and water)

Carbon tetrachloride: 6D, USCG (air and water)

Castor beans: USCG (water)

Castor pomace: USCG (water)

Caustic potash, dry: 6D, USCG, HM-57

Cesium metal: 6D

Chlordane: 6D

Chloric acid: 6D

Chloroacetic acid, solid: 6D, HM-57

Chlorobenzol: USCG

Chloroform: 6D, USCG (air and water)

Chlorophenyl trichlorosilane: 6D, HM-57

Chloroplantinic acid: 6D (air)

2-chloropropene: 6D

Chromic acid mixture, dry: petition

Chromic fluoride, solid: 6D, HM-57

Chromic fluoride solution: 6D, HM-57

Coal tar dye, liquid: HM-57

Coconut meal pellets: USCG (water)

Combustible liquid, n.o.s.: 6D, USCG, HM-102

Compounds, cleaning, liquid (containing phosphoric, acetic, etc.): HM-57

Copper chloride: 6D (air)

Copra: USCG (water)

Cotton: USCG (water)

Cotton batting: USCG (water)

Cotton seed hull fiber or shavings: USCG (water)

Cotton wadding: USCG (water)

Cotton waste: USCG (water)

Creosote, coal tar: 6D, USCG

Crotonic acid: HM-57

Crotonylene: 6D

DDT: 6D (air)

Decahydronaphthalene: USCG

Diacetone alcohol: 6D

Diazinon: 6D (air)

Dibromodifluoromethane: 6D (air)

Dichloracetic acid: 6D, HM-57

Dichloroacetyl chloride: 6D, HM-57

Dichlorobenzene, ortho: 6D (air)

Dichlorobenzene, para: 6D (air)

Dichlorodifluoroethylene: 6D (air)

Dichloroisopropyl ether: HM-57

Dichlorometane: 6D (air)

Dichloropentane: USCG

2,4-Dichlorophenoxy acetic acid: 6D (air)

Dichlorophenyl trichlorosilane: 6D, HM-57

Dichloropropene-dichloropropane mixture: HM-57

Dichloropropene and propylene dichloride mixture: HM-57

Dieldrin: 6D (air)

Di (2-ethyl hexyl) phosphoric acid: HM-57

Diethyl ketone: 6D

Dihydropyran: 6D

Diisopropylamine: 6D

Diisopropylethanolamine: HM-57

Diisopropylether: 6D

2,3-Dimethylbutane: 6D

Dimethyl carbonate: 6D

1,4-Dimethyl cyclohexane: 6D

Dinitro cyclohexylphenol: 6D (air)

Dioxane: 6D

Diphenylmethylbromide, solid: 6D, HM-57

Divinyl ether: 6D

Dye intermediate: HM-57

Ethyl acrylate: 6D

Ethyl benzene: 6D

Ethyl borate: 6D

Ethyl bromide: 6D

Ethyl butyl acetate: USCG

Ethyl butyl ether: 6D

Ethyl butyraldehyde: 6D

Ethyl butyrate: 6D

Ethyl chloroacetate: 6D

Ethyl crotonate: 6D

Ethylene chlorohydrine: 6D

Ethylene diamine: 6D

Ethylene dibromide: 6D (air and water)

Ethylene glycols: USCG

Ethyl hexaldehyde: USCG

Ethyl lactate: USCG

Ethyl propionate: 6D

Ethyl silicate: USCG

Excelsior: USCG (water)

Exothermic ferrochrome: USCG (water)

Exothermic ferromanganese: USCG (water)

Exothermic silicon chrome: USCG (water)

Feed, wet, mixed: USCG (air and water)

Felt, waste: USCG (water)

Ferric chloride, solid: 6D (air)

Ferric chloride solution: HM-57

Ferrophosphorus: USCG (water)

Ferrosilicon: 6D, USCG (air and water)

Fibers: USCG (water)

Fish meal: USCG (water)

Fluoboric acid: 6D, HM-57

Fuel oil: USCG

Fumaryl chloride: 6D, HM-57

Garbage tankage: USCG (water)

Germane: Petition

Hay: USCG (water)

Hay, loose, wet, or damp: USCG (water)

Heptachlor: 6D (air)

Hexachloroethane: USCG (water)

Hexadiene: 6D

Hexaldehyde: USCG

Hexamethylene diamine, solid: 6D, HM-57

Hexanoic acid: HM-57

Hydrogen, cryogenic liquid: Notice on cryogenic liquids soon to be issued by the Board.

Hydrogen selenide: 6D, petition

Hypochlorite solution, containing not more than 7% available chlorine: 6D (air)

Iminobispropylamine: HM-57

Iodine pentafluoride: 6D, HM-57

Iron oxide, spent: 6D (air)

Isobutyl acetate: 6D

Isobutylamine: 6D

Isobutyric acid: HM-57

Isobutyric anhydride: HM-57

Isopentanoic acid: HM-57

Isopropyl acid phosphate, solid: 6D, HM-57

Isopropylamine: 6D

Isopropyl nitrate: 6D

Jute: USCG (water)

Kapok: USCG (water)

Kerosene: USCG

Lead dross: USCG (water)

Lead peroxide: petition

Lead sulfate, solid: 6D, HM-57

Lindane: 6D (air)

Lithium acetylide-ethylene diamine complex: 6D

Lithium borohydride: 6D

Lithium nitrate: 6D

Magnesium aluminum phosphide: 6D

Magnetized materials: 6D (air)

Malathion: 6D (air)

Manganese dioxide: 6D (air)

Mercury, metallic: 6D (air)

Mesityl oxide: USCG

Metal borings, turnings, etc.: USCG (water)

Methyl acrylate: 6D

Methyal: 6D

Methyl amyl acetate: USCG

Methyl amyl ketone: USCG

Methyl butene: 6D

Methyl butyrate: 6D

Methyl chloroform: 6D (air)

Methyl cyclohexane: 6D

Methyl cyclopentane: 6D

Methyl dichloroacetate: 6D, HM-57

Methyl ethyl pyridine: HM-57

Methyl furan: 6D

Methyl methacrylate monomer, uninhibited: Petition

Methyl norbornene dicarboxylic anhydride: 6D, HM-57

Methyl pentadiene: 6D

Methyl pentane: 6D

Methyl propionate: 6D

Methyl propyl ketone: 6D

Methyl sulfide: 6D

Mining reagent: HM-57

Mipafox: 6D (air)

Molybdenum pentachloride: 6D (air)

Monoethanolamine: HM-57

Naphthalene: 6D, USCG (air and water)

Nitrogen, cryogenic liquid: Notice on cryogenic liquids soon to be issued by the Board.

Nitromethane: Board proposal in this docket.

Oakum: USCG (water)

Oiled material: 6D, USCG (air and water)

Organic phosphate compound, liquid (extremely toxic): HM-51

Organic phosphate compound, mixture, dry (extremely toxic): HM-51

Other regulated material, Group A, n.o.s.: 6D

Other regulated material, Group B, n.o.s.: 6D

Other regulated material, Group D: Board proposal in this docket.

Oxygen, cryogenic liquid. Notice on cryogenic liquids soon to be issused by the Board.

Paper, scrap: USCG (water)

Paraldehyde: USCG

Pentachlorophenol: 6D

Perfluoro-2-butene: 6D (air)

Pesticide, water reactive, etc.: USCG (water)

Petroleum coke: USCG (water)

Phencapton: 6D (air)

Phenylene diamine, meta or para: 6D (air)

Phosphine: 6D, petition

Phosphorus heptasulfide: 6D

Phosphorus trisulfide: 6D

Photographic flash lamps: USCG (water)

Polystyrene beads: 6D

Potassium dichromate: 6D (air)

Potassium fluoride solution: 6D, HM-57

Potassium hydrogen fluoride solution: 6D, HM-57

Potassium hydrogen sulfate, solid: 6D (air)

Potassium hydroxide, dry: 6D, HM-57

Potassium metabisulfite: 6D (air)
Propionaldehyde: 6D
Propionic acid: 6D, HM-57
Propionic acid solution: HM-57
Propionic anhydride: HM-57
Propyl acetate: 6D
Propylamine: 6D
Propyl chloride: 6D
Propylene diamine: HM-57
Propyl formate: 6D
Pyroforic solid, n.o.s.: Board proposal in this docket.
Radioactive material, fissile, n.o.s.: Board proposal in this docket.
Rags, scrap: USCG (water)
Road oil: USCG
Rosin: USCG (water)
Rubber curing compound: USCG (water)
Rubidium metal: 6D
Rubidium metal, in cartridges: 6D
Rust preventive coating: USCG
Sawdust: USCG (water)
Selenic acid: 6D, HM-57
Soda lime: 6D, HM-57
Sodium aluminate, solid: 6D (air)
Sodium dichromate: 6D (air)
Sodium fluoride, solid: 6D (air)
Sodium fluoride solution: 6D, HM-57
Sodium hydrogen sulfate, solid: 6D (air)
Sodium hydrogen sulfate solution: 6D, HM-57
Sodium hydrogen sulfite, solid: 6D, HM-57
Sodium hydroxide, dry: 6D, HM-57
Sodium metabisulfite: 6D (air)
Sodium methylate, alcohol mixture: HM-57, petition
Sodium monoxide: 6D, HM-57
Sodium pentachlorophenate: 6D (air)
Sodium phenolate, solid: HM-57
Sodium phosphide: 6D
Stannic phosphide: 6D
Stannous chloride: 6D (air)
Styrene monomer: 6D, USCG

Sulfur, solid: USCG (water)
Sulfurous acid: 6D, HM-57
Tetrachloroethane: 6D, USCG (air and water)
Tetrachloroethylene: 6D (air)
Tetraethylene pentamine: HM-57
1,2,3,6-tetrahydrobenzaldehyde: HM-57
Tetra hydrofuran: 6D
Tetramethyl ammonium hydroxide: HM-57
Tetramethyl methylene diamine: 6D (air)
Textile treating compound mixture, liquid: HM-57
Thioglycolic acid: 6D, HM-57
Thiram: 6D (air)
2,4-toluene diamine: 6D (air)
2,4-toluene diisocyanate: 6D
Toluene sulfonic acid: HM-57
Trichloroacetic acid, solid: 6D, HM-57
Trichloroacetic acid, solution: 6D, HM-57
Trichloroethylene: 6D (air)
Triethylamine: 6D
Trimethyl acetyl chloride: 6D, HM-57
Twisted jute packing: USCG (water)
Uranium hexafluoride, low specific activity, etc.: Board proposal in this docket.
Uranium hexafluoride, fissile, etc.: Board proposal in this docket.
Uranium metal, pyroforic: Board proposal in this docket.
Uranyl nitrate hexahydrate solution: Board proposal in this docket.
Valeric acid: HM-57
Valeryl chloride: 6D, HM-57
Vinyl ethyl ether, inhibited: 6D
Vinyl isobutyl ether: 6D
Water reactive solid, n.o.s.: Board proposal in this docket.
Waxes, liquid: USCG
White acid, etc.: 6D, HM-57
Yeast: 6D (air)
Zinc chloride solution: 6D, HM-57
Zirconium hydride: 6D
Zirconium tetrachloride, solid: HM-57

3. Exemptions. Section §172.101, column 5, proposes to eliminate packaging exemptions for certain materials. The list of these materials is as follows:

Acetone cyanohydrin
Acetonitrile
Acrylonitrile
Allyl alcohol
Allyl chloride
Aluminum hydride
Aluminum phosphide
Arsenic trichloride
Arsine
Boron trifluoride
Cesium, metal
Chloric acid
Chlorine
Chlorophenyl trichlorosilane

2-chloroprene
Dichlorophenyl trichlorosilane
Ethylene diamine
Dihydropyran
Dilsopropyl ether
Dimethyl sulfide
Ether, (ethyl)
Hexadiene
Hydrobromic acid, more than 49% strength
Hydrogen peroxide solution, over 52% hydrogen perioxide
Hydrogen sulfide
Hydrofluosilicic acid
Isopropylamine

Lithium acetylide-ethylene diamine complex
Lithium borohydride
Lithium nitride
Magnesium aluminum phosphide
Mesityl oxide
Methylal
Methyl butene
Nicotine, liquid
Organic phosphate liquid, n.o.s. (25% or less)
    (extremely toxic)
Parathion mixture, liquid
Pentane

Propylamine
Propyl chloride
Rubidium metal
Sodium phosphide
Stannic phosphide
Strychnine
Tetraethyl dithio pyrophosphate mixture, liquid
Tetraethyl pyrophosphate mixture, liquid
Tetrahydrofuran
Vinyl ethyl ether
Zirconium hydride

The Board is of the opinion that these materials present a level of hazard in transportation such that specification packaging should be a requirement for this shipment, including small quantities. Considering the packaging authorized and the properties of these materials which indicate they are highly flammable, extremely toxic, extremely corrosive, highly reactive, or highly flammable and toxic, the Board believes this proposal to be warranted.

4. Passenger carrying aircraft. For the same reasons, the following materials which are now permitted to be transported on passenger carrying aircraft, are proposed to be not permitted for transportation on such aircraft.

Acetaldehyde
Acetone cyanohydrin
Acrylonitrile
Aluminum phosphide
Arsenic trichloride
Benzyl chloride
Crude nitrogen fertilizer solution
Cyclopentane
Dihydropyran
Diisopropyl ether
2,3-Dimethylbutane
Dimethyl sulfide
Dioxane
Dioxolane
Divinyl ether
Ether, (ethyl)
Ethyl methyl ether
Ethyl nitrite
Gas drips, hydrocarbon
Gasoline
Hexadiene
Hydriodic acid
Isopentane
Isopropylamine
Mercaptan mixture, aliphatic (combustible liquid)
Methylal

Methyl butene
Methyl formate
Methyl furan
Methyl pentadiene
Methyl proprionate
Methyl propyl ketone
Methyl sulfide
Monoethylamine
Neohexane
Nicotine, liquid
Organic phosphate, liquid (extremely toxic)
Organic phosphate compound mixture, liquid,
    n.o.s. (25% or less) (extremely toxic)
Parathion mixture, liquid
Polystyrene beads
Propylamine
Propyl chloride
Propylene oxide
Strychnine
Sulfur trioxide
Tetraethyl dithio pyrophosphate mixture, liquid
Tetraethyl pyrophosphate mixture, liquid
Tetrahydrofuran
2,4-toluene diisocyanate
Vinyl ethyl ether
Zirconium hydride

5. Passenger carrying rail car. Another important change is the specificity of what is permitted to be carried in a passenger carrying rail car. Presently there are limitations on rail carriers in baggage service. The proposed rule would prescribe what hazardous materials may be transported in a passenger rail car. The present regulations are silent in this respect and only address themselves to what is permissible in rail baggage service. As this constitutes a new area of regulation, the Board, for the time being, simply is proposing to apply the limitations for passenger-carrying aircraft. Under the circumstances this is considered to be a reasonable interim approach while the whole subject of the carriage of hazardous materials on passenger vehicles is evaluated. The Board expects to more fully cover this area in future rule making regarding hazardous materials on surface passenger vehicles. Meanwhile, so as not to unduly expand the areas of rule making in this docket, this interim approach has been adopted.

6. Transportation aboard vessels. In column 7, the stowage and segregation requirements for water transportation of hazardous materials are proposed to be changed from what is now found in 46 CFR 146 to provide consistency with the International Maritime Dangerous Goods Code developed by the Intergovernmental Maritime Consultative

Organization (IMCO). For further discussion of the stowage and segregation changes, see that section of the preamble dealing with Part 176.

7. Labeling and classification. The list of Hazardous Materials (§ 172.101) sets forth the U.S. classification and label requirements for each material in columns 2 and 3. A problem is that the classification and label requirements for all the materials are not completely consistent with the UN system. For shipments by water, greater consistency is becoming a necessity. Vessels under many foreign flags carrying packaged hazardous materials in compliance with the UN system and calling at the United States ports are becoming a frequent occurrence as more and more maritime nations begin to adopt the IMCO recommendations for the carriage of hazardous materials by vessels. The Board believes that some means must be provided to insure that the UN system of classification and labeling is permitted aboard these vessels while within the navigable waters of the United States. To achieve this end, the UN (also IMCO) classification and labeling requirements for each product listed in § 172.101 are set forth in column 4, if known. The regulations proposed would permit a carrier by water to transport hazardous materials classified and labeled in accordance with either the United States or UN systems. The present U.S. Coast Guard regulations in 46 CFR 146.02–10 and 146.02–11 permit the labeling and packaging of certain import, export, and transiting shipments to be in accordance with foreign requirements. Consequently, this recognition is not really a change in philosophy or regulation.

8. Docket HM-51. Review of § 172.101 will also reveal certain proposed changes relating to Docket HM-51. For the purposes of rulemaking, Docket HM-51 is superseded by this docket and the matters for rulemaking under that docket are now incorporated in this docket. The definitions as proposed in the Second Advance Notice in Docket No. HM-51 (36 FR 2934) are incorporated herein with one slight modification suggested by a commenter regarding particle sizes of dusts. See § 173.326a in this docket. A very important matter to note is that the list of hazardous materials does not include a number of items listed in C.A.B. No. 82, Air Transport Restricted Articles Tariff No. 6D. As stated in HM-57, Notice No. 73-6 (38 FR 24915) the Board wishes to eliminate discrepancies between the Tariff and the list in the Hazardous Materials Regulations.

Based on the evaluation of the information available to the Board, it is of the opinion that many of the toxic materials listed in the Tariff are properly classed and proposes hereby to add them to § 172.101. In an effort to ascertain that the materials listed in § 172.101 as proposed herein meet the highly toxic or irritating material criteria, the Board requests that additional data be made available to it.

The Board is primarily interested in reviewing data representing results of the tests or criteria prescribed in § § 173.326, 173.326a, and 173.326b. Accordingly the Board will add those appropriate items as listed below, with packaging as prescribed for highly toxic materials or irritating materials, not specifically provided for, § § 173.345, 173.364, or 173.384, as pertinent.

As an ancillary action, after the data has been reviewed and classification determinations have been completed through this rulemaking, the Department will undertake to notify the Civil Aeronautics Board of all discrepancies. In this manner, it is hoped that Tariff 6D and the DOT Hazardous Materials Regulations will be made more uniform thereby facilitating compliance with all regulations.

This list of materials proposed to be added is as follows: ("Irr" indicates irritating material; "Toxic" indicates the Board has some information indicating the material may not be highly toxic but it is not entirely conclusive and needs verification):

Acetylene tetrachloride
Allyl isothiocyanate: Irr
Antu
Benzadine base
Benzyl cyanide
Benzylidene chloride: Toxic
Beryllium chloride
Beryllium hydroxide
Beryllium metal, flake, or powder
Beryllium oxide
Beryllium sulfate
Brombenzyl cyanide, liquid: Irr
Butyl isocyanate: Irr
Carbonyl fluoride
Chloroacetone, stabilizer: Irr
Chloronitrobenzene, meta or para, solid
Chloronitrobenzene, ortho, liquid
Cyclohexyl isocyanate: Irr
Dichloroethyl sulfide

Dichlorovinylchloroarsine
Dichlorovos (DDVP)
Diethyl sulfate: Toxic
Dimefox
Dinitroaniline, liquid
Dinitrocresol
Dinitrotoluenes, liquid: Toxic
Dinoseb
Diphenylchloroarsine, solid: Irr
Endrin
Ethyl bromoacetate
Ethyl chloroacetate
Ethyl dichloroarsine
Ethyl isocyanate: Irr
Hydrogen selenide
Isodrin
Lead acetate: Toxic
Methyl chloroacetate
Methyl isocyanate: Irr

Methyl isothiocyanate: Irr
Naphthylamine: Toxic
Nickel arsenate
Nitroaniline
Nitrotoluene, liquid: Toxic
Nitrotoluene, solid: Toxic
Osmium tetroxide
Oxalic acid, solid: Toxic
Oxalic salts, solid: Toxic
Pentachloroethane: Toxic
Phenylcarbylamine chloride
Phenyldichloroarsine
Phenylisocyanate: Irr
Phenyl mercuric acetate, solid
Phenyl mercuric hydroxide, solid
Phenyl mercuric nitrate, solid
Pindone
Potassium cuprocyanide
Propyl isocyanate: Irr
Schradan

Selenic acid, solid
Silver arsenite
Sodium arsanilate
Tetrachlorodinitroethane
Thallium acetate
Thallium bromide
Thallium carbonate
Thallium chloride
Thallium hydroxide
Thallium iodide
Thallium monoxide
Thallium nitrate
Thallium peroxide
Thallium sesquichloride
Thallium sulfide
Toluidine, liquid
Trichloroacetyl chloride
Vanadium trichloride
Zinc phosphide

9. Change of classifications. In the proposed amended listing of hazardous materials, a number of materials in the present list are proposed to be changed classification. These proposed changes are a result either of Docket HM-103 which necessitates a philosophy that materials be classed according to the hazards they present in a transportation incident, or of the proposed revision of § 173.2 which prescribes the proper classification when a material meets more than one hazard classification. The list of materials for which the classification is proposed to be changed is as follows:

Acetyl chloride
Allyl alcohol
Allyl chlorocarbonate
Bromine pentafluoride
Bromine trifluoride
Chloropicrin and nonflammable, nonliquefied
    compressed gas mixture
Cyanogen chloride
Cyanogen gas
Diethyl dichlorosilane
Dimethyl sulfate
Ethyl chloroformate
Ethyl chlorothioformate
Ethylene imine
Fluorine
Hexaethyl tetraphosphate and compressed gas
    mixture
Hydrazine anhydrous
Hydrocyanic acid, liquefied, solution, etc.
Hydrogen peroxide
Insecticide, liquefied gas, containing extremely or
    highly toxic material

Iodine pentafluoride
Methyl bromide and nonflammable, nonliquefied
    compressed gas mixture
Methyl chloroformate
Nickel chloroformate
Nickel carbonyl
Nitrating (mixed) acid
Nitrating (mixed) acid, spent
Nitric acid
Nitric acid, fuming
Nitric oxide
Nitrobenzol
Nitrogen dioxide
Nitrogen peroxide
Nitrogen tetroxide
Nitrogen tetroxide-nitric oxide mixture
Parathion and compressed gas mixture
Pentaborane
Perchloric acid
Phosgene
Phosphoric anhydride
Phosphorus pentachloride

It should be noted that packaging requirements remain the same as present. The changes are limited to classification and labeling. Of course, this would also change the shipping paper description requirements. Editorially, the packaging sections would be appropriately relocated or amended as necessary, in the final amendment.

VIII. PACKAGING REQUIREMENTS

In this docket and in Docket HM-103 published elsewhere in this FEDERAL REGISTER, poisonous materials would no longer be designated as Class A or B poisons but rather as Extremely or Highly Toxic Materials.

Materials would be reordered in priority of care in packaging and handling according to their degree of toxicity.

Because persons associated a level of hazard distinction in definition for the previous Class A and Class B, (when the definitions quantitatively did not make such a distinction), these persons have mistakenly associated "Class A poison" with Extremely Toxic and "Class B poison" with Highly Toxic. This association should not be made. The extremely and highly toxic distinctions are proposed to be established to better relate the integrity of the packaging to the level of hazard of the material packaged. Several materials formerly identified as Class B poisons are now proposed to be identified as extremely toxic materials. Their specification packaging is not proposed to be changed because the Board believes the presently authorized packaging is adequate.

The following is a list of such materials:

Chloropicrin, absorbed
Chloropicrin liquid
Methyl bromide and chloropicrin mixture, liquid

Motor fuel antiknock compound
Perchloro methyl mercaptan

However, the Board believes that other Class B poison materials now proposed to be identified as extremely toxic materials, should not be authorized in some of the presently specified packaging. Appearing below is a list of these materials and, as a guide only, a summary description of the packaging that would no longer be authorized.

Acetone cyanohydrin: 5C, 17C, 17E, 37B, 10A, 10B, 10C, 11A, 11B, 12B with several glass bottles, 12D, 15A, or 16B with glass bottles, 15C, 16A, 19A, tank cars, tank trucks, 1A, 1D, 1E, 21C, 42C, 42D, 15P, 22C, 37P, 12A, 29, 42C, and 12P.

Arsenic trichloride: Same as acetone cyanohydrin.

Cyanogen bromide: Glass or earthware inside containers.

Hydrocyanic acid solutions: 11A, 11B

Nicotine, liquid: Same as acetone cyanohydrin.

Organic phosphate compound liquid (extremely toxic): 17C, 17E, 21C, 37A.

Organic phosphate compound mixtures, n.o.s., dry: 17C, 17H, 37A, 21C for over 15% by weight.

Organic phosphate compound mixtures, n.o.s., liquid: 17C, 17E, 37B, 21C, 37A.

Parathion, liquid: 17C, 17E, 21C, 37A.

Parathion mixture, dry: 17C, 17H, 37A, 21C for over 15% by weight.

Parathion mixture, liquid: 17C, 17E, 37B, 21C, 37A.

Strychnine: 5A, 6B, 17E, 17H, 37A, 37B, 10A, 10B, 10C, 11A, 12A (12B or 12C) with glass bottles and some other inside packagings, 15A and 15B with certain inside packaging, 15C, 16A, 19A, 18B, 22A, tank cars, 21C, and 37P.

Tetraethyl dithio pyrophosphate, liquid: 17C, 17E, 21C, 37A.

Tetraethyl dithio pyrophosphate mixture, dry: 17C, 17H, 37A, 21C for over 15% by weight.

Tetraethyl dithio pyrophosphate mixture, liquid: 17C, 17E, 37B, 21C, 37A.

Tetraethyl pyrophosphate, liquid: 17C, 17E, 21C, 37A.

IX. PART 173—SHIPPERS

Many changes are proposed in this part, the great majority being of an editorial nature. However, the Board recognizes that in the publication of such a voluminous document proposed changes might be unnoticed. It desires to highlight all proposed changes as clearly as possible.

Therefore, under these circumstances, the Board has determined that a preamble covering this part section by section would be the best approach to avoid any oversights. The proposed changes are therefore highlighted as follows:

Section 173.1. Editorial clarification. Proposed deletion of unnecessary language.

Section 173.2. Editorial when viewed with the notice in Docket HM-103.

Sections 173.3, 173.4, and 173.5. Editorial.

Section 173.6. Proposed new requirements covering the preparation of shipments of hazardous materials for transportation by aircraft. Most of these proposed requirements are derived from C.A.B. No. 82, Air Transport Restricted Articles Tariff No. 6-D and the International Air Transport Association (IATA) Restricted Articles Regulations.

Section 173.7. Editorial.

Section 173.9. Editorial. Also, proposed new requirements have been added in paragraph (d) to give added flexibility in using certain packagings of foreign manufacture. This proposed change is based on a petition from a domestic chemical manufacturer.

Section 173.18. Proposed new requirements for consolidation of aircraft shipments. These requirements are derived from rules in C.A.B. No. 82 Air Transport Restricted Articles Tariff No. 6D.

Section 173.21. Editorial. Also, proposed new requirements regarding the forwarding of Bureau of Explosives reports on cigarette lighters to the Department.

Sections 173.22, 173.23, 173.24, and 173.26. Editorial.

Section 173.27. Editorial changes caused by the incorporation of 14 CFR Part 103 in these regulations.

Section 173.28. Editorial.

Section 173.29. The Board is proposing to revise § 173.29 in its entirety to require that packagings and tanks which have not been cleaned or purged of all hazardous materials residue be carried in the same manner as if they were filled with hazardous materials. It has been brought to the Board's attention on several occasions that present regulations do not require identification of the potential hazards of materials such as parathion in so-called empty containers. The Board believes that packagings containing residual quantities of hazardous material should be subject to the same requirements as those that are filled. However, a provision has been added to the shipping paper regulations proposed in Docket No. HM-103 authorizing addition of the word "Empty" to a shipping description when packagings have not been cleaned and purged.

Sections 173.30, 173.31, 173.32, 173.33, and 173.34. Editorial.

Sections 173.51, 173.52, 173.54, 173.57, 173.59, 173.60, 173.61, 173.62, 173.63 and 173.64. Editorial.

Section 173.65. Proposed limitation to present exemption which would not permit transportation of small quantities of explosive chemicals described as drugs or medicines, by express rail car or aircraft.

Section 173.66. Proposed deletion of provision based on obsolete reference to an emergency.

Sections 173.67, 173.68, 173.69, 173.70, 173.71, 173.72, 173.73, 173.74, 173.75, 173.76, 173.77, 173.78, 173.79, and 173.80. Editorial.

Sections 173.86 and 173.87. Editorial.

Sections 173.89, 173.90, 173.91, 173.92, 173.93, 173.94, and 173.95. Editorial.

Section 173.101. Editorial.

Sections 173.103, 173.104, 173.105, 173.106, 173.107, 173.108, 173.109, 173.110, 173.111, 173.112, 173.113, and 173.114. Editorial.

Section 173.116. Editorial. Also, proposed addition of reference temperature of 130° F. regarding filling.

Section 173.116a. Editorial. To set forth the applicability of the U.S. Coast Guard regulations when transportation by water is involved for liquids having a flash point above 80° F. (Tagliabue open-cup).

Section 173.118. Proposed changes in requirements regarding "exempt" quantities of flammable liquids. "Exemptions" would now be limited to specification packaging. By reclassification to ORM-D, exemption from other regulations would also be effected under certain circumstances.

Section 173.118a. Proposed exemption provisions for combustible liquids.

Section 173.119. Editorial. Also, proposed packaging limitations as presently set forth in Tariff 6-D and the IATA Restricted Articles Regulations.

Section 173.119a. Proposed added combustible liquids packaging requirements which apply when such materials are transported by aircraft or passenger vessel.

Section 173.119b. Proposed added packaging requirements which apply when certain other combustible liquids are transported by aircraft or passenger vessel.

Section 173.120. Proposed simplification of requirements applicable to vehicles being transported by rail or highway.

Section 173.121. Editorial.

Section 173.122. Editorial. Also, proposed marking requirement for tank cars as originally introduced in Docket HM-101, Notice 72–5 (37 FR 7104).

Section 173.123. Editorial.

Section 173.124. Editorial. Also, proposed marking requirement for tank cars as originally introduced in Docket HM-101, Notice 72–5 (37 FR 7104).

Section 173.125. Editorial. Also, proposed deletion of provisions based on obsolete reference to an emergency. Proposed limitation on packaging authorized aboard an aircraft based on standards of the IATA Restricted Articles Regulations.

Sections 173.126 and 173.127. Editorial.

Section 173.128. Editorial. Also, proposed incorporation of requirements from 46 CFR 146 and the IATA Restricted Articles Regulations. Proposed deletion of provisions based on obsolete reference to an emergency. Proposed deletion of certain labeling and marking exemptions.

Section 173.129. Editorial. Also, proposed limitations on packaging authorized aboard aircraft based on the IATA Restricted Articles Regulations. Proposed deletion of certain labeling and marking exemptions.

Section 173.130. Proposed to authorize reclassification of refrigerating machines containing a limited amount of a flammable liquid to ORM-D.

Section 173.132. Editorial. Also, proposed incorporation of requirements from 46 CFR 146 and the IATA Restricted Articles Regulations. Proposed deletion of provisions based on obsolete reference to an emergency. Proposed deletion of certain labeling and marking exemptions.

Section 173.133. Editorial. Also, proposed deletion of provisions based on obsolete reference to an emergency.

Sections 173.134, 173.135, 173.136, 173.137, and 173.138. Editorial.

Section 173.139. Editorial. Also proposed marking requirement for tank cars as originally introduced in Docket HM-101, Notice 72-5 (37 FR 7104).

Sections 173.140 and 173.141. Editorial.

Section 173.143. Editorial.

Section 173.144. Editorial. Also, proposed limitation on packaging authorized aboard aircraft based on the IATA Restricted Articles Regulations. Proposed deletion of certain labeling and marking exemptions.

Section 173.145. Editorial. Also, proposed limitation on packaging authorized aboard aircraft based on the IATA Restricted Articles Regulations.

Section 173.146. Editorial.

Section 173.147. Proposed change in requirements regarding "exempt" quantity of methyl vinyl ketone. "Exemption" would now be limited to specification packaging. By reclassification to ORM-D, exemption from other regulations would also be effected under certain circumstances.

Sections 173.148 and 173.149. Editorial.

Section 173.149a. Proposed addition of requirements for the shipment of nitromethane.

Section 173.149b. Proposed incorporation of requirements from 46 CFR 146 and Tariff 6-D for formaldehyde.

Section 173.150. Editorial.

Section 173.151. Editorial. Proposed separation of oxidizing materials and organic peroxides as two separate classes of hazardous materials.

Section 173.151a. Proposed addition of an organic peroxide definition.

Section 173.153. Proposed changes in requirements regarding "exempt" quantities of flammable solids, oxidizing materials, and organic peroxides. "Exemption" would now be limited to specification packaging. By reclassification to ORM-D, exemption from other regulations would also be effected under certain circumstances.

Sections 173.154, 173.154a, 173.155, 173.156. Editorial.

Sections 173.158, 173.159, 173.160, and 173.161. Editorial.

Section 173.162. Editorial. Proposed changes in requirements regarding "exempt" quantity of charcoal. "Exemption" would now be limited to specification packaging. By reclassification to ORM-D, exemption from other regulations would also be effected under certain circumstances.

Section 173.163. Editorial.

Section 173.164. Proposed addition of "chromic acid mixture, dry" based on Manufacturing Chemists Association petition.

Sections 173.165, 173.166, 173.167, 173.168, 173.169, 173.170, 173.171, and 173.172. Editorial.

Sections 173.174 and 173.175. Editorial.

Section 173.176. Editorial. Also, proposed changes in requirements regarding exemptions for matches. Exemptions would now be limited to packaging. By reclassification to ORM-D, exemption from other regulations would also be effected under certain circumstances.

Section 173.177. Editorial. Also, proposed deletion of provision relating to obsolete reference to an emergency.

Sections 173.178, 173.179, 173.180, and 173.181. Editorial. Also, proposed deletions based on petiton stating that slow burning (nonflammable) film does not meet flammable solid definition.

Section 173.182. Editorial. Based on Docket HM-104, Notice No. 72-10 (37 FR 16108), note would be added for calcium nitrate. Also, proposes to delete exemptions from labeling and marking, except those provided in § 173.153, if the material meets oxidizing materials definition.

Sections 173.183, 173.184, and 173.185. Editorial.

Section 173.186. Editorial. Also, proposed deletion of provision relating to obsolete reference to an emergency.

Section 173.187. Editorial.

Section 173.188. Editorial. Also, proposed deletion of provision relating to obsolete reference to an emergency.

Section 173.189. Editorial.

Section 173.190. Editorial. Also, proposed marking requirements for tank cars as originally introduced in Docket HM-101, Notice 72-5 (37 FR 7104).

Sections 173.191, 173.192, 173.193, and 173.194. Editorial.

Section 173.195. Editorial. Also, proposed deletion of provision relating to obsolete reference to an emergency. Proposes to incorporate text of § 173.196.

Section 173.196. Editorial.

Sections 173.197, 173.197a, 173.198, 173.199, 173.200, 173.201, 173.202, and 173.203. Editorial.

Section 173.204. Editorial. Also, proposed limitation on packaging authorized aboard aircraft based on IATA Restricted Articles Regulations. Proposed deletion of provision relating to obsolete reference to an emergency.

Sections 173.205, 173.206, 173.207, 173.208, 173.209, 173.210, 173.211, 173.212, 173.213, 173.214, and 173.216. Editorial.

Section 173.217. Editorial. Also, proposed deletion of certain labeling and marking exemptions.

Section 173.218. Editorial.

Section 173.220. Editorial. Also, proposed deletion of certain labeling and marking exemptions.

Sections 173.221 and 173.222. Editorial.

Section 173.223. Editorial. Also, proposed deletion of certain labeling and marking exemptions.

Section 173.224 and 173.225. Editorial.

Section 173.226. Editorial. Also, proposed change in requirements regarding "exempt" quantity of powdered thorium metal. "Exemption" would be limited to specification packaging. By reclassification to ORM-D, exemption from other regulations would also be effected under certain circumstances.

Sections 173.227 and 173.228. Editorial.

Section 173.229. Editorial. Also, proposed changes in requirements regarding "exempt" quantities of chlorate and borate mixtures or chlorate and magnesium chloride mixtures. "Exemption" would be limited to specification packaging. By reclassification to ORM-D, exemption from other regulations would also be effected under certain circumstances.

Sections 173.230, 173.231, 173.232, 173.233, 173.234, 173.235, 173.236, 173.237, 173.238, and 173.239. Editorial.

Section 173.240. Editorial. Changes are to recognize proposed criteria for materials corrosive to aluminum proposed to be covered as an ORM-B classification. ORM-B, n.o.s. materials would be subject to restrictions only when intended to be transported by air.

Section 173.241. Editorial. Also proposed addition to reference temperature of 130° F. regarding filling.

Section 173.242. Editorial.

Section 173.244. Editorial. Also, proposed change in requirements regarding exemptions for corrosive materials. Exemptions would be limited to specification packaging. By reclassification to ORM-D, exemptions from other regulations would also be effected under certain circumstances.

Section 173.245. Editorial. Also, proposed incorporation of certain packaging limitations aboard aircraft as set forth in Tariff 6-D and the IATA Restricted Articles Regulations.

Sections 173.245a, 173.245b, and 173.246. Editorial.

Sections 173.247 and 173.248. Editorial. Also, proposed incorporation of packaging limitation aboard aircraft as set forth in the IATA Restricted Articles Regulations.

Section 173.249. Editorial. Also, proposes to delete certain older specification tank cars. Proposes changes in requirements regarding exemptions for certain corrosive liquids. Exemptions would be limited to specification packaging. By reclassification to ORM-D, exemptions from other requirements would also be effected.

Section 173.250. Editorial. Also, proposed simplification of requirements governing the transportation of automobiles.

Section 173.251. Editorial.

Section 173.252. Editorial. Also, proposed to require marking of tank cars as originally introduced in Docket No. HM-101, Notice 72.5 (37 FR 7104).

Section 173.253. Editorial. Also, proposed deletion of provisions relating to obsolete reference to an emergency.

Sections 173.254 and 173.255. Editorial.

Sections 173.256, 173.257, and 173.258. Editorial. Also, proposed incorporation of certain packaging limitations aboard aircraft based on requirements of the IATA Restricted Articles Regulations.

Section 173.259. Editorial.

Section 173.260. Editorial. Proposed deletion of exemption governing batteries shipped by rail.

Section 173.261. Editorial. Also, proposed changes in requirements regarding exemptions for fire-extinguisher charges which are corrosive. Exemptions would be limited to specification packaging. By reclassification to ORM-D, exemptions from other regulations would also be effected.

Section 173.262. Editorial. Also, proposed incorporation of certain packaging limitations aboard aircraft based on requirements of the IATA Restricted Articles Regulations.

Section 173.263. Editorial. Also, proposed incorporation of certain packaging limitations aboard aircraft based on requirements of the IATA Restricted Articles Regulations. Proposed deletion of exemption from labeling and marking for hydrochloric acid of not over 20% strength unless packaged in accordance with §173.244 for ORM-D.

Section 173.264. Editorial. Also, proposed deletion of provisions relating to obsolete references to an emergency. Proposed to require marking of tank cars as originally introduced in Docket No. HM-101, Notice 72-5 (37 FR 7104).

Section 173.265. Editorial. Also, proposed incorporation of certain packaging limitations aboard aircraft based on requirements of the IATA Restricted Articles Regulations.

Section 173.266. Editorial. Also, proposed incorporation of certain packaging limitations aboard aircraft based on requirements of the IATA Restricted Articles Regulations. Proposed to require marking of tank cars as originally introduced in Docket No. HM-101, Notice 72–5, (37 FR 7104).

Section 173.267. Editorial.

Section 173.268. Editorial. Also, proposed to require marking of tank cars as originally introduced in Docket No. HM-101, Notice 72–5 (37 FR 7104).

Section 173.269. Editorial. Also, proposed incorporation of certain packaging limitations aboard aircraft based on requirements of the IATA Restricted Articles Regulations.

Sections 173.270 and 173.271. Editorial.

Section 173.272. Editorial. Also, proposed deletion of labeling and marking exemptions for sulfuric acid of not over 25 percent concentration unless packaged in accordance with § 173.244 for ORM-D. Proposed incorporation of certain packaging limitations of Tariff 6-D and the IATA Restricted Articles Regulations aboard aircraft.

Section 173.273. Editorial. Also, proposed to require marking of tank cars as originally introduced in Docket No. HM-101, Notice 72–5 (37 FR 7104).

Sections 173.274 and 173.275. Editorial.

Section 173.276. Editorial. Also, proposed incorporation of packaging limitation aboard aircraft based on requirements of the IATA Restricted Articles Regulations.

Section 173.277. Editorial. Also, proposed incorporation of packaging limitations aboard aircraft of Tariff-6D and the IATA Restricted Articles Regulations. Proposed changes in requirements regarding exemptions for hypochlorite solutions. Exemptions in this section would be limited to specification packaging. By reclassification to ORM-D exemptions from other regulations would also be effected.

Section 173.278. Editorial. Also, proposed incorporation of packaging limitation aboard aircraft based on the requirements of the IATA Restricted Articles Regulations.

Section 173.279. Editorial. Also, proposed deletion of exemptions from labeling and marking for anisoyl chloride unless packaged in accordance with § 173.244 for ORM-D.

Sections 173.280, 173.281, 173.282, 173.283, 173.284, and 173.285. Editorial.

Section 173.286. Editorial. Proposed changes to the requirements on exemptions for chemical kits. Exemptions would be limited to specification packaging. By reclassification to ORM-D, exemptions from other regulations would also be effected.

Sections 173.287 and 173.288. Editorial. Also, proposed incorporation of packaging limitations aboard aircraft based on IATA Restricted Articles Regulations.

Section 173.289. Editorial. Also, proposed deletion of authorization for Bureau of Explosives approval. Proposed incorporation of packaging limitations aboard aircraft based on IATA Restricted Articles Regulations.

Section 173.290. Editorial.

Section 173.291. Editorial. Also, proposed incorporation of packaging limitations aboard aircraft based on the IATA Restricted Articles Regulations.

Sections 173.292, 173.293, and 173.294. Editorial.

Section 173.295. Editorial. Also, proposed incorporation of packaging limitation aboard aircraft based on the requirements of the IATA Restricted Articles Regulations.

Sections 173.296 and 173.297. Editorial.

Section 173.298. Editorial. Also, proposed incorporation of packaging limitation aboard aircraft based on the requirements of the IATA Restricted Articles Regulations.

Sections 173.299, 173.299a, 173.300. Editorial.

Section 173.301. Editorial. Also, proposed addition of specific limitations on handling of cylinders containing an extremely toxic gas.

Sections 173.302, 173.303, 173.304, and 173.305. Editorial.

Sections 173.306 and 173.307. Editorial. Also, proposed changes in requirements regarding exemptions for certain compressed gases. Exemptions would be limited to packaging. By reclassification to ORM-D, as set forth in proposed § 173.307, exemptions from other regulations would also be affected under certain circumstances. Paragraphs 173.306 (c) and (f) are shown only for information. They are subject to rulemaking in Docket HM-106, Notice 72–2 (38 FR 7470).

Section 173.308. The Board proposes to prescribe certain limitations on the shipment of cigarette lighters or other similar devices charged with fuel.

Section 173.314. Proposed marking of certain tank cars as originally presented in Docket No. HM-101, Notice 72–5 (37 FR 7104).

Sections 173.315 and 173.316. Editorial.

Section 173.325. Editorial.

Sections 173.326, 173.326a, and 173.326b. Proposed new definitions for Extremely Toxic, Highly Toxic, and Irritating materials. These definitions are based on comments received in the two Advance Notices of Proposed Rulemaking issued under Docket HM-51 (35 FR 8831 and 36 FR 2934).

Section 173.327. Editorial. Also, proposed additional general requirements regarding filling. Proposed tank car marking requirements as originally presented in Docket No. HM-101, Notice 72–5 (37 FR 7104).

Sections 173.328, 173.329, 173.330, 173.331, 173.332, and 173.333. Editorial.

Section 173.334. Editorial. Also, propose to require approval of valves by the Department in place of the Bureau of Explosives.

Section 173.335. Obsolete. Proposed to be deleted.

Sections 173.336, 173.337, 173.338, and 173.343. Editorial. Section 173.343 replaced by § 173.326a.

Section 173.344. Editorial. Proposed specific temperature reference for filling.

Section 173.345. Editorial. Also, proposed deletion of exemptions from labeling and marking for poisonous materials.

Section 173.346. Editorial. Also, proposed incorporation of packaging limitations from Tariff 6-D and the IATA Restricted Articles Regulations for aircraft.

Section 173.347. Editorial. Also, proposed deletion of provision relating to obsolete reference to emergency.

Sections 173.348 and 173.349. Editorial. Also, proposed incorporation of packaging limitations from Tariff 6-D and the IATA Restricted Articles Regulations for aircraft.

Section 173.350. Editorial.

Section 173.351. Editorial. Also, proposed deletion of obsolete packaging and incorporation of performance testing for glass packagings for hydrocyanic acid solutions.

Section 173.352. Editorial. Also, proposed incorporation of packaging limitations from IATA Restricted Articles Regulations for aircraft.

Section 173.353. Editorial. Also, proposed additional limitations on packaging containing compressed gases.

Sections 173.354, 173.355, and 173.356. Editorial.

Section 173.357. Editorial. Also, proposed deletion of provision relating to obsolete reference to emergency.

Section 173.358. Editorial. Also, proposed deletion of certain packagings for materials classed as Extremely Toxic.

Section 173.358a. Editorial. Also, proposed adjustment of former § 173.358 to recognize those materials that are classed as Highly Toxic and provide more flexibility in authorized packaging than is provided for Extremely Toxic materials.

Section 173.359. Editorial. Also, proposed deletion of certain packaging for materials classed as Extremely Toxic. Proposed deletion of exemption from labeling and marking for certain mixtures.

Section 173.359a. Editorial. Also, proposed adjustment of former § 173.359 to recognize those materials that are classed as Highly Toxic and provide more flexibility in authorized packaging than is provided for Extremely Toxic materials.

Sections 173.360, 173.361, 173.362, and 173.363. Editorial.

Section 173.364. Editorial. Also, proposed deletion of exemption from labeling and marking for Highly Toxic solids. Note: throughout the following sections, the use of multi-wall paper bags is prohibited from carriage on aircraft.

Section 173.365. Editorial. Also, proposed deletion of provision relating to obsolete reference to emergency. Proposed incorporation of packaging limitations of the IATA Restricted Articles Regulations for aircraft.

Section 173.366. Editorial.

Section 173.367. Editorial. Also proposed deletion of provision relating to obsolete reference to emergency. Proposed incorporation of packaging limitations of the IATA Restricted Articles Regulations for aircraft.

Section 173.368. Editorial.

Sections 173.369 and 173.370. Editorial. Also, proposed incorporation of packaging limitations of the IATA Restricted Articles Regulations for aircraft. Proposed deletion of exemption from labeling and marking for cyanides, cyanide mixtures, and carbolic acid.

Section 173.371. Editorial. Also, proposed addition of reference temperature regarding the form of dinitrobenzol for application of packaging requirements.

Sections 173.372, 173.373, and 173.374. Editorial.

Section 173.375. Editorial. Also, proposed deletion of provision relating to obsolete reference to emergency.

Section 173.376. Editorial.

Section 173.377. Editorial. Also, proposed deletion of certain packagings for mixtures classed as Extremely Toxic. Proposed to incorporate packaging limitations of the IATA Restricted Articles Regulations for aircraft.

Section 173.377a. Editorial. Also, proposed adjustment of former § 173.377 to recognize those materials that are classed as Highly Toxic and provide more flexibility in authorized packaging than is provided for Extremely Toxic materials.

Section 173.377b. Proposed deletion of exemptions from labeling and marking for certain mixtures.

Section 173.379. Editorial. Also, proposed deletion of glass or earthenware packaging for cyanogen bromide.

Section 173.381. Editorial. For definition, see § 173.326b. Also, proposed additional general packaging requirements that are common to most hazardous materials.

Section 173.382. Editorial. Proposed clarification of the requirements for packagings when a compressed gas mixture is used. See § 173.305. Under present circumstances, persons may believe that ordinary drums are authorized as packaging for certain compressed gases.

Section 173.384. Obsolete. The Board believes this material is no longer being shipped.

Sections 173.385, 173.386, 173.387, and 173.388. Editorial.

Sections 173.389 through 173.399. Editorial.

Section 173.426. Proposed requirements for truck bodies and trailers transported by rail car, regarding handling of fumigated rolling stock.

Section 173.432. Editorial. Also, proposed inclusion of flammable compressed gases in prohibition against loading into tank cars from motor vehicles or drums.

Section 173.500. Proposed new definitions for ORM-A, B, C, and D materials. These ORM-A, B, and C definitions are now in use and are contained in Tariff 6-D and the IATA Restricted Articles Regulations.

Section 173.505. New section.

Section 173.510. Proposed requirements which contain general packaging rules for all ORM substances.

For §§ 173.605 through 173.1085, see the listing of hazardous materials in this preamble that are proposed to be added to the Hazardous Materials Regulations. In that list, the source of the regulation is indicated. All the materials covered by these sections are either presently subject to some form of restriction for air or water transportation, or both, as indicated or the Board has received a petition for their inclusion. Otherwise, no new materials have been added by the Board to these categories. The packaging proposed is generally what is now required by Tariff 6-D or the IATA Regulations for aircraft, or 46 CFR 146.27–100 for vessels.

In addition, a change is being proposed to § 173.115(b) that would affect the classification of materials with any flash point if they are thermally unstable. In this change, it is proposed to classify a thermally unstable liquid material as a flammable liquid even though its flash point is over 100° F.

X. PART 174—CARRIERS BY RAIL

This part is proposed to be reorganized. The Board proposes to combine present Parts 174, 175 and 176 into one Part 174, at the same time making several changes of substance. The table below sets forth the new section numbers followed by section references to where the regulations are now located in 49 CFR. In addition, changes of significant substance are highlighted.

§ 174.1 (§ 174.500). Editorial.

§ 174.3 (§ 174.501). Editorial.

§ 174.5 (§ 174.504). Editorial.

§ 174.8 (§ 174.501). Editorial. Also proposes requirement to inspect all placarded cars at inspection points.

§ 174.9 (§ 174.596). No change.

§ 174.10 (§ 174.598). No change.

§ 174.11 (§ 174.505). Editorial.

§ 174.12 (§ 174.576). Editorial.

§ 174.14 (§ 174.582). Editorial.

§ 174.16 (§ 174.564). Editorial.

§ 174.18 (§ 174.588(g) and (h)).

§ 174.20 (§ 174.575). Editorial.

§ 174.22 (§ 174.589(a) 174.560 Note 1). Updating. Some definitions are now found in § 170.30. Others, considered unnecessary have been deleted. Important to note is the proposed definition for "train."

§ 174.24 (§ 174.510). Editorial. Also, proposes new requirements concerning possession of the paper and position in train of each placarded car.

§ 175.25 (§ 174.589(f)). Editorial. Also, proposes to expand the application of this regulation to toxic gases.

§ 174.27 (§ 174.511). Editorial.

§ 174.45 (§ 174.506). Editorial.

(§ 174.511). Editorial.

§ 174.47 (§ 174.580, 174.588(c)). Editorial.

§ 174.49 (§ 174.578). Editorial.

§ 174.50 (§ 174.594). Proposed to be revised to delete certain non-regulatory language and improve requirements regarding any leaking tank cars containing hazardous materials.

§ 174.55 (§ 174.532(a), (1), (6), (8), (9), (10) and § 174.586(d) and (e)). Proposes to add several new requirements for the handling of hazardous materials in general. Some of these proposed requirements formerly were limited to certain classes of hazardous materials.

§ 174.57 (§ 174.566(c)). Editorial.

§ 174.59 (§ 174.547). Editorial. Also proposes new requirements with respect to some details of placard replacement.

§ 174.61 (§ 174.533). Editorial. Also proposes changes in present requirements and adds containers on flat cars.

§ 174.63 (§ 174.534). Editorial. Also proposes an additional requirement regarding impact speeds to be withstood.

§ 174.67 (§ 174.561). Editorial. Proposes to update some requirements and make certain recommendations regulatory requirements, namely: (a) (1), (b) introductory, (b) (2), (e) and (n).

§ 174.69 (§ 174.562). Editorial.

§ 174.81 (§ 174.538). Editorial. Also proposes to add "corrosive solids" to segregation requirements.

§ 174.83 (§ 174.589(c) (d)). Editorial. Also proposes new rule regarding impacting of certain cars.

§ 174.84. Proposes new requirements concerning flat cars carrying placarded trailers or containers.

§ 174.85 (§ 174.589(b)). Editorial.

§ 174.87 (§ 174.589). Editorial.

§ 174.88 (§ 174.589(g)). Editorial.

§ 174.89 (§ 174.589(m)). Editorial.

§ 174.90 (§ 174.589(h)). Editorial.

§ 174.95 (§ 174.589(m)). Editorial.

§ 174.100 (§ 174.502). Editorial.

§ 174.101 (§ 174.526). Editorial.

§ 174.102 (§ 174.527). Editorial.

§ 174.103 (§ 174.588(a), (b), (d), (e), (f), and (g)).

§ 174.104 (§ 174.525). To be handled by separate rule making action.

§ 174.105 (§ 174.581). Editorial.

§ 174.106 (§ 174.503). Editorial.

§ 174.107 (§ 174.577). Editorial.

§ 174.109 (§ 174.585). Editorial.

§ 174.110 (§ 174.591). Editorial.

§ 174.112 (§ 174.529). Editorial.

§ 174.114 (§ 174.590). Editorial.

§ 174.115 (§ 174.530). Editorial.

§ 174.200 (§ 174.532(b)). Editorial.

§ 174.201 (§ 174.532(i)). Editorial.

§ 174.204 (§ 174.560). Editorial.

§ 174.208 (§ 174.579). Editorial—Also proposed to be amended to include truck bodies or trailers on rail cars.

§ 174.280 (§ 174.532(m)). Proposes to expand application to all poisonous materials even though they are not so classed, in this case extremely or highly toxic materials which are classed as gases.

§ 174.290 (§ 174.532(L)). Editorial.

§ 174.300 (§ 174.532(a), and (d)). Editorial.

§ 174.304 (§ 174.560(a)). Editorial.

§ 174.380 (§ 174.532(m)). Proposes to expand application to all poisonous materials even though they are not so classed, in this case classed as flammable liquids.

§ 174.410 (§ 174.532(g)). Editorial.

§ 174.450 (§ 174.592, 174.593). Editorial.

§ 174.480 (§ 174.532(m)). Proposes to expand application to all poisonous materials even though they are not so classed, in this case classed as flammable solids.

§ 174.510 (§ 174.532(k)). Editorial.

§ 174.515 (§ 174.566(b)). Editorial.

§ 174.580 (§ 174.532(m)). Proposes to expand application to all poisonous materials even though they are not so classed, in this case classed as oxidizing materials.

§ 174.600 (§ 174.532(l) (9)). Proposes to extend requirement to all poisonous materials.

§ 174.615 (§ 174.566 (a) and (b)). Editorial.

§ 174.625. Editorial.

§ 174.680 (§ 174.532(m)). Editorial. Also, limitation now refers to labeled toxic materials.

§ 174.700 (§ 174.532(j) and 174.586(h)). Editorial.

§ 174.715 (§ 174.566(d)). Editorial.

§ 174.750 (§ 174.588(c)). Editorial.

§ 174.800 (§ 174.532(h), (h) (1), 174.532(f), 174.586(g)). Editorial.

§ 174.810 (§ 174.532(h) (3) and (4)). Editorial.

§ 174.812 (§ 174.532 (h) and (h) (2)). Editorial—Also, the recommendatory language is proposed to be made mandatory.

XI. PART 175—AIRCRAFT OPERATORS

It is proposed to place the requirements pertaining to the transportation of hazardous materials by aircraft in new Part 175. Part 103 of Title 14, Code of Federal Regulations would be canceled if this new part is adopted. Concurrently, the FAA Administrator will issue appropriate amendments in Part 91 of the Federal Aviation Regulations to reflect this proposed consolidation.

A number of substantive changes to the regulations are proposed pertaining to the acceptance, loading, and transportation of hazardous materials by civil aircraft. Sections of particular importance are as follows:

Section 175.1. If adopted as proposed, the scope of Part 175 will be confined to requirements that must be observed by aircraft operators. Shipper requirements pertaining to carriage by air are contained in Parts 172 and 173. The FAA will continue to exercise jurisdiction over shippers when shipments are offered for carriage by air.

Section 175.5. Aircraft leased to and operated by region nationals outside the United States are exempt from the regulations proposed herein. Otherwise, the regulations continue to be applicable to the loading and transportation of hazardous materials in any civil aircraft in the United States and in civil aircraft of United States registry anywhere in air commerce.

Section 175.10. Contains exceptions indicating the kinds of materials that are not covered by the regulations in the part. Of significance is the addition to the present regulations of provisions pertaining to aircraft parts and equipment to include supplies and replacement items if authorized or required to be carried aboard an aircraft for its operation and to include certain articles for passenger consumption and use. This section also deals in part with proposed Notice 73–13 (38 FR 10157; published April 25, 1973). The proposal is to permit the carriage of medicinal and toilet articles in crewmember or passenger baggage including carry-on baggage within certain quantity limitations.

Section 175.20. Sets forth conditions requiring aircraft operators to train personnel in the handling and carriage of hazardous materials by reference to the training program and manual requirements in certain parts of 14 CFR.

Section 175.30. Sets forth conditions for acceptance of shipments for transportation by aircraft. It specifies the quantity limitations, shipping papers, certification, and labeling requirements by referencing the appropriate subparts of proposed New Part 172.

Section 175.35. Contains proposed requirements for the shipping papers which must be carried aboard aircraft transporting hazardous materials.

Section 175.40. Specifies that aircraft operators must keep labels on hand and that they must replace lost or detached labels.

Section 175.45. This section deals, in part, with proposed Notice 73–17 (38 FR 149637; published June 7, 1973). The proposal is to permit certificate holders under Parts 121, 127, and 135 to report certain incidents involving dangerous articles to the FAA District Office holding the carrier's operating certificate and charged with the overall inspection of its operation.

Section 175.50. Sets forth the conditions pertaining to deviations from the provisions in the Part. Of significance is the addition of the proposed requirement that safety instructions be written.

Section 175.75. States the limitations on the quantities of hazardous materials permitted aboard aircraft. Note that it is proposed to change the quantity limitations from 50 pounds net weight to 65 pounds gross weight. The proposed increase takes into account the weight of packaging material. The same applies to the weight of compressed gas cylinders which has been increased from 150 lbs. net to 300 lbs. gross. Gross weights are the standard commercial measurement for revenue purposes and the Board believes they should be used in place of net weight of content measurements for the purposes of these limitations. Net weights are appropriate for purposes of specifying the limitations on the amount of a material which may be placed into a package. It should also be noted that no limitation is proposed on the number of packages for "Other Regulated Materials" aboard an aircraft. Another significant proposed change is to consider a reusable transport container in the same category as an inaccessible cargo pit or bin for the purpose of carrying hazardous materials.

Section 175.85. This section deals with cargo location aboard an aircraft. Note that paragraph (b) requires that materials authorized for cargo-only aircraft must be accessible to a crewmember during flight and that hazardous materials, other than magnetized material, must be inaccessible to persons other than crewmembers on passenger-carrying aircraft.

SUBPART C

Contains specific regulations that would apply to particular kinds of materials according to their classification. Included are requirements pertaining to self-propelled vehicles, the carriage of gasoline, kerosene, or aviation fuel in aircraft, the carriage of flammable liquids in cargo aircraft in the State of Alaska, special requirements for poisons, and special requirements pertaining to radioactive materials.

Section 175.205. This section has been added to propose safety requirements for vehicle ferry operations.

Section 175.220. This section is being proposed to permit the transportation in the State of Alaska in cargo-only aircraft of certain flammable liquids in quantities not to exceed 55 U.S. gallons. Such carriage will be permitted only under certain conditions set forth in the text of the section, one of which is that another means of transportation is impractical.

Persons reviewing proposed new Part 175 should compare it with existing Part 103, Title 14, Code of Federal Regulations. Also, they should study the list of hazardous materials proposed § 172.101 since it contains significant regulatory proposals pertaining to the kinds and quantities of hazardous materials permitted to be transported aboard civil aircraft. These changes are highlighted in that portion of the preamble dealing with Part 172.

XII PART 176—CARRIERS BY WATER

Persons interested in a detailed section by section review of how the regulations in 46 CFR Part 146 were affected by the transfer of these regulations in 49 CFR Part 176 should review Marine Safety Council Public Hearing Agenda, C6—249, June 1, 1973. (The June date is for record purposes only and is not indicative of its date of issue.) This publication, CG—249, will be automatically distributed to those organizations who receive Coast Guard reprints. An errata will also be distributed to account for the changes to Part 176 since the printing of CG—249. A copy of this publication may be obtained by interested persons not on the mailing list for reprints by a letter request to Commandant (G—MHM), U.S. Coast Guard, Washington, D.C. 20590. This preamble is intended to simply highlight the more important changes that are being proposed regarding the Hazardous Materials Regulations as they apply to carriers by water.

The objectives of the transfer of the 46 CFR Part 146 regulations are summarized as follows:

1. To achieve a better regulations format by including all the DOT Hazardous Materials in one Code of Federal Regulations, except for bulk transportation by water and military explosives regulations.

2. To simplify by reorganization the regulations that are being transferred to 49 CFR Part 176.

3. To completely revise the stowage and segregation requirements to provide as much consistency as possible with the International Maritime Dangerous Goods Code developed by the Intergovernmental Maritime Consultative Organization (IMCO).

4. To eliminate the separate shipper requirements originally contained in 46 CFR 146 by integrating them completely with the shipper requirements set forth in 49 CFR Part 173.

5. To revise and update certain handling requirements for hazardous materials.

6. To establish regulations for barge-carrying vessels.

7. To revise the requirements for roll-on/roll-off vessels to make the United States requirements consistent with new requirements being adopted by IMCO.

8. To revise the requirements for stowage and segregation of land-sea containers and portable tanks carried aboard vessels.

9. To add requirements pertaining to the existing regulations for the stowage of hazardous materials within land-sea containers.

10. To revise the requirements for the inspection of hazardous materials aboard vessels.

11. To modify the requirements for ferry vessels.

12. To modify the requirements for jettisoning hazardous materials.

13. To modify the requirements regarding reports concerning hazardous materials leakage incidents.

14. To require that the dangerous cargo (hazardous materials) manifest be kept on or near the bridge house of vessels, in a special place.

15. To change the name of the classification "Hazardous Articles" to "Other Regulated Material," abbreviated ORM.

16. To revise shipper requirements for "No Label Required" items to allow for consistency with the rail and highway regulations.

17. To revise requirements for the Dangerous Cargo Manifest to allow use of IMCO's correct technical name.

1. Changes relating to format. Presently the requirements for the shipment of hazardous materials by water are published in a separate Title of the Code of Federal Regulations (46 CFR) from those published for the same materials shipped by highway and rail (49 CFR). For hazardous materials (other than in bulk quantity), the requirements for shippers in both titles are almost identical. However, publication in two titles leads to much duplication, creating an unnecessary duplication of administrative effort by government and a need by the public to keep abreast of two series of regulatory developments and changes. Understandably, this separation and the fact that each title (CFR) has a distinctive format, has lead to an important number of obvious inconsistencies between the two titles. The treating of these inconsistencies has resulted in significant time and effort spent by the public and government to alleviate undesired interruptions in the smooth flow of goods between modes of transportation. In a number of instances these administrative inconsistencies have created confusion whereby shipments have been delayed during transit.

With the continuing need for regulation development and consequent change, and with an emphasis on consistent regulations for intermodal shipments, this situation is completely undesirable from the standpoint of the public in knowing and understanding the regulations and of the government in efficiently managing a consistent hazardous materials regulatory program. The proposed incorporation of 46 CFR 146 in 49 CFR 170–189 would provide a more concise code of regulations for administration by the Board and thereby be more understandable and easier to follow by shippers, carriers, and other interested members of the public.

The incorporation of 46 CFR 146 (except § 146.29) into 49 CFR 170–189 necessarily would change the present format of the U.S. Coast Guard Dangerous Cargo Regulations. Generally speaking, the proposal consists of placing the detailed requirements for the water transportation of hazardous materials in a new Part 176 of 49 CFR, and revising the current list of hazardous materials in 49 CFR 172.5 to accommodate certain U.S. Coast Guard requirements. The shipper requirements would be in 49 CFR 173.

The specific stowage location for each material (i.e., on deck or below deck) would appear in a column 7 of the list (172.101). Also appearing in this column, would be any special requirements for a given hazardous material. Presently this information appears in column (2) and (4) through (7) of Tables A through K of 46 CFR 146. For a further explanation of the proposal to revise the list of hazardous materials see that portion of this preamble covering Part 172 and the proposed § 172.100.

The information that now appears in Subparts 146.01 through 146.27 of 49 CFR (exclusive of Tables A through K and the Table for Radioactive Materials) would be relocated in a new 49 CFR Part 176. Information relating to shippers would be integrated into 49 CFR Part 173 (most of the information is now found in this part, the regulations in 46 CFR Part 146 being a duplication).

2. Changes relating to content. The Board desires to create as complete consistency as possible between the regulations of rail, highway, air, and water, and similar requirements of other countries. Although many present requirements of the various modes of transportation in the United States and in other countries are similar, major differences exist. However, the U.S. Coast Guard believes it is necessary in water transportation to resolve these differences, and that they can be resolved without an inordinate amount of difficulty. This notice proposes to accomplish this as one of its objectives.

a. Packaging. Presently, in several cases, 46 CFR Part 146 and 49 CFR 173 authorize different outside packagings for the same hazardous materials. In some instances, the differences are warranted due to the additional hazards presented by the relatively confined conditions aboard vessels at sea. The regulations in 49 CFR Part 173 are proposed to be amended to recognize these situations. In other cases, the differences are not considered justifiable but are truly the product of outdated (46 CFR Part 146 not having been revised when 49 CFR Part 173 was changed) or incompletely or erroneously stated requirements in 46 CFR Part 146. To correct this situation, the U.S. Coast Guard proposes to authorize the same outside packagings for water transportation as for surface transportation, with certain exceptions as set forth in Part 173 in this document and as outlined below. The U.S. Coast Guard is not considering adoption of the IMCO packaging requirements in 49 CFR Part 173 at this time. Notable packaging exceptions are as follows:

i. Portable tanks, other than those DOT—51 and 60 specification tanks presently authorized by 46 CFR 146 for certain materials, would not be permitted for gases or liquids.

ii. Tank cars, motor vehicle tank trucks, and other bulk transporters, other than those presently authorized by 46 CFR 146 for certain materials, would not be permitted.

iii. Certain other individual packagings would not be permitted as specifically set forth in 49 CFR 173. (These exceptions would be very few.)

b. Stowage. This notice proposes adoption of the stowage requirements of the IMCO Dangerous Goods Code. Most maritime countries, including the United States, are participating in the continuing development of this Code and many are presently using it for their national regulations. Consequently, the adoption of the Code's standards by the United States would result in much consistency with the maritime shipping regulations of many other countries.

The current regulations in 46 CFR Part 146 differ in basic philosophy from those of IMCO. A major shift in the thinking of United States carriers by water regarding stowage is called for by this proposal for adoption of the IMCO stowage system. The regulations in 46 CFR Part 146 have been based on a principle that "on deck" stowage is safest. Of primary importance in the early stages of development of this theory was the ease of access to hazardous cargo, the fact that flammable or toxic vapors (in the event of leakage) would not be confined, and the positioning of hazardous cargo for jettisoning in case of an accident. On the other hand, though the IMCO Code recognizes that on deck stowage is necessary for certain extremely hazardous materials, its safety philosophy is that "below deck" stowage is more advisable, when possible, because it is most important to achieve the highest degree of physical protection possible for hazardous cargo to reduce the potential for outside influences to cause a hazardous materials incident.

Although these two philosophies are quite divergent, the U.S. Coast Guard believes that the additional features built into the IMCO requirements (such as stipulating accessible stowage when "under deck" stowage is used), present an overall system which promotes a completely adequate, and perhaps higher, level of safety.

The IMCO system with its three stowage terms "on deck," "under deck," and "under deck away from heat," together with the accessibility requirement, is significantly simpler than the numerous stowage terms now in use in 46 CFR Part 146. This simplicity should lead to easier understanding and improved compliance thereby creating safer conditions aboard vessels transporting hazardous materials.

The stowage requirements for portable tanks and containers are proposed to be revised to preclude "under deck" stowage for hazardous materials required to be stowed "on deck" only (46 CFR 146.07–40). The regulations now provide an exemption for tanks and containers from the "on deck" only stowage requirement for some hazardous materials. The U.S. Coast Guard believes this is not in the best interests of safety. Also, this exemption seriously conflicts with the standards of IMCO.

c. Segregation. The segregation requirements for break bulk stowage outlined in the IMCO Code are proposed for adoption in this notice. The present requirements in 46 CFR Part 146 regarding segregation are quite extensive and complex and there are more exceptions than general rules regarding their use. On the other hand, the U.S. Coast Guard believes the IMCO system is very simple and there are few exceptions to the general rules.

IMCO has under consideration a recommended Code for the segregation of hazardous materials in highway vehicles aboard roll-on/roll-off vessels. This Code has been approved by the responsible subcommittee of IMCO. All that remains to be obtained is the approval of the General Assembly. In anticipation of the formal adoption of this Code, it is being proposed in this notice.

The segregation requirements for containerships and break bulk vessels which handle containers and portable tanks (46 CFR 146.07) are very general and in some cases not clear in their application. Some classes of hazardous materials may be stowed in the same hold separated by only one container, whereas, on a break bulk vessel, packages containing the same hazardous materials must be separated by a hold. These are materials where the intermingling under accident conditions could very significantly increase the magnitude of an existing accident and complicate any emergency action. The U.S. Coast Guard, therefore, believes that the segregation requirements for such materials in containers are in need of improvement and proposes major changes to the segregation requirements for containerships.

This notice also proposes requirements for the segregation of barges on board barge-carrying vessels. The segregation requirements in 46 CFR 146 are not directly applicable to this type of vessel. In developing the requirements proposed in this notice for these vessels, the primary objective pursued was to achieve an equivalent degree of safety as is proposed herein for break bulk vessels. The U.S. Coast Guard acknowledges that barge carrying vessels may afford a higher level of safety for the carriage of hazardous materials, in which case a less rigorous segregation philosophy could be considered. However, due to the fact this type of vessel has been operating only a short time, the U.S. Coast Guard does not believe it has sufficient justification in the interests of safety to depart from the approach it has assumed.

Hazardous materials on barges are not required to be in compliance with the segregation requirements in 46 CFR Part 146 that apply to break bulk vessels. The U.S. Coast Guard believes that this situation poses undue hazards in the handling of hazardous materials aboard barges and proposes to revise the regulations to make it clear that barges are not exempt from these segregation requirements. This proposal was originally set forth in the FEDERAL REGISTER of March 20, 1971 (36 FR 5400) but has been delayed in further development pending this major proposal regarding 46 CFR Part 146.

d. Other carrier requirements. 46 CFR Part 146 contains no detailed requirements for the proper stowage of hazardous materials within containers. Subpart 146.07 stipulates that hazardous materials loaded into containers must be properly secured but gives no specific guidance. In the case of segregation within a container, the requirements are not completely clear. This notice proposes specific requirements for the stowage of hazardous materials in containers. Also, since there is no statement exempting containers, the U.S. Coast Guard advises that it would consider the standard segregation requirements as being applicable to stowage of hazardous materials within containers.

e. Requirements for the carriage of bulk solids. The various subparts of 46 CFR 146 contain requirements for the carriage of solids in bulk (materials received without mark or count). These requirements will be contained in a new part of subchapter 0 in Title 46 and will be published in a separate document at a later date.

46 CFR Subparts 146.01 through 146.28 will be deleted in their entirety if the proposals contained in this document and the future proposal to include in subchapter 0 of Title 46 the carriage of solids in bulk are adopted.

The Coast Guard will hold a hearing on April 9, 1974 at 9:30 a.m. in room 2230 of the Nassif Building, Department of Transportation, 400 Seventh Street, SW., Washington, D.C. Interested persons are invited to attend the hearing and present oral or written statements on this proposal. It is requested that anyone desiring to attend the hearing notify the Commandant (G—CMG), U.S. Coast Guard, Washington, D.C. 20590.

VIII. PART 177—CARRIERS BY PUBLIC HIGHWAY

No major changes are proposed for Part 177 except to §§ 177.817 and 177.823 which relate to the proposals made in Docket HM-103. These proposals are subject to adoption contingent on adoption of the corresponding proposals made in that Docket.

SHIPPING PAPERS

a. Section 177.817. The Board proposes to revise § 177.817 to relate the shipping paper requirements to Subpart C of Part 172 as proposed in Docket No. HM-103. The revised section contains only those requirements that are specifically applicable to motor carrier operations. It should be noted that the proposed revision to this section (1) clarifies the requirements pertaining to shippers' certificates, (2) provides that materials identified as ORM—A, B, C, or D are not subject to the shipping paper requirements for transportation by highway unless such shipments are accepted for subsequent transportation by air or water, and (3) proposes to clarify the possession requirements for shipping papers with the recommendation that shipping papers be placed in a pouch or other device that is mounted on the inside of the door to the left of the driver's position. Comments as to the practicality of making such a location a mandatory requirement in the future are invited. New paragraph (d) in the revised section proposes to require that shipping papers pertaining to hazardous materials be arranged so they will appear first upon examination during transportation. This is in order to make such documents readily available should incidents requiring their immediate examination occur during transportation. In considering this requirement, commenters should acquaint themselves with the increased reliance placed on shipping papers for the specific identification of certain hazards, certain kinds of LCL and LTL shipments as proposed in Docket HM-103. Also included is a recommendatory provision indicating that the driver of a motor vehicle should offer the shipping papers in his possession to emergency personnel for their examination.

MARKING AND PLACARDING MOTOR VEHICLES

b. Section 177.823. The Board is proposing to revise § 177.823 to make reference to the requirements contained in new Part 172 as proposed in Docket No. HM-103. In addition, the revised section contains two proposed exceptions from the placarding requirements, the first pertaining to a vehicle escorted by a representative of a state or local government, and the second when a carrier has received permission from the Department. These provisions are proposed to provide some flexibility in the placarding requirements in order to facilitate the movement of vehicles during emergencies.

XIV. CONCLUSION

The Board has under development a plan whereby it would completely restructure the major Parts of 49 CFR Parts 170 to 189. This would permit more orderly arrangement of the present regulations and any new regulations published. This plan could have been placed into effect in this notice, however, it was the opinion of the Board that it would have made this docket extremely more difficult to review for the public. It is proposed, in the final amendment, to generally reorganize the regulations as follows:

Part 170 General Information, Regulations, and Definitions.
Part 171 Rule-Making Procedures of the Hazardous Materials Regulations Board.
Part 172 List of Hazardous Materials and Hazardous Materials Communications Regulations.
Part 173 Shippers; General Requirements for Shipments and Packagings.
Part 174 Carriers by Rail.
Part 175 Aircraft Operators.
Part 176 Carriers by Water.
Part 177 Carriers by Public Highway.
Part 178 Shipping Container Specifications.
Part 179 Specifications for Tank Cars.
Part 180 (Vacant).

Part 181 Explosives.
Part 182 Gases.
Part 183 Flammable and Combustible Liquids.
Part 184 Flammable Solids.
Part 185 Oxidizing Materials and Organic Peroxides.
Part 186 Toxic Materials, Irritating Materials, and Etiologic Agents.
Part 187 Radioactive Materials.
Part 188 Corrosive Materials.
Part 189 Other Regulated Materials (ORM) A, B, C, and D.
Parts 181 to 188 would contain the specific packaging requirements for hazardous materials which are now located in Part 173.

The Board welcomes any comments on this planned reorganization.

In order to assist interested persons in their efforts to understand fully the proposals made in this Notice, the Board has scheduled a meeting which will be open to the public. The meeting will be for the purpose of answering questions and providing explanations of the proposals contained in this Notice. Statements relative to the merits of the proposals in this Notice may not be made at the meeting, but must be submitted in writing.

The meeting will be held on February 13, 1974, in the Departmental Auditorium located on Constitution Avenue between 12th and 14th Streets, NW., Washington, D.C. beginning at 10:00 a.m.

Commenters are requested to make their comments in a manner that will clearly identify the particular matters on which they are commenting. Unless comments are general in nature pertaining to the entire Notice, it is requested that each paragraph of comments be identified in the following manner:

"Part 176 We think _____" or "Section 172.503 We believe _____"

Also, those commenters submitting more than two pages of comments are requested to submit six copies in order to facilitate their handling by the Board.

(39 F.R. 3022, January 24, 1974)

[49 CFR Parts 172, 173]
[Docket No. HM-112; Notice 74-7]
TRANSPORTATION OF HAZARDOUS MATERIALS
Classification and Packaging of Organic Peroxides

On January 24, 1974, the Hazardous Materials Regulations Board ("the Board") proposed a consolidation of the hazardous materials regulations now found in 14 CFR Part 103, 46 CFR Part 146, and 49 CFR Parts 170 to 179. See Docket HM-112, Notice No. 73-9, 39 FR 3021. In this notice of proposed rule making, the Board stated that the consolidation of the organic peroxide regulations would be the subject of separate rule making to be introduced in a notice before action was completed on Notice 73-9.

The Board had hoped that its organic peroxides proposal would include changes to the present listings and packings now prescribed in the regulations cited above. These changes were to be based on extensive work done under the auspices of the United Nations Committee of Experts on the Transport of Dangerous Goods. It now has become obvious to the Board that the results of this work will not be ready to permit processing in a timely fashion in Notice 73-9. As a result, the Board has decided to withhold any new proposals for listing and packaging of organic peroxides until a later date.

The listings for organic peroxides shown in Notice 73-9 are hereby withdrawn. Instead, the proposals set forth herein complete the presentation in Docket HM-112 of the Board's intent on handling of these materials until the United Nations work is completed. Consequently, the specific item descriptions and packagings proposed in this notice are essentially a restatement of the present descriptions and packagings in 46 CFR Part 146 and 49 CFR Parts 172 and 173. Also, succinic acid peroxide, and organic peroxides not otherwise specified would not be permitted on passenger-carrying aircraft. It is also proposed to change § 173.119(m) to cover flammable liquids which are also highly toxic.

Flammable liquids which are also corrosive or organic peroxides would continue to be covered as is presently provided in § 173.119(m).

It should be noted also that § 172.602, in Docket HM-103, Notice No. 73-10, 39 FR 3163, would require that any shipper wishing to ship an organic peroxide not assigned a hazard information number in § 172.101 would be required to obtain a written designation of the hazard information number from the Department before making the initial shipment.

(39 F.R. 14519, April 24, 1974)

[Docket No. HM-112; Notice No. 73-9]
HAZARDOUS MATERIALS
Consolidation of Regulations and Miscellaneous Proposals; Public Hearing
Regarding Transportation Aboard Aircraft

Notice is hereby given that the Hazardous Materials Regulations Board ("the Board") will hold a public hearing beginning at 9:30 a.m. on February 10, 1975 in Room 300, Federal Office Building 10A (commonly referred to as the FAA Building) located at 800 Independence Avenue, SW., Washington, D.C. to receive comments from interested persons on the present regulations pertaining to the transportation of hazardous materials aboard aircraft or the need to issue new regulations which might improve the protections afforded the traveling public and aircraft crews.

The present regulations pertaining to the transportation of hazardous materials aboard aircraft are found in Part 103 of Title 14, Code of Federal Regulations, and by references contained therein, Title 49 Code of Federal Regulations, (Parts 170-189). The basic design of the system of regulation found in Part 103 is the incorporation of the partial exemption (specification packaging, marking, and labeling) sections in Title 49 for transportation of hazardous materials aboard passenger-carrying aircraft and the requirements pertaining to shipment via rail express for cargo-only aircraft. The system has been basically the same for more than 20 years without substantial revision.

On January 24, 1974, the Board published a notice of proposed rule making (39 FR 3022) under Docket HM-112 proposing a new Part 175 under Title 49 to replace Part 103 of Title 14. Also proposed were a number of revisions to the new list of hazardous materials (§ 172.101) and certain sections of Part 173 which, if adopted, would no longer permit the transportation of certain materials aboard passenger-carrying aircraft. However, the Board did not propose a complete revision of the regulations in this regard based on an evaluation of each material or generic classification listed.

In comments dated August 30, 1974, the Airline Pilots Association (ALPA) indicated its " ... basic policy and recommendation that hazardous materials should be banned from passenger-carrying aircraft except for those items which are medically necessary for the good of the population, dry ice for the prevention of perishable goods, and magnetic material when packaged and stowed under the appropriate regulations. . . . " In addition, during their meeting held November 18 to 27, 1974, the ALPA Board of Directors adopted a resolution calling for (1) discontinuing the transportation of all hazardous materials aboard passenger-carrying aircraft except for certain radiopharmaceuticals, magnetized materials, and dry ice, and (2) limiting the hazardous materials to be carried in cargo-only aircraft to only those materials presently authorized aboard passenger-carrying aircraft.

In light of the foregoing, the Deputy Secretary of Transportation established a task force to make a complete and informed review of the hazardous materials presently being moved in air commerce. The task force is made up of representatives of the Federal Aviation Administration and the Office of Hazardous Materials and is under the direction of the Director of the Office of Hazardous Materials. The task force has been directed to address the points in the ALPA Resolution and to examine those hazardous materials being carried on passenger-carrying and cargo-carrying aircraft with a view toward taking those materials that could be moved by surface transportation off aircraft when no justification for movement in air commerce can be shown. The purpose of this hearing is to assist the task force in obtaining the views of interested members of the shipping industry, the public, and the transportation industry.

This hearing will focus on the materials presently authorized to be transported aboard aircraft. It is contemplated that a second public hearing will be held on the operating requirements that possibly should be imposed in addition, or as an alternative, to those presently specified or proposed in Docket HM-112; Notice 73–9. The time, location and details pertaining to the hearing on operating requirements will be announced in a later issue of the FEDERAL REGISTER.

Commenters are also advised that this proceeding does not include those matters pertaining to radioactive materials covered by FAA Docket No. 3668; Notice 74–18 (39 FR 14612) published April 25, 1974 and those on implementation of regulations necessary to accomplish compliance with section 108 of the Hazardous Materials Transportation Act (Pub. L. 93–633).

The departmental task force will prepare a report following consideration of the comments presented at the hearing, or in writing. That report will be made part of this docket and may serve as the basis for further rule making.

In preparation of views for presentation, commenters should consider the potential hazards of materials presently authorized aboard passenger-carrying aircraft, cargo-only aircraft, or both, taking into account (1) the quantity authorized, (2) the prescribed packaging, (3) the nature and degree of the potential hazard, i.e., toxicity, flammability, corrosivity, pyrophoricity, explosivity, etc., (4) the potential for commingling with other materials having incompatible characteristics and (5) any other factor that should be considered relative to the safety of the passengers, crew, or operation of an aircraft. Commenters may also present views on the necessity that certain identified materials be permitted aboard aircraft in the future in the public interest even though this notice contains no specific proposal to change the regulations to prohibit the transportation of any particular material via aircraft.

Presiding at the hearing will be the Director of the Office of Hazardous Materials as a designated representative of the Board Member for the Federal Aviation Administration in accordance with 49 CFR 170.31(b). Any person who wishes to make an oral statement at the hearing should notify the Director in writing or preferably by telephone or telegram providing his name, address, telephone number, and the approximate time needed for his presentation. The notification should be provided on or before February 4, 1975 and addressed to Director, Office of Hazardous Materials, Department of Transportation, Washington, D.C. 20590 (202–426–0656).

Interested persons not desiring to make oral presentations are invited to give their views in writing. Communications should identify the docket number and be submitted in duplicate to the Director at the above address by February 20, 1975.

A transcript of the hearing will be made and anyone may purchase a copy of the transcript from the reporter. A copy of the transcript and copies of all comments received will be available for examination by interested persons at the Office of the Secretary, Hazardous Materials Regulations Board, Room 6215 Trans Point Building Second and V Streets, SW., Washington, D.C., both before and after the closing date for comments.

(40 F.R. 4329, January 29, 1975)

[Docket No. HM-112; Notice No. 73–9]
TRANSPORTATION OF HAZARDOUS MATERIALS
ABOARD AIRCRAFT
Notice of Hearing

Notice is hereby given that the Hazardous Materials Regulations Board ("the Board") will hold a public hearing beginning at 9:30 a.m. on February 20, 1975 in Room 310, Federal Office Building 10A (commonly referred to as the FAA Building) located at 800 Independence Avenue, SW., Washington, D.C. to receive comments from interested persons on present regulations pertaining to the operating requirements applicable to carriers by aircraft when they transport hazardous materials and the need for any amendments to those regulations which would improve the protections afforded the traveling public and aircraft crews.

The present operating requirements pertaining to the transportation of hazardous materials aboard aircraft are in Part 103 of Title 14, Code of Federal Regulations. Training requirements pertaining to hazardous materials are in Parts 121 and 135 of the same title. On January 24, 1974, the Board published a notice of proposed rule making (39 FR 3022) under Docket HM-112 proposing among other things a new Part 175 under Title 49, Code of Federal Regulations to replace Part 103 of Title 14. Included in that notice were several proposals pertaining to operating requirements including changes in documentation procedures.

On January 29, 1975, a notice was published in the FEDERAL REGISTER (40 FR 4329) announcing a public hearing to be held on February 10, 1975 at the location stated above. That hearing will focus on the materials presently authorized to be transported aboard aircraft. The hearing announced in this notice will focus on the operating requirements that possibly should be imposed in addition, or as an alternative, to those presently specified or proposed in Docket HM-112; Notice 73-9.

Commenters are also advised that this proceeding does not include those matters pertaining to radioactive materials covered by FAA Dockets Nos. 13668 and 14249.

The notice published on January 29 also discussed a Departmental task force which will prepare a report following consideration of the comments presented at the hearing scheduled for February 10 and the hearing announced herein or submitted in writing in connection with either hearing. The report will be made part of this docket and may serve as the basis for further rule making.

In preparation of comments, commenters should consider (1) the training requirements presently specified in 14 CFR, Parts 121 and 135 and the need for modification thereof; (2) the documentation requirements and informational needs of crew members prescribed in 14 CFR 103.25 and proposed under Docket HM-112 (39 FR 3022) for new § 175.35, or the need for modifications thereof; (3) the purpose and benefit of the requirement that all hazardous materials (of the type and quantity authorized aboard cargo only aircraft) be accessible to crew members during flight; (4) the type and quantity of emergency gear and apparatus, such as fire extinguishers, that should be kept aboard aircraft used to transport hazardous materials; (5) the possibility of requiring registration of shippers, carriers and package manufacturers relative to the transportation of hazardous materials via aircraft; (6) the use of the gross weight of a package or the net weight of its contents to determine quantity and stowage limitations aboard aircraft and (7) any other operational factor that should be considered relative to the safety of the passengers, crew, or operation of an aircraft.

The Director of the Office of Hazardous Materials acting as the designated representative of the Board Member for the Federal Aviation Administration in accordance with 49 CFR 170.31(b) will preside at the hearing. Any person who wishes to make an oral statement at the hearing should notify the Director in writing or preferably by telephone or telegram providing his name, address, telephone number, and the approximate time needed for his presentation. The notification should be provided on or before February 18, 1975 and addressed to Director, Office of Hazardous Materials, Department of Transportation, Washington, D.C. 20590. ((202) 426-0656)

Interested persons not desiring to make oral presentations are invited to give their views in writing. Communications should identify the Docket number and be submitted in duplicate to the Director at the above address by February 28, 1975.

A transcript of the hearing will be made and anyone may purchase a copy of the transcript from the reporter. A copy of the transcript and copies of all comments received will be available for examination by interested persons at the Office of the Secretary, Hazardous Materials Regulations Board, Room 6215 Trans Point Building, Second and V Streets, S.W., Washington, D.C., both before and after the closing date for comments.

(40 F.R. 5386, February 5, 1975)

[Docket No. HM-103; HM-112; Amdt. Nos. 171-32,
172-29, 173-94, 174-26, 175-1, 176-1, 177-35]
CONSOLIDATION OF HAZARDOUS
MATERIALS REGULATIONS

On January 24, 1974, the Hazardous Materials Regulations Board (the Board) published two proposals identified as Docket No. HM-103; Notice No. 73-10 (39 FR 3164) and Docket No. HM-112; Notice No. 73-9 (39 FR 3022). HM-103 proposed to adopt a Hazard Information System to enhance the identification and communication of hazards during transportation and made other proposal pertaining to shipping papers, package marking, package labeling, and transport vehicle placarding. HM-112 proposed to consolidate the air, water, and surface transportation hazardous materials regulations of the Department into one volume of the Code of Federal Regulations, 49 CFR Parts 100–189. In addition, HM-112 proposed several other miscellaneous amendments.

The reasons for the proposals under HM-103 and HM-112 were explained in the preamble to these dockets.

Interested persons were invited to participate in these rule-making proceedings. Since the closing date of the comment period for Dockets HM-103 and HM-112, the Department of Transportation has established the Materials Transportation Bureau (the Bureau) which has been delegated the rule-making authority previously exercised by the Hazardous Materials Regulations Board.

In consideration of many comments received on implementing a hazard information system, and to give the public an opportunity to evaluate other hazard information systems, the Board published Notice 73–10, Docket HM-103, on June 25, 1975 (40 FR 26687), terminating its original proposal under that docket establishing a two-digit number to identify and communicate hazards during transportation. The remainder of the docket pertaining to shipping papers, package marking, package labeling, and transport vehicle placarding was retained although modified to some degree based on recommendations received. All comments received that pertain to the elements of the HM-103 Notice retained by the Board have been given full consideration by the Bureau before it decided on the amendments made herein. Of the many comments received, several were not specifically directed to the contents of Docket HM-103 pertaining to the elements retained by the Board. Such comments will be given separate consideration and may be the subject of future rule-making action.

Some of the most significant amendments under HM-103 are the adoption of a uniform vehicle placarding system, a uniform marking system for cargo tanks, portable tanks, and tank cars, and an improved format for shipping paper entries. These amendments and the package labels are the major elements of a system the Board had proposed for better identification of hazards in transportation. Many commenters had objected to the complexity of the placarding system that had been proposed in the Notice. The Bureau reconsidered the matter, believes some of the objections valid and has simplified considerably the implementation of the placarding system.

The Bureau believes that the action taken to place the shipping paper, marking, labeling, and placarding requirements in a separate Part will simplify the effort by the shippers and carriers in locating the pertinent requirements. Comments addressed to the changes retained by the Bureau under Docket HM-103 are discussed under the respective Part designations with references from 49 CFR, 14 CFR, and 46 CFR, parenthetically noted.

All comments received on Docket HM-112 have been given full consideration by the Bureau before a decision was made on the amendments contained herein. Of the many comments received, a number were not specifically directed to the content of Docket HM-112. Such comments will be given separate consideration and may be the subject of future rule-making action. In many cases, amendments have been introduced which are purely editorial in nature, such as the substitution of "subchapter" for "chapter", rewording for clarification, etc. In each case where a substantive change is not evident, the amendment has been identified as "no substantive change". In addition, references to rail express shipments and to conditions related to "the emergency" have been deleted. In the case of rail express, this service is no longer available. References to "the emergency" were adopted during the Second World War. These provisions which were temporary relaxations in the regulations are no longer appropriate. These changes have been identified. Another general change which appears in these amendments is in the language identifying authorized carriers to encompass common, contract and private carriers since the safety provisions of these regulations are applicable to all carriers. Also, references are made to §172.330 which provides marking requirements for packagings to improve hazard communication.

Some of the most significant amendments under HM-112 have resulted from the incorporation of the U.S. Coast Guard Regulations, 46 CFR Part 146, and Federal Aviation Regulations, 14 CFR Part 103, into 49 CFR Parts 170–177. In this consolidation, the Bureau has replaced the long-standing acceptance of rail express criteria as the basis for determining the rules applicable to aircraft. In its place, the Bureau has adopted the approach of evaluating commodities in terms of their direct implications to aircraft operations as a means of establishing more realistic standards for the transportation of hazardous materials in air commerce. For consistency, the Bureau has decided to apply this same approach in classifying materials transported by vessel that were previously classified as "Hazardous Articles". In Notice 73–9 there were proposed regulations for the shipment of flammable and combustible liquids aboard vessels (§173.116a) that were not consistent with regulations of these same materials shipped via other modes of transportation. This inconsistency involved the different methods and temperature ranges used in determining flashpoints of flammable and combustible liquids. In order to correct this disparity, another notice (Notice 73–9A) was published under this docket on September 8, 1975, that revised the flammable and combustible liquid definition criteria to be uniform for all modes of transport. A hearing was held on October 1, 1975, and all of the written comments filed in response to the notice supported the change to standardize the definitions. In order to expedite implementation of uniform criteria, the action was transferred to Docket HM-133 and published December 31, 1975 (40 FR 60030) with certain corrections published on February 23, 1976 (41 FR 7497). The amendments published herein reflect those changes except certain matters related to the original proposals made under Docket HM-112 on January 24, 1974.

To address the specific additional requirements for transportation by air or water, the concept of "Other Regulated Materials" ORM-A, B, and C was introduced to include those materials identified as "Hazardous Articles" when transported by water, "Other Restricted Articles" when transported by air, and materials corrosive only to aluminum.

On September 8, 1975, an extension of the effective date was published (40 FR 41527) pertaining to materials corrosive only to aluminum. It was stated that disposition of the classification of these materials would be in the amendments made herein. Since these amendments class such materials as ORM-B, no further action under Docket HM-57 is necessary and it is hereby terminated so far as materials corrosive only to aluminum are concerned. HM-112 also proposed an additional category, ORM-D, to provide certain exceptions for limited quantity packages of hazardous materials. Included in this category are those items identified as "Consumer Commodities". It is the Bureau's opinion that the packaging of those hazardous materials available to the general public in retail outlets provides an adequate level of safety in the transportation environment except to a limited degree that requires their identification when carried aboard aircraft. Accordingly, ORM-D materials are not required to be labeled and they are not required to be identified on shipping papers except where offered for transportation aboard aircraft.

In light of the Bureau's conclusions regarding ORM-D materials, the Board's proposed deletion of the marking and labeling exceptions for hazardous materials packaged in limited quantities has been modified. Limited quantities will continue to be excepted from labeling (except when shipped by air) and specification packaging requirements, but they must now be marked in accordance with the marking requirements. However, with regard to poisons, the Bureau believes that positive identification of the material and its associated hazards is essential particularly to preclude the combination in transportation of poisons and foodstuffs. Accordingly, it has been decided to implement the requirement for both marking and labeling of Poison B materials as was proposed in the notice.

Several comments on HM-112 suggested that the Bureau consider the establishment of degrees of hazard within each classification, such as "Highly Corrosive" in comparison to "Corrosive", etc. While the suggestion has merit, the difficulties of establishing parameters for such gradations within a classification and the necessarily arbitrary selection of transition points lead the Bureau to conclude that such proposals require further study and should not be adopted at this time. In addition, Docket HM-112 proposed extensive revisions to Parts 170.171. In particular, rulemaking procedures in Part 170 would have been shifted to Part 171 and general information and regulations in Part 171 would have been redesignated Part 170. In consideration of comments received, as well as the recent replacement of Part 170 in its entirety by a new Part 102, which prescribes procedures applicable to all of the Bureau's rule-making activities (40 FR 31767, July 29, 1975) and a new Part 107, Subpart B, which prescribes exemption procedures applicable to all hazardous materials regulations (40 FR 48466, October 15, 1975), the Bureau has decided to retain the present designation of Part 171 and to delete many of the proposed changes. Comments addressed to the remaining changes are discussed under the respective section designations with the HM-112 Notice designations parenthetically noted.

The basic purpose of these amendments is to achieve editorial clarity and uniformity in existing regulations. Substantive changes are primarily related to matters of identification and classification of certain materials moving in commerce.

In accordance with section 102(2) (c) of the National Environmental Policy Act (Pub L. 91–190, 42 U.S.C. 4321 et seq.) the Bureau has evaluated the environmental impact of these amendments. On the basis of that evaluation, the Bureau has concluded that this regulatory consolidation is not a major Federal action significantly affecting the quality of the human environment and, accordingly, that the preparation of an environmental impact statement is not required.

By finalizing the proceedings in Dockets HM-103 and HM-112 through the issuance of the amendments set forth below, the Materials Transportation Bureau is completing the first major phase of its continuing effort to improve, update, and simplify the regulations governing the shipment and transportation of hazardous materials. The next phase, which the Bureau intends to embark upon as soon as practicable, will be a recodification of the entire body of hazardous materials regulations. Whereas the principal purpose of the HM-103/112 phase has been the consolidation and substantive improvement of specific segments of the regulations, the purpose of recodification phase will be to rearrange and revise the language of the regulations in comprehensible and accessible form to reduce inconsistency, redundancy and obsolescence. In an effort to minimize the number of future recodification language changes to be made in the parts now being restated in their entirety (i.e. Parts 171, 172, 174, 175, and 176), numerous non-substantive changes in terminology and style have been incorporated into those parts.

In the remarks which follow, an attempt has been made to address each section affected by this amendment. The extent of the change from current regulations is summarized, any modification to the changes as proposed in the notices is indicated and the Bureau's response to comments received is noted.

PART 171—GENERAL INFORMATION, REGULATIONS, AND DEFINITIONS

171.1 (HM-112 § 170.1) Revised to set forth purpose and scope of the regulations in subchapter.

171.2 New section setting forth the application of Department of Transportation Regulations to the transportation of hazardous materials.

171.6 (HM-112, § 170.6) The proposed amendments to the requirements and conditions for special permits have been deleted in view of the revocation of § 171.6 by the exemption procedures prescribed in new Part 107 published under Docket HM-127 in the FEDERAL REGISTER on October 15, 1975 (40 FR 48466).

171.7 (HM-112, § 170.7) Additional matter incorporated by reference. The incorporations of the IATA and IMCO material was not proposed but is being adopted in this amendment because the Bureau considers these materials pertinent. Several comments recommended reference to the current edition of any standard to avoid obsolescence. The Bureau recognizes the value of such an approach but considers it impractical since prior review of applicable standards would not be possible and unacceptable requirements would result. Also, recognizes footnotes as regulations.

171.8 (HM-112, § 170.30) A new section of definitions and abbreviations. In response to comment, the Bureau has incorporated in this section all of those definitions which appear to be applicable to one or more modes of transportation. New definitions have been added and others rewritten for clarification.

171.9 Existing 171.9 on Vessel Stores is deleted as unnecessary; rules of construction are adopted under same section designation.

171.11 Paragraph (c) with respect to shipper compliance with Federal Railroad Emergency Orders has been deleted. By petition dated July 16, 1975, the Hazardous Materials Advisory Committee (HMAC) of the Transportation Association of America took exception to new regulations published on December 12, 1974, under Docket HM-123 (39 FR 43310) that require shippers to comply with the provisions of Federal Railroad Emergency Orders as those orders pertain to shippers. Emergency Order No. 5, issued on October 25, 1974, imposed particular handling requirements for certain tank cars and a special notation on shipping papers. Since a provision pertaining to the identification of tank cars has been added to new section 172.203, the Bureau agrees with HMAC that section 171.11(c) and section 173.5(c) (contains language identical to section 171.11) should be deleted.

171.12 (HM-112, §§ 170.12 and 173.9) Clarification of the requirements for international shipments to recognize applicability of the U.S. regulations to import and export shipments as well as to shipments being transferred in a U.S. port area and recognizes certain IMCO provisions. Makes provision for marking packages with specification identifications when they meet the requirements of Part 178. This previously was authorized only for export shipments by § 173.9(c). It should be noted that paragraph (b) applies only to the IMCO class and label, not to the shipping description, marking and packaging of imported or exported hazardous materials. The Bureau believes that these newly adopted provisions will facilitate the movement of materials in international commerce.

171.14, 171.15, and 171.16 (HM-112, §§ 170.14, 170.15, and 170.16) No substantive change.

PART 172—LIST OF HAZARDOUS MATERIALS AND HAZARDOUS MATERIALS COMMUNICATIONS REGULATIONS

Part 172 has been completely revised.

172.1 Sets forth the purpose and scope of Part 172.

172.3 Sets forth the applicability of Part 172.

172.100 (HM-112, § 172.100) Explains the purpose and use of the Hazardous Materials Table. The increased number of people coming into contact with the regulations for the first time has prompted the Department to attempt a more elaborate explanation of the purpose and use of the Hazardous Materials Table. Much difficulty has been experienced by many individuals in attempting to use this table. The evaluation of material descriptions and the selection of the proper shipping name are probably the most common problems encountered. Once the correct class and shipping name has been determined further use of the table is, we believe, quite straightforward.

172.101 (HM-112, § 172.101) The Hazardous Materials Table was proposed with a somewhat modified format from that in current use and the table has been further modified for this amendment based on comments received. The United Nations class and label column which appeared in the proposed table has been deleted in the amended version. The presence of the UN class and label designation in the commodity table was felt to be undesirable since these are not regulatory but only information items. There were several requests for the addition of the UN class and commodity numbers to the table.

We believe that this has merit from the standpoint of preparation of international shipments, however, due to space limitations it was not feasible to include these items in the commodity table. It is anticipated that an auxiliary table containing this information along with some other helpful information may be prepared at a later date. The content of this table, i.e., its regulatory nature, will determine whether it is included as part of the Code or as an addition to the index which is currently published as a separate non-regulatory document. Other comments concerning the table indexed by specific column or material or shipping name description are as follows:

172.101 Column 2. Hazardous material descriptions and proper shipping names. Major comments suggested the addition or deletion of certain entries. IATA submitted a large list of commodity names or descriptions for addition to the table in order that the international tariff and the United States Regulations would be more nearly uniform. It is believed that the addition of a number of these names has merit, however, it was felt that the addition of a specific material to the table with the subsequent packaging and labeling requirements, should not be done without affording the opportunity for public comment on these entries. Additional entries made at this time were very limited and were of a type generally devised to eliminate some confusion that existed concerning the hazard class of a solution or device. Potential additions and deletions to the Table of Hazardous Materials is a subject of continual review. When appropriate, the public will be given an opportunity to comment on such changes. Substantive changes to the requirements pertaining to entries in column 1 are discussed under comments addressed to specific commodities as follows:

Acetaldehyde. Because of its high volatility, low flashpoint, wide flammability limits (4–6 percent in air) and low ignition temperature, is no longer permitted exceptions for limited quantities.

Acetone Cyanohydrin. Due to its low decomposition temperature with evolution of extremely toxic gases is fully regulated regardless of the quantity being shipped and is forbidden on passenger-carrying aircraft.

Acetylene. Due to its inherent instability and subsequent requirements for stabilization for shipment is fully regulated regardless of the quantity being shipped and is forbidden on passenger-carrying aircraft.

Acrylonitrile. In addition to being poison by skin absorption, may polymerize with the evolution of heat and is flammable even in dilute solutions with water. This material is therefore considered sufficiently hazardous that it is completely regulated regardless of the quantity shipped and is forbidden on passenger-carrying aircraft and passenger vessels.

Allyl Alcohol. Previously classified as a Poison B is now classified as a flammable liquid requiring both a flammable liquid and poison label. In keeping with this change, the maximum quantity permitted in one package by air has been adjusted to those normally permitted flammable liquids.

Aluminum Chloride. Has been deleted as a specific entry due to the uncertainty that exists concerning its correct classification. Comments were received to the fact that this material in the dry state is pyrophoric in which case this material would be classed as a flammable solid. Previous comments in HM-57 addressed its skin corrosivity and metal corrosivity. Additional tests have been requested on this material and future consideration will be given to addition of this material to the table.

Aluminum Powder. Is believed to be sufficiently hazardous in a finely divided uncoated form to warrant classification as a flammable solid. This is an asterisk entry, meaning that not all forms are regulated. With the present qualitative definition of a flammable solid, some judgement, based on experience with the material must be used for classification of the powdered material. When a quantitative definition of a flammable solid becomes effective the classification of all flammable solids will be subject to reevaluation.

Ammonium Hydroxide (with not more than 44 percent Ammonia); Ammonia Solution (with 44 percent or more ammonia in water); Ammonia Solution (with not more than 44 percent ammonia in water) See Ammonium hydroxide; and Crude Nitrogen Fertilizer Solution (more than 25.4 p.s.i.g.). In the current 49 CFR these commodities are listed as asterisk entries, without the parenthetical qualifying phrase which describes the concentration of ammonia. In the nonflammable compressed gas class as entries the current "*Aqua ammonia solution containing anhydrous ammonia" has been replaced with a clearer descriptive name (Ammonia solution). The asterisk has been removed, and the qualifying phrase added "with 44 percent or more ammonia in water". According to vapor pressure tables of pure ammonia water solutions, at 44 percent ammonia the vapor pressure exceeds 40 p.s.i.a. at 70° F. which is one definition of a nonflammable compressed gas. Crude nitrogen fertilizer may contain compounds other than ammonia and water which contribute to the vapor pressure of the solution, therefore, the qualifying phrase "(more than 25.4 p.s.i.g.)" was added in place of the asterisk. The definition of these commodities remains unchanged. The introduction of a new entry of ammonium hydroxide in the corrosive class is to identify solutions which it is believed in the past were shipped as corrosive liquids, n.o.s. The minimum concentration of 10 percent was proposed based on past exemptions for mineral acids and alkalies in concentrations less than 10 percent. Corrosivity data establishing the minimum concentration that should be regulated as corrosive has not yet been obtained, therefore the entry for ammonium hydroxide has been made an "asterisk" entry with no minimum concentration specified.

Ammonium Perchlorate. For many years it has been the practice to ship certain grades of ammonium perchlorate as Class A Explosives. Title 46 CFR recognized this hazard and listed ammonium perchlorate of 15 micron particle size or less as Explosive A. There is some disagreement over the particle size at which the explosive characteristics become significant. This disagreement may be based to some extent on the method used to measure the particle size and particle size distribution. The effect of moisture on the sensitivity has not been addressed in either the testing or the classification. The particle size description has therefore been removed in the Hazardous Materials Table and the entry asterisked with a reference to Explosive A. The entry of Ammonium perchlorate with the class listed as oxidizer is also an asterisk entry, referring to its potential as an Explosive Class A, i.e., ammonium perchlorate must be classed either as an Explosive A or an Oxidizer.

Ammonium Permanganate is sufficiently sensitive to initiation that shipment is forbidden in several foreign countries. It is believed that exceptions should no longer be permitted for this material. It is believed that this material is sufficiently hazardous that it should be completely regulated regardless of the quantity shipped and is forbidden in air transport or on passenger vessels.

Ammonium Sulfide. One shipper reports that this material as a 50 to 55 percent solution in water will not only flash but will sustain combustion for an appreciable period of time, depending upon the temperature of the solution. Ammonium sulfide solutions were thus classified as a flammable liquid as an asterisk entry rather than ORM-A as originally proposed.

Amyl Acetate. There are several isomers of this compound, and several technical grades. Several comments were received on whether or not this material should be classed as a flammable or as a combustible liquid because of the wide range of flash points recorded. This problem is not unique with this compound but occurs with a number of other materials and the class was determined from similar reasoning, i.e., the available flash points were reviewed and the majority of flash points were less than 100° F. Amyl acetate was thus classed as a flammable liquid. This does not prevent a shipper with a material of this general description, with a flash point of 100°F. or more, from shipping the material as a combustible liquid n.o.s.

Beryllium Compounds. Toxicity data were submitted on several beryllium compounds showing these material did not meet the definition of a poison according to 49 CFR, therefore an asterisk was added to this entry.

Bromochloro Methane. Toxicity data were submitted on this material indicating no need for regulation, therefore, it was removed from the commodity table.

Butyl Acetate. Comments were received to the effect that the flash point of the several isomers of butyl acetate span the 73° F. range and therefore packaging requirements are different for different isomers. Similar comments were received for amyl acetate and hexaldehyde. The Bureau believes that the actual flash point of the material being shipped should be used to determine its classification and the packaging requirements for this commodity. If the flash point spans the 100° F. point and classification as either flammable liquid or combustible liquid is possible, the material being shipped should be classed and named accordingly. Butyl acetate is listed only as a flammable liquid. If the material being shipped has a flash point greater than 100° F. it should be shipped as combustible liquid, n.o.s., as paint thinner, combustible liquid, or some other entry appropriate for a combustible liquid.

Calcium Carbide. Packaging requirements, based on current industry practices were submitted and were considered by the Bureau to be preferable to the general packaging requirements for a flammable solid n.o.s., thus a new packaging section was added for calcium carbide.

Combustible Liquids. There were numerous comments concerning the proposed reduction in maximum quantity of combustible liquids permitted by air from 55 gallons to 15 gallons. A re-evaluation of this problem, considering the fact that the minimum flash point of a combustible liquid has been specified as 100° F. and also that there are no regulations for combustible liquids in containers less than 110 gallons, has resulted in the Bureau removing any restriction on the maximum quantity of combustible liquids by air.

Divinyl Ether. Due to its high volatility, low flash point, and anesthetic properties, is no longer permitted exceptions for limited quantities.

Ethyl Ether. Exceptions for limited quantities have been deleted for ethyl ether because of its low flash point, low boiling point and anesthetic properties. The Bureau believes it should be forbidden aboard passenger-carrying aircraft.

Ethyl Methyl Ether. Due to its high volatility, low flash point, and anesthetic properties, is no longer permitted exceptions for limited quantities.

Ethyl Nitrite. Due to its high volatility (low boiling point) low flash point, and low explosion temperature (194° F.) is no longer permitted exceptions for limited quantities.

Fluorine. The preamble to HM-112, Notice 73–9 listed fluorine as one of the commodities for which the classification had been changed. The table (172.101) listed fluorine without any change in the class. This amendment makes the change to the nonflammable gas class as intended, however for hazard identification the labels specified are POISON and OXIDIZER. These were felt to be more meaningful than the nonflammable compressed gas label.

Gas Mixtures. Carbon dioxide-oxygen mixtures and helium-oxygen mixtures will be tested when a quantitative definition of an oxidizer gas has been accepted, and the material classed accordingly. Until that time the correct classification of oxygen containing small amounts of carbon dioxide or helium remains the responsibility of the shipper.

Hexamethylene Imine. U.S. Department of Health, Education, and Welfare, Toxic Substances List of 1974 gives an oral LD50 or 33mg./kg. for rats. Therefore, a poison label was added.

Hydrazine. Was proposed to be reclassed as flammable liquid in the notice on the basis that it is a thermally unstable material. For the amendment, hydrazine has been broken into two classes depending on concentration: 64 percent or more is classed as a flammable liquid (concentrations from 64 percent up to 100 percent hydrazine are used as monopropellant explosives) and less than 64 percent is classed as corrosive material and listed as an asterisk entry. The packaging reference remains unchanged and is the same for both classes.

Hydrochloric and Hydrofluoric Acids. In the notice it was proposed that the maximum quantity in one package of either of these acids when carried aboard passenger aircraft be limited to one quart. Experimental evaluation of corrosion rates of dilute solutions of these two acids made subsequent to the notice show corrosion rates of 0.02 to 0.035 inches per week with 7075-T-6 aluminum. Restricting the total amount of acid per package does not provide an adequate level of safety when the number of packages permitted aboard the aircraft is not limited in view of the high corrosion rates of these materials. Therefore the amendment prohibits the shipment of these two acids aboard passenger-carrying aircraft.

Hydrogen Peroxide. In high concentrations is a high energy monopropellant which can decompose with explosive violence under the right stimuli. Therefore this amendment, as did the proposal, does not permit exceptions for limited quantities of hydrogen peroxide in concentrations greater than 52 percent.

Mesityl Oxide. Data submitted by two commenters indicated no need to regulate this material as a poison B, therefore the requirement for a poison label has been deleted.

Monoethyl Amine. Due to its high volatility, low flashpoint and corrosive characteristics, exceptions for limited quantities are no longer permitted for this material.

Nitric Acid. There was some objection to the proposed reclassification of nitric acid as an oxidizer material, on the basis that it has always been classed as a corrosive material. It is believed that the Logan Airport crash, November 1973 (NTSB-AAR-74-16) fully demonstrates the oxidizing characteristics of nitric acid. In addition, future development of the regulations along more technical lines (rather than tradition) for such items as mixed loading charts and selection of cushioning and absorbent materials would best be served by identifying nitric acid in its more concentrated forms as an oxidizer. Since the Bureau has no data indicating at what concentration the oxidization characteristics of this material is significant, it has relied on a technical bulletin (MCA Chemical Safety Data Sheet SD-5), which lists 52 percent HNO_3 as the most dilute commercial grade and the current packaging requirements recognize 40 percent HNO_3 as a suitable breakpoint. Concentrations greater than 40 percent are therefore classed as oxidizer and concentrations 40 percent or less are classed as corrosive. This breakpoint will be re-evaluated when a quantitative definition of an oxidizer is developed.

Nitro-Carbo-Nitrate. A comment requested that a new hazard class "Blasting Agents" be added for this and similar materials. The Bureau believes that this is outside the scope of this rulemaking and will defer consideration of the request to a later date.

Nitro Aniline. The Bureau believes that a movement toward technically correct chemical names with indexing based more on family names or structure rather than alphabetically using prefixes such as, mono, ortho, para, etc., is desirable. The shipping requirements for ortho and para nitroaniline are the same, and there is no apparent basis for distinguishing between these two isomers. The main entry has therefore been changed to "nitroaniline" with alternate shipping names of Orthonitroaniline and paranitroaniline. It is anticipated that the alternate shipping names will eventually be deleted. A similar fate may be anticipated for similar entries. Deletion of the mono prefix is also anticipated at some future date.

Organic Peroxides. In the notice it was proposed to regulate all organic peroxides by name and permit no n.o.s. entries for this class. Packaging requirements were not included in the notice as they were not available at that time, and have not yet been obtained. Therefore, all new entries for organic peroxides have been deleted, except for those published in the HM-112 notice published on April 24, 1974 and the organic peroxides currently listed in the regulations (including n.o.s. entries) have been returned to the hazardous materials table § 172.101.

Organic Phosphate and Organic Phosphorus Compounds. This amendment adds a new entry in a hazardous materials table for organic phosphorus compounds. Though not proposed, this addition became necessary when it was brought to our attention that based on the chemical structure of the compound molecule, such materials as Phorate and Thimet are not technically organic phosphates. These materials have been traditionally treated, regulated, described and shipped as organic phosphates by the majority of the shippers who recognize the extreme hazard involved with these materials and the fact that the emergency response is similar or the same as that for organic phosphates. However, recent incidents occurring with this material in transportation demonstrate that some less knowledgeable shippers have been describing this material as poisonous liquid or solid n.o.s. with the corresponding lesser packaging requirements. The Bureau believes that it was the intention of the regulations to include these compounds as organic phosphates. However, since there seems to be some uncertainty as to the accuracy of this description, the hazardous material table is being modified to accurately describe and specify the requirements for shipment of organic phosphorus compounds and mixtures.

Vinyl Ethyl Ether. As proposed, this amendment no longer permits exceptions for limited quantities of this material due to its high volatility, low flashpoint, and anesthetic properties.

Zirconium Picramate. Wet with at least 20 percent water had been listed as an oxidizing material with corresponding section references. This amendment changes the class of this material to flammable solid. The Bureau believes this more accurately reflects the hazard presented by the material and is more appropriate for mixed loading compatability considerations. However, no change has been made in the packaging requirements or quantity limitation with respect to this material.

172.101 Column 3. Hazard Class. In the transfer of a commodity from one class to another, objection was taken to the use of packaging references in the former class. It is felt that if suitable packaging is referenced that there is no need to change it. Recodification of these regulations in the near future will place the packaging references within the part or subpart assigned that class.

Subdivision of Compressed Gases. Objection was taken to the proposed subdivision of compressed gases into either flammable or nonflammable gases. The proposed deletion of the poison gas class was considered a major error. The identification of compressed gases with poison as an auxiliary hazard and use of the standard poison label was also found objectionable. These objections were partially responsible for the return, in this amendment, to the currently recognized Poison A and Poison B designations and identification system.

Those materials currently identified as poison A materials have been identified in the table as Poison A and are required to carry a Poison Gas label and placard. Other gases that have a lesser poison hazard but which are still considered poison, according to the definition of Poison B, are required to carry the poison label in addition to the flammable or nonflammable gas label unless specifically stated otherwise for a specific gas (see fluorine). When a more quantitative definition of oxidizer material is available, it is possible that some additional changes will be made in the hazard class of certain materials.

Poisonous or Toxic Materials. A number of comments were received concerning the uncertainty between extremely and highly toxic definitions as well as the confusion over the differences between irritants and ORM-A materials. It was suggested that a return to the old nomenclature as something that was currently understood would be appropriate. It was also pointed out that the hazard class and the wording on the label and placard were different. Recently there has been considerable discussion with OSHA and the Consumer Product Safety Commission over the criteria for evaluating the poison hazard of a material. Although the proposed definition was derived after considerable discussions with the National Academy of Sciences and HEW several years ago, the current thinking is toward somewhat modified versions of these definitions. International discussions on this subject (UN and IMCO) have also indicated some need to modify the proposed definitions. It was therefore decided to return to the current definitions of Poison A and Poison B, and Irritant, and leave the definition of ORM-A essentially the same as that of ORA-A as defined by IATA. A notice will be prepared on this subject for public comment as soon as possible. New names for these hazard classes will also be considered at that time. Any material currently listed as Poison A is again listed as Poison A in the Hazardous Materials Table and any material proposed in HM-112 as Extremely Toxic has been restored to its previous classification.

ORM Entries for Other Regulated Materials, Group A; Other Regulated Materials, Group B; Other Regulated Materials, Group C; and Other Regulated Materials, Group D have been deleted from the table and replaced with the simple entry, ORM-A, ORM-B, ORM-C and ORM-D. The quantitative definition of ORM-A has been deleted and will be reconsidered concurrently with the notice on the poison materials definition to be prepared as soon as possible. The definition of ORM-A becomes essentially the same as the definition used by IATA for ORA-A. ORM-D has been limited to consumer commodities in this amendment. However, the ORM-D class may be expanded in the future to include such things as diagnostic materials, chemical kits used for routine chemical analysis, etc.

Assignment of HI Numbers. A number of comments were submitted concerning the assignment of HI numbers to specific commodities. Because the proposal on the HI System has been terminated (see 40 FR 26687, June 21, 1975), these comments are not being addressed in this rulemaking. However, it should be noted that any data submitted has been filed as part of the background data on the appropriate commodity. The number of submissions of actual data was very few. Most comments consisted of qualitative expressions such as " . . . based on our experience of many years in handling this material we believe that the HI number should be 30 rather than 31." Such information is of little value in a data file.

Heat of Dilution Factor. Identification of this potential hazard has been deleted from the hazardous materials data or characterization requirements because of the withdrawal of the Hazard Information System (see 40 FR 26687, June 25, 1975). Thus comments directed toward this factor are no longer pertinent.

172.101 Column 4. Label(s) Required. With reversion to the previous Poison A and Poison B classes, the use of the Poison Gas label is also required and changes have been made in this column to reflect the changes required for this revision. The requirement for all poisons (Poison A and B) to bear the poison label may require double labeling where a hazard with a higher ranking order exists, and thus the requirement for a poison label has been added where current data indicate the material meets the Poison B definition. Compressed gases not currently designated as Poison A materials but which meet the inhalation criteria of a Poison B are required to bear the Poison label in addition to the flammable gas, nonflammable gas or oxidizer label. Phosphorus pentasulfide and sesquisulfide were proposed in the notice as requiring a Flammable Solid, Dangerous When Wet and Poison label. Objection was taken to the use of three labels. The proposed requirement for the poison label has not been adopted based on the description of a "Dangerous When Wet" material as one which when in contact with water liberates a flammable or poison gas in dangerous quantities. Because the Dangerous When Wet label implies a potential poison hazard on exposure of the material to water, the proposed additional requirement for a Poison label is considered unnecessary.

Column 5. Exceptions. The Transportation Safety Act of 1974 (Pub. L. 93–633) defines exemptions in the same manner that we have defined special permits for many years. Therefore the word "exemptions" as it has been used in all previous regulations must be replaced with some alternate wording in order that these two types of relief from the regulations are not confused. The word chosen for use in this table is "exceptions". What formerly were referred to as exempt quantities are now identified as "Limited Quantities", in the specific sections referenced in the Hazardous Materials Table.

In the notice of proposed rulemaking it was proposed that exceptions (exemptions) for limited quantities be deleted for a number of commodities that were listed by name in the regulations. These exceptions are being removed on the basis of flash point, boiling point, stability, and narcotic or anesthetic effects.

Also, exceptions should apply only to those items specifically identified as safe in limited quantities, and should not apply to n.o.s. entries or general descriptions of materials such as "Compound, cleaning . . ." The Bureau intends to continue the evaluation of exceptions for all hazardous materials as soon as possible.

Column 6. Packaging Requirements have been separated from exceptions for limited quantities. It is believed that this makes the section references more meaningful.

Column 7. Maximum quantity in one package aboard aircraft. There were a number of objections to the reduction of combustible liquids aboard aircraft from 55 gallons to 15 gallons. In view of the lack of general restrictions applied to combustible liquids, (except for bulk shipments), any restriction as to maximum quantity of combustible liquids on aircraft, unless specifically designated in § 172.101, has been deleted. Rail express has been deleted from this table as obsolete since rail express no longer exists.

Column 8. Water shipments. Some objection was taken to the loss of additional descriptions of hazardous materials that currently appear in 46 CFR. The Bureau felt that the space required for these descriptions was not justified. These descriptions will be maintained as part of the background files by the Department. It is recommended that those individuals currently holding copies of 46 CFR keep them for future reference to identify more specifically such items as "garbage tankage."

172.200 (§ 173.427(a)) Sets forth the applicability of Subpart C. Several commenters recommended the use of a separate shipping document for listing hazardous materials. Consequently, a Notice of Public Hearing was published in the FEDERAL REGISTER in Volume 39, No. 238, December 10, 1974, Docket No. HM-103; Notice 73–10A for comment on a proposed Hazardous Materials Manifest as a replacement for the shipping paper. Several commenters at the hearing on February 11, 1975, presented objections to the increased cost of preparing a separate Hazardous Materials Manifest, and presented several ways in which the shipping paper could be prepared mechanically to meet the requirements. The Bureau agrees with these comments and has not adopted the proposal to require a Hazardous Materials Manifest.

172.201(a) (1) (HM-103, § 172.201) Prescribes the procedure for listing hazardous materials on a shipping paper when materials not classed as hazardous are also listed. Based on the comments received and the presentations at the February 11, 1975 hearing, the Bureau reconsidered the proposed requirement and provided for alternate methods of listing the hazardous materials; either list the hazardous materials first, or highlight hazardous materials entries by entering them in a color which contrasts with all other entries on the shipping paper, or place an "X" in a column with a colored border placed before the proper shipping name. The Bureau believes this will permit expeditious location of hazardous materials entries and still permit sequential listings for planned load discharging. The basis for this requirement was discussed extensively in the notice and in an advance notice of proposed rulemaking published on June 27, 1972 (37 FR 12660).

172.201(a)(2) (HM-103, § 172.201) Provides for shipping paper entries to be legible and to be printed (manually or mechanically) in English. Several commenters suggested that the requirement not exclude manually printed entries. The Bureau agrees and has made such provisions.

172.201(a) (3) (HM-103, § 172.201) Provides for shipping paper entries to contain no codes or abbreviations. Several commenters suggested that codes improve shipping paper entries. The Bureau does not agree. In accidents involving hazardous materials in transportation, emergency response personnel must be able to readily identify the hazards involved and at times no means other than the shipping paper is available. The Department has many examples of unreadable codes on shipping papers and examples where the reverse side of a shipping paper which contained the code explanation was not reproduced.

172.201(a) (4) (§ 173.427(a); HM-103, § 172.201(b)) Permits inclusion on the shipping paper of additional information concerning the material provided the information is not inconsistent with the required description. Several commenters recommended deleting the word "shipping" following the word "additional" in the notice. The Bureau agrees and provides accordingly.

172.202 (§ 173.427(a); HM-103, § 172.202) Provides for the required description of hazardous materials on the shipping paper. Changes to the existing rule are minor in that the classification is not required for an n.o.s. entry when, except for the authorized abbreviation "n.o.s." the proper shipping name and the classification are identical. Also, the sequence of the entries for a proper shipping name and classification is prescribed thereby precluding use of code references or split entries. Several commenters recommended that the requirements in the notice pertaining to the size of the print in the classification be deleted as being too restrictive; others recommended the word "immediately" in § 172.202(c) of the notice be deleted as being too restrictive. The Bureau agrees to these requested deletions and these amendments reflect that agreement.

172.203 (§ 173.427; HM-103, § 172.203) Sets forth requirements for shipping paper entries in addition to those required in § 172.202.

172.203(a) (§ 171.6(a) (2); HM-103, § 172.203(a)) Provides for identifying on the shipping paper a shipment made under a DOT exemption by entry of the exemption number in association with the proper shipping name and classification. No substantive change.

172.203(b) Provides for identifying a "Limited Quantities" shipment on the shipping paper. This was not in the notice, however, the Bureau considers it to be a minor requirement, yet one that is needed to identify shipments of hazardous materials that present a limited hazard in transportation because of the "form" or "quantity" in which they are packaged. Because of the limited hazard, the Bureau has excepted these materials from specification packaging and labeling requirements except when offered for transportation by air. Also, the proposal would have required the shipping paper to contain the entry "No label required" under certain conditions. The use of the term "limited quantities" replaces the requirement for the shipping paper entry "No label required" which has been misleading in the past so far as air transportation is concerned.

172.203(c) (§ 173.427(a) (1); HM-103, § 172.203(g)) Prescribes the "Blasting caps" entry on the shipping paper. No substantive change.

172.203(d) Provides for identifying on the shipping paper the placarding exception for corrosive materials that are not corrosive to skin.

172.203(e) (HM-103, § 172.203(j)) Provides for the required entry of the word "EMPTY" on the shipping paper for certain packagings and permits the optional entry of the word "EMPTY" for other packagings. Several commenters recommended this optional entry be deleted or be changed to read "EMPTY—Last Contained—". The Bureau agrees that the suggested wording is appropriate and permits its use as an alternate to the word "EMPTY".

172.203(f) (FAR § 103.3; HM-103, § 172.203(c)) Prescribes the "Cargo-only Aircraft" entry for air shipments not authorized aboard passenger-carrying aircraft. No substantive change.

172.203(g) (§ 174.584; HM-103, § 172.203(d)) Prescribes additional shipping paper entries applicable to rail shipments.

172.203(g) (1) and (2) (§ 174.584; HM-103, § 172.203(d)) Provides for entering the placard notation and endorsement on the shipping paper. Several commenters suggested that the placard notation requirements be deleted as being redundant and unnecessary. The Bureau believes that this notation must be retained to permit carrier personnel to replace missing or destroyed placards to enhance safety in transportation. No substantive change.

172.203(g) (3) (§ 174.584(g)) Provides for the entry "EMPTY" on shipping papers for certain rail shipments. No substantive change.

172.203(g) (4) (§ § 171.11, 173.5; HM-123) Provides for a handling precaution entry on the shipping paper for DOT 112A and 114A tank cars without head shields that contain a flammable compressed gas. This was not in the notice; however, the Bureau believes this to be the appropriate action to incorporate the provisions of FRA E.O. No. 5 into the shipper part of the regulations as recommended in a petition for reconsideration to regulations adopted under Docket HM-123 (39 FR 43310).

172.203 (h) (§ 173.427 and HM-113) Provides for additional shipping paper entries for certain shipments by highway in Specification MC 330 and MC 331 cargo tanks. No substantive change.

172.203(h) (1) (§ 173.427(a) (3) and HM-113) Provides for an additional shipping paper entry for anhydrous ammonia in Specification MC 330 and MC 331 cargo tanks consistent with new requirements adopted under HM-113 (40 FR 24902).

172.203(h) (2) (§ 173.427(a) (4); HM-103, § 172.203(i)) Provides for an additional shipping paper entry for liquefied petroleum gas in Specification MC 330 and MC 331 cargo tanks. No substantive change.

172.203(i) (1) (46 CFR § § 146.05–12(f), 146.05–13(e) and 146.06–20; HM-103, § 172.202(a) (4) The Bureau has added in this section certain additional shipping paper requirements for domestic and export shipments by water consistent with presently prescribed requirements in the 46 CFR references.

172.203(i) (2) (46 CFR § 146.05-12(f) (5); HM-103, § 172.203(e)) Provides for adding the technical name to the shipping paper for export water shipments of certain materials described by an n.o.s. entry in § 172.101 as previously required in § 146.05-12. To insure uniformity of shipping paper requirements for export shipments, the words "chemical name" have been changed to "technical name".

172.203(j) (§ 173.427(a) (5); HM-103, § 172.203(k)) Prescribes additional shipping paper entries for shipments of radioactive materials. The more significant changes from the existing rule are the requirements for identifying the transport index for each package bearing RADIOACTIVE YELLOW-II or RADIOACTIVE YELLOW-III labels; for identifying the fissile class or stating "Fissile Exempt" as appropriate, and the authorization for the use of packagings approved by the U.S. Nuclear Regulatory Commission or the Energy Research and Development Administration and under certain conditions, by the applicable International Atomic Energy Agency Certificate of Competent Authority.

172.204(a) (§ 173.430; HM-103, § 172.204(d) (2)) Provides for manual or mechanical signing of the certification. Not a new requirement, but a clarification of the method of signing. No substantive change.

172.204(b) (§ 173.430(c); HM-103, § 172.204(a)) Provides for certification exceptions for shipments by highway in a cargo tank supplied by the carrier, and by highway during transportation by a private carrier. One commenter recommended paragraph (b) (2) of § 172.204 be reworded for clarity. The Bureau agrees and has reworded the paragraph essentially as it was in 49 CFR, 173.430(c).

172.204(c) (§ 173.430(b); FAR 103.3(e); HM-103, § 172.204; HM-112, § 175.30)) Provides for certification for shipments by air. One significant change from the existing rule pertains to the requirements for two copies of the certification, and another pertains to air shipment of radioactive materials as mandated by § 108 of the Hazardous Materials Transportation Act (Title I of Pub. L. 93-633, January 3, 1975).

172.204(c) (1) (HM-103, 172.204(c)) Provides for alternative wording of the certification for hazardous materials shipped by air including a portion of the IATA certificate. This exact language was not in the notice since IATA has since revised the language of its certificate. While certain portions of the IATA certificate have not been included as part of the alternate certificate, the Bureau believes the objectives of the certification requirement will be achieved through use of the portions adopted.

172.204(c) (2) (14 CFR 103.3(b); HM-112, § 175.30) Prescribes the requirement for executing in duplicate the certification for air shipments consistent with existing requirements.

172.204(c) (3) (HM-103, § 172.204(b) (1)) Provides for additional certification for shipments by passenger and cargo-only aircraft. Existing § 173.430(b) (1) required this added certification for passenger aircraft only. One commenter recommended the proposed paragraph be reworded for clarity because the phrase "permitted hazardous material" was confusing. The Bureau agrees and has reworded the paragraph accordingly.

172.204(c) (4) (FAR § 103.3(e); Pub. L. 93-633) Provides for additional certification for radioactive materials authorized aboard passenger-carrying aircraft. This provision was not in the notice, but published as a regulation in the FAR reference on April 17, 1975, to become effective on May 3, 1975, in accordance with the mandate of § 108 of the Hazardous Materials Transportation Act (Title I of Pub. L. 93-633, January 3, 1975).

172.204(d) (§ 173.430(a); HM-103, § 172.204(d)) Specifies who may sign a certification and how the signature may be accomplished. Significant changes from the existing rule provide for the signature to be legible and accomplished manually or mechanically by a "principal, officer, partner, or employee of the shipper or his agent."

172.300 (§ 172.401; HM-103, §§ 172.300, 172.302(a) (1)) Sets forth general requirement for the marking of shipping names on packages. Several commenters recommended that "cylinders" be added to this section. A cylinder is a packaging and need not be individually identified.

172.302 (46 CFR, § 146.05–15(e) (1); HM-103, § 172.302(f)) Provides for marking export water shipments of a material described by an "n.o.s." entry in § 172.101. Several commenters recommended that this requirement be deleted since the Hazard Information (HI) number and package label would provide adequate hazard identification. The proposal concerning HI numbers was withdrawn (see 40 FR 26687, June 25, 1975), and the Bureau believes that the marking requirement from the 46 CFR reference must be retained to enhance safety in transportation in export water shipments. In response to comments, this amendment limits the requirement for marking the technical names of components of mixtures to the "technical name of the hazardous materials giving the mixture its hazardous properties."

172.304 (§ 173.401(a) (1); HM-103, § 172.302(d) and § 172.305(f); HM-101, § 173.31(a) (6)) Prescribes general marking requirements. Significant changes from the existing rule are that the prescribed marking must be durable, in English, and that it may not be near any other marking (such as advertising) that could substantially reduce its effectiveness. A commenter requested clarification on whether a product label was considered to be a "marking or advertisement". It is the Bureau's position that a product label that contains precautionary information consistent with the intent of the DOT regulations would not be precluded by this restriction.

172.306 (§§ 173.401(b) and 177.816; HM-103, § 172.302(e)) Provides for marking the name and address of the consignee on the package and specifies the exceptions to this requirement. One commenter recommended that the exceptions be extended to packages in a freight container. The Bureau agrees, has made such provisions, and has added a clarification of the intent of the provision as pertaining to "carload lot", "truckload lot" and "freight container load".

172.308 (§ 173.400(a); HM-103, § 172.302(c)) Provides for the use of certain abbreviations in markings. No substantive change.

172.310 (HM-111, § 173.401(f)) Provides for additional markings for packages containing radioactive materials. This requirement proposed under Docket HM-111, was published after notice and public comment on December 31, 1974, and became effective on March 31, 1975 (39 FR 45238).

172.312 (§ 173.401(c); HM-103, § 172.302(h)) Provides for markings on certain packages containing liquid hazardous material to indicate the orientation of the inside packaging. One commenter recommended that specification containers 6D, 21P, 37M and 27P containing liquid hazardous materials be excepted from the "This Side Up" marking requirement because the intent of the package orientation marking is to identify the top of a package that has been overpacked. The Bureau agrees and has provided for this exception. This section also provides for the use of arrows to augment the required package orientation marking. This was not in the notice, however, it was recommended by a commenter and the Bureau believes the use of arrows will contribute to safety. To prevent confusion and make the use of arrows effective, this amendment provides that arrows on packages containing hazardous material may not be used for purposes other than to indicate the correct package orientation.

172.316 (HM-103, § 172.302) Provides for marking an outside packaging containing a material classed as ORM by identifying the ORM immediately following or below the proper shipping name. Those classed as ORM-A and C must be marked ORM-A and ORM-C, as appropriate. Those classed as ORM-B must be marked ORM-B unless corrosive only to aluminum when wet, and then must be marked ORM-B—KEEP DRY to indicate the precautions needed to prevent corrosive action. Since ORM-D materials essentially are consumer commodities packaged in limited quantities, the Bureau believes that, as recommended by several commenters, the marking of the proper shipping name and the appropriate ORM are adequate except that those ORM-D packages intended or offered for transportation by air must be marked ORM-D-AIR to indicate they meet the requirements of § 173.6. Although not contained in the notice, this amendment provides for attaching the marking to an ORM package with a tag when circumstances prevent the application of the marking to the package surface.

Also, the Bureau has added a provision authorizing the marking on a package containing a material classed as ORM to be accepted in lieu of the certification required on shipping papers in those instances wherein shipping papers are not required. This was not proposed in the notice, but the Bureau believes this will accomplish the requirements to enhance safety without requiring additional paperwork.

172.326 (§ 173.401(a) (1); HM-103, § 172.305(f)) Provides for marking the proper shipping name on the head and one side of each portable tank when appropriate or on two opposing sides. The Bureau agrees with commenters who recommended one marking be placed on an "operating side" of a portable tank, and has made such provision. Based on one commenter's recommendation, the Bureau reconsidered the proposed definition of portable tanks and has excluded multi-unit-tank-car-tanks. The most significant change from existing rules is the requirement that the marked name of contents on each portable tank must properly identify the actual hazardous material the portable tank contains.

172.328 (§ 177.823(b); HM-103, § 172.308) Provides for markings on cargo tanks. The most significant change from existing rules is that for gases, the marking must be the proper shipping name or an appropriate common name.

172.330 (HM-101, § 173.31(a) (6); HM-103, § 172.310) Provides for markings on tank cars including the marking visibility and the location of the marking on both sides of the tank car near the stencilled DOT specification marking. Also requires that markings indicate the actual hazardous material the tank car contains. The most significant changes from existing rules are the provisions indicated above and the fact that tank car marking requirements are implemented by references in Part 173 of the regulations as originally proposed under Docket HM-101 (37 FR 7104).

172.400 (§ 172.402(b); HM-103, § § 172.400(a) and 172.402) Provides for general labeling requirements, and exceptions thereto. The most significant changes to existing requirements are: (a) The establishment of a maximum size for freight containers wherein labels are required instead of placards; (b) the exclusion of packages containing materials classed as ORM from labeling; and (c) modification of the labeling prohibition in existing § 173.404(b).

In accordance with the CGA (Compressed Gas Association) proposal in response to the notice, the Bureau has adopted in this section a change to CGA Pamphlet C-7, Appendix A ("A Guide for the Preparation of Precautionary Markings for Compressed Gas Containers") which will identify one additional hazard when appropriate for a nonflammable or a flammable compressed gas. Adoption of the amended standard was proposed in Notice 73–10A (39 FR 43091). The standard has since been revised to remove references to hazard information numbers. Commenters recommended that a hazardous material classed as a "Poisonous compressed gas" be authorized under the provisions of the CGA Pamphlet C-7, Appendix A. This recommendation is not adopted because the Bureau believes that materials classed as Poison A must be labeled because of the special requirements applicable to them. Other comments recommended that packages in freight containers and palletized loads be excepted from labeling requirements. The Bureau believes that the hazards must be identified for proper handling and as an assistance to emergency response personnel during incidents involving hazardous materials in transportation. This was discussed in the preamble to Docket No. HM-28 (Removal of Label Exemptions) and the following language from the preamble is considered appropriate: "In the common carrier sector, even when truckload lots are involved, it is essential that a uniform and consistent labeling system be maintained in order to assure its maximum effectiveness. Also, there are occurrences which necessitate the 'breaking up' of truckload lots even though not intended at the time of shipment." (See 36 FR 9068, May 19, 1971.) The Bureau believes the same thinking applies to palletized loads and freight containers.

172.401 (§ § 173.404 and 173.86; HM-103, § 172.401) Specifies labeling that is prohibited by the regulations in this subchapter and provides for the use of certain labeling on packages in import and export shipments. The Bureau agrees with the commenters who recommended that proposed § 172.401(a) be clarified to permit the shipment of packages of hazardous and nonregulated materials (unless otherwise prohibited) in the same outside packaging, and has reworded the paragraph accordingly. Several commenters discussed the problem a carrier could encounter in determining whether or not a labeled package offered for transportation contained regulated material for which the label was appropriate. The intent of the regulation is to prevent a carrier from "knowingly" accepting for transportation an improperly labeled package. The carrier is not expected to open a package to inspect its contents. The certification of the shipper is to be accepted unless inspection or other information leads the carrier to believe the labeling may be in error. The Bureau agrees with the commenters who recommended that proposed § 172.401 be reworded to better identify the exceptions therein and has rewritten the section accordingly. The most significant change from existing rules is the labeling exception for a package containing a sample of a material being shipped to a laboratory for testing to determine its hazard characteristics.

172.402 (§ § 173.402 and 173.388; HM-103, § 172.402) Provides additional labeling requirements for specific materials including those offered for air transport authorized only on cargo aircraft. The most significant new requirements apply to the labeling of a material meeting the definitions of more than one class, one of which is Poison B, and the labeling of packages containing samples for laboratory examination. The multiple labeling is adopted as proposed in the notice.

The Bureau agrees with the commenter's recommendation that specific requirements be placed in this section for the CARGO AIRCRAFT ONLY and MAGNETIZED MATERIAL labels that have been used in the past, but not required by 49 CFR. Several commenters recommended deletion of the requirement for multiple labeling which they considered unnecessary with the use of the Hazard Information (HI) numbers to identify additional hazards. Since the HI numbers have been withdrawn from the HM-103 proposal, the Bureau believes it is essential to use labels to identify certain significant hazards in transportation. Because of the significance of the potential hazard in transporting poisons with foodstuffs, the Bureau believes that a poison that is an additional hazard must be identified, and has made such provisions based on a commenter's recommendations. One commenter recommended that provisions be made to cover the possibility of a shipper accidentally "overlabeling" a sample being sent to a laboratory for classification. This amendment makes such provisions because the Bureau believes that when a material must be sent to a laboratory for classification the shipper must classify it to the best of his ability based on the defining criteria in this subchapter, the hazard precedence specified in § 173.2 of this subchapter, and his knowledge of the material. It then must be properly packaged, marked, and labeled as required for the hazards determined by the shipper. Subsequent laboratory tests of the sample may or may not be in agreement with the shipper's initial determination of hazard classification.

172.403 (§ § 173.399, 173.402(a) (10); HM-111; § 173.416; HM-103, § 172.403) Prescribes specific labeling requirements for radioactive materials. One commenter mistakenly interpreted the language in this section and confused RADIOACTIVE YELLOW-II with Fissile Class II Category. No substantive change.

172.404 (§ 173.403; HM-103, § 172.404) Provides for label requirements on mixed packing. The Bureau agrees with commenters' recommendations and has inserted the word "hazardous" to indicate the intent of the regulation that the labeling on the outside packaging must identify the hazards contained therein. Also, the Bureau substituted the term "outside container" for "outside packaging" for consistency with the definitions. The wording of this section was changed by the Bureau from that in the notice to allow for the withdrawal of the Hazard Information numbers and to be more specific in identifying the pertinent references in Part 173.

172.405 Provides for permissive substitution for the word "OXYGEN" for "OXIDIZER" on the label for oxygen.

172.406 (§ 173.404(a), § 173.402(a); HM-103, § 172.405) Provides for the label placement requirements on packages and for exceptions thereto. The more significant changes to existing rules are: (a) The requirement to place the label near the marked name of contents; (b) the optional use of a tag under certain conditions to attach labels; (c) the requirement to place multiple labels next to each other; (d) the requirement for placing at least two labels on certain packages; and (e) the requirement under certain conditions for placing labels on a contrasting color background or having an outer border line on the label.

Several commenters recommended this section be rewritten for clarity particularly concerning (a) the use of a tag on a cylinder; (b) the dotted line border on labels; and (c) affixing labels near the closure on a freight container. The Bureau has rewritten the section accordingly and based on commenter's recommendations the Bureau added provisions for (a) attaching a label by means of a tag to a package having an irregular surface to which a label will not adhere; (b) optional placement of a solid or dotted line outer border on a label; and (c) placement of two of the required labels on packages of radioactive materials. The latter requirement was inadvertently omitted from the notice, however, it requires no more than that already required by existing § 173.402(a) (10).

One commenter suggested that the use of only one label on a package would be less confusing to emergency services personnel. The Bureau believes that for certain significant hazards, two labels identifying those hazards are required so that personnel encountering a spill or fire involving such material will have a better opportunity to make an informed judgment as to actions required.

172.407 (§ 173.404; HM-103, § 172.411 through § 172.450) Prescribes the label specifications. These basically are the labels prescribed in existing § 173.405 through § 173.422 of 49 CFR, and the MAGNETIZED MATERIAL and CARGO AIRCRAFT ONLY labels which are presently used for air shipments. The most significant change from the existing specifications is that the required color for each label must be within the prescribed color tolerances as was proposed in the notice. The Bureau does not agree with the commenter's recommendation that the label color specifications not be mandatory. Part of the hazard identification and communication system is the label color. The color is particularly significant for those who cannot read, and the proper differentiation between colors is significant to those persons that are partially color blind.

A maximum tolerance for each color is allowed while color separation is maintained. This will permit easier compliance while preventing confusion of colors such as red and orange, yellow and orange, and blue and green. Several label manufacturers have forwarded label samples to the Bureau for examination for compliance with requirements as to size, format, and color. The majority of those submitted met the specifications and were within color tolerances. One printing ink manufacturer stated that it was part of his service, at no cost to the customer, to mix the inks to meet the color standard established by the customer.

Several commenters recommended tags be permitted to attaching labels to cylinders. This amendment adopts that recommendation. The Bureau agrees with a commenter's recommendation that the outer border of a label be a solid line or a dotted line when an outer border is required, and has made such provisions in § 172.406(d).

172.411 through 172.450 (§ 173.405 through § 173.422 and § 172.119(i); HM-103, § 172.411 through § 172.450) Specifies the use of sixteen labels and references the requirements for additional labels such as the bung label. The most significant additions to the existing labeling requirements are the MAGNETIZED MATERIAL and CARGO AIRCRAFT ONLY labels. These labels which are adopted by DOT are presently used for air shipments but were not contained in 49 CFR. The EMPTY label specification was not in the notice, but was in existing § 173.420 and the Bureau placed it in this amendment to continue its use as has been provided for in existing § 173.29. The specifications pertaining to the minimum size, shape, content, design, and color of each label remains as presently required in 49 CFR.

172.500(a) (§ § 177.823(a), 174.547; HM-103, § 172.500(a)) Sets forth the applicability of Subpart F which pertains to the placarding of motor vehicles, rail cars, portable tanks and freight containers.

172.500(b) (HM-103, § 172.500(b)) Sets forth exceptions to the placarding requirements, namely, that the provisions of Subpart F do not apply to materials classed as etiologic agents or as ORM-A, B, C, or D.

172.502(a) (§ § 177.823(b) (1), 177.823(d); HM-103, § 172.501(a)) Provides that placards may not be displayed for commodities that are not classified as hazardous materials, and that placards which are incorrect for a material may not be displayed. By this amendment placards may now be displayed on a transport vehicle or freight container that contains quantities of hazardous materials excepted from placarding requirements if the placards properly represent the hazardous material contained therein.

172.502(b) (§ 174.546(d); HM-103, § 172.501(a) (2)) Provides that signs or other devices which may be confused with placards specified in this subpart may not be affixed to or displaced on transport vehicles, portable tanks, or freight containers.

172.504(a) (§ § 177.823(a) (1), 174.540 through 174.542; HM-103, § 172.502(a)) Sets forth the required placards for classes of hazardous materials and certain materials by name. Many commenters objected to the complexity of the Hazard Information System proposed in Docket HM-103 and the resultant placarding system which they felt to be so complex as to be unworkable, and confusing to the extent that it would be a detriment to safety. Because the proposal with regard to the Hazard Information System has been withdrawn (see 40 FR 26687, June 25, 1975), this amendment adopts a placarding system without numerical hazard identifiers.

172.504(a) (§ 174.540 to § 174.552, § 177.823(a); HM-103, § 172.502(a) and Table (1)) Tables (1) and (2) specify the placarding required for motor vehicles and rail cars carrying any quantity of certain hazardous materials, including multiple placards for certain radioactive materials. The notes to Table (2) provide, among other things, for alternate placards, under certain conditions, for combustible liquids and flammable solids and the permissive use of the OXYGEN placard. Significant changes from existing rules are the requirements for placarding any quantity of a flammable solid requiring the DANGEROUS WHEN WET label, multiple placards for certain radioactive materials, and for placement of the rail POISON GAS and EXPLOSIVES A placards on a white background.

172.504(b) (HM-103, § 172.502(b)) Provides for the use of the DANGEROUS placard when no more than an aggregate gross weight of 5,000 pounds of one class of material is loaded at one loading facility into a freight container or transport vehicle. The basis for this new requirement is to more specifically identify the potential hazards of materials in mixed loads when more than a specified volume is loaded at one point, while taking into account the potential for confusion should such a requirement be imposed on all LTL pickups and deliveries.

172.504(c) (§ 177.823; HM-103, § 172.502(b)) Provides for a placarding exception in highway transportation and in TOFC or COFC rail service when motor vehicles and freight containers contain less than 1,000 pounds aggregate gross weight of certain materials covered by Table (2).

172.506 (§ 177.823(a); HM-103, § 172.503(a)) Prescribes the responsibility for the placarding of motor vehicles. Many commenters objected to the proposal in Docket HM-103 which requires the person offering the hazardous material to give to the highway carrier the placards required for the material, or to affix such placards if a representative of the highway carrier was not present. Many commenters stated that this requirement would be unworkable in actual practice and presented problems as to responsibility for correct placards, especially with mixed loads. The Bureau agrees and has changed the regulation to require the person who offers the material to give the required placards to the highway carrier and assigns to the highway carrier the responsibility to affix the correct placards. The requirement for the shipper to give a motor carrier the proper placards for the material he offers for transportation is a significant change from existing rules. Motor carriers will continue to be responsible for any placarding required by the aggregation of shipments at their terminals, or different pickup or delivery points. However, the regulation provides an exception if the carrier's vehicle is equipped with the required placards and they are displayed at the time the material is offered for transportation.

172.508 (§§ 174.548 and 174.549; HM-103, § 172.503(b)) Requires the shipper to affix the appropriate placards to each rail car containing the material he has offered for transportation.

172.510(a) (§ 174.540; HM-103, § 172.513(c) Prescribes a square background for the EXPLOSIVES A and POISON GAS rail placards. This allows for better identification of rail cars containing those materials which, by regulation, may not be uncoupled or cut off while in motion and which may not be struck by other cars or coupled with more force than needed to complete coupling.

172.510(b) (§ 173.119(h); HM-103, § 172.502(b)) Requires a DOME placard be affixed on domed tank cars transporting certain flammable liquids.

172.510(c) (§ 174.562; HM-103, § 172.527) Requires the display of EMPTY placards to communicate the potential hazards of so-called empty cars that have not been cleaned or purged or reloaded with a material not subject to the regulations.

172.510(d) (§ 174.579, § 173.426) Requires the FUMIGATION placard on each freight container and transport vehicle offered for rail transportation which contains lading that has been fumigated or treated with poisonous liquid, solid, or gas.

172.512 (§ 174.549, § 177.823; HM-103, § 172.504) Provides for placarding freight containers, including a provision for placarding freight containers having a capacity of less than 640 cubic feet when offered for transportation by air. It takes into account intermodal movements, including those for which the 1,000 pound placarding exception does not apply.

172.514 (§§ 174.541 and 177.823; HM-103, § 172.505) Provides for placarding of cargo tanks and portable tanks. In addition, placarding is required on empty portable tanks having a capacity of 110 gallons or more and cargo tanks unless reloaded with materials not subject to the regulation or sufficiently cleaned of residue or purged of vapor to remove the potential hazard. Not included are so-called "ton tanks" meeting specifications 106 or 110 due to their design and the manner in which they are handled and loaded.

172.516 (§§ 177.823(a), 174.549; HM-103, § 172.508) Sets forth visibility requirements for placards on transport vehicles, portable tanks, and freight containers including horizontal display, and the maintenance of visibility and legibility of placards. Also sets forth methods of attachment of placards. Several commenters discussed possible difficulties that may be encountered in displaying placards under various conditions. In consideration of these comments, the Bureau has provided for the placards to be made of various types of materials and for alternate attachment methods provided the format, color and legibility of the placard is maintained.

172.519 through 172.558 (Part 174, Subpart C; § 177.823; HM-103, §§ 172.511 through 172.550) Sets forth the specifications for each placard. The significant changes to existing specification are: (a) the placards are the same for all modes of transportation; (b) all hazardous material placards have similar dimensions and general appearance; (c) all except the DANGEROUS placards basically are enlargements of the labels; and (d) in addition to the improved format, the OXYGEN, CHLORINE, FLAMMABLE SOLID and ORGANIC PEROXIDE placards are adopted to improve the identification and communication of the hazards of materials in transportation.

PART 173—SHIPPERS

173.1 Clarification, deleting unnecessary language.

173.2 Amended to include a priority listing of hazard classification to be followed in cases involving multiple hazards. As a result of several comments, explosives required to be classed by the Department of Defense have been excepted from the requirements of this section. The listing which appeared in Notice HM-112 has been modified to take into account the comments of several interested parties. It should be further pointed out that this priority listing is intended to direct the shipper's attention to such things as more restrictive packaging requirements and does not purport to classify materials on the basis of hazard potential of unpackaged material.

173.3 Amended to incorporate the provisions of former §§ 173.1, 173.3, 173.4, 173.5 and 173.6.

173.4 Deleted.

173.5 Deleted.

173.6 This section now sets forth the general requirements for shipment of hazardous materials by air. The requirements are deemed necessary to provide adequate safety in the transportation of hazardous materials by air. Many comments were received regarding the proposed provisions. Several comments objected to the 4-foot drop test contending that this did not represent actual transportation conditions. Since this is intended as a bench mark for the determination of packaging integrity, the Bureau does not consider these objections valid, and has retained this requirement. Several comments objected to the outage requirements for liquid containment on the grounds that packagings over 110 gallons capacity could perform satisfactorily with a less restrictive limitation. The Bureau recognizes the merit of these comments and has revised this provision accordingly. Other comments expressed concern that cylinders might not be considered as "packaging." In view of the definition provided in § 171.8 for packaging, which includes cylinders as a type of container, this difficulty should not arise. Editorial changes throughout this section have been made in response to comments received.

173.7 Changed to reference the U.S. Energy Research and Development Administration.

173.9 Deleted as unnecessary; provisions incorporated in § 171.12.

173.10 Deleted as unnecessary in view of §§ 171.15 and 171.16.

173.18 Proposed § 173.18 has been deleted because in certain respects it is unnecessary and, in its entirety, could be considered in conflict with § 175.30(b) which requires that aircraft operators inspect packages containing hazardous materials.

173.21 Paragraph (a) has been amended for clarification, and paragraph (b) adds a requirement regarding the forwarding of Bureau of Explosives reports on cigarette lighters to the Department. A comment objected to the designation of the B of E as the examining laboratory. Lacking laboratory facilities, it is the Bureau's position that the Department must rely upon the recognized capabilities of the B of E. The use of other laboratories may be considered in future rule-making actions.

173.22, 173.23 and 173.24 No substantive changes.

173.25 Clarifies the intent of orientation markings, "THIS SIDE UP" or "THIS END UP" to indicate position of closures of inside packages.

173.26 No substantive change. Deletes reference to rail express.

173.27 Pertaining to containers, the limitations on the maximum quantities of hazardous materials permitted in packages offered for transportation by aircraft must conform with the provisions of § 172.101 of this subchapter. Incorporates provisions of 14 CFR.

173.28 Paragraph (n) has been redesignated paragraph (o). A new paragraph (n) has been added to provide for reuse of single-trip packaging for ORM materials and other materials not subject to specification packaging requirements.

173.29 Many comments objected to the revisions proposed in the Notice HM-112 pertaining to shipment of empty packagings having a capacity of 110 gallons or less. The Bureau recognizes the impact which might result from such revisions. Accordingly, the matter will be addressed in a future rule-making action. However, the Bureau has adopted requirements pertaining to empty packagings having a capacity greater than 110 gallons because the Bureau believes there is sufficient quantity of material in an "empty" tank to warrant regulation.

173.30 Revised to include references to the loading and unloading requirements from 46 CFR and 14 CFR for hazardous materials transported aboard vessels or aircarft.

173.31, 173.32, and 173.33 Incorporate provisions of 46 CFR. Several comments suggested changing the word "Testing" to "Retesting." The Bureau considers the word "Qualification" to include testing and retesting requirements and has made the appropriate changes.

173.34 No substantive changes.

173.51 The language of this section formerly indicated a prohibition of transportation of certain explosives by common carriers. Due to the hazards presented by the materials covered by this section, the Bureau takes the position that this prohibition must be applicable to all types of carriers including private and contract carriers.

173.52, 173.54, 173.56, 173.57, 173.58, 173.59 and 173.60 Deletes reference to rail express and clarifies certain marking and labeling provisions.

173.61 No substantive changes.

173.62, 173.63, 173.64 Deletes reference to rail express.

173.65 A change to exemption provisions, deletes authorization for shipment by rail express.

173.66 Deletes obsolete reference to emergency.

173.67, 173.68, and 173.69 Deletion of references to rail express.

173.70, 173.71, 173.72, 173.74, 173.75, 173.76, 173.77 and 173.78 Deletes reference to rail express, and eliminates types of carriage, as discussed earlier in this preamble for § 173.51.

173.79 and 173.80 Deletes reference to rail express.

173.87 No substantive changes.

173.89 and 173.90 Deletes reference to rail express.

173.91 and 173.92 No substantive changes.

173.93 Deletes reference to rail express, adds provisions applicable to cargo-only aircraft from 14 CFR.

173.94 No substantive changes.

173.95 Deletes reference to rail express.

173.101 No substantive changes.

173.103 Deletes reference to rail express.

173.104, 173.105, 173.106, 173.107, 173.108 and 173.109 No substantive changes.

173.110 Deletes reference to rail express.

173.111, 173.112, and 173.114 No substantive changes.

173.115 Paragraph (f) has been deleted in light of the definition provided in § 171.8. The proposed amendment to § 173.115 pertaining to a thermally unstable material is withdrawn pending further study of the defining criteria for such material. However, the Bureau intends to use the provisions of § 173.115 (g) adopted under Docket HM-102 (40 FR 22263) when it determines that a material should be made subject to the regulations due to its thermally unstable properties.

173.116 In Notice HM-112 packagings were required to be provided with sufficient outage to be less than liquid full at 130°F. This reference temperature was selected as the highest likely to be encountered in transportation in the United States. Many comments objected to this reference temperature on the grounds that no recognition was given to insulated tanks, heat sink capacity of large liquid volumes, etc. The Bureau agrees and has limited this requirement to packagings with capacities of 110 gallons or less.

173.116a Deleted. This proposal is withdrawn in light of proposals adopted under Notice 73-9a (40 FR 4537).

173.118 In Notice HM-112, it was proposed in paragraph (a) to change the requirements pertaining to "Exempt" quantities of flammable liquids. In light of changes to these amendments pertaining to the ORM-D class, the Bureau has decided not to proceed with the changes to this section which would require labeling of "exempt" quantities except for shipments offered for transportation by air. However, the Bureau believes that in order to provide proper hazard communication they should not be exempt from marking requirements and has revised the regulation accordingly.

The Bureau believes that materials presenting hazards identified by more than one class should not, in the interests of safety, be allowed the exceptions specified in paragraph (b) of this section, and revised the paragraph accordingly.

173.118a A new section on exception provisions for combustible liquids incorporating requirements from 173.118 and provisions relating to air and water shipments has been added. Many comments were directed toward a "new" requirement for marking tank cars with the proper shipping name. The only such marking requirement is for tank cars containing combustible liquids when exported by vessel.

173.119 Subparagraphs (a) (1), (b) (7), (k) (3), (m) (1), (m) (5) and (m) (8) have been revised to prohibit use of certain packagings for the transportation of flammable liquids aboard aircraft since the Bureau does not consider them suitable for such service. Also, a provision has been added to subparagraph (a) (3) restricting use of drums specified therein to capacities of 5 gallons or less. Paragraph (b) has been revised for the same reasons stated for the revisions to § 173.118(b). Paragraph (1) has been deleted as unnecessary. Paragraph (m) has been amended to recognize the additional hazards of organic peroxides and poison B liquids. Section 173.119(a) and (b) have been amended to reflect this change. Several comments of an editorial nature have been recognized.

173.119a and 173.119b Deleted as unnecessary because of the provisions of § § 173.118 and 173.118a.

173.120 Simplification of requirements applicable to vehicles being transported by rail or highway.

173.121 Deletes reference to rail express. 173.122 References § 172.330, 173.123 No substantive changes. 173.124 References § 172.330.

173.125 Deletes obsolete reference to emergency, and prohibits the use of certain packagings aboard aircraft since the Bureau does not consider them suitable for such service.

173.126 Deletes references to rail express. 173.127 No substantive changes.

173.128 Incorporates requirements of 46 CFR and prohibits the use of certain packaging aboard aircraft since the Bureau does not consider them suitable for such service. Deletes obsolete reference to an emergency and removes marking and labeling exceptions. Many comments objected to the deletion of marking and labeling exceptions on the basis of safety experience over the past several years. Notwithstanding, in view of the necessity for an improvement in the communication of hazard information, the Bureau believes that these marking and labeling requirements are essential.

173.129 Prohibits the use of certain packagings aboard aircraft since the Bureau does not consider them suitable for such a service and deletes certain labeling and marking exceptions. Many comments objected to the deletion of marking and labeling exceptions. For the reasons cited in § 173.128, the Bureau believes these deletions necessary.

173.130 Deletes marking exception for refrigerating machines containing a limited amount of flammable liquid to improve hazard information communication. However, in response to comments, the Bureau agrees that retention of the labeling exception is justified, except that this exception is not applicable to transportation by air.

173.132 Incorporates provisions of 14 CFR and prohibits the use of certain packagings aboard aircraft since the Bureau does not consider them suitable for such service and deletes marking and labeling exceptions. See discussion of § 173.128.

173.133 Deletes obsolete reference to an emergency. 173.134 Deletes reference to rail express.

173.135 and 173.136 Incorporate provisions of 14 CFR, and in § 173.138 includes diethyl dichlorosilane based on reclassification.

173.137 No substantive change. 173.138 Deletes reference to rail express. 173.139 References § 172.330. 173.140 and 173.141 No substantive changes. 173.143 Deletes reference to rail express. 173.144 See discussion for § 173.129. 173.145 Prohibits the use of certain packagings aboard aircraft since the Bureau does not consider them suitable for such use. 173.146 No substantive change.

173.147 Deletes marking exception for limited quantities of methyl vinyl ketone to improve hazard information communication, however, in response to comments, the Bureau agrees that retention of labeling exception is justified, except that this exception is not applicable to transportation by air.

173.148 and 137.149 No substantive change.

173.149a Adds packaging provisions for shipments of nitromethane to recognize addition to Hazardous Materials Table.

173.149b This proposed section withdrawn since the Bureau believes formaldehyde to be properly classed as ORM-A.

173.150 No substantive change.

173.151 and 173.151a These provide a separate classification and definition for an oxidizer and an organic peroxide to recognize different hazard characteristics. Several comments were directed toward the definition of an oxidizer in comparison to that provided under the Hazard Information System in HM-103. Because the proposal with respect to the Hazard Information System has been withdrawn from HM-103, these comments are not pertinent. A comment recommended establishment of standard tests for classification of organic peroxides. This suggestion will be addressed in future rulemaking.

173.153 Deletes marking exception for certain quantities of flammable solids, oxidizers and organic peroxides to improve hazard information communication. However, in response to comments, the Bureau agrees that retention of the labeling exception is justified, except that this exception is not applicable to transportation by air.

173.154 No substantive change. 173.154a, 173.155, and 173.158 No substantive changes. 173.159 Deletes reference to rail express, incorporates provisions of 46 CFR.

173.160 and 173.161 No substantive changes.

173.162 Deletes marking exception for charcoal to improve hazard information communication. However, in response to comments, the Bureau agrees that retention of the labeling exception is justified, except that this exception is not applicable to transportation by air. Also deletes reference to rail express. Removes authorization for the use of bags aboard aircraft, since the Bureau does not consider them suitable for such service.

173.163 No substantive changes.

173.164 Addition of "Chromic acid mixture, dry" in response to Manufacturing Chemists Association's petition, and clarification of gross weight restriction.

173.165, and 173.166 No substantive changes. 173.167 Deletes reference to rail express. 173.168 No substantive change. 173.169, 173.170, 173.171, 173.172, and 173.174 Deletes reference to rail express. 173.175 No substantive change.

173.176 Deletes marking exception for matches, strike-on-box, book and card, (when packaged in outside wooden or fiberboard boxes with nonflammable material) to improve hazard information communication. However, the Bureau feels that retention of the labeling exception is justified except that this exception is not applicable to transportation by air. Also deletes reference to rail express. The word "exposed" has been deleted because this material is considered hazardous regardless of whether or not it has been exposed.

173.177 Deletes obsolete reference to an emergency.

173.178 Deleted as no longer necessary and replaced by a new section 173.178 to prescribe packaging requirements for calcium carbide to recognize its addition to the Hazardous Materials Table. Several comments objected to a requirement that calcium carbide be packaged in accordance with § 173.154 citing transportation safety experience over several years with other packagings. The Bureau agrees with these comments and, since the exclusion of moisture is the primary safety consideration in handling calcium carbide, has prescribed water-tight containment as a requirement for transportation.

173.179, 173.180 and 173.181 Deleted as no longer necessary. 173.182 Deletes labeling exception for nitrates to improve hazard information communication.

173.183 and 173.184 No substantive change. 173.185 Deletes reference to rail express.

173.186 Deletes obsolete reference to an emergency and reference to rail express.

173.187 No substantive change. 173.188 Deletes obsolete reference to an emergency. 173.189 No substantive changes.

173.190 References § 172.330, deletes reference to rail express and authorizes air transportation. Incorporates provisions of existing § 173.232 with respect to transportation of tank cars containing residual phosphorous.

173.191, 173.192, 173.193 and 173.194 No substantive changes. 173.195 Editorial, and deletes obsolete reference to an emergency. 173.196 Deletes reference to rail express. 173.197 Deletes reference to rail express. 173.197a and 173.198 No substantive changes. 173.199 and 173.200 Deletes reference to rail express. 173.201, 173.202, and 173.203 No substantive changes.

173.204 Deletes obsolete reference to an emergency. Prohibits the use of certain packagings aboard aircraft since the Bureau does not consider them suitable for such service.

173.205 No substantive changes.

173.206 Adds packaging requirements for several commodities to recognize additions to Hazardous Materials Table, and retains labeling exception for lithium or rubidium metal in cartridges.

173.207 and 173.208 No substantive changes.

173.209, 173.210, and 173.211 Deletes reference to rail express. 173.212 No substantive change. 173.213 Deletes reference to rail express, 173.214 and 173.216 No substantive changes.

173.226 Deletes marking exception for thorium metal, powdered, to improve hazard information communications, however, the Bureau feels the retention of the labeling exception is justified except that this exception is not applicable to transportation by air.

173.227 and 173.228 No substantive changes. 173.229 Deletes marking exception for chlorate and borate mixtures or chlorate and magnesium chloride mixtures to improve hazard information communication, however, the Bureau feels the retention of the labeling exception is justified except that this exception is not applicable to transportation by air.

173.217 Deletes marking exceptions for certain commodities to improve hazard information communication. However, the Bureau feels that retention of labeling exceptions is justified, except that this exception is not applicable to transportation by air.

173.218 and 173.219 No substantive changes.

173.220 Deletes marking exceptions for magnesium or zirconium to improve hazard information communications. However, the Bureau feels the retention of the labeling exceptions is justified, except that this exception is not applicable for transportation by air.

173.221 and 173.222 No substantive changes.

173.223 Deletes marking exception for peracetic acid to improve hazard information communications. However, the Bureau feels the retention of the labeling exception is justified, except that this exception is not applicable to transportation by air.

173.224 No substantive change.

173.225 Adds packaging requirements for phosphorous trisulfide and phosphorous heptasulfide to recognize additions to Hazardous Materials Table.

173.226 Deletes marking exception for thorium metal, powdered, to improve hazard information communications, however, the Bureau feels the retention of the labeling exception is justified except that this exception is not applicable to transportation by air.

173.227 and 173.228 No substantive changes. 173.229 Deletes marking exception for chlorate and borate mixtures or chlorate and magnesium chloride mixtures to improve hazard information communication. However, the Bureau feels that the retention of the labeling exception is justified except that this exception is not applicable to transportation by air.

173.230 and 173.231 No substantive changes.

173.232 Former provisions of this section are incorporated in §173.190. A new §173.232 has been added to provide for the transportation of aluminum powder in recognition of the hazard characteristics of the material.

173.233, 173.234, 173.235 and 173.236 No substantive changes. 173.237 Adds packaging requirements for chloric acid to recognize addition to Hazardous Materials Table. 173.238 Deletes reference to rail express. 173.239 and 173.239a No substantive changes.

173.240 Recognizes deletion of materials which are corrosive only to aluminum. Such materials will be classed as ORM-B and regulated for air transportation only. The Bureau believes that these materials present a hazard limited to air transportation.

173.241 Secifies reference temperature of 130° F. for filling requirement.

173.242 No substantive change. 173.244 Deletes marking exception for certain quantities of corrosive materials to improve hazard information communication. However, in response to comments, the Bureau agrees that the retention of labeling exception is justified except that this exception is not applicable to transportation by air. One comment pointed out the deletion of glass bottles as authorized inside containers for corrosive liquids. The Bureau agrees that these are suitable containers and has modified the section accordingly. Prohibits the use of certain packagings aboard aircraft since the Bureau does not consider them suitable for such service. Several comments contended that the wording of §173.245(b) prohibits the use of plated, clad or lined tanks. The Bureau agrees and has deleted the word "entirely" to permit such use.

173.245a, 173.245b, and 173.246 No substantive changes. 173.247, and 173.248 Prohibits the use of certain packagings aboard aircraft since the Bureau does not consider them suitable for such service.

173.247a No substantive change. 173.249 Deletes marking exception for certain materials to improve hazard information communication. However, in response to comments, the Bureau agrees that the retention of labeling exception is justified except that this exception is not applicable to transportation by air. Several comments objected to the proposed deletion of certain specification tank cars and packagings. The Bureau considers these objections valid and has retained the tank cars as accepted packaging. The packagings deleted by the amendment to this section are now provided for in § 173.245. Specifically identifies certain additional materials with similar hazard characteristics and makes several editorial changes.

173.250 Simplification of requirements for transportation of automobiles and other vehicles equipped with wet storage batteries.

173.251 No substantive change. 173.252 References the marking requirements of § 172.330, prohibits the use of certain packagings aboard aircraft since the Bureau does not consider them suitable for such service. 173.253 Deletes obsolete reference to an emergency.

173.254, and 173.255 No substantive changes. 173.256, 173.257, and 173.258 Prohibits the use of certain packagings aboard aircraft since the Bureau does not consider them suitable for such service.

173.259 No substantive change. 173.260 Removes exception for carload quantities of electric storage batteries (wet) by rail. Experience demonstrated that shipment of these types of batteries in tightly closed rail cars pose an unacceptable hazard.

173.261 Deletes marking exception for fire extinguisher charges which are corrosive to improve hazard information communication. However, the Bureau feels that retention of the labeling exception is justified, except that this exception is not applicable to transportation by air.

173.262 Prohibits the use of certain packagings aboard aircraft since the Bureau does not consider them suitable for such service.

173.263 Deletes obsolete reference to an emergency. Prohibits the use of certain packagings aboard aircraft since the Bureau does not consider them suitable for such service. Deletes marking exceptions for certain corrosive liquids to improve hazard information communications, however, the Bureau feels that retention of the labeling exception is justified, except that this exception is not applicable to transportation by air.

173.264 Deletes obsolete reference to an emergency and also references the marking requirements of § 172.330. Incorporates provisions of 46 CFR.

173.265 Prohibits the use of certain packagings aboard aircraft since the Bureau does not consider them suitable for such service.

173.266 Prohibits the use of certain packagings aboard aircraft since the Bureau does not consider them suitable for such service and references the marking requirements of § 172.330.

173.267 Deletes reference to rail express, and prohibits the use of certain packagings aboard aircraft since the Bureau does not consider them suitable for such service.

173.268 Deletes reference to rail express and references the marking requirements of § 172.330.

173.269 Prohibits the use of certain packagings aboard aircraft since the Bureau does not consider them suitable for such service.

173.270, and 173.271 No substantive changes. 173.272 Deletes marking exception for certain concentrations of sulfuric acid to improve hazard information communication. However, the Bureau feels that retention of labeling exception is justified, except that this exception is not applicable to transportation by air. Prohibits the use of certain packagings aboard aircraft since the Bureau does not consider them suitable for such service.

173.273 References the marking requirements of § 172.330. 173.274, and 173.275 No substantive changes.

173.276 Prohibits the use of certain packagings aboard aircraft since the Bureau does not consider them suitable for such service.

173.277 Prohibits the use of certain packagings aboard aircraft since the Bureau does not consider them suitable for such service. Also deletes marking exceptions for hypochlorite solutions to improve hazard information communication. However, the Bureau feels that retention of labeling exception is justified, except that this exception is not applicable to transportation by air.

173.278 Prohibits the use of certain packagings aboard aircraft since the Bureau does not consider them suitable for such service.

173.279 Deletes marking exception for anisoyl chloride to improve hazard information communications. However, the Bureau feels that the retention of labeling exception is justified, except that this exception is not applicable to transportation by air.

173.280 Deletion of diethyl dichlorosilane to reflect its reclassification from a corrosive to a flammable liquid. The packaging requirements for this material are now provided for in § 173.135. Prohibits the use of certain packagings aboard aircraft since the Bureau does not consider them suitable for such service.

173.281, 173.282, and 173.283 No substantive changes. 173.286 Proposed changes deleted. No substantive change. 173.287, and 173.288 Prohibits the use of certain packagings aboard aircraft since the Bureau does not consider them suitable for such service.

173.289 Deletes authorization of Bureau of Explosives approval. Prohibits the use of certain packagings aboard aircraft since the Bureau does not consider them suitable for such service.

173.290 No substantive change. 173.291 Prohibits the use of certain packagings aboard aircraft since the Bureau does not consider them suitable for such service.

173.292, 173.293 and 173.294 No substantive changes. 173.295 Prohibits the use of certain packagings aboard aircraft since the Bureau does not consider them suitable for such service.

173.296 and 173.297 No substantive changes. 173.298 Prohibits the use of certain packagings aboard aircraft since the Bureau does not consider them suitable for such service.

173.299, 173.299a and 173.300 No substantive changes. 173.301 Proposed changes deleted. No change from current 49 CFR. 173.302 and 173.303 No substantive changes. 173.304 Proposed changes deleted. No substantive change.

173.306 Deletes marking exception for certain compressed gases to improve hazard information communication. However, in response to comments, the Bureau agrees that retention of the labeling exception is justified, except that this exception is not applicable to transportation by air. Also, in response to several comments, the word "receptacle" has been replaced by "container", and in 173.306(d) (2) a provision has been introduced to authorize transportation of vehicles with fuel tank emptied and securely closed. In response to many comments, the maximum capacity for containers excepted from marking and labeling has been increased to 50 cubic inches. The Bureau believes this change does not affect the transportation safety of this material. Other comments directed to this section and concerning the inclusion of pneumatic accumulators as well as a revision of the pressure and volume criteria for determining requirements may be considered in separate rulemaking.

173.307 New section excepting from the requirements of this subchapter, certain compressed gases which pose no hazard during transportation.

173.308 New section setting forth requirements for shipment of cigarette lighters or similar devices charged with fuel to recognize the hazard peculiar to these devices.

173.314 References the marking requirements of § 172.330. 173.315, and 173.316 No substantive changes.

173.325, 173.326, 173.343, and 173.381 Retains definitions that were proposed to be deleted for poisonous gas or liquid, Poison A, poisonous liquids or solid, Poison B and irritating material. In response to comments on this rulemaking, the Bureau has agreed that "Poisonous materials" is more meaningful in communication than is "Toxic materials" and has made this change. In consideration of comments to Docket HM-51, the Bureau had proposed in this notice to introduce a new classification approach for toxic materials. Several comments objected on the grounds that this change would require extensive procedural changes and merits further study. It is the intent of the Bureau to proceed with new rulemaking action on poisonous materials at the earliest possible date. This rulemaking will take into account the comments received and the recent developments in the classification of poisonous materials by other government agencies as well as international organizations. One of the major aims of the future rulemaking will be to propose classifications that will not be in conflict with those of other U.S. or international agencies.

173.327 Specifies reference temperature for 130° F. for filling requirement. Recognition is given for packagings of greater than 110 gallons capacity to be excepted from these requirements as discussed regarding § 173.116. Also references marking requirements of § 172.330.

173.328, 173.329, 173.330, 173.331, and 173.332 No substantive changes. 173.333 Incorporates provisions of 46 CFR. 173.334 Requires approval of cylinder valves by the Department in place of the Bureau of Explosives. 173.335 Obsolete and deleted. The Bureau understands that the police grenades covered by this section are no longer shipped. 173.336 Incorporates provisions of 46 CFR. 173.337 No substantive change. 173.338 Obsolete and deleted. The Bureau understands that the commodities covered by this section are no longer shipped. 173.343 See discussion for 173.325.

173.344 Specifies reference temperature of 130° F. for filling requirement. Recognition is given for packaging of greater than 110 gallons capacity to be excepted from these requirements as discussed regarding § 173.116.

173.345 Deletes marking and labeling exceptions for poisonous liquids to improve hazard information communication. 173.346 Prohibits use of certain packagings aboard aircraft since the Bureau does not consider them suitable for such service.

173.347 Deletes obsolete reference to an emergency. 173.348 and 173.349 Prohibits use of certain packagings aboard aircraft since the Bureau does not consider them suitable for such service. 173.350 No substantive change. 173.351 Adds performance tests for glass packagings to assure adequacy of containment.

173.352 Prohibits use of certain packagings aboard aircraft since the Bureau does not consider them suitable for such service.

173.353 Deletes obsolete reference to an emergency and incorporates provisions of 46 CFR. 173.354, 173.355, and 173.356 No substantive changes. 173.357 Proposed changes deleted. Deletes obsolete reference to an emergency and incorporates provisions of 46 CFR. No other change from current 49 CFR.

173.358 Proposed section 173.358a deleted. See discussion for § 173.326. 173.359 Proposed change from poison to toxic deleted. Marking and labeling exceptions are deleted for certain Poison B materials to improve hazard information communication. Proposed section 173.359a deleted. See discussion for § 173.326.

173.360, 173.361, 173.362, 173.362a, and 173.363 No substantive changes. 173.364 Proposed change from poison to toxic deleted. Marking and labeling exceptions are deleted for certain Poison B materials to improve hazard information communications.

173.365 Prohibits the use of certain packagings aboard aircraft since the Bureau does not consider them suitable for such service. 173.367 Prohibits the use of certain packagings aboard aircraft since the Bureau does not consider them suitable for such service.

173.368 No substantive change. 173.369 and 173.370 Prohibits the use of certain packagings aboard aircraft since the Bureau does not consider them suitable for such service and deletes marking and labeling exceptions for certain Poison B materials to improve hazard information communication.

173.371 Adds reference temperature for determining physical form of dinitrobenzol to establish packaging requirements. 173.372, 173.373 and 173.374 No substantive changes. 173.375 Deletes obsolete reference to an emergency. 173.376 No substantive change.

173.377 Proposed change deleted, except marking and labeling exceptions are deleted for certain Poison B materials to improve hazard information communication, and prohibits the use of certain packagings aboard aircraft since the Bureau does not consider them suitable for such service. Proposed sections 173.377a and 173.377b deleted. See discussion for § 173.326.

173.379 No substantive change. 173.381 See discussion for § 173.326. Also sets forth additional packaging requirements to improve transportation safety. In response to comments, the Bureau does not believe that the proposed requirement for packages to be "Hermetically closed" is more restrictive than the present requirement for packages to be "tightly closed." Therefore, this amendment adopts the proposed language.

173.382 Clarification of packaging requirements for compressed gas mixture with irritating materials, and specifically identifies certain additional materials with similar hazard characteristics and establishes packaging requirements for those materials.

173.384 Deleted as obsolete. The Bureau understands that the material covered by this section is no longer being shipped.

173.385, 173.386, and 173.387 No substantive change. 173.389 through 173.398 No substantive change. 173.399 Deleted. The labeling provisions of this section have been incorporated in new § 172.403. 173.400 through 173.417, and 173.420 through 173.422 Deleted. The provisions of this section have been incorporated in new § 172.423. 173.425 Deleted because the regulations are no longer necessary.

173.426 Editorial clarification. 173.427, 173.430, and 173.431 Deleted. See new § 172.204.

173.432 Add flammable compressed gas to those materials prohibited from being loaded into tank cars from motor vehicles or drums so as to reduce hazards in handling.

173.500 New section. Definitions of ORM-A, B, C and D materials.

173.501 New section on applicability of regulations to "Other Regulated Materials." 173.505 New section. Exceptions for "Other Regulated Materials." 173.510 New section. Sets forth general packaging requirement for "Other Regulated Materials." In response to comments, recognition has been provided for outage for packagings of over 110 gallons capacity to be excepted from these requirements. 173.605 through 173.1085 New sections added to set forth packaging requirements for materials added to the Hazardous Materials Table based on 46 CFR, and individual petitions received by the Bureau. A definition of magnetized material has been added in § 173.1020 in response to comments.

173.1050 In response to comments indicating that photographic flash bulbs do not pose a hazard during transportation, the Bureau has deleted this proposed section from this amendment.

173.1200 A new section setting forth the requirements for ORM-D materials, providing packagings and quantity limitations for certain commodities included in this class.

PART 174—CARRIERS BY RAIL

This part deals with the requirements for rail carriers engaged in the transport of hazardous materials. Comments received are addressed in a manner that will follow the format of Part 174. The section numbers for Part 174 are followed by the former section number from Title 49 in parenthesis.

174.1 (174.500) This section sets forth the purpose and scope of Part 174.

174.3 (174.501) Much of the material in former section 174.501 is covered elsewhere in the revised regulations and therefore has been deleted from this section. Since the material which was not deleted speaks entirely to shipments which are not in proper condition for transport, the heading has been changed to "Unacceptable Hazardous Materials Shipments."

174.5 (174.504) No substantive change. However, the section recognizes exception from shipper's certificate requirements for private carriage by rail.

174.7 (174.500) This section sets forth the responsibility for compliance with the requirements of Part 174.

174.8 (174.501 174.589 174.598) This section incorporates some of the material mentioned as being deleted in connection with § 174.3. It also requires an inspection of all placarded cars at points where trains are required to be inspected. The Bureau believes that this requirement will contribute substantially to safety without placing an unreasonable burden on the rail carriers. Several comments were received recommending that inspection by DOT of "methods of manufacture, packing, and storage" be restricted to those items only which affect safety in transportation. This recommendation has been incorporated in the wording of this section.

174.9 (174.596) No substantive change. Several comments were received relating to the "card". As recommended by these comments, the "card" has been revised to permit showing that either the tank, the safety valve, or both, is overdue for retest and the correct section reference has been inserted.

174.10 (174.598) No substantive change. Several commenters objected to some provisions of this inspection requirement. As adopted, § 174.10 is unchanged from existing § 174.598(a) because the Bureau believes that the provisions of this section are necessary in the interest of safety.

174.11 (174.505) No substantive change. One comment was received requesting clarifying wording to indicate that shipments can be made from Canada if they comply with applicable Canadian Transport Commission Regulations. This has been incorporated.

174.12 (174.576) No substantive change.

174.14 (174.582) No substantive change. The proposed wording "flammable extremely toxic gas" has been deleted and replaced by "poison gas" in accord with the Bureau's response to comments as discussed in regard to § 173.325 of this subpart.

174.16 (174.564) No substantive change. Section 174.16 reads the same as existing § 174.564 since the Bureau believes these provisions are necessary in the interest of safety. 714.18 (174.588) No substantive change. 174.20 (174.575) No substantive change.

174.22 Some of the definitions proposed in this section have been placed in § 171.8 as suggested by several commenters. Other commenters objected to the changed definition of "train". The Bureau has reviewed these objections but is aware that some rail carriers do not use "markers" on trains, with the result that the existing definition of "train" does not have general applicability. In a court decision, it was suggested that the definition of a train be clarified and a definition be developed that would provide for a more uniform understanding and enforcement. This revised definition will define all trains, with or without markers, including "transfer trains."

174.24 (174.510) No substantive change. Rail carriers are not required to tender shipping papers for ORM materials. As suggested by several commenters, the requirement for showing the placement of each placarded car has been deleted from this section and placed in § 174.26 (numbered as § 174.25 in the Notice). Likewise "shipping paper" has been defined by reference to Subpart C of Part 172.

174.25 (174.584) Expanded to take into account the new placards required under Part 172. The requirements relating to an initial switching operation have been expanded because the Bureau believes this is necessary to provide hazard information communication. Proposed § 174.25 has been renumbered § 174.26 and a new § 174.25 has been incorporated to read consistent with existing § 174.584 which was eliminated inadvertently from the Notice. It requires no additional burden and is consistent with an existing regulation. The "Dangerous" placard endorsement requirement for combustible liquid shipments has been removed since placement in the train of such shipments is not specified. In addition, the placard notation has been amended to read "Combustible".

174.26 (174.589) (HM-112, § 174.25) Revised section that includes a notice to train crews regarding "Poison" placarded cars in addition to "Explosives" placarded cars. The Bureau believes that the degree of hazard presented by "Poison" placarded cars is comparable to that presented by cars placarded "EXPLOSIVES A." Deletion of "Notice to Crews" for Class B Explosives has been made. One comment was received regarding the possibility of this section applying to switching. The section indicates in Paragraph (a) that it applies only to trains. Another commenter is concerned that chlorine laden rail cars would be subject to "Notice to Crews" requirements. Deletion of the "HI number" references removes chlorine from this requirement.

Commenters requested that a "Notice to Crews" be used to indicate the location of all placarded cars in a train and that this "Notice" may be used in lieu of a consist. The Bureau believes that a requirement for notification to train crews of the position of any placarded cars is essential for hazard information communication, but that it is not necessary that all of their information be on the "Notice to Crews." The consecutively numbered "Notice to Crews" is reserved for the materials considered to have maximum hazards and the information on the other cars permitted to be on another document such as a train consist.

174.27 (174.511) These provisions now appear in Part 172.

174.33 (174.507, 174.547) No substantive change. One comment was received objecting to the requirement that rail carriers have an adequate supply of labels. This provision currently is in section § 174.507 as is the placard supply requirement existing in § 174.547. The Bureau believes these requirements necessary to ensure hazard information communication. 174.45 (174.506) No substantive change. A commenter requested that reports be made to the Bureau of Explosives. This section does not preclude the submission of these reports to the Bureau of Explosives. 174.47 (174.580, 174.588) No substantive change. 174.49 (174.578) No substantive change.

174.50 (174.594) The revision deletes certain non-regulatory language and makes the provisions of the section applicable to any leaking tank cars since the Bureau believes that these requirements should apply to all hazardous materials rather than a certain few.

174.55 (174.532) Incorporates several new requirements for the handling of hazardous materials in general. The Bureau believes in the interest of safety that some of these requirements which were formerly limited to certain classes of materials must be expanded to cover all classes. Reference to the Bureau of Explosives procedures has been made recommendatory rather than mandatory as proposed in the notice since the Bureau believes that other methods may be equally satisfactory.

174.56 (174.532) No substantive change. This section separately sets forth specific requirements for the loading of drums containing hazardous materials into rail cars. 174.57 (174.566) No substantive change. 174.59 (174.589) No substantive change. As a result of several comments, the proposed requirement to allow movement of rail car without proper placards and car certificates only by obtaining approval from the Department has been deleted to facilitate short movements in wrecking operations. 174.61 (174.533) No substantive change except to include these tanks in COFC service. This section has been amended to authorize these tanks in COFC service only when approved by the Federal Railroad Administrator because the Bureau believes that adequate tie-downs are essential for safe transportation. Several comments were received concerning center-of-gravity greater than 98 inches. As a result, this matter has been deleted and will be revised and incorporated into a test procedure to be used by the Federal Railroad Administrator in evaluating proposed methods of securement. Also, the wording in this section has been revised so as to make its provisions consistent with § 174.61. Bureau of Explosives Pamphlet 6C is referenced as a recommended method, rather than as a mandatory method.

174.67 (174.561) In several places recommendations have been made mandatory because the Bureau believes these requirements are essential to safety in transportation. Several commenters recommended that this section be revised in accordance with a proposal submitted by the Manufacturing Chemists' Association. Because the MCA proposal is considered substantial and was received by the Department after the publication of Docket HM-112, it will be considered in future rulemaking activities.

174.69 (174.562) No substantive change. As recommended by several commenters, the requirement to remove "commodity cards" has been deleted. 174.81 (174.538). No substantive change. Several comments were received requesting that "in transportation" be added to paragraph (a) so as to delineate DOT requirements from those of other agencies. The scope of Parts 170-189 is clearly defined as transportation regulations and such a reference in this section would be duplicative. 174.83 (174.589(d)) No substantive change. Several comments were received concerning the added expense of "shoving to rest" certain tank cars such as those transporting chlorine. The section has been revised to clearly denote the need for such handling for cars placarded "EXPLOSIVES A" and "POISON GAS" as is required by current regulations.

174.84 (174.589(c)) No substantive change. This section is an extension of the provisions in section 174.589(c). Although two comments were received objecting to this section, the Bureau believes that returning these requirements is necessary for adequate safety in switching flat cars loaded with placarded trailers or containers. 174.85 (174.589(e)) No substantive change, except that an editorial change has been made to indicate that it applies only to cars placarded "EXPLOSIVES A."

174.86 (174.589(1)) No substantive change. 174.87 (174.589(m)) Incorporates provisions in proposed §§ 174.87, 174.89, 174.90, 174.93 and 174.95 that deal with placing of placarded cars in passenger and mixed trains. Paragraphs (a) and (a) (1) are expanded to include the placement of all placarded cars except those requiring the COMBUSTIBLE placard. The Bureau believes this necessary in the interest of safety. Paragraph (a) (2) is expanded to include all hazardous materials requiring labels for reasons of hazard information communication. In accordance with one comment, paragraph (a) (2) has been changed to say "requiring" rather than "bearing" so as to clarify the meaning of the requirement.

174.88 (174.589(g)) Several comments were received recommending that the "middle of the block" requirement be deleted. The Bureau agrees and has deleted paragraph (b) (2) in its entirety.

174.89 (174.589(n)) As proposed, this section now prohibits positioning a car placarded "Radioactive Materials" next to any other placarded car (other than one placarded "Combustible" or "Radioactive Materials"), an engine, occupied caboose, or carload of undeveloped film.

174.90 (174.589(h), (k)) No substantive change. 174.91 (174.589(i)) The existing train placement exception for tank cars placarded DANGEROUS when moved in a train consisting entirely of DANGEROUS placarded tank cars has been eliminated since the Bureau believes the use of buffer cars to be essential for safety in transportation.

174.92 (174.589(j)) No substantive change. 174.93 New section which sets forth a requirement that at least one buffer car be placed between an empty placarded tank car and the engine or caboose. Since "empty" tank cars may contain a substantial quantity of product, the Bureau believes that this provision is necessary for adequate safety.

174.95 (174.589) The proposed revisions of this section have been incorporated in § 174.87. 174.100 (174.502) No substantive change. 174.101 (174.526) No substantive change. Numerous comments were received requesting Bureau of Explosives Pamphlets No. 6 and 6A not be made mandatory. This has been done since the Bureau believes that other methods may be equally satisfactory. 174.102 (174.527) No substantive change. 174.103 (174.588) No substantive change. One comment was received recommending that paragraph (c) (2) be amended to read " . . . competent, who has been duly certified by the Department, or is willing . . . " The Bureau does not wish to develop a certification program for distressed explosive shipments, preferring to rely upon the rail carriers and the Bureau of Explosives for proper disposition.

174.104 (174.525) No substantive change however, incorporates amendments published under Docket HM-114. 174.105 (174.581) No substantive change. 174.106 (174.503) No substantive change. 174.107 (174.577) No substantive change. 174.109 (174.585) No substantive change.

174.110 (174.591) Revised to require placarding regardless of the presence of a responsible employee because the Bureau believes that even an occasional absence of the appropriate placard would reduce safety.

174.112 (174.529) No substantive change. One comment was received mentioning that Class B Explosive laden box cars would require either steel floors or spark shields. This is the intent and is incorporated in the final rule. Reference to Bureau of Explosives' Pamphlet Nos. 6 and 6C has been made recommendatory rather than mandatory since the Bureau believes that other methods may be equally satisfactory. 174.114 (174.590) No substantive change.

174.115 (174.530) Deletes the "no placards required" provision from existing rule. Under this amendment, FLAMMABLE placards are required for cars containing Class C explosives. The Bureau believes this placarding is necessary because these materials do present a serious fire hazard in transportation. One comment was received saying that the intent of the section is obscure. We have reviewed this section which was § 174.530 and believe that the intent to specify loading requirements for Class C explosives is clearly stated. Reference to Bureau of Explosives' Pamphlet Nos. 6 and 6C has been changed to recommendatory rather than mandatory since the Bureau believes that other methods may be equally satisfactory.

174.200 (174.532) Revised to apply only to shipments of flammable gases. Special handling requirements for flammable liquids are now set forth in § 174.300 of this amendment.

174.201 (174.532) No substantive change. Several comments were received regarding 106A and 110A tanks being subject to this requirement. In order to accommodate these suggestions, the heading to this section has been changed to "Compressed gas cylinders" eliminating reference to ton multi-unit tank car tanks.

174.204 (174.560) No substantive change except revised to apply only to compressed gases. Requirements for tank cars containing flammable liquids or Poison A materials are now set forth in § 174.304 and § 174.600 respectively.

174.208 (174.579) Incorporates provisions that were proposed in § 174.625. Revised to include cars loaded with truck bodies and trailers. The Bureau believes that these present hazards just as great as those presented by cars when they have been fumigated.

174.280 (174.532) A section prohibiting any package of gaseous material bearing a poison label from being transported in the same vehicle with edible materials.

174.290 (174.532) No substantive change. One comment was received requesting revision of § 174.290(a) (7) to reference Department of Defense procedures and specifications. The reference has been incorporated in this amendment. Existing requirements in § 174.532 (1) (8-13) were inadvertently not included in the Notice, but have been provided for in this amendment.

174.300 (174.532) Sets forth special handling requirements for flammable liquids that were formerly provided in § 174.532. One commenter mentioned the inconsistency that TOFC units can have certain heating and refrigerating devices whereas rail cars may not. The Bureau is aware of this difference but believes no amendment is necessary since, to our knowledge, the type of equipment in question is not utilized in rail service.

174.304 (174.560) Sets forth tank car delivery requirements for flammable liquids that were formerly provided in § 174.560. No substantive change.

174.380 (174.532) A section prohibiting any package of flammable liquid bearing a poison label from being transported in the same vehicle with edible materials.

174.410 (174.532) No substantive change. 174.450 (174.592, 174.593) No substantive change. 174.480 (174.380) A section prohibiting any package of flammable solids bearing a poison label from being transported in the same vehicle with edible materials.

174.510 (174.532) No substantive change. 174.515 (174.566) No substantive change except revised to apply only to potassium permanganate. Requirements for cleaning cars which contained poisons are now set forth in § 174.615 of this amendment.

174.580 (174.532) A section emphasizing that, regardless of class, any package bearing a poison label must not be transported in the same vehicle with edible materials.

174.600 (174.560) Sets forth special handling requirements for Poison A materials (regardless of flash point) that formerly appeared in § 174.560.

174.615 (174.566) Sets forth requirements for the cleaning of cars that contained poisonous materials. These requirements formerly appeared in § 174.566.

174.625 (174.579) The provisions of this proposed section have been incorporated in § 174.208.

174.680 (174.532) Sets forth the general prohibition against transporting any package bearing a poison label in the same vehicle with edible materials. No substantive change.

174.700 (174.532, 174.586) No substantive change. One editorial comment was received and the words "continuously occupied" are now used to clarify the intent of the requirement in paragraph (c) of this section. 174.715 (174.566) No substantive change. One comment was received regarding responsibility for cleaning cars. This comment, considered beyond the scope of the HM-112 Notice, may be the subject of future rulemaking. 174.750 (174.588) No substantive change. 174.800, 174.810, 174.812, (174.532, 174.586) No substantive change. The reasons for consolidating the requirements regarding corrosive materials into these sections were to facilitate use of the regulations. In § 174.810, wet electric storage batteries are prohibited from being loaded in refrigerator cars or "plug-door" type cars because experience has demonstrated that shipment of these type of batteries in such cars poses an unacceptable hazard.

PART 175—CARRIAGE BY AIRCRAFT

This part deals with the carrier requirements for aircraft operators engaged in the transport of hazardous materials which were formerly in 14 CFR Part 103. Comments received are addressed in a manner that will follow the format of Part 175.

One comment objected to the absence of definitions in this part. This objection has handled by consolidating all definitions in § 171.8.

175.1 New section describing the purpose and scope of Part 175. 175.3 New section on the acceptability of hazardous materials shipments for transportation by air.

175.5 Applicability provision transferred from 14 CFR Part 103. One commenter recommended that this section disclaim any applicability of the International Air Transport Association (IATA) Regulations. This comment was not within the scope of the proposed rulemaking and cannot be made part of this rulemaking.

175.10 Sets forth exceptions from this part for certain commodities formerly covered in 14 CFR Parts 103, 121, and 134. Exceptions for radioactive materials formerly in 14 CFR Part 103 are now provided for in § 173.7 of this subchapter.

175.20 Requirements with regard to who must comply with applicable regulations are transferred from 14 CFR Parts 121 and 135.

175.30 Specifies the responsibilities of aircraft operators with respect to accepting hazardous materials shipments for transportation. Requirements are transferred from 14 CFR Part 103. Several comments suggested editorial changes which were adopted. Also, one comment recommended that the amendment be changed to preclude aircraft operators from accepting hazardous materials shipments prepacked by shippers in transport containers. Because this section requires the operator of an aircraft to inspect the "outside container" (see § 171.8 of this amendment) in which a hazardous material is packaged before placing the material aboard an aircraft, the Bureau believes that the concern of the commenter has been answered. However, an express prohibition against tendering prepacked transport containers may be considered in future rulemaking.

175.33 Requirements for notifying pilot-in-command are transferred from 14 CFR Part 103.

175.35 New section setting forth shipping paper requirements for air transportation to provide hazard information communication. 175.40 New section covering labeling maintenance and replacement requirements for aircraft operators so as to maintain hazard information communications.

175.45 Requirements for reporting hazardous materials incidents are transferred from 14 CFR Part 103.

175.75 Quantity limitation requirements are transferred from 14 CFR Part 103. Several comments objected to the proposal that the quantity limitations be changed from 50 pounds net weight to 65 pounds gross weight for hazardous materials loaded in an inaccessible cargo pit or bin and from 150 pounds to 300 pounds for nonflammable compressed gases loaded under the same conditions. The Bureau agrees that gross weight limitation would not clearly identify the amount of hazardous materials that may be contained in a package and that the net weight limitation should be used. The amendment has been changed to reflect these recommendations.

175.78 Stowage compatibility requirements are transferred from 14 CFR Part 103. One comment suggested this section be changed to reference the compatibility-of-cargo requirements for rail and highway currently contained in 49 CFR. This suggestion was not adopted because the Bureau believes that the limitations imposed which require corrosive materials not be stowed in a position that will allow contact with a package of flammable solids, oxidizers, or organic peroxides are sufficient to permit an adequate level of safety.

175.79 New section setting forth loading and storage requirements for packages that are marked for package orientation to minimize the possibility of leakage.

175.85 Cargo location requirements are transferred from 14 CFR Part 103. Three comments objected to the accessibility and safeguarding portions of this proposal. The Bureau believes these requirements to be necessary in the interest of safety. The exception to the accessibility requirement embodied in paragraph (b) of this section was adopted under Docket HM-128 (see 40 FR 58284, December 16, 1975).

175.90 Requirements with respect to damaged or leaking packages are transferred from 14 CFR Part 103.

175.305 (HM-112, § 175.205) New section setting forth requirements for transporting self-propelled vehicles to provide for the safe transportation of those vehicles when their fuel tanks are not completely drained.

175.310 (HM-112, § 175.210; HM-128) Requirements are transferred from 14 CFR Part 103. 175.320 (HM-112, § 175.220; HM-128) Requirements are transferred from 14 CFR Part 103. 175.630 Special requirements for poisons are transferred from 14 CFR Part 103. 175.700 Special requirements on radioactive materials are transferred from 14 CFR Part 103. 175.710 Special requirements for fissile Class III radioactive materials are transferred from 14 CFR Part 103.

PART 176—CARRIAGE BY VESSEL

This part deals with the requirements for water carriers engaged in the transport of hazardous materials which were formerly in 46 CFR Part 146. The section numbers for Part 176 are followed by the section number from Title 46 CFR in parenthesis. Comments received are addressed in a manner that will follow the format of Part 176.

The following sections were deleted and not reproduced in Part 176 because the existing provisions of 49 CFR 100–199 cover their substance:

46 CFR 146.01–1 through 6, 146.01–8 through 11, 146.02–3, 146.02–8, 146.02–14(e), 146.02–18, 146.02–19, 146.02–25, 146.02–30, 146.03–3, 146.03–6, 146.04–1 through 5, 146.05, 146.06–8, 146.07–10(b) and (c), 146.07–15, 146.08–25, 146.08–31, 146.19–1 through 30, 146.19–100, 146.20–1 through 13, 146.20–21(a), (b) and (d), 146.20–100, 146.20–200, 146.20–300, 146.21–1, 146.21–65 through 100, 146.22–1 and 3, 146.22–25, 146.22–100 and 200, 146.23–1, 146.23–30, 146.23–100, 146.24–1 through 25, 146.24–100, 146.25–1 through 15, 146.25–55 and 60, 146.25–100, 200 and 300, 146.26–1, 146.26–100, 146.27–1, 146.27–25, and 146.27–100.

The requirements for military explosives (46 CFR 146.29) are being retained in Title 46 and were not subject to any changes under HM-112.

The provisions of 46 CFR 146.09–7 and 8 were deleted as the Bureau believes the packagings therein are no longer in use.

The following sections are informative in nature rather than regulatory and hence are deleted:

46 CFR 146.02–21, 146.19–80, 146.21–10, 146.24–80, 146.25–70, and 146.26–5.

The following sections were deleted as being no longer necessary with the new format of the regulations:

46 CFR 146.03–1, 146.03–5, 146.03–8 through 10, 146.03–13 through 14, 146.03–16 through 23, 146.03–25 through 31, 146.03–33, 146.03–35, 146.03–37, 146.03–39, 146.06–3, 146.06–4, 146.06–7, 146.07–1(a), 146.07–35, 146.21–15(b), 146.22–5(a), 146.22–35, 146.23–5, 146.23–35, 146.27–5, 146.27–10, and 146.28.

The provisions for the carriage of bulk, solid hazardous materials were moved to 46 CFR Part 148.

The provisions of the following sections are addressed in § 176.83:

46 CFR 146.06–9, 146.10–5, 146.19–40, 146.20–16, 146.22–15, 146.23–25, 146.24–55, 146.25–45, 146.26-20, and 146.27–15.

176.1 (46 CFR 146.02–1) No substantive change. Several comments were received which recommended clarification of the term "vessel." This was achieved in § 171.8, devoted to definitions for this subchapter.

176.3 New section setting forth requirements for acceptability of shipments.

176.5 (46 CFR 146.02-2 146.01-7 146.02-4) No substantive change. One comment recommended provisions be provided to direct interested persons to the regulations appropriate to the carriage of hazardous materials in bulk. An additional paragraph was added to § 176.5 to accommodate this reference. Another comment suggested that specific reference be made to 46 CFR 146.29 for regulations pertinent to the shipment of military explosives. A specific reference was added to § 176.5 to reference these regulations. 176.9 (46 CFR 146.06–5) No substantive change.

176.11 (46 CFR 146.02–9 146.06–6 146.02–10, 146.02–1) Provisions covering packages of combustible liquids formerly proposed in HM-112, § 176.5 are incorporated in this section. Combustible liquids in containers of 110 gallons or less are not subject to these regulations when shipped aboard passenger vessels. Safety concerns do not warrant the regulation of those quantities of combustible liquids as their only hazard is that of additional fuel to an existing fire. A comment was received which suggested a change in wording to insure clarity. This section has been changed to recognize IMCO requirements only. This change was made to greatly increase safety control while facilitating enforcement as well as compliance.

176.13 (46 CFR 146.02–5 146.02–4) Removes reference to shipper compliance as this requirement appears in Part 172.

176.15 (46 CFR 146.02-6, 146.07-30) No substantive change. 176.18 (46 CFR 146.02-6a) No substantive change. 176.22 (46 CFR 146.03-4 146.03-6 146.03-7 146.03-11 146.03-12 146.03-15 146.03-24 146.03-32 146.03-34 146.03-38 146.03-40 146.07-1) No substantive change. See § 171.8 for stowage terms—these terms utilized by IMCO have been adopted. Several comments were received which recommended changes to the definitions as proposed. The definitions have been transferred to § 171.8 as a result of comments received and certain additions, deletions or changes have also been made as a result of the comments received. 176.24 (46 CFR 146.06-1) 146.08-15(a) No substantive change.

176.27 (46 CFR 146.05-11 146.07-20 146.08-20(a) 146.08-20(c)) No substantive change. The "Certificate" statement required by this section is not being repeated in this section; as one comment stated, spelling out the statement would be redundant since Part 172 addresses certification.

176.30 (46 CFR 146.06-12 146.06-15 146.06-20 146.05-14 146.06-14) No substantive change except allows optional use of IMCO classification. In response to comments received the stowage location provided hazardous materials is being added to this section. Including this information on the dangerous cargo manifest is an existing requirement but was inadvertently omitted from the notice. Another comment indicated that the label applied to the hazardous materials should be noted on the dangerous cargo manifest. This is not considered necessary since the classification of the material will identify the label required.

176.31 (46 CFR 146.06-12) No substantive change. 176.33 (46 CFR 146.06-10) No substantive change. 176.36 (46 CFR 146.02-22) No substantive change. 176.39 (46 CFR 146.02-12 146.02-13) Reference to "vapor proof" lighting devices has been changed to "explosion proof" in light of comments received. This section has also been modified as a result of comments to accommodate situations where physical inspection of cargo is not possible. Changed to specify inspection intervals in order to provide better guidance as to intent of regulations and to facilitate safety. Temperature recording requirements dropped as being unnecessary given regular interval inspections. No other substantive changes.

176.45 (46 CFR 146.02-15 146.02-20(c)) No substantive change except for wording in § 176.45(b) reflecting regulations published elsewhere in 46 CFR in regard to pollution. Requirements of § 146.02-15(c) are now contained in § 176.710. One comment suggested that this section prohibit emergency hot work repair except when resulting from incidents involving hazardous materials. Emergency hot work is covered by § 176.54 and the authorization of such work is not dependent upon accidents or incidents involving hazardous materials. Another comment suggested that a marine chemist should be consulted following an incident involving hazardous materials in order that proper procedures be followed. Such a requirement may not always be applicable. The requirement to notify the District Commander is sufficient as a general requirement.

176.48 (46 CFR 146.02-15, 146.02-35, 146.20-51, 146.24-75) No substantive 164.20-51, 146.24-75) No substantive 15(b), 146.20-49(a), 146.24-70) No substantive change. A comment was received that suggested that a marine chemist be required to evaluate all damaged or leaking packages of hazardous materials. Damaged or leaking packages are not permitted for transportation unless repaired to the satisfaction of the master of the vessel. Further specifics in this section are inappropriate. This section does not preclude the vessel's master from seeking appropriate expertise.

176.52 (46 CFR 146.02-16) No substantive change. One comment received on this section requested that twenty-four hours advance notice be provided to the water carrier by the shipper, of the exact nature of the cargo. A requirement of this sort would be unenforcible. This amendment, as was proposed, requires the information on the cargo be supplied at the time of delivery, but does not preclude the shipper and carrier from exchanging necessary information at an earlier time. Another comment recommended that this subpart prohibit shippers from offering hazardous materials under false or deceptive names, and prohibit the carrier from knowingly transporting such material. Part 173, shipper requirements, prohibits the shipper from offering hazardous materials under a false or deceptive name. Since Part 176 is devoted to water carrier requirements this prohibition will not be restated. The word "knowingly" is added to the prohibition on water carriers transporting such materials.

176.54 (46 CFR 146.02-20) No substantive change. Several comments were received stating that the prohibition on burning and welding were restrictive. In consideration of the comments received, this section has been modified to allow repairs involving welding and burning when authorized by the Coast Guard Captain of the Port. Emergency repairs to the vessel's main propulsion plants or auxiliaries will be authorized without prior approval. Another comment suggested that a certified marine chemist be consulted prior to the authorization of "hot work." Since the authorization to conduct "hot work" may be granted by the Coast Guard Captain of the Port, consultation with a certified marine chemist will remain at the discretion of the Captain of the Port.

176.57 (46 CFR 146.02-17, 146.08-15(b), 146.08-20(b)) No substantive change. 176.58 Restrictions previously set forth for explosives in regard to requiring use of "good housekeeping" procedures now apply to all hazardous materials.

176.60 (46 CFR 146.06-11) No substantive change. 176.63 (46 CFR 146.03-34) No substantive change other than adopting IMCO terms. In response to a comment, the "on-deck" stowage situation for hazardous materials, when stowed in a deck house, is being amplified to require that the deck house be vented to the atmosphere by a permanent structural opening such as a door, hatch, companionway, manhole, etc.

176.69 Adopts general requirements for stowage from IMCO regulations. The general stowage requirements, that "under deck" stowage should be used when authorized and that hazardous materials, except those classed as ORM must be stowed in an accessible manner, prompted several comments. It was pointed out that "under deck" stowage is not always desirable for certain hazardous materials and that where "on deck" and "under deck" stowage is authorized the final decision should be left to the master of the vessel. Since the requirement for "under deck" stowage states that this stowage should be used when authorized, the utilization of "on deck" stowage is not prohibited. Several comments indicated that accessible stowage is not always possible on all types of vessels such as container vessels. The requirements to stow all hazardous materials in an accessible manner has been modified to reflect comments received and to recognize specialized vessels.

176.72 Adopts several new handling requirements for hazardous materials designed to help ensure the integrity of the packages during handling. Sets forth general requirements for proper and safe handling of hazardous materials and expands stowage requirements to apply to all classes of hazardous materials. Several comments requested that the proposed requirement to limit deck loads of hazardous materials to 50% of the deck space be deleted. This requirement is being retained since additional "under deck" stowage is being provided hazardous materials, and since the limitation to 50% of the total deck space is an IMCO recommendation. The wording in this subpart has been modified for those portions that refer to firefighting equipment and the securing of break bulk cargo on deck, as a result of comments received. A comment suggested that only flammable hazardous materials be restricted from stowage within twenty-five feet of an operating or embarkation point of a lifeboat. It is felt that all hazardous materials present an increased hazard at the operating and embarkation points to lifeboats, especially during an emergency situation, therefore, this restriction is adopted as proposed.

176.74 (46 CFR 146.19–35(g) 146.20–19 146.27–20 146.25–40 146.24–30 146.22–5(b)) No substantive change (176.74(c) is an IMCO requirement).

176.76 (46 CFR 146.07–1 146.07–5 146.07–40 146.07–25 146.08–1(a) 146.08–10) The regulations have been changed to provide for compliance with IMCO container loading requirements including specific requirements for securing the load within a container and consistent intermodal placarding and marking requirements. One comment suggested alternative separation requirements between "reefer" units and containers of flammable liquids and gases. The separation requirement has been modified as a result of the comments received. A comment suggested that where the lading is contained within an intermodal container or vehicle body only that portion of the lading which consists of explosives or other dangerous materials must be entirely contained within the intermodal container or vehicle body without overhang or projection. The proposal is being modified to accommodate this comment. Several comments pointed out that to require "void" spaces to be filled with dunnage when hazardous materials are loaded in containers and vehicles would require the placement of dunnage in spaces created when barrels or other circular packages are loaded within an intermodal container or vehicle body.

This was not the intent of the proposal and as suggested by the comment the requirement to use dunnage is being applied to "slack spaces" in the load. Many comments recommended that the requirement to stow dry cargoes over liquid cargoes should not be a mandatory requirement since consideration should be given to the density of the particular cargoes involved as well as the packaging. This requirement has been modified to provide for situations where judgment would dictate a different stowage situation. Comments received on the weight limitation for portable tanks suggested that the entry in this section be consistent with a similar requirement in §173.32, and one comment suggested that the maximum gross weight be authorized at 55,000 pounds. The proposal has been modified to incorporate the comments received. Comments received on the carriage of rail cars and highway vehicles by vessels suggested that barges be added as an authorized vessel, however, the proposed definitions for trainships and trailerships would include barges and a specific mention of barges would be unnecessary. The requirement for securing packages within a transport vehicle or intermodal container has been modified to eliminate the need for vertical restraint when shifting of the load has been prevented. Several comments were received which requested the inclusion of Class A Explosives to the requirements governing the transportation of hazardous materials in highway vehicles, railroad vehicles, and intermodal containers.

Class A Explosives were specifically excluded from transportation under the requirements of this section because the existing regulations prohibit the transportation of Class A Explosives in railroad or highway vehicles and intermodal containers without prior approval from the Commandant of the U.S. Coast Guard. The Coast Guard has tasked the Chemical Transportation Industry Advisory Committee with developing regulations for the transportation of Class A Explosives by vessel in railroad or highway vehicles and intermodal containers. These regulations will be issued as a separate rulemaking. One comment requested that the requirement for securing dunnage to the floor of a vehicle or container, when the cargo consists of dense material or heavy packages, be deleted because the requirement was too vague to be enforceable. This requirement is being retained in order to provide guidance with respect to the proper bracing and securing of dense or heavy cargoes within a vehicle or container. The determination of when a cargo is of sufficient density or weight that the dunnage should be secured to the floor will be left to the judgment of the shipper and the Coast Guard Captain of the Port. The guidance provided by this requirement will insure a secure stow for heavy and dense materials. Several wording changes were suggested by comment to this subpart and have been adopted in an effort to maintain clarity.

176.77 New requirements to cover barge carrying vessels. The regulations did not adequately address this form of carriage of hazardous materials as it did not previously exist.

176.78 (46 CFR 146.09-15, 146.20-35, 146.19-70, 146.21-57, 146.22-7, 146.23-13, 146.24-27, 146.25-43, 146.26-35, 146.27-35) No substantive change. Several of the existing sections that have been incorporated in this new section have been modified as they apply to fork-lift trucks as a result of comments received. The definition for an "LP" designated unit has been changed to agree with the definition of a "G" designated unit with the distinction being the fuel utilized. The section pertaining to the load back rest for fork-lift trucks has been reworded for clarity. The safety standard for the forks has been modified to conform closer to a performance standard. The section referring to tire guards has been reworded to eliminate uncertainty. The section referring to the steering mechanism of fork-lift trucks is not being changed. Although a comment was received to modify this section, as proposed, the section adequately addresses the intended steering mechanism requirements. The section pertaining to truck capacity has been modified in order to avoid confusion. In addition, two other sections in this subpart which refer to the requirement of this section for portable fire extinguishers used during fork-lift truck operations have been revised to better identify the size and type extinguisher and the approval required. The approval, however, is not so specific as to preclude fire extinguishers found on foreign vessels as long as the extinguisher is equivalent.

176.79 (46 CFR 146.09) No substantive change. 176.80 Segregation requirements for hazardous materials based on recommended IMCO provisions.

176.83 (46 CFR 146.20-90, 146.19-40, 146.21-30, 146.22-10, 146.22-15) Adopts IMCO stowage criteria except for explosives. The segregation chart in this section has been modified to correct certain inaccuracies which appeared in the proposal and were pointed out by several comments. Comments received on the segregation term "away from" stated that the term needed to be defined by a limiting distance in order to preclude interpretation of this segregation term. A suggestion of eight feet for "away from" separation is not being adopted. The separation required by this term is being defined as ten feet in order to provide consistency with IMCO recommendations and is being further modified when the cargoes requiring this separation are in containers or vehicles. A comment was received that supported the proposal as it relates to the non-application of segregation requirements for ORM materials. That proposal is adopted by this amendment except for specific requirements for certain materials. A comment was received which suggested that the segregation term "Separate by a complete cargo space or hold from" be qualified by a minimum distance when "on deck" stowage is used. This is not considered necessary since this amendment, as proposed, states that when "on deck" stowage is used for cargoes requiring this segregation that a corresponding longitudinal distance is required for separation. Another comment suggested that UN compatibility groupings be used for developing segregation for explosives during transportation. This suggestion may be the subject of a future rule-making as it is outside the scope of this rulemaking.

176.88 (46 CFR 146.08) No substantive change. 176.89 (46 CFR 146.08-40) No substantive change. 176.90 (46 CFR 146.08-45) No substantive change. 176.91 (46 CFR 146.08-45) No substantive change. A comment received on this section suggested that the term "motorboat" requires a definition if the intent of this section is to cover diesel powered craft. This was not the intent. 176.92 (46 CFR 146.08-55). No substantive change. 176.93 (46 CFR 146.08-55 146.08-35(c)) No substantive change. 176.95 (46 CFR 146.10-2 146.10-4) No substantive change. 176.96 (46 CFR 146.10-3) No substantive change. 176.98 (46 CFR 146.10-6) No substantive change. 176.99 (46 CFR 146.10-50 146.10-6) No substantive change. 176.100 (46 CFR 146.20-85 146.20-87) No substantive change except existing 300 pound or less net weight exemption has been deleted to be compatible with 33 CFR 126.17. A comment received indicated that this section would ultimately require four permits for certain movements of Class A Explosives and that the existing regulations require only two in similar circumstances. The proposed regulations are basically the same as the existing requirements and will require one permit for the transportation of Class A Explosives by the water mode except for situations where the specific movement is not authorized by the regulations, then two permits will be required.

176.105 (46 CFR 146.20-31, 146.20-35, 146.20-43, 146.09-12, 146.29-29) No substantive change. 176.110 (46 CFR 146.20-49) No substantive change. 176.115 (46 CFR 146.20-15, 146.20-19) No substantive change. 176.120 (46 CFR 146.20-29) Amended to recognize the use of modern hatch cover designs. 176.125 (46 CFR 146.20-29) No substantive change. A comment to this section stated that this proposal would be too restrictive for the loading of explosives when intermodal containers secured on deck contained only non-regulated cargo. The three foot limitation for deck cargo, over which explosives must pass, is directed to break bulk vessels and not to container vessels. The requirement is based on an existing requirement and is being retained, however, the requirement is being reworded to specifically identify the break bulk vessel in order to avoid interpretation.

Another comment suggested that the "three feet" be deleted and that the height of the hatch coaming or bulwark railing be used as the determining factors for deck loads over which explosives must pass. The requirement is being retained as proposed since the height of the deck cargo is not being restricted to three feet when the height of the hatch coaming or bulwark is higher. However, a limitation of three feet is required when the height of the hatch coaming or bulwark is less than one foot.

176.130 (46 CFR 146.20-37) Comments to this section suggest that vertical restraint should not be required when the shape of the packages, stuffing pattern and bracing prevent shifting of the load. It was not proposed to make a substantive change to this section. However, based on the comments received and satisfactory experience gained, this section has been modified to eliminate vertical restraint requirements where shifting of the load is prevented.

176.135 (46 CFR 146.09-1) No substantive change. 176.138 (46 CFR 146.09-2 146.20-33) No substantive change. Comments to this section suggested that the construction of magazines be based on IMCO requirements. The suggestion is valid and the Coast Guard is considering the incorporation of IMCO recommendations for the transportation of explosives via vessels in a future rulemaking.

176.141 (46 CFR 146.09-3) No substantive change. A comment addressing § 176.138 and this section suggested that the sheathing requirement and the decking requirement be eliminated since the requirement is very costly and may actually serve no real purpose. The requirements are being retained since they are based on safety considerations and elimination cannot be accomplished while providing an equivalent alternative. The comment cited, as example, the military ammunition ships which do not use wood sheathing or decking, however, these vessels are specially designed vessels for the transportation of military explosives. The incorporation of nonsparking metal dunnage as used on the military ammunition ships would be far costlier than the requirements presently required for commercial vessels. Changes to these sections will be a matter of future rule-making.

176.144 (46 CFR 146.09-4) No substantive change. Comments to this section suggest that the ventilation of magazines can better be accomplished by eliminating the sheathing requirements and incorporating the recommendations of IMCO. This is an existing requirement and is being retained, however, the incorporation of IMCO recommendations for explosives may be a matter of a future rulemaking. 176.147 (46 CFR 146.09-5) No substantive change.

176.150 (46 CFR 146.09-6) No substantive change. Comments suggested clarification of the size limitation placed on portable magazines. The proposed portable magazine size specification limited the stowage of explosives in a portable magazine to 100 cubic feet plus 10%, the 10% being added as an allowance for error on the basic limitation of 100 cubic feet. In response to the comments, the capacity of a portable magazine is limited to no greater than 110 cubic feet. 176.155 (46 CFR 146.20-16) No substantive change. 176.156 (46 CFR 146.20-23) No substantive change. 176.157 (46 CFR 146.20-17) No substantive change. 176.158 (46 CFR 146.20-21) No substantive change. 176.159 (46 CFR 146.20-25) No substantive change.

176.163 (46 CFR 146.09-11, 146.20-37) No substantive change. Two comments were received which stated that grounding need not be required on conveyors constructed of aluminum or other non-sparking material when used for the handling of explosives. This requirement is being retained since it is an accepted safety measure when working with explosives, is an easy measure to perform, and is based on existing requirements. 176.165 (46 CFR 146.20-39) No substantive change. 176.167 (46 CFR 164.20-41) No substantive change. This section has been modified as a result of comments received to provide for the use of non-sparking tools or tools covered with a non-sparking material. 176.169 (46 CFR 146.20-39) No substantive change. 176.171 (46 CFR 146.20-45) No substantive change. 176.173 (46 CFR 146.20-47) No substantive change. 176.177 (46 CFR 146.20-53 146.20-55 146.20-59 146.20-63 146.20-65 146.20-67 146.20-69 146.20-71 146.20-73 146.20-75 146.20-77 146.20-79 146.20-81 146.20-83) No substantive change except for purposes of enhancing safety, the authorization to use oil or chemical burning lamps or lanterns has been deleted. Several comments were received requesting changes to the requirements for magazine vessels, however, only minor rewording has been made for clarification. The requirements for explosives are not being amended by this rulemaking as it is inappropriate to make such substantive changes without affording interested persons an opportunity to comment.

176.200 (46 CFR 146.24-30 146.24-50) No substantive change. Adopts IMCO criteria for stowage of poison gas away from foodstuffs. 176.205 (46 CFR 146.24-35) No substantive change except for allowing "below deck" stowage of some flammable compressed gases in accordance with IMCO. The only comment to this section suggested that flammable compressed gases be limited to "on deck" when stowed on board a vessel. In consideration of the proposed incorporation of IMCO stowage criteria for these materials and the additional safety considerations required when flammable compressed gases are stowed "under deck" this section is retained as proposed. 176.210 (46 CFR 146.24-30) No substantive change. 176.220 (46 CFR 146.24-65) No substantive change. 176.225 (46 CFR 146.24-60) No substantive change. 176.305 (46 CFR 146.21-35) 146.21-40 146.21-45 146.21-25 146.21-15 146.26-25 146.21-50) No substantive change. 176.315 (46 CFR 146.21-20 146.26-20) No substantive change. Clarification and combines treatment of hazards of combustible and flammable liquids. 176.320 (46 CFR 146.21-25) No substantive change. 176.325 (46 CFR 146.21-55 146.26-15) No substantive change. 176.328 (46 CFR 146.27-30) Comment to this section suggested that when motor vehicles are carried in containers, the container cannot always be open for inspection purposes and that relief from inspection be provided containerized cargo. This comment and others relating to the problem of inspecting containers on board container vessels has been resolved in § 176.39, "Inspection of cargo."

176.329 (46 CFR 146.21-60 146.26-30) This proposed section, which made no substantive changes to existing requirements, has not been adopted. Exemptions for potable spirits, adopted under Docket HM-102 (40 FR 22263, May 22, 1975), are now provided for in § 173.118(c) of this subchapter.

176.331 Adopted per IMCO. Requires stowage separation of packages (containing flammable liquid) bearing Poison labels from foodstuffs.

176.400 (46 CFR 146.22-15). No substantive change. Comments to this section suggested rewriting the requirements in order to preclude the stowage of oxidizers and organic peroxides together. This section is retained unchanged since the Segregation Table § 176.86(b), stipulates that these materials must be "separate from" each other. This segregation term requires stowage in separate holds.

176.405 (46 CFR 146.22-20) No substantive change. 176.410 (46 CFR 146.22-15, 146.22-30, 146.20-23, 145.22-15) No substantive change. Comments to this section suggested that the first retardant bulkhead requirement be eliminated when ammonium nitrate or nitro carbo nitrate are stowed within the hold with explosives and the recommendations of IMCO be adopted for this stowage situation. This requirement is being retained since major changes to the regulations involving the transportation of explosives will be considered under a separate rulemaking.

176.415 (46 CFR 146.22-30, 146.22-40) No substantive change. 176.419 Adopted per IMCO. Requires stowage separation of packages (containing flammable solids or oxidizers) bearing Poison labels from foodstuffs.

176.600 (46 CFR 146.25-45) Changed to adopt IMCO stowage criteria. Requirements for separation of foodstuffs and Poison A and Poison B materials. Comment to this section requested that "separate from" stowage be provided to poisons from materials known to be for food contact application as well as from foodstuffs. This requested additional requirement is not being adopted because it would create an enormous area for interpretation and would be practically unenforceable.

176.605 (46 CFR 146.25-50) No substantive change. 176.700 (46 CFR 146.19-35) No substantive change. 176.710 (46 CFR 146.19-50, 146.02-13(b)) No substantive change. In response to a comment, a cross reference has been added for reporting requirements to § 171.16. 176.715 (46 CFR 146.19-50) No substantive change. 176.800 (46 CFR 146.23-10, 146.23-15, 146.23-25) No substantive change. 176.805 (46 CFR 146.23-20) No substantive change.

176.900 (46 CFR 146.27-25) No substantive change except specifying what "adequately dunnaged off the bulkhead" means. Specific requirements for cotton have been deleted because of the change in hazard classes of cotton to a flammable solid. Several wording changes have been made to this section in response to comments received. The paragraph which provides for a wooden bulkhead, when cotton or other fibers are stowed adjacent to a boundry bulkhead subject to heat, has been modified to provide for bulkheads which are permanently insulated. The paragraph which provides for particular fixed fire smothering equipment has been modified to allow other approved fixed fire smothering systems. The paragraph requiring tarpaulins to be fitted over hatch openings, so as to provide a tight hold, has been modified to recognize hatch covers which when fitted in place provide the required tighthold.

176.901 New section (part of proposed § 176.900) setting forth requirements for stowage of cotton or fibers with resin or pitch. 176.902 New section (part of proposed § 176.900) setting forth the requirements for stowage of cotton or fibers with vegetable, animal, or resin oil. 176.903 New section (part of proposed § 176.900) setting forth requirements for stowage of cotton or fibers with coal.

176.904 New section (part of proposed § 176.900) setting forth requirements for stowage of cotton or fibers with synthetic nitrate of soda.

Sections of 46 CFR dealing with the requirements for the transportation of limited quantities of hazardous materials do not appear in Part 176 because these provisions are found in Part 173.

PART 177—CARRIAGE BY PUBLIC HIGHWAY

This part deals with the requirements for highway carriers engaged in the transport of hazardous materials. Comments received are addressed in a manner that will follow the format of Part 177.

177.800 through 177.813 No substantive change.

177.815 Deletion of certain duplicative requirements regarding labeling which are set forth in Part 173.

177.816 Deletion of a duplicative marking requirement which is set forth in Part 173.

177.817 Several comments expressed opposition to the prohibition against carriers accepting for transport a hazardous material accompanied by an improperly prepared shipping paper. Application of such a prohibition, they stated, is difficult because of the complexity of the shipper paper requirements and the problem of educating the drivers so that they are aware of these requirements. They further stated that burden for compliance with shipping paper requirements should rest with the shipper. The Bureau agrees that shippers must be responsible for properly completing shipping papers, and the regulations so require. However, in order for the motor carrier industry to comply with carrier operating requirements, such as placarding, carrier personnel must be able to recognize when hazardous materials are being offered for transportation, what class of material is being offered, and in what quantities. This information must be obtained from the shipping papers at the time the shipment is tendered to the carrier in order for the carrier to initiate compliance with the carrier operating requirements. Therefore, a modified version of the proposal has been retained in this amendment.

Several comments objected to the proposed provision that hazardous materials be listed first on the carrier's shipping documents which cover both hazardous and non-hazardous materials (by reference to Subpart C of Part 172). Since carrier shipping documents are usually prepared in the same order as the shipper's original shipping paper, and the shipper must list hazardous materials first, the Bureau does not agree that this proposal will be disruptive to the motor carrier industry, or that it will impose on that industry any new burdens. Therefore, the proposed change with revised language is being adopted by this amendment.

In the notice, the Board proposed to clarify the possession requirements for shipping papers carried by a driver. Included in the proposal was a recommendation that shipping papers be placed in a pouch or other device that is mounted on the inside of the door to the left of the driver's position. The purpose of this proposal was to standardize shipping paper locations so that emergency response personnel would be better able to find such documents when the need arises. The industry's opinion was that the pouch could interfere with the driver's ability to use the various controls on the vehicle and would be an irritation to the driver's left leg as he actuated the clutch pedal. They also expressed an objection on the basis of the cost of purchasing and installing the pouches. In addition, the motor carrier industry objected to the proposal that the shipping paper be on the driver's person when he is away from the vehicle. They cited possible loss or destruction of the paper due to this additional handling, and the problem emergency response personnel would have in locating a driver should an incident occur at the vehicle. The Bureau believes the motor carrier industry's comments have some merit, especially those comments relating to the problem of locating a driver when he is away from his vehicle. While the Bureau still believes there is a definite need for standardization of shipping paper location on a motor vehicle while hazardous materials are in transit, it does not want the location requirements to be so restrictive as to not allow motor carriers any choice of location for the documents. Therefore this amendment leaves the carrier with some discretion on choice of location, while maintaining a meaningful degree standardization.

It was also proposed that hazardous materials shipping papers, when carried with other shipping papers, be arranged so that they will appear first upon examination of the papers. The motor carrier industry stated that such a requirement would be disruptive to the normal practice of arranging shipping papers in the order in which deliveries are to be made. As an alternative, the industry suggested that the Department list alternative systems, allowing carriers the choice of selecting the system most suitable to their operation. The Bureau agrees and this amendment provides accordingly.

It was also suggested that the Department should " . . . establish a Recommended Practice urging motor carriers to establish systems that will provide quick access to hazardous materials shipping papers." (Emphasis added.) Since the shipping papers are often the only means of specifically identifying the hazardous materials, these documents must be readily available and identifiable should incidents, requiring their immediate examination, occur during transportation. Not requiring an identification system could mislead emergency response personnel, upon rapid examination of shipping documents, into believing hazardous materials are not on a vehicle. For this reason, the Bureau does not agree that an identification system for shipping papers in the driver's possession should only be recommended.

To insure that a line of communication is established between a driver and emergency response personnel, this amendment requires that shipping papers be readily available to emergency response personnel. Such action was deemed necessary in order to enhance the safety of emergency response personnel and the general public.

177.819, 177.821 and 177.822 No substantive change.

177.823 The Board proposed two conditions where in an emergency a vehicle containing hazardous materials could be moved without the required markings and placards. Many comments suggested that the alternatives proposed were too restrictive and could prevent the immediate movement of a vehicle as necessary for handling of an emergency. The Bureau agrees and has added a third provision under which a vehicle could be moved when necessary, in an emergency, to protect life or property.

177.824 & 177.834 through 177.840 No substantive change.

177.841 In the preamble to Notice 73-9, the Board explained the proposed change to paragraph (d) in § 177.841 regarding the loading of poisons with food stuffs, and invited comments as to whether or not there should be some relaxation of the restriction and, if so, how it should be accomplished. Many comments suggested that the overpack unit approved under Special Permit No. 6869 should be adopted as a means of accomplishing some relaxation of the restriction. In addition, the American Trucking Associations, Inc. (ATA) proposed a "pallet overpack" as a means of safely loading poisons in the same vehicle with foodstuffs.

The Bureau does not believe that sufficient experience data has been obtained at this time for use of the Special Permit No. 6869 overpack to justify its adoption under § 177.841. The Bureau will continue to monitor experience data obatined from holders of the permit as well as experience gained from the segregation requirements recognized by the United Nation's Committee of Experts on the Transport of Dangerous Goods. As additional experience data is obtained, the Bureau will consider reopening rule making on the subject. Also, the Bureau will consider permitting use of the "pallet overpack" suggested by ATA if submitted under the prescribed exemption procedures established by the Department.

   177.842, 177.843, 177.848, 177.853 through 177.861 & 177.870 No substantive change.

(41 F.R. 15972, April 15, 1976)

SUMMARY • Docket No. HM-113 • *This was a proposal to control on an interim basis, the stress corrosion problem in DOT MC 330 and 331 tank trucks used to transport anhydrous ammonia, pending development of a final solution.*

[49 CFR Parts 173, 177, 178]
[Docket No. HM-113; Notice No. 73–11]
MC 330 AND MC 331 CARGO TANKS
Stress Corrosion Cracking

   The Hazardous Materials Regulations Board is considering amendments to certain sections of the Department's Hazardous Materials Regulations, which pertain to shipments of anhydrous ammonia in MC 330 and MC 331 cargo tanks. These sections are: §§ 173.33, 173.315, 173.427, 177.817, 177.824, and 178.337.

   In 1968 the Department first sought to prevent or reduce stress corrosion cracking in these tanks by amendments to regulations (HM-5, 33 FR 2389, 7493). Required thereunder was immediate inspection of all MC 330 and MC 331 cargo tanks, in order to determine need for repair and to insure product retention integrity. When shipped in quenched and tempered tanks, anhydrous ammonia had to either be inhibited with 0.2% water by weight or be at least 99.995% pure. Any tank, new or used, which had been opened for inspection, test, or repair—or any tanks which had been used for other than ammonia—had to be purged of air before loading.

   In 1971 the Department became concerned with indications that the stress corrosion cracking problem remained unsolved. On November 20, 1971, a Notice of Public Meeting was issued requesting participation by persons ". . . having knowledge concerning the existence or extent of stress corrosion cracking, and methods by which it can be prevented . . . ." (36 FR 22192). The meeting was held on December 14, 1971. Considerable evidence substantiated that the cracking problem either remained or had reappeared. There was even evidence of instances where cracking was so extensive as to require that cargo tanks be scrapped.

   The meeting produced varying opinions as to the cause of the cracking. Since then, the National Association of Corrosion Engineers (NACE) has studied the problem and has filed a petition for rulemaking, seeking amendments to the rules designed: (1) To provide interim measures to reduce the probability of cracking; (2) to provide detection of and control over stress corrosion cracking occurring during the interim; and (3) to measure the effectiveness of the recommended requirements until the matter is further researched.

   Believing that the NACE petition warrants prompt consideration of action, the Board issues this Notice. The rules proposed are for the most part based upon the information in the NACE petition.

   The proposals would add certain tank inspection and test requirements; require notice to the Bureau of Motor Carrier Safety when a tank is removed from service; further restrict the use of MC 330 or MC 331 cargo tanks constructed from quenched and tempered steel by precluding their use for transporting anhydrous ammonia when a water inhibitor is not added; require that shippers monitor and periodically check the amount of water inhibitor added in order to assure close adherence to the 0.2% water-by-weight shipping requirement, when quenched and tempered tanks are used; specify the types of water to be added as an inhibitor; require shipping paper notations showing the more restricted use for tanks of quenched and tempered steel construction; specify certain requirements for MC 331 tanks regarding post-weld heat treatment; and require a carrier which offers to sell or lease a MC 330 or MC 331 cargo tank to provide a copy of the results of any required tests to each prospective purchaser or lessee.

(39 F.R. 1059, January 4, 1974)

SUMMARY • Docket No. HM-114 • *Established new requirements governing selection, preparation, inspection, certification, and equipment on railroad box cars used to transport Class A explosives.*

[49 CFR Part 174]
[Docket No. HM-114; Notice No. 74–1]
RAIL CARS USED TO TRANSPORT CLASS A EXPLOSIVES
Selection, Preparation, Inspection, Certification, and Loading

   The Hazardous Materials Regulations Board is considering amendment of § 174.525 which prescribes the requirements for selection, preparation, inspection, certification and loading of railroad cars used to transport Class A explosives.

As a result of recent rail accidents and incidents involving Class A explosives, the Federal Railroad Administration (FRA) issued Emergency Order No. 3 on August 9, 1973, to supplement the Hazardous Materials Regulations (38 FR 21952). This Emergency Order provides that each car transporting Class A explosives must be equipped with certain "low-sparking" type of brake shoes and all brake shoes on the car must be of the same and proper type and design, in safe and suitable condition for service, and comply with prescribed wear limits. In addition, the Order provides that the car must be equipped with a continuous steel sub-floor or metal spark shields of prescribed dimensions. However, if the car is not equipped with prescribed steel sub-floor or metal shields, the car may be used to carry Class A explosives only if it is inspected at intervals and in the manner set forth in the Emergency Order.

On November 2, 1973 the Association of American Railroads (AAR) filed a request for modification of Emergency Order No. 3 or, in the alternative, for review as provided in section 203 of the Federal Railroad Safety Act of 1970 (45 U.S.C. 432). Some of the modifications requested by the AAR deal with matters that are included in this Notice of Proposed Rule Making. They are included in this notice to afford an opportunity for public participation in their resolution. Upon completion of the rule-making proceeding initiated by this notice, FRA intends to terminate Emergency Order No. 3.

Although the accidents involving Class A explosives which occurrd on the Southern Pacific Transportation Company at Roseville, California on April 28, 1973, and at Benson, Arizona on May 24, 1973, are still under investigation, the FRA believes that § 174.525 must be amended to eliminate potential fire hazards on rail cars used to transport Class A explosives. These hazards result from overheated friction journal bearings, overheated and "sparking" brake shoes, and the presence of combustible material on the undersides of cars.

Interested persons are invited to give their views on these proposals. Communications should identify the docket number and be submitted in duplicate to the Secretary, Hazardous Materials Regulations Board, Department of Transportation, Washington, D.C. 20590. Communications received on or before March 31, 1974, will be considered before final action is taken on these proposals. All comments received will be available for examination by interested persons at the Office of the Secretary, Hazardous Materials Regulations Board, Room 6215, Buzzards Point Building, Second and V Streets S.W., Washington, D.C., both before and after the closing date for comments. The proposals contained in this notice may be changed in light of the comments received.

In addition to assure that all interested persons have an opportunity for oral presentation, the FRA will conduct a public hearing commencing at 10 a.m., on March 21, 1974, in Room 2545, Federal Building, 650 Capitol Mall, Sacramento, California.

The purpose of this public hearing is to obtain information to assist the FRA in developing a final rule in this proceeding, not to determine the cause of circumstances surrounding any of the recent rail accidents or incidents involving hazardous materials which are still under investigation.

The hearing will be an informal not a judicial or evidentiary type of hearing. There will be no cross-examination of persons making statements. An FRA staff member will make an opening statement outlining the matter set for hearing. Interested persons will then have an opportunity to present their oral statements. At the completion of all oral statements those persons who wish to make rebuttal statements will be given the opportunity to do so in the order in which they made their initial statement. Additional procedures for conducting the hearing will be announced at the hearing. Interested persons may present oral or written statements at the hearing. All statements will be made a part of the record of the hearing and be a matter of public record. Persons who wish to make oral statements at the hearing should notify the Office of the Chief Counsel, Federal Railroad Administration, Room 5101, Nassif Building, 400 Seventh Street SW., Washington, D.C. 20590, before March 14, 1974 stating the amount of time requested for their initial statement.

The proposed changes in Paragraph (b) of § 174.525 are described below.

Subparagraph (1). It is proposed to delete the words "when available" and "on other". The first deletion would make absolute the present conditional specifications contained in the subparagraph. The second is clarifying in nature.

Subparagraph (3). It is proposed to substitute "holes" for "loose boards", add "doors" and substitute "which may hold fire from sparks" for "liable to hold sparks and start a fire". The first two changes are merely clarifying in nature while the third change is proposed both for clarification and to conform with the language of subparagraph (4).

Subparagraph (4). It is proposed to delete "or broken boards" to conform with similar changes in other subparagraphs.

Subparagraph (6). It is proposed to amend this subparagraph to require that after December 31, 1975, each car used to transport Class A explosives must be equipped with roller bearings, and to amend the present first sentence of this subparagraph to reflect this proposal by substituting "The roller bearings or journal boxes, and the trucks" for "The journal boxes and trucks."

Overheating of friction journal bearings often resulting in open flames from burning oil and pads, is recognized as a major hazard in railroad operations. Since roller bearings are much less likely to overheat and even less likely to generate open flames if they should overheat, virtually all of the new freight cars placed in service as well as older cars rebuilt in recent years are equipped with roller bearings. At present, approximately one-half of the national freight car fleet is equipped with roller bearings. In these circumstances, FRA believes that cars carrying Class A explosives should be required to be equipped with roller bearings.

Subparagraph (11). The FRA proposes to redesignate existing subparagraph (11) as subparagraph (13) and to add a new subparagraph (11). The proposed new subparagraph provides that after December 31, 1974, each car carrying Class A explosives must be equipped with high-friction composition brake shoes and brake rigging designed for these shoes and that until then the car must be equipped with either high-friction composition brake shoes or high-phosphorous brake shoes and brake rigging designed for the type of brake shoe used. Proposd subparagraph (11) would also require all brake shoes on the car to be of the same type and in safe and suitable condition for service. High-friction composition brake shoes would be required to have a minimum thickness of three-eighths inch and high-phosphorous brake shoes, of one-half inch.

Sparks generated by contact between brake shoes and wheels during braking of trains present a serious fire potential which assumes critical dimensions when a car is carrying Class A explosives. Cast iron brake shoes produce a heavy shower of sparks during braking which could ignite any combustible material under the car. High-phosphorus brake shoes are much less susceptible to this sparking effect but since they are made of metallic material, they do produce some sparks during heavy braking. High-friction composition shoes normally generate almost no sparks. Low-friction composition brake shoes also generate practically no sparks. However, because only a very small portion of the nation's freight car fleet is equipped with low-friction composition brake shoes, this type of brake shoe is virtually unknown to many railroad maintenance employees and is not carried in stock by many railroads. Consequently, there is a strong possibility that worn or missing low-friction composition brake shoes may be improperly replaced with high-friction composition brake shoes thereby creating serious fire and safety hazards. Mixed types of brake shoes on a car and worn-out brake shoes are also hazardous.

Subparagraph (12). The FRA proposes to redesignate existing subparagraph (12) as subparagraph (14) and to add a new subparagraph (12). The proposed new subparagraph provides that a car carrying Class A explosives must have either a metal sub-floor with no combustible material exposed beneath the car or have metal spark shields extending from the center sill to the side sills and from each end sill to at least twelve inches beyond the extreme treads of the inside wheels of each truck. The spark shields must be tightly fitted against the sub-floor so that no vacant space to catch sparks or combustible material is exposed. The new subparagraph also provides that the metal sub-floor or spark shields may not have an accumulation of oil, grease or other debris which could support combustion.

In recent demonstrations using a static wheel dynamometer at speeds up to 45 m.p.h. and blowers to simulate the actual railroad environment, slivers of brake shoe material became embedded in cracks in wood placed at car sub-floor height above the test wheel, at distances of more than thirty-six inches beyond the center of the axle in the direction of rotation. In these demonstrations, radiant heat equivalent to that radiated by an overheated wheel, charred wood subflooring protected by a tightly-fitted metal shield but did not cause the wood to burn. Particles of brake shoes deposited in a catchpan at ballast level continued to glow for minutes. Accordingly, metal shielding of the area above each truck is necessary to prevent fire caused by heat radiated from an overheated wheel or by burning fragments of brake shoe material becoming lodged in wood sub-flooring. This shielding is still necessary even when a car is equipped with high-friction composition brake shoes because in the event of "sticking brakes" or sustained heavy braking, the resin in the composition material may ignite and burn freely causing the brake shoe to disintegrate and freely-burning fragments to be propelled and lodged against the bottom of the car. This shielding will also minimize fire hazards resulting from high-friction composition brake shoes being mistakenly replaced with cast iron brake shoes, a not uncommon occurrence.

Subparagraphs (13) and (14). In these subparagraphs which presently are numbered (11) and (12), the term "qualified inspector" is proposed to be substituted for "competent employee." This change is proposed to describe more precisely the person required to examine, inspect and certify cars used to transport Class A explosives.

In addition, a number of changes are proposed in paragraph (c) of § 174.525.

Subparagraph (1). The term "qualified inspector" is proposed to be substituted for "competent employee" to conform with proposed subparagraphs (13) and (14) of paragraph (b).

Subparagraph (3). The FRA proposes to delete "or to the side of wooden cars between car initials and the car door". As a result, all car certificates would be required to be attached to the fixed placard boards which are now standard equipment on freight cars. Also, the text of Certificate No. 1 would be changed to become a general certification that the car complies with the requirements of the recently issued FRA Freight Car Safety Standards (38 FR 32224) as well as those of this part pertaining to cars used to transport Class A explosives.

Pursuant to the provisions of Section 102(2) (c) of the National Environmental Policy Act (42 U.S.C. 4321 et seq.), the FRA has considered the requirements of that Act concerning Environmental Impact Statements and has determined that the amendments proposed in this notice would not have a significant impact upon the environment. Accordingly, an Environmental Impact Statement is not necessary and will not be issued with respect to the proposed amendments.

This notice is issued under the authority of sections 831–835 of Title 18, United States Code, and section 9 of the Department of Transportation Act (49 U.S.C. 1657).

(39 F.R. 4668, February 6, 1974)

[Docket No. HM-114; Amdt. No. 174-24]
PART 174—CARRIERS BY RAIL FREIGHT
Rail Cars Used To Transport Class A
Explosives

This amendment prescribes standards to eliminate potential fire hazards resulting from overheated friction journal bearings, overheated and sparking brake shoes, and the presence of combustible material on the undersides of cars used to transport Class A explosives. It establishes new requirements for selection, preparation, inspection, certification and loading of these railroad cars.

On February 6, 1974, the Hazardous Materials Regulations Board published a notice of proposed rulemaking (NPRM), Docket No. HM-114; Notice No. 71-1 (39 FR 4668), which proposed this amendment. The reasons for this amendment were discussed in that notice of proposed rulemaking. Interested persons were invited to comment and several comments were received by the Board. In addition, a public hearing was held on March 21, 1974, to provide interested persons an opportunity to present information orally to assist the Board in developing a final rule in this proceeding. All written comments received and those made at the public hearing have been fully considered by the Board. The interest shown and the views expressed are appreciated by the Board.

The major issues raised in these comments involve bearings, brake shoes, and spark shields. These issues were discussed at length in the NPRM and are also discussed separately below.

Bearings. One commenter strongly opposed requiring all box cars used to transport Class A explosives to be equipped with roller bearings. He noted that solid bearing cars travel an average of one million miles before occurrence of an overheated journal or "hot box" and that only four percent of these result in reportable accidents. This average mileage figure is expected to improve as more cars are equipped with stabilized bearings under an industry specification which requires the owning railroad to stabilize bearings whenever it disassembles a truck. Because of the expected improvement in the ratio of overheated journals and freight car miles operated, the present requirement that trucks and journals be inspected before a car is loaded with Class A explosives, and the substantial investment of the railroad industry in "hot box" detectors, this commenter contended that there is no factual basis for excluding solid bearing box cars from Class A explosives service. Any type of bearing can fail for any number of reasons, including poor design, bad maintenance and undetected structural flaws. Since the primary cause of solid bearing failure is inadequate servicing and any existing service-related defects will be corrected during the required preloading inspection, he concluded that cars equipped with solid bearings should remain acceptable for the transportation of Class A explosives.

One commenter stated that if the proposed roller bearing requirement were adopted it would create a serious shortage of cars available to transport class A explosives.

After carefully considering these arguments, the Board still believes that safety considerations require that all box cars used to transport Class A explosives be equipped with roller bearings after December 31, 1975. Cars transporting Class A explosives should be equipped with the safest journal bearings available because a major cause of derailments due to equipment failures is the overheating of journals. In its report entitled "Journal Failure Report" dated October 1972, the FRA established that in the years 1968, 1969 and 1970, the failure ratio of plain bearings to roller bearings was 11.4, 8.1 and 5.8, respectively. The report projected that the failure rate of plain bearings would probably level off at about three times the failure rate of roller bearings. By January 1, 1976, sufficient time will have elapsed since issuance of Emergency Order No. 3, by the Federal Railroad Administration (FRA) on August 9, 1973 (38 FR 22172), for car owners and railroads to locate and equip a sufficient number of cars to transport Class A explosives. Approximately one-half of the national rail car fleet is equipped with roller bearings and all new cars placed in service are equipped with roller bearings.

Brake Shoes. None of the commenters opposed the proposed elimination of cast iron brake shoes. However, two commenters vigorously opposed elimination of high phosphorous brake shoes on cars used to transport Class A explosives. One commenter submitted extensive test data to support its contention that with respect to the fire hazard, the high phosphorous shoe coupled with the Association of American Railroads standard spark shield, is the safest material developed for braking railroad freight cars.

Both commenters also stated that the sparking of high phosphorous shoes under heavy sustained braking promptly stops when the brakes are released. On the other hand, it was noted that, while high friction composition brake shoes do not spark under these conditions, they may ignite and burn freely and will continue to burn freely with an open flame after the brakes have been released.

Both commenters recognized the hazards inherent in the possible misapplication of standard metal brake shoes instead of high phosphorous brake shoes. One commenter relied upon a program for positive identification of high phosphorous brake shoes which is now underway within the railroad supply industry to resolve this problem. The other commenter indicated that the high phosphorous brake shoe may soon replace the cast iron brake shoe and that this would eliminate any possibility for misapplication of metal shoes.

FRA has carefully considered these comments but still believes that the exclusive use of high-friction composition brake shoes on cars transporting Class A explosives is necessary to assure safety. Sparks resulting from the friction of a brake shoe wearing on a wheel tread provide a high potential for ignition of any exposed combustible material. The high-friction composition brake shoe has a paractically zero sparking effect.

The risk of high friction composition shoes igniting and burning is rather remote since combustion occurs only under the most severe and sustained braking conditions. Although the high phosphorous type brake shoe exhibits a dramatically reduced tendency for sparking compared to the common cast iron shoe, it is still a metallic material which can produce sparks under certain braking conditions.

Spark Shields. Two commenters opposed the spark shields proposed by FRA. They contended that composition brake shoes alone provide sufficient protection on cars not equipped with spark shields and that high phosphorous shoes on cars equipped with the smaller AAR standard spark shields provide a sufficient degree of safety.

FRA does not agree. The criteria requiring spark shields is the flammability of material exposed to ignition from truck effects such as overheated journals, dragging equipment, and sparks from braking. Although the high-friction composition brake shoe has minimal sparking characteristics, there is no positive assurance that metal type brake shoes will not be substituted in error. Although measures have been developed recently to prevent misapplication of cast iron shoes on brake heads intended for none other than high-friction composition type brake shoes, it will be years before these measures are implemented and become totally effective. In the interim, and pending availability of sufficient cars with all-metal sub-flooring, the security of Class A Explosives demands the protection of the larger FRA spark shields.

Several commenters suggested that the FRA require the placement of at least one spacer car not containing hazardous materials regulated under 49 CFR Parts 170–189 between cars of explosives. This suggestion and a number of other suggestions which were beyond the scope of a notice of proposed rulemaking are being studied by the FRA and may be the subject of future rulemaking proceedings.

Several commenters requested clarification of the term "qualified inspector" in the proposed amendment. Accordingly, FRA has changed this term to "qualified person designated under § 215.15."

(39 F.R. 41365, November 27, 1974)

SUMMARY • Docket No. HM-115 • *This was a rulemaking regarding a new specification tank truck for cryogenic liquids, and other requirements for the transportation of these materials by rail and highway.*

[Docket No. HM-115; Notice No. 74–3]
[49 CFR Parts 172, 173, 177, 178, 179]

CRYOGENIC LIQUIDS

The Hazardous Materials Regulations Board (the "Board") is considering amending §§ 172.5, 173.29, 173.33, 173.300, 173.304, 173.314, 173.315, 173.316, 177.817, 177.840, and 179.102–4, and adding new §§ 178.338 and 179.102–18. These proposed amendments and additions provide for the transportation of cryogenic liquids in certain packagings and establish a new DOT specification MC 338, for a cargo tank constructed for use in transporting certain cryogenic liquids.

These proposals are based, in part, on a petition by the Compressed Gas Association, Inc. ("CGA"). The CGA pointed out, in its petition for rule making addressed to the Board, that large quantities of cryogenic liquids are being transported under many special permits authorizing "cryogenic" cargo tanks of various designs. The CGA submitted to the Board a proposed specification for the construction of cryogenic cargo tanks which the Board is identifying herein as "MC 338." This proposed specification is based on numerous designs for cryogenic tanks which have been utilized for several years with a reportedly satisfactory safety record. The CGA did not, however, include in its proposed specification all the requirements that the Board considers necessary for the transportation of cryogenic liquids.

Therefore, the Board is proposing to modify the CGA proposal in several respects, the major changes being as follows:

1. Filling densities. It is proposed that each cargo tank have no less than 2 percent outage below the inlet of any safety relief valve under conditions of incipient opening. The Board does not believe that it should permit loading which results in a pressurized condition during transportation where use of safety valves would routinely be depended upon to operate to prevent rupture of the containment vessel. In addition, a set-to-discharge pressure for relief valves on DOT-4L cylinders has been specified to make the filling density limitations meaningful.

2. Liquid oxygen and aluminum. The Board proposes to prohibit the use of aluminum in pumps and in inner vessels on cargo tanks which are used to transport liquefied oxygen. The Board belives that it does not presently have sufficient information to determine whether adequate safety levels are maintained when aluminum materials are used in contact with liquefied oxygen.

3. ASME Code. The Board proposes that the ASME Code be followed in construction of cryogenic cargo tanks. The CGA proposal limits reference to the Code only to materials used in the construction of cryogenic tanks.

4. "G" loadings. The Board has not agreed with the petitioner's statements regarding design calculations for external supports for tanks without frames and for supports of internal tanks. The Board believes that similar requirements for MC 331 cargo tanks are equally applicable to cryogenic tanks.

5. Aluminum jacket. The CGA proposed that aluminum be prohibited in the construction of a jacket for a cargo tank used to transport flammable cryogenic liquids except under restrictive conditions. The Board proposes to completely prohibit the use of aluminum jackets with cargo tanks transporting flammable cryogenic liquids and to extend this prohibition to cargo tanks transporting liquefied oxygen because of the properties of these materials in a fire situation.

6. Manholes. Although the CGA proposal does not include a requirement for manholes in the inner tank, the Board proposes that such manholes be required becaue of past experience pertaining to repairs of existing cargo tanks and the modifications that have been necessary. Providing manholes will ease access into the inner vessel and protect against degradation of the thermal integrity of the tank.

7. Remote control shut-off valves. The CGA recommends the use of remotely controlled shut-off valves with each filling and discharge valve for flammable cryogenic liquids. The Board agrees with this proposal but believes similar requirements are appropriate for poisonous ladings to preclude a large spill of these materials as well.

On October 16, 1971, the Board published a notice of proposed rulemaking, Docket No. HM-91 (36 FR 20166), proposing regulations governing the shipment of certain Cold Compressed Gases in Tank Cars. Several comments received raised valid objections to some of the proposals contained therein. In addition, since that time the Board has experienced difficulties with certain types of tank cars and has serious doubts concerning the validity of the assumptions upon which that notice was based. Therefore, Docket No. HM-91 is hereby withdrawn. If the Board decides to pursue these matters at a future date (other than ones related to vinyl fluoride and hydrogen chloride), a new docket will be opened. Those portions of Docket No. HM-91 relating to vinyl fluoride and hydrogen chloride are hereby reintroduced in this rulemaking proposal for further consideration.

(39 F.R. 7950, March 1, 1974)

[Docket No. HM-115]
CRYOGENIC LIQUIDS
Change of Status From Notice of Proposed Rule Making To Advance Notice of Proposed Rule Making; Comments Date Postponed

On March 1, 1974, the Hazardous Materials Regulations Board published Notice 74-3 (39 FR 7950) proposing amendments to the Department's hazardous materials regulations that pertain to the transportation of cryogenic liquids. The Board has decided to convert Notice 74-3 to an advance notice of proposed rule making in order that it may consider several technical points in greater detail, after which it will publish a new notice under Docket HM-115.

In consideration of the foregoing, Notice 74-3 has been redesignated by the Board as an advance notice of proposed rule making.

(39 F.R. 32624, September 10, 1974)

[Docket No. HM-115]
CRYOGENIC LIQUIDS
Change of Status From Notice of Proposed Rule Making To Advance Notice of Proposed Rule Making; Comments Date Postponed

On March 1, 1974, the Hazardous Materials Regulations Board published Notice 74-3 (39 FR 7950) proposing amendments to the Deaprtment's hazardous materials regulations that pertain to the transportation of cryogenic liquids. The Board has decided to convert Notice 74-3 to an advance notice of proposed rule making in order that it may consider several technical points in greater detail, after which it will publish a new notice under Docket HM-115.

In consideration of the foregoing, Notice 74-3 has been redesignated by the Board as an advance notice of proposed rule making.

(39 F.R. 32624, September 10, 1974)

[49 CFR Parts 171, 172, 173, 174, 176, 177, 178,179]
[Docket No. HM-115; Notice No. 79–3]

CRYOGENIC LIQUIDS

The generation of cryogenic, or extremely low temperature, materials on a commercial scale began in the late 1920's and production has steadily increased since that time. Recent increases in shipments have been sharper due to the use of cryogens as an energy source and in new scientific and industrial processes. While in large part cryogens are either used at the generation point or are converted and moved as gas via pipeline, significant amounts are transported in packages. Transporting gases in cryogenic form allows for the movement of large quantities of product in a relatively small space since a liquefied gas occupies only approximately 1/600th the volume of the atmospheric pressure gas. And using the maintenance of low temperature rather than high pressure to keep a gas liquefied allows for significant weight reduction in the container.

ADVANCE NOTICE OF PROPOSED RULEMAKING

To date the regulation of cryogenic liquids under the Hazardous Materials Regulations has been done on a piecemeal basis that has resulted in a lack of completeness and uniformity. In 1974, to begin the process of rectifying this situation, MTB's predecessor issued an advance notice of proposed rulemaking (ANPRM) pertaining to the transportation of cryogenic liquids (39 FR 7950, March 1, 1974; 39 FR 32624, September 10, 1974). The ANPRM provided for packaging, shipping and carriage requirements applicable to flammable or pressurized cryogenic liquids. This notice of proposed rulemaking continues many of the provisions contained in the ANPRM, incorporates some changes made as a result of comments to the ANPRM and, in several areas, extends the scope of the ANPRM.

Under present regulations those cryogens that are flammable or pressurized cannot be transported in bulk (with the exception of hydrogen in tank cars) except when under an exemption issued by DOT. Also, nonflammable, nonpressurized cryogenic liquids, which make up the vast majority of container shipments are presently not regulated under the Hazardous Materials Regulations, except in the water mode.

Under this notice, the general scheme of regulation proposed in the ANPRM for flammable or pressurized cryogens is continued, with significant changes noted in this preamble. Additionally, this notice goes beyond the ANPRM in several areas, most notably in the following proposals:

1. The extension of jurisdiction to encompass bulk transportation of flammable cryogenic liquids in intrastate commerce by highway;

2. The requirement that shippers of flammable cryogenic liquids in portable tanks, cargo tanks or tank cars and carriers of flammable cryogenic liquids in cargo tanks by highway, file registration statements with DOT.

3. The provision for rail shipment of ethylene in the cryogenic form;

4. The requirement that drivers of vehicles used to transport flammable cryogenic liquid by cargo tanks be provided specific training.

5. The imposition of regulatory requirements on the transportation of nonflammable, nonpressurized cryogenic liquids.

A total of 46 comments to the ANPRM were received. The comments were generally favorable to the concept of a rulemaking with regard to cryogenic liquids, but the various comments took issue with aspects of the ANPRM. The significant comments and their resolution are discussed below.

COMMENTS ON OTHER THAN THE CARGO TANK SPECIFICATION

Comments were received suggesting various changes to the definition of "cryogenic liquids." After a thorough search of available technical materials, the MTB has determined that there is no single, universally-accepted, definition of the term. The MTB is concerned with transportation safety which does not necessarily require using definitions that are the most scientifically accurate. In addressing safety concerns, however, it was determined that a modification of the ANPRM definition of "cryogenic liquid" was appropriate. Whereas the ANPRM defined the term relative to the temperature of the material at the time of loading and used the relatively high temperature of $-40°$ F. as cutoff, this notice defines "cryogenic liquid" in terms of the boiling point of the material and uses $-130°$ F. as a cutoff. This accomplishes the following:

1. Carbon dioxide, nitrous oxide and vinyl fluoride, whose bulk shipment is presently provided for as liquefied gases, are excluded from the definition;

2. Ethane and hydrogen chloride, anhydrous, while currently authorized for bulk shipment only under exemption, have physical properties similar to carbon dioxide, nitrous oxide and vinyl fluoride and are not cryogens under this notice. This notice proposes to allow their bulk shipment in insulated, non-cryogenic tank cars and cargo tanks.

3. The ANPRM specifically proposed to except nonflammable, nonpressurized cryogens from regulation. That exception is deleted in this notice. While proposed section 173.320 exempts these materials from some of the re-

quirements applicable to other cryogenic liquids, the MTB is proposing that certain requirements apply due to the extreme thermal hazard these low temperature materials pose to human tissue and, in the case of vessels, to shipboard structures.

In response to comments received, the MTB has further refined and clarified the filling densities and design pressures for DOT 4L cylinders. The new proposal provides more flexibility in filling depending on the maximum start-to-discharge pressure and also provides for some increase in fill density to more accurately reflect the satisfactory results of current practice. However, the MTB does not agree that data applicable to 4L cylinders is also applicable to larger containers such as the proposed MC338 cargo tank. This is primarily due to the fact that the cargo tank is not designed to vent. The filling densities in the ANPRM for the cargo tank have been reviewed, but only minor modifications have been made, as appropriate.

Some commenters wanted the references in section 173.316 of the ANPRM to the cryogenic liquids at the time when it is offered for transportation or when transportation begins to instead refer to the liquid at the time the container is filled. While there is usually no distinction in the meanings of the various terms in relation to the condition of the material inside the container, this is not the case with cryogenic liquids. MTB agrees with the comments when cryogenic liquids are being considered in terms of holding time and has made the necessary changes reflected in sections 173.318 and 173.319 of this notice.

The MTB received comments requesting that the so-called atmospheric gases (i.e. nitrogen, oxygen, argon and helium) in the form of pressurized cryogens be permitted to vent from cargo tanks during transportation. Some commenters compared these gases to the exhaust gases of the tractor unit and also pointed out the fact that the air we breathe is composed of 79% nitrogen and 21% oxygen. The MTB, however, is disinclined to permit promiscuous venting of pressurized cryogens. The design of the cargo tanks for these commodities is predicated on a holding time value that precludes the need for venting. These same design parameters determine the filling density of the tank. In the past, the great preponderance of these nonflammable cryogens have been transported by rail and highway in a nonpressurized, and hence nonregulated, form. This notice proposes to allow the venting of the atmospheric gases and helium, during transportation as a cryogen, if the venting occurs at pressures less than 25.3 psig.

Several commenters questioned the ANPRM requirement that cargo tanks be subjected to periodic retests at a pressure of 1.5 times the design pressure of the tank. The commenters felt that that pressure is unsafe and that the ratio of test pressure to design pressure should be 1.25 in order to avoid cyclic failure of the inner tank. The MTB firmly believes, however, that the 1.5 ratio retest pressure is necessary to establish the continued integrity of the cargo tank, and that pressure vessels designed and built according to the proposed DOT specification (which closely follows the ASME Code) will not experience cyclic failure at this pressure. It is possible that the data supplied by these commenters related only to non-ASME Code vessels. It should be noted that tank cars are not subject to this type of retest requirement. This is because of the required presence of a high grade vacuum (75 microns or less of mercury) at the time the tank car is offered for transportation. Since the vacuum is checked before each trip a constant log will indicate any leakage in the inner tank or outer jacket. Cargo tanks may or may not be vacuum insulated, thus there is no similar requirement for shipments in cargo tanks and greater reliance must be placed on the pressure integrity of the containment vessel.

COMMENTS ON THE CARGO TANK SPECIFICATION

The new specification for a cargo tank for use in transporting cryogenic liquids, DOT specification MC 338, was proposed in the ANPRM. Numerous comments were received in response to this specification, and certain revisions have been made. While most of the proposed revisions are for the purpose of clarification and consistency, others are more substantive and are addressed in the following paragraphs.

A suggestion was made by a commenter that the simple combustion test is sufficient for determining the combustion-sustaining characteristics of insulation for cargo tanks used to transport oxygen. The MTB agrees with this suggestion and has incorporated it into this notice.

Some commenters stated the insulation on cargo tanks often is covered by a metal jacket that is lapped, but not sealed, and that this construction has given satisfactory service on cargo tanks where an inner material is used to seal the insulation. The MTB agrees with this position and has allowed for it in the proposal.

Comments were received recommending the deletion of the requirement that mixing devices be installed for venting cargo tanks transporting flammable ladings. It was pointed out that such devices have always been optional equipment and are available only for hydrogen. The MTB agrees that this deletion can be made, since the restrictions imposed on cargo tank loading and one way travel time provided adequate assurance that venting will not take place during transportation.

The Hazardous Materials Regulations require impact tests to be performed on many packaging materials and the MTB believes that they should be required for packaging materials used in cargo tanks where the lading temperature is extremely cold. However, because of available data on the high ductile strength of aluminum at cold temperatures, the MTB proposes to exclude it from the required impact tests.

This notice incorporates a change suggested by one commenter that all tank nozzle-to-shell and nozzle-to-head welds have full penetration. The MTB believes that this is a reasonable requirement that will contribute to the safety of the tank.

The Compressed Gas Association recommended that certain packaging materials be excluded from the requirement for manholes. They pointed out that both stainless steel and aluminum, in extensive low-temperature service, have demonstrated significant corrosion resistance. The added heat leak and other difficulties associated with manways outweigh, in their judgement, any safety benefits or convenience resulting from their use with these packaging materials. The MTB agrees with the recommendation in part, and has specified the packaging materials to be used in the construction of cargo tanks for which manways are not required. In addition, for the purpose of clarification, the manholes or manways which are required are proposed to be located at the rear, or on the rear, or on the rear head of the tank.

The MTB agrees with the suggestion of numerous commenters that any liquid connection to a pressure building coil should be required to have a shut-off valve, and has made this change to increase safety and reliability.

Some commenters suggested that other cryogens in addition to nitrogen be authorized for use in the holding time tests. The MTB can see no reason for restricting these tests to the use of nitrogen only and proposes to authorize that the tests be performed using any cryogenic liquid having a boiling point at atmospheric pressure equivalent to the coldest design service temperature of the tank.

The Compressed Gas Association requested that the holding time marked on the tank be permitted to be less than that to which the tank is actually rated by test under section 178.338-9. Such a practice could be economically beneficial if a tank with a long rated holding time is used on short hauls. By allowing a marked rated holding time less than the rated holding time, maintenance of thermal integrity costs could be reduced without compromising safety. MTB is proposing to include this suggestion in section 178.388-9.

The proposed MC 338 specification anticipates two basic designs. The first design encompasses an inner tank that not only contains the cargo, but also acts as the main structural member of the tank. The insulation in this design is affixed to the inner tank and shrouded with a metal jacket. The second design encompasses an inner tank supported within an outer tank that provides a vacuum envelope and the main structural strength of the assembled cargo tank. In both designs the structural members must be designed to withstand identical "g" loadings, with identical safety factors. And in the second design this is true for both the inner tank and outer tank structural members. A number of comments were directed to this second design and pointed out that in an accident the outer tank will be able to contain the inner tank because the insulation material between the two tanks will absorb some of the energy of any acceleration forces, and therefore the inner tank structural members should be permitted to be designed to lower "g" loadings. The commenters suggested that the inner tank members be designed, with a safety factor of four, to 1½ "g" in the vertical upward, longitudinal and lateral directions and 2 "g" in the vertical downward direction versus the ANPRM's proposed 2 "g" longitudinal and lateral and 3 "g" vertical. This notice continues to propose the higher values for "g" loadings.

The Compressed Gas Association recommended that appropriate gauging devices be authorized as a primary control for filling cargo tanks. They stated that certain gauging devices have proven safe and accurate for large containers and are often the only means available to accurately determine fill level. The MTB concurs in this opinion, and has removed the mandatory verification by weight requirement for cargo tanks from this notice.

Based on a recommendation by the Compressed Gas Association, the MTB proposes to require each MC 338 tank vehicle owner to obtain photographs, pencil rubs, or other facsimiles of the required cargo tank nameplates. In support of their recommendation, the Association maintained that it is both advisable and customary in the industry for the owner to obtain this information.

ALUMINUM CARGO TANKS IN OXYGEN SERVICE

Until recently the MTB has prohibited the use of aluminum in certain packages that come in contact with cryogenic oxygen. At one time this prohibition extended to gaseous oxygen. Several years ago, however, aluminum high pressure cylinders under exemption were permitted to contain gaseous oxygen and certain clean bore, portable tanks have been permitted to transport oxygen under an exemption. Also, there are many aluminum cargo tanks currently in use transporting nonpressurized, thus nonregulated, cryogenic oxygen.

In the past MTB has stated, in response to public queries, that the general prohibition on aluminum use in cryogenic oxygen service is not due to any inherent incompatibility between aluminum and oxygen. Rather, it is based on concern regarding the potential hazard resulting from the mixture of contaminants with oxygen in an aluminum cargo tank either designed in such a way as to allow for areas of contaminant entrapment, or improperly cleaned of contaminants after manufacture and prior to use, or both.

The National Aeronautics and Space Administration has collected and published standards for oxygen equipment cleaning procedures. Similarly, the Compressed Gas Association has developed a new standard for oxygen service cleaning equipment.

The proposals contained in the ANPRM prohibited the use of aluminum for an inner vessel used to transport oxygen. Numerous comments opposed this position pointing to the relatively successful industry use of large aluminum tanks in oxygen applications under exemption and in nonregulated use. Commenters also discussed the refined cleaning procedures currently available.

In view of these past applications and the recently developed cleaning standards promulgated by the Compressed Gas Association, the MTB is proposing in this notice to allow the use of aluminum for MC 338 tanks in oxygen service. By making this proposal, MTB is able to show the public what specific regulatory controls may be used to permit the use of aluminum in the containment of oxygen, cryogenic liquid. Based on comments received MTB will decide whether or not to go forward with this proposal.

MTB is interested in public comment on this proposal reflecting on whether aluminum tanks should be permitted for cryogenic oxygen service at all and, if this practice can be safely permitted, whether the proposed regulatory controls are adequate. MTB is particularly interested in comments on CGA Pamphlet G-4.1, "Cleaning Equipment for Oxygen Service," which is incorporated into the cleanliness standard proposed in section 178.338–15, and its adequacy for aluminum tanks in oxygen service. A copy of this publication is available for review in the public docket. Besides the cleanliness standard in section 178.388–15, the important sections of the proposal relating to the use of aluminum tanks in oxygen service are 178.338–16(d), which requires that after inspection and testing the interior cleanliness of the tank must be verified, 178.338–1(c), which requires that the tank be designed and constructed so as to preclude the entrapment of foreign material, and 178.338–16(b), which prohibits the use of certain weld inspection methods for tanks intended for oxygen service.

ECONOMIC CONSIDERATIONS

While many special permits (now called exemptions) for the bulk transportation of pressurized cryogenic liquids were in effect prior to 1979, the terms of these exemptions were not uniform, and to a certain extent were incomplete. Since 1970, however, the regulatory thrust, has been to enhance and standardize these terms and conditions. As a result of this, design features and performance requirements have been more closely delineated and quantified, with resulting modifications and redesign of such cargo tanks.

Since all shippers and carriers involved with the transportation of flammable or pressurized cryogenic liquids are legally required to be operating under exemption procedures, and since these exemptions generally reflect the scope and nature of the regulations set forth in this notice, the MTB has determined: (1) that the economic costs of complying with the proposed specification requirements have been or should have been already incurred; and (2) that consequently the economic impact does not represent a net significant impact on industry. The MTB also concludes that the proposed regulations covering the transportation of nonflammable, nonpressurized cryogenic liquids will have no measurable impact on shipper/carrier operating revenues or profit margins, or an end-use cost to consumers.

However, in response to several commenters who stated that potentially severe economic impacts may be sustained by individual firms—especially small, marginal operators—in complying with this revised proposal, the MTB invites, and will seriously consider, the views and comments of all interested parties as to how their revisions will affect phases of their operating and investment cost schedules, utilization of equipment, and all other relevant expense categories.

The MTB wishes to emphasize, however, that this information will not be considered as a basis for providing exemptions or relief from any subsequent or final regulations that may be promulgated. The purpose of the requested information is to enable the MTB to set in perspective the nature and extent of the expected improvement in public safety which this proposal is designed to achieve.

There are approximately 52 exemptions for cryogenic cargo tanks and tank cars currently in effect. There are approximately 22 exemptions for portable tanks. This notice does not cover the portable cryogenic cargo tanks. The current exemptions will continue to remain in effect until the petitioner allows them to expire. All original exemptions authorized within the last 30 months or so for cargo tanks and rail cars should comply fully with the proposal contained in this notice. Exemptions which originated prior to this time would be in varying degrees of compliance with the proposals contained in this notice and the feasibility of modifying the container to conform would have to be ascertained on an individual basis. This is provided for in this notice.

IMPACT OF NOTICE ON EXISTING EXEMPTIONS

This notice contains some proposed relaxed requirements and specifications that are at variance with current exemptions in some cases. Such proposals originated in large part as a result of the exemption appeal process, but were not resolved in that forum. Rather, these issues, raised in the exemption process, were determined by the Associate Director for Hazardous Materials Regulation to be matters of such general applicability and future effect as to warrant being made the subject of ruelmaking. Therefore, in accordance with 49 CFR 197.109(e) the issues have been incorporated in this rulemaking. No further action on these matters will be accomplished in the exemption process.

REVIEW BY SECTIONS

The following is an analysis and explanation, by section, of the more significant features of this regulatory proposal.

Section 171.1. This section would be amended to clarify the scope of the Department's regulations and to make them applicable to the bulk transportation of flammable cryogenic liquids in intrastate commerce by highway. The authority for this extension of regulatory jurisdiction is found in sections 103 and 105 of the Hazardous Materials Transportation Act (49 U.S.C. 1802, 1804). The Department may issue regulations for the safe transportation in commerce of hazardous materials and "commerce" includes any trade, traffic, commerce, or transportation that affects interstate trade, traffic, commerce, or transportation. Bulk quantities of flammable cryogenic materials, when transported by motor vehicle in intrastate commerce, affect interstate trade, traffic, commerce, or transportation because of the inherent risk presented by such material to people and property in interstate commerce on the same highways.

Section 171.7. This section incorporates certain industry publications by reference, most importantly for the purposes of this rulemaking a Compressed Gas Association pamphlet entitled "Cleaning Equipment for Oxygen Service."

Section 171.8. Defines "filling density" and "SCF" (standard cubic foot). For "filling density" the existing locations of the definitions for compressed gases in various containers are noted, as are the locations of the definitions proposed in this notice applicable to cryogenic liquids in various containers. The proposed definitions of filling density for cryogenic liquids reflect the need to allow for shrinkage in cargo tank and tank car capacity resulting from cryogenic temperatures.

Section 172.101. It is proposed that the Hazardous Material Table be revised to reflect the definition of a cryogenic liquid to be established in this rulemaking. All "cryogenic liquids" would have that term reflected in their proper shipping names. Additionally an entry in the Table would be provided for ethane-propane mixtures and the hydrogen chloride entry would become hydrogen chloride, anhydrous.

For the cryogenic liquids, the proposed additions and changes to this section prescribe maximum quantity limitations for the air mode and various other restrictions for water shipments. The weight limitations for aircraft shipments were determined based on the properties of each material. Relatively inert materials are authorized on both passenger and cargo aircraft, but in differing amounts depending on the material and type of aircraft. Flammables and oxidizers are forbidden on both clases of aircraft. Substantive comments are solicited on the proposed weight limits and restriction.

Section 172.203. A notation is proposed for the shipping papers for DOT 113 tank cars indicating that these cars may not be humped or cut off while in motion.

Section 172.328. The word "compressed" would be deleted from paragraph (c) to ensure that cargo tank marking requirements would apply to non-pressurized cryogens.

Section 172.504. Table 2 of the general placarding requirements would be amended by correcting the existing reference to "Nonflammable gas (fluoride)" to "Nonflammable gas (fluorine)." Also, the reference to oxygen would be changed to reflect the "cryogenic liquid" terminology in this docket and the present Note 2 to the Table would be deleted to reflect the fact that the placarding requirement must be complied with under this proposal whether the oxygen, cryogenic liquid is pressurized or not.

Section 173.5. It is proposed to require that shippers of flammable cryogenic liquids in portable tanks, cargo tanks or tank cars file registration statements with MTB. This proposal implements the authority provided in section 106(b) of the Hazardous Materials Transportation Act (49 U.S.C. 1805(b)), and consistent with that provision a registration statement would only be required once very two years, beginning in 1980, and a "registration window" of two months duration would be provided. The information obtained through the registration statements would enable MTB to ascertain who is shipping these materials, the location of facilities, warranting periodic inspections, and the number and types of portable tanks, cargo tanks and tank cars used.

Section 173.29. An addition would be made to this section covering empty packagings previously used to ship flammable cryogenic liquids. All residual cryogenic liquid would be required to be removed. This requirement would not apply to DOT 4L cylinders.

Section 173.31. Paragraph (a) of this section would be revised to provide for tank cars now transporting cryogenic liquids under exemption to be examined to see if the appropriate DOT specifications are or can be met. If not, MTB must be notified. Since only one additional cryogenic liquid, ethylene, could be transported by rail under this proposal, only tank cars currently carrying ethylene under exemption would be affected by this provision. This section also requires the monitoring of the daily pressure rise in Class 113 cars and if certain limits are exceeded requires the successful completion of one of two alternative retests prior to the next shipment. The reference in Retest Table 1 to the DOT-113A175W car would be deleted since this car is no longer used, the Table provisions for the DOT-113A60W car would be made consistent with the requirements in this section, and the DOT-113C120W car would be added to the Table.

Section 173.33. The primary changes proposed for this section involve the inclusion of cargo tanks used to carry cryogenic liquids into the existing provisions in this section. A provision, like that in 173.31, would require holders of exemptions permitting cargo tanks to carry cryogenic liquids to attempt to bring the tanks into compliance with the new DOT MC 338 specification. This would include performing a holding time test. A provision in paragraph (d) would required that a MC 338 tank be examined after each shipment to determine that "the actual holding time is not significantly different from the marked rated holding time."

Section 173.300. A definition would be added for "cryogenic liquid" and the definition of "filling density" would be amended to indicate the different sections that apply to cryogenic liquids.

Section 173.304. Atmospheric cryogens would be removed from this section (they are proposed to be placed in § 173.316) and the redesignation of hydrogen chloride to hydrogen chloride, anhydrous would be reflected.

Section 173.314. Hydrogen chloride, anhydrous would be added as an authorized compressed gas for shipment by tank car. This material may currently be shipped by tank car only under DOT exemption. The entry for vinyl fluoride inhibited in the Table in paragraph (c) has been revised to provide a range of allowable filling densities. The maximum density has been increased from that currently permitted, based on proven safe practice under exemption, and the safety provided by the new special requirements paragraph in this section and in section 179.102-4. Note 17 to the Table in paragraph (c) of this section has been revised. The existing note 17 improperly addresses construction requirements in a part directed towards shippers. Proposed paragraph (g) would establish special requirements for hydrogen chloride, anhydrous and vinyl fluoride, inhibited applicable to shippers. The material in current Note 17 applicable to tank construction would be addressed for vinyl fluoride, inhibited within proposed section 179.102-4.

Section 173.315. The proposals would provide for the shipment of ethane, ethane-propane mixtures and hydrogen chloride, anhydrous in insulated MC-331 cargo tanks. These materials can presently be shipped in cargo tanks only under exemption. Proposed revisions to paragraphs (c) and (h) establish outage and gauging device requirements for tanks containing these materials.

Section 173.316. This section would be extensively revised. Presently it applies only to liquefied hydrogen and to shipments in both tank cars and cylinders. As proposed this section would provide requirements for all cryogenic liquids permitted to be shipped in cylinders, i.e., argon, nitrogen, oxygen and hydrogen. Various paragraphs in this section would establish general requirements, requirements for pressure controlling valves, and specification cylinder requirements and filling limits for each commodity. A range of fill densities for all the commodities except hydrogen would be permitted, based on various pressure control valve settings.

Section 173.318. This totally new section proposes shipper requirements for cryogenic liquids in cargo tanks. Many of the provisions in this section are the result of successful past practices with shipments under exemption, other reflect requirements applicable to existing nonregulated cryogenic cargo tanks. The provisions in this section establish general requirements; safety relief device requirements; weight of lading requirements (to be established either by weighing or by the use of authorized gauging device); a two percent outage requirement; a temperature of lading requirement to ensure that the tank will not vent within its marked holding time; and maximum permitted filling densities. Argon, helium, nitrogen, oxygen, carbon monoxide, hydrogen, methane and natural gas would be authorized for shipment in an MC 338 cargo tank and a range of maximum permitted filling densities is provided depending on the setting of a pressure control device.

Section 173.319. This new section proposes requirements for cryogenic liquids in tank cars. Currently section 173.316(a)(1) sets forth shipper requirements for hydrogen, the only cryogenic liquid now authorized by regulation for shipment in tank cars. It is proposed to move the current 173.316 requirements for hydrogen to this section, and to authorize the tank car shipment of ethylene. This section would (1) set forth general requirements primarily involving loading, (2) require at least 0.5 percent outage and an absolute pressure in the annular space of less than 75 microns of mercury, (3) require the temperature of the cryogenic liquid at the time of filling be such that the tank will not vent in less than the holding time established pursuant to section 179.400-4, and (4) establish the maximum permitted filling density and maximum start-to-discharge pressure for the two commodities.

Section 173.320. Argon, helium, nitrogen and oxygen transported in packages such that the pressure will not exceed 25.3 psig (i.e. nonpressurized) are proposed to be excepted from the packaging requirements of the Hazardous Materials Regulations applicable to cryogenic liquids. This section sets out the general exception and then indicates those requirements from which these materials are not excepted.

Section 174.25. A minor change is proposed to reflect the new term "oxygen, cryogenic liquid."

Section 174.67. Paragraph (a) would be amended to correct a mistake. Currently only section 174.200 is referenced whereas all of subpart F of Part 174 should be.

Section 174.83. Paragraph (b) would be revised to add tank cars carrying flammable cryogens to those transporting explosive A or poison gas and which may not be cut off in motion, coupled with excessive force or struck by any car moving under its own momentum.

Section 174.204. It is proposed to extend the existing delivery requirements applicable to tank cars containing anhydrous ammonia, liquefied hydrocarbon gas and liquefied petroleum gas to apply to cars containing cryogenic liquids as well.

Section 176.76. Paragraph (h) would be added to require that a cryogenic liquid in a cargo tank aboard a vessel must be in Specification MC 338 tank, even if the cryogenic liquid is nonflammable and nonpressurized and generally excepted from container requirements by section 173.320. Also valves, fittings or piping (exclusive of the tank itself) which come in contact with lading would not be allowed to be made of aluminum on an MC 338 tank carrying any cryogenic liquid aboard a vessel. Finally, this provision requires the time between loading and unloading of the cargo tank to not exceed the marked rated holding time calculated pursuant to section 178.338–9(c).

Section 177.816. The MTB is proposing that the driver of each vehicle used to transport a flammable cryogenic liquid in a cargo tank receive formal training at least every 24 months. Included would be training pertaining to requirements of the Hazardous Materials Regulations applicable to flammable cryogenic liquids; requirements of the Federal Motor Carrier Safety Regulations applicable to drivers; the properties and potential hazards of the flammable cryogenic liquids being transported; instructions on the operating characteristics, emergency features and loading limitations of the transport vehicle; and the procedures to be followed in case of accident or other emergency. It is proposed that written records of training be kept and that the driver be issued a certificate of training. Recognizing the potential risks presented by flammable cryogenic liquids when they are transported in large quantities in cargo tanks, the MTB believes the proposed training requirement are appropriate and necessary.

Section 177.818. For the same reasons stated in section 177.816, the MTB is proposing that special instruction be carried with the shipping papers on each vehicle transporting a flammable cryogenic liquid in a package exceeding 125 gallons water capacity. The special instructions would include general precautions, manual venting instructions, emergency procedures, and the names and telephone numbers of persons to be contacted in case of emergency or accident.

Section 177.824. Paragraph (e) would be revised to cover the MC 338 cargo tank in this retest requirements section. The proposal references the detailed testing requirements set forth in the MC 338 specification.

Section 177.825. A registration requirement for carriers of flammable cryogenic liquids is proposed, similar to that proposed in section 173.5 for shippers of flammable cryogenic materials. The reasons and authority for the proposed registration requirements are the same for both sections.

Section 177.840. This section would be revised to provide carrier requirements for the transportation of cryogenic liquids. Paragraph (h) would require drivers to be knowledgeable in the handling of the cryogenic material being transported, to manually vent under certain conditions, and to log, for each shipment, the cargo tank pressure at various specified times. Paragraph (1) would require the time between loading and unloading of the tank at its intended delivery point to not exceed the one-way travel time calculated under section 173.388–9(c) and for distribution service modified by paragraph (j) of this section. Paragraph (j) would explain how to determine one-way travel time for cargo tanks in distribution service. Paragraph (k) would set out the requirements that must be met before a cryogenic cargo tank would be considered empty. Paragraph (1) would require a specified breathing apparatus for drivers of vehicles transporting carbon monoxide, cryogenic liquid.

Section 178.57–13. This section would be revised to reference paragraphs (b) and (c) of section 173.316. This corrects the present situation under which pressure controlling valve requirements in 173.316 are not referenced in the 41 specification and also reflects the fact that the requirements in present paragraph (b)(2) of section 173.304 are proposed to be incorporated in section 173.316.

Section 178.337–11. It is proposed to have the existing MC 331 discharge opening rquirements applicable to flammable liquids, flammable compressed gas and anhydrous ammonia made applicable to hydrogen chloride, anhydrous for which bulk shipment would be authorized under this notice.

Section 178.338. This section proposes a cargo tank specification, MC 338, for the carriage of cryogenic liquids. This specification was also contained in the ANPRM in substantially the same form in which it is presently presented. The following paragraphs explain how the tank specification that appears in this notice differs from the one that appeared in the ANPRM.

Section 178.338–1 is reworded to provide a more orderly description of the cargo tank and its design factors and form. Insulation requirements are now described in terms of a performance standard.

Aluminum is now excluded from the requirement for impact testing in section 178.338-2 due to its ductility at low temperatures.

In section 178.338–3, Note 1 of paragraph (b) has been deleted and its contents are now part of the text of paragraph (b).

Paragraph (f) of Section 178.338–4 requires full penetration on nozzle to shell and nozzle to head welds.

There has been only minor rewording of section 178.388-5 and a change to note a referenced section's new location in this proposal.

Tanks constructed of certain materials are excepted from the manhole requirement in section 178.338-6. Also, this section now specifies a location requirement for the manhole.

Paragraph (a) of section 178.338-7 limits the product drainage requirements to tanks intended to carry products that are flammable. The rest of the section is reworded for clarity.

In Section 178.338-8, the wording is clarified and various section number references have been corrected. Also, for a pressure-building coil, this section now requires that a valve at the liquid connection to the coil must be provided.

In section 178.338-9, devoted to holding time, a number of changes have taken place. The use of a cryogenic liquid other than nitrogen is now allowed in performing the holding time test. An optional holding time test is permitted for tanks made to the same design as a tank that has been subjected to the full holding time test. The term "marked rated holding time" is introduced, which permits a tank, based on its intended use and possible economic benefits to the operator, to have its specification plate marked with a lesser holding time than the tank is sucessfully tested for. A revised and more relaxed one-way travel time definition is provided for tanks with a marked rated holding time in excess of 72 hours.

Paragraph (d) is added to section 178.338-10 and requires that every part of the loaded cargo tank must be at least 14 inches above level ground (exclusive of wheel assemblies).

The reference in paragraph (c) of section 178.338-11 to poisonous ladings has been removed since none of the cryogens proposed to be authorized for the MC 338 tank are poisonous.

There is only minor rewording to section 178.338-12.

Paragraph (f) of section 178.338-13 in the NAPRM, which referenced section 178.338-18(c) regarding design weight of lading used in determining loadings, is deleted.

In section 178.338-14, the ANPRM requirement in the second sentence of paragraph (a)(4) is removed since it merely desribes how to carry out the first sentence and is unnecessary. The second sentence of paragraph (c) in the ANPRM is now contained in section 178.338-7(b).

The cleanliness standard in section 178.338-15 now requires that tanks constructed for oxygen service be cleaned in accordance with Compressed Gas Association Pamphlet G-4.1 entitled "Cleaning Equipment for Oxygen Service."

The ANPRM's restriction on the use of the pneumatic test to certain applications is removed from section 178.338-16. Also, for tanks constructed in accordance with Part UHT of the ASME Code, it is now required for pneumatic test, as well as hydrostatic tests, that the test pressure be twice the design pressure. The ANPRM's provision for weld inspection using specified nondestructive testing other than radiography is removed. It is proposed that all welds on the cargo tank shell or heads be radiographed.

Corrections are made to the references in section 178.33817.

Section 178.338-18 is reworded to more clearly establish the separate nameplate and specification plate requirements.

Section 178.338-19 now requires that the manufacturer furnish, beside the tank manufacturers data report, a photograph, pencil rub or other facsimile of the nameplate and specification plate. The section is also reworded to more clearly indicate the requirements if a tank is manufactured in two or more stages.

Section 179.102. This section sets forth special commodity requirements for pressure tank car tanks.

Section 179.102-4 deals with vinyl fluoride, inhibited and the proposed changes would incorporate construction details now improperly in Note 17 to the Table in section 173.314, establish a cold temperature requirement for cars to be used to transport this material, require specific types of steel and impact tests for construction, raise the lowest possible pressure setting of the safety vent, require certain specified appurtenances and finally, have the adequacy of the insulation determined on performance oriented basis rather than solely on an engineering standard basis.

Section 179.102-7. This new section sets forth special commodity requirements for hydrogen chloride, anhydrous. Presently, this material cannot be shipped in bulk, except under exemption. This section would be based on the requirements currently in the exemptions. Many of the requirements would be similar to those proposed for vinyl fluoride, inhibited.

Section 179.400. This section would be restructured to be more performance standard oriented rather than engineering standard oriented and rearranged in logical sequence to progress from the conception of the tank to final construction.

Section 179.401. This section would be revised to delete the DOT 113A175-W car which is believed to be obsolete and to add the DOT 113C120-W car. This section also incorporate the use of the DOT 133C120-W car for ethylene service. Heretofore cryogenic ethylene has only been permitted to be shipped in bulk under exemption.

(44 F.R. 12826, March 8, 1979)

49 CFR Parts 171, 172, 173, 174, 176, 177, 178, and 179
[Docket No. HM-115, Amdt. Nos. 171-74, 172-82, 173-166, 174-43, 176-17,
177-60, 178-77, 179-32]

CRYOGENIC LIQUIDS

Information collection requirements contained in this regulation have been approved by the Office of Management and Budget under the provisions of 44 U.S.C. Chapter 35 and have been assigned the following numbers: OMB #2137-0541 (Shipper's or Carrier's Registration Statements, §§ 173.11 and 177.826), OMB #2137-0542 (Cargo Tank Pressure and Temperature Records, § 177.840), OMB #2137-0539 (Special Instructions for Cryogenic Liquids, § 177.818), OBM #2137-0540 (Flammable Cryogenic Training Records, § 177.816), and OMB #2137-0017 (Cargo Tank Certification Record Requirements, §§ 178.338-2, 178.338-4, 178.338-9).

This rulemaking has a long history. Prior to today's publication the following notices have been published in the **Federal Register,** and a public hearing held, with regard to Docket HM-115:

(1) On March 1, 1974, the Hazardous Materials Regulations Board, MTB's predecessor, published a Notice of Proposed Rulemaking (NPRM) (39 FR 7950).

(2) On June 19, 1974, the Board extended the comment period (39 FR 21166).

(3) On September 10, 1974, the Board converted the NPRM into an Advance Notice Of Proposed Ruelmaking (ANPRM) and further extended the comment period (39 FR 32624).

(4) On March 8, 1979, MTB published a NPRM which, while in general terms continuing the regulatory scheme proposed in the 1974 ANPRM, made significant changes in several areas. Readers who wish more information on these changes are referred to the preamble discussion in the NPRM (44 FR 12826). This notice also announced a public hearing.

(5) On April 5, 1979, corrections and changes were made to the NPRM (44 FR 20461).

(6) On April 17, 1979, MTB held a public hearing on the NPRM.

(7) On June 21, 1979, additional changes were made in the NPRM and the deadline for filing comments was extended (44 FR 36211).

The purpose of this rulemaking is discussed in the March 8, 1979 preamble. Briefly, some of the transportation subject to this rule has historically been permitted only under an exemption (called in years past a special permit). This exemption program, provided for in § 107(a) of the Hazardous Materials Transportation Act (49 U.S.C. 1806(a)), has required those wishing to transport these commodities to come to MTB for specific approval of the packaging (and how the packaging is to be maintained), the hazardous materials to be transported in the packaging and the conditions of transportation. Exemptions create burdens on shippers, carriers, packaging manufacturers and the MTB itself and should be replaced by regulation as soon as appropriate. To the extent that the regulations have not provided for the carriage of these materials, these burdens have been necessary in order to accommodate commerce and, at the same time, ensure public safety. This rule will alleviate the difficulties engendered under the exemption program and the standardization achieved by issuing these rules will contribute to improved overall safety in the transportation of these materials.

Another major impact of this rule is the application of certain regulatory requirements to nonflammable cryogenic gases, like nitrogen and helium, transported in nonpressurized form. These commodities have previously been unregulated by MTB except when transported by vessel. However, MTB believes that the extreme thermal hazards these materials pose in their cryogenic form warrant the imposition of the limited requirements contained in this rule. This will be more fully discussed below.

Approximately 30 comments were received on the proposals contained in the NPRM. All comments, including late submissions, have been fully considered by MTB during the development of this final rule. Several well-defined aspects of the NPRM were the subject of most of the commentary. These significant issues are discussed below by subject. Following the subject-by-subject review is a Review by Section which briefly discusses each section of the rule and the significant changes that have been made since the NPRM.

USE OF ALUMINUM

The NPRM proposed that aluminum be permitted to be used to fabricate the inner vessel of an MC-338 cargo tank, but that the jacket surrounding the insulation of these containers, when used in oxygen service, must be made of steel. This proposal was based on the adequacy of recently developed industry cleaning standards. The NPRM included a discussion of the use of aluminum in certain packages that come in contact with cryogenic oxygen and specifically asked for public comment concerning this issue (44 FR 12828).

Several commenters supported the use of aluminum in the fabrication of the outer jacketing, as well as the inner vessels, of packagings used for the transportation of cryogenic oxygen. The commenters mentioned the excellent low temperature characteristics of aluminum and the fact that the oxygen does not react adversely with aluminum. In further support of this position, one commenter noted that MTB was planning to issue a proposal for an aluminum cylinder specification for the transportatin of oxygen and other gases. (The final rule was published in the **Federal Register** on December 24, 1981, under Docket HM-176, Specification and Usage Requirements for DOT-3AL Seamless, Aluminum Cylinders (46 FR 62452)).

MTB agrees with these comments to the extent that it proposed, and now adopts in this rule, requirements permitting the use of inner tanks constructed of aluminum to transport cryogenic oxygen, if proper precautions are taken to ensure that the packaging is cleaned of all foreign matter. In order to ensure the cleanliness of cargo tanks, MTB is adopting the cleaning standards contained in Compressed Gas Association (CGA) Pamphlet G-4.1. However, MTB strongly believes that it should not permit the use of aluminum jackets on cargo tanks used to transport a flammable cryogen or pressurized, cryogenic oxygen. In a fire situation any escape of these materials would serve to greatly intensify the fire. Considering that aluminum loses most of its strength at 500° F. and melts at 1200° F., it follows that the protection afforded by aluminum is far less than that afforded by steel, which retains a great deal of its strength at aluminum's melting point and which does not melt until heated to approximately 2600° F. Notwithstanding the good safety experience of nonregulated, nonpressurized cryogenic oxygen cargo tanks, as well as some early exemption (special permit) tanks, and in the absence of any meaningful tests on cargo tanks demonstrating the survivability of aluminum in a fire environment, MTB has concluded that aluminum outer jackets do not provide an acceptable level of safety in cargo tanks used to transport cryogenic oxygen, at pressures in excess of 25.3 psig, or flammable cryogens.

As proposed in the NPRM, this rule prohibits the use of aluminum valves with rubbing or abrading aluminum internal parts in packagings transporting cryogenic oxygen. The reason for this prohibition is the possibility that the heat generated by abrading parts, together with the formation of feather-tipped projections creates a potential for combustion. The same prohibition was also proposed for flammable cryogens; however, the final rule, based on comments and MTB's reevaluation of potential consequences resulting from release in a fire, this prohibition has been extended to any aluminum valve, pipe or fitting.

PRESSURE RELIEF DEVICE SYSTEMS

Several commenters suggested that the requirements contained in CGA Pamphlets S-1.1, S-1.2 and S-1.3 provide adequate protection for cargo tanks from overpressurization. One commenter contended that a single relief device system that is sized using the CGA standards is adequate. The commenter believes that the proposal in the NPRM, that there be a primary and secondary pressure relief device system of equal capacity, was excessive and unwarranted. This contention has also been raised in applications for exemption. MTB has consistently maintained that spring loaded relief valves are mechanical and therefore subject to malfunction, and that in an accident situation a tank may be in an inverted position discharging liquid and therefore relieving pressure at a much lower rate than when discharging vapor. There is also a significant body of analytical and experimental data that raises questions as to the adequacy of the CGA valve sizing due to several assumptions made in the formulás, such as the heat conduction values of the insulation and the total tank area considered as exposed to fire, which provide lower calculated capacities than MTB believes is necessary to assure an acceptable level of safety in a fire situation. Therefore, MTB considers that the primary and secondary relief capacities prescribed in the NPRM are necessary to assure an acceptable level of protection in transportation and they have been retained in this rule at § 173.318(b).

One commenter objected to the proposed marking of each valve with a rated pressure equal to or exceeding the tank design pressure at the coldest temperature expected to be encountered (proposed § 178.338(b)(2)). The commenter believed that such marking conflicts with requirements found to be acceptable by the ASME Code and various piping codes. MTB agrees and has determined that this marking is not necessary and therefore has revised this paragraph to require only that each valve must be constructed and rated for a pressure equal to or exceeding the tank design pressure at the coldest temperature to be encountered. Additional proposed market requirements in § 173.318(b)(5)(ii), for pressure relief devices remain unchanged in this final rule.

MINIMUM OUTAGE AND FILLING DENSITIES

One commenter implied that the proposed minimum outage requirement was too restrictive for pressurized nonflammable cryogenic liquids. The commenter expressed that the minimum outage for pressurized nonflammable cryogenic liquids should be 0.5%. MTB believes that a 2.0% minimum outage is appropriate for pressurized nonflammable as well as flammable cryogenic liquids. The greater minimum outage requirement will significantly reduce the likelihood that a cryogenic liquid will vent as a liquid through a pressure relief device. Given the fact that thermal hazards are posed by the release of these extremely cold liquid materials, whether they are flammable or not, this rule retains the proposed outage requirements.

Section 173.318(d) has been revised to exclude helium from the outage requirement because at relieving pressures it is no longer a liquid.

The commenter also recommended that the filling densities proposed in § 173.318(f) be changed to be consistent with his minimum outage recommendation. Since MTB will continue to use a minimum outage of 2.0%, the filling density changes are not necessary.

VENTING, HOLDING TIME, TRIP MONITORING, AND EQUILIBRATION
OF CRYOGENIC LIQUIDS

In the NPRM venting, holding time and trip monitoring were addressed in proposed §§ 173.33(d)(1)(ii), 173.318(e), 177.840(h), (i), and (j), and 178.338–9, and these requirements would have applied to all cryogenic liquids. Several commenters strongly urged that these requirements only apply to flammable cryogenic ladings. MTB has looked again at the issue of enroute venting and agrees with the commenters that enroute venting of nonflammable atmospheric cryogens and helium presents little hazard. Except for the holding time test contained in the MC-338 cargo tank specification, these requirements now only apply to flammable cryogenic liquids. Each MC-338 cargo tank must have a rated holding time to be considered a specification cargo tank, whether it is used to transport flammable or pressurized nonflammable cryogenic liquids.

Equilibration, which was not addressed in the NPRM, is permitted at § 177.840(i) and (j) of this rule. The concept of equilibration was advocated by several commenters. Essentially it represents a relaxation of the holding time requirement for flammable cryogens by allowing one-way travel time to be redetermined after a full equilibration of the cargo tank is performed. Equilibrations may only be performed at a facility that handles, and is therefore familiar with, flammable cryogenic liquids. Allowing such equilibrations will permit longer hauls to be made safely.

LIQUEFACTION TEMPERATURE

There has been some confusion regarding the term "liquefaction temperature" which was used at several points in the NPRM in specifying design service requirements for packagings (see §§ 173.315(a) Table, Note 11, 178.338–1(c)(2)). The term was used without specifying the pressure at which the liquefaction temperature was to be established. This was an unintended omission because the liquefaction temperature of a cryogenic liquid or cold refrigerated gas varies with pressure. When transported in a pressurized condition the temperature is higher than at one atmosphere.

Several commenters called MTB's attention to this problem. (In the ANPRM at § 178.338–1(b)(2) the liquefaction temperature had been stated as being at one atmosphere.) Some commenters who favor design service temperatures based on liquefaction temperatures under pressurized shipping conditions indicated their belief that the omission seemed to support their position. One commenter indicated that it was his understanding that liquefaction temperature was to be determined at one atmosphere. It was MTB's intention that the liquefaction temperature be determined as indicated by this latter commenter.

Despite the omission in the NPRM, in the exemption process for cryogenic liquids and refrigerated gases such as ethane-propane mixes, MTB has, with but a single exception, specified that design service temperature be based on the liquefaction temperature at a pressure of one atmosphere, and MTB is continuing with the limitation in this rule. If a packaging has a design service temperature based on the liquefaction temperature of the lading under pressurized conditions and there is a sudden rapid release of the pressure for one reason or another, the release could result in temperatures colder than the design temperature of the packaging.

The reference pressure of one atmosphere has been added in the rule at §§ 173.315, Note 11, and 178.338–1(c)(2) to make the meaning of the term "liquefaction temperature" absolutely clear.

JACKET DESIGN

Both the ANPRM and the NPRM contained lengthy formulas, details of welding and reinforcing ring construction, etc. for cargo tank evacuated jackets, which were patterned from rail car specifications. The June 21, 1979 **Federal Register** publication of changes to the NPRM altered this approach, at § 178.338-1(f), by simply referring to specific paragraphs of the ASME Code. One industry commenter was critical of this approach asserting that the ASME charts could be misapplied and/or are in error. As best MTB can determine, the commenter's point is that using ASME Appendix V charts in designing a tank for 7.5 p.s.i. minimum external pressure, which based on a safety factor of four should produce a collapsing pressure of 30 p.s.i., will theoretically result in a collapsing pressure of less than 30 p.s.i. as a result of a reduction in safety factors by ASME in 1977. MTB has dealt with this concern by revising § 178.338–1(f)(1) to reference a critical collapsing pressure of 30 p.s.i. rather than a design pressure of 7.5 p.s.i. This gives the same result as was intended in the NPRM and at the same time avoids any possible ambiguity that may result from the use of the ASME charts. This is consistent with the original Compressed Gas Association (CGA) proposal in CGA-341.

POSTWELD HEAT TREATMENT

One commenter recommended that in the case of tanks constructed in accordance with Part UHT of the Code, postweld heat treatment be required by MTB only when required by the ASME Code (§ 178.338–2(e)). If postweld heat treatment were only conducted when required by the ASME Code, most of these cargo tanks would not be postweld heat treated since the cargo tank thickness is less than the thickness that the ASME Code rquires be post-weld heat treated. However, MTB believes that the postweld heat treatment of these thinner materials is necessary to provide maximum ductility to resist the vibrations and stresses inherent in the highway environment. Experience over many years with the use of postweld heat treatment in cargo tanks has been favorable and MTB feels there are important safety considerations justifying the requirement with regard to MC-338 tanks.

PRESSURE TESTING

The NPRM, at § 178.338–16, proposed that newly constructed tanks be subjected to a pressure test at 1 ½ times design pressure, except that for tanks constructed of UHT steel a test pressure of two times design pressure was proposed. Several commenters urged that a test pressure of only 1 ¼ times design pressure be required. MTB believes that a test to 1 ½ times design pressure is necessary to prove tank integrity. One industry group asserted that such a requirement precludes pneumatic testing, which is necessary for tanks designed to carry large volumes of low density material, such as hydrogen or helium, and for which testing with dense material such as water presents difficulties. To the extent that a manufacturer does not wish to perform a pneumatic test, MTB does not believe this point is meritorious with regard to initial manufacturer's testing which can be performed while a tank is still on the fabrication floor where it can be supported over sufficient tank area to avoid distortion from water weight. MTB believes the general test pressure requirement of 1 ½ times design pressure ensures tank integrity and represents sound design and engineering practice in light of the fact that the highest setting for a relief device is 150 percent of design pressure.

The two times design pressure requirement for tanks constructed in accordance with Part UHT of the ASME Code is also being retained. It is not anticipated that this material will be used frequently in the construction of MC-338 cargo tanks; however, to the extent that it is used, MTB believes that the particular composition of this material and the process by which it is manufactured justify the higher test pressure requirements. This is consistent with other MTB requirements that exceed ASME Code requirements in several areas (such as in § 178.337 for MC-331 cargo tanks) when UHT steels are used.

One commenter recommended that the retest requirement at proposed § 173.33(d)(2) be eliminated for MC-338 cargo tanks with evacuated jackets. The commenter stated that such tanks in cryogenic liquid service are not subject to internal or external corrosion, and that in effect the tank is under constant "leak test" because the slightest leak into the vacuum space would be readily detected. MTB does not agree that this retest should be eliminated. The retest, at higher that service pressure, is necessary to ensure the continued integrity of the tank.

MTB does agree with industry commenters that the proposed retest at 1 ½ times design pressure would, for purposes of personnel safety, require hydrostatic testing and that this not only poses structural problems for hydrogen and helium tanks, but creates problems in drying and preventing contaminants from entering the tank. Since the cryogenic materials are noncorrosive and since the tank has been subjected to a pressure integrity test of 1 ½ times design pressure at time of manufacture, MTB believes it is appropriate, without compromising safety, to allow the retest to be performed at 1 ¼ times design pressure. A procedure for performing pneumatic tests is provided at § 178.338–16(b).

DESIGN LOADINGS

MTB received several comments concerning the "g" loading requirements. Concern was expressed regarding the different methods contained in the NPRM for defining the loadings (i.e., by "g" loadings and also as a function of weight of the tank). In order to clarify the support and anchoring requirements, MTB has revised § 178.338-13 by expressing all design forces as a function of the static loading imposed in the cargo tank and its attachments when the cargo tank is filled to the design weight of lading. These design forces apply to the tank and jacket as well as the support system.

One commenter urged that the proposed design forces be reduced. While present DOT exemptions (not specify-ing MC-331 construction) do not specify design forces for the tank or jacket of the vacuum-insulated cargo tank, MTB believes that it is necessary to specify some minimum design force considerations. It should be noted that the static loading requirements for the inner tank of a vacuum-insulated cargo tank were reduced from those proposed in the original NPRM by the June 21, 1979 publication modifying the NPRM. At the time this change was made, MTB stated that it would "not compromise the safety of the cargo tank because the outer tank, which is the structural member used in place of a motor vehicle frame, is proposed to be designed to "g" loadings of three vertical and two longitudinal and lateral." (44 FR 36211.)

As for the outer tank (generally referred to as a jacket in this rule) of a vacuum-insulated cargo tank, the same commenter believes the static loading requirements should be reduced from three vertical and two in all other directions to two vertical and 1½ in all other directions. This is the same loading requirement that was proposed (and is here made final) for the inner tank and its suspension members. MTB does not believe the commenter's recommended design forces are adequate. However, upon reevaluation, MTB does believe that the vertical loading requirement of three, proposed in the NPRM can be safely reduced since, for tanks with evacuated jackets, any dynamic forces are dampened by virtue of the double wall construction. This rule, therefore, imposes static loading requirements for the jacket of an MC-338 cargo tank of two in all directions. This agrees with the industry standard expressed in CGA-341.

Nonvacuum-insulated MC-338 cargo tanks are very similar in construction to MC-331 tanks with support members attached directly to the cold liquid tank. The NPRM proposed static loadings for this type of MC-338 tank that were the same as the required loadings for the MC-331. MTB believes the construction of this tank justifies the loading requirement of three in the vertical downward direction since those dampening factors that supported the reduction in vertical loading for the evacuated jacket are not present in this type of construction. There is also no logical reason for having a lower vertical loading requirement for this tank than for the MC-331 tank. For this tank, the proposed loadings in the NPRM are retained in the final rule.

NONPRESSURIZED ATMOSPHERIC GASES AND HELIUM

The MTB is adopting regulations that will apply to nonpressurized atmospheric gases and helium. In the NPRM, MTB stated as a basis for its proposed regulation of nonpressurized atmospheric gases and helium that " . . . certain requirements apply due to the extreme thermal hazard these low temperature materials pose to human tissue and, in the case of vessels, to shipboard structures."

Several commenters took strong exception to this specific proposal. Generally, they questioned the value of shipping papers and incident reports in dealing with the "so-called" thermal hazard of these cryogens. Also, commenters took exception to a statement in the regulatory evaluation for the proposal that indicated such materials could be considered corrosive from a technical standpoint. A commenter suggested that MTB's actual motivation was to obtain a reliable record (through incident reporting) of experience in transportation of such materials for purposes of more extensive regulation in the future. This commenter suggested such a record could be developed based on existing incident and accident reports without additional regulation. Further, he asserted that the existing safety record leads inevitably to the conclusion that such a "thermal hazard" does not, in fact, exist.

Another commenter noted that the ANPRM issued on March 1, 1974, and September 10, 1974 (39 FR 7950; 39 FR 32624) did not include these materials. He further indicated that MTB's proposal to regulate these materials due to the extreme thermal hazards these low temperature materials pose is tantamount to a finding by MTB, which the commenter felt had not been adequately supported, that, pursuant to 49 U.S.C. 1803 (enacted after the ANPRM was issued), the transportation of these cryogens may pose an unreasonable risk to health and safety or property.

MTB finds that the quantity and form of all materials subject to this final rule as cryogenic materials may pose an unreasonable risk to health and safety or property when transported in commerce due to the extreme thermal hazard they pose to human tissue. However, MTB acknowledges that the unreasonable risk potential presented by these materials because of their low temperature is not as significant as the risks posed by many other hazardous materials and, therefore, only a limited number of regulations should apply to their transportation.

If a liquid cryogen were to come into contact with any part of a human body, irreversible damage to tissue could be caused unless the contact is of short duration. Thsi fact is verified in industry literature. While the industries known by MTB to engage in shipment and transportation of these materials have a good safety record, MTB concludes these materials are hazardous materials for purposes of transportation in commerce and that some, but not all, of the hazard communication regulations and the incident reporting requirements should apply to them.

Other cryogens are regulated because they pose risks in addition to their extreme thermal hazard. Principal among these risks are flammability and toxicity (e.g., carbon monoxide). A lesser degree of risk is posed by atmospheric gases and helium when transported at pressures greater than 25.3 psig. These pressurized materials are presently required to be carried in cargo tanks under conditions specified in exemptions because they are, by definition, compressed gases. The principal reason for a higher level of regulation of these materials is to assure that the integrity of their packagings is sufficient to prevent failure at the pressures the packagings can be expected to encounter during transportation. However, in terms of immediately available energy, there is no significant difference in the risk presented by a tank containing a cryogenic atmospheric gas at a pressure above 25.3 psig versus a tank containing a cryogenic atmospheric gas at a pressure below 25.3 psig. (MTB does not draw the same conclusion relatively to tanks only partially filled since they have a greater immediate available energy potential when their relief valves are set at higher discharge pressures.)

Therefore, MTB concludes that the principal risk presented by packagings filled with atmospheric gases and helium is the thermal hazard these low temperature materials pose to human tissue and to shipboard structures and that specification packagings are not required for these materials when transported by motor vehicle or railroad at pressures at or below 25.3 psig.

It must be noted that the absence of accident data cannot be conclusive as to whether or not a material is a hazardous material. There are a number of hazardous materials currently subject to the regulations (e.g., spent nuclear fuel) for which there is no accident data indicating fatalities, injuries, or serious property damage as a result of accidents in transportation. If MTB were to agree that a material can not and should not be regulated unless there is accident data concerning transportation of the material in the form and manner to be regulated. MTB would be agreeing that it could only react to adverse experience in deciding if a material may pose an unreasonable risk when transported in commerce. The legislative history of the Hazardous Materials Transpotation Act clearly establishes that a strictly reactive posture was not the intent of Congress.

In this final rule, MTB is adopting requirements pertaining to the identification of atmospheric gases and helium when they are carried in packagings at pressures at or below 25.3 psig and incident reporting requirements consistent with those specified for other hazardous materials. The communications requirements relate to the preparation of shipping papers and the marking of packagings including cargo tanks and tank cars. Except for oxygen, MTB is not adopting requirements pertaining to labeling of packages and the placarding of vehicles and rail cars (except for carriage aboard aircraft and vessels). With the exception of oxygen, MTB does not believe that placarding and labeling are necessary for these materials. The identification of the materials by marking will be sufficient for communication of the risk they present during transportation and will allow for improved emergency response through use of identification numbers. MTB believes the risk presented by oxygen warrants the implementation of placarding and labeling requirements for this material (see NFPA Pamphlet 53M entitled, "Fire Hazards in Oxygen-Enriched Atmospheres 1979", for information concerning accidents involving oxygen, including cryogenic oxygen).

THE IMPACT OF THIS RULEMAKING ON EXISTING EXEMPTIONS

A number of cargo tanks are currently being operated under exemption, carrying pressurized and/or flammable cryogenic liquids. The NPRM proposed that where practicable these tanks be modified, remarked and rerated to MC-338 Specifications. Where not praticable, the NPRM proposed that the exemption holder so advise MTB, giving the reasons. The NPRM did not address what MTB would do after being so notified. The same process was proposed for rail cars currently transporting ethylene, cryogenic liquid under exemption. Exemption holders were concerned about the impact of the rule on outstanding exemptions and the extent of modifications to these existing tanks, and attendant expense, that MTB would require in an effort to bring them into compliance with the new specification requirements.

MTB has decided to completely resolve this issue by allowing existing cargo tanks and tank cars authorized to transport a cryogenic liquid or certain cold refrigerated gases under an outstanding exemption to be remarked and rerated as specification tanks without undergoing modification. MTB believes this "grandfathering" of existing tanks is necessary to avoid potential severe economic consequences to some exemption holders and can be justified from a safety point of view because of the thorough technical review involved in the exemption process, notwithstanding the fact that certain aspects of certain exemptions may differ from this final rule. However, no new construction under these exemptions may be initiated after January 1, 1984.

Under this rule, the owner or the person using a cargo tank or tank car complying with a DOT exemption listed below should remove the DOT exemption number stenciled on the cargo tank or tank car and stamp the identification plate as specified by § 173.31(a)(8), § 173.33(b)(2), or § 173.33(b)(3), as applicable, according to the proper specification. A copy of the applicable exemption in effect on December 31, 1983, must be retained by the owner or operator, if not the owner, for a tank remarked as a DOT specification packaging.

The DOT exemptions affected by this rulemaking are as follows:

Exemptions—Tank Cars

4717	6231	7491
5736	6392	
5792	6825	

Exemptions—MC-330 or MC-331 Cargo Tanks

5062	7744	8443
6939	7957	8476
6978	8336	8504
7632	8442	

Exemptions—MC-338 type Cargo Tanks

2587	4554	5959	6464	7192	8404
2708	4760	6039	6536	7207	8583
2805	5186	6111	6545	7444	8753
3367	5196	6113	6571	7494	8763
3648	5322	6173	6738	7513	8768
4108	5365	6197	6755	7600	8778
4399	5413	6205	6768	7603	8805
4400	5485	6218	6802	7849	8889
4404	5825	6243	6919	7911	8894
4490	5852	6403	6923	8286	8954
4497	5954	6432	7025	8393	

Exemptions-Cylinders

Cylinders marked "DOT E-6668", "DOT E-8404" or cylinders marked "DOT SP-6668" prior to issuance of the exemption may be marked "DOT-4L" as specified by § 173.23(D) of the rule.

REVIEW BY SECTION

Readers are reminded that this Review discusses only significant comments on the proposals in the NPRM and changes made to the NPRM in this Final Rule. For those provisions that are unchanged, readers are referred to the preamble discussion in the original NPRM (44 FR 12826, March 8, 1979) and subsequent correction documents (44 FR 20461, April 5, 1979; 44 FR 36211, June 21, 1979).

Section 171.1. There is no change, except that the proposed language in § 171.1(a)(i) is now contained in § 171.1(a)(iii) due to amendments to the Hazardous Materials Regulations subsequent to the issuance of the NPRM. Due to the length of time between the NPRM and this Final Rule, provisions have been relocated within a section in a number of instances. Throughout the remainder of this preamble, only significant relocations will be commented on.

Section 171.7. In § 171.7, paragraph (c)(3) is revised to reflect the new address of the Compressed Gas Association and, although they were not a part of the NPRM, paragraph (d)(3)(v) is revised to recognize the 1980 edition of CGA Pamphlet S-1.2 and paragraph (d)(12) is revised to recognize the 1978 edition of Federal Standard H-28 which superseded National Bureau of Standards Handbook H-28. The purpose of these and other changes is to recognize the most up-to-date standards so that users of these references will not have to hold superseded editions. MTB has examined the new standards and for the purpose of this incorporation by reference, recognition of the newer editions does not substantively affect the Hazardous Materials Regulations.

Section 171.8. The proposed definition of "filling density" is changed by including a separate entry for hydrogen, cryogenic liquid, in cylinders. The separate filling density reference for hydrogen, cryogenic liquid, in cylinders is needed for clarity. The separate requirement for hydrogen existed at proposed § 173.316(c)(3) of the NPRM, however the lack of a separate reference in § 171.8 may have led to some confusion. This change will avoid that. Also, MTB has added definitions of "atmospheric gases", "BTU", "cryogenic liquids", and "NPT". These added definitions are also to add clarity. The definition of "atmospheric gases" is especially important since under certain circumstances this rule excepts these materials from certain requirements.

Section 172.101. Five commenters addressed the proposed changes to the Hazardous Materials Table, including the Air Transport Association (ATA) and the Compressed Gas Association (CGA).

ATA's comments included the following suggestions: (1) to add "air, cryogenic liquid", "neon, cryogenic liquid", and "cryogenic liquid, n.o.s." to the Hazardous Materials Table (the Table); (2) to forbid pressurized argon, helium, and nitrogen in cryogenic form from transportation by passenger aircraft; and (3) to authorize the use of cylinders for all cryogenic liquids listed in the Table. In the final rule, MTB has amended the Hazardous Materials Table as suggested by ATA to add "neon, cryogenic liquid". However, MTB does not agree that the descriptions "air, cryogenic liquid" and "cryogenic liquid, n.o.s." should be added to the Table. MTB is unaware of any exception from the practice whereby the various cryogenic liquid fractions are separated and shipped in the form of liquid oxyen, liquid nitrogen, etc. Concerning the prohibition proposed by the ATA on the transportation of cryogenic argon, helium and nitrogen aboard passenger carrying aircraft, ATA did not provide any explanation of its concerns that malfunctioning of a pressure controlling valve would result in operation of a safety relief valve. If for any reason, including a blockage in the pressure controlling valve, there is pressure rise in the cylinder, MTB believes that the safety valve should function at its set-to-discharge pressure. MTB's decision to authorize these materials aboard passenger carrying aircraft is consistent with the decision made by the Dangerous Goods Panel of the International Civil Aviation Organization (ICAO). MTB agrees with ATA that the use of cylinders should be allowed and, accordingly, has authorized these materials to be transported by air in DOT-4L cylinders or in accordance with ICAO requirements under § 171.11.

ATA also proposed that three separate entries be provided in the Table for each nonflammable cryogen. The entries would designate the cryogen as pressurized, low pressure and nonpressurized. MTB does not agree with ATA that three separate entries should be listed for the cryogenic liquids. The basis for ATA's recommendation was that separate entries, rather than the exception proposed at § 173.320, would clarify the applicable requirements for the varying pressures. MTB takes the position that the use of a single entry in the Table, with the exceptions referenced in column 5(a), is consistent with the format of the Table and is well understood by persons affected by the regulations. Should MTB conclude at some future date that some means are needed to differentiate among the various categories of these materials, it will consider the matter in a future rulemaking.

ATA also expressed concern that cryogenic liquids, such as helium and neon, having boiling points colder than the freezing point of air should not be permitted in air transportation when in Dewar flasks because of the possibility the vents may freeze with solid air thereby allowing internal pressure buildup MTB is aware that these packagings have been transported aboard passenger aircraft in the past with no reports of adverse experience that would justify such a prohibition. ATA provided insufficient data as to why this prohibition should be imposed and MTB is not aware of any data or information that supports ATA's position. MTB's decisions not to prohibit the transportation of Dewar flasks containing authorized cryogenic liquids by aircarft is consistent with ICAO. MTB's response is the same with regard to ATA's suggestion that, as another alternative, MTB should consider banning all air transportation of cryogenic liquids regardless of pressure in the packaging.

Another concern of ATA's was that the provisions in proposed § 173.320 do not clarify the fact that such shipments offered for transportation by aircraft are subject to shipping papers, marking, labeling and other requirements pertaining to carriage aboard an aircraft in Part 175. MTB agrees and has revised § 173.320 by removing the reference to aircraft.

Maximum net quantity limits aboard aircarft (the Table, column (6)) were recommended by ATA based on the pre-December 31, 1982 listings in the International Air Transport Association (IATA) Restricted Articles Regulations and the Air Transport Association's Tariff 6-D. CGA's comments also addressed quantity limits. CGA suggested quantity limitations for argon, helium, and neon (CGA, as well as ATA, suggested that neon be added to the Table) that exceeded those in the NPRM. A quantity limitation of 550 pounds of argon by cargo-only aircraft was recommended to allow the shipment of the "150 liter, industry standard container". A similar quantity limitation was recommended for neon. CGA recommended that the quantity limitation of 300 pounds of helium by cargo-only aircraft be increased to 1100 pounds to be consistent with present air shipments in standard containers that include a 1000 gallon container. CGA did not suggest any change in the proposed passenger aircraft limitation of 100 pounds for argon or helium and suggested the same limitation be provided for the new neon entry. MTB agrees with the intent of CGA's suggestions relative to maximum quantity per package aboard an aircraft, however, MTB has modified the suggestion by adopting the limitations used by ICAO for cryogenic liquids, rather than those of CGA, ATA or IATA. MTB believes usage of these quantity limitations will facilitate transportation of these materials moving in international commerce by air without compromising safety. The maximum net quantity in one package by cargo-only aircraft of helium, neon, and nitrogen is 1,100 pounds and the maximum net quantity of argon is 300 pounds. The quantity limitation in one package of these gases by passenger aircraft is 100 pounds.

Several commenters addressed the proposed stowage requirements for water shipments, which in several instances were more stringent that the stowage curently allowed. Specifically, argon, cryogenic liquids and helium, cryogenic liquid were proposed to be allowed "on deck" stowage, and oxygen, cryogenic liquid was proposed to be allowed "on deck" stowage on a cargo vessel, but forbidden on a passenger vessel. (Present regulations allow helium and the cryogenic forms of argon and oxygen to be stowed on deck or under deck. Argon and oxygen when stowed under deck must be "under deck away from heat.") After further consideration of the proposed stowage requirements, the Coast Guard, which assisted MTB in the preparation of these requirements, agreed to the stowage requirements contained in the present regulations except for that which permits oxygen, cryogenic liquid, to be stowed "under deck" on a passenger vessel. This rule does, however, permit oxygen, cryogenic liquid to be stowed on deck on a passenger vessel and it does not appear there will be any hardship created by this requirement.

Three commenters also objected to the proposed provision prohibiting hydrogen, cryogenic liquid, on board a cargo vessel. The commenters argued that the "5" in column 7(a) should be changed to "1" for consistency with other flammable cryogenic liquids which have similar properties. MTB and the Coast Guard do not agree with the commenters because hydrogen with its wide flammability limits poses a greater potential hazard than other flammable cryogenic liquids. Therefore, the suggested change has not been made.

A change has been made in column (7)(b) for gaseous hydrogen from "54" to "4". The reason for this second change is to correct a typographical error since the number "54" had no meaning and the entries "5" and "4" were conflicting. Retaining only the number "4" allows the water shipment aboard passenger vessels of limited quantities of the material.

ATA and CGA were not in favor of including the term "cryogenic liquid" as part of the proper shipping name. ATA, as discussed earlier, supported use of the descriptions "low pressure", "non-pressurized" and "pressurized". CGA recommended that the term "pressurized liquid" be used for consistency with the terminology in existing regulations. MTB believes the "cryogenic liquid" terminology more readily conveys the thermal hazard posed by these materials to emergency personnel and transport workers. Therefore, the term "cryogenic liquid" is being retained as part of the proper shipping description in the final rule.

MTB has added identification numbers to the new entries being included in the Hazardous Materials Table. This is in accordance with the final rule issued under HM-126A (which was completed after the NPRM for this action (45 FR 34560; May 22, 1980)). The new cryogenic liquid descriptions are preceded by an "NA" prefix.

Another major change since the NPRM is the addition of certain entries designated as "liquid (refrigerated)". These entires cover materials that, while not meeting the definition of a cryogenic liquid, are gases at atmospheric pressure and ambient temperatures and are transported in cold liquid form. MTB has added these materials to this rulemaking in order to eliminate a number of exemptions. Because of the grandfathering of packagings covered by existing exemptions, the addition of these materials to the Table and related requirements will not add to the burden of current exemption holders. In fact, the addition of these materials to the Table eliminates the need for an exemption to transport these harzardous materials.

Finally, several commenters requested that the current proper shipping name of hydrogen chloride be retained rather than modified by the description "anhydrous", as proposed in the NPRM. The commenters pointed out that the use of "hydrogen chloride" to describe anhydrous material is universally known in the industry and adding the word "anhydrous" is unnecessary. As requested by these commenters, MTB has allowed the current name, "hydrogen chloride", to remain.

Section 172.203. Three commenters pointed out that the notation on shipping papers used to alert rail carriers that certain tank cars may not be humped or cut off while in motion should be applicable to DOT 113D tank cars. MTB agrees with the commenters. Omission of the DOT 113D tank car was an oversight. MTB has revised paragraph (g)(3) to clarify that the notation is applicable to any class DOT 113 tank car.

Section 172.328. The parenthetical language "(including a cryogenic liquid)" has been added to clarify the fact that this provision applies to those materials.

Section 172.504. Instead of deleting Note 2 to Table 2 and renumbering the remaining Notes, as proposed in the NPRM, the text to Note 2 is removed and Note 2 is reserved.

Section 173.11 (proposed as § 173.5). Several commenters objected to the proposed requirements in this section for shippers of flammable cryogenic liquids in portable tanks, cargo tanks and tank cars, and the proposed requirements in § 177.825 (now § 177.826) for carriers of flammable cryogenic liquids in cargo tanks by highway, to file registration statements with MTB. Three of the commenters opposed the requirements as unnecessary. One of the commenters pointed out that registration is not required for other hazardous materials, that there is no evidence that it is needed here, and that such a requirement is contrary to the intent of the Paperwork Reduction Act of 1980 (44 U.S.C. 3501–3520). Three commenters objected to the provision in the proposal that would require the registration statement to be acknowledged by MTB prior to shipment of flammable cryogenic liquids since it was believed this would result in unnecessary and unreasonable delays. A different commenter suggested that the words "cargo tank" should be included in paragraph (a)(5) for consistency with the intent stated in the preamble. The remaining commenter suggested that the requirement should not apply to shippers of tank cars since it was not a requirement in the exemptions and the requested information would not enhance safety.

MTB disagrees with the commenters who believe that there is no need for registration of shippers and highway carriers of flammable cryogens. The registration requirement adopted in this rule will provide important information which will enable MTB to identify and locate shippers and carriers of these very hazardous materials for periodic inspections of their operations. For the same reason, MTB has not excepted shippers of tank cars from the requirement.

MTB does agree with the commenters' concern regarding the potential time lapse prior to MTB acknowledgement. It is MTB's intention that the registration statements be processed in a timely manner. Accordingly, as suggested by one of the commenters, MTB has revised the requirement to allow the registration statements to be filed by certified mail with a return receipt requested. The signed copy of the returned certified mail receipt will serve as MTB's acknowledgement of the registration statement. A similar change has been made to § 177.826.

Additional changes to this section since the NPRM, include the requirement that registration filings be with the Associate Director for Hazardous Materials Regulation, rather than the Associate Director for Operations and Enforcement. This is not a substantive change and merely represents a reorganization of functions within MTB. The section has been reorganized and the commenters suggestion that cargo tanks be added in proposed paragraph (a)(5) has been adopted (the corollary paragraph in the final rule is (b)(4)). Finally, the dates on which registration begins have been revised to be consistent with the effective date of this rule.

Section 173.23. A new paragraph (d) is added to permit remarking of cylinders covered by a DOT exemption, as discussed under the heading "Impact of This Rulemaking on Existing Exemptions."

Section 173.29. Proposed paragraph (f)(3) would have prohibited venting of any gas, other than an atmospheric gas, remaining in a tank car previously used to transport a cryogenic liquid. Upon review, MTB believes the same purpose is served by the present requirement of paragraph (c)(2) which requires that an empty tank car that has been used to transport a hazardous material must be shipped in the same manner as when it contained a greater quantity of material unless it has been cleaned, purged or reloaded with a material not subject to the regulations. Therefore, proposed paragraph (f)(3) has not been incorporated into the final rule.

Section 173.31. Proposed paragraph (a)(6) (paragraph (a)(8) herein) is revised as discussed under the heading "Impact of this Rulemaking on Existing Exemptions" to permit remarking of tank cars covered by a DOT exemption. Additionally, the provision has been changed to allow the operator as well as the owner to do the remarking.

A new paragraph (a)(9) is added to provide that tank cars built to Specifications DOT-113A175W, DOT-113C60W, DOT-113D60W, or DOT-113D120W may continue in service, but no new construction of these tank cars is authorized. No new construction of any of the tanks have been authorized under exemption since 1973. The only tank cars authorized for new construction are specification DOT-113A60W (for hydrogen) and specification DOT-113C120W (for ethylene).

Proposed paragraph (c)(13)(i) is revised, as suggested by two commenters, to clarify that a class DOT-113 tank car is not required to be retested unless the pressure rise in the tank exceeds 3 psi per day. If the pressure rise exceeds 3 psi, the tank must be properly retested prior to being returned to service. The NPRM language was not as clear and referred only to 113A cars. Application of this provision to all DOT-113 cars was intended. The correction of this oversight represents a relaxation of the proposed requirements. Two commenters to paragraph (c)(13)(i) stated that the requirement to monitor the cars for pressure rise should only apply to cars transporting flammable cryogens. MTB notes that at the present time the only cryogens authorized for shipment in tank cars are flammable and hence although the language of this section does not limit application to flammable cryogens, §172.101 does.

In paragraph (c)(13)(ii)(B), the proposed evaporation rate retest has been replaced with a calculated heat transfer rate (CHTR) retest to more accurately reflect the performance tests specified for class DOT-113 tank cars operating under exemption and to be consistent with a change made to proposed §179.400-4 and published in this rule. The CHTR retest provides a method for qualification of a tank car insulation system that can be more readily accomplished than that which was proposed in the NPRM.

Paragraph (c)(iii) has been rewritten to clarify the fact that successfully completing either of the retest prescribed in (c)(ii) will allow a car to be placed back in service.

In this section, as in a number of others, the term "safety relief valve" has been replaced by the term "pressure relief valve." While both terms are generally understood by industry to refer to the same device, MTB believes that "pressure relief valve" is the more accurate term and MTB, in line with certain industry groups, is using this terminology in this rule. MTB has not attempted to change all references in the Hazardous Materials Regulations where the term "safety relief valve" is used. For purposes of the Hazardous Materials Regulations the terms have the same meaning.

Retest Table 1, which until this rule followed paragraph (d)(4), now follows paragraph (c)(13) and has been expanded to include tank cars formerly under DOT exemption that are to be remarked in accordance with paragraph (a)(8) of this section. Additionally, the DOT-113A175W car, proposed to be deleted, has been left in the Retest Table to meet the needs of a commenter who indicated that tank cars of this type have been reconditioned and returned to service. In response to another comment, the footnote references have been corrected. Also, the vapor tight pressure for the DOT-113A175W tank car has been reduced to 80 percent of the start-to-discharge pressure of the safety relief valve. This change is consistent with the vapor tight pressure valves used for other DOT class tank cars. Two other commenters recommended that a new footnote "t" be added following the table to cover the retest of alternate safety relief valves when installed. Such a footnote is not necessary, however, because §173.31(c)(13)(v) prescribes a retest that is applicable to any alternate safety relief device.

Section 173.33. Several changes to this section have been discussed previously in the preamble in the subject-by-subject review: (1) Cargo tanks currently used to transport cryogenic liquids under exemption will be "grandfathered" and will not have to be modified or have the inability to make such modifications explained to MTB (proposed §§173.33 (b)(2) and (b)(2)(iii); (2) the holding time verification (proposed §173.33(d)(1)(ii)) now only applies to cargo tanks containing flammable cryogenic liquids rather than all cryogenic liquids; and (3) the required retest pressure for an MC-338 tank (proposed §173.33(d)(2)) has been reduced from 1½ times maximum allowable working pressure to 1¼ times that figure.

Additionally, the section has been revised to require cargo tanks currently used to transport certain cold, refrigerated gases under exemption to be stamped with the MC-330 or MC-331 specification identification followed by the exemption number. This serves to "grandfather" these tanks, thereby removing them from the exemption program (although the exemption number will remain with the tank for its service life to substantiate the authorized use of the tank).

The pneumatic testing method in paragraph (d)(2) no longer specifies that a soap and water solution or other suitable material must detect leaks through the presence of foaming or bubbling. Now any "equally sensitive method" is acceptable. This change provides greater flexibility and recognizes that there may be a number of acceptable methods for detecting leaks.

Another change allows either the owner or operator of a tank currently operating under exemption to perform the remarking. The NPRM spoke in terms of the exemption "holder". This change recognizes the fact that frequently a tank may be under long-term lease to an operator who has exclusive use and possession of the tank. This person is authorized to perform the remarking, thereby reducing the possibility that a tank would have to be taken out of service so that the "holder" could perform the remarking.

The names "carbon dioxide" and "nitrous oxide" are removed and replaced by "carbon dioxide, liquid (refrigerated)" and "nitrous oxide, liquid (refrigerated)" each time they appear in this section.

Section 173.300. A commenter suggested that the definition of a cryogenic liquid in proposed paragraph (f) be changed to specify a reference pressure at 14.7 psig for "technical clarification." MTB believes a reference pressure would clarify the provision and, for consistency with the liquefaction temperature, has specified a pressure of one atmosphere.

In the NPRM, the definition for filling density in paragraph (g) would have been redesignated as paragraph (h) and amended to indicate the different sections applicable to cryogenic liquids. However, MTB has removed present paragraph (g) in the final rule since a definition of filling density has been added to § 171.8.

Section 173.304. The proposed Table to paragraph (a)(2) has been changed. "Hydrogen chloride" is no longer being deleted and added as "Hydrogen chloride, anhydrous." As noted in the discussion of § 172.101, the name of this material is no longer being changed. However, the name "carbon dioxide, liquefied" is being removed and the name "carbon dioxide" is added.

As noted in the April 5, 1979 addition to the NPRM, paragraph (b) has been revised to clarify filling limits for vinyl fluoride, inhibited.

Section 173.314. Several commenters suggested that tank cars containing hydrogen chloride, liquid (refrigerated) should be marked "HYDROGEN CHLORIDE" for consistency with the requirement for chloride materials which have similar hazards. MTB agrees and has added the requirement in paragarph (b)(6). In addition, in the Table to paragraph (c), the maximum pressure for tank cars containing hydrogen chloride, liquid (refrigerated) when offered for transportation is increased from 80 psig to 90 psig, based on the concerns of several commenters that increased time be allowed for adequate inspection of tank cars. The effect of the change is minimal and will have no adverse effect on safety. Also, the entries to the Table have been changed, as noted in discussions of other sections, to reflect different shipping names for carbon dioxide, liquid (refrigerated) and hydrogen chloride, liquid (refrigerated).

Section 173.315. MTB believes the reference to the marking requirement in the introduction text to this paragraph is unnecessary and, therefore, it has been deleted. In the Table to paragraph (a), the obviously incorrect filling density for hydrogen chloride, liquid (refrigerated) of "10.3 percent" is changed to "103.0 percent". Also, MTB is authorizing the shipment of refrigerated ethane, ehtane-propane mixture, and hydrogen chloride, liquid (refrigerated), in insulated MC-338, as well as MC-331, cargo tanks. This eliminates an unnecessary restriction in the NPRM that precluded the use of an MC-338 tank for materials for which a less heavily insulated MC-331 tank was authorized.

Note 11 to the Table has been extensively revised. The NPRM proposed that each tank have a design service temperature no warmer than the material's liquefaction temperature. As discussed earlier in the preamble under the heading "Liquefaction Temperature," the term "liquefaction temperature" has been replaced with the more readily understood term "boiling point at one atmosphere" to eliminate possible ambiguity. The requirement in present Note 11 that a cargo tank must be designed for a service temperature no higher than minus 100° F. which was inadvertently omitted in the NPRM, has been retained.

The holding time requirement has been rewritten for clarity, but without changing the substance of the requirement.

The last sentence in proposed Note 11 would have required that, before being transported in an empty condition, a cargo tank be drained, vented or blown down sufficiently to prevent venting while en route. In view of the existing requirement that an empty cargo tank last used to transport a hazardous material must be transported in the same manner as when it contained a greater quantity of material. MTB does not believe the provision is necessary and therefore it has not been included in the final rule.

A requirement has been added to Note 11 that a cargo tank used to transport a flammable gas must have an outer steel jacket. The reasons are discussed earlier in this preamble under the heading "Use of Aluminum."

Finally, the names "carbon dioxide" and "nitrous oxide" are removed and replaced by the names "carbon dioxide, liquid (refrigerated)" and "nitrous oxide, liquid (refrigerated)," respectively, each time they appear in this section.

Section 173.316. Proposed paragraph (a)(3) has been changed so that the steel jacket requirement does not apply to cylinders transporting cryogenic oxygen. This change was urged by several commenters and also responds to petitions submitted by CGA dated July 3, 1975, and January 19, 1977. This change is supported by the satisfactory safety record of aluminum jacketed cylinders in cryogenic oxygen service under DOT E-6668, and the fact that prior to the granting of this exemption, in 1972, extensive fire tests were performed with satisfactory results. These factors, as well as the quality of material in each package, distinguish cylinders from cargo tanks and support the position adopted in this final rule that aluminum jackets be authorized for cylinders, but not for cargo tanks, used to transport cryogenic oxygen. See the earlier preamble discussion under the heading "Use of Aluminum."

Proposed paragraph (a)(4) has been revised to limit its applicability to cylinders used to transport cryogenic oxygen. Flammable cryogens are now addressed in a new paragraph (a)(5), which prohibits an aluminum valve, pipe or fitting on any cylinder used to transport a flammable cryogenic liquid. This prohibition was supported by the commenters and is necessary, due to the low melting point of aluminum, in order to preclude an increase in the consequences of a fire involving a cylinder containing a flammable cryogenic liquid.

One commenter suggested that pressure relief devices on cylinders should comply with CGA Pamphlet S-1.1, entitled "Pressure Relief Device Standards, Parts I—Cylinders for Compressed Gases." MTB adopted CGA Pamphlet

S-1.1 under an earlier rulemaking to require that pressure relief devices on cylinders comply with CGA standards (Docket HM-163E, 46 FR 22194, April 16, 1981). The requirement is contained in § 173.34(d).

Hydrogen has been excepted from applicability of the filling density definition in paragraph (c)(1) and a separate definition has been added as Note 1 to the Table in (c)(3). This was an oversight in the NPRM since the Table to (c)(3) clearly stated that filling density was based on cylinder capacity at minus 423° F.

As suggested by commenters and discussed in the explanation for § 172–101, cryogenic helium and neon are authorized for shipment in cylinders in this final rule and this is reflected by their addition to paragraph (c)(2).

Section 173.318. The proposed requirement to allow aluminum valves without rubbing or abrading aluminum parts have been retained in paragraph (a)(4) for cargo tanks used to transport cryogenic oxygen. However, a prohibition on an aluminum valve, pipe or fitting being used on a cargo tank used to transport a flammable cryogenic liquid has been placed in a new paragraph (a)(5). The reason for this change is the same as for the similar change made to § 173.316.

A new paragraph (b)(1)(ii) has been added to clarify that each pressure relief valve must be designed and constructed for a pressure equal to or exceeding the tank design pressure at the coldest temperature to be encountered. New paragraph (b)(5)(iii) contains a requirement to mark the set-to-discharge pressure of each pressure control valve. This requirement is consistent with that contained in § 173.315 for other cargo tanks.

In the Table to proposed paragraph (f)(3), the design service temperature for methane or natural gas was −320° F. This was in error. The correct design service temperature specified in the final rule is −260° F.

The operational requirement, proposed in §§ 178.338–9(c) and 178.338–18(d), specifying that a cargo tank containing a flammable cryogenic liquid must be marked with the one-way travel time (OWTT) for the material contained in the cargo tank has been deleted from the specification section. The requirement is now, more appropriately, in new paragraph 173.318(g). Additional comments and changes to this section have been addressed earlier in this preamble under the headings "Pressure Relief Device Systems," "Minimum Outage and Filling Densities," and "Venting, Holding Time, Trip Monitoring and Equilibration of Cryogenic Liquids."

Section 173.319. The requirements in this section have been revised to clarify that they apply only to flammable cryogens.

MTB received conflicting comments on the use of scales to determine the amount of cryogenic liquids loaded into a tank car as proposed in paragarph (a)(2). MTB has decided to delete the required use of scales for hydrogen cars, but has retained the method for cars containing other cryogenic liquids. The exception for hydrogen is made due to the extremely low density of this material and the fact that even a reasonably accurate scale capable of weighing a rail tank car would likely have a plus or minus error that would make it unrealiable as a method of checking the amount of hydrogen that has been loaded.

The requirement in proposed paragraph (a)(3) to notify the Bureau of Explosives (B of E) when a tank car has not been received by the consignee after 20 days was opposed by three commenters. One of these commenters pointed out that a tank car operating between such places as Texas and Alaska would exceed 20 days in transit on every shipment. The other two commenters argued that the design of cars for flammable cryogenic ladings provides for a 30 day holding time and therefore taking action after 20 days is premature. After considering the matter, MTB has decided to retain the notification requirements. MTB and the Federal Railroad Administration believe that, for tank cars transporting flammable cryogens, tracing should begin after 20 days in transit to ensure that a car will be located and routed to reach its destination within the holding time of the car.

Paragraph (c) has been changed to relate filling temperature to pressure rise, instead of the holding time, for consistency with changes made to § 179.400–4. These changes recognize that it is not current practice to conduct holding time tests on tank cars in cryogenic liquid service.

Finally, the permitted filling densities in the table in paragraph (d)(2) have been modified for ethylene to accommodate tank cars currently operating under a DOT exemption that under this rule will be remarked in accordance with § 173.31(a)(1). The DOT-113A175W car has been added to the authorized specifications for hydrogen as noted in the preamble discussion of § 173.31.

Section 173.320. The applicability of the exception provided by this section is expanded to include all atmospheric gases and to make it clear that the regulations pertaining to cryogenic liquids do not apply to refrigeration systems during transportation by motor vehicle, railcar, or vessel. The wording has also been clarified to emphasize that in order to qualify for this section, the design and construction of the packaging, not the thermal protection, must limit the pressure to 25.3 psig.

The exception has also been expanded to exclude atmospheric gases (except oxygen) and helium in qualifying packaging from the labeling and placarding requirements in Part 172. For carriage aboard aircraft, the provisions of § 173.11 apply; therefore, there is no reference to aircarft in § 173.320.

The issue of imposing any requirements on the shipment of atmospheric gases and helium is discussed earlier in this preamble under the heading "Nonpressurized Atmospheric Gases and Helium."

Part 174. Except for minor reorganization of § 174.204(a)(2), the sections in this Part are amended and revised as proposed in the NPRM.

Section 176.76. A number of changes have been made to this section. Several commenters recommended that this section be expanded to provide for cargo tanks formerly under a DOT exemption that have been remarked in accordance with § 173.33(b)(2). MTB agrees and this change has been made in paragraph (h)(1).

The reference, in proposed paragraph (h)(1) to § 173.318(a)(4) is now treated more clearly by restating the applicable requirements in paragraph (h)(2). Proposed paragraph (h)(2) is now (h)(3).

In line with other provisions regarding holding time, proposed paragraph (h)(2) (now paragraph (h)(3)) has been changed so that it now only applies to flammable cryogens. See the discussion under the heading "Venting, Holding Time, Trip Monitoring and Equilibration of Cryogenic Liquids" earlier in this preamble. Several commenters correctly pointed out that proposed paragraph (h) does not provide for tank cars containing a cryogenic liquid to be transported by vessel. This was an oversight on MTB's part. Accordingly, a new paragraph (h)(4) has been added to allow shipment of cryogenic liquids by vessel in DOT-113 or (for nonpressurized atmospheric gases and helium) AAR 240W tank cars. Because of the extreme thermal hazard cryogenic liquids pose to shipboard structures, a requirement has been added in new paragraph (h)(4) to require portable tanks, cargo tanks and tank cars containing cryogenic liquids to be stowed "on deck" only.

Section 177.816. The NPRM proposed that drivers of cargo tanks used to transport a flammable cryogenic liquid be required to receive formal training, at least once every 24 months, on the proper handling of the particular flammable cryogen being transported in the particular type of cargo tank. Included in the proposal were requirements that written records of training be kept by the carrier and that the driver be issued a certificate of training to be carried on his person while operating the motor vehicle. Three comments were received concerning this section. The Interstate Commerce Commission supported a detailed, structured curricula. The second commenter suggested that corresponding revisions should be made to § § 391.11, 391.51, and 397.19 of the Federal Motor Carrier Safety Regulations, and the third commenter opposed the proposal as redundant and as an unnecessary regulatory burden that would do nothing to enhance safety in view of § 177.800. This commenter also questioned whether the driver would be placed out-of-service if he did not have the certificate of training in his possession during a DOT roadside inspection.

This final rule adopts a driver training requirement because MTB believes that the program will increase a driver's knowledge and general awareness of the applicable Hazardous Materials Regulations, and his knowledge of the hazard of cryogenic liquids and the handling and operating characteristics of the particular vehicle used to transport the material. This requirement reflects good business practice and helps to ensure that this type of training will be performed at regular intervals.

In response to the comments, MTB notes that the training presently required by § 177.800 is general in nature and does not cover the details proposed, and now contained, in § 177.816, which is specifically applicable to flammable cryogenic liquids. Adoption of a structured curricula as suggested by the ICC, is outside the scope of this rulemaking. If MTB determines in the future that there is a need for a structured driver training program, it will be made the topic of a future rulemaking.

The major change to the NPRM is that a certificate of training is no longer required. MTB reevaluated the burdens and benefits of such a requirement and now believes that requiring the paperwork to be completed and then retained by the driver is unnecessary in view of the fact that the information can be obtained from the carrier's files.

In order to reduce paperwork burdens, MTB has reduced the period of time that a carrier must keep the driver's record of training after a driver has left the carrier's employ from three years to 90 days.

Paragraph (a) is changed to indicate that when interchange operations are involved, only the originating carrier is subject to this section.

Section 177.818. One commenter recommended that the provisions of this section be limited to transportation of flammable cryogenic liquids. The omission of the word "flammable' preceding "cryogenic liquid" in this section was in error. The omission has been corrected.

Section 177.824. This section is adopted as proposed in the NPRM.

Section 177.826. MTB has revised this section along the lines of § 173.11 to allow carriers to file registration statements by certified mail with a return receipt serving as MTB's acknowledgment. One commenter expressed concern that certain information contained in the registration statement, such as the number and type of vehicles in use by a carrier, may be considered proprietary and should be protected from public release. For years, each motor carrier has filed with the Federal Highway Administration's Bureau of Motor Carrier Safety (BMCS) a listing of all MC-330 and MC-331 cargo tanks the carrier has in service (49 CFR § 177.824(f)). MTB does not believe that the type of information that must be provided under this section is any more proprietary than that provided to BMCS on the MC-330 and MC-331 cargo tanks. However, if a carrier believes that its information is entitled to confidentiality under the Freedom of Information Act (5 U.S.C. 552), or is the type of material referred to in 18 U.S.C 1905, the carrier is entitled to request confidential treatment under the terms of 49 CFR 107.5.

In paragraph (c), the initial filing period has been revised to be consistent with the effective date of this amendment.

Section 177.840. Paragraph (h) has been restructured for clarity.

Paragraphs (i) and (j) have been rewritten to allow for equilibrations. This is discussed earlier in the preamble under the heading "Venting, Holding Time, Monitoring, and Equilibration of Cryogenic Liquids."

MTB believes paragraph (j) of the NPRM which specified a 50 percent reduction in one-way travel time for cargo tanks used in distribution service (peddle runs) is no longer necessary in light of the new provision on equilibration specified in §177.840 and the modified provisions to allow venting of nonflammable cryogens as specified in §173.320. Accordingly, it has been deleted.

Paragraph (k) of the NPRM, which discussed empty cargo tanks has also been removed, as discussed previously in this preamble with regard to §173.29, MTB believes that this proposal was unnecessary in light of the present requirement in §173.29(c)(2).

Paragraph (l) of the NPRM is now paragraph (k) and has been clarified to indicate that the apparatus must be approved by the National Institute of Occupational Safety and Health.

Section 178.57. Several changes have been made to this section that were not addressed in the NPRM. These changes, however, reflect an expanded use of the DOT-4L cylinder and in most instances are necessary to conform to other changes that have been made in this rule.

In §178.57-2, The maximum service pressure has been revised to more accurately reflect the maximum pressures authorized in Part 173 for cryogenic liquids during transportation. In paragraph (c) of this section, design service temperatures are specified for argon, helium, neon, nitrogen and oxygen, in addition to hydrogen. The temperatures specified are the same as those previously required under paragraph (c) except for helium and neon, which were not previously authorized for transportation in DOT-4L cylinders.

Section 178.57-5 has been changed to reflect the fact that for the carriage of certain commodities an aluminum outer jacket is authorized. See the discussion of this in the preamble discussion of §173.316.

Section 178.57-8 has been revised to provide for aluminum as well as steel jackets on DOT-4L cylinders. A steel jacket on a non-evacuated insulation system may be no less than 0.060 inch thick, as prescribed by present regulations. When an aluminum jacket is used on a non-evacuated insulation system, it may be no less than 0.070 inch thick. This is consistent with existing exemptions. Paragraph (c) also provides that when a steel or an aluminum jacket is used on vacuum insulated cylinders, the jacket must be designed to a minimum collapsing pressure of 30 psi. The 30 psi minimum collapsing pressure will provide greater flexibility than the specification of a minimum thickness and is better suited to controlling the wall thickness of an evacuated jacket to provide resistance to collapsing. Paragraph (d) of the same section has been added to ensure that the requirement in §178.57-20(a)(4) is complied with at the time of manufacture.

In §178.57-10, the definition of "P" has been rewritten to indicate that the pressure test need not be hydrostatic and to clarify that this figure is to be expressed in psi.

Section 178.57-12 now provides that the fitting, boss or pad provided for each opening may be "integral" in lieu of "securely attached". The reference in paragraph (a)(2) has been changed from "American Standard taper pipe threads" to "NPT". This change, with the associated definition in §171.8, helps pinpoint the reference without making a substantive change. Paragraph (a)(3) has been rewritten for clarity. As previously worded it was unclear whether the required inertness and leakage prevention characteristics referred to the threads or the gasket. Although the answer may have been obvious, it has now been stated correctly.

Section 178.57-13 was addressed in the NPRM and it is adopted here with the added rquirement that flow capacity meet the industry standards in CGA Pamphlet S-1.1.

In §178.57-20, the references to "service temperature" in paragraph (a)(2) have been changed to "design service temperature", without substantive effect. An identical change has been made in paragraph (a)(4) along with minor rewriting for clarity. The examples shown in the present paragraph (a)(4) have been moved to paragraph (a)(5). Paragraphs (a)(8) and (a)(9) have been added to required special orientation, instructions and, when appropriate, a marking which identifies a cylinder with an aluminum jacket.

Section 178.57-21 has been revised to recognize that an aluminum jacket is authorized for nonflammable cryogens and therefore the materials of construction for the cylinder and outer jacket are now specified separately.

Section 178.57-22 is being changed to make very minor changes to correct improper uses of plurals, to change "service temperature" to "design temperature" and to move the chemical analysis table to the proper location.

Section 178.337. MTB has revised §178.337-1(a) for clarity and to authorize MC-331 cargo tanks to be constructed with aluminum. If aluminum is used for the inner tank, the tank must be insulated. When an insulated tank is intended to be used to transport a flammable gas, a steel jacket is required. Although not addressed in the NPRM, MTB believes these modifications are justified in view of its decision to allow aluminum in the construction of MC-338 cargo tanks under similar conditions. Section 178.337-1(e) has been revised to recognize the changes made to §§173.315(a) Table Note 11 and 178.337-1(a).

Section 178.337–11(c) has been revised to include the term "NPT" as a reference standard for the discharge openings. Also the reference to hydrogen chloride has been changed to be consistent with the new proper shipping name "hydrogen chloride, liquid (refrigerated)."

Section 178.338–1. An additional reference has been added in paragraph (a)(2) of § 178.338–1 to clarify the design service temperature. See the preamble discussion under the heading "Liquefaction Temperature", for further discussion.

A commenter requested that paragraph (c)(1) be revised to require each tank to be "designed, constructed, and stamped . . . ." A tank constructed to the MC-338 specification is required to be marked by stamping, embossing, or other means, as prescribed by § 178.338–18. Therefore, MTB believes adding the word "stamped" to this section is not appropriate. Also, the commenter requested that paragraph (c)(1) allow closing welds to be made with non-removable backing strips. Construction of such a cargo tank is not prohibited by the regulations provided the tank is designed and constructed to allow washing of the interior surface, as prescribed by paragraph (c)(3), and meets the cleanliness requirements contained in § 178.338–15. Therefore, no change is necessary. As suggested by the same commenter, MTB has made a minor revision in the lower design pressure value specified in paragraph (c)(1) from 25 psig to 25.3 psig for consistency with § 173.320. The word "cargo" has been removed, in the reference to "cargo tank" in paragraph (c)(3), for consistency with the definition of tank in § 178.338–1(b).

In paragraph (e), a commenter requested that MTB specify a thickness in inches, in addition to the gauge values specified for the minimum metal thicknesses, in order to ensure proper calculations. However, the gauge-to-inch values suggested by the commenter were not in accordance with gauge thicknesses prescribed in the Table to § 173.24(c). The commenter did not explain why the values in § 173.24 were unsuitable. MTB has now specified minimum thicknesses, in inches, based upon the minimum thicknesses prescribed in § 173.24(c).

The design parameters for evacuated jackets in paragraph (f) were revised in the June 21, 1979 notice. In that notice, MTB acceded to commenter's requests and incorporated some of CGA's recommendations for a vacuum jacket contained in CGA-341 which, to a large extent, references the ASME Code. In its comments to the NPRM, as modified, the CGA recommended adoption of § 178.338–1(f) as initially proposed in the March 8, 1979 NPRM and expressed its view that the June 21, 1979 revision to paragraph (f)(2) was unsuitable. After considering these comments, and for reasons discussed under the heading "Jacket Design" earlier in this preamble, MTB has adopted the proposals for evacuated jackets contained in the June 21, 1979 notice, but has specified that the minimum collapsing pressure for evacuated jackets must be a least 30 p.s.i. This is consistent with standards in Pamphlet CGA-341.

Section 178.338–2. A commenter objected to the proposed requirements in paragraph (a) that jacket material must be in accordance with the ASME Code and to the additional postweld heat treatment required for cargo tanks constructed of UHT materials in proposed paragraph (e). The commenter stated that ASME material has never been mandatory for evacuated jacket material. This commenter cited the narrow temperature range for proper heat treatment of 5%, 8% and 9% nickel steels as the reason for opposing the proposed requirements for UHT materials. These same objections were raised to proposed § 178.338–4(a).

MTB is not aware of any test or experience data on the suitability of cargo tanks constructed of UHT materials that have not been postweld heat treated. Also, MTB is not aware of any cargo tank constructed of 5% or 8% nickel steels in pressurized cryogenic liquid service. Furthermore, there is no authorized cargo tank constructed of 9% nickel steels in pressurized cryogenic liquid service under exemption that MTB can use as the basis for obtaining experience data. The one cryogenic cargo tank constructed of a modified 9% nickel steel being operated under exemption has not been postweld heat treated, but it is mounted on a flat-bed trailer; therefore, it is not subject to full torsional forces and dynamic shock loadings normally encountered during transportation. In the absence of such data, MTB strongly believes the requirements proposed in the NPRM are justified in order to fully account for the dynamic forces encountered in the transportation environment. These forces are not considered by the ASME Code, which basically establishes standards for stationary pressure vessels. Furthermore, MTB believes that anyone who fabricates with UHT material is capable of properly heat treating these materials. Other UHT steel pressure vessels require postweld heat treatment. Therefore, the proposed requirements are retained in the final rule.

Section 178.338–3. A commenter requested that MTB incorporate the more lenient metal thickness requirements of the ASME Code. Since, as noted above, ASME designs do not consider the dynamics of the transportation environment. MTB is adopting the proposal contained in the NPRM.

Section 178.338–4. One commenter recommended that proposed § 178.338–4(c), which deals with the intersection of nozzles, supports, and other welds with longitudinal welds in the tank and load bearing jacket, be deleted. The commenter believes that the ASME Code adequately addresses welding details. However, MTB believes that the prohibition on such intersections, other than by the welds of load rings or stiffening rings, contributes to overall tank integrity at negligible additional cost. Therefore, MTB is retaining § 178.338–4(c) and, along the same lines, is adding language to proposed § 178.338–13(d) (now at § 178.338–13(a)), specifically providing for fillet weld discontinuity in the attachment of supports and bumpers, in order to reinforce the § 178.338–4(c) requirement.

Section 178.338–5. This section is adopted as proposed in the NPRM.

Section 178.338–6. One commenter recommended that MTB delete the requirement for a manhole on cargo tanks in oxygen service. The manhole requirement was proposed as § 178.338–6 in the June 21, 1979 **Federal Register** publication amending the NPRM. The commenter stated that oxygen is not a corrosive and that by following the proposed tank cleaning requirements the possibility of contamination of the tanks would be reduced to a level that would eliminate the need for a manhole. The commenter also believes that a manhole would provide a means of future contamination by allowing for an internal inspection without proper follow-up cleaning procedures. MTB believes that contamination of oxygen tanks during construction could pose a serious safety problem and therefore, in order to assure that the proper degree of cleanliness has been accomplished, the tank must be internally inspected after construction and prior to final closure. At the time of tank manufacture, a manhole provides the surest means of ensuring that the tank is contaminant-free prior to final closure. This assurance of cleanliness is crucial in tanks intended for oxygen service.

After the tank has been in service, the presence of a manhole allows for subsequent thorough internal examination. MTB believes the benefits to be gained by having the manhole outweigh any risk that a subsequent examination will cause contamination. The cleanliness requirements are geared to ensuring that after a tank has been reentered, and prior to subsequent reuse, that an oxygen compatible environment, i.e., no contamination, is reestablished.

Section 178.338–7. A commenter recommended that proposed paragraph (a), providing for complete drainage of flammable ladings, be revised to remove the implication that a pipe must drain down from the lowest point of the tank. MTB agrees and has revised the paragraph accordingly. The requirement for the closure of openings in a tank in proposed paragraph (b) is addressed in §§ 178.338–8, 178.338–10 and 178.338–11; therefore, proposed paragraph (b) has not been adopted.

Section 178.338–8. Paragraph (a) has been reworded to specifically require compliance with § 173.318(b) rather than merely referring to that section. The wording in paragraph (b)(2) has been simplified without a change in meaning. In paragraph (b)(5), at the suggestion of one commenter, MTB is allowing a check valve to be used on MC-338 cargo tanks. Allowing the use of a check valve provides an alternative that meets the intent of MTB's intitial proposal in that it provides for complete, positive closure. Further discussion of this section is contained under the heading "Pressure Relief Device Systems" earlier in the preamble.

Section 178.338–9. In response to proposed paragraph (b), a commenter pointed out that it is impossible to maintain a liquefied gas at a temperature corresponding to its boiling point at atmospheric pressure, as proposed in the NPRM. The commenter maintained that "in the absence of thermal stratification, the fluid would maintain itself at a saturation temperature corresponding to the lowest back pressure that can be maintained stable after filling the vessel." MTB agrees with the commenter. Accordingly, MTB has specified the boiling point at a reference pressure of one atmosphere and has revised the requirements to reflect a more practical pressure as suggested by the commenter. The requirement appears in paragraph (b)(1) of the final rule.

In proposed paragraph (b)(1) (paragraph (c) of this final rule), a commenter recommended that the holding time obtained in the optional test be required to be not less than 90% of the marked rate holding time, rather than requiring test results be within 10% of the original test. The commenter also recommended that MTB specify the Normal Evaporation Rate (NER) at not more than 110% of the original test. MTB has incorporated the suggested change to the optional test regimen in paragraph (c), but has not included the NER test (which was not defined by the commenter). This test has been mentioned by various industry members in the past and MTB has always requested that the test be defined. Further, MTB believes a test should be performed to validate the use of the NER test by comparing the results of the NER to the currently specified holding time tests. Without this information, MTB is unable to evaluate the value of the NER versus the holding time test.

Proposed paragraph (c) has been revised for clarity and it appears as paragraph (b)(2) in the final rule. The one-way travel time requirements have been moved to § 173.318(g).

Section 178.338–10. A commenter recommended certain changes be made in proposed paragraph (a) to specify applicability of the collision damage protection requirements to lines on a tank which, if damaged, could result in the loss of the lading, and to lines which connect to the safety relief devices. MTB has incorporated the intent of the comment into the final rule. In paragraph (c), the word "cargo" is added, as suggested by a commenter, when referring to the tank. MTB is not incorporating a comment suggesting a specific bumper height requirement. The NPRM contained a performance-oriented requirement ("adequate to protect all valves and fittings . . . ,") that is adopted in this final rule.

Section 178.338–11. A commenter suggested that the first sentence in paragraph (b) be revised to read "Each liquid filling and liquid discharge line. . . . " MTB agrees that the word "liquid" should be added since the lines being referred to in this paragraph are not those which discharge vapor. This change makes the meaning clearer and it has also been incorporated into paragraphs (b) and (c).

As proposed in paragraph (c) in the NPRM, the location of valve seats on cargo tanks was dependntt on whether the tank was vacuum insulated or not. A commenter disagreed with the proposal as it related to tanks with evacuated jackets. The commenter pointed out that, based on its information, the construction of vacuum jacketed valves does not facilitate location of the valve seat inside the jacket. The commenter argued that attempts to comply with the proposed requirement will result in vacuum and maintenance problems. As part of its analysis of the comment, MTB reviewed the outstanding DOT exemptions for vacuum insulated cargo tanks designed to carry flammable cryogens. This review revealed that none of these cargo tanks had the construction specified in the NPRM. After reevaluating the matter, MTB agrees with the commenter that for evacuated jacket construction it is not necessary to specify valve seat location inside the jacket and therefore this provision has been changed to require the valve to be as close to the tank as practicable.

The wording regarding the remotely controlled valves in proposed paragraph (c) as been changed in response to a comment. For consistency with other requirements in paragraph (c), the references in proposed paragraph (c)(1) and (c)(2) to internal shut-off valves have been changed to remotely controlled shut-off valves. Paragraph (c)(1) has also been changed to make clear that the thermal means of closure need not be a fusible element and that equivalent devices are acceptable.

Section 178.338-12. A commenter objected to the proposal on yield section and suggested that it be deleted. The commenter indicated that it had been reported to him that yield sections are hazardous and voiced the opinion that the risk of catastrophic failure appears to exceed the hazard which the yield section was intended to eliminate. The yield section is designed to break under strain without affecting the product retention capabilities of valves on a tank, and there is no evidence that safety has been compromised because of such a requirement in the past. Therefore, MTB has incoporated the requirement in the rule. The term "shear section" rather than "yield section" is being used since MTB believes the former term more accurately describes the mode of failure.

Section 178.338-13. This section has been reorganized. Paragraph (d) in the NPRM is now paragraph (a). Paragraph (b) in the NPRM is essentially continued as paragraph (b), however paragraph (c) in the NPRM, which referred to paragraph (b), has been deleted as unnecessary. Paragraph (e) in the NPRM is now paragraph (c). Paragraph (a) in the NPRM which dealt with cargo tanks that were "not permanently attached to or integrated with" the vehicle chassis has been deleted since it is unnecessary. MTB is not aware of any cargo tank designed or fabricated in this fashion that is used in cryogenic service.

Changes in static loading requirements in this section have been made and these are discussed earlier in the preamble under the heading "Design Loadings."

Section 178.338-14. The only change made to this section since the NPRM is in paragraph (a)(1) where the requirement that the device indicate the maximum permitted liquid level now specifies that this indication be "at the loading pressure." MTB believes this improves the clarity of this provision without making a substantive change.

Section 178.338-15. The only change to this section is that the closure that is now specified is that of the "manhole of the tank" instead of the "manway or the tank" specified in the NPRM. This change is necessary to conform to the requirements in § 178.338-6, which were first proposed in the June 21, 1979 amendment of the NPRM (§ 178.338-15 was not corrected at that time.)

Section 178.338-16. The discussion of testing requirements in this section is discussed earlier in this preamble under the heading "Pressure Testing."

Paragraph (a) has been reorganized so that the basic test pressure requirements need not be stated twice, once under "hydrostatic test" and once under "pneumatic test". Now the basic test requirements have been set out as paragraph (a) and the additional requirements applicable to pneumatic testing are set forth in paragraph (b).

It was proposed in paragraph (b) that "all welds in or on the cargo tank shell or heads shall be radiographed. . . ." One commenter recommended that jacket welds not be required to be radiographed and that additional methods of weld inspection be authorized for the internal tank. MTB agrees that jacket welds need not be radiographed because weld integrity is established by other means, such as the thermal integrity test. Howver, MTB believes that radiographic inspection of all tank heads and shell welds is necessary to ensure the pressure integrity of the tank. The revised requirement, which is at § 178.338-16(c) of the rule, does not preclude the use of other inspection methods in addition to that required by this rule, so long as any deposits or other contamination from such methods are removed prior to final tank closure. This incorporates a change to the NPRM that was made in the June 21, 1979 **Federal Register** publication.

Paragraph (c) in the NPRM (paragraph (d) herein) has been changed by removing the proposed requirement that any non-mechanical cutting undertaken in defect repair required the qualification of the cutter, the welder and the combination of cutting and welding. These requirements are covered by the ASME Code.

Section 178.338-17. This section has been adopted as proposed in the NPRM.

Section 178.338-18. A commenter recommended that the nameplate be affixed to the left side rather than to the right (curb) side of the vehicle. No reason was provided for the suggested change, which would modify a long standing requirement. The recommendation is not adopted.

The commenter also stated that it is unnecessary and redundant to require two plates. The matter of duplicate plates is particularly important. The ASME Code requires that the nameplate be attached to the Code vessel and be visible after insulation is applied to the Code vessel. If duplicate plates were not used, an opening would have to be made in the insulation system to expose the nameplate. Such an opening would adversely affect the efficiency of the insulation system. Further, paragraph UG-119(e) of the Code contains provisions for installation of a duplicate nameplate on the insulation jacket. This duplicate plate must contain all of the information found on the original plate, including the Code symbol. Therefore, MTB is retaining the requirement in this final rule.

Finally, the commenter favored a ¼ inch size letter requirement in place of the proposed ⅜ inch lettering for the identification plates. MTB believes ⅜ inch letters provide more legible markings than smaller sizes without imposing undue burdens on manufacturers and therefore has retained the proposed requirement in this rule.

Only minor changes have been made in this section. Paragraphs (b)(8) and (b)(9) in the NPRM have been placed in reverse order in this rule. Paragraph (b)(9) has been expanded to specify that for a cargo tank used to transport several cryogenic liquids only one MRHT need be marked on the specification plate. The MRHT's for the additional commodities may be placed adjacent thereto.

The NPRM proposal for paragraph (d) has been deleted from this section and is now the introductory pargraph of § 173.318(g).

Section 178.338–19. Certain minor changes have been made to this section. A commenter recommended to delete the words "including the ASME Code" in proposed paragraph (a) because "the manufacturer will provide the U-1 form as required by the Code." These forms are not always available. A person has only to review the exemption requests received by MTB to realize the significant number of instances wherein an applicant has a tank made to ASME Code, but where no drawings, nameplates, or U-1 forms are available. Therefore, the proposed requirement has been retained in the final rule. Paragraph (a) has been restructured and it now specifies that the manufacturer shall furnish the required documents to the owner.

Paragraph (b) has been changed to specifically require the manufacturer of each stage to not only furnish a certificate covering his work to any succeeding manufacturer, but also to pass along any certificates received from earlier manufacturers. This will ensure that at the end of staged construction the final manufacturer will possess certificates covering the entire construction. Another change requires that all these certificates be furnished to the owner.

Paragraph (c) has been changed so that upon change of ownership, the old owner does not retain the original documents for one year as proposed in the NPRM, but need only retain photocopies for that time. This change allows the original documents to be transferred to the new owner.

Section 179.100–7. In § 179.100–7(a), the reference entry "ASTM-A537–70, Grade A" is corrected to read "ASTM-A537-80, Class 1." The erroneous reference to Grade A has appeared in several editions of Title 49 CFR. This change correcting the error is also made at various other places in § § 179.102-4 and 179.102-17.

Section 179.102–1. In § 179.102–1 the heading and the introductory test to paragraph (a) have been amended to read "carbon dioxide, liquid (refrigerated)."

Section 179.102–4. Several commenters recommended that proposed paragraph (b) be revised to allow the manway, nozzles and anchorlegs to be fabricated with stainless steel, which would not be required to be impact tested. The commenters supported their position on the grounds that stainless steel can substantially reduce the heat flow into the tank, and thereby increase the holding time and improve transportation safety. MTB believes the recommendation has merit and has provided for the use of ASTM A240 Type 304, 304L, 316, or 316L stainless steel. The use of ASTM Specification A516 and A537 materials are authorized, as proposed in the NPRM.

Several commenters objected to the requirement in proposed paragraph (c) that insulation material must be "self-extinguishing." The commenters argued that even though most insulations do not support combustion, the materials are not technically "self-extinguishing" as the term would indicate. Two commenters indicated that they were unsure how the term would be defined. MTB agrees with the commenters that the use of the term "self-extinguishing" without further definition would be confusing and that presently approved materials are not in fact "self-extinguishing." After reevaluating this proposal, MTB has decided to remove the "self-extinguishing" requirement and require that the insulation be of approved material.

Commenters pointed out that in proposed paragraph (d) there is no need to require dual safety relief systems for vinyl fluoride. Two commenters indicated that a rupture disc on a tank of flammable gas can be considered a poor safety practice. MTB agrees with the commenters and has deleted the proposed requirement for a safety vent. In proposed paragraph (d) the word "piped" is changed to "directed", as suggested by two commenters, for consistency with § 179.102-3(a)(2), which requires openings in the protective housing cover for relief valve discharge.

In paragraphs (f) and (h), the use of a thermometer well or a pressure gage has been made optional rather than mandatory. This option offers a shipper flexibility in the manner in which compliance with the provisions of Note 17 to the Table in § 173.314(c) is determined.

Finally, paragraph (1) has been revised to define more clearly the acceptance standards for welds. The welds must meet the acceptance standards contained in W11.06 of AAR specifications for Tank Cars, Appendix W.

Section 179.102–17. Paragraphs (b), (c), (d), (f) and (m) have been changed in the same manner, and for the same reasons, as their counterparts in § 179.102–4 ((b), (c), (d), (f) and (l)). In addition, as suggested by commenters, paragraph (d) has been revised to allow a pressure relief device to be trimmend with monel "or other approved material" and to change "teflon coated monel" to read "fluoronated hydrocarbon polymer coated monel" since the word "teflon" is a registered trademark of E. I. du Pont de Nemours and Company.

Section 179.400. The nomenclature used to identify tank cars was questioned by two commenters. In the final rule, "Class DOT-113" is used to refer inclusively to specifications DOT-113A, DOT-113C, DOT-113D, etc. When a specific design is being referred to, it will be identified by a particular specification, for example "DOT-113A60W". This procedure conforms to that stated in section 2-2(b) of AAR Specifications for Tank Cars.

Section 179.400-1. Except for an editorial change, this section is adopted as proposed in the NPRM.

Section 179.400-2. This section is adopted as proposed in the NPRM.

Section 179.400-3. This section has been restructured for clarity. Additionally, a change was made to correct an error in the NPRM which indicated that the tank must have "heads designed concave to pressure." While this is correct for the inner tank heads, it is wrong with regard to the jacket heads. Jacket heads must be designed convex to pressure and this necessary change has been made in the final rule.

Section 179.400-4. The section heading and the requirements in § 179.400-4 are revised. The performance standard proposed in the NPRM has been modified, as suggested by commenters, to allow nitrogen to be used as the test medium rather than the actual lading which may be a flammable gas. In addition, the formulas for heat transfer rates have been defined and clarified.

Section 179.400-5. Two commenters to this section requested that MTB allow the use of nickel steel in the construction of DOT-113D tank cars. MTB is not authorizing nickel steels for newly constructed Class DOT-113 tank cars becuse this material does not have adequate impact properties when subject to temperatures as low as the design service temperature authorized for these tank cars.

Section 179.400-6. Two commenters requested that in proposed paragraph (b) the words "and stresses" be deleted since the AAR Specifications for Tank Cars specifies "loads". This has been done. An updated section of AAR Specifications for Tank Cars is also referenced.

Section 179.400-7 (proposed § 179.400-8). MTB has deleted the reference to "approved contour" of tank heads since all heads specified meet this requirement.

Section 179.400-8 (proposed § 179.400-7(a) through (d)). In proposed paragraphs (a) and (b) (which are also paragraphs (a) and (b) herein), the inference that 2:1 or 3:1 ellipsoidal heads are required for the inner tank has been removed by replacing the word "the" with the word "any". Paragraph (d) has been reworded for consistency with the wording in the other paragraphs. Also, the jacket head thickness of ½ inch provides head puncture resistence equivalent to that required of certain other classes of tank cars.

Section 179.400-9 (proposed § 179.400-7(e)). The formula in paragraph (b) (proposed paragraph (e) (1)), which specifies the width of the jacket plate on each side of the stiffening ring, has been corrected. Also as suggested by commenters, MTB has reworded § 179.400-9(c) (proposed as § 179.400-7(e)(2)) to allow differing structural shapes to be credited for stiffening of the outer jacket with external pressure, and to add a provision to require external closed rings be provided with a drain opening to reduce corrosion of the stiffening ring.

Sections 179.400-10, 179.400-12 through 179.400-15, 179.400-18, 179.400-21 through 179.400-23 and 179.400-26. Except for minor corrections and editorial changes, requirements in these sections have been adopted as proposed in the NPRM. Most of the changes were suggested by commenters. The sections have been renumbered in the sequence normally followed in designing a tank car:

Present	Proposed
179.400-10	179.400-7(f)
179.400-12	179.400-10
179.400-13	179.400-11
179.400-14	179.400-12
179.400-15	179.400-13
179.400-18	179.400-16
179.400-21	179.400-19
179.400-22	179.400-20
179.400-23	179.400-21
179.400-26	179.400-24

Section 179.400-11 (proposed as § 179.400-9). Two commenters suggested in proposed paragraph (a) that "access opening" be changed to "opening" to eliminate the implication that the opening is a manway, and to revise the

weld procedures to recognize difficulties in performing a fusion double welded butt joint for certain circumferential closing joints in the cylindrical portion of the outer jacket. MTB agrees with the commenters and has revised paragraphs (a) and (b) accordingly.

Section 179.400–16 (proposed as § 179.400–14). Except for an editorial change, the requirements are adopted as proposed in the NPRM. Two commenters has suggested that the provision in the last sentence in paragraph (b), stating that a cutting torch "may not be used", be revised by using the wording "must not". The term "may not", as used in this sentence, means no person is authorized or permitted to use a cutting torch on the welded closure. This use of the term is consistent with the rule of construction set forth at § 171.9(b)(4).

Section 179.400–17 (proposed as § 179.400–15). Two commenters brought to MTB's attention that the proposed requirements for vacuum insulated loading and unloading lines and insulated shut-off valves in paragraph (a)(1) should apply only to DOT-113A60W tank cars (which are designed only for hydrogen). Other DOT-113 tank cars are designed for the warmer cryogens and the requirements are unnecessary. MTB agrees and has made an appropriate revision.

The commenters also requested that the proposed provision in paragraph (a)(3) requiring vapor phase blowdown line discharge to be directed upward and away from operating personnel be deleted. The commenters indicated that the vapor blowdown valve is used only when the tank car is hooked up to a closed system and that "blowdown flare may be hazardous to personnel." Despite the fact that in normal operations the line will only be opened to a closed system, MTB agrees with the commenters about the hazard of blowdown flare but believes if a flare occurs, for whatever reason (including inadvertently opening the wrong valve), the flare should be directed away from operating personnel. For this reason, MTB has retained this requirement in the final rule.

Section 179.400–19 (proposed as § 179.400–17). Proposed paragraph (b) of this section contained a requirement that a tank car be equipped with a connection for a liquid gage and, in addition, a fixed length dip tube and a vapor phase pressure gage. As recommended by two commenters, the word "liquid" has been removed before the word "gage" and all of paragraph (b) has been reorganized to clarify that a car is required to be equipped with only one of the two alternative methods provided in paragraph (b)(1) to determine the quantity of liquid lading in the car.

Section 179.400–20 (proposed as § 179.400–18). Paragraph (c)(4) has been revised to include a pressure controlling and mixing device on DOT-113A60W tank car, as required in the past by § 179.400–18(c)(3). The requirement in § 179.400–18(c)(3) that the device must prevent venting of certain gas mixtures was inadvertently omitted in the proposal and has been added in paragraph (c)(4)(iii) of the final rule.

Section 179.400–24 (proposed as § 179.400–22). Three commenters took exception to the proposal in paragraph (b) to prohibit marking, stenciling or stamping on shells or heads of inner tanks. One of the commenters pointed out the provision would conflict with requirements for identification markings of plate thickness, material, and welds contained in § AAR.15 (now contained in § AAR 5.1.4) and in Appendix W, § W10.04, of the AAR Specifications for Tank Cars. MTB believes that the structural integrity of a tank at cold temperatures should not be compromised by stamping; however, MTB agrees that there is no valid reason to preclude the marking or stenciling of a tank and these prohibitions have been removed from the final rule. The requirements in the final rule are consistent with present requirements in the AAR Specifications for Tank Cars.

Section 179.400–25 (proposed as § 179.400–23). Two commenters recommended that in proposed paragraph (a) MTB reference the standard stenciling requirements of Appendix C of the AAR Specifications for Tank Cars. MTB believes the commenters' suggestion has merit and has incorporated the change into the final rule at § 179.400–25.

Three commenters maintained that the requirement for marking the name of the hazardous material for which the tank was designed on a tank car is covered by § 173.319 and, therefore, proposed paragraph (b) is unnecessary. MTB agrees with the commenters. Proposed paragraph 173.319(a)(4)(iv) (which has been adopted into the final rule) requires a tank car be marked with the name of the material contained in the tank car during transportation. Therefore, proposed paragraph (b) has been removed. The remaining paragraphs have been redesignated as paragraphs (b), (c), (d), and (e) in this final rule.

Exceptions were taken, in proposed paragraph (c), to the use of the words "minimum loading temperature", on the basis that "design service temperature" was used elsewhere in the proposal to indicate the same thing. MTB agrees with the commenters and has made the change in the final rule.

Finally, exception was taken to the proposals in paragraphs (c) and (d) on the basis that these requirements would conflict with a suggestion by the commenters that the filling density volume be marked on the tank. This suggestion was tied to a suggested revision to the definition of filling density in § 173.300 of the NPRM. MTB has not adopted this revision to the definition of filling density in § 173.300 and therefore, has not incorporated these additional changes suggested here.

Section 179.401–1. Except for minor changes in terminology due to changes elsewhere in the rule, this section is adopted as proposed in the NPRM.

(48 F.R. 27674, June 16, 1983)

49 CFR Parts 171, 172, 173, 176, 178, and 179
[Docket No. HM-115, Amdt. Nos. 171–174, 172–82, 173–166, 176–17, 178–77, 179–32]
Cryogenic Liquids, Revisions

On June 16, 1983, MTB published a final rule in the FEDERAL REGISTER under Docket No. HM-115 (48 FR 27674). MTB received 18 petitions for reconsideration to certain provisions of the final rule. A majority of the requested changes were contained in a petition submitted by the Compressed Gas Association (CGA).

Several petitioners objected, among other issues, to changes to proper shipping names and identification number prefixes to certain entries of the Hazardous Materials Table (the Table), in §172.101, and requested the effective date of the final rule be postponed. MTB believed those issues warranted immediate handling so that changes could be included in the 1983 edition of Title 49, Code of Federal Regulations, Parts 100–199. Therefore, MTB separated those issues from other issues raised in the petitions and handled them in a document which was published in the FEDERAL REGISTER on November 1, 1983 (48 FR 50440). In that document, MTB postponed the mandatory effective date of the final rule until October 1, 1984. MTB also revised the proper shipping names for the cryogenic liquids and cold form gases to include the international descriptor, "refrigerated liquid" and the identification number prefix was changed from "NA" to "UN". Entries for cryogenic liquid were designated "(cryogenic liquid)", in the Table, to distinguish those gases from cold form gases such as carbon dioxide, nitrous oxide, and hydrogen chloride. For compressed gases, MTB provided for continued use of the descriptions as presently found in the HMR as well as for the optional use of the international descriptions which include the word "compressed". For example, "Argon, compressed" and, for domestic transportation only, "Argon" are acceptable descriptions. Other substantive issues raised in the petitions are addressed in this document.

Several issues raised by the petitioners are addressed in earlier preamble discussions. For additional information, readers are referred to preamble discussions which appeared in the notice of proposed rulemaking (NPRM) (44 FR 12826, March 8, 1979), and related correction documents (44 FR 20461, April 5, 1979; 44 FR 36211, June 21, 1979), the final rule (48 FR 27674, June 16, 1983) and the correction and revision document of November 1, 1983 (48 FR 50440).

TANKS OPERATING UNDER DOT EXEMPTIONS

Under the final rule, the owner or person using a cargo tank or tank car under, "and in compliance with," a DOT exemption issued before October 1, 1984, if required to remove the DOT exemption number stenciled on the cargo tank or tank car and stamp the identification plate, as specified by §173.31(a)(8) or §173.33(b)(2), with the proper specification.

Several petitioners pointed out that the phrasing of §§173.31(a)(8) and 173.33(b)(2) implies that tank cars and cargo tanks must continue to be used in conformance with the terms of the exemptions. This is not MTB's intention. Tank cars and cargo tanks which are remarked as specification packagings cease to be governed by their previous exemption. Instead, they are subject to the applicable requirements, conditions, and limitations prescribed in the HMR. Sections 173.31(a)(8) and 173.33(b)(2) are revised for clarity. Section 173.33(b)(2) is revised for consistency with §§173.31(a)(8) and 173.33(b)(3).

Sections 173.31(a)(8), 173.33(b)(2) and (b)(3) require the owner or the operator, if not the owner, to retain a copy of the exemption that was in effect on September 30, 1984. It is not MTB's intention to require renewal of an exemption for the purpose of having a valid exemption on September 30, 1984. Also MTB did not specify where the exemption must be retained. The rule is revised to require that the exemption in effect at the time a tank car or cargo tank is remarked as a DOT specification packaging be retained on file during the period the tank car or cargo tank is in service. MTB does not agree with a petitioner who suggested that it is necessary for a copy of the exemption to be carried with each cargo tank as was required under the exemption. However, this does not prevent any person from carrying a copy of the exemption on a cargo tank.

After October 1, 1984, an exemption affecting a cargo tank or tank car of a type covered by the final rule will not be renewed unless the holder of the exemption submits information to the Associate Director for Hazardous Materials Regulation stating the reason why the tank does not qualify for remarking as a specification packaging.

All applicable DOT exemptions are listed in the preamble on page 27678 of the final rule. Other exemptions affected by the rule are as follows:

Exemptions—MC-338 type Cargo Tanks

E-7227
E-8602 (Model HL 1920M)
E-8644

Exemptions—Class DOT-105 Tank Cars
  E-3992
Exemptions—MC-330 or MC-331 Cargo Tanks
  E-6215
  E-8199

PRESSURE RELIEF DEVICE SYSTEMS

CGA and several other petitioners took exception to the requirement for a primary system of one or more spring loaded pressure relief valves and, except for tanks in carbon monoxide service, a secondary system of one or more frangible discs. CGA requested that the requirement be revised to permit cargo tanks to be equipped with a primary system consisting of spring loaded or pilot-operated pressure relief valves and a secondary system consisting of frangible discs or pressure relief valves. CGA maintains that a complete blowdown of certain ladings may present a greater hazard than controlled relief of the hazardous material through a spring loaded pressure relief valve. After further consideration, MTB agrees, in part, with CGA and other petitioners. MTB has revised § 173.318(b)(1)(i) to provide for a primary consisting of one or more pressure relief valves and a secondary system of one or more frangible discs or pressure relief valves. The pressure relief valves of the primary and secondary systems may be any type of pressure relief valve designed to automatically open and close at predetermined pressure. This option on use of frangible discs does not apply to the secondary system on cargo tanks in carbon monoxide service which are required to be equipped only with pressure relief valves.

CGA and other petitioners requested revision of subparagraph 173.318(b)(1)(viii) which contains a requirement that any shut-off valve or device that interferes with the proper operation of a pressure control valve must be designed and installed so that the cargo tank may not be operated for transportation purposes when the pressure control valve operation is impeded.

In its comments to the requirement, CGA stated:

The present wording would require an interlock so that the vehicle could not be operated if the pressure control operation is impeded. This would lead to an unsafe condition at the time of final unloading of a flammable refrigerated liquid. Even though the liquid has been completely drained and the gas pressure has been reduced to atmospheric pressure on the return trip, there is the hazard of venting flammable gas if the pressure is controlled by the low pressure road valve rather than by the higher set pressure relief valve. This is because the refrigeration heat sink of liquid is no longer present to absorb the constant incoming steady heat leak. The heat leak instead goes into warming the residual cold gas, and the gas pressure can rise quite rapidly as a result, possibly exceeding the road relief valve setting before the return is completed. Current industry practice is usually to transfer from the road relief valve to the pressure relief valve on the return trip.

The present wording also precludes the provision for multiple deliveries at increasingly higher pressure levels between deliveries without venting gas to the lowest pressure level at which the cargo tank was loaded. This is contrary to a number of present exemptions that allow this type of operation. Such exemptions include E-2708, E-4490 and E-7192.

In addition to this revision to Section 173.318(b)(1)(viii), corresponding revisions should be made to Section 173.318(g), 173.318(g)(3), 177.840(i), 177.840(j), 177.840(k), 177.840(k)(3), 178.338-9(a), 178.338-9(a)(1), 178.338-9(b)(1), 178.338-9(b)(2), and 178.338-18(b)(9).

On further consideration, MTB agrees that the provision in subparagraph 173.318(b)(1)(viii) may require that the residual lading be reduced to impracticable levels at final unloading in order to prevent venting when pressure is controlled (limited) by the pressure control valve. MTB also agrees that the interlock requirement would preclude multiple deliveries without venting appreciable quantities of lading at each delivery point. Accordingly, subparagraph (b)(1)(viii) is removed. Remaining subparagraphs (ix) and (x) are renumbered as subparagraphs (viii) and (ix), respectively. Paragraph (g)(3) is removed and a revision is made to the introductory text of paragraph 173.318(g) to permit the display of more than one one-way-travel-time (OWTT) marking on a cargo tank. CGA's other requests for reconsideration are denied since they are not necessary with the removal of subparagraph 173.318(b)(1)(viii).

CGA requested a revision to § 173.318(b)(2)(i) to exclude cargo tanks in atmospheric gas (except oxygen) service and helium service from the requirement of primary and secondary pressure relief device systems of equal capacities. CGA maintained that MTB made distinctions in other sections between nonflammable ladings versus flammable and oxygen ladings based on the fact that atmospheric gases (except oxygen) and helium do not intensify a fire in fire exposure incidents. Thus, CGA asserts that it is unnecessary to apply the redundancy for flow capacity based on fire conditions for both the primary and secondary systems for atmospheric gases (except oxygen) and helium. MTB agrees and grants the request for reconsideration by revising subparagraph (b)(2)(i) to allow cargo tanks used in atmospheric gas (except oxygen) and helium service to be equipped with the primary system only.

CGA requested that the secondary system have a minimum total capacity at a pressure not to exceed 120% of the tank design pressure in place of 150% as prescribed by § 173.318(b)(2)(iii). CGA maintained that the change in the setting would provide a greater margin of safety. MTB believes the change is unnecessary and the request for reconsideration is denied. Section 173.318(b)(2)(iii) specifies that the pressure of the secondary system may not exceed 150 percent of the tank design pressure. Therefore, a pressure at 120 percent of the tank design pressure is permitted. MTB specified the secondary system at a minimum total capacity of 150 percent to allow the secondary system to function after the primary system which relieves at a pressure of 120% of the tank design pressure. MTB believes these systems should operate in sequence to provide for a controlled release of the lading.

CGA requested revision of the requirement that the primary system of pressure relief valves must have a liquid flow capacity equal to or exceeding the maximum rate at which the tank is to be filled at a pressure not to exceed 120% of the tank design pressure in subparagraph (b)(2)(iv). CGA maintained that a tank filled by pumping equipment which is capable of producing pressures in excess of the design pressure of the tank may be equipped with a by-pass on the pump discharge or other suitable method to prevent accumulation of pressures in the tank in excess of 120% of the tank design pressure. MTB does not agree and the request for reconsideration is denied. MTB believes that the design and construction of the primary pressure relief valves should be capable of sustaining a flow capacity at pressures not to exceed 120% of the tank design pressure during filling operations. CGA provided no information on the adequacy or fail-safe function of a by-pass on a discharge pump or other special controls that will prevent excessive pressure build-up in tanks used for cryogenic liquids. Therefore, no change is being made to the provision.

Section 171.7. MTB is adding in paragraph (d)(5) certain ASTM Standards which are referenced in §§ 173.316(a)(4), 173.318(a)(4) and 178.338–2(a).

Section 171.8. A petitioner requested that the temperature reference in the definition of "SCF" (Standard Cubic Foot) be changed from 60°F. to 70°F. for consistency with the U.S. industry standard contained in CGA Pamphlet P-11, "Metric Practice Guide" and the temperature used to define a compressed gas in § 173.300. MTB does not agree and the request for reconsideration is denied. The term "SCF" defines the standard conditions used to determine the relieving capacity of pressure relief devices. These standard conditions of 60°F. and 14.7 psia are presently contained in the HMR and are consistent with those used by CGA for determining and sizing pressure relief devices in CGA Pamphlets S-1.1 and S-1.2. No change is made in the definition.

Section 172.101. A petitioner stated that the provision for ethylene, refrigerated liquid to be stowed "below deck" on cargo vessels is unsafe and is inconsistent with stowage requirements applicable to other flammable cryogenic liquids. MTB agrees with the petitioner and grants the request for reconsideration by removing the "3" in column 7(a) of the Table.

A petitioner objected to the provision prohibiting the transportation of hydrogen, refrigerated liquid on a cargo vessel. The petitioner argued that the prohibition on hydrogen, refrigerated liquid is inconsistent with requirements that apply to other flammable cryogenic liquids, such as natural gas and carbon monoxide, and that the "light density of hydrogen vapor and up-and-away venting provide an adequate margin of safety." The petitioner argued that there is exemption experience to support transporting cargo tanks and portable tanks containing hydrogen, refrigerated liquid on a cargo vessel "on deck". MTB and Coast Guard, which assisted MTB in the preparation of the final rule, maintain that because of its very wide flammability range and the fact that it burns with an invisible flame, hydrogen poses a greater potential hazard than other flammable cryogenic liquids. MTB considers it necessary to apply special safety controls for hydrogen when transported on board a cargo vessel or a case-by-case basis by exemption. To allow transportation of hydrogen under regulations of general applicability would not assure adequate safety and, therefore, the request for consideration is denied.

A petitioner requested that the quantity limitation in one package of argon, refrigerated liquid be increased from 300 pounds to 1,100 pounds by cargo aircraft for consistency with the quantity limitation authorized for nitrogen, neon, and helium, and for consistency with the quantity limitation for argon, refrigerated liquid adopted by the Dangerous Goods Panel of the International Civil Aviation Organization (ICAO). MTB agrees with the petitioner that the quantity limitation should be consistent with that recommended by ICAO. MTB is granting the request for reconsideration by revising the Table to provide for 1,100 pounds of argon, refrigerated liquid to be transported by cargo aircraft.

Section 173.23. A petitioner correctly pointed out that cylinders meeting the DOT-4L specification are not required to be retested and, therefore, the schedule for remarking cylinders manufactured under DOT E-6668 or E-8404 should be changed. MTB grants the request for reconsideration by revising paragraph (e) to require the cylinders be remarked "DOT-4L" by January 1, 1986. (This requirement appeared as paragraph (d) in the rule and was redesignated paragraph (e) under Docket HM-189 which was published in the **Federal Register** on November 1, 1983; 48 FR 50444.)

Section 173.31. Two petitioners took exception to the prohibition in paragraph (a)(9) against new construction of DOT-113D120W tank cars made with nickel alloy steel inner tanks which are authorized under DOT exemption. One of the petitioners maintained that there is no technical reason or unsatisfactory exemption experience to support prohibiting new construction of DOT-113D120W tank cars. The other petitioner alleged that MTB based its decision on disallowing new construction of DOT-113D120W tank cars merely on the fact that there has been no new construction of the tank car since 1973. MTB agrees, in part, with both petitioners. MTB conducts continuing reviews of packagings authorized for use in the HMR to remove specifications which are no longer being manufactured. MTB does not believe these efforts would be well-served by providing for new construction of a tank car in the HMR when there is no evidence of demand for its construction. Therefore, the petitioner's request for reconsideration is denied. However, because of the satisfactory safety record of existing DOT-113D120W tank cars, MTB believes continued use of existing tank cars should be authorized.

The Association of American Railroads (AAR) pointed out that requirements for the retest of the alternate pressure relief valve on DOT-113D120W tank cars were omitted in the final rule. MTB is revising subparagraph (c)(13)(v) to correct this oversight and specify the same test procedure as is required for DOT-113C120W tank cars.

Three petitioners pointed out that new § 173.314(c) authorizes DOT-105A600W tank cars for hydrogen chloride service, but does not provide for DOT-105 tank cars in hydrogen chloride service that are authorized under DOT E-3992. MTB agrees with the petitioners. Omission of existing tank cars, built with ASTM a 212B steel to low temperature ASTM A300 testing qualifications, under DOT E-3992 was an oversight. MTB grants the petitioners' request for reconsideration by adding a new paragraph (a)(10) to authorize continued use of these tank cars.

Section 173.33. Changes to this section are addressed earlier in this preamble under the heading "Tanks Operating Under DOT Exemptions".

Section 173.300. CGA requested that the definition of "cryogenic liquid" in paragraph (f) be removed and a new definition for "refrigerated liquid" be added to read: "A refrigerated liquid is a cold liquefied gas which, when charged into an insulated transport container, cannot be held indefinitely due to vaporization or pressure rise caused by heat transfer from the surroundings." CGA also requested that the descriptor "cryogenic liquid" be changed to "refrigerated liquid" each time it appears in the HMR. MTB is denying the request for reconsideration because CGA's suggested definition provides no distinction between the so-called "cold form gases", such as carbon dioxide, nitrous oxide, hydrogen chloride and vinyl chloride, which are not regulated as cryogenic liquids.

CGA also suggested a second alternative to adding the above definition of "refrigerated liquid". The alternative provided for adding a sentence at the end of the present definition of "cryogenic liquid" to read: "A material meeting this definition is described as a 'Refrigerated liquid' in Part 172 of this subchapter". MTB agrees and grants the request for reconsideration. In the November 1 correction document, MTB authorized the international descriptor, "refrigerated liquid", to be a part of the proper shipping name for cryogenic liquids and the cold form gases. The cryogenic liquid descriptions were specifically identified in italics in the Table to distinguish the cryogenic gases from the cold form gases. Therefore, at the end of the definition for cryogenic liquid, MTB is adding a clarification that materials meeting the definition are described, in part, as "..., refrigerated liquid (cryogenic liquid)" in the Table.

MTB is revising the definition of a cryogenic liquid to clarify that these materials may not meet the definition of a compressed gas in paragraph (a).

Section 173.314. MTB is revising the entry for vinyl fluoride in the table in § 173.314(c) to continue the applicability of Note 23. Note 23, as amended under Docket HM-175 (49 FR 3468, January 27, 1984), requires each class 105 tank car built after August 31, 1981, to conform to specification 105J. Tank cars built before September 1, 1981, with a capacity exceeding 18,500 gallons and used to transport flammable gases are required to be retrofitted by December 3, 1986, to conform to specification 105J.

The AAR and another petitioner requested that paragraph (g)(2) be revised by adding a provision that appears in DOT E-3992 that requires tank cars in hydrogen chloride service to be weighed when full and when empty. Prior to offering an empty tank car for transportation, the car must be emptied below three percent of weight of the original load. The requirement is similar to Rule 35 of the Uniform Freight Classification. MTB is considering addressing tank cars containing a residue of a hazardous material in a proposed rule in the future and, therefore, the request for reconsideration is denied. Upon consideration, MTB also believes the requirement that the pressure in an empty tank car may not exceed 70 psig is unnecessary in view of requirements in § 173.29(c). Accordingly, paragraph (g)(2) is removed and paragraph (g)(3) is redesignated paragraph (g)(2).

Section 173.316. Two petitioners objected to a provision in paragraph (a)(4) prohibiting cylinders in oxygen service from having aluminum valves or fittings with internal rubbing or abrading aluminum parts which may come in contact with cryogenic oxygen. One petitioner believed it was MTB's intention to apply the provision prohibiting rubbing or abrading aluminum parts to cargo tanks in oxygen service and not to cylinders in oxygen service. Both petitioners maintained that safety experience has been satisfactory in using "... an anodized aluminum body with an internal anodized aluminum piston ...."

MTB believes that internal rubbing or abrading aluminum parts which may come in contact with cryogenic oxygen must not be used in any cylinder used to transport cryogenic oxygen. The prohibition is needed because of the potential for ignition and subsequent rapid burning of aluminum when subject to fire engulfment temperatures, to friction heat from abrasion, or high oxygen flow velocities over surfaces with sharp projections or abrupt directional changes. However, MTB agrees with the petitioners that anodized aluminum has a lower friction coefficient than non-anodized aluminum. Therefore, MTB is granting the request for reconsideration by revising paragraph (a)(4) to allow the use of rubbing or abrading aluminum parts that have been anodized in conformance with ASTM Standard B 580 in cylinders used in oxygen service. A similar change is made to § 173.318(a)(4) for cargo tanks in oxygen service.

A petitioner requested that paragraph (b) be revised by referencing § 173.304(b)(2) for requirements on pressure control valves. MTB agrees with the petitioner that the paragraph should be clarified. However, MTB would be in error to reference paragraph 173.304(b)(2) since it was removed in the final rule. The requirements pertaining to pressure control valves on cylinders which appeared in paragraph 173.304(b)(2) are contained in CGA Pamphlet S-1.1. These requirements are made applicable by § 173.34(d), which incorporates CGA Pamphlet S-1.1. For clarity, MTB is revising paragraph (b) by replacing the words "pressure control valve" with the words "pressure control system" in the paragraph heading and text.

MTB is revising the introductory text of paragraph (c) to clarify that DOT-4L cylinders containing a cryogenic liquid must be transported in the vertical position.

Two petitioners requested that the table in paragraph (c)(2) be amended by adding additional filling densities to allow for pressure control valve settings at 1 ¼ times a marked service pressure of 500 psi for DOT-4L cylinders. MTB received data supporting filling densities at settings of 450, 540, and 625 psig from one petitioner. The petitioner argued against reducing pressure control valve settings on DOT-4L cylinders by 15 psi. The petitioner contends: "The control valve pressure settings in the table represent ranges of pressure. Thus, if a control valve setting of 235 psig for a vacuum insulated DOT-4L200 cylinder were required ($200 \times 1.25 = 250 - 15 = 235$), the value of the filling density of 295 psig would be used because an entry for 235 psig does not exist." Also, the petitioner argued that "[i]t is possible to have a cryogenic 4L cylinder without a vacuum jacket in which case the control valve setting, as per paragraph 173.304(b)(2), is one and one-fourth times the service pressure without subtracting the 15 psi." MTB agrees and grants the request for reconsideration by revising the table to add additional pressure control valve settings. The settings must be in conformance with paragraph 173.316(c)(2) for the named gases and § 173.34(d), which incorporates CGA Pamphlet S-1.1. Paragraph 5.9.3 of CGA Pamphlet S-1.1 specifies that a pressure control valve setting must be set 15 psi lower than 1 ¼ times the marked service pressure on DOT-4L cylinders insulated by a vacuum.

Petitioners requested that the filling density entry for nitrogen at a pressure control valve setting at "295" be revised by removing "69" and adding "68". MTB agrees and grants the request for reconsideration.

Section 173.318. Two petitioners urged MTB to reconsider the requirement in paragraph (a)(3)(i) which prohibits the use of aluminum outer jackets on cargo tanks in oxygen service. The petitioners argued that MTB's position on this matter for cargo tanks is inconsistent with action taken by MTB in allowing aluminum jackets on oxygen cylinders, that the reasons used by MTB to justify allowing aluminum jackets on cylinders can be used also to support aluminum jackets on cargo tanks, and that the operating experience of aluminum jacketed non-specification cargo tanks in oxygen service has been excellent for over 50 years. Neither petitioner submitted any test data on cargo tanks demonstrating the survivability of aluminum in a fire environment which was a significant factor in MTB's decision to allow aluminum jackets on cylinders in oxygen service. MTB strongly believes that aluminum as a material of construction for the cargo tank jacket must not be used because it loses strength and melts at much lower temperatures than steel in a fire situation. Increase influx of heat and the attendant pressure buildup resulting from loss of jacket integrity would accelerate the rate of oxygen release and intensify the fire. A steel jacketed tank's relative survival time in fire engulfment is over two times that of an aluminum jacketed tank, as was discussed by MTB in the preamble of the final rule under the heading "Use of Aluminum" (48 FR 27674). The request for reconsideration is denied. However, as discussed above under § 173.316, MTB is revising paragraph (a)(4) to allow the use of aluminum parts that have been anodized in accordance with ASTM Standard B 580 on cargo tanks in oxygen service.

MTB is relaxing the provision in paragraph (a)(5) to allow use of aluminum valves, pipes and fittings external to the jacket provided no lading is retained in these parts during transportation.

See preamble discussion in this document under the heading "Pressure Relief Device Systems" for changes made to the provisions on pressure relief valves in paragraph (b).

A petitioner requested that the words "pressure control valve" be deleted in subparagraph (b)(1)(iii) because a pressure control valve is not a pressure relief device. The petitioner's request for reconsideration is denied. MTB believes that when a cargo tank is filled to the pressure setting of the pressure control valve, the presure control valve acts as a pressure relief device to relieve pressure. The paragraph is revised for clarity. Also, subparagraph (b)(1)(iii) is revised to reference requirements in flow capacities in subparagraph (b)(2)(i).

A petitioner requested that subparagraph (b)(5)(ii) be revised by deleting the word "actual" preceding the words "discharge rate" and that the words "of free air" be added immediately following "(SCFM)". The petitioner stated that the changes would permit the flow capacity to be marked using the standard flow rating method. MTB agrees and grants the petitioner's request for reconsideration.

A petitioner requested that paragraph (g) be revised to allow for the display of more than one one-way-travel-time (OWTT) marking on the tank when it is used to transport a cryogenic liquid at different pressure levels, MTB agrees and grants the request for reconsideration by revising the introductory text of paragraph (g) and paragraph (g)(3) to allow more than one OWTT marking on a cargo tank.

Section 173.319. AAR recommended that the word "flammable" be deleted in paragraph (a)(4) thereby making the requirements applicable to all cryogenic liquids transported by rail. AAR did not explain why it believed atmospheric gases and helium which are transported by rail at pressures less than 25.3 psig should be regulated to an extent greater than specified in § 173.320. MTB is denying the request for reconsideration because it is outside the scope of this rulemaking. Further consideration will be given to the matter upon receipt of a petition for rulemaking.

Section 173.320. A petitioner requested that MTB add a provision requiring Dewar flasks be equipped with a suitable pressure relief device when used for helium or neon, refrigerated liquid at pressures below 25.3 psig. The petitioner maintained that the neck of the Dewar flask may freeze with solid air thereby allowing internal pressure buildup and rupture of the packaging. MTB is denying the request for reconsideration because it is outside the scope of this rulemaking. Further consideration will be given to the matter upon receipt of a petition for rulemaking. Further, shippers are reminded that it is their responsibility to determine the suitability of packagings in conformance with § 173.24.

Paragraph (b) is removed and redesignated paragraph (g) in § 176.11. MTB takes this opportunity to clarify in a new paragraph (b) that atmospheric gases and helium at pressure below 25.3 psig may be offered for carriage aboard an aircraft in conformance with § 171.11.

Section 176.11. Paragraph 173.320(b) which excepts atmospheric gases used in a refrigeration system from regulation by vessel is redesignation paragraph 176.11(g).

Section 176.76. A petitioner requested that paragrpah (h)(2) be revised for clarification by adding the words "during transportation" immediately after the words "cryogenic liquid". The petitioner's request for reconsideration is denied because the introductory text to paragraph (h) makes it clear that the regulations apply to cryogenic liquids transported by vessel.

Section 178.57-2. Two petitioners requested that the maximum authorized service pressure on DOT 4L cylinders be continued at 500 psi in place of 360 psi as specified in the final rule. MTB agrees and grants the request for reconsideration by specifying a pressure at 500 psi to correspond with the additional filling densities authorized in the table in § 173.316(c)(2).

Section 178.57-13. A petitioner requested revision of this section to reference § 173.304(b)(2) for requirements on pressure control valves. The request for reconsideration is denied because § 173.304(b)(2) which contained requirements on pressure control valves on DOT-4L cylinders was removed under the final rule. The requirements previously contained in § 173.304(b)(2) are contained in CGA Pamphlet S-1-1, which is incorporated by reference in § 173.34(d). The last sentence in § 173.57-13 containing an incorrect reference to CGA Pamphlet S-1.1 for requirements on flow capacity of relief devices is removed.

Section 178.57-20. A petitioner requested revision of paragraph (a)(9) to allow the letters "AL" to be added immediately following the specification markings in place of stamping the words, "ALUMINUM JACKET", on the jacket. The petitioner maintained that the two-letter marking appropriately identifies aluminum jacketed cylinders and is less expensive. The petitioner also contended that the material of construction of the jacket may not be known at the time of manufacture of the inner containment vessel (cylinder) and, therefore, marking the jacket material designation on the cylinder should not be required under paragraph (b). MTB agrees and grants the request for reconsideration by revising paragraphs (a)(9) and (b) accordingly.

Section 178.57-22. A petitioner requested a revision of the information required in the inspector's report to clarify that the materials of construction of the inner container must conform to paragraph (a) of § 178.57-21. MTB agrees and grants the request for reconsideration.

Section 178.337-11. The National LP-Gas Association and another petitioner objected to the requirement in paragraph (c) permitting liquid or vapor discharge openings sized at 1 ¼ NPT to be equipped with an excess flow valve and a manually operated external valve. The petitioners maintain discharge openings sized at 1 ¼ inches are better protected by a remotely controlled internal shut-off valve. MTB revised the paragraph under the final rule due to claims of limited availability of internal valves sized at 1 ¼ inches. However, MTB has since confirmed that the 1 ¼ NPT internal valve is readily available. MTB is granting the petitioners' request for reconsideration by revising paragraph (c) to require that MC-331 cargo tanks must be equipped with internal valves on vapor or liquid discharge openings that are 1 ¼ NPT or larger in size after September 30, 1984.

Section 178.338-1. A petitioner requested that 22 gauge stainless steel in place of 20 gauge stainless steel be allowed for construction of non-evacuated jackets. The petitioner stated that 22 gauge steel offers the same protection as 20 gauge steel, is less costly and adds less weight. A review of exemptions reveals that many of the older exemptions authorized 22 gauge stainless steel jacket and MTB has no record of incidents caused by puncture or the influx of moisture. Therefore, MTB is granting the request for reconsideration by authorizing stainless steel jackets having a minimum thickness of 22 gauge.

A petitioner agreed to the requirement, in paragraph (f)(1), of a 30 psi critical collapsing pressure for evacuated jackets but took exception to the requirement that jacket heads, shell and stiffening rings must be designed in accordance with the ASME Code. The petitioner maintained that the ASME does not provide a minimum collapsing pressure format and, therefore, references to the ASME Code should be deleted. MTB agrees and grants the request reconsideration by removing the references.

Section 178.338–2. A petitioner objected to the requirement that the jacket material of a MC-338 cargo tank be in conformance with the ASME Code as being too restrictive and that it eliminates presently used materials. The petitioner argued that ASME materials are intended primarily for pressure vessels subjected to internal pressure and that the availability of the sheet materials is extremely limited. The petitioner requested that paragraph (a) be revised to allow evacuated jackets to be constructed of ASME materials or materials meeting ASTM specifications A 242, A 441, A 514, A 572, A 588, A 606, A 607, A 633, A 715. MTB agrees with the petitioner's request for reconsideration and has made the change.

Two petitioners objected to the requirement, in paragraph (c), for impact testing of all tank material, except aluminum. One petitioner stated that impact testing is not necessary on materials when not required by the ASME Code, especially for stainless steels, such as Type 304 stainless steel. MTB does not agree. The ASME Code basically establishes standards for stationary pressure vessels and it does not consider the dynamic forces encountered in the transportation environment. In order to assure adequate strength and toughness of the materials throughout the range of service temperatures encountered, the petitioner's request for reconsideration is denied.

Section 178.338–3. A petitioner requested that paragraph (a) be revised to specify a minimum thickness of not less than 0.090-inch for the tank. The petitioner contends that 0.090-inch thick stainless steel permits a tank design pressure of 40 psi and is approximately 40 percent thicker than the ASME minimum thickness for stainless steel. The present requirement specifies a thickness of not less than ⅛ or 0.125-inch.

Several exemptions for vacuum insulated cryogenic cargo tanks authorize the use of a stainless steel inner tank of 0.110-inch thickness. These tanks with pressure control valves set below 25 psig are used for atmospheric gasses and, therefore, are not specification regulated except when transported by vessel. There has been no adverse experience reported on the operation of these tanks.

There is a thickness threshold, particularly in large diameter tanks, below which distortions from welding and handling are likely to occur, and where reasonable shape rigidity is compromised. Even though reinforcing members are attached to provide rigidity in thin wall vessels, a point is reached where any attachment disturbs the ideal tank contours and provides a source for fatigue stresses. MTB has not been provided an anlysis of these factors and, therefore, a minimum thickness threshold has not been convincingly established. MTB must assume, lacking an engineering and safety analysis, that the minimum thickness should be in the vicinity of 0.125-inch based on experience in this thickness. Considering the experience with 0.110-inch thickness, the fact that the inner tank is well protected and is not subjected to any corrosive atmosphere, and the fact that the strength must meet the dynamic force requirements of § 178.338–3(b), the petitioner's request for a minimum thickness of 0.090-inch is denied. However, MTB belives 0.110-inch minimum thickness for the inner tank of a vacuum insulated cargo tank is acceptable and is revising paragraph (a) accordingly.

Section 178.338–4. A petitioner requested revision of paragraphs (a) and (f) to remove the requirement that welds in evacuated jackets be in conformance with the ASME Code. MTB takes the position that the evacuated jacket is a load bearing member and should have acceptable welds. Therefore, MTB believes these welds should meet recognized standards in the ASME Code and MTB is denying the petitioner's request for reconsideration. However, MTB is revising paragraph (a) to remove a duplicative requirement that all undercutting in shell and head material must be repaired as specified in the ASME Code. Paragraph (f) is revised to remove the duplicative requirement to paragraph (a) that all joints must be in accordance with the ASME Code.

Section 178.338–6. A petitioner requested that paragraph (c) be revised to allow location of a welded manhole on the front head of an MC-338 cargo tank. The petitioner argued that no strength reduction would occur due to required reinforcement of openings in the tank. In light of the petitioner's comment and upon further consideration, MTB agrees and grants the petitioner's request for reconsideration. The rationale for the original requirement, developed from a detailed study of an accident involving an MC-331 cargo tank, was that the design and location of the bolted manhole cover assembly in the front tank head allowed the manhole assembly to transmit accident impact loadings that caused failures in the tank head and shell. Most manholes used in MC-338 cargo tank are welded manholes fabricated nearly flush with the tank shell and located beneath the insulation jacket. Because such designs are unlikely to transmit and concentrate accident impact loads as occurred in the MC-331 cargo tank failure, MTB has decided that it is not necessary to restrict the location of such manholes. However, a manhole with a bolted closure when impacted is likely to transmit and concentrate accident loads into the tank. For this reason, MTB continues to prohibit manholes with bolted closures on the front head of MC-338 cargo tanks.

Section 178.338–9. A petitioner requested that MTB add a procedure for determining heat transfer rate and hold time requirements similar to that used for class DOT-113 tank cars. MTB agrees and grants the request for reconsideration by adding a new paragraph (c)(3) containing alternate procedures for determining the heat transfer rate and holding time of cargo tanks used in nonflammable cryogenic liquid service.

Section 178.338–10. A petitioner stated that the term "ultimate strength" is obsolete and should be replaced with the term "tensile strength". MTB agrees and grants the petitioner's request for reconsideration by revising paragraphs (b) and (c) accordingly. Similar changes are made in § 178.338–13.

Section 178.338–12. A petitioner stated that a shear section may be of questionable value outboard of valves located forward of the tandem, but has no useful purpose if the valves are within a rear cabinet forward of, and protected by, the bumper. MTB agrees that protection of valves provided by the bumper arrangements should be recognized and MTB is granting the request for reconsideration by revising the section, as suggested by the petitioner.

Section 178.338–13. In comments on paragraph (c), a petitioner stated that increased tensile strengths of materials at operating temperatures should be defined using values contained in the ASME Code. The petitioner also pointed out that the higher strength that materials have at low temperatures should not be recognized in applications where the material may not be at the low temperature. MTB agrees and grants the petitioner's request for reconsideration by revising paragraph (c) accordingly.

Section 178.338–14. A petitioner requested revision of the last sentence in paragraph (a)(3) by replacing the parenthetical words "(percent outage)" with the words "(water capacity in pounds)". The petitioner stated that a setting expressed as a percentage does not reflect the actual outage for loading conditions and may be misleading. It is MTB's position that if a fix-length dip tube or trycock line gauging device is used to establish the maximum permitted liquid level at the loading pressure, it must be designed to assure conformance with the maximum permitted filling density prescribed in § 173.318. Therefore, after further consideration, MTB believes the requirement specifying the type of setting is unnecessary and it is removed. Accordingly, the petitioner's request for reconsideration is denied since it is unnecessary with the removal of paragraph (a)(3).

One petitioner objected to the placement of the pressure gauge on the tank jacket but provided no substantive data to justify removal of this requirement from paragraph (b). Therefore, the request for reconsideration is denied.

Also, a petitioner requested that the requirement on orifices in paragraph (c) be revised to limit applicability to tanks in flammable cryogenic liquid service, and to remove trycock lines from the restriction of openings not greater than 0.060 inch diameter. The petitioner maintained that larger openings are needed for trycock lines to ensure proper operation. MTB agrees with the petitioner in both cases and the requests for reconsideration are granted. The requirements are limited to tanks in flammable cryogenic service, and openings for trycock lines, if provided, may be no larger than ½-inch nominal pipe size.

Section 178.338–16. Paragraph (a) is revised to remove the requirement that the material of construction for the evacuated jacket must be in conformance with the ASME Code. This revision is consistent with the changes in § 178.388–2(a) to allow ASTM materials, as requested by a petitioner.

Section 178.388–18. A petitioner requested that the requirement in paragraph (a) be revised to permit ¼-inch lettering in place of ⅜-inch lettering on nameplates. MTB believes ⅜-inch letters provide more legible markings at negligible cost. The petitioner's request for reconsideration is denied.

A petitioner stated that in paragraphs (b)(1) and (2) the abbreviation "veh." is unnecessary and should be removed, in paragraph (b)(5) the "certificate date" is unnecessary as the "date of manufacture" is sufficient, in paragraph (b)(8) the correct abbreviation for weight is "wt." and not "wgt.", and in paragraph (b)(9) the word "cryogen" is not defined.

The petitioner's first two requests for reconsideration are denied. MTB believes the abbreviation "veh." is needed to clarify that the vehicle manufacturer is the final manufacturer of a portion of the vehicle, such as the tank or jacket. The "certificate date" is the date that the completed cargo tank is certified as conforming to all applicable requirements of the MC-338 specification as prescribed in § 178.338–19(a), and because it may differ from the manufacture date, it is retained. Relative to the petitioner's latter two requests for reconsideration, MTB agrees "wt." is the acceptable abbreviation for weight and revises paragraph (b)(8) accordingly. In paragraph (b)(9), the term "cryogen" is replaced with the term "cryogenic liquid".

Section 179.102–1. In response to petitioners' request for reconsideration, MTB is revising paragraph (a)(6) to remove the requirement that the tank anchor-to-tank shell fillet welds must be examined by radioscopy. A similar revision is made to §§ 179.102–(l) and 179.102–17(m).

Sections 179.102–4 and 179.102–17. MTB is revising paragraph 179.102–4(a) to incorporate an amendment adopted under Docket HM-175 (49 FR 3468, January 27, 1984 which requires that each specification 105 tank car built after August 31, 1981, be in conformance with specification 105).

Three petitioners requested revisions to paragraphs 179.102–4(b) and 179.102–17(b) to clarify that stainless steel is not authorized for use as the material of construction for the tank. MTB agrees and grants the requests for reconsideration by revising the two paragraphs.

Several petitioners objected to the requirement, in paragraphs 179.102–4(g) and 179.102–17(g), permitting the installation on a tank car of a gaging device if it is a fixed length dip tube. The petitioners pointed out that most tank cars are equipped with a closed magnetic level gaging device and the use of these gaging devices should be continued as they are also authorized under DOT E-3992. MTB agrees and grants the petitioners' request for reconsideration be revising the paragraphs to permit gaging devices that are approved by the AAR Committee on Tank Cars. The term "gaging device" is used in place of the term "gauging device" for consistency with the usage of this term in Part 179.

A petitioner requested that, in paragraphs 179.102–17(d) and (i), the term "fluorinated hydrocarbon polymer" be removed and replaced with the more specific term "PTFE". MTB agrees and grants the petitioner's request for reconsideration. However, the term "polytetrafluoroethylene" is used in place of its abbreviation.

Another petitioner objected to the restriction in paragraph 179.102–4(i) that precludes use of steels containing certain elements in tank cars used in vinyl fluoride service. Of principal concern is the restriction against aluminum and copper because of their presence in the type of steel used in the construction of valves. The petitioners' request for reconsideration is denied. MTB will not change the restriction until compatability data that specifically relates to vinyl fluoride are developed and reviewed since vinyl fluoride is known to be reactive with certain alloys.

Petitioners took exception to the requirement, in paragraphs 179.102–4(j) and 179.102–17(k), that the jacket of a tank car be stenciled with the words, "COLDEST LADING TEMPERATURE". The petitioners requested that the present wording of "MINIMUM OPERATING TEMPERATURE" continue to be authorized. One petitioner stated that "MINIMUM OPERATING TEMPERATURE" is more meaningful for the design and operating condition of the tank; whereas, "COLDEST LADING TEMPERATURE" may be misunderstood as being the temperature to the lading at any given time. MTB agrees and grants the petitioners' request for reconsideration by revising the paragraphs to permit continued use of the present marking.

Petitioners requested removal of the requirement, in paragraphs 179.102–4(1) and 179.102–17(m), that tank anchor-to-tank shell fillet welds must be examined by radioscopy. The petitioners maintained that radioscopy is not used to examine tank car fillet welds. MTB agrees and grants the petitioners' request for reconsideration by revising the paragraphs. A similar chnage is made in § 179.102–1(a)(6) for tanks in carbon dioxide, refrigerated liquid service.

Section 179.400–4. A petitoiner requested revision of paragraph (a)(1) and the expression "q" in paragraph (a)(5) by adding "of water capacity" immediately following "Btu/day/lb." MTB agrees and grants the request for reconsideration.

Section 179.400–8. A petitioner indicated that in paragraph (c) the formula for minimum thickness should read "$t = PL/8SE (3 + \sqrt{L/r})$. MTB disagrees with the petitioner. In the November 1 publication, MTB corrected the formula to read "$t = [PL(3 + \sqrt{(L/r)})]/(8SE)$". In the corrected formula, only the term "(8SE)" is the denominator and the term "(L/r)" is the square root expression.

A petitioner requested revision of paragraph (d) to allow the minimum wall thickness of the outer jacket head to be ½ inch "before forming" in place of the required ½ inch "after forming". The requirement that jacket heads be at least ½ inch thick is intended to provide head puncture resistance and is equivalent to the requirement for head shields on certain other classes of tank cars which are used to transport flammable gases. Therefore, the petitioner's request for reconsideration is denied.

Section 179.401–1. Editorial changes have been made to certain entries in the table to § 179.401–1.

(49 F.R. 24306, June 12, 1984)

49 CFR Parts 173, 177, 178, and 179
[Docket No. HM-115, Amdt. Nos. 173–180, 177–68, 178–62, 179–36]
Cryogenic Liquids; Corrections and Revisions

This document corrects typographical errors, omissions, discrepancies, and provides answers to questions of general interest received by MTB with respect to those publications. Also, an alternate configuration for the pressure relief device system on cargo tanks used in atmospheric gas (except oxygen) and helium service is authorized.

Because the amendments adopted herein clarify and correct certain provisions of the final rule and impose no new regulatory burden on any person, notice and public procedure are unnecessary and these amendments may be made effective without the customary 30 day delay following publication.

MTB has determined that this rule, as promulgated, is not a "major rule" under terms of Executive Order 12291 or significant under DOT implementing procedures (44 FR 11034). A final regulatory evaluation and environmental assessment was not prepared as the amendments herein are not substantive changes to the final rule.

Based on limited information available concerning size and nature of entities likely to be affected by these amendments, I certify that these amendments will not have a significant economic impact on a substantial number of small entities.

The following is a section-by-section summary of the amendments.

Section 173.10. Paragraphs (a) and (e) are amended for consistency with § 174.204(a)(2) which extended applicability of the delivery requirements to flammable cryogenic liquids.

Section 173.23. Paragraph (e) is amended for consistency with § 178.57–20(a)(9) which, as amended in the June 12, 1984 document, requires a DOT 4L cylinder with an aluminum jacket to be marked "AL."

Section 173.33. In paragraph (b)(2), as amended in the June 12, 1984 document, the parenthetical expression in the first sentence is corrected by adding the word "plate" preceding the word "placed". The fourth sentence is corrected by removing the word "of" and adding the word "or".

Section 173.314. In paragraph (c), the section reference is corrected to read "§ 173.10" in place of "§ 173.432".

Section 173.316. In paragraph (c)(2), as amended in the June 12, 1984 document, the table is corrected by reinstating an omitted filling density at "230 psig" and the design service temperature of oxygen is corrected to read "−320" in place of "−452".

Section 173.318. In paragraphs (b)(1)(i) and (b)(2)(i), MTB is authorizing use of an alternate pressure relief system on cargo tanks used in atmospheric gas (except oxygen) and helium service consisting of at least one pressure relief valve and, if needed, additional pressure relief valves or frangible discs with a combined relieving capacity sized for fire conditions. Present requirements provide that the primary pressure relief system must consist of one or more pressure relief valves. MTB believes that relaxation on the type of pressure relief devices authorized allows the industry greater flexibility in the selection of pressure relief devices and may remove certain operational restraints and reduce hardware costs without any reduction in the level of safety. MTB is making corresponding changes in paragraph (b)(2)(iii) to specify the minimum total capacity for the alternate pressure relief system, and in paragraph (b)(2)(iv) to allow the alternate system to have a liquid flow capacity equivalent to the liquid loading flow rate at a pressure not to exceed 150 percent of the tank design pressure in place of the present 120 percent limit.

In paragraph (e), the parameters for the temperature of a flammable cryogenic liquid in a cargo tank are relaxed to allow a setting that corresponds with the start of travel condition in place of the temperature at loading.

In paragraph (f)(3), the filling density entry for hydrogen at 17 psi is amended by removing "6.5" and adding "6.6". This higher filling density is presently authorized by exemption.

In paragraph (g), as amended in the June 12, 1984 document, MTB expanded the one-way-travel-time (OWTT) marking to include display of the pressures used when a tank is partially unloaded at one or more locations and the control of tank pressure is transferred from the pressure control valve to the pressure relief valve. However, MTB failed to provide an exception from these additional markings for tanks unloaded at one location. MTB is revising paragraph (g) to correct the oversight. Additionally, MTB has received several inquiries on whether the use of an OWTT and the corresponding rated holding time based on loading conditions other than the liquefaction temperature are authorized. MTB takes the position that such OWTTs have always been authorized by § 173.318(e) and, therefore, no change is required to the final rule. Also see preamble discussion to § 177.840.

Section 173.320. The introductory text in paragraph (a) provides an exception from requirements of the subchapter for atmospheric gases and helium used in process systems, such as a refrigeration system, and during loading and unloading operations. Paragraph (a) is revised to clarify that this exception applies even though the pressure may exceed 25.3 psig under these conditions.

Section 172.330(a)(1) requires a tank car to be marked with the proper shipping name or authorized common name of the material contained in the tank car when required by Part 173 or 179. The exceptions specified in § 173.320 do not impose any requirement to mark a proper shipping name on a tank car. However, it does require tank cars to be marked with identification numbers in conformance with § 172.330(a)(2).

Section 177.840. Paragraph (i) is revised to clarify that the cargo tank may not be placed in transportation unless the pressure of the flammable cryogenic liquid lading is equal to or less than that used to determine the marked rated holding time in order to prevent premature venting.

Section 178.338–9. The printing of the formula in paragraph (c)(3)(i) is corrected.

Section 178.338–16. The first sentence in paragraph (a) is amended by removing the word "cargo" preceding the word "tank". This change is made for consistency with the definition of "tank" in § 178.338–1(b).

Section 179.102. In the June 12, 1984 document, MTB granted a petition for reconsideration by revising §§ 179.102–1(a)(6), 179.102–4(1) and 179.102–17(m) to remove the requirement that tank anchor-to-tank shell filled welds must be examined by radioscopy. However, a requirement that the tank anchor-to-tank shell fillet welds must meet the acceptance standards of AAR specifications for Tank Cars, Appendix W, paragraph W11.06 was continued. It has been brought to MTB's attention that the referenced standard pertains to radiotaped welds and, therefore, is inappropriate. MTB agrees and has corrected the oversight.

Section 179.400–8. The printing of the formula in paragraph (c) is corrected.

(49 F.R. 42733, October 24, 1984)

SUMMARY • Docket No. HM-116 • *Amendments intended to clarify the regulations regarding shipment of samples of explosives; authorizing the Energy Research and Development Administration to test and classify their own explosives.*

Hazardous Materials Regulations Board
[49 CFR Parts 172, 173]
[Docket No. HM-116; Notice No. 74-5]
TRANSPORTATION OF HAZARDOUS MATERIALS
Proposed Classification of New Explosives
and Shipment of Samples of Explosives

The Hazardous Materials Regulations Board (the Board) is considering amendment of § 173.86 of the Department's Hazardous Materials Regulations. The proposal has three purposes:

(1) To permit the U.S. Atomic Energy Commission (USAEC) to examine, classify, and approve for its own activities new explosives as being safe for transportation;

(2) To restructure the current regulations contained in § 173.86 for clarification; and

(3) To provide a shipping name for samples of explosives shipped under § 173.86a.

The U.S. Atomic Energy Commission has petitioned the Board to amend § 173.86 to include the USAEC as an entity authorized to examine, classify, and approve for its own activities, new explosives for shipment. The USAEC states that it operates Government-owned laboratories and research facilities which are equivalent in experience and capability to the laboratories of the Bureau of Explosives and the Department of Defense for purposes of examining and testing explosives. Since the USAEC facilities are often used by the Department of Defense to examine, classify, and approve new explosives pursuant to Department of Defense authority under § 173.86, the Board believes that it is in the public interest to authorize the USAEC to perform these functions for its own explosives.

In addition, it is the Board's opinion that § 173.86 of the Hazardous Materials Regulations in its current format is a section of the regulations difficult to understand. The Board believes that the difficulty exists because the section covers both "shipments of samples of explosives," and "classification of new explosives" in an interchangeable manner, when in actuality they should be treated separately. For clarification, the Board proposes to restructure current § 173.86 into two separate § § 173.86 and 173.86a.

Proposed § 173.86 addresses itself only to the classification of new explosives. Included in this section would be the definition of a new explosive, and the particular agency who must examine, classify, and approve a new explosive as being safe for transportation.

Proposed § 173.86a addresses itself to the shipment of samples of explosives for laboratory examination. Included in the proposed sections would be a definition of a "sample for laboratory examination," and the packaging, marking, and labeling requirements for shipping samples for laboratory examination.

Since samples of explosives for laboratory examination are authorized for transportation under certain specified conditions, the Board proposes to change § 172.5 to require specific identification of these materials as samples when shipped under § 173.86a.

In accordance with section 102 of the National Environmental Policy Act (Pub. L. 91-90, (42 U.S.C. 4321 et seq.)) the Board has considered the environmental impact of this proposal. It has determined that the changes proposed in this notice would not have a significant impact upon the environment. Accordingly, it considers that an Environmental Impact Statement is not necessary and it does not intend to issue such a statement with respect to this proposed amendment.

(39 F.R. 12261, April 4, 1974)

PART 172—LIST OF HAZARDOUS MATERIALS
PART 173—SHIPPERS
[Docket No. HM-116; Amdt. Nos. 172-30, 173-95]
Classification of New Explosives and Shipment of Samples of Explosives

The purpose of these amendments to Parts 172 and 173 of the Department of Transportation's Hazardous Materials Regulations is to:

1. Permit the Energy Research and Development Administration (ERDA), formerly a part of the U.S. Atomic Energy Commission, to examine, classify, and approve for its own activities new explosives as being safe for transportation;

2. Restructure and clarify the regulations contained in section 173.86 pertaiing to new explosives, and

3. Provide a shipping name for samples of new explosives shipped under section 173.86.

Interested persons have been afforded the opportunity to participate in the making of these amendments by commenting on a notice of proposed rule making (Notice No. 74-5) published April 4, 1974 (39 FR 12261). All commenters who responded to the proposed change in section 173.86(b) supported authorizing ERDA to classify and approve new explosives made by or under the supervision of ERDA. ERDA operates government-owned laboratories and research facilities which are equivalent in experience and capabilities to the laboratories of the Bureau of Explosives and the Department of Defense for examining and testing explosives. Because of their expertise in this area, this Bureau has included ERDA as an authorized agency for evaluating new explosives.

Several commenters objected to the proposed definition of a "new explosive" in section 173.86(a). They felt the definition was still unclear and submitted suggestions for the definition of a "new explosive." With minor changes, the Bureau adopted the recommendations and has added a more complete definition of a "new explosive."

One commenter objected to the proposed change in section 173.86(c), that ignition elements be removed from samples of explosives before shipment. The commenter felt the requirement, if interpreted literally, would mean that the leg wire and bridge wire would have to be removed from electric blasting caps. Although no commenter mentioned it, a literal interpretation would also mean that the fuses would have to be removed from many types of fireworks. The Bureau agrees that this would be an impractical and unnecessary requirement and has withdrawn that portion of the proposal.

Two commenters stated that it is often necessary to move new, unclassified explosives from plants to laboratories or field test sites for functional evaluations. If the new explosives do not perform as expected, the manufacturer may not pursue their development any further, and the Bureau believes it would be an unnecessary expense to send samples to one of three agencies authorized to approve them if these explosives will not be used commercially and move in transportation. The Bureau, therefore, has added a new paragraph (e) to section 173.86 to allow transport of these unclassified explosives for testing purposes under special conditions.

(41 F.R. 15013, April 9, 1976)

PART 172—LIST OF HAZARDOUS MATERIALS
PART 173—SHIPPERS
[Docket No. HM-116; Amdt. Nos. 172-30, 173-95]
Classification of New Explosives and Shipment of Samples of Explosives

The purpose of these amendments to Parts 172 and 173 of the Department of Transportation's Hazardous Materials Regulations is to:

1. Permit the Energy Research and Development Administration (ERDA), formerly a part of the U.S. Atomic Energy Commission, to examine, classify, and approve for its own activities new explosives as being safe for transportation;

2. Restructure and clarify the regulations contained in section 173.86 pertaining to new explosives, and

3. Provide a shipping name for samples of new explosives shipped under section 173.86.

Interested persons have been afforded the opportunity to participate in the making of these amendments by commenting on a notice of proposed rule making (Notice No. 74-5) published April 4, 1974 (39 FR 12261). All commenters who responded to the proposed change in section 173.86(b) supported authorizing ERDA to classify and approve new explosives made by or under the supervision of ERDA. ERDA operates government-owned laboratories and research facilities which are equivalent in experience and capabilities to the laboratories of the Bureau of Explosives and the Department of Defense for examining and testing explosives. Because of their expertise in this area, the Bureau has included ERDA as an authorized agency for evaluating new explosives.

Several commenters objected to the proposed definition of a "new explosive" in section 173.86(a). They felt the definition was still unclear and submitted suggestions for the definition of a "new explosive." With minor changes, the Bureau adopted the recommendations and has added a more complete definition of a "new explosive."

One commenter objected to the proposed change in section 173.86a(c), that ignition elements be removed from samples of explosives before shipment. The commenter felt the requirement, if interpreted literally, would mean that the leg wire and bridge wire would have to be removed from electric blasting caps. Although no commenter mentioned it, a literal interpretation would also mean that the fuses would have to be removed from many types of fireworks. The Bureau agrees that this would be an impractical and unnecessary requirement and has withdrawn that portion of the proposal.

Two commenters stated that it is often necessary to move new, unclassified explosives from plants to laboratories or field test sites for functional evaluations. If the new explosives do not perform as expected, the manufacturer may not pursue their development any further, and the Bureau believes it would be an unnecessary expense to send samples to one of three agencies authorized to approve them if these explosives will not be used commercially and move in transportation. The Bureau, therefore, has added a new paragraph (e) to section 173.86 to allow transport of these unclassified explosives for testing purposes under special conditions.

(40 F.R. 15013, April 9, 1976)

[Docket No. HM-117; Amdt. Nos. 171–44, 173–127, 178–54]

SPECIFICATION 1M GLASS CARBOY IN EXPANDED POLYSTYRENE PACKAGING AND CANCELLATION OF CERTAIN OBSOLETE SPECIFICATION PACKAGINGS

On June 14, 1974, the Hazardous Materials Regulations Board published a notice of proposed rulemaking, Docket HM-117, Notice 74–8 (39 FR 20805), which proposed these amendments. The background and the basis for incorporating these changes in the regulations were discussed in that notice. Interested persons were invited to submit their views in writing and consideration has been given to all comments received relating to matters within the scope of the notice. The primary drafters of this document are Marlo E. Gigliotti, Technical Division, and Edward T. Mazzullo, Standards Division, Office of Hazardous Materials Regulation and George W. Tenley, Jr., Office of the Chief Counsel, Research and Special Programs Administration.

SHIPPER RESPONSIBILITY

In the Notice, the Materials Transportation Bureau (MTB) proposed to amend §173.22 to clarify the shipper's responsibility for compliance with requirements involving specification packagings. This proposal is not addressed in this rulemaking, since its provisions were incorporated in Docket HM-134 (41 FR 38175, September 9, 1976).

EXPANDED POLYSTYRENE PACKAGING

In the Notice, the term "polystyrene" was used. In these amendments, the term "expanded polystyrene" now appears in place of "polystyrene." This change has been adopted because polystyrene which is not expanded (or expandable), means a rigid; high density material that has been injection molded into form, or extruded into sheet and thermoformed into shape. It is unlike "expanded polystyrene" which is cellular (foamed), low density material that is produced by applying heat to expandable polysytrene beads. The MTB believes it is important that a differentiation be made between polystyrene and expanded polystyrene because a packaging made of polystyrene is usually brittle (unless modified by rubber or polymers) and may have certain relatively poor mechanical properties. However, packaging made of expanded polystyrene is lightweight, strong, and has excellent energy and shock absorbing characteristics.

In §178.17, the heading of the proposed specification 1M has been redesignated to read: "Non-reusable glass carboy in non-reusable expanded polystyrene packaging." This heading has been changed to indicate that the glass carboy is not to be reused for hazardous materials. The prohibition on reuse will preclude the carboys from being abused since they will no longer be removed from the original protective expanded polystyrene packaging and placed inside another expanded polystyrene packaging. This position, accepted by the MTB, was suggested by a commenter who noted that glass containers are subject to breakage and to weakening as a result of possible abuse prior to reuse. An informal survey of the present exemption (special permit) holders indicates that glass carboys are normally not reused.

Many comments were received concerning the proposal in §178.17–3(b)(3), relating to construction requirements of the expanded polystyrene packaging. The majority of the commenters believe that the density requirement can be safely reduced from the proposed minimum two pounds per cubic foot without detracting from the ability of the expanded polystyrene to adequately protect a glass carboy during performance of the free fall drop tests prescribed in §178.17–5(b)(1).

The MTB believes, based on the testing conducted in this regard, that an optimum density for the expanded polystyrene packaging is 1.7 pounds per cubic foot (pcf), and, therefore, has revised §178.17–3(b)(3) to require a density for the expanded polystyrene ranging from 1.7 pcf minimum to 2.0 pcf, maximum.

The MTB has also revised §178.17–4(b)(1) to clarify the requirements for closing an expanded polystyrene packaging with tape. Also, §178.17–4(b)(2) has been added to provide an alternate means of preparing an expanded polystyrene package for shipment, i.e., by closing with nonmetallic strapping meeting prescribed specifications.

GLASS CARBOY

One commenter noted that while §178.17–3(a)(1) would require the glass carboys to be "properly annealed," there is no definition of what constitutes a properly annealed glass carboy. It was recommended that an annealing requirement be specified and a method to determine the level of annealing be included. The MTB agrees. Sections 178.17–3(a)(2) and 178.17–5(a)(1) have been revised to include a requirement that the real temper of the glass carboy be not greater than 5 when tested in accordance with the examination method contained in the American Society for testing and Materials publication ASTM C148–77, Method A. This publication is incorporated by reference in § 171.7 as part of this rulemaking.

One commenter objected to the proposal in § 178.17–3(a)(2) that the top lip of each carboy must be smooth and even, by stating that the requirement is unclear and not capable of being measured. While the purpose of the requirement was to eliminate potential fracture points, the MTB agrees with the commenter and has deleted this requirement.

A commenter responding to the proposal made in § 178.17–3(a)(4) contended that the establishment of the maximum amount of glass which can be used in connection with the manufacture of the carboys should be deleted because this requirement is not necessary for the glass carboy to meet the minimum performance requirements of the specification. The MTB agrees and has deleted the maximum tolerance reference, while preserving the minus tolerance limit. These requirements appear in this amendment as § 178.17–3(a)(3).

One commenter recommended deletion of the proposed change in § 178.17–3(a)(5) regarding the minimum glass thickness requirements. The commenter submitted the following statement:

> The proposed wall thickness requirement of 0.075-inch is not necessarily or directly related to the glass container's performance capability. Some of the carboys subjected to the proposed four-foot drop test had wall thicknesses substantially below 0.075-inch and yet did not break. Further, it is not technically feasible to determine by mechanical or electronic means the minimum wall thickness in 6.5-gallon capacity glass carboys. The wall thickness of such carboys can be determined only by destructive testing, i.e., cutting the container.

The MTB agrees with the comments and has removed § 178.17–3(a)(5) from the proposal.

In § 178.17–5(a), the pressure test requirements for the glass carboy have been substantially revised. The MTB disagrees with one commenter that an internal pressure test is neither a necessary nor significant test of a carboy's performance capabilities. The MTB does, however, agree with the commenter's contentions that the proposed internal pressure and hydrostatic pressure tests are different methods of measuring the same capabilities, that the proposed hydrostatic pressure test requiring destruction of at least one-half of one percent of the carboys produced is unnecessarily harsh, and that the requirement to test one out of every two hundred carboys produced is not suited to modern, high volume, production methods. The MTB feels that safety considerations are provided for by requiring a pressure test to be conducted on one carboy from each mold at least every eight hours. In addition, provision has been made for tests using either an instantaneous or a sustained pressure, and for either destructive or non-destructive testing.

In § 178.17–5(b) drop test requirements have been revised to place the responsibility for drop testing on the manufacturer of the expanded polystyrene packaging, to provide periodic testing every 4 months instead of every 6 months and to provide rejection criteria when the packaging fails a periodic drop test. Since the primary purpose of the drop test is to evaluate the ability of the expanded polystyrene packaging to protect the inside glass carboy, the MTB feels that the manufacturer of the expanded polystyrene packaging should be made responsible for this testing. A requirement to perform drop tests at least every 4 months is considered more meaningful than the 6 month interval originally proposed because it should insure at least one periodic test during a production year which could conceivably be less than 6 months long. If packaging defects (such as improper density, bead adhesion, molding, or design) are discovered during periodic drop testing, it is the MTB's opinion that defects in packagings produced prior to the unsuccessful tests will be detectable by examining the packagings using visual inspection, physical measurements, or other testing as appropriate. Therefore, a requirement has been placed on the person performing the drop tests to, upon discovery of a packaging defect, examine packagings on hand produced prior to an unsuccessful test for the defect, and to inform packaging users of defective packagings which have been released.

The MTB on its own initiative has made two changes pertaining to §§ 178.17–1 and 178.17–6. In § 178.17–1(d) and § 178.17–6(c) provisions have been included to allow a glass carboy meeting the requirements of § 178.4 (DOT-1D) to be used with the DOT-1M expanded polystyrene packaging. This option has been provided so that unused DOT-1M glass carboys already in the distribution system (e.g., under exemptions DOT-E 5615 and DOT-E5526) can be used in the shipment of authorized hazardous materials.

On January 9, 1979, personnel from both the MTB and private industry were witness to tests of completed packages made as prescribed by the new Specification 1M. One purpose of the test was to demonstrate the effectiveness of two different types of expanded polystyrene packagings currently being produced. Both types of packagings passed the tests prescribed in § 178.17–5(b)(1) without leakage from or breakage of any of the tested glass carboys.

RESPONSIBILITY FOR COMPOSITE PACKAGING

The provisions of §§ 173.22(a) and 173.24(d) place responsibility on the shipper for compliance with the requirements of Part 178 over which the shipper has control. With regard to the new specification 1M, a general requirement has been added in § 178.17–1(b) to require each glass carboy to fit snugly in its expanded polystyrene packaging. It is felt that this requirement,.in addition to the provisions of §§173.22(a) and 173.24(d), clearly places responsibility on the shipper for the compatibility of the individual components of the composite packaging and eliminates the need for registering assemblers. Therefore, the proposal to have assemblers of the composite packaging register with the MTB has been deleted.

DELETION OF OBSOLETE SPECIFICATIONS

Prior to publication of the Notice, the MTB had information indicating that specification packagings 1B, 1C, 1E, 28, 28A, 31, 34B, and 43A were no longer being manufactured or used. In the Notice, the MTB proposed to delete the specifications for these packagings from Part 178 and to remove the authorizations for their use wherever they appear in Part 173. Since no comments were received supporting retention of the specification packagings and their authorizations for use, the Bureau has determined that the packagings are obsolete. Therefore, in keeping with the objectives to simplify and clarify the regulations, the appropriate deletions have been made to Parts 173 and 178 in this amendment.

(44 F.R. 14194, March 12, 1979)

SUMMARY · Docket No. HM-117 · *This docket authorized a new DOT specification 1M 6-½ gallon glass carboy encased in expanded polystyrene and requirements governing its use.*

[49 CFR Parts 173, 178]
[Docket No. HM-117; Notice No. 74–8]
SPECIFICATION 1M GLASS CARBOY IN POLYSTYRENE PACKAGING AND CANCELLATION OF CERTAIN OBSOLETE SPECIFICATION PACKAGINGS
Notice of Proposed Rule Making

The Hazardous Materials Regulations Board is considering amendment of Parts 173 and 178 of the Department's Hazardous Materials Regulations to provide a specification for glass carboys in polystyrene packaging, to authorize use of this packaging with various commodities, and to clarify a shipper's responsibility for compliance with specifications involving certain packagings.

The Manufacturing Chemists Association and a number of holders of outstanding special permits have petitioned the Board to amend the regulations to provide for a new specification for a glass carboy in polystyrene packaging and to authorize use of this packaging with various commodities. This proposal identifies the new packaging as specification DOT-1M.

The Board believes that the assignment of responsibility for compliance with the Department's regulations involving composite packaging specifications must be clearly set forth. In Docket No. HM-69; Amendment No. 178–20 (36 FR 16579), the Board established a manufacturer's registration system for the purpose of identifying manufacturers of containers and to clearly establish responsibility for compliance with the specification. The proposed specification includes, in addition to identification of the manufacturer of each component, identification of the person who assembles the completed unit constituting the specification to clearly specify responsibilities for composite packagings. The Board proposes that each assembler obtain a resgistration number for identification on the exterior packaging. The assembler's registration number will contain the letter "A" immediatley following the numerals. This identification will be in addition to the polystyrene packaging manufacturer's registration number if he is not the assembler.

In addition, the Board proposes an amendment of §173.22 to clarify shipper responsibility for compliance with specifications when the specifications prescribe functions to be performed by the shipper as described in Parts 173 and 178.

The Board has available information which indicates that certain specification packagings are obsolete. Therefore, by this notice, the Board is proposing to cancel packaging specifications 1B, 1C, 1E, 28, 28A, 31, 34B, and 43A from Part 178 and also proposes to remove the authorizations for their use wherever they appear in Part 173. Any person using one of the above-listed specification packagings who requests that it be continued in the regulations for either construction or use should provide the Board with information concerning its use, including the number being used and the type of use.

(39 F.R. 20805, June 14, 1974)

SUMMARY • **Docket No. HM-118** • *This rulemaking is a proposal to classify expandable polystyrene resin as a flammable solid. It is also proposed to amend the flammable solid definition.*

[49 CFR Parts 172, 173]
[Docket No. HM-118; Notice No. 74–9]

EXPANDABLE POLYSTYRENE RESIN AND THE
DEFINITION OF A FLAMMABLE SOLID
Notice of Proposed Rule Making

The Hazardous Materials Regulations Board ("the Board") of the Department of Transportation proposes to (1) amend Part 172 of the Hazardous Materials Regulations to specify that expandable polystyrene containing either a flammable liquid or flammable gas is classed as a flammable solid and (2) amend the definition of a flammable solid in § 173.150.

By letter dated January 8, 1954, the Chief Inspector, Bureau of Explosives, Association of American Railroads, expressed his view that expandable polystyrene is not classed as a "dangerous article under the I.C.C. Regulations; since it does not fall within the scope of any of the definitions contained therein." He went on to state that, "Any hazard which the material might have will be minimized by the use of tight containers, and any containers which will satisfactorily prevent the escape of vapors under normal conditions will be acceptable."

The Board agrees with the letter statement expressed above. However, the Board believes the statement is also applicable to thousands of other materials presently subject to the Department's regulations. Also, the statement had no binding effect since the Chief Inspector had stated earlier that it was his view that the material was not classed as a "dangerous article." The situation here is the same as that mentioned in the preamble of the Notice to Docket HM-102 (37 FR 11898) stating that a material must be classed as a hazardous material before it may be made the object of regulatory requirements.

The opinion of the Chief Inspector was not challenged until 1973 when the Department took the position that the expandable polystyrene shipped by one manufacturer was classed as a flammable solid, n.o.s. That particular manufacturer has petitioned the Board for a special permit for certain packaging waivers. Since receipt of the petition, the Board has been contacted by several manufacturers of expandable polystyrene presenting different views on how the material should be handled under the regulations. The Board believes this matter should be handled by a public rule making procedure being fully aware that millions of pounds of this material are shipped annually and that the definition set forth in § 173.150, when literally construed, may not address the potential hazard of expandable polystyrene.

According to information available to the Board, pentane is the most common glowing (expanding) agent in the expandable polystyrene shipped at the present time. Pentane has a flash point of minus 40° F. or lower, is flammable when mixed with air at 1.5 to 7.8 percent by volume, and is 2½ times heavier than air. The percentage of pentane in expandable polystyrene is usually 5 to 8 percent by weight. Open burning tests have been conducted on expandable polystyrene, and the Board takes no issue with the conclusion that the material does not burn so vigorously and persistently as to create a serious transportation hazard. However, the Board does believe the material poses a potential hazard in transportation due to the possible emission of pentane (or other flammable material) vapors into confined areas. The Board belives the material should be classed as a flammable solid so that the potential hazard will be communicated by labeling and placarding, and approprite packagings specified.

By letter dated April 12, 1974, the Assistenat General Counsel to the Society of the Plastics Industry, Inc. (SPI) stated the following:

"Initially, we wish to observe that any regulation adopted by the Department with respect to this material should be limited to expandable polystyrene resin. The designation "expandable" signifies that resin into which pentane has been introduced, the presence of which generates the interest of the Department of Transportation in regulating this commodity. Additionally, "resin" is preferable to "beads" for purposes of identification in the Department's regulations.

"As we discussed, the members of SPI have transported hundreds of millions of pounds of expandable polystyrene resins during a period of more than a decade without incurring a single reported incident of fire or explosion. Any danger which does exist is the result of the accumulation of pentane gas which bleeds from the resins and leaks from the container into an enclosed environment which lacks ventilation. Any such danger can thus be alleviated by adequate venting during storage/transit or prior to unloading and further by a ban upon exposure to open flame or lighted cigarettes. Once any accumulated gas has thus been dissipated, the resin itself cannot be considered to be a flammable solid within the definition of 49 C.F.R. § 173.150. Identification of the resin as a flammable solid is thus a misnomer and further fails to adequately identify the hazard presented.

"It is the recommendation of SPI and its members that the Department of Transportation not require expandable polystyrene resin to be identified as a flammable solid but rather that a specific rule section be adopted which (1) prohibits transportation of this material in refrigerator-type equipment and (2) requires the shipper to provide warning identification labels to be applied to the transportation equipment advising of the requirement to ventilate that equipment for a period of ten minutes before unloading and further advising of a prohibition upon smoking or open flame in the vicinity of such equipment. Such notification will, we respectfully submit, give adequate warning of the minor hazard which may be presented by this commodity."

Concerning the first paragraph quoted above, the Board agrees to use the word "resin" in place of the word "beads" but is proposing that the word "polystyrene" appear first as the key word for alphabetical listing in § 172.5. Concerning the statement in the second quoted paragraph about the excellent experience in shipping this material to the present time, the Board has no information to the contrary and the accident experience, or lack of experience, involving this material (which is not presently the subject of the Department's incident-reporting requirements) is not the basis for this rule making action.

Concerning the recommendation that expandable polystyrene resin not be classed as a flammable solid, the Board does not agree for the reason stated earlier in this preamble, a material must be classed as a hazardous material before it may be made the object of regulatory requirements. Further, the suggested prohibition against the use of refrigerated equipment raises the question as to what other types of semi-airtight equipment should be considered or whether ventilation equipment should be specified and what type of criteria utilized. Furthermore, if this approach is followed, it may be necessary to examine the criteria pertaining to the transportation of all materials capable of emission of flammable vapors. The Board believes such an undertaking would be considerable in scope and length. The Board also takes the same view pertaining to special markings to be applied to transport vehicles. If such markings are specified when expandable polystyrene resin is transported, perhaps they should also be considered for other materials if the present FLAMMABLE (for highway) DANGEROUS (for rail) placards, or those proposed in Docket HM-103 (39 FR 3164) are considered to be insufficient in communicating the potential hazards of flammable materials in transportation. The Board does not agree with the alternatives suggested by SPI at this time but will consider carefully all views presented before deciding on a future course of action.

The Board believes the definitions in the regulations for each of the classifications should be specified in quantitative terms when it is practicable and feasible to do so. Its efforts to date have not been easy (e.g., Docket HM-57 for corrosive materials). However, there are situations when exclusive reliance upon quantitative definitions could preclude the regulation of materials, such as polystyrene resin, that should be made the subject of safety requirements. If the Board is to carry out its assigned responsibilities in an adequate manner, there should be sufficient flexibility provided in its regulations to enable the Department to regulate materials that pose a potential hazard in transportation. Therefore, the Board is proposing to amend § 173.150 by adding a reference to § 172.5 in order to classify as flammable solid materials that present potential hazards associated with other substances in the flammable solid class. Additions to the list in § 172.5 would be made in accordance with the rulemaking procedures specified in 49 CFR Part 170. This same procedure is presently used by the U.S. Coast Guard to identify materials as "Hazardous articles" under 46 CFR 146.27-1.

Concerning packaging, the Board will consider recommendations for packagings presently used that are not provided for in § 173.154.

With respect to the proposal made in Docket HM-112, Notice 73-9 (39 FR 3021) regarding § 173.101, under the description "polystyrene beads," that proposal remains effective insofar as it pertains to the Hazard Information Number assignment, the passenger-carrying rail car and aircraft limitations, and the vessel stowage requirements. Otherwise, that proposal is modified as presented in this notice.

(39 F.R. 25235, July 9, 1974)

SUMMARY · Docket No. HM-119 · *This docket established a new DOT specification 35 open-head, non-reusable polyethylene drum (pail), and propose requirements governing its use.*

[Docket No. HM-119; Notice No. 74-10]

POLYETHYLENE DRUM PACKAGING
Proposed Specification 35

The Hazardous Materials Regulations Board is considering an amendment of Parts 173 and 178 of the Department's hazardous materials regulations to provide a specification for non-resuable, molded, open-head polyethylene drums, to authorize use of this packaging with certain dry and paste materials, and to provide for the use of a large capacity plastic drum in § 173.245b.

This proposal contains a specification for non-reusable, molded, open-head polyethylene drums which is based on the satisfactory shipping experience reported to the Board under special permits. Some of these special permits have been in effect for over five years.

The proposal also reflects the authorization for use of the drum for the dry and paste materials that were permitted under the provisions of the special permits.

The present packagings authorized for corrosive solids not specifically provided for in § 173.245b include the use of a plastic drum of not over six gallons capacity. The Board has proposed to amend the plastic drum packagings in § 173.245b by increasing the volumetric capacity to seven gallons in order to permit use of the specification drum covered by this proposal.

(39 F.R. 27693, July 31, 1974)

[Docket No. HM-119; Amdt. Nos. 171–37, 173–106, 178–42]
POLYETHYLENE DRUM PACKAGING
Specification 35

Interested persons were afforded the opportunity to participate in the making of these amendments by a notice of proposed rulemaking. Docket No. HM-119, Notice 74–10, published in the FEDERAL REGISTER (39 FR 27693) on July 31, 1974.

Two commenters suggested that the proposals in § 173.217 be extended to include liquid material meeting the corrosive material definition. This has not been done because such action is beyond the scope of the notice of proposed rulemaking, involving hazards outside the consideration of what was proposed there.

Several commenters expressed concern relative to the proposed polyethylene material specifications. The comments pertained to the property of melt index and the specification of a value of 6.0 maximum for that property. The commenters contended that melt index (which is a melt flow property measured under low shear conditions) is not an appropriate property, by itself, to insure the toughness of polyethylene and that the measure of safety of the drums will be provided by the performance tests in § 178.16–13. These commenters further observed that the significance of melt index to the strength of molded polyethylene varies depending on the scale on which it is determined and that molecular weight distribution (i.e., whether broad or narrow) of the polyethelene, with respect to melt index, is also a factor. The Bureau agrees and has withdrawn its proposal for a melt index specification.

The specifications for density range and percent elongation have been changed. The values now specified (0.941 to 0.965 and 75% minimum, respectively) more adequately apply to the grade of polyethylene that is used in the injection molding of these drums.

Although it was not specifically proposed, the Bureau believes that a better means to identify the polyethylene used in the molding of the drum is necessary. For this reason, the Bureau is requiring drum manufacturers to have, and retain, certain data relating to the material properties of the molding resin, at the start of production. The data, which may be obtained from the material supplier or determined by the drum manufacturer, must include values for density, melt index, molecular weight, molecular weight distribution, izod impact, tensile impact, environmental stress crack resistance, brittleness temperature, and flexural modulus. The data must indicate the method used to obtain the numerical values and include all parameters (e.g., method of preparing these specimens; number of specimens tested; speed of testing; type of test specimens and dimensions; etc.) relative to test conditions. (ASTM methods may be used where applicable.) It is further required that these material properties data, for each lot of molding resin processed into drums, be recorded and retained—along with a 100-gram sample of the resin and at least three drums molded from the resin—for at least one year by each molder at each producing plant.

Several commenters objected to the requirement for ultraviolet (UV) light protection being mandatory and contended that the drums are nonreusable, have a short life and are seldom stored outdoors or subjected to prolonged exposure to direct sunlight. The Bureau agrees and has made inclusion of UV light protection in the polyethylene an optional provision.

Two commenters, responding to the proposal on the rated (marked) capacity of the drum, suggested use of the specification of minimum-maximum volume percentages, as prescribed in other DOT specifications. The Bureau agrees and has incorporated this change into the amendment.

Other commenters responding to the drum closure proposal in § 178.16–10(a) suggested the type of steel head to be used and further recommended that the use of gaskets not be mandatory. The Bureau believes the comments have merit and has included alternate specifications for a steel head and a provision that gaskets are required only when necessary to meet performance tests.

Several commenters recommended changes in the procedure for the drop test prescribed in § 178.16–13(a)(1), in order to clarify the requirement for conditioning the drums and contents to 0° F., and to delete the requirement for the drums to be rolled following the drop tests. The Bureau agrees and has clarified the wording in the specification. The purpose of temperature conditioning is to ensure that drum and contents are at or below 0° F. at the commencement of testing. A choice as to means of achieving this result is up to the tester. In addition, the Bureau has omitted the requirement of rolling the drums following drop tests, as this technique is only meaningful when applied to test containers filled with water.

Numerous commenters responded to proposed § 178.16–13(a)(2), regarding the vibration test, and recommended its exclusion from the proposed specification. The Bureau believes that the vibration test is an important aspect of the specification, and the requirement has been retained. However, the test procedure has been rewritten for clarification. Also, criteria for acceptance, rejection, and penalty have been included in the procedure so that it agrees with the drop test in these respects.

Several commenters responded to proposed § 178.16–13(a)(3) and either (1) objected to the requirement for a static compression test at 130° F., or (2) suggested changes to the testing procedure. The Bureau does not agree to the deletion from the specification of a static compression test that is performed on stacked, fully loaded drums which are at or above a temperature of 130° F. at the start of the test. Changes in the test procedure, in keeping with the comments received, have been made and, as in the other performance tests, acceptance/rejection/penalty criteria are included.

Many commenters in responding to proposed § 178.16–16(a) pointed out that for an injection molding process the drop testing of one container for each lot of 1,000 produced per machine is excessive. The Bureau agrees that the reproducibility of design in items produced by the injection molding process, when compared to the blow molding process, has a far greater continuity and repeatability. Therefore, the Bureau has accepted the suggestion that the specification be changed to decrease the test requirement to one container for every 10,000 drums molded by each machine. Also, an alternative is provided which requires that drop tests be conducted once every 24 hours of uninterrupted production from each machine; or at the resumption of production from a machine after shut down for more than one hour.

One commenter requested a change in the proposed marking requirement (§ 178.16–19(a)), i.e., allowing characters ¼-inch in height in place of ½-inch. The Bureau agrees this comment has merit and has modified the proposal accordingly.

For clarity or for editorial reasons, a number of other changes which did not appear in the 1974 notice have been incorporated into the specification. Those changes include the addition of ASTM references in § 171.7, authorization of increased net weight in § 173.245(b)(6), a single Table of Sections entry for § 178.16, specificaiton of lug placement for steel top heads, and clarification of drum orientation during drop tests.

(42 F.R. 36261, July 14, 1977)

SUMMARY • Docket No. HM-120 • *This docket proposed new switching requirements for handling placarded rail cars carrying hazardous materials. Action withdrawn.*

[49 CFR Part 174]
[Docket No. HM-120; Notice No. 74–11]
FREIGHT CARS
Switching of Placarded "Dangerous"

The Hazardous Materials Regulations Board is considering amendment of § 174.589, of Title 49 Code of Federal Regulations, which prescribes the requirements for handling placarded freight cars carrying hazardous materials.

As a result of preliminary findings concerning the catastrophic tank car explosions at Decatur, Illinois, on July 19, 1974, and Wenatchee, Washington, on August 6, 1974, the Board believes that § 174.589 should be amended to prevent further occurrences. Although the National Transportation Safety Board has not yet completed its investigation of these accidents nor determined their probable cause, it appears that the Decatur accident may have occurred as a result of rough handling and that rough handling of freight cars placarded "Dangerous" during switching operations may have contributed to the Wenatchee accident.

The proposed changes in § 174.589 are described below:

Paragraph (c). It is proposed to expand the provisions of this section to include freight cars placarded "Dangerous." This would prohibit the uncoupling or cutting off of these cars while they are in motion, the striking of these cars by other cars moving under their own motion, and the coupling of these cars with more force than is necessary to complete the coupling but in no case at a speed of more than 4 m.p.h.

Paragraph (d). It is proposed to delete this paragraph as surplusage since all freight cars placarded "Dangerous" would be required to be handled in accordance with the provision of paragraph (c).

Pursuant to the provisions of section 102(2)(c) of the National Environmental Policy Act (42 USC 4321, et seq.), the Board has considered the requirements of that Act concerning Environmental Impact Statements and has determined that the amendments proposed in this notice would not have a significant impact on the quality of human environment within the meaning of the Act. Accordingly, an Environmental Impact Statement is not necessary and will not be issued with respect to the proposed amendments.

(39 F.R. 29197, August 14, 1974)

[49 CFR Part 174]
[Docket No. HM-120; Notice No. 74–11]
SWITCHING OF FREIGHT CARS PLACARDED
"DANGEROUS"
Notice of Withdrawal of Proposal

On August 14, 1974, the Hazardous Materials Regulations Board published a notice of proposed rule making, Docket No. HM-120; Notice No. 74–11 (39 FR 29197). This notice proposed changes in section 174.589 that would prohibit the uncoupling or cutting off of cars placarded "DANGEROUS" while they were in motion, the striking of such cars by other cars moving under their own motion, and the coupling of these cars with more force than is necessary to complete the coupling, but in no case at the speed of more than four miles per hour.

At the request of commenters, the date for filling comments was extended from September 20, 1974 to October 30, 1974 (39 FR 33808, October 22, 1974).

Subsequent to the issuance of Notice No. 74–11, the Federal Railroad Administration published FRA Emergency Order No. 5 (39 FR 38230, October 30, 1974). The Emergency Order States:

ORDER

"In addition to the requirements of Parts 170–189 of Title 49 of the Code of Federal Regulations governing the transportation of hazardous materials, effective 12:01 a.m., October 27, 1974, a railroad may transport flammable compressed gas in DOT 112A and DOT 114A uninsulated tank cars that are not equipped with head shields prescribed by the Hazardous Materials Regulations Board in Docket HM-109, Amendment No. 5, 173–83, 179–15 published in the July 31, 1974, issue of the FEDERAL REGISTER (39 FR 27572), 49 CFR 179.100–23, only under the following conditions:

"(a) DOT specification tank cars 112A and 114A that are not equipped with head shields required by 49 CFR 179.100–23, transporting flammable compressed gas requiring dangerous placards shall not be cut off in motion. No car moving under its own momentum shall be allowed to strike any DOT 112A or 114A tank car containing flammable compressed gas placarded dangerous, that is not equipped with head shields required by 49 CFR 179.100–23, nor shall any such car be coupled into with more force than is necessary to complete the coupling.

"(b) The shipping papers required by 49 CFR 174.510 for loaded tank cars containing flammable compressed gas with placarded dangerous must carry the notations: "DOT 112A" or "DOT 114A" and "must be handled in accordance with FRA E.O. No. 5."

"(c) Railroad employees must be informed of the presence of these cars and instructed to handle them in accordance with the requirements of this order.

"A civil penalty of not less than $250 nor more than $2500 will be assessed for each violation of this order.

"An opportunity for review of this order is provided in accordance with Section 554 of Title 5 of the United States Code.

"Issued in Washington, D.C. on October 25, 1974."

More than forty comments were received in response to HM-120; Notice No. 74–11. An analysis of the comments indicated that a large number of commenters wanted further study to be made with some interim precautionary measures being taken with the cars that have caused the problems. Several commenters suggested that the Federal Railroad Administration take action on the tank cars that have been involved in most of the serious accidents—the 112A and 114A uninsulated tank cars transporting flammable compressed gas. In addition, commenters suggested that tests be conducted to determine safe coupling speed(s) and the efficiency of hump yard retarders. The Federal Railroad Administration believes that the commenters have raised valid points which require further study and that Emergency Order No. 5 covers the specific problem of the DOT 112A and 114A uninsulated tank cars transporting flammable compressed gas.

It should be noted that the head shield retrofit program for the DOT 112A and 114A tank cars will, when completed, eliminate the need for Emergency Order No. 5.

In view of the foregoing, the Office of Hazardous Materials Operations (OHMO) has determined that action on the proposed amendments is not appropriate at this time, and that Docket No. HM-120; Notice No. 74–11 should be withdrawn. The withdrawal of this proposal, however, does not preclude the OHMO from issuing similar notices in the future or commit the OHMO to any course of action.

(41 F.R. 16661, April 21, 1976)

SUMMARY • Docket No. HM-122 • *Clarification of regulations in 49 CFR Parts 171 and 173 to the effect that all hazardous materials regulations are applicable to private carriers by highway.*

[Docket No. HM-122; Amdt. Nos. 171–26, 173–88]
PART 171—GENERAL INFORMATION AND REGULATIONS
PART 173—SHIPPERS
Application of Regulations to Private and
Contract Carriers by Motor Vehicle

The purpose of these amendments to the Hazardous Materials Regulations is to make it clear once again that, except as otherwise specifically provided in their text, the regulations are applicable to contract and private carriers by motor vehicle who operate in interstate or foreign commerce and to shippers who utilize the services of those carriers.

It has come to the attention of the Board that, in spite of clear indications in the Regulations that they are applicable to transportation by private and contract motor carriers (see, e.g., notes preceding § 173.1) and court decisions and administrative interpretations to the same effect, there have been instances in which private carriers have contended that the rules in the Hazardous Materials Regulations are inapplicable to them. The contentions have been founded upon the fact that both § 171.1(a) and § 173.1(a) make reference to "common" carriers.

The adjective "common," as used therein, was intended to apply only to rail movements. As noted above, it has always been the intention of hte Board to apply the regulations to all carriers who fall within the ambit of the Explosives and Combustibles Act, 18 U.S.C. 831–835, and to shippers who use the services of those carriers. As 18 U.S.C. 831 makes clear, the term "carrier" includes "any person engaged in the transportation of passengers or property by land, as a common, contract, or private carrier."

In consideration of the above, the Board is revising § 171.1 and amending § 173.1(a) to make it clear that the regulations apply to contract and private carriers by motor vehicle and to shippers who use the services of those carriers.

Since these amendments are interpretative and do not alter the substance of the rules, notice and public procedure thereon are unnecessary.

(39 F.R. 42366, December 5, 1974)

SUMMARY • Docket No. HM-123 • *This docket establishes the clear requirement that shippers by rail must comply with any applicable requirements of Emergency Orders issued by the Federal Railroad Administrator.*

[Docket No. HM-123; Admt. Nos. 171–27, 173–89]
PART 171—GENERAL INFORMATION AND REGULATIONS
PART 173—SHIPPERS
Shippers by Rail Freight

Emergency Order No. 5 was issued by the Federal Railroad Administration (FRA) on October 25, 1974 (39 FR 38230–10–30–74), with an effective date of 12:01 a.m. October 27, 1974. This required the railroads to handle all DOT 112A and 114A tank cars (without head shields) transporting flammable compressed gas in the same manner as Class A poisons and Class A explosives during switching operations. To provide an additional means for railroad employees to identify these cars, E.O. No. 5 required that shipping papers for these tank cars contain the following notations: "DOT 112A" or "DOT 114A" as appropriate and "Must be handled in accordance with FRA E.O. No. 5."

In order that a railroad will be notified by the shipper that a tank car he is tendering is one requiring special handling in accordance with FRA E.O. No. 5, the Hazardous Materials Regulations are being amended to require the shipper to comply with all FRA emergency orders. Therefore, a shipper of a DOT 112A or 114A (without head shields) containing flammable compressed gas will be required to put the following notation on the shipping papers: "DOT 112A" or "DOT 114A" as appropriate, and "Must be handled in accordance with FRA E.O. No. 5."

As a situation exists which demands immediate adoption of these regulations in the interest of public safety, it is found that notice and public procedure hereon are impractical and good cause exists for making these amendments effective in less than 30 days.

(39 F.R. 43310, December 12, 1974)

SUMMARY • Docket No. HM-124 • *This docket pertains to alternative cargo tank outlet valve protection, in place of otherwise prescribed shear sections.*

[Docket No. HM-124; Notice No. 75–2]
BOTTOM OUTLET VALVES ON MC-312 CARGO TANKS
Proposed Rulemaking

The Hazardous Materials Regulations Board (the Board) is considering amending § 178.343–5(b) of the Hazardous Materials Regulations which specifies that outlet valves must be protected by a shear section.

The Board was petitioned by the Truck Trailer Manufacturer's Association to provide alternative means of protecting outlet valves on motor vehicles to assure against the accidental escape of liquid cargoes. They suggested the change for two reasons: (1) The change would provide for an alternative protection of the valve equivalent to the protection offered by the presently required valve shear section, and (2) this change will allow the use of valves unavailable with shear section.

After reviewing the data submitted in support of the petition, the Board has concluded that the petitioner's contention may have merit.

(40 F.R. 13316, March 26, 1975)

[Docket No. HM-124; Amdt. 178–38]

PART 178—SHIPPING CONTAINER SPECIFICATIONS
Bottom Outlet Valves on MC 312 Cargo Tanks

The purpose of this amendment to § 178.343–5(b) of the Hazardous Materials Regulations is to provide an alternate means of protecting bottom outlet valves on MC 312 cargo tanks to assure against the accidental escape of liquid cargoes. Section 178.343–5(b) presently requires this protection be provided by the use of bottom outlet valves equipped with a shear section.

On March 26, 1975, the Hazardous Materials Regulations Board published a notice of proposed rulemaking, HM-124; Notice No. 75-2, in the FEDERAL REGISTER (40 FR 13316) which would permit the use of bottom outlet valves without shear sections on MC 312 cargo tanks if protection for the valves is provided by suitable guards capable of absorbing a concentrated horizontal force of at least 8,000 pounds.

A number of commenters suggested that the proposed alternate method be revised to require that the bottom outlet valves be located in the protective envelope provided by the suspension sub-assembly of the cargo tanks. The Bureau believes it overly restrictive to allow the suspension subassembly to serve as the only alternate to a shear section. As proposed, § 178.343–5(b) (1) (ii) will permit the suspension sub-assembly or any other method which will meet the strength criteria to serve as valve protection. For this reason, § 178.343–5(b) (1) (ii) remains as proposed in the notice.

One commenter agreed that bottom outlet valve protection can be provided by the use of guards; however, it was suggested that the guards provide protection at least equivalent to the afforded top outlets for overturn protection. The Bureau does not have data available at this time to prescribe strength requirements for guards on MC 312 cargo tanks other than those that are presently presented for the MC 306 and MC 307 cargo tanks (see 49 CFR 178.340–8(d) (1) (ii). In addition, the Bureau is not aware of any data that would indicate that the present requirements are not adequate. Therefore, the strength criteria of guard protection for bottom outlet valves on MC 312 cargo tanks will remain as proposed.

Another commenter pointed out that the word "emergency" is not used in the present wording of § 178.343–5(b) (1). It is recognized that this section deals with bottom outlet product piping and not necessarily emergency valve piping. Therefore, the word "emergency" is deleted from the proposed regulations.

(41 F.R. 3869, January 27, 1976)

SUMMARY · Docket No. HM-125 · *A proposed prohibition against new construction of DOT 112A and 114A uninsulated pressure tank car tanks for use in the transportation of hazardous materials, on the basis of accident experience with such tank cars. See also Docket Nos. HM-63, HM-103, and HM-144.*

[49 CFR Parts 173, 179]

[Docket No. HM-125; Notice No. 75–4]

PROHIBITION OF NEW CONSTRUCTION
OF SPECIFICATION DOT-112A AND 114A UNINSULATED
PRESSURE TANK CAR TANKS
Transportation of Hazardous Materials

The Hazardous Materials Regulations Board is considering amending §§ 173.314 and 179.101 of the Department's Hazardous Materials Regulations to prohibit new construction of Specification DOT-112A and 114A uninsulated pressure tank car tanks for use in the transportation of hazardous materials.

Since 1958 there have been a series of disastrous railroad accidents involving these uninsulated pressure tank cars transporting liquefied flammable compressed gases. During these accidents tank cars were readily punctured, enormous quantities of gas were released and ignited. The flames in turn overheated adjacent, similarly laden, tank cars causing them to rupture, burn and "rocket" for considerable distances. One such disastrous accident occurred at Laurel, Mississippi, on January 25, 1969, when 13 of 15 LP Gas laden 112A tank cars violently ruptured and rocketed causing the death of three townspeople, and injuries to over 30 others, and property damage totalling millions of dollars. On June 21, 1970, at Crescent City, Illinois, rupturing tank cars of the 112A Specification resulted in injuries to approximately 10 persons; and again on October 1971, in Houston, Texas, ruptured tank cars of a flammable compressed gas caused the death of one fireman and injured approximately 35 firemen and newsmen on the scene. Many other accidents have occurred which have presented severe hazardous conditions for the public and emergency response personnel. On February 11, 1974, near Oneonta, New York, ruptured and overheated 112A specification tank cars caused injuries to over 50 persons, primarily firefighters who were endeavoring to control the fires.

At Decatur, Illinois, on July 19, 1974, a DOT 114A tank car was punctured in a switching accident resulting in a gas-air blast that injured 130 people and killed seven. Again at Houston, Texas, on September 21, 1974, a switching accident resulted in a similar explosion that killed one person and injured 150 persons.

A summary of the accidents reported to DOT, covering accidents involving DOT Specification 112A and 114A uninsulated tank cars from January 1, 1969, to date, reveals that there were 192 accidents involving 434 DOT 112A or 114A cars, which resulted in 68 head punctures, 13 shell punctures, 59 tank ruptures and a total of 156 had loss of lading. Twenty-three persons were killed and 936 injured as a result.

Prior to the introduction of 112A and 114A uninsulated tank cars liquefied compressed gases were transported exclusively in Specification 105A insulated tank cars. It is estimated that approximately 22,000 112A and 114A uninsulated pressure tank cars are in service today, compared to the approximately 33,000 Specification 105A insulated tank cars in hazardous materials service. Although there are less 112A/114A than 105A tank cars in service, the RPI-AAR Tank Car Safety Research and Test Project reports (for 1965–1970) indicate that more than twice as many 112A/114A tank cars were damaged, as compared to 105A tank cars, and that five times as many 112A/114A tank cars ruptured.

The relative utilization for both the 112A/114A tank cars and the 105A tank cars has been considered. The utilization factor was considered in order to determine whether the higher utilization of 112A/114A tank cars, as compared to 105A tank cars, would serve to explain the higher rate of accidents being experienced by the 112A/114A tank cars. Data, submitted in connection with the audited carload waybill sampling project of the FRA, was reviewed to provide information concerning the relative utilization of both types of cars. The information obtained from this data did not support the concept that the higher utilization of 112A/114A tank cars would serve to explain the higher rate of accidents being experienced by the 112A/114A tank cars.

FRA accident data accumulated during the period of January 1, 1969, through December 31, 1974 indicates that:

	112A/114A	105A
Number of accidents reported to FRA . . .	193	101
Number of cars derailed and/or damaged.	434	213
Number of cars sustaining a head puncture.	68	13
Number of cars sustaining a shell puncture without a head puncture.	13	5
Number of cars ruptured without puncture.	59	8
Number of tanks sustaining partial or total loss of hazardous lading. . . .	156	39
Number of persons killed as a result of tanks being punctured or ruptured...	23	0
Number of persons injured	936	151

Available data has shown 105A tank cars to be substantially superior to 112A/114A tank cars from a safety standpoint. (Many of the 105A tank car ruptures occurred as a result of chemical detonation of the lading rather than by a weakening of the shell due to overheating, or puncture.) In fact, extremely hazardous commodities such as Motor Fuel Anti-Knock Compound and Ethylene Oxide, which may detonate if exposed to high temperatures, are required to be carried in insulated tank cars.

As a result of unintentional release of liquefied petroleum gas from 112A/114A tank cars during train accidents, the Railway Progress Institute and the Association of American Railroads have undertaken a multi-million dollar joint research program to evaluate the conditions leading up to tank car puncture and rupture (tank "rocketing"), and to develop improvements to eliminate this safety problem.

The Federal Railroad Administration and the Hazardous Materials Regulations Board have already taken regulatory action to improve tank car safety by requiring protective head shields on DOT Specification 112A and 114A uninsulated tank car heads to reduce the incidence of head punctures on tank cars carrying liquefied compressed gases (HM-109). In addition, FRA is closely monitoring tank car operations. The Administrator issued FRA Emergency Order No. 2 (37 FR 28311) prohibiting further use of certain uninsulated pressure tank cars found to have a structural inadequacy which results in tank shell cracking and possible leakage of hazardous lading. These cars were not permitted to be returned to service until this structural inadequacy has been corrected by modification of these cars (39 FR 2124).

In an effort to improve tank car safety by way of regulatory operating practices, the FRA issued Emergency Order No. 5 on October 25, 1974, prohibiting the free switching of all 112A and 114A tank cars transporting flammable compressed gas. By requiring controlled coupling of such tank cars, the FRA has moved to decrease the danger of head punctures in train yard operations.

The Federal Railroad Administration is also undertaking research efforts in this area to improve tank car safety. The FRA program is a multi-faceted research program aimed at determining the failure modes of these type tank cars and solutions to eliminating these safety problems. Theoretical and experimental studies of one-fifth scale and full size tank cars to determine and evaluate their performance in fires are underway. Torching as well as enveloping fires are being studied and various insulating materials are being evaluated to develop specifications for "thermal shielding" to prevent tank overheating and "rocketing." Fire tests have shown that the application of a "thermal shield" (insulation) will extend the survival time of a propane laden tank car subject to a pool fire from twenty-five minutes to ninety minutes. This extra time will give fire-fighters a better chance to cool the tank car surface and prevent impingement of the tank and an explosion. In addition, the "thermal shield" reduces the severity of those tank explosions which do occur.

Likewise, metallurgical analysis is also being conducted by the FRA on the materials used in the construction of tank cars. The objective of this study is to develop new specifications for tank shell steel and better design attachments to the tank shell. Theoretical structural stress analysis is being performed to obtain a better understanding of a tank car as a structure.

Also, the relief capacity of spring loaded safety relief devices is being measured in both the vapor and in the liquid stages to determine the actual flow capacity of existing tank car relief devices. The existing equations for sizing relief valve flow may be inadequate for uninsulated pressure tank cars. This testing will lead to new design criteria for sizing of the relief valves used for these cars. Additionally, a new FRA-AAR/RPI test facility will afford industry a place to test and certify new safety relief devices.

While the Board is aware of these research efforts and agrees that significant results will be developed from them, the Board is concerned that existing uninsulated pressure tank cars built to DOT specifications present a great public danger of propane tank car fires and subsequent explosions. The Board recognizes that removing all 112A and 114A tank cars from this service would intensify the national energy crisis and severely restrict the rail movement of fuels, fertilizers, chemicals, and liquefied compressed gases. The Board also realizes that improved operating practices, headshields and thermal shielding will greatly reduce the hazards inherent in 112A and 114A tank cars as currently constructed. However, the Board feels that new construction of 112A and 114A tank cars should be prohibited until a new design or type of car is developed. Accordingly, the Board proposes to prohibit new construction of Specification DOT 112A and 114A tank cars for the transportation of hazardous materials after December 31, 1975.

Pursuant to the provisions of section 102(2) (c) of the National Environmental Policy (42 U.S.C. 4321 et seq.), the Board has considered the requirements of that Act concerning Environmental Impact Statements and has determined that the amendments proposed in this notice would not have a significant impact upon the environment. Accordingly, an Environmental Impact Statement is not necessary and will not be issued with respect to the proposed amendments.

(40 F.R. 17853, April 23, 1975)

[49 CFR Parts 173, 179]
[Docket No. HM-125; Notice No. 75-4]
PROHIBITION OF NEW CONSTRUCTION
OF SPECIFICATION DOT-112A AND 114A UNINSULATED
PRESSURE TANK CAR TANKS
Withdrawal of Notice

Notice 75-4 in Docket HM-125 was published by the Hazardous Materials Regulations Board and developed in response to several rail accidents involving DOT specification 112A and 114A tank car tanks, which had resulted in loss of life and substantial property damage. However, subsequent to issuance of Notice 75-4, the Bureau proposed more extensive amendments concerning modification of construction standards and criteria for the use of such tank car tanks (Docket HM-144; Notice 76-12; 41 FR 52324).

Notice 76-12 was the basis for the amendments published by the Bureau on September 15, 1977. These amendments addressed the concerns expressed in Notice 75-4 by requiring for all 112A and 114A tank car tanks: (1) bottom and top shelf couplers; (2) headshields or protective headjackets on those tank car tanks used to transport anhydrous ammonia and flammable gases; and (3) thermal protection on those tank car tanks used to transport flammable gases. In light of this action, the preamble to the final rule stated:

> As was indicated in the notice, with the promulgation of standards and specifications upgrading existing specification 112 and 114 so as to improve design and construction, the Bureau will withdraw Notice 75-4, under Docket HM-125.
>
> Accordingly, the Bureau hereby withdraws Notice 75-4 under Docket No. 125.

(43 F.R. 3598, January 26, 1978)

SUMMARY • Docket No. HM-126 • *A DOT docket on hazard information systems, to assist the agency in evaluating the merits of each with a view to subsequent proposal of a uniform multimodal system, compatible with international requirements and meeting the primary needs of fire and other emergency services. Also see Docket No. HM-103.*

[49 CFR Parts 170–189]
[Docket No. HM-126]
HAZARD INFORMATION SYSTEMS
Request for Comments

The Hazardous Materials Regulations Board (the Board) is soliciting comments concerning the merits of various Hazard Information Systems (HI-Systems). The Board believes there are deficiencies in the hazard communication requirements of its regulations and that a HI-System may be necessary to provide for the adequate communication of hazards of materials in transportation. Some of the benefits that could be derived through implementation of a HI-System are:

1. Recognition of multiple hazards of individual materials;
2. Recognition of multiple hazards of mixed ladings; and
3. Provision of sufficient information whereby fire fighting and other emergency response personnel can acquire immediate information to make better informed judgments on how to handle emergency situations.

By publication of a notice in this issue of the FEDERAL REGISTER (40 FR 26687) the Board has announced termination of further consideration of the use of two digit numbers to identify the hazards of materials during transportation under Docket No. HM-103; Notice 73–10. The Board also stated that there is still sufficient need for development and implementation of an effective hazard information system and made reference to the hazard information systems and criteria included in this advance notice for public comment.

The Board believes that in analyzing various HI-systems it is obvious that, even though they are intended to achieve the same end result, i.e., convey emergency response information, they pursue several different routes or philosophies by attempting to provide (1) information concerning the hazards of a material, (2) information concerning the hazards of a material in combination with the degree of risk involved; and (3) information concerning the actions to be taken during an emergency response situation.

The systems discussed in this notice incorporate these basic philosophies to different degrees and for reasons that can be attributed, at least in part, to the economic, social, and/or industrial structures of a particular nation, and the authority or interest advocating them. The differences in these philosophies are both subtle and obvious, and they lead to analytical difficulties in choosing criteria by which to evaluate the various systems.

The Board believes that the following criteria, though not necessarily inclusive, are factors to be considered in the evaluation of hazard information systems:

1. Capability of the general public to recognize the existence of the immediate dangers presented by a material;
2. Presentation of information in a manner so that the general public will be able to accurately transmit basic information to emergency response personnel;
3. Compatibility, intermodally and internationally;
4. Capability of application to both bulk and non-bulk shipments;
5. Capability of functioning without use of manual or other subsidiary documents;
6. Capability to meet the needs of emergency response personnel, carriers, shippers, and the general public;
7. Capability of integration with documentation, packaging, and vehicle identification requirements to help insure accuracy;
8. Capability of implementation without undue economic burden on shippers and carriers; and
9. Capability of indicating degree of hazard.

It is the Board's position that any alpha/numeric/symbolic hazard information system adopted in the future be compatible with and adaptable to the placards it adopts under Docket HM-103.

The following existing or proposed hazard information systems are summarized in this notice:

1. The previously proposed DOT HI System.
2. The NFPA 704M System.
3. The RID/ADR System.
4. The HAXCHEM System.
5. A proposal by Union Carbide Corporation (similar to a proposal submitted by Air Products and Chemicals Corporation).
6. A proposal by the U.S. Coast Guard.
7. The Canadian System (rail).
8. A proposal by the International Air Transport Association.
9. A proposal by Pennwalt Corporation.

THE DOT HAZARD INFORMATION SYSTEM

This system is based on two digit hazard information numbers on shipping papers, package labels and vehicle placards as hazards communicators. A two digit hazard information number would identify the primary and additional (if any) hazards of a material. The first digit of a hazardous information number is the United Nations class number for the material and the second digit indicates whether there are significant additional hazards. A zero indicates there are no significant additional hazards.

There are approximately 59 hazard information numbers to identify materials with hazards ranging from a single hazard, such as "Nonflammable Gas" with the HI number 20, to five hazards for a material that is a "Flammable Solid" that is also "Pyrophoric, Poisonous, Water Reactive and Corrosive" which would have the HI number 47.

Defining criteria for each hazard and a specified precedence of hazards for the assignment of hazard information numbers have been developed so a material meeting a specific defining criteria will always be assigned the same hazard information number—in most cases without review by a government agency. A shipping paper containing the shipping name and classification of a hazardous material would also contain its hazard information number. A package label and a vehicle placard would bear the hazard information number in the lower corner.

An important element of the system is an Emergency Response Manual which consists of a card for each hazard information number to identify the expected health, fire and explosion hazards for each material. These cards also contain suggested responses for use by emergency response personnel during the first 10 to 30 minutes of an incident involving hazardous materials in transportation as well as suggested first aid actions.

NFPA 704 SYSTEM

The system identifies the hazards of the material in terms of three principal categories, namely "health", "flammability", and "reactivity" (instability); and indicates the order of severity numerically by five divisions ranging from "Four (4)", indicating a severe hazard, to "Zero (0)", indicating no special hazard.

The information is presented by a spatial system of diagrams of a diamond shape signal divided into four segments with "health" always being on the left; "flammability" at the top; and "reactivity (instability)" on the right. Color categories for backgrounds are blue for "health" hazard, red for "flammability", and yellow for "reactivity (instability."

A fourth space in the diagram should be used to indicate unusual reactivity with water, and is indicated by placing the letter "w" with a line through the center (w̶) in the space. The space may also be used to indicate other additional information such as pressurized vessels, radioactivity, proper fire extinguishing agent, or protective equipment required in case of fire or other emergencies.

The system for ranking degrees of hazard is based on relative rather than absolute values. The system prescribes the number to be used in each category by describing the effects of the material, in order of severity, in each category, i.e., health, flammability, and reactivity.

RID/ADR SYSTEM (EUROPE)

The RID/ADR System applies only to bulk transport by rail and highway in Europe. It requires that in addition to a placard on the format of the U.N. label, tank vehicles and rail cars must display an orange colored plate 30 cm high and 40 cm wide; this plate must have shown on it in black, two numbers, one above the other. The top number may be either two or three digits and may or may not be preceded by the letter "X". This is the Hazard Identification number. The second number is the United Nations Serial Number of the particular commodity.

The first figure of the Hazard Identification number indicates the primary hazard as follows:

2. Compressed Gas.
3. Flammable Liquid.
4. Flammable Solid.

5. Oxidizer Material or Organic Peroxide.
6. Toxic Material.
8. Corrosive Material.

The second and third digits indicate secondary or tertiary hazards respectively as follows:

0. No Additional Hazard.
1. Explosion Risk.
2. Gas May Be Given Off.
3. Flammability Risk.
5. Oxidizer Risk.

6. Toxic Risk.
8. Corrosive Risk.
9. Risk of Violent Reaction from Spontaneous Decomposition.

When the first and second digits are the same, an intensification of the primary hazard is indicated e.g., 33 indicates a highly flammable liquid, 66 indicates a very dangerous toxic subtance, etc. A refrigerated gas is indicated by a hazard identification number of 22. The number 42 would indicate a flammable solid which may give off a gas upon contact with water. The letter "X" preceding a hazard identification number indicates that water should not be applied to the commodity.

The hazard information number is established by the governmental authorities for each material transported in bulk.

HAZCHEM SYSTEM (UNITED KINGDOM)

In an attempt to give to emergency services information which will enable them to act independently of reference books and instructions, the "Hazchem System" was developed. The basic principle of "Hazchem" is that it gives direct information on the action to be taken by firemen and policemen and requires no interpretation of information on hazards. The United Kingdom accepts the system for international highway and railroad traffic.

The specific hazard identification panel is fire resistance and has an orange reflectorized background on which appears two numbers. The top number gives information on the hazards to be expected and the lower number is the United Nations number by means of which the substance can be precisely identified.

The "Hazchem" code gives information under the following headings:

1. Firefighting media.
2. Personal protection.
3. Explosive risk.

4. Spillages.
5. Evacuation.

The "Hazchem" scale gives information on the firefighting media to be used by the use of the numbers 1, 2, 3, or 4, and information on personal protection, explosive risk and spillage action by the letters PRST and WXYZ. Where the policeman or fireman should consider the possibility of initiating the evacuation of an area the letter E is added. Each fireman and policeman would carry a durable card showing the "Hazchem" scale enabling him to translate letters on a transport vehicle into direct action, simple first aid measures appear on the reverse side of the card.

UNION CARBIDE PROPOSAL

This proposal involves placarding, labeling and documentation to convey hazard information. The placard would be rectangular in shape and would be in two levels. The upper level would display the word "DANGEROUS" in 4-inch high letters. The lower level would consist of 6-inch squares, each bearing a hazard symbol within a diamond, accompanied by appropriate language associated with the hazard symbol. The system contemplates the addition or removal of symbol squares, either by the use of sliding rail holders or self-adhesive squares, as commodities are placed in or removed from the placard vehicles. For empty unpurged vehicles, the "DANGEROUS" placard would be left visible, with the word "EMPTY" displayed on the lower level.

Another proposed placard is diamond shaped similar to the placards presently used by rail carriers. This placard is proposed for both rail and highway and consists of the word "DANGER" displayed diagonally across the placard. The hazard symbols would be attached to the placard as proposed for the rectangular placard. The reverse side of the placard would bear the words "DANGEROUS-EMPTY", for use on empty unpurged rail cars or motor vehicles.

The labeling system proposed by Union Carbide would make use of the present DOT labels used singly for commodities with a single hazard only, and multiple labels for dual or triple hazard materials. As an alternative, single labels with imprinted multiple symbols for dual hazard and triple hazard materials could be utilized.

For shipping paper identification, Union Carbide proposes that for each hazardous material entry, some distinctive letters, such as "HAZ", "DOT", or "HM" be shown preceding the description. In addition, all of the hazard classifications applicable to each material would be shown.

An emergency response manual is suggested by Union Carbide as a means of furnishing supplemental information to emergency personnel. The manual would be alphabetically indexed and cross referenced to all hazards and combinations thereof. The manual would be keyed to the language obtained from the shipping documents, placards or labels.

PROPOSAL BY THE U.S. COAST GUARD

The United States Coast Guard proposed a Hazard Information System as an alternative to the HI System propose in HM-103. In this approach, the UN label or a placard incorporating the basic design of the UN label is to identify the primary hazard. A two digit number is applied to the lower quadrant of the label or placard.

The first digit, a number from 1 to 5, indicates the relative degree of the primary hazard (i.e., the hazard identified by the label or placard). The first digit of the hazard information number relates to the primary hazard as follows:

1. Low degree of hazard such as: Combustible liquids with flashpoint over 141° F; solid Class B poison, poisonous by ingestion only; corrosive solid, corrosive to metals only.

2. Intermediate degree of hazard such as: Combustible or flammable liquid with flashpoint between 73–141° F; liquid Class B poison, poisonous by ingestion only; corrosive liquid, corrosive to metals only.

3. High degree of hazard such as: Flammable liquid with flashpoint between −18° F and 73° F; Class B poisons, poisonous by inhalation or skin absorption; corrosive materials, corrosive to skin.

4. Extremely high degree of hazard such as: Flammable liquids with flashpoints below −18° F or Reid vapor pressure above 27 PSIA with flashpoint below 73° F; Class A poisons; pyroforic material.

5. Evacuate area, do not attempt to contain fire or spill.

The second digit would indicate the secondary hazard of the commodity based on the U.N. Class Number as follows:

2d digit:	Meaning
0	No secondary hazard.
1	Thermal instability hazard.
2	Hazard resulting from gas evolution.
3	Flammable.
4	Flammable solid—dangerous when wet.
5	Oxidizer.
6	Poison.
7	Radioactive.
8	Corrosive.

When no hazard information number has been published for a particular commodity the shipper could determine a hazard information number based on the above criteria. Prior to the initial shipment of the commodity, the shipper would submit the number assigned plus supporting data to the Department of Transportation for approval and subsequent publication in the FEDERAL REGISTER. The shipper could ship the commodity under the hazard number he has determined until such time as the final determination of the number is published by the Department.

Placards on vehicles containing mixed lading bearing different hazard information numbers would bear the hazard information number derived as follows:

1. The first digit would be the highest first digit of all the hazard information numbers on the commodities within the vehicle.

2. The second digit would be one of the second digits of all the hazard information numbers on commodities within the vehicle based on the following precedence of severity of secondary hazard: 1, 6, 3, 5, 8, 4, 2, 7.

This system could be summarized on the reverse side of the required shipping documents.

CANADIAN SYSTEM HAZARD INFORMATION—EMERGENCY RESPONSE FORM, CARLOADS

Carload, trailerload, and containerload shipments of dangerous commodities, as defined in regulations, originating in Canada would be accompanied by a Hazard Information—Emergency Response Form which would be furnished by the shipper to the carrier and which would accompany the car, trailer, or container from the shipper's siding to the consignee's siding.

When multi-unit shipments of a single dangerous commodity are made from one shipper at point of origin to one consignee at one destination, one only Hazard Information—Emergency Response Form would accompany each such shipment, and a list of the car numbers would be shown thereon.

In the case of compartmentized tank cars loaded with more than one regulated commodity, a separate response form would be required for each commodity and its location in the car indicated in the appropriate block provided on the form.

The Emergency Response Form would include information on the potential hazards of the commodity under the headings:

Fire

Explosion

Health

Immediate Action Information would be included under the headings:

General

Fire

Spill or Leak

First Aid

The shipper modifies the potential hazard and immediate action information sections where necessary by adding or deleting instructions applicable to the particular commodity being shipped.

In addition, the response form provides the following information:

1. Placard endorsement.
2. Car initials and number.
3. Consignee.
4. Destination.
5. Routing.
6. Proper shipping name.
7. Classification.
8. Placard notation.
9. Date shipped.
10. Shipper.
11. Shipping point.
12. Weight or volume.
13. Shipper's certificate.
14. Emergency telephone number.

IATA SYSTEM

The system identifies the hazards of the material by a two digit number, with the first digit, which corresponds to the United Nations Class number identifies the main danger; the second digit, which is an arbitrary figure, indicates subsidiary risks. The second digit would be zero (0) if there were no subsidiary risk.

The system uses letters for the second digit in lieu of numbers to indicate tertiary hazards.

The information is presented in a spatial system of diagrams on a rectangular shaped placard divided into three segments, with the primary hazard number always on the bottom left, the secondary hazard number or numeral always on the bottom right. The system provides for a barred "W" in the top portion to indicate reactivity with water. Background colors indicate the following: Orange for first group of each class, white for the second group, and green for the third group, with blue provided for the barred "W".

The system suggests placing the appropriate danger labels next to the placard to make the system more explicit.

The system calls for a "simple rigid card" to be carried by emergency response personnel concerning actions which should be taken from the hazard communication.

PROPOSAL BY THE PENNWALT CORPORATION

The system involves labeling, placarding, and documentation as a means of hazard communicators, and is predicated upon the establishment of an "Order of Importance" listing of hazard classifications in which one classification precedes in degree of severity, those which follow it. The assignments are as follows:

1. Explosive materials.
2. Compressed gases.
3. Flammable liquids, combustible liquids, flammable gases.
4. Flammable solids.
5. Oxidizing agents.
6. Poisonous materials.
7. Radioactive materials.
8. Corrosive materials.
9. Miscellaneous materials not covered by any other classification but of sufficiently dangerous character that some means of warning should be displayed.

The corresponding primary hazard number would be printed in the space provided on labels of design similar to the DOT HI System. If the product presents multiple hazards, numbers corresponding to each of the additional hazards would also be printed, with the primary hazards, in the space provided.

Placards are of the general size and shape of present placards required for the various modes of transportation by DOT regulations, with the appropriate label for the material, the shipping name of the material, the corresponding hazard number(s), and the primary classification.

Documentation contains a column, of contrasting background, whose use is restricted solely to entry of the hazard number(s) for each product listed, with the shipping name and classification.

In order to assist interested persons in their efforts to understand the various hazard information systems mentioned in this Notice, copies of all material available to the Board concerning specific systems will be made available for review or will be furnished to persons requesting such additional information. Requests should be addressed to: Chief, Regulations Division, Office of Hazardous Materials, Department of Transportation, Washington, D.C. 20590.

Interested persons are invited to give their views on these hazard information systems or other hazard information systems not discussed in this Notice. In addition, comments are invited relative to the evaluation criteria contained herein or any other criteria that should be considered. Communications should identify the docket number and be submitted in duplicate to the Secretary, Hazardous Materials Regulations Board, Department of Transportation, Washington, D.C. 20590. Communications received on or before November 5, 1975 will be considered before further action is taken. All comments received will be available for examination by interested persons at the Office of the Secretary, Hazardous Materials Regulations Board, Room 6215 Trans Point Building, Second and V Streets, SW., Washington, D.C., both before and after the closing date for comments.

(40 F.R. 26688, June 25, 1975)

[49 CFR Part 172]
[Docket No. HM-126A: Notice No. 79–9]

DISPLAY OF HAZARDOUS MATERIALS IDENTIFICATION NUMBERS; IMPROVED EMERGENCY RESPONSE CAPABILITY

This Notice proposes the adoption of an identification numbering system for hazardous materials based primarily on the system adopted for worldwide use by the United Nations Committee of Experts on the Transport of Dangerous Goods. Use of the identification numbering system will provide for improved identification of hazardous materials and the communication of information concerning the handling of such materials when they are involved in transportation accidents. Display of identification numbers would be required in connection with descriptions on shipping papers, on packages of 110 gallons capacity or less, and would be required on orange panels affixed to portable tanks, cargo tanks and tank cars. The MTB believes the use of identification numbers will provide an improved capability to quickly obtain and transmit accurate information about hazardous materials involved in accidents in order to obtain information necessary to deal with such accidents. Another element the MTB believes to be important in providing an improved emergency response capability is a manual it has under development. It will associate identification numbers with brief, concise instructions that will assist emergency personnel during the first minutes of a hazardous materials accident.

Many expressions of interest concerning the need to improve hazardous materials emergency response capabilities have been received from members of Congress, the National Transportation Safety Board, the Secretary of Transportation's Hazardous Materials Task Force, state officials, and most important, fire, police and other emergency service organizations. The MTB believes that the improvements proposed in this Notice will, if adopted, provide the basis for an improved emergency response capability that is not presently available through direct use of technical names; e.g., hexadecyltrichlorosilane (UN 1781), to identify hazardous materials and accurately and quickly communicate information about them.

The primary drafters of this Notice are Alan I. Roberts and Lee E. Metcalfe of the Materials Transportation Bureau.

BACKGROUND

On June 25, 1975, the Hazardous Materials Regulations Board (HMRB) published in the **Federal Register** (40 FR 26688, Docket HM-126) a request for public comment on hazard information systems. The request listed nine factors to be considered in the evaluation of a hazard information system:

1. Capability of the general public to recognize the existence of the immediate dangers presented by a material;
2. Presentation of information in a manner so that the general public will be able to accurately transmit basic information to emergency response personnel;
3. Compatibility, intermodally and internationally;
4. Capability of application to both bulk and non-bulk shipments;
5. Capability of functioning without use of a manual or other subsidiary documents;
6. Capability to meet the needs of emergency response personnel, carriers, shippers, and the general public;
7. Capability of integration with documentation, packaging, and vehicle identification requirements to help insure accuracy;
8. Capability of implementation without undue economic burden on shippers and carriers; and
9. Capability of indicating degree of hazard.

The request for public comments went on to describe nine different systems that either had been proposed to the Department or were known to be supported by one or more segments of the affected industry or emergency response community. An exception was the DOT hazardous materials information system which was previously subject of a rulemaking proposal under Docket HM-103. The MTB has considered all of these systems and the comments that were received in response to the request, and is now of the view that none are acceptable for adoption. The following is a summary of some of the reasons for this conclusion:

1. At least two of the systems would employ a preassigned before shipment numbering technique to provide specific action information or an indication of the level of hazard (Hazchem, NFPA 704). Although preassignment of numbers presumably would be made by MTB, commenters have not recommended any criteria except for flammability. Use of preassigned numbers would impose substantial delays on the introduction of new materials into commerce and the extensive effort and resources necessary for implementation would make such a system unworkable.

2. Several of the systems would employ numbers or words based primarily on the hazards or properties of a material (HI, RID/ADR, USCG, IATA, Union Carbide, Air Products, Pennwalt). The MTB believes that reliance on the identification of hazards alone is no longer a feasible approach, that specific identification of many materials is essential to the accomplishment of an improved emergency response program. Communication of hazards is important in the initial phases of an accident, and appropriate first contact handling information can be based on hazards; however, many accident situations dictate the need for more detailed information on how to deal with specifically identified materials. This is particularly so in accidents involving combinations of hazardous materials as is frequently the case in major train derailments.

3. One system would call for shippers to furnish the carrier an emergency response from that would move with each rail car (Canadian Transport Commission). While the basic idea has merit, the MTB considers that (1) such a system would result in the production of emergency response information widely varying in content based on subjective judgments of different shippers, except where industry associations might prepare response information for

common materials; (2) the enforceability of such a requirement without detailed Government guidelines would be questionable; and (3) while such a program may be workable for bulk shipments and full loads, the huge amount of paperwork involved would make its implementation impracticable for less-than-carload shipments.

4. One system would call for display of the order of severity in three risk categories—health, flammability, and reactivity (NFPA 704). The MTB and its predecessor have examined the utility of such a system on several occasions. However, no criteria have been suggested for the health and reactivity presentation; therefore, qualitative evaluations must be made in assigning levels of hazard (hence, the preassignment factor discussed in 1 above). In addition, this proposed system places too many hazard factors into three fields, thereby precluding appropriate assessment. Examples are the assignment of health hazard level three to chlorine, liquid oxygen, sodium hydroxide (lye), hydrogen sulfide, and tetraethyl-lead, and the assignment of reactivity level one to calcium oxide (quicklime), antimony pentachloride and chromic acid. The kind of hazard and the extent of the hazard posed by each of these materials is very different from that posed by the others that share the same assigned number. The appropriate action required to handle accidents involving these materials also is quite different in each case. Consequently such a system is not appropriate for use in a transportation emergency response situation.

The MTB believes that determination of the degree of hazard beyond an examination of the properties of a material requires an assessment of the environment and circumstances at the scene of an accident. In most instances, a preliminary determination of risk can be made based on the information obtained at the scene of an accident. If the initial assessment reveals that a potentially serious situation exists; e.g., a large spill from a tank vehicle placarded POISON and bearing an identification number, then, beyond that basic initial response information provided in the proposed manual, a call should be made to the central response information center to obtain expert advice in fully assessing the level of risk based on the accident environment; i.e., topography, wind direction, proximity to habitations, configuration of the accident, etc., and the properties and quantity of the spilled material in relation thereto. It is not conceivable that all the factors that must be considered in determining risk can be transmitted directly from placards on transport vehicles. Therefore, the identification numbering system has been selected as the best means to assist in quicker and more accurate determinations of risk.

The MTB extends its appreciation to the many individuals companies, and organizations for their considerable efforts in preparing and submitting comments in reply to the Request for Public Comment. A number of comments were accepted and previously adopted under Docket HM-112, and others not directly related to the proposals in this Notice will be considered in a future rulemaking action.

Of all the comments received in response to the Request, the MTB finds those presented by the Association of American Railroads (AAR) particularly supportive of the concepts on which the proposals in this Notice are based. For this reason, we restate a large portion of the AAR comments:

THE BASIC REQUIREMENTS OF A HAZARD INFORMATION SYSTEM

The communication of hazard information is a very difficult task. It must be capable of use by many different groups, each with a different need for information, and it must be usable in an emergency environment without requiring the use of potentially unavailable supplementary materials. The information which it transmits must be easily recognized, quickly understood and accurately communicated.

WHO NEEDS A HAZARD INFORMATION SYSTEM?

The most general answer to this question is that anyone who may potentially come in contact with a transportation accident involving hazardous materials needs an information system in order to be able to make intelligent decisions about the safety of his own life and property and about the lives and property of others.

The general public is the broadest and most inclusive subcategory of people with a need. This group includes both those who are at the scene of an incident and know it and those who are in proximity and are not aware of it. As a rule of thumb, it is probable that no system of hazard information will protect those who are not able to immediately view the sight of the incident except as it alerts those who are in proximity and then pass along the message.

Carrier and shipper personnel are the next group of people with a need for a hazard information system. They may be considered to have certain minimal training in either necessary response countermeasures or in the ability to communicate the fact that an incident involving hazardous materials has taken place.

Emergency response personnel, generally, are taken to be the police and fire departments serving the area in which the hazardous incident takes place. Despite the wide variation in training and competence of such personnel, they are assumed, at a minimum, to be familiar with routine fire fighting and evacuation procedures and, at a more sophisticated level, may be able to apply advanced levels of countermeasure technology.

Expert emergency response teams consist of those with special training in dealing with hazardous commodities. They may be in the employ of shippers and manufacturers, carriers, and governmental agencies. They consist of that group with the highest possible degree of knowledge about the specific commodities involved in particular incidents.

WHAT IS REQUIRED OF A HAZARD INFORMATION SYSTEM?

Each of the four groups outlined above has different needs for a system of hazard information.

The general public should be made aware immediately that a hazard exists and should be able to communicate the very basic facts of that hazard to experts with a higher level of knowledge. These basic facts include the location of the incident, the type of vehicle involved, the occurrence of leakage, fire or explosion and, if vehicle markings are seen, transmission of the fact of such markings or of the wording on them.

Carrier and shipper personnel should be able to communicate all of the information possessed by the general public and, in addition, they should be able to identify the nature of the hazard involved. By having access to the shipping papers, the carrier and shipper personnel will be able to pass along detailed information concerning the commodities involved to their superiors and to the first emergency units to arrive at the scene. In some instances, carrier and shipper personnel information such as that provided by CHEMTREC.

[Emergency] response personnel require immediate information on the nature of the hazards of the materials involved and any special handling requirements. They should be able to use the information obtained from lower levels of response sophistication to access more detailed data to assist them in handling the specific product involved under the circumstances prevailing at the time of the accident.

Expert emergency response teams need to be able to use the information obtained at the scene and from lower level response groups to provide the most detailed and sophisticated direction for the handling of the incident. They should also be able to assist in proper cleanup and disposal activities following the immediate minimumization of threats to life and property.

SUMMARY

Having analyzed the abilities and needs of those who will have contact with any hazard information system, its essentials are easily stated:

• It must be immediately recognizable as a system describing the existence of a hazard.

• It must be readily understandable by anyone who comes in contact with it.

• The information gained by anyone who comes in contact with it must be readily communicable to all others with a need to know.

• For those with a greater degree of ability, it must be indicative of a response pattern applicable to the particular incident.

• It must be uniform among all transportation modes.

• It must be capable of accommodating new hazardous commodities and, for higher levels of sophistication, it would be desirable that it be capable to easy computer implementation.

THE COMPONENTS OF THE HAZARD INFORMATION SYSTEM
DEVELOPED BY THE AAR

In this section of the AAR's comments, the primary component parts of our hazard information system will be discussed and, wherever possible, related to the needs of various parties as recognized above.

SHIPPING PAPERS

The railroad industry believes that primary reliance must be placed on the shipping papers for information on the name of the product involved and its hazards. Historically, the shipping paper has been seen as the only single source of complete identifying information which travels with a shipment of hazardous materials. Because of all the important exemptions which appear in the regulations for the labeling and placarding of hazardous materials, the complete identification of particular shipments can be accomplished only through the shipping papers which accompany it. The recognition of deficiencies in the current use and content of shipping papers has led to the development of recommendations by the AAR for their correction.

Currently, shipping papers identify only the required classification, that is, the "primary hazard" of the material. This is no longer sufficient; the regulations must be amended to require the inclusion of all major hazards ["major hazards" are those recognized to the tertiary level] for dangerous commodities. This will mean that a shipment of, for instance, sulphuric acid must now be listed on [the] shipping paper as "SULPHURIC ACID, corrosive material, poisonous liquid, Class B."

To make the presence of hazardous materials obvious on the shipping paper, the proper shipping name and all hazards must be listed first and consecutively in each commodity's freight description. The order of materials (regulated or non-regulated) as listed on the shipping paper can be random if this previously described format is employed. The uniform presentation of information on shipping papers will enable transportation and emergency response personnel to readily identify hazardous materials from their descriptions and classifications.

In order to hasten the ability of on-scene personnel to contact those with expert technical information about the commodities involved, an emergency telephone number must be added to the shipping document. This telephone number must provide, on a 24-hour basis, immediate access to emergency assistance from the shipper, a designated industry response group, CHEMTREC, The Bureau of Explosives, or whatever source is deemed by the shipper to be best. It is of the utmost importance that both the immediate and long-term response to a hazardous materials incident be initiated as quickly as possible. In view of the potential for reducing injury, danger to property and adverse environmental consequences, the requirement for an emergency telephone number on shipping documents, to be placed there by the shipper, is a minimal burden.

The railroads of the United States have established a numerical Standard Transportation Commodity Code (STCC) for identifying hazardous materials. The use of STCC numbers is required by mutual railroad agreement and the number is exhibited on waybills covering all shipments. Recent efforts by the STCC Technical Committee of the AAR have resulted in the development of a new series of STCC numbers whose purpose is to specifically identify discrete hazardous materials. This "49-series" was made effective January 1, 1976 and is the vehicle for hazardous materials management which the transportation industry has long needed.

STCC numbers are 7-digit combinations assigned to specific commodities based on hazard characteristics. Within the 7-digit STCC number a means is provided to allow the quick identification of primary, secondary and, in some cases, tertiary hazards, whether or not hazards subordinate to the primary hazard are now recognized by DOT. The 49-series begins with the first two digits, 49, as an identifier for hazardous materials. The next three digits identify the potential hazards of the materials and the final two digits identify the specific commodity....

It is important to note that many materials properly shipped under a general classification, i.e., "flammable liquids, n.o.s.", have now been given specific 49-series STCC numbers based on the specific tariff description currently used. Their movement in transportation is well documented, and therefore, these commodities, as listed in STCC Tariff 1-D (or subsequent revisions), should henceforth be listed in § 172.5 as proper shiping names. The railroads request that the Hazardous Materials Regulations require the shipper to insert the STCC numbers on all shipping papers tendered with goods for transportation. Only [the] shipper knows enough about the materials he produces and ships to ensure the accurate assignment of STCC numbers.

The STCC Code is designed to be flexible and to allow for future expansion to include both new commodities and new combinations of [hazard] classification; a section is already reserved in the STCC Code to designate substances which are environmentally hazardous as defined by the Environmental Protection Agency.

The precise logic which has been employed to create the 49-series has resulted in numbers whose utility far surpasses those proposed by other organizations including MCA and the United Nations. The middle three digits provide a complete set of numbers to reveal all major potential hazards of each product and this unique feature does not exist in any other system of which the AAR is aware....

The AAR insists that no secondary, untried system of "hazard numbers" be implemented. The 49-series STCC Code is now in existence and in use and the next revision of the Bureau of [Explosives] publication "Emergency Handling of Hazardous Materials in Surface Transportation" will contain a complete listing of all commodities in the 49-series code and the appropriate hazard information for handling emergency situations involving these commodities. A cross reference index will be supplied to locate commodity information if only the 49-series STCC number is known; in addition, a growing number of rail carriers will be able to provide hard copy information about the commodities involved by entering the STCC numbers of commodities involved in incidents into their computers.

Out of an abundance of caution, it must be emphasized that the STCC Code does not make the railroads' system just another "numbers game" Hazard Information System. STCC numbers allow precise commodity identification and verification, allow computer accessing for commodity information, provide a means for the retention of hazardous materials movement statistics and assist in the computer generation of shipping documents. They are not intended for use by the general public and, thus, require no first-level-of-response decoding. Their primary purpose is fulfilled at the shipper/carrier interface and in the ability of the shipper or carrier to make precise commodity information known to emergency personnel....

EMERGENCY ACTION GUIDE

A unique feature of the AAR Hazard Information System is the Emergency Action Guide for Hazardous Materials Incidents....

The purpose of the Emergency Action Guide is to indicate, primarily for carrier personnel, an initial response pattern to be taken to save lives and prevent injuries. Such an initial response will usually not extend beyond the first fifteen minutes or so following an accident and, with this in mind, the Guide identifies the basic nature of the hazard for each classification of material and describes the "ground rules" for initial emergency action. The Guide, while not suggested as a mandatory requirement, is proposed to be included as part of a trainman's timetable for relevant operations and could be displayed by poster in cabooses and locomotives. When reproduced on durable material, and perhaps folded, such a guide could also be carried in truck cabs, fire engines, and police vehicles. Recommended procedures for using the Guide would include instructions to those to whom it is issued to read it in advance of an emergency in order that, when an emergency occurred, the particular applicable section of the Guide could be easily found.

A basic purpose of the Emergency Action Guide is to fill the gap between the accident and the receipt of information from the party responding to the emergency telephone number and, even if the Guide was not part of the information carried on local community emergency vehicles, it could be available from the train crew, or from the truck driver, to assist the first emergency units on the scene.

The above is not a complete reproduction of the AAR's comments, nor is the copyrighted AAR Emergency Action Guide reproducible in this notice. Sample pages submitted by the AAR may be reviewed in the docket file. However, the quoted comments are representative of AAR's viewpoints concerning matters covered by this Notice. The MTB agrees that primary reliance must still be placed on shipping papers for information relative to "non-bulk" shipments, and that while the display of information on shipping papers has considerably improved following the adoption of regulations under Docket HM-112, further improvements are necessary. The MTB agrees that an identification number should be assigned to hazardous materials for rapid communication of information and employment in the emergency response system. The MTB agrees with the AAR that an emergency action guide or emergency response manual should be employed in the improved system. However, the MTB does not fully agree with the AAR's recommendations on how to acomplish improvements in these three areas.

THE IDENTIFICATION NUMBERS

The MTB is proposing that the UN numbering system be used as the basis for implementation of an improved emergency response system. For a number of years, the United Nations Committee of Experts on the Transport of Dangerous Goods has been assigning identification numbers to specific chemicals and generic descriptions for "dangerous goods." This system is becoming more and more recognized on a worldwide basis. It is referenced in the Dangerous Good Code of the International Maritime Consultative Organization and is under consideration for adoption in the forthcoming standards to be implemented by the International Civil Aviation Organization. This system also is referenced in various procedural documents under development pertaining to international trade documentation. The MTB recently was advised by its Canadian Government counterpart, Transport Canada, that UN identification numbers will be used within Canada in connection with shipments of hazardous materials and that the numbers will be the basis for the Canadian emergency response system.

Several important factors were considered by the MTB in deciding to propose the use of the basic UN numbering system for improved emergency response information:

1. The numbers are assigned by Governmental authorities under the aegis of the Economic and Social Council of the United Nations and have the same meaning throughout worldwide commerce.

2. They have been assigned specifically to identify hazardous materials (dangerous goods) and are intended to have no other meaning or use. Thus, their formulation and application is not dictated by any other competing or overriding considerations.

3. They are assigned, for the most part, to materials requiring separate recognition that are shipped in commercial quantities. Therefore, only four digits are necessary. Other chemical identifier systems have as many as nine digits. The fewer the digits in an identification number, the greater its reliability in an emergency response system employing radio and telephone communications.

4. Numbers are assigned on the basis of the next open number without regard to the particular chemical properties of a material, thereby, avoiding any problems in the availability of numbers for future assignments. Availability of numbers almost inevitably would be a problem if the internal arrangement of digits was meant to have some special meaning; e.g., a chemical grouping or properties assignment. The use of identification numbers will (1) serve to verify descriptions of chemicals; (2) provide for rapid identification of materials when it may be inappropriate or, even worse, confusing to require display of lengthy chemical names on vehicles; (3) assist in preparation of correct documentation when language translation is necessary; and (4) be an aid in speeding communication of information on materials from accident scenes and the return of more accurate emergency response information. For example, the UN number for acetone is 1090, while the number for acetone cyanohydrin in 1541—a distinction of great importance in conveying appropriate emergency response information.

It is proposed to list UN or U.S. identification numbers in a new column 3a of § 172.101. The number will be preceded by "UN" if the description preceding it is exactly the same or sufficiently similar to the international description (as displayed in the IMCO Code). If the description in § 172.101 is significantly different but addresses the same material as an UN entry, it will be given the same number but will be preceded by "NA" in the United States and Canada. A four-digit number beginning with "9" indicates there is no corresponding UN description (specific or generic) for the material assigned that number. This is the mechanism the MTB and Transport Canada propose to use to assign identification numbers to materials, such as hazardous substances, that have not as yet been assigned numbers by the United National Committee of Experts.

The AAR is to be complimented on its attempts to implement a hazardous materials identification numbering system. Its efforts in this regard have, in MTB's opinion, led the way towards providing an improved emergency response capability for hazardous materials. This Standard Transportation Commodity Code (STCC) system, in the absence of any established alternative, is being used in U.S. rail transport with increased frequency. Nevertheless, the MTB believes that what it is proposing in this Notice is preferable to the STCC system as a national, multimodal system for the following reasons:

1. While the 49 STCC series is not to be used for economic purposes, by necessity its formulation was constrained by the need for it to "bridge" into other STCC series numbers established for freight rating/statistical purposes. The following STCC listings illustrate this point:

Specified name in STCC tariff 1–G	STCC number
Combustible liquid n.o.s. (asphalt tile plasticizer)	49 154 08
Combustible liquid n.o.s. (plasticizers, paint, lacquer or varnish)	49 154 10
Combustible liquid n.o.s. (belt dressing)	49 154 16
Combustible liquid n.o.s. (electric wire saturating or finish compound, paraffin base).	49 154 18
Combustible liquid n.o.s. (bay rum)	49 154 20
Combustible liquid n.o.s. (perfumery)	49 154 22
Combustible liquid n.o.s. (after shave lotions)	49 154 24
Combustible liquid n.o.s. (polishing wheel cement thinner)	49 154 26

It is obvious that many STCC entries would serve no useful purpose for improved emergency response. There are a number of entries in Section 172.101 that are end-use terms, a practice which should not be expanded unless proven necessary.

2. If the descriptions in § 172.101 are greatly expanded, as suggested by the AAR, the undesirable result would be reduced accuracy and rapidity of access to information that is essential.

3. The STCC identification numbering system is published in a tariff which is on file with the Interstate Commerce Commission. Additions or deletions are made by, or under the supervision of, representatives of the railroad industry. Neither the transfer of all the listings for hazardous materials in the STCC Tariff nor incorporation of the tariff by reference is appropriate for publication in the Department's Hazardous Materials Regulations. If the AAR believes there are materials inappropriately described by generic descriptions in § 172.101, it should petition the MTB to describe each of them by technical name, thereby affording public participation in a rulemaking proceeding.

4. Recognition of the STCC system is limited to the United States and even within the United States its acceptance and use is further limited—primarily to the shipment of materials by railroad. It is not recognized on a worldwide basis, nor is it likely to be in light of the now-established UN numbering system and the growing application of that system.

DISPLAY OF THE IDENTIFICATION NUMBER

The MTB proposes to require the display of the identification number, as listed in proposed column 3a of § 172.101, on all shipping papers as the third element of a basic description (immediately following the hazard class entry) and on the outside of packagings of 110-gallons capacity or less (immediatley following the prescribed shipping name of contents). It also proposes to require that the number be displayed on orange panels affixed to portable tanks, cargo tanks, and tank cars. While not a part of this proposal, the MTB will consider in the future the need to expand the required display of the orange panel to other than bulk shipments of materials.

The MTB believes that the display of the identification numbers must be as uniform as possible to enable ready recognition by emergency response personnel. Further, the MTB recognizes that, with display of the identification numbers on vehicles, it may not be necessary to expand the name marking requirements for tank cars and cargo tanks—an alternative of questionable benefit when complex chemical names are involved. The display of the numbers on the orange panel will lead to rapid identification of contents. If an identification number addresses a generic description, and such a description is considered inadequate for a particular material, the MTB believes the material should be described in § 172.101 by technical name with an identification number assigned to it following an appropriate rulemaking proceeding. The MTB recognizes that it may have to consider a number of additions and changes to the list of hazardous materials in the future. Concerning the marking of identification numbers on packages, two major factors were considered. First, freight personnel often became the first contact "emergency" personnel when spills and leaks occur. The MTB visualizes that they will be able to make use of the manual in the same manner as emergency response personnel. Sole dependence on complex chemical names without reference to an identification number may lead to an erroneous response. This same view applies to emergency response personnel coming into contact with packages directly in vehicles, on freight docks, or elsewhere. The second factor is the value of the identification number in verifying the shipping information displayed on documents with the information displayed on packages to preclude error.

THE MANUAL

One of the principal comments the HMRB received in response to its proposals under Docket HM-103 was the view that the proposed system (HI) placed heavy reliance on a manual that the HMRB suggested would be placed in each of the 392,000 emergency vehicles (fire, police, etc.) in the Nation. At approximately the same time as the initial proposal under HM-103, the Office of Hazardous Materials initiated distribution of its "Emergency Services Guide for Selected Hazardous Materials" which provides guidance during the initial phases of an emergency for 45 selected materials that are transported in bulk quantities. To date, more than 880,000 copies of the Guide have been distributed to fire, police, and other emergency services organizations without charge. This fact, plus information that good use has been made of the manual, reinforces the MTB's view that a manual should be a viable part of this improved emergency response effort. It will serve to provide immediate emergency response information, especially important during the first minutes when many judgements and decisions must be made.

The MTB has undertaken the development of a manual of approximately 100 pages that will address virtually all hazardous materials transported in commerce. The manual is being developed on the principle that its statements of information and recommended actions must be concise and clearly understood. In examining other manuals, it is evident that there are only a limited number of recommendations (instructions) that can be given in a manual even though the recommendation may address hundreds of materials. Rather than write a manual with separate instructions for each material—a method that would be unduly burdensome, expensive, and unnecessary—the MTB is planning the manual according to the following format. At the beginning of the manual there will be two lists of virtually all domestic and international shipping names for hazardous materials. The first list will be arranged in numerical order of identification numbers, and the second list in alphabetical order. In both lists, there will be a page reference number associated with each entry. The numbered pages, approximately 60 to 70, will contain initial emergency response information and instructions. Many materials calling for the same initial emergency response action will be covered by the same page of information and instructions. The information pages are being developed in a style similar to those contained in the "HI" manual proposed under Docket HM-103.

The MTB believes that, in many instances, the manual will be all that is needed at the scene of an incident to terminate a minor event or to keep a minor event from becoming a major event of great consequence. However, when a situation requires further assistance, the manual refers the user to CHEMTREC (Chemical Transportation Emergency Center) and other organizations to obtain more detailed information and the help of technical experts.

Other features of the manual under development include (1) display of certain chemical descriptions in red to indicate when evacuation may be necessary; (2) evacuation tables related to small and large spills of the materials listed in red; (3) a page devoted to the display of placards with an indication that the recommendation on a particular page should be followed relative to each placard if, for some reason, shipping papers are not available or the identification number is not displayed; and (4) a typical shipping paper highlighting the description of a hazardous material and its identification number.

EXCLUSIONS

The proposals in this Notice do not apply to any material for which its proper shipping description does not have an associated identification number displayed in column 3a of §172.101. Included in this category would be Consumer commodities, ORM-D and classes A and B explosives.

Consumer commodities properly classed ORM-D due to their quantity and form are not considered to pose a risk sufficient to warrant the application of these requirements. Classes A and B explosives, while under consideration for inclusion in the future, will have to be reviewed in greater detail to determine if a large number of identification numbers are necessary to make more finite distinctions in their risks or if a complete restructuring of the explosives descriptions would be appropriate. The MTB believes that, at the present time, it is desirable to use the "A" and "B" references for explosives presently displayed on shipping papers and placards for direct application in the manual and for communications.

OTHER ACTIONS

This rulemaking proposal is considered to be a matter of high priority to the MTB. Therefore, other proposals pertaining to the hazardous materials communications regulations will be dealt with in two separate rulemaking actions. Included in these actions will be proposals to (1) permit the use of certain international descriptions (IMCO and forthcoming ICAO descriptions emanating primarily from the UN recommendations); (2) require further description for certain n.o.s. entries; (3) prescribe additional labeling requirements and permit use of certain newly adopted international labels; (4) require a special marking on portable tanks, cargo tanks and tank cars containing unodorized LP gas; and (5) address a number of petitions pertaining to shipping papers, certifications and placarding.

IMPLEMENTATION: PRIORITIES AND COSTS

The MTB views the timing of the implementation of requirements proposed in this Notice as the major feature of its cost. This view is based on factors such as the present inventories of preprinted shipping documents, premarked (name of contents) inventories of packagings, and the necessary training and education associated with the implementation of the procedures necessary to comply with the regulations proposed herein if they are adopted. Therefore, the MTB believes it should make a statement concerning its contemplated implementation of the regulations proposed in this Notice.

The highest priority will be placed on implementation of documentation and identification (the orange panel) requirements applicable to transportation of hazardous materials in portable tanks, cargo tanks and tank cars. The MTB believes the mandatory compliance date should be no more than one year after the date of publication of the final rule.

The second highest priority will be placed on implementation of the documentation requirements for hazardous materials in packaging (other than bulk) excluding those, such as Consumer commodities, ORM-D, that will not be assigned identification numbers. The MTB believes the mandatory compliance date should be no longer than one year after the date of publication of the final rule.

The lowest priority will be assigned to packaging (less than 110-gallon capacity) markings. The MTB believes that two years should be provided for implementation or, as an alternative, a "grandfather clause" should be provided for packagings in stock prior to the effective date of the new regulations.

Concerning the display of identification numbers on orange panels, the MTB has obtained preliminary cost figures from a major supplier of labels and placards and has determined that, depending on the quantities ordered, the costs of prenumbered panels would be approximately as follows: Panels made of tag board for one time use— 10 to 25 cents per panel; panels made of extruded vinyl for short term use—20 to 45 cents per panel; panels made of cast vinyl with plastic coating having an estimated life of five years—38 to 55 cents per panel. The vinyl panels would be adhesive backed to facilitate attachment.

The MTB expects the greatest costs will be incurred in the display of the orange panels on cargo tanks operated by common/contract carriers because of the high frequency in the change of usage of cargo tanks in this particular segment of the transportation industry. This fact is recognized in the existing regulations relative to the marking of the name of contents of cargo tanks. However, the MTB believes the added cost associated with implementation of these proposed requirements is warranted in the interest of providing an improved emergency response capability and is preferable to the alternative of extending the present tanks, as is presently required for portable tanks.

Specific views are invited concerning the proposed implementation of the regulations proposed in this Notice. Commenters should provide factual information concerning inventories of preprinted documents, adjustments in ADP programs, packaging turnover, and training. Broad generalized statements will not be beneficial to the MTB in its efforts to make a serious determination of how the rules are to be implemented, if they are adopted.

In further regard to implementation, the MTB was asked recently if any of the present regulations would preclude application of orange panels such as are proposed in this rulemaking. The MTB's response indicated there is no existing regulation that would prohibit the display of the United Nations identification number after the class entry on a shipping document, nor the display of an orange panel bearing an identification number.

The MTB originally considered that a rulemaking action of this type should be the subject of an advance notice of proposed rulemaking in order to solicit public comments relative to its development. On further consideration, the MTB has decided that a notice of proposed rulemaking is appropriate in order to expedite implementation since the proposals contained herein are confined to one basic issue; i.e., the display of an identification number, with other matters pertaining to improved communications being made the subject of separate rulemaking actions.

CONCLUSION

The MTB believes (1) adoption of the hazardous materials identification system proposed in this Notice will be a vital initial step in improving emergency response capabilities; (2) that, while the system can function without reliance on a manual, a manual of the type it has under development will significantly enhance emergency response capabilities, particularly during the first minutes of an emergency; (3) the proposed system is compatible intermodally and internationally and will be applicable to bulk and non-bulk shipment to an extent considered necessary; and (4) the proposed system is capable of integration with existing documentation, packaging and vehicle identification requirements that will insure the accuracy of information transmitted to and from emergency response personnel.

(44 F.R. 32972, June 7, 1979)

[49 CFR Part 172]
[Docket No. HM-126A; Notice No. 79–9]
Display of Hazardous Materials Identification Numbers; Improved Emergency
Response Capability; Descriptions for Organic Peroxides; Extension
of Comment Period

The MTB is developing a rulemaking proposal for future publication that will pertain to organic peroxides. It will be primarily addressed to packaging and special shipping requirements. Considering its recent proposal under this Docket pertaining to the display of hazardous materials identification numbers to provide an improved emergency response capability, the MTB believes that, during the interim, it is necessary that most of the organic peroxides that may be shipped in commerce be separately identified and assigned individual identification numbers. This will provide the capability to give separate recognition to the different risks posed by organic peroxides, such as (1) the differing degrees of thermal sensitivity; (2) violence of thermal decomposition; (3) susceptibility to ignition by friction; (4) flammability; and (5) corrosivity.

Approximately 135 organic peroxide entries would be added to the Hazardous Materials Table. Although packaging would not be listed for each new organic peroxide entry, each is cross-referenced to an entry that is already in the Table that has packaging and other requirements which are applicable to the new entry. Each of the new entries has an identification number which would be entered with the name of the material on shipping papers and packages. The identification numbers would be used as a basis for referencing appropriate emergency response information. Certain United Nations and IMCO entries contain concentrations that are greater than those authorized by the DOT regulations and thus would not be acceptable for transportation in the higher concentrations. For example, diisopropylbenzene hydroperoxide has a maximum of 72 percent in solution in the IMCO entry whereas only 60 percent peroxide is authorized in the entry in Section 172.101. Therefore, only a maximum of 60 percent peroxide may be offered for transportation or transported under the DOT regulations.

Paragraph (b)(5) of Section 172.100 would be revised to establish a requirement for entering on a shipping paper, and marking on the package, the technical name of each organic peroxide offered for transportation. While the proper shipping name derives from the technical name entry for each organic peroxide, the organic peroxide entry referenced by the word "see" would continue to contain the requirements for columns 4, 5, 6, and 7 in Section 172.101 until a complete rulemaking proposal for organic peroxides is developed, proposed, and adopted. In light of the proposals made in this Notice, and in consideration of a number of requests for further time to comment on Notice No. 79–9, the MTB is extending the comment period on that Notice to coincide with the closing date for comments on this Notice.

(44 F.R. 43858, July 26, 1979)

49 CFR Parts 172, 173
[Docket No. HM-126B; Notice No. 79–14]
Improved Descriptions of Hazardous Materials for Emergency Response

The MTB published a notice of proposed rulemaking in the FEDERAL REGISTER on June 7, 1979 (44 FR 32972; Docket No. HM-126A; Notice No. 79–9), proposing the adoption of a numerical identification system for hazardous materials transported in commerce. In the July 26, 1979, FEDERAL REGISTER (44 FR 43858; Docket No. HM-126A; Notice No. 79–9 and 44 FR 43864; Docket HM-171; Notice No. 79–11) supplemental notices were published proposing to adopt the numerical identification system for organic peroxides, and to authorize the optional use of United Nations' shipping descriptions and identification numbers for certain hazardous materials in place of the descriptions required by existing DOT regulations. The objective of the proposals in this notice is to augment the previous proposals by adding to the Hazardous Materials Table certain entries necessary to improve the identification of the hazards of many materials. The MTB believes these improved hazardous materials identifications are essential to the successful accomplishment of an emergency response system which will be accessed by means of identification numbers.

The MTB has been requested by the Environmental Protection Agency to consider requiring the identification of each n.o.s. entry on shipping papers and package markings by the technical name of the hazardous material. This would permit more accurate identification of the material for emergency response actions. This is already a requirement for export shipments by vessel and the MTB agrees that safety would be enhanced by such a requirement since more specific information would be immediately available for use in emergency response actions. For a mixture containing two or more hazardous materials, at least two of the components which contribute most to the hazards of a material would be required to be identified. However, the MTB does not propose to apply this requirement to hazardous materials authorized to be described and shipped as Limited Quantities.

The Hazardous Materials Table does not contain a generic description applicable to all pesticides (i.e., pesticide, liquid or solid, n.o.s.). Addition of a generic description for pesticides to the Table would not provde sufficient information to identify the type of pesticide and, consequently, it would be difficult to specify appropriate action to be taken in the event of an accident involving spillage or exposure. Conversely, it would be virtually impossible to list each pesticide by name and possible formulation. The MTB believes an appropriate approach would be to identify and describe pesticides by chemical groups based on their chemical structures. This approach would enable first aid and medical advice to be linked to such groups. To this end, fifteen groups of pesticides have been identified which the MTB has proposed for addition to the Hazardous Materials Table. Within each of the fifteen groups, there would be three separate entries, which would distinguish the form (i.e., liquid or solid), and for liquids would distinguish the hazard class (i.e., flammable liquid or poison B liquid). Thus, a total of forty-five descriptions would be added to the Table to identify pesticides by chemical structure, form, and hazard class. The MTB estimates that these forty-five descriptions would apply to more than ninety percent of the pesticides transported.

Also, the MTB is proposing the addition of eight generic n.o.s. entries addressing multiple hazards. These multiple hazard entries consist of such n.o.s. descriptions as Corrosive liquid, poisonous, n.o.s., Flammable liquid, corrosive, n.o.s. and Oxidizer, corrosive liquid, n.o.s. The MTB believes these new entries will provide improved identification of a number of hazardous materials in association with the additional labeling requirement proposed for the entries in column (4) of the Hazardous Materials Table.

In addition to assigning identification numbers to the hazardous materials, the MTB believes that certain additional shipping paper entries would be beneficial to emergency response personnel and carriers. Specifically the entries being proposed are the phrase "Water Reactive" for a material required to be labeled FLAMMABLE SOLID and DANGEROUS WHEN WET; and the word "Poison" for a material required to bear a POISON label, classed other than as a Poison B and not otherwise identified as a poisonous material on the shipping paper. The MTB agrees with the Association of American Railroads recommendation that such a warning be added to the shipping paper entry for certain materials required to be labeled FLAMMABLE SOLID and DANGEROUS WHEN WET to quickly identify the material as having a potential of being water reactive during an emergency. The American Trucking Associations, Inc., petitioned the MTB for a rule change to add the word "Poison" to a shipping paper to assist the carrier in complying with § 177.841(e). The MTB agrees that it would permit carrier personnel who load vehicles to be aware of the POISON label and to plan loads accordingly. The same situation would exist for the rail carrier when such references as § § 174.280, 174.380, 174.480, 174.580 and 174.680 are considered, for the air carrier when considering § 175.630, and for carriage by vessel when § § 176.331 and 176.600 are considered. Also, such an entry would assist quick identification of a poison hazard during an emergency.

Dichloropropene and propylene dichloride mixture was placed in the Hazardous Materials Table as a Corrosive material under Docket HM-57 (38 FR 35467; December 28, 1973). However, a review of current references, including the United Nations "Transport of Dangerous Goods" and the IMCO "Dangerous Goods Code", indicate the hazard class of Flammable liquid is more appropriate. The National Fire Protection Association, in its manual entitled "Fire Hazard Properties of Flammable Liquids, Gases and Volatile Solids," indicates the flash point of the first named material in the mixture as 95° F. and the other as 60° F. thus, changing the hazard class of the mixture to Flammable liquid would reflect the flammable nature of this mixture and more appropriately describe the hazard that would be important in emergency response.

Also, the MTB proposes to revise the heading and paragraph (a) of § 173.352 to include Cyanide solutions, n.o.s. classed as a Poison B, UN 1935, which would be added to the Hazardous Materials Table even though not shown in the § 172.101 Table in this notice. The omission of this material from the Table came to MTB's attention shortly before publication of this notice. Due to the manner in which MTB programs and retrieves this Table from automatic data processing equipment, reprogramming the Table would have inordinately delayed publication of this notice. However, despite the fact that the entry Cyanide solutions, n.o.s. does not appear in the formal proposal, MTB is proposing that it be added to the Hazardous Materials Table and hereby gives notice of such proposal. The MTB believes the packagings authorized by § 173.352 for sodium cyanide or potassium cyanide are more appropriate for cyanide solutions, n.o.s. than the general packagings that would otherwise by authorized for this material under § 173.346 for a poisonous liquid, n.o.s.

Fuel, aviation turbine, engine is now in the Table as a Flammable liquid and the MTB proposes to provide an additional entry for it as a Combustible liquid. The MTB has been informed that having this fuel properly identified on the shipping paper will help the aviation industry insure that the correct fuel is being delivered for use in the operation of aircraft. Aviation turbine engine fuel shipped as Fuel oil or as Combustible liquid, n.o.s. for use in aircraft apparently leaves a degree of uncertainty about the actual identity of the material. The MTB believes that the cost for this relatively minor change in documentation would be far outweighed by even a small improvement in aviation safety.

The MTB proposes that §173.151a(a)(3) be revised to permit continued classification of a hazardous material according to its predominant hazard when it contains an organic peroxide without placing an asterisk before each organic peroxide entry. It is possible that when certain stabilizing diluents are added to certain organic peroxides the predominant hazard is that of the diluent rather than the organic peroxide.

A number of other additions and changes to the Hazardous Materials Table are proposed based on petitions for rulemaking and other sources. Proposed additional entries include Propargyl alcohol which is flammable and poisonous; Chloroprene, uninhibited which would be listed as forbidden (uninhibited chloroprene may polymerize spontaneously so as to cause dangerous evolution of heat); and Chloroprene, inhibited which is a Flammable liquid. Further, additional entries proposed are: Alcoholic beverage; Benzidine; Bromochloromethane; Calcium hypochlorite, hydrated; Chlordane (Flammable liquid); Furan; Morpholine; Morpholine, aqueous mixture; Paraldehyde; Pinene; and 1,1,1, Trichloroethane. Packaging reference revisions are proposed for the Compressed gas, n.o.s.; Refrigerating machine and Strychnine, solid entries while an additional label requirement is proposed for the three Hydrogen peroxide entries.

A comment to Docket HM-126A was received which indicated that if a longer comment period had been available, an evaluation of the assignment of identification numbers would have been made for submission with the comment. Since identification numbers have been proposed for assignment to hazardous materials in Docket HM-126A (44 FR 32972; June 7, 1979), Docket HM-126A Supplement (44 FR 43858; July 26, 1979), and this Docket, such comments may be submitted in response to this notice, if they were not provided in earlier comments.

(44 F.R. 65020, November 8, 1979)

49 CFR Part 172
[Docket No. HM-126C; Notice 84–3]
Required Use Of Emergency Response Guidebooks and Material
Safety Data Sheets

The National Transportation Safety Board (NTSB) has recommended that the Department of Transportation determine, by mode of transportation, the feasibility of requiring comprehensive product-specific emergency response information such as MSDS's for hazardous materials moving in bulk quantities. The American Trucking Associations, Inc. (ATA) has petitioned MTB for a rule requiring placement of Emergency Response Guidebooks in certain transportation facilities. Comments that relate to the ATA petition have been received from other parties. This notice solicits comments on the potential benefits and consequences of required use of the ERG and/or MSDS's to communicate information on the hazards of materials while they are moving in commerce.

This notice contains a substantial amount of material that is directly quoted. Primary sources of the quoted materials are as follows:

NTSB—National Transportation Safety Board, 800 Independence Avenue, SW., Washington, D.C. 20594, James E. Burnett, Chairman

ATA—American Trucking Associations, Inc., 1616 P Street, NW., Washington, D.C. 20036, Robert A. Hirsch, Attorney and Richard M. Doyle, Hazardous Materials Specialist

IBT—International Brotherhood of Teamsters, Chauffeurs, Warehousemen and Helpers of America, 25 Louisiana Avenue, NW., Washington, D.C. 20001, R. V. Durham, Director, Department of Safety and Health

WTA—Wyoming Trucking Association, Inc., 109 Rancho-Avenue, Casper, Wyoming 82602, Larry E. Meredith, Managing Director.

The following are also referenced in this notice:

CIS—NIH/EPA Chemical Information System, CIS User Support Group, Computer Sciences Corporation, P.O. Box 2227, Falls Church, VA 22042, Katherine Noble, Project Manager

CHEMTREC—Chemical Transportation Emergency Center, Chemical Manufacturers Association, 2501 M Street, NW., Washington, D.C. 20037, Joe J. Mayhew, Director

NFPA—National Fire Protection Association, Batterymarch Park, Quincy, MA 02269, Robert W. Grant, President

AAR—Association of American Railroads, Washington, D.C., Thomas Phemister, Director, Bureau of Explosives

ERG—Emergency Response Guidebook, Materials Transportation Bureau, Research and Special Programs Administration, U.S. Department of Transportation, Washington, D.C. 20590, Alan I. Roberts, ERG Project Manager

As background to its Safety Recommendation I-83-2, issued November 29, 1983, NTSB stated the following:

About 11:00 a.m., e.s.t., on October 13, 1982, an eastbound tractor/cargo-tank semitrailer, owned and operated by Matlack, Incorporated, overturned when its driver took evasive action to avod a head-on collision with a westbound pickup truck with another pickup truck in tow that crossed the centerline on State Route 299 approximately one-fourth mile west of Odessa, Delaware. The tank-trailer contained 5,600 gallons of divinylbenzene (DVB), 150 gallons of which leaked from the tank through a clean-out cap and a pressure relief device in the dome. As a result of the accident, five persons were treated for injuries at a local hospital; four (including the two Matlack drivers) were released and one was admitted for further treatment. In addition, 48 emergency response persons were treated and released for respiratory problems and skin rashes associated with exposure to the DVB.

Police officers were notified of a highway accident but were not informed that a hazardous material was involved. Upon arrival, police and ambulance crews devoted their activities to site security and first-aid to the crash victims.

The first arriving police officers reviewed the shipping papers and then returned them to the driver. The shipping papers described the cargo as "5,600 gallons of COMBUSTIBLE LIQUID, not otherwise specified (n.o.s.) (Divinylbenzene, 55, Inhibited) NA 1993." The truck was properly placarded in accordance with Department of Transportation regulations.

Approximately 100 emergency response personnel responded to the accident, but none of them had either previous experience or formal training for handling a hazardous materials transportation accident.

About 1 hour after the crash, emergency response persons began complaining of respiratory and skin problems, and 48 of them were evacuated from the accident site and taken to a hospital for treatment, including the police officers who initially examined the cargo's shipping papers. About the same time, Matlack's drivers were also transported to a hospital and carried the shipping papers with them.

The emergency response personnel at the site knew the name of the cargo, but initially were unable to obtain information on its potential hazards and on the emergency response procedures to follow. When Matlack officials arrived on-scene and discovered that the shipping papers were not in the truck cab, a police official called the hospital where the drivers were being treated to obtain a verbatim reading of the papers. The papers confirmed the name of the cargo, but they did not contain emergency response guidance. The hospital treating the 48 emergency response persons did not have medical treatment information on DBV and substituted the medical treatment prescribed for benzene exposure.

The primary, on-scene reference material on hazardous materials was the Hazardous Materials Emergency Response Guidebook (Guidebook) 1980 edition, published by the Materials Transportation Bureau, Research and Special Programs Administration, U.S. Department of Transportation. Divinylbenzene is not one of the hazardous materials listed by name in the Guidebook.

Witnesses stated that emergency personnel looked for divinylbenzene in the Guidebook and, upon discovering that it was not listed, followed the response guidelines prescribed for divinyl ether—the only Guidebook entry with the term "divinyl." The entry for divinyl ether refers the reader to Guide 30, which first describes the material as a poison which may be fatal through inhalation, oral intake, or skin absorption, and second, capable of producing a spreading, flammable vapor. Information contained on the truck's placards (I.D. #1993, U.N. hazard class #3, and a flame symbol over a red background) directs the emergency response personnel to Guide 26 in the Guidebook; however, witnesses reported that the placards, although undamaged and unobscured, were not used to identify the cargo during the early stages of the incident. The appropriate guide, #26, first describes the various commodities in this group, including DVB, as capable of burning, and second, of producing vapors which may cause dizziness or suffocation as well as skin and eye irritation. The difference in the primary risk described in the two guides would explain the limited caution exercised by emergency response personnel.

The absence of shipping papers, the failure to observe placards, and the misuse or misunderstanding of the Hazardous Materials Emergency Response Guidebook reportedly combined to increase uncertainty among public officials, to protract the incident both in time and scope, and to lead to a reduced level of cooperation between emergency response personnel and carrier representatives. The lack of experience and training of those responding to this accident in handling hazardous materials and the lack of emergency response information specific to the transported hazardous material are by no means unique in highway accidents throughout the Nation.

Progress has been made over the last decade in providing information to emergency response groups on hazardous materials involved in accidents; however, critical information available to first arriving emergency personnel is still limited in many respects, especially for n.o.s. products. The Department of Transportation regulations require that hazardous materials shipments be placarded and accompanied by shipping documents. In addition, the Department has expended considerable time and effort in developing, updating, and disseminating nearly one-half million copies of its Hazardous Materials Emergency Response Guidebook for use by emergency response personnel. However, this Guidebook (for sound practical reasons) and the required shipping documents lack physical property data, medical treatment guidance, environmental precautions, and detailed hazard conditions which are specific to the commodity in transit. These data sources give emergency responders general information on the potential hazards during the fist 20–30 minutes into the accident. In most cases, however, additional references must be identified, quickly accessed, and used to determine the hazards and remedies for the specific commodity being transported.

Specific emergency guidance information is often not readily available when n.o.s. shipments are involved in accidents. For example, the I.D. number on the placard in this incident was "1993" which identifies divinylbenzene also applies to 17 other commodities or groups of commodities listed in the tables of 49 CFR 172. In addition, n.o.s. commodities are less likely to be included in commonly used emergency response guides than are specified commodities. Divinylbenzene, for example, in addition to not being listed in DOT's Hazardous Materials Emergency Response Guidebook, is not listed in the U.S. Coast Guard's Chemical Hazards Response Information System (CHRIS) manual, the Association of American Railroad's Emergency Handling of Hazardous Materials in Surface Transportation Guide, or the National Institute for Occupational Safety and Health/Occupational Safety and Health Administration (NIOSH/OSHA) "Pocket Guide to Chemical Hazards"—some of the most widely used guides. Divinylbenzene is listed in the Fire Protection Guide on Hazardous Materials, published by the National Fire Protection Association.

The Safety Board is aware that the development and distribution to emergency responders of an all purpose data sheet for every hazardous material subject to transportation would be a formidable and expensive task. Fortunately, information of this type already exists for most chemicals manufactured in this country. One source is the Material Safety Data Sheet which, in two pages, provides the manufacturer's name, physical property data, medical treatment, fire and explosion hazards, environmental protection guidance, protective measures, and other hazard information—all specific to the chemical in question. The OSHA recommends that a Material Safety Data Sheet be available at the workplace for each hazardous material which is handled there. An OSHA official reported that it is "rare" to find a manufacturing or shipping facility which does not follow this recommendation. The Safety Board is also aware that these documents are commercially available from a variety of sources. Moreover, one commercial service collects chemical and protective action data from a variety of government and industry sources and provides hard-copy information via a telephone-computer link to subscribers of the service.

The product-specific information available from such sources would be an asset to emergency response personnel if it were to accompany the shipping papers. Such information is widely used and readily available to shippers and manufacturers and could be supplied by carriers at a minimal cost per shipment. The Safety Board believes that use of this type of information as a part of the hazardous materials shipping documents which greatly benefit the effective handling of emergencies involving bulk shipments of hazardous materials. The ranking emergency response official at the Odessa incident obtained a divinylbenzene Materials Safety Data Worksheet on the second day of the emergency. According to the official, had the data sheet been available at the outset, considerable time would have been saved in identifying the cargo, the health effects from exposure to the hazardous materials, and the type of emergency activities necessary to respond to the accident.

Therefore, the National Transportation Safety Board recommends that the U.S. Department of Transportation, Research and Special Programs Administration: Determine, by mode of transportation, the feasibility of requiring comprehensive product-specific emergency response information such as Materials Safety Data Sheets, to be appended to shipping documents for hazardous materials transported in bulk quantities, giving particular attention to the early emergency response problems posed by n.o.s. commodities in transit. For those modes of transportation for which a positive determination results, incorporate necessary requirements into Title 49 of the Code of Federal Regulations. (Class 11, Priority Action) (I-83-2)

. . . .

The following basic health threat information is stated for divinylbenzene in NFPA's Fire Protection Guide on Hazardous materials (NFPA 49) Seventh Edition: LIFE HAZARD—Moderately toxic by inhalation. Eye and respiratory irritant. Effect on skin unknown but probably little, if any.

The following basic health threat information is taken from the Material Safety Data Sheet prepared by the manufacturer of the divinylbenzene involved in the accident referenced by the NTSB: TOXOCITY—Moderately toxic by inhalation. Irritating to eyes and respiratory tract. Effect on skin unknown, but considerd to be negligible.

The following basic health threat information is taken from Guide 26 of the 1980 Emergency Response Guidebook (DOT-P 5800.2): HEALTH HAZARDS—Vapors may cause dizziness or suffocation. Contact may irritate or burn skin and eyes. Fire may produce irritating or poisonous gases.

The following is the initial information on the basic health threat of divinylbenzene that is provided to a caller by CHEMTREC: Inhalation of vapors can cause irritation to eyes and respiratory tract. Contact with liquid can irritate skin and eyes.

The following is the basic health threat data for divinylbenzene contained in the CIS system: DIRECT CONTACT—Potentially [irritating] With Prolonged Contact Skin, Eyes. GENERAL SENSATION—Possible Dizziness Or Drowsiness From Vapors. Mild Eye Irritation But No Corneal Damage. Disagreeable Odor. ACUTE HAZARD LEVEL—Moderate Inhalative Toxicant. CHRONIC HAZARD LEVEL—Moderate Inhalation Hazard When Exposed Chronically At Sublethal Concentrations. Prolonged Skin Contact May Cause Irritation. DEGREE OF HAZARD TO

PUBLIC HEALTH—Moderated Inhalative Hazard From Both Acute And Chronic Exposures. 1.5–2 Cupsful Estimated Lethal Dose to Man [by ingestion].

By letter dated October 26, 1983. ATA petitioned for amendment to the Hazardous Materials Regulations as follows:

. . . .

Pursuant to the provisions of 49 CFR 106.31, ATA hereby petitions U.S. Department of Transportation (DOT) to require, by rule, motor carriers involved in the transportation of hazardous materials to maintain a copy of the Emergency Response Guidebook (Guidebook), DOT P5800.2 at each motor carrier facility where hazardous materials shipments are loaded or unloaded from vehicles.

In its introduction, the Guidebook explains that it "was developed for use by firefighters, police and other emergency response officials as a guide for initial actions to be taken to protect themselves and the public when . . . handl[ing] incidents involving hazardous materials." Although the Guidebook was designed primarily for use at the location of incidents occurring on highways or on railroads, the introduction goes on to explain DOT's belief that the Guidebook should also be of value in handling incidents occurring at terminal facilities. The motor carrier industry concurs.

Without question, the Guidebook provides essential information about each of the materials classified by DOT as hazardous. This information is set forth in the Guidebook in a comprehensive, yet efficient and practical format. The Guidebook transforms the DOT's numerical hazardous (sic) identification system into an effective emergency response system, by providing information that can be quickly located, easily understood, and utilized by any response personnel in any emergency.

Experience has shown that the majority of hazardous materials incidents in the motor carrier industry have occurred at terminal locations; that incidents, in general, occur infrequently; and, that most of these have involved minor spills. Nonetheless, the benefits which emergency response personnel have realized at highway and rail incidents by using the Guidebook—a fact recently testified to by the International Association of Fire Chief[1]—should be no less great for incidents occurring at terminals. The Guidebook should prove invaluable during the initial phases of incident response, and it should facilitate both prompt and effective first aid to any terminal worker who may accidentally come into contact with a hazardous material.

At present, many motor carriers have already taken the initiative by voluntarily making the Guidebook available to their terminal workers. However, a DOT requirement that it be made available at all terminal locations will be of industry-wide significance, benefiting the public at the same time.

The DOT's promulgation of this rule will address another industry concern as well. Currently, a growing number of jurisdictions have already required, or are considering requiring, motor carriers to make material safety data sheets available at their terminals.

The necessity that motor carriers maintain and make available a material safety data sheet for each hazardous material that may in the course of a year pass through their terminals poses a substantial burden upon carriers. It does so without realizing benefits in addition to those which the Guidebook already can achieve. The essential emergency response information, and first aid and health risk information contained in each is virtually the same. However, the contents and format of material safety data sheets is not standardized. It can and, indeed, does vary with each manufacturer preparing one. Further, not every chemical manufacturer currently prepares data sheets for its products, nor is a data sheet prepared for every material manufactured. In addition, most data sheets are keyed to the product's trade name rather than to the DOT's proper shipping name. When it is considered that the motor carrier workers, for whom the material safety data sheets are purportedly intended, are already familiar with the DOT proper shipping name, the placarding and labeling requirements, and all of the other relevant DOT regulations, it seems clear that the material safety data sheets would not be nearly as functional as the Guidebook, especially in an emergency setting. At the same time, the noninclusion of highly technical data in the Guidebook—such as freezing and boiling points; vapor density; vapor pressure; specific gravity; and viscosity—makes the Guidebook more suitable for effective use by lay persons.

Accordingly we urge DOT to adopt this proposed regulations, and that it rule as inconsistent, or otherwise prohibit, the requirements of State and local governments that motor carriers must maintain and make available at their terminal locations material safety data sheets or similar informational guides (excluding the Guidebook) for hazardous materials being transported in commerce.

. . . .

By letter dated November 7, 1983, ATA supplemented its petition as follows:

. . . .

---

[1]Proposed Amendments to the Hazardous Materials Transportation Act: Hearing on S. 1108, Title IV, Before the Senate Committee on Commerce, Science and Transportation, 96th Cong. 1st Sess. (1983).

On October 26, 1983, American Trucking Associations, Inc. (ATA) petitioned the Materials Transportation Bureau (MTB) regarding the DOT's Emergency Response Guidebook. We asked MTB to adopt a rule requiring "motor carriers involved in the transportation of hazardous materials to maintain a copy of the . . . Guidebook . . . at each motor carrier facility where [such] shipments are loaded or unloaded from vehicles," and to "rule as inconsistent, or otherwise prohibit, the requirements of State and local governments that motor carriers must maintain and make available at their terminal locations material safety data sheets or similar informational guides (excluding the Guidebook) for hazardous materials being transported in commerce."

The purpose of this letter is to supplement our October 26th petition, by clarifying the scope of the relief we requested therein.

As we discussed in our petition, a growing number of State and local governments are requiring that motor carriers maintain and make available to their terminal workers specific information pertaining to chemical and physical properties of and emergency response and first aid information for each hazardous material motor carriers transport. Generally, this information is being required of carriers in the form of a "material safety data sheet."

Pursuant to these State and local laws, a motor carrier's more of several enumerated activities. These activities have included: "transporting"; "distributing"; and "handling." Additionally, many jurisdictions also regulate "storage."

Without question, motor carriers must perform one or a combination of these activities ("transport", "distribute", "handle"), in the normal course of transporting hazardous materials in commerce. Semantics notwithstanding, as a practical matter, such activities (including loading and unloading) embrace the integral functions of a transportation movement.

For the same reason such temporary stoppages in transit as the coming to rest of a container (non-bulk) of a hazardous material in a carrier's terminal while it awaits loading into a vehicle, or the temporary parking in a carrier's terminal area of a fully or partially loaded tank truck prior to its outbound movement should be defined as "transportation." As such, both the activity and the location where they occur should be subject to the Department of Transportation's jurisdiction. These types of temporary cessations in the transportation movement should not be classified as "storage."[2]

We want to emphasize that, in petitioning for this rule, the motor carrier industry's intent is not to avoid its responsibility to provide pertinent safety information concerning hazardous materials to its workers. We believe, however, that such information, covering the host of DOT-regulated hazardous materials, is already available through the Guidebook, where it is provided in a format which is provably more effective and efficient than that of material safety data sheets.

. . . .

By letter dated December 29, 1983, WTA stated the following:

. . . .

The Wyoming Trucking Association, Inc. endorses the petition filed with your Department by the American Trucking Association, Inc., to require motor carriers involved in the transportation of hazardous materials to maintain a copy of the emergency response guidebook at each facility where hazardous materials are loaded or unloaded from vehicles.

The ATA petition requests that the DOT pre-empt state and local requirements for material data sheets.

Wyoming has many points where hazardous materials are loaded and unloaded, but as a bridge state many more loads cross with closed doors.

. . . .

By letter dated January 27, 1984, the acting Chief Counsel, Research and Special Programs Administration (RSPA) stated the following in response to the WTA letter:

. . . .

As the ATA notes in its petition, and as you reiterate in your letter, one purpose of such a requirement is to overcome or forestall the imposition by state or local governments of requirements that motor carriers maintain material safety data sheets for each hazardous material received or shipped at each terminal.

In accepting your letter, and docketing it as a comment on the ATA petition, I wish to point out that the acceptance of the ATA petition, or any subsequent rulemaking arising out of it, would not in and of itself represent

[2]The United States Environmental Protection Agency agrees with this proper distinction between "transportation" and "storage." Pursuant to its regulation, 40 CFR 263.12, a carrier's holding of a waste in a specification container for a period of ten days or less at any one location does not constitute storage.

the preemption of any current or future state or local requirement. Under provision of the Hazardous Materials Transportation Act (HMTA) (49 U.S.C. 1812) and the Regulations of the MTB (49 CFR 107.201–107.225), the preemption of a state or local requirement occurs upon a finding that the requirement is inconsistent with the HMTA or a regulation issued thereunder. The administrative process (or, if a party challenging the state or local requirement elects, the judicial process) is separate and distinct from the rulemaking process invoked by the ATA petition. Consequently, the question of the preemption of state and local requirements mandating the use of safety data sheets, would not be relevant to any rulemaking action that might arise from the ATA petition.

. . . .

By letter dated January 13, 1984, the IBI stated the following:

. . . .

It has come to our attention that the American Trucking Association (ATA) petitioned the Materials Transportation Bureau (MTB) on October 26, 1983, to require motor carriers involved in the transportation of hazardous materials to maintain a copy of the DOT Emergency Response Guidebook at each facility where hazardous shipments are loaded and unloaded. In it's petition (P-922), ATA requested that the DOT requirement preempt state and local requirements for Material Safety Data Sheets (MSDS).

While we support the effort to require motor carriers to maintain copies of the Emergency Response Guidebook, we are concerned about DOT preemption of state and local laws requiring the maintenance of MSDS at transportation facilities. We believe that a DOT regulation requiring the use of the guidebook cannot preempt a state or local law requiring a MSDS because each document provides significantly different information, and therefore a different benefit, to employees involved in the transportation of hazardous materials.

The DOT Emergency Response Guidebook was developed in 1980 for use during the initial stages of a transportation emergency. The guidebook classifies hazardous materials by shipping name and provides acute health hazard information. While MSDS do offer information on acute health hazards and emergency action, unlike the DOT guidebook, however, MSDS provide information on chronic and long latency health effects from exposure to a hazardous material. The differences between the Guidebook and the MSDS can best be illustrated by examining the treatment of a particular hazardous material, benzene.

Benzene solvent is produce in billion gallon quantities per year. Scientific data strongly suggest that benzene is a human carcinogen. The DOT guidebook lists the health effects of benzene in terms of acute health effects. The guidebook states that benzene: "Vapors may cause dizziness and suffocation. Contact may irritate or burn skin and eyes. Fire may produce irritating or poisonous gases. Runoff from fire control or dilution may cause pollution".

The chronic or long latency health effects associated with benzene exposure are not mentioned in the DOT handbook.

Similar to the guidebook, an MSDS for benzene would note the acute effects linked to benzene exposure. In contrast, however, an MSDS would describe chronic and long latency effects of benzene exposure in terms of leukemia, lymphatic and hematopoietic cancer. We believe that workers transporting hazardous materials have a right to be informed of the chronic health effects associated with exposure to shipped materials as well as the acute health effects. The DOT guidebook alone does not offer complete health hazard information.

In addition to the discrepancy betweeen the guidebook and the MSDS in terms of the completeness of chronic and long latency health effects information, the documents differ in their treatment of chemical mixtures. MSDS generally include health hazard information on all chemicals present in concentrations greater than 1% in a hazardous chemical mixture. In contrast, the guidebook provides health effects information for the major constituent of a chemical mixture, leaving the other minor constitutents unnamed and unaddressed. We believe that the health effects associated with exposure to all constituents of a chemical mixture should be made known to employees transporting hazardous materials. This can only be accomplished through the use of an MSDS.

In summary, MSDS generally provide more thorough health hazard information in terms of chronic health effects and chemical mixture information than the DOT guidebook. On balance, the DOT guidebook is particularly useful in emergency incidents. Since the documents are useful for different purposes, we feel it is inappropriate to substitute the guidebook for an MSDS. Instead, we recommend that the shipper be required to maintain both the guidebook and MSDS at all facilities involved in the transportation of hazardous materials. Clearly, the benefits of employee access to complete health hazard information on hazardous materials in transportation outweigh any burden placed on the carrier to obtain and maintain the guidebook and MSDS documents, documents readily available from DOT and chemical manufacturers/shippers, respectively.

. . . .

In its comments on proposals made under Docket HM-126A (44 FR 32972; June 7, 1979), the IBT stated the following:

. . . .

More than any other group, Teamster members bear the brunt of the inherent dangers involved in the transportation of hazardous materials. Teamster members package, ship, transport, and receive a major share of all hazardous materials moving in interstate (and intrastate) commerce. Transportation workers stand alone as the vital first link in dealing with hazardous materials incidents. Their actions in the first minutes following an incident may well determine if that incident will remain minor or result in a major catastrophe.

This NPRM recognizes, but does not act upon the fact that transportation workers must be adequately prepared and equipped with the knowledge necessary to prevent this type of catastrophe. If the proposed identification system was accompanied by an adequately prepared, and evenly distributed, Emergency Response Manual," it would go far in improving the emergency response capabilities at all levels of response, including the initial "person at the scene" (i.e., the transportation worker), the responding emergency service personnel, and the special assistance personnel who may be required.

A properly distributed "Emergency Response Manual" would also go far in correcting one of the most neglected aspects of hazardous materials transportation, the health and safety of the transportation workers themselves. Before transportation workers can protect other potential victims, they must first be able to protect themselves. This requires a knowledge of the nature of the hazard and explicit information on how to respond in those first critical minutes. An uninformed, incapacitated, or dead transportation worker will be of little or no use to other potential victims or later, to the arriving emergency service personnel.

. . . it is imperative that the proposed identification system be released only if it is in conjunction with a cross-referenced "number vs. technical name" table, and only if MTB can assure that all likely recipients of emergency calls have such a table at hand. A more practical alternative, and one much more appropriate to the overall purpose of the system, namely, improved emergency response capability, would be the concurrent release and distribution of the cross-referenced "Emergency Response Manual," which MTB has already largely developed. Furthermore, this approach would be more cost-effective in that the necessary cross-referenced table would not have to be published and distributed twice (i.e., once with the number identification rule, and once with the future "Manual").

Acknowledged throughout the text of this NPRM (HM-126) is the fact that the first minutes of a hazardous materials incident are the most critical. Also undisputed is the fact that the actions of the transportation worker at the scene in those critical minutes may mean the difference between a minor incident and a major catastrophe. Based on this, we most strongly urge that distribution of the Manual to certain personnel be made mandatory in the final rule. Mandatory distribution might be achieved in the following manner (for trucking personnel):

1. Operators of vehicles transporting hazardous materials:

a. Single Commodity—The driver would receive only a single "response guide" from the Manual (with the emergency phone number) for the single type of hazardous material being transported.

b. Mixed Commodity—The driver would receive "response guides" from the Manual for any type of hazardous material which he may haul as well as the "identification number vs. guide number" cross-reference table at the front of the Manual. A driver would not need the lengthy alphabetical listings of all the technical names versus the identification and guide numbers.

2. Shipping, receiving, the warehouse personnel—copies of the full Manual, with cross-referenced lists and all the Emergency Response Guides, would be required in central locations and at any office likely to receive a hazardous materials emergency call.

Similar requirements would, of course, be required for other modes of transportation and specialists in those industries should be consulted as to specific needs.

The material quoted above is provided as background to this notice which is a solicitation of facts and viewpoints rather than a proposal to take any specific action. MTB solicits comment on the following:

1. What material specific information is provided by a MSDS that would mitigate the potential consequences of a discharge beyond the type of information provided by the ERG and CHEMTREC, and how quickly would that information be needed? In commenting, please take into account that the information on file at CHEMTREC is based on MSDSs provided by manufacturers and that CHEMTREC can provide information that is not contained in the ERG e.g., flash-point, boiling point, flammable limits, and vapor density. Also, CHEMTREC has access to shippers and the CIS for more detailed information on hazardous materials. If comments are presented concerning the value of TLV (threshold limit value) data, it is requested that supporting information be provided in support of how such data (TWA—time weighted average; STEL—short-term exposure limit; C—ceiling) can be effectively applied in the transportation environment. For example, what type of monitoring equipment could be reliably used to make an assessment of a spill area? Should MTB imply that confidence may be placed in use of such equipment? Up to the present time, it has been MTB's opinion that this approach would not be appropriate; therefore, current ERG guidance for any cargo (not only regulated hazardous materials) is "Move And Keep People Away From Inci-

dent Scene; Do Not Walk Into Or Touch Any Spilled Material; Avoid Inhaling Fumes, Smoke and Vapors Even If No Hazardous Materials Are Involved; Do Not Assume That Gases Or Vapors Are Harmless Because Of Lack Of Smell".

2. (a) Should DOT consider discontinuing distribution of the ERG in favor of MSDSs accompanying shipments of hazardous materials? (b) Should consideration of MSDSs be limited to bulk shipments as suggested by NTSB? In commenting, please consider the possibility of undesirable results in applying both systems to transportation, e.g., the different language contained in basic health threat information (as demonstrated above for divinylbenzene) as well as differing response information. In preparing for issuance of this notice, MTB reviewed 29 CFR 1915.97 relative to preparation of U.S. Department of Labor Form OSHA 20 and the Occupational Safety and Health Administration's (OSHA) final rule amending 29 CFR Part 1910 (48 FR 53280; November 25, 1983). The information specified for inclusion in MSDSs (§ 1910.1200(g)) does not require manufacturers and importers to use standard language for either the communication of risk or the mitigation of risk. To a significant degree, this is overcome by training (§ 1910.1200(h)) required to be given by employers in Standard Industrial Classification Codes 20 through 39. (c) To what extent could and should DOT rely on training of emergency response and transportation personnel in use of MSDS information rather than the ERG, taking into account that more than 180,000,000 shipments of hazardous materials are made annually in the United States?

3. If, following review of the comments on this notice, MTB decides to propose a mandatory placement of ERG's in transportation facilites: (a) How should MTB describe (define) those facilities in the regulations? (b) Should ERG's be required in vehicles used to transport hazardous materials, as suggested by IBT? (c) What would be the means of acquisition of the ERG's? (d) How much time should be provided for acquisition and implementation? (e) Could such a requirement be implemented without having an effect on necessary revisions and updates of the ERG? (f) In order for MTB to assess the cost of such a program in a regulatory analysis, how many vehicles (including rail), vessels, aircraft, and terminal facilities would be subject to such a requirement (taking into account the last quoted paragraph of IBT's comments above)? The following information is provided for background: There were 750,000 copies of the 1980 ERG (DOT 5800.2) delivered, without charge, by MTB to emergency response (and associated) organizations between 1981 and 1984 and more than 200,000 obtained from commercial sources. More than 600,000 copies of the 1984 ERG (DOT 5800.3) have been distributed by MTB since December 1, 1983. While DOT's distribution costs have been less than $1.00 per copy, the charge at the Government Printing Office (GPO) for the ERG was set at $7.00 per copy. Four commercial sources of the ERG base their prices on quantities ordered. It is contemplated that the ERG will be revised and redistributed at three year intervals if the program is continued following this proceeding.

4. (a) Is there another way to deal with " ... emergency response problems posed by n.o.s. commodities ... " as discussed by NTSB in Recommendation I-83-2? On May 22, 1980 MTB published a final rule under Docket HM-126B (preamble page—45 FR 34565) setting forth requirements for more specific identification of poisons, including those covered by n.o.s. entries in § 172.101. The purpose of the rule, which is set forth in § 172.203(k), is to make identification of poisons more specific for immediate response purposes. (b) Should MTB consider expanding the requirements to hazardous materials of all classes? Commenters should note that the present rule does not require the technical names of compounds or principal constituents if the entry on a shipping paper (in association with the n.o.s. entry coming from § 172.101) is a name in the NIOSH Registry (RTECS—Registry of Toxic Effects of Chemical Substances) which contains more than 59,000 substance entries. The reason for providing this option is the problem emergency response personnel could have in dealing with long and complex chemical names (with dozens of letters and numbers in some cases) and the fact that RTECS is a component of the NIH/EPA CIS computer system that may be accessed by CHEMTREC at any time specific identification of a material is necessary. At the time the rule was promulgated. MTB had determined that it was only essential for materials meeting the definition of a class B poison (regardless of class precedence). Also, a different rule for identification of hazardous substances in mixtures was issued at the same time under Docket HM-145B. (c) What would be the burden of such a requirement? and (d), Can or should such a requirement be construed as deriving the same benefit as possession of a MSDS during transportation?

(49 F.R. 10048, March 16, 1984)

SUMMARY · MTB-1 · *Public notice of the establishment of the Materials Transportation Bureau as successor to the Hazardous Materials Regulations Board; establishment of rule making procedures in 49 CFR Part 102; supersedes HM-1.*

[MTB Docket No. 1]
ESTABLISHMENT OF MATERIALS TRANSPORTATION BUREAU AND RULEMAKING PROCEDURES

The purpose of this amendment is to restructure Chapter I of Subtitle B of Title 49 of the Code of Federal Regulations to reflect the establishment of the Materials Transportation Bureau within the Department of Transportation and to establish rulemaking procedures applicable to the activities of that Bureau, including those pertaining to hazardous material and pipeline safety.

This amendment makes a series of technical changes to conform the overall format of Chapter I to the new organizational structure and to provie an arrangement that will accommodate future regulatory issuances. Primary among such future regulatory issuances will be the consolidation, reorganization and reissuance of the hazardous materials regulations now appearing in 14 CFR Part 103, 46 CFR Part 146, and 49 CFR Parts 170 through 179. To facilitate this effort, a separate subchapter is being established in advance.

It also establishes a new Part 102 which describes the procedures applicable to the Bureau, including the Offices of Hazardous Materials Operations and Pipeline Safety Operations, in prescribing public rules and provides for appropriate participation by interested persons. All pending rule-making proceedings initiated by the Hazardous Materials Regulations Board under Part 170 procedures or by the Office of Pipeline Safety under Part 5 procedures will be continued by the Bureau under new Part 102 procedures.

The new Part 102 provides for general notices of proposed rule making, to be published in the FEDERAL REGISTER, except in cases where the Director finds that notice is impractical, unnecessary, or contrary to the public interest, and except for interpretive rules, general statements of policy, and rules relating to Bureau organization, procedure, or practices. The authority to conduct rule-making proceedings and to issue final rules may be delegated to other officials in the Bureau. Delegations authorizing the Directors of Hazardous Materials Operations and Pipeline Safety Operations to conduct rule-making proceedings within their respective areas of response, but not including the issuance of final rules, are set forth in Appendix A of the new part.

The new part also provides for the consideration of petitions for rule making, petitions for reconsideration of adopted rules, and petitions for extension of time to comment on notices of proposed rule making.

The opportunity for an informal hearing is a required procedure with respect to certain rule-making functions administered by the Bureau. Even when they are not required by statute, hearings may be held as a supplementary fact-finding procedure, whenever it is considered necessary or desirable. Any hearing, whether or not required to be held by statute, will be nonadversary, with no formal pleadings and no adverse parties, and any resultant rule will not necessarily be based exclusively on the record of the hearing.

All final rules will be published in the FEDERAL REGISTER, unless, in accordance with section 552(a) of Title 5, United States Code, actual and timely notice has been given to all persons subject to it.

Since this amendment relates to Bureau organization, procedures, and practices, notice and public procedure thereon is unnecessary and it may be made effective in less than 30 days after publication in the FEDERAL REGISTER.

(40 F.R. 31767, June 29, 1975)

SUMMARY • Docket No. HM-127 • *Amendments establishing the Materials Transportation Bureau's exemption or waiver procedures under the Hazardous Materials Transportation Act, replacing the special permit program and waivers formerly issued by the FAA.*

[Docket No. HM-127]
EXEMPTIONS FROM HAZARDOUS
MATERIALS REGULATIONS
Interim Procedures

Scope. These procedures apply to all applications for exemptions, including waivers, deviations or special permits, and all renewals or extensions thereof, with respect to the following bodies of hazardous materials regulations:

1. 14 CFR Part 103 governing transportation of dangerous and magnetized materials by aircraft.

2. 46 CFR Part 146 governing transportation or storage of explosives or other dangerous articles or substances and combustible liquids on board vessels.

3. 49 CFR Parts 172 through 179 governing the transportation and shipment of hazardous materials by rail and highway.

Background. Effective July 7, 1975, the authorities to issue, modify, or revoke the regulations governing the transportation of hazardous materials by air, highway, rail and water and bulk transportation (other than ship's stores and supplies and the bulk transportation of hazardous materials which are loaded or carried on board a vessel without benefit of packages or labels and received and handled by the vessel without mark or count) and the authorities to issue, modify, or revoke exemptions therefrom, previously exercised by the Federal Aviation Administration, the Federal Highway Administration, the Federal Railroad Administration and the Coast Guard, were consolidated and transferred to the newly established Materials Transportation Bureau. The Materials Transportation Bureau has the responsibility for developing and issuing Department-wide procedural regulations for the implementation of section 107 of the Hazardous Materials Transportation Act (Pub. L. 93–633), which establishes significant new procedural requirements in connection with administrative handling and issuance of hazardous materials exemptions. These new requirements include public proceedings, the issuance of FEDERAL REGISTER notices and the opportunity for public comment on nearly all applications for exemptions.

In addition to the new procedural requirements established by the Hazardous Materials Transportation Act, the United States District Court for the District of Columbia has issued an order requiring, among other things, immediate implementation of similar procedural requirements. The United States Court of Appeals for the District of Columbia has granted a temporary stay of the order. The Government has requested an extension of that stay pending appeal of the order. The Government's request was based on a commitment that the new Materials Transportation Bureau would issue procedural regulations by October 15, 1975, fully implementing the new requirements. In this connection it is anticipated that a notice of proposed rule making will be issued on or about July 30, followed by a public comment period of 30 to 45 days.

In view of the centralization and transfer of the authority to issue exemptions and the plan for implementation of the new procedures, it is necessary to establish interim procedures which will minimize the confusion that applicants for exemptions are likely to experience during the conversion period and will avoid unnecessary disruptions in essential movements of goods covered by such exemptions.

General Procedures. Except in the case of air transportation applications requiring priority treatment, applications for exemptions, waivers, deviations, or special permits which would have been submitted for processing under 14 CFR 103.5, 14 CFR 11.25 (with respect to the requirements of 14 CFR Part 103), 46 CFR 146.02–25, or 49 CFR 170.13–15 should continue to be prepared in accordance with those sections. However, they should be filed with the Director, Office of Hazardous Materials Operations, Materials Transportation Bureau, Washington, D.C. 20590, for his issuance or denial with a duplicate copy filed as indicated in 14 CFR or 46 CFR, as appropriate.

Air Transportation Applications Requiring Priority Treatment. The Federal Aviation Administration, acting in accordance with the procedures in 14 CFR 103.5, will continue to receive and act on applications submitted under that section for authority to deviate from the requirements of 14 CFR Part 103 when such a deviation is necessary for the expeditious transportation of articles for the protection of life or property.

Termination Date: These interim procedures are cancelled, effective October 16, 1975.

(40 F.R. 31976, July 30, 1975)

[49 CFR Parts 107, 170]
[Docket No. HM-127; Notice No. 75–7]
HAZARDOUS MATERIALS REGULATIONS
Proposed Exemption Procedures

The purpose of this notice is to propose an amendment to Title 49, Code of Federal Regulations, to establish a new Part 107 "Procedures" and place therein a new Subpart A "Exemptions". Future subparts to be included in Part 107 would cover sanctions, registration, and preexemption matters. The regulations now being proposed would prescribe procedures to be followed in applying for and the processing of applications for exemptions from the Materials Transportation Bureau's regulations governing the transportation of hazardous materials which, except for bulk transportation by water and certain ships' supplies, govern the movement of such materials by any mode of transportation.

Section 107 of Pub. L. 93–633, enacted in January 1975, sets up certain procedural requirements for granting of administrative relief from the regulations controlling the transportation of hazardous materials to be issued under section 105 of that law. Under § 107, applicants must establish on the basis of a safety analysis to be submitted by them as part of their application that a level of safety will be achieved which will (1) be equal to or higher than that reached by following the regulations, or (2) if there is no existing level of safety, be consistent with the public interest and the policy of Pub. L. 93–633. Unless it is shown that an emergency requires otherwise, each application is to be made the subject of a FEDERAL REGISTER notice and made available for public inspection and comment before action is taken to grant or deny it. The maximum period for an exemption is two years. Any application for renewal must be processed in the same manner as an original application.

An exemption, as that term is used in section 107 and in the proposed regulations, includes all forms of administrative relief from the requirements of substantive regulations. Until July 7, 1975, when the authority to issue such relief was centralized in the Materials Transportation Bureau, it was granted by one of a combination of four of the Department of Transportation's operating administrations (U.S. Coast Guard, Federal Aviation Administration, Federal Railroad Administration, or Federal Highway Administration) under various labels including waiver, exemption, deviation and special permit. Moreover, the relief was requested, processed and granted or denied under multiple procedural schemes. (See 14 CFR 11.25, 14 CFR 103.5, 46 CFR 146.02–25 and 49 CFR 170.13.) For example, a person seeking relief from a regulation in 14 CFR Part 103 pertaining to air commerce could (1) petition the Federal Aviation Administration for an exemption under 14 CFR 11.25, (2) apply to the Federal Aviation Administration (or in an emergency to a Flight Standards District Office) for authority for deviation, or (3) petition the Hazardous Materials Regulations Board (consisting of representatives of the four operating administrations) for a special permit for a waiver or exemption. The proposed regulations would be the exclusive body of procedures for all such future administrative relief actions.

The proposed regulations would describe in general terms, what must be included in applications for exemptions and specify that all applications are to be filed with the Bureau's Office of Hazardous Materials Operations. Except in those situations in which an applicant asserts that an existing emergency requires priority processing, applications would have to be filed at least 120 days before the proposed effective date. When an applicant asserts that an existing emergency requires priority processing, and supplies supporting facts and reasons, he would have the choice of applying in the normal fashion by written submission or he could apply by telephone with a follow-up written filing within 15 days.

Upon receipt of an application it would be reviewed for completeness and docketed if complete or returned to the applicant if incomplete. If it is asserted by the applicant that there is an existing emergency, an immediate examination of the supporting facts and reasons would be conducted and a determination made as to whether or not an emergency justifying priority processing did in fact exist. If determined that it did not exist, the applicant would be so notified and the application would be processed on a routine basis. If determined that an emergency did exist, the application would be processed on a priority basis in which case the usual public proceeding (FEDERAL REGISTER notice and opportunity for public comments) could be omitted.

The bases upon which determinations of existing emergencies would be made are set forth in proposed § 107.9.

Renewal applications would be submitted in the same manner as original applications and would be required to meet the same requirements. Under the proposed regulations the Bureau would publish a notice of each application in the FEDERAL REGISTER (probably a weekly or bi-weekly compilation in digest form) and afford an opportunity to comment. If the application is granted, notice of the granted, would also be published.

No public hearing, argument or other formal proceeding would be held on an application before its disposition. Any interested person could, upon written request, appear before an appropriate official of the Bureau to discuss an application or its denial. Applications, including their related safety analyses, and public comments would be available for inspection in the Bureau's public docket room.

An exemption would be terminated upon a determination, after notice to the holder, that it was no longer consistent with the public interest; that it is no longer needed because of a change in the regulations; or that it was granted on the basis of false, fraudulent or misleading representation or information. Likewise, an exemption could be suspended upon a determination that it was not being exercised in accordance with its terms or that new information requires it to be amended to adequately protect life and property.

The proposed effective date is October 16, 1975.

Interested persons are invited to submit views and comments on the proposal. A public hearing will be held for that purpose at 9:30 a.m. on August 26, 1975, in the third floor auditorium of Federal Office Building 10A (commonly referred to as the FAA Building) located at 800 Independence Avenue, SW., Washington, D.C. Interested persons not desiring to present oral presentations are invited to submit their comments in writing. Comments should refer to the docket number and be submitted to: Docket Section, Materials Transportation Bureau, U.S. Department of Transportation, Trans Point Building, Washington, D.C. 20590. All comments received before the close of business on September 12, 1975, will be considered, and will be available in the docket for examination both before and after the closing date. Comments received after the closing date and too late for consideration will be treated as suggestions for future rule making.

(40 F.R. 32758, August 4, 1975)

[Docket No. HM-127]
**HAZARDOUS MATERIALS
TRANSPORTATION**
Establishment of Exemption Procedures

On August 4, 1975, the Office of Hazardous Materials Operations published in the FEDERAL REGISTER (49 CFR 32758) a notice of proposed rule making in which it proposed a new set of procedures to be followed in applying for and in the processing of requests for exemptions from the Department of Transportation's regulations governing the transportation of hazardous materials. These procedures would implement the statutory requirements of section 107 of the Hazardous Materials Transportation Act (Title I of Pub. L. 93–633).

As provided in that notice of proposed rule making, interested persons were given until the close of business on September 12, 1975, to submit comments and a public hearing was held in Washington, D.C., on August 26, 1975. In response to the considerable interest expressed by persons shipping and transporting hazardous materials via aircraft in or to Alaska, a second public hearing was scheduled for Anchorage (40 FR 37247, August 26, 1975) and held there on September 2, 1975. The Materials Transportation Bureau has given due consideration to the comments received and has adopted the proposal with the modifications set forth below.

GENERAL ARRANGEMENT

As proposed the exemption procedures would have been Subpart A of Part 107. The Bureau has decided to move them to Subpart B and use Subpart A for general procedural provisions applicable to the hazardous materials program. A section on definitions and one covering request for confidential treatment of documents are included in this rule making. Other general procedural provisions will be added to Subpart A as they are developed. For the convenience of those who may wish to make a direct comparison between the proposed provisions and the final provisions, a table of comparative section numbers is set forth below:

Proposed §	Final §
107.1	107.101.
107.3	107.101.
107.5	107.103, 107.105, and 107.113.
107.7	107.107.
107.9	107.115.
107.11	107.109.
107.13	107.117.
107.15	107.119.
107.17	107.123.

PROCESSING PROCEDURES

Several comments addressed the proposed provision that would have allowed any interested person to meet informally with a Bureau official to discuss an exemption application or the action taken on an application. For the most part those comments expressed concern that this would allow the presentation of adverse information and arguments without opportunity for rebuttal. Some of the comments went on to suggest remedial safeguards, which tend to be formal, adversary and time-consuming.

The Bureau shares the concern expressed in those comments. It was this concern that accounts for the requirement in the proposed regulations which would have required a memorandum of all such meetings to be placed in the public docket. After reflecting on the comments, the Bureau is less than convinced as to the adequacy of the memorandum to docket provision.

For these reasons and in light of the opportunity being provided for interested persons to present their views on individual exemption applications through the public procedure process being implemented, the Bureau has decided not to adopt the proposed informal meeting provision. All comments received by the Bureau regarding an application will be available in the docket to other interested persons who comment, the applicant or any other person for such rebuttal as any of them may desire to submit.

Recommendations asking that all interested persons who comment on an application be required to serve copies of their comments on the applicant and those suggesting that the regulations should prescribe the form and style of comments are rejected as totally inconsistent with the statutory requirements governing the opportunity that must be offered the public for expressing their views informally on each application. As noted in § 107.123, the applicant or any other member of the public may obtain copies of any docketed document.

Several comments suggested that the regulations should set the public comment period on applications at 30 days. While the Bureau feels that 30 days will probably be the appropriate comment period for many, if not most, applications, it does not believe that the flexibility allowed by statute in this regard should be negated by a rigid regulation.

In finalizing these procedures, the Bureau is transferring into Appendix B the standard conditions previously made applicable to special permits by 49 CFR 171.6 where they will be applicable to all exemptions to be issued under these new procedures. Based on the recommendations of several commentors, it has also included in Appendix B and is making applicable to all future exemptions for carriage of hazardous materials by aircraft, a restatement of the standard terms previously contained in 14 CFR 103.5.

The proposed provision stating that the Director, OHMO, in acting on an application "may initiate rule making . . . in addition to or in lieu of granting or denying the application" drew diverse comments. Suggestions were made that the words "or in lieu of" should be dropped; that exemptions should automatically become regulations after two years or after having been once renewed; and that all existing exemptions as well as applications should be reviewed with a view to rule making. The Bureau appreciates the need for codifying into permanent regulations those exemption-tested concepts proven to be safe. (See Docket HM-128, 40 FR 45197, October 1, 1975). It feels, however, that this objective is not to be achieved by prescribing a regulation commanding it to occur. Rather, its attainment is a function of administrative effort and realistic, properly balanced procedures. For these reasons, the Bureau is adopting this provision as proposed.

PROCESSING TIME

Comments from a number of holders of special permits and existing exemptions and from others who are potential future applicants for exemptions contested the proposed provision which would have required all applications (except those seeking priority treatment on the basis of an existing emergency) and all applications for renewals to be submitted at least 120 days before the requested effective date or expiration date. Those comments suggested shortening the period to 45, 60, 75, 80, or 90 days. In a related comment it was recommended that should the Bureau adopt the mandatory 120-day advance filing date, it should also allow for priority processing of an application which may not qualify as an emergency but for which there are compelling reasons for expedited handling after the public comment period has closed.

As discussed elsewhere in this preamble, a new section has been added governing all emergency exemptions. Also, as discussed elsewhere, the mandatory lead time of 120 days for renewal applications has been reduced to 60 days. With respect to original applications, the Bureau feels that, as a general proposition, it must have the 120-day period available to conduct the statutorily required public proceedings and to evaluate the application, its safety analysis and the public comments thereon. This is not to say that the Bureau is going to consume the full 120 days in the processing of each application. Each will be processed as expeditiously as practicable.

In view of these comments and considerations, § 107.103(b)(10) and (c) pertaining to non-emergency, original applications have been changed so as to provide the following. The provision dealing with the 120-day advance filing date has been modified from mandatory to advisory by changing the word "must" to "should". A provision has been added stating that applications are processed in the order received unless the Director, OHMO, is persuaded by information the applicant may submit in his application that priority processing is called for after the public proceeding on the application is completed.

Several comments asserted that the proposed provision stating that the administrative review of applications for completeness and conformity would be made within 30 days after its receipt should be changed by reducing that period or making it run concurrently with the public comment period.

As discussed elsewhere, the period has been reduced to 15 days for renewal applications. With respect to original applications, as in the case of the total 120-day processing period, the Bureau feels that this 30-day period must be available when the volume of new applications, complexity of particular applications and similar administrative considerations so require. Although the Bureau is adopting the 30-day figure as proposed, it should be noted that § 107.107 does not state that the Bureau is going to hold each application for 30 days before publishing the required Federal Register notice. That notice will be filed with the Federal Register for publication as quickly as the Bureau is satisfied that an application is complete and in conformity with the requirements of § 107.103(b). It is anticipated that, generally, such filings will be made well before the thirtieth day. What § 107.107 does inform the applicant is that if his application is not returned within 30 days, he can assume that it has been found complete and in conformity with § 107.103(b) and that the public comment process has been initiated.

RENEWALS

A number of comments focused on those parts of the proposal which would have required applications for renewals to be submitted and processed in the identical manner as applications for initial issuance. There were three major points made by commentors in this regard. First, they asserted that an applicant for renewal should not be required to resubmit what is frequently extensive technical data which the Bureau already has on file. Second, they expressed the opinion that since that data had already been through the Bureau's evaluation and approval process, it should not require the Bureau 120 days to review it on renewal. And third, they pointed out that holders of exemptions (and other existing forms of administrative relief from the hazardous materials regulations) which will expire between the proposed effective date (October 16, 1975) and 120 days thereafter may be technically precluded from the opportunity to obtain a renewal without a lapse because, under existing procedures, they would not have applied 120 days in advance of the expiration date.

Modifications have been made in response to each of these three lines of comments. The principal change has been to provide a separate section governing the filing of applications for renewal (§ 107.105) and limiting the applicability of the proposed section on applications to applications for initial issuance (§ 107.103). The Hazardous Materials Transportation Act requires that "[e]ach person applying for . . . an exemption or renewal shall, upon application, provide a safety analysis . . . to justify the grant of such exemption." The Bureau agrees that no useful purpose will be served by refiling of data which is already part of the public record and believes that the statutory objective can best be served in the case of renewal applications by requiring the applicant to review his earlier submissions and certify their continued accuracy and applicability and update that data as necessary. In all cases, a renewal application should be accompanied by a report on the applicant's activities covered by the exemption since its issuance, including all accidents or incidents relating to those activities. New § 107.105(a) so provides. The Bureau also agrees that review of renewal applications should not be as time-consuming as evaluation of initial applications even though public notice and comment proceedings are required in both cases. Accordingly, new § 107.105(b) provides, in advisory rather than mandatory terms, that applications for renewals should be submitted at least 60 days before the expiration date of the exemption.

In conjunction with this shortening of the renewal processing period from 120 to 60 days, the administrative review period provided for in § 107.107 has also been shortened from 30 to 15 days and a new § 107.105(c) has been added to expressly provide for the continuation of an exemption in the unlikely event that processing of a timely filed renewal application is not completed before the scheduled expiration date. These changes do not, of course, preclude earlier submissions. Moreover, renewal applications containing requests for amendments will be processed in the same manner as original applications and therefore should be submitted accordingly.

So that applicants for renewals who file during the first 60 days after the effective date of these procedures will not be prejudiced by the new requirements governing the contents of renewal applications, § 107.105(d) has been added to provide that renewal applications received during that period will be processed if they meet the content requirements under the Department's procedures in effect immediately prior to the effective date of these new procedures. The processing of such applications will, of course, include the public notice and opportunity for public comment steps described in § 107.109(a).

PROCESSING OF EMERGENCY APPLICATIONS

The vast majority of those who commented on the emergency exemption portion of the proposed regulation were concerned with its application to air commerce. The comments of shippers and carriers alike spoke with favor about the timely and efficient treatment they had received under 14 CFR 103.5 when faced with a pressing need for relief from the regulations. While the Bureau agrees that there are many well-tested and worthwhile features to the 14 CFR 103.5 procedures, some of which are being adopted in these regulations, it must be recognized that 14 CFR 103.5 allowed two bases for administrative relief from the hazardous materials regulations applicable to air commerce. One basis was "emergency" which as an overall concept is transferable to these regulations for implementing § 107 of the Hazardous Materials Transportation Act. The other basis was the impracticability of other forms of transportation. Regardless of the logic and soundness of this second basis, it is not by itself recognized by § 107 of the Act as a legitimate basis for the Bureau declaring that an emergency exists. The necessity to eliminate this second basis for emergency exemptions has caused great concern by interests in Alaska which has traditionally accounted for more than 75% of all relief granted under 14 CFR 103.5.

After reviewing the case-by-case history of actions taken under 14 CFR 103.5, the Bureau, on September 26, 1975, initiated rule making (40 FR 45197, October 1, 1975) based on demonstrated favorable safety experience thereunder. It is expected that such regulations will eliminate the need for several classes of reoccurring emergency exemptions for Alaska and other remote areas where a cargo-only aircraft is the only practicable means of transportation. To allow for finalization of this related proposed rule making, the Bureau is adopting certain transition procedures set forth in § 107.125 which will enable it to respond to the special situation in Alaska. To ensure that the needs of the citizens of Alaska are properly served during the transition period which the Bureau expects to be completed by January 16, 1976, the FAA and the Bureau have arranged for all essential exemption activities and decisions to be made in Alaska. For example, the "official designated by the Director, OHMO," to perform certain Bureau functions under § 107.125 will be stationed in the FAA Regional Office in Anchorage.

The Bureau feels that these steps, when fully completed, together with the modifications made in these regulations, will result in an accommodation of the well-articulated needs of persons in Alaska and at the same time fully satisfy the procedural requirements prescribed by the Hazardous Materials Transportation Act.

In response to recommendations that applications for emergency exemption be treated separately from general applications, the Bureau has grouped all provisions relating to the application for and processing of emergency exemptions into a separate distinct section (§ 107.113). In so doing and in response to related comments, provision has been made for making application through FAA District Offices in the case of air commerce and for 24-hour telephone numbers for the other modes of transportation.

DETERMINATION OF EXISTING EMERGENCY

The proposed criteria for determinations as to whether or not an emergency exists evoked a wide range of comments. At one extreme was a recommendation that the proposed criteria for making determinations be converted to flat declarations that an emergency does in fact exist when, in the view of an applicant, any of the described conditions occur (i.e., risk to life or property or the chance of serious economic loss). At the other extreme was an assertion that an emergency exists only if there is "an imminent risk of a substantial injury to human health, welfare or life itself which is not outweighed by the public's statutory right to know of and participate in the pending exemption proceeding." The author of the latter comment would further restrict his narrow concept of emergency by providing that "no relief should be available where it appears that the applicant himself has induced or provoked the alleged emergency by unnecessarily delaying his filing." To deny an applicant the means to abate a danger to his own "health, welfare or life" is an unreasonable penalty to impose for late filing of an application. Such a penalty is unconscionable when, as in most such cases, the danger is to the health, welfare or life of innocent third parties rather than that of a dilatory applicant.

In between these extremes were suggestions that express recognition should be given to cost/benefit considerations and seasonal movement of products such as agricultural chemicals, and that lack of other forms of transportation should be considered to be an emergency authorizing the use of aircraft along the lines of present 14 CFR 103.5. A few commentors stated that there was a need for the criteria to be more specific, particularly with regard to the term "serious economic loss". One such commentor sought specificity as to whose economic loss (e.g., shipper, carrier, consignee, general public) is to be considered under the criteria. Another commentor asserted that the criteria were not sufficiently specific to inform him as to how he could frame an application guaranteed to qualify it for emergency treatment. Another commentor complained that an emergency had not been "totally defined."

One commentor stated that there appeared to be no reason for the parenthetical expression in the protection of life and property criteria which excludes "the hazardous materials to be transported" from the class of property for which an emergency exemption can be sought. This exclusion was proposed because the Bureau means to limit emergency determinations under that criterion to situations in which there is an urgent need for the hazardous material concerned to be (1) delivered elsewhere in order to alleviate a condition posing a threat to life or property, or (2) moved from its present location in order to protect life or property from the hazards the material may present.

The commentor who would limit "emergencies" to situations involving risk to health, welfare or life on the theory that the governing statute (§ 107(d) of the Hazardous Materials Transportation Act) so requires, reads into the statute words of limitation that simply are not there. Those who seek specificity to precisely cover a particular factual situation would have the Bureau so narrow the criteria as to risk freezing out other legitimate emergency situations that surely will arise.

Several comments concerned the manner in which the emergency determination authority should be exercised under the proposed criteria. Although it does not consider it necessary or appropriate for inclusion in the regulations, the Bureau finds considerable merit in one commentor's admonition that "the finding that an emergency exists must result from a balancing of all of the relevant information available to the Department." The Bureau intends to do precisely this in making emergency determinations, particularly those which will be made under the "serious economic loss" criteria of § 107.115(b). While the Bureau fully anticipates that its emergency determinations under the "serious economic loss" criteria will nearly always be limited to situations in which the hazardous material concerned needs to be delivered elsewhere to prevent serious economic loss, it recognizes also the possibility of that infrequent instance when a manifest injustice or absurdity could result if the criteria is literally limited to needed deliveries.

Various elements of the Department of Defense (DOD) expressed the view that certain of their shipments of hazardous materials which require exemptions when transported by commercial carriers should be entitled to emergency exemptions in the interest of national defense. Two U.S. Army commentors recommended that such emergency exemptions should be granted for "shipments to be made by or for the DOD in support of the national defense program, when certified by the DOD as essential and critical." The Naval Sea Systems Command requested a "grandfather clause for DOD Special Permits in order that the transportation of DOD weapons systems/components will not be disturbed."

The Bureau does not find authority in law which would authorize it to adopt any of the DOD proposals for grandfather clauses or DOD certifications. The responsibility vested in the Secretary of Transportation by § 107(d) of the Hazardous Materials Transportation Act to determine that an emergency exists must be carried by him or by one of his subordinates within his Department. It cannot be transferred horizontally to another Executive Department. In addition, the Bureau believes that those determinations can only be made case-by-case on the basis of existing circumstances.

Comments from the Air Force question the need for requiring them to reapply biennially for an exemption issued in 1961 for an indefinite period. The provisions of §§ 107(a) and 114(b) (2) of the Hazardous Materials Transportation Act are controlling on this point. Section 114(b) (2) operates to terminate the Air Force exemption and any other similar indefinite exemptions on January 4, 1977, unless renewed before that date, in accordance with regulations issued under § 107 of the Act. That section, under which the current regulations are being issued, does not allow any exemption or renewal thereof to be issued for more than a two-year term. As their comments suggest, a properly framed petition for rule making based on their satisfactory safety experience under the exemption they now hold may well be a more satisfactory way for the Air Force to proceed. Other elements of DOD may also find a well-reasoned petition for rule making preferable to repeated applications for exemptions. It should be recognized that transportation of hazardous materials aboard DOD's own vessels, vehicles and aircraft does not require either rule making or exemptions since such operations are not considered to be in commerce.

After considering the various points advanced by the commentors on the criteria for determining whether an emergency exists, the Bureau finds itself in concurrence with the commentor who took the position that:

"The regulations implementing the exemption power should . . . refrain from specific definition of an "emergency", for the very nature of emergencies is their unforeseen timing and character. The need for expedited treatment as an emergency matter is best left to the judgment and discretion of the Materials Transportation Bureau and its staff, to determine on a case-by-case basis as each situation arises. Attempts at definition of an indefinable concept will only serve to frustrate the equitable exercise of this power, by boxing it into criteria that fail to accommodate every situation that will be encountered."

APPEALS

Several comments noted the lack of a specific appeal procedure for those applicants whose applications may be denied. One of the commentors went on to suggest that perhaps the reconsideration procedures in Part 102 pertaining to rule making should be available for appealing exemption denials. The other commentors recommended the addition of specific appeal procedures to this body of exemption regulations. The Bureau believes this latter approach to be preferable and therefore has added a new § 107.121 expressly providing for the appeal of various actions taken under these exemption regulations to the Director of the Materials Transportation Bureau whose decisions will be administratively final.

CONFIDENTIAL INFORMATION

A few commentors criticized the proposed regulations for not being specific as to the disposition of documents for which requested confidential treatment is denied by the Bureau. Several of those commentors expressed the view that an applicant faced with an adverse determination on his request for confidentiality should have the option of withdrawing his application. The Bureau agrees with both of these views. Therefore, it has established more comprehensive procedures governing requests for confidential treatment. These procedures in new § 107.3 will apply to all documents submitted with respect to hazardous materials, not just applications for exemptions. They provide for notice to an applicant when his request for confidentiality is denied and an opportunity for him to respond or withdraw his application before the Bureau discloses the information. Section 107.117 concerning withdrawal of a pending application has been modified to expressly allow an applicant to recover the contested documents if he withdraws his application before it is finally determined. These changes do not go so far as to allow the return of such documents after an application has been finally denied as was suggested by one commentor. The Bureau is not prepared to authorize withdrawal of documents once they have served as a basis for a completed official action on an application for exemption, be it an approval or a denial.

Also, in response to comments, the Bureau has modified § 107.124(b), pertaining to what the Bureau makes available for public inspection, by deleting the misleading reference to materials "not relevant to the petition" and by adding a citation to the Department of Transportation's Freedom of Information Act regulations.

PARTIES TO EXEMPTIONS

A large number of comments made the point that a procedure should be established for extending the terms of an exemption granted to one person to other persons in like circumstances without requiring a complete duplication of the various steps and evaluations performed with regard to the original application. The arguments in support of this view are in many respects similar to those which justified simplification of the renewal process.

The procedures suggested by the commentors were for the most part analogous to the "registration" concept that has been employed for the past few years by the Hazardous Materials Regulations Board under its special permit program. Under that program, once the efficacy of a proposal susceptible of being performed by persons in addition to the applicant was established and a special permit (i.e., exemption) issued, the Board would allow other persons to "register" (i.e., become a co-holder) under that special permit.

The Bureau agrees that provision should be made for a similar process under these new exemption procedures. However, rather than merely codifying the earlier procedures, the Bureau has decided to incorporate certain modifications to bring them in line with the intent and purpose of the Hazardous Materials Transportation Act and in particular § 107 thereof. Since § 106 of the Act uses the term "registration" to describe an entirely different concept, the term "party to an exemption" has been adopted. New § 107.111 sets forth the requirements for applying for status as a party to an exemption and describes the Bureau's processing thereof. By filing an application to become a party to an exemption, the applicant constructively adopts as his own the technical and safety information submitted by the applicant for the exemption and if he is granted status as a party to the exemption he is bound by the limitations and conditions that apply to the initial holder of the exemption and will be identified separately as a holder on the exemption documents issued to him.

NATIONAL TRANSPORTATION SAFETY BOARD COMMENTS

On September 12, 1975, the closing day for comments on the proposed exemption regulations, the Chairman of the National Transportation Safety Board (NTSB) filed as comments to the docket an advance copy of two formal NTSB recommendations subsequently delivered to the Secretary of Transportation on September 25, 1975.

Both of the NTSB's recommendations stem from its conclusion that "the proposed exemption procedures do not fulfill the intent of Section 107 of the Hazardous Materials Transportation Act, which calls for 'a safety analysis as prescribed by the Secretary to justify the grant of such exemption.'" The NTSB does not believe that the information required by proposed § 107.5(b) (4)–(7) and (9) [§ 107.103(b) (4)–(7) and (9) in these final regulations] will result in a clear presentation of specific safety concerns and does not constitute a safety analysis. In its view each applicant should be required to "prepare a formal safety analysis statement which would—

"(1) Identify the ways persons could be injured with respect to the quantity and form of the materials to be transported,

"(2) Identify the specific risks for which the applicant considers it necessary to establish safety control measures, based on § 107 (9) (i) and (ii) of the proposed exemption procedures [§ 107.103(b) (9) (i) and (ii) of these final regulation], and

"(3) Describe measures which would eliminate these risks." In addition to assuring that such formal statements would cause applicants to focus on safety problems, the NTSB believes that the mass of data that would be derived could be used as base data in future risk analyses.

Having expressed these views, the NTSB proceeded to recommend to the Secretary of Transportation that he

"(1) Prescribe the content and form for a safety analysis statement to accompany applications for exemptions to the Materials Transportation Bureau's regulations. (Recommendation HM-75-1) (Class I).

"(2) Revise proposed 49 CFR 107.5(b) (9) to require submission of a safety analysis statement, in the form prescribed by the Secretary of Transportation, to support the applicant's belief that his proposed exemption will achieve the level of safety specified in 49 CFR 107.5(b) (9) (i) and (ii). (Recommendation HM-75-2) (Class 1)."

The provisions of the Bureau's proposed procedures cited by the NTSB require an applicant to prepare (1) a detailed technical description of his proposal, (2) a quantitative and qualitative chemical analysis of the material concerned, (3) an analysis of all related shipping experience and accident experience, (4) a statement of the special transportation controls needed for the mode of transportation proposed to compensate for any increased risks that would be encountered should the exemption be granted, (5) a schedule of events under the proposal, and (6) a statement setting forth the applicant's analysis of why he believes his proposal will achieve a level of safety at least equivalent to that provided by the regulations or, if there is no regulatory standard, will adequately protect against risks to life and property which are inherent in the transportation of hazardous materials. These were the items of information and the analyses that the Bureau considered necessary for it to properly evaluate a proposal. Notwithstanding the construction assigned to the term "safety analysis" and the intent imputed to § 107 of the Hazardous Materials Transportation Act by the NTSB, the Bureau is of the firm belief that the information gathering and analytical requirements which it proposed with respect to applications for exemptions fulfills the "intent" of § 107 and will provide the Bureau with the information it needs to evaluate the proposals and establish the proper regulatory safeguards in these cases in which an exemption is granted.

In finalizing these regulations, the Bureau has modified items (7) and (9) in the list of required application contents in light of the NTSB comments. Item (7) has been amended to require an applicant to identify increased risks likely to result if an exemption is granted and specify the safety control measures necessary to compensate for them. Item (9), which requires a statement from the applicant as to why he believes his proposal will achieve the required statutory level of safety, has been amended to require that statement to cover the safety control measures proposed by the applicant. These changes, as the NTSB suggested, should help assure that applications focus on the safety problems which need to be considered.

The Bureau, however, cannot fully agree with the NTSB that each applicant should be required to "identify the ways persons could be injured with respect to the quantity and form of the materials to be transported." The NTSB approach applied literally would mean that a recent applicant seeking Bureau approval for a different (and what may well be a better) technique for applying glue in the fabrication of several different styles of hazardous material specification fiberboard boxes would have been confronted with an overwhelming task. One style of the fiberboard boxes alone is used to carry hundreds of different hazardous materials. Under the NTSB proposal, the applicant would have been required to identify the ways persons could be injured with respect to each of those hundreds of hazardous materials. While it is undoubtedly true that "the data derived from this procedure could be used as base data in future risk analysis", it is more likely that the applicant would have abandoned the effort. It is the Bureau's view that the risks to be identified and addressed by the applicant, by those who choose to comment on the application, and by the Bureau staff, are those risks that would arise as a direct result of granting the exemption. In rejecting this part of the NTSB's suggested changes, the Bureau does not mean to give the impression that it finds the suggestion totally without merit. In particular cases, the Bureau foresees requiring an applicant to supply the full range of information which the NTSB would require for all cases. The obtaining of such information on a case-by-case basis is clearly provided for in § 107.109(b) [proposed § 107.11(b)], which may have been overlooked in the formulation of the NTSB's comments.

Recommendation HM-75-1 calls for "a safety analysis statement to accompany applications for exemptions to the Materials Transportation Bureau's regulations." In addition to regulations pertaining to hazardous materials, the Bureau also prescribes and administers regulations under the Natural Gas Pipeline Safety Act of 1968. Although it would appear that the NTSB intended to include those regulations within the coverage of Recommendation HM-75-1, exemptions from those regulations are beyond the scope of this rule making and are governed by a different statutory standard.

Except as stated above, the Bureau is satisfied that the proposed regulations, modified as described in this preamble, reflect and accommodate the NTSB's Recommendations HM-75-1 and HM-75-2. The Bureau also believes that through the public notice and comment procedures being established, the NTSB will be afforded new opportunities to apply its insight and expertise to the matter of the transportation of hazardous materials in commerce.

OTHER MATTERS

Two comments addressed the proposed requirement that applications state the composition and percentage of each chemical which is the subject of an exemption application. Both commentors felt that information on traces or insignificant amounts need not be included in an application. One commentor would set the floor at 5%. The Bureau understands and appreciates the commentor's point. While it is prepared to follow a general practice of accepting applications which provide the specified information with respect to all components which make up 1% or more of a mixture or solution, the Bureau believes that making this practice a fixed rule may, on occasion, induce an applicant to omit essential information.

Section 107.109(c) has been modified to accommodate suggestions that an applicant whose application is denied should be given the reasons for the denial.

Comments on the proposed termination and suspension provisions asserted that an exemption should not be subject to suspension for failure of the holder to adhere to its terms unless those terms are "repeatedly violated". The Bureau believes that such a change would effectively negate any therapeutic effect that is othewise likely to result from the establishment of this sanction. A related suggestion stated that an immediate amendment rather than suspension is the appropriate administrative action to be taken when new information shows that an exemption does not adequately protect against risks to life and property. The Bureau believes that this might be so in some cases. In others, a suspension pending actual determination of an appropriate amendment may be necessary. It was to provide for this flexibility that the proposed suspension provision in question was cast in discretionary terms. The Bureau sees no reason to change it.

One commentor questioned the legality of giving packaging manufacturers, reconditioners, and other similarly situated persons the right to apply for exemptions under the proposed regulations. The commentor stated that through legislative oversight such persons were not expressly mentioned in § 107 of the Hazardous Materials Transportation Act as being potential applicants for exemptions. The commentor also correctly pointed out that a bill (S. 2024, 94th Cong.) on this subject has been introduced in the Senate. That bill had its origins in the Bureau which is of the view that its enactment would merely clarify the matter and that legislative validation of the questioned class of potential applicants is not required. A person's right to petition an agency for relief from a regulation of that agency which directly affects that person is so well established as to be beyond question.

Several editorial adjustments have been made in response to comments and to be consistent with the changes discussed elsewhere in this preamble.

EFFECTIVE DATE

Since these amendments establishing new exemption procedures and making related changes to existing regulations are procedural rather than substantive and because of the need for immediate public guidance with respect to the new exemption procedures, they are being made effective in less than 30 days after publication of the FEDERAL REGISTER. As proposed in the notice of proposed rule making issued on July 30, 1975 (40 FR 32758, August 4, 1975), these amendments become effective on October 16, 1975.

RELATED CHANGES TO OTHER TITLES

Elsewhere in this edition of the FEDERAL REGISTER, 14 CFR 103.5 is being revoked and 46 CFR 146.02-25 is being amended to conform with the adoption of these new exemption procedures.

(40 F.R. 48466, October 15, 1975)

[Docket No. HM-127; Amdt. No. 107–2]
PART 107—PROCEDURES
Domestic Agents for Non-Residents of the United States

The primary purpose of these amendments to the Materials Transportation Bureau's procedural regulations is to require non-residents of the United States who apply for exemptions from the hazardous materials transportation regulations to designate resident agents in the United States upon whom service of process may be made. Applications filed by non-residents after the effective date of these amendments will have to include a designation of the applicant's agent in the United States.

In addition, the standard conditions applicable to all exemptions are being amended to make it clear that the exemption number notation required to be on shipping papers need not be followed immediately by the entries required by 49 CFR 173.427. It is sufficient that the exemption number notation is made in such a manner as to be clearly identified with the other required entries for the item concerned.

Since these amendments relate solely to Bureau procedures and practices, notice and public procedure thereon is unnecessary.

(41 F.R. 7509, February 18, 1976)

[Docket No. HM-127; Amdt. No. 173–93]
PART 173—SHIPPERS
Use of Hazardous Materials Packagings Authorized Under Exemptions

As a feature in the exemption procedures (49 CFR 107.101–107.125) which became effective on October 16, 1975 (40 FR 48466), the Materials Transportation Bureau established a means whereby a person who seeks the same administrative relief as the holder of a particular exemption can become a party to that exemption (49 CFR 107.111). The rights, duties and obligations of a party to an exemption are the same as those of the basic holder of the exemption. In addition to this group of persons whose interests are identical to those of a basic exemption holder, there are two other groups of persons whose activities are affected to a lesser extent by certain exemptions which pertain to packagings for hazardous materials. These are (1) persons who purchase and use packagings manufactured by an exemption holder in accordance with that exemption, and (2) persons who receive from an exemption holder and distribute hazardous materials which have been packaged by the exemption holder in accordance with that exemption. Under the new exemption procedures, the Bureau has been receiving applications to be made "parties" from all three groups.

In the process of evaluating an application for an exemption concerning packaging—whether it concerns a type not previously approved or the use of an approved type for a purpose not previously approved—the Bureau considers the full transportation life of the packaging. Consequently, if the Bureau issues an exemption authorizing the applicant to produce and market a new packaging, any limitations or special conditions which the Bureau has concluded should be imposed on the use of that packaging are expressly stated in the exemption. Similarly, if the Bureau issues an exemption authorizing a shipper to package a hazardous material in a packaging that is at variance with what is required by the regulations, the exemption includes a statement as to whether or not the hazardous materials may be reshipped in the original packaging by distributors, agents, purchasers and the like. The exemption also specifies whether or not the packaging is reusable.

Under the "special permit" process which preceded the new exemption procedures, any shipper who merely wished to purchase and use a packaging manufactured under a special permit and each person who received and wished to reship a hazardous material packaged in accordance with a special permit was required to follow the same procedure as a person who wished to exercise the same authority and perform all of the same functions as the basic special permit holder, i.e., obtain a copy of the special permit and formally register under that permit number with the Office of Hazardous Materials Operations. In converting this "registration" process to the new "parties" procedures, the primary focus was on those prospective applicants who would be seeking the full authority of some-one else's exemption. Consequently, becoming a "party" is considerably more involved than was "registering".

The Bureau believes that the new public notice and comment procedures whereby a person may become a full party to an exemption through the constructive adoption of the original applicant's submission are both necessary and appropriate in the case of an applicant who seeks the same status as the holder of the exemption. However, it is not of the same opinion with respect to the incidental users and distributors of packaging materials covered by an exemption. The principal safety objective in their case is to clearly impose upon them a duty to limit their use of the packaging, insofar as transportation is concerned, to the purposes for which it was found qualified during the exemption evaluation process. So long as copies of the packaging exemption which already contain the necessary specifics are made available to users, any additional registration or other formal process accompanying the delivery of those copies is wasteful and of no value in terms of safety.

Accordingly, the Bureau is adding a new § 173.22a to the regulations governing shippers of hazardous materials explicitly setting forth the requirements governing non-exemption holders who ship hazardous materials in exempt packaging which they have received or purchased in the course of business. Under the terms of the new exemption procedures, such packaging will, of course, be plainly and durably marked "DOT-E" followed by the applicable exemption number (49 CFR Part 107, Subpart B, Appendix B).

Since this amendment grants relief and imposes no additional burden on the persons affected, I find that notice and public procedure thereon are impracticable and that good cause exists for making it effective in less than 30 days after publication in the FEDERAL REGISTER.

(41 F.R. 3477, January 23, 1976)

SUMMARY · Docket No. HM-128 · *Amendments to title 14 CFR (now 49 CFR Part 175) providing for transportation of certain hazardous materials heretofore not permitted aboard small, single-pilot aircraft where other means of transportation are not available or are impracticable; intended particularly for Alaskan transportation but not limited to that geographical area.*

[14 CFR Part 103]
[Docket No. 128; Notice No. 75–9]
CARRIAGE OF HAZARDOUS MATERIALS ABOARD AIRCRAFT
Notice of Proposed Rule Making

The Materials Transportation Bureau (MTB) is considering a series of amendments to Part 103 which would codify into that body of permanent regulations authority which in the past has been granted through the granting of administrative relief from various regulatory restrictions. They were granted by the Federal Aviation Administration on a case-by-case basis, to transport, subject to specific terms and conditions, certain materials on cargo-only aircraft when there was no other practicable means of transportation.

Each proposed amendment is based on the experience and favorable record of safety associated with the carriage of the material concerned over the last several years under exemptions or authorizations to deviate from the existing requirements of Part 103.

ACCESSIBILITY ON SINGLE PILOT, SMALL CARGO-ONLY AIRCRAFT

Section 103.31(b) of Title 14 CFR requires hazardous materials acceptable only for cargo aircraft to be carried in a location accessible to a crewmember in flight. Compliance with this regulation requires the presence of at least two crewmembers aboard the aircraft, even though only one person may be required to fly it. Materials that are not accessible to a crewmember in flight are subject to the quantity limitations prescribed for inaccessible materials in § 103.19(a) and (c). As a consequence, the utilization of a small, cargo aircraft capable of operation by a single pilot is severely handicapped by the regulation due to its payload limitations and the expense of adding an additional crewmember.

The restriction imposed by § 103.31(b) bars the use of a small, single pilot aircraft to transport materials such as gasoline and other flammable liquids to remote communities, isolated sites of exploration teams, and other facilities located in areas not served by ground transportation or where roads can only be used during certain months, unless some administrative relief from that restriction is granted.

For a number of years the FAA, acting under the provisions of 14 CFR 103.5, has issued authorizations for small, single pilot cargo-only aircraft to deviate from the accessibility requirements of § 103.31(b) to make deliveries of essential hazardous materials within the State of Alaska and other remote areas when other means of transportation were not practicable or in emergencies.

In view of the excellent safety record of operations involving the carriage of hazardous materials in small aircraft pursuant to the conditions and limitations prescribed in those authorizations, the MTB proposes to amend § 103.31(b) by relieving small, single pilot, cargo-only aircraft from the accessibility requirements of that paragraph while being used to transport hazardous materials to places which cannot be supplied by other means of transportation. The MTB believes these small aircraft operations can be conducted under the proposed amendment at a level of safety equivalent to that otherwise achieved through compliance with Part 103. Section 103.19(a) and (c) which also deals with accessibility would also be amended to reflect the amendment to § 103.31(b).

DOT SPECIFICATION 17E CONTAINERS FOR FUEL

Section 103.33(c)(1) of Title 14 CFR allows certain limited supplies of fuel to be carried by small passenger-carrying aircraft and helicopters in Alaska and other remote areas, in metal containers that are either DOT Specification 2A containers of not more than 5 gallons capacity, each packed in DOT Specification 12B fiberboard boxes, in one of three DOT specification wooden boxes, or in a non-specification wooden box at least ½-inch thick. Section 103.33(c)(2) allows the use of any 10-gallon container of at least 28-gauge metal, if packed in one of the three DOT specification wooden boxes, or the ½-inch wooden box.

The Specification 2A container is required to be constructed of 28-gauge metal (0.0129 inch minimum thickness). A DOT Specification 17E container of 5-gallon capacity is required to be constructed of 24-gauge metal (0.0209 inch minimum thickness). Thus, a 5-gallon 17E is more than 60% thicker than the Specification 2A. A 24-gauge container is more resistant to puncture than a 28-gauge container by an order of 800 inch-pounds to 600 inch-pounds. It is MTB's conclusion that a 24-gauge 17E drum, alone, is at least equivalent in integrity to a 28-gauge Specification 2A container packed in a Specification 12B fiberboard box. Accordingly, MTB proposes to amend § 103.33(c) by adding DOT Specification 17E containers of not more than 5 gallons capacity as a packaging authorized for use under that section.

ACCUMULATED EXPERIENCE UNDER EXEMPTIONS AND AUTHORIZATIONS TO DEVIATE

Section 103.9 provides that no person may carry any dangerous material in a cargo-only aircraft except those that: (1) are specified in 49 CFR 172.5 as acceptable for shipment by rail express; (2) do not exceed the maximum quantity for each outside container specified in 49 CFR 172.5 for rail express; and (3) are packaged, marked, and labeled as specified in 49 CFR Part 173 for shipment by rail express.

Over the past several years the need to deliver a number of particular commodities classified as hazardous materials to remote places in Alaska and elsewhere has given rise to the development of sets of special limitations and conditions for allowing those commodities to be transported by the only available means of transportation (i.e., cargo-only aircraft) in quantities in excess of the standard limitations prescribed for rail express in § 172.5. As a result, considerable experience has been gained and the techniques for safe transportation of these larger quantities of essential commodities have been perfected.

Therefore, the MTB proposes to add a § 103.37 to Part 103 expressly authorizing cargo-only aircraft operating under special limitations and conditions designed to assure a high level of safety, to deliver to places not served by other practical means of transportation certain hazardous materials which the MTB believes have been demonstrated through the FAA's exemption and deviation authorization experience to be fully capable of being safely transported.

EXPLOSIVES FOR USE IN BLASTING OPERATIONS

To meet the need for explosives to perform essential blasting operations and to conduct geological testing activities at remote locations, it has been necessary for exemptions and authorizations to deviate from the rail express prohibitions relating to Explosives A. In each case, the carriage of the explosives has been subject to specific requirements to assure a high level of safety. Air cargo-only transportation of commercial explosives has been performed under these controlled conditions for avalanche control, firefighting in wilderness areas, tunnel and other major earth-moving construction in areas inaccessible by surface transportation, and oil and other mineral exploration and extraction activities in remote areas.

Therefore, the MTB proposes to incorporate into the permanent body of regulations governing the transportation of hazardous materials the authority to transport explosives for blasting operations as the exclusive cargo on cargo-only aircraft to remote places. Blasting caps would be authorized for carriage on separate flights under the same conditions or with other non-hazardous cargo when placed in special packaging designed and constructed to contain the explosive force of the blasting caps should they be initiated.

FLAMMABLE LIQUIDS IN 55-GALLON CONTAINERS

Gasoline and certain other flammable liquids, as defined in 49 CFR 173.115(a), are limited for rail express and thus also for cargo-only aircraft to a maximum quantity of 10 gallons for each outside container by 49 CFR 172.5.

A Special Federal Aviation Regulation (SFAR), No. 28, was issued on March 28, 1974 (39 FR 12337, published April 5, 1974), to permit the carriage of flammable liquids, other than pyroforic liquids, in cargo-only aircraft within the State of Alaska in quantities that exceed the maximum quantity limitations of 49 CFR 172.5 but are not in excess of 55 gallons per outside container. As set forth in the preamble to SFAR No. 28, the principal reason for its adoption was to meet the demand for flammable liquids in areas of Alaska where other means of transporting larger quantities are unavailable or impracticable.

This demand was met for a number of years prior to issuance of that SFAR and since its expiration in March of this year through the issuance of deviation authorizations under § 103.5.

In addition to Alaska, a number of requests for deviation authorizations to carry flammable liquids in quantities in excess of the limitations of 49 CFR 172.5 via cargo-only aircraft to remote places elsewhere in the United States (primarily in the Pacific Northwest) have been granted during recent years. A review of operations under SFAR No. 28 and the related deviation authorizations indicates that no accidents or incidents have been recorded as a result of these operations.

Therefore, the MTB proposes to incorporate into the permanent body of regulations governing the transportation of hazardous materials the authority to transport gasoline and certain other flammable liquids used primarily for heating purposes by cargo-only aircraft in 55-gallon or smaller drums to remote places.

FLAMMABLE LIQUIDS IN INSTALLED BULK TANKS

The carriage of flammable liquids such as gasoline in bulk tanks, the installation of which has been approved under a supplemental type certificate has been permitted, pursuant to the exemption authority in Part 11 of the Federal Aviation Regulations (14 CFR Part 11) under certain limited circumstances. This means of transporting large quantities of flammable liquids has been employed for several years to supply the needs of isolated villages, exploration teams, Alaskan pipeline related operations, and other facilities not served by ground transportation or only seasonally served.

In view of these facts, the MTB proposes to authorize the carriage of certain flammable liquids to remote places where there are no other means of transportation in supplemental type certificate approved bulk tank installations subject to certain conditions developed and perfected through the exemption process experience. These conditions and limitations would, for the most part, govern the loading and unloading and carriage of liquids in the approved bulk tanks.

Interested persons are invited to submit views and comments on the proposal. A public hearing will be held for that purpose at 9:30 a.m. on October 23, 1975, in the third floor auditorium of Federal Office Building 10A (commonly referred to as the FAA Building) located at 800 Independence Avenue SW., Washington, D.C. Interested persons not desiring to present oral presentations are invited to submit their comments in writing. Comments should refer to the docket number and be submitted to: Docket Section, Materials Transportation Bureau, U.S. Department of Transportation, Trans Point Building, Washington, D.C. 20590. All comments received before the close of business on November 6, 1975, will be considered, and will be available in the docket for examination both before and after the closing date. Comments received after the closing date and too late for consideration will be treated as suggestions for future rule making.

To the extent the proposals made herein may be adopted, the MTB contemplates combining them with those it adopts in new Part 175 of 49 CFR proposed under Docket HM-112 (39 FR 3022, January 24, 1974).

(40 F.R. 45197, October 1, 1975)

[Docket HM-128; Amdt. No. 103–27]

PART 103—TRANSPORTATION OF DANGEROUS ARTICLES AND MAGNETIZED MATERIALS
Carriage of Certain Hazardous Materials on Cargo-Only Aircraft as Only Means of Transportation

The purpose of these amendments to 14 CFR Part 103 is to authorize certain materials, the air transportation of which is otherwise prohibited or restricted, to be transported by cargo-only aircraft subject to special handling and operational controls when other means of transport are not practicable.

On October 1, 1975, the Materials Transportation Bureau published a notice of proposed rulemaking on this subject (40 FR 45197) inviting public comments and announcing a public hearing for October 23, 1975, in Washington, D.C. All comments received have been given due consideration.

As a result of comments received, the following changes which the Bureau believes compatible with the basic proposal have been made in the proposed amendments in addition to language changes for clarification:

1. The introductor clause in § 103.33 is reworded to (1) make that section equally applicable to small aircraft and helicopters operating in remote areas of the United States outside Alaska as well as within Alaska, and (2) include other flammable liquids, such as alcohol, used as fuel.

2. In the table in § 103.37(a), the entry pertaining to electric blasting caps (more than 1000) Class A explosives is revised to allow them to be carried on the same aircraft with non-hazardous cargo if they are packaged in IME 22 cap containers.

3. In the table in § 103.37(a), the entry pertaining to electric blasting caps (1000 or less) is revised to allow them to be carried in IME 22 or MC 201 cap containers on the same aircraft with other hazardous cargo, except Class A or Class B explosives.

4. In the table in § 103.37(a), the entry pertaining to high explosives Class A is revised to allow them to also be carried on the same aircraft with certain other commercial blasting agents which are similar to high explosives, are difficult to initiate, and are presently authorized for carriage aboard cargo-only aircraft.

5. In § 103.37(b)(2) requiring advance permission from airports where hazardous materials covered by new § 103.37 are to be loaded or unloaded or where the aircraft is to land enroute, a new provision is added in recognition of the possibility of the aircraft being diverted to an alternate airport.

6. In § 103.37(b)(6), the written instructions to be provided to the pilot of a cargo-only aircraft scheduled to carry any of the special hazardous materials shipments authorized by these amendments are modified to include the name of the airport official[s] who have granted the required approval for the use of his airport facilities.

7. The proposed prohibition in § 103.37(c)(4)(ii) against smoking and flame or spark producing devices within 50 feet of aircraft loading or unloading flammable or combustible liquid bulk tanks is transferred to paragraph (b)(9) of that section so as to make the prohibition applicable to all loadings and unloadings of hazardous materials covered by new § 103.37.

The Bureau has not adopted a number of other recommendations for the reasons stated:

1. It was recommended that electric blasting caps in IME 22 or MC 201 cap containers be allowed to accompany any other hazardous or non-hazardous cargo. As stated above, the proposed amendment has been modified to allow Class A shipments (1000 or more) with other non-hazardous materials and Class C shipments (less than 1000) with hazardous materials exclusive of Class A or B explosives as well as with non-hazardous cargoes.

The Bureau is considering a broad review and possible revision of the regulations pertaining to the classification and packaging of blasting caps for all modes of transportation. Until that effort is completed, the Bureau is not prepared to issue a regulation that would authorize large quantities of blasting caps (i.e., Class A) to be transported in the same aircraft with other hazardous materials such as flammable liquids, corrosive liquids, or oxidizing materials. The Bureau's primary concern in this regard is the possible adverse interaction that could occur even considering the high integrity of the IME 22 container system.

2. It was also recommended that Class A high explosives be allowed to accompany non-hazardous materials. The Bureau believes that shipments of these explosives should be made on an exclusive basis receiving the full attention of the loading, off-loading, and airborne cargo handling personnel, the only exception being those closely related materials which have been added to the high explosives entry in the § 103.37(a) table. These materials, as pointed out by several commentors, are also used in blasting operations and are generally handled and treated with the same consideration as high explosives. Any further relaxations in this regard can only be considered and the related circumstances evaluated on a case-by-case basis.

(40 F.R. 58284, December 16, 1975)

SUMMARY • Docket No. HM-130 • *Amendments to title 46 CFR to increase the maximum authorized gross weight to 55,000 pounds for portable tanks in marine transportation.*

[46 CFR Part 146]
[Docket No. HM-130]
[CGD 74-292]
CARRIAGE OF PORTABLE TANK
Proposed DOT Specification Packaging

The Coast Guard is considering amending the dangerous cargo regulations to allow the carriage of a portable tank having a gross weight of 55,000 pounds or less.

Written comments. Interested persons are invited to participate in this proposed rulemaking by submitting written data, views, or arguments on the proposal contained in this document to the Executive Secretary, Marine Safety Council (G-CMC/82), Room 8234, U.S. Coast Guard Headquarters, 400 Seventh Street, SW., Washington D.C. 20590. (Telephone 202-426-2477). Each person submitting comments should include his name and address, identify the notice (CGD 74-292), any specific wording recommended, and reasons for any recommended changes. Comments received will be available for examination by interested persons in Room 8234, Department of Transportation, Nassif Building, 400 Seventh Street, SW., Washington, D.C. Copies will be furnished upon payment of fees prescribed in 49 CFR 7.81.

Public hearing. The Coast Guard will hold a hearing on July 1, 1975 at 9:30 a.m. in Room 8334, Department of Transportation, Nassif Building, 400 Seventh Street, SW., Washington, D.C. Interested persons are invited to attend the hearing and present oral or written statements on this proposal. It is requested that anyone desiring to attend the hearing notify the Executive Secretary of the time needed for his presentation at least ten days before the day the public hearing is held. Written summaries or copies of oral presentations are encouraged.

Closing date for comments. All communications received before July 16, 1975 will be evaluated before final action is taken on this proposal. The proposed amendment may be changed in the light of comments received.

Under the authority of 46 U.S.C. 170 (11), the Commandant of the Coast Guard may exempt any vessel or class of vessels from any provisions of the Dangerous Cargo Act (R.S. 4472, as amended; 46 U.S.C. 170) or any regulations or parts of regulations established under that Act when he finds that a vessel, route, area of operation, condition of the voyage, or other circumstances render the application of the Act or regulation unnecessary for the purposes of safety.

On June 11, 1960 (25 FR 5236), 46 CFR 146.05-5 increased the gross weight of DOT specification portable tanks from 8,000 pounds to 20,000 pounds. Since 1960, the Commandant, U.S. Coast Guard, has permitted, under the authority of 46 U.S.C. 170(11), the carriage by vessels of DOT specification tanks that exceed the 20,000 pound limitation after finding that this carriage secured the necessary provisions against the hazards of health, life, limb, and property, the stated objectives of the Dangerous Cargo Act. No report of accidents due to the increased weight of the tanks has been received by the Coast Guard. Because of the favorable experience in the permitted carriage, the Liquid and Bulk Tank Division of Fruehauf Corporation has petitioned the Coast Guard to amend the regulations by increasing the gross weight of DOT specification tanks from 20,000 pounds to 55,000 pounds.

In consideration of that petition the Coast Guard proposes to amend 46 CFR Part 146 as follows:

1. By revoking § 146.05-5(h) which increased the gross weight limitation for DOT specification tanks from 8,000 pounds to 20,000 pounds.

2. By striking the figures "8,000" and "20,000" wherever they appear in §§ 146.21-100, 146.23-100, 146.24-100, and 146.25-200 and inserting "55,000" in place thereof.

3. By striking the figure "20,000" in the second and third sentences in § 146.07-1(b)(4) and inserting the figure "55,000" in place thereof.

The Coast Guard has determined that the proposal in this document would have no foreseeable significant impact on the quality of the human environment. An environmental assessment with a negative declaration has been drafted. Copies of this draft may be obtained in Room 8306, Coast Guard Headquarters, Washington, D.C. 20590. Interested persons are invited to comment on this draft statement.

(40 F.R. 24532, June 9, 1975)

[Docket No. HM-130; Amdt. CGD 74-292]

PART 146—TRANSPORTATION OR STORAGE OF EXPLOSIVES OR OTHER DANGEROUS ARTICLES OR SUBSTANCES AND COMBUSTIBLE LIQUIDS ON BOARD CARGO VESSELS
Portable Tank Gross Weights

The purpose of these amendments to Part 146 of Title 46, Code of Federal Regulations, is to allow the carriage of a portable tank having a gross weight of 55,000 pounds or less on board a vessel.

Interested persons have been afforded an opportunity to participate in this rulemaking by notice of proposed rulemaking that was published in the FEDERAL REGISTER on June 9, 1975 (40 FR 24532), and by public hearing that was held on July 1, 1975. Ten written comments were received, all of which endorsed the proposed amendment. One of the commenters suggested that paragraphs §§ 146.21-15(b), 146.26-10(b) and 146.26-20(f) should also be amended by striking the figures "20,000" wherever they appear and inserting "55,000" in place thereof. This suggestion has been incorporated in the amendment.

Compliance with these amendments before the effective date is considered by the Bureau as compliance with existing regulations.

(40 F.R. 52027, November 7, 1975)

SUMMARY • Docket No. HM-131 • *Proposed changes to the DOT regulations to require operators of aircraft to specially inspect and monitor all radioactive materials transported by air, by use of radiation detection instruments.*

[14 CFR Part 103]
[Docket No. HM-131; Notice No. 75-10]
TRANSPORTATION OF DANGEROUS ARTICLES AND MAGNETIZED MATERIALS
Proposed Inspection and Monitoring Requirements for Radioactive Materials

The Materials Transportation Bureau is considering amending Part 103 of the Federal Aviation Regulations to require aircraft operators to perform certain inspection and monitoring of radioactive material shipments.

On April 25, 1974, a notice of proposed rule making (Docket No. 13668; Notice No. 74-18; 39 FR 14612) on this same subject was published by the Federal Aviation Administration (FAA). That proposal was finalized, with certain changes, on February 4, 1975, as Amendment No. 103–23, to have an effective date of June 30, 1975.

Numerous petitions were subsequently received by FAA requesting an extension of the June 30, 1975, effective date, citing unclear specifications as to the radiation monitoring instrument needed to perform the required monitoring of radioactive materials packages, stating that the criteria of plus or minus 20 percent accuracy was deficient, in the absence of specifying a range limit. The petitioners further stated that even when more definitive instrument specifications have been developed and published, that the instrument manufacturers and suppliers would require further time to supply the proper equipment to the aircraft operators, who, in turn, would need further time to properly train their cargo handling personnel in the use of such instruments.

The petitioners also stated that, as published, the rule was unclear as to which type of radiation, i.e., gamma, beta, alpha, or neutron, or whether all four types were to be monitored. If the latter were to be the case, they stated that more than one type of instrument at each monitoring station would be required, since no single instrument was commercially available which had the capability of suitably detecting all of the types of radiation.

Several petitioners also requested reconsideration of the requirement for monitoring each package after its removal from the aircraft and prior to its next departure. They pointed out that the "planeside monitoring," which would be required as a result, presented serious, if not impossible, operational difficulties due to the nature of airline operations which involve short duration turnarounds and enroute stops, and are affected by factors such as weather conditions. They further pointed out that the monitoring of off-loaded packages could best be performed after transfer to the freight terminal. They argued that if such monitoring detected radiation or contamination in excess of some stated levels and the aircraft had already departed, monitoring of the aircraft could be performed at its next stop.

After consideration of the merits of the petitions, the FAA extended the effective date for compliance with the radiation monitoring requirements to January 1, 1976 (Amendment 103–25, Docket No. 14530, 40 FR 26673). FAA also announced at the same time that it had instituted a study to develop more realistic specifications for the radiation monitoring instrument.

Since March 1975, the Office of Hazardous Materials Operations of the Materials Transportation Bureau, in cooperation with the FAA, the Civil Aero-medical Institute, the Transportation Systems Center, and the U.S. Nuclear Regulatory Commission have studied the problem with the objective of clarifying the technical specification and use of the radiation monitoring instrument. Before arriving at a proposed clarification, it was the consensus of the above group that the objective of the monitoring had to be clearly identified. It recognized that there were two possible objectives in monitoring, i.e., to detect levels of radiation or contamination resulting from the unusual and unlikely loss of shielding or breach of containment or to verify that the shipper's assigned transport index (T.I.) was in compliance. It was clearly recognized that either objective would dictate significantly different instrument specifications, and also that no single type of instrument could adequately detect all types of radiation. The group agreed that the principal objective of radiation monitoring by air carriers be to detect the radiation hazard situation and not to verify compliance of the T.I. The group also agreed that from the practical standpoint, gamma radiation was the most significant potential problem, recognizing that very few packages emitting only neutron radiation are transported, and also recognizing that alpha and beta radiation do not present an external radiation hazard. It was further recognized that monitoring for external gamma radiation alone might not in every case detect alpha or beta contamination from a leaking package.

The proposals herein afford an opportunity for interested persons to comment on what the Materials Transportation Bureau considers to be an appropriate and realistic instrument specification, as well as a practicable proposal for the application of the monitoring requirement.

Because these proposals on radiation monitoring are substantially different from the monitoring requirements finalized on February 4, 1975, and because the January 1, 1976, effective date for those requirements will pass before the proposals herein are finalized, the requirements of paragraph (d)(3) of section 103.3 and paragraphs (c), (d), and (e) of section 103.23, finalized on February 4, 1975 (Amdt. 103–23), have been revoked. (See Dkt. 13668 in this issue of the FEDERAL REGISTER.)

These proposals are substantially different from those finalized on February 4, 1975, in the following respects:

ACTION LEVEL

As proposed herein, the "action level" in performing the radiation monitoring would be 15 milliroentgens per hour (mr/hr). As published in § 103.23(d)(2) (Amdt. 103–23), the aircraft operator would have been required to verify that the measured T.I. was in agreement with that as assigned to the package label by the shipper or zero in the case of white labeled packages. The "action level" of 15 mr/hr is being proposed on the basis of what the Bureau believes to be a reasonable value for the average measurement error that might be experienced in relation to the maximum transport index allowed to be assigned by the shipper to a package carried aboard an aircraft. The proposal recognizes the objective stated above that the principal purpose of the radiation monitoring is to detect those unlikely situations involving levels of gamma radiation which indicate a loss of shielding or a breach of containment. This is consistent with the Bureau's position that verification of transport indexes assigned by shippers is not the principal purpose of monitoring by air carriers. The proposed "action level" of 15 mr/hr also will be more conducive to the utilization of "fixed" radiation monitors in an automated scanning "pass by" type system in those stations handling significant numbers of packages. Such systems can also be very effective in reducing potential exposure to any package handling personnel who may routinely handle the monitoring and processing of radioactive shipments. It should be understood that in citing the 15 mr/hr "action level", it is not intended to imply that such a level of radiation would be "acceptable" with respect to the shipper's requirements. No change is being proposed with regard to the maximum transport index of 10 which applies to the shipper of radioactive packages. Rather, the proposed aircraft operator monitoring is intended to provide an additional or backup safeguard.

INSTRUMENT SPECIFICATION

More definitive operating characteristics are being provided, specifying the required operational range, percent efficiency, energy range sensitivity, battery check capability, maximum response time, and nonsaturation feature at high levels of radiation.

"OFF-LOAD" INSPECTION AND MONITORING

As proposed herein, a visual inspection of each package would be required after off-loading each package from an aircraft, prior to its departure. In the event this inspection reveals suspected leakage or damage to the package integrity, radiation monitoring would have to be performed immediately. After the visual inspection and transfer of the package into a freight terminal, and before release to another transport mode or to the consignee, the radiation monitoring would have to be carried out. In the event that the off-loaded package is to be transferred to another air carrier for carriage aboard an aircraft, the Bureau interprets the provisions of proposed § 103.23(c)(1) to require that the new air carrier monitor the package in accordance with § 103.23(d) before placing it in an aircraft.

The effective date for the amendments proposed herein would be six months after their publication. This would recognize the lead time necessary for manufacturers and suppliers of radiation monitoring equipment to deliver equipment to air carriers and for air carriers to train cargo handling personnel in the operation and use of the equipment.

(40 F.R. 57688, December 11, 1975)

[49 CFR Part 175]
[Docket No. HM-131; Notice No. 75–10]
CARRIAGE BY AIRCRAFT
**Proposed Inspection and Monitoring Requirements for Radioactive
Materials—Withdrawal of Notice**

On December 11, 1975, the Materials Transportation Bureau (MTB) published Docket No. HM-131, Notice No. 75–10 in the FEDERAL REGISTER (40 FR 57688). This notice modified an earlier Federal Aviation Administration (FAA) rulemaking action which prescribed inspection requirements to be carried out by air carriers for hazardous materials shipments. The FAA notice (Docket No. 13668) was published on April 25, 1974, (39 FR 14612), issued with certain revisions as an amendment on February 4, 1975, (40 FR 5140), and was to have become effective March 7, 1975. Among the requirements were specific monitoring procedures to be followed, including specifications for the radiation monitoring equipment to be used. As a result of numerous comments, the monitoring requirements for radioactive materials packagings were deleted from the FAA amendment, and Docket No. HM-131 was published by MTB for the purpose of clarifying the instrument specifications and implementing the monitoring requirements. The comment period for Docket No. HM-131 expired on February 17, 1976.

Strong objections have been received regarding the impositions upon air carriers caused by the requirements proposed in Docket No. HM-131. Several carriers and carrier associations have pointed out the additional costs which would be incurred in the procurement of the required instruments, and in the training of personnel to carry out the monitoring operations. They questioned the feasibility of training personnel to the level of competency required. Also, many shippers objected to the delays in the transporting of their materials which could be caused by the new requirements. Numerous carriers and shippers contended that responsibility for compliance with the restrictions on maximum permitted radiation levels would more appropriately rest with the shipper, and that the carrier should be allowed to rely upon the shipper's certification, except in cases involving apparent damage or leakage.

Alternatives to the requirements proposed in Docket No. HM-131 were suggested by some commenters. They included central monitoring stations operated by a Federal agency, or the registration of shippers of radioactive materials.

Therefore, after thoroughly considering the comments received, the MTB is withdrawing its proposals under Docket No. HM-131, Notice No. 75–10 for the following reasons:

1. The proposed requirement could result in increased exposure to cargo handlers, particularly since many carriers assign relatively few of their personnel to handling such activities, and the monitoring operation would extend the period of time during which an individual is subject to exposure.

2. Since the publication of the FAA notice on April 25, 1974, implementation of Section 108 of the Hazardous Materials Transportation Act (P.L. 93–633) has restricted the carriage of radioactive materials to those used or intended for use in research, or medical diagnosis or treatment. This substantially reduces the likelihood of inadvertent exposures to the public.

3. A Notice of proposed rulemaking appears elsewhere in this issue of the FEDERAL REGISTER which, in response to recommendations from the Nuclear Regulatory Commission (NRC), proposes amendments to reduce the maximum radiation level permitted for packages of radioactive materials aboard passenger aircraft, and would increase the required separation distance between passengers and radioactive cargo. The changes discussed in paragraph 2 represent significant increases in the Federal regulatory control of the carriage of radioactive materials by aircraft. If these changes, together with the visual inspection requirements now specified, are eventually used in conjunction with the proposed changes discussed in this paragraph, then it is the judgment of the MTB that these measures will increase safety in the air transportation of radioactive materials more effectively than would the monitoring requirements proposed in Docket No. HM-131.

An additional consideration is the possibility that the medical use of radio-pharmaceuticals could be interrupted as a result of: (1) delays in handling of the materials due to the monitoring requirements; and (2) possible increased transportation costs due to the costs of the proposed instrument and personnel training requirements.

(42 F.R. 37426, July 21, 1977)

SUMMARY • Docket No. HM-132 • *Amendment of the regulations to authorize shipment of asphalts in non-DOT specification tank trucks equivalent to DOT Specification MC 306.*

[Docket No. HM-132; Amdt. No. 173–92]
PART 173—SHIPPERS
Asphalts in Cargo Tanks

A petition has been received requesting reclassification of these materials as combustible liquids, however, no justification for this request based on safety was submitted. The Bureau believes that the materials should be identified as flammable liquids not only because of their flash points but also due to the fact that they are usually shipped and maintained at temperatures at or above their flash points during transportation.

The Bureau has received a number of requests for exemptions for continued shipment of these materials in a type of cargo tank similar to a MC 306 cargo tank. Of the several tanks authorized for the carriage of flammable liquids, a specification MC 306 provides the minimum safety features believed necessary to accommodate such cargo. The cargo tanks apparently in widespread use, which are covered by these exemption requests, are manufactured by several firms and meet the basic structural requirements for a MC 306 cargo tank except for manhole design, venting, and emergency flow control.

Tanks used for the transport of hot asphalt are well-insulated double-shell tanks. This construction increases their resistance to impact and puncture over that of standard MC 306 tanks. The vapor pressures of asphalt materials are quite low. One asphalt, RC-70, which has a flash point in the 80° F. range, has a boiling range of approximately 350 to 600° F. Since asphalt materials do not reach these temperatures during transportation, the need for venting is minimized. Also, the high viscosity of these materials makes the valving requirements specified for a MC 306 cargo tank impractical.

(40 F.R. 59598, December 29, 1975)

SUMMARY • Docket No. HM-134 • *Public notice of DOT's intention to adopt existing Hazardous Materials Regulations under the new Hazardous Materials Transportation Act, thereby making those former regulations subject to the broader jurisdiction and higher penalty provisions of the new statute.*

[14 CFR Part 103]
[46 CFR Parts 64 and 146]
[49 CFR Parts 171–179]
[Docket No. HM-134; Notice No. 76–2]
HAZARDOUS MATERIALS REGULATIONS
Notice of Proposed Rule Making

The Materials Transportation Bureau of the Department of Transportation proposes to amend certain of the hazardous materials regulations for which it has responsibility to expressly reflect a reissuance of those regulations under the authority of the Hazardous Materials Transportation Act (Title I of Pub. L. 93–633). To accomplish that purpose the Bureau proposes to revise the authority citations and, where necessary, the applicability of the hazardous materials regulations in 14 CFR Part 103, 46 CFR Parts 64 and 146, and 49 CFR Parts 171–179.

On January 3, 1975, the Hazardous Materials Transportation Act (HMTA) was signed into law. It was the declared policy of Congress in enacting the HMTA " . . . to improve the regulatory and enforcement authority of the Secretary of Transportation to protect the Nation adequately against the risks to life and property which are inherent in the transportation of hazardous materials in commerce."

Some of the provisions of the HMTA that more clearly reflect the declared policy accomplish the following:

1. Broaden the definition of commerce to include transportation which affects interstate transportation;

2. Provide for a broader application of hazardous materials regulations in certain geographical locations;

3. Provide for Federal pre-emption of inconsistent state and local regulations and law;

4. Extend the Secretary's authority to impose civil penalties to violations committed in the rail and highway modes;

5. Remove statutory restriction on the Secretary's authority to centralize Department of Transportation regulatory activities relating to the safe transportation of hazardous materials by the various modes; and

6. Extend the Secretary's regulatory authority to cover the manufacturers of packages and containers to be used in the transportation of hazardous materials.

Congress recognized that prior to January 3, 1975, much had already been done with respect to providing for the safe transportation of hazardous materials but was convinced of the necessity to bring those previous actions into conformity with the purposes and provisions of the HMTA. Section 114(b)(2) of the HMTA reads in part, "The Secretary shall take all steps necessary to bring orders, determinations, rules, and regulations into conformity with the purposes and provisions of this title as soon as practicable . . . "

This proposal to revise the citations of authority and, where necessary, the applicability of certain hazardous materials regulations in Titles 14, 46, and 49 of the Code of Federal Regulations, to expressly reflect a reissuance of those regulations under the authority of the HMTA is one step in meeting that Congressional mandate.

Although the Bureau considers reissuance of the hazardous materials regulations under the authority of the HMTA to be mandatory and therefore leaving the Bureau without discretion in the matter, the Bureau wants the affected public to be aware of the pending reissuance and desires that interested persons submit written views and comments with respect to matters which they regard as being affected by that reissuance.

Comments by interested persons should refer to the docket number and be submitted in duplicate to the Section of Dockets, Office of Hazardous Materials Operations, Department of Transportation, Washington, D.C. 20590. All comments received before the close of business on April 1, 1976, will be considered, and will be available for examination in Room 6213 Trans Point Building, 2100 Second Street, SW., Washington, D.C. both before and after the closing date. Comments received after the closing date and too late for consideration will be treated as suggestions for future rule making.

(41 F.R. 9188, March 3, 1976)

[Docket No. HM-134; Amdt. Nos. 171–34, 172–32, 173–100, 174–27, 175–2,
176–2, 177–37, 178–40, 179–17]
HAZARDOUS MATERIALS REGULATIONS
Reissuance

On March 3, 1976, the Materials Transportation Bureau (MTB) published a notice of proposed rulemaking in the Federal Register identified as Docket No. HM-134; Notice No. 76-2. In that notice the MTB proposed to amend certain of the hazardous materials regulations for which it has responsibility to expressly reflect a reissuance of those regulations under the authority of the Hazardous Materials Transportation Act (Title I of Publ. L. 93–633) (HMTA). To accomplish that purpose the MTB proposed to revise the authority citations and, where necessary, the applicability of the hazardous materials regulations in 14 CFR Part 103, 46 CFR Parts 64 and 146, and 49 CFR Parts 170–179.

The basis for the March 3 proposal is found in § 114(b)(2) of the HMTA:

"The Secretary shall take all steps necessary to bring orders, determinations, rules, and regulations into conformity with the purposes and provisions of this title as soon as practicable . . ."

This reissuance, one step in meeting the mandate of section 114(b)(2), provides the necessary legal connection being made between the hazardous materials regulations and the HMTA so that the provisions of the HMTA and the authorities vested in the Secretary of Transportation by the HMTA can be implemented in relation to those regulations.

By virtue of the reissuance, a number of HMTA provisions will come into operation in such a way as to modify the overall application of the hazardous materials regulations. Specifically, reissuance of the regulations under the HMTA:

Extends the MTB's regulatory authority to the manufacturers of packages and containers that are represented, marked, certified, or sold for use in the transportation of hazardous materials;

Preempts inconsistent State and local regulations and law unless the State or local political subdivision concerned applies to the MTB for and is granted a determination that the inconsistent law or regulation is not preempted;

Extends to the concerned operating elements of the Department of Transportation authority to impose civil penalties for violations committed in the rail and highway modes;

Significantly increases the criminal fines for all violations of the hazardous materials regulations;

Provides, as additional enforcement tools, authority for the administrative issuance of compliance orders and judicial issuance of injunctions and awarding of punitive damages; and

Makes the hazardous materials regulations as fully applicable in Puerto Rico, the Virgin Island, Samoa and Guam as they already are in the 50 States and the District of Columbia.

In addition to these specific changes which will become operational on the effective date of this reissuance, there is another significant feature of the HMTA yet to be exercised. In prescribing the range of transportation which may be regulated through the issuance of hazardous materials regulations, § 103(1) of the HMTA includes transportation which affects interstate transportation. Clearly, the scope of this new regulatory authority as described by § 103(1) of the HMTA, is broader than that which has been exercised under 18 U.S.C. 834 as limited by the definition of "interstate and foreign commerce" in 18 U.S.C. 831. The Bureau now contemplates exercising his expanded HMTA authority through individual rulemaking proceedings which it would initiate as the need for extending the hazardous materials regulations to particular intrastate situations affecting interstate commerce come into focus. Meanwhile the classes of shippers and transporters who are subject to the reissued hazardous materials regulations remain the same as immediately prior to the reissuance.

Although the MTB considered the proposed reissuance to be mandatory and therefore leaving the Bureau without discretion in the matter, interested persons were invited to participate in the HM-134 rulemaking by submitting comments with respect to matters which they regarded as being affected by the proposed reissuance. All comments received on Docket HM-134 have been given full consideration by the Bureau before a decision was made on the amendments contained herein. Of the many comments received, a number have been determined to be outside the scope of the HM-134 proposal. Such comments have been or will be given separate consideration and may be the subject of future rulemaking action. The remaining comments, including those offering full support for the proposed reissuance, concerned themselves with the following:

Lack of adequate notice in HM-134 as to specific substantive changes to existing regulations.

Retaining certain existing citations of authority to the regulations.

Deferring the proposed reissuance until the amendments under Dockets HM-112 and HM-103 are finalized.

Deferring the proposed reissuance pending recodification and simplification of the regulations.

DOT jurisdiction over non-commercial as well as commercial transportation.

Preemption of State and local laws and regulations.

An allegation that the reissuance of 49 CFR 173.115(a) under the terms of the proposal would be in violation of Sections 104 and 105(a) of the HMTA.

The need to clarify what have been called ambiguities in the Shipping Container Specifications of Part 178 of 49 CFR before reissuing those specifications.

The issues raised by these comments will be addressed separately in the discussion that follows.

ADEQUACY OF NOTICE

Several comments expressed the attitude that the proposed reissuance was merely a technical or housekeeping amendment and that HM-134 could not legally be used as the vehicle to make any substantive changes in the nature, extent and effect of the hazardous materials regulations.

The Materials Transportation Bureau does not consider the amendments to be made under the proposed reissuance to be merely technical or housekeeping in nature. The Congressional mandate of Section 114(b) (2), to which the proposal was in response, envisions much more than that. As previously stated, the reissuance will result in the necessary legal connection being made between the hazardous materials regulations and the HMTA so that the provisions of the HMTA and the authorities vested in the Secretary of Transportation by the HMTA can be implemented in relation to those regulations. The proposal gave notice of this result when it was stated that, "This proposal to revise the citations of authority and, where necessary, the applicability of certain hazardous materials regulations . . . , to expressly reflect a reissuance of those regulations under the authority of the HMTA is one step in meeting" the "Congressional mandate" of Section 114(b) (2) of the HMTA. This amendment will do no more than what was proposed in the March 3rd notice.

RETENTION OF EXISTING AUTHORITY CITATIONS

Because the Transportation of Explosives Act (18 U.S.C. 831–835) has not been repealed, one comment urged that the existing authority citations to that Act in 49 CFR Parts 171–179 be retained. The Bureau believes that to do so would not be consistent with the statutory mandate of section 114(b)(2) of the HMTA which requires the hazardous materials regulations to be brought into "conformity with the purposes and provisions of [the HMTA] as soon as practicable."

RELATIONSHIP TO HM-112 AND HM-103

Subsequent to the March 3rd notice, the amendments under Dockets HM-112 and HM-103 were published (41 FR 15972, April 15, 1976). Those amendments became effective July 1, 1976, and resulted in a consolidation of hazardous materials regulations for all modes in 49 CFR. The consolidation revised and relocated to 49 CFR certain hazardous materials regulations previously in 14 and 46 CFR. The action resulted in the revocation of 14 CFR Part 103 and a substantial reduction of 46 CFR Part 146 (41 FR 15972, 41 FR 28116). Therefore, amendments to authority citations in 14 CFR Part 103 proposed by this rulemaking are unnecessary and amendments to authority citations in 46 CFR Part 146 need not be as extensive as originally anticipated.

The comments requesting the Bureau to defer action on the proposed reissuance until final action was taken on outstanding hazardous material Dockets HM-103 and HM-112 have been overtaken by events. Since that time those dockets have been finalized and became effective July 1, 1976 (41 FR 15972). The fact that mandatory compliance with various amendments under Dockets HM-103 and HM-112 is not required until after July 1, 1976, is not considered by the Bureau to be a valid reason for further delay in finalizing HM-134. The regulations as amended by HM-103 and HM-112 set forth the currently applicable standards. A violation of any of those standards subjects the violator to the applicable penalty provided by statute unless (1) the action constituting the violation is an action authorized in the effective date provision under Dockets HM-103 and HM-112, as amended at 41 FR 26014, to be performed in the manner provided by the regulations in existence on June 30, 1976, and (2) that action does in fact comply with the regulations as they existed on June 30, 1976.

RECODIFICATION

One comment urged the Bureau to consider completely recodifying the existing hazardous material regulations prior to their reissuance under the HMTA. In light of the Congressional mandate embodied in §114(b)(2) of the HMTA and a recodification that may take a year or more to complete, the Bureau must reject this suggested approach.

By finalizing the proceedings in Dockets HM-103 and HM-112 the Materials Transportation Bureau completed the first major phase of its continuing effort to improve, update, and simplify the regulations governing the shipment and transportation of hazardous materials. The next phase will be a recodification of those hazardous materials regulations over which the Bureau has authority. Whereas the principal purpose of HM-103 and HM-112 was the consolidation and substantive improvement of specific segments of the regulations, the purpose of the recodification phase will be to rearrange and revise the language of the regulations in a more comprehensible and accessible form to reduce inconsistency, redundancy and obsolescence. The changes made in 49 CFR Parts 171, 172, 174, 175 and 176 by Dockets HM-103 and HM-112 have already instituted many of the changes that would be ordinarily achieved by a pure codification. Therefore, future codification efforts will be directed towards 49 CFR Parts 173, 177, 178 and 179.

Even if it were prepared to initiate formal codification rulemaking focusing on Parts 173 and 177–179, the Bureau believes that first a reasonable period should be allowed for the major changes effected by HM-103 and HM-112 to season and for persons affected by those changes to become accommodated to them.

NONCOMMERCIAL TRANSPORTATION

One comment read as follows:

"The statute [HMTA] does not limit the DOT's jurisdiction to commercial transportation. Historical experience in administering the hazardous materials regulations, however, supports the conclusion that concerns for public safety do not require regulations governing personal transportation in personal vehicles. [It is recommended] that this conclusion be reflected in regulations adopted under HM-134, thereby continuing the traditional limitation of coverage to commercial transportation."

While, as pointed out earlier, the action being taken by the Bureau in this proceeding does not expand the coverage of the hazardous materials regulations to any class of shippers or transporters who are not already subject to those regulations, it should be recognized that the express policy of Congress is enacting the HMTA was "to improve the regulatory and enforcement authority of the Secretary of Transportation to protect the Nation adequately against the risks to life and property which are inherent in the transportation of hazardous materials in commerce." For the Bureau to do as this commentor suggests and circumvent that declared policy by attempting to divest itself of the responsibility and jurisdiction authority placed in it is neither appropriate nor wise. The Bureau must remain unhampered in its ability to act, within its now expanded jurisdictional authority, in the best interest of health and safety. In exercising its new broader jurisdiction in future rulemakings, the Bureau, of course, is not precluded from giving consideration to what this commentor calls "traditional limitations of coverage" or to the practical and economic impact that is likely to accompany actual exercises of this broadened jurisdiction.

PREEMPTION

Another series of comments addressed the issue of Federal preemption as provided for by § 112 of the HMTA. Some of these comments presented a pro or con position on the concept of preemption. Because preemption of inconsistent State and local laws and regulations is Congressionally mandated, a discussion of the advantages or disadvantages of preemption in this rulemaking is moot. Appearing elsewhere in this edition of the FEDERAL REGISTER are procedural regulations for the full implementation of section 112.

Other comments made the unfounded accusation that the Materials Transportation Bureau had proposed the reissuance solely to seek an advantage in current litigation where preemption is a central issue. Though coincidence might lend support to that belief, it is regrettable that lack of confidence in the Bureau's concern for public safety led to that distorted conclusion. In February of this year the Bureau was asked by the Seante Subcommittee on Transportation why it did not, upon enactment of the HMTA on January 3, 1975, immediately move to reissue the existing hazardous materials regulations under the authority of the HMTA. The Bureau responded:

"The consolidation rulemaking identified as Docket No. HM-112 was published in January 1974. DOT efforts with regard to the HM-112 rulemaking began in 1971. By July 1975, when the reorganization within DOT to implement the provisions of the HMTA were completed, a vast amount of time by both the Department and the affected public had already been invested in the HM-112 rulemaking. Because of that time and because further delay of the rulemaking would have major consequences on persons associated with all modes of transportation, the MTB considered finalizing of HM-112 to be its foremost priority.

Upon completion of HM-112, it was the MTB's intention to immediately move to reissue the consolidated regulations under the authority of the HMTA and develop procedural regulations to implement certain provisions of the Act."

In accord with that stated intention, when the Bureau committed itself to submitting the final HM-112 rulemaking to the FEDERAL REGISTER on April 1, 1976, the Bureau published HM-134 on March 3, 1976, and at the same time continued its development of procedural regulations to implement the preemption and enforcement provisions of the HMTA.

The remaining comments with regard to preemption ran the spectrum from full support of the reissuance and its effect on the preemption provision of the HMTA; to suggesting that until efficient procedures for making preemption determinations have been established, State and local regulations in existence on the effective date of the HMTA be presumed to be not preempted unless the Bureau makes a determination to the contrary; to urging the MTB to reject any proposal to amend its regulations which would result in the preemption of any inconsistent State and local regulations and law. These last two comments reveal a basic misunderstanding of the relationship between reissuance and the premption provisions of the HMTA. The Bureau believes a discussion to clarify this misunderstanding is in order.

The relationship between the authority of the Federal Government and that of State and local governments with respect to the safe transportation of hazardous materials, was specifically addressed by Congress in section 112 of the HMTA.

In Senate Report No. 93–1192 accompanying S. 4057, which was eventually enacted as the HMTA, the Committee on Commerce in discussing section 112 stated:

"The Committee endorses the principle of Federal preemption in order to preclude a multiplicity of State and local regulations and the potential for varying as well as conflicting regulations in the area of hazardous materials transportation."

The express provision of preemption in section 112 of the HMTA provides that, "Except as provided in subsection (b) of this section, any requirement of a State ... which is inconsistent with any requirement set forth" in the Act "or in a regulation issued under" the HMTA "is preempted." Section 112(a)).

Section 112(b) provides that a State requirement not consistent with any requirements set forth in the HMTA or in a regulation issued under the HMTA is not preempted if upon the application of an appropriate State agency, the Secretary determines, in accordance with procedures prescribed by regulation, that such requirement (1) affords an equal or greater level of protection to the public than is afforded by the requirements of the HMTA or of regulations issued under the HMTA, and (2) does not unreasonably burden commerce. The contemplated procedural regulations appear elsewhere in this edition of the FEDERAL REGISTER.

In order to fully implement the provisions of section 112(a), reissuance of existing hazardous materials regulations under the authority of the HMTA is necessary. Without this reissuance, section 112 would be inoperative with respect to those regulations. The procedures for State and local governments and individuals to seek administrative determinations under sections 112(a) and 112(b) with regard to State and local laws and regulations appearing elsewhere in this edition of the FEDERAL REGISTER are effective on the same date this reissuance becomes effective.

Additionally, Congress did not intend to limit section 112 to a prospective application. The section applies to any requirement . . . which is inconsistent. In addition, the language of section 112(a) is self operative, i.e., "inconsistent" State requirements are, by law, preempted. The procedures published elsewhere in this edition of the FEDERAL REGISTER merely provide the opportunity to receive an administrative ruling as to consistency or inconsistency (as well as the procedures to seek non-preemption determinations under section 112(b). The fact that such an administrative ruling had not been issued under section 112 with respect to a particular requirement of a State or political subdivision will carry no implication as to the consistency or inconsistency of that requirement with the Act or any regulations issued under the Act.

PROCEDURAL REQUIREMENTS OF HMTA

Several comments alleged that the proposed reissuance of 49 CFR 173.115(a) under the terms of the proposal would be in violation of sections 104 and 105(a) of the HMTA.

Section 173.115(a) defines a flammable liquid for purposes of transportation as any liquid having a flashpoint below 100° F, (37.8° C) with certain exceptions. Because the material polyester resin has a flashpoint below 100° F, it is classed as a flammable liquid. The basis for the definition of flammable liquid was the subject of various rulemaking actions and will not be rediscussed here. Participants to those rulemakings have since sought to have polyester resin exempted from the flammable liquid class even though it met the definition. Every available procedure has been used to gain this relief including a still to be completed appeal from the denial of the exemption request under the procedures of 49 CFR 107.121. These same participants now seek to argue their case by alleging procedural inadequacies in this rulemaking.

They first claim that by reissuing § 173.115(a) under the authority of the HMTA, the Bureau proposes to classify polyester resin as a flammable liquid without having made the statutorily required finding that it poses a threat to life or property. They assert that such action is in violation of section 104 of the HMTA which reads in part as follows:

"Upon a finding by the Secretary, in his discretion, that the transportation of a particular quantity and form of material in commerce may pose an unreasonable risk to health and safety or property, he shall designate such quantity and form of material . . . as a hazardous material (emphasis supplied)."

The Bureau takes this opportunity to point out that section 104 permits the Secretary to designate a material as hazardous if it may pose an unreasonable risk to health and safety or property. The section does not require the Secretary to go beyond a finding that the material is capable of posing such a risk. The merits of the previous rulemaking with respect to the flammable liquid definition have been discussed at length in Docket HM-102 and are summarized in the discussion which follows. Although nearly all of the significant activity in Docket HM-102 predated the HMTA, the procedures used to reach the conclusions of that rulemaking are in full accord with the new requirements of section 104.

These same commentors also assert that the reissuance of § 173.115(a) violates section 105(a) of the HMTA, in that the Bureau did not provide "an opportunity for informal, oral presentation" on the proposal. It was under Docket HM-102 that the Hazardous Materials Regulations Board (predecessor of the MTB) published the currently applicable definition of flammable liquids found in 49 CFR 173.115 (4 CFR 22263, May 22, 1975). A historic review of the efforts that lead to that publication will also lead to a better understanding of the Bureau's reaction to this assertion with regard to section 105(a) of the HMTA.

The first notice of proposed rulemaking relative to Docket HM-102 appeared in the FEDERAL REGISTER on February 27, 1968, under Docket HM-3 (33 FR 3382). Subsequent to that notice meetings were held with industry and trade associations such as Manufacturing Chemists Association, the National Fire Protection Association, the National Paints and Coatings Association, the American Petroleum Institute, the Society of the Plastics Industries, Inc., and the American Society for Testing and Materials. Of concern was the appropriate definition for flammable liquids considering the transportation environment, and the need for compatibility with the flammable liquid definitions of other government agencies. The recommended test method specified for evaluating the flammability hazard of liquids was discussed extensively with these groups.

In 1969, a contract (DOT-OS-00007) was established with the Bureau of Mines for an evaluation of the Department's hazard classification system and test methods. This resulted in the publication of a report in April 1970 titled, "Recommendation of Flash Point Method for Evaluation of Flammability Hazard in the Transportation of Flammable Liquids." The report contained an analysis of the flammability hazard presented in transportation; and recommended a definition for flammable liquids, and a method of test as a criterion for this definition. An additional contract (DOT-OS-00038) was let in 1969 for the study of the transportation environment. An extensive report was published under this contract in October 1971 by the General American Transportation Corporation titled, "A Survey of Environmental Conditions Incident to the Transportation of Materials." Much of this study was devoted to the temperature environment to be encountered in transportation. Based on these two studies, the proposed definition of a flammable liquid was considered to be somewhat conservative with respect to the maximum temperature limit to be encountered in transportation.

Representatives of the Consumer Product Safety Commission, and the Department of Labor's Occupational Safety and Health Adminstration (OSHA) were specifically asked to participate in the discussions and evaluation of the classification problem. On May 29, 1971, OSHA published its regulations for Flammable and Combustible Liquids (29 CFR 1910.106 effective date August 31, 1971). They specify a definition for flammable liquids identical with the definition proposed by this Department. A public hearing was held by OSHA concerning this standard at which time the desirability of uniformity of definition between the Department of Labor and Department of Transportation was discussed.

As a result of these many efforts, modifications of the proposed amendment were made at various times. The more important modifications included:

1. Deletion of the open cup as an alternate test method. [HM-67 (35 FR 18534), December 5, 1970]

2. Additional requirements for combustible liquids and the addition of certain exemptions for flammable liquids newly regulated under this amendment. [HM-102 (30 FR 2768), January 24, 1974]

3. Addition of a Setaflash tester as an alternate test method and additional exemptions for certain specific commodities. [HM-102 (FR 22263), May 22, 1975]

Four additional notices appeared in the FEDERAL REGISTER during this time period. These were concerned primarily with an extension of the time period for consideration of this rulemaking. Extensive files on this subject are open to the public. The Bureau maintains a public reading room and welcomes visits by any interested party.

Regardless of the statutory authority for a particular proposed regulation, when the Bureau issues a notice of proposed rulemaking, it schedules a public hearing on its own motion if it believes that the rulemaking is likely to be materially advanced by supplementing the written comments with oral presentations. In the case of proposed hazardous materials regulations under section 105(a) of the Hazardous Materials Transportation Act, the Bureau is fully cognizant of the "opportunity for informal, oral presentation" provision which distinguishes that section. We do not, however, subscribe to the view that "a hearing is now guaranteed by law." We perceive the true state of the law on this matter to be somewhat less absolute. Even where a statute goes further than section 105(a) and states in unqualified terms that a hearing shall be held, a hearing is not required if the only reasons asserted for holding it are legally insufficient. Dyestuffs and Chemicals, Inc. v. Flemming, 271 F. 2d 281 (8th Cir. 1959), cert. denied, 362 U.S. 911 (1960). In accord is the § 102.27(a) provision of the Bureau's procedural regulations which provides that a petition for a hearing will be granted only if the petitioner shows good cause for the hearing.

While the Bureau routinely schedules one or more informal public hearings on all of its significant hazardous materials rule-making proposals, as it did in HM-102 which resulted in the change to the definition of "flammable liquid", there is no compelling legal or policy reason for the Bureau to hold hearings on those occasions when there is nothing material to be heard.

CLARIFICATION OF AMBIGUITIES IN REGULATIONS BEING MADE APPLICABLE TO MANUFACTURERS

Several comments on the proposed reissuance received from container manufacturers and associations representing these manufacturers asserted that reissuance of 49 CFR Part 178 should be deferred until certain clarifications are made to the provisions of that Part.

Prior to the enactment of the HMTA, the DOT lacked any direct jurisdiction over manufacturers of packages and containers that are used for transporting hazardous materials in commerce. Therefore, detailed regulations spelling out the requirements for containers were necessarily addressed to a class of persons subject to DOT jurisdiction (i.e. shippers). Congress saw this as a serious weakness that frustrated the DOT's efforts to adequately regulate for the safe transportation of hazardous materials. Congress cured this weakness by vesting in the Secretary of Transportation, under various provisions of the HMTA, regulatory and enforcement authority over those manufacturers. One goal of this reissuance is to implement that new authority, therefore, necessitating certain changes to Parts 178 and

179 to delineate the new responsibilities of container manufacturers resulting from the reissuance of those parts under the HMTA.

To accomplish that purpose, Parts 178 and 179 have been amended to provide that any person performing a function prescribed by either part must perform that function in accordance with the applicable part. The Bureau is specifically requiring that the manufacturer mark the packaging or container with the DOT specification. That mark will be understood to certify compliance by the manufacturer, that the functions performed by the manufacturer, have been performed in compliance with the applicable part.

Because of the DOT's historical lack of direct jurisdiction over container manufacturers, Parts 178 and 179 presently prescribe many functions with regard to packagings and containers that are, by their nature, performed after the package or container has left the control of the manufacturer. To insure that a packaging or container meets all the applicable specification requirements when ultimately used in the transport of hazardous materials, new § 178.0-2 asks that the manufacturer of a packaging or container inform each person to whom that packaging or container is transferred of any specification requirements which have not been met at the time of transfer. In addition, § 173.22(a) has been revised to clarify the shipper's responsibility for compliance with packaging and container specifications. The revised paragraph requires that when a shipper performs a function covered by, or having an effect on a specification requirement of Part 178 or 179, the shipper must perform that function in accordance with the specification. (This revision to § 173.22 was proposed under Docket HM-117 published on June 14, 1974 (39 FR 20805).

As a result of market forces, i.e. shippers required by regulation to transport hazardous materials in containers meeting the specifications of Parts 178 and 179, container manufacturers have produced their product in compliance with those specifications. One commenter states, "we manufacture our [container] according to DOT specification . . . or exemption. All tests are also performed to the same specification."

Thus, while proclaiming past voluntary compliance with DOT container manufacture and testing specifications, these manufacturers are now heard to say that the same specifications are so vague and questionable that proposed reissuance would put a very unfair burden on them by imposing the risk of civil penalties and criminal prosecution. They assert that such burdens should not be placed on the manufacturer until the vagueness and ambiguity are removed from the specifications.

The Bureau is not persuaded by these comments to defer reissuance of 49 CFR Parts 178 and 179 under the authority of the HMTA. Several reasons lead to this conclusion:

Container manufacturers' own past practices under the "shipper" regulations with regard to manufacture and testing provide guidance as to the purpose and meaning of the regulations governing the design and construction of containers;

Container manufacturers' long exposure to the past practices of the DOT with regard to the interpretation and enforcement of manufacture and testing specifications with respect to shippers also provide guidance as to future DOT practices with regard to manufacturers;

As in the case of any Federal compliance or enforcement action, any container manufacturer who may be accused of non-conforming to a regulation which he considers vague or ambiguous is entitled to assert that as a defense; and

Container manufacturers have the same rights and opportunities to request interpretations of and changes to the regulations as are currently exercised with great frequency by the shippers and carriers who have long been directly subject to the Federal hazardous materials transportation regulations.

(41 F.R. 38175, September 9, 1976)

SUMMARY • Docket No. HM-135 • *Authorized pneumatic retesting of specifications MC 330 and 331 cargo tanks and addressed requirements on manufacturer's certification for these tanks; also allowed placarding of certain motor vehicles otherwise exempt from placarding, when those vehicles are going to be moved by rail or water and would have to be placarded in those modes.*

[Docket No. HM-135; Amdt. Nos. 173–96, 177–36]

PART 173—SHIPPERS
PART 177—SHIPMENTS MADE BY WAY OF COMMON, CONTRACT, OR PRIVATE CARRIERS BY PUBLIC HIGHWAY
Certification of Cargo Tanks and Placarding of Motor Vehicles

The purpose of these amendments to the Hazardous Materials Regulations of the Department of Transportation is to: (1) Permit pneumatic testing of Specifications MC 330 and MC 331 cargo tanks; (2) clarify the manufacturer's certificate retention requirements for Specification MC 330 cargo tanks, and (3) permit the placarding of motor vehicles containing shipments of less than 1,000 pounds of certain hazardous materials when such shipments are part of an intermodal movement by motor, water, or rail.

A petition has been received from Racon, Inc., requesting that § 173.33(e) be amended to permit pneumatic retesting of Specifications MC 330 and MC 331 cargo tanks used exclusively for certain refrigerant gases. The petitioner states that hydrostatic retesting causes a rust deposit in these tanks which must be removed by sandblasting before the tanks may be returned to service. Such a procedure, petitioner claims, creates a destructive action which affects the integrity of the tank.

The Bureau believes the petition has merit, and further believes that the choice of using a pneumatic retest method should be available to all users of Specifications MC 330 and MC 331 cargo tanks, regardless of the commodities transported. Since section 177.824 presently permits a choice of retest methods for other specification cargo tanks, this amendment will give all cargo tank users the choice of retest method.

On December 2, 1974, Docket No. HM-110; Amendment Nos. 173–87 and 177–31 (39 FR 41741) was published by the Hazardous Materials Regulations Board which among other things added a new § 177.814 entitled "Retention of manufacturer's certificate and retest reports," requiring that each user of a cargo tank retain a copy of the tank manufacturer's certificate and all records from retesting the cargo tank. Section 177.814 referred to provisions in the specifications whereby a motor carrier could certify a cargo tank in place of a manufacturer's certification. The Board failed to recognize however, that the specifications for MC 330 and MC 331 cargo tanks do not provide for certification by other than the manufacturer of the cargo tank since these tanks are built according to the ASME Code, and only the tank manufacturer can certify compliance with the Code requirements. Therefore, § 177.814 is being changed to recognize this distinction by excepting specifications MC 330 and MC 331 tanks from carrier certification.

It has been brought to the Bureau's attention by a petition from the National LP-Gas Association, that the specification for MC 330 cargo tanks did not require a manufacturer's certification. Instead a manufacturer's data report was required to indicate compliance with the ASME Code under which the tank was constructed. The petitioner points out that users of specification MC 330 cargo tanks cannot comply with § 177.814 because certificates were not required for these tanks, and because the users cannot test the tanks to determine if in fact they were built to the specification. Therefore, petitioner asks that § 177.814 be amended to provide that users of specification MC 330 tanks can copy the information imprinted on the identification plate and ASME data plate permanently attached to the tank, and retain this information in place of the original manufacturer's data report when such report is not available. The Bureau believes the petition has merit and is amending § 177.814 accordingly.

Section 177.823 presently prohibits the placarding of cargo tanks and motor vehicles containing less than 1,000 pounds of a hazardous material except for explosives, Class A and Class B; poisons, Class A and certain radioactive materials. Since the regulations of the U.S. Coast Guard and the Federal Railroad Administration require the placarding of containers and trailers containing any amount of these materials, shipments are often frustrated when moving between highway and water or highway and rail. In order to facilitate the ease of intermodal movement of hazardous materials, the Bureau is amending the highway placarding requirement to permit placarding for less than 1,000 pounds when the motor vehicle or cargo-carrying container has a prior or subsequent movement by water or rail.

Since these amendments will allow a retesting procedure that will have the effect of enhancing the integrity and safety of certain cargo tanks and because these amendments will provide for consistency between various Departmental regulations and remove an unwarranted frustration on the intermodal movement of hazardous materials, the Materials Transportation Bureau finds that notice and public procedure thereon are impracticable and unnecessary.

In addition, because these amendments are a relaxation of the existing rules and place no additional burden on any person, they are being made effective in less than 30 days after publication in the FEDERAL REGISTER.

(41 F.R. 17735, April 28, 1976)

SUMMARY • Docket No. HM-138 • *This critical rule making established the preemption and enforcement procedures to be followed by the Materials Transportation Bureau in administering regulations under the Hazardous Materials Transportation Act. Also see Docket No. HM-134.*

[Docket HM-138; Amdt. No. 107–3]
PART 107—PROCEDURES
Preemption and Enforcement Procedures

The purpose of these amendments to 49 CFR Part 107 is to (1) establish procedural regulations that implement the preemption provisions of section 112 of the Hazardous Materials Transportation Act (Title I of Pub. L. 93–633), (2) prescribe procedures to be followed by the Materials Transportation Bureau (MTB) in carrying out its enforcement responsibilities under sections 109, 110, and 111 of the same Act, and (3) add several general procedural provisions covering the MTB's hazardous materials public docket room, service of documents and subpoenas.

AMENDMENTS TO GENERAL PROVISIONS

The list of definitions in § 107.3 is expanded to include a definition of "person" which covers all of the commonly recognized classes of legal entities regularly doing business, as well as individuals. A definition of "respondent" is added for use in connection with the new Subpart D enforcement procedures. A definition of "State" is being incorporated verbatim from section 103(5) of the Hazardous Materials Transportation Act.

A new § 107.9 is added setting forth a general description of the materials available for public inspection and copying in the MTB's hazardous materials public docket room.

A new § 107.11 prescribes the means for effecting service of documents for purposes of the various hazardous materials procedural regulations.

A new § 107.13 specifies who has authority to issue subpoenas under section 109(a) of the Hazardous Materials Transportation Act and prescribes the procedures for their issuance and service, the payment of witness fees, the handling of motions to quash or modify, and their enforcement.

PREEMPTION

In enacting section 112 of the Hazardous Materials Transportation Act, the Congress endorsed the principle of Federal preemption in order to preclude a multiplicity of State and local regulations and the potential for varying, as well as conflicting, regulations in the area of hazardous materials transportation (S. Rep. No. 93–1193, 93rd Cong., 2d Sess. 37 (1974)). The pertinent provisions of section 112 read as follows:

Sec. 112. (a) General.—Except as provided in subsection (b) of this section, any requirement of a State or political subdivision thereof, which is inconsistent with any requirement set forth in this title, or in a regulation issued under this title, is preempted.

(b) State Laws.—Any requirement of a State or political subdivision thereof, which is not consistent with any requirement set forth in this title, or in a regulation issued under this title, is not preempted if, upon the application of an appropriate State agency, the Secretary determines, in accordance with procedures to be prescribed by regulation, that such requirement (1) affords an equal or greater level of protection to the public than is afforded by the requirements of this title or of regulations issued under this title and (2) does not unreasonably burden commerce. Such requirement shall not be preempted to the extent specified in such determination by the Secretary for so long as such State or political subdivision thereof continues to administer and enforce effectively such requirement.

Section 112(a) operates to preempt and nullify requirements of States and their political subdivisions pertaining to the transportation of hazardous materials which are inconsistent with Federal requirements. The determination as to whether a State or local requirement is consistent or inconsistent with a Federal statute or Federal regulations is traditionally judicial in nature. It is to be expected that conflicts between State and local hazardous materials requirements and the Hazardous Materials Transportation Act and related regulations will be the subject of future litigation. The Bureau, however, does not see the courts as the exclusive arbitrators of the inconsistency questions that may be posed by § 112(a). Therefore, to facilitate the operation of section 112(a), procedures are established in new Subpart C providing States and their political subdivisions and persons affected by requirements of States or political subdivisions the opportunity to seek administrative rulings as to the consistency or inconsistency of any particular State or political subdivision requirement. Also established are the procedures required by section 112(b) by which States or political subdivisions may obtain formal determinations that their inconsistent requirements are not preempted.

The new Subpart C prescribes means for obtaining inconsistency rulings and preemption determinations with respect to existing hazardous materials requirements of States and political subdivisions of States. These procedures are not a vehicle for receiving pre-enactment approval by the MTB. The Bureau does not hold itself out as legislative drafting or editorial service nor will it provide legal research service for State or local governments or legislators. It will, however, upon application, review State and political subdivision requirements which have been officially adopted by a State or political subdivision.

In defining the term "political subdivision" used in section 112 of the Act, the Bureau has taken the view that it includes not only the agencies of a state government, as well as municipalities and other local governments, but also entities created by one or more States or municipalities.

Section 107.203 invites applications for inconsistency rulings from States, political subdivisions, and persons affected by hazardous materials requirements of States and political subdivisions. It also specifies what each applicant must submit as part of his application. Section 107.205 requires private applicants to serve a copy of their application on the State or political subdivision concerned and gives the State or political subdivision 45 days in which to comment. Irrespective of who files an application, the Office of Hazardous Materials Operations (OHMO) may serve notice on additional parties or publish notice of the application in the FEDERAL REGISTER and invite interested persons to file written comments. Under § 107.207 the Director, OHMO, exercises broad investigative authority to fully develop the information necessary for his evaluation of the application. Section 107.209 reserves to the Director, OHMO, the authority to review and issue an inconsistency ruling on his own motion, invoking as necessary the same procedural mechanisms as are available in the case initiated by an application. Paragraph (c) of this section sets forth criteria employed by the Director OHMO, in making his ruling. These criteria comport with the test for conflicts between Federal and State statutes enunciated by the Supreme Court in Hines v. Davidowitz, 312 U.S. 52 (1941). Each ruling is issued in writing and is served on the applicant and others who may be involved in the proceeding. When major issues are involved or the ruling has broad application, it may be published in the FEDERAL REGISTER in addition to being served on interested individuals.

Section 107.211 provides a 30-day appeal period for any person aggrieved by a particular ruling.

Section 107.215 invites States and political subdivisions (not private parties) to submit applications for non-preemption determinations. As specified in paragraph (b) of this section, each application must set forth the facts and the applicant's rationale and arguments in support of the non-preemption determination sought.

Section 107.217 requires a State or political subdivision applying for a preemption determination to serve notice of that application on those persons which the State or political subdivision can reasonably identify as being affected by the requested determination. The OHMO may add to the service list or publish notice in the FEDERAL REGISTER, or both. All persons upon whom notice is served or who are identified in the FEDERAL REGISTER notice will have the opportunity to submit written comments. Under § 107.219, as in the case of inconsistency rulings, the Director, OHMO, has broad procedural authority to fully develop the necessary information for making his determination. This includes the authority to require supplemental submissions from the applicant. The Director may dismiss an application without prejudice if there is insufficient information to make his determination; if the applicant fails to provide requested additional information; or if the applicant fails to provide notice to affected persons. The Director will only consider a request for a preemption determination if the inconsistency of the State or political subdivision requirement in issue has been affirmatively fixed by the order of a court of competent jurisdiction, by a § 107.209 administrative ruling that has become final, or through the express acknowledgement of inconsistency by the applicant. When the OHMO has received all of the information necessary for reaching a determination, all participants are so notified and, if a formal determination is not issued within 90 days, § 107.223 affords the applicant the same opportunity for appeal as in the case of a denial.

Section 107.221, which provides for the issuance of a written determination, specifies the principal factors considered by the Director, OHMO, in reaching that determination. Whether a particular inconsistent State or political subdivision requirement unreasonably burdens commerce is a question of fact and requires a balancing of the State's or political subdivision's interest in public health and safety against the national interest in maintaining a free flow of commerce. The factors listed in § 107.221(b), no single one of which is dispositive in determining whether a State or political subdivision requirement creates an unreasonable burden, provide in combination the test by which that determination is to be made. Such have been the factors employed by the Supreme Court in deciding whether various State transportation safety requirements impose an unreasonable burden on interstate commerce. See e.g. South Carolina State Highway Department v. Barnwell, 303 U.S. 177 (1938); Southern Pacific v. Arizona, 325 U.S. 761 (1945); Bibb v. Navajo Freight Lines, 359 U.S. 520 (1959). Section 107.221 also provides for service of each preemption determination upon the applicant and other interested persons.

As in the case of inconsistency rulings, a 30-day appeal period for preemption determinations is provided by § 107.225. An appeal may be filed by any person aggrieved by a particular preemption determination.

ENFORCEMENT

New Subpart D describes the investigative procedures and enforcement authorities of the MTB with respect to its assigned area of responsibility for enforcement of the hazardous materials regulations. It also sets forth the MTB's procedures governing the imposition of sanctions with respect to violations of those regulations.

Section 107.301 describes the division of enforcement responsibilities among the operating elements of the Department of Transportation, including the MTB. The MTB exercises its hazardous materials enforcement responsibility through the OHMO. As stated in § 107.303, Subpart D covers the OHMO segment of this enforcement responsibility. Section 107.305 describes the investigative procedures followed by OHMO inspectors.

Under separate headings the remainder of Subpart D describes the various enforcement sanctions available to the OHMO and with respect to those which are primarily administrative to nature (i.e., compliance orders and civil penalties) sets forth procedures governing the processing of individual actions. The choice of which enforcement action or combination of actions to be initiated in a given case is, of course, a matter of administrative discretion with the OHMO acting under the policy direction of the Director, MTB. The four types of enforcement sanctions available to the OHMO are compliance orders, injunctive actions, civil penalties, and criminal penalties.

Sections 107.307 through 107.321 set forth the procedures applicable to the issuance of compliance orders under section 109(a) of the Hazardous Materials Transportation Act. Sections 107.431 through 107.359 set forth the procedures applicable to civil penalties assessed under section 110(a) of the Hazardous Materials Transportation Act. In many respects of the compliance order and civil penalty procedures are similar. Both are initiated by the issuance of a notice of probable violation with the respondent having 30 days in which to reply. Both provide the respondent with an opportunity to present a rebuttal to the allegations, in writing or orally, or a combination of both. They also provide means for settlement through informal compromise. In addition, both provide the respondent with options as to the degree of formality with which the matter is to be processed.

After receiving a notice of probable violation indicating that the OHMO is considering issuance of a compliance order, the respondent may (1) elect not to contest the allegations, (2) propose the negotiation of a consent order, or (3) contest the allegations. If a consent order cannot be worked out or the respondent contests the allegations, a hearing is held before an official who has the authority to dismiss the case or issue an order directing compliance. His order may be appealed within 20 days after service.

In a situation where an ongoing or impending violation poses a risk requiring corrective steps be taken for the protection of public health and safety without delay, an order directing immediate compliance can be issued under § 107.319. Although prior notice and opportunity for a hearing procedure do not apply in such a situation, the order is subject to administrative appeal.

Section 107.331 and 107.333 reflect generally the authority contained in section 111(a) and (b), respectively, of the Hazardous Materials Transportation Act. These provisions provide for injunctive relief and punitive damages for violations of the hazardous materials regulations and for the enforcement of compliance orders. When time permits the Department of Justice will bring an action on behalf of the MTB in the appropriate U.S. District Court. However, when the situation involves an imminent hazard the Director, MTB, may bring the action on his own motion in the appropriate U.S. District Court.

When a respondent receives a notice of probable violation indicating that the Director, OHMO, is considering assessing a civil penalty, the respondent may, within 30 days of service, (1) pay the amount of the preliminary assessment, (2) make an informal response denying the allegations in whole or in part and offering explanatory information, or (3) request a hearing. The filing of an informal response provides the opportunity for informal conference and possible compromise of the case. If a hearing is requested it will be conducted before an OHMO official who may dismiss the case or issue an order assessing a civil penalty. His order may be appealed within 20 days after service. In any case in which a civil penalty is assessed, the factors listed in § 107.359 are considered.

At any time after the issuance of a notice of probable violation in a civil penalty case and before it is referred to the Department of Justice for collection, the civil penalty can be compromised and settled by payment of the amount agreed upon in compromise.

Section 107.371 reflects the criminal penalty provided for in section 110(b) of the Hazardous Materials Transportation Act. Section 107.373 describes generally the procedures followed by the OHMO and the MTB with respect to possible criminal violations.

Since these amendments relate to practices and procedures of the MTB and its Office of Hazardous Materials Operations, notice and public procedure thereon is not necessary. Because these amendments are intimately related to action being taken in Docket HM-134 appearing elsewhere in this edition of the FEDERAL REGISTER, these amendments are being made effective concurrently with those in HM-134. However, because the MTB contemplates a review of the procedures established by these amendments after they have been in operations for at least six months and since it is desirous of public participation in that review, interested persons are invited to submit such comments as they may desire with respect to the new preemption and enforcement provisions of new Subparts C and D of 49 CFR Part 107 to the Docket Section, Materials Transportation Bureau, U.S. Department of Transportation, Trans Point Building, Washington, D.C. 20590. All comments received before the close of business on June 1, 1977, will be considered during the review, and will be available in the docket for examination, both before and after that date.

To be given particular attention during the review will be the civil penalty hearing procedures set forth in § 107.355 which the Bureau recognizes may be more formal and burdensome than necessary to fulfill the statutorily required "opportunity for hearing". Accordingly, the Bureau is particularly desirous of receiving comments with respect to this specific subject.

(41 F.R. 38167, September 9, 1976)

49 CFR Parts 107, 171, and 173
[Docket No. HM-138A; Notice No. 81-6]
Enforcement Procedures and Related Miscellaneous Proposals

On September 9, 1976, the MTB published Amendment No. 107-3 (41 FR 38167) prescribing, as pertinent here, the procedures to be followed by the MTB in carrying out its enforcement responsibilities under sections 109, 110, and 111 of the Hazardous Materials Transportation Act (HMTA) (49 U.S.C. 1808, 1809, and 1810). Because the provisions adopted therein related to the practices and procedures of the MTB, Amendment 107-3 was issued as a final rule without prior notice and comment.

However, the MTB stated in the preamble its intention to review the procedures based on experience gained in their operation and invited public participation in that review through the submission of comments. No written comments were ever received in response to that solicitation.

Notwithstanding the absence of public comments in response to the requirements adopted in 1976, the MTB has had extensive dealings with persons subject to its hazardous materials enforcement jurisdiction through individual enforcement cases. This experience, together with the ongoing evaluation of the effectiveness, efficiency, and fairness of the hazardous materials enforcement program, form the bases for the proposals made in this notice of proposed rulemaking. Many of the changes proposed herein are merely editorial in nature, others seek to remove redundancies, and the remainder actually alter the substance of, or add to, existing requirements. With respect to due process protections existing in current requirements, the proposals made herein would not detract from them, but rather add thereto. There follows an analysis by section of those proposed changes which would alter the substance of current requirements or add new ones.

PART 107: SUBPART A—GENERAL PROVISIONS

Section 107.13. Two minor changes would be made in paragraphs (a) and (h) to clarify who may issue a subpoena and the related area of how a subpoena is modified or quashed. Paragraph (a) would be amended to substitute a reference to an "official presiding over a hearing" for the current reference to "the MTB official designated to preside over a hearing." This change would reflect the fact that, although not required by statute to do so, it has been an MTB practice to obtain the services of non-MTB personnel (i.e., Administrative Law Judges) to preside over hearings.

The proposed change to paragraph (h) would make it consistent with paragraph (a) by correctly identifying to whom a subpoena recipient should apply in order to have a subpoena modified or quashed. As proposed, an application to modify or quash would be sent to the official who issued the subpoena.

SUBPART B—EXEMPTIONS

Section 107.109. This proposal would add as a basis for denying an application for exemption false statements, misrepresentations, or omissions of material fact used to support the application. This authority in the Associate Director for HMR to terminate further consideration of application for the stated reasons, would be consistent with current authority under § 107.119 allowing the Associate Director for HMR to terminate existing exemptions for the same reasons. By providing this authority at the application stage, administrative time and money can be saved in these cases where misrepresentation is a problem.

Section 107.119. Two important proposals made in this Notice relate to the enforcement of the terms and conditions of an exemption. This section contains proposals regarding the impact of enforcement actions against exemption holders and parties to exemptions on the right to continue activities under the exemption. Associated with this proposal is a proposal in § 107.307 providing for initiation of an enforcement case for failure to comply with the terms of an exemption or an approval.

Under the § 107.119 proposal, the Associate Director, Office of Hazardous Materials Regulation (OHMR), could suspend the authority to operate under an exemption during the pendency of an enforcement action against the holder thereof or party thereto. This would apply when action is brought against the holder or party in cases where the exemption itself is involved as well as in cases where it is not. In a like manner, the Associate Director, OHMR could terminate an exemption when the holder or party has been found to be in violation in accordance with the enforcement procedures in Subpart D, or when the holder or party submitted false information in its application, or if it misrepresented or failed to reveal a material fact.

The § 107.307 proposal is based on the premise that a failure to comply with the terms or conditions of an exemption or approval in effect renders the exemption or approval a nullity, and places the holder or party in the position he would have been in absent the exemption or approval, i.e., he is required to comply with the underlying regulations to which the exemption or approval was addressed. In addition to the proposals herein, language would be added to each exemption and approval to reflect this procedure. In addressing exemption and approval enforcement in this manner, the respondent would be able to avail itself of the procedural protections of the enforcement procedures.

SUBPART D—ENFORCEMENT

Section 107.229. Because establishing knowledge is fundamental to a finding of violation under the HMTA and these regulations, it is proposed to add a definition of the term "knowledge" or "knowingly".

As developed by the Supreme Court in United States v. International Minerals and Chemical Corporation, 402 U.S. 558 (1971), knowledge, as proposed herein, means that a person who is engaged in activity subject to the HMTA and the hazardous materials regulations is presumed to be aware of that statute and those regulations, as applicable to the particular activity in question. Actual knowledge that a given act is a violation of a particular requirement is not required, and the concept includes, with respect to the facts which establish a violation, what the person should have known in the proper exercise of its responsibilities. This definition is of course subject to the limitation prescribed in section 110(a) of the HMTA that excepts from liability for imposition of a civil penalty an employee who acts without knowledge (thus placing the onus solely on the employer).

In addition, it is proposed to add a definition of the term "investigation" which would encompass the inspection activity of the MTB authorized under section 109(c) of the HMTA, as well as the investigation authority granted under section 109(a). This addition is considered appropriate because the majority of the MTB hazardous materials field effort involves compliance inspections.

Compliance Orders and Civil Penalties. To the extent possible, the requirements and procedures applicable to both civil penalty actions and compliance order actions would be merged to avoid redundancy.

Section 107.307. One of the more significant substantive changes proposed in this Notice, and reflected in the merged civil penalty and compliance order provisions, is the authority of the MTB's Office of Operations and Enforcement (OOE) to seek in one notice of probable violation, both a civil penalty and a compliance order. Although the OOE has historically had this authority under the HMTA, it had never been implemented in the regulations, and thus has never been exercised. The provision for a dual notice of probable violation is proposed as an enforcement tool in those cases where the OOE has reason to believe that either a civil penalty or a compliance order alone would not achieve the desired level of compliance.

Section 107.309. A new section would be added to Subpart D covering use by the OOE of warning letters as an enforcement tool. Although this form of enforcement imposes no sanctions, its use is appropriate where an inspection reveals that a person's compliance status is generally good and that any probable violations noted are minor in nature and clearly present an aberration to an otherwise sound program.

The OOE has used the warning letter for approximately three years, and its use has demonstrated its effectiveness. Although the letter does not require a response most persons who receive one submit information explaining the cause of the problems observed or demonstrating why the observed conduct was not in violation. This form of communication between the OOE and regulated persons serves the dual purposes of informing those persons of deficiencies and also educating them as to the requirements of the Hazardous Materials Regulations applicable to their operations.

Section 107.311(c). The change proposed in this paragraph would allow the OOE to amend a notice of probable violation already issued to a respondent. This provision would enable the OOE to take action on new information that it may discover, as in the case of an inspection of a separate facility of the respondent, without having to issue a new notice. Should the OOE amend a notice of probable violation under this proposal, the respondent would be given 30 days to respond and could treat the amended notice as an initial notice for the purpose of choosing his response option (i.e., informal, hearing, or payment/compliance).

Section 107.317(b). This proposal would require a respondent, as part of his request for an informal conference, to state which of the allegations made by OOE in the notice of probable violation he admits and which ones he denies, as well as the issues the respondent will raise at the conference. This proposal is designed to identify at the outset what the conference will entail and thus enable a fuller discussion of salient points.

Section 107.319(a). The proposal in this section providing for an Administrative Law Judge (ALJ) to preside over hearings convened at the request of a respondent under proposed § 107.313(a)(3), reflects the current practice of the MTB. Although it has been determined in an opinion by the DOT General Counsel that a hearing of the type provided for in 5 U.S.C. 554 (formerly, the Administrative Procedure Act) is not required by the HMTA, the MTB has provided APA-type hearings before an ALJ and believes it is in the interest of administrative due process to continue to do so. Since the DOT does not maintain a staff of ALJs, it is necessary for the Chief Counsel of RSPA to obtain them on a case-by-case request to the Office of Administrative Law Judges, Office of Personnel Management, or through contractual arrangement with retired or otherwise inactive ALJs. Consequently, minor delays can be expected between the time of the respondent's request and the assignment of an ALJ.

It should be noted that a request for a hearing would have to be made by a respondent in both civil penalty and compliance order cases. Under current § 107.315, a hearing is automatically invoked whenever a respondent challenges an allegation in a notice of probable violation proposing a compliance order. There is no reason why the compliance order case should differ from a civil penalty case, and the MTB believes that the affirmative election for a hearing is the better procedure.

Section 107.321(a). Although the proposals in this Notice do not prescribe a full complement of procedural requirements governing the conduct of a hearing, the MTB does believe, and has proposed, that testimony offered during a hearing should be oral. With respect to other procedural requirements governing the proceeding, the ALJ would have discretion to impose whatever format he chose.

Section 107.325. Under the current regulations there are two deficiencies relating to appeals that need to be corrected. One, relating to the respondent, is dealt with in paragraph (b) and provides that in cases not involving a hearing the respondent may appeal an order of the Associate Director for OOE to the Director, MTB. Although not currently provided in the regulations, this right of appeal has been provided to each respondent since the inception of the hazardous materials enforcement program of the MTB.

The other deficiency is the failure of the regulations to permit the OOE to appeal an adverse decision of an ALJ to the Director of MTB, thereby placing OOE in a procedural position inferior to the respondent. Accordingly, paragraph (a) permits either party to an enforcement hearing to appeal an adverse final order of an ALJ.

In addition, paragraph (c) provides basic requirements that an appeal must conform to, and is designed to assist the Director of MTB in making a thorough and expeditious decision.

Finally, a new paragraph (f) would make failure to comply with a term or condition of a compliance order a basis for OOE initiating an enforcement case.

Section 107.327. This section would prescribe the mechanisms governing an offer in compromise by the respondent in both civil penalty cases and compliance order cases. An offer of an amount in compromise of a civil penalty proposed or assessed would stay the running of any response period then outstanding. This proposal would merely formalize current practice. If accepted, an amount in compromise would constitute a full satisfaction of the civil penalty and would have no effect on the finding of a violation. Thus, the compromise would go only to the amount of the penalty, and not to the underlying violations on which the penalty is based.

The proposals concerning compliance orders in this section are designed to be more specific than current language. For example, under current language there is no requirement that the respondent identify the facts or proposed compliance order terms and conditions he challenges. The proposal would require a statement as to both. In addition, language would be added in paragraph (b)(3) establishing an administrative cause of action for failure to comply with the terms of an executed consent order (this provision would thus be consistent with the proposed requirement in § 107.325 that would make failure to comply with the terms of a compliance order a basis for the MTB to initiate an enforcement case).

Section 107.331. An amendment to paragraph (b) is necessary in order to make clear that civil penalties for up to the $10,000 maximum provided in the HMTA, may be assessed for each violation of a requirement relating to the manufacture, fabrication, marking, maintenance, reconditioning, repair, or testing of a container or package. In this context, each violation means each container found to have been in violation of an applicable requirement.

Section 107.333. Although no changes are proposed in the statutorily prescribed assessment criteria, this rulemaking provides an opportunity to state the MTB position regarding use of these criteria (currently prescribed in § 107.359). It has been the practice of the MTB, and this practice would continue under these proposals, to use the assessment criteria merely as general guidelines in establishing the preliminary civil penalties proposed in the Notice of Probable Violation. The MTB believes this to be appropriate since at the Notice stage no assessment of a final civil penalty has been made.

In addition, with respect to the assessment criteria relating to the economic impact of a civil penalty on a respondent, the MTB believes that that is a matter best raised by the respondent if that factor is to have the greatest effect on penalty mitigation. Only the respondent can gauge, and provide pertinent and current information concerning, its ability to pay and its ability to continue in business in the face of the civil penalty proposed in the notice of probable violation or assessed under an order.

PART 171

The definition of "person" in § 171.8 would be amended in this proposal to correct an error that occurred in the final rule consolidating the Hazardous Materials Regulations (HM-112, 41 FR 15995, April 15, 1976). In that rule, where "person" was first defined, the MTB's predecessor excluded governmental agencies from the definition to reflect the fact that the Hazardous Materials Regulations have never been applied directly to regulate customary governmental activity. Such activity as the transportation, for governmental purposes, of hazardous materials by an agency of the Federal, State or local governments is not regulated. However, the act of offering such materials to a commercial carrier for transportation has always been considered subject to the regulations. This fact is witnessed by the existence of exceptions for certain Federal activities (see, for example, 49 CFR 173.7). Similarly, a governmental agent boarding a commercial passenger aircraft with a hazardous material in his possession is employing a commercial activity subject to the regulations.

Where governmental functions may be mixed with commercial activity, such as where an employee of a State university transports a hazardous material, the facts of the specific activity will have to be examined to determine whether the transportation is for a governmental purpose and therefore not subject to the regulations.

As noted previously, § 171.2 would be revised to indicate that not only do the specific requirements prescribed in subchapter C apply to the activities of persons engaged in hazardous materials transportation, but also the requirements imposed through exemptions and approvals issued under the Hazardous Materials Regulations. This proposal is a necessary adjunct to the language proposed in § 107.307 (discussed above in connection with § 107.119) establishing the violation of the term of an exemption, approval, or order as a basis for asserting the enforcement jurisdiction of section 110 of the HMTA.

In addition, it is proposed to add language to § 171.2, to clarify the fact that DOT markings and designations may not be used on any container, or in connection with any shipment, that is not in full compliance with all applicable provisions in the Hazardous Materials Regulations. Thus, for example, this language is intended to make clear the fact that a drum reconditioner may not place his drum reconditioner's registration number bearing the letters DOT, on a drum which is not a fully complying DOT specification container (including all required embossment markings).

(46 F.R. 47091, Sept. 24, 1981)

49 CFR Parts 106, 107, 171, and 173
[Docket No. HM-138A; Amdt. Nos. 106–4, 107–11, 171–70, 173–161]
Exemption and Enforcement Procedures
and Related Miscellaneous Provisions

These amendments are based on proposals made in Notice 81-6 (46 FR 47091; September 24, 1981). A total of 21 commenters responded to the Notice, representing 10 trade associations, 7 corporations, 3 Federal agencies, and one state agency. The majority of the comments were well developed and in several cases offered alternative approaches or language. Consequently, several important changes have been made to the proposals in the Notice. These changes take the form of additional clarifying information, or a return to previous provisions in lieu of specific proposals. The following section-by-section analysis discusses the comments, the changes made in response to them, and reason for not making some of the changes suggested.

PART 107: SUBPART A—GENERAL PROVISIONS

Section 107.13. One commenter suggested that the reference in paragraph (a) to the "official designated to preside over a hearing" be changed to include a reference to an administrative law judge. The change has not been made since § 107.13 is a general section which includes other proceedings, in addition to enforcement hearings, which would not necessarily be conducted by an administrative law judge.

The same commenter stated that paragraph (h) should provide that an application to quash or modify an administrative subpoena issued under § 107.13 be submitted to an official other than the person who issued the subpoena. MTB does not agree. The person who issued the subpoena is in the best position to correctly and expeditiously weigh the reasons to quash or modify against the reasons on which he or she relied in issuing the subpoena.

SUBPART B—EXEMPTIONS

Section 107.109. In responding to paragraph (c)(1), three commenters expressed concern that MTB's efforts to cull out at the application stage exemption requests based on false or misleading statements could lead to denials where the statements or omissions were the result of inadvertency or were so inconsequential as to have no bearing on the merits of the application. MTB agrees that the proposed language could be misunderstood and has adopted the change suggested by two commenters that the words "false" and "misleading" be qualified by the word "materially," so that an exemption will be denied under paragraph (c)(1) only when the information in question relates directly and substantially to the requirements prescribed in § 107.103.

Section 107.119. Of all the proposals in the Notice, none received more adverse comments than those relating to exemption amendment and suspension (paragraph (b)(2)) and exemption termination (paragraph (c)(4)). After reviewing the 11 comments received, in light of the intent of the proposals and the enforcement needs of MTB, § 107.119 has been changed substantially from the proposals in the Notice.

Recognizing, as the comments point out, the potential unfairness and operational hardship that could result from action against an exemption holder based on enforcement actions not necessarily involving the exemption or not yet resolved, proposed paragraphs (b) and (c) have been redone to substantially reflect current language. A new paragraph (e) has been added to provide the Associate Director for HMR with the enforcement tool which MTB still believes is necessary for oversight of operations conducted under exemptions. As adopted, paragraph (e) permits the Associate Director for HMR to refer an exemption to the Associate Director for OOE for initiation of an enforcement case based on that exemption. If the enforcement case results in a finding of violation against the holder or a party to the exemption, the Associate Director for HMR could use that finding as a basis for amending, suspending, or terminating the exemption as to the exemption holder or party involved.

This approach should be an appropriate response to those commenters who felt that paragraphs (b) and (c) would have permitted the Associate Director for HMR to take action against one exemption for alleged failure to comply with other exemptions or other unrelated regulatory provisions.

SUBPART D—ENFORCEMENT

Section 107.229. Although only four comments were addressed to the proposed definition of "knowledge" or "knowingly," they reflect generally a disagreement either as to the need for the definition or the approach used. Two commenters stated that basing the definition in part on United States v. International Minerals and Chemical Corporation, 402 U.S. 558 (1971), was improper because, as one commenter argued, under that case ignorance could be a defense, or as stated by the other commenters, the Hazardous Materials Transportation Act was not involved in the International Minerals case.

Notwithstanding those comments opposed to the definition, MTB believes that it is appropriate to include a definition of this critical concept in the enforcement procedures. However, MTB recognizes that modification of the proposed definition is necessary. Accordingly, these amendments to Part 107 adopt a definition of these key terms which is simpler than that proposed while still addressing a point raised in the vast majority of enforcement proceedings held to date, i.e., the respondent's inadvertency and lack of intent to violate the regulations.

Section 107.301. Although the proposed change to paragraph (e) was not discussed in the preamble to the Notice, and was only provided to describe the jurisdiction of the various modal administrations in DOT having hazardous materials enforcement responsibilities, three commenters opposed its adoption. The proposal was included to reflect the fact that MTB has an overlapping and secondary jurisdiction over shipments which move by more than one mode of transportation. Although the range of this jurisdiction is theoretically large due to the fact that a large number of hazardous materials shipments move by more than one mode of transportation, in fact and in practice the overlap is quite small, and through coordination with other DOT elements, does not burden regulated persons unreasonably. In response to the comments, and to clarify the enforcement jurisdiction of the MTB, this section has been changed to reference the delegation of authority cited in 49 CFR Part 1 (§ 1.53).

Section 107.305. One commenter contended that paragraph (a) relating to the authority of OOE to conduct investigations was overboard in its statement of OOE's investigative jurisdiction. In linking the actual extent of OOE's investigative jurisdiction with OOE's enforcement jurisdiction, the commenter states that OOE cannot investigate what it is not authorized to enforce. Although paragraph (a) was intended to reflect the investigation-enforcement relationship suggested by the commenter, MTB agrees that the clarity of this relationship could be improved. Accordingly, paragraph (a) has been emended to make specific reference to the delegations cited in 49 CFR Part 1.

Another suggestion by the same commenter concerning termination of an investigation under paragraph (d) also has been adopted. As amended, paragraph (d) requires OOE to notify a person who has been the subject of an investigation when the investigation is terminated.

Four comments were received concerning the confidentiality provision of paragraph (e). Three of those comments stated that the protection afforded respondents during the pendency of a case is undermined by the statement that confidentiality will be granted "unless otherwise determined by the OOE." Despite the fact that this language is currently in paragraph (e) and was thus not a part of the Notice, MTB agrees with the commenters that the provision should be amended to remove any reference to a discretion in OOE to deny confidentiality of information or persons identified in an investigation. However, it should be noted that OOE could be directed by a court in a given case to reveal information or identities granted confidentiality under paragraph (e).

Finally, MTB rejects the suggestion of one commenter that reference should be made in paragraph (e) to 18 U.S.C. 1905 concerning criminal sanctions for government employees who unlawfully disclose information arising out of investigations such as those authorized by § 107.305. It is not appropriate to place in these procedural requirements a citation to sanctions against Federal employees where such requirements apply independently of the enforcement program of MTB.

Section 107.307. MTB's reference in paragraph (a) to violation of an approval as a basis for enforcement action was criticized by two commenters. Arguing that failure to comply with the requirements of an approval is in essence noncompliance with the underlying regulation requiring the approval, these commenters contend that a person does not violate the approval and thus cannot be the subject of an enforcement action for an alleged failure to comply with an approval. MTB has reevaluated this proposal in light of these comments and believes that there is merit to their position. Recognizing that there is a fundamental difference between exemptions and approvals, MTB is persuaded that they cannot be treated in the same manner. As noted by one commenter, an approval is merely an extension of the regulation that establishes it, and as such the responsibility to comply is based on the regulation not the approval. Accordingly, it is appropriate as recommended by these commenters to delete the reference to approvals where appropriate throughout the proposals, and these changes have been adopted herein. However, it is important to note that failure to comply with a condition or term prescribed in an approval would be considered failure to comply with the underlying regulation requiring the approval, and could be the basis for an enforcement action.

Another commenter objected to paragraph (a) because "The revision essentially permits OHMR (sic) to assess a civil penalty or issue an order directing compliance without first taking the time to evaluate the nature of the violation." The point raised by this comment does not address a revision to paragraph (a), but rather deals with existing language. In addition, the commenter is incorrect in its view of what paragraph (a) authorizes the Associate Director for OOE to do. The paragraph merely authorizes the conduct of proceedings by which appropriate sanctions are determined and applied. Section 110(a) of the Hazardous Materials Transportation Act (HMTA) (49 U.S.C. 1809(a)) mandates the consideration of several assessment criteria before an order imposing a civil penalty may be issued. It is the proceeding which establishes the record on which a sanction is based.

Section 107.309. Only one commenter expressed any reservation concerning the proposal dealing with the issuance of warning letters. The commenter contends that the proposal is deficient because it provides no guidance to regulated persons as to circumstances under which warning letters would be issued in lieu of compliance orders or civil penalties. The preamble discussion of this proposal in the Notice gave an example of a typical situation in which a warning letter is frequently used. MTB believes that the preamble language is sufficient to describe generally how warning letters are used, and accordingly, has not made any changes to the proposal.

Section 107.311. One commenter recommended that this section be amended to include language requiring issuance of notices in a timely manner. The commenter states that delays between inspection and issuance of notices can exceed a year, and that such delays are per se prejudicial to the respondent. MTB disagrees strongly that delay between inspection and notice is "per se prejudicial." Clearly if a delay does adversely affect a respondent's ability to respond, that fact would be a relevant part of the respondent's defense to the allegations in the notice. At the conclusion of every inspection, the person inspected is put on notice during an exit interview as to any probable violations noted and the possible sanctions that could result. Given the small staff available to conduct inspections and the technical nature of the subject matter, delays are inevitable, but the OOE strives to keep them to a minimum. Accordingly, the suggested change to § 107.311 has not been made.

Another commenter argued that under paragraph (c), amendments to notices should be allowed only where the new information relates directly to the allegations in the original notice. In all other cases, OOE should have to issue a new notice based on new allegations. MTB believes this recommendation has merit, and paragraph (c) has been changed accordingly.

Section 107.313. As noted in the preamble to the Notice, one of the primary features of the proposals was elimination of redundancies between the procedures applicable to compliance orders and civil penalties. One aspect of that consolidation was the combining in one enforcement case of both a civil penalty and a compliance order. However, that proposal in paragraph (b) was objected to by a commenter who argued that even if such action was legal (which the commenter doubted), use of the proposed standard "where appropriate" was unclear and unreasonable. Furthermore, the commenter argued that in no event "should it be deemed 'appropriate' to assess a civil penalty if the notice of probable violation has not proposed such penalty."

The decision as to which enforcement sanction to apply in a given matter can only be determined on a case-by-case basis. The responsibility of MTB in each case is to provide an adequate basis in the notice of probable violation to inform the respondent of the nature of the allegations and sanctions selected so he can respond effectively. As to the commenter's second point, the intent of the proposal was to provide for a double sanction in the notice, and MTB agrees that it should be made clear that in no case should a double sanction be applied at the order stage without having first been proposed at the time of the notice (or amended notice under § 107.311). Accordingly, a clarification has been added to paragraph (b).

Another commenter, noting the variation in enforcement procedures between the five administrations in DOT having enforcement responsibility under the HMTA, contends that the enforcement procedures proposed in Notice 81–6 do not make it clear that failure to request a hearing under this section and § 107.319 constitutes a waiver of the right to request a hearing. The commenter is correct in stating that any reply other than under paragraph (a)(3) and § 107.319 constitutes a waiver of the right to request a hearing. To clarify this fact, express waiver language has been added to paragraph (b).

Section 107.315. The only comment addressed to this proposal argued that it was deficient because it failed to deal with the possibility that a respondent may wish to admit some probable violations while denying others. MTB believes that a respondent in this position could, and is encouraged to, respond under §§ 107.313(a)(2) and 107.317 by making an informal response, or under §§ 107.313(a)(3) and 107.319 by requesting a hearing. Under either course, the case would proceed only with respect to those probable violations put at issue by the respondent.

Section 107.317. Although not specifically identifying the proposal, one commenter apparently opposed the provision that would require a respondent, as part of its request for an informal conference, to state which allegations are admitted and which are denied, as well as the issues that will be raised at the conference. The commenter stated that this requirement would frustrate what the commenter described as "settlement conferences."

It should be noted that the informal conference is not intended to be a settlement conference, although the discussion of that point could be relevant. Rather, the informal conference is an aspect of the informal response that is designed to allow an effective and less costly alternative to a hearing. However, MTB does believe that this purpose could be served without a need for the respondent to state which allegations are admitted or denied, and that proposal has been deleted. The proposal to require a statement of the issues to be raised has been retained, however.

Finally, pursuant to a comment received, language has been added to paragraph (c) enabling a respondent to request that an informal conference be held by telephone.

Section 107.319. As proposed in the Notice, paragraph (b)(2) of this section contained a typographical error which stated that issues in a case that was to be the subject of hearing had to be admitted or denied. Under current language to which no proposed change was to be made, paragraph (b)(2) requires that a respondent requesting a hearing "state with respect to each allegation (made in the Notice) whether it is admitted or denied."

In commenting on this erroneous proposal, one commenter pointed out the burden it would place on a respondent at the notice stage of the proceeding. Because the comment deals primarily with what the commenter believes to be an unfair demand for detail at the earliest stage in the proceeding, it is relevant to paragraphs (b)(2) and (b)(3) which do require the respondent to "state with particularity the issues to be raised by the respondent at the hearing." MTB believes that there is merit to the points raised by the commenter to the extent they suggest changes that would enable MTB to know generally with what issues the hearing will deal, while at the same time not prejudicing the respondent's case at the hearing. Accordingly, the language of paragraphs (b)(2) and (b)(3) has been modified to require the respondent to identify which allegations of violation, if any, it admits, generally what issues it will raise at the hearing, and provides that issues not raised in the hearing request will not be barred from presentation later.

Section 107.321. In response to the points made by one commenter, two important changes have been made to the procedural requirements of this proposal. In order to achieve the consistency between cases, cited by the commenter as being valuable in establishing precedents and providing a more uniform record, paragraph (b) has been modified to require that all hearings are held in accordance with the Federal Rules of Civil Procedure and the Federal Rules of Evidence. Authority has been granted the Administrative Law Judge (ALJ) to modify their application where the ALJ determines modification is necessary in a given case. As noted by the commenter, this change will avoid problems that have arisen where non-DOT ALJs, with varying backgrounds, use procedures with which they are familiar but which may appear awkward for practitioners who are not familiar with them.

Another change recommended by the commenter, and adopted herein, is deletion of the requirement that all testimony be given orally. To the extent that written testimony can contribute to a more orderly process, more effective determination of the issues and facts in a case, and a better record for appellate review, its use should not be discouraged. Accordingly, the procedures have been amended to remove the provision, and thereby leave to the ALJ and parties the determination as to the type of testimony to be used.

Proposed paragraph (c) has been modified to remove, as superfluous, the reference to OOE offering information "necessary to fully inform the presiding officer. . . . " Reference to OOE having the burden of proof is all that is necessary in addressing its responsibility to provide information in support of its allegations.

Another commenter suggested that the phrase "in defense of the allegations" appearing in paragraph (d) be changed to reflect the fact that the respondent is not providing a defense of the allegations, but rather is disputing those allegations. MTB agrees with this point and an appropriate change to paragraph (d) has been made.

Section 107.323. One commenter noted that paragraph (a) states that the decision by an ALJ in a hearing proceeding represents "the final decision" in the proceeding. The commenter questions how the decision of the ALJ can be considered final when an appeal to the Director, MTB is available to both parties under proposed § 107.325. The commenter's point is well taken and an appropriate change to paragraph (a) has been made to remove the reference to "final decision."

The same commenter noted what it considered other inappropriate language in paragraph (a). The commenter contends that the reference to "the relief sought by the OOE" should be changed to more accurately reflect what OOE seeks to obtain through the proceeding. MTB agrees that the language should be clarified, and accordingly, the word "relief" has been changed to "sanction."

Section 107.325. The proposal to permit the OOE to appeal an adverse ruling by an ALJ to the Director, MTB was objected to by three commenters. Basing their opposition primarily on the view that since OOE is in effect a "prosecutor," and should not, therefore, be able to appeal an adverse ruling, these commenters argue that if an appeal is to be granted due process dictates that the appeal be to someone other than the person who has ultimate responsibility over OOE's program.

As to the contention that OOE should not be able to appeal an adverse ruling by an ALJ, MTB does not agree that due process considerations preclude this right of appeal. The enforcement program of the OOE is clearly civil in nature—both in practice and as contemplated by Congress in passage of the HMTA, and consequently, OOE is not precluded from appealing adverse rulings.

The commenter's argument against OOE's right to appeal to the Director, MTB, however, is sound. To avoid, to the degree practicable, the appearance of a conflict of interest, these amendments adopt the approach used by the Federal Railroad Administration, and permit an aggrieved party to appeal an adverse decision of an ALJ to the administrator, Research and Special Programs Administration. This action elevates consideration of issues on appeal to the highest level in the agency and represents a desired separation between the office presenting the case and the ultimate decision maker on appeal. In non-hearing cases, the Director, MTB would continue to decide all appeals.

Another commenter argued that the proposal in paragraph (f) to make violation of the terms of a compliance order the basis for an enforcement proceeding, was contrary to the HMTA. A review of the HMTA and the comment confirms the correctness of the commenter's argument, and the proposal has been deleted accordingly. Also, the identical language in § 107.327(b)(3) has been deleted.

The same commenter objected to language in proposed paragraph (e) authorizing the Director, MTB to maintain an order in effect despite an appeal to the order if he determines that action to be necessary "in the public interest." MTB agrees with the commenter that standards for making that decision should be stronger, and has adopted the commenter's suggestion that it be based on a finding of "immediate danger to the public."

Section 107.327. Two commenters objected to the failure of the compromise procedures to provide for settlement of an enforcement proceeding with no admission by the respondent of violation. One commenter argued that without a provision for settlement resulting in no finding of violation, amicable resolutions of enforcement proceedings will be frustrated and respondents will be required to bear the expense of fighting allegations of violations to conclusion. The other commenter contended that without a provision for settlement accompanied by no finding of violation, respondents will be unreasonably exposed to "private challenges based upon an admission of guilt."

To the extent that these arguments establish a need for providing for the settlement of cases without a finding of violation, MTB believes they have merit. In fact, settlements of the type contemplated by the commenters have been reached in a few cases. However, these amendments retain the concept of compromise currently in use in MTB's hazardous materials enforcement program. In the vast majority of cases to date, the facts upon which the allegations of violation have been based have not been in dispute. As such, the commission of those violations is pertinent under the assessment criteria in the HMTA and the HMR in subsequent cases involving the same respondent.

On the other hand, there have been cases where a settlement involving only the payment of money by the respondent is the most reasonable solution. Consequently, these amendments adopt a specific provision for settlement without a finding of violation. However, discretion remains with the Associate Director for OOE to reject settlement offers, and this discretion would most likely be exercised in cases where only the amount of civil penalty is in dispute.

In response to comments, two other changes have been made to this section. The proposal that a compromise offer be in the form of a certified check or money order submitted prior to an agreement accepting the offer has been deleted, since the language of new paragraph (a)(1) now speaks to the compromise offer being initiated by either respondent or OOE. The requirement that the respondent include, as part of a proposed consent agreement, an acknowledgement that the notice of probable violation may be used to construe the terms of the compliance order resulting from the agreement, has been deleted as being unnecessary as a rule of general applicability.

Section 107.329. Although not representing a new proposal in Part 107, this section was objected to by one commenter who contended that provision in the HMR for a compliance order for immediate compliance has no legal basis in the HMTA. Viewing § 107.329 as a means of addressing imminent hazard situations through the administrative process, the commenter argues that the HMTA provides for imminent hazard situations in section 111(b) (49 U.S.C. 1810(b)) by requiring DOT to proceed judicially, either on its own motion or through the Attorney General. Consequently, this section, which enables the waiving by OOE of the administrative procedures otherwise applicable to compliance order proceedings, represents, argues the commenter, a denial of due process.

MTB agrees that, with respect to cases involving imminent hazard situations as defined in section 111(b) of the HMTA, use of § 107.329 would operate to abrogate the due process rights of the respondent by denying it the right to be heard. However, § 107.329 was designed to reach cases falling short of the imminent hazard standard, but which are of sufficient potential safety impact to warrant extraordinary treatment.

Notwithstanding the purpose and intent of § 107.329, experience does not indicate a need to retain it. With the range of legal and equitable remedies available under the HMTA, particularly the imminent hazard provisions of section 111(b), as implemented in § 107.341, all factual situations likely to occur are provided for. Accordingly, § 107.329 has been deleted.

Section 107.331. (§ 107.329, as adopted). The proposal to amend the maximum penalty provision applicable to container manufacturers, reconditioners, repairers, and retesters received numerous adverse comments due to the interpretation given the proposal in the preamble. Although the amendment to this section merely adopts the language of section 110(a) of the HMTA, the preamble noted that " . . . each violation means each container found to have been in violation of an applicable requirement."

A common thought expressed in most of the comments was that the above view was not possible under the HMTA, since Congress has applied the concept of a continuing violation only to violations involving shippers and carriers of hazardous materials and not to the manufacturers, rebuilders, reconditioners, and retesters of the containers in which those materials are transported. Consequently, those commenters argue that it would be unlawful for MTB to cite, for example, each of 200 cylinders in a given lot for a violation common to that lot, e.g., improper marking. To do so would be to cite a continuing violation.

MTB has reviewed thoroughly the questions posed by the commenters regarding this issue, particularly the contention that to cite each container in a lot would be to cite a continuing violation, and although not agreeing totally with the analyses offered by the commenters, does believe that adjustments to its interpretation are warranted.

Moreover, these changes will not, based on experience gained under the HMTA, frustrate the purposes of the enforcement policy, since the actual amendment merely reflects the language of the HMTA.

Throughout the container specifications in Part 178, there are numerous requirements for sample testing where one container, or a portion of one container, is authorized to represent the compliance status of all other containers of the same lot. Similarly, there are certain testing requirements which must be performed at regular intervals, e.g., every four months or at the start or restart of production. Violations involving groups of containers which are permitted to be certified, marked, and sold based on representative functions being performed on a certain number of samples of those groups, are considered as a single violation. Thus, if a leakage test for 17E drums under § 178.116–13 is improperly performed and constitutes the basis for a violation, only one violation has been committed regardless of how many 17E drums of a given size and type are produced during the four months covered by the leakage test. Similarly, the physical test requirements of § 178.37-16, if improperly performed, would result in one violation for each type of test and not for each of the 200 cylinders for which the sample (coupon) is representative.

In stating this enforcement position on the issue, it is MTB's intent to reflect a level of regulation appropriate to observed enforcement needs. If in the future it should appear that those needs require the expanded civil penalty authority possible under section 110 of the HMTA, that authority would be asserted through additional rulemaking.

PART 171

Section 171.2. Several changes have been made in the proposals in this section based on the views of a commenter. First, as noted in the discussion of § 107.307, the references to approvals have been deleted, since, as the commenter pointed out, an approval is merely an aspect of an underlying substantive regulation, and any failure to comply with the terms of an approval would constitute a failure to comply with the underlying regulation.

Secondly, paragraph (d)(4) has been deleted as being vague. Should cases of misleading marking involving marks other than those described in subparagraphs (d)(1), (d)(2), and (d)(3) arise, such marks could be identified in subsequent rulemaking. Finally, paragraph (e) has been deleted as superfluous because the provisions of paragraphs (c) and (d) adequately address container manufacturers, reconditioners, repairers, fabricators, and retesters.

Section 171.8. The proposal to amend the definition of person to include governmental entities received several comments, each of them opposed to the provision as proposed. The two Federal agencies that responded to the expansion of the definition both argued that since the HMTA did not define the term person, the regulations cannot create a definition. Furthermore, since the current definition in § 171.8 was carried over from predecessor regulations adopted pursuant to 18 U.S.C. 831 et seq., where the statutory definition did not include governmental entities, there is no basis in law to include them in this rulemaking. In light of the comments received, and based on a review of interpretative difficulties that have arisen over the absence of a definition of "person" in the HMTA, the MTB has decided to withdraw the proposal to adopt one in this rulemaking, and pursue instead a statutory amendment to the HMTA. This effort will commence with a proposal in the Department's legislative initiatives submitted to the 98th Congress.

Miscellaneous changes. In addition to the amendments discussed above, two other changes have been adopted in this rulemaking. These changes, neither of which appeared in the Notice, are designed to facilitate activities under the HMR and impose no regulatory burden on persons operating thereunder.

Section 171.19 has been amended to delete the termination date applicable to approvals and authorizations issued by the Bureau of Explosives of the Association of American Railroads. Section 171.19 was adopted pursuant to MTB's phased program to withdraw preexisting approvals and authorizations issued by the Bureau of Explosives, and recognized the need to provide for transition between the elimination of such approvals and authorizations and the issuance thereof by MTB. However, it now appears that numerous Bureau of Explosives approvals and authorizations were issued without termination dates. Consequently, to withdraw them on December 31, 1984, would place a prohibitive administrative burden on MTB, and a concomitant burden on affected industries. Accordingly, in lieu of the current termination dates. Consequently, to withdraw them on December 31, 1984, would place a prohibitive administrative burden on MTB, and a concomitant burden on affected industries. Accordingly, in lieu of the current termination date, MTB has adopted all written approvals previously issued by the Bureau of Explosives as if they had been issued by MTB. This action in no way precludes MTB from taking future rulemaking action to terminate specific classes of approvals and authorization if it determines that conditions of safety and administrative efficiency warrant. It must be stressed that as conditions to the conduct of operations under such preexisting approval, the approval must be in writing, be available for inspection by personnel from the Department on demand, and must have been issued without a termination date. If any or all of these conditions are not met, the approval is a nullity and operations under the invalid "approval" would subject the person conducting them to all applicable sanctions under the HMTA. Furthermore, it should be noted that these approvals are viewed as equivalent to MTB issued approvals from an enforcement standpoint, and enforcement action may be taken against a person who fails to comply with all terms and conditions prescribed in the approval.

Another change, to § 173.22, has been made to enable a shipper to rely on exemption or approval markings in discharging its responsibility under that section to determine that a packaging or container is in compliance with an applicable specification in Part 178 or 179, or, as adopted herein, an approval or exemption. This change is necessary to clearly articulate the shipper's duty, while enabling it to rely on the most immediate evidence of compliance. Absent this change, the shipper would be technically in violation if he used a packaging or container, that bore no specification marking under Parts 178 and 179 (this does not abrogate any exception authorized by this subchapter). It should be noted, however, that compliance with these requirements, and reliance on the actions of others permitted by this section do not in any way relieve a person from complying with all other applicable requirements or conditions of use.

In addition, the shipper may also rely on specification markings which were, at the time of manufacture of the container, the appropriate markings, although at the time of the offering for transportation are no longer in effect.

(48 F.R. 2646, January 20, 1983)

[49 CFR Part 178]
[Docket No. HM-140; Notice No. 76–6]
SHIPPING CONTAINER SPECIFICATIONS
Notice of Proposed Rulemaking; DOT Specification 17H Modification

The Materials Transportation Bureau (MTB) is considering an amendment to Part 178 of the Department's Hazardous Materials Regulations pertaining to the minimum thickness of the removable head of a 55-gallon capacity, DOT specification 17H steel drum. The Bureau proposes to change the minimum thickness of the removable head from 16-gauge to 18-gauge.

The Bureau had previously received two applications requesting exemptions from the minimum thickness requirements specified for certain components of a DOT-17H steel drum. One application requested the use of an 18-gauge removable head on a 55-gallon capacity, DOT-17H steel drum. The specification requires a 16-gauge removable head for a drum of this capacity. The other application requested the use of a 20-gauge body and bottom head on a 30 gallon capacity, DOT 17H, steel drum. The specification requires an 18-gauge body and bottom head for a drum of this capacity. The applications were denied because adequate justification was not provided.

The Bureau is proposing a reduction in the minimum thickness of the removable top head of a 55-gallon capacity, DOT-17H drum because a further review of the performance test results and of the specification thickness requirements of the heads of the drum indicate that further study is warranted.

One feature of the DOT 17H specification, as well as other DOT packaging specifications, that is not related to a performance test is puncture resistance. This aspect is addressed only by gauge thickness specifications. In the case of the 55-gallon capacity drum, the bottom head may be constructed of 18-gauge steel. Therefore, the Bureau believes there is no basis for the requirements of the removable head to be continued at 16-gauge for purposes of puncture resistance. Accordingly, the Bureau proposes to change the specification requirements of the 55 gallon capacity, DOT-17H drum to authorize the same gauge for the removable head that is specified for the bottom head.

The proper mating of removable heads to drums is essential to assure compliance with specification requirements. Such mating, adequate to prevent leakage, is accomplished by the assembly of the removable head to the drum body in conjunction with the use of a gasket and a locking ring.

The Bureau anticipates comments expressing concern that a 12-gauge bolted ring with drop forged lugs and a ⅝-inch bolt or an approved equally efficient closure, as specified in § 178.118–8, will not be suitable in conjunction with an 18-gauge removable head. This is a valid concern; however, such a concern may also be applicable to a 16- or 14-gauge removable head. The Bureau would welcome a discussion of this subject in the comments it receives to this notice.

(41 F.R. 41713, September 23, 1976)

[Docket No. HM-140; Amdt. No. 178–44]
[49 CFR]
PART 178—SHIPPING CONTAINER SPECIFICATIONS
DOT Specification 17H Modification

On September 23, 1976, the Materials Transportation Bureau (MTB) published a Notice of Proposed Rulemaking, Docket No. HM-140; Notice 76–6 (41 FR 41713) which proposed this amendment. The background and the basis for incorporating this change into the regulations was discussed in that notice. Interested persons were invited to submit their comments by the close of business on November 30, 1976. Primary drafters of this document are Darrell L. Raines, Office of Hazardous Materials Operations, Regulations Division, and Douglas A. Crockett, Office of the Assistant General Counsel for Materials Transportation Law.

There was a total of 8 comments received regarding this proposed amendment. Four were in complete agreement.

One commenter did not recommend approval until sufficient data are obtained to show that overall integrity of the drum will not be affected. The MTB has witnessed satisfactory testing of 55-gallon DOT-17H drums, with an 18-gauge removable head, which were equipped with a new type of closure that was approved by the Bureau of Explosives and assigned Approval Number BA 2058. This particular closure appeared to be an improvement over the standard closure now required by 49 CFR 178.118-8 for 55-gallon drums.

One commenter recommended approval of the 18-gauge cover provided the closure system described in BA 2058 is used. This same commenter expressed concern about the reuse of an all 18-gauge drum without any specific stipulation in the regulation as to the closure system. The MTB believes that the additional requirements of footnote 3 will eliminate the concern expressed by this commenter.

One commenter expressed his opposition to the proposed change by stating

... Common sense suggests that a steel drum manufactured with a lighter gauge metal cannot be as safe a shipping container as a heavier gauge. He recommended that the Bureau develop and adopt more discriminatory tests. In addition, this commenter recommended that a reconditioned drum have a cover of original manufacture and not permit an 18-gauge converted cover. Another commenter strongly suggested that footnote 3 not be amended as proposed, but instead, the MTB should add a footnote (4) to read "18-gauge authorized provided the drum is fitted with the closure specified in Bureau of Explosives' Authorization BA 2058.

In view of the comments received and upon further consideration, the MTB believes that the 18-gauge removable head should only be authorized upon demonstration of successful tests and manufacturers' registration with OHMO. This amendment is effective upon publication in the FEDERAL REGISTER, rather than 30 days thereafter, since it is a relaxation of existing restrictions on the construction of DOT Specification 17H drums.

(42 F.R. 61464, December 3, 1977)

[49 CFR Part 172]
[Docket No. HM-141]

COLOR CODING OF COMPRESSED GAS PACKAGES
Advance Notice of Proposed Rule Making

The Materials Transportation Bureau (Bureau) has been asked to consider amending Part 172 of the Hazardous Materials Regulations as they apply to the identification of compressed gases in cylinders by means of a color coding system (system) applied to the individual compressed gas cylinder (cylinder). There have been numerous bills considered by both houses of Congress in the past several years to establish this system. The purpose of the system is to provide safety for any person within the sphere of hazard surrounding the cylinder whether the cylinder is in transportation, use, or storage. Further, it was anticipated that the system would protect any person using the gas from the cylinder by identifying the gas contained in the cylinder.

To develop a Notice of Proposed Rulemaking, certain information is required and, therefore, the Bureau is providing this opportunity for comment from the public for the development of a system of color coding compressed gas packagings. Comments should be addressed to the following subjects:

(1) How useful is a color coding concept?

(2) Should all types of compressed gas packagings be included in the system such as cylinders, cargo tanks, tank cars, and portable cargo tanks?

(3) Should there be a volume limitation for packagings related by the system?

(4) Should this volume limitation be expressed by volume of gas contained or water capacity in the packaging?

(5) Should one of the following standards be adopted:

(a) The Defense Supply Agency Standard Interim F11GT/162, Appendix B, Reference Drawing Group B;

(b) American National Standards Institute Standard ANSI Z48.1-1954 (R1971) Method of Marking Portable Compressed Gas Containers to Identify the Material Contained;

(c) Color coding by use of a four color code system to identify the UN numbering system for compressed gases, this color code system is to be similar to that used by the Electronic Industry Association Standard RS-359, August 1968 (Reaffirmed August 1972). This color coding numbering system could be accomplished by painting the entire packaging in a four color combination or by use of a 2-inch × 12-inch pressure sensitive material which has four color bands 3 inches wide;

(d) Single color code system for each compressed gas;

(e) Metal stamping or embossing name of package contents on the shoulder of the package or markings similar to those specified in CGA Phamphlet C-7, Appendix A; or

(f) Paint stencil or silk screen name of contents in one location longitudinal to the package sidewall?

Since 49 CFR, Parts 172-176, already contains requirements for: (1) shipping papers, way bills, switching orders, and other billing, (2) marking of portable tanks, (3) marking or placarding rail cars and motor vehicles, and (4) labeling of cylinders, substantiation should be provided in the comments for establishing the fact that the proposed color coding system selected will significantly enhance safety above and beyond that provided by the current regulations and further will fully meet the necessary overall safety needs of the public. Inasmuch as it is now required that economic impact evaluations are to be made prior to rulemaking, the public should indicate the anticipated costs of any identification system that might be selected from those mentioned by the Department or proposed by the commenter.

(41 F.R. 43188, September 30, 1976)

[49 CFR Part 172]
[Docket No. HM-141]
COLOR CODING OF COMPRESSED GAS PACKAGES
Termination of Docket

An advance notice of proposed rulemaking (41 FR 43188, Sept. 30, 1976) was published by the Materials Transportation Bureau (MTB) in response to requests and to bills considered by both Houses of Congress in the past several years seeking the establishment of a system of color codes for compressed gas packages. The purpose of a color code would be to provide safety for any person within the sphere of hazard surrounding a compressed gas package, whether the package is in transportation, use, or storage, by facilitating ready identification of the gas contained in the package.

Most commenters objected to the concept of a color code system because:

(1) Some degree of color blindness affects a significant segment of the population (variously described as 10 percent to 15 percent of the working population).

(2) The great multiplicity of gases and the almost infinite number of mixtures (sometimes including up to five different gases) would require a complex color code system.

(3) The complexity of such a system would require a reference source that would be unwieldy in size and possibly of limited usefulness.

(4) The continual development of new mixtures would require frequent updating of the reference source.

(5) The variable appearance of color pigments under different light sources, and the possible fading of pigments, might result in substantial confusion.

(6) The many years of transition, while older compressed gas packages of varying color combinations are emptied and returned to a source for filling and application of a new color code, would be a major source of confusion.

(7) Labels, placards and/or markings already provide identification of each gas by means of color, symbol and text.

Comments from organizations and individuals associated with the delivery of health care services questioned the value of modifying existing medical gas color codes even though medical gases constitute only a small part of the total number of gases and gas mixtures in use at the present time. The American Association for Respiratory Therapy addressed color coding as being probably the least important mechanism for safe handling of compressed gases.

The National Society for the Prevention of Blindness, Inc., strongly disagreed with the premise that any color code system would enhance safety for personnel involved in loading, handling or storage. This society viewed color codes as impractical and possibly dangerous.

Various military organizations suggested use of a number of standards dependent on the intended use of compressed gas packages, such as diving, medical, or general application. These commenters did not address fire fighting or emergency response uses of compressed gases. However, MTB notes that the color red is widely used at present to identify fire extinguishers that may contain any of a number of different gases, such as carbon dioxide, nitrogen, or various refrigerant gases.

Few comments were received from persons concerned with fire protection. The Fire Equipment Manufacturers' Association, Inc., requested to be excluded from any color code system that might be established. The association was concerned with the cosmetic and esthetic acceptance of portable fire extinguishers in both the workplace and the home. Further, the association remarked that for industrial use, a color coding system already is employed to identify certain agents for special fire application. This system was not explained further.

The Florida State Fire Marshal's Office limited comment to liquefied petroleum gas transportations safety and stated that changing the color of packagings used to transport this material could lead to confusion with resultant danger to the public. Similarly, the California Division of Industrial Safety suggested that extending the use of a color code system to shipping containers would confuse people accustomed to differing color code systems in use within their workplace. The Safety Department of the University of Wisconsin further supported this view by stating that color markings would be more detrimental than advantageous and cannot be substituted in any conceivable way for a label.

The docket also includes a letter from the Deputy Assistant Secretary of the Department of Labor (DOL) to a

Member of the House of Representatives relating to color code legislation. The letter explains that with the advent of many sophisticated mixtures of exotic gases, color coding alone became too complex to assure the necessary recognition of hazards. A vendor's catalog is cited as containing a list of over 200 gas mixtures, including a mixture of carbon monoxide, helium, oxygen, argon, and nitrogen. The letter also states that the Occupational Safety and Health Administration is working with the Department of Transportation and a number of other Federal agencies in an attempt to standardize labeling requirements, including the use of color, symbol and legend. (The MTB published new standards for labeling requirements including the use of color, symbol and legend on April 15, 1976 (41 FR 15972), docket HM-103/112. The DOL currently has this and other standards under consideration.)

The MTB wrote the DOL, on October 15, 1976, requesting statistics on deaths or injuries involving compressed gas cylinders where the cause of death or injury was the injured or deceased's lack of knowledge as to the gas contained in the cylinders. The DOL injury data system is based on investigations where one fatality, or five hospital admissions occur. In subsequent communication, MTB was informed that no such deaths or injuries had been reported.

The single accident situation cited by the primary and virtually single proponent of the color coding concept occurred on his company's premises in 1955 and involved a compressor test in which oxygen was inadvertently used rather than an inert gas. In his statement, he suggests the infrequency of such accidents by saying, "The amazing thing to us is that we can find no case so far where this type of accident has occurred in the past." Similarly, the MTB has not found one case that has occurred since. While the proponent has stated that "we know of no safety precautions we could have taken, except an actual analysis of all gases, which could have prevented this disaster," the facts as presented by the proponent indicate that insufficient planning has led to an exhaustion of suitable supplies of gases to be used for the tests. To hasten delivery of the needed gas the company made an uncustomary pickup rather than waiting for delivery by fully trained personnel. This can be considered the primary causal factor in the unfortunate incident. The proponent stated September 14, 1955, that "we (the company) are taking every possible step to make certain that such substitution cannot occur again." The MTB notes that the precautions taken by the proponent's company have been successful in the subsequent 23 years, which suggests that color coding is not necessary to prevent death and injury.

The proponent of the color code system maintains that only six colors are specified by MIL-STD-101B. However, the DOD lists 115 various color combinations of four or more colors, representing only a small number of the gas mixtures possible. The coding of cylinders in the six basic colors referred to may adequately prevent the particular kind of accident referred to by the proponent, but it would not prevent the inadvertent mixture of gases within a grouping that could lead to adverse reactions in a laboratory or industrial environment. As pointed out earlier in this document, commenters indicate that even the military does not have only one color code system, but several systems patterned for specific uses.

The staff conclusions of the MTB on the major economic and public safety issues involved in the adoption of a uniform, nationwide color code system for compressed gas cylinders are quoted below:

A uniform, nation-wide system of color coding compressed gas cylinders as a means of preventing serious cylinder accidents in the normal, routine environment characterizing the workplace or households is felt to be of little or marginal value as a safety measure; and the adoption of such a system may increase, rather than decrease, serious accidents involving such cylinders.

The costs of adopting a uniform, nation-wide system of color coding compressed gas cylinders would not be negligible. With an estimated 110 million cylinders in existence, and a unit conversion cost of $1 to cover such expenses as conversion of current workplace procedures, new safety training requirements, the development of new owner identification systems, sandblasting and repainting cylinders, and so forth—the cost of the new system could easily exceed $110 million. When viewed against the expected marginal benefits to be derived from it, the color coding system does not appear to be the most cost-effective way to improve the public safety associated with the transportation, handling and storage of compressed gas cylinders.

The use of color code systems for cylinders in emergency response situations where rapid reaction time is of vital importance to save life or limb (as in hospitals, or in the military) does have an obvious and demonstrated value. However, a color code system for the purpose of improving emergency response/reaction times involving (after-the-fact) transportation accidents is of less obvious value, since cylinders are usually stored or packed together in a manner that allows only a few cylinders to be taken in at a glance; and since shipping paper requirements provide a much more rapid means for identifying cargo content.

The complete staff summary of "The Economic and Public Safety Implications of a Uniform Nationwide Color Code System for Compressed Gas Cylinders," is available for public review in the Dockets Branch, Office of Program Support, Materials Transportation Bureau, Department of Transportation, Washington, D.C. 20590. Primary drafters of this document are Paul H. Seay, Technical Division, and Douglas A. Crockett, Standards Division, Office of Hazardous Materials Regulation, MTB.

Based on the comments received, and the conclusions reached, docket HM-141 is hereby terminated without further action.

(43 F.R. 46050, October 5, 1978)

[49 CFR Part 173]
[Docket No. HM-142]
ETIOLOGIC AGENTS
Advance Notice of Proposed Rule Making

The Materials Transportation Bureau (MTB) is considering amending 49 CFR Part 173, as it applies to the transportation of etiologic agents, to extend the coverage of those regulations to a number of presently excluded substances.

The Hazardous Materials Transportation Act calls for a regulatory program applicable to materials which, when transported in commerce, pose an unreasonable risk to health, safety, or to property. However, under present MTB regulations, only those etiologic agents known to be hazardous to humans are regulated. In addition, although they are subject to certain regulatory requirements of Department of Health, Education, and Welfare, cultures of etiologic agents in quantities of 50 milliliters or less, diagnostic specimens, and biological products are also excluded from the MTB's regulations governing the transportation of hazardous materials. The MTB is concerned that these gaps in its present regulatory scheme may be leaving unaddressed some rather substantial risks associated with the transportation of many of those excluded etiologic agents. To provide the MTB with a more comprehensive basis for a future proposal to amend the regulations, the MTB request comments on the following specific areas of interest:

1. DEFINITION OF ETIOLOGIC AGENT

(a) Is the definition of etiologic agent given in 49 CFR 173.386 adequate?

(b) Should the definition be expanded to include agents which are harmful to plants and animals?

(c) Should the definition be expanded to include biological materials (such as recombinant DNA) used in or derived from genetic studies?

2. EXCEPTIONS

(a) Should etiologic agents in quantities of 50 milliliters or less (per outside packaging) be further regulated by the MTB as to packaging, marking, and labeling?

(b) Should the MTB, when determining the quantity of etiologic agent below which regulation is unnecessary, use a system which takes into account the potency, i.e., the toxigenicity or virility of the agent (similar to the system used for poisons)?

(c) Should the MTB establish more specific regulatory requirements for diagnostic specimens and what should these be?

3. LABELING REQUIREMENTS

(a) Should a small size label, consistent with the general label format for other hazardous materials, be adopted to accommodate use of small packages for etiologic agents?

(b) Alternately, should a minimum package or overpack size be established to enhance safety by making it less likely for the package to become lost during shipment?

4. TRANSPORTATION OF IMPORTED SHIPMENTS
OF ETIOLOGIC AGENTS

(a) To what extent do prevailing practices regarding transportation of imported shipments of etiologic agents, or suspected etiologic agents particularly diagnostic specimens, pose a health or safety risk?

(b) What, if any, kind of monitoring or clearance procedures are necessary to adequately control perceived risks attributable to transportation of imported shipments of etiologic agents?

Comments are welcome on these questions, as well as any additional recommendations for enhancing the safety in transportation of etiologic agents.

(41 F.R. 52086, November 26, 1976)

[49 CFR Parts 172, 173, 174, 176, 177]
[Docket No. HM-143; Notice No. 76–11]
BLASTING AGENTS
Proposed Rule Making

The purpose of this notice of proposed rulemaking is to propose the following amendments to Parts 172, 173, 174, and 176 of the Department's Hazardous Materials Regulations:

1. Remove the shipping name Nitro carbo nitrate;
2. Add a new shipping name, Blasting agent, n.o.s. and a new class, Blasting agent;
3. Provide packagings for Blasting agents; and
4. Provide a new label and a new placard for Blasting agents.

The Department of Transportation's Hazardous Materials Regulations do not now include a definition of a blasting agent. A material used for blasting must be classified as one of three classes—Class A explosive, Class B explosive, or Oxidizer (nitro carbo nitrate). Neither the Class B explosive nor the Oxidizer classification is appropriate for many blasting agents.

On April 19, 1972, the Institute of Makers of Explosives petitioned the then Hazardous Materials Regulations Board to create a new hazard class called "Blasting Agents."

The inclusion of a blasting agent description and hazard class will contribute to increased safety in transportation because some materials now shipped as nitro carbo nitrates (oxidizing materials) also present a potential explosive hazard.

Both the Mining Enforcement and Safety Administration (MESA) and the Bureau of Alcohol, Tobacco, and Firearms (BATF) publish definitions of blasting agents. MESA bases its storage requirements on the classification of an explosive as determined by this Department. Many materials used for blasting which would be considered blasting agents by MESA and BATF must be classed as Class B explosives under the DOT regulations. MESA requires magazine storage for DOT Class B explosives, but does not require magazine storage for materials identified as blasting agents.

In an effort to resolve these problems and to bring the DOT regulations into closer conformity with the regulations of MESA and BATF, the Materials Transportation Bureau (MTB) is proposing to incorporate a definition of a blasting agent into the DOT regulations. This definition is essentially the same as the statutory definition in the explosives laws administered by BATF (18 U.S.C., Section 841(e)) with certain additions which MTB considers necessary to achieve an acceptable level of safety in transportation.

The MTB considers blasting agents to be very insensitive explosives and is proposing that they be subject to the requirements of Section 173.86 which prescribes shipping requirements for new explosives. The MTB is also proposing a blasting agent label and placard.

Blasting agents would not be subject to specification packaging requirements. In addition, the MTB is proposing to delete the description, nitro carbo nitrate, from the regulations since all materials now so described would be included in the Blasting agent, n.o.s. description. A reasonable time would be provided for the change in the description of those materials now identified as nitro carbo nitrates to be revised to the blasting agent description.

The proposal requires more tests and spells out more detailed testing than are now required for Class A and Class B explosives because;

1. The packaging requirements for blasting agents are less restrictive than those for Class A and Class B explosives and, therefore, additional testing is considered necessary to clearly establish the basis for regarding a particular substance as a blasting agent; and

2. The Materials Transportation Bureau regulatory plan includes several changes such as rewriting the present explosive regulations. Some of the tests required for blasting agents (and possibly other tests not delineated here) could be incorporated in the proposed regulations. It is desirable to publish a blasting agent definition as expeditiously as possible, since the complete revision of the explosive section will be published in a future notice of proposed rulemaking.

The 212° F. temperature specified in the differential thermal analysis test is not found in any of the present regulations. It was chosen because many blasting agents contain appreciable quantities of water which can be affected at or above this temperature.

(41 F.R. 52083, November 26, 1976)

[49 CFR Parts 172, 173, 174, and 176]
[Docket No. HM-143; Notice 76–11]
BLASTING AGENTS
Public Conference

On November 26 1976, the Materials Transportation Bureau (MTB) published in Docket HM-143 a Notice of Proposed Rulemaking (41 FR 52083) which proposed the following amendments to Parts 172, 173, 174, and 176 of the Department's Hazardous Materials Regulations:

1. Remove the shipping name, Nitro Carbo Nitrate (NCN); 2. Add a new shipping name, Blasting Agent, n.o.s., and a new class, Blasting Agent; 3. Provide packagings for blasting agents; and 4. Provide a new label and a new placard for blasting agents.

On July 25, 1977, Monsanto Company petitioned MTB for an informal conference on blasting agents, in accordance with 49 CFR § 102.25. Monsanto objected to the proposed rule on the general ground that the proposal would create a new hazard class based not upon the intrinsic characteristics, kind, or degree of hazard presented by the material but upon the material's end use. In support of their petition for a public conference, Monsanto argued that NCN is an oxidizer with an excellent safety record and it was wrong in principle to include in a single class materials which differ greatly in the degree and kind of hazard simply because they may be used for the same purpose. Further, Monsanto believes if the facts were fully understood the proposal would be either dropped or modified to retain the oxidizer classification for NCN. Monsanto asserted that there is widespread opposition to the proposal and that a public conference could explore the objections of other interested parties in addition to permitting Monsanto to explain its position and answer questions from the Office of Hazardous Materials Operations.

(42 F.R. 43416, August 29, 1977)

[Docket No. HM-143; Amdt. Nos. 172–48,
173–124, 174–34, 175–8, 176–7, 177–45]
BLASTING AGENTS
Final Rules

The Office of Hazardous Materials Operations published a notice of proposed rulemaking under docket HM-143 on November 26, 1976 (41 FR 52083; Notice 76–11). At the request of one of the major producers of nitro carbo nitrates, a public conference was held on September 23, 1977.

The principal drafters of this document are Charles W. Schultz, Technical Services Branch, Office of Hazardous Materials Regulation, and George W. Tenley, Office of the Chief Counsel, Research and Special Programs Administration.

Comments on the notice were received from a total of twenty-one organizations and persons. All comments have been carefully considered and the significant comments are discussed as they apply to the sections appearing in the notice.

Section 172.101. Hazardous Materials Table: One commenter objected to the n.o.s. following the blasting agent description because he said it implies that there are other blasting agents which are specifically named. The Materials Transportation Bureau (MTB) does not agree since it is adding the shipping name, ammonium nitrate-fuel oil mixtures to the table and contemplates that other specific descriptions for blasting agents will be added in the future.

Two commenters said that the description nitro carbo nitrate (NCN) is a more specific name than blasting agent and recommended that the shipping description NCN be retained. The MTB believes that this description should no longer be used since it has been so closely associated with the Oxidizer class for many years and has only been recognized as a description by other agencies because it is contained in the DOT regulations. Also the term is considered so vague that its continued use would cause considerable confusion in light of the new test criteria being adopted in this amendment for classification purposes. In consideration of the fact that a very large volume of the materials to be classed as blasting agents are ammonium nitrate-fuel oil mixtures, a shipping description is being provided for such mixtures, as an alternative to the Blasting agents, n.o.s., description, if they meet the definition and test procedures specified for blasting agents.

Concerning the commenters' argument that such materials be retained in the Oxidizer class, the MTB acknowledges the good shipping experience of these materials in the past. However, they do present a potential explosive hazard, though very insensitive to fire and shock, and this potential risk should be recognized through appropriate classification and resultant labeling and placarding. Therefore, the MTB does not agree that such materials should be retained in the Oxidizer class.

Section 172.411. Explosive A, Explosive B, Explosive C and Blasting Agent labels: One commenter stated that the colors prescribed for the blasting agent label associated these products with explosives. He also said "The label . . . doesn't contain the Hazard Symbol . . . a direct contradiction to OHM's reasoning for symbols as directed in HM 103–112." The MTB considers blasting agents to be very insensitive explosives and, as stated in the preamble to the notice, the reason no symbol is displayed is due to the fact that blasting agents present a much lower level of hazard in terms of their detonation potential than Class A and Class B explosives for which an "exploding bomb" is displayed on the label. This is analogous to the labeling system recommended by the United Nations where no hazard symbol is required on labels for Division 1.5 materials.

Section 172.504. General placarding requirements: Two commenters pointed out that no provisions had been made indicating under what conditions the blasting agent placard will be required. Blasting agents will be placed in Table 2 immediately following Class C explosives.

Section 172.524. Blasting Agents placard: The MTB considers that a BLASTING AGENT placard rather than an OXIDIZER placard is necessary to alert emergency personnel that an explosion is possible if blasting agents are involved in large fires.

Five organizations commented on the proposed placards for blasting agents. The main criticisms were that the placards are too small and of the wrong color and shape. The blasting agent placard is similar in these respects to those required for explosives and it would be incongruous to require a larger placard for a material having less potential hazard than Class A and Class B explosives. If one of the placards for explosives or blasting agents is changed, all would have to be changed, which is beyond the scope of this rulemaking. The MTB, however, will be initiating a proceeding during 1979 to evaluate overall experience with the new placard system adopted under Docket HM-103/112 in 1976 and review all related questions and concerns including those raised in this rulemaking. The Bureau recognizes the concerns raised by those commenters objecting to addition to a new placard. Offsetting this is the fact that virtually all of the proponents of this rulemaking, the shippers of blasting agents, will be required to placard rail cars and freight containers, and will be required to give BLASTING AGENT placards to motor carriers if their vehicles are not equipped with such placards. Also, the Bureau is not making compliance with this amendment mandatory for approximately eight months following its publication to allow sufficient time for the making of necessary adjustments in shipper/carrier programs.

Section 173.86. New explosives and blasting agent definitions; approval and notification: Three commenters objected to placing blasting agents in this section, recommended that the entire section be moved to a new § 173.52a, and that changes be made to the section. While § 173.86 has been located in the part of the regulations governing Class A explosives for many years, it has always been applicable to all classes of explosives. The proposals to move it to another section will be considered in a future rulemaking action involving recodification.

Section 173.96 (now 173.114a) Blasting agents. (a) Definition of blasting agent. Six organizations commented on the proposed definition of a blasting agent.

One commenter stated that the proposed definition was not really a definition because it does not specify physical properties or chemical composition and that a definition should not be based on a negative test. The present definition of nitro carbo nitrate (NCN), to which the commenter apparently does subscribe, also does not specify any physical properties and addresses chemical composition in only the most general terms. The NCN definition also incorporates a negative test—i.e., the material may not detonate when tested with a number 8 blasting cap. The physical properties and chemical compositions of materials which will be included in the blasting agent description are of such great variety that it would be impossible to describe them in a regulation of any reasonable length.

This same commenter objected to the phrase " . . . very little probability of initiation to explosion or of transition from burning to detonation under conditions incident to transportation. . . . " It was further stated that "the railroad industry firmly believes that MTB and the commodity manufacturer must be certain that a material will not detonate before it is classed as nondetonable." This appears to be a rather paradoxical comment since his opening comments indicated that the overall proposal " . . . is neither wise nor necessary." Presently, NCNs are transported as oxidizing materials and are known to be detonable under certain conditions. They are designed to function by detonation and would be of no commercial value if they were nondetonable.

Two commenters objected to the fact that the definition did not include a prohibition against ingredients which are explosives as defined in the regulations. The MTB believes that such a prohibition is neither necessary nor desirable. There are many explosive formulations, which contain no explosive ingredients, which are far more hazardous than those containing such ingredients. An obvious example of this is explosives composed of chlorates and organic materials compared with commercial dynamite which contains nitroglycerin. The former are so hazardous and unpredictable that they are no longer produced as commercial explosives in the United States while dynamite has been used commercially for many years and is still being used in some applications.

Several commenters were concerned that the ingredients in blasting agents containing explosives might separate out during storage. Any separation of components of the blasting agents which occurs in the packaging is a localized change in composition and character of the material and the product may not be offered for transportation unless it has been established that the separation will not result in an increased level of hazard.

One commenter expressed concern about the behavior of blasting agents containing explosive ingredients in a fire. The rule requires that the largest commercial package (up to 200 kg) be subjected to a fire test. Any instability under these conditions should be detected in this test.

Four commenters objected to placing blasting agents in § 173.96 which is under the heading CLASS B EXPLOSIVES; DEFINITIONS. The commenters interpreted this as implying that the MTB was considering blasting agents to be similar to Class B explosives. There was no intent by the MTB to relate blasting agents to Class B explosives since such a relationship would be inappropriate in most instances.

One commenter suggested that blasting agents be placed in Subpart E of Part 173. This subpart covers flammable solids, oxidizers and organic peroxides. Another commenter suggested that blasting agents be placed in a new § 173.114a. The MTB believes that the hazards of blasting agents in transportation are more closely associated with those of explosives than with flammable solids and oxidizing materials and is placing the entry in a new § 173.114a until the codification efforts are completed.

Several commenters stated that the proposed definition appeared to restrict blasting agents to those materials used in the mining industry. The definition has been reworded to correct this impression.

Section 173.96(b) (now 173.114a(b)) Tests: There were many comments on the proposed tests. There was almost unanimous objection to the card-gap test. Tests conducted by the manufacturers of blasting agents and by the U.S. Bureau of Mines have demonstrated that the card-gap test described in DOD TB 700–2 (May 19, 1967) is not suitable for most materials which would be described as blasting agents. Tests conducted at the U.S. Bureau of Mines also showed that a meaningful card-gap test, the blasting cap test, and the bullet (projectile) test all gave the same results as far as potential transportation hazards are concerned. Therefore, the rifle bullet test and the card-gap test are not being adopted. Tests also established that the lead witness cylinder, presently described in Note 1 to § 173.53 is a more sensitive indicator of detonation than the detonating cord proposed in the notice. Since the use of the detonating cord witness might allow some materials to be classed as blasting agents which would not pass the blasting cap test if a lead witness cylinder were employed, the blasting cap test has been modified to require the lead cylinder witness. Several commenters said that the 500 gram test portion proposed in the thermal sensitivity test is too large for practicable testing. The MTB agrees and has reduced the sample size to 50 grams.

Section 173.96(c)(2) (now 173.114a(c)(2)). There were many objections to the positive statement that blasting agents may not be transported in portable tanks, cargo tanks or tank cars. The MTB believes that this prohibition is necessary for safety in transportation because:

1. The largest package required to be tested in a fire under the rule is 200 kg. The results of a test on this quantity do not necessarily indicate what would happen if a 3000 gallon tank of the same material was involved in a fire. In certain instances, the MTB might want to require additional testing before authorizing the transport of blasting agents in bulk.

2. The initiation to explosion of high energy liquids in bulk is not completely understood. The MTB might require special testing of a highly fluid blasting agent before allowing bulk transport.

Section 174.81 (Amended). This paragraph has been changed to place blasting agents after Class C explosives in the loading and storage table.

Section 175.78 Stowage compatibility of cargo. One commenter noted that there was nothing in the air carrier regulations to prevent blasting agents from being loaded in contact with or close proximity to special fireworks and railroad torpedoes. This prohibition has been added.

Section 175.320 Cargo-only aircraft; only means of transportation.

The table has been amended to replace "oxidizing materials" and "Nitrocarbonitrate" with blasting agent, n.o.s. The reason for this is that nitrocarbonitrates are now classed as oxidizers and the restrictions applying to nitrocarbonitrate must now be applied to blasting agents. This change was overlooked in the notice.

Section 176.83 (a) and (b): A new entry, blasting agents, n.o.s., has been added to Table I under item 17.

Blasting Agents has been added after Explosives C in Table II.

Section 177.848(a): This paragraph has been changed to place blasting agents after Class C explosives in the loading and storage table.

(43 F.R. 57898, December 11, 1978)

[Docket No. HM-143; Amdt. Nos. 172–48, 173–124, 174–34,
175–8, 176–7, 177–45]
49 CFR Parts 172, 173, 174, 175, 176, 177
Blasting Agents

On December 11, 1978, the MTB published a final rule under Docket HM–143 in the Federal Register (43 FR 57898), establishing the new shipping descriptions of Blasting Agent, n.o.s., Ammonium nitrate-fuel oil mixture, and a new hazard class, Blasting Agents. Since this publication, the MTB has received three petitions for reconsideration in accordance with the provisions of 49 CFR 106.35.

One petitioner stated that the amendment creates a new hazard class " . . . which is inconsistent with all other regulatory schemes all over the world."; that the basis of this new hazard class is the end use of the materials rather than their intrinsic characteristics; that no need or justifiable evidence was presented by any participant in the proceedings to warrant the establishment of the new class; that establishment of the new class makes all training material somewhat obsolete; and that there are twelve other alleged errors in the Docket.

The MTB is not aware of any governmental regulatory class scheme, except those of the United States and Canada, which officially recognizes the term "nitrocarbonitrate." The "Transport of Dangerous Goods" (ST/SG/AC.10/1/Rev. 1, Page 153) as published by the United Nations does not include the term "nitrocarbonitrate" in its index and further implies that this term is used only in North America. The materials called nitrocarbonitrates are included in Explosives, Blasting, Type B. Also, nitrocarbonitrate is not a recognized shipping name under the Inter-governmental Consultative Organization (IMCO) regulations. Because of this, the MTB believes that the term "blasting agent" is more in line with international regulatory schemes.

As to the statement that the new hazard class is based on end use rather than intrinsic characteristics, it is the MTB's opinion that the phrase " . . . a material designed for blasting . . ." constitutes only a limited part of the definition of a blasting agent and is included only to aid in the identification of the type of materials being addressed by these regulations. The fundamental part of the definition is the test criteria and these tests do evaluate the intrinsic characteristics, kind, and degree of hazard of these materials. The statement that there was no need or justification for establishing a new class of materials was addressed in the preamble of Docket HM-143 (43 FR 57898) in the last paragraph under Section 172.101. The MTB's opinion has not changed since this publication.

A petitioner's statement that the new class makes training materials somewhat obsolete may be addressed to any new rulemaking. It is the MTB's opinion that training materials must reflect the regulations and be based on the regulations, rather than regulations being based on training materials.

Another petitioner stated that the cost of compliance with the required tests would place an extremely high and unjustifiable burden on the industry. The petitioner also asserted that there existed large inventories of materials prepared in compliance with the requirements for nitrocarbonitrates, that supplies of preprinted packaging existed, and that compliance with the August 15, 1979, effective date would not be practicable since a longer period of time is necessary to deplete existing stocks of completed packagings and packaging supplies. The MTB has reviewed the test requirements for blasting agents and believes that such tests are necessary. However, for products consisting of only "prilled ammonium nitrate and fuel oil," the MTB has determined that only the blasting cap sensitivity test need be passed. Based on figures obtained from industry, this will eliminate the requirement to perform approximately 10,000 tests for this type of blasting agent with a significant cost savings to industry. Also included in this amendment is an allowance for materials presently described, offered and transported as nitrocarbonitrate to continue to be shipped in accordance with the regulations in effect on August 14, 1979, until December 31, 1980.

Another petitioner requested that manufacturers of blasting agents be allowed to perform the required tests, that the classing of blasting agents be done by DOT, and that the effective date be extended to August 15, 1981. Upon further consideration, it is the MTB's opinion that materials which contain only prilled ammonium nitrate and fuel oil may be tested by individual manufacturers with the results being forwarded to the MTB. According to industry figures, these materials comprise about 85% of materials included in the blasting agent classification. Testing of the remaining materials must be conducted by one of the designated agencies, with approvals for the class and packaging being issued by the Associate Director for Operations and Enforcement of the MTB. The MTB believes that extension of the effective date is not required in light of the allowance for materials presently shipped as nitrocarbonitrate.

(44 F.R. 31180, May 31, 1979)

[49 CFR Parts 173, 179]
[Docket No. HM-144; Notice No. 76-12]

TRANSPORTATION OF HAZARDOUS MATERIALS
Shippers; Specification for Pressure Tank Car Tanks

As a result of a series of serious railroad accidents involving pressure tank cars transporting hazardous materials, the Materials Transportation Bureau is considering amending Parts 173 and 179 of the Hazardous Materials Regulations to modify the specifications for uninsulated pressure tank car tanks (112 and 114 specifications) so as to improve design and construction of new and existing cars.

BACKGROUND

On March 15, 1976, a "Petition for Advance Notice of Proposed Rulemaking ... to amend 49 CFR Part 179, Subpart C; 49 CFR Part 173; Docket No. HM-125, Notice 75-4; and Docket No. HM-109, Amendment Nos. 173-83 and 179-5" was submitted to the Bureau by the Railway Progress Institute Committee on Tank Cars. The petitioner (representing the five principal tank car builders and lessors in the United States: ACF Industries, General American Transportation Corporation, North American Car Corporation, Pullman Leasing Company and Union Tank Car Company) stated that: This petition requests significant changes in the regulations which will improve the safety of transportation of flammable compressed gases and anhydrous ammonia in railroad tank cars.

The petition seeks the following:

A. Amendment of 49 CFR Part 179 to add specifications for two new DOT class tank cars. These cars would be "thermally" shielded counterparts of DOT Class 112A and 114A cars. Thermal shield systems could be of any type (e.g., coating or insulation with jacket) that qualifies thermally. If a jacket is used, a ½-inch thick jacket head would be used in lieu of a tank head shield.

B. Amendment of 49 CFR Part 173 to authorize the use of these two new specification tank cars for the transportation of all products currently authorized in 112A and 114A tank cars.

C. Amendment of 49 CFR 173.314 to prohibit the transportation of flammable gases and anhydrous ammonia:

1. In DOT Class 112A and 114A tank cars built after the date the specifications proposed in "A" are published, and

2. In DOT Class 112A and 114A tank cars after six years from the date that the specifications proposed in "A" are published.

D. Withdrawal of Docket HM-125 which proposed to prohibit new construction of DOT Class 112A and 114A tank cars.

E. Amendment of the tank head shield specifications (49 CFR 173.314 and 179.100–23):

1. To extend the date for equipping Class 112A and 114A tank cars with such shields from December 31, 1977, to December 31, 1979;

2. To delete the requirements for head shields in DOT Class 112A and 114A tank cars built new after the date that the two specifications proposed in "A" are published; and

3. To modify certain of the head shield design requirements.

Several other interested persons have addressed one or more of these subjects in commenting on notices of proposed rulemaking (particularly in HM-109 and HM-125), or in related correspondence. In establishing this new Docket and issuing this notice of proposed rulemaking, the Bureau intends to consolidate its rulemaking activity for pressure tank cars that pertain to upgrading the existing specifications 112 and 114 to improve their design and construction. After this has been accomplished, Docket HM-125 proposing to prohibit new construction of 112A and 114A tank cars will be terminated.

Pressure tank cars transporting hazardous materials have been involved in accidents and caused concern since the adoption of the first "pressure" specification on January 1, 1918. However, since 1969 there has been a growing concern due to an increase in the number of pressure tank cars involved in derailments during which they have lost their lading under violent, catastrophic conditions. According to information reported to the Department, from January 1, 1969, through December 31, 1975, there have been 519. 112 and 114 pressure tank cars in derailments of which 168 lost some, or all of their lading. These occurrences have caused 18 deaths, 832 injuries, and 45 major evacuations involving more than 40,000 persons.

As a result of analyzing these accidents, the National Transportation Safety Board (NTSB) has issued several recommendations regarding pressure tank cars used to transport hazardous materials, particularly liquefied flammable gases. On October 6, 1969, the NTSB issued Recommendation NTSB-69-R-29 which called for prototype tank cars to be thoroughly tested under the full scope of accident conditions known to be encountered in service and for the development and implementation of suitable regulations to correct any identified deficiencies.

On January 24, 1971, the NTSB issued Recommendation NTSB-71-R-9, calling for a revision of the specifications for the construction of new tank cars. Other NTSB recommendations have been issued recommending that existing and new pressure tank cars be upgraded to provide a greater level of safety.

Considerable research has been performed by the Department through the Federal Railroad Administration in conjunction with the U.S. Army Ballistics Research Laboratory, the Association of American Railroads, the Railway Progress Institute and the Railroad Tank Car Safety Research and Test Project Committee, in analyzing the problem of puncture and rupture of pressure tank cars involved in an accident environment. Twenty-five reports have been written and placed in the Public Docket. Most of these reports can be obtained from the National Technical Information Service (NTIS), Springfield, Virginia 22151. A list of these reports is in Appendix A to this notice.

Additional references to research performed concerning pressure tank car problems is contained in Railroad Research Information Service Special Bibliography dated October 1976, pages 351–379 (PB-258-066).

Due to the catastrophic nature of accidents involving pressure tank cars, the Bureau believes that promulgation of improved design and construction standards for new cars and for retrofitting such improvements on existing cars at the earliest opportunity is essential to assure safety. Based upon the results of the research programs being conducted by the Federal Railroad Administration and industry, performance standards for puncture resistance from impacts and thermal protection from fire exposure are being proposed in this Notice.

PROPOSAL

A new § 179.105 entitled "Special Requirements for Specification 112 and 114 Tank Cars" is proposed to be added in Part 179 of the regulations. This section provides new specifications for improving the safety of these tank cars. It contains a requirement that within six months after the effective date of the final rule, all new specification 112 and 114 tank cars are to be built equipped with "shelf couplers," a tank head puncture resistance system, a thermal protection system and a safety relief valve of adequate capacity to protect each thermally insulated tank.

Previously built specifications 112 and 114 tank cars shall be required to be similarly equipped in accordance with the following schedule:

1. Either shelf couplers or a tank head puncture resistance system within one year after the effective date of the rule;

2. Notwithstanding "1", shelf couplers within two years after the effective date of the rule; and

3. Thermal protection and tank head puncture resistance systems with adequate safety relief valve capacity within four years after the effective date of the rule.

In order to assure compliance with the requirements for thermal protection and head puncture resistance within the four-year period, it is further proposed that each car owner be required, as a minimum, to so equip its previously built 112 and 114 tank cars in accordance with a prescribed schedule. This schedule requires that 20 percent of each owner's tank cars be equipped during the first year, 30 percent the second year, 30 percent the third year and the final 20 percent the fourth year. This schedule takes into account production start-up problems during the first year when arrangements must be made for shop space and production techniques must be refined. In addition, it recognizes the difficulties likely to be experienced during the fourth year of locating, removing from service, and re-equipping the remaining cars in the fleet which traditionally have been the most difficult to locate and remove from service. The end result would be that after four years, all previously built 112 and 114 tank cars used to transport compressed gases would be equipped with shelf couplers, a tank head puncture resistance system, an adequate relief valve and a thermal protection system.

THERMAL PROTECTION

Analyses of accidents involving uninsulated pressure tank cars by both the Federal Railroad Administration and industry (including shippers, tank car builders and railroads) recognize the need to establish a standard for thermal protection. The Federal Railroad Administration in cooperation with the industry conducted pool fire tests at the U.S. Army Ballistics Research Laboratory at White Sands, New Mexico. Also, at a torch facility located at the Pueblo Test Center, extensive testing was conducted to obtain thermal evaluation of numerous promising thermal protection candidates in several forms. Both small plate sample and full scale tank cars were subjected to the torching environment. Based on these tests, information is available to specify a performance standard for thermal protection for pressure tank cars. In proposed § 179.105–4, two tests are specified for qualifying thermal protection systems. One is a pool fire for a time period of 100 minutes, and the other is a torch fire for 30 minutes.

Calculations based on the results of full scale pool fire tests conducted at White Sands, New Mexico, indicate that all of the liquid lading in a thermally protected tank having a nominal capacity of 33,600 gallons will be vented when exposed to a pool fire of 100 minutes duration. Previous experimental tests and computations have shown that the severity of a failure is directly related to the amount of liquid lading present at the time of failure. If no liquid lading remains, the possibility of rupture is remote. Accordingly, 100 minutes has been selected as the duration for the pool fire test to qualify proposed thermal insulation systems, and a description of the qualifying test procedure is included. Evidence indicates that systems incorporating "coating" of insulating materials or insulating materials encased in a steel jacket can qualify under this test procedure. Likewise, based upon torching tests conducted at the Pueblo Test Center, a torch fire test requirement is specified. During the Pueblo Tests is was calculated that a tank car will empty its liquid contents within 30 minutes through a hole in its shell, resulting from the penetration and withdrawal of a coupler head. For this reason, 30 minutes has been selected as the prescribed minimum duration of the torch test.

A simulated torch fire test is described as are methods for qualifying proposed thermal insulation systems in the torching environment. Again, tests indicate that systems incorporating a "coating" of insulating materials and insulating materials encased in a steel jacket can qualify and are available for use.

TANK HEAD PUNCTURE PROTECTION

Another major area of concern to the Bureau has been protection of tank heads from punctures, particularly punctures caused by vertical disengagement of couplers on adjacent cars. Proposed § 179.105–5 establishes criteria for protecting the tank head from puncture. These criteria are based upon analyses of accidents and impact tests involving tank head punctures in which tank cars loaded close to their rail load limit of 263,000 pounds have impacted at speeds of up to 18 miles per hour.

Three options are proposed to afford adequate tank head puncture resistance:

1. Installation of a protective head shield system that meets the requirements of existing § 179,100–23;

2. Installation of a specified steel jacket head having a minimum thickness of ½ inch; or

3. A tank head puncture resistance system with the capability of withstanding specified impacts without loss of lading based upon a performance requirement.

COUPLERS

Impact tests recently performed by the Federal Railroad Administration at the Pueblo Test Center have demonstrated that the use of shelf couplers in addition to the application of tank head puncture resistance systems, effectively lessens the possibility of tank punctures by constraining vertical disengagements of couplers or causing a coupler head to break away thereby preventing it from acting as a ram. The retrofit schedule for head puncture resistance systems for previously built cars is proposed to extend over a four-year period. The Bureau believes that the impact resistance that can be realized from the relative ease of application of shelf couplers can and should be achieved much more quickly. For this reason, proposed § 179.105–6 would require the installation of specifically designated shelf E couplers, F top shelf couplers, or other couplers approved by the Federal Railroad Administrator, within one year on 112 and 114 tank cars not equipped with head shields. In this connection, the Bureau notes that in August 1974 the Association of American Railroads petitioned for a requirement that shelf couplers be applied to all 112 and 114 pressure tank cars within one year.

SAFETY RELIEF VALVES

Tests conducted by the Federal Railroad Administration indicate that existing safety relief valves installed on uninsulated 112 and 114 tank cars may not provide sufficient relief capacity under extreme fire accident conditions. However, these tests have demonstrated that if thermal protection is applied to a tank, the existing valves provide sufficient relief capacity. Section 179.105–7 would require that newly built and retrofitted cars having thermal protection be equipped with the same capacity safety relief valves currently required on noninsulated 112 and 114 tank cars.

MARKING REQUIREMENTS

Section 179.105–8 provides revised stencilling requirements for indentifying 112 and 114 tank cars equipped with thermal protection systems. The Bureau believes this is necessary to assist in identifying cars equipped with thermal and tank head puncture resistance systems and the type of systems applied.

TANK CAR APPROVAL

The regulations proposed in this notice do not contain any requirement for "approval" by the AAR Committee on Tank Cars as do many of the existing Part 179 provisions, since the Bureau believes the addition of thermal protection and tank head puncture protection can be properly achieved by compliance with the proposed standards without the imposition of "approval" requirements.

CANADIAN TANK CARS

In § 179.105–1, paragraph (c) is being proposed to require that after four years after the effective date of the final rule, 112 and 114 tank cars built to specifications promulgated by Canadian Transport Commission (formerly the Board of Transport Commissioners for Canada) and used to transport compressed gases in the United States must also be equipped in accordance with the same special requirements as United States built and owned specification 112 and 114 tank cars. Because of the catastrophic consequences of accidents involving 112 and 114 tank cars, the Bureau believes that all such cars used in the United States to transport compressed gases must be equipped as proposed in this notice within four years after the effective date of the rule.

PART 173

A revision to § 173.31(a) (3) is proposed so as to enable new and retrofitted 112 and 114 tank cars stencilled with "T" and "J" to be used in the same manner as corresponding tank cars stencilled "A" and "S."

In § 173.314, the Table in paragraph (c) has a Note 23 which now provides that after December 31, 1977, 112 and 114 tank cars used to transport compressed gases must be equipped with protective head shields. The Bureau proposes to modify this requirement so as to require either protective head shields or shelf couplers on these cars within one year after the effective date of the final rule. If the tank car has head shields, shelf couplers are required to be installed within two years after the effective date. Also, the change would require all such tank cars to be equipped with thermal protection and tank head puncture resistance systems within four years after the effective date.

In order to maintain editorial consistency between the new proposed § 179.105 requirements and existing requirements in other sections of Part 173 and 179, the Bureau will issue conforming changes in § § 173.8, 179.5, 179.14, 179.100–4, 179.100–15, 179.100–21, 179.101–1, and 179.103 in the final rule.

The Bureau has evaluated this proposal in accordance with the policies of the Department of Transportation as published in the April 16, 1976, issue of the FEDERAL REGISTER (41 FR 16200) and believes that the proposed changes in this notice will result in substantial reductions in property loss and damage, and are otherwise warranted from the standpoint of public safety.

The estimated minimum capital investment necessary to implement the requirements proposed in this notice relative to existing tank cars is $5,000 per tank car. This figure does not include the installation of head shields since they are presently required by an earlier amendment to § 173.314. For new tank cars, the minimum cost is estimated to be $4,200 in additional capital investment per car, based on an estimated 500 new cars that will be placed into service each year. Therefore, the minimum cost of implementing the requirements proposed in this notice will be $100,000,000 for the estimated 20,000 existing tank cars to be retrofitted and the additional annual investment for 500 new cars will be $2,100,000 (current dollars). Based on these data, the average annual sum of capital to be invested over the four-year period would be $27,100,000 if the minimum requirements proposed herein are adopted.

On the benefit side, the Bureau believes that the foregoing costs will be offset not only by reductions in the number of accidents involving property loss and damage, but also by the magnitude of dollar losses sustained. This does not take into account the social benefits—and to the extent they can be quantified, the economic benefits—in public safety that will be derived by significantly reducing the number of deaths, injuries and evacuations that have characterized the accident experience of 112/114 tank cars in the past. Accident data for calendar years 1969–1975 indicates that 519 tank cars were involved in derailments and 168 of these cars lost some or all of their lading. These occurrences resulted in 18 deaths, 832 injuries and 45 major evacuations involving more than 40,000 persons. Four of these accidents resulted in losses estimated as totaling more than $100,000,000.

(41 F.R. 52324, November 29, 1976)

[Docket No. HM-144; Amdt. Nos. 173-108, 179-19]
PART 173—SHIPPERS—GENERAL REQUIREMENTS FOR SHIPMENTS AND PACKAGINGS
PART 179—SPECIFICATIONS FOR TANK CARS
Shippers; Specification for Pressure Tank Car Tanks

These amendments are the result of the joint efforts of the Federal Railroad Administration (FRA) and the Materials Transportation Bureau (the Bureau). In accordance with internal DOT procedures, the FRA has developed the substantive provisions of these amendments for review and issuance by the Bureau. Accordingly, further information concerning substantive provisions of these amendments may be obtained from the above contact.

BACKGROUND INFORMATION

On November 19, 1976, as a result of a series of serious railroad accidents involving uninsulated pressure tank cars (built to specifications 112 and 114) transporting hazardous materials, the Materials Transportation Bureau issued a Notice of Proposed Rulemaking, Docket No. HM-144; Notice No. 76-12 (41 FR 52324). The purpose of that Notice was to elicit public comment on a proposed rule to improve the design and construction of new and existing 112 and 114 tank cars. Specifically, the Notice proposed that a new Section 179.105 entitled "Special Requirements for Specification 112 and 114 Tank Cars" be added to Part 179 of the regulations. This section would prescribe new specifications for improving the safety of these cars. The Notice would have required that all newly built 112 and 114 tank cars, be equipped with "shelf couplers," a tank head puncture resistance system, a thermal protection system and a safety relief valve of adequate capacity to protect each thermally insulated tank.

Also, the Notice proposed that existing 112 and 114 tank cars be retrofitted according to the following schedule:
1. Either shelf couplers or a tank head puncture resistance system be installed within one year after the effective date of the rule;
2. Notwithstanding "1," shelf couplers be installed within two years after the effective date of the rule; and
3. Thermal protection and tank head puncture resistance systems with adequate safety relief valve capacity be installed within four years after the effective date of the rule.

In order to assure compliance with the requirements for thermal protection and tank head puncture resistance within the four-year period, an annual completion schedule was also proposed.

The reasons for these proposals were discussed in considerable detail in the Notice. Interested persons were invited to participate in the rulemaking proceeding through the submission of written comments. Fifty-three submissions were received and have been fully considered by the Bureau in the development of this final rule.

Subsequent to the issuance of the Notice, three serious railroad accidents occurred involving 112 and 114 tank cars.

On November 26, 1976, at Belt, Montana, the Burlington Northern, Inc., had a train derailment. Two persons were killed, six persons were seriously injured and fifteen others were treated for injuries when twenty-four freight cars derailed. One of the derailed cars was CGTX 64226, a 112A tank car, loaded with approximately 31,000 gallons of propane. The tank sustained a tank head puncture, began to release its contents and subsequently ruptured. A second 112A tank car, CGTX 64141, loaded with butane was subjected to the fire environment. Approximately two hours after the accident this tank car ruptured from the heat exposure.

On February 20, 1977, in Dallas, Tex., an Atchison, Topeka and Santa Fe Railway freight train derailed. In the derailment, UTLX 38355, a 112A tank car loaded with 32,437 gallons of propane, sustained a tank head puncture near the base of the head. The escaping propane ignited and the resulting torching flame impinged upon and heated GATX 97359, another 112A tank car, which contained 30,321 gallons of isobutane. After about forty minutes of fire impingement, GATX 97359 exploded violently. The tank separated into three major parts. Fortunately, no injuries resulted from this accident, but the estimated third-party damage has been set at $3,500,000.

On March 16, 1977, at Love, Ariz., an Atchison, Topeka and Santa Fe Railway train derailed. Eight propane laden tank cars were involved:

ACFX 17359—112A340W.   RTMX 3487—105A300W.
ACFX 17355—112A340W.   RTMX 3526—105A300W.
RTMX 3515—105A300W.   RTMX 3532—105A300W.
ACFX 17358—112A340W.   RTMX 3492—105A300W.

All RTMX tank cars had ½-inch jacket heads and insulation.

ACFX 17359 sustained a puncture in the "belly" of the tank. The spilled contents burned. ACFX 17355 ruptured as a result of flame impingement. RTMX 3515 sustained a tank head puncture caused by a wheel cutting through the jacket head and the tank head. The contents ignited and burned. ACFX 17358 sustained a small head puncture. The contents spilled out through this hole and burned; also burning of contents occurred at the safety relief valves. RTMX 3487 which was exposed to fire impingement burned its contents at the safety relief valve. The tank did not rupture; the contents were released through the safety relief valve. The remaining three tank cars, RTMX 3526, 3532 and 3492, sustained no appreciable fire damage.

In the opinion of the Bureau, the tank head punctures sustained by CGTX 64226 at Belt, Montana, and UTLX 38355 at Dallas, Tex., would have been prevented had these cars been equipped with a tank head puncture resistance system. If neither tank car tank head had been punctured, there would have been no spill of product, no fire, and no resulting tank ruptures. No loss of life, nor serious injury would have occurred at Belt, Mont., and very little third-party property damage would have ensued at Dallas, Tex.

The Love, Ariz., accident is more difficult to analyze. The estimated speed at the time of the derailment was 48 miles per hour. The tank tear in ACFX 17359 occurred in the tank shell; tank head protection would not have prevented this tank puncture. ACFX 17355 ruptured due to heat exposure while RTMX 3487 did not rupture. It released its contents through its safety relief valve. Insulation appears to have assisted RTMX 3487 in resisting the adverse affects of fire exposure. Although RTMX 3515 was equipped with a ½-inch jacket head, it sustained a tank head puncture. The high derailing speed of 48 miles per hour appears to have given sufficient energy to a car wheel so that it could puncture both the ½-inch jacket head and the tank head. However, since much of the wheel's energy was dissipated in penetrating the jacket head, it is the Bureau's opinion that the ½-inch steel jacket kept the tank head hole to a minimum. This reduced the amount of fire in the area of this car.

As a result of analyzing comments received, two significant changes have been made in the final rule.

1. Specification 112 and 114 tank cars used to transport hazardous liquids (such as flammable or poisonous liquids) and nonflammable gases other than anhydrous ammonia (such as "fluorocarbon" gases), need only be equipped with shelf couplers. Such cars will continue to be designated as 112A/114A tank cars.

2. Specification 112 and 114 tank cars used to transport anhydrous ammonia need only be equipped (or retrofitted) with a tank head puncture resistance system and shelf couplers; thermal protection is not required. Such cars will be designated as 112S/114S tank cars. Several other changes have been made in the final rule. These changes and comments are discussed in the "Section by Section Analysis" which follows.

SECTION BY SECTION ANALYSIS
SECTION 173.31 QUALIFICATION, MAINTENANCE, AND USE OF TANK CARS

The purpose of amending paragraph (a) (3) is to authorize the use of classes DOT-112T and 112J tank cars having equal or higher marked test pressure when classes DOT–112A and 112S are prescribed, and similarly to authorize the use of classes DOT-114T and 114J tank cars having equal or higher marked test pressures when classes DOT-114A and 114S are prescribed. No specific comments on this change were received; the amendment is being adopted as proposed.

In the notice of proposed rulemaking, §§ 179.105–2(a) (4) and 179.105–3(a) (1) and (2) would require that each newly built 112 and 114 tank car be equipped and each previously built 112 and 114 be retrofitted, with a coupler restraint system. Furthermore, the Notice proposed that such a system be retrofitted within one year on cars not equipped with tank head puncture resistance systems and within two years on all other 112 and 114 cars. The requirement for retrofitting existing cars has been placed in a new paragraph (a) (5) of § 173.31 so as to clearly indicate that it applies to all 112 and 114 tank cars no matter how used, while newly built tank car requirements remain in § 179.105–2(a) (4).

Several comments were received indicating that it appeared that under the proposal most 112 and 114 tank cars would have to be retrofitted with a coupler restraint system within one year and that it was doubtful that the application of approximately 40,000 shelf couplers (two per tank car) could be accomplished in one year. The Bureau agrees and the retrofit period has been extended to June 30, 1979.

SECTION 173.314 REQUIREMENTS FOR COMPRESSED GASES IN TANK CARS

Currently, note 23 to the table in paragraph (c) states:

Specification 112A or 114A tank cars used for transportation of compressed gases must be equipped with protective head shields after December 31, 1977. See § 179.100–23 for head shield specification.

Note 23 appears in the Table after specifications 112 and 114 for anhydrous ammonia and flammable gases (such as butadiene, LPG, vinyl chloride, etc.). The notice proposed to change this requirement to:

. . . either protective head shields or shelf couplers after (one year after effective date), shelf couplers after (two years after effective date); and thermal protection and tank head puncture resistance systems after (four years after effective date) . . .

The final rule separates the requirements for 112 and 114 tank cars used to transport flammable compressed gases from those used to transport anhydrous ammonia. These requirements are being placed in two notes. Note 23 requires specification 112 and 114 tank cars used for the transportation of flammable compressed gases to be equipped with thermal protection and tank head puncture resistance systems by January 1, 1982.

Note 24 covers anhydrous ammonia cars. Many commenters indicated thermal protection did not appear to be necessary to improve safety on 112 and 114 tank cars transporting nonflammable compressed gases such as "fluorocarbons" and anhydrous ammonia. The Bureau concurs, particularly since accident records maintained by the Federal Railroad Administration show no incidents of thermal rupture of a 112/114 tank car when transporting nonflammable compressed gases. However, due to the toxicity of anhydrous ammonia and the fact that FRA accident records show deaths and injuries caused by the tank head puncture of 112/114 tank cars transporting anhydrous ammonia, new Note 24 requires installation of tank head puncture resistance systems (such as headshields) by January 1, 1982, for 112/114 tank cars transporting anhydrous ammonia. However, tank head puncture resistance systems are not required on cars used to transport other nonflammable compressed gases.

SECTION 179.105 SPECIAL REQUIREMENTS FOR SPECIFICATION 112 AND 114 TANK CARS

This new section sets forth the new requirements for newly built and for retrofitting previously built 112 and 114 tank cars.

SECTION 179.105–1 GENERAL

This new section sets forth three requirements:

1. Tanks built under specification 112 and 114 must meet the requirements of §§ 179.100, 179.101 and when applicable §§ 179.102 and 179.103.

2. AAR approval is not required for changes in nor additions to specifications 112 and 114 tank cars necessary to comply with § 179.105.

3. 112 and 114 tank cars built to specifications promulgated by the Canadian Transport Commission that are not equipped as described in § 179.105 may not be used to transport compressed gases in the United States after December 31, 1981.

No comments were received pertaining to "1" or "3," but many comments were received regarding deletion of "AAR Approval ("2")."

Commenters requested that approval by the AAR Committee on Tank Cars be required for changes in or additions to 112/114 tank cars necessary to comply with § 179.105. It was stated that the railroad "Interchange Rules" require a "Certificate of Construction" before a tank car may move in interchange service. Modifications and additions would still be required to be approved by the Committee. In addition, many commenters expressed belief that the Committee's expertise is essential to assure that modifications to these tank cars are performed properly.

The Bureau recognizes that the existing car owner/rail carrier approval system which is set forth in the AAR "Interchange Rules" may be continued by the AAR Tank Car Committee and that its approval for interchange may, therefore, be required by industry for all additions, modifications and repairs performed to comply with § 179.105. However, the Bureau does not believe that this approval need be imposed by regulation. These standards adopted for improved tank car safety are augmented by specific design criteria (such as specified couplers, head shield designs and thermal protection system), thereby affording tank car owners sufficient guidance to perform the modifications and additions required by this rule. For these reasons, the Bureau has deleted the requirement for AAR Tank Car Committee approval in this rule.

SECTION 179.105-2 NEW CARS

The notice proposed that all 112 and 114 tank cars built after six months the effective date be equipped with:
1. A thermal protection system (§ 179.105-4); 2. A tank head puncture resistance system § 179.105-5; 3. A safety relief valve meeting the requirements of § 179.105-7; and 4. A coupler restraint system (§ 179.105-6).

Based upon comments received, the Bureau has decided to establish three types of 112 and 114 tank cars. This decision has been alluded to earlier in this preamble. Accordingly, § 179.105-2 has been revised as follows:

1. Newly built 112A and 114A tank cars are authorized to transport hazardous liquids and nonflammable compressed gases, other than anhydrous ammonia. Each is required to be equipped with a coupler restraint system that meets the requirements of § 179.105-6.

2. Newly built 112S and 114S tank cars are authorized to transport anhydrous ammonia as well as commodities authorized to be transported in 112A/114A tank cars. Each is required to be equipped with a tank head puncture resistance system that meets the requirements of § 179.105-5 as well as a coupler restraint system that meets the requirements of § 179.105-6.

3. Newly built 112T, 112J, 114T and 114J tank cars are authorized to transport flammable and non-flammable compressed gases, including anhydrous ammonia, and hazardous liquids. Each is required to be equipped with all four safety measures: Thermal protective system (§ 179.105-4), tank head puncture resistance system (§ 179.105-5), coupler restraint system (§ 179.105-6) and a safety relief valve that meets the requirements of § 179.105-7.

Several commenters suggested that the use of a coupler restraint system such as "shelf couplers" and the perpetuation of FRA Emergency Order No. Five would obviate the need for installation of a tank head puncture resistance system. FRA Emergency Order No. Five orders railroad carriers to handle 112 and 114 tank cars not equipped with head shields which are transporting flammable compressed gases in switch yards by "shoving to rest." It was issued as a result of three serious rail switching accidents in which tank cars not equipped with head shields sustained tank head punctures as a result of overspeed impacts. Release of the flammable gas lading and subsequent ignition caused deaths, injuries and substantial property damage. The intent of the Emergency Order was to provide an interim safety measure until all 112 and 114 tank cars transporting flammable gases were equipped with tank head shields (January 1, 1978), and not be applied permanently.

As a result of testing performed at the Transportation Test Center, Pueblo, Colo., it has been demonstrated that for some overspeed switching impacts, shelf couplers will prevent some tank head punctures. For other impacts under differing conditions, shelf couplers were not effective in preventing tank head puncture, but head shields were effective and did prevent punctures. And, under certain test impact conditions involving more than one tank car, a combination of both shelf couplers and head shields were needed to prevent tank head puncture. These test results were summarized at the FRA public briefing held on December 8, 1976.

Tank head punctures occur in train derailments as well as in switching mishaps. Emergency Order No. Five has no effect on train derailment conditions. Therefore, the Bureau has concluded that for tank cars carrying anhydrous ammonia and flammable gases, both a coupler restraint system and a tank head puncture resistance system are necessary to prevent tank puncture in derailments as well as in switch yard accidents. However, for tank cars carrying products having less volatility such as hazardous liquids, and gaseous products having non-toxic, nonflammable properties, the consequence of puncture and product release is not as serious. Therefore, the application of a coupler restraint system will afford adequate public safety. When both a tank head puncture resistance system and a coupler restraint system are applied to all 112 and 114 tank cars used to transport flammable gases, the need for Emergency Order No. Five will end.

179.105-3 PREVIOUSLY BUILT CARS

In the notice of proposed rulemaking, the Bureau proposed that previously built (existing) 112 and 114 tank cars be retrofitted in a four-year time period with a thermal protective system and a tank head puncture resistance system.

Based upon comments received, the Bureau has modified this requirement. The rule requires a tank head puncture resistance system to be retrofit installed on 112S, 112T, 112J, 114S, 114T and 114J tanks cars, but this system is not required on 112A and 114A tank cars. Likewise, thermal protection must be retrofit installed on 112T, 112J, 114T and 114J tank cars, but it is not required on 112A, 112S, 114A and 114S tank cars.

As proposed in the Notice, a coupler restraint system is required to be retrofit installed on all 112 and 114 tank cars by July 1, 1979; for clarity this requirement is stated in § 173.31(a) (5) as well as § 179.105-3(a).

Each car owner is required to install thermal protection and tank head puncture resistance systems in conformance to the following schedule:

1. "Lead time" until January 1, 1978.
2. Twenty percent of cars owned by January 1, 1979.
3. Fifty percent of cars owned by January 1, 1980.
4. Eighty percent of cars owned by January 1, 1981.
5. All cars owned by January 1, 1982.

Many commenters said that they believed that this schedule did not provide adequate time to perform a retrofit installation program of this magnitude. Several requested a six-month "lead time;" essentially this has been granted. Several requested a retrofit schedule of a five, or six-year time period, in lieu of the proposed four-year period. This extended retrofit time period was carefully considered by the Bureau. However, due to the serious catastrophic consequences which can result from a single accident involving uninsulated, non-head shielded tank cars transporting flammable compressed gases, or involving non-head shielded tank cars transporting anhydrous ammonia, the Bureau believes that it is imperative to have retrofit installation completed as soon as is practicable.

Other factors evaluated by the Bureau in making this decision were:

1. Current regulations contained in § 173.314(c) require head shield installation by January 1, 1978. This requirement is now being phased in over an additional 4-year period.

2. Anhydrous ammonia cars (112S and 114S) will not require thermal protection, thus reducing the magnitude of the retrofit program.

3. Cars for "fluorocarbon gases" and hazardous liquids (112A and 114A) will not require thermal protection nor tank head protection, thus likewise reducing the magnitude of the retrofit program.

For these reasons, the 4-year schedule has been retained.

SECTION 179.105–4 THERMAL PROTECTION

The purpose of this section is to establish performance standards and also testing procedures to verify compliance with these performance standards for thermal protection systems to be applied to 112T, 112J, 114T and 114J tank cars. This section contains six paragraphs:

(a) Performance standard; (b) Test verification; (c) Simulated pool fire test; (d) Simulated torch fire test; (e) Analysis; and (f) Exterior tank color.

Paragraph (a) in the Notice proposed a requirement that:

Each specification 112 and 114 tank car shall be equipped with a thermal protection system that prevents the release of any of the car's contents (except release through the safety relief valve) when subjected to:
(1) A pool fire for 100 minutes, and
(2) A torch fire for 30 minutes.

Several commenters suggested that the "performance standards" be replaced by "test requirements" or "qualification tests," or supplemented by "design specifications." These comments pertained not only to thermal protection but also to tank head puncture resistance (§ 179.105–5) and, to a lesser extent, to "coupler restraint" (§ 179.105–6). The Bureau has considered those options but still believes that the performance standards are the best approach because they provide incentives for innovation. Also, the requirement that tank cars covered by these specifications retain the level of protection specified throughout their service life, eliminates the need for detailed maintenance procedures.

The performance standards prescribed are individual requirements which must be demonstrated under simulated pool and torch fire environments. It is not intended that an undamaged tank car be capable of withstanding a 100-minute pool fire and a 30-minute torch fire; and, to satisfy these requirements, one insulated test plate need not be subjected to both tests. Compliance may be achieved by subjecting an insulated test plate to the pool fire test and a similar test plate equipped with the same thermal protection system to the torch fire test. A tank car having an acceptable thermal system which is involved in a severe accident in which the tank and the system sustain considerable physical damage may not be capable of surviving a 100-minute pool fire and a 30-minute torch fire. The thermal protection system may not eliminate all tank car ruptures, but when properly maintained most thermal ruptures will be prevented. Damaged or deteriorated systems must be repaired or replaced before the car is again used.

Several commenters recommended that additional fire testing and in-service testing of thermal protection systems be conducted before final rulemaking. The Bureau recognizes that additional testing can always be said to provide additional data, but it does not concur in the need for additional testing in this case. The Bureau believes that the extensive series of fire tests conducted by FRA and RPI/AAR demonstrate not only the utility and practicality of a thermal shield system, but also provide sufficient data for reliable cost projections. Furthermore, the Bureau believes that the successful in-service use of thermal shields to protect aerospace hardware and stationary compressed gas tanks, in addition to considerable railroad tank car experience, demonstrate the reliability of several competitive types of thermal shields. The Accelerated Life Test (ALT) program, being conducted by FRA, RUI/AAR, and several shippers, has provided adequate evidence that at least four reliable thermal shield systems are readily available (see discussion below). The Bureau has sought to achieve a condition of adequate protection at minimum cost and sees no reason why additional thermal shield systems cannot be developed.

Most commenters indicated that the extensive testing programs conducted by FRA with the assistance of the RPI/AAR were beneficial in analyzing the problems encountered with uninsulated pressure tank cars and developing solutions to these problems. However, one commenter offered the opinion that DOT sponsored tank car tests were not run to gather information regarding a wreck environment, but were designed to make the tank car look bad in support of some preconceived theories. This commenter used as an example the White Sands Missile Range fire tests which were purported to be "about as far from a wreck environment as could be devised." The Bureau does not agree with this assessment and believes that the tests conducted were reasonable (although not necessarily conservative) simulations of tank car accident scenarios. It should be noted that the time to rupture of an uninsulated tank car in a White Sands test was 24 minutes, whereas in analyzing rail accidents, the RPI/AAR has found that in at least twenty-three instances, the times required to rupture tank cars engulfed in accidental fire were less than 24 minutes. The Bureau, therefore, feels that these fire tests were reasonable simulations of wreck environments.

A similar comment was that the two full-scale fire tests conducted at White Sands were run under controlled conditions and yet variables were not controlled. Several instances were cited. The commenter stated that in the non-insulated test the car contained 3,200 gallons more propane than the car in the thermally coated test and that the initial propane temperature was lower in the thermally coated test than in the uninsulated test. The Bureau has analyzed these differences and has concluded that they did not significantly affect the test results and that the differences were partially compensating. For example, a greater initial volume of propane in the thermally coated test would have increased the time to rupture and conversely, a higher initial propane temperature would have decreased the time to rupture. The commenter noted that the propane composition was not reported for the thermally coated test. On the basis of the temperature-pressure data at the start of each test, the Bureau has concluded that there was no significant difference in the propane used in the two tests. Also, the "commercial propane" used in both tests was supplied from the same source. The commenter noted that in the uninsulated test, the temperatures and pressures reported do not match the temperatures and pressures expected. The Bureau has reviewed the temperature-pressure data from the uninsulated test and has concluded that the experimental pressure-temperature data correlate reasonably well with theoretical data. The commenter mentioned that in both tests pure propane temperature-pressure relationships were used to estimate missing data ignoring important factors of superheat, supercooling, compressibility, and the influence of impurities. The Bureau has reviewed the effects mentioned and has concluded that the procedures used in both tests were adequate.

One commenter criticized the quality of some of the reports listed in Appendix A of Notice No. HM-144. To support his position, the commenter used several quotes from "Reference 9" of the Notice. The Bureau does not concur with the assessment of this commenter and believes that the quotes used are taken out of context. For example, the commenter used the following quote from "Reference 9," page 39, first paragraph under B: "Unfortunately, no useful liquid level data were recorded." This quote only referred to direct liquid level measurements. This report also described an indirect method that was used to measure liquid level. Another misleading quote was from "Reference 9," page 39, third paragraph under VI.A: "One or more sign reversals occurred in the recorded emf values during the test, indicating that the data recorded are not reliable." This latter quote was only in reference to a particular, small group of thermocouples. The majority of thermocouples did not experience any sign reversals and were reliable.

Paragraph (b) in the Notice established the method of verifying by testing that the thermal protection system meets the performance standard. Several commenters noted that the Bureau did not identify thermal systems which are deemed acceptable and do not require further testing. The commenters recommended that either a list of approved thermal systems be presented or sufficient time be allowed for the testing of new systems. The Bureau believes there is merit to these comments. Accordingly, the following list specifies thermal protection systems that do not require test verifications under § 179.105-4(b) based upon successful simulation testing conducted by FRA. This list is not intended to be all inclusive, and systems that may be submitted to the Bureau in the future and which are shown to meet the test specifications in § 179.105-4 will also be excepted from the test verification. Information concerning the systems listed below as well as any which may be excepted from verification in the future, is available for inspection in the Docket Section, Room 6500, Transpoint Building, 2100 Second Street, SW., Washington, D.C. 20590.

1. One inch minimum thickness Deltaboard (12 pounds per cubic foot, 15 pounds per cubic foot) encased in an 11-gauge steel jacket. Manufacturer, (Deltaboard) Rockwell Manufacturing Company, Leeds, Alabama.

2. The tank car external surface is prepared by sandblasting to remove all existing paint, primer, grease and loose material. 2 mils (dry) of Thermolag primer-351 are applied to the clean surface. 165 mils (dry) of Thermolag 330-1 subliming compound is next applied to the primed surface. 5 mils (dry) of Thermolag topcoat 350 is applied to the subliming coating. Manufacturer TSI-Inc., St. Louis, Missouri.

3. The tank car external surface is prepared by sandblasting to remove all existing paint, primer, oil, grease, and

loose materials. 3 mils (dry) of primer (Military Standard MIL-P-52192B) are applied to the clean surface. Chicken wire (1″ hexagonal, 22 gauge) is next attached to the primed surface. 180 mils (dry) of Chartex 59 thermal coating is then applied. 3 mils (dry) of a topcoat (AMBERCOAT 75) is then applied. Manufacturer (Topcoat)-Ameron, Brea, California. Manufacturer (Chartek 59), Avco, Lowell, Massachusetts.

4. The tank car external surface is prepared by sandblasting to remove all existing paint, primer, grease and loose material. .7 mils (dry) of primer (a 1:1 ratio by volume of 513–003 base component and 9110x350 activator component) is applied to the clean surface. 235 mils (dry) of thermal shield coating (a nominal 5:1 ratio by volume of 821x359 base component and 9110x407 activator component) is next applied to the primed surface. 2 mils (dry) of topcoat (a 2:1 ratio by volume of 821x317 base component and 9110x376 activator component) is applied to the thermal shield material. Manufacturer, De Soto, Inc., Des Plaines, Illinois.

Paragraph (c) proposed a testing simulation for the "pool fire" performance standard. This paragraph prescribed in detail the pool fire test environment and the method of testing the thermal protection system in that environment. In a similar manner, paragraph (d) proposed a testing simulation for the "torch fire" performance standard. The torch fire test environment and the method of testing in that environment were specified. Many comments were submitted about both proposals.

Several commenters have questioned the availability of facilities for conducting the thermal performance tests. It should be noted that the Transportation Test Center facility is available under stipulated user agreements for conducting any of the prescribed tests in § 179.105–4. Interested parties should contact the Transportation Test Center Director for information on use of the facility. The test facility is not unique; it uses standard components and technology. Thus, it is the Bureau's belief that the test facility and the test procedures themselves can be readily duplicated.

Some commenters requested a reduction in the pool fire time criteria on the basis that tank cars do not fail at 800° F., but fail at a higher temperature, e.g. 1050°F. In addition, some commenters contend that the 100-minute figure for pool fires is not theoretically consistent with the 30-minute requirement for the torch fire environment. In setting forth performance standards for thermal protection systems in terms of both pool fire and torch fire exposure criteria, the Bureau intended to ensure that the thermal protection system would retain the required thermal capacity with a safety factor over the range of exposures encountered. The pool fire environment, which involves interactive safety relief valve action and more extensive exposure of the tank car exterior, is considered the prime performance requirement. The 100-minute, 800° F. stipulations provide safety margins. Both full scale and simulated pool fire tests have shown that there are available thermal protection systems which can meet this pool fire criteria. Given the fundamental pool fire performance standards, the torch fire requirement is designed to ensure the adequacy of the thermal protection system in another commonly encountered, but not necessarily more severe fire environment. The Bureau does not feel that it is necessary to increase the torch resistance time in order to be theoretically consistent with the pool fire time specifications. The total performance standard is composed of the two fire environment elements and the Bureau is convinced that when viewed as a whole, the requirements not only adequately cover the scope of experienced exposures but can be met by several currently available products.

One commenter agreed with the 40 plus or minus 10 mph flame velocity requirement for the torch fire test for nonjacketed systems, but contended that it was not necessary for steel jacketed systems. The Bureau stipulated the torch fire criteria to include evaluation of erosive effects as the commenter rightfully concluded. The Bureau sees no reason why it should prejudge how a jacketed and a nonjacketed system might differ, and accordingly the specifications treat all systems equally. Each must successfully withstand the same environment and be tested under identical conditions.

Several commenters have recommended that smaller plate sizes be allowed for testing thermal protection systems. One commenter also requested that small circular plates be allowed. The Bureau's intent in requiring large test specimens is to evaluate an entire thermal shield system, including attachments. By requiring these plates, the phenomena of edge effects, inhomogenities, heat paths due to attachment requirements, etc., are minimized.

Paragraph (e) requires that the entire tank car surface be analyzed to assure that it will achieve the performance standards for thermal protection.

Some commenters requested a clarification of what analysis is required by this paragraph. Several commenters suggested that certain structures (e.g., ladders) be exempted from this requirement. One commenter requested that a maximum of 350 square inches be exempted from this requirement. The Bureau's intent in requiring this analysis is to ensure that those portions of the tank car shell that are not covered by the thermal protection system do not pose an unacceptable safety hazard. In other words, there must be equivalent thermal resistance in these areas. In calculating the thermal resistance in these areas, the structural strength of the tank and attachments may be used to demonstrate adequate thermal resistance.

One commenter recommended that the requirement in § 179.101–1(a), Note 4, for white paint on specifications 112 and 114 compressed gas tank cars be eliminated if thermal protection systems are installed. The Bureau agrees. Paragraph (g) has been added to § 179.105–4 and it states that 112 and 114 tank cars equipped with thermal protection need not be painted white.

SECTION 179.105–5 TANK HEAD PUNCTURE RESISTANCE

The purpose of this section is to establish performance criteria and testing standards to verify compliance with the performance criteria for tank head puncture resistance systems to be applied to 112S, 112T, 112J, 114S, 114T and 114J tank cars. The section consists of three paragraphs:

(a) Performance standard;

(b) Test verification; and

(c) Tank head puncture resistance test.

Paragraph (a) in the notice proposed a requirement that each tank car be capable of sustaining, without loss of contents, coupler-to-tank head impacts within the area of the tank head described in § 179.100–23 (approximately the lower half of the head) at relative car speeds of 18 miles per hour.

These test conditions were developed as a result of analyzing accident data compiled by the Federal Railroad Administration which was used in promulgating MTB Docket HM-109, Tank Head Shields. Also, data derived from coupler impact tests at the Transportation Test Center were used in verifying the specific test criteria.

Several commenters stated that they questioned the need for tank head protection on shelf coupler-equipped tank cars. The Bureau does not concur; its reasons have been set forth under the discussion of § 179.105–2.

Paragraph (b) requires test verification by full-scale testing to the performance standard or by use of the "alternate test procedures" set forth in (c). However, test verification is not required if the car owner elects to install: 1. Protective head shields (§ 179.100–23); or 2. Full tank head jackets of at least ½-inch steel.

One commenter discussed MTB Docket HM-109; notice 75-3, which proposed to permit a hand brake bracket to pass through a hole in the head shield so that the bracket could be mounted on the tank head rather than on the shield. The Bureau withdrew that notice on the basis of comments indicating that the requirement for adding a ⅜-inch thick steel pad on the tank head to support the bracket would be costly and the bracket reinforcement would locally rigidize the area. This could cause poor tank steel impact resistance. The Bureau has not re-opened this matter in this docket since it was disposed of in Docket HM-109.

Likewise, commenters suggested certain changes to the specific head shield specifications contained in § 179.100–23. The Bureau believes that these suggestions, which were raised in one form or another under proceedings in HM-109, were adequately handled in that docket.

Paragraph (c) describes the test protocol to be followed in verifying a tank head puncture resistance system. One commenter questioned the requirement that the coupler of the ram car be perpendicular to the impacted car upon impact. This commenter contended that in reality, the ram car coupler would usually be at a lesser angle and in most cases would strike the impacted car a glancing blow. The Bureau agrees that the ram car coupler will often be at a lesser angle to the impacted car in actual impacts. However, impacts can occur (and indeed have been observed in the FRA/RPI/AAR Switchyard Impact Program) in which the ram car coupler is perpendicular to the impacted car. Since the perpendicular impact is the most severe situation, the Bureau believes it should be used in the test procedure. Also, having the ram car coupler strike the impacted car at some other angle would unduly complicate both conducting the test and interpreting the data.

SECTION 179.105–6 COUPLER VERTICAL RESTRAINT SYSTEM

In the notice, § 179.105–6 proposed standards and specifications for coupler vertical restraint systems. The purpose of such systems is to resist vertical disengagement of coupled couplers so as to reduce tank head puncture. The section has been somewhat reorganized in this rule from the way that the paragraphs were published in the notice, but the basic content remains the same. These paragraphs are captioned:

(a) Performance standard; (b) Test certification and approval (entitled test verification in the notice); (c) Coupler vertical restraint tests (proposed as paragraph (d) in the notice); and (d) Listing of approved couplers (proposed as paragraph (c) in the notice).

Most commenters endorsed the proposal to apply a coupler restraint system to 112 and 114 tank cars. Several suggested that § 179.14 entitled *Tank Car Couplers* be revised to include "E-shelf" and "F-shelf" couplers in the list of couplers approved by the Federal Railroad Administrator. Since the notice did not address couplers on all tank cars (§ 179.14) but rather was limited to couplers on 112 and 114 tank cars, this suggestion is not being adopted in this proceeding. However, it is being considered by the FRA and the Bureau and may be adopted in the future.

Some commenters questioned the use of buff loads in the required testing. They indicated that it is a less severe test of the coupler's ability to avert coupler disengagement than a test without buff loads. Some commenters have also questioned the extreme difficulty in conducting tests with the required buff loads in existing facilities. In requiring buff loads, it was the Bureau's intent to insure that introduction of potentially higher levels of vertical loads in combination with the range of feasible buff loads did not produce other undesirable failures. The 2,000-pound buff load was meant to provide full buff engagement of the couplers while the vertical strength was tested. The intent of the specified 725,000-pound buff application with vertical loading was to insure that the coupler would not fail as a result of the combination of stresses. The Bureau is satisfied that the lower level buff load will adequately test the vertical restraint system and sees no reason to enter into testing requirements of other portions of the coupler system in this specification, particularly in view of the difficulty in applying the higher levels of buff loading in conjunction with the vertical loads testing. Accordingly, paragraphs 179.105-6(a) and (c) (3) have been revised.

In accordance with information furnished by the AAR, F-top shelf couplers are designated SF70CHT and SF70-CHTE in this amendment.

SECTION 179.105-7 SAFETY RELIEF VALVES

Section 179.105-7 in the notice proposed to require the relieving or discharge capacity of safety relief valves on thermally protected cars to be at least the same as on non-insulated tank cars. The effect of this proposal would have been to permit existing safety relief valves to be retained on cars retrofitted with thermal protection and to require the same safety relief valve capacity on newly built 112 and 114 thermally protected tank cars.

Several commenters recommended the development of a formula which takes into account the insulating effect of thermal protection. Other commenters either thought that the valve size could be reduced with increases in thermal protection or wanted an explanation of the Bureau's thinking to assist design engineers.

The Bureau, after a thorough reexamination, has confirmed that the proposed safety relief valve requirement is correct for the *minimum* thermal protection requirements of the specification. This safety relief valve requirement is consistent with the 100-minute, 800° F. pool fire performance standard and other overall system considerations in protecting the tank car from premature rupture. In response to the commenter concerns, the Bureau is permitting a modified sizing equation to reflect the contribution of additional or higher thermal insulation properties of the cars covered under this specification. The application of a modifying factor to the established uninsulated tank equations, prescribed in section A8.01 of Appendix A of the "AAR Specifications for Tank Cars," supports the previously published result at the minimum required thermal insulation level, and further allows determination of appropriately reduced safety relief capacities at higher than minimum levels of thermal protection.

Thus, the rule permits a reduction in the safety relief valve capacity on a thermally insulated car in proportion to the total number of minutes the tank is protected in the pool-fire test as related to the 100-minute standard. However, owners may continue to use the current safety relief valves on retrofitted and on newly built cars if they so desire.

SECTION 179.105-8 STENCILING

Mandatory stenciling reflecting the installation of tank head puncture resistance and thermal protection systems is prescribed in §179.105-8. The rule differs from the notice in that the provision is retained for 112S and 114S tank cars, e.g., 112 and 114 tank cars equipped with a tank head puncture resistance system, but not equipped with a thermal protection system.

Several commenters suggested that instead of using alternative letters in place of the "A" in the specification (in other words instead of using 112J in lieu of 112A to indicate a 112 car having jacketed thermal insulation), the Bureau use new specification numbers such as 122A and 124A. The Bureau has not adopted this suggestion. Since head shield equipped cars are required to be stenciled 112S and 114S, and over 600 cars have been so stenciled, the Bureau believes that continuation of this system to embrace thermal protection systems is logical. New specification numbers would necessitate additional regulatory wording in §173.31 as well as in other sections of Part 179, for example, §179.100. The use of differing letters indicating specific applied systems will accomplish the same identification function in an easier manner.

Some commenters stated that the "A" is a "spacer" and persons do not expect to obtain information from this letter. Tank car specification 103 and proposed specification 113 use letters to denote information about the car design. For example, a DOT103CW tank car has a stainless steel tank and a proposed DOT113C120W tank car is designed to be capable of handling cold temperature product loadings such as encountered with liquefied natural gas.

For these reasons, the stenciling system proposed in the notice is being retained along with the current requirement for 112S and 114S stenciling.

DISCUSSION OF OTHER COMMENTS

Specification 105 Tank Cars. Several commenters mentioned that many DOT specification 105 tank cars are used to transport the same products as are transported in 112 and 114 tank cars. The commenters believe that the 105 tank car may not have as good thermal and tank head puncture resistance protection as is being specified for the

112T/J and 114T/J cars. This matter is beyond the scope of this docket. Therefore, the Bureau will consider the matter of safety standards for specification 105 tank cars and may initiate rulemaking in the future.

Tank Car Steel. One commenter stated that a report by Dr. W. S. Pellini entitled, "Fracture Properties of Tank Car Steels—Characterization and Analysis" was not part of the references cited in the notice of proposed rulemaking. The report was not included because the report was not available to the Bureau at the time of the publication of the notice. However, the views of Dr. Pellini were known to the Bureau and were used in evaluating the overall tank car problem. On February 24, 1975, Dr. Pellini gave a presentation on tank car steels to representatives of DOT and industry. Also, on several occasions Dr. Pellini has discussed his views on tank car steels with DOT staff members. Based upon the work conducted by the National Bureau of Standards, the Battelle Columbus Laboratories, and Dr. Pellini, the Bureau concluded that existing tank car steels, are adequate. Accordingly, the Bureau did not propose any change to existing tank car steel specifications in this proceeding.

Docket HM-125. As was indicated in the notice, with the promulgation of standards and specifications upgrading existing specification 112 and 114 so as to improve design and construction, the Bureau will withdraw notice No. 75-4, under Docket HM-125.

Economic Impact. Several commenters took exception to the estimated cost projections stated in the notice. The principal objection was the use of minimum imposed costs rather than maximum possible costs. The use of minimum costs is considered to be the only practicable means for calculation of the economic impact of a rule. Any other calculations would of necessity be based on the anticipated decisions of car owners as to the options they choose to comply with the rule. However, the costs for protective head shields are included as suggested by several commenters even though they were required by an earlier rule (Amendment No. 179-15, 39 FR 27572, July 30, 1974). Also, updated information on the costs for couplers and the purging of cars has been used in the revised calculations. Adjustments have also been made to take into account that certain of the requirements proposed in the notice are not included in this rule.

The implementation of this rule will require a cash outlay of $107.9 million in 1976 dollars.

The following summarizes the per unit tank car investment costs of the rule and the number of tank cars thought to be affected.

Car type and utilization	Additional protection	Minimum cost	Number of tank cars
(112T/J, 114T/J) flammable gases, anhydrous ammonia, non-flammable gases, hazardous liquids.	Thermal, head, and couplers	$6,900	15,300
(112S, 114S) anhydrous ammonia, nonflammable gases, hazardous liquids.	Head and couplers	1,900	2,700
(112A, 114A) nonflammable gases, hazardous liquids	Couplers	500	2,000

The Bureau believes that the foregoing costs will be offset not only by reductions in the number of accidents involving property loss and damage, but also by the magnitude of dollar losses sustained. This does not take into account the social benefits—and to the extent they can be quantified, the economic benefits—public safety that will be derived by significantly reducing the number of deaths, injuries and evacuations that have characterized the accident experience of 112 and 114 tank cars. Since 1969, more than 500 of these tank cars have been involved in derailments of which more than 170 of these cars lost some or all of their lading. These occurrences resulted in 20 deaths, 855 injuries, and 45 major evacuations involving more than 40,000 persons. Four of these accidents resulted in estimated losses of more than $100,000,000.

The Bureau considers that the requirements set forth in this rule represent a cost-effective solution to the safety problems presented by 112 and 114 tank cars over the past several years.

(42 F.R. 46306, September 15, 1977)

[Docket No. HM-144; Notice No. 78-5]
[49 CFR Parts 173 and 179]
SHIPPERS: SPECIFICATION FOR PRESSURE
TANK CAR TANKS

This Notice is the result of the joint efforts of the Federal Railroad Administration (FRA) and the Materials Transportation Bureau (the Bureau). In accordance with internal DOT procedures, the FRA has developed the substantive proposals of this Notice for review and issuance by the Bureau. Accordingly, further information concerning substantive provisions of this Notice may be obtained from the above contact.

BACKGROUND INFORMATION

Emerging Need for Expedited Retrofit. On September 15, 1977, the Bureau published in the FEDERAL REGISTER (42 FR 46306) a final rule concerning specifications for tank cars which included the following timetable:

1. Existing specification 112 and 114 tank cars used to transport flammable gases were to be retrofitted with thermal and tank head protection (such as a "head shield") over a 4-year period ending on December 31, 1981.

2. Existing specification 112 and 114 tank cars used to transport anhydrous ammonia were to be retrofitted with tank head protection (such as a head shield) over a 4-year period ending on December 31, 1981.

3. All specification 112 and 114 tank cars were to be equipped with special couplers designed to resist coupler vertical disengagements. These couplers were to be retrofitted on all cars by July 1, 1979.

The recent major accidents at Pensacola, Fla., on November 9, 1977, at Waverly, Tenn., on February 22, 1978, and at Lewisville, Ark., on March 29, 1978, in combination with an incident of apparent vandalism near Youngstown, Fla., on February 26, 1978, have again focused attention on measures to improve the safety of rail transportation of hazardous materials. In the decade prior to the issuance of these new tank car safety requirements, under Amendments 173–108 and 179–19, 20 persons were killed because of accidental lading release from specification 112 and 114 tank cars. However, in the 6 months following the issuance of the rule, 17 additional persons have been killed.

While it is not possible to prevent the release of dangerous products in all situations, the severity and variety of circumstances relating to the occurrence of recent accidents have pointed out the need to take all feasible steps to protect the public against potential major disasters involving the transportation of flammable gases, anhydrous ammonia, and other hazardous materials. In particular, attention has been directed toward the possibility of accelerating the retrofit timetable for 112 and 114 tank cars.

On March 15, 1978, the Transportation and Commerce Subcommittee of the House Committee on Interstate and Foreign Commerce conducted hearings on railroad safety matters which had come to national attention as a result of the incidents which had occurred at Pensacola, Waverly, and Youngstown. At this hearing, the National Transportation Safety Board (NTSB) stated that it believed that with a strong sustained effort the special couplers and head shields could be installed on all 112 and 114 tank cars by late in December 1978.

On March 20, 1978, a second hearing was conducted jointly by the Subcommittee on Federal Spending Practices and Open Government and the Subcommittee on Civil Service and General Services of the Senate Committee on Governmental Affairs. At this hearing the NTSB reiterated its position regarding the acceleration of the retrofit schedule. After reviewing the testimony, the subcommittees requested that the FRA consider revising the retrofit schedule.

Further, on April 4–6, 1978, the National Transportation Safety Board conducted a special hearing in which a major focus was the timetable for the retrofit installation of the 112 and 114 tank car safeguards. At the conclusion of that hearing, its Chairman stated that the NTSB had determined that shelf couplers and tank head protective shields should and could be installed on all 112 and 114 tank cars by the end of 1978.

On April 7, 1978, the FRA conducted a special safety inquiry into the retrofit timetable for 112 and 114 uninsulated pressure tank cars. The purpose of this special inquiry was to obtain sufficient information to enable the FRA to determine whether the existing tank car retrofit schedule could be accelerated. The FRA received pertinent manufacturing, maintenance and cost data pertaining to this retrofit program from persons representing the National Transportation Safety Board, railroad carriers, tank car shippers, tank car owners, tank car builders, and coupler manufacturers.

Data submitted in the FRA special safety inquiry, together with other information available to the Department of Transportation, have made it possible to describe more accurately the problems associated with the retrofit process and to fashion a revised retrofit schedule which will improve the safety of specification 112 and 114 tank cars as quickly as possible without creating major economic disruptions. The balance of this Notice will describe the affected tank car pool and retrofit plans which have been made with respect to these cars, summarize the major obstacles to acceleration of the retrofit program, and outline the basic rationale underlying the proposed new schedule.

Number of Tank Cars and Retrofit Elections. As a result of the special safety inquiry and other information received, the following summarizes the current 112 and 114 tank car pool.

The Universal Machine Language Equipment Register (UMLER), which is maintained by the Association of American Railroads, lists a total of 22,228 DOT and Canadian Transport Commission (CTC) specification 112 and 114 tank cars and 105 individual reporting marks covering these tank cars as of April 11, 1978. Included in this UMLER listing are United States, Canadian and Mexican owned tank cars and car owners (UMLER lists one Mexican owner with fifty tank cars).

Based upon UMLER information and information received from United States tank car owners, the number of DOT specification 112 and 114 tank cars currently does not exceed 20,400 and the number of United States owners is fewer than 100.

Data submitted to the FRA indicate that approximately 3,400 of these 112 and 114 tank cars will be dedicated to anhydrous ammonia service. These tank cars will require "head shields," but not thermal protection, and will be retrofitted to DOT specifications 112S and 114S. Approximately 700 of these tank cars have already been equipped with head shields.

Approximately 2,000 of these tank cars are used to transport vinyl chloride monomer, a flammable compressed gas, on essentially an exclusive basis. Because weight is a critical factor, it is expected that these tank cars will be retrofitted with systems having the least additional weight, e.g., a "spray-on" thermal protection with separate head shields. Consequently, these tank cars will be retrofit converted to DOT specifications 112T and 114T.

Owners of an additional 2,000 specification 112 and 114 tank cars used in flammable gas service such as for transporting propane appear to have elected to use the "spray-on" thermal protection and separate head shields, thereby retrofit converting to DOT specifications 112T and 114T.

Another group of approximately 500 of these 112 and 114 tank cars will be used exclusively in non-flammable gas and hazardous liquids services. These tank cars will require only a shelf coupler retrofit.

Owners of the remaining 112 and 114 tank cars (approximately 12,500) have elected or are expected to use a jacketed insulation with integral tank head protection and will be retrofit converting their cars to DOT specifications 112J and 114J.

Relative Difficulty of Retrofit Tasks. As described above, specification 112 and 114 tank cars used in various services will be subject to the application of various retrofit "packages." All 112 and 114 cars are required to be equipped with shelf couplers, and that task is not integrally related to any other part of the process—either with regard to car availability or the mechanical steps involved. Therefore, both the existing retrofit program and the program proposed by this Notice treat the application of shelf couplers as a matter separate from the application of tank head protection and thermal protection.

The head protection and thermal protection tasks present a more complicated problem. The rationale of the existing schedule contemplated that these two elements of the retrofit would likely be accomplished in most cases as a single process so as to hold down costs and out-of-service time and minimize unfavorable impacts on the transportation of essential products.

In the case of the jacketed retrofit, which will evidently be used for the vast majority of cars requiring both protective devices, existing techniques of application will continue to mandate a unified retrofit process. However, the "spray-on" thermal protection method in combination with a "head shield," which is expected to be employed for roughly 4,000 cars, is capable of separation into two retrofit stages.

The NTSB and others have identified shelf couplers and head protection as those measures requiring most urgent attention. Shelf couplers, as discussed below, should not present a major problem based on recently developed information.

Representatives of the major tank car companies, in testimony before the FRA special safety inquiry, made statements supporting the conclusion that the complete retrofit program could probably be accomplished in a three-year period by utilizing extra shifts and withdrawing additional cars from service at any given time. However, these witnesses warned that a significant reduction of allowed time below three years could upset plans already established for the orderly accomplishment of the retrofit and could actually delay the final overall completion of the retrofit tasks.

The FRA and the Bureau have attempted to evaluate what reductions might be possible in the time allowed to complete the application of tank head protection. In doing so, it has been necessary to consider two factors as they apply to each of the retrofit packages ("S," "T," "J").

The first factor is car availability. That is, given a proposed regulatory deadline, how many cars would be removed from service at any given time? Can these cars be made available for retrofit in an orderly manner?

The second factor is capacity. That is, do the affected parties have reasonable access to the necessary plant, equipment, skilled labor and any other components necessary to do the job?

In addition to the two factors bearing on feasibility, the effect of various proposed deadlines on retrofit elections has been considered. Most particularly, the FRA and the Bureau have given some weight to the superior protective qualities of the jacketed retrofit package. Any new regulatory deadlines which might require the immediate application of head protection would have the likely effect of discouraging the use of the jacketed retrofit, since the unitary process requires more shop time and can be accomplished at fewer facilities.

Thus, the proposed schedule outlined below emphasizes the completion of retrofit tasks which are more easily accomplished with less out-of-service time at a greater number of potential facilities. Although it is proposed to accelerate the timetable for the unitary jacketed retrofit, an effort has been made to leave undisturbed the elections which have already been made concerning the use of that approach.

PROPOSED SCHEDULE

Shelf Coupler Application. Based upon information gathered from coupler manufacturers, tank car owners and tank car shippers, it appears that shelf couplers can be applied to all 112 and 114 tank cars not later than December 31, 1978. An adequate supply of these couplers is or soon will be available, and application is not difficult. Such application can be performed at any location having a light duty crane. Railroad repair facilities ("rip tracks") on major tank car shipping routes are able to assist in applying these couplers. Accordingly, it is proposed to amend section 173.31(a)(5) to require retrofit installation of shelf couplers not later than December 31, 1978. Since the proposed accelerated coupler retrofit schedule would not result in additional "shopping," or significant "out-of-service" time, this change in schedule should not result in any appreciable change in retrofit cost.

Non-Jacketed Thermal Protection with Separate Tank Head Protection (Specifications 112T and 114T). As stated, it appears that approximately 4,000 specification 112 and 114 tank cars will be equipped with nonjacketed, "spray-on" thermal protection and separate tank head protection ("T" retrofit package). These cars when retrofitted will be specification 112T and 114T tank cars. Due to the urgency of placing tank head protection on these cars at the earliest possible time, it is proposed to amend section 179–105–3(d) to require that:

1. All tank head protection (head shields) be applied not later than December 31, 1979; and
2. Thermal, "spray-on" coating be applied not later than December 31, 1980.

Since this change in schedule could result in as many as 50 percent of these tank cars (e.g., the tank cars originally scheduled for retrofit in 1980 and 1981) having to be out-of-service twice (once for "head shield" application and once for thermal protection application) additional retrofit costs could occur. It was indicated at the FRA special safety inquiry that each such retrofit application could remove the car from service for up to 45 days. Since these non-retrofitted tank cars have an average monthly rental of $300, the overall maximum additional cost would be $900,000 (e.g., 2,000 tank cars × $300/mo. × 1½ mo.). As noted below, 45 days is a relatively high estimate.

Although some participants in the FRA special safety inquiry suggested that "head shields" could be applied by not later than the end of 1978, the Bureau believes that such a drastic compression is not feasible.

Considerable concern exists among some parties as to the methods of retrofitting head shields to the tank cars. Several persons have questioned whether the "trapezoidal" head shield can be adequately attached to the tank car draft sill. Nine specification 112 tank cars were equipped with trapezoidal type head shields and fatigue tested at the Transportation Test Center at Pueblo, Colo. As of March 24, 1978, these head shields had been subjected to an average of 248 coupling impacts (ranging in speed from 4 to 10 miles per hour) and approximately 100,000 miles of over the road service. No fatigue problems were detected. Also, another type of head shield consisting of a half tank car head was installed on each end of one tank car. As of the same date, these two head shields were subjected to 248 coupling impacts and approximately 78,000 miles of over the road service. Again, no fatigue problems were detected. This testing indicates that no fatigue problems should occur when the head shield is attached to the tank car using proper welding techniques and a sound attachment design.

However, the welded attachment of all of these head shields to the tank cars was performed under controlled conditions. Most shield designers and manufacturers indicated that this welding operation was the critical factor and needed to be performed by highly skilled welders under controlled conditions in enclosed shops in order to avoid a risk of failure during train operations and consequent serious derailment. Since this retrofit application can result in a significant out-of-service period, the reduction in the supply of tank cars which would result from compressing this schedule to any greater degree could cause severe economic difficulty.

Tank-Head Protection without Thermal Protection (Specifications 112S and 114S). It appears that approximately 3,400 specification 112 and 114 tank cars will be dedicated to the transportation of anhydrous ammonia. These cars, which are required to be equipped with tank head protection ("head shields") ("S" retrofit package), will when retrofitted be specification 112S and 114S tank cars. Again, due to the urgency of placing tank head protection on these cars at the earliest possible time, it is proposed to amend section 179–105–3(d) to require that this tank head protection be applied not later than December 31, 1979.

It appears that such a change in schedule will not result in any appreciable increase in retrofit costs.

As was indicated in the discussion of the application of head shields to tank cars being retrofitted to the 112T and 114T specifications, suggestions have been made that head shield application be completed by the end of 1978. These tank cars are used exclusively to store and transport anhydrous ammonia. Due to the prolonged cold weather, most of these cars will not be available for retrofitting until early July and will be needed to store manufactured anhydrous ammonia beginning in early September. Any significant out-of-service disruption could result in a severe fertilizer shortage in the spring of 1979. For this reason, it appears that a second year (1979) will be required to perform this retrofit if significant disruption is to be avoided.

Jacketed Insulation with Integral Tank Head Protection (Specifications 112J and 114J). Of the roughly 20,400 specification 112 and 114 tank cars subject to the retrofit requirements of HM-144, approximately 12,500 are planned to be retrofitted with a jacketed insulation system incorporating integral tank head protection ("J" retrofit package). These cars when retrofitted will be specification 112J and 114J tank cars. Several major tank car builders have indicated that these cars could be completely retrofitted not later than December 31, 1980 and our analysis supports this conclusion. Accordingly, it is proposed to amend section 179.105–3(d) to require this retrofit operation to be performed so that:

1. Twenty-five percent of these tank cars owned by each tank cars owner be retrofitted not later than December 31, 1978;

2. An additional 40 percent of these tank cars owned by each tank car owner be retrofitted not later than December 31, 1979; and

3. An additional 35 percent of these tank cars owned by each tank car owner be retrofitted not later than December 31, 1980.

Likewise, based upon statements made at the FRA safety inquiry as well as other information received, it is believed that this proposed acceleration of the retrofit schedule should not result in any appreciable increase in retrofit costs.

Consideration has been given to requiring either total completion of this type of retrofit at an earlier date or increasing the percentage of tank cars required to be retrofitted during 1978 and 1979. Since this type of retrofit requires considerable ability in metalsforming and insulation application, only a few tank car repair shops have the existing capacity to perform this work. Construction of additional plant capacity would consume considerable time, while use of new car construction shops could cause severe tank car shortages and cause economic problems for many petroleum and chemical shippers and users. More importantly, any additional compression could cause critical out-of-service problems during the heating and fertilizing seasons, resulting in insufficient fuel during the winter and insufficient fertilizer in the spring. For this reason, as well as considering shop facility capacity, it appears that this retrofit schedule would cause the least overall economic disruption while achieving a more rapid implementation of the safety standards.

Availability of Cars During the Retrofit Period. Without question the most serious constraint facing the FRA and the Bureau in the development of a compressed timetable has been the availability of pressure tank cars to perform essential transportation services. Witnesses at the FRA special safety inquiry indicated that the pressure tank car fleet is fully utilized during much of the year either to carry or to store fuels, fertilizer and industrial chemicals. This testimony is consistent with other information available to the Department of Transportation. Therefore, the FRA and the Bureau have attempted to fashion the proposed new retrofit schedule in a way which is intended to minimize disruptions in service. However, it is recognized that the compression of the program into a shorter time period may result in localized shortages of essential products. Comment is specifically solicited, therefore, on the following analysis of out-of-service time and the consequences of that analyses for users of the products transported and stored in 112 and 114 tank cars.

Application of a shelf coupler is a relatively simple operation requiring not more than a total elapsed time of one-hour per tank car using a two or three man crew and a light duty crane. The difficulty arises in having the appropriate pair of shelf couplers at the proper location so as to be ready for application to a specific tank car. However, this is a problem which is solvable through proper planning. In terms of total out-of-service time, coupler retrofit can cause a tank car to be "out-of-service" for a time period of up to one day. This one-day time period is caused by switching the tank car to and later from a "repair" or "work" track. Since many 112 and 114 tank cars will have to be moved to repair tracks for other purposes prior to the end of the year, this impact should not be significant. Through the exercise of proper initiative, couplers may also be applied at major shipping points without any out-of-service time attributable to the application of the couplers.

Application of "head shields," "spray-on" thermal protection and jacketed insulation systems require the tank car to be shipped to a repair facility. Shippers, car owners and tank car lessors agreed that a time period of from twelve to fifteen days is required to move a tank car from an unloading point to a repair shop and that a like period of time is required to move a tank car from a repair shop to a loading point. Estimates of the time required to perform the retrofit operations and related maintenance ranged from twelve to thirty days. This includes provision for preinstallation operations. An average period of fifteen days appears to be realistic. Thus, to total out-of-service time estimates range from 36 to 60 days. An average out-of-service time of 45 days is used in the following analyses. However, some time credit must be assigned to the fact that during this 45-day period the empty tank car has moved from the consignee's unloading facility to the shipper's loading facility. A ten-day time period would be the minimum average time required for this empty movement were not retrofit or maintenance shopping involved. Accordingly, the net retrofit out-of-service time chargeable to this program has been determined at 35 days (five weeks) for each shop cycle.

In order to determine the effect of out-of-service time, it is assumed that the major retrofit program will begin about July 1, 1978. Thus, there will be approximately five 5-week cycles in 1978, 10 such cycles in 1979 and 10 additional such cycles in 1980. Allowance for plant vacations and possible holiday interruption is taken into account by using a fifty, rather than a fifty-two week year.

The effect of this five-week retrofit cycle on approximately 2,000 vinyl chloride tank cars being converted to specifications 112T and 114T can be analyzed.

1. Under the existing retrofit schedule, fifty-percent (1,000) of these tank cars were to be retrofitted with "spray-on" thermal insulation and "head shields" not later than December 31, 1979.

2. Under the proposed accelerated retrofit schedule all 2,000 of these tank cars would have to be retrofitted with "head shields" by that date.

3. Therefore, at least 1,000 vinyl chloride monomer tank cars already have been scheduled for total retrofit not later than December 31, 1979; and thus, not more than 1,000 such tank cars will require two shoppings, one shopping between the present date and the end of 1979 for application of "head shields," and one shopping during 1980 for the application of "spray-on" thermal protection. By careful planning, some owners should be able to complete additional cars in a single shopping.

4. 2,000 tank cars will be out-of-service for a five-week retrofit cycle between the present date and December 31, 1979, with fifteen such cycles. This means that an average of 133 (2,000 tank cars divided by fifteen cycles) will be out-of-service at any one time due to retrofit applications being performed during the time period of July 1, 1978, through December 31, 1979.

5. A maximum of 1,000 tank cars will require retrofit installation of "spray-on" thermal protection during 1980. This means that an average of 100 (1,000 tank cars divided by ten cycles) will be out-of-service at any one time due to the retrofit applications being performed during 1980.

In the same manner, the effect of this five-week retrofit cycle on the approximately 2,000 specification 112 and 114 tank cars transporting liquefied flammable gases which are being converted to specifications 112T and 114T can be analyzed as follows:

An average of 133 tank cars will be out-of-service at any one time during the time period of July 1, 1978, through December 31, 1979; and an average of 100 tank cars will be out-of-service at any one time during 1980.

Likewise, the effect of this five-week retrofit cycle on the approximately 3,400 dedicated anhydrous ammonia tank cars being converted to specifications 112S and 114S can be analyzed.

1. Approximately 700 of these tank cars have been converted or built to specifications 112S and 114S.

2. Approximately 2,700 of these tank cars must have "head shields" retrofit installed by December 31, 1979.

3. With fifteen such cycles, this means that an average of 180 of these tank cars will be out-of-service during any one cycle for the time period of July 1, 1978, through December 31, 1979.

The proposed accelerated retrofit schedule would require that the approximately 12,500 specification 112 and 114 tank cars being converted to specifications 112J and 114J be retrofitted according to the following schedule: 25 percent in 1978; and additional 40 percent in 1979; and an additional 35 percent 1980.

Thus, during the time period of July 1, 1978, through December 31, 1978, there would be five, five-week retrofit cycles. Approximately 3,125 (25 percent of 12,500) tank cars would require retrofit shopping during this time period. Approximately 625 (3,125 tank cars divided by 5 cycles) such tank cars would be out-of-service at any one time during July 1, 1978, through December 31, 1978.

During 1979, approximately 5,000 (40 percent of 12,500) of these tank cars would require retrofit shopping. Approximately 500 (5,000 tank cars divided by 10 cycles) such tank cars would be out-of-service at any one time during the year.

During 1980, approximately 4,375 (35 percent of 12,500) of these tank cars would require retrofit shopping. Approximately 438 (4,375 tank cars divided by 10 cycles) such tank cars would be out-of-service at any one time during the year.

In summary, this analysis shows that under the requirements of the proposed retrofit schedule an average of 848 tank cars (4.2 percent) will be out-of-service at any one time between July 1, 1978, and December 31, 1980. Average units out-of-service for individual years are (a) 1,071 tank cars during 1978, (b) 946 tank cars during 1979, and (c) 638 tank cars during 1980. Greater impacts may be experienced within individual categories of service. These numbers represent an overall lower percentage than that estimated by the tank car companies. Since the analysis assumes an even flow of cars through the shops the number of cars actually withdrawn from service at any given time may be higher or lower.

Since most of the tank car builders indicated that retrofit operations will be performed at facilities other than their principal new car fabrication facilities, and since current production of tank cars of all types is considerably less than total capacity, additional new pressure tank car construction could ease shortages occurring during the retrofit period.

Canadian 112 and 114 Tank Cars. Approximately 2,000 specification 112 and 114 tank cars have been constructed to specifications promulgated by the Canadian Transport Commission (CTC) and are used principally in Canada. However, approximately 80 percent of these CTC specification 112 and 114 tank cars transport hazardous commodities on the United States railroad network at some time. Accordingly, it is proposed to amend § 179.105–1(c) to require shelf couplers on all such CTC tank cars transporting hazardous materials in the United States not later than December 31, 1978, and require total retrofit not later than December 31, 1980.

Compliance. In order to assist in monitoring compliance with the HM-44 retrofit schedule, a separate Notice of Proposed Rulemaking is being developed. This Notice will propose requirements for car owner reporting of retrofit plans and accomplishments.

Economic Impact. In analyzing the effect of accelerating the retrofit schedule as proposed in this Notice of Proposed Rulemaking, the FRA and the Bureau have attempted to identify additional costs resulting from compression of the schedule. A specific possible increased cost of $900,000 has been identified for non-jacketed thermal protection and separate tank head protection application. Other additional costs are not now identifiable in definitive terms. However, the Bureau recognizes that compliance with the compressed retrofit schedule proposed in this Notice will result in some additional costs such as overtime payments, second and third shift differential payments, and possible premium payments for components. Also there may be additional transportation costs associated with "double shopping" of a small number of DOT specification 112T and 114T tank cars, as well as some additional labor cost. It is the belief of the Bureau that such additional costs will be only a small percentage of the cost of the initial rule and that the benefits to public safety and industry of accelerating the retrofit of these safety features will far outweigh any additional cost. Commenters are requested to submit cost information pertinent to this proposal.

(43 F.R. 20250, May 11, 1978)

[49 CFR Part 179]
[Docket No. HM-144; Notice No. 78–8]

SPECIFICATION FOR PRESSURE TANK CAR TANKS: COMPLIANCE REPORTING

The Materials Transportation Bureau published on September 15, 1977, a final rule establishing additional safety requirements for DOT Specification 112 and 114 tank cars (42 FR 46306). The requirements included improved couplers, tank head protection and thermal protection. Tank car owners were afforded a 4-year period to complete the application of required protective systems to cars built prior to January 1, 1978. On May 11, 1978, the Bureau published a notice of proposed rulemaking (NPRM) to shorten the period allowed for retrofit of these cars, 43 FR 20250. This rulemaking is proposed as a means of facilitating the implementation of the proposed retrofit program.

Information received at a Federal Railroad Administration (FRA) special safety inquiry of April 7, 1978, indicated that a substantial shortening of the retrofit period was possible but would intensify both logistical and car availability problems. The "logistical" problems described by witnesses at the safety inquiry relate to the diversion of cars from their normal service in a controlled, incremental fashion to assure the full utilization of retrofit plant capacity. Problems of car availability will be directly impacted by the success of the tank car companies and private car shops in scheduling a phased retrofit of the fleet.

The Bureau is hopeful that tank car owners will take adequate measures to assure the phased completion of the retrofit without creating short-term critical shortages of 112 and 114 cars. The tank cars constitute a substantial portion of existing pressure tank car capacity for the transportation of certain essential fuels, fertilizers, and industrial chemicals. Due to the high demand for such equipment for use in transportation and as temporary storage, it is expected that tank car owners will have to make careful plans to assure completion of the program by the proposed regulatory deadlines.

Neglect by a tank car owner or owners to establish an adequate pace of retrofit could result in a failure to meet regulatory deadlines. Since it is the policy of the Bureau not to grant exemptions from the regulatory deadlines, it is possible that serious shortages of cars could exist on one or more of the regulatory deadlines as a result of an accumulation of unequipped cars which would be prohibited from use in transportation. It will, therefore, be necessary for the FRA, which is responsible for enforcing the tank car regulations, to monitor closely the manner in which tank owners comply with the regulatory deadlines. In the event it appears that any tank car owner has failed to establish a program leading to the timely completion of the retrofit tasks, FRA may find it necessary to institute compliance order proceedings under 49 CFR Part 209 (42 FR 56742; October 28, 1977) or take other, appropriate legal action.

The reporting rule proposed in this notice will provide the FRA with the information necessary to carry out its enforcement mission. In the judgment of the Bureau, the information requested does not go beyond the basic kind of data which tank car owners would have to develop in the normal course of business to facilitate compliance with the substantive regulations. Any cost directly attributable to the reporting requirement would, therefore, be limited to preparation of correspondence.

The proposed rule would not require the use of any standard form. Reporting requirements would lapse after completion of the retrofit process and submission of a final report.

FRA has estimated that fewer than 100 tank car owners would be required to submit reports under the rule. A separate amendment to the regulations would define tank car owner to mean a person whose reporting mark appears on the car. The definition will assure that the individual or business which is responsible for the control and maintenance of the car is also responsible for seeing that the car is equipped with the required safety devices.

Section 179.105-9. The Bureau proposes to establish basic reporting requirements under a new section 179.105-9. The section would require four basic kinds of reports.

Initial report. Paragraph (a) would require each tank car owner to make an initial report to FRA not later than September 30, 1978, providing specific information concerning the type of retrofit package which will be employed, and describing the progress already made to comply with the retrofit schedule.

Quarterly report. Paragraph (b) would require each tank car owner to provide a quarterly update of the progress made in applying head protection, thermal protection, and improved couplers.

Final report. Paragraph (c) would require each tank car owner to certify in a final compliance report, the completion of the retrofit program.

Report on change in status. Paragraph (d) would require the reporting, in connection with the quarterly submission, of any material event bearing on the responsibility of any person with respect to the accomplishment of the retrofit tasks. The purpose of this provision is to assure that responsibility for compliance can be fixed on the appropriate person and to provide an explanation for any irregularities caused by the transfer or destruction of any cars.

Section 179.105-1(d). The Bureau proposes to define "tank car owner," as that term is used in connection with requirements for specification 112 and 114 tank cars, to mean any person whose assigned reporting mark appears on the tank car. The reporting mark system is used in the railroad industry as a basis for identifying effective responsibility for and control of rolling stock. For practical compliance purposes, then, the "tank car owner" described by the proposed definition is also the person who "marks, maintains, reconditions, repairs, or tests" a tank car within the meaning of section 105 of the Hazardous Materials Transportation Act (49 U.S.C. 1804).

(43 FR 24865, June 8, 1978)

[Docket No. HM-144; Amdt. Nos. 173-117, 179-23]
PART 173—SHIPPERS—GENERAL REQUIREMENTS FOR SHIPMENTS AND PACKAGINGS
PART 179—SPECIFICATIONS FOR TANK CARS
Shippers; Specification for Pressure Tank Car Tanks

These amendments are the result of the joint efforts of the Federal Railroad Administration (FRA) and the Materials Transportation Bureau (the Bureau). In accordance with internal DOT procedures, the FRA has developed the substantive provisions of these amendments for review and issuance by the Bureau. Accordingly, further information concerning substantive provisions of these amendments may be obtained from the above contact.

BACKGROUND INFORMATION

Emerging Need for Expedited Retrofit. On September 15, 1977, the Bureau published in the FEDERAL REGISTER (42 FR 46306) a final rule concerning specifications for tank cars which included the following timetable:

1. Existing specification 112 and 114 tank cars used to transport flammable gases were to be retrofitted with thermal protection and tank head protection (such as "head shield") over a 4-year period ending on December 31, 1981.

2. Existing specification 112 and 114 tank cars used to transport anhydrous ammonia were to be retrofitted with tank head protection (such as a head shield) over a 4-year period ending on December 31, 1981.

3. All specification 112 and 114 tank cars were to be equipped with special couplers designed to resist coupler vertical disengagements. These couplers were to be retrofitted on all cars by July 1, 1979.

Major accidents at Pensacola, Fla., on November 9, 1977, at Waverly, Tenn., on February 22, 1978, and at Lewisville, Ark., on March 29, 1978, in combination with an incident of apparent vandalism near Youngstown, Fla., on February 26, 1978, focused attention on measures to improve the safety of rail transportation of hazardous materials.

As a result of these accidents and subsequent hearings conducted by Congress, the National Transportation Safety Board and the FRA, the Bureau published a notice of proposed rulemaking, Docket No. HM-144; Notice No. 78-5 (43 FR 20250; May 11, 1978). The purpose of that notice was to elicit public comment on a proposed rule to accelerate the time schedule for the retrofit program specified in this docket under amendments numbered 173-108 and 179-19.

The reasons for adopting a shortened retrofit program were discussed in considerable detail in the notice. Interested persons were invited to participate in the rulemaking proceeding through the submission of written comments.

Thirty-nine submissions were received and have been fully considered by the FRA and the Bureau in the development of this final rule.

In response to comments received, four changes have been made in the final rule.

1. In § 173.314 paragraph (c) table, the date in note 24 has been revised. This date is now December 31, 1980, and is consistent with the date in note 23.

2. In § 179.105–1 paragraph (c)(1) has been revised. The date for coupler retrofit for Canadian-owned tank cars moving under load in the United States has been extended 3 months to March 31, 1979, and permission granted for "empty" return movement to Canada of these nonequipped cars until July 1, 1979.

3. In § 179.105–1 paragraph (c)(2) has been revised to make the requirements applicable to Canadian-owned 112 and 114 tank cars consistent with those applicable to United States owned 112 and 114 tank cars when the retrofit program has been completed.

4. In § 179.105–3 paragraph (d)(3)(i) the percentage of "J" retrofits that must be completed by December 31, 1978, has been established at 20 percent. This is the percentage required by amendment 179–19.

SECTION-BY-SECTION ANALYSIS

Section 173.31 Qualification, Maintenance, and Use of Tank Cars. The notice proposed to reduce the retrofit time period for application of shelf couplers by 6 months from June 30, 1979, to December 31, 1978. Several commenters suggested that, although adequate supplies of shelf couplers might be fabricated by the end of 1978, they doubted that 100 percent of the cars could be equipped by that date. Logistical problems were cited. Several suggestions were offered:

1. Retain the June 30, 1979, date;

2. Use March 31, 1979, as the required date; and

3. Permit "empty" tank cars not equipped with shelf couplers to be transported after December 31, 1978 for either 3 or 6 additional months.

Information available to the FRA and the Bureau indicates that an adequate supply of shelf couplers will be manufactured during 1978 for retrofitting all of the 112 and 114 tank cars. However, it is recognized that getting the couplers to the tank cars may present problems. Shelf couplers manufactured during the last quarter of 1978 will need to be installed rapidly and the railroads have indicated that they will assist in this retrofit installation. They have designated specific repair facilities located along the major 112 and 114 tank car routes to stock these couplers and perform the retrofit. Likewise, in order to reduce this logistical problem of getting the tank car and the shelf couplers together, several car owners are utilizing portable installation equipment and installing their shelf couplers at shipper and consignee facilities.

While an "empty" tank car does not present the same degree of hazard as does a loaded tank car, an "empty" tank car can transport up to 1,000 gallons of product. Puncture of a car containing 1,000 gallons of propane or anhydrous ammonia could cause a serious hazardous materials accident. Tank cars, either in storage or after being unloaded, can be equipped with shelf couplers using portable equipment at the storage and unloading sites.

For these reasons the FRA and the Bureau believe that the December 31, 1978, date is sound and consistent with the goal of upgrading the safety in using these tank cars as quickly as is possible. It is recognized that compliance will require the close cooperation of the coupler suppliers, tank car owners, tank car users, and the railroads but all evidence indicates that this retrofit program can be completed by December 31, 1978, if these parties make a diligent effort.

Section 173.314 Requirements for Compressed Gases in Tank Cars. Several commenters recommended that in note 24, the date be December 31, 1980, so as to be consistent with the date in note 23 and so as to enable tank cars scheduled for the jacketed ("J") retrofit to be used in anhydrous ammonia service in 1980 while awaiting such retrofit. The FRA and the Bureau concur. Accordingly, note 24 has been amended to prohibit the shipment of anhydrous ammonia in 112 and 114 tank cars not equipped with head shields after December 31, 1980.

Section 179.105–1 General. Paragraph (c) of this section covers specification 112 and 114 tank cars manufactured to specifications promulgated by the Canadian Transport Commission (CTC). Amendment 179–19 required CTC specification 112 and 114 tank cars operating in the United States to comply with the DOT special requirements not later than December 31, 1981. The subsequent notice proposed to require shelf couplers installation not later than December 31, 1978, and require all other retrofitting not later than December 31, 1980.

Information received from Canadian owners and shippers, as well as the CTC, indicates that some 3,300 specification 112 and 114 tank cars were built to CTC specifications. The majority will be retrofitted to the "J" specification.

Two Canadian respondents questioned the December 31, 1980, deadline. They stated that this deadline might cause lessors to take tank cars out of "international" service (service between Canada and the United States) for up to 2 years. The FRA and the Bureau believe that this entire retrofit program must be completed by December 31, 1980, so as to assure adequate safety in the transportation of liquefied flammable gases and anhydrous ammonia in these tank cars. For this reason, the December 31, 1980, date is adopted.

Considerable comments were made concerning the December 31, 1978, shelf coupler retrofit deadline.

Canadian commenters reminded the FRA and the Bureau that under the December 31, 1981, deadline the Canadian tank car owners had approximately 4 years to install shelf couplers. Under the proposed revised deadline, this retrofit must be accomplished within less than 1 year for CTC specification tank cars used in "international" service. Commenters indicated that while approximately 80 percent of the CTC cars were in "international" service, the Canadian goal is 100-percent application of shelf couplers to these cars to provide maximum flexibility in utilization.

Canadian coupler manufacturers have just begun to produce type "E" shelf couplers and indicate that quantity production of type "F" shelf couplers will not begin before December 1978. Canadian tank car owners indicate that approximately 70 percent of their 112 and 114 tank cars require type "E" shelf couplers and 30 percent require type "F" shelf couplers.

This information on Canadian shelf coupler requirements and availability is new. It was not developed at the FRA special safety inquiry held in April and was not available to either the FRA or the Bureau. In light of these Canadian comments and since the proposed revised retrofit schedule compresses the Canadian-owned tank car coupler requirements from 4 years to 1 year, paragraph (c) has been modified:

1. The deadline for loaded CTC specification 112 and 114 tank cars moving in the United States to be equipped with shelf couplers is not later than March 31, 1979; and

2. The deadline for "empty" CTC specification 112 and 114 tank cars moving in the United States returning to Canada is by July 1, 1979.

Although "empty" tank cars present a hazard, the FRA and the Bureau recognize the difficulties in retrofit installing Canadian-manufactured shelf couplers on Canadian-owned tank cars while these cars are in the United States. In addition to logistical problems, such an operation can present unique "international" problems such as customs taxation. To reduce this type of difficulty, it has been decided to extend "empty" tank car movement rights to Canadian tank cars for 3 months so that they may return to Canada for shelf coupler retrofit.

It is recognized that this schedule is tight, but the FRA and the Bureau believe that promulgation of this schedule is essential to upgrade safety when CTC specification 112 and 114 tank cars are traveling in the United States. Further, the FRA and the Bureau are aware that the Canadian railroads have indicated a willingness to assist in the coupler retrofit program and have established a procedure for changing couplers while the tank cars are in rail transportation. Also, portable installation operations can be utilized in Canada.

Another problem mentioned by Canadian commenters was with paragraph (c)(2) of this section. In the notice this paragraph indicated that all compressed gases being transported in CTC 112 and 114 tank cars moving in the United States after December 31, 1980, would have to have such tank cars equipped with thermal protection and tank head puncture resistance. The effect of this requirement would be to place more stringent regulations on CTC 112 and 114 tank car shipments than those imposed on similar DOT specification tank car shipments. This difference was not intended. Accordingly, paragraph (c)(2) in § 179.105–1 has been amended to indicate that after December 31, 1980, CTC specification 112 and 114 tank cars:

1. Transporting flammable compressed gases in the United States shall have the prescribed thermal protection and tank head puncture resistance; and

2. Transporting anhydrous ammonia in the United States shall have the prescribed tank head puncture resistance.

Section 179.105–3 Previously Built Cars Paragraph (A). Paragraph (a) of this section covers retrofit installation of shelf couplers. As was stated under the analysis of § 173.31, several commenters stated that they believed that not all of the specification 112 and 114 tank cars could have shelf couplers retrofit installed by not later than December 31, 1978. For the reason stated in the analysis of § 173.31, the FRA and the Bureau believe that the December 31, 1978, deadline can be met with diligent effort by coupler suppliers, tank car owners, tank car users, and the railroads. Further, the FRA and the Bureau believe that complete coupler retrofit by the end of 1978 is necessary in order to quickly upgrade safety when these cars are being used to transport hazardous materials.

Paragraph (D). Paragraph (d) mandates specific retrofit schedules for performing the "S," "T," and "J" retrofits.

The "S" Retrofit Schedule. The notice proposed to require complete retrofitting of all tank cars being converted to the "S" specification by December 31, 1979. One commenter stated that he doubted that all of the head shields could be retrofit installed by the end of 1979. He is concerned with out-of-service time for his anhydrous ammonia cars as well as extra transportation costs associated with empty tank car movements to retrofit repair facilities. Another commenter recommended that all head shields be applied by the end of 1978.

In the notice, the FRA and the Bureau indicated their reasons for selecting the December 31, 1979, deadline. This reasoning included the fact that due to the prolonged cold 1978 spring, many anhydrous ammonia cars were not ready for retrofitting until July and these tank cars will again be needed to store manufactured anhydrous ammonia beginning in early September. The FRA and the Bureau believe that by extending the date through 1979 so as to include two summer periods, the estimated 3,400 tank cars exclusively dedicated to anhydrous ammonia service can be retrofitted with head shields.

The question of who will pay the empty tank car transportation costs is a matter to be resolved between the tank carowners and the railroads.

The "T" Retrofit Schedule. In the notice, paragraph (d)(2) proposed for "T" retrofitting that the deadlines for applying the tank head puncture resistance system ("head shields") be December 31, 1979, and for applying thermal protection be December 31, 1980. Considerable comment was received on this proposed revision of the schedule as well as the text of the preamble. Comments addressed to the statement in the preamble concerning " . . . the superior protective qualities of the jacketed retrofit package" will be discussed later in this analysis.

Many comments were received suggesting that the "T" retrofit schedule be the same as the "J" retrofit schedule. In the opinion of the FRA and the Bureau the "T" retrofit, particularly the application of "head shields," is easier to perform than the "J" retrofit. The jacketed method must be accomplished as a unitary process with the tank head protection, jacket shell and insulation being applied at one shop. This work can be performed at only a few locations.

At the National Transportation Safety Board special hearing on April 4, 1977, it was demonstrated that a "head shield" could be installed in approximately 94 minutes. While this example is a most specific case it demonstrates that "head shields" can be installed relatively quickly and easily as compared to full jackets and insulation.

The FRA and the Bureau believe that it is essential to install tank head puncture resistance as quickly as is possible, and that with a determined effort by industry all of these head shields can be applied by the end of 1979.

One commenter recommended that the "head shield" installation part of the "T" retrofit be required to be completed by the end of 1978. The FRA and the Bureau believe that compressing the retrofit deadline to the end of 1978 for "head shield" application would result in a considerable number of tank cars being out-of-service during the first quarter of 1979, just when they are needed the most. By establishing this deadline at December 31, 1979, this retrofit installation can be expeditiously completed with a minimum of out-of-service time.

Several commenters stated that they believed that it is unrealistic to state that "T" type retrofitting would be done in two stages; first, application of "head shields," and, second, application of thermal protection. Instead, these commenters believe that, as a practical matter, these two operations will be done at the same time. These amendments do not preclude such action. However, opportunity is being afforded to tank car owners to perform these two applications at different times. Obviously, the sooner that the entire retrofit is completed, the sooner the tank car will have the completed safety features detailed under this docket.

The "J" Retrofit Schedule. In the notice, paragraph (d)(3) proposed that the deadline for performing the jacketed retrofit be compressed from 4 years to 3 years and that the cumulative percentage of tank cars required to be completed at the end of each of the 3 years be 25 percent, 65 percent, and 100 percent, respectively.

A considerable number of comments were received concerning the completion percentage for 1978 (the first year). In amendment 179–19 this percentage was 20 percent; in notice No. 78–5 the percentage was proposed to be increased to 25 percent. Comments were received to the effect that plans and commitments had been made based upon the 20-percent figure. Further, construction of one new facility for performing the jacketed retrofit is not yet complete. However, these commenters believed that the second and third year requirement of 65 percent and 100-percent retrofit could be attained.

The FRA and the Bureau concur that the increased requirement for 25 percent completion by December 31, 1978, may not be attainable. Therefore, the 20-percent figure published under amendment 179–19 has been retained. The December 31, 1979, figure of 65 percent and the December 31, 1980, figure of 100 percent which were proposed in notice No. 78–5 appear to be attainable. Therefore, these percentages have been adopted as proposed in the notice.

Comments Concerning Alleged "Superior Protective Qualities" of Jackets. Several commenters have questioned the references in the notice of proposed rulemaking and subsequent departmental statements to the "superior protective qualities" of the jacketed retrofit. The point which was intended by these statements was that the presence of steel jacketing provides additional, if limited, protection against puncture or pressure vessel failure in an accident environment. This conclusion was based on the Department's extensive experience with the performance of steel jacketed insulated tank cars in actual service over a number of years.

Subsequent to the issuance of the notice, a major tank car company conducted tests of the particular thermal coating which it had selected for use in the retrofit program. The results of those tests were submitted for consideration in relation to the present rulemaking. The tests were designed to evaluate the extent to which the particular thermal coating might assist in preventing puncture and weakening of the tank shell. Under the test protocol employed, the resulting data indicated that the thermal coating provided protection at least equivalent to that afforded by a conventional jacketed system.

The FRA and the Bureau believe that the development of this new data underscores the validity of statements made by various commenters to the effect that no official preference should have been expressed in this rulemaking action for any particular system of protective devices which can be shown to meet the minimum thermal and tank head puncture resistance performance standards established by the substantive regulations.

It is the position of the Department that the retrofit should go forward as quickly as is feasible, with each tank carowner making such elections as the owner may deem appropriate in light of overall safety considerations. As pointed out by several commenters, disruptions in retrofit elections will result in a delay of the overall retrofit process. However, it remains true that the jacketed method requires somewhat more time, must be accomplished as a unitary process, and can be installed at only a few locations. For these reasons, the final rule, like the proposed rule, distinguishes between the "J" and "T" retrofits with respect to deadlines for the application of tank head protection.

Compliance Reporting. One commenter recommended that tank carowners be required to periodically report their retrofit progress and that this information be published periodically in the FEDERAL REGISTER.

On June 8, 1978, the Bureau published a notice of proposed rulemaking under docket No. HM-144 covering "Compliance Reporting" (notice No. 78-8; 43 FR 24865; June 8, 1978). The purpose of that notice is to elicit public comment on a proposal requiring DOT specification 112 and 114 tank carowners to provide a listing of those tank cars and report progress toward completion of retrofit plans. Accordingly, this comment will be considered when analyzing the response to notice 78-8.

Withdrawal from Service Compared to Retrofitting. One commenter indicated that he planned to withdraw his 112 and 114 tank cars from service by a combination of conversion (to DOT specification 111) and scrapping. He desired relief from the complete retrofit schedule provided he withdrew tank cars at the rate prescribed by the "J" retrofit schedule. While the FRA and the Bureau understand the problems being encountered by this tank carowner in converting his existing small capacity (12,000–15,000 gallons) uninsulated pressure tank cars to "economic" tank cars, it is believed that early installation of shelf couplers and speedy retrofit conversion of tank cars, according to their intended use, is essential in order to attain an adequate level of safety. Accordingly, no relief is granted to this commenter, or any other carowner, to substitute "withdrawal" from service for retrofitting.

Likewise, another owner of a very few number of 112 tank cars desires relief from the shelf coupler deadline because he is endeavoring to sell or otherwise dispose of these tank cars. Since the purpose of this provision is to achieve safety through the application of coupler vertical separation restraint, and since it is not directed to any one or group of owners, but is instead directed to all tank cars, no exception to the shelf coupler deadline is being granted. It should be noted that 112A and 114A tank cars equipped with shelf couplers may be used for the transportation of nonflammable compressed gases (except anhydrous ammonia) and hazardous liquids without further safety modification.

Waiver of FRA Periodic Inspection Deadline. One respondent requested a waiver of the FRA periodic inspection requirements for his hazardous materials laden tank cars. While it is not a part of this rulemaking, the FRA and the Bureau consider the periodic freight car inspection to be an important method of effecting an eventual overall reduction of railroad accidents. It is believed that both the FRA periodic inspection program and the HM-144 retrofit program are essential to safety and both can be carried out at the same time. However, since the railroad safety requirements are administered by the FRA, requests for waiver should be addressed to that Administration.

Effect of Strikes, Etc. One commenter has notified the Department that his retrofit activities are at a standstill because of a strike at two of his facilities. Another commenter indicates difficulty due to construction delay of a new facility.

The FRA and the Bureau appreciate that problems develop in any safety program. However, the need to provide for public safety outweighs acquiescence to these problems, and it is believed that solutions are available. It will be the policy of the Department in this retrofit matter not to issue waivers nor exemptions, but rather to assure that these regulations are adhered to in the manner and on the dates prescribed.

Economic Impact. In analyzing the effect of accelerating the retrofit schedule the FRA and the Bureau have attempted to identify additional costs resulting from compression of the schedule. A specific possible increased cost of $900,000 has been identified for nonjacketed thermal protection and separate tank head protection application. Other additional costs are not identifiable in definitive terms, and commenters did not present definitive information on specific costs to be incurred solely as a result of this accelerated schedule. However, the Bureau recognizes that compliance with the compressed retrofit schedule contained in this amendment will result in some additional costs such as overtime payments, second and third shift differential payments, and possible premium payments for components. Also there may be additional transportation costs associated with "double shopping" of a small number of DOT specification 112T and 114T tank cars, as well as some additional labor costs. It is the belief of the FRA and the Bureau that such additional costs will be only a small percentage of the cost of the initial rule and that the benefits to public safety and industry of accelerating the retrofit of these safety features will far outweigh any additional cost.

(43 F.R. 30057, July 13, 1978)

[Docket No. HM-144; Amdt. No. 179-24]
PART 179—SPECIFICATIONS FOR TANK CARS
Specification for Pressure Tank Car Tanks: Compliance Reporting

The Materials Transportation Bureau published on September 15, 1977, a final rule establishing additional safety requirements for DOT specification 112 and 114 tank cars (42 FR 46306). The requirements included improved couplers, tank head protection and thermal protection. Tank car owners were afforded a four-year period to complete the application of required protective systems to cars built prior to January 1, 1978. On July 13, 1978, the Bureau published a supplemental amendment shortening the period allowed for retrofit of these cars (43 FR 30057). This rulemaking is issued as a means of facilitating the implementation of the retrofit program.

On June 8, 1978, the Bureau published a notice of proposed rulemaking, Docket No. HM-144; notice No. 78-8 (43 FR 24865). The purpose of that notice was to elicit public comment on a proposed rule which would define "tank car owner" and require owners of DOT specifications 112 and 114 tank cars to provide:

1. A listing of cars owned;
2. Plans of the owner regarding retrofit; and
3. Quarterly reports concerning the owner's efforts to meet established compliance deadlines.

The reasons for considering these compliance reporting requirements were discussed in considerable detail in the Notice. Six submissions were received and have been fully considered by the Federal Railroad Administration (FRA) and the Bureau. After analysis of these comments it has been determined that the rule should be issued in the form proposed.

Three commenters opposed implementation of this reporting requirement. These commenters are shipper-owners of specification 112 and 114 tank cars. Two of these respondents assured DOT that their retrofit activities would be completed within the time limits prescribed. The third respondent suggested that time extensions might be requested to complete the retrofit program. These three believed that implementation of the compliance reporting requirements would be an unnecessary burden on tank car owners.

However, two commenters including one tank car shipper-owner favored the compliance reporting as a method to inform the FRA and the Bureau of retrofit status.

The FRA and the Bureau believe that it is necessary to monitor compliance with the regulatory deadlines. As was stated in the Notice, neglect by a tank car owner or owners to establish an adequate pace of retrofit could result in a failure to meet regulatory deadlines. Such failure could result in an accumulation of unequipped cars which would be prohibited from use in transportation. It will, therefore, be necessary for the FRA, which is responsible for enforcing the tank car regulations, to monitor closely the manner in which tank owners comply with the regulatory deadlines. In the event it appears that any tank car owner has failed to establish a program leading to the timely completion of the retrofit tasks, FRA may find it necessary to institute compliance order proceedings under 49 CFR Part 209 (42 FR 56742; October 28, 1977) or take other, appropriate legal action. The reporting rule contained in this Amendment will provide the FRA with the information necessary to carry out its enforcement mission.

One commenter suggested that use of the "UMLER" (Universal Machine Language Equipment Register) might afford FRA and the Bureau with adequate information. The "UMLER" keeps information concerning the tank car specification of every listed car. However, FRA experience with "UMLER" is that information is at least ninety-days behind events; and "UMLER" would provide no information regarding "shelf coupler" retrofit (which is not a specification change). Not would "UMLER" indicate owners' plans. Because of the compressed retrofit time schedule and our commitment to rapid completion of this important phase of the retrofit program, it is believed that this reporting system is necessary to obtain the needed information on an accurate, timely basis.

Additionally, one commenter recommended that the quarterly reports received from tank car owners be summarized and published in the FEDERAL REGISTER until full retrofit compliance is met. It is believed that by having these reports available for public inspection, adequate availability of information is afforded regarding the status of the retrofit program. Accordingly, it is not planned to publish quarterly summaries.

One respondent who owns specification 112 and 114 tank cars under more than one reporting mark requested that for purposes of this rule and compliance with the retrofit schedule his several reporting marks be treated as a unit. The FRA and the Bureau concur provided such a request is included when the information required by section 179.105-9(a) is submitted. However, leased tank cars cannot be considered as "owned" by lessees.

One commenter requested that departmental attention be focused on the cause of railroad accidents and that expeditious movement of liquefied petroleum gas not be disrupted. The FRA is taking all feasible action to address the causes of train accidents but believes that compliance reporting is also necessary. Also, the FRA and the Bureau believe that monitoring retrofit compliance will help to assure adequate tank car capacity for the transportation of liquefied flammable gases and anhydrous ammonia.

(43 F.R. 39792, September 7, 1978)

49 CFR Parts 170, 171, 172, 173, 174, 175, 176, 177, 178, 179, 180, 181, 182, 183, 184, 185, 186, 187, 188, 189] [Docket No. HM-145]
ENVIRONMENTAL AND HEALTH EFFECTS MATERIALS
Advance Notice of Proposed Rulemaking

In issuing this advance notice of proposed rulemaking, the Materials Transportation Bureau (MTB) is giving notice that it is considering whether new or additional transportation controls are necessary for classes of materials presenting certain hazards to humans and to the environment and which are not generally subject to the existing Hazardous Materials Regulations (HMR). The MTB is particularly interested in receiving views on the practicality and need for transportation controls on materials whose potential release during or incident to transportation may result in an unreasonable risk to property, the environment, or to human health and safety as has been determined through exposure in the work place or exposure by environmental accumulation.

This action is in response to recommendations from other organizations who have expressed a desire for the MTB to take more effective steps to deal with certain unregulated materials.

Comments by: March 14, 1977.

Addressed to: Docket Section, Office of Hazardous Materials Operations, Department of Transportation, Washington, D.C. 20590. Comments should reference Docket No. HM-145. It is requested that comments be submitted in five copies.

BACKGROUND

A number of public and private organizations and environmental agencies have expressed to MTB the view that the MTB should consider establishing transportation controls to deal with materials which are not regulated or are only partially regulated by the U.S. Department of Transportation's (DOT) HMR, transportation of which may pose certain hazards that the DOT previously has not formally recognized. The Natural Resources Defense Counsel, the General Electric Company, the National Tank Truck Carriers, the National Maritime Safety Association, the U.S. Environmental Protection Agency (EPA), and the Occupational Health and Safety Administration (OSHA) of the Department of Labor have expressed various concerns with the transportation of materials that may cause or contribute to the incidence of cancer, birth defects, genetic changes, environmental damage, and other effects, some poorly understood, and which in the past have been regulated, if at all, primarily because of other more easily recognized hazard characteristics. Such materials are referred to herein as "environmental and health effects materials." The MTB is considering the development of rules to deal with the transportation of a variety of environmental and health effects materials, to incorporate a systematic approach to identification of the kinds of hazards that might require attention, identification of materials that pose such hazards, and evaluation of the appropriateness of regulating such materials in transportation. Any such action would be based on Section 104 of the Hazardous Materials Transportation Act of 1974 (Pub. L. 93–633, 88 Stat. 2156) which authorizes the Secretary of Transportation to designate as a hazardous material any material the transportation of which in a particular quantity and form "may pose an unreasonable risk to health and safety or property ... "

EXISTING DOT REGULATIONS

Historically, the DOT has established its regulatory control upon properties of materials that pose a significant potential hazard to humans from acute exposures. The program to minimize this hazard has been primarily directed at controlling the handling of the materials and was further confined to the circumstances of the hazardous materials transportation activity. This philosophy has led to the development of a series of regulations found in Title 49 of the Code of Federal Regulations. These regulations define the classes of hazardous materials and list materials contained in the classes (49 CFR 172.101).

Present DOT definitions of classes of materials regulated as hazardous are found in 49 CFR Part 173. Definitions dealing primarily with toxic effects, found in Subpart H therein, include those of Poison A (§ 173.326), Poison B (§ 173.343), Irritating materials (§ 173.381), Etiologic agents (§ 173.386) and Radioactive materials (§ 173.389). The existing definitions are generally limited in scope by reliance on testing criteria that may not provide adequate consideration of the risks that transporting some materials may have on health or environmental effects. Some of these limitations in the transportation regulations can be recognized as: (a) Not listing as HMR, those materials which when directly exposed to man over a prolonged period of time (month to years) effect his health; (b) not listing as HMR, those materials which when discharged into the environment pose imminent and substantial danger to public health or welfare, including, but not limited to, fish, shellfish, wildlife, shorelines and beaches, or (c) not listing as HMR, those materials which when found in man's food, water, or air may endanger his health. These risks have been addressed to some extent by agencies outside this Department.

ACTIONS OF OTHER AGENCIES

In connection with possible modification of existing MTB classification criteria, the MTB may consider partial or full adoption of criteria, and lists of materials identified thereunder, which have been developed for specific purposes by other agencies. This approach has been employed in this Department's definition of etiologic agents, 49 CFR 173.386, which rely on identification of such agents by the Department of Health, Education, and Welfare.

The EPA has proposed rules under section 311 of the Federal Water Pollution Control Act (33 U.S.C. 1321) which identify 306 materials as hazardous substances, based upon their toxicity to aquatic, mammalian, and plant organisms, as well as their potential for entering the navigable waters of the United States (see Appendix A).

The OSHA of the Department of Labor has published a list of materials it considers to be human carcinogens (see Appendix B). The selection criteria used recognizes effects of chronic occupational exposure which may be quite remote in time from the onset of exposure. OSHA has also proposed rules governing occupational exposure to asbestos (see Appendix B), which would include controls on asbestos handling incident to transportation. The Inter-governmental Maritime Consultative Organization is actively concerned with possible hazards associated with health effects of asbestos particles released during transportation.

The Organization for Economic Cooperation and Development has issued a decision of the Council on Protection of the Environment by Control of Polychlorinated Biphenyls (PCB's), which was adopted at its 315th meeting in Paris, France, February 13, 1973, and which recommended that member countries require labeling and specification packaging for the transport of PCB's. Both the EPA and the U.S. Department of State have indicated concern over the health effects of these materials founded, in part, upon the PCB's levels found in the fisheries of the Great Lakes, certain foodstuffs, and in the milk fat of nursing mothers in several States. In Section 6 of the Toxic Substances Control Act (Pub. L. 94–469, October 11, 1976) Congress has directed EPA to prescribe methods of marking and disposal of PCB's and has completely banned manufacture and distribution of these materials within two and one-half years of the effective date of the Act, subject to exception by the EPA Administrator.

LEGISLATION

Additional mechanisms, either existing or in development, which address health or environmental effects of various materials, may exist at both the Federal and State level. Such programs as can be identified may be considered by the MTB in evaluating any action it may take. State programs pertaining to the transportation of materials called hazardous wastes are of particular interest.

Recent Federal legislation includes the previously mentioned Toxic Substances Control Act which provides EPA with authorization to require pre-market evaluation of new chemicals, as well as evaluation of some presently known materials. Although full implementation of this Act by EPA is some time off, activities of EPA and industries regulated under the Act may provide a great deal of information concerning environmental and health effects materials.

Title III of the Resources Conservation and Recovery Act of 1976 (Pub. L. 94–580, October 21, 1976) directs the EPA Administrator to develop criteria for identifying hazardous wastes and a list of such wastes to be subject to EPA regulatory control. Any proposed or existing hazardous waste transportation control activities using specific packaging, labeling, and shipping documents are of interest in the MTB's evaluation of environmental and health effects materials.

REQUEST FOR COMMENT

To assist the MTB in its examination of the possible need for further identification and control of enviromental and health effects materials moving in commerce, comments on the following subjects would be useful:

1. Whether or not additional regulation of environmental and health effects materials in transportation is needed and why. If so,

2. What sort of human health effects should be considered.

3. What sort of environmental effects should be considered.

4. What criteria should be used to ascertain effects and identify materials. The MTB is concerned that duplication of research efforts carried out by other agencies be avoided as far as possible and is interested in the suitability of considering lists of materials identified by other agencies as having adverse environmental or health effects.

5. Whether modifications to existing DOT hazardous material classifications, or establishment of new classes, would best accommodate the identified environmental and health effects materials.

6. What sort of transportation controls may be needed for identified environmental and health effects materials.

Presently available controls include specification of the physical containment necessary for transportation of a hazardous material, as well as systems to insure adequate communication of information on the material and its hazards to persons handling the material while it is in transportation and to persons responding to an emergency. Degree of control generally reflects the intensity of a given hazard. Should packaging controls be necessary, performance standards rather than specification standards may be considered.

7. With regard to hazardous waste, what classification system may be used to clearly identify mixtures as opposed to single compound materials; what packagings may be appropriate for transportation; and how existing transportation documentation can be used to cover transport of hazardous wastes from the generator (shipper) to the disposer (consignee).

8. Should new or additional transportation controls be necessary, what the impact on affected industries may be, and what a reasonable implementation schedule would be. The MTB is specifically concerned with avoiding costs which are not essential to the maintenance of transportation safety, and obtaining cost data to determine whether an inflation impact statement will be required.

9. Should new or additional transportation controls be necessary, whether the preparation of an environmental impact statement will be required.

10. Any other matters relevant to the identification and control in transportation of environmental and health effects materials, or to the need therefore, including the need for uniformity in the applicability of such safety regulations as might be developed under this docket to the various modes of transportation.

PROGRAM PLAN

If rulemaking is determined appropriate, under this docket, the MTB may consider a limited revision of the hazard classification; develop a list of substances; and provide a discussion for the basis of their selection. In addition, this effort may include consideration of regulatory requirements pertaining to communications, packaging, handling, and personnel training.

The MTB will be reviewing any comments received to answer questions outlined above and with a view to establishing selection criteria and rationale which would indicate specifically: (a) What types of toxicological data are meaningful; (b) in what context should these data be used; and (c) what degree of risk may be viewed as acceptable under what given conditions. Certain testing requirements may be established by the MTB to address: (a) The potential biological threat of a material; and (b) the probable occurrence of that threat during transportation.

The materials included in the EPA Hazardous Substances List and the OSHA list of carcinogenic chemicals, which are not presently regulated by the MTB in the Code of Federal Regulations, Title 49, are contained in Appendix A and B of this advance notice. These lists are provided as example lists of materials only and interested parties may wish to include in their comments specific reference to these listed materials as appropriate.

If sufficient interest in expressed in comments, an informal hearing on this subject will be held in Washington, D.C., no earlier than February 7, 1977. The time, location, and agenda of the hearing, if required, will be published in the FEDERAL REGISTER.

(41 F.R. 53824, December 9, 1976)

[49 CFR Parts 171, 172, 173, 174, 175, 176, 177]
[Docket No. HM-145A; Notice No. 78–6]
TRANSPORTATION OF HAZARDOUS WASTE MATERIALS
Proposed Provisions; Hearing

The RCRA (Pub. L. 94–580), which amends the Solid Waste Disposal Act (42 U.S.C. 3251 and following), directs the Environmental Protection Agency (EPA) to promulgate standards to be applied to persons generating, transporting, or treating, storing, or disposing of hazardous waste materials. The RCRA specifically directs EPA to promulgate regulations establishing standards applicable to transporters of hazardous wastes, regarding—

(1) Recordkeeping concerning such hazardous waste transported, and their source and delivery points;

(2) Transportation of such waste only if properly labeled;

(3) Compliance with the manifest system . . . ; and

(4) Transportation of all such hazardous waste only to the hazardous waste treatment, storage, or disposal facilities which the shipper designates on the manifest form to be a facility holding a permit issued under this subtitle (42 U.S.C. 6923).

The RCRA also requires EPA to publish regulations for generators of hazardous waste, regarding—

(1) Recordkeeping practices [to] identify the quantities of such waste [and] the constituents thereof . . . ;

(2) Labeling practices for any containers used for the storage, transport or disposal [to identify the waste];

(3) Use of appropriate containers for such hazardous waste;

(4) Furnishing of information on the general chemical composition of such hazardous waste to persons transporting, treating, storing, or disposing of such wastes;

(5) Use of a manifest system to assure that all such hazardous waste generated is designated for treatment, storage, or disposal in . . . facilities . . . for which a permit has been issued [except on-site facilities] . . . ; and

(6) Submission of reports to the Administrator [concerning quantities of hazardous waste generated and its disposition] [42 U.S.C. 6922].

Such transporter regulations as EPA may issue are required by the RCRA to be consistent with DOT regulations under the HMTA (title I of Pub. L. 93–633). However, EPA is authorized to recommend to the Secretary of Transportation changes to existing DOT regulations such as the designation of additional materials as hazardous under the HMTA.

EPA intends to propose generator standards and all other required hazardous waste regulations under subtitle C during the summer of 1978, as they are developed. On April 28, 1978, EPA proposed standards applicable to transporters of hazardous waste and also announced the June 20 public hearing previously noted (43 FR 18506). EPA stated at that time that the comment period for all EPA subtitle C proposals would remain open for at least 60 days beyond the last date of publication. For the same reasons, the deadline for comment on this proposal, to be announced later by DOT, will be at least 60 days after the last notice published by EPA (45 days' notice will be given). Since the RCRA provides 6 months after publication of final transporter standards before they can become effective, 6 months will be provided between publication of any final DOT rule on hazardous wastes and a mandatory effective date (except for waste materials that are hazardous materials under the existing DOT Hazardous Materials Regulations).

The regulations for transportation of hazardous wastes that are proposed herein are largely duplicative of the EPA transporter standards proposal of April 28. Some of the references to 40 CFR part 250 may have to be changed, as those references are now based on the numeration shown in unpublished EPA working drafts of generator standards that may be altered before publication as an EPA notice of proposed rulemaking or final rule. Depending on the results of the DOT proposal, EPA eventually may publish jointly with DOT, modify their own proposal, or adopt forthcoming DOT regulations. EPA and DOT intend to enforce jointly any DOT regulations governing transportation of hazardous waste. The RCRA also provides for State authorization to establish and administer a State hazardous waste program in lieu of an EPA program (42 U.S.C. 6926), so that most of the functions the RCRA assigns to EPA ultimately may be carried out by States authorized for that purpose by EPA.

EPA and DOT held a joint hearing on October 26, 1977, at Des Plaines, Ill., to consider the possible development by DOT of regulations under the HMTA to achieve the goals of the RCRA regarding transporters of hazardous waste. Public comment received as a result of the joint hearing expressed concern with the possibility that compliance with duplicative EPA and DOT regulations could cause inefficiency or confusion and strongly supported the prospect of development of joint EPA-DOT regulations for this purpose. The Materials Transportation Bureau (MTB) believes that the DOT Hazardous Materials Regulations are capable of being modified under the HMTA to address the transportation hazards of waste materials and that the RCRA states the need for such a modification. Development of the necessary regulations under the HMTA can result in essentially a single set of DOT regulations applicable to the transportation of hazardous materials which include hazardous wastes. Suitable DOT regulation of hazardous waste transportation will eliminate most of the additional and possibly duplicative regulations that would otherwise have to be issued by EPA. However, commenters should note that this DOT proposal is not identical to the April 28 EPA transporter proposal. Some differences of substance exist, and it is probable that some transporter requirements will be published by EPA, notwithstanding any final hazardous waste rules that DOT may issue.

Waste materials differ from virgin materials in several respects that impact transportation. The physical, chemical and biological characteristics of a waste are likely to be more difficult to establish in transportation emergencies than the characteristics of virgin materials which are often known by the commercial uses to which they are put. The mixing of waste materials from different sources increases the difficulty of predicting and communicating the hazard of the resulting mixture. In addition, the shipper of a waste material may have little economic interest in monitoring the transportation and delivery of the waste to a treatment, storage or disposal site. In the case of hazardous waste, the added costs of proper disposal (for example, high temperature incineration, or deposit in a permitted landfill) may result in the use of less expensive methods of disposal that adversely affect public health and safety. Transportation responsibility is a key element to ensuring proper off-site disposal.

In addressing these problems, the RCRA relies heavily on use of a hazardous waste manifest which resembles hazardous materials shipping papers required by DOT. Unlike hazardous materials shipping papers, the manifest is intended to provide a record of the transportation and delivery for disposal, treatment or storage of a hazardous waste. The accountability provided by the hazardous waste manifest extends to intrastate activity and is intended to ensure transportation to a proper facility prepared to accept the particular hazardous waste consigned to it.

Existing DOT Hazardous Materials Regulations generally do not apply to intrastate highway carriers or shippers. The distinction between inter- and intrastate transportation is much less pronounced in transportation by other modes. Both forms of transportation are regulated in air commerce, most rail carriers operate as interstate carriers, and movement of hazardous materials on the navigable waters of the United States is subject to Coast Guard enforcement activities. In highway carriage, however, there are substantial intrastate carriage operations not presently subject to the Hazardous Materials Regulations.

The HMTA defines "commerce" to include interstate commerce and intrastate transportation that affects interstate commerce (HMTA, § 103(1); see also HM-134, 41 FR 38175, September 9, 1976). The fact that the RCRA applies to all waste transporters, regardless of whether interstate commerce is directly involved, amounts to a finding that intrastate commerce in hazardous wastes affects interstate commerce. The necessity of assured delivery to a permitted disposal facility, as against possible diversion of shipments to improper disposal sites, requires regulation of intrastate movements. To the generator/shipper, or to the carrier of a hazardous waste, it may not be clear whether a given shipment is being offered for interstate or intrastate transportation, since the accompanying shipping paper may show alternate consignee facilities. This uncertainty also may hamper enforcement efforts, if the proposed DOT amendments restricted their application solely to interstate transportation. In view of this, the proposal herein would apply to both interstate and intrastate transportation of hazardous wastes by all modes.

The proposed amendments:

(1) Would apply to the offering, transportation and delivery, both interstate and intrastate, of hazardous waste materials.

(a) Government agencies. Existing Hazardous Materials Regulations do not apply to Federal, State or local governments that carry hazardous materials as a part of a governmental function, using government employees and vehicles, but do apply to such agencies that offer hazardous materials for transportation by common or contract carriers. The DOT proposal would apply similarly. Note that the EPA transporter proposal would apply to all governmental agencies, regardless of the nature of their activities.

(b) Commonwealth of the Northern Mariana Islands. Due to differences in statutory definitions, the DOT proposal, unlike the EPA transporter proposal, would not apply to the Northern Marianas (49 U.S.C. 1802(5); 42 U.S.C. 6903(31)).

(c) On-site disposal. Carriers subject to existing Hazardous Materials Regulations are regulated when traveling on or across any public highway, and the DOT proposal would similarly apply, subject to some EPA exceptions. The EPA transporter proposal would define "on-the-site" treatment, storage or disposal, to which transporter standards would not apply, to include contiguous sites separated only by public or private right[s]-of-way (see EPA proposal, § § 250.30(b) and 250.31(h)).

(2) Would amend only the DOT Hazardous Materials Regulations. The Federal Motor Carrier Safety Regulations (49 CFR Parts 390–397) would not be extended to include intrastate carriers, even though those regulations were recently incorporated by reference in the Hazardous Materials Regulations at 49 CFR 177.804. Also, this proposal does not address bulk shipment of hazardous materials by vessel, which is governed by Coast Guard regulations at 46 CFR Parts 30–40, 64, 98, 148, and 151.

(3) Would define "hazardous waste" to include presently recognized hazardous materials when shipped as waste, unless not recognized by the EPA as a hazardous waste. In view of their expected identification or listing under the RCRA as hazardous waste, the definition also would include materials shipped as waste, such as soil contaminated with polychlorinated biphenyls, which are not now recognized as hazardous materials. The definition would result in the designation of all hazardous wastes as hazardous materials, although some hazardous materials shipped as waste might not be hazardous wastes. Hazardous wastes would become a subset of hazardous materials, classed either by a specific entry in the Hazardous Materials Table, or by inclusion in a new ORM-E class (hazardous waste, n.o.s.).

(4) Would prohibit the offering, transportation or delivery of hazardous wastes to sites not sanctioned by the RCRA. With narrow exceptions, a hazardous waste, under this proposal, could be transported only to a site permitted by the EPA or by an authorized State for treatment, storage or disposal of that waste (exceptions include household refuge and small amounts of some wastes that may be excepted by EPA).

(5) Would require additional information on shipping papers. Additional information required to complete a RCRA manifest would have to be entered on hazardous materials shipping papers, or the manifest itself could be used as the shipping paper, provided it meets DOT requirements.

(6) Would allow a greater latitude for use of nonrefillable and reconditionable containers for some hazardous waste shipments.

(7) Would establish a stringent definition of when a container (under 110 gallons) is "empty" for purposes of the DOT Hazardous Materials Regulations generally, with a general placarding exception, and a particular hazardous waste marking exception. Many shipments of "empty" packagings containing residues of a hazardous material would not generally require placarding, and packagings containing a hazardous waste residue would not have to be marked "waste" if the residue is otherwise adequately identified under the Hazardous Materials Regulations.

(8) Would require carriers to submit a telephone report for any improper hazardous waste discharge, and to submit additional details on the written hazardous materials incident report (form F5800.1) presently required by DOT. Unlike the EPA transporter proposal (§ 250.37(a)), this proposal does not address emergency situations. Existing DOT rules concerning carriers (e.g., 49 CFR 177.823 and .854) and emergency exemptions (49 CFR 107.113)

should suffice. Moreover, the DOT proposal does not assign responsibility for cleaning up a spill but rather leaves it to be assigned by State or local law. Note that the EPA transporter proposal would require the carrier to clean up a spill or to take such action as is directed by an appropriate agency (§ 250.37(c)), and would limit reports to accidental discharges (§ § 250.31(f); 250.37(b)).

(9) Would require carriers to retain for three years a copy of each manifest or other document certifying delivery, and would require receipting signatures for shipments of hazardous waste transferred between carriers in any mode (compare the EPA transporter proposal, § 250.35(b)).

(10) Would preempt inconsistent State and local requirements. The HMTA provides in section 112 that any requirement of a State or its political subdivision which is inconsistent with the HMTA or with regulations issued thereunder is preempted. If this proposal is issued as a final rule under the HMTA, it will preempt inconsistent State and local requirements. The proposal contains a provision which specifically contains a provision which specifically addresses preemption of different or additional packaging, marking, labeling and placarding requirements, certain reporting requirements, and certain shipping paper requirements. Note that the RCRA provides for the establishment of authorized State hazardous waste management programs after EPA review, and that a waiver of preemption is possible under the HMTA for qualified State requirements (see 49 CFR Part 107).

(11) Would not result in a major national economic impact, nor in a major increase in costs or prices for carriers generally, industries that ship hazardous waste, levels of government, geographic regions or specific elements of the population. Required use of specification packaging would be new only to intrastate hazardous waste carriers and their shippers not now required by State or local law to use DOT specification packaging for hazardous materials. This proposal would not require specification packaging for waste materials not presently regulated as hazardous materials. The basis for this conclusion is discussed in somewhat greater detail in an economic evaluation prepared by MTB, and is based in part on a preliminary EPA study entitled "The Economic Impact on the Hazardous Waste Transportation Regulations (section 3003 of the RCRA)" which was prepared by A. D. Little, Inc. Both documents are available for examination in the dockets section at the address previously shown.

(12) Do not require preparation of an Environmental Impact Statement. The transportation rules proposed herein do not directly concern disposal methods the RCRA requires for hazardous wastes and are not expected to have a significant impact on the environment.

Primary drafters of this proposal are Lee Metcalfe, Office of Hazardous Materials Operations, and Douglas Crockett, Office of the Chief Counsel, Research and Special Programs Administration.

DETAILED REVIEW BY SECTIONS

Section 171.3 is the basic provision regarding the offering, transportation and delivery of hazardous waste. The proposed rules would apply to all transportation, including intrastate transportation.

The identification of a material as a hazardous waste may result from a shipping paper entry or from actual knowledge by the offerer or transporter. A carrier is not obligated to examine each component of a waste shipment to determine whether it contains a hazardous waste, but he has knowledge that he is being offered a hazardous waste if a material is shown on shipping papers as a hazardous material and is offered for treatment, storage or disposal, or if the carrier, due to the nature of his carriage operation, for example, has adequate reason to be aware that a material so offered is hazardous.

Under paragraph (b), any motor carrier transporting a hazardous waste for which an accompanying manifest is required would have to display the same identification marking on his vehicle that is required by the Federal Motor Carrier Safety Regulations. Note that the EPA transporter proposal would limit this requirement to vehicles that must be placarded or that are carrying more than 1,000 lbs. of a hazardous waste (§ 250.38(b)).

A hazardous waste could be delivered only to a permitted facility unless the exception in paragraph (c)(2) applies. Certain small shipments that EPA anticipates may be shipped without a manifest could go to facilities not holding permits under subtitle C of the RCRA, provided each shipment is offered and transported separately from, and not consolidated with, other hazardous wastes, but note that DOT shipping papers may still be necessary even in the absence of a manifest. Note also that carriers who consolidate individual hazardous waste shipments for delivery to a treatment, storage or disposal site might be subject to EPA requirements as generators of hazardous waste, a circumstance that would occur if different wastes are consolidated so as to change the identity of the resulting composite waste material. Even under existing DOT regulations, if a carrier mixes hazardous materials after acceptance from a shipper in a way that changes the hazards involved, the resulting hazardous material cannot be transported except in compliance with DOT regulations concerning that new material.

Where small shipments not under a manifest are consolidated, and the identity of the resulting mixture is not different from its constituent components, a manifest is not expected to be required for transportation of that mixture (although shipping papers may be), but a manifest may be required by the permitted consignee facility to which the shipment is delivered.

Household refuse and household septic tank pumpings would be excepted from the subchapter by paragraph (d), so that neither the offer of such materials nor their transportation would be regulated by DOT.

Paragraph (e) would state the MTB's intent and opinion regarding preemption of certain categories of State hazardous waste regulations that concern transportation. The Hazardous Materials Regulations generally preempt State and local requirements which are inconsistent with requirements contained in those regulations, a fact that would be equally true of the proposal if published as a final rule. Paragraph (e) is intended to clarify some aspects of the preemptive effects of the proposal. State or local action not addressed in paragraph (e) would still be pre-empted if inconsistent with requirements of this proposal should it become a final rule under the HMTA.

Section 171.8. Section 171.8 contains proposed definitions related to hazardous waste. Note that the proposed definition of "hazardous waste" is limited to materials that are subject to EPA requirements that are to be specified in 40 CFR Part 250.

Section 171.17. Section 171.17 proposes reporting requirements in addition to those presently specified in §§ 171.15 and 171.16, although compliance with proposed § 171.17 would satisfy the requirements of § 171.15. Telephonic contact with the National Response Center, U.S. Coast Guard, or an EPA regional official at an appropriate regional office would be required when any discharge of a hazardous waste to the environment occurs during transportation. Excluded from the DOT requirement would be telephonic contact with a discharge occurs within the facilities of a shipper or consignee. (However, the discharge of a substance designated as hazardous by EPA under 40 CFR Part 116 into or upon the navigable waters of the United States is reportable under the Clean Water Act, Pub. L. 92–500, section 311(b)(5)). This proposed exclusion would have no effect on the telephonic notification requirement specified in § 171.15. Under the proposal, it is possible that two telephonic notifications would be required. Paragraph (b) of this section specifies the information to be supplied to the official to whom the report is made, and paragraph (c) specifies reporting requirements in addition to those presently required to be filed on DOT Form F5800.1 under § 171.16. The additional information that would be required may necessitate the filing of supplemental reports if the disposition of a discharged material is not known by the reporter within the 15-day reporting time frame. State reporting requirements for hazardous wastes are limited to duplication of Federal requirements except for immediate notification (see proposed § 171.3).

Section 172.101. Present §§ 172.100 and 172.101 would be combined into a revised and amended § 172.101 in order that the Hazardous Materials Table and the language introducing the table be contained in one section of the regulations. Paragraphs (b) (2) and (3) would be modified to exclude from single mode applicability those materials which are hazardous wastes and to subject those materials to transportation requirements regardless of the mode of transportation involved. Paragraph (c)(8) would be added to specify that a proper shipping name include the word "waste" when the material described is a hazardous waste subject to EPA's requirements in 49 CFR Part 250. Use of this method to appropriately prescribe the proper shipping name for a hazardous waste alleviates the necessity for the addition of hundreds of proper shipping name entries to the Hazardous Materials Table. Only one new shipping name "Hazardous waste, n.o.s.," would be added to the Hazardous Materials Table; the other italicized entries are for cross-reference and referral purposes.

Section 172.200. Section 172.200(b) would be amended to remove the ORM exceptions to the shipping paper requirements when a material being offered or transported is a hazardous waste. Also, an additional entry pertaining to ORM-E materials would be included.

Section 172.201. Section 172.201(c) would be added to require both the name and address of the shipper and the consignee on shipping papers for hazardous wastes. These entries are essential to the establishment of appropriate controls relative to the origin and destination of hazardous wastes.

Section 172.203. Section 172.203 would include a new paragraph (j) requiring the inclusion of the EPA specified name for a hazardous waste when the proper shipping name for a material is specified in § 172.101 by other than its technical name.

Section 172.205. A new § 172.205 would be added concerning the carriage and disposition of hazardous waste manifests to be required by the EPA in 40 CFR 250.22. Note that paragraph (a) requires a receipting signature whenever a shipment is transferred between carriers in any mode (compare EPA proposal, § 250.35(b)). Paragraph (b) provides that a hazardous waste manifest may be used as the shipping paper for a material if it contains all the information required by Part 172, Subpart C.

Section 172.300. Section 172.300 would be amended to specifically reference § 173.29 regarding empty packagings and a paragraph (b) would be added which would require packages of hazardous waste to be marked as required by EPA.

Section 172.306. Section 172.306 would be amended by adding a new paragraph (b) requiring certain packages to be marked with both the name and address of a shipper of a hazardous waste, thereby excluding the option afforded in paragraph (a) of the section which permits the name and address of the shipper or the consignee to be displayed. This provision would parallel proposed § 172.201(c).

Section 171.316. Section 172.316 would be amended to include ORM-E materials in the package marking requirements for ORM's.

Section 172.504. Section 172.504 would be amended by adding a new paragraph (d) to exclude certain packagings containing only the residue of hazardous materials from consideration in determining the applicability of the placarding requirements.

Section 173.2. Section 173.2 would be amended to add ORM-E to the "order of hazards" listing for classification purposes.

Section 173.28. Section 173.28 would be amended to consolidate three paragraphs that contain restrictions pertaining to containers marked NRC and STC which are reused. Paragraph (n) would be amended to include a reference to ORM-E materials regarding reuse of STC marked packagings and a new paragraph (p) would be added to permit the reuse of NRC or STC marked specification packagings for one-time shipments of hazardous wastes under certain specified conditions. Note the first condition stipulates that the material must be packaged "in accordance with this part"; therefore, a flammable liquid, n.o.s. would be packaged in accordance with § 173.119. This reuse authorization for hazardous wastes would not permit any deviation from the packaging requirement of Part 173 except as specifically stated in § 173.28. Proposed paragraph (p)(5) would permit the collection of certain STC marked packagings for reconditioning and reuse in accordance with the regulations if they can be properly requalified under the provisions specified in paragraph (m) of the section.

Section 173.29. Section 173.29 would be amended to require, with certain exceptions, a packaging that contains the residue of a hazardous material to be offered for transportation in the same manner as required when it previously contained a greater quantity of a hazardous material. A similar proposal was made under Docket HM-112 (39 FR 3022, 3095) on January 24, 1974. A number of commenters on that proposal requested that the "empty container" proposal for packagings having capacities of 110 gallons or less be removed from the HM-112 rulemaking action and considered in a separate rulemaking in order to afford commenters additional time to submit appropriate recommendations concerning the handling of so-called "empty packagings." Pursuant to the requests of several commenters, the matter was removed from that rulemaking action; however, no comments or recommendations have been received since that time. In specifying appropriate regulations for hazardous materials and hazardous wastes in connection therewith, the Bureau believes it is essential to deal with the subject of so-called "empty packagings" containing the residues of hazardous materials. Also proposed paragraph (c) would specify that a packaging being discarded would, without any qualification, be considered to contain a hazardous waste if it contains the residue of a hazardous material. The term "[place where it is] to be discarded" does not refer to those facilities where packagings themselves are reprocessed or reconditioned for further use but does include long-term storage facilities. The Bureau is fully aware of the ramifications of its proposed amendments to § 173.29 and solicits constructive comments relative to alternative regulatory requirements that would achieve the same level of safety as proposed in this rulemaking.

Section 173.118a. Section 173.118a would be amended to exclude a combustible liquid from the 110 gallons or less exception when it is a hazardous waste subject to 40 CFR Part 250. Paragraph (b)(1) would be amended to include a reference to hazardous waste manifests. Paragraph (b)(5) would be amended to reference the special reporting requirements in proposed § 171.17.

Section 173.389. Section 173.389 would be amended to restate the definition of radioactive materials to clarify the fact that the definition applies only for purposes of the Hazardous Material Regulations.

Section 173.500. Section 173.500 would be amended to clarify the definition of ORM materials. This clarification is essential to implementation of the ORM-E class which is included in new paragraph (a)(5). Note that the ORM-E definition includes hazardous wastes subject to the regulations of the EPA in 40 CFR Part 250, but is designated so as not to exclude other materials that may be included within this class at a future time.

Section 173.505. Section 173.505(a) would be revised for clarification of the limited quantity exceptions of ORM-A, B, and C materials, since the basic function of the regulations presently stated is to exclude certain materials from specific packaging requirements when they are offered for transportation in packages as Limited Quantities. The proposal is a more coherent restatement of this function.

Section 173.510. Section 173.510 would be amended to exclude the basic packaging requirements of that section from the exceptions specified in § 173.505 and a new paragraph (5) would be added requiring that transport vehicles used to transport ORM materials be free from leaks and that all openings must be securely closed. Of significance is the proposed prohibition against use of open-top freight containers and transport vehicles for bulk shipments. This prohibition would apply to the use of open or tarp-covered dump trucks for the transport of hazardous wastes or any other hazardous material in bulk. Commenters opposed to such a prohibition are invited to submit suggestions concerning appropriate controls for tarp-covered vehicles that would assure compliance with § 173.24.

Section 173.1300. A new Subpart I would be added to Part 173 to address ORM-E materials and a new § 173.1300 would be added to address hazardous wastes, n.o.s. No specific packaging requirements are being proposed in this rulemaking for such materials other than a reference to the basic requirements for ORMs in § 173.510 and Subpart A to Part 173. For example, if a hazardous waste is to be offered for transportation by air, the requirements of § 173.6 in Subpart A would apply. It may be necessary to add specific packaging requirements for certain hazardous wastes (not in any other class according to § 173.2) in the future when they are specifically identified. Paragraph (b) provides an exception under certain conditions.

MODAL PARTS

Sections 173.24 and 176.11 would be amended to exclude hazardous wastes from the exceptions specified for ORM materials.

Sections 174.45, 175.45, 176.48, and 177.807 would be amended to reference proposed new § 171.17 concerning the reporting of discharges involving hazardous wastes.

Section 177.823(b) would be added to include a requirement for motor vehicles transporting hazardous wastes to be marked as specified at 49 CFR 397.21 (b) and (c).

(43 F.R. 22626, May 25, 1978)

49 CFR Parts 171, 172, 173, 174, 176, 177
[Docket Nos. HM-118, 126A, 126B, 145A, 145B, 159, and 171; Amdt. Nos. 171–
53, 172–58, 173–137, 174–37, 176–11, 177–48]
Identification Numbers, Hazardous Wastes, Hazardous Substances,
International Descriptions, Improved Descriptions, Forbidden Materials,
and Organic Peroxides

This action by the Materials Transportation Bureau (MTB) consolidates several related rulemakings into one final rule. By "related", MTB means that in many instances, the same sections of the Hazardous Materials Regulations are affected by the different rulemakings covered by this final rule. The notices of rulemakings containing the proposals, identified by docket number, Federal Register publication, date of publication and title are as follows:

1. Docket HM-118, Notice No. 74–9 (39 FR 25235, July 9, 1974), Expanded Polystyrene Resin and the Definition of a Flammable Solid.

2. Docket HM-145A, Notice No. 78–6 (43 FR 22626, May 25, 1978), Transportation of Hazardous Waste Materials.

3. Docket HM-145B, Notice No. 79–2 (44 FR 10676, February 22, 1979), Transportation of Hazardous Substances.

4. Docket HM-126A, Notice No. 79–9 (44 FR 32972, June 7, 1979), Display of Hazardous Materials Identification Numbers; Improved Emergency Response Capability.

5. Docket HM-126A (additional proposal), Notice No. 79–9 (44 FR 43858, July 26, 1979), Descriptions for Organic Peroxides.

6. Docket HM-159, Notice No. 79–12 (44 FR 43861, July 26, 1979), Forbidden Materials.

7. Docket HM-171, Notice No. 79–11 (44 FR 43864, July 26, 1979), Use of United Nations Shipping Descriptions.

8. Docket HM-126B, Notice No. 79–14 (44 FR 65020, November 8, 1979), Improved Descriptions of Hazardous Materials for Emergency Response.

This consolidated publication of final regulations pertaining to the subjects covered by the dockets identified above was requested by many commenters responding to the different proposals. MTB agrees that all the referenced proposals should be acted upon in one body of final regulations so that persons affected by these new and revised regulations may plan their future business activities relative to training, development and acquisition of shipping documents, the marking of packages, and the development of procedures to comply with the revised incident reporting and newly adopted identification requirements. However, MTB is not able to be fully responsive to those commenters who requested that all of the amendments covered by this action be made effective on the same date (in several comments, the date suggested was two to three years from the date of publication of final regulations). With the exception of regulations pertaining to hazardous wastes, hazardous substances, empty packagings, and certain organic peroxides, more than one year is being provided for the implementation of procedures to comply with the requirements adopted in this action; in fact approximately three years is being provided for compliance with the packaging marking requirements. Instead of specifying a set of lengthy and complicated effective dates in this preamble, the effective date for compliance with each regulation that is effective after November 20, 1980, is set forth in a specific regulation associated with each new or revised requirement. Sections containing compliance dates after November 20, 1980, are § 172.101 (j)(k); § 172.200(c); § 172.203(k); § 172.300(c)(3); § 172.324(b); § 172.336(c) (6) and (7); and § 172.402(a)(10). The principal requirements that become effective on November 20, 1980 (with certain exceptions) pertain to the transportation of hazardous wastes, hazardous substances, certain forbidden materials (organic peroxides), and empty packagings.

Also bearing on the matter of effective dates, is the requirement for compliance by MTB with the Federal Reports Act of 1942 and procedures administered thereunder by the Office of Management and Budget (OMB) relating to prior clearance of recordkeeping requirements imposed by Federal regulatory action. Prior OMB clearance is required with respect to the provisions adopted herein which impose recordkeeping or report preparation requirements.

MTB will inform the public through notification in the Federal Register when OMB clearance of these requirements has been received. It is anticipated that this clearance process will be completed prior to November 20, 1980, the earliest of the effective dates prescribed herein.

It should be noted that most of the materials that this rule indicates by name to be "Forbidden" materials in column 3 of the Table in § 172.101 are and have been "Forbidden" materials in the past under geneal prohibitions.

The listing of these materials by name, and the effective dates specified for these amendments, does not change the present "Forbidden" status of these materials if they were not authorized to be offered for transportation prior to this publication.

Concerning special requirements pertaining to hazardous wastes and hazardous substances, it is important to note that those requirements do not apply unless a material is a hazardous waste or a hazardous substance (or both) according to the definitions in § 171.8. This determination is separate from determining if a material is otherwise a hazardous material.

This preamble is structured to provide a discussion of the comments received and MTB's action relative to significant matters pertaining to the individual rulemaking proposals under captions identifying each docket. The termination of Docket HM-118 is covered by the discussion of Docket HM-159 relating to forbidden materials. The Review by Sections also contains discussions of many comments received in response to the various proposals.

DOCKET HM-126A—DISPLAY OF HAZARDOUS MATERIALS
IDENTIFICATION NUMBERS

The amendments under this Docket require the display of identification numbers on shipping papers and packages in association with proper shipping names and the display of identification numbers on orange panels or placards affixed to portable tanks, cargo tanks, and tank cars. The numbering system is based on the system adopted for worldwide use by the United Nations Committee of Experts on the Transport of Dangerous Goods. The purpose of these amendments is to improve the capability of emergency personnel to quickly identify hazardous materials and ensure the accurate transmission of information to and from the scenes of accidents involving hazardous materials. The identification numbers will also enable emergency response personnel to gain quick access to immediate response information in a guidebook that will be widely distributed by the Department in the near future.

MTB received more than 250 comments in response to the proposal, a majority of which expressed complete support for the proposal. Approximately 50 comments supported the use of the identification numbering system generally, but did not agree with the proposed extent to which the display would be required. There were 12 comments received supporting other types of systems; the Standard Transportation Commodity Codes, Chemical Abstract Service (CAS) numbers, the National Fire Protection Association 704 system, and the HAZCHEM system which is used in the United Kingdom. With the exception of the CAS numbers, each of the alternative systems suggested was discussed in the preamble to the Notice. The CAS registry system, while not discussed in the Notice, was given consideration by MTB and discussed in the regulatory evaluation for the Notice as follows:

The Chemical Abstract Service (CAS) registry system for chemicals was given consideration. This system is presently in use at the Environmental Protection Agency (EPA) pursuant to EPA Order 2800.2 dated June 30, 1975. Each chemical substance is assigned a unique, multi-digit identifying number. There are an estimated 4,500,000 chemical registrations presently on file at CAS and there are an estimated 375,00 entries being added to the registry annually.

Examination of this system raises several significant problems relative to its being employed as the basis for an emergency response program. A principal concern would be the use of numbers having up to nine digits to convey emergency response information to and from the scene of hazardous materials transportation accident. The use of more than four digits to convey specific identification information would not be suitable for such a purpose, and considering the variable number of digits under such a system, it would likely be counterproductive. There is no need for such a large number of specific identifications of chemicals to accomplish an appropriate emergency response program. With such a large number, it would be impractical to publish a manual [for on-scene use] indexing the shipping name of each material. . . . Also, the shipping community would be faced with a tremendous burden and expense in stocking identification mechanisms for display on vehicles, making appropriate entries on shipping papers and packages, and determining the assignment of appropriate numbers to mixtures and solutions of the materials.

The distinction between this system and the UN number assignment system previously discussed is the fact that the UN assigns numbers to recognize under separate identifications those materials that are shipped in large volumes, such as nitric acid, or to separately identify materials having certain properties that would call for their being specifically identified without regard to the volume of such materials being transported. An example in the latter category is the specific identification of all organic peroxides that are commercially produced and shipped. It is recognized that it would be of benefit to all concerned with the regulation and transportation of chemicals if a common code were promulgated for all purposes, including toxic substances control and emergency response actions. This is one of the rationale[s] behind a forthcoming recommendation of the Toxic Substances Strategy Committee to the President, which states: "All the research and regulatory agencies concerned with toxic substances control should be required to use the CAS Registry Numbers as a uniform chemical identification system in all their files and proceedings." While the CAS registry system may be appropriate as a mechanism for the identification and control of the introduction of toxic substances into commerce, it would not be appropriate for emergency response purposes and would be detrimental to the proper implementation of such a program.

As stated in the Notice, the HAZCHEM and NFPA 704 systems would require a preassigned before-shipment numbering technique to provide specific action information or an indication of the level of hazard. One commenter, responding to MTB's statement that no criteria have been established for the (NFPA 704) health and reactivity presentations, stated: "In fact, over a thousand chemicals have been already rated and assigned and can be used as a basis for rating similar materials. NFPA #704 is designed for user development of qualitative criteria. Further, if quantitative criteria need to be developed, it can be done using existing procedures." In the Notice, MTB indicated its concern over the qualitative approach by illustrating the application of the same health hazard number to materials such as liquid oxygen and hydrogen sulfide, etc., and believes that this kind of assignment would not provide a sound basis for the communication of emergency response information nor for the assessment of risk at a transportation accident. Concerning the development of quantitative criteria, MTB agrees this could be accomplished. However, more than three basic fields of display and more than four levels of hazard would be required. Considering the millions of variables presented by hazardous materials in transportation, MTB believes that such a system would be severely impeded by the difficulties that would be encountered in seeking its effective implementation.

Several commenters took exception to the views expressed by MTB relative to use of Standard Transportation Commodity Code (STCC) numbers for the identification of hazardous materials. They indicated their intent to continue use of STCC numbers for their own purposes, including emergency response activities, and saw no reason why MTB could not implement requirements for the identification of hazardous materials based on this existing system. MTB stated its reasons in the Notice for not proposing the use of STCC numbers and understands fully the desire of the railroad industry to use the STCC system. MTB views that were expressed in the Notice (44 FR 32976) concerning the STCC system remain unchanged relative to the appropriate implementation of an identification numbering system for hazardous materials.

A commenter stated: "Our final concern is that only one identification system be adopted nationally and internationally. In the US, the railroads already have their STCC code for identification of hazardous materials. Will the DOT proposal eliminate this requirement? If not, shipping papers and containers will have two sets of numbers— the DOT's and the railroad's." MTB does not intend to take any action prohibiting use of the STCC system, nor any other economic/statistical system, such as the Brussels Nomenclature. These systems were designed for purposes other than identification of hazardous materials. While the 49 series STCC numbers are separately assigned to hazardous materials, the railroad industry has adopted a "bridge" for 49 series numbers to the basic STCC system in order that correct rates and statistics can be generated according to different basic freight classifications. MTB is not aware of any requirement that shippers provide STCC numbers on bills of lading or display of STCC numbers on packages, as suggested by the commenter. A railroad waybill generally contains a special block which appears above the space provided for the description of articles or commodities. This block would contain identifying STCC numbers, not only for hazardous materials, but also for machinery, lumber, and other nonhazardous commodities. The identification number required by DOT will be displayed as part of the basic hazardous material description and MTB sees no potential for conflict between the two systems.

Five commenters expressed the view that no change to the present hazardous materials identification system is necessary. One commenter suggested that MTB continue to work to improve the emergency response capabilities of its hazard information system, based on existing regulations, through increased activity in the training of emergency response personnel. Further, the commenter placed great emphasis on the potential for mistakes in the display of identification numbers. MTB acknowledges that there may be some errors in entering identification numbers on shipping documents. It is for this reason that the suggestion of another commenter that the requirement for proper shipping names on the shipping papers be dropped is not adopted, and is one of several reasons for MTB's adoption of the requirement that identification numbers be displayed on packagings of 110-gallon capacity or less, as well as larger packagings. The Notice mentioned the value of an identification number in verifying the shipping information displayed on documents with the information displayed on packages. The proper display of shipping names will assist in verification of an identification number when doubt exists. Further, the Emergency Response Guidebook has been revised during its development by MTB to include a complete alphabetical index in addition to the numerical index for hazardous materials. However, in a study performed for MTB, using members of the Baltimore Police Department as test subjects, the number identification approach to accessing the response information in the draft Emergency Response Guidebook produced a lower error rate than did the use of the shipping name identification approach (2% vs 10% errors). Also, accessing the response information through use of identification numbers was twice as fast as accessing through use of shipping names. While the potential for clerical error in transferring identification numbers to carrier documents is real, this potential is outweighed by the potential for miscommunication of many complicated names of different hazardous materials shipped in commerce. Therefore, MTB is adopting the identification numbering system basically as proposed in the Notice. MTB agrees that there should be increased activity in the training of emergency response personnel. However, it is also obligated to implement adequate regulatory requirements leading to the quick and accurate communication of hazardous material information, which cannot be accomplished, as suggested by these commenters, through use of the system in existence up to the time of this rule.

MTB stated in the Notice that the adoption of the identification numbering system it proposed will "... provide the basis for an improved emergency response capability that is not presently available through direct use of technical names; e.g., hexadecyltrichlorosilane (UN 1781), to identify hazardous materials and accurately and quickly communicate information about them." The basic UN numbering system, and associated North American (NA) identification numbers for materials not appropriately covered by the UN, was selected by MTB as the basis for an improved emergency response information system because (1) the numbers are assigned by governmental authorities and easily incorporated by the Department in its regulations; (2) all of the identification numbers preceded by the letters "UN" will have the same meaning throughout worldwide commerce; (3) identification numbers have been assigned specifically to identify hazardous materials and have no other intended meaning or use; (4) their formulation and application are not dictated or driven by economic considerations for freight classification, rate-making, or statistical purposes; (5) they are assigned, for the most part, to materials requiring separate recognition that are shipped in commercial quantities, thereby precluding the need for more than four digits; and (6) the identification numbers are assigned on the basis of the next open number without regard to the particular chemical properties or end-use of a material, thereby avoiding any problems in the validity of numbers for future assignments.

Further, MTB stated that use of identification numbers for hazardous materials will (1) serve to verify descriptions of chemicals; (2) provide for rapid identification of materials when it might be inappropriate or confusing to require the display of lengthy chemical names on vehicles; (3) aid in speeding communication of information on materials from accident scenes and in the receipt of more accurate emergency response information; and (4) provide a means for quick access to immediate emergency response information in the Emergency Response Guidebook (manual) that will be distributed by MTB.

Several commenters objected to the proposal that identification numbers be displayed on packages in association with required shipping names. One commenter stated that his organization does not recommend that emergency services personnel enter vehicles to determine the identification number for hazardous materials and that reliance should be placed on identification numbers displayed on shipping papers. MTB agrees that emergency services personnel should not place themselves unnecessarily in jeopardy by entering vehicles for the purpose of identifying hazardous materials if they can identify the materials by other means. However, there are circumstances which require entry into vehicles during or following emergencies. In the Notice, MTB stated the following (44 FR 32976):

Concerning the marking of identification numbers on packages, two major factors were considered. First, freight personnel often become the first contact "emergency" personnel when spills and leaks occur. The MTB visualizes that they will be able to make use of the manual in the same manner as emergency response personnel. Sole dependence on complex chemical names without reference to an identification number may lead to an erroneous response. This same view applies to emergency response personnel coming into contact with packages directly in vehicles, on freight docks, or elsewhere. The second factor is the value of the identification number in verifying the shipping information displayed on documents with the information displayed on packages to preclude error.

While this quoted language clearly makes reference to the possibility that emergency personnel could come into contact with packages directly in vehicles, this factor was given undue emphasis by the commenter. Other factors support MTB's action. For example, according to the incident reports on file at MTB, a large percentage of the total of hazardous materials incidents occurs during freight handling operations and not in vehicles. MTB believes that identification numbers on packages will be of significant use in mitigating the consequences of spills occurring during freight handling operations when shipping papers are not always immediately available.

A commenter questioned the benefit that could be derived from the use of identification numbers by the public. He stated, "To expect the general public to notice, or be knowledgable of, the numerical system is beyond realism." The prime objective of the proposals and the adopted rule is improved emergency response information. MTB has a continuing program to educate industry and emergency response personnel. It is also hoped that the new system will be given much publicity by industry and the emergency services organizations. The present placarding system serves as a hazard alerting system for the benefit of the general public and emergency services. With the added display of identification numbers on tanks, MTB visualizes that there may be many instances where members of the general public, even if not generally knowledgeable of the system, would be able to convey to the emergency services by telephone the identification numbers they see on orange panels or placards.

One commenter, a trade association of chemical manufacturers, expressed support for the display of identification numbers on shipping papers and stated its conclusion "... that [identification] numbers of portable tanks, cargo tanks and tank cars can contribute to the safe evaluation and initial handling of an emergency." However, the commenter recommended "... the modification of existing hazardous materials placards to incorporate the four-digit number in the center section of placards." In support of this recommendation, the commenter stated:

A dual placarding system is unnecessary in that the present placard format can be modified to incorporate the four-digit numbers in the center section of the placards in a size and display that meets the visibility objectives of the MTB and emergency response services. In training personnel to comply with operating procedures, the more simplistic a procedure, the higher the degree of compliance. Thus a one step procedure, keeping it simple, will receive a higher degree of correct implementation than a two step procedure.

The proposed application of ID panels on portable tanks, cargo tanks and tank cars establishes a double process to select, match and apply a set of placards and a set of ID panels. Training personnel to meet present placarding requirements is difficult enough because of turnover, absenteeism and multiplicity of products at a loading station. The more complex an operation—the more confused employees are apt to become. We are convinced a dual placarding requirement would cause a sharp rise in placarding errors, thereby reducing the increase in safety sought by this docket. Such errors would create an added threat to emergency response personnel. . . .

In our view the concept of applying adhesive backed ID panels to vehicles and tanks is not practical. CMA member companies have had less than desirable experience with adhesive placards in that under certain weather conditions they are next to impossible to apply and removal after use is often extremely difficult and costly. Therefore, it is likely that if transport vehicles are required to display ID panels, as proposed in the docket, the vehicles would need to be equipped with ID panel holders. Costs to install ID panels and costs on labor for dual placarding is an unwarranted and unnecessary expense, and is inflationary and wasteful.

The integration of ID numbers into the placards provides two other significant advantages. It directly associates product identity for emergency response with the other hazard precautions communicated on the container, i.e. the placard. The display of the placard in a holder assures that the identification numbers have not fallen off the vehicle and avoids confusion that would set in if the ID number on the vehicle was different than the placard required for the ID panel. CMA's proposal permits use of existing vehicle placard holders.

The MTB in HM Docket No. 126 dated June 25, 1975, stated—"It is the Board's position that any alpha/numeric/symbolic hazard information system adopted in the future be compatible with and adaptable to the placards it adopts under Docket HM-103." CMA's proposal is consistent with this MTB objective, and conversely the ID panel is not. CMA has developed a reasonable one-step process that will accomplish the objectives of MTB in the docket with no lowering of the level of visibility.

The commenter also proposed the use of plain placards for display of identification numbers addressing materials that are not otherwise subject to the placarding requirements (e.g., hazardous substances to be covered by 9000 series identification numbers).

Except for the special consideration that must be given to Class A poison gases and radioactive materials and their associated placards, POISON GAS and RADIOACTIVE, MTB agrees with many of the points raised by the commenter and has modified the rule to allow the display of identification numbers on placards as an alternative to their display on orange panels. Even though a number of commenters supported the display of identification numbers on placards as the only means of display, there may be circumstances when it would be appropriate to display identification numbers on orange panels (e.g., for international shipments or the identification of hazardous substances), as was proposed in the Notice.

Included in the same comments were suggested alternative methods for the display of "class" words on placards. Of necessity, there would be a substantial reduction in the size of the letters in the words, thereby lessening the benefit of their display on placards. The commenter suggested the words be placed immediately above the identification numbers (letters approximately ¾″ high), or in the bottom triangle below the identification numbers (letters ranging in size from approximately ¾″ to 1 ¼″ high). In addressing this matter, MTB was faced with one of its most difficult decisions relative to the rulemaking actions covered by this publication.

In deciding to allow display of identification numbers on placards, as an alternative to their display on orange panels, MTB decided to address not only the potential problem raised by a reduction in the size of "class" words, but also the need for words on placards and future problems relative to their display in a single language. MTB has received inquiries concerning bilingual displays in English/Spanish (for example, Puerto Rico is declared by law to be a State for the purposes of these regulations) and English/French (for example, it is anticipated that there will be increasing interest in bilingual requirements relative to commerce between the U.S. and Canada).

MTB has concluded that "class" words have substantially reduced letter size, will be of little benefit to emergency services in recognizing the kinds of hazards they may be dealing with that also has concluded that, with two exceptions, the removal of "class" words from placards will not significantly affect the ability of emergency services personnel to recognize hazards based solely on the hazard alerting presentation and format of a placard; i.e., size, shape, color, and pictograph.

One exception is the distinction between liquids and gases. The loss of this distinction, which would result from the removal of "class" words, is counterbalanced by the requirements that gases be identified by name on tanks and that the international class number (2 for gases) be displayed at the bottom of placards.

A second exception involves the fact that there are operational considerations relative to placement of cars in trains based on the type of placard displayed. The car placement rules in Subpart D of Part 174 specify restrictions on the location of cars containing Class A poison gases and radioactive materials in a train solely on the basis of the placard type displayed; i.e., POISON GAS or RADIOACTIVE. There are also references to COMBUSTIBLE placards in Part 174 that provide exceptions to the car placement rules when combustible liquids, rather than flammable liquids or gases, are being carried. (However, an error in recognizing a placard bearing identification numbers identifying a combustible liquid would not cause a violation of the car placement rules.)

After considering these problems, MTB is promulgating a final rule that removes the requirement that "class" words be displayed on placards, except when a POISON GAS or RADIOACTIVE placard is required. In those two instances, the "class" words must be on a placard in the same manner as presently required. This will necessitate that identification numbers in those two instances be displayed using orange panels without the option of placing the numbers on the placards. Also, in recognition of the fact that combustible liquids are subject to fewer constraints than flammable liquids and gases (e.g., car placement requirements by rail and use of tunnels by highway), this rule specifies that the bottom of placards for combustible liquids, when identification numbers are placed thereon, shall be white rather than red. This distinction will facilitate the identification of combustible liquids in tanks.

MTB is adopting a requirement that, as a condition relative to the display of identification numbers on placards, the international class designation for hazardous materials be displayed at the bottom of a placard. This kind of display will overcome the lack of distinction between gases and liquids, since a number "2" will be required on the bottom of a placard for a gas. Admittedly, it will take some time for the emergency services to become fully familiar with this type of designation. As a step to improve familiarity, a description of each of the international class numerical designations is contained in the Emergency Response Guidebook. Also, a special display has been added to the Guidebook to illustrate a placard bearing identification numbers with a number "2" in the bottom triangle. The caption reads: "A number 2 at the bottom of a placard without any name means that the material in the tank is a gas. See the next page for the meaning of other numbers at the bottom of placards." However, this is not an initial step, since MTB has been distributing placard and label charts to emergency service personnel and other interested persons for more than 7 years (more than 1½ million copies to date). These charts show the international class designations on labels and placards.

Denial of Petition. By petition dated March 6, 1979, the Association of American Railroads (AAR) requested ". . . that the MTB broaden the scope of HM-145A and HM-145B . . . by proposing the adoption of a requirement that shippers include on the shipping papers the code numbers for 'hazardous materials,' 'hazardous substances,' and 'hazardous wastes' through use of the 49-Series of the Standard Transportation Commodity Code Tariff." The petition pointed out the need for the code numbers to give shippers and carriers the ability to make precise commodity information known to emergency personnel.

MTB agrees with the AAR concerning the need for a numerical coding system to assure the communication of precise commodity information; however, for the reasons stated in Notice 79-9 (44 FR 32972) under Docket HM-126A and earlier in this preamble, and following full consideration of all comments received on its proposal, MTB has decided that a numbering system based on the United Nation's system will be used. Therefore, the AAR petition, to the extent that it request MTB to propose use of the 49 Series of the Standard Transportation Commodity Code Tariff, is hereby denied.

HM-126A—DESCRIPTIONS FOR ORGANIC PEROXIDES

An additional proposal under Docket HM-126A contained a listing of each organic peroxide (with identification number) that may be shipped in commerce in order that the different kinds of risks presented by these materials may be recognized during implementation of emergency response procedures. These differing risks include (1) thermal sensitivity; (2) violence of thermal decomposition; (3) susceptibility to ignition by friction; (4) flammability; and (5) corrosivity. Approximately 135 organic peroxide entries were proposed to be added to the Hazardous Materials Table.

One commenter requested that di-n-propyl peroxydicarbonate 87% maximum, di-sec-butyl peroxydicarbonate 77% maximum, and di-(2-ethylhexyl) peroxydicarbonate 77% maximum be added to the Table. These materials were proposed (and adopted) to be in the Table as technical by pure materials, and, therefore, these entries would adequately cover those peroxides in the concentrations requested by the commenter.

One commenter objected to the proposed requirement that the technical names of the organic peroxides be used as proper shipping names. The reasons given were that such complicated names do not assist emergency personnel in recognizing a hazard quickly, misspellings and errors are more likely on shipping papers and the required use of these names would place a burden on domestic producers of organic peroxides that would be out of proportion to the volume of organic peroxides imported or exported. MTB does not agree with these objections. The use of the technical names of organic peroxides has been required for many years by the International Maritime Dangerous Goods Code and by § 172.203(j)(2) for water shipments without the difficulties noted by the commenter. No other commenters objected to the use of the technical names as the proper shipping names.

Another commenter said that in some instances the hazard of the material in which an organic peroxide was dissolved might completely overshadow the hazard of the peroxide. MTB agrees with this comment and has changed the wording in § 173.15a(a)(3) to authorize organic peroxide solutions to be classed as other than an organic peroxide.

Two commenters pointed out that the adoption of the United Nation(UN) nomenclature for organic peroxides would automatically bring some peroxide formulations under regulation which have previously not been subject to regulation in the United States. This situation can be handled under the provisions of § 173.151a(a)(4) which provides for data to be submitted to MTB indicating that a particular peroxide formulation does not present a hazard in transportation and, based on the data submitted, that product may be shipped as not subject to the regulations.

One commenter suggested that a listing of the peroxides be deferred until complete packaging and shipping requirements have been developed. The reason given was that the cross-referencing to a general entry can lead to errors and inconsistencies. MTB believes that it is necessary to publish the technical names of the peroxides at this time due to the development of the Emergency Response Guidebook. MTB acknowledges that there were some errors in the listings and cross references that appeared in the Notice of Proposed Rulemaking, and these errors have been corrected in this final rule.

A commenter noted that the proposed list of organic peroxides contained several entries in which the concentrations of peroxide solutions were lower than those presented in the UN recommendations. It was urged that the U.S. had representatives at the UN meetings when the international concentrations were agreed upon and took no exception to them following detailed review and consideration of data presented. MTB agrees with this commenter and has changed the concentrations of those peroxides in question to agree with the concentrations proposed in the UN recommendations.

One commenter noted that the organic peroxide, n.o.s. entries and two specific peroxides listed by the UN did not appear in the Notice. The listing of proposed entries in the Notice was in addition to the entries already appearing in the Table. The entries cited by this commenter already appear in the Table and, therefore, were not included in the Notice.

A commenter suggested that the organic peroxide nomenclature be brought into conformance with the latest proposals by the UN Editorial Committee. MTB agrees with this suggestion and the entries in this final rule have been modified to reflect the latest UN accepted nomenclature.

A commenter suggested that, since there are two Organic peroxide, liquid or solution, n.o.s. entries—one classed as an Organic peroxide and the other as a Flammable liquid—§ 172.100 should be amended to identify this difference. MTB does not agree that this difference needs to be specifically addressed. These entries presently appear in the Hazardous Materials Table and no specific explanation of these entries is provided. It remains the responsibility of the shipper to determine the proper class of the hazardous material being shipped, to determine a proper shipping name in accordance with that class and then ship that material in compliance with the applicable regulations. This final rule does not alter this responsibility.

A commenter referenced his Docket HM-126A comments to the comments he made in response to Docket HM-159 in regard to a 9% active oxygen content for certain organic peroxides. MTB's response to this commenter is contained in that portion of this preamble addressing Docket HM-159.

Several commenters noted that neither the UN nor North America (NA) indicator was placed before the identification numbers for the organic peroxides in the proposal. This final rule places either a UN or NA indicator, as appropriate, before the identification numbers for organic peroxide entries in the Table.

Docket HM-126B—IMPROVED DESCRIPTIONS OF HAZARDOUS
MATERIALS FOR EMERGENCY RESPONSE

The amendments under this Docket require more specific identification of certain hazardous materials, mainly in four areas, as follows:

1. New generic descriptions for pesticides are being added to the Table in § 172.101. These descriptions identify and describe pesticides by chemical groups based on their chemical structures. Each of the 15 groups is divided into liquids and solids. The liquids are further divided to distinguish between materials that are classed as Flammable liquids and those classed as Poison B.

2. If the technical name of the principal poisonous constituent of a material is not identified in a shipping name, it must be identified in association with the basic description on a shipping paper.

3. If a shipping name or class name for a poisonous material does not indicate the material is a poison, the word "Poison" must be shown on the shipping paper. Also, for water reactive materials (other than Water reactive solid, n.o.s.), the words Dangerous When Wet must be shown.

4. Nine n.o.s. descriptions addressing multiple hazards are added to the Table.

Certain other amendments under the Docket are addressed in the Review by Sections.

MTB received many comments expressing support for the proposals it made under this Docket, with one major exception—the display of technical names after n.o.s. descriptions. The principal arguments against this proposal were that (1) many of the technical names that would be required would be lengthy and complex; (2) such information would be of little or no value in an emergency; (3) the four-digit identification number, tied to emergency response information in a manual or provided by CHEMTREC, would be sufficient for emergency response purposes; (4) the purpose of the additional information stated in the preamble of the Notice for this Docket was in conflict with statements referencing lengthy technical names in the preamble to Docket HM-126A; and (5) such a requirement would require the disclosure of proprietary information.

Upon completion of its review of the comments on its proposal pertaining to technical names, MTB was compelled to fully reconsider the proposal. At the outset, MTB acknowledges that it erred in not explaining the proposal in sufficient detail in the Notice and that one sentence in the preamble did not sufficiently discuss the distinction between the benefits of the proposed additional descriptive information and the benefits from using identification numbers. The sentence which read, in part, "MTB agrees that safety would be enhanced by such a requirement since more specific information would be immediately available for use in emergency response actions," should more accurately have stated, "MTB agrees that safety would be enhanced by such a requirement since more specific information would be available for first-aid and clean-up actions immediately following the initial phases of an emergency."

Upon further consideration, MTB concludes that the requirement for more specific information should relate only to poisonous materials. If some form of technical identification is not provided in the shipping name for a material that is poisonous, neither a manual nor an emergency response information center can overcome this shortcoming when it comes to providing appropriate first-aid measures beyond initial actions. For example, if a shipping paper contains only the description "Poison B liquid", no specific response information can be provided as to appropriate antidotal procedures within a sufficient time interval. Even if a response center were able to reach the shipper of the material, considerable time would be lost in attempting to ascertain the specific identification of the poisonous material in the vehicle involved in the accident of concern, unless the shipper introduces only products of the same chemical group into commerce.

MTB has concluded that this kind of immediate information would be of less importance for materials such as flammable liquids and corrosives. Also, MTB believes that it has become necessary to distinguish materials that cause death by systemic poisoning from those that result in death due to corrosive destruction of tissue. While this distinction is not made for consumer products, MTB believes it should be made in regard to the transportation of hazardous materials.

MTB has dropped proposed § 172.203(j) and has merged the requirement for the identification of poisons into paragraph (k), entitled "Poisonous materials." A provision has been added to exclude consumer commodities and chemical groups identified in shipping names. The "technical" identification would not be required for limited quantities because of difficulties that may be encountered in shipping untested samples for laboratory examination. A major feature is a provision allowing use of names contained in the National Institute for Occupational Safety and Health (NIOSH) Registry of Toxic Effects of Chemical Substances or more commonly, NIOSH Registry. The NIOSH Registry contains many short and specific names which are cross-referenced to technical names. The NIOSH Registry will be kept on file at CHEMTREC. Shippers desiring the addition of their materials to the NIOSH Registry should contact The Editor, Registry of Toxic Effects of Chemical Substances, National Institute for Occupational Safety and Health, 4676 Columbia Parkway, Cincinnati, Ohio 45226, Telephone: (513) 684–8317. By allowing alternative use of names in the NIOSH Registry or identification by chemical groups, MTB believes the claimed problem with potential disclosure of proprietary information has been alleviated.

The proposed requirement that the technical names of all n.o.s. described materials be marked on packages has not been adopted. While MTB agrees that many product labels, in particular those for pesticides, contain this information, this is not the case for all poisonous materials. However, MTB is aware of labeling proposals presently under development by the EPA and the Occupational Safety and Health Administration (OSHA), which are much more extensive than those proposed by MTB under Docket HM-126B. For this reason, MTB believes it appropriate to drop the marking portion of its proposal in light of the contemplated actions of EPA and OSHA in order to avoid any conflicts or unnecessary duplication of requirements.

Many commenters supported the proposed and adopted requirement that the word Poison be added to shipping descriptions when not included in a shipping name or class description for a material meeting the DOT poison criteria. Also, most commenters supported the required display of information for water reactive materials, except they recommended Dangerous When Wet rather than Water Reactive as being more appropriate in communicating the risk. MTB agrees and the requirement has been adopted accordingly.

A commenter correctly pointed out that MTB failed to include Oxidizer, poisonous liquid or solid in the Notice. This is contained in § 172.402(a)(3). As noted earlier, appropriate entries have been included in the amendment. Also, commenters pointed out MTB's failure to address labeling for multiple hazards in § 172.402 in addition to the proposed dual labeling specified for the eight generic n.o.s. entries in the Table. One commenter suggested that dual labeling not be required if the identification numbering system is adopted. MTB does not agree. MTB views the DOT required labels to be a hazard-alerting portion of the hazard information system and, in terms of incident prevention and immediate knowledge of risk, an essential element of the system. In adopting the dual labeling requirements for the generic entries, MTB agrees that similar treatment should be given to other materials having the same hazards. Therefore, a change is being made to § 172.402 accordingly.

DOCKET HM-145A—TRANSPORTATION
OF HAZARDOUS WASTE MATERIALS

The amendment under this Docket provide for the proper identification of hazardous waste materials for transportation and to ensure that such wastes ultimately are delivered to predetermined designated facilities through implementation of certain record-handling requirements. The requirements apply to hazardous waste carried in either interstate or intrastate (with one exception) transportation. The hazardous waste regulations have been coordinated with the Environmental Protection Agency's (EPA) development of a national hazardous waste regulatory program as mandated by § 3003 of the Resource Conservation and Recovery Act (RCRA) (42 U.S.C. 6923).

MTB's Office of Hazardous Materials Operations published a Notice of Proposed Rulemaking regarding the transportation of hazardous wastes under Docket HM-145A on May 25, 1978 (43 FR 22626; Notice 78–6). The preamble to the Notice provided an extensive explanation of the proposed regulations from the standpoint of what is expected in implementing RCRA requirements; the differences in the jurisdiction of EPA and DOT; and the overall objective of the proposed requirements. Rather than repeating that discussion here, interested readers are referred to the earlier document.

As a part of this rulemaking proceeding, six EPA-DOT joint public hearings were held in various parts of the United States. Virtually all the comments made at the hearings were addressed to EPA proposals, except that many commenters expressed the view that MTB should issue the regulations pertaining to the transportation of hazardous wastes. MTB agrees with those views insofar as the requirements pertain to carrier activities. Further, MTB is adopting a limited number of requirements that apply to generators (shippers) of hazardous wastes in addition to existing applicable hazardous materials regulations. EPA has recently published regulations, to be codified at 40 CFR Part 262 (45 FR 12722, February 26, 1980), applicable to generators of wastes. The EPA regulations are extensive in scope and apply to a generator's nontransportation activities as well as to the offering of wastes for transportation. MTB believes that it will be necessary for shippers to be fully cognizant of EPA's regulations, which include requirements relating to record-handling, storage limitations, and special package markings, since many of these requirements are inappropriate for MTB to adopt under the Hazardous Materials Transportation Act (HMTA) (49 U.S.C. 1801–1812).

EPA also has published regulations, to be codified at 40 CFR Part 263 (45 FR 12737, February 26, 1980), applicable to transporters (carriers) of hazardous wastes (hereinafter referred to as wastes). With two exceptions, however, the EPA has acknowledged (at 40 CFR 263.10) that a carrier complying with DOT requirements, as amended by this rule, applicable to transportation of wastes, will be considered in compliance with corresponding EPA requirements even though those requirements may be stated differently. One exception is EPA's requirement for carrier identification numbers, which must be obtained as a one-time requirement; the second exception is EPA's requirements for the cleanup of spills or discharges. For the convenience of carriers, the former exception is cited and the latter quoted in notes to § 171.3 of this rule.

Throughout this rulemaking proceeding, MTB worked closely with EPA in the joint development of appropriate transportation requirements. This rule is being promulgated under the HMTA to address the carriage of hazardous wastes, which may pose an unreasonable risk to health and safety or property when transported in commerce. Consistent with the regulations adopted by EPA, MTB is adopting requirements primarily to ensure that hazardous wastes are properly identified to carriers and that they are delivered to predetermined designated facilities. Proper identification of wastes is essential in order to implement the transportation aspects of a "cradle to grave" hazardous waste tracking system.

One commenter, referring to proposed § 171.3(e)(3) (which is § 171.3(c)(3) of this final rule), stated that the proposal "[does] not uphold the preemption of inconsistent state or local requirements with regard to additional requirements on shipping papers" and, therefore, stands in opposition to Section 112 of the HMTA and the preemption procedures in 49 CFR Part 107, Subpart C. Another commenter was concerned about possible regulation by the State or locality to which a waste shipment is destined and claimed that the specific preemption language in the proposal did not go as far as Section 112 of the HMTA would allow.

Both comments misconstrue the proposal. The term "inconsistent" as used in the HMTA describes the type of state and local transportation safety regulation that is preempted by requirements under the HMTA. Section 171.3(c) lists and classifies as "inconsistent" certain areas of possible State and local regulatory actions pertaining to hazardous wastes which MTB believes would be disruptive of the national uniformity required in the identification of hazardous materials (including wastes) in transportation. Section 171.3(c) does not list all the conditions under which it might view a State or local law as "inconsistent," nor does it apply to non-transportation requirements that may be imposed by State or local law. The final rule says that MTB considers that State and local requirements pertaining to certain aspects of the transportation of wastes are inconsistent with DOT requirements if they apply differently from or in addition to them. Section 171.3(c) is a declaration of intent that State and local requirements in this safety area be uniform to encourage compliance by shippers and carriers of hazardous wastes and to avoid unnecessary regulatory impediments to the reliable transportation and delivery of these materials.

In a related matter, a recently issued MTB inconsistency ruling (IR-2) (44 FR 75566, December 20, 1979) expresses a view of HMTA preemption in other regulatory areas similar to that expressed in this final rule. While preemption is customarily a judicial matter, the statement in § 171.3(c) may reduce the need for subsequent dispute resolution.

In the final rule, the reference to additional requirements for shipping papers (including hazardous waste manifests) by states and localities to which a shipment is consigned has been dropped. When a hazardous waste manifest is being carried, as required by this rule, it is considered a shipping paper, and the language in § 171.3(c) applies. Any agency of a state which requires waste disposal facilities to obtain information in addition to or differing from that required by EPA and DOT to be contained in manifests must specify some means other than shipping papers (including hazardous waste manifests) for those waste facilities to obtain that additional or different information from generators. MTB believes that a nationwide standard for manifests is an important factor in accomplishing the safe and effective delivery of wastes to designated facilities without delays due to the manner of documentation.

After the comment period on this docket had closed, representatives of environmental protection departments of several Northeastern States, most notably New Jersey, expressed concern to Departmental representatives that the final rule might contain language preempting certain aspects of in-place State hazardous waste management programs. They were especially concerned about State requirements for information about individual waste shipments that may be in addition to that required by DOT's shipping paper requirements.

The language of the final rule does not preclude State requirements for such additional information with regard to wastes generated or disposed of within that State. However, as mentioned earlier, § 171.3(c) does not state all the conditions under which a State or local law might be preempted. State or local requirements applicable to hazardous waste transportation that are not addressed by § 171.3(c) may nonetheless be found preempted under administrative procedures found at 49 CFR Part 107, Subpart C.

Several commenters discussed the proposed changes to § 173.28 which addresses the reuse of containers (packagings). One commenter suggested that proposed paragraph (n) be revised to make it consistent with existing § 173.28(i) and to make it clear that NRC (nonreusable container) marked specification packagings may be used to transport the indicated hazardous materials. MTB agrees and has modified the rule accordingly. Another commenter suggested that STC (single trip container) marking requirements be removed from the regulation. While this suggestion may have merit, it would require action that is considerably beyond the scope of this rulemaking since each packaging specification in Part 178 that references such a marking would have to be modified.

One commenter expressed strong objection to proposed § 173.28(p) which proposed to permit the reuse of NRC- and STC-marked packagings for shipments of wastes under specified conditions. The same commenter concurred subsequently in the submission of another commenter who recommended that the requirement be restated to allow the reuse of NTC- and STC-marked packagings for wastes if they are "reconditioned . . . in accordance with this part." Since there are no reconditioning requirements for NRC-marked packagings, these commenters may be suggesting allowing their reuse for wastes, as proposed in the Notice, while the more substantial STC-marked drums would have to be reconditioned before use for waste shipments. MTB does not agree with this point of view. It is more likely, however, that the commenters oppose allowing NRC- and STC-marked packagings to be used for waste shipments unless they are reconditioned. For the one-time use authorized by this final rule, MTB does not believe reconditioning to be necessary for either STC- or NRC-marked packagings.

The same commenter pointed out the important contribution of the drum reconditioning industry to the saving of national resources and energy. MTB acknowledges the significant contribution of the drum reconditioning industry to savings in energy and resources. However, NRC-marked packagings are not presently covered by reconditioning requirements, and MTB believes they should be put to appropriate use in waste disposal activities under limited circumstances and that there is no logical reason to exclude STC-marked drums from similar use. Shipments of hazardous wastes in reused STC- and NRC-marked drums are subject to special requirements: (1) they must be moved under manifests; (2) discharges must be reported and cleaned up; (3) deliveries are restricted to designated facilities; (4) the adopted rule limits transportation to highway; (5) packages must be held 24 hours after filling for inspection (a requirement that does not apply to reconditioned drums); (6) and packages may not be loaded or unloaded by common carrier personnel, who would not normally be as familiar or trained to deal with a waste material as would the employees of private and contract carriers. This last fact is relevant to the objection of another commenter who suggested common carriers should not be treated differently. On a number of occasions, MTB has recognized the distinction between private (and at times contract) carriers and common carriers in the application of regulations and the issuance of exemptions (see, for example, § 173.29 in this rule and § 173.316(a)(2)(vi)) and believes such a distinction is appropriate in this case. MTB considers that the adoption of this rule will contribute to saving energy and national resources since many packagings that are used to bring materials into industrial concerns may now be reused for wastes and it will be unnecessary for new or reconditioned drums to be used in their place.

Several commenters addressed MTB's proposed revision of § 173.29 dealing with empty packagings. One commenter challenged MTB's statement in the Notice preamble which indicated that "it is essential to deal with the subject of so-called 'empty packagings' containing the residues of hazardous materials" (43 FR 22629). The commenter suggested that MTB did not have data to justify this action. By a separate and subsequent submission, this commenter petitioned for rulemaking to delete § 173.29(e) as it pertains to required removal, obliteration, or covering of labels when packagings are transported in open-top or flat bed vehicles. Another commenter considered the proposal to be outside the scope of this proceeding and suggested that it should only be addressed to wastes. MTB is in substantial disagreement with these comments. However, MTB does agree that the rulemaking petition has merit since the deletion of § 173.29(e) was proposed in the Notice and is adopted here.

The revision to § 173.29 addresses all so-called empty packagings having a capacity of 110 gallons or less—not only drums. The residues in a packaging may pose an unreasonable risk to health and safety or property regardless of whether or not the packaging is being transported for recovery of residue, reconditioning, refilling, or disposal. Even small quantities of certain poisonous residues can be lethal upon contact with the skin. Residues of corrosives can cause severe burns. Residues of some flammable liquids can generate vapors that can be explosive when mixed in certain volumes with air. The commenter is correct in suggesting that MTB does not have accident data to support this requirement, for the simple reason that so-called empty containers have never been subject to the reporting requirement of § 171.15 or § 171.16, MTB believes it is appropriate to take action based on the fact that hazardous materials currently are subject to regulation as such when in smaller quantities than are often contained in so-called empty packagings. One commenter suggested that a drum "having less than 1% of the marked capacity of the container, should be deemed 'empty'," This would mean that in excess of two quarts of residue could remain in a 55-gallon drum and be excepted from the rule. Furthermore, the commenter did not suggest how the 1% or less would be determined. MTB does not agree with this suggestion because of the hazards these quantities may pose. Concerning the same commenter's petition to allow prescribed hazardous materials labels to remain on a packaging containing the residues of hazardous materials, MTB agrees and has adopted the requirement as proposed in the Notice. Labels, as required for filled packagings, must remain on so-called empty packagings unless the packagings have been cleaned and purged of all residues.

MTB does not consider the adoption of rules for all so-called empty packagings, and not just those being discarded as waste, to be outside the scope of this rulemaking action. MTB's intention to do this was expressed and thoroughly discussed in the Notice. Obviously, the status of packagings containing the residues of hazardous materials is often subject to change. While it may have been originally intended that they be transported for recovery or reuse, it could be decided at any point in the cycle that they be discarded. MTB is unable to perceive any rational approach that would call for a distinction in how a hazard is communicated (i.e., through labels, markings, and shipping papers) merely due to a difference in destination. Therefore, it was essential that the entire matter be dealt with in this rulemaking action (the first action addressing the subject since it was removed from Docket HM-112 in 1976), to assure uniform application of the requirement to all so-called empty packagings.

The hazardous waste regulations being issued under Docket HM-145A constitute a significant regulation under the terms of Executive Order 12044 and DOT implementing procedures (44 FR 11034; February 26, 1979). In accordance with regulation evaluation requirements contained in those documents, MTB invites public comments, responses and reactions to this rule during the first year of its implementation. Comments should be titled "Hazardous Waste Rules" and addressed to: Dockets Branch, Materials Transportation Bureau, U.S. Department of Transportation, Washington, D.C. 20590. It is requested that five copies of any comments be submitted. Comments received prior to November 20, 1981, will be considered.

At the conclusion of the comment period, MTB will reexamine the rule based on the comments received. Subsequently, MTB will publish a notice in the Federal Register indicating what, if any, action it is taking or proposing to take based on its reexamination.

HM-145B—TRANSPORTATION OF HAZARDOUS SUBSTANCES

This rule provides for the identification of hazardous substances when a "reportable quantity" is contained in a package. The identification will be accomplished by requiring that the name of the hazardous substance and the letters "RQ" be placed on shipping papers and displayed on packages in association with the descriptions for hazardous substances. Also adopted is a new § 171.17 specifying the reporting requirements for discharges of hazardous substances into or upon the navigable waters or adjoining shorelines.

The Notice proposed to identify a hazardous substance as any material that is subject to the EPA regulations found in 40 CFR Parts 116 and 117, and to require the identification of hazardous substances when loaded in reportable quantities at any one loading location, considering all packages loaded at that site. If final rules were adopted as proposed, there would have been no qualification or exception relative to the aggregation of packages in vehicles other than limiting their application to aggregate quantities loaded at one loading location. Also, there would have been no qualification or exception for small quantities of these materials in mixtures or solutions. The Notice also proposed the entry of a statement on each shipping document specifying a notification requirement, including a telephone number to be called in the event of a discharge. The notification requirement set forth in proposed § 171.17 specified that a report would be required when any amount of a hazardous substance was discharged during transportation, if a reportable quantity of a hazardous substance was present in a vehicle, as indicated by the identification statement on the shipping paper. The proposal also would have required special markings on portable tanks, cargo tanks, and tank cars for the identification of hazardous substances. Significant changes have been made to the proposal.

Several commenters stated that the DOT would exceed its authority under the Hazardous Materials Transportation Act (HMTA) 49 U.S.C. 1801-1812) by adding the ORM-E hazard class to cover hazardous substances which, they stated, are not hazardous materials that meet the definition of any existing hazard class. Other commenters stated that MTB holds no authority to promulgate regulations pertaining to materials identified as hazardous substances by another Government agency because Section 104 of the HMTA requires the Secretary of Transportation or his delegate, per 40 CFR 1.53(b)(1) and paragraph 5 of Appendix A to Part 1, to make a finding that "a particular quantity and form of material in commerce may pose an unreasonable risk to health and safety or property . . . " while in the past MTB has primarily dealt with the materials that may pose acute hazards to persons or property, addressing materials that may pose chronic hazards to health through the environment is well within DOT's authority under the HMTA. The pertinent language in Section 311 of the Clean Water Act (33 U.S.C. 1321), EPA's authority for designating hazardous substances, is contained in paragraph (b)(4), which provides for determination of "those quantities of . . . any hazardous substances the discharge of which may be harmful to the public health or welfare. . . ." Clearly, many of the risks involved from the transportation of hazardous materials relate to the possibility of unintentional release, and such releases may involve discharges into the navigable waters of the United States. To the extent that EPA has designated certain substances in specified quantities as potentially harmful, it is appropriate for MTB to designate those quantities of those materials as hazardous materials under the HMTA. Moreover, should MTB not take this action, it would be left to EPA to fill the void covering the transportation of those hazardous substances not reached by DOT regulations. Such a split in regulatory coverage would be inefficient and a hinderance to all concerned.

Several commenters suggested that the definition of a hazardous substance and a reportable quantity should be redefined to agree with EPA and that Part 117 of EPA's regulations pertaining to hazardous substances should be incorporated into the DOT regulations as an appendix. MTB agrees that there is a need to redefine hazardous substances. However, in recognition of the distinction between transportation operations and fixed facilities, MTB believes it should adopt a definition for hazardous substances under the HMTA that recognizes the unique features of our transportation system while sufficiently accomplishing the purposes and intent of § 311 of the Clean Water Act. Therefore, MTB has revised its definition of a hazardous substance by addition of two major features that were not included in the Notice. The first feature is a limitation of the application of DOT's regulations to reportable quantities of a hazardous substance contained in one package (see the definition of a package in § 171.8); in fact, the concept of a reportable quantity in a single package is now contained in the definition of "hazardous substance" in § 171.8. The second feature is a limitation of the application of DOT's regulations with regard to mixtures and solutions containing materials identified by the letter "E" in § 172.101. Certain concentrations of these materials will be excluded from the applicability of DOT's regulations pertaining to hazardous substances. This is being accomplished by the inclusion of a table in the definition. For example, if the reportable quantity for a certain hazardous substance is 1,000 pounds, less than a 2 percent concentration by weight of that material in a mixture or solution will *not* be subject to DOT's regulations as a hazardous substance. Further, the 2 percent or greater concentration (by weight) of that material must result in a reportable quantity being contained in one package to be subject to DOT's regulations as a hazardous substance. Concerning discharge notifications, MTB has modified the requirement consistent with the suggestions of many commenters who stated that MTB was proposing a rule that was considerably beyond the scope of the purpose and intent of § 311 of the Clean Water Act. MTB proposed the rule in that fashion because it anticipated considerable difficulty on the part of transport workers in establishing when reportable quantities were discharged. MTB erred in overstating the proposed reporting requirement for that purpose and agrees that the discharge reporting requirement should be closely aligned with the statutory reporting requirement.

A number of commenters stated that transportation employees would not always be able to make notifications of discharges directly and that many communications would of necessity be through carrier personnel other than the operators of vehicles. MTB has modified the rule to recognize such a situation by stating that the person in charge of the transport vehicle, etc., shall make the notification directly or indirectly through the carrier. The prime responsibility still rests with the person in charge, however, the final rule grants flexibility in the method of notification. MTB believes this is consistent with the intent of § 311 of the Clean Water Act. A new paragraph (e) has been added to § 171.17 to require the carrier to make a notification if the person in charge of a vehicle is incapacitated or otherwise unable to make a notification. For instance, there may be a situation when the driver of a motor vehicle is severely injured in an accident and would not be able to make the required notification. In such a case, an official of the carrier operating the vehicle would be required to make the notification as soon as he has knowledge that a hazardous substance (i.e. a reportable quantity) has been discharged.

Many commenters objected to the proposed requirement for a notification statement on shipping papers, stating that such a statement was not necessary to accomplish the purposes of the proposed rulemaking. Several commenters stated a simple identification of a hazardous substances in a shipment would be sufficient to implement the reporting requirements and that carrier employees could be educated to know and understand the meaning of the identification. MTB agrees with these commenters and has decided that the letters "RQ", entered on shipping papers and packages (other than portable tanks, cargo tanks and tank cars), in association with required descriptions, will be sufficient to implement the identification of hazardous substances. This decision also addresses the concern expressed by several commenters concerning the unique display of the letter "E" as was proposed in the Notice.

Many commenters discussed the significant impact of the proposed regulations on consumer commodities expressing the view that, if adopted, the effective use of the ORM-D hazard class would be essentially negated and that the concept of limited quantities would be destroyed. MTB believes that this concern has been overcome by the manner in which it has defined hazardous substances for purposes of application of the hazardous materials regulations.

While the regulations for hazardous substances adopted by DOT in these amendments are closely related to those of the EPA found in 40 CFR Part 117, they are not identical in their applicability. EPA's regulations require, without qualification, notification of discharges of reportable quantities of hazardous substances under conditions specified in § 311(b)(5) of the Clean Water Act and, as specified in the notice, provisions of 40 CFR 117.21. The DOT regulations require notification when a reportable quantity is discharged from a package (e.g., a drum, cargo tank motor vehicle or tank car) that is marked or documented as containing a reportable quantity.

Persons who do not have knowledge that a reportable quantity of a hazardous substance has been discharged are not required by EPA to make notifications. MTB has been advised by EPA that it will not bring civil or criminal suit for failure to make notification when such notification is not required by DOT's regulations, unless it can be shown that there was actual knowledge that a reportable quantity was discharged.

Shippers and carriers must also bear in mind that compliance with the provisions of this rule do not relieve them from possible civil liability under § 311 resulting from discharges of reportable quantities. This liability relates to the discharge itself and removal costs and is addressed in regulations promulgated by the EPA and the U.S. Coast Guard.

HM-159 (HM-118)—FORBIDDEN MATERIALS

This rule adds the names of materials to the Hazardous Materials Table that MTB considers to be too hazardous to be permitted in commercial transportation. Also, the rule adds N-methy-N'-nitro-N-nitrosoguanidine as a flammable solid and adds a new § 173.179 prescribing packaging requirements for this material. Changes have been made to §§ 173.21 and 173.51 pertaining to forbidden materials and packaging.

A total of 28 comments were received in response to the Notice. All commenters were in general agreement with the proposal to add certain materials considered to be too hazardous for commercial transportation to the Table. A commenter presented data indicating that 2,6-dichloro-4-nitrophenol is not a forbidden material. MTB agrees with the data presented and this material has been removed from the list. The same commenter also stated that some iodoso compounds might be considered forbidden, but that others would not be in this category. Pending further detailed investigation of these substances, the proposed entry of iodoso and iodoxy compounds (dry) has been removed from the list of named forbidden materials.

A comment from a manufacturer of organic peroxides suggested that the term "active oxygen" would be better than "available oxygen" for those entries in the Table containing such limitations. MTB agrees with the commenter and the term "active oxygen" has been substituted for the term "available oxygen."

Two air carrier associations concurred in the proposed § 172.100(d) and suggested that the idea contained in that section be applied to all commodities in the Table. Based on these comments, MTB has reviewed the entire use and meaning of the asterisk in the Table and has decided to eliminate the asterisk, thus allowing a shipper to determine if a specific material should be regulated under the hazard class identified in the Table. However, in conjunction with this change, it is necessary for MTB to introduce a new symbol into the table in order to identify those materials which have been designated as hazardous materials of a particular class, whether or not they meet the definition for the hazard class in which they have been designated. If a shipper wishes to ship a formulation of a material identified by the plus (+) symbol as non-regulated, or in a class other than that specified in the Table, he must supply the Associate Director, Office of Hazardous Materials Regulation, MTB with data which establishes that it does not present a hazard in transportation, or presents a different hazard than that which is listed in the Table.

Another commenter objected to the limiting of peroxyacetic acid to 40% by weight instead of the 43% authorized in the UN Recommendations and the IMCO regulations. There are several current organic peroxide entries which have different concentration limits in the UN recommendations and the DOT Hazardous Materials Regulations. Since the concentration limits in the UN Recommendations were agreed to by the U.S. delegations to the UN meetings, MTB has revised the limits in the Table to agree with those in the UN Recommendations.

Several commenters objected to all or parts of the proposed revision of § 173.21. Some organic peroxide manufacturers objected to the use of 130° F. in paragraph (a)(2) because their interpretation was that this temperature is the minimum decomposition temperature for which refrigeration would be required. They argued that they have shipped certain organic peroxides with decomposition temperatures of 120° F. without refrigeration for years and, also, that the UN Recommendations use 122° F as the decomposition or polymerization temperature below which refrigeration is required for the most active organic peroxides. MTB considers 130° F. the maximum temperature that could be expected during transportation, and there are many sections in the regulations which reference 130° F. The fact that an organic peroxide, or any other material, decomposes below 130° F. does not necessarily mean that it must be stabilized or refrigerated. The paragraph states in part "with an evolution of a dangerous quantity of heat or gas. . . ." If the decomposition or polymerization does not create a hazard in transportation, the provisions of the paragraph do not apply regardless of the decomposition temperature of the material. Therefore, the temperature reference of 130° F. has been maintained in the final rule.

Several commenters objected to the fact that § 173.21(a)(2) did not contain a statement concerning the time a material would have to be exposed to the 130° F. temperature in order to be considered forbidden from transportation. MTB agrees that this is a weakness in the proposed wording and has altered the wording of the rule to reference two test methods. The test methods are: ASTM E-487 "Standard Method of Test for Constant Temperature Stability of Chemical Materials" and the Organic Perioxide Producers' Safety Division (OPPSD) "Self Accelerating Decomposition Test (SADT)." Several commenters expressed concern that this paragraph does not make it clear that approvals issued by the Bureau of Explosives would be continued in effect until an orderly transition to approval by the Associate Director for Operations and Enforcement, MTB could be accomplished. MTB acknowledges the validity of the objection and has included a clarification statement in the rule which references § 171.19.

Several commenters stated that proposed § 173.21(a)(3) was too vague and it was suggested that a phrase be added such as ". . . e.g., the release of flammable vapor in such quantities that an explosive mixture would be created within the transport vehicle." MTB agrees that this objection, and wording to the effect of that suggested has been included in the rule.

On July 9, 1974, the Hazardous Materials Regulations Board, the predecessor of MTB, published a Notice a Proposed Rulemaking under Docket No. HM-118 proposing to list and classify polystyrene resin, expandable, containing a flammable liquid or gas, as a flammable solid. No final action was taken on this rulemaking proposal. Section 173.21(a)(3) of this rule will forbid the offering for transportation of packages which evolve a dangerous quantity of flammable gas or vapor from a material not otherwise subject to the regulations. MTB believes this prohibition is sufficient to preclude the type of potential hazard which was the concern addressed by the Hazardous Materials Regulations Board in its proposal under Docket No. HM-118. Therefore, the proposals under Docket No. HM-118 are hereby terminated.

Several commenters said that § 173.21(a)(4) needed clarification. The objections were based on the fact that there was no definition of detonation and that there is no recognized test method for determining whether detonation has occurred in a package as a result of a thermal stimulus. In response to the first objection, MTB has included a definition of detonation in the final rule. The second objection is not correct. There are three tests specified in the regulations for determining whether a packaged material detonates as a result of a thermal stimulus. One of these is described in § 173.88(g), Note 2. Another method is found in DOD TB 700-2 (May 19, 1967) which is referenced in § § 173.86(b) (2) and (3). Both of these test methods have been in the DOT regulations for many years and have been used extensively on both military and commercial materials to determine whether a detonation will occur in a package exposed to a thermal stimulus. While both test methods were designed for testing propellants, they can be and have been used to test other hazardous materials. The third test method is described in § 173.114a(b)(6) and may be used in evaluating whether a detonation has occurred. MTB has considered it inadvisable to reference these methods in this rule because such a reference could suggest that a chemical manufacturer who is not familiar with testing explosives should attempt to perform these tests. This type of testing should be done only by personnel who are well versed in the testing of explosives and this fact has been stated in the rule.

Docket HM-171—USE OF UNITED NATIONS SHIPPING DESCRIPTIONS

The amendments under this Docket authorize the use of United Nations shipping descriptions and identification numbers for certain hazardous materials in place of the descriptions required by existing DOT regulations. These amendments are intended to facilitate the international transportation of hazardous materials and to minimize the economic burdens imposed on shippers by the multiplicity of package markings and shipping paper descriptions now required for compliance with both domestic and international requirements. In addition, the amendments provide optional stowage locations for hazardous materials when transported by vessel. The optional stowage locations authorized are those provided for the particular hazardous materials in the International Maritime Dangerous Goods (IMDG) Code published by the Inter-Governmental Maritime Consultative Organization (IMCO).

A number of comments were received which expressed complete support for the proposal. In general, the supporting commenters endorsed the proposal since it would eliminate costly redundancy in shipping paper descriptions and packaging markings. One supporting comment is quoted since it provides some quantification of the importance of the international transportation of chemicals to our economy:

Shipments of chemicals and allied products were valued at $126.5 billion in 1978. The export activity continued to be strong in 1978 with the value of all chemical exports totaling $12.62 billion, an increase of 16.7 percent over 1977. While the imports of chemicals also increased, the favorable balance of trade in the chemical area increased from $5.84 billion in 1977 to $6.19 billion in 1978, a gain of 6 percent. In the future, these shipments are expected to increase and will be affected by international regulations to a greater degree.

Several comments were received expressing opposition to the proposal. It should first be noted that many of the issues raised concerned the use of IMCO classifications and labeling for certain hazardous materials. Although limited to import and export shipments in the present regulations, this authorization has been a provision of the DOT regulations since adoption of amendments under Docket No. HM-112 in 1976. For this reason, MTB believes it is reasonable to assume that shipper and carrier personnel should, in the execution of their responsibilities in the preparation and acceptance of shipments, already have gained a basic familiarity regarding the use of IMCO classifications and labels as an alternative to the class and labels as an alternative to the class and labels prescribed for certain hazardous materials in § 172.101.

The fundamental argument raised in opposition to the proposed amendments is that the existence of an optional hazardous materials list will, in the words of one commenter, have a "chaotic effect" on the regulated industries, particularly on the rail and motor carrier industries, because it would complicate the regulations. MTB agrees that the provision of options to various requirements increases the volume of regulations and, to a certain degree, their complexity. In spite of this fact, experience has shown that such regulatory provisions are essential if the regulations are to be effective without unnecessarily burdening industry. For example, it could be argued that the hazardous materials placarding requirements could be vastly simplified by eliminating the "DANGEROUS" placard and certain exceptions to the placarding requirements, and simply requiring that appropriate placards be displayed for each hazardous materials transported regardless of quantity. Such simplification is obviously not in the best interests of the regulated industries and would undoubtedly be declared totally unacceptable by the very commenters who oppose the amendments under Docket HM-171. MTB believes that these amendments will do much to enhance safety by minimizing redundant, conflicting and confusing shipping paper and package marking requirements. Under the current practice of incorporation of IMCO classification and labeling provisions by reference, it is difficult for rail or motor carrier personnel to determine compliance with these provisions. The optional list will eliminate confusion and errors on the part of carrier personnel by making this information readily available to them in § 172.102.

A number of objections to the proposed amendments were raised on the basis of placarding implications. One commenter expressed concern that rail carrier personnel would be unable to replace missing placards with placards conforming to IMCO labels. The proposal did not address the use of placards conforming to IMCO specifications either in place of, or in addition to, DOT placards. This same commenter also incorrectly interpreted the existing § 171.12(b) as allowing freight containers to which IMCO enlarged labels have been affixed in place of DOT specification placards, to be transported by railway, motor vehicle or aircraft in the United States.

Several commenters suggested that use of the optional list will make it impossible for carrier personnel to verify the correctness of placarding and various operational measures such as segregation. Still other commenters requested clarification of the relationship between offering a material under its classification in § 172.102, and the DOT placarding requirements. both placarding and segregation, as well as other operational considerations such as car placement, are dictated by the classification of the particular hazardous material. It is for this reason that, when a material is offered under its IMCO class, the name of the DOT class which most closely corresponds to that IMCO class must be shown as part of the description on shipping papers. This indication is currently required by § 171.12(b) and was subsequently proposed for inclusion in § 172.102. When a shipper offers a hazardous material under the class provided in § 172.102, it is that class which governs all applicable transportation requirements regardless of what the classification of the material might be under § 172.101. Since the DOT class name corresponding to that IMCO class will be written out in the basic description, carrier personnel need only verify that the class shown on the shipping papers is, in fact, the class shown for that description in § 172.102, and that placarding, segregation, etc., have been accomplished as required for the class under which the material is offered.

A commenter noted that the classification of certain hazardous materials was different in § 172.101 and § 172.102. It was precisely this observation which prompted the authorization currently contained in § 171.12(b) which accommodates these classification differences so that it is not necessary for shippers to relable their export shipments.

The same commenter observed that the proper shipping name, marking, labeling and placarding may be different for different shipments of the same material from the same shipper. This is a valid observation that could also be made regarding existing authorizations contained in § 171.12(b). In the period since § 171.12(b) was adopted in 1976, MTB has not been informed that this potential difference in description, classification and labeling provisions for the same material offered by a particular shipper has created any difficulties. In addition, MTB believes that by extending the authorization to use optional descriptions, labeling and classification to domestic as well as international transportation, individual shippers will tend to use one of the two options for all shipments of a particular material rather than prepare some shipments according to § 172.101 and other shipments according to § 172.102.

One commenter, who represents a large group of motor carriers, maintained that "... to allow a second type of shipping description to be utilized by shippers in the U.S., in place of those specified in existing rules which have been in place only 2-½ [sic] years is wrong" because it invites noncompliance and results in expensive retraining of personnel, and summarized the proposal as "... another case of the international shipment tail-wagging the domestic dog." MTB questions the implication that expensive and extensive retraining of carrier personnel will be necessitated by the adoption of Docket HM-171. In general, carrier personnel will have to be informed that, should they not find in § 172.101 a hazardous materials description which appears on a shipping paper, they should check to see if it appears in § 172.102. If it does not, the shipment is in non-compliance. If it does appear, they may then proceed to verify the correctness of the classification, labeling, placarding, etc. For this reason, MTB does not believe this change will result in a necessity for extensive retraining of personnel.

This same commenter offered a National Transportation Safety Board (NTSB) report on non-compliance with the hazardous materials regulations in support of this argument that adoption of the optional list would further complicate the regulations and detract from transportation safety. MTB does not accept this argument for reasons previously stated and would draw the attention of this commenter to the fact that NTSB has supported this proposal as being "...responsive, in part, to the Safety Board's Recommendation I-78-71 dated January 17, 1978, in that it provides for the use and cross-reference of IMCO shipping descriptions."

One commenter suggested that the need for an optional hazardous materials list could be eliminated for import/export shipments by retaining existing § 171.12(b) and by allowing the IMCO description for a material to appear on shipping papers and requiring that the description and class of the material appearing in § 172.101 must precede the IMCO proper shipping name and class on the shipping papers. Adoption of this suggestion would be more restrictive than current regulations in that, in addition to a requirement to always show the proper shipping name from § 172.101, the class prescribed for the materials in § 172.101 would also have to be indicated. Under the existing regulations, the IMCO classification may be used in place of the DOT class for import/export shipments of certain hazardous materials. For this reason, and because this suggestion would do nothing to eliminate redundant shipping paper description and package marking requirements, the suggestion is not adopted. This commenter also maintained that authorization to use IMCO descriptions for purely domestic transportation was "totally unacceptable" at this time because the IMCO system is obscure to carrier and emergency response personnel. MTB notes that water carrier personnel have been successfully using the IMCO system for several years and that, as indicated in numerous comments, this segment of the carrier industry does not agree with the statement made by this commenter. Furthermore, as previously stated, one of the benefits of the optional list, as opposed to the existing practice of incorporation of IMCO by reference, is that it removes the "obscurity" of the IMCO system by placing the necessary information in the hands of all carrier personnel. MTB assumes also that carrier personnel have been properly trained to deal with the current authorization for use of IMCO descriptions, classes and labels for import/export shipments. With the provision of this information in § 172.102 readily available for use by all carrier personnel, MTB can see no valid reason why the option should not be extended to domestic as well as import/export shipments.

A commenter noted that certain materials have different identification numbers assigned under § 172.101 and § 172.102, and that this difference could thwart emergency response efforts. These differences result from the fact that IMCO lags several years behind the UN in adoption of UN Recommendations. The identification numbers assigned in § 172.101 have been based on the most recent UN Recommendations while those indicated in § 172.102 are those identification numbers assigned to materials in the current edition of the IMCO Code. It is essential that § 172.102 be maintained consistent with the IMCO Code to prevent frustration of international shipments. These occasional number differences should not affect response actions since the Emergency Response Guidebook has been designed to function with identification numbers contained in either list.

A request was received to include in the regulations a list of countries who have adopted the IMCO Code. MTB does not believe it appropriate to include such a list in the regulations. This information, is, however, available from MTB upon request.

THE EMERGENCY RESPONSE GUIDEBOOK

Notice 79-9, under Docket HM-126A (44 FR 32972), included a discussion of MTB's proposed distribution of a manual. This document is now entitled "Emergency Response Guidebook" (ERG). Its development is completed and it is expected that its production will begin within forty-five days after publication of these amendments. The format of the ERG is essentially the same as was discussed in the Notice.

The MTB is completing its planning for distribution of the ERG, without charge, to emergency response entities throughout the United States. MTB has been advised by two firms that they plan to make commercial distribution of the ERG.

Since the ERG contains materials closely associated with the regulations published herein, it bears a copyright provision authorizing its reproduction, without modification, without further permission from DOT.

The MTB expresses its appreciation to the many individuals, organizations and businesses that assisted in the development of the ERG. The many exchanges of ideas and information resulted in the development of a much improved document. MTB recognizes that further improvement will be suggested and may be warranted; therefore, it contemplates that the ERG will be republished in approximately two years.

CORRECTIONS

MTB anticipates that a limited number of errors will be discovered upon review of the amendments in this publication; e.g., a printing error in the pound or kilogram entries for "E" identified materials in the Table to § 172.101. MTB plans to handle corrections in one Federal Register publication before September 1, 1980. Any person discovering an error may contact the individuals named earlier in this preamble, directly by telephone or by letter. A distinction should be made between discovery of errors and taking exception (disagreeing) with MTB's decisions concerning substantive matters in the amendments. Substantive matters will be handled in accordance with 49 CFR 106.35 and 106.37.

REVIEW BY SECTIONS

Section 171.1. Section 171.1 is revised from its proposed scope in Dockets HM-145A and HM-145B to specify the applicability of the Hazardous Materials Regulations with regard to the transportation of hazardous wastes and hazardous substances by intrastate motor carriers. No such distinction is considered necessary relative to transportation by rail car, aircraft or vessel since the nature of such modes makes their operations subject to DOT's regulations without regard to the intrastate or interstate nature of individual shipments. The revision of this section was not proposed in the Notice. However, this does not constitute a substantive change since the new § 171.1 language merely includes hazardous waste and restates the jurisdictional scope originally proposed in § 171.3, which was discussed in the Notice to Docket HM-145A preamble. In anticipation of applications to EPA for interim authorizations to manage State hazardous waste programs pursuant to procedures specified in 40 CFR Part 123, MTB has excluded intrastate hazardous waste motor carriers from the application of the regulations in States holding interim authorizations (See Section 3006 of the RCRA).

Section 171.2. Two minor amendments are made to this section to reduce the references to § 171.12 and to add a reference to § 176.11. The provisions of current § 171.12(b) are transferred to the new § 172.102. These changes were proposed in Docket HM-171.

Section 171.3. Section 171.3 contains the basic requirements pertaining to the offering, transportation, and delivery of hazardous wastes, generally as proposed in Docket HM-145A. It should be noted that the applicability of specific hazardous waste requirements is based on a determination that a hazardous waste manifest is required according to EPA's regulations in 40 CFR Part 262. Under paragraph (b), a motor carrier may not transport a hazardous waste for which a manifest is required unless the carrier is identified on the vehicle in the manner prescribed by 49 CFR 397.21, one of the Federal Highway Administration's Federal Motor Carrier Safety Regulations, or 49 CFR 1058.2, a regulation of the Interstate Commerce Commission. This requirement applies when a manifest is required, regardless of the quantity of wastes transported. Paragraph (b)(2) requires compliance with the manifest requirements, and paragraph (b)(3) specifies the limitations on the delivery of wastes. A note is added following paragraph (b) emphasizing the fact that penalties exist for discharging hazardous waste at other than designated facilities. Paragraph (c) (§ 171.3(e) of the Notice) addresses those actions of a state, or its political subdivision, considered inconsistent with DOT's Hazardous Materials Regulations. A change has been made to proposed § 171.3(c)(3) that was not contained in the Docket HM-145A Notice to clarify the fact that hazardous waste manifests are considered to be shipping papers when they are being carried aboard transport vehicles. Paragraph (d) specifies the conditions under which an official may authorize the removal of a waste without the preparation of a manifest when a discharge occurs during transportation. Note 1 to this section provides advisory information to shippers and carriers that they are required by EPA's regulations to obtain identification numbers. Note 2 quotes EPA's regulation pertaining to the clean up of hazardous wastes discharged during transportation. With the addition of these two notes, and the amendments to the regulations set forth in this action, EPA has stated in its rules (see 40 CFR 263.10; 45 FR 12743, February 26, 1980) that carriers (transporters) of hazardous wastes will know of the hazardous waste transportation requirements applicable to their operations by reading DOT's Hazardous Materials Regulations (and Note 2 to § 171.3) without the necessity of reading EPA's corresponding regulations.

As published, § 171.3 differs from the version shown in the Notice. Most of the textual change eliminates unnecessary duplication of requirments proposed and adopted elsewhere in the rule. The reference to 49 CFR 397.21 has been adjusted to reflect the necessity of displaying the required vehicle markings regardless of whether or not placarding is necessary, a fact to which the scope of § 397.21 otherwise keys. Proposed, but not adopted, was a lengthy reference to an exception for certain shipments not subject to EPA hazardous waste manifest requirements. This exception has been broadened to include all shipments not subject to EPA manifest requirements and therefore not incorporated into the DOT definition of "Hazardous waste" in § 171.8 of this rule.

To preclude anticipated problems. MTB has provided for the substitution of equivalent specification drums for hazardous wastes that may be impractical to package in the authorized specification drums.

Section 171.7. In § 171.7 paragraphs (c)(27), (c)(28), (d)(5)(ix), (d)(20), (d)(21) and (d)(22) are added. Paragraph (d)(20) is added to recognize the latest edition of the NIOSH Registry of Toxic Effects of Chemical Substances as an acceptable source of common names, chemical names and trade names that may be used in place of a technical name to meet the requirements of § 172.203(k). Paragraph (d)(21) is added to provide proper reference to the UN Recommendations on the Transport of Dangerous Goods which had been cited in § 172.519 in Docket HM-103. The other additions to this section provide citations to obtain test methods used to determine thermal stability (as addressed in § 173.21). An explanation of the addition of these test references can be found in the portion of the preamble of this document addressing Docket HM-159.

Section 171.8. Section 171.8 contains definitions related to hazardous wastes and hazardous substances, revised somewhat from those proposed in Dockets HM-145A and HM-145B. Note that the definition of "Hazardous waste" is limited to materials that are subject to EPA hazardous waste manifest requirements specified in 40 CFR Part 262. Added to the definition is a qualification concerning manifests so that carriage of hazardous wastes in States holding interim authorizations from EPA, in accordance with 40 CFR Part 123, is not excluded from the application of DOT's Hazardous Materials Regulations (other than carriage by intrastate motor carriers—See Section 171.1(a)(3)). As stated earlier, MTB believes that a nationwide standard for manifests is an important factor in accomplishing the safe and effective delivery of wastes to designated facilities. This factor is particularly important relative to the operations of carriers transporting materials in interstate commerce. Note also that for the purposes of the DOT regulations, a material is a hazardous substance based on certain conditions that are set forth in the definition. Any hazardous material that may also be a hazardous substance under DOT regulations is identified in the Table by the letter "E" along with the reportable quantity for that material. When a material identified by the letter "E" is in a mixture or solution, the material is evaluated against the percentage concentration by weight in paragraph (d) to determine whether it falls within the definition. If a pure material is involved, or a mixture or solution with letter "E" materials which equal or exceed the paragraph (d) concentrations, the final step is to determine if it is a hazardous substance. For the purposes of DOT regulations, a material identified by the letter "E" becomes a hazardous substance when a reportable quantity (indicated in pounds and kilograms following the proper name) is contained within one package, or if not packaged, within one transport vehicle. For a mixture, each material must be separately evaluated since a hazardous substance determination must be made for each letter "E" material in a mixture. Thus, a package with a net weight of 20 pounds and containing a mixture of material "A" having an RQ of 10 pounds and "B" having an RQ of 100 pounds could, depending on the concentration, contain a hazardous substance with regard to "A", but not "B." Different percentages of an RQ for each substance in a mixture would *not* be aggregated. Note well that letter "E" materials that are not classed as ORM-E's, are hazardous materials even if they are not hazardous substances.

Section 171.12. Paragraph (b) of this section is no longer necessary since its provisions are contained in § 172.102 as proposed in Docket HM-171. Therefore, this paragraph is replaced by an appropriate cross reference to § 172.102.

Sections 171.15 and 171.16. It was proposed, in Dockets HM-145A and HM-145B, to set forth the hazardous waste and hazardous substance discharge reporting requirements in a new § 171.17. However, EPA has proceeded to align its hazardous waste discharge reporting requirements closely with those presently set forth in DOT's Hazardous Materials Regulations. Therefore, a new provision is not necessary and § 171.15 is modified to make it clear that hazardous wastes are covered by the reporting requirements, and § 171.16 is modified to require that a copy of the hazardous waste manifest (or other document used in place of the manifest) be attached to the report, thereby ensuring the availability of information such as generator and carrier identification numbers. The only other change to the present DOT reporting requirement is the required entry of an estimate of the amount of waste removed from the site of a discharge, the name and address of the facility to which it was taken, and the manner of disposition of any unremoved waste. This information is to be entered in the "Remarks" portion of the presently required report. MTB added the toll call telephone number to § 171.15 and 171.17 for reporting hazardous waste and hazardous substance discharges from location where toll free 800 numbers do not apply, such as Hawaii, Alaska and Puerto Rico.

Section 171.17. Section 171.17 provides a reporting requirement for the discharge of a hazardous substance that is closely aligned with the notification requirement of § 311 of the Clean Water Act. The final rule requires notification when the carrier determines that a hazardous substance in transportation has been discharged into the navigable water or adjacent shorelines. This is a lesser reporting requirement than was contained in notice to Docket HM-145B in which it was proposed that each discharge of a letter "E" material be required to be reported, regardless of the location of the discharge and whether it was a sufficient quantity to constitute a hazardous substance.

Section 172.101. Present §§ 172.100 and 172.101 are combined into a revised and amended § 172.101 in order that the Hazardous Materials Table and the language introducing the Table may be contained in one section of the regulations. Indications of changes are to the language presently appearing at § 172.100, which as a result of this rule will appear at § 172.101. Paragraph (a) has been revised to accommodate this change.

The introductory text to paragraph (b) is revised to eliminate the asterisk (*), add the plus (+), and to include the letter "E."

Paragraph (b)(1) is revised to acknowledge the new plus symbol (+) in column 1 for material in column 2. The plus (+) was added to this section to identify materials for which the hazard class is fixed, without regard to whether the material meets the definition of the hazard class assigned. The asterisk (*) is no longer referenced in this paragraph or in column 1 of the Table. Justification for the removal of the asterisk, is discussed in this preamble text dealing with HM-159. An alternate shipping name and hazard class for materials may be authorized by the Office of Hazardous Materials Regulation, MTB upon submission of data justifying such reclassification and description changes.

Paragraphs (b)(2) and (b)(3) are modified to exclude from any limitation of requirements to single mode applicability those materials which are hazardous wastes or hazardous substances, both of which are subject to certain regulations regardless of the mode of transportation involved. Paragraph (b)(4) is added to include the meaning of the letter "E" used in column 1 of the Table. The letter "E" identifies materials named in column 2 that may be hazardous substances depending on quantity contained in a single package, in which case they are subject to applicable regulations regardless of the transporting mode. It is important to note that although a hazardous material may be reclassed according to its hazard(s), if a hazardous substance does not meet the definition of any other hazard class, it must be reclassed as such (ORM-E) and shipped in accordance with the applicable regulations for hazardous substances.

The introductory text to paragraph (c) is revised editorily to reflect the authorized modification of a proper shipping name as specified in paragraphs (b)(4), (c)(10), (c)(11), (c)(12) and (c)(13).

Paragraph (c)(4) has been amended by adding a recommendatory sentence indicating that the sequence of each entry as shown in column 2 of the Table is the preferred sequence for marking and shipping paper requirements. The use of the preferred sequence will better enable emergency response personnel to locate the material when using the Emergency Response Guidebook since it is this sequence which appears in the Guidebook. Paragraph (c)(5) is revised, as proposed in Docket HM-126A, to require the technical name to be used as the proper shipping name for all "see" references for Organic peroxides. This is discussed in the Docket HM-126A preamble.

Paragraph (c)(6) contains a provision to limit the application of the DOT Hazardous Materials Regulations concerning hazardous materials identified as poisons. Under this limitation, a hazardous material having the word poison or poisonous in the proper shipping name must be considered to be a poison only if tests indicate death resulted from systemic poisoning rather than by corrosive destruction of tissue. MTB believes this limitation is appropriate and will provide adequately for emergency response information.

Paragraph (c)(9) is added, as proposed in Docket HM-145B, to explain the addition of the reportable quantity to the Table as an italicized entry in parentheses following the proper shipping name of potentially hazardous substances in §172.101. This makes the reportable quantity for each of these materials readily available for reference so a determination can be made as to whether a hazardous substance is being offered for transportation.

Paragraph (c)(10) is added, as proposed in Docket HM-145A as §172.101(c)(8), to specify that a proper shipping name include the word "waste" when the material described is a hazardous waste subject to EPA's requirements in 40 CFR Part 262. Use of this method to appropriately prescribe the proper shipping name for a hazardous waste alleviates the necessity for the addition of hundreds of proper shipping name entries to the Table. Only one new shipping name "Hazardous waste, liquid or solid, n.o.s." is added to the Table for hazardous waste.

Paragraph (c)(11) is added to aid in improving the identification of hazardous materials for emergency response personnel. To reduce the use of n.o.s. entries, MTB is authorizing the use of the name of the hazardous material in certain solutions and mixtures as part of the proper shipping name. This new procedure provides a partial answer to questions such as "What is Acetone under DOT regulations?" by allowing a mixture of a hazardous material and non-hazardous materials to be identified by the proper shipping name of the hazardous material, providing the hazard class remains unchanged. Thus, an Acetone, water and mineral oil solution that is a Flammable liquid may be identified as Acetone solution. Flammable liquid instead of Flammable liquid, n.o.s.

Paragraph (c)(12) is added to provide a means for identifying hazardous substances when they are not identified in a proper shipping name. This could occur through the use of the proper shipping name Hazardous substance, liquid or solid (as appropriate), n.o.s., which was requested by commenters for mixtures or solutions of hazardous substances. MTB was provided with an example of a mixture of Toxaphene and Xylene with a flash point above 200° F. consisting of a hazardous substance (i.e., any letter "E" material that has equalled or exceeded a reportable quantity in one package). Therefore, the entry would be Hazardous substance, liquid, n.o.s. This would be followed by the hazard class, ORM-E, the identification number for Hazardous substance, liquid, n.o.s., which is NA9188, and the identifier, RQ. In association with the basic description, the name of the hazardous substance that equals or exceeds a reportable quantity in one package must be identified.

Paragraph (c)(13) contains a provision that was proposed in Docket HM-145B in paragraph (b)(1)(iii) and that has been revised by MTB for clarity. It also contains a requirement for consideration of forbidden materials and those materials identified with the plus (+) symbol when determining a hazard class for a material that no longer meets the defining criteria for the class shown for the material in the Table. For example, Xylene is identified with the letter E and is classed as a Flammable liquid. However, in certain mixtures, the flash point could be raised to between 100° and 200° F. and, if shipped in a cargo tank or tank car, could equal or exceed Xylene's 1,000 pound reportable quantity. In this case, it would meet the definition of a hazardous substance in §171.8 and its proper shipping name would be Combustible liquid, n.o.s. It may not be shipped as Xylene under the paragraph (c)(11) rule because its hazard class has changed.

It is important to note that although a hazardous substance may be reclassed according to its hazard(s), if it does not meet the definition of any other hazard (and since by definition it is a reportable quantity), it must be reclassed as an ORM-E.

Paragraph (d) is revised as proposed in Docket HM-159 to clarify the intent of the term "Forbidden" as it appears in column 3 of the Table. The prohibition as "Forbidden" does not apply if the materials are diluted, stabilized or incorporated in devices and are classed in accordance with the definitions of hazardous materials contained in Part 173.

New paragraph (e) describes the addition of column 3(a) containing the identification numbers as proposed in Docket HM-126A. Present paragraphs (f), (g), (h), and (i) were formerly paragraphs (e), (f), (g), and (h) and contain no revisions.

MTB added paragraph (j) to establish a general grandfather provision for shipping paper entries and package markings for materials packaged before the effective date of an amendment to the Table that changes a proper shipping name or hazard class of a material in the Table.

Except for hazardous wastes and hazardous substances, if the proper shipping names or hazard classes of materials are changed by this rule, such changes are not mandatory prior to July 1, 1981, as stated in paragraph (k) of this section.

Hazardous Materials Table: The Hazardous Materials Table is amended to identify potentially environmentally hazardous materials through the use of the letter "E." These letter "E" materials include hazardous substances designated by the EPA under 40 CFR Part 117 and not presently subject to the DOT regulations (classed as ORM-E), as well as presently regulated hazardous materials that have been designated by the EPA as hazardous substances. These latter materials are in two major categories. One is materials that are now listed by name in § 172.101, and the other is materials not now identified by name but which are now regulated under the n.o.s. listings in § 172.101. A total of 368 entries in the Hazardous Materials Table are identified as potentially hazardous substances. A total of 369 had been proposed, but EPA dropped Calcium oxide and Calcium hydroxide from their hazardous substance list and MTB did likewise. MTB added Aluminum sulfate, solid, to the ORM-E list. Aluminum sulfate, solution, remains as an ORM-B. Guthion is changed to a "see" entry referenced to the entry Azinphos methyl because Guthion is a registed trade name.

Fourteen potentially hazardous substances are identified as ORM-A. This is based on the chemical, physical and other comparable properties of the compounds. The properties of these compounds are such that each compound can cause extreme annoyance or discomfort to passengers and crew in the event of leakage during transportation. Based on data provided by commenters, Xylenol was transferred from the ORM-E list to the ORM-A list and the following five ORM-A's were transferred to the ORM-E list:

2,4-Dichlorophenoxyacetic acid ester
2,4,5-Trichlorophenocyacetic acid amine
2,4,5-Trichlorophenoxyacetic acid ester
2,4,5-Trichlorophenoxyacetic acid salt
2,4,5-Trichlorophenoxypropionic acid ester

Ninety-nine materials are classed as ORM-E. This is based on the EPA designation of certain materials as hazardous substances on March 13, 1978 (42 FR 10494), and the fact that according to MTB's evaluation they do not meet the defining criteria of any other hazard class. In addition to the five materials transferred from ORM-A to ORM-E as discussed above, Aluminum sulfate, solid, was added to the ORM-E's and Heptachlor and Vanadium pentoxide were reclassed from Poison B to ORM-E based on data provided by commenters. Also, the following changes in class assignments were made by MTB based on data submitted by commenters: Xylenol was reclassed from an ORM-E to ORM-A; Dodecylbenzenesulfonic acid was reclassed from an ORM-E to a Corrosive material; Ethylenediamine was reclassed from a Flammable liquid to a Corrosive material; and Sodium hydrosulfide, solid, was reclassed from a Flammable solid to a Corrosive material.

Although there appears to be a discrepancy between the number of newly identified materials in this proposal and the number of materials in the EPA list of hazardous substances, the materials in this proposal are those covered in the EPA list. The difference in the number of materials results from the necessity of identifying in § 172.101 the different forms, mixtures or solutions of a material for proper regulation. For example, "Aldrin" appears once in the EPA list and six times in § 172.101. Two materials, Calcium oxide and Calcium hydroxide, were dropped from the hazardous substance list by EPA. Therefore, neither is identified in the Table as a potentially hazardous substance; however, Calcium oxide remains as an ORM-B as it was before the EPA hazardous substance list was developed.

In addition to the changes to the Table to accommodate the potential hazardous substances, MTB made changes to improve packaging or to more accurately describe a hazardous material based on data provided by commenters.

One commenter was concerned that MTB did not include a generic description for a flammable liquid that is also toxic. Dual hazards presented by some materials were addressed subsequently in HM-126B. Among others, the proper shipping name Flammable liquid, poisonous, n.o.s. with identification number UN 1992 was proposed. With this addition to the Table, both toxic and non-toxic flammable liquids are now covered by generic descriptions.

The discrepancy between identification numbers for Uranium Hexafluoride, fissile and Uranium Hexafluoride, low specific activity appearing in proposed HM-126A (NA9173 and NA9174, respectively) and the draft Emergency Response Guidebook (UN2926 and UN2927, respectively) has been corrected. The correct identification numbers are NA9173 and NA9174.

One commenter petitioned that morpholine, a flammable liquid, should be added by name to the Table. Morpholine appears by name in the UN Recommendations on the Transport of Dangerous Goods (UN2054), but not in the DOT Regulations. Since Morpholine is not in the Table by name, it must presently be described as Flammable liquid, n.o.s. MTB agrees and is adding the proper shipping name Morpholine to the Table, classed as Flammable liquid and assigned identification number UN2054.

One commenter was concerned that Pine oil was assigned identification number UN1272. Pine oil is listed by name in the Table and classed as a Combustible liquid. The UN Recommendations on the Transport of Dangerous Goods also lists Pine oil by name but classes this material as Flammable liquid. Since the upper flash point limit for the UN flammable liquid hazard class overlaps with the lower flash point limit of DOT's combustible liquid hazard class, and the emergency response for most flammable liquids and combustible liquids is identical, MTB believes that the appropriate identification number for Pine oil is UN1272.

One commenter suggested that the wrong identification number had been assigned to several proper shipping names and that the "UN" or "NA" prefix assigned to some identification numbers was wrong. The commenter submitted recommendations to correct these errors. Upon review, MTB has changed certain identification numbers and prefixes.

Another commenter stated that Hydrofluoric acid, anhydrous (UN1052) and Hydrogen fluoride (NA1790) are the same material and should be assigned the same identification number. MTB agrees with the commenter that UN 1052 is the correct identification number for both materials and has changed the Table accordingly.

Another commenter was concerned that the prefixes in the identification numbers assigned to certain cross referenced proper shipping names do not agree. For example, identification number NA1133 ws assigned to Cement, liquid, n.o.s. but identification number UN1133 was assigned to Adhesive, n.o.s. which is cross referenced to Cement, liquid, n.o.s. In order to resolve the difficulty of assigning identification numbers to cross referenced proper shipping names, MTB is deleting identification numbers from certain proper shipping names that are cross referenced to another shipping description. This does not apply to organic peroxides because both entries are pertinent.

Various commenters reported that the Table has been affected by other final rules that have been published recently. MTB is well aware of this and has made necessary changes in the Table to keep it up to date.

Several commenters reported that certain potentially hazardous substances were classed incorrectly and submitted data to support their position. Additionally, these and other commenters stated that requirements and/or limitations for certain of these materials were wrong or inconsistent with those for materials having similar properties. Based on the comments and MTB's evaluation, changes that affect hazard class, packaging, quantity in one package, etc., have been made.

At the request of a commenter, the proper shipping name Metal alkyl solution, n.o.s., classed as Flammable liquid, has been added to the Table. Addition of this entry, provides a better description for nonpyrophoric solutions of metal alkyls in flammable solvents than Flammable liquid, n.o.s. which is currently used. Also, the commenter stated that the emergency response for an accident involving metal alkyls in flammable solvents differs somewhat from that for a flammable liquid. Technical adjustments have been made to the descriptions for formaldehyde so the proper shipping name correctly describes the material as Formaldehyde solution. The hazard class depends on the size of the packaging, and the identification number assigned depends on the flash point of the formaldehyde solution. An alternate shipping name is Formalin, not Formalin solution. The section for specific packaging requirements for Nicotine, liquid referenced in the Table has been changed from § 173.358 to relect the correct reference, § 173.346.

MTB added Gasohol to the Table to reflect the increasing use of the gasoline-ethyl alcohol mixtures as a vehicle fuel and to permit an emergency response reference for it.

MTB has revised the entry for Aluminum sulfate, based on commenter recommendation, to include both the solid form and solutions. Aluminum sulfate, solid is classed as ORM-E, whereas Aluminum sulfate solution is classed as ORM-B.

One commenter expressed concern that Zinc phosphide had been incorrectly classed as a Flammable solid. Data submitted by the commenter indicated that Zinc phosphide meets the definition of a Poison B material and, although Zinc phosphide reacts with water, its reaction rate is so slow that liberated gases are dissipated before a hazard develops. MTB agrees and has reclassed Zinc phosphide as a Poison B.

One commenter took exception to the classification of Heptachlor as Poison B. Data he submitted included oral toxicity for analytical reference standard grade (99.8%) and technical grade (74%) heptachlor, and inhalation and skin absorption toxicity for technical grade heptachlor. Also, skin corrosion data was submitted. MTB agrees with the commenter that the data does not support classification of heptachlor as a poison B or corrosive material. Hence, Heptachlor has been reclassed as ORM-E.

One commenter pointed out that Dichlorvos is a Poison B liquid. However, the applicable sections for packaging referenced in column 5 of the Table (i.e., § 173.364 and § 173.365) are for a Poison B solid. The packaging references for Dichlorvos, therefore, are corrected to § 173.345 and § 173.346, respectively.

Two commenters stated that they ship Dodecylbenzenesulfonic acid as a corrosive liquid. The properties are similar to an alkanesulfonic acid, which is listed by name in the Table and classed as a corrosive material. MTB agrees with these commenters and has reclassed Dodecylbenzesulfonic acid as Corrosive material from ORM-E based on the data provided.

One commenter from a company that manufacturers agricultural chemicals objected to the inclusion of Guthion as a proper shipping name. Guthion is the trade name for one of the company's registered pesticides. The commenter stated that azinphos methyl is the common name and should be the proper shipping name. As a compromise, MTB has added Azinphos methyl as an alternate name for Guthion.

A commenter stated that Zinc hydrosulfite does not meet the definition for a Flammable solid and that the material is not hazardous in transportation. The UN Recommendations on the Trasnport of Dangerous Goods lists Zinc hydrosulfite in Class 9—Miscellaneous dangerous goods. Goods listed in this class are not considered to be dangerous when transported by rail or motor vehicle. Thus, MTB has reclassed Zinc hydrosulfide as an ORM-A.

One commenter stated that Vanadium pentoxide had been improperly proposed to be classed as Poison B. Data he submitted indicate that the oral toxicity (LD_{50}) of Vanadium pentoxide is greater than 50 mg/kg. MTB review supports this conclusion, therefore, Vanadium pentoxide has been reclassed as an ORM-E.

Sodium hydrosulfide, solid, is reclassed as a Corrosive material. MTB has classed this material as a Flammable solid in the proposal based on information obtained from the United Nations Recommendations on the Transport of Dangerous Goods. However, data submitted by a commenter was sufficient for MTB to conclude that Sodium hydrosulfide, solid, is not a Flammable solid but is corrosive to skin.

Several commenters stated that pesticides identified and described by chemical groups based on their chemical structures provide sufficient information to specify appropriate action to be taken in event of an accident involving spillage or leakage. The 45 new descriptions for 15 pesticide groups identify the type of pesticide and enable first aid and medical advice to be linked to these pesticides. MTB agrees with the commenters that the n.o.s. modifier should be deleted from these descriptions. Coupled with other action taken in this rulemaking, this eliminates the proposed requirement to provide the technical name of the pesticide when a proper shipping name includes the chemical element or group. MTB now estimates that these descriptions cover at least 95 percent of the pesticides transported. Pesticides not described by technical names or family groups in the Table can be described by general or generic descriptions such as Insecticide, liquid, n.o.s.; Compound, tree *or* weed killing, liquid; Flammable liquid, poisonous, n.o.s. The use of a generic name or general description to describe a pesticide requires that the technical name of the pesticide be included as part of the shipping description. These commenters argued that the Table already contains descriptions for organic phosphates and the addition of three new descriptions for organophosphorus pesticides is not necessary. MTB believes that these new descriptions provide vital information on both the chemical structure and end use of the material. Many organophosphorus materials in commerce are not pesticides. Thus, the new organophosphorus pesticide descriptions do not apply to these materials. However, the descriptions for organic phosphates may apply.

A pesticide in the 15 chemical groups that does not meet the defining criteria for a flammable liquid and/or poison, may meet the definition of another hazard class and have to be described by a shipping name appropriate to that class. For example, an organotin pesticide in a liquid formulation that does not meet the definition of a flammable liquid or poison, may be a corrosive liquid. In this case, the proper shipping name for that organotin pesticide would be Corrosive liquid, n.o.s.

Nitrogen trifluoride was proposed to be listed in the Table as Forbidden in the notice to Docket HM-126B. MTB has made the necessary corrections based on commenter suggestions. Nitrogen trifluoride is classed as Nonflammable gas with identification number UN2451.

Two commenters objected to the proposed addition of Pinene to the Table classed as Flammable liquid without an asterisk in column 1, since an asterisk denotes that a material may or may not be regulated under the class shown depending on whether or not the commodity meets the definition of the class listed for that entry. One commenter stated that tests conducted indicate that pinene has a closed cup flash point range between 99° and 100° F. With a flash point below 100° F., a material meets the definition of a Flammable liquid. With a flash point at or above 100° F. and below 200° F., a material meets the definition of a Combustible liquid. According to MTB data, pinene has two isomers. Alphapinene has a flash point of 91° F. Betapinene has a flash point of 117° F. The flash point of pinene containing an isomeric mixture falls between 91° and 177° F. and depends on the percentage of each isomer present. Since this rulemaking deletes all asterisks from the Table, pinene, when classed as a Flammable liquid, would be described as Pinene. When classed as a Combustible liquid, pinene would be described as Combustible liquid, n.o.s.

A commenter pointed out that "Alcoholic beverages", classed as Flammable liquid, in containers having a rated capacity of one gallon or less, are not subject to the hazardous materials regulations per § 173.118(c). Thus, the proposed one quart net quantity per package limitation for passenger carrying aircraft is wrong. MTB agrees and column 6(a) in the Table has been changed to read "See § 173.118(c)." This commenter also pointed out that the correct identification number is UN1170 and not NA1987.

Since the proper shipping name Engine, internal combustion has been proposed, one commenter recommended that Motor, internal combustion be deleted from the Table stating: "The motor receives its power from an outside source. The engine develops power internally." MTB does not dispute the commenter's argument, however, the description has been retained. The terms "motor" and "engine" have become synonymous in the automobile industry. MTB seriously doubts that motor companies in this industry would consider changing their names to engine companies.

A commenter objected to the proposed requirement to label hydrogen peroxide solutions (up to and including 52% peroxide) with a corrosive label to identify the secondary hazard. This commenter stated that "the noncorrosiveness for less than 52% is an industry fact." Based on the data presented, MTB has deleted the requirement for a CORROSIVE label on hydrogen peroxide solutions containing not more than 52% peroxide.

Comments were received concerning the new entry and requirements for Calcium hypochlorite, hydrated. The description has been revised to include in italicized print "(minimum 5.5% but not more than 10% water, and more than 39% available chlorine)". The material is a potential hazardous substance and has been so designated by an "E" in column 1 of the Table. The associated RQ is 100/45.4. Specific packaging requirements are referenced to § 173.217. This section contains packaging requirements for similar type compounds. The statement "keep cool and dry" has been added in column 7(c) of § 172.101.

Stowage requirements have been changed to authorize both "on deck" and "below deck" locations on board cargo vessels and passenger vessels for certain potential hazardous substances. The proposed regulation in Docket HM-145B authorized only "below deck" locations which were unduly restrictive.

EPA has changed the reportable quantity (RQ) for Calcium hypochlorite from RQ-10/4.54 to RQ-100/45.4. This change has been incorporated into the entry for Calcium hypochlorite mixture in the Table.

Several hazardous materials that contain one or more potential hazardous substances were not properly identified in the HM-145 proposal. The materials are identified now by an "E" in column 1. The RQ assigned to these materials is based on the RQ of the potential hazardous substance. If two or more potential hazardous substances are present, the lower/lowest RQ value is listed. For example, Nitrating acid (RQ-1000/454) is a mixture containing Nitric acid (RQ-1000/454) and Sulfuric acid (RQ-5000/2270). The other materials in this category that are identified in the Table as potential hazardous substances are Chlorosulfonic acid-sulfur trioxide mixture (RQ-1000/454); Hypochlorite solution (RQ-100/45.4); Methyl bromide and ethylene dibromide mixture, liquid (RQ-1000/454); Nitrating acid (RQ-1000/454); Nitrating acid, spent (RQ-100/454); Nitrohydrochloric acid (RQ-1000/454); Nitrohydrochloric acid, spent (1000/454); Sodium nitrite mixed with potassium nitrate (RQ-100/45.4); Sodium nitrite mixture (RQ-100/454), and White acid (RQ-5000/2270).

Section 172.102. A new § 172.102 is added as proposed in Docket HM-171. This section contains the Optional Table as well as the text necessary to explain the table and implement its use.

Paragraph (a) of this section sets forth the basic purpose of the Optional Table which provides hazardous materials descriptions, classification, labeling and vessel stowage requirements which may be used for certain hazardous materials as an alternative to the corresponding requirements provided in § 172.101. However, materials subject to the DOT regulations that are not considered dangerous under IMCO recommendations must be transported in accordance with the applicable DOT regulations. This exclusion has been included to insure that it is clearly understood that materials such as a combustible liquid with a flash point greater than 141° F. and less than 200° F. (in packagings with a capacity exceeding 110 gallons), which are not considered dangerous according to IMCO definitions are subject to all applicable DOT requirements.

A statement is also included in this paragraph to clarify the fact that many of the materials shown in the Optional Table are not subject to the DOT regulations and that their inclusion in the Optional Table does not constitute a designation of the material as a hazardous material. Only materials (1) designated as hazardous materials in § 172.101, including hazardous wastes and hazardous substances; (2) identified as forbidden in § 172.101; or (3) covered by the prohibition specified in § 173.21 or § 173.51, are subject to the DOT regulations. Entries for materials not designated as hazardous in § 172.101 are retained in the Optional Table to alert persons who may be engaged in importing or exporting such materials that the materials may be considered hazardous under widely applied international standards and to provide basic guidance relative to the classification and labeling of these materials in international transport.

One commenter suggested that proposed § 172.102 should be amended to recognize the fact that materials not regulated by DOT may be described on shipping papers by the IMCO proper shipping name and hazard class, and the package marked and labeled as provided in IMCO. MTB believes this change is unnecessary. Section 172.401, concerning prohibited labeling, specifically authorizes labels prescribed by IMCO to be applied to packages even though the material may not be considered hazardous under the DOT regulations. Regarding shipping paper descriptions and package markings, the DOT regulations do not prohibit description and marking as prescribed by IMCO in the case of materials not regulated by DOT. It is however, suggested that in such cases the shipping papers bear a notation indicating that the particular material is not subject to regulation under the DOT Hazardous Materials Regulations. The same commenter suggested that the proposed § 172.102 be revised to include a specific authorization to allow IMCO placards to be affixed to portable tanks and freight containers in addition to any placards required by the DOT regulations. MTB believes this comment has merit and it will be addressed in a future rulemaking.

Paragraph (b) of § 172.102 specifies conditions under which the description, class or label(s) provided in the Optional Table may be used rather than the DOT description, class or label(s) respectively. Class A and B explosives and radioactive materials are excluded from application of the provisions of § 172.102. Therefore, in order for a shipper to determine if he may use the Optional Table he must first establish the hazard class of the hazardous material under consideration in accordance with all applicable requirements of the DOT regulations. This is particularly important in the case of explosives where the hazard class may not necessarily be established solely by the shipper. Once the shipper has classed the hazardous material as provided in the DOT regulations, has determined that the material is not a Class A or B explosive or a radioactive material, and is not a forbidden material, he may then proceed to use the Optional Hazardous Materials Table if he so desires.

One commenter suggested that a symbol be introduced in Column (1) of the Optional Table to indicate to a shipper that a particular material listed is not subject to the DOT Hazardous Materials Regulations. Another commenter suggested that a symbol be introduced in Column (1) to indicate to a shipper that a particular material is considered a hazardous substance under the Hazardous Materials Regulations. MTB believes that neither of these amendments is necessary and that adoption of these amendments could result in improper use of the Optional Table. As stated in § 172.102(a), designations of materials as hazardous materials are made only in § 172.101. Therefore, it is always necessary to determine whether a material is regulated, and, if regulated, the appropriate description and class for the material provided in § 172.101, before using the Optional Table.

Several commenters suggested future inclusion of Explosives A and B entries in the Optional Table. Although MTB is not prepared to recognize IMCO description, classification, and labeling provisions for these materials for transportation in the United States at this time, this information has been included in § 172.102 with the letter "N" indicated in Column (1) adjacent to each entry.

The conditions in the existing § 171.12(b) under which the IMCO class and label(s) may be used when a hazardous material is transported by air, highway, or rail have been retained in § 172.202, with the exception of the condition that limited the application of that paragraph to import, export or transiting shipments. The IMCO shipping name may be used only when the material conforms to all additional defining or limiting conditions prescribed for the description in the appropriate schedule in the IMCO Code. Individual IMCO Code schedules often contain criteria or additional information which limit the applicability of a particular description, and MTB believes that these additional provisions must be observed in selecting an IMCO shipping description from § 172.102. The use of an IMCO shipping name is also made conditional upon inclusion of the UN number shown for the entry (if any) in the Optional Table immediately after the required class entry in the shipping papers. This is required not only to insure consistency with the United Nations standards for transport documentation, but also to enhance emergency response capabilities.

Paragraph (e) of § 172.102 requires that the description for a material designated as a potential hazardous substance in § 172.101 and offered for transportation as a hazardous substance (i.e., a reportable quantity in a single package) must be augmented by the technical name of the substance if that name does not appear in the optional shipping description. This is to insure that hazardous substances do not lose their basic identity when a shipper chooses to use a shipping description from the Optional Table in place of the name that would be otherwise required by § 172.101.

One request for clarification of the intent of paragraph (e) was received. The provisions of this paragraph would only apply to a material that is transported as a hazardous substance under § 172.101, but is offered for transportation under a description in the Optional Table. Materials designated as potential hazardous substances are identified by the letter "E" in Column (1) of the Hazardous Materials Table in § 172.101.

One commenter proposed that entries be inserted in the Optional Table in connection with IMCO's provisions for limited quantity shipments. MTB considers the addition of these entries inappropriate in that they are not considered proper shipping names under the IMCO Code.

Paragraphs (f) through (l) explain the content of Columns 1 through 7, respectively, of the Optional Table. Column 1 contains the letter "N" adjacent to certain entries. This indicates, as explained in paragraph (f), that the particular shipping description, class and label(s) shown in § 172.102 are not acceptable alternatives to the applicable DOT requirements in § 172.101 and, therefore, may not be used. This prohibition is imposed only when MTB believes the IMCO description and/or classification appearing in § 172.102 will not adequately communicate the hazard(s) of the material in all modes of transport.

Column 2 of the Optional Table lists the proper shipping names contained in the IMCO Code. The basic format of entries and methods of selection and presentation of proper shipping names on shipping documents and packages are identical to those in § 172.101. As previously discussed, entries that are contained in the IMCO Code which describe materials that could only be classified as Class A or B explosives or Radioactive materials under DOT regulations are proceeded by the letter "N." Also omitted from the Optional Table are a limited number of entries from the IMCO Code which:

(1) Are not included in the UN Recommendations and have no UN number assigned;

(2) Are not "n.o.s." entries that require addition of the technical name of the hazardous material; and,

(3) In the opinion of MTB, are not sufficiently explicit to permit appropriate response measures to be initiated in the event of an incident.

Examples of such entries are the IMCO shipping names "Acaricides" and "Nematocides." Hazardous materials falling within such descriptions will, therefore, be transported under the next most appropriate description in § 172.102 (such as Poisonous liquid, n.o.s., in the case of the above examples), or under the appropriate description in § 172.101.

Column 3 of the Optional Table sets forth the IMCO hazard class or division for the material as appropriate. Paragraph (h) of § 172.102 includes a brief definition of each of the IMCO hazard classes and divisions and refers the user to the IMCO Code for more detailed definitions.

A comment was received which proposed that a table be included in this section providing the name of the DOT class which most closely corresponds to each IMCO class or division. MTB believes this information would be helpful and, rather than include this information in tabular format, has revised the proposed paragraph (f) of § 172.102, which is now paragraph (h), to include the name of the DOT class(es) most closely corresponding to each IMCO class or division in parentheses following the name of that class or division. MTB believes that this addition should satisfy the concerns of another commenter who proposed that a column be added to the Optional Table to provide this information for each entry.

Column 4 of the Table indicates the United Nations number assigned to the material, if any. In certain cases where no UN number has been assigned to a particular material, MTB has inserted in Column 4 the UN number of the appropriate generic, or "n.o.s.," entry under which the material would be included. In such cases, the UN number listed has been shown in parentheses and is the required identification number if the optional description is used.

Column 5 specifies the labels to be applied to a package containing the hazardous material. Specifications for the labels may be either as provided in the DOT regulations or in the IMCO Code. The label described as "St. Andrew's Cross" in this column refers to the label specified for Division 6.1, Packaging Group III materials in the UN and IMCO recommendations.

Column 6, which is for informational purposes only, provides the packaging group assigned to the material in the IMCO Code. An explanation of the meaning and purpose of this grouping system, as well as the grouping criteria developed for certain hazard classes, is presented in the recommendations prepared by the United Nations Committee of Experts on the Transport of Dangerous Goods. These recommendations, entitled "Transport of Dangerous Goods," may be obtained from the United Nations bookstores in New York or Geneva, Switzerland. MTB believes that a number of individuals involved in the international transportation of hazardous materials have gained sufficient working knowledge of the grouping system to merit the inclusion of packaging groups in the Optional Table.

A number of comments were received which concerned the implications of these amendments relative to packaging. As indicated in the Notice of Proposed Rulemaking, these amendments do not in any way affect the packaging currently required for hazardous materials under the DOT regulations and the Packaging Group column was included primarily for informational purposes. It should be noted, however, that the packaging group of a material is necessary information under IMCO labeling recommendations in determining the appropriate label to be applied to materials of Division 6.1. For this reason, the packaging group column in the optional list has been retained notwithstanding one comment which suggested deletion of the column. One commenter proposed that this column be expanded to include the IMCO Code page number for the particular material to provide further packaging guidance. This proposal was not adopted since, as previously indicated, this portion of the rules is not intended to address specific packaging provisions, and because there is a possibility that the four digit IMCO Code page number could be confused with the identification number. Several commenters expressed the opinion that MTB should explore means of addressing differences between IMCO and DOT packaging requirements. This comment is not within the scope of this rulemaking and will have to be addressed in a future rulemaking.

Column 7 sets forth the vessel stowage requirements for the hazardous materials as provided in the IMCO Code. Although § 172.101 was revised with the publication of Docket HM-103/112 to include IMCO stowage requirements to the maximum extent possible, the differences between shipping descriptions for certain hazardous materials in the DOT regulations and the IMCO Code made it impossible to include the IMCO stowage requirements for all hazardous materials. MTB believes that consistency between the DOT and IMCO stowage requirements is necessary to insure that vessels loaded in United States ports will not be in violation of the stowage requirements in force in those nations that have incorporated the IMCO Code into their national regulations, and so that vessels of those nations will not be in violation of U.S. requirements. Inclusion of the majority of the IMCO shipping descriptions in the Optional Table will make it possible to authorize the use of IMCO stowage when hazardous materials are transported under an appropriate IMCO description. The meanings of numbers used to designate acceptable stowage locations are explained in paragraph (k). The numbers used in §172.102 retain the same meaning assigned to them in §172.101; however, the explanations of the meanings of these numbers have been revised in an effort to provide greater clarity.

One commenter suggested that the maximum transport temperature specified for organic peroxides in Column 7(c) of the Optional Table should be replaced with the appropriate emergency and control temperatures recently adopted by the UN. In order to maintain consistency with existing IMCO vessel stowage requirements, these revisions have not been made. It is envisioned that these changes will be made to §172.102 when IMCO adopts amendments to the Code to replace maximum transport temperature with control temperature and emergency temperature.

Several requests were received to include various hazardous materials in the Optional Table which have been assigned UN numbers but have not yet been included by IMCO in the Code. MTB believes it would be premature to include such entries in the Optional Table at this time since they are not yet authorized by IMCO for use in international transportation. It is anticipated that many of these entries will be adopted by IMCO in the near future and that, thereupon, the entries will be added to the Optional Table.

A number of commenters took issue with the classification or labeling provided for certain materials in the IMCO Code and requested either that the Optional Table be modified or that appropriate amendment proposals be submitted to the UN or IMCO. No such changes have been incorporated in the Optional Table since this would result in inconsistencies with IMCO recommendations. MTB believes such action would be contrary to the intent of this rulemaking. However, the issues raised by these commenters will be considered and amendment proposals may be submitted to the UN or IMCO, if appropriate.

A number of modifications have been made to the Optional Table. Through a careful editorial review and as a result of several comments received, errors in a number of entries were identified. These errors have been corrected to insure that entries in the Optional Table are consistent with the corresponding entry in the IMCO Code. In addition, IMCO Code Amendment 16–78, which became effective March 1, 1980, has been incorporated into the Optional Table to assure that it reflects the latest IMCO Code provisions. MTB envisions that the Optional Table will be amended upon publication of each IMCO Code amendment so that it will always contain the most up-to-date information.

One commenter suggested that the stowage locations specified for Alkaline corrosive liquids, n.o.s.; Corrosive liquids, n.o.s.; and Corrosive solids, n.o.s. be revised to agree with those provided for the same entries in §172.101. Since the IMCO Code allows stowage of these materials as specified by the competent authority, MTB agrees that the proposed change is entirely appropriate and has revised the stowage location designations accordingly.

An objection was raised to the identification numbers indicated for the entries Sodium fluoride, solution and Silicofluorides, solid, n.o.s. on the basis that neither the IMCO Code nor the UN Recommendations specifies an identification number for those entries. Numbers for these materials were, however, adopted by the UN Committee at its Tenth Session (December 1978) and will appear in the next edition of the UN Recommendations. The numbers contained in the proposal have, therefore, been retained. The same commenter suggested that the letter "N" be inserted before the entry "Hydrazine, anhydrous and solutions containing less than 36% water by weight" because the class and labels provided for these hydrazine solutions in the Optional Table disagree with the DOT class and labels for these materials and because he considered the DOT classification to be adequate. MTB believes that use of the IMCO class and labels for these solutions will not result in a derogation of safety in transportation. The insertion of the letter "N" before entries solely because the class and label(s) under IMCO may be different from those provided in §172.101 is contrary to the purpose and intent of adopting the Optional Table. The suggestion has, therefore, not been incorporated into the Optional Table.

A number of requests were received to add certain shipping descriptions to the Optional Table which appear as proper shipping names in §172.101 and are also acceptable alternate descriptions for the materials in the IMCO Code. Such descriptions have been included in Roman type in the Optional Table with a cross reference to the entry which appears as the primary description for the material in the IMCO Code. Other comments requested the addition to the Optional Table of certain shipping descriptions which appear in §172.101 but not in the IMCO Code. MTB does not consider it appropriate to add such entries because they are not recognized by IMCO for international shipments.

One commenter noted several discrepancies between the entries for "Fishmeal" or "Fishscrap" in the Optional Table and those in the IMCO Code. The entries in the Optional Table have been revised to agree with those in the IMCO Code.

Section 172.200. Section 172.200(b) is revised as proposed in Dockets HM-145A and HM-145B, to remove the

ORM exceptions to the shipping paper requirements when a material being offered or transported is a hazardous waste or a hazardous substance. The wording of the proposal has been simplified in the final rule, without changed effect. The proposed entry for ORM–E materials has been deleted since the exception proposed in the Notice is unnecessary under the final rules.

Section 172.201. Paragraphs (a)(1)(ii), (a)(1)(iii), and (a)(4)(i) of § 172.201 are revised by MTB to accommodate the revision to § 172.202(a)(3) which requires the identification number preceded by "UN" or "NA", as appropriate, to be entered as the third element of the basic description, as proposed in Docket HM-126A. Also, MTB added a provision to paragraph (a)(1)(iii) to authorize the entry of "RQ" in the "HM" column in place of the "X" to identify the entry as representing a hazardous substance. This was recommended by several commenters, and MTB concurs. As proposed in Docket HM-171, paragraph (a)(4)(i) is amended to allow the optional insertion of the entries "IMCO" or "IMCO Class" in the hazardous materials description on the shipping papers. MTB believes that certain shippers may desire to include these entries to clarify the fact that a hazardous material is being offered under the IMCO hazard class, particularly when this hazard class difers from that provided for the material in § 172.101. A proposal that the entry "IMCO" be allowed to appear immediately before the proper shipping name has not been adopted since it is MTB's belief that the proper shipping name should appear first in the basic hazardous materials description.

Section 172.202. Paragraphs (a)(1) and (a)(2) are revised by MTB to clarify the fact that the entries in § 172.102 are optional. Also, paragraph (a)(2) is revised to reduce some of the shipping paper entries. Whenever entries from the Optional Table are used for domestic shipments, § 171.102 applies. A number of commenters expressed concern that the proposal would allow unrestricted mixing of DOT and IMCO shipping descriptions, classification and labeling which could result in confusion and suggestd that this paragraph be amended to prevent such unrestricted mixing. MTB agrees with these comments and has amended § 172.202 to require that the proper shipping name, class and identification number for a material appearing on the shipping paper must be taken either entirely from § 172.101 or entirely from § 172.102, and has amended § 172.400 to insure that the package labeling is consistent with the proper shipping name market on the package.

Various methods were suggested by commenters for insuring that some indication is included in the shipping paper description when an entry from § 172.102 is being utilized. MTB believes that such an indication is already provided, since the class of a material is always expressed numerically in Column (3) of the Optional Table and, therefore, would have to be indicated in the same manner on the shipping papers. This means that a numerical indication of the class on the shipping papers will serve as a direct indication that the entry under which the material is offered is taken from § 172.102.

Paragraph (b) is revised for clarity and to provide an example since the rule change to paragraph (a)(2)(ii) made the previous example, Corrosive liquid, n.o.s., incorrect. It should be noted, as paragraph (b) now indicates, that the basic description now consists of three elements: the proper shipping name, the hazard class, and the identification number. However, technical names may be required to be entered after the proper shipping name. These requirements were proposed in Dockets HM-126A and HM-145B.

Some shipping paper entries are required to be made "in association with" the basic description. The term "in association with" means that the additional entry may follow the complete description for a hazardous material in any reasonable format, as long as it is clearly part of the entry.

The requirement to enter the basic description in a prescribed sequence, with certain exceptions specified, does not preclude the use of a shipping paper format with columns. However, the basic description sequence must be maintained, with authorized exceptions.

Paragraph (b) is revised to show the addition of the identification number to the basic description.

One commenter suggested that a provision be added which would require the indication of flashpoint on shipping papers in order to assist water carriers in planning vessel stowage. MTB believes that, in general, the indication of the appropriate IMCO division number for flammable liquids sufficiently specifies flashpoint for stowage purposes, and that the relatively few instances where the stowage of hazardous materials of other classes is dependent on flashpoint would not justify a requirement to show flashpoint in all cases. However, as authorized by § 172.201(a)(4), such an entry may be added after the basic description.

Section 172.203. The deletion of paragraph (i)(2), the redesignation of paragraph (i)(3) as (i)(2), and the addition of paragraphs (j), (k), and (l) were proposed in HM-126B. However, paragraph (i)(2) has been revised and not deleted. General shipping paper requirements for hazardous substances are consolidated in paragraph (c). Paragraph (e) is revised to include required hazardous substance shipping papers entries for "empty" packagings. MTB revised paragraph (d)(1)(ii) to authorize the entry of a generic chemical description for the chemical form of a Radioactive material. This provision was not proposed, but is included by MTB as part of this final rulemaking. It is a relaxation of present rules.

Paragraph (i)(2) is revised to provide for the identification of at least two hazardous components of certain mixtures and solutions for shipping paper entries for export shipments by vessel. Based on a considerable number of comments, the general requirement for such an identification of components for all n.o.s. entries as was proposed in §172.203(j) of Docket HM-126B has been deleted. Paragraph (i)(2) also contains two significant exceptions from the two component requirement: First, for an n.o.s. material, other than a mixture that meets the definition of more than one hazard class, if the description identifies the name of the chemical element or group that is primarily responsible for the material being classed as it is, no additional component needs to be identified. Second, for an n.o.s. material which is a mixture of materials of different hazard classes and meets the definition of more than one hazard class, if the description contains the name of the chemical element or group responsible for one of the hazard classes, only the technical name of the component(s) not identified must be entered in parentheses.

Paragrph (j) is added to require the "Dangerous When Wet" entry on shipping papers for material labeled "DANGEROUS WHEN WET." In Docket HM-126B, it was proposed that "Water Reactive" be entered on the shipping paper, but commenters pointed out that the hazard should be identified, and MTB agrees.

Several commenters stated that a blanket requirement for entering the technical name for all n.o.s. entries, as was proposed in §172.203(j) of Docket HM-126B, would contribute little to safety and yet be a fairly costly requirement. MTB agrees and as a final rule in paragraph (k) of §172.203 the requirement has been limited to materials that are poisonous, but are not identified as such by their basic description. The compliance date for the identification of poisons on the shipping paper is July 1, 1981, but such entries may be made voluntarily after July 1, 1980.

It was proposed in Docket HM-145A to require the EPA name for the waste to be included in parentheses whenever the DOT shipping name did not include the technical name of the material. Since publication of that Notice, EPA has decided to use DOT shipping nomenclature. Consequently, there are no additional EPA names and this provision has been eliminated from the final rule.

Section 172.204. Paragraph (b)(1) is revised to exclude hazardous waste from the exceptions from certification. This revision is necessary for consistency with EPA's requirement for certification of hazardous waste shipments. This provision was overlooked in DOT's May 25, 1978, Notice and has been added to the final rule. Based on a petition, MTB is authorizing a permissive revision to the certification so it will read "This is to certify that the herein-named materials. . . ." This will permit a shipper to place the certification beside or above the hazardous materials entries if desired. This authorization is contained in a new note following the certification. The previous note was deleted by MTB because it was no longer pertinent.

Section 172.205. A new §172.205 is added as proposed in Docket HM-145A concerning the carriage and disposition of hazardous waste manifests which are required by EPA in 40 CFR Part 262. It is important to note that no hazardous waste subject to the manifest requirement may be offered, transported, transferred or delivered without a manifest with signatures and dates for receipt of the waste. The wording in this section of the final rule differs from that proposed in the Notice to more clearly state the requirements and to reduce the potential compliance costs to persons affected. There are specific requirements pertaining to the hazardous waste manifest relative to acceptance for transportation, delivery to a designated facility or a place outside the United States, and the retention of copies by the shipper and carrier(s) for three years from the date of receipt by the initial carrier. The format of a manifest is not specified in referenced EPA regulations; therefore, any document that contains all the required entries, signatures and dates is acceptable. A manifest that contains all the information required on a shipping paper by §172.202 and §172.203 may be used as a shipping paper. Paragraph (h) allows use of the present railroad waybill system to document the delivery of hazardous wastes. However, a representative of the railroad receiving a waste shipment from a generator or a motor carrier must date and sign a copy of the manifest. The person delivering the hazardous waste to the initial rail carrier must send a copy of the dated and signed manifest to a representative of the designated facility. This provision does not apply to an intermodal transfer following transportation by a rail carrier.

Section 172.300. Section 172.300 has been revised by MTB to reflect the package marking requirements resulting from the proposals in Dockets HM-126A and HM-126B, and to reflect suggestions from commenters. Marking requirements proposed in this section in Docket HM-145A for waste have been dropped since they are adequately covered elsewhere in this rule. The proposed HM-145A package marking requirement is now covered by EPA marking and certification requirements, and the proposed additional §172.306 requirement in Docket HM-145A was deleted by MTB since a companion proposal was dropped by EPA.

Several commenters suggested that clarification should be provided relative to the applicability of the provisions of §172.102 to the marking of packagings with the name of contents. It is the intent of MTB that the proper shipping name in §172.102 may be marked on packagings. To eliminate any confusion in this regard, § 172.300 has been amended to include a specific reference to §172.102. One petitioner recommended §172.300 be revised to include an exclusion from the identification number marking requirements for Limited Quantity packages. MTB agrees and has included such a provision. The compliance date for displaying identification numbers on packages is July 1, 1983. Voluntary compliance may begin after June 30, 1980, and is encouraged.

Section 172.316. Section 172.316 is amended as proposed in Dockets HM-145A and HM-145B, to include ORM-E materials in the package marking requirements for ORM's.

Section 172.324. The requirement to enter the letters RQ in association with the proper shipping name on certain packages is a slight modification from the requirement proposed in Docket HM-145B, as a result of commenters' suggestions.

The compliance date for marking RQ on packages of hazardous substances is July 1, 1983. Voluntary compliance may begin after June 30, 1980, and is encouraged.

Section 172.326. Based on commenters' suggestions, MTB has revised for clarity the wording from that proposed in HM-126A for the portable tank marking requirement, and has added paragraph (e) to provide for portable tanks that are "empty" of previously contained hazardous materials. This later provision was omitted from the HM-126A proposal, however, it incorporates "empty" packaging concepts previously addressed elsewhere, but more appropriately contained in this section. A basic provision for "empty" packagings is contained in §173.29 as presented in this final rulemaking.

Commenters recommended that consideration be given to revising the portable tank marking requirements by differentiating between those tanks having a capacity of 1000 gallons or more and those having a capacity of less than 1000 gallons. MTB agrees and has made separate provisions in paragraph (a) for the two sizes indicated. Paragraph (d) requires a motor vehicle or rail car transporting portable tanks to display identification markings if those markings on the portable tank are not visible. This change to §172.326 was not proposed, but it is necessary to carry out the basic intent of the indentification marking requirements.

Section 172.328. Several comments were received describing the difficulties that would be encountered in complying with the identification number marking requirement with a separate orange panel for each material being transported in a portable tank, cargo tank or tank car, as proposed in Docket HM-126A. MTB has revised the proposed requirements in this final rule by accepting a suggestion to permit application of an identification number on an appropriate square-on-point vehicle placard as an alternative to the orange panel. This is provided for in §§172.332 to 172.338.

Section 172.330. Changes were proposed to §172.330 in Docket HM-126A to provide for displaying identification numbers on cargo tanks transporting hazardous materials. Based on commenter suggestions, MTB has made minor editorial revisions to clarify the provisions for rail cars and for multi-unit tank car tanks, with no change in scope. Also, based on commenter suggestions, MTB has added a provision which clarifies the marking requirements for empty tank cars and multi-unit tank car tanks.

The required compliance date for displaying the appropriate identification number on tank cars and multi-unit tank car tanks is July 1, 1983. On a voluntary basis, compliance may begin after June 30, 1980, and is encouraged.

Section 172.332. Identification number marking specifications were proposed in Docket HM-126A. However, MTB has reduced the overall physical dimensions slightly to eliminate unneeded area. Based on numerous comments, MTB has revised the requirements somewhat, and in §172.336 has limited the application with regard to: (1) cargo tanks transporting fuel oils or distillate fuels; (2) multi-compartmented cargo tanks and tank cars; and (3) multi-unit tank car tanks. Also, based on comments, an alternative method for displaying the identification number on a hazard warning placard is authorized in §172.334. Paragraphs (a) and (d) of §172.332 contain the basic description of the orange panel and paragraph (b) prescribes the type of material required for its fabrication. Because metal or plastic was not prescribed for placards, the question has frequently been asked as to whether such material could be used. The answer with regard to placards is also appropriate with regard to these panels: Any material may be used as long as it is at least as durable as the tag board specified in the rule. Based on suggestions from commenters, MTB authorized, in paragraph (c), the name and hazard class of the material transported to be shown on the orange panel in letters no larger than 18 point type.

The intent of the requirement in paragraph (e) that the identification number on an orange panel be displayed in proximity to any required placard(s) was questioned by some commenters. If placards are required, any required orange panels with the identification number must be located in proximity to the placards, so they may be readily visible to emergency response personnel. Some hazardous materials do not require placarding, but an identification number may be required, for example on materials classed as ORM-E.

Section 172.334. This section provides an alternative method for displaying identification numbers. Basically, it provides for 3½ inch black number on a white background to be displayed near the center of the placard. This display must be on the appropriate placard for the material, and it will necessarily replace or essentially cover the words on the placard. The name and hazard class of the material may be shown in the top outer border of the space bearing the identification number in letters no larger than 18 point type when an identification number is displayed on a placard.

Paragraph (b) contains a black and white illustration, in miniature, of the identification number for Acetone displayed on a FLAMMABLE placard. As illustrated, the hazard class number 3 is in reverse print. If this alternative method for displaying the identification number is used and the identification number block is affixed over or in place of the hazard word on the placard, in this case FLAMMABLE, the hazard class number in the lower corner may be a black number on a white background, if not already present as a 1¾ inch numeral in reverse print.

Section 172.336. Sections 172.336 and 172.338 were not in the proposal, but MTB separated the identification number marking requirements into compatible areas and renumbered the sections accordingly. Paragraph (a) provides for permissive identification number marking on full load shipments by transport vehicles and freight containers. This is expected to enhance safety and emergency response effort and may be desired by some shippers. Paragraph (b) establishes a specification for a white square-on-point configuration background for displaying an identification number under the alternative provision (alternative to the orange panel) for materials that are not authorized or required to be placarded. It is important to note that this white display background containing an identification number is not a placard. Paragraph (c) contains exceptions to the identification number display requirements. These exceptions vary from total exception, under certain circumstances, for nurse tanks, to partial exception, as for the forward end of cargo tank trucks and cargo tank semi-trailers. Limited exceptions are also provided for multi-compartmented cargo tanks and tank cars as well as certain cargo tanks transporting distillate fuels such as fuel oil. Several commenters pointed out the problems created with multi-compartmented cargo tanks and tank cars and a partial exception is provided. MTB believes that with this exception the basic intent of identifying the materials adequately for emergency response is accomplished and, at the same time, certain burdens are relieved and possible confusion avoided. This is true as well in the other instances where exceptions are provided.

Paragraph (c) contains the July 1, 1981, compliance date for displaying the identification number on portable tanks, cargo tanks, and tank cars. For multi-unit tank car tanks the compliance date is July 1, 1983. On a voluntary basis, the identification number may be displayed on these tanks any time after the July 1, 1980, effective date.

Section 172.338. The proposal did not contain a provision for replacement of lost or missing identification numbers, however, several commenters identified this as a potential problem area needing resolution. This section provides for this occurrence.

Section 172.400. Paragraph (b)(3) is revised to add the words "freight container load" in the sequence of "carload or truckload shipments." This change updates a regulation that was published before freight containers were a common form of transportation and allows MTB to eliminate an exemption (under 49 CFR Part 107, Subpart B). As requested by several commenters, a new paragraph (d) is added to this section and paragraph (a) is revised to insure that packages marked only with a proper shipping name from the Optional Table also will be labeled in accordance with that Table. This reflects the requirement in § 172.202 that the entire basic description and the label requirement must be taken from either the § 172.101 Table or the § 172.102 Optional Table.

Section 172.402. Paragraphs (a)(5) through (a)(9) are added to reflect additional multiple labeling requirements that are established as proposed in Docket HM-126B. Initially, it was considered adequate to have the multiple labeling reflected in Column (4) of the Table, but commenters recommended that, for consistency, it be entered in this section. MTB agrees and has so provided. The compliance date of July 1, 1983, is contained in paragraph (a)(10). Voluntary compliance may begin any time after the July 1, 1980, effective date.

Section 172.407. Paragraph (h) is amended for consistency with § 172.102(h) as it was proposed in Docket HM-171. This rule authorizes IMCO specification labels in all cases except Explosives A and Explosives B. This change clarifies that, except for the specifications for color tolerance which must meet DOT requirements, labels may meet either DOT or IMCO specifications, except that a foreign language text alone is authorized only on import shipments. In addition, to eliminate an exemption issued to the Department of Defense to authorize additional text on labels as required by the country of destination, MTB is adding a provision to paragraph (h) of this section authorizing such an addition.

Section 172.503. Section 172.503 is added to provide a reference from the placarding rules to the identification number marking alternative authorized in § 172.334.

Section 172.504. Table 2 in paragraph (a) is revised to provide for using the DANGEROUS placard for Class C explosives. This has been authorized under exemption DOT-E 7902 which was issued after the applicant pointed out that in case of fire involving Class C explosives, emergency response personnel could be injured when taking routine actions applicable to the FLAMMABLE placard without checking to determine what was involved in the fire. A normal precaution for emergency response personnel when observing the DANGEROUS placard is to try to determine the materials involved by obtaining shipping papers or through other available means. Also, changes are made to Table 2 to eliminate the requirement for affixing the DANGEROUS placard for Class C explosives, the BLASTING AGENTS placard for Blasting agents or the OXIDIZER placard for Nitrocarbonitrate, if the freight container or transport vehicle is transporting Class A or B explosives and is appropriately placarded for Class A or Class B explosives. Further provisions are made for affixing only the FLAMMABLE GAS placard when a motor vehicle is transporting Nonflammable gas and Flammable gas. Although these placarding changes had not been proposed, they provide relief from some of the existing rules, eliminate an outstanding exemption and at the same time provide adequate warning for the materials involved. Also, a new paragraph (d) is added to exclude certain packagings containing only the residue of hazardous materials from consideration in determining the applicability of the placarding requirements.

Section 172.519. Paragraph (d) is revised as suggested by commenters to increase the size of the UN hazard class number display on placards. Where such a display had been permissive, it now becomes mandatory in certain situations when required by this subchapter, such as under § 172.334. Paragraph (f) is added to authorize a variance in the placard specification so the alternative identification number marking requirement can be accomplished.

The alternative, provided by § 172.334, authorizes the display of identification numbers on the appropriate placard for the hazardous material being transported.

Section 173.2. Paragraph (a)(16) is added to include ORM-E in its proper order of hazard.

Section 173.21. The title is revised and the text is amended as proposed in HM-159 to provide better guidance on materials or packaging conditions that are not acceptable in transportation. The term "Forbidden materials" in the context of this section is new and clarification of the application of the term is provided. Section 173.21 applies to any material considered to be forbidden and is not limited to materials falling within established hazard classes. Included in the revision of this section is a prohibition against the offering of packages that evolve a dangerous quantity of flammable gas or vapor released from a material not otherwise subject to the regulations, e.g. the release of flammable blowing agent vapors from a manufactured product in such quantities that an explosive mixture would be created within the transport vehicle. Under this final rule, each refrigeration method, when used as a means of stabilization, must be approved by the Associate Director for Operations and Enforcement. This change is in accord with the approval authority withdrawals from the Bureau of Explosives presently being handled by amendments published under Docket HM-163. Several commenters objected to the fact that proposed paragraph (a)(2) did not contain any statement concerning the time a material would have to be exposed to the 130° F. temperature in order to be considered forbidden from transportation. MTB agrees that it is a weakness in the proposed wording and has altered the wording to reference two test methods. The test methods are: ASTM E-487 "Standard Method of Test for Consent Temperature Stability of Chemical Materials" and the Organic Peroxide Producers' Safety Division (OPPSD) "Self Accelerating Decomposition Test (SADT)." Several commenters expressed concern that this paragraph does not make it clear that approvals issued by the Bureau of Explosives would be continued in effect until an orderly transition to approval by the Associate Director for Operations and Enforcement could be accomplished. MTB acknowledges this objection and has included a clarification statement in the rule referencing § 171.19.

Section 173.28. Section 173.28 is amended to consolidate three paragraphs that contain restrictions pertaining to containers marked NRC or STC. Paragraph (n) is amended to include a reference to ORM-E materials regarding reuse of STC-marked packagings and a new paragraph (p) is added to permit the reuse of NRC or STC specification packagings for one-time shipments of hazardous wastes under certain specified conditions. Note the first condition stipulates that the material must be packaged "in accordance with this Part". For example, a Flammable liquid, n.o.s., must be packaged in accordance with § 173.119. This reuse authorization for hazardous wastes does not permit any deviation from the packaging requirements of Part 173 except as specifically stated. This final rule differs from the Docket HM-145A proposal to reflect input from commenters.

Section 173.29. Section 173.29 is amended, as proposed in Docket HM-145A, to require, with certain exceptions, a packaging that contains the residue of a hazardous material to be offered for transportation in the same manner as required with it previously contained a greater quantity of a hazardous material. However, there are significant exceptions in paragraph (a) concerning marking, placarding, shipping papers, and stowage.

Paragraph (a)(3)(ii) excepts from shipping paper requirements the transportation by contract or private carrier of certain "empty" packagings containing a hazardous materials residue when the purpose of the transportation is to reuse or recondition the packaging. This was not proposed in the Notice but was requested by several commenters. The exception recognizes the fact that private and contract carriers who perform this transportation are familiar with the hazards and packagings involved in the transportation of these materials.

Section 173.51. Section 173.51 is revised as proposed in the notice to Docket HM-159 to make provisions for additional coverage of forbidden materials.

Section 173.118a. Section 173.118a is amended to exclude a combustible liquid from the 110 gallons or less exception when it is a hazardous waste subject to 40 CFR Part 262. Thus, a combustible liquid that is a hazardous waste and is offered for transportation in a packaging having a capacity of 110 gallons or less must be shipped as a waste combustible liquid and all regulations pertaining to the transportation of waste materials apply. This was not proposed, but is added by MTB based on a comment that pointed out the omission. Paragraph (b)(1) is amended to include a reference to hazardous waste manifests. Paragraph (b)(2) is revised by MTB to eliminate a conflict between the identification number marking requirements for portable tanks, cargo tanks and tank cars in the Docket HM-126A proposal and the exception authorization in paragraph (b). Paragraph (b)(5) is revised to include the hazardous substance discharge reporting requirements of § 171.17.

Section 173.151a. As proposed in Docket HM-126B, paragraph (a)(13) is revised to permit continued classification of a hazardous material according to its predominant hazard when it contains an organic peroxide without having to place a plus before each organic peroxide entry. It is possible that when certain stabilizing diluents are added to certain organic peroxides the predominant hazard is that of the diluant rather than the organic peroxide.

Section 173.154. Based on a petition, MTB is adding "Calcium hypochlorite, hydrated" to the Hazardous Materials Table as a proper shipping name with specification packaging referencing § 173.217. In order to eliminate confusion, the reference in § 173.154(a)(20) to hydrated calcium hypochlorite is deleted.

Section 173.179. As proposed in Docket HM-159, § 173.179 is added to prescribe packaging for N-methyl-N'-nitro-N-nitrosoguanidine, which is added to the Table as a Flammable solid.

Section 173.182. The introductory text to paragraphs (a) and (b) are amended as proposed in Docket HM-145B to provide appropriate packaging for the following materials that have been identified by EPA as hazardous substances: Beryllium nitrate, Cupric nitrate, Ferric nitrate, Mercuric nitrate, Nickel nitrate, and Zirconium nitrate.

Section 173.217. Section 173.217 is amended based on a petition for rulemaking requesting that "calcium hypochlorite hydrated" be added to the Table, with a packaging reference to §173.217. MTB is in agreement with the data presented in the petition and has added the entry.

Section 173.352. The heading and paragraph (a) are revised to include Cyanide solutions, n.o.s. classed as a Poison B, UN 1935, which is added to the Table as proposed in Docket HM-126B. MTB believes the packagings authorized by §173.352 for sodium cyanide or potassium cyanide are more appropriate for Cyanide solutions, n.o.s. than the general packagings that would otherwise be authorized for this material under §173.346 for a poisonous liquid, n.o.s.

Section 173.364. Paragraph (a) is revised to provide certain exceptions for Poisonous solid Limited Quantities that are similar to those authorized for Poisonous liquid Limited Quantities. This was an apparent omission from the Docket HM-112 rulemaking and provides relief from certain regulations for shipments of these materials.

Section 173.389. Section 173.389 is amended as proposed in Docket HM-145A to restate the definition of radioactive materials to clarify the fact that the definition applies only for purposes of the Hazardous Materials Regulations. This clarification is necessary since EPA regulations address materials having lower levels of radiation.

Section 173.500. Section 173.500 is amended to clarify the definition of ORM materials. This clarification is essential to implementation of the ORM-E class which is included in new paragraph (b)(5). Note that the ORM-E definition includes hazardous wastes subject to the regulations of the EPA in 40 CFR Part 262, and hazardous substances as defined in §171.8. Except for the amendment of Note 1, which resulted from a comment about the apparent conflict between the hazardous waste requirements and the exception for combustible liquids in certain packagings, the final rule is as was proposed in Dockets HM-145A and HM-145B.

Section 173.505. Paragraph (a) is revised to acknowledge a restriction on the ORM exceptions in that §173.21 applies to any hazardous material offered for transportation. As adopted, the provision differs from the HM-145A proposal in format, but the content remains the same.

Section 173.510. Section 173.510 is amended to exclude the basic packaging requirements from the exceptions specified in § 173.505, and a new paragraph (a)(5) is added requiring that transport vehicles used to transport ORM materials must have discharge openings securely closed. This is a significant change from the HM-145A proposal which would have precluded the use of open-top vehicles. Numerous comments were received describing procedures for effectively using tarps to cover dump trucks and other open-top vehicles. MTB believes that many of the comments have merit and has revised the requirement accordingly. MTB added a note in paragraph (a)(1) to inform shippers that EPA has prescribed packaging for certain PCB's for storage for disposal.

Section 173.1300. A new Subpart O is added to Part 173 to address ORM-E materials, and a new §173.1300 is added to address Hazardous waste, liquid or solid, n.o.s., and Hazardous substance, liquid or solid, n.o.s. These two entries resulted from Dockets HM-145A and HM-145B. No specific packaging requirements are specified in this rule for such materials other than a reference to the basic requirements for ORM's in §173.510. For example, if a hazardous waste is to be offered for transportation by air, the additional requirements of §173.6 in Subpart A apply. It may be necessary in the future to add specific packaging requirements for certain hazardous wastes or hazardous substances if problems relative to the safe transportation of materials in this class (ORM–E) are identified.

Section 174.24. This section is amended to exclude hazardous substances and hazardous wastes from the exceptions specified for ORM materials.

Section 174.25. Paragraph (b)(6) is added to require that the letters "RQ" be entered on the shipping paper either before or after the basic description for hazardous substances. Although a similar requirement was proposed in Docket HM-145B in §172.203, it was not proposed for this section. Several suggestions were made during a joint EPA/MTB hearing on hazardous waste that the identifier be entered after the basic description and that it contain the reportable quantity for the material. MTB agrees in part and has authorized the placement of the identifier "RQ" to be entered either before or after the basic description. Thus, under §172.201(a)(4), if it is placed after the basic description, the shipper may enter the designated reportable quantity. MTB agrees with commenters who suggested provisions be added to require identification on a shipping paper of a hazardous substance residue in a tank car and has added such a provision to the end of paragraph (c).

Section 174.45, 175.45, 176.48 and 177.807. As proposed in Dockets HM-145A and HM-145B, §§174.45, 176.48 and 177.807 are revised to provide for the reporting of the discharges of hazardous wastes and hazardous substances. As a result of a change in the EPA reporting requirements for hazardous waste, reporting can be accommodated through a slight revision to §§171.15 and 171.16. However, §171.17, modified somewhat from the Docket HM-145B proposal, is in the final rule for the reporting of the discharge of a hazardous substance. Based on several comments that suggested hazardous substance discharge reports be limited to reportable quantity discharges (incorporated in the definition of hazardous substance) into or upon navigable waters, MTB deleted the §175.45 proposal because it appears unlikely that such a discharge would occur from an aircraft.

Section 176.11. This section is amended to exclude hazardous substances and hazardous wastes from the exceptions specified for ORM materials.

Section 177.823. In Docket HM-145A MTB had proposed a marking requirement for motor vehicles used for transporting hazardous waste. However, since the requirement appears to duplicate a requirement in § 171.3, the § 177.823 requirement is dropped.

(45 F.R. 34560, May 22, 1980)

ENVIRONMENTAL PROTECTION AGENCY
40 CFR Parts 123, 260 and 262 Hazardous Waste Management System: General Standards for Generators of Hazardous Waste, State Hazardous Waste Program Requirements.

Note: This document contains a correction published on Monday March 8, 1982.

I. AUTHORITY

This proposed rule is issued under the authority of sections 2002, 3001 through 3007, and 3010 of the Solid Waste Disposal Act, as amended by the Resource Conservation and Recovery Act of 1976 (RCRA), as amended, 42 U.S.C. 6912, 6921 through 6927, 6930.

II. BACKGROUND INFORMATION

A. History of the Manifest. On February 26, 1980, EPA established a manifest system to assure that hazardous waste designated for delivery to an offsite treatment, storage or disposal facility actually reached its destination. The central element of that system is the "manifest", a control and transport document that accompanies the waste from its point of generation to its point of destination.

Although EPA considered requiring a uniform manifest form when it developed its regulations, the Agency decided to require that specific information accompany the waste. (45 FR 12728–9, February 26, 1980) At that time, the regulated community was already required by the Department of Transportation to use a shipping paper for the transportation of hazardous materials. DOT's regulations allow industry to use a shipping paper format of their choosing for the required information.

The information requirements for the DOT shipping paper and the manifest were similar; thus, EPA concluded that a shipping paper could be used to satisfy RCRA manifest requirements if additional information required by EPA was included. By not requiring a specific form, EPA's intent was to provide the regulated community with the option of adapting their existing DOT shipping papers to function as manifests or designing their own forms to fulfill specific needs.

B. Manifest Implementation Problems. Since the introduction of the Federal manifest system, there has been a proliferation of manifests as various States decided to develop and print their own forms. At least 21 States require generators to use specific manifest forms, often with varying additional information requirements. The current system has caused two major problems.

First, the lack of uniformity in the manifests required by States has created a burden for both generators and transporters. Currently, a transporter carrying hazardous waste may be required to carry the manifest of each State in which he travels in order to comply with its manifest requirements. Failure to carry a particular State's manifest may delay or prevent shipments from reaching their destination or subject the transporter to a State enforcement action. Under these conditions, a generator may be required to go through the costly and inefficient procedure of filling out several manifest forms with duplicative information in order to ensure that the waste shipment reaches the designated facility.

Second, the lack of uniform requirements prevents generators with plants in more than one state from standardizing their manifesting procedures. This prevents multistate corporations from achieving efficiency in their information collection activities.

C. State and Industry Involvement. In order to examine the feasibility of developing a uniform manifest document to solve these problems, EPA and DOT asked two organizations representing the States and the regulated community to submit comments to EPA. The State group, the Association of State and Territorial Solid Waste Management Officials (ASTSWMO) and the industry group, the Hazardous Materials Advisory Council (HMAC), each developed recommendations concerning the content and use of a uniform manifest and submitted them to EPA and DOT in March of 1981. EPA and DOT reviewed the recommendations, prepared a draft manifest form and met with the ASTSWMO and HMAC committees in July of 1981. During ASTSWMO's September, 1981 national meeting, its members reviewed the Uniform Hazardous Waste Manifest form and subsequently submitted further comments to EPA. A discussion of ASTSWMO's latest recommendations is included in Part IV.

D. EPA and DOT Joint Rulemaking. Since the current problems associated with the manifest involve both DOT and EPA, the Agencies have worked together to devise a regulatory solution. Therefore, several amendments are being proposed in this area.

EPA is proposing to amend 40 CFR Part 262 to require generators to use the Uniform Hazardous Waste Manifest form in order to meet manifest requirements. Amendments to 40 CFR Part 123 would make use of the form a requirement for State interim and final authorization. DOT is proposing to amend its regulations to require that transporters of hazardous waste comply with EPA's proposed amendments and to clarify that any State law or regulation requiring a different or additional manifest is inconsistent with DOT's Hazardous Materials Regulations.

The effect of these amendments is twofold. First, the use of a nationally uniform manifest would be required for all offsite transport of hazardous waste. Secondly, no State could require a transporter to carry additional information on or with the manifest. Thus a transporter could not be held legally liable for failure to carry a State's particular manifest form.

Neither the EPA nor DOT proposed amendments would prohibit a State from requiring additional information concerning each shipment of hazardous waste from the generator or treatment, storage or disposal facility. For example, a State may require that certain disposal related data be present at the facility before the facility accepts the waste. The proposed amendments do not preclude the transporter from voluntarily carrying such information but do prohibit any State from requiring him to do so.

This regulatory approach is authorized by the Hazardous Materials Transportation Act (HMTA), 49 USC 1801 *et. seq.*, and RCRA. HMTA expressly preempts any State or local requirement that is inconsistent with HMTA or the regulations issued thereunder. As stated in its preamble, DOT believes that national uniformity is necessary in this area to avoid a patchwork of differing State requirements inconsistent with the HMTA regulations. In addition, section 3006(b) of RCRA authorizes EPA to disapprove any State application for final authorization to administer and enforce the Federal hazardous waste program if a State program is inconsistent with the Federal or other State programs. As explained in the next section, EPA proposes to exercise the section 3006(b) authority in this area to require national consistency.

E. State Authorization under RCRA. EPA is proposing to change the Part 123 State program requirements to require any State applying for Phase I or Phase II of interim authorization or final authorization to require the use of the Uniform Hazardous Waste Manifest form. The revised manifest requirement in § 123.34 will be applicable to all States applying for final authorization. The revised manifest requirements in § 123.128 will be applicable to any State which applies for any component of Phase II of interim authorization which is announced after the effective date of today's amendments (see 40 CFR 123.122(d)(2)). These changes will assure that a uniform format is used for manifest information, that State programs will be in compliance with DOT regulations and that, in the case of final authorization, the consistency requirements in 3006(b) of RCRA are met.

The DOT preemption regulations operate independently of RCRA and apply to all States, whether or not they have received interim or final authorization. Thus, DOT's broader preemptive authority will extend to all authorized States and will ensure a comprehensive solution to the manifest problem.

F. Implementation Dates. EPA and DOT are proposing common effective dates for the proposed amendments. Since DOT regulations apply independently of RCRA regulation, coordination is necessary for consistent implementation of the uniform national manifest.

After considering the alternative of phasing in the use of the uniform manifest form, EPA and DOT have chosen to require a uniform effective date. This approach was selected in order to avoid confusion concerning the compliance date. Use of the form will be required for all transportation of hazardous waste 180 days after the publication of the final rule. This effective date was chosen in order to provide quick relief from the current situation and also to provide States and industry with time to implement the use of the new form.

EPA has received comments requesting that the effective date allow the States time to update their data management systems with a minimum of disruption to their current operating procedures. The Agency solicits comments from both States and industry on whether the proposed effective date allows sufficient time to implement the form.

As stated in a previous section, DOT requirements would apply to all hazardous wastes transportation in States with and without interim authorization. Accordingly, all State manifests will be subject to Federal preemption at the time DOT's amendments become effective.

III. THE UNIFORM HAZARDOUS WASTE MANIFEST

EPA, with DOT's agreement, proposes the Uniform Hazardous Waste manifest with the following features:

Form Design. EPA incorporated ideas from several existing State and industry manifest forms in order to design a uniform manifest that would satisfy the majority of users. The form itself is the standard paper size of 8½″ by 11″ to allow for easy filing. The form was also designed to facilitate the processing of data from it into automated data processing systems. Therefore, all the information which would be entered by a data machine operator was placed on the right hand side of the form. The left hand margin indicates what information is filled in by the generator, transporter, and the designated facility. The form also contains three inches of horizontal space for listing the proper DOT shipping name.

The manifest document is designed to contain an optional continuation sheet (EPA form 8700–22A) on which the generator may list both additional hazardous wastes and transporters. This eliminates the need for industries, notably laboratories with many different waste streams, to fill out multiple separate manifests for one shipment.

Instructions for filling out the Uniform Hazardous Waste Manifest are included in today's **Federal Register.** While the instructions are required to be followed, EPA is not mandating that they be included with the form. Additional special instructions or emergency information may be printed outside the margin or on the back of the form at the user's discretion.

Information Requirements. In developing the Uniform Hazardous Waste Manifest, EPA has transferred its existing information requirements onto a standard form. EPA has not increased its requirements except for one minor item: the inclusion of the treatment, storage and disposal facility's telephone number. The Agency believes that the inclusion of the phone number will aid the transporter in contacting the facility in the event of an emergency or delay.

Two existing information requirements, the manifest document number and the EPA assigned generator identification number have been combined to form a unique manifest document number. This manifest document number will allow each generator to manifest up to 100,000 shipments before repeating a number. Because every generator has a separate EPA ID number, each document number will be unique and therefore easily identifiable within the hazardous waste management system. Further, the newly defined manifest document number reemphasizes the Agency's belief that the generator should assign the number for each shipment of waste.

The proposed form contains two optional spaces. The first is the hazardous waste number space in which the EPA or State hazardous waste identification number can be filled in. This is not an information requirement. Rather, it is a convenient feature of the form to aid the States and industry in their recordkeeping tasks. The second optional space is entitled "Special Handling Instructions." The generator may use this space to record extra instructions, comments, or an alternate designated facility's address in the event an emergency prevents delivery to the primary designated facility.

A third optional space for the vehicle identification number is being considered. During the drafting of the uniform manifest, EPA and DOT debated whether or not to include the vehicle identification requirement. It was decided not to include that requirement in this proposal for practical reasons. During the transport of one hazardous waste load, several vehicles may be used if shipments are consolidated or the equipment fails. The generator, who is responsible for filling out the manifest, may be unaware of these possible changes and therefore could not be reasonably expected to provide such information on the manifest. For this reason the vehicle identification number is not being proposed as a Federal information requirement. DOT and EPA would appreciate comments on whether or not to include an optional space for the vehicle ID number on the form.

The Agency has also included a manifest discrepancy indication space on the form. EPA regulations require owners or operators of facilities to note on each copy of the manifest any significant discrepancies between the quantity or type of hazardous waste designated on the manifest and the quantity or type of hazardous waste a facility actually receive (40 CFR 264.72, 265.72). When recording significant discrepancies on the Uniform Hazardous Waste Manifest, the TSDF owner or operator will be required to use the discrepancy indication space.

Form printing and number of copies. The Federal government will not print or distribute uniform manifest forms. Generators may print their own forms or obtain them from commercial form design companies and, in this way, retain the flexibility of selecting manifest packages that fit their needs.

For example, a generator may decide to vary the number of copies in a manifest package depending upon where he ships his waste. EPA requires that the manifest consist of two copies for the generator, one for each transporter and one for the designated facility. In addition to the EPA requirement, some States mandate that a manifest copy be sent to them at the initiation and completion of each shipment. In order to easily comply with the State requirement, the generator can obtain a manifest package that includes the number of copies to fulfill the Federal requirement plus additional copies for the State. Or the generator may decide to photocopy legible copies to satify tracking requirements.

IV. REQUEST FOR ADDITIONAL MODIFICATIONS BY STATES

As explained in section II C, the draft uniform manifest was reviewed by ASTSWMO's Hazardous Waste Task Group during its September 1981 national meeting. The Task Group supported the form's information requirements and format but recommended that it be modified to suit the broader needs of the various State manifest systems.

ASTSWMO requested that EPA include these latest recommendations of its Task Group in this preamble. EPA is aware of the concerns of the States and is taking this opportunity to obtain comments on the various State recommendations in the hope of designing a manifest form that fits the needs of all its users.

The recommendations advocate that EPA allow the inclusion of additional State information on the form, modify its design to allow for a larger special handling instruction box, and eliminate the continuation sheet. In addition, the issue of State printing and distribution of the uniform manifest form was raised.

Additional State Information. The States have the authority to request information from treatment, storage or disposal facilities and generators. Some States, through ASTSWMO, have indicated that they desire information concerning each hazardous waste shipment beyond that required on the manifest. The amendments proposed by EPA and DOT would prohibit States from requiring additional information to be carried with the waste. However, it is within a State's authority to require that such information be present at the designated facility as a condition of acceptance of the waste. Some States have indicated that they will require such a condition in order to obtain this information.

Therefore, rather than require the generator to fill out both a manifest and a State form, the States are proposing that the uniform manifest be redesigned to contain optional spaces for the inclusion of State information. ASTSWMO's recommendations include the following:

• Add spaces for State identification numbers under the designated EPA identification number spaces for generators, transporter and treatment, storage and disposal facilities.

• Add a space within the Descrepancy Indication box for the recording of codes which indicate the treatment, storage or disposal processes utilized in handling the hazardous waste.

• Modify the design of the form to allow both the weight *and* volume of the waste to be easily recorded. Current Federal regulations only require the weight or volume to be listed.

The Agency recognizes the need on the part of the States to obtain certain information for the efficient management of their hazardous waste programs. The adoption of ASTSWMO's suggestion could benefit industry by providing the opportunity of meeting all Federal and State requirements on one form. However, EPA believes that the addition of State information spaces would complicate the form making it more difficult for the generator to comply with Federal manifest requirements. Generators would once again be subject to dealing with varying information requirements since individual States would require different information be filled in as a condition of acceptance of the waste at the facility.

The purpose of the Uniform Hazardous Waste Manifest form is to ease the regulatory burden on generators and transporters by providing a uniform format for information necessary for the transportation of hazardous waste. In designing the form, the Agency has striven for uniformity but is well aware of the advantages of a more flexible form which contains optional spaces. In order to design a form that best suits the needs of its users, the Agency invites comments on the preferability of a strictly uniform manifest form (i.e., no optional spaces included) or form that could contain additional State information.

Special Handling Instructions Box. In order to provide room for State information, ASTSWMO suggested reducing the number of waste listing spaces on the manifest from six to four. This additional space could then be used to maximize the area of the Special Handling Instructions box. The Agency solicits comments on whether four spaces are sufficient to list the majority of hazardous waste shipments on one manifest and if the need exists to maximize the area on the Special Handling Instructions box.

Elimination of the Continuation Sheet. The States also indicated that the addition of a continuation sheet to the manifest will significantly complicate their data management systems. As noted above, the primary reason for the inclusion of a continuation sheet was to eliminate the need for generators, notably laboratories with many different waste streams, to fill out multiple manifests for one shipment. In its latest series of recommendations, ASTSWMO has advocated that the Department of Transportation create classification codes for various groups of laboratory waste streams, thereby eliminating the need for a continuation sheet. Creation of such codes would require rulemaking action on the part of DOT. EPA believes that the continuation sheet is a necessary component of the uniform manifest form in the absence of such rulemaking. Comments concerning classification codes for laboratory waste should be adressed to DOT.

State Printing and Distribution of Manifest Forms. A final ASTSWMO recommendation advocates the expansion of the top margin of the uniform manifest form to provide room for States to print their names and State assigned document number. Further, some States have indicated a preference to print and distribute the manifest forms because it allows them to print their own emergency instructions. In addition, ASTSWMO assets that State document numbers are necessary for the effective use of any automated State manifest tracking system.

EPA and DOT are concerned that the implementation of this recommendation could result in situations in which the generator may be required to fill out multiple manifests for one shipment. A generator shipping his waste interstate could be required by his State to use its printed manifest form and also be required to use the destination State's form as a condition of acceptance of the waste by the TSD facility. This type of interference with the transport of hazardous waste is what the uniform manifest form was designated to prevent. Comments on the desirability of State printing and distribution of forms are requested.

V. CLASSIFICATION

EPA has determined that today's proposed rule will not result in: an annual effect on the economy of $100 million or more; a major increase in costs or prices for consumers, individual industries, Federal, State, or local government agencies, or geographic regions; or significant adverse effects on competition, employment, investment, productivity, innovation, or the ability of United States based enterprises to compete in domestic or export markets. Today's action is expected to reduce the current burden on the regulated community. Therefore, today's proposed rule is not subject to the major rule provision of the Executive Order 12291 and a regulatory impact analysis is not required.

The Uniform Hazardous Waste Manifest is subject to the OMB clearance requirements of the Paperwork Reduction Act of 1980. The Agency has already received clearance for the information requirements of the manifest system. If, on the basis of comment, the Agency revises the form to include additional information requirements that would increase the information burden of the regulatory community, the form will be resubmitted to OMB for clearance.

Furthermore, this proposed rule, if adopted, would not have a significant impact on a substantial number of small entities, thereby triggering the requirements of the Regulatory Flexibility Act. The required use of a Uniform Hazardous Waste Manifest will reduce the burden faced by regulated hazardous waste generators and transporters, including small businesses as defined by that Act. Since the cost to industry due to the manifest form is realized largely in terms of labor costs and labor is a large percentage of small business's cost, the uniform manifest should provide small businesses with significant relief. Accordingly, I certify pursuant to 5 U.S.C. § 601 *et seq.* that the proposal, if promulgated, will not have a significant economic impact on small entities.

This proposal regulation was submitted to the Office of Management and Budget for review as required by Executive Order 12291. Any written comments from OMB to EPA and any EPA response to those comments are available for public inspection at the Office of Solid Waste Docket, Room 2636, U.S. EPA, M St. SW, Washington, D.C. 20460.

(47 F.R. 9336, March 4, 1982)

[49 CFR Parts 171, 172, 173, 174, 175, 176, 177]
[Docket No. HM-145B; Notice No. 79-2]

TRANSPORTATION OF HAZARDOUS SUBSTANCES

When Congress enacted the Federal Water Pollution Control Act (33 U.S.C. 1251 *et seq.*) (FWPCA) it declared in subsection 311(b)(1) of the FWPCA that it is the policy of the United States that there should be no discharges of oil or hazardous substances into or upon the navigable waters of the United States, adjoining shorelines or into or upon the waters of the contiguous zone. Section 311(b)(2) requires the Administrator of the EPA to designate as hazardous substances such elements and compounds which, when so discharged, "present an imminent and substantial danger to the public health or welfare including, but not limited to, fish, shellfish, wildlife, shorelines and beaches." Section 311(b5) of the FWPCA requires that any person in charge of a vessel or an onshore or offshore facility is to report discharges of certain quantities of these hazardous substances. Failure of a person in charge to make such report as soon as he has knowledge of such discharge subjects him to criminal penalties. Section 311 of the FWPCA also has numerous provisions concerning civil liability of owners and operators for the discharge, cleanup and removal of oil and hazardous substances.

Section 311(1) of the FWPCA authorizes the President to delegate administration of Section 311 of the FWPCA to appropriate heads of Federal departments and agencies. It also directs each such department and agency to use, whenever appropriate, the services of other departments and agencies, so as to avoid duplication of effort. By Executive Order 11735 of August 7, 1973, the President assigned some duties under Section 311 of the FWPCA to EPA, some others to DOT, and still others to both.

The Clean Water Act of 1977 (Pub. L. 95-217) enacted December 27, 1977, amended Section 311 of the FWPCA in many details. Among other things, it increased the maximum liability for removal and mitigation costs. The Clean Water Act did not change reporting requirements or penalties for failure to report other than to extend the applicability to the Outer Continental Shelf in the case of persons otherwise subject to the jurisdiction of the United States.

On March 13, 1978, EPA published in the FEDERAL REGISTER (43 FR 10474) a set of regulations, including 40 CFR Part 116, which designated 271 substances as hazardous in accordance with Section 311 of the FWPCA. On the same day, EPA proposed an additional 28 compounds to be so designated. It is expected that before the rules proposed herein become effective, all of the 28 proposed compounds will have been added to the list of 271.

The March 13, 1978, FEDERAL REGISTER also contained 40 CFR Parts 117-119, in which EPA classified hazardous substances in various respects, i.e., removability, reportable quantities and potential penalties. These regulations have never become effective because of delayed effective dates, court orders, and finally the passage of Pub. L. 95-567 in November of 1978. That law voided some of the provisions on which the proposed regulations were based. Accordingly, 40 CFR Parts 117-119 are being withdrawn. EPA, however, proposes to retain regulations to be renumbered 40 CFR Part 117, which spell out the minimum quantity of designated hazardous substance which, if discharged, must be reported.

IMPACT OF EPA'S RULES ON THE TRANSPORTATION INDUSTRY

According to Section 311 of the FWPCA, it is the person in charge of a vessel or an onshore or offshore facility who must report immediately the discharges of designated hazardous substances in quantities which may be harmful. Section 311(a)(10) of the FWPCA defines an onshore facility to include, but not be limited to, motor vehicles and rolling stock.

In practice, it is expected that, for example, the manufacturer of a chemical substance will have no difficulty identifying the substance or that the substance has been designated by EPA to be hazardous. But the master of a vessel carrying a hazardous substance, even though the loading plan may show what is being carried, may not be familiar with the EPA designations. Also, as has been pointed out by representatives of the railroad and trucking industries, a common carrier may be required to carry a mixture identified under a commercial trademark or a chemical product not sufficiently descriptive to permit even an expert to determine the chemical identity of the cargo. The transportation industry has expressed concern that a person in charge of a truck or a train may be subject to criminal penalties for failing to report a discharge in spite of the absence of positive information identifying the cargo as falling under the EPA regulations.

DOT-EPA COORDINATION

Coordination of DOT and EPA activities to minimize or abate environmental pollution by the transportation industry is not new. The FWPCA itself assigns some duties to the Administrator of EPA, others to the Secretary of the department in which the Coast Guard is operating, i.e. DOT. Executive Order 11735 similarly divides responsibilities between these two agencies. However, to simplify reporting of a hazardous substance discharge, the Coast Guard's National Response Center has been designated as the single reporting point to which discharges of hazardous substances are to be reported. This response center then will alert the proper office, Coast Guard or EPA, to respond to the particular events.

ADVANCE NOTICE OF PROPOSED RULEMAKING—HM-145

On December 9, 1978, DOT published HM-145 as an advance notice of proposed rulemaking (41 FR 53824) covering the subject of this proposal. All comments received concerning this advance notice have been given full consideration in relation to the rules proposed herein. A number of commenters expressed concern over jurisdictional matters such as EPA's apparent infringement on DOT's authority and the question of overlapping jurisdiction. Also, several commenters indicated that DOT should not attempt to develop criteria for such materials and that the regulations should not be applicable to materials in quantities of less than 110 gallons per package.

Under the rules proposed herein, the transportation of substances designated as hazardous pursuant to the FWPCA would not be subject to a completely new set of regulations but an existing body of transportation regulations would be extended using DOT's authority under the Hazardous Materials Transportation Act (HMTA) (49 U.S.C. 1801 *et seq*.). In the handling of discharge incidents, EPA or DOT is required by Section 311 of the FWPCA to respond. This is still true whether or not any transportation regulations are adopted under this rulemaking. Concerning alleged jurisdictional infringement, it should be understood that EPA and DOT are each attempting to accomplish their statutory responsibilities, and the Materials Transportation Bureau (MTB) believes that it is not a matter of infringement but rather the simple question of whether DOT or EPA should administer the regulations pertaining to the transportation of materials that are hazardous substances and which also may be considered to pose an unreasonable risk to health and safety or property.

Concerning the development of criteria, the MTB agrees with the commenters who suggested that DOT should not attempt to develop the criteria for materials that are subject to the FWPCA unless they fall within the realm of the existing defining criteria for materials presently designated as hazardous materials. The MTP believes the EPA has both the expertise and the technical resources necessary to deal with the determination and designation of those materials which should be considered for inclusion in the reporting requirement mandated by the FWPCA. Despite EPA's role in this proposal, it should be understood that the authority for the proposed requirements is the HMTA.

Concerning the suggested 110-gallon breakpoint, it has been determined by EPA that certain materials pose a significant risk at much smaller quantities, depending on the toxicity of the material under consideration. Therefore, the suggestion has not been adopted in this proposal.

RELATIONSHIP TO DOCKET HM-145A—HAZARDOUS WASTES

It is expected that many interested persons may be concerned about the relationship between this proposed rulemaking and the proposals made for hazardous wastes under Docket HM-145A on May 25, 1978 (43 FR 22626). In order to maintain a cooperative relationship with EPA and, recognizing the difficulties of timing the implementation of two major actions within that agency, the MTB's proposals for hazardous wastes and hazardous substances are being dealt with separately. However, it should be recognized that many of the proposed changes are common to both proposals and that, barring errors and omissions, there should be no overlaps or conflicts if both actions are completed as proposed. This is the case even if the implementation dates of final regulations are different.

PROPOSED APPLICABILITY OF REGULATIONS

It is proposed to make the regulations applicable to all hazardous substances, regardless of quantity, when they are listed by the EPA as subject to the FWPCA. Thus, any hazardous substance would be readily identifiable from shipping paper entries regardless of amount actually carried on a transport vehicle. This proposal also would require discharge reporting to be *based* not on actual amounts released, but on amounts loaded at one loading point, so that transporters would not have to keep a running total in multiple pickup and delivery operations, and would not have to estimate spill size to decide whether an immediate report is required. It is the view of the MTB that determinations of what constitutes a reportable discharge should be kept as simple as possible, consistent with the purposes of the HMTA, in order to encourage compliance and to facilitate effective enforcement. Therefore, the MTB is proposing that no immediate report of a release is required unless a reportable quantity is loaded at one location, e.g., a shipper's facility or freight terminal, and that fact noted on the shipping paper by quantities associated with the descriptions and by the notification statement. The accumulation of packages of a particular substance loaded in less than reportable quantities at more than one loading site would not be a reportable quantity under this proposal.

Also, in making the reportable quantity determination at one loading site, different substances would not be added together to make that determination. In other words, as long as a reportable quantity of a single hazardous substance is not loaded at one location, a carrier would not be required to report a release under proposed § 171.17. However, in determining a reportable quantity, concentrations of a material are not required to be shown on a shipping paper and may not be considered in making that determination, i.e., ten pounds of a 50% mixture of a hazardous substance would be treated the same as ten pounds of the pure hazardous substance. Similarly, once the presence of a reportable quantity in transportation is signalled on the shipping paper(s), a discharge of *any* hazardous substance from that transport vehicle, aircraft, vessel or facility to the environment requires a report to the National Response Center on all hazardous substances in the transport vehicle aircraft, vessel or in the area of the discharge if the discharge occurs in a facility and the source of the discharge cannot readily and safely be determined for the report.

This would preclude the necessity of a transport worker (e.g., truck driver or warehouseman) making a determination of the actual quantity discharged in deciding when a discharge report is required.

Although serious consideration was given to requiring different materials to be added on a proportional basis in determining a reportable quantity, the related computation, paperwork and difficulty of enforcement does not appear to be justified, especially since all the packages are unlikely to be simultaneously involved in a release. Instead, the MTB is proposing a simpler system which is equivalent in overall stringency. Comment is specifically solicited on whether the adding of different substances to determine unity is desirable in calculating a reportable quantity.

It is proposed to make the regulations in this proposed rulemaking applicable to transportation in both intrastate and interstate commerce consistent with the applicability of the FWPCA and DOT's authority under the HMTA.

REVIEW BY SECTIONS

Section 171.1. It is proposed to revise § 171.1 to specify the applicability of the proposed regulations to the motor vehicle transportation of hazardous substances by intrastate carrier. No such distinction is considered necessary relative to transportation by rail car, aircraft or vessels since such operations are subject to DOT's regulations without regard to the intrastate or interstate nature of individual shipments.

Section 171.8. Section 171.8 contains proposed definitions of "EPA," "hazardous substance," and "reportable quantity."

Section 171.17. Section 171.17 proposes a reporting requirement in addition to those presently specified in § 171.15 and 171.16, although compliance with proposed § 171.17 would satisfy the requirements of § 171.15(a) and (b). Notification to the National Response Center, U.S. Coast Guard would be required when any discharge of a hazardous substance occurs during transportation, if a reportable quantity is present as indicated by a shipping paper having the statement required by § 172.203(j)(2).

Section 172.101. Present §§ 172.100 and 172.101 would be combined into a revised and amended § 172.101 in order that the Hazardous Materials Table and the language introducing the Table would be contained in one section of the regulations. The introductory sentence to Paragraph (b) would be modified to include the letter "E" as an identifier in column 1 for material in column 2 that has been identiied by the EPA as a hazardous substance and to subject those materials to transportation requirements by all modes. The letter "E" would be required to precede the proper shipping name on each shipping paper, package and name of the material in each portable tank, cargo tank or tank car for identification purposes during the transportation of any hazardous substance.

Paragraph (b)(1) would be revised to provide for changing the hazard class of certain materials from that specified in column 3 of the Hazardous materials Table. When a shipper determines that, because of a manufacturing process or other reason, a material is changed so that its basic hazard(s) is changed, a material would be reclassed and offerd for transportation under the provisions of the new class. It is important to note that although a hazardous substance could be reclassed according to its hazard(s), if it did not meet the definition of any other hazard it would be reclassed as an ORM-E. It it were reclassed to any hazard class except ORM-E, it would be shipped under its proper shipping name within that class with the name of the hazardous substance entered in parentheses after the proper shipping name. However, it if were reclassed as an ORM-E, it would retain its original proper shipping name since there is no ORM-E, n.o.s. or hazardous substance, n.o.s. proposed.

Paragraphs (b)(2) and (3) would be modified to exclude from single mode applicability those materials which are identified by the letter "E" in column 1 and to subject those materials to transportation requirements regardless of the mode of transportation involved.

Paragraph (c)(8) would be added to explain the addition of the reportable quantity as an italicized entry in parentheses following the proper shipping name of hazardous substances in § 172.101. This would make the reportable quantity for each hazardous substance readily available for reference and for determination of when the notification statement on a shipping paper is required.

Hazardous Materials Table: The Hazardous Materials Table would be amended to include hazardous substances designated by the EPA under 40 CFR Part 117 and not presently subject to the regulations. Also, the Table would be amended to identify presently regulated hazardous materials that have been designated by the EPA as hazardous substances. These latter materials would be in two major categories. One would be materials that are now listed by name in § 172.101, and the other would be materials not now identified by name but which are now regulated under the n.o.s. listings in 172.101. Of the 165 materials being added by name, 45 are presently regulated under a n.o.s. listing, such as Beryllium fluoride which is regulated under the listing of Beryllium compound, n.o.s. The total number of materials that would be identified as hazardous substances under this proposal would be 359. These are listed and discussed in this preamble individually or in groups to identify the criteria for designating each as a hazardous material.

Although there appears to be a discrepancy between the number of newly identified materials in this proposal and the number of materials in the EPA list of hazardous substances, the materials in this proposal are those covered in the EPA list. The difference in the number of materials results from the necessity to identify in § 172.101 the different forms, mixtures or solutions of a material for proper regulation. For example, "Aldrin" appears once in the EPA list and six times in § 172.101.

Hazard Class Determinations. The additions to § 172.101 to accommodate the hazardous substances would consist of 92 materials classed as ORM-E, 45 materials that are now regulated under n.o.s. listings which would be identified by name, and 28 that would be classed as ORM-A or ORM-B. Provisions are also made in § 172.101 to identify 194 materials currently being regulated by name that EPA has designated as hazardous substances.

Other Regulated Materials—ORM-A. Nineteen hazardous substances would be identified as ORM-A. This is based on the chemical, physical and other comparable properties of the compounds. The properties of these compounds are such that each compound can cause extreme annoyance or discomfort to passengers and crew in the event of leakage during transportation. These ORM-A materials are listed below:

Ammonium carbamate	Toxaphene
Ammonium carbonate	Trichlorfon
Ammonium oxalate	Trichlorophenol
Captan	2,4,5-Trichlorophenoxyacetic acid
Chlorpyrifos	2,4,5-Trichlorophenoxyacetic acid amines
2,4-Dichlorophenoxyacetic acid esters	2,4,5-Trichlorophenoxyacetic acid esters
Maleic acid	2,4,5-Trichlorophenoxyacetic acid salts
Maleic anhydride	2,4,5-Trichlorophenoxypropionic acid
Paraformaldehyde	2,4,5-Trichlorophenoxypropionic acid esters
TDE	

Other Regulated Materials—ORM-B. Nine hazardous substances would be identified as ORM-B. This is based on the chemical, physical and other comparable properties of the compounds. The properties of these compounds are such that each compound is capable of causing significant damage to a transport vehicle, aircraft or vessel from leakage during transportation. These ORM-B materials are listed below:

Aluminum sulfate	Ammonium fluoborate
Ammonium bisulfite, solid	Ammonium silicofluoride

Calcium hydroxide
Ferrous chloride, solid
Lead fluoborate

Lead fluoride
Zirconium sulfate

Other Regulated Materials—ORM-E. Ninety-two materials would be classed as ORM-E. This is based on the EPA designation of certain materials as hazardous substances on March 13, 1978 (42 FR 10494), and the fact that according to our evaluation they do not meet the defining criteria of any other hazard class. These ORM-E materials are listed below:

Adipic acid
Ammonium acetate
Ammonium benzoate
Ammonium bicarbonate
Ammonium chloride
Ammonium chromate
Ammonium citrate
Ammonium sulfamate
Ammonium sulfite
Ammonium tartrate
Ammonium thiocyanate
Ammonium thiosulfate
Antimony trioxide
Benzoic acid
n-Butyl phthalate
Cadmium acetate
Cadmium bromide
Cadmium chloride
Calcium chromate
Calcium dodecylbenzenesulfonate
Chromic acetate
Chromic sulfate
Chromous chloride
Cobaltous bromide
Cobaltous formate
Cobaltous sulfamate
Cupric acetate
Cupric oxalate
Cupric sulfate
Cupric culfate, ammoniated
Curpic tartrate
Dicamba
Dichlorbenil
Dichlone
Dinitrotoluene
Diquat
Diuron
Dodecylbenzenesulfonic acid
Ethylenediaminetetraacetic acid
Ferric ammonium citrate
Ferric ammonium oxalate
Ferric fluoride
Ferric sulfate
Ferrous ammonium sulfate
Ferrous sulfate
Fumaric aid

Isopropanolamine didecylbenzenesulfonate
Kelthane
Kepone
Lead acetate
Lead iodide
Lead stearate
Lead sulfide
Lead thiocyanate
Lithium chromate
Mercaptodimethur
Methoxychlor
Naled
Napthenic acid
Nickel ammonium sulfate
Nickel chloride
Nickel hydroxide
Nickel sulfate
Nitrophenol
Nitrotoluene
Pentachlorophenol
Polychlorinated biphenyls
Potassium chromate
Propargite
Pyrethrins
Quinoline
Resorcinol
Sodium chromate
Sodium dodecylbenzenesulfonate
Sodium phosphate, dibasic
Sodium phosphate, tribasic
Strontium chromate
Triethanolamine dodecylbenzensulfonate
Vanadyl sulfate
Xylenol
Zinc acetate
Zinc ammonium chloride
Zinc borate
Zinc bromide
Zinc carbonate
Zinc chloride, solid
Zinc fluoride
Zinc formate
Zinc phenolsulfonate
Zinc silicofluoride
Zinc sulfate
Zirconium potassium fluroide

The 45 materials from various n.o.s. classes would be identified within the flammable liquid, combustible liquid, flammable solid, oxidizer, poison B, radioactive material, and corrosive material hazard classes as follows:

Flammable Liquids/Combustible Liquids. Three hazardous substances would be identified as flammable liquids and one as a combustible liquid. Closed cup flash points, except for epichlorohydrin, were obtained from the literature. Based on the open cup flash point of 105° F. reported for epichlorohydrin, the MTB estimates the closed cup flash point to be 95° F. These liquids and their respective closed cup flash point are listed below:

Material	Flash Point (°F.) CC.
Benzonitrile	167
Dichloropropene	95
Epichlorohydrin (estimated)	95
Ethylenediamine	93

Flammable Solids. Three hazardous substances would be identified as flammable solids. This is based on the chemical and physical properties of the compounds, and the fact that similar compounds (e.g., sodium phosphide) are classed as flammable solids. Sodium hydrosulfide, solid is listed as a "substance liable to spontaneous combustion" in the United Nations document entitled Transport of Dangerous Goods. These flammable solids are listed below:

Sodium hydrosulfide, solid
Zinc hydrosulfite
Zinc Phosphide

Oxidizers. Seven hazardous substances would be identified as oxidizers. Each compound is a nitrate and has been subject to the regulations under the proper shipping name "Nitrate, n.o.s." In addition, five of the nitrates are listed by name as oxidizing substances in the United Nations document entitled *Transport of Dangerous goods.* These oxidizers are listed below:

Beryllium nitrate
Cupric nitrate
Ferric nitrate
Mercuric nitrate

Nickel nitrate
Zinc nitrate
Zirconium nitrate

Poison B. Seventeen hazardous substances would be identified as Poison B materials. Data on oral toxicity using rats (orl-rat LD50: mg/kg) was obtained from the National Institute for Occupational Safety and Health (NIOSH) *Registry of Toxic Effects of Chemical Substances* (1977 Edition) for 12 compounds. Data on toxicity by skin absorption using rabbits (skn-rbt LD50: mg/kg) was obtained from the NIOSH Registry for dichlorovos. Three compounds are currently subject to the Poison B regulations as "n.o.s." entries. Arsenic trisulfide is covered by the proper shipping name "Arsenical compound, n.o.s. solid". Beryllium chloride and beryllium fluoride are covered by the proper shipping name "Beryllium compound, n.o.s." No data is available for selenium oxide. However, based on chemical and physical properties and the toxicity of other selenium compounds (e.g., sodium selenite), it is the MTB's opinion that selenium oxide meets the criteria for this hazard class. These poison B materials are listed below:

Material	Toxicity (orirat LD 50: mg/kg)
Arsenic trisulfide	—
Beryllium chloride	—
Beryllium fluoride	—
Carbofuran	5
Coumaphos	16
Dichlorovos (skn-rbt)	107
Disulfoton	10
Endosulfon	18
Endrin	3
Ethion	13
Guthion	16
Heptachlor	40
Mevinphos	4
Mexacarbate	14
Selenium oxide	—
Sodium selenite	7
Vanadium pentoxide	10

Corrosive Material. Ten hazardous substances would be identified as corrosive materials. This is based on the chemical and physical properties of the compounds and the fact that several similar type compounds are classed as corrosive materials. These corrosive materials are listed below:

Ammonium bisulfite solution

Ammonium hydrogen fluoride, solid

Antimony tribromide, solid

Antimony tribromide solution

Antimony trifluoride, solid

Antimony trifluoride solution

Cresol

2,2-Dichloropropionic acid

Ferrous chloride solution

Hexachlorosyclopentadiene

Sodium bifluoride, solid

Sodium bifluoride solution

Sodium hydrosulfide solution

Radioactive Material. One hazardous substance, uranyl acetate, would be identified as a radioactive material. This compound has been subject to the regulations under one of several proper shipping names for n.o.s. entries for radioactive materials (e.g., Radioactive material, n.o.s.).

Section 172.200. Section 172.200(b) would be amended to remove the ORM exceptions as to the shipping paper requirements when a material being offered or transported is a hazardous substance.

Section 172.202. Section 172.202 would be amended to provide for identifying the quantity of a hazardous substance in pounds or kilograms on a shipping paper so a reportable quantity can be calculated.

Section 172.203. Section 172.203 would include a new paragraph (j) requiring the inclusion of the letter "E" within parentheses, brackets or a circle on the shipping paper immediately before the proper shipping name for each material identified in §172.101 as a hazardous substance. Also a notification statement would be required to be placed on the shipping paper anytime a transport vehicle, aircraft, vessel, or freight container is loaded with a reportable quantity of any hazardous substance at any one loading site, transported by vessel.

Section 172.316. Section 172.316 would be amended to include ORM-E materials in the package marking requirements for ORMs.

Section 172.324 would be added and §§172.326, 172.328 and 172.330 would be revised to provide for including the letter "E" in a circle immediately before the proper shipping name on each package containing a material identified in §172.101 as a hazardous substance.

Section 173.2. Section 173.2 would be amended to add ORM-E to the "order of hazards" listing for classification purposes.

Section 173.118a. Section 173.118a would be amended to exclude from the 110 gallons or less exception a combustible liquid when it is a hazardous substance identified in §172.101.

Paragraph (b)(5) would be amended to reference the special reporting requirements in proposed §171.17.

Section 173.500. Section 173.500 would be amended to clarify the definition of ORM materials. This clarification is essential to the implementation of the ORM-E class which is included in new paragraph (a)(5). Note that the ORM-E definition would include hazardous substances subject to 40 CFR Part 117 but is stated so that other materials may be included within this class at a future time.

Section 173.505. Section 173.505(a) would be revised for clarification of the exceptions for ORM-A, B, and C materials packaged in limited quantities since a basic function of the regulations is to exclude certain materials from specific packaging requirements when packages containing limited quantities of these ORMs are offered for transportation. This proposal more coherently restates this function.

Section 173.510. Section 173.510 would be amended to more clearly set forth the basic packaging requirements applicable to ORMs and a new subsection (b)(5) would be added requiring that transport vehicles used to transport ORM materials be free from leaks and have all openings securely closed. The proposed prohibition against use of open-top freight containers and transport vehicles for bulk shipments is significant. This prohibition would apply to the use of open or tarp-covered dump trucks for the transport of hazardous substances. Commenters opposed to such a prohibition are invited to submit suggestions concerning appropriate controls of tarp-covered vehicles that would assure compliance with §173.24.

Section 173.1300. A new Subpart O would be added to Part 173 to address ORM-E materials generally and a new §173.1300 would be added to address hazardous substances. No specific packaging requirements are being proposed in this rulemaking for such materials other than a reference to the basic requirements for ORMs in §173.510. For example, if a hazardous substance is to be offered for transportation by aircraft, the requirement of §173.6 in Subpart A would apply since Subpart A is referenced for such shipments in §173.510.

Modal Parts: Sections 174.24 and 176.11 would be amended to exclude hazardous substances from the exceptions specified for materials classed as ORMs.

Sections 174.45, 175.45, 176.48 and 177.807 would be amended to reference the proposed new §171.17 concerning the reporting of discharges involving hazardous substances.

(44 F.R. 10676, February 22, 1979)

49 CFR Part 172
[Docket No. HM-145C Amdt. No. 172–66]
Listing of Hazardous Materials

BACKGROUND

Section 306(a) of the Comprehensive Environmental Response, Compensation, and Liability Act (CERCLA) requires that, within 90 days after the date of enactment of the Act, each substance that is listed or designated as a hazardous substance under the Act shall be listed as a hazardous material under the Hazardous Materials Transportation Act (HMTA). Section 306(b) provides that common and contract carriers shall not be liable under CERCLA for releases of hazardous substances prior to the effective date of the listing of that substance as a hazardous material unless it is demonstrated that the carrier has actual knowledge of the identity or nature of the substance.

The purpose of these provisions is twofold: First, to assure coordination of the implementation of CERCLA (as it relates to transportation) with the administration of the HMTA so as to avoid regulatory inconsistencies and overlaps; and, second, to provide reasonable notice, through the HMTA regulatory system, to transporters of hazardous substances that they are subject to the liability and other provisions of CERCLA.

LISTINGS

The purpose of this final rule is to fulfill the requirements of Section 306(a) of CERCLA by listing as hazardous materials those substances that EPA has determined to be "hazardous substances," as defined in Section 101(14). That definition incorporates six lists of substances, five of which have been developed under other statutory authorities:

1. Section 311(b)(2)(A) of the Federal Water Pollution Control Act (FWPCA);
2. Section 3001 of the Solid Waste Disposal Act;
3. Section 307(a) of the FWPCA;
4. Section 112 of the Clean Air Act; and
5. Section 7 of the Toxic Substances Control Act (TSCA).

The sixth list is comprised of substances for which authority to designate is granted to EPA in Section 102 of CERCLA.

The listing in this rule does not include:

1. Substances listed under Section 311(b)(2)(A) of the FWPCA. These substances were incorporated into the Hazardous Materials Table on May 22, 1980, (45 FR 34560) and are currently covered by the Hazardous Materials Regulations. It is therefore unnecessary to repeat them in this listing.

2. Substances under Section 7 of the TSCA. No substances have yet been designated under this authority.

3. Substances designated under Section 102 of CERCLA. No substances have yet been designated under this authority.

The listing in this rule includes substances designated under Section 307(a) of the FWPCA, Section 3001 of the Solid Waste Disposal Act, and Section 112 of the Clean Air Act. It should be noted that many of these substances either are already listed in the Hazardous Materials Table or meet an existing hazard class definition and are currently subject to the Hazardous Materials Regulations. Today's listing indicates by asterisk (*) those materials that were listed as hazardous substances in the Department's May 22, 1980, final rule. With respect to the other substances listed, reference should be made to the existing regulations to determine their applicability to those substances.

Effect of Listings. This rule meets the requirement of Section 306(a) of CERCLA since hazardous substances, as defined in Section 101(14) of CERCLA, are listed as hazardous materials under the Hazardous Materials Transportation Act. It does not, however, extend the applicability of the Department's Hazardous Materials Regulations to any materials that are not already covered by those regulations. For example, if in the past shipping papers were not required by 49 CFR 72.200 for a material included in this new listing, they will not now be required as a result of the listing. Specifically, the Department is not at this time incorporating these materials into the list of "hazardous substances," as defined in the Hazardous Materials Regulations (49 CFR 171.8), nor is the Department assigning reportable quantities (RQs) for purposes of the Hazardous Materials Regulations.

Section 102 of CERCLA provides that pending establishment by EPA of a different quantity, the RQ for all hazardous substances shall be one pound. The Hazardous Materials Regulations provide that shipping papers must be issued for all shipments of hazardous substances (as defined in § 171.8) that equal or exceed their reportable quantity. Therefore, the effect of listing the materials covered by this action as "hazardous substances," as defined by the Department, and assigning an RQ of one pound would be to vastly increase the number of shipments requiring shipping papers under the Hazardous Materials Regulations. For example, every shipment of galvanized steel containing more than one pound of zinc would require a hazardous materials shipping paper. This result would not promote the purposes of CERCLA, and it would be contrary to the Department's goal of minimizing paperwork requirements.

At such time as EPA exercises its authority under Section 102 to establish RQs for particular substances, the Department will determine the appropriateness of listing those substances as "hazardous substances" and assigning those RQs to them.

It should be noted that, as discussed above, some of the materials listed in this rule are already designated as hazardous substances in the Hazardous Materials Table, and RQs for these materials have already been assigned.

Section 102 of CERCLA provides that all materials in today's listing that have not been assigned an RQ shall have an RQ of one pound pending establishment of a different quantity by EPA. Section 103 (a) and (b) of CERCLA requires that all releases of an RQ of a hazardous substance into the environment be reported to the National Response Center. While the Department does not currently contemplate changing its general incident reporting requirements (49 CFR 171.15 and 171.16). EPA is currently developing a notice explaining how it will implement the requirements of Section 103 (a) and (b) of CERCLA. Shippers and transporters of materials listed in this rule should contact EPA (Mr. H. D. Van Cleave, Acting Director, Emergency Response Division (WH-548), Office of Hazardous Emergency Response, U.S. EPA, 401 M Street, SW., Washington, D.C. 20460, (202) 245-3045) for additional information regarding these reporting requirements.

With respect to those materials listed in this rule that are not already covered by the Hazardous Materials Regulations, the Department is aware that, since shipping papers are not required for these materials, carriers will not always be aware that CERCLA's release notification requirements apply to them. While this uncertainty is unfortunate, it is far preferable to the imposition of extensive new shipping paper and other regulatory requirements that would be necessary to provide carriers with certainty. Furthermore, this uncertainty is an interim consequence of the enactment of CERCLA; as EPA establishes RQs for these materials, the Department will begin to incorporate them into the Hazardous Materials Table as "hazardous substances" and assign RQs to them, at which point the Department's requirements of the Hazardous Materials Regulations will apply.

With respect to the release notification requirements for hazardous substances, (as defined in 49 CFR 171.8), it should be noted that, in addition to the requirements of 49 CFR 171.17 for releases "into or upon the navigable waters or adjoining shorelines," Section 103 (a) and (b) of CERCLA requires the reporting of such releases into the "environment," which is defined broadly to include surface water, ground water, land surface, and ambient air.

(46 F.R. 17738, March 19, 1981)

49 CFR Parts 171, 172, and 173
[Docket No. HM-145E, Advance Notice]
Reportable Quantity of Hazardous Substance

BACKGROUND

The Comprehensive Environmental Response, Compensation, and Liability Act of 1980 (CERCLA), also known as the "Superfund" law, is the major Federal legislation designed to address the need for a comprehensive system to respond to releases of hazardous substances into the environment and to impose liability for then on responsible persons (42 USC 9601 et seq.). While a principal focus of concern under CERCLA has been releases from hazardous waste disposal facilities, the law also applies extensively to the transportation of hazardous substances. (See, for example, 42 USC 9601(9)).

The comprehensiveness of CERCLA is indicated by the breadth of the definition of the term, "hazardous substance," which includes substances designated by the Environmental Protection Agency (EPA) under six separate statutory authorities:

1. Section 311 of the Clean Water Act (CWA);
2. Section 3001 of the Solid Waste Disposal Act;
3. Section 307(a) of the CWA;
4. Section 112 of the Clean Air Act;
5. Section 7 of the Toxic Substances Control Act; and
6. Section 102 of CERCLA.

In order to coordinate the implementation of CERCLA with the Department's administration of the Hazardous Materials Transportation act (HMTA), Congress adopted section 306)a) of CERCLA (42 U.S.C. 9656), which provides that each substance that is listed or designated as a "hazardous substance" under the Act shall be listed as a "hazardous material" under the HMTA. Section 306(b) provides that, with certain exceptions, carriers of CERCLA "hazardous substances" shall not be held liable under the CERCLA liability scheme until after the effective date of the listing of the released substance as a "hazardous material" in accordance with section 306(a).

MTB Regulations: At the time of the adoption of CERCLA, a large number of CERCLA "hazardous substances" were already subject to the Department's Hazardous Materials Regulations (HMR), as developed and administered by the Department's Materials Transportation Bureau (MTB). Those substances can be divided into three categories. First, substances that meet MTB's hazard class definitions, contained in 49 CFR Part 173, are fully regulated as hazardous materials. For example, cyanide, while it is a CERCLA "hazardous substance" because it is designated under section 307 of the CWA, is already fully regulated under the HMR because it meets the hazard class definition for a Class B Poison (49 CFR 173.343).

Second, with regard to CERCLA "hazardous substances" that had been designated under section 311 of the CWA, those substances that were not already subject to the HMR had been subjected to them by a final rule issued by MTB on May 22, 1980, the purpose of which was to coordinate the administration of the HMTA with EPA's administration of the CWA (45 FR 34560). The effect of that rule was that, when such a substance is transported in a package containing a quantity of the substance which equals or exceed the "reportable quantity" (RQ) for that substance, as determined by EPA, the shipment must conform with certain shipping paper and package marking requirements (49 CFR 173.1300). If a quantity of a hazardous substance equal to or exceeding its RQ is releases, the release must be immediately reported (49 CFR 171.17). Thus, for purposes of the HMR, "hazardous substance" is defined, in effect, as a material designated by EPA under section 311 of the CWA, but only when it is transported in a package containing at least the RQ of that substance. These "hazardous substances", along with their RQ's, were incorporated into the Hazardous Materials Table at 49 CFR 172.101.

Third, with regard to CERCLA "hazardous substances" that has been designated under section 3001 of the Solid Waste Disposal Act, commonly known as the Resource Conservation and Recovery Act (RCRA), i.e., "hazardous wastes," MTB issued a final rule, simultaneously with the rule discussed above, to coordinate the administration of the HMTA with EPA's administration of RCRA (45 FR 34560). That rule incorporated EPA's "cradle-to-grave" manifest system and other requirements for the transportation of hazardous wastes into the HMR (49 CFR 171.3), and defined the term, "hazardous waste," to mean, for purposes of the HMR, any material subject to the EPA manifest system (49 CFR 171.8). One effect of this definition is that shipments by shippers qualifying for EPA's small generator exemption (generally, those generating less than 1,000 kilograms of hazardous waste per month) of materials not satisfying a DOT hazard class definition are not presently subject to the HMR (See 40 CFR 261.5).

MTB Implementation of CERCLA. Section 306(a): On March 19, 1981, MTB issued a final rule, listing all CERCLA "hazardous substances" as "hazardous materials" under the HMTA, in accordance with section 306(a) (46 FR 17738). In issuing that rule MTB had examined the issue of whether, in addition to "listing" the substances as hazardous materials, MTB should apply the Hazardous Materials Regulations (HMR) to shipments of those materials that were not already subject to them.

In section 102(b) of CERCLA, borrowing the concept of "reportable quantities" from the CWA, Congress assigned a statutory RQ of one pound to all CERCLA "hazardous substances" (except those for which EPA had already assigned RQs under section 311 of the CWA). Section 102(a) of CERCLA authorized EPA to set different RQs, as appropriate. Therefore, in adopting the rule in accordance with section 306 of CERCLA, MTB faced the issue of whether to treat all CERCLA "hazardous substances" as "hazardous substances" under the HMR, and thereby to impose shipping paper and package marking requirements on all shipments of packages of CERCLA "hazardous substances" equal to or exceeding the statutory RQ of one pound. Since the lists of CERCLA "hazardous substances" contain many common materials that are generally not regarded as hazardous in transportation (e.g., copper, zinc, and lead), and since the increased regulatory and paperwork burden would have been very substantial, MTB decided not to impose those requirements at that time.

MTB indicated, however, that it would reexamine the issue when EPA set RQs for the substances as authorized by section 102(a) of CERCLA: "At such time as EPA exercises its authority under section 102 to establish RQs for particular substances, MTB will determine the appropriateness of listing those substances as "hazardous substances and assigning those RQs to them" (46 FR 17738).

Subsequently, MTB received petitions for reconsideration of the rule from the National Tank Truck Carriers, Inc., the American Trucking Association, and the American Association of Railroads, three national trade associations representing common carriers of hazardous materials. All of the petitioners objected to MTB's failure to impose shipping paper requirements on shipments of CERCLA "hazardous substances" in packages equal to or exceeding the statutory RQ of one pound. They argued that, since section 306(b) of CERCLA exempted carriers form liability under that act prior to the effective date of the listing required by section 306(a). Congress must have intended that carriers be given actual notice that they are transporting listed materials in order to be subject to liability under CERCLA, and that the HMR shipping paper requirements under that ACT prior to the effective date of the listing required by section 306(a), Congress must have intended that carriers be given actual notice that they are transporting listed materials in order to be subject to liability under CERCLA, and that the HMR shipping paper requirements were intended as the means for providing that notice. They argued further that it was inappropriate to subject carriers to the liability and reporting and other requirements of CERCLA unless a means for notifying them of that fact were established.

On November 30, 1981, MTB published a denial of these petitions (46 FR 58086). MTB concluded that, while section 306(a) of CERCLA required the listing of CERCLA "hazardous substances" as "hazardous materials" under the HMTA, Congress had not expressed an intent to affect the Department's long-standing discretion under section 105 of the HMTA to determine whether, and to what extent, to regulate hazardous materials. MTB also restated its position that, given the tremendous paperwork burden that would be required to provide carriers with the notice they sought, it would be inappropriate to impose shipping paper requirements at that time, but that, when EPA adjusted the RQs, MTB would again examine the issue.

EPA Notice of Proposed Rulemaking: On May 25, 1983, EPA published a Notice of Proposed Rulemaking (NPRM) in which it proposed to exercise its authority under section 102(a) of CERCLA to adjust RQ's for CERCLA "hazardous substances" (48 FR 23552). MTB submitted comments to EPA on their notice which are provided as appendix A to this notice. Briefly, EPA has proposed RQ adjustments for 387 out of the 696 CERCLA "hazardous substances." The remainder are still under evaluation by EPA, and adjustments to them will be proposed when the evaluations have been completed. With regard to those substances for which EPA has completed its evaluations, it proposes either to retain the RQ of one pound or to establish a new RQ of 10, 100, 1000, or 5000 pounds. Interested persons are referred to the preamble of the NPRM for detailed discussion of the proposed adjustments and to the proposed rule itself for the proposed RQs.

REGULATORY CONSIDERATIONS AND REQUEST FOR COMMENTS

Executive Order 12291: In keeping with its commitment in issuing the rulemaking in accordance with section 306 of CERCLA, MTB is initiating the process of determining whether to incorporate EPA's adjusted RQ's into the HMR and to subject the CERCLA "hazardous substances" to regulation as "hazardous substances" under the HMR. In making that determination, MTB is subject to the provisions of Executive Order 12291, secton 2 of which provides:

In promulgating new regulations, . . . all agencies, to the extent permitted by law, shall adhere to the following requirements:
(a) Administrative decisions shall be based on adequate information concerning the need for and consequences of proposed government action;
(b) Regulatory action shall not be undertaken unless the potential benefits to society for the regulation outweigh the potential costs to society;
(c) Regulatory objectives shall be chosen to maximize the net benefits to society;
(d) Among alternative approaches to any given regulatory objective, the alternative involving the least net cost to society shall be chosen; and
(e) Agencies shall set regulatory priorities with the aim of maximizing the aggregate net benefits to society, taking into account the condition of particular industries affected by regulations, the condition of the national economy, and other regulatory actions contemplated for the future. (46 FR 13193).

Controlling Paperwork Burdens: By a final rule issued on March 31, 1983 (48 FR 13666), the Office of Management and Budget (OMB) promulgated regulations under the Paperwork Reduction Act of 1980 that impose additional requirements on agencies considering the issuance of regulations that would require the "collection of information" (5 CFR Part 1320). As broadly defined in those regulations, the term, "collection of information" includes the shipping paper and package labeling requirements under the HMR (5 CFR 1320.7(c)(2)). Therefore, in considering whether to subject CERCLA "hazardous substances" to those requirements, MTB is required to comply with those regulations.

In part, those regulations provide that agencies may not impose information collection requirements without first obtaining OMB approval (5 CFR 1320.4(a)). Those regulations also provide:

To obtain OMB approval of a collection of information, an agency shall demonstrate that it has taken every reasonable step to ensure that:
(1) The collection of information is the least burdensome necessary for the performance of the agency's functions to comply with legal requirements and achieve program objectives;
(2) the collection of information is not duplicative of information otherwise accessible to the agency; and
(3) The collection of information has practical utility . . . (5 CFR 1320.4(b)).

Regulatory Alternatives: MTB has determined that the following are the principal regulatory alternatives available to it when EPA issues its final rule to adjust RQs:
1. MTB may issue a rule to incorporate all CERCLA "hazardous substances" into the Hazardous Materials Table as "hazardous substances" under the HMR, applying the RQ for each substance in effect at that time (including the statutory RQ of one pound for those substances for which EPA has not concluded its evaluations).
2. With regard to those CERCLA "hazardous substances" for which EPA has completed its evaluations and established RQs, MTB may issue a rule to incorporate those substances into the Hazardous Materials Table as "hazardous substances" under the HMR, but withhold further action on all other CERCLA "hazardous substances" until EPA has completed its evaluations and established RQs for them.

3. MTB may withhold further action until EPA has completed its evaluations and established RQs for all CERCLA "hazardous substances."

4. As a variation of Alternative 1, MTB may issue such a rule with regard only to those substances for which MTB possesses adequate information concerning the need for and consequences of such a rule, and for which the potential benefits outweigh the potential costs.

5. As a variation Alternative 2, MTB may issue such a rule with regard only to those substances for which MTB possesses adequate information concerning the need for and consequences of such a rule, and for which the potential benefits outweigh the potential costs.

6. As a variation of Alternative 3, MTB may issue such a rule with regard only to those substances for which MTB possesses adequate information concerning the need for and consequences of such a rule, and for which the potential benefits outweigh the potential costs.

7. MTB may do nothing. This would mean that the only hazardous substances regulated by the HMR would be those already listed in the Hazardous Materials Table.

8. MTB may decline to apply the HMR to hazardous substances by removing those already listed in the Hazardous Materials Table and by not adding any others. This would return DOT to its traditional role of safety regulation under HMTA.

Categories of Substances: In order to analyze these alternatives in accordance with F.O. 12291 and 5 CFR Part 1320, MTB requires additional information regarding the need for and consequences of adopting the EPA adjusted RQs, the associated costs and benefits, and the relative burdens and the practical utility of the various alternatives. For this purpose, CERCLA "hazardous substances" can be divided into four categories:

1. Those which are already subject to the HMR because they fall within one or more of the hazard classes defined in 40 CFR Part 173.

2. Those which are already subject to the HMR because they have been designated under section 311 of the CWA and are, therefore, incorporated into the HMR as "hazardous substances."

3. Those which are already subject to the HMR because they have been designated under section 3001 of RCRA and are, therefore, incorporated into the HMR as "hazardous wastes."[1]

4. Those which are not currently subject to the HMR.

With regard to Category 1 substances, the impact of the adoption of RQs under the regulatory alternatives generally appears to be relatively minor because most of these materials are already fully regulated under the HMR. The principal change would be that shipping papers would be required to display "RQ" and certain other information (49 CFR 172.203(c)). However, there may be important exception to this generalization. For example, materials classified as ORM-A under the HMR are subject to regulation only when transported on aircraft and/or vessels unless they are hazardous wastes in which case they are regulated by all modes. Designation of these materials as "hazardous substances" under the HMR would subject them to regulation when transported by other modes, as well. For example, tetrachloroethylene, a common drycleaning solvent classified under the HMR as an ORM-A, is a CERCLA "hazardous substance" because it is designated under both section 307 of the CWA and section 3001 of RCRA. EPA is still assessing the effects of this substance, and has proposed retaining the statutory RQ of one pound at least until those assessments are completed. Therefore, if MTB were to adopt that RQ, currently unregulated shipments of tetrachloroethylene by highway and rail in packages exceeding one pound would become subject to the HMR.

With regard to Category 2 substances, the adoption of RQs under the regulatory alternatives would only affect substances for which EPA has adjusted the RQ from that which had been assigned under section 311 of the CWA. For example, EPA has proposed to adjust the RQ for pentachlorophenol, a common wood preservative, from ten pounds to one pound. Thus, if pentachlorophenol were frequently transported in packages containing more that one pound, but less than ten pounds, of the substance, MTB's adoption of an adjusted RQ of one pound could result in a significant increase in the number of shipments subject to the HMR. On the other hand, EPA has proposed to increase the RQs for some of these substances, which might result in a reduction of the number of shipments subject to the HMR. It should be noted that, as discussed above, the original purpose of MTB's adopting RQ's for these materials was to coordinate its program with that of EPA, and that that purpose would no longer be served if DOT were to retain the current RQs after EPA has changed them.

With regard to Category 3 substances, as discussed above, only shipments of hazardous waste that are subject to EPA's manifest requirements are currently subject to the HMR. Therefore, application of the HMR to all shipments of hazardous waste in packages containing more that the RQ for that waste could result in a substantial increase in the number of shipments subject to the HMR. For example, assuming that packages contain at least the RQ, ship-

---

[1]There is considerable overlap among these first three categories. For example, sodium cyanide is regulated under the HMR as a poison B, it has been designated as a "hazardous substance" under section 311 of the CWA, and, when carried for disposal, it is a "hazardous waste" under section 3001 of RCRA. Considerations of this overlapping is essential to determine the impact of the regulatory alternatives on any given material.

ments originating with shippers who qualify for EPA's small generator exemption would become subject to the HMR.

Finally, with regard to Category 4 substances, the impact of the adoption of RQs under the regulatory alternatives would be similar to the impact on Category 3 substances that are not subject to EPA's manifest requirements. For example, diethylphthalate (other than waste diethylphthalate) is currently not subject to the HMR. However, EPA has proposed an RQ for diethylphthalate of 100 pounds. Therefore, under the regulatory alternatives, all shipments of diethylphthalate in packages exceeding 100 pounds would become subject to the HMR.

Questions to be Addressed by Comments: The above discussion of the potential impacts of the regulatory alternatives is based on preliminary observations. By this advance notice, MTB solicits more specific information from the public in order to conduct the analyses required by E.O. 12291 and 5 CFR Part 1320. Specifically, MTB solicits responses to the following questions:

1. What is the anticipated frequency of shipments that are not currently subject to the HMR, but that would become subject to them if MTB were to adopt RQs under the regulatory alternatives?

2. What new cost would the adoption of RQs under the regulatory alternatives impose on shippers and carriers of CERCLA "hazardous substances?"

3. What is the anticipated frequency of releases from these shipments that would exceed the RQ, and what is the likelihood of clean-up efforts resulting from the reporting of these releases?

4. What are the anticipated environmental benefits of such clean-up efforts?

5. With regard to these releases, what is the likehood that clean-up would occur even if MTB did not adopt one of the regulatory alternatives?

6. What would be the effect of the adoption of RQs under the regulatory alternatives on international commerce?

7. What would be the effects of the adoption of RQs under the regulatory alternatives on the potential for liability and on the insurability of shippers and carriers of CERCLA "hazardous substances?"

8. Some CERCLA "hazardous substances" are hazardous only in certain forms. For example, while lead and other heavy metals are designated as CERCLA "hazardous substances," they are hazardous only when they occur in very small particles. Therefore, EPA has proposed to exempt from reporting releases of these metals except for particles that are less than 100 micrometers in diameter. Are there other CERCLA "hazardous substances" that are hazardous only in certain forms to which MTB could similarly limit the applicability of the HMR?

9. If MTB were to extend the applicability of its shipping paper and package marking requirements through the adoption of RQs under the regulatory alternatives, would the increased frequency of their use tend to diminish their effectiveness as hazard warnings?

10. What other factors should MTB consider in determining the need for and consequences of the regulatory alternatives?

11. What other factors should MTB consider in determining the potential benefits and the potential costs to society of the regulatory alternatives?

12. What other information would be of value to MTB in conducting the analyses required by E.O. 12291?

13. To assist MTB in fulfilling the requirement of 5 CFR Part 1320, which of the regulatory alternatives, necessary for the proper performance of the agency's function, would be the least burdensome?

14. What is the "practical utility", as that term is defined at 5 CFR Part 1320.7(9), of the "collection of information" that would result from the adoption of RQs under the regulatory alternatives?

Appendix A—MTB Comments on EPA NPRM "Notification Requirements; Reportable Quantity Adjustments"
In response to EPA's Notice of Proposed Rulemaking (NPRM), entitled, "Notification requirements; Reportable Quantity Adjustments" (48 FR 23552), the Materials Transportation Bureau (MTB) of the U.S. Department of Transportation submits the following comments.

At the outset, MTB notes and concurs with EPA's expressed intention "to work with DOT to develop a coordinated (sic) and integrated set of regulations so that shippers and carriers of hazardous substances will be subject to only one set of regulations" (48 FR 23560). This intention comports fully with the longstanding policy of both agencies to work together in areas of shared regulatory responsibility to ensure that regulatory duplications, overlaps, inconsistencies, and gaps do not occur.

In accordance with this policy, MTB has previously expressed its commitment to examine the question of whether to subject CERCLA "hazardous substances" to additional regulation under the Hazardous Materials Transportation Act (HMTA) at such time as EPA exercises its authority under section 102(a) of CERCLA to adjust the reportable quantities (RQs) for those substances. The enclosed Advance Notice of Proposed Rulemaking (ANPRM) is MTB's initial step in conducting this examination, and any comments that EPA may have on that ANPRM will be greatly appreciated and fully considered.

Very briefly, the primary purpose of the ANPRM is to solicit information necessary to conduct the analyses that MTB is required to perform by Executive Order (E.O.) 12291 before issuing regulations to subject CERCLA "hazardous substances" to additional regulation under the HMTA. As EPA's contractor, ICF, Inc., has observed, "the extension of the HMR (MTB's Hazardous Materials Regulations) to CERCLA hazardous substances is a discretionary action, and . . . the effects of that extension (such as shipping paper costs) should be attributed to DOT's regulatory action, not EPA's adjustment regulation."

While MTB fully concurs with this conclusion, since EPA has the authority under CERCLA to designate "hazardous substances" and to establish RQs for them, EPA's actions, in effect, establish the starting point for MTB's considerations. Therefore, those actions clearly have a very substantial influence on the ultimate effect of any action by MTB to subject CERCLA "hazardous substances" to additional regulation under the HMR. For example, if a given CERCLA "hazardous substance," is frequently transported in packages containing 15 pounds of the substance, EPA's determination that the RQ for that substance should be 10 pounds, rather than 100 pounds, would result in many more shipments of the substance, being subject to the HMR if MTB were to extend the applicability of the HMR to that substance. Therefore, although E.O. 12291 properly places on MTB the ultimate responsibility to examine the impacts of extension of the HMR, it would be appropriate for EPA, in the process of adjusting RQs, to consider the influence of its actions on those impacts.

An excellent example of EPA's apparent consideration of these impacts is the treatment in the NPRM of metals that are CERCLA "hazardous substances" because they are designated as "toxic pollutants" under section 307 of the CWA. Recognizing that these metals (such as lead, copper and zinc) are hazardous only in the form of very small particles, EPA stated that "no reporting of releases of massive forms of these substances is required if the diameter of the pieces of the substance released is equal to or exceeds 100 micrometers" (48 FR 23601). This limitation of the form of the substances makes it much more practicable for MTB to subject these substances to regulation under the HMR; large numbers of innocuous shipments containing these metals in other forms would not be subject to regulation.

Similar limitations may also be appropriate for other CERCLA "hazardous substances." For example, asbestos, while demonstrably hazardous in some forms, in a commonly used industrial material that is frequently transported as a part of products that pose little risk, such as brake linings and asbestos-cement pipes. To subject shipments of these products to the HMR because they contain a quantity equal to or greater than the RQ for asbestos (currently one pound) would very probably not be cost-effective. However, if EPA were to limit the forms of asbestos that are subject to the RQ to those forms that present the hazards for which asbestos was designated a CERCLA "hazardous substance" (e.g., unbonded particles), application for the HMR only to those forms of asbestos would be more feasible.

Apart from these general comments relating to coordination between EPA and MTB, MTB has several specific comments relating to the section of the preamble of the NPRM entitled, "Additional criteria considered but not currently used for adjusting reportable quantities." First, EPA states, "until passage of CERCLA, not all releases of CERCLA hazardous substances have been uniformly subject to DOT reporting requirements" (48 FR 23567). This statement implies that, *since* passage of CERCLA, all releases of those substances have been uniformly subject to DOT reporting requirements. As discussed in the enclosed ANPRM, and in the rulemaking actions cited therein, DOT has not extended application of the HMR, including the reporting requirements, to CERCLA "hazardous substances" that were not already subject to them.

Second, MTB disagrees with EPA's suggestion that "Release Potential" might be an appropriate criterion on which to base adjustments to RQs. As EPA states elsewhere in the NPRM:

> This rulemaking proposes adjustments to the statutory RQs based upon specific scientific and technical criteria which correlate with *the possibility of hazard or harm upon the release of a substance in a reportable quantity.* These revised RQs, therefore, enable the Agency to focus its resources on those releases which are *most likely to pose potential threats to public health and welfare and the environment* (48 FR 23560) (emphasis added).

The likelihood of release of a substance is irrelevant to these considerations; the "possibility of hazard or harm" and the "potential threats to public health and welfare and the environment" caused by a given release are the same whether the release was likely or unlikely to occur. Therefore, the need for EPA to be notified and to undertake reponse is likewise the same.

With regard to the specific factors identified by EPA in that section, MTB also disagrees with the suggestion that "Transportation Mode" or "Packaging and Containerization" might be appropriate factors for consideration in determining the likelihood of releases. With regard to transportation mode, the NPRM states, "If some hazardous substances are generally shipped by a transportation mode that exposes them to a particularly high risk of large releases, the RQ may be reduced. . ." (48 FR 23567). It is, of course, the purpose of the HMR to assure that all hazardous materials are transported, by whatever mode, in a manner that does not pose an unreasonable risk of release. Generally, each mode poses equivalent risks of release. In transportation between two specific points it may be possible to determine that one mode presents a lower risk than another, but that conclusion could be established only after a detailed risk analysis and would apply only to that particular situation. For example, while a risk analysis might demonstrate that, for transportation between New Orleans and Houston, barging would pose a minutely lower risk than rail, that conclusion would obviously be absurd for transportation between Denver and Phoenix.

Further, other factors, such as condition of equipment, track, or roadway, have a much greater influence on transportation risk than does mode of transportation.

With regard to "packaging and containerization," CERCLA "hazardous substances" that are subject to the packaging requirements of the HMR are required to be packaged in such a way that releases will not occur during the normal course of transportation. In the event of an accident involving properly packaged materials, the type of packaging will generally have little effect on the likelihood or quantity of release. With regard to package size, MTB agrees with EPA that there is likely to be a correlation between the size of package and the size of release. As discussed previously, to the extent that EPA can establish RQs so that only releases from packages of a sufficient size to pose a substantial threat are subject to reporting requirements, application of the HMR will be more practicable. Again, however, the relevant concern is the severity of the release, rather than its likelihood.

(48 F.R. 35965, August 8, 1983)

49 CFR Parts 171, 172, and 173
[Docket No. HM-145E; Notice No. 86–3]
Reportable Quantity of Hazardous Substances

I. BACKGROUND

Section 306(a) of CERCLA requires the Secretary of Transportation to list all "hazardous substances" as defined by CERCLA as "hazardous materials" under the Hazardous Materials Transportation Act (HMTA). By a final rule issued on March 19, 1981, Docket No. HM-145C [46 FR 17738]. RSPA fulfilled this requirement by amending 49 CFR 172.101 to include a list of these materials immediately following the Hazardous Material (Table).

The definition of "hazardous substance" as defined in section 101(14) of CERCLA includes substances designated under section 307(a) and section 311 of the Federal Water Pollution Control Act (Clean Water Act or CWA), section 3001 of the Solid Waste Disposal Act (commonly known as the Resource Conservation and Recovery Act (RCRA)), and section 112 of the Clean Air Act (CAA). With the exception of substances designated under section 311 of the CWA, CERCLA also established the reportable quantities for hazardous substances from these sources at one pound and gave EPA the authority to adjust the size of the reportable quantity by regulations. While listing all of the CERCLA hazardous substances in the HMR, RSPA did not apply regulations under the HMTA to those substances that were not already subject to the HMR. If such action were taken, there would have been a vast increase in the number of shipments regulated under the HMR, many of them for relatively innocuous materials. During the incorporation of the CERCLA list of hazardous substances under HM-145C, RSPA indicated that when the EPA exercised its authority under section 102 of CERCLA to adjust the RQs for those substances, RSPA would again examine the question of whether to subject these substances to regulation under the HMTA.

In order to develop a coordinated regulatory program, RSPA has worked closely with EPA's Emergency Response Division on the designation of hazardous substances and adjustment of reportable quantities (RQs) since CERCLA was enacted.

EPA Action. In a Notice of Proposed Rulemaking (NPRM) issued on May 25, 1983 (48 FR 23552), EPA proposed to adjust the RQ level for 387 of the 698 CERCLA "hazardous substances". In this NPRM, EPA listed for the first time the "hazardous substances" designated by section 101(14) of CERCLA. This NPRM also discussed in detail the CERCLA notification requirements, the methodology and criteria used by the EPA to adjust the reportable quantity levels, and the RQ adjustments proposed under section 102 of CERCLA and section 311 of the CWA.

RSPA commented on this NPRM on August 8, 1983 (48 FR 35965). In the comments, RSPA concurred with EPA's expressed intention "to work with DOT to develop a coordinated (sic) and integrated set of regulations so that shippers and carriers of hazardous substances will be subject to only one set of regulations" [48 FR 23560]. RSPA in turn expressed a commitment to examine to what extent CERCLA "hazardous substances" should be regulated under the HMR.

By Final Rule published in the Federal Register on April 4, 1985 [50 FR 13456], EPA exercised its authority under CERCLA and adjusted the RQs for 340 of the CERCLA hazardous substances. RSPA is responding to this EPA action by issuance of this NPRM.

DOT Action. As previously stated, EPA published an NPRM on May 25, 1983 [48 FR 23552]. This NPRM proposed to adjust the RQ levels for 387 of the 698 CERCLA hazardous substances. In response to the EPA NPRM, DOT published an Advance Notice of Proposed Rulemaking (ANPRM) on August 8, 1983 [48 FR 35965]. This ANPRM considered the regulatory impact that the incorporation of CERCLA hazardous substances into the HMR would place on shippers and carriers. Basically, the ANPRM examined the extent the HMR should apply to CERCLA hazardous substances. In this ANPRM, RSPA proposed eight regulatory alternatives as possible ways of incorporating the CERCLA hazardous substances into the HMR. These alternatives are repeated as follows:

Proposed Regulatory Alternatives:

1. RSPA may issue a rule to incorporate all CERCLA "hazardous substances" into the Table as "hazardous substances" under the HMR, applying the RQ for each substance in effect at that time (including the statutory RQ of one pound for those substances for which EPA has not concluded its evaluations).

2. With regard to those CERCLA "hazardous substances" for which EPA has completed its evaluations and established RQs, RSPA may issue a rule to incorporate those substances into the Table as "hazardous substances" under the HMR, but withhold further action on all other CERCLA "hazardous substances" until EPA has completed its evaluations and established RQs for them.

3. RSPA may withhold further action until EPA has completed its evaluations and established RQs for all CERCLA "hazardous substances".

4. As a variation of Alternative 1, RSPA may issue a rule with regard only to those substances for which RSPA possesses adequate information concerning the need for and consequences of such a rule, and for which the potential benefits outweigh the potential costs.

5. As a variation of Alternative 2, RSPA may issue a rule with regard only to those substance for which RSPA possesses adequate information concerning the need for and consequences of such a rule, and for which the potential benefits outweigh the potential costs.

6. As a variation of Alternative 3, RSPA may issue a rule with regard only to those substances for which RSPA possesses adequate information concerning the need for, and consequences of such a rule, and for which the potential benefits outweigh the potential costs.

7. RSPA may do nothing. This would mean that the only hazardous substances subject to the HMR would be those already listed in the Table.

8. RSPA may decline to apply the HMR to hazardous substances by removing those already listed in the Table and by not adding any others. This would return DOT to its traditional role of safety regulation under the HMTA.

Comments on Proposed Alternatives. Thirty-three commenters addressed these alternatives in their comments. The majority of commenters felt that RSPA should further regulate hazardous substances. Commenters taking this position either Alternative No. 1 or No. 2. Both of these alternatives are similar because they propose to regulate CERCLA hazardous substances under the HMR. The difference between these two alternatives is the manner in which hazardous substances are adopted into the HMR.

Under Alternative No. 1, all CERCLA hazardous substances would be regulated under the HMR following issuance of a final rule by RSPA. Those substances whose reportable quantity level had not been adjusted would still be regulated under the HMR at the statutory RQ of one pound. Under Alternative No. 2 only those substances that had EPA adjusted RQs would be regulated under the HMR. Those substances whose RQ values had not been adjusted would be listed but not regulated. RSPA would consider subjecting these substances to regulation when their RQs had been adjusted by EPA.

While each of the proposed regulatory alternatives provides a method for dealing with CERCLA hazardous substances, RSPA believes that the adoption of CERCLA hazardous substances into the HMR under Alternative No. 2 is the most effective course of action. This alternative complies with the Paperwork Reduction Act by controlling the additional paperwork burdens associated with regulating CERCLA hazardous substances. Review of the proposed alternatives indicates that while the choice of one of the other alternatives may be feasible, their selection is not warranted. Selecting Alternative No. 3, 4, 5, or 6 would require RSPA to either delay issuance of a rule on hazardous substances until EPA takes further action or require that RSPA issue a rule on hazardous substances based solely on RSPA's evaluation of these substances. RSPA does not believe that either approach is acceptable.

Alternative No. 7 proposes that RSPA do nothing. RSPA would continue to only regulate those substances that are already in the Table. From the comments RSPA received to this alternative, carriers are concerned that pursuing this course of action would ignore their need for additional information to be shown on shipping papers. This additional information is believed necessary to alert them to the fact that they are carrying hazardous substances and to inform them that if a "reportable" amount is spilled, a report is required by CERCLA. If RSPA were to take no further action, there would be no identification of new hazardous substances on shipping papers. For this reason, RSPA does not believe that selection of this alternative is warranted.

The selection of Alternative No. 8 would mean that RSPA would decline to regulate hazardous substances under the HMR. Choosing this alternative, RSPA would be acting on the premise that the regulatory controls which are currently in place for hazardous substances are adequate and that regulation under the HMR is not necessary. RSPA does not believe this. There is a need for carriers to have information about hazardous substances. If a carrier is aware that a hazardous substance is being transported and a spill occurs in a reportable quantity, the carrier can make the required spill report. Also, shippers should know as soon as possible what the final RQs are for hazardous substances, so they can finalize their instructions to shipping and operating personnel as well as to their carriers. They can also make any needed changes to their shipping documentation and training materials.

Considering the tremendous paperwork burden associated with the adoption of Alternative No. 1, and the fact that the selection of Alternative No. 2 will keep the HMR consistent with the EPA regulations, RSPA believes that Alternative No. 2 is the most practical way of satisfying these information needs.

It is RSPA's position that DOT has discretionary authority under Section 105 of the HMTA to determine whether, and to what extent, to regulate hazardous materials. (A detailed discussion of this position was provided in the ANPRM.) With regard to the CERCLA hazardous substances for which EPA has issued a final rule adjusting their reportable quantities, RSPA has decided to propose to add them all to the hazardous materials table.

Questions Posed in the ANPRM. RSPA requested that commenters to the ANPRM respond to fourteen questions regarding CERCLA hazardous substances. Specifically, these questions were aimed at giving RSPA an idea of the potential impact attributable to each of the proposed regulatory alternatives, and to assist RSPA in making the analyses required by Executive Order (E.O.) 12291 and 5 CFR Part 1320. Each question is repeated herein:

1. What is the anticipated frequency of shipments that are not currently subject to the HRM, but that would become subject to them if RSPA were to adopt RQs under the regulatory alternatives?

2. What new costs would the adoption of RQs under the regulatory alternatives impose on shippers and carriers of CERCLA "hazardous substances"?

3. What is the anticipated frequency of releases from shipments that would exceed the RQ and what is the likelihood of clean-up efforts resulting from the reporting of these releases?

4. What are the anticipated environmental benefits of such clean-up efforts?

5. With regard to these releases, what is the likelihood that clean-up would occur even if RSPA did not adopt one of the regulatory alternatives?

6. What would be the effect of the adoption of RQs under the regulatory alternatives on international commerce?

7. What would be the effect on the adoption of RQs under the regulatory alternatives on the potential for liability and on the insurability of shippers and carriers of CERCLA "hazardous substances"?

8. Some CERCLA "hazardous substances" are hazardous only in certain forms. For example, while lead and other heavy metals are designated as CERCLA "hazardous substances" they are hazardous only when they are in very small particles. Therefore, EPA has proposed to exempt from reporting releases of these metals except for particles that are less than 100 micrometers in diameter. Are there other CERCLA "hazardous substances" that are hazardous only in certain forms to which RSPA could similarly limit the applicability of the HMR?

9. If RSPA were to extend the applicability of its shipping paper and package marking requirements through the adoption of RQs under the regulatory alternatives, would the increased frequency of their use tend to diminish their effectiveness as hazard warnings?

10. What other factors should RSPA consider in determining the need for and consequences of the regulatory alternatives?

11. What other factors should RSPA consider in determining the potential benefits and the potential costs to society of the regulatory alternatives?

12. What other information would be of value to RSPA in conducting the analyses required by E.O. 12291?

13. To assist RSPA in fulfilling the requirement of 5 CFR Part 1320, which of the regulatory alternatives, necessary for the proper performance of the agency's function, would be the least burdensome?

14. What is the "practical utility", as that term is defined at 5 CFR Part 1320.7(q), of the "collection of information" that would result from the adoption of RQs under the regulatory alternatives?

Comments on Questions Posed in the ANPRM. Of the thirty-three comments received, only eight commenters addressed any or all of these questions. A brief synopsis of their responses is provided herein. A detailed discussion of the comments received to these questions is available for review in the regulatory evaluation and supporting documentation located in the public docket file of HM-145E.

RSPA received a few comments to the question concerning the impact the adoption of RQs would have on international commerce. These comments indicated that by increasing the number of CERCLA hazardous substances subject to the HMR, there may be an increase in the number of materials that would be regulated in international commerce. Commenters stated that new requirements for hazardous substances have the potential to be burdensome and confusing to shippers in other countries. No substantive evidence was provided to back up this claim. Since the adoption of the first hazardous substances into the HMR in Docket HM-145B on May 22, 1980, [45 FR 34560] and November 10, 1985, [45 FR 74640] RSPA has not been made aware of any major problems occurring relative to the import or export of hazardous substances. Secondly, DOT possesses very limited knowledge of the volume of hazardous commodities which are imported and exported annually. Without knowing the number of hazardous substances that will be incorporated into the HMR and the number of shipments of these materials that are being imported into or exported from the United States, RSPA has no way of knowing what effect the adoption of RQs under one of the proposed regulatory alternatives would have on international commerce. This question was posed by RSPA in the AMPRM because RSPA had no quantifiable data on these shipments. RSPA had hoped commenters would furnish this data. They did not.

Only a few comments were received concerning the effect the adoption of RQs would have on the liability and insurability of shippers and carriers of CERCLA hazardous substances. Most of the comments RSPA received on this question pointed out that should any of the regulatory alternatives be adopted, there would virtually be no change in the liability or insurability of shippers and carriers of these substances. One commenter, representing an association comprised of carriers, thought the full adoption of hazardous substances under the HMR would greatly improve the liability and insurance situation of carriers. This commenter contended that if carriers had better, more complete knowledge that a hazardous substance was being transported, they would exercise greater caution when transporting these shipments. As a result, insurance rates for transporting these substances might stabilize. However, another commenter contended that a shipper's and carrier's potential for liability might be increased substantially should the CERCLA hazardous substances be adopted, due to the increase in the number of regulated shipments. No supportive documentation was furnished by either commenter to substantiate their positions.

In response to the question regarding whether there are other CERCLA hazardous substances that are only hazardous in certain forms, RSPA only received one comment. The commenter expressed concern about lead oxide and lead silicate having an RQ of one pound under the EPA rule and stated that even though these substances are less toxic than other substances with higher RQs, EPA has failed to develop a specific RQs for them. This commenter urged RSPA to develop substantial supportive documentation on all substances which present an imminent hazard and threat to the environment before imposing any regulations on them.

RSPA is sympathetic to this commenter's concern, however, EPA is the only agency which has the authority under CERCLA to adjust hazardous substance RQs. RSPA's authority under the HMTA is to determine which materials pose an unreasonable risk to health and safety or property when transported in commerce and to regulate them at an appropriate level. The EPA develops a substantial supportive record for a named substance before an RQ level is established. In the final rule dated April 4, 1985, [50 FR 13456] EPa stated that they have deliberately decided not to establish RQs for the many broad, generic classes of organic and metallic compounds. Although not specifically listed by name, lead oxide and lead silicate would be included in one of these broad, generic classes. EPA has also determined that the notification requirements under CERCLA apply only to those specific compounds whose RQs are listed in the EPA Table, the List of Hazardous Substances and Reportable Quantities (40 CFR 302.4).

Comments made to the question concerning what other factors RSPA should consider in determing the need for and consequences of the regulations stressed two points. One, RSPA should consider the effect that each of the regulatory alternatives would have on small businesses. Secondly, RSPA should consider a carrier's liability and the potential environmental damage that could result if a hazardous substance is spilled and the appropriate response is not initiated because the substance was not properly reported.

RSPA is required to give full consideration to the impact any regulatory action it takes would have on small businesses. With regard to carrier's liability, CERCLA requires that all spills of a CERCLa hazardous substance which occur in a reportable quantity be reported to the National Response Center. The incorporation of CERCLA hazardous substances into the HMR will promote better identification of these materials, leading to greater compliance with the reporting requirements and lessening the potential for environmental damage.

Concerning the question about what new costs shippers and carriers would incur should CERCLA hazardous substances be adopted under one of the regulatory alternatives, most commenters stated that shipment costs would increase proportionally with the increase in the number of shipments. No quantitative data was provided to support this conclusion. RSPA acknowledges that costs may increase if the number of shipments subject to the HMR increases. Of course, any increase in costs is dependent on many factors ranging from the degree of hazard posed by the materials being shipped to the volume of materials and the number of shipments being made. With the exception of Alternative No. 7, selection of any of the alternatives may entail additional costs. This is expected. These costs result from compliance with the documentation, packaging, marking, labeling and placarding requirements of the HMR. E.O. 12291 requires that no regulatory action shall be undertaken unless it can be demonstrated that the potential benefits to society which the proposed regulation provides outweigh its potential costs. (A detailed evaluation of the economic impact of this regulation can be found in Appendix B of the regulatory evaluation).

In response to the question concerning the anticipated frequency of shipments currently not subject to the HMR and the cost increase that would occur if these shipments become subject to the HMR, more than half of the commenters believed the number of regulated shipments would increase. Most of the commenters thought this increase would be substantial, however, no quantitative data was furnished. These comments support RSPA's contention in the ANPRM that if certain categories of CERCLA hazardous substances are brought under the umbrella of the HMR, there could be a significant increase in the number of shipments subject to the HMR. When a material becomes subject to the HMR, shippers incur costs by complying with documentation and package marking requirements. Carriers become subject to additioinal requirements such as loading requirements, driving and insurance rules and reporting requirements. These requirements, of course, would not be new ones for shippers and carriers of those substances who are already subject to the HMR.

II. PROPOSED RULE

The primary purpose of this notice of proposed rulemaking is to incorporate CERCLA hazardous substances into the HMR as proposed under Alternative No. 2. RSPA proposes three actions. First, RSPA proposes to add to the Table those CERCLA hazardous substances which have EPA established final RQs. RSPA will take no action at this time on unevaluated hazardous substances as these will be assessed by RSPA after EPA establishes their final RQ's. When final RQs are established for these substances, RSPA will consider adding them to the Table. The changes proposed in this NPRM to incorporate certain CERCLA hazardous substances into the HMR should enhance the carriers' awareness, through the shipping paper identification, that they are transporting hazardous substances.

The second action RSPA is proposing is to revise the definition of a hazardous substance in §171.8 to include hazardous wastes which are not specifically listed by the EPA as a hazardous waste, but which exhibit an EPA characteristic of ignitability, corrosivity or reactivity (ICR). RSPA believes that revision of the definition for a hazardous substance to include ICR materials is necessary because EPA has stated that when a reportable quantity of any ICR waste is released (spilled), the release is reportable under CERCLA.

The third action RSPA is proposing is to change the hazardous substance discharge notification requirements of 49 CFR 171.17. EPA pointed out in its comments to the ANPRM, HM-145E, that the language of this section applies only to discharges of a hazardous substance which occur in a reportable quantity into navigable waters or upon adjoining shorelines. EPA stated "while this language was written to conform to the requirements of the Clean Water Act, it is not consistent with the multimedia reporting requirements of CERCLA which apply to releases to air, land and water". RSPA agrees with EPA's statement and is proposing to change the provisions of §171.17 to either make the discharge notification requirements applicable to all media or to remove the discharge notification requirements from the HMR. As an option, RSPA could include a reference to the notification requirements of CERCLA in the HMR. For purposes of this rule, the terms "release" and "environment" have the same meaning as specified by EPA in its definition of these terms in 40 CFR 302.3.

The RQs for certain hazardous substances which were previously incorporated into the Table (HM-145B; 45 FR 34560 and 45 FR 74640) have been adjusted by EPA. RSPA proposes to change the RQ for each of these hazardous substances to correspond to the adjusted final RQ set by EPA.

In its final rule, EPA adjusted the RQs of 340 hazardous substances, including 21 waste streams. However, EPA's rule included many hazardous substances from Section 311 of the CWA whose "adjusted" final RQ is the same as its statutory RQ. These substances were previously incorporated into the Table under HM-145B. Therefore, the number of hazardous substances addressed in this rule is less than 340. These hazardous substances would be listed in the Table by name followed by the RQ and identified in column 1 of the Table as a hazardous substance by the symbol "E". The proposed changes in the Table affect certain sections in Part 173.

Unlisted Hazardous Substances:

A. *ICRE Wastes.* Under the Resource, Conservation and Recovery Act (RCRA), ICRE wastes are wastes which are not specifically listed by the EPA as hazardous wastes, but which possess characteristics of ignitability, corrosivity, reactivity or extraction procedure (EP) toxicity (ICRE). These characteristics are defined by EPA in 40 CFR 261.20–261.24. In the EPA final rule of April 4, 1985 [50 FR 13456], EPA states that releases of non-designated wastes which exhibit ICR characteristics are reportable under section 103(a) of CERCLA. The RQ for non-designated substances which are ICR wastes is 100 pounds. That rule states: "Substances exhibiting the characteristic of extraction procedure (EP) toxicity are not at issue here, because the chemicals at which the EP toxicity test is aimed are all specifically designated as hazardous under section 302.4 of today's regulation." The RQs for wastes exhibiting the characteristic of EP toxicity are listed in the table at 40 CFR 302.4 and are keyed to the contaminant (hazardous substance) on which the chracteristic of EP toxicity is based. According to the EPA, when a reportable quantity of any of these unlisted ICRE wastes is released (spilled), the release is reportable under CERCLA.

Section 102 of CERCLA authorizes EPA to designate hazardous substances over and above those designated by Congress in the statute, and to establish reportable quantities for them. Using this authority, EPA plans to designate about 500 new hazardous substances in the next five years. In addition, EPA proposes to add generic groups of materials to the hazardous substance list. These generic groups include those undesignated substances which exhibit a characteristic of ignitability, corrosivity or reactivity (ICR). ICR materials are discussed in EPA's Final Rule (50 FR 13460 April 4, 1985) and are derived from EPA's RCRA regulations found in 40 CFR 261.21–261.23. The characteristics which determine whether a material is an ICR material are the same as or similar to RSPA's definitions for the hazard classes of Flammable liquid, Combustible liquid, Flammable gas, Oxidizer, Corrosive material, Explosive A, Explosive B, Flammable solid, ORM-E and RSPA's criteria for forbidden materials.

Based on EPA's interpretation that ICRE wastes are reportable under CERCLA, RSPA proposes to take the following actions to include these substances in the HMR. First, RSPA proposes to amend the hazardous substance definition in §171.8 to include ICR wastes (materials exhibiting an EP toxic characteristic are not included here because the constituents making them toxic are individually listed). This will insure that ICR wastes are included as hazardous substances in the HMR. Secondly, RSPA proposes to include instructions in the introductory language to the Table which explain the procedure to follow when selecting the proper shipping name for ICR wastes. This is necessary because ICR wastes may or may not meet the definition of a DOT hazard class other than ORM-E. For those wastes the RSPA is proposing to require that the proper shipping name include in parenthesis the EPA ICR characteristic exhibited by the waste so that the EPA ICR characteristic which makes the waste hazardous is identified. RSPA also proposes to amend the shipping paper requirements to require the applicable ICR characteristic to be included as part of the proper shipping name. Further, RSPA proposes to amend the marking requirements to require the applicable ICR characteristic to be displayed on packagings of 110 gallons or less.

B. ICRE substances. EPA also proposed in a NPRM issued on April 4, 1985 [50 FR 13514], a 100-pound RQ for releases of nondesignated substances which exhibit the RCRA characteristics of ICR. Substances exhibiting the characteristic of EP toxicity are not at issue because the substances at which the EP toxicity test are aimed are all designated by EPA. Under the EPA proposal, any spill of an ICRE substance (whether it is a waste or not) is reportable under CERCLA if the material (1) exhibits an ICRE characteristic and (2) is spilled in a reportable quantity. This proposal has not been finalized by the EPA, and RSPA does not intend to take any action on these substances until the EPA proposal becomes final. If the EPA proposal becomes final for all ICRE substances, not just wastes, RSPA will propose to amend the HMR in the same manner as we are proposing for ICRE wastes. Both the definition and the introductory language to the Table for a hazardous substance would be amended to include ICRE substances. The RSPA seeks comments on suggested methods to follow when incorporating both ICRE wastes and ICRE substances into the HMR. Also the RSPA solicits comments on the problems shippers and carriers foresee should these materials be incorporated into the HMR.

When activities designating CERCLA hazardous substances are complete, there will be approximately 1400 specifically named substances (approximately 900 designated by the statute and 500 to be designated by EPA within the next five years). This estimate may be conservative because it does not include any non-designated substance which exhibits an ICR characteristic (materials exhibiting an EP toxic characteristic are not included here because the constituents making them toxic are individually listed). Therefore, it appears that within five years almost all of the hazardous materials regulated by DOT will also be hazardous substances, since the Table presently contains only about 2400 proper shipping names, including both the specific and generic (n.o.s.) entries. The significance associated with the use of a special notation (the "E" in column 1 and the "RQ" in column 2) in the Table to identify hazardous substances for reporting purposes may be lost if virtually all of the hazardous materials are subject to the notification requirements at various RQs. In addition, RSPA may not be able to identify and class many of these materials in the Table. ICR substances could fall within as many as eight different hazard classes other than ORM-E. A similar problem exists with the incorporation of RCRA waste streams which is proposed in this NPRM. If RSPA is unable to class a material, it may be inappropriate to list it in the Table. There is certainly the argument to be made that adding numerous substances in the ORM-E class to the listing of acutely hazardous materials in the hazardous materials Table reduces the level of safety by diluting the Table to include a large number of substances that do not pose special transportation hazards. This could reduce shippers' and carriers' attention to safety by imposing comparable restrictions on substances which differ significantly in hazards.

RSPA solicits comments on the desirability of using some other system of identifying hazardous substances other than the one which presently exists and is proposed in this notice (i.e., listing a material in the Table with special notation as noted above). Other options might include establishing a separate list of all hazardous substances with their RQs. Such a list would not contain either a column for the hazard class (which, in many cases RSPA is unable to assign) or specific packaging requirements. Another option might be to simply require that all spills of chemical materials be reported, in effect consolidating all of the existing and possible reporting requirements for spills into a single, simple rule for shippers and carriers.

RSPA solicits comments on the desirability of removing the requirement in the HMR to report spills of hazardous substances. This is presently contained in § 171.17. Section 103 of CERCLA contains explicit reporting requirements for discharges of hazardous substances. Also, both the U.S. Coast Guard and EPA have rules which require reporting of discharges of these same materials (See 33 CFR 153.201 and 40 CFR 117.21 and 302.6). Referencing the existence of these reporting requirements in the HMR might be better than having another reporting requirement. RSPA believes the penalties under EPA's CERCLA rules and Coast Guard's CWA rules for failure to report a hazardous substance discharge are adequate and there may be no need to require further reporting for hazardous substance discharges in the HMR. An example of how this could be done may be found in Note 2 at the end of § 171.3.

III. REVIEW BY SECTIONS

Section 171.8. The definition of a hazardous substance would be revised to include wastes which exhibit a RCRA characteristic of ICR.

Section 171.17. Paragraph (a) would be revised to expand the "environment" (i.e., air, land and water) into which a discharge may occur and for which a discharge notification is required. Other options include removing this requirement from the HMR or referencing the CERCLA reporting requirements in the HMR.

Section 172.101. Paragraph (c)(14) would be added to cover specific waste streams that have been designated as hazardous substances by EPA and whose hazard class assignment may change from shipment to shipment because of variations in the composition of the waste stream. RSPA proposes to add 21 waste streams to the Table, all classed as ORM-E. These waste streams would not be listed by their EPA names. They would be added as follows: (1) Those waste streams whose hazard characteristics meet the definition of the ORM-E hazard class only would be assigned the proper shipping name "Hazardous waste, liquid *or* solid, n.o.s.", as appropriate. Included as part of the proper shipping name would be the EPA hazardous waste number (RCRA waste number) assigned to the waste stream (e.g., F003). Thus, a waste stream identified by EPA as "F003" containing the following spent nonhalogenated solvents and the still bottoms from the recovery of these solvents: (a) Xylene, (b) Acetone, (c) Ethyl acetate, (d) Ethylbenzene, (e) Ethyl ether, (f) Methyl isobutyl ketone, (g) n-Butyl alcohol, (h) Cyclohexanone, (i) Methanol, when classed as ORM-E and shipped in liquid form in a reportable quantity, would be described as "RQ, Hazardous waste, liquid, n.o.s. (F003) ORM-E, NA9189", (2) those waste streams having hazard characteristics that meet the definition of a hazard class other than ORM-E would be classed accordingly and assigned the proper shipping name (either a specific name or an n.o.s. entry) that most appropriately describes the waste stream. The EPA hazardous waste number assigned to the waste stream "F003" that also meets the definition of a flammable liquid would be classed as a flammable liquid. This waste stream when shipped in a reportable quantity would be described as "RQ, Waste flammable liquid, n.o.s. (F003), Flammable liquid, UN1993". If RSPA were to adopt the EPA descriptions, there would be several DOT proper shipping names which would contain more than ten words.

Paragraph (c)(15) would be added to cover hazardous substances which are waste materials that are not specifically listed under RCRA and exhibit an EPA characteristic of ignitibility, corrosivity or reactivity (ICR). Because of variations in the composition of the wastes being shipped, selection of the proper shipping name and assignment of a hazard class to these wastes may change from shipment to shipment depending on their hazards. For this reason, RSPA proposes to require that the proper shipping name include in parenthesis the EPA ICR characteristic which makes the waste hazardous under RCRA. Selection of a proper shipping name for these wastes, when shipped in a reportable quantity, would be as follows: (1) Those wastes whose hazard characteristics only satisfy the definition of the ORM-E hazard class would be assigned the proper shipping name "Hazardous waste, liquid, *or* solid, n.o.s.", as appropriate, followed by the specific EPA ICR characteristic in parenthesis; (2) those wastes having hazard characteristics that meet the definition of a hazard class other than ORM-E and also have an EPA ICR characteristic would be assigned the proper shipping name (either a specific name or an n.o.s. entry) that most accurately describes the waste; for example, if a waste meets the definition of a flammable liquid and the waste material is not specifically listed in the Table by name, the proper shipping name would be "Waste flammable liquid, n.o.s. (EPA ignitibility)"; (3) those wastes which are specifically listed by name in the Table and exhibit an EPA ICR characteristic would be assigned the proper shipping name specifically listed for that waste. For example, the proper shipping name for Methyl acetate which is a waste would be "Waste methyl acetate (EPA ignitibility)".

Hazardous Materials Table. The Table would be amended to add 31 hazardous substances designated by the EPA under 40 CFR Part 302. These substances are not presently subject to the HMR unless they are also hazardous wastes shipped on a Uniform Hazardous Waste Manifest. The EPA has assigned an unqualified "final RQ" to these hazardous substances.

The Table would also be amended to identify hazardous materials which are presently regulated that have been designated by the EPA as hazardous substances. These materials make up two major categories. One category would include materials that are now listed by name in the Table and the other would include materials not now identified in the Table by name but which are now regulated under the n.o.s. listings in the Table. Of the 111 materials being added by name, 51 are presently regulated under a n.o.s. listing, such as methyl isobutyl ketone which is regulated as a flammable liquid under the proper shipping name of "Flammable liquid, n.o.s." or under an end use listing, such as (densensitized) nitroglycerin which is regulated as a class A explosive under the proper shipping name of "High explosive, liquid". The total number of materials that would be identified as hazardous substances under this proposal would be 189. These are listed and discussed in this preamble individually or in groups to identify the criteria used for designating each as a hazardous material.

The Table would be amended to change the RQ of several entries based on EPA's adjustment of the RQ of 45 hazardous substances that were designated previously under section 311 of the CWA. The RQ for 15 hazardous substances was raised and the RQ for 30 hazardous substances was lowered.

The Table would be amended to add the "E" designator and RQ to certain mixtures that contain one or more hazardous substances to identify the hazardous substances(s) present in the mixture. The "E" designator and RQ level would also be added to certain n.o.s. descriptions for materials that may contain a hazardous substance.

Finally, the Table would be amended to make corrections to certain listings to align them with the corresponding international description or to indicate that the description is not an exact match with a similar international description. These corrections would be made in conjunction with the adjustments being made to various materials in the Table that are affected by the EPA rule on hazardous substances.

Although there may appear to be a discrepancy between the number of newly identified materials in this proposal and the number of materials in the EPA list of hazardous substances, the materials in this proposal are those covered in the EPA list. The difference in the number of materials results from the necessity to identify in the Table the different hazard classes, forms, mixtures or solutions of a material for proper regulation.

Hazard Class Determinations. The additions to the Table to accommodate hazardous substances would consist of 31 materials classed as ORM-E, 51 materials that are now regulated under n.o.s. listings which would be identified by name, and 29 materials that would be classed as ORM-A. Provisions are also made in the Table to identify 51 materials that are currently being regulated by name in the HMR. EPA has designated these materials as hazardous substances.

Other Regulated Materials—ORM-A. Twenty-nine hazardous substances would be classed as ORM-A, based on the chemical, physical and other comparable properties of the materials. The properties of the materials are such that each material can cause extreme annoyance or discomfort to passengers and crew of a transport vehicle in the event of leakage during transportation. Three of these materials, each marked with an asterisk, would be classed as combustible liquids when packaged in containers having a capacity exceeding 110 gallons. The ORM-A materials are listed below:

Acrylamide
*Benzylidene chloride
Bis(2-chloroethoxy) methane
Bromoform
4-Bromophenyl phenyl ether
p-Chloroaniline
4-Chloro-m-cresol
Chlorodibromomethane
2-Chlorophenol
4-Chlorophenyl phenyl ether
Dibromomethane
*m-Dichlorobenzene
Dichlorobromomethane
2,4-Dichlorophenol
2,6-Dichlorophenol

Dimethoate
Dimethyl phthalate
4,6-Dinitro-o-cyclohexylphenol
1,1-Dimethyl-2-phenylethanamine
Hexachloropropene
*Isophorone
Malononitrile
Methapyrilene
1,4-Naphthoquinone
Strontium sulfide
1,2,4,5-Tetrachlorobenzene
2,3,4,6-Tetrachlorophenol
1,2,4-Trichlorobenzene
2,4-Xylenol

Note that dinitrocyclohexylphenol is presently listed in the Table, but only the 4,6-dinitro-o-cyclohexylphenol isomer has been designated by EPA as a hazardous substance. Also, xylenol was previously designated as a hazardous substance (RQ-1000/454) by EPA, however, EPA has designated 2,4-dimethylphenol as a hazardous substance with a "final RQ" of 100/45.4. Another name for 2,4-dimethylphenol is 2,4-xylenol, which is a xylenol.

Other Regulated Materials—ORM-E. Thirty-one materials would be classed as ORM-E. The materials include 10 specific substances and twenty-one waste streams. Their classification is based on the EPA designation of certan materials as hazardous substances with an unqualified "Final RQ" on April 5, 1985 [50 FR 13456], and the fact that according to our tentative evaluation they do not meet the defining criteria of any other hazard class. These ORM-E materials are listed below:

Butyl benzyl phthalate
2-Chloronaphthalene
Diethyl phthalate
Di-n-octyl phthalate
Maleic hydrazide
N-Nitrosodiphenylamine
Pronamide

Resperine
Silver
Trichlorofluoromethane
EPA RCRA waste (stream) numbers F003, F007, F008,
    F009, F010, F011, F012, K014, K023, K024, K036,
    K037, K044, K045, K047, K071, K083, K093, K094,
    K103, K106.

Each waste stream would be described as Hazardous waste, liquid *or* solid, n.o.s. followed by a specific RCRA waste number and the assigned RQ. Because of various components, fractions, concentrations and properties, the hazards associated with a particular waste stream may be greater than anticipated. Each waste stream should be examined carefully. Comments concerning the hazards of these waste streams are welcome. Recommendations for hazard class assignment with supporting data are encouraged.

The 51 materials now described by the various generic n.o.s. descriptions would be identified within the Flammable liquid, Combustible liquid, Poison A, Poison B and Corrosive material hazard classes as follows:

Flammable Liquids/Combustible Liquids. Nine hazardous substances would be identified as Flammable liquids and six as Combustible liquids. Closed cup flash points were obtained for these substances from the literature. Three of the combustible liquids, each marked with an asterisk, would be classed as ORM-A when packaged in containers of 110 gallons or less. These liquids and their respective closed cup flash point are listed below:

Material	Flash point (°F.) CC.
Acetophenone	180
*Benzylidene chloride	198
2-Chloroethyl vinyl ether	61
Cyclohexanone	116
*m-Dichlorobenzene	146
1, 1-Dichloroethane	22
Ethyl methacrylate	60
*Isophorone	184
Isopropylbenzene	115
Methacrylonitrile	54
Methyl isobutyl ketone	56
2-Nitropropane	82
1, 3-Pentadiene	−20
2-Picoline	79
Propionitrile	43

Poison B. Thirty-two hazardous substances would be identified as poison B materials. Data on oral toxicity using rats (orl-rat LD_{50}: mg/kg) and toxicity by skin absorption using rabbits (skn-rbt LD_{50}: mg/kg) was obtained from the National Institute for Occupational Safety and Health (NIOSH) *Registry of Toxic Effects of Chemical Substances* (RTECS) (1981–82 Edition) for 27 compounds. Soluble cyanide salts not identified by name would be covered by two existing but modified cyanide descriptions and a new cyanide description that specifically addresses inorganic cyanides, n.o.s. Toxicity data for specific salts of dinitro-o-cresol are listed in the RTECS. No data is available for the remaining three hazardous substances. However, based on chemical and physical properties and the toxicity of similar compounds, it is the RSPA's opinion that these hazardous substances meet the criteria for this hazard class. These Poison B materials are listed below:

Material	Toxicity (LD_{50}: mg/kg)	
	Oral-rat	Skin-rbt
1-Acetyl-2-thiourea	50.0	
Aldicarb	0.9	200
5(Aminomethyl)-3-isoxazolol	45.0	
4-Aminopyridine	20.0	
Ammonium vanadate	18.0	
Chloracetaldehyde	23.0	67
2-Chlorophenyl thiourea	4.6	
3-Chloropropionitrile	50.0	
Cyanides (soluble cyanide salts), not elsewhere specified		
O, O-Diethyl S-methyl dithiophosphate		
Diethyl-p-nitrophenyl phospate	1.8	
O, O-Diethyl O-pyrazinylphosphorothioate	3.5	
Diisopropyl fluorophosphate	6.0	
4,6-Dinitro-o-cresol	10.0	
4,6-Dinitro-o-cresol salt		
Dinoseb	25.0	80
2,4-Dithiobiuret	5.0	
Endosulfan sulfate		
Endothall	38.0	
Endrin aldehyde		
Famphur	35.0	
Fluoroacetamide	6.0	
Isodrin	7.0	

(cont.)

Material	Toxicity (LD$_{50}$: mg/kg)	
	Oral-rat	Skin-rbt
Methomyl	17.0	
Naphthylthiourea (alpha)	6.0	
Octamethlpyrophosphoramide	5.0	
Osmium tetroxide	14.0	
N-Phenylthiourea	3.0	
Potassium silver cyanide	21.0	
Thiofanox	8.5	39
Thiosemicarbazide	0.9	
Warfarin	3.0	

Corrosive material. Three hazardous substances would be identified as corrosive materials. This is based on the chemical and physical properties of the compounds and the fact that several similar type compounds are classed as corrosive materials. These corrosive materials are listed below:

Benzene sulfonyl chloride
1,4-Dichloro-2-butene
Phthalic anhydride

Poison A. One hazardous substance, carbonyl fluoride, would be classed as poison A. Carbonyl fluoride is a toxic, nonflammable, colorless, irritating gas with a pungent odor. Inhalation toxicity data listed in the RTECS are as follows: inhalation-rat LC$_{50}$: 360 ppm/1 hr. Using this data, the value in milligrams per liter is calculated to be LC$_{50}$: 0.97 mg/L.

Proposed reportable quantity changes for certain hazardous substances. Based on action taken by EPA, the RQ of the following entries would be changed (raised or lowered) as indicated:

Entry	Present	Change
Acetic acid solution	RQ-1000/454	RQ-5000/2270
Acetic acid, glacial	RQ-1000/454	RQ-5000/2270
Acetic anhydride	RQ-1000/454	RQ-5000/2270
Ammonium fluoride	RQ-5000/2270	RQ-100/45.4
Ammonium sulfide solution	RQ-5000/2270	RQ-100/45.4
Amyl acetate	RQ-1000/454	RQ-5000/2270
Aniline	RQ-1000/454	RQ-5000/2270
Antimony potassium tartrate, solid	RQ-1000/454	RQ-100/45.4
Antimony trioxide	RQ-5000/2270	RQ-1000/454
Benzonitrile	RQ-1000/454	RQ-5000/2270
n-Butyl phthalate	RQ-100/45.4	RQ-10/4.54
Calcium carbide	RQ-5000/2270	RQ-10/4.54
Calcium hypochlorite, hydrated	RQ-100/45.4	RQ-10/4.54
Calcium hypochlorite, mixture	RQ-100/45.4	RQ-10/4.54
Dichlobenil	RQ-1000/454	RQ-100/45.4
1,1-Dichloropropane	RQ-5000/2270	RQ-1000/454
1,3-Dichloropropane	RQ-5000/2270	RQ-1000/454
Dinitrobenzene	RQ-1000/454	RQ-100/45.4
Dintrophenol solution	RQ-1000/454	RQ-10/4.54
Ethylenediamine	RQ-1000/454	RQ-5000/2270
Furfural	RQ-1000/454	RQ-5000/2270
Hydrofluoric acid solution	RQ-5000/2270	RQ-100/45.4
Hydrogen fluoride	RQ-5000/2270	RQ-100/45.4
Kelthane	RQ-5000/2270	RQ-10/4.54
Malathion	RQ-10/4.54	RQ-100/45.4
Mercaptodimethur	RQ-100/45.4	RQ-10/4.54
Methylamine, anhydrous	RQ-1000/454	RQ-10/4.54
Methylamine, aqueous solution	RQ-1000/454	RQ-100/45.4

Entry	Present	Change
Methyl methacrylate monomer, inhibited	RQ-5000/2270	RQ-1000/454
Methyl methacrylate monomer, uninhibited	RQ-5000/2270	RQ-1000/454
Mevinphos	RQ-1/0.454	RQ-10/4.54
Mevinphos mixture, dry	RQ-1/0.454	RQ-10/4.54
Mevinphos mixture, liquid	RQ-1/0.454	RQ-10/4.54
Nitrogen dioxide, liquefied	RQ-1000/454	RQ-10/4.54
Nitrogen tetroxide, liquefied	RQ-1000/454	RQ-10/4.54
Nitrophenol	RQ-1000/454	RQ-100/45.4
Phosgene	RQ-5000/2270	RQ-10/4.54
Phosphorus oxychloride	RQ-5000/2270	RQ-1000/454
Phosphorus trichloride	RQ-5000/2270	RQ-1000/454
Prophlene dichloride (*1,2-Dichloropropane*)	RQ-5000/2270	RQ-1000/454
Propylene oxide	RQ-5000/2270	RQ-100/45.4
Pyrethrins	RQ-1000/454	RQ-1/0.454
Quinoline	RQ-1000/454	RQ-5000/2270
Resorcinol	RQ-1000/454	RQ-5000/2270
Sodium	RQ-1000/454	RQ-10/4.54
Sodium, metal dispersion	RQ-1000/454	RQ-10/4.54
Sodium, metal liquid alloy	RQ-1000/454	RQ-10/4.54
Sodium fluoride, solid	RQ-5000/2270	RQ-1000/454
Sodium fluoride, solution	RQ-5000/2270	RQ-1000/454
2,4,5-Trichlorophenoxyacetic acid	RQ-100/45.4	RQ-1000/454
2,4,5-Trichlorophenoxyacetic acid amine	RQ-100/45.4	RQ-5000/2270
2,4,5-Trichlorophenoxyacetic acid ester *or* salt	RQ-100/45.4	RQ-1000/454
Vinyl acetate	RQ-1000/454	RQ-5000/2270
Zirconium potassium fluoride	RQ-5000/2270	RQ-1000/454

The above list of entries in the Table whose RQ would be changed is not all inclusive. For example, no attempt was made to list the potassium sodium alloys where these alloys are subject to the HMR as they apply to hazardous substances because sodium is a hazardous substance whose RQ was adjusted by EPA from RQ-1000/454 to RQ-10/4.54. Potassium is not a hazardous substance. No attempt was made to list mixtures, etc. However, the proposed changes for the entries not listed above should be obvious upon careful review of the proposed changes to the Table.

Some of the chemicals designated by EPA as hazardous substances are rather obscure. Very little information was available to help RSPA assess the hazards of these chemicals. With outside help, bits and pieces of information were obtained that allowed RSPA to assign a hazard class to all but one chemical . . . ethylenebis-(dithiocarbamic acid), CAS Registry No. 111–54–6, which has an EPA assigned RQ-5000/2270 and RCRA Waste Number U114. To the best of our knowledge, ethylenebis(dithiocarbamic acid) is a nonisolated intermediate produced *in situ* in the manufacture of certain pesticides (e.g., Maneb and Zineb). Comments concerning the hazards and hazard class assignment of ethylenebis(dithiocarbamic acid) are requested. Ethylenebis(dithiocarbamic acid) does not appear in the list of proposed changes to the Table. If there is no new information concerning this material provided in the comments to this rule, RSPA proposes to class this material as an ORM-E.

The hazard class assigned to zinc bromide would be changed from ORM-E to corrosive material. Information received by RSPA indicates that zinc bromide in both solid form and aqueous solution is corrosive to skin. The existing entry for zinc bromide would be revised to cover the material when shipped in solid form. A new entry would be added to cover zinc bromide when shipped as a solution (i.e., zinc bromide, solution).

Section 172.203. Paragraph (c)(3) would be added to require that the applicable EPA ICR characteristic be included on the shipping paper as part of the proper shipping name.

Section 172.324. Paragraph (c) would be added to require that the proper shipping name, including the applicable ICR characteristic, be shown on each packaging having a rated capacity of 110 gallons or less.

Section 173.202. In conjunction with a proposed change in the Table to align a description with the international proper shipping name, the word sequence in "Sodium potassium alloy (liquid)" would be changed to "Potassium sodium alloy (liquid)". Accordingly, the title of the section and paragraph (a) would be revised.

Section 173.206. In conjunction with two proposed changes in the Table to align the descriptions with international proper shipping names, "Sodium, metallic" would be changed to "Sodium" and the description "Sodium potassium alloy (solid)" would be changed to "Potassium sodium alloy (solid)". For sodium, the title of the section and paragraphs (a), (a)(3), (a)(10), (b) and (c) would be revised accordingly. For potassium sodium alloy (solid), the tital of the section and paragraphs (a) and (a)(10) would be revised accordingly.

Section 173.326. In conjunction with a proposed change in the Table to delete the entry for "Nitrogen peroxide, liquid" because it is not an international proper shipping name and it is a duplication of the entry for "Nitrogen tetroxide, liquid", the description in paragraph (a)(10) would be changed from "Nitrogen peroxide (tetroxide)" to "Nitrogen tetroxide".

Section 173.336. For the considerations stated in §173.326 and the proposed revision of the entry "Nitrogen tetroxide, liquid" to "Nitrogen tetroxide, liquefied" and the entry "Nitrogen dioxide, liquid" to "Nitrogen dioxide, liquefied" the descriptions in both the title of the section and paragraph (a) would be revised accordingly.

Section 173.347. In conjunction with the proposed change in the Table to align the description for "Aniline oil" with the international proper shipping name, the title of the section and paragraphs (a), (c)(1) and (d) would be changed to "Aniline".

Section 173.373. In conjunction with a proposed change in the Table to include the three isomers (ortho, meta and para) of nitroaniline in the description, the title of the section and paragraphs (a), (a)(4) and (a)(5) would be revised accordingly.

Section 172.655. In conjunction with a proposed change in the Table to align the description "Naphthalene *or* naphthalin" with the international proper shipping name, the title of the section and paragraphs (a), (b) and (c) would be changed to "Naphthalene, crude *or* refined."

(51 F.R. 22902, June 23, 1986)

49 CFR Parts 171 and 172
[Docket No. HM-145F, Amdt. Nos. 171–90, 172–108]
Hazardous Substances

I. BACKGROUND

On October 17, 1986, the President signed into law the Superfund Amendments and Reauthorization Act of 1986 (Pub. L. 99–499), which made several important changes to the Comprehensive Environmental Response, Compensation, and Liability Act of 1980 (CERCLA). Section 202 of Pub. L. 99–499 amended section 306 of CERCLA to require that the Secretary list *and regulate* hazardous substances listed or designated under section 101(14) of CERCLA as hazardous materials within thirty days of enactment of the Amendments. RSPA is today publishing a final rule under Docket HM-145F to fulfill this requirement.

RSPA has been considering incorporating CERCLA hazardous substances into the Hazardous Materials Regulations (HMR, 49 CFR Parts 171–179) under Docket HM-145E, and published both an advance notice of proposed rulemaking (ANPRM, 49 FR 35965, August 8, 1983) and a notice of proposed rulemaking (NPRM, 51 FR 22902, June 23, 1986) dealing with these issues. The Superfund Amendments of 1986 have overtaken most of the issues presented in these two notices. In this final rule, RSPA has selected the most practical method of listing and regulating hazardous substances in order to comply with the statutory deadline. A few issues remain, such as whether or not to remove the hazardous substance discharge notification requirement found at 49 CFR 171.17 from the HMR. These issues will be dealt with in the future under Docket HM-145E. Issues raised in HM-145E which are dealt with in this final rule will not be raised again under Docket HM-145E.

Today's rule includes a list of current hazardous substances with their reportable quantities (RQs), furnished by the U.S. Environmental Protection Agency (EPA). This list appears in an Appendix to §172.101 (Appendix) which replaces the CERCLA List. In addition, the rule contains amendments which apply the HMR to these hazardous substances. It is RSPA's intention to make changes from time to time to the list of hazardous substances or their RQs in the Appendix as adjustments are made by EPA.

The listing of hazardous substances and application of the HMR to them in this rule is being done differently from procedurs which have been required until now. Formerly, hazardous substances were integrated into the Hazardous Materials Table (Table) found at 49 CFR 172.101. This listing of hazardous substances in the Table contained the normal complement of·entries (i.e., proper shipping name, hazard class, identification number, required labels, packaging requirements, quantity limitations aboard aircraft, and entries for water shipments), and in addition, for each hazardous substance, the notation "E" was placed in column 1 with an italicized "RQ" notation in column 2 following the proper shipping name which in turn was followed by two numbers: the reportable quantity in both pounds and kilograms. When a hazardous substance was shipped (and by definition, at least a reportable quantity of the material had to be present in the package), both the shipping paper and the package (unless it was a bulk package (greater than 110 gallons)) had to bear the notation "RQ". This was to alert persons that a reportable quantity of a hazardous substance was present, and, should a spill occur, require the spill to be reported to the National Response Center (NRC).

In this final rule, RSPA is not integrating hazardous substances into the Table, but is placing them in a separate list, along with the reportable quantity for each substance, in the Appendix. In fact, this rule removes the special notations ("E" in column 1 and "RQ" and quantities in column 2) for hazardous substances presently in the Table. Where the hazard class for the hazardous substance is "ORM-E" (Other Regulated Material, category E), which means that that material does not meet any DOT hazard class definition except ORM-E, and is regulated only because it is a hazardous substance, the entire entry is removed from the Table. All specified hazardous substances listed and regulated under the HMTA are now found in the separate list in the Appendix. The only generic hazardous substance entry now present in the Table is "Hazardous substance, liquid or solid, n.o.s." The hazard class is ORM-E and the identification (ID) number is NA9188. One other ORM-E entry remains in the table, i.e., "Hazardous waste, liquid *or* solid, n.o.s."; ID number NA 9189. Materials which are designated hazardous substances and which satisfy a DOT hazard class other than ORM-E will be listed in both the Table and the Appendix, but the special notations no longer appear in the Table. This rule also removes the "CERCLA List" that appears after the Table and replaces that list with the Appendix.

RSPA has chosen this approach because of the great difficulty RSPA has with classing these materials. In order to place them in the Table, RSPA must determine their proper hazard class, using the hazard class definitions found in the HMR. RSPA has not been able to do this in a timely fashion due to the inherent differences in the technical and programmatic approach to these materials taken by RSPA and EPA on whose initial determinations RSPA relies.

EPA adjusted the reportable quantities of a number of hazardous substances in a final rule published on April 4, 1985 (50 FR 13456). Following this rule, RSPA published the notice of proposed rulemaking (NPRM) under Docket HM-145E. In the NPRM, RSPA proposed to integrate approximately 200 of these hazardous substances into the Table. Although RSPA has information from EPA on the physical, chemical, and toxicological properties of those materials, this NPRM was not published until June 23, 1986. This was due to the difficulty in determining the proper hazard class for the materials because they were either not suited to the established process for hazardous materials classification or because many of them were relatively obscure materials. In some cases DOT was not even able to establish the physical state (solid, liquid, or gas) for the materials designated by EPA. Given the size of this problem and the short time available to issue regulations in accordance with Pub. L. 99–499, RSPA has decided to abandon this approach and let shippers, who should know the properties of their materials, determine their proper shipping names, hazard classes, and the correct identification numbers. To do this, a shipper has the Table with its specific and generic entries, the hazard class definitions contained in Part 173, and the list of hazardous substances, including their RQ's, as designated by EPA in the Appendix. Under the HMR it has always been the responsibility of the shipper to class each material for shipment (except for explosives which require prior laboratory testing), and that responsibility remains in this final rule.

RSPA is aware that this approach will create some inconsistencies in the application of the regulations. For example, asbestos is presently regulated as an ORM-C, but the regulations only apply to asbestos that has commercial value, not waste asbestos. The packaging for commercial asbestos is specified at § 173.1090. However, asbestos is on the EPA list of hazardous substances at a reportable quantity of one pound, and this applies to all asbestos, commercial and waste, provided it is in a friable (loose) form. Therefore, under this rule commercial asbestos is regulated as an ORM-C, with packaging specified at § 173.1090, and waste asbestos is regulated as an ORM-E, with packaging specified at § 173.1300. This inconsistency occurs because of the statutory mandate in the Superfund Amendments to regulate all hazardous substances. RSPA will undertake regulatory action in the near future to correct this and other inconsistencies. Because the determination of the appropriate degree of regulation is discretionary, unlike today's action which is based on a statutory mandate, the future rulemaking will provide for notice and comment. Interested persons should withhold their comments until that notice is published.

Other than the expanded list of hazardous substances and the relocation of hazardous substances from the table to the Appendix, the regulatory requirements remain essentially the same. The shipper will have to determine the hazard class and proper shipping name for the material and the authorized packaging for the material using the Table and the packing authorizations contained in Part 173. When a hazardous substance is present in a shipment (i.e., there is a reportable quantity or more of the designated material in the package), the shipping paper entry must contain the notation "RQ". This requirement is unchanged. When the proper shipping name does not contain the name of the constituents which make the material a hazardous substance, that information must be added in association with the basic description. This requirement is also unchanged. In the case of waste streams, RSPA is requiring the use of the EPA waste number instead of the entire narrative waste stream description. the EPA waste number for the waste stream must be entered in association with the proper shipping name. In the case of a hazardous substance which satisfies one of the EPA "ICRE" hazardous waste characteristics of ignitibility, corrosivity, reactivity, or extraction procedure toxicity (EP toxicity), the requirement for additional information must be satisfied by using the letters, "EPA" followed by the word "ignitibility", or "corrosivity", or "reactivity", or "EP toxicity", as appropriate, in association with the basic description.

Procedures for marking non-bulk packagings (those of 110 gallons or less) also remain essentially the same. The "RQ" notation is required when a hazardous substance is present and if the proper shipping name does not include the constituent or constituents which make the material a hazardous substance, that information must be added in association with the proper shipping name. As is the case with shipping papers, when the hazardous substance is a waste stream or a waste material exhibiting an EPA "ICRE" characteristic, the additional identifying information required in the marking in association with the proper shipping name must be the waste stream number or, for the ICRE materials, the letters "EPA" and the word "ignitibility", or "corrosivity", or "reactivity", or "EP toxicity" as appropriate.

The regulatory action in this final rule is mandated by statute, and, for this reason, with one exception, RSPA is not affording persons by § 172.101(j) which allows up to one year after a change in the Table to use up stocks of preprinted shipping papers and to ship packages that were marked prior to the change. The exception is that RSPA is allowing preprinted shipping papers to be used and previously marked packages of hazardous substances to be transported if prepared in conformance with the requirements for hazardous substances prior to January 1, 1987. For example, shipping papers for a hazardous substance which read: "RQ, Adipic acid, ORM-E, NA9077", may be used until exhausted or until January 1, 1988, whichever comes first. After exhaustion or one year, such a shipment would have to be described as: "RQ, Hazardous substance, solid, n.o.s., ORM-E, NA9188, (adipic acid)". This also applies to marked packages. However, if the reportable quantity for the material has changed and the shipping paper entry or package marking does not reflect the reportable quantity as it appears in the Appendix in this rule, the shipment does not qualify for the exception in § 172.101(j) and must comply with the new requirements after January 1, 1987.

II. REVIEW BY SECTIONS

Section 171.8. The definition of a hazardous material is revised to specifically include hazardous substances. The definition of a hazardous substance is revised to reference a new Appendix to § 172.101 which follows the Hazardous Materials Table (Table) at § 172.101. This Appendix replaces the CERCLA List currently shown and contains all hazardous substances and their reportable quantities. Reference to petroleum products has been removed from the hazardous substance definition since the determination of what materials should be designated as hazardous substances rests with EPA. Reference to "or in one transport vehicle if not packaged" has been removed since RSPA considers vehicles to be packagings when they are the primary means of containment (i.e., are used to transport materials in bulk).

Section 171.11. The wording of (d)(1)(i) of this section is amended to require the display of the waste stream number or "EPA" and the applicable ICRE characteristic on shipping papers.

Section 171.12a. The wording of (a)(3)(i) of this section is amended to require the display of the waste stream number or "EPA" and the applicable ICRE characteristic on shipping papers.

Section 172.101, Preamble. Paragraph (b) is revised to eliminate all references to the letter "E" in the Table. Subparagraph (c)(9) is revised to remove reference to "E" and "reportable quantity" and to add provisions for selecting proper shipping names for hazardous substances.

Section 172.101, Hazardous Materials Table. The Table is revised by removing the letter "E" from Column 1 of the Title heading and all places where it appears in Column 1 of the Table. All RQ designations and quantities are removed from the descriptions in Column 2 of the Table (for example, "*(RQ-100/454)*"). The Table is revised by removing all entries for hazardous substances which only meet the definition of the ORM-E hazard class, with the exception of the generic entry "Hazardous substance, liquid *or* solid, n.o.s.". This includes removing the following five entries with "See" references to certain hazardous substances classed as ORM-E:

(1) 2,4-D ester. See 2,4-Dichlorophenoxyacetic acid ester;
(2) EDTA. See Ethylenediaminetetraacetic acid;
(3) PCB. See Polychlorinated biphenyls;
(4) 2,4,5-T amine, ester, or salt. See 2,4,5-Trichlorophenoxyacetic acid, amine, ester, or salt;
(5) 2,4,5-TP ester. See 2,4,5-Trichlorophenoxypropionic acid ester.

The entry "Hazardous waste, liquid or solid, n.o.s." remains in the Table and continues to bear the ORM-E hazard class designation. Hazardous substances meeting only the DOT hazard class definition for ORM-E appear in the new Appendix to § 172.101, along with all of the other CERCLA hazardous substances. Certain hazardous substances which satisfy the definition of a DOT hazard class other than ORM-E remain in the Table and also appear in the new Appendix. However, the "E" symbol, "RQ", and quantities no longer appear in the Table for these materials.

Section 172.101, Appendix. The CERCLA List is removed and replaced by an Appendix entitled "List of Hazardous Substances and Reportable Quantities." The appendix lists those materials which are hazardous substances as listed or designated under Section 101(14) of CERCLA.

Section 171.102. Paragraph (e) of this section is amended to require display of the waste stream number or "EPA" and the applicable ICRE characteristic on shipping papers.

Section 172.203. Paragraph (c) is amended to reference the new Appendix of hazardous substances which follows the Table and to require that hazardous substance constituents be shown in parentheses in association with the basic description, if the proper shipping name does not identify the hazardous substance constituents as shown in Appendix A to § 172.101. A new sentence is added to this section to require that a waste stream number or "EPA" and the applicable ICRE characteristic be shown, instead of the name of the constituent from the Appendix, in parentheses on the shipping paper in association with the basic description for those waste materials which are either waste streams or exhibit an ICRE characteristic.

Section 172.324. Paragraph (a) is revised to require the name of a hazardous substance constituent to be shown as a package marking, if the proper shipping name does not identify the hazardous substance constituent, as shown in the Appendix to § 172.101. Paragraph (b) is revised to require that all packages of 110 gallons or less that contain waste streams or waste exhibiting ICRE, characteristics, be marked in association with the proper shipping name with the waste stream number or "EPA" and the appropriate ICRE characteristic in parentheses. Existing paragraph (b) is redesignated as paragraph (c).

(51 F.R. 42174, November 21, 1986)

49 CFR Parts 171 and 172
[Docket No. HM-145F, Amdt. Nos. 171–90, 172–108]
Hazardous Substances; Corrections

November 21, 1986, RSPA amended the Hazardous Materials Regulations (HMR) by incorporating into the HMR, as hazardous materials, all substances designated as hazardous substances under the Comprehensive Environmental Response, Compensation and Liability Act of 1980 (CERCLA). This action was necessary to comply with the Superfund Amendments and Reauthorization Act of 1986. In the final rule, hazardous substances and their reportable quantities (RQs) were listed in an Appendix to § 172.101. In addition, the final rule contained amendments making the HMR applicable to these hazardous substances. The effective date of that final rule was January 1, 1987. However, RSPA published an amendment on December 24, 1986 (51 FR 46672) which extends that effective date to July 1, 1987, to afford shippers sufficient time to comply with the rule. This amendment corrects the errors which appear in the regulatory text of that rule.

Paragraph (d)(1)(i) of § 171.11 is revised to clearly indicate the correct format for adding additional descriptive information when the proper shipping name does not include the name of the hazardous substance.

Sections 171.12a(a)(3) and 172.102(e) are revised to reference applicable description requirements for hazardous substances in §§ 172.203(c) and 172.324. This corrects certain grammatical errors and clarifies the application of requirements. Editorially, paragraph (c) of § 172.101 is amended by removing the reference contained in this paragraph to paragraph (b)(4), since the symbol "E" no longer appears in Column 1 of the Hazardous Materials Table. Further, the spelling of the word "ignitibility" is changed everywhere it appears in the regulatory text so that the spelling of the word is consistent with EPA's spelling (i.e., ignitability).

In the Appendix to § 172.101 which begins on page 42177, paragraphs 2, 3 and 4 of the introductory text to the List of Hazardous Substances and Reportable Quantities are revised for clarity and to correct certain errors which appeared upon publication. The portion of paragraph 2 which appears on page 42178 is revised to include reference to "K numbers" since waste streams are referenced by both "F" and "K" numbers.

Several changes are made to the List of Hazardous Substances and Reportable Quantities. A few of the reportable quantities are changed (either raised or lowered) because of incorrect entries in the original list. Many of the broad generic categories of materials which appear on the list in upper case letters are removed from the list because there are no RQ's assigned to these categories. Also removed from each of these entries are the two asterisks (**) which reference a footnote that stated no RQ is being assigned to that particular generic or broad class. These entries and the footnote are removed because, by definition, a material must have a reportable quantity to be a hazardous substance. Several adjustments are made to certain entries on the list by revising, removing, or adding either the entire line entry or a portion of the entry. Some entries on the list are rearranged so they appear in correct alphabetical sequence.

The symbols "*" and "@" are deleted from certain entries which appear on the list because either the exact name of the hazardous substance does not appear in the § 172.101 Hazardous Materials Table or the name of the synonym for the hazardous substance which RSPA added is inappropriate. The footnote at the end of the list which is referenced by the symbol "***" is removed because it refers to EPA requirements. This symbol is also removed from the entries "RADIONUCLIDES", "Ferric dextran" and "Iron dextran". The symbol "#" which appears on the list after certain "F" and "K" numbered wastes and at the end of the list as a footnote is removed because it is inappropriate. The footnote represented by the symbol "¢" is revised to state explicitly that solid metals which are in pieces whose particle size is larger than 100 micrometers (0.004 inches) are not hazardous substances under the HMR. For convenience, the appendix to § 172.101 is reprinted in its entirety.

On page 42195, paragraph (c) of § 172.203 is revised to clarify when the name or names of hazardous substance constituents must appear in parentheses on the shipping paper in association with the basic description. On this same page, § 172.324 is revised for ease of understanding and to plainly state when the name or names of a hazardous substance constituent must be marked in parentheses on a package having a capacity of 110 gallons or less. Changing each of these paragraphs removes the limitation contained in the final rule that made the requirement for additional information apply only to mixtures or solutions. It now applies to shipments of pure materials as well.

The use of EPA waste numbers to identify waste streams ("F" and "K" numbers) was discussed in the preamble to the final rule (page 42175, column 2). For waste streams, the EPA waste number must be entered on shipping papers in association with the basic description (not . . . "in association with the proper shipping name." . . . as originally stated). The EPA waste number for the waste stream must also be marked on non-bulk packagings (those of 110 gallons or less) in association with the proper shipping name.

Since the final rule authorized the use of "F" and "K" numbers to identify waste streams, many people have inquired about the acceptability of using "D" numbers to identify EPA unlisted hazardous wastes which exhibit "ICRE" characteristics. Upon consideration of these comments, RSPA agrees that the use of "D" numbers should be an authorized alternative to showing the letters "EPA" and the applicable ICRE characteristic. Therefore, RSPA is revising the appropriate sections of the rules text to allow the use of the terms "EPA ignitability" or "EPA corrosivity" or "EPA reactivity" or "EPA EP toxicity", as appropriate *or* use of the corresponding "D" number, as appropriate, on shipping papers in association with the basic description and as marking on non-bulk packagings (those of 110 gallons or less) in association with the proper shipping name.

ADMINISTRATIVE NOTICES

Because the amendments adopted herein were mandated by the Superfund Amendments and Reauthorization Act of 1986 (Pub. L. 99–499, October 17, 1986), it has been determined that notice and public procedure are contrary to the public interest. No determinations have been made under the Regulatory Flexibility Act (5 U.S.C. 601, *et seq.*).

Under the terms of "DOT Regulatory Policies and Procedures" (44 FR 11034, February 26, 1979), since these amendments are part of an emergency rulemaking governed by a short-term statutory deadline, no determination has been made as to whether it is "significant".

I certify that these amendments do not require preparation of an environmental impact statement under the National Environmental Policy Act (49 U.S.C. 4321, *et seq.*).

Although the provisions of Pub. L. 99–499 provide insufficient time for RSPA to perform the required analyses and make required findings under the applicable statutory, regulatory, and executive authorities, the agency is aware that amendments of such broad applicability may produce significant impacts on industry segments, a substantial number of which may be small enterprises.

Because RSPA's role in regulating hazardous substances is directly tied to EPA's ongoing hazardous substances responsibility, primarily through the agency's determination of reportable quantities, amendments will be made to HMR as necessary to satisfy the intent of Congress expressed in Pub. L. 99–499.

(52 F.R. 4824, February 17, 1987)

[49 CFR Parts 173 and 178]
[Docket No. HM-146; Notice No. 76–13]
SHIPPING CONTAINERS
Extension of Service Life of DOT 3HT Cylincers

The Materials Transportation Bureau (MTB) is considering amending §§ 173.34 and 178.44 of the Department's Hazardous Materials Regulations as they pertain to shipments of certain nonflammable, non-poisonous compressed gases by extending the service life of DOT 3HT specification cylinders from 15 years to 24 years.

DOT 3HT cylincers are made of high strength steel with the high operating wall stresses. Because of these high stresses, the authorized service life of the 3HT cylinder is specifically limited. Authorized service life in the original DOT 3HT specification was limited to 4380 pressurization (equivalent to a 12-year life based on once-a-day pressurization) or 12 years, whichever occurred first. In 1970 while retaining the 4380 pressurization limitation, the 12 years was extended to 15 years by Docket HM-31, Amendment Nos. 173–19 and 178–19 (35 FR 5331, March 31, 1970). That docket provided for consideration of further extension if sufficient justification was provided.

This proposal is based on the petitions by the Air Transport Association, Inc. (ATA), dated July 17, 1975, and Compressed Gas Association, Inc. (CGA), dated March 6, 1973. These Associations have submitted the results of a comprehensive test program performed on cylinders selected at random representing eight existing 3HT cylinder sizes. The test cylinders selected had been in service for an average of 14.2 years. The test results show that:

1. Each cylinder of the 26 cylinders subjected to a pressure cycling test passed the 10,000 cycles required for new cylinders.

2. Of 24 cylinders that were pressurized to burst, one cylinder failed to withstand the specified minimum burst pressure requirement for new cylinders (2.22 times the service pressure). This cylinder burst at 2.11 times the service pressure. Considering that this cylinder has been in service for 14 years and has successfully passed other prescribed tests, such as the hydrostatic test, magnetic particle inspection, and visual inspection, this minimal deviation in burst pressure is not considered significant.

3. The test regimen, consisting of but not limited to complete internal and external visual inspection, magnetic particle inspection, hydrostatic testing, cycling and physical testing, was followed and all results obtained were satisfactory.

The ATA, in its supporting data, showed that the field service recharging frequency of these cylinders was estimated to be not more than once in 10 days. Based on the 4,380 pressurization allowed, this is equivalent to 120 years service life.

The MTB agrees that the petitions have merit. In addition, the MTB on its own initiative is proposing to require that the cylinder be permanently marked with its rejection elastic expansion (REE). The reason for this proposal is to make the retester aware of this requirement peculiar only to the DOT 3HT specification. The MTB proposes to amend certain sections in the regulations, covering cylinder rejection due to excessive elastic expansion, as follows:

1. For new cylinders, the REE instead of the original elastic expansion (EE) must be marked on the cylinder.

2. For existing cylinders, the REE must be determined and then marked on the cylinder near the original EE prior to the next retest.

(41 F.R. 54958, December 16, 1976)

[Docket No. HM-146; Amdt. Nos. 173–112, 178–45]
PART 173—SHIPPERS—GENERAL REQUIREMENTS
FOR SHIPMENTS AND PACKAGINGS
PART 178—SHIPPING CONTAINER SPECIFICATIONS
Extension of Service Life of DOT 3HT Cylinders

On December 16, 1976, the Materials Transportation Bureau (MTB) published a Notice of Proposed Rulemaking, Docket HM-146, Notice 76–13 (41 FR 54958) which proposed these amendments. The background and basis for incorporating these amendments were discussed in that notice. Interested persons were invited to give their views prior to the closing date of March 22, 1977. Primary drafters of this document are Jose B. Pena, Office of Hazardous Materials Operations, Technical Service Branch, and Douglas A. Crockett, Office of the Chief Counsel, Research and Special Programs Directorate.

Several commenters suggested that the marking of the rejection elastic expansion on cylinders not yet marked as of the effective date of this rule be marked before the next retest date so that these cylinders need not be removed from service immediately. The MTB agrees and accordingly includes the recommendation in this rule.

Another commenter pointed out that some cylinders manufactured to DOT Special Permit 5967 in the past were incorrectly marked with "DOT 3HT" in addition to "SP 5967". Cylinders made under SP 5967 are not DOT 3HT cylinders and the service life extension provided in this amendment for 3HT cylinders does not apply to SP 5967 cylinders, even if those cylinders are marked as 3HT cylinders. Such cylinders can be identified by the marking "SP 5967" which will appear on the shoulder of the cylinder. Any cylinder so marked, regardless of any other markings that may also appear, is not a DOT 3HT cylinder, and the service life extension provided herein does not apply. This comment is made here to notify shippers, carriers or other uses of 3HT cylinders that improperly marked cylinders do exist and must not be confused with DOT 3HT cylinders.

(42 F.R. 63644, December 19, 1977)

PART 173—SHIPPERS—GENERAL REQUIREMENTS
FOR SHIPMENTS AND PACKAGINGS
Carbon Dioxide Transported Aboard Aircraft

The purpose of this amendment to Part 173 of the Department's Hazardous Materials Regulations is to except from the coverage of those regulations carbon dioxide in solid form, when used as a refrigerant to preserve materials for medical diagnostic or treatments purposes. Regulations applicable to all carbon dioxide in solid form were published in the FEDERAL REGISTER on April 15, 1976 (41 FR 15972) under Docket HM-103/112. No comments were received which objected to the requirements imposed by this docket publication. Recently, the Materials Transportation Bureau has become concerned that these regulations may be disrupting certain shipments of essential medical supplies and specimens for critical medical diagnoses since carbon dioxide in solid form is used extensively for the preservation of materials being shipped by aircraft for medical diagnostic and treatment purposes. The Bureau recognizes that certain hazards are associated with carbon dioxide when shipped by air. It was for this reason that carbon dioxide in solid form was placed under regulation. IT is the Bureau's opinion that on balance the benefits which accrue from the rapid movement of these materials outweigh the limited risks involved. Therefore, to the extent the regulations restrict the timely performance of essential medical services and treatment, some adjustment to the regulations is called for.

This Bureau plans to review the total effect of its regulations as they apply to carbon dioxide in solid form. Meanwhile, in consideration of the desirability of not disrupting the movement of these materials by aircraft, the regulations are amended in that regard. Because this action does not impose any burden on the general public, public proceedings are unnecessary.

(42 F.R. 5059, January 27, 1977)

[Doc. No. HM-148; Amdt. No. 172–33]

PART 172—HAZARDOUS MATERIALS TABLE
AND HAZARDOUS MATERIALS
COMMUNICATIONS REGULATIONS
Correction of ORM Entries

On December 30, 1976 (41 FR 57018), the Materials Transportation Bureau (MTB) published its last major correction to amendments appearing under Docket HM-103/112 on April 15, 1976 (41 FR 15972). The shipping names ORM-A n.o.s. and ORM-B n.o.s. were shown in the Hazardous Materials Table in the April 15 publication as applicable only to transportation by air and water, shipment of such materials by other modes being not subject to the regulations. However, ORM-A n.o.s. and ORM-B n.o.s. materials were proposed in a notice appearing on January 24, 1974 (39 FR 3022) to apply solely to air shipment, and it was the intent of that notice that any material meeting either ORM-A or ORM-B definition would only be regulated when shipped and transported by air unless otherwise indicated by an individual entry specifically identifying the material.

That intent is evident from an examination of the 1974 notice, which indicates that the ORM-A and ORM-B classes were derived from C.A.B. No. 82, Air Transport Restricted Articles Tariff 6-D and from the International Air Transport Association Restricted Articles Regulations (39 FR 3025). The text proposed regulation of both n.o.s. entries (39 FR 3076) as well as most specifically identified ORM-A and ORM-B materials by air only. Some specifically identified materials (e.g., Chloroform (ORM-A) and Calcium Oxide (ORM-B) were proposed to be regulated by water as well as air.

The ORM-B class was in part also derived from a previously established rulemaking under Docket HM-57 which dealt generally with materials having corrosive effects on steel, aluminum and human skin. On September 8, 1975 (40 FR 41527), MTB noted that its "disposition of the classification of materials corrosive only to aluminum will be in the amendments . . . under Docket No. HM-112." In both the previous January 24, 1974 HM-112 notice and the subsequent April 15, 1976 HM-112 amendments, references proposed and implemented in 49 CFR 173.240 (Corrosive Material; definition) excluded aluminum corrosion and the April 15 preamble observed that materials corrosive only to aluminum "will be classed as ORM-B and regulated for air transportation only" (41 FR 15984).

The necessary correction to the ORM-A n.o.s. and ORM-B n.o.s. entries in the Table, consisting of a removal of the "W" in column 1 of each entry, was overlooked in HM-103/112 publications and is being made herein. As this amendment relieves a previously stated requirement and does not impose any burden on the general public, public proceedings are unnecessary.

(42 F.R. 7139, February 7, 1977)

[49 CFR Parts 172,175]
[Docket No. HM-149; Notice 77–2]

AIR TRANSPORTATION OF SMALL QUANTITIES
OF MATERIALS EXHIBITING VERY LOW LEVELS
OF RADIATION
Proposed Exemption Renewal

Effective until May 3, 1977, small quantities of radioactive materials otherwise qualifying for marking, labeling and packaging exceptions under 49 CFR 173.391 are excepted by § 175.10(a) (6) from the remaining requirements of Subchapter C when transported by air.

The § 175.10(a) (6) exception was published two years ago (49 FR 5168, February 4, 1975; 40 FR 17141, April 17, 1975) as 14 CFR 103.1(c) (4), excepting such small quantities from application of Part 103. It was republished on April 15, 1976 (41 FR 15972) as an exception only to requirements of 49 CFR Part 175. On December 30, 1976 (41 FR 57018), the exception was modified to except such air shipments from all of the requirements of Subchapter C. That amendment was made necessary by uncertainty over the effect on the exception of other parts of Subchapter C of 49 CFR, uncertainty which arose as a result of the transfer of the exception from 14 CFR to 49 CFR.

Conforming with § 107 of the Act (49 U.S.C. 1806) governing exemptions, the § 175.10(a) (6) exception, like its predecessor 14 CFR 103.1(c) (4), is limited to a two-year life unless reexamined and renewed. As was the case with 14 CFR 103.1(c) (4) which became effective on May 3, 1975, § 173.10(a) (6) will terminate unless renewed on May 3, 1977. The MTB proposes to renew it on a finding that renewal is consistent with the public interest and the policy of the Act. Renewal is proposed with the following modifications.

Upon renewal, the text of the exception for radioactive materials qualifying under § 173.391(a) will appear as a new paragraph designated § 175.10(b). The effect of this change will be to require shipping papers and shippers' certifications for air shipment of materials which are not a component part of an instrument or manufactured article. Air shipments of instruments and manufactured articles containing small amounts of radioactive materials will continue to be excepted by § 175.10(a) (6) from the requirements of the subchapter, including shipping paper requirements. The difference in proposed treatment of such materials reflects the fact that low-level radioactive materials incorporated in instruments (timepieces, electronic tubes, smoke detectors, and similar devices) are component parts of those items and are in a nondispersible form.

The text of proposed § 175.10 (a) (6) and (b) is updated to reference § 173.391 as of the effective date of renewal, because that section has remained essentially unchanged since the exception was first published in 1975. The reference to § 175.10(a) (6) in § 172.204(c) (4), concerning certification as to intended use, is proposed to be changed to read "§ 175.10(b)", and the parallel reference to § 175.10(a) (8) is proposed for deletion, since articles addressed in § 175.10 (a) are presently not subject to Subchapter C.

(42 F.R. 16459, March 28, 1977)

[Docket No. HM-149; Amdt. Nos. 172–35, 175–4]
PART 172—HAZARDOUS MATERIALS TABLE AND HAZARDOUS MATERIALS COMMUNICATION REGULATIONS
PART 175—CARRIAGE BY AIRCRAFT
Air Transportation of Limited Quantities of Low-Level Radioactive Materials; Exemption Renewal

By notice (42 FR 16459, March 28, 1977) the MTB proposed to renew the expiring exemption found at 49 CFR 175.10 (a) (6). Twelve comments were received which may be grouped as follows:

(1) Objection to imposition of the proposed shipping paper requirements for § 173.391(a) materials, and request for a public hearing on that subject.

(2) Objection to continuation of surface shipping paper requirements for § 173.391 (b) and (c) materials in light of the proposed continuance of air shipping paper exceptions for § 173.391 (b) and (c) materials;

(3) Request for amendment of § 172.203(d)(1)(iii) to allow alternative use of a statement of maximum permitted activity in place of the actual activity per package;

(4) Request for a permanent regulation instead of an exemption renewal;

(5) General support for the March 28 proposals.

There were not received any objections to basic continuation of the exemption. Given the public interest in matters related to the exemption, such as shipping paper requirements for surface transportation as well as for transportation by air, the MTB has decided to limit final action under this docket to the exemption itself and to consider the question of shipping paper requirements and the request for a public hearing as separate matters under a new docket. A notice of proposed rulemaking for that purpose will appear in a subsequent issue of the FEDERAL REGISTER.

The exemption itself is authorized by § 107(a) of the Hazardous Materials Transportation Act of 1974 (Title I of Pub. L. 93–633; 49 U.S.C. 1806(a)) and necessitated by § 108(b) of that Act (49 U.S.C. 1807(b)). It is predicated on the very limited hazards posed by those materials meeting the criteria of § 173.391 (a), (b), and (c). Because the existing exemption, which relieves a restriction stated in the Act and in § 175.700(d), will expire on May 3, 1977, an effective date of less than 30 days following this publication is necessary to avoid disrupting exempted shipments. Continuation of the exemption will have a negligible environmental impact and will not impose any additional costs on shippers, carriers or consumers.

(42 F.R. 22366, May, 3, 1977)

[Docket No. HM-151, Amdt. Nos. 171–36, 172–37]
PART 171—GENERAL INFORMATION, REGULATIONS, AND DEFINITIONS
PART 172—HAZARDOUS MATERIALS TABLE AND HAZARDOUS MATERIALS COMMUNICATIONS REGULATIONS
Label and Placard Colors; Hazard Numbers

A color standard for label and placard colors was published as a final rule under Docket No. HM-103/112 on April 15, 1976 (41 FR 15972), compliance with which became mandatory on January 1, 1977.

This standard, proposed in 1974 under Docket No. HM-103 (39 FR 3164, January 24, 1974), involved two series of color charts provided by DOT that display standard colors. The colors on the charts are also numerically described in Appendix A to Part 172 by certain technical specifications (Munsell notations). The visual display on each chart incorporates a degree of latitude, or tolerance, to account for variations in printing materials and processes and was intended to serve as a visual control on label and placard colors, while the Munsell notations were provided to ensure constancy and reproducibility of the Color Tolerance Charts.

However, the manner in which §§172.407(d), 172.519(e) and Appendix A to Part 172 are stated makes it appear that the regulatory standard is the Munsell description rather than the visual display on the Color Tolerance Charts. Use of the numeric Munsell description as a standard could necessitate an instrumented examination of label and placard colors, relegating the Color Tolerance Charts to serving as a visual representation of the specified Munsell descriptors. Since an instrumented color analysis is beyond the practical capacity of many label, placard and packaging manufacturers, and many if not most shippers, some correction to the published standard is required. Moreover, field inspections such as those conducted by the Bureau of Motor Carrier Safety, as a practical matter, cannot include color instrumentation. Inspectors will use the Color Tolerance Charts and judge compliance by visual comparison between those charts and label and placard colors.

Accordingly, this rulemaking restates the color standard to establish as the controlling standard the colors displayed on the Color Tolerance Charts. The Munsell notations, a technical description, have been retained in Appendix A to Part 172 to ensure accurate reproduction of the charts. In this restatement, the weathering and fadeometer tests have been withdrawn from Appendix A and placed in §§172.407 and 172.519. As applicable to labels, the weathering test has been modified to take into account the practical limitations of packaging materials upon which labels are affixed or printed. Also, advisory references to two standards adopted by the American Society for Testing and Materials are included to illustrate what is meant by the fadeometer test requirement. Any fadeometer test that is a recognized standard procedure may be used, and either a wet or dry method may be selected. Appendix A is restated for clarity in its entirety, but the Munsell notations are themselves unchanged. Changes appear only in heading, footnotes, and the format of the Chroma column in Table I.

A limited exception to the required use of the Color Tolerance Charts has been included for labels printed before March 1, 1979, directly onto the surface of packagings. The costs and technical problems of printing with close color tolerances on packaging surfaces such as fiberboard, which may be both porous and pigmented, will require further evaluation. The Office of Hazardous Materials Operations will publish a notice outlining in some detail the factors bearing on a possible resolution and soliciting public comment.

As an additional matter, the MTB, acting on a petition concerning §172.407 (g)(3), is amending that provision. Section 172.407(g) allows the United Nations and Intergovernmental Maritime Consultative Organization hazard class number to be displayed in the lower corner of a label, but the number "must be . . . (a)pproximately 0.25-inch (6.3 mm.) high." That requirement, in §172.407(g) (3), is too vague to be useful. Labels have sufficient space to allow display of a hazard number up to one-half inch (12.7 mm.) in size, and §172.407(g) (3) is being amended to reflect more accurately that practical limitation.

This change is a relaxation of existing requirements and is not expected to impose any additional costs or burdens on the public, industry or government, or to have any significant environmental or economic impact. In view of that, and because the existing mandatory standards may be causing unnecessary compliance difficulties at the present time, public notice and comment are being dispensed herewith and the change is being made effective in less than 30 days after publication in the FEDERAL REGISTER.

(42 F.R. 34283, July 5, 1977)

[Docket No. HM-151A, Amdt. No. 172-50]
PART 172—HAZARDOUS MATERIALS TABLE AND HAZARDOUS MATERIALS COMMUNICATIONS REGULATIONS
Label and Placard Colors

A color standard for label and placard colors was published as a final rule under Docket No. HM-103/112 on April 15, 1976 (41 FR 15972), compliance with which became mandatory on January 1, 1977. That standard, proposed in 1974 under Docket No. HM-103 (39 FR 3164, January 24, 1974) and amended by Docket No. HM-151 (42 FR 34283, July 5, 1977), involved two series of color charts provided by DOT that display standard colors. The colors on the charts are also numerically described by Tables 1 and 2 of Appendix A to Part 172 through certain technical specifications (Munsell notations). The visual display on each chart incorporates a degree of latitude, or tolerance, to account for variations in printing materials and processes and to serve as a visual control on label and placard colors, while the Munsell notations were provided to ensure constancy and reproducibility of the Color Tolerance Charts.

In 1976 it became necessary to suspend that standard so far as it applied to labels printed directly onto the surfaces of a packaging, because such surfaces could not be controlled for printing purposes without great expense. Fiberboard, for example, is both pigmented and porous, with the result that pigments applied to it are absorbed and their visual effects modified. The suspension terminates on March 1, 1979, at which time the previously published standard applies. During the suspension, several trade associations concerned with aspects of packaging manufacture have examined existing conditions and have recommended to the Materials Transportation Bureau (Bureau) that the previously published standard be modified to account for the difficulties encountered in printing on kraft and fiberboard.

This amendment contains a color standard recommendation that was developed during the past 18 months by representatives of the container Corporation of America, the Fibre Box Association and the Paper Shipping Sack Manufacturing Association. The Bureau has reviewed this recommendation, finds it reasonably consistent with the existing standard, and adopts it. To produce this recommended standard, inks were selectively obtained and mixed and an improved quality control was used in the printing processes. Although reproduction of the previously published DOT color tolerances was the object of this industry effort, the printing processes and the porosity and pigmentation variations of kraft paper and fiberboard surfaces necessitate new color tolerances. Similar difficulties have caused these associations to advise that they are not able to produce color tolerance charts for visual reference for compliance. Consequently, the color tolerances developed in this manner are represented in Table 3 by Commission internationale de L'Eclarlage (C.I.E.) coordinates only. DOT will use the existing color charts for compliance purposes, although resort to instrumentation will be necessary where an initial visual reference indicates that the colors of a label printed on a packaging surface are improper.

To assist those who may prefer instrumentation to ensure compliance with color requirements for placards and for labels which are not printed on packaging surfaces, the C.I.E. coordinates have been added to the Munsell notation for the existing Color Tolerance Charts (see new Tables 1 and 2). This will result in C.I.E. coordinates being available for instrumentally evaluating all hazardous materials warning labels and placards.

Although the work described above was performed with variations of kraft and fiberboard surfaces, the Bureau has made the standards applicable to labels printed on any packaging surface, regardless of the substrate. Kraft paper and fiberboard represents the most difficult case, but other situations may exist wherein the manufacturer's inability to select a printing substrate, due to structural requirements and economic factors affecting packaging construction, may result in deviations from the previously published standard. The temporary two-year suspension of the previously published standard has been extended to July 1, 1979, to allow time for packaging manufacturers to modify their processes to bring them into compliance with the new standard published in this document. However, they have a choice of compliance with either Table 1, 2, or with Table 3.

Since the previously published standard would otherwise apply to labels printed on packaging surfaces after March 1, 1979, the less restrictive standards published in this document constitute a relaxation of existing requirements. This amendment is not expected to impose any additional costs or burdens on the public, industry or government, or to have any significant environmental or economic impact. The amendment in fact results from petitions received from the fiberboard and kraft paper packaging industries. In view of the above, the Bureau finds that public notice is unnecessary. However, comments are welcome and should be sent to the address indicated earlier.

(44 F.R. 9756, February 15, 1979)

[49 CFR Part 175]
[Docket No. HM-152; Notice No. 77–6]

CARRIAGE BY AIRCRAFT
Requirements for Radioactive Materials

In July of 1974, the U.S. Atomic Energy Commission (AEC) transmitted to the Federal Aviation Administration (FAA) of the Department of Transportation several recommendations regarding the transportation of radioactive materials aboard civil aircraft ("Recommendations for Revising Regulations Governing the Transportation of Radioactive Material in Passenger Aircraft," July, 1974, on public file in the Section of Dockets, Office of Hazardous Materials Operations, 2100 2nd Street, SW., Washington, D.C.). These recommendations have been under review and have been the subject of discussions between the staffs of the two agencies and the successors to the AEC, the U.S. Nuclear Regulatory Commission (USNRC) and the Energy Research and Development Administration (ERDA). The MTB has evaluated these recommendations and the several discussions held thereon, and believes that they provide a basis for the proposals in this document to reduce the radiation exposure to persons aboard aircraft transporting radioactive materials.

The proposed rules would revise § 175–700, applicable only to passenger-carrying aircraft, to restrict the carriage of radioactive materials packages required to bear a Radioactive Yellow-III label to those with a transport index of 3.0 or less. Additionally, in order to insure the least amount of potential exposure to passengers, the proposed rules would require each radioactive material package required to bear a Radioactive Yellow-II or radioactive Yellow-III label to be stowed on the floor of the cargo compartment of the aircraft. Furthermore, a package required to bear either of those labels could be carried on a passenger-carrying aircraft only if the radioisotope it contains has a radioactive half-life that does not exceed 30 days. Exceptions to the half-life restriction would be provided for radioactive materials that are susceptible to rapid chemical deterioration (such as those requiring dry ice refrigeration), those having a half-life exceeding 10^8 years (such as natural or depleted uranium), and certain export or import shipments as specifically approved by the Director, OHMO.

A new § 175.701 is proposed, setting forth minimum spacing distances between people or animals and packages of radioactive materials carried aboard passenger-carrying aircraft. This section would replace the required separation distances contained in existing § 175.700.

The proposed new § 175.701 would permit the aircraft operator to develop a system of predesignated areas for the stowage of packages of radioactive materials aboard passenger-carrying aircraft. The specific details of the proposed use of such a "spacing out" system by an aircraft operator would be required to be approved by the Director, MTB. Under this proposal, a system of predesignated areas would be approved by the Director if it were designed to assure that: (1) the packages are placed in each predesignated area in accordance with § 175.701 (a); and (2) the predesignated areas are laterally separated from each other by at least four times the applicable distance specified in the table in § 175.701 (b) (2) as measured in accordance with § 175.701(b) (1). These proposals are intended to preclude any radiation level "peaking" from the cumulative effect of radiation emitted from each predesignated area.

Proposals to amend § § 175.75(a) (3) and 175.702 would provide for an increase in the amount of radioactive material permitted to be carried aboard a cargo-only aircraft, and would set forth the requirements for stowage in such situations. Current § 175.75(a) (3) limits the maximum quantity of radioactive materials that may be carried aboard an aircraft to an amount that totals a transport index of 50. It is proposed to amend § 175.75(a) to increase the maximum amount that may be carried aboard a cargo-only aircraft to a total transport index of 200. More specifically, under proposed § 175.702, when the total transport index does not exceed 50, the separation distance requirements applicable to passenger-carrying aircraft would apply to cargo-only aircraft. However, when the transport index of all packages exceeds 50, the proposal would require a minimum separation distance of 30 feet (9 meters). Additionally, in such cases, groups of packages would be limited to a transport index of 50, with each group separated from every other group by not less than 20 feet (6 meters). When packages of fissile radioactive materials are being carried, the total transport index for any aircraft would be limited to a maximum of 50, rather than 200, to assure nuclear criticality safety.

A new § 175.703 is proposed to incorporate the existing requirements of § 175.700 for separation of radioactive materials packages from undeveloped film. The new section would also provide conditions for overpacking or "bagging" or properly marked and labeled packages of radioactive materials within an outer enclosure such as a heavy gauge plastic bag or a fiberboard box. Present requirements for labeling and transport index determinations do not address this situation. The proposed procedures would specify the conditions for such use.

(42 F.R. 37427, July 21, 1977)

49 CFR Parts 173 and 175
[Docket No. HM-152; Amdt. Nos. 173–136, 175–13]
Requirements for Transportation of Radioactive Materials

On July 21, 1977, a notice of proposed rulemaking (Docket HM-152; Notice 77–6) was published in the Federal Register (42 FR 37427) announcing the Materials Transportation Bureau (MTB) intention of further restricting the transportation of radioactive materials aboard civil aircraft. These proposed changes to the regulations were prompted by a report titled "Recommendations for Revising Regulations Governing the Transportation of Radioactive Material In Passenger Aircraft" which was prepared by the U.S. Atomic Energy Commission (AEC) and transmitted to the Federal Aviation Administration (FAA) of the Department of Transportation (DOT) in July, 1974. The principal recommendation of this report is to reduce by approximately one-half the maximum permissible radiation level at seat height to 2 millirem per hour and the average radiation dose rate to 1 millirem per hour. In its subsequent discussions with the FAA and the successors of the AEC, the U.S. Nuclear Regulatory Commission (NRC) and the Energy Research and Development Administration (ERDA), the MTB determined that the proposed rules were necessary to attain a greater level of safety for passengers and crew members of passenger-carrying aircraft without unduly subjecting ground service personnel and crews of cargo-only aircraft to the threat of increased exposure to radiation.

Comments received in response to the notice of proposed rulemaking were evaluated on the basis of their: (1) applicability to this particular rulemaking: (2) effectiveness in helping to reduce radiation levels and (3) reasonableness of the methods by which this objective is to be realized. Comments were received from 37 different sources representing the views of air carriers and air carrier associations, organizations of airline employees, producers and associations of producers of nuclear materials, consumer interest groups, private individuals and various Federal agencies. The points raised by these commenters were generally reflective of the special interest each party foresaw as being impacted by such a rule change. The comments were very useful in preparing this final rule.

The most significant difference between the final rule and the proposed rulemaking is the absence of § 175.700(a)(5) which would have imposed restrictions for the transportation of radioactive materials based upon their half-life or susceptibility to rapid chemical deterioration. This particular issue drew the greatest amount of response with nearly one-third of the commenters objecting to it. Commenters pointed out that restricting radioactive materials according to half-life would not in and of itself be effective in reducing the level of radiation exposure since the prescribed limits will be effectively maintained through adherence of package transport index and distance separation factors. The MTB agrees with this conclusion and has therefore decided to eliminate the proposed restriction in this rulemaking.

One commenter who represents an international corporation which ships 20,000 radioactive shipments per month complained that "these regulations appear to be directed at our particular class of shipper for one mode of transportation." This commenter contended that "if a 'potential hazard' exists for air transportation, it would also exist for land and water transportation." Although the transportation of radioactive materials by modes other than air is not addressed in this docket, it must be pointed out that the short half-life of radiopharmaceuticals requires rapid delivery such as that provided by aircraft, particularly passenger-carrying aircraft. This rapid delivery requirement has resulted in approximately 800,000 packages of radioactive materials being transported by passenger-carrying aircraft in 1975 ("Final Environmental Statement on the Transportation of Radioactive Material by Air and Other Modes," Dec. 77, NUREG-0170, pp. 1–11,16), as a consequence of this increased shipping activity the annual population dose from direct radiation exposure has risen beyond levels which are not "as low as reasonably achievable (ALARA)." This is not the case with radioactive materials being transported in passenger-carrying motor vehicles, rail cars and vessels, and similar regulatory actions are not warranted for these modes at the present time. The contention that the proposed regulations would be discriminatory to a particular class of shipper by establishing certain requirements relating to the use of overpacks is well taken. Consequently, the proposal has been broadened to permit the consolidation of radioactive materials packages by persons other than the original shipper, with the condition that their determination of the T.I. for the overpack be made by addition of the individual package T.I.'s and not actual measurements.

In response to the commenter who pointed out a discrepancy which would exist between the air and highway modes through separate requirements for the labeling of overpacks, the amendment provides an exception in § 173.393(r) permitting a single label to be applied on nonrigid overpacks as well as the use of the term "mixed" on this label. Under the proposal, "mixed radioactive materials" was the proposed description for use on a label affixed to an overpack containing different radionuclides. In this final rule, the description has been changed to "mixed" because the terms "radioactive" and "contents" already appear as part of the label and because of the limited availability of space on the label. Although the proposed rule did not make reference to any changes in Part 173, the MTB subsequently recognized that the requirements originally contained in proposed § 175.703(b) were more appropriate to shippers than to air carriers. Therefore, § 173.393 has been amended to reflect these requirements. In this way also, handling, marking, and labeling requirements for packages of radioactive materials contained in overpacks are now addressed for all modes of transportation.

One commenter who objected to any requirement for the additional labeling of clear plastic overpacks argues correctly that other hazardous materials in similar transparent overpacks are not subject to this requirement; however, for the benefit of cargo handlers the MTB believes that the presence of a label which specifies the composite T.I. is valuable in helping to reduce exposure time which would otherwise be spent in making a close examination of the individual packages. The hazards associated with radioactive materials dictate that standard procedures which apply to hazardous materials generally are not always adequate in reducing unnecessary risks. Therefore, this requirement has been included in the revised regulation.

A comment urging the MTB to reconsider its decision on maximum radiation levels at seat height did not contain new information to support a change from the rule as proposed. As mentioned earlier, the MTB is adopting a standard which is based on a maximum of 2 millirem per hour and an average 1 millirem per hour at seat height. This is a 50 percent reduction of the previously authorized limit. While it is obvious that the reduction of any radiation exposure is desirable, the imposition of a lower limit has not been shown to be of a significant benefit commensurate with its cost. There is a cutoff point where benefits begin to diminish very rapidly when additional measures in the form of increased shielding, lower transport index limitations, and distance separation factors are applied. As it was pointed out in the "Assessment of the Environmental Impact of the FAA Proposed Rulemaking Affecting the Conditions of Transport of Radioactive Materials on Aircraft" (BNWL-B-421) the question then is what is a reasonably achievable exposure limit and package T.I. limit? The MTB believes that the present data indicate these amended limits are as low as reasonably achievable.

Many other comments were submitted in response to Notice 77–6, however, their content was not considered useful in meeting the objectives of this rulemaking of reducing radiation levels and population exposure rates associated with the transportation of radioactive materials by air. Expanding this rulemaking to include the substance of these comments was determined to be inappropriate at this time.

The following is a section by section summary of the revised regulations which address particular comments contained in the docket.

PART 175—CARRIAGE BY AIRCRAFT

Section 175.75: Specifies T.I. limits for passenger-carrying and cargo-only aircraft. Several commenters objected to the 200 T.I. limitation for cargo-only aircraft contending that the MTB "has failed to demonstrate that there is a compelling need for lifting the current restriction" and that most aircraft are not physically capable of safely handling such large volumes. It should be pointed out that the 200 T.I. is a quantity limit which applies to the largest of aircraft. Small and intermediate aircraft will naturally be restricted to smaller cargo storage areas. The MTB has purposely increased the T.I. limit so that shippers and carriers might be encouraged to divert radioactive materials from passenger-carrying flights, thereby reducing the annual population dose.

Section 175.85: Changes the reference in paragraph (d) to read "§175.701." Several carriers sought an exception for radioactive materials in paragraph (b) which would permit their stowage in an inaccessible location on cargo-only aircraft. This particular item has already been specifically addressed in another rulemaking, Docket No. HM-168; Amdt. 175–11, which appeared in the Federal Register on January 31, 1980 (45 FR 6946).

Section 175.700: Provides regulations specific to radioactive materials on passenger-carrying aircraft. As mentioned earlier, the proposed restriction of radioactive materials by half-life has been dropped from the final rule. In paragraph (a)(3) the wording has been revised to permit loading not only on the floor of the cargo compartment but on the floor of an airfreight container as well. This action is being taken in response to one commenter who pointed out the added safety benefits afforded by containerized cargo. A number of commenters suggested additional requirements which would have specified a free clearance of at least 20 inches from the topmost surfaces of all packages to the nearest surface of the partition separating the cargo compartment from the passenger compartment, and stowage in the rear most practicable position in the aircraft. In the opinion of the MTB, this is simply another means of meeting the objective of reducing radiation exposure levels and it does not appear necessary to impose such particular requirements since package T.I. limits and distance separation factors provide the level of safety desired while still allowing carriers the opportunity to comply in a manner which is most appropriate to their operations.

Another commenter not able to find a specific provision covering the shipment of a "large radioactive source" recommended retention of the wording previously contained in §175.700(c). Present constraints in the regulations under §173.391 limit the carriage of radioactive materials by passenger-carrying aircraft to only those materials which are intended for use in, or incident to, research, or medical diagnosis or treatment and thus already have the effect of limiting most packages to quantities considerably below permissible limits. Also, with the introduction of a 3.0 T.I. package limit, large quantity packages would be prohibited to be shipped by passenger-carrying aircraft. This same commenter also noted the absence of paragraph (d) concerning the limitation of radioactive materials to those intended for use in, or incident to, research, or medical diagnosis or treatment. This oversight has been corrected in redesignated paragraph (c).

Section 175.701: Specifies separation distances on passenger-carrying aircraft. A number of commenters supported a complete revision of this section which would tend to concentrate packages of radioactive materials in the rear of the aircraft at shorter separation distances. This would have resulted in an average radiation level in the passenger compartment of 0.5. millirem per hour, but this average would be at the expense of particular passenger seating areas being subjected to a maximum radiation level of 5 millirem per hour. Since this plan does not meet the primary objectives of this rulemaking it was not considered desirable.

With regard to paragraph (b)(1), one commenter suggested that the words "these packages" be substituted for the phrase "each individual package" as found in the second sentence. The MTB agrees that the intent and meaning is more clearly related and the suggestion has been adopted. Also, in response to the oversight pointed out by one commenter the words "or predesignated area" have been added to the table contained in paragraph (b)(2).

A great deal of criticism was received in response to the separation distances prescribed in paragraph (b). Commenters argued that such a system is too complex to be workable in real world conditions especially when one considers variables such as short loading times, other hazardous materials on board, the presence of animals in cargo compartments and general cargo already loaded on the floor of the aircraft. Another commenter argued that a total transport index (TTI) of 3.0 to 10.0 be assigned to each aircraft cargo compartment taking into account the aircraft size. None of these suggestions were sufficiently justified to be incorporated into the final rule. The provisions of §175.701(b) in this rulemaking are not a substantive change of existing regulations but rather a regulatory refinement which will more evenly distribute the packages of radioactive materials so that lower levels of exposure will be realized. To assign a TTI to each cargo compartment would unnecessarily restrict the amount of radioactive materials that may be carried while not commensurately reducing the maximum allowable exposure levels.

One commenter suggested that "the time is opportune to eliminate requirements to apply separation distances relating to animals" as this requirement is not one of the considerations of the International Atomic Energy Agency

(IAEA) Regulations on the Safe Transport of Radioactive Materials. Considering that the effects of radiation exposure are also damaging to animals and they in turn represent an element of property which is subject to protection from hazardous materials transported in commerce, it is the opinion of the MTB that current separation requirements remain unchanged.

The concept of "predesignated areas" also drew the attention of numerous commenters. One carrier protested the intervention of the DOT in the carrier's prerogative for utilizing available space. The carrier claimed this would amount to unnecessary government regulation and delay the implementation of changes prompted by aircraft modification, seasonal traffic flows or other influence from outside sources. It should be pointed out, however, that this is a voluntary election which the carrier is free to make and is only offered as an alternate means of safely transporting radioactive materials by passenger-carrying aircraft. To the commenter who objected to this proposal as being "not for safety but to increase the amount of radioactive packages which could be carried on passenger-carrying aircraft", the MTB notes that the utilization of predesignated areas must also insure an equivalent level of safety. In a similar manner the MTB rejects the lateral separation factor of 2 rather than 4 which was proposed by the same carrier who objected to the very idea of predesignated areas. There exists a threat of radiation level "peaking" from the additive effect of radiation emitted from each predesignated area when a factor of 2 is applied, possibly resulting in unacceptably high exposure levels.

Section 175.702: Comments received in response to this section were of a cursory nature. Some commenters questioned the 200 T.I. limit for cargo-only aircraft claiming that it is unrealistic since most aircraft can't handle more than 50 T.I. However, in order to provide an incentive which would be effective in helping to reduce the demand by shippers and carriers of radioactive materials for space on passenger-carrying aircraft, the 200 T.I. limit is considered reasonable for aircraft which are able to meet all separation requirements.

One commenter, who is a frequent shipper of radioactive materials, expressed a concern over the ability of specialized carriers of small parcels to comply with the distance separation requirements, as these carriers frequently use smaller aircraft in their operations. The problem is not a new one occasioned by the introduction of this section, and in fact one of the carriers mentioned by this commenter is presently operating under authority of exemption number E-7060 which provides relief from the 50.0 T.I. limitation, while requiring a documented radiation protection program for carrier personnel. Additionally, the kind of operations permitted by the terms of this exemption were proposed for authorization in Docket HM-166B; Notice 79–8 (44 FR 29503) but, due to numerous adverse comments, the proposal was deleted from the amendment. The provisions in this amendment are considered by MTB to be sufficient to meet the needs of shippers without unduly jeopardizing safety.

A comment also addressed the distance separation requirements established in paragraph (b)(2) (i) and (iii). The representative of a foreign flag air carrier sought DOT compatibility with those regulations currently set forth by the International Air Transport Association (IATA). Basically, this would provide a system of steps between 50 and 200 T.I. with corresponding increases in separation distance beginning at 15 feet, 4 inches (4.65 meters) and progressing to 28 feet, 10 inches (8.75 meters). While the MTB is interested in consistency with international standards for the safe and smooth flow of goods, the restriction imposed in this amendment is not seen as a burden to commerce especially when one considers the 200 T.I. ceiling and the relative ease with which most cargo-only aircraft operating in international service would be able to handle such a load, in compliance with the DOT minimum.

Section 175.703. The proposals in this section (now § 173.393(r)) drew the overwhelming majority of adverse comments. Specifically, the commenters objected to the proposals for compression testing, marking and labeling of overpacks. Commenters contended that there were inconsistencies in the use of overpacks for packages of radioactive materials when compared to those for other hazardous materials. One commenter pointed out that marking and labeling of clear plastic overpacks is currently excepted by § 173.25(a) when the markings and labeling of the inside packages are visible. The commenter viewed the added requirements as an example of over-regulation by the government. Despite the commenters arguments, the MTB does not believe that there was sufficient support provided for the points raised to make a revision of the rule as proposed. As was discussed earlier in this document the relationship of time is a critical factor in the accumulation of a dose of radiation, and the availability of a label(s) with the aggregate T.I. entered thereon is seen as an effective means for reducing the time spent by ground handlers in determining the activity of the package for proper placement in the aircraft, surface transport vehicle or storage area.

One commenter argued that subjecting overpacked packages of radioactive materials in nonspecification packagings to Type A container test requirements for compression was highly impractical. The commenter further stated that "The plastic bag overpack which we used is placed over a variety of box shapes and sizes. Therefore, each shipment prepared would technically have to be tested 24 hours prior to movement." This commenter also suggested that the limited quantity of radioactive materials permitted in non-specification packagings is a sufficient safeguard in and of itself to eliminate the need for a performance standard greater than the standard requirements for all packages presently called for in § 173.24. The MTB agrees with these comments especially since most overpacks would be composed to Type A packages which are permitted, in part, because of their ability to withstand heavy loads. Accordingly, this proposed requirement in paragraph (b)(2)(iv) has been eliminated from the final rule.

Another commenter, objecting to the proposed restriction in paragraph (b)(2)(viii) which would have prohibited the consolidation of packages from more than one original shipper, claimed the proposal was confusing since it didn't specify who was to be considered the original shipper; that is the manufacturer, distributor, or central hospital serving satellite facilities. This proposed restriction has been removed from the final rule since it would have eliminated some safety benefits, such as reduced radiation levels achieved by shielding from surrounding packages. To address the comment directly, it should be noted that any person initiating a shipping paper is considered to be the original shipper.

Another commenter was concerned about the proposed option of entering the words "mixed radioactive materials" on the label of overpacks and shipping papers in place of specifically identifying the particular radionuclides. This proposal has been modified in part with the present requirement for specifying each package and its radioactive materials contents by activity, physical and chemical forms, transport index and the like on the shipping papers being retained. However, considering that each individual package label already specifies the particular radionuclide(s) contained therein, and considering further the limited space available for such information, use of the generic description "mixed" is determined to be sufficient on labels applied on overpacks and this element of the proposed regulation has been adopted. The required information available from shipping papers was recognized as being too valuable to emergency response personnel and other interested persons to justify deletion of the requirement. To avoid possible confusion in completing the labeling requirements for an overpack containing packages of radioactive materials, guidance has been taken by specifying that the number of curies entries must be a cumulative total of all such similarly labeled packages contained therein. For the purposes of these regulations the T.I. marked on the label of an overpack may be used in calculating maximum vehicle loading limitations and distance separation requirements.

A commenter suggested that the label entry "mixed radioactive materials" would be in conflict with parallel requirements found in the Official Air Transport Restricted Articles Tariff No. 6-D (CAB No. 82) which specify that each radionuclide must be identified on the Yellow-II and Yellow-III labels. This difference has been avoided by changing the sense of proposed § 175.703(b)(2)(i) from a mandatory requirement to a permissive use of the word "mixed" in the contents entry on the label.

In this section also, one commenter recommended compatibility with the IATA separation requirements from undeveloped film. As no evidence was presented which would support this commenter's desire to promote uniform standards while adequately protecting other property at the same time, the more restrictive distance separations imposed by this section have been retained. However, the units of measure have been revised to include meters.

It should be noted that this amendment does not include the previously authorized option contained in § 175.710(c)(3) applicable to the carriage of Fissile Class III radioactive materials. As the MTB has received no requests for approval of procedures other than those specified in subparagraphs (c)(1) and (2) it was determined that retention of this approval system was unnecessary.

(45 F.R. 20097, March 27, 1980)

[Docket No. HM-157; Amdt. No. 177-39]

PART 177—CARRIAGE BY PUBLIC HIGHWAY
Incorporation of the Federal Motor Carrier Safety Regulations by Reference

This amendment is consistent with the policy of reissuing regulations formerly issued under the Explosives and Other Dangerous Articles Act (EODAA) (18 U.S.C. 831-34) so that regulations will be effective under the new HMTA. Such reissuance is performed pursuant to the direction in the HMTA to bring all rules and regulations into conformity with the purposes and provisions of that Act as soon as practicable, Pub. L. 93-633, section 114(b)(2). The Federal Motor Carrier Safety Regulations (FMCSR) are currently applicable to carriers of hazardous materials pursuant to 49 CFR 397.2 which is issued jointly under the Interstate Commerce Act (ICA), 49 U.S.C. 304, and the EODAA, 18 U.S.C. 831-835. The effect of this amendment is merely to make civil penalties and other enforcement tools of the HMTA applicable to those hazardous materials carriers already subject to Parts 390-397. Because this amendment merely reissues, under new authority, regulations already in effect, notice and comment are unnecessary. For the same reason, this amendment is effective upon publication.

As the FMCSR are being incorporated by reference, the intent, scope of application and preemptive effects of the FMCSR, as reissued under the HMTA, are unchanged. The Department does not intend for this action to alter the categories of persons subject to the FMCSR, to alter the substance of those regulations, or to preempt State or local law not preempted by the FMCSR before incorporation into Part 177. However, §§ 397.3 and 397.9 are not being reissued under the HMTA at this time. Reissuance of both sections is being deferred pending further review, and in the meantime both sections will continue, as in the past, to be enforceable with criminal penalties provided by the statutes under which they were originally promulgated.

(43 F.R. 4858, February 6, 1978)

[49 CFR Parts 172 and 173]
[Docket No. HM-159; Notice No. 79–12]
Forbidden Materials

On February 23, 1978, the MTB published an Advance Notice of Proposed Rulemaking (43 FR 7449) concerning materials which are believed to be too hazardous to be permitted in commercial transportation. The Advance Notice included four lists of materials and requested that the public comment on the following three questions:

1. Should the Hazardous Materials Table be the consolidated central location for the listing of forbidden materials by chemical name or should that listing be placed in a separate section?

2. What, if any, additional materials should be identified in the regulations as forbidden?

3. Are there any materials listed in this notice which do not meet the regulatory criteria making them a forbidden material? If so, identify these materials and explain why they should not be considered forbidden materials.

A total of fifty-three comments were received and evaluated. Only one commenter was opposed to having a list of forbidden materials. The reasons for this opposition were that no list could be complete, the absence of a specific chemical from the list would imply that it is not forbidden, and there is no need for a list because the regulations provide criteria for prohibiting certain materials from being transported. The MTB disagrees and believes that all known materials considered to be too hazardous for transportation should be included in a list. This has been done previously, however, the list has not been as extensive as the list presently proposed.

All other commenters were in favor of incorporating forbidden materials in Title 49, Code of Federal Regulations (49 CFR). Thirteen commenters stated that these materials should only be placed alphabetically in only 49 CFR 172.101 based on the fact that there should only be a single source list for all hazardous materials. Four commenters suggested that a separate list be provided in some other section of the regulations. This was based on the belief that a separate section would be easier to use and would more easily identify these materials. Five commenters stated that the forbidden materials should be put in both 49 CFR 172.101 and another section. The basis for this position is that the commenters felt that all materials should be included in the Table in § 172.101 but that the list of forbidden materials also be included in a separate section so that persons could more easily determine which materials are forbidden without a complete review of the Table in § 172.101. The MTB believes that placing the names of forbidden materials only in § 172.101 is better than the other two alternatives because: (1) A person using the regulations should start at the Hazardous Materials Table and if it is noted that a material is forbidden he does not have to look any further; (2) A person using the regulations could possibly overlook the forbidden materials if they were in a separate section; and (3) Placing the materials in both § 172.101 and another section results in unnecessary duplication of regulations, causes confusion, and does not contribute appreciably to safety.

Two commenters were concerned that if a material was shown as forbidden this would mean that solutions of that material or devices containing that material would also be forbidden. This is not the intent of the MTB and this is made clear in the proposed change to § 172.100.

Two commenters stated that certain triazoles have properties which would indicate they are forbidden but other triazole compounds do not have such properties. Pending further detailed investigation into these chemicals, triazoles are being removed from the proposed list. The same situation exists with triazones which were also deleted from the proposal.

One commenter submitted reports from the Bureau of Explosives (B of E) which classed the material, Bis 2-fluoro-2,2-dinitro ethylformal, (FEFO), as a Class A explosive. The MTB is in agreement with the report and, therefore, this material has been deleted from this proposed list as a forbidden material.

One commenter suggested that the material, nitroisobutanetriol trinitrate, be added to the list and another commenter stated that the material, t-butoxy-carbonylazide, should be added. Based on the information submitted on each of these materials, they have been added as forbidden materials. Two commenters recommended that the concentration of ketone peroxides be expressed in terms of active oxygen, rather than percentage of peroxide, and that the active oxygen content of these materials be limited to 9 percent. The MTB agrees with the data submitted and has incorporated such changes in this notice.

Twenty-five commenters opposed forbidding the transportation N-methyl-N-nitrosoguanidine because it is a very important reagent in cancer and mutagenic research. The MTB does not believe that the product should be shipped under § 173.65(d) which provides essentially no regulation. A proposal has been made for shipping limited amounts of this chemical in packagings recommended by the B of E. The MTB is proposing to class this material as a flammable solid when packaged in accordance with B of E recommended packagings.

In the Advance Notice it was proposed to list by name the forbidden explosives now appearing in § 173.51. The MTB has reconsidered this proposal and is now proposing to include two new entries in § 172.101 which are referenced to 173.51. These are "Forbidden Explosives" and "Explosives, forbidden." In this proposal, 173.51 has been rewritten to make it clearer and concise. The major proposed changes in this section include: the inclusion of most of the fireworks with explosives because fireworks are classed as explosives: the revision of the present entry "fireworks containing copper sulfate and a chlorate" to include any acidic metal salt and a chlorate due to the fact that the hazard of spontaneous combustion is not limited only to copper sulfate and a chlorate; and the inclusion of devices in an effort to be consistant with other sections of 49 CFR governing explosives which also include devices.

"Forbidden materials," with a reference to §173.21, is a proposed new entry which did not appear in the Advanced Notice. Section 173.21 would be amended for clarification. This section applies to any material considered to be forbidden and is not limited to materials falling within established hazard classes. Included in the proposed revision of this section is a prohibition against the offering of packages that evolve a dangerous quantity of flammable gas or vapor released from a material not otherwise subject to the regulations, e.g. the release of flammable blowing agent vapors from a manufactured product in such quantities that an explosive mixture would be created within the transport vehicle. It is also proposed that each refrigeration method, when used as a means of stabilization, be approved by the Associate Director for Operations and Enforcement. This change is in accord with the approval withdrawals presently being handled by amendments published under Docket HM-163.

This proposed rulemaking, which would prohibit the transportation of certain materials known to be susceptible to accidental detonation in a fire (other than an explosive), is responsive to Recommendation No. 3 in the National Transportation Safety Board's report (No. NTSB-RAR-76-1) on the explosion which occurred in Wenatchee, Washington on August 6, 1974.

(44 F.R. 43861, July 26, 1979)

[Docket No. HM-160; Amdt. No. 172–47,
173–123, 174–33, 175–7, 177–44]

PART 173—SHIPPERS—GENERAL REQUIREMENTS FOR SHIPMENTS AND PACKAGINGS

Transportation of Asbestos; Revision of Amendment No. 173–123;
Effective Date Extension

On December 4, 1978, the MTB published a final rule under Docket HM-160 in the FEDERAL REGISTER (43 FR 56664). Since this publication, the MTB has received several petitions for reconsideration in accordance with the provisions of 49 CFR 106.35. The petitions requested reconsideration of the provisions and/or extension of the effective date of the final rule. This document will incorporate methods of shipment which were identified in the notice of proposed rulemaking (43 FR 8562, March 2, 1978) and also those which were included in the final rule. These amendments represent minimum safety requirements and are intended to reduce the risks to public health associated with the generation of unacceptable airborne concentrations of asbestos that my result from packaging and handling of asbestos shipments in commercial transportation.

Two petitioners based their petitions on the fact that the final rule contained a provision requiring bags and other non-rigid containers of asbestos to be palletized and unitized by some method such as shrink-wrapping in plastic film or wrapping in fiberboard secured by strapping. It was noted that this requirement was not included in the notice of proposed rulemaking (43 FR 8562), thus making comments on this requirement impossible during the normal comment period for the proposed rulemaking. Petitioners also posed the question of whether or not freight containers, rail cars, etc., constituted rigid, airtight packagings as required in §173.1090(d)(1). It was stated that if such containers were not included in this provision, all shipments of bags or non-rigid containers would be required to be palletized and unitized according to the provisions of §173.1090(d)(2), and that this requirement would impose great hardship on the asbestos industry and on shippers of large volumes of asbestos who normally ship using exclusive use vehicles and rail cars. It was also indicated that neither the equipment nor facilities exist at the present time to achieve compliance with the palletizing and unitizing requirement of the final rule by the published effective date. The MTB has determined that freight containers and, probably, motor vehicles and rail cars would not satisfy the requirements of §173.1090(d)(1).

By allowing the use of unitized pallet loads as identified in the final rule, the MTB intended to recognized less restrictive handling requirements for bagged asbestos than those that would have been required by the "loading by consignor/unloading by consignee" approach. However, it was not the intent of the MTB to eliminate the more restrictive consignor/consignee approach. Therefore, §173.1090(d) is being revised to allow the option of either the consignor/consignee approach as identified in the original proposed rulemaking with inclusion of an exclusive use provision or the unitized pallet approach using bags or other non-rigid packagings as required by the final rule.

One petitioner noted that a method of shipment of asbestos via water was the use of slings which are shrink-wrapped or stretch-wrapped and transported in the hold of a vessel without the use of pallets. It was the petitioner's contention that the use of pallets would increase the incidents of unintentional release of asbestos due to the interaction of the pallets against the bags which are unitized by the slings. It was noted that pallets were used in all instances except when placed in the hold of the transport vessel. Given the lack of detailed data on the amount of asbestos fibers released in transportation and the circumstances and cause for such release, the MTB is in general agreement that increased unintentional releases may be likely if pallets were used under the method identified in the petition.

Therefore, § 173.1090(d) is being revised to allow slings in loads that are shrink-wrapped or stretch-wrapped to be transported by water without the use of pallets. Future monitoring of hazardous materials incident reports will assist the MTB in determining the safety and efficiency of this and other methods for shipment of asbestos.

One petitioner suggested that the terms "pallet" and "palletized" be defined in the rulemaking. The MTB does not intend to publish a definition of pallet or palletized. It is the MTB's opinion that any rigid platform or board upon which goods may be placed for transportation would meet the requirements when unitizing a load of bags or other non-rigid packagings.

Several petitioners cited a need for the MTB to define the term "dust and sift proof". For the purposes of this amendment, the MTB considers dust and sift proof to mean packagings which are constructed so as to prevent the release of their contents either through materials of construction, seams, or closures during conditions normally incident to transportation.

One petitioner requested that the use of gluing of bags into a unit be allowed as an alternative to the unitizing methods identified in the final rule. It is the MTB's opinion that the use of shrink-wrapping or other similar methods of enclosure assist not only in unitizing a pallet load of bags or other non-rigid packagings, but also assist in the prevention of airborne asbestos contamination of individuals involved in the transportation of asbestos. Simply gluing these packagings together to form a unit would not provide this added measure of safety to which the final rule addresses itself. Therefore, gluing of bags into a unit is not being included as an alternative unitizing method.

One petitioner requested that quantities of less than 2,000 pounds, net weight, per vehicle be excepted from the palletizing and unitizing requirement. It is MTB's opinion that the palletizing and unitizing requirement does not unreasonably restrict the shipment of asbestos in any quantity. This requirement is necessary to provide a minimum level of safety.

Several petitioners requested an extension of the effective date of the final rule. The effective date has been extended to allow five months for compliance as originally intended by the December 4, 1978 publication.

(44 F.R. 18673, March 29, 1977)

[49 CFR Parts 172, 173, 174, 175, 176, 177]
[Docket No. HM-160; Notice No. 78-3]
TRANSPORTATION OF ASBESTOS

I. BACKGROUND—PRODUCTION/CONSUMPTION/TRANSPORTATION
PATTERNS

Asbestos is a generic term used to describe a number of naturally occuring fibrous, hydrated mineral silicates. It includes chrysolite, crocidolite, amosite, anthophyllite asbestos, tremolite asbestos, and actinolite asbestos. These are all asbestiform minerals which, when crushed, produce asbestos, fibers with various chemical and physical properties.

Asbestos fibers are generally characterized by high tensile strength and flexibility, and favorable chemical resistance, heat and frictional properties. Certain grades of asbestos can be spun and woven, while others can be laid and pressed to form paper, or used for structural reinforcement of materials such as cement, plastic, asphalt and tile. The asbestos content of these latter products ranges from 5 to 15 percent by weight.

Although asbestos is adaptable to more than 2,000 uses, the construction industry accounts for nearly two-thirds of the United States (U.S.) asbestos fiber consumption. The remaining 33 percent is utilized in a myriad of industrial and consumer products.

During the ten-year period ending 1976, the total amount of new asbestos introduced into the U.S. transportation system averaged approximately 1,700,000 short tons annually. Slightly more than half this tonnage consists of crude or milled asbestos fibers, with the remainder consisting of asbestos contained in products manufactured in whole or in part from asbestos. There is, of course, a lag between the time a shipment of asbestos fibers enters the transportation system and the time that shipment enters the transportation system as a manufactured product.

Growth in the utilization and transportation of asbestos is expected to slowly increase at a rate of about 2 percent per year.

During the entire history of the asbestos industry in the United States, domestic sources have been able to meet only a small percentage of U.S. requirements. Roughly 90 percent of the total U.S. industrial demand for all grades and types of asbestos fibers is thus dependent on foreign imports—with Canada being the major source of supply. In 1975, 93 percent of U.S. imports came from Canada with the next largest suppliers being the Republic of South Africa (3 percent) and the U.S.S.R. (2 percent). Eight other countries shipped smaller amounts of asbestos to the U.S. in 1975.

The transportation pattern for asbestos is characterized by three distinct stages:

In the first stage, asbestos ore is transported from the mine site to a milling plant where the ore is crushed and processed into fibers.

Of the five mills operating in the U.S. as of late 1976, three are located at the mines, but the other two are 32 and 52 miles distant. Transportation is reported to be by open-hopper vehicles such as dump trucks.

The second stage is characterized by the shipment of crude or milled asbestos fibers from the mills (mostly foreign) to industries that use asbestos in the products they manufacture. The largest industry in this category (SIC 3292) as of 1972 consisted of 142 establishments primarily engaged in the manufacture of asbestos textiles, asbestos building materials, asbestos insulating materials for covering boilers and pipes, and other products composed wholly or chiefly of asbestos fibers. Other industries receiving asbestos fibers are mainly industries engaged in the production of asphalt felts and coatings (SIC 2952), hard surface floor coverings (SIC 3996), gaskets (SIC 3293), and paper products (SIC 2261).

These industries generally receive their shipments of asbestos fibers in pressure packed, five-ply paper or woven vinyl bags weighing about 100 lbs. per bag, with the bags glue-locked to each other and shrink-wrapped to a pallet (wrapped with a film of plastic which is then shrunk). In the aggregate, rail shipments handle about 80 percent of all milled asbestos fibers entering the U.S. transportation system, with merchant vessel shipments accounting for the remaining 20 percent. Packaging for ocean transportation, both for imported and exported asbestos fibers, is changing to containerization. All shipments, for example, from the Union of South Africa have been reported to be containerized. Although rail shipments of asbestos fibers are not yet being containerized, the majority of such shipments are made in sealed, railroad box cars which are routed for the most part direct to the asbestos products manufacturing industries.

The third stage in the transportation of asbestos involves the shipment of manufactured products made wholly or in part from asbestos. Approximately 61 percent of the asbestos used in manufactured products is firmly imbedded or "locked in" such products as floor tiles, asbestos cement pipes and sheets, floor products and plastics. These products generate less airborne fibers than asbestos manufactured products that are friable or in powder form, although at the present time, it is not known whether the transportation of products in either category under current conditions presents an unreasonable risk to the public health and safety. As defined in the Department of Commerce's 1972 Census of Transportation, approximately 92 percent of asbestos manufactured products are shipped by motor vehicle, with rail and all other modes accounting for about 7 percent and 1 percent, respectively.

II. GENERAL EFFECTS OF AIRBORNE ASBESTOS EXPOSURE/EVIDENCE OF ACCIDENTAL RELEASE OF ASBESTOS FIBERS

The MTB believes it to be firmly established that asbestos, in its several commercial forms, poses serious health hazards to individuals subject to long term exposure to airborne asbestos concentrations. As noted in the 1972 preamble of the Occupational Safety and Health Administration (OSHA) standard on asbestos (37 FR 11318): "No one has disputed that exposure to asbestos of high enough duration is causally related to asbestosis and cancers . . " Recent new evidence, however, as reported by OSHA, not only tends to confirm this finding but also suggests that serious potential health risks are involved with even relatively low-level, brief or intermittent exposure to airborne asbestos concentrations. Although there is no detailed information available on the amount of asbestos fibers released in transportation, the MTB believes that, in consideration of the carcinogenic and other health hazards associated with asbestos, there is a sufficient basis for establishing regulatory control of asbestos in transportation.

III. RELATIONSHIP TO HM-145

On December 9, 1976, the MTB published an Advanced Notice of Proposed Rulemaking (41 FR 53824) in Docket No. HM-145 entitled "Environmental and Health Effects Materials." In that Notice, the MTB announced that it was considering whether new or additional transportation controls are necessary for certain classes of materials which are not generally subject to the existing Hazardous Materials Regulations.

A large number of comments were received in Docket HM-145. The MTB has concluded that a considerable amount of time and effort is still needed in staff evaluation of these comments before it will be in a position to issue a notice or notices of proposed rulemaking for environmental and health effects materials, either on a comprehensive or on a selective basis.

Several comments on HM-145, however, were specifically directed to the idea that transportation regulatory controls for asbestos be established as soon as possible, with the suggestion that asbestos be addressed on an individual basis, rather than writing until the eventual resolution of Docket HM-145. The MTB agrees with the urgency of the views expressed on this matter, and therefore asbestos is being treated separately under this proposed rulemaking.

IV. QUANTITATIVE VERSUS QUALITATIVE STANDARDS

In determining that there is a need for transportation controls on asbestos, the MTB has considered the desirability and practicality of utilizing quantitative or qualitative emission criteria or some combination of the two, either as developed by the MTB or as developed by other agencies.

Currently, quantitative permissible exposure limits for asbestos, as promulgated by the Environmental Protection Agency (EPA) have not been established for environmental or nonoccupational settings. The EPA's standard for airborne asbestos emissions falls under its "no visible emissions" criterion. This is in contrast to the quantitative criteria established by OSHA for airborne concentrations of asbestos fibers in occupational or worksite conditions. The criteria of OSHA consist of an 8-hour time weighted average standard, and a maximum "ceiling concentration" standard. Both standards are to some extent based on the ability of current devices to measure asbestos airborne concentrations in a systematic, meaningful manner, and are not directly based on any causal or threshold relationship between the standards and the probability of contracting an asbestos induced disease.

The MTB believes that its proposed non-specification packaging standards as applied to the transportation of asbestos fiber is an effective and efficient means of precluding potential problems associated with asbestos airborne emissions occurring during transportation; and that they are consistent with the standards of both EPA and OSHA.

V. SCOPE AND IMPACT

The standards as herein proposed would only apply to the transportation of what are generally regarded as milled or crude asbestos fibers, but would exclude asbestos contained in a natural or artificial binding-material and manufactured products containing asbestos.

In light of the regulatory controls already in existence or under consideration by other federal agencies, and until such time as the MTB has more specific and concrete information that the normal packaging and handling of these forms of asbestos is such as to create unreasonable asbestos exposure problems, the MTB does not believe their specific regulation in transportation is warranted.

In reviewing the potential inflationary and economic impacts associated with the proposed rule, the MTB has determined that such impacts will be minimal. Based on the foregoing discussion of the production and transportation pattern for asbestos shipments, it is clear that the only aspect of that pattern which might experience a cost impact pertains to import and export shipments by merchant vessel. This follows from the fact that the proposed rule would generally require all such shipments to be containerized in contrast to packaging alternatives now available to shippers. However, the cost-differential between these alternative cargo handling methods is not only very small, but a significant portion of such shipments are already containerized, or being containerized.

As proposed herein, a new paper shipping name "Asbestos" would be added to the list of hazardous materials in 49 CFR 172.101. The proposed classification for "asbestos" would be as an ORM-C (Other Regulated Material, Group C).

(43 F.R. 8562, March 2, 1978)

[Docket No. HM-160; Amdt. Nos. 172–47,
173–123, 174–33, 175–7, 176–6, 177–44]
TRANSPORTATION OF ASBESTOS

On March 2, 1978, a notice of proposed rulemaking (HM-160; Notice 78–3) was published in the FEDERAL REGISTER (43 FR 8562) stating that the MTB was planning to exercise regulatory control over the transportation of asbestos. Specific regulatory requirements were proposed for the control of certain forms of asbestos (e.g., milled or crude asbestos fibers). No requirements were proposed for asbestos fibers which are immersed or fixed in a natural or artificial binder material, or for manufactured products containing asbestos. Interested persons were invited to participate in the rulemaking proceeding through submission of written comments on the proposal to the MTB. All submissions, including late submissions, that were received on the proposal were fully considered by the MTB in the development of this final rule.

NEED TO REGULATE THE TRANSPORTATION OF ASBESTOS

Several commenters felt that the MTB had failed to establish a need to regulate the transportation of asbestos. One of the commenters suggested that there was no need for the proposed regulatory control of asbestos in transportation because the "methods and procedures now in use for the packaging and transport of asbestos meet the requirements of Part 173.23(A)(sic) of the Transportation Act, that is 'under conditions normally incident to transportation there will be no significant release of the hazardous materials to the environment' and 'the effectiveness of the packaging will not be substantially reduced . . . (t)he proposal contains no documentation to justify additional regulation." This commenter, while apparently believing that asbestos is a hazardous material, was incorrect in suggesting that asbestos is currently regulated by the MTB; or in suggesting that the purpose of Notice 78–3 was to justify the additional regulation by the MTB of asbestos in transportation. The transportation of asbestos is not now regulated by the MTB. It was precisely the purpose of Notice 78–3 that it should be. If, as the commenter suggests, the transportation of asbestos is now "in compliance with pertinent provisions of the Transportation Act," this rulemaking action will formalize and insure in a uniform and systematic manner that this is the case.

Another commenter stated that Notice 73–8 did "not establish a foundation for regulation, in that it does not document, or even allege for that matter, the actual release of fiber during the transportation of asbestos." As was pointed out in Notice 78–3, the MTB has "no detailed information on the amount of asbestos fibers released during transportation." The MTB does not now regulate asbestos, and has not therefore systematically collected accident data on the amounts of asbestos released in transportation or data on the frequency of such accidents. Most asbestos fiber, however, is currently shipped in bags, and it is undeniable that these bags can and do break, or can be and are being torn or punctured, with a consequent release of some or all of the bag contents. It can be speculated, moreover, that if all of the 750,000 tons of asbestos annually shipped in the United States were packaged in, as one commenter states, the "standard package" of a 100-pound bag; and if as little as one-tenth of one percent of these bags were damaged in transportation during the year (one out of a thousand) and if on the average 1 percent of the contents of the bags so damaged were released, the total amount of asbestos released per year would equal about 7.5 tons. These calculations give a general idea of the magnitude of asbestos fiber that would be released, given a 99.9 percent efficiency factor for "bag integrity" in transportation, and a 99.0 percent efficiency factor in minimizing the amount of asbestos released given a tear in the bag. The rather evident fact that asbestos has been accidentally released during transportation has not been contradicted by anything submitted to the public docket on this rule-making action. One commenter, for example, in discussing the use of open-bed trailers with side racks and tarpaulins to transport asbestos stated that there is no evidence that the use of such trailers "has contributed to bag breakage and the release of airborne concentrations of asbestos fiber." The Asbestos Information Association, an incorporated nonprofit organization representing 51 firms in the United States and Canada engaged in the manufacture or processing of asbestos-containing products and the mining/milling of asbestos fibers, stated that with "the very large volume of asbestos shipped, occasional container damage may occur."

Although several commenters who discussed this matter do not contend that asbestos has not been released in transportation, they generally are of the view that the amounts that are being released are not significant or of a sufficient amount to pose an unreasonable risk to public health. The MTB does not agree; it believes that the amounts of asbestos fibers that are being released now, or would be released in the future, in the absence of these amendments, *may* pose an unreasonable risk to health.

Several commenters were concerned with the statement appearing in Notice 78–3 that "asbestos in its several commercial forms, poses serious health hazards to individuals subject to long term exposure to airborne asbestos concentrations." One commenter stated that "not all long-term exposures to airborne concentrations pose any health hazards . . . ." Another commenter suggested that the statement needed "more explicit definition" and that "reference should have been made to unanswered questions within the scientific community concerning mineral type, fiber size and smoking in the asbestos-cancer relationship." One commenter stated that there is a dose-response relationship between exposure to asbestos and disease causation, and that this conclusion is supported by an OSHA statement from its June 7, 1972 preamble to its standard for exposure to asbestos dust (37 FR 11318). The OSHA statement is that: "No one has disputed that exposure to asbestos of high enough *intensity and long enough* duration is causally related to asbestosis and cancers" (emphasis added). Although the MTB had also quoted this statement in Notice 78–3, the words underlined for emphasis had been inadvertently omitted. Under these circumstances, some commenters apparently felt that the MTB was asserting the view that because, according to some commenters, asbestos is ubiquitous, long term exposure to ambient levels of asbestos fibers poses serious health hazards to all people, without regard to their occupational or paraoccupational status. It was not the intention of the MTB to assert this view. That there are or can be "undisputed grave consequences from exposure to asbestos" (37 FR 11318) does not depend on the questioned conclusiveness of the evidence reported by OSHA (40 FR 47652) regarding the potential health hazards posed by low-level, brief or intermittent exposure to asbestos. The MTB relies on the foregoing FEDERAL REGISTER references for the general view that exposure to asbestos may pose an unreasonable risk to the public.

SECTION 173.1090(a) AND (b)

Several commenters stated that there are certain mineral ores, ore concentrates and milled mineral products which may have trace amounts of asbestos, or minor amounts of asbestos occurring as contaminants. They suggested that these materials presented no risk to property and little, if any, risk to public health and safety in transportation. Moreover, since the packaging requirements proposed in Notice 78–3 applied to only certain kinds of asbestos, namely milled or crude asbestos fibers produced by an asbestos mill, they further suggested that only "commercial asbestos fibers" be defined as a hazardous material.

The MTB recognizes that there are certain mineral ores, ore concentrates and milled mineral products, as well as other products, that contain certain amounts of asbestos, and that the commercial value of these minerals or products is not dependent on their asbestos content. The specific requirements in these amendments for the control of asbestos fibers in transportation do not apply to such materials or products, nor do they apply to asbestos as a waste product[1] or as a contaminating trace element. The amendments apply only to asbestos in its several commercial forms since it is those forms of asbestos that have been firmly established as posing serious health hazards to individuals. A new paragraph has been added which would define commercial asbestos as any material or product containing asbestos that has commercial value because of its asbestos content, and appropriate modifications have been made in the amendments to reflect this clarification. This new paragraph is identified in this amendment as paragraph (b) (paragraphs (b) and (c) in the notice are now paragraphs (c) and (d), respectively).

One commenter recommended that the scope of Notice 78-3 be amended to include, in addition to asbestos fibers, "all mineral and man-made (fibers) which have been identified by U.S. Government agencies as being carcinogenic and which may pose serious health risk." On December 9, 1976, the MTB published an Advance Notice of Proposed Rulemaking (41 FR 53824) in Docket No. HM-145 entitled "Environmental and Health Effects Materials." In that Notice, the MTB announced that it was considering whether new or additional transportation controls are necessary for certain classes of materials which are not generally subject to the existing Hazardous Materials Regulations. The question of whether all mineral and manmade fibers, which have been identified by U.S. Government agencies as being carcinogenic and which pose an unreasonable risk to public health, should be controlled in transportation will be considered in terms of the further development and resolution of the issues associated with Docket HM-145. Notice 78-3 however, pointed out that a large number of comments were received in Docket HM-145, and that a considerable amount of staff evaluation of these comments was still required before it would be possible to issue a notice or notices of proposed rulemaking for environmental and health effects materials, either on a comprehensive or on a selective basis.

SECTION 173.1090(c)(1)

Several commenters objected to the reference made to metal or fiber drums to illustrate the rigid packaging alternative for asbestos fibers. These commenters stated that the asbestos industry has not developed the technology to use this type of packaging alternative; that available technology is not transferable to the use of metal or fiber drums; and that, among other things, the use of this alternative could generate far greater airborne concentrations of asbestos than packaging and shipping practices currently in effect. As one commenter pointed out:

> Commercial asbestos is fluffy. It is difficult to pack this material in a rigid container, and, because the fiber would gradually compact during shipment, it would be difficult to remove it for introduction into the manufacturing process. It would also be extremely cumbersome, if not impossible, to empty rigid containers effectively and rapidly into hoods designed for bags. Spillage would no doubt occur and workers would be unnecessarily exposed to fibers.

Another commenter recommended that a DOT Specification 56 portable tank be included in the amended rule as an acceptable package "for the transportation of asbestos-type products." This commenter stated that "with the use of equipment designed for the purpose, the D.O.T. 56 package can be readily filled or emptied without release of any product dust to the atmosphere or contact with the product by the operator." Another commenter insisted that only metal drums and not fiber drums were acceptable for the transportation of asbestos fibers. These commenters apparently lost sight of the fact that proposed § 173.1090(c)(1) does not "mandate," as one commenter suggested, or even encourage the use of rigid, airtight packaging such as metal or fiber drums or even portable tanks. It provides an alternative method of shipping commercial asbestos fibers. As was indicated in Notice 78-3, the MTB believes that its proposed non-specification packaging standards as applied to the transportation of commercial asbestos is an effective and efficient means of precluding potential problems associated with asbestos airborne emissions occurring during transportation; and that they are consistent with the standards of the EPA and the OSHA. Some of the commenters however were also apparently unaware that the transportation standards for the control of asbestos are designed to be comprehensive in nature such that, once the standards are promulgated, commercial asbestos cannot be packaged and transported in any matter not specified in the amendments. If under more advanced technology the use of rigid, airtight packaging would lessen the likelihood of airborne asbestos emissions associated with bag breakages under current industry wide non-uniform non-standardized packaging practices, then it is necessary that alternative transportation standards be available so as not to preclude the development and utilization of such

---

[1]Under Docket HM-145A (43 FR 22626, May 25, 1978), new standards and procedures were proposed for the transportation of hazardous waste materials. That proposal would include waste asbestos if so identified by EPA under Section 3001 of the Solid Waste Disposal Act as amended by the Resource Conservation and Recovery Act (Pub. L. 94–580).

technology. Although the public record on Notice 78-3 contains statements that the asbestos industry is seeking to improve the technology involved in the shipment and handling of commercial asbestos so as to minimize the possibility for the accidental release of such asbestos incident to transportation, it is by no means certain that the pace of such technological improvements is rapid enough or that the best, economically feasible technology is being considered. However, the classification of asbestos as an ORM-C will, for the first time, require the submission of incident reports to the MTB by carriers of any unintentional release of asbestos during transportation, and enable the MTB to monitor the safety performance record associated not only with the transportation alternatives available under current technology as provided for by these amendments, but also with any improvements in that technology.

For these reasons, the substance of proposed § 173.1090(c)(1) is being retained but modified to reflect an even broader range of permissible rigid, airtight packaging alternatives. This section now is identified in this amendment as § 173.1090(d)(1) because of the addition of new paragraph (b).

SECTION 173.1090(c)(2)

Proposed paragraph (c)(2) of Notice 78-3 covered the transportation alternative of shipping commercial asbestos in bags when in closed freight containers, motor vehicles, or rail cars that were loaded by the consignor and unloaded by the consignee. Several commenters noted that, unless reliance was placed on using the rigid, airtight packaging alternative provided in the proposal, this alternative would preclude the shipment of asbestos fibers by open-bed trailers. One commenter noted that there is "no evidence to indicate that the use of open-bed trailers with side racks and tarpaulins has contributed to bag breakage and the release of airborne concentrations of asbestos fiber." Another commenter noted that the type of bag permitted by proposed paragraph (c)(2) was not specified, and that the shipper could package asbestos in burlap bags, or very thin paper or polyethylene bags which could permit asbestos fibers to be easily released into the air during transit. Another commenter was concerned with "small volume users of asbestos and customers who, from time to time, require sample shipments for trial production runs of a few hundred pounds," and who under § 173.1090(c)(2) would be forced to acquire the exclusive use of a railcar or highway trailer, or rely on the alternative provided by § 173.1090(c)(1).

Given the lack of detailed data on the amount of asbestos fibers released in transportation and the circumstances and causes for such release, the MTB is in general agreement with the thrust of these comments; accordingly, a new paragraph (d)(2) recognizes less restrictive handling of bagged asbestos than was proposed.

SECTIONS 174.840, 175.640, 176.906, 177.844

In these Sections, Notice 78-3 had proposed that, incident to its transportation, asbestos must be loaded, handled, and any asbestos contamination removed, in a manner that will prevent occupational exposure to airborne asbestos particles (emphasis added).

Some commenters objected to the word "prevent," believing that this word was intended to mean completely precluding the possibility of an accident occurring in which asbestos fibers would be released; or completely isolating people involved in the transportation, loading and unloading of asbestos from exposure to asbestos fibers from whatever source such fibers were generated. One commenter pointed out that with "the very large volume of asbestos shipped, occasional container damage may occur." Another commenter pointed out, although in a somewhat contradictory fashion, that since "asbestos is ubiquitous," therefore "airborne levels of asbestos fibers can be present in any place of employment, regardless of whether or not asbestos or products containing known quantities of asbestos are handled" (emphasis added). The Asbestos Information Association in its comments stated that "asbestos is ubiquitous, and there are no workplaces where there is zero occupational exposure to asbestos" (original emphasis). If Notice 78-3 was not as clear as it might have been on this point, it is only necessary to say that the basic purpose of these amendments is to minimize the exposure to airborne asbestos particles accidentally released during or incident to transportation; and appropriate changes to Parts 174, 175, 176, and 177 have been made to reflect this purpose.

ORM-C CLASSIFICATION

Notice 78-3 proposed that the classification for "asbestos" would be as an ORM-C, (Other Regulated Material, Group C). Several commenters were uncertain and concerned about the marking requirements associated with ORM-C classifications. One commenter noted that the designation ORM-C would "carry no meaningful warning to the person handling or opening the package." Another noted that the present regulations of the Occupational Safety and Health Administration (OSHA) on labeling requirements for asbestos convey much more information than an ORM-C marking requirement. These commenters were apparently not completely familiar with the marking requirements associated with ORM-C designated materials. The ORM-C marking not only warns when a package contains hazardous material, but it is also a certification by the person offering the package for transportation that the material is properly described, classed, packaged, marked, and labeled (when appropriate) and in proper condition for transportation according to applicable regulations of the Department. Neither function precludes or preempts OSHA labeling requirements or creates "contradictory regulatory requirements for labeling" as one commenter suggested. For these reasons, no changes have been made with respect to any marking requirements for asbestos packages.

ECONOMIC/INFLATIONARY IMPACT

In reviewing the potential economic and inflationary impacts associated with the final rule, the MTB has determined that such impacts will be minimal. Based on the comments received, and the consequent modification of Notice 78–3, the only economic costs associated with final amendment pertain to the reporting requirements to be submitted to MTB on the accidental releases of commercial asbestos fibers during or incident to transportation. The absolute annual magnitude of these costs will be, of course, a function of the total number of incident reports that are submitted on accidental releases of asbestos fibers; but in view of the undisputed grave consequences from exposure to asbestos fibers, these reporting requirements will not impose an unnecessary burden on the economy, on individuals, or on public and private organizations.

(43 F.R. 56664, December 4, 1978)

[49 CFR Part 173]
[Docket No. HM-162; Notice No. 78–9]
SHIPPERS—GENERAL REQUIREMENTS
FOR SHIPMENTS AND PACKAGINGS
Metric Equivalence for Quantity Limitations

By petition dated February 7, 1977, the Manufacturing Chemists Association (MCA) recommended revision of section 173.26(a) of the Department's Hazardous Materials Regulations to facilitate conversion to metric measurements in the transportation of hazardous materials. The MCA stated that this change would permit the conversion of any hazardous materials package to metric measurements and that such a change would provide shippers and packaging manufacturers with the necessary latitude to convert to more practicable capacities measured in metric units, such as are now provided for by the regulations of the International Air Transport Association and the Inter-Governmental Maritime Consultative Organization. The MCA petition is similar to an earlier petition of the International Air Transport Association containing the rationale that the 10-percent increase in the net quantity per package (dry measure) for import and export shipments would have a negligible effect on safety, since the packaging requirements otherwise would be the same.

With the exception of an exclusion pertaining to packagings having large volumes, the Bureau agrees with the petitioners and believes that adoption of the changes proposed herein (1) will have no adverse effect on the safe transportation of hazardous materials; (2) will be of considerable assistance to shippers converting to systems of metric measurement for both domestic and international purposes; and (3) will not impose any additional costs on packaging manufacturers or shippers since use of the provisions of § 173.26 is optional.

The second sentence in the proposed change states "Specification packagings must be marked to indicate the use of metric measurements and must be tested accordingly." An illustration of compliance with this proposed requirement for a DOT-17E drum would be,

"DOT-17E STC ABC 18–220L–78" and a corresponding change in the quantity of water used in the drop test based on a rated capacity of 220 liters.

(43 F.R. 28216, June 29, 1978)

[Docket No. HM-162; Amdt. No. 173–122]
PART 173—SHIPPERS—GENERAL REQUIREMENTS
FOR SHIPMENTS AND PACKAGINGS
Metric Equivalence for Quantity Limitations

A Notice of Proposed Rulemaking on this subject was published in the FEDERAL REGISTER on June 29, 1978 (43 FR 28216). The Notice was based on a petition received from the Manufacturing Chemists Association requesting revision to section 173.26(a) of the Department's Hazardous Materials Regulations to facilitate conversion to metric measurements in the transportation of hazardous materials. In setting forth the proposal, the Materials Transportation Bureau (the Bureau) expressed the view that the changes proposed would have no adverse effect on the safe transportation of hazardous materials and would be of considerable assistance to shippers converting to systems of metric measurement for both domestic and international purposes, and that the proposed change to the regulations, if adopted, would not impose any additional costs on packaging manufacturers or shippers since use of the provisions of section 173.26 is optional.

The comments received ranged from full support of the proposed amendment, to total objection. Many commenters pointed out that the symbol (rather than abbreviation) for a milliliter is mL rather than ml, and a change has been made accordingly. Several commenters failed to note that application of section 173.26(a) is permissive rather than mandatory. One commenter stated: "Although 1 quart may be rounded to 1 litre, in fact, 1 quart equates to .946 352 9 litre (Standard for Metric Practice ASTM E 380–76E). Therefore, all present U.S. liquid volume measures will be increased by 5.7 percent. This percent increase is unacceptable for packages exceeding 1 gallon." The commenter did not indicate why the increase is unacceptable and apparently missed the point that the application of the regulation is optional. If his particular industry group finds its use unacceptable, they should not use it. The same commenter said:

> This stated "increase" in measure will put the petroleum industry in conflict with, among others, CFR-49 178.116–2 which states: "Minimum actual capacity of containers shall be not less than rated (marked) capacity plus 4 percent." I select this "rated capacity" since the DOT-17E specification was the illustration used in Notice No. 78–9. But this will hold for nearly all rated capacities over the current regulation which reads: "1 gallon for liquids and 10 pounds for solids." As the size of the container increases, the absolute percent will remain constant but the actual discrepancy in gallons will increase. Under accepted conversion standards, there would be 5.9 gallons less in a container at 110 U.S. gallons than the stated metric equivalent under the proposed rule making.

It is not clear what the accepted conversion standards are that are alluded to by the commenter. However, if his concern is in regard to a reduction in safety because of a change in outage, it should be noted that four percent of 220 L is 8.8 L while four percent of 208 L (approximately 55 gallons) is 8.32 L. Therefore, a 220 L drum will require greater minimum outage than a 208 L drum.

The commenter did not illustrate an unsafe condition that would be created by the proposal. However, we accept his comment for further study concerning the validity of all the outage requirements considering recently adopted filling restrictions set forth in various shipping sections of the regulations, such as section 173.116(b) for flammable liquids.

Two commenters, associated with aviation, expressed opposition to the proposal based on the confusion that would be created by the use of metric measurements and the possibility that it would contribute to possible overloading of aircraft. In responding to the comments, we assume the commenters are referring primarily to the information contained in the shipping documents. For many years, the regulations have authorized the display of metric quantity measurements on shipping documents. At the present time, section 172.202(a)(4) does not preclude weight being entered in kilograms and volume and liters, and many shipments, particularly those in international commerce, are presently shipped and documented using metric measurement. The Bureau believes that it has become necessary for all persons involved in commerce to become acquainted with metric measurements whether hazardous materials are involved or not. Consistent with this view, the Department of Transportation is entering into an effort to familiarize its enforcement personnel with the metric equivalents that are authorized by this amendment.

One commenter pointed out that the conversion factor should not be applicable to radioactive materials since different conditions could be produced relative to the original analysis and would, therefore, require new criticality and radiation evaluations. The commenter has raised a point that was overlooked in preparation of the proposal. It was not intended that § 173.26 apply to regulations containing specific conversions such as § 173.396, or § 178.24a which specifies the requirements for DOT Specification 2E. Therefore, the word "only" has been added to the first sentence of the rule in two places to limit its application to those regulations containing only limitations specified in U.S. liquid measures or avoirdupois weight.

A commenter recommended that the Bureau clarify that the purposes of this rulemaking are to make it possible to manufacture, mark, and test "metric capacity" containers and state that the new rule does not mandate that a 5-gallon pail must be made to hold 20 liters nor 55-gallon drums made to hold 220 liters. The commenter's request has merit. A literal interpretation of the proposal may result in a conclusion that the only basis for utilization of metric quantities would be at specifically one liter per quart or 500 grams per pound. This was not intended. It was the Bureau's intent that metric equivalents up to the limitations specified in the proposal may be utilized and the first sentence of the rule has been modified to indicate that metric units may be substituted on an equivalent basis and up to and including one liter per quart and 500 grams per pound. This same commenter requested that the language of § 173.26 be adjusted to reflect that changes in steel thicknesses are not required for the equivalent metric sizes. The Bureau does not consider such an additional provision to be necessary since there is no implication in the rule to indicate such a requirement. For example, § 178.116–6 contains a table indicating marked capacities in gallons. This amendment permits the conversion of the gallons entries to metric equivalents on the basis of one liter per quart. Therefore, the 10-gallon entry may be considered to read "40 liters" without any change to the minimum thicknesses specified. Tables of this type were taken into account during development of the proposal. The Bureau considers that the performance tests specified are sufficient to maintain overall containers integrity and that puncture resistance continues to be accomplished through the specification of minimum thicknesses.

The United States Environmental Protection Agency submitted comments stating:

> Administratively, the change is small and EPA certainly encourages the utilization of the metric system whenever and wherever feasible. However, without more evidence than is presented in the notice of proposed rulemaking. EPA must seriously question the wisdom of this proposal from a public health and environmental quality viewpoint.
>
> In the existing regulations the maximum quantities eligible for substitution are 1 gallon of liquids and 10 pounds of solids. The allowable 5 and 10 percent, respectively, increases allowed by conversion to metric measurements at these levels would indeed be insignificant. However, when the maximum limits are raised to 110 gallons of liquids and 1,000 pounds of solids, the difference between the United States measurement system and the metric system conversions specified becomes significant, viz., 6 gallons for liquids and 100 pounds for solids. Considering the variety of hazardous materials and containers eligible under this regulation it does not appear to be a simple matter. In the event of an accidental release and under certain circumstances it seems possible that such increases could significantly affect public health and the environment. EPA believes that more thought and analysis needs to be completed prior to proceeding further under this docket.

The Bureau does not agree that further thought and analysis is needed concerning this rulemaking. At its inception, full consideration was given to the increases being authorized and the preamble to Notice 78–9 contained a reference to the Bureau's agreement with the petitioner that such a provision would have a negligible effect on safety. The Bureau continues to agree with the petitioner and believes that the adoption of the amendment contained herein will have no significant effect on the safe transportation of hazardous materials. It should be noted that when metric measurements are used, packagings must be tested accordingly. This means that a DOT Specification 17E drum marked "220 L" must meet the test requirements specified in section 178.116–12 at 98 percent of its increased capacity and not based on 55 gallons, or approximately 208 liters capacity. The same holds true for other specifications for which test requirements are specified. The Bureau is satisfied that the conversion limitations permitted by this amendment will not have any significant impact on safety from the standpoint of public health or environmental quality, considering all the limitations and conditions specified in the regulations.

(43 F.R. 56043, November 30, 1978)

[Docket No. HM-163; Amdt. Nos. 171–41, 173–119, 178–49]
BUREAU OF EXPLOSIVES
Withdrawal of Certain Delegations of Authority

Upon the formation of the Department of Transportation in 1967, the functions of the Interstate Commerce Commission (ICC) relative to formulating regulations for the safe transportation of explosives and other dangerous articles (hazardous materials) pursuant to 18 U.S.C. 831–835 were transferred to the Secretary of Transportation by section 6(e)(4) of the Department of Transportation Act (49 U.S.C. 1657(e)(4)). The hazardous materials regulations have been administered by the Department of Transportation since that time.

As a matter of routine, the ICC had delegated certain functions to a nongovernmental body, the B of E. Such delegations were made pursuant to paragraph (e) of 18 U.S.C. 834, which provided that:

> (e) In the execution of sections 831–835, inclusive, of this chapter the Commission may utilize the service of carrier and shipper associations, including the Bureau for the safe transportation of explosives and other dangerous articles . . .

The Hazardous Materials Transportation Act (title I of Pub. L. 93–633) (HMTA), enacted January 3, 1975, under which authority the hazardous materials regulations were reissued effective January 3, 1977, does not contain any reference to this practice of Federal reliance on authority delegated to the B of E. The HMTA directs the Secretary of Transportation to "take all steps necessary to bring . . . regulations into conformity with (its) purposes and provisions . . . as soon as practicable. . . ." (§ 114(b)(2)). Reissuance of the hazardous materials regulations was undertaken to comply with this direction. Withdrawal of delegations to the B of E is being undertaken for the same reason.

Over the past several years, a number of the delegations have been deleted or assumed by the Department in conjunction with various specific rulemaking projects, including the recent docket HM-103/112 amendments (41 FR 15972, 41 FR 40614, and 41 FR 57018) which consolidated all of the Department's hazardous materials regulations into 49 CFR Parts 100–199.

These amendments constitute the first action in an overall phased program to withdraw all of the delegations of authority to the B of E in 49 CFR Parts 100–199. The amendments herein constitute a withdrawal of some of the functions previously delegated and an assumption thereof by the Materials Transportation Bureau (MTB). The MTB has engaged the Department's Transportation Systems Center (TSC), Cambridge, Mass., to perform all of the functions addressed herein under the direction and supervision of the MTB, with the exception of § 173.34(e)(1). The delegations to the B of E remaining after this action will be dealt with in future rulemaking actions.

In the case of functions herein assumed by the MTB, or to be performed by the TSC, as well as those to be considered in future rulemaking, it is necessary to provide for continuity and continued effectiveness of existing B of E registrations for a specified transition period. Accordingly, a new section 171.18 is being added to provide for continued effectiveness of any registration that is valid at the time MTB assumes that registration function. Such registrations continue to be valid until they lapse in the normal course, with one exception in § 173.34(e)(1), or are otherwise restricted by regulation. "Registration" (e.g. § 178.1–4(a)), a clerical function, is distinct from the filing of "reports" (e.g. § 178.1–9(f)), also a clerical function, but one for which renewal is unnecessary. Both functions are distinct from "approval," a judgmental function to be treated in a subsequent notice of proposed rulemaking (except for § § 173.34(e)(1) and 173.21(d), treated herein).

In the course of evaluating B of E delegations, the MTB expects to propose amendments to many of the sections requiring departmental action in order to establish substitute performance criteria, or to simplify, clarify, or delete such requirements. The amendments in this document may be subject to further rulemaking.

The function in section 173.34(e)(1) herein withdrawn from the B of E will be performed on an interim basis by persons qualified under section 173.300a. Section 173.34(e)(1) presently requires B of E approval of equipment, used in periodic retesting of DOT specification compressed gas cylinders. Persons qualified under section 173.300a are presently authorized to verify inspections and tests required in the construction of new cylinders. Because the knowledge and skills required for inspection and test verifications on new cylinders are clearly adequate for inspection of retesting equipment, the MTB intends to rely on persons so qualified to inspect retest equipment until such time as other criteria may be developed. However, an existing backlog of applications for equipment inspection under section 173.34(e)(1) requires immediate attention. During this interim period, anyone wishing to arrange for inspection under section 173.34(e)(1) should contact the MTB.

(43 F.R. 36445, August 17, 1978)

[Docket No. HM-163A; Amdt. No. 171–45]
PART 171—GENERAL INFORMATION, REGULATIONS, AND DEFINITIONS
Approvals/Authorizations Issued by the Bureau of Explosives

On August 17, 1978, the Materials Transportation Bureau published Docket No. HM-163; Amdt. Nos. 171–41, 173–119, 178–49 (43 Fr 36445), as the first action in an overall phased program to withdraw all of the delegations of authority to the B of E in 49 CFR Parts 100–199. The reasons for the action taken as well as those to be considered in future rulemakings were clearly stated in the preamble to the above referenced amendment and will not be repeated here.

The MTB realizes that it is necessary to provide for continuity and continued effectiveness of existing B of E approvals and authorizations. Accordingly, a new section 171.19 has been added to provide for continued effectiveness of any approval that is valid at the time MTB assumes or abolishes that function. Any approval or authorization with a valid expiration date will continue in effect until that expiration date but not beyond December 31, 1984. Any approval or authorization that was issued by the B of E without an expiration date or with an expiration date after 1984, will automatically expire on December 31, 1984.

When the July 20, 1978, final rule to Docket HM-121 (43 FR 31138) withdrew certain authority previously delegated to the B of E, no public comment had been received on existing B of E authorization subject to that amendment. Since that time, MTB has been notified of the existence of certain authorizations which are affected by the July amendment. Because of the need for immediate action to avoid unintended impacts on the holders, this amendment is issued without prior notice and is effective immediately. However, comments are still solicited and should be addressed as previously indicated. Comments will be considered in subsequent publications in Docket HM-163.

(44 F.R. 18027, March 26, 1979)

[49 CFR Part 178]
Docket No. HM-163B; Notice 79–7
Shipping Container Specifications; Withdrawal of Certain Bureau of Explosives Delegations of Authority

On August 17, 1978, the Materials Transportation Bureau published Docket No. HM-163; Amdt. Nos. 171–41, 173–119, 178–49 (43 FR 36445). These referenced amendments constituted the first action in an overall phased program to withdraw all of the delegations of authority to the B of E in 49 CFR Parts 100–199. The MTB will continue to use the service and expertise of the B of E laboratory for the testing of explosives and other hazardous materials. Results of tests performed by the B of E will be forwarded to the MTB for review and final disposition. The preamble to the above referenced amendments clearly stated the reasons for the action taken as well as those to be considered in future rulemaking. In view of the above referenced preamble, repeating it again in this notice is not deemed necessary.

The Bureau realizes that it is necessary to provide for continuity and continued effectiveness of existing B of E approvals and authorizations for a specified transition period. Accordingly, a new section 171.19 was added earlier under Docket No. HM-163A (Amdt. No. 171-45) to provide for continued effectiveness of subject approvals and authorizations until their expiration dates or until December 31, 1984, whichever is earlier.

These proposed changes should have little or not economic impact on the private sector, consumers, State or local governments since these proposals would merely require approval from MTB instead of the B of E. Also, in some instances the requirement for MTB to receive certain reports and to witness certain tests would be deleted.

For simplicity, all paragraphs affected by these proposed amendments, which now read the same, have been grouped together with the present wording and the proposed amendment. Comments regarding this format will be appreciated and useful for future rulemaking actions under Docket HM-163.

(44 F.R. 26772, May 7, 1979)

49 CFR Part 178
[Docket No. HM-163B; Amdt. No. 178-59]
Withdrawal of Certain Bureau of Explosives Delegations of Authority

On May 7, 1979, the Materials Transportation Bureau (MTB) published a Notice of Proposed Rulemaking, Docket HM-163B; Notice 79-7 (44 FR 26772) which proposed these amendments. The background and the basis for incorporating these amendments into the regulations were discussed in that notice. Interested persons were invited to give their views prior to the closing date of June 30, 1979.

The Bureau received only four comments on Notice 79-7, all of which were favorable to the proposed changes.

One commenter, in addition to expressing his approval of the format used for Notice 79-7, recommended the deletion of the last portion of proposed §178.236-2(e), §178.237-2(e), §178.238-2(e), and §178.239-2(e) which reads, "in accordance with your former freight classification (UFC) Rule 40, or National Motor Freight Classification (NMFC) item 200." The commenter further stated that the strength of the bags is already well-covered in the specifications mentioned, and the tests required are all performance tests which will adequately prove the performance of the bag without reference to the freight classification data. In view of the above, and upon further considerations, the Bureau agrees that reference to Rule 40 of the Uniform Freight Classification or Item 200 of the National Motor Freight Classification is unnecessary.

The second commenter suggested that the reference to the Associate Director for HMR and Associate Director for OE be changed to the Director, MTB because of frequent organizational changes within MTB. The Bureau agrees that the possibility of an organizational change affecting the Associate Director for HMR and Associate Director for OE is greater than a change affecting the Director, MTB. However, the Bureau maintains that our regulations should specify the office within the Bureau that is charged with a particular function. For this reason, the rule change is being incorporated as proposed in Notice 79-7.

The third comment received was a brief statement expressing approval of the format in which HM-163B and HM-166A notices of proposed rulemaking were prepared.

The last commenter stated that there appears to be a contradiction between §§178.118-6(a)(2) and 178.118-8(b). Docket HM-140 revised footnote 3 of §178.118-6(a) to authorize the use of a 18-gauge removable head on 55-gallon DOT Specification 17H steel drums under certain specified conditions. The change being adopted by this rulemaking, in §178.118-8(b), does not authorize the use of an 18-gauge head unless specifically approved by the Associate Director for OE. On the other hand, when a 14-gauge or 16-gauge head is used, other types of closing devices are authorized if they perform without failure under the tests required by §§178.118-12 and 178.118-13, and a record of such tests is retained during the period the closure is in use.

Although these amendments do not add anything that was not proposed in the notice, it is anticipated that there may be questions regarding the sentence "Equally efficient means of testing may be authorized upon approval by the Associate Director for OE" which appears in most of the paragraphs entitled "Leakage test." Prior to these amendments the B of E was authorized to approve alternate methods of testing. Two fairly well known methods approved by the B of E are identified as the T-Zone test and the Pocket tester. Use of the Pocket tester is limited to drums of approximately 5-gallons capacity. These previously approved test methods will continue in effect until December 31, 1984, in accordance with 49 CFR 171.19. However, this does not mean that the Associate Director for OE will approve these same test methods in the future. For test methods other than those required in the regulations or approved earlier by the B of E, an approval must be obtained in writing from the Associate Director for OE.

The MTB intends to publish a notice of proposed rulemaking at a later date requesting comments on the best way to incorporate into the regulations those approvals and authorizations that were issued by the B of E for alternate leakage test methods. The MTB intends to complete action on the proposed rulemaking in sufficient time to avoid further extension of the December 31, 1984 deadline.

(44 F.R. 66197, November 19, 1979)

49 CFR Parts 171, 173
[Docket No. HM-163D, Notice No. 79–15]
Withdrawal of Certain Bureau of Explosives Delegations of Authority

On August 17, 1978, the Materials Transportation Bureau published Docket No. HM-163; Amdt. Nos. 171–41, 173–119, 178–49 (43 FR 36445). These referenced amendments constituted the first action in an overall program to withdraw all of the delegations of authority to the B of E in 49 CFR Parts 100–199.

On March 26, 1979, the MTB published Docket No. HM-163A; Amdt. No. 171–45 (44 FR 18027) to recognize certain approvals and authorizations issued by the B of E.

On May 7, 1979, the MTB published Docket No. HM-163B; Notice 79–7 (44 FR 26772) proposing to withdraw or cancel certain delegations of authority to the B of E in Part 178 of 49 CFR. The final rule is expected to be published in the very near future.

Docket No. HM-163C; Amdt. Nos. 171–50, 173–132, 178–57 (44 FR 55577) was published on September 27, 1979, to transfer from the Transportation Systems Center, Cambridge, Massachusetts, to the Bureau's Associate Director for Operations and Enforcement the responsibility for: (1) Approving cigarette lighters or other ignition devices; (2) registering container manufacturers' marks or symbols; and (3) receiving and maintaining reports required to be filed in connection with hazardous materials shipping containers and packagings.

The MTB plans to continue use of the service and expertise of the B of E laboratory for the testing of explosives and other hazardous materials. However, consideration will be given to the use of additional laboratories, such as the Bureau of Mines, when acceptable arrangements can be made. Results of tests performed by the B of E will be forwarded to the Associate Director for Operations and Enforcement, Materials Transportation Bureau, Washington, D.C. 20590 by the applicant for review and final disposition. The preamble to the August 17, 1978, amendment clearly stated the reasons for the action taken as well as those to be considered in future rulemaking. In view of the above referenced preamble, repeating it again in this notice is not deemed necessary.

These proposed changes should have little or no economic impact on the private sector, consumers, State or local governments since these proposals would merely require the final approval to be granted by the Associate Director for Operations and Enforcement instead of the B of E. In some instances the requirement for B of E examination and approval by MTB would be deleted.

(44 F.R. 67476, November 26, 1979)

49 CFR Parts 171, 173, 174, 177
[Docket No. HM-163-D; Amdt. Nos. 171–54; 173–138; 174–38; 177–49]
Hazardous Materials Regulations; Withdrawal of Certain Bureau of Explosives
Delegations of Authority

On November 26, 1979, the Materials Transportation Bureau (MTB) published a Notice of Proposed Rulemaking, Docket HM-163D; Notice 79–15 (44 FR 67478) which proposed these amendments. The background and the basis for incorporating these amendments into the regulations were discussed in that notice. Interested persons were invited to give their views prior to the closing date of January 15, 1980.

The MTB received eight comments on Notice 79–15.

The main objections received were in reference to § 171.20 and § 173.86. The objections were (1) no time limitation on the approval response from the Associate Director for OE after an application for approval has been submitted, (2) no mention of an appellate review in the event that the Associate Director for OE denies an approval, and (3) the economic hardship and excessive time delay that would occur if the present authority now delegated to the Department of Defense and the Department of Energy was withdrawn.

In response to the first objection, the MTB has and will continue to rely on the expertise and recommendations of the B of E. Therefore, we do not visualize the need to incorporate a time period for the Associate Director for OE to respond to an approval request at this time. All applications received for approval will be processed as expeditiously as possible. If actual practice dictates the need for a time limit at a later date, the MTB will consider the issuance of a notice of proposed rulemaking for public comment.

In reference to the second objection, § 171.20 has been revised by adding paragraph (c) to allow any applicant to file an appeal with the Director, MTB in the same manner as provided in § 107.121 for an exemption.

The proposed changes in § 173.86 were not intended to disrupt or change the present authority delegated to the Department of Defense and the Department of Energy. Therefore, § 173.86(b) has been revised to require OE approval only on those items examined by the B of E.

Two paragraphs in Part 174 and three paragraphs in Part 177 have been revised and included in this rulemaking to coincide with similar changes made in Part 173. The changes proposed for § 173.34(d) and § 173.303(a) have been withdrawn from this rulemaking and will be republished in a separate notice of proposed rulemaking in the near future. In addition to § 173.34(d) and § 173.303(a) the MTB believes that the only remaining delegation of authority to the B of E in Parts 173, 174, 177 and 178 that has not been changed is § 177.821(e). The MTB will include these three proposed changes in the same notice.

(45 F.R. 32692, May 19, 1980)

[49 CFR Part 177]
[Docket No. HM-164; Advance Notice]
CARRIAGE BY PUBLIC HIGHWAY
Highway Routing of Radioactive Materials; Inquiry

I. SCOPE OF THIS DOCKET

A. Background. On April 20, 1978, the MTB published an opinion (43 FR 16954) concerning the legal relationship between section 175.111 of the New York City health code and regulations issued by DOT under the Hazardous Materials Transportation Act (HMTA, Title I of Pub. L. 93–633). Section 175.111 of the city's health code prohibits the transportation in or through the city of most commercial shipments of radioactive materials. The HMTA is the basic Federal legislation under which the transportation safety of hazardous materials, including radioactive materials, is regulated. In the opinion, MTB concluded that HMTA routing authority is sufficient to preempt State and local highway routing requirements (see HMTA, §§ 105, 112; 49 U.S.C. 1804, 1811), but that because a routing requirement has not yet been established under the HMTA, that act does not at present preempt section 175.111 of the city's health code.

This municipal safety requirement, and other similar requirements imposed by State and local jurisdictions elsewhere, affect interstate commerce. In some cases local requirements may so vary from one another as to be incompatible. In other cases they may impose significant additional responsibilities on shippers, carriers, or neighboring jurisdictions. Existing State and local requirements for highway carriers of various radioactive materials now restrict use of bridges, tunnels, and roads otherwise open to public use. Local jurisdictions have also imposed requirements for permit fees, advance notice, escorts, and specified times of travel. In many cases, these local restrictions are associated with local responsibilities for emergency response or for traffic control (such as the establishment of truck routes). This rulemaking will examine the transportation safety aspects of highway routing of radioactive materials. The examination will include consideration of routing decisions now being made by carriers and the methods by which those decisions are made. The rulemaking will examine the safety effects of existing and possible Federal, State, and local highway routing controls, including effects of actions by one State or locality on another.

Only highway routing of radioactive materials will be considered in this docket. This does not rule out the possible future consideration of materials in other hazard classes and other modes of transportation. However, highway transportation, of all four modes of transportation, offers the largest number of routing possibilities and the greatest access to population centers. When highway carriers transport radioactive materials, they now face immediate and significant disparities in safety requirements imposed by State and local jurisdictions.

B. Safety. Both DOT and the Nuclear Regulatory Commission (NRC) share responsibility for insuring use of safe methods of preparing and transporting radioactive materials. DOT regulations pertain to packaging, labeling and marking, placarding and shipping paper entries, keyed to the radiation hazard of the material being transported (49 CFR parts 170–178, especially §§ 173.7(b), 173.389–.398 and parts 390–397, especially part 397). Complementary NRC regulations, pertaining to packaging of certain radioactive materials, are found at 10 CFR part 71. In addition NRC regulations in 10 CFR part 73 concern the physical security of special nuclear materials, at both fixed facilities and while in transportation.

An existing DOT regulation generally addresses highway routing of hazardous materials (49 CFR 397.9(a)), including radioactive materials, when carried in substantial quantities. Section 397.9 was issued under statutes that predate the HMTA (18 U.S.C. 834 and 49 U.S.C. 304), and states:

§ 397.9 Routes. (a) Unless there is no practicable alternative, a motor vehicle which contains hazardous materials must be operated over routes which do not go through or near heavily populated areas, places where crowds are assembled, tunnels, narrow streets, or alleys. Operating convenience is not a basis for determining whether it is practicable to operate a motor vehicle in accordance with this paragraph.

. . . .

Another DOT regulation expressly recognizes State and local traffic regulation (49 CFR 397.3). Section 397.3 approves those State and local requirements which concern the mechanics of driving and handling vehicles. Those

State and local requirements are roughly comparable to Federal requirements in 49 CFR part 392. Section 397.3 states:

§ 397.3 State and local laws, ordinances, and regulations. Every motor vehicle containing hazardous materials must be driven and parked in compliance with the laws, ordinances, and regulations of the jurisdiction in which it is being operated, unless they are at variance with specific regulations of the Department of Transportation which are applicable to the operation of that vehicle and which impose a more stringent obligation or restraint.

A third regulation, issued under the HMTA, approves certain hazardous materials restrictions imposed on the use of tunnels by State or local authority (49 CFR 177.810). Section 177.810 states:

§ 177.810 Vehicular tunnels. Nothing contained in parts 170–189 of this subchapter shall be so construed as to nullify or supersede regulations established and published under authority of State statute or municipal ordinance regarding the kind, character, or quantity of any hazardous material permitted by such regulations to be transported through any urban vehicular tunnel used for mass transportation.

Sections 397.3 and 397.9, and section 177.810(a), taken together, reflect the fact that routing of highway traffic in hazardous materials has been a matter left primarily to State and local regulation, and the principle that such State and local regulation should not have the actual effect of altogether forbidding highway transportation between any two points, even where other modes of transportation are available. These provisions constitute the present posture of DOT highway routing policy.

In addition to these provisions, there are also a number of publications available, concerning radioactive materials transportation, which will be considered in this docket. The list below is not inclusive:

(1) Final Environmental Statement on the Transportation of Radioactive Material by Air and Other Modes (NUREG-0170), U.S. Nuclear Regulatory Commission, Office of Standards Development, December 1977 (available from the National Technical Information Service for $12).

(2) Lippek and Schuller, Legal, Institutional, and Political Issues in Transportation of Nuclear Materials at the Back End of the LWR Nuclear Fuel Cycle, September 30, 1977 (Battelle Human Affairs Research Centers, 4000 Northeast 41st Street, Seattle, Wash. 98105).

(3) Transport of Radioactive Material in the United States (NUREG-0073), U.S. Nuclear Regulatory Commission, Office of Standards Development, May 1976 (single copies may be obtained by writing to Division of Technical Information and Document Control, U.S. Nuclear Regulatory Commission, Washington, D.C. 20555).

(4) Environmental Survey of Transportation of Radioactive Materials to and from Nuclear Power Plants (WASH–1238), U.S. Atomic Energy Commission, Directorate of Regulatory Standards, December 1972 (copies available from the National Technical Information Service for $7.25).

In addition, the Nuclear Regulatory Commission has contracted for a generic environmental assessment on transportation of radioactive materials near or through large densely populated areas. Results of this effort will be considered as they become available.

The items listed are available for public inspection in the MTB dockets room. Copies may be obtained from the publishing agencies or, where indicated, from the National Technical Information Service, Springfield, Va. 22161 (payment to NTIS should be enclosed).

C. The need for consistent rules. Consistency among Federal, State, and local transportation requirements affects both efficiency and safety in transportation. For highway transportation, differences in regulatory requirements may affect safety in a number of ways, such as—

(1) Routes used may not be the best available;

(2) Confusion resulting from differences in locally enforced rules may result in noncompliance with either Federal or local rules;

(3) Rerouting that results from a locally imposed rule may have unconsidered effects on other localities, especially on their emergency responsibilities.

However, regulatory uniformity may not be always desirable or possible, due to local transportation conditions and the emergency responsibilities of local authorities. There are therefore practical limits on the possible scope of uniform or exclusive HMTA routing requirements that might be developed in this docket.

II. SOME POSSIBLE REGULATORY ALTERNATIVES

Four alternatives are outlined below, to illustrate several procedures which might be used to regulate highway routing of radioactive materials. MTB is not proposing to employ any of the alternatives. They are outlined merely as illustrations of available HMTA authority. As illustrations, they reflect differences in State and local decision-making participation, differences in cost to governments, business, and consumers, and differences in judgment as to the necessity for additional Federal scrutiny of radioactive materials carriage by highway. The first three alternatives are probably in ascending order of stringency, cost, and degree of DOT rulemaking scrutiny. A draft regulatory evaluation, available for inspection in the public docket, tentatively concludes the implementation of the regulatory examples below would probably not have major economic consequences under Executive Order 12044.

A. Require compliance by radioactive materials highway carriers with a general routing rule to be established by MTB. The test of 49 CFR 397.9 might serve as a model for development of a general routing requirement (variations would require an exemption under part 107). Specific route approval or licensing of highway carriers would not be necessary or possible.

B. Require each highway carrier to be licensed only for variance from radioactive materials routes permitted under a generally applicable MTB routing rule, but permit voluntary licensing. Alternative B, a partial licensing scheme, would have many of the features of alternative C, a full licensing scheme, outlined below. However, alternative B would involve the establishment of a general Federal routing rule under which much or most highway carriage of radioactive materials would occur, with specific route approval required only for carriage operations that depart from the general rule. Both the general rule, as well as any specific route approvals, might consider, in addition to actual routes, matters such as carrier fitness, travel times, and availability of alternate methods of transportation other than highway carriage. The general rule, or a specific route approval, would be sufficient authority for highway carriage operations conducted in compliance with applicable Federal requirements, and State and local requirements not consistent with those Federal requirements would be preempted.

This alternative could also provide for specific route approval, when justified, on a voluntary basis upon application by a carrier, or as a requirement upon application from a State or local government. Specific route approval would be used primarily for situations involving unusual local conditions or routes involving substantial controversy.

C. Require each highway carrier to be licensed for each radioactive material route. This alternative would require each highway carrier to obtain prior MTB approval of any route to be used in the transportation of radioactive materials. The carrier might file proposed routes supported by a statement of safety and jurisdictional considerations. Public comment would be solicited. If the carrier's proposal were accepted by MTB, it would authorize carrier operation under the plan for a certain term, perhaps 2 years. Plan approval would preempt State and local requirements not consistent with it, but could make federally enforceable those State and local requirements affecting the carrier which are consistent with the plan. In some cases, special locally imposed requirements might be expressly incorporated into the plan by the carrier or MTB.

It would be necessary to establish some general criteria by which route plans could be judged. As in alternative B, matters which might be examined could include carrier fitness, travel times, and availability of alternate methods of transportation. Such criteria additionally would be useful to carriers in preparing plans, and to State and local governments in administering their highway regulatory programs.

At the end of the term, a carrier could file for renewal. At that time his safety record, and conditions affecting his performance, could be evaluated, again by a public process. Under some circumstances, and subject to procedural considerations, the carrier's plan approval could be revoked or modified before the term had run.

This alternative would make it impossible to move a designated radioactive material by highway unless the route used were previously approved by MTB. Consequently, existing routing practices would have to be phased out gradually, to reduce confusion and commercial disruption. The mechanics of this alternative resemble those of the process now used by MTB in issuing exemptions. Implementing this alternative may require substantial administrative resources.

D. Invite the Nuclear Regulatory Commission to consider routing restrictions for its licensees. The Nuclear Regulatory Commission addresses routes used to transport special nuclear materials (10 CFR part 73) and has the authority to consider routing in both regulatory and licensing proceedings.

III. REQUEST FOR COMMENT

Comment is solicited on the preceding discussion and on the questions below.

Should radioactive materials be subject to more stringent Federal highway routing requirements than now imposed by 49 CFR 379.9?

(A) If so—

(1) What types, quantities and forms of radioactive materials should be considered?

(2) What benefits might be achieved?

(3) What factors in addition to population density and highway conditions should be considered in connection with routing? Should those factors include such things as emergency response training for drivers, special equipment, or the operating convenience and efficiency of the carrier? Should these factors be considered in place of routing?

(4) How would additional Federal rules impact State and local regulatory programs, or emergency response capabilities? To what extent is greater uniformity in State and local requirements desirable, and to what extent achievable through Federal rulemaking?

(5) What kind of Federal rule is desirable? Is a generalized DOT requirement preferable to a procedure that entails an individual DOT examination of some or all routes? Do local conditions affecting route selection necessitate individual Federal examination? If detailed examination of highway routes is necessary, by what procedures should it be accomplished?

(6) What additional costs may be involved if new routing rules are developed and implemented? How are those costs likely to affect shippers, carriers, Federal, State, and local governments, utilities, and the public?

(B) If not—

(1) What are the likely costs and benefits of taking no action?

(2) Do existing disparties between State and local rules concerning highway carriage of radioactive materials need to be harmonized? If so, how?

A hearing will be held to consider views on this advance notice, at a time and place to be subsequently announced. Drafters of this document are Douglas A. Crockett, Office of Hazardous Materials Regulation, MTB, and George W. Tenley, Office of the Chief Counsel, Research and Special Programs Administration.

Commenters are advised that section 105(b) of the HMTA requires DOT to consult and cooperate with the Interstate Commerce Commission before issuing any regulation with respect to the routing of hazardous materials.

(43 F.R. 36492, August 17, 1978)

49 CFR Parts 173 and 177
[Docket No. HM-164; Notice No. 80-1]
Highway Routing of Radioactive Materials

I. HISTORICAL BACKGROUND

In 1976, truck shipments of irradiated reactor fuel (spent fuel) from Brookhaven National Laboratories' Long Island facility were interdicted by an amendment to the New York City Health Code. The Health Code amendment had the practical effect of banning most commercial shipments of radioactive materials in or through the City. Associated Universities, Inc., which operates Brookhaven National Laboratories, asked DOT whether that ordinance was preempted by Federal transportation safety requirements issued under the Hazardous Materials Transportation Act (HMTA) (49 U.S.C. 1801 *et seq.*). On April 20, 1978, DOT published an Inconsistency Ruling (43 FR 16954) in which it viewed the City's Health Code amendment as an extreme routing requirement intended to protect the very dense urban population found inside the City. DOT concluded that the HMTA could preempt local requirements such as New York City had implemented, but because highway routing authority had not yet been exercised under the HMTA, the City's health code was not preempted by HMTA requirements.

A number of other State and local governments have either passed, or proposed, legislation that severely restricts transportation of certain radioactive materials through their jurisdictions. These actions do not seem to be based on the relative significance of previous accidents involving radioactive materials transportation. The information available to DOT through the Department's Hazardous Materials Incident Reporting System, to which carriers report incidents involving any release of a hazardous material in transportation, or any suspected radioactive contamination, indicates that radioactive materials transportation has a good safety record. In 1977 the DOT estimated that 2.5 million packages of radioactive materials were being transported by all modes yearly. This estimate closely approximates the 2.19 million packages reported in the study "Final Environmental Statement on the Transportation of Radioactive Material by Air and Other Modes" (December 1977) (NUREG 0170) (p. 1–18) as being shipped in 1975. From 1971, when the reporting system was established, until August 1979, a total of 463 incident reports were received involving radioactive materials (0.5% of the total reports received). In comparison, approximately 45,000 incident reports were received which involve flammable liquids (51% of the total). Of the 463 reports filed since 1971 involving radioactive materials, 323 concerned highway transportation, and of this number approximately 275 were reports of minor or suspected contamination to the container and/or transport vehicle due to improperly prepared shipments. The more severe of the reported highway incidents involved vehicle accidents which resulted in packages of radioactive materials being burned, thrown from the vehicle, or rolled on by the vehicle. These events occurred in about 15% of the reported incidents. Examples of such incidents reported last year include:

(1) The January 10 collision near Morristown, Tennessee of a truck tractor and flat-bed trailer carrying 5 cylinders, each containing 6800 pounds of radioactive material fissile, n.o.s. (Uranium Hexafluoride UF_6) into the rear end of a tank truck. The crash resulted in the total loss of the truck power unit and personal injuries to the driver. The cylinders however, remained intact and the trailer sustained very limited damage. The load was returned to Oak Ridge, Tennessee using another power unit. No loss of contents or increased radiation levels were detected.

(2) A single vehicle accident on March 22 involving a truck tractor and enclosed semi-trailer carrying 54 steel drums of 55 gallon capacity, each containing approximately 810 pounds of Radioactive Material, LSA, n.o.s. (yellow-cake). In this incident the vehicle was travelling on a portion of I-235 near Wichita, Kansas. The shoulder of the road was composed of soft dirt due to a recent excavation required for the construction of an interchange. Travelling at a speed of 50–52 MPH the right rear wheels went into the soft shoulder on the right side of the road. When the driver attempted to steer the truck back onto the road, the truck began to swerve to the left, overturned, and landed across the road on its right side. As a result of the accident, 51 drums came through the roof of the trailer and

scattered as far as 100 yards from the truck in the direction the truck was initially travelling. About 1800 pounds of the 43,782 pounds of yellowcake was spilled. Cleanup operations and recovery of the yellowcake required 9 days to complete. This incident resulted in personal injuries to the driver but no radiological damage occurred to personnel and essentially none to the environment.

(3) The loss of a package of radiopharmaceuticals (radioactive yellow-III label) from the rear of a local delivery truck on August 16 onto a city street in Des Moines, Iowa. The package weighing 29 pounds consisted of a lead shielded generator (Molybdenum 99/Technetium 99) and glass vials of a sterile saline solution. Extensive damage was incurred by the package from the wheels of passing motor vehicles resulting in the scattering of its contents. While several of the glass vials were broken the generator itself was not damaged to the point of releasing its contents, nor was there an increase in radiation levels.

None of these or any of the radioactive materials incidents reported to date resulted in radiological health consequences as severe as the consequences reported sometimes to result from the behavior of flammable liquids in transportation accidents. Nonetheless, it seems likely that State and local interest in radioactive materials transportation will continue. Reasons for this interest involve qualitative differences between transportation hazards posed by radioactive materials and transportation hazards posed by other materials.

Transportation accident risk and estimates of population doses from normal accident-free transportation for radioactive materials have been made in NUREG 0170 and in the preliminary report "Transport of Radionuclides in Urban Environs: A Working Draft Assessment" (May 1978, SAND77–1927) (Urban Environs Draft) (both documents are available for review in the public docket). Those estimated risks are within the magnitudes of other socially accepted risks, such as evidenced in highway traffic fatality rates.

Public concern with radioactive materials transportation, however, is more profound than those estimates would suggest is justified. In part this concern reflects the distinction between risks which are likely to be concentrated and similar risks spread over differing times and locations. The annual death rate from passenger car accidents, for example, usually is perceived as less catastrophic than major aircraft accidents, although far more people die in automobile accidents. This distinction may reflect the perceived limits of society to deal with catastrophic occurrences.

Discomfort from a lack of public familiarity with radiation hazards also increases the likelihood of local responses to radioactive materials transportation risks. Accident risk, for example, may be expressed in such unfamiliar terms as numbers of latent cancer fatalities, early deaths or morbidities, and genetic effects. Unlike other hazardous materials, radioactive materials present an impact during accident-free, or normal, transportation. This impact, called normal dose, results from the fact that under normal circumstances, some small amounts of radiation penetrate the outer surfaces of most packages of radioactive materials. Normal dose is very small, but it is statistically significant in terms of the overall impacts that result from radioactive materials transportation.

Radiation hazards themselves are comprised of a number of phenomena. A radioactive material may be solid, liquid, or gaseous, and thus may or may not easily be dispersed in a transportation accident. A radioactive material may be ingested or absorbed selectively and retained in plant, animal, and human tissues for varying lengths of time due to the basic chemical and physical characteristics of the different radioactive materials as well as the nature of the tissues. A person also can be exposed to radiation by being near an exposed radiation source. Radiation ordinarily cannot be detected except by instrumentation, unlike the well understood flammability hazard of such materials as gasoline.

Radiation health effects are not widely understood but include genetic effects and latent cancer, conditions which may not be manifested until many years after exposure (which may not be recognized at the time it occurs). A thorough understanding of radiation and its known health effects requires a significant degree of technical knowledge. Other materials possess similar hazards, but the combination of these characteristics in the case of radioactive materials has produced a degree of public concern which has affected actions taken or being considered by State and local governments.

II. DISCUSSION OF PUBLIC COMMENTS

In August 1978, DOT issued an advance notice of proposed rulemaking (43 FR 36492, August 17, 1978) opening this docket and asking for public comment to assist in deciding whether rules to govern highway routing of radioactive materials should be developed and proposed, and if so, what the rules should say. The advance notice did not propose any action but asked for comment on whether any action should be taken by DOT. Over 550 comments were received, falling principally into six groups.

A. Individuals; Public Interest and Environmental Organizations. This group comprises almost 70% of all comments received and falls into two subgroups:

(1) Individuals and organizations opposed to the transportation of nuclear materials or Federal involvement in local affairs. These commenters made two major points: local laws which are stricter than Federal regulations should be allowed to stand, and radioactive materials, particularly spent fuel, are inherently dangerous and should not be

transported through heavily populated areas. One commenter urged MTB to adopt a full licensing scheme to apply to shipments involving a large number of curies (a unit of radioactivity) with an expressly reserved right in State and local governments to impose stricter standards. This commenter suggested banning large curie shipments from urban areas with population densities above 10,000 persons per square mile.

(2) Individuals and organizations favoring wider Federal preemption of State and local laws. These commenters stressed the excellent transportation safety record of radioactive materials and urged that additional requirements not be imposed. Many commenters in this group asked MTB to adopt a general routing rule which would specifically preempt unnecessary local restrictions that impede commerce.

B. State Governments and Political Subdivisions. Views were expressed by approximately 19 States, 7 counties and 10 cities or towns. Several States endorsed existing DOT requirements and supported a general routing rule such as that found at 49 CFR 397.9(a). Most commenting States appear to favor a general routing rule with provision for some State input. Most States also appear to be interested in obtaining more information on the types, quantities, and forms of radioactive materials shipped, and the routes actually used. Local governments, on the other hand, generally opposed any type of Federal interference with local laws and ordinances. Commenters from both urban and rural counties, as well as from cities, generally opposed transportation of radioactive materials through their jurisdictions.

C. Motor Carrier Industry. Commenters in the motor carrier industry were concerned with inconsistent State and local laws. The American Trucking Associations, Inc., (ATA) suggested that MTB establish a general routing rule which would give carriers some degree of flexibility within certain guidelines to use their own discretion over choice of routes. To provide for State input, ATA suggested that MTB prioritize highways for routing purposes by characteristics that States could use in determining specific routes within their jurisdictions. ATA also suggested the use of a "circuity limit" to establish maximum rerouting distances that could be required by States under this scheme. Finally, ATA states that any such routing requirements should be keyed to vehicles carrying sufficient amounts of radioactive materials to require placarding. (When certain amounts of any hazardous material are carried in a motor vehicle, DOT requires that a placard, or warning sign, be affixed to the vehicle. For radioactive materials, the placard bears the word "RADIOACTIVE" and an appropriate symbol.)

D. Shippers of "Low-Level" Radioactive Materials and Other Hazardous Materials. This group includes commenters representing manufacturers, users, and shippers of radiopharmaceuticals, medical and industrial isotopes, and other "low hazard" radioactive materials. It also includes shippers concerned with possible future routing controls on other hazardous materials (a matter beyond the scope of this docket). These commenters generally saw little reason to impose more stringent rules, but felt that if such rules were to be imposed, low-level radioactive materials should be excepted because of their time-critical nature (many medical radioisotopes lose their radioactivity over a relatively short period of time), low transport hazard, and medical/research value. Suggestions ranged from excepting all Type A quantity (from 0.001 to 1,000 curies of material per package, depending on the material) and limited quantity packages (small amounts otherwise generally excepted from DOT specification packaging, marking and labeling requirements) to excepting all non-placarded shipments.

E. Shippers of Large Quantity or "High-Level" Radioactive Materials. This group primarily includes shippers or shipper organizations associated with the nuclear power industry. Although there were only nine commenters in this category, one commenter represented 24 electric utility companies which are operating 39 nuclear power generators and planning the construction of 61 new generators. This commenter maintained that routing controls applying only to radioactive materials cannot be justified on the basis of safety alone, but that the proliferation of local restrictions on transportation justify the imposition by MTB of a general routing requirement to preempt State and local requirements. One commenter suggested a general rule that would require avoidance of heavily populated areas when possible, would provide for "voluntary licensing" of carriers for specific routes, and would permit State and local governments to seek an order from MTB prohibiting transportation of certain radioactive materials over specific routes.

F. Bridge and Turnpike Authorities. Comments were received from bridge and turnpike authorities, and from the International Bridge, Tunnel and Turnpike Association. These commenters expressed concern that their facilities might become part of a "designated hazardous materials route" established by MTB and pointed out that such action might raise their insurance rates.

III. REGULATORY BACKGROUND

A. Synopsis of Proposed Rule. The proposal presented in this publication would establish a general rule which would apply to any motor vehicle carrying radioactive materials requiring placarding. The general rule would require such a vehicle to be operated on a route that presents a risk to the fewest persons unless there is not any practicable alternative highway route available or unless it is operated on a "preferred" highway as subsequently defined. Subject to this provision, the motor vehicle would have to be operated on a route which minimizes transit times, so as to minimize unnecessary exposure. The carrier would be responsible for notifying the driver of the presence of radioactive materials in the shipment and for indicating generally the route to be followed.

A second, additional and more specific rule would apply to any motor vehicle transporting a package containing a large quantity of radioactive materials, as defined by existing DOT regulations. Such a motor vehicle would be required to operate on "preferred" highways, defined as any highway approved for that purpose by an appropriate State agency, and any Interstate highway for which an equivalent substitute has not been provided by such State agency. The vehicle would operate in accordance with a written route plan prepared by the carrier before departure. State agencies could designate preferred highways, after consultation with local jurisdictions, based on the policy of an overall minimization of radiological and nonradiological impacts of both normal transportation and transportation accidents. When necessary, a motor vehicle containing a large quantity of radioactive materials could operate away from preferred highways under the provisions of the general rule. The driver of a motor vehicle containing a large quantity package would be required to receive specific training. Each shipper of a large quantity package would be provided by the carrier with a copy of the written route plan, which the shipper would file with MTB (except for irradiated reactor fuel covered by NRC requirements). The filed route plans would be used by MTB to provide data on routes, amounts and shipment frequencies for use in State and local emergency response planning. Information on the movements of irradiated reactor fuel would be available after the MTB received this information from the NRC.

The specific large quantity rule would require use of an Interstate urban circumferential or bypass route to avoid cities if available, instead of an Interstate through route, notwithstanding a minor transit time increase. For cities with Interstate through routes without Interstate circumferential or bypass routes, a State could designate any available circumferential or bypass route if it is essentially equivalent in performance or design to an Interstate circumferential or bypass situated in some other urban location.

B. Existing DOT Requirements for Transport of Radioactive Materials. This document focuses on routing and related operational controls for highway transportation of radioactive materials. Existing provisions in the DOT Hazardous Materials Regulations address required packaging and related transportation controls, which constitute the primary safety measures in radioactive materials transportation. A brief summary of those existing rules follows.

Packaging for radioactive materials transportation is based on amount, kind, and physical form of the radioactive material to be transported. Each radionuclide is assigned to a Transport Group, of which there are seven that are ordered to reflect the various radionuclides' degree of radiotoxicity and relative hazard in transportation. For each Transport Group, two quantity limits are established which define Type A and Type B quantities, for which Type A and Type B packaging then is prescribed. If the radionuclide is in "special form" rather than "normal form", quantity limits for Type A and B quantities are larger, because materials in special form are difficult to disperse, either because of the inherent properties of the materials (such as a solid metal) or because the materials are specially prepared (as through encapsulation).

In most cases, a warning label must be applied to each package of radioactive material. The kind of label required depends on the radiation dose rate at or near the surface of the package. The dose rate, in turn, is determined by the type of packaging and shielding used within the package, and by the type and quantity of radionuclides present in the package. There are three labels which may appear on a package of radioactive materials: White I, Yellow II, and Yellow III. The amount of surface radiation allowed for each type of label is identified subsequently in the discussion of radioactive materials covered by this rulemaking. It is sufficient to state here that any vehicle which carries a package labeled Yellow III must show the radioactive material placard on all four sides of the transport vehicle. In addition, all vehicles which carry Fissile Class 3 (certain fissile radioactive materials which require special transportation arrangements for that reason) and large quantity packages must be placarded regardless of the dose rate of the package.

Three other terms that affect packaging are "limited quantity", "low specific activity" (LSA), and "large quantity". Limited quantities of radioactive materials are small amounts, such as may be found in certain manufactured articles (instruments, electronic tubes). Limited quantities of the various radionuclides also are defined generally by an activity limit in millicuries or curies associated with each Transport Group. Such amounts are excepted from many transportation controls, such as requirements for specification packaging, marking of the shipping name on the package, and labeling the package for a radiation hazard.

LSA materials are materials that contain very little radioactivity per unit weight. Uranium ore, for example, may be shipped as LSA. These materials frequently are shipped in large volume shipments and are transported in Type A packaging unless moved in an exclusive use vehicle (*i.e.,* where a single shipper alone uses the vehicle and all loading and unloading occurs under the direction of the shipper or the consignee, a practice through which larger shipments are permitted).

"Large quantity" amounts of radioactive materials are defined by Transport Group and vary from a minimum of 20 or more curies (for materials such as plutonium, Transport Group I) to 50,000 or more curies (certain radioactive gases, Transport Groups VI and VII). Large quantity amounts must be shipped in Type B packaging, most of which require approval for that purpose, prior to use, by the Nuclear Regulatory Commission.

The distinction between Type A packaging and Type B packaging is significant. In addition to having adequate radiation shielding, Type A packaging is designed to withstand normal transportation conditions as simulated by tests described in the Hazardous Materials Regulations: exposure to the equivalent of extreme climatic conditions;

and drop, penetration, compression and vibration tests representing other conditions encountered in normal transportation. Type B packaging, on the other hand, often must be heavily shielded and is designed to withstand extreme accident conditions as simulated by a 30-foot drop onto an unyielding surface; a 40-inch drop onto the end of a pointed steel bar; exposure to a temperature or fire of 1,475° F. for 30 minutes; and submersion in three feet of water for eight hours.

In the vast majority of possible accidents experimental work has indicated that in the event of an accident a release of 0.1 percent of the contents would be a reasonable assumption for Type A packages. On the basis of general handling experience it is further assumed that the actual intake of radioactive material into the body by a person coming into contact with air or surfaces contaminated by such a release is unlikely to exceed 0.1 percent of the amount released from the package. Thus, it is unlikely that any one person would ingest more than one-millionth of the maximum allowable package contents in the event of an accidental release. Stated differently the Type A package quantity limitations are such that an intake of one-millionth of the maximum allowable package contents would not result in a radiation dose to any organ in the body exceeding internationally accepted limits; nor a radiation level of 1 rem per hour at 10 feet from the unshielded contents.

Type B packaging, in a severe transportation accident, would be expected to survive without any significant release of its contents. Spent fuel assemblies, for example, are shipped by highway as large quantity shipments in massive packagings (casks) that may be five in diameter, fifteen feet long and weigh up to 35 tons. Casks are practically impervious to small-arms fire and small explosive charges.

In a highway accident near Oak Ridge, Tennessee, on December 8, 1970, a spent fuel cask was thrown more than 100 feet when a truck driver while negotiating a wide turn lost control after swerving to avoid another vehicle. Although the driver was killed in the impact, there was no release of spent fuel or increase in radiation. Spent fuel casks of an earlier design also have been subjected to destructive testing simulating severe, high speed highway and rail accidents. The casks survived with only minor damage that would have posed little or no risk to the public if the events had been real rather than simulated.

Associated with irradiated fuel and present during its transportation by highway are certain decay gases and volatile fission products along with the essentially solid materials. Given a set of circumstances in which the cask is subjected to extreme crushing forces of 200,000 pounds and a subsequent fire of 1875°F. for 2 hours duration, estimates have been made of the resulting radiological consequences. In Section 5–6 of NUREG 0170 some of these "worst-case" shipment scenarios were considered. One such hypothetical case involves a shipment of spent fuel being transported through a high-density urban area (15,444 people per square kilometer). It was hypothesized that if such an incident were to occur, 100% of the gaseous and volatile materials would be released as an aerosol and then dispersed into the atmosphere where wind currents and other weather conditions would influence both the area and degree of radioactive contamination. Under these particular circumstances it is estimated that the contaminated area would require evacuation for 10 days and the cost of clean-up, lost incomes and temporary living expenses would amount to $200 million (1975). Radiological health consequences are estimated to be minimal with no early or latent cancer fatalities. While an event such as this is likely to occur only once in 3 billion years, the data is significant when weighing its risk against other risk levels which are determined to be acceptable. Extreme incidents which involve the release of as little as 1% of the solids as an aerosol would have extremely serious consequences. Such an incident, however, is likely only once in 25 billion years and is thought by MTB not to warrant undue concern. A more typical high speed collision and fire in a highway accident is not likely to result in extensive radiological injuries or damage from the presence of either Type A, Type B or large quantity packages of radioactive materials.

C. Normal and Accident Exposure Resulting From Transport of Radioactive Materials. This proposal was developed after consideration of impacts from both transportation accidents and accident-free (normal) transportation. Accident risk includes both radiological risks and nonradiological risks (such as impact damage in a motor vehicle collision). Normal transportation is considered principally from the radiological standpoint of normal population dose. Nonradiological impacts of normal transportation are considered secondarily and largely consist of the costs associated with motor vehicle operation (such as fuel use).

Normal dose is the amount of radiation exposure received generally by persons who come near packages of radioactive materials during accident-free transportation, such as package handlers, truck crews, pedestrians and other passers-by. Normal dose usually is expressed in terms of rems (Roentgen Equivalent in Man, a measure of biological damage from radiation) or units thereof. The term "person-rem" is used to express total (integrated) population dose. The normal dose from a package of radioactive materials is dependent upon the amount of radiation emitted through the package surfaces, which is described by the Transport Index (usually a measure of radiation at three feet from the package surface). Essentially all packages of radioactive materials, from small Type A packages to spent fuel casks, emit at least small amounts of radiation even when in compliance with all Federal packaging requirements. The amount of radiation exposure received by the population as normal dose is proportional to the time during which exposure occurs. It declines at least geometrically with distance from the package. A longer trip means a longer period of exposure which results in greater normal doses to truck crews (drivers) and also may mean greater doses to the surrounding population. In highway transportation, the dose received by the truck crew is the largest single component of normal dose that can be changed by modifying transportation practices. The health effects discussed in this publication are those predicted by a health effects model used in NUREG 0170. Commenters wishing to address the validity and degree of certainty associated with that health effects model will find a brief discussion in NUREG 0170 on p. 3–11.

In NUREG 0170, the impact of normal dose from radioactive materials transportation is summed up in the following way for all modes of transportation.

> The estimated total annual population dose [from radioactive materials transportation] is 9,790 person-rem in 1975 and 25,400 person-rem in 1985. This dose has the same general characteristics as other chronic exposures to radiation such as natural background. The predicted result of public exposure to this radiation is approximately 1.19 latent center fatalities and 1.7 genetic effects in 1975 and 3.08 latent cancer fatalities and 4.4 genetic [effects] in 1985. When the value of 9,790 person-rem may seem large, it is small when compared with the [forty million] person-rem received by the total U.S. population in the form of natural background radiation . . . [T]he average annual individual dose [from radioactive materials transportation] is approximately 0.5 [millirem], which is a factor of 300 below the average individual dose from background radiation. [p. 4–49]

Total accident risk is an estimate that combines both the chance that an accident will occur and the probable consequences if it does. Total risk sums both radiological consequences and nonradiological consequences. Accident risk from radiological hazards depends on a variety of factors, but principally on the severity and rates of accidents on the roads traveled (other factors contribute to the accident probability, such as driver training and vehicle condition) and on the density and proximity of the population along the route. All else being equal, unsafe highways, long trips and dense populations near the highways result in higher accident risks. Accident risk also includes the nonradiological hazards, such as the injuries and damage that may be realized in any motor vehicle accident. Nonradiological accident risks generally appear to be much greater than radiological accident risks, but the prediction of radiological accident risks involves more variables than nonradiological accident risks and therefore is less confident.

Regarding radiological risk from potential transportation accidents in all modes of transportation, NUREG 0170 estimates that

> The accident risk for the 1975 level of shipping activity . . . is very small: roughly 0.005 additional [latent cancer fatalities] per year, or one additional [latent cancer fatality] every 200 years, plus an equal number of genetic effects. This number of [latent cancer fatalities] is only 0.3% of those resulting from normal transport population exposures.

> . . . .

> The projected accident risk in 1985 is . . . about 3.5 times the 1975 risk, but is still very small in comparison to the [latent cancer fatalities] resulting from normal transport.

> . . . .

> The principal nonradiological impacts are those injuries and fatalities resulting from accidents involving vehicles used exclusively for the transport of radioactive materials. The number of expected annual nonradiological fatalities [in 1975] is almost 50 times greater than the expected number of additional [latent cancer fatalities] resulting from radiological causes [in transportation accidents] but is less than one fatality every five years. [pp. 5–52, 53]

D. Related Factors Affecting Route Selection Under Proposal. In view of statistics showing lower accident rates and reduced travel times in travel on Interstate highways, this proposal favors use of the Interstate System. MTB believes that in most cases this policy will produce the most significant transportation safety impact reduction and it offers a clear standard for compliance and enforcement purposes. However, the policy is modified by two other considerations which should be kept in mind by persons reviewing this proposal.

First, for reasons of cargo security discussed later in this document, the Nuclear Regulatory Commission (NRC) recently established interim physical security rules (44 FR 34466, June 15, 1979) for transportation of irradiated reactor fuel (spent fuel). Those rules include the following requirements for NRC licensees who ship spent fuel:

(a) Advance notice to and approval from the NRC for each shipment of spent fuel.

(b) Advance arrangements with law enforcement agencies along the route for emergency assistance.

(c) Use of routes that avoid heavily populated areas where practicable, and additional protective measures approved by the NRC where that is not possible.

(d) A trained escort accompanying each shipment.

(e) Motor vehicles that are equipped with radiotelephone and CB radio communications equipment and that are capable of being immobilized.

(f) Procedures for coping with threats and physical security emergencies.

The security of spent fuel in transit was a major concern to commenters in the 1978 hearing on the advance notice of proposed rulemaking in this docket and in the hearing in 1977 regarding the inconsistency ruling on the New York City Health Code amendment. Development of the current DOT proposal reflects existing arrangements between DOT and NRC wherein NRC exercises responsibility for any necessary physical security requirements during transportation. The DOT proposal is therefore directed at reducing impacts associated with normal and accident situations arising in transportation, while NRC is concerned with preventing malicious or deliberate release of radioactive materials. The DOT proposal, however, would extend the NRC physical security requirements to nonlicensee shippers, such as the Department of Energy.

Second, the proposal acknowledges that some local conditions may justify special routes for shipments of large quantity packages. One such condition is expressly recognized in the proposal and concerns cities which have an Interstate direct route and an Interstate (or equivalent) circumferential or bypass route. The proposal also provides for State action to establish or modify routes for carriers of large quantity packages.

The benefit of routing that avoids cities, or heavily populated areas generally, is difficult to predict, but involves a trade-off between the increased impacts due to longer shipment distances and the decreased impacts due to avoiding dense populations. Avoidance of heavily populated areas is a requirement that currently applies to all shipments of hazardous materials by motor vehicle if the amounts are sufficient to require placarding:

> Unless there is no practicable alternative, a motor vehicle which contains hazardous materials must be operated over routes which do not go through or near heavily populated areas, places where crowds are assembled, tunnels, narrow streets, or alleys. Operating convenience is not a basis for determining whether it is practicable to operate a motor vehicle in accordance with this paragraph. [49 CFR 397.9(a)].

Requiring motor vehicles to avoid heavily populated areas usually will increase trip distance and travel time. For the transportation of radioactive materials, under some circumstances those increases can result in an increased normal dose. If use of less safe highways or increased travel times are necessary to avoid heavily populated areas, accident risk also may be increased. The extent of the safety benefit that might result from motor vehicles avoiding heavily populated areas (such as a possible decrease in normal dose or in accident consequences) is influenced by factors such as differences in population densities, effectiveness of local emergency planning, physical features and weather conditions along the various routes that might be used and the times and days they are used. These factors are site-specific and hard to generalize on a national scale except on a statistical basis.

Some generalizations, however, can be made. Because of their lower accident rates and greater efficiency, use of Interstate highways usually will result in fewer accidents and in reduced travel times. Given equivalent roadways, routing radioactive materials carriers on longer Interstate circumferential roads, with adjoining populations that are less dense than those adjoining a shorter Interstate through route, usually will increase normal truck crew dose and the probability of an accident but usually will decrease total normal dose and accident consequences. The possible reduction in radiological accident consequences in such a situation depends on variable factors including population distribution in the area and meteorological conditions which can affect the movement of airborne debris.

Differences exist between Interstate routes through and around a city. A circumferential Interstate route may have a higher average speed and lower accident rate than an Interstate through route, but the accidents may be more severe. Because of the cost and availability of land, the greater access requirements, the design standards of some urban freeways may be less than optimal and possibly less than those of a suburban circumferential Interstate highway. Data from NUREG 0170 and recent traffic accident statistics indicate that routing to avoid cities may offer a slight reduction in overall radiological risk, but at the probable expense of a greater number of fatalities and injuries resulting from an increase in traffic accidents associated with increased distances. However, even though the resultant increase in nonradiological fatalities appears to be larger than the decrease in radiological fatalities anticipated, the difference is small in terms of absolute numbers (a difference of possibly one fatality every 100 years at 1985 levels of shipping activity). There also is necessarily more uncertainty in the prediction of radiological consequences from transportation than in the prediction of traffic fatalities, due to the number of variables involved, so a conservative approach also suggests circumferential routing.

There also are sound administrative reasons to require that Interstate circumferential and bypass routes be used. Circumferential routing around cities is more consistent than direct routing with requirements that apply to other hazardous materials transported by highway (49 CFR 397.9(a)).

The proposed required use of circumferential routes by large quantity carriers, however, is predicated on the safety and efficiency of transportation on Interstate highways. Where other highways are designated to establish an urban circumferential route, they should offer the same advantages as comparable Interstate circumferentials. For the designation of preferred highways other than urban circumferentials, the proposal would assume an evaluation of all factors pertinent to reducing the impacts of highway transportation of radioactive materials, rather than the abbreviated method of relying on the similarity of the preferred routes to Interstate highways. State action is more fully discussed later in this document.

From a regulatory standpoint, consideration must be given to the need for requirements which are efficient and comprehensible, which encourage compliance and which can be enforced. The term "heavily populated areas", not used in the proposal, is disfavored for this reason. Instead, an attempt has been made to state the routing factors which would be used for placarded vehicles, and to state that the carrier would be responsible for acting to ensure those factors are observed in the operation of its motor vehicles. MTB also must consider the extent to which State and local site-specific participation can be useful in establishing or modifying routes used by highway carriers of radioactive materials.

IV. ANALYSIS OF PROPOSED RULE

A. *Radioactive Materials Subject to Routing Requirements.* The proposal in this notice is based on the type of radioactive material shipped and the quantity (activity) per shipment. Essentially there are three transportation situations that would require different treatment under this proposal (see table "Examples of Radioactive Materials Under Proposal"):

(1) Packages for which the carrier is not required to placard his vehicle would be excepted from any routing restrictions. These packages comprise the majority of all radioactive materials shipped and include packages excepted from labeling or bearing the White I or Yellow II radioactive material label as a result of a relatively low radiation dose rate at or near the package surface (see CFR 172.403). A package is excepted from labeling under certain conditions if it contains limited quantities of radionuclides (identified in 49 CFR 173.391(a)), manufactured articles (clocks, smoke detectors, or electronic tubes) which contain limited quantities of radioactive materials, or certain other manufactured articles (identified in 49 CFR 173.391(c)). Also excluded from labeling are some low specific activity (LSA) radioactive materials when shipped in an exclusive use motor vehicle (see 49 CFR 173.392).

A radioactive White I label is required on all other packages which have a dose rate measuring up to 0.5 millirem per hour at any point on the external surface of the package (excluding Fissile Class II or III or large quantity radioactive materials). A radioactive Yellow II label is required on any package measuring more than 0.5 millirems but not more than 50 millirems per hour at any point on the external surface of the package, and not exceeding one millirem per hour at three feet from any point on the external surface of the package, (*i.e.,* the Transport Index may never exceed 1.0 for these packages). A wide range of radioactive materials thus would be excepted from any routing requirement since they are either excepted from labeling or carry the White I or Yellow II label and thus are excepted from placarding.

(2) Packages for which placarding is required would be subject to a general routing requirement. This category of packages includes those requiring a Yellow III label or containing Fissile Class III materials or a large quantity of radioactive material. Also, any package which measures more than 50 millirem per hour at any point on the package surface or which exceeds one millirem per hour at three feet from any point on the external surface of the package (i.e., the Transport Index is greater than 1.0; see 49 CFR 173.389(i)) requires placarding. The proposal would require all such packages, if not transported on an Interstate or specially designated highway, to be transported so as primarily to risk exposure to the least number of people and secondarily to minimize travel times.

Many commercial shipments of radioactive materials fall within this category. For example, many medical-use shipments, both Type A and B quantities, require a Yellow III label and must be placarded. Medical isotopes used for scanning procedures in hospitals such as Tc-99M, Au-198 or I-131 are occasionally packaged such that the Transport Index exceeds 1.0. Isotopes used for teletherapy and medical research such as Co-60 and Cs-137 usually require a Yellow III label. Many industrial-use shipments would also fall into this category. Isotopes such as americium, berylium, Cs-137, and Kr-85 are used by the well-logging industry to determine properties of rock formations. Ir-192 and Co-60 are used in radiography to measure structural integrity of welded joints. Isotopes which are used in industrial gauging devices include Ra-226, Sr-90, Am-241 and others. Many of these industrial isotopes would require a Yellow III label when packaged according to accepted practice.

In short, radioactive materials subject to the general routing requirement in proposed § 177.825(a) include any packaged radionuclide, regardless of quantity, which has a Transport Index of 1.0 or greater.

(3) Shipments of packages containing a large quantity of radioactive materials (defined at 49 CFR 173.389(b)), including spent fuel, would be subject to additional Federally imposed restrictions as well as the possibility of Federally recognized State restrictions. This category includes the most toxic radionuclides, which are found in Transport Groups I and II, when shipped in quantities over 20 curies per package as well as larger quantities in the other Transport Groups. Included in Transport Groups I and II are many shipments of nuclear fuel cycle material, plutonium, polonium, mixed fission products, some isotopes of uranium, and certain commonly shipped isotopes such as Am-241, Ra-226, and Sr-90. A large number of shipments of materials in the first two Transport Groups are already subject to stringent physical security requirements during transportation established by the NRC. Special nuclear materials, potential theft targets which include many shipments of plutonium and the uranium isotopes U-233 and U-235, as well as spend fuel, a possible terrorist target, when shipped by NRC licensees are subject to the physical security requirements in 10 CFR Part 73.

EXAMPLES OF RADIOACTIVE MATERIALS UNDER PROPOSAL

Radioactive Materials Not Subject to Proposal	Radioactive Materials Subject to General Rule	Radioactive Materials Subject to Additional Restrictions
Shipments for which carrier not required to placard vehicle	Shipments which require carrier to placard vehicle-any Type A or Type B package with radiation levels such that Yellow III label required	Shipments which include any Large quantity package

Radioactive Materials Not Subject to Proposal

1. Packages excepted from labeling

   –Smoke detectors
   –Clocks
   –Manufactured articles
   –Very small quantities of radionuclides

2. Packages with either White I or Yellow II label

   –Most radiopharmaceuticals used in hospitals

3. Low Specific Activity (LSA) in Exclusive Use Vehicles

   –Limited to uranium ores and thorium ores unconcentrated

Radioactive Materials Subject to General Rule

1. Medical and research use isotopes

   –Mo-99, Au-198, and I-131 used for scanning procedures
   –Small source Co-60 and Cs-137 for research activities

2. Industrial use isotopes

   –Ir-192 and Co-60 used for radiography

   –Cs-137 and Kr-85 used in well-logging industry
   –Ra-226, Sr-90 and Am-241 used for industrial gauging devices

3. Low level radioactive wastes

   –Packages LSA in exclusive use vehicles

Radioactive Materials Subject to Additional Restrictions

1. Over 20 curies of most toxicradionuclides — Transport Groups I and II

   –Americium
   –Plutonium[1]
   –Polonium
   –Spent reactor fuel[1]

2. Any other radionuclide depending on the number of curies in package

	No. of Curies
–Transport Group III –	over 200
–Transport Group IV –	over 200
–Transport Group V –	over 5,000
–Transport Group VI-VII –	over 50,000
–Special Form –	over 5,000

   –Example – Large source (10,000 curies) Co-60 used for teletherapy or large source Cs-137 (100,000 curies) used for medical research

[1]Subject also to NRC physical security requirements in 10 CFR Part 73.

BILLING CODE 4910–60–C

B. General Routing Requirement. The general routing scheme contained in proposed § 177.825(a) would require placarded vehicles carrying radioactive materials first to avoid areas posing hazards to large numbers of people and as a subordinate consideration to operate over routes selected to reduce time in transit. Consideration of "time in transit" includes a prudent evaluation of delays that may result from potential occurrences such as anticipated bad weather. Either of two exceptions permit variance from the condition that selected routes avoid population exposure: (1) when a practicable alternative highway route is not available, or (2) when the motor vehicle is operated on preferred highways under conditions set out in proposed § 177.825(b)(1). The criteria for determining when a "practicable alternative" highway route is not available are the same as those considered to apply under the existing hazardous materials routing rule in 49 CFR 397.9(a): operating necessity and safety. Operating necessity includes such factors as access to origin and destination points, and necessary fuel and repair stops. Safety includes considerations such as adverse weather and roadway conditions, but does not include travel time which is subordinate to the requirement to avoid population exposure. In no case is the operating convenience of the carrier a valid consideration. The second exception from the requirement relies on motor vehicle operation over routes that are intended for large quantity shipments under proposed § 177.825(b).

The requirement that transit time be minimized poses particular problems in multiple stop operations. This is because the number of possible routes between any two points and the number of possible sequential combinations of various stops theoretically can be a very large number. Consequently, for purposes of compliance with the proposed rule, it would be sufficient that a motor vehicle operator choose only the probable quickest route to his next stop, although any more efficient method of selecting routes to reduce transit time may be used.

The proposed general rule would apply only to motor vehicles which are required to be placarded. There are three reasons for this choice. First, hazardous materials placards are highly visible and easily observed by Federal, State and local enforcement authorities. Second, placarding itself is not required for most radiopharmaceuticals, industrial isotopes and other low-hazard radioactive materials. These materials are shipped in large numbers of packages and may be manufactured as well as used in the same urban area. They would be extremely difficult to control by routing requirements. Third, the existing routing rule in 49 CFR 397.9(a) applies to placarded motor vehicles. Carriers as well as enforcement authorities are familiar with the existing connection between placarding and routing control, a fact that should improve initial compliance with any final rules issued in this docket.

Commenters have suggested that cities with a population density of 10,000 to 12,000 persons or more per square mile should be avoided by radioactive materials carriers. The MTB has not used the term "heavily populated area". It does not appear practical to define it as a function of population densities or absolute population figures. The term is vague and its purpose difficult to enforce. In its interim rule on physical security of spent fuel, NRC uses census figures which are publicly available. That program, however, involves specific route approval from NRC for security reasons, which MTB does not consider justified in dealing with normal exposure and the possibility of accidents. A route restriction for placarded highway carriers based on a specific population figure would require an easily accessible, authoritative and highly detailed source of population information. Census figures usually are based on political boundaries, total populations, and total land areas. These figures do not distinguish uneven population distributions within a particular jurisdiction. Use of total jurisdictional figures (or population density figures averaged over a given jurisdiction) may result in unnecessary avoidance of entire jurisdictions or permitted transit through localized areas of high population density within a jurisdiction.

The proposal would require the carrier to affirmatively ensure that routes are selected to minimize the number of persons that may be exposed to a radiological risk. This is the basic goal to which any prohibition of travel in heavily populated areas would aim. Further comment on this is welcome.

C. Special Restrictions on Shipments of Large Quantities, Such as Spent Fuel Shipments. The large quantity package has been selected as the cutoff point for additional requirements presented in this proposal concerning required use of preferred highways, route plans and driver training. MTB recognizes that a substantial argument can be made for choosing some other cutoff point or for not using any such distinction at all, particularly in light of the NUREG 0170 estimates that for all modes of transportation, large quantity packages account for only about 2% of the normal population dose and 37% of the latent cancer fatalities expected to result from transportation accidents (1985 projection, NUREG 0170, pp. 4-44, 5-34). However, large quantity packages generally travel 30% to 50% farther per shipment than Type B and Type A packages (NUREG 0170, p. A-13). Large quantity packages are estimated to have comprised about 378 out of a total of 1.3 million packages of radioactive materials shipped by truck in 1975. NUREG 0170 projects that 1,911 large quantity packages will be shipped in 1985 out of a total of 3.5 million radioactive materials packages shipped by truck (pp. A-11, A-21, 22), although an estimate of 600 large quantity packages would reflect reduced spent fuel shipments and absence of recycled plutonium shipments in the NUREG 0170 model for 1985. However, it is quite possible that the estimates for large quantity shipments for both years may be several times the stated estimates, due to the manner in which the information was gathered.

Of all the radioactive materials packages shipped, only large quantity packages pose even a remote risk of extraordinary or catastrophic accident consequences. Requiring specified routes for large quantity packages would add to the public certainty as to the location and nature of these unusual risks and permit more rational emergency response planning for remote events that nonetheless may require substantial planning efforts.

MTB thinks that the hazards associated with other than large quantity packages do not warrant requiring them to be routed on preferred highways (Interstate and State-designated highways) and that the enforcement, compliance and possibly economic costs of such a requirement could be substantial. By not requiring all placarded motor vehicles to operate on preferred highways, the proposed general rule acknowledges the pronounced differences between large quantity shipments and other placarded shipments, the fact that the annual volume of all placarded shipments is large, and that a substantial part of those shipments may involve local multiple-stop delivery operations. Although the proposal would nor require all placarded motor vehicles to operate on preferred highways, that result is encouraged, subject to the carrier's judgement, since some circuitous travel and questions about population exposure may be avoided by an election to travel on a preferred highway. Required use of preferred highways, however, would be limited to motor vehicles that transport a large quantity package.

(1) Type of roadway. The type of roadway on which radioactive materials would be transported was thought by commenters from all groups to be a prime consideration in any routing requirement. MTB is proposing that large quantity shipments of radioactive materials be restricted to carriage only on a preferred highway. A preferred highway would be defined as any specific highway designated as a preferred highway by an appropriate State-wide agency, and any Interstate highway for which substitute is not provided by such agency.

Interstate routes. An Interstate highway is an expressway usually with fully controlled access which is part of the 42,500-mile Interstate highway system as designated by Congress. However, the term as used in this proposal includes roadways which also are designated "temporary" Interstates.

The Interstate System is part of the Federal-aid primary system connecting principal cities of the United States. Interstate highways would be defined in the text of this proposal as preferred highways because the Interstate System is built to exacting and generally uniform specifications and offers the safest and often most direct routes available.

Statistics published by the Department's Federal Highway Administration for 1976 indicate that the possibility of an accident involving a fatality or injury on an Interstate highway is as little as 25% of what it is on a non-Interstate highway ("Fatal and Injury Accident Rates on Federal-Aid and Other Highway Systems/1976", September 1978). Since 1967 when such statistics first became available, the fatal accident rate (fatal accidents per 100 million vehicle miles) for Interstate highways consistently has been very much lower than the rate for non-Interstate highways. These figures suggest that travel on Interstate highways significantly reduces the probability of an accident. Consequently, in the absence of State action, MTB believes any vehicle carrying a large quantity shipment generally should be routed via Interstate highways. Restricting large quantity radioactive materials carriers to the Interstate System also is one of the few alternatives determined in NUREG 0170 to be cost-effective, because it substantially reduces overall normal population dose (p. 6–12).

State-designated routes. MTB believes that States may be able to offer useful refinements, particularly in view of the fact that State and local agencies also bear the basic emergency response duties and costs. The proposal would recognize action by appropriate State agencies to designate non-Interstate public roads as preferred highways, and to remove the preferred status of an Interstate highway if an equivalent route is provided. Permissible State action is further discussed later in this document under the heading "Guidelines for State regulation."

A motor carrier who is required to transport a large quantity package on a preferred highway, or a motor carrier of other radioactive materials packages who voluntarily uses a preferred highway would be required to use the most direct preferred highway and would not be required to evaluate population densities. However, in the absence of State action to the contrary, a carrier would be required to use an Interstate or other preferred circumferential or bypass route in favor of an Interstate route through a city. This position represents a compromise between considerations of normal population dose including that of motor carrier personnel, possible accident exposure and the need for uniform and efficient compliance and enforcement.

Exceptions. The motor vehicle would be authorized to leave or travel off preferred highways when necessity or safety considerations dictate and when necessary to travel from shipment origin to the nearest preferred highway and from a preferred highway to the shipment destination. Necessary food, rest, fuel, service and repair stops would be permitted. Any travel on nonpreferred highways would still be subject to the general rule stated in paragraph (a) of proposed § 177.825 including required routing to limit the number of persons potentially exposed to risks.

In the proposed rule, MTB has not attempted to answer the question of how far out of the way a carrier must go to access and use a preferred highway. It would be preferable that the question be answered by State agencies by means of designating additional preferred highways to account for situations wherein an unreasonable amount of circuitous travel may result from carriers accessing the Interstate highways of the State. However, MTB is considering several possible methods of establishing a limit on the circuity that a carrier must accept to access a preferred highway. Two possible rules, which differ in their effect, have been examined. One rule would generally state that a carrier need not increase travel distance more than 25% to access a preferred highway, measuring from points selected by the carrier. This approach has some effects which are much less than optimal. A second rule, which is more precise, would permit the use of a formula to select routes that include non-preferred highways: for each possible route, total mileage on non-preferred highways would be increased by 25% and added to mileage on preferred highways. The route with the smallest mileage sum, computed in that fashion, would be used. Both rules might be offered in the alternative, at the option of the carrier. For enforcement purposes, a violation could be shown only by a demonstration that neither rule was followed.

The proposal, as drafted, would rely on the mutual interests of carriers and State agencies to produce local accommodations on questions concerning access to preferred highways. Comment is solicited on this point.

Placards. MTB is giving serious consideration to proposing the required use of a distinctive mark or logo on radioactive hazard warning placards to permit the ready recognition of motor vehicles carrying large quantity packages. Under existing rules, large quantity packages could be identified only by examining the shipping paper or the package markings.

One method under discussion would involve the use of the placard background presently required for certain railcars (49 CFR 172.510, 172.527). MTB believes use of some such device to distinguish motor vehicles carrying large quantity packages may be necessary and solicits comment on this issue.

(2) Route plans. A motor carrier transporting a large quantity of radioactive materials would have to prepare a route plan complying with the provisions of paragraph (b) in proposed § 177.825. A similar requirement now applies to carriers of Class A explosives (see 49 CFR 397.9(b)). The route plan would be supplied to both the shipper and the driver of the vehicle, in most cases before departure. The shipper's copy, for nonexclusive use shipments, could be provided later by mail. The plan would contain specific information concerning the route selected, and emergency telephone numbers for each State traversed. DOT believes that it would be preferable to rely on a single telephone number to access all emergency responses and is considering possible methods of achieving this result. For the purposes of this proposal, however, the text used indicates the basic intent: that the carrier be prepared in advance to contact State emergency response personnel immediately in the event of an accident. The State police in many cases may be the appropriate agency for summoning emergency response assistance.

The shipper would file a copy of each route plan submitted by the carrier with the MTB within 90 days of the date a large quantity shipment of radioactive materials is accepted for transportation. NRC licensees who already are required to provide this information to NRC under physical security requirements would be excepted from filing, since that information will be available to DOT. The MTB intends to make shipment information in accumulated route plans accessible to State agencies for emergency response planning and is considering several possible methods of providing this service. For shipments made under physical security requirements, however, some restrictions on release of information may have to be observed to avoid compromising that security.

(3) *Driver training requirements.* This proposal would apply a driver training requirement to motor carriers transporting large quantity packages of radioactive materials. The training would include instruction to the driver every two years on the Hazardous Materials Regulations pertaining to radioactive materials, the Federal Motor Carrier Safety Regulations (49 CFR Parts 390–397) applicable to operation of the motor vehicle, the hazards and characteristics of large quantities of radioactive materials, emergency features or other special characteristics of the vehicles to be used to transport those materials, and any emergency procedures to be followed in the event of an accident. The training would be evidenced by a certificate in the driver's qualification file and on his person during transportation. The driver training proposal was derived from a proposal now under development concerning drivers of tank trucks. A similar proposal also appears in Docket HM-115 (44 FR 12826, 12842, March 8, 1979) regarding drivers of certain tank trucks carrying flammable cryogenic liquids. For planning purposes, MTB is assuming that training would not exceed 20 hours a year for new drivers and would involve written training materials and written examination. The actual extent of training would be subject to the carrier's judgement and the driver's previous training.

D. Cargo Security. Spent fuel is the most widely recognized large quantity of radioactive materials routinely shipped. For that reason, spent fuel casks could become the target of terrorist activity, although the likelihood of a successful act of sabotage that breaches a spent fuel cask and disperses its contents may be quite small. The NRC recently established new interim physical security procedures in 10 CFR 73.37 for the shipment by its licensees of spent fuel. Those procedures are intended to remain until current studies of the ability of spent fuel casks to withstand acts of sabotage are completed. The MTB has reviewed the interim procedures and believes they will provide adequate physical protection for spent fuel shipments.

Because physical requirements under the NRC's rules may conflict with the DOT highway routing proposal made herein, paragraph (b)(4) of proposed § 177.825 would permit variation from the proposed rule's requirements if necessitated by security requirements under the NRC's rules. This provision also would permit variation for security reasons under previously established NRC rules applicable to special strategic nuclear materials.

Since the NRC interim safeguards' rules only apply to NRC licenses, such as operators of commercial nuclear generating stations, MTB is proposing to require shipments of spent fuel by nonlicensees to be made in accordance with general requirements approved by MTB as being essentially equivalent to the NRC requirements. Some shipments made by contractors of the Department of Energy, such as Brookhaven National Laboratories, and possible contractors of the Department of Defense, may be subject to this provision.

In accordance with the DOT-NRC memorandum of understanding, the NRC has primary responsibility for physical security requirements. The MTB believes it is doubtful that terrorist acts would be directed against small source nonfissile isotopes, because of the small radiological consequences involved, and does not see a need for physical security requirements for such shipments. The NRC now is examining the possible need for physical protection of large source nonfissile isotopes and smaller quantities of special nuclear material during transportation. The MTB will await NRC judgment in this matter before considering any further action regarding physical security.

E. Guidelines for State Regulation. The result of stringent local regulation of highway carriers of radioactive materials has been described by some commenters to this docket as a "burden" on commerce. It is the MTB's view that the existence of a burden on commerce imposed by a State of local requirement is relevant to rulemaking responsibilities under the HMTA so far as it may affect transportation safety. The HMTA does not necessarily exclude State and local regulation of highway carriers of hazardous materials, nor is that result desirable. However, the HMTA does provide adequate preemptive authority to ensure that the Act and regulations issued under its authority are effective as intended.

The MTB believes it is important that State and local views be considered in routing decisions. There is, however, an obvious difficulty in permitting local governments to exercise what amounts to a veto power over interstate commerce. A small jurisdiction which does not directly benefit from shipping activities within its borders will often find attractive the option of diverting traffic into neighboring jurisdictions, with concomitant safety impacts in those jurisdictions. Local safety rules that are excessively stringent may produce counterproductive safety impacts and possible violation of Federal requirements in the transportation of improperly identified shipments. A balance is needed in routing decisions between local knowledge of local conditions and the wider demands of safety in interstate commerce. The proposed rule, for this reason, encourages routing participation by State and local governments through an agency with State-wide jurisdiction that would be accessible to all those persons that may be affected by routing decisions. The proposal reflects the current MTB view that a greater degree of uniformity in rules affecting radioactive materials transportation by highway is needed and that unless necessary to ensure the physical security of the cargo, as previously discussed (or otherwise justified by exemption or waiver of preemption), any State or local requirement that amounts to a transportation ban on highway carriage of radioactive materials is not reasonable.

The term "State agency with Statewide enforcement authority" is used in the proposal to describe those State agencies that may designate non-Interstate highways as preferred highways and disapprove (and thus terminate) the defined preferred status of a segment of an Interstate highway for which the State agency has provided an alternate and equivalent preferred highway. The term "agency" is intended to describe an entity (including a common agency of more than one State, such as one established by interstate compact) which is authorized to use State legal process to impose and enforce routing requirements on carriers of radioactive materials without regard to intrastate jurisdictional boundaries. This description would exclude, for example, a bridge authority unless that authority also is empowered to impose and enforce such rules concerning radioactive materials transportation on State highways generally. This description would not exclude the possibility of more than one agency in a single State sharing responsibility for designating preferred highways.

Reliance on routing designation by agencies with State-wide authority may pose particular problems for cities and for agencies which operate under interstate compacts and which have responsibility for areas with defined jurisdictional boundaries. For this reason, State action establishing a preferred highway must be preceded by consultation with affected local jurisdictions. A route modification to bypass a major city, for example, would require consultation with that city and with any impacted adjacent jurisdictions. A route modification that impacts jurisdictions in another State would require consultation with those jurisdictions. Also, bridge, tunnel and turnpike authorities would rquire action by a State-wide agency in order to restrict passage of radioactive materials carriers on an Interstate or other preferred highway. Note that the provision in 49 CFR 177.810, which saves for such agencies the right to restrict hazardous materials transportation generally, would be modified to reflect this part of the proposal. Commenters may wish to propose other methods of dealing with the problem of providing a forum for State routing decisions which permits all interests affected by such decisions to participate in the decision process.

Under the proposal, an appropriate State-wide agency would be able to take the following actions.

Designation or modification of preferred highways other than Interstate highways. The goal in designating a preferred highway would be an overall reduction in both radiological and nonradiological impacts from transportation of large quantity packages. Basic criteria for this goal would include:

(1) Normal radiological impacts—including radiation exposure to drivers, cargo handlers, persons in other vehicles and pedestrians, occurring during normal, accident-free transportation.

(2) Normal nonradiological impact—including costs to carriers and shippers, and other impacts of motor vehicle operation such as vehicle emissions and traffic congestion.

(3) Radiological accident impact—including injuries, deaths, property damage, cleanup costs, and costs of emergency response preparedness.

(4) Nonradiological accident impact—including deaths, injuries, and property damage.

This State agency action would be predicated on the results of a technical safety review of available routing choices. It would be prudent for the State agency to document the process.

Modification of the preferred status of Interstate highways. The preferred status of an Interstate highway could be removed as part of an action based on the above-stated criteria only if the continuity of the Interstate System would be maintained by designation of a preferred highway which is essentially equivalent.

Urban circumferentials and bypasses. The proposal would require an Interstate circumferential or bypass route to be used in favor of an urban Interstate through route. Where an urban Interstate through route exists without an Interstate circumferential or bypass route, an abbreviated designation process could be used by a State agency to establish a non-Interstate circumferential or bypass as the preferred route. In this situation, an urban Interstate through route could be replaced by any circumferential or bypass route which is equivalent to other urban Interstate circumferentials or bypass routes elsewhere in either design standards or performance (*i.e.,* actual traffic flows and accident rates).

Continuity must be maintained for Interstate highways, but for non-Interstate preferred highways, continuity would be a safety factor which might not be as important as other safety considerations. However, where a preferred highway would direct traffic to a State's boundary, jurisdictions in the next State which would be impacted by the traffic must be consulted and the impacts considered as part of the designation process. A State boundary, in other words, may define the limits of a State agency's authority, but it does not define the limits of the impacts which must be considered in exercising that authority.

Cargo security and the possibility of sabotage or deliberate release of radioactive materials from a large quantity package are not directly considered in the designation of preferred highways. As previously mentioned, under the current division of responsibilities between DOT and NRC, an accounting for these factors is an NRC responsibility which is discharged through NRC physical security requirements in 10 CFR Part 73 (or the equivalent under this proposal for non-NRC licensees) for which an allowance is made in this proposal. Those requirements, which now apply to shipments of spent fuel as well as special nuclear material, involve the NRC in approving routes and other countermeasures selected to reduce threats to the physical security of the cargo.

V. ALTERNATIVES NOT PROPOSED

A. Intrastate Carriers. The HMTA provides authority to regulate intrastate commerce that affects interstate commerce (49 U.S.C. 1802(1)(B)). The existing Hazardous Materials Regulations do not apply to purely intrastate carriers, that is, carriers whose business does not involve them at any time in the transportation of materials in interstate commerce. Intrastate carriers operate only within a State and do not carry materials in transportation whose origin or destination points are not within the State. As a practical matter, such carriers would be most likely to be used in local pickup and delivery services, warehouse distribution and so forth. Intrastate carriers of radioactive materials are regulated by State law and further controlled by requirements expressed through conditions imposed by the NRC on its licensees. Those conditions include provisions which are identical to requirements imposed on interstate carriers by DOT. Regulation of the routes used by intrastate carriers of large quantity radioactive materials shipments was considered but not proposed because of the primarily local character of such transportation, and the very limited number of such shipments likely to move by intrastate carrier. States are free, at the present time, to establish routing controls for intrastate carriers. Future action by MTB will be considered if new information warrants.

B. Other Modes and Other Hazardous Materials. Interest has been expressed in routing considerations applicable to rail carriers, in view of the amounts of spent fuel the railroads eventually may be called upon to carry. Rail operations, however, differ significantly from highway operations and rail routing raises a separate set of issues. Also, the routing choices available in rail operations with regard to populated or congested areas are considerably more limited than in highway transportation.

The MTB does not rule out the development of highway materials other than radioactive materials, especially for hazardous materials shipped in bulk by highway. It is not practical, however, to attempt to deal with this subject in this docket. A study currently is being conducted for the Federal Highway Administration which eventually may provide a basis for developing general hazardous materials highway routing criteria.

C. Full Licensing of Carriers. Both registration and licensing of highway carriers of large quantities of radioactive materials were considered. With the route plan requirements proposed, however, ready identification of carriers would be possible without registration. Moreover, carriers already are subject to safety and reporting requirements under the Federal Motor Carrier Safety Regulations, and this proposal would require specialized driver training. MTB sees little additional advantage in requiring registration or licensing and has not proposed to require either.

D. Transport Group Limitation. Instead of referencing packages containing a large quantity of radioactive materials as the key to required use of preferred highways. MTB considered referencing large quantities in Transport Groups I and II only. Those transport groups include the most toxic radionuclides which are defined as large quantity when shipped in packages containing more than 20 curies. However, in view of the substantial amounts of other transport groups that can be carried in individual packages, it was felt that use of the large quantity cutoff for this purpose was justified without reference to transport groups. MTB would be interested in suggestions as to other feasible cutoff points. Note that reference to transport groups and to large quantity is proposed to be eliminated in a scheduled revision of the DOT and NRC rules concerning radioactive materials (HM-169, 44 FR 1852, January 8, 1979; 44 FR 23266, April 19, 1979; 44 FR 47966; August 16, 1979; 44 FR 60771, October 22, 1979). Consequently, if the large quantity cutoff is retained, it may be expressed in terms of the A_2 values proposed in that rulemaking rather than transport groups. MTB also solicits views on whether special form materials should be treated separately. In this proposal, large quantity, in addition to specified amounts in each of the transport groups, means 5,000 or more curies of any material in special form. Under the HM-169 proposal, A_1 values also could be used for special form materials.

VI. EXPECTED ENVIRONMENTAL AND ECONOMIC IMPACTS

The primary operational effect of this proposal would be to encourage use of the Interstate System by carriers of radioactive materials. Although carriers transporting packages containing a large quantity of radioactive materials are generally required to use either the Interstate System or State-designated preferred highways, carriers transporting packages containing lesser quantities are likely also to tend to use the Interstate and preferred highways especially in areas of heavy population, if this proposal is implemented. Overall radiological effects of this proposal would include a very slight reduction in total latent cancer fatalities attributable to normal dose in 1985 and a lesser reduction in the annual latent cancer fatality accident risk (based on NUREG 0170 projections). Some additional reduction in radiological consequences may result from State designation of preferred highways. A slight increase in nonradiological consequences may result from routing on preferred urban bypass or circumferentials. Overall, environmental impacts should be negligible.

Economic costs are expected not to exceed $330,000 annually under 1985 levels of shipping activity and mostly would consist of carrier costs for driver training and route plan preparation and filing. This estimate, however, does not include possible additional insurance costs to State and local bridge and tunnel authorities on the Interstate System or on highways that may be designated by future State action as preferred highways. At present, MTB lacks any quantitative data on this subject. Commenters are encouraged to provide any available estimates.

Because of the level of costs anticipated and the limited potential for environmental impact, the MTB does not consider the preparation of an environmental impact statement or a regulatory analysis necessary for this proposal. A more detailed examination of costs and environmental impacts is available in the draft regulatory evaluation and environmental assessment which may be obtained from the Dockets Branch at the address indicated at the beginning of this notice. Because this proposal varies from the highway routing requirement at 49 CFR 397.9(a), at the time a final rule is published, some further adjustment to § 397.9(a) is contemplated to avoid any conflict.

(45 F.R. 7140, January 31, 1980)

49 CFR Parts 171, 172, 173, 177
[Docket No. HM-164, Amdt. Nos. 171–59, 172–64, 173–143, 177–52]
Radioactive Materials; Routing and Driver Training Requirements

I. BACKGROUND

The history of these amendments is summarized in the Notice of Proposed Rulemaking (NPRM) of January 31, 1980 (45 FR 7140). Individuals interested in this docket should review that publication as well as the Advance Notice of Proposed Rulemaking (ANPRM) of August 17, 1978 (43 FR 36492) since references are made to both documents. To set the context for the present discussion, however, the most important background items relating to these amendments are briefly summarized here.

In 1976, truck shipments of irradiated reactor fuel (spent fuel) from Brookhaven National Laboratories' Long Island facility were interrupted by an amendment to the New York City Health Code. The Health Code amendment had the practical effect of banning most commercial shipments of radioactive materials in or through the City Associated Universities, Inc., which operates Brookhaven National Laboratories, asked DOT whether that ordinance was preempted by Federal transportation safety requirements issued under the Hazardous Materials Transportation Act (HMTA) (49 U.S.C. 1801 et seq.). On April 20, 1978, DOT published an Inconsistency Ruling (43 FR 16954) in which it viewed the City's Health Code amendment as an extreme routing requirement intended to protect the very dense urban population found inside the City. DOT concluded that the HMTA could preempt local requirements such as New York City had implemented, but because highway routing authority had not yet been exercised under the HMTA, the City's Health Code was not preempted by HMTA requirements. Since this ruling a number of other State and local governments have either passed, or proposed, legislation that severely restricts transportation of certain radioactive materials through their jurisdictions.

The Department of Transportation subsequently published the ANPRM entitle "Highway Routing of Radioactive Materials; Inquiry" in August, 1978. The public was invited to comment on the need and possible methods for establishing routing requirements pertaining to highway carriers of radioactive materials under the HMTA. A public hearing was held in conjunction with the ANPRM on November 29, 1978 in Washington, D.C. The Department received over 550 comments from a broad cross-section of the public including representatives from State and local governments, public interest and environmental organizations, the motor carrier industry, the shipping industry, bridge and turnpike authorities, Federal agencies, and Congressional officials, in addition to the many individual citizen comments. Based upon these comments and the Department's own judgment an NPRM was published on January 31, 1980.

The NPRM set out specific proposals for routing certain types of radioactive materials shipped by highway, and driver training requirements. The stated purpose of those proposals was to reduce the possibility of exposure and inadvertent releases in normal and accident situations in transportation, and to clarify the scope of permissible State and local actions. The four month public comment period scheduled in the NPRM was subsequently extended to five months. The Materials Transportation Bureau (MTB) conducted seven public hearings from late March to early June of 1980. The seven hearings were held in Philadelphia, Atlanta, Chicago, Denver, Seattle, Boston and New York. In addition, MTB conducted public meetings in Akron, Ohio, Eugene, Oregon and Union City, California. The Department has received and reviewed over 1,000 public comments on the January 31 notice. In addition, over 1,600 pages of transcripts from the seven public hearings have been reviewed as well as statements made at the three public meetings.

Because of the great interest generated by these routing proposals and because of the many and varied issues involved, DOT has decided to include an extended discussion of public comments as a supplement to the final rules document. Although principal comments are discussed in this preamble, inclusion in the docket of a supplementary discussion of public comments allows the Department to provide more detailed responses than would be practical in the preamble. For those readers interested in public comments on the Advance Notice, a summary is provided on pages 7141 and 7142 of the January 31 NPRM. In addition to the supplement of public comments, all individual public comments and all public hearing transcripts are available for inspection at the address previously listed.

Other essential background information covered by the NPRM includes an analysis of the existing DOT safety program for the transportation of radioactive materials, DOT accident experience with nuclear material transportation, a technical discussion of projected public risk from the transport of radioactive materials in the United States, the Nuclear Regulatory Commission (NRC) physical security program for shipments of spent nuclear fuel by its licensees, DOT's interrelationship with NRC's transport requirements, and an extensive discussion of the proposed routing and training requirements. The present document will reference some of this information. However, those discussions will not be restated here except as they relate to substantive public comments.

II. GENERAL DISCUSSION

The Department of Transportation has examined the transportation of radioactive materials exhaustively since issuing the ANPRM nearly two and a half years ago. This process has included the review of over 1600 public comments and 2000 pages of transcripts from public hearings in addition to a number of risk assessment studies on the subject. On the basis of these comments, documented risk studies and past accident experience for radioactive material transport, the Department has concluded that the public risks in transporting these materials by highway are too low to justify the unilateral imposition by local governments of bans and other severe restrictions on the highway mode of transportation. Other modes of transport generally do not appear to offer alternatives which clearly lower public risks to the extent that use of the highway mode should be substantially restricted. DOT also believes, however, that these currently low risks wil be further minimized by the adoption of driver training requirements and provisions of a method for selecting the safest available highway routes for carriers of large quantity radioactive materials, as accomplished in this rule.

The estimated low risks in transporting radioactive materials also support the belief that the present packaging requirements are adequate to protect the public. A detailed discussion of DOT's packaging requirements was presented in the NPRM. As was clearly pointed out in the proposed rules, this rulemaking is not an examination of packaging requirements, the adequacy of which is assessed by DOT and NRC on a continuing basis. There has been no new documented evidence presented during the public comment process to show that the current packaging requirements result in unacceptable risks to the public.

Many commenters question the need for these routing rules and some view them as nothing more than a method of accommodating the transportation requirements of the nuclear power industry. Some maintain that State and local restrictions have been applied mostly to nuclear fuel cycle shipments such as spent fuel and have not frustrated shipments of radiopharmaceuticals or other "necessary" small quantity radioisotopes. They suggest that DOT's stated intention of providing uniformity and consistency, at least in part, to ensure shipments of needed nuclear medical materials is based on an invalid perspective.

The Department has examined many of the local restrictions for radioactive material transportation and continues to believe that many result in unnecessary restrictions on the transportation of all types of radioactive materials, including non-fuel cycle materials. Some public comments support this. For example, the Society of Nuclear Medicine presented the following comments at the Chicago hearing on April 3, 1980.

> The Society has great concern for the proliferation of State and local statutes and ordinances enacted to control the transportation of radioactive materials into, through, and out of these jurisdictions. It can be stated that this non-uniformity of controls in the transportation of these medical necessities constitutes one of the most rapidly increasing and serious impediments to nuclear medicine health care delivery with which we are faced. Thus, the Society views with favor those portions of this docket which will provide for uniformity of regulation, on a national basis, while still providing for adequate state and local input in the implementation of the final rule.

The Society points out that over 3,300 medical centers, hospitals and clinics in the U.S. are engaged in nuclear medicine, and it estimates that one out of every two patients admitted to hospitals require some type of "nuclear medicine procedure."

The Petroleum Equipment Suppliers Association, a trade association representing 251 companies that supply goods and services to over 10,000 companies engaged in petroleum drilling and production, point out that the relatively small quantities of industrial isotopes which its members ship are often covered by State and local restrictions:

Increasingly, state and municipal governments are enacting routing restrictions and prohibitions and requirements for pre-notification and escorts. The rising tide of these regulations and ordinances threaten not only to burden interstate commerce involving the use of radioactive sources by (oil and gas well) service companies, but to actually destroy the ability of these companies to provide these services. Without these services and exploration and production of oil and gas in this country will effectively cease.

DOT remains firm in its belief that the impact of piecemeal State and local restrictions on the transportation of all radioactive materials, including non-fuel cycle materials, signifies a need for nationally consistent routing rules.

It is also the Department's determination that public safety can be improved through a nationally uniform rule that ensures the use of available highway routes that are known to be safe for large quantity radioactive materials. In developing this rule, three basic conclusions underlie the approach taken:

(1) Route selection should be based on some valid measure of reduced risk to the public,

(2) Uniform and consistent rules for route selection are needed from both a practical and safety standpoint, and

(3) Local views should be carefully considered in routing decisions since routing is a site-specific activity unlike other transport controls such as marking and packing.

With respect to the first conclusion, DOT is of the opinion that an assessment of risk to the public should include a consideration of both normal radiological exposure which is inherent in the transportation of radioactive materials as well as a consideration of potential accidents which could result in additional radiological exposure. Further, an assessment of risk to the public from accidents involving large quantity radioactive materials should include a balanced consideration of factors which affect both the likelihood of an accident as well as the consequences.

Many commenters seem to be concerned only with consequence—particularly high consequence accidents involving large quantity radioactive materials in a heavily populated urban center. Local authorities, for example, are concerned with postulated "worst-case" accidents because of a fear that their emergency response capabilities are insufficient for such hypothetical catastrophes. The Department, also, is concerned with such events and is mindful of the large economic consequences estimated for such hypothetical events by a recent draft environmental assessment completed for the NRC by Sandia National Laboratories (Transportation of Radionuclides in Urban Environs: Draft Environmental Assessment", July, 1980). These estimates relate to a scenario which assumes the worst credible accident for certain truck shipments of spent fuel and polonium in densely populated urban areas. One could conclude from the study that a way to lower the possibility of such high consequences is to reroute the shipments away from urban areas entirely. However, the study also indicates that this may not be the best alternative if one considers overall risks to the public, since routes that avoid the urban areas may have much higher accident rates which increase the chance of a severe accident occurring in the first place. It is DOT's opinion that public policy for the routing of radioactive materials should be based not only upon a concern for worst-case accident consequences, but also upon all other factors which contribute to the overall risk involved in transporting large quantity radioactive materials. This policy is embodied in this rulemaking by requiring use of Interstate highways which generally have much lower accident rates than other roadways, while at the same time requiring that cities be avoided where possible by using either Interstate beltways or State-designated bypass routes to minimize the possibility of worse-case accidents.

With respect to the second conclusion, DOT recognizes the need to balance local and national interests in providing for uniformity and consistency in routing. DOT is providing a national framework for highway routing of radioactive materials within which State and local concerns can be addressed. This framework is needed because of the current patchwork of conflicting State and local routing requirements. It is recognized that there may be local situations which are so unusual that they cannot be adequately accommodated within this framework. These situations can be called to the attention of the Department through existing administrative channels that may involve either special or general rulemaking. However, because of the role of the State governments in designating routes and the nature of the routing guidelines being provided to the States which stress the participation of the local governments, DOT does not expect such situations to be numerous.

The third conclusion, which concerns the need for local input in routing decisions, also serves as a basis for the routing rules developed under this rulemaking. Routing as a safety control for the transport of any hazardous material is different from the more traditional safety controls such as packaging, package marking, vehicle placarding and loading. Routing is largely a site-specific activity which cannot be entirely accommodated at the Federal level. Therefore, DOT is encouraging a decentralized decision-making process in this area within a Federally-provided regulatory framework. The Department believes that in the interest of uniformity and safety, it is both appropriate and practical for many routing decisions to be made at the State level. The fifty State governments are in a better position than the Federal government to respond to local concerns and likewise are in a better position than the 23,000 or so local jurisdictions to consider overall safety impacts from routing decisions. The ensure adequate consideration of local viewpoints, DOT believes an advisory group primarily composed of local officials should be established in each State to periodically review the effectiveness of the State/local consultation (discussed in more detail elsewhere in this document).

III. FEDERAL/STATE/LOCAL ROLE IN ROUTING
RADIOACTIVE MATERIALS

The Hazardous Materials Transportation Act grants DOT the authority to regulate the transportation of hazardous materials. Among other things, section 105(a) of the HMTA specifically identifies routing as one form of regulation that the Secretary may deem necessary and appropriate for the safe transportation of hazardous materials. Before the issuance of the ANPRM for Docket HM-164, the Department had not implemented routing regulations for any hazardous material under this clear authority granted by the HMTA. A general routing provision does exist at 49 CFR 397.9 providing guidance to carriers and drivers of placarded motor vehicles. That provision predates the issuance of the HMTA and has not yet been adopted in regulations issued under the authority of that Act.

However, a number of actions by State and local governments relating to a specific hazardous material (radioactive material) and a specific mode of transportation (highway) have raised the question of whether more specific Federal routing requirements should be issued. The DOT must consider the overall safety impact of piecemeal, uncoordinated local actions on hazardous material transportation. The ANPRM and the NPRM made clear the Department's intention to consider only routing requirements for radioactive materials shipped by highway, the focus of most State and local actions, rather than undertake a comprehensive regulatory proceeding to consider all classes of hazardous materials and all modes of transportation. The fact that this proceeding considers only one hazard class and one mode does not rule out future Federal actions for other hazardous materials and other modes of transportation.

By issuing these regulations the Department has made the determination that routing requirements can improve safety—not only by providing for the use of the safest highway routes, but also by addressing the safety impacts of narrowly conceived local actions. In order to fulfill the mandate on hazardous material routing, it is DOT's responsibility to set out a national framework within which legitimate local concerns can be addressed. To establish this framework the DOT has the authority to make the basic decision as to what radioactive materials pose a significantly serious risk such that routing controls are necessary, and how these materials should be routed. The Department has made these decisions in this rulemaking and a brief synopsis now follows.

First, a general routing rule is established for all radioactive material shipments by highway which require a warning placard. These include many of the thousands of shipments of radiopharmaceuticals, industrial isotopes, and low-level wastes that are made annually. The general rule emphasizes that the carrier choose routes which minimize radiological risk by considering such factors as population, accident rates, and transit time.

Second, special requirements apply to motor vehicles transporting large quantity packages of radioactive materials. These requirements include preferred routing, written route plans and driver training certification. Preferred routes are identified as Interstate highways and State-designated routes.

The Interstate highway system lays the basic Federal framework for providing safe and efficient routes for large quantity radioactive materials. Accident rates along these roadways are sharply lower than on any other type of roadway. Several studies also support the safety and efficiency of the Interstate highway system for the carriage of hazardous materials. In comments to Docket HM-164, the NRC developed a hypothetical case study of routing alternatives using information generated by NUREG 0170 ("Final Environmental Statement on the Transportation of Radioactive Materials by Air and Other Modes", December, 1977). Both the NRC case study and NUREG 0170 are discussed extensively in the NPRM and in the Final Regulatory Evaluation and Environmental Assessment prepared in support of this document. The case study clearly shows that use of Interstate highways generally result in lower radiological risks from the transportation of radioactive materials. Also, pilot tests were conducted for the Federal Highway Administration to apply routing criteria developed for all hazardous materials ("Development of Criteria to Designate Routes for Transporting Hazardous Materials by Highway", July, 1980). These tests were performed with the help of local officials in Nashville, Tennessee and Seattle, Washington and the results clearly demonstrate the advantages of the Interstate highways as compared to other roadways in minimizing risks associated with hazardous material transportation.

Carriers of large quantity radioactive materials are required to use Interstate beltways when possible to avoid city centers. Carriers are allowed off the Interstate system only to follow a State-designated route; in a documented case of emergency; to obtain necessary fuel or vehicle repairs; or to travel to and from a pick-up or delivery site not located on an Interstate System highway.

The Department believes that use of Interstate highways ensures a safe route of travel for large quantity radioactive materials. However, the Department recognizes the limitations of relying solely on the Interstate System and, as already mentioned, the inherent site-specific nature of routing. There is a clear need for a mechanism to accommodate these factors. Several examples serve to point this out:

1. Most points of origin and destination for large quantity radioactive materials shipments are not located on interstate highways. Additional safety benefits may be realized if access routes between the Interstates and these points are designated by the State.

2. The low accident rate associated with interstate highways is based on a national average. DOT recognizes there

are situations where accident rates will be higher for a particular segment of an Interstate than for a nearby alternate route.

3. The accident rate is not the only important element to consider in assessing risk to the public—one must also consider the consequences of a serious accident, even though the probability of that accident may be small. Therefore, the population along the route of travel should also be considered. Since Interstate highways serve to connect population centers, the benefits of using an Interstate highway with its lower accident rate going through a city should be carefully examined and compared with the benefits of using a more circuitous, secondary road around the city.

4. Use of the Interstate highway system may necessitate circuitous travel resulting in some increase in normal radiological exposure and, in some cases, higher accident risks. More direct non-Interstate routes may exist which could provide greater safety to the public.

The task which confronted DOT in this rulemaking was to provide for a more site-specific analysis to resolve these situations while at the same time maintaining national uniformity and a safe, viable transport system for nuclear materials.

Many commenters feel that local governments should be responsible for routing within their jurisdictions. First, they argue that local governments have the primary responsibility for protecting the health and safety of their citizens and therefore should determine if routes through their jurisdictions are acceptable. It is the town, city or county which provides initial emergency response to protect health and property in the event of an accident. Secondly, they argue that route selection is a site-specific process and that local officials are the most knowledgeable of local roads and local conditions. However, DOT sees serious problems from both a practical and safety standpoint associated with placing ultimate routing authority with each of the 23,000 local jurisdictions in the country.

Local jurisdictions are inherently limited in perspective with respect to establishing routing requirements. While the Department recognizes that local governments are accountable only to their own citizens, such a limited accountability has some undesirable effects. For example, a routing restriction in one community may have adverse safety impacts on surrounding jurisdictions. Also, some communities in determining that they do not have the appropriate expertise or manpower to perform a routing analysis, may find attractive the option of completely prohibiting the transport of radioactive materials through their jurisdictions. This has already happened in some cases. Uncoordinated and unilateral local routing restrictions placed on carriers of radioactive materials would simply not be conducive to safe transportation. There is a clear need for national uniformity and consistency.

DOT believes that the role of State governments is the key for ensuring that the safest highway routes are used by carriers of large quantity radioactive materials. A State government has a much broader perspective than local governments since it is charged with providing for the safety and welfare of all its communities. The safety impacts of a routing decision on all communities within the State can be assessed.

There are a number of other advantages to the exercise of route designation authority at the State level. States have the capability to incorporate local input directly into their routing analyses through existing State administrative and lawmaking procedures. At the same time States have the capability of working with the Federal government and are familiar with implementing regulations under a variety of Federal programs. States often have the greater manpower and technical training necessary to perform a routing analysis which adequately considers all factors related to public risk. For example, many States exercise authority under the NRC's Agreement State Program to regulate possession and use of certain source and by-product nuclear materials. Many States have radiation safety officials as well as knowledgeable transportation officials available to collaborate on a routing analysis.

States not only have the capability to consider local viewpoints on route selection, but also can address concerns of tunnel, turnpike and bridge authorities. The Department does not seek to force the use of all such facilities for nuclear material transportation. Rather this rulemaking establishes a system by which the State can consider the use of these facilities on the basis of overall risk to the public. A State government, after a careful evaluation of the total risks to the public, may conclude that a safer route is available and that certain facilities should be avoided.

Many commenters have reservations about the role of the States and the efficacy of the State route designation process. Probably the greatest reservation is shown by local officials who are concerned that the States may not actively pursue local interests before routes are designated. The State Planning Council on Radioactive Waste Management submitted comments supporting the concept of State-designating routing. However, the Council, composed of State and local officials, strongly encouraged DOT to "develop appropriate mechanisms and procedures to enable local participation in routing decisions."

The Department also wants to ensure that local communities have input into the State route selection process. DOT believes that the key to incorporation of local viewpoint into routing decisions is the cooperation between State and local governments before designation of routes. The Department has considered establishing specific guidelines for States of follow to ensure a formalized procedure for local consultation. However, there is great difficulty associated with this approach given the variations in organizational structure and administrative processes from State to State.

Instead, the Department is taking two steps to ensure that consideration is given to local viewpoints. First, the final rules contain a general requirement that the States consult with affected local jurisdictions before establishing a preferred route. DOT believes that the States must adequately consider local input, especially in light of the routing guidelines which necessitate the accumulation of local data relating to accident rates, population characteristics and other information that would require local cooperation.

However, the Department also understands that reasonable differences of opinion may exist in this sensitive area. As a result, DOT believes that each State should establish an advisory group composed largely of city and county officials. The purpose of the group would be to meet periodically, recommend to the State appropriate methods of consulting with local jurisdictions, and review the effectiveness of those measures in actual practice. Such State advisory groups would provide a valuable oversight function that should help to continually improve the State routing program.

State officials commented that the preferred routing system places a large burden on State governments and requested clear guidance from DOT on routing decisions. The Department also believes this to be extremely important in the interests of both national uniformity and safety. As a result, DOT is preparing a publication entitled "Guidelines for Selecting Preferred Highway Routes for Large Quantity Shipments of Radioactive Materials" ("DOT Guidelines") which is discussed in more detail elsewhere in this preamble.

IV. PRENOTIFICATION AND TIME-OF-DAY RESTRICTIONS

An extremely large number of commenters favored some type of requirements relating to prenotification and time-of-day controls. The Department notes that most State officials strongly endorsed these measures. In light of these public comments the Department has carefully reconsidered both types of controls.

Prenotification. A number of reasonable arguments have been made in support of prenotification: To aid the State in its route designation activity; to ensure better enforcement by utilizing State and local enforcement personnel in addition to Federal inspectors; and to more rapidly facilitate emergency response capability in case of vehicular accident. Prenotification on a case-by-case basis for all shipments of radioactive materials would result in a severe burden not only on shippers and carriers but also on the governmental units receiving this voluminous information with a doubtful increase in safety. Many commenters agreed with DOT, except for shipments of certain high-level radioactive materials. In most cases, the desire for prenotification by State and local officials centers around spent fuel and certain other nuclear waste materials.

On June 30, 1980, Congress enacted legislation (section 301 of the NRC Authorization Act, Pub. L. 96–295) directing the NRC to develop regulations which will require its licensees to provide State governments with advance notification for certain shipments of nuclear wastes. The NRC issued an NPRM on this matter on December 9, 1980 (45 FR 81058), proposing to require prenotification for licensee shipments of all wastes required to be shipped in Type B packaging, which includes spent fuel. The NRC has asked for the public to comment on the NPRM before March 9, 1981. Since these proposals would apply to a substantial number of shippers and carriers regulated by DOT, a discussion of the proposed requirements bears on the issue of prenotification raised in comment on the proposals for highway routing made in this docket by DOT.

In its NPRM, NRC proposed two sets of prenotification requirements. One set of proposed requirements concerns shipments of spent fuel in quantities greater than 100 grams mass. Such shipments are large quantity shipments subject to the routing requirements established by DOT in this docket when transported by highway. This treatment of spent fuel separately from other nuclear wastes is necessary because spent fuel shipments are also subject to physical security requirements which the NRC has imposed to guard against theft and sabotage. Information concerning exact schedules used in spent fuel shipments therefore must be considered sensitive. In the NPRM, the NRC proposes to require licensees to notify the governor of each State through which a shipment will pass at least four days before arrival at the State boundary. The notification would identify the shipper, carrier, receiver, the material to be transported, and the times of departure from origin and arrival at the State boundary. The licensee would have to immediately notify the State governor if the transportation schedule changes by more than six hours.

The confidentiality of information concerning the exact schedule of such sensitive shipments (i.e. dates and times of shipments) would have to be protected by the governor's office as if it were national security information (see proposed 10 CFR 73.21 in the NRC NPRM). Although treated as confidential, the information could be passed on to local officials as long as it is transferred under the security conditions described by NRC in its proposal. Other shipment information would not be considered confidential. Confidential information could be declassified ten days following the departure of the shipment (or the last shipment in a series) from the State.

The second set of prenotification requirements proposed by NRC in its NPRM would apply to any other nuclear wastes that are required to be shipped in Type B packaging. This category of materials includes large quantity radioactive waste shipments which also are subject to the routing system established in this docket. The NRC would require advance notice of shipment to the governor at least four days before the beginning of an estimated seven-day period of departure from the shipment origin. Information to be supplied would include the point of origin, the estimated seven-day period or periods of arrival both at the State boundary and at the shipment destination, and a point of contact for schedule changes. Prenotification information for nuclear wastes, other than spent fuel as described previously, would not be considered sensitive information and the State governor would not have to protect its confidentiality. The NRC estimates that over 24,000 waste shipments, including spent fuel, will be subject to these advance notice requirements annually, although only a small portion will be large quantity shipments.

The NRC prenotification proposals would not apply to two particular groups of large quantity radioactive materials shipments. First, nonlicensee shipments of nuclear waste, primarily those in support of DOE research and development activities, are not covered by the NRC prenotification proposals. Second, radioactive materials that are not waste products (primarily large source teletherapy shipments and possibly some other large source medical and industrial isotopes) also are not covered by the NRC prenotification proposals.

Further, there remain some unanswered questions concerning the nature of a prenotification system—what specific materials should be covered, how early the advance notice should be given, how the State or local governments would handle what may be voluminous paperwork, and what information is necessary. Congress has provided an indication of what is appropriate in this controversial area and the NRC is considering proposals which will not be made final for some time.

Another recent development also may prove useful to DOT in determining the efficacy of a prenotification system. The Puget Sound Council of Governments (PSCOG) is conducting a study in prenotification for certain materials as part of a comprehensive regional study of hazardous materials transportation under contract to DOT. PSCOG will present its findings on the effectiveness and practicality of advance notice to DOT in early 1981.

Two other facts also should be noted. First, the NRC intends to publish an atlas of all highway routes that have been approved for shipment of spent fuel. This information therefore will be publicly available to all State and local governments and other interested parties. Second, the existing NRC physical security program for spent fuel requires confidential notification and coordination with affected local officials (local law enforcement agencies) concerning approved routes.

In light of these considerations, DOT has decided not to take final action at this time concerning prenotification. In order to prevent a possibly severe inconsistency betwen NRC and DOT transportation requirements, the DOT will have to wait at least until final rules are issued for NRC licensees before undertaking a rulemaking proceeding to consider specific prenotification requirements for other types of large quantity shipments. In its further consideration of prenotification, DOT will also consider the role of escort vehicles provided by State or local governments. This subject is addressed later in this document in the general discussion of the preemptive effects of Docket HM-164.

Time-of-day restrictions. Many commenters are also strongly in favor of some kind of time-of-day restriction for nuclear material transportation. Again, most commenters are concerned with high-level nuclear wastes and spent fuel. There are practical as well as safety problems associated with uncoordinated time restrictions. For example, it has been estimated that the average shipment distance for a large quantity package of radioactive materials is approximately 2,200 kilometers. This implies travel through a large number of State, county and municipal jurisdictions. Even if the various time restrictions for these jurisdictions were known in advance by the carrier, delays enroute could be numerous. Some commenters argue that the delays caused by certain time restrictions are justified on the basis of the increased accident risks which exist during rush hour traffic in an urban area. However, the Department must also consider the added risks of normal radiological exposure accruing to the vehicle driver and bystanders at any temporary delay site. This may be a more important consideration from the standpoint of overall public risk, especially when one considers that several temporary delays could occur for each shipment. Also, there may be additional security problems related to the temporary delay of spent fuel shipments.

The Department does see some need for a coordinated effort to carefully examine the transportation of large quantity materials during periods of heavy rush hour travel in large urban areas. DOT believes that the States can address this situation as part of their route designation program by providing for suitable alternative routes to avoid certain heavily traveled highways during peak travel times. This would amount to a time of day restriction on certain highways, but would not require the hazardous material be unnecessarily delayed in one area.

V. OTHER TRANSPORTATION CONTROLS RELATED TO ROUTING

The notice of proposed rulemaking also addressed a number of other State and local actions generally related to the routing of radioactive materials. This included not only prenotification and time-of-day restrictions, but also escort requirements, restrictions pertaining to special personnel or equipment and any other action which would have the effect of unnecessarily limiting the transportation of radioactive materials through a jurisdiction. For the most part, DOT views these transport controls differently from the site specific nature of routing in one important aspect. These requirements are not directly related to characteristics that are peculiar to a specific geographical location. With the possible exception of the previously mentioned prenotification and time-of-day restrictions, the Department does not believe that public safety concerning the transportation of radioactive materials can be measurably improved by such State and local actions.

The Department has noted that the rationale supporting the need for various State and local actions often involves concerns in three areas: the adequacy of the emergency response system for hazardous material transportation; questions over liability for nuclear materials involved in highway accidents; and doubts over the effectiveness of the Federal enforcement of regulations. As a result, many citizens, as well as some State and local officials, believe that additional controls at the State and local level are justified, no matter how fragmented they may be. The Department does not subscribe to this philosophy. Even in cases where criticism may be justified, piecemeal State and local action instituted because of a concern over these issues and limiting the carriers' ability to function would not solve the problems. In fact, steps are now being taken by DOT and other Federal agencies to improve Federal, State and local capabilities in these critical areas.

With respect to emergency response, the Department of Transportation has prepared a comprehensive training program for responding to radioactive material transportation accidents. This training program "Handling Radioactive Materials Transportation Emergencies" is directed to "first-on-the-scene" emergency service personnel such as local fire, police and ambulance organizations. The comprehensive training package consists of slides, tapes, student workbooks and instructor guides. It is a simple and straight-forward instruction kit to provide local and State personnel with a basic understanding of the subjects of radiation and associated hazards, packaging required for nuclear material, transportation regulations, protective measures and procedures, and planning and preparedness for transportation accidents. DOT has been coordinating the development of this training program for the past two years with emergency service personnel as well as State and local officials. This 6 to 8 hour training package supplements the 20 hour training program already available to emergency response personnel responding to other hazardous material transportation emergencies. The entire training program will be distributed to governors of each State upon request.

A booklet entitled "Response to Radioactive Materials Transportation Accidents" is also nearing completion. It was distributed as an interim edition in the spring of 1980 and the response from State radiation control program directors and emergency management authorities has been very favorable. It is intended to provide local emergency response authorities with basic information on the first steps to take at the scene of an accident until the arrival of State or other radiological response teams.

The Federal Emergency Management Agency (FEMA) is the agency primarily responsible for coordinating Federal assistance to State and local governments that are developing plans for responding to radiological accidents at both fixed nuclear facilities and at the scene of transportation accidents. FEMA has taken a number of steps toward this end. Recently proposed rules (45 FR 42341) were published on procedures and criteria for reviewing and approving the adequacy of State and local plans and preparedness. FEMA has also established the Federal Radiological Preparedness Coordinating Committee (FRPCC) consisting of a number of separate Federal agencies including DOT. This committee is coordinating all Federal assistance and guidance to various State and local agencies for developing and testing emergency response plans. The FRPCC responsibilities in this area include the following:

—Establish policy and guidance to other Federal agencies
—Develop preparedness criteria
—Provide direct assistance to State and local governments
—Review and approve State radiological emergency plans and preparedness
—Implement a program of public education
—Develop and manage an emergency response training program including field test exercises materials
—Issue guidance for radiation instrumentation systems.

The Department of Transportation is providing assistance to FEMA in the preparation of Federal guidance to State and local governments for use in developing the transportation portions of radiological emergency response plans. DOT will also assist FEMA in its review and approval of State and local plans and in the evaluation of exercises to test those plans.

In support of this effort, a Federal interagency task force was recently organized. The task force, with participation by State and local authorities, is preparing an important planning document "Guidance for Developing State and Local Radiological Emergency Response Plans for Transportation Accidents." Federal agencies including DOT, NRC, FEMA, DOE and the Environmental Protection Agency have collaborated on this effort to provide State and local authorities with guidelines to develop effective response plans. A preliminary guidance document will be published in the **Federal Register** for public review and comment during the first quarter of 1981.

A committee composed only of State and local officials has been organized to provide direct input into activities conducted by this task force. The Interorganizational Advisory Committee, composed of State civil defense and radiation control authorities and local emergency management officials, should prove to be an effective sounding board for planning and guidance documents developed by the task force.

It should also be noted that the routing scheme established by this docket will enhance State and local emergency response planning. The International Association of Fire Chiefs, in its comments to Docket HM-164, states:

> . . . we fully support Docket HM-164, Highway Routing of Radioactive Materials, for the following reasons:
> 1. Some nation-wide method for the routing of radioactive truck shipments is necessary. For each local jurisdiction to impose

specific routing requirements would present an untenable situation. However, under the proposed regulations, each state would establish the routing after reviewing local input. The key here is to require local jurisdiction input.

2. The requirements that the carrier file a route plan with MTB is very important. In this way MTB will be able to provide data on routes, amounts, and shipment frequencies. This data will then be used by the local fire departments for their emergency response planning guides.

Questions over the adequate availability of funds to reimburse local jurisdictions and individuals affected by nuclear transportation accidents seem to be another impetus to various State and local actions. Final responsibility for nuclear transportation accidents really depends upon accident specific really depends upon accident specific factors and will usually be settled in the courts. Some of the factors affecting financial responsibility include the nature of the accident itself, the shipper or carrier involved, the type of radioactive material involved and the geographic location of the accident. For most types of radioactive materials the extent of financial liability and the types of costs to be reimbursed would be determined by the applicable State tort law.

If the origin or destination of the radioactive material is an indemnified facility such as a nuclear power plant, the provisions of the Price-Anderson Act (42 U.S.C. 2210) assure a source of funds to cover certain personal injury and property damage claims. The law extends to persons other than the licensee, such as the carrier, who may be liable for an accident. Insurance coverage up to $560 million per accident is provided by a combination of licensee private insurance policies and indemnity agreements between the licensees and the NRC.

The Federal Highway Administration (FHWA) is now in the process of determining appropriate levels of financial responsibility for motor carriers of hazardous materials. On July 1, 1980, the President signed the Motor Carrier Act of 1980 (Pub. L. 96–296) into law. Section 30 of the Act, among other things, establishes minimum levels of financial responsibility for motor carriers transporting hazardous materials in interstate or intrastate commerce (applicable to vehicles with a gross weight rating of 10,000 pounds or more). The purpose of section 30 is to assure the public that a motor carrier maintains an adequate level of financial responsibility sufficient to satisfy most claims covering public liability, property damage and environmental restoration.

The minimum levels set in the Act include $5 million for each vehicle operated by carriers of large quantity radioactive materials and certain other hazardous materials. DOT has unlimited authority to adjust this level upward and may also adjust downward to not less than $1 million for each vehicle for an initial two-year period.

The FHWA's Bureau of Motor Carrier Safety (BMCS) issued an ANPRM (Docket No. MC-94, 45 FR 57676) entitled "Minimum Levels of Financial Responsibility for Motor Carriers" on August 28, 1980. The purpose of the notice is to obtain public comments and data and to eventually make any necessary adjustments to the minimum levels scheduled by Congress to go into effect on July 1, 1981.

Many commenters have also suggested that doubts about Federal enforcement efforts have resulted in increased State and local regulatory activities. The major criticism of commenters to this docket is that the preemptive effect of DOT's routing rules will eliminate or frustrate enforcement efforts at the State and local level. It is contended that State and local enforcement is needed to supplement the Federal inspection effort.

Although it is clear that this rulemaking will preempt certain State and local actions, DOT does not believe this will reduce enforcement efforts at any level. States have been increasingly active in the enforcement of Federal highway safety and hazardous material transport regulations. Many States have adopted the Federal Motor Carrier Safety Regulations and the Federal Hazardous Materials Regulations as strongly encouraged by DOT. Most States already have enforcement systems in place to carry out the provisions of these regulations. A number of States have initiated substantial hazardous material training programs for law enforcement and other personnel. DOT has provided training to State and local personnel at its Transportation Safety Institute in Oklahoma City. Such State-level enforcement activities will not be hampered by these final rules. In fact, it is DOT's contention that enforcement, particularly at the State level, will be enhanced by the States routing function provided by this rulemaking.

At the Federal level, the Department's BMCS has the primary responsibility for ensuring compliance with the Hazardous Materials Regulations by motor carriers. BMCS is now authorized 210 hazardous material or safety specialists in the field and expects additional positions next fiscal year. BMCS is now administering a four-State demonstration program which funds approximately 100 additional State inspectors. Also, pending before Congress is the Commercial Motor Vehicle Safety Act which, if enacted, would authorize a 50-State grant program that could result in a total of 2,200 State inspectors for motor carrier safety. Moreover, the NRC's enforcement staff of over 100 inspectors is directing its inspection efforts increasingly toward the transportation activities of their licensees. This will enhance the overall enforcement program particularly for transporters of nuclear fuel-cycle materials.

A number of commenters note that penalties were not mentioned in the January 31 NPRM and suggest the need for such. Penalties for violation of radioactive materials transportation requirements under the HMTA are the same as prescribed for other hazardous materials. Civil penalties may include a maximum fine of $10,000 for the occurrence of each violation for each day. Criminal penalties may include a fine and imprisonment up to $25,000 and five years. Civil and criminal penalty actions can be taken against container manufacturers as well as shippers and carriers of radioactive materials. In addition, the States provide for civil and criminal penalties under their own legislation and the levels vary from State to State.

The Department believes that much is being done in the areas of emergency response planning and training, carrier financial responsibility, and regulatory enforcement. Furthermore, both local and State expertise have been solicited to help in the process of strengthening various programs. DOT certainly recognizes the legitimate concern and acknowledges the expertise of State and local officials in these areas. However, independently applied restrictions which frustrate the ability of a motor carrier to safely and expeditiously move nuclear materials are not the proper approach to enhance over-all public safety. It is DOT's opinion that State and local concerns can be more adequately satisfied under programs coordinated at the Federal level which incorporate State and local viewpoints.

VI. PREEMPTIVE EFFECT OF DOCKET HM-164

Because of the extensive nature of the Part 177 amendments, the relationship among the levels of regulation of the different categories of radioactive materials, and the need for an understandable interface between Federal and State regulation of radioactive materials transportation, DOT believes that certain regulatory actions by State and local governments should not be taken. To explain this view, DOT sets out its policy on the relation of State and local regulation to the Federal requirements in Part 177 in a new appendix to that part. An appendix appears to be a more appropriate method of stating this policy than the regulatory text used in the January 1980 notice of proposed rulemaking, and an appendix permits a more extensive discussion of the policy. The section-by-section analysis appearing later in this preamble details the specific reasons for the policy. Some general issues will be discussed here.

The structure of the amendments to Part 177 accommodates State regulation of carriers' routes in defined circumstances, as well as some limited local regulation. Briefly, an appropriate State-wide agency may designate routes for motor vehicles transporting large quantity radioactive materials. Local governments, if permitted by State law, may exclude such motor vehicles from locations from which they are excluded by Part 177 or by State action consistent with Part 177. For placarded vehicles carrying lesser quantities of radioactive materials, both State and local governments may adopt § 177.825(a) verbatim. Section 177.825(a), established in this rulemaking, requires a carrier to consider certain information in route selection and to provide general guidance to the motor vehicle operator as to routes used. While State regulation is circumscribed as regards routes used by such carriers, adoption of § 177.825(a) will permit a State to directly enforce that provision without necessary recourse to Federal enforcement personnel. The same purpose is served by the limited local regulation permitted for placarded carriers of both large quantity and less than large quantity shipments. Routing restrictions for unplacarded motor vehicles are not necessary. The preemptive effects of the final rules in this docket are intended to occur at the effective date of the rules.

The basic justification for publishing a statement concerning the preemptive effects of Docket HM-164 was questioned by many commenters. The HMTA expressly preempts State and local requirements that are "inconsistent" with HMTA requirements, both the law itself and regulations issued under it. DOT has previously established procedures to permit it to interpret the HMTA's preemptive effects when so requested by State or local governments, or by other interested persons. These procedures, codified in Part 107 of 49 CFR, offer a less expensive alternative for resolving preemptive issues than litigation although such issues are ultimately judicial in nature. It is apparent that new rules which deal extensively with matters of regulatory concern to State and local governments, such as those published in this Docket, will necessitate guidance from DOT as to the preemptive effects on State and local authority. DOT believes that this guidance will be considerably more useful if provided, as far as possible, before the rules become effective. The Part 177 appendix is intended to serve this purpose.

Underlying the appendix are several conclusions about the Federal-State relationship in the area of radioactive materials transportation. First, as expressed in the Part 107 preemption procedures, DOT believes that "inconsistent", as used in the HMTA, refers to State and local rules that directly conflict with HMTA requirements, and also to those that are "an obstacle to the accomplishment and execution" of the HMTA (§ 107.209(c)(2)). Therefore, the policy statement in the appendix concerns characteristics of State and local regulatory activity that are necessary to effect, or to avoid hindering, accomplishment of the goals and purposes of the Part 177 amendments. Those amendments balance complementary national, State and local interests in regulating motor carriers to ensure that public health and safety are served by Federal, State and local rules that are widely applied and understood and that are based on a comprehensive examination of factors affecting radioactive materials transportation safety.

This rulemaking does not delegate Federal authority to regulate motor carriers, a fact that has been misunderstood by many commenters. The rules published in Docket HM-164 define and make Federally enforceable the use of Interstate System highways for carriers of large quantity radioactive materials. They also make Federally enforceable those routes designated by appropriate State agencies, based on DOT's own determination that such routes, if derived from an adequate safety analysis like the "DOT Guidelines" are likely to result in a further reduction of radiological risk that is reliable and reasonably related to the costs of evaluating, enforcing and using selected routes. Further, DOT has concluded that route designations that do not meet the conditions outlined in the Part 177 appendix are unreliable tools for minimizing radiological risk, may result in unconsidered safety impacts, may unnecessarily burden commerce, and generally result in a confused patchwork of safety regulation that is not conducive to compliance.

In the appendix, DOT has not attempted to specify in detail the process to be used by a State agency in route designation except in two respects. A safety analysis as described must be performed to ensure reliable results, and the designating State agency must consult with affected local or neighboring State jurisdictions. State consultation with affected local jurisdictions is necessary to ensure that the information used to perform a safety analysis is the best available. It is important, for this reason, that the consultative process between the State routing agency and local governments be both substantive and thorough.

In considering this need, DOT has concluded that an appropriate method for effecting the consultative process should include public notice and opportunity for comment, public hearing when appropriate, and direct notice to affected local jurisdictions. To ensure that these processes are adequate, DOT also believes that a standing advisory body consisting largely of local officials who are concerned with routing issues should be establish in each State to recommend to the State appropriate consultative methods and to evaluate the effectiveness of those methods in actual use. This is particularly important in States that are likely to impose frequent routing decisions or to deal with particularly controversial issues. An *ad hoc* advisory body may suffice in States that are unlikely to take frequent routing action. For example, a State that expects only limited traffic in large quantity shipments on an acceptable Interstate route may wish to conduct an initial review of the routes of travel using an advisory body convened for that specific purpose. Another consideration related to the State-local consultative process concerns routing actions which local governments believe should be taken within their jurisdictions. A local jurisdiction which requests State action, for example to shift traffic from an urban segment of Interstate highway, should identify potential alternate routes to the appropriate State routing agency and state why those other routes may be a better choice for routing large quantity shipments. A State advisory body might be able to provide a useful preliminary evaluation of local requests of this kind and to identify any need and possible methods for further State-local consultation.

Commenters also raised questions about the effect of this rulemaking on the local authority of Indian tribes. DOT believes that, where an Indial tribe has effective routing authority similar to that exercised by a counterpart State agency, it should be exercised as described in the Part 177 appendix. Tribal regulatory authority over motor carriers must exist separately from the Part 177 amendments, since those amendments do not delegate any such authority. The source of tribal authority may differ from that of State authority in that tribal authority is recognized by treaty or Acts of Congress. Consequently, it is possible that limits on tribal authority may occur as a result of Federal law other than the HMTA. Rather than a question of HMTA preemption, tribal routing authority may involve a question of the proper relationship between the HMTA and other Federal law. In specific situations, it may be necessary to examine other Federal law to determine the practical limits on tribal authority to impose routing controls on motor vehicles carrying radioactive materials. In the Part 177 amendments, DOT is treating Indian tribes as it treats States. DOT recognizes, however, that specific factual and legal circumstances may differ from those that affect State authority and is prepared to examine these circumstances on an individual basis, as the need is shown.

DOT's decision against required use of escort vehicles is discussed in the section-by-section discussion of the new appendix to Part 177. However, an obvious relationship exists between prenotification and the voluntary provision of escort vehicles by jurisdictions through which a large quantity shipment may pass. DOT intends to examine situations where an escort might be provided voluntarily by a local jurisdiction, under circumstances in which the presence of an escort is not a precondition to passage through the jurisdiction, and in which the transport vehicle is not delayed at the jurisdictional boundary. Escort vehicles in some cases may also be provided by shippers of spent fuel under the existing NRC physical security program for transit through some heavily populated local jurisdictions. In view of this, DOT intends to examine the possible impact of such voluntary, locally provided escort services on the DOT routing rules, existing NRC physical security rules and proposed NRC prenotification rules.

VII. SECTION-BY-SECTION DISCUSSION OF FINAL RULES

Summary of Changes from NPRM. There are several important changes from the proposals issued in the NPRM based upon the Department's review of the public comments. First, new provisions are added to Part 172 to aid shippers, carriers and enforcement personnel in the identification of radioactive materials shipments which are subject to the preferred routing system. These provisions include a new shipping paper entry and a white placard background applying only to shipments involving a large quantity package of radioactive materials.

Secondly, new definitions for "State routing agency," "preferred route", and "State-designated route" are added to the regulations. These definitions are added to answer questions concerning the appropriate routing agency designated by the States and the manner by which States exercise their authority to designate preferred routes.

The wording of both the general routing rule (proposed § 177.825(a)) and the preferred routing rule (proposed § 177.825(b)) have been modified somewhat. Although the effect of the general routing rule remains the same, the criteria for the carriers to use in selecting a route has been revised to make the rule more manageable and enforceable. Several points concerning the preferred routing rule may not have been clear in the NPRM and should be emphasized.

It is important to emphasize that the final rule establishes the Interstate highway system as a self-functioning Federally prescribed routing network capable of providing for the safe movement of nuclear materials even if the States choose not to designate routes. However, because the level of safety provided through use of Interstate highways may be improved by site-specific evaluations, DOT believes that the States should be extended as much flexibility as possible in their route designation process. For example, the final rule does not require a carrier to use Interstate beltways or bypass routes when other routes have been designated by the States as substitutes. The States can consider the need for circumferential routes to avoid urban areas on a more site-specific basis in their own routing analyses. The beltway provision still applies to carriers using Interstate System preferred routes when the States have not designated another route. This flexibility is consistent with the routing guidelines being developed for the States.

Another change to the preferred routing rule is the reference to the DOT Routing Guidelines as criteria for States to use in designating preferred routes. As will be covered in more detail in the next section, the guidelines will provide the States with a clear, step-by-step procedure for performing a routing analysis that is both more understandable and flexible than the criteria presented in the January 31 NPRM.

The last major change between the proposed and final rule involves inconsistency between Federal and State/local transportation requirements. Proposed paragraph (d) of § 177.825 has been deleted. Instead of addressing this topic in the routing rule itself, DOT has chosen to include an expanded discussion of DOT policy in a separate appendix to Part 177 as mentioned previously.

The remainder of the final rules are basically unchanged from the NPRM, except for redesignation of certain paragraphs. The following section-by-section discussion provides a synopsis of DOT's rationale for each section including reference to substantive public comments. A more detailed discussion of public comments is provided in the previously mentioned docket supplement.

§ 171.7 Incorporation of State routing guidelines by reference. The publication "Guidelines for Selecting Preferred Highway Routes for Large Quantity Shipments of Radioactive Materials" (DOT Guidelines) is incorporated by reference in § 171.7. Repeated reference has been made to the need for State and local involvement in routing decisions on the one hand, and the need for uniformity and consistency of those decisions on the other. Many commenters, particularly State officials, support the preferred routing system for large quantity nuclear material to accommodate this goal, but only if DOT provides clear and practical guidelines for use by State authorities. The DOT Guidelines are intended to fulfill this function.

In developing the guidelines, the Department has drawn upon two recent research projects. The first is a study completed for the FHWA entitled "Development of Criteria to Designate Routes for Transporting Hazardous Materials by Highway". This research project involved a study of all factors which contribute to the selection of highway routes for all hazardous materials classes. The most important factors related to the lowering of public risk are then selected as the basic criteria upon which an agency should base its highway routing decisions. Although the Department does not consider this generic research to be final, the study does establish a methodology which can be useful after further refinements are made relating to the particular class of hazardous materials for which routing is to be evaluated.

With this in mind, the Department initiated another research project to develop routing criteria oriented specifically to the peculiar characteristics of radioactive materials transportation. This study is being conducted for the Materials Transportation Bureau and is titled "Guidelines for Selecting Routes for Highway Shipments of Large Quantity Radioactive Materials". The routing guidelines developed thus far provide flexibility to the appropriate State and tribal routing authorities, either to designate the use of an Interstate highway and provide an alternative, or to identify other appropriate routes. Further refinements in the guidelines are expected after the completion of pilot tests to be conducted with the help of two State governments in January 1981. It is expected that the guidelines will be published and made available to State agencies shortly thereafter.

Another important element of the guidelines relates to recommendations for the soliciting of local input into routing decisions. It should be noted that the routing guidelines provide for substantial local input in themselves. Much of the data necessary to perform the routing analysis will be generated from local sources: accident rates, population statistics, conditions of roadways, emergency response capabilities, property values, evacuation capabilities, and location of facilities such as schools and hospitals which require special consideration. Nevertheless, the Department believes it essential that the State specifically provide for a process of consultation with appropriate local authorities.

§ 171.8 Definitions. A number of commenters suggested that DOT specifically identify the agency in each State that would have the authority to designate preferred routes. As stated previously, the Department has no authority to do so. The designation of routes for large quantity radioactive materials is an authority which only the States can exercise for themselves. Each State has legal and organizational peculiarities relating to the regulation of radioactive material transportation. Often, authority is divided among various agencies within the same State. Consequently, each State should determine for itself the appropriate routing agency within the general definition established by § 171.8.

The definition of "State routing agency" includes interstate compacts and appropriate Indian tribal authorities (see the discussion of § 177.825(b) relating to Indian lands). As specifically mentioned in the NPRM, this definition excludes a bridge/tunnel/turnpike authority unless that authority also is empowered to impose such rules concerning radioactive materials transportation on State highways generally. Routes designated by a State routing agency may be enforced by that agency, or by any other appropriate State agency. This definition may apply to more than one agency in a single State sharing responsibility for designating preferred highways.

Two other definitions are added. The first is the definition of a "preferred route". A preferred route includes "State-designated routes" which is also defined in § 171.8. A definition for State-designated routes is necessary to clearly show the criteria the State must follow in establishing preferred routes: application of DOT routing guidelines or an equivalent routing analysis, prior consultation with affected local agencies, and coordination with adjoining States to ensure continuity of routes.

§ 172.203(d)(1)(iii) Shipping papers. For identification and enforcement, a requirement is added to § 172.203 to require the shipper to enter "Large quantity" as part of the hazardous material description on the shipping paper. This will alert the carrier that he has received a package of radioactive materials for which routing controls are required and that a route plan must be prepared.

§ 172.507 and § 172.527 Placarding. Vehicle identification requirements are added to Part 172 to require a white background for the RADIOACTIVE warning placard. The white background will aid enforcement personnel to distinguish between large quantity shipments and other placarded shipments for which preferred routing is not required.

Public comments strongly favored some method of distinguishing between vehicles which contain large quantity packages and vehicles which do not contain large quantity packages but which still require the RADIOACTIVE warning placard. DOT considered several methods of accomplishing this. The white placard background is determined to be the most passive system considering effectiveness and cost of implementation. The white background system has been used for some time to distinguish certain hazardous materials shipped by rail for the purpose of car handling.

§ 173.22 (b) and (c) Shipper's responsibility for physical security, and filing of route plans. Without change from the proposals in the NPRM, the Department is adding provisions to § 173.22(b) to require shippers of irradiated reactor fuel (spent fuel) to provide physical protection under either a plan now required by the NRC (see "Physical Protection of Irradiated Reactor Fuel in Transit", 45 FR 37399, June 3, 1980, and 10 CFR Part 73) or a plan approved by MTB. Also, a provision is added to § 173.22(c) to require shippers of a large quantity package of radioactive materials to file a copy of the route plan prepared for that shipment within 90 days following the shipment with DOT. The Department intends to consolidate the information contained in the route plans and supply it to interested parties. For further discussion of route plans and physical security see the discussion of § 177.825(c) and § 177.825(e), respectively.

§ 177.810 Tunnels. Section 177.810 is revised to except radioactive materials from requirements that restrict their transportation through urban vehicular tunnels used for mass transportation. An informative sentence is also added which directs carriers to § 177.825. This action is being taken to facilitate achievement of the basic objective of the general routing rule to minimize radiological risk and to allow the States flexibility to designate preferred routes for large quantity shipments. The States, in exercising their routing prerogative under this rule, may determine through their routing analysis that a safer route exists which does not require the use of tunnels and other such facilities. In that case, the States may reimpose restrictions for large quantity radioactive materials.

Many commenters questioned the rationale behind the exception for radioactive materials in § 177.810 as opposed to restrictions for other hazardous materials. The State of California, which retains control over the shipment of hazardous materials through its tunnels, held that it is imperative that the State maintain the flexibility to prohibit such transportation. The Maryland Department of Transportation objected to the proposed revision of § 177.810 and took the position that any vehicle required to display the RADIOACTIVE placard should not be permitted to traverse an urban vehicular tunnel used for mass transit. DOT does not believe that this is necessarily the case from a health and safety standpoint. Traditional locally imposed restrictions on tunnel traffic frequently focus on explosives and flammable gases, for which the confinement provided by a tunnel may act to exacerbate the risk. In cases involving radioactive materials, the fact of confinement does not operate to increase overall risk.

For large quantity shipments, it is DOT's position that tunnel restrictions should not be based merely on the nature of the facility but on the overall risks between available routes, and that such restrictions should be imposed only by an agency with State-wide responsibilities that permit adequate consideration of other alternative routes. Thus, use restrictions on tunnels and similar facilities should not be determined solely by facility operators, but rather their use should be available for consideration as possible alternatives in the State procedures leading to route selection. The amendment to § 177.810 is necessary for States to be able to evaluate the site-specific risks involved over various routes without being hampered by locally imposed constraints which may be counterproductive. One proper factor that a State agency would consider in route designation is the potential property damage to the tunnel itself in the event of an accident.

In the absence of a State routing agency's action to review the status of tunnels and similar facilities located within its jurisdiction, a large quantity carrier will generally be limited to such facilities that are part of an Interstate System highway. Other placarded carriers could use such facilities only after considering the safety factors specified in new § 177.825(a).

§ 177.825(a) General routing rule. Paragraph (a) of this section is adopted with some change in wording from that proposed. The basic objective of the general routing rule remains the same: the carrier must examine all available highway routes and choose a route that minimizes radiological risk to the public. In making this determination, the carrier must consider available information on the most important factors which contribute to the minimization of radiological risk. These factors are identified in the final rule as population, accident rates of available highways, transit time, and the time of day and day of week during which the shipment occurs.

The NPRM also included such factors as terrain, physical features, weather conditions, and effectiveness of local emergency planning. These factors have been deleted from § 177.825(a) of the final rule for various reasons. The influence of terrain and physical features on public risk from transportation is largely accounted for by considering accident rates of the alternative roadways. It is not believed that these factors should be singled out for special consideration by the carrier since they are only two factors which contribute to overall highway accident rates. Weather condition is a factor over which the carrier has no control, has little advance knowledge of, and could often change during actual transportation. Determining the effectiveness of local emergency planning would be a difficult burden to place on the carrier in light of the subjective judgement that would be necessary and the lack of available information to the carrier. It is the Department's belief that effective emergency response planning is an activity that all communities should be involved with. As already discussed, DOT and FEMA are collaborating to provide an emergency response training and preparedness program to achieve this end. Economic factors such as property values have not been included because they generally follow population density and are not otherwise readily available to carriers.

The last major change to the general routing rule involves the replacement of the term "risk radiological exposure to the fewest persons" with "minimized radiological risk." Risk minimization is the basic goal to be achieved. Certainly limiting exposure to the fewest people possible is one element of reducing overall radiological risk, but it is not the only consideration.

Many commenters reviewing this section took exception to what they called the non-specific, unquantifiable criteria carriers and drivers must evaluate in choosing a route which will minimize radiological risk. There was general agreement that placarded vehicles carrying other than large quantity packages of radioactive materials should not be forced to comply with the very specific routing rules established for those shipments. However, no one offered a more acceptable rule to govern general routing requirements. While most of those persons commenting on this section considered the lack of precise, measurable factors to be an advantage which carriers could use to operate vehicles at their own discretion, the American Trucking Association (ATA) expressed its concern over the rule's implication that only one possible route could qualify. The ATA went on to state that, given the dynamic state of affairs of the prescribed criteria, the optimum route could vary even during the course of actual transportation, and carriers would find themselves subject to the whim and fancy of respective State and local governments in issuing citations for unacceptable route selection.

DOT does not expect that any of the suggested actions of carriers or compliance personnel will occur with such frequency that the value of the rule as a general statement of meaning or intent will be diminished, especially in light of the improved wording of the rule. For clarification purposes DOT does acknowledge that more than one route could qualify as an acceptable alternative and it is not incumbent on the carrier or driver to make detailed calculations in selecting the most appropriate route.

The public interest group Rural America was alarmed by DOT's emphasis on routing vehicles carrying such materials in a manner that might affect the health and safety of small towns and rural people. Such a policy, they said, reflects the Department's failure to recognize the needs and rights of populations residing in rural areas, and they see in the rule a discriminatory stance regarding sparsely populated areas. In directing carriers to select routes which minimize radiological risk DOT does not agree that it is merely shifting a burden from one group of persons to another, although it is true that population density is one factor the carrier must consider. Rather, DOT expects to see a decrease in the amount of exposure to all persons in the general population.

The Department once again would like to point out that this general routing rule applies to thousands of shipments involving relatively low-hazard radiopharmaceuticals, and other medical and industrial isotopes. These shipments often involve multiple pickups and deliveries, interchanges with other modes of transportation, and the comingling of radioactive materials with non-hazardous materials on the same vehicle. A general requirement to accommodate a great number of shipments in such a complex transportation environment will necessarily involve some vagueness. The rule is intended to guide motor carriers by specifying important factors to consider in evaluating a number of available routes.

§ 177.825(b) Preferred Routes for Large Quantity Radioactive Materials. In the notice of proposed rulemaking DOT discussed its reliance on the Interstate System of highways as being the primary roadways over which radioactive materials shipped under a route plan are to be carried. The general designation as preferred highways is, therefore, granted to these highways based upon an overall performance rating with respect to lower accident rates and their capacity for reducing transit times. For the most part, public comment expressed support for this proposal as well as the related provision which allows States the prerogative to modify the preferred status of Interstate highways and designate other roads as acceptable alternatives.

Some commenters argued that specific segments of the Interstate System are not as safe as statistics indicate for the system as a whole and that DOT should not make such widespread designations without performing a mile by mile review of the roadway. The NPRM recognized that each mile of the entire 42,500 miles of Interstate highway is not so consistent in design engineering or accident history that there would be an even correlation of the system's parts equal to that of the whole. That is one of the reasons why an option is extended to the States which enables them to modify the preferred status of those segments for which there is a more acceptable alternative. As a basic system, however, even in the absence of State action, the Interstate System highways are well-suited for the use required by the rule. It also serves as a measure for use by the States in their designation of some additional highways which provide an essentially equivalent or greater level of safety. This basic system of highways as primary routes also supports emergency response planning by increasing the confidence of planners in their knowledge of routes of travel.

The requirement that carriers of large quantity radioactive material packages use an Interstate circumferential or bypass route around a city was generally recognized by commenters as a reasonable precaution. This requirement did not, however, receive unanimous approval.

One commenters suggested that the use of beltways would not automatically result in the avoidance of all heavily populated areas. The City of Baltimore expressed its opinion that during peak-hour traffic patterns, it may be less hazardous to direct shipments over an Interstate through route rather than over a beltway and wanted this option left open to the States in their modification of Interstate highways and designation of other preferred routes. Comments from the State of Massachusetts pointed to situations where some metropolitan areas have multiple beltways and they feared that the rule as proposed might allow for routing over the shorter circumferential route, even though a second route, with superior design standards and lower population density, is available.

In response to these comments, DOT must reaffirm its belief that packages of large quantity radioactive materials can be transported over any Interstate highway, and most other comparable routes, with a confident level of safety. However, this does not imply that reasonable routing rules should not be imposed by State governments which increase this level of confidence. Consequently, in applying a rule which addresses the broad national interest DOT has chosen to direct carriers to use urban Interstate circumferential beltways in the belief that, when considering both normal and accident conditions of radioactive materials transportation, an aggregate benefit will be realized. States are encouraged to exercise their option to designate other streets and highways as preferred routes and to modify the status of Interstate highways. Such action, if justified, could include the direction of traffic onto Interstate through routes or onto a specific Interstate bypass. Each of the above referenced comments regarding beltways, then, would seem to be satisfied through a responsible exercising of the State's prerogative to designate routes and modify the status of Interstate highways. The guidelines developed by DOT to assist the States in their selection of preferred highways, and for similar use by local units of government in their consultation with the States, is also an effective means by which comprehensive, safety related routing decisions can be made.

In commenting on which radioactive materials should be restricted to preferred highways few persons took exception to the choice of large quantity packages. As a matter of fact there was widespread agreement, among those persons acknowledging the need to transport radioactive materials, that the Interstate System of highways and equivalent roads are the most appropriate routes for large quantity packages. As pointed out in the NPRM, Docket HM-169 will probably eliminate the term "Large Quantity". For routing purposes, some multiple of A_2 values (see the discussion on package curie limits in Docket HM-169, 44 FR 1852, January 8, 1979) will very likely be used to identify radioactive material packages now described as large quantity packages in § 173.389.

Several Indian organizations expressed a concern that the NPRM failed to recognize "the unique legal status of Indian tribal governments and tribally-owned lands." Specifically they contended that Indian tribes are, in effect, quasi-sovereign governments possessing rights of self-government under the terms of various treaties with the Federal government. As such, organizations such as the Council of Energy Resource Tribes (CERT) maintain that Indian tribes have the same prerogative as State governments to designate preferred routes for large quantity radioactive materials across tribal lands.

Commenters from Indian organizations support their arguments from the legal standpoint that DOT's preemptive authority may be limited by tribal ownership rights. CERT contends that:

. . . Indian tribes do not lose title to the land on which State or interstate highway rights-of-way are obtained through negotiated agreements between the tribes and the State government. Thus, a tribe may not have relinquished its right to restrict the use of the easement for a purpose that the tribe feels endangers the health and safety of its people. The DOT may not have the authority to preempt such tribal restrictions because the Hazardous Materials Transportation Act does not expressly apply to Indian lands.

Indian commenters did not voice an objection to the transportation of radioactive material across their lands per se. The commenters were oriented toward allowing the route-designation option with the Indian tribes the same as for the States. It is pointed out that many Indian reservations are located near mining or milling activities associated with nuclear materials as well as Federal disposal sites for radioactive waste materials. Further, many Indian lands are crossed or are in proximity to highways used for transportation of all types of nuclear materials.

The applicability of the HMTA to Indian tribal lands will depend on the specific facts and laws involved. Generally, however, DOT does recognize the special status of Indian tribal governments in the Federal system. Accordingly, the final rules allow Indian tribal governments to exercise routing authority in a similar manner as provided for the State governments. This is accomplished by including appropriate Indian tribal authorities in the definition of "State routing agency" in § 171.8.

While the Interstate System of preferred highways will permit the transport of radioactive materials between any two points, DOT recognizes that in some instances this may involve an excessive amount of time and mileage thereby reducing the overall effectiveness of the safety objectives intended by this rulemaking. However, rather than prescribing an arbitrary numerical percentage increase against which carriers would have a blanket approval to use non-Interstate System highways, DOT believes that the States are fully competent to deal with such actual cases as they arise and will respond to them in an appropriate fashion. It is anticipated that particular situations which involve a regular flow of materials will come before the State in the form of requests or petitions from carriers seeking the designation of preferred highway for a certain non-Interstate highway. Considering the key role played by the States in designating routes, it is believed that this approach is the most reasonable method to address circuitous travel that may result occasionally from use of the Interstate System. Also, this will likely result in the selection of a route based on a documented measure of public risk rather than one based on an arbitrary percentage figure. It is expected that the States, in considering the approval or denial of the carrier request or petition, will perform a routing analysis similar to that prescribed by the DOT Routing Guidelines. DOT will reevaluate the final rules in the first year after they become effective and will consider whether or not they need to be modified to provide other methods of dealing with circuitous travel.

One final point on the State designation of routes should be made. Commenters have questioned whether such a State-designate route would be established on a shipment by shipment basis or be a generic route established to handle shipments on a continuing basis. DOT is of the opinion that the application of the DOT routing guidelines or some other equivalent routing analysis by a State routing agency would be sufficient to establish preferred routes for routine use by carriers of large quantity radioactive material packages. State-designated routes are not considered to be shipment specific routes except under unusual, one-time-only shipment situations (see Section VI.D. of Appendix A to Part 177).

§ 177.825(c) Route Plans. An essential component of the final rule is the route plan prepared by the carrier or its designated representative. This document must be prepared by carriers of large quantity packages in compliance with the preferred routing system established in § 177.825(b). A similar requirement already exists for carriers transporting packages of Class A or Class B explosives. Admittedly, there are a great number of variables to be considered in route planning when one looks at the aggregate of total packages, multiple shipping locations, and widespread destinations. However, for any particular shipment the routing possibilities are somewhat limited by the safety criteria established by DOT and the practical alternatives such as available roadways. Accordingly, DOT does not foresee any severe administrative burdens being required of carriers beyond their capacity to perform, nor does it expect that carriers will be indiscriminate in their selection of routes. Certainly DOT recognizes the interest of shippers in routing decisions and expects that they will be very influential in the final selection. However, carriers remain the party with ultimate responsibility for compliance with § 177.825(c) and they are cautioned to carefully evaluate any route plan submitted for their adoption by other parties.

The proposal to require the preparation and filing of route plans for large quantity radioactive materials packages drew a considerable amount of public comment. For the most part, persons who would be the beneficiaries of the information contained therein supported the proposed requirement while shippers who would be responsible for the administrative filing of the route plans seriously questioned the need for this information by DOT or any other unit of government. The objections can be synopsized as follows: that the States have not expressed any interest in such data; that DOT seems to want the data only for purposes of passing it on to the States; and that the filing of such information could lead to serious problems related to proprietary information as well as security.

In answer to these comments, DOT fails to agree that the States are not interested in the data which can be extracted from written route plans, that proprietary information would not be protected, or that the potential for sabotage would increase by any noticeable degree. Quite to the contrary, DOT is of the opinion that the States and other units of government extending all the way to cities and towns have expressed a very affirmative desire to share completely in the accessibility of detailed information contained in the route plan. Their motives in obtaining such data appear to be in fulfilling their role in compliance and emergency response preparedness activities related to protection of the local public health and safety. Many of these jurisdictions suggest a requirement for duplicative filing of route plans with all interested units of government. Such a burdensome filing requirement has not been adopted, and DOT believes it can meet the needs of local government through its periodic reports and answers to specific inquiries regarding any of the reportable information.

With respect to proprietary information and security, information which DOT can be expected to release on these shipments will deal with statistical accounting of package contents, routes used, identification of origins and destinations and the like. Effectively this is no more information than is currently available to those who wish to monitor the shipping activities of the relatively few facilities at which large quantity radioactive materials packages are handled. Also, any information for which confidential treatment is requested and justified may be protected from disclosure under 49 CFR 107.5. DOT remains firm in its belief that the requirements for preparation and filing of written route plans are reasonable and necessary.

There was an almost unanimous call from State and local officials, as well as interested persons, seeking information contained in the route plan prior to the actual transfer of the radioactive materials. These requests will be satisfied in part by the previously mentioned NRC rulemaking which will necessitate the prenotification of any interested State in which spent fuel or a Type B waste shipment is to be transported.

Other commenters interested in the specific form and substance of DOT's reports to the States requested clarification and updating as to how this information will be provided and to what agency. The agencies of primary interest in these reports is expected to be those organizations in the various States which are empowered to designate preferred highways. Consequently, they will be the principal addressees. In addition, copies will be furnished to the Office of the Governor of each State, to the tribal governments, and to the extent possible any other organization or interested party specifically identified by any of the aforementioned. All other persons would be free to inspect these reports in the Offices of MTB, or may acquire copies of them.

§ 177.825(d) Driver training. DOT has added one provision to the route plan requirement that requires the carrier to submit a supplement to an original route plan when the carrier is forced to deviate from the route plan for emergency or other reasons. The supplement must be submitted to the shipper within 30 days following the deviation and must document the reason for the deviation and the route actually used. This supplement is required when the carrier must leave the preferred route temporarily even in cases to access rest fuel or vehicle repair stops unless the facility used is actually located along the preferred route.

Requirements pertaining to driver training and certification are incorporated in these final rules with only minor changes from that proposed. These requirements are redesignated as § 177.825(d) (see § 177.825(b)(3) and § 177.825(C) in the NPRM). The large majority of commenters favored some type of driver training requirements for operators of vehicles carrying large quantity radioactive materials. Most of the criticism of the driving training requirements involves the extent of training to be required and the method of ensuring that adequate training is provided.

Many commenters maintain that training should not be left to the discretion of the carrier and that the carrier training program should be inspected and certified by DOT. Others commented that the proposed training was not specific enough. Some commenters also expressed their belief that DOT should be responsible for establishing the entire training program, addressing the minimum number of hours required and details on the actual content of training materials.

On the other hand, some shippers and carriers criticized the proposed training requirements as unnecessary and, in some cases, duplicative of existing training requirements. Also, it was maintained that truck drivers should not be expected to become experts on hazardous materials regulations or on the properties and hazards of radioactive materials. There was a feeling that the additional cost of providing driver training just for the transportation of one particular type of hazardous material could result in the loss of some transportation service for these materials.

In response to these criticisms it should first be mentioned that the driver training requirements are based on similar proposed requirements for drivers transporting another hazardous material. Docket HM-115 (44 FR 12826, March 8, 1979) proposed training for drivers of certain tank trucks carrying flammable cryogenic liquids. The Department's intention in Docket HM-164 has been to develop an effective driver training program that is consistent with that for a cryogenic liquids and possibly for other types of hazardous materials in the future.

The current DOT stance on hazardous materials driver training, as established by HM-115, is to require that training be provided for the material involved and that the training program be implemented within the general guidelines provided by the Department. Any driver training requirement must be able to accommodate the many variables involved in hazardous materials transportation such as: the different materials and different associated hazards; the varying level of knowledge and experience of truck drivers; and the wide difference in the effectiveness of various methods of training. For this reason, it is believed that the driver training requirement must be of a general nature and that it is the Department's role to set out the major requirements which allow the flexibility to develop an individualized training program that will accomplish the safety objectives desired. It is not believed that DOT certification of the individual driver training program is needed at this time. Compliance with the driver training requirement for large quantity radioactive materials, just as for any other hazardous materials requirement under the HMTA, will be the subject of safety inspections conducted by MTB, BMCS and various State enforcement personnel. The enforcement of the driver training program will also be aided by the requirement that the driver be furnished with a certificate stating that such training has been provided.

In response to comments from carriers and shippers, DOT believes that driver training for radioactive materials is necessary as a reasonable precaution for large quantity shipments and that truck drivers would not have to become regulation specialists in order to comply with the training objectives. Further, costs necessary to establish a training program should not be high or result in scarcity of service. To some extent, DOT agrees with the contention that some of the proposed requirements duplicate existing training requirements in the Hazardous Materials Regulations. This is true of the requirements proposed in the NPRM (§ 177.825(c)(1) (ii) and (iv)) relating to the motor carrier safety regulations and the operating and handling characteristics of the vehicle. Existing §§ 177.804 and 397.1 now require that drivers be familiar with motor carrier safety regulations, including those in Part 397 for hazardous materials. Minimum requirements for all truck drivers, including provisions relating to the operations of the motor vehicle, are addressed in Part 391 "Qualifications of Drivers". Consequently, the proposed training requirements relating to these areas have not been included in the final amendments.

§ 177.825(e) Physical security requirements for spent fuel. Paragraph (e) is added to this new section to incorporate the requirements proposed in § 177.825(b)(4). The effect of this paragraph is to require motor carriers to transport shipments of irradiated reactor fuel in compliance with a physical protection plan established by the shipper. These plans, approved by DOT or NRC, may sometimes involve transportation requirements different from those specified in § 177.825 but designed to assure at least an equal measure of protection to public safety, and take precedence over the other rules published in § 177.825. Shipments affected by this paragraph include those made by an NRC licensee, and consignments from the DOD and DOE transported by for-hire carriers (except defense-related shipments accompanied by personnel specifically designated by or under the authority of those agencies to preserve national security). A number of commenters expressed their disapproval with the provisions of this regulation which effectively designates NRC as the lead agency for matters involving transportation security for spent fuel.

While the responsibility for prescribing physical protection requirements applicable to special nuclear materials and highly irradiated spent fuel offered for transportation by NRC licensees has been relegated to the NRC, through the memorandum of understanding (MOU) currently in effect between DOT and NRC (44 FR 38690, July 2, 1979), DOT believes that the Hazardous Materials Regulations should contain a specific rule which requires those shippers not otherwise licensed by NRC to comply with safeguards designed to ensure physical security of spent fuel.

DOT recognizes that a considerable amount of "for hire" transportation of spent fuel is performed under security arrangements in support of operations conducted by the DOD and DOE. In the case of shipments escorted by personnel specifically designated by or under the authority of those agencies, for the purpose of national security, a broad exception is granted in §§ 173.7(b) and 177.806(b) which frees common and contract carriers from compliance with the Hazardous Materials Regulations.

This exception was issued with the understanding that it could be revised at some subsequent date if time and experience demonstrated the need. In the more than 30 years that this exception has been in force the DOT is not aware of any instance where the public health and safety have been jeopardized because of shipper or carrier noncompliance with the specific requirements of the Hazardous Materials Regulations. The DOT, therefore, is not inclined to remove the exception at this time since the original conditions of issuance still remain.

The proposal that DOT more closely regulate packages of large quantity radioactive materials shipped by or under the direction of the DOD and DOE attracted a great deal of interest and comment. Some commenters were surprised to learn that, in addition to the exceptions for national security in §§ 173.7(b) and 177.806(b), shipments transported by the military and other government agencies, using their own personnel and transport vehicles, are not subject to the Hazardous Materials Regulations and urged the inclusion of such agencies as regulated carriers. Others followed this topic by indicating that military shipments can and do have accidents, and could pose a grave threat to the communities through which they travel.

The question of DOT jurisdiction and authority over such governmental transportation activities was most recently discussed by MTB in its Docket HM-145A, Notice No. 78-6 (43 FR 22626, May 25, 1978), Transportation of Hazardous Waste Materials. In that document DOT restated its determination not to exercise its authority over Federal, State or local government agencies that carry hazardous materials as a part of a governmental function, using government employees and vehicles. The Department believes that such transportation continues to be conducted in a responsible manner. Also, no new information has come to the attention of DOT regarding the actual occurrence of serious incidents involving hazardous materials transported by this class of carriers. Therefore, it is the opinion of DOT that an extension of its regulations to the degree sought by these commenters is unnecessary at this time.

A matter closely related to the above involves shipments made by governmental agencies through common or contract carriers without escorts provided by such agencies. Essentially, these shipments must be in general compliance with DOT's requirements for safe transportation. Certain exceptions, however, do permit DOD and the Bureau of Alcohol, Tobacco and Firearms (ATF) to make shipments of hazardous materials in packagings not otherwise prescribed by the regulations. Of particular concern to the matter at hand is the treatment of physical security controls applicable to unescorted shipments of spent fuel made by or on behalf of the DOD or DOE. Highly irradiated spent fuel elements pose identical biological and radiological risks regardless of their origins; be it the reactor vessel of an electrical power plant, nuclear submarine or research facility. Other factors also remain relatively constant. For instance, highways retain their same characteristics regardless of who uses them, spent fuel casks are of the same basic designs, and in many cases it is quite conceivable that the carrier, vehicle and driver used to transport shipments for an NRC licensee one week would subsequently be employed by a DOD or DOE contractor to perform a similar service. The same conclusions that justify requiring a licensee to provide physical protection in compliance with a plan established under regulations prescribed by the NRC apply to others who ship spent fuel.

Consequently, the final rule is adopted in the same form as proposed, thereby requiring the respective departments (unless they perform the transportation with their own vehicles) to either submit copies of their physical protection plans to MTB for approval, or, when necessary to preserve the national security, provide an escort of personnel specifically designated by or under their authority. Shipments of irradiated reactor fuel by DOE in support of its research and development activities are not generally considered by DOT to be carried out to preserve national security (as opposed to defense-related shipments made by both DOD and DOE) and are therefore subject to this Department's regulations.

A number of commenters criticized the exception for physical security of spent fuel shipments and some even expressed their belief that it merely allows spent fuel to be shipped under the cloak of secrecy and security thereby avoiding DOT safety rules. It is difficult for DOT to follow the logic of this contention when one considers that the NRC security rules are much more stringent than the DOT safety rules proposed for large quantity radioactive materials. (The NRC physical security program is discussed on page 7144 of the HM-164 NPRM). Nevertheless, this may be a moot point in the near future. DOT has been notified by NRC that NRC licensees shipping spent fuel may be required to follow DOT's preferred routing system, including the use of State-designated routes. The licensees would be relieved of the requirement to obtain prior route approval from the NRC as long as they use preferred routes. In addition, the licensees would have to continue to adhere to all other requirements in the NRC security program including continuous monitoring of shipment, communication with local law enforcement agencies, vehicle immobilization features, escorts, and prenotification to both the NRC and possibly to State governors.

Part 177 Appendix. A new appendix is added to Part 177. It sets out DOT policy and advice on how State and local governments can exercise their own authority over motor carriers in a manner that will be consistent with rules in Part 177 concerning radioactive materials. Sections I and II are introductory. Sections III, IV and V discuss the three categories of radioactive materials shipments, previously addressed in the preamble, which depend on whether or not the motor vehicle transporting the material is required by Part 172 to be placarded, and if so, whether the material is a large quantity radioactive material. Section VI concerns radioactive materials generally.

Sections I and II—Section I states the purpose of the appendix. Section II defines "routing rule" for purposes of the appendix. Emergency action by State or local authorities to deal with immediate threats to public health and safety, as where a highway is impassable, is not a routing rule. Also, the definition excludes rules of the road that apply to vehicles without regard to the hazardous nature of their cargo. "Routing rule" does refer to governmental action that so affects or burdens commerce as to selectively redirect hazardous materials traffic.

Section III—discusses State and local rules that affect motor vehicles transporting large quantity radioactive materials.

State rules. A State cannot make transportation between two points impossible by highway. The radiological risks in transporting large quantity radioactive materials by highway are small and total preclusion of shipments cannot be justified on that basis. A prohibition on use of Interstate System highways is justified only where an equivalent alternate route is specified that offers risk minimization at least equal to the forbidden Interstate segment. Because of their average accident rate and usual design features, Part 177 requires use of Interstate System highways unless a safer route is designated by an appropriate State routing agency after consulting with local jurisdictions and evaluating the actual routes involved.

The fact that a route may be designated for use on a temporary basis for a limited time does not invalidate a demonstrated safety benefit and is encouraged. For example, if justified by safety analysis, a State agency can designate alternate routes in support of time-of-day restrictions in congested areas. A State agency might specify a safer route to be used instead of an Interstate System highway segment or instead of a State-designated route during periods of peak local traffic.

Criteria in Section III.A.2. of the appendix describe necessary features of preferred highways designated by States. One criterion is that preferred routes are designated by a State agency with authority under State law to impose its routing rules anywhere in the State. The rules must be similarly enforceable by State authority anywhere in the State although not necessarily by the same agency. One State agency, for example, could impose routing rules that are enforced by the State police. The State agency must be able to exercise this authority on all public roads in the State regardless of the boundaries of local jurisdictions such as cities and counties, or special authorities such as operate toll roads. This broad authority is necessary for two reasons. First, neither the appendix nor Part 177 delegates regulatory authority over motor carriers. State law must provide that basic regulatory authority. Second, the State agency must be able to consider *any* public highway in its route selection process with knowledge that the lowest risk route may be selected and its use enforced.

A local jurisdiction is not likely to consider all the routing options that affect it and is not normally responsible for considering the impacts of its own rules on other jurisdictions. Similar problems can occur at the State level. The total number of State agencies concerned with transportation of radioactive materials, however, is considerably more limited than the number of local jurisdictions that conceivably might exercise routing authority, a factor which reduces the potential for confusion and enhances compliance.

A closely related criterion in Section III.A.2. specifies that route selection by a State agency be preceded by consultation with affected jurisdictions. Impacts of routing decisions must be considered regardless of the jurisdiction in which they may occur. Affected jurisdictions will include such entities as cities and counties, and may also include neighboring States. Where neighboring States are affected, the impacted local jurisdictions there must be consulted, preferably through a similar State-wide agency. Local jurisdictions know local conditions that affect, or may be affected by traffic in hazardous materials. Without consideration of local views on such matters as accident rates, risk minimization efforts are hampered.

The criterion does not specify the consultation process, although some local governments in commenting asked that the process be spelled out. DOT believes that the rulemaking process used, like the basic rulemaking authority of a State agency, is largely a matter of State law. To ensure reliable results, however, it would be appropriate to provide public notice, opportunity to comment, and a hearing if justified (as in informal Federal rulemaking) and to individually notify and request comments from those local jurisdictions which can be identified as likely to be affected by the routing decisions under consideration.

DOT also believes that each State should establish an advisory group composed largely of city and county officials. The purpose of the group would be to meet periodically, recommend to the State appropriate methods for consulting with local jurisdictions, and review the effectiveness of those measures in actual practice. Such State advisory groups would provide a valuable oversight function that should help to continually improve the State routing program. DOT views adequate, substantive local consultation as essential to State route designations. State routing rules that are not preceded by adequate local consultation are unreliable and inconsistent with the Part 177 amendments established in this docket. A failure in local consultation will jeopardize the enforceability of State route designations for large quantity carriers.

Another criterion in Section III.A.2. specifies that the State designation is preceded by a comparative risk analysis of possible routes. A comparative analysis is essential to ensure that risk is indeed being minimized. The "DOT Guidelines" provide a basic analytical technique that may be used to minimize radiological risk. A more sensitive analysis based on that technique also is acceptable.

Local rules. Local governments may regulate the routes of carriers of large quantity radioactive materials, but only in support of State and Federally designated preferred highways. Local prohibition of motor vehicles transporting large quantity radioactive materials is consistent with Part 177 only so far as the vehicles' presence is forbidden by Part 177 (or by State route designations consistent with Part 177). On the other hand, Part 177 presumes that no local routing rules will apply to motor vehicles on preferred highways where Federal and State route designation is exclusive, or to vehicles at locations off the preferred highways under circumstances permitted by Part 177 (*e.g.* a fuel or repair stop).

Sections IV and V—Section IV concerns rules that apply to placarded motor vehicles which do not contain a large quantity of radioactive material. Section V concerns rules that apply to unplacarded vehicles.

A State or local routing rule that attempts to regulate placarded vehicles not transporting large quantity radioactive materials is consistent with Part 177 only if the rule is identical to § 177.825(a). The language of that section is by necessity general. Since uniform application is intended (in part to aid compliance), § 177.825(a) should not be subject to interpretations that vary between jurisdictions. Local variations in the language of § 177.825(a) would invite varying interpretation and application of the rule. Section 177.825(a) is intended to be nationally uniform. More stringent regulation of placarded motor vehicles is not necessary given the hazard level involved, and will impose unnecessary burdens on commerce that do not provide a reasonable safety benefit.

A State or local routing rule that attempts to regulate radioactive materials that are permitted by Part 177 to be carried in an unplacarded vehicle is not consistent with that part. Such rules are unnecessary, given the very limited hazard involved.

Section VI—Section VI concerns a variety of other State and local rules that are associated with routing rules.

State and local rules cannot conflict with physical security requirements imposed by the NRC. Part 177 permits a carrier to vary from its requirements if necessary to comply with the NRC physical security program. By making NRC physical security rules enforceable under the HMTA, DOT intends that State and local rules also permit necessary variances.

State or local rules that require special personnel, equipment or escort are not consistent with Part 177. Precautions of this nature are taken under NRC rules to ensure the physical security of spent fuel shipments, with which local or State rules may conflict. Their imposition for transportation safety alone serves little purpose and poses serious difficulties for carriers. The existence of State and local requirements for special equipment may effectively dictate the continuous use of the equipment in all jurisdictions. Varying requirements between jurisdictions pose additional problems that may necessitate equipment changes and delays *en route,* or avoidance of an otherwise desirable route. Containment and packaging equipment are themselves exclusively set by Federal regulations. Special personnel and escort requirements pose similar problems. State and local escort requirements in particular are a source of delay in transportation if the escort is not required for the entire journey. Whether an escort vehicle is provided by the carrier or by a local jurisdiction, if presence of an escort vehicle is a condition of entering the jurisdiction, the transport vehicle is likely to have to stop at jurisdictional boundaries to establish communication with the escort vehicle. It also is likely that delay will result from the early arrival of a transport vehicle or the late arrival of an escort vehicle.

Earlier in this document, DOT stated its intention to further consider requirements for prenotification to State governments of large quantity shipments, following completion of an NRC rulemaking on prenotification for shipments of nuclear waste. Because the voluntary provision of escort vehicles by local governments is closely related to prenotification issues, such voluntary escort services will also be reconsidered at that time.

Shipping paper entries and other hazard warning devices bear little special relationship to local safety problems. In fact, the utility of such measures heavily depends on their universal recognition. Variations in hazard warning devices dilute the effectiveness of those required by Parts 172 and 177, which are understood nationally and internationally, and may hamper emergency response.

State and local requirements for filing route plans or other documents containing shipment specific information pose a potential for unnecessarily delaying motor vehicles. In many cases such requirements are redundant with Federal requirements concerning safety and security. They are not likely to measurably enhance local emergency response capabilities. When applied at a State or local level, they are likely to result in an inefficient use of emergency preparedness resources.

Accident reports imposed by State or local governments that are necessary to ensure immediate emergency assistance are consistent with Part 177. Accident reports required at a later time are duplicative of requirements of Part 177 (which references §§ 171.15 and 171.16). Reports submitted to DOT are publicly available, and States may make prior arrangements for DOT to provide them with copies of incident reports as they become available. The appendix does not concern general accident report requirements, such as a State requirement that any motor vehicle accident involving injury or substantial property damage be reported to the State police during a stated period following the accident.

Prenotification was discussed previously in the preamble. Prenotification requirements by State and local governments, if found necessary, will be established in a nationally uniform manner. Unless DOT reaches and acts on a conclusion that prenotification rules are necessary, beyond those Congress has directed NRC to impose on certain radioactive wastes, independent State and local prenotification requirements are not consistent with Part 177.

Lastly, because of the importance of expediting radioactive materials shipments, due to the risk and added normal dose attendant to delay, other forms of State and local regulation that affect motor carriers of radioactive materials should not result in unnecessary delay (see § 177.853(a)). A delay is unnecessary unless it is required by an exercise of State or local regulatory authority over a motor vehicle that so clearly supports public health and safety as to justify the safety detriment and burden on commerce caused by the delay (such as in an emergency).

§ 397.9 Routing for hazardous materials. The Bureau of Motor Carrier Safety is revising 49 CFR 397.9 of the Federal Motor Carrier Safety Regulations in amendments published elsewhere in this **Federal Register** issue. This will direct the motor carrier's attention to the new routing requirements for radioactive materials in § 177.825. The amendment is needed to prevent an inconsistency between routing provisions required for radioactive materials in § 177.825 and those required for other hazardous materials in § 397.9(a).

VIII. ENVIRONMENTAL AND ECONOMIC IMPACT

DOT has prepared a Final Regulatory Evaluation and Environmental Assessment (DOT Assessment) in support of these final rules (copies may be obtained from the Dockets Branch previously cited). It is clear from the available technical information referenced in the Assessment that radiological risks in transporting radioactive materials resulting from both normal exposure and accidents are very low. Even if one allows that the risk estimates developed by these technical risk studies are underestimated by an order-of-magnitude, the projected overall risks from the transportation of radioactive materials would still be extremely low. Furthermore, one cannot ignore historical accident experience which is shown to be quite good for radioactive material transportation when compared to other hazard classes by MTB's incident reporting system (1971–1980). Although historical experience by itself may not necessarily be the best method of projecting future events, the low historical accident rates do tend to support the research conclusions that the risks in transporting nuclear material by highway are low.

The primary operational effect of these rules is to require or encourage use of the Interstate System by carriers of radioactive materials. Although carriers transporting packages containing a large quantity of radioactive materials are generally required to use either the Interstate System or State-designated preferred highways, carriers transporting packages containing lesser quantities are likely also to tend to use the Interstate and preferred highways especially in areas of heavy population.

Adoption of the preferred routing system which utilizes Interstate System highways and State-designated highways for large quantity radioactive materials is determined to be the appropriate course of action for routing. This alternative has the potential for the greatest safety impact and is feasible and cost effective considering the marginal safety benefits involved.

DOT agrees that "high consequence" accidents in densely population urban areas should be of great concern, but not to the extent that public policy on hazardous material routing should be formulated solely on the basis of avoiding such "worst case" accidents. For example, the high consequence estimates of the 1980 Urban Environs Study referenced earlier may be reduced substantially by avoiding the city but overall public risk may actually increase if the carrier is forced to use poor, secondary and circuitous rural roads. Nevertheless, it is clearly a reasonable precaution to minimize the possibility of "worst case" accidents by requiring use of a circumferential Interstate highway if it is available. If one is not available, a State may conduct a routing analysis to examine availability of other routes for comparison with the Interstate through route.

Overall radiological effects of this rule include a reduction in total latent cancer fatalities attributable to normal dose and a lesser reduction in the annual latent cancer fatality accident risk (based on NUREG 0170 projections). Some additional reduction in radiological consequences should result from State designation of preferred highways. A slight increase in nonradiological consequences may result from routing on preferred urban bypasses or circumferentials. Overall, environmental impacts should be negligible.

Economic costs are expected not to exceed $330,000 annually under 1985 levels of shipping activity and mostly would consist of costs for driver training and route plan preparation and filing. Some additional cost may result from the new placard background and shipping paper requirements for large quantity shipments. Also, this estimate does not include possible additional insurance costs to State and local bridge and tunnel authorities. MTB requested, but did not receive, any quantitative data on this subject.

Because of the level of costs anticipated and the limited potential for environmental impact, MTB does not consider the preparation of an environmental impact statement or a regulatory analysis necessary for these amendments. As mentioned, a more detailed examination of costs and environmental impact is available in the Final Regulatory Evaluation and Environmental Assessment.

DOT intends to conduct an evaluation of the final rule a year following its effective date. This evaluation will consider the rule's efficacy as regards public health and safety, and its actual effects on carriers, and State and local jurisdictions, with particular attention to any difficulties that have appeared during the rule's implementation. As part of that evaluation, notice will be published to solicit public comment, and a direct solicitation of comments will be made to the States and to interested groups such as the National Governors Association and the National League of Cities. As previously indicated, DOT also will be reexamining prenotification as well as its relation to escort vehicles voluntarily provided by State or local governments sometime following publication (in early 1981) of final NRC prenotification rules for nuclear waste shipments. An advance schedule for both preceedings will appear in a future **Federal Register** publication of the DOT Regulations Agenda.

Work on the DOT Guidelines, referenced herein, will be continued and the document is expected to be released in the first half of 1981, following pilot tests early in the year.

(**46 F.R. 5298, January 19, 1981**)

49 CFR Part 172
[Docket No. HM-166F; Advance Notice]
Limited Quantities of Radioactive Materials

I. BACKGROUND OF REGULATIONS

Ever since the general consolidation of the Hazardous Materials Regulations (HMR) was accomplished under Docket HM-103/112 (41 FR 14972, April 15, 1976), an inconsistency has existed between the regulations applicable to aircraft and those applicable to the other modes insofar as they pertain to limited quantities of radioactive materials and radioactive devices. As that consolidation was a major revision of the HMR wherein the requirements for carriers by aircraft and vessel were included in the comprehensive set of regulations already applicable to carriers by rail and public highway, it was not possible to resolve all of the philosophical differences represented by the various modes. For the sake of expediency, it was determined that the general implementation of the consolidated HMR should not be unnecessarily delayed by varying requirements which reflect the legitimate differences professed by the modal administrations regarding an acceptable level of safety. A thoughtful consideration of issues such as the topic at hand was judged to be a more prudent course and is expected to result in regulations that assure a proper degree of protection for public health and safety without unduly burdening shippers or carriers.

At the present time all packages containing limited quantities of radioactive materials or radioactive devices transported by any mode are excepted from specification packaging, marking, and labeling, and are further excepted from the provisions of § 173.393 pertaining to general packaging and shipping requirements applicable to other radioactive materials. These exceptions are consistent with those provided for limited quantities of hazardous materials belonging to most other hazard classes based upon the limited consequences that could be expected when they are involved in incidents. Hazard classes which do not provide exceptions based upon a limited quantity include those belonging to the explosives group and poison A materials. While the exception from package marking does not apply to most other hazard classes, it should be noted that in the case of dispersible radioactive materials the outside of the inner container must bear the marking "Radioactive."

Analysis of the limited quantity exceptions for radioactive materials as they apply to each of the modes reveals the following differences in regulatory control:

Rail: (a) A carrier may not accept for transportation a package containing a limited quantity of radioactive materials unless it has received a properly certified shipping paper (see § 174.24).

(b) A detailed hazardous materials incident report must be filed with MTB in the event of an unintentional release or other reportable circumstance (see §§ 171.15 and 171.16) and any contamination resulting from a release must be cleaned up (§ 174.750(a)).

Air: (a) Excepted from *all* requirements of the HMR, including shipping paper provisions and hazardous materials incident reports (see § 175.10(a)(6)).

Water: (a) Excepted from *none* of the requirements of Part 176, and therefore must have proper shipping papers (see § 176.24).

(b) A detailed hazardous materials incident report must be filed with MTB in the event of an unintentional release or other reportable circumstance (see §§ 171.15 and 171.16) and any contamination resulting from a release must be cleaned up (§ 176.710).

Highway: (a) A carrier may not transport a package containing a limited quantity of radioactive materials unless it is accompanied by a properly prepared shipping paper (see § 177.817).

(b) A detailed hazardous materials incident report must be filed with MTB in the event of an unintentional release or other reportable circumstance (see §§ 171.15 and 171.16) and any resulting contamination must be cleaned up (§ 177.861).

It can be seen that the span of control over these materials ranges all the way from being practically negligible when transported by aircraft to very extensive when transported by vessel. MTB believes that the inherent risks associated with the transportation of these materials by each mode are not sufficiently different to justify this disparity. Consequently, this ANPRM seeks public comment from shippers, carriers, emergency response personnel and other interested persons in helping to resolve these differences, or otherwise support their continued existence based upon a technical review of the regulations with consideration given to the nature, form and quantities of radioactive materials involved.

II. CURRENT REGULATORY ACTIVITIES

In Docket HM-169, Notice No. 79-1 (44 FR 1852, January 8, 1979) the MTB proposed a general revision of the HMR as they apply to radioactive materials to make them more compatible with international standards. Although a considerable amount of comment was received with respect to limited quantities, most of it addressed specific requirements such as the proposed elimination of the marking exception. Other commenters suggested that the all encompassing exception applicable to aircraft should be extended to the other modes. Although useful, the information in that Docket does not provide MTB with a complete set of data for use in making a thorough safety analysis for these materials by all modes.

On November 23, 1979, MTB published a notice of receipt of an application for exemption—8300-N (44 FR 67267). In this application United Parcel Service is seeking an exemption from the requirements for shipping papers when limited quantities of radioactive materials are to be transported by rail or over the public highways. Once again the MTB received public comment urging favorable action in this area but still it appears that even with the addition of these comments, and the data provided therein, the Bureau is not sufficiently informed to resolve the broader issues addressed in this inquiry.

In the area of international transportation regulations, the MTB is aware of current proposals to the "Technical Instructions for the Safe Transport of Dangerous Goods by Air" in which the International Civil Aviation Organization (ICAO) would treat limited quantities of radioactive materials as essentially unregulated commodities. To qualify for this exception the radioactive materials would have to meet a definition of limited quantity equivalent to one of those proposed in Docket HM-169, be packaged in accordance with general requirements applicable to all radioactive materials, and except for articles manufactured from natural or depleted uranium or natural thorium and empty packages, contain the marking 'Radioactive' so that it is visible upon opening the package. These materials could then be offered for transportation without an accompanying detailed shipping paper. Instead the shipper need only indicate the presence of these hazardous materials by entering a specified phrase—for example, "excepted radioactive material"—on whatever shipping document accompanies the shipment. These proposals seem to evolve from present operating practices long since adopted by international air carriers with apparently no adverse impact on health or safety. To the extent that air carriers and certain international officials believe the public health and safety are adequately protected by these procedures considering the very small quantities of radioactive material involved, the MTB believes that it is worth investigating their applicability to other modes as well.

One of the functions a shipping paper provides is to make detailed information available to emergency response or cleanup personnel responding to an accident. In this regard, it has been alleged by some shippers and carriers that the information is not imperative due to the very small quantities of radioactive materials that may be shipped this way. Additionally, this detailed information is available or can be obtained from the consignor and the need to provide this information on the shipping papers has been questioned. Consequently, the MTB is interested in determining if the detailed description required by § 172.203 is *necessary* for adequate response to accidents, considering the limited hazard of these materials and other methods which are available for obtaining this information in a timely manner.

Another area for consideration is the marking requirements for these materials. There is an important interface between shipping papers and marking as they relate to:

(a) recognition that a hazardous material is being shipped;

(b) identification of the material being shipped;

(c) proper handling and stowage of the materials involved; and

(d) appropriate action in the event of an accident.

Therefore, the MTB is also seeking comments of how the marking requirements may need to be modified if the shipping paper requirements are changed.

III. REQUEST FOR COMMENT

Comment is solicited on the preceding discussion and on the following questions. Do the requirements presently contained in the HMR, applicable to the transportation of limited quantities of radioactive materials, provide an appropriate degree of regulation to adequately protect the public health and safety?

(a) If so—

(1) How do the transportation conditions of the various modes differ to justify diversity of regulatory control?

(2) Can the exception from package marking requirements be supported to show that protection of the public health is not being jeopardized?

(3) In the case of intermodal transfers, do the more restrictive regulations impose an unwarranted economic burden without providing a commensurate increase in safety?

(4) Does the lack of an incident reporting requirement for limited quantities of radioactive materials transported by aircraft significantly diminish the effectiveness of the DOT's accident analysis system?

(b) If not—

(1) How should the regulations be revised?

(2) Do the hazards associated with all limited quantity radioactive materials and devices pose such a low risk that the MTB can remain confident in this exception, or should certain radionuclides, forms, etc. be excluded from limited quantity exceptions?

(3) What would be the approximate cost/benefit of any suggested change?

(4) Will there be an adverse impact on emergency response activities if detailed shipping paper requirements are waived for rail, water, and highway shipments of limited quantities?

(5) Do the marking requirements need to be modified if the detailed shipping paper requirement is waived for rail, water and highway shipments? If so, how?

(45 F.R. 80843, December 8, 1980)

49 CFR Parts 172, 173, and 175
[Docket No. HM-166F; Notice No. 81–8]
Transportation of Limited Quantities of Radioactive Materials and Devices

On December 8, 1980, MTB published an advance notice in the **Federal Register** (45 FR 80843) calling for comments on the need for, or possible elimination of, certain regulatory requirements applicable to the transportation of radioactive materials and radioactive devices in limited quantities. That notice identified the glaring inconsistency which has existed between shipments transported by air versus those transported by any of the surface modes, ever since the Hazardous Materials Regulations (HMR) were consolidated in 1976. Rules proposed in this notice are based upon (1) public comments received in response to the previously cited publication, (2) an assessment of risks inherent in the transportation of these radioactive materials, (3) consideration of risks inherent in each of the modes, (4) an evaluation of hazardous materials incidents reported since 1971, and (5) a comparative analysis of radioactive materials and materials belonging to other hazard classes, with respect to the relaxed requirement for transportation of small quantity packages and the favorable safety records generally achieved by each class. A discussion of pertinent issues and comments received in response to the advance notice follows.

I. ADEQUACY AND SUITABILITY OF CURRENT REGULATIONS

A review of comments generally confirms MTB's own assessment that requirements for the most frequently used modes (highway and air) are not consistent, and may be excessive for surface modes and too relaxed for air transportation. In its comments, however, 3M Static Control Systems expressed the opinion that requirements for transportation by air do assure protection of public health and safety. 3M points out further that the International Atomic Energy Agency's Regulations for the Safe Transport of Radioactive Materials exempt qualifying packages from regulation by all modes in a fashion similar to MTB's exception for air transportation, and it proposes that the HMR be amended to reflect these less restrictive requirements for shipments transported by highway, rail, and water as well. This opinion and recommendation was shared by several other commenters and supported by a claim from Hoffman-La Roche, Inc., that tens of thousands of such packages were safely transported for them in 1980 with only three known incidents occurring. Of these, none involved release of radioactive materials and the internal containers were simply repackaged and returned for disposition.

A. Technical Requirements. Data supplied in comments filed by Miles Laboratories, Inc. and others support MTB's earlier conclusion that activity limits are so low as to present no significant risk to public health. That conclusion holds true not only for the vast majority of packages that contain a small percentage of the maximum permissible activity limit, but for theoretical packages containing one-hundred percent of the authorized activity limit as well, based on generally accepted release fractions and intake rates. Other commenters, like the American College of Radiology, agree that the present regulatory limit of external radiation levels not exceeding 0.5 millirem per hour at the package surface (2.0 millirem per hour for exclusive use shipments) present no radiation danger to persons handling the packages, even if deformed by damage. No commenter to this docket indicated either a need or desire to revise the activity limits applicable to limited quantities of radioactive material, radioactive devices, or packages containing more than one radioactive device. Also, MTB believes that current limits adequately provide for the public safety, regardless of the mode in which the packages are transported.

1. Classification with other hazardous materials. Although not discussed in the advance notice or in comments to the docket, MTB believes that consideration must be given to reordering the precedence of hazards listed in § 173.2 to downgrade limited quantity radioactive materials to a level more appropriate to their actual risk. While actual incidents are not documented, the HMR have been criticized for a "loophole" which some persons contend allows flammable liquids and corrosive liquids containing trace amounts of radioactive material to be transported aboard aircraft as completely unregulated materials. To correct this situation, MTB is proposing that radioactive materials in limited quantities be separated from the major classification and downgraded to a position between "corrosive material (solid)" and "irritating materials."

2. Packaging. Only one commenter responding to the advance notice addressed the subject of container integrity. The Lawrence Berkeley Laboratory, while acknowledging the low risks associated with limited quantity radioactive materials, suggests measures be taken to preclude any incidental leakage of dispersible radioactive material (from the inner container) in the form of a liquid or an alpha-emitting solid. To achieve this, they propose that DOT 2N metal cans of the sealed or friction-lid type when used as the inner container should be able to withstand atmospheric pressure differentials and the dropping or crushing incurred in minor accidents. They further propose that items too large or not practical for limited quantity radioactive material packaging be shipped as low specific activity radioactive materials. These suggestions have not been included in the proposed rule, since MTB is satisfied that general requirements for packages in §§ 173.6 and 173.24 and provisions of § 173.91 already provide an adequate level of regulatory control.

B. Administrative Requirements. Although MTB received widespread agreement on its standards for the more critical elements of transportation safety (i.e. packaging, quantity limits, and external radiation levels), the same cannot be said for those requirements which address the communication of hazard warning information (i.e. shipping papers and package markings). As previously indicated, several commenters believe an acceptable level of safety is being achieved in the air mode, and since air is generally considered to be the most critical mode, they imply that requirements for detailed shipping papers, incident reporting and the like for limited quantity radioactive materials transported by surface modes are superfluous.

1. Shipping papers. It seems worthy to note that every commenter responding to MTB's inquiry, "In the case of intermodal transfers, do the more restrictive regulations impose an unwarranted economic burden without providing a commensurate increase in safety?," noted that hazardous materials shipping papers are a reason for frustration of shipments or impose a significant economic burden. The frustration of shipments reportedly occurs on occasions when motor carriers interline packages to air carriers. Air carrier personnel sometimes become suspicious when they observe hazardous materials shipping papers in the motor vehicle driver's possession and are then asked to accept the packages without similar documentation. The absence of DOT shipping papers, it is explained, often leads to unnecessary delays while pertinent regulations are researched and, as a result, packages of radioactive materials requiring delivery in a timely manner fail to be loaded on scheduled flights.

Conversely, if a shipper seeks to avoid such delays by preparing a hazardous materials shipping paper to accompany packages during air transportation, it may incur additional freight charges attributed to hazardous materials. Miles Laboratories, Inc. cite an example of increased transportation charges amounting to $6.00 per shipment whenever they ship via Federal Express. That surcharge is applied whenever the HMR prevent Federal Express from transporting packages in local pick-up and delivery service which is incidental to its air operations, unless the packages are accompanied by detailed shipping papers. Miles Laboratories, Inc. claims its air transportation costs are increased by at least $27,000.00 per year as an indirect result of MTB's requirement to describe packages of limited quantity radioactive materials in detail on shipping papers, when offered for transportation in the highway mode. Considering that projections for the year 1985 estimate over 800,000 packages of limited quantity radioactive materials and radioactive devices will be transported by air, the aggregate cost imposed through the hazardous materials surcharge is substantial. In fact, it is estimated that savings in excess of $1 million per year may be realized by shippers if carriers follow MTB's lead in the deregulation of these materials.

Although there is an identifiable cost associated with the preparation and distribution of shipping papers, most commenters apparently chose to ignore this incidental and relatively small administrative cost of regulatory compliance, in favor of emphasizing the more direct costs. However, calculated savings of more than 15,000 person hours per year is possible with elimination of the shipping paper requirements, as proposed herein.

Without exception, every comment filed in response to MTB's query on possible adverse impact to emergency response activities, if detailed shipping paper requirements are waived for surface modes, very boldly proclaim that such activities would not suffer, and some in fact suggest that the impact may be positive. The latter conclusion is based upon several commenters' assessments that the risks presented by limited quantity radioactive materials are so low as to not even warrant notification of traditional emergency response personnel such as firefighters, but instead rely on in-house safety personnel to take appropriate measures which further reduce the already low risk. These commenters reason that calling on emergency service personnel, who are trained and equipped to handle acutely hazardous incidents, is a poor utilization of their resources. However, it was not explained how the absence of detailed shipping papers would alleviate any overreaction of this sort.

While acknowledging that detailed shipping papers for packages of limited quantity radioactive materials and radioactive devices provide benefits which only increase public safety by a slight margin, MTB believes that an indication of the presence of a radioactive material must be communicated in some general fashion if a damaged or a stray package is discovered during transportation. Consequently, the proposed rules contain a new provision for qualifying packages under the limited quantity exception which specifies that the shipper must furnish a written notice on or with the package which reads "Radioactive material, limited quantity, n.o.s., UN 2910" *or* "Radioactive device, n.o.s., UN 2911", as appropriate, followed by the statement "This package meets all requirements of 49 CFR 173.391 for limited quantity radioactive materials."

2. Package markings. Unlike the general agreement reached by most commenters on the need for and value of detailed shipping papers, package markings involve a more diverse range of opinions. Those persons in favor of maintaining the status quo point to the long history of safety in transportation for packages of limited quantity radioactive materials and radioactive devices. They contend that the absence of external markings has not resulted in any mishandling of packages to the extent that there was ever a serious threat to public health or the environment and, therefore, question the justification for a new requirement at this time. In addition, they also worry that markings which include the word "radioactive" may in fact delay the otherwise speedy delivery of these packages by unnecessarily alarming carrier personnel.

In their comments, the Lawrence Berkeley Laboratory supports a requirement for marking packages as "Radioactive material, limited quantity, n.o.s." However, their concern does not seem to be so much with the communication of hazard warning information as it is with expediting delivery. MTB agrees that some small benefits in the form of reduced normal dose may be gained by minimizing the period of time these radioactive materials packages spend in transit. It does not, however, believe that such a package marking would have the effect of shortening transit times.

United Parcel Service indicated support for the elimination of detailed shipping paper requirements but suggested instead inclusion of those presently required entries as package markings. In this way they believe sufficient information would be available to properly handle damaged packages. Obviously this approach goes far beyond all other existing requirements for package markings and its relative merits appear dubious while the burden on shippers would be considerable.

In consideration of the above it is the determination of MTB that current marking requirements prescribed in § 173.391(a)(4) provide an adequate level of safety without unduly burdening shippers and that regulation should, therefore, remain unchanged.

3. Incident reporting. One of the principal means available to MTB for assessing the effectiveness of the Hazardous Materials Regulations is the incident reporting system. Information accumulated in that system over the past ten years suggests that limited quantity radioactive materials and radioactive devices are being safely transported. In fact, while it is estimated that several million such packages have been transported during this past decade, MTB has records on only fourteen reported incidents involving these materials. However, since air carriers are presently excepted from reporting requirements when limited quantity radioactive materials are involved in an accident, MTB must acknowledge that it does not have total confidence in its data and resulting conclusions. (Note.—Of the fourteen incident reports discussed above, more than half were submitted by carriers operating in the air mode, even though they are not required to do so.) As no carriers responding to the advance notice chose to address the matter of incident reports, MTB is of the opinion that present requirements applicable to the surface modes are reasonable and necessary and should be extended to include air carriers as well. Appropriate revisions are proposed for §§ 173.391 and 175.10. A conservative estimate of 10 person-hours per year (10 reports at one hour each) is the additional paperwork burden which MTB believes would be imposed if the proposed rule change is adopted. The burden would be shared by approximately 73 of the more than 340 for-hire air carriers now operating in the U.S. MTB solicits specific comments on the accuracy of its estimates for the actual number of additional Hazardous Material Incident Reports which may be required, the time required for their preparation, and the affected population. Carrier estimates of the number of incidents occurring within their system over the past several years and not reported because of the current exception from §§ 171.15 and 171.16 would be most useful.

4. Shipping descriptions. In its formulation of these proposed rules, MTB was once again made aware of problems faced by carriers who transport radiopharmaceuticals. They claim that the proper shipping names "Radioactive material, n.o.s." and "Radioactive material, limited quantity, n.o.s." trigger responses by Federal, State, and local enforcement personnel which are quite often inappropriate to any risks associated with these materials. After confirming that the carrier is not transporting particularly objectionable materials, it is generally allowed to proceed but only after a sometimes lengthy delay. These carriers reason that if the DOT proper shipping name more clearly identified the materials by their intended use, such problems would be greatly reduced without compromising safety. As MTB's experience with materials belonging to the hazard classes flammable liquid, corrosive material, poison B, and others reflects no adverse effects in transportation which can be attributed to their use for, and description as, "Medicines, n.o.s.," it is proposing in § 172.101 to introduce a similar proper shipping name, "Radiopharmaceuticals, n.o.s.", for hazard class radioactive material.

(46 F.R. 61908, December 21, 1981)

49 CFR Parts 171 and 173
[Docket No. HM-166-I; Notice No. 81-2]
Transportation of Liquefied Petroleum Gas In Intrastate Commerce

Since passage of the Hazardous Materials Transportation Act (HMTA) of 1974 (49 USC 1801 et seq.) the MTB has encouraged the adoption of the Hazardous Materials Transportation Regulations 49 CFR Parts 170 to 179) by the States in order to promote uniformity in safety regulation throughout the nation. Certain areas of transportation safety demand a strong, predominant Federal role. In the HMTA's Declaration of Policy and in the Senate Committee language reporting out what became § 112 of the HMTA, Congress indicated a desire for uniform national standards in the field of hazardous materials transportation and, with the HMTA, gave the Department of Transportation the authority to promulgate those standards. Although the HMTA has not totally precluded State or local action in this area, it is the MTB's opinion that, to the extent possible, Congress intended to make such State or local action unnecessary.

It has come to the attention of the MTB that the adoption by individual States of the Hazardous Materials Transportation Regulations has created an anomalous situation in certain States for certain cargo tank owners and operators. DOT regulations require cargo tanks for LPG to be constructed in compliance with either DOT Specification MC-330 or MC-331. However, a number of cargo tanks not subject to DOT regulations (nor ICC regulations prior to 1967) have been constructed and used in intrastate commerce for many years. While they were manufactured in accordance with certain consensus standards and were otherwise qualified for use, they do not meet the standards now required in DOT regulations. The result of a State's adoption and enforcement of DOT regulations is to immediately require that all cargo tanks in that jurisdiction comply with DOT specifications without provision for an adequate transition period.

MTB also has been advised of a difficulty encountered by a carrier based in Nevada. For a number of years, this carrier operated only small cargo tank trucks (commonly referred to as "bobtails") in intrastate commerce. Due to a change in business conditions, it became necessary for the carrier to acquire a cargo tank semitrailer (meeting DOT requirements) for carriage of LPG from California to its base in Nevada. Upon entering interstate operations, all of the carrier's operation, including operation of the small cargo tanks, came under DOT jurisdiction. The MTB believes that appropriate relief should be provided to remedy a situation that may not be uncommon and believes the conditions proposed in this NPRM, in association with allowing use of non DOT specification cargo tanks, assure an adequate level of safety for the transportation of LPG in small cargo tanks during the transition period.

This proposal is limited in its applicability to intrastate commerce, including a cargo tank operated by a motor carrier that may operate other motor vehicles in interstate commerce.

The proposed revision would allow the continued use of a cargo tank for transportation of LP gas that is not marked according to Specification MC-330 or MC-331, provided it (1) is marked and conforms to the edition of the ASME Code in effect when it was manufactured; (2) has a minimum design pressure of 250 psig; (3) has a capacity of 3500 gallons or less; (4) was manufactured prior to January 1, 1981; (5) conforms to NFPA Pamphlet 58; (6) has been inspected and tested in accordance with §173.33 as specified for Specification MC-330 or MC-331; and (7) it is operated in conformance with the regulations except the specification requirements.

The procedure proposed in this NPRM will allow the continued safe use of cargo tanks constructed in conformance with the ASME Code when a State upgrades its regulatory program by adopting the Hazardous Materials Transportation Regulations, as well as allowing continued use of such tanks for local shipments by interstate carriers. MTB has been advised by industry representatives that all new tanks are being manufactured in compliance with DOT specifications; therefore, new construction after January 1, 1981, is not covered by this NPRM.

It is also proposed to update the reference in §171.7(d)(6) to Pamphlet 58 of the National Fire Protection Association since this is the edition currently available from that organization.

The MTB has determined that this proposed regulation will not, if promulgated, have a significant economic impact on a substantial number of small entities.

If this proposed regulation is not adopted, there will be a serious economic hardship on small LPG carriers because their nonspecification cargo tanks will no longer be authorized for transportation of LP gas in several States. New DOT specification tanks would have to be purchased and delivery to LPG customers would be severely disrupted.

(46 F. R. 27146, May 18, 1981)

49 CFR Parts 171 and 173
[Docket No. HM-166I; Amdt. Nos. 171–64, 173–53]
Transportation of Liquefied Petroleum Gas in Intrastate Commerce

On May 18, 1981, the MTB published a notice of proposed rulemaking under Docket No. HM-166I; Notice No. 81–2 (46 FR 27146), which proposed an amendment to allow the continued use of certain nonspecification cargo tanks for the transportation of LPG in intrastate commerce.

Since passage of the Hazardous Materials Transportation Act (HMTA) of 1974 (49 USC 1801 et seq.) the MTB has encouraged the adoption of the Hazardous Materials Transportation Regulations (49 CFR Parts 170 to 179) by the States in order to promote uniformity in safety regulation throughout the nation. However, the adoption of the Department's Hazardous Materials Regulations has created a few problems for some cargo tank owners and operators in certain States. DOT regulations require cargo tanks for LPG to be constructed in compliance with either DOT Specification MC-330 or MC-331. However, a number of cargo tanks now subject to DOT regulations (nor ICC regulations prior to 1967) have been constructed and used in intrastate commerce for many years without incident. While they were manufactured in accordance with certain consensus standards and were otherwise qualified for use, they do not meet the standards now required by DOT regulations. The result of a State's adoption and enforcement of DOT regulations is to immediately require that all cargo tanks in that jurisdiction comply with DOT specifications without provision for an adequate transition period.

The MTB received six comments on Notice No. 81-2. Two commenters stated that they supported the proposed amendments. However, they thought that the DOT should broaden the proposal to include interstate use of the nonspecification tanks. The MTB does not concur in the use of cargo tanks having a design pressure of less than 250 psig in interstate commerce. Three reasons for this denial are (1) it goes far beyond on what was proposed in the Notice; (2) it would be unfair to all of the carriers and owners who have purchased DOT specification equipment, and (3) it would be setting a precedent for the use of nonspecification cargo tanks in interstate commerce which would denigrate the value and validity of a nationally uniform system establishing the level of safety required for cargo tanks used to transport flammable gases such as propane.

Two comments recieved supported the proposed amendments with certain exceptions. First, they recommended that proposed paragraph § 173.315(k) be revised to include reference to the API-ASME Code. One commenter recommended that a new paragraph (i) be added to § 173.315(k) to read "Tanks designed and constructed in accordance with paragraphs U-68 or U-69 of the 1949 and earlier editions of the ASME Code and having a design pressure of 200 psi may be used provided that they comply with the other provisions of 173.315(k) of this subchapter. Such tanks may be rerated at a working pressure 25 percent in excess of the design pressure for which the tank was originally constructed, and if rerated shall be marked as follows: "Re-rated working pressure . . . psi." For purposes of setting safety relief valves and pressure control valves, and for establishing maximum and minimum design pressures, the rerated working pressure shall be considered as the equivalent of the design pressure as defined in these regulations." One commenter questioned use of the words "ASME certificate" in § 173.315(k)(3). One commenter later withdrew his recommendation regarding reference to the API-ASME Code. The MTB does not concur that specific reference be made to the joint API-ASME Code. Inasmuch as the DOT (ICC prior to 1967) specification cargo tanks have referenced only the ASME Code for design, construction, and inspection requirements, it is not considered appropriate to include a reference to the API-ASME Code. Tanks designed and constructed in accordance with the ASME Code which have a design pressure less than 250 psig must be rerated to a working pressure of not less than 250 psig before entering service. The minimum design pressure which DOT is willing to accept for a cargo tank used to transport LPG is 250 psig.

One commenter stated that although the intention of the proposal is to "allow continued use of nonspecification cargo tanks for the transportation of LPG in intrastate commerce," proposed section 173.315(k)(6) goes far beyond the current situation. This commenter further stated that the Department has required, and enforced, the use of DOT specification cargo tanks by interstate carriers regardless of the inter/intra-state nature of the commerce. Since the proposal would permit a deterioration of the present safety situation, it is not believed to be in accordance with Congress' intent regarding "uniform national standards." Finally, this commenter recommended eliminating the words "including its operation by a motor carrier otherwise engaged in interstate commerce" from proposed § 173.315(k)(6) and provide the "grandfather" exception to only intrastate carriers. The MTB does not agree with this commenter because application of the "grandfather" exception only to intrastate carriers would not alleviate the problems faced by a carrier whose status has changed from intrastate carrier to interstate carrier.

The last comment received was from the Hazardous Substances Transportation Board (HSTB) of the Pennsylvania Department of Transportation which concurred in part with the proposed amendments. However, they recommended that documentation be required to be carried on each vehicle to establish the fact that the cargo tank complies with the requirements of § 173.315(k). Reasons cited by the HSTB for such documentation were to facilitate the highway enforcement program of the various States and to prevent unnecessary disruption of service by enforcement officials. While MTB recognizes that a requirement for the documentation recommended by the HSTB would somewhat ease the enforcement burden on both State and Federal enforcement personnel, this benefit is outweighed by the recordkeeping burden placed on the motor carrier to maintain a copy of these documents in each vehicle at all times. It is the policy of the Federal government to reduce, not add to the paperwork burdens to the regulated community. However, if the carrier elects to carry this documentation with the vehicle, it may facilitate inspection and prevent delays by enforcement officials.

Except for minor editorial changes in § 173.315(k)(1), (3), (k)(6), and a new (k)(7) no other changes have been made to Notice No. 81-2.

The MTB has determined that this regulation is consistent with Section 2 of Executive Order 12291, and is a non-major rule under the terms of that Order. Pursuant to the Regulatory Flexibility Act, this rule will not result in a significant economic impact on a substantial number of small entities because its effect is to eliminate a burdensome restriction on certain carriers of LPG.

(47 F. R. 7242, February 18, 1982)

49 CFR Part 173
[Docket No. HM-166K, Notice No. 81-7]
Transportation of Anhydrous Ammonia in Intrastate Commerce

Since passage of the Hazardous Materials Transportation Act (HMTA) of 1974 (49 U.S.C. 1801 et seq.) the MTB has encouraged the adoption of the Hazardous Materials Regulations (HMR) (49 CFR Parts 170 to 179) by the States in order to promote uniformity in safety regulations throughout the nation. Certain areas of transportation demand a strong predominant Federal role. In the HMTA's Declaration of Policy and in the Senate Committee language reporting out what became section 112 of the HMTA, Congress indicated a desire for uniform national standards in the field of hazardous materials transportation, and, with the HMTA, gave the Department of Transportation the authority to promulgate those standards. Although the HMTA has not totally precluded State or local action in this area, it is the MTB's opinion that to the extent possible, Congress intended to make such State or local action unnecessary.

On May 22, 1980, the MTB promulgated a rule designating certain hazardous materials as hazardous substances under the HMTA and assigned them reportable quantities (RQ's). (See FR Volume 45, No. 101, Thursday, May 22, 1980, page 34560–34705.) In that rulemaking, DOT asserted its authority over the intrastate shipment of hazardous substances by motor carrier and made the provisions of the HMR apply to the carriage of these substances. Anhydrous ammonia was one of the materials designated a hazardous substance with a reportable quantity of 100 lbs, and its transport in this or larger amounts per package anywhere in the United States was made subject to the HMR, including specification containers.

It has come to the attention of the MTB that the adoption by individual States of the Hazardous Materials Transportation Regulations and the assertion of DOT's authority over the intrastate transport of hazardous substances has created an anomalous situation in certain states for certain cargo tank owners and operators. DOT regulations require cargo tanks for anhydrous ammonia to be in compliance with either DOT Specification MC-330 or MC-331 at a design pressure of 265 psig. However, a number of cargo tanks not subject to DOT (nor ICC regulations prior to 1967) have been constructed and used in intrastate commerce for many years. While they were manufactured in accordance with certain consensus standards, including the ASME Code, and were otherwise qualified for use, they do not meet, or were not marked and certified to meet, the standards now required by DOT regulations. The result of a state's adoption and enforcement of DOT regulations and the designation of anhydrous ammonia as a hazardous substance is to immediately require that all cargo tanks in that jurisdiction comply with DOT specification without an adequate transition period.

This proposal is limited in its applicability to intrastate transportation by highway in jurisdictions where the use of these cargo tanks has been permitted in the past. It includes intrastate operations by a motor carrier that may operate other motor vehicles in interstate commerce.

The proposed revision would allow the continued use (in states where such use is permitted), of a cargo tank for the transportation of anhydrous ammonia that is not marked according to Specification MC-330 or MC-331 or one that has a design pressure of 250 psig, provided it (1) is marked and conforms to the edition of the ASME code in effect when it was manufactured; (2) has a minimum design pressure of 250 psig; (3) was manufactured prior to January 1, 1981; (4) is painted white or aluminum; (5) has been inspected and tested in accordance with § 173.33 as specified for Specification MC-330 or MC-331; (6) is operated exclusively in intrastate commerce (including its operation by a motor carrier otherwise engaged in interstate commerce); and (7) is operated in conformance with the regulations except the specification requirements.

The MTB has determined that this proposed regulation will not, if promulgated, have a significant economic impact on a substantial number of small entities.

If this regulation is not adopted, there may be a serious economic hardship on small anhydrous ammonia carriers because their nonspecification cargo tanks will no longer be authorized for transportation of anhydrous ammonia in several states. New DOT specification tanks would have to be purchased and deliveries to users of anhydrous ammonia could be severely disrupted.

(46 F. R. 47099, September 24, 1981)

49 CFR Part 173
[Docket No. HM-166K, Amdt, No. 54]
Transportation of Anhydrous Ammonia in Intrastate Commerce

On Thursday, September 24, 1981, the Materials Transportation Bureau published a Notice of Proposed Rulemaking, Docket No. HM-166K, Notice No. 81-7 [46 FR 47099], which proposed to amend the Hazardous Materials Regulations (HMR) to allow the continued use of certain specification and nonspecification cargo tanks in intrastate anhydrous ammonia service in States which historically have permitted, or never prohibited, the use of these cargo tanks for that service.

The Notice was based in part on a petition from the fertilizer industry to allow the continued use of these cargo tanks in order to avoid hardship both to farmers using anhydrous ammonia for fertilizer and the fertilizer industry. Comments on the Notice were solicited with a closing date of November 23, 1981.

A total of eleven comments were received, ten from private individuals or companies and one from the Pennsylvania Department of Transportation (Pennsylvania DOT). Based on the comments received, the proposals in the Notice, with one revision, are being incorporated as amendments to the HMR. The following is a summary of the comments received on the Notice.

Of the ten comments received from individuals or companies, nine requested that the amendments proposed in the Notice be adopted. None of these comments went into detail on the proposed amendments or recommended any changes to them. All of these comments stressed the hardship that would fall on the agricultural community and the fertilizer industry which services it if the amendments were not adopted. The tenth commenter protested the elimination of the use of State approved cargo tanks in the first place.

The Pennsylvania DOT concurred in part in the amendments proposed in the Notice. However, they recommend that documentation be required to be carried on each vehicle to establish the fact that the vehicle had complied with the requirements contained in the proposed amendment. Reasons cited by the Pennsylvania DOT for such documentation were to make enforcement of the HMR by State authorities easier and to prevent unnecessary disruption of service by enforcement officials. While MTB recognizes that a requirement for the documentation suggested by the Pennsylvania DOT would somewhat ease the enforcement burden of both State and Federal enforcement personnel, this benefit is out-weighed by the recordkeeping burden placed on the carrier to maintain a copy of these documents in each vehicle at all times. It is the policy of the Federal government to reduce, not add to the paperwork burden of the regulated community. However, if the carrier elects to carry this documentation with the vehicle, it may facilitate inspections and prevent delays by enforcement officials.

As stated in the summary of the Notice it was the intention of the proposal to authorize the use of specification and nonspecification cargo tanks with a design service pressure of 250 psig in the carriage of anhydrous ammonia in intrastate service. However, due to oversight, the proposed change dealt only with nonspecification cargo tanks. Cargo tanks built to Specification MC-330 or MC-331 with a design service pressure of 250 psig were omitted from the proposed rule change. This omission has been corrected by including Specification MC-330 and MC-331 cargo tanks in Note 17.

(47 F.R. 7244, February 18, 1982)

49 CFR Parts 171, 172 and 173
[Docket No. HM-166L; Notice No. 82-1]
Regulation of Consumer Commodities; Paint and Paint Related Material
Adhesive, n.o.s.

This document is the thirteenth in a series of notices and amendments designed to reduce regulatory burdens by incorporating changes in the Hazardous Materials Regulations based on either petitions for rulemaking submitted in accordance with 49 CFR 106.31 or on MTB's own initiative. The MTB published its first notice of proposed rulemaking under Docket HM-166 on November 30, 1978 (43 FR 56070).

This Notice is based on a petition from the National Paint and Coatings Association, Inc. (NPCA) and upon MTB's own initiative. The changes proposed herein would authorize (1) the use of the word "Paint" or the words "Paint related material" as proper shipping names because several of the descriptions used currently for shipping names in § 172.101 are obsolete or are seldom used to describe today's paint products; and (2) expand the coverage of the consumer commodity category for flammable liquids by lowering the flash point limitation for one gallon inside containers specified in § 172.1200(a)(1)(iii) from 73° F. to 20° F. ("Flash point" means the minimum temperature at which a substance gives off flammable vapors which in contact with spark or flame will ignite.) This latter provision will apply to all consumer commodities, although MTB believes the greatest benefit will be derived by the paint industry.

The NPCA has stated that approximately 400 million cans of paint shipped in 1980 were subject to the Department's Hazardous Materials Regulations and that more than 300 million cans were of the one-gallon size. Approximately 100 million one-gallon cans shipped in 1980 had a flash point between 20° F. and 73° F. Therefore, if the proposal is adopted, labels will not be required on approximately 25 million cases of paint for transportation by highway, rail and vessel, resulting in a savings to shippers of .5 million dollars. Also it is estimated that adoption of this proposed rulemaking will eliminate the need for shipping papers on approximately one-half million individual shipments moving by highway, rail and vessel and will result in annual savings of approximately one million dollars to shippers and carriers of paint and paint related material alone.

As indicated in Docket No. HM-36A; Notice No. 80-5 (45 FR 40628) June 16, 1980, the consequences of incidents involving paint and paint related material shipped in small packagings have been minor, involving only a few minor injuries and no deaths. In view of the limited risk presented by these materials in packagings of one gallon or less, the MTB is proposing to lower the specified flash point from 73° F. to 20° F. in § 173.1200 for materials which are flammable and are eligible to be reclassed as ORM–D and offered for shipment as consumer commodities. This proposed change does not apply to carriage aboard aircraft because these flammable materials, at the increased volume proposed by this Notice, should continue to be subject to inspection as required by § 175.30(b) and (c) before they are loaded aboard passenger carrying aircraft (if authorized) and cargp-only aircraft. Due to the fire hazard should there be any leakage from packages, MTB believes this distinction between carriage aboard aircraft and surface transportation is necessary.

The MTB is proposing to delete the entry "Adhesive, n.o.s. *See* Cement, liquid, n.o.s." in § 172.101 and add "Adhesive, n.o.s." classed as Flammable liquid and Combustible liquid. Several inquiries have been received concerning the identification number assigned to the proper shipping name "Cement, liquid, n.o.s." and the alternate names "Adhesive, n.o.s," and "Cement, adhesive, n.o.s." which reference the primary entry. The identification number assigned is NA1133. The "NA" prefix was assigned because "Cement, liquid, n.o.s." does not match the description listed in the United Nations Recommendations for the Transport of Dangerous Goods. Identification numbers with the "NA" prefix are associated with descriptions that are not recognized for international shipments, except to Canada. This has caused problems because a single description cannot be used for both domestic and international shipments of hazardous materials. Assigning the identification number UN1133 to the proper shipping name and class, "Adhesive, n.o.s., Flammable liquid" and "Adhesive, n.o.s., Combustible liquid," should alleviate the problem.

This proposed rulemaking will result in a substantial savings of time and money to all shippers of paint, paint related material, and other flammable liquids, that may be described as "Consumer commodities" because of the reduction in labeling and related paperwork.

(47 F.R. 4538, February 1, 1982)

49 CFR Parts 171, 172 and 173
[Docket No. HM-266L; Amdt. Nos. 171–72, 172–79, 173–163]
Regulation of Consumer Commodities; Paint and Paint Related Material

On Monday, February 1, 1982, the Materials Transportation Bureau (MTB) published a Notice of Proposed Rulemaking (NPRM) Docket Number HM-166L (47 FR 4538) which addressed paints and paint related materials. The NPRM proposed to reduce the number of shipping names associated with paint in the Hazardous Materials Table (49 CFR 172.101) from approximately 28 to 7. In addition, the NPRM proposed to relax certain shipping requirements for paint and paint related material by allowing a flammable liquid with a flash point higher than 20° F. to be shipped as "Consumer commodity," ORM–D when in inside packagings of one gallon or less. At present, the Hazardous Materials Regulations (HMR) restrict the volume of flammable liquids having flash points below 73° F. being shipped as "Consumer commodity" to one quart. The effect of such change would be to allow four one gallon metal cans of paint in fiberboard boxes to be shipped without requiring that they be labeled or accompanied by shipping papers except when carried aboard aircraft.

MTB received a total of 37 comments in response to the NPRM. While the paint manufacturing industry and carriers generally favored the proposal, persons interested in fire protection strongly opposed those portions of the notice which would have allowed the increased quantity of paint with a flash point below 73° F. to be shipped as "Consumer commodity, ORM–D". There was little opposition to consolidation of shipping names and much support for it. Fourteen comments received from industrial firms that manufacture or ship paints and adhesives support the NPRM without exception. One manufacturer did want the shipping names "varnish" and "enamel" retained because products with these names are used to coat electrical wires, a use which most people do not associate with "paint." The purpose of shipping names in the hazardous materials table is not to pinpoint the ultimate use of a product with great exactitude, but rather to provide a standardized format which succeeds in communicating the basic properties, or kinds of hazardous materials in transportation.

In addition to industrial firms, seven trade associations, representing paint producers, carriers, and shippers supported the NPRM.

MTB received comments from four carriers. Two supported the NPRM. One air carrier expressed concern that the relaxed requirements would exclude shipments by air and would cause confusion and inadvertent violation of the regulations because shippers may not know that one portion of a journey might be accomplished by air, requiring shipping papers. Since the package would not be labeled and marked so the contents could be identified, a violation might ensue if the package was shipped by air with no shipping papers. A rail carrier expressed concern that the proposed shipping names would not identify whether the hazardous material was a hazardous substance identified under the Comprehensive Environmental Response, Compensation and Liability Act (CERCLA or "Superfund"). The identification of CERCLA hazardous substances is addressed at length in a previous MTB publication (see Docket No. HM-145C, 46 FR 17738, March 19, 1981). It is extremely unlikely that a hazardous substance, as presently defined in the HMR, in packagings addressed in this NPRM would be a constituent of paint in sufficient quantity to constitute a reportable quantity (RQ). If a hazardous substance were present in sufficient quantity, the marking provisions of § 172.324 would apply and the package would have to be marked with the name of the hazardous substance and the letters "RQ".

If a package contains a material which is listed in the CERCLA List (§ 172.101) but which is not a hazardous material or "hazardous substance" as presently defined in § 171.8, that material is not subject to the requirements of the HMR regardless of whether it is a "Consumer commodity" or not. This issue is discussed at length in Docket HM-145C and this final rule has no effect on it.

The Air Transport Association (ATA) expressed concern that the NPRM did not include all paint related items that appear in the ICAO Technical Instructions, specifically paint driers and thinners. One purpose of the NPRM was to reduce the number of shipping names associated with paint, including thinners, driers, removers and reducers. MTB feels that these materials can all be safely shipped under the shipping names "Paint" or "Paint related material" with separate entries for the flammable liquid, combustible liquid and corrosive material hazard classes.

MTB received comments from the fire departments of 7 municipalities, two from Members of the International Associations of Fire Chiefs, one from a fire protection engineer, and one from the National Fire Protection Association, all opposing the relaxation of shipping requirements (the use of the Consumer commodity, ORM-D hazard class) for flammable liquids as proposed in the NPRM. Comments from the fire departments and fire chiefs opposed relaxation of the communications requirements (labels and shipping papers) associated with shipments under the ORM-D hazard class. They expressed the opinion that the absence of labels and shipping papers would increase the danger to fire service personnel, or the general public, or both. The fire protection engineer expressed the view that the present regulations are consistent with National Fire Protection Association (NFPA) requirements and OSHA regulations and if the NPRM became final, the HMR would no longer be consistent. He also thought that there would be precedence for opening up the Consumer commodity, ORM-D hazard class to other flammable liquids which are not paints.

Comments received frm the NFPA expressed the view that hazards at warehouses storing paints would be greatly increased. The comments stated that DOT labels and markings on outside containers are used for purposes of material classification of flammable liquids into various NFPA subclasses based on flash point and that this classification system is vital to the nationally recognized and widely used Flammable and Combustible Liquids Code, NFPA 30–1981, and without it the NFPA maintained there could be severe "fire overloading" of storage and warehousing facilities.

An evaluation of the merits of the comments reveals some concern for relaxation of shipping requirements, as proposed in the NPRM, is justified. It is true that accident data compiled from incident reports do not reveal a serious fire problem with paint as it is now shipped, however, an argument can be raised that this condition exists because of the adequacy of the present regulations and that relaxation of requirements would produce more problems than benefits. Prior to issuing the notice, MTB did not fully consider the problems that could arise if paints were inadvertently shipped by air as pointed out by the air carrier, or the nontransportation impacts of the proposals, such as the storage classification problem pointed out by the NFPA. In addition, while the comments from fire service personnel did not provide any factual data to support their concerns for increased risk to fire fighters and the public, the opinions of fire protection professionals should be given further consideration before action is taken. Because of these factors MTB has decided to withdraw those portions of the NPRM which would allow paints with flash points between 20° F. and 73° F. to be shipped as "Consumer commodity, ORM-D".

The NPRM explained the reasons why a change was needed for the entry "Adhesive, n.o.s. *See* Cement, liquid, n.o.s." However, it has been noted that "liquid, n.o.s." is not a part of the proper shipping name for the entry "Adhesives" in the United Nations Recommendations for the Transportation of Dangerous Goods and "liquid, n.o.s." is not a part of the proper shipping name for the entry "Cement" in the IMDG Code. For these reasons, "liquid, n.o.s." has been deleted from both shipping names in the § 172.101 Table. Also, § 173.132 has been changed accordingly.

(48 F.R. 17094, April 21, 1983)

49 CFR Part 173
[Docket No. HM-166M; Amdt. No. 173–55]
Reinstatement of Department of Energy Approval Authority for Radioactive
Materials Package Designs

On December 11, 1980, the MTB published Docket No. HM-56; Amdt. No. 106–3, 107–8, 171–58, 172–63, 173–142, 174–39, 175–18, 176–12, 177–51, 178.64 (45 FR 81570) which made numerous miscellaneous changes to 49 CFR. Item Number 39 of the referenced docket changed the designations "U.S. Atomic Energy Commission" and "USAEC" to read "U.S. Nuclear Regulatory Commission" and "USNRC" each time they appeared in the following sections and section headings: § 173.393a(a), (a)(1), (a)(2), (a)(3) and (a)(5), § 173.394(b)(3) and (c)(2), § 173.395(b)(2) and (c)(2), § 173.396(b)(4) and (c)(3). Prior to the above changes, the DOE and the NRC as successors to the Atomic

Energy Commission had approval authority under the above referenced sections. Although the final rule was not intended to impose burdens upon any person, it did have an impact on the ongoing programs of DOE. The DOE has stated that their energy, space, medical, industrial and waste programs would meet with lengthy delays if all package designs were required to pass through the approval process of the NRC and these delays could severely limit the effectiveness of the DOE nuclear program.

In view of the strict procedures the DOE requires to be followed to certify its own package designs for radioactive materials, the MTB agrees that DOE packaging requirements and evaluation techniques which demonstrate compliance with safety standards equivalent to those contained in 49 CFR Parts 100 to 177 and 10 CFR Part 71 are sufficient to protect the public health and safety.

Since this amendment is only reinstating an approval authority that was in 49 CFR prior to Docket HM-56 and does not impose additional requirements, public notice has not been provided and this amendment is effective without delay.

(47 F.R. 7243, February 18, 1982)

49 CFR Parts 172 and 173
[Docket No. HM-166-O; Notice No. 82-4]
Deletion of Certain Commodity Entries

This notice proposes the deletion of 171 entries from the Hazardous Materials Table, § 172.101. The number of entries covered by this proposal according to class is as follows:

Combustible liquid	29
Corrosive material	10
Flammable gas	2
Flammable liquid	30
Flammable solid	25
Forbidden	2
Nonflammable gas	2
ORM-A	1
ORM-C	18
Oxidizer	2
Poison B	4
Cross Entries or Informational	46

The MTB has reviewed the properties of ten materials identified by Code 3 in the proposed regulatory text, and believes that they present only a minimal hazard in transportation. In almost all cases, the commodities identified by Code 3 do not meet the definition of a hazardous material. However, should this notice be promulgated as proposed, it would remain the shippers responsibility to determine whether the material to be offered for transportation meets the definition of a hazardous material. If it does the shipper must classify and ship the material according to hazard class, in compliance with all applicable regulations. With respect to any of the entries proposed for deletion, the material may still be regulated as a hazardous material even though the entry has been deleted. In that case a shipper would use the specific proper shipping name which is most appropriate, or an n.o.s. description if no specific name is appropriate.

This proposal is part of a continuing effort by the MTB to eliminate unnecessary regulation and to reduce the volume of regulation when appropriate. If this proposal is promulgated as an amendment, it will have the effect of reducing the size of the Hazardous Materials Table, thus making it easier to use.

(47 F.R. 24157, June 3, 1982)

49 CFR Parts 172 and 173
[Docket No. HM-166-O; Amdt. Nos. 172-86, 173-169]
Deletion of Certain Commodity Entries

On June 3, 1982 the MTB published Notice No. 82-4 (47 FR 24157) under Docket HM-166-O which proposed to delete 171 entries from the Hazardous Materials Table, § 172.101, and to remove or amend sections in Part 173 of the Hazardous Materials Regulations (HMR) which are associated with some of the entries which were proposed for deletion. The public comment period ended on August 2, 1982. The MTB received 40 comments to HM-166-O. The comments were carefully considered by the MTB. On the basis of comments received, some of the entries proposed to be removed are retained. The MTB believes that the result of this amendment will be to reduce the volume of regulation, making the HMR easier to use and less burdensome on the regulated public. The following is a discussion of the most significant comments.

The majority of the commenters were either in favor of the proposal, or in favor of the proposal except as it affects the specific proper shipping name of a material offered for transportation by the commenter. Five commenters were against the proposal in its entirety.

One commenter contends that the MTB failed to specify an alternative proper shipping name to be used in the event that a shipping name is deleted. The MTB disagrees. It is the shipper's responsibility to identify and classify the hazardous materials that are being offered for transportation. In order for the MTB to specify an alternate proper shipping name to an entry that is being removed, information would have to be supplied regarding the properties of the material to determine if they meet the definition of a hazard class. For example, the proper shipping name "Rubber curing compound" is one of the entries being removed from the Table. In order for the MTB to suggest an alternate proper shipping name for this commodity, technical data would have to be supplied, e.g., does the commodity meet the definition of a flammable solid.

One commenter suggested that by eliminating certain proper shipping names and by substituting the generic equivalent, the MTB would be creating problems for emergency response personnel. The MTB does not agree. The use of a vague proper shipping name does not increase the ability of emergency response personnel to do their job.

One commenter suggested that if commodities were deleted from the Table and not deleted from international regulations, this would create a problem of dual descriptions for shippers that ship a commodity both domestically and internationally. ICAO descriptions are presently accepted for descriptions of domestic shipments by air and IMO descriptions as listed in § 172.102 are permitted if part of the transportation of the material is by vessel. Also, it should be pointed out that, since § 172.102 was adopted in 1980, no petition has been received to broaden the applicability of that section to domestic surface shipments.

Several commenters objected to the deletion of the proper shipping names; "Drugs, n.o.s."; "Medicines, n.o.s."; and "Cosmetics, n.o.s." The arguments set forth by these commenters are that the use of a proper shipping name such as "Flammable liquid, n.o.s.", for these commodities could be confusing to the carrier and might frustrate transportation. While the MTB does not agree that these entries are more appropriate than a more specific entry (including a generic class n.o.s. entry), they are retained in the Table at this time.

One commenter objected to the removal of the proper shipping name "Gas drips", on the basis that the name gas drips conveys an important message to emergency response personnel. The commenter pointed out that gas drips are testing samples from pipelines, and have a distinctive mercaptan smell. The shipping name gas drips conveys this information to emergency response personnel and could prevent a possible overreaction in an incident where the mercaptan smell is present. The MTB agrees with the commenter and the proper shipping name, "Gas drips" is retained in the Table.

Three commenters objected to the removal of the proper shipping name "Acid, sludge" on the basis that it is not a vague name, that it is a term that is widely understood in industry and, therefore, should be retained in the Table. The MTB disagrees. The three commenters who objected to the deletion of this proper shipping name also disagreed as to what constitutes acid, sludge. Several commenters also objected to the deletion of the proper shipping name, "Resin, solution" on similar grounds, that it is a term widely understood in industry and, therefore, it should be retained in the Table. Again the commenters disagreed as to what constitutes resin solution, and MTB has removed the entry because it is vague. Also, the entry "Resin, solution", Flammable liquid which was inadvertently left out of the notice, has been removed.

Two commenters suggested that the informational entry, "Film, photographic", should be retained because it is useful. The MTB disagrees. Safety film is not considered a hazardous material. The type of film that does meet the definition of a hazardous material is the nitrocellulose based film which is no longer manufactured. In a comment to Docket number HM-8, Notice 71-13, the Eastman Kodak Company stated that it had not produced nitrate based film since 1949. It is true that some nitrate based film may be in transportation to and from film archives; however, the proper shipping name, "Film (*nitrocellulose*)" is being retained in the Hazardous Materials Table. The MTB believes that thirty-four years is a long enough period of time to inform the shipping public that safety film is not subject to the provisions of the Hazardous Materials Regulations.

Several commentes objected to the removal of the following proper shipping names:

Entry	*Hazard Class*
Coal tar distillate	Combustible liquid
Coal tar distillate	Flammable liquid
Coal tar light oil	Combustible liquid
Coal tar light oil	Flammable liquid
Coal tar naptha	Combustible liquid
Coal tar naptha	Flammable liquid
Coal tar oil	Combustible liquid
Coal tar oil	Flammable liquid

The commenters point out that these shipping names are widely used in industry and, therefore, should be retained. The MTB agrees that the proper shipping name, "Coal tar distillate" should be retained, however, the other proper shipping names in the coal tar group should be deleted because Coal tar distillate is descriptive of all these commodities.

Several commenters objected to the removal of the following entries:

Entry	Hazard class
Lighter fluid	Flammable liquid
Naptha distillate	Combustible liquid
Naptha distillate	Flammable liquid
Naptha solvent	Combustible liquid
Naptha solvent	Flammable liquid
Petroleum naptha	Combustible liquid

The basis for these objections is that they are commonly used shipping names and, therefore, should be retained in the Table. Upon consideration of these comments, the MTB has retained the entries for "Naphtha" and for "Petroleum naphth" in the Table because it believes that these proper shipping names are descriptive of all of these commodities.

One commenter objected to the deletion of the proper shipping names, "Hydrocarbon gas, liquefied" and "Hydrocarbon gas, nonliquefied", on the basis that they are more descriptive than an n.o.s. entry. The MTB believes that liquefied petroleum gas is the same as hydrocarbon gas, liquefied. However, since there is not an entry for petroleum gas, nonliquefied, the MTB believes it is appropriate to retain the entries for hydrocarbon gas.

Several commenters objected to the deletion of the entry, "Denatured alcohol" because of the possible confusion between denatured alcohol and other types of alcohol, including potable alcohols. The MTB agrees with these commenters and has retained the proper shipping name, "Denatured alcohol" in the Table. However, the MTB believes that the proper shipping name, "Rum, denatured", should be deleted because the proper shipping name "Denatured alcohol" includes "Rum, denatured".

One commenter objected to the deletion of the proper shipping name "Potassium fluoride solution", because a description under the n.o.s. category would not be adequate to provide proper product definition due to specific properties of that commodity. The MTB agrees, the proper shipping name "Potassium fluoride solution" is retained in the Table.

One commenter objected to the deletion of the entry, "Crude oil", because the proper shipping name is widely used in industry and is more appropriate than an n.o.s. entry. The MTB agrees with the commenter and has retained the entry in the Table.

Several commenters objected to the deletion of the proper shipping names, "Boiler compound, liquid" and "Water treatment compounds". The bases for these objections are that they are commonly used names and that there are packaging exceptions for these commodities which the commenters fear would be lost. The MTB believes that these proper shipping names are too vague to be useful, however, the packaging exceptions which exist for these commodities are retained in Part 173.

Several commenters expressed concern that changing proper shipping names would result in economic loss due to the preprinting of shipping papers. This situation has been taken into account in §172.101(j)(2). Stocks of preprinted shipping papers and package marking may be continued in use, in the manner previously authorized, until depleted or for a one-year period, whichever is less. The effective date of this rulemaking is one year from publication of this amendment, therefore, a two-year period is provided for depletion of existing stocks of preprinted package markings and shipping papers.

(48 F.R. 52306, November 17, 1983)

49 CFR Part 172 and 173
[Docket No. HM-166-O; Amdt. No. 172–61, 173–174]
Deletion of Certain Commodity Entries

On November 17, 1983, MTB published a final rule in Docket HM-166-O [48 FR 52306) which removed certain entries (proper shipping names) from the Hazardous Materials Table (HMT) § 172.101, effective September 30, 1984. The MTB received eleven petitions for reconsideration of that rule. The following is a summary of the petitions.

One petitioner complained that removing entries from the HMT results in inconsistencies between the two tables found in §§ 172.101 and 172.102. MTB wrote to this petitioner asking for further clarification of that complaint. In response, the petitioner modified the petition by urging MTB to make the rule effective only after international bodies delete the same entries from their list of shipping names. The petitioner contends that requiring different proper shipping names for domestic and international shipments leads to confusion on the part of shippers.

MTB does not agree that the effective date of the final rule should be postponed until international bodies have deleted the same entries from their lists. MTB has submitted a paper to the United Nations Committee of Experts on the Transport of Dangerous Goods recommending that the entries removed from the HMT be removed from the list in Chapter 2 of its Recommendations. Indeed many of those names appear on the international lists only because the international bodies drew heavily on the HMT in their initial choice of descriptions.

Several of the materials removed from the HMT do not satisfy the definition of any specific DOT hazard class. They were regulated in a class (ORM–C) that is not recognized by international bodies. That situation can cause greater confusion for shippers, both domestically and internationally, than inconsistency between the two tables.

The other ten petitioners objected to the deletion of the proper shipping name "Resin solution", Flammable liquid. In the preamble to the final rule, it was stated that the entry "Resin solution", Flammable liquid was inadvertently left out of the notice and that proper shipping name also was being removed. Although MTB still believes that the proper shipping name is vague, the entry is restored to the HMT because of the procedural error that did not give an opportunity for public comment on its removal. Several of these petitioners objected also to the removal of the entry "Resin solution", Combustible liquid. This entry was in the notice of proposed rulemaking, Notice No. 82–4 (47 FR 25157) and public comment was received and addressed in the preamble to the final rule. The petitioners did not furnish adequate justification to support a change in that amendment.

The entry "Road asphalt" was not removed in the final rule but was inadvertently removed from § 173.131. This document corrects that error by adding "Road asphalt" in the heading and text of § 173.131. Also, the entry "Paper scrap" was removed from the HMT but the corresponding section in Part 173 was not removed. That oversight is corrected in this document by removing § 173.1075.

(49 F.R. 14353, April 11, 1984)

49 CFR Part 173
[Docket No. HM 166P; Notice No. 82–8]
Radiation Level Limits for Exclusive Use Shipments of Radioactive Materials

I. Basis for Proposed Rule

I. Background. The radiation level limitations for exclusive use, or full-load, shipments of radioactive materials have been a part of domestic and international regulatory standards for many years. MTB published a notice of proposed rulemaking (NPRM), Docket HM-169, in the **Federal Register** (44 FR 1852) on January 8, 1979, which includes a broad historical relationship between DOT and the International Atomic Energy Agency (IAEA) regulatory standards. Without extensive elaboration about radiation levels for exclusive-use shipments, HM-169 proposed changes which include restricting packages to 1000 millrem per hour at any point on the surface rather than 1000 millirem per hour at one meter from any point on the package surface. Other minor proposals in HM-169 related to external radiation levels for exclusive-use shipments are mainly nonsubstantive, rewordings of existing provisions.

Subsequent to issuance of the NPRM in HM-169, there have been a number of requests for interpretation of § 173.393(j) with respect to "open vehicles" and "closed vehicles" operating under exclusive-use conditions. The meaning and application of § 173.393(j) have been difficult to interpret for purposes of compliance and enforcement. It was determined by both the Department of Transportation and the Nuclear Regulatory Commission (NRC) in meeting with representatives of the nuclear industry that even the wording of the changes proposed in Docket HM-169 does not adequately clarify the issues, and further consideration is needed.

Another area of confusion in interpretation involves the relationship between § 173.393(j) and § 177.842(a). Some carriers and shippers erroneously concluded that the exception in § 177.842(a) from the total transport index limitation of 50 per vehicle for packages complying with § 173.393(i) is applicable to their operation, provided the radiation levels on and around their vehicle are within the limits of § 173.393(j) (2), (3) and (4). That misinterpretation stems from a failure to consider the first sentence in § 173.393(j) as presently written, which excludes packages meeting limits of § 173.393(i).

2. Applications Concept. The provisions of § 173.393(j) are intended to apply to special conditions which consider radiation safety aspects of the shipment and the resulting radiation exposures. When package radiation levels exceed the limits of § 173.393(i), the package may be shipped if the shipper assumes certain carrier responsibilities and makes arrangements with the carrier to maintain the required exclusive-use conditions, wherein radiation levels of the vehicle and not just the packages are considered.

Similarly, shippers make arrangements with carriers for exclusive-use shipments of some packages of radioactive materials that are classified as fissile or low specific activity. For the more common nonexclusive use shipments, the primary mechanisms for controlling radiation exposure of carrier personnel are: (1) package radiation limits specified in § 173.393(i), (2) the total transport index of the packages in the vehicle, and (3) their separation distance from normally occupied spaces. However, when reasonable radiation level reduction efforts do not achieve § 173.393(i) limits, packages may be transported in exclusive-use vehicles with radiation exposure controlled by limiting radiation levels on and near the vehicles, as well as the package, and by the imposed exclusive-use shipment controls.

Basically, the proposed radiation level limitations for packages and vehicles in §173.393(j) would apply to shipments of radioactive materials only when the package surface radiation level or transport index limits exceed the §173.393(i) limits and some other method of controlling radiation exposure is necessary.

3. Package Limits. The existing limitation of 1000 millirem per hour at one meter from the surface of the package is considered to be unsatisfactory. IAEA standards in effect since 1973 have set a limit of 1000 millirem per hour at the surface of the package, and the United States is the only major industralized country that has not yet adopted the IAEA standards. A soon to be issued final rule under HM-169 is expected to impose the limit of 1000 millirem per hour at the package surface.

Without a package surface limitation, it is hypothetically possible for a package of minimum dimensions (4″ × 4″ × 4″) to have radiation levels near 400 rem per hour at the surface and still meet a limit of 1000 millirem per hour at one meter. In the interest of safety for carrier personnel, emergency services personnel, and the public, the 1000 millirem per hour at the surface is considered to be a more appropriate limit. At the present time, most exclusive-use shipments are probably restricted by vehicle radiation level limits rather than by package radiation level limits.

If a packaging design is such that it cannot be offered for transport as a package in full compliance with the HMR because the surface radiation level is too high, then the activity of its contents must be lowered, or the unit must be provided with additional shielding. The new configuration must satisfy all requirements of a package before it may be offered for transportation.

4. Accessible Surface Limits. It is proposed that external radiation level limits for vehicle and/or package be established with respect to accessibility by personnel during transport. Establishing limits with respect to readily accessible surfaces should eliminate most interpretative difficulties and discrepancies present in existing provisions for "open" and "closed" vehicles in the current regulations. The basic concern is to establish a maximum rate of exposure that might be received by any transport worker, or general public personnel, and that maximum rate need not be different for "open" and "closed" vehicles.

Controls established by the shipper for exclusive-use transport of a package with surface levels above 200 millirem per hour or a transport index above 10 would, in many cases, include the use of a "closed transport vehicle" (see §173.389(q)) to achieve radiation levels on external surfaces at or below 200 millirem per hour. In some cases, in lieu of a permanently enclosed vehicle, the shipper may make arrangements with a carrier to use a flatbed trailer or other "open" vehicle and convert it to a "closed transport vehicle" by means of permanent or temporary personnel barriers. This conversion must be within the basic carrier safety requirements (such as tie downs, blocking and bracing, materials integrity, etc). As proposed, the package radiation level limitation for a package within such a personnel barrier would be 1000 millrem per hour at the surface. All readily accessible external surfaces of the vehicle or the barrier could not exceed the 200 millrem per hour limit. Another example of such an arrangement is an "open" vehicle which employs barriers that do not result in the vehicle being converted to a "closed transport vehicle," but still it achieves radiation levels which do not exceed the 200 millrem per hour limit at readily accessible surfaces. An example of this is the use of shielded outer packages such as casks which are commonly used to transport drums of low specific activity (LSA) waste.

It is emphasized that the 200 millirem per hour is a maximum not a goal. In keeping with the principle of "as low as reasonably achievable" (ALARA), the radiation levels at all accessible surfaces (sides, ends, top, and bottom) should be kept ALARA.

In the existing regulations, the absence of a stated radiation level limit at the vertical planes projected from the lateral edges of a package being carried on an open transport vehicle is also believed to be a significant omission. The proposed revision should eliminate the problem and reduce misinterpretations.

5. Two Meters From Accessible Surfaces Limits. The proposed limitation of 10 millirem per hour at any point two meters from any accessible external surface of the vehicle or load constitutes no significant change from existing rules for most conventional closed vehicles such as vans, but it would restrict slightly the presently allowed radiation levels for packages with surface levels less than 200 millirem per hour that are transported on vehicles like flatbeds. At present, the 10 millirem per hour limit is at the plane 2 meters from the lateral edges of the flatbed. In cases such as a package on an "open" flatbed, the proposed amendment would set the 10 millirem per hour limit at 2 meters from the package surface. The proposed revision also excludes the 10 millirem per hour limitations at 2 meters from the top and underside of vehicle surfaces. This is a practical consideration since it is uncommon for persons to be 2 meters above or below accessible surfaces of exclusive-use vehicles, and it would be an unreasonable burden in most facilities to obtain radiation measurements 2 meters below the underside of a vehicle and 2 meters above the top surface. The proposed revision would eliminate all reference to vertical planes projected from vehicle surfaces. Instead, the 10 millirem per hour limit would refer directly to readily accessible external surfaces.

The radiation level limitation at two meters is intended to control radiation exposure to personnel not associated with the shipment, such as people in other vehicles moving along side the shipment, persons nearby when the vehicle is stopped temporarily, or persons refueling or servicing the vehicle. In combination with the proposed accessible surface limitation, these revisions should improve controls on radiation exposure to members of the general public.

6. Occupied Spaces Limit. The proposed revision does not change the maximum radiation level for locations normally occupied during transport. The only change is to state more clearly the intended conditions under which a private carrier is excepted from the 2 millirem per hour limit. It was previously assumed that all personnel of a private carrier involved in transporting radioactive material under exclusive-use provisions would be operating under a regulated radiation safety program. Since this is not always the case, a qualifying requirement to assure radiation safety would be added to the regulations.

It is assumed that the carrier's personnel responsible for its regulated radiation safety program will assure that all exposures are kept as low as reasonably achievable, monitor operations, and establish limits in line with appropriate radiation safety requirements.

This 2 millirem per hour limit does not apply to vehicles carrying radioactive materials packages under conditions controlled by the total transport index limit per vehicle and separation distance requirements between normally occupied areas and the nearest package. The 2 millirem per hour is not an unreasonably low upper limit for occupied spaces. It is noted that the existing transport index and separation distance tables in Parts 174 and 177 do not always result in such a limit. MTB plans to propose amendments in the near future which will assure better radiation exposure control via transport index and separation distance criteria.

7. Designated Agent Requirements. The carrier responsibilities assumed by the shipper for exclusive-use shipments include controlling certain aspects of the shipment from point of origin to destination.

As stated in the regulations, instructions provided by the shipper to the carrier for maintaining exclusive-use conditions must be included with the shipping papers. Part of the requirements in the definition of "exclusive-use" in § 173.389(o)(2) refer to initial, final, and intermediate loading and unloading by the consignor, consignee, or designated agent. Under these proposals, it would be made clear that a designated agent must have radiological expertise appropriate for handling the radioactive material being transported for the shipper. The assurance that the designated agent has the appropriate radiological capabilities must be the responsibility of the shipper when he establishes exclusive use controls.

Consignors and consignees in nearly all cases are licensed by the NRC or a State agency, to possess, use, or transfer the radioactive materials in a consignment. To be licensed, they must demonstrate necessary capabilities for handling the radioactive materials authorized by their license. It would be inappropriate for the HMR to require all initial, intermediate, and final loading and unloading of exclusive-use shipments by the consignor, consignee or their designated agent and then be silent on requirements for designated agents. As a minimum, the radiological capabilities should include knowledge of methods and procedures for minimizing radiation exposures when handling radioactive materials shipments, and the ability to recognize and control radiological emergencies that could occur in handling the consignment. The designated agents radiation protection practices should conform to recognized national standards on such matters as probable annual radiation exposure and use of personnel radiation dosimetry equipment.

This requirement for radiological capabilities should complement radiation level limitations in assuring safety for persons involved with the shipment and the general public, for exclusive-use shipments.

(47 F.R. 44356, October 7, 1982)

49 CFR Part 172, 173 and 175
[Docket No. HM-166Q and HM-166F; Notice No. 10]
Exceptions for Small Quantities of Hazards Materials

Background. On July 10, 1980, MTB published Docket No. HM-139C (45 FR 46419), Conversion of Individual Exemptions to Regulations of General Applicability, which contained the withdrawal of a previous proposal (45 FR 18994) that would have authorized a standardized packaging for small quantities of specified hazardous materials. MTB made an attempt in that Docket to create a standardized packaging, based on three DOT exemptions (7755, 7921, 8116), which were to be referenced in § 173.4. The resulting comments were in such disagreement that MTB decided to withdraw the proposal, and announced that it would address the issue of a general exception for analytical standards separately at a later date.

The Scientific Apparatus Makers Association (SAMA), which represents leading firms engaged in the design, manufacture, and distribution of high technology instruments, petitioned MTB to reconsider several of SAMA's suggested approaches for resolving the issues surrounding a standardized packaging for small quantities of certain hazardous materials. Pursuant to the SAMA petition, MTB has carefully considered eight DOT exemptions (7755, 7921, 8116, 8285, 8292, 8423, 8581, and 8658) which authorize several different packaging techniques for small quantities, and concludes that all of the affected exemptions can be accommodated under the standardized packaging proposed herein.

Basic requirements. MTB is proposing in this notice to grant significant relief from the Department's Hazardous Materials Regulations under specified conditions. The proposal applies only to small quantities of flammable liquids, flammable solids, oxidizing materials, organic peroxides, corrosive materials, Poison B. ORM A, B and C, and limited quantity radioactive materials which also meet the definition of these other hazard classes. The quantity limitation per inner receptacle would be 25 milliliters for liquids and 25 grams for solids except, in the case of poisons, the allowable quantity would be based on the LD 50 value of the material. Each inner receptacle would be packed in a secondary packaging with sufficient cushioning material and with a material that would absorb the entire contents of each receptacle containing a liquid. The secondary packaging would be securely packed in an outside package meeting the basic requirements of § 173.24. The package would have to be capable of passing drop tests consistent with those specified for materials of packing Group I in the U.N. Recommendations for the Safe Transport of Dangerous Goods and a compression test similar to the test required by § 173.398(b)(3)(v).

The package would not have to be marked with the shipping name of the material or bear a hazard warning label. However, the shipper would be required to place inside the package a statement certifying compliance with the proposed new section. Also, the name and address of the shipper would have to be included in the statement and displaced on the outside of the package. The exception would be limited to the person who prepares the package for shipment; therefore, repacking or other changes to the packaging would not be authorized under the exception.

While the proposed rule would authorize a wide range of hazardous materials to be placed in such a package, materials assigned certain identification numbers in §§ 172.101 and 172.102 would not be authorized under the exception unless specifically approved by the Associate Director for HMR. This limitation is believed necessary due to the significant risks presented by the materials identified in the list. For example, it may be necessary for persons seeking approval to use a rigid secondary packaging such as a high density polyethylene box, in order to preclude crushing and possible release of a strong oxidizer, even though the basic provisions of the proposed rule are rather conservative in assuring an acceptable level of safety for its application.

In subsequent correspondence supporting its petition, SAMA stated it "... estimates that if the Department of Transportation adopts our petition to deregulate small quantities of materials previously designated 'hazardous' by DOT, a savings to industry would approach the $25 million level (a conservative figure). This figure includes shipping costs, direct labor, and an estimate of overhead." While MTB agrees that considerable savings would result from the adoption of the proposal, it must be emphasized that this proposal cannot be characterized as "deregulation" but rather a significant reduction in the level of regulation that will apply to significant number of shipments due to their quantity and manner in which they will be offered for transportation.

Requirements for radioactive materials. Although MTB is presently considering requirements for the transportation of limited quantity radioactive material proposed in Docket HM-166F; Notice No. 81-8 (46 FR 61908, December 21, 1981) regulatory changes proposed for these materials in this Docket are significantly broader and it was determined they should be open for public comment. This proposal addresses only those limited quantities of radioactive materials which also meet the definition of one or more of the other hazard classes addressed in § 173.3, since it is reasoned that shippers of limited quantity radioactive materials not meeting the definition of another hazard class would elect to comply with the requirements of § 173.391(a). To qualify for this general exception from the HMR, these packages would first have to meet activity and external dose rate limitations, and other criteria specified in § 173.391(a). To further assure an acceptable level of safety these materials would also be subject to the quantity limits appropriate to the additional hazard.

Renewal of exemption for limited quantity radioactive materials. Though normally handled in Docket No. HM-149, MTB is taking this opportunity to propose the renewal for two years of the limited exemption found at 49 CFR 172.204(c)(4), 175.10(a)(6), and 175.700(c) for air transportation of small quantities of materials exhibiting very low levels of radiation.

Conforming with Section 107 of the Hazardous Materials Transportation Act (49 U.S.C. 1806) governing exemptions, the exemption in the sections cited above is limited to a two-year life unless reexamined and renewed. The exemptions were last renewed under Docket HM-149C (46 FR 24184, April 30, 1981). The legal background and regulatory history of these exemptions are discussed in a preceding notice of proposed rulemaking (42 FR 16459, March 28, 1977). As the exemptions are due to expire on May 2, 1983, MTB proposes to renew them if, following receipt and review of comments, it finds that the renewal is consistent with the public interest and safety.

(47 F.R. 54130, November 15, 1982)

49 CFR Parts 172, 173 and 175
[Docket No. HM-166Q, and HM-166F; Amdt. Nos. 172-83, 173-167, 175-28]
Exceptions for Small Quantities of Hazardous Materials; and Limited
Quantities of Radioactive Materials

Background. These amendments are based on proposals made in Notice No. 10 (82-10) Docket HM-166Q and HM-166F (47 FR 51430, November 15, 1982), and Notice No. 81-8, Docket HM-166F (46 FR 61908, December 12, 1981). This consoldiation of the separately issued notices into a single final rule is considered desirable since each of the proposals is affected by the other. A discussion of each notice and the comments received in response thereto follows.

A. Notice No. 81–8, Limited Quantities of Radioactive Materials. In this notice, MTB proposed adoption of certain rules applicable to the transportation of limited quantity radioactive materials and radioactive devices. That notice proposed changes to eliminate certain regulatory inconsistencies which have existed between shipments transported by aircraft and those transported by the surface modes ever since the HMR were consolidated in 1976. These amendments to the HMR are based on the relatively low hazards associated with limited quantity radioactive materials and radioactive devices when compared with other hazardous materials and implement requirements which will provide an adequate level of safety in transportation.

1. General. Response to this Notice of Proposed Rulemaking (NPRM) was received from twenty commenters, of which shippers of radioactive materials in limited quantities and radioactive devices comprised a majority. Comments were received from one carrier (United Parcel Service) and two carrier associations (Air Transport Association and Radiopharmaceutical Shippers and Carriers Conference) having particular interests in the proposed rule because of its potential impact. In addition, comments were received from the University of Chicago, the University of Rochester Medical Center, and the Society of Nuclear Medicine—a specialty medical society of physicians, medical scientists, and technologists engaged in the practice of nuclear medicine. Nearly half the comments received fully support the rules as proposed. While no commenters look exception to the basic objectives established in the NPRM, there are several areas in which a consensus on how best to achieve the desired results was not reached. A brief discussion of specific comments follows.

2. Regulations for transportation by aircraft. The Air Transport Association (ATA) took exception to MTB's proposal in that ATA believes the proposed requirements would be inconsistent with the International Civil Aviation Organization's (ICAO) "Technical Instructions for the Safe Transport of Dangerous Goods by Air". Specifically, ATA feels that if these materials are hazardous, they should be fully regulated and the pilot notified of their presence. If the materials are not hazardous, then ATA feels they should not be regulated. The basic requirements for these materials are identical in the two sets of regulations. MTB recognizes that the DOT rules provide options not allowed by ICAO at this time, but MTB is hopeful that a similar rule might be adopted by ICAO in the future.

3. Quasi shipping papers. Although the proposal to eliminate the detailed hazardous materials shipping paper and certification requirements received strong support from most commenters, some objections were raised regarding the new requirement for communicating the presence of radioactive materials in limited quantities. The objections center on logistics relative to the notice which commenters predict will arise as shippers and carriers seek to comply with the revised rules rather than any possible increase of risks in transportation.

Mallinckrodt, Inc. contends that a provision which permits the notice to be enclosed in the package is such that, for a carrier to be satisfied that the shipment does comply, it will necessitate opening the package. Another commenter, Beckman Instruments, Inc., believes that the presence of such a notice will in many cases result in a refusal, particularly by air carriers, to carry the package. MTB considers neither of these possibilities so serious, however, as to require a change in the rule as proposed. The rule is purposely flexible so that compliance with the requirement for written notification may be achieved by whichever means the shipper determines most appropriate. As common carriers by aircraft do not presently require shippers to identify excepted packages by external markings or hazard warning entries on the air waybill, MTB does not believe they will be so inclined following publication of this rule, except through implementation of the ICAO requirements. Since the ICAO requirement is one option contained in the final rule, this can be worked out by the shippers and carriers involved.

The National Bureau of Standards of the Department of Commerce sees the requirement for an additional written notice on their multipurpose shipping form as being very inconvenient. As an alternative, they propose that the shipper's certification required by § 172.204 also be acceptable. While the commenter's proposal has merit, it could not be effectively implemented without also retaining the descriptions "radioactive material, instruments *and* articles" and "radioactive material, limited quantity, n.o.s." as proper shipping names. With the availability of those proper shipping names, a shipper could then elect to describe its packages in the manner presently required for surface mode shipments. Essentially this would require the shipper to make a determination of which option is most acceptable; one that involves the simple statement provided in § 173.421–1(a), or another that requires, as a minimum, entries for the proper shipping name, identification number, total quantity, the name of each radionuclide, a description of the physical and chemical form, the activity contained in each package, and the shipper's certification. MTB believes an overwhelming majority of shippers would elect the former method and for that reason has decided not to provide for the suggested alternative which could generate confusion.

Associated with comments received on the proposal to require written notification as to the package contents are counter-proposals from several sources which suggest that limited quantity radioactive materials be subject to package marking requirements presently applicable to all other hazard classes. Presently, there is basic agreement between the U.S. and international regulations in not requiring package markings and MTB still agrees with reasoning which supports not requiring external markings on limited quantity radioactive materials and radioactive instruments and articles. There are no specific precautions that a carrier need take when transporting these materials under normal conditions. If the packages are involved in an accident or are lost, their radioactive nature is communicated by: (1) Marking on the inner packaging (§ 173.421(d)); and (2) the information transmitted on the "notice enclosed in or on the package, included with the packing list or otherwise forwarded with the package" (§ 173.421–1(a)). MTB believes this immediately available information adequately alerts personnel of the hazard which is present in the event of a mishap.

Comments filed on behalf of the Radiopharmaceutical Shippers and Carriers Conference called attention to a redundancy when the description "radioactive material, limited quantity, n.o.s." is used with the required statement "This package meets all requirements of 49 CFR 173.391 (*now § 173.421*) for limited quantity radioactive materials." Those comments infer that redundant information in shipping descriptions is unnecessary and should not be required. MTB agrees that this information can be combined in a single statement and still meet the intended purposes. As a result, the required statement now reads "This package conforms to conditions and limitations specified in 49 CFR 173.421 for Radioactive Material, limited quantity, n.o.s., UN 2910, 49 CFR 173.422 for Radioactive Material, Instruments *and* Articles, UN 2911; or 49 CFR 173.424 for Radioactive material, articles manufactured from natural *or* depleted uranium or natural thorium, UN 2909", as appropriate.

4. Hazard ranking. The proposal to separate radioactive material in limited quantities from the broad class of radioactive materials appearing in § 173.2 and position it between "corrosive material (solid)" and "irritating materials" was met with widespread approval. However, E.I. du Pont de Nemours and Company (Du Pont) expressed its concern over materials which also meet the definition of a higher order hazard class being subject to specification packaging, marking, labeling shipping paper, and placarding requirements regardless of the quantity of the other hazardous material present and, therefore, the degree of hazard contained. Du Pont went on to say that compliance with the proposed requirements, if unaltered, would increase its costs by $750,000 annually due to increases in packaging, freight and administrative expenses while providing no commensurate increase in safety. The comments do not contain data which support Du Pont's assertion that transportation-related costs would increase significantly. However, the comments seem to suggest that the extremely small number of incidents involving limited quantity radioactive materials as reported to MTB, could not, in a cost-benefit analysis, justify even a slight percentage increase of those costs. Du Pont's comments on this topic close by suggesting that any final rule pertaining to the reordering of the precedence of hazards table include an exception for limited quantity radioactive materials meeting the definition of a higher order hazard, in quantities equal to or less than one (1) pint (liquid or solid) and, alternatively, require such materials to be subject to the requirements of § 173.421 as amended by this rulemaking.

Since it is the intent of these amendments to grant regulatory relief to commonly shipped materials which have demonstrated an outstanding history of safe transportation. MTB agrees with Du Pont's recommendation that provisions be made to continue the broad exception from the regulations for limited quantity radioactive materials which also meet the definition of certain other hazard classes. To accomplish this, it was necessary to include radioactive materials in § 173.4 and withdraw the original proposal of reordering radioactive materials in limited quantities as a separate description in § 173.2(a). These materials are now identified as an exception in § 173.2(b)(5). Following those actions, it was then necessary to specify in § 173.421–2 requirements which preserve the previous exceptions from specification packaging, marking, and labeling for limited quantity radioactive materials meeting the definitions of certain other hazard classes, without jeopardizing safety by completely disregarding the other hazard. In § 173.421–2, two instances are considered in which limited quantity radioactive materials that also meet the definition of an additional hazard class may be offered for transportation.

The first instance (§ 173.421–2(a)) pertains to materials in hazard classes other than ORM-A, B, and C and combustible liquids in packagings having a rated capacity of 110 gallons or less. This rule directs shippers to class the material by the non-radioactive material hazard class and to prepare packages for shipment in accordance with provisions applicable to that other class.

Section 173.421–2(b) pertains to materials meeting the definition of an ORM-A, B, or C, and combustible liquids in packagings having a rated capacity of 110 gallons or less. In these instances, if the material is a hazardous waste or hazardous substance *or* if it is to be transported in a mode appropriate to the ORM class, the radioactivity is subrogated to the status of being the subsidiary hazard and the shipper is directed to class the material in the other hazard class. If, however, the material is not a hazardous waste or hazardous substance *and* the material is offered for transportation in a mode to which requirements of the HMR pertaining to the specific material and hazard class do not apply, the shipper is required to class it a radioactive material.

For packages not classed radioactive material, an indication of the presence of radioactive materials is communicated through a requirement that the shipper enter the statement "Limited quantity radioactive material" on the shipping paper in association with the basic description.

5. "Radiopharmaceuticals" Proper Shipping Name. The proposal to adopt a shipping description which distin-

guishes radioactive materials used for medical purposes from those used in power production, industrial radiography, and other non-medical applications drew three comments opposing the entry as compared to eight comments favoring the proposal.

In support of this proposal are a variety of shipper and carrier organizations which claim that the description "radiopharmaceuticals" should benefit the medical community, transportation personnel, emergency response teams, and the radiopharmaceutical industry without increasing the risks to public health and safety in transportation. Such benefits are generally thought by these commenters to be derived from:

(1) A more accurate indication of the risk involved in the transport of and, when necessary, an appropriate response to incidents involving radiopharmaceuticals; (2) more expeditious and less costly shipments of health care products; (3) fewer frustrated shipments; and (4) a more accurate characterization of the nature of these materials which improves general understanding without loss of awareness of the radioactive hazard involved.

A comment from the University of Rochester Medical Center expressed opposition to the proposed name because it is viewed as being misleading, if the intent of this change is to distinguish between the more hazardous long-lived radioisotopes and the shorter-lived isotopes used in nuclear medicine which are assumed to be less hazardous. The commenter goes on to say that establishing another name for radioactive material would be confusing to transportation workers and would probably not accomplish the intended goal of speedy delivery.

The ATA response to this proposal suggests the new proper shipping name is not consistent with United Nations (UN), International Civil Aviation Organization (ICAO), and International Atomic Energy Agency (IAEA) standards which ATA understands to be an on-going goal of MTB. Accordingly, ATA recommends that MTB first introduce this proper shipping name to the UN and IAEA. In addition, since this proposal appears less restrictive and inconsistent with the ICAO technical instructions ATA objects to its adoption as a final rule.

Finally, comments filed by Du Pont expressed the view that the proposed proper shipping name would not provide for the breadth of relief intended and does not address the underlying cause of carrier delays resulting from frequent compliance checks by Federal, State, and local enforcement personnel, in Du Pont's opinion, the underlying cause is a lack of general understanding of the actual levels of hazards present as described by existing proper shipping names which would be more appropriately addressed by providing immediate advice to emergency response personnel with specific information as to the nature of the hazard present. Du Pont continues by claiming their experience indicates the proposed name would apply to only a limited number of products, and though achieving some desirable benefits, that gain would be negated by the additional burden of dual descriptions for international air transportation of these products due to the resulting dissimilar descriptions currently specified by countries which have adopted the Restricted Articles Regulations of the International Air Transport Association. The comment concludes by requesting that MTB withdraw the proposed addition of the proper shipping name "Radiopharmaceuticals, n.o.s." from its final rulemaking while retaining the existing descriptions.

Most of the comments supporting adoption of the description focused on the perceived benefit of more rapid delivery. It is contended that carrier personnel would be able to expedite the movement of these materials without undue surveillance since these products are associated with human health care.

Considering the controversy arising from this proposal MTB has decided not to adopt it since this Docket is designed to address only those issues which are not controversial and which may be handled in an expeditius manner. With the option available for shippers to voluntarily add additional information to the packages and shipping papers, it is thought best to avoid the proper shipping name "Radiopharmaceuticals, n.o.s." at this time.

B. Notice No. 10, Exceptions for Small Quantities of Hazardous Materials. In this NPRM it was proposed to grant significant relief from the Department's Hazardous Materials Regulations for the transportation of small quantities of flammable liquids, flammable solids, oxidizers, organic peroxides, corrosive materials. Poison B, ORM-A, B and C, and limited quantity radioactive materials which also meet the definition of any of these other hazard classes. The relief proposed was dependent on conformance to newly proposed performance packaging requirements, and specified restrictions. As a result of a petition filed by the Scientific Apparatus Makers Association (SAMA), MTB proposed what it believed to be an acceptable standarized packaging for the shipment of these materials. MTB also believes that DOT exemptions (6971, 7755, 7921, 8116, 8285, 8292, 8423, 8581, and 8658), which presently authorize the use of several different packaging techniques, may be eliminated by the proposed rule.

Codification of the rule originally proposed to be in § 173.3(d) has been changed to § 173.4 Exceptions for small quantities.

These proposals were made on the basis of favorable shipping experience achieved under exemption which demonstrate that certain innovative techniques are both safe and effective. This rulemaking tailors the terms of those specific exemptions cited above into a rule of general applicability.

1. General. MTB received 42 comments from industry, trade associations and one Federal agency relative to Notice No. 10. The ATA objects to small quantities of hazardous materials moving in air transport without identification and suggests that incompatible materials might be shipped together. In general, ATA supports the ICAO rules and feels there should be no exception for small quantities.

MTB believes that the approach toward the exception of small quantities of hazardous materials is correct. Furthermore, it is believed that adequate safeguards are in place to prevent the likelihood of incompatible materials from posing an unacceptable risk during air transportation. MTB is currently recommending that ICAO adopt an exception similar to this rule.

Most of the comments support the proposal to standarize the packagings for small quantities of hazardous materials with only a few minor revisions. Basically, comments concern: (1) Raising the quantity limitations (for liquids and solids) per inner receptacle; (2) increasing the LD_{50} value for Poison B materials; and (3) allowing the certification to be placed on the "outside" of the package.

2. Addition of Materials Belonging to the Combustible Liquid, Flammable Gas and Non-flammable Gas Hazard Classes. One commenter recommended the inclusion of flammable gases, nonflammable gases, and combustible liquids to the list of authorized materials. MTB has not added these gases because it would be beyond the scope of the NPRM. Combustible liquids, other than those which are hazardous wastes or hazardous substances, in quantities of 110 gallons or less are not subject to the regulations, therefore, the inclusion of this hazard class is unnecessary.

3. Increased Quantity Limits. Nearly one-third of the commenters suggested that the quantity limitation per inner receptable as proposed in paragraph (d)(1) (i) and (iii) of § 173.3 be increased from the proposed 25 milliliters for liquids and 25 grams for solids to at least 100 milliliters for liquids and 100 grams for solids (except for poisons) as "these are the quantities requested by customers".

MTB has increased the maximum permissible quantity limitation per inner receptacle to 30 milliliters for liquids and 30 grams for solids. MTB prefers keeping the quantity limit at amounts which it considers reasonable from a safety standpoint and yet adequate to meet the needs of shippers.

Over a third of the commenters suggested that MTB reconsider the quantity limitation for Poison B materials as proposed in paragraph (d)(1)(iii) of § 173.3 and clarify the wordings in the paragraph. Some commenters believe that the LD_{50} should be increased anywhere from 5 times to 1000 times the LD_{50} value. Most of the commenters believe that the more stringent packaging requirements would provide more than adequate safety protection for Poison B materials and, in the event of an inner receptable failure during transportation, the contents would be taken-up by the absorbent material surrounding it, thereby greatly mitigating the potential for exposure of a harmful quantity.

MTB agrees with the commenters' assertions relative to the high integrity of the packagings, but believes an increase of only 20 times the LD_{50} value to be a more sound approach toward minimizing the risk potential of packages containing Poison B materials. MTB believes that by limiting the maximum quantity per inner receptacle, toxic materials, such as parathion, may be safety transported without posing a significant risk to cargo handlers and emergency response personnel. This paragraph has also been revised for clarification.

4. Miscellaneous Comments to Proposed § 173.3. One commenter suggests that the "closure" requirement as proposed in paragraph (d)(3) of § 173.3 should apply to air shipments only. MTB disagrees. The closure requirement was a critical element in MTB's consideration to propose a broad exception for "small quantity" shipments when transported by all modes and would provide an extra margin of safety that supports the adoption of the exceptions.

Two commenters suggested that MTB reword the compression requirements proposed in paragraph (d)(6)(i) of § 173.3 for clarity. MTB has revised this paragraph as a few words were erroneously omitted in the NPRM.

One commenter requested clarification as to whether all five drop tests proposed in paragraph (d)(6)(ii) of § 173.3 must be performed. *All* tests must be conducted to establish the performance of the package. However, the five tests need not be performed on the same package. The paragraph is revised by replacing the word "any" with the word "each".

Two commenters questioned the six-foot drop proposal. MTB believes the six-foot drops are necessary to assure an acceptable level of safety for shipments moving under a broad exception from the communication regulations.

A comment suggests that the words "or 173.25" as proposed in paragraph (d)(7) of § 173.3 be deleted. MTB agrees, noting that in this instance, the potential hazard would be the violation of § 173.21. The packagings (cushioning material), as proposed, must be capable of absorbing the entire content of the inner receptacle. Section 173.21 effectively forbids the offering of incompatible materials together in the same package or overpack.

Two commenters believe the 65 pound weight limit as proposed in paragraph (d)(8) of § 173.3 is unnecessarily restrictive. MTB disagrees. The same requirement currently exists for materials classed ORM–D which contain many of the basic materials addressed in this rulemaking. The 65 pound gross weight limitation per package contributes to safe handling and limits the aggregate risk presented by an individual package; similar benefits should be achieved with the new packaging without being unnecessarily restrictive to shippers or consignees. In addition, MTB has removed the reference "29.48 kilograms" from the paragraph. Persons wishing to use the metric units may refer to § 173.26.

Several commenters opposed the requirement of placing the certification "inside" the package as proposed in paragraph (d)(9) of § 173.3 and contend that it should appear on the outside where it may be seen without opening the package. MTB has reconsidered its approach on this issue and has modified the rule.

Paragraph (b) of § 173.4 contains the same date that is specified in § 173.421–1(b)(2). This was done so that both exemptions from the legislative prohibition to transport radioactive materials on passenger-carrying aircraft could be evaluated for renewal at the same time.

5. Limited Quantities of Radioactive Materials on Passenger-Carrying Aircraft. See **Federal Register** issued May 2, 1983 (48 FR 19719).

6. Editorial amendment. Paragraph (g) of § 172.101 is revised to reflect a reference to § 171.3, 173.3, 173.4 and 173.5 since each contains certain exceptions in addition to those that were referenced in paragraph (g).

(48 F.R. 30132, June 30, 1983)

49 CFR Part 173
[Docket No. HM-166Q; Amdt. No. 173–167]
Exceptions for Small Quantities of Hazardous Materials; Correction

The MTB is correcting the final rule for small quantities of Poison B materials appearing in § 173.4 (a)(1)(iii) because the Poison B hazard class, as defined in § 173.343, does not require the determination of an LD_{50} (or LC_{50}) or even an approximate lethal dose (ALD). The MTB emphasizes that the DOT toxicity tests are "limit tests" (i.e., pass/fail) adapted to meet DOT transportation safety requirements. Since "LD_{50} tests" are *not* required, the final rule must be corrected to remove the reference to "LD_{50} value" in association with § 173.343 which was inadvertently included in the final rule.

On further consideration MTB believes that a single limit of one gram (based on twenty times the breakpoint of 50 milligrams per kilogram of body weight for oral toxicity specified in § 173.343) is appropriate for the exceptions for poisons provided by § 173.4. Any calculation to determine the maximum quantity of a Poison B material authorized per inner receptacle is meaningless without a specific toxicity value and MTB believes that it is inappropriate to call for additional tests to be conducted in order to gain the benefits provided by the exception.

(49 F.R. 19025, May 4, 1984)

40 CFR Parts 171, 172, 173, and 175
[Docket No. HM-166-S, Notice No. 84–10]
Magnetized Material

Under the provisions of 49 CFR 173.1020, a material is considered to be "magnetized" and subject to the requirements of the Hazardous Materials Regulations (when transported by aircraft) when it has a magnetic field strength of 0.002 gauss or greater at a distance of 7 feet from any point on the surface of the package, or which is of such mass that is could affect aircraft instrumentation, particularly magnetic compasses. Furthermore, a material with a measurable magnetic field greater than 0.00525 gauss, when measured from any package surface at a distance of 15 feet, must be shielded to reduce the reading to a level that is no greater than 0.00525 gauss before being offered for transportation by aircraft.

This notice is in response to a petition for rulemaking filed with the MTB by the Motor Vehicle Manufacturers Association (MVMA), under the provisions of 49 CFR 106.31. The MVMA petitioned the MTB to deregulate certain materials, such as auto fenders and other automobile parts, which may meet the lower criteria in § 173.1020 for magnetized material, but pose little or no transportation hazard because they do not affect aircraft instrumentation.

The MTB believes that the current rules on magnetic materials are obsolete due to improvements in the technology of aircraft instrumentation over the past 30 years. Modern aircraft use electronic compasses with magnetic compasses as backups. The sensors for the magnetic backup compasses of modern aircraft are located sufficiently far away from cargo bays so that the possible marginal magnetic properties of metal objects such as automobile parts will not cause a measurable deflection effect on the compass. It is normal procedure for pilots to check the aircraft's magnetic compass and electronic compass against the runway heading before takeoff.

In order to perform an accurate test for magnetism at the lower gauss limit currently specified in § 173.1020 (0.002 gauss), a test would have to be conducted away from any possible magnetic sources, using a gauss meter which costs about $1,500.00. To conduct this test, large metal objects which are to be offered for transportation, and which are not intentionally magnetic, but have acquired magnetic properties during their manufacture or because of their orientation, would have to be tested in a controlled environment. If the test shows that the metal objects are above the 0.002 gauss level at a distance of 7 feet, under current rules they must be labeled with the MAGNETIZED MATERIAL label, marked "ORM-C", and shipped as a hazardous material. The metal objects must be reloaded and brought to the airport for transportation by aircraft. When the metal objects are moved, the magnetic field created by the mass of those objects may change due to the new placement of the objects. When they are loaded aboard the aircraft, the magnetic field may again change, due to placement of the object aboard an aircraft. Consequently these tests at low gauss levels are not repeatable and the necessity for and benefit of regulating metallic ladings at low levels at low levels of magnetism is questionable.

The MTB believes that testing metal objects for magnetism is not necessary, other than those deliberately fabricated with magnetic properties. Marginal magnetic properties of other ladings are of such low levels that the probability of deflecting aircraft insruments is negligible. This notice proposes to eliminate the 0.002 gauss threshold and to continue to forbid the transportation of magnetized material over the 0.00525 gauss level at a distance of 15 feet from the package surface for carriage aboard aircraft. The MTB believes that the rules proposed in this notice, if adopted, would not reduce the level of air safety and would relieve shippers and carriers of burdens imposed by undue regulation.

The MTB will propose that the standards for magnetized material in the International Civil Aviation Organization (ICAO) Technical Instructions be changed to reflect these proposed amendments if a final rule is issued as proposed. Should these amendments be adopted as a final rule, the Federal Aviation Administration (FAA) is considering the publication of an Advisory Circular for aircraft built before 1955 to assure that the proper procedure for stowage of magnetized material is followed to avoid affecting instruments on aircraft which might not have compasses with remote sensors.

(49 F.R. 37438, September 24, 1984)

49 CFR Parts 171, 172, 173 and 175
[Docket No. HM-166-S, Amdt. Nos. 171–84, 172–102, 173–195, and 175–34]
Magnetized Material

On September 24, 1964, RSPA published a notice of proposed rulemaking in the FEDERAL REGISTER, (Notice 84–10), (49 FR 37438). That notice proposed to amend the regulations governing the transportation of magnetized materials aboard aircraft. The notice was published in response to a petition for rulemaking submitted by the Motor Vehicle Manufacturers Association (MVMA), RSPA proposed to deregulate certain materials, such as automobile fenders and other automobile parts, which may meet the lower magnetic criteria in 49 CFR 173.1020 for magnetized material (0.002 gauss or greater at a distance of 7 feet from any point on the surface of the package). The notice also proposed to eliminate the *Magnetized Material* labeling requirement, and the ORM–C marking for packagings which have a gauss level of 0.00525 gauss or less at a distance of 15 feet, and to forbid transportation of materials by aircraft, which as packaged, have a gauss level of over .00525 gauss at a distance of 15 feet.

As stated in the notice, RSPA believes that the current rules on magnetized materials are obsolete and fail to recognize improvements in the technology of aircraft instrumentation over the past 30 years that substantially prevent most magnetized materials from having an adverse effect on the operation of instruments.

In response to Notice 84–10, RSPA received 21 written comments. The respondents included shippers of magnetized materials, the Airline Pilots Association (ALPA), the Air Transport Association of America (ATA), and the U.S. Air Force (USAF). Of those commenters expressing an opinion on the overall merits of the proposal, all commenters were in favor of the proposal except the ALPA, the ATA and the USAF.

The ALPA expressed concern that while individual shipments may not contain sufficient magnetic force to affect aircraft instrumentation, multiple shipments aboard an aircraft may affect instrumentation. Operating information which is required to be furnished to pilots of aircraft (operating under rules contained in 14 CFR Parts 121 and 135) include cockpit checklists. Cockpit checklists include making certain that instruments are working properly. If multiple quantities of individual shipments, each containing nonregulated amounts of magnetized material are stowed aboard an aircraft so as to affect the instrumentation, the problem would become apparent as the pilot performs the pre-flight check. If as a result the instruments are diverted, corrective action must be taken before takeoff. RSPA believes that such occurrences are highly unlikely because of the remote positioning of magnetic flux detectors in modern aircraft. No test results or technical justification were submitted in support of ALPA's position.

The ATA commented that a number of its members were in favor of the proposed rule and one member was against the proposal, stating that in recent years DC-8 aircraft were twice affected by materials with magnetic properties. No information or documentation was supplied regarding the details of these incidents, and there was no indication as to whether the materials were properly or improperly transported under the provision of the HMR.

The USAF commented in opposition to the proposed rule without providing technical details.

RSPA and the Federal Aviation Administration (FAA) believe that these rules will not reduce the level of air safety and will relieve a burden of undue regulation on shippers and carriers. To assure the proper stowage of cargo aboard aircraft which might not have compasses with remote sensors or aircraft having compass master units located within the fuselage, the FAA is publishing an Advisory Circular to provide information relevant to the preparation and loading of magnetic materials for shipment in civil aircraft.

RSPA is delaying the effective date of this rule for 90 days to allow petitions for reconsideration to be submitted by interested parties. Commenters who can provide test results or technical justification may petition RSPA for reconsideration following the procedure in § 106.35.

It should be noted that this rule amends the rules for air transportation of hazardous materials under the provisions of 49 CFR and does not affect the rules under the Technical Instructions for the Safe Transport of Dangerous

Goods by Air, published by the International Civil Aviation Organization (ICAO). However, RSPA will recommend that the ICAO Technical Instructions be amended accordingly.

(50 F.R. 48419, November 25, 1985)

49 CFR Part 173
[Docket No. HM-166V, Amdt. No. 173–199]
Hazardous Materials; Uranium Hexafluoride

On November 18, 1986, RSPA published a final rule under Docket HM-166V, entitled "Hazardous Materials: Uranium Hexafluoride", amendment numbers 172–107 and 173–198 at 51 FR 41631. These amendments strengthened the hazardous materials regulations (HMR, 49 CFR Parts 171 through 179) applying to the transport of UF_6. Specifically the amendments required that containers for UF_6 be constructed, inspected, tested, cleaned and repaired in accordance with a consensus standard. ANSI Standard 14.1–1982. In addition, the amendments placed limitations on filling cylinders and transporting of filled cylinders.

Subsequent to publication of the final rule, RSPA learned that the new requirements had disqualified a large number of existing UF_6 cylinders, thereby causing serious problems for shippers of this material. In response to five petitions for reconsideration, RSPA on December 24, 1986, published at 51 FR 46674 a revision of the final rule, which both delayed, until June 30, 1987, the requirement that cylinders be constructed in accordance with ANSI Standard 14.1–1982 and modified the packaging requirement by authorizing not only ANSI Standard 14.1–1982 but previous editions of that standard. The revision also announced that a public meeting would be held on March 2, 1987, to discuss issues related to RSPA's rules on the transport of UF_6. The revision did not alter or delay requirements for cleaning cylinders.

On March 2, 1987, the public meeting was held on the U.S. Department of Transportation. Commenters cited potential disruption in defense and civilian nuclear activities if approximately 50,000 UF_6 packagings were removed from transportation by the new regulation.

In the final rule, cleaning procedures specified in Appendix A of ANSI Standard 14.1–1982 were incorporated in 49 CFR 173.420(a)(1) and made applicable to the cleaning of new and used packagings for UF_6. During the March 2, 1987 meeting, several attendees noted that the cleaning procedures in Appendix A are not suitable for cleaning used packagings. Specifically, the Appendix A procedures require that hydrostatic testing be performed prior to cleaning the interior. The commenters believed that introduction of water into a packaging containing a residue of UF_6 during hydrostatic testing could cause a reaction between UF_6 and the water. This reaction produces hydrogen fluoride (HF), a corrosive material which may damage the interior of the packaging and could potentially be released and injure persons in the area. RSPA agrees with these commenters and this emergency final rule is being issued to change § 173.420(a)(1) to reference cleaning procedures in ANSI Standard N14.1–1982, which contains procedures for cleaning both new and used packagings, and to delete the reference to Appendix A of the standard, which is suitable for new packagings only.

Other issues discussed in the March 2, 1987 meeting are under consideration by RSPA and not addressed in this final rule.

Under the provisions of section 553 of the Administrative Procedure Act, agencies are permitted to issue a rule in final form when notice and public procedure are impracticable, unnecessary or contrary to the public interest. This final rule eliminates an erroneous and potentially hazardous regulatory requirement and delay of its implementation is believed to be contrary to the public interest. Therefore, I find, under 5 U.S.C. 553, that notice and public procedures on the rule and delay in its effective date are contrary to the public interest.

(52 F.R. 7581, March 17, 1987)

[49 CFR Parts 107, 173, 176, 178]
[Docket No. HM-167; Notice No. 78-121]
INTERMODAL PORTABLE TANKS
Proposed Packaging Specifications

The portable tank has become an increasingly popular method of packaging hazardous materials for international transport. The rapid increase in the use of portable tanks for the international transport of hazardous materials can be attributed, in part, to the advent of containerized transport. When mounted in a frame with dimensions conforming to those of a standard freight container, a portable tank can be readily transported aboard a containership. The loading and unloading of these ships is greatly facilitated through use of the modern cargo handling equipment normally employed in containerized transport, and since the tank can be directly transferred from ship to transport vehicle, lost time in the land-sea interface is minimized. In addition, assuming sufficient demand for a product exists, the portable tank provides an economic advantage over shipping hazardous materials in smaller packages by reducing packaging and handling costs. Safety in handling is enhanced through use of these tanks since direct exposure of operating personnel to the materials being transported is minimized.

Recognizing the inherent advantages of transporting hazardous materials in portable tanks and the increasing popularity of this method of transport, IMCO's Subcommittee on the Carriage of Dangerous Goods undertook the preparation of a recommendatory standard governing the design and construction of portable tanks intended for the international carriage of hazardous materials by sea. The purpose of this effort was to promote the harmonization of the various national requirements for the design, construction and use of portable tanks in order to facilitate the international transport of hazardous materials in portable tanks. The United States took an active part in the development of these recommendations and, in 1972, an amendment to IMCO's International Maritime Dangerous Goods (IMDG) Code was published which included the newly developed standards for portable tanks.

Shortly after the initial publication of IMCO's standards for portable tanks intended for use in the marine mode, the United Nations Committee of Experts on the Transport of Dangerous Goods began development of a recommendatory standard for portable tanks which would be acceptable for multimodal transport. The first phase of work on this standard was completed in 1974 with the Committee's adoption of a recommendation for the design and construction of a portable tank suitable for the carriage of flammable liquids by both land and sea modes. Subsequent to this initial effort, additional standards were developed by the committee which addressed the carriage of other classes of hazardous materials in these tanks.

While the portable tank standard initially adopted by the Committee of Experts was considered a suitable standard for tanks to be transported by both land and sea modes, it was nevertheless substantially different from that developed by IMCO in that it was based to a great extent on existing European road and rail requirements. Recognizing the need for the harmonization of these two standards, the UN Committee of Experts and IMCO's Subcommittee on the Carriage of Dangerous Goods have worked closely over the past two years in an effort to minimize the difference between their two recommendations. These efforts have been, for the most part, successful and the harmonized standard resulting from this work, as well as experience gained through the Department's exemption program, has provided the basis for the portable tank specifications proposed in this notice.

Exemptions authorizing the use of portable tanks similar in design and construction to those proposed in this notice have been sought and, where appropriate, granted for over ten years. The exemptions were necessary because the Department's regulations failed, with the exception of DOT Specification 60 portable tank, to provide for portable tanks designed specifically for the transportation of flammable, corrosive, and poisonous and other liquid hazardous materials. The exemptions have been granted for portable tanks generally fulfilling the requirements of both a DOT specification for cargo tanks and earlier versions of the international standards.

The initial effort towards the development of DOT portable tank specifications conforming to the IMCO standards was undertaken by the Chemical Transportation Advisory Committee (CTAC), which is a committee providing advice and guidance to the Commandant of the Coast Guard regarding the transport of hazardous materials in the marine mode. This work was begun shortly after the initial publication of IMCO's portable tank recommendations and the initial draft of the IM 100 and IM 101 portable tank specifications was completed by 1974. However, as a result of several major regulatory projects underway at that time, in particular the consolidation of the Hazardous Materials Regulations under Dockets HM 112 and HM 103, the draft specifications prepared with the assistance of the CTAC were never published as a Notice of Proposed Rulemaking, and, due to the changes in the IMCO standards stemming from the harmonization effort with the UN, the draft required extensive revision to insure conformance with the latest IMCO and draft UN standards.

As the UN/IMCO portable tank standards gain ever increasing worldwide acceptance, it has become imperative for the United States to recognize portable tanks constructed to these standards in order to facilitate the international transport of hazardous materials in portable tanks. The two most predominant types of international portable tanks are those designated as IMCO Type 1 and Type 2 tanks. These are low pressure tanks designed specifically for the carriage of hazardous liquids with vapor pressures less than 43 pounds per square inch, absolute, at 150° F. The Type 1 tank must have a maximum allowable working pressure (MAWP) of 25 pounds per square inch, gauge, or greater, while the Type 2 tank must have a MAWP of less than 25 pounds.

Existing DOT regulations currently contain no comparable specifications for portable tanks intended solely for the carriage of flammable, corrosive, poison and various other hazardous liquids and, as a result, the transportation of these liquids within the United States in tanks conforming to IMCO Type 1 or Type 2 standards can presently be authorized only by exemption. The ever increasing demand to transport hazardous materials in these types of tanks, resulting from the general worldwide recognition of the IMCO standards, has necessitated the issuance of dozens of such exemptions. Therefore, by introducing the IM 100 and IM 101 specifications (conforming to the IMCO Type 2 and Type 1 tanks respectively), regulatory recognition of these portable tank types, which are all ready being widely used in the United States under exemption, is provided. This will serve not only to facilitate the international transport of hazardous materials in portable tanks, but also to eliminate the time constraints and economic burdens inherent in the exemption process.

One of the goals of the efforts underway internationally to develop standards such as the UN/IMCO portable standards is to eliminate a duplicity or redundancy of approval efforts by various national authorities through implementation of widely recognized and mutually acceptable international standards. Implementation of the portable tank specifications proposed in this notice will serve to accomplish this by affording the American importer or exporter the opportunity of shipping in a DOT specification portable tank that is compatible with the accepted international portable tank standards. To promote the acceptance of these tanks without the necessity of additional inspections by foreign administrations in whose jurisdictions the tanks will also be used, the MTB believes it is necessary that the construction and testing of the tanks be supervised by a disinterested party. This notice proposes that certain qualified, disinterested persons may be designated as approval agencies and that such agencies may approve Specification IM 100 and IM 101 tanks. It is anticipated that many of the parties who will apply for designation as approval agencies will be organizations already recognized by foreign governments to issue approvals for portable tanks on behalf of those governments. The MTB believes that this action will minimize the total number of authorities who must approve a tank to be used in international transport in that one approval agency may be able to issue approvals acceptable to many national administrations and for various modes of transport, with only a single inspection. This should reduce the cost and time required to move these multimodal tanks in international commerce.

The primary drafter of this document is LT Edward A. Altemos, USCG, Office of Hazardous Materials Regulation. The following is an analysis and explanation, by section, of the more significant features of this regulatory proposal.

Section 107.3. This section defines "Approval Agency".

Sections 107.401–407. These sections set forth the procedures whereby the Associate Director for Hazardous Materials Regulation (HMR), MTB, may designate qualified persons to approve certain DOT specification packagings on his behalf. The procedure for filing an application for designation as an approval agency would be given as well as the information required to be included in the application. Section 107.403 would provide the criteria by which the Associate Director for HMR selects individuals or organizations to act as approval agencies. The criteria are designed to permit selection of any organization, foreign as well as domestic, that is technically competent for the purpose and free from undue influence by those involved with the fabrication, ownership or movement of the packages it will be called upon to evaluate. The required action on the part of the Associate Director for HMR in processing each application is provided as well as the standard conditions which are a part of each designation issued. Finally the procedures by which the Associate Director for HMR, MTB, may withdraw such designations, or by which an approval agency may voluntarily terminate its designation, would be set forth. The designation could only be withdrawn if:

(1) The application submitted by an approval agency contains misrepresentation of facts relative to the organization;

(2) The approval authority fails to comply with the terms of its designation; or,

(3) The approval authority is incompetent.

Generally, before the Associate Director for HMR could withdraw a designation as an approval agency, the organization concerned must be informed of the reasons for which the withdrawal has been undertaken and must be accorded an opportunity to demonstrate or achieve compliance in the deficient areas.

Section 173.32a. This section would provide the basic procedures by which Specification IM 100 and IM 101 portable tanks could be approved. Applications for approval of portable tanks would be submitted by either the owner of the tanks or the manufacturer. Each application would be required to include all engineering drawings and calculations necessary for the approval agency to ensure that the tank design complies in all respects with the appropriate specification. An incomplete application would be returned to the applicant within thirty days of its receipt by the approval authority with an indication of the reasons for which the application has been deemed to be incomplete. If an application is complete the approval agency would review the design and witness all required tests before issuing the approval certificate. The approval authority would maintain a set of the approved drawings and calculations for each tank design it has approved, as well as a copy of each approval certificate it issues, for a period not less than 15 years. In addition, a copy of each approval issued would be forwarded to the MTB. If an application for approval is denied, the approval authority would inform the applicant of the reasons for which the approval was denied. Denial of an application for approval could be appealed to the Associate Director for HMR.

Existing portable tanks, which are being operated under DOT exemption and which are substantially in conformance with either the IM 100 or IM 101 specifications, could be designated as a specification tank by the Associate Director for HMR. Because many of the portable tanks currently being operated under DOT exemption were designed to comply with either the IMCO standard or the DOT MC 306 or 307 specifications, the MTB believes that a number of exemptions could be designated as IM 100 or IM 101 specification tanks as appropriate.

Finally, procedures are proposed governing the review and approval of modifications to existent Specification IM 100 and IM 101 portable tanks. Requirements for retesting of modified tanks and the reissuance of approval certificates, if appropriate, are proposed.

Section 173.32b. This section proposes periodic retesting and reinspection of Specification IM 100 and IM 101 portable tanks. Tanks would be pressure retested at intervals of not more than five years and spring loaded pressure relief devices would be retested at intervals of not more than two and one-half years. In addition each tank would be visually reinspected at least once every two and one-half years. It is proposed that any tank which becomes seriously deteriorated or which is damaged to an extent which may adversely affect the tank's ability to retain its contents, must be pressure retested. All tests and inspections are to be witnessed by one of the authorized approval agencies and records of all tests and inspections conducted would be retained by both the owner of the tank and by the witnessing approval agency.

Section 173.32c. This section proposes basic requirements for the use of Specification IM 100 and IM 101 portable tanks. The majority of the requirements in this section have been derived from § 173.32 and consist of general requirements governing the use of portable tanks. However, one paragraph has been included which would permit the Associate Director for HMR to authorize for transportation in Specification IM 100 or IM 101 portable tanks certain hazardous materials which are not specifically authorized in Part 173. It is anticipated that Part 173 would be periodically updated to reflect the authorizations issued in this manner.

Section 173.116. Outage requirements are proposed for portable tanks containing flammable liquids. The filling relationship proposed is intended to limit the loading of a tank such that the tank is not more than 98 percent full by volume at a temperature of 122° F.

Sections 173.119 through 173.630. Various amendments are proposed to authorize the carriage of hazardous materials in Specification IM 100 or IM 101 portable tanks. The specific materials authorized have been taken either from existing exemptions under which portable tanks have been operating or from the lists of hazardous materials suitable for carriage in such tanks as developed by the UN Committee of Experts or IMCO. The MTB is fully aware that many hazardous materials not proposed for carriage in Specification IM 100 or IM 101 portable tanks by this notice are suitable for carriage in such tanks. Commenters are therefore urged to suggest amendments to Parts 173 that would authorize additional materials for carriage in these tanks. Such proposals should include sufficient data relative to the properties of the materials to permit the MTB to make a determination regarding the acceptability of the material for bulk transportation as well as the type of tank which should be required.

The maximum allowable working pressures required for tanks carrying particular materials have, in most instances, been specified according to the pressure ratings suggested for these materials in the international standards. In general, these standards make use of four basic pressure ratings for portable tanks. These pressure ratings approximately correspond to the maximum allowable working pressures of 14.2 psig, 25 psig, 38 psig and 58 psig which are utilized extensively throughout this notice. In other cases maximum allowable working pressures proposed for particular materials in this notice have been based upon successful experience gained in transporting the material in similar tanks under exemption. In general, higher working pressure tanks are required for more dangerous materials or for materials transported under the generic shipping descriptions from which the severity of the hazards of the particular materials cannot be readily ascertained. Similarly, other requirements such as increased minimum shell thickness or prohibition of bottom outlets, are based upon the severity of hazard of the material. For hazardous materials possessing significantly high vapor pressures, the maximum allowable working pressures proposed are no less than the sum of the vapor pressure of the material at 150° F. plus an allowance of 5 psig for dynamic loadings. The 150° F. reference temperature has been taken from the UN/IMCO recommendations.

Section 176.340. A new § 176.340 is proposed which lists the types of portable tanks acceptable for the carriage of combustible liquids in the water mode. Except for the Specification IM 100 and IM 101 portable tanks, all other tanks listed are currently authorized for carriage of combustible liquids under 46 CFR 90.05-35. The MTB believes that there will be benefit to the user of the regulations if this list is reproduced in the regulations dealing solely with the transportation of packaged hazardous materials. Finally, it is proposed that portable tanks of types other than those listed may be used provided they are approved by the Commandant (G-MHM).

Section 178.270. Proposes general design and construction requirements applicable to both Specification IM 100 and IM 101 portable tanks. These tanks are intended for the carriage of liquids having a vapor pressure less than 43 psia at a temperature of 150° F. and are basically equivalent to the IMCO Type 2 and Type 1 tanks respectively.

A basic requirement proposed for all tanks is that they be designed to retain contents under all normal conditions incident to transportation. In addition, they would be required to have a cross section that is capable of being analyzed, either mathematically or experimentally, to insure that the maximum stress levels prescribed will not be exceeded in normal service. It is proposed that the center of gravity of all filled tanks be approximately centered within the points of attachment of lifting devices to insure that the loads imposed during lifting are evenly distributed throughout all lifting attachments. Insulation, when installed, would be jacketed to prevent damage or other conditions which would reduce its effectiveness. All tanks would be constructed of steel and the chemical and physical properties of the materials of construction would meet the minimum requirements established in Section VIII, Division 1 of the Boiler and Pressure Vessel Code of the American Society of Mechanical Engineers (ASME Code).

Maximum permissible stress levels for tanks, at the hydrostatic test pressure and under specified dynamic loadings, are proposed in § 178.270-4. Some difficulty exists in adhering to the IMCO and draft UN standards in this regard in that both of these recommendations contain a provision that tanks should be constructed to a recognized pressure vessel code but then proceed to specify maximum permissible stress levels that are inconsistent with the

ASME Code which is the most widely recognized pressure vessel design code used in the United States. In particular, the ASME Code limits the maximum permissible stress at test pressure to the lower of 93.75 percent of the specified minimum yield strength or 37.5 percent of the specified minimum tensile strength, while the international standards limit stress at the test pressure to the lower of 75 percent of the specified minimum yield strength (which for austenetic steels would be determined at the 1.0 percent offset) or 50 percent of the specified minimum tensile strength. The ASME Code stress levels insure, at all times, a minimum factor of safety of 4:1 against ultimate strength (at the maximum allowable working pressure) while the UN/IMCO stress levels would, for certain materials, permit a minimum 3:1 factor of safety against ultimate strength at that pressure. For the purpose of this notice, stress levels conforming to those prescribed in the ASME Code are proposed. The MTB recognizes, however, that inconsistency between domestic regulations and the international standards of a matter as fundamental as maximum permissible stress levels could, conceivably, adversely affect free movement of portable tanks in international commerce. For this reason, commenters are specifically requested to comment on the relative merits of both approaches to limitation of stress levels (i.e. ASME Code method and UN/IMCO method) devoting particular attention to:

(a) The potential that safety would be compromised if a 3:1 factor of safety was to be applied in certain cases;

(b) Any potentially significant barriers to trade that could result from the use of the stress level limitations proposed herein that are inconsistent with those in the international standards; and,

(c) The extent to which tanks constructed to a 3:1 factor of safety are actually in use throughout the world and reports of relevant transportation experience concerning such tanks.

Minimum thickness for heads and shells are proposed in § 178.270-5. The values specified assume tank construction is of mild steel which has been defined as a steel with a guaranteed minimum tensile strength of 52,500 pounds per square inch and a guaranteed minimum percentage elongation of 27, as recommended in the UN/IMCO standards. For tanks constructed of other than mild steel as defined, the required minimum thickness could be reduced by use of a mathematical equivalence relationship. The tensile strength of the material of construction as determined through actual testing could be used in place of the guaranteed minimum tensile strength tabulated for the material in the ASME Code.

In § 178.270-6 it is proposed that the stress in tank supports, frameworks or lifting attachments should not exceed 80 percent of the specified minimum yield strength of the materials of construction under specified conditions of dynamic loading. Furthermore, framework on tank containers would fully comply with the requirements of 49 CFR Parts 451 and 452 (43 FR 16948-16951, April 20, 1978) issued to implement the International Convention for Safe Containers.

Requirements for joints in tank shells are proposed in § 178.270-7. All joints would be made by fusion welding. Joint preparation, welding procedures and efficiencies would be as prescribed in the ASME Code. If they are not ASME Code qualified, welders would have to be qualified by the approval agency using the procedures provided in the ASME Code.

General requirements concerning the protection of valves and accessories are proposed in § 178.270-8.

In § 178.270-9 it is proposed that all tanks be provided with inspection openings of adequate size to allow for complete internal inspection. For tanks with a capacity exceeding 500 gallons a manhole of specified minimum dimensions would have to be provided.

Tanks not fitted with vacuum relief devices would be required by § 178.270-10 to be designed to withstand a positive external pressure differential of 6 pounds per square inch. When vacuum relief valves are installed, tanks would have to be designed to withstand an external pressure not less than the set pressure of the relief device.

Requirements for pressure and vacuum relief devices are proposed in § 178.270-11. The requirements for the number and type of relief devices required as well as the pressure setting of these devices are based on the provisions of the UN and IMCO portable tank standards. It is proposed that each portable tank be fitted with at least one pressure relief device and that at least one spring loaded pressure relief device be provided for tanks with capacities of 500 gallons or greater. General requirements governing the location, arrangement and construction of relief devices are proposed in § 178.270-11(b). The spring loaded relief device required on tanks with capacities of 500 gallons or greater, or frangible discs fitted on tanks with lesser capacities that have no spring loaded device installed, would be required to function at a pressure of 125 percent of the maximum allowable working pressure. Additional frangible pressure relief devices installed to insure adequate venting capacity in fire situations, would be required to function at a pressure of 150 percent of the maximum allowable working pressure. Any fusible element installed would be required to have a fusing temperature of not more than 250° F. Vacuum relief devices, if fitted, would be required to open at a nominal overpressure of not less than 3 pounds per square inch.

Requirements for minimum total venting capacity of relief devices are proposed in § 178.270-11(d). A minimum relieving capacity of one standard cubic foot of air per minute per 30 square feet of tank area is prescribed for each spring loaded pressure relief device. In addition, the total required venting capacity for all installed pressure relief devices is presented in a tabular format as a function of tank surface area. The values for venting capacities proposed were obtained using the venting capacity relationship presented in Pamphlet S-1.2, published by the Compressed Gas Association, assuming hexane as the fluid vented. It is further proposed that the required venting capacity determined according to the proposed table for insulated tanks be permitted to be reduced in proportion to the efficiency of the installed insulation subject to a maximum reduction to 25 percent of the venting capacity specified in the table. General requirements relating to the survivability of insulation under fire conditions are also proposed. These requirements, as well as those proposed for markings on pressure relief devices, are as provided in the latest IMCO portable tank standards.

Requirements for the construction and arrangement of valves, piping and other accessories are proposed in § 178.270-12. It is proposed that all tank nozzles be fitted with manually operated stop valves with the exception of nozzles installed in the vapor space which are closed by a blank flange during transport. Valves on filling and discharge connections must be fitted as close to the shell as practicable and filling and discharge connections must be fitted with a secondary means of closure such as a blank flange. It is proposed that valves be rated at a pressure not lower than the maximum allowable working pressure of the tank, and that they be closed by a clockwise motion of the handwheel when screwed spindles are employed. Internal stop valves, when installed, would be required to be self-closing and to be fitted entirely within the shell or the welded discharge flange. Piping would be required to have a bursting strength not less than four times the maximum allowable working pressure of the tank and to be adequately supported to prevent damage due to jarring or vibration. It is proposed that all nozzles and shell penetrations be designed in accordance with the ASME Code. Glass or other easily destructible gauging devices would be prohibited.

Requirements for the initial test and inspection of specification IM 100 IM 101 portable tanks are proposed in § 178.270-13. Each tank would be required to be subjected to a pressure test at a pressure of 150 percent of the maximum allowable working pressure of the tank. Leakage, undue distortion, or other conditions that indicate a weakness which might, in the opinion of the witnessing approval authority, render the tank unsafe for transportation services, would constitute failure of the test. Internal heating coils, if installed, would be required to be hydrostatically tested to a pressure not less than the greater of 200 psig or 150 percent of the rated pressure of the coils.

A prototype tank of each portable tank design would be required to undergo inertial restraint tests to insure that the design and construction of the tank supports is adequate to preclude separation of the tank from the framework or skids which would secure the portable tank to the transport vehicle or vessel, under the inertial loads encountered in transportation. Furthermore, a prototype of any tank design which is to be authorized for rail transport would be required to pass a series of rail impact tests based upon tests prescribed by the Association of American Railroads.

Requirements for marking certain information on a metal identification plate are proposed in § 178.270-14. These markings would be in addition to those currently required by § 172.326. Information required to be shown is based upon that suggested in the IMCO and UN standards. Additional information required by other national authorities or international organizations would be permitted to be displayed on the same metal plate.

Section 178.271. Section 178.271 proposes additional requirements applicable only to Specification IM 100 portable tanks. The maximum allowable working pressure of these tanks would be less than 25 psig but not less than 14.2 psig. Except as limited or modified in the proposed specification, IM 100 portable tanks with pressure greater than or equal to 15 psig would be designed and constructed in accordance with the ASME Code. It would not be required that these tanks be inspected by an ASME Code inspector or that a Manufacturer's Data Report (ASME Form U-1) be issued for the tanks. Provisions are proposed by which the approval agency may authorize a reduction in the minimum shell thickness required under § 178.270-4 provided additional protection against puncture of the tank is provided. Guidelines concerning the acceptable means of providing additional external protection are proposed.

Section 178.272. Section 178.272 proposes additional requirements applicable only to specification IM 101 portable tanks. It is proposed that the maximum allowable working pressure of these tanks be 25 psig or greater. In addition such Specification IM 101 portable tanks would be designed and constructed in accordance with the ASME Code except as limited or modified in the specification. It would not be required that these tanks be inspected by an ASME Code inspector or that a Manufacturer's Data Report (ASME Form U-1) be issued for the tanks.

(43 F.R. 58050, December 11, 1978)

49 CFR Parts 107, 171, 172, 173, 174, 176, 177, and 178
[Docket No. HM-167; Amdt. Nos. 107-8, 171-60, 172-65, 173-144,
174-40, 176-13, 177-53, 178-65]
Intermodal Portable Tanks

On December 11, 1978, the MTB published a notice of proposed rulemaking (NPRM) under Docket HM-167, Notice No. 78-12 (43 FR 58050) which proposed to authorize the use of two new packaging specifications for portable tanks. Specifications for these proposed intermodal (IM) portable tanks were based primarily on international standards. Interested persons were invited to participate in the rulemaking process, and all comments received were given full consideration by the MTB.

The majority of the commenters were in agreement with the MTB that the requirements to be adopted under

HM-167 should be compatible with standards in the "Recommendations on Multimodal Tank Transport" adopted by the UN Committee of Experts, and standards specified in the IMCO Code.

Based on these comments and the desire of the MTB to harmonize HM-167 with international standards, the IM portable tank specifications are designated as IM 101 and IM 102 (instead of the proposed IM 100 and IM 101) to correspond with the similarly designed and constructed IMCO Type 1 and IMCO Type 2 portable tanks. In addition, measurements in the final rule are specified in metric and nonmetric units to eliminate possible discrepancies in conversions.

A major effort has been made to harmonize this rule with the UN Recommendations by use of an IM Tank Table patterned after the lists of dangerous goods permitted for transport in portable tanks which appear in the IMCO and UN recommendations. Prior to use of an IM tank for transportation of a hazardous material, the person offering the material must determine from the Table if the tank is authorized for the material and, if so, what special requirements apply. Some of the requirements are maximum allowable working pressure, pressure relief device configuration, and bottom outlet configuration for various materials authorized for transportation in an IM portable tank.

THE IMPACT OF RULEMAKING ON EXISTING EXEMPTIONS

It is the MTB's intention that, to the maximum extent possible, existing intermodal portable tanks authorized for use under an outstanding exemption be covered under this rule thereby eliminating the need for the exemption. The MTB has decided, however, that in order to be allowed to be re-marked as a specification IM portable tank, an existing intermodal tank must, as a minimum, conform to the basic provisions of this rule.

Accordingly, each owner or manufacturer of an intermodal portable tank which is in service under a DOT exemption or was constructed under a DOT exemption on or before May 1, 1981, should examine the tank and the tank drawings to determine if the tank meets the requirements of an IM 101 or IM 102 specification portable tank. In order to be marked and used as an IM 101 or IM 102 tank, the tank must be modified, re-rated and re-marked as specified by § 173.32(d) herein according to the proper specification by May 1, 1983. The following tanks may be re-marked as a DOT specification tank:

1. A tank in full compliance with an IM 101 or IM 102 specification.

2. A tank that is modified and brought into full compliance with an IM 101 or IM 102 specification. (These modifications may include but are not limited to, the resetting, replacement, or the addition of pressure relief devices).

3. A tank which, while otherwise in conformance with an IM 101 or IM 102 specification, has its pressure relief device sited on the top of the tank within 12 degrees of the top longitudinal centerline, provided the inlet of each pressure relief device is in the vapor space of the tank.

4. A tank constructed of austenitic stainless steels which, while otherwise in conformance with the IM 102 specification (e.g. § 178.270–5), has an absolute minimum equivalent head and shell thickness of not less than 3.0 mm (0.118 inches).

5. A tank with an outside diameter greater than 1.8 m (5.9 feet) constructed of other than the reference mild steel which, while otherwise in conformance with the IM 101 specification, has a minimum shell thickness not less than the value derived from the formula in § 178.270–5(c) based on a required thickness of the reference steel equal to 6 mm (0.236 inches).

The requirement for location of pressure relief devices along the top longitudinal centerline is to preclude discharges of liquid lading to an extent considered practical. This requirement will increase the level of safety at no additional cost for newly constructed portable tanks. The MTB has sought, in relaxing the location requirement for pressure relief devices on existing portable tanks, to reduce required modifications while maintaining safety.

For the minimum wall thickness, the UN/IMCO standards specify nonequivalent metric and nonmetric values of 3.00 mm/0.125 inches for IMCO Type 2 portable tanks and 6.00 mm/0.250 inches for IMCO Type 1 portable tanks. In the final rule, for new construction, the MTB has equated the more conservative nonmetric value to its equivalent metric value (i.e., 3.16 mm/0.125 inches and 6.35 mm/0.250 inches). For existing tanks, however, the MTB has determined that a tank under exemption built to the UN/IMCO metric standards provides adequate safety, and absent other areas of nonconformance, tanks meeting these standards may be re-marked as DOT specification tanks.

The owner of a portable tank under exemption which is remarked as an IM 101 or IM 102 portable tank is required to forward to the Associate Director for HMR and to retain at his principal place of business a written report containing the following information:

1. A statement certifying that each tank or series of identical tanks manufactured to a single design is in compliance, except as authorized herein, with the applicable IM 101 or IM 102 specification (§ § 178.270, 178.271, 178.272).

2. The identification of the person certifying the portable tank.

3. The applicable DOT exemption number and the serial number of each tank covered by the report.

4. A summary of the modifications made to the tank to bring it into conformance with the IM 101 or 102 specifications including authorized deviations.

Hazardous materials offered for transportation in tanks which are re-marked as specification tanks, are subject to the authorizations, conditions and limitations of this final rule without regard to previous authorizations, conditions and limitations under an exemption.

After September 1, 1981, any exemption affecting a portable tank of a type covered by this final rule will not be renewed upon its expiration unless the owner or manufacturer of the tank has submitted information to the Associate Director for Hazardous Materials Regulation stating a valid reason why the tank cannot be brought into compliance with the requirements of this amendment.

AAR'S COMMENTS ON THE DESIGN, TEST, AND USAGE OF IM
PORTABLE TANKS FOR RAIL SERVICE

The Association of American Railroads (AAR) made numerous comments and recommendations with respect to the design and construction of IM portable tanks as authorized for rail transport. Because these comments relate only to rail transportation they are discussed collectively rather than in the review by sections.

AAR suggested that "AAR.600 Specifications For Acceptability of Tank Containers in COFC Service" (AAR.600) be used as a guideline for IM portable tanks in rail service. AAR pointed out that HM-167 should recognize, as does AAR.600, both the rigors of normal transportation and the extra measures of performance necessary in accident situations. The MTB and the Federal Railroad Administration (FRA), which assisted MTB in the preparation of this rule, agree with the AAR's concept of considering both normal and accident environments in setting the minimum design requirements for packagings, and these unique and severe aspects were fully considered in this rulemaking.

MINIMUM DESIGN PRESSURE

AAR contended that the railroad industry would not accept tanks with a design pressure of less than 35 psig. This would, in effect, prohibit all IM 102 and a large number of IM 101 portable tanks from rail service. The MTB believes AAR's comment results from a misunderstanding of the permitted usage of IM 101 and IM 102 portable tanks. Unlike many DOT specification packagings which are authorized for an entire generic group of hazardous materials such as flammable liquids, IM portable tanks are authorized only for particular materials or for a limited generic group of materials according to design pressure and other characteristics of the tank. For example, an IM 102 portable tank with a design pressure of 15 psig when used for flammable liquids would be limited to flammable liquids, having no other hazards, with a vapor pressure below 9.2 psia at 149° F., and a flash point at 32° F. or above. Thus, a tank with a relatively low working pressure is authorized only for hazardous materials of relatively low hazard for which the tank provides an adequate level of safety. In addition, in most cases, the design of specification IM 101 and IM 102 portable tanks is controlled by the minimum thickness requirement as discussed below. Therefore, the MTB does not agree with the AAR recommendation that the working pressure of tanks authorized for rail transport must be 35 psig or higher.

Minimum Tank Wall Thickness. The AAR recommended that the minimum tank wall thickness acceptance criteria adopted in the final rule be equivalent to the thickness of plates criteria specified in AAR.600. AAR.600 requires for all tank sizes used to transport a hazardous material, regardless of the hazard severity of that material, a tank wall thickness of:

(1) ⅜ inch (0.375 inches, 9.5 mm) carbon steel; or,

(2) 8 MSG (0.1644 inches, 4.2 mm) austenitic stainless steel.

The AAR supported these wall thickness requirements by stating "Tanks in hazardous materials service via rail must have added integrity to withstand possible accidents and the ⅜" mild carbon steel thickness of AAR.600 is a calculated equivalent, taking necessary safety factors into account. The selection of 8 gauge for a minimum stainless wall is based on a comparison of the puncture data of carbon steel and stainless steel." The AAR comments did not specifically address the various minimum tank wall thickness requirements proposed in the NPRM.

In the final rule, based on consideration of the size of the tank, the tank material and the hazard properties of the material being shipped, the MTB has adopted the following minimum tank wall thickness requirements:

• For large diameter (greater than 5.9 feet) IM 101 portable tanks fabricated of austenitic stainless steel, the minimum wall thickness required is approximately 0.188 inches (depending on the properties of the material of construction) and therefore is greater than the 0.164 inches recommended by the AAR. These IM 101 portable tanks are similar in design to the majority of the IMCO type 1 portable tanks.

• For large diameter IM 101 tanks fabricated of mild steel, the minimum wall thickness required is 0.250 inches. This value for mild steel is related to the minimum wall thickness required for austenitic stainless steel by the equation in new §178.270-5(c) which was adopted from the UN Recommendations, such that the puncture resistances for these materials are equivalent. Thus, for mild steels, the MTB believes the 0.250 inches minimum thickness requirement, even though less than the 0.375 inches recommended by the AAR, is comparable to the thickness for austenitic stainless steels supported by the AAR.

• For small diameter (5.9 feet or less) IM 101 portable tanks and for IM 102 (IMCO type 2) portable tanks, the MTB has adopted in this rule minimum tank thickness requirements that are consistent with the UN recommendations and the IMCO Code. These thickness requirements are in accord with other regulations calling for packaging integrity proportional to both the severity of hazard and the quantity of the material being transported in a packaging. As discussed earlier with respect to the minimum design pressure requirements, IM 102 tanks are limited to the carriage of hazardous materials of relatively low hazard.

The MTB does not believe that the AAR has sufficiently supported its view concerning an inadequacy in the minimum wall thickness requirements proposed in the NPRM and has not adopted the AAR's recommended minimum thickness requirements.

DYNAMIC SHOCK LOADING

The AAR recommended that the tank and frame design loads for IM portable tanks authorized for rail be 2 "g" vertical combined with 3.5 "g" longitudinal and 1.5 "g" lateral for consistency with the loads encountered in rail transportation. The MTB agrees that the severity of longitudinal shock loading in the rail mode—which can exceed 3.5 "g" during switching operations and certain over-the-road operations such as start, stop, slack run-out and slack run-in—must be considered in order to provide an adequate level of safety through the entire life of an IM portable tank. However, the MTB and the FRA believe that the most efficient method of dealing with this shock load environment with respect to the total multi-modal transportation system is not through increased structural requirements on the specification IM 101 and IM 102 portable tanks which would result in increased unit cost, tare weight, and less efficient transportation of hazardous materials. Rather, the MTB and the FRA believe IM 101 and IM 102 portable tanks, when shipped by rail, should be shipped under conditions approved by the Associate Administrator for Safety, FRA, including a requirement that rail cars have end-of-car cushioning or its equivalent. Such cushioning reduces over-the-road shock to below 2 "g" longitudinal and substantially reduces coupling shocks. The requirement for FRA approval of IM portable tanks has been added in § 174.63. In addition, a requirement that IM portable tanks not be coupled with excessive force has been added in § 174.84 to reduce high longitudinal shocks.

SAFETY VENTS

The AAR pointed out that the current regulations do not authorize the use of safety vents (frangible discs) for flammable liquids shipped by tank car. The MTB agrees that such a requirement is valid for all bulk shipments of flammable liquids via rail and, accordingly, has added such a requirement for IM portable tanks in § 173.32c.

Also, in consideration of the fact that the AAR prohibits tank cars from having gravity actuated vacuum vents, the MTB has added a requirement that any vents or valves on a portable tank must be designed to provide total containment of the hazardous material in an overturn accident situation.

PORTABLE TANKS IN TOFC SERVICE

The AAR expressed the view that the TOFC shipment of IM portable tanks is unsafe for two reasons. First, the AAR believes that there is an increased risk associated with TOFC securement, i.e., the securement of the portable tank to the motor vehicle chassis and the securement of the motor vehicle chassis to the flatcar. Second, the AAR believes that "... the combined center of gravity for the flatcar, chassis, and container is approximately 139″ and this grossly exceeds the 98″ maximum center of gravity for freight cars allowed by paragraph 2.13, AAR Specification M-1002."

Even though the AAR did not submit any data, calculations, or test results to support its position, the MTB feels that there may be some merit in the AAR's views. In order to obtain sufficient data to evaluate TOFC service, all sections of this rule pertaining to TOFC service have been removed and will become a part of a new rulemaking action (HM-177) entitled "Transportation of Hazardous Materials in TOFC Service". The MTB has scheduled a public hearing for February 25, 1981, at which time it will receive data, calculations and test results pertaining to the transportation of hazardous materials in TOFC service. MTB is particularly interested in receiving data, calculations, and test results showing the effectiveness of TOFC securement and the safety effect of a high center of gravity on TOFC service. The notice of the public hearing appears elsewhere in this issue of the Federal Register.

REVIEW BY SECTIONS

The following is an analysis and explanation by section of differences between this final rule and the NPRM. Persons interested in significant features of this rule that were not changed from the NPRM are referred to the discussion in the NPRM. Additionally, this review contains a discussion of substantive comments received in response to the NPRM.

Section 107.3. The definition of "approval agency" has been clarified and a definition of "competent authority" has been added.

Sections 107.401–107.405. These sections have been completely revised. Under Part 107, Subpart E, approval agencies will be designated as such by the MTB rather than subjected to an approval process as proposed in the NPRM. It is MTB's desire that IM portable tanks approved by an approval agency designated by MTB be acceptable to other governments in the implementation of their requirements, including international standards. Agencies will be designated by the MTB if they meet the requirements and conditions specified by § 107.402.

Approval agencies may be commercial enterprises or agencies of other governments. One commenter stated " . . . on the subject of Approval Agency . . . rather than leave open to any number of acceptable approving agencies, I feel consideration should be given to another approach. I suggest that those agencies previously qualified, assuming they would still qualify, be designated as 'Approval Agencies'. On those others who apply and are qualified I believe it would be advantageous to limit the number to those who can in combination serve the need of the field but not saturate with its attendent possible problems." The MTB believes any person meeting the required qualifications should be designated as an approval agency. No effort will be taken to restrict the number of qualified applicants seeking designation as approval agencies or to impede competition in the field.

Another commenter objected to excluding IM tank manufacturers and owners from eligibility as approval agencies. The commenter contended that self-certification would not compromise safety nor jeopardize compliance. The commenter further argued that DOT presently allows manufacturer certification of certain other packagings. The MTB believes IM portable tanks should be certified by an independent inspection agency because IM portable tanks will be authorized for a broad range of hazardous materials and will be exposed to significantly different modal environments. An additional reason for the designated approval agency process is to establish a basis for their acceptability to other national governments. It is anticipated that domestic and foreign approval agencies recognized by foreign governments will apply for designation as approval agencies under this final rule. Therefore, the number of agencies necessary to approve a tank used in international transport will be minimized in that one approval agency may be able to issue approvals acceptable to many national administrations and for various modes of transport thereby reducing the burdens encountered by IM tank manufacturers.

One other commenter questioned whether in proposed § 107.404(a)(5) (which is § 107.405(b) in this final rule), it was the MTB's intent to require notification every time an inspector is hired or fired by the competent authority and all changes in managerial structure, geographic area of service, clients, etc. The commenter contended that the requirement should be revised to require notification to the Associate Director, Office of Hazardous Materials Regulation (OHMR) of only substantive changes in the information submitted in the application for designation as an approval agency. The MTB agrees with the commenter and an applicant is required by the final rule to submit considerably less information in an application than was proposed in the Notice. The MTB believes all required information is essential and is needed to evaluate an agency's continued qualifications to perform the applicable packaging approval function.

The procedures proposed in § 107.406 whereby the Associate Director for the HMR may suspend or terminate a designation granted under this subchapter are adopted with a number of modifications in § 107.405 of this rule. It should be noted that § 107.405(a)(3) specifies that failure of a competent authority to recognize qualified designated approval agencies domiciled in the United States may be the basis for suspension or termination of a designation made by the Associate Director for HMR of a foreign approval agency. MTB believes this provision is necessary to assure that IM tank manufacturers will not be unnecessarily burdened with required use of different inspection agencies serving the same purpose. Obviously, if the United States recognizes the approval agency designated by another government, equal treatment should be expected of that government relative to designated approval agencies domiciled in the United States.

Section 171.7. Paragraphs (c) and (d) of this section have been amended to reflect a reference appearing in § 178.270–3 to the "ISO 82–1974(e) Steel-Tensile Testing."

Section 171.8. Definitions are added for "IM Tank Table" "intermodal portable tank" or "IM portable tank", "outage" or "ullage" and "p.s.i." or "psi."

Section 172.203. A new paragraph (1) has been added to require that any material described on shipping papers by an n.o.s. entry in § 172.101 or § 172.102 and offered for transportation in an IM portable tank have the technical name, component, chemical element or group contributing to the hazard or hazards of the material shown in parentheses.

Section 173.32a. Based on the comments the MTB received on this section several changes were made.

A commenter indicated that it appears that if an approved agency supervises the manufacture of portable tanks, it cannot remain an independent party. MTB agrees and the duties of an approval agency have been revised to reflect that these duties include functions such as review of tank designs and the witnessing of all required tests but do not extend to actually supervising the manufacture of the tanks.

This same commenter suggested that, in paragraph (a), provision to be made to permit an owner or a manufacturer to submit test data on tank designs to reduce the need for certain calculations and to provide consistency with § 178.270–2(c). Another commenter recommended that three sets of all engineering data, rather than two sets, be submitted to an approval agency (as required by § 107.404(b)) in order that the third set could be retained by a field inspector who may even be located in a different geographical location than the main office of the agency. The MTB agrees with both commenters and has made the recommended changes.

Paragraph (d) has been revised, as discussed under the section titled "Impact of This Rulemaking on Existing Exemptions", to permit re-marking of certain portable tanks covered by a DOT exemption.

The MTB has revised proposed paragraph (e) (paragraph (f) herein) in agreement with a commenter who pointed out that only modifications affecting conformance with § 178.270, that is, the structural integrity of the tank, its support structure or its ability to retain lading, should require prior approval by an approval agency and not minor cosmetic modifications. Also, as suggested by a commenter, an owner or a manufacturer desiring modification of a tank may use another approval agency if the initial approval agency is no longer operating or is not available.

A new paragraph (g) has been added to contain procedures whereby the Associate Director for HMR may terminate an approval certificate if:

1. Information upon which the approval was based is fraudulent or substantially erroneous; or
2. Termination of the approval is necessary to adequately protect against risk to life and property.

However, except in emergency situations, before the Associate Director for HMR may terminate an approval certificate, the owner or manufacturer and the approval agency must be informed of the reasons for the termination and be provided with an opportunity to comment or to achieve compliance.

Section 173.32b. A commenter objected to use of the term "hydrostatic" and suggested that it be replaced with the word "pressure." The commenter contended that hydrostatic applies to the use of water only. While historically the term may have applied in practice to the use of water, most technical references apply the term to any liquids. Therefore, the MTB has not changed its use of the term and any suitable liquid may be used. However, requirements on hydrostatic testing in paragraph (a)(1) have been revised based on a comment that inspection of a tank for corrosion and dents while it is under pressure may not be safe. In addition, requirements in paragraph (a)(2) have been revised to require that spring loaded pressure relief valves must be removed from a tank and tested at least every two and one-half years.

Also, based on objections from several commenters of the cost and inconvenience of using an approval agency to perform the visual inspection, the MTB is permitting the 2½ year interval visual inspection to be satisfied by an owner or his agent or, as proposed in the NPRM, by an approval agency. The MTB has revised these requirements to require that a visual inspection be performed at the time a hydrostatic test is due.

A new paragraph (c) has been added to inform owners of portable tanks meeting the definition of a container (49 CFR 450.3(3)) that these tanks must be offered for international transportation in accordance with the applicable requirements of Parts 450 through 453 of this title. The applicable provisions of these parts are primarily limited to the container frame and attachments for portable tanks.

Based on several comments, in paragraph (d), an alternate location has been provided for test markings. The markings for the hydrostatic test and the visual inspection test required by this section may be marked on the side of a tank near the identification plate, or may be stamped on the identification plate.

The MTB also agrees with a commenter that relief devices in deteriorated or damaged portable tanks that are being repaired or retested, need not be retested unless there is reason to believe that the relief devices are damaged or deteriorated. However, in either case, the valves must be removed from the tank and visually inspected for damage or deterioration. The provisions on damaged or deteriorated portable tanks (proposed paragraphs (c) and (d)) have been consolidated in paragraph (e).

The recordkeeping requirements for this section are contained in a new paragraph (f).

Section 173.32c. The MTB has made several modifications to the basic requirements governing the use of specification IM 101 and IM 102 portable tanks proposed in this section.

A new requirement has been added in paragraph (c) of this section to prohibit the filling of an IM portable tank for which the prescribed periodic test and inspection under § 173.32b has become due. The MTB believes this requirement is justified in order to ensure proper maintenance of tanks. The requirement does not apply to any tank filled prior to the test due date.

Two commenters objected to the 55,000 pound gross weight limitation for IM portable tanks in the proposal as being too restrictive when compared to the ISO limit of 67,000 pounds for a 40 foot van. The MTB agrees with the commenters and has deleted the restriction. However, the MTB has added a requirement in paragraph (i) of the rule to specify that in no case may an IM portable tank be loaded to a weight that exceeds the maximum gross weight specified on the tank identification plate. The 55,000 pound gross weight restriction also has been removed from § 173.32(a)(2).

In paragraph (j), the MTB has specified for IM portable tanks a minimum filling density of 80% by volume to limit dynamic instability that would result from lesser filling densities.

Also, a new paragraph (k) has been added to clarify that there may not be any leakage of material through a frangible disc or tell-tale device. For additional discussion, see the section review of § 178.270–11.

A large number of commenters, responding to the proposal for rear-end protection, expressed their views that the bumper requirements specified in 49 CFR 178.340–8 are not necessary for portable tanks and would place a severe economic burden on shippers. The commenters expressed the view that (1) the bumper requirements in § 178.340–8 are intended to protect the piping under cargo tanks in rear under-ride accidents, (2) that this is not a danger with portable tanks which are carried on trailers and have no exposed piping which could be damaged in under-ride accidents, and (3) the ISO frame and the head thicknesses used in IM portable tanks would provide greater inherent impact resistance when compared to cargo tanks. The MTB agrees with the commenters and has deleted the proposed bumper requirements. However, the MTB has added a requirement in paragraph (1) to preclude overhang or projection of any part of an IM portable tank when loaded on a highway or rail transport vehicle, to assure the protection provided by the vehicle itself.

The procedures proposed in paragraph (d) authorizing the Associate Director for HMR to approve, under certain conditions, hazardous materials for transportation in IM portable tanks have been deleted and placed, with modifications, in new § 173.32d.

Section 173.32d. This section contains requirements whereby a person may request the Associate Director for HMR for approval to add a material to the IM Tank Table or to delete an entry from the Table. The procedures for filing an application and the required information to be included in the application are contained in the preface to the IM Tank Table. Note that paragraph (d) specifies that additions to the Table have interim status until opportunity is provided for public comment on each addition.

Section 173.116. Several commenters contended that the proposal relative to outage in paragraph (i) does not relate to UN, IMCO, or DOT cargo tank requirements and, therefore, should not be adopted. The MTB agrees with the commenters and has deleted the proposed outage requirement. See § 173.32c for requirements on filling density.

Section 173.118a. Except for an editorial change, this section was adopted as proposed in the notice.

Section 173.119 through 173.630. These sections contain various amendments to authorize carriage of hazardous materials in specification IM 101 and IM 102 portable tanks. The MTB received many comments requesting authorization to ship various hazardous materials not proposed in the notice in IM 101 and IM 102 specification portable tanks. Many of these comments dealt with specifying tank constructional features necessary to safely ship various hazardous materials. The MTB has adopted the use of the IM Tank Table to provide a clearer and more concise format for the large number of hazardous material involved and to provide harmony with international standards. The MTB carefully reviewed all the comments on hazardous materials, the UN Recommendations and IMCO Code, and the existing regulations in setting the standards in the IM Tank Table. Some materials discussed in various comments are still under review and may be added to the IM Tank Table at a later date.

Sections 174.63 and 174.84. The MTB has revised the heading to § 174.63 to include IM portable tanks and has added a new paragraph (d) to provide for COFC service under conditions approved by the Associate Administrator for Safety, FRA. Section 174.84 has been revised to provide that flat cars carrying IM portable tanks may not be coupled with excessive force. Also, see discussion of AAR's comments above.

Section 176.340. This section is adopted as proposed in the notice.

Section 177.834. This section is revised to provide that IM portable tanks may not be stacked on each other or placed under other freight during transportation by motor vehicle.

Section 178.270–1. The MTB agrees with the commenters on the proposal to this section that a vapor pressure of 43 psia at a temperature of 122° F. instead of 150° F. should be used in order to provide agreement with the criteria used for the IMCO Type 1 and Type 2 portable tank, and accordingly has revised paragraph (a). In addition, the MTB notes that this criteria prohibits the shipment of several flammable liquids (ethylene oxide, etc.), which are considered as a gas in most international regulations, in specification IM 101 tanks. This presents no conflict with DOT regulations, however, since DOT requires the shipment of these materials in high pressure tanks.

Section 178.270–2. The MTB disagrees with the commenter who stated that other acceptable experimental methods for stress analysis of a tank should be permitted to be approved by an approval agency, instead of the Associate Director for HMR, to expedite handling of such approvals. Section UG-101 of the ASME Code, which is the basis for the approval, is adaptable and should be used. Any other experimental methods need to be approved by the MTB to centralize data on such methods and to minimize duplicative review efforts.

Section 178.270–3. Sections 178.270–3 and 178.270–4 are interrelated and as a result of the incorporation of comments received, the MTB has revised both sections. Several commentors expressed confusion with respect to the materials authorized in proposed § 178.270–3 for the construction of IM 101 and IM 102 intermodal portable tanks and objected to the restriction that only ASME Code specified materials would be authorized. It is the intent of the MTB to limit the materials of construction to steel. The MTB has replaced the word "steel" with the words "carbon and alloy steels" to clarify that all types of steel manufactured to a recognized national code and meeting the specified criteria are authorized for the construction of IM 101 and IM 102 portable tanks. Materials other than steel, such as aluminum, nickel, and monel, are not authorized.

The MTB has also, in § 178.270–3, expanded the authorization for steels and the criteria for the determination of the maximum allowable stress value to permit the use of non-ASME materials by using ASME criteria for the determination of the maximum allowable stress value for the actual steel used. Two methods of deriving the maximum allowable stress values are offered. The first method allows the maximum allowable stress value to be based on the actual measured yield and tensile strengths of the group of plates used to fabricate the tank shell. The yield and tensile strength value is limited to not greater than 120 percent of the minimum values at 93° C. (200° F.) specified in the national standard used to manufacture the steel. This limiting value was chosen because yield and tensile strengths may vary in a plate. The MTB feels that 120 percent is the maximum safe variation from the guaranteed minimum national standard value. The second method allows the maximum allowable stress value to be based on the specified minimum yield and tensile strengths at 93°C. (200° F.) specified by the national standard used to manufacture the material.

Several commenters stated that the evaluation of the maximum allowable stress value at 300° F. is excessive and is not specified in either the UN Recommendations or IMCO Code. Upon review the MTB agrees that 300° F. is excessive. However, good design practice, the UN Recommendations and the IMCO Code state that in choosing a material and in determining the wall thickness for a tank, the maximum and minimum filling or working temperatures should be taken into account. Therefore, since the working pressure is based on a product vapor pressure at 149° F. for all IM 101 and IM 102 portable tanks and tank shell temperatures over the ullage volume have been measured to be in the range of 200° F, the MTB believes it is reasonable to require the stress evaluation temperature to be at least 200° F. and has amended §178.270-3 accordingly.

A commenter stated that a gauge length of L/D=4 for the tensile test specimen should replace the L/D=5 in §178.270-3. The MTB disagrees. A gauge length of five is used to provide harmony with the UN Recommendations for multimodal tank transport. In addition, the relationship $[L_0 = 5.65 \, (S_0)^{1/2}]$ between specimen gauge length and cross-sectional area has been added to clarify that both bar and strip specimens are authorized. Also, the MTB has specified usage of the procedures in ISO 82-1974(e) Steels-Tensile Testing. These requirements should cause no undue burden on the domestic IM portable tank industry.

Section 178.270-4. In the NPRM the MTB discussed the inconsistency between the UN/IMCO standards and the ASME requirements regarding maximum allowable stress levels for a tank. The MTB requested comments on the relative merits of both systems and also requested specific comment on the following three issues:

(a) The potential that safety would be compromised if a 3:1 factor of safety was to be applied in certain cases;

(b) Any potentially significant barriers to trade that could result from the use of the stress level limitations proposed in the NPRM that are inconsistent with those in the international standards; and,

(c) The extent to which tanks constructed to a 3:1 factor of safety are actually in use throughout the world and reports of relevant transportation experience concerning such tanks.

The MTB received 11 comments on the maximum allowable stress value. The commenters included foreign tank manufacturers, domestic and foreign shippers and carriers, and an approval agency. The opinions expressed by the commentors fell into three groups: those supporting a 3:1 factor of safety; a 4:1 factor of safety; and a 3:1 factor of safety provided adequate safety is demonstrated.

One of the six commentors who supported the 3:1 factor of safety proposed other control measures which would in effect result in a tank with a factor of safety greater than 4:1. Cost was the principal reason for support of the 3:1 factor of safety. Other reasons cited were the adequacy of the 3:1 factor of safety and the adverse affect of inconsistent stress levels on the free movement of the tanks in international trade.

Unfortunately, the MTB did not receive any analysis or data to support either the 4:1 or 3:1 factor of safety. Thus, the MTB evaluated the two systems based on available data and the physical requirements of the two systems to define the safety and economic effects of the inconsistencies between the two systems.

For austenitic stainless steel, the ASME Code limits the maximum allowable stress at test pressure to the lower of 93.75 percent of the specified minimum yield strength determined at the 0.2 percent offset. Similarly, international standards limit stress at the test pressure to the lower of 75 percent of the specified minimum yield strength at the 1.0 percent offset. When the maximum allowable stresses for these two systems are compared for specific materials, the stress is found to be essentially equal and controlled by the 75 percent of the yield strength for the UN/IMCO. In effect, the result of using either system will be a tank having approximately the same shell thickness. The MTB believes that it is advantageous to be in harmony with UN Recommendations and the IMCO Code when possible. Therefore, for austenitic stainless steels, the MTB is adopting the UN/IMCO method as well as the ASME method for specifying the maximum allowable stress levels.

For carbon and low alloy steels, the ASME Code limits the maximum allowable stress at test pressure to the lower of 93.75 percent of the specified minimum yield strength or 37.5 percent of the specified minimum tensile strength. This method utilizes a 0.2 percent offset. The ASME Code stress levels insure at all times a minimum factor of safety of 4:1 against ultimate strength (at the maximum allowable working pressure): whereas, the UN/IMCO stress levels would permit a range in the factor of safety from approximately 4:1 to approximately 3:1 against the ultimate strength at that pressure.

In the UN Recommendations and the IMCO Code, and as the MTB proposed in the NPRM, the determination of the tank wall thickness is a function of minimum wall thickness requirements as well as the maximum allowable stress requirements. In evaluating the respective influence of each of these parameters on the side walls of a tank, the MTB found that in almost all cases the minimum wall thickness requirements exceed the thickness required by hoop stress. In most of the applications for exemption received by the MTB for IM portable tanks, the "fixed minimum" wall thickness of the cylindrical shell portion as established by the IMCO Code regulations and the UN Recommendations exceed the thicknesses required by ASME. Thus, in the side walls of the tanks, the minimum thickness requirement controls the required wall thickness and the 4:1 vs 3:1 factor of safety has little effect on the tank design.

The MTB believes that IM portable tanks must be capable of resisting the bending, torsion and shear stresses created by the cyclic application of loadings. To adequately resist such forces, careful consideration must be given to the structural reinforcements in those vital areas where the mechanically induced stresses, in addition to pressure induced stresses, are significant factors. Most tank failures result from fracture starting at a flaw, crack, or stress concentration. The more common locations of stress concentrations are at openings and attachments, and it is in such areas that it is particularly important to maintain the margin of strength and safety—through added reinforcement—provided by the ASME Code. In addition, the IM portable tank specifications and the corresponding international standard do not prescribe heat treatment or radiographic examination of welds which are important requirements in tank designs having a factor of safety below 4:1. Considering all factors in the design of the IM 101 and IM 102 portable tanks, a tank designed with a 3:1 factor of safety offers no significant economic advantage in cost over a tank designed with a 4:1 factor of safety. For these reasons the MTB has adopted the maximum allowable stress values for carbon and low alloy steel as proposed.

Section 178.270-5. A new paragraph (d) in § 178.270-5 has been added to provide greater compatibility with the UN Recommendations and to provide a better definition of the minimum wall thickness requirements for portable tanks used to transport certain hazardous materials for which the IM Tank Table requires greater thickness than the standard values specified in § 178.270-5(b). In this case, the specified minimum shell and head thickness is based on the steel defined in paragraph § 178.270-5(a) and on a tank diameter of 1.8m (5.9 feet). For other materials of construction and for different diameters, the required wall thickness varies as defined by the formula (that is, thinner wall thicknesses for stronger materials of construction, thicker walls for larger diameter tanks).

The metric and nonmetric equivalencies specified for tank shell thickness in the UN Recommendations and IMCO Code are not equivalent, but are based on the closest standard plate thickness. In the final rule, minimum shell thicknesses are specified for the reference mild steel. These thicknesses must also be used as the basis for the calculation of the equivalent minimum shell thickness for other steels. The MTB has based the minimum allowed shell thickness on the larger of the standard plate thicknesses specified in the UN Recommendations and the IMCO Code.

Section 178.270-6. A new paragraph (b) has been added to require that IM portable tanks designed for international transportation must be in accordance with 49 CFR Parts 450 through 453.

Section 178.270-7. The MTB has revised § 178.270-7 to clarify the requirements for welded joints in the tank shells. The ASME Code has been referenced for weld procedures and welder performance. Additionally, the MTB agrees with several commenters that the requirement for all longitudinal welds to be in the upper half of the tank shell is too restrictive and is not practical for low pressure tanks constructed of stainless steel, which is the predominant material used for IM portable tanks. Therefore, the MTB has deleted proposed paragraph 178.270-7(b).

Section 178.270-8. Except for an editorial change, this section is adopted as proposed in the notice.

Section 178.270-9. The MTB, in agreement with a commenter, has reduced the minimum size requirements for inspection openings. In addition, for clarity, the MTB has added the requirement that each inspection opening must be located above the liquid level in any tank. This requirement is implied by both the UN Recommendation and the IMCO Code.

Section 178.270-10. A commenter recommended that the phrase "and in any case at least three pounds per square inch" be added at the end of § 178.270-10(b) to provide protection from tank collapse due to vacuum and to provide better compatibility with the UN Recommendations. The MTB agrees and has added the phrase "and in any case at least 0.21 bar (3 p.s.i.)" to the end of paragraph 178.270-10(b).

Section 178.270-11. This section had been completely revised. To provide closer harmony with the UN Recommendation and the IMCO Code in proposed paragraph 178.270-11(a)(1) (paragraph (a)(1) herein), the cut-off point for when a portable tank must be equipped with a spring-loaded pressure relief device has been changed from a capacity of "500 gallons or more" to "more than 500 gallons." Thus, a 500-gallon tank is excluded from the requirement for a spring-loaded pressure relief device.

Two commenters argued that a certain amount of flexibility should be provided in the siting of the pressure relief devices specified in proposed paragraph (b)(2) (paragraph (b)(1) in this rule). The MTB agrees with the necessity for flexibility in the siting location but believes it is necessary to place a boundary on the allowable siting locations. A requirement for the siting pressure relief valve inlets, in the top center of the tank has been added with a specified limitation.

Two commenters pointed out that the proposed requirement in paragraph (b)(4) that the valve disc in a spring-loaded pressure relief device must be free to turn on its seats has practical drawbacks, particularly when flame-traps and anti-ice devices are installed. The MTB agrees that these drawbacks would offset any intended safety benefits and has deleted the requirement. Proper functioning of the valves is better assured by periodic valve inspection and retesting.

Two other commenters pointed out that the use of a frangible disc on the inlet side of a spring-loaded pressure relief valve for the transportation of highly corrosive materials was not addressed in the NPRM. The added requirements in paragraph (a)(3), recommended by one commenter, are based on the UN Recommendations and the IMCO Code. If a frangible disc is inserted in series with a pressure relief valve, the space between the disc and the valve must be provided with a suitable tell-tale indicator, pressure gage, needle valve, try cock, etc. to permit detection of frangible disc leakage as a result of pinholing or disc rupture. A frangible disc will not burst at the intended internal tank pressure should pressure build up in the space between the disc and the pressure relief valve. In an extreme case in which the tank is subject to a very rapid pressure rise as for example in the case of a fire, the disc may not rupture below the tank test pressure. Any possibility of pressure build up in the region between the disc and the safety relief valve would be eliminated if an open vent is used. However, such a vent would result in the release of hazardous vapors in the event of a premature failure of a frangible disc. Therefore, any tell-tale indicator for the space between the frangible disc and the safety relief valve must be designed to preclude loss of any hazardous material during transportation. After filling and prior to transportation, the tell-tale indicator must be inspected to determine that the disc is vapor tight. Instructions for the use of the tell-tale indicator must be provided in the vicinity of the indicator at all times.

A commenter stated that in proposed paragraph 178.270-11(d) (paragraph (c) herein) the primary pressure relief device setting be related to the product (total containment pressure) and not the tank (maximum allowable working pressure—MAWP) in order to remove conflict with European land mode maximum allowable relief valve settings. The setting of primary relief valves on the basis of the maximum allowable working pressure is the method used in the UN Recommendations and the IMCO Code. This method uses total containment pressure as only one of the criteria used to authorize a hazardous material in a particular tank and allows great flexibility in the commodities authorized for shipment in a particular tank. In the case where the total containment pressure is sustantially below the MAWP, a greater margin of product containment is provided. Thus, the MTB believes that a change to the proposal is not necessary.

A commenter argued that in proposed paragraph (d)(1) (also (d) herein) one standard cubic foot of air per minute per 30 square feet of exposed tank area was too small a value for a minimum relief valve capacity. The MTB agrees and has revised this paragraph to include: first, a minimum valve size and capacity for each spring-loaded relief device, and second, the minimum total relief valve capacity for each tank as specified in the MC 307 cargo tank specification. This criteria has been used in the evaluation of portable tank exemption applications for many years.

Proposed paragraph 178.270-11(d)(2) (paragraph (d)(2) herein) has been revised for clarity to provide that, regardless of the pressure relief device type and design, the pressure in a tank, even in a fire situation, may never exceed the test pressure of the tank. The minimum cross sectional area requirement for vacuum relief valves was deleted from § 178.270-11(d)(1) and placed in § 178.270-11(d)(4).

Two commenters objected to usage of the tabular requirement in proposed paragraph (d)(3) for the minimum total emergency vent capacity as too restrictive and not in full harmony with the UN Recommendations and the IMCO Code. After review of these requirements, the MTB agrees with the commenters and has added as an alternative to the table, a formula based on the Compressed Gas Association's Pamphlet S-1.2 to allow the minimum emergency vent capacity to be sized for a specific hazardous material or group of materials intended for transportation in a particular IM portable tank. When this approach is used, the tank approval certificate must specify those hazardous materials that may be transported in the tank based on the reduced vent capacity.

One commenter questioned the justification for the greatly increased values in Table 1, Minimum Emergency Vent Capacity, as compared with the values stated in current regulations for cargo tanks in § 178.342-5. The MTB has three reasons for using the values given in Table 1: (1) Table 1 is identical with the corresponding table given in both the UN Recommendations and in the IMCO Code; (2) the table is designed to provide a safe total venting capacity for all hazardous materials transported in IM 101 and IM 102 intermodal portable tanks; and, (3) because of the size and intermodal nature of IM 101 and IM 102 portable tanks, the probability of these tanks being in a total fire engulfment situation is higher than for a cargo tank. Therefore, the MTB has determined that the total vent capacities required in Table I are fully justified.

Several parties recommended that the MTB eliminate the high temperature insulation requirement and authorize the use of readily available insulating materials in proposed paragraph (d)(3) (paragraph (d)(5) herein). The 1200° F requirement for the insulation and jacket materials is in full harmony with the corresponding UN Recommendations and IMCO Code. The MTB feels it is necessary to maintain this harmony for the IM 101 and IM 102 intermodal portable tanks. Also, since the total venting requirement for the tank is substantially reduced when the requirements of (f)(4) are met, it is appropriate that the thermal integrity requirements remain high. In addition, the MTB wishes to clarify that when the effects of insulation are not used to reduce the required total venting capacity, only the provisions of § 178.270-2(e) are applicable to the insulation. The MTB has also redefined the coefficient "F" in proposed paragraph (d) (paragraph (d) in this rule) using the same relationship used by both the UN Recommendations and the IMCO Code.

Upon further consideration the MTB has revised proposed § 178.270–11(e)(2) (now paragraph (d)(6)) to authorize flow rating at 110% of the MAWP.

Section 178.270–12. Two commenters recommended that the location for internal discharge valves be expanded to include "within its companion flange." Also, they recommended the addition of a requirement for a shear section outboard of the internal discharge valve. The MTB feels the recommendations are valid and are in harmony with both DOT and international requirements and they have been adopted in paragraphs § 178.270–12(d) and (e).

Section 178.270–13. The § 178.270–13(d) heading has been revised and the MTB has provided that a tank identical in design, except of a smaller size, to a previously tested and approved tank need not be prototype qualification tested.

Section 178.270–14. A requirement has been added in this section that the relief valve settings must be included on the identification plate of an IM portable tank to avoid confusion during inspection and testing. The provisions concerning the identification plate have been amended by the change of entries and the size of the lettering and by the addition of metric units to provide harmony with the UN Recommendation and the IMCO Code. The location for the marking of the date and identification of the witnessing or performing party for both the last visual inspection and the last hydrostatic test has been made optional to provide operational flexibility. The markings may be on the tank or on the metal identification plate in accordance with the provisions of § 173.32b(d).

Sections 178.271 and 178.272. The IM 101 and IM 102 portable tanks must be designed and constructed in accordance with the ASME Code, with the few exceptions noted in the specification. Inspection and supervision of testing, however, must be performed by an approval agency and not ASME. Therefore, the MTB has added the statement "ASME certification and stamp not required" to both § 178.271 and § 178.272 for clarity.

Also, concerning effective dates, the MTB must comply with the Federal Reports Act of 1942 and procedures administered thereunder by the Office of Management and Budget (OMB) relating to prior clearance of recordkeeping requirements imposed by Federal regulatory action. Prior OMB clearance is required with respect to the provisions adopted herein which impose recordkeeping or report preparation requirements.

MTB will inform the public through notification in the Federal Register when OMB clearance of these requirements has been received. It is anticipated that this clearance process will be completed prior to the effective date prescribed herein.

(46 F.R. 9880, January 29, 1981)

49 CFR Part 174
[Docket No. HM-167; Amdt. No. 174–40A]
Intermodal Portable Tanks—"Trailer-on-Flatcar Service"

On January 29, 1981, the MTB published the final rule on Intermodal Portable Tanks under Docket HM-167 (46 FR 9880) which authorized the use of two new specification intermodal portable tanks. One of the principal commenters to the NPRM under Docket HM-167 was the AAR who submitted several comments concerning the design, test, and usage of IM portable tanks when carried in rail service. The AAR comments were discussed in the preamble to the final rule (46 FR 9881–9883).

The AAR took exception to the proposed authorization for the transportation of portable tanks in TOFC service. The MTB and the FRA carefully reviewed the matter and decided that there may be merit in AAR's comments. As a result, TOFC service was prohibited in the final rule and MTB opened Docket HM-177 entitled "Transportation of Hazardous Materials in TOFC Service." In addition, the MTB held a public hearing on February 25, 1981 (46 FR 8055), in Rosemont, Illinois, at which a representative of the AAR stated he did not believe that tank containers were being transported at high centers of gravity (higher than 98 inches above rail) in TOFC operations. Subsequent to the hearing, a petition for reconsideration of the final rule issued under Docket HM-167 was received from the AAR removing its objection to authorized transportation of portable tanks in TOFC service. However, the AAR objected to the requirement of obtaining FRA approval for container-on-flatcar (COFC) service. Rather, the AAR proposal supported the use of industry standards to monitor TOFC and COFC service. The MTB is not aware of the existence of a formal industry standard specifying conditions for TOFC and COFC transportation of hazardous materials in intermodal portable tanks. Therefore, the MTB will keep Docket HM-177 open to receive proposals for the adoption of such a standard. However, as an interim measure until proposals are received and considered during a rulemaking proceeding, FRA's Associate Administrator for Safety will approve the conditions for TOFC and COFC service for IM portable tanks, as is required for DOT 51, 56, and 57 portable tanks and cargo tanks (See § § 174.61(c) and 174.63(b)).

As stated in the preamble to the final rule (46 FR 9880, page 9888), the Office of Management and Budget (OMB) must clear the report preparation and recordkeeping requirements adopted in the rule. Acting under the Paperwork Reduction Act of 1980, OMB has approved the specified reporting and recordkeeping requirements; Applications for Designation as an Approval Agency (OMB Approval No. 2137–0008), Applications for Approval of an IM Portable Tank (OMB Approval No. 2137–0011), Certification Reports for an IM Portable Tank under DOT Exemption (OMB Approval No. 2137–0012), Proposed Addition (or Change) to IM Tank Table (OMB Approval No. 2137–0013), and Manufacturing Data and Test Reports (OMB Approval No. 2137–0014).

(46 F.R. 24185, April 30, 1981)

[49 CFR Parts 107, 171, and 175]
[Docket No. HM-168; Notice No. 78–13]
HAZARDOUS MATERIALS ABOARD AIRCRAFT

This notice addresses four issues involving the transportation of hazardous materials aboard aircraft. The principal drafters of this document are Alan I. Roberts and Lee E. Metcalfe of the Office of Hazardous Materials Regulation and George W. Tenley of the Office of Chief Counsel, Research and Special Programs Administration.

Applicability of Part 175. The issue concerning the definition of "public aircraft" has existed for a number of years and basically surrounds the term "used exclusively." The MTB is aware that the matter has been the subject of litigation on at least one occasion and that there has been much confusion concerning the applicability of its regulations in this area. Therefore, the MTB proposes to remove the references to "civil aircraft" and "public aircraft" in Appendix B to Part 107, and §§ 171.8, 175.1, and 175.5 and, in lieu thereof, to specifically except government-owned and operated aircraft and those nongovernment-owned aircraft that are under the exclusive direction and control of a government for a period of not less than 180 days as specified in a written contract or lease. The exception would not extend to government-owned and operated aircraft used for commercial purposes.

In determining whether a government does in fact have "exclusive direction and control" over a given aircraft, the MTB believes that the government must, under the terms of the contract or lease, exercise at least the following responsibilities:

1. Approval of crew members and the determination that they are qualified to operate the aircraft (as established by documents showing compliance with pertinent Federal Aviation Regulations);

2. Determining the airworthiness and maintenance of the aircraft (as established by documents showing compliance with pertinent Federal Aviation Regulations); and

3. Dispatching the aircraft, including times of departure, airports to be used, and the type and amount of cargo to be carried.

These criteria are proposed for inclusion in the proposed revision to § 175.5, and the MTB believes they are necessary to assure the safety of an operation not otherwise conducted under the jurisdiction of the Hazardous Materials Regulations.

By specifying a minimum time period, the MTB intends to exclude aircraft operated on behalf of governments for short periods from the exception and to make such operations subject to the Department's Hazardous Materials Regulations. By adoption of the proposed revision to § 175.5 without reference to either "civil" or "public" aircraft, the MTB believes the matter will be clarified to the benefit of all concerned and that the proposal is consistent with responsibilities assigned to the Department by the Hazardous Materials Transportation Act.

Aircraft Leased to Foreign Nationals. The MTB is proposing to remove the exception in § 175.5 pertaining to aircraft of United States registry under lease to and operated solely by foreign nationals outside the United States. This action is being proposed in recognition of obligations imposed upon the United States through its membership in the International Civil Aviation Organization (ICAO) and its commitment as a signatory to the Chicago Convention. Chapter 3.5, Annex 6, Part 1 of the Convention provides that explosives and other dangerous articles, except where necessary for operation, navigation, or safety of an aircraft, may only be carried if their carriage is approved by the aircraft's state of registry and they are in conformance with the regulations of that state. Consequently, in the absence of a demonstrated need to retain the exception based on considerations of safety, United States obligations as imposed by the Convention dictate that the exception be deleted. Therefore, the MTB is proposing that the operations of such aircraft be subject to the same requirements as those applicable to U.S. flag carriers.

Reports of Discrepancies. The Airline Pilots Association (ALPA) has petitioned the MTB to amend the Department's Hazardous Materials Regulations to require aircraft operators to report to the Federal Aviation Administration (FAA) any hazardous materials package that is not prepared for shipment in accordance with Parts 172 and 173 of the regulations. The National Transportation Safety Board (NTSB) made a similar proposal through its Safety Recommendation A-26, issued March 26, 1974, as a result of its investigation of an accident which occurred at Logan Airport in Boston, Massachusetts, on November 3, 1973. Subsequent to the NTSB recommendation, the Department's Hazardous Materials Regulations pertaining to the transportation of hazardous materials aboard aircraft were completely revised in new Part 175. The MTB believes that new § 175.3 entitled "Unacceptable Hazardous Materials Shipments" and new § 175.30 entitled "Accepting Shipments" pertaining to the acceptance of shipments are responsive in part to the NTSB recommendation. However, no discrepancy reporting requirement was adopted at that time. The term "discrepancy" has been selected to preclude the connotation of prejudgment as to the existence of a violation, and the proposed reporting requirement would only apply following the acceptance of a shipment for carriage aboard aircraft. This is to preclude the burdensome reporting of attempts to obtain acceptance for carriage when carrier personnel readily recognize nonconforming shipments and refuse to accept them for carriage. The MTB is proposing to keep the reporting requirement as uncomplicated as possible while accomplishing its intended purpose and believes that its adoption will significantly improve safety through improved communication between aircraft operators and FAA personnel.

Accessibility. The MTB is proposing to amend Section 175.85 to specify the circumstances under which a hazardous material will be considered in a location accessible to a crew member during flight. This a matter that has been at issue for several years since no definition of the term "accessible" has been provided by regulation in the past. This lack of definition has resulted in a number of different and conflicting interpretations, and the MTB believes that the matter should be resolved by the amendment proposed in this rulemaking. ALPA has petitioned for a much lengthier elaboration of the accessibility requirement, but the MTB believes that the brief description of what it considers to be accessible would result in the same limitations ALPA proposes. For example, ALPA suggests language pertaining to accessibility to hazardous materials on pallets or in containers. The proposed rule would require that the material be loaded so that it can be seen and handled. ALPA recommends that a hazardous materials package be placed so that a crew member will have unobstructed access from at least one side of each hazardous material package from a walkway (isle) and that there be sufficient space available to permit isolation of the package if necessary. The proposed rule would require that the material be loaded in such a manner that it can be separated from other cargo during flight. ALPA suggests that hazardous materials packages may be loaded one on top of the other provided each package is accessible from at least one side. The proposed rule would not preclude stacking of packages so long as they can be seen, handled, and separated from other cargo during flight.

The MTB believes that there are several kinds of hazardous materials that should not be subject to the accessibility requirement of §175.85. These are materials that would pose no significant risks to the structural integrity of an aircraft and may be safer to carry in inaccessible locations that are not in proximity to crew members. The MTB proposes to exclude the following from the accessibility requirement: radioactive materials; poison B, liquids or solids (except those labeled FLAMMABLE); and irritating materials.

For purposes of clarity, the latter portion of §175.85(b) pertaining to the operation of single pilot, cargo-only aircraft would be placed in a revised paragraph (c) and the regulation presently found in paragraph (c) would be transferred to new paragraph (f). No substantive changes to these regulations are proposed.

(43 F.R. 57928, December 11, 1978)

49 CFR Part 175
[Docket No. HM-168; Amdt. No. 175–11)
Aircraft Accessibility Requirements for Radioactive Materials

On December 11, 1978, the MTB published a Notice of Proposed Rulemaking (Docket HM-168; Notice 78–13; 43 FR 57928) which contained proposals to amend certain regulations pertaining to the transportation of hazardous materials aboard aircraft. One proposal in Notice 78–13 was to exclude certain classes of materials from the accessibility requirements of §175.85. Even though the final rule under Docket HM-168 will be published in the near future, the MTB believes it should immediately address the issue of accessibility requirements for radioactive materials. This amendment is concerned solely with an exclusion from accessibility requirements for radioactive materials.

Six commenters in Docket HM-168 specifically addressed the proposal to except certain classes of materials from accessibility requirements and all comments were favorable. Therefore, the proposal is considered noncontroversial and, with respect to radioactive materials, is adopted in this amendment. To effect the change, the phrase "Except for radioactive materials, . . . " has been added to the introductory sentence of paragraph (b) of §175.85. The MTB has elected to proceed expeditiously in this matter for several reasons. The requirement that radioactive materials be accessible is not in keeping with good radiation protection principles, the MTB's stated objectives regarding radioactive materials, nor the proposed (42 FR 37427) segregation requirements in Docket HM-152.

Good radiation protection principles dictate that exposures be kept "As Low As Reasonably Achievable" (ALARA) and it is reasonable to load packages of radioactive materials as far as practical from crewmembers. Requiring accessibility has negligible safety value since there is little or nothing a crew could or should do to mitigate leakage from a radioactive materials package. The MTB is working to incorporate the ALARA principle into its regulations and this rulemaking is a step in that direction. Also, the requirements proposed in Docket HM-152 for the segregation of radioactive materials packages would be difficult to achieve on modern aircraft if the packages were required to be accessible.

Furthermore, recent actions by Congress, the Nuclear Regulatory Commission and the Department of Energy require that plutonium be carried on aircraft only in a highly crash resistant packaging (PAT-1). Due to certain crash associated conditions, specifically longitudinal crush, the PAT-1 package is restricted to the aftermost location of an aircraft. This requirement, when coupled with the previous requirement for accessibility of the package, essentially thwarted the carriage of plutonium by air.

(45 F.R. 6946, January 31, 1980)

49 CFR Parts 171, 175
[Docket No. HM-168; Amdt. Nos. 171–55; 175–15]
Hazardous Materials Aboard Aircraft

On December 11, 1978, the MTB published a notice of proposed rulemaking (Docket HM-168, Notice 78-13; 43 FR 57928) which addressed four issues involving the transportation of hazardous materials aboard aircraft. The MTB's actions and significant public comments concerning the proposals contained in the Notice are discussed in the following paragraphs.

Applicability of Part 175. There has been much confusion concerning the applicability of the Hazardous Materials Regulations (HMR) to nongovernment-owned aircraft which are "used exclusively" by a government. In order to lessen the possibility of noncompliance due to any misunderstanding, the MTB has determined that it is necessary to clarify the applicability of the HMR in this area. This amendment prescribes the requirements which define exclusive direction and control (by a government) of nongovernment-owned aircraft to provide a clear distinction between the applicability and inapplicability of regulations issued under the Hazardous Materials Transportation Act (HMTA) (49 U.S.C. 1801 et seq.).

The proposal contained in Notice 78–13 has been adopted with certain revisions. The changes include a shortening of the prescribed minimum lease period from 180 days to 90 days and an editorial revision of § 175.5 for the purpose of clarity.

The proposal to delete the word "civil" from Appendix B of Part 107 was handled in the final rule to Docket HM-166B (45 FR 13087: February 28, 1980).

Four comments submitted by certain Federal agencies criticized the proposal to require that aircraft be chartered or leased for a minimum of 180 days as a condition of exclusive use. Two of the commenters requested periods of 30 days or 60 days on the premise that the shorter periods are more in keeping with their current utilization of leased aircraft. Upon review of submitted data concerning aircraft utilization, the MTB has reduced the minimum lease period to 90 days. It is believed that this change will cover the majority of these government leasing arrangements, thus easing the burden of compliance on the agencies involved. It is the MTB's opinion that a stipulated leased period of less than 90 days would be inconsistent with the intent of this rulemaking action.

Two of the critical commenters suggested that any minimum lease period is unacceptable. One suggested that any shipment accompanied by a government courier or custodian should be excepted from the HMR. The MTB disagrees with this suggestion. The accompaniment of a shipment by a government representative is, of itself, irrelevant with regard to determining the applicability of the HMR to the shipment. The other commenter contended that the operations of a government agency are not in commerce and hence not subject to the HMR. The MTB disagrees with this contention. For transportation by aircraft, a government's hazardous materials shipments are subject to the HMR (1) when the government uses aircraft (civil or private) for commercial purposes (as in the case of certain foreign airlines which operate within the United States and are owned by foreign governments); and (2) when a government uses a nongovernment-owned aircraft which is not established as being under its exclusive direction and control.

One commenter suggested that provision be made to except government-owned, contractor-operated aircraft from compliance with the HMR. The MTB does not believe it necessary to incorporate the suggested provision. If government-owned aircraft are operated exclusively within the government's own sphere of activities for non-commercial purposes, regulations issued under HMTA do not apply so such operations. Non-commercial operations are conducted by certain Federal, state and local government agencies and by government subdivisions such as state universities.

Aircraft Leased to Foreign Nationals. action is taken in recognition of an obligation of the United States under the Chicago Convention. Chapter 3.5, Annex 6, Part 1 of the Convention provides that explosives and other dangerous articles, except where necessary for operation, navigation or safety of an aircraft, may only be carried if their carriage is approved by the aircraft's state (nation) of registry and they are in conformance with the regulations of that state.

Prior to this amendment, Part 175 did not apply to aircraft of United States registry which were under lease to and operated by foreign nationals outside the United States. Historically, the exception was included in the regulations in recognition of the difficulty the U.S. Federal Aviation Administration has had in obtaining compliance with Part 175, and other parts of the HMR which are incorporated by reference in Part 175, in those operations. Unfortunately, the language of the exception did not include a requirement that an equivalent body of regulations must be complied with in lieu of Part 175 to insure that United States obligations under Annex 6 of the Chicago Convention would be met. Determinations of equivalency pose significant problems due to differing requirements. Moreover, there could be many instances where insufficient time would be available in a leasing transaction to compare the HMR with the requirements and prohibitions of another state.

To remedy the situation, the exception in §175.5 has been revised to stipulate that the HMR do not apply to United States registered aircraft under lease to and operated by foreign nationals outside the United States if (1) hazardous materials are carried in accordance with the pertinent regulations of the state of the foreign operator, and (2) the materials are not forbidden aboard aircraft by §172.101. The amendment on this subject represents a substantial revision of the proposal contained in Notice 78-13 which would have required full compliance with the HMR by foreign operators. The revision is based on the merits of several comments to the docket which addressed the inability of foreign operators to comply with requirements of the HMR. Commenters contended that, in many instances, compliance with the HMR would place them in conflict with national or international regulations applicable to the operations of foreign operators. Further, they contended that it would be difficult, if not impossible, to obtain compliance with the HMR by foreign shippers offering hazardous materials for shipment totally outside the United States. The MTB believes these comments have merit and the revision which has been adopted alleviates these difficulties while satisfying United States obligations under the Chicago Convention.

While the MTB has no assurance that the regulations of the nation of a foreign operator will always achieve a level of safety equivalent to that provided by the HMR, it believes the amendment to be the best and most practical solution to the problem. Several comments addressed the difficulty of enforcing any requirement which would apply to foreign operators. Enforcement could, admittedly, be difficult in some instances. However, the requirement is considered necessary and there are means available to the FAA for legal recourse against U.S. lessors, foreign operators, and their agents. Since many of the materials prohibited for air transportation by the HMR are also banned by other governments, it is not anticipated that there will be significant problems involving non-compliance with the prohibitions. It is also anticipated that U.S. lessors may desire to stipulate compliance with the new provision as part of their leasing agreements.

It should be noted that the International Civil Aviation Organization is presently drafting a new Annex entitled "Safe Transport of Dangerous Goods by Air" and that this Annex will be considered by the MTB when it is issued. It is anticipated that §175.5 will require further amendment at that time.

Reports of Discrepancies. This requirement is based on a petition from the Air Line Pilots Association (ALPA), and a similar proposal contained in National Transportation Safety Board (NTSB) Safety Recommendation A-74-26, to require that an aircraft operator report each instance a package offered to the operator for transport by air is discovered not in compliance with the HMR. The proposal contained in Notice 78-13 has been adopted with certain revisions.

The NTSB, in it comments on Docket HM-168, contended " . . . that none of the MTB's actions to date have corrected the deficiency in the existing system which permits a shipper to seek acceptance of a shipment by other carriers once it has been refused by a carrier because of noncompliance with Federal standards." The NTSB further stated "The Safety Board urges the MTB to take action to insure that carriers are required to hold a shipment and its papers until the FAA is notified and the shipment is corrected." Although the MTB agrees with the spirit of this recommendation, the HMTA does not grant the MTB authority to confer upon private individuals (aircraft operators and carriers) the authority to confiscate (in effect) the property of another individual based on presumed, though not proven, noncompliance with the HMR.

In addition to what the MTB considers to be legal impediments to this recommendation, there are practical impediments as well. It would be extremely difficult for FAA enforcement personnel to deal with discrepancy notifications involving potential shipments prior to their acceptance by aircraft operators. A shipment containing a hazardous material must be offered to the carrier in accordance with the regulations. An offering occurs when (1) the package is presented, (2) the shipping paper is presented, (3) the certification is executed, and (4) the transfer of the package and shipping paper is completed with no further exchange (written or verbal) between the shipper and aircraft operator, as usually evidenced by the departure of the shipper. At this point, it is clear that the operator has accepted the shipment and the shipper has removed himself from a final opportunity to take corrective action that would preclude a violation of the HMR relative to transportation of hazardous materials aboard aircraft. Absent a clear showing that the package was offered for transportation as illustrated above, the MTB doubts that an effective enforcement action could be taken against persons who appear at airline cargo terminals seeking to ship packages of hazardous materials.

Based on impediments to implementation of a more comprehensive reporting requirement, the requirement which has been adopted limits required reporting to shipment discrepancies which are discovered subsequent to acceptance of the shipment for transportation and limits "reportable" discrepancies to those discrepancies which are not detectable as a result of proper examination by a person accepting the shipment under the acceptance criteria of §175.30.

This notification requirement will facilitate the timely investigation by FAA personnel of shipment discrepancies involving situations where inside containers do not meet prescribed packaging or quantity limitation requirements and where packages or baggage are found to contain hazardous materials after having been offered and accepted as other than hazardous materials.

Accessibility. The purpose of this change is to define the term "accessible" as used in § 175.85(b) and to except certain materials from accessibility requirements. The definition is considered necessary because there has been confusion in the past over the meaning of the term "accessible." Certain materials have been excepted from accessibility requirements because they pose no significant risks to the structural integrity of an aircraft and may be safer to carry in inaccessible locations that are not in proximity to crewmembers.

The proposal contained in Notice 78-13 has been adopted with revisions. It should be noted that radioactive materials were excepted from accessibility requirements prior to this amendment in Docket HM-168, Amendment 175-11 (46 FR 6946) published on January 31, 1980. The reasons for this action are explained in the preamble to Amendment 175-11.

The types of materials which are excepted from accessibility requirements have been expanded to include not only radioactive, poison B and irritating materials but also etiologic agents. The change is based on the merits of comments by ALPA to the effect that infectious, disease producing agents should be carried as far away from the flight crew as possible. The MTB has revised the proposed exception to permit packages suitable for passenger-carrying aircraft to be transported inaccessibly on cargo-only aircraft in quantities exceeding the quantity limitation (50 lb. net weight per inaccessible cargo compartment or freight container) imposed by § 175.75(a)(2). Without this provision packages containing quantities permitted on passenger-carrying aircraft would be subject to more restrictive location requirements on cargo-only aircraft than packages of identical materials containing larger, cargo-only aircraft, quantities.

The exception from accessibility requirements provided for small, single pilot, cargo-only aircraft being used where other means of transportation are impracticable or not available has been retained in this amendment but has been relocated from paragraph (b) to paragraph (c) in § 175.85.

Based on the merits of comments by an aircraft operator in Alaska, an additional exception from the accessibility requirements of § 175.85(b) has been provided. This exception applies to cargo-only aircraft being operated where other means of transportation are impracticable or not available and will permit operators of such aircraft to deviate from accessibility requirements to the extent that they may use alternate cargo stowage procedures where such procedures are reviewed prior to implementation and approved in writing by an appropriate FAA field office. The exception addresses those situations in which large packages are carried in quantities necessitating the stowage of some of the packages in locations on the aircraft where they cannot be readily handled and separated from other cargo during flight, such as when aircraft loads of 55 gallon drums of hazardous materials are carried.

The above-mentioned changes have necessitated a format revision of § 175.85. Paragraph (b) now contains only the requirement that cargo be carried accessibly. Former paragraph (c) has been redesignated paragraph (f) and a new paragraph (c) has been added which contains certain exceptions to the requirements of §§ 175.85(b) and 175.75(a).

Comments to the docket indicated some misunderstanding of the accessibility requirements. First, this final rule is a clarification, rather than a significant revision, of accessibility requirements. The MTB has never considered it acceptable to load materials acceptable only for cargo-only aircraft in inaccessible locations on aircraft such as inaccessible cargo compartments, inaccessible freight containers or, for packages located in accessible cargo compartments, in locations making them inaccessible. Second, this final rule does not preclude the use of either pallets or freight containers. Pallets may be used, within the context of § 175.85(b), if at least one side or end of each palletized package containing hazardous materials is visible and it is possible to handle and separate packages during flight, if necessary.

In order to render palletized packages visible and separable, it may not be feasible to load pallets to full capacity. However, lost pallet capacity may be minimized by leaving open areas in the middle of pallets so that interior packages are rendered visible, or by using packages of other than hazardous materials as interior packages on pallets.

With regard to freight containers, such containers may be used within the context of § 175.85(b), if the containers are loaded in a manner that permits inspection of their contents during flight, and if each package of hazardous materials contained therein is loaded accessibly, that is, it can be seen, handled and separated from other cargo during flight. By the same token, nonstructural containers and insulating or padding materials are not precluded from use as long as covered packages can be readily inspected, handled and separated from other cargo during flight.

One commenter has suggested that a definition of "accessible" proposed in the ICAO Annex 18 entitled "Safe Transport in Dangerous Goods by Air" be made part of the HMR. The MTB will give consideration to Annex 18 in its entirety at a future date. Because of this, and because of differences between the ICAO proposed definition and the definition proposed in Notice 78-13, a verbatim adoption of the ICAO definition would not be appropriate at this time.

(45 F.R. 35329, May 27, 1980)

49 CFR Part 172
[Docket No. HM-171; Amdt. No. 68]
Use of United Nations Shipping Descriptions

The purpose of the Optional Hazardous Materials Table is to facilitate the international transportation of hazardous materials and to minimize a duplicity of package marking and labeling requirements, and shipping paper descriptions, that would be required to comply with both domestic and international standards. In order to satisfy this intent, it is essential that the Optional Hazardous Materials Table be maintained in alignment with the provisions of the applicable international standards.

The Optional Hazardous Materials Table, in its present form, is reflective of the provisions of the IMCO Code including all amendments through Amendment 16–78. Recently, IMCO published Amendments 17–79 and 18–79 to the IMCO Code, and established June 1, 1981, as the world-wide implementation date for these amendments. Consequently, the MTB is revising the Optional Hazardous Materials Table to maintain the necessary alignment with the IMCO Code and is establishing June 1, 1981, as the effective date for revision of the table.

Amendments 17–79 and 18–79 to the IMCO Code include changes to or deletion of approximately 360 existing hazardous materials entries and the addition of approximately 260 new entries. Because of the optional nature of § 172.102, as well as the large number of changes necessitated by the recent IMCO amendments (many of which are minor in nature), the MTB does not consider it necessary to itemize each change being made to the Optional Hazardous Materials Table. However, to facilitate identification of new or modified entries, such entries have been identified by the inclusion of an asterisk (*) in Column (1) of the table. A copy of the amended optional table has been provided to the Association of American Railroads for use in their efforts to develop a hazardous materials table which consolidates the tables currently appearing in §§ 172.101 and 172.102.

Since this rule does not impose mandatory additional requirements, notice and procedure thereon are considered unnecessary.

(46 F.R. 29392, June 1, 1981)

49 CFR Parts 171, 172, 176, and 178
[Docket No. HM-171B; Amdt. Nos. 171–67, 172–75, 176–14, 178–73]
Use of United Nations Shipping Descriptions

The purpose of the Optional Table is to facilitate the international transportation of hazardous materials and to minimize a duplicity of shipping paper descriptions and package marking and labeling requirements that would be required to comply with both domestic and international standards. In order to satisfy this intent, it is essential that the Optional Table be maintained in alignment with the provisions of the applicable international standards as close as practical.

The Optional Table, in its present form, is reflective of the provisions of the IMDG Code including all amendments through Amendment 18–79. Recently, IMO published Amendment 19–80 to the IMDG Code, and established November 1, 1982, as the world-wide implementation date for the amendment. Consequently, MTB is revising the Optional Table to maintain the necessary alignment with the IMDG Code, but is establishing October 1, 1982, as the effective date for revision of the Optional Table in order to insure that the updated version of the Optional Table appears in the October 1, 1982, reprinting of the Code of Federal Regulations. Users are, therefore, cautioned that the amended entries in the Optional Table are not considered effective by IMO until November 1, 1982.

Amendment 19–80 to the IMDG Code includes changes to or deletion of approximately 140 existing hazardous materials entries and the addition of approximately 60 new entries. In addition, this amendment corrects several errors and omissions in the existing Optional Table.

On May 22, 1982, the name of the Inter-Governmental Maritime Consultative Organization (IMCO) was changed to the International Maritime Organization (IMO). Also, since IMO identifies its International Maritime Dangerous Goods Code by the term "IMDG Code." MTB is changing its term for this code from "IMCO Code" to "IMDG Code." Accordingly, where applicable, these changes are being incorporated into the regulations.

This is the third publication of a final rule under Docket Number HM-171. The first rule under this docket was published in the **Federal Register** on May 22, 1980 (45 FR 34560). The second rule under this docket which is designated as HM-171A was published in the **Federal Register** on June 1, 1981 (46 FR 29392). The final rule published herein is designated HM-171B. Succeeding rules published under docket HM-171 will be designated in alphabetical sequence.

Since this rule does not impose mandatory additional requirements, notice and procedure thereon are considered unnecessary. Further, given the need to provide the widest possible dissemination of the changes to the Optional Table and October 1, 1982 schedule for republication of the affected title of the Code of Federal Regulations, I find that good cause exists for making this rule effective in fewer than 30 days after its publication in the **Federal Register.**

Again, affected persons are reminded that the amended entries are not considered effective by IMO until November 1, 1982.

(47 F.R. 44466, October 7, 1982)

49 CFR Parts 171 and 172
[Docket No. HM-171C; Amdt. Nos. 171–75, 172–84]
Optional Hazardous Materials Table; Use of United Nations
Shipping Descriptions

The purpose of the Optional Table is to facilitate the international transportation of hazardous materials and to minimize a duplicity of shipping paper descriptions and package marking and labeling requirements that would be required to comply with both domestic and international standards. Consequently, it is essential that the Optional Table be maintained in alignment with the provisions of the applicable international standards to the maximum extent possible.

The Optional Table, in its present form, is reflective of the provisions of the IMDG Code including all amendments through Amendment 19–80. Recently, IMO published Amendment 20–82 to the IMDG Code, and established December 1, 1983, as the worldwide implementation date for the amendment. Consequently, MTB is revising the Optional Table to maintain the necessary alignment with the IMDG Code, and is also updating the matter incorporated in §171.7 to include a reference to Amendment 20–82. The effective date for these amendments is established as December 1, 1983, in order to conform to the implementation date set by IMO.

Amendment 20–82 to the IMDG Code includes changes to, or deletion of, approximately 480 existing hazardous materials entries and the addition of approximately 85 new entries. In addition, this amendment corrects several errors and omissions in the existing Optional Table. Because of the large number of changes to the table it is being republished in its entirety with all new or amended entries indicated by an asterisk in Column (1). These asterisks do not, however, constitute a part of the amendment.

Since this rule does not impose mandatory additional requirements, notice and procedure thereon are considered unnecessary.

(48 F.R. 50234, October 31, 1983)

49 CFR Part 172
[Docket No. HM-171D, Amdt. No. 172–105]
Transportation of Hazardous Materials; IMO Proper Shipping Names

Section 172.102 of the Hazardous Materials Regulations (HMR, 49 CFR Parts 171–179) contains the Optional Hazardous Materials Table. The origin of this table is the International Maritime Dangerous Goods Code (IMDG Code) of the International Maritime Organization (IMO). The table contains entries for proper shipping names, hazard classes, identification numbers, labels, packing groups, and vessel stowage requirements. The HMR allows shippers of international shipments which are transported by vessel to use the entries in the §172.102 Table to describe and label their shipments where the description (proper shipping name, I.D. number, label or hazardous class) in the §172.101 Table differs from the IMO description.

The §172.102 Table has not been kept current with the changes in the IMDG Code. IMO Amendment 22–84 became effective on July 1, 1986. Neither Amendment 22–84 nor the previous Amendment, 21–83, has been entered in the §172.102 Table. Together these two amendments contain about 200 changes or additions to the IMO list which are not reflected in the current §172.102 Table. Export shipments which are dispatched using descriptions in the Optional Table may not be in compliance with the IMDG Code. The effect of this non-compliance could be that the shipment would be frustrated by authorities en route to destination.

In order to remedy this situation, RSPA today is issuing an emergency final rule which will amend §172.102 to allow the Director of the Office of Hazardous Materials Transportation to issue an approval of a hazardous material description (on an interim basis) which is not in the Optional Hazardous Material Table but which is in a current edition of the IMDG Code, to be used as if it were in the Optional Table.

This regulation is a rule of agency procedure. Its effect is not to alter the rights or interest of parties, but simply to permit affected parties to seek approval from the Director of the Office of Hazardous Materials Transportation to use a shipping description listed in the IMDG Code but not listed in the Optional Table. Without this procedural change, a party wishing to do so would have to use a more cumbersome and time-consuming exemption procedure. As a procedural rule, this regulation is exempted from the notice and comment requirements of the Administrative Procedure Act (5 U.S.C. 553(b) (A)).

This final rule is made effective immediately upon issuance, rather than the typical 30 days following publication. The Administrative Procedure Act permits issuance of an immediately effective final rule for "good cause" (5 U.S.C. 553(d) (3)). This regulation is made effective immediately because amendments to the IMDG effective July 1 create approximately 200 shipping descriptions not listed in the Optional Table. Until such time as the Hazardous Materials Regulations are amended, it is essential to have an expeditious administrative means of harmonizing the two sets of requirements. Requiring affected parties to go through the existing exemption process would slow commerce and place an unreasonable economic burden on the parties. For example, RSPA is aware of currently pending shipments of materials that could be seriously delayed in the absence of this rule's procedural mechanism. Consequently, RSPA has determined that good cause exists for making the rule effective immediately upon issuance.

(51 F.R. 25639, July 15, 1986)

49 CFR Parts 173 and 178
[Docket No. HM-172; Notice No. 80-2]
Marking and Record Retention Requirements for Cylinders

The cylinder specifications in Part 178 of the Hazardous Materials Regulations require that each cylinder be marked with a serial number and an identifying symbol of the maker. Numbers and symbols of the purchaser or the user are also permitted by the applicable cylinder specifications to be marked on a cylinder if the maker's symbol appears on the cylinder near the original test date. The MTB believes that deletion of the references in Part 178 to the purchaser's or user's markings in each cylinder specification would eliminate confusion between the required markings of the maker and these optional markings of the purchaser or user. The deletion of the purchaser's and user's markings from the individual cylinder specifications would not prohibit a maker from marking on cylinders a series of numbers specified by an original purchaser, so long as duplication of numbers used by a particular maker does not occur. The maker's serial number may be any number conceived by or suggested to the maker as long as that maker uses it only once, so that the number, in association with the maker's mark, is unique. Markings of the purchaser or user would continue to be permitted on a cylinder as additional information under the provisions of § 173.34(c)(1).

The maker's serial number and an identifying symbol required to be marked on a cylinder, together with the original test date, provide the only means of tracing a cylinder to its original test reports. The original test reports, in many cases, are needed for proper retest, repair, or rebuilding of a cylinder as these functions must be performed by a process similar to that used during original manufacture. The test report may be the only source for obtaining this information. Because of the invaluable use of such required markings to locate the test reports, the MTB believes these markings should not be changed. Accordingly, this proposal would delete § 173.34(c)(3)(ii) and thereby prohibit a cylinder owner from changing the manufacturer's serial number and his identifying symbol marked on a cylinder. An approval is not required for changes in optional markings.

On September 27, 1979, the MTB published in the **Federal Register** an amendment (Docket HM-163C, 44 FR 55577) to reassign certain approval, registration and recordkeeping responsibilities from the Transportation System Center of the Research and Special Programs Administration to MTB's Associate Director for Operations and Enforcement (OE). Certain of these responsibilities relative to the submission of cylinder reports are discussed below.

Under current requirements, it is the duty of the inspector to furnish complete reports to the purchaser, the maker, and the Associate Director for OE. Also, if a cylinder has been repaired or rebuilt, test reports as required in the original cylinder specification must be furnished to these same parties. The submission of test reports to all purchasers as required by present cylinder specifications is not necessary. For example, a purchaser of a cylinder intended for personal use may not have a need for such reports. However, should a purchaser want the reports, he should have access to them. This proposal would correct this situation by revising the applicable cylinder specifications to require the furnishing of such reports to the purchaser only if they are requested.

Although the cylinder specifications presently require the inspector to submit test reports to the purchaser, the maker and the Associate Director for OE, there is no requirement that any of these parties retain these records. The MTB believes that due to the long term use of cylinders, these reports should be permanently retained by the inspector and the maker of the cylinder. MTB has incorporated these provisions into these proposals. The requirements for submission of test reports to the MTB would be deleted.

(45 F.R. 9960, February 14, 1980)

49 CFR Parts 173 and 178
[Docket No. HM-172; Amdt. No. 173-156, 178-70]
Marketing and Record Retention Requirements for Cylinders

A notice of proposed rulemaking covering cylinder marking requirements and proposing to adopt test report retention requirements for certain cylinders was published under Docket HM-172 on February 14, 1980 (45 FR 9960). The current cylinder specifications in Part 178 of the Hazardous Materials Regulations require that each cylinder be marked with a serial number and an identifying symbol of the maker. Numbers and symbols of the purchaser or the user are also permitted if the maker's symbol appears on the cylinder near the original test date.

The cylinder specifications also contain a requirement that complete reports be sent to the cylinder maker, purchaser, and MTB. However, there is no requirement that anyone retain the report.

MTB believes that deletion of the references in Part 178 to the purchaser's or user's markings in each cylinder specification eliminates confusion between the required markings of the maker and the optional markings of the purchaser or user.

All cylinder specifications contain a prohibition against duplication of serial numbers. The deletion of the purchaser's and user's markings from the individual cylinder specifications does not prohibit a maker from marking on cylinders a series of numbers specified by an original purchaser. The maker's serial number may be any number conceived by or suggested to the maker as long as that maker uses it only once so that the number, in association with the maker's mark, is unique.

Several comments to this proposal were received in response to the Notice. Three commenters suggested that MTB establish a register of cylinder owners as a provision for ownership control. This is outside the scope of the notice. A registry of cylinder owners would not appreciably improve transportation safety while requiring a large outlay of funds by MTB for development and a large annual investment of funds for maintenance of such a system.

One commenter recommended that the revision to § 173.34(c)(1) authorize additional markings to accompany the required markings. This is authorized by the wording of the revision; however, the specified marking may not be altered or rearranged.

Several comments were received concerning the retention of test reports. Based on a consideration of these comments, MTB has decided to establish a 15 year test report retention period for all cylinders considered in the notice, except for DOT Specification 3HT cylinders which are installed aboard aircraft. The DOT Specification 3HT cylinder test reports must be maintained for the cylinder's authorized service life. MTB is including the 4BA, 4AA480 and 4BW specifications which were omitted from the Notice. This will provide for consistency among similar specifications. MTB believes that the 15 year test report retention period is sufficient to monitor the durability of cylinders including those having no prescribed retest period.

Three commenters suggested that it would be advantageous for MTB to receive summary reports. MTB believes this has merit, however, this was not proposed in the Notice of Proposed Rulemaking, would result in additional costs to the public as well as DOT, and is not in accordance with efforts to reduce the paperwork burden on the public.

One commenter indicated that a requirement for the inspector and the maker to retain the test reports would increase costs. Inspections by MTB of cylinder manufacturing facilities have revealed that it is a common practice of independent inspection agencies and manufacturers to retain test reports as proof of performance of their responsibilities. MTB believes that by formalizing the test report retention requirements that have been a common industry practice, a savings will result to the industry through the elimination of the reporting requirements that are cancelled by this rulemaking.

One commenter recommended that a restriction similar to that specified by § 178.45-17(e) for the DOT Specification 3T cylinder be made applicable to other cylinders. This restriction prohibits a cylinder from being marked DOT-3T unless specific requirements are met. While MTB feels that this restriction is appropriate, it should be noted that § 178.0-2(a) and (b) contain similar provisions for all specification packagings, including cylinders, and thus the prohibition in each specification would be superfluous.

One commenter suggested that regulatory control over the quality of cylinders being produced would be lost or greatly decreased unless MTB registered all cylinders. MTB does not agree. MTB believes the quality of cylinders can be maintained only through the cooperation of the cylinder industry and an effective surveillance program.

One commenter suggested that the transfer of cylinder ownership should be accompanied by the addition of the new owner's marks as an ownership guarantee. MTB believes that owner's marks on cylinders would not necessarily be effective in protecting owners against misuse of their cylinders because the presence of a symbol on a cylinder does not prove ownership. MTB does believe that the conveyance of a cylinder bearing the unique manufacturer's mark and serial number when supported by proper documentation can be used as an indication of ownership.

SPECIFIC REQUIREMENTS

Paragraph (c)(1) of § 173.34 is revised to clarify the authority for placing markings other than required markings on cylinders. This paragraph also prohibits indentations from being made in the sidewall of a cylinder unless specifically permitted in the applicable specification for the cylinder. Paragraph (c)(3) is revised to combine paragraph (i) into (c)(3) since paragraph (ii) is deleted as proposed.

Except for the cylinders which are identified by lot numbers, the specifications are amended to specify that the required cylinder markings must consist of the maker's symbol and a serial number.

Except for the cylinders for which there is no requirement for report submittal, the specifications are amended to change the inspector's duties to require the cylinder inspector to submit reports required by the cylinder specifications to the cylinder maker and, upon request, to the purchaser. This amendment also establishes a general requirement that the inspector retain the cylinder test reports for 15 years from the original test date of the cylinders. For the DOT Specification 3HT cylinders, the inspector must retain the cylinder test reports for the authorized service life of the cylinders.

Except for the cylinders for which there is no requirement for report submittal, the cylinder specifications are amended to establish a general requirement for the manufacturer of the cylinders under these specifications to retain the cylinder test report for 15 years from the original test date of the cylinders. However, the manufacturers of the DOT Specification 3HT cylinders must retain the cylinder test reports for the authorized service life of the cylinders.

(47 F.R. 16183, April 15, 1982)

49 CFR Part 173
[Docket No. HM-172A, Advance Notice]
Marking Owner Symbols on Compressed Gas Cylinders

By a petition dated December 15, 1982, CGA requested a revision to an amendment published under Docket HM-172 revising § 173.34, *inter alia* subparagraph (c)(1) of that section (Amdt. 173–153; 47 FR 16183). Most of the petition is quoted as follows:

Proposed Revision. The Compressed Gas Association (CGA) requests revision of 49 CFR 173.34(c)(1) as amended, to include the following italicized language:

(1) Additional information such as owner symbols, not affecting the markings prescribed in the applicable cylinder specification, may be placed on the cylinder. Only one owner symbol may be displayed on each cylinder. Such symbols may be removed or changed only by the owner or with the owner's written permission. No identation may be made in the sidewall of the cylinder unless specifically permitted in the applicable specification.

Note.—For the purpose of avoiding duplicate owner symbols, the Compressed Gas Association maintains a registry of such symbols. See CGA Pamphlet C-16.

CGA has a vital interest in maintaining a registry of owner symbols. As phrased, the Note following the revised section would simply advise regulated parties of the existence of the registry, and would inform them of our intention to avoid confusing duplication of symbols by different owners.

Purpose of the revision. Section 173.301(b) of today's regulations requires that "A container charged with a compressed gas must not be shipped unless it was charged by or with the consent of the owner of the container."

There is a strong safety basis for this provision. Cylinders commonly remain in service for many decades. The excellent safety record in the transportation of gases over such time periods is derived in significant part from the careful charging, periodic retesting, and maintenance functions performed by the cylinder owner or on his behalf. The cylinder, as a reused container kept within the control of the owner, remains the owner's responsibility, and safety in transportation is enhanced as a result.

Compliance with this provision would be greatly facilitated by the two requirements that we ask you to insert in Section 173.34. The appearance of multiple owner symbols on the same cylinder obviously would frustrate the intent of Section 173.301(b). We request, therefore, that the current owner be made more readily identifiable by having Section 173.34 declare that only one owner symbol at a time may appear on a cylinder.

The second requested requirement would provide assurance that the symbol of an owner would not be removed without his permission. Section 173.301(b) recognizes the owner's responsibility for cylinder maintenance and record keeping, and the safety value of those functions. This information provides historical continuity in use of the container, for the sake of safety in transportation and use. Haphazard and unlimited alteration of owner symbols without owner permission would nullify the benefit of the current requirement to seek the owner's permission before shipping a recharged cylinder.

Other benefits of the revision. Clarification of the ownership of cylinders also provides benefits in the environmental area, facilitating rapid identification of interested parties in the event of improper discard of a cylinder. In situations involving civil liability as well, more ready determination of ownership will be beneficial to the public.

Conclusion. CGA's registration system of owner symbols will be maintained as long as the requested language appears in title 49 CFR. The registration system will be maintained without expense to DOT, and it will not entail any use of government personnel.

Registering a symbol under the CGA system will not be required in Title 49, and therefore no regulatory impact, small business, paperwork reduction, environmental impact, or information gathering issues are involved. The proposed requirements pertaining to appearance of only a single owner's symbol, and no changing of owner symbols without owner permission, merely facilitate existing requirements and also impose no new obligations on any parties requiring regulatory impact analyses.

. . . . .

Subsequent to receipt of the petition, CGA provided MTB a draft copy of its Pamphlet C-16 as adopted by CGA, but not yet formally printed. A copy is available for examination in the public docket under HM-172A.

Interested persons are encouraged to present their views on the CGA petition described above. Of particular value would be comments addressing the following questions concerning the CGA petition:

1. What constitutes an owner's symbol and how should the term be defined?

2. Should a distinction be drawn between symbols stamped into the metal of a cylinder (shoulder or footring, when permitted) and those that are non-permanent e.g., painted logos, trademarks or tradenames?

3. If a rule is adopted, as proposed by CGA, what impact would it have on owners of existing cylinders?

4. In view of the advantages discussed by CGA in ensuring that the current owner of a cylinder be readily identifiable, should there be a mandatory requirement that an owner symbol be placed on a cylinder?

5. What actions, if any, should be required of a carrier, shipper, or a freight forwarder who finds more than one owner symbol on a cylinder?

Commenters are not limited to responding to the questions raised above and may submit any facts and views consistent with the intent of this notice. In addition, commenters are encouraged to provide comments on "Major rule" considerations under terms of Executive Order 12291, "significant rule" considerations under the DOT regulatory procedures (44 FR 11034), potential environmental impacts subject to the Environmental Policy Act, information collection burdens which must be reviewed under the Paperwork Reduction Act, and economic impact on small entities subject to the Regulatory Flexibility Act.

(49 F.R. 14405, April 11, 1984)

49 CFR Part 173
[Docket No. HM-172B; Notice No. 84–12]
Cylinder Retester Identification Procedures

On August 17, 1978, MTB amended § 173.34(e)(1) to require, as pertinent here, that:

> Each person who represents that he performs retesting [of cylinders] in accordance with [paragraph (e)] must have a retester's identification number issued by the MTB or have a valid Bureau of Explosives approval issued prior to August 17, 1978. Such persons holding a B or E approval without a termination date must obtain a retester's identification number from MTB before December 31, 1979. [Docket HM-163; 43 FR 36445; August 17, 1978.]

This action was part of an overall program to withdraw all delegations of authority previously made to the Bureau of Explosives (B of E). At the time, MTB believed that the cylinder retest verification and approval functions being performed by the B of E could be performed by persons qualified as independent inspection agencies as provided in § 173.300a under the direction of MTB. MTB believes the soundness of its actions relative to cylinder retesting have been substantiated by the favorable experience of the independent inspection program since August 1978. It is primarily this experience which is the foundation of the proposals in this notice.

To assure the continuity of the cylinder oversight responsibility, MTB developed procedures and requirements for retesters to assure that cylinder hydrostatic retesting would be performed by qualified retesters. These procedures and requirements are available to prospective cylinder retesters upon request from MTB.

The cylinder retest approval procedures require a prospective applicant to request, in writing, an application form and a list of participating independent inspection agencies approved under § 173.300a. The applicant must make arrangements to have an independent inspection agency perform an inspection of his cylinder retest facility. Based upon a favorable recommendation from the independent agency, and other data demonstrating that the applicant has capabilities in both personnel and equipment to retest cylinders, the applicant is issued a cylinder retester identification number by MTB. Each retester identification number is valid for five years. Conditions contained in each document granting a retester identification number include the following:

a. Limits on the types of cylinders that may be retested.

b. A requirement that testing must be performed by or in the presence of a designated hydrostatic test operator whose performance has been witnessed by the independent inspection agency, or one who has been qualified in a program acceptable to MTB.

c. A requirement that a copy of the document which grants the retester his identification number must be maintained adjacent to the testing apparatus, and a copy of the document must be signed by the facility manager and returned to MTB within 20 days of receipt by the facility.

d. A requirement that the retester notify MTB within 20 days of any change in management, equipment, or designated testers.

Requirements for notification of changes in management pertain only to those management employees directly responsible for supervision of the retesting. Depending on the size and organization of some companies, MTB believes it may be advantageous for some retesters to maintain a centralized personnel listing of all testing personnel throughout the company. Consequently, in place of notification, a retester may be permitted to maintain a current listing, with qualifications, of each person who conducts retesting by making a written request and securing written acceptance by MTB. Similarly, notification of changes in equipment is required only for replacement of a component which would cause the cylinder hydrostatic retest equipment to be different from that observed and recommended by the independent inspection agency.

In addition, the retester is required to request authorization from MTB to retest cylinders other than those upon which the inspector's report was based.

The above description of the current retester qualification program is intended to serve two purposes. First, it is presented to offer notice that the program exists and to explain its operation, so as to avoid expanding the already voluminous provisions of § 173.34(e) through codification of these procedures therein. It should be noted that a general recodification of § 173.34 is being developed by MTB, and this action is consistent with that effort. Secondly, it is necessary to describe the current program and procedures in order to provide a basis for the additional requirements proposed below.

It is proposed to enable MTB to specify certain special conditions in addition to the standard conditions, based on its review of the applicant's retest program and the completed record of the independent inspection agency. If special conditions were specified, the applicant would be given 20 days to submit written comments on those conditions to MTB. Pending a decision by MTB on the applicant's request for adjustments, the applicant could proceed with its retesting program if it complied with all conditions contained in the issuance document, including those special conditions being challenged.

The other proposed new provision (which would appear as an amendment to § 173.34(e)) is that upon the successful completion of each retest, the retester would be required to stamp the cylinder with his identification number and the date of retest. MTB believes this proposal is important because it would identify the person who performed the test. The application of the retest markings could not adversely affect other required markings on the cylinder.

In proposing to adopt the procedures and requirements discussed above, and to give full effect to the program generally, it is necessary to propose a change to § 173.300a to reflect the fact that independent inspection agencies approved thereunder would be eligible to perform inspections in functional areas other than the currently stated limitation manufacturer's testing under Part 178. As proposed, an approved independent inspection agency could perform inspection functions as required by any provision in the Hazardous Materials Regulations (49 CFR Parts 171–189) for which it has been qualified.

MTB requests that persons commenting on this notice include any information that they may have concerning the difficulties and burdens experienced by currently authorized cylinder retesters in meeting the conditions of their registration. MTB also requests that any special problems concerning corporate organizations (i.e., centralized programs or decentralized programs) be brought to our attention.

(49 F.R. 39177, October 4, 1984)

49 CFR Part 173
[Docket No. HM-172B, Amdt. No. 173-194]
Cylinder Retester Identification Procedures

On October 4, 1984, RSPA published a Notice of Proposed Rulemaking (NPRM) in the **Federal Register** (49 FR 39177). The NPRM proposed certain procedures requiring a cylinder retester to mark a retested DOT cylinder with an identification number (issued by RSPA). The proposed identification number would be placed between the month and year of the retest date on each cylinder retested. The NPRM also proposed additional changes to permit an approved independent inspection agency to perform inspection functions as authorized by any provision in the Hazardous Materials Regulations (HMR) (49 CFR Parts 171–172) for which it has been qualified. This final rule contains amendments to the OHMT which are based on the proposals contained in the NPRM and the comments received in response to that NPRM. Interested persons should refer to Notice 84–12 (49 FR 39177) for detailed background information.

RSPA received 32 written comments in response to Notice 34–12, most of which came from manufacturers of fire fighting equipment. Comments were also received from two trade associations, the Air Transport Association (ATA), and The Chlorine Institute, Inc. The majority of the respondents were supportive of the proposal.

A discussion of the significant comments and amendments adopted in this final rule follows:

Three commenters expressed their concern that cylinder logos (an identifying statement) which each of their companies have used for years to identify their own cylinders, and which are recognized in their own geographic areas, would no longer be allowed; one of the three commenters suggested that the use of a symbol or logo should be allowed as an option. RSPA disagrees. RSPA views logos, to which several companies subscribe, as only being recognizable in small geographic areas. There are situations where two or more companies located in different parts of the country are using the same logos. In the case of symbols, there is no easy way to distinguish a retester's symbol from those of cylinder owners, users, manufacturers or inspectors. In addition, RSPA is unable to register retester's logos on its automatic data processing equipment.

The ATA, which represents carriers that contract with independent agencies for the hydrostatic retesting of cylinders, and four other commenters expressed concern that the proposed markings will require additional space on the cylinder shoulder, and more work for retesters because of the additional required markings. RSPA disagrees.

During the past few years, RSPA has observed the proposed identification system in use both domestically and internationally. A number of retesters have been marking their cylinders in the manner proposed in the NPRM, using one die containing all characters of the identification number. The one die concept has produced legible markings and the process takes less of the retester's time than the application of single character dies. RSPA has observed the one die concept for marking cylinders undergoing retest, and found that the amount of space needed for marking the retested cylinders only increased slightly.

The Chlorine Institute (CI), a trade association which represents chlorine producers and packagers engaged in packaging chlorine into cylinders, took strong exception to the proposed marking procedures. They stated that chlorine is distributed primarily by companies that, for the most part, conduct their own hydrostatic tests. The CI argued that these companies maintain meticulous cylinder retest records and maintain complete control and accountability for their own cylinders. Therefore, these members do not support the proposal on the basis that the identification requirements would serve no useful purpose. RSPA disagrees. Although the chlorine industry may be doing an excellent job of retesting and keeping track of their cylinders, there is a larger number of users that are not willing or able to be as meticulous. Furthermore, RSPA takes the position that retest records do not facilitate compliance oversight unless there is traceability from the cylinder to the records.

(50 F.R. 46054, November 6, 1985)

49 CFR Parts 173 and 179
[Docket No. HM-174; Notice No. 80–6]
Shippers: Specifications for Tank Cars

ACCIDENT EXPERIENCE

At the time the Department of Transportation commenced its review of specifications for pressure tank cars, there had been a series of disastrous railroad accidents involving rail transportation of flammable compressed gases, toxic compressed gases and other hazardous materials. Most of these accidents involved uninsulated pressure tank cars built to the DOT Specifications 112 and 114.

The Federal Railroad Administration (FRA) accident data accumulated during the period of January 1, 1969, through December 31, 1974, had indicated:

	DOT specification	
	112/114	105
Number of accidents reported to FRA	193	101
Number of cars derailed and/or damaged	434	213
Number of cars sustaining a head puncture	86	13
Number of cars sustaining a shell puncture without a head puncture	13	5
Number of cars ruptured without puncture	59	8
Number of tanks sustaining partial or total loss of hazardous lading	156	39
Number of persons killed as a result of tanks being punctured or rupturing	23	0
Number of persons injured as a result of tanks being punctured or rupturing	836	151

On the basis of this accident data, the Department determined that non-retrofitted 112/114 tank cars presented a greater threat to the public safety than the 105 tank cars. However, 105 tank cars have been involved in a number of train accidents over the past 25 years which dramatize the importance of assuring that these tank cars are also equipped with the best safety protection that is feasible. For accidents prior to 1972, a comprehensive analysis was made by the Railway Progress Institute and the Association of American Railroads (AAR) as part of the Railroad Tank Car Safety Research and Test Project (RA-01-2-7), Phase 01, Report on Summary of Ruptured Tank Cars Involved in Past Accidents, a copy of which has been placed in the docket. The report identified a significant number of accidents in which 105 tank cars were punctured or ruptured due to thermal input. Among the accidents described were the following:

On April 23, 1963, at Bradtsville, Pennsylvania, a derailment resulted in the overturning of a 105A300 car carrying LPG causing an LPG leak and fire. This fire caused the violent rupture of an overturned 112A400W car also carrying LPG. Ten minutes later, the 105 car also violently ruptured, hurling sixty percent of the tank 900 feet.

On December 13, 1964, at West Columbus, Ohio, a 105A100W car carrying ethylene oxide punctured and a fire ensued. Another 105A100W car carrying ethylene oxide subsequently ruptured.

On August 22, 1967, at Texarkana, Texas, a 105A300W tank car carrying butadiene was engulfed in a fire and subsequently ruptured.

On May 27, 1968, at Cotulla, Texas, an intense fire resulted from the puncture of two 105A300W cars carrying vinyl chloride. Two 105A100W tank cars carrying ethylene oxide ruptured after about one hour's exposure to this fire. An end of one of the cars was hurled 300 feet.

On September 11, 1969, a derailment at Glendora, Mississippi, resulted in the head puncture of a 105A200W car carrying vinyl chloride and the shell puncture of a 105A200W car also carrying vinyl chloride. The car with the head puncture "torched" a 105A300W car carrying vinyl chloride. This latter car subsequently ruptured, hurling one half of the tank 600 feet and the other half 200 feet.

Two recent accidents demonstrate the potential consequences of release of product from these cars. On February 26, 1978, near Youngstown, Florida, an Atlanta and Saint Andrews Bay train derailed when joint bars were intentionally removed from the rail. During the derailment, a 105 tank car containing chlorine was punctured in the bottom of the tank shell. Eight persons died and 138 were injured as a result of contact with chlorine gas that settled in the area near the derailment.

On April 8, 1979, near Crestview, Florida, a Louisville and Nashville train derailed 25 cars containing hazardous materials. At least five 105 tank cars released some product during that derailment, including a 105A500W tank car that released chlorine from a puncture in the tank shell and a 105A300W tank car containing anhydrous ammonia that split into several pieces and rocketed. Due to the release of several types of hazardous materials, over 4,000 people were evacuated from the surrounding area.

For the reasons discussed in the section by section analysis of sections 179.100-23 and 179.106-5, the systems proposed in this proceeding are not directed at the prevention of damage to tank shells (as distinguished from tank heads) such as occurred at Youngstown and Crestview. However, these accidents illustrate that 105 tank cars are vulnerable to loss of lading through mechanical damage. The human and economic consequences that result from such instances may be substantial.

Other recent accidents further illustrate that the safety systems mandated for 112/114 tank cars are also relevant to the design of 105 tank cars. On June 16, 1977, two Missouri Pacific trains collided at Neelyville, Missouri, causing the lower tank head of a 105A300W tank car containing vinyl chloride to be punctured by the coupler of the adjoining car. On March 16, 1977, a 105A300W tank car transporting butane was punctured in the tank head by a freight car wheel when in Atchinson, Topeka & Santa Fe train derailed at Love, Arizona. In neither of the foregoing accidents was the damaged car equipped with shelf couplers or HM-144 head protection.

On September 8, 1979, a Southern Pacific train derailed near Paxton, Texas. Two 105A300W cars lost their lading in that accident. One of those cars contained isobutylene and was apparently breeched by fire, although it may have sustained damage to a tank head during the derailment. The other car, containing ethylene oxide, ruptured violently as a result of exposure to a "pool fire" fueled by flammable liquids also released during the accident. Neither 105 car was equipped with shelf couplers, HM-144 head protection, or HM-144 thermal protection.

PRIORITY ACTION FOR 112 AND 114 TANK CARS

Since the Specification 112 and 114 tank cars were determined to present a more serious threat to public safety, the Bureau and the FRA decided to assign first priority to improving the construction and maintenance standards applicable to those cars. It was further decided that after these 112 and 114 cars had been structurally upgraded, the Bureau and FRA would then consider a revision of the standards applicable to the 105 tank car to provide a level of safety comparable to that of the improved 112/114 tank cars.

Accordingly, on September 15, 1977, the Bureau issued amendments Nos. 173-108 and 179-19 (42 FR 46306). In summary, these amendments required:

1. Existing and newly built specification 112 and 114 tank cars used to transport flammable gases such as propane, vinyl chloride and butane to have both thermal and tank head protection.

2. Existing and newly built specification 112 and 114 tank cars used to transport anhydrous ammonia to have tank head protection (such as a head shield).

3. All specification 112 and 114 tank cars to be equipped with special couplers designed to resist coupler vertical disengagements (shelf couplers).

The retrofitting of couplers and of head shields on tank cars that transport anhydrous ammonia has been completed. Approximately ninety percent of the 112 and 114 tank cars used to transport propane and other liquefied flammable gases have been retrofitted with tank head puncture and thermal protection. The remaining ten percent (approximately 1,700) of these tank cars will have their retrofit completed by the end of this year.

Now that the 112 and 114 tank car retrofit program is in its final stage, the Bureau and FRA believe that the 105 tank car should now be addressed. This Notice of Proposed Rulemaking proposes to do so in a manner consistent with amendment Nos. 173-108 and 179-19.

Additionally, in developing this Notice, the Bureau and the FRA have determined that two related safety items should be addressed:
1. The need for full tank head puncture resistance; and
2. Application of shelf couplers to all existing and newly built DOT specification tank cars.

NATIONAL TRANSPORTATION SAFETY BOARD RECOMMENDATIONS

On November 22, 1978, the National Transportation Safety Board (NTSB) issued "Recommendation R-78-58." It stated: "Require that top and bottom shelf couplers be installed on all DOT 105 tank cars as soon as possible (Class I, Urgent Action)." On March 12, 1980, the NTSB expanded this "recommendation" to suggest that the DOT extend Federal requirements for top and bottom shelf couplers to all tank cars which carry hazardous materials and extend requirements for shelf couplers, head shields and thermal protection to type 105 cars when they are newly manufactured or rebuilt.

FRA SAFETY INQUIRY

On April 13, 1978, the Federal Railroad Administration conducted a Special Safety Inquiry into Improved Safety Standards for Insulated Pressure Tank Cars. Testimony was heard from representatives of the NTSB, the AAR, shipping industries, tank car builders and leasing companies, and railroad operating unions. There was general agreement on the concept of completing the retrofit of the 112 and 114 tank cars before beginning any retrofit of 105 tank cars. Also, there was general agreement that shelf couplers should be retrofitted on 105 tank cars after the coupler retrofit on 112 and 114 tank cars was completed. Differing opinions were expressed as to the need for the further retrofitting of 105 tank cars.

Subsequent to that hearing, the Bureau and FRA determined that all new tank cars are being equipped with shelf couplers and that most existing 105 tank cars are also being so equipped when coupler repairs or replacements become necessary. Also, as a result of that hearing and information received subsequently from car builders, FRA believes that most of the 6,000 newly built 105 tank cars that are utilized in flammable gas service have been equipped with ½-inch jacket heads and high temperature insulation.

CHARACTERISTICS AND USE OF 105 TANK CARS

There are approximately 27,000 U.S. and Canadian owned Specification 105 tank cars, of which approximately 24,000 are built to DOT specifications and 3,000 are built to Canadian Transport Commission (CTC) specifications. The U.S. fleet consists of approximately 1,400 aluminim 105 tank cars and 22,600 steel tank cars. Until 1973, most 105 tank car tanks had capacities ranging from 10,000 gallons to 20,000 gallons. Since 1973, FRA estimates that more than 6,000 DOT specification 105 tank cars have been manufactured that have capacities ranging from 25,000 to 34,000 gallons.

Report FRA/ORD-80/60, entitled "105A Tank Car Fleet Characterization Study," is included in the docket and contains additional information on the variations in age, structural designs, capacities, thermal insulations, and other characteristics of 105 tank cars.

Many DOT Specification 105 tank cars are used to transport the same hazardous commodities as are transported in the 112 and 114 specification tank cars. In addition, the 105's are used to carry other hazardous materials such as chlorine, ethylene oxide, butadiene, hydrocyanic acid, motor fuel anti-knock compounds, poisons and combustible/flammable liquids and solids.

All 105 specification tank cars have some amount of thermal insulation and all have steel jacket coverings of varying thicknesses. Most tank heads and shell thicknesses on the 105 cars are greater than those on the 112 cars. However, it is estimated that at least 8,000 of the existing 105 cars that carry the same commodities as 112 and 114 tank cars do not have the equivalent level of puncture and thermal protection mandated for the 112 and 114 tank cars.

RETROFIT OF TANK HEAD AND THERMAL PROTECTION

Elsewhere in this issue of the Federal Register, the Bureau and the FRA are issuing an Advance Notice of Proposed Rulemaking soliciting comments concerning the retrofit application of tank head and thermal protection systems to existing 105 tank cars and other DOT Specification tank cars that are used to transport the same hazardous materials as 105 cars.

SECTION-BY-SECTION ANALYSIS

§ 179.31 Qualification, maintenance, and use of tank cars. The proposed amendment of paragraph (a)(3) of Section 173.31 would authorize the use of class DOT-105J cars that have equal or higher marked test pressure than the test pressure for the prescribed 105A tank car. This is proposed in order to provide authorization to use the 105J tank car under current provisions of Part 173.

Paragraph (a)(6) would require shelf couplers to be installed on all 105 tank cars by December 31, 1981. Since there are approximately 24,000 specification tank cars in the United States fleet, the total number of shelf couplers required to be installed would be approximately 48,000 (two per tank car). However, the Bureau and the FRA believe that approximately 6,000 of these cars already are equipped with shelf couplers and that shelf couplers can be applied on the remaining 18,000 tank cars within the proposed time period without disrupting the flow of vital commodities transported by these 105 tank cars. Shelf couplers are easily installed. Moreover, the Bureau and the FRA believe that the retrofit installation of shelf couplers on the 105 tank cars can quickly improve the safety performance of these cars at minimal cost. At the FRA Special Safety Inquiry, industry representatives agreed that these special couplers assist in keeping cars in line and preventing tank punctures.

Paragraph (a)(7) would require shelf couplers to be retrofit installed on all other DOT specification tank cars by not later than December 31, 1984. The Bureau and FRA estimate that there are approximately 135,000 DOT Specification tank cars. Of this number, approximately 24,000 have been built to specification 105 and approximately 18,000 to specifications 112 and 114. These tank cars are covered by proposed paragraph (a)(6) and existing paragraph (a)(5). Thus, paragraph (a)(7) would apply to approximately 93,000 cars. However, the Bureau and FRA estimate that approximately 20,000 of these tank cars are already equipped with shelf-couplers retrofit installed within four years based upon industry's experience with the HM-144 shelf coupler retrofit programs.

§ 179.14 Tank car couplers. The proposed deletion of existing paragraphs (a)(1)(2) and (4) would remove the authority to apply non-shelf F-Style couplers to new tank cars. The remaining couplers specified in paragraphs (3) and (5) are top and bottom shelf E-style and top shelf F-style couplers. Thus, the effect of this proposed change would be to require shelf couplers on all 105 tank cars built after December 31, 1980. The Bureau and the FRA understand that current practice is to install shelf couplers on all new tank cars.

§§ 179.100-23 and 179.106-5 Head shields. The proposed change in Section 179.100-23 would require that all new DOT Specification 105, 112 and 114 tank cars built after December 31, 1980, be equipped with a tank head puncture resistance system providing protection for the entire tank head, rather than only the lower half of the tank head. The purpose of this requirement is to assure that new tank cars will be designed to provide the maximum feasible protection against tank head mechanical damage in a derailment environment. The proposed addition of paragraph (c) in Section 179.100-23 would authorize continued use of 112 and 114 tank cars equipped to present HM-144 requirements.

The rule issued in Docket No. HM-144 governing the application of safety systems to DOT Specification 112/114 uninsulated pressure tank cars (Section 179.105-5) required only the lower half of the tank head to receive protection. This requirement was based on analysis, research and testing conducted in the early 1970's, and represented the best judgment of what was prudent and feasible at the time the requirements of Docket No. HM-109, the predecessor to Docket No. HM-144, were promulgated (39 FR 27572; July 30, 1974). Docket No. HM-144 added coupler restraint systems and thermal protection to the requirements for retrofit of 112/114 tank cars, producing an overall system of safety protection that renders these tank cars highly resistant to product loss in a derailment environment.

The HM-144 requirements, then, represented a very satisfactory approach to the protection of pressure tank cars. Nevertheless, recent accidents have illustrated that human and economic losses resulting from individual accidents may dramatically exceed the levels previously anticipated. In addition, at least three tank car companies have incorporated full tank head protection into their designs for the retrofit of 112/114 tank cars and the construction of some 105, 112 and 114 tank cars. This voluntary initiative by private industry has demonstrated both the economic and technical feasibility of providing full tank head protection.

These developments have caused the Bureau and the FRA to reconsider the issue of new pressure tank car construction with respect to protection against mechanically-caused failure of the pressure tank. Puncture, tearing, or critical scoring of a pressure tank equipped to HM-144 specifications can occur in at least three modes:

First, the tank shell may be damaged in a derailment involving significant forces. Roughly one out of ten instances of major product loss involves shell penetration. The application of material (such as insulation or jacketing) on the exterior of the shell may provide limited protection. However, it appears not to be currently feasible to provide impact resistance on the tank shell comparable to that required for tank heads. The additional weight associated with shell shielding materials and support structure, coupled with the gross weight on rail limits, would reduce the product-carrying capacity of these cars. The reduction in capacity, in turn, would increase the cost of transporting these products by rail.

The second tank failure mode is penetration of the required head protection. An extreme derailment or a high force impact between groups of cars could result in failure of the tank head protection system. However, extensive testing and recent rail accident experience have demonstrated that the likelihood of such failure is very small; and any further effort to strengthen the system would face the same limits of practicability discussed above.

The third failure mode is penetration of the tank above the level protected by the required head shield. Prior to application of shelf couplers on 112/114 cars, roughly one out of ten mechanically-caused failures of those cars occurred in that manner. The derailment at Pensacola, Florida, November 9, 1977, for instance, involved a puncture just above the area that would likely have been protected by a shield covering the lower half of the tank head. Failures of 105 cars, while less frequent, have followed the same basic pattern as 112/114 cars with respect to mechanical damage resulting in product loss.

While application of shelf couplers will tend to reduce the likelihood that objects will strike the top half of pressure tank car heads, it is certain that some such instances will occur. Indeed, a retrofitted 112 tank car was punctured in the top portion of the tank head when on January 14, 1980, at Ridgefield, Washington, a Burlington Northern train derailed. The release of anhydrous ammonia through the puncture resulted in two deaths, while derailment forces in that accident may have been sufficient to overcome the protection in accord with that taken under the 112 and 114 tank car safety program.

CANADIAN TANK CARS

In proposed Section 179.106–1, paragraph (c) would require that 105 tank cars built to specifications promulgated by the CTC transporting hazardous materials in the United States must also be equipped in accordance with the same special requirements and time constraints as United States built and owned specification 105 cars. Because of the potential catastrophic consequences of accidents involving 105 tank cars, the Bureau and FRA believe that all such cars used to transport hazardous materials in the United States must be so equipped.

Specifically, existing CTC specification 105 tank cars would be required to be equipped with shelf couplers by not later than December 31, 1981, if used to transport hazardous materials in the United States. Likewise, each new CTC specification tank car built after December 31, 1980, would be required to be equipped with shelf couplers, a tank head puncture resistance system, a thermal protection system, and a large capacity safety relief valve in the same manner as new DOT specification 105 tank cars if it is used to transport hazardous materials in the United States.

NEW TANK CAR REQUIREMENTS

Proposed Section 179.106–2 contains four new safety requirements applicable to new 105 tank cars constructed after December 31, 1980. These four safety features are identical to those now required on newly constructed 112 and 114 tank cars.

Coupler Vertical Restraint System. Each new 105 tank car would be required to be equipped with a coupler vertical restraint system (shelf couplers). These couplers have demonstrated an ability to reduce tank and running gear damage under certain rail accident conditions. Further, AAR Interchange Rules have required such couplers on all new tank cars since January 1, 1978.

Tank Head Puncture Resistance System. Each new 105 tank car would be required to have a tank head puncture resistance system (head shields) similar to that proposed for new 112 and 114 tank cars. A review of recent accidents involving 112 and 114 tank cars equipped with tank head protection confirms that this protection is effective in preventing tank head punctures. According to FRA accident records, no 112 nor 114 tank car has sustained a tank head puncture in the area protected by the head shield. There have been three reported tank head punctures in areas not protected by the head shield. The Bureau and FRA believe that applying full tank head protection to 105 pressure tank cars will materially improve the rail transportation safety of liquefied compressed gases and other highly hazardous liquids being carried in these cars.

Thermal Protection. Although 105 tank cars are required to be insulated with a material capable of controlling product temperature in the transportation environment, there is no current requirement that this insulation protect the tank from overheating in a fire environment. All specification 112 and 114 tank cars transporting flammable liquefied compressed gases are now required to have high temperature thermal protection to protect tank in a fire environment. The Bureau and the FRA believe that addition of a high temperature thermal requirement to the current insulation requirement on 105 tank cars is necessary to assure the use of the best available materials for new construction. The Bureau and FRA are aware that such insulating materials have been installed on many 105 tank cars.

Safety Relief Valves. Tests conducted by the FRA in conjunction with the rulemaking contained in MTB Docket HM-144 indicate that the safety relief valves installed on uninsulated 112 and 114 tank cars might not provide sufficient relief capacity under extreme fire accident conditions. However, these tests also demonstrated that if thermal protection were added to tanks, the capacity demands on these valves would be reduced. Accordingly, Section 179.105-7 was issued to require that newly built and retrofitted 112 to 114 cars that have thermal protection be equipped with the same capacity safety relief valves that were required on noninsulated 112 and 114 tank cars.

For these reasons, it is being proposed that newly built 105 tank cars that have thermal protection also be equipped with the larger capacity safety relief valve which was initially developed for uninsulated pressure tank cars.

PREVIOUSLY BUILT CARS

Proposed Section 179.106-3 would require the retrofitting of shelf couplers on all existing 105 tank cars. As has been stated previously under the discussion of Section 179.31, the Bureau and FRA believe that rapid retrofit installation of shelf couplers on all 105 tank cars not already so equipped is essential from the standpoint of safety. The supply of shelf couplers is sufficient to permit the retrofitting of the approximately 18,000 DOT 105 tank cars not so equipped within twelve months.

STENCILLING

In order that shippers, carriers and others may easily identify tank cars having the various described safety features, proposed Section 179.106-4 would require 105 tank cars newly built in accordance with Section 179.106 to be stencilled 105J. This stencilling will provide for easy identification of the tank car's safety features and facilitate compliance with the loading and handling regulations.

(45 F.R. 48671, July 21, 1980)

49 CFR Parts 173 and 179
[Docket No. HM-174; Amdt. Nos. 173-145, 179-27]
Shippers; Specifications for Tank Cars

These amendments are the result of the joint efforts of the Federal Railroad Administration (FRA) and the Materials Transportation Bureau (MTB). In accordance with internal Department of Transportation (DOT) procedures, the FRA has developed the substantive provisions of this rule for review and issuance by the MTB.

The MTB proposed a series of revisions in a notice published on July 21, 1980 (45 FR 48671). Interested persons were requested to submit their views. Comments received were from individual shippers, shipper organizations, a railroad organization, a rail labor organization, the National Transportation Safety Board (NTSB), and tank car manufacturers. All of the comments have been carefully reviewed and fully considered during the formulation of the final rule set forth in this document.

With the exception of shelf couplers, the FRA and the MTB deliberately separated new car construction requirements under this rulemaking action from retrofit matters under Docket HM-175. This action allows the MTB to clearly state that the decisions reached in HM-175 are independent of the decisions that may be reached in HM-175.

DISCUSSION OF COMMENTS

General. Several commenters expressed the opinion that the MTB was mandating changes without sufficient accident analysis. One commenter stated that a derailment accident history comparison between 112/114 and 105 tank cars for the period 1965 through mid-1979 shows that, on the basis of car-year exposure, the 105 car as a group is less vulnerable to head puncture, shell puncture, fitting damage, rupture, and lading loss than other tank car types. Although the source of this data is not stated, it apparently came from a study of the 105 tank car population and accident data published by the Railway Progress Institute and the Association of American Railroads (Report No. RA-17-1-43; August 1980). It should be recognized that conclusions based on the car accident data are dependent upon how the data are statistically normalized to reflect, among other things, that more than twice as much flammable gas is transported in 112/114 tank cars than in 105 tank cars.

In analyzing accident data over the last 25 years, the FRA has concluded that 105 tank cars have been involved in a number of train accidents with consequences similar to 112 and 114 tank cars dramatizing the importance of assuring that these tank cars are equipped with a level of safety protection consistent with the risk.

Several commenters also expressed the opinion that the MTB was mandating changes without sufficient testing of Specification 105 tank cars. Some commenters discussed the detailed testing of 112/114 tank cars and suggested that similar testing of 105 cars be performed prior to mandating changes in 105 tank cars. Over the last 10 years, the FRA has built test facilities and conducted numerous tests in cooperation with various industry groups. Researchers investigated the capability, feasibility and even the practical aspects of life cycle durability of tank car safety improvement options. An extensive portion of the resulting findings relate directly to puncture resistance, thermal protection and safety valve systems regardless of the particular application to a tank car type, whether it be a 112, 114 or 105.

The thrust of many commenters' arguments seems to be that the MTB should defer applying the HM-144 performance standards to the 105 tank cars until it determines the degree to which current 105 tank car designs meet those standards. The FRA and the MTB are confident that they have adequate information to proceed with this final rule without delay because:

(1) the data base resulting from earlier tests and experience with DOT Specification 112 and 114 tank cars is appropriate;

(2) in terms of the commenters' concerns, this rule applies only to new tank cars that, except for one additional commodity, will carry the same commodities covered in HM-144; and

(3) it is unrealistic to expect that all variations of the 105 tank car designs can or need to be tested as systems.

Furthermore, the FRA research program is not intended to identify all feasible options that satisfy the performance specifications promulgated in Docket HM-144 and which are being extended in this rule. The supply industry has the necessary expertise to develop any new options that they feel may be more cost effective than the existing options being used on 112 and 114 tank cars. Indeed, at the present time the FRA and the Railway Progress Institute are using FRA facilities to test various combinations of jacketed systems and thermal coatings. The FRA and the MTB believe that sufficient analysis and testing, including full scale testing of 112 tank cars, has been conducted in order to proceed with changes in new 105 tank car requirements. Some 105 tank cars which meet the head and thermal protection requirements of this rule are being built presently. Moreover, as was noted by many of the commenters, there is a great diversity of 105 tank car designs. Therefore, the FRA and the MTB believe that it would be a prohibitive burden to require that each 105 tank car design be subjected to full scale fire, impact, and valve testing. However, FRA has facilities at the Transportation Test Center where appropriate testing as previously established in HM-144 can be performed by any tank car builder or owner at reasonable expense.

Several commenters suggested that the DOT should be more concerned with the causes of rail accidents, such as poor track maintenance and operational problems, rather than mandating changes to 105 tank cars. FRA has research, regulatory, and Federal assistance programs underway to improve track maintenance, equipment maintenance and operating practices. In addition, the FRA recently completed a study, requested by Congress, on the relationship of the size, weight, and length of rail cars to the safety and efficiency of rail transportation that points the way for further improvements in freight car design. However, these efforts will not eliminate all accidents. FRA and the MTB believe that although the risk to the public from hazardous materials will be reduced by these efforts, there is still a need to improve the safety of tank cars that carry certain hazardous materials.

Many commenters gave examples of why commodities should be separately treated with respect to thermal and tank head protection. They believe it is not necessary to add safety requirements to tank cars used to transport certain commodities, for example, carbon dioxide. This particular commodity is not toxic and will not support a fire. Many commenters supported commodity specific tank car requirements in a general way and some provided more specific recommendations, such as:

—gives its acquiescence to the present HM-144 thermal and tank head protection systems only for flammable gases in new specification 105 cars as this acknowledges the reality of current car builder practices.

—agrees that new construction of 105 cars for these commodities should incorporate the same puncture and thermal protection requirements intended for 112 and 114 cars for transporting the same commodities.

There is substantial justification to limit added safety features only to tank cars transporting commodities that need extra protection as was prescribed by the HM-144 amendment.

Although there are administrative and operational advantages in specifying uniform safety protection requirements which would apply to every new 105 tank car, the MTB agrees with those commenters who suggested continuing the specific commodity and class designation approach of HM-144.

The information assembled in this proceeding has persuaded the MTB that higher levels of 105 tank car protection are called for with respect to the same kinds of commodities that earlier prompted the additional HM-144 requirements for 112 and 114 tank cars—flammable gases and anhydrous ammonia—plus one additional commodity having characteristics which approximate those of flammable gas—ethylene oxide. That information does not provide comparable justification for extending those requirements to 105 tank cars carrying other hazardous commodities. However, because FRA and MTB remain concerned with the adequacy of tank car puncture resistance and thermal protection for other hazardous commodities, we will continue to examine this question (e.g. HM-175) and initiate corrective regulatory action as necessary.

SPECIFIC COMMENTS AND ANALYSIS OF MAJOR ISSUES

The following is a summary of the comments received and an explanation of the revisions made by the MTB in response to those comments.

Shelf Coupler Retrofit (§ 173.31). As proposed in the NPRM, paragraph (a)(6) of § 173.31 would require a coupler vertical restraint system (shelf couplers) to be installed on all 105 tank cars by December 31, 1981, and paragraph (a)(7) of § 173.31 would require the system on other DOT specification tank cars by December 31, 1984. The commenters supported overwhelmingly the idea that all 105 tank cars should be equipped with shelf couplers and noted that the requirement could be made effective immediately for new 105 tank car construction since it is already the practice. The only issues raised involved the time frame and priorities for the retrofit installation of couplers.

A majority of commenters requested that the final rule allow 18 months for retrofitting 105 tank cars. Several of these commenters noted that it is approximately 18 months from the publication of the NPRM (July 21, 1980) until the proposed date for retrofitting 105 tank cars (December 31, 1981), apparently presuming that the MTB intended an 18-month retrofit period. The specific reasons for requesting 18 months included perceived problems of availability of the couplers and potential disruption of commerce due to shopping. The National Transportation Safety Board called for the expedited installation of shelf couplers on 105 tank cars, but declined to suggest an appropriate interval.

As to the other DOT specification tank cars, there was a similar general agreement that retrofit installation of shelf couplers is warranted. However, several commenters believe that the requirement should extend only to those other DOT specification tank cars that carry hazardous materials. On the other hand, other commenters stated that shelf couplers should be required to be installed on all new or rebuilt freight cars. There were differences among the commenters about priorities for retrofitting these tank cars as well as the appropriate time period to complete the process. The suggested interval ranged variously from an unspecified "expedited" basis to 48 months, 54 months, 60 months, 72 months, 78 months, 84 months, and even 108 months. The reasons advanced for time extensions included differing estimates as to: (1) the number of cars involved, (2) the time required to locate, move and retrofit the cars, and (3) the availability of couplers. In addition, some commenters suggested that nonplacarded cars be given additional time beyond the December 31, 1984, proposed date. Other commenters noted that whatever interval is chosen, the retrofit should focus first on those cars actually carrying hazardous materials; and one commenter would accord priority to cars of 22,000 gallons or more.

One commenter believes that FRA underestimated the size of the total tank car fleet. However, AAR's Yearbook of Railroad Facts shows 178,069 tank cars in service at the end of 1979. FRA estimates that about 75 percent of the total tank car fleet carries placaded hazardous materials during all or part of its life. The 75 percent equates to the 135,000 DOT specification tank cars used as the starting figure in the economic evaluation. The estimate is supported by FRA analysis of tank car shipments and UMLER file data.

Based on analyses of total cars, performance on retrofits under HM-114 and coupler manufacturer capabilities and assurances, the MTB has set a February 28, 1982, completion date for the 105 tank car retrofit and a February 28, 1985, completion date for the retrofit of the other tank cars. The latter date provides a period of approximately 48 months for the effective date of the regulation. The four-year retrofit period is consistent with known industry capability and the established safety value of shelf couplers. Shelf coupler availability is not a limiting factor.

In the shelf coupler retrofit program for Specification 112 and 114 tank cars, it is estimated that more than 16,000 cars were equipped within six months. In the 112/114 tank car retrofit, arrangements were made with railroads and private shops to provide for application of couplers to many cars, with minimum delays, along major hazardous materials routes. Similar arrangements would be possible for the retrofit program required by this final rule. Other cars can be equipped during normal cyclical maintenance at the home shop. Although such a measure is not likely to be necessary given the experience of the 112/114 tank car retrofit, field application of couplers could be made if necessary.

As indicated in the NPRM, the FRA and the MTB estimate that of approximately 24,000 Specification 105 tank cars, 18,000 have not yet been equipped with shelf couplers. Of those tank cars bearing specifications other than 112/114 or 105, approximately 73,000 remain to be equipped.

The FRA and the MTB have established the key priority with respect to order of retrofit by requiring that all 105 tank cars be equipped during the first year. It would be both unnecessary and disruptive to specify a detailed order of retrofit for the remaining fleets based on car size, commodity carried, or annual mileage. The MTB believes that industry will utilize its specialized knowledge to assure that the tank cars carrying the most hazardous materials are retrofitted first. The incentive for industry to support such a program is economic. The incremental cost to retrofit tank cars carrying the most hazardous material first is minimal, if any, since the cars must be fitted within a limited time period under this rule. Industry will prefer under these conditions to achieve the greatest risk reduction. The benefit to industry is a decline in the potential of a serious accident and the accompanying costs. This approach by the MTB uses the free market system to get the best safety performance at the least cost to government and industry.

At the same time, the flexibility afforded by the final rule will permit intelligent planning by industry based on car availability and routine maintenance intervals. The FRA and the MTB believe that this flexibility will assure completion of the retrofit at an earlier date than would be the case if shippers and car owners were required to manage the logistics of equipping multiple groups of cars according to a rigid schedule.

Cars previously built to ICC or DOT specifications that are not in placarded hazardous materials service are not subject to this retrofit requirement unless and until they are placed in such service (see 49 CFR 179.1). However, shippers are cautioned that shelf couplers are "safety appurtenances" for which inspection will be required following the completion date of the respective retrofit periods (see 49 CFR 173.31(b)). Also, couplers may be changed at any time due to damage in the service environment; therefore, it is imperative that coupler type be ascertained at the time of loading to assure compliance with the regulations.

Compliance Reporting. Many commenters seemed to assume that a reporting system for the coupler vertical restraint retrofit is necessary, although none was proposed in the NPRM. The FRA and the MTB believe that it would be useful to measure compliance and are considering issuing an NPRM to require annual reports covering the DOT specification tank cars to be retrofitted by February 28, 1985. A suitable reporting procedure would help to measure progress and ensure that the deadline is met.

REQUIREMENTS FOR SPECIFIC COMMODITIES IN TANK CARS

Sections 173.124, 173.314, and 173.354. These sections have been amended to require that certain new 105 tank cars meet the special requirements of § 179.106. Section 173.314 has also been amended to clarify that certain new and previously built 112 and 114 tank cars are required to meet the special requirements of § 179.105. The purpose of these changes is to alert readers of Part 173 to the changes in Part 179. Section 173.314 has been further amended to correct typographical errors in the table. These typographical errors occurred in the entries for difluoroethane; dimethylamine, anhydrous; monomethylamine, anhydrous; methyl chloride; trimethylamine, anhydrous; and liquified petroleum gas (pressure not exceeding 300 pounds per square inch at 105 degrees F).

FULL TANK HEAD PUNCTURE RESISTANCE SYSTEM VERSUS LOWER
HALF SYSTEM (§ 179.100–23)

As proposed in the NPRM, § 179.100–23 would require that each end of a DOT Specification 105, 112, and 114 tank car built after December 31, 1980, be equipped with a tank head puncture resistance system that covers the entire tank head. This was not proposed because of any inadequacy of the HM-144 tank head puncture resistance standard (lower half of the tank head). Indeed, the NPRM clearly stated that the ". . . HM-144 requirements represented a very satisfactory approach to the protection of pressure tank cars." Rather, full head system was proposed on the basis that ". . . human and economic losses resulting from individual accidents may dramatically exceed the levels previously anticipated." However, the dramatically higher costs only occur if there is an accident. The majority of commenters opposed the proposed full tank head system on the basis that the FRA did not identify any accident where a car equipped to the HM-144 standard (shelf couplers and half head) had failed to protect the tank head. The FRA and the MTB agree that there is not to date any specific accident data demonstrating the HM-144 tank head protection system is inadequate. The FRA and the MTB also agree that there is not to date any clearly identifiable additional margin of safety provided by a full tank head puncture resistance system that would warrant Federally mandating the full tank head protection system.

Several commenters representing major groups did support a full tank head puncture resistance system. Their comments did not contain an analysis of what additional protection would be provided by a full head system or any accident history of HM-144 equipped cars indicating a failure of the HM-144 system. In the absence of definitive accident data, and in light of benefits attributed by the NTSB and other commenters to the combination of half head protection in conjunction with shelf couplers, the FRA and the MTB do not believe it is appropriate to impose rigid Federal requirements for a full tank head puncture resistance system. Accordingly, the MTB is not requiring full head protection for 105, 112, and 114 tank cars as proposed in the NPRM, but is instead extending the same HM-144 requirements to the 105 tank cars defined in § 179.106–2. Consequently, editorial changes in the title and text have been made in the final rule to clarify that this section is an alternative requirement for all tank cars required to satisfy the head puncture resistance requirements of § 179.105–5.

Even though not required by this rule, the FRA and the MTB note with approval some evidence of evolving voluntary industry practice to provide full head protection.

§ 179.106 SPECIAL REQUIREMENTS FOR SPECIFICATION 105
TANK CARS

§ 179.106–1 General. The 105 tank car special requirements are set forth in § 179.106. Several commenters objected to paragraph (b) of § 179.106–1. Paragraph (b) provides that AAR approval is not required for changes or additions to Specification 105 tank cars for compliance with § 179.106. The FRA and the MTB recognize that the existing car owner/rail carrier approval system which is set forth in AAR "Interchange Rules" may be continued by the AAR Tank Car Committee and that its approval for interchange may, therefore, be required by industry for all additions, modifications and repairs performed to comply with § 179.106. However, the FRA and the MTB to not believe that this approval needs to be imposed by regulation. These standards adopted for improved tank car safety are augmented by specific performance oriented design criteria (such as specified couplers, head shield design and thermal protection systems) thereby affording tank car owners sufficient guidance to perform the modifications and additions required by this rule. For these reasons the MTB had not included a requirement for AAR Tank Car Committee approval in the rule.

New Car Requirements (§ 179.106–2). The requirements for new 105 tank cars are set forth in § 179.106–2. The requirements for coupler vertical restraint systems have previously been discussed. The analyses of the comments relating to the tank head puncture resistance systems, the thermal protection systems and the safety relief valve requirements are discussed separately.

The MTB has decided to allow more time before newly built tank cars must comply with this section. It has become apparent from comments submitted that the NPRM's effective compliance date of January 1, 1981, might cause unreasonable delays in the delivery of tank cars already ordered. The FRA and the MTB recognize the problems associated with lead times in construction procurements. The rule provides a six-month period from the effective date to the time when a newly built tank car must comply with this section. This period will give adequate time for car orders to be filled by the builder in accordance with this rule. In prescribing the September 1, 1981, date the FRA and the MTB considered, but rejected, numerous suggestions that the rule be based upon the date ordered. One commenter stated: "Because of shop backlogs of up to two years ... any changes in specifications must be referenced to car order date rather than car built date." The FRA and the MTB decided that a "date ordered" basis would lead to delays in installing the safety systems of up to two years and confusion in identifying those newly built cars which must comply with the rule. It is worthwhile to mention that FRA has been advised that many new 105 tank cars that will carry flammable gases are already being constructed in compliance with the tank head and thermal requirements of this rule.

Tank Head Puncture Resistance System (§ 179.106–2). Several commenters supported full tank head puncture resistance requirements for all newly constructed 105, 112, and 114 tank cars. Several other commenters supported the HM-144 standard for head protection (lower half of the tank head) on all newly constructed 105 tank cars. One commenter supported the full head requirement for new 105 tank cars, while offering no opinion regarding the 112 and 114 tank cars. Most commenters supported commodity differentiation and were not opposed to the principle of mandating HM-144 standards on those 105 tank cars that carry the same commodities as the 112 and 114 tank cars (flammable gases and ammonia). One commenter noted that the industry has voluntarily installed head protection on 105 tank cars carrying flammable gases for several years.

The majority of commenters, however, were opposed to requiring either full or HM-144 equivalent head protection on all new 105 tank cars without regard to the commodity being carried. These commenters noted that commodity differentiation was an integral part of HM-144 requirements applicable to 112 and 114 tank cars. According to these many commenters, the wide variety of commodities carried in the 105 tank cars and the attendant cost of providing an all encompassing level of protection precludes mandating the same head and thermal protection system for every 105 tank car.

Other objections to the proposed tank head requirements for 105 tank cars were raised. Some commenters reiterated that the accident record indicates that the 105 tank car is superior to the 112 and 114 tank cars in its ability to survive an accident environment. Hence, they contend that there is not a similar justification for the additional requirements as there was in HM-144. In addition, a number of commenters stated that the incremental benefit of shelf couplers reduces the safety benefit of a tank head protection system to an unacceptably small level.

The MTB is extending HM-144 head puncture resistance requirements to new 105 tank cars that will carry the HM-144 commodities and ethylene oxide, notwithstanding the allegedly better safety record of 105 cars when compared to the unretrofitted 112 and 114 cars. A relatively better overall safety record is not at all surprising since 105 tank cars have some insulation and varying degrees of additional tank head puncture resistance. While the thermal insulation and head protection systems of many 105 tank cars do not meet the HM-144 standard, nevertheless, as a group, 105 tank cars do provide varying degrees of additional protection over the unretrofitted 112 and 114 tank cars. Having established a specified level of tank head puncture resistance and thermal requirements in HM-144 for certain commodities carried by 112 and 114 tank cars, the MTB has no hesitation about utilizing that same standard for 105 tank cars carrying those same commodities.

The FRA and the MTB do not agree with the argument that shelf couplers provide an adequate level of safety that eliminates the need for tank head protection. Essentially the same issue was raised and rejected in the HM-144 proceedings. Tests performed as early as 1976 at the Transportation Test Center in Pueblo, Colorado, demonstrated that shelf couplers will prevent tank head punctures during some overspeed switching impacts. However, for other impacts under differing conditions, shelf couplers were not fully effective in preventing tank head punctures while half head shields were effective in preventing most punctures. It was also found that a combination of shelf couplers and half head shields was needed to prevent tank head punctures over the range of realistic impact conditions.

The FRA and the MTB have concluded that certain newly built 105 tank cars need a coupler restraint system and a tank head puncture resistance system. This dual protection, required for 112 and 114 tank cars in 1977, will significantly reduce tank punctures in derailments and switch yard accidents.

HIGH TEMPERATURE THERMAL PROTECTION

§ 179.106–2 New Cars. The level of thermal protection proposed for 105 tank cars is from § 179.105–4 (HM-144 thermal protection standard). Almost all the commenters opposed the NPRM proposal for thermal protection on all newly built DOT 105 tank cars. More than one-half of all commenters said that if thermal protection were to be required for all DOT 105 tank cars without regard to commodity, the rule should be deferred pending additional testing and data.

Several other objections were raised on various points. Most of these were aimed at the cost consequences of requiring added safety systems of marginal benefit for the transport of commodities where, these commenters contend, the accident history does not justify additional safety features.

The FRA has reviewed the accident history and has not found any justification for not requiring the same level of thermal protection in 105 tank cars when they carry the identical hazardous commodities as 112 and 114 cars. On the other hand, there are some commodities presently authorized in 105 tank cars that pose a lower risk in fire environments.

The MTB has revised the NPRM proposal so that the final rule formally extends the thermal protection standards of § 179.105–4 to 105 cars transporting flammable gases and ethylene oxide. Ethylene oxide is included because it has properties comparable to flammable gases. Ethylene oxide has a very low flash point (less than 0 degrees F) and does not need oxygen for combustion. It is flammable over an unusually wide range of mixtures with air, from 2 percent through 100 percent. Additionally, it barely misses the temperature/pressure relationship for being classified as a flammable gas. Its vapor pressure is 38.5 psi absolute at 100 degrees F, which is extremely close to the pressure criterion of 40 psi absolute at 100 degrees F that is used to define a flammable gas under DOT regulation (49 CFR 173.300). (The UN recommendations and IMGO Code both classify ethylene oxide as a flammable gas.)

The MTB recognizes that some existing 105 tank cars have thermal protection systems that may already meet the thermal protection requirements. DOT has previously approved various thermal protection systems and maintains a list of those approved systems. Tank cars built with approved systems are excepted from the test verification requirements of paragraph (b) of § 179.105–4. Information on these systems is available in the Dockets Branch, Room 8426, Nassif Building, 400 Seventh Street, SW., Washington, D.C. 20590.

The MTB has established a September 1, 1981, date for the thermal protection system requirement. The six-month period after the effective date of this rule is included for the reasons discussed in the tank head puncture resistance section.

Safety Relief Valves (§ 179.106–2). Most commenters objected to the proposal for the larger flow capacity safety valve for all commodities authorized to be carried in DOT Specification 105 tank cars. Since the final rule for the larger safety valve applies only to those DOT Specification 105 tank cars which carry flammable gases and ethylene oxide, the justification for the larger valve is the same as that given in HM-144.

In summary, extensive research, conducted both before and after the rulemaking under HM-144, has indicated that:

(1) Since rail cars often overturn in accidents, the controlling condition in sizing for pressure relief is the liquid flow or upset car condition and not exclusively the vapor flow criterion used prior to HM-144; and

(2) Existing valve sizing equations underestimate the total heat flux inputs which can occur in accident environments.

Accordingly, the MTB has modified § 179.106–2 to specify that revised valve sizing is applicable only for new 105 tank cars carrying flammable gases and ethylene oxide. For the commodities covered by HM-144, valves with sufficient capacity have been satisfactorily used in extensive 112/114 tank car service and pose no real installation obstacles for new Specification 105 tank cars. As with the tank head puncture resistance system and the thermal protection system, MTB has established a September 1, 1981, date for the revised safety valve requirement.

§ 179.106–3 Previously Built Cars. This section requires the retrofitting of shelf couplers on all existing 105 tank cars by February 28, 1982. The issues have been discussed under § 173.31.

§ 179.106–4 Stencilling. Several commenters recognized the concept proposed in the NPRM for using the letter "J" to indicate full tank head and thermal protection as logical. They went on to recommend a broader system to comprehend the several DOT 105 tank car designs already in service and to anticipate the possible regulatory changes that may affect some existing cars. For example, the following nonconflicting letters were suggested: "A" standard jacket head; "S" for ½ inch half high head shield; "T" for ½ inch half high head shield plus nonjacketed high temperature thermal protection; "U" for ½ inch half high head shield plus high temperature thermal protection under metal jacket; "H" for ½ inch full head shield; "K" for ½ inch full head shield and nonjacketed high temperature thermal protection; and "J" for ½ inch full head shield plus thermal protection under metal jacket. These commenters further offered that this scheme would facilitate recordkeeping for DOT 105 tank cars.

The FRA and the MTB do not agree that an elaborate lettering system that includes the variety of existing car designs is necessary at this time. Additional car categories may become necessary in the future because of further regulatory actions, but MTB does not believe it is appropriate to anticipate what those actions might include. Accordingly, the final rule adopts the letters A, S, and J for three categories of 105 tank cars. It provides an identification system that is consistent with the 112/114 tank car identification system.

OTHER DISCUSSION

Economic Impact. The FRA included an economic evaluation for the docket when the NPRM was issued. That evaluation included cost figures for full head shields on all newly built 105, 112 and 114 tank cars. It also included cost figures for shelf couplers, thermal protection and safety valves as specified by HM-144 on all newly built 105 tank cars. The final rule requires that newly built 105 tank cars carrying flammable gases have lower half head protection, thermal protection and safety valves. The rule also requires shelf couplers, lower half head protection, thermal protection and safety valves for newly built 105 tank cars carrying ethylene oxide. Finally, the rule requires shelf couplers and lower half head protection for newly built 105 tank cars carrying anhydrous ammonia. These changes reduce the scope of the rule and the overall industry cost. The MTB believes that the benefits identified in the earlier analysis will not be significantly reduced despite the reduced scope of the final rule since the commodities included in the final rule are the ones that have historically resulted in costly accidents. Accordingly, the MTB believes another economic evaluation is not warranted. A new economic evaluation taking into account the adjustments made in the final rule would continue to show that this regulation will not have a major adverse economic impact on industry, the public or government.

Several commenters expressed concern that the proposed safety modifications would add to the tank car weight. These commenters were concerned that the added weight would reduce the amount of commodity that could be transported in the car. This weight sensitive concern is not significant because of the limited scope of the final rule. FRA estimates that only a very small percent of the total volume of all hazardous commodities transported by railroads would be affected.

Beyond general expressions of negative cost/benefit from treating all 105 tank cars the same and from requiring full head shields, the commenters provided very little specific cost data. After a thorough review of initial calculations in the economic evaluation prepared for the NPRM, the FRA and the MTB conclude that the original estimates are accurate.

Finally, as previously mentioned, one commenter who did not provide supporting details, argued that the number of cars needing shelf couplers is much greater than the MTB estimate. The FRA has reexamined this issue. Based on the best data to which it has access, the FRA has found that the initial estimate is reasonably accurate for establishing that a four-year period provides sufficient time to complete the shelf coupler retrofit without severe economic penalty.

(46 F.R. 8005, January 26, 1981)

49 CFR Part 179
[Docket No. HM-174; Amdt. No. 179–27A]
Specifications for Tank Cars

MTB received four petitions for reconsideration of the final rule issued in Docket HM-174 (46 FR 8005, January 26, 1981). While each petition contained its own views, the primary area of reconsideration concerned the safety valve sizing requirements. The petitioners did not submit any new information to support their views on the adequacy of the safety valve sizing methods presently used for specification 105 tank cars. MTB and FRA have reviewed the supporting material contained in Dockets HM-144 and HM-174 for flammable gases. Several petitioners recommended that MTB continue to use the CGA-AAR type valve sizing equations which had been used for 40 years prior to the adoption of the safety valve sizing requirement of HM-144. One petitioner pointed out that MTB used this "traditional" formula in Docket HM-167, Intermodal (IM) Portable Tank Specifications. IM portable tanks, which are generally transported as single units, are used to transport hazardous materials that are liquids at ambient temperatures and pressures. This is quite different from a flammable gas that is compressed to liquefy it for transportation and storage. Additionally, it is not uncommon to find several tank cars involved in a train derailment thereby increasing the opportunity for one damaged tank car to supply fuel for a fire that could cause other tank cars to rupture. Therefore, MTB and FRA believe that the safety valve sizing method used for 105 tank cars should not be the same as the method used for IM portable tanks.

One petitioner took exception to the requirements for head and thermal protection. The commenter contends that the application of top and bottom shelf couplers affords sufficient protection to jacketed/insulated 105A tank cars. The commenter provided no new information to support this view beyond the factors which were previously considered during the HM-174 rulemaking proceeding. Therefore, no change in the head and thermal protection requirements is being adopted.

One petitioner expressed concern over the inclusion of ethylene oxide in the same category as propane. Although ethylene oxide does not meet the DOT definition for a flammable gas, MTB included ethylene oxide in the final rule because of its similar properties. Ethylene oxide is so close to being a flammable gas that the UN Recommendations and the IMCO Dangerous Goods Code classify it as a flammable gas. The petitioner suggested that the use of a larger safety valve may decrease the level of safety by reducing the effectiveness of the protective inert gas blanket and also expressed the view that, once auto-ignition occurs, the internal pressure of the tank car makes little difference.

The final rule added ethylene oxide as an additional commodity to the list of commodities previously covered in HM-144. Although the notice of proposed rulemaking in HM-174 proposed to require the larger safety valve on all newly constructed 105 tank cars and, hence, on cars built to carry ethylene oxide, the special focus on ethylene oxide as a commodity subject to the requirement did not occur until the final rule stage. It appears to MTB that once the focus turned to ethylene oxide, genuine concerns, albeit speculative ones at this point, began to develop about the impact of the larger valve for ethylene oxide because of its unique characteristics. Ethylene oxide is a flammable liquid which is toxic and corrosive. Once ignited, ethylene oxide will burn inside a tank car without additional oxygen

MTB and the FRA are not persuaded by the scant information in petitions that the larger safety valve for ethylene oxide is less safe. Neither is MTB nor FRA persuaded that the safety benefits attributable to a larger valve are irrelevant for cars carrying ethylene oxide. However, MTB is extending the compliance date for the safety valve sizing requirement on specification 105 tank cars used to transport ethylene oxide to September 1, 1982, to afford the full AAR tank car committee and other interested parties an opportunity to study the question of safety valve sizing for ethylene oxide and to submit the results of any studies for review and consideration. MTB requests that any new information relating to this matter be submitted no later than June 1, 1982.

(46 F.R. 42678, August 24, 1981)

49 CFR Parts 173 and 179
[Docket No. HM-174; Amdt. Nos. 173–172, 179–34]
Shippers; Specifications for Railroad Tank Cars

On January 26, 1981 (46 FR 8005), MTB issued a final rule establishing certain construction standards for DOT specificaton 105 tank cars built to carry specified hazardous materials. The construction standards include a safety valve sizing requirement for those tank cars built to carry ethylene oxide. The final rule required that each DOT specification 105 tank car used to transport ethylene oxide and constructed after August 31, 1981, must have a safety valve sized in accordance with 49 CFR 179, 105-7.

MTB received several petitions for reconsideration of the final rule. These petitions addressed, among other things, the safety valve sizing requirement for tank cars used to transport ethylene oxide. The petitioners argued that the larger safety valve for ethylene oxide would be less safe because of that commodity's peculiar characteristics. They also argued that the valve sizing equation in the rule should not be applied because ethylene oxide is a liquid while the equation is designed for gases.

While MTB and the Federal Railroad Administration (FRA) were not persuaded that these arguments were adequately supported, the compliance date was extended from September 1, 1981, until September 1, 1982 (46 FR 42678), then from September 1, 1982, until September 1, 1983 (47 FR 38697), and finally from September 1, 1983, until March 1, 1984 (48 FR 39630).

The extensions were granted to permit the AAR Tank Car Committee (AAR Committee) and other interested parties an opportunity to study the question of safety valve sizing for ethylene oxide and to submit the results for review and consideration. During the past two years, the AAR Committee conducted an extensive study of safety valve sizing. A final report was furnished to MTB and FRA earlier this year and has been placed in the Docket.

AAR SAFETY RELIEF VALVE REPORT

The AAR Report is a comprehensive study of safety valve sizing and of the railroad accident environment. The report does address ethylene oxide safety valve sizing, but the overall analysis and conclusions of the AAR Report have equal validity for other materials. The AAR Report does not attempt to make the case, originally offered to justify the extension of the compliance date, that ethylene oxide should be treated differently as a result of peculiar chemical or physical properties. Rather, the AAR Report attempts to make the case that the safety relief valves currently used on DOT specification 105 and 111 tank cars transporting ethylene oxide and liquefied flammable gases (LFG) are adequately sized for railroad accident fire environments when used in conjunction with high temperature thermal insulation that meets the requirements of 49 CFR 179.105-4.

FRA and MTB have determined that the data and analysis do not support the conclusion that the current valves are adequately sized and, therefore, will not further extend the compliance date for equipping new construction ethylene oxide tank cars with large capacity safety relief valves. However, as a result of the extensive analyses by

AAR and FRA on the relationship of thermal insulation and safety valve sizing, adequate data are available to establish an optional approach that will permit continued use of the current valves on new ethylene oxide tank cars if additional thermal insulation is added.

Summary Technical Analysis of the AAR Report. The AAR Report discusses the functions of safety relief valves and design considerations for these valves; summarizes the equations that govern the flow of nonflashing liquids, saturated vapors, and flashing liquids through valves; reviews 49 technical papers dealing with various aspects of safety relief valves; reviews the results from selected pool fire simulation tests; and summarizes the results of subcooled water flow tests of two safety relief valves. The report also includes a description of computer programs developed by the AAR Committee to analyze the effects of fire engulfment on a tank car. The programs predict the temperatures and pressures of the tank car and the flow rate through the valves. The programs can treat both the case of the tank car remaining upright and the case of its being overturned. The programs can take account of a wide variety of fire environments, loading conditions, and insulation properties. The report includes selected case studies using the computer programs which purport to show the adequacy of the safety relief valves used on existing propane and ethylene oxide tank cars.

FRA and MTB Assessment of the AAR Report. The FRA and MTB assessment of the AAR Report, which has been placed in the docket, is summarized below:

The AAR Report confirms two key findings of a 1970 research study sponsored by FRA: (1) The overturned tank car accident scenario, in which the safety valve releases liquid rather than vapor, should be considered in the sizing of relief valves, and (2) the thermal conductance of insulation systems should be estimated at elevated temperatures. The report also contains useful information on design considerations and functions of safety relief valves.

The AAR Report makes two general recommendations with which MTB and FRA concur. First, the AAR recommends basing the allowable tank pressure on a percentage of the tank test pressure. Second, the AAR recommends that either (a) the theoretical burst strength of the tank not exceed the tank pressure within a reasonable time or (b) if the theoretical burst strength does not exceed the tank pressure, the tank should be empty when this occurs.

The computer program developed by the AAR for propane tank cars could be a useful guide for sizing relief valves. The computer program developed by the AAR for ethylene oxide greatly simplifies the physical phenomena; despite this simplification, that program could also be a useful guide for sizing relief valves if the results are cautiously interpreted.

MTB and FRA's primary reservations about the AAR Report are the assumptions made concerning (a) the heat flux from the fire to the tank car, (b) the effective thermal conductance of the jacket insulation in a fire, and (c) the allowable test pressure.

The AAR Report assumes that the tank car is only one-fourth engulfed in the fire and, therefore, that the heat flux to the tank is about 8,000 BTU/(hr ft^2). In a study sponsored by FRA and conducted by IIT Research Institute (IITRI), it was found that a heat flux of about 25,000 BTU/(hr ft^2) is necessary to correlate the data obtained in full-scale fire tests of liquefied petroleum gas tank cars. The IITRI results are also consistent with a fire data analysis sponsored by FRA and conducted by Cornell Aeronautical Laboratory.

The AAR Report assumes a value of 2.3 BTU/(hr ft^2 °F) for the effective thermal conductance of the jacket insulation satisfying the requirements of 49 CFR 179.105–4. The report does not explain how this 2.3 value was derived. However, in separate correspondence, the AAR has stated that they based this 2.3 value on a preliminary evaluation by IITRI. In this preliminary evaluation, IITRI assumed small heat losses from the back of the test plates used in pool fire simulation tests. In a later evaluation of these tests, IITRI has revised the plate heat loss estimates and has obtained a conductance value of 4.0.

The AAR has also submitted a report dated May 27, 1983, and entitled "Thermal Conductances of Fire Protection Insulations for Tank Cars." This report concludes that, based on laboratory thermal conductivity measurements, two thermal shield systems that passed the 800°F/100 minute pool fire simulation test would result in an overall system conductance of 2.3 BTU/(hr ft^2 °F) in a fire environment. MTB and FRA believe that these laboratory tests have not been demonstrated to be an acceptable simulation of railroad pool fires and that the DOT pool fire simulation tests are the best available simulation.

The AAR Report uses an allowable tank pressure of 120 percent of the tank test pressure. MTB and FRA believe it is illogical to specify a tank test pressure for undamaged, unheated tank cars and then optimistically assume that tank cars in an accident environment can withstand 120 percent of tank test pressure.

Using these extremely optimistic assumptions, the AAR Report concludes that the safety valves on existing propane and ethylene oxide tank cars are adequate. Using the more fully analyzed and conservative safety assumptions of the IITRI study, the MTB and FRA conclude that the existing valves are not adequate. MTB and FRA conclude that the safety valve/thermal protection combinations mandated in the final rules for Dockets HM-144 (42 FR 48306) and HM-174, and proposed in the NPRM for Docket HM-175 (48 FR 16188), are justified. However, based on the

IITRI calculations, MTB and FRA are allowing an option whereby additional thermal protection can be provided so that the ethylene oxide safety valve currently used can continue to be used.

Optional Approach for Ethylene Oxide. For pressure tank cars, DOT requires (49 CFR 178.100–15) that the total safety relief valve discharge capacity must be sufficient to prevent building up pressure in the tank in excess of 82½ percent of the tank test pressure or 10 psi above the start-to-discharge pressure, whichever is higher. There are similar requirements (49 CFR 179.200–18, 49 CFR 179.300–15, 49 CFR 179.400–18, and 49 CFR 179.500–12) for other types of tank cars. (The Department permits certain pressure tank cars carrying certain compressed gases to have a valve discharge capacity that would allow a build up of pressure to 90 percent of the tank test pressure. (49 CFR 179.102–11). The DOT safety valve requirements for rail tank cars are similar to the valve requirements for stationary tanks in the American Society of Mechanical Engineers (ASME) Code Section VIII, Division 1. The ASME Code requires that, if a tank may be exposed to a fire or other unexpected external source of heat, the pressure-relieving capacity must limit the pressure to 80 percent of the hydrostatic test pressure.

In the DOT regulations cited above, the fire characteristics and accident situations that the safety valves must protect against are not explicitly stated. In 1973 FRA conducted a full-scale fire test of a DOT specification 112A 340W tank car, without thermal protection, filled with propane. In this test the tank car was upright, the safety valve was discharging vapor, and the car was engulfed in a pool fire. The maximum pressure attained was 357 psig (105 percent of the test pressure).

A study by the Railway Progress Institute and AAR of eight DOT specification 112 and 114 tank cars without thermal protection that were involved in actual railroad fires indicates that in several of those cases the tank car pressure apparently exceeded the DOT pressure specifications. In an analytical study sponsored by FRA and conducted by Cornell Aeronautical Laboratory, it was concluded that "the consequences of inadequate relief capacity—overpressure—as a contributor to car failure could be effectively masked by evidence of fire and mechanical damage. Common post-accident test ... will not reveal this condition." It was further concluded in the study that "reputed observations at derailment sites of relief flow from cars which subsequently ruptured indicated that the flow may have been substantially reduced from that anticipated for a fully-opened valve."

To mitigate the problems with inadequate safety relief valve capacity, the Department undertook research and then initiated regulatory action. In 1973 FRA sponsored a full-scale pool fire test of an upright DOT specification 112 tank car filled with propane and equipped with thermal protection and a large capacity safety valve. In this test the maximum pressure reached was 320 psig (94 percent of the tank test pressure), and the thermal protection used was less than that required in 49 CFR 179.105–4. MTB and FRA believe that had the tank car in this test been equipped with the 49 CFR 179.105–4 thermal protection, the pressure would not have exceeded the permissible pressure limits.

Based on the IITRI calculations discussed above, MTB and FRA believe that the standard safety valve/thermal protection requirements promulgated in the final rules of Dockets HM-144 and HM-174 and proposed in the NPRM of Docket HM-175 will result in compliance with 49 CFR 179.100–15 or 49 CFR 179.102–11 for tank cars that remain upright in railroad fires.

Research sponsored by FRA and conducted at Cornell Aeronautical Laboratory, U.S. Air Force Rocket Propulsion Laboratory, the University of Maryland, and IITRI demonstrates that, for the same fire environment, the valve sizing requirements for a tank car that is overturned and therefore discharging liquid from the safety valve will usually be greater than the valve sizing requirements for a tank car that is upright. In promulgating past valve sizing requirements, MTB and FRA believed (and still believe) that strict compliance with the requirements of 49 CFR 179.100–15 or 49 CFR 179.102–11 for overturned tank cars in fires would lead to unreasonably large relief valves. The IITRI calculations indicate that the large capacity safety valve/thermal protection will result in satisfactory safety for overturned ethylene oxide tank cars. For 105A 100W or 111A 100W ethylene oxide tank cars the maximum tank pressure will slightly and temporarily exceed the prescribed pressure limit of 85 psig, but the predicted pressures quickly recede to below the 85 psig level. For 105A 200W and 105A 300W ethylene oxide tank cars there are no predicted pressure problems using the large capacity safety valve/thermal protection.

FRA and MTB have analyzed the IITRI calculations to determine if an acceptable option can be developed to allow the use of 1100 standard cubic feet per minute (scfm) (at 85 psig) safety relief valves on new ethylene oxide tank cars. The IITRI calculations indicate that if the thermal protection were increased, the safety performances would be satisfactory. These thermal protection/safety valve combinations would result in excessive pressures, but by the time these pressures are reached, the fires should be greatly diminished in intensity or under control. In addition, the use of the additional thermal protection would provide more protection in torch fires and less chance of an autoignition of ethylene oxide in torch and pool fires than would the standard thermal protection/safety valve combination. MTB and FRA believe these benefits compensate for the higher pressures.

MTB and FRA have selected a minimum value of 550° F in the 100 minute pool fire test for the thermal protection required if the small 1100 scfm valve is used on ethylene oxide tank cars. The IITRI calculations indicate no clear cut break point in the range between 500° F and 600° F in terms of dramatic safety differences. MTB and FRA believe that the 550° F valve will provide a margin of safety to compensate for the simplifying assumptions made in the IITRI calculations.

SECTION-BY-SECTION-ANALYSIS

Section 173.124 Ethylene oxide. Paragraph (a)(5)(ii) of § 173.124 is revised to clarify that DOT specification 105 tank cars built after August 31, 1981, and before March 1, 1984, are not required to have a safety valve sized in accordance with § 179.106–2(c)(4). The several extensions of the compliance date for the large capacity safety relief valve for new construction ethylene oxide cars (49 CFR 179.102(a)(9)) could create confusion about the application of § 173.124(a)(5)(ii). The language change does not result in any substantive change, but merely clarifies the relationship between § 173.124 and § 179.102–12.

Section 179.105–7 Safety relief valves Current § 179.105–7 establishes safety valve sizing requirements. The requirements are applicable to DOT specification 105 tank cars by virtue of § 179.106–2(c)(4), which requires that DOT specification 105J tank cars be equipped with a safety relief valve that meets the requirements of § 179.105–7. Thus, the § 179.105–7 safety valve sizing requirements will apply to DOT specification 105 tank cars built after February 29, 1984, to transport ethylene oxide since those tank cars must be constructed in accordance with specification 105J.

The amendment in § 179.105–7 in this document provides an optional method to meet the safety valve sizing requirement. In addition to sizing the valve according to the formula prescribed in section A8.02 of Appendix A of the AAR Specifications for Tank Cars applicable to compressed gases in non-insulated tanks, a valve with a capacity of at least 1100 scfm at 85 psi may be used in conjunction with a thermal protection system that will pass the pool fire simulation tests prescribed in § 179.105–4(d) with none of the thermocouples on the uninsulated scale of the steel plate indicating a temperature in excess of 550° F. According to the industry, an 1100 scfm valve is the size currently used on ethylene oxide tank cars.

Note that the option is not related to date of construction. Thus, any DOT specification 105 tank car used to transport ethylene oxide could be modified to specification 105J. Similarly, cars currently being built may be fitted with the option and marked as DOT specification 105J cars.

Note also that revised § 179.105–7 includes DOT specification 111 tank cars. Inclusion of specification 111 tank cars does *not* impose any burden. Section 179.105–7 does not require any specification 111 tank car to be equipped with a valve sized according to § 179.105–7. The requirement for any specification 111 tank car to have a valve sized in accordance with § 179.105–7 would result only from a separate regulatory action where that issue is fully addressed. Inclusion of the reference to specification 111 tank cars means only that the option of using additional thermal insulation instead of a larger capacity safety relief valve would be available for a specification 111 tank car in the same way as it is available for specification 105 tank cars.

(49 F.R. 3473, January 27, 1984)

49 CFR Parts 173 and 179
[Docket No. HM-175]
Shippers: Specifications for Tank Cars

The Bureau and the Federal Railroad Administration (FRA) feel that previously completed regulatory actions for new and existing DOT Specification 112 and 114 tank cars together with regulatory actions now in process for DOT Specification 105 tank cars will significantly alleviate the consequences of major accidents involving hazardous commodities. Thus far, there has been no recorded accident in which the HM-144 mandated puncture or thermal protection systems have failed. Nevertheless, these safety improvement actions and proposed actions have not directly addressed several safety concerns in the total system of rail transport of hazardous commodities.

The Bureau and FRA now propose to collect additional information which will allow a comprehensive evaluation of the need, means, and cost to:

1. Extend the specified puncture and thermal protection levels of DOT Specification 112 and 114 tank cars (HM-144) to existing DOT Specification 105 tank cars that transport the same commodities as 112 and 114 tank cars;

2. Extend the specified puncture and thermal protection levels of DOT Specification 112 and 114 cars (HM-144) to existing DOT Specification 105 tank cars that transport other hazardous commodities such as ethylene oxide, butadiene, poisons, and combustible and flammable liquids or solids.

3. Extend the specified puncture and thermal protection levels of the DOT Specification 112 and 114 tank cars (HM-144) to other new and existing DOT Specification tank cars that carry the same commodities as DOT Specification 105 cars, e.g., DOT Specification 111 tank cars.

The rules promulgated by the Bureau in HM-144 were formulated as performance standards, setting minimum levels of protection from impact and from fire for flammable compressed gasses and anhydrous ammonia carried in DOT Specification 112 and 114 tank cars. It can be argued that these performance standards should be the minimum standards for all tank cars carrying those products. It can be further argued that the same level of protection should be afforded for other equally hazardous commodities that pose a similar degree of risk. The Bureau and the FRA believe that the issues raised by these arguments need to be more fully explored and analyzed.

Unlike the DOT Specification 112/114 tank cars, the DOT Specification 105 tank cars cannot be treated as a uniform single group. They are composed of many sub-groupings which differ from each other in terms of shell and jacket thickness, insulating system properties, structural features, type and capacity of pressure relief systems, fittings, pressure rating and amount of the various ladings they transport. Generally, the sub-groupings exist because each has been designed to accommodate a specific, rather narrow set of hazardous commodities. Even within each sub-grouping, there are differences in design due to evolutions over the past forty years or because of the exercise of manufacturing options.

Although 105 tank cars carry the same or equally hazardous commodities as 112 and 114 tank cars, some of the sub-group 105 tank car designs provide less puncture resistance and thermal protection than 112 and 114 cars. The deficiencies were acknowledged during the 112 and 114 regulatory proceedings and in subsequent hearings and testimony to Congress.

In a notice of proposed rulemaking (NPRM) published elsewhere in this issue of the **Federal Register,** the Bureau is proposing to require the equivalent tankhead and thermal protection specified for 112 and 114 tank cars in HM-144 to be provided on newly built 105 tank cars. The proposed rule would also require that shelf couplers be installed on all 105 tank cars by December 31, 1981, and installed on all other DOT specification tank cars by December 31, 1984. Retrofit of existing tank cars for tank head and thermal protection is covered by this notice.

While there does not appear to be any major technical obstacles to retrofitting those 105 cars which are identified as needing additional head and thermal protection, there may be unique problems in assessing the degree of protection now possessed when compared to the performance standards of HM-144 and the magnitude of economic burdens associated with retrofitting, retiring, or changing the usage of certain existing 105 cars.

This ANPRM contains several subject areas in which the Bureau and FRA are soliciting additional facts from the public, railroads, shipping industries, tank car builders, leasing companies, railroad operating unions, and other involved safety interest groups and associations. The additional information which is collected will be used in resolving the issues in retrofitting existing 105 cars and dealing with other DOT Specification cars which carry the same hazardous commodities.

The NPRM contains data on some major accidents in which DOT Specification 105 cars have been involved. It is clear that several catastrophic accidents have occurred involving 105 cars and that both commodities covered under HM-144 and other hazardous commodities not covered under HM-144 have been released. Ethylene oxide and chlorine are examples of commodities that are not handled in 112 and 114 cars but pose a serious threat and have a documented accident history. In addition, DOT Specification 111 cars may also carry hazardous commodities, ethylene oxide for example. The Bureau is especially interested in obtaining more detailed information on major accidents related to 105 tank cars and other DOT Specification tank cars that carry the same hazardous commodities as 105 cars. Flammable compressed gases, anhydrous ammonia, chlorine, nitrosyl chloride, sulfur dioxide, sulfuryl fluoride, acrolein, pyrophoric liquids, metallic sodium, hydrofluoric acid, hydrocyanic acid, nitrogen tetroxide, motor fuel antiknock compounds, butadiene, cryogenic liquids and ethylene oxide are specific commodities for which data are desired. Specific written comments are requested for the following questions and topic areas.

1. Please provide details of accidents experienced with the commodities noted above according to DOT Specification car and sub-group tank car design characteristics.

1.1	Date of each accident.	1.17	Type(s) of underframe(s).
1.2	Location—State, nearest terminal, milepost, operating railroad.	1.18	Capacities in U.S. gallons.
1.3	Summary narrative of significant events.	1.19	Lading(s).
1.4	Deaths due to release of hazardous material.	1.20	Safety valve type(s).
1.5	Injuries due to release of hazardous material.	1.21	Safety valve setting(s), PSIG (start to discharge and full opening).
1.6	Dollar damage estimate.	1.22	Type of damage sustained:
1.7	People evacuated.		1.22-1 Head punctures (location).
1.8	Times of significant events.		1.22-2 Shell punctures (location).
1.9	Car number(s).		1.22-3 Fittings.
1.10	ICC/DOT specification type of car(s).		1.22-4 Rupture due to fire exposure.
1.11	Date built.		1.22-5 Burn hole.
1.12	Type and thickness of jacket(s).		1.22-6 Crack initiation location and propagation.
1.13	Type and thickness of insulation materials.		1.22-7 Amount of product released.
1.14	Thickness of tank head(s); lower and upper halves.		1.22-8 Distance pieces hurled.
1.15	Thickness of tank shell(s).		1.22-9 Distance covered by vapor cloud.
1.16	Tank and jacketing materials—type(s) of steel.		

The Bureau of FRA have estimated the protection levels of various subgroupings of DOT Specification 105 cars as compared to the levels mandated for 112 tank cars. The criteria utilized for the puncture and thermal protection assessment gave credit for shell thickness, jacket thickness and insulation characteristics. The following table is a summary of the estimate made, and illustrates one possible scheme for rating the comparative protection levels of existing 105 tank cars.

2. How should the existing level of head and thermal protection be determined for the various sub-groups of 105 tank cars? For example, should a point value be given for each type and thickness of material?

3. What process should be employed to identify and mark each car in the fleet according to its level of head and thermal protection?

4. Is a different level of head and thermal protection needed for hazardous commodities other than those carried in 112 and 114 tank cars? If so, what level and why for the particular commodity? How should these selected cars be identified and marked?

5. Comment on degree to which table is an accurate summary of the existing DOT 105 fleet in terms of characteristics selected for grouping the cars, the number of cars in the sub-groups, and the protection level indicated.

6. Please provide a breakdown of 105 cars owned by you according to appropriate sub-groupings and characteristics as in the table. Please provide similar information for 105 cars used but not owned by you. Please identify the cars by reporting mark and car number.

Based on surveys and subsequent assessments, the Bureau and FRA believe that at least 8000 existing 105 tank cars that carry the same hazardous commodities as the 112 and 114 tank cars would have to be retrofitted to bring them up to level of protection reasonably equivalent to that prescribed in HM-144.

7. How many 105 tank cars owned by you would be (a) retired, (b) displaced to other service, or (c) retrofitted if HM-144 performance levels for head and thermal protection were mandated?

8. What would be the reasons and economic consequences or retirement of certain cars in lieu of retrofitting? What consequences from changing the usage rather than retrofitting? Please provide specific information on the age and size of the cars which would be strong candidates for retirement or a change in usage.

9. What effect would decisions to retire or change the usage of 105 tank cars have on new car procurements?

10. What is your assessment of the technical feasibility of retrofitting the various sub-groups of 105 cars and other DOT Specification cars that carry the same commodities as the 105 cars?

11. What is your estimate of the cost of retrofitting a given 105 car? Please specify the sub-group and relevant car characteristics (e.g., capacity) that you base your overall estimate upon and identify the specific cost elements. Also, identify the type of protection system employed for purposes of the estimate, e.g., spray on insulation or jacketed insulation. Finally, include a cost estimate for the out of service time and other cost factors not included above.

12. What should be the retrofit priorities and what time frames would be reasonable? Please specify the basis for your priorities and time periods.

13. Which of the current relief valves are adequate? To what degree can relief valves or discs be modified?

14. Are there any peculiar problems or impacts unique to your situation or due to the fact that you may be a small business? Finally, there are several issues of more general applicability for which the Bureau and FRA are soliciting information.

15. What methods or processes should be utilized to determine that a given tank head or thermal protection system meets or exceeds a specific performance level?

16. What requirements, procedures, and methods should be utilized for car stencilling?

17. What should be the reporting requirements for monitoring the progress of any mandated retrofit program?

18. What operational changes might be considered in lieu or retrofitting (humping restrictions, train make-up requirements, dedicated train service, special routing, special inspection procedures, on-board detection systems, speed restrictions)?

(45 F.R. 48668, July 21, 1980)

49 CFR Parts 173 and 179
[Docket No. HM-175; Notice No. 83–1]
Specifications for Railroad Tank Cars Used To Transport
Hazardous Materials

At the time DOT commenced its review of specifications for pressure tank cars in the early 1970's, there had been a number of serious railroad accidents involving rail transportation of flammable compressed gases, toxic compressed gases and other hazardous materials. Most of these accidents involved uninsulated pressure tank cars of large capacity (over 18,500 U.S. gallons) built to specification 112 and 114.

Since the specification 112 and 114 tank car shipments of hazardous material were determined to present a more serious threat to public safety, MTB and the Federal Railroad Administration (FRA) assigned first priority to improving the construction and maintenance standards applicable to those cars. It was further decided that after these specification 112 and 114 tank cars had been structurally upgraded, the MTB and FRA would then consider a revision of the standards applicable to specification 105 tank cars.

Accordingly, on September 15, 1977, MTB published a final rule in Docket HM-144 (42 FR 46306). In summary, the rule requires that:

1. Existing and newly built specification 112 and 114 tank cars used to transport flammable gases such as propane, vinyl chloride and butane have both thermal protection (large capacity safety relief valves and high temperature thermal insulation) and tank head protection (such as a head shield);

2. Existing and newly built specification 112 and 114 tank cars used to transport anhydrous ammonia have tank head protection; and

3. All specification 112 and 114 tank cars be equipped with special couplers designed to resist coupler vertical disengagements (shelf couplers).

After the upgrading of specification 112 and 114 tank cars was substantially completed, MTB initiated rulemaking for specification 105 tank cars. On January 26, 1981, MTB published a final rule in Docket HM-174 (46 FR 8005) affecting new construction of specification 105 tank cars. That final rule also includes a shelf coupler requirement applicable to all specification tank cars. The rule requires that:

(1) Specification 105 tank cars built before March 1, 1981, be retrofitted over a one-year period ending on February 28, 1982, with a coupler vertical restraint system equivalent to that required on specification 112 and 114 tank cars;

(2) All other specification tank cars built before March 1, 1981, be equipped with a coupler vertical restraint system equivalent to that required on specification 112 and 114 tank cars over a four-year period ending on February 28, 1985;

(3) After February 28, 1981, newly built specification 105 tank cars be equipped with a coupler vertical restraint system equivalent to that required on specification 112 and 114 tank cars;

(4) After August 31, 1981, newly built specification 105 tank cars transporting flammable gases, anhydrous ammonia and ethylene oxide be equipped with a tank head puncture resistance system equivalent to that required on certain specification 112 and 114 tank cars (S, T, and J cars);

(5) After August 31, 1981, newly built specification 105 tank cars transporting flammable gases and ethylene oxide be equipped with high temperature thermal insulation equivalent to that required on certain specification 112 and 114 tank cars (T and J cars); and

(6) After August 31, 1981, newly built specification 105 tank cars transporting flammable gases and ethylene oxide be equipped with safety relief valves sized according to the requirements for specification 112 and 114 tank cars.

On July 21, 1980, the same day the notice of proposed rulemaking in Docket HM-174 was issued, MTB also issued an advance notice of proposed rulemaking (ANPRM) in Docket HM-175 (45 FR 48668). That notice sought additional information to allow an evaluation of the need, means, and cost to extend the specified puncture and thermal protection levels of specification 112 and 114 tank cars to:

1. Existing specification 105 tank cars used to transport the same hazardous materials permitted in specification 112 and 114 tank cars;

2. Existing specification 105 tank cars used to transport other hazardous materials such as ethylene oxide, butadiene, poisons, and combustible and flammable liquids or solids; and

3. All other new and existing specification tank cars used to transport the same hazardous materials permitted in specification 105 tank cars, e.g., specification 111 tank cars.

Comments received in response to the ANPRM and the substantial body of information developed in Dockets HM-144 and HM-174 have been considered in developing the rule proposed in this notice. In all, thirty-eight comments to the ANPRM were submitted to MTB. The depth of coverage in response to the eighteen questions posed by MTB ranged from one or two-page documents limited to a single concern, general observations, or a blanket endorsement of some other commenter's submittal, to documents of more than forty pages covering each of the major issue areas.

QUESTION #1: ACCIDENT DATA

In the ANPRM, MTB requested details of accidents involving a broad range of hazardous materials according to specification tank car and designated subgroup tank car design characteristics. The tank car subgroup designs were specified in the ANPRM.

Ten responses contained some accident data. Most of the comments were directed at individual commodities or limited movements. One commenter supplied detailed accident data on all major hazardous material rail accidents since 1958. Other commenters supplied accident data on particular hazardous materials or tank car fleet basis. Although the great majority of tank cars included in these data carried flammable gases or anhydrous ammonia, other hazardous materials also were mentioned. These other hazardous materials included ethylene oxide, chlorine, motor fuel anti-knock compound, nitrogen fertilizer solution, refrigerants/dispersants/fluorocarbons, carbon dioxide, sulfur dioxide, ethyl ether, metallic sodium, anhydrous hydrofluoric acid, benzene, acrolein, hydrogen sulfide, vinyl acetate, flammable liquids, propylene oxide, and anhydrous hydrochloric acid. The comments received in this area did not contradict or significantly add to MTB's own data base.

QUESTION #2: EXISTING PROTECTION LEVELS

The MTB requested comments on how the levels of head and thermal protection should be ascertained for the various subgroups of existing specification 105 tank cars. Specifically, MTB asked if a point value system reflecting each type and thickness of material should be established.

Most commenters either directly stated or inferred that the levels of needed protection cannot be established for other hazardous materials in the same way as for flammable gases. They contend that safety requirements should be based on the characteristics of the hazardous material being transported and the risks inherent in its transportation by rail. Some commenters noted that the Railway Progress Institute (RPI), the Association of American Railroads (AAR), and FRA were conducting scale model testing and that specific methodologies should await final test results. Several commenters stated that full scale testing was needed to determine the degree of protection provided to such hazardous materials by various specification tank car designs. One commenter recommended that objective engineering formulas from FRA test data be developed to relate the physical and thermal characteristics of the head/tank material, insulation material, jacket material, and tank design.

Overall, MTB did not receive any analyses or approaches to evaluate the levels of puncture and thermal protection that the various specification 105 tank cars now possess. The commenters tended to avoid this question on the grounds that it was not feasible, without additional testing, to determine the levels of protection now provided by the various specification 105 tank car designs.

QUESTION #3: IDENTIFICATION AND MARKING

The MTB requested information on the processes that might be employed to identify and mark each car in the fleet according to the level of head and thermal protection it possesses.

Most commenters answered this question in the same way as for Question #16 on stenciling. They recommend that the current stenciling methods and the alphanumeric system to identify particular car/commodity/protection for specification 112/114 and new specification 105 tank cars be extended to all specification 105 tank cars.

QUESTION #4: COMMODITY SPECIFIC PROTECTION

The MTB solicited comments on whether a different level of thermal and head protection is needed for hazardous materials other than those carried in specification 112 and 114 tank cars. For those hazardous materials which may need different levels of protection, MTB also requested information on what the levels should be and the reasons for the protection level selected. In addition, MTB asked for suggestions on how cars carrying these hazardous materials should be identified and marked.

The commenters who responded to this question believed that hazardous materials should be evaluated individually, with many requiring different or lesser levels of protection than required for the hazardous materials addressed in HM-144. Several commenters offered specific examples of hazardous materials that warrant special protection consideration. Several commenters felt that the existing requirements are already in proportion to protection needs; e.g., the authorized tank car is often required to be constructed to a higher strength specification than the vapor pressure of the hazardous material at ambient temperature warrants.

Commenters generally restricted their opinions to products and car types with which they were directly associated. A variety of reasons were offered in support of various positions.

Two commenters noted that there has never been an incident involving loss of product from a tank car transporting metallic sodium and the metallic sodium is shipped at low vapor pressure. Therefore, they conclude that there is no need for head and thermal protection of tank cars used to transport metallic sodium.

One commenter felt that there might be a justification for a heavier head jacket for cars carrying pyrophorics. The same commenter claimed that there was no justification for retrofit of motor fuel anti-knock compound cars because the disaster potential was much less than for some of the other hazardous materials transported in specification 105 tank cars.

Several commenters identified flammable gases, ethylene oxide, chlorine, and certain poisons as commodities that may be candidates for additional protection. One commenter believed that flammable liquids and combustible liquids have characteristics that do not require the same protection level as do flammable or poisonous gases.

Another commenter cited ethylene oxide, chlorine, and acrolein as hazardous materials that may need additional protection. This commenter believed that the HM-144 thermal protection requirement would not be appropriate for chlorine and acrolein. It was noted that the HM-144 requirement is designed to prevent tank steel from exceeding 800° F for a specific time during exposure to torch or pool fires, whereas chlorine reacts with steel at about 500° F and acrolein polymerizes at about 400° F. Another commenter noted the steel-chlorine reaction problems and suggested that future testing should be directed at producing high temperature insulating systems to improve the thermal resistant performance of chlorine cars. This commenter noted that, for chlorine cars, there have been more shell punctures than head punctures. Therefore, the commenter concluded that if a thicker tank head or jacket is necessary, then consideration should be given to requiring thicker material for the entire tank. One commenter identified chlorine, vinyl chloride, and isobutylene as the most viable hazardous materials for tank car retrofitting. Another commenter pointed to ethylene oxide cars as the cars most needing retrofit.

No consistent or rigorous criteria for ranking hazardous material in order of the need for additional protection levels evolved from the comments submitted.

QUESTION #5: ACCURACY OF MTB ESTIMATES

The MTB included in the ANPRM a table of car groupings and estimated protection levels to stimulate comments. It is entitled, "Summary of Estimated Protection Provided by Broad 105 Car Classification Sub-Groupings." Comments on the applicability and accuracy of the table were solicited.

No in-depth analyses or detailed comments pertaining to the table were supplied by the commenters. Several commenters felt that the table represented a reasonable breakdown of the specification 105 tank car fleet. Some suggested that the breakdown of the specification 105 tank car fleet in the table be compared with the results of a study being done by Dynatrend for FRA and/or with the results of Phase 17 of the RPI/AAR Tank Car Project. A more detailed breakdown of the specification 105 car fleet (e.g., hazardous material, by car size, and by car accident history) was recommended by others. Most commenters, however, only addressed the number of cars in each subgroup, not the estimated protection levels of the subgroups.

The commenters generally felt that the two columns offered by the MTB on the estimated status of the tank head and thermal protection of the subgroups were subjective. Many stated that additional testing was needed to quantify the existing levels of head and thermal protection by car design. One commenter felt that the table was misleading for anhydrous ammonia since the table indicates that specification 105 anhydrous ammonia cars have significantly less thermal protection than HM-144 requirements, even though high temperature thermal protection is not required in HM-144 for specification 112 and 114 tank cars used to carry anhydrous ammonia.

QUESTION #6: FLEET CHARACTERISTICS

The MTB asked that tank car owners and users provide breakdowns of specification 105 tank cars owned or used according to appropriate subgroupings and characteristics in the table published in the ANPRM.

The limited data submitted did not permit MTB to compile a more accurate or comprehensive characterization of the specification 105 tank car fleet.

QUESTION #7: RETIREMENT, DISPLACEMENT, RETROFIT CONSIDERATIONS

The MTB requested that tank car owners estimate how many specification 105 tank cars would be retired, displaced, or retrofitted if HM-144 performance levels were mandated for the entire specification 105 tank car fleet.

Generally, it appears that the responses to this question were made from varying assumptions as to what the final rule requirements would be. Hence, it is impossible to draw any firm conclusions on the number of cars that would be displaced, retired, or retrofitted if a retrofit requirement for the entire 105 tank car fleet was established.

QUESTION #8: REASONS AND CONSEQUENCES OF RETIREMENT OR DISPLACEMENT OF TANK CARS

Comments were requested about the reasons for and economic consequences of the retirement of certain cars in lieu of retrofitting. Comments were also requested about the reasons for and the economic consequences of changing car usage rather than retrofitting. The MTB further requested that commenters provide specific information on the age and size of cars which would be considered as strong candidates for retirement or for a change in usage.

Most commenters answered this question in a way which suggested that actual "real world" decisions would have to await a clearer expression of regulatory intent. Various rather general individual opinions were advanced. One commenter stated that cars 28 years or older would be retired because the retrofit cost could not be fully depreciated over the remaining life of those cars. Another commenter stated that cars listed in the ANPRM Table as subgroup C (LPG) would either be scrapped, or converted to a non-pressure class if a demand for other use could be found for those cars. Another commenter identified small tank cars (approximately 11,000 gallon size) as being strong candidates for a change in usage or retirement.

QUESTION #9: NEW CAR PROCUREMENTS

Comments were requested on the effect that tank car retirement or usage changes would have on new car procurements.

Most comments were not detailed or specific. It was difficult to determine a consistent or overall direction of the replies. Some commenters stated that all of the retired/displaced cars would be replaced, while one said that none of the retired/displaced cars would be replaced.

The MTB did not obtain new information from the comments which significantly aided in estimating the impact that the regulatory changes would have on new car procurements.

QUESTION #10: TECHNICAL FEASIBILITY OF RETROFITTING

Assessments of the technical feasibility of retrofitting specification 105 cars were solicited by MTB.

Most of the commenters who responded to this question did not advance technical doubts as to the feasibility of accomplishing the potential retrofit. There were strong opinions that it was not economically practical (because of high costs) to retrofit tank cars by replacing existing valves with larger safety valves. There were also general objections to a retrofit based on the belief that an unretrofitted specification 105 tank car is safer than the unretrofitted specification 112 and 114 tank cars built prior to HM-144.

Even though no significant retrofit obstacles were identified, several commenters observed that because of the growth in diameter of tanks, insulation thickness was restricted by clearance dimensions along railroad rights-of-way. Other commenters considered retrofitting as feasible from a purely technical aspect, but thought that additional testing would be necessary to resolve certain design questions. Several commenters expressed uncertainties concerning the length of time that external coatings applied to specification 105 car jackets could be expected to meet high temperature thermal protection requirements without losing mechanical integrity in severe railroad operational environments.

QUESTION #11: RETROFIT COST ESTIMATES

The MTB requested estimates of the prospective costs of retrofitting given types of specification 105 tank cars. The MTB asked that commenters specify the subgroup and relevant car characteristics (e.g., capacity) upon which their estimates were based; that the specific cost elements be isolated; and that the type of protection system be identified. In addition, cost estimates for out-of-service time and other cost factors were requested.

The nature and the amount of the estimates supplied by the commenters varied over a wide range. In some cases, it was not clear what cost elements were embraced by the totals.

QUESTION #12: RETROFIT PRIORITIES AND TIMEFRAMES

The ANPRM solicited comments concerning what retrofit priorities and timeframes might be reasonable, and the basis for the priorities and timeframes advocated.

Several of the commenters, while registering doubts about the justification of a retrofit, stated that any retrofit priority should be based on the characteristics of the hazardous materials, with the timeframe related to the need for retrofit, the number of cars affected, and the availability of shop space. One commenter recommended that the retrofit priorities should be based upon the Environmental Protection Agency's Hazardous Substances Categories, and that the number of cars known to be in each of these priority categories would then provide a basis for arriving at reasonable retrofit timeframes. Several commenters believed that retrofit priorities should be based on the characteristics of the hazardous material and car size while one commenter believed that retrofit priorities should be based upon a reconciliation of the level of hazard inherent in the hazardous material and the degree of protection already existing in the cars used. Several commenters suggested priority be given to the HM-144 commodities (flammable gases and anhydrous ammonia).

QUESTION #13: SAFETY RELIEF VALVES

The MTB asked for comments on the adequacy of current safety relief valves and the extent to which relief valves or discs can be modified on existing cars.

The large majority of commenters felt that a requirement should not be imposed that would mandate replacement of existing valves with larger ones. Although one commenter believed that larger safety valves are of benefit and recommended continued use wherever feasible, many other commenters expressed the opinion that the relief valves currently used on specification 105 tank cars are adequate. The only commenter who discussed directly the feasibility of modifying relief valves or discs stated that existing relief valves on specification 105 tank cars cannot be modified to obtain a larger capacity.

The primary concern of the commenters is the large cost that would be involved if installation of a large capacity safety valve requires cutting a larger hole in the tank. However, the commenters did not indicate the number of cars which would require the cutting of a larger hole.

QUESTION #14: PECULIAR PROBLEMS OR IMPACTS

The MTB requested information on problems or impacts unique to a commenter's situation, including whether the owner or shipper was a small business.

The general flavor of the comments was that those small businesses owning or operating only a few cars would tend to suffer the most if a comprehensive retrofit were to be required. As in answers to several other questions, commenters seemed to be concerned that an arbitrary requirement might be imposed which would be expensive to comply with and provide little real direct benefits in return.

QUESTION #15: PROOF OF PERFORMANCE LEVELS

The MTB requested opinions on the methods or processes that should be utilized to determine the degree to which given tank head or thermal protection systems meet or exceed a specified performance level.

Several commenters favored complete full-scale testing to prove the effectiveness of proposals in new areas. One commenter suggested that both full-scale puncture and drop hammer tests be performed for systems not previously tested. Several commenters endorsed the FRA thermal test program which was then underway as a step in the right direction.

QUESTION #16: STENCILLING

The MTB asked for comments on the requirements, procedures and methods that might be utilized in stenciling or labeling the tank cars.

Most commenters felt that current methods for tank car stenciling could be easily augmented to advise the shippers of the degree of protection provided by specific car types.

QUESTION #17: REPORTING NEEDS

The MTB asked for comments on the reporting requirements for monitoring the progress of any mandated retrofit program.

Most commenters stated that if a reporting requirement is considered necessary then a system similar to that used for HM-144 should be adopted. Several commenters recommended a quarterly reporting schedule, while others recommended a semi-annual report schedule. One commenter suggested that industry should submit to DOT a timetable to monitor progress on a quarterly basis and that DOT should require reports by exception, i.e., from only those who are not meeting the submitted schedule.

QUESTION #18: OPERATIONAL SUBSTITUTES

The MTB asked for comments on operational changes that might be instituted in lieu of retrofitting and still provide acceptable protection levels.

Several commenters suggested that DOT should await recommendations of the Systems Safety Analysis Group of the Inter-industry Task Group. Other commenters suggested that a rule, similar to FRA Emergency Order No. 5, should be adopted. Several commenters believed that operational changes are a more costly solution, especially since they are recurring costs, and should not be required.

DISCUSSION

The MTB and FRA have examined the comments received in respect to actual accident trends, previous regulatory actions, testing results, technical evaluations, traffic statistics and other pertinent information in the possession of MTB and FRA. As a part of this examination, several options were developed and evaluated. A comprehensive benefit-cost analysis was employed to assess the relative merits of each major option.

The provisions of this NPRM are discussed in relationship to each of the three areas of consideration listed in the ANPRM.

The first area addressed in the ANPRM was the issue of extending the specified puncture and thermal protection levels of specification 112 and 114 tank cars (HM-144) to existing specification 105 tank cars that transport the same commodities as specification 112 and 114 tank cars.

In assessing the advisability of regulatory action in this area, the MTB and FRA note that:

(a) Any specification 112 tank car, 114 tank car, or 105 tank car built after August 31, 1981 must have a specified puncture protection level, thermal protection level and safety valve sizing as appropriate, when used to transport flammable gases and the toxic non-flammable gas, anhydrous ammonia;

(b) The safety improvement effectiveness of these protection requirements has been established through operating experience;

(c) Almost all of the flammable gas and anhydrous ammonia rail movement volume in specification 105 tank cars is carried in newer cars having capacities in excess of 18,500 gallons;

(d) Other than having jacketed ambient temperature insulation, the basic design of the large capacity specification 105 tank cars is essentially similar to the specification 112 and 114 tank cars prior to being retrofitted in accordance with HM-144;

(e) Requiring retrofit of the safety valves on existing large capacity (over 18,500 U.S. gallons) specification 105 tank cars will not require cutting a larger hole in the tank since these cars have a manway cover with an 18 inch diameter; and

(f) The benefit-cost ratio for the retrofit of these large specification 105 tank cars (numbering about 3,000) is positive.

The FRA has determined that the retrofit of specification 105 tank cars would be cost beneficial, provided that the retrofit requirement is limited to those cars having capacities exceeding 18,500 U.S. gallons and used to transport those hazardous materials that historically have an unfavorable accident history and disaster potential (i.e., LPG, ethylene oxide, and anhydrous ammonia). The retrofit proposed in this NPRM requires protection systems that are identical or substantially similar to those that have been extremely successful when applied to other tank cars transporting these commodities. The selection of large capacity specification 105 tank cars for retrofitting is consistent with the previous tank car retrofit rulemaking (HM-144) since the specification 112 and 114 tank cars previously required to be retrofitted were large capacity cars.

A large capacity safety valve is proposed because the combination of thermal insulation and larger safety valve provides the best level of safety. The large safety valve minimizes the product left in the tank at the point of tank failure and provides additional benefits in the case where the car is overturned and must vent liquid as well as vapor.

The cost involved in requiring a large capacity safety valve, as was required in HM-144 and HM-174, is not excessive since MTB and FRA are proposing a rule limited to tank cars with a volume over 18,500 gallons. The MTB and FRA believe that most, if not all, of these cars already have a manway of at least 18 inches in diameter, i.e., large enough to be equipped with the large capacity valve. Thus, no changes in the tank structure, would be necessary. At most, a new manway cover plate will be necessary.

For those tank cars with a manway cover having a major dimension or diameter of less than 18 inches, MTB and FRA have proposed an alternative system that would allow the use of additional thermal protection in conjunction with existing safety valves. With this system, there should be no liquid remaining in the tank car at the moment of failure in those incidents in which the tank car remains upright. Under this option, there could be situations where the tank car pressures slightly exceed the pressures prescribed in §179.100-15 and §179.102-11; however, the pressures would not exceed the tank test pressure. A technical analysis explaining the safety validity of this option, and the consistency of the option with safety criteria established in HM-144, is in the docket.

The MTB and FRA are interested in determining the number of tank cars, if any, that have a manway cover with a major dimension or diameter of less than 18 inches. Therefore, it is requested that tank car owners who have cars with a manway cover having a major dimension or diameter of less than 18 inches provide the following information in their comments to the docket:

(a) The reporting mark(s) of the car(s);

(b) the manway cover diameter or major dimension, layout, and arrangement, including the location of all valves and fittings;

(c) a description of the existing insulation system on the tank car; and

(d) a list of all hazardous materials that are transported in the particular car(s).

At the present time and with the information available, MTB and FRA do not believe a rule requiring a retrofit of existing specification 105 tank cars having capacities less than 18,500 gallons carrying the identified hazardous materials is warranted on a cost/benefit basis. Many of these cars are nearing the end of their service life. Hence, the cost of retrofit might not be recovered in the remaining tank car life. More importantly, these smaller capacity cars have a lower utilization rate, reducing their exposure to potential accident situations. Finally, their smaller capacity presents a smaller safety risk should they be involved in an accident.

The second area of consideration for potential rulemaking listed in the ANPRM was to extend the specified puncture and thermal protection levels of specification 112 and 114 tank cars (HM-144) to existing specification 105 tank cars that transport other hazardous materials such as ethylene oxide, butadiene, poisons, and combustible and flammable liquids or solids.

Beyond some general mention of certain hazardous materials, the commenters did not support the extension of thermal and puncture requirements as applied to flammable and toxic poisons to "other" hazardous materials carried in existing specification 105 tank cars. The MTB and FRA have reviewed the safety record and the threats posed by all of the hazardous materials carried in specification 105 tank cars. Using the same rationale as was used in the final rule in HM-174, the MTB and FRA believe that the characteristics and threats posed by ethylene oxide are so close to materials classes as flammable gas that the same retrofit requirements should be adopted for tank cars used to transport this material. The expected benefits are the same as those identified for flammable gases.

The FRA is unable to justify similar retrofit requirements for specification 105 tank cars carrying other hazardous materials. This is true even though hazardous materials such as chlorine, acrolein, and motor fuel anti-knock compounds have properties that pose a high risk in accidents. However, these hazardous materials differ from flammable gases in that the identical level of thermal protection may not provide the same duration to failure or alleviate the potential catastrophic consequences to the same degree. Moreover, these hazardous materials are often carried in cars with different and better relative protection levels than tank cars authorized for flammable gases.

In regard to thermal protection, the more dangerous of the other hazardous materials undergo rapid decomposition or polymerization at elevated temperature, but neither the failure mechanisms nor countermeasures are as yet adequately tested or understood. For example, in the case of chlorine, it has not been determined that there have been any accident consequences that would definitely have been avoided if high temperature thermal protection identical to that required for flammable gases had been in place. Moreover, unlike flammable gases, a thermally induced rupture of a chlorine car would not increase the fire or continue a chain reaction spread of fire engulfment.

The MTB and FRA also considered increased puncture resistance for existing cars carrying hazardous materials such as chlorine, motor fuel anti-knock compound, and sulfur dioxide. The MTB and FRA are not convinced that an increased puncture resistance requirement is justified based on accident experience and the current protection levels built into the cars authorized to transport these materials. The primary basis for this view is the fact that these hazardous materials are required to be shipped in tank cars with pressure ratings in excess of that needed to contain these products. Therefore, these cars already have some increased degree of built-in puncture resistance. Although this tank head puncture resistance may not be equivalent to HM-144/HM-174 performance levels, the safety record of these cars is such that MTB and FRA cannot now justify their retrofit or redesign to achieve an incremental amount of additional protection. It should also be noted that all specification 105 tank cars carrying hazardous materials are now required to have shelf couplers (per HM-174), which diminishes the probability of coupler-inflicted tank punctures.

With the exception of ethylene oxide, MTB did not find sufficient threats to safety, nor cost-benefit justification, for proposing an extension of thermal and head protection for tank cars to materials other than those addressed in dockets HM-144 and HM-174.

The final area of consideration identified in the ANPRM suggested that there may be safety benefits in regulatory action to extend the specified puncture and thermal protection levels of the specification 112 and 114 tank cars (HM-144) to other new and existing specification tank cars that carry the same hazardous materials as specification 105 tank cars, e.g., certain specification 111 tank cars.

The MTB and FRA have examined the hazardous materials transported in both specification 105 tank cars and specification 111 tank cars. There are approximately 300 specification 111A100W4 tank cars that are authorized to move flammable gases and ethylene oxide, the hazardous materials determined by MTB and FRA to warrant additional protection. The size of this fleet will not increase, because on October 9, 1981 the Association of American Railroads (AAR) specified that:

(a) After October 1, 1981, no specification 111A100W4 tank cars may be newly built for or converted to ethylene oxide service. Existing specification 111A100W4 tank cars in ethylene oxide service must comply with the coupler restraint requirements of 49 CFR §179.105–6 by March 31, 1982; and

(b) After October 1, 1981, each class of tank car converted to ethylene oxide service must comply with Class 105J coupler restraint, head shield, and thermal protection requirements for new construction.

However, for the same reasons as previously stated with respect to the large (over 18,500 U.S. gallons) specification 105 tank cars, MTB and FRA believe that the retrofit of the large specification 111 tank cars carrying ethylene oxide and flammable gases is cost beneficial.

In order to be consistent with the AAR actions and avoid unnecessary complexity, the proposed rule does not authorize any newly constructed specification 111 tank cars for ethylene oxide or flammable gas service. New construction of specification 111 tank cars is not needed since specification 105 tank cars may be constructed to carry these commodities.

SECTION-BY-SECTION ANALYSIS

Section 173.124 Ethylene oxide. It is proposed to amend paragraph (a)(5) of §173.124 to require that each specification 105 tank car built before September 1, 1981, with a capacity in excess of 18,500 U.S. gallons, shall conform to specification 105J when transporting ethylene oxide after December 31, 1986. (Specification 105 tank cars built after August 31, 1981 are currently required to conform to specification 105J when transporting ethylene oxide.) Requiring a specification 105J tank car would mean that existing specification 105 tank cars would have to be retrofitted with high temperature thermal protection, tank head protection, and large safety valves by December 31, 1986 if they are to continue in ethylene oxide service.

It is also proposed to amend paragraph (a)(5) to require that each specification 111 tank car, with a capacity in excess of 18,500 U.S. gallons, conform to specification 111J when transporting ethylene oxide after December 31, 1986. Thus, each existing specification 111 tank car would have to be retrofitted with high temperature thermal protection, tank head protection, and large safety valves by December 31, 1986 in order to continue to carry ethylene oxide.

Section 173.314 Requirements for compressed gases in tank cars. It is proposed to amend this section to require that existing DOT specification 105 tank cars (those built prior to September 1, 1981) used to transport anhydrous ammonia, and with a capacity exceeding 18,500 U.S. gallons capacity, be retrofitted by December 31, 1986 with lower tank head protection, *i.e.,* conform to specification 105S. It is further proposed that existing specification 105 tank cars with a capacity exceeding 18,500 U.S. gallons, used to transport flammable gases, be retrofitted with thermal protection, head protection, and large safety valves by December 31, 1986, *i.e.,* conform to specification 105J. It is also proposed to require that each specification 111 tank car with a capacity exceeding 18,500 U.S. gallons, used to transport flammable gases, shall conform to specification 111J.

Section 179.102–12 Ethylene oxide. It is proposed to amend this section to require that existing specification 105 tank cars (built prior to September 1, 1981 and with a capacity exceeding 18,500 gallons) used to transport ethylene oxide be retrofitted with high temperature thermal protection, tank head protection, and large safety valves by December 31, 1986 if they are to continue in ethylene oxide service.

Section 179.105–7 Safety relief valves. It is proposed to add a new paragraph (c) which would allow the use of smaller safety valves if the thermal protection exceeds the minimum thermal protection required in §179.105–4.

Section 179.106–1 General. It is proposed to amend this section by requiring that existing specification 105 tank cars manufactured to the specifications of the Canadian Transport Commission conform to the same standards prescribed for DOT specification 105 tank cars.

Section 179.106–3 Previously built cars. It is proposed to amend this section by establishing performance requirements for specification 105S and 105J tank cars build before September 1, 1981. The proposed requirements for the 105S and the 105J tank cars in this section are identical to the requirements in §179.106–2 for new cars.

Section 179.200–27 Alternative requirements for tank head puncture resistance systems. It is proposed to add this section to clarify that specification 111 tank cars may utilize a head shield as prescribed in §179.100–23 instead of meeting the puncture resistance requirements in §179.105–5.

Section 179.202–18 Ethylene oxide. It is proposed to add a new paragraph(a) (10) in §179.202–18 to require that each specification 111 tank car used after December 31, 1986 for the transportation of the ethylene oxide, with a capacity exceeding 18,500 U.S. gallons, conform to class 111J.

Section 179.203 Special requirements for specification 111 tank cars. It is proposed to add a new section setting out special requirements for specification 111 tank cars that is parallel to §179.106 for specification 105 tank cars.

(48 F.R. 16188, April 14, 1983)

49 CFR Parts 173 and 179
[Docket No. HM-175; Amdt. Nos 173–173, 179–35]
Specifications for Railroad Tank Cars Used To Transport
Hazardous Materials

In the early 1970's DOT commenced its review of specifications for pressure tank cars. There were a number of serious railroad accidents involving rail transportation of flammable compressed gases, toxic compressed gases, and other hazardous materials. Most of these accidents involved uninsulated pressure tank cars of large capacity (over 18,500 U.S. gallons) built to specifications 112 and 114.

Since the specification 112 and 114 tank car shipments of hazardous material were determined to present a more serious threat to public safety, MTB and the Federal Railroad Association (FRA) assigned first priority to improving the construction standards applicable to those cars. It was further decided that after these specification 112 and 114 tank cars had been structurally upgraded, the MTB and FRA would consider a revision of the standards applicable to the specification 105 tank cars.

Accordingly, on September 15, 1977, MTB published a final rule in Docket HM-144 (42 FR 46306). In summary the rule requires that:

(1) Existing and newly built specification 112 and 114 tank cars used to transport flammable gases such as propane, vinyl chloride, and butane have both thermal protection (large capacity safety relief valves and high temperature thermal insulation) and tank head protection (such as a head shield);

(2) Existing and newly built specification 112 and 114 tank cars used to transport anhydrous ammonia have tank head protection; and

(3) All specification 112 and 114 tank cars be equipped with special couplers designed to resist coupler vertical disengagement (shelf couplers).

After the upgrading of specification 112 and 114 tank cars was substantially completed, MTB initiated rulemaking for specification 105 tank cars. On January 26, 1981, MTB published a final rule in Docket HM-174 (46 FR 8005) affecting new construction of specification 105 tank cars. The rule requires that:

(1) Specification 105 tank cars built before March 1, 1981, be retrofitted over a one-year period ending February 28, 1982, with a coupler vertical restraint system equivalent to that required on specification 112 and 114 tank cars;

(2) After February 28, 1985, all other specification tank cars be equipped with a coupler vertical restraint system equivalent to that required on specification 112 and 114 tank cars;

(3) After February 28, 1981, newly built specification 105 tank cars be equipped with a coupler vertical restraint system equivalent to that required on specification 112 and 114 tank cars;

(4) After August 31, 1981, newly built specification 105 tank cars transporting flammable gases, anhydrous ammonia, and ethylene oxide be equipped with a tank head puncture resistance system equivalent to that required on certain specification 112 and 114 tank cars (S, T, and J cars);

(5) After August 31, 1981, newly built specification 105 tank cars transporting flammable gases and ethylene oxide be equipped with high temperature thermal insulation equivalent to that required on certain specification 112 and 114 tank cars (T and J cars); and

(6) After August 31, 1981, newly built specification 105 tank cars transporting flammable gases and ethylene oxide be equipped with safety relief valves sized according to the requirements for specification 112 and 114 tank cars.

On July 21, 1980, the same day the notice of proposed rulemaking in Docket HM-174 (45 FR 48671) was issued, MTB also issued an advance notice of proposed rulemaking (ANPRM) in Docket HM-175 (45 FR 48668). That notice sought additional information to allow an evaluation of the need, means, and cost to extend the specified puncture and thermal protection levels of specification 112 and 114 tank cars to:

(1) Existing specification 105 tank cars used to transport the same hazardous materials permitted in specification 112 and 114 tank cars;

(2) Existing specification 105 tank cars used to transport other hazardous materials such as ethylene oxide, butadiene, poisons, and combustible and flammable liquids or solids; and

(3) All other new and existing specification tank cars used to transport the same hazardous materials permitted in specification 105 tank cars, *e.g.*, specification 111 tank cars.

After analyzing the comments received in response to the ANPRM and comprehensively evaluating the costs and benefits of a variety of potential regulatory options, MTB issued a notice of proposed rulemaking (NPRM) in HM-175 on April 14, 1983 (48 FR 16188). The NPRM, which is substantially the same as the final rule, proposed the following requirements:

(1) After December 31, 1986, specification 105 tank cars built before September 1, 1981, that have a capacity exceeding 18,500 U.S. gallons and are carrying a flammable gas, anhydrous ammonia, or ethylene oxide would have to be equipped with lower half tank head protection (such as a head shield);

(2) After December 31, 1986, specification 105 tank cars built before September 1, 1981, that have a capacity exceeding 18,500 U.S. gallons and are carrying a flammable gas or ethylene oxide would have to be equipped with:

(a) High temperature thermal insulation; and

(b) Safety relief valves sized according to the requirements for specification 112 and 114 tank cars; and

(3) After December 31, 1986, specification 111 tank cars that have a capacity exceeding 18,500 U.S. gallons and are carrying flammable gas or ethylene oxide would have to be equipped with: (a) Lower half tank head protection; (b) high temperature thermal insulation; and (c) safety relief valves sized according to the requirements for specification 112 and 114 tank cars.

A total of 25 comments in response to the NPRM were received, representing a diverse group of interested persons. The commenters include the Association of American Railroads (AAR), major chemical and petroleum companies, tank car leasing companies, manufacturers of high temperature insulating materials, shippers, several fire chiefs and one state association of fire chiefs, numerous trade associations of shippers, tank car owners, and the National Transportation Safety Board (NTSB).

A number of commenters generally endorsed the proposed rule since it would increase the level of safety in the transportation of the affected hazardous materials. One such commenter, NTSB, urged that consideration be given to further rulemaking to address other hazardous materials and the smaller capacity specification 105 tank cars (under 18,500 U.S. gallons) transporting liquified flammable gases (LFG), anhydrous ammonia, and ethylene oxide. In that regard, FRA and MTB will continue to evaluate the need for new rules which exceed the current requirement that all tank cars transporting a hazardous material, after March 1, 1985, be equipped with shelf couplers. As indicated in the preamble to the notice of proposed rulemaking in this docket, retrofitting the smaller capacity specification 105 tank cars does not appear to be justified on a benefit/cost basis.

Several commenters, including the AAR and a major ethylene oxide shipper/tank car owner, supported the proposed tank head and thermal insulation requirements, but opposed the requirement to retrofit with a large capacity safety relief valve.

A number of commenters opposed the proposed rule on the belief, without explanation, that the costs exceed the benefits. Other commenters who opposed the rule in whole or in part or who believed that the rule is premature, more fully explained the basis of their objections. These objections also related generally to the cost/benefit issue, principally by challenging the accuracy of both the cost estimates (cost of retrofit and number of cars involved) and the benefit estimates (effectiveness rate, statistical base for accident frequency, and impact of prior rulemakings). No commenters disputed the technical feasibility of retrofitting tank cars with the safety systems proposed. Apart from a question about the need for a large capacity safety relief valve in addition to high temperature thermal insulation (800° F material in the simulation pool fire test), no commenters disputed the belief that tank head and thermal protection would improve safety; some questioned how much improvement would be achieved.

Several commenters stated that FRA's $12,000 cost estimate to retrofit a tank car is too low. They suggested a $14,000 to $15,000 range as reflecting the true current cost. FRA agrees that the $12,000 cost in the economic impact analysis developed for the NPRM is lower than the current cost. That analysis which is comprehensive and detailed,was begun at the time of the 1980 ANPRM and uses 1980 dollars for *both* the costs and benefits. The final economic impact analysis includes an updating to 1983 dollars for *both* costs and benefits. In 1983 dollars, the cost of retrofitting a car is estimated to be approximately $14,000. Adopting a conservative approach, the same inflation factor (Department of Labor's Consumer Price Index) was applied on the benefit side, even though many of the constituent parts of the benefit side were rising at a faster pace, *e.g.,* medical costs.

The result of the update is that the benefit/cost ratio remains the same and would still be highly favorable (1.42) even if a $15,000 retrofit cost figure were used.

Some commenters felt that FRA had not adequately considered the cost of valve changes in its estimation of retrofit costs. FRA is not convinced that its earlier cost estimate is too low. However, under the final rule, appropriate thermal protection may be achieved either by a given level of insulation and the larger safety valve or by a greater level of insulation without any changes to the existing valve. In its initial analysis (Economic Impact Analysis of the Retrofit of 105 and 111 Tank Cars Carrying Hazardous Materials, Exhibits FRA-10,300, FRA-10,310) FRA showed that the cost of the two options is the same. According to the figures used in that study, which represent costs in 1980 dollars, it would cost $6,000 to provide the required level of thermal protection. If the larger valve is installed, the valve cost per car is $1,000 and the installation cost is $5,000. If the smaller valve is not replaced, $6,000 worth of insulation must be installed. In the proposed rule, the option was limited to those cars with a manway less than 18 inches in diameter. The extra insulation option, therefore, was not available for most of the cars to be retrofitted. This limitation is not in the final rule.

Several of the commenters believe that FRA's estimate of retrofit costs is too low. This is based on the commenters' estimates of the number of cars required to be retrofitted. However, these commenters did not provide FRA with details as to the source of their information or the methods by which they calculated their estimates. FRA can only reiterate the methodology it employed in arriving at the estimate of 3,206 cars requiring retrofit.

FRA based its estimates on two tank car studies—"Characteristics of 103, 104 and 111 tank Cars," Arthur D. Little, Incorporated, 1981, and "Tank Car Study, Task 9—Additional Data Analysis," Dynatrend Incorporated, 1981. The starting point was the 21,378 type 105 and 111 cars identified in the Dynatrend study as being involved in the transportation of the subject hazardous materials. FRA then excluded those cars with a capacity under 18,500 U.S. gallons (15,688 cars). Of the remaining 5,690 cars, 2,616 were found to have adequate safety features, leaving 3,072 cars which would require the retrofit. Forty-six of these cars were judged to have a value too low to justify the cost of the retrofit, and were therefore assumed to be retired or placed in other service. (The cost of purchasing new cars to replace those to be retired is included in the total cost estimate of the retrofit program.)

Several opponents of the proposed rule thought FRA and MTB gave inadequate consideration in its estimation of the benefits to be derived from the additional safeguards to the protection afforded by the double shelf couplers. We do not agree. In preparing its estimation of the incidents/accidents which could be prevented by the proposed requirements, FRA reduced the expected benefits to eliminate those benefits attributable to double shelf couplers. In other words, the estimated HM-175 benefits are in addition to the benefits expected from the couplers.

According to a study by the AAR (Phase 02 Report on Effectiveness of Shelf Couplers, Head Shields and Thermal Shields, Supplement AARR-482, August 20, 1981), the HM-144 retrofit of Class 112 (114) tank cars with double shelf couplers, head shields, and thermal protection has proven to be approximately 95 percent effective in preventing head punctures and ruptures due to fire. Since the HM-144 retrofit for 112/114 cars is very similar to the proposed retrofit for 105/111 cars, this effectiveness rate could have been used in our HM-175 analysis if double shelf couplers had not already been installed on 105 and 111 cars. FRA believes 80 percent to be a reasonable estimate of effectiveness for thermal head protection beyond that protection provided by the couplers. This effectiveness rate would have to drop to 51.5 percent before there would be a break even benefit/cost ratio, even ignoring the chances of a catastrophic event.

Indeed, the estimate of benefits does not include the possibility that a major catastrophic event may be prevented, even though at least one catastrophic accident involving a 105/111 car carrying hazardous materials would be likely in the absence of the HM-175 requirements. Had the analysis included the likelihood of a catastrophic accident, the benefit/cost ratios of all alternatives would have increased substantially. This is especially so in light of the fact that HM-175 deals with the larger (over 18,500 U.S. gallon) tank cars which, because of both size and rate of utilization, would be most likely to be involved in a catastrophic event.

Several commenters recommended that this rule be postponed to permit further testing and evaluation of the performance of existing specification 105 tank cars in accident situations. MTB and FRA do not concur with this recommendation. Since 1970, FRA has sponsored an extensive research program on the performance of flammable gas tank cars in simulated accident situations. The Railway Progress Institute (RPI) and the AAR have also sponsored their own research program on flammable gas tank cars, and FRA, the RPI, and the AAR have collaborated on many projects.

Most of the tests in these programs utilized specification 112/114 tank cars. However, because of the similarities between flammable gas specification 112/114 cars and flammable gas specification 105 cars, FRA does not believe that the results would have been significantly different.

FRA did conduct fire simulation tests of representative insulation systems used in existing specification 105 tank cars and concluded that, while these systems provide more protection than is present on uninsulated specification 112/114 cars, the protection is much less than now required by 49 CFR 179.105-4 for specification 112/114 cars. Section 179.105-4 requires that a thermal shield provide sufficient protection so that the back face temperature of a test plate not exceed 800° F in a 100 minute pool fire simulation. By contrast, most of the existing insulation systems on specification 105 cars reached 800° F in only 20–40 minutes of testing. While the time period is longer than for the uninsulated 112 and 114 tank cars prior to being retrofitted, the difference is not considered significant enough to reduce the benefits expected from a high temperature thermal retrofit of specification 105 tank cars.

Instead of repeating its past testing program on flammable gas tank cars to address specification 105 tank cars, FRA plans to focus its limited research resources on determining what, if any, additional protection is required for materials and tank cars not addressed in Dockets HM-144, HM-174, and HM-175.

Several commenters recommended that no changes in safety valve sizes on existing specification 105 tank cars be mandated until MTB and FRA analyzed an AAR report entitled, "A Study of Pressure Tank Car Safety Relief Valve Sizing Requirements." MTB and FRA have analyzed that report, both for this rulemaking and in the related docket, HM-174. A detailed assessment of the AAR report is in the docket. A detailed discussion of the report is included in the preamble to Amendment No. 173–172, 179–84, Docket HM-174, which is published in today's Federal Register.

Based on our analysis of the AAR report and on an independent study of safety valve sizing sponsored by the FRA and conducted by the IIT Research Institute (IITRI), MTB and FRA conclude that the valves on existing 105 and 111 tank cars carrying flammable gases and ethylene oxide are not adequate even if the tank cars are equipped with a thermal protection system that results in a maximum temperature of 800°F in a 100-minute simulated pool fire test. However, based on the IITRI calculations, MTB and FRA are allowing an option whereby additional thermal protection can be provided so that the currently sized safety valves on all existing specification 105 and 111 tank cars carrying flammable gases or ethylene oxide may continue in use.

One commenter requested that an exception be made for its specification 105 anhydrous ammonia tank cars that have thick (3/8″ or 1/2″) head jackets. These jackets apparently do not satisfy the requirements of either §179.100–23 or §179.105–5. The commenter provided insufficient information (e.g., type of steel used for the jacket) to enable MTB and FRA to evaluate the merits of this request.

SECTION-BY-SECTION ANALYSIS

Section 173.124 Ethylene Oxide. Paragraph (a)(5) of §173.324 is amended to require that each specification 105 tank car built before September 1, 1981, with a capacity in excess of 18,500 U.S. gallons, conform to specification 105J when transporting ethylene oxide after December 31, 1986. (As a result of earlier actions taken in Docket HM-174, specification 105 tank cars built after August 31, 1981, are currently required to have tank head and high temperature thermal insulation when transporting ethylene oxide.) Requiring a specification 105J tank car for ethylene oxide means that by December 31, 1986, existing specification 105 tank cars is excess of 18,500 U.S. gallons must be retrofitted with high temperature thermal protection, tank head protection, and larger capacity safety relief valves (or additional thermal protection).

Paragraph (a)(5) is also amended to require that each specification 111 tank car, with capacity in excess of 18,500 U.S. gallons, conform to specification 111J when transporting ethylene oxide after December 31, 1986. thus, by December 31, 1986, each existing large capacity specification 111 tank car in ethylene oxide service must be retrofitted with high temperature thermal protection, tank head protection, and larger safety valves (or additional thermal protection). A new subparagraph (a)(5)(v) is added to specify that specification 111 tank cars built after March 1, 1984 are not permitted to transport ethylene oxide.

Section 173.314 Requirements for Compressed Gases in Tank Cars. This section is amended to require that existing specification 105 tank cars (those built prior to September 1, 1981) used to transport anhydrous ammonia, and with a capacity exceeding 18,500 U.S. gallons capacity, be retrofitted by December 31, 1986, with lower half tank head protection, i.e., conform to specification 105S. The final rule further requires that existing specification 105 tank cars with a capacity exceeding 18,500 U.S. gallons, used to transport flammable gases, be retrofitted by December 31, 1986 to conform to specification 105J. Consistent with the proposed rule, the final rule also requires that by December 31, 1986, each specification 111 tank car with a capacity exceeding 18,500 U.S. gallons, used to transport flammable gases, shall conform to specification 111J. In response to a comment by AAR, the final rule includes a sentence in "Note 23" that provides that specification 111 tank cars built after March 1, 1984 are not authorized to transport flammable gases.

Section 179.102–12 Ethylene Oxide. Section 179.102–12 is amended to require that each existing specification 105 tank car (built prior to September 1, 1981, and with a capacity exceeding 18,500 gallons) used to transport ethylene oxide be retrofitted by December 31, 1986, with high temperature thermal protection, tank head protection, and a safety valve sized in accordance with §179.105–7 if it is to continue in ethylene oxide service. The safety valve sizing requirement means that either a large capacity valve or additional high temperature insulation must be installed.

Section 179.105–7 Safety Relief Valves. Section 179.105–7 is amended by adding paragraph (d) to permit continued use of the currently installed valve on specification 105 cars transporting flammable gases if the thermal protection exceeds the minimum thermal protection required in §179.105–4. This provision as proposed has been modified in several respects. First, the performance requirement for the additional thermal protection has been revised to require that, in the simulation pool fire tests required in §179.105–4, none of the thermocouples on the uninsulated side of the steel plates indicate a plate temperature in excess of 550° F (instead of 540° F in the proposed rule). This minor change reflects additional data developed by FRA after publication of the notice of proposed rulemaking.

Second, all existing specification 105 and 111 tank cars carrying flammable gases and ethylene oxide (instead of only those cars with a manway cover of less than 18 inches diameter as proposed) will be allowed to use the option of additional thermal protection in lieu of a larger capacity safety relief valve. This change has been made to give tank car owners additional flexibility in satisfying the safety objectives of this rulemaking.

Third, the option to use additional thermal insulation instead of using a larger capacity safety relief valve is limited in paragraph (d) to cars transporting flammable gases. This has been done because today's amendment to the final rule in Docket HM-174 provides a similar option specifically developed for cars transporting ethylene oxide. Under that provision (§179.105–7(c)), the use of 550° F material permits a safety valve sized with a flow capacity as low as 1100 scfm at 85 psi, which corresponds to the currently utilized valve.

Fourth, the final rule permits the use of the currently installed valve if the additional thermal insulation is provided, rather than the use of a valve sized in accordance with the formula for compressed gases in insulated tanks. This change has been made to make clear that the currently installed valves are acceptable. The change has also been made because most valves on the tank cars are sized with a capacity greater than the minimum capacity required under the formula. FRA wants to make clear that the valve capacity may not be reduced below its current level. (Nothing in the rule would preclude increasing the capacity of the safety relief valve.)

Section 179.106–1 General. Section 179.106–1 is amended to require that existing specification 105 tank cars manufactured to the specifications of the Canadian Transport Commission conform to the same standards prescribed for DOT specification 105 tank cars.

Section 179.106–3 Previously Build Cars. Section 179.106–3 is amended to establish performance requirements for specification 105S and 105J tank cars built before September 1, 1981. The requirements for the 105S and 105J tank cars in this section are identical to the requirements in §179.106–2 for new cars.

Section 179.200–27 Alternative Requirements for Tank Head Puncture Resistance Systems. This section is added to clarify that specification 111 tank cars may utilize a head shield as prescribed in §179.100–23 instead of meeting the puncture resistance requirements in §179.105–5.

Section 179.202–18 Ethylene Oxide. Paragraph (a)(10) is added in §179.202–18 to require that each specification 111 tank car used after December 31, 1986, for the transportation of ethylene oxide, with a capacity exceeding 18,500 U.S. gallons, conform to class 111J. Paragraph (a)(11), though not included in the NPRM, is also added. It specifies that specification 111 tank cars built after March 1, 1984, are not permitted for the transportation of ethylene oxide. This addition merely reflects the previous action of the AAR Tank Car Committee and is added in response to the AAR's comment.

Section 179.203 Special Requirements for Specification 111 Tank Cars. The final rule adds §179.203 which sets out special requirements for specification 111 tank cars that parallels section 179.106 for specification 105 tank cars.

One change from the proposed rule is the deletion of the words "before October 1, 1981" from paragraph (d). Paragraph (d) requires that specification 111 tank cars built to specifications promulgated by the Canadian Transport Commission must be equipped in accordance with §179.203–2 by December 31, 1986. The change means that all specification 111 cars over 18,500 U.S. gallons transporting flammable gases or ethylene oxide, after December 31, 1986, must conform to specification 111J. Also, a paragraph (e) is added to specify that specification 111 tank cars built after March 1, 1984 are not permitted for the transportation of flammable gases or ethylene oxide.

(49 F.R. 3468, January 27, 1984)

[49 CFR Parts 173 and 179]
[Docket No. HM-175]
Specifications for Tank Cars; Response to Petitions

MTB received three petitions for reconsideration of the final rule issued in Docket HM-175 (49 FR 3468. Jan. 27, 1984). The petitioners are Dow Chemical Company (Dow), Mallard Transportation Company (Mallard), and the AAR.

The final rule in HM-175 made changes in the construction and maintenance standards for certain railroad tank cars used to transport hazardous materials. The changes are as follows:

(1) After December 31, 1986, DOT specification 105 tank cars built before September 1, 1981, that have a capacity exceeding 18,500 U.S. gallons and are carrying a flammable gas, anhydrous ammonia, or ethylene oxide must be equipped with lower half tank head protection (such as a head shield);

(2) After December 31, 1986, DOT specification 105 tank cars built before September 1, 1981, that have a capacity exceeding 18,500 U.S. gallons and are carrying a flammable gas or ethylene oxide must be equipped with either: (a) High temperature thermal insulation (800° F material) and safety relief valves sized according to the requirements for specification 112 and 114 tank cars, or (b) high temperature thermal insulation (550°F material) and currently installed safety relief valves; and

(3) After December 31, 1986, DOT specification 111 tank cars that have a capacity exceeding 18,500 U.S. gallons and are carrying a flammable gas or ethylene oxide must be equipped with lower half head protection and either (a) high temperature thermal insulation (800° F material) and safety relief valves sized according to the requirements for specification 112 and 114 tank cars, or (b) high temperature thermal insulation (500°F material) and currently installed safety valves.

Since the main concerns of each petitioner varied from the other petitioners, each petition was separately considered.

Dow's Petition.   Dow submitted a one-page petition for reconsideration stating that "the requirements and compliance schedule of subject rulemaking are unreasonable and premature." The basis of or the "unreasonableness" of the final rule according to Dow, is that an estimated 40 cars of its affected fleet of 883 cars would be out of service for retrofitting at any given time during the retrofit period. This five percent average reduction in its available fleet during the next several years is unreasonable in Dow's view.

MTB and FRA are not persuaded by Dow's argument on the reasonableness of the final rule. First, Dow did not submit data indicating the utilization rate of its affected fleet. Thus, there is no evidence that the five percent reduction in the available fleet would present any actual problem.

Second, based on national traffic statistics covering the entire tank car fleet transporting the materials covered by the final rule, there appears to be substantial traffic volume fluctuations on a month-to-month basis. MTB and FRA believe that Dow should be able to schedule retrofitting during the periods of low traffic so as to substantially reduce or eliminate any adverse impact. A more complete analysis of the impact of traffic volume fluctuations is included in an economic evaluation of the petitions for reconsideration, which is in the docket.

Third, even assuming a marginal reduction in Dow's available fleet during the retrofit period, there was no information included in the petition to enable MTB and FRA to weigh the potential adverse impact to Dow as compared to added safety benefits of a prompt retrofit schedule.

Finally, even if more complete information from Dow indicated that the retrofit schedule presented a serious problem for Dow, as the owner of a major portion of the cars affected by the final rule, the proper way to proceed would be to address Dow's specific needs and not to revise the basic rule.

Dow's second contention is that the rule is premature. Dow argues that there are not any approved 550° F thermal protection systems nor any off-the-shelf large capacity valves designed for ethylene oxide. MTB and FRA are not persuaded by this contention. As is often the case, a specific requirement creates the necessary market for product testing and development. Subsequent to Dow's petition, MTB published a revised list of excepted thermal protection systems (49 FR 33524, Aug. 23, 1984). The list included five thermal protection systems that meet the 550° F standard. With respect to a large capacity valve for ethylene oxide, MTB and FRa are not aware of any bona fide request to a valve manufacturer for the construction of such a valve. If a timely order is made and a valve cannot be manufactured within the retrofit period, MTB and FRA will consider an extension of the deadline.

Accordingly, Dow's petition for reconsideration is denied.

Mallard's Petition.   Mallard submitted a two-page petition. Mallard's basic contention is that the final rule is excessively costly for Mallard to comply with. However, the petition did not attempt to rebut FRA's extensive benefit/cost analysis included in the docket. Nor did Mallard argue that the rule as a whole is not beneficial. Rather, the petition alleges that the Mallard Transportation Company is a small business under the Regulatory Flexibility Act (RFA) and that it would be financially hurt by the rule. Mallard's petition states "we will spend monies that will never be recovered."

MTB is denying Mallard's petition for several reasons. First and foremost is that, whatever the ultimate merits of Mallard's contention of unreasonable economic harm, the alleged economic harm is peculiar to Mallard and thus is not a basis for revising the rule generally.

Moreover, neither MTB nor FRA believe Mallard has yet made a case of significant economic injury to it as a small business. First, Mallard is a leasing company that owns approximately 220 tank cars, not an inconsequential asset base. While the petition does not provide enough economic information about Mallard Transportation Company to reach a final determination, it is not clear that Mallard would qualify as a small business under the RFA.

Second, the basic purpose of the RFA is to provide special treatment for small business in those cases where uniform treatment of all business, regardless of size, would actually produce disproportionate burdens on small businesses that may adversely affect competition in the marketplace, discourage innovation, or restrict improvements in productivity. This is not the case with the final rule in Docket HM-175 since the cost burdens imposed are not related to the size of the business. Rather, the cost burdens are purely marginal in nature because they are directly proportional to the number of relevant cars owned by a company. A more complete discussion of this issue is included in an economic evaluation of the petitions for reconsideration that is in the docket.

Finally, Mallard has not shown the degree of adverse economic impact to enable MTB and FRA to assess that impact against the safety benefits attributable to the retrofit. Thus, while it is true that Mallard may have to spend approximately $150,000 to retrofit 10 cars, there is insufficient data to determine the potential hardship that the expenditure would cause. MTB and FRA can consider further Mallard's individual situation at such time as additional information is provided.

AAR's Petition.   The AAR submitted a 44-page petition for reconsideration (including attachments). The petition addresses the single issue of safety valve sizing. In addition to the 44-page petition itself, the AAR's analysis involves references to numerous studies, computer programs, and technical reports involving hundreds of pages of highly technical material. The discussion in this notice of the AAR's petition, therefore, is summary in nature. A technical analysis prepared to FRA of the AAR's petition for reconsideration in Docket HM-175 is entered in the docket.

The disagreement between the AAR and the Department of Transportation (DOT) concerning safety valve sizing is longstanding. The AAR contested the valve sizing approach adopted in Docket HM-144 (42 FR 46306, Sept. 15, 1977) for DOT specifications 112 and 114 tank cars. AAR restated its objections in Docket HM-174 (46 FR 8005, Jan. 26, 1981), which involves new construction of DOT specification 105 tank cars. As a result of the AAR's petition for reconsideration of the final rule in HM-174, MTB postponed the compliance date for installing the large capacity safety relief valve on the new construction of DOT specification 105 tank cars built to transport ethylene oxide from September 1, 1981, until March 1, 1984. During that period the AAR prepared a comprehensive study of safety valve sizing. At the same time, FRA was continuing its longstanding research effort on the safety valve sizing issue.

The contentions raised by AAR in its study submitted to the docket in HM-174 have been previously addressed by MTB and FRA. A summary of the MTB and FRA position is included in the preamble discussion to the amendment of the final rule in HM-174 published on January 27, 1984 (49 FR 3473) and a detailed response is included in the docket. The amendment of the final rule was made in response to the AAR's petition for reconsideration in Docket HM-174.

The petition for reconsideration in Docket HM-175 is essentially a request to address once again the AAR's contentions addressed in the HM-174 rulemaking. (The AAR's petition for reconsideration of the final rule in HM-175 also requested another reconsideration of the actions taken in HM-174. The procedural validity of the request need not be addressed since resolution of the technical issues as it affects Docket HM-175 effectively disposes of the identical technical issues in Docket HM-174.) Indeed, the AAR's petition does not raise new arguments about the safety valve sizing issue, but it does contain additional data and analysis in support of the arguments raised in its earlier study.

MTB and FRA thoroughly reviewed the AAR's petition for reconsideration in HM-175 and conclude that it does not contain data or analysis that could cause a change in the conclusions reached in responding to the AAR's petition for reconsideration of the final rule in HM-174. The longstanding disagreement reflects the technical complexity involved in the question of safety valve sizing. It also reflects the reality that totally clear cut answers to the many subcomponents of the analytical framework do not exist. Extrapolation from limited data, mathematical simplification of complex physical phenomena, use of data based on experiments involving an entirely different scale (laboratory testing as opposed to full-scale testing), and other analytical difficulties characterize the process of determining the appropriate valve size.

While the AAR and FRA have "nits" to pick about each other's computer program and analytical approach, the critical differences reflect differing judgments about how to deal with uncertainty in the data and about what constitutes the proper level of safety. The fundamental difference between FRA and the AAR continues to be the fire environment that tank cars should be expected to withstand. The AAR petition proposes that tank cars only be required to withstand what the AAR denotes as "uncontrolled fires," whereas FRA believes that they should withstand more severe fires, what the AAR denotes as "catastrophic fires." Similarly, FRA and the AAR differ on whether there is a potential for total tank fire engulfment (FRA) or only a one quarter portion of the tank engulfed (AAR).

Obviously, FRA and the AAR continue to have an honest disagreement, reflecting both a differing assessment of research and technical literature in the field, and a different determination of the appropriate margin of safety. One thing is clear. As recently as ten years ago, before the adoption of the safety criteria in issue (800° F high temperature thermal insulation *and* a large capacity safety relief valve, or 550° F insulation), it was not uncommon for railroad tank cars transporting flammable gases to rupture violently as a result of being exposed to fire. The consequences of a thermally induced rupture of such a car can be catastrophic in terms of loss of life and property damage. Since adoption of the safety criteria, beginning in Docket HM-144 and now including Docket HM-174 and Docket HM-175, that accident experience has been virtually eliminated. While the accident reduction might have occurred without requiring a large capacity safety relief valve in addition to high temperature thermal insulation (800° F material), it is far from certain that the reduction would have occurred.

Since it is our view that the proposal of the AAR petition to amend the final rule to size safety valves in accordance with the AAR's study pose unnecessary and unacceptable safety risks, the petition is denied.

(49 F.R. 43963, November 1, 1984)

49 CFR Parts 100 through 199
[Docket HM-177]
Public Hearing and Request for Comment on Trailer-on-Flatcar
Transportation of Hazardous Materials

On December 11, 1978, an NPRM was published in the Federal Register (43 FR 58050) under Docket HM-167. Comments were received from the Association of American Railroads (AAR) concerning TOFC service for the transportation of hazardous materials in intermodal tanks. The AAR believes that there is an increased risk associated with TOFC securement, i.e., the securement of the portable tank to the motor vehicle chassis and the securement of the motor vehicle chassis to the flatcar.

Additionally, the AAR believes that ".... the combined center of gravity for the flatcar, chassis and container is approximately 139″ and this grossly exceeds the 98″ maximum center of gravity for freight cars allowed by paragraph 2.1.3, AAR Specification M–1002."

Even though the AAR did not submit any data, calculations, or test results to support its position, the MTB believes the AAR's views should be given further consideration before it makes a final decision concerning the transportation of tank containers in TOFC service. Also, the MTB recognizes that, in view of the AAR's references to a 98″ maximum center of gravity, the entire matter of transportation of hazardous materials in TOFC service should be examined to determine if a rulemaking proposal should be initiated under this Docket. This examination should include semitrailers (vans) and freight containers mounted on chassis as well as tank containers.

MTB is particularly interested in obtaining comments and information concerning the following factors that should be addressed in relation to TOFC operations.

(1) The current manner in which TOFC rail cars and other car types having center of gravities (when loaded) in excess of 98″ are handled to ensure adequate safety. What special requirements and/or procedures are imposed?

(2) The extent which supplemental snubbing and/or hydraulic stabilizers can improve the dynamic performance (car roll angle, side bearing loading, spring motion, vertical loading fluctuation) of high center of gravity cars. If such control devices are effective, can the center of gravity limitation be raised and still have the same level of safety performance?

(3) The contribution of track cross level variations and/or distance between rail joints in causing or exaggerating rock and roll in high center of gravity cars, and in TOFC loads in particular.

(4) Evaluations as to the effectiveness of operational changes (i.e., speed restrictions, humping limitations, route selection, etc.) in countering the adverse effects of high center of gravity loaded cars. Is there a set of operating conditions wherein high center of gravity loads, including intermodal tanks can be safely moved in TOFC service? At what additional cost?

(5) The trade offs and options which are important factors in determining and setting center of gravity restrictions and/or limits. To what degree is the hazard of the cargo a controlling consideration?

(6) Beyond center of gravity influences, the other factors which must be taken into account when assessing the safety of movement by TOFC. What components of operation are unique to TOFC service?

(7) The extent to which improvement in securement, end of car cushioning, better loading/unloading methods, etc., can reduce concern for the safety of hazardous materials in TOFC service. Can the securement of the portable tank to the chassis and the chassis to the flatcar be made adequate for a realistic railroad environment? What combination of improvements can make such TOFC service safe?

(8) An enumeration of special requirements which are recommended for transport of hazardous materials in TOFC service but which are not applicable for general TOFC movements. What additional requirements can be justified for the transport of hazardous materials? For example, should stacking of certain packagings (e.g. double decking of drums) be prohibited?

(9) The past shipping experience with hazardous material movement in TOFC service. Aside from incidents (involving unintentional releases) reported to MTB, what has been the accident history vs. the total number of shipments made? Do some railroads tend to have more problems related to such movements than others?

(10) Testing which has been performed, or could be performed to measure the current safety level of hazardous materials in TOFC service, and which could be used to evaluate countermeasure improvements. What are the results of past testing? What are the recommendations for additional testing to prove or disprove various contentions? How should such testing be performed and who should do the testing?

Interested persons are invited to participate in the hearing. Persons intending to present oral statements for the record should advise the information contact mentioned earlier in this Notice.

While unsupported views and opinions will be accepted, information as requested above in the form of data, calculations or concerning accident experience and test results would be most useful. In particular, the MTB invites the AAR to provide data and calculations supportive of its 98″ maximum center of gravity limitation.

(46 F.R. 8055, January 26, 1981)

49 CFR Parts 100 through 199
[Docket No. HM-177]
Trailer-on-Flatcar Transportation of Hazardous Materials

On January 26, 1981, MTB published in the **Federal Register** a notice announcing a public hearing and soliciting comments, data, and test results on Trailer-on-Flatcar (TOFC) securement and the effect of a high center of gravity on the safe transportation of hazardous materials in TOFC service (46 FR 8055). In the Notice, MTB requested that commenters address certain specific factors in relation to TOFC operations, including semitrailers (vans) and freight containers mounted on a chassis.

The Notice was in response to comments concerning the securement and center of gravity effect that were received from the Association of American Railroads (AAR) under Docket HM-167 "Intermodal Portable Tanks," and AAR's petitions for reconsideration of the issuance and renewal of certain DOT exemptions authorizing portable tanks in TOFC service. In its comments addressing an application for exemption, AAR stated that it believed there was an increased risk associated with TOFC securements, i.e., the securement of the portable tank to the motor vehicle chassis and the securement of the motor vehicle chassis to the flatcar. In addition, AAR stated that " . . . the combined center of gravity for the flatcar, chassis, and container is approximately 139″ and this grossly exceeds the 98″ maximum center of gravity for freight cars allowed by paragraph 2.1.3, AAR Specification M-1002." AAR did not submit any calculations or test data to support its position. However, MTB felt that perhaps some of AAR's assertions had merit. Therefore, MTB prohibited TOFC service in the final rule in Docket HM-167 as an interim measure, and made TOFC service a separate new rulemaking action under Docket HM-177.

Three persons testified at the public hearing under Docket HM-177 which was held in Rosemont, Illinois on February 25, 1981. A representative of AAR stated that AAR took no exception to TOFC operations for intermodal portable tanks, retracting its previous position. The other two commenters did not present any technical information directly related to the factors listed in the Notice. Six written comments were received to the docket. In general, the commenters who addressed the factors listed in the Notice agreed that the combined center of gravity of a flatcar, chassis and portable tank does not exceed the 98″ maximum specified for interchange service. Furthermore, the commenters indicated that they had no shipping experience that would indicate that "rock and roll" and double-interfacing securement systems presented any safety problem. No comments have been received since the April 2, 1981, closing date.

Subsequent to the public hearing, AAR submitted comments withdrawing its opposition to TOFC operations for intermodal portable tanks under Docket HM-177 and requested dismissal of the docket. After AAR withdrew its opposition, MTB could see no reason for not permitting TOFC service. An amendment subjecting TOFC service to the same approval requirement as required for container-on-flat car (COFC) service was published in the Federal Register under Docket HM-167 on April 30, 1981 (46 FR 24185).

In view of the foregoing, MTB concludes that adoption of a formal standard to further limit or restrict hazardous materials transportation in TOFC service is not justified. Docket HM-177 is hereby terminated.

(47 F.R. 30800, July 15, 1982)

49 CFR Part 173
[Docket HM-178; Advance Notice]
Definition of Flammable Solid

The flammable solid class has within it the most diversified types of hazardous materials of any of the hazard classes defined in the Department's Hazardous Materials Regulations. For this reason, the MTB is considering dividing the class into several groups and is proposing test methods and criteria which will enable a shipper to determine whether a material he wishes to ship falls within one of those groups. The principal reason for this advance notice of proposed rulemaking is to request comments from interested persons as to the adequacy of the definition criteria under consideration in this ANPRM including the reproducibility of the results of the tests in Appendix B. Specific questions on which the MTB would like to receive comments and meaningful data are:

1. Are there any additional types of materials which should be included in the flammable solids class (e.g., should the temperature in (g) be lowered to include materials such as molten sulfur)? Should some of the proposed groups be eliminated?

2. If the definition in (g) is adopted, certain molten metals would become subject to the regulations. What type of packaging controls should the MTB consider for such materials. (The MTB has been contacted by officials of two States concerning the regulation of molten materials).

3. Current §171.8, as pertinent here, defines "Water reactive material (solid)" as a material, that on contact with water, will evolve flammable or toxic gases in dangerous quantities. Is there any substance known which, on contact with water, will evolve a toxic nonflammable gas? If no such material is known to exist, water reactive materials would be included in paragraph (f) of the proposed definition.

4. Are there any consensus standard test methods which could be used in place of, or in addition to, the suggested methods?

For tests, such as the bacterial action or fermentation test, the MTB does not have specific information regarding sample sizes and temperature limitations. It is, therefore, requested that persons with experience with this type of spontaneous heating provide the MTB with pertinent information regarding development of such criteria.

The MTB anticipates that a number of highly competent and qualified experts will, upon review of Appendix B, consider the proposed methods for testing of flammable solids to be less than representative of currently available technology. This is intentional. The MTB has attempted to develop methods that would not require the acquisition of expensive and complicated test equipment. For example, metal drums could be used as test chambers and ovens with minor modifications.

This advanced notice of proposed rulemaking requests comments from interested persons regarding the definition of a flammable solid. Such comments may be used for future rulemaking purposes.

In consideration of the foregoing, the MTB is considering the issuance of a proposal to revise the flammable solid definition as follows:

§173.150 Flammable solid: definition. For the purpose of this subchapter, a "Flammable Solid," is any solid material, including gels and pastes, other than one classed as an explosive or a blasting agent, which is described in the following paragraphs:

(a) Pyrophoric solids which ignite when exposed to moist air at or below 55° C (130° F).

(b) Solids subject to spontaneous heating by reaction with oxygen and which contain unsaturated oils or other easily oxidizable substances.

(c) Solids subject to spontaneous heating by fermentation or bacterial action and which self-heat due to the action of bacteria or other organisms.

(d) Readily ignitable solids which are easily ignited and burn so vigorously and persistently as to create a hazard in transportation.

(e) Solids which can be ignited by friction.

(f) Solids which in contact with water evolve flammable gases.

(g) Solids or molten materials shipped at (elevated) temperatures exceeding 315° C (600° F), which can cause ignition of combustible materials.

Tests to be used to evaluate the above descriptions are found in Appendix B to this Part.

APPENDIX B: METHODS FOR TESTING FOR FLAMMABLE SOLIDS

Pyrophoric Solids. At least on pound of material, in the particle size in which it will be shipped, shall be placed in an apparatus where the temperature, humidity and rate of air flow can be controlled. Air at 55° C (130° F) and having a relative humidity of 50 percent (± 5 percent) shall be passed at 8 kilometers per hour through the apparatus containing the material. A material that ignites in one hour in this test is classed as a flammable solid and considered a pyrophoric material.

Solids Subject to Heating by Reaction With Oxygen. The material, other than a gel or paste, in the particle size in which it will be shipped, shall be placed in a constant temperature oven having a test chamber with a volume of not less than 120 liters. The temperature control of the oven shall be set so as to maintain a temperature of 55° C (130° F) ±3° C in the oven when it is empty. Fifty kilograms of the material to be tested, or an amount occupying 50 percent of the volume of the test chamber, whichever is less, shall be placed in a holder within the oven. The holder shall be made of wire mesh with a mesh size no smaller than necessary to contain the materials and shall be supported in the geometrical center of the oven so as to clear its bottom and sides by no less than 5 centimeters. Air shall be introduced into the bottom of the test chamber at a rate, in liters per hour, no greater than 3 percent, nor less than 2 percent, of the volume of the test chamber. If the temperature at the center of the sample rises 20° C (36° F) or more within 7 days of initiation of the test, the material is classed as a flammable solid.

Solid subject to Heating by Fermentation or Bacterial Action. The temperature control of an oven is set to maintain a temperature of 37.8° C (100° F) when the oven is empty. One pound of material under test shall be in the oven for 7 days. If the temperature at the center of the sample rises 20° C (36° F) or higher during this period, the material is classed as a flammable solid.

Readily ignitable solids which burn vigorously and persistently. Ease of ignition shall be determined by attempting to ignite 10 grams of the material under test, arranged in a conical (when practical) pile with:

1. Mechanical (metal) sparks
2. Electrostatic sparks (0.006 joules delivered from a 0.002 to 0.004 microfarad capacitor)
3. A small flame source such as a match.

A material which can be ignited by one of these sources must be considered readily ignitable. Whether the material burns vigorously and persistently shall be determined as follows: when 50 ml of water are applied to a 50 g of a material which is burning and the water produces no effect or increases the rate of burning, that material is considered to burn so vigorously and persistently as to create a hazard in transportation. A material which is readily ignitable and burns vigorously as defined in one of the above tests is classes as a flammable solid.

Solids Which Can Be Ignited by Friction. A solid which is as sensitive or more sensitive to friction than phosphorous pentasulfide is classed as a flammable solid. Tests may be conducted on any commercial friction tester or using a laboratory constructed apparatus which gives comparable results.

Solids Which Emit Flammable Gases in Contact With Water. Twenty-five grams of the substance being tested are placed in a gas generator and treated with 50 ml of water. Any material which evolves a flammable gas at a rate exceeding one ml per gram per hour must be classed as a flammable solid.

(46 F.R. 25492, May 7, 1981)

49 CFR Part 173
[Docket HM-179 Advance Notice]
Definition of Oxidizer

For purposes of definition, oxidizers have been divided into two groups, liquids and solids. Test methods and criteria are proposed which will enable a shipper to determine whether a material he wishes to ship meets the definition of an oxidizer. The principal reasons for this advance notice of proposed rulemaking are to request comments from interested persons as to the adequacy of the proposal and to solicit comments concerning alternative methods. Relative to the latter purpose, MTB would like to know if there are any consensus standard test methods which could be used in place of, or in addition to, the suggested methods?

The MTB anticipates that a number of highly competent and qualified experts will, upon review of Appendix B, consider the proposed methods for testing of oxidizers to be less than representative of currently available technology. This is intentional. The MTB has attempted to develop methods that would not require the acquisition of expensive and complicated test equipment.

In consideration of the foregoing, the MTB is considering the issuance of a proposal to revise §173.151 and add an Appendix C to Part 173 of Title 49, Code of Federal Regulations as follows:

§173.151 Oxidizer; definition. For the purpose of this subchapter, an "Oxidizer" is a material which may cause the ignition of combustible materials without the aid of an external source of ignition or when mixed with combustible materials, increases the rate of burning of these materials when the mixtures are ignited. Oxidizers are divided into two general groups:

(a) Solid Oxidizers. A solid oxidizer is a solid substance which accelerates the burning rate of dry wood sawdust as much or more than ammonium persulfate, when tested in accordance with the method described in Appendix C of this part.

(b) Liquid oxidizers. A liquid oxidizer is a liquid substance which when tested in accordance with the method described in Appendix C of this part, will initiate a fire in the test container.

APPENDIX C: METHODS FOR TESTING FOR OXIDIZERS

Solid Oxidizers. The test method for solid oxidizers as prescribed in paragraphs 1 through 5 measures the potential of a solid substance to increase the burning rate of a combustible material when the two are intimately mixed.

1. Ammonium persulfate shall be the reference material. It shall pass through a No. 100 Sieve in the U.S. Sieve Series (ASTM-E-11-61) and shall contain less than 5% by weight of water.

2. Wood sawdust shall be the combustible material in this test and shall pass through a NO. 50 Sieve in the U.S. Sieve Series (ASTM E-11-61) and shall contain less than 5% by weight of water.

3. A five pound mixture of ammonium persulfate and wood sawdust shall be prepared in a 1 to 1 ratio by weight. A second five pound mixture of the material to be tested, in the particle size in which it will be shipped, and wood sawdust shall be prepared in a 1 to 1 ratio by weight. Both of these mixtures should be mixed as thoroughly as possible.

4. The two mixtures shall be formed into conical piles on incombustible surfaces and the edge of both piles ignited (at one place) simultaneously with laboratory burners or flares. The ignition sources shall be applied until piles are well started and then removed.

5. A substance which causes the sawdust to burn at a rate equal to or greater than the burning rate of the ammonium persulfate mixture shall be classed as an oxidizer.

Liquid Oxidizers. The test method for liquid oxidizers as prescribed in paragraphs 1 through 5 measures the potential of a liquid substance to initiate a fire when brought into contact with combustible materials.

1. The test shall be conducted in a box $2' \times 2' \times 2'$ constructed of natural wood (not plywood) having sides not more than ½″ thick. A ¼″ diameter hole shall be drilled at the horizontal center of each of the vertical side 4 inches from the bottom edge. A one inch diameter hole shall be drilled in the center top of the box.

2. The box shall be packed tightly with wood excelsior containing less than 5% of water and the top of the box properly secured in place.

3. Two liters of the liquid under test shall be poured into the box through the hole in the top of the box and, at the same time, the box shall be tipped slightly in different directions so as to distribute the liquid as uniformly as possible throughout the excelsior.

4. The box and its contents shall be allowed to stand for two hours.

5. A substance which causes the ignition of the excelsior shall be classed as a liquid oxidizer.

(46 F.R. 31294, June 15, 1981)

49 CFR Parts 172 and 174
[Docket HM-180 Advance Notice]
Placarding of Empty Tank Cars

MTB requests comments from interested persons concerning a petition to eliminate the requirement for display of EMPTY placards on tank cars. The petition is quoted as follows:

PETITION FOR RULEMAKING

Pursuant to the procedures specified in Part 106 of your regulations, the International Association of Fire Chiefs (IAFC) hereby petitions for removal of all references to EMPTY placards set forth in Title 49, Code of Federal Regulations, Parts 172 and 174.

Your regulations presently require than an EMPTY placard be displayed on a tank car that has been emptied but not cleaned and purged of all hazardous residues. This means that a significant quantity of a hazardous material may remain in a tank car while the placards displayed on the car indicated that it is empty. An EMPTY placard (or sign) may be useful to emergency personnel in making assessments in how to deal with accidents if the tank car bearing such a message is truly empty (cleaned and purged). However, THE PRESENT SYSTEM IS MISLEADING AND DANGEROUS and should be terminated.

We believe that many qualified experts will agree that, in many instances, a nearly empty tank car presents a potentially greater danger than a filled car. If the Federal Railroad Administration were to conduct a full-scale fire test on a 32,000-gallon propane car which is empty according to the law but full of propane vapor and possibly a few gallons of liquefied propane, what would be the result? If a car similarly containing chlorine were breached in an accident, what would be the result? Keep in mind the cars just described would be placarded EMPTY during their transportation.

We recognize that placards are used by the railroad industry to comply with the car placement and other hauling requirements of your regulations. However, we maintain that the system could be modified to accomplish those objectives and still provide for proper communication of the risks presented by the hazardous material in tank cars. For example, the words on shipping papers now reading "EMPTY Last contained . . ." could be changed to read "NEAR EMPTY" in recognition of the need for brevity in the entries on shipping documents. Some other indication on the placard could also be developed which would not give the false impression that the EMPTY one does.

An additional argument is our contention that the placarding system should be consistent. Tank trucks must remain placarded when empty unless cleaned and purged of all hazardous material residue or reloaded with a nonhazardous material, while rail tank cars can use the EMPTY placard with residue and vapors present. It is difficult for us to have an effective and comprehensive training program for hundreds of thousands of firefighters when major inconsistencies, such as demonstrated in this petition, ar permitted by your agency to exist.

Based on its evaluation of the comments received in response to this advance notice, MTB may issue a Notice of Proposed Rulemaking proposing to delete or modify the requirements for display of EMPTY placards.

(46 F.R. 37953, July 23, 1981)

49 CFR Parts 172, 173, and 174
[Docket No. HM-180 Notice No. 84–6]
Placarding of Empty Tank Cars

Prior to November 1927, the hazardous materials regulations required that when lading was removed from tank cars, placards be removed. The Interstate Commerce Commission issued an order in Docket 3666 on August 1, 1927, authorizing after November 1, 1927, on a voluntary basis, use of the "DANGEROUS-EMPTY" placard. It was a requirement at that time that each loaded tank car containing an "Inflammable" (liquid), a "Corrosive Liquid", "Compressed Gas," or "Poisonous" (liquid) had to have displayed on each end and each side the appropriate placard. Upon removal of the tank car contents (except residue) these placards had to be removed. The "DANGEROUS-EMPTY" placard as a voluntary display continued in use from November 1, 1927 to July 14, 1959 (Order 39, 24 FR 5641), when the "DANGEROUS-EMPTY FLAMMABLE POISON GAS" placard was made mandatory for tank cars containing residual Flammable Poison Gas. In July 1962 (Order 55, Docket 3666, July 6, 1962) a "POISON GAS-EMPTY" placard was established and display was required on each tank car containing the residue of a Poison Gas. Otherwise, the use of "DANGEROUS-EMPTY" placards on tank cars remained voluntary until July 1, 1977, the effective date of new placarding requirements under Docket HM-103 (41 FR 16131, April 15, 1976).

Rulemaking under Docket No. HM-103 (41 FR 15972) established requirements for placarding each transport vehicle and freight container with placards generally resembling the United Nation's hazard warning labels for dangerous goods in transportation. EMPTY placard requirements were established for tank cars, but different requirements were established for cargo tanks. The placards on a cargo tank motor vehicle must remain when it is empty unless it has been cleaned and purged of hazardous residue and vapor. Since 1976, comments have been received from emergency response personnel about the confusion caused by the two placarding systems for an "empty" cargo tank and an "empty" tank car. Also, comments have been received from rail carrier personnel about missing or lost tank car placards. Tank car placards generally are made of tagboard with the EMPTY display being printed on the reverse side of the hazard warning placard. These tagboard placards are loosely held in placard holders and may blow out or be removed. They must be replaced at an additional expense by carriers.

In June 1981, the International Association of Fire Chiefs (IAFC) petitioned for a rule change which was quoted by the MTB in an Advance Notice of Rulemaking, Docket No. HM-180 (46 FR 37953, July 23, 1981). The IAFC petition stressed the difference between the placarding requirements for "empty" cargo tanks and tank cars, and stated that use of the EMPTY placards on tank cars is "misleading and dangerous." Further, IAFC stated that cargo tanks and tank cars should be placarded in a consistent manner, that is, both should remain placarded when emptied unless cleaned and purged of all residue and vapor or reloaded with another material.

Most of the 52 comments received on the advance notice were from representatives of emergency services and industry. One was received from the Association of American Railroads and three were from rail carriers. Five were from city, state and federal agencies concerned with safe transportation of hazardous materials.

Approximately one-third of the comments from industry favored the retention of the EMPTY placard, stating that it is beneficial to emergency response personnel. Others believe the EMPTY placard is beneficial to the rail carriers in car placement activities. The majority of the rail carrier comments were in favor of retaining the EMPTY placard for car placement reasons. One large rail carrier, however, presented an opposing position indicating that computer generated instructions provide for the makeup of a train beginning with the initial loaded switch from the shipper at origin through the spotting for unloading at destination and the return of the empty tank car. Therefore, car placement and train makeup is not dependent upon the determination of the empty or loaded status of rail cars from the placards.

One industry commenter stated that in the course of business they receive "empty" tank cars and return "empty" tank cars between company facilities. This commenter supported the IAFC petition in that the EMPTY placard gave an erroneous message. They recommended, as a replacement, the hazard placard for the material with the word "RESIDUAL" in the lower triangle of the placard. This would continue the use of display panel with the hazard warning placard on one side but with RESIDUAL on the reverse side. This commenter also suggested the shipping paper notation be changed from "EMPTY" and "EMPTY LAST CONTAINED" to "RESIDUAL." MTB believes this recommendation to change the shipping papers description has merit.

Another chemical manufacturer recommended:

> [T]hat all references to the "EMPTY" placard for rail car be removed and the regulations be worded such that rail cars, containing residue or vapor of a hazardous material, must remain placarded with the proper placard required when the car was loaded. Cars containing non-hazardous materials or cars which have been cleaned of residue and purged of vapor would not be placarded, thereby precisely determining whether a hazard exists. For a placarded car, the shipping paper could be a reliable indicator of whether the rail car was loaded or empty.
>
> It has been our experience that there has always been confusion and potential for error in the use of the "EMPTY" placards. When empty cars return to our plants, we notice that the cars come back in one of four ways; (1) correctly placarded with the "EMPTY" placard, (2) still placarded as in the loaded movement, (3) a combination of loaded and "EMPTY" placards, or (4) with one or all placards missing. As can be seen, the use of the "EMPTY" placard is a hit-or-miss proposition whereby you depend on the consignee to reverse the loaded placard. We are sorry to say that not all consignees are as conscientious in complying with the regulations as we attempt to be.

One large chemical manufacturer who commented had submitted a petition (P-819) in April 1981, before the date of the publication of the advance notice, recommending that placarded tank cars remain placarded unless reloaded with another material or cleaned and purged of hazardous material. This petition contained the following as justification for the requested rule change:

> Requiring tank cars containing a residue of a hazardous material to be placarded in the same manner as when they contained a greater amount of the material will fully alert those handling the car of potential dangers. Other empty bulk containers (cargo tanks and portable tanks) are handled this way, and we are not aware that this has been a problem to anyone. Further, requiring the same placard for empty and full cars will encourage permanent placarding for those cars in dedicated service: our experience with non-permanent placards indicates they are frequently lost in transit. Anticipating that many (most) shippers will be displaying the DOT's identification number on placards, it is more important than ever that the correct placards be put on the car and that they stay there.

Elimination of the empty placard will simplify the regulations and reduce their burden by:

• Reducing the number (and cost) of placards kept in inventory (i.e., one style placard will do the job for all bulk containers).

• Eliminating the need for changing or reversing placards after tank cars are unloaded.

• Reducing confusion of whether a tank car is empty and needs to have placards changed or reversed.

MTB also believes that a large chemical manufacturer who ships large quantities of metallic sodium UN1428 has identified a specific deficiency related to the EMPTY FLAMMABLE SOLID W placard. This petitioner stated, in part, the following:

> A substantial volume of sodium, metal in tank cars that require the display of a flammable solid W placard. When shipped full the significant water reactive hazard of this commodity is noted on the placard; however, when the placard is reversed showing the word "empty" instead of the symbol "W", this critical hazard is not adequately identified to emergency personnel. The heel in an empty, sodium metal tank car constitutes like hazard as when full; it is extremely dangerous when exposed to minute quantities of water.

Nearly half of the commenters, and this included the vast majority of the comments from emergency services, recommended the removal of the EMPTY placard and the use of the same placard for a loaded and an "emptied" tank car. These commenters generally expressed the belief that the basic placard would provide adequate warning for initial emergency response action. Follow-on actions could be determined from the complete identification of the contents of involved tank car by checking the shipping papers. Further and more detailed actions could be planned from information provided on a consist and from outside sources after specific tank cars had been identified.

MTB believes display of EMPTY placards on tank cars containing hazardous materials is not appropriate for communicating risk and that safety would be enhanced if the placarding system does not differentiate between loaded tank cars and those containing a residue of a hazardous material. Placement of tank cars of hazardous materials, whether loaded or containing a residue, could be accomplished and verified through documentation.

A cost comparison with the present placarding system revealed that using reusable vinyl placards and leaving tank cars placarded as when filled, unless their service is changed or they are cleaned and purged, would result in average annual savings of approximately $1.4 million in placarding costs.

PROPOSED RULE CHANGES

Paragraph (e) of §172.203 would be revised to change the shipping paper entry for empty packagings and empty portable tanks, cargo tanks, tank cars and multi-unit tank car tanks that contain the residue of a hazardous material to include in the description the word RESIDUAL instead of the word EMPTY.

Footnote 4 to Table 2 in §172.504 would be revised to eliminate reference to the EMPTY placard. The second sentence of Footnote 4 prohibits display of the EMPTY COMBUSTIBLE placard. This prohibition would not be needed if the EMPTY placard is eliminated.

Paragraphs (a) and (c) of §172.510 would be revised to eliminate references to the EMPTY placards. The amended paragraph (c), in addition, would prescribe requirements for assuring that an emptied tank car containing the residue of a hazardous material is properly placarded as when it was loaded.

Section 172.525 and its accompanying paragraph (c)(10) in Appendix B to Part 172 which contain the specifications for the EMPTY placards would be removed.

Paragraph (a)(3) of §173.190 prescribes EMPTY FLAMMABLE SOLID placarding requirements for tank cars containing Phosphorus, white or yellow residue. The proposed change to eliminate the EMPTY placard would eliminate the need for this requirement in this section because the placarding requirements for Phosphorus are based on its hazard class and are given in §172.504 and §173.25. Therefore, MTB proposes to remove this placarding requirement from §173.190(a)(3) since each empty tank car containing a hazardous material residue would retain its placards under this proposed rule change.

The final entry in the placarding notation table in §174.25 which applies to empty tank cars would be revised to remove the exception pertaining to any tank car that had contained a combustible liquid. Also, §174.25 would be revised to remove the exception pertaining to tank cars that had contained combustible liquids and to change the shipping paper description for emptied tank cars that contain the residue of hazardous materials from "EMPTY" to "RESIDUAL" to better indicate the hazard.

Paragraph (e) of §174.50 would be revised to remove the term "empty" and reword the requirement for clarity to indicate that no open-flame light may be brought near any leaking placarded tank car.

Sections 174.69 would be revised to remove the requirements for removing, replacing or reversing placards on empty tank cars. Also, a requirement would be added making the person who is responsible for removing the lading from a tank car responsible for assuring it is properly placarded before it is offered for transportation, if it contains the residue of a hazardous material.

Sections 174.87, 174.89, 174.90, 174.91, 174.92 and 174.93 concerning the placement of placarded empty tank cars have been reviewed. MTB does not believe that the proposed changes to the tank car placarding requirements would adversely affect the car placement requirements.

(49 F.R. 32090, August 10, 1984)

49 CFR Parts 171, 172, 173, and 174
[Docket No. HM-180; Amdt. Nos. 171–82, 172–98, 173–189, 174–47]
Placarding of Empty Tank Cars

I. BACKGROUND

In June 1981, MTB received a petition from the IAFC which stated that the use of the EMPTY placard on tank cars is "misleading and dangerous" because it implies a lack of hazard even though, in many instances, an "empty" tank car containing a residue of a hazardous material presents a potentially greater danger than a tank car that is filled with the material. For this reason, the IAFC petition requested that MTB eliminate the requirement for display of EMPTY placards on tank cars. In response to this petition, MTB published on June 23, 1981, an advance notice of proposed rulemaking (ANPRM) (Docket HM-180, 46 FR 37953). This ANPRM reiterated the IAFC petition verbatim. Based on MTB's evaluation of the comments which they received on the ANPRM, a notice of proposed rulemaking (NPRM) was prepared by MTB and published on August 10, 1984, under Docket HM-180 (49 FR 32090). After reviewing the comments received to that NPRM, MTB has prepared this final rule (FR).

The petition from the IAFC alluded in a general way to a potentially hazardous condition. Since residues in a tank car do not provide the cooling capacity of the liquid in tank cars which are fully loaded, tank cars which only contain residues may present a risk of violent rupture in a shorter time than fully loaded cars when in a fire. Under fire impingement conditions, the actions of emergency response personnel involved with a tank car which contains only the residue of a liquefied compressed gas or a flammable liquid may be different from the actions which they might take in responding to a tank car which contains a full load of materials. Upon consideration of the substantive points made in the IAFC petition, MTB prepared an ANPRM.

The ANPRM (46 FR 37953) requested comments from interested persons on eliminating the requirement for display of the EMPTY placard. Evaluation of the comments MTB received to the ANPRM showed that almost half of the commenters favored use of the same placard for a "loaded" and an "emptied" tank car. Other commenters favored retention of the EMPTY placard or use of a placard with the word "residual" in place of the word "empty".

Based on MTB's evaluation of the comments received to the ANPRM, a notice of proposed rulemaking (NPRM) was prepared by MTB and published under Docket No. HM-180 (49 FR 32090). In this NPRM, MTB again proposed to amend the placarding regulations by eliminating the EMPTY placard. This amendment would have required that a tank car which contained only the residue of a hazardous material be placarded with the same hazard warning placard required when the tank car contained a greater quantity of the materials. Although this proposal would increase an individual's awareness, as far as knowing the type of hazardous material contained in the tank car, emergency response personnel would still have no rapid method of determining whether a tank car involved in a fire was loaded or contained only the residue of the material. One rail carrier cited, in his comments to the NPRM, an accident in which a tank car which contained only the residue of anhydrous ammonia exploded in twenty minutes after being subjected to a fire.

With this in mind and after further consideration of comments, MTB prepared this FR. MTB has concluded that a new RESIDUE placard for tank cars is needed to accurately communicate the appropriate warning to emergency response personnel. Further, to make the hazard communication information on the shipping paper consistent with that of the placard, MTB is revising the shipping paper entry from "EMPTY: Last Contained . . . " to "RESIDUE: Last Contained . . . ". It should be noted that the Association of American Railroads stated in their comments to the NPRM that "the shipping paper requirements for empty tank cars which last contained hazardous materials do not apply to tank cars which held combustible liquids." This is incorrect. Under the Provisions of §174.25(c) the shipping paper (billing) is not required to show the words "Empty" or "Empty: Last Contained", but a shipping paper must be prepared and accompany the shipment.

II. RESPONSE TO COMMENTS

MTB received forty-eight comments to the NPRM: 30 were from the chemical industry and nine were from rail carriers. The remaining nine comments came from firefighters, a State Department of Transportation, the National Transportation Safety Board (NTSB) and the Department of Defense.

Two commenters recommended that the words "residual" or "residue" be defined if used in place of the word "empty" on the placard. None of these terms are defined in the HMR. The NTSB recommended that the maximum quantity of a hazardous material that may be moved in an "empty" tank car be specified. MTB agrees with these recommendations and is adding, in §171.8, a definition for "residue" which includes, for tank cars only, a quantitative limitation of three percent or less of the tank car's capacity. Historically, the Department's Federal Railroad Administration has considered a tank car to be "empty" when the residue remaining in the tank car does not exceed three percent of the weight of the car's last loaded movement. This operational definition is derived from Rule 35 of the Uniform Freight Classification Tariff. MTB believes that the quantitative definition for the word "residue" adopted in this rule is consistent with the FRA definition, is easy to understand and will enhance safety by providing emergency response personnel with accurate information regarding the maximum quantity of hazardous material that may remain in the tank car. Since this definition parallels the definition in use today, negligible costs should be incurred by this clarification of terminology. Eight commenters indicated that the word "residue" or "residual" should not be substituted for the word "empty" on the shipping paper because the name used for shipping several materials that are not subject to the HMR contain the word "residue" or "residual." MTB reviewed the words "residue", "residual", and the "residuum" as listed in the Uniform Freight Classification (UFC) 6000 and concluded that of the 47 line entries containing these words, five have names that could be confused with the description of hazardous materials. However, because of the format of the description of a hazardous material on the shipping paper, i.e., "RESIDUE: Last Contained Sulfuric acid, Corrosive material, UN 1830", there should be no confusion with UFC entries.

Five chemical industry commenters and five rail carriers, all of whom use tank cars to transport hazardous materials, requested that the present requirements for the EMPTY placard not be changed. When making up trains, the EMPTY placard is used operationally by rail carrier personnel to help them identify the correct tank cars when they are switching, humping and sorting cars. The rail carriers made the point that the EMPTY placard contributed to the safety of those rail employees involved in rail car switching, placement, and humping operations. MTB believes the RESIDUE placard can be used in rail car operations in the same manner as the EMPTY placard is now used.

Several commenters who use tank cars regularly emphasized the fact that the residues of some hazardous materials present a greater hazard during a fire than tank cars which contain bulk loads of those same materials. MTB believes this evaluation is correct. Three safety or emergency response organizations and six chemical industry commenters recommended adoption of a RESIDUE or RESIDUAL placard to replace the present EMPTY placard. Eighteen commenters supported the MTB proposal contained in the NPRM to use the same placard on both loaded tank cars and tank cars which only contain the residues of hazardous materials. Fifteen of these comments were from the chemical industry and the other three were from safety or emergency response personnel.

III. DISCUSSION

Notwithstanding a rail carrier's reliance on the EMPTY placard for certain operational procedures (which they say also enhances safety), MTB must consider hazard warning placards in light of their contribution to emergency response as well. Based on the comments received and the knowledge gained by MTB from the comments MTB concludes that emergency response considerations overwhelmingly favor the use of the RESIDUE placard and that the final rule should be revised from what was proposed in the NPRM. Further, MTB believes that the shipping paper entries for a tank car which contains only the residue of a hazardous material should show the same hazard warning as the revised placard. Therefore, the rule is changed accordingly. The placarding requirements are revised from EMPTY to RESIDUE for a tank car which contains only the residue of a hazardous material. This rule change should resolve the problems emergency response personnel were having with the EMPTY placard, and still indicate a difference between a tank car that is loaded and one that contains only the residue of hazardous material. Revision of the placard will also improve hazard communication by removing the black triangle from the top of the placard, thus allowing display of the hazard symbol.

There should be no significant difference between the purchase price of the EMPTY placard presently required by the regulations and the RESIDUE placard that will be required by this amendment. In addition, the cost of placing the placards on the tank cars would be the same. A year is given from the publication date of this rule for the change over to the RESIDUE placard, during which time on-hand stocks of the EMPTY placard should be depleted.

IV. REVIEW BY SECTIONS

1. Section 171.8 is revised by adding a definition for "Residue". This definition was not proposed in the notice, but commenters to the HM-180 Notice requested that "Residue" be defined. From this definition one can determine the maximum quantity of hazardous material that is contained in a tank car placarded with the RESIDUE placard.

2. Paragraph (e) of §172.203 is revised to change the shipping paper entry for empty packagings and empty portable tanks, cargo tanks, tank cars and multi-unit tank car tanks that contain the residue of a hazardous material to include in the description the word RESIDUE instead of the word EMPTY. The description will be preceded by the words "RESIDUE: Last Contained . . . ".

3. Footnote 4 to Table 2 in §172.504 is revised to change the placard name from EMPTY to RESIDUE.

4. Paragraphs (a) and (c) of §172.510 are revised to change the special placarding provisions for rail from "POISON GAS—EMPTY" to "POISON GAS—RESIDUE" in paragraph (a) and to change the placard name from EMPTY to RESIDUE in the title of paragraph (c) and in the third line of paragraph (c). Paragraph (c)(1) is also reworded for clarity.

5. Section 172.525, and accompanying paragraph (c)(10) in Appendix B to Part 172 which contains the specifications for the EMPTY placard, have the placard name changed from EMPTY to RESIDUE.

6. Paragraph (b)(3) of §173.190 prescribes the EMPTY-FLAMMABLE SOLID placarding requirements for tank cars which contain the residue of white or yellow Phosphorus. Since this final rule changes the placard name from EMPTY to RESIDUE, the placarding requirement for Phosphorus in 173.190(b)(3) is changed to FLAMMABLE SOLID-RESIDUE.

7. Paragraph (c) of 174.25 is amended by revising the paragraph to make the requirement consistent with the shipping paper requirements in Subpart C to Part 172.

8. Paragraph (e) of §174.50 is revised for clarity by rewording the requirement that no open-flame light may be brought near any placarded tank car that is leaking.

9. Section 174.69 is revised to restate the requirement that the person who is responsible for removing the lading from a tank car is also responsible for assuring that if the tank car contains the residue of a hazardous material it is properly placarded before it is offered for transportation.

10. The section title and the text §174.93 are revised for consistency and clarity to change the car placement requirements for EMPTY placarded tank cars to reflect the change to the RESIDUE placard.

(50 F.R. 39005, September 26, 1985)

49 CFR Parts 171, 172, and 174
[Docket No. HM-180, Notice No. 86–1]
Placarding Tank Cars Which Contain Hazardous Material Residue

1. BACKGROUND

On September 26, 1985, the RSPA published a final rule (Docket HM-180; 50 FR 39005), which concerned the placarding of empty tank cars. In the final rule, RSPA changed the placarding and shipping paper requirements for tank cars that contain residues of hazardous materials. Also contained in the final rule was a quantitative definition for the word "residue."

In response to the final rule, RSPA received eight petitions for reconsideration under 49 CFR 106.35. Each of the petitioners objected to the definition of residue which was contained in the final rule. Specifically, concern was expressed about that portion of the definition which states:

"Residue of a hazardous material, as applied to the contents of a tank car (other than DOT Specification 105 or 110 tank cars), means a quantity of material not greater than 3 percent of the cars marked volumetric capacity."

One of the petitioners, Shell Oil Company, requested that RSPA reconsider the quantitative wording of the residue definition. Shell, as well as the other petitioners, contend that the public was given no opportunity to comment on the residue definition and that it is not operationally possible to comply with the definition. Commenters state that there is no accurate method of determining exactly how much material remains in a tank car after unloading.

Although the RSPA does not necessarily agree with certain of the factual statements made in the petition submitted by Shell, we believe their comments on the final rule concerning the residue definition represent a consensus of the concerns expressed by the other petitioners. For this reason, the substantive points of the Shell petition are quoted as follows:

"There is no practical method of determining precisely how much material remains in a tank car after unloading, considering tank car design and the limitations of facilities at many unloading locations.

Some gauging tapes are designed to gauge the liquid level at the top of the car rather than at the bottom.

Devices used in pressure cars don't reach to the bottom of the car and therefore can't measure the remaining liquid which may be more or less than 3% depending on the pumping effort expended.

Some receivers unload butane and other compressed gases by heating a small amount of the liquid and pumping the resulting vapor through the vapor line to force out the liquefied gas. The remaining vapors occupy 100% of the marked volumetric capacity of the car. Although for all practical purposes, the car contains only the residue of a hazardous material, it does not meet the definition in Section 171.8 and presumably would be considered a loaded car.

Viscous liquids such as resins and regulated asphalts also present a problem since there is no accurate method of measuring the clingage on the tank walls.

Weighing cars before they leave the unloading location is not a viable alternative since most facilities don't have track scales. Installing scales at all locations receiving hazardous material tank cars would be prohibitively expensive. Moving a car from the unloading point to a railroad scale to determine its status could result in a violation, if for example, a car displayed residue placards and the scale weight indicated it contained more than 3% of the volumetric capacity. The additional switching would also be costly in terms of time and money.

The definition of what constitutes an empty car does not enhance safety but merely increases the potential for unintentional violations of the regulations. The 3% measure is basically an economic issue and is already covered in Rule 35 of the Uniform Freight Classification.

There is no need for a specific percentage figure to be included in the residue definition since fire fighters would follow the same procedure regardless of the exact amount of the residue. Shell suggests the following definition be added to §171.8 in lieu of the one in the final rule.

"Residue means the hazardous material remaining in a packaging, including a tank car, after its contents have been emptied and before the packaging is refilled or cleaned and purged of vapor to remove any potential hazard."

II. RESPONSE TO COMMENTS

In response to the petitions, the RSPA proposes to redefine "residue" by limiting it to liquids (excluding vapor and gases) and expanding the quantitative limitation from 3% to 4% with a tolerance of plus or minus one percent in order to provide shippers greater flexibility in their determination of compliance with the rule. The RSPA is taking this action because of the problems that may be encountered in accurately determining the amount of residue which remains in an unloaded tank car. Aside from weighing the car, which may be impractical because most unloading facilities don't have track scales, there appears to be no accurate method of measuring the precise amount of residue which remains in a tank car. However, RSPA believes that a quantitative requirement must be adopted in order to make the rule effective.

The public is invited to submit any substantive comments or information which they may have concerning the "residue" definition proposed in this notice. Comments should, as a minimum, address the maximum amount of residue which, from a safety standpoint, could be left in a tank car which is placarded with the RESIDUE placard.

In addition to the petitions addressed to the definition of residue, the RSPA also received comments stating that it was inconsistent and confusing to require a COMBUSTIBLE placard to remain on a tank car which only contains the residue of a combustible liquid. Currently, tank cars which contain residue of hazardous material may be placarded with the appropriate RESIDUE placards.

The only exception to this is for tank cars which contain combustible liquid residue. These cars must continue to display the COMBUSTIBLE placard. Upon further consideration of this point, the RSPA agrees that this is inconsistent and may cause confusion. For this reason the RSPA believes that a RESIDUE placard should be required on tank cars which contain combustible liquid residues. Therefore, the RSPA proposes to remove the second sentence in footnote 4 to Table 2 of §172.504 and paragraph (c) of §172.510. Revision to both of these sections is necessary to require the display of a RESIDUE placard on a tank car which contains residue of a combustible liquid. Further, RSPA proposes to revise §172.525 to require the lower triangle of the combustible RESIDUE placard to be white and to require and word "RESIDUE" to be black on this same placard. This change is necessary to identify those tank cars which contain combustible residue so they are not subjected to the train placement requirements. In addition, it is proposed to reinstate the first sentence of footnote 4 to Table 2 of §172.504 as it appeared prior to Amendment No. 172–98 (50 FR 39005).

For clarification, the RSPA proposes to revise paragraph (c) of §172.334. This revision was inadvertently omitted from Docket HM-196, the final rule that addressed the Packaging and Placarding Requirements for Liquids Toxic by Inhalation. This revision would prohibit the display of identification numbers on subsidiary placards such as the POISON placard required by §172.505. The RSPA proposes to revise §172.525 to authorize the display of identification numbers on the RESIDUE placard or orange panel in accordance with the marking requirements contained in Subpart D of Part 172. The RSPA also proposes to revise the Table in paragraph (a)(2) of §174.25 to clarify the shipping paper requirements for tank cars which contain combustible liquids.

Editorially, the RSPA proposes to amend §172.525(a)(2) to include a reference to §172.544 so there is a citation in this section which refers to the RESIDUE placard for combustible liquids. Further, the RSPA proposes to amend paragraph (c) of §174.25 to clarify exactly when the letters "RQ" must be entered on the shipping paper and to remove the combustible liquid exception from the description requirements.

Also, §174.93 would be amended to except tank cars which contain combustible liquid residue from the train placement requirements. This is consistent with the provision for loaded tank cars which contain Combustible liquids. In the case of a tank car which contains the residue of a hazardous material, the letters "RQ" are required to be shown on the shipping paper when the quantity of the residue is equal to or exceeds the reportable quantity level and the residue or a constituent of the residue is a hazardous substance. Comments are requested on the changes proposed to §§172.334, 172.504, §172.510, §172.525, §174.25 and §174.93.

The RSPA is also interested in receiving comments on the placarding requirements for tank cars which carry residues of materials which exhibit inhalation toxicity. After April 30, 1986 tank cars loaded with these materials are required to display POISON placards as well as the placards required in §172.504. When the tank car is unloaded, each of these placards would have to be changed to RESIDUE placards. The RSPA is interested in receiving suggestions on optional placarding methods that could be employed when placarding tank cars which contain residue of materials which exhibit inhalation toxicity.

III. REVIEW BY SECTIONS

1. It is proposed to revise §171.8 by expanding the quantitative limits of the "residue" definition and to restrict the application of the definition to liquids.

2. It is proposed to revise paragraph (c) of §172.334 to prohibit the display of an identification number marking on a subsidiary placard.

3. It is proposed to reinstate the first sentence of footnote 4 to Table 2 of §172.504 as it appeared prior to Amendment No. 172.08 and to remove the second sentence which prohibits a RESIDUE (EMPTY) placard on a tank car containing Combustible liquid.

4. It is proposed to revise paragraph (c) of §172.510 by removing the prohibition against using a RESIDUE placard for combustible liquid residues.

5. It is proposed to revise §172.525 so a reference to the RESIDUE placard for combustible liquids is shown.

6. It is proposed to revise §172.525 to require the lower triangle of the RESIDUE placard for combustible liquid residues to be white and to require the word "RESIDUE" to be black on the combustible liquid RESIDUE placard.

7. It is also proposed to revise §172.525 to authorize identification numbers to be displayed on RESIDUE placards or on orange panels in association with RESIDUE placards.

8. It is proposed to revise §174.25 for clarification.

9. It is proposed to revise §174.93 to except tank cars which contain combustible liquid residue from train placement requirements.

(51 F.R. 9079, March 17, 1986)

49 CFR Parts 171, 172 and 174
[Docket No. HM-180, Amdt. Nos. 171–88 172–104 and 174–60]
Placarding of Tank Cars Which Contain Hazardous Material Residue;
Disposition of Petitions for Reconsideration

I. BACKGROUND

On March 17, 1986, RSPA published a notice of proposed rulemaking pertaining to disposition of petitions for reconsideration (51 FR 9079). This notice was prepared in response to eight petitions for reconsideration which were filed as a result of a final rule issued under Docket HM-180 that was published on September 26, 1985 (50 FR 39005). That final rule amended by the HMR by changing the placarding and shipping paper requirements for "empty" tank cars which contain residues of hazardous materials. Under the final rule of Docket HM-180, a quantitative definition of what constitutes a residue was adopted. Further, the applicable regulations in Parts 172 and 174 were also revised to reflect other amendments which were made in that final rule.

In the March 17, 1986 notice, RSPA proposed to redefine "residue" by restricting the applicability of the definition to liquids and expanding the quantitative limitation from 3% to 4% with a measurement tolerance of plus or minus one percent. RSPA believed that by raising the percentage of residue which may remain in a tank car and providing a tolerance of plus or minus one percent, shippers should have less difficulty in complying with the rule. RSPA also believed that adopting a quantitative limitation was important in order to make the rule effective. RSPA invited the public to submit comments concerning the expanded "residue" definition. RSPA requested that comments, as a minimum, address the maximum amount of residue which can safely remain in a tank car placarded with the RESIDUE placard.

In addition to requesting comments regarding the definition of "residue", RSPA stated in the notice that some of the comments received from the petitioners pointed out how inconsistent it was to require tank cars which contain combustible liquid residue to remain placarded as full loads. Currently, tank cars which contain residue of hazardous material must display the appropriate RESIDUE placards unless (1) the tank car contains the residue of a combustible liquid, or (2) the tank car is reloaded with a non-hazardous material, or (3) the tank car is sufficiently cleaned of residue and purged of vapor to remove any potential hazard. Tank cars which contain residue of a combustible liquid must continue to display COMBUSTIBLE placards. In view of the comments received by the petitioners and to promote consistency in the regulations, RSPA proposed in the notice to require the use of RESIDUE placards on those tank cars which contain combustible liquid residue. RSPA also proposed to amend §174.93 so that tank cars containing combustible liquid residue would be excepted from the train placement requirements.

RSPA proposed to revise paragraph (c) of §172.334 to prohibit the display of identification numbers on subsidiary placards such as the POISON placard required by §172.505. Commenters were also asked to address the placarding requirements for tank cars that carry residues of hazardous materials which meet the criteria specified in the new §173.3a pertaining to inhalation toxicity. As of May 1, 1986, tank cars loaded with hazardous materials which exhibit that criteria must display the POISON placards required by §172.505 as well as the primary placards required by §172.504. When unloaded, all tank cars containing residues, except explosives, poison gas, or radioactive material, must have all placards, including the POISON placards required by §172.505, reversed or changed, as appropriate, to RESIDUE placards. RSPA requested comments and suggestions on alternate or preferred placarding methods that could be employed when placarding tank cars that contain a residue of materials which exhibit inhalation toxicity. RSPA also proposed to make other changes to the regulations for clarification and consistency.

II. RESPONSE TO COMMENTS MADE TO THE NOTICE

RSPA received twenty four comments to the NPRM. Comments were received from various chemical companies, several oil companies, two railroads and their association and the Hazardous Materials Advisory Council. A large number of these commenters were opposed to RSPA adopting a quantitative definition for "residue". In addition to these comments, a few of the commenters urged RSPA to develop a separate and distinct placard for those materials which exhibit inhalation toxicity. Several commenters stated that the size of the word "RESIDUE" on the RESIDUE placard was too large and was obscured by the placard holder. Based on the regulatory changes which were being made in the rule, commenters requested a one year extension of the mandatory compliance date of October 1, 1986. Discussion of these points as well as RSPA response to each of these comments follows.

Definition of Residue. RSPA proposed on March 17, 1986 (51 FR 9079) to redefine "residue". RSPA expanded the quantitative limitation in the definition contained in the final rule from 3% to 4%, allowing a tolerance of ±1%. Our intent was to provide shippers greater flexibility in complying with the rule. Although RSPA recognized that it would be difficult to accurately determine exactly the amount of residue which remains in a tank car after unloading, RSPA believed that safety would be enhanced by adopting a quantitative definition. At the very least, emergency responders would have a benchmark on which to base their decisions, should an incident occur that involved a tank car which contained only the residue of a hazardous material.

The majority of comments RSPA received stated that there was no need for a quantitative definition for "residue". Commenters stated that aside from it being difficult to determine exactly how much residue remains in an unloaded tank car, firefighters would follow the same procedures regardless of the amount of residue in a tank car. One commenter stated that it is too difficult to establish a volumetric dividing line between "full" and "residue" for safety purposes, because the hazard posed by a given material varies both by the hazard class of the material and the quantity of material present. A few commenters also suggested that RSPA eliminate the plus or minus allowance from the definition of "residue" because there was no need for the definition to contain such an allowance.

Based on the comments received concerning the definition of "residue" and recognizing the difficulty in determining the exact amount of residue which remains in a tank car once it is unloaded, RSPA has decided to adopt the definition for "residue" which was proposed in the comments submitted by I.E. duPont de Nemours. The new definition does not quantitatively specify the amount of residue which may remain in a tank car which displays the RESIDUE placard. The new definition for "residue" states that "Residue means the hazardous material remaining in a packaging, including a tank car, after its contents have been unloaded to the maximum extent practicable and before the packaging is either refilled or cleaned of hazardous material and purged to remove any hazardous vapors".

Special Placement Requirements for Placards. Three commenters proposed that RSPA establish special placement requirements for placards, i.e. standardizing the location of the primary and subsidiary placards. Commenters stated that such a system is necessary because of the multiple placarding requirements of HM-196 and the new Canadian regulations which require the use of multiple placards for certain commodities. It was felt that with these new requirements, emergency response efforts may be hampered by confusion between the primary and secondary hazards. Therefore, it was suggested by these commenters that the primary hazard placard always be displayed to the left of any required subsidiary placard.

RSPA believes that imposing a special placard placement requirement may be necessary if problems are encountered discriminating between the primary and subsidiary placard, however, RSPA questions the need at this time to establish such a system. We have not been made aware of any problems or incidents that have taken place which can be attributed to confusion occurring between the placement of the primary and subsidiary hazard placards. Therefore, RSPA is not imposing any additional placement requirements on primary and subsidiary placards in this rule.

Placarding Materials Which Exhibit a Poison-Inhalation Hazard. For those materials which exhibit a Poison-inhalation hazard (§173.3a), several commenters suggested that RSPA develop a separate and distinct placard or marking which accurately communicates to the public that such a hazard is present. It was stated that while training and education of emergency response personnel is helpful, there will still be confusion under the current system as to whether or not a material exhibits an inhalation hazard. This is essentially true for materials in tank cars which satisfy the definition for a Poison B and also exhibit a Poison-inhalation hazard. Under HM-196, duplication of POISON placards is not required. Therefore, use of the primary POISON placard satisfies the subsidiary placarding requirement of §172.505. Commenters pointed out that unless there is a method developed to easily identify those materials which exhibit a Poison-inhalation hazard, especially when Poison B materials are present, the emergency responders will have no way of knowing if a material presents a Poison-inhalation hazard.

RSPA acknowledges that establishment of a separate or distinct placard or marking for materials which exhibit an inhalation hazard may be necessary, however, RSPA believes that establishment of such a placard is beyond the scope of this rulemaking. Further, in developing a new, distinct placard for materials which exhibit an inhalation hazard, a consideration must be given to the format of labels as well as the potential for conflict with existing international requirements. For these reasons. RSPA believes that it would be inappropriate to establish such a placard in this rulemaking.

Lettering Size of the Word "RESIDUE" on the Placard. RSPA received three comments which stated that the size of the lettering of the word "RESIDUE" on the placard should be changed in 1 inch rather than 1½ inches because the word "RESIDUE" was obscured by the placard holder, RSPA agrees with these commenters, RSPA never intended for the word "RESIDUE" to be partially underneath the lower cross bar of the placard holder. This is why both §172.525 and Appendix B to Part 172 only require the letters to be "approximately" 1½ inches (40mm) high. Further, since the RESIDUE placard is the only placard which has lettering in the lower triangle of the placard, RSPA did not believe that the RESIDUE placard could be confused with any other placard. Nevertheless, in view of the comments RSPA received on this point, and to ensure by that the word "RESIDUE" appears on the placard clearly and unobscured by the placard holder, RSPA is changing the size of the lettering for the word "RESIDUE" to 1 inch.

RSPA realizes that since the use of RESIDUE placards with 1 ½ inch lettering has been authorized since November 1, 1985, there may be stocks of these placards on hand. Therefore, RSPA is authorizing the use of RESIDUE placards that have 1 ½ inch lettering until July 1, 1987. On that date, the use of RESIDUE placards with 1 inch lettering will be mandatory.

Combustible Residue Placard. In this notice, RSPA proposed to require the use of RESIDUE placards on tank cars which contain residues of combustible liquids. Two commenters suggested that if a COMBUSTIBLE RESIDUE placard is required, it should be different from the other RESIDUE placards so RESIDUE placards for combustible liquids would not be confused with other RESIDUE placards (especially the FLAMMABLE RESIDUE placard). RSPA does not believe that taking such action is necessary because the bottom triangle of the COMBUSTIBLE RESIDUE placard will always be white with the word "RESIDUE" in black letters. The bottom triangle of the other RESIDUE placards (including the FLAMMABLE RESIDUE placard) will be black with the word "RESIDUE" in white letters. With these differences, there should not be any confusion between the COMBUSTIBLE RESIDUE placard and the other RESIDUE placards. Therefore, no change has been made to the COMBUSTIBLE RESIDUE placard that was proposed in the notice. Its use is required on tank cars which contain the residues of combustible liquids.

The Norfolk Southern Corporation and the Association of American Railroads contended that in the past tank cars which contained residues of combustible liquids required no placards. RSPA disagrees. Prior to the promulgation of HM-18, tank cars containing combustible liquids were required to be placarded with COMBUSTIBLE placards and shipping papers were required. When unloaded, these tank cars were still required to display COMBUSTIBLE placards and have accompanying shipping papers. This is supported by the provisions of both §172.501(c) and §174.25(c). These sections prohibited the display of EMPTY placards on tank cars which contain only the residue of a combustible liquid and prohibited the use of the words "Empty" or "Empty: Last Contained" on the shipping papers of those tank cars which last contained combustible liquids. These sections did not state that the use of COMBUSTIBLE placards and shipping papers were no longer required. RSPA understands that the rail industry may have interpreted these exceptions to imply that no placards or shipping papers were required for unloaded (empty) tank cars which last contained combustible liquids. This interpretation was incorrect. RSPA finds nothing in the regulations or the administrative record which indicates that tank cars which contain combustible liquid residues are currently excepted from the requirements to have shipping papers and to display the COMBUSTIBLE placards required by §172.504 (see also §173.118a). Nevertheless, to promote consistency in the regulations and to enhance safety, the requirements contained in this rule specify that a tank car which contains the residue of a combustible liquid must have shipping papers and the basic description on the shipping papers must include the words "RESIDUE: Last Contained". Further, tank cars which contain combustible liquid residue must display RESIDUE placards.

Mandatory Compliance Date. Several commenters requested that RSPA extend for one year the mandatory compliance date for the rules promulgated in HM-180. Specifically, the Norfolk Southern Corporation urged RSPA to defer all of the RESIDUE placard requirements and corollary shipping paper requirements. They stated that this extension was needed because of the uncertainty as to the status of the RESIDUE placarding requirements until the issuance of the March 17, 1986 notice, the possible creation and issuance of a new, more unique COMBUSTIBLE RESIDUE placard, and the change that must be made to the size of the lettering of the word "RESIDUE" on the RESIDUE placard. They also stated that this extension was needed to allow ample time for training and compliance planning.

RSPA believes that delaying the effective date of the rule for one year is unnecessary. The only change being made to shipping papers by the rule other than changing the word "empty" to "residue", is that §174.25(c) now requires the shipping papers (billing) for tank cars which contain the residue of a combustible liquid to contain the words "RESIDUE: Last contained . . . ". Previously, §174.25(c) did not require the shipping paper (billing) for tank cars containing combustible liquid residue to show the words "Empty" or "Empty: Last Contained". As we have previously stated, there has always been a requirement for shipping papers to accompany all tank cars which contain residues. Further, RSPA does not believe there is a need to establish a more unique COMBUSTIBLE RESIDUE placard. Regarding the change being made to the specification (size) of the word "RESIDUE" on the placard, RSPA believes adequate time (more than a year) is being provided for depletion of on-hand stocks of RESIDUE placards with 1 ½ inch lettering and procurement of new placards. Although changes are being made to the definition of "residue" by eliminating the quantitative levels specified, this new definition should not impose any additional operational requirements on shippers and carriers. In addition, RSPA believes that providing a grace period of one year before the mandatory compliance date of July 1, 1986 provides ample time for shippers and carriers to deplete their on hand stocks of RESIDUE placards with 1 ½ inch lettering. Therefore, the amendments contained in this rule are effective October 1, 1986. However, compliance with the regulations as amended herein is authorized immediately. Use of RESIDUE placards with 1½ inch lettering is authorized until July 1, 1987.

Subsidiary Risk Placard. RSPA also received two comments which stated that the subsidiary risk placard should not be reversed to a RESIDUE placard when a tank car is unloaded. In effect, both commenters believed that the ability to communicate the hazard of inhalation toxicity would be lost by subjecting materials regulated under HM-196 to the residue placarding requirements of HM-180. One of these commenters pointed out that it would be possible to use permanent adhesive placards for the supplementary POISON placards if these placards permanently read POISON. It was stated that if the supplementary placards are required to be changed or reversed to RESIDUE placards, it will be necessary to install additional placard holders. Both commenters stressed how important it is to communicate the unique hazard of materials which exhibit an inhalation toxicity hazard (§173.3a).

RSPA believes that the subsidiary risk placard required by §172.505 should be reversed or changed to a RESIDUE placard when a tank car has been unloaded and only contains residue. This will communicate with greater certainty the fact that only the residue of a material remains in the tank car and will ensure that the RESIDUE placard requirements for all materials remain consistent.

III. REVIEW BY SECTIONS

1. Section 171.8 is revised by amending the definition of "residue". The definition, as amended, does not specify quantitatively the amount of hazardous material residue which may remain in a tank car placarded with RESIDUE placards.

2. Paragraph (c) of §172.334 is revised to prohibit the display of an identification number marking on a subsidiary placard.

3. The first sentence of footnote 4 to Table 2 of §172.504 is reinstated as it appeared prior to Amendment No. 172–98 and the second sentence of footnote 4 is removed so that a RESIDUE placard must be displayed on a tank car which contains the residue of a combustible liquid.

4. The exception provided for combustible liquid residues in paragraph (c)(1) of §172.510 is removed for consistency. Tank cars which contain residue of a combustible liquid are now required to display RESIDUE placards. For clarity, paragraphs (c)(2) and (c)(3) are redesignated as (c)(1) and (c)(2).

5. Paragraphs (a)(1) of §172.525 and (c)(10) of Appendix B to Part 172 are revised by changing the size of the letters in the word "RESIDUE" on the placard from 1½ inches (40mm) to 1 inch (25mm).

6. Paragraph (a)(2) of §172.525 is revised to authorize identification numbers to be displayed on RESIDUE placards or on orange panels in association with RESIDUE placards.

7. The first sentence which follows paragraph (a)(2) of §172.525 is amended to include a reference to §172.544 so that a reference is provided to the COMBUSTIBLE placard. Also, the sentences which follow paragraph (a)(2) and precede paragraph (b) of §172.525 are codified and designated as a new paragraph (a)(4).

8. Paragraph (a)(3) is added to §172.525 and requires the lower triangle of the RESIDUE placard for combustible liquid residues to be white and the word "RESIDUE" to be shown in black letters on the COMBUSTIBLE-RESIDUE placard.

9. Section 174.25 is revised for clarification.

10. Section 174.93 is revised to except tank cars which contain combustible liquid residue from the train placement requirements.

(51 F.R. 23075, June 25, 1986)

49 CFR Parts 171, 172, 173, and 178
[Docket No. HM-181, Advance Notice No. 82–3]
Performance-Oriented Packagings Standards

BACKGROUND

The Hazardous Materials Regulations (HMR), issued under authority of the Hazardous Materials Transportation Act, are found in the Code of Federal Regulations (CFR), Title 49, Subchapter C, which is comprised of Parts 171 through 179. Subchapter C occupies approximately one thousand and fifty pages of the CFR. The two largest parts of Subchapter C are Part 173, "SHIPPERS—GENERAL REQUIREMENTS FOR SHIPMENTS AND PACKAGINGS," of about three hundred pages, and Part 178, "SHIPPING CONTAINER SPECIFICATIONS," of about four hundred and fifty pages.

Basically, to use the HMR, a shipper locates in the Hazardous Materials Table (49 CFR 172.101) the hazardous material to be shipped. Column 5(b) of the Table refers to the appropriate packaging section of the shipper's requirements (49 CFR Part 173) for that material. Part 173, addressed primarily to shippers, lists the DOT packagings authorized for that material and includes a reference to the appropriate section of Part 178 (Packaging Specifications) where the specification for those packagings are found. Part 178, addressed primarily to container manufacturers, contains detailed construction specifications for a wide variety of packagings.

As new materials with hazardous characteristics have been developed and produced, and as advances in technology have produced new packaging methods, the number of specification packagings has grown until at present they occupy, as previously noted, about four hundred and fifty pages of the CFR. The specification packaging sections of Part 178 cover packagings required for various hazardous materials and range from small polyethylene bottles to cargo tanks. However, the major portion of Part 178 is devoted to specifications for non-bulk packagings, i.e., authorized capacities of 110 gallons or less, and includes approximately 93 specifications for carboys, drums, barrels, boxes, cases, trunks, tubes, bags, and various sorts of containers designed to be enclosed by larger containers. Not included in these 93 specifications are those covering cylinders for compressed gases, and packagings designed solely for radioactive or explosive materials, none of which is addressed in this proposal.

The detailed packaging specifications must be complied with by persons desiring to either manufacture packagings, or ship hazardous materials in authorized packagings. This includes requirements for: Material, thickness, fastenings, capacity, coatings, openings, joining, carrying devices, and miscellaneous other construction requirements. Much of the information is given in great detail and is repetitious. For example, there are fourteen specifications for wooden boxes. Most specifications list the acceptable types of wood from which lumber must be used to construct the box and this list may be repeated in the next specification for a similar, but slightly different box. In addition to the types of acceptable wood being specified, the thickness and width of the boards, the kind and dimension of nails, and the spacing of the nails in joining the box may also be specified.

In sharp contrast to this system of detailed specifications for container construction, is a system of performance-oriented packaging standards that has been developed in the form of Recommendations by the United Nations Committee of Experts on the Transport of Dangerous Goods (U.N. Recommendations). These standards address the same types of non-bulk containers (drums, barrels, boxes, bags, carboys, and inside containers or receptacles) as do the DOT specifications. Typically, these standards have general requirements for materials, construction, and a maximum size. For example, the UN 1A1 steel drum must have welded seams if it is to be used to carry liquids, must have welded or mechanically seamed chimes, the diameter of the opening may not exceed 7 cm (2.75 inches), and the drum may not exceed 450 liters (118.88 gallons) capacity. There is an additional requirement for rolling hoops if the drum capacity is greater than 60 liters (15.85 gallons). Aside from these very general construction requirements, the strength and integrity of the drum is established by a series of performance tests which the drum must pass before it is authorized for the carriage of hazardous materials; hence the term "performance-oriented" packaging standards. For drums, the principal tests are a drop test, i.e., a drum, filled and prepared as if for shipment, is allowed to free fall on to a level unyielding surface without spilling its contents, and a stacking test. In addition, hydraulic and leakage tests are prescribed if the drum is intended to transport liquids. The height specified for the drop test in the UN system is determined by the "Packing Group" of the hazardous materials to be transported. The UN system divides hazardous materials into three "Packing Groups" depending on their relative hazards. Packing Group I consists of very dangerous materials, Packing Group II involves materials considered to present a moderate degree of danger, and Packing Group III addresses materials presenting only minor danger. The steel drum, for example, would have to survive a drop from a height of 1.8 meters (5.91 feet) if it were to carry a material in Packing Group 1, 1.2 meters (3.94 feet) for Group II, and 0.8 meters (2.62 feet) for Group III materials (if the specific gravity of the hazardous materials to be shipped in the drum does not exceed 1.2). Within the broad general construction requirements given in the performance-oriented standard, the drum manufacturer is free to exercise its design and production ingenuity to produce a packaging which is both cost effective, as determined by the marketplace, and safe, as determined by the performance standards which it must meet.

PURPOSE OF THE ANPRM

The Materials Transportation Bureau (MTB) is considering adopting a set of performance-oriented packaging standards based on the U.N. Recommendations. The goals of this effort are: (1) Simplification of the Hazardous Materials Regulations for both bulk and non-bulk shipments; (2) a significant reduction in the volume of the regulations; (3) provision for greater flexibility in the requirements for the design and construction of hazardous materials packaging in order to recognize technological advancements in packaging; (4) the promotion of safety in transport through the use of better packaging; (5) a reduction in the need for exemptions; and, (6) the facilitation of international commerce, including commerce between the United States and Canada.

The simplification of the HMR is the most important goal of this proposal. The existing regulations have their origin in the early efforts of the transportation industry to protect its personnel and equipment from materials with dangerous properties. Some container specifications go back to the early days of this century. The regulations were developed in a piecemeal fashion with recurring adjustments or "fixes" to take care of particular problems. Individuals subject to the HMR have complained that they are so complicated in both content and organization as to be almost unusable except by someone with extensive training or experience. In addition, the complexity of the HMR may penalize the small businessman, whether shipper or carrier, who has insufficient hazardous materials business to justify employing an expert in the regulations. Industry is not the only place where the complexity of the HMR places a burden. A staff of eight hazardous materials specialists at MTB spend approximately half of their time giving explanations of the HMR either in writing or over the telephone. As already indicated, one of the principal reasons for this Advance Notice of Proposed Rulemaking (ANPRM) is to demonstrate how both shipper requirements and the container specifications can be greatly simplified by the adoption of a set of performance-oriented packaging standards.

Another reason for considering a shift to performance-oriented packaging standards is to provide greater flexibility in the requirements for the design, construction, and use of packagings having capacities of less than 450 liters. The current packaging specifications are very detailed and are not readily adaptable to innovations in packaging technology. Consequently, the use of an improved packaging which does not meet a detailed specification would be prohibited until a specification is amended or an exemption is issued. This imposes a costly burden on both the industry and government with an uncertain benefit in increased safety.

In addition, international trade would be facilitated by removing as an obstacle to trade the selected shipping and packaging requirements of the HMR. Through both law and policy, it has been decided that standards-related activities shall not be a barrier to trade. Title IV of the Trade Agreements Act of 1979 (Public Law 96–39) addressing Technical Barriers to Trade (Standards), states in pertinent part:

> No Federal agency may engage in any standards-related activity that creates unnecessary obstacles to the foreign commerce of the United States. . . . Each Federal agency, in developing standards, shall take into consideration international standards and shall, if appropriate, base the standards on international standards. . . . Each Federal agency shall, if appropriate, develop standards based on performance criteria, such as those relating to the intended use of a product and the level of performance that the product must achieve under defined conditions, rather than on design criteria, such as those relating to the physical form of the product or the types of material of which the product is made.

Within the next five years, our trading partners in other countries are expected to adopt non-bulk packaging requirements based on the UN Recommendations. Without a change in DOT's present regulations, non-bulk packages of hazardous materials shipped in conformance with the HMR would probably not be acceptable in those countries, and in imported shipments prepared only in accordance with the U.N. Recommendations would not be acceptable here. Since the U.S. has a very favorable balance of trade in chemicals (many of which are hazardous materials), the U.S. may sustain a net loss from any reduction in trade caused by barriers such as non-reciprocal packaging and shipping requirements. In order to avoid such a reduction in trade, DOT will have to adopt the UN system, or maintain two systems—the existing regulations for domestic shipments and a UN based system for international shipments. The second alternative is highly undesirable since it would add to the volume of the HMR and require additional government resources to maintain the two systems.

The final goal of this project, the reorganization of the bulk packaging requirements of the HMR, would represent an attempt to simplify and rationalize those portions of the HMR applying to packagings and packages having a capacity greater than 450 liters.

ORGANIZATION OF THE ANPRM

This ANPRM consists of three parts: a revised Hazardous Materials Table, performance-oriented packaging standards, and revised shipper requirements which assign those standards to particular hazardous materials. In other words, the packaging requirements for particular hazardous materials are given in the shipper's requirements (Part 173) by referencing the approved packaging in the Part 178 standards.

The performance-oriented non-bulk packaging standards in this ANPRM would be codified in Part 178 as Subpart L (§178.500 to §178.525). Subpart M of Part 178 (§178.600 to §178.608) would contain the performance tests for non-bulk packagings.

In addition to the performance-oriented packaging standards covering drums, barrels, jerricans, boxes, bags and composite receptacles, Subpart L would also contain definitions and terms, the alpha-numeric grouping used as identification codes, some general construction requirements, and the requirements for marking non-bulk packagings used in the U.N. system.

The requirements proposed in Subpart L are essentially the same as the general recommendations on packagings found in Chapter 9 of the U.N. Recommendations. It should be noted that Chapter 9 is currently in the process of being updated by the Committee of Experts, and the provisions contained in this advance notice reflect the majority of the changes proposed to be incorporated into the UN Recommendations as of March 1982. It is envisaged that, should a Notice of Proposed Rulemaking be published, the packaging standards proposed would reflect the most current text of Chapter 9 of the UN Recommendations. In addition, changes are proposed to make the text of the

UN Recommendations mandatory ("should" has been changed to "shall") and, where the language is ambiguous or unclear, changes have been made for the sake of clarity.

Not addressed in this ANPRM are packagings used exclusively to transport explosives or radioactive materials, or the specifications for cylinders. Regulations applying to explosives, radioactive materials and compressed gases are undergoing separate reviews.

Part 173 of the HMR assigns packaging requirements, or exceptions from those requirements, for hazardous materials. If, for example, a shipper is interested in ascertaining the packaging requirements for a hazardous material, it is first necessary to find the appropriate proper shipping name in the Hazardous Materials Table (49 CFR 172.101), and then look to column 5 entitled "Packaging." Paragraph references in column 5 are in Part 173, where both the packaging requirements for the particular hazardous material as well as the applicable exceptions, if any, are found.

Because of the many entries in the Hazardous Materials Table (approximately 2700, divided into twenty-two hazard classes), it is not practical in this ANPRM to republish the entire Hazardous Materials Table (§172.101) and shipper requirements (Part 173) to reflect the incorporation of the performance-oriented packaging standards. Instead, a sample hazard class, "Flammable liquid," was chosen to demonstrate how the Hazardous Materials Table and the shipping requirements in Part 173 could be rewritten to accommodate the proposed new packaging standards for both bulk and non-bulk packagings.

The flammable liquid hazard class was chosen for this demonstration because it is a complex one, specifying many different packagings for a long list of materials with varying characteristics, and is probably the "worst case" among all of the hazard classes in its difficulty for reconstructing the HMR to accommodate performance-oriented packaging standards and the revised bulk requirements.

For purposes of demonstration, this ANPRM includes a revised Hazardous Materials Table (§172.101) with all entries for the hazard class "Flammable liquid." The columns bear the same headings as the current Hazardous Materials Table. In general, proposed changes have been made only in entries in columns 5(b) (specific packagings requirements) and 6(a) and 6(b) (maximum net quantity in one package for carriage aboard passenger-carrying aircraft or railcar); however, a limited number of descriptions in column (2) would be revised, and two new descriptions would be added.

With two exceptions (mercaptan mixtures and methyl vinyl ketone), references in column 5(a) to "exceptions from specific packaging requirements," which do not refer to specification packaging, were not changed and the existing references to non-specification packaging remain the same. With the exceptions noted above, sections which deal with non-specification packaging are not included in the shipper's requirements of this demonstration because they will not change.

References in column 5(b) (specification packaging), refer to the proposed shipper requirements which in turn refer to the performance-oriented packaging standards for non-bulk packages proposed in Subpart L of Part 178 or bulk packaging requirements in Parts 178 and 179.

Columns 6(a) and 6(b) would contain the maximum quantities of flammable liquids, in liters, permitted aboard passenger-carrying and cargo-only aircraft, respectively. The quantities indicated in this proposal reflect, to the maximum extent possible, quantity limits established in the Technical Instructions for the Safe Transport of Dangerous Goods by Air, published by the International Civil Aviation Organization (ICAO). This would be a change from the existing Hazardous Materials Table.

The proposed sections on shipper requirements (Part 173) of this ANPRM consist essentially of: (1) Five non-bulk packing methods contained in paragraphs §173.119 (a) through (e), (2) one general bulk packing method contained in paragraph §173.119(f), (3) thirteen specific bulk packing methods contained in paragraphs §173.119 (g) through (s), followed by (4) §173.120 through §173.129 which specify both bulk and non-bulk requirements for particular hazardous materials. Sections 173.130 through 173.149a would be deleted and reversed for possible future use.

The general philosophy employed in the development of the five general non-bulk packing methods is that hazardous materials presenting a similar hazard in transport should be packaged in the same manner. For the purposes of the ANPRM, "similar" means materials of the same UN Packing Group. The general criteria applied by the UN Committee in assigning flammable liquids to particular Packing Groups are:

Packing Group I—Boiling point less than or equal to 35° C.
Packing Group II—Boiling point greater than 35° C and flashpoint less than 23° C.
Packing Group III—Boiling point greater than 35° C and flashpoint greater than or equal to 23° C.

However, in this connection, it should be noted that when assigning Packing Groups to specific materials, the UN Committee of Experts may deviate from these general criteria on the basis of the actual transportation experience. This general philosophy of like packaging for materials of similar hazard has not been applied consistently in existing DOT's regulations. Numerous instances exist where the packaging for two materials of virtually identical hazard is different, as is the case of ethyl mercaptan and butyl mercaptan. This may result in the shipper being forced to use a less desirable (from the point of view of user preference rather than from inherent safety) packaging, or alternatively, requesting a DOT exemption.

A second, and somewhat related general principle, is incorporated into the proposed non-bulk packing methods. This general principle is that any packing which is considered suitable for the transport of a hazardous material by virtue of the hazards and physical properties of the material should be authorized for the material. This principle also has not been consistently applied in the existing DOT regulations. For example, a fiberboard box with glass or earthenware inner receptacles may be authorized for a particular material whereas a wooden box with glass or earthenware inner receptacles is not. Such omissions cannot be justified on the grounds of safety and are most often the result of the piecemeal fashion in which the regulations have evolved. In general, the packagings authorized in the regulations reflect what industry is currently using. However, as industry practice changes and packaging technology advances, the types of packagings suitable to transport hazardous materials have not been updated appropriately. MTB believes that authorizing all types of packaging suitable for the transport of particular hazardous materials will vastly increase the flexibility of industry to utilize the most desirable packaging available while, at the same time, eliminating the burden of exemptions on both industry and MTB without adversely affecting safety.

With the exception of transportation by aircraft, the maximum capacities of inner receptacles of combination packagings will no longer be specified. MTB believes that the size of the inner receptacle should no longer be the controlling factor since the entire non-bulk packaging will have to pass the prescribed performance tests. For transportation by aircraft, certain combination packagings would not be authorized and limitations would be prescribed for the capacities of inner receptacles. This would ensure the maximum possible consistency with the corresponding provisions in the ICAO Technical Instructions.

The following is a description of the five no-bulk Packing Methods found in §173.119. Paragraph (a) of §173.119 provides packagings authorized for the general class of flammable liquids corresponding to UN Packing Group I. Packages identified in this paragraph must meet Packing Group I performance tests—the most severe.

Paragraph (b) of §173.119 is also for flammable liquids which fall under UN Packing Group I, but, because of certain incompatibilities or differences in quantities allowed aboard aircraft, they are not included in the general category (173.119(a)) and instead are treated separately. This paragraph contains a table of authorized packagings suitable for particular hazardous materials which are identified by their UN or NA identification number.

Paragraph (c) of §173.119 like §173.119(a) is a generalized category of flammable liquids, but in this instance they are less volatile and fall under Packing Group II. The packages for this group of materials need only meet Packing Group II performance tests.

As paragraph (c) of §173.119 corresponds to §173.119(a), for Group II materials, so paragraph (d) of §173.119 corresponds to §173.119(b), for Group II materials. In other words, §173.119(d) is for flammable liquids falling under Packing Group II which have specific requirements or restrictions either in the material or design of the packaging, or the quantity allowed aboard aircraft. As in §173.119(b), §173.119(d) contains a list of suitable packagings in tabular form with the hazardous materials identified by their identification numbers.

Paragraph (e) of §173.119 is for flammable liquids even less volatile than the two preceding categories, and classified in the UN system as falling under Packing Group III. There are no special requirements for this group of materials, so they are all able to be accommodated within a single paragraph. Packages specified in this paragraph need only pass the Packing Group III performance test requirements. This paragraph contains several provisions not found in the four preceding paragraphs. The first, in §173.119(e)(3), provides for the use of maximum net quantity limits per package allowed on aircraft by the ICAO Technical Instructions, provided that packagings tested and marked as meeting Packing Group III performance standards are utilized. Should a shipper elect not to use Packing Group III performance tested packagings, the current DOT exception from specification packaging for materials with a flash point above 23°C (73°F) would be retained and appear in §173.119(e) (4). However, in such cases the maximum net quantity of 4 liters per packing allowed aboard passenger carrying aircraft, as provided in the existing DOT regulations, would still apply.

Following the five paragraphs on non-bulk packaging (§173.119 (a) through (e)) is a single paragraph (§173.119(f)) consolidating the general bulk packaging requirements for flammable liquids. Paragraph §173.119(f) is divided into four subparagraphs covering: (1) Tank cars, (2) cargo tanks, (3) intermodal portable tanks and (4) portable tanks other than intermodal. The remainder of §173.119 consists of thirteen paragraphs (§173.119(g) through (s)) each covering the bulk packaging requirements of either a single material or a class of materials with similar properties (for example the mercaptans). The non-bulk packaging requirements of these thirteen materials or classes are sufficiently non-specific so that they may be covered by the general non-bulk packing methods found in paragraph §173.119 (a) through (e), but their requirements for bulk shipments are sufficiently unique to require the separate bulk packaging paragraphs.

The remainder of the proposed Part 173 consists of §173.120 through §173.129 with each section devoted to particular materials. This is necessary because the unique nature of the material (such as automobiles) or particular hazardous properties (such as ethylene oxide) make it impossible to cover either their bulk or non-bulk shipping requirements under the general paragraphs in §173.119. However, two of these sections (§173.128 covering mercaptans and §173.129 covering methyl vinyl ketone) only contain packaging exemptions, and thus general packaging methods would apply.

The rearrangement of the bulk packaging requirements would provide an opportunity for the transfer of operational and maintenance requirements for specific packagings from sections covering individual materials into sections covering operation and maintenance (§§173.31, 173.32 and 173.33) where they belong; for example, the dome placard requirements now contained in §173.119(h) could be moved to §173.31. This would remove many redundant entries. Although these changes do not appear in this ANPRM they could be included in a future notice if this project proceeds to that stage.

Viewed in its entirety, this ANPRM, proposes: (1) A revised Hazardous Materials Table (§172.101) for the "Flammable liquid" hazard class; (2) revised shipper requirements for flammable liquids (Part 173); (3) a new Subpart L to Part 178 containing performance-oriented non-bulk packaging standards based on the UN Recommendations; and, (4) a new Subpart M to Part 178 containing the required performance tests for the packagings specified in Subpart L. For example, column 5(b) of the proposed Hazardous Materials Table refers shippers to flammable liquids to either §173.119 or to one of the sections in Part 173 dealing with particular flammable liquids. The non-bulk packaging requirements of the vast majority of flammable liquids would be covered by one of the five paragraphs in §173.119 ((a) or (b) for Group I materials, (c) or (d) for Group II materials, and (e) for Group III materials) while the bulk requirements would be covered in §173.119(f). The marking requirements and construction standards for the non-bulk packagings referred to in Part 173 are presented in Subpart L of Part 178 (§§178.500 through 178.525), and the performance tests that those packagings must meet are given in Subpart M of part 178 (§§178.600 through 178.608).

If the concept set forth in this ANPRM is adopted as a final rule, this proposal would eventually eliminate the following 93 DOT specification packagings: Carboys, jugs in tubs, and rubber drums:

1A, 1D, 1X, 1EX, 1H, 1K, 35, 1M, 34; Inside containers and linings: 2T, 2C, 2D, 2U, 2E, 2F, 2G, 2TL, 2J,2K, 2L, 2M, 2N, 2R, 2S, 2SL; Metal barrels, drums, kegs, cases, trunks, and boxes: 5, 5A, 5B,5C, 5K, 5L, 5M, 5P, 6B, 6C, 6J, 6D, 42B, 42D, 17C, 17E, 17F, 17H, 37K, 37A, 37B, 37P, 37M, 37C, 13A, 32A, 32B, 32C, 32D, 33A; Wooden barrels, kegs, boxes, kits, and drums: 10B, 14, 15A, 15B, 15C, 15D, 15E, 15X, 15P, 16A, 16B, 16D, 19A, 19B, 18B, 22A, 22B, 22C; Fiberboard boxes, drums, and mailing tubes: 12B, 12C, 12D, 12E, 12A, 12P, 12R, 21C, 21P, 29; and Bags, cloth, burlap, paper or plastic: 36A, 36B, 36C, 44B, 44C, 44D, 44E, 45B, 44P. In addition, MTB estimates that more than 300 pages of text in the Code of Federal Regulations would be eliminated.

It is important to emphasize that the proposed shift to performance-oriented non-bulk packagings does not represent a total change in philosophy from the standards imposed under existing DOT specifications. Although termed "specifications" and heavily oriented toward detailed construction requirements, most of the current DOT packaging standards are, in reality, a combination of construction specifications and performance standards. Many, and perhaps most, of the existing DOT specification non-bulk packagings would successfully pass the performance tests prescribed in this notice and could be marked with the appropriate proposed markings. Therefore, although the required markings on non-bulk packagings would change after a suitable transition period, existing types and designs of packagings could continue to be manufactured and used in the future if considered desirable.

TRANSITION

In order to facilitate changeover from the present DOT specification system to the UN performance-oriented system, a transition period would be specified to allow packaging manufacturers to design, retool, and begin production of the new non-bulk packagings, and to allow packaging manufacturers and shippers to exhaust their existing stock. MTB solicits comments on the length of the transition period which would begin after issuance of a final rule and continue until only performance-oriented non-bulk packagings are permitted.

MTB envisions that, upon publication of a final rule, the DOT packaging specifications listed above would no longer appear in Title 49 even though their new construction and continued use would be authorized throughout the transition period. The performance-oriented non-bulk packaging standards would replace the existing packaging specifications in the CFR.

Likewise, the revised Part 173 would replace the existing shipper requirements. With fewer packagings authorized, Part 173 would also undergo a significant reduction in the number of pages of the HMR. However, more important than the size of the reduction of the regulations is the fact that with a reduction of a number of authorized non-bulk packagings the shipper requirements will be greatly simplified. This should not only ease the burden on the regulated public in complying with the regulations, it should also reduce the burden on government through an expected decrease in the number of requests for interpretation and exemption received annually.

INDUSTRY CONTRIBUTION

The development of this proposed format for the presentation of packaging requirements in Part 173 benefitted greatly from the assistance rendered by a group of industry packaging experts working under the joint auspices of the Chemical Packaging Committee (CPC) of the Packaging Institute and the chemical Manufacturers Association (CMA). This group developed a series of sample formats for the presentation of the packaging requirements in Part 173 which proved to be of considerable assistance to MTB in the preparation of this ANPRM. Although the format suggested in this document is a compilation of views expressed by both industry and government and intended to be responsive to the needs of both, the views presented by the joint CPC/CMA group were invaluable to MTB. The interest, dedication and foresightedness of this group is clearly recognized and very much appreciated by MTB.

PROPOSED REVISIONS

In order that interested parties may better understand this proposal and have methodology on which to comment, the ANPRM contains, as a demonstration, the following proposed changes to the HMR:

Section 171.8 would be revised by adding definitions for "bulk" and "non-bulk" packagings.

In §171.12, paragraph (c) would be revised to provide that hazardous materials being imported into, exported from, or transiting the U.S. may be transported in packagings tested and marked in compliance with the UN Recommendations or equivalent national international regulations of countries outside the U.S., provided the package is otherwise offered, accepted and transported in accordance with the DOT regulations. This means, for example, that a steel drum (1A1) tested and marked in compliance with the UN Recommendations or equivalent international (e.g., IMCO, ICAO, RID, ADR) or national regulations under the conditions established by the Competent Authority of a foreign country, could be used in the transportation of an import, export, or transiting shipment in the U.S., provided a 1A1 steed drum is authorized in Part 173 for the transport of the particular hazardous material and the shipment is otherwise handled in accordance with the HMR.

The Hazardous Materials Table in §172.101 would be extracted and amended for the hazard class "Flammable liquid" to reflect performance-oriented non-bulk packaging standards based on the UN Recommendations and revised bulk shipping requirements.

Subpart A (General) of Part 173 is not included for purposes of this demonstration. However, it is anticipated that in the event this ANPRM leads to a Notice of Proposed Rulemaking (NPRM), §173.6 of Subpart A (Shipment by Air) would be revised to reflect the latest requirements of ICAO.

Subpart B (Preparation of Hazardous Materials for Transportation) of Part 173 would be amended to reflect general requirements in accordance with those contained in Chapter 9 of the UN Recommendations appropriate as shipper responsibilities, and to reflect specific packaging requirements for the flammable liquid hazard class of hazardous materials based on the performance-oriented packaging standards contained in the UN Recommendations and on the Packing Group for the specific materials.

Subpart D (Flammable, Combustible and Pyrophoric Liquids; Definitions and Preparation) of Part 173 would be amended, as a demonstration in this ANPRM, to reflect the packaging requirements and performance-oriented non-bulk packaging standards of the UN Recommendations.

A new Subpart L would be added to Part 178 contained definitions, markings, general packaging requirements and packaging standards for non-bulk packagings based on the U.N. Recommendations.

A new Subpart M would be added to Part 178 containing the performance tests prescribed for the packagings given in Subpart L.

MTB for the purpose of this ANPRM, proposes to amend Part 178 to include both the performance-oriented packaging standards and the required tests for packagings associated with those standards contained in Chapter 9 of the UN Recommendations.

(Ed.—Docket No. HM-181 is a major on-going DOT rule making, proposing the domestic implementation of many international requirements. Be sure to determine the current status of this rule making, which is referenced throughout the *Red Book*.)

(47 F.R. 16268, April 15, 1982)

49 CFR Parts 173 and 178
[Docket No. HM-182; Advance Notice]
Specifications for and Use of Specification 17E Steel Drums

On April 19, 1979, PPG Industries, Incorporated, filed an application for an exemption from 49 CFR 173.119(a)(3) to allow the shipment of certain flammable liquids having a flashpoint of 20°F or less in 55 gallon DOT 17E steel drums with a triple-seamed chime construction and having a body thickness of 20 gauge in place of the minimum body thickness of 18 gauge required by the regulations. PPG supported this request by citing test results which, the applicant held, proved that the drum could pass all current DOT test requirements for an all 18 gauge DOT 17E steel drum. The request was assigned number 8187-N and published for comment in the **Federal Register** on May 21, 1979 (44 FR 29551).

After a thorough study of the application and the supporting information provided, as well as consideration of all public comments submitted relative to the application, the Associate Director for Hazardous Materials Regulations advised PPG Industries that the request was denied. An appeal of this denial was filed in accordance with 49 CFR 107.121 and was subsequently denied by the Director, Materials Transportation Bureau. The basis for denial of the appeal as well as the original request, was *inter alia,* that the subject raised in the application was of such general applicability and future effect that it could not appropriately be considered as an exemption and could only be adequately evaluated through a rulemaking proceeding (49 CFR 107.109(e)).

On January 1982, a petition for rulemaking was filed by Inland Steel Container Company (Inland) which requested amendment of the DOT Hazardous Materials Regulations to allow the use of a triple-seamed steel drum meeting all requirements of DOT Specification 17E except that the minimum thickness of body and head sheets would be reduced from that currently prescribed. The Inland petition is similar in many respects to the PPG petition; however, it is much broader in scope in that it applies to all hazardous materials for which a Specification 17E drum is currently authorized. The Inland petition is supported by a number of letters submitted by major shippers of hazardous materials.

Because of the increasing general interest on the part of shippers and steel drum manufacturers in the use of triple-seamed drums complying with DOT Specification 17E with the exception of reduced body and/or head thicknesses, MTB believes it is now appropriate to solicit views on this subject from all segments of the public, looking towards a possible amendment of the Hazardous Materials Regulations authorizing the use of such drums. In order to afford the public ample opportunity to consider the merits of such action, MTB is providing in this publication a reproduction of the salient points of the Inland petition. The Inland petition proposes specific amendments and provides supporting information as follows:

SUBSTANCE OF THE PETITION

It is proposed that existing regulations be revised to authorize the shipment of certain commodities in steel drums which conform in all respects except steel thickness to Specification 17E (§178.116),and which are manufactured with triple-seamed chimes. Specifically it is proposed that:

(1) Commodities that are presently authorized in Part 173 to be shipped in drums of 55 gallon capacity or less, constructed with head sheets of 18 gage steel and body sheets of 20 gage steel, be authorized in triple seamed drums constructed of 20 gage steel throughout, and

(2) Commodities that are presently authorized through §173.119(a) to be shipped in drums of 55 gallon capacity or less constructed of 18 gage steel throughout, be authorized in triple-seamed drums constructed with head sheets of 18 gauge steel and body sheets of 20 gauge steel.

To accomplish this purpose, the following revisions are proposed:

§178.116–6. This paragraph to be modified by appending footnote 3 to the 18 gauge minimum shown in the "Head sheet" column of the table for 55 gallons maximum capacity drums. The footnote to read as follows: "20 gage authorized, provided heads are seamed to bodies by a process which results in chimes with seven overlapping layers formed from the parent and body steel."

§173.119(a)(3). This paragraph to be modified by revising the third sentence to read as follows: "Drums with a marked capacity in excess of 30 gallons must be constructed of 18 gage body and head sheets, except that 20 gage is authorized for the body sheets if heads are seamed to bodies by a process which results in chimes with seven overlapping layers formed from the parent head and body steel."

BACKGROUND OF THE PROPOSAL

The Specification 17E drum in the 55 gallon size, which is the subject of this petition is well-known as the most widely used packaging for hazardous liquid substances, where the degree of hazard falls within a range from relatively low to moderately severe. Because it has become highly standardized in dimension and construction, it is also the accepted packaging for non-hazardous liquids. It is, in fact, a world standard packaging for the transport of a vast variety of liquid substances in industrial and commercial trade.

U.S. production of new drums that conform to the requirements of this specification was approximately 22.5 million units in 1980. Of these, 9.0 million were 18 gage throughout and 11.5 million were made with 20 gage bodies and 18 gage heads (20/18). In addition, approximately 2.0 million drums of the same type of construction, but made of 20 gage steel throughout were produced.

Because of the considerable market for this type of packaging, the design of the drum and the processes for its manufacture have become highly developed. Drums are produced today on automated lines operating under controlled conditions at rates up to 900 drums per hour. Product uniformity and quality is not only a requirement for shipper satisfaction, it is essential to the efficient operation of these facilities.

From a product design standpoint, there have been two significant developments in recent years that have led to the improved drum performance. The first of these, the drop panel bottom, significantly reduced the incidence of vibration induced bottom failures. The second, and the one which is the basis for this petition, is the triple seam.

Initial development of what has come to be known as the triple seam chime construction, was done in Europe in the late 1960's in response to a perceived need to improve drum performance under the conditions which are simulated in a drop test. Several designs were produced and in fact still available, but all designs have tended to approach a standard in which a tight chime with seven overlapping layers is produced. This construction is attainable uniformly under well-controlled, high speed production conditions.

Triple seamed drums were introduced in the U.S. in 1978 by means of license agreements with European manufacturers. Since that time, the drum, in a wide variety of gage combinations, has become standard in Europe, and is being produced broadly in the U.S. by some 14 manufacturers.

In Europe, triple seam chime construction has resulted in a dramatic shift in the mix of drum types. Prior to 1970, the standard drum was 1.25 mm, with some production of 1.0mm drums and .625mm chime reinforced Monostress. Since that time, the 1.0mm drum, along with a .9/1.0 version, has become the most common. Additionally, drums are being produced in greater quantity in .825mm and .75mm versions, and a triple seamed Monostress drum is being used in a combination .6/.825mm version.

In the U.S. triple seamed drums are being produced in 18, 20/18 and 20 gage versions in quantity, and in 16 gage on a more limited basis.

It should be noted that while steel is purchased in the U.S. to a given decimal with no negative tolerance, the practice in Europe is different. Steel is purchased there with both plus and minus tolerances. Therefore, steel stated as 1.0mm can be expected to vary both above and below that thickness by 2 to 4%; whereas, steel purchased in the U.S. to a .0324″ minimum decimal (20 gage) can be expected to average from 2 to 4% heavier. Appendix I illustrates the relationship between European metric steel thicknesses and U.S. gages.

In summary, triple seam chime construction is now a highly developed production technique; it is the accepted method for a very significant segment of word drum production; and it has enabled manufacturers to produce lighter, less costly drums. The reasons for this transition are presented in the following section on performance.

PERFORMANCE CAPABILITY OF TRIPLE SEAMED DRUMS

Triple seaming is a process by which a drum head is joined to drum body by mechanically rolling up circumferential flanges on the two pieces into seven layer chime. This construction contrasts with previous double seaming practice which results in a five layer chime. Appendix II pictorially compares the two types of chimes in cross-section.

A triple seamed chime as a result of the additional thickness and interlocking of the layers has superior performance characteristics when compared to a double seamed chime. It is more resistant to denting and other concentrated impact damage; it performs in vastly superior manner when subjected to the more dispersed crushing force of a fall from a height; it has better resistance to the unrolling force exerted by severe internal pressure; and its tight construction is inherently more leakproof.

As a result of this performance capability, drums manufactured with triple seamed chimes in a given steel gage meet and exceed the performance standards set for heavier gage drums. This is clearly demonstrated in the tests conducted for Inland by Gaynes Laboratories.

In the tests three drum styles were subjected to the series stipulated in the United Nations Recommendations on the Transportation of Dangerous Goods for packagings suitable for the carriage of Packing Group I substances with a specific gravity 1.2 or less. This level exceeds in severity that specified in DOT specification 17E and includes all of the tests required for that specification.

The three styles tested were 55 gallon 18 gage, 55 gallon 20/18 gage and 55 gallon gage. The drums were manufactured with triple seamed chimes, and except for the nonspecification gage of the 20 gage drums, were made in accordance with the construction requirements in DOT specification 17E.

The U.N. tests conducted were the 1.8 meter (6′) diagonal and flat drop tests, 250 kpa (36.2 psi) internal hydrostatic pressure test, 8 meter (26′) stacking test and 30 kpa (4.3 psi) air pressure leak test. Supplemental tests included the DOT specified 7 psi air pressure leak test, a 40 psi internal hydrostatic pressure test and a one hour, 285 rpm vibration test on a table having a 1″ displacement. There were no failures in any of the tests conducted.

Copies of the test reports prepared by Gaynes Laboratories are included in Appendices III, IV and V.

STATEMENT OF SHIPPER INTEREST

Accompanying this petition are several letters from shippers, addressed to the Office of Hazardous Materials, which set forth an interest in the adoption of our proposed regulation change. The interest stated is based on the following reasons:

1. Shippers have had successful experience with the use of 20 gage and 20/18 gage triple seamed drums for several years and in numerous shipments. Some of this experience is described in the shippers' letters.

2. Triple seamed drums in test experience perform in superior fashion when compared with double seamed drums.

3. The ability to use lighter weight drums in transport offers an opportunity to reduce costs. The following weight* comparisons are noted:

55 gallon 18 gage—43.5#
55 gallon 20/16 gage—36#
55 gallon 20 gage—33#

*Based on theoretical weight at DOT minimum decimal. Actual weights will be somewhat higher.

4. The use of triple seamed drums is an enhancement to transportation safety. This construction method offers the opportunity for steel thickness reduction without compromising safety in transport.

CONCLUSION

Inland Steel Container believes that for the reasons which have been discussed in the body of this petition, a regulatory change is warranted. Further, it is believed that this is consistent with the position that DOT has supported in international discussions promoting performance oriented specifications for packagings.

Approval of the change would provide benefits to U.S. business in cost reduction and in improved competitive position in world trade.

APPENDIX I.—COMPARISON OF U.S. AND EUROPEAN STEEL THICKNESSES FOR STEEL DRUMS

[DOT thicknesses]

U.S. gage	DOT nominal		DOT minimum	
	Inches	Millimeters	Inches	Millimeters
24	0.0239	0.607	0.0209	0.531
22	.0299	.759	.0269	.683
20	.0359	.912	.0324	.823
19	.0418	1.062	.0378	.960
18	.0478	1.214	.0428	1.087
16	.0598	1.519	.0533	1.354

[European Ordering Thicknesses]

Nominal		U.S. Manufacturers gage range
Inches	Millimeters	
0.0246	0.625	24
.0295	.75	22
.0354	.90	20
.0394	1.00	19
.0492	1.25	18
.0591	1.50	16

For the sake of brevity, Appendix II of the Inland petition, which presents a cross-sectional diagram of both a triple-seamed chime and a standard double-seamed chime, has not been reproduced in this ANPRM. In addition, since the tests and test results presented in Appendices III, IV and V of the Inland petition were summarized in the body of the petition, they have not been reproduced in this ANPRM. Copies of each of these appendices are available for review in the docket file.

The statement of shipper interest referred to in the Inland petition were filed by the following shippers of hazardous materials: Monsanto Company, PPG Industries, Inc., Economics Laboratory, Inc., Nalco Chemical Company.

Copies of these statements are available for review in the docket file.

In connection with this ANPRM and the Inland petition for rulemaking, it should be noted that applications for exemptions, again concerning the minimum thickness of DOT 17E drums, have been received for consideration under 49 CFR 107.103. These applications request authorizations to manufacture, mark and sell a steel drum with head sheets of a thickness below the minimum thickness prescribed but are otherwise in full compliance with Specification 17E. The following applications are being reviewed:

1. Mauser-Werke G. m.b.H., West Germany. Top and bottom heads will be made of 1mm (0.0366 inch minimum) steel in place of 18 gauge (0.0428 inch minimum) steel presently required. For materials authorized to be packaged in DOT-17E, 20/18 gauge drums.

2. G. Shonung Co., West Germany. Top and bottom heads will be 1 mm (0.0366 inch minimum) steel in place of 18 gauge (0.0428 inch minimum) steel presently required. For materials authorized to be packaged in DOT-17E drums.

3. Natico, Inc., USA. Top and bottom heads would be 20 gauge (0.0324 inch minimum) steel in place of 18 gauge (0.0428 inch minimum) steel presently required. For materials authorized to be packaged in DOT-17E drums.

Comments are solicited on the amendments requested in the Inland petition and on the following questions:

1. What has been the transportation experience in the use of both all 20 gage triple-seamed drums and 20/18 gage triple-seamed drums for hazardous (when authorized) and nonhazardous materials;

2. What would be the safety implications, if any, particularly with respect to resistance to puncture, if the reduced body and head thicknesses were authorized under the conditions proposed by Inland; and,

3. What would be the economic benefits and consequences associated with adoption of the Inland petition with regard to the purchase of new and used drums.

(47 F.R. 25167, June 10, 1982)

49 CFR Parts 173, 177, and 178
[Docket No. HM-183, Advance Notice No. 82–5]
Design, Maintenance, and Testing of MC-306 Cargo Tanks

The MC-306 type cargo tank is the major highway transport vehicle used for the movement of flammable and combustible liquids. The MC-306 type cargo tank category includes MC-300, MC-301, MC-302, MC-303, and MC-305 cargo tanks. Statistics indicate that when this type cargo tank is in an accident, a high incidence of product leakage occurs in an overturn situation.

In a 1975 BMCS analysis of incident and accident reports resulted in a decision to formalize a "Tank Integrity Program." The program's main objective was to determine how cargo tank incident causation could be identified and mitigated. A two-phase program was undertaken to accomplish this objective. Phase I of the program called for a review of existing research and a thorough analysis of multi-source accident bases. The Phase II effort was to provide for crash testing of cargo tanks and was to be predicated on Phase I results.

A contract to perform Phase I was awarded to Dynamic Science, Inc., Phoenix, Arizona. The contractor reviewed existing research and accident data, conducted field investigations and evaluated current specifications. The results of this review were inconclusive because the existing accident data were not sufficiently comprehensive.

Accordingly, the contract was modified to have the contractor conduct tests to determine if the current tank designs provided adequate protection to prevent leakage of cargo in overturn situations. Three tests were performed to complete this task: (1) A static vertical guard loading test; (2) a static horizontal pipe loading test; and (3) a tipover test. These tests were conducted using MC-305 and MC-306 cargo tanks.

The static vertical guard loading test was conducted on a 1971 MC-306 cargo tank. Major leaks developed at all hatch cover vent/check valves, at one hatch cover seal, and at one discharge vent valve when the vehicle was rotated upside down with only 10 percent of the full load in the tank. After sealing these leaks, the actual vertical roof loading test was conducted at two times its load weight with no subsequent leakage or damage to the tank structure.

Static horizontal pipe loading tests were conducted on both a 1971 MC-306 and a 1966 MC-305 cargo tank. The pipe elbows failed at the shear section, as designed. A hairline crack developed at a weld on the MC-306 cargo tank resulting in a slight cargo leakage. The valves located upstream of the shear section on the MC-305 cargo tank were unseated and resulted in a major cargo release.

Tipover tests were also conducted on both the MC-305 and MC-306 cargo tanks. There was considerable damage to the MC-305 cargo tank shell but no leakage from it. All of the dome covers leaked, however, and some leaks approached a rate of 15 gallons per minute. The partitions between compartments appeared to have broken, thereby permitting mixing of compartment contents. Major damage occurred on the right front corner of the MC-306 cargo tank shell. A 4-inch weld split along the front bulkhead-to-shell-seam. This permitted cargo leakage at the rate of 60 gallons per minute. Leakage would have occurred from the vents and valves, if these openings had not been deliberately sealed. The compartment partitions also broke through in this test.

The results of the tipover tests indicate a need to improve the Specification MC-306 cargo tank standards to reduce the likelihood of leakage in overturn accidents. Particular attention must be given to leakage from valves, vents, and manhole covers.

For a more detailed description of testing procedures, results and recommendations, see "Analysis of Cargo Tank Integrity in Rollovers" (contract number DOT-FH-11-9193). This document is available to the public through the National Technical Information Service (NTIS), NTIS Accession No. PB 279506.

The results of the Phase I testing indicated that while there are certain integrity problems with the cargo tank shell, the primary source of a product leakage was from cargo tank openings. It was determined that the problem of product leakage from cargo tank openings was an area which warranted immediate attention. Consequently, Phase II, which was to be primarily aimed at testing cargo tank integrity, was postponed and the effort was focused on the cargo tank opening problem.

The contract to perform Phase II was awarded to Dynamic Science, Inc., Phoenix, Arizona, in October 1978. The contract had four objectives: (1) To assess present maintenance practices and requalification requirements as they affect a cargo tank's continuing product retention capability; (2) to assess existing specifications for manhole covers, fill covers, and other product retention items and identify specific items which represent potential leakage points in overturn accidents; (3) develop test procedures and engineering drawings for a simulator capable of testing manhole covers and other product retention devices in overturn situations; and (4) develop engineering recommendations to improve cargo tank product retention capabilities that can be incorporated into the cargo tank specification and qualification requirements of the Department of Transportation.

In order to assess carrier maintenance practices, a survey of 10 geographically separated carriers was performed. The survey included both large and small operations of five common and five private carriers. The basic findings of the survey were: (1) very little maintenance is done on critical components such as manholes, high capacity vents and breather vents; (2) scheduled maintenance frequency was an extreme variable; (3) structural maintenance was reported to be the most difficult, but components provided the most maintenance problems; (4) manholes, valve operators, adapters, and internal valves were identified as the tank components that required the most attention, repair, and replacement effort; and (5) shop inspection and repair systems were formally established and well supported by internal files which revealed that most maintenance is directed to power units and cargo tank running gear.

The carrier survey revealed that those components requiring the most maintenance effort had little maintenance performed on them. Field and laboratory testing was used to identify those cargo tank components that would be involved in preventing leakage in overturn accidents. Sixty-one cargo tanks with a total of 187 compartments were tested. The tanks were pneumatically tested to satisfy the requirements of 49 CFR 177.824 except that the test pressure was limited to one (1) psi to prevent damage to the tank structure. The tests were performed with the internal valves open and then with the internal valves closed. These tests identified leaks for all compartment system components except breather vents which were made inoperative in the manner required by 49 CFR 177.824(d)(1)(ii) when pressure tests are required. The primary sources of leakage were the manhole assembly, internal valve, high capacity vent, liquid level sensor, weld and shell cracks, vapor recovery should, cleanout opening and discharge outlet, adapter and manifold. Approximately 80% of the total leaks had top sources and 20% had bottom sources. The majority of the leaks in the manhole assembly were in the filler cover, fusible plug and dome cover.

Since breather vents were rendered inoperative during the field tests, no performance data could be obtained. Breather vents are usually located in the manhole filler cover which has been identified as a primary leakage source in overturn accidents. It was, therefore, necessary to test them under representative overturn conditions in order to accomplish a complete cargo tank compartment evaluation. A total of 119 breather vents of 16 different types were tested. The average leakage through the vent device was determined to be a steady stream of between ⅛ to ½ inch in diameter.

The field and laboratory testing provided data which indicated the sources from which leakage could be expected in cargo tank overturns. The static testing performed did not reveal the forces that accompany or induce leakage. Dynamic test data were necessary to determine the impact environment which produces leakage. To obtain the dynamic test data, a cargo tank compartment overturn simulator was designed and constructed. The simulator is capable of providing repeatable 90° and 180° tests without sustaining structural deformation. Two series of tests were performed using the overturn simulator.

The first series of tests were performed in the 90° overturn mode. The tests showed that significant leakage occurred on initial impact. Generally, test fluid was released on impact through the pressure actuated vent in the filler assembly. In liquid spray form, the test fluid covered an area approximately 15 feet to either side and above the simulator and 15–18 feet ahead of the manhole cover. For this series of tests, the average peak pressure at the manhole on impact was 15.6 psig.

The second test series consisted of both 90° and 180° overturns and included a fire test. The test results for this test series were identical to those of the first test series for all manhole assemblies with pressure actuated vents in the filler cover. The fire test was in a 90° overturn simulation with gasoline as the test fluid. On impact, the gasoline formed the spray pattern described above and ignition occurred at 838 milliseconds after impact. The resulting fireball had a maximum height of 21.1 feet, a maximum depth of 11.8 feet and a maximum width of 20.6 feet. Average temperatures recorded during the four seconds after impact were 1217° F at the manhole and 325° F fifteen feet in front of the manhole.

A meeting was held on February 19–21, 1980, to brief industry representatives on program results and obtain their input on possible regulatory changes which would cover the production and repair of cargo tanks and maintenance and operation of cargo tanks. The twenty-five people attending this meeting represented cargo tank and tank component manufacturers, carriers, repair agencies and trade organizations. In general, there was concurrence in changes that would result in overall safety and uniform practices. There was mostly nonconcurrence of changes that would require new designs or increased technical performance characteristics.

For more detailed information on Phase II results and recommendations see "Cost-Effective Methods of Reducing Leakage Occurring in Overturns of Liquid Carrying Cargo Tanks" and "Reducing Leakage Occurring in Overturns of Liquid Carrying Cargo Tanks" (Contract number DOT-FH-11-9494). These documents are available to the public through the National Technical Information Service (NTIS), NTIS Nos. PB-82-199936 and PB 82–198243, respectively.

Comments are solicited concerning the views, findings and recommendations of the contractor in the reports on Phases I and II cited above.

In view of the foregoing discussion on MC-306 type cargo tanks, comments are solicited on the following questions:

1. What design performance changes in manhole closures and venting devices are necessary to achieve the overturn integrity now required for MC-306 type cargo tanks?

2. Can the existing MC-306 type cargo tank fleet be retrofitted with improved manhole assemblies (manhole closure with or without PAV) without requiring changes in conventional 16 and 20 inch openings in compartment structures? If yes, please provide an estimate of the cost of installation per compartment.

3. Is it possible to remove pressure-actuated venting (PAV) from manhole fill covers?

4. Would a requirement for visual inspection prior to each loading applicable to manhole closures, vents, valves and piping improve the cargo retention capability of MC-306 type cargo tanks?

5. Should 49 CFR 177.824 be revised to require that MC-306 type cargo tanks be pressure (pneumatic or hydrostatic) tested? if so, at what intervals?

6. Please provide an estimate of the cost of visual inspection and a pressure test (pneumatic and hydrostatic) on a MC-306 type tank.

7. Are the skills required to test, inspect, and verify the integrity of these cargo tanks within the capabilities of currently employed carrier maintenance personnel?

8. What methods are presently used by cargo tank manufacturers to ensure that component parts are in compliance with the applicable DOT regulations?

9. Should the scope of this Docket be expanded to address the design and construction of the MC-306 cargo tank in its entirety?

(47 F.R. 27876, June 28, 1982)

49 CFR Parts 171, 172, 173, 176, 177, 178 and 180
[Docket Nos. HM-183, 183A; Notice No. 85–4]
Requirements for Cargo Tanks

This proposal includes provisions—

1. To require that all manufacturers of cargo tanks hold a current ASME certificate of authorization.

2. To require that each cargo tank designed with an internal design pressure of 15 psig or greater be "constructed and certified in conformance with the ASME Code", and each cargo tank with an internal design pressure less than 15 psig be "constructed in accordance with the ASME Code".

3. To require that all new specification cargo tanks be certified by an Authorized Inspector who is commissioned by the National Board of Boiler and Pressure Vessel Inspectors (National Board).

4. To require that ring stiffeners on a cargo tank be of a design that can be visually inspected.

5. To authorize the use of external self-closing stop valves in place of internal self-closing stop valves in certain circumstances.

6. To require that the strength of connecting structures on a multitank cargo tank be equal to that required of the cargo tank motor vehicle.

7. To specify minimum standards for the strength and size for a manhole on all new cargo tanks.

8. To require retrofit of any manhole closures not conforming to the prescribed strength requirement, within five years from the effective date of the final rule.

9. To specify the accident damage protection required for cargo tank motor vehicles.

10. To specify in Parts 173 and 177 the relationship between the cargo tank and its lading to guide manufacturers and shippers.

11. To clarify that the prescribed minimum thickness for the tank shell and heads excludes materials added for cladding, lining or corrosion allowance.

12. To specify the parameters to be considered in determining the effective stresses on a cargo tank.

13. To clarify that a remote means of closure for all internal or external self-closing stop valves is required.

14. To require that on all cargo tanks constructed after (the effective date of the final rule) all pressure relief devices be reclosing, except a frangible disc may be used in series with a reclosing pressure relief device.

15. To revise the MC 307 and MC 312 cargo tank specifications to provide for the manufacture of vacuum-loaded cargo tanks.

16. To specify a minimum design pressure of 15 psig for Specification MC 312 cargo tanks.

17. To require that all specification cargo tanks be pressure retested.

18. To require that all specification cargo tanks be visually inspected every year.

19. To require that the shell and head of an unlined cargo tank in a service corrosive to tank metal be thickness tested at least once every two years.

20. To specify certain additional safety control measures for a cargo tank used to transport a lading having more than one hazard class.

21. To require that a cargo tank used to transport a poison B material and certain hazardous materials having multiple hazards have a minimum design pressure of 25 psig.

22. To require that a cargo tank inspector or tester meet certain minimum knowledge and experience qualifications.

23. To require that major repairs on cargo tanks be performed by a facility that is a holder of an ASME U stamp, a National Board R stamp or be witnessed and certified by an Authorized Inspector.

24. To require that an owner of a cargo tank used in the transportation of hazardous materials retain certain records.

I. BACKGROUND

Cargo tanks are used for the bulk transport by highway of vast quantities of liquid and gaseous hazardous material essential to the support of our Nation's economy. This transportation over our Nation's highways represents a significant potential collective risk to the public and requires a continuing examination of safety control measures to ensure that the risk is minimized.

Each year MTB receives a large number of incident reports on the release of hazardous materials as a result of cargo tank motor vehicle collisions, overturns and loading/unloading accidents. As a result of these reports, BMCS and MTB jointly initiated in 1975 a long term research and development program. The objective of the research has been to evaluate the records of cargo tank accidents, reexamine existing HMR pertaining to cargo tanks, and document typical industry practices with respect to all DOT cargo tank specifications, except the recently adopted MC 338 cryogenic cargo tank. The industry practices portion of the research program included the examination of industry design, fabrication, operation, inspection, maintenance, test, and repair practices and procedures.

The initial focus of the program was on the MC 306 type cargo tank which is the major highway transport vehicle used to transport flammable and combustible liquids, such as gasoline and fuel oils. For the purpose of the research, the MC 306 type cargo tank included the MC 306 and its predecessors, the MC 300, MC 301, MC 302, MC 303, and MC 305 cargo tanks. Research findings showed that the MC 306 type cargo tank is highly susceptible to leakage and presents a substantial fire risk when the tank is involved in overturn accidents. Based on these findings, MTB and BMCS issued an advance notice of proposed rulemaking (ANPRM), which was published in the Federal Register under Docket HM-183, on June 28, 1982 (47 FR 27876). In the ANPRM, MTB and BMCS solicited comments on the advisability of revising the HMR covering MC 306 type cargo tanks to reduce the risk of release of lading from the cargo tank in overturn accidents. Comments to Docket HM-183 are included in the subject and the section-by-section discussions herein.

The second cargo tank type studied was the specification MC 331 and its predecessor, the MC 330. The MC 331 type cargo tank is used to transport liquefied gases, such as ammonia (NH_3) and liquefied petroleum gas (LPG). Because of the high pressure integrity required to contain these gases and the high potential risks associated with their release, cargo tanks used to transport these gases are designed, constructed and certified to the American Society of Mechanical Engineers (ASME), Boiler and Pressure Vessel Code. The ASME Code prescribes requirements for manufacturer qualification, manufacturing quality control and independent inspection and certification. Research findings confirmed that, while these cargo tanks are for the most part adequately designed and constructed, in some cases they are not being properly maintained and requalified. External corrosion which caused leaks, particularly in hidden areas such as the fifth wheel and stiffening rings, frequently went undetected. A large percentage of the pressure relief valves failed the basic operating tests. The study also found a substantial amount of stress-corrosion cracking, a condition that may severely degrade the integrity of a cargo tank. The regional geographical distribution of cargo tanks showing evidence of stress corrosion cracking suggests that shoppers in certain locales are offering hazardous materials for transportation in cargo tanks when those materials are not compatible with the tank material. The research contractor, Dynamics Sciences, Inc., pointed out that such practices were most probably induced by improper quality control procedures for the lading. Additionally, HMR contain no requirement pertaining to the qualification of cargo tank testers and research findings indicate that cargo tank tests are not always being performed by persons who are qualified to perform those functions.

The third group of cargo tanks studied was the MC 307 type cargo tank including its predecessor, the MC 304; and the MC 312 type cargo tank including its predecessors, the MC 310 and MC 311. The MC 307 and MC 312 type cargo tanks are used to transport high vapor pressure flammable liquids, poisonous materials, corrosive materials, and hazardous materials meeting two or more of these hazard classes. The variety of hazardous materials transported in these cargo tanks is substantial. Like the MC 331 type cargo tank, the MC 307 and MC 312 type cargo tanks with higher design pressures are required to conform with the ASME Code. As with the other types of cargo tanks, research findings have shown that poor maintenance, repair and requalification of these cargo tanks are major problems. Corrosion (external and internal) resulting from the road environment, lading spillage and chemical reaction between the lading and the tank materials, is a predominant problem. Another serious problem that has been identified is leakage of lading during loading and unloading operations due to malfunctioning valves.

MTB and BMCS believe these research findings have identified many shortcomings, inconsistencies and ambiguous requirements that currently exist in the cargo tank regulations. These deficiencies have resulted in situations, in which improper procedures have been used (intentionally as well as unintentionally) in the construction, operation, repair, and testing of cargo tanks.

MTB and BMCS agree with the research findings that cargo tank integrity and therefore safety in the transportation of hazardous materials in cargo tanks could be improved by revising and clarifying the HMR. This proposal is intended to accomplish that goal as follows:

(1) Sections 173.33, 177.814 would be completely revised. The requirements dealing with the continuing qualification, maintenance, and periodic testing of cargo tanks contained in those sections would be removed and placed in proposed Part 180, Subpart E. Other requirements pertaining to the manufacturing of cargo tanks contained in those sections have been removed and placed in the appropriate specifications in Part 178.

(2) Sections 178.340, 178.341, 178.342, and 178.343 would be revised for clarity. Identical requirements applicable to MC 306, MC 307, and MC 312 cargo tanks would be contained in the general section, §178.340. The sections in the individual specifications, i.e., §§178.341, 178.342, 178.343, would be arranged to correspond with the sections contained in §178.340. for example, §178.340–10 contains requirements on pressure relief devices applicable to MC 306, MC 307 and MC 312 cargo tanks. Any specific requirement applicable to an individual specification appears in §178.341–10 for MC 306 cargo tanks, §178.342–10 for MC 307 cargo tanks, and §178.343–10 for MC 312 cargo tanks.

In the past years, MTB has received in excess of 25 petitions for rulemaking, granted in excess of 40 exemptions, and issued numerous telephonic and written interpretations on regulations pertaining to cargo tanks. In preparing this NPRM, MTB and BMCS have considered all the comments to Docket HM-183, the applicable outstanding petitions for rulemaking interpretations issued by the MTB and BMCS, and the recommendations made by the research contractors, and the National Transportation Safety Board (NTSB), as well as some found in general literature.

Both MTB and BMCS have participated in the development and issuance of the research contracts, the ANPRM and this NPRM. In the interest of brevity "we" is used hereinafter in referring to "MTB and BMCS."

II. SPECIFICATION DESIGN AND CONSTRUCTION REQUIREMENTS

A. Cargo Tank Overturn Integrity. Research conducted by the States of Michigan and California and DOT has shown that failures of the tank shell, manhole closures and pressure relief valves occur frequently in cargo tank overturn accidents. In a substantial number of instances, these failures resulted in serious leakage, sometimes resulting in fires. These research studies showed that leakage resulted from tank shell puncture, tank shell rupture, weld failure, manhole closures being blown off or badly deformed and, thereby, not reseating, and failure of pressure relief valves and vents.

As an example, a December 1981 study by the California Highway Patrol titled "California Tank Truck Accident Survey" provides insight into the problem. The report analyzed 131 tank truck accidents that occurred from February 1, 1980 to January 31, 1981. The report states: "Spills occurred in 79 percent of the overturns and 16 percent of the nonoverturn spills." With respect to frequency of fires, the study states: "Fires occurred in 22 percent of the spill accidents and two percent of those without spills" and "Forty-four (44) percent of the fires were with cargoes of gasoline, whereas this type of cargo was being carried in only 23 percent of all tank truck accidents."

With respect to the source of spills, the report states: "Most of the spills for which the source was reported occurred through cracks, ruptures, and punctures (52 percent). Another 44 percent occurred through dome covers, vents, or safety valves. The domes also leaked or came open in 11 of the 33 cases of ruptures or punctures of the tank."

The identification of manhole closures as a major source of spills in overturn accidents has resulted in many manufacturers developing a manhole closure of greater integrity. Many cargo tank owners have equipped their new cargo tanks and, in some cases, retrofitted existing cargo tank with these higher integrity manhole closures. Further, the Truck Trailer Manufacturers Association (TTMA) has adopted a recommended practice, RP#61–82, that all manhole closures be capable of withstanding a minimum static pressure of 36 psig without leakage or permanent deformation. We are proposing that all manhole closures on construction of new cargo tanks be designed and tested to a minimum static pressure of 36 psig without leakage or permanent deformation. Because of the severe consequences of manhole closure failures and the fact that cargo tanks may remain in service for 30 years or more, we believe that the retrofitting of manhole closures on existing cargo tanks is necessary. To minimize the economic impact of any retrofitting, we are proposing that the retrofitting be accomplished over a five-year period, with at least 20% of the affected cargo tanks being retrofitted each year. We believe a five year period will permit scheduled replacement that can coincide with normal tank testing, repair, and component replacement, thereby minimizing the cost of retrofit. Based on comments to Docket HM-183 and the research program, the estimated cost of such a retrofit ranges between $20 and $250 per manhole when included as part of scheduled maintenance and testing.

Substantially increasing the tank wall thickness is often proposed as a countermeasure to tank puncture and rupture. We believe that a cargo tank should retain its lading in an overturn on the roadway if the cargo tank does not strike a substantial obstacle. Many cargo tanks have survived such accidents without loss of lading. However, we believe that requiring a substantial increase in cargo tank wall thickness to eliminate tank puncture or rupture would constitute such a major action that more detailed accident analyses, engineering analyses and the consideration of costs and benefits should be studied before such an action is initiated. Although we are not proposing an increase in wall thickness in this rulemaking, we are proposing greater manufacturing quality controls on cargo tanks, improved accident damage protection and more comprehensive and systematic consideration of cargo tank overturn integrity. Discussion of these issues are found later in this preamble.

B. Design Pressure of Cargo Tanks. We are aware that the present regulations do not clearly state a specific relationship between the design pressure of the cargo tank and the properties of the lading. Further, the regulations do not specify a consistent relationship between the design pressure and operational factors such as loading and unloading pressures and the pressure of gas paddings. The lack of such a specific set of relationships can result in: (1) The release of a hazardous material under normal transportation conditions; and (2) the uncontrolled release of a hazardous material as a result of an overturn accident even when the cargo tank is not severely damaged.

As an example, the MC 306 cargo tank has a design pressure of not less than the static pressure generated by the lading, and the pressure relief devices are set-to-discharge at 3 psig. Section 173.119 permits the transportation in cargo tanks of a flammable liquid with a vapor pressure of 16 psia (1.3 psig) at 100° F. Some of these ladings have a vapor pressure as high as 5.8 psig at 115° F. Thus such a lading would continuously vent under normal transportation conditions. Further, in an overturn accident the pressure relief device is required to withstand both the vapor pressure and the static pressure of the lading. For a typical MC 306 cargo tank the static pressure of the lading at the bottom of an inverted vehicle is in the range of 1 to 2 psig. The resulting pressure on a pressure relief device set-to-discharge at 3 psig could be as high as 7.4 psig. This condition would result in the uncontrolled release of the flammable liquid lading through the pressure relief system of an inverted cargo tank that is not in motion (static condition) even though the system is undamaged and operating as designed. Such potential releases are unacceptable and are not limited to MC 306 cargo tanks. MC 307 and 312 cargo tanks are vulnerable under similar conditions. MC 307 cargo tanks often transport ladings protected by a gas pad at or near the tank's design pressure. The pressure from this pad, in addition to the vapor pressure and static pressure from the lading, could result in an uncontrolled release of the lading in an overturn accident. The MC 312 cargo tank specification has no minimum design pressure limit and yet such a cargo tank is often used to transport high density ladings that can produce substantial static pressures.

We believe that such conditions may present an unacceptable risk. Accordingly, we are proposing to specify both design and shipper requirements for relating the cargo tank design pressure to both operational factors and lading physical properties. Included would be a design reference temperature of 115° F. for liquid hazardous materials transported in cargo tanks, the same as presently specified for liquefied gases in uninsulated cargo tanks. The proposal would require that the design pressure be greater than or equal to the largest of the following:

(1) The minimum pressure prescribed in the individual specification;

(2) The pressure prescribed for the lading in Part 173;

(3) 120 percent of the sum of the vapor pressure of the lading at 115° F., the pressure of any gas pad (including air) in the ullage space, and the maximum static pressure exerted by the lading; or

(4) The maximum pressure used to load or unload the lading.

The design pressure requirement would apply to cargo tanks constructed after the effective date of the final rule and could result in an increase in the design pressure of some cargo tanks which are used for the transportation of certain materials. We believe, however, that the design pressure of cargo tanks used for most ladings, including gasoline, will not be significantly affected by this necessary change.

Under proposed §173.33, shippers would be required to verify that the design pressure of the cargo tank is adequate for their hazardous material. In some cases, certain hazardous materials will no longer be authorized for transportation in the lower pressure cargo tanks. We believe this change will affect a relatively small percentage of hazardous materials being transported in cargo tanks and will not have a substantial effect on most shippers and carriers.

We are aware that:

(1) Many MC 305 and 306 cargo tanks have been manufactured and marked with a design pressure of zero psig in non conformance with §178.341–1(b) which states, "The design pressure of each cargo tank shall be not less than that pressure exerted by the static head of the fully loaded tank in the upright position";

(2) The present MC 312 cargo tank specification authorizes a design pressure of zero psig;

(3) The MC 300, 301, 302, 303, and 305 cargo tank specifications do not address design pressure and do not require the marking of the design pressure; and

(4) The MC 310 and 311 cargo tank specifications require the marking of maximum allowable working pressure (MAWP) and maximum working pressure (MWP), respectively, but not the design pressure.

Notwithstanding these marking and manufacturing inconsistencies, each MC 300, 301, 302, 303, 304, 305, 306, 310, 311 and 312 cargo tank has been pressure tested to at least 3 psig. Accordingly, proposed §173.33 would not preclude the use of a cargo tank having a design pressure of less than 3 psig or without a marked design pressure, nor would it preclude the remarking of the marked design pressure on a cargo tank.

Specifically, we are proposing that:

(1) Any existing MC 306 type or MC 312 cargo tank whose pressure relief system is set-to-discharge at 3 psig may be marked with a design pressure not greater than 3 psig;

(2) Any existing MC 300, 301, 302, 303, 304, 305, 306, 307, 310, or 312 cargo tank may be marked or remarked with a design pressure based on the design requirements of proposed §178.340–1(l);

(3) MC 300, 301, 302, 303, 305, 306 and 312 cargo tanks with no marked design pressure or a design pressure of less than 3 psig may be used for authorized ladings where the largest pressure derived from the design pressure requirements in proposed §173.33 is less than or equal to 3 psig; and

(4) MC 310 and MC 311 cargo tanks may be used for authorized ladings where the largest pressure derived from the design pressure requirements in proposed §173.33 is less than or equal to the marked MAWP and MWP, respectively.

C. Accident Damage Protection. We believe the cargo tank design should take into consideration; (1) The need to protect the structural integrity of the head, shell, and piping in collision and overturn accidents; (2) the puncture, abrasion, and crush resistance of the head and shell; and (3) the distribution of loads transmitted into the head and shell by projections, fittings, appurtenances and protection devices. We have observed cargo tanks where little or no consideration was given to the distribution of accident loads from roll-over guards into the tank shell. The involvement of these cargo tanks in accidents in some cases has resulted in the puncture or tearing of the tank shell and the subsequent loss of lading. To clarify the requirement that the cargo tank motor vehicle has to be considered as a system with respect to accident damage protection, it is proposed that accident damage protection devices or projections be designed to distribute any potential accident load into the tank shell so that the stress resulting from the specified loads in combination with the stresses resulting from maximum design pressure during transportation do not result in a stress in the tank shell greater than 75 percent of the ultimate strength of the tank material at that point. We believe that any projections from a tank shell, fitting, or closure that retain lading under any tank orientation should reasonably be able to withstand the forces to which they may be subjected due to collision with other vehicles or objects or in cargo tank rollover or else be adequately protected. Accordingly, in the proposal on accident damage protection, both general and specific requirements are prescribed where we believe the present requirements have allowed questionable design practices.

D. Vacuum-Loaded Cargo Tanks (Waste Tanks). MTB received two petitions (P-870, P-883) requesting the addition of provisions to the HMR covering the transportation of hazardous waste materials in vacuum-loaded cargo tanks. The petitions requested that MTB add a new vacuum-loaded cargo tank specification based on the present MC 307 or MC 312 cargo tank specification modified to allow (a) the use of external self-closing valves that are protected by the truck's suspension and frame in place of internal self-closing valves, (b) the use of circumferential reinforcements placed much farther apart than prescribed in the regulations and compensated for by a thicker cargo tank shell, (c) design and construction with a minimum internal design pressure of 25 psig and external design pressure of 15 psig because of the partial vacuum developed in the tank.

We have included these provisions in the proposed MC 307 and MC 312 cargo tank specifications. We are not proposing to restrict transportation in MC 307 and MC 312 vacuum-loaded cargo tanks to waste materials. We believe that the structural integrity of these cargo tanks would be adequate for other hazardous materials and that there is no significant difference in the risks associated with the transportation of waste materials and other types of hazardous materials. If adopted in a final rule, these provisions would eliminate the need for over 40 exemptions authorizing the use of vacuum-loaded cargo tanks.

E. Cargo Tank Manufacturer Qualification, Quality Control, and Certification. The present DOT certification system for cargo tanks, with the exception of ASME Code cargo tanks, allows the manufacturer to certify that a cargo tank conforms with all requirements of the appropriate specification. The manufacturer's technical knowledge and skill, integrity and product liability are some factors that assure the cargo tank purchaser and the public that the cargo tank is in conformance with the HMR and adequate for the safe transport of hazardous materials. The HMR provides no specific criteria for the assessment of any of these factors. Most cargo tank manufacturers exhibit great knowledge, skill and integrity. On the other hand, however, a number of manufacturers have demonstrated very limited knowledge and skill with respect to matters such as stress analyses, welding, metallurgy, recognized good design and quality control practices, and the HMR.

In contrast, a plumber or an electrical contractor involved in the construction of a building in most cases has to demonstrate knowledge and skill with respect to his trade including knowledge of local regulations. Further, a plumber's or electrician's work is generally inspected by a State, city, county, insurance inspector or an independent engineering firm to assure the building is adequately constructed, and provides an acceptable level of safety for the public. The construction of boilers and pressure vessels used in buildings are similarly regulated. We believe that the risk to the public from the transport of bulk quantities of hazardous materials in cargo tanks over our streets and highways is no less than that presented by a building's plumbing, electrical systems and boilers or pressure vessels. For this reason, we believe that a qualification system is necessary for cargo tank manufacturers and repair facilities. Further, we believe that an independent inspector is needed to assure cargo tank design construction, and test quality. Pursuant to this goal, we are proposing that all cargo tank manufacturers hold a current ASME certificate of authorization and that each cargo tank be certified in accordance with the ASME Code and the appropriate specifications by an Authorized Inspector holding a valid commission from the National Board of Boiler and Pressure Vessel Inspectors (National Board). The National Board "is organized for the purpose of promoting greater safety to life and property by securing concerted action and maintaining uniformity in the construction, installation, inspection and repair of boilers and other pressure vessels and their appurtenances, thereby assuring acceptance and interchangeability among Jurisdictional Authorities responsible for the administration and enforcement of the various sections of the American Society of Mechanical Engineers (ASME) Boiler and Pressure Vessel Code". Its inspectors are employees of states or cities or work for an "Authorized Inspection Agency," such as an insurance company.

Most States in the United States and every Province in Canada now require pressure vessels having a design pressure of 15psig or greater to be designed, fabricated, and inspected in conformance with the ASME Code by a National Board Authorized Inspector. Presently, the HMR require ASME Code construction on MC 330, MC 331 and MC 338 cargo tanks, MC 307 cargo tanks with a design pressure in excess of 50 psig, and MC 312 cargo tanks that are unloaded with a pressure in excess of 15 psig. This proposal would align the Federal standards for cargo tanks having a design pressure of 15 psig or greater with the system of quality assurance used by most of the States. For cargo tanks built with a design pressure of 15 psig and greater, both the manufacturer and the Authorized Inspector must certify on the cargo tank certificate that the cargo tank is constructed and certified in conformance with the appropriate MC specification and the ASME Code.

Cargo tanks built with a design pressure less than 15 psig are below the lower boundary of applicability defined by the ASME Code. We are proposing to use the ASME Code for tanks in this lower pressure range in order to gain the many benefits of the ASME Code. Because ASME makes no provisions for certification and stamping of such tanks, we are proposing that both the manufacturer and the Authorized Inspector certify on the cargo tank certificate that the cargo tank meet the appropriate cargo tank specification and, thereby, "constructed in accordance with the ASME Code".

We realize that the cargo tank specifications contained in this proposal would differ in some respect from the requirements in the ASME Code. These differences result because a cargo tank is a highway transport vehicle and is subject to different accident scenarios and dynamic shock loadings during transportation. These forces are not considered by the ASME Code, which basically establishes standards for stationary pressure vessels. Therefore, we have provided for these additional factors, which are unique to the road transport of hazardous materials.

We recognize also that at the present time many Authorized Inspectors are not knowledgeable about the requirements contained in the DOT cargo tank specifications. It is very important that an Authorized Inspector be knowledgeable of the specification requirements to insure the adequacy of the overall cargo tank motor vehicle. To achieve this goal, we intend to work in combination with the National Board to develop a qualification process for Authorized Inspectors desiring to certify DOT specification cargo tank motor vehicles.

III. INCREASED TESTING AND INSPECTION

A major problem uncovered during the research program was that maintenance and testing of cargo tanks is inadequate. This is particularly true where the requirements for cargo tank retesting and reinspection are concerned. The requirements in §177.824 are generally assumed to apply to the testing and inspection of repaired cargo tanks rather than to an ongoing test and inspection program. Several commenters to Docket HM-183 stated that it was not clear whether §177.824 applied to cargo tank repairs or to an ongoing test and inspection program.

In reviewing the existing regulations concerning the maintenance, testing and inspection of cargo tanks, we have found that there are no well defined requirements for an ongoing cargo tank maintenance program. We have placed too much reliance on the provisions of §173.24 which states "For specification containers, compliance with the applicable specifications in Parts 178 and 179 of this subchapter shall be required in all details, except as otherwise provided in this subchapter" for the continuing qualification, maintenance and testing of cargo tanks. This reliance on §173.24(d) has led many shippers and carriers who are not fully aware of all the requirements of the HMR to think that no ongoing cargo tank maintenance program is required. Several omissions regarding the requirements of the HMR which have surfaced are:

(1) Other than a 2 year visual inspection, there is no requirement for an ongoing cargo tank maintenance system;

(2) There is no requirement for cargo tanks to be subject to a formalized and systematic maintenance program; and

(3) Cargo tank corrosion is not adequately addressed in the present HMR.

We agree with both the research findings and commenters that the maintenance, inspection and testing of cargo tanks are inadequate. Further, we believe that the present requirements are inadequate to properly protect the public and are therefore unacceptable. This conclusion is supported by the enactment in October 1984 of Pub. L. 98-554 (Motor Carrier Safety Act of 1984), specifically section 210(b), which states:

> The Secretary shall, by rule, establish Federal standards for inspections of commercial motor vehicles and retention by employers of records of such inspections. Such standards shall provide for annual or more frequent inspections of commercial motor vehicles unless the Secretary finds that another inspection system is as effective as an annual or more frequent inspection system. For purposes of this title, such standards shall be deemed to be regulations issued by the Secretary under section 206.

While not a part of the Hazardous Materials Transportation Act, the authority for the HMR, this law is an example of the public's and Congress' understanding of the importance placed on the maintenance and inspection of parts and accessories necessary for safe operation of a commercial motor vehicle, including a cargo tank motor vehicle. It is obvious that maintenance and inspection of a cargo tank containing a hazardous material during transportation is even more critical to public safety than a commercial motor vehicle used to transport food products. Accordingly, we are proposing requirements for annual and other periodic inspections as well as a requirement for a formal tank maintenance program for cargo tanks. These proposed requirements would be in addition to the requirements in the FHWA Federal Motor Carrier Safety Regulations governing maintenance and inspection of equipment.

The present §177.824(a) requires inspection and retesting of all authorized specification cargo tanks, except for a cargo tank having a capacity of 3,000 gallons or less used exclusively for flammable liquids. There is little basis in terms of safety for such an exception. Such cargo tanks are built to the same specifications, operated under similar conditions, and operated over the same roads with the same hazardous materials as typical 4,000 to 9,000 gallon cargo tanks motor vehicles. Although smaller in capacity than a typical cargo tank motor vehicle, a vehicle transporting 3,000 gallons of flammable liquid may pose a risk to the public in an accident. Further, we believe that periodic maintenance, inspection, and retest of any vehicle transporting hazardous material must be an integral part of any responsible operator's safety management program. For these reasons, we are proposing that all cargo tank motor vehicles be subject to test and inspection requirements.

In order to ensure that all DOT specification cargo tanks undergo a formalized maintenance program, we have expanded and placed all the requirements for the maintenance and retesting of cargo tanks presently contained in §§173.33 and 177.824 in Subpart E to Part 180. Subpart E of Part 180 would be entitled, "Cargo Tanks: Qualification, Maintenance and Use." Proposed §180.407(a) Table shows the specific type of testing and inspection required for each DOT Specification cargo tank. The table also specifies the test and inspection interval for various cargo tank configurations. Other provisions contained in this subpart are procedures for the test, inspection, repair and modification of a cargo tank. To ensure that all work is properly performed, requirements for the qualification of persons performing and certifying this type of work are also proposed.

Part 180 would contain all requirements applicable to persons who perform functions relating to maintenance and continuing qualification of packagings, such as the prescribed inspections, testing, reconditioning and repair requirements. Persons performing these functions include retesters, reconditioners, approval and inspection agencies, shippers and carriers. We believe that consolidating all the qualification, maintenance and use requirements in one part of the HMR will result in regulations that are easier to understand. Also, this action removes test and inspection criteria from regulations applicable to shippers in Part 173 and to carriers in Parts 174-177. An outline of the subparts that would be contained in part 180 and the present sections containing these requirements is as follows:

Part 180—Continuing Qualification and Maintenance of Packagings:

Subpart A—General.
Subpart B—Non-bulk packagings (except cylinders); Qualification and maintenance (§173.28).
Subpart C—Cylinders; Qualification and maintenance (§173.34).
Subpart D—Portable Tanks; Qualification and maintenance (§173.32, 173.32a, 173.32b, 173.32c).
Subpart E—Cargo Tanks; Qualification and maintenance (§173.33).
Subpart F—Tank Cars; Qualification and maintenance (§173.31).

Only the requirements contained in subpart E are addressed in this proposal.

IV. TRANSPORTATION OF HAZARDOUS MATERIALS IN CARGO TANKS

A. Flammable Liquids or Pyrophoric Liquids. We propose to eliminate several discrepancies pertaining to the transportation of flammable liquids and pyrophoric liquids in cargo tanks. In this proposal, §173.119 and certain other sections in Subpart D are revised to clarify and reiterate the following:

(a) MC 310, MC 311 and MC 312 cargo tanks that are used in flammable liquid service must be equipped with pressure relief devices having venting capacities adequate for flammable liquid ladings. We are proposing that pressure relief system on these cargo tanks when used in flammable liquid service must be equivalent to a pressure relief system required for a MC 307 cargo tank.

(b) Any cargo tank with bottom outlets would have to be equipped with self-closing stop valves capable of being remotely operated by both thermal and mechanical means as required for MC 306 and MC 307 cargo tanks.

(c) Under the present §173.119(m), there is no requirement for safeguards to assure the containment integrity of a cargo tank used to transport flammable liquids with secondary hazards. Based on a petition (P-607), parameters would be provided to allow the use of certain cargo tanks for a lading with more than one hazard class. In the proposal, a cargo tank used to transport a flammable liquid that is also corrosive must be constructed of, or lined with a material that is compatible with the lading or be constructed with added shell thickness for a corrosion allowance. A cargo tank transporting a flammable liquid that is also a poison B liquid must have a design pressure of 25 psig. These design parameters agree with the requirements of subparts F and H of the HMR.

(d) The use of air pressure to load/unload a cargo tank containing a lading having a flash point of 20° F. closed cup or less would be prohibited. We believe the unloading procedure may pose a potential risk of fire should an air enriched atmosphere at or near the flammability range of the flammable liquid lading be created in the vapor space of the tank.

(e) The use of non-reclosing pressure relief devices would be prohibited except when such devices are arranged in series with a spring loaded pressure relief valve. Nonreclosing pressure relief devices present the risk of a continuous release of lading as compared to a partial release to relieve pressure. Such a continuous release of a flammable lading in the event of a fire could produce catastrophic results.

Sections that would be affected by these revisions are as follows: 173.119, 173.135, 173.145.

B. Flammable Solids, Oxidizers and Organic Peroxides. We believe that certain flammable solids, oxidizers and organic peroxides have hazardous characteristics similar to flammable liquids and pyrophoric liquids. Therefore, we are proposing that cargo tanks equipped with bottom outlets when used to transport these hazardous materials have self-closing stop valves capable of being remotely operated by both thermal and mechanical means as required for other cargo tanks used in flammable liquid service.

Sections that would be affected by this revision are as follows: 173.154, 173.190, 173.224.

C. Corrosive Materials. Requirements for corrosive materials in cargo tanks would be standardized. We are proposing that a cargo tank used to transport a corrosive material be constructed of, or lined with a material that is compatible with the lading or have added shell thickness for a corrosion allowance. Further, cargo tanks equipped with bottom outlets when used to transport corrosive materials would be required to be equipped with self-closing stop valves that are capable of being remotely operated by mechanical means.

Sections that would be affected by this revision are as follows:

173.245, 173.247, 173.249, 173.250a, 173.252, 173.253, 173.254, 173.255, 173.257, 173.262, 173.263, 173.264, 173.266, 173.267, 173.268, 173.271, 173.272, 173.273, 173.274, 173.276, 173.280, 173.289, 173.292, 173.294, 173.295, 173.296, 173.297.

D. Poison B Materials. Requirements for poison B materials in cargo tanks would be standardized. Given the substantial acute and environmental risks associated with the release of a poisonous material, we are proposing that a cargo tank used to transport a poison B material have a design pressure of 25 psig. Any cargo tank equipped with bottom outlets used to transport a poison B liquid would be required to have self-closing stop valves capable of being remotely controlled by both thermal and mechanical means. The type of cargo tank authorized to transport poison B solids would be based on the specific properties of these materials.

Sections that would be affected by these revisions are as follows:

173.346, 173.347, 173.352, 173.358, 173.359, 173.369, 173.373, 173.374

V. OTHER ISSUES

Question eight in Docket HM 183 asked "Should the scope of this docket be expanded to address the design and construction of the MC 306 cargo tank in its entirety?" "No!", with some qualification was the answer given by all manufacturers and operators of MC 306 cargo tanks. The commenters stated that the basic design of the MC 306 cargo tank was adequate and major changes were not justified. On the other hand, several commenters stated that gains in safety and productivity are probable in a major redesign of cargo tanks. These commenters stated, and we agree, that to achieve such gains would require considering the cargo tank motor vehicle as a system and, further, the highway and vehicle as part of a transportation system. Such an effort involves vehicle size and weight laws and vehicle design standards and is a large, long term undertaking.

A number of cargo tank operators commented that more uniform requirements and a greater level of enforcement are necessary to stimulate improvement in cargo tank maintenance. We agree with these commenters. Aiding enforcement personnel by removing the many regulatory shortcomings, inconsistencies and ambiguous requirements is a goal of this NPRM.

Additionally, the DOT has been active in fostering adoption of the Federal regulations by States and supporting them in their enforcement efforts. MTB's State Hazardous Materials Enforcement Development program provided monies to States to adopt and enforce the HMR. Concommitant with this program is the training of enforcement personnel in the technical aspects for the regulations dealing with the transport of hazardous materials. BMCS also has an ongoing program, titled the Motor Carrier Safety Assistance Program, whose objective is to substantially increase commercial motor vehicle enforcement programs by providing development and implementation grants to States in return for their agreeing to adopt Federal safety programs, including the Federal Motor Carrier Safety Regulations and the HMR, or compatible requirements, and establishing enforcement policies and programs. Again, many State personnel have and are being trained in enforcement procedures and regulatory compliance requirements.

New training courses in cargo tank regulations have been developed by BMCS in an effort to provide better training of enforcement personnel. Emphasis areas dealing with such requirements as minimum shell thickness measurements during inspections provide for uniform compliance on the part of industry.

Many commenters stated the driver qualifications and trainings are as important to controlling safety as packaging integrity. We agree that driver qualification and training are critical to the safe transport of any commodity and particularly hazardous materials. The strengthening of driver qualification and training requirements will be addressed in future rulemaking actions.

REVIEW BY SECTION

The following is an analysis and explanation, by section, of the more significant features of this regulatory proposal.

Section 171.3. This section would permit the continued use of a non DOT specification cargo tank for transportation of a hazardous waste provided it is operated exclusively by an intrastate motor carrier in a State where its use is permitted by that State.

Section 171.7. This section would incorporate the latest edition of the ASME Code, as requested by petitioners (P-794, P-824); extend applicability to include Sections II and V of the ASME Code; reflect a change in the reference numbering of the American Welding Society (AWS) Code from 3.0 to 2.1 (and a change in address); incorporate additional ASTM and CGA standards; and add references to certain publications of the Rubber Manufacturers Association and the American Society for Non-Destructive Testing.

Section 171.8. Definitions of "Authorized Inspector" and "Authorized Inspection Agency" would be added to clarify who is qualified to perform inspections as required by the HMR. The definition of "cargo tank" would be revised to clarify that the term includes all attached appurtenances, reinforcements, fittings and closures. A definition of "cargo tank motor vehicle" would be added for clarity and to provide for consistent use of terminology when referring to the transport vehicle.

Section 172.203. This section would remove the reference "173.315(a)(1)" and insert in its place "173.315(a)" for consistency with a change made in that section.

Section 173.33. This section would be completely revised and entitled "Hazardous materials in cargo tank motor vehicles". It would contain only general shipper requirements for the use of cargo tanks transporting hazardous materials. For example, this section would provide a means for a shipper to determine if the design pressure of a cargo tank is adequate for a particular commodity. Present requirements pertaining to commodities, cargo tank design, qualification, maintenance and use of cargo tanks would be revised and placed as appropriate in Parts 173, 177, 178 and 180. Highlights of other proposed changes are:

(1) MC 300, MC 301, MC 302, MC 303, MC 305, MC 306 and MC 312 cargo tanks with no marked design pressure or marked with a design pressure of less than 3 psig may be used for authorized ladings where the largest pressure derived under paragraph (d) is less than or equal to 3 psig.

(2) MC 310 and MC 311 cargo tanks may be used for authorized ladings where the largest pressure derived from paragraph (d) is less than or equal to the marked MAWP or MWP respectively.

(3) "MC 306 type" cargo tanks maybe upgraded to the MC 306 cargo tank specification; MC 304 cargo tanks to the MC 307 specification; and "MC 312 type" cargo tanks to the MC 312 specification.

(4) Non-reclosing pressure relief devices are prohibited on a cargo tank used in flammable liquid service unless such device is in series with a spring loaded pressure valve.

(5) A cargo tank may not be loaded with a hazardous material if a dangerous reaction may occur in the tank between a contaminant or a residue and the new lading.

(6) No two or more materials, the mixing of which may produce an unsafe condition, may be transported in a cargo tank motor vehicle.

Sections 173.119–173.374. These sections would be revised as discussed earlier in this preamble under the heading

"Transportation of Hazardous Materials in Cargo Tanks." In addition, certain other changes would be made for clarity.

Section 173.131 would remove the reference to 178.340-10; 178.341-4 and 178.341-5 and insert in their place references to §§178.340-14 and 178.340-16, 178.341-10 and 178.341-11, as applicable.

Section 173.250a would delete paragraph (a)(2) as unnecessary since these requirements are contained in paragraph (a)(1).

Section 173.252 would clarify that the ¾ inch minimum thickness prescribed for the tank shell and head is exclusive of any lining, cladding or corrosion allowance.

Section 173.272 would clarify that the temperature of the lading may not exceed the design temperature of the cargo tank.

Section 173.289 would delete paragraph (a)(4) as unnecessary since the requirement is contained in paragraph (a)(1).

Section 173.292 would delete paragraph (a)(2) as unnecessary since the requirement is contained in paragraph (a)(1).

Section 173.294 would clarify that the tank must be made of, or lined with, pure nickel or stainless steel.

Section 173.295 would clarify that cargo tanks made of steel are authorized for stabilized benzyl chloride only. The requirements in paragraph (a)(10) pertaining to cargo tanks made of nickel would be removed and placed in paragraph (a)(9). Paragraph (a)(10) would be reserved.

Section 173.315 would remove references to §173.33 and add the applicable reference to Part 180. Additionally, the present provisions of §§173.33(f)(7)(8) and (9) would be removed and placed in §§173.315(h)(4), 173.315(n) and 173.315(o), respectively.

Section 173.318 would add the present requirements contained in §173.33(d)(1)(ii) as new §173.318(g)(4).

Section 176.76. The requirement in §173.33(a)(1) authorizing a cargo tank motor vehicle containing a hazardous material to be transported aboard a vessel would be removed and placed in 176.76(b)(1).

Sections 177.800, 177.801 and 177.802. Sections 177.800 and 177.801 would be simplified and revised for clarity. Section 177.802 would be removed and a new section governing the inspection of carrier facilities and records would be added. These inspection requirements are similar to those applicable to rail and air carriers.

Section 177.814. The existing provisions on record retention and reporting requirements of this section would be removed and placed in §180.417. A new §177.814 would reference those requirements.

Section 177.824. The existing provisions on the inspection and retest of cargo tanks of this section would be removed and placed in Part 180 of this sub-chapter. A new §177.824 would reference those requirements.

Section 177.840. Paragraph (f) would inform carriers of certain equipment testing requirements applicable to the transportation of chlorine that are contained in §173.315(o).

Section 178.337. This section would add those requirements presently found in §173.33 which pertain to the manufacture of a DOT Specification MC 331 cargo tank.

In §178.337-1, paragraph (a) would remove a reference to §173.33(i) and insert in its place a reference to paragraph (e). Paragraph (e) would remove a reference to §173.33(i). The requirements presently contained in §173.33(i) would be added in new paragraph (e)(2).

Section 178.337-2(c) would remove the reference to §173.33(g)(1) and add those requirements presently contained in §173.33(g)(1). Also a provision to allow the use of properly joined aluminum baffles would be added.

Section 178.337-3 would revise the minimum thickness formula contained in paragraph (b), as discussed earlier in this preamble under §178.340-3.

Section 178.337-4(b) would require that welding procedure and welder performance tests be performed in accordance with the ASME Code.

Section 176.337-6(a) would be revised to require manholes on all cargo tanks.

Section 178.337-8(b) would remove the references to §173.33(f)(9), (h)(4) and (5). The special requirements for chlorine angle valves contained in paragraph (b) would be placed in §178.337-9(b).

Section 178.337-9 would contain requirements for piping, pipe fittings, pressure devices, hoses and other pressure parts. New paragraph (b)(7) would incorporate certain special requirements for chlorine cargo tanks now found in §§173.33 and 178.337-8.

Paragraph (d)(1) would incorporate the provision now found in §173.33(j) permitting mounting a refrigeration unit on a motor vehicle.

Section 178.337-11(a) would remove all references to §173.33(h) and add those requirements presently contained in §173.33(h) through (h)(2) and certain provisions presently contained in §178.337-11(c). A new paragraph (a)(4) would incorporate certain special requirements for chlorine excess flow values that are presently contained in§173.33(h)(4). Paragraph (b) would remove the reference to §173.33(h)(3) and add the requirements contained in that paragraph. Section 178.337 would require that cable linkages for the internal valve be corrosion resistant.

Section 178.337-14(b) would remove the reference to §173.33(f)(7) and add the requirements contained in that paragraph.

Section 178.337-15 would remove the references of §173.33(f) (6) and (10) and add the requirements contained in those paragraphs.

Section 178.338. This section would be revised by adding those requirements presently found in §173.33 which pertain to the manufacture of a DOT Specification MC 338 cargo tank.

Section 178.338-8(b) would remove the reference to §173.3(f) and add certain applicable requirements contained in §173.33 (f) through (f)(5).

Section 178.338-17 would remove the references to §§173.33(f)(6) and 173.318(a)(4) and add the requirements now contained in those paragraphs.

Section 178.340. This section contains the general design and construction requirements applicable to MC 306, MC 307 and MC 312 cargo tanks. This section would add a table of headings covering §§178.340-1 through 178.340-14 for easier reference of the requirements contained in these sections.

Section 178.340-1. In this section, over 30 terms would be defined for clarity. New terms would be used whenever existing terms are found inappropriate. For example, "fail safe" device would replace the term "shear section" to permit the use of any device designed to fail in order to protect a major part. This proposal considers each "compartment" a separate cargo tank. Similarly, a multiple compartment tank is a series of connected cargo tanks. Thus, the term "compartment" will no longer be used.

This section would specify the ASME Code as well as the appropriate specification as the basis for cargo tank motor vehicle design and construction. Each cargo tank with a design pressure of 15 psig or greater would have to be "constructed and certified in conformance with the ASME Code" and code stamped. Each cargo tank with a design pressure of less than 15 psig would have to be "constructed in accordance with the ASME Code", but may not bear on ASME code stamp. In either case, the design and construction techniques would be the same and the design and construction would be certified by an Authorized Inspector. This procedure will provide shippers, cargo tank motor vehicle owners, and insurance carriers with greater assurance that each specification cargo tank motor vehicle conforms to specification requirements.

Based on satisfactory experience under the exemption program and several petitions (P-537, P-768, P-870, P-883), vacuum-loaded cargo tanks made to the MC 307 and MC 312 cargo tank specifications would be authorized.

Design parameters for connecting structures would provide for the over-all structural integrity of the cargo tank motor vehicle based, in part, on NTSB recommendations H-83-25 through -30.

The structural requirements specified for connecting structures are the same as those specified for a cargo tank with the exception of pressure.

Section 178.340-2. Requirements for construction materials, including the thickness of these materials, would be contained in this section. All construction materials for the cargo tank shell, heads, bulkheads and baffles would be of a metal suitable for building pressure vessels in conformance with the ASME Code. This provision, in response to a petition for rule change (P-870) and satisfactory exemption experience, would expand the types of construction materials authorized and, thereby, eliminate the need for an exemption for cargo tanks made of metallic materials other than steel or aluminum.

Two petitioners (P-310, P-452) requested that the minimum thicknesses for shells, heads, bulkheads and baffles be expressed in decimals in place of gauge. These petitions are denied as unnecessary since the proposed specifications would require conformance with the ASME Code.

The present requirement that cargo tanks in corrosive service must be designed for a 10-year service life would be removed. We believe this requirement restricts the economic optimization of the tank design and does not provide assurance that the tank shell thickness is greater than the prescribed minimum thickness. We are proposing that cargo tanks may be designed for any service life. An acceptable level of safety would be maintained by new inspection and thickness monitoring requirements specified in proposed subpart E, of Part 180. Tanks with shells below the minimum thickness must be repaired or the specification plate must be removed.

Section 178.340-3. Requirements for the structural integrity of cargo tanks would be contained in this section. The shell thickness and cargo tank structure would conform with the structural requirements prescribed in this section, in addition to the minimum shell and head thickness requirements in the individual specification and the accident damage protection requirements in §178.340-8. Further, the cargo tank manufacturer must consider all structural loading and damage that would result from accidents, such as an overturn on the highway.

The proposal contains a requirement that stresses resulting from specific operating loadings be evaluated in the design of a cargo tank in place of the present general loading requirement. We are proposing to specify the same loadings presently prescribed for the MC 331 cargo tank. We believe the MC 331 loading requirements adequately take into account the forces actually encountered in transportation. For frameless cargo tank motor vehicles, where the tank shell serves as the vehicle frame, the basic structural integrity of a cargo tank would be calculated using a specific formula. The proposed formula is an improved version of that specified in §178.337-3 of the MC 331 cargo tank specification. The proposed formula was developed as a part of the MC 331 cargo tank integrity study by the research contractor, Dynamic Sciences, Inc., and its industry advisors. The proposed formula includes shear stresses resulting from internal pressure, a factor not included in the present formula in §178.337-3. For cargo tanks mounted on a frame or built with integral structural supports, the stress analysis appropriate to the tank and structural configuration would be made using the required loadings. Since conformance with the ASME Code would be required, the present 5:1 safety factor would be relaxed to the 4:1 safety factor allowed by the ASME Code.

Section 178.340–4. Requirements for the joints in the tank would be contained in this section. Joint preparation, welding procedures and welder performance are as prescribed in the ASME Code. This Code includes supervision of welding, information on proper type and size of electrodes used, number of passes, etc., and in addition, require that work drawings indicate details and tolerances as well as verification of welder competence all of which are necessary to assure the structural integrity of the tank.

Since conformance with the ASME Code would be required, the confusing requirements of optimum fabrication techniques and efficiency determination would be removed. Any welding technique and its corresponding allowable weld efficiency permitted by the ASME Code would now be acceptable. Conformance with the ASME Code would also obviate the need for the compliance test in addition to ASME Code requirements.

Section 178.340–5. Manhole requirements would be contained in this section. Under the proposed rule, a manhole closure would have a structural strength much greater than that presently prescribed. A minimum manhole size would also be prescribed. Additional comments and changes to this section are addressed earlier in this preamble under the heading "Cargo Tank Overturn Integrity".

Section 178.340–6. Supports and anchoring requirements would be contained in this section. The minimum dynamic loadings specified for the cargo tank would be applicable to the tank supports and anchoring systems. See earlier preamble discussion under §178.340–3.

Section 178.340–7. Circumferential reinforcement requirements would be contained in this section. These requirements are relaxed to permit, under certain conditions, the unreinforced portion of the shell to exceed 60 inches. This relaxation is based on a petition (P-537) and the satisfactory experience of cargo tanks made to this requirement under exemption. This section would also prohibit any type of reinforcement that precludes visual inspection of the tank shell or head. The placement of reinforcement rings over circumferential shell joints would be prohibited.

Section 178.340–8. Requirements for protection from accident damage would be contained in this section. This proposal is organized into both general requirements and individual requirements for the following accident protection zones: the bottom damage protection zone (the lower ⅓ of the tank), the roll over damage protection zone (the upper ⅔ of the tank), and the rear-end tank protection zone (area subject to rear-end or backing collisions).

The general requirements for accident damage protection would require a cargo tank, its piping, closures and valves that may contain lading to be designed and constructed to minimize the potential for loss of lading resulting from an accident. Projections from the cargo tank shell that retain lading such as domes and sumps would be required to be designed to minimize the possibility of the loss of lading in an accident, be constructed of a material having a strength equivalent to that of the tank shell, and have a thickness at least equal to that specified by the appropriate specification. Any projection that extends more than two inches from the tank shell would be required to be protected from accident damage. To provide for maximum design freedom and to minimize cargo tank cost, projections that have a strength of at least 125 percent of the requirements specified for the appropriate accident damage protection device would not be required to have accident damage protection. Additionally, such projections may be considered as accident damage protection devices for the protection of fittings, piping, etc.

a. Bottom damage protection. Bottom damage protection devices are intended to protect any outlet, projection, sump, or piping located in the bottom damage protection zone from accidental damage such as a collision with another vehicle or with a road side structure, such as a guard rail.

The present specifications require the use of internal self-closing stop valves to provide lading retention in collision accidents. Vacuum-loaded cargo tanks, operating under DOT exemption, are authorized to have external self-closing stop valves which are protected by bottom accident damage protection. These cargo tanks have demonstrated a high level of integrity. The bottom accident damage protection used on vacuum-loaded cargo tanks consists of the vehicle frame, rear wheels, suspension system and rear end tank protection. Because of the excellent safety record of vacuum-loaded cargo tanks, we are proposing to allow external self-closing stop valves with bottom accident damage protection as an alternative to the internal self-closing stop valves on all MC 307 and MC 312 cargo tanks.

Accident data from studies of under-ride accidents indicate that a significant percentage of under-ride accidents occur at or near highway speed limits. Based on these studies, we are proposing that the bottom accident damage protection system must be capable of absorbing or deflecting an energy of 275,000 foot-pounds based on the ultimate strength of the material. This is equivalent to the impact of a 4,000 pound automobile at a speed of 50 miles per hour or the impact of a 80,000 pound truck backing into a stationary structure at 10 miles per hour. This impact energy is applied to the bottom damage protection device at any point and from any direction, i.e. front, rear, side or bottom, over an area not greater than 6 square feet.

Also, requirements would be relaxed to allow the use of valve protection features other than a shear section to provide greater flexibility for cargo tank manufactures and operators.

b. Rollover damage protection. The most common highway accident involving the loss of cargo tank lading is a rollover. Such accidents generally result in damage to the upper ⅔ of the cargo tank. The most vulnerable areas are the cargo tank sides, top and front head. Present requirements allow the lateral strength of rollover damage protection devices to be only ¼ the strength required normal to the shell. In a rollover accident, the rollover damage protection system can receive lateral loads that equal or exceed the normally applied load. For this reason, we are proposing that the rollover damage protection system on each cargo tank motor vehicle be designed for a lateral load equivalent to twice the gross weight of the loaded cargo tank motor vehicle. If more than one rollover protection device is used, each device must be designed for a lateral load no less than one-half the gross weight of the loaded cargo tank motor vehicle.

c. Rear-end tank protection. A petitioner (P-452) maintains that the present rear bumper requirements serve two functions. First, as required by §178.340–8 the bumper must protect the cargo tank and any tank component that may retain lading from damage as a result of a collision with another vehicle or with a structure during backing. Second, as required by 49 CFR 393.86 the bumper serves as a rear-end under-ride protection device to protect occupants of any vehicle that may collide with the rear-end of the cargo tank. The petitioner requested that the regulations be revised to separate the two functions.

We agree and are proposing to relax the requirement for rear-end protection by allowing a rear-end tank protection device which may be different from the rear-end under-ride protection device. This would be accomplished by removing the height restriction for the location of the rear-end tank protection. The placement of the rear-end tank protection device would protect the cargo tank and any piping, fittings, etc., from collision damage. The strength requirements for the rear-end tank protection device would remain unchanged. The rear-end under-ride device must conform with the design and strength requirements specified in 49 CFR 393.86.

MTB has received several petitions (P-904, P-911, P-923) for rule change requesting that rear-end tank protection be required only on the rearmost unit of a "double" cargo tank motor vehicle configuration. The petitioners argued that the present requirement adds cost and weight to the cargo tank configuration with no safety benefits. We do not agree. We believe that the forward unit of a "double" is vulnerable to rear-end tank damage particularly in turning maneuvers. This vulnerability increases in proportion to the length of the draw bar between the cargo tank units. The forward unit of a "double" is at times operated without the protection afforded by the rear units. Operation of such a forward unit, whether with a full load or with only residual lading presents an unacceptable risk. Furthermore, we believe that the removal of the location restriction on rear-end tank protection devices, as proposed in this rule, would eliminate the need for a heavy supporting structure and minimize the cost and weight of the protection device. For these reasons, these petitions are denied.

Section 178.340–9. This section would contain and clarify the present requirements relative to pumps, piping, hoses, connections, etc. contained in §§178.340–8(d) (3), (4), (5) and (6). It would also contain a provision that hose couplings must be designed for a burst pressure of not less than 120 percent of the design burst pressure of the hose.

A requirement would be added that loading/unloading and charging lines be of sufficient strength or be protected by a fail-safe device in order to prevent damage to the cargo tank that could result in loss of lading from any forces applied by loading/unloading or charging lines attached to the cargo tank. This should prevent the uncontrolled loss of lading from the cargo tank should the cargo tank motor vehicle be moved while the loading/unloading or charging lines are still attached to the facility's tanks. Finally, a provision would be added to authorize the use of nonmetallic piping, valves or connections located outboard of the lading retention system.

Section 178.340–10. Pressure relief device requirements would be contained in this section. The proposal differs from existing regulations as follows:

1. Provisions for possible arrangements, types, location and pressure setting of pressure relief devices are expanded and revised for clarity. This section contains requirements currently found in §§178.341–4, 178.342–4 and 178.343–4.

2. Research and experiments have shown that cargo tanks can experience short duration pressure surges of up to 50 psig in a rollover. This has been shown to cause the release of about two gallons of lading through the pressure relief devices on a typical MC 306 cargo tank. The proposal clarifies the fact that the present regulations require pressure relief systems to be capable of withstanding a pressure surge without leakage of lading and yet be capable of operating when there is a sustained pressure rise in the tank.

3. We are not aware of a valid reason for a normal vent valve set-to-discharge at 1 psig since, in many rollover conditions, the static head of the lading would open and operate such a valve. Additionally, the vapor pressure of some ladings exceeds 1 psig in transportation resulting in continuous venting. Therefore,we are proposing to prohibit the use of normal vent valves.

4. We propose to revise the procedures used to establish the rated flow capacity for pressure relief devices. The proposal requires that a manufacturer of a pressure relief device certify that the pressure relief device model (design, size and set pressure) is designed, tested and meets the requirements contained in this section and the applicable cargo tank specification. Each pressure relief device model would be flow capacity tested prior to its first use. The rated flow capacity for each pressure relief device model would be based on testing of at least three prototype

pressure relief devices. The marked rated flow capacity of the pressure relief device would not be greater than 90 percent of the average value for the devices tested.

Additionally, we are proposing that an Authorized Inspector witness and approve the flow capacity test for each pressure relief device model and sign the manufacturers certification for that model.

5. We believe that the risk from hazardous material transportation is substantially reduced when packaging is designed to retain the lading in non catastrophic accidents and to minimize the quantity of lading released when release is inevitable. However, in a cargo tank accident, particularly an overturn followed by a fire, the functioning of a frangible disc or a fusible element would result in the release of a substantial quantity of lading. A reclosing pressure relief device on the other hand would minimize the quantity of lading released. Further, we believe that frangible disc and fusible elements particularly in low pressure applications are much more likely to fail as a result of impact and liquid surge than reclosing pressure relief devices. Accordingly, for all cargo tanks constructed after the effective date of the final rule, we are proposing that all pressure relief devices be reclosing, except a frangible disc may be used in series with a reclosing pressure relief device.

Section 178.340-11. Requirements for tank outlets would be contained in this section. In this proposal, the present requirements contained in §§178.341-5, 178.342-5 and 178.343-5 are revised to provide practical guidelines for the design of cargo tank openings, outlets and their attached piping, connections and appurtenances.

Requirements for remote closure of a cargo tank loading/unloading outlet would be clarified. Each cargo tank self-closing stop valve used for loading and/or unloading would be required to have a remote means of closure. When used to transport a flammable liquid, pyrophoric liquid, oxidizer or Poison B liquid, a cargo tank would be required to be equipped with a self-closing stop valve capable of being remotely operated by both thermal and mechanical means. These requirements presently are contained in §173.119 but have not been consistently required for all flammable liquids and other hazardous materials presenting similar risks. Self-closing stop valves with mechanical means of remote closures would be authorized for corrosive materials.

Based on a petition (P-537) and on exemption experience, requirements for internal outlet self-closing stop valves would be relaxed to allow the use of an external self-closing stop valve in certain cases. See the preamble discussion of §178.340-6 on accident damage protection.

Provisions regarding top outlets would be clarified. An opening in the top of a cargo tank that is securely closed during transportation with a welded or bolted blank flange or by a threaded plug would not be considered an outlet. Such openings would be considered projections and if they extend more than two inches from the tank shell or head accident damage protection would be required. See proposed §178.340-8(a)(1).

Section 178.340-12. Gauging device requirements would be contained in this section. Existing requirements for gauging devices are applicable only to MC 307 cargo tanks. These requirements would be revised and made applicable to MC 306 and MC 312 cargo tanks. As proposed, any method that measures the maximum permitted liquid level or amount of the lading to an accuracy of 0.5 percent would be permitted. Sight glass gauging devices would continue to be prohibited.

Section 178.340-13. This section would contain the general pressure testing requirements presently found in §§178.341-7, 178.342-7 and 178.343-7. This section would also contain a general pressure test procedure which emphasizes that a cargo tank must be subjected to a prescribed test pressure to assure the pressure integrity of the cargo tank.

The proposal authorizes pneumatic pressure testing for all DOT Specification MC 306, MC 307 or MC 312 cargo tanks. If pneumatic testing is used, the cargo tank must be inspected for leakage or other signs of defects at the inspection pressure specified in the applicable specification. The pneumatic inspection pressure specified would be lower than test pressure due to the potential hazards involved in a pneumatic pressure test.

Section 178.340-14. This section would revise and clarify the cargo tank marking requirements presently contained in §178.340-10. The required markings will all be found on the cargo tank nameplate or the cargo tank motor vehicle specification plate. It would also require that all markings must be in English.

Since all DOT Specification MC 306, MC 307 and MC 312 cargo tanks will either be "constructed and certified in conformance with the ASME Code" or "constructed in accordance with the ASME Code", each cargo tank nameplate would be required to contain all the information required by the ASME Code. In addition, the nameplate would also be required to contain other markings such as cargo tank design pressure, cargo tank test pressure, cargo tank design temperature range, material specification (shell and heads), minimum shell and head thickness, and maximum design density of the lading.

The marking of the cargo tank design pressure would be required because shippers of hazardous materials must be able to determine if a cargo tank has adequate lading retention capability for the commodity to be shipped. The cargo tank design temperature range marking would make shippers and carriers aware of the temperature range at which the cargo tank can be safely operated. Operating the cargo tank at temperatures outside this range might affect the strength and physical properties of the cargo tank's material of construction. Marking the minimum shell thickness for the cargo tank top, side and bottom, would acknowledge the industry practice, especially in MC 306 cargo tank construction, of building a cargo tank with thicker top and bottom plates and a thinner sidewall. Marking the maximum design density of the lading on the nameplate should prevent the loading of a cargo tank with a lading which is heavier than the design density of the cargo tank.

Additionally, each cargo tank motor vehicle would be required to have a specification plate providing safety and operational data. The specification plate would contain such information as exposed surface area, maximum loading and unloading rates, heating system design pressure and temperature, lining material, the name of the cargo tank and cargo tank motor vehicle manufacturer and the cargo tank motor vehicle certification date.

The marking of the exposed surface area would provide a means of determining the required venting capacity. The marking of the maximum loading and unloading rates and the heating system design pressure and temperature would provide shippers and carriers with important operational information about the cargo tank motor vehicle. For example, this information could prevent the use of a heating medium which is warmer than the design temperature of the system.

The proposal allows the combination of the nameplate and specification plate on an uninsulated cargo tank. In this case, the combination plate would be welded or brazed to the cargo tank. For an insulated cargo tank, the nameplate would be required to be welded or brazed to the cargo tank and the specification plate would be welded, brazed, or riveted to the insulation jacket or to an integral supporting structure of the cargo tank motor vehicle. The ASME Code requires that the nameplate be visible after insulation is applied. Additionally, paragraph UG-119(f) of the ASME Code contains provisions for installation of a duplicate nameplate on the insulation jacket. The use of duplicate plates is also important on ASME Code cargo tanks if the specification plate is lost or becomes illegible. The nameplate which is on the cargo tank would serve to verify ASME Code construction and would be enable the owner to replace the specification plate.

Section 178.340–15. This section would revise and clarify the certification requirements presently contained in §178.340–10. The proposal requires that in addition to a responsible official for the manufacturer that an Authorized Inspector would also be required to sign a certificate certifying that the cargo tank or cargo tank motor vehicle is designed, fabricated, tested and completed in conformance with the application specification.

Additionally, the proposal requires that the manufacturer of each stage of a cargo tank motor vehicle construction not only must furnish a certificate covering its work to any succeeding manufacturer, but also must pass along any certificates received from earlier manufacturers. This will ensure that at the end of staged construction, the final manufacturer will possess certificates covering the entire construction process. The proposal also would require that the manufacturer furnish the owner with all certificates and other documentation required by this section.

Section 178.341. Individual specification requirements applicable to an MC 306 cargo tank motor vehicle are contained in this section.

The MC 306 type cargo tank is a low pressure cargo tank used mostly for the transportation of low vapor pressure flammable liquids. The research contractor estimates that there are approximately 102,500 cargo tanks of all specifications in hazardous materials service and that approximately 57 percent (57,900) of these are MC 306 type cargo tanks. The MTB incident data base for the years 1980 through 1983 shows that MC 306 type cargo tanks accounted for about 88 percent of the deaths reported for cargo tanks of known type in hazardous material service. We believe that this data depicts the combined high risk potential of the highly flammable hazardous materials transported and the comparatively low integrity of MC 306 type cargo tanks relative to other specification cargo tanks. MC 306 type cargo tanks have a thin shell and heads that require additional reinforcement to accommodate the prescribed environmental loads.

Because we do not have sufficient data to assess the suitability of the basic shell integrity of MC 306 type cargo tanks in accidents or to evaluate various countermeasures such as increased shell and head thickness and improved roll stability, we have not proposed any revision of the regulations on this issue. This proposal, however, does address the matching of hazardous material properties with tank design parameters.

Presently there are no requirements for the design pressure specified for the MC 306 cargo tank except that the tank must be designed for a "Design Pressure" not less than the static head of the fully loaded tank in the upright position. Several petitioners (P-562, P-827) requested revision of the requirement that the tank shell be designed to withstand a pressure not less than that exerted by the static head of the lading. The petitioners claim that the design pressure of a typical MC 306 cargo tank is zero. We disagree. At present, MC 306 cargo tanks must have a design pressure of not less than the static head of a fully loaded tank and be pressure tested at 3 psig. The low capacity normal vent is set-to-discharge at one psig. The emergency pressure relief devices for fire conditions are set-to-discharge at 3 psig and are flow rated at 5 psig. The unloading relief protection system is designed to prevent pressure in the tank from exceeding 3 psig. The tank is in effect normally operated at up to 3 psig and in an emergency situation at 5 psig which is 166 percent of the test pressure.

MC 306 cargo tanks should be designed for a test pressure of 5 psig pressure at which the pressure relief valve is rated. Section 173.119 permits MC 306 cargo tanks for ladings with a vapor pressure of 16 psia at 100°F. or 1.3 psig. Some of these ladings have a vapor pressure of 5.8 psig at 115° F, the reference design temperature. For a typical MC 306 cargo tank of elliptical cross section, minor axis of 61 inches, typical ladings can produce a static pressure of 1.5 to 2.0 psig. For example, pentane should be transported in a cargo tank with a design pressure of at least 8.7

psig [1.2 times (vapor pressure at 115° F = 5.8 psig; plus the static head = 1.4 psig for a tank 60 inches high)]. Given the potential magnitude and the variability of these properties over the vast number of hazardous materials authorized, we believe a design pressure of not less than 3 psig is necessary. A maximum design pressure of 14.9 psig is being proposed for the MC 306 cargo tank.

The pressure design parameters for a tank are generally based on the vapor and static pressure of the lading and the loading/unloading pressure. For most cargo tanks, the tank test pressure is 150 percent of the design pressure and the emergency relief devices are set-to-discharge and flow rated at not lower than the design pressure and not above the test pressure. This philosophy assures that the emergency relief devices would not operate under normal transportation conditions and yet during an emergency condition would operate at less than or equal to a pressure for which the tank has been proof-tested.

Based on this discussion, we propose to revise the MC 306 specification as follows:

1. A minimum design pressure of 3 psig would be specified for MC 306 cargo tanks. This design pressure is consistent with the setting presently used for both loading/unloading and emergency relief devices.

2. The maximum design pressure for MC 306 cargo tanks would be increased to 14.9 psig. This is a relaxation of the present requirements and will be consistent with §173.119 which allows the MC 306 for a lading with a vapor pressure of 16 psig at 100° F. and takes into account that some ladings will have higher vapor pressures at 115° F.

3. A test pressure of 150% of the design pressures, and in no case less than 5 psig, would be required for MC 306 cargo tanks. This is an extension of the present requirement of flow rating the relief devices at 5psig (1.66 times 3 psig). We believe this 1.66 ratio is adequate for a 3 psig design pressure and is compatible with the presently used operating and emergency venting pressures.

4. The minimum set-to-discharge pressure for any pressure relief valve would be the design pressure (3 psig minimum). The pressure relief valve must close at not less than the design pressure.

5. For a cargo tank designed to be loaded or unloaded with the dome cover closed, the cargo tank must be equipped with a vacuum relief device to limit the vacuum to 1 psig and a pressure relief valve to limit the tank pressure to design pressure based on the product transfer rate marked on the cargo tank specification plate. This change allows greater flexibility for cargo tanks having a design pressure greater than 3 psig.

6. The use of the existing relief valve for normal venting conditions (set-to-discharge at 1 psig) would be prohibited. This relief valve has been shown by research to be a source of leakage in a cargo tank rollover and allows the release of lading under normal transportation conditions. The release of lading through these valves does not conform to the no leakage requirement in §178.341–4(d)(2) of the present regulations. The leakage in a rollover may be due to static pressure plus vapor pressure of lading acting on the device.

Certain other revisions to the MC 306 cargo tank motor vehicle specification would also be made. The use of rupture discs would continue to be prohibited in cargo tanks intended for a flammable liquid lading. As stated earlier in the preamble discussion to §178.340–5, the required strength of the manhole would be increased to that used by most of the tank manufacturers.

Section 178.342. Individual specification requirements applicable to a MC 307 cargo tank motor vehicle are contained in this section. Based on exemption experience, construction of a vacuum-loaded cargo tank motor vehicle designed in conformance with the MC 307 cargo tank motor vehicle specification would be authorized.

Petitioners (P-262, P-327) have requested a revision that would change the maximum allowable stress of ⅓ of the ultimate strength of the metal used to ¼. This suggested revision has been incorporated by the adoption of the ASME Code.

Section 178.343. Individual specification requirements applicable to a MC 312 cargo tank motor vehicle are contained in this section. Based on exemption experience, construction of a vacuum-loaded cargo tank motor vehicle with an internal design pressure of 25 psig and an external design pressure of 15 psig designed in conformance with the MC 312 cargo tank specification would be authorized. Given the substantial acute and environmental risks associated with the release of corrosive materials, we are proposing a minimum design pressure of 15 psig for MC 312 cargo tank motor vehicle.

Section 180 Subpart E. The requirements for the continuing qualification, maintenance and periodic testing of MC 300 through MC 312, MC 330, MC 331 and MC 338 cargo tank motor vehicles presently found in §§173.33, 177.814 and 177.824 would be contained in this new subpart.

Section 180.405. This section would contain general requirements for the use of cargo tank motor vehicles. The requirements are based on the regulations presently found in §173.33 or have been developed as a result of the research programs described earlier in the preamble. The requirements would provide for the continued use of— an existing cargo tank motor vehicle made to an obsolete specification, cargo tank motor vehicles conforming with and used under a DOT exemption whose provisions have been incorporated into the HMR, and DOT Specification cargo tank motor vehicles having no marked design pressure or a marked design pressure of less than 3 psig (see earlier preamble discussion titled "Design Pressure of Cargo Tank"). These requirements would assure that the level of safety of each cargo tank motor vehicle would be in accordance with the intent of the regulations.

As stated earlier in the preamble discussion to §178.340–5, research findings have shown that in a rollover condition manhole covers frequently fail or are deformed, causing lading leakage. To correct this problem, proposed §180.405(g) would require a retrofit on certain cargo tank manhole closures within 5 years of the effective date of the final rule with at least 20 percent of the affected cargo tanks being retrofitted each year. Each manhole closure would be required to be capable of withstanding a static internal pressure of at least 36 psig or the cargo tank test pressure, whichever is greater, without leakage or permanent deformation.

Presently §173.33(b)(2) authorizes MC 304 cargo tanks to have pressure relief devices and outlets which conform with DOT Specification MC 307 cargo tanks. We can find no valid reason why other DOT Specification cargo tanks made to an obsolete specification cannot be similarly modified. This proposal would allow MC 300, MC 301, MC 302, MC 303 and MC 305 cargo tanks to have pressure relief devices and outlets which conform with MC 306 cargo tanks, MC 310 and MC 311 cargo tanks to have pressure relief devices and outlets which conform with MC 312 cargo tanks, and MC 330 cargo tanks to have pressure relief devices and outlets which conform with the MC 331 cargo tanks.

Section 180.407. This section would contain all the requirements for the test and inspection of DOT Specification cargo tank motor vehicles. It would contain the general requirements presently in §§173.33 and 177.824, in addition to the proposed new requirements which have been developed as a result of the research program. Among the revisions to the general requirements is a clarification of the test requirements and the required test and inspection frequency in tabular format. The proposed test frequency would be consistent with and justified by the research program and associated recommendations and findings. Additionally, this section would emphasize the need to have trained and experienced inspectors.

Paragraph (d) would contain the requirements for an annual external visual inspection and testing. This section would contain the present periodic testing and inspection requirements contained in §177.824(b) and the following new requirements:

(a) The annual external visual inspection would be expanded to include DOT Specification MC 330, 331 and 338 cargo tank motor vehicles based on the need demonstrated by the research studies;

(b) Provisions on the inspection and testing of remote closure devices and the inspection of other appurtenances and accessories (that the research studies have shown should be inspected) would be added.

Paragraph (e) would provide the conditions under which an annual internal visual inspection is required. This section would contain some of the present requirements contained in §177.824(b).

Paragraph (f) would contain the retesting requirements on all lined cargo tanks. These are proposed new requirements which resulted from the research program and NTSB recommendations.

Paragraph (g) would specify the pressure retest requirements for all cargo tanks. This section would contain the requirements presently found in §§177.824(c), (d) and 1773.33(d), and the following revisions or additions:

(a) Specify the retest pressure for all cargo tanks in tabular form.

(b) Specify that each cargo tank must be pressurized to test pressure, except that the inspection for leaks and damage at design pressure would be permitted if the cargo tank is pneumatically retested.

(c) Require that pressure bearing parts of a cargo tank heating system be hydrostatically tested at a minimum of one and one-half times the heating system design pressure once every five years. The test pressure is based on several petitions for rulemaking (P-262, P-316, P-327).

(d) Require that each pressure retest be witnessed and certified by an Authorized Inspector. Additionally, each cargo tank motor vehicle would be required to have an external and internal visual inspection performed by an authorized inspector every five years as part of the pressure retest. To minimize the economic impact of pressure retesting all affected DOT specification cargo tanks, we are proposing that at least 20% of each owner's affected cargo tanks be retested each year.

(e) Establish a requirement that each MC 330 and MC 331 cargo tank constructed of QT steel be subjected to periodic internal fluorescent magnetic particle reinspection when used for the transport of liquefied petroleum gas or any lading that may cause stress corrosion cracking, the same as is now required for cargo tanks in ammonia service. Research on the integrity of MC 330 and MC 331 cargo tanks shows that the incidence of stress corrosion cracking is increasing and not limited to cargo tanks in anhydrous ammonia service. We believe requiring fluorescent magnetic particle reinspection is necessary for cargo tanks that transport any material that may cause stress corrosion cracking, except those constructed of other than part UHT materials that are postweld heat treated. We are specifically soliciting comments and rationale as to whether this proposed requirement should be extended to cargo tanks made of steel that are not postweld heat treated and to cargo tanks of less than 3500 gallons capacity without a manhole.

(f) Paragraph (h) would require an annual leakage test on all cargo tanks. This would be a new requirement based on the results of the research studies which found numerous leaks in the majority of cargo tank compartments tested. In order to limit cost and equipment down time, the test prescribed in the Environmental Protection Agency's "Determination of Vapor Tightness of Gasoline Delivery Tank Using Pressure-Vacuum Test" is an acceptable alternative test.

(g) Paragraph (i) would require thickness testing of cargo tank motor vehicles under certain conditions. This requirement is based on the need shown by analysis of accidents involving cargo tank motor vehicles in corrosive service. This requirement is consistent with a rule-related notice that was published in the Federal Register (48 FR 15127; April 7, 1983) requiring the owner to assure himself that the cargo tank motor vehicle conforms at all times with the specification under which it was constructed.

Thickness testing would be required when:

1. Visual inspection at any time indicates significant corrosion.

2. A cargo tank motor vehicle is unlined and is in corrosive service or is exposed to a corrosive environment capable of affecting the structural integrity of the cargo tank motor vehicle.

Section 180.409. This section would establish minimum qualification standards for any persons performing or witnessing a test or inspection required in §180.407. Specific qualification standards are prescribed for each required test or inspection. This section would be included to assure that all cargo tanks are tested and inspected by experienced and qualified personnel who are familiar with both the cargo tank motor vehicle specifications and proper testing methods. This section addresses a finding in the cargo tank research program that showed some testers are unqualified.

Section 180.411. This section would establish pass-fail criteria for each required test and inspection contained in §180.407. The establishment of pass-fail criteria would ensure that the results of each required test and inspection in §180.407 would be evaluated in the same way.

Section 180.413. This section would contain the repair, replacement of appurtenances, and modification requirements presently prescribed in §§173.33-(d)(11), (e) and 177.824, and the following revisions:

1. Repair requirements for MC 300 through MC 305, MC 310 and MC 311 cargo tanks would be specified. Such requirements are merely implied in the existing regulations.

2. All cargo tank repairs would be required to be performed in a facility having a current certificate of authorization from the ASME for Section VIII (Division 1), having a current certificate from the National Board, or under the direct supervision of an Authorized Inspector.

3. Modification, stretching, and rebarreling parameters would be specified and qualifications for facilities performing such work would be prescribed. The facility requirements would be identical to those specified for repairs.

Section 180.415. This section would contain the marking requirements presently contained in §177.824(h) plus a new requirement to identify the type of test or inspection performed. This new requirement would enable shippers, carriers and enforcement personnel to readily determine if a cargo tank motor vehicle has been tested or inspected as required by proposed §180.407.

Section 180.417. This section would contain the reporting and record retention requirements presently prescribed in §§173.33 and 177.824, and the following revisions:

1. A requirement that each owner of a cargo tank motor vehicle retaining in its files a certificate or manufacturer's data report certifying that the cargo tank motor vehicle is in conformance with the specification under which it was constructed. This proposed requirement has been added based on the results of the research program which indicated that a significant number of owners did not have these documents in their possession. Additionally, the research program showed that most owners did not have drawings or calculations for their equipment. This information will be of great assistance to the owner and other persons in verifying that a cargo tank conforms with the regulations. If a manufacturer's certificate is unavailable, owners of non-ASME Code stamped cargo tanks manufactured before the effective date of the final rule would be permitted to certify their equipment when supervised by an Authorized Inspector, owners of ASME Code stamped cargo tanks would be able to obtain a copy of the manufacturer's data report from the National Board or copy the information contained on the cargo tank nameplate and ASME Code plate. In both cases, the owner and the Authorized Inspector would be required to sign the certificate stating that the cargo tank fully conforms to the appropriate specification.

2. This section would expand the information required on the cargo tank test or inspection report presently required in §177.824. The additional information will make readily available to all concerned parties exactly what type of testing or inspection was performed along with the results of each test or inspection. This proposed requirement was based upon the recommendations of the research program and comments to Docket HM-183 recommending close monitoring of compliance with the HMR.

3. Reporting requirements for MC 330 and MC 331 cargo tanks carrying liquefied petroleum gas or any other material that may cause stress corrosion cracking would be specified. This is presently required for MC 330 and MC 331 cargo tanks in anhydrous ammonia service. We believe that this requirement is justified because research has shown that the incidence of stress corrosion cracking is increasing and is not limited to cargo tanks in anhydrous ammonia service.

Comments

MTB and BMCS are requesting that interested persons submit constructive comments, together with supporting data, for or against the rules proposed in this notice. The submission of general comments without supporting data or documentation will not assist MTB and BMCS in the development of a final rule. In order to fully consider the impact of this complex and technical proposal on public safety and the regulated industry, commenters are strongly encouraged to provide substantive data, calculations and test results to support their views. Commenters are requested to make their comments in a manner that will clearly identify the particular matters on which they are commenting. Unless comments are general in nature pertaining to the entire Notice, it is requested that each paragraph of comments be identified in the following manner:

"Part 176. We think —" or "Section 172.503 We believe —".

MTB and BMCS are particularly interested in receiving constructive comments in the following areas:

(1) What would be the incremental cost and benefits to the public and to cargo tank motor vehicle purchasers of requiring ASME Code construction of new cargo tank motor vehicles?

(2) What would be the incremental cost and benefits of requiring the use of National Board Authorized Inspectors and requiring that an Authorized Inspector certify that a cargo tank motor vehicle conforms with the applicable DOT specification?

(3) What would be the incremental cost and benefits of requiring that repairs to a cargo tank be performed by the following: a cargo tank manufacturer holding an ASME "U" stamp; a repair facility holding a "R" stamp; or under the direct supervision of an Authorized Inspector who certifies the repair as being acceptable?

(4) What would be the incremental savings on insurance to cargo tank manufacturers and cargo tank motor vehicle operators on cargo tank motor vehicles built to the ASME Code and certified by an Authorized Inspector?

(5) What would be the incremental savings realized by authorizing the use of external self-closing stop valves as proposed in §178.340–11(a)?

(6) What would be the incremental savings realized if the rear end tank damage protection requirements as proposed in §178.340–8(d) were adopted?

(7) Should we require National Board registration for all new construction of cargo tank motor vehicles?

(8) Should we require that an Authorized Inspector perform and approve or witness and approve all the tests and inspections specified in Part 180 of this proposal?

(9) What would be the impact of no longer authorizing the use of MC 300, MC 301, MC 302, MC 303, or MC 305 cargo tank motor vehicles for transporting hazardous materials?

(10) Should we limit the service life of a cargo tank motor vehicle in hazardous materials service? If so, what should this service life limit be?

(11) Are there any hazardous materials presently authorized in MC 306 type cargo tanks that are too hazardous to be transported in such tanks?

(12) Is it likely that a lined cargo tank with external insulation could have corrosion that significantly reduces its structural integrity yet would still pass the prescribed pressure test? If so, what test or inspection procedures could be used to identify and prevent this condition?

(13) Should MTB required special construction, test, operational or restricted lading authorizations on a cargo tank fitted with a very large closure (e.g., an openable rear head) located below the normal liquid level of a full cargo tank?

(14) In addition to answers to the above questions, MTB solicits comments on the overall cost effectiveness (and the incremental cost change) if the rules proposed in this Notice are adopted.

(50 F.R. 37766, September 17, 1985)

49 CFR Part 173
[Docket No. HM-183B; Notice No. 86–6]
Rear Bumpers on Cargo Tank Truck

Section 178.340–8(b) has been in effect since December, 1967 and similar bumper requirements have been in effect since the early 1940's for previously manufactured specification cargo tanks. This section requires that all cargo tanks must be protected by the use of rear bumpers. However, a large number of cargo tank trucks (commonly called bobtails) used in combination with cargo tank full trailers have been manufactured without rear bumpers. The number of units manufactured without rear bumpers is estimated to be approximately 3500. These combination units are used primarily for the transportation of gasoline, fuel oil and other petroleum distillate products.

As a result of accidents and incidents involving the transportation of hazardous materials in cargo tanks, increased emphasis has been placed on cargo tank compliance for the past few years. This increased emphasis, combined with research efforts evaluating the integrity of all specification cargo tanks and increased adoption and enforcement of the Hazardous Materials Regulations (HMR's) by individual States, has resulted in disclosure of violations of the HMR's with respect to cargo tank operation and manufacture. One area specifically identified was the lack of rear bumpers on cargo tank trucks operated in and out of combination with cargo tank full trailers.

In March, 1983, the California Highway Patrol (CHP) requested an interpretation regarding the rear bumper requirements for a three-axle cargo tank truck towing a two-axle cargo tank trailer. A letter of interpretation was issued by the BMCS, in coordination with the RSPA, to the CHP in April, 1983, which stated:

> Section 178.340–8(b) specifically requires that "every" cargo tank be provided with a rear bumper. In the example sited two cargo tanks are present, the cargo tank attached to the towing unit and the cargo tank attached to the trailer. Therefore, both cargo tanks must be equipped with rear bumpers as required by this section.

Similar letters of interpretation were issued to carriers and manufacturers during 1983. Subsequently, the CHP issued a directive stating that no enforcement actions would be taken regarding the absence of rear bumpers due to petitions for rulemaking filed with the RSPA by industry representatives requesting relaxation of the rear bumper requirements for "doubles" and a pending rulemaking action regarding cargo tanks by the DOT. However, in December 1983 the CHP Cargo Tank Advisory Committee, comprised of the CHP, State Fire Marshal, proprietary carriers, for-hire carriers, petroleum companies, tank manufacturers and DOT, was advised that the present requirement for rear bumpers on all cargo tanks was in effect and enforceable by the BMCS whether or not the CHP withheld enforcement.

The RSPA and BMCS published a joint rulemaking regarding manufacture, testing and in-use requirements for cargo tanks (HM-183, 183A) on September 17, 1985. In this NPRM, the petitions for rule change requesting that rear end tank protection be required only on the rearmost unit of a "double" cargo tank motor vehicle configuration were denied stating:

> . . . The petitioners argued that the present requirement adds cost and weight to the cargo tank configuration with no safety benefits. We do not agree. We believe that the forward unit of a "double" is vulnerable to rear-end tank damage particularly in turning maneuvers. This vulnerability increases in proportion to the length of the draw bar between the cargo tank units. The forward unit of a "double" is at times operated without the protection afforded by the rear units. Operation of such a forward unit, whether with a full load or with only residual lading presents an unacceptable risk. . . .

As part of the administrative proceedings on Docket HM-183, 183A, two public hearings and two public meetings were conducted. At the public hearing in Burlingame, California, held in December, 1985, the DOT again stated that rear bumpers are required on all cargo tanks and that those without rear bumpers are in violation of the regulations. It remains our opinion that rear bumpers are required on all cargo tank motor vehicles. These rear bumpers provide protection to the tank and associated piping in the event of a rear-end collision.

Several enforcement cases involving the lack of a rear bumper on cargo tanks were initiated in 1985. Subsequent to the NPRM and enforcement cases, representatives of the affected industry requested a meeting with the RSPA and BMCS. These representatives indicated that if immediate compliance was required, the economic impact of removing all cargo tanks that are not in compliance would be harmful to the economy. Additionally, it was stated that such an action would also serve to threaten public safety in that enough petroleum products might not be delivered to support public or private transportation as well as emergency response units. It was also stated that the lack of enforcement of this requirement for more than 40 years fostered the belief by manufacturers and cargo tank operators that such cargo tanks complied with the regulations.

We do not concur with the argument that lack of enforcement indicated acceptance of cargo tanks manufactured without rear bumpers. It is our opinion that the regulations requiring rear bumpers is quite clear and that cargo tanks manufactured without the rear bumpers are in violation of the regulations. Additionally, the Federal Motor Carrier Safety Regulations (FMCSR's) (49 CFR Parts 305 through 399) require rear end protection on each motor vehicle. Therefore, the rear bumper is used to comply not only with the HMR's, but also, with the FMCSR's.

However, we do acknowledge that strict enforcement of the rear bumper requirement could cause hardship both for motor carriers and the public in general. Because of the potential hardship, we are proposing to allow a 36 month time period for cargo tank operators to bring their units into compliance. By allowing this time period, little, if any, interruption of petroleum product delivery should occur. This proposal would also allow motor carriers the ability to bring into compliance portions of their fleets on a periodic basis, thus eliminating the potential for removing all non-complying units at a single time.

It should be noted, however, that the proposed 36 month compliance period applies only to those units that are operated in "double" combinations. If a cargo tank truck is operated without the cargo tank full trailer attached, a rear bumper is mandatory. Operation of the cargo tank truck without a rear bumper would be a violation of the regulations and would be subject to enforcement and penalty actions.

Alternative. Industry representatives have indicated that they know of no incidents that have occurred due to a lack of a rear bumper on the cargo tank truck. Additionally, the Truck Trailer Manufacturers Association has requested a grandfathering of existing cargo tank trucks from the rear bumper requirements and suggested a modification to the regulations which would require such cargo tank trucks to be operated only in combination when no bumper is present.

In order to fully assess the requirement for a rear bumper on these combination units, we are requesting information regarding accident history and other pertinent comments regarding the need for a rear bumper. Additionally,

OHMT and BMCS are requesting that interested persons submit constructive comments, together with supporting data, for or against the rules proposed in this notice. The submission of general comments without supporting data or documentation will not assist OHMT and BMCS in the development of a final rule. OHMT and BMCS are particularly interested in receiving constructive comments in the following areas:

(1) What would be the incremental costs of requiring a rear bumper to be installed on presently non-complying units?

(2) Presently, a cargo tank manufacturer is required to certify that the cargo tank is manufactured in accordance with all applicable requirements. The manufacturer of the cargo tank may not know if the cargo tank truck is to be operated in combination with a cargo tank full trailer. What method of certification would be necessary for the cargo tank manufacturer to assure that the cargo tank truck complies only when operated in combination with a cargo tank full trailer?

(3) What marking should be required to be displayed on the cargo tank truck to indicate that a bumper is required when it is not being operated in combination?

(4) Should an existing cargo tank truck be grandfathered, while newly manufactured cargo tanks be required to be equipped with a rear bumper?

(5) Does tow bar length have any effect on safety, particularly in cornering maneuvers where the cargo tank truck could be struck from the rear?

(6) How would compliance with the bumper strength requirements be met when a temporary bumper is installed on the cargo tank truck when not operated in combination with a cargo tank full trailer? What is the likelihood that the temporary bumper may or may not be installed properly?

(7) How frequently are cargo tank trucks, without rear bumpers, operated not in combination with a full trailer?

(51 F.R. 28605, August 8, 1985)

49 CFR Parts 171, 172, 173 and 175
[Docket No. HM-184; Amdt. Nos. 171-66, 172-74, 173-158 and 175-23]
**Implementation of the ICAO Technical
Instructions**

In consideration of a perceived need to improve the overall safety of the transportation of hazardous materials on a worldwide level, the International Civil Aviation Organization (ICAO) Accident Investigation and Prevention Divisional meeting in 1974 adopted Recommendation 7/1 citing a need to study all aspects relating to the transportation of hazardous materials aboard aircraft. In late 1974, this recommendation was adopted by the ICAO Air Navigation Commission (ANC) and the ICAO Secretariat was directed to study the matter. Principal regulations currently in force throughout the world were reviewed and ICAO contracting states and interested international organizations were requested to provide information on the regulations they followed and comments on the need for increased international cooperation in this area. All responses received indicated that such a need did exist.

In light of the conclusions drawn from this study, the ANC concluded that ICAO should undertake the development of international standards and guidance material concerning the transportation of hazardous materials by air. By early 1976 a study group established by the ANC had prepared a draft set of Standards and Recommended Practices (SARPS) and supporting technical instructions. The SARPS, when adopted, would become an annex to the Convention on International Civil Aviation (Chicago Convention). Because the SARPS covered only the basic principles relating to the safe transport of hazardous materials by air, they were to be supported by a technical document providing all relevant details. To insure maximum uniformity with existing international standards for the transport of hazardous materials by other modes, the draft SARPS and Technical Instructions were based on the Recommendations prepared by the United Nations Committee of Experts on the Transport of Dangerous Goods and the regulations issued by the International Atomic Energy Agency (IAEA) and also took appropriate account of other existing standards for the transportation of hazardous materials by air.

In March, 1978, the ANC established the Dangerous Goods Panel (DGP), and directed it to continue work on the development of the SARPS and Technical Instructions. In February 1981, the DGP concluded its efforts and forwarded the final draft of the SARPS and supporting Technical Instructions for the Safe Transport of Dangerous Goods by Air (the Technical Instructions) to the ICAO council for review and approval. On June 26, 1982, the Council adopted the SARPS as Annex 18 to the Chicago Convention. Under the provisions of the Chicago Convention, Annex 28 would become effective (voluntary compliance authorized) on January 1, 1983, and applicable (compliance required) on January 1, 1984, unless more than half of the 150 contracting governments registered disapproval by October 26, 1981. By the latter date, 33 contracting states had registered unqualified support of Annex 18. Two states, the United States and Switzerland, registered a partial approval of the Annex. No states disapproved of the Annex.

The partial disapproval of the Annex by the United States stemmed from certain provisions in the Annex which incorporate the Technical Instructions by reference and thereby lend to the Technical Instructions the character of an annex. This, in effect, requires that all contracting states insure full compliance with all details of the Technical Instructions. While the United States fully supported the basic principles espoused in the Annex, the vast majority of which are consistent with existing provisions in the Hazardous Materials Regulations, it was not in a position to mandate compliance with all details of the Technical Instructions, for this reason, the United States advised that ". . . the United States wishes to register disapproval of paragraph 2.2.1 of Annex 18. Furthermore, the United States wishes to register disapproval of Chapters 3 through 9 of the Annex to the extent that the provisions in these Chapters directly incorporate the provisions of the Technical Instructions." It is important to note that the reasons behind the partial disapproval of Annex 18 by the United States relate more to the administrative and legal difficulties under United States law that would be associated with mandating compliance with the Technical Instructions than to a general concern for the adequacy of the Technical Instructions from the point of view of safety in the air transport. Indeed, notwithstanding the United States' partial disapproval of Annex 18, the Materials Transportation Board (MTB) believes they can provide a basis for the safe transport of hazardous materials by air and will do much to facilitate the international transport of hazardous materials by air.

It is clear that by January 1, 1984, compliance with the ICAO Technical Instructions will be mandated by virtually all ICAO member states, which include all the industrialized nations in the world and all of the major trading partners of the United States. Furthermore, as a result of a decision by the International Air Transport Association (IATA) Restricted Articles Board (RAB) to implement the ICAO Technical Instructions effective December 31, 1982, shippers will be unable to have shipments accepted for international transportation after that date unless the shipment complies with the provisions of the ICAO Technical Instructions. For these reasons, the MTB believes it is in the interest of both shippers and carriers of hazardous materials by air to take action to amend the Hazardous Materials Regulations to recognize the ICAO Technical Instructions. Such action would also be consistent with the United States' obligations as a signatory government to the Chicago Convention.

It should be emphasized that this Notice of Proposed Rulemaking is responsive, in part, to a petition submitted by the Air Transport Association of America requesting that the Hazardous Materials Regulations be amended to recognize the ICAO requirements for the transport of hazardous materials. The petition notes the necessity and urgency of recognition of the ICAO Technical Instructions through the Hazardous Materials Regulations and suggests that there is a general belief (shared by shippers, airline pilots and air carriers, alike) that the ICAO rules ". . . evidence well organized, technically accurate and workable standards for worldwide transportation of hazardous materials." Because this petition so accurately sets forth the need to take action to amend Hazardous Materials Regulations in order to take account of the ICAO requirements, the MTB believes there is merit in quoting the substance of that petition in this notice:

The Air Transport Association's Cargo Committee, member air carrier's Heads of Cargo divisions and the Restricted Articles Board (member air carrier's representatives from the fields of engineering, chemistry, cargo services, training and safety), take this opportunity to urgently petition the Bureau, pursuant to 49 CFR 106.31, regarding 49 CFR recognition of the ICAO rules and dangerous goods list emanating from Annex 18 and its Technical Instructions for the Safe Transport of Dangerous Goods by Air.

At recent air carrier, Air Line Pilot and other meetings, air carrier representatives, pilots and shippers have noted that the new ICAO rules evidence well organized, technically accurate and workable standards for world-wide transportation of hazardous materials. DOT-MTB personnel at these meetings have stated that the United States Government has an obligation under the Chicago Convention to recognize the newly developed ICAO rules, even though it was reported that our government had partially disapproved certain paragraphs in Annex 18. This registration of disapproval, however, was only to the extent that the 'Technical Instructions' were incorporated into the Annex itself.

We realize that our government will also find it necessary to file 'differences' with the ICAO organization such as, quantity limitations on aircraft and hazardous substances and wastes, to mention a few areas which immediately come to mind.

This brings us to the point of urgency, and the fact that time is of the essence, as it will be permissive for the rest of the world to use the ICAO dangerous goods rules by the end of 1982.

In reality, these rules will take effect for all international traffic moving on the world's scheduled airlines on December 31, 1982, when the IATA Dangerous Goods Regulations—24th (ICAO) Edition becomes effective. Hazardous Materials traffic moving into, out of and through the U.S. will by necessity need to be in a position to abide by these dangerous goods rules.

It is with this in view that we respectfully request DOT-MTB to take all immediate steps possible to provide rulemaking which will implement the ICAO rules into U.S. regulations as an 'alternative' to 49 CFR (with U.S. differences), including the simultaneous publishing of an 'optional' hazardous materials table which will provide for the use of the ICAO Dangerous Goods List (with U.S. differences) for both domestic and international traffic.

Such immediate rulemaking will permit all air carriers the option of utilizing the ICAO rules for acceptance, handling, on line transportation and interline transportation concurrent with their implementation.

Also, such immediate rulemaking will provide the FAA's Office of Civil Aviation Security with the needed time to prepare for the alternative use of ICAO rules by the nations shippers and transporters, as well as other nations shippers and transporters.

We are extremely desirous of as smooth a transition as is possible with respect to the introduction of the ICAO rules. It is our obvious fear that ill-timed introduction to these rules could cause great confusion in our industry. This confusion could provide for unnecessary embargos of hazardous materials/dangerous goods traffic, the disruption of transportation and the opportunity for misrepresentation of such traffic, thus, defeating our ever present concern for safety.

The MTB is proposing to amend the Hazardous Materials Regulations to allow, under certain conditions and with certain limitations, hazardous materials packaged, marked, labeled, classified and described and certified on shipping papers as provided in the ICAO Technical Instructions to be offered, accepted and transported by aircraft within the United States and aboard aircraft of United States' registry anywhere in air commerce. Furthermore, certain amendments are being proposed to Part 175 to align the requirements for the loading and handling of hazardous materials aboard aircraft with those of the Technical Instructions in order to insure that United States' aircraft operating in foreign countries, or that foreign aircraft operating in the United States, will not be in violation of the requirements for loading and handling applied in the country in which the aircraft is operating.

If certain of the requirements imposed on operators by the ICAO Technical Instructions are considered to be overly restrictive, or MTB is unable at this time to justify them on the grounds of safety, no corresponding amendment to Part 175 is being proposed. In such cases it would be the responsibility of the air carrier to insure compliance with any appropriate provisions of the Technical Instructions when operating in a country that mandates compliance with the Technical Instructions. In other cases, where the implementation of a provision in the ICAO Technical Instructions would result in the derrogation of an existing DOT requirement that the MTB believes should be maintained in the interest of safety, no amendment to Part 175 is proposed. In such cases ICAO will be officially notified, upon conclusion of this rulemaking action, that the United States is maintaining the more stringent requirement so that an appropriate exception can be noted in the ICAO Technical Instructions.

The following is an analysis of this proposal by section which provides the background behind the proposed changes:

Section 171.2. Paragraph (a) of this section would be amended to include a cross-reference to a new §171.11 which contains an authorization to use certain provisions of the ICAO Technical Instructions in place of the corresponding provisions in the Hazardous Materials Regulations. The cross reference is similar to those currently contained in paragraph (a) to the sections authorizing compliance with the IMCO Code.

Section 171.11. A new §171.11 would be added. This section details the extent to which a shipper and air carrier may comply with certain provisions of the ICAO Technical Instructions as an alternative to the corresponding provisions in the DOT Hazardous Materials Regulations. In particular, the section would, subject to certain conditions, allow hazardous materials to be transported by aircraft, and by motor vehicle incident to transportation by aircraft if, under the ICAO Technical Instructions, the hazardous material is properly:

(1) Packaged,

(2) Marked,

(3) Labeled,

(4) Classified,

(5) Described and certified on shipping papers,

(6) Otherwise in the required condition for shipment, and

(7) Within the quantity limits prescribed for transportation by either passenger or cargo aircraft, as appropriate.

The conditions imposed on the alternative use of the ICAO Technical Instructions would be set forth in paragraph (d) of the section. The special requirements concerning hazardous waste and hazardous substances are necessitated by the statutory mandate to insure appropriate safeguards for these materials during transportation. The additional conditions proposed in paragraphs (d)(4) and (d)(5), two of which apply only to transportation by motor vehicle incident to air transportation, are considered by the MTB to be necessary for appropriate emergency response in the event of an incident involving the materials.

Section 171.7. This section would be amended to incorporate the 1983 edition of the ICAO Technical Instructions by reference and to provide information relative to the source for obtaining the Technical Instructions.

Section 171.8. This section would be revised to incorporate a definition for the abbreviation "ICAO" and for the term "Unit load device". The definition proposed for "Unit load device" is the definition of that term in the ICAO Technical Instructions.

Section 172.401. This section would be amended to exempt from the prohibited labeling provisions any package labeled in accordance with the ICAO Technical Instructions regardless of the mode of transport of the package.

Section 172.446. Paragraph (a) and the facsimile of the Magnetized Material label would be revised by deleting the text "MAGNETIZED MATERIAL LABEL" in the lower right corner of the border of the label. This is being proposed so that the DOT specifications for this label will be consistent with the ICAO specifications for the same label. A new paragraph (c) would be added to allow the continued use of labels meeting the existing DOT specifications that are in stock as of January 1, 1983, until those stocks are depleted.

Section 172.448. Paragraph (a) and the facsimile of the Cargo Aircraft Only label would be amended by replacing the text "DANGER-PELIGRO" with the word "DANGER" and by removing the text "CARGO AIRCRAFT ONLY LABEL" in the lower right corner of the label. These changes are being proposed so that the DOT specifications for this label will be consistent with the ICAO specifications for the same label. A new paragraph (c) would be added to allow the continued use of labels meeting the existing DOT specifications that are in stock as of January 1, 1983, until those stocks are depleted.

Section 173.250. The reference to §175.305 would be changed to §175.10 because the provisions governing the transport by aircraft of wheelchairs with wet electric storage batteries would now appear in §175.10 rather than in §175.305.

Section 175.3. The specific references to Parts 172 and 173 would be deleted to enable a carrier to accept a shipment prepared in accordance with the new §171.11.

Section 175.10. One exception currently appearing in this paragraph would be modified and six new exceptions would be added to insure that all the hazardous materials excepted from regulation under the ICAO Technical Instructions will also be excepted from application of the DOT Hazardous Materials Regulations when carried aboard aircraft. Paragraph (a)(5) would be amended to impose a ten pound gross weight limit on the amount of small arms ammunition that a passenger or crew member may carry in checked baggage. This change is proposed to make this exception consistent with the corresponding exception in the ICAO Technical Instructions. Several new exceptions would be added to deal with alcoholic beverages, perfumes or colognes carried by the operator for sale aboard the aircraft or carried by passengers or crew as carry-on baggage; and with carbon dioxide, solid (dry ice) used in food or beverage service aboard the aircraft or by passengers to refrigerate perishables in carry-on baggage. Consistent with their presentation in the ICAO Technical Instructions, the recently published DOT amendments which provide exceptions for the carriage of electrically powered wheel chairs will now be shown in this section. Several exceptions which currently appear in this section but not in the ICAO Technical Instructions, and which apply primarily to domestic aircraft operations, are not effected by these amendments.

Section 175.30. Paragraph (a) of this section would be amended to allow an operator to accept shipments of hazardous materials if described, certified on shipping papers, labeled, and marked in accordance with §171.11. However, freight containers would still be required to be placarded or labeled in accordance with Subpart F of Part 172. Paragraph (b) would be amended to make the preloading package and overpack inspection requirements consistent with ICAO requirements by introducing the concept of "Unit load device". In addition, specific references to the DOT requirements for overpacks which appear in §173.25 and §175.393(r) would be removed to permit acceptance of overpacks that are prepared in accordance with the provisions of the ICAO Technical Instructions. Finally, a new paragraph (e), concerning the pre-acceptance inspection of overpacks, would be added which would be consistent with the IACO requirements concerning inspection of overpacks.

Section 175.33. The provisions for notification of pilot-in-command would be revised to make the information required to be provided more consistent with that required in the ICAO Technical Instructions. It would be required that the notification be provided to the pilot "as early as practicable" prior to departure. Hazardous materials shown on the notification would be permitted to be described by their proper shipping name, hazard class and identification number as provided in the ICAO Technical Instructions provided any additional descriptions required by §171.11 appear on the shipping papers. A new requirement that the notification include the total number of packages of each material and their total quantity or transport index, as appropriate, is added to be consistent with ICAO requirements. Certain other entries required by ICAO to appear on the notification have not been proposed for inclusion in this section because the MTB does not believe that the increased regulatory burden associated with their inclusion can be justified on the grounds of safety; however, this information may be added to the notification at the discretion of the operator. Finally, a new paragraph (b) would be added to require that a copy of the notification be carried aboard the aircraft during flight.

Section 175.35. This section would be deleted. The MTB believes there is no reason to carry the shipping papers aboard the aircraft if a copy of the notification to pilot-in-command is aboard.

Section 175.75. This paragraph would be amended by deleting the specific reference to Part 172 in paragraph (a)(i) in order to allow the carriage of hazardous materials prepared in accordance with §171.11 and by including references in paragraph (a)(3) to the existing single package maximum transport indices for passenger and cargo only aircraft.

Section 175.78. This section would be amended to reflect the minimum segregation requirements for hazardous materials of various classes as provided in the ICAO Technical Instructions. The MTB believes these changes are necessary to insure that uniform loading requirements will be applied to United States' registered aircraft operating overseas as well as to foreign registered aircraft operating in the United States. In addition, the MTB believes that these provisions would represent an improvement over the current requirements of this section which specify segregation requirements only for corrosive materials. For example, §175.78(a) currently would allow flammable liquids and oxidizers to be stowed next to or in contact with each other. This could result in a potentially dangerous situation in the event of leakage of the packages.

Section 175.79. Paragraph (a) would be amended by removing the provision concerning securing of packagings. This provision will be transferred to a new §175.81 concerning securing of cargo in order to provide greater consistency with the manner of presentation of these requirements in the ICAO Technical Instructions. Paragraph (b) would be amended to clarify that packagings fitted with both top and side closures need not be stored and loaded with the side closures up. This clarification will make paragraph (b) consistent with the corresponding provision in the ICAO Technical Instructions.

Section 175.81. A new §175.81 concerning securing of packages would be added to consolidate the general requirement for securing previously contained in §175.79 and the specific securing provisions for packages containing radioactive materials previously contained in §175.85(d).

Section 175.85. Paragraph (a) of this section would be amended to clarify the conditions under which hazardous materials may be carried on the main deck of a passenger (combi) aircraft, in order to provide consistency with the ICAO Technical Instructions. Paragraph (b) would be amended to eliminate the use of the word "accessible" through introduction of the ICAO definition of "accessible" directly into the loading requirement. The ICAO definition of "accessible" is, for all practical purposes, identical to that currently used in the section, except the criterion that a crew member be able to separate the package containing cargo-aircraft-only material from other cargo would apply only when size and weight of the package permit. The third sentence in the existing text of paragraph (b) of this section has been removed because the same text already appears in paragraph (c)(3) of this section. Finally, the list of hazardous materials in paragraph (c)(1) which need not be carried in a location that is accessible to a crew member in flight would be revised to be consistent with the corresponding provision of the ICAO Technical Instructions.

Section 175.88. A new §175.88 concerning inspection of unit load devices would be added consistent with the provisions of the ICAO Technical Instructions. This section would require that a unit load device be inspected prior to loading aboard an aircraft to make certain it is free from evidence of damage to, or leakage from, any hazardous materials it may contain. This is consistent with current requirements that packages and overpacks be inspected for evidence of damage or leakage prior to loading on an aircraft.

Section 175.90. This section would be completely revised to add certain provisions contained in the ICAO Technical Instructions to the existing provisions concerning damaged shipments. A provision would be added to require that packages and overpacks be inspected for damage or leakage upon removal from an aircraft or unit load device and that the area in which a unit load device was stowed aboard an aircraft be inspected, upon removal of the unit load device, for evidence of leakage of, or contamination from, the dangerous goods carried. If such evidence is found, the aircraft compartment must then be inspected for contamination and any contamination removed. In addition, provisions would be added providing actions to be taken in the event that a package containing etiologic agents is found to be damaged or leaking.

Section 173.305. Paragraph (b) of this section would be deleted since the provisions governing the transport of wheelchairs with wet electric storage batteries would now appear in §175.10.

Section 175.320. A new paragraph (b)(10) would be added to prohibit the international transportation of hazardous materials under this section unless permission is first obtained from the countries of origin, destination, transit and overflight as required by the ICAO Technical Instructions.

Section 175.630. Several changes are proposed to this section to maintain consistency with corresponding requirements in the ICAO Technical Instructions. The applicability of the section would be expanded to cover etiologic agents as well as poisons. Poisons and etiologic agents, however, would be allowed to be transported aboard an aircraft in the same compartment with foodstuffs, feed or other edible material provided the poisons or etiologic agents and the foodstuffs, feed or other edible materials are loaded in separate unit load devices that are not stowed adjacent to each other aboard the aircraft. Finally, the provision requiring that the area in which a package bearing a poison label was stowed aboard the aircraft be visually inspected after removal of the package would be deleted. This would be replaced by the new provision in §175.90 requiring that the package be inspected for evidence of damage or leakage upon removal from the aircraft and that the aircraft itself need only be inspected if the package shows evidence of damage or leakage.

Section 175.701. Six of the separation distances prescribed in the table in paragraph (b)(2) of this section would be revised to align the required separation distances with those provided in the ICAO Technical Instructions.

Section 175.705. A new §175.705 would be added to require a periodic check for radiological contamination of aircraft used routinely for the transport of radioactive materials. If the check revealed the aircraft to have contamination above prescribed levels, it would have to be taken out of service. This section is proposed to provide greater consistency with the provisions of the ICAO Technical Instructions.

(47 F.R. 33295, Aug. 2, 1982)

49 CFR Parts 171, 172, 173, and 175
[Docket No. HM-184; Amendment Nos. 171–69, 172–77, 173–160, 175–25]
Implementation of the ICAO Technical Instructions

On August 2, 1982, the MTB published a notice (Docket HM-184; Notice 82–9) in the Federal Register (47 FR 33295) which requested public comment on the need to amend the Hazardous Materials Regulations (HMR) in order to take account of the ICAO Technical Instructions which become effective on January 1, 1983. Background concerning the implementation of the ICAO Technical Instructions on a worldwide basis pursuant to Annex 18 of the Convention on International Civil Aviation (Chicago Convention), and the potential implications of their implementation with respect to hazardous materials shipments being imported into or exported from the United States by air, were discussed in the preamble to that notice.

Under these amendments hazardous materials will be allowed to be shipped, under certain conditions and limitations specified in §171.11, when packaged in accordance with the ICAO Technical Instructions. In July, 1982, a working group of the ICAO Dangerous Goods Panel met to review in detail the packaging provisions for hazardous materials in the ICAO Technical Instructions. In the course of this review, some difficulties regarding the compatibility of certain hazardous materials with authorized packagings were discovered and corrected. These corrections will be incorporated into the 1984 edition of the ICAO Technical Instructions. In the interim, the MTB must emphasize to users of the ICAO Technical Instructions that under the packaging provisions of the Technical Instructions there is a general requirement that any packaging material in direct contact with a hazardous material must be resistant to any chemical or other action of the hazardous material that could reduce the effectiveness of the packaging. This provision appears in paragraph 1.1.3 of Part 3, Chapter 1 of the ICAO Technical Instructions. Therefore, no packaging is considered to comply with the ICAO Technical Instructions unless this general requirement is fulfilled. Shippers of hazardous materials must be cognizant of this, and the other general packing requirements in the ICAO Technical Instructions, when determining the acceptability of packagings for use pursuant to the ICAO Technical Instructions.

Twenty-two commenters responded to Notice 82-9. Based on the comments received, the proposals contained therein are being adopted with certain changes, as final amendments to the HMR. All commenters expressed general support for the proposals. Other significant comments and the actions taken thereon are discussed by general subject area, or in the analysis by section, in the following paragraphs.

A number of commenters submitted general comments related to differences between the HMR and the ICAO Technical Instructions regarding classification, labeling, marking and identification numbers assigned to particular hazardous materials. One commenter went so far as to suggest that MTB replace the Hazardous Materials Table in §172.101 with the ICAO Table. The MTB is aware of these differences, and it is because of these differences that the MTB believed it necessary to propose amendments that would allow optional compliance with the ICAO Technical Instructions. However, because of the intermodal nature of the basic DOT classification system and the Hazardous Materials Table in §172.101, and in view of the relatively small quantity of hazardous materials transported by air as compared to the other modes of transport, the MTB does not believe that amendments to §172.101 or the basic DOT classification, marking and labeling system can be supported solely on the basis of conflicts with the ICAO Technical Instructions. Therefore, comments of that nature received in response to Notice 82-8 are considered to be outside of the scope of this rulemaking. Interested persons desiring to propose amendments to the DOT system of classification or to §172.101 are invited to petition for such changes.

One commenter felt that, under the proposals, a shipper or carrier "... would be required to adhere to the stricter of the DOT or ICAO standards, in those circumstances in which the requirements under the two systems differ." This statement is incorrect since, to the extent that the ICAO Technical Instructions have been incorporated by reference into §171.11, compliance with the ICAO Technical Instructions with regard to packaging, marking, labeling, classification, description and certification on shipping papers and quantity limitations would be permitted as an alternative to compliance with the corresponding requirements in the HMR.

Two commenters discussed the potential recognition of the International Air Transport Association (IATA) Dangerous Goods Regulations in the HMR. One suggested that the IATA regulations should be allowed to be used as the "source" of the ICAO Technical Instructions by incorporating the IATA regulations by reference into the HMR. The other commenter noted that only the ICAO standards should be recognized since ICAO is an intergovernmental body and, therefore, the ICAO Instructions would not include commercial variations that could be adopted by IATA. Having already noted several differences between the ICAO Technical Instructions and the 24th edition of the IATA Dangerous Goods Regulations, the MTB agrees with the latter commenter. Therefore, the IATA Dangerous Goods Regulations will not be incorporated by reference in the HMR.

A commenter expressed concern for the potential impact of the notice on domestic air transportation, especially with respect to ORM-D materials. He urged MTB to make every effort to preserve existing classifications such as ORM-D "... in the accommodation of international codes that are as yet untested." The MTB appreciates the concerns of this commenter and notes that under the amendments published in this document the existing DOT classification system, including ORM-D, is preserved, and that hazardous materials may be offered and accepted for transportation by air in accordance with this classification system.

One air carrier requested that MTB adopt the 1973 edition of the International Atomic Energy Agency (IAEA) regulations concerning the transport of radioactive materials. This has already been proposed under Docket No. HM-169 and a final rule is expected to be published in the near future. Therefore, no changes have been made in this amendment as a result of the comment.

MTB's attention was drawn to the fact that there are many hazardous materials and UN numbers listed in the ICAO Technical Instructions that are not listed in the Emergency Response Guide (ERG). The MTB is aware of this situation and consideration will be given to adding these entries to the next edition of the ERG. Also, the Chemical Transportation Emergency Center (CHEMTREC), which is referenced in the ERG, will be notified concerning the UN numbers that are not presently listed.

In addition to these general comments, a number of commenters suggested specific changes to individual sections of the proposed rule. These comments, and the actions taken in response to them, are summarized in the following review by sections:

Section 171.2. No change to the proposed rule.

Section 171.7. An error in the proposed rule concerning the numbering of the paragraphs to be added has been corrected.

Section 171.8. No change to the proposed rule.

Section 171.11. Several questions have been raised concerning the applicability of this section to transportation by motor vehicle. It is intended that this section apply to the transportation by motor vehicle of hazardous materials that have been or will be transported by air. Transportation by motor vehicle need not be immediately incident to the air transportation. For example, a consignee receiving hazardous materials transported by air under the provisions of §171.11 may reconsign the shipment to another party by motor vehicle under the provisions of this section. In such cases, it is not intended that the shipment be repackaged, remarked or relabeled to conform to DOT requirements applying solely to carriage by motor vehicle in domestic transportation.

When shipments of hazardous materials are transported by motor vehicle under the provisions of §171.11, it is intended that the requirements of Subpart F of Part 172 concerning placarding of the motor vehicle will still apply. However, in such cases the placard for the DOT class required by §171.11 (d)(4)(i) that is shown on the shipping paper must be applied. This will be the placard for the DOT class most closely corresponding to the ICAO class. In order to clarify the intent that placarding is required for transportation by motor vehicle, paragraph (d)(4) of this section has been revised to make specific reference to Subpart F of Part 172. In addition, in order to assist emergency response and enforcement personnel, a new paragraph (d)(4)(iii) has been added recommending that shipping papers used for transportation by motor vehicle contain an indication that the shipment is being made pursuant to §171.11.

Several commenters were not clear as to whether the prohibition in proposed paragraph (c) applied only to materials prohibited from being offered or accepted for transportation, as indicated by the word "Forbidden" in Column (3) of the Hazardous Materials Table in §172.101, or whether it also applied to materials forbidden for transportation aboard passenger or cargo carrying aircraft as indicated in Column (6) of the table. Paragraph (c) has been reworded to clarify that it applies only to those materials forbidden from being offered or accepted for transportation by any mode, as indicated by the word "Forbidden" in Column (3) of the Table.

Four commenters responded to the proposed §171.11(d)(1) suggesting that the HMR exempt hazardous substances from regulation when transported by air. Due to the requirement of Section 306 (a) of the Comprehensive Environmental Response, Compensation and Liability Act of 1980 (PL 96–510) regarding the transportation of hazardous substances, the MTB cannot provide for such an exemption in the HMR and no change has been made concerning the applicability of the HMR to hazardous substances in air transport. In response to another comment, the wording of paragraph (d)(1)(i) has been revised to make it more consistent with the wording of paragraph (d)(1)(ii).

Two commenters requested clarification regarding the materials to which paragraph (d)(3) would apply. Examples of such materials include hazardous substances, hazardous wastes and combustible liquids with a flashpoint above 60.5° C when shipped in packagings of greater than 110 gallons capacity. A proposal that these materials be listed in paragraph (d)(3) has not been adopted because the MTB believes such a listing to be unnecessary.

A number of comments were received concerning proposed paragraphs (d)(4) and (d)(5). Two commenters suggested that neither of these paragraphs were necessary for safety and that they imposed an administrative burden and should, therefore, be deleted. The MTB disagrees and believes that, on the basis of comments submitted under Docket No. HM-126, there is widespread support for inclusion of this information on shipping papers. Two commenters suggested that paragraph (d)(5) should be altered so that it applies to materials meeting the criteria for ICAO Division 6.1 packing Group I or II. Because of the differences between these criteria and the DOT definition of a Poison B, this suggestion has not been incorporated into the final rule. Finally, two commenters felt that inclusion of the word "Poison" on shipping papers for air transport was redundant since the ICAO class number "6.1" would already be shown as either the classification or as a subsidiary risk and suggested that addition of the word "Poison" should only be required for transportation by motor vehicle. The MTB agrees with this comment and has amended paragraph (d)(5) accordingly.

In response to three comments received, proposed paragraph (d)(6) of this section has been amended to clarify the fact that ICAO class or division numbers are not considered abbreviations and may, therefore, be used. This paragraph has been redesignated paragraph (d)(7) in the final rule and, as a result of a number of questions raised concerning the applicability of the proposed §171.11 to shipments of radioactive materials, a new paragraph (d)(6) has been included to highlight the special requirements that would apply to shipments of radioactive materials.

Section 172.101. A section reference in the Hazardous Materials Table to the exception for transporting wet electric storage batteries with steel chairs has been revised as a consequence of changes made elsewhere in this amendment.

Section 172.401. No change to the proposed rule.

Section 172.446 and 172.448. Two comments were received objecting to the proposal to allow depletion of stocks existing on January 1, 1983, of the Magnetized Material and Cargo Aircraft Only labels, particularly for international transportation. The MTB believes that it is unreasonable to require immediate change to a new label format and, therefore, has retained the provision in question.

Section 172.504. Although no change was proposed to this section, two commenters suggested modifications to the placarding requirements contained therein. One commenter suggested that a separate paragraph concerning the placarding of aircraft freight containers should be included to avoid any differences with the ICAO system. The MTB believes such a paragraph is unnecessary since the provision of paragraph (b) of this section, which allows such freight containers to be labeled in accordance with §172.406(e)(3), provides for consistency with the ICAO system. The second commenter proposed that this section be modified to allow "... the details of the hazardous class of materials contained within a container to be entered on the pallet tag. ..." This suggestion has not been incorporated into the final rule since it would be inconsistent with current ICAO provisions which specifically require the display of hazard class labels on freight containers, as opposed to "details of the hazardous class of materials" on pallet tags.

Section 175.3. No substantial change to the proposed rule.

Section 175.10. Two commenters objected to the change proposed to paragraph (a)(5) of this section, one noting that the 10 pound limitation on small arms ammunition in checked baggage would be inconsistent with section 107(c), of the Hazardous Materials Transportation Act (49 U.S.C. 1806(c)). The MTB agrees with these commenters and has withdrawn the proposed amendment to this paragraph.

Two commenters noted that, since toilet articles are already addressed in paragraph (a)(4) of this section, the probable intent of proposed paragraph (a)(16) of this section was to allow perfumes and colognes purchased through duty free sales to be carried as carry-on baggage. They requested that this intent be clarified in the text of the paragraph. The MTB agrees, and paragraph (a)(16) has been reworded accordingly.

A commenter suggested that the proposed paragraph (a)(17) be combined with existing paragraph (a)(13) since both deal with exceptions for dry ice. The MTB believes it is preferable to leave the two provisions in separate paragraphs since the former paragraph addresses carriage of dry ice as a refrigerant for cargo and provides certain marking requirements whereas the latter paragraph concerns dry ice for use in cabin service or as a refrigerant in carry-on baggage. The commenter also requested that the figure four pounds in paragraph (a)(17) be changed to two kilograms to insure consistency with the ICAO Technical Instructions. The MTB believes this is unnecessary since the existing §173.26 allows the use of two kilograms as the metric equivalent to four pounds.

At the request of a commenter, paragraph (a)(19) has been revised to include a cross reference to the definition of non-spillable battery that is contained in §173.260(d).

Section 175.30 A suggestion to revise paragraph (b) of this section to provide that the preloading inspection may be conducted by an agent of the operator has not been incorporated in the final rule. The MTB believes this is unnecessary since an agent of the operator is authorized to perform the inspection required in this paragraph. Although no change was proposed to paragraph (d) of this section, a commenter requested that "refrigerating machines" be added to the list of hazardous materials excepted from the provisions of §175.30(a) and (b). The MTB considers this request to be outside of the scope of this rulemaking and no change has been made to this paragraph.

A commenter suggested deletion or revision of proposed paragraph (e)(1) of this section in order to improve clarity. The MTB believes the proposed text is clear and, in order to provide consistency with the text appearing in the ICAO Technical Instructions, has retained the proposed text. The commenter also suggested that the word "packagings" should be replaced by "packages" in subparagraphs (e)(2) and (e)(3). The MTB agrees with this change with regard to subparagraph (e)(2), but believes that "packagings" is the correct term in paragraph (e)(3) and has retained the proposed text of that subparagraph.

Section 175.33. A commenter suggested minor revisions to the proposed text of subparagraphs (a)(2), (4) and (5) of this section in order to improve clarity. The MTB has retained the text as proposed for paragraph (a)(2) in order to maintain consistency with similar text in §172.203, but has added the words "(if applicable)" in subparagraph (a)(4) and changed the word "may" to "must" in subparagraph (a)(5), as requested by the commenter. In addition, subparagraph (a)(5) has been editorially revised to clarify its intent.

Section 175.35. Two commenters noted that the proposal to delete this section is inconsistent with the ICAO Technical Instructions which require a copy of the shipping paper to accompany the shipment. The MTB agrees with these commenters and §175.35 is retained.

Section 175.75. Five commenters requested that the quantity limitations imposed in subparagraphs (a)(2) of this section be deleted. Because no change was proposed to this paragraph in the notice, the MTB considers the deletion of this paragraph to be outside of the scope of this rulemaking and the paragraph has, therefore, been retained. Nevertheless, the MTB believes that there may be merit in considering the deletion or amendment of this limitation provided evidence supporting such action is supplied and that full public participation in this action is insured. Consequently, petitions to delete or amend §175.75(a)(2), with full supporting information, are invited for consideration in a separate rulemaking action.

Section 175.78. Two commenters noted an inconsistency between the segregation specified in Table 1 and that required under the ICAO Technical Instructions, and requested that Table 1 be modified to remove this inconsistency. The MTB agrees that Table 1 should be consistent with ICAO requirements and the Table has been modified accordingly. Another commenter suggested that "Blasting Agents" be added to Table 1 with the same segregation required for oxidizers. The MTB agrees that Blasting Agents should be added to this Table, but since they would be classified by ICAO under Division 1.5, they have been added to the entry for explosives in the Table. As a consequence of this action, existing paragraph (b) is unnecessary and has been removed.

Section 175.79. No change to the proposed rule.

Section 175.81. No substantal change to the proposed rule.

Section 175.85. A commenter requested that paragraph (c)(1)(v) of this section, which provides an exception from the accessibility requirements for certain flammable liquids, be amended to include the upper limit of liquid flashpoint to which the exception applies. The MTB believes this is unnecessary since the upper liquid flashpoint limit to which provisions concerning flammable liquids apply is already clearly stated in the definition of a flammable liquid contained in §173.115. The commenter also requested clarification as to whether the list in paragraph (c)(1) pertains only to the hazard class of materials or whether it also applies to subsidiary risks. It is intended that this list pertain only to hazard class and not to subsidiary risks.

Another commenter suggested that paragraph (c)(1)(v) be modified by including and exception for flammable liquids meeting the definition of another hazard class in order to make this provision consistent with the corresponding provision of the ICAO Technical Instructions. The MTB agrees with this comment and has excepted flammable liquids meeting the definition of another hazard class from the provisions of paragraph (c)(1)(v). This commenter also requested that the list of materials in paragraph (c)(1), that are excepted from the accessibility requirement, be expanded to include ORM materials, noting that this would be consistent with the proposed inclusion of ICAO Class 9 materials in the list. The MTB agrees with this comment in principle, and has added ORM-A, C, D and E materials to the list. ORM-B has not been added since these materials would be classified as corrosives by ICAO and as such would be subject to the accessibility requirement.

Section 175.88. No substantial change to the proposed rule.

Section 175.90. No substantial change to the proposed rule.

Section 175.305. No change to the proposed rule.

Section 175.320. A commenter requested clarification as to whether proposed new paragraph (b)(10) should also be included in Part 107. The MTB believes this is unnecessary since Part 107 deals with the DOT approval of a shipment while the proposed new paragraph deals with a potential requirement for foreign approval of a shipment. The same commenter queried whether proposed paragraph (b)(10) could, in some cases, be more stringent than ICAO by requiring a foreign approval for the shipment of certain materials which are not subject to regulation under the ICAO Technical Instructions. Upon reexamination, the MTB believes this could be the case. Therefore, the wording of paragraph (b)(10) has been revised to make it precautionary.

Section 175.630. A commenter supported the proposed amendment to this section which would limit the ability to transport etiologic agents in the same compartment with foodstuffs, feed or other edible material. However, he suggested that this rule would be undermined unless the MTB prohibited the transportation of etiologic agents in mail bags or required that postal authorities appropriately label mail bags containing etiologic agents. The MTB believes such action to be unnecessary since diagnostic specimens, biological products and etiologic agents in quantities of less than 50 milliliters are excepted by paragraph 173.386(d) provided certain packaging requirements are met.

Section 175.701. No change to the proposed rule.

Section 175.705. No change to the proposed rule.

(47 F.R. 54817, Dec. 6, 1982)

49 CFR Parts 173 and 178
[Docket No. HM-185; Notice No. 82-7]
Standards for Polyethylene Containers

On June 21, 1979, MTB published a notice (44 FR 36211) in the Federal Register which announced a public meeting and solicited comments pertaining to the feasibility of establishing standards for polyethylene containers used as packagings for hazardous materials. A public meeting was held on July 24, 1979, and in response to substantial public interest a subsequent meeting was held on November 13, 1979 (announced in 44 FR 58767). Both meetings took place in Washington, D.C. These meetings were non-evidentiary and as such no transcripts of the proceedings were made. However, prepared statements of speakers and other written comments submitted in response to the notices are available for review in the docket file. Items discussed at the meetings included the following:

1. A presentation of the National Bureau of Standard's current views relative to stress cracking, permeation and compatibility of polyethylene packagings used for hazardous materials. Particular attention was focused on the development of methods by which these phenomena can be predicted without direct testing on a case by case basis or, alternatively, appropriate test criteria for determining compatibility of materials with polyethylene.

2. Reuse provisions for polyethylene packagings, particularly with regard to shipment of Poison B liquids and solids in polyethylene containers.

3. Use of polyethylene packagings for flammable liquids, with discussion of the tendency of polyethylene to pick up a static charge and associated risks.

4. Shipping experience of shippers using polyethylene containers for hazardous materials.

This notice of proposed rulemaking is based upon oral and written comments from the public in response to the two aforementioned notices, MTB's own rulemaking initiatives and four petitions for rulemaking submitted by representatives of the plastics industry. Subjects addressed in this notice are discussed in the following paragraphs.

I. BACKGROUND

It is proposed to revise Specification 34 for polyethylene drums. Maximum authorized capacity, as specified in §178.19–3, would be increased from 30 gallons to 55 gallons. A minimum container wall thickness of 0.125 inches would be required, with 0.090 inches thickness authorized in corners and undercuts. In §178.19–7, a compression weight load of 2400 pounds would be prescribed for conduct of the static compression test. These changes are based on the terms of existing exemptions for 55 gallon polyethylene drums and on the merits of a petition for rulemaking submitted by the Society of the Plastics Industry (SPI).

Authorizations for use of 30 gallon specification polyethylene drums are currently in the Hazardous Materials Regulations (HMR) in §§173.245, 173.263, 173.264, 173.265, 173.266, 173.272, 173.276, 173.277 and 173.288. These sections would be revised to permit use of a 55 gallon drum. In addition, based on successful shipping experience under the terms of exemptions, authorizations for use of Specification 34 would be added to §§173.125, 173.247, 173.256, 173.257, 173.269, 173.271, 173.287, 173.348, 173.349, 173.357, 173.361 and 173.362a.

In §178.19–7, it is proposed to increase the time interval between periodic retests from four months to one year, in order to reduce the burden imposed by frequent retests. Conduct of drop tests (both at ambient temperature and at 0°Fahrenheit) would be modified to require drops (of four feet in height) onto both the side and bottom of drums, in addition to the current requirement for dropping drums diagonally onto the top chime, in order to be compatible to those requirements currently contained in polyethylene drum exemptions. Specification 34 would be editorially revised in its entirety for clarity. One change would substitute the words "polyethylene drum" where the specification currently references "polyethylene container" or "polyethylene container for use without overpack."

Another proposal would apply not only to Specification 34, but also to all other specification polyethylene containers. It involves elimination of specification requirements relative to the physical composition and characteristics of polyethylene resins used in the molding of containers and increased reliance on performance requirements as opposed to detailed design and construction requirements for achieving a high level of container integrity. In Appendix B to Part 178 and in the various specifications for polyethylene containers in Part 178, tables which specify properties (i.e., melt index, density, tensile strength, etc.) for polyethylene resins would be deleted as would the requirement in §178.16–4 pertaining to retention of data related to melt index and density. This change would provide manufacturers flexibility in making innovative changes to their manufacturing processes without the need for their applying for relief from specified properties and is responsive to two petitions for rulemaking and MTB's own rulemaking initiatives.

It is essential that a high level of packaging integrity for polyethylene containers be maintained in the absence of specified resin properties. The performance requirements found in the individual specifications are indicative of the ability of a container to withstand the rigors of the transportation environment without release of its contents, barring abnormal abuse. Performance requirements would be emphasized by editorially revising each polyethylene container specification in Part 178 in order to clarify that each container manufactured under provisions of a specification must be capable of withstanding without failure the performance tests prescribed in that specification. Also, it would be required that polyethylene resins used to mold containers be "not previously used" so as to prohibit the use of polyethylene resins made from reprocessed containers which were previously used for the shipment of any material. This prohibition is not intended to prohibit reprocessing of either excess from the molding process or unused containers.

MTB has permitted the use on certain polyethylene drums under its exemptions program for a number of years. Authorizations for packaging specific hazardous materials in these drums have been limited to materials which have been tested for permeation and compatibility with the polyethylene packaging. This approach has been necessary because of difficulties in predicting permeability and other effects on polyethylene when exposed to different ladings. Ordinarily, after sufficient experience with a new packaging has been accumulated through the exemptions program, a packaging is considered for inclusion in the HMR as an authorized container. It is proposed to amend Specification 34 to accommodate polyethylene drums of 55 gallon capacity which are presently authorized by exemptions. For the purpose of revising the specification, MTB believes it is necessary to rely on a continuation of the approach which is used in the exemptions program, i.e., the polyethylene packaging should be examined in connection with each specific material proposed to be transported therein. Further, it is proposed that this approach be made applicable to the packaging of hazardous materials in any polyethylene packaging, in order to alleviate an existing deficiency in the HMR. Therefore, new test criteria are proposed for use in determining chemical compatibility and rates of permeation. A generalized requirement in § 173.24(c)(9) stipulates that a polyethylene packaging must be compatible with its lading. Section 173.24 would be amended to prescribe maximum permissible rates of permeation. Proposed rates are 0.5% for extremely toxic poisonous materials, defined as those materials having an oral toxicity of less than 20 mg/kb (LD_{50}, oral rat) or dermal toxicity of less than 80 mg/kg (LD_{50}, dermal rabbit) and a maximum of 2.0% for all other hazardous materials.

A method to be used in determining chemical compatibility and rates of permeation would be added as Appendix B of Part 173. The test method prescribed testing of the specific hazardous material in the polyethylene packaging in which shipment is intended. The package, as prepared for shipment, is stored at elevated temperatures of 130° F for 90 days or 140° F. for 14 days (at the tester's option) and examined for evidence of chemical incompatibility and rate of permeation. The proposed method is based on requirements in exemptions for polyethylene containers. It should be noted that the proposal establishes a standard and does not impose a specific requirement to test. It would not be necessary to test each combination of hazardous material and polyethylene packaging. Many combinations have already been tested under the exemptions program or have successful shipping histories. Untried combinations of materials and packagings would require testing only if there were no basis for making a reasonable determination as to compatibility and permeation rates in the absence of such testing.

Under the current reuse provisions of § 173.28, polyethylene containers which have been used for shipment of Poison B liquids and solids may be emptied and reused for the shipment of other materials, both hazardous and nonhazardous. MTB has been petitioned by SPI to limit the reuse of such containers to Poison B liquids and solids, i.e., such containers would remain in "dedicated service" as packagings for Poison B materials. The petition requests that a skull and crossbones symbol and the warning "Contains Poison, Limit Reuse for Poison Only" be required as a permanent marking on polyethylene containers used as outside packagings, or as a stencilled marking on outside packagings other than polyethylene which hold inside polyethylene containers, in those instances where Poison B materials are to be packaged. The petition addresses the concern that the poison hazard may remain in empty polyethylene containers, even after cleaning. MTB agrees with the need to address this concern, but is of the opinion that imposition of the economic burden which would be associated with permanent embossment of warnings on polyethylene containers is not justifiable. As an alternative, MTB proposes to amend § 173.28(d) to require that polyethylene containers once used as packagings for Poison B materials be limited to Poison B materials for reuse. If poison labels appear on such containers, the labels would be maintained in a legible condition until the containers are disposed of or destroyed. Permanent embossment of the skull and crossbones symbol or warning statements could be performed on a voluntary basis. It is believed that the MTB's proposal addresses the safety issue without imposing substantial costs on shippers or container manufacturers.

It is proposed to eliminate paragraphs (a) and (b) of § 173.23 from Part 173. Paragraph (a) permits continued use of Specification 5B, 6J and 37A metal drums manufactured prior to March 18, 1964, having inside Specification 2S, 2SL, 2T or 2TL polyethylene liners. Paragraph (b) permits use of certain polyethylene containers for use without overpack manufactured prior to September 5, 1966, and marked ICC-34. It is believed that these containers are obsolete and, therefore, the provisions of § 173.23(a) and (b) are no longer needed. The remaining paragraph (c) would be redesignated paragraph (a).

MTB is aware that there are differences between changes proposed in this rulemaking for specification packagings and changes suggested in Docket HM-181; Advance Notice No 82–3 (47 FR 16268) which envisions the deletion of specification packagings and adoption of international performance-oriented packaging standards. Commenters are reminded that this notice of proposed rulemaking is intended to authorize use of certain polyethylene packagings in a relatively short period of time whereas Docket HM-181 is an advance notice in which changes are set forth for demonstration purposes and, even if promulgated, would not become final in the near future.

II. REQUEST FOR COMMENTS

MTB invites comments on all aspects of the proposed rule and the issues discussed in this preamble. Of particular interest are comments addressed to the following issues.

1. MTB is proposing to add authorizations for use of Specification 34 based on shipping experience acquired under the exemptions program for the following materials: alcohol, n.o.s.; thionyl chloride; compound, cleaning, liquid (corrosive material); electrolyte (acid) and alkaline battery fluid; perchloric acid; phosphorous oxychloride;

sulfuric acid of greater than 95 percent to not over 100.5 percent concentration; chromic acid solution; arsenic acid; carbolic acid, liquid; chloropicrin and chloropicrin mixtures; aldrin mixtures; and, dinitrophenol solutions. MTB requests comments concerning other materials or classes of material which may be suitable for shipment in Specification 34 based on established shipping experience, evidence of chemical compatibility, or their similarity to the aforementioned materials or to those materials for which Specification 34 is currently authorized.

2. MTB estimates that the recordkeeping requirement of § 178.19-7(d) will impose an annual burden of 4 hours on each of approximately 25 manufacturers, for a total annual burden of 100 hours. Comments are requested as to the validity of this estimate.

3. It is proposed to delete specifications for polyethylene resins used to mold containers. MTB believes that packaging integrity would be maintained by reliance on performance requirements and proposed requirements pertaining to permeation and chemical compatibility. Comments are requested concerning the need, if any, from the standpoint of safety for specifying the characteristics of polyethylene resins.

4. MTB believes that requirements proposed in this notice for determining chemical compatibility and rates of permeation (proposed § 173.24(d) and Appendix B to Part 173) for hazardous materials packaged in polyethylene containers, will correct an existing deficiency in the HMR and are necessary in order to achieve a high level of safety with regard to the transportation of hazardous materials in polyethylene packagings. If promulgated, the new requirements would have an impact on both shippers of hazardous materials and polyethylene container manufacturers, some of whom are small businesses. It is believed the regulation would not have a significant economic impact on a substantial number of small businesses under the criteria of the Regulatory Flexibility Act. MTB requests comments concerning estimates of degree of impact on small businesses, in terms of economic cost and numbers and types of businesses affected, proposals for practicable alternatives to supplement or replace the test method proposed as new Appendix B to Part 173 and estimates regarding the economic costs (or savings) attributable to such alternatives.

5. It is proposed (in proposed § 173.28(d) to require "dedicated service" for polyethylene containers used to package Poison B materials. The proposal would impose a general requirement that polyethylene containers, once used to package Poison B materials, be reused only for poison B materials, be used only for Poison B materials and a specific requirement that Poison B labels remain on such containers, when so required. The proposal is based on the belief that residue remaining in such containers after initial use poses a hazard both to persons handling, cleaning or refilling the used, empty containers and, in subsequent use, to persons who come into contact with supposedly nonpoisonous materials which become contaminated with poisonous residues. MTB requests comments on practicable alternatives to its proposal, to include estimates of associated costs and benefits.

6. It is proposed to delete paragraphs (a) and (b) of § 173.23 because it is believed that the containers authorized therein are obsolete. The metal drums authorized in § 173.23(a) are all at least 18 years old and the polyethylene drums authorized in § 173.23(b) are all at least 15 years old. If any shippers are still using these containers, comments are requested from them concerning numbers and types of containers used, shipping experience and period of time necessary to deplete any existing stocks.

7. Exemptions potentially affected by this rulemaking are as follows: DOT-E 6637, E 6700, E6726, E 6800, E6883, E 6986, E 7035, E 7072, E 7082, E 7220, E 7502, E 7788, E 7888, E 7933, E 7940, E 8051. Comments and suggestions for eliminating any, or all, of these exemptions, or portions thereof, are requested.

III. SECTION-BY-SECTION SUMMARY OF PROPOSED CHANGES.

Section 173.23. Paragraphs (a) and (b) would be deleted to eliminate obsolete provisions, paragraph (c) would be redesignated paragraph (a).

Section 173.24. Paragraph (c)(9) would be deleted, paragraph (d) would be redesignated (e) and a new paragraph (d) would be added specifying rates of permeation and chemical compatibility requirements.

Section 173.28. The section title would be revised for clarity and paragraph (d) would be revised to limit reuse of polyethylene containers used to package Poison B materials.

Section 173.125. Authorization for use of Specification 34 would be added as paragraph (a)(7).

Section 173.245. Paragraph (a)(26) would be revised to remove the 30 gallon limitation for Specification 34.

Section 173.247. Authorization for use of Specification 34, for thionly chloride only, would be added as paragraph (a)(20).

Section 173.256. Authorization for use of Specification 34 would be added as paragraph (a)(8).

Section 173.257. Authorization for use of Specification would be added as paragraph (a)(13).

Section 173.263. Paragraph (a)(28) would be revised to remove the 30 gallon limit for Specification 34.

Section 172.264. Paragraph (a)(18) would be revised to remove the 30 gallon limit for Specification 34.

Section 173.265. Paragraph (d)(6) would be revised to remove the 30 gallon limit for Specification 34.

Section 173.266. Paragraph (b)(8) would be revised to remove the 30 gallon limit for Specification 34.

Section 173.269. Authorization for use of Specification 34 would be added as paragraph (a)(7).

Section 173.271. Authorization for use of Specification 34 would be added as paragraph (s)(20).

Section 173.272. Paragraph (g) would be revised to authorize use of Specification 34 for sulfuric acid in concentrations of 95 percent to 100.5 percent and paragraph (i)(9) would be revised to remove the 30 gallon limit for Specification 34.

Section 173.276. Paragraph (a)(10) would be revised to remove the 30 gallon limit for Specification 34.

Section 173.277. Paragraph (a)(6) would be revised to remove the 30 gallon limit for Specification 34.

Section 173.287. Authorization for use of Specification 34 would be added as paragraph (b)(9).

Section 173.288. Paragraph (e) would be revised to remove the 30 gallon limit for Specification 34.

Section 173.348. Authorization for use of Specification 34 would be added as paragraph (a)(5).

Section 173.349. Authorization for use of Specification 34 would be added as paragraph (a)(4).

Section 173.357. Authorization for use of Specification 34 would be added as paragraph (a)(4).

Section 173.361. Authorization for use of Specification 34 would be added as paragraph (a)(4).

Section 173.362a. Authorization for use of Specification 34 would be added as paragraph (a)(3).

Appendix B to Part 173. A test method for determining chemical compatibility and rates of permeation for hazardous materials in polyethylene containers wold be added as Appendix B.

Section 178.16. In § 178.16–1, paragraph (c)(1) would be added to emphasize performance requirements and, in § 178.16–4, paragraph (a) would be revised to delete specifications for polyethylene resins and requirements for retaining data concerning melt index and density.

Section 178.19. The section would be revised in its entirety in order to clarify its language, increase authorized capacity to 55 gallons, delete specifications for polyethylene resins, revise test procedures and emphasize performance requirements.

Section 178.21. Paragraph (b) of § 178.21–1 would be added to emphasize performance requirements; paragraph (a) of § 178.21–3 would be revised to add a requirement that polyethylene resin may not have been used previously and Note 1 which follows this paragraph would be deleted to remove specifications for polyethylene resins.

Section 178.24. The specification title and the title and text of § 178.24–1 would be revised for clarity with a performance requirement added as § 178.24–1(c): a requirement that polyethylene resin may not have been used previously would be added and specifications for polyethylene resins would be deleted in § 178.24–2(a).

Section 178.24a. In § 178.24a–3, paragraph (c) would be deleted and paragraph (a) would be revised to add a requirement that polyethylene resin not be previously used and to delete specifications for polyethylene resins.

Section 178.27. In § 178.27–1, Note 1 would be deleted to remove specifications for polyethylene resins, paragraph (a) would be revised to add a requirement that polyethylene resin may not have been used previously and paragraph (b) would be added to emphasize performance requirements.

Section 178.35. The section title and § 178.35–1 would be revised for clarity; emphasis on performance requirements would be added as § 178.35–1(c); and in § 178.35–2, specifications for polyethylene resins would be deleted and a requirement that polyethylene resin may not have been used previously would be added.

Section 178.35a. The section title and § 178.35a–1 would be revised to delete specifications for polyethylene resins, require that resins may not have been used previously and emphasize performance requirements.

Appendix B to Part 178. Appendix B, entitled "Specifications for Plastics," would be deleted in its entirety.

(47 F.R. 37592, Aug. 26, 1982)

49 CFR Parts 173 and 178
[Docket No. HM-185; Amdt. Nos 173–176, 178–79]
Standards for Polyethylene Packagings

On August 26, 1982, MTB published a notice of proposed rulemaking (Notice 82–7) in the Federal Register (47 FR 37592). In the notice, MTB proposed adoption of certain rules applicable to polyethylene packagings used for hazardous materials. This document contains changes to the Hazardous Materials Regulations (HMR) based on the proposals in Notice 82–7 and the merits of comments received from the public in response to the notice. The interested reader is referred to Notice 82–7 for additional background information.

MTB received 32 comments to Notice 82–7 from 25 respondents. Respondents represented 12 chemical shippers, six container manufacturers, one individual and six associations. The majority of commenters expressed support for the proposals, while offering comments and suggestions for improving specific aspects of the rulemaking. Only one commenter opposed the rulemaking, citing the lack of fire resistance standards in Specification 34 as a reason why MTB should not increase the authorized capacity for Specification 34 polyethylene drums from 30 to 55 gallons. Several commenters suggested changes to the proposals without expressing either support or opposition. Amendments adopted in this final rule and significant comments are discussed in the following paragraphs.

Deletion of Obsolete Requirements, "Grandfather" Provision for Exemption Packagings (§ 173.23). There were no objections from commenters concerning deletion of obsolete specifications in paragraphs (a) and (b) of § 173.23. The change is made as proposed. Based on the merits of several comments, a new paragraph (a) is added. It provides that polyethylene drums which are manufactured and marked in accordance with various DOT exemptions may be used in place of Specification 34 drums, for those materials for which Specification 34 drums are authorized. The exemption packaging must conform to all Specification 34 requirements, with the exception of the specification marking requirement and must be marked "DOT-34" in characters at least one half inch in height to identify it as an authorized package. Without this provision, these exemption polyethylene drums would be authorized for continued use only if their manufacturers periodically applied for, and received, renewal of the applicable exemptions. This amendment should result in a substantial savings to polyethylene drum manufacturers and shippers.

PERMEATION AND COMPATIBILITY CRITERIA (§ 173.24, APPENDIX A TO PART 173)

In Notice 82-7, MTB proposed standards for hazardous materials packaged in polyethylene packagings with regard to compatibility between the packaging and its lading and permeation of lading through the container. Eighteen commenters specifically addressed this proposal, most offering suggestions for revising the proposed standard.

One commenter suggested that the permeation and compatibility issue should be deferred for consideration under Docket HM-181, entitled "Performance-Oriented Packagings Standards" (ANPRM published April 15, 1982; 47 FR 16268) because of the complexity of the issue and its effect on polyethylene packagings. It is apparent that permeation and compatibility criteria may be essential to implementation of a performance oriented packaging system as envisioned in Docket HM-181, particularly if plastics other than polyethylene are permitted for construction of packagings for hazardous materials. However, MTB sees no reason for deferring action on the issue since the new standards will facilitate the use of 55 gallon polyethylene drums other than under the terms of exemptions and will fulfill an existing need for permeation and compatibility criteria applicable to polyethylene packagings.

The proposal in Notice 82-7 is adopted with changes based on the merits of comments and the MTB's own initiative. Maximum permissible rates of permeation of 0.5% for Poison B materials and 2.0% for other hazardous materials are specified. Three time and temperature combinations are specified for the test procedure: 14 days at 60° C (140° F), 28 days at 50° C (122° F) and 180 days at 18° C (64° F) or higher. The language of paragraph (d) of § 173.24 is revised to clarify that the permeation and compatibility criteria are standards rather than a requirement to test, and the paragraph is revised editorially. The test procedure in Appendix B is revised to incorporate the suggestions of commenters. These changes and the reasons for them are discussed in greater detail in the following paragraphs.

The permeation and compatibility criteria which were proposed in Notice 82-7 are revised by using the language "Each polyethylene packaging . . . must be capable of withstanding without failure the procedure specified in Appendix B . . ." rather than the proposed language "The procedure specified in Appendix B . . . shall be followed . . . ." This change responds to the requests of several commenters that MTB clarify that the criteria represent standards rather than a requirement to test, except in those instances where the specific packaging authorization in Part 173 requires testing. This change permits shippers to make a reasonable determination of compliance with the standard, in many instances without the need for actual testing, based on: (1) Successful shipping experience, (2) industry generated data concerning permeation and compatibility, (3) the shippers' knowledge of the material, or (4) testing performed on similar products. Actual testing is required for specifically named materials unless such testing has been performed for materials packaged in exemption packagings. The language is also revised so that approval by the Associate Director for Hazardous Materials Regulation is required only in certain instances such as use of test procedures other than that specified in Appendix B or use of packagings and materials which exceed the prescribed permeation rates but are known to be safe for transport.

One commenter contended that the permeation and compatibility criteria should not apply to small quantities of materials which are permitted to be shipped in nonspecification packagings, claiming a hardship would be imposed on users of such packagings. Another commenter contended that the criteria should not apply to inside plastic receptacles used in combination packagings. MTB disagrees with these commenters on the basis that incompatibility or permeation of hazardous material through a nonspecification packaging or an inside receptacle may be potentially just as hazardous as incompatibility or permeation through a specification packaging. Therefore, the adopted criteria apply to all polyethylene packagings and receptacles used for hazardous materials.

With regard to the proposed criteria, several commenters suggested a 180 day test period at ambient temperature as an alternative to the two proposed time-at-temperature combinations of 54.4° C (130° F) for 90 days and 60° C (140° F) for 14 days. This suggestion is consistent with the permeation criteria proposed in Docket No. HM-181 and is adopted in this final rule. Commenters suggested that testing for 14 days at 60° C (140° F) yields results roughly equivalent to testing for 180 days at ambient temperature. They contended that the time-at-temperature test of 90 days at 54.4° C (130° F) is not equivalent to either of the aforementioned time-at-temperature combinations. A test at 50° C (122° F) for 28 days was suggested as an appropriate alternative. This latter time-at-temperature combination is referenced in several industry standards for testing permeation and compatibility. MTB agrees with this suggestion and the final rule provides three alternative time-at-temperature combinations: 14 days at 60° C (140° F), 28 days at 50° C (122° F) and 180 days at 18° C (64° F) or higher.

Several commenters objected to the use of toxicity levels in the proposed criteria which differ from those found in 49 CFR 173.343. Commenters alleged that the proposed toxicity levels (i.e., oral toxicity of less than 20 mg/Kg (LD50, oral rat) and dermal toxicity of less than 80 mg/Kg (LD50, dermal rabbit)) are arbitrary and may require testing beyond that currently required to determine if a material meets Poison B criteria because of either oral or dermal toxicity (i.e., oral toxicity of 50 mg/Kg or less (LD50, oral rat) and dermal toxicity of 200 mg/Kg or less (LD50, dermal rabbit)). Based on the merits of these comments, MTB is revising the toxicity levels to be consistent with § 173.343, so that the lower permissible rate of permeation of 0.5 percent applies to hazardous materials which meet Poison B criteria. The higher rate of permeation of 2 percent applies to all other hazardous materials.

Several commenters proposed alternate rates of permeation ranging from 1 percent to 10 percent on a yearly basis. As indicated in the preceding paragraph, MTB is adopting maximum permissible rates of permeation of 0.5 percent for Poison B materials and 2.0 percent for other hazardous materials, determined over the time period of the test, which is roughly equivalent to 1 percent and 4 percent, respectively, per year. It is believed that these rates provide an acceptable level of safety, as evidenced through limited application under various polyethylene container exemptions, without being overly restrictive as to the types of materials which may be packaged in polyethylene. One commenter contended that a general prohibition against hazardous conditions during transportation and handling is a greater inducement to safety than setting specific permeation limits. Another suggested that the shipper of hazardous material should be responsible for determining an appropriate rate of permeation. MTB disagrees with these commenters and believes that both a general prohibition against hazardous conditions and specified maximum permissible rates of permeation are necessary to provide regulatory guidance and achieve an acceptable level of safety.

The test procedure which was proposed for adoption in the notice is revised based on the merits of comments. Three samples are specified for conducting the permeation and compatibility test. The term "rate of permeation" is clarified and expressed as the percentage loss of hazardous material content during the time period over which the test is conducted. The format and language of the procedure are revised for clarity. One commenter suggested that a static compression test be required (after the storage test and prior to the drop test) and that test containers be cut apart for internal inspection after completion of testing. Another commenter proposed rejection criteria based on loss of tare weight during conduct of the storage test. These suggestions may have merit but since they go beyond the scope of testing proposed in Notice 82-7, they are not adopted in this final rule.

Several commenters recommended that MTB incorporate by reference Packaging Institute, USA (PI/USA) Test Procedure T4101-80. MTB has reviewed this test procedure and believes it to be unsatisfactory for incorporation by reference due to the use of recommendatory language and subjective failure criteria. It also appears to go beyond the scope of testing provided in Notice 82-7, in that it requires testing of both test coupons and whole containers and provides a limitless array of time and temperature combinations. The test procedure for whole containers (Part B) is roughly equivalent to the procedure adopted in this final rule, although PI/USA T4101-80 does not prescribe maximum permissible rates of permeation. Two commenters proposed environmental stress crack resistance (ESCR) test procedures as base level tests for determining if different types and designs of polyethylene containers are suitable as packagings for hazardous materials. Although the commenters differed in their recommendations, the test procedures generally involve placing an environmental stress crack agent into a polyethylene container for a specified period of time at a specified temperature. The container passes the test if it does not crack or leak. Another commenter proposed a compatibility test that is conducted at three different temperatures (0° C (32° F), 23.8° C (75° F) and 37.7° C (100° F)) over a one year period by immersing test coupons of polyethylene, cut from the intended packaging, in the hazardous material which is to be packaged. Measurements of coupon material thickness and tensile strength are made at the beginning and end of the test period. Failure is defined as more than 5% loss of either material thickness or tensile strength. Although both of these proposed tests may have merit, MTB believes that they are beyond the scope of the proposals contained in Notice 82-7 and they are not adopted in this final rule.

MARKING OF POLYETHYLENE PACKAGINGS USED FOR POISONOUS MATERIALS (§ § 173.24, 173.28)

In Notice 82-7, it was proposed to amend § 173.28(d) to require that polyethylene packagings used for Poison B materials be limited to Poison B materials in any subsequent reuse. Notice of this limitation was proposed to be accomplished by means of maintaining Poison labels on such containers. This proposal was based on a petition from the Society of the Plastics Industry (SPI) for a permanent marking on such containers which would include a skull and crossbones symbol and the warning "Contains Poison, Limit Reuse for Poison Only." Both proposals have the objective of warning reusers of containers of potential hazards. MTB initially believed that its proposal was a more cost effective means of accomplishing the safety objective. Comments addressed to this proposal appear to represent a consensus of polyethylene drum manufacturers and shippers and overwhelmingly favor a permanent marking requirement instead of the proposed provision for maintenance of Poison labels. Commenters accurately point out that it is difficult to maintain labels on empty packagings, particularly when the packagings undergo a reconditioning process. Implicit in these comments is that it is neither cost effective, nor effective from a safety standpoint, to require maintenance of labels instead of a permanent marking. Several marking alternatives were suggested: the SPI recommendation; just the skull and crossbones symbol; the terms "Poison", "Contains Poison", and "Poison-Do not re-use except for Poisons". MTB believes these comments have merit and has adopted a permanent marking requirement in this final rule. MTB believes that a minimal marking requirement using the word "Poison" is adequate for conveying notice of a potential hazard and is adding the requirement in §173.24(d) for the permanent marking of polyethylene packagings used for poisonous materials. The word "Poison" must be marked in lettering of at least ¼ inch in height. Additional text or symbols may be included with the required marking. A one year delay in the effective date of the marking requirement is provided to ease any burden that might arise from the new marking requirement. Paragraph (d) of §173.28 is revised to clarify that permanent markings should not be removed. Also, a new paragraph (i) is added to §173.28 which recommends that polyethylene packagings previously used for poisons be reused for poisons or hazardous wastes only.

One commenter suggested that all polyethylene packagings used for hazardous materials should be permanently marked in order to identify those packagings that have been in hazardous materials service. Although the proposal has some merit, such a requirement may not be justifiable in terms of cost, is controversial and is not adopted in this final rule.

One commenter supported the proposed limitation on reuse of polyethylene containers but suggested the limitation not apply to containers used as packagings for hazardous wastes destined for disposal. MTB agrees with this suggestion and has provided an exception in §173.28(i) to permit for hazardous wastes, when authorized, regardless of whether the hazardous waste is a poisonous material.

AUTHORIZATION FOR USE OF SPECIFICATION 34 POLYETHYLENE DRUMS (PART 173) AND ELIMINATION OF EXEMPTIONS

Twelve commenters addressed the proposed Part 173 authorizations for use of Specification 34 polyethylene drums. There was strong support for revising the authorizations for use currently in the HMR (§§173.245, 173.263, 173.264, 173.265, 173.266, 173.272, 173.276, 173.277 and 173.288) to permit use of a 55 gallon maximum capacity Specification 34 instead of the previously authorized maximum capacities of 30 gallons or less. With the exception of §173.245, these sections are changed as proposed in Notice 82-7. Because of the likelihood of previously untried corrosive liquids being shipped in Specification 34 drums under the packaging authorization in §173.245 (corrosive liquids, n.o.s.), MTB believes it is essential that compatibility between lading and container be established prior to first shipment. Therefore, the authorization for use in §173.245 (a)(26) is revised both to permit use of a 55 gallon Specification 34 and to require the shipper to assure conformance with the permeation and compatibility criteria of §173.24(d), prior to first shipment. This will necessitate testing in most instances. However, if testing has previously been performed for a given material and packaging by the shipper, the container manufacturer or a third party, then it needs to be repeated.

MTB has proposed adding authorizations for use to 12 packaging sections, as follows: §§173.125, 173.247, 173.256, 173.257, 173.269, 173.271, 173.287, 173.348, 173.349, 173.357, 173.361 and 173.362a. There were no objections to the proposed changes to §173.125 (alcohol, n.o.s.) and §173.257 (alkaline and acid battery fluid) and these are changed as proposed.

One commenter pointed out that arsenic acid (§173.348), carbolic acid (§173.349), chloropicrin and chloropicrin mixtures (§173.357), aldrin mixtures (§173.361) and dinitrophenol solutions (§173.362a) are often dissolved in carrier solutions containing surfactants which may have deleterious effects on polyethylene. The commenter suggested that compatibility between lading and packaging should be established for these poisonous materials by actual testing prior to first shipment. MTB shares this commenter's concerns and, therefore, the final rule requires the shipper to assure conformance with the permeation and compatibility criteria of §173.24(d) prior to the first shipment for these five materials. It should be noted that testing has been performed for many of these materials as a condition for their being authorized for shipment in specification 34-type drums under various exemptions and, therefore, may not have to be repeated.

Several commenters suggested that perhaps the other materials should remain under exemption because of compatibility problems or inadequate shipping experience. The materials addressed are thionyl chloride; compound cleaning liquids containing more than 30% hydrofluoric acid, perchloric acid, phosphorous oxychloride, chromic acid solutions and chloropicrin. MTB notes that all of these materials have successful shipping experience under exemption and that there has been over one year of additional experience acquired since comments were submitted to Notice 82-7. Based on this experience MTB does not believe it is necessary to prohibit use of Specification 34 for these materials. However, since it appears that these materials may require a higher degree of care in their packaging, MTB is requiring that the shipper assure conformance with the permeation and compatibility criteria of §173.24(d) prior to first shipment. The requirement does not preclude use of test data generated by someone other than the shipper, such as by the drum manufacturer or another shipper, or the use of test data acquired under the terms of exemptions.

In Notice 82–7, MTB requested comments concerning materials or categories of material other than those proposed which may be suitable for shipment in Specification 34 drums. Several commenters recommended adding authorizations to Part 173 for use of Specification 34 for the following materials, contending that successful shipping experience has been acquired under the terms of exemptions. These materials are: flammable liquids with flash points above 20° F. (§173.119); paint (§173.128); organic peroxides, including those classed flammable liquid (§173.119, 173.221–223); hydrobromic acid not over 49% (§173.262); poisonous liquids, n.o.s. (§173.346); and cyanide solutions (§173.352). These materials are currently authorized in polyethylene drums under the terms of approximately 30 exemptions.

MTB agrees with commenters that some successful shipping experience has been demonstrated. However, it is noted that in some instances these materials generally have not been shipped under exemption for as long a period of time as the materials proposed in Notice 82–7 and that under provisions of many exemptions, materials are specifically identified to and acknowledged in writing by MTB's Office of Hazardous Materials Regulations (OHMR) prior to first shipment. Under this latter provision OHMR has required that permeation and compatibility testing be performed prior to first shipment. In order to achieve an equivalent level of safety as that provided under the exemptions program, MTB believes it is necessary to require the shipper to assure conformance to the permeation and compatibility criteria of §173.24(d) prior to first shipment. Authorizations for use of Specification 34 are adopted in this final rule, for those materials named in the preceding paragraph, with this stipulation. One commenter recommended an across-the-board authorization to use Specification 34 for any compatible material. In the absence of a demonstration of transportation safety through shipping experience gained under the terms of exemptions or international regulations, MTB is reluctant to expand authorizations for use beyond those discussed above and, therefore, no action is taken on the commenter's recommendation.

There are approximately 700 named materials (including trade names, generic names and "not otherwise specified" descriptions) authorized to be packaged in Specification 34 and 34-style polyethylene drums under the provisions of 46 DOT exemptions which are affected by this rulemaking.

The 46 exemptions are listed as follows:
 DOT-E: 6397, 6637, 6700, 6726, 6787, 6800, 6883, 6986, 7035, 7062, 7072, 7082, 7220, 7249, 7502, 7538, 7682, 7788, 7888, 7933, 7940, 8051, 8067, 8188, 8197, 8247, 8301, 8339, 8389, 8468, 8488, 8498, 8499, 8537, 8585, 8709, 8780, 8823, 8888, 8896, 9014, 9054, 9115, 9119, 9252, 9253.

MTB has not conducted a detailed review of these exemptions to determine which authorizations for use or exemptions may be rendered unnecessary by this final rule. Some of these exemptions have many materials authorized (for example, two exemptions authorize 118 and 210 materials, respectively) and a one by one review of the material authorizations, to determine which might be eliminated, was not feasible. It is anticipated that exemption holders will not apply for renewal of those exemptions rendered obsolete upon expiration of the exemptions. Also, in some instances it may not be feasible for a polyethylene drum manufacturer to maintain an exemption if it is only needed for one or two materials. It appears probable that at least 13, and possibly as many as 42, exemptions will be rendered obsolete. Shipments under the remaining exemptions will be significantly reduced due to the packaging authorizations adopted in the final rule, but some exemptions will probably be continued in effect because they will still contain those packaging authorizations for materials such as low flash point flammable liquids which are not adopted in the final rule. In addition, three of the exemptions may be continued in effect because they authorize container thicknesses of less than 0.125 inches in specific locations for a drum of 55 gallons capacity whereas this final rule specifies a minimum wall thickness of 0.125 inches.

CONTAINER SPECIFICATION REQUIREMENTS (PART 178)

Twelve respondents submitted comments addressed specifically to the proposed changes to Part 178. Comments indicated support for the proposals, particularly with regard to the elimination of requirements relative to the physical composition and characteristics of polyethylene resins used in the manufacture of containers and the increase of the maximum authorized capacity of Specification 34 to 55 gallons. Detailed specifications for polyethylene resins are removed from §§178.16, 178.19, 178.21, 178.24, 178.24a, 178.27, 178.35, 178.35a and Appendix B to Part 178. The proposed editorial revision to the aforementioned sections to emphasize performance requirements, is adopted in the final rule. Also the word "molded" has been removed from the title of Specification 34, in recognition of other manufacturing techniques which may be used for the manufacture of polyethylene drums.

It has been determined that the vibration test procedure which appears in §§178.19–7, 178.21–3, 178.24–7, 178.27–4, 178.35–5, 178.35a–4 and 178.133–11 is technically incorrect in referring to "an amplitude of one inch."

Such an amplitude imposes a more rigorous vibration test than is intended, and is not practicable. The correct terminology, as found in § 178.16–13(a)(2), is "a vertical double-amplitude (peak-to-peak displacement) of one inch." This error is corrected in this final rule in the aforementioned sections. Since this change imposes no additional requirements and is a relaxation of an existing requirement, prior notice is deemed unnecessary.

Changes to requirements for packaging testing and test record retention are being considered under Docket HM-181 with regard to all non-bulk specification packagings other than cylinders. One commenter has suggested that the changes in testing proposed in Notice 82–7 should be postponed for consideration under Docket HM-181. MTB notes that the test requirements proposed in Notice 82–7 are different from those envisioned under Docket HM-181 and that any changes to test requirements made in this final rule may be rendered obsolete by future rulemaking action under Docket HM-181. As there is no pressing need to change the existing requirements from the standpoint of safety, MTB is deferring action on this issue for consideration under Docket HM-181. Therefore, § 178.19 is revised in part, rather than in its entirety as proposed, in order to implement the increase in capacity to 55 gallons, emphasize performance requirements, correct the language of the vibration test and make other editorial corrections. In Notice 82–7, MTB proposed minimum wall thickness of 0.125 inches for Specification 34 polyethylene drums of greater than 15 gallons capacity, with a thickness of 0.090 inches permitted in corners and undercuts. Several commenters contended that it is in the best interests of safety not to provide a reduction of material thickness in corners and undercuts. MTB agrees with this contention and the final rule requires a minimum wall thickness of 0.125 inches throughout the container. One commenter suggested that Specification 34 be revised to authorize rated capacities of up to 210 gallons. MTB considers this suggestion to be beyond the scope of this rulemaking.

It was proposed in Notice 82–7 to require that polyethylene resins used to mold containers be unused so as to prohibit the use of polyethylene resins made from reprocessed containers which previously might have been used as hazardous materials packagings. The proposal was supported by commenters. However, several commenters suggested that the proposal be revised to permit use of reground materials (such as excess molding material or off-specification containers) from an ongoing manufacturing process. MTB agrees with this suggestion and the language in §§ 178.16–4(a), 178.19–2(a), 178.21–3(a), 178.24–2(a), 178.24a–3(a), 178.27–1(a), 178.35–2(a) and 178.35a–1(d) is revised accordingly.

SECTION-BY-SECTION SUMMARY OF CHANGES

Section 173.23. Paragraph (b) is deleted to eliminate obsolete provisions and paragraph (a) is revised to permit use of DOT-34 style exemption polyethylene drums wherever Specification 34 drums are authorized for use.

Section 173.24. Paragraph (c)(9) is deleted, paragraph (d) is redesignated paragraph (e) and a new paragraph (d) is added specifying rates of permeation and chemical compatibility requirements for hazardous materials packaged in polyethylene, and requiring the permanent marking of polyethylene packagings used for poisonous materials with the word "Poison" in letters at least ¼ inch in height.

Section 173.28. Paragraph (d) is revised to clarify that permanent markings should not be removed from used packagings and paragraph (i) is added to recommend that polyethylene packagings not be reused for anything other than poisons or hazardous wastes if the packagings were previously used for poisonous materials.

Section 173.119. Authorizations for use of Specification 34 are added as paragraphs (b)(11) and (m)(19). The authorizations necessitate that the shipper assure conformance with the requirements of § 173.24(d) prior to first shipment.

Section 173.125. Authorization for use of Specification 34 is added as paragraph (a)(7).

Section 173.128. Authorization for use of Specification 34 is added as paragraph (a)(5). The authorization necessitates that the shipper assure conformance with the requirements of § 173.24(d) prior to first shipment.

Section 173.221. Authorization for use of Specification 34 is added as paragraph (a)(12). The authorization necessitates that the shipper assure conformance with the requirements of § 173.24(d) prior to first shipment.

Section 173.222. Authorization for use of Specification 34 is added as paragraph (a)(6). The authorization necessitates the conduct of permeation and compatibility testing, in accordance with § 173.24(d) prior to first shipment.

Section 173.223. Authorization for use of Specification 34 is added as paragraph (a)(7). The authorization necessitates that the shipper assure conformance with the requirements of § 173.24(d) prior to first shipment.

Section 173.245. Paragraph (a)(26) is revised to remove the 30 gallon limitation for Specification 34 and to require that the shipper assure conformance with the requirements of § 173.24(d) prior to first shipment.

Section 173.247. Authorization for use of Specification 34 is added as paragraph (a)(21). The authorization necessitates that the shipper assure conformance with the requirements of § 173.24(d) prior to first shipment.

Section 173.256. Authorization for use of Specification 34 is added as paragraph (a)(9). The authorization necessitates that the shipper assure conformance with the requirements of § 173.24(d) prior to first shipment.

Section 173.257. Authorization for use of Specification 34 is added as paragraph (a)(13).

Section 173.262. Authorization for use of Specification 34 is added as paragraph (a)(5). The authorization necessitates that the shipper assure conformance with the requirements of § 173.24(d) prior to first shipment.

Section 173.263. Paragraph (a)(28) is revised to remove the 30 gallon limit for Specification 34.

Section 173.264. Paragraph (a)(18) is revised to remove the 5 gallon limit for Specification 34.

Section 173.265. Paragraph (d)(6) is revised to remove the 30 gallon limit for Specification 34.

Section 173.266. Paragraph (b)(8) is revised to remove the 30 gallon limit for Specification 34.

Section 173.269. Authorization for use of Specification is added as paragraph (a)(8). The authorization necessitates that the shipper assure conformance with the requirements of § 173.24(d) prior to first shipment.

Section 173.271. Authorization for use of Specification 34 is added as paragraph (a)(1). The authorization necessitates that the shipper assure conformance with the requirements of § 173.24(d) prior to first shipment.

Section 173.272. Paragraph (g) is revised to authorize use of Specification 34 for sulfuric acid in concentrations of 95 percent to 100.5 percent and paragraph (i)(9) is revised to remove the 30 gallon limit for Specification 34.

Section 173.276. Paragraph (a)(10) is revised to remove the 30 gallon limit for Specification 34.

Section 173.277. Paragraph (a)(6) is revised to remove the 30 gallon limit for Specification 34 and to authorize the use of vented closures for Specification 34.

Section 173.287. Authorization for use of Specification 34 is added as paragraph (b)(9). The authorization necessitates that the shipper assure conformance with the requirements of § 173.24(d) prior to first shipment.

Section 173.288. Paragraph (e) is revised to remove the 5 gallon limit for Specification 34.

Section 173.346. Authorization for use of Specification 34 is added as paragraph (a)(6). The authorization necessitates that the shipper assure conformance with the requirements of § 173.24(d) prior to first shipment.

Section 173.348. Authorization for use of Specification 34 is added as paragraph (a)(5). The authorization necessitates that the shipper assure conformance with the requirements of § 173.24(d) prior to first shipment.

Section 173.349. Authorization for use of specification 34 is added as paragraph (a)(4). The authorization necessitates that the shipper assure conformance with the requirements of § 173.24(d) prior to first shipment.

Section 173.352. Authorization for use of Specification 34 is added as paragraph (a)(8). The authorization necessitates that the shipper assure conformance with the requirements of § 173.24(d) prior to first shipment.

Section 173.357. Authorization for use of Specification 34 is added as paragraph (a)(4). The authorization necessitates that the shipper assure conformance with the requirements of § 173.24(d) prior to first shipment.

Appendix B to Part 173. Appendix B is added to provide a procedure for determining chemical compatibility and rates of permeation.

Section 178.16. In § 178.16-1, paragraph (c) is added to emphasize performance requirements and, in § 178.16-4, paragraph (a) is revised to delete specifications for polyethylene resins and requirements for retaining data concerning melt index and density.

Section 178.19. The section title and § 178.19-1 are revised for clarity and to emphasize performance requirements, § 178.19-2 is revised to delete specifications for polyethylene resins, § 178.19-3 and paragraph (c)(2) of § 178.19-7 are revised to authorize a container capacity of 55 gallons and paragraph (c)(1) of § 178.19-7 is revised to correct deficiencies in the vibration test.

Section 178.21. Paragraph (b) of § 178.21-1 is added to emphasize performance requirements, the introductory text of paragraph (a) of § 178.21.3 is revised to add a prohibition against using a polyethylene resin that has been used previously, paragraph (c)(4) of § 178.21-3 is revised to correct deficiencies in the vibration test and Note 1 which follows paragraph (a) of § 178.21-3 is deleted to remove specifications for polyethylene resins.

Section 178.24. The specification title and the title and text of § 178.24-1(c) are revised for clarity and to emphasize performance requirements; prohibition against using a polyethylene resin that has been used previously is added and specifications for polyethylene resins are deleted in § 178.24-2(a); and § 178.24-7(a)(3) is revised to correct deficiencies in the vibration test.

Section 178.24a. In § 178.24a-3, paragraph (c) is deleted and paragraph (a) is revised to add a prohibition against using a polyethylene resin that has been used previously and to delete specifications for polyethylene resins.

Section 178.27. In § 178.27-1, Note 1 is deleted to remove specifications for polyethylene resins, paragraph (a) is revised to add a prohibition against using a polyethylene resin that has been used previously and paragraph (b) is added to emphasize performance requirements. Paragraph (a)(3) of § 178.27-4 is revised to correct deficiencies in the vibration test.

Section 178.35. The section title and § 178.35-1 are revised for clarity; emphasis on performance requirements is added as § 178.35-1(c); in § 178.35-2, specifications for polyethylene resins are deleted and a prohibition against using a polyethylene resin that has been used previously is added; and paragraph (a)(4) of § 178.35-5 is revised to correct deficiencies in the vibration tests.

Section 178.35a. The section title and § 178.35a-1 are revised for clarity, to delete specifications for polyethylene resins, require that resins may not have been used previously and to emphasize performance requirements. Paragraph (a)(3) of § 178.35a-4 is revised to correct deficiencies in the vibration test.

Section 178.133. Paragraph (b) of § 178.133-11 is revised to correct deficiencies in the vibration test.

Appendix B to Part 178. Appendix B, entitled "Specifications for Plastics," is deleted in its entirety.

(49 F.R. 24684, June 14, 1984)

49 CFR Parts 172 and 173
[Docket HM-187; Notice No. 83–2]
Requirement for Small Arms Ammunition

On June 30, 1982, the Sporting Arms and Ammunition Manufacturers Institute, Inc. (SAAMI) petitioned the Office of Hazardous Materials Regulation (OHMR) to authorize the transportation of small arms ammunition classed as ORM-D instead of Class C explosive. The Materials Transportation Bureau (MTB) considers the SAAMI request to have merit but to be too broad in scope because small arms ammunition includes such devices as tear gas cartridges, tracer cartridges for machine guns, and seat ejector cartridges for aircraft. However, MTB is proposing that limited types of small arms ammunition which are used in rifles, shotguns, and pistols be classed as ORM-D. The SAAMI proposal suggested an additional "Small arms ammunition" entry in 49 CFR 172.101. The entry "Small arms ammunition" with the hazard class ORM-D is proposed to be added to the hazardous materials table.

Years of experience in shipping small arms ammunition indicate that these devices present only a very small hazard in transportation and this has been confirmed by extensive tests conducted by SAAMI. The MTB has reviewed a documentary film of the SAAMI tests and arrangements can be made for this film to be viewed by any interested person. Most of the weight in a small arms shipment is in the projectiles and the shell or cartridge cases. There is only a relatively small amount of propellant explosive present. For example, in shot shells the weight of the propellant powder is only about 4% and in rifle cartridges only about 8% of the total weight. There is no significant hazard from the shot or bullets in a fire because they are not accelerated unless the ammunition is fired in a weapon.

Information furnished by SAAMI indicates that savings to their members on shipments by one motor carrier alone would be approximately $100,000 per year. Although MTB does not have specific data on the numbers of shipments by other shipper (such as distributors) and by other carriers it is believed that the annual savings on these shipments would also be substantial. MTB solicits comments from interested parties on the cost savings and burden reduction associated with this proposal, including specific data to quantify these savings.

Under existing regulations, small arms ammunition is not required to be in specification packaging and is not required to be labeled. This proposal would permit a different marking on the packages and would eliminate the need for shipping papers for surface transportation.

(48 F.R. 24146, May 31, 1983)

49 CFR Parts 172 and 173
[Docket HM-187; Amdt. Nos 172–82, 173–175]
Requirement for Small Arms Ammunition

On May 31, 1983, a notice of proposed rulemaking (Docket HM-187; Notice No. 83–2) was published in the Federal Register (48 FR 24146) announcing a proposal by the Materials Transportation Bureau (MTB) to add the hazardous materials description and proper shipping name entry "Small arms ammunition" under the hazard class ORM-D. The basis for MTB's action was a petition from the Sporting Arms and Ammunition Manufacturers Institute, Inc. (SAAMI). On June 30, 1982, SAAMI petitioned the Office of Hazardous Materials Regulation (OHMR) to authorize the transportation of small arms ammunition classed as ORM-D rather than Class C explosive. Although the MTB recognized in the Notice the merit of the SAAMI request, it was considered too broad in scope because of the wide variety of items that would be included under the category of small arms ammunition such as tear gas cartridges, tracer cartridges for machine guns, and seat ejector cartridges. Therefore, MTB noted that for the purposes of the rulemaking it was including only certain types of small arms ammunition used in rifles, shotguns, and pistols.

Twenty-three comments were received in response to the Notice. These comments were evaluated on the basis of their applicability to this particular rulemaking and their merit. Of the comments received, over half of the commenters firmly supported the addition of the optional entry "Small arms ammunition", classed as ORM-D in the hazardous materials table of 49 CFR 172.101 for domestic shipments. Most of these commenters pointed to that in their many years of experience in shipping small arms ammunition there have been relatively few incidents and no injuries that have arisen as a result of small arms ammunition posing a hazard in transportation.

Small arms ammunition contains only a relatively small amount of propellant explosive in proportion to its total weight. It will not sustain burning without additional fuel. The negligible hazard presented by packages of small arms ammunition has been confirmed by extensive tests conducted by SAAMI. In these tests, a total of 111 cases of sporting ammunition containing 145,500 rounds of the most popular types and brands of shotgun shells, rimfire cartridges, centerfire pistol and revolver cartridges, and centerfire rifle cartridges were consumed in four different tests. The tests included burning a frame building containing sporting ammunition, burning packed ammunition in an open area, burning packed sporting ammunition enclosed in an fire-resistant structure, and subjecting packed ammunition to severe shock. These series of tests confirmed the fact that mass detonation of sporting ammunition in a fire is not probable and was not evidenced in any of the tests.

It was found that even under extreme conditions of heat and confinement, there was no indication of either mass detonation or explosion. These tests also confirmed that there is a very limited "projection" hazard from a fire involving sporting ammunition. Where projection occurred, the materials with the highest velocity were the primer caps which, because of their nonaerodynamic shape and light weight, traveled short distances with low velocity. It was found that adequate protection would be provided if the usual protective clothing (including face mask) is worn by fire protection personnel.

As was mentioned in the notice of proposed rulemaking (NPRM), MTB also reviewed a documentary film of the SAAMI tests produced in cooperation with the Fire Prevention Bureau of the City of Chicago. The MTB believes this film accurately depicts the very limited hazard that is present when transporting small arms ammunition. In addition to this film and the SAAMI tests, six separate burn tests were conducted by the City of Fridley, Minnesota, Fire Department. These burn tests used ammunition furnished by members of SAAMI and included shotgun shells, centerfire rifle and pistol cartridges and rimfire cartridges. The tests were conducted to duplicate situations which fire fighters and emergency response personnel might encounter. These tests confirm the SAAMI's position that the fire fighting techniques currently in use by most of the nation's fire fighters can be used to effectively and safely extinguish fires involving sporting ammunition.

The Department of Defense (DOD) expressed specific concern about the shipment of small arms ammunition overseas. They erroneously thought this proposal was applicable to all small arms ammunition shipments, and would require the remarking of all of their small arms stock on hand. Because of this misinterpretation, DOD requested that MTB initiate action with the various international bodies concerned with the movement of hazardous materials, to permit them to ship small arms ammunition overseas without the requirement for remarking or packaging. Since the transport of these materials as Class C Explosives will still be permitted and display of internationally required markings is not precluded, such action is not necessary to accomplish the intended purpose of this rulemaking. Class C Explosive as a hazard class for small arms ammunition is not being terminated, but rather ORM-D as an option for shipping certain types of ammunition is being provided.

In addition to DOD's concern, several comments were received from representatives of organizations and associations contending that reclassifying small arms ammunition from Explosive C to ORM-D would cause major problems for emergency response personnel due to the lack of a shipping paper requirement and the changes that would occur in the marking of shipping documents. Some commenters felt it was imperative that water carriers be notified via the shipping paper that small arms ammunition is fully regulated for international transport by vessel. They contend that without this notification, shipboard personnel would have no knowledge that a small arms ammunition shipment was being made and, in case of an emergency, emergency response personnel would have no way of knowing where the ammunition is stowed on the vessel. These commenters contend that when ammunition is offered for shipment by vessel as an ORM-D, it might not be declared under the International Maritime Organization's Dangerous Goods Code where the goods normally would be classed as explosives having a U.N. division of 1.4. They also contend that the lack of shipping papers and the change in marking requirements would reduce the tight control over the commodity which may lead to not only improper stowage of these materials on vessels, but increase the potential for problems, delays, and penalties for carriers and importers at overseas ports.

Under the regulations as they now exist, shipping papers indicate to water carriers that small arms ammunition is a regulated item in the water mode. MTB believes that these concerns are unwarranted because the addition of the entry "Small arms ammunition" as an ORM-D will not prohibit a shipper from using the original classification of small arms ammunition as a Class C explosive, nor does it waive the requirements of any international regulation with which an international shipper may have to comply. Regarding the point that the lack of shipping papers and change in marking requirements would reduce the tight control over the commodity and lead to the ammunition not being identified to an ocean carrier as being subject to International Maritime Organization (IMO) rules, MTB believes that this problem is no different than the problems involving other materials that are regulated differently by IMO and DOT. MTB is constantly involved in striking a balance between a strong desire for compatibility with international standards and establishing appropriate levels of regulation for materials in domestic commerce.

A similar comment concerning the need for shipping papers was received from the County of Ventura Fire Protection District of Camarillo, California, stating that shipping papers should be required because of the potential for the release of toxic gases when certain types of small arms ammunition are subjected to heat and detonation. They stated that as a result of toxic gases being released, nitrogen compounds can be released in large quantities along with amines and other gases which could cause pulmonary edema along with other physical symptoms and lead to the deterioration of vital body functions. In response to these comments, MTB doesn't believe the toxic products of combustion that are present in small arms ammunition fires will be any greater than those toxic gases that would be released during a fire involving a large number of materials that are not regulated as hazardous materials. It is for this reason that the new 1984 issue of the DOT Emergency Response Guidebook contains explicit precautionary instructions for emergency service personnel to be followed when they approach the scene of an accident involving any cargo (not only regulated hazardous materials.)

This same commenter made reference to the accident record of those shippers transporting small arms ammunition, suggesting that it leaves something to be desired. A review of hazardous materials incidents reported to the

MTB involving shipments of small arms ammunition revealed that over the last decade there have been no fires, explosions, or hazardous situations reported that were a result of the transportation of small arms ammunition. The majority of these incidents involved broken packagings which permitted individual cartridges to spill out. These were simply collected and repackaged. MTB believes that this record confirms that the transportation of certain types of small arms ammunition poses only a minimal hazard. This belief is supported by correspondence from a representative of a major ammunition manufacturer who states that in shipping his products domestically and internationally for over 29 years there has only been one accident in which his product was involved in a fire, and there were no injuries or deaths as a result of that accident.

This same ammunition manufacturer provided cost data showing that as a result of this final rule there could be a reclassification made to the freight class rating of certain types of small arms ammunition by the National Motor Freight Classification board which could possibly result in a transportation cost savings in excess of $1,000,000 for the industry. This is a potential cost saving in excess of that suggested by SAAMI. Information furnished by SAAMI indicated that savings to their members on shipments by one motor carrier alone would be approximately $100,000 per year. Although MTB solicits comments from interested parties on the cost savings and burden reduction associated with this rule, only these two estimates were received. MTB does believe that these figures indicate that the cost savings and burden reduction associated with this rule may be substantial.

The IAFC and two other commenters also proposed that placards be required for small arms ammunition and other class C explosive shipments. MTB believes that the minimal hazard posed by small arms ammunition classed as ORM-D material does not warrant the placarding of vehicles. Therefore, this suggested change is not adopted.

One commenter representing an ammunition manufacturer supported the addition of the entry "Small arms ammunition", but proposed that this classification include ammunition for revolvers and industrial 8 gauge ammunition. MTB believes this to be a reasonable request based on the fact that ammunition for revolvers (a type of pistol) is considered to already be included under this rule, and industrial 8 gauge ammunition is considered to pose no greater hazard in transportation than the other classes and types of ammunition under this rule. Therefore, these changes are adopted in this final rule.

The Institute for Legislative Action of the National Rifle Association was in general agreement with SAAMI's proposal, but suggested that the proposed § 173.1201 be amended by adding the word "projectile" after the description "detonating explosive" and by increasing the caliber for rifle and pistol ammunition from 45 caliber to 50 caliber. The MTB believes that the addition of the word "projectile" may serve to clarify the applicability of this section and for this reason adopts this addition in the text of this section. MTB also believes that increasing the caliber of ammunition in this section from 45 to 50 caliber is acceptable, and would not present any significant additional hazard. For this reason, this change is also adopted.

One commenter representing the Air Transport Association expressed as his chief concern the fact that the ORM-D classification for small arms ammunition does not provide for weight limitations when carried in inaccessible cargo compartments on aircraft. MTB believes that placing gross weight limitations on the number of packages permitted in an inaccessible cargo compartment is unnecessary, as packages of small arms ammunition, ORM-D, pose no greater hazard than other ORM-D materials which are not subject to such limitations, and ORM-D shipments by air will still be required to be accompanied by shipping papers. This same commenter suggested the use of a marking such as "1.4S, Small arms ammunition", in addition to the marking ORM-D to enhance identification of such shipments in case of fire in any location (storage, unit load device, etc.). MTB has not adopted this suggestion; however, there is nothing to preclude a shipper from displaying 1.4S on packages, if they comply with international standards (including competent authority approval) for that class and division. In their concluding comment, this same air carrier association stated that this proposal, although not controversial, was not directed toward a commonality with the International Civil Aviation Organization (ICAO) Regulations. MTB's response to this comment is the same as stated above relative to international shipments by vessel.

Based on the comments received and considering the testing programs that confirm the limited risk of certain types of small arms ammunition, MTB believes that the addition of small arms ammunition under the ORM-D hazard class is justified. Therefore, the proposal contained in Notice No. 83-2 is revised in accordance with the foregoing discussion and for editorial clarity and is adopted in this final rule.

(49 F.R. 21933, May 24, 1984)

49 CFR Parts 100–179, Ch. I
[Docket No. HM-188]
Transportation of Hazardous Materials Between
Canada and the United States

Until the mid 1970's, the regulations of the CTC (formerly the Board of Transport Commissioners for Canada) were, with few exceptions, identical to those found in the HMR. It was due to this regulatory compatibility that trans-border shipments of hazardous materials moved without confusion on the part of shippers and carriers as to the applicability of regulatory requirements of each country. This may not be the present situation since CTC's and DOT's regulations for the safe transportation of hazardous materials differ in a number of significant ways.

The CTC regulations are entitled "Regulations for the Transportation of Dangerous Commodities by Rail" (TDCR) and are prescribed by "General Order No. 1974–1-Rail" of the Canadian Transport Commission dated July 31, 1974. The following is stated in general notice as a forward to the TDCR:

> These regulations are applicable to dangerous commodities transported over railways subject to the jurisdiction of the Commission.

Paragraph 9(a) of § 173.8 of the HMR reads as follows:

> Except for hazardous wastes and hazardous substances, shipments of hazardous materials which conform to the regulations of the Canadian Transport Commission (formerly the Board of Transport Commissioners for Canada), may be transported from the point of entry in the United States to their destination in the United States, or through the United States en route to a point in Canada. Empty rail tank cars may be transported in conformity with Canadian Transport Commission regulations from point of origin in the United States to a point of entry into Canada.

Considering the applicability of CTC's TDCR and the paragraph from the HMR quoted above, so-called reciprocity in regulations exist at the present time only in regard to transportation of hazardous materials (dangerous commodities) by railroad and only to materials that are subject to both CTC and DOT regulations (e.g., there is no CTC regulation presently applicable to combustible liquids; therefore, the provisions of § 173.8 do not apply and combustible liquids must be transported in conformance with the HMR).

Most recent major revisions to CTC's regulations became effective on May 1, 1982. Many of the revisions reflect conformance to international standards based on recommendations issued by the United Nations Committee of Experts on the Transport of Dangerous Goods. The following discussions do not constitute a comprehensive analysis of CTC's regulations; they are provided only to illustrate some of the differences between CTC and DOT regulations (two numbers before a decimal in a section citation denote a CTC rule, e.g., § 71.1, whereas three numbers before a decimal denote a DOT rule contained in 49 CFR, e.g., § 171.1):

1. The list of dangerous commodities in § 72.5 most closely aligns with the list of the International Maritime Dangerous Goods Code (IMDG Code) and differs in many respects from the list in § 172.101. For example, numerous explosives descriptions not given in § 172.101 are listed in § 72.5. The list if virtually identical to DOT's optional hazardous materials table in § 172.102 and quite similar to the International Civil Aviation Organization's Technical Instructions for the Safe Transport of Dangerous Goods by Air, use of which is permitted by § 171.11.

2. A number of materials are classed differently in § 72.5 than they are in § 172.101. Also, the international class numbering system is used in § 72.5 rather than the class words in § 172.101. Examples are: (1) Aluminum hydride bears class number 4.2 which, according to § 72.1, means a substance "liable to spontaneous combustion" whereas, in § 172.101, it is classed as a Flammable solid; (2) Ammonia, an hydrous and Hydrogen chloride bear class number 2.3 which, according to § 72.1, means "Poison gases" whereas, in § 172.101, each of these materials is classed as a nonflammable gas; and (3) according to § 72.1, high-strength nitric acid, e.g., a 72% concentration, bears class number 8 for corrosives, and according to § 172.101, nitric acid at this concentration is classed as an oxidizer.

3. Section 73.427 sets forth CTC's requirements for shipping papers. Included is a provision that the class number of a material be used rather than a class word(s) required by § 172.202 in referencing § 172.101. Use of class numbers alone is not generally permitted by the HMR for imported shipments moving by rail or highway; therefore, for basic descriptions of hazardous materials on shipping papers appearing in the United States, only shipments by rail coming from Canada are presently permitted (by § 173.8) to have classes identified on shipping papers by numbers in place of class words.

4. Except for placards for Explosives and Poison Gas, placards for railcars specified in § 74.548 are wordless enlarged UN labels bearing class numbers in the bottom corner. For example, the only distinction between a FLAMMABLE GAS and a FLAMMABLE LIQUID placard is class number 2 or 3, respectively, in the bottom corner. DOT requires class words on placards except when identification numbers are permitted and displayed. In § 74.548a, CTC not only requires identification numbers to be displayed on bulk packagings, as is required by DOT, but on every carload, container load, or trailer load of hazardous materials requiring a placard. Except for division 2.3 (poison gas), which is addressed in § 74.548b, § 74.548a requires identification numbers to be displayed only on placards and does not permit optional use of orange panels.

5. The CTC regulations contain no provisions for the transport of consumer commodities as provided by the HMR.

6. Paragraph 73.9(c) permits, with certain exceptions, the transport into the United States of hazardous materials prepared in accordance with the IMDG Code, provided they are transported in closed freight containers and that the DOT certification statement appears on the shipping papers. The HMR do not generally permit the transport by rail of hazardous materials packaged or placarded in accordance with the IMDG Code without regard to the corresponding requirements in the HMR.

7. The CTC regulations contain no listing of forbidden materials, as listed in § 172.101, and certain provisions appear in § 173.21 for which there is no corresponding provision in § 73.21. For example, § 173.21(d) forbids §... materials (other than those classed as explosives) which will detonate in a fire". The CTC regulations contain no such general prohibition. Other examples of materials forbidden for transport under the HMR, but permitted by the CTC regulations, include "new" explosives not approved in accordance with § 173.86 and Methyl ethyl ketone peroxide containing more than 9 per cent active oxygen.

MTB believes that it should make every reasonable effort to recognize shipments coming into the United States in conformance with CTC regulations (and future Transport Canada regulations) in consideration of the fact that (1) CTC fully recognizes shipments moving into Canada under DOT regulations (§ 73.8), and (2) more than $3 billion worth of hazardous materials (dangerous commodities) move annually between our countries.

Commenters are invited to address any potential safety impacts being encountered or contemplated as a result of the present "reciprocal" regulatory provisions of § 73.8 and § 173.8 or that may result from Transport Canada requirements (a number of which, if adopted as proposed in the Canada Gazette on June 19, 1982, would be the same as those of CTC). Of particular concern to MTB are those potential safety impacts that may be related to emergency response actions because of several fundamental differences in communications requirements. This concern may be offset by the fact that both regulatory systems use identification numbers assigned to materials based on the worldwide UN System. It is these identification numbers which provide rapid access to emergency response information in the U.S. Emergency Response Guidebook and Canada's Emergency Response Guide for Dangerous Goods.

Commenters are encouraged to discuss the value of this materials identification numbering commonality in offsetting other differences in light of the wide dissemination of the Guidebook and its growing use by fire, police, and other emergency response entities in the United States.

MTB wishes to emphasize that the purpose of this Advance Notice of Proposed Rulemaking is to solicit comments concerning safety impacts due to differences in regulations pertaining to the safe transportation of hazardous materials. It is not intended to address the merits of CTC's regulations (or those proposed by Transport Canada); nor is it intended to serve as a forum for such a purpose.

(48 F.R. 20255, May 5, 1983)

49 CFR Parts 171 and 173
[Docket No. HM-188B, Notice No. 85–2]
Transportation of Hazardous Materials Between
Canada and the United States

On February 6, 1985, Transport Canada published new multi-modal regulations for the transport of dangerous goods (hazardous materials) in Part II of the Canada Gazette. The regulations are officially titled "Regulations respecting the handling, offering for transport and transporting of dangerous goods" or simply the "Transportation of Dangerous Goods Regulations", issued pursuant to the provisions of the Transportation of Dangerous Goods Act of July 17, 1980. For the purpose of this notice, these regulations are referred to as the "TDG Regulations". Certain parts of these regulations were effective at the time of publication, other parts became effective on April 8, 1985, but the majority of the regulations, and particularly those dealing with specific transport requirements as opposed to administrative matters, are scheduled to enter into force on July 1, 1985. Copies of the TDG Regulations may be obtained from the Canadian Government Publications Center, Supply Services Canada, Ottawa, K1A 0S9, Canada, at a cost of $18 (Canadian) per copy.

On March 27, 1985, the Embassy of Canada delivered a note to the Department of State which formally requested that the United States take steps to amend the DOT HMR to grant reciprocal recognition to the TDG Regulations in order to facilitate the transport of hazardous materials between the United States and Canada. A specific proposed text for a revised 49 CFR 173.8 was attached to this note. Because the note summarizes certain aspects of the TDG Regulations, as well as underscoring the need to facilitate hazardous materials movements between Canada and the United States, the note, and the attached suggested text of § 173.8 are reproduced here for information.

The Embassy of Canada presents its compliments to the Department of State and is pleased to advise the Department that final regulations, issued under the authority of the Transportation of Dangerous Goods Act of 1980, were published in the Canada Gazette, Part II, on February 6, 1985. These regulations facilitate the northbound movement of dangerous goods between Canada and the United States and the Canadian authorities hereby request that the corresponding U.S. regulations be modified to facilitate the southbound movement of equivalent goods.

The Canadian Regulations parallel closely provisions in Title 49 of the United States Code of Federal Regulations, the International Maritime Code for Dangerous Goods (IMDG), and UN Recommendations for the Transport of Dangerous Goods.

Included in these regulations, all of which will be in force by July 1, 1985, are provisions to facilitate trade entering Canada from the United States. The new regulations on bilateral trade (called "transborder shipments" in the regulations) cover transportation by road, rail and water if in home voyage Class II, and stipulate that, with limited exceptions, shipments complying with

49 CFR shall be acceptable in Canada. Under Parts IV and V of the new Canadian regulations, goods in all nine United Nations Classes, except Class 1 or Divisions 3 or 4 of Class 2, are regulated in a manner that facilitates compliance with both the Canadian and U.S. Regulations. Thus, trade from the U.S.A. to Canada is facilitated without undue delay and without the additional cost of repackaging, relabelling, replacarding and redocumenting at the border. Equivalent acceptance of Canadian Regulations to the first destination in the U.S.A. by appropriate amendment of 49 CFR would allow the current extensive and mutually beneficial hazardous goods trade between Canada and the United States to continue.

The Canadian authorities therefore request that the appropriate United States authorities proceed expeditiously with pertinent amendments to the hazardous materials provisions of 49 CFR which currently recognize conformity with the Canadian Transport Commission's Dangerous Commodity Regulations for rail transport as being equivalent to compliance with 49 CFR. The amendments would recognize conformity with the Transportation of Dangerous Goods Regulations for all modes of transport as being equivalent to conformity with 49 CFR.

Should the required amendments to 49 CFR not be in place by July 1, 1985, there is a distinct risk of disruption in the significant dangerous goods trade between our two countries. Once implemented, however, the amendments would permit the flow of transborder trade involving dangerous goods to proceed safely under adequately controlled conditions.

The Embassy wishes to emphasize that Canadian Regulations provide for the same level of safety as U.S. Regulations, and are almost identical with Regulations under the IMDG Code, which is presently recognized as acceptable for goods in transit to first destination in the United States. Furthermore, Canadian Regulations are closer to international (U.N.) Recommendations that are United States Regulations, and over the past few years the United States has altered its own Regulations to meet U.N. standards more closely.

The Embassy understands that the Office of Hazardous Materials Regulation of the United States Department of Transportation agrees in principle that 49 CFR should be amended as suggested above. The Embassy therefore requests that the State Department bring to the attention of the Department of Transportation the urgency of proceeding with a rulemaking proceeding (sic) to institute such amendments. To facilitate this process, the Canadian authorities have prepared the attached draft of a proposed amendment. (It should be noted that the reference in this text to both the Transportation of Dangerous Goods Regulations and to Canadian Transport Commission Regulations provides for those regulatory requirements which are not addressed at this time in the Transportation of Dangerous Good (sic) Regulations but which are required for rail shipments.) The Embassy would appreciate the State Department's providing to the appropriate USA authorities a copy of this text on which they may wish to draw in drafting the pertinent amendments to 49 CFR.

TEXT OF PROPOSED AMENDMENT TO SECTION 173.8 OF 49 CFR

Section 173.8 Canadian shipments and packagings.

(a) For all dangerous goods other than those classified as Class 1 or Divisions 3 or 4 of Class 2 under the Transportation of Dangerous Goods Regulations made pursuant to the Transportation of Dangerous Goods Act, 1980:

(1) Shipments of hazardous materials entering the United States from Canada or empty rail cars which contain residues of hazardous materials, that are being returned to Canada, which conform with the Regulations of the Government of Canada pursuant to the Transportation of Dangerous Goods Regulations and in addition, in the case of rail shipments, conform with the Canada Transport Commission Regulations for the Transportation of Dangerous Commodities by Rail for those requirements of the Canadian Transport Commission not addressed by the Transportation of the point of entry in the United States to their destination in the United States or through the United States en route to a destination in Canada: or, in the case of empty rail cars containing a residue of hazardous material, from their point of unloading in the United States to a destination in Canada.

(b) For dangerous goods classified Class 1 or Division 3 or 4 of Class 2 under the Transportation of Dangerous Goods Regulations:

(1) Shipments of hazardous materials which conform in Safety Marks and in Shipping Name to the Regulations of the Government of Canada pursuant to the Transportation of Dangerous Goods Regulations (under the Transportation of Dangerous Goods Act) and in all other requirements to either 49 CFR or the Transportation of Dangerous Goods Regulations and the Canadian Transport Commission Regulations for the Transportation of Dangerous Commodities by Rail, may be transported from the point of entry in the United States to their destination in the United States, or through the United States en route to a point in Canada. Empty rail tank cars may be transported in conformity with the Transportation of Dangerous Goods Regulations and with Canadian Transport Commission Regulations from point of origin in the United States to point of entry in Canada.

(c) Except as specified in 173.301(i) specification packagings made and maintained in full compliance with the corresponding specifications prescribed by the Railway Transport Committee of the Canadian Transport Commission (formerly the Board of Transport Commissioners for Canada), in its Regulations for the Transportation of Dangerous Commodities by Rail, and marked in accordance therewith (e.g., BTC, CTC, etc.) may be used for the shipment of hazardous materials within the United States.

While issue could be taken with a number of statements in this note, particularly with regard to the extent that the reciprocity provisions contained in the TDG Regulations will facilitate shipments entering Canada from the United States, it must be emphasized that this is not the purpose of this notice. The MTB believes that achieving the maximum level of reciprocity between the TDG Regulations and the HMR is both necessary and beneficial to both the United States and Canada for a number of reasons. However, it must be borne in mind that the purpose of these regulations is to insure safety in the transport of hazardous materials and, in the event of an incident or accident, to permit the nature of the hazards of the materials involved to be readily identified to emergency response personnel. The latter purpose can only be realized through extensive training efforts of personnel involved in the handling of hazardous materials and in response to hazardous materials incidents. In order for such training to be effective, it is essential that the salient points of the hazardous materials regulations (e.g. labeling, placarding and shipping paper description requirements) remain relatively stable, and that, when significant changes to these fundamental requirements are introduced, their introduction is a gradual process allowing sufficient time for retraining. Therefore, the MTB considers that the purpose of this notice is to explore and solicit comment on the extent to which recognition can be accorded to the TDG Regulations without seriously jeopardizing the hazard warning and emergency response systems based on the existing HMR.

ANALYSIS OF THE TDG REGULATIONS

In order to adequately assess the potential safety implications associated with recognition of the TDG Regulations, it is first necessary to examine the differences between them and the HMR. While a complete analysis and description of these differences in this notice is impracticable, it is possible to highlight some of the fundamental differences to facilitate the development of comments. Undoubtedly, many commenters will desire to conduct a far more extensive comparison individually. The following brief description of the evolution of the TDG Regulations, as well as some of the more significant differences between those regulations and the HMR, is provided to help stimulate comment.

Until the mid-1970's the regulations of the CTC (formerly the Board of Transport Commissioners for Canada) were, with few exceptions, identical to those found in the HMR. It was due to this regulatory compatibility that transborder shipments of hazardous materials moved without confusion on the part of shippers and carriers as to the applicability of regulatory requirements of each country, and that broad "reciprocal" recognition was accorded to the CTC Regulations in § 173.8. However, several years ago changes were made to the CTC Regulations that caused them to differ significantly in many respects from the HMR.

At that time, in recognition of the increasing number of "land bridge" shipments and import shipments arriving in Canada in conformance with the provisions of the International Maritime Organization's (IMO) International Maritime Dangerous Goods Code (IMDG Code), the CTC Regulations were substantially revised to replace the then existing proper shipping names, and the classification, labeling and placarding systems with those provided in the IMDG Code, which, in turn, differ only in minor respects from those in the Recommendations of the United Nations Committee of experts on the transport of Dangerous Goods (U.N. Recommendations).

To assess the potential safety implications of these newly introduced differences between the CTC Regulations and the HMR in order to determine if the broad reciprocal recognition accorded the CTC Regulations through § 173.8 was still appropriate, the MTB published an advance notice of proposed rulemaking on May 5, 1983, under Docket No. HM-188 (48 FR 20255) and also conducted a public hearing on the matter on June 2, 1983. Numerous comments were received which, for the most part, stressed the importance of maintaining reciprocal regulatory recognition of the Canadian regulations and supported the fact that, although the new description, classification, labeling and placarding requirements of the CTC Regulations differed in many respects from the corresponding provisions of the HMR, there was not an adverse effect on safety by continuing to permit shipments to enter the United States from Canada when in conformance with CTC Regulations. One of the principle reasons for this belief was the commonality to both systems of the UN number as a means for specific hazardous materials identification. As a result of this action, it was finally concluded that there was no need from the point of view of transport safety to rescind the broad recognition of the CTC Regulations in § 173.8, and the section has remained with only one relatively minor amendment until this time. In this context, it is important to note that the reciprocity in regulation exists at the present time only in regard to transportation of hazardous materials (dangerous commodities) by railroad and only to materials that are subject to both CTC and DOT regulations (e.g. there is no CTC regulation presently applicable to cumbustible liquids: therefore, the provisions of § 173.8 do not apply and combustible liquids must be transported in conformance with the HMR).

Unlike the CTC Regulations, the new TDG Regulations apply to all modes of transport. On the other hand, the new regulations are similar to the CTC Regulations in that both employ the basic description, classification, labeling and placarding requirements that are provided in the U.N. Recommendations and IMDG Code. Since the CTC requirements have been permitted for several years under § 173.8 from the point of view of safety, it would not appear to be a radical departure to extend recognition of this method of description, classification, labeling and placarding to all modes of transport through recognition of the TDG Regulations. This would particularly appear to be true since the UN system is already widely employed in the marine and air modes in the United States through regulatory recognition of the IMDG Code and the International Civil Aviation Organization's (ICAO) Technical Instructions for the Safe Transport of Dangerous Goods by Air, respectively. It should also be noted that this system of hazardous materials description, classification, labeling and placarding has been included in the Optional Hazardous Materials Table (§ 172.102) of the HMR since 1980, and, with certain exceptions has been permitted for the rail and highway movement of hazardous materials that are in the course of being imported or exported by vessel. Nevertheless, it would appear appropriate to highlight some of the more fundamental differences between the TDG

Regulations and the HMR. Once again, it must be emphasized that the following discussions do not constitute a comprehensive analysis of the TDG Regulations. They are provided only to illustrate some of the differences between the TDG Regulations and the HMR.

1. The list of dangerous goods in the TDG regulations most closely aligns with the list of the U.N. Recommendations. ICAO Technical Instructions and the IMDG Code and differs in many respects from the list in § 172.101. For example, numerous descriptions not given in § 172.101 are listed in the TDG Regulations. This list is similar to DOT's optional hazardous materials table in § 172.102.

2. A number of materials are classed differently in the TDG Regulations than they are in § 172.101. Also, the international class numbering system is used in the TDG Regulations rather than the class words as in § 172.101.

3. In addition to classifying a number of materials differently, the TDG Regulations include a new class of "Corrosive Gases" (Division 4 of Class 2) which does not exist in the DOT classification system. Included in this new class are a total of nine gases including such gases as Anhydrous ammonia and Chlorine which are classified as Non-flammable gases under the HMR. The Corrosive gas class is not included in the U.N. Recommendations. IMDG Code of ICAO Technical Instructions and these gases would be required to be labeled with a label consisting of a white square on point with a black gas cylinder in the upper half. The corrosive gas placard is simply an enlarged version of the label.

4. Under the TDG Regulations, the class number of a material is used rather than a class word(s) required by § 172.202 in referencing § 172.101. Use of class numbers alone is not generally permitted by the HMR for imported shipments moving by rail or highway: therefore, for basic descriptions of hazardous materials on shipping papers appearing in the United States, only shipments by rail coming from Canada are presently permitted (by § 173.8) to have classes identified on shipping papers by numbers in place of class words.

5. Except for placards for Explosives and Poison Gas, placards specified in the TDG Regulations are wordless enlarged UN labels bearing class numbers in the bottom corner. For example, the only distinction between a Flammable Gas and a Flammable Liquid label in class number 2 or 3, respectively, in the bottom corner. Except in cases where the identification numbers are permitted on placards, DOT requires that the class words be displayed.

There is one additional general point regarding the TDG Regulations that is important to note. While it is envisioned that eventually the TDG Regulations will address all aspects of the transport of dangerous goods by all modes, they are not at this time complete. While they do apply to all modes of transport, they do not currently address all aspects of dangerous goods transport. For example, the TDG Regulations contain provisions applicable to all modes for classification, labeling, placarding, marking of packages and preparation of shipping papers, but they do not, at this time, contain regulations on packaging. To fill in these "gaps", the existing modal regulations (e.g. the CTC Regulations) will remain in place. However, the TDG Regulations will, to the extent that they address a particular aspect of the transport of dangerous goods, supersede the existing modal regulations. For example, the classification, labeling, placarding, package marking and preparation of shipping papers for a rail shipment in Canada will be governed by the TDG Regulations while the other aspects of transport such as packaging and car placement, not addressed in the TDG Regulations, will continue to be governed by the CTC Regulations.

AMENDMENTS TO THE HMR

In light of the publication of the TDG Regulations, and of the Note transmitted by the Canadian Embassy, the MTB is proposing to add a new § 171.12a to the HMR which would allow, with certain exceptions and limitations, hazardous materials to be transported into the United States from Canada in conformance with the TDG Regulations. The MTB carefully studied the text suggested by the Canadian government in the Embassy Note, but concluded that the broadly worded text did not fully take into account some of the safety consequences of recognizing the TDG Regulations, nor certain special controls the MTB must exercise, e.g., with respect to the transport of hazardous substances and hazardous wastes. Therefore, the MTB is proposing a text which it considers more appropriately reflects the degree of recognition that should be given the TDG Regulations.

It is proposed that the revised regulations for Canadian shipments be included in a new § 171.12a rather than in § 173.8 where the existing reciprocity provisions appear. The MTB believes that Part 171 is a more appropriate location for this provision because the proposed section deals with transport requirements in general, rather than just packaging or shipper requirements. As a consequence, the existing § 173.8 would be removed and reserved.

It will be noted that the proposed § 171.12a only addresses transport by rail and highway. This is due to the fact that the MTB believes it unnecessary to include reciprocity provisions for the air and marine modes since the HMR already incorporates by reference the ICAO Technical Instructions and the IMDG Code. Under the TDG Regulations, air transport in Canada is governed by the ICAO Technical Instructions. Since § 171.11 of the HMR already permits compliance with the Technical Instructions, the MTB believes there is no need to address shipments arriving from Canada by air in the proposed § 171.12a. Similarly, in the marine mode, the TDG Regulations, as supported by the Canadian Coast Guard regulations issued pursuant to the Canada Shipping Act, require, with the exception ot a home-trade voyage Class II, compliance with the IMDG Code. Since §§ 171.12, 172.102 and 176.11 already generally permit compliance with the IMDG Code, the MTB considers it unnecessary to address marine mode transport in the proposed § 171.12a.

Paragraph (a) of the proposed § 171.12a contains general permission for shipments entering the United States from Canada by rail and highway to be classified, labeled, placarded, marked and described and certified on a shipping paper in accordance with the TDG Regulations. Certain exceptions to this general authorization are contained in paragraph (b) of § 171.12a.

For some hazardous materials allowed to be transported in conformance with the TDG Regulations, paragraph (a) would require certain additional information to appear on shipping papers and package markings. For instance, the letters "RQ" would have to appear on shipping papers and in package markings, when appropriate, in order to trigger the necessary reporting requirements in the event of the release of a hazardous substance.

The exceptions to the general authorization to comply with the TDG Regulations, as set forth in paragraph (b), would be forbidden materials and packagings, explosives and materials classified as "Corrosive gases" under the TDG Regulations. The reason for excluding forbidden materials and packages from the provisions of paragraph (a) is considered self-evident. Explosives have been excluded, with the exception of allowing the use of the labels and placards required by the TDG Regulations which would contain the phrase "explosive A, B or C" as appropriate, because of the substantial differences in the classification systems and because of the heavy reliance of thousands of local ordinances on the present DOT classifications for explosives. The gases classified as "Corrosive gases" under the TDG Regulations have been excepted from the provisions of paragraph (a) because of the anticipated difficulties in the emergency response area of introducing, in a short time frame, classifications, labels and placards not heretofore known to emergency response personnel. On the basis of comments received in response to this notice, the MTB would be prepared to recognize the Corrosive gas classification, label and placard if it appears that this would not adversely affect the ability of emergency response organizations to respond to transport emergencies.

Paragraph (C) of the proposed § 171.12a is identical to the existing paragraph 173.8(b). Since the CTC specifications for packagings will be retained in effect for the time being, the MTB believes that the provisions currently contained in § 173.8(b) should be retained.

Paragraph (d) proposes to continue to give recognition to the CTC Regulations to the extent that they will still apply to rail shipments in Canada (i.e., to the extent they are not superseded by the TDG Regulations). Since there exists at this time no national regulations in Canada governing the transport of hazardous materials by road, the proposed paragraph (e) would require that shipments entering the United States from Canada by highway under the provisions of § 171.12a(a) be otherwise transported in accordance with the HMR.

Commenters are invited to address any potential safety impacts contemplated as a result of the proposed "reciprocal" regulatory provisions of § 171.12a. Of particular concern to MTB are those potential safety impacts that may be related to emergency response actions because of several fundamental differences in communications requirements. This concern may be offset by the fact that both regulatory systems use identification numbers assigned to materials based on the worldwide U.N. system. It is these identification numbers which provide rapid access to emergency response information in the U.S. Emergency Response Guidebook and Canada's Emergency Response Guide for Dangerous Goods. Commenters are encouraged to discuss the value of this materials identification numbering commonality in offsetting other differences in light of the wide dissemination of the Guidebook and its growing use by fire, police, and other emergency response entities in the United States.

(50 F.R. 23026, May 30, 1985)

49 CFR Parts 171, 172, 173, 174, 176, 177 and 179
[Docket No. HM-188B, Amdt. Nos. 171–83, 172–100, 173–191, 178–48, 176–22, 177–66, 179–39]
Transportation of Hazardous Materials Between
Canada and the United States

On February 6, 1985, the Canadian Government (specifically Transport Canada) published new multi-modal regulations for the transport of dangerous goods (hazardous materials) in Part II of the Canada Gazette. The regulations are officially titled "Regulations Respecting the Handling, Offering for Transport and Transporting of Dangerous Goods" or simply the "Transportation of Dangerous Goods Regulations", and have been issued pursuant to the provisions of the Canadian Transportation of Dangerous Goods Act of July 17, 1980. For the purpose of this final rule, these regulations are referred to as the "TDG Regulations". The TDG Regulations apply to the transport of dangerous goods within Canada, including shipments destined for the United States and those entering Canada from the United States. Certain parts of these regulations were effective at the time of publication, other parts became effective on April 8, 1985, but the majority of the regulations, and particularly those dealing with specific transport requirements as opposed to administrative matters, became effective on July 1, 1985. Owing to the problems concerning transportation of hazardous materials between the United States and Canada addressed in this final rule, the TDG Regulations were amended prior to becoming effective to permit, in general, shipments of hazardous materials moving between the United States and Canada to continue to be transported in accordance with the requirements of HMR until October 31, 1985.

On May 30, 1985, the MTB published a notice of proposed rulemaking in the FEDERAL REGISTER under Docket No. HM-188B (50 FR 23036) which proposed to amend the HMR to permit transportation of hazardous materials from Canada to the United States in accordance with the TDG Regulations subject to certain conditions and limitations. This proposal was published in response to a note delivered to the Department of State by the Embassy of Canada which formally requested that the United States take steps to amend the HMR to grant recognition to the TDG Regulations in order to facilitate the transport of hazardous materials between Canada and the United States. A public hearing was held on these proposals on June 27, 1985, and three persons offered oral comments at that time. A complete transcript of that hearing is in the public docket.

In addition to inviting comments on the proposed amendments to the HMR, and on certain specific questions posed in the notice of proposed rulemaking, the MTB raised the following questions for comment in the public hearing:

1. What are the safety implications of recognizing the corrosive gas (Division 2.4) classification in the TDG Regulations with respect to the labeling and placarding of shipments entering the United States? Would the ID number on the corrosive gas placard be adequate to ensure appropriate emergency response? Could such a placard be specifically authorized for transportation within the United States on shipments destined for Canada since TDG Regulations do not recognize the DOT placarding required for gases classed as corrosive gases by the TDG Regulations?

2. What would be the safety implications of recognizing the Canadian shipping paper requirements for explosives shipments entering the United States if the papers also contained the DOT proper shipping name and hazard class of the explosives?

3. Should the explosives labels and placards required by the TDG Regulations for shipments to the United States, which the MTB has proposed to recognize, also be allowed for transportation within the United States on shipments destined for Canada, since the TDG Regulations do not recognize the DOT explosives labels and placards?

4. Information currently available indicates that the Canadian Transport Commission will amend its regulations effective July 1, 1985, to fully incorporate the TDC Regulations. The existing §173.8 of the DOT regulations fully recognizes shipments entering the United States in conformance with the CTC regulations. This means that effective July 1, shipments could legally enter the United States in full conformance with the CTC regulations, as modified through incorporation of the TDG Regulations (i.e., with TDG required shipping papers, labels and placards (including the "Corrosive Gas" label and placard)). Does this situation present an unacceptable risk from the safety standpoint? Should DOT take immediate action to amend §173.8 to prevent this situation?

A total of sixteen comments, in addition to those offered at the hearing, were received in response to this notice of proposed rulemaking and the additional questions raised at the hearing. The following is an analysis, by section, of the comments and any changes made in the final rule as compared to the proposed rule. Aside from recognition of the CORROSIVE GAS placard specified in the TDG Regulations, which was discussed as an option for consideration in the notice, certain minor nonsubstantive changes have been included in this final rule consistent with changes to the TDG Regulations which became effective on July 1, 1985.

Section 171.7. This section is being amended to include the TDG Regulations, as amended as of July 1, 1985, in the matter incorporated by reference. While this was not proposed in the notice, the MTB feels that is necessary in order to more precisely specify those TDG Regulations that are being recognized under the HMR.

Section 171.12a. There was general support by commenters for this section as proposed, although one commenter, the International Association of Fire Chiefs (IAFC), expressed concern about the proposed rule because hazardous materials entering the United States from Canada ". . . would be permitted to be identified through the Canadian system." This is, of course, the principal reason for publication of the proposed rule. The MTB believes it would impose tremendous and unreasonable burdens to require that shipments entering the United States from Canada fully conform to the HMR. This would certainly be a drastic departure from past philosophy concerning recognition of Canadian regulations. The MTB believes that a total rejection of the TDG Regulations cannot be supported on safety grounds and has, therefore, adopted this section. However, the section appearing in this final rule differs in certain respects from that proposed as a result of the introduction of changes taking account of comments, to improve its presentation, and to correct certain omissions in the proposal that came to light through the continuing discussions with Transport Canada.

Paragraph (a) of §171.12a has been expanded to include packaging authorized under the TDG Regulations. While the TDG Regulations do not currently address packaging in general, they do prescribe packaging for specific types of hazardous materials such as limited quantities and consumer commodities. Because of the similarity of the packagings prescribed to that required under HMR for the same types of materials, the MTB believes that these TDG authorized packagings should be recognized. In addition, the reference to hazardous materials "certified" on a shipping paper in accordance with the TDG Regulations that appeared in the proposal has been removed since the

TDG Regulations do not require the use of a shipper's certification. Several commenters objected to the fact that the proposal would recognize the use of class numbers rather than class names in shipping papers. In view of the fact that §173.8 has permitted this practice for rail shipments entering the United States from Canada for several years, and because the DOT Emergency Response Guidebook, and the pocket version of the guidebook recently made available for distribution to drivers by a major trade association, contain a table explaining the meaning of these class numbers, the MTB has made no change to the proposed rule as a result of this comment.

The paragraph (a)(2)(i) that appeared in the proposal is now considered by the MTB to be superfluous and it has, therefore, been removed, with the proposed paragraphs (a)(2)(ii) and (a)(3) being combined to improve clarity. As a result, the proposed paragraphs (a)(4), (a)(5), and (a)(6) are now designated (a)(3), (a)(4) and (a)(5), respectively. One commenter suggested that the proposed paragraph (a)(5) be modified to require that only one hazardous waste manifest be required by providing that waste shipments destined for Canada be made using the Canadian hazardous waste manifest, while shipments destined for the United States be made under the United States manifest. The uniform hazardous waste manifest currently required in the United States by DOT and EPA regulations was developed through years of intensive effort and after thorough coordination with the states. Because of the impact that the change suggested by this commenter would have regarding the use of this uniform hazardous waste manifest within the United States; the MTB has not adopted this suggestion. However, discussions are now underway between the United States Environmental Protection Agency and their Canadian counterparts, and the concerned states in an attempt to reach an agreement whereby such a reciprocal recognition of hazardous waste manifests may be possible in the future.

Finally, a provision has been added to paragraph (a)(5) to require that only "UN" or "NA" numbers may be used and that "PIN" numbers provided for in the TDG Regulations are not acceptable. This action was taken in response to a recent amendment to the TDG Regulations which no longer requires the "PIN" prefix for consignments transported to the United States. One commenter suggested that this paragraph be modified to require English text on labels and placards when text is required or present. It should be noted that when text is required on labels and placards by the TDG Regulations, both the English and the French texts are required. Also, for a number of years, the HMR have recognized either foreign text or no text on labels conforming to IMO or UN standards (see §172.407), as well as "wordless" placards under CTC Regulations. For these reasons, the suggested change to require English text of labels and placards has not been adopted in the final rules.

Paragraph (b)(2) of§171.12a has been changed considerably from that appearing in the proposal. In view of the general support for this section, and of the lack of any specific opposition to the proposed use of the explosives labels and placards prescribed in the TDG Regulations for shipments to the United States, the MTB has expanded the permissive use of these labels and placards to include shipments originating in the United States destined for Canada. Since the TDG Regulations do not recognize the DOT explosives labels and placards for shipments entering Canada, the MTB believes this action will reduce the burdens on shippers and carriers or explosives by eliminating the necessity to change labels and placards at the border or for dual labeling and placarding. The MTB has also decided to add a provision to paragraph (b)(2) which recognizes shipping papers for explosives prepared in accordance with the TDG Regulations provided the shipping paper also includes the letters "DOT" followed by the DOT proper shipping name and hazard class of the explosives. This is similar to the recognition accorded shipping papers for explosives prepared in conformance with the HMR granted to United States, shipments entering Canada by the TDG Regulations. The MTB has included this provision to minimize the need for preparation of dual sets of shipping papers. Comment was specifically solicited on both of the preceding additions to this paragraph at the public hearing, and there was no negative comment regarding these suggested provisions.

The provision excluding any recognition of the TDG Regulations regarding the transport of Class 2, Division 4 materials (i.e., "Corrosive gases") that appeared in paragraph (b)(3) of the proposal has been removed. The notice specifically raised the issue of the recognition of the corrosive gas class and indicated that, on the basis of comments received, the MTB would be prepared to recognize the Corrosive gas classification, label and placard if it appeared that this would not adversely effect the ability of emergency response organizations to respond to transport emergencies.

The majority of commenters supported the recognition of the corrosive gas class. Reasons indicated by commenters for this support included:—The presence of the UN number on placards and shipping papers which permits appropriate emergency response to be initiated through use of the DOE Emergency Response Guidebook.

—The presence of the UN number on placards and shipping papers which permits appropriate emergency response to be initiated through use of the DOE Emergency Response Guidebook.

—The requirement in paragraph 5.38(1) of the TDG Regulations to mark the name of the gas on each road and rail vehicle used to transport corrosive gases in bulk.

—The danger of delay and disruption of shipments resulting from the need to relabel and replacard, and to prepare new shipping papers for, consignments of corrosive gases and the increased risks and exposures associated with such delays and disruptions.

—The risk of commission of serious and dangerous errors in the course of replacarding, relabeling and redocumenting shipments and in transcribing UN numbers and other descriptive information.

—The opinion that the corrosive gas classification more accurately describes and communicates the actual hazards of these gases than does the DOT "Nonflammable gas" hazard class.

—The belief that the commonality of the gas cylinder pictogram to both the Corrosive gas and Nonflammable gas labels and placards overrides the color differences, and still enables emergency responders to recognize that a shipment contains a Corrosive gas.

—The presence of the name of the gas on shipping papers providing for easy identification of the gas.

—The speed and effectiveness of organizations and systems such as CHEMTREC (Chemical Manufacturer's Association), CANUTEC (Transport Canada) and CHLOREP (Chlorine Emergency Plan of the Chlorine Institute) in providing advice and assistance to emergency response personnel in the event of transport incidents.

—The serious facilitation and economic burdens that would be imposed on shippers and carriers by the need to change labels, placards and shipping papers at border crossings if the Corrosive gas classification is not recognized.

Those opposed to recognition of the corrosive gas class generally cited lack of familiarity on the part of emergency response personnel with this class, label and placard, and the belief that such recognition would undermine the regulatory uniformity of the HMR and of emergency response aids such as the DOT Emergency Response Guidebook, as the reasons for their opposition. Several commenters specifically noted the need to include reference to the corrosive gas class in the DOT Emergency Response Guidebook before this class is recognized for transportation of these gases within the United States.

The existing § 173.8 has permitted, for several years, the gases now classed as corrosive gases by the TDG Regulations to enter the United States by rail when classed, labeled and placarded "Poison gas" in accordance with the CTC Regulations. This was permitted, after soliciting public comment (Docket No. HM-188), on the basis that the UN number on the placard and in shipping papers provided sufficient information to initiate appropriate emergency response in the event of an accident. The MTB is unaware of any adverse experience as a result of these Canadian shipments. The MTB believes there is little difference between this situation and the use of a "Corrosive gas" placard with the appropriate UN number. Certainly, at any considerable distance it would be difficult to distinguish between the "Corrosive gas" and "Poison gas" placards, and response actions would be based on the UN number indicated on the placard. In addition, as noted by several commenters, it must be borne in mind that paragraph 5.38(1) of the TDG Regulations requires each road or rail vehicle used for the bulk transportation of a corrosive gas be marked on each side with the name of that gas in letters not less than 102 mm (4 inches) high. The value of this marking in aiding the identification of the contents in the case of an accident cannot be underestimated.

After consideration of the comments regarding recognition of the corrosive gas classification, and on the basis of the recent past experience with shipments made under the CTC classification of "Poison gas", the MTB believes that recognition of the corrosive gas class will not have a serious adverse effect on emergency response to accidents and this final rule permits shipments to enter the United States from Canada in conformance with the TDG Regulations concerning corrosive gases. In order to address the concerns expressed by several of the commenters, the MTB will undertake to distribute to each of the emergency response organizations to which the DOT Emergency Response Guidebook was sent, appropriate information concerning the gases classed as corrosive gases under the TDG Regulations including examples of the placard employed for such gases. The MTB believes that recognition of the corrosive gas class will not jeopardize emergency response actions, but will eliminate the costly burdens that would have been imposed if it were necessary to change labels and placards at the border. For the same reasons, and in response to a comment, the MTB has included a new paragraph (c) in § 171.12a that did not appear in the proposal, permitting the corrosive gas labels and placards (which are required for transport within Canada even for shipments originating in the United States) to be used in the United States for shipments destined for Canada provided an indication is included in the shipping papers that the placards/labels applied have been used for the purpose of transportation to Canada. As a result of the inclusion of this additional paragraph, proposed paragraphs (c), (d) and (e) are redesignated (d), (e) and (f), respectively.

The proposed paragraph (e) has been revised by adding a provision allowing the return of empty cargo tanks to Canada in conformance with the TDG Regulations. While the notice of proposed rulemaking proposed that, based on the existing provisions of § 173.8, this practice be permitted for empty rail tank cars, a similar provision for empty cargo tanks was inadvertently omitted.

Sections 172.401 and 172.502. Although not originally proposed, these sections have been amended to specifically provide that the display of any label or placard required by the TDG Regulations is not considered a prohibited display. The MTB believes this is necessary to permit the display of labels and placards required by the TDG Regulations, such as the label and placard specified for Class 9 materials, that may otherwise be considered to be a prohibited display within the United States. In this context it is noted that it is not required that a carrier maintain any display of labels or placards not required by the HMR. Therefore, the amendment of these sections will not impose any additional responsibility or burden on any person, but will permit labels and placards such as those required by the TDG Regulations for Class 9 materials to be applied in the United States when a shipment is destined for Canada.

Section 173.8. This section is being removed as proposed. There were no comments submitted regarding this proposed action.

Section 174.11. One commenter noted that as a result of the removal of § 173.8, it would be necessary to amend § 174.11 to reflect the addition of the new § 171.12a and the incorporation of the new TDG Regulations. The MTB concurs, and has amended the section accordingly.

Sections 174.59 and 177.823. One commenter requested that § 174.59 be amended to allow railroads to replace lost Corrosive gas placards with the appropriate placard required by the HMR on the basis that both railroad and emergency response personnel would be familiar with the placards required by the HMR and that it would be both unreasonable and unnecessary to expect United States' railroads to maintain supplies of the Canadian placards. The MTB agrees with the suggestion and has amended § 174.59 accordingly. A similar provision in § 177.823 has been similarly amended.

Section 171.12, 173.314, 176.11, 177.805, 179.105–1 and 179.106–1. References to § 173.8, which has now been removed, have been revised to refer to the new § 171.12a.

(50 F.R. 41516, October 11, 1985)

49 CFR Parts 171, 172, 173, 174, 175, 176, and 177
[Docket No. HM-191; Notice No. 84–4]
Classification of Detonating Board, and Packaging of Detonators

I. BACKGROUND

A. Cordeau Detonant Fuse

1. General. Currently, cordeau detonant fuse (detonating cord) is classed as a class C explosive. This classification includes manufactured articles, like detonating cord, which contain class A explosives, or class B explosives, or both, as components but in restricted quantities. The class C explosive hazard classification of detonating cord takes into account inherent properties, such as its resistance to heat, fire, impact, shock, pressure, and stray electric, currents, which minimize the potential for detonation. The class C explosive hazard class typically addresses materials which, if initiated, will cause little or no explosive destruction beyond the limits of the outside shipping container, and which will not cause similarly packaged materials in close proximity to mass detonate.

Charging with the explosive materials (usually pentaerythrite tetranitrate (PETN) or cyclotrimethylene-trinitramine (RDX)) is currently limited to 400 grains per linear foot. However, there are no limitations on: (1) The length of cord which may be shipped in a single package, or (2) the aggregate quantity of explosive material contained in a package. This results in potentially hazardous situations during transportation, since it is very predictable that the entire contents of a package of detonating cord will explode, if a detonation occurs inside the package. Detonation of one package of detonating cord is likely also to cause the mass explosion of similar packages in a shipment, as well as that of other class A or class B explosives or blasting agents which may be included in a shipment.

Although detonating cord has a demonstrated history of no exploding when subjected to normal conditions of transportation, the possible consequences of its exploding are considered by MTB to be so severe as to require shippers and carriers to take additional precautions which serve to minimize risks to life and property. Reclassification of detonating cord as a class A explosive would effectively assure that regulatory requirements applicable to that hazard class are sufficient to provide an adequate level of safety in transportation.

The MTB is not alone in this assessment. Although the type of detonating cord being produced today has been transported with relative safety for nearly 50 years, a petition for rulemaking filed by the IME, the safety association of the commercial explosives industry in the United States and Canada, whose members produce a major portion of the commercial explosives materials transported in the United States, suggests that the present classification of detonating cord as a class C explosive permits an excessive amount of the high explosive ingredient to be transported under rules which are principally concerned with the fire hazard of materials rather than the potential for exploding.

2. Reaction of Detonating Cord in a Transportation Incident and Testing for Sensitivity. The MTB has information on one serious incident involving detonating cord. The incident occurred on a pier in Brooklyn, N.Y. in 1956. It resulted in the death of 10 persons, injuries to another 247 persons, and property damage estimated at $10 million. An explosion occurred within 25 minutes of the sounding of the first alarm of a raging fire in a covered pier which reportedly contained 37,000 pounds of detonating cord. The blast destroyed a 10,000 square foot section of the pier,

caused major destruction of buildings and equipment in a 1000 food radius, and caused broken glass damage in a one mile radius.

Destructive testing of detonating cord demonstrates its relative insensitivity. A major test of this material performed at the Aberdeen Proving Grounds, Aberdeen, Md. in the 1950's is said to have demonstrated that detonating cord will burn rather than detonate when subjected to the influence of fire and reasonably confined conditions. Other tests involving a 10-pound weight dropped from a height of 16 feet also failed to detonate the material upon impact.

When detonating cord is tested against the criteria specified in § 173.53(c) it does demonstrate the characteristics of a class A explosive. The number 8 test blasting cap will consistently detonate samples of detonating cord. While it can be argued that other class C explosives will also detonate under such testing, detonating cord is probably the only material in that class which may pose a risk to detonation of other similar packages in close contact; thereby creating potentially serious threats to life and property.

3. Other Factors Influencing Reclassification. The MTB acknowledges the relatively good history of safe transportation for detonating cord over the nearly 50 years that it has been classified class C explosive. However, a reevaluation of the potential risk of this material resulted in MTB's decision to propose in this notice a change in its hazard class C to class A. Foremost, in this decision, aside from the threat of mass explosion, is the inadequacy of requirements pertaining to the communication of hazard warnings. Operational requirements pertaining to areas like loading and storage, routing, parking, and attendance and surveillance were also influential in this determination.

(a) Communication of Hazard. The DOT requirements pertaining to the communication of hazard warnings for detonating cord include: (1) Shipping papers which identify the proper shipping name, hazard class, and total quantity; (2) package markings which display the proper shipping name followed by the notation "Handle Carefully"; (3) labeling of the package with the EXPLOSIVE C label; and (4(placarding of transport vehicles with the "DANGEROUS" placard, under certain conditions. These warnings are intended to benefit the safety of operating personnel employed by shippers and carriers, emergency response personnel who may be called upon in the event of an incident, and the general public. There are, however, instances when the communications may be inadequate.

For one, it should be pointed out that when detonating cord is transported by a motor carrier and there is less than 1000 pounds of the material on the transport vehicle, a hazard warning placard is not required. Consequently, the required warnings are limited to the shipping paper, and packaging markings and labels, all of which may be inaccessible during a serious emergency. This may result in some persons believing that the vehicle does not contain any hazardous material. Others may approach the vehicle using standard emergency response procedures which generally do not consider the threat of mass explosion. In either event personnel would be exposed to a potential danger greater than they would, or should be required to, expect. MTB believes there should be an appropriate hazard warning on the transport vehicle to advise emergency response personnel of the risk presented by any significant quantity of detonating cord.

Secondly, since the DANGEROUS placard is a general purpose instrument to communicate basis hazard warnings, it cannot be solely relied upon to direct emergency service personnel to the most appropriate response for detonating cord. For instance, a motor vehicle carrying detonating cord and another hazardous material such as a poison B, a corrosive material, or a flammable solid, may be placarded DANGEROUS when less than 5000 pounds of either class has been loaded therein at one loading facility. The DANGEROUS placard, however, is not an appropriate hazard warning for materials which may mass explode.

(b) Operational Requirements. Since detonating cord is currently included in the class C explosive hazard class it is not subject to many requirements which are designed to assure a level of safety commensurate with the risk presented by class A explosives. While the long history of relatively safe transportation of detonating cord may indicate that operational controls currently applied are sufficient, MTB believes the margin of safety is simply too thin. Controls which apply to materials in the class A explosive hazard class are both reasonable and appropriate for significant quantities of detonating cord.

The following is a listing of some of the more significant requirements which would provide increased safety in transportation, which applied to detonating cord.

(1) Switching rules which prohibit the free movement of cars, and requirements for a buffer car when in a rail yard (§ 174.83).

(2) Placement rules for cars in rail yards and sidings (§ 174.85).

(3) Train placement rules (§ § 174.88 and 174.90).

(4) General requirements which prohibit transportation aboard aircraft (§ 172.101).

(5) Detailed requirements for loading and unloading class A explosives on vessels (§ 176.105).

(6) Attendance and surveillance of motor vehicles. (§ 397.5).

(7) Selection of routes (§ § 174.105, 175.320(b)(8), and 397.9).

4. Miscellaneous Proposals. In its petition for rulemaking, the IME suggests that shippers of detonating cord be permitted an exception to classify the material class C explosive whenever the net total explosive weight of detonat-

ing cord on a vehicle is less than fifty pounds. MTB believes an adequate level of safety in transportation would be realized when a small quantity of detonating cord is transported under rules applicable to class C explosives. However, the IME petition is considered unnecessarily broad. Consequently, the proposed rule puts certain limitations on the material, carriers, and modes of transportation to which the class C explosives classification would apply.

One variation from the IME petition is to use the gross weight of the package containing detonating cord rather than the net explosive weight. This is being done primarily after considering the release of energy that is possible with 50 pounds of a material like PETN, and to establish a practical basis for determination of compliance. To further provide for an acceptable level of safety, the explosive content of detonating cord shipped under the exception would be limited to 100 grains per linear foot.

Another variation from the IME petition is MTB's decision to limit the exception to shipments being transported by: (1) motor vehicles which are operated by private carriers, and (2) private vessels. All other carriers transporting detonating cord would be required to handle it as a class A explosive.

The IME petition also recommends that detonating cord be restricted from transportation with detonators, with certain exceptions. Specifically, the petition states:

(a) "Explosive A" detonating cord should be restricted from transportation with "Explosive C Detonators" except when the IME 22 Box is utilized.

(b) "Explosive C" detonating cord and "Explosive C Detonators" may be transported without restriction.

(c) Both "Explosive A" and "Explosive C" detonating cord should be restricted from transportation with "Explosive A Detonators."

Except for article (b), MTB agrees that the suggested restrictions are necessary. The presence of any detonators and detonating cord on the same motor vehicle is considered to be unacceptable if the IME Standard 22 container is not used. Appropriate changes are, therefore, proposed in the loading and storage charts to §§ 174.81, 176.83 and 177.848.

B. Detonators

1. General. Beginning January 1, 1985, all detonators and detonating primers, including those approved prior to January 1, 1980, must be prepared for shipment and transported in accordance with amendments to the Hazardous Materials Regulations (HMR) adopted under Docket HM-161 (44 FR 70721, December 10, 1979). One of the more significant changes adopted under that Docket is the criteria used in determining the hazard class of detonators and detonating primers.

Currently, certain blasting caps (detonators) approved for transportation prior to January 1, 1980 may be classed class A explosive or class C explosive depending on the actual number of devices. If there are 1000 or less, the hazard class is class C explosive. If there are more than 1000, the hazard class is class A explosive. It was determined in HM-161 that classification based on the number of devices is not an appropriate method. The final rule adopted in that Docket, therefore, requires that detonators be evaluated on their ability to undergo limited propagation. If detonators undergo limited propagation in the shipping package, the class is class C explosive. Limited propagation means that if one detonator near the center of the shipping package is exploded, the aggregate weight of explosives, excluding ignition and delay charges, in this and all additional detonators in the outside packaging that explode may not exceed 25 grams.

Detonators which mass detonate in the shipping package may not be classed as a class C explosive. Mass detonate means that more than 90 percent of the devices tested in a package explode practically simultaneously. This method of classification is considered more acceptable from the point of view of safety in transportation.

As shippers and carriers begin to comply voluntarily with the revised regulations prior to the January 1, 1985 mandatory compliance date, they realize that the long-standing authorization for transporting certain blasting caps with high explosives on the same motor vehicle is no longer permitted. This is a consequence of the IME Standard 22 being specifically addressed to the transportation of class C detonators. At the present time 1000 or less detonators may be carried in an IME Standard 22 container to compartment without an outside DOT specification wooden or fiberboard box. This currently permitted under terms of exemption numbers E-5243 and E-6984. However, the exemptions do not include those detonators which are electric blasting caps that are classed class A explosive.

On January 14, 1983, the IME filed a petition for rulemaking to amend the HMR in order to preserve an operation which has a demonstrated record of safety in transportation. In essence, the petition seeks to:

• Continue use of the IME No. 22 container or compartment as a method of transporting detonators with class A or B explosives in the same manner as has been in use since June 30, 1973.

• Apply that same method to the combined transportation of detonators and blasting agents.

• Make certain editorial corrections to the regulations.

2. Performance of IME Standard 22 Container in Tests and Accidents. Fire tests with the IME Standard 22 container demonstrate conclusively that the package remains intact following the detonation of many blasting caps.

In the 12 years since the IME Standard 22 container or compartment was first used, there are only three reported incidents that involve this container. In these three cases, each of which involved total destruction of the motor vehicle, it is believed that the IME Standard 22 container provided the level of protection for which it is designated. A brief description of each incident follows:

On April 5, 1979, the driver of a motor vehicle that was loaded with explosives and traveling on Route 52 in Keystone, West Virginia noticed a fire in the rear of the truck bed. He pulled the truck over to the side of the road and tried to extinguish the fire. When the driver realized the fire was out of control, he began warning people in the surrounding area. An explosion occurred shortly thereafter. The truck was carrying 1,000 pounds of dynamite (high explosive, class A explosive), and 500 blasting caps (class C explosive) in an IME Standard 22 container. A number of unexploded blasting caps, unburnt instruction sheets and pieces of the fiberboard box which held the caps were found in the area of the blast. It is believed that the initial detonation occurred in the dynamite outside the IME Standard 22 container.

On January 29, 1981, the driver of a motor vehicle that was loaded with explosives and traveling on Highway 53 near Shelbyville, Kentucky lost control of the vehicle and it overturned. The gas tank ruptured and the vehicle began to burn. The uninjured driver was able to get out of the vehicle and stop traffic. An explosion occurred shortly thereafter. The truck was carrying 8,500 pounds of ammonium nitrate-fuel oil mixture (blasting agent), seven pounds of explosive boosters (class A explosive), and 2,000 electric blasting caps (class A explosive) in an IME Standard 22 container. A number of unexploded blasting caps were found in the area of the blast. It is believed that the initial detonation occurred in one of the materials outside of the IME Standard 22 container.

On June 1, 1981, the driver of a motor vehicle that was loaded with explosives and traveling on Highway 211 north of Cheyenne, Wyoming noticed a fire in the rear of the vehicle. He stopped the vehicle and went for help. By the time emergency personnel arrived the vehicle was engulfed in flames. Because the vehicle was loaded with 1,950 pounds of a high explosive (class A explosive) and 78 electric blasting caps, the area was secured and the fire was allowed to burn itself out. There was no explosion. As a result of the fire, the vehicle and its load were completely destroyed. However, the blasting caps, which were in an IME 22 container, did not mass detonate. That IME 22 container, though fire damaged, was intact following the incident and did provide the protection for which it was designed.

3. Conclusion. Based on its assessment of the IME Standard 22 container, and the history of safe transportation under exemption E-6984, MTB is satisfied that an adequate level of safety would be maintained if the HMR were amended to permit certain blasting caps, class A explosives, to be transported in an IME Standard 22 container on the same motor vehicle with other high explosives. Suggested actions contained in the IME petition for rulemaking are proposed in this notice, though not in the same format.

II. REVIEW BY SECTIONS

General References to IME Standard 22 throughout the HMR address not only the manner of design and construction of the container or compartment but the permitted uses and requirements specified in that publication as well.

Section 171.7 would be revised to incorporate the most recent edition of the IME Standard 22. A copy is available for review in the docket file for this rulemaking action and is filed as part of the original document.

Section 172.101 would be amended by removing the proper shipping name "cordeau detonant fuse" and adding the proper shipping name "detonating cord."

Section 173.53 would be amended by adding a definition of detonating cord at paragraph (w).

Section 173.66 Paragraph (c) would be revised to permit the carriage of certain detonators (blasting caps) by motor vehicle without first being packed in an outside specification wooden or fiberboard box. Instead, the detonators would have to be transported in an IME Standard 22 container or compartment. The total number of detonators would be limited to 1,000 per motor vehicle. An additional requirement is proposed to prohibit materials from being loaded on top of the IME container or immediately outside the door of an IME compartment so that venting may occur as intended by the design.

Paragraph (e) would be amended by adding subparagraph (3) to permit use of an IME Standard 22 container or compartment as an alternative to the specification wooden and fiberboard boxes ordinarily required for class A detonators which conform to requirements pertaining to their design and limited explosive content.

Section 173.81 would be added to specify packaging requirements for detonating cord. The proposed packaging requirements are similar to those now specified in § 173.104 for cordeau detonant fuse.

Paragraph (c) would permit the classification of 50 pounds or less of detonating cord not exceeding 100 grains of explosive per linear foot as a class C explosive. That classification would be permitted when carriage is performed by a private carrier only.

It is recognized that paragraph (c), as proposed, is more restrictive than the rule change petitioned by the IME. However, this is in keeping with MTB's desire to limit the population at risk to the smallest possible number, and then further limiting that segment to those persons (i.e.,private carrier personnel) who are most aware of the hazards associated with the material, and who are trained to handle such a material. Interested persons who believe this proposal is too restrictive should explain fully in their comments why a less restrictive rule should be adopted. For example, it has been suggested that common carrier service may be essential to small businessses in delivering detonating cord in small lots to their customers. A number of common carriers specialize in transporting explosives while others have little or no experience in transporting explosives other than class C explosives. What distinction, if any, can MTB make by rule that would limit carriage to carriers qualified to handle explosives that have the potential to mass detonate? Data, such as a safety analysis, relevant shipping and accident experience, and cost estimates which may be evaluated by MTB should accompany comments filed in response to this proposed restriction.

Section 173.100(d) would be removed and reserved since detonating cord is proposed to be described in § 173.81.

Section 173.103 would be amended by adding a requirement in paragraph (c)(4) that each inside packaging containing detonators, class C explosives, must be marked "class C explosives". This marking is considered necessary since use of the IME Standard 22 container or compartment, and paragraph (c)(2), otherwise permit their transportation without first being packed in a specification wooden or fiberboard box. Also, since an IME Standard 22 container or compartment that is an integral part of the vehicle body, or is permanently attached to a motor vehicle, is not required to be marked or labeled, there is no communication of this hazard, except through the shipping paper.

Section 173.104 would be amended by removing references to the shipping description cordeau detonant fuse.

Section 174.81 would be amended by adding detonating cord to the list of materials identified in row b of the table (e.g. high explosives and propellant explosives, class A), and by removing cordeau detonant fuse from row 8.

The letter "X" would be added at the intersection of rows d and 10 to prohibit the rail transportation of detonators and blasting agents on the same transport vehicle.

Row 7a, headed "Detonators, detonating primers", would be added to the group of class C explosives. These articles would be prohibited from transportation with other explosives, blasting agents and poisonous gases; the same as class A detonators and detonating primers.

The heading of row 10 would be revised to include ammonium nitrate-fuel oil mixtures.

Footnote 1 and references to it within the table would be removed since class C detonators and detonating primers are proposed to be separately identified in row 7a.

Reference to footnote 2 in row 13 would be added at row e since its applicability is also to ammunition for cannon.

Footnote 5 would be revised to remove references to blasting agents and ammonium nitrate-fuel oil mixtures. These materials were included with oxidizers (row 12) prior to establishment of the blasting agents hazard class. However, as they are now included with blasting agents in row 10 the footnote's reference should be limited to ammonium nitrate, fertilizer grade.

Section 175.320 would be amended by changing a description in the table in paragraph (a) from cordeau detonant fuse to detonating cord.

Section 176.83 would be amended by adding detonating cord, class A explosive to the list of materials identified in row 2 of the table (e.g. high explosives and propellant explosives class A), and by removing cordeau detonant fuse from row 8.

Detonators and detonating primers (class C explosives) would be added to row 14a. These articles would be prohibited from loading and stowage with other explosives and blasting agents; the same as class A detonators and detonating primers.

All detonators would be prohibited from loading and stowage with blasting agents.

Detonating fuses would be prohibited from loading and stowage with blasting agents.

Detonating cord would be prohibited from loading and stowage with initiating explosives, all detonators and detonating primers, detonating fuses, and special fireworks and railway torpedoes.

Section 177.835(g) would be revised to add blasting agents and detonating cord, Class C explosive as other materials that are prohibited from being transported on the same motor vehicle with detonators or detonating primers. The exceptions in paragraphs (g)(1) and (g)(2) would, however, permit blasting agents to be transported on the same motor vehicle with detonators under certain conditions.

Paragraph (g) would be revised editorially to clarify that detonators may be transported on the same motor vehicle with detonating primers.

Section 177.848 would be amended by adding detonating cord to the list of material identified in row b of the table (e.g. high explosives and propellant explosives, class A), and by removing cordeau detonant fuse from row 8.

Row 7a, headed "Detonators, detonating primers" would be added to the group of class C explosives. These articles would be prohibited from transportation with other explosives, blasting agents and poisonous gases; the same as class A detonators and detonating primers. An exception is made for shipments transported under provisions of § 177.835.

Row 8a, headed "Detonating cord", would be added to the group of class C explosives. Detonating cord would be prohibited from transportation with initiating explosives, all detonators and detonating primers, detonating fuzes, and special fireworks and railway torpedoes. An exception is made for shipments transported under provisions of §§ 173.81(c) or 177.835.

The heading of row 10 would be revised to include ammonium nitrate-fuel oil mixtures.

Reference to footnote 1 at coordinates d-9, d-11, d-12, d-13 and d-14 would be removed since its applicability is to class C detonators.

All detonators would be prohibited from transportation with blasting agents. An exception is made for shipments transported under provisions of § 177.835.

Footnote 1 would be revised to apply to class A as well as class C detonators.

Reference to footnote 2 in row 13 would be changed from row d to row e since its applicability is to ammunition for cannon and not detonators.

Footnote 5 would be revised to remove references to blasting agents and ammonium nitrate-fuel oil mixtures. These materials were included with oxidizers (row 12) prior to establishment of the blasting agents hazard class. However, as they are now included with blasting agents in row 10 the footnote's reference should be limited to ammonium nitrate, fertilizer grade.

(49 F.R. 20873, May 17, 1984)

49 CFR Parts 171, 172, 173, 174, 175, 176 and 177
[Docket No. HM-191; Amdt. Nos. 171–82, 172–95, 173–182, 174–46, 175–33, 176–20, 177–64]
Classification of Detonating Cord, and Packaging of Detonators

1. BACKGROUND

On May 17, 1984, MTB published a notice of proposed rulemaking (Notice 84-4) in the Federal Register (49 FR 20873). That notice proposed the reclassification of cordeau detonant fuse from class C explosive to class A explosive, and a change in its proper shipping name to "cord, detonating flexible". In addition, the notice proposed changes to permit carriage of detonators, class A explosive, within an IME standard 22 container under rules which, until now, applied to detonators, class C explosive, only. This final rule contains amendments to the Hazardous Materials Regulations (HMR) based on proposals in Notice 84–4 and the merits of comments filed in response to that notice. Interested persons should refer to Notice 84–4 for detailed background information.

In response to Notice 84–4, MTB received 109 written comments. The respondents represent a variety of interests. There are 97 users of explosives, of which 66 are engaged in the oil well service industry. Seven commenters are dealers of explosive materials and two commenters are manufacturers of explosives. Comments were also received from the IME, the U.S. Army and the International Fire Service Training Association (IFSTA).

Each commenter accepted (though some did so reluctantly) the proposal to reclassify large quantities of detonating cord as a class A explosive. In addition, each commenter, with the exception of IFSTA, favored the proposed classification for small quantities of detonating cord as a class C explosive. However, each commenter, save one, opposed the proposal that would restrict the transportation of detonating cord classed as a class C explosive to private carriage.

With respect to that part of the proposal addressing the packaging of detonators, the commenters generally favored the proposed rules. However, three commenters requested that detonators packaged for transportation prior to January 1, 1985, be authorized for shipment in accordance with the regulations in effect on October 31, 1979. That authorization is requested to extend through December 31, 1985.

A discussion of significant comments and amendments adopted in this final rule follows.

A. Detonating Cord

1. Classification. A majority of commenters took exception to MTB's assumption that most packages of detonating cord will not behave in a manner characteristic of class C explosives. Several commenters suggested MTB develop evidence which supports such speculation before making extensive changes in the regulation of detonating cord.

While MTB does not have hard data to support this reclassification, it notes the petition initiating this rulemaking docket was filed by the IME whose member companies produce a major portion of the commercial explosives materials transported in the United States. Their expertise in this area is unquestioned and it is irresponsible for MTB to ignore or unduly delay that request for reclassification.

Though some commenters do not agree that detonating cord ia a class A explosive, they reluctantly accept that classification. However, those commenters qualified their acceptance of the class A explosive hazard class on the condition that transportation of relatively small quantities of detonating cord may continue as class C explosives. This is of particular concern to wireline operators in the oil well service industry who recommended an increase in the proposed maximum permitted gross weight from 50 pounds to 100 pounds.

Upon reconsideration of the limited risk posed by relatively small quantities of detonating cord, and in response to comments received, MTB is raising the maximum permitted gross weight of packages of detonating cord that may be offered for transportation or transported as class C explosives to 100 pounds.

Other commenters believe the classification of detonating cord must be determined on the basis of its performance under test conditions, and not by the assignment of a weight limit. The following comments from the Department of Army are very specific in that regard.

> As pointed out in 49 CFR the most important consideration in determining DOT class is the risk posed to detonation of other similar packages in close contact, if one of the packages is detonated. The Army is presently testing some current packages of detonating cord, in accordance with the requirements of TB 700–2/NAVSEAINST 8020.8/TO 11A-1-47/DLAR 8220.1, to determine their proper DOT class. It is possible that these tests will determine that our current detonating cord should be DOT class B or C. In addition, new packages containing less cord per package have been proposed and will be tested when design is complete. It is our design intention that these new packages will not pose an explosive propagation hazard and will qualify as DOT class C.

> .   .   .   .   .

> DOT classes should be based on the actual hazard that each package (defined as packaging materials and contents per 49 CFR 171.8) poses. Different quantities of the same item packaged in different ways may present different hazards. In the case of detonating cord, a DOT class A assignment could be presumed and assignment of DOT class B or class C specifically permitted for DOD when test results justify such assignment to specific packaged configurations. There is ample precedent for the assignment of different DOT classes based on different hazards for similar items.

MTB is very much interested in the results of this proposed testing and encourages the design of packages that do not pose an explosive propagation hazard. The development of such packages is beneficial to safety in transportation. In addition, shippers and consignees of detonating cord may also benefit by the reduced transportation charges normally assessed for class C explosives, as compared to transportation charges for a comparable quantity of class A explosives. The United Nations (U.N.) Recommendations currently provide for classification of detonating cord in class 1.1D and 1.4D (essentially equivalent to DOT's class A and C explosive hazard classes, respectively) and the design of packages that do not mass explode allows MTB to maintain a consistency with the U.N.

Exception is taken to DOD's suggestion that some packages of detonating cord may be classified as a class B explosive. The class B explosive hazard class is peculiar to materials which, in general, function by rapid combustion rather than detonation. The explosive compounds used in detonating cord will detonate. Thus, the class B explosive hazard class is not appropriate.

Considering some shippers are confident they may design packages of detonating cord that react to adverse environments in a manner characteristic of the class C explosive hazard class, this final rule adopts a new entry in the Hazardous Materials Table for "Cord, detonating flexible, class C explosive". The safety benefits and economic incentives derived from having a set of regulations that permits the class C explosive hazard class as a possible alternative to the class A explosive hazard class will, hopefully, stimulate shippers to further reduce threats to health and property.

2. Modal restrictions. MTB proposed that detonating cord not exceeding 100 grains per linear foot, and not exceeding 50 pounds may be transported only by a private carrier, operating in the highway or water mode, under requirements typically applicable to the class C explosive hazard class. In all other cases, the notice proposed that detonating cord shall be offered for transportation and transported as a class A explosive. There was much resistance to that proposal.

For the most part, opposition came from shippers and users of detonating cord. One large shipper states "(b)illions of feet of detonating cord have been transported as a class C explosive by for-hire carriers in the last fifty years and, to our knowledge, there have been no transportation incidents that suggest that the transportation of detonating cord, class C explosive, should be limited to private motor carriers." The commenter went on to recommend "... the maximum quantity offered for shipment as a class C explosive be changed to 100 lbs. gross weight. This is consistent with the original IME proposal since, by weight, the explosive content of packaged detonating cord is slightly less than 50 percent." This call for a 100 pound gross weight limitation was echoed by 86 commenters. A single commenter suggested "(a) more realistic approach would be to allow 250 lbs. of detonating cord to be shipped class C by common carrier."

The IME reiterated this good safety record for detonating cord, but then provided MTB with information on a previously unreported transportation incident near Colstrip, Montana in which 100,000 feet of cord detonated during a fire in the cargo compartment of a motor vehicle operated by a private carrier. That incident occurred in 1974. It resulted in a total loss of the vehicle and contents, but there were no deaths or injuries.

Each of the commenters from the oil and gas well perforating industry complained of the difficulty and expense they might incur if suppliers are unable to ship detonating cord as a class C explosive via common carrier. Tariff restrictions imposed by common carriers are claimed to make the cost of transportation of small quantities of class A explosives prohibitive. Also, the small user is frequently several hundred miles distant from the supplier and the costs associated with private carriage thereby become prohibitive.

One perforating company with operations throughout the world filed the following comments:

Many of these wells are located in parts of the world where storing large quantities of explosives is very difficult and sometimes impossible. It is imperative, therefore, that these locations have the ability to transport small quantities of detonating cord by air. Approximately 90 percent of all our explosives shipments are sent by cargo or passenger aircraft.

The present proposal to classify detonating cord as a class A explosive would prohibit us from transporting it on cargo and passenger aircraft. This proposal, therefore, would greatly impact the well service industry's ability to provide timely service to the oil industry.

The commenters have shipping statistics which show detonating cord may be safely transported by private and for-hire carriers. However, they do not go on to explain fully why a rule less restrictive than proposed should be adopted for each class of for-hire carriers. In the absence of data which shows conclusively that small quantities of detonating cord, when detonated, will not significantly affect transportation safety; MTB believes it must restrict transportation of detonating cord to modes limiting the population at risk to the smallest possible number and, to the extent possible, to those persons who regularly accept the risks inherent with the transportation of hazardous materials.

In consideration of comments received, MTB is adopting amendments which are less restrictive than proposed. Specifically, the gross weight of packages containing detonating cord, class A explosive which may be offered for transportation or transported as class C explosives is raised to 100 pounds. In addition, carriage is permitted via each mode of transportation. However, detonating cord, offered for transportation or transported under provisions of § 173.81(c) may not be carried aboard a passenger-carrying aircraft, except for a passenger-carrying aircraft used during that flight to serve a remote site, like an off-shore drilling unit, by shuttling operating personnel and carrying their necessary supplies and equipment.

3. Miscellaneous comments. Several commenters addressed problems associated with the transportation by vessel of class A explosives to locations such as Alaska, Hawaii and Guam. They correctly indicate that class A explosives are prohibited from many ports, and ports which do handle class A explosives impose charges which greatly exceed the charges in effect for class C detonating cord. The problems addressed by these commenters are significantly reduced by MTB's increasing the maximum permitted weight from 50 pounds to 100 pounds, and permitting carriage by common carriers.

A commenter in the oil well service industry complained the reclassification of detonating cord, as a class A explosive, would adversely affect operations which require use of jet perforating guns. The commenter states:

Many oil field applications of detonating cord also involve the use of other class C materials. For example, oil and gas wells are almost universally completed using shaped charges conveyed in a hollow steel cylinder called a perforating gun. These charges are normally ballistically linked using a length of detonating cord. As both materials are currently classified as class C explosives, they can be assembled at a central location and transported together to remote well locations simply and economically. Reclassification of detonating cord as a class A explosive would require separate transportation of these materials with assembly of the perforating gun at the remote well site. This would thus substantially increase the cost of providing this service while compounding operational difficulties.

The commenter is mistaken. Since the jet perforating gun is a separately identified explosive article, the shipper must offer it for transportation and transport it under rules applicable to the completed article, and not on the bases of regulations applicable to each of its components. The HMR currently provide for classification of jet perforating guns as a class A explosive or a class C explosive, depending on the total amount of explosive contents of the guns being transported per motor vehicle. Thus, this rule does not affect the classification of a perforating gun.

B. Detonators

1. Packaging in IME Standard 22 Container. Several commenters took exception to current and proposed requirements associated with the transportation of detonators in an IME 22 container. One comment from the Department of the Army reads as follows:

The proposal to amend 49 CFR 173.103 to require inside containers of DOT class C detonators to be so marked in order to permit use of the IME Standard 22 container creates an untenable situation because 173.103 and 173.66 combine to base DOT class on the efficiency of the total packaged configuration, including inside, intermediate, and outside packagings. If DOT class C is determined, based on inside, intermediate, and outside packaging, and the intermediate and outside packaging is then removed, there is doubt about the propriety of marking "class C explosive" on what has thereby become a DOT class A package. We would recommend that some other administrative method be selected to achieve the intended result since use of methods such as that proposed are usually reasons for DOT to impose sanctions on shippers.

Intermediate and outside packagings do influence the classification of detonators. However, as indicated in Notice 84-4, more than 12 years experience with the IME-22 container in the transportation of detonators indicates its performance in accident and fire situations is excellent. Furthermore, during that period, use of the IME-22 container was extended to detonators meeting criteria for a class A explosive under recently revised rules. Based on that experience, MTB is confident requirements pertaining to use of the IME-22 container provide an adequate level of safety.

Although the commenter states "(w)e would recommend that some other administrative method be selected to achieve the intended result . . .", MTB notes no such recommendation was provided. The record does not, therefore, support the commenter's suggestion that the regulations be revised to prohibit transportation of detonators which are not packaged in a intermediate or specification outside packaging.

Twenty-five commenters objected to the proposed requirement that an IME-22 box shall be used when detonators and detonating cord are transported on the same motor vehicle. The following comment is representative of this group.

I disagree with DOT's position that . . . restricting the transportation of class C detonating cord and class C detonators from carriage on the same vehicle unless the detonators are packed in an IME-22 box would not cause "significant impact on a substantial number of small entities." My company is definitely a small business entity and this proposal would severely limit my capability to compete in oil field wireline service industry.

. . . . .

I believe it is very important to small wireline businesses that you continue to allow . . . properly packaged class C detonators (packaged in accordance with 49 CFR 173.103(d)(1) and 173.103(d)(2), to be transported on the same vehicle with class C detonating cord. If you do so, I'll be able to remain competitive in my business and you will have performed your duty to insure the safe transportation of detonating cord and the safety of the general public.

These commenters contend an equivalent level of safety may be achieved by using packages of detonators suitable for transportation aboard passenger-carrying aircraft. Packages conforming to § 173.103(d) may not exceed 50-pounds gross weight. In addition, those packages must be so designed that if one device near the center of a package is detonated, no other device in the package will detonate and there will be no communication of detonation from one package to another. There is, then, strong bases for the packaging option suggested by these commenters.

In consideration of the above, § 177.835(g) is amended to permit transportation of detonators, class C explosive, packaged to conform with § 173.103(d), on the same motor vehicle with detonating cord, class C explosive.

2. Detonators packaged prior to January 1, 1985. Several commenters took note of the impending date for mandatory compliance with requirements promulgated under Docket HM-161 (44 FR 70721, December 10, 1979). The following chronology tracks events which result in a problem for these commenters.

• HM-161 adopted a rule which did away with the classification of blasting caps on the basis of quantity (i.e., more than 1000 are class A and 1000 or less are class C).

• HM-161 adopted a rule which determines hazard class on basis of limited propagation (class C), explosion of more than 25 grams of the aggregate gross weight of explosives, excluding ignition and delay charges, in the outside packaging (class A) and mass detonation (class A).

• HM-161 provides in § 173.66(G) a transition period of 5 years for shippers to have their blasting caps approved as class C explosives. (Jan. 1, 1980–Dec. 31, 1984).

• To the surprise of most shippers, there is difficulty developing packages that meet criteria for class C explosive for a majority of their detonators. Therefore, shippers continue to classify on the basis of quantity.

• A principle reason that shippers continue to mark and label their packages "blasting caps, class C explosive' is the limitation in the 1979 edition of the IME Standard 22 permitting class C blasting caps only.

• IME filed petition for rulemaking P-891 on January 14, 1983. That petition seeks to also permit transportation of class A detonators in an IME Standard 22 container.

The IME submitted the following comment:

In order to preserve an operation which has a demonstrated record of safety in transportation, the explosives industry awaited action on its petitions which are covered in HM-191 to clarify HM-161. Therefore, packaging certain blasting caps as detonators was delayed.

There will be a substantial inventory of blasting caps in the field as of January 1, 1985. In order to deplete this inventory, IME respectfully requests that material packaged prior to January 1, 1985 continue to be shipped under existing markings and regulations until January 1, 1966 at which time all remaining inventory will be relabeled.

The Department of the Army commented by stating:

The proposal would require reaccomplishing standard emergency resupply packages of munitions items, which are configured to support a declared national emergency. Repackaging and reworking present inventories would increase packaging and transportation costs. . . .

. . . . .

Recommend a grandfather clause be incorporated to allow Department of Defense to package, mark, and ship present stocks of this material in accordance with the regulations prior to implementation of this ruling. This will allow Department of Defense to deplete our inventory without costly repackaging and remarking.

While a five year transition period provided shippers adequate time to reclassify packages of detonators, they were hampered by the restriction in IME Standard 22 which limits use of that container to blasting caps that are class C explosives. With adoption in this final rule of a revised edition of the IME Standard 22 permitting class A detonators also, there is no longer a reason to mark packages of detonators "blasting caps", class C explosive. However, as adoption of that revised edition of IME Standard 22 comes at the eleventh hour, MTB recognizes a considerable number of packages marked and classed under the grandfathered rules are stored in explosives magazines. Considering that blasting caps have been safely transported under that description and class for well over 50 years, MTB believes benefits gained by remarking the packages are disproportionate to costs. Section 173.66(g), therefore, is revised to permit the transportation of detonators, packaged prior to January 1, 1985, in accordance with regulations in effect on October 31, 1979. This provision terminates on December 31, 1985. Each package offered for transportation after December 31, 1985 must conform with rules in effect at the time of shipment.

3. IME Standard 22. For consistency with amendments proposed in Notice No. 84–4, the IME submitted a later revision to their Standard Library Publication (SLP) No. 22 which specifies conditions that must be met before detonating cord, class C explosive, may be transported on the same motor vehicle with detonators. Essentially, the revision permits transportation of class C detonating cord and detonators in a manner already in effect for high explosives and detonators. It is agreed the revision is necessary. Therefore, MTB is incorporating by reference IME SLP No 22 revised as of January 1, 1985.

II. REVIEW BY SECTIONS

General References to IME Standard 22 throughout the HMR address not only the manner of design and construction of the container or compartment but the permitted uses and requirements specified in that publication as well.

Section 171.7 is revised to incorporate the January 1, 1985 edition of IME Standard 22. A copy is available for review in the docket file for this rulemaking action.

Section 172.101 is amended by removing the proper shipping name "cordeau detonant fuse" and adding two entries with the proper shipping name "Cord, detonating flexible" (class A and class C).

Section 173.53 is amended by adding a definition of detonating cord at paragraph (w).

Section 173.66 Paragraph (c) is revised to permit the carriage by motor vehicle of certain detonators (blasting caps) that are not packed in an outside specification wooden or fiberboard box. Instead, the detonators shall be transported in an IME Standard 22 container or compartment.

Paragraph (e) is amended by adding subparagraph (3) to permit use of an IME Standard 22 container or compartment as an alternative to specification wooden and fiberboard boxes ordinarily required for class A detonators.

Paragraph (g) is amended by extending by one year the period during which detonators may be transported under rules in effect on October 31, 1979.

Section 173.81 is added to specify packaging requirements for detonating cord, class A explosive. The packaging requirements are similar to those specified in §173.104 for class C detonating cord.

Paragraph (c) permits transportation of 100 pounds or less of detonating cord not exceeding 100 grains of explosive per linear foot as a class C explosive.

Section 173.100(d) is revised to redefine detonating cord. The expanded definition specifies the approving agency shall assure that a detonation in one package will not communicate detonation to adjacent packages.

Section 173.103 is amended by adding a requirement in paragraph (c)(4) that each inside packaging containing detonators, class C explosives, shall be marked "class C explosives". The Department of the Army did not concur in this proposal. However, the commenter did not provide support for its nonconcurrence.

The requirement that inside packagings for detonators be marked "class C explosive" is essential to safety in transportation. Absent this marking requirement, intermediate and end-use shippers could mistakenly remove unmarked inner packages of class A detonators containing an excessive amount of explosive material and transport them in an IME-22 container as a class C explosive. As safety requirements pertaining to transportation of class A explosives are more stringent, the regulations must consider the worst possible case. Consequently, the shipper may take advantage of the less stringent regulations pertaining to class C explosives, but only upon a positive endorsement that the detonators otherwise qualify as class C explosives.

Twenty-six commenters requested that MTB permit use of packages which conform to requirements of §173.103(d), as an alternative to the IME-22 container. Thus, paragraph (d) is amended by adding a requirement that each package of detonators suitable for transportation aboard a passenger-carrying aircraft shall have a marking which assures motor carrier personnel those detonators may be transported with hazardous materials normally requiring segregation or separation. The required marking states "This package conforms to conditions and limitations specified in 49 CFR 173.103(d)." Currently, the air carrier's need to know this information is satisfied by the required shipping paper entry "Cargo aircraft only". However, as the package now has special significance in the highway mode, motor carriers must see an endorsement which assures that otherwise incompatible materials are not transported on the same vehicle.

Section 173.104 is revised to give formal notice that each outstanding approval for cordeau detonant fuse, class C explosive is amended by changing the hazard class from class C explosive to class A explosive, and by changing the proper shipping name from cordeau detonant fuse to cord, detonating flexible. To facilitate an orderly transition, a period of approximately 18 months is granted during which detonating cord, for which an approval was issued prior to January 1, 1985, may be transported subject to conditions of the approval and in accordance with requirements of the HMR in effect on December 31, 1984.

Section 174.81 is amended by: adding detonating cord to the list of materials identified in row b of the table (e.g. high explosives and propellant explosives, class A), removing cordeau detonant fuze from row 8, and adding detonating cord at row 8a. Detonating cord (class A and C) is prohibited from transportation in the same transport vehicle with initiating explosives, all detonators and detonating primers, detonating fuzes, and special fireworks and railway torpedoes.

The letter "X" is added at the intersection of rows d and 10 to prohibit the transportation of detonators and blasting agents on the same transport vehicle.

Row 7a, headed "Detonators, detonating primers", is added to the group of class C explosives. These articles are prohibited from transportation with certain other explosives, blasting agents and poisonous gases.

Row 8a, headed "Detonating cord", is added to the group of class C explosives. Detonating cord is prohibited from transportation with initiating explosives, detonators and detonating primers, detonating fuzes, and special fireworks and railway torpedoes.

the heading of row 10 is revised to include ammonium nitrate-fuel oil mixtures.

Footnote 1 and references to it within the table are removed since class C detonators and detonating primers are now separately identified in row 7a.

Reference to footnote 2 is added at the intersection of rows 13 and e since its applicability is also to ammunition for cannon.

Footnote 5 is revised to remove references to blasting agents and ammonium nitrate-fuel mixtures. Those materials were included with oxidizers (row 12) prior to establishment of the blasting agents hazard class. However, as they are now included with blasting agents in row 10 the footnote's reference is limited to ammonium nitrate, fertilizer grade.

Section 175.320 is amended by changing a description in the table in paragraph (a) from cordeau detonant fuse to detonating cord. This specific authorization to transport detonating cord a cargo only aircraft supersedes the general restriction in column (6)(b) of the Hazardous Materials Table.

Section 176.83 is amended at Table I by: adding detonating cord, class A explosive to the list of materials identified in row 2 of the table (e.g. high explosives and propellant explosives class A); removing cordeau detonant fuze from row 15; and adding detonating cord at row 15a. Detonating cord (class A and C) is prohibited from transportation in the same hold or compartment with initiating explosives, all detonators and detonating primers, detonating fuzes, and special fireworks and railway torpedoes.

Detonators and detonating primers (class C explosives) are added at row 14a. These articles are prohibited from loading and stowage with other explosives and blasting agents; the same as class A detonators and detonating primers.

All detonators are prohibited from loading and stowage with blasting agents.

Detonating fuzes are prohibited from loading and stowage with blasting agents.

Section 177.835(g) is revised to add blasting agents and detonating cord, Class C explosive as materials that may not be transported on the same motor vehicle with detonators or detonating primers. However, exceptions in paragraph (g) permit transportation of detonators and blasting agents on the same motor vehicle. Paragraph (g) is revised to permit transportation of detonators packaged to conform with requirements specified in § 173.103(d) with detonating cord, class C explosives.

Paragraph (g) is further revised to clarify that detonators may be transported on the same motor vehicle with detonating primers.

Section 177.848 is amended by adding detonating cord to the list of materials identified in row b of the table (e.g. high explosives and propellant explosives, class A), and by removing cordeau detonant fuze from row 8.

Row 7a, headed "Detonators, detonating primers" is added to the group of class C explosives. These articles are prohibited from transportation with other explosives and blasting agents; generally the same as class A detonators and detonating primers. An exception is made for shipments transported under provisions of § 177.835(g).

Row 8a, headed "Detonating cord", is added to the group of class C explosives. Detonating cord is prohibited from transportation with initiating explosives, all detonators and detonating primers, detonating fuzes, and special fireworks and railway torpedoes. An exception is made for shipments transported under provisions of §§ 173.81(c) or 177.835(g).

The heading of row 10 is revised to include ammonium nitrate-fuel oil mixtures.

Reference to footnote 1 at coordinates d-9, d-11, d-12, d-13 and d-14 is removed since requirements applicable to class C detonators are now specified in row 7a.

Footnote 1 is revised to apply to class A detonators, and class C detonators transported with class C detonating cord.

Reference to footnote 2 in row 13 is changed from row d to row e since its applicability is to ammunition for cannon and not detonators.

Footnote 5 is revised to remove references to blasting agents and ammonium nitrate-fuel oil mixtures. These materials were included with oxidizers (row 12) prior to establishment of the blasting agents hazard class. However, as they are now included with blasting agents in row 10, the footnote's reference is limited to ammonium nitrate, fertilizer grade.

(50 F.R. 798, January 7, 1985)

49 CFR Part 175
[Docket No. HM-192]
Quantity Limitations Aboard Aircraft

The provisions of 49 CFR 175.75(a)(2) impose a limitation of fifty pounds net weight on the quantity of hazardous materials, permitted to be carried aboard a passenger aircraft, that may be carried in an inaccessible manner aboard an aircraft. Historically, the intent of this provision was to preclude large quantities of hazardous materials being aggregated in any one stowage location aboard an aircraft, thereby reducing the possibility of an uncontrollable event should an incident occur.

On December 6, 1982, the MTB published a notice of proposed rulemaking in the FEDERAL REGISTER under Docket No. HM-184 (47 FR 33295) which requested public comment on the need to amend the HMR in order to take account of the International Civil Aviation Organization's (ICAO's) Technical Instructions for the Safe Transport of Dangerous Goods by Air (ICAO Technical Instructions), which were to become effective on January 1, 1983. Although this notice did not propose an amendment to § 175.75(a)(2), five commenters proposed the deletion of this paragraph on the basis that no corresponding provision existed in the ICAO Technical Instructions. Because no change was proposed to this paragraph in the notice, the MTB considered the deletion of this paragraph to be outside of the scope of that rulemaking and the paragraph was, therefore, retained. Nevertheless, the MTB indicated that there may be merit in considering the deletion or amendment of this limitation provided evidence supporting such action is submitted and that there would be full public participation in a rulemaking proceeding. Consequently, petitions to delete or amend § 175.75(a)(2), with full supporting information, were invited for consideration in a separate rulemaking action.

On April 15, 1983, Japan Air Lines Company, LTD., (JAL) submitted a petition for rulemaking which requested the deletion of § 175.75(a)(2) contending that the quantity limitation was arbitrary, unjustifiable and inconsistent with other provisions of Part 175 as well as inconsistent with the ICAO Technical Instructions. In order to afford the public ample opportunity to consider the merits of such action, the MTB is providing in this publication the following reproduction of the salient points of the JAL petition:

"5. The 50 pound restriction prescribed in 14 CFR 175.75(a)(2) (sic) is arbitrary inasmuch as no justification has ever been articulated in support of this limitation. Significantly, no such "per aircraft" weight limitation has ever been imposed with respect to "Other Regulated Material" (ORM) as defined in 49 CFR 173.500, including consumer commodities containing hazardous materials otherwise subject to regulation, which are classed as ORM-D items. It seems patently inconsistent and unsupportable to dictate that an aircraft may carry no more than 50 pounds of a given material packaged and prepared for transportation in accordance with the strict requirements of Part 175 while permitting the same aircraft to carry an unlimited quantity of that same material contained in "consumer commodities" and probably not packaged as securely as the non-ORM items. Moreover, no reasoning is given in support of the determination to permit no more than 50 pounds of hazardous materials (and 150 pounds of non-flammable compressed gas) in an inaccessible cargo compartment, while an unlimited quantity of hazardous material may be carried in accessible cargo compartments.

6. Another inconsistency results from the fact that the quantity limitation applies only to passenger-carrying aircraft. Shipments of hazardous material that are acceptable for carriage on passenger aircraft would be subject, even without these per-aircraft overall weight limitations, to much more stringent individual quantity and packaging requirements than those applicable to hazardous materials transported on cargo-only aircraft. It seems rather inconsistent, therefore, to impose these per-aircaft limitations on materials already packaged in accordance with the rigid MTB standards for passenger-carrying aircraft, while not applying any such per-aircraft limitations to goods carried on cargo-only aircraft which need not be packaged in accordance with the more stringent standards.

7. Furthermore, the ICAO Technical Instructions for the Safe Transport of Dangerous Goods by Air, which has received virtually worldwide acceptance and application, contains no such per aircraft limitation, and to our knowledge there has never been any incident attributable to the transportation in a single aircraft of quantities of hazardous materials in excess of the limitation prescribed in 49 CFR 175.75(a)(2). Although the U.S. is not obliged to replicate the ICAO Standards in its own regulations, the benefits of worldwide uniformity were recognized, and indeed, the ICAO Technical instructions were in large measure adopted, in the final rule issued in Docket HM-184, 47 FR 54817, of 6 December 1982. In keeping with the philosophy of worldwide uniformity and consistency, JAL submits that it is appropriate for the U.S. regulations to be brought into conformity with the ICAO technical instruction in this regard. Indeed, teh current lack of uniformit may increase the risks incident to the through transportation of hazardous commodities. Since aircraft operating between countries that adhere to the ICAO technical instructions may well carry quantities of hazardous materials in excess of the MTB limitations, it becomes necessary to off-load such excess materials at an enroute station before the aircraft departs for the United States. Aside from the dangers incident to the additional handling of the hazardous materials in the off-loading and re-loading process, backlogs of hazardous materials are frequently caused to accumulate, which can greatly increase the risk of a catastrophic accident occurring, particularly in the many parts of the world where the climate is exceptionally hot and/or humid, and the warehouse facilities are rather poor. If U.S. regulations were brought into conformity with the ICAO technical instructions in this regard, the necessity of off-loading and storing excess hazardous cargo at these intermediate points, and the unnecessary additional safety risks attendant to these procedures, could be avoided.

8. In summary, the per-aircraft quantity limitations presently set forth in 49 CFR 175.75(a)(2) are arbitrary, inconsistent, and may actually serve to undermine the overall objective of the MTB scheme of regulations, which is to ensure maximum safety in the transport of hazardous goods by air. The promulgation of any such standards should be done only on the basis of a comprehensive analysis of the justification for, and broader ramifications of, such action, and can most effectively be undertaken through close coordination with ICAO."

Although the MTB does not necessarily agree with the statements made in the JAL petition, the substance of the petition is, without question, of broad and general applicability and consequently of interest to many parties. For this reason, comments are solicited on the amendments requested in the JAL petition, on the evidence offered in support of that petition and on the following questions:

1. What has been the transportation experience in areas outside of the United States where no corresponding aircraft quantity limitations are imposed?

2. What would be the safety implications, if any, if the JAL petition were granted?

3. What would be the economic benefits and consequences associated with adoption of the amendment proposed by JAL?

4. If instead of removing § 175.75(a)(2), certian classes (e.g., Poison B, liquids or solids) or sub-classes (e.g., Flammable liquids with a flashpoint above 73° F and no subsidiary risks) of hazardous materials were to be excepted from these quantity limitations, what hazard classes or sub-classes could be safely excepted and why?

5. If instead of removing 175.75(a)(2), the 50 pound limitation were replaced by a higher quantity limit, what quantity limit would be practicable and why?

(49 F.R. 13717, April 6, 1984)

49 CFR Part 175
[Docket No. HM-192]
Quantity Limitations Aboard Aircraft

A total of 28 comments were received in response to the advance notice of proposed rulemaking. Commenters expressed widely varying opinions regarding the action proposed in the JAL petition, from full agreement that § 175.75(a)(2) be removed from the HMR to opposition to making any change to the existing provisions. Other commenters proposed that the quantity limitations in § 175.75(a)(2) be increased. Because this matter is so controversial, the Air Line Pilots Association (ALPA) requested that a public hearing be held on the matter. The MTB agrees with ALPA that this is both an important and controversial matter, and believes that any change to the existing provisions of § 175.75(a)(2) must be carefully examined. Accordingly a public hearing will be held concerning this matter. In addition, the MTB believes it is important to summarize the comments that have been received in response to the advance notice of proposed rulemaking in order that certain matters raised by various commenters may be further considered at the hearing. The MTB will also accept additional written comments on matters raised in this document.

Six comments were received from emergency response organizations (e.g. fire departments and fire service training organizations). All of these commenters opposed any change to § 175.75(a)(2), alleging that removal of these quantity limitations could result in a catastrophic loss of life. One of these commenters termed any consideration of deleting the existing quantity limitation as "ludicrous". It should be noted that none of these commenters provided any detailed information to support these opinions, nor did any attempt to answer the specific questions posed by the MTB in the advance notice of proposed rulemaking. These questions had been posed in an attempt to gather information in order to determine the merits of the JAL petition.

One comment was received from a foreign government. The Director General of Civil Aviation of Portugal supported deletion of the provisions of § 175.75(a)(2) on the basis that such action would not be prejudicial to safety and that it would create further uniformity with the International Civil Aviation Organization (ICAO) Regulations in Annex 18 of the Convention on International Civil Aviation and the ICAO Technical Instructions for the Safe Transport of Dangerous Goods by Air. However, no information was provided to support the contention that removal of these quantity limitations would not be prejudicial to safety.

Comments were received from nine chemical shippers or shipper related organizations. All of these commenters supported some modification to the existing quantity limitations, with eight suggesting the removal of the limitations. Many of these commenters cited distribution and marketing difficulties that had arisen from these regulations such as the following example:

> American Hoechst Corporation divisions, subsidiaries and our parent facility have experienced, as a result of this regulation, marketing disadvantages with unnecessary handling and packaging difficulties which, in many cases, has defeated the purpose of using air freight service in the first place.

. . . .

> Current packaging standards, whether performance or specification criterion, when complied with offers sufficient control to transport hazardous material safety as proven by the lengthy service from responding air lines to this docket.

In its comments, Dow Chemical U.S.A. cited similar difficulties, and also attempted to outline the origin of the "50 pound" weight restriction. The salient points of the Dow comment are as follows:

> 1. This is to advise that the Dow Chemical Company supports the elimination of the present 50 lb. weight restriction for hazardous materials that currently applies to inaccessible cargo compartments on passenger aircraft. The rule served it's purpose years ago when most passenger aircraft had baggage and cargo compartments that were easily accessible to a crew member; and there was a very limited experience with transporting hazardous materials by air. Today not a single passenger aircraft operated by the Trunk Air Carriers have in-flight access to any of the baggage or cargo compartments. As a result, it creates an undue and unnecessary burden upon carrier and shipper alike.
> 2. The original rule stems from the early history when cargo and passenger aircraft had both accessible and inaccessible cargo and baggage compartments. The DC-3's had an in-flight accessible compartment behind the crew and one that was inaccessible in-flight behind the passenger compartment. The DC-4, DC-6 and DC-7 as well as the Connies and the Strata Cruisers all had the larger accessible compartment behind the crew as well as the belly compartments that were inaccessible during flight for all practical purposes. Some had trap doors or hatches that could be removed but they were primarily for mechanical inspection and normally used on the ground.
> 3. The limit of 50 lbs. was believed to be the heaviest weight that a cargo agent could physically handle with any degree of care. It also was high enough to cover an aircraft battery which the air carriers frequently transported as company material. Since there were other cargo compartments for the larger shipments the 50 lb. limit did not necessarily create a problem for either the shipper or the carrier.
> 4. The world headquarters of Dow's Pharmaceutical Division, Merrell Dow, is located in Cincinnati, Ohio. At one time they used a central purchasing plan that sourced their global manufacturing points on Cincinnati for certain of their raw materials and expensive drugs, many of which were regulated by DOT. The purpose of the central sourcing was for better quality control and to buy in large quantities at a lower price. In most cases the materials would move from Cincinnati to international points by air. Since Cincinnati had little or no all-cargo aircraft, the distribution system relied heavily upon the passenger aircraft and packaged accordingly. This was especially true for destinations like South Africa.
> In the past, shipments were small and it went fairly smoothly, but as production increased the shipments became larger. As an example, Cherry Extract. Due to its flash point it is shipped as a flammable. On shipments of 100 to 200 pounds they were not only separated in compartments, but split up between aircraft, often being separated from the restricted article certificate and causing undue delays. Reducing the size of the shipment to 50 lbs. helped but also increased the costs.

. . . .

> 5. In researching hazardous material incidences that have been reported to DOT and discussing the matter with knowledgeable air carrier personnel, we are unable to find a record of any hazardous material incidents aboard an aircraft that was caused by a large quantity of hazardous material. The control of the potential hazard of the material is in the packaging, not in the quantity aboard the aircraft in any one compartment.

The Council for the Safe Transportation of Hazardous Articles (COSTHA) supported some change to the existing quantity limitations, citing successful transportation experience with hazardous materials classified as ORM-D Consumer Commodities to which the quantity limitations of § 175.75(a)(2) do not apply. The following extract of the COSTHA comment outlines the suggested modification to the existing quantity limitations:

> For a number of years, COSTHA participants have been shipping Consumer Commodities ORM-D-AIR via aircraft, without being subject to the 50-pound limitation. This transportation has been free of any undue passenger or crew hazard exposure. This verifies that certain packaging and quantity restrictions imposed by the rules result in safe transportation.
> Without addressing what higher quantity would be equally safe, it seems reasonable to conclude that similar quantities of

similar materials, otherwise classed, will provide the same degree of safety for passengers and crew alike. There is such a category under the U.S. regulations commonly referred to as "limited quantity".

There are a few exceptions to this equivalency generalization, but except for pure gases in quantities over 4 fluid ounces, the only material difference would seem to be the overall ORM-D gross package weight limit of 65 pounds. Pure flammable gases and non-flammable gases may need to be considered as a separate category.

The following points should be considered:

1. Experience that has been reported by foreign commenters to the docket (Question 1).

2. The obvious lack of safety implications for permitting the same items as now permitted by another name (Question 2).

3. The marked increase in efficiency and timeliness in moving goods because of the greater availability of scheduled airlines, thereby avoiding the delays that shippers frequently experience (Question 3).

4. The known outstanding experience of moving ORM-D materials by air without significant risk (Question 4).

5. And using the established 65-pound package limit now recognized by the DOT Hazardous Materials Regulations and ICAO TI Packing Note 910 (actually 55.1 pound) (Question 5).

There is a basis to remove immediately the total quantity limit by substituting a package size restriction, and a hazard risk level. The level could be that set by the current U.S. limited quantity categories.

While this solution is not entirely satisfactory because it does not correlate directly to the international regulatory system, the general risk level (better described as the "lack-of-significant-risk" level) might be adequately reflected by choosing UN Group II and III materials and quantities for passenger aircraft as given in the ICAO TIs. While it is true that the ICAO quantities would result in larger packaging for some materials than the 65-pound limit suggests, it is equally true that under U.S. experience many of these materials are moving safely in commerce, some as unregulated, and have been for many years. Safety in their movement is more related to methods of packaging than a packaging size having its origins in the ancient regulations for railway express cars.

The major advantage to such an approach is that it is not U.S.-regulation oriented. It uses existing international criteria. It would seem to be a reasonable compromise for at least the initial step.

We request, therefore, a formal rule making proposal at least using the U.S. limited quantity levels. During these considerations very serious study should be given to considering alternatively the use of ICAO passenger aircraft quantity limits or Group II and III limitations on package sizes, rather than aircraft total quantities. Under this proposal, the 50-pound quantity limitation would continue to apply to Group I substances where they are authorized aboard aircraft.

Eleven comments were received from air carriers and air carrier organizations. All of these commenters supported either the removal of the quantity limitations or an increase in the quantity limits. The International Air Transport Association (IATA) and seven foreign air carriers submitted comments fully supporting the JAL petition to remove any limitation on the quantity of hazardous materials permitted to be carried in an inaccessible location aboard a passenger aircraft. The majority of these commenters provided specific opinions regarding the questions posed in the advance notice of proposed rulemaking. The comments submitted by Air France are typical of these comments and, although the comments are relatively lengthy, the MTB believes there is merit in reproducing those comments in this document. The five questions referred to in the Air France comments are those posed by the MTB in the advance notice.

In response to the reference advance notice of proposed rulemaking, AIR FRANCE wishes to submit the following comments:

(a) We fully support the statements in the JAPAN AIR LINES Co., Ltd., petition dated April 15, 1983, quoted in the reference Docket.

(b) With regard to the questions raised in the Docket itself:

(1) What has been the transportation experience in areas outside of the United States where no corresponding aircraft quantity limitations are imposed?

As most international airlines serving airports located on United States territory, AIR FRANCE had for the past two decades the experience of simultaneously operating:

—Flights to, from, or through a U.S. airport, where the quantity limitations per aircraft hold or compartment in § 175.75(a)(2) were applied, and

—Flights not serving a U.S. airport, where no such quantity limitations were applied and only the quantity limitations were applied and only the quantity limitations per package in (up to 1982) International Air Transport Association (I.A.T.A.) Regulations and (from 1983 on), International Civil Aviation Organization (I.C.A.O.) Technical Instructions for the Safe Transport of Dangerous Goods by Air were enforced, with no total aircraft or compartment quantity limitation.

We registered during this long period no evidence of either more incidents or more potentially hazardous ones on the international flights, as compared to the flights serving an airport on U.S. territory, which were—and still are—performed in compliance with the requirements of § 175.75(a)(2).

Further, our records of all incidents or abnormalies concerning carriage of dangerous goods (hazardous materials) associated with AIR FRANCE flights, held since 1972, include no case where the total quantity (number of packages) per aircraft, or cargo compartment, or unit load device was identified as a factor which might eventually have increased the risk.

(2) What would be the safety implications, if any, if the J.A.L. petition were granted?

To the best of our knowledge, this would introduce no adverse safety implications. On the contrary, positive safety improvement effect could be expected, insofar as past experience demonstrates safety regulations are best and most universally complied with when they are systematically the same for all flights. Exceptions, and rules with too many variants, have consistently been found more difficult to enforce. The currently prevailing international situation, whereby in accordance with I.C.A.O. Technical Instructions there is no aircraft quantity limitation on most routes, but there is one for any shipment to, from or through an airport located in the United States or on U.S. territory, constitutes an additional complexity which may have an adverse effect on safety.

(3) What would be the economic benefits and consequences associated with adoption of the amendment proposed by J.A.L.?

The currently prevailing situation under 49 CFR 175.75(a)(2) is primarily detrimental to the U.S. general public, insofar as U.S. shippers or consignees may not benefit from the possibility of shipping any significant amount of hazardous cargo on passenger aircraft flights: They are in practice limited to the use of freighter aircraft flights, noticeably less frequent and available to a significantly lesser number of international destinations.

Since there is no evidence to demonstrate it contributes to a higher safety, this rule therefore seems to constitute an unwarranted restraint on international commerce by air, primarily detrimental to United States citizens or companies.

Deletion of the rule would result in more equal competition between the U.S. and foreign chemical industries on worldwide markets, as well as more equal competition between U.S. and foreign air carriers on international routes not touching an airport located on U.S. territory. Past experience has shown that removing such restrictions on fair competition usually results in traffic development beneficial to the shipping public as well as the airline industry at large.

(4) If, instead of removing § 175.75(a)(2), certain classes ... of hazardous materials were to be excepted from these quantity limitations, what hazard classes or sub-classes could be safely excepted and why?

Based on the findings of the I.C.A.O. group of international expects as reflected in I.C.A.O. Technical Instructions for the Safe Transport of Dangerous Goods by air, all classes or sub-classes of hazardous materials (dangerous goods) can safely be exempted from aircraft quantity limitations, with the exception of:

—United Nations Class 7, Radioactive Materials, where a maximum quantity of 50 Transport Indexes per aircraft should remain applicable in accordance with International Atomic Energy Agency (I.A.E.A.) rules.

(5) If, instead of removing § 175.75(a)(2), the 50 pound limitation were replaced by a higher quantity limit, what quantity limit would be applicable and why?

Except as provided for under comment No. 4) above in accordance with I.C.A.O. Technical Instructions, we believe 50 pound (or 150 pounds of non flammable compressed gas) to be an arbitrary limit, and it does not, to the best of our knowledge, seem possible to substantiate this value or any higher set value in terms of safety. This is because safety, in the context of international I.C.A.O. regulations as well as 49 CFR is based on quantity limitations per package in relation with stringent packaging requirements, with the intent of making each individual package harmless once all specified requirements have been complied with. The actual safety problem is to ensure every single package is totally harmless: If it is not, it should be deemed unacceptable on passenger aircraft, where it may not be accessible during flight in the event of an incident. If it is, then having 2, 3 or in identical packages, all meeting the safety requirements per package, will not change the safety risk.

The Air Transport Association of America (ATA) reported that its member airlines are in basic agreement that the quantity limitations imposed by § 175.75(a)(2) are too restrictive, although they are divided as to whether to remove them or raise them, with one carrier stating that they are satisfied with the present rule and quantities. However, they noted that only international carriers can provide transportation experience in areas outside of the United States where no corresponding aircraft quantity limitations are imposed. The ATA went on to make the following observations regarding the origin of this limitation, problems encountered by carriers as a result of the limitation and a suggested interim measure to increase the permitted quantities in order to help the carriers as a result of the limitation and a suggested interim measure to increase the permitted quantities in order to help to resolve the problems that have been encountered:

Certain air carriers have reported that, in complying with § 175.75(a)(2), it has forced multi-piece air freight shipments of hazardous materials of the same class to be split, requiring loading in multiple ULD's and cargo compartments, and/or movement on numerous aircraft over a period of days, and in certain instances, requiring routing to different transfer points. This all provides for additional handling, and exposure of the pieces in the shipment to an increased possibility of incompatible loading and damage. In this situation, additional paperwork is also required, i.e., extra copies of dangerous goods declarations, pilot notifications, and separate manifests for each additional flight.

In considering this issue, one has to ponder the question of how the 50 pounds of hazardous materials (and 150 pounds of nonflammable compressed gas) in an inaccessible cargo compartment, found its way into the regulations. It appears that it may have been inherited from passenger rail car and rail express car quantities established many years ago, perhaps by the old Railway Express Agency (REA).

This also makes it all the more difficult to recommend practicable quantity limits. Perhaps an approach applying modification to § 175.75(a)(2) could be introduced to increase the quantity from 50 pounds to 300 pounds, and 150 pounds of non-flammable compressed gas to 500 pounds, for a period, such as, one year from the date of an interim rule. Close evaluation of the increased quantities could be made during and at the end of this period. Further determination could then be made with respect to the issuance of a final rule.

The Flying Tiger Line was the only United States air carrier to submit comments on the notice separate from those submitted by the ATA. Flying Tigers expressed reservations concerning the complete removal of the quantity limitations in question and indicated a belief that " ... statements and/or petitions relating to experience can be misleading. The United States of America has imposed a Hazardous Incident Reporting Procedure (49 CFR 171.15

and 175.16) for many years. This same requirement does not exist worldwide, which suggests information received by DOT-MTB may be partially self-serving rather than complete." Flying Tigers went to support an increase in the present quantity limitations along the lines suggested by the ATA. The Flying Tigers comments also made the following observations of general interest, and posed certain additional questions:

We further believe that the introduction of the ICAO Technical Instructions January 1, 1983, (acceptable to DOT when used in accordance with 49 CFR 171.11) permitted numerous quantity increases per existing packages on passenger and cargo aircraft. While we support the introduction and acceptance of ICAO Technical Instructions, it is suggested there is inadequate history to insure the same past levels of safety have been maintained. Carriers, and carrier organizations continue to file exceptions suggesting some form of disapproval with current regulations, and the level of safety provided by same.

There are obvious economic benefits to shippers and passenger air carriers that could be derived from either a relaxation of § 175.75(a)(2), or if the limitations were removed. The questions which arise are primarily based on removal of all quantity limitations, and the impact on safety. You may wish to consider the following:

1. Current quantity limitations present minimal problems to carriers insuring non-compatible Hazardous Material is properly separated in accordance with 49 CFR 175.78 (Table a). Removal of limitations can result in mixing non-compatible Hazardous Material due to aircraft space constraints. This problem may be compounded due to aircraft ground time at a facility, (further compounded by other flight activity, which must be completed in conjunction with flight prior to departure). This degree of danger is an unknown factor, which can only be determined by number and quantity of non-compatibles on a given flight and the potential reaction based on contact.

2. Aircraft configuration, (B747 vs B747 Combi, etc.) should be considered. Should a combi-aircraft be permitted to transport an unlimited quantity of hazardous materials on the main cargo deck of a passenger aircraft? What is the potential degree of danger to passengers in the event of incident? Does this impede crew ability to respond to emergency?

Note: We believe this same incident occurring in a belly compartment can be more easily addressed by crews and presents a lesser degree of danger to passengers.

Sabena Belgian World Airlines also expressed some reservations regarding the total removal of these quantity limitations, and suggested a revision of § 175.75(a)(2) similar in many respects to that proposed by the ATA and Flying Tigers. In addition, the Sabena comments contained a number of important observations of a general nature, particularly with respect to the need to load hazardous materials on the main deck of combi-aircraft because the existing quantity limitations so severely restrict the loading of hazardous materials in the inaccessible underfloor holds. The following comments by Sabena are of particular interest:

In order to meet the present quantity limitations to, from or via the USA, we (SABENA) are obliged to load most of hazardous materials permitted on passenger aircraft in the main deck cargo compartment of our combi aircraft (these compartments are fully accessible Class B cargo compartments), but the adequacy of main deck Class B cargo compartments for transport of hazardous materials has been questioned by some parties. It has been considered that the loading of hazardous materials in main deck cargo compartments should not be encouraged, because combi main deck holds must be kept ventilated at all times. However, in theory, any fire in aircraft underfloor holds would be suppressed by oxygen starvation.

We believe that the present restriction of § 175.75(a)(2) does not recognize the principle that when hazardous materials are properly packaged, they no longer constitute any appreciable degree of hazard. We prefer to see more stringent packaging regulations where necessary with the complete elimination of quantity restrictions per aircraft (other than for radioactive materials). In fact, this was done in the ICAO Technical Instructions where specification packagings for hazardous materials have been required for transport on passenger-carrying aircraft as well. In this connection, we disagree with point 6 of the JAL petition which states: 'Shipments of hazardous materials that are acceptable for carriage on passenger aircraft would be subject ... to much more stringent individual ... packaging requirements than those applicable to hazardous materials transported on cargo aircraft.' We believe that the possibility of ruptured packagings does in fact exist with the non-specification packagings presently permitted for carriage on passenger aircraft, but that this possibility is remote with the specification packagings required for carriage on cargo aircraft or required by the ICAO Technical Instructions for carriage on passenger aircraft.

Considering the above, we feel that § 175.75(a)(2) should be retained for hazardous materials permitted to be carried aboard passenger-carrying aircraft when such materials are packed in non-specification packagings, but that § 175.75(a)(2) should not apply when hazardous materials are packed in marked specification packagings as provided in the ICAO Technical Instructions. This principle would be reviewed if and when ICAO adopts provisions for limited quantities of dangerous goods. Nevertheless, since no incidents have been reported in air transport, as stated above, a certain relaxation of the present restrictions of § 175.75(a)(2) seems desirable, and it is suggested that the quantity limitation be increased from 50 pounds to 300 pounds. On the other hand, we see no need to limit the quantity of non-flammable compressed gases, in view of the very specialized type of packagings (cylinders) used for these materials.

In addition to the general suggestion to raise the quantity limitation to 300 pounds, Sabena also proposed in their comments to exempt certain hazardous materials with a limited level of hazard from the provisions of § 175.75. Specifically, Sabena proposed that § 175.75 be amended to read as follows:

Section 175.75 Quantity limitations aboard aircraft.

(a) ...

(1) ...

(2) More than 300 pounds net weight of hazardous material permitted to be carried aboard passenger-carrying aircraft—

(i) ...

(ii) ...

(iii) ...

(3) ...

(b) No limitation applies to the number of packages of the following materials aboard an aircraft:

(i) Hazardous materials in marked specification packagings as provided in the ICAO Technical Instructions,

(ii) Non-flammable compressed gases,

(iii) Small-arms ammunition or Explosives of ICAO Division 1.4, compatibility group S,

(iv) Flammable liquids with a flashpoint above 90°F. (32°C.) that do not meet the definition of another hazardous class,

(v) Combustible liquids subject to the requirements of this subchapter,

(vi) Substances of ICAO Packing Group III in Division 6.1,

(vii) Materials of IACO Class 9, and

(viii) ORM Materials.

While not ruling out the possibility of inceasing the quantity limitations in § 175.75(a)(2), ALPA in their comments emphasized that such action should only be taken after careful study to insure that flight safety is in no way compromised. ALPA also indicated that testing should be done in order to assess the effects of hazardous materials releases not only in inaccessible compartments, but in accessible compartments as well. The following extract from the ALPA comments summarizes their views on this matter.

In summary, ALPA has reservations on the dilution of the safety aspects of 49 CFR by the proposed elimination of the limitations imposed by § 175.75(a)(2) solely to bring it in line with the ICAO Technical Instructions. We would strongly recommend that any consideration toward a reduction of the valuable safety quantity limitations of § 175.75(a)(2) be based on testing of inaccessible cargo compartments containing hazardous materials under actual flight conditions to assess their capability to withstand safely the possible problems created by the effects of the materials being carried in the compartments. These tests should include, but not be limited to, fire, toxic leaks, corrosive spills, the effects of high ground ambient temperatures, excessive humidity conditions, and the effects of explosive decompression. While it is recognized that this ANPRM is addressing only inaccessible cargo compartments, we would recommend that the same type of testing be accomplished for accessible cargo compartments. With the worldwide acceptance and use of the combi aircraft, the same issues will again have to be addressed.

We cannot support any reduction of the safety aspects of 49 CFR 175.75(a)(2) based purely on simplifying airline procedures or for economic gain. The fact that the possibility of a catastrophic ground accident concerns the petitioner, JAL, should certainly indicate that unlimited quantities of hazardous materials in inaccessible cargo compartments may have even more catastrophic results during flight.

Since this subject has become so controversial, we would recommend that a public hearing be scheduled.

As previously indicated, owing to the diversity of views on the question of limitations on the quantity of hazardous materials that may be carried in an inaccessible location aboard a passenger aircraft, the MTB agrees with the ALPA suggestion that a public hearing be scheduled relative to this matter. At this hearing the MTB desires to receive further information and constructive comments on the questions raised in the advance notice of proposed rulemaking, comments and further information regarding the general matters raised in the comments highlighted in this document and comments on the following specific questions:

1. What will be the difference in the effects of fire, toxic leaks, or corrosive spills if an incident occurs in an inaccessible compartment as compared to an accessible compartment?

2. What is the relative hazard of transporting unlimited quantities of hazardous materials in an accessible location aboard passenger aircraft (e.g. on the main deck of a combi aircraft), as compared to carrying the same quantities in inaccessible locations, and why? Should the scope of this docket be expanded to address the transport of hazardous materials in accessible locations that may afford less safety than inaccessible locations?

3. If the MTB proceeds with a rulemaking to increase the quantities permitted to be carried in an inaccessible location, should a distinction in the quantities permitted be made on the basis of the classification of the compartment (i.e. C, D or E) in which hazardous materials are to be loaded?

4. Should the use of unit load devices be considered as a condition for permitting an increase in the quantities of hazardous materials permitted to be transported in inaccessible locations?

5. What are the merits of the approaches suggested by COSTHA, ATA and Sabena to modifying the existing quantity limitations?

(50 F.R. 6013, February 13, 1985)

49 CFR Part 173
[Docket HM-193, Notice No. 84–8]
**Tritium and Carbon-14; Low Specific Activity Radioactive Materials
Transported for Disposal**

A. BACKGROUND

The requirements of § 173.425 address most shipments of low-level radioactive waste transported from NRC or Agreement State licensees to licensed disposal facilities. Medical, biomedical, and related research institutions generate relatively large volumes of tritium and carbon-14 contaminated wastes that meet the definition of low specific activity radioactive material (§ 173.403(n)(4)(iii)). Much of the waste from these institutions is several orders of magnitude below the maximum activity level limit established for low specific activity radioactive materials. However, they still exceed the statutory definition of radioactive materials which includes any material having a specific activity greater than 0.002 microcuries per gram of material (49 U.S.C. 1807).

Most scintillation media wastes also meet the definition of a flammable liquid and are suspected to be carcinogens as well. Animal carcasses and tissues are not classified as hazardous materials per se but their disposal is often times handled in the same manner as hazardous materials. The flammability of the very low specific activity scintillation media is considered by MTB to present a greater hazard in transportation than their radiotoxicity. This proposal, therefore, would require that very low specific activity scintillation media be packaged, marked, labeled and otherwise prepared for shipment and transported on the basis of their flammability or another acute hazard, if present. Animal carcasses and tissues containing low levels of tritium or carbon-14 which do not meet the definition of another hazard class could be transported as materials not subject to the HMR.

B. NRC RULE CHANGE

The NRC investigation of problems associated with these low activity wastes from the biomedical community resulted in rules documents published in the **Federal Register** on October 8, 1980 (45 FR 67018) for the proposed rule, and March 11, 1981 (46 FR 16230) for the final rule. As adopted, the new Section 20.306 allows licensees greater latitude in the disposal of certain wastes containing low concentrations of tritium and carbon-14. In essence, if the specific activity of animal carcasses and tissues and liquid scintillation media are not greater than 0.05 microcuries per gram, they may be disposed of without regard to the radioactive nature of the materials. When compared to other radionuclides, the fundamentally lower radiation hazards of tritium and carbon-14 allow these low activity wastes to be disposed of safely when emphasis is placed on the other hazardous or noxious properties presented by the materials.

C. RADIOLOGICAL HAZARDS DURING TRANSPORT

Existing provisions of the HMR require the shipper to consider all hazards associated with a material when packaging and offering a hazardous material for transportation. Although the subject materials meet the definition of radioactive material for purposes of transportation, the radiological consequences to personnel and the environment in the event of release during transportation are considered to be extremely small. The potential risks associated with possible gradual buildup of activity at disposal locations over a period of time was considered in the NRC rulemaking. The potential risk of a buildup of activity as a result of transportation activities is even less.

(49 F.R. 33469, August 23, 1984)

49 CFR Part 107
[Docket No. HM-194; Amdt. No. 107–13]
Designation of Testing Agencies; United Nations Packagings

BACKGROUND

On October 12, 1984, MTB published a notice of proposed rulemaking (Notice No. 84-13) in the **Federal Register** (49 FR 40056). That notice proposed the adoption of a procedure by which MTB may designate third-party packaging testing agencies for the purpose of certifying conformance of packaging designs to U.N. standards. This final rule contains amendments to the Hazardous Materials Regulations (HMR) based on the proposals in Notice 84–13 and the merits of comments filed in response to that notice. Interested persons should refer to Notice 84–13 for additional background information.

In response to Notice 84–13, MTB received 35 written comments. The respondents include international exporters of hazardous materials, packaging manufacturers and packaging testing organizations. Of those commenters expressing an opinion on the overall merits of the proposal, 28 are in favor, two gave conditional support and one commenter strongly opposed the idea. The two commenters giving conditional support fail to see a need for implementing the proposed procedure. However, they indicated a willingness to support a procedure which is not keyed to "third-parties". Numerous commenters requested that the rule also permit the designation of "in-house" laboratories.

CLARIFICATION

The title of this docket is revised by replacing the word "Laboratories" with the word "Agencies" since the qualifications of a person performing the certification function is the prime factor in accomplishing the purpose of this rulemaking. Obviously, one qualification in the ability to determine the adequacy of test equipment for performance of a necessary function in his own facility (laboratory) or, for example, in the facility of a packaging manufacture.

ACKNOWLEDGMENT OF PROBLEM

The comments do not reflect a universal agreement that the regulatory requirements for international road and rail transport within Europe (ADR and RID, respectively) will require that packagings conforming to U.N. standards be certified by MTB or a designated approval agency. Some commenters are convinced this will occur. Other commenters contend the ADR/RID countries will come to recognize that self-certification of performance-oriented packagings standards is just as reliable as the long-accepted policy for self-certification of specification packagings.

One company which sees the ADR/RID requirement as a serious threat is Monsanto. Currently, Monsanto is having containers tested in Europe. The commenter complains "this procedure is time consuming and expensive and results in an adverse effect on exports". Another company, Olin Chemicals, takes the position that it must assume the status quo of the ADR/RID package testing requirements and, therefore, supports the proposal as a short term solution. However, Olin is one of several commenters who believe, in the long term, ADR/RID countries can be convinced that self-certification is an acceptable alternative.

The principal argument raised by commenters who fail to recognize the ADR/RID requirements as a problem for U.S. shippers centers on Paragraph 9.7.1.1. of Chapter 9 of the U.N. Recommendations. That Paragraph essentially leaves testing procedures to the discretion of the competent authority. Thus, PPG Industries, Inc. (PPG) and others contend ADR/RID countries must respect the determination of the U.S. Competent Authority on whichever system is selected.

Like most commenters to the Docket, MTB must assume that requirements promulgated by the ADR/RID countries will be fully implemented on the effective date of May 1, 1985, and it would not be appropriate to presume that U.S. shippers enjoy some privilege regarding compliance with these amendments not accorded ADR/RID member states after May 1, 1985. Consequently, the MTB believes a potential problem exists, and a positive step should be taken to minimize the negative impacts should this potential problem be realized.

ADEQUACY OF THE PROPOSAL

Most commenters viewed the proposal as an acceptable response to the potential problem. It is characterized as a straightforward approach to a clearly defined problem; i.e., ADR/RID countries require that tests be performed by a person independent of the packaging manufacturer and the proposal provides that inspections may be conducted by third-party inspection agencies approved by the Associate Director for HMR. While some commenters argue that a non-governmental inspection agency will nevertheless be unacceptable, other commenters see evidence that this concept is taking hold in some ADR/RID countries.

Writing in opposition to the proposed rule, PPG expresses doubt that ADR/RID countries will accept the findings of approved agencies. To PPG, it is obvious that MTB can satisfy the ADR/RID countries only by testing and approving packagings in its own right.

Several other commenters cite evidence that suggests a willingness by the ADR/RID countries to accept certifications which are not issued by the national competent authority directly. For instance, although The Netherlands maintains its own national testing laboratories, it recently set a precedent by conferring authorization to conduct tests leading to packaging approval on TOPA, an independent government-approved laboratory. Similarly, the competent authority of the United Kingdom authorizes bodies qualified under the National Testing Laboratory Accreditation Scheme in undertake tests leading to packaging approval.

There is a diversity of opinion on how a packaging approval system should operate. Most commenters emphasize that the self-certification process used in the U.S. today should continue, and that whichever method develops for international traffic it should not prejudice certifications for domestic traffic. However, Natico, Inc., a drum manufacturer, concluded its comments by stating "(T)hird party testing and certification for the transportation of hazardous materials should be made mandatory."

While the Notice was specific to designation of independent third-party testing laboratories, there was some discussion of other possibilities. The National Bureau of Standards (NBS) advised MTB of its National Voluntary Laboratory Accreditation Program (NVLAP) and suggested this alternative is better than developing a separate system. NBS currently recognizes a series of nine different laboratory accreditation programs (LAP). Some of these cover the testing of insulating materials, concrete, carpeting, and paint and paper products. Accreditation is established through a demonstration, to technical experts working for NBS, of the applicant laboratory's proficiency. As each LAP looks more to the competency of a laboratory than its independence, there are many in-house testing laboratories which enjoy accreditation and apparent recognition by certain countries belonging to the International Laboratory Accreditation Conference. Thus, the concept of self-certification is given greater credibility through the accreditation program.

The NVLAP program was only recently brought to the attention of MTB and did not receive consideration prior to development of the notice of proposed rulemaking for this docket. However, the design of the NVLAP system appears to satisfy several interests. In its position as a facilitator, NBS would relieve MTB of most of the day-to-day functions necessary in the administration of an approvals program for testing agencies. Also, as the accreditation in existing LAP's includes many "in-house" laboratories, the request of numerous commenters to drop the "third-party" provision from the applicant's qualifications statement might be satisfied. The NVLAP system, however, goes considerably beyond the scope of Notice 84–13 and MTB doubts that any rulemaking would be necessary for implementation of such a program since self-certification is already permitted by the HMR.

USE OF IN-HOUSE TESTING FACILITIES

Some shippers, who wish to maintain the status of their own testing laboratories, contend they meet MTB's objectivity test when testing packaging materials produced for them by an unrelated packaging manufacturer. To them, this is a form of quality assurance that adds another dimension to packaging integrity. For instance, when it wishes to ship materials in steel drums, 3M believes it can test drums made by any number of manufacturers with the same degree of objectivity as a third-party testing laboratory, but at a considerable savings of time and money. Packaging manufacturers who wish to maintain the status of their testing facilities argue very convincingly about the competency of their personnel and equipment, but their objectivity is not as apparent as that of a shipper testing a manufacturer's packaging.

It seems that a large number of commenters wrongly interpreted the requirement of § 107.402(b)(5). While that requirement specifies the agency must perform its functions independent of the manufacturers and owners of the packaging concerned, it does not prevent the agency from witnessing physical tests at a shipper or packaging manufacturer's facility. In fact, the requirements of § 107.402 were first applicable to the certification of specification intermodal portable tanks, and § 173.32a(b)(3) clearly indicates that witnessing a manufacturer's or owner's testing of tanks is acceptable. Nothing in today's final rule intends to modify that practice. Consequently, each in-house testing facility which possesses the necessary equipment and personnel may conduct the required tests, but the packaging certification must be issued under authority of a third-party designated for that purpose by the Associate Director for HMR.

QUALIFICATIONS FOR APPROVAL AGENCIES

Wyle Laboratories took exception to the proposed rule's absence of detailed criteria by which testing laboratories must operate. The commenter suggested quality-control criteria, such as traceability of calibration standards, calibration frequency and documentation, and standardization of procedures, would assure conformance to the packaging's certified level of performance.

Testing required by the U.N. Recommendations is very basic and may be adequately conducted with basic instruments. Currently, most packaging testing is performed by manufacturers and, to this date, there are no indications their instruments (pressure gauges, micrometers, scales, etc.) are so poorly maintained that MTB should establish specific requirements which assure greater accuracy. It is doubtful that independent agencies are any less competent in their knowledge or equipment.

ECONOMIC IMPACT

Mallinckrodt, Inc., while supporting the proposed rule only as a necessary means to avoid frustration of its export shipments, complained manufacturers and shippers of small packages (bottles inside fiberboard boxes) will bear a disproportionate burden of costs resulting from this rulemaking. Whereas steel drums are limited to a relatively small number of designs and capacities, the number of different packaging consisting of fiberboard boxes with inside glass or plastic bottles is extremely large. Thus, a greater amount of time and money is required to demonstrate that each packaging conforms to the performance standards. Due to the shortness of time available for filing comments, the commenter was unable to present data supporting this claim; however, MTB does not challenge its validity except to note that there are provisions in the U.N. Recommendations that allow the selective testing of packaging that differ only in minor respects from a tested type, and specifically provide some relief for the use of various inner packaging in a combination packaging without requiring retesting of the completed packaging. Another point Mallinckrodt wished to make is that testing would have to be performed twice; first to the satisfaction of the shipper or manufacturer developing the package, and secondly to the satisfaction of the designated certification agency. This also results in increased costs of time and money. To preclude what it considers unnecessary costs, Mallinckrodt requested that MTB push for acceptance by the ADR/RID countries of the self-ceritifcation process currently used in the U.S.

There is no disputing the commenter's statement. However, if MTB were to do nothing, international shippers would be in the even worse position of possibly having to incur additional costs by having their packaging approved in a foreign country. As indicated by other commenters, that option is probably the most undesirable.

MAJOR RULEMAKING

One commenter, PPG, took exception to MTB's determinaiton that this is not a major rule. The commenter states:

First, this proposal could have significant economic impact on PPG and other shippers. PPG has approximately 350 detailed packaging specifications that, if this proposal were adopted, may have to be "voluntarily tested to demonstrate adequacy." The cost of this testing could easily reach one million dollars. Secondly, this proposal, if adopted, would completely change the packaging approval process in the United States. It would become apparent to ADR/RID countries that the U.S. approves of third party certification of packaging, and it would be required for all shipments to these countries.

The MTB made it quite clear the Notice represents a possible solution to a problem that will likely affect many shippers, including PPG. An overwhelming amount of support for that proposal came from the regulated community. In addition, MTB clearly indicated in the Notice that the requirements in the HMR pertaining to U.N. performance-oriented packaging standards (§ 171.2 and § 178.0–3) can be met through self-certification. Consequently, affected persons are free to choose the option which they believe is most appropriate to their circumstances. MTB emphasizes that the result of the adoption of changes to procedural rules by this amendment does not impose a new mandatory burden on any person.

One of the criteria specified in Executive Order 12291 regarding a "major rule" refers to a regulation that is likely to result in "(s)ignificant adverse effects on . . . the ability of United States-based enterprises to compete with foreign-based enterprises in domestic or export markets." The MTB believes this rule may facilitate the ability of U.S.-based enterprises to compete in export markets.

SUMMARY OF AMENDMENTS

Section 107.401 is revised to expand its scope to include certifications issued for packagings conforming to standards appearing in the UN Recommendations on the Transport of Dangerous Goods. Paragraph (b) is added to clearly indicate that designated agencies share authority with MTB. Accordingly packaging manufacturers and shippers may ask the Associate Director for HMR to provide a packaging ceritification. In addition, the affected party may appeal to the Associate Director for HMR an adverse determination made by a designated agency.

Section 107.402 is amended in paragraphs (b)(3), (b)(4)(ii) and (b)(6) by expanding the scope of packagings covered in this section to include packagings conforming to U.N. Recommendations.

Section 107.404 is amended to indicate a designated agency issues a certification, rather than an approval certificate which is appropriate only to serially numbered intermodal portable tanks.

Subpart E, with the amendments discussed above and other conforming changes, is presented in its entirety for clarity.

(50 F.R. 10060, March 13, 1985)

49 CFR Part 107
[Docket No. HM-194; Notice No. 84–13]
Designation of Testing Laboratories; United Nations Packagings

BACKGROUND

Many nations that regulate packagings for the transportation of hazardous materials are adjusting their regulatory systems to recognize the performance-oriented packaging standards by the United Nations (UN) in Chapter 9 of the Recommendations prepared by the United Nations Committee of Experts on the Transport of Dangerous Goods (UN Recommendations). Such a proposal is under consideration in the United States under Docket No. HM-181 (47 FR 16268, April 15, 1982).

Individual nations and groups of nations are engaged in this effort. In addition, UN-affiliated organizations such as the Internationl Maritime Organization (IMO) and the International Civil Aviation Organization (ICAO) are adjusting their dangerous goods codes to conform with Chapter 9 standards.

Within Europe, the regulatory bodies for road and rail (ADR and RID, respectively) are very advanced in this process, and UN packaging will be required in Europe for shipments of all flammable liquids (to 141° F), corrosive materials and poisons as early as May 1, 1985, unless authorized transitional packaging is used. The new packagings require prescribed UN markings. Unlike certain other regulatory bodies, such as ICAO, which "grandfathered" all existing packaging specifications, the RID and ADR grandfather clause would not accommodate U.S. packaging specifications unless they are tested and marked in accordance with the previous RID and ADR requirements, or with the UN Recommendations.

Of particular concern in our proposed adoption of this procedure is the fact that most European nations will soon require that new packagings not only be tested and marked on accordance with the UN Recommendations, but that they be approved by their own governments on the basis of design testing conducted by a government laboratory or by third-party testing laboratories recognized by those governments. Consequently, those same governments currently either have their own certifying laboratories, or are recognizing third-parties to conduct testing, for the purpose of issuing government approvals of packagings. Should governments refuse to accept packagings marked and self-certified as conforming with UN standards by U.S. shippers or packaging manufacturers, as provided for in 49 CFR 178.0–3, it may become necessary for U.S. shippers or packaging manufacturers to send empty packaging overseas for testing and approval. In order to attempt to avoid such a situation, MTB believes it must provide a viable alternative to packaging self-certification for U.S. packaging manufacturers and shippers that will be more akin to the approval procedures that are, or apparently will be, employed in Europe.

Historically, manufacturers of DOT specification packaging (except for certain cylinders and intermodal portable tanks) have been authorized to engage in testing and self-certification. MTB has no current plan to require third-party testing, or testing by MTB-designated laboratories on a mandatory basis in association with implementation of standards addressed by Chapter 9 of the UN Recommendations; nor does MTB currently intend to require registration or approval of packagings which are successfully tested and certified by a third-party laboratory.

A testing laboratory has filed a petition for rulemaking for establishment of a procedure to provide U.S. Competent Authority recognition of its facility and this rulemaking is in response to that petition. The MTB proposes to adopt amendments to 49 CFR Part 107 whereby qualified testing laboratories can be designated by the U.S. Competent Authority and, therefore, may provide independent certification of conformance with UN Recommendations for each shipper and packaging manufacturer who chooses to seek such certification. This is a strictly voluntary procedure—there is no proposed MTB requirement that such laboratories be used, nor is such a requirement contemplated at this time.

Shippers and packaging manufacturers seeking a recognized third party testing laboratory already face an extremely tight schedule. Delay in rulemaking could mean that the May 1, 1985 deadline is unattainable if anticipated problems become real at that time. Therefore, because of potential difficulties with acceptance of packagings in Europe as of May 1, 1985, it is important that the MTB consider implementation of this program as quickly as possible.

SUMMARY OF PROPOSED AMENDMENTS

Section 107.401 would be revised to expand its scope to include certifications issued for packagings conforming to standards appearing in the UN Recommendations on the Transport of Dangerous Goods. In addition, a new paragraph would be added to clearly indicate that authority delegated to approved agencies is shared with MTB. Accordingly, packagings manufacturers and shippers may apply for certification directly to the Associate Director for HMR, or appeal an adverse determination by a designated approval agency.

Section 107.402 would be amended in paragraphs (b)(3), (b)(4)(ii) and (b)(6) to expand the scope of packagings covered by this section to include those conforming to UN Recommendations.

Section 107.404 would be amended to indicate that a designated approval agency which examines and tests packagings conforming to UN Recommendations will issue a certification, rather than an approval certificate, which is appropriate only to intermodal portable tanks.

(49 F.R. 40056, October 12, 1986)

49 CFR Parts 172 and 173
[Docket No. HM-195]
Reclassification of Special Fireworks

On October 26, 1984, the United States Display Fireworks Association (USDFA) filed a petition for rulemaking under the provisions of 49 CFR § 106.31. The petition is published verbatim in this notice. MTB's publication of the

USDFA's petition as an Advance Notice of Proposed Rulemaking does not constitute a decision by MTB to undertake a rulemaking action on the substance of the petition. This Advance Notice is issued solely to obtain comments on the merits of the petition from interested parties as one aspect of its decision on whether to proceed with rulemaking.

LIST OF SUBJECTS

49 CFR Part 172

Hazardous materials transportation. Labeling. Packaging and containers.

49 CFR Part 173

Hazardous materials transportation. Packaging and containers.

Petition for Rulemaking. Proposed by the United States Display Fireworks Association

I. Introduction *A. Preamble.* As compared to fireworks used by the general public and display projectiles used for military purposes, special (or display) fireworks are most commonly used on the Fourth of July and at public events and are handled by professional pyrotechnic managers or semi-professional personnel. This is an industry that is characterized by small, family owned businesses, often dating back seven to ten generations to Italy.

These professional artisans developed public display pyrotechnics in the United States based on age-old technologies, practices and methods. The tools and materials of this industry segment have changed little over the last two hundred years. Nonetheless, the industry has, in the main, complied with the recent trends toward more government regulation and have done so willingly. Yet a point has come where federal regulations governing transportation of special fireworks products threaten the livelihood of businesses and degrade rather than promote certain attendant safety factors. Such demands of cost and time beyond what can be reasonably expected in the public interest often seriously hinder the quality of service to users and customers and threaten to drive these small companies further into unprofitability as well as to discourage new generations and new entrants of people into the field. One result of this trend is the importation of more foreign made fireworks products. In addition, more domestic manufacturers are having difficulty taking delivery on foreign and domestic materials needed to manufacture safe display pyrotechnics which have now become synonymous with the birthday of our country.

We respectfully ask that the Secretary of Transportation provide regulatory relief for this industry and heed its safety recommendations.

B. Purpose of Petition. In accordance with the provisions in Part 106 of 49 C.F.R. the U.S. Display Fireworks Association proposes a rulemaking for the establishment of amendments to the present rules contained in Parts 172.101–173.100. The purpose of the proposed rulemaking is to seek relief from regulatory over-classification of the transportation of display fireworks and to enhance safety in transporting products on public roads by differentiating between special and display fireworks and reclassifying display articles under Class C regulations.

C. Reason for Petition. The companies comprising the display fireworks industry make up a unique segment of the pyrotechnics industry. These mostly family owned, small businesses use long established artisan oriented techniques for producing and using special fireworks.

Because (sic) of federal regulations these companies can no longer ship products by means of air freight or rail. Now, however, enforcement of Department of Transportation regulations has effectively precluded economic shipment by common carrier trucks. Regulatory requirements caused by classification of display fireworks as Class B result in truck companies demanding a "mini-rate" charge which is economically prohibitive. Certain other regulations often make shipment, even by company owned vehicles, cost prohibitive because of required additional employees attendant in the trucks during stops where rural safe havens are not readily available. From a safety standpoint, required placards announcing "B-Explosive" flatly invite vehicle breakins and theft of fireworks which are then unsafely and improperly ignited (without a projection tube).

In summary, the reason for this petition is threefold: (1) make truck common carrier transportation economically available to the display fireworks industry; (2) make company vehicle transportation economic and time efficient; and (3) enhance safety by the elimination of placards entitled "B-Explosives".

D. Definitions. Display Fireworks: special fireworks that are not military in use and whose purpose is for typical, local display events which are prepared and activated by trained personnel

In this petition, special fireworks and display fireworks may be used interchangeably.

Company Operated Vehicle: a roadway vehicle owned or operated by the manufacturer or distributor of display fireworks who is not a contract transportation agent or a common carrier

Manufacturer: a domestic person or company who makes or modifies or sells or activates display fireworks articles

Distributor: a domestic person or company who sells or activates display fireworks

Foreign Display Fireworks: articles which are originally manufactured outside the United States and that are not modified or operationally enhanced by a domestic manufacturer

The Industry: the segment of the pyrotechnics and special fireworks industry in the United States that manufactures and distributes display fireworks

II. Proposed Rulemaking A. *Amendment of Specific Sections.* As experienced by manufacturers and distributors of display fireworks. Part 173.88(d) effectively constitutes an over-classification of display fireworks. In order to achieve relief from the requirement that a motor vehicle "must be attended at all times by its driver or a qualified representative of the motor carrier that operates it", it is recommended that special fireworks, as defined in Part 173.88(d); be redefined so as to separate display fireworks articles from military projectiles (for which no relief is sought) by reclassifying display fireworks under Class C of Part 173.100 of 49 C.F.R.

This change also requires amending Part 172.101 to add to a listing of "Fireworks, display" so that Column 3 reads "Class C explosive", Column 4 reads "Explosive C", Column 5(a) reads "none", Column 5(b) reads "173.100", Column 6(a) reads "Forbidden", (6)(b) reads "150" and Columns 7(a) and (b) read "1.2" with 7(c) reading "Passenger vessels in metal lockers only". Inclusion of display fireworks under Class C would apply to the transportation of display fireworks in vehicles operated by fireworks manufacturers and distributors, and by common carriers or contract carriage operators.

These amendments reclassify display fireworks as Class C but, as a result of Part 172.101, 6(a) prohibit the transport of such material by passenger carrying aircraft or railcar. This restriction is in industry recognition of the safety needs peculiar to those modes of transportation and to continue those overriding safety measures which should remain unmodified.

B. *Assessment of Proposed Rulemaking.* 1. Safety Considerations. On February 16, 1984, the applicant provided a Statement of Safety Considerations For The Transportation of Class B, Special Fireworks (Appendix A) to, and at the request of, the Office of the Secretary. Other safety considerations are detailed as follows.

a. Record: The absence of any on-the-road accident involving explosion or ignition of special fireworks is clear demonstration that the industry is capable of exercising a high degree of safety in transporting its products. Obviously, this safety record demonstration is operational in nature and provides imperical evidence of safety over the period of time since DOT and other Federal agencies have been keeping data and records.

b. Professional Managers and Technicians: Because of the pyrotechnic nature of special fireworks, the personnel involved in transporting products in manufactuer and distributor operated vehicles are almost always trained, technical artisans. Rarely do industry members employ vehicle drivers only to haul products to customer sites. When common carriers are involved in shipping special fireworks, packages are adjusted in size and weight so that such products can be transported without the presence of professional managers. Therefore, safety is enhanced by such professional personnel who are normally involved in special fireworks handling.

c. Safety Comparison: In comparing the transportation safety record of display fireworks and petroleum products (which are not required to have vehicle attendance at all times), the relative operational safety of display fireworks is generally greater. For example, a load of gasoline can immediately ignite upon contact of spark or flame. This is not particularly true of display fireworks. This comparison demonstrates and suggests an over-classification of display fireworks as Class B.

d. Placard Related Safety: The special fireworks industry is unfortunately plagued by breakin and robbery of company owned vehicles because of both placarding reading "B-Explosive" and company names that include the word "fireworks". Many companies have not used the name "fireworks" because of this reason.

By reclassification of display fireworks as Class C, placards announcing explosives will be eliminated which will further enhance safety by stopping theft and the unprofessional use of stolen fireworks.

2. Indirect Considerations. There have been occurrences (sic) where domestic, special fireworks manufacturers and distributors have found that federal transportation regulations have tended to restrict sale and delivery of products to customers in certain geographical areas. Therefore, there has been some tendency in the domestic industry in recent years not to market its products as widely and to the full array of domestic users it once did. This industry retreat is not unrelated to the problems and the often excessive costs of complying with Class B regulations which confront this particular industry.

At the same time, and not unrelated to adjustments in the marketplace, there has been a rise in the importation and use of foreign made display fireworks (sic) which typically are less safe than domestic products. For various reasons, unenhanced and unmodified foreign projectiles have tended to be sold directly (and without the benefit of modifications) to domestic users, many of whom once purchased from domestic sources.

Increasingly, there is a tendency for customers to buy more available, cheaper and unsafe foreign fireworks. Foreign government subsidization of fireworks manufacturing and export and the fact that product liability is usually not enforceable and, finally, the use of less "lifting powder" all make foreign products cheaper to U.S. customers who often can not get timely and economical delivery of safe bona fide products from domestic manufacturers due to what becomes regulatory barriers. There is a demonstrable record of mishaps and unsafe operation of such foreign special fireworks, the latest of which was premature detonation of firewoks that occurred in the Washington metropolitan area on the Fourth of July, 1984. This information is offered because of the accompanying safety degradation that may be of significance to the Department.

3. Regulatory Compliance. It has been proven that it is difficult to comply with certain Class B regulations and still profitably and safely deliver special fireworks to users. For example, approximately two thirds of most transportation of display fireworks takes place in areas where no safe havens exist. Therefore, it often requires greater driving time and transportation logistics to safely transport products than if certain regulatory requirements for display fireworks did not exist. In some respects this increased "exposure time" degrades transportation safety.

4. Regulatory Precedent. The applicant believes that this petition is similar in nature to Docket HM-187, Amendment Numbers 172-92 and 173-175, a petition of the Sporting Arms and Ammunition Manufacturers Institute. DOT acted favorably on that petition for reasons such as successful transportation history and minimal hazard. We believe this rulemaking is a precedent which provides certain justification to grant the request under this petition.

5. Operational Relief and Business Benefits. The applicant has assessed the operational problems posed by the referenced regulations against what specific and minimal amendments are necessary to give relief from the encumbrances which are typically experienced from over-classification of display fireworks. Those amendments are specified in Section II, A of this petition. The following are real and operational relief and benefits which will result from amending regulations.

a. Cost and Time: An amendment under this petition will relieve the inflation of employee and operational time cost which have threatened the economic viability of the display fireworks industry and the probable success of its business future.

b. Service to the Public: The applicant states that its ability to safely serve the public would be significantly enhanced by the removal of display fireworks from Class B Hazardous Material Transportation Regulations thereby allowing both company and common carrier vehicles to deliver fireworks to the customer in less time and with less exposure to circumstances which pose safety hazards.

c. Placards Regulations: Removal of Class B placarding requirements through reclassification of display fireworks will enable the industry to more successfully avoid vehicle breakins and thus protect against illegal detonation of fireworks articles by thieves.

6. Legislative History and Congressional intent. The petitioner references the hearing record of S. 1933, The Federal Railroad Safety Act of 1969, as an example of the intent of Congress to provide relief as is recommended under this petition. Title III (Hazardous Materials Control) of the Act created a hazardous materials technical staff to oversee all transportation of dangerous materials. The Department of Transportation testified in these hearings on cost/benefit in which transportation of explosives is cited. The Department noted that "The regulations should minimize the hazard to the public, within the limits of economic feasibility."

7. Supportive Data. An industry survey of the Association's eighty members confirmed the restrictive impact of current Class B regulations on the transportation of display fireworks. The forty-five responses to the survey establish a substantial assessment pattern. Federal regulations have brought changes in transporting display fireworks that render members unable to satisfactorily serve customers. Two-thirds of the companies do a significant portion of business in rural areas where there are no readily available safe havens. Although two-thirds of the companies are over twenty years old, no one has reported or ever known of a highway accident resulting in fire or explosion of display fireworks in transport. Members unanimously believe manufacturers and distributors, as well as common carriers, can safely transport display fireworks without the application of the regulatory requirements occasioned by classification under Class B Explosives. These professionals have historically shown their understanding transportation safety and their ability to self regulate transportation of products in the industry.

III. CONCLUSION

The applicant respectfully requests a rulemaking and approval of this petition based on Congressional intent, regulatory purpose, enhancement of safety in the public interest and the need to maintain economic integrity of the display fireworks industry.

APPENDIX A

1. History—There has never been accidental explosion of Class B fireworks during transportation—according to an informal survey of professionals in the Class B industry.

2. Shipping Containerization—All Class B products are packed tightly in lined, waterproof cartons and banded for safety purposes.

3. Manufacturer Operated Vehicles—All trucks are locked and have metal-lined exteriors and woodlined interiors.

4. Ignition Safety Properties—Class B products will only ignite by application of direct flame to ignition devices. Ignition will not occur from high temperature or impact alone.

5. Transportation Safety Comparison—It is commonly held by industry personnel that Class B products, in transit, are less subject to explosion than gasoline or petroleum products in transit.

(49 F.R. 45627, November 19, 1984)

49 CFR Parts 172 and 173
[Docket No. HM-195]
Reclassification of Special Fireworks

On November 19, 1984, RSPA, published in the **Federal Register** an ANPRM soliciting comments on the merits of a petition filed by the USDFA to reclassify special (or display) fireworks from class B explosives to class C explosives (49 FR 45627). These fireworks are commonly used for public display on the Fourth of July and other special events. The USDFA summarized their reasons for the petition as follows: "(1) Make truck common carrier transportation economically available to the display fireworks industry, (2) make company vehicle transportation economic and time efficient; and (3) enhance safety by the elimination of placards entitled "B Explosives". The petition was published verbatim in its entirety in the ANPRM.

RSPA received over 90 comments on the ANPRM. The majority of the commenters, representing fire and safety emergency response agencies, objected to the changes proposed in the petition. Most of these commenters expressed concern over the downgrading of certain safety controls for motor vehicles transporting special fireworks. They expressed strong opposition to allowing the display of DANGEROUS placards in place of EXPLOSIVE placards and no placards on motor vehicles containing less than 1,000 pounds of special fireworks. They objected to elimination of the requirements for the attendance and surveillance of vehicles, special restrictions on parking the vehicle in certain areas, and the preparation of route plans. Many commenters attributed the excellent safety record for transporting special fireworks to these safety and expressed concern for the safety of firefighters responding to incidents involving fireworks if the safety controls are downgraded.

Commenters in favor of the petition expressed their concern over the deteriorating financial posture of the explosive industry resulting from Federal, State and local regulations, competition from explosive importers, and higher transportation rates, insurance premiums, raw materials and labor costs. One commenter maintained that present state-of-the-art materials and methods being used by fireworks manufacturers have "dissolved" the safety concerns experienced by earlier manufacturers.

On March 13, 1985, RSPA notified the USDFA of a preliminary determination that the petition should be denied based on RSPA's review of the petition and the comments received in response to the petition, and afforded the USDFA 30 days to provide additional information to support their petition or to make any comments on the comments received in response to their petition. The USDFA filed a letter dated April 12, 1985, responding to the comments received to the petition and again requesting some regulatory relief.

In order to asssit RSPA in making a determination on whether some regulatory relief may be warranted on a selective basis for special fireworks, RSPA requested the U.S. Bureau of Mines to conduct testing of assorted special fireworks packed in various packaging configurations. The testing was conducted on May 28–31, 1985. Testing procedures conformed to the UN Test Series 6, which is used to determine how explosives react when involved in a fire or explosion. In addition, a special test exposing a truck partially loaded with 500 pounds of special fireworks to an external fire source was conducted to determine whether this quantity of special fireworks would explode violently or just burn. Test results showed that the special fireworks functioned primarily by rapid combustion and therefore, are properly classed as class B explosives.

In a letter dated September 16, 1985, the USFA withdrew its petition without making any comment. Based on a review of the comments received in response to publication of the petition in the ANPRM, testing results, and the USDFA's withdrawal of their petition, Docket HM-195 is hereby withdrawn.

(Ed.—Also see latest activity in Docket No. HM-181)

(51 F.R. 4405, February 4, 1985)

49 CFR Parts 172 and 173
[Docket No. HM-196, Notice No. 85]
Packaging and Placarding Requirements for Liquids Toxic by Inhalation

NEED FOR ACTION

On December 3, 1984, a discharge of a material identified as methyl isocyanate (MIC) occurred at the pesticide plant of Union Carbide (India) in Bhopal, India. More than two thousand people died as a result of the discharge.

On December 19, 1984, the Chairman of the National Transportation Safety Board (NTSB) addressed a letter to the Administrator of the Research and Special Programs Administration (RSPA), urging the Department to reexamine its system of hazard identification and classification, and to update it in accordance with current technology in order to raise the minimum level of protection provided in the Hazardous Materials Regulations. The NTSB letter is as follows:

Dear Ms. Douglass: The December 3, 1984, release of methyl isocyanate (MIC) from a manufacturing plant at Bhopal, India, resulted in a tragedy of monumental proportions. It is difficult to accept the fact that a material, whose primary hazard as classified by the Department of Transportation (DOT) is its flammability, would cause such widespread death due to its toxicity.

Because of its continuing concern about deficiencies in the DOT's hazard identification and classification system, as described in the Safety Board's Safety Report, "Status of Department of Transportation's Hazardous Materials Regulatory Program," (NTSB-SR-81-2) and more recently, as discussed at its July 26–27, 1983, public hearing on the safety of railyards in populated areas, the Board has compiled toxicity and other data for these materials and information about the safety measures taken by Union Carbide and others in their transportation. Our review of this information indicates that there is an urgent need to improve the manner in which toxic materials are classified and to raise the minimum levels of protection required by federal regulations in transporting these materials.

Although MIC is classified by DOT regulations as flammable liquid just as is gasoline,[1] Union Carbide's handling of this material reflects more fully the true hazard it presents. For example, without limits on the quantity per container and in accordance with DOT regulations, Union Carbide or anyone else, could elect to ship this material by rail in the least protected of DOT specification tank cars (ARA, 103, 104, and 111); instead, Union Carbide requires MIC to be transported in DOT Specification 115 tank cars which are double-walled and insulated and have stainless steel tanks limited in capacity to 8,000 gallons. Similarly, containers offering greater than the required protection are used by Union Carbide for highway shipments. Moreover, Union Carbide does not allow the transportation of this material by highway with other materials on the same vehicle and, based on indepth studies, it specifies the routing of all its shipments to assure minimum exposure to the public of this material during its transportation. These increased safeguards for all shipment of MIC are possible because Union Carbide is the only U.S. manufacturer and is able to control fully its distribution; however, this is not true of other materials which pose similar toxic threats in the event of a transportation emergency.

The DOT system for identifying and classifying the hazards of materials is the outgrowth of a system developed over the years by industry. In developing the system, industry primarily used accident experience to make judgments about the hazard posed by a material and about the adequacy of packaging methods to minimize the potential for releases of material during transportation. Also controlling industry's assessment of the types and degree of the hazards posed by materials were its consideration only of acute threats to life, its limitation of concern to the safety of people in the immediate area of an accident, and its belief that accidents almost always would involve a fire. Since DOT's inheritance of this hazard classification system in 1967, an overall, objective assessment using current technology, has not been made to determine its continued adequacy for identifying fully the hazards posed to public safety and health when materials are released as a result of transportation accidents.

The tragedy of Bhopal, resulting in the deaths of more than 2,000 people, involved the release of material from a tank containing about 3,750 gallons of MIC: Fire was not involved and the DOT material classification provided no inference that such a release posed a major threat to public safety. In an attempt to understand why this release of MIC produced results similar to those normally associated with Class A Poisons and why this hazard was not identified by the DOT's system for classifying the hazards of materials, the DOT's requirements for identifying toxic hazards were reviewed. The table below which lists various materials and selected properties of materials was developed by the Safety Board. The table includes toxic materials shipped under several DOT classifications so as to compare the lethal properties of Class A Poisons with those of materials in other classifications. As can be seen from this data, the property which most distinguishes Class A Poisons from others is their higher vapor pressures (all are gases as opposed to liquids). The L_{60} values, while not directly comparable, show all listed materials to be lethal at concentrations significantly below the lower flammability limit. For example, MIC can be lethal at 5 parts per million (ppm) yet does not reach its lower flammable limit until there exists a 53,000 ppm concentration.

Although not specifically acknowledged by the DOT's definition, there is a relationship between the standard vapor pressure, lethality, and boiling point of materials classed as Class A Poisons. This relationship recognizes the natural tendency of toxic materials to vaorize into the air. However, it does not consider the ability of other toxic materials, when heated thermally or chemically, to vaporize as readily as Class A Poisons under standard conditions. Moreover, the definition of Class A Poisons establishes no standards or tests for determining which materials constitute a threat sufficient to be included in this classification. Furthermore, the criteria established for identifying Class B Poisons contain no upper limits on toxicity such that materials exceeding a specified toxicity would be classes as a Class A Poison and be protected during transportation at the level specified for Class A Poisons. Stated otherwise the DOT hazard classification does not consider the possible site-specific hazards to public safety and health of materials in accident environments. As can happen when materials are involved in transportation accidents, it appears that the vaporization rate of MIC at Bhopal was increased by heat generated by a chemical reaction causing the release of lethal but nonflammable vapors which were distributed widely by air currents.

The hazard identification and classification system must identify completely the hazards posed to life and health by each material during normal transportation and during emergencies because this knowledge influences greatly decisions made about the level of protection required for containers used in transporting materials and influences public safety protection measures which are instituted when materials are released during transportation. The DOT first was cautioned in 1969 about deficiencies in its hazard classification system by the National Academy of Science (NAS) in its report, "A Study of Transportation of Hazardous Materials: A Report to the Office of Hazardous Materials of the U.S. Department of Transportation." Because the recommendations made in the NAS report were not implemented by the DOT and because similar deficiencies have been identified in

---

[1]If a material is both flammable and meets the criteria for Poison B materials, it must be classified as flammable according to the requirements of 49 CFR 173.2.

Material	DOT class	Lse¹(ppm/time)	Boiling point (°F)	Vapor pressure (mm)	Flammable limits (percent)
Methyl isocyanate	Flammable liquid	5/4 hrs	102	348	5.3–26.
Toluene disocyante	Poison B	10/4 hrs	484	0.024	0.9–9.5.
Phosgene	Poison A	150/10 min	47	1,180	Nonflammable.
Acrolein	Flammable liquid	150/10 min	125	214	2.8–31.
Hydrogen cyanide	Poison A	150/30 min	79	620	5.6–40.
Cyanogen chlonde	Poison A	118/30 min	55	1,010	6.6–32.
Epichlorohydrin	Flammable liquid	250/4 hrs	239	13	3.8–21.
Nitric acid (red furring)	Corrosive liquid	49/4 hrs	185	103	Noncombustible.

¹Lethal concentration—the concentration at which 50 percent of the animals (generally rodents) die when exposed for the time specified.

accident investigations since 1972, the Safety Board has made several recommendations (R-72-44, I-76-3, R-80-12, I-81-6, and I-81-14) calling for improvements in the DOT hazard identification and classification system as well as for improvements in packaging requirements for specific hazardous materials.

One recommendation of particular importance in light of the tragedy at Bhopal is R-80-12. That recommendation called for an examination of speciality products and Class A Poisons to determine if the toxicity hazard of materials transported in DOT Specification 111 tank cars was sufficient to require the protection afforded by head shields and thermal insulation. In the Federal Railroad Administration's July 14, 1982, response to this recommendation, the Safety Board was advised that the toxicity hazards of products transported in DOT Specification 111 tank cars were being reviewed as a part of actions being taken in rulemaking Docket HM-175 and that the benefit/cost analysis for HM-175 had been completed. The FRA committed itself to including the results of the review of other products shipped in DOT Specification cars as well as the review of the benefit/cost analysis in the final action taken on Docket HM-175. Based on this commitment, the board acted on October 1, 1982, to close R-80-12 as acceptable action. On January 27, 1984, final action was taken on Docket HM-175; that action did not include an assessment of the hazards posed to public safety and health based on the toxicity of materials.

The Safety Board continues to urge that early attention be given by the DOT to reexamination of its hazard identification and classification system. However, the tragedy at Bhopal is another reminder of the need for immediate action by the DOT is identify materials that, during accident conditions, can present toxic threats to public safety and health similar to those demonstrated in the recent release of MIC. Many questions about the toxicity of materials now unanswered by DOT's hazard identification and classification system must be answered to determine which flammable liquids, Class B Poisons, corrosives, and other materials, can pose life-threatening hazards during accident conditions as we now know MIC can. For example, the properties listed in the above table indicates that acrolein poses hazards similar to those MIC. We believe these materials can be indentified expeditiously through a study of additional materials-specific properties to assess the volatility of the materials based on their vapor pressures and boiling points. In this way the hazards posed by the materials when fire does not result during accidents as well as the relationship to published toxicity data on materials can be related [sic] to tranportation environments.

The Safety Board encourages you to pursue, as a priority action, the identification of those materials now being transported that, during transportation emergencies, can pose life-threatening hazards to the public. The results of this effort then should be used to adopt, on an emergency basis, necessary changes in DOT's regulations concerning the transportation of those products found to possess hazards similar to MIC.

Respectfully yours,
Jim Burnett,
Chairman.

The Department believes there is merit in the basis concerns raised by NTSB—in particular, the points addressing inhalation risks due to the volatility of toxic liquids and the need for immediate action relative to the packaging of such materials. This NPRM addresses toxic liquids that have significant volatility, their packaging, and improved communication of their presence in transport vehicles.

BACKGROUND

As mentioned by the NTSB, the present system for identifying and classifying the poisonous (toxic) hazards of materials has its basis in recommendations made to the Interstate Commerce Commission prior to transfer of regulatory responsibilities to DOT in 1967. However, the following background information on recommendations and efforts to improve the classification system for toxic (poisonous) materials begins with the National Academy of Sciences (NAS) report in 1969 which is mentioned in the NTSB letter.

Since the NAS Report has been cited on a number of occasions, particularly that portion dealing with the classification of hazardous materials, the report of the Panel that addressed the subject is included as an appendix to this notice. There were four appendices to the NAS Report; three discussed general approaches to classification and the fourth addressed test methods for flashpoint. No new criteria were suggested for determination of inhalation risks taking into account the volatility of materials. Relevant to this NPRM is a portion of Appendix II-A addressing health hazards which reads as follows:

APPENDIX II-A.—SUGGESTED APPROACHES TO CLASSIFICATION

Health Hazards. The health hazards of materials being transported are characterized by their acute effects on human health according to the subcategories that follow. Note that the subject of mechanical trauma has not been considered in this classification. Consideration should be given, but is not included here, to the problem of the evolution of toxic gases during fires.

.   .   .   .

Systemic Hazards. Degree 1. Use standard definitions for toxic substances by inhalation, ingestion, and absorption through the skin as set forth in the proposed revision of USDA Interpretation 18, Item 18, published in the **Federal Register** on April 4, 1969.

Degree 2. Use standard definitions for highly toxic substances (poisons) by inhalation, ingestion, and absorption through the skin given in the FHSA regulations, except that an LD_{50} or LC_{60} shall supplant the single dosage to 10 animals. This is in keeping with test methods recommended by USDA regulations, Interpretation 18, and NAS-NRC Report 1138.

Irritant Gases and Vapors, Dusts, and Mists Hazards. Degree 1. As considered here, these substances make reference to reversible local irritant effects on eyes, nose, and throat exclusive of systemic effects. An irritant action must be determined by human experience since animal tests are not presently available. Lachrymatory action on the eye and sternutators are also included in this category.

.   .   .   .

On June 6, 1970, the following appeared in the **Federal Register** [Docket HM-51; 35 FR 8831] relative to inhalation hazards:

ADVANCE NOTICE OF PROPOSED RULEMAKING

On August 21, 1968 (33 FR 11862), the Hazardous Materials Regulations Board announced a plan to revise the regulations governing the transportation of hazardous materials. That document announced the intention to issue notices of proposed rule making in at least four areas, including, "classification and labels", and invited public help in developing the basic regulatory principles to guide the Board in revising the regulations.

The Board is planning to consider, in the near future, a proposal for classification tests for poisonous materials. To assist the Board in that consideration, the public is invited to express its views on the health hazard classification tests proposed herein. This document is not a proposal to change the regulations. It is an effort to get public participation early in the rule-making process.

The present definitions of poisonous materials contain specific testing criteria only in the case of class B poisons. There are no criteria now provided for class A poisons or irritating materials (including test gases). As a result, the public cannot practically rely upon those definitions to determine when the Federal regulations apply. In order to correct that situation, the Department proposes to adopt testing criteria for those latter two categories.

The National Research Council-National Academy of Sciences assisted the Department in developing these test criteria. In addition, the testing procedures and benchmarks used by the Departments of Agriculture and Health, Education, and Welfare have also been considered to ensure harmony between the regulatory standards of the several Federal agencies having jurisdiction in this area (see, for example, § 191.1 of the regulations of the Department of Health, Education, and Welfare, 21 CFR Part 191, and § 362.116 of the regulations of the Department of Agriculture, 7 CFR Part 362).

Types of health hazards. The health hazards of materials being transported are characterized by their acute effects on human health. Hazards to be considered are:

Systemic hazards.

Contact hazards.

Irritant hazards.

Systemic hazards exist when materials are capable of causing harmful effects through inhalation, ingestion, or absorption through the skin.

.   .   .   .

Hazard degrees. Degrees of hazard are ranked according to the potential severity of the hazard to people. The establishment of hazard degrees is necessary in order to establish packaging criteria reflecting the potential severity of the damage if a product should escape from its packaging during transportation. This potential must be taken into account in the design and integrity of packaging used in the shipment of the toxic products. The major categories and criteria are as follows:

Extremely dangerous poisons. Materials would be classified as extremely dangerous poisons if, on short exposure, they could cause deaths or major residual injury to humans. In the absence of adequate data on human toxicity, a material would be presumed to be extremely poisonous to humans if it falls within any one of the following categories when tested on laboratory animals:

.   .   .   .

(2) Inhalation. Any material that has an LC_{30} of 75 parts per million by volume or less or 0.75 milligrams per liter by volume or less of vapor, mist or dust when administered by continuous inhalation for 1 hour or less to both male and female rats, each weighing between 200 and 300 grams. If the material is administered to the animals as a dust or mist, more than 90 percent of the particles available for inhalation in the test must have a diameter of 10 microns or less.

. . . . .

Toxic materials. Materials would be classified as toxic if on short exposure they could cause serious temporary or residual injury to humans. In the absence of adequate data on human toxicity, a material would be presumed to be toxic to humans if it falls within any one of the following categories when tested on laboratory animals:

. . . .

(2) Inhalation. Any material that has an LC_{50} of more than 75 parts per million by volume but not more than 200 parts per million or more than 0.75 milligram but not more than 2 milligrams per liter of vapor, mist, or dust when administered by continuous inhalation for 1 hour or less to both male and female rats, each weighing between 200 and 300 grams. If the product is administered to the animals as a dust or mist, more than 90 percent of the particles available for inhalation in the test must have a diameter of 10 microns or less.

. . . .

On February 12, 1971, the following appeared in the **Federal Register** [Docket-51; 36 FR 2934] relative to inhalation hazards:

SECOND ADVANCED NOTICE OF PROPOSED RULEMAKING

On June 6, 1970, the Hazardous Materials Regulations Board published an Advance Notice of Proposed Rule Making Docket No. HM-51 (35 FR 8831), inviting public assistance in developing regulatory principles for the classification of certain hazardous materials on the basis of their health hazards.

The comments received generally related to toxicity test procedures, classification, and degrees of toxicity.

Toxicity test procedures. Most commenters agreed that toxicity test procedures should be uniform among regulatory agencies, noting even minor variations by DOT could be confusing. Apprehension was displayed concerning the use of tests and other criteria which were not developed specifically for the transportation environment.

. . . .

Degrees of toxicity. There was no common opinion expressed in this area. One group of commenters suggested retaining only one toxic category as Poison B, leaving the poison A category for gases only, and possibly placing some quantitative benchmarks on this category. Others agreed in principle with the designation of various degrees but suggested modifications.

. . . .

The present definitions of poisonous materials only contain specific testing criteria or guidelines for Class B poisons. There are no criteria or sufficiently descriptive guidelines for Class A or Class C poisons. Consequently, the public may encounter difficulty in relying solely on those definitions to determine the applicability of the regulations. In order to improve this situation, the Board proposes to adopt testing criteria wherever possible and better descriptive guidelines for all toxic materials covered by the Department's regulations.

The National Research Council-National Academy of Sciences assisted the Department in developing these test criteria. In addition, the testing procedures and hazard degrees used by the Departments of Agriculture and Health, Education, and Welfare were considered to insure harmony among the regulatory standards of Federal agencies having jurisdiction with respect to health hazards of chemicals.

The health hazards of materials being transported are proposed to be characterized by their acute effects on human health. The hazards considered are systemic hazards and irritant hazards. Systemic or internal hazards exist when materials, if inhaled, ingested, or absorbed through the skin can have harmful effects on organs and tissues other than at the site of contact.

. . . .

Degrees of hazard would be ranked according to the potential severity of the hazard to people. The establishment of hazard degrees is necessary in order to establish packaging criteria reflecting the potential severity of the damage if a product should escape from its packaging during transportation. The major categories and criteria which would be proposed are as follows:

Extremely toxic substances. Materials would be classified as extremely toxic substances if, on short exposure, they could cause death or major residual injury to humans. In the absence of adequate data on human toxicity, a material would be presumed to be extremely toxic to humans if it fell within any one of the following categories when tested on laboratory animals, according to the U.S. Department of Agriculture test procedures described under Title 7, Chapter 3, § 362.8 of the Federal Regulations.

. . . .

(2) Inhalation: Any material that has an LC_{50} of 50 parts per million or less by volume of a gas or vapor, or 0.50 milligrams or less of mist or dust per liter of air when administered by continuous inhalation for 1 hour to both male and female white rats (young adults). If the material is administered to the animals as a dust or mist, more than 90 percent of the particles available

for inhalation in the test must have a diameter of 10 microns or less, provided the Department finds it reasonably foreseeable that such concentrations could be encountered by man.

. . . . .

Highly toxic materials. Materials would be classified as highly toxic if, on short exposure, they could cause serious temporary or residual injury to humans. In the absence of adequate data on human toxicity, a material would be presumed to be highly toxic to humans if it fell within any one of the following categories when tested on laboratory animals, according to the U.S. Department of Agriculture test procedures described under Title 7, Chapter 3, § 362.8 of the Code of Federal Regulations.

. . . . .

(2) Inhalation: Any material that has an LC_{50} of more than 50 parts per million by volume of gas or vapor but not more than 200 parts per million or more than 0.50 milligram, but not more than 2 milligrams of mist or dust per liter of air when administered by continuous inhalation for 1 hour or less to both male and female white rats (young adults). If the product is administered to the animals as a dust or mist, more than 90 percent of the particles available for inhalation in the test must have a diameter of 10 microns or less provided the Department finds that it is reasonably foreseeable that such concentrations could be encountered by man.

On January 24, 1974, DOT published extensive proposals under HM-112 combining actions under a number of dockets, including HM-51. Included in the rulemaking was proposed adoption of a new placarding system, improved packaging for air shipments and standardized shipping paper requirements, and new definitive classification criteria for extremely and highly toxic materials. The proposals in the notice pertaining to inhalation risks were as follows:

§ 173.326 Extremely toxic materials; definition.
(a) For the purpose of this subchapter, a substance is considered to be an extremely toxic material if it falls within any one of the following categories when tested on laboratory animals according to the test procedures described in this paragraph:

. . . . .

(2) Inhalation. Any material that has an LC_{50} or 50 parts per million or less by volume of a gas or vapor, or 0.50 milligram or less of mist or dust per liter of air when administered by continuous inhalation for 1 hour to both male and female white rats (young adults). If the material is administered to the animals as a dust or mist, more than 90 percent of the particles available for inhalation in the test must have a diameter of 10 microns or less, provided it is reasonably foreseeable that such concentrations could be encountered by man in transportation.

. . . . .

§ 173.326a Highly toxic materials; definition.
(a) For the purpose of this subchapter, a substance is considered to be a highly toxic material if it falls within any one of the following categories when tested on laboratory animals according to the test procedures described in this paragraph:

. . . . .

(2) Inhalation. Any material that has an LC_{50} or more than 50 parts per million by volume of gas or vapor but not more than 200 parts per million or more than 0.50 milligram, but not more than 2 milligrams of mist or dust per liter of air when administered by continuous inhalation for 1 hour or less to both male and female white rats (young adults). If the product is administered to the animals as a dust or mist, more than 90 percent of the particles available for inhalation in the test must have a diameter of 10 microns or less provided it is reasonably foreseeable that such concentrations could be encountered by man in transportation.

. . . . .

The comments received in response to the three notices generally reflected opposition indicating (1) no demonstrated need for change, (2) conflict with definitions of other agencies, (3) differences with international standards, (4) proliferation of sublabelling elements, and (5) increased freight rates. There were several comments relative to volatility, but not in a positive sense. A typical comment was as follows:

. . . . .

Inhalation

A liquid could have an LC_{50} of 75 ppm under laboratory test conditions but present a negligible hazard in transportation because of low vapor pressure. Similarly, a solid could be highly toxic if tested in a highly divided dust form but be shipped as particles too large to penetrate into the lungs. To be valid, this classification must embody the concept of likelihood of test concentrations actually existing in the field. This concept appears generally in statutory codes. Thus, the Federal Hazardous Substances Act says, "—provided such concentration is likely to be encountered by man when the substance is used in any reasonably foreseeable manner."

Accordingly, we recommend that the first sentence in this paragraph be revised by adding something similar to the following: "—provided such concentration is likely to be encountered by man under any reasonably foreseeable conditions in normal transportation."

The recommendation quoted above was included in the second ANPRM. No positive recommendations were received concerning a means to address the volatility of liquids in association with the LC_{50} values.

On April 15, 1976, as part of the preamble to the Final Rule under Docket HM-112 (41 FR 15976), DOT stated the following:

. . . .

Poisonous or Toxic Materials. A number of comments were received concerning the uncertainty between extremely and highly toxic definitions as well as the confusion over the differences between irritants and ORM-A materials. It was suggested that a return to the old nonmenclature as something that was currently understood would be appropriate. It was also pointed out that the hazard class and the wording on the label and placard were different. Recently there has been considerable discussion with OSHA and the Consumer Product Safety Commission over the criteria for evaluating the poison hazard of a material. Although the proposed definition was derived after considerable discussions with the National Academy of Sciences and HEW several years ago, the current thinking is toward somewhat modified versions of these definitions. International discussions on this subject (UN and IMCO) have also indicated some need to modify the proposed definitions. It was therefore decided to return to the current definitions of Poison A and Poison B, and Irritant, and leave the definition of ORM-A essentially the same as that of ORA-A as defined by IATA. A notice will be prepared on this subject for public comment as soon as possible. New names for these hazard classes will also be considered at that time. Any material currently listed as Poison A is again listed as Poison A in the Hazardous Materials Table and any material proposed in HM-112 as Extremely Toxic has been restored to its previous classification.

. . . .

The "current thinking" alluded to was primarily related to the potential of the vapors of toxic materials to cause harm as a result of discharges during transportation, usually expressed in terms of boiling point or vapor pressure. MTB now considers its decision to terminate the proposed definitions in anticipation of development of improved methodology to be unfortunate because of the lengthy delay in bringing the matter to an appropriate resolution. If, as proposed in 1974, the rule had been adopted, materials such as MIC would have been classed Extremely Toxic Materials because of the precedence table proposed in § 173.2 (Hm-112; 39 FR 3094) which would have given "Extremely toxic liquid or solid" a classification precedence over "Flammable liquid".

According to the proposed rule, under HM-112, the packaging for any extremely toxic material covered by an n.o.s. entry would have been very restricted according to § 173.328 (39 FR 3115) unless the necessary safety control measures for a material were addressed in a separate section of the regulations by rulemaking.

While MTB was awaiting further international action on resolution of the definitions pertaining to inhalation toxicity, action was taken to improve the communication requirements for hazardous materials in transportation. On May 22, 1980, a Final Rule was issued under HM-126 (45 FR 34560) requiring, as relevant to this NPRM, that (1) dual hazards of materials addressed by n.o.s. (not otherwise specified) entries in § 172.101 (49 CFR) be recognized by new shipping names, e.g., "Flammable liquid, poisonous, n.o.s.", (2) under § 172.203(k), a shipping paper contain, in association with the shipping name specified for a material, its technical or NIOSH Registry name, for improved identification in emergencies; (3) the word "Poison" be displayed on a shipping paper in association with the description and class when the description and class do not indicate that a material is a poison; and (4) UN/NA numbers be displayed on shipping papers and packagings for direct reference to emergency response information, including DOT's Emergency Response Guidebook. While the new requirements provided substantial improvement in identifying risks, they did not provide for packagings of high integrity for materials described as "Flammable liquid, poisonous, n.o.s." posing a substantial risk due to their volatility; therefore, the new shipping entry referenced § 173.119(m) for packaging without special regard to materials posing inhalation hazards as opposed to those posing oral and dermal hazards.

At the international level, work began as early as 1974 on the development of new criteria for toxic materials that would not only consider acute toxicity or inhalation (LC_{50}), but also the ability of material to reach a dangerous concentration in the event of a discharge.

The United Nations (UN) Committee of Experts on the Transport of Dangerous Goods had been aware for some time that a system of classification based solely on the LC_{50} of materials does not always characterize the actual hazard presented by materials in transport. What the Committee considered necessary was a method of classification and packaging grouping that more accurately reflects the probability of poisoning by considering the volatility of a material as well as its toxicity.

The first proposal to do this, submitted to the UN Committee of Experts by the delegation of the Soviet Union in December of 1974, proposed that the relative inhalation hazard of materials be assessed through determination of the material's "toxic point". The "toxic point" of a material was defined as the temperature at which the vapor concentration of a material reached its LC_{50}. Although this concept appeared sound initially, it soon became evident that the toxic point method had some practical drawbacks, relating in particular to its heavy reliance on determination of the vapor pressure of a material at a number of different temperatures in order to accurately determine its toxic point.

In response to the criticisms expressed by some members of the Committee of Experts regarding the "toxic point" method, the U.S. representative with assistance of representatives from U.S. industry began to examine alternate approaches to the problem.

In May of 1977, the United States delegation submitted an alternate proposal to the UN Committee proposing use of a material's normal boiling point as an indicator of relative volatility, rather than vapor pressure. The United Kingdom delegation proposed a third method for consideration. This method made use of LC_{50} as an indicator of acute inhalation toxicity, and used a "volatility" factor as an indicator of the potential of the substance to reach lethal concentration in event of a spill. The "volatility" of the substance was defined as the saturated vapor concentration of the substance measured at 20°C.

The Committee of Experts carefully assessed the merits of each of the three methods and, unable to decide on the use of one method to the exclusion of the other two, and recognizing the need to address the problem of inhalation risks in transport at the earliest possible time, the Committee decided, at its Tenth Session in 1978, to adopt all three methods for publication in the next edition of the UN Recommendations. After publication of these methods and criteria in the UN Recommendations, they were implemented by the International Maritime Organization (IMO) through Amendment 17–79 to the International Maritime Dangerous Goods Code (IMDG Code). The IMDG code is the basic standard governing the international transportation of hazardous materials by sea. As shippers and carriers began to work with the new methods, it soon became evident that the use of the three methods was causing some confusion and that it would be best to settle on a single method. For this reason, the United States delegation proposed to the UN Committee that a special meeting be held to re-examine this question in an attempt to arrive at a single method. The Committees agreed, and a meeting was held in Hartford, Connecticut, in October 1981. A number of the member governments of the UN Committee participated in the meeting, as well as the World Health Organization, the Hazardous Materials Advisory Council, and the European Council of the Federation of Chemical Manufacturers.

Following an exchange of views, the group agreed to recommend to the Committee that the following three principles be used as the framework for determining inhalation toxicity for vapours:

(a) Toxicity should be represented by LC_{50} (1 hour, rat).

(b) Inhalation potential should be represented by Saturated Vapor Concentration at a reference temperature of 20°C.

(c) A system combining the above two factors should be developed, allowing substances to be placed in order of their overall inhalation toxicity risk.

In the months that followed, the UN Committee continued work on the development of appropriate grouping criteria on the basis of these principles. Finally, at its Twelfth Session in December 1982, the UN Committee adopted revised criteria for assessment of the inhalation hazard of materials, making use of LC_{50} and "volatility" (i.e., the saturated vapor concentration at 20°C.).

The first international organization to implement the new method and criteria was the International Civil Aviation Organization (ICAO). This was accomplished with the publication of the first edition of the ICAO Technical Instructions for the Safe Transport of Dangerous Goods by Air in 1983. These Technical Instructions, published pursuant to Annex 18 to the Convention on International Civil Aviation (Chicago Convention), are binding regulations for the international transport of hazardsous materials by most governments that are participants in the Chicago Convention.

Since January 1, 1983, the HMR (§ 171.11) have contained provisions that incorporate by reference the ICAO Technical Instructions. Under these provisions, shippers of dangerous goods may offer shipments, domestically and internationally, by air as required by the ICAO Technical Instructions with certain exceptions. It is currently estimated that in excess of 80 percent of dangerous goods transported by air within the United States are now being transported in accordance with the ICAO Technical Instructions rather than in accordance with the detailed provisions of the DOT Hazardous Materials Regulations. In this context, it should be noted that under the ICAO Technical Instructions any material which falls into Packing Group I by virtue of its toxic inhalation hazard is forbidden for transport aboard passenger and cargo-only aircraft.

The International Maritime Organization (IMO) has also implemented the new method for assessing the inhalation risk of toxic materials with the publication of Amendment 21–83 to the IMDG Code, which became effective on January 1, 1985. Since 1976 the DOT Hazardous Materials Regulations have permitted transportation of hazardous materials being imported into or exported from the United States in accordance with IMDG Code classifications. Because virtually all hazardous materials transported by sea must be transported in accordance with the IMDG Code, these materials must now be offered for transportation by sea in conformance with the improved UN inhalation assessment methods.

At approximately the same time that the UN Committee of Experts began considering the revision of criteria for classification and grouping of toxic materials presenting an inhalation risk, work was also begun on the development of a scheme for determination of the precedence of hazards of substances possessing multiple hazards. The intent of such a scheme is to establish a procedure for determining which of the hazards presented by a material would be considered the primary hazard and, therefore, establish the hazard class of the material.

In December of 1978, the Committee of Experts adopted such a scheme for determination of hazard precedence which was subsequently published in the UN Recommendations. Since that time, this scheme has been adopted both by IMO and ICAO. One of the principal concerns of the Committee during the development of this scheme was to insure that it took proper account of materials that present a serious risk of poisoning due to inhalation of vapors. Under this scheme, a liquid, other than an organic perioxide or radioactive material, is classified as a poison regardless of the level of any other risk (e.g., flammability, corrosivity, etc.), if it meets the criteria established for the Packing Group I inhalation toxicity which is the same criteria prepared for § 173.3a in this NPRM. The UN precedence scheme for classification will be fully addressed by MTB under Docket HM-181.

Taking into account the preceding background information, MTB believes the Group I criteria that is presently in effect for international transportation can and should be used as the basis for implementing improved transportation safety requirements within the United States for volatile toxic liquids. Further, MTB believes this action should be initiated immediately because action on matters to be addressed by HM-181 will not be completed in the near future.

PROPOSED AMENDMENTS TO PART 172

§ 172.203(k)(4)—MTB proposes to add a new subparagraph that will require an additional description on a shipping paper reading "Poison-Inhalation Hazard" for any liquid hazardous material having a saturated vapor concentration at 20°C (68°F) equal to or greater than ten times its LC_{50} value if that value is 1000 parts per million or less. It should be stressed that this proposed requirement would apply to any liquid material (e.g., acrolein) meeting this criteria, not only materials subject to "n.o.s." packing requirements.

§ 172.504(c)—MTB proposes to revise the sentence at the end of the paragraph to exclude materials subject to new § 172.505 from the 1000 pound placarding exception provided for motor vehicles and freight containers.

§ 172.505—A new section would be added to the placarding rules requiring POISON placards, in addition to placards required by § 172.504, to be displayed on each motor vehicle, rail car and freight container that contains any quantity of a material required to be identified by a "Poison-Inhalation Hazard" description on a shipping paper according to § 172.203(k)(4).

The MTB believes the inhalation risks presented by materials meeting the criteria proposed for § 172.203(k)(4) are significant to such a degree that communication of their nature and presence is necessary without exception.

PROPOSED AMENDMENTS TO PART 173

§ 173.3a—MTB is proposing to add a new § 173.3a to Part 173 that will address the inhalation risks of liquid materials that are currently classed as Flammable, Corrosive, Oxidizer, Poison, or Organic Peroxide, and whose packagings are specified in sections that contain "n.o.s." (not otherwise specified) packaging requirements. This section would not apply to material (e.g., acrolein which have specific packaging prescribed in regulations other than general n.o.s. packagings. Matters relating to these materials will be addressed under Docket HM-181.

The proposed criteria for § 173.3a address the principal factors involved in the potential hazard presented by volatile toxic materials. These are the fundamental toxicity of material, as expressed by an LC_{50} value, and the probability that such a concentration will evolve in the atmosphere above spilled material. This latter factor is directly proportional to the vapor pressure of the liquid and is expressed as the saturated vapor concentration.

The LC_{50} is the concentration of the material in air which is most likely to cause death in 50 percent of both male and female albino rats within 14 days after a continuous exposure of one hour. For purposes of this section, the value is expressed in milliliters per cubic meter or parts per million (ppm). Provision is made for using LC_{50} data much of which is currently available in published literature instead of conducting tests involving large numbers of animals.

The saturated vapor concentration is the maximum concentration of vapor in air which is produced when the vapor is in equilibrium with the liquid at a temperature of 20° C (68° F). This value is also expressed in milliliters per cubic meter (ppm).

The criteria proposed are those published in the United Nations' (UN) Recommendations of the Committee of Experts on the Transport of Dangerous Goods for materials which require Group I packaging because of high inhalation toxicity. UN Recommendations specify that a material with an LC_{50} of more than 1000 ppm does not require Group I packaging due to inhalation hazards. A material is subject to Group I packaging when the LC_{50} is 1000 ppm or less and the saturated vapor concentration is ten or more times the LC_{50} value. While MTB is proposing to use the UN criteria for Group I as a basis for this proposed rule, it is not proposing to authorize use of the packagings for Group I materials specified in Chapter 9 of the UN Recommendations. The packaging proposed in this NPRM is the same as specified for Poison A materials under the current HMR with a provision for material-specific approval of other packagings based on a determination of equivalency to packagings prescribe for Poison A materials or suitable packagings specifically prescribed for certain hazardous materials in other classes (e.g., acrolein).

It is relevant to add here mention of a collateral issue. During the past two years, we have received more than 1800 letters, including more than 100 from members of Congress, protecting or questioning the use of animals in stating our toxicity criteria. Most of the letters required individual responses explaining that our present regulations do not require specific LC_{50} or LD_{50} data, but a determination as to whether a material has a specified toxicity at a certain breakpoint. In other words, each of the present tests is a limit test requiring no more than 10 laboratory animals rather than hundreds suggested in the letters. MTB recognizes that this NPRM proposes use of specific LC_{50} data and the problems we may face in responding to numerous protests by electing to use such data. However, we believe our public safety responsibilities outweigh the concerns expressed by opponents to use of LD_{50} or LC_{50} data and, unless an equivalent and acceptable procedure for determination of inhalation toxicity is provided as an alternative, we firmly believe LC_{50} must be used for the purposes of the new safety control measures proposed for new § 173.3a as well as the improved communication requirements proposed for shipping papers and placarding. In order to minimize testing, however, there are provisions in the proposed rule allowing conversion of 4-hour LC_{50} data to 1-hour LC_{50} data, and use of LC_{50} data contained in published tests.

COMMENTS

Interested persons are invited to submit constructive comments on the rules proposed in this notice. MTB does not solicit comments on the technical merits of the NTSB letter (e.g., the lack of a reference temperature in the fifth column of the table in the letter). Comments are solicited on the merits of the basic issue raised by NTSB which is the purpose of this rulemaking action, i.e., improved packaging of violatile liquids that present significant toxicity risks, and improved communication of the presence of those risks during transportation.

Earlier in this preamble there is mention of comments received in response to Dockets HM-51 and 112 concerning conflicts with other agencies. Commenters are invited to point out any conflicts that could be encountered relative to the requirements of other agencies if a final rule is adopted as proposed in this NPRM. Such a consideration should take into account the fact that this NPRM is limited to proposed changes affecting shipping papers, placarding and use of packagings.

MTB requests data concerning materials that may be affected by the rules proposed in this notice. Of particular interest would be the technical names of materials affected and any additional costs that will be encountered in changing to packagings that would be required, if a final rule is adopted as proposed for § 173.3a.

(50 F.R. 5270, February 7, 1985)

49 CFR Parts 172 and 173
[Docket No. HM-196; Amdt. Nos. 172–99 and 173–190]
Packaging and Placarding Requirements for Liquids Toxic by Inhalation

As a result of a release of a hazardous material identified as methyl isocyanate at a pesticide plant in Bhopal, India, on December 3, 1984, the National Transportation Safety Board (NTSB) requested that the Department reexamine its system of hazardous materials identification and classification, and to update it in accordance with current technology in order to raise the minimum level of protection provided in the Hazardous Materials Regulations. On February 7, 1985, the MTB published a Notice of Proposed Rulemaking, Docket No. HM-196; Notice No. 85–1 (50 FR 5270) proposing special packaging and communication requirements for certain poisonous liquids based on their potential inhalation hazards. An extension of time to file comments was published in the **Federal Register** on March 13, 1985 (50 FR 10088).

The MTB received forty-five comments regarding Notice No. 85–1. Most of the commenters were basically supportive of MTB's efforts to establish a higher level of safety for the materials addressed by the Notice. However, practically all of the commenters had specific concerns and comments regarding the proposed changes.

Three of the comments received were general in nature and did not recommend any specific changes to the regulations. Four commenters were in favor of MTB issuing a second notice of proposed rulemaking, incorporating knowledge gained from comments submitted to the current docket. MTB believes this rule is too important to delay further and does not agree that another notice of proposed rulemaking is necessary or appropriate.

Several commenters were concerned about the lack of a provision that will allow continued shipping of those materials that are presently in the transportation system. MTB agrees that sufficient time must be allowed in order for the shippers of materials affected by this final rule to bring their practices into conformance therewith. In § 173.3a(d), a priority has been established for compliance first in regard to shipments in bulk packagings (May 1, 1986) with compliance for non-bulk packagings five months late (October 1, 1986). This should provide ample time for shippers to implement the requirements of this rule.

A major concern expressed by several commenters pertained to the application of the proposal to small or limited quantities of materials. MTB agrees that this final rule should not apply to materials packaged in primary containment units of one liter or less. An exception is the labeling requirement specified in § 172.402(a)(10) which is consistent with the POISON labeling requirement for all limited quantities that meet the definition of the Poison B class.

MTB believes the present shipping paper requirements for limited quantities, the POISON label which must be displayed, and small quantities of material per primary containment unit (inside package or container) justify the exclusion of limited quantity packages from the application of this final rule. In addition, § 173.4 is amended to authorize an exception of one gram quantities of liquids that are toxic by inhalation with the exception of those not authorized under § 173.4(a)(1)(iii).

One commenter suggested that the words "Poison-Inhalation Hazard" be included as part of the label which would be affixed to packages containing such materials. The commenter stated that this information would provide visibility to those having contact with the package. MTB agrees in principle with this suggestion and has amended § 172.301 to require the words "Inhalation hazard" in association (near) the required label(s) on packages. Excluded are one liter quantities as discussed above.

Several commenters stated that when "Poison by Inhalation" is a subsidiary risk identifier, the U.N. hazard class number located in the bottom quadrant of the placard should not be required. Also, the four-digit ID number should not be an integral part of the subsidiary risk "POISON" placard. MTB agrees with these commenters, and neither the display of U.N. hazard class numbers nor the four-digit ID numbering requirements have been changed by this rule.

MTB does not agree with the one commenter who recommended that § § 176.30(a)(6), 176.30(b), 176.74, and 176.83 be amended. Paragraph (k)(4) of § 172.203 requires the words "Poison Inhalation Hazard; to be entered on the shipping paper. Since § 176.30(a)(6) and § 176.30(b) requires the information to be the same as required by § 172.203, repeating the same requirement in Part 176 is redundant. Also, special attention in § 176.74, 176.76 and 176.83 is not considered necessary in light of the requirement specified in § 176.24.

A majority of the commenters recommended that § 172.101 Table be amended to identify those materials that are subject to this rule. It is apparent that many of these commenters believe that the burden for such a determination should rest fully on MTB. Such a view raises fundamental questions concerning the basic structure of the hazardous materials transportation scheme of regulation which has been in use more than 75 years i.e., a material is, or is not, subject to regulation according to classification criteria (e.g., § 173.115 for flammable liquids) or special criteria (e.g., § 173.4 for special exceptions). It has been estimated that more than 30,000 different chemicals (including compounds and mixtures or formulations) are shipped in commerce subject to the HMR and most are not listed by name in § 172.101. In most cases it is the criteria (or descriptive definitions in certain cases) that shippers must use to determine whether materials offered for transportation are subject to the HMR.

MTB construes some of these comments as endorsing a system of preclearance, i.e., notification of MTB when a new material is to be introduced into commerce. This would be before the first shipment in order for MTB to acknowledge the material by listing, or other means, based on the data provided by the shipper concerning the material. As a matter of practicality, this option is not viable based on the present staffing in MTB to exercise its HMR program nor would it be a desirable imposition on shippers of hazardous materials.

Several commenters suggested that special requirements in § 173.3a would be overlooked if special identifications were not provided in the § 172.101 Table for each material affected by the rule. MTB is concerned, and somewhat confused, by this view. There are a number of special requirements not specifically addressed in the Table. For example, there are special packaging requirements for shipment by aircraft in § 173.6. There are special prohibitions in § 173.21. There are also special exceptions provided in the regulations that are not addressed in the Table, e.g., § 173.3 for use of "salvage drums" and § 173.4 for small quantities. Also, it appears that several commenters based their comments on the Table alone without consideration of the rules in § 172.101 which introduce the Table and its applicability. In order to provide added clarification concerning use of the Table, a new sentence is added at the end of § 172.101(a) emphasizing the existence of other requirements in Parts 171 (e.g., § 171.12 for imported packages), Part 172, and Subparts A and B of Part 173. This emphasis also includes the applicability of new § 173.3a.

Two commenters suggested that MTB create a new hazard class for "Toxic by Inhalation" equal in status to the other hazard classes. MTB does not believe that adoption of a new hazard class is necessary to accomplish the purpose of this rule, consistent with the proposals set forth in the NPRM. As stated in the preamble of the NPRM, the entire classification scheme will be considered under Docket HM-181. In the meantime, in MTB's opinion, the U.N. criteria for inhalation toxicity hazard are the most appropriate ones for the purpose of this rulemaking.

Several commenters expressed concern about the mechanics involved in obtaining approvals for new packagings that may be required. Also, they expressed concern about the workload and time it would take to obtain an approval. MTB intends to give priority treatment to requests for approval—in particular those presenting data usable for comparison with materials addressed by specific packaging provisions in Part 173 (other than n.o.s. packagings). In addition, elimination of packagings of one liter or less from the packaging requirements will relieve the approval burden to some degree. Also, the priority specified for implementation, as specified in § 173.3a(d), will serve to distribute the approvals burden over a longer period of time.

A few commenters stated that the Poison A packaging is too restrictive and that the proposed rules "go too far" and that they fail to consider the success of current practices. MTB recognizes that there are other packagings which have been used for several years that can be safely used for materials that are toxic by inhalation. For example,

Specification 51 portable tanks and DOT-5 series drums are packagings that have an excellent safety record. Such packagings will be fully considered for approval by MTB pursuant to § 173.3a(a)(3).

One commenter stated that: (1) The MTB's proposed categorization of these materials appears to be more restrictive than that permitted by ICAO; (2) cargo aircraft shipment should be permitted for "Poison-Inhalation-Hazard" materials, particularly for small quantities (up to 1 liter or 1 kg); and (3) for small quantity research items, shippers should have the option of assuming that an item is a "Poison-Inhalation-Hazard" without actually having the LC_{50} data. The answers to this commenter are: (1) ICAO's criteria of whether a material is forbidden, or if permitted, the quantities permitted are based on a basic philosophy summarized in Table S-2-7 in the supplement of the Technical Instruction. Without printing the Table, the general rule is that any 6.1, Group I liquid, that is in Group I by virtue of inhalation hazard, is forbidden on both passenger and cargo aircraft. It is true that ICAO permits cargo aircraft shipment of some materials which may be subject to this rule because of their inhalation hazard. However, based on our participation in ICAO deliberations, we are certain that ICAO would have listed these as forbidden/forbidden if they had known that the material presented such a hazard, because the general rules in Table S-2-7 would have been applied. This was not done because ICAO has no way to tell from the UN listing of a material that it has been placed in Group I because of an inhalation hazard as opposed to an oral or dermal hazard. Once data on the inhalation hazard of these materials is available, we believe ICAO will forbid them on cargo aircraft. Methyl isocyanate will be forbidden on cargo aircraft with publication of the 1986 edition of the ICAO Technical Instructions; (2) MTB did not propose to change column (6) of the § 172.101 Table for materials subject to this rule; and (3) in § 172.402(h) provisions are already provided for shipment of samples for laboratory analysis.

Several commenters suggested that MTB establish certain reference sources for obtaining published LC_{50} data to limit the scope of the required literature search. Some of the commenters went on to suggest that the current edition of the NIOSH's "Registry of Toxic Effects of Chemical Substances (RTECS)" be used as the reference source. While MTB agrees that RTECS is a recommended source for obtaining LC_{50} data and uses it as the principle reference source, MTB does not wish to limit the reference sources to a few publications. We will accept use of credible LC_{50} data from any published reference. To avoid causing unnecessary confusion, the wording in § 173.3a(c)(4) is amended to reflect this view.

The recommendation that the definition of Poison A in § 173.326 be amended is not adopted. MTB does not find any immediate need to amend § 173.326. This section was not addressed in the proposal, nor did we receive any constructive suggestions on how it should be amended.

Three commenters suggested that a distinction (or clarification) be made between systemic poisoning and corrosive poisoning (poisoning due to destruction of tissue). This is not an easy task. As a safety issue, it is the end result that matters, not the precise mechanism by which the results are incurred. Therefore, MTB considers "Poison-Inhalation Hazard" to include both systemic and corrosive poisoning. The same commenters raised the question of how to convert LC_{50} data based on other than one hour exposure tests into one hour exposure values. They went on to suggest that for systemic poisoning the conversion factor should be based on the equation: Total dose = dosage × length of exposure. For example, LC_{50} values based on 4 hour exposure should be converted to one hour value by multiplying the LC_{50} (4 hour) value by 4, not 2 as proposed in the NPRM. They indicate that the same conversion factor (or straight line conversion) is not applicable to the LC_{50} values due to corrosive poisoning. MTB agrees with the reasoning for corrosive poisoning but disagrees with the 4 hour conversion factor for systemic poisoning. As stated in the NPRM, the criteria for inhalation toxicity came from the UN and is a result of several years of intense work in which the U.S. (including industry) participated. Without any thorough evaluation, it is not prudent to arbitrarily create new criteria which certainly will cause more problems. The MTB is aware of the controversy and difficulties in using a conversion factor of two to convert an LC_{50} value based on 4 hour exposure to an LC_{50} value based on one hour exposure. This method is even more difficult to apply to LC_{50} values based on exposure times less than one hour or longer than 4 hours. However, the majority of LC_{50} data published are either based on one hour or 4 hour exposure times. All things considered, the UN criteria remains most appropriate for the purpose of this rulemaking. With regard to corrosive poisoning. MTB's position is that the only meaningful LC_{50} value is that obtained with one hour exposure time. MTB knows of no meaningful conversion method.

One commenter suggested that more exact test parameters be established to promote uniformity of LC_{50} testing. The same commenter recommended the use of the test procedure described in the Organization for Economic Cooperation and Development (OECD) for Acute Inhalation Toxicity. MTB has reviewed the OECD procedure and agrees with the commenter that, with minor modification, the OECD's procedure be used when conducting LC_{50} testing. The OECD procedure requires at least a four hour exposure period which is not as appropriate for transportation as for other situations. For transportation purposes the exposure time need not be greater than one hour and § 173.3a(c)(1) reflects this view.

Four commenters suggested that the definition of "Saturated Vapor Concentration" and the method of calculating it from vapor pressure data be elaborated on for clarification. MTB agrees with the suggestions and has amended §§ 173.3a(b)(2) and 173.3a(c)(2) accordingly.

More than one hundred chemicals were mentioned by the commenters as possibly being subject to this rule. MTB has reviewed those chemicals mentioned, using RTECS and other available literature, and has identified at least 36 that are considered to be subject to this rule. They are—

Acetone cyanohydrin	Methyl bromide
Acrolein, inhibited	Methyl chloroformate
Allyl alcohol	Methyl chloromethyl ether
Allylamine	Methyl hydrazine
Bromine trifluoride	Methyl isocyanate
n-Butylisocyanate	Monochloroacetic acid, liquid
Chlorine trifluoride	Nickel carbonyl
Chloracetonitrile	Nitric acid, red fuming
Chloropicrin	t-Octylmercaptan
Crotonaldehyde	Pentaborane
Dimenthyl hydrazine, unsymmetrical	Phosphorus oxychloride
Ethyl chloroformate	Phosphorus trichloride
Ethyl isocyanate	Propinitrile
Ethylene chlorohydrin	n-Propyl chloroformate
Ethyleneimine	Tetramethoxy silane
Isopropyl chloroformate	Tetranitromethane
Mesitylene	Titanium tetrachloride
Methacrylonitrile	Trimethoxy silane

Among these chemicals, eleven are not specifically listed by name in the § 172.101 Table and would be shipped using generic n.o.s. proper shipping names such as "Flammable liquid, n.o.s.", or "Poison B liquid, n.o.s." etc. The remaining 25 chemicals in the list are specifically listed by name in the § 172.101 Table. Four of them refer to § 173.119, and two of them refer to § 173.346, ad the packaging requirements. MTB considers those packaging requirements to be deficient for reasons described in the NPRM. To remedy this, Column 5(b) of the § 172.101 Table has been amended by adding § 173.3a respectively for those six chemicals to require more restrictive packaging requirements. These amendments are not meant to imply that other materials are not subject to this rule. Also, the reason for leaving the § 173.119 or § 173.346 packaging requirements in the Table is to provide packaging requirements for mixtures and solutions of these chemicals which do not meet the inhalation hazard criteria of this rule (see § 172.101(c)(11)).

The following is a section-by-section summary of the amendments:

AMENDMENTS TO PART 172

Section 172.101. A sentence, which did not appear in the Notice, is added to paragraph (a) to inform users of the regulations that not all requirements of general applicability are found in the references in the Hazardous Materials Table. A reference to § 173.3a has been added in column (5)(b) of the Hazardous Materials Table for 6 materials to inform shippers that these materials may not be packaged in all of the packagings provided in § 173.119 and § 173.346. However, those packagings may be suitable for certain mixtures or solutions of these materials that pose risks lower than concentrations making them subject to this rule;

Section 172.203(k)(4). The reason for adding this paragraph was discussed in the Notice. This section has been changed because commenters informed MTB that the original wording was ambiguous;

Section 172.301(a). This paragraph has been amended to require packagings over one liter and no greater than 110 gallons capacity to be marked "Inhalation Hazard" in association with the required label(s);

Section 172.402(a)(10). MTB is adding a new subparagraph requiring display of POISON labels, in addition to any other label required, for packages containing materials meeting the criteria specified in § 173.3a(b)(2);

Section 172.504(c). The revised sentence in this section has been changed slightly, for clarity, from that proposed in the Notice;

Section 172.505. In agreement with the suggestions of several commenters, a provision has been added to indicate that duplication of POISON placards is not required nor display of UN class numbers at the bottom of additional placards.

AMENDMENTS TO PART 173

Section 173.3a. A subparagraph has been added to § 173.3a(a)(2) to except materials addressed in paragaraph (b)(1) and (b)(2) of this section from the packaging requirements of (a)(1) and (a)(3) of the section when packaged in basic containment units having a rated capacity of one liter or less.

Some commenters said the wording in (b)(2) of this section was not clear and they were unable to tell whether "that value" referred to the LC_{50} value or the saturated vapor concentration. The wording has been changed to make it clear that it is the LC_{50} value.

Paragraph (c)(1) has been changed to incorporate a reference to the procedure of the Organization for Economic Cooperation and Development (OECD) as was requested by one commenter.

Paragraph (c)(2) has been expanded to provide more detail on the method of calculating the saturated vapor concentration from the vapor pressure of a material at 20°C, as was suggested by some commenters.

It was pointed out by one commenter that the use of a multiplying factor to convert an LC_{50} based on a 4 hour exposure to an LC_{50} equivalent to a one hour exposure is not valid for a material which causes death by direct pulmonary effect, as opposed to one which acts by systemic poisoning. A clarification has been included in (c)(3).

Paragraph (c)(4) has been changed to mention the RTECS as a source of LC_{50} data.

Paragraph (c)(5) has been added to authorize the use of a limit test instead of a precise LC_{50} determination when no data are available in the literature or when the data in the literature are questionable. This provision will reduce the number of test animals that must be used to accomplish the purpose of this rule.

Paragraph (d) has been added to specify a compliance date for bulk packagings, a later compliance date for non-bulk packaging, and to allow two years for determination of applicability based on a 48 hour rather than 14 day observation period.

(50 F.R. 41092, October 8, 1985)

49 CFR Part 174
[Docket No. HM-197; Notice No. 85-2]
Shippers; Use of Cargo Tanks, Portable Tanks, IM Portable Tanks, and Multi-
Unit Tank Car Tanks in TOFC and COFC Service

The use of multi-modal freight containers to transport both hazardous and nonhazardous materials by rail is a common practice in the United States and internationally. However, the use of portable tanks, IM portable tanks, and multi-unit tank car tanks, which are all tank designed for use in more than one mode of transportation, is generally prohibited in rail transportation in trailer-on-flatcar (TOFC) and container-on-flatcar (COFC) service. (See 49 CFR 174.61 and 174.63). An identical prohibition against TOFC/COFC transportation of hazardous materials applies to cargo tanks. (See 49 CFR 174.61.) Cargo tanks are tanks designed for highway use, but which potentially could be used in TOFC/COFC service.

Notwithstanding the general prohibition, the current regulations provide that specific approval can be given for tank TOFC and COFC service under conditions approved by the FRA. Four approvals for TOFC and COFC service of DOT specification 51, IM 101 and IM 102 tanks have been granted in the past several years. Recently, both the number of requests and the scope of the requests for approval of tank TOFC/COFC service have expanded. MTB and FRA believe this trend will continue as shippers perceive benefits in using containers (tanks) capable of use in several modes of transportation.

MTB and FRA believe that more complete safety criteria for TOFC and COFC service of tanks transporting hazardous material need to be established. Identification and development of appropriate safety criteria are needed, whether to guide a case-by-case approval process or to establish regulatory standards. It is appropriate to establish safety criteria at this time since, in many cases, they can be implemented without adversely affecting established transportation practices.

Tank TOFC/COFC service of hazardous materials involves many of the same safety issues as transportation in a traditional railroad tank car (single-unit tank car tank). These safety issues include pressure relief, identification, special commodity requirements, and special handling requirements. However, TOFC/COFC service involves other or different safety issues. In particular, securement of the container to the flatcar or in the highway trailer, and securement of the highway trailer to the flatcar are concerns unique to TOFC/COFC service. MTB and FRA believe that tank TOFC/COFC service of hazardous materials has its place in the overall transportation system. However, neither MTB nor FRA is now convinced that tank TOFC/COFC would provide the same level of safety in all instances as a traditional railroad tank car given the wide range of hazardous materials involved and the differences in tank design. The adequacy of securement of a trailer to the flatcar and the trailer's potential vulnerability in TOFC service are areas of most interest to FRA.

In order to identify and develop safety criteria for tank TOFC/COFC service of hazardous materials, MTB and FRA request all interested parties to address the topic areas listed below.

1. Securement and cushioning of trailers and containers. E.g., are the Association of American Railroads' specifications M-952-82, M-943-80, M-931-83, and M-1002 (paragraph 600-19 of the January 1, 1983 revision,) adequate minimum safety standards?

2. Surge prevention. E.g., would a minimum filling density requirement provide adequate protection?

3. Special handling requirements. E.g., are train placement restrictions similar to, or identical with, those in 49 CFR §§ 174.91 to 174.93 necessary or appropriate?

4. Special commodity requirements. E.g., should thermal protection and puncture resistance requirements be established for tanks in TOFC/COFC service transporting materials such as flammable gases, ethylene oxide, and anhydrous ammonia?

5. Commodity restrictions. E.g., should certain hazardous materials currently authorized in the various tanks under consideration be forbidden in TOFC service, COFC service, or both types of service?

6. Lading transfer. E.g., should transfer of lading from a tank while in TOFC/COFC be prohibited or would safety standards similar to, or identical with, those in 49 CFR 173.10, 177.834, 177.837, and 179.67 provide an adequate level of safety?

7. Identification. E.g., what marking should be required on a tank that indicates it is authorized for TOFC/COFC service and that identifies its special features (pressure relief devices, bottom outlets, special structural features)?

8. Pressure relief. E.g., should all tanks in TOFC/COFC service transporting a flammable liquid be required to have safety relief valves in lieu of, or in addition to, safety vents?

9. System performance test requirements. E.g., how many, if any, impact tests of a complete TOFC or COFC system are needed to demonstrate the adequacy of the system?

Commenters are not limited to responding to the questions raised above and may submit any facts and views consistent with the intent of this notice. In addition, commenters are encouraged to provide comments on "major rule" considerations under terms of Executive Order 12291, "significant rule" considerations under the DOT regulatory procedures (44 FR 11034), potential environmental impacts subject to the Environmental Policy Act, information collection burdens which must be reviewed under the Paperwork Reduction Act, and economic impact on small entities subject to the Regulatory Flexibility Act. A transcript of the hearing will be made and placed in the public docket.

(50 F.R. 18278, April 30, 1985)

49 CFR Part 172
[Docket No. HM-198; Notice No. 86–6]
Molten Sulfur; Molten Materials Generally

The National Transportation Safety Board (NTSB) has recommended that the RSPA: (1) Regulate molten sulfur and, as appropriate, other molten materials, as hazardous materials, (2) prescribe packaging and handling standards, and (3) incorporate information relating to the hazards of these materials into warning devices and publications available to emergency responders and others involved in the transportation of molten materials.

As background to its Safety recommendation I-85-19, issued on August 12, 1985, the NTSB stated the following:

About 11:50 a.m., P.s.t., on January 19, 1985, a tractor with two tank trailers, operated by Cal Tank Lines, struck the concerete median barrier of the southbound lanes of Interstate 680 on the Benicia-Martinez bridge in Benicia, California. The trailers, carrying molten sulfur, overturned into the northbound lanes. One trailer was destroyed by ensuing fires, and the other was breached in several places. The molten sulfur splashed onto vehicles traveling in the northbound lanes as well as onto the roadway and its shoulders. The sulfur was ignited by an undetermined source and burned for approximately 3 hours. The driver of the truck and the driver of one of the vehicles in the northbound lanes died, and 26 persons were taken to local hospitals; 3 persons were admitted. Persons were evacuated from the area near the accident site, and the roadway was closed for 15 minutes.

Firefighters reported that when they arrived visibility was extremely poor due to a heavy white smoke. In their haste to attend the injured, and because bystanders appeared to be suffering no ill effects from the smoke, the initial responders carried out rescue operations without donning any protective breathing apparatus. These firefighters later were treated for breathing difficulties related to vapors from the burning material. When the fire chief arrived, he tried to identify the cargo by looking for placards, but there were none on the trailers. After the injured had been sent to the hospital, the firefighters turned their attention to dealing with the material spilled from the trailers. Firefighters, now in chemical protective suits, attempted to plug the holes in one of the trailers using wooded plugs, but they were unsuccessful because the molten material ignited the plugs. At the same time, other firefighters were hesitant to apply extinguishants to the burning material on the ground since they did not know what the material was.

About 1:15 p.m., two firefighters approached the cab of the truck and found a waybill and other papers on the ground outside the tractor. Using information from these papers, the carrier was contacted, and at 1:30 p.m. the material was identified as molten sulfur. By that time, several additional persons had been sent to the hospital suffering from either contact burns due to the molten sulfur or inhalation of its combustion products. Even after the firefighters learned the identity of the material, they had difficulty finding information on how to handle the emergency and how to treat those injured in the emergency response guidebooks they had available. Ultimately, the fire chief finally was able to find limited information on handling molten sulfur in the U.S. Department of Transportation's (DOT) 1984 Emergency Response Guidebook.

The molten sulfur was a causal factor in the two deaths and in most of the injuries involved in this accident. When firefighters arrived, the truck driver was alive, but trapped in the cab of his truck. Firefighters attempted to extricate the driver but were forced to retreat due to the heat from the burning sulfur. Sometime after 11:30 p.m., the driver's body was removed from the cab of the truck. The coroner's report listed the cause of the driver's death as "inhalation of fire and smoke with asphyxiation." The other fatality was splashed by molten sulfur as the tanks climbed the barrier. He died 3 days later of thermal burns. Many of those injured as a result of this accident suffered irritation of the mucous membranes. Sulfur dioxide, a combustion product of sulfur, produces this effect.

The NTSB continued:

While the temperature at which molten sulfur is transported is not sufficient to ignite most combustibles, its elevated temperature presents a hazard nevertheless, as this accident involving the deaths of 2 persons and injury to 26 others and substantial property damage demonstrated. The Safety Board is concerned that there may be other unregulated molten materials in the transportation system which also might cause severe casualties involving persons, damage to property, and major disruption to communities.

Therefore the National Transportation Safety board recommends that the Research and Special Programs Administration:

Regulate molten sulfur and, as appropriate, other molten materials, as hazardous materials, prescribe packaging and handling standards, and incorporate information relating to the hazards of these materials into warning devices and publications available to emergency responders and others involved in the transportation of molten materials. (Class II, Priority Action) (I-85-19).

Classify as priority action on the proposed rulemaking in Docket HM-178 regarding the definition of a flammable solid, and establish a timetable for its completion. Include in the final rule test requirements and clear, objective criteria for shippers to identify those materials included in this hazard class. (Class II, Priority Action) (I-85-20).

RSPA has conducted an analysis of the inherent hazards involved in the transportation of molten materials to determine whether these materials are adequately regulated to provide safe transportation. As part of this analysis, RSPA reviewed the results of a separate investigation of the Benicia, California accident conducted by the National Fire Protection Association that was discussed in the January 1986 issue of Fire Journal. Analysis of the investigation report showed the following:

1. Detailed emergency response information on molten sulfur was not immediately available because molten sulfur is not subject to regulation as a hazardous material.

2. Firefighters had difficulty confirming the nature of the cargo.

3. Visibility at the site was severely limited as a result of fog and dense vapors of sulfur dioxide.

RSPA agrees with the NTSB's assessment that molten sulfur should be regulated as a hazardous material. RSPA believes that analysis of the investigation of this incident and other incidents involving molten sulfur justifies the regulation of this material as a flammable solid, even though it may not be in a solid state during transportation. This action is necessary so that emergency response personnel will have sufficient initial warning information to assist them in handling this kind of incident. RSPA has not determined if specific packaging standards are necessary. In this notice RSPA is proposing that molten sulfur be subject only to the general packaging requirements contained in § 173.24. This section contains general packaging requirements for the transportation of all hazardous materials.

This notice also solicits comments on the nature and scope of the transportaiton of molten sulfur and other molten materials that are currently not regulated as hazardous materials. Evaluation of these comments will aid RSPA in determining whether other molten materials should be regulated, and if specific packaging standards are necessary for the safe transportation of molten sulfur and other molten materials. RSPA will evalute these comments and then determine if it will propose further amendments to the HMR.

In this notice, RSPA proposes to specifically list molten sulfur in the Hazardous Materials Table (HMT) in § 172.101, and classify the material as a flammable solid. Solid sulfur is currently regulated domestically as ORM-C only for shipments by vessel. Approximately 6 million long tons of molten sulfur are shipped domestically by highway and rail each year. Sulfur, in a molten state, is not currently listed as a hazardous material in the HMT. However, molten sulfur is listed in the Optional Hazardous Materials Table, § 172.102, reflecting regulated status in the International Maritime Dangerous Goods (IMDG) Code. (This table is used for import and export shipments by vessel.) Furthermore, the U.S. Coast Guard presently regulates bulk vessel movements of molten sulfur in Subchapter O of 46 CFR. If the rules proposed in this notice are adopted, molten sulfur would be treated in the same manner as in the U.N. Recommendations. However, since it is unclear whether or not molten sulfur meets the current flammable solid definition contained in § 173.150, the RSPA is proposing to include a plus mark (+) associated with the entry in the HMT. This would indicate that molten sulfur is subject to the regulations as a flammable solid whether or not the material meets the definition of the hazard class contained in § 173.154.

The definition of a flammable solid is less than precise. The NTSB pointed this out in the background to its Safety Recommendations I-85-19 and 1-85-20. A more precise flammable solid definition is under study by the United Nations Committee of Experts on the Transportation of Dangerous Goods (U.N.). Despite the current efforts to improve the definition, the U.N. recommendations classify molten sulfur as a flammable solid.

In addition to proposing to regulate the transportation of molten sulfur, which would have the effect of requiring shipping papers, marking, labeling, and transport vehicle placarding, the RSPA is soliciting comments on the following questions in order to determine the need for further regulation of molten sulfur and other molten materials as appropriate:

1. What molten materials, not currently regulated by DOT, are being transported?

2. At what temperatures are these materials transported?

3. What modes of transportation are being used for the transportation of molten materials?

4. What packagings (bulk and/or nonbulk) are being used for the transportation of molten materials?

5. What is the minimum/maximum quantity of molten materials being transported in any one transport vehicle?

6. What are the hazards associated with the transportation of specific types of molten materials?

7. Should shipping papers, markings, and placarding be required for shipments of all or specific molten materials?

8. Are DOT packaging standards, including specification packagings, necessary for the safe transportation of molten materials?

9. What packaging design criteria are used to minimize the loss of lading as a result of a collision or overturn?

10. Does industry have, or is industry working on, packaging standards for the transportation of any molten material?

11. What accident data is available to demonstrate the suitability of presently used packagings for molten materials?

(51 F.R. 42114, November 21, 1986)

APPENDIX D

Pertinent Hazardous Materials Statutes

HAZARDOUS MATERIALS TRANSPORTATION ACT
49 U.S.C. 1801 et seq., 88 Stat. 2156, P.L. 93–633.

DECLARATION OF POLICY

1801. It is declared to be the policy of Congress in this title to improve the regulatory and enforcement authority of the Secretary of Transportation to protect the Nation adequately against the risks to life and property which are inherent in the transportation of hazardous materials in commerce. (Sec. 102 of P.L. 93–633).

DEFINITIONS

1802. As used in this title, the term—

1. "commerce" means trade, traffic, commerce, or transportation, within the jurisdiction of the United States, (A) between a place in a State and any place outside of such State, or (B) which affects trade, traffic, commerce, or transportation described in clause (A);

2. "hazardous material" means a substance or material in a quantity and form which may pose an unreasonable risk to health and safety or property when transported in commerce;

3. "Secretary" means the Secretary of Transportation, or his delegate;

4. "serious harm" means death, serious illness, or severe personal injury;

5. "State" means a State of the United States, the District of Columbia, the Commonwealth of Puerto Rico, the Virgin Islands, American Samoa, or Guam;

6. "transports" or "transportation" means any movement of property by any mode, and any loading, unloading, or storage incidental thereto; and

7. "United States" means all of the States. (Sec. 103 of P.L. 93–633.)

DESIGNATION OF HAZARDOUS MATERIALS

1803. Upon a finding by the Secretary, in his discretion, that the transportation of a particular quantity and form of material in commerce may pose an unreasonable risk to health and safety or property, he shall designate such quantity and form of material or group or class of such materials as a hazardous material. The materials so designated may include, but are not limited to, explosives, radioactive materials, etiologic agents, flammable liquids or solids, combustible liquids or solids, poisons, oxidizing or corrosive materials, and compressed gases. (Sec. 104 of P.L. 93–633.)

REGULATIONS GOVERNING TRANSPORTATION
OF HAZARDOUS MATERIALS

1804. **(a) GENERAL**—The Secretary may issue, in accordance with the provisions of section 553 of title 5, United States Code, including an opportunity for informal oral presentation, regulations for the safe transportation in commerce of hazardous materials. Such regulations shall be applicable to any person who transports, or causes to be transported or shipped, a hazardous material, or who manufactures, fabricates, marks, maintains, reconditions, repairs, or tests a package or container which is represented, marked, certified, or sold by such person for use in the transportation in commerce of certain hazardous materials. Such regulations may govern any safety aspect of the transportation of hazardous materials which the Secretary deems necessary or appropriate, including, but not limited to, the packing, repacking, handling, labeling, marking, placarding, and routing (other than with respect to pipelines) of hazardous materials, and the manufacture, fabrication, marking, maintenance, reconditioning, repairing, or testing of a package or container which is represented, marked, certified, or sold by such person for use in transportation of certain hazardous materials.

(b) **COOPERATION.**—In addition to other applicable requirements, the Secretary shall consult and cooperate with representatives of the Interstate Commerce Commission and shall consider any relevant suggestions made by such Commission, before issuing any regulation with respect to the routing of hazardous materials. Such Commission shall to the extent of its lawful authority, take such action as is necessary or appropriate to implement any such regulation.

(c) **REPRESENTATION.**—No person shall, by marking or otherwise, represent that a container or package for the transportation of hazardous materials is safe, certified, or in compliance with the requirements of this Act, unless it meets the requirements of all applicable regulations issued under this Act. (Sec. 105 of P.L. 93–633.)

HANDLING OF HAZARDOUS MATERIALS

1805. (a) CRITERIA.—The Secretary is authorized to establish criteria for handling hazardous materials. Such criteria may include, but need not be limited to, a minimum number of personnel; a minimum level of training and qualification for such personnel; type and frequency of inspection; equipment to be used for detection, warning and control of risks posed by such materials; specifications regarding the use of equipment and facilities used in the handling and transportation of such materials; and a system of monitoring safety assurance procedures for the transportation of such materials. The Secretary may revise such criteria as required.

(b) **REGISTRATION.**—Each person who transports or causes to be transported or shipped in commerce hazardous materials or who manufactures, fabricates, marks, maintains, reconditions, repairs, or tests packages or containers which are represented, marked, certified, or sold by such person for use in the transportation in commerce of certain hazardous materials (designated by the Secretary) may be required by the Secretary to prepare and submit to the Secretary a registration statement not more often than once every two years. Such a registration statement shall include, but need not be limited to, such as a person's name; principal place of business; the location of each activity handling such hazardous materials; a complete list of all such hazardous materials handled; and an averment that such person is in compliance with all applicable criteria established under subsection (a) of this section. The Secretary shall by regulation prescribe the form of any such statement and the information required to be included. The Secretary shall make any registration statement filed pursuant to this subsection available for inspection by any person, without charge, except that nothing in this sentence shall be deemed to require the release of any information described by subsection (b) of section 552 of title 5, United States Code, or which is otherwise protected by law from disclosure to the public.

(c) **REQUIREMENT.**—No person required to file a registration of statement under subsection (b) of this section may transport or cause to be transported or shipped extremely hazardous materials, or manufacture, fabricate, mark, maintain, recondition, repair, or test packages or containers for use in the transportation of extremely hazardous materials, unless he has on file a registration statement. (Sec. 106 of P.L. 93–633.)

EXEMPTIONS

1806. (a) GENERAL.—The Secretary, in accordance with procedures prescribed by regulation, is authorized to issue or renew, to any person subject to the requirements of this title, an exemption from the provisions of this title, and from regulations issued under section 105 of this title, if such person transports or causes to be transported or shipped hazardous materials in a manner so as to achieve a level of safety (1) which is equal to or exceeds that level of safety which would be required in the absence of such exemption, or (2) which would be consistent with the public interest and the policy of this title in the event there is no existing level of safety established. The maximum period of an exemption issued or renewed under the section shall not exceed 2 years, but any such exemption may be renewed upon application to the Secretary. Each person applying for such an exemption or renewal shall, upon application, provide a safety analysis as prescribed by the Secretary to justify the grant of such exemption. A notice of an application for issuance or renewal of such exemption shall be published in the Federal Register. The Secretary shall afford access to any such safety analysis and an opportunity for public comment on any such application, except that nothing in this sentence shall be deemed to require the release of any information described by subsection (b) of section 552 of title 5, United States Code, or which is otherwise protected by law from disclosure to the public.

(b) **VESSELS.**—The Secretary shall exclude, in whole or in part, from any applicable provisions and regulations under this title, any vessel which is excepted from the application of section 201 of the Ports and Waterways Safety Act of 1972 by paragraph (2) of such section (46 U.S.C. 391a (2)), or any other vessel regulated under such Act, to the extent of such regulation.

(c) **FIREARMS AND AMMUNITION.**—Nothing in this title, or in any regulation issued under this title, shall be construed to prohibit or regulate the transportation by any individual, for personal use, of any firearm (as defined in paragraph (4) of section 232 of title 18, United States Code) or any ammunition therefore, or to prohibit any transportation of firearms or ammunition in commerce.

(d) **LIMITATION ON AUTHORITY.**—Except when the Secretary determines that an emergency exists, exemptions or renewals granted pursuant to this section shall be the only means by which a person subject to the requirements of this title may be exempted from or relieved of the obligation to meet any requirements imposed under this title. (Sec. 107 of P.L. 93–633.)

TRANSPORTATION OF RADIOACTIVE MATERIALS
ON PASSENGER-CARRYING AIRCRAFT

1807. (a) GENERAL—Within 120 days after the date of enactment of this section, the Secretary shall issue regulations in accordance with this section and pursuant to section 105 of this title, with respect to the transportation of radioactive materials on any passenger-carrying aircraft in air commerce, as defined in section 101 (4) of the Federal Aviation Act of 1958, as amended (49 U.S.C. 1013 (4)). Such regulations shall prohibit any transportation of radioactive materials on any such aircraft unless the radioactive materials involved are intended for use in, or incident to, research, or medical diagnosis or treatment, so long as such materials as prepared for and during transportation do not pose an unreasonable hazard to health and safety. The Secretary shall further establish effective procedures for monitoring and enforcing the provisions of such regulations.

(b) **DEFINITION.**—As used in this section, "radioactive materials" means any materials or combination of materials which spontaneously emit ionizing radiation. The term does not include materials in which (1) the estimated specific activity is not greater than 0.002 microcuries per gram of material; and (2) the radiation is distributed in an essentially uniform manner. (Sec. 108 of P.L. 93–633.)

POWERS AND DUTIES OF THE SECRETARY

1808. (a) GENERAL.—The Secretary is authorized, to the extent necessary to carry out his responsibilities under this title, to conduct investigations, make reports, issue subpoenas, conduct hearings, require the production of relevant documents, records, and property, take depositions, and conduct, directly or indirectly, research, development, demonstration, and training activities. The Secretary is further authorized, after notice and an opportunity for a hearing, to issue orders directing compliance with this title or regulations issued under this title; the district courts of the United States shall have jurisdiction, upon petition by the Attorney General, to enforce such orders by appropriate means.

(b) **RECORDS.**—Each person subject to requirements under this title shall establish and maintain such records, make such reports, and provide such information as the Secretary shall by order or regulation prescribe, and shall submit such reports and shall make such records and information available as the Secretary may request.

(c) **INSPECTION.**—The Secretary may authorize any officer, employee, or agent to enter upon, inspect, and examine, at reasonable times and in a reasonable manner, the records and properties of persons to the extent such records and properties relate to—

1. the manufacture, fabrication, marking, maintenance, reconditioning, repair, testing, or distribution of packages or containers for use by any person in the transportation of hazardous materials in commerce; or
2. the transportation or shipment by any person of hazardous materials in commerce.

Any such officer, employee, or agent shall, upon request, display proper credentials.

(d) **FACILITIES AND DUTIES.**—The Secretary shall—

1. establish and maintain facilities and technical staff sufficient to provide, within the Federal government, the capability of evaluating risks connected with the transportation of hazardous materials alleged to be hazardous;
2. establish and maintain a central reporting system and data center so as to be able to provide the law-enforcement and firefighting personnel of communities, and other interested persons and government officers, with the technical and other information and advice for meeting and emergencies connected with the transportation of hazardous materials; and
3. conduct a continuing review of all aspects of the transportation of hazardous materials in order to determine and to be able to recommend appropriate steps to assure the safe transportation of hazardous materials.

(e) **ANNUAL REPORT.**—The Secretary shall prepare and submit to the President for transmittal to the Congress on or before May of each year a comprehensive report on the transportation of hazardous materials during the preceding calendar year. Such report shall include, but need not be limited to—

1. a thorough statistical compilation of any accidents and casualties involving the transportation of hazardous materials;

2. a list and summary of applicable Federal regulations, criteria orders, and exemptions in effect;
3. a summary of the basis for any exemptions granted or maintained;
4. an evaluation of the effectiveness of enforcement activities and the degree of voluntary compliance with applicable regulations;
5. a summary of outstanding problems confronting the administration of this title, in order of priority; and
6. such recommendations for additional legislation as are deemed necessary or appropriate. (Sec. 109 of P.L. 93–633.)

PENALTIES

1809. (a) CIVIL.—(1) Any person (except an employee who acts without knowledge) who is determined by the Secretary, after notice and an opportunity for a hearing, to have knowingly committed an act which is a violation of a provision of this title or of a regulation issued under this title, shall be liable to the United States for a civil penalty. Whoever knowingly commits an act which is a violation of any regulation, applicable to any person who transports or causes to be transported or shipped hazardous materials, shall be subject to a civil penalty of not more than $10,000 for each violation, and if any such violation is a continuing one, each day of violation constitutes a separate offense. Whoever knowingly commits an act which is a violation of any regulation applicable to any person who manufactures, fabricates, marks, maintains, reconditions, repairs, or tests a package or container which is represented, marked, certified, or sold by such person for use in the transportation in commerce of hazardous materials shall be subject to a civil penalty of not more than $10,000 for each violation. The amount of any such penalty shall be assessed by the Secretary by written notice. In determining the amount of such penalty, the Secretary shall take into account the nature, circumstances, extent and gravity of the violation committed and, with respect to the person found to have committed such violation, the degree of culpability, any history of prior offenses, ability to pay, effect on ability to continue to do business, and such other matters as justice may require.

(2) Such civil penalty may be recovered in an action brought by the Attorney General on behalf of the United States in the appropriate district court of the United States or, prior to referral to the Attorney General, such civil penalty may be compromised by the Secretary. The amount of such penalty, when finally determined (or agreed upon in compromise), may be deducted from any sums owed by the United States to the person charged. All penalties collected under this subsection shall be deposited in the Treasury of the United States as miscellaneous receipts.

(b) CRIMINAL.—A person is guilty of an offense if he willfully violates a provison of this title or a regulation issued under this title. Upon conviction, such person shall be subject, for each offense, to a fine of not more than $25,000, imprisonment for a term not to exceed 5 years, or both. (Sec. 110 of P.L. 93–633.)

SPECIFIC RELIEF

1810. (a) GENERAL.—The Attorney General, at the request of the Secretary, may bring an action in an appropriate district court of the United States for equitable relief to redress a violation by any person of a provision of this title, or any order or regulation issued under this title. Such district courts shall have jurisdiction to determine such actions and may grant such relief as is necessary or appropriate, including mandatory or prohibitive injunctive relief, interim equitable relief, and punitive damages.

(b) IMMINENT HAZARD.—If the Secretary has reason to believe that an imminent hazard exists, he may petition an appropriate district court of the United States, or upon his request the Attorney General shall so petition, for an order suspending or restricting the transportation of the hazardous material responsible for such imminent hazard, or for such other order as is necessary to eliminate or ameliorate such imminent hazard. As used in this subsection, an "imminent hazard" exists if thre is substantial likelihood that serious harm will occur prior to the completion of an administrative hearing or other formal proceeding initiated to abate the risk of such harm. (Sec. 111 of P.L. 93–633.)

RELATIONSHIP TO OTHER LAWS

1811. (a) GENERAL.—Except as provided in subsection (b) of this section, any requirement, of a State or political subdivision thereof, which is inconsistent with any requirement set forth in this title, or in a regulation issued under this title, is preempted.

(b) STATE LAWS.—Any requirement, of a State or political subdivision thereof, which is not consistent with any requirement set forth in this title, or in a regulation issued under this title, is not preempted if, upon the application of an appropriate State agency, the Secretary determines, in accordance with procedures to be prescribed by regulation, that such requirement (1) affords an equal or greater level of protection to the public than is afforded by the requirements of this title or of regulations issued under this title and (2) does not unreasonably burden commerce. Such requirement shall not be preempted to the extent specified in such State or political subdivision thereof continues to administer and enforce effectively such requirement.

(c) OTHER FEDERAL LAWS.—The provisions of this title shall not apply to pipelines which are subject to regulation under the Natural Gas Pipeline Safety Act of 1958 (40 U.S.C. 1671 et seq.) or to pipelines which are subject to regulation under Chapter 39 of title 18, United States Code. (Sec. 112 of P.L. 93–633.)

TRANSPORTATION OF EXPLOSIVES ACT
18 U.S.C. 831–835, P.L. 86–710 (Repealed)

831. DEFINITIONS As used in this chapter—

Unless otherwise indicated, "carrier" means any person engaged in the transportation of passengers or property, by land, as a common, contract, or private carrier, or freight forwarder as those terms are used in the Interstate Commerce Act, as amended, and officers, agents, and employees of such carriers.

"Person" means any individual, firm, copartnership, corporation, company, association, or joint-stock association, and includes any trustee, receiver, assignee, or personal representative thereof.

"For-hire carrier" includes common and contract carriers.

"Shipper" shall be construed to include officers, agents, and employees of shippers.

"Interstate and foreign commerce" means commerce betwen a point in one State and a point in another State, between points in the same State through another State or through a foreign country, between points in a foreign country or countries through the United States, and commerce between a point in the United States and a point in a foreign country or in a Territory or possession of the United States, but only insofar as such commerce takes place in the United States. The term "United States" means all the States and the District of Columbia.

"State" includes the District of Columbia.

"Detonating fuzes" means fuzes used in military service to detonate the explosive charges of military projectiles, mines, bombs, or torpedoes.

"Fuzes" means devices used in igniting the explosive charges of projectiles.

"Fuses" means the slow-burning fuses used commercially to convey fire to an explosive combustible mass.

"Fusees" means the fusees ordinarily used on steamboats, railroads, and motor carriers as night signals.

"Radioactive materials" means any materials or combination of materials that spontaneously emit ionizing radiation.

"Etiologic agents" means the causative agent of such diseases as may from time to time be listed in regulations governing etiologic agents prescribed by the Interstate Commerce Commission (Secretary of Transportation) under section 834 of this chapter.

832. TRANSPORTATION OF EXPLOSIVES, RADIOACTIVE MATERIALS, ETIOLOGIC AGENTS, AND OTHER DANGEROUS ARTICLES

(a) Any person who knowingly transports, carries, or conveys within the United States, any dangerous explosives, such as and including, dynamite, blasting caps, detonating fuzes, black powder, gun-powder, or other like explosives, or any radioactive materials, or etiologic agents, on or in any passenger car or passenger vehicles of any description operated in the transportation of passengers by any for-hire carrier engaged in interstate or foreign commerce, by land, shall be fined not more than $1,000 or imprisoned not more than one year, or both; and, if the death or bodily injury of any person results from a violation of this section, shall be fined not more than $10,000 or imprisoned not more than ten years, or both: Provided, however, That such explosives, radioactive materials, or etiologic agents may be transported on or in such car or vehicle whenever the Interstate Commerce Commission (Secretary of Transportation) finds that an emergency requires an expedited movement, in which case such emergency movements shall be made subject to such regulations as the Commission (Secretary) may deem necessary or desirable in the public interest in each instance: Provided further, That under this section it shall be lawful to transport on or in any such car or vehicle, small quantities of explosives, radioactive materials, etiologic agents, or other dangerous commodities of the kinds, and in such amounts, and under such conditions as may be determined by the Interstate Commerce Commission (Secretary of Transportation) to involve no appreciable danger to persons or property: And provided further, That it shall be lawful to trasnport on or in any such car or vehicle such fusees, torpedoes, rockets, or other signal devices as may be essential to promote safety in the operation of any such car or vehicle on or in which transported. This section shall not prevent the transportation of military forces with their accompanying munitions of war on passenger-equipment cars or vehicles.

(b) No person shall knowingly transport, carry, or convey within the United States liquid nitroglycerin, fulminate in bulk in dry condition, or other similarly dangerous explosives, or radioactive materials, or etiologic agents, on or in any car or vehicle of any description operated in the transportation of passengers or property by any carrier engaged in interstate or foreign commerce, by land, except under such rules and regulations as the

Commission (Secretary) shall specifically prescribe with respect to the safe transportation of such commodities. The Commission (Secretary) shall from time to time determine and prescribe what explosives are "other similarly danger-ous explosives," and may prescribe the route or routes over which such explosives, radioactive materials, or etiologic agents shall be transported. Any person who violates this provision, or any regulation prescribed hereunder by the Interstate Commerce Commission (Secretary of Transportation) shall be fined not more than $1,000 or imprisoned not more than one year, or both; and, if the death or bodily injury of any person results from a violation of this section, shall be fined not more than $10,000 or imprisoned not more than ten years, or both.

(c) Any shipment of radioactive materials made by or under the direction or supervision of the Atomic Energy Commission or the Department of Defense which is escorted by personnel specially designated by or under the authority of the Atomic Energy Commission or the Department of Defense, as the case may be, for the purpose of national security, shall be exempt from the requirements of sections 831–835 of this chapter and the rules and regulations prescribed thereunder. In the case of any shipment of radioactive materials made by or under the direc-tion or supervision of the Atomic Energy Commission or the Department of Defense, which is not so escorted by specially designated personnel, certification upon the bill of lading by or under the authority of the Atomic Energy Commission or the Department of Defense, as the case may be, that the shipment contains radioactive materials shall be conclusive as to contents, and no further description shall be necessary or required; but each package, receptacle, or other container in such unescorted shipment shall on the outside thereof be plainly marked "radioac-tive materials," and shall not be opened for inspection by the carrier.

833. MARKING PACKAGES CONTAINING EXPLOSIVES AND OTHER DANGEROUS ARTICLES

Any person who knowingly delivers to any carrier engaged in interstate or foreign commerce by land or water, and any person who knowingly carries on or in any car or vehicle of any description operated in the transportation of passengers or property by any carrier engaged in interstate or foreign commerce, by land, any explosive, or other dangerous article, specified in or designated by the Interstate Commerce Commission (Secretary of Transportation) pursuant to section 834 of this chapter, under any false or deceptive marking, description, in-voice, shipping order, or other declaration, or any person who so delivers any such article without informing such carrier in writing of the true character thereof, at the time such delivery is made, or without plainly marking on the outside of every package containing explosives or the dangerous article the contents thereof, if such marking is required by regulations prescribed by the Interstate Commerce Commission (Secretary of Transportation), shall be fined not more than $1,000 or imprisoned not more than one year, or both; and, if the death or bodily injury of any person results from the violation of this section, shall be fined not more than $10,000 or imprisoned not more than ten years, or both.

834. REGULATIONS BY INTERSTATE COMMERCE COMMISSION (SECRETARY OF TRANSPORTATION)

(a) The Interstate Commerce Commission (Secretary of Transportation) shall formulate regulations for the safe transportation within the United States of explosives and other dangerous articles, including radioacive materials, etiologic agents, flammable liquids, flammable solids, oxidizing materials, corrosive liquids, compressed gases, and poisonous substances, which shall be binding upon all carriers engaged in interstate or foreign commerce by land or water.

(b) The Commission (Secretary) of its (his) own motion, or upon application made by any interested party, may make changes or modifications in such regulations, made desirable by new information or altered condi-tions. Before adopting any regulations relating to radioactive materials the Interstate Commerce Commission (Secre-tary of Transportation) shall advise and consult with the Atomic Energy Commission.

(c) Such regulations shall be in accord with the best-known practicable means for securing safety in transit, covering the packing, marking, loading, handling while in transit, and the precautions necessary to deter-mine whether the material when offered is in proper condition to transport.

(d) Such regulations, as well as all changes or modifications therof, shall, unless a shorter time is specified by the Commission (Secretary), take effect ninety days after their formulation and publication by the Commission (Secretary) and shall be in effect until reversed, set aside, or modified.

(e) In the execution of sections 831–835 inclusive, of this chapter the Commission (Secretary) may utilize the services of carrier and shipper associations including the Bureau for the Safe Transportation of Explosives and other Dangerous Articles, and may avail itself (himself) of the advice and assistance of any department, commission, or board of the Federal Government, and of State and local governments, but no official or employee of the United States shall receive any additional compensation for such service except as now permitted by law.

(f) Whoever knowingly violates any such regulation shall be fined not more than $1,000 or imprisoned not more than one year, or both; and, if the death or bodily injury of any person results from such violation, shall be fined not more than $10,000 or imprisoned not more than ten years, or both.

835. ADMINISTRATION

(a) The Interstate Commerce Commission (Secretary of Transportation) is authorized and directed to administer, execute, and enforce all provisions of sections 831–835, inclusive, of this chapter, to make all necessary orders in connection therewith, and to prescribe rules, regulations, and procedure for such administration, and to employ such officers and employees as may be necessary to carry out these functions.

(b) The Commission (Secretary) is authorized to make such studies and conduct such investigations, obtain such information, and hold such hearings as it (he) may deem necessary or proper to assist it (him) in exercising any authority provided in sections 831–835, inclusive, of this chapter. For such purposes the Commission (Secretary) is authorized to administer oaths and affirmations, and by subpoena to require any person to appear and testify, or to appear and produce documents, or both, at any designated place. No person shall be excused from complying with any requirement under this paragraph because of his privilege against self-incrimination, but the immunity provisions of the Compulsory Testimony Act of February 11, 1893 (49 U.S.C. 46), shall apply with respect to any individual who specifically claims such privilege. Witnesses subpoenaed under this subsection shall be paid the same fees and mileage as are paid witnesses in the district courts of the United States.

(c) In administering and enforcing the provisions of sections 831–835, inclusive, of this chapter and the regulations prescribed thereunder the Commission (Secretary) shall have and exercise all the powers conferred upon it by the Interstate Commerce Act, including procedural and investigative powers and the power to examine and inspect records and properties of shippers to the extent that such records and properties pertain to the packing and shipping of explosives and other dangerous articles and the nature of such commodities.

NOTE: The Department of Transportation Act (Public Law 80–570, approved and effective October 15, 1966) transferred to and vested in the Secretary of Transportation all functions, powers, and duties of the Interstate Commerce Commission, and of the Chairman, members, officers, and offices thereof, under the Act quoted above.

FEDERAL AVIATION ACT OF 1958
49 U.S.C. 1301 et seq.

1471.(a)(1) CIVIL PENALTIES AND POSTAL OFFENSES

(a)(1) Any person who violates (A) any provisions of titles III, V, VI, VII, or XII of this Act, or any rule, regulation, or order issued thereunder, . . . shall be subject to a civil penalty of not to exceed $1,000 for each such violation . . . , except that the amount of such civil penalty shall not exceed $10,000 for each such violation which relates to the transportation of hazardous materials. The amount of any such civil penalty which relates to the transportation of hazardous materials shall be assessed by the Secretary, or his delegate, upon written notice upon a finding of violation by the Secretary, after notice and an opportunity for a hearing. In determining the amount of such a penalty, the Secretary shall take into account the nature, circumstances, extent, and gravity of the violation committed and, with respect to the person found to have committed such violation, the degree of culpability, any history of prior offenses, ability to pay, effect on ability to continue to do business, and such other matters as justice may require. . . .

1472. CRIMINAL PENALTIES

(h)(1) In carrying out his responsibilities under this Act, the Secretary of Transportation may exercise the authority vested in him by section 105 of the Hazardous Materials Transportation Act to provide by regulation for the safe transportation of hazardous materials by air.

(2) A person is guilty of an offense if he willfully delivers or causes to be delivered to an air carrier or to the operator of a civil aircraft for transportation in air commerce, or if he recklessly causes the transportation in air commerce of, any shipment, baggage, or other property which contains a hazardous material, in violation of any rule, regulation, or requirement with respect to the transportation of hazardous materials issued by the Secretary of Transportation under this Act. Upon conviction, such person shall be subject, for each offense, to a fine of not more than $25,000, imprisonment for a term not to exceed 5 years, or both.

DANGEROUS CARGO ACT
46 U.S.C. 170

170. REGULATION OF CARRIAGE OF EXPLOSIVES OR
OTHER DANGEROUS ARTICLES ON VESSELS

(1) Vessel defined. The word "vessel" as used in this section shall include every vessel, domestic or foreign, regardless of character, tonnage, size, service, and whether self-propelled or not, on the navigable waters of the United States, including its Territories and possessions, but not including the Panama Canal Zone, whether arriving or departing, or under way, moored, anchored, aground, or while in drydock; it shall not include any public vessel which is not engaged in commercial service, nor any vessel subject to the provisions of section 391a of this title, which is constructed or converted for the principal purpose of carrying inflammable or combustible liquid cargo in bulk in its own tanks: *Provided,* That the provisions of subsection (3) of this section shall apply to every such vessel subject to the provisions of section 391a of this title, which is constructed or converted for the principal purpose of carrying inflammable or combustible liquid cargo in bulk in its own tanks.

(2) **Passenger-carrying vessel defined.**   The phrase "passenger-carrying vessel" as used in this section, when applied to a vessel subject to any provision of the International Convention for Safety of Life at Sea, 1929, means a vessel which carries or is authorized to carry more than twelve passengers.

(3) **Transportation, etc., of certain explosives prohibited.**   It shall be unlawful knowingly to transport, carry, convey, store, stow, or use on board any vessel fulminates or other detonating compounds in bulk in dry condition, or explosive compositions that ignite spontaneously or undergo marked decomposition when subjected for forty-eight consecutive hours to a temperature of one hundred and sixty-seven degrees Fahrenheit, or compositions containing an ammonium salt and a chlorate, or other like explosives.

(4) **Transportation, etc., of certain high explosives on passenger-carrying vessels prohibited; exceptions.**   It shall be unlawful knowingly to transport, carry, convey, store, stow, or use on board any passenger-carrying vessel any high explosives such as, and including, liquid nitroglycerin, dynamite, trinitrotoluene, picrates, detonating fuzes, fireworks that can be exploded en masse, or other explosives susceptible to detonation by a blasting cap or detonating fuze, except ships' signal and emergency equipment, and samples of such explosives (but not including liquid nitroglycerin) for laboratory or sales purposes in restricted quantities as may be permitted by regulations of the Commandant of the Coast Guard established hereunder.

(5) **Same; non-passenger-carrying vessels.**   It shall be unlawful knowingly to transport, carry, convey, store, stow, or use on board any vessel other than a passenger-carrying vessel, any high explosive referred to in subsection (4) of this section except as permitted by the regulations of the Commandant of the Coast Guard established hereunder.

(6) **Transportation, etc., of other explosives or other dangerous articles; exceptions.**   (a) It shall be unlawful knowingly to transport, carry, convey, store, stow, or use (except as fuel for its own machinery) on board any vessel, except one specifically exempted by paragraph (b) of this subsection, any other explosives or other dangerous articles or substances, including inflammable liquids, inflammable solids, oxidizing materials, corrosive liquids, compressed gases, poisonous articles or substances, hazardous articles, and ships' stores and supplies of a dangerous nature, except as permitted by the regulations of the Commandant of the Coast Guard established hereunder: *Provided,* That all of the provisions of this subsection relating to the transportation, carrying, conveying, storing, stowing, or use of explosives or other dangerous articles or substances shall apply to the transportation, carrying, conveying, storing, stowing, or using on board any passenger vessel of any barrels, drums, or other packages of any combustible liquid which gives off inflammable vapors (as determined by flashpoint in open cup tester as used for test of burning oil) at or below a temperature of one hundred and fifty degrees Fahrenheit and above eighty degrees Fahrenheit.

(b) This subsection shall not apply to—

(i) vessels not exceeding fifteen gross tons when not engaged in carrying passengers for hire;

(ii) vessels used exclusively for pleasure;

(iii) vessels not exceeding five hundred gross tons while engaged in fisheries;

(iv) tugs or towing vessels: *Provided, however,* That any such vessel, when engaged in towing any vessel that has explosives, inflammable liquids, or inflammable compressed gases on board or on deck, shall be required to make such provisions to guard against and extinguish fire as shall be prescribed by the Commandant of the Coast Guard;

(v) cable vessels, dredges, elevator vessels, fireboats, icebreakers, pile drivers, pilot boats, welding vessels, salvage and wrecking vessels;

(vi) inflammable or combustible liquid cargo in bulk: *Provided, however,* That the handling and stowage of any inflammable or combustible liquid cargo in bulk shall be subject to the provisions of section 391a of this title.

(7) **Regulations for protection against hazards created by explosives or other dangerous articles.**   In order to secure effective provisions against the hazards of health, life, limb, or property created by explosives or other dangerous articles or substances to which subsection (3)—(4), (5) or (6) of this section apply—

(a) The Commandant of the Coast Guard shall by regulations define, describe, name, and classify all explosives or other dangerous articles or substances, and shall establish such regulations as may be necessary to make effective the provisions of this section with respect to the descriptive names, packing, marking, labeling, and certification of such explosives or other dangerous articles or substances; with respect to the specifications of containers for explosives or other dangerous articles or substances; with respect to the marking and labeling of said containers; and shall accept and adopt for the purposes above mentioned in this subsection such definitions, descriptions, descriptive names, classifications, specifications of containers, packing, marking, labeling, and certification of explosives or other dangerous articles or substances to the extent as are or may be established from time to time by the Interstate Commerce Commission insofar as they apply to shippers by common carriers engaged in interstate or foreign commerce by water. The Commandant of the Coast Guard shall also establish regulations with respect to the marking, handling, storage, stowage, and use of explosives or other dangerous articles or substances on board such vessels; with respect to the disposition of any explosives or other dangerous articles or substances found to be in an unsafe condition; with respect to the necessary shipping papers, manifests, cargo-stowage plans, and the description and descriptive names of explosives or other dangerous articles or substances to be entered in such shipping documents; also any other regulations for the safe transportation, carriage, conveyance, storage, stowage, or use of explosives or other dangerous articles or substances on board such vessels as the Commandant of the Coast Guard shall deem necessary; and with respect to the inspection of all the foregoing mentioned in this paragraph. The Commandant of the Coast Guard may utilize the services of the Bureau for the Safe Transportation of Explosives and Other Dangerous Articles, and of such other oragnizations whose services he may deem to be helpful.

(b) The transportation, carriage, conveyance, storage, stowage, or use of such explosives or other dangerous articles or substances shall be in accordance with the regulations so established, which shall, insofar as applicable to them, respectively, be binding upon shippers and the owners, charterers, agents, masters, or persons in charge of such vessels and upon all other persons transporting, carrying, conveying, storing, stowing, or using on board any such vessels any explosives or other dangerous articles or substances: *Provided,* That this section shall not be construed to prevent the transportation of military or naval forces with their accompanying munitions of war and stores.

(c) Nothing contained in this section shall be construed to relieve any vessel subject to the provisions of this section from any of the requirements of title 52 (secs. 4399 to 4500, inclusive) of the Revised Statutes or acts amendatory or supplementary thereto and regulations thereunder applicable to such vessel, which are not inconsistent herewith.

(d) Nothing contained in this section shall be construed as preventing the enforcement of reasonable local regulations now in effect or hereafter adopted, which are not inconsistent or in conflict with this section or the regulations of the Commandant of the Coast Guard established hereunder.

(e) The United States Coast Guard shall issue no permit or authorization for the loading or discharging to or from any vessel at any point or place in the United States, its territories or possessions (not including Panama Canal Zone) of any explosives unless such explosives, for which a permit is required by the regulations promulgated pursuant to this section, are packaged, marked, and labeled in conformity with regulations prescribed by the Interstate Commerce Commission under section 835 of Title 18, and unless such permit or authorization specifies that the limits as to maximum quantity, isolation and remoteness established by local, municipal, territorial, or State authorities for each port shall not be exceeded. Nothing herein contained shall be deemed to limit or restrict the shipment, transportation, or handling of military explosives by or for the Armed Forces of the United States.

(8) **Masters, owners, etc., required to refuse unlawful transportation of explosives or other dangerous articles.** Any master, owner, charterer, or agent shall refuse to transport any explosives or other dangerous articles or substances in violation of any provisions of this section and the regulations established thereunder, and may require that any container or package which he has reason to believe contains explosives or other dangerous articles or substances be opened to ascertain the facts.

(9) **Publication of, hearings on, and effective date of proposed regulations.** Before any regulations or any additions, alterations, amendments, or repeals thereof are made under the provisions of this section, except in an emergency, such proposed regulations shall be published and public hearings with respect thereto shall be held on such notice as the Commandant of the Coast Guard deems advisable under the circumstances. Any additions, alterations, amendments, or repeals of such regulations shall, unless a shorter time is authorized by the Commandant of the Coast Guard, take effect ninety days after their promulgation.

(10) **Tendering explosives or other dangerous articles for shipment without divulging true character or in violation of section.** It shall be unlawful knowingly to deliver or cause to be delivered, or tender for shipment to any vessel subject to this section any explosives or any other dangerous articles or substances defined in the regulations of the Commandant of the Coast Guard established hereunder under any false or deceptive descriptive name, marking, invoice, shipping paper, or other declaration and without informing the agent of such vessel in writing of the true character thereof at or before the time such delivery or transportation is made. It shall be unlawful for any person to tender for shipment, or ship on any vessel to which this section applies, any explosives or other dangerous articles or substances the transportation, carriage, conveyance, storage, stowage, or use of which on board vessels is prohibited by this section.

(11) **Exemption of vessels from section or regulations when compliance unnecessary for safety.** The Commandant of the Coast Guard may exempt any vessel or class of vessels from any of the provisions of this section or any regulations or parts thereof established hereunder upon a finding by him that the vessel, route, area of

operations, conditions of the voyage, or other circumstances are such as to render the application of this section or any of the regulations established hereunder unnecessary for the purposes of safety: *Provided,* That except in an emergency such exception shall be made for any vessel or class of vessels only after a public hearing.

(12) **Agencies charged with enforcement.** The provisions of this section and the regulations established hereunder shall be enforced primarily by the Coast Guard of the Department of the Treasury; which with the consent of the head of any executive department, independent establishment, or other agency of the Government, may avail itself of the use of information, advice, services, facilities, officers, and employees thereof (including the field service) in carrying out the provisions of this section: *Provided,* That no officer or employee of the United States shall receive any additional compensation for such services, except as permitted by law.

(13) **Detention of vessels pending compliance with section and regulations; penalty for false swearing.** Any collector of customs may, upon his own knowledge, or upon the sworn information of any reputable citizen of the United States, that any vessel subject to this section is violating any of the provisions of this section or of the regulations established hereunder, by written order served on the master, person in charge of such vessel, or the owner or charterer thereof, or the agent of the owner or charterer, detain such vessel until such time as the provisions of this section and of the regulations established hereunder have been complied with. If the vessel be ordered detained, the master, person in charge, or owner or charterer, or the agent of the owner or charterer thereof, may within five days appeal to the Commandant of the Coast Guard, who may, after investigation, affirm, set aside, or modify the order of such collector. If any reputable citizen of the United States furnishes sworn information to any collector of customs that any vessel, subject to this section, is violating any of the provisions of this section or of the regulations established hereunder, and such information is knowingly false, the person so falsely swearing shall be deemed guilty of perjury.

(14) **Violation of section or regulations; penalty; liability of vessel.** Whoever shall knowingly violate any of the provisions of this section or of any regulations established under this section shall be subject to a criminal penalty of not more than $2,000 or imprisoned not more than 5 years, or both, for each violation. In the case of any such violation on the part of the owner, charterer, agent, master, or person in charge of the vessel, such vessel shall be liable for the penalty and may be seized and proceeded against by way of libel in the district court of the United States in any district in which such vessel may be found.

(15) **Same; increased penalty in event of personal injury or death.** When the death or bodily injury of any person results from the violation of this section or any regulations made in pursuance thereof, the person or persons who shall have knowingly violated or caused to be violated such provisions or regulations shall be fined not more than $10,000 or imprisoned not more than ten years, or both.

(16) **Transportation of motor vehicles carrying gasoline, etc., penalty for violations.** The transportation by vessels of gasoline or any other inflammable or combustible liquid or inflammable gas when carried by motor vehicles using the same as a source of their own motive power, or motive power for driving auxiliaries forming a part of the vehicle, shall be lawful under the conditions as set forth in the regulations established by the Commandant of the Coast Guard under this section: *Provided, however,* That the motor or motors in any vehicle be stopped immediately after entering the said vessel, and that the same be not restarted until immediately before said vehicle shall leave the vessel after said vessel has been made fast to the wharf or ferry bridge at which she lands. All other fire, if any, in such vehicle shall be extinguished before entering the said vessel and the same shall not be relighted until after said vehicle shall leave the vessel: *Provided further,* That the Commandant of the Coast Guard, may, by regulation, permit the operation on board vessels of motive power for driving auxiliaries forming a part of motor vehicles, under such conditions as he may deem proper: *Provided further,* That any owner, charterer, agent, master, or other person having charge of a vessel shall have the right to refuse to transport motor vehicles the fuel tanks of which contain gasoline or other inflammable or combustible liquid or inflammable gas used as a source of power for the vehicle or its auxiliaries: *Provided further,* That the owner, motor carrier, and operator of any such vehicle in which all fires have not been extinguished or the motor or motors stopped as required by this subsection or regulations established thereunder, and the owner, charterer, agent, master, or person in charge of the vessel on which such vehicle is transported, shall each be liable to a penalty of not more than $500, for which the motor vehicle and vessel respectively, shall be liable: *And provided further,* That a violation of this subsection shall not subject any person to the penalty provided in subsection (14) or (15) of this section.

(17) (A) Any person (except an employee who acts without knowledge) who is determined by the Secretary, after notice and an opportunity for a hearing, to have knowingly committed an act which is a violation of any provision of this section, or of any regulation issued under this section, shall be liable to the United States for a civil penalty of not more than $10,000 for each day of that violation. The amount of such civil penalty shall be assessed by the Secretary by written notice. In determining the amount of such penalty, the Secretary shall take into account the nature, circumstances, extent, and gravity of the violation committed and, with respect to the person found to have committed such violation, the degree of culpability, any history of prior offenses, ability to pay, effect on ability to continue to do business, and such other matters as justice may require.

(B) Such civil penalty may be recovered in an action brought by the Attorney General on behalf of the

United States, in the appropriate district court of the United States or, prior to referral to the Attorney General, such civil penalty may be compromised by the Secretary. The amount of such penalty, when finally determined (or agreed upon in compromise), may be deducted from any sums owed by the United States to the person charged. All penalties collected under this subsection shall be deposited in the Treasury of the United States as miscellaneous receipts.

170a. SAME; USE BY VESSELS OF LAUNCHES, LIFEBOATS, ETC., EMPLOYING COMBUSTIBLE FUEL; REGULATIONS.

Nothing contained in section 170 of this title shall prohibit the use by any vessel of motorboats, launches, or lifeboats equipped with engines using an inflammable or combustible fuel, nor shall anything contained in said section prohibit such motorboats, launches, or lifeboats from carrying such inflammable or combustible fuel in their tanks: *Provided,* That no such inflammable or combustible fuel for the engines of such motorboats, launches, or lifeboats shall be carried except as may be prescribed by regulations of the Commandant of the Coast Guard: *Provided further,* That the use of such lifeboats shall be under such regulations as shall be prescribed by the Commandant of the Coast Guard.

170b. SAME; APPROPRIATIONS.

There are authorized to be appropriated such sums of money as may be necessary to carry out the provisions of sections 170–170b, 391a, 402, 414, and 481 of this title and sections 382–385 of Title 18. (Oct. 9, 1940, ch. 777, § 8, 54 Stat. 1028.)

RESOURCE CONSERVATION & RECOVERY ACT
42 U.S.C. 6921, 90 Stat. 2806, P.L. 94–580, as amended.

DEFINITIONS
6903. As used in this Act

. . .

(5) the term "hazardous waste" means a solid waste, or combination of solid wastes, which because of its quantity, concentration, or physical, chemical, or infectious characteristics may—

(A) cause, or significantly contribute to an increase in mortality or an increase in serious irreversible, or incapacitating reversible, illness; or

(B) pose a substantial present or potential hazard to human health or the environment when improperly treated, stored, transported, or disposed of, or otherwise managed.

(6) the term "hazardous waste generation" means the act or process of producing hazardous waste.

. . .

(12) the term "manifest" means the form used for identifying the quantity, composition, and the origin, routing, and destination of hazardous waste during its transportation from the point of generation to the point of disposal, treatment, or storage.

. . .

(27) The term "solid waste" means any garbage, refuse, sludge from a waste treatment plant, water supply treatment plant, or air pollution control facility and other discarded material, including solid, liquid, semisolid, or contained gaseous material resulting from industrial, commercial, mining, and agricultural operations, and from community activities, but does not include solid or dissolved material in domestic sewage, or solid or dissolved materials in irrigation return flows or industrial discharges which are point sources subject to permits under section 402 of the Federal Water Pollution Control Act, as amended (33 U.S.C. 1342), or source, special nuclear, or byproduct material as defined by the Atomic Energy Act of 1954, as amended (42 U.S.C. 2011 et seq.).

IDENTIFICATION AND LISTING OF HAZARDOUS WASTE

6921. (a) Criteria for identification or listing. . . . (T)he Administrator (of the Environmental Protection Agency) shall, after notice and opportunity for public hearing, and after consultation with appropriate Federal and State agencies, develop and promulgate criteria for identifying the characteristics of hazardous waste, and for listing hazardous waste, which should be subject to the provisions of this subtitle, taking into account toxicity, persistence, and degradability in nature, potential for accumulation in tissue, and other related factors such as flammability, corrosiveness, and other hazardous characteristics. Such criteria shall be revised from time to time as may be appropriate.

STANDARDS APPLICABLE TO GENERATORS OF HAZARDOUS WASTE

6922. (a) In general. . . . (A)fter notice and opportunity for public hearings and after consultation with appropriate Federal and State agencies, the Administrator shall promulgate regulations establishing such standards, applicable to generators of hazardous waste identified or listed under this subtitle, as may be necessary to protect human health and the environment. Such standards shall establish requirements respecting—

(1) recordkeeping practices that accurately identify the quantities of such hazardous waste generated, the constituents thereof which are significant in quantity or in potential harm to human health or the environment, and the disposition of such wastes;

(2) labeling practices for any containers used for the storage, transport, or disposal of such hazardous waste as will identify accurately such waste;

(3) use of appropriate containers for such hazardous waste;

(4) furnishing of information on the general chemical composition of such hazardous waste to persons transporting, treating, storing, or disposing of such wastes;

(5) use of a manifest system and any other reasonable means necessary to assure that all such hazardous waste generated is designated for treatment, storage, or disposal in, and arrives at, treatment, storage, or disposal facilities (other than facilities on the premises where the waste is generated) for which a permit has been issued as provided in this subtitle or pursuant to title I of the Marine Protection, Research, and Sanctuaries Act (33 U.S.C. 1411 et seq.). . . .

(b) Waste minimization. Effective September 1, 1985, the manifest required by subsection (a)(5) shall contain a certification by the generator that—

(1) the generator of the hazardous waste has a program in place to reduce the volume or quantity and toxicity of such waste to the degree determined by the generator to be economically practicable; and

(2) the proposed method of treatment, storage, or disposal is that practicable method currently available to the generator which minimizes the present and future threat to human health and the environment.

STANDARDS APPLICABLE TO TRANSPORTERS
OF HAZARDOUS WASTE

6923. (a) Standards. . . . (A)fter opportunity for public hearings, the Administrator, after consultation with the Secretary of Transportation and the States, shall promulgate regulations establishing such standards, applicable to transporters of hazardous waste identified or listed under this subtitle, as may be necessary to protect human health and the environment. Such standards shall include but need not be limited to requirements respecting—

(1) recordkeeping concerning such hazardous waste transported, and their source and delivery points;

(2) transportation of such waste only if properly labeled;

(3) compliance with the manifest system referred to in 42 U.S.C. 6922(a)(5); and

(4) transportation of all such hazardous waste only to the hazardous waste treatment, storage, or disposal facilities which the shipper designates on the manifest form to be a facility holding a permit issued under this subtitle, or pursuant to title I of the Marine Protection, Research, and Sanctuaries Act.

(b) Coordination with regulations of the Secretary of Transportation. In case of any hazardous waste identified or listed under this subtitle which is subject to the Hazardous Materials Transportation Act (49 U.S.C. 1801 et seq.), the regulations promulgated by the Administrator under this section shall be consistent with the requirements of such Act and the regulations thereunder. The Administrator is authorized to make recommendations to the Secretary of Transportation respecting the regulations of such hazardous waste under the Hazardous Materials Transportation Act and for additional materials to be covered by such Act.

(c) Fuel from hazardous waste. . . . (A)fter opportunity for public hearing, the Administrator shall promulgate regulations establishing standards, applicable to transporters of fuel produced (1) from any hazardous waste identified or listed under 42 U.S.C. 6921, or (2) from any hazardous waste identified or listed under 24 U.S.C. 6921 and any other material, as may be necessary to protect human health and the environment. Such standards may include any of the requirements set forth in paragraphs (1) through (4) of subsection (a) as may be appropriate.

. . .

EXPORT OF HAZARDOUS WASTE

6938. (a) In general. . . . (N)o person shall export any hazardous waste identified or listed under 42 U.S.C. 6921 unless

(1) (A) such person has provided the notification required in subsection (c) of this section,

(B) the government of the receiving country has consented to accept such hazardous waste,

(C) a copy of the receiving country's written consent is attached to the manifest accompanying each waste shipment, and

(D) the shipment conforms with the terms of the consent of the government of the receiving country required pursuant to subsection (e), or

(2) the United States and the government of the receiving country have entered into an agreement as provided for in subsection (f) and the shipment conforms with the terms of such agreement.

(b) **Regulations.** . . . (T)he Administrator shall promulgate the regulations necessary to implement this section. Such regulations shall become effective 180 days after promulgation.

(c) **Notification.** Any person who intends to export a hazardous waste identified or listed under this subtitle . . . shall, before such hazardous waste is scheduled to leave the United States, provide notification to the Administrator. Such notification shall contain the following information:

(1) the name and address of the exporter;

(2) the types and estimated quantities of hazardous waste to be exported;

(3) the estimated frequency or rate at which such waste is to be exported, and the period of time over which such waste is to be exported;

(4) the ports of entry;

(5) a description of the manner in which such hazardous waste will be transported to and treated, stored, or disposed in the receiving country; and

(6) the name and address of the ultimate treatment, storage or disposal facility.

. . .

(f) **International agreements.** Where there exists an international agreement between the United States and the government of the receiving country establishing notice, export, and enforcement procedures for the transportation, treatment, storage, and disposal of hazardous wastes, only the requirements of subsections (a)(2) and (g) shall apply.

(g) **Reports.** . . . (A)ny person who exports any hazardous waste identified or listed under this subtitle shall file with the Administrator no later than March 1 of each year, a report summarizing the types, quantities, frequency, and ultimate destination of all such hazardous waste exported during the previous calendar year. . . .

COMPREHENSIVE ENVIRONMENTAL RESPONSE, COMPENSATION, AND LIABILITY ACT
42 U.S.C. 9601 et seq., 94 Stat. 2767, P.L. 96–510, as amended

DEFINITIONS
9601. For purpose of this title, the term—

. . .

(14) "hazardous substance" means (A) any substance designated pursuant to section 311(b)(2)(A) of the Federal Water Pollution Control Act (33 U.S.C. 1321(b)(2)(A)), (B) any element, compound, mixture, solution, or substance designated pursuant to section 102 of this Act (42 U.S.C. 9602), (C) any hazardous waste having the characteristics identified under or listed pursuant to section 3001 of the Solid Waste Disposal Act (42 U.S.C. 6921) . . . , (D) any toxic pollutant listed under section 307(a) of the Federal Water Pollution Control Act (33 U.S.C. 1317(a)), (E) any hazardous air pollutant listed under section 112 of the Clean Air Act (42 U.S.C. 7412), and (F) any imminently hazardous chemical substance or mixture with respect to which the Administrator (of the Environmental Protection Agency) has taken action pursuant to section 7 of the Toxic Substances Control Act (15 U.S.C. 2606). The term does not include petroleum, including crude oil or any fraction thereof which is not otherwise specifically listed or designated as a hazardous substance under subparagraphs (A) through (F) of this paragraph, and the term does not include natural gas, natural gas liquids, liquefied natural gas, or synthetic gas usable for fuel (or mixtures of natural gas and such synthetic gas);

. . .

(18) "onshore facility" means any facility (including, but not limited to, motor vehicles and rolling stock) of any kind located in, on, or under, any land or nonnavigable waters within the United States;

. . .

(20)(A) "owner or operator" means (i) in the case of a vessel, any person owning, operating, or chartering by demise, such vessel, (ii) in the case of an onshore facility or an offshore facility, any person owning or operating such facility, and (iii) in the case of any abandoned facility, any person who owned, operated, or otherwise controlled activities at such facility immediately prior to such abandonment. Such term does not include a person, who, without participating in the management of a vessel or facility, holds indicia of ownership primarily to protect his security interest in the vessel or facility;

    (B) in the case of a hazardous substance which has been accepted for transportation by a common or contract carrier and except as provided in 42 U.S.C. 9607(a)(3) or (4), the term "owner or operator" shall mean such common carrier or other bona fide for hire carrier acting as an independent contractor during such transportation, (ii) the shipper of such hazardous substance shall not be considered to have caused or contributed to any release during such transportation which resulted solely from circumstances or conditions beyond his control;

    (C) in the case of a hazardous substance which has been delivered by a common or contract carrier to a disposal or treatment facility and except as provided in 42 U.S.C. 9607(a)(3) or (4), (i) the term "owner or operator" shall not include such common or contract carrier, and (ii) such common or contract carrier shall not be considered to have caused or contributed to any release at such disposal or treatment facility resulting from circumstances or conditions beyond its control;

. . .

    (22) "release" means any spilling, leaking, pumping, pouring, emitting, emptying, discharging, injecting, escaping, leaching, dumping, or disposing into the environment (including the abandonment or discarding of barrels, containers, and other closed receptacles containing any hazardous substance or pollutant or contaminant), but excludes . . . ;

. . .

    (26) "transport" or "transportation" means the movement of a hazardous substance by any mode, including pipeline, and in the case of a hazardous substance which has been accepted for transportation by a common or contract carrier, the term "transport" or "transportation" shall include any stoppage in transit which is temporary, incidental to the transportation movement, and at the ordinary operating convenience of a common or contract carrier, and any such stoppage shall be considered as a continuity of movement and not as the storage of a hazardous substance;

. . .

    (28) "vessel" means every description of watercraft or other artificial contrivance used, or capable of being used, as a means of transportation on water. . . .

LIABILITY

9607. (a) Notwithstanding any other provisions or rule of law, and subject only to the defenses set forth in subsection (b) of this section—

    (1) the owner and operator of a vessel or a facility,

    (2) any person who at the time of disposal of any hazardous substance owned or operated any facility at which such hazardous substances were disposed of,

    (3) any person who by contract, agreement, or otherwise arranged for disposal or treatment, or arranged with a transporter for transport for disposal or treatment, of hazardous substances owned or possessed by such person, by any other party or entity, at any facility owned or operated by another party or entity and containing such hazardous substances, and

    (4) any person who accepts or accepted any hazardous substances for transport to disposal or treatment facilities or sites selected by such person,

from which there is a release, or a threatened release which causes the incurrence of response costs, of a hazardous substance, shall be liable for—

    (A) all costs of removal or remedial action incurred by the United States Government or a State not inconsistent with the national contingency plan;

    (B) any other necessary costs of response incurred by any other person consistent with the national contingency plan;

    (C) damages for injury to, destruction of, or loss of natural resources, including the reasonable costs of assessing such injury, destruction, or loss resulting from such a release; and

    (D) the costs of any health assessment or health effects study carried out under 42 U.S.C. 6904.

The amounts recoverable in an action under this section shall include interest on the amounts recoverable under subparagraphs (A) through (D). . . .

    (b) There shall be no liability under subsection (a) of this section for a person otherwise liable who can establish by a preponderance of the evidence that the release or threat of release of a hazardous substance and the damages resulting therefrom were caused solely by—

    (1) an act of God;

    (2) an act of war;

(3) an act or omission of a third party other than an employee or agent of the defendant, or than one whose act or omission occurs in connection with a contractual relationship, existing directly or indirectly, with the defendant (except where the sole contractual arrangement arises from a published tariff and acceptance for carriage by a common carrier by rail), if the defendant establishes by a preponderance of the evidence that (a) he exercised due care with respect to the hazardous substance concerned, taking into account the characteristics of such hazardous substance, in light of all relevant facts and circumstances, and (b) he took precautions against foreseeable acts or omissions of any such third party and the consequences that could foreseeably result from such acts or omissions; or (4) any combination of the foregoing paragraphs.

. . .

TRANSPORTATION

9656. (a) Each hazardous substance which is listed or designated as provided in 42 U.S.C. 9601(14) shall, . . . at the time of such listing or designation, . . . be listed and regulated as a hazardous material under the Hazardous Materials Transportation Act.

(b) A common or contract carrier shall be liable under other law in lieu of 42 U.S.C. 9607 for damages or remedial action resulting from the release of a hazardous substance during the course of transportation which commenced prior to the effective date of the listing and regulation of such substance as a hazardous material under the Hazardous Materials Transportation Act, or for substances listed pursuant to subsection (a) of this section, prior to the effective date of such listing: *Provided however,* that this subsection shall not apply where such carrier can demonstrate that he did not have actual knowledge of the identity or nature of the substance released.

TOXIC SUBSTANCES CONTROL ACT
15 U.S.C. 2601 et seq., 90 Stat. 2003, P.L. 94–469, as amended.

REGULATION OF HAZARDOUS CHEMICAL SUBSTANCES AND MIXTURES

2605. (a) Scope of regulation. If the Administrator (of the Environmental Protection Agency) finds that there is a reasonable basis to conclude that the manufacture, processing, distribution in commerce, use, or disposal of a chemical substance or mixture, or that any combination of such activities, presents or will present an unreasonable risk of injury to health or the environment, the Adminstrator shall by rule apply one or more of the following requirements to such substance or mixture to the extent necessary to protect adequately against such risk using the least burdensome requirements:

(1) A requirement (A) prohibiting the manufacturing, processing, or distribution in commerce of such substance or mixture, or (B) limiting the amount of such substance or mixture which may be manufactured, processed or distributed in commerce . . . (or an measures short of such requirements).

INDEX